Chemistry

FIFTEENTH EDITION

Jason Overby

College of Charleston

McGraw Hill

CHEMISTRY

Published by McGraw Hill LLC, 1325 Avenue of the Americas, New York, NY 10019.

This book is printed on acid-free paper.

4 5 6 7 8 9 CPI 29 28 27 26 25

ISBN 978-1-266-81744-1
MHID 1-266-81744-1

Cover Image: *3DStach/Shutterstock*

mheducation.com/highered

About the Author

Jason Overby/McGraw Hill

Jason Overby was born in Bowling Green, Kentucky, and raised in Clarksville, Tennessee. He received his B.S. in chemistry and political science from the University of Tennessee at Martin. After obtaining his Ph.D. in inorganic chemistry from Vanderbilt University, Jason conducted postdoctoral research at Dartmouth College.

Since joining the Department of Chemistry and Biochemistry at the College of Charleston, South Carolina, Jason has taught a variety of courses ranging from general chemistry to advanced inorganic chemistry. He is also interested in the integration of technology into the classroom, with a particular focus on adaptive learning. Additionally, he conducts research with undergraduates in inorganic and organic synthetic chemistry as well as computational organometallic chemistry.

In his free time, he enjoys boating, bowling, and cooking. He is also involved with USA Swimming as a nationally certified official. He lives in South Carolina with his wife Robin and two daughters, Emma and Sarah.

Contents in Brief

Contents

Martin McCarthy/theasis/E+/Getty Images

Measurement and the Properties of Matter 1

zoom-zoom/iStock/Getty Images

Atoms, Ions, and Molecules 40

NASA

Mass Relationships in Chemical Reactions 79

Leo Matyushkin/Shutterstock

Reactions in Aqueous Solutions 124

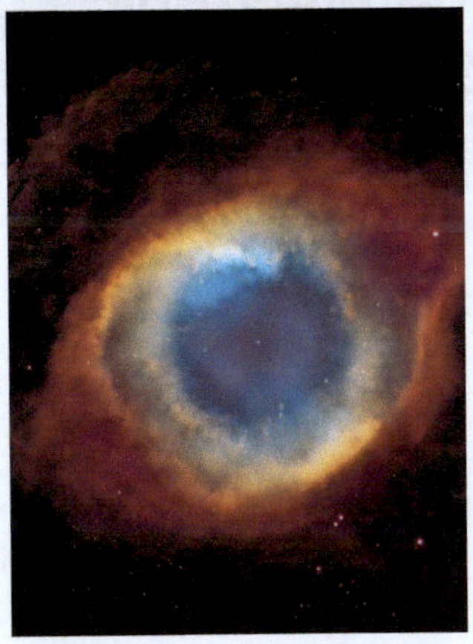

STScI/NASA

Gases 175

ThamKC/iStock/Getty Images

Thermochemistry 230

Andrew Dunn/Alamy Stock Photo

Quantum Theory and the Electronic Structure of Atoms 273

welcomia/Shutterstock

Periodic Relationships Among the Elements 328

Jason Overby

Compounds and Bonding 370

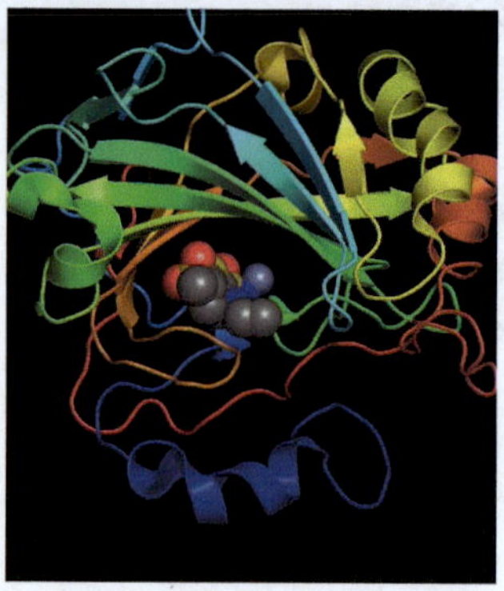

Courtesy of Jason Overby

Structure and Bonding Theories 418

Anna Kireieva/Shutterstock

Intermolecular Forces and Liquids and Solids 470

MADDRAT/Shutterstock

Physical Properties of Solutions 527

Per-Andre Hoffmann/Image Professionals GmbH/Alamy Stock Photo

Chemical Kinetics 567

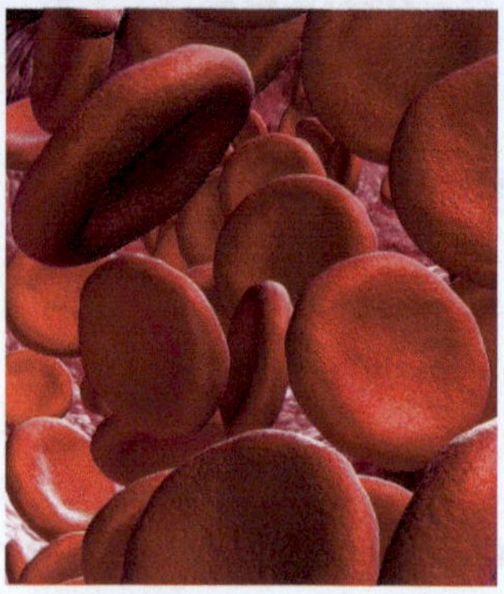

yodiyim/Shutterstock

Chemical Equilibrium 629

stevemendenhall/RooM/Getty Images

Acids and Bases 672

Michael McCoy/Science Source

Acid-Base Equilibria and Solubility Equilibria 725

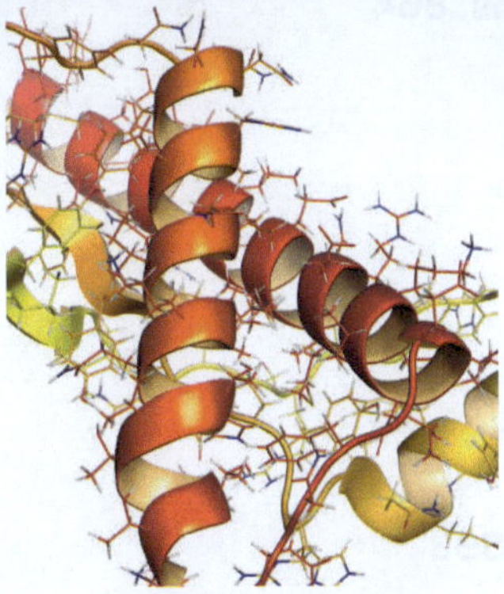

Pieter Van de Vijver/molekuul/123RF

Entropy, Gibbs Energy, and Equilibrium 777

Gene Chutka/E+/Getty Images

Electrochemistry 815

xenotar/iStockphoto/Getty Images

Nuclear Chemistry 863

View Stock/age fotostock

Chemistry in the Atmosphere 904

yongyut rukkachatsuwa/Shutterstock

Metallurgy and the Chemistry of Metals 934

Science Photo Library/Getty Images

Nonmetallic Elements and Their Compounds 963

Akkaradet Bangchun/Shutterstock

Coordination Chemistry 1001

Amith Nag Photography/Moment/Getty Images

Organic Chemistry 1030

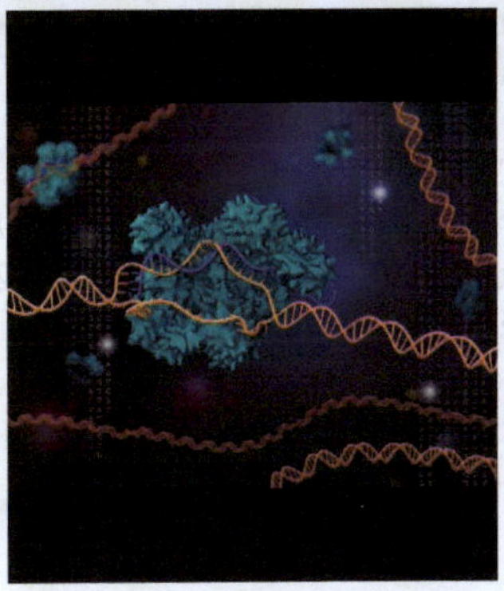

Meletios Verras/Shutterstock

Synthetic and Natural Organic Polymers 1062

List of Applications

The opening sentence of this text is, "Chemistry is an active, evolving science that has vital importance to our world, in both the realm of nature and the realm of society." Throughout the text, Chemistry in Action boxes and Chemical Mysteries (digital only) give specific examples of chemistry as active and evolving in all facets of our lives.

Chemistry in Action

Chemical Mystery (digital only)

List of Videos

The following videos are correlated to *Chemistry*. Within the chapter are icons letting the student and instructor know that a video is available for a specific topic. Videos can be found in the eBook.

Preface

In the fifteenth edition of *Chemistry,* the long history of focusing on a solid background of chemical concepts while providing an appreciation of the importance of chemistry in our daily lives continues unabated. Helping both instructors and students in attaining these goals requires a solid logical flow of material, so we believe that it is our duty to present a wide range of chemical topics with consistent presentation. We focus on balancing theory with application by demonstrating various principles with useful examples whenever possible.

As in earlier editions, we strive to present a text with clear and concise explanations that clarify difficult concepts but are detailed enough that students will gain a solid foundation as they progress through the chemistry curriculum. We always encourage feedback from all instructors and students.

ALEKS Data Parity with *Chemistry,* 15th Edition

A major source of frustration for instructors and students alike over the past two decades has been the lack of data coordination between a print text, eBook, and an online resource. Often it is not clear what is a true accepted value for an important chemical item (e.g., standard enthalpy of formation, acid ionization constant, or standard electrode potential for a half-reaction) as different sources often give a multitude of differing numbers. With the fifteenth edition of *Chemistry*, no longer will students be confused about which particular chemical value to employ in a calculation when using ALEKS. The data between the print text, eBook, and ALEKS are now parallel with one another. Numerous tables, figures, and problems throughout the text have been updated to mirror the data in ALEKS, thereby presenting unified chemical data.

Learning Objectives and Summary of Concepts and Facts

In the previous edition, a new listing of Learning Objectives was added to each chapter. To better help instructors and students focus on relevant topics and ideas in a given section, we are providing Learning Objectives at the beginning of each section followed by a Summary of Concepts & Facts at the end of each section. We believe that this perfectly introduces the pertinent idea(s) to be covered and finishes with a reinforcement of those ideas at the end of a section.

13.2 Rate Laws

Learning Objectives

- Summarize reaction order and provide examples of a zeroth-, first-, and second-order reaction rate law.
- Write the rate law of a reaction given experimental data.
- Express the units of the rate constant *k* for a reaction.

Summary of Concepts & Facts

- The rate law expresses the relationship of the rate of a reaction to the rate constant and the concentrations of the reactants raised to appropriate powers. The rate constant *k* for a given reaction changes only with temperature.
- Reaction order is the power to which the concentration of a given reactant is raised in the rate law. Overall reaction order is the sum of the powers to which reactant concentrations are raised in the rate law. The rate law and the reaction order cannot be determined from the stoichiometry of the overall equation for a reaction; they must be determined by experiment. For a zero-order reaction, the reaction rate is equal to the rate constant.

Student Hot Spots

In the electronic version of this text, all the learning resources for the Student Hot Spots are readily available.

Student Hot Spot

Student data indicate you may struggle with significant figures. Access your eBook for additional Learning Resources on this topic.

STUDENT HOT SPOT

Student data indicate you may struggle with this content. View the following video, "Converting Element Mass into Moles and Atoms," to make sure you understand the concept before moving on.

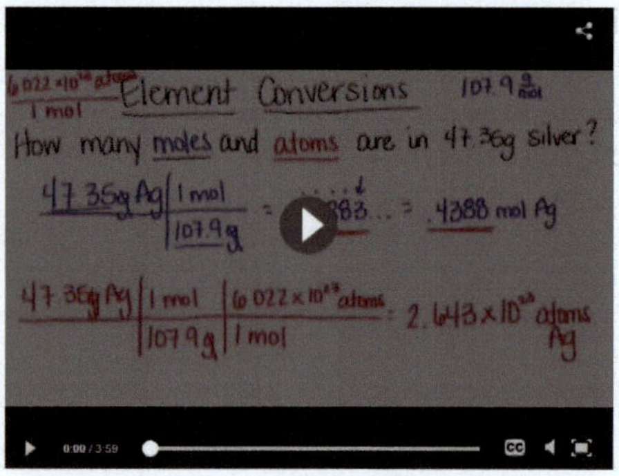

Further, access to student results has guided the editing of content in many chapters. While many of the changes are subtle, some are more comprehensive. The ability to edit based on real-time assessment data from students is the new paradigm for textbook authoring. Undoubtedly this changes how we provide and enhance learning materials for our students in the future!

Chapter Summary

A Chapter Summary is provided at the end of every chapter with a succinct overview of all sections in the chapter.

These are intended to be a reinforcement of the important ideas a student will have seen in the chapter. These can easily be used as study guides as well.

Chapter Summary

Molecular Geometry Molecular geometry refers to the three-dimensional arrangement of atoms in a molecule. For relatively small molecules, in which the central atom contains two to six bonds, geometries can be reliably predicted by the valence-shell electron-pair repulsion (VSEPR) model. This model is based on the assumption that chemical bonds and lone pairs tend to remain as far apart as possible to minimize repulsion. (Section 10.1)

Dipole Moments In a diatomic molecule, the difference in the electronegativities of bonding atoms results in a polar bond and a dipole moment. The dipole moment of a molecule made up of three or more atoms depends on both the polarity of the bonds and molecular geometry. Dipole moment measurements can help us distinguish between different possible geometries of a molecule. (Section 10.2)

Valence Bond Theory Valence bond theory is a quantum mechanical description of bonding that assumes electrons in a molecule occupy atomic orbitals of the individual atoms and these atomic orbitals overlap to form a covalent bond. Valence bond theory states that stable molecules form when the potential energy of the system of atoms has decreased to a minimum. Diatomic molecules can be easily explained by this minimization of potential energy with changes in distance between reacting atoms. (Section 10.3)

Hybridization of Atomic Orbitals Hybridization is the quantum mechanical description of chemical bonding. Atomic orbitals are hybridized, or mixed, to form hybrid orbitals. These orbitals then interact with other atomic orbitals to form chemical bonds. Various molecular geometries can be generated by different hybridizations. The hybridization concept accounts for the exception to the octet rule and also explains the formation of double and triple bonds. (Sections 10.4, 10.5)

Molecular Orbital Theory Molecular orbital theory describes bonding in terms of the combination of atomic orbitals to form orbitals that are associated with the molecule as a whole. Molecules are stable if the number of electrons in bonding molecular orbitals is greater than that in antibonding molecular orbitals. We write electron configurations for molecular orbitals as we do for atomic orbitals, using the Pauli exclusion principle and Hund's rule. (Sections 10.6, 10.7, 10.8)

Art Program and Design

For this edition, the art program was further revised to impart a more modern look and enhance visibility as well as provide better ADA compliance for vision-impaired students. Clear graphics are a vital component of the student learning process and as such, all molecular models, graphs, periodic tables, and other figures have been updated with a new look and color scheme. In some instances, illustrations have been replaced with scientifically accurate photographs for enhanced chemical context. Many chapter-opening photographs have been updated for new insights into various chemical topics and applications.

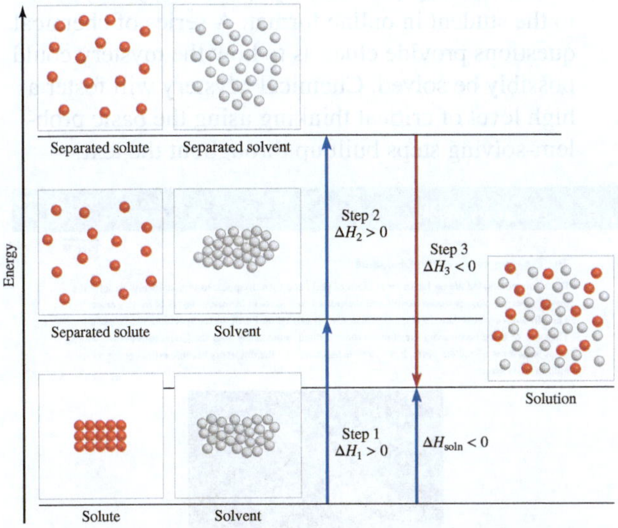

Problem Solving

The development of problem-solving skills has always been a major objective of this text. The two major categories of learning are shown next.

Worked examples follow a proven step-by-step strategy and solution.

- **Problem statement** is the reporting of the facts needed to solve the problem based on the question posed.
- **Strategy** is a carefully thought-out plan or method to serve as an important function of learning.
- **Solution** is the process of solving a problem given in a stepwise manner.
- **Check** enables the student to compare and verify with the source information to make sure the answer is reasonable.
- **Practice Exercise** provides the opportunity to solve a similar problem to become proficient in this problem type. The margin note lists additional similar problems to work in the end-of-chapter problem section.

End-of-Chapter Problems are organized in various ways. Each section under a topic heading begins with Concept Review questions followed by Problems. The Additional Problems section provides more problems not organized by section, followed by the problem type Interpreting, Modeling, & Estimating.

Many of the examples and end-of-chapter problems present extra tidbits of knowledge and enable the student to solve a chemical problem that a chemist would solve. The examples and problems show students the real world of chemistry and applications to everyday life situations.

Visualization

Graphs and Flow Charts are important in science. In *Chemistry,* flow charts show the thought process of a concept and graphs present data to comprehend the concept. A significant number of Problems and Review of Concepts & Facts, including many new to this edition, include graphical data.

Molecular art appears in various formats to serve different needs. Molecular models help to visualize the three-dimensional arrangement of atoms in a molecule. Electrostatic potential maps illustrate the electron density distribution in molecules. Finally, there is the macroscopic to microscopic art helping students understand processes at the molecular level.

Photos are used to help students become familiar with chemicals and understand how chemical reactions appear in reality.

Figures of apparatus enable the student to visualize the practical arrangement in a chemistry laboratory.

Study Aids

Setting the Stage

Each chapter starts with the Chapter Outline and a brief opening discussion of the chapter.

Chapter Outline enables the student to see at a glance the big picture and focus on the main ideas of the chapter.

Tools to Use for Studying

Useful aids for studying are plentiful in *Chemistry* and should be used constantly to reinforce the comprehension of chemical concepts.

Worked Examples along with the accompanying Practice Exercises are very important tools for learning and mastering chemistry. The problem-solving steps guide the student through the critical thinking necessary for succeeding in chemistry. Similar problems in the end-of-chapter problems section are listed at the end of the examples, enabling the student to apply new skill to other problems of the same type. Answers to the Practice Exercises are listed at the end of the chapter problems in the print text, or immediately following the Practice Exercises in the eBook.

Key Equations are highlighted within the chapter, drawing the student's eye to material that needs to be understood and retained. The key equations are also presented in the chapter summary materials for easy access in review and study.

Chapter Summary provides a quick review of concepts presented and discussed in detail within the chapter.

Key Words are a list of all important terms to help the student understand the language of chemistry.

Testing Your Knowledge

Review of Concepts & Facts lets students pause and check to see if they understand the concept presented and discussed in the section. Answers to the Review of Concepts can be found in the Student Solution Manual.

End-of-Chapter Problems enable the student to practice critical thinking and problem-solving skills. The problems are broken into various types:

- By chapter section. Starting with Review Questions to test basic conceptual understanding, followed by Problems to test the student's skill in solving problems for that particular section of the chapter.
- Additional Problems use the knowledge gained from the various sections and/or previous chapters to solve the problem.
- Interpreting, Modeling, & Estimating problems teach students the art of formulating models and estimating ballpark answers based on appropriate assumptions.

Real-Life Relevance

Interesting examples of how chemistry applies to life are used throughout the text. Analogies are used where appropriate to help foster understanding of abstract chemical concepts.

End-of-Chapter Problems pose many relevant questions for the student to solve. Examples include: Why do swimming coaches sometimes place a drop of alcohol in a swimmer's ear to draw out water? How does one estimate the pressure in a carbonated soft drink bottle before removing the cap?

Chemistry in Action boxes appear in every chapter on a variety of topics, each with its own story of how chemistry can affect a part of life. The student can learn about the science of scuba diving and nuclear medicine, among many other interesting cases.

CHEMISTRY *in Action*

The Search for the Higgs Boson

In this chapter, we identify mass as a fundamental property of matter, but have you ever wondered: Why does matter even have mass? It might seem obvious that "everything" has mass, but is that a requirement of nature? We will see later in our studies that light is composed of particles that do not have mass when at rest, and physics tells us under different circumstances the universe might not contain *anything* with mass. Yet we know that *our* universe is made up of an uncountable number of particles with mass, and these building blocks are necessary to form the elements that make up the people to ask such questions. The search for the answer to this question illustrates nicely the process we call the scientific method.

Current theoretical models tell us that everything in the universe is based on two types of elementary particles: bosons and fermions. We can distinguish the roles of these particles by considering the building blocks of matter to be constructed from fermions, while bosons are particles responsible for the force that holds the fermions together. In 1964, three different research teams independently proposed mechanisms in which a field of energy permeates the universe, and the interaction of matter with this field is due to a specific boson associated with the field. The greater the number of these bosons, the greater the interaction will be with the field. This interaction is the property we call mass, and the field and the associated boson came to be named for Peter Higgs, one of the original physicists to propose this mechanism.

This theory ignited a frantic search for the "Higgs boson" that became one of the most heralded quests in modern science. The Large Hadron Collider at CERN in Geneva, Switzerland (described in Chapter 19), was constructed to carry out experiments designed to find evidence for the Higgs boson. In these experiments, protons are accelerated to nearly the speed of light in opposite directions in a circular 17-mile tunnel and then allowed to collide, generating even more fundamental particles at very high energies. The data are examined for evidence of an excess of particles at an energy consistent with theoretical predictions for the Higgs boson. The ongoing process of theory suggesting experiments that give results used to evaluate and ultimately refine the theory, and so on, is the essence of the scientific method.

Illustration of the data obtained from decay of the Higgs boson into other particles following an 8-TeV collision in the Large Hadron Collider at CERN.
Thomas McCauley/Lucas Taylor, CERN/Science Source

On July 4, 2012, scientists at CERN announced the discovery of the Higgs boson. It takes about 1 trillion proton–proton collisions to produce one Higgs boson event, so it requires a tremendous amount of data obtained from two independent sets of experiments to confirm the findings. In science, the quest for answers is never completely done. Our understanding can always be improved or refined, and sometimes entire tenets of accepted science are replaced by another theory that does a better job explaining the observations. For example, scientists are not sure if the Higgs boson is the only particle that confers mass to matter, or if it is only one of several such bosons predicted by other theories.

But over the long run, the scientific method has proven to be our best way of understanding the physical world. It took 50 years for experimental science to validate the existence of the Higgs boson. This discovery was greeted with great fanfare and recognized the following year with a 2013 Nobel Prize in Physics for Peter Higgs and François Englert, another one of the six original scientists who first proposed the existence of a universal field that gives particles their mass. It is impossible to imagine where science will take our understanding of the universe in the next 50 years, but we can be fairly certain that many of the theories and experiments driving this scientific discovery will be very different than the ones we use today.

Chemical Mystery (eBook only) poses a mystery case to the student in online format. A series of chemical questions provide clues as to how the mystery could possibly be solved. Chemical Mystery will foster a high level of critical thinking using the basic problem-solving steps built up throughout the text.

CHEMICAL *MYSTERY*

Page 38

The Disappearance of the Dinosaurs

Dinosaurs dominated life on Earth for millions of years and then disappeared very suddenly. To solve the mystery, paleontologists studied fossils and skeletons found in rocks in various layers of Earth's crust. Their findings enabled them to map out which species existed on Earth during specific geologic periods. They also revealed no dinosaur skeletons in rocks formed immediately after the Cretaceous period, which dates back some 65 million years. It is therefore assumed that the dinosaurs became extinct about 65 million years ago.

Chuck Eckert/Alamy Stock Photo

Page 39

Among the many hypotheses put forward to account for their disappearance were disruptions of the food chain and a dramatic change in climate caused by violent volcanic eruptions. However, there was no convincing evidence for any one hypothesis until 1977. It was then that a group of paleontologists working in Italy obtained some very puzzling data at a site near Gubbio. The chemical analysis of a layer of clay deposited above sediments formed during the Cretaceous period (and therefore a layer that records events occurring *after* the Cretaceous period) showed a surprisingly high content of the element iridium (Ir). Iridium is very rare in Earth's crust but is comparatively abundant in asteroids.

Instructor and Student Resources

ALEKS®

ALEKS (Assessment and LEarning in Knowledge Spaces) is a web-based system for individualized assessment and learning available 24/7 over the Internet. ALEKS uses artificial intelligence to accurately determine students' knowledge and then guides them to the material that they are most ready to learn. ALEKS offers immediate feedback and access to ALEKSPedia—an interactive text that contains concise entries on chemistry topics. ALEKS is also a full-featured course management system with rich reporting features that allow instructors to monitor individual and class performance, set student goals, assign/grade online quizzes, and more. ALEKS allows instructors to spend more time on concepts while ALEKS teaches students practical problem-solving skills. And with ALEKS 360, your student also has access to this text's eBook. Learn more at **www.aleks.com/highered/science**

Instructors have access to the following instructor resources:

- **Art** Full-color digital files of all illustrations, photos, and tables in the book can be readily incorporated into lecture presentations, exams, or custom-made classroom materials. In addition, all files have been inserted into PowerPoint slides for ease of lecture preparation.
- **Animations and Videos** Numerous full-color animations and videos illustrating important concepts and processes are also provided. Harness the visual impact of concepts in motion by importing these files into classroom presentations or online course materials.
- **Accessible PowerPoint Lecture Outlines** Ready-made presentations that combine art and lecture notes are provided for each chapter of the text.
- **Computerized Test Bank** Over 3,000 test questions that accompany *Chemistry* are available for creating exams or quizzes.
- **Instructor's Solutions Manual** This supplement contains complete, worked-out solutions for *all* the end-of-chapter problems in the text.

ALEKS® Virtual Labs

McGraw Hill Virtual Labs is a must-see, outcomes-based lab simulation. It assesses a student's knowledge and adaptively corrects deficiencies, allowing the student to learn faster and retain more knowledge with greater success. First, a student's knowledge is adaptively leveled on core learning outcomes: Questioning reveals knowledge deficiencies that are corrected by the delivery of content that is conditional on a student's response. Then, a simulated lab experience requires the student to think and act like a scientist: recording, interpreting, and analyzing data using simulated equipment found in labs and clinics. The student is allowed to make mistakes—a powerful part of the learning experience! A virtual coach provides subtle hints when needed, asks questions about the student's choices, and allows the student to reflect on and correct those mistakes. Whether your need is to overcome the logistical challenges of a traditional lab, provide better lab prep, improve student performance, or make your online experience one that rivals the real world, Learn-Smart Labs accomplishes it all.

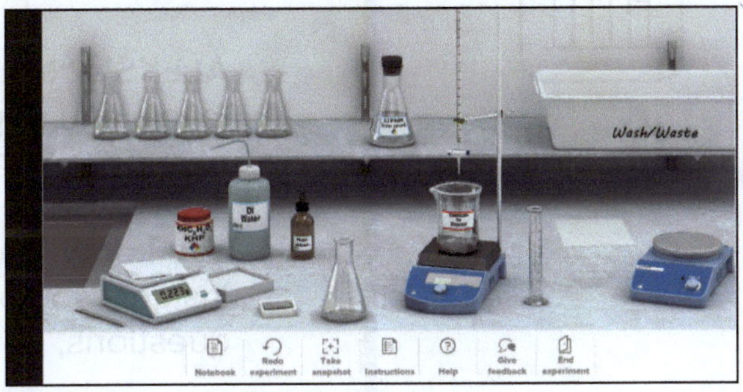

McGraw Hill

McGraw Hill Create™

With **McGraw Hill Create,** you can easily rearrange chapters, combine material from other content sources, and quickly upload content you have written, like your course syllabus or teaching notes. Find the content you need in Create by searching through thousands of leading McGraw Hill textbooks. Arrange your book to fit your teaching style. Create even allows you to personalize your book's appearance by selecting the cover and adding your name, school, and course information. Order a Create book and you'll receive a complimentary print review copy in three to five business days or a complimentary electronic review copy (eComp) via email in minutes. Go to www.mcgrawhillcreate.com today and register to experience how McGraw Hill Create empowers you to teach *your* students *your* way. www.mcgrawhillcreate.com

Mc Graw Hill **ALEKS**®

Break down barriers and build student knowled

Students start your course with varying levels of preparedness. Some will get it quickly. Some won't. ALEKS is a course assistant that helps you meet each student where they are and provide the necessary building blocks to get them where they need to go. You determine the assignments and t content, and ALEKS will deliver customized practice until your students truly get it.

Experience the ALEKS Difference

Easily Identify Knowledge Gaps

Gain greater visibility into student performance so you know immediately if your lessons clicked.

- **ALEKS's "Initial Knowledge Check"** helps accurately evaluate stude levels and gaps on day one, so you know precisely where students a at and where they need to go when they start your course.
- **You know when students are at risk of falling behind** through ALEK Insights so you can remediate—be it through prep modules, practice questions, or written explanations of video tutorials.
- **Students always know where they are,** know how they are doing, a can track their own progress easily.

Gain More Flexibility and Engagement

Teach your course your way, with best-in-class content and tools to immerse students and keep them on track.

- **ALEKS gives you flexibility** to assign homework, share a vast library curated content including videos, review progress, and provide stud support, anytime anywhere.
- **You save time** otherwise spent performing tedious tasks while havin more control over and impact on your students' learning process.
- **Students gain a deeper level of understanding** through interactive and hands-on assignments that go beyond multiple-choice question

Student Solutions Manual

Students will find answers to the Practice Exercises and detailed solutions for selected problems from the text in the Student Solutions manual. In addition, there are problem-solving strategies and tutorial solutions that surround each chapter's most important topics and problem types.

Acknowledgments

We would like to thank the following reviewers of the fourteenth edition:

Daniel Ashley, *Spelman College*
Mufeed Basti, *NC A&T State University*
Ken Brown, *Winston-Salem State University*
Nicole Christofield, *Ball State University*
Robin De Roo, *Cypress College*
Becky Gee, *University of California, Berkeley*
Sujata Guha, *Tennessee State University*
Lisa Hibbard, *Spelman College*
Harvey Hou, *Alabama State University*
Sapna Jain, *Alabama State University*
Steven LaiHing, *Oakwood University*
Zhihai Li, *Ball State University*
Abdul K. Mohammed, *North Carolina Central University*
Bhanu Priya Viraka Nellore, *Mississippi Valley State University*
Joshua M. Patterson, *Delaware State University*
Seong Seo, *Albany State University*
Thiagarajan Soundappan, *Navajo Technical University*
Jesse Tye, *Ball State University*

We have benefited much from discussions with our colleagues at Williams College and the College of Charleston, and correspondence with instructors here and abroad.

It is a pleasure to acknowledge the support given us by the following members of McGraw Hill: Executive Portfolio Manager Ian Townsend, Senior Product Developer Mary Hurley, Senior Director of Digital Content Dr. Shirley Hino, Senior Marketing Manager Cassie Cloutier, Lead Core Content Project Manager Laura Bies and Assessment Content Project Manager Ron Nelms, Lead Designer David Hash, and Senior Content Licensing Specialist Lori Hancock.

A Note to the Student

General chemistry is commonly perceived to be more difficult than most other subjects. There is some justification for this perception. For one thing, chemistry has a very specialized vocabulary. At first, studying chemistry is like learning a new language. Furthermore, some of the concepts are abstract. Nevertheless, with diligence you can complete this course successfully, and you might even enjoy it. Here are some suggestions to help you form good study habits and master the material in this text.

- Attend classes regularly and take careful notes.
- If possible, always review the topics discussed in class the same day they are covered in class. Use this book to supplement your notes.
- Think critically. Ask yourself if you really understand the meaning of a term or the use of an equation. A good way to test your understanding is to explain a concept to a classmate or some other person.
- Do not hesitate to ask your instructor or your teaching assistant for help.

The fifteenth edition tools for *Chemistry* are designed to enable you to do well in your general chemistry course. The following guide explains how to take full advantage of the text, technology, and other tools.

- Before delving into the chapter, read the chapter *outline* and the chapter *introduction* to get a sense of the important topics. Use the outline to organize your note taking in class.
- At the end of each chapter you will find a summary of facts and concepts, the key equations, and a list of key words, all of which will help you review for exams.
- Definitions of the key words can be studied in context on the pages cited in the end-of-chapter list or in the glossary at the back of the book.
- Careful study of the worked-out examples in the body of each chapter will improve your ability to analyze problems and correctly carry out the calculations needed to solve them. Also take the time to work through the practice exercise that follows each example to be sure you understand how to solve the type of problem illustrated in the example. The answers to the practice exercises appear at the end of the chapter in your print text, following the questions and problems, or immediately following the practice exercises in your eBook. For additional practice, you can turn to similar problems referred to in the margin next to the example.
- The questions and problems at the end of the chapter are organized by section.
- The appendices show a list of important figures and tables with page references. This index makes it convenient to quickly look up information when you are solving problems or studying related subjects in different chapters.

If you follow these suggestions and stay up-to-date with your assignments, you should find that chemistry is challenging, but less difficult and much more interesting than you expected.

—Jason Overby

CHAPTER

1

Measurement and the Properties of Matter

A scanning tunneling microscope probes individual small molecules when they adsorb on graphene, a single-atom thin sheet of carbon atoms.
Martin McCarthy/theasis/E+/Getty Images

CHAPTER OUTLINE

Chemistry is an active, evolving science that has vital importance to our world, in both the realm of nature and the realm of society. Its roots are ancient, but as we will see, chemistry is every bit a modern science.

We will begin our study of chemistry at the macroscopic level, where we can see and measure the materials of which our world is made. In this chapter, we will discuss the scientific method, which provides the framework for research not only in chemistry but in all other sciences as well. Next, we will discover how scientists define and characterize matter. Then we will spend some time learning how to handle numerical results of chemical measurements and solve numerical problems. In Chapter 2, we will begin to explore the microscopic world of atoms and molecules.

1.1 Chemistry: The Study of a Science for the Twenty-First Century

Chemistry is *the study of matter and the changes it undergoes.* Chemistry is often called the central science because a basic knowledge of chemistry is essential for students of biology, physics, geology, ecology, and many other subjects. Indeed, it is central to our way of life; without it, we would be living shorter lives without automobiles, electricity, computers, and many other everyday conveniences.

Although chemistry is an ancient science, its modern foundation was laid in the nineteenth century, when intellectual and technological advances enabled scientists to break down substances into ever-smaller components and consequently to explain many of their physical and chemical characteristics. The rapid development of increasingly sophisticated technology throughout the twentieth century has given us even greater means to study things that cannot be seen with the naked eye. Using computers and special microscopes, for example, chemists can analyze the structure of atoms and molecules—the fundamental units on which the study of chemistry is based—and design new substances with specific properties, such as drugs and environmentally friendly consumer products.

化学

The Chinese characters for chemistry mean "the study of change."

It is fitting to ask what part the central science will have in the twenty-first century. Almost certainly, chemistry will continue to play a pivotal role in all areas of science and technology. Before plunging into the study of matter and its transformation, let us consider some of the frontiers that chemists are currently exploring (Figure 1.1). Whatever your reasons for taking general chemistry, a good knowledge of the subject will better enable you to appreciate its impact on society and on you as an individual.

Compared with other subjects, chemistry is commonly believed to be more difficult, at least at the introductory level. There is some justification for this perception; for one thing, chemistry has a very specialized vocabulary. However, even if this is your first course in chemistry, you already have more familiarity with the subject than you may realize. In everyday conversations we hear words that have a chemical connection, although they may not be used in the scientifically correct sense. Examples are "electronic," "quantum leap," "equilibrium," "catalyst," "chain reaction," and "critical mass." Moreover, if you cook, then you are a practicing chemist! From experience gained in the kitchen, you know that oil and water do not mix and that boiling water left on the stove will evaporate. You apply chemical and physical principles when you use baking soda to leaven bread, choose a pressure cooker to shorten the time it takes to prepare soup, add meat tenderizer to a pot roast, squeeze lemon juice over sliced pears to prevent them from turning brown or over fish to minimize its odor, and add vinegar to the water in which you are going to poach eggs. Every day we observe such changes without thinking about their chemical nature. The purpose of this course is to make you think like a chemist, to look at the *macroscopic world*—the things we can see, touch, and measure directly—and visualize the particles and events of the *microscopic world* that we cannot experience without modern technology and our imaginations.

At first, some students find it confusing that their chemistry instructor and textbook seem to be continually shifting back and forth between the macroscopic and microscopic worlds. Just keep in mind that the data for chemical investigations most often come from observations of large-scale phenomena, but the explanations frequently lie in the

Figure 1.1 *(a) Ethanol for fuel use is produced from corn in industrial facilities. (b) Production of photovoltaic cells, used to convert light into electrical current. (c) Computer memory storage chips take advantage of the ability to repeatedly write and erase data on the nanoscale. (d) The automation of protein sequencing provides valuable insight into the structure of important biomolecules.*
(a): Julie Kendall/E+/Getty Images; (b): MS Mikel/Shutterstock; (c): Nor Gal/Shutterstock; (d): Nicolas/E+/Getty Images

unseen and partially imagined microscopic world of atoms and molecules. In other words, chemists often *see* one thing (in the macroscopic world) and *think* another (in the microscopic world). Looking at the rusted nails in Figure 1.2, for example, a chemist might think about the basic properties of individual atoms of iron and how these units interact with other atoms and molecules to produce the observed change.

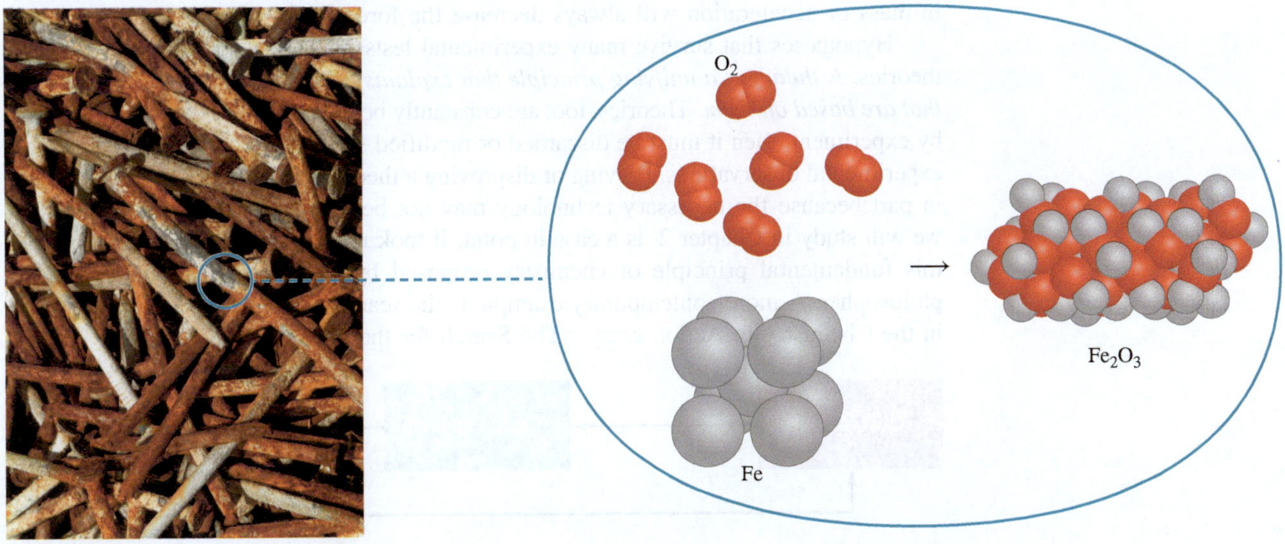

Figure 1.2 *A simplified molecular view of rust (Fe_2O_3) formation from iron (Fe) atoms and oxygen molecules (O_2). In reality, the process requires water, and rust also contains water molecules.*
B.A.E. Inc./Alamy Stock Photo

1.2 The Scientific Method

Learning Objectives

- Summarize the steps of the scientific method.
- Distinguish between qualitative and quantitative data.

All sciences, including the social sciences, employ variations of what is called the ***scientific method,*** *a systematic approach to research.* For example, a psychologist who wants to know how noise affects people's ability to learn chemistry and a chemist interested in measuring the heat given off when hydrogen gas burns in air would follow roughly the same procedure in carrying out their investigations. The first step is to carefully define the problem. The next step includes performing experiments, making careful observations, and recording information, or *data,* about the system—the part of the universe that is under investigation. (In the examples just discussed, the systems are the group of people the psychologist will study and a mixture of hydrogen and air.)

The data obtained in a research study may be both ***qualitative,*** *consisting of general observations about the system,* and ***quantitative,*** *comprising numbers obtained by various measurements of the system.* Chemists generally use standardized symbols and equations in recording their measurements and observations. This form of representation not only simplifies the process of keeping records but also provides a common basis for communication with other chemists.

When the experiments have been completed and the data have been recorded, the next step in the scientific method is interpretation, meaning that the scientist attempts to explain the observed phenomenon. Based on the data that were gathered, the researcher formulates a ***hypothesis,*** *a tentative explanation for a set of observations.* Further experiments are devised to test the validity of the hypothesis in as many ways as possible, and the process begins anew. Figure 1.3 summarizes the main steps of the research process.

After large amounts of data have been collected, it is often desirable to summarize the information in a concise way, as a law. In science, a ***law*** is *a concise verbal or mathematical statement of a relationship between phenomena that is always the same under the same conditions.* For example, Sir Isaac Newton's second law of motion, which you may remember from high school science, says that force equals mass times acceleration ($F = ma$). What this law means is that an increase in the mass or in the acceleration of an object will always increase its force proportionally, and a decrease in mass or acceleration will always decrease the force.

Hypotheses that survive many experimental tests of their validity may evolve into theories. A ***theory*** is *a unifying principle that explains a body of facts and/or those laws that are based on them.* Theories, too, are constantly being tested. If a theory is disproved by experiment, then it must be discarded or modified so that it becomes consistent with experimental observations. Proving or disproving a theory can take years, even centuries, in part because the necessary technology may not be available. Atomic theory, which we will study in Chapter 2, is a case in point. It took more than 2000 years to work out this fundamental principle of chemistry proposed by Democritus, an ancient Greek philosopher. A more contemporary example is the search for the Higgs boson discussed in the Chemistry in Action essay, "The Search for the Higgs Boson."

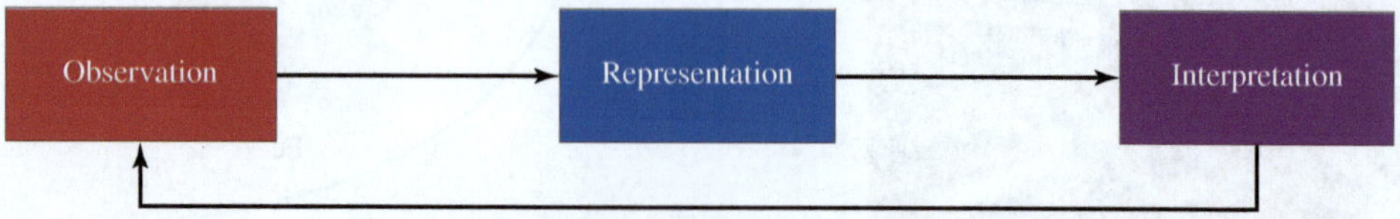

Figure 1.3 *The three levels of studying chemistry and their relationships. Observation deals with events in the macroscopic world; atoms and molecules constitute the microscopic world. Representation is a scientific shorthand for describing an experiment in symbols and chemical equations. Chemists use their knowledge of atoms and molecules to explain an observed phenomenon.*

CHEMISTRY *in Action*

The Search for the Higgs Boson

In this chapter, we identify mass as a fundamental property of matter, but have you ever wondered: Why does matter even have mass? It might seem obvious that "everything" has mass, but is that a requirement of nature? We will see later in our studies that light is composed of particles that do not have mass when at rest, and physics tells us under different circumstances the universe might not contain *anything* with mass. Yet we know that *our* universe is made up of an uncountable number of particles with mass, and these building blocks are necessary to form the elements that make up the people to ask such questions. The search for the answer to this question illustrates nicely the process we call the scientific method.

Current theoretical models tell us that everything in the universe is based on two types of elementary particles: bosons and fermions. We can distinguish the roles of these particles by considering the building blocks of matter to be constructed from fermions, while bosons are particles responsible for the force that holds the fermions together. In 1964, three different research teams independently proposed mechanisms in which a field of energy permeates the universe, and the interaction of matter with this field is due to a specific boson associated with the field. The greater the number of these bosons, the greater the interaction will be with the field. This interaction is the property we call mass, and the field and the associated boson came to be named for Peter Higgs, one of the original physicists to propose this mechanism.

This theory ignited a frantic search for the "Higgs boson" that became one of the most heralded quests in modern science. The Large Hadron Collider at CERN in Geneva, Switzerland (described in Chapter 19), was constructed to carry out experiments designed to find evidence for the Higgs boson. In these experiments, protons are accelerated to nearly the speed of light in opposite directions in a circular 17-mile tunnel and then allowed to collide, generating even more fundamental particles at very high energies. The data are examined for evidence of an excess of particles at an energy consistent with theoretical predictions for the Higgs boson. The ongoing process of theory suggesting experiments that give results used to evaluate and ultimately refine the theory, and so on, is the essence of the scientific method.

Illustration of the data obtained from decay of the Higgs boson into other particles following an 8-TeV collision in the Large Hadron Collider at CERN.
Thomas McCauley/Lucas Taylor, CERN/Science Source

On July 4, 2012, scientists at CERN announced the discovery of the Higgs boson. It takes about 1 trillion proton–proton collisions to produce one Higgs boson event, so it requires a tremendous amount of data obtained from two independent sets of experiments to confirm the findings. In science, the quest for answers is never completely done. Our understanding can always be improved or refined, and sometimes entire tenets of accepted science are replaced by another theory that does a better job explaining the observations. For example, scientists are not sure if the Higgs boson is the only particle that confers mass to matter, or if it is only one of several such bosons predicted by other theories.

But over the long run, the scientific method has proven to be our best way of understanding the physical world. It took 50 years for experimental science to validate the existence of the Higgs boson. This discovery was greeted with great fanfare and recognized the following year with a 2013 Nobel Prize in Physics for Peter Higgs and François Englert, another one of the six original scientists who first proposed the existence of a universal field that gives particles their mass. It is impossible to imagine where science will take our understanding of the universe in the next 50 years, but we can be fairly certain that many of the theories and experiments driving this scientific discovery will be very different than the ones we use today.

Scientific progress is seldom, if ever, made in a rigid, step-by-step fashion. Sometimes a law precedes a theory; sometimes it is the other way around. Two scientists may start working on a project with exactly the same objective, but will end up taking drastically different approaches. Scientists are, after all, human beings, and their modes of thinking and working are very much influenced by their background, training, and personalities.

The development of science has been irregular and sometimes even illogical. Great discoveries are usually the result of the cumulative contributions and experience of many workers, even though the credit for formulating a theory or a law is usually given to only one individual. There is, of course, an element of luck involved in scientific discoveries, but it has been said that "chance favors the prepared mind." It takes an alert and well-trained person to recognize the significance of an accidental discovery and to take full advantage of it. More often than not, the public learns only of spectacular scientific breakthroughs. For every success story, however, there are hundreds of cases in which scientists have spent years working on projects that ultimately led to a dead-end, and in which positive achievements came only after many wrong turns and at such a slow pace that they went unheralded. Yet even the dead ends contribute something to the continually growing body of knowledge about the physical universe. It is the love of the search that keeps many scientists in the laboratory.

Summary of Concepts & Facts

- The study of chemistry involves three basic steps: observation, representation, and interpretation. Observation refers to measurements in the macroscopic world; representation involves the use of shorthand notation symbols and equations for communication; interpretations are based on atoms and molecules, which belong to the microscopic world.
- The scientific method is a systematic approach to research that begins with the gathering of information through observation and measurements. In the process, hypotheses, laws, and theories are devised and tested.

Review of Concepts & Facts

1.2.1 Which of the following statements is true?
(a) A hypothesis always leads to the formulation of a law.
(b) The scientific method is a rigid sequence of steps in solving problems.
(c) A law summarizes a series of experimental observations; a theory provides an explanation for the observations.

1.2.2 A student collects the following data for a sample of an unknown liquid. Which of these data are qualitative measurements and which are quantitative measurements?
(a) The sample has a volume of 15.4 mL.
(b) The sample is a light yellow liquid.
(c) The sample feels oily.
(d) The sample has a mass of 13.2 g.

1.3 Measurement

Learning Objectives

- Use SI unit prefixes.
- Employ measurements such as mass, volume, and density in calculations.
- Perform conversions among the different temperature scales.

The study of chemistry depends heavily on measurement. For instance, chemists use measurements to compare the properties of different substances and to assess changes resulting from an experiment. The measurements chemists make are often used in calculations to obtain other related quantities. Different instruments enable us to measure a substance's properties: The meterstick measures length or scale; the buret, the pipet, the graduated cylinder, and the volumetric flask measure volume (Figure 1.4); the balance measures mass; the thermometer measures temperature. These instruments provide measurements of ***macroscopic properties,*** which *can be determined directly.* ***Microscopic properties,*** *on the atomic or molecular scale, must be determined by an indirect method,* as we will see in Chapter 2.

A measured quantity is usually written as a number with an appropriate unit. To say that the distance between New York and San Francisco by car along a certain route is 5166 is meaningless. We must specify that the distance is 5166 kilometers. The same is true in chemistry; units are essential to stating measurements correctly.

SI Units

For many years, scientists recorded measurements in *metric units,* which are related decimally—that is, by powers of 10. In 1960, however, the General Conference of Weights and Measures, the international authority on units, proposed a revised metric system called the ***International System of Units*** (abbreviated ***SI,*** from the French *Système Internationale d'Unites*). Table 1.1 shows the seven SI base units. All other units of measurement can be derived from these base units. Like metric units, SI units are modified in decimal fashion by a series of prefixes, as shown in Table 1.2. We will use both metric and SI units in this book.

Measurements we will utilize frequently in our study of chemistry include time, mass, volume, density, and temperature.

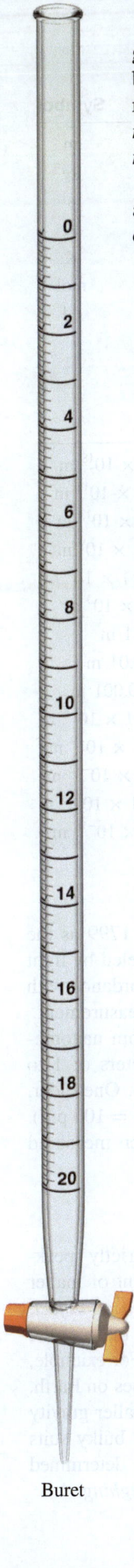

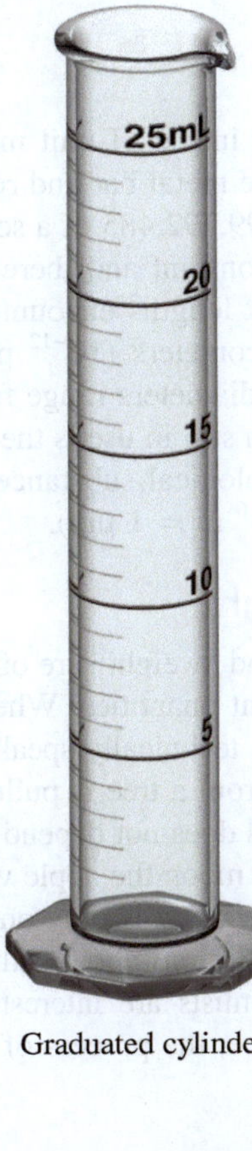

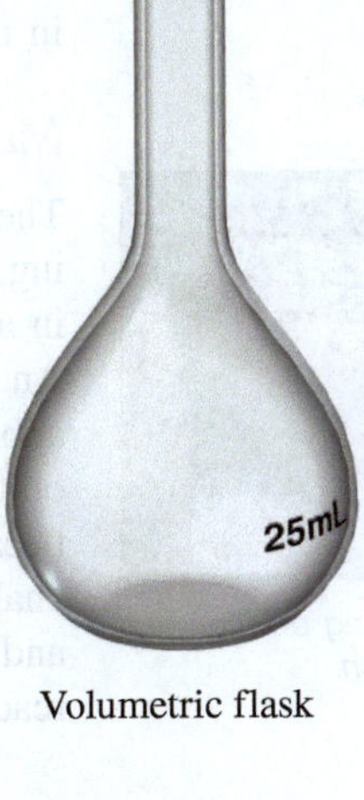

Figure 1.4 *Some common measuring devices found in a chemistry laboratory. These devices are not drawn to scale relative to one another. We will discuss the uses of these measuring devices in Chapter 4.*

Table 1.1 SI Base Units

Base Quantity	Name of Unit	Symbol
Length	meter	m
Mass	kilogram	kg
Time	second	s
Electrical current	ampere	A
Temperature	kelvin	K
Amount of substance	mole	mol
Luminous intensity	candela	cd

Table 1.2 Prefixes Used with SI Units

Prefix	Symbol	Meaning	Example
exa-	E	1,000,000,000,000,000,000, or 10^{18}	1 exameter (*E*m) = 1×10^{18} m
peta-	P	1,000,000,000,000,000, or 10^{15}	1 petameter (Pm) = 1×10^{15} m
tera-	T	1,000,000,000,000, or 10^{12}	1 terameter (Tm) = 1×10^{12} m
giga-	G	1,000,000,000, or 10^{9}	1 gigameter (Gm) = 1×10^{9} m
mega-	M	1,000,000, or 10^{6}	1 megameter (Mm) = 1×10^{6} m
kilo-	k	1,000, or 10^{3}	1 kilometer (km) = 1×10^{3} m
deci-	d	1/10, or 10^{-1}	1 decimeter (dm) = 0.1 m
centi-	c	1/100, or 10^{-2}	1 centimeter (cm) = 0.01 m
milli-	m	1/1,000, or 10^{-3}	1 millimeter (mm) = 0.001 m
micro-	μ	1/1,000,000, or 10^{-6}	1 micrometer (μm) = 1×10^{-6} m
nano-	n	1/1,000,000,000, or 10^{-9}	1 nanometer (nm) = 1×10^{-9} m
pico-	p	1/1,000,000,000,000, or 10^{-12}	1 picometer (pm) = 1×10^{-12} m
femto-	f	1/1,000,000,000,000,000, or 10^{-15}	1 femtometer (fm) = 1×10^{-15} m
atto-	a	1/1,000,000,000,000,000,000, or 10^{-18}	1 attometer (am) = 1×10^{-18} m

Length

Length is measured in the SI unit meter (m). The meter was defined in 1799 as the length of a prototype metal bar and redefined in 1983 as the distance traveled by light in a vacuum in 1/299,792,485 of a second. This definition is fixed in accordance with the speed of light constant and thereby reduces the uncertainty in the measurement.

The majority of lengths encountered in chemistry typically range from nanometers (10^{-9} m) to picometers (10^{-12} pm). Proteins commonly have diameters of 1 to 3 nm while atomic diameters range from 50 to 250 pm (0.05 to 0.25 nm). One older, non-SI unit of length still in use is the angstrom (1 Å = 10^{-10} nm = 0.1 nm = 100 pm). For comparison, biological substances such as cells and bacteria are often measured in micrometers (10^{-6} m = 1 μm).

Mass and Weight

An astronaut jumping on the surface of the moon.
NASA

The terms "mass" and "weight" are often used interchangeably, although, strictly speaking, they are different quantities. Whereas mass is a measure of the amount of matter in an object, ***weight,*** technically speaking, is *the force that gravity exerts on an object.* An apple that falls from a tree is pulled downward by Earth's gravity. The mass of the apple is constant and does not depend on its location, but its weight does. For example, on the surface of the moon the apple would weigh only one-sixth what it does on Earth, because the moon's gravity is only one-sixth that of Earth. The moon's smaller gravity enabled astronauts to jump about rather freely on its surface despite their bulky suits and equipment. Chemists are interested primarily in mass, which can be determined readily with a balance; the process of measuring mass, oddly, is called *weighing.*

The SI unit of mass is the *kilogram* (kg) but the smaller gram (g) is often more convenient.

$$1 \text{ kg} = 1000 \text{ g} = 1 \times 10^3 \text{ g}$$

Previously, the kilogram was defined in terms of a particular object (Figure 1.5), a cast cylinder of platinum and iridium. As of 2019, the kilogram has been redefined as a value related to a universal physical constant (the Planck constant; see Section 7.2).

Figure 1.5 *The prototype kilogram is made of a platinum-iridium alloy. It is kept in a vault at the International Bureau of Weights and Measures in Sèvres, France. In 2007 it was discovered that the alloy has mysteriously lost about 50 μg!*

Jacques Brinon/AP Images

Volume

Volume is an example of a measured quantity with derived units. The SI-derived unit for volume is the *cubic meter* (m^3). Generally, however, chemists work with much smaller volumes, such as the cubic centimeter (cm^3) and the cubic decimeter (dm^3):

$$1 \text{ cm}^3 = (1 \times 10^{-2} \text{ m})^3 = 1 \times 10^{-6} \text{ m}^3$$
$$1 \text{ dm}^3 = (1 \times 10^{-1} \text{ m})^3 = 1 \times 10^{-3} \text{ m}^3$$

Another common (but non-SI) unit of volume is the liter (L). A ***liter*** is *the volume occupied by one cubic decimeter.* One liter of volume is equal to 1000 milliliters (mL) or 1000 cm^3:

$$\begin{aligned} 1 \text{ L} &= 1000 \text{ mL} \\ &= 1000 \text{ cm}^3 \\ &= 1 \text{ dm}^3 \end{aligned}$$

and one milliliter is equal to one cubic centimeter:

$$1 \text{ mL} = 1 \text{ cm}^3$$

Figure 1.6 compares the relative sizes of two volumes. Even though the liter is not an SI unit, volumes are usually expressed in liters and milliliters.

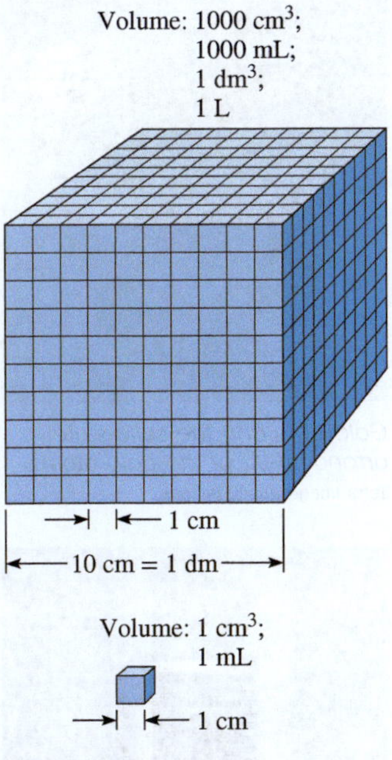

Figure 1.6 *Comparison of two volumes, 1 mL and 1000 mL.*

Density

Density is another measured quantity with derived units. The equation for density is

$$\text{density} = \frac{\text{mass}}{\text{volume}}$$

or

$$d = \frac{m}{V} \tag{1.1}$$

where *d, m,* and *V* denote density, mass, and volume, respectively. Because density is a property that does not depend on the quantity of mass present, for a given substance the ratio of mass to volume always remains the same. In other words, *V* increases as *m* does. Density usually decreases with temperature.

The SI-derived unit for density is the kilogram per cubic meter (kg/m^3). This unit is awkwardly large for most chemical applications. Therefore, grams per cubic centimeter (g/m^3) and its equivalent, grams per milliliter (g/mL), are more commonly used for solid and liquid densities. Because gas densities are often very low, we express them in units of grams per liter (g/L):

$$1 \text{ g/cm}^3 = 1 \text{ g/mL} = 1000 \text{ kg/m}^3$$
$$1 \text{ g/L} = 0.001 \text{ g/mL}$$

Table 1.3 Densities of Some Substances at 25°C

Substance	Density (g/cm^3)	Substance	Density (g/cm^3)
Hydrogen*	0.0000899	Aluminum	2.70
Graphene aerogel†	0.00016	Diamond	3.5
Air*	0.0013	Iron	7.9
Ethanol	0.79	Lead	11.3
Water	1.00	Mercury	13.6
Graphite	2.2	Gold	19.3
Table salt	2.2	Osmium‡	22.6

*Measured at 1 atmosphere and 273 K.
†Graphene aerogel is the least dense solid known.
‡Osmium (Os) is the densest element known.

Table 1.3 lists the densities of several substances.

Examples 1.1 and 1.2 show density calculations.

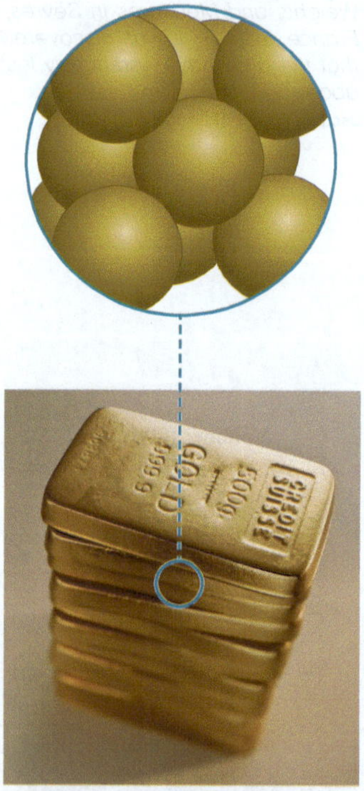

Gold bars and the solid-state arrangement of the gold atoms.
Tetra Images/Getty Images

Example 1.1

Gold is a precious metal that is chemically unreactive. It is used mainly in jewelry, dentistry, and electronic devices. A piece of gold ingot with a mass of 301 g has a volume of 15.6 cm^3. Calculate the density of gold.

Solution We are given the mass and volume and asked to calculate the density. Therefore, from Equation (1.1), we write

$$d = \frac{m}{V}$$

$$= \frac{301\ \text{g}}{15.6\ \text{cm}^3}$$

$$= 19.3\ \text{g/cm}^3$$

Practice Exercise A piece of platinum metal with a density of 21.5 g/cm^3 has a volume of 4.49 cm^3. What is its mass?

Similar problems: 1.9, 1.10.

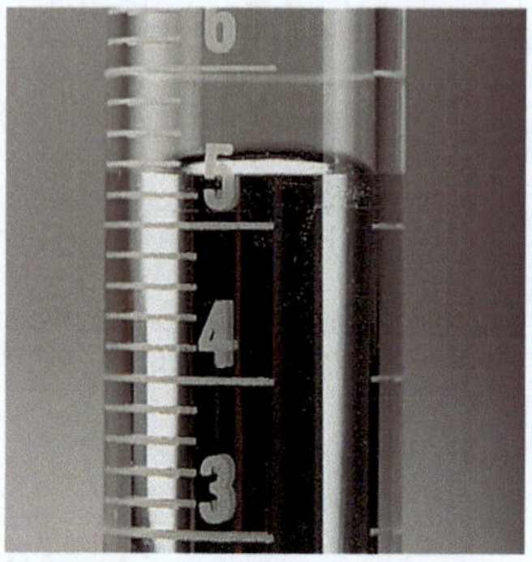

Mercury.
Stephen Frisch/McGraw Hill

Example 1.2

The density of mercury, the only metal that is a liquid at room temperature, is 13.6 g/mL. Calculate the mass of 5.50 mL of the liquid.

Solution We are given the density and volume of a liquid and asked to calculate the mass of the liquid. We rearrange Equation (1.1) to give

$$m = d \times V$$

$$= 13.6\ \frac{\text{g}}{\cancel{\text{mL}}} \times 5.50\ \cancel{\text{mL}}$$

$$= 74.8\ \text{g}$$

Practice Exercise The density of sulfuric acid in a certain car battery is 1.41 g/mL. Calculate the mass of 242 mL of the liquid.

Similar problems: 1.9, 1.10.

Temperature Scales

Three temperature scales are currently in use. Their units are °F (degrees Fahrenheit), °C (degrees Celsius), and K (kelvin). (Note that the Kelvin scale does not have the degree sign.) The Fahrenheit scale defines the normal freezing and boiling points of water to be exactly 32°F and 212°F, respectively. The Celsius scale divides the range between the freezing point (0°C) and boiling point (100°C) of water into 100 degrees.

As Table 1.1 shows, the ***kelvin*** is *the SI base unit of temperature:* It is the *absolute* temperature scale. By absolute we mean that the zero on the Kelvin scale, denoted by 0 K, is the lowest temperature that can be attained theoretically. This means that temperatures expressed in kelvins can never be negative. On the other hand, 0°F and 0°C are based on the behavior of an arbitrarily chosen substance, water. Figure 1.7 compares the three temperature scales.

The older definition of kelvin is 1/273.16 of the temperature of the triple point of water (exactly 0.01°C or 32.018°F) (see Section 11.9). Like the meter in 2019, the kelvin is now defined in terms of a universal physical constant (the Boltzmann constant, see Section 17.2).

The size of a degree on the Fahrenheit scale is only 100/180, or 5/9, of a degree on the Celsius scale. To convert degrees Fahrenheit to degrees Celsius, we write

$$?°\text{C} = (°\text{F} - 32°\text{F}) \times \frac{5°\text{C}}{9°\text{F}} \tag{1.2}$$

The following equation is used to convert degrees Celsius to degrees Fahrenheit:

$$?°\text{F} = \frac{9°\text{F}}{5°\text{C}} \times (°\text{C}) + 32°\text{F} \tag{1.3}$$

Both the Celsius and the Kelvin scales have units of equal magnitude; that is, one degree Celsius is equivalent to one kelvin. Experimental studies have shown that absolute zero on the Kelvin scale is equivalent to −273.15°C on the Celsius scale. Thus, we can use the following equation to convert degrees Celsius to kelvin:

$$?\ \text{K} = (°\text{C} + 273.15°\text{C})\frac{1\ \text{K}}{1°\text{C}} \tag{1.4}$$

We will frequently find it necessary to convert between degrees Celsius and degrees Fahrenheit and between degrees Celsius and kelvin. Example 1.3 illustrates these conversions.

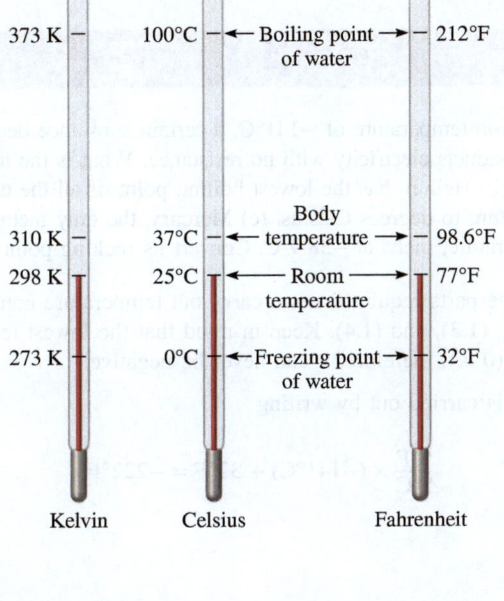

Figure 1.7 *Comparison of the three temperature scales: Celsius, Fahrenheit, and absolute (Kelvin) scales. Note that there are 100 divisions, or 100 degrees, between the freezing point and the boiling point of water on the Celsius scale, and there are 180 divisions, or 180 degrees, between the same two temperature limits on the Fahrenheit scale. The Celsius scale was formerly called the centigrade scale.*

CHEMISTRY *in Action*

The Importance of Units

In December 1998, NASA launched the $125 million Mars Climate Orbiter, intended as the red planet's first weather satellite. After a 416-million-mile journey, the spacecraft was supposed to go into Martian orbit on September 23, 1999. Instead, it entered the Martian atmosphere about 100 km (62 mi) lower than planned and was destroyed by heat. The mission controllers said the loss of the spacecraft was due to the failure to convert English measurement units into metric units in the navigation software.

Engineers at Lockheed Martin Corporation who built the spacecraft specified its thrust in pounds, which is an English unit. Scientists at NASA's Jet Propulsion Laboratory, on the other hand, had assumed that thrust data they received were expressed in metric units, as newtons. Normally, pound is the unit for mass. Expressed as a unit for force, however, 1 lb is the force due to gravitational attraction on an object of that mass. To carry out the conversion between pound and newton, we start with 1 lb = 0.4536 kg and from Newton's second law of motion,

$$\begin{aligned}\text{force} &= \text{mass} \times \text{acceleration}\\ &= 0.4536\text{ kg} \times 9.81\text{ m/s}^2\\ &= 4.45\text{ kg m/s}^2\\ &= 4.45\text{ N}\end{aligned}$$

because 1 newton (N) = 1 kg m/s^2. Therefore, instead of converting 1 lb of force to 4.45 N the scientists treated it as 1 N.

The considerably smaller engine thrust expressed in newtons resulted in a lower orbit and the ultimate destruction of the spacecraft. Commenting on the failure of the Mars mission, one scientist said: "This is going to be the cautionary tale that will be embedded into introduction to the metric system in elementary school, high school, and college science courses till the end of time."

Artist's conception of the Martian Climate Orbiter.
NASA

The Chemistry in Action essay, "The Importance of Units," shows why we must be careful with units in scientific work.

A magnet suspended above a superconductor that is cooled below its transition temperature by liquid nitrogen.
ktsimage/iStockphoto/Getty Images

Example 1.3

(a) Below the transition temperature of −141°C, a certain substance becomes a superconductor, meaning it conducts electricity with no resistance. What is the temperature in degrees Fahrenheit? (b) Helium has the lowest boiling point of all the elements at −452°F. Convert this temperature to degrees Celsius. (c) Mercury, the only metal that exists as a liquid at room temperature, melts at −38.9°C. Convert its melting point to kelvins.

Solution These three parts require that we carry out temperature conversions, so we need Equations (1.2), (1.3), and (1.4). Keep in mind that the lowest temperature on the Kelvin scale is zero (0 K); therefore, it can never be negative.

(a) This conversion is carried out by writing

$$\frac{9°\text{F}}{5°\text{C}} \times (-141°\text{C}) + 32°\text{F} = -222°\text{F}$$

(Continued)

(b) Here we have

$$(-452°F - 32°F) \times \frac{5°C}{9°F} = -269°C$$

(c) The melting point of mercury in kelvins is given by

$$(-38.9°C + 273.15°C)\frac{1\ K}{1°C} = 234.3\ K$$

Practice Exercise Convert (a) 327.5°C (the melting point of lead) to degrees Fahrenheit; (b) 172.9°F (the boiling point of ethanol) to degrees Celsius; and (c) 77 K, the boiling point of liquid nitrogen, to degrees Celsius.

Similar problems: 1.12, 1.13, 1.14.

Summary of Concepts & Facts

- SI units are used to express physical quantities in all sciences, including chemistry.

Review of Concepts & Facts

1.3.1 The density of platinum is 21.45 g/cm^3. What is the volume of a platinum sample with a mass of 11.2 g?

1.3.2 The melting point of adamantane is 518°F. What is this melting point in kelvins?

1.3.3 The density of copper is 8.94 g/cm^3 at 20°C and 8.91 g/cm^3 at 60°C. This density decrease is the result of which of the following?
(a) The metal expands.
(b) The metal contracts.
(c) The mass of the metal increases.
(d) The mass of the metal decreases.

1.4 Handling Numbers

Learning Objectives

- Employ scientific notation when handling very large or very small numbers.
- Apply significant figure rules in calculations.
- Discriminate between accuracy and precision.
- Differentiate between systematic errors and random errors.

Having surveyed some of the units used in chemistry, we now turn to techniques for handling numbers associated with measurements: scientific notation and significant figures.

Scientific Notation

Chemists often deal with numbers that are either extremely large or extremely small. For example, in 1 g of the element hydrogen there are roughly

$$602,200,000,000,000,000,000,000$$

hydrogen atoms. Each hydrogen atom has a mass of only

$$0.00000000000000000000000166 \text{ g}$$

These numbers are cumbersome to handle, and it is easy to make mistakes when using them in arithmetic computations. Consider the following multiplication:

$$0.0000000056 \times 0.00000000048 = 0.000000000000000002688$$

It would be easy for us to miss one zero or add one more zero after the decimal point. Consequently, when working with very large and very small numbers, we use a system called *scientific notation.* Regardless of their magnitude, all numbers can be expressed in the form

$$N \times 10^n$$

where N is a number between 1 and 10, and n (the exponent) is a positive or negative integer (whole number). Any number expressed in this way is said to be written in scientific notation.

Suppose that we are given a certain number and asked to express it in scientific notation. Basically, this assignment calls for us to find n. We count the number of places that the decimal point must be moved to give the number N (which is between 1 and 10). If the decimal point has to be moved to the left, then n is a positive integer; if it has to be moved to the right, n is a negative integer. The following examples illustrate the use of scientific notation:

(1) Express 568.762 in scientific notation:

$$568.762 = 5.68762 \times 10^2$$

Note that the decimal point is moved to the left by two places and $n = 2$.

(2) Express 0.00000772 in scientific notation:

$$0.00000772 = 7.72 \times 10^{-6}$$

Here the decimal point is moved to the right by six places and $n = -6$.

Keep in mind the following two points. First, $n = 0$ is used for numbers that are not expressed in scientific notation. For example, 74.6×10^0 ($n = 0$) is equivalent to 74.6 (recall that any number to the power zero is equal to one). Second, the usual practice is to omit the superscript when $n = 1$. Thus, the scientific notation for 74.6 is 7.46×10 and not 7.46×10^1.

Next, we consider how scientific notation is handled in arithmetic operations.

Addition and Subtraction

To add or subtract using scientific notation, we first write each quantity—say, N_1 and N_2—with the same exponent n. Then we combine N_1 and N_2; the exponents remain the same. Consider the following examples:

$$(7.4 \times 10^3) + (2.1 \times 10^3) = 9.5 \times 10^3$$

$$(4.31 \times 10^4) + (3.9 \times 10^3) = (4.31 \times 10^4) + (0.39 \times 10^4) = 4.70 \times 10^4$$

$$(2.22 \times 10^{-2}) - (4.10 \times 10^{-3}) = (2.22 \times 10^{-2}) - (0.41 \times 10^{-2}) = 1.81 \times 10^{-2}$$

Multiplication and Division

To multiply numbers expressed in scientific notation, we multiply N_1 and N_2 in the usual way, but *add* the exponents together. To divide using scientific notation, we divide N_1 and N_2 as usual and subtract the exponents. The following examples show how these operations are performed:

$$(8.0 \times 10^4) \times (5.0 \times 10^2) = (8.0 \times 5.0)(10^{4+2})$$
$$= 40 \times 10^6$$
$$= 4.0 \times 10^7$$
$$(4.0 \times 10^{-5}) \times (7.0 \times 10^3) = (4.0 \times 7.0)(10^{-5+3})$$
$$= 28 \times 10^{-2}$$
$$= 2.8 \times 10^{-1}$$
$$\frac{6.9 \times 10^7}{3.0 \times 10^{-5}} = \frac{6.9}{3.0} \times 10^{7-(-5)}$$
$$= 2.3 \times 10^{12}$$
$$\frac{8.5 \times 10^4}{5.0 \times 10^9} = \frac{8.5}{5.0} \times 10^{4-9}$$
$$= 1.7 \times 10^{-5}$$

Significant Figures

Except when all the numbers involved are integers (for example, in counting the number of students in a class), it is often impossible to obtain the exact value of the quantity under investigation. For this reason, ***significant figures,*** which are *the meaningful digits in a measured or calculated quantity,* are used to indicate the margin of error in a measurement. When significant figures are used, the last digit is understood to be uncertain. For example, we might measure the volume of a given amount of liquid using a graduated cylinder with a scale that gives an uncertainty of 1 mL in the measurement. If the volume is found to be 6 mL, then the actual volume is in the range of 5 mL to 7 mL. We represent the volume of the liquid as (6 ± 1) mL. In this case, there is only one significant figure (the digit 6) that is uncertain by either plus or minus 1 mL. For greater accuracy, we might use a graduated cylinder that has finer divisions, so that the volume we measure is now uncertain by only 0.1 mL. If the volume of the liquid is now found to be 6.0 mL, we may express the quantity as (6.0 ± 0.1) mL and the actual value is somewhere between 5.9 mL and 6.1 mL. We can further improve the measuring device and obtain more significant figures, but in every case, the last digit is always uncertain; the amount of this uncertainty depends on the particular measuring device we use.

Figure 1.8 shows a modern balance. Balances such as this one are available in many general chemistry laboratories; they readily measure the mass of objects to four decimal places. Therefore, the measured mass typically will have four significant figures (for example, 0.8642 g) or more (for example, 3.9745 g). Keeping track of the number of significant figures in a measurement such as mass ensures that calculations involving the data will reflect the precision of the measurement.

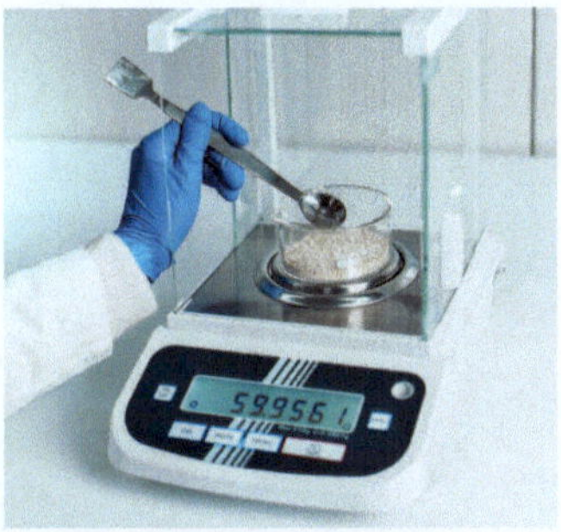

Figure 1.8 *A Fisher Scientific A-200DS Digital Recorder Precision Balance.*
Predrag Popovski/Moment/Getty Images

Guidelines for Using Significant Figures

We must always be careful in scientific work to write the proper number of significant figures. The following rules determine how many significant figures a number has:

1. Any digit that is not zero is significant. Thus, 845 cm has three significant figures, 1.234 kg has four significant figures, and so on.
2. Zeros between nonzero digits are significant. Thus, 606 m contains three significant figures; 40,501 kg contains five significant figures; and so on.
3. Zeros to the left of the first nonzero digit are not significant. Their purpose is to indicate the placement of the decimal point. For example, 0.08 L contains one significant figure, 0.0000349 g contains three significant figures, and so on.

4. If a number is greater than 1, then all the zeros written to the right of the decimal point count as significant figures. Thus, 2.0 mg has two significant figures, 40.062 mL has five significant figures, and 3.040 dm has four significant figures. If a number is less than 1, then only the zeros that are at the end of the number and the zeros that are between nonzero digits are significant. This means that 0.090 kg has two significant figures, 0.3005 L has four significant figures, 0.00420 min has three significant figures, and so on.
5. For numbers that do not contain decimal points, the trailing zeros (that is, zeros after the last nonzero digit) may or may not be significant. Thus, 400 cm may have one significant figure (the digit 4), two significant figures (40), or three significant figures (400). We cannot know which is correct without more information. By using scientific notation, however, we avoid this ambiguity. In this particular case, we can express the number 400 as 4×10^2 for one significant figure, 4.0×10^2 for two significant figures, or 4.00×10^2 for three significant figures.

Student Hot Spot

Student data indicate you may struggle with significant figures. Access your eBook for additional Learning Resources on this topic.

Example 1.4 shows the determination of significant figures.

Example 1.4

Determine the number of significant figures in the following measurements: (a) 394 cm, (b) 5.03 g, (c) 0.714 m, (d) 0.052 kg, (e) 2.720×10^{22} atoms, (f) 3000 mL.

Solution (a) Three, because each digit is a nonzero digit. (b) Three, because zeros between nonzero digits are significant. (c) Three, because zeros to the left of the first nonzero digit do not count as significant figures. (d) Two. Same reason as in (c). (e) Four. Because the number is greater than one, all the zeros written to the right of the decimal point count as significant figures. (f) This is an ambiguous case. The number of significant figures may be four (3.000×10^3), three (3.00×10^3), two (3.0×10^3), or one (3×10^3). This example illustrates why scientific notation must be used to show the proper number of significant figures.

Practice Exercise Determine the number of significant figures in each of the following measurements: (a) 35 mL, (b) 2008 g, (c) 0.0580 m^3, (d) 7.2×10^4 molecules, (e) 830 kg.

Similar problems: 1.21, 1.22.

A second set of rules specifies how to handle significant figures in calculations.

1. In addition and subtraction, the answer cannot have more digits to the right of the decimal point than either of the original numbers. Consider these examples:

$$\begin{array}{rl} 89.332 & \\ +\ 1.1 & \longleftarrow \text{one digit after the decimal point} \\ \hline 90.432 & \longleftarrow \text{round off to } 90.4 \end{array}$$

$$\begin{array}{rl} 2.097 & \\ -\ 0.12 & \longleftarrow \text{two digits after the decimal point} \\ \hline 1.977 & \longleftarrow \text{round off to } 1.98 \end{array}$$

The rounding-off procedure is as follows: To round off a number at a certain point we simply drop the digits that follow if the first of them is less than 5. Thus, 8.724 rounds off to 8.72 if we want only two digits after the decimal point. If the first digit following the point of rounding off is equal to or greater than 5, we add 1 to the preceding digit. Thus, 8.727 rounds off to 8.73, and 0.425 rounds off to 0.43.

2. In multiplication and division, the number of significant figures in the final product or quotient is determined by the original number that has the *smallest* number of significant figures. The following examples illustrate this rule:

$$2.8 \times 4.5039 = 12.61092 \longleftarrow \text{round off to } 13$$

$$\frac{6.85}{112.04} = 0.0611388789 \longleftarrow \text{round off to } 0.0611$$

3. For logarithms, the number of significant figures in the answer is determined by the number of significant figures after the decimal point. This rule applies to both natural logarithms (ln) and base-10 logarithms (log). The following examples illustrate this rule:

$$\ln(29.23) = 3.3751956 \longleftarrow \text{round off to } 3.3752$$

$$-\log(0.0593) = 1.2269453 \longleftarrow \text{round off to } 1.227$$

4. When taking the natural or base-10 antilogarithm of a number, the number of significant figures in the answer is the same as the number of significant figures to the right of the decimal point in the original measurement. The following examples illustrate this rule:

$$e^{4.876} = 1.311051929 \times 10^2 \longleftarrow \text{rounds off to } 1.31 \times 10^2$$

$$10^{-3.32} = 4.7863009 \times 10^{-4} \longleftarrow \text{rounds off to } 4.8 \times 10^{-4}$$

5. Keep in mind that *exact numbers* obtained from definitions or by counting numbers of objects can be considered to have an infinite number of significant figures. For example, the inch is defined to be exactly 2.54 centimeters; that is,

$$1 \text{ in} = 2.54 \text{ cm}$$

Thus, the "2.54" in the equation should not be interpreted as a measured number with three significant figures. In calculations involving conversion between "in" and "cm," we treat both "1" and "2.54" as having an infinite number of significant figures. Similarly, if an object has a mass of 5.0 g, then the mass of nine such objects is

$$5.0 \text{ g} \times 9 = 45 \text{ g}$$

The answer has two significant figures because 5.0 g has two significant figures. The number 9 is exact and does not determine the number of significant figures.

Example 1.5 shows how significant figures are handled in arithmetic operations.

Student Hot Spot

Student data indicate you may struggle with handling significant figures in calculations. Access your eBook for additional Learning Resources on this topic.

Example 1.5

Carry out the following arithmetic operations to the correct number of significant figures: (a) 12,343.2 g + 0.1893 g, (b) 55.67 L − 2.386 L, (c) 7.52 m × 6.9232, (d) 0.0239 kg ÷ 46.5 mL, (e) 5.21×10^3 cm + 2.92×10^2 cm, (f) ln(6.43), (g) $10^{-8.92}$.

Solution In addition and subtraction, the number of decimal places in the answer is determined by the number having the lowest number of decimal places. In multiplication and division, the significant number of the answer is determined by the number having the smallest number of significant figures.

(a)
$$\begin{array}{r} 12{,}343.2 \text{ g} \\ +\quad 0.1893 \text{ g} \\ \hline 12{,}343.3893 \text{ g} \end{array} \longleftarrow \text{round off to } 12{,}343.4 \text{ g}$$

(b)
$$\begin{array}{r} 55.67 \text{ L} \\ -\ 2.386 \text{ L} \\ \hline 53.284 \text{ L} \end{array} \longleftarrow \text{round off to } 53.28 \text{ L}$$

(Continued)

(c) $7.52\ \text{m} \times 6.9232 = 52.06246\ \text{m} \longleftarrow$ round off to 52.1 m

(d) $\dfrac{0.0239\ \text{kg}}{46.5\ \text{ml}} = 0.0005139784946\ \text{kg/mL} \longleftarrow$ round off to 0.000514 kg/mL or 5.14×10^{-4} kg/mL

(e) First we change 2.92×10^2 cm to 0.292×10^3 cm and then carry out the addition $(5.21\ \text{cm} + 0.292\ \text{cm}) \times 10^3$. Following the procedure in (a), we find the answer is 5.50×10^3 cm.

(f) $\ln(6.43) = 1.8609745 \longleftarrow$ round off to 1.861

(g) $10^{-8.92} = 1.2022644 \times 10^{-9} \longleftarrow$ round off to 1.2×10^{-9}

Practice Exercise Carry out the following arithmetic operations and round off the answers to the appropriate number of significant figures: (a) 26.5862 L + 0.17 L, (b) 9.1 g − 4.682 g, (c) 7.1×10^4 dm $\times$ 2.2654×10^2 dm, (d) 6.54 g ÷ 86.5542 mL, (e) $(7.55 \times 10^4\ \text{m}) - (8.62 \times 10^3\ \text{m})$, (f) log(109.3), (g) $10^{17.234}$.

Similar problems: 1.23, 1.24.

The preceding rounding-off procedure applies to one-step calculations. In *chain calculations*—that is, calculations involving more than one step—we can get a different answer depending on how we round off. Consider the following two-step calculations:

First step: $A \times B = C$

Second step: $C \times D = E$

Let's suppose that A = 3.66, B = 8.45, and D = 2.11. Depending on whether we round off C to three or four significant figures, we obtain a different number for E:

Method 1	**Method 2**
$3.66 \times 8.45 = 30.9$	$3.66 \times 8.45 = 30.93$
$30.9 \times 2.11 = 65.2$	$30.93 \times 2.11 = 65.3$

However, if we had carried out the calculation as $3.66 \times 8.45 \times 2.11$ on a calculator without rounding off the intermediate answer, we would have obtained 65.3 as the answer for E. Although retaining an additional digit past the number of significant figures for intermediate steps helps to eliminate errors from rounding, this procedure is not necessary for most calculations because the difference between the answers is usually quite small. Therefore, for most examples and end-of-chapter problems where intermediate answers are reported, all answers, intermediate and final, will be rounded.

Accuracy and Precision

In discussing measurements and significant figures, it is useful to distinguish between *accuracy* and *precision*. ***Accuracy*** tells us *how close a measurement is to the true value of the quantity that was measured.* ***Precision*** *refers to how closely two or more measurements of the same quantity agree with one another* (Figure 1.9).

(a) (b) (c)

Figure 1.9 *The distribution of holes formed by darts on a dart board shows the difference between precise and accurate. (a) Good accuracy and good precision. (b) Poor accuracy and good precision. (c) Poor accuracy and poor precision.*

The difference between accuracy and precision is a subtle but important one. Suppose, for example, that three students are asked to determine the mass of a piece of copper wire. The results of three successive weighings by each student are

	Student A	**Student B**	**Student C**
	1.964 g	1.970 g	2.000 g
	1.971 g	1.972 g	2.002 g
	1.978 g	1.968 g	2.001 g
Average value	1.971 g	1.970 g	2.001 g

The true mass of the wire is 2.000 g. Therefore, Student B's results are more *precise* than those of Student A (1.970 g, 1.972 g, and 1.968 g deviate less from 1.970 g than 1.964 g, 1.971 g, and 1.978 g from 1.971 g), but neither set of results is very *accurate.* Student C's results are not only the most *precise,* but also the most *accurate,* because the average value is closest to the true value. Highly accurate measurements are usually precise, too. On the other hand, highly precise measurements do not necessarily guarantee accurate results. For example, an improperly calibrated meterstick or a faulty balance may give precise readings that are in error.

The terms *accuracy* and *precision* are related to two types of error encountered in measurements. ***Systematic errors*** *occur in a predictable manner, leading to measured values that are constantly off from the true value.* Systematic errors are normally found as part of the experimental procedure caused by imperfect instrument setup or methods of observation and can be corrected with repeated measurements. Accurate measurements have low systematic error associated with them.

Random errors are *not predictable, leading to measured values that vary greatly from the true value.* Random errors are present in every experiment and affect every experiment in different ways. Such errors might result from fluctuations in an instrumental measurement or an incorrect interpretation of an instrument's reading. Precise measurements have low random error associated with them. However, it is quite possible to have high random error and still determine an accurate average result.

Summary of Concepts & Facts

- Numbers expressed in scientific notation have the form $N \times 10^n$, where N is between 1 and 10, and n is a positive or negative integer. Scientific notation helps us handle very large and very small quantities.
- Significant figures provide meaningful information about measured or calculated values. They are useful in determining the margin of error in a measurement.
- Accuracy and precision are useful in evaluating data. Accuracy reflects the closeness of a value to its true value while precision indicates the closeness of individual measurements to one another. These two terms are related to systematic errors (errors that occur in a predictable way) and random errors (errors that are unpredictable).

Review of Concepts & Facts

1.4.1 Give the length of the pencil with proper significant figures according to the two rulers you use for the measurement.

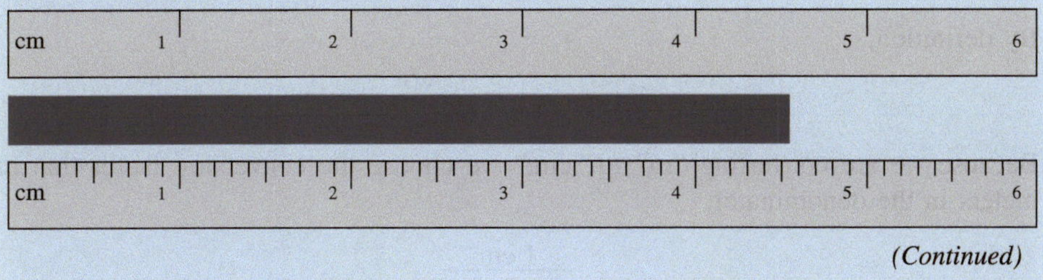

(Continued)

1.4.2 A student measures the density of an alloy with the following results: 10.28 g/cm^3, 9.97 g/cm^3, 10.22 g/cm^3, 10.15 g/cm^3, 9.94 g/cm^3. How should the average value for the density be reported?

1.4.3 Four mass measurements of a metal cube were made using a laboratory balance. The results are 24.02 g, 23.99 g, 23.98 g, and 23.97 g. The actual mass of the metal cube is 25.00 g. Are the mass measurements accurate? Are the mass measurements precise?

1.5 Dimensional Analysis in Solving Problems

Learning Objectives

- Manipulate conversion factors in dimensional analysis problems.
- Solve problems utilizing dimensional analysis.

Careful measurements and the proper use of significant figures, along with correct calculations, will yield accurate numerical results. But to be meaningful, the answers also must be expressed in the desired units. We use *dimensional analysis* (also called the *factor-label method*) to convert between units in solving chemistry problems. Dimensional analysis is based on the relationship between different units that express the same physical quantity. For example, by definition 1 in = 2.54 cm (exactly). This equivalence enables us to write a conversion factor as follows:

$$\frac{1 \text{ in}}{2.54 \text{ cm}}$$

Because both the numerator and the denominator express the same length, this fraction is equal to 1. Similarly, we can write the conversion factor as

$$\frac{2.54 \text{ cm}}{1 \text{ in}}$$

which is also equal to 1. Conversion factors are useful for changing units. Thus, if we wish to convert a length expressed in inches to centimeters, we multiply the length by the appropriate conversion factor.

$$12.00 \cancel{\text{in}} \times \frac{2.54 \text{ cm}}{1 \cancel{\text{in}}} = 30.48 \text{ cm}$$

We choose the conversion factor that cancels the unit inches and produces the desired unit, centimeters. Note that the result is expressed in four significant figures because 2.54 is an exact number.

Next let us consider the conversion of 57.8 meters to centimeters. This problem can be expressed as

$$? \text{ cm} = 57.8 \text{ m}$$

By definition,

$$1 \text{ cm} = 1 \times 10^{-2} \text{ m}$$

Because we are converting "m" to "cm," we choose the conversion factor that has meters in the denominator,

$$\frac{1 \text{ cm}}{1 \times 10^{-2} \text{ m}}$$

and write the conversion as

$$? \text{ cm} = 57.8 \cancel{\text{m}} \times \frac{1 \text{ cm}}{1 \times 10^{-2} \cancel{\text{m}}}$$
$$= 5780 \text{ cm}$$
$$= 5.78 \times 10^3 \text{ cm}$$

Note that scientific notation is used to indicate that the answer has three significant figures. Again, the conversion factor 1 cm/1 × 10^{-2} m contains exact numbers; therefore, it does not affect the number of significant figures.

In general, to apply dimensional analysis we use the relationship

$$\text{given quantity} \times \text{conversion factor} = \text{desired quantity}$$

and the units cancel as follows:

$$\cancel{\text{given unit}} \times \frac{\text{desired unit}}{\cancel{\text{given unit}}} = \text{desired unit}$$

In dimensional analysis, the units are carried through the entire sequence of calculations. Therefore, if the equation is set up correctly, then all the units will cancel except the desired one. If this is not the case, then an error must have been made somewhere, and it can usually be spotted by reviewing the solution.

A Note on Problem Solving

At this point you have been introduced to scientific notation, significant figures, and dimensional analysis, which will help you in solving numerical problems. Chemistry is an experimental science and many of the problems are quantitative in nature. The key to success in problem solving is practice. Just as a marathon runner cannot prepare for a race by simply reading books on running and a pianist cannot give a successful concert by only memorizing the musical score, you cannot be sure of your understanding of chemistry without solving problems. The following steps will help to improve your skill at solving numerical problems.

1. Read the question carefully. Understand the information that is given and what you are asked to solve. Frequently it is helpful to make a sketch that will help you to visualize the situation.
2. Find the appropriate equation that relates the given information and the unknown quantity. Sometimes solving a problem will involve more than one step, and you may be expected to look up quantities in tables that are not provided in the problem. Dimensional analysis is often needed to carry out conversions.
3. Check your answer for the correct sign, units, and significant figures.
4. A very important part of problem solving is being able to judge whether the answer is reasonable. It is relatively easy to spot a wrong sign or incorrect units. But if a number (say, 9) is incorrectly placed in the denominator instead of in the numerator, the answer would be too small even if the sign and units of the calculated quantity were correct.
5. One quick way to check the answer is to round off the numbers in the calculation in such a way so as to simplify the arithmetic. The answer you get will not be exact, but it will be close to the correct one.

Student Hot Spot

Student data indicate you may struggle with conversion factors. Access your eBook for additional Learning Resources on this topic.

Example 1.6

A person's average daily intake of glucose (a form of sugar) is 0.0833 pound (lb). What is this mass in milligrams (mg)? (1 lb = 453.6 g.)

(Continued)

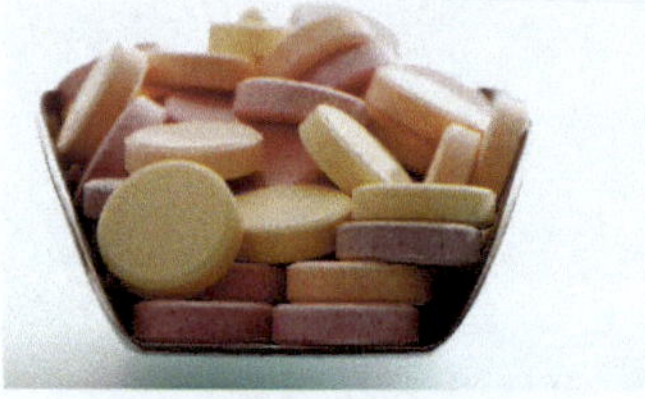

Glucose tablets can provide people with diabetes with a quick method for raising their blood sugar levels.

Foodcollection/Getty Images

Strategy The problem can be stated as

$$? \text{ mg} = 0.0833 \text{ lb}$$

The relationship between pounds and grams is given in the problem. This relationship will enable conversion from pounds to grams. A metric conversion is then needed to convert grams to milligrams (1 mg = 1×10^{-3} g). Arrange the appropriate conversion factors so that pounds and grams cancel and the unit milligrams is obtained in your answer.

Solution The sequence of conversions is

$$\text{pounds} \rightarrow \text{grams} \rightarrow \text{miligrams}$$

Using the following conversion factors

$$\frac{453.6 \text{ g}}{1 \text{ lb}} \text{ and } \frac{1 \text{ mg}}{1 \times 10^{-3} \text{ g}}$$

we obtain the answer in one step:

$$? \text{ mg} = 0.0833 \text{ lb} \times \frac{453.6 \text{ g}}{1 \text{ lb}} \times \frac{1 \text{ mg}}{1 \times 10^{-3} \text{ g}} = 3.78 \times 10^4 \text{ mg}$$

Check As an estimate, we note that 1 lb is roughly 500 g and that 1 g = 1000 mg. Therefore, 1 lb is roughly 5×10^5 mg. Rounding off 0.0833 lb to 0.1 lb, we get 5×10^4 mg, which is close to the preceding quantity.

Practice Exercise A roll of aluminum foil has a mass of 1.07 kg. What is its mass in pounds?

Similar problem: 1.33.

As Examples 1.7 and 1.8 illustrate, conversion factors can be squared or cubed in dimensional analysis.

Example 1.7

A helium tank has a volume of 275 L. What is the volume in m^3?

Helium is a commonly used gas for inflating balloons.

Dmitry Lobanov/123RF

Strategy The problem can be stated as

$$? \text{ m}^3 = 275 \text{ L}$$

How many conversion factors are needed for this problem? Recall that 1 L = 1000 cm^3 and 1 cm = 1×10^{-2} m.

Solution We need two conversion factors here: one to convert liters to cm^3 and one to convert centimeters to meters:

$$\frac{1000 \text{ cm}^3}{1 \text{ L}} \text{ and } \frac{1 \times 10^{-2} \text{ m}}{1 \text{ cm}}$$

Because the second conversion deals with length (cm and m) and we want volume here, it must therefore be cubed to give

$$\frac{1 \times 10^{-2} \text{ m}}{1 \text{ cm}} \times \frac{1 \times 10^{-2} \text{ m}}{1 \text{ cm}} \times \frac{1 \times 10^{-2} \text{ m}}{1 \text{ cm}} = \left(\frac{1 \times 10^{-2} \text{ m}}{1 \text{ cm}}\right)^3$$

This means that 1 cm^3 = 1×10^{-6} m^3. Now we can write

$$? \text{ m}^3 = 275 \text{ L} \times \frac{1000 \text{ cm}^3}{1 \text{ L}} \times \left(\frac{1 \times 10^{-2} \text{ m}}{1 \text{ cm}}\right)^3 = 0.275 \text{ m}^3$$

(Continued)

Check From the preceding conversion factors you can show that 1 L = 1×10^{-3} m^3. Remember that when a unit is raised to a power, any conversion factor you use must also be raised to that power. Therefore, a 275-L storage tank would be equal to 275×10^{-3} m^3 or 0.275 m^3, which is the answer.

Practice Exercise The volume of a room is 1.08×10^8 dm^3. What is the volume in m^3?

Similar problem: 1.38(d).

Example 1.8

Liquid nitrogen is obtained from liquefied air and is used to prepare frozen goods and in low-temperature research. The density of the liquid at its boiling point (−196°C or 77 K) is 0.808 g/cm^3. Convert the density to units of kg/m^3.

Strategy The problem can be stated as

$$? \text{ kg/m}^3 = 0.808 \text{ g/cm}^3$$

Two separate conversions are required for this problem: g ⟶ kg and cm^3 ⟶ m^3. Recall that 1 kg = 1000 g and 1 cm = 1×10^{-2} m.

Solution In Example 1.7 we saw that 1 cm^3 = 1×10^{-6} m^3. The conversion factors are

$$\frac{1 \text{ kg}}{1000 \text{ g}} \quad \text{and} \quad \frac{1 \text{ cm}^3}{1 \times 10^{-6} \text{ m}^3}$$

Finally,

$$? \text{ kg/m}^3 = \frac{0.808 \cancel{\text{g}}}{1 \cancel{\text{cm}^3}} \times \frac{1 \text{ kg}}{1000 \cancel{\text{g}}} \times \frac{1 \cancel{\text{cm}^3}}{1 \times 10^{-6} \text{ m}^3} = 808 \text{ kg/m}^3$$

Check Because 1 m^3 = 1×10^6 cm^3, we would expect much more mass in 1 m^3 than in 1 cm^3. Therefore, the answer is reasonable.

Practice Exercise The density of the lightest metal, lithium (Li), is 5.34×10^2 kg/m^3. Convert the density to g/cm^3.

Similar problem: 1.39.

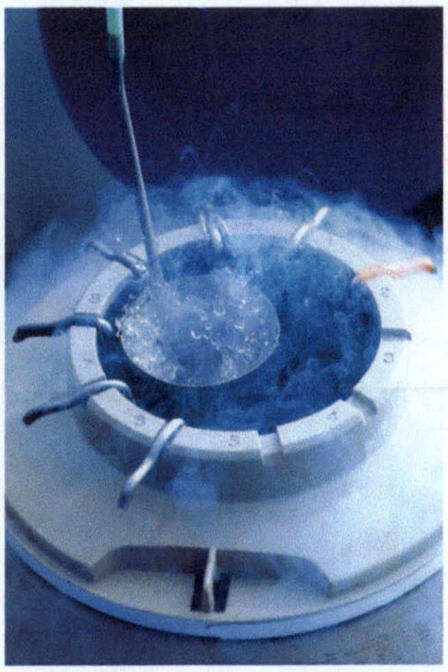

Liquid nitrogen is used for frozen foods and low-temperature research.

Science Photo Library/Alamy Stock Photo

Summary of Concepts & Facts

- Dimensional analysis is an effective method for conducting calculations. In this method, units are used as connective pieces and ultimately cancel to leave only the final unit(s) required in the answer.

Review of Concepts & Facts

1.5.1 What is the correct answer with the appropriate number of significant figures and units for the following energy conversion calculation?

$$3.54 \times 10^{12} \text{ erg} \times \frac{1.000 \text{ J}}{1.000 \times 10^7 \text{ erg}} \times \frac{1.000 \text{ cal}}{4.184 \text{ J}} \times \frac{1 \text{ kcal}}{1000 \text{ cal}} =$$

1.5.2 What is the volume in L of a 1.24-kg sample of benzene (density = 0.876 g/mL)?

1.5.3 The heat of combustion of TNT is 14.5 MJ/kg. What is this heat of combustion in J/g?

1.6 Real-World Problem Solving: Information, Assumptions, and Simplifications

Learning Objective

- Estimate approximate answers to problems without exact numerical solutions.

In chemistry, as in other scientific disciplines, it is not always possible to solve a numerical problem exactly. There are many reasons why this is the case—for example, our understanding of a situation is not complete or data are not fully available. In these cases, we must learn to make an intelligent guess. Such approaches are sometimes called "ball-park estimates," which are simple, quick calculations that can be done on the "back of an envelope." As you can imagine, in many cases the answers are only order-of-magnitude estimates.†

In most of the example problems that you have seen so far, as well as the questions given at the end of this and subsequent chapters, the necessary information is provided; however, to solve important real-world problems such as those related to medicine, energy, and agriculture, you must be able to determine what information is needed and where to find it. Much of the information you might need can be found in the various tables located throughout the text, and a list of tables and important figures is given in the Index of Important Figures and Tables. In many cases, however, you will need to go to outside sources to find the information you need. Although the Internet is a fast way to find information, you must take care that the source is reliable and well referenced. One excellent source is the National Institute of Standards and Technology (NIST).

To know what information you need, you will first have to formulate a plan for solving the problem. In addition to the limitations of the theories used in science, typically assumptions are made in setting up and solving the problems based on those theories. These assumptions come at a price, however, as the accuracy of the answer is reduced with increasing simplifications of the problem, as illustrated in Example 1.9.

Example 1.9

A modern pencil "lead" is actually composed primarily of graphite, a form of carbon. Estimate the mass of the graphite core in a standard No. 2 pencil before it is sharpened.

Strategy Assume that the pencil lead can be approximated as a cylinder. Measurement of a typical unsharpened pencil gives a length of about 18 cm (subtracting the length of the eraser head) and a diameter of roughly 2 mm for the lead. The volume of a cylinder V is given by $V = \pi r^2 l$, where r is the radius and l is the length. Assuming that the lead is pure graphite, you can calculate the mass of the lead from the volume using the density of graphite given in Table 1.3.

Solution Converting the diameter of the lead to units of cm gives

$$2\text{ mm} \times \frac{1\text{ cm}}{10\text{ mm}} = 0.2\text{ cm}$$

which, along with the length of the lead, gives

$$V = \pi\left(\frac{0.2\text{ cm}}{2}\right)^2 \times 18\text{ cm}$$
$$= 0.57\text{ cm}^3$$

(Continued)

†An order of magnitude is a factor of 10.

Rearranging Equation (1.1) gives

$$m = d \times V$$
$$= 2.2 \frac{g}{cm^3} \times 0.57^3$$
$$= 1 \text{ g}$$

Check Rounding off the values used to calculate the volume of the lead gives $3 \times (0.1 \text{ cm})^2 \times 20 \text{ cm} = 0.6 \text{ cm}^3$. Multiplying that volume by roughly 2 g/cm^3 gives around 1 g, which agrees with the value just calculated.

Practice Exercise Estimate the mass of air in a ping-pong ball.

Similar problems: 1.41, 1.42.

Considering Example 1.9, even if the dimensions of the pencil lead were measured with greater precision, the accuracy of the final answer would be limited by the assumptions made in modeling this problem. The pencil lead is actually a mixture of graphite and clay, where the relative amounts of the two materials determine the softness of the lead, so the density of the material is likely to be different than 2.2 g/cm^3. You could probably find a better value for the density of the mixture used to make No. 2 pencils, but it is not worth the effort in this case.

Summary of Concepts & Facts

- Many problems can be approximated by using simplifications or assumptions in solving the problem. However, the accuracy of the estimated answer is reduced with increased simplification of the problem.

Review of Concepts & Facts

1.6.1 Estimate how far (in feet) light travels in 1 nanosecond.

1.7 Classifications of Matter

Learning Objectives

- Classify matter as substances or mixtures.
- Differentiate between homogeneous and heterogeneous mixtures.
- Categorize substances as elements or compounds.

We defined chemistry in Section 1.1 as the study of matter and the changes it undergoes. ***Matter*** is *anything that occupies space and has mass.* Matter includes things we can see and touch (such as water, earth, and trees), as well as things we cannot (such as air). Thus, everything in the universe has a "chemical" connection.

Chemists distinguish among several subcategories of matter based on composition and properties. The classifications of matter include substances, mixtures, elements, and compounds, as well as atoms and molecules, which we will consider in Chapter 2.

Substances and Mixtures

A ***substance*** is *a form of matter that has a definite (constant) composition and distinct properties.* Examples are water, ammonia, table sugar (sucrose), gold, and oxygen.

(a) (b)

Figure 1.10 *(a) The mixture contains iron filings and sand. (b) A magnet separates the iron filings from the mixture. The same technique is used on a larger scale to separate iron and steel from nonmagnetic objects such as aluminum, glass, and plastics.*
Ken Karp/McGraw Hill

Substances differ from one another in composition and can be identified by their appearance, smell, taste, and other properties.

A ***mixture*** is *a combination of two or more substances in which the substances retain their distinct identities.* Some familiar examples are air, soft drinks, milk, and cement. Mixtures do not have constant composition. Therefore, samples of air collected in different cities would probably differ in composition because of differences in altitude, pollution, and so on.

All mixtures are classified as either homogeneous or heterogeneous. When a spoonful of sugar dissolves in water we obtain a ***homogeneous mixture*** in which *the composition of the mixture is the same throughout.* If sand is mixed with iron filings, however, the sand grains and the iron filings remain separate (Figure 1.10). This type of mixture is called a ***heterogeneous mixture*** because *the composition is not uniform.*

Any mixture, whether homogeneous or heterogeneous, can be created and then separated by physical means into pure components without changing the identities of the components. Thus, sugar can be recovered from a water solution by heating the solution and evaporating it to dryness. Condensing the vapor will give us back the water component. To separate the iron-sand mixture, we can use a magnet to remove the iron filings from the sand, because sand is not attracted to the magnet [see Figure 1.10(b)]. After separation, the components of the mixture will have the same composition and properties as they did to start with.

Elements and Compounds

Substances can be either elements or compounds. An ***element*** is *a substance that cannot be separated further into simpler substances by chemical methods.* To date, 118 elements have been positively identified. Most of them occur naturally on Earth. The others have been created by scientists via nuclear processes, which are the subject of Chapter 19 of this text.

For convenience, chemists use symbols of one or two letters to represent the elements. The first letter of a symbol is *always* capitalized, but any following letters are not. For example, Co is the symbol for the element cobalt, whereas CO is the formula for the carbon monoxide molecule. Table 1.4 shows the names and symbols of some of the more common elements. The symbols of some elements are derived from their Latin names—for example, Au from *aurum* (gold), Fe from *ferrum* (iron), and Na from *natrium* (sodium)—whereas most of them come from their English names.

Atoms of most elements can interact with one another to form compounds. Hydrogen gas, for example, burns in oxygen gas to form water, which has properties

Table 1.4 Some Common Elements and Their Symbols

Name	Symbol	Name	Symbol	Name	Symbol
Aluminum	Al	Fluorine	F	Oxygen	O
Arsenic	As	Gold	Au	Phosphorus	P
Barium	Ba	Hydrogen	H	Platinum	Pt
Bismuth	Bi	Iodine	I	Potassium	K
Bromine	Br	Iron	Fe	Silicon	Si
Calcium	Ca	Lead	Pb	Silver	Ag
Carbon	C	Magnesium	Mg	Sodium	Na
Chlorine	Cl	Manganese	Mn	Sulfur	S
Chromium	Cr	Mercury	Hg	Tin	Sn
Cobalt	Co	Nickel	Ni	Tungsten	W
Copper	Cu	Nitrogen	N	Zinc	Zn

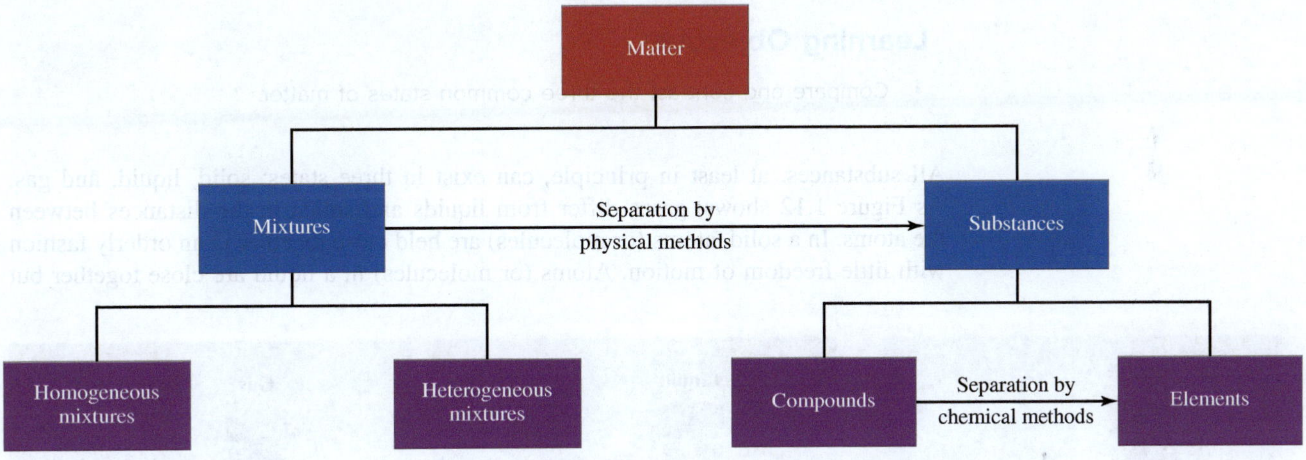

Figure 1.11 *Classification of matter.*

that are distinctly different from those of the starting materials. Water is made up of two parts hydrogen and one part oxygen. This composition does not change, regardless of whether the water comes from a faucet in the United States, a lake in Mongolia, or the ice caps on Mars. Thus, water is a ***compound,*** *a substance composed of atoms of two or more elements chemically united in fixed proportions.* Unlike mixtures, compounds can be separated only by chemical means into their pure components.

The relationships among elements, compounds, and other categories of matter are summarized in Figure 1.11.

Summary of Concepts & Facts

- Chemists study matter and the changes it undergoes. The substances that make up matter have unique physical properties that can be observed without changing their identity and unique chemical properties that, when they are demonstrated, do change the identity of the substances. Mixtures, whether homogeneous or heterogeneous, can be separated into pure components by physical means.
- The simplest substances in chemistry are elements. Compounds are formed by the chemical combination of atoms of different elements in fixed proportions.

Review of Concepts & Facts

1.7.1 Which of the following diagrams represent elements and which represent compounds? Each color sphere (or truncated sphere) represents an atom. Different colored atoms indicate different elements.

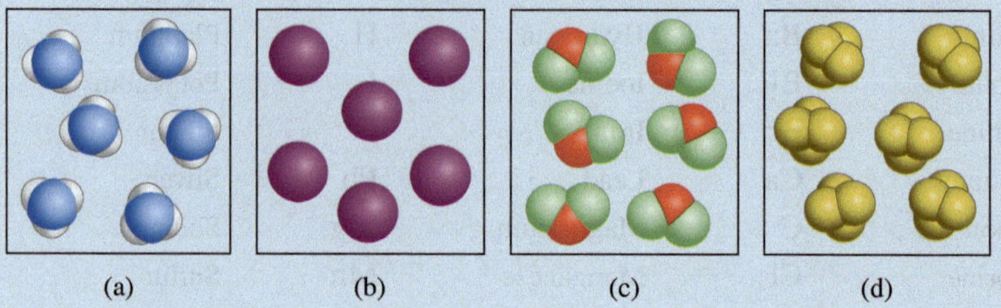

1.8 The Three Common States of Matter

Learning Objective

- Compare and contrast the three common states of matter.

All substances, at least in principle, can exist in three states: solid, liquid, and gas. As Figure 1.12 shows, gases differ from liquids and solids in the distances between the atoms. In a solid, atoms (or molecules) are held close together in an orderly fashion with little freedom of motion. Atoms (or molecules) in a liquid are close together but

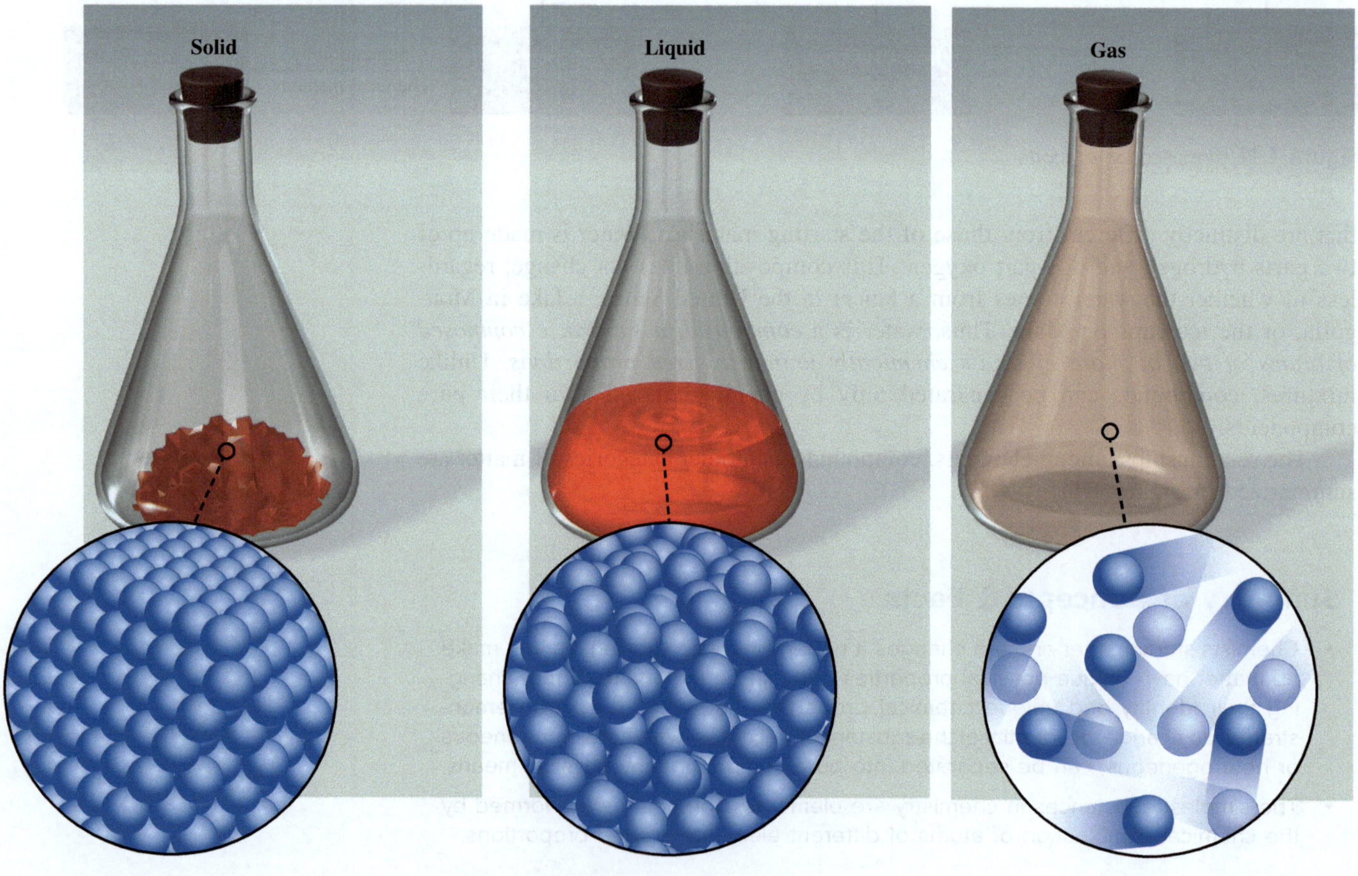

Figure 1.12 *Microscopic views of a solid, a liquid, and a gas.*

Figure 1.13 *The three states of matter. A Bunsen burner changes ice into water and steam.*
Charles D. Winters/Timeframe Photography/McGraw Hill

are not held so rigidly in position and can move past one another. In a gas, the atoms (or molecules) are separated by distances that are large compared with the size of the atoms (or molecules).

The three common states of matter can be interconverted without changing the composition of the substance. Upon heating, a solid (for example, ice) will melt to form a liquid (water). (The temperature at which this transition occurs is called the *melting point.*) Further heating will convert the liquid into a gas. (This conversion takes place at the *boiling point* of the liquid.) On the other hand, cooling a gas will cause it to condense into a liquid. When the liquid is cooled further, it will freeze into the solid form. Figure 1.13 shows the three states of water. Note that the properties of water are unique among common substances in that the molecules in the liquid state are more closely packed than those in the solid state.

While most matter does exist in one of the three common states, there are many additional forms of matter, including plasma and Bose-Einstein condensate (BEC). A plasma behaves very much like a gas and can expand indefinitely but consists of high-energy particles, including neutral atoms, electrons, and ions (charged particles derived from atoms). A plasma is strongly influenced by electrical and magnetic fields even though the plasma has an overall charge of zero. Examples of plasma outside laboratory environments include plasma screen televisions, fluorescent lights, and lightning. It is theorized to be the most common form of matter in the universe as stars such as the sun consist of plasma. The three common states—gas, liquid, and solid—along with plasma are considered as the fundamental states as they are easily observable.

The BEC was first predicted in 1924 by Satyendra Bose and Albert Einstein. Its existence was confirmed in 1995 when gaseous rubidium atoms were cooled to near absolute zero (170 nanokelvins). The kinetic energy of the atoms at this temperature is low enough that atoms condense together to form the equivalent of a "superatom" with many interesting properties, including superfluidity. The BEC allows scientists to study quantum mechanics experimentally in the laboratory.

Under extreme conditions, many other states of matter are known to exist. Other low-temperature states of matter include fermionic condensates, Rydberg molecules, the quantum Hall state, photonic matter, and dropletons. At very high temperatures, states of matter known to exist include electron-degenerate and neutron-degenerate matter. Interestingly, degenerate matter has been shown to have densities with nearly 10 trillion times that of the most dense substances observed on Earth.

Summary of Concepts & Facts

- All substances, in principle, can exist in three states: solid, liquid, and gas. The interconversion between these states can be affected by changes in temperature.

Review of Concepts & Facts

1.8.1 An ice cube is placed in a closed container. On heating, the ice cube first melts and the water then boils to form steam. Which of the following statements is true?
(a) The physical appearance of the water is different at every stage of change.
(b) The mass of water is greatest for the ice cube and least for the steam.

1.9 Physical and Chemical Properties of Matter

Learning Objectives

- Appraise properties of matter as either chemical or physical.
- Identify properties of matter as either extensive or intensive.

Substances are identified by their properties as well as by their composition. Color, melting point, and boiling point are physical properties. A ***physical property*** *can be measured and observed without changing the composition or identity of a substance.* For example, we can measure the melting point of ice by heating a block of ice and recording the temperature at which the ice is converted to water. Water differs from ice only in appearance, not in composition, so this is a physical change; we can freeze the water to recover the original ice. Therefore, the melting point of a substance is a physical property. Similarly, when we say that helium gas is lighter than air, we are referring to a physical property.

On the other hand, the statement "Hydrogen gas burns in oxygen gas to form water" describes a ***chemical property*** of hydrogen. *To observe this chemical property of hydrogen we must carry out a chemical change*—in this case, burning the hydrogen gas in oxygen. After the change, the original chemical substance, the hydrogen gas, will have vanished, and all that will be left is a different chemical substance—water. We *cannot* recover the hydrogen from the water by means of a physical change, such as boiling or freezing.

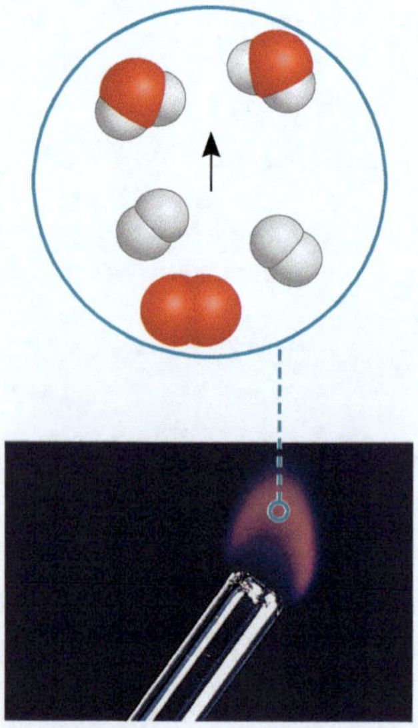

Hydrogen burning in air to form water.
Ken Karp/McGraw Hill

Every time we hard-boil an egg, we bring about a chemical change. When subjected to a temperature of about 100°C, the yolk and the egg white undergo changes that alter not only their physical appearance but their chemical makeup as well. When eaten, the egg is changed again, by substances in the body called *enzymes.* This digestive action is another example of a chemical change. What happens during digestion depends on the chemical properties of both the enzymes and the food.

All measurable properties of matter fall into one of two additional categories: extensive properties and intensive properties. The measured value of an ***extensive property*** *depends on how much matter is being considered.* ***Mass,*** *which is the* *quantity of matter in a given sample of a substance,* is an extensive property. More matter means more mass. Values of the same extensive property can be added together. For example, two copper pennies will have a combined mass that is the sum of the masses of each penny, and the length of two tennis courts is the sum of the lengths of each tennis court. ***Volume,*** defined as *length cubed,* is another extensive property. The value of an extensive quantity depends on the amount of matter.

The measured value of an ***intensive property*** *does not depend on how much matter is being considered.* **Density,** defined as *the mass of an object divided by its volume,* is an intensive property. So is temperature. Suppose that we have two beakers each containing 100 mL of water at 25°C. If we combine them to give 200 mL of water in a larger beaker, the temperature of the combined quantities of water will be still be 25°C, the same as it was in two separate beakers. The density of the combined quantities of water will also be the same as the original quantities. The temperature and the density of water do not depend on the amount of water present. Unlike mass, length, and volume, temperature and other intensive properties are not additive.

Summary of Concepts & Facts

- Characterizing a substance requires knowing its physical properties, which can be observed without changing its identity, and chemical properties, which can be demonstrated only by chemical changes.
- Any measurable property can be further characterized as intensive or extensive. Intensive properties do not depend on the amount of matter present while extensive properties are affected directly by the amount of matter present.

Review of Concepts & Facts

1.9.1 The diagram in (a) shows a compound made up of atoms of two elements (represented by the blue and red spheres) in the liquid state. Which of the diagrams in (b)–(d) represents a physical change and which diagrams represent a chemical change?

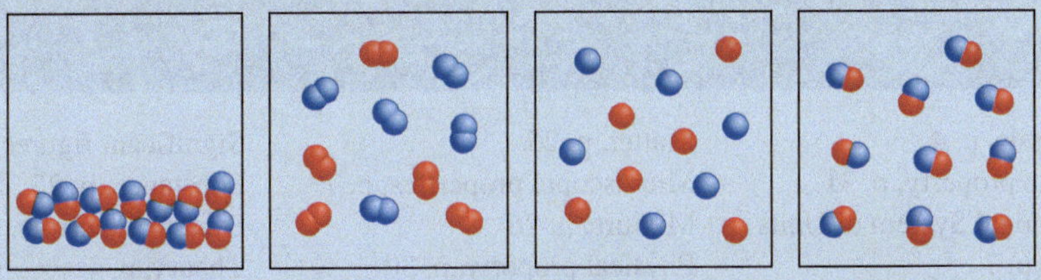

(a) (b) (c) (d)

1.9.2 Determine which of the following properties are intensive and which are extensive.

(a) The hardness of diamond is 10 on the Mohs scale.
(b) The melting point of water is 0°C.
(c) A cube of lead has an edge length of 2.5 cm.

Chapter Summary

The Study of Chemistry Chemistry is the study of the properties of matter and the changes it undergoes. Elements and compounds are substances that take part in chemical transformation. (Sections 1.1 and 1.2)

Measurements and Units Chemistry is a quantitative science and requires measurements. The measured quantities (for example, mass, volume, density, temperature) usually have units associated with them. The units used in chemistry are based on the International System (SI) of units. (Section 1.3)

Handling Numbers Scientific notation is used to express large and small numbers, and each number in a measurement must indicate the meaningful digits, called significant figures. (Section 1.4)

Dimensional Analysis A simple and effective way to perform chemical calculations is dimensional analysis. In this procedure, an equation is set up in such a way that all the units cancel except the ones for the final answer. (Section 1.5)

Real-World Problem Solving Solving real-world problems frequently involves making assumptions and simplifications. (Section 1.6)

Common States of Matter Chemists classify matter based on its composition and properties. These include substances and mixtures. Mixtures may be heterogeneous or homogeneous. Substances may be either elements or compounds. The three common states of matter are solid, liquid, and gas. However, many other states of matter are known at the extreme ranges of temperature (very cold and very hot). (Sections 1.7 and 1.8)

Physical and Chemical Properties To characterize a substance, we need to know its physical properties, which can be observed without changing its identity, and chemical properties, which can be demonstrated only by chemical changes. (Section 1.9)

Key Equations

$d = \frac{m}{V}$ (1.1)	Equation for density
$?°C = (°F - 32°F) \times \frac{5°C}{9°F}$ (1.2)	Converting °F to °C
$?°F = \frac{9°F}{5°C} \times (°C) + 32°F$ (1.3)	Converting °C to °F
$? K = (°C + 273.15°C)\frac{1 K}{1°C}$ (1.4)	Converting °C to K

Key Words

Accuracy, p. 18
Chemical property, p. 30
Chemistry, p. 2
Compound, p. 27
Density, p. 31
Element, p. 26
Extensive property, p. 30
Heterogeneous mixture, p. 26
Homogeneous mixture, p. 26
Hypothesis, p. 4
Intensive property, p. 31
International System of Units (SI), p. 7
Kelvin, p. 11
Law, p. 4
Liter, p. 9
Macroscopic properties, p. 7
Mass, p. 30
Matter, p. 25
Microscopic properties, p. 7
Mixture, p. 26
Physical property, p. 30
Precision, p. 18
Random errors, p. 19
Qualitative, p. 4
Quantitative, p. 4
Scientific method, p. 4
Significant figures, p. 15
Substance, p. 25
Systematic errors, p. 19
Theory, p. 4
Volume, p. 31
Weight, p. 8

Questions & Problems

Red numbered problems solved in Student Solutions Manual

1.2 The Scientific Method

Review Questions

1.1 Explain what is meant by the scientific method.

1.2 What is the difference between qualitative data and quantitative data?

Problems

1.3 Classify the following as qualitative or quantitative statements, giving your reasons. (a) The sun is approximately 93 million miles from Earth. (b) Leonardo da Vinci was a better painter than Michelangelo. (c) Ice is less dense than water. (d) Butter tastes better than margarine. (e) A stitch in time saves nine.

1.4 Classify each of the following statements as a hypothesis, a law, or a theory. (a) Beethoven's contribution to music would have been much greater if he had married. (b) An autumn leaf gravitates toward the ground because there is an attractive force between the leaf and Earth. (c) All matter is composed of very small particles called atoms.

1.3 Measurement

Review Questions

1.5 Name the SI base units that are important in chemistry. Give the SI units for expressing the following: (a) length, (b) volume, (c) mass, (d) time, (e) energy, (f) temperature.

1.6 Write the numbers represented by the following prefixes: (a) mega-, (b) kilo-, (c) deci-, (d) centi-, (e) milli-, (f) micro-, (g) nano-, (h) pico-.

1.7 What units do chemists normally use for density of liquids and solids? For gas density? Explain the differences.

1.8 Describe the three temperature scales used in the laboratory and in everyday life: the Fahrenheit scale, the Celsius scale, and the Kelvin scale.

Problems

1.9 Bromine is a reddish-brown liquid. Calculate its density (in g/mL) if 586 g of the substance occupies 188 mL.

1.10 The density of methanol, a colorless organic liquid used as solvent, is 0.7918 g/mL. Calculate the mass of 89.9 mL of the liquid.

1.11 Convert the following temperatures to degrees Celsius or Fahrenheit: (a) 95°F, the temperature on a hot summer day; (b) 12°F, the temperature on a cold winter day; (c) a 102°F fever; (d) a furnace operating at 1852°F; (e) −273.15°C (theoretically the lowest attainable temperature).

1.12 (a) Normally the human body can endure a temperature of 105°F for only short periods of time without permanent damage to the brain and other vital organs. What is this temperature in degrees Celsius? (b) Ethylene glycol is a liquid organic compound that is used as an antifreeze in car radiators. It freezes at −11.5°C. Calculate its freezing temperature in degrees Fahrenheit. (c) The temperature on the surface of the sun is about 6300°C. What is this temperature in degrees Fahrenheit? (d) The ignition temperature of paper is 451°F. What is the temperature in degrees Celsius?

1.13 Convert the following temperatures to kelvin: (a) 113°C, the melting point of sulfur, (b) 37°C, the normal body temperature, (c) 357°C, the boiling point of mercury.

1.14 Convert the following temperatures to degrees Celsius: (a) 77 K, the boiling point of liquid nitrogen, (b) 4.2 K, the boiling point of liquid helium, (c) 601 K, the melting point of lead.

1.4 Handling Numbers

Review Questions

1.15 What is the advantage of using scientific notation over decimal notation?

1.16 Define significant figure. Discuss the importance of using the proper number of significant figures in measurements and calculations.

Problems

1.17 Express the following numbers in scientific notation: (a) 0.000000027, (b) 356, (c) 47,764, (d) 0.096.

1.18 Express the following numbers as decimals: (a) 1.52×10^{-2}, (b) 7.78×10^{-8}.

1.19 Express the answers to the following calculations in scientific notation:

(a) $145.75 + (2.3 \times 10^{-1})$
(b) $79{,}500 \div (2.5 \times 10^{2})$
(c) $(7.0 \times 10^{-3}) - (8.0 \times 10^{-4})$
(d) $(1.0 \times 10^{4}) \times (9.9 \times 10^{6})$

1.20 Express the answers to the following calculations in scientific notation:

(a) $0.0095 + (8.5 \times 10^{-3})$
(b) $653 \div (5.75 \times 10^{-8})$
(c) $850{,}000 - (9.0 \times 10^{5})$
(d) $(3.6 \times 10^{-4}) \times (3.6 \times 10^{6})$

1.21 What is the number of significant figures in each of the following measurements?

(a) 4867 mi
(b) 56 mL
(c) 60,104 tons
(d) 2900 g
(e) 40.2 g/cm^3
(f) 0.0000003 cm
(g) 0.7 min
(h) 4.6×10^{19} atoms

1.22 How many significant figures are there in each of the following? (a) 0.006 L, (b) 0.0605 dm, (c) 60.5 mg, (d) 605.5 cm^2, (e) 960×10^{-3} g, (f) 6 kg, (g) 60 m.

1.23 Carry out the following operations as if they were calculations of experimental results, and express each answer in the correct units with the correct number of significant figures.

(a) 5.6792 m + 0.6 m + 4.33 m
(b) 3.70 g − 2.9133 g
(c) 4.51 cm × 3.6666 cm
(d) $(3 \times 10^{4}$ g + 6.827 g)/(0.043 cm^3 − 0.021 cm^3)

1.24 Carry out the following operations as if they were calculations of experimental results, and express each answer in the correct units with the correct number of significant figures.

(a) 7.310 km ÷ 5.70 km
(b) $(3.26 \times 10^{-3}$ mg) − $(7.88 \times 10^{-5}$ mg)
(c) $(4.02 \times 10^{6}$ dm) + $(7.74 \times 10^{7}$ dm)
(d) (7.8 m − 0.34 m)/(1.15 s + 0.82 s)

1.25 Three students (A, B, and C) are asked to determine the volume of a sample of ethanol. Each student measures the volume three times with a graduated cylinder. The results in milliliters are: A(87.1, 88.2, 87.6); B(86.9, 87.1, 87.2); C(87.6, 87.8, 87.9). The true volume is 87.0 mL. Comment on the precision and the accuracy of each student's results.

1.26 Three apprentice tailors (X, Y, and Z) are assigned the task of measuring the seam of a pair of trousers. Each one makes three measurements. The results in inches are X(31.5, 31.6, 31.4); Y(32.8, 32.3, 32.7); Z(31.9, 32.2, 32.1). The true length is 32.0 in. Comment on the precision and the accuracy of each tailor's measurements.

1.5 Dimensional Analysis in Solving Problems

Problems

1.27 Carry out the following conversions: (a) 22.6 m to decimeters, (b) 25.4 mg to kilograms, (c) 556 mL to liters, (d) 10.6 kg/m^3 to g/cm^3.

1.28 Carry out the following conversions: (a) 242 lb to milligrams, (b) 68.3 cm^3 to cubic meters, (c) 7.2 m^3 to liters, (d) 28.3 μg to pounds.

1.29 The average speed of helium at 25°C is 1255 m/s. Convert this speed to miles per hour (mph).

1.30 How many seconds are there in a solar year (365.24 days)?

1.31 How many minutes does it take light from the sun to reach Earth? (The distance from the sun to Earth is 93 million mi; the speed of light = 3.00×10^8 m/s.)

1.32 A jogger runs a mile in 8.92 min. Calculate the speed in (a) in/s, (b) m/min, (c) km/h. (1 mi = 1609 m; 1 in = 2.54 cm.)

1.33 A 6.0-ft person weighs 168 lb. Express this person's height in meters and weight in kilograms. (1 lb = 453.6 g; 1 m = 3.28 ft.)

1.34 The speed limit on parts of the German autobahn was once set at 286 kilometers per hour (km/h). Calculate the speed limit in miles per hour (mph).

1.35 For a fighter jet to take off from the deck of an aircraft carrier, it must reach a speed of 62 m/s. Calculate the speed in miles per hour (mph).

1.36 The "normal" lead content in human blood is about 0.40 part per million (that is, 0.40 g of lead per million grams of blood). A value of 0.80 part per million (ppm) is considered to be dangerous. How many grams of lead are contained in 6.0×10^3 g of blood (the amount in an average adult) if the lead content is 0.62 ppm?

1.37 Carry out the following conversions: (a) 1.42 light-years to miles (a light-year is an astronomical measure of distance—the distance traveled by light in a year, or 365 days; the speed of light is 3.00×10^8 m/s), (b) 32.4 yd to centimeters, (c) 3.0×10^{10} cm/s to ft/s.

1.38 Carry out the following conversions: (a) 70 kg, the average weight of a male adult, to pounds; (b) 14 billion years (roughly the age of the universe) to seconds (assume there are 365 days in a year); (c) 7 ft 6 in, the height of the basketball player Yao Ming, to meters; (d) 88.6 m^3 to liters.

1.39 Aluminum is a lightweight metal (density = 2.70 g/cm^3) used in aircraft construction, high-voltage transmission lines, beverage cans, and foils. What is its density in kg/m^3?

1.40 Ammonia gas is used as a refrigerant in large-scale cooling systems. The density of ammonia gas under certain conditions is 0.625 g/L. Calculate its density in g/cm^3.

1.6 Real-World Problem Solving: Information, Assumptions, and Simplifications

Problems

1.41 What is the mass of one mole of ants? (*Useful information:* A mole is the unit used for atomic and subatomic particles. It is approximately 6×10^{23}. A 1-cm-long ant weighs about 3 mg.)

1.42 How much time (in years) does an 80-year-old person spend sleeping during his or her life span?

1.43 Estimate the daily amount of water (in gallons) used indoors by a family of four in the United States.

1.44 Bowling alleys generally stock bowling balls from 8 to 16 lb, where the mass is given in whole numbers. Given that regulation bowling balls have a diameter of 8.6 in, which (if any) of these bowling balls would you expect to float in water?

1.7 Classifications of Matter

Review Questions

1.45 Give an example for each of the following terms: (a) matter, (b) substance, (c) mixture.

1.46 Give an example of a homogeneous mixture and an example of a heterogeneous mixture.

1.47 Give an example of an element and a compound. How do elements and compounds differ?

1.48 What is the number of known elements?

Problems

1.49 Give the names of the elements represented by the chemical symbols Li, F, P, Cu, As, Zn, Cl, Pt, Mg, U, Al, Si, Ne. (See Table 1.4 and the list of The Elements with Their Symbols and Atomic Masses.)

1.50 Give the chemical symbols for the following elements: (a) cesium, (b) germanium, (c) gallium, (d) strontium, (e) uranium, (f) selenium, (g) neon, (h) cadmium. (See Table 1.4 and the list of The Elements with Their Symbols and Atomic Masses.)

1.51 Classify each of the following substances as an element or a compound: (a) hydrogen, (b) water, (c) gold, (d) sugar.

1.52 Classify each of the following as an element, a compound, a homogeneous mixture, or a heterogeneous mixture: (a) water from a well, (b) argon gas, (c) sucrose, (d) a bottle of red wine, (e) chicken noodle soup, (f) blood flowing in a capillary, (g) ozone.

1.8 The Three Common States of Matter

Review Questions

1.53 Identify each of the diagrams shown here as gas, liquid, or solid.

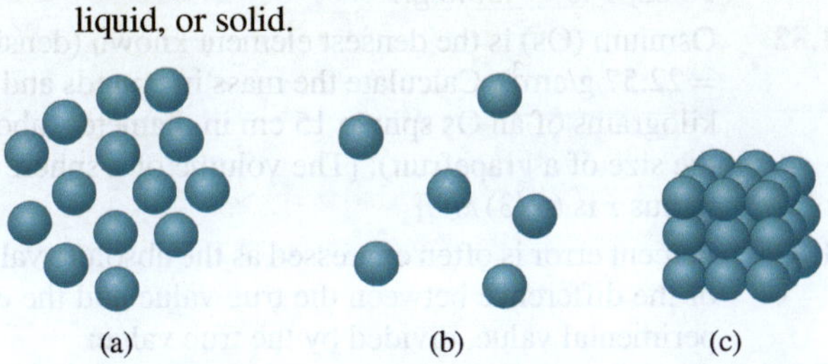

1.54 Explain how the distances between particles typically change with different states of matter.

1.9 Physical and Chemical Properties of Matter

Review Questions

1.55 Using examples, explain the difference between a physical property and a chemical property.

1.56 How does an intensive property differ from an extensive property? Which of the following properties are intensive and which are extensive? (a) length, (b) volume, (c) temperature, (d) mass.

Problems

1.57 Do the following statements describe chemical or physical properties? (a) Oxygen gas supports combustion. (b) Fertilizers help to increase agricultural production. (c) Water boils below 100°C on top of a mountain. (d) Lead is more dense than aluminum. (e) Uranium is a radioactive element.

1.58 Does each of the following describe a physical change or a chemical change? (a) The helium gas inside a balloon tends to leak out after a few hours. (b) A flashlight beam slowly gets dimmer and finally goes out. (c) Frozen orange juice is reconstituted by adding water to it. (d) The growth of plants depends on the sun's energy in a process called photosynthesis. (e) A spoonful of table salt dissolves in a bowl of soup.

Additional Problems

1.59 Give one qualitative and one quantitative statement about each of the following: (a) water, (b) carbon, (c) iron, (d) hydrogen gas, (e) sucrose (cane sugar), (f) table salt (sodium chloride), (g) mercury, (h) gold, (i) air.

1.60 Which of the following statements describe physical properties and which describe chemical properties? (a) Iron has a tendency to rust. (b) Rainwater in industrialized regions tends to be acidic. (c) Hemoglobin molecules have a red color. (d) When a glass of water is left out in the sun, the water gradually disappears. (e) Carbon dioxide in air is converted to more complex molecules by plants during photosynthesis.

1.61 In 2008, about 95.0 billion pounds of sulfuric acid were produced in the United States. Convert this quantity to tons.

1.62 In determining the density of a rectangular metal bar, a student made the following measurements: length, 8.53 cm; width, 2.4 cm; height, 1.0 cm; mass, 52.7064 g. Calculate the density of the metal to the correct number of significant figures.

1.63 Calculate the mass of each of the following: (a) a sphere of gold with a radius of 10.0 cm [the volume of a sphere with a radius r is $V = (4/3)\pi r^3$; the density of gold = 19.3 g/cm^3], (b) a cube of platinum of edge length 0.040 mm (the density of platinum = 21.4 g/cm^3), (c) 50.0 mL of ethanol (the density of ethanol = 0.798 g/mL).

1.64 A cylindrical glass bottle 21.5 cm in length is filled with cooking oil of density 0.953 g/mL. If the mass of the oil needed to fill the bottle is 1360 g, calculate the inner diameter of the bottle.

1.65 The following procedure was used to determine the volume of a flask. The flask was weighed dry and then filled with water. If the masses of the empty flask and filled flask were 56.12 g and 87.39 g, respectively, and the density of water is 0.9976 g/cm^3, calculate the volume of the flask in cm^3.

1.66 The speed of sound in air at room temperature is about 343 m/s. Calculate this speed in miles per hour. (1 mi = 1609 m.)

1.67 A piece of silver (Ag) metal weighing 194.3 g is placed in a graduated cylinder containing 242.0 mL of water. The volume of water now reads 260.5 mL. From these data calculate the density of silver.

1.68 The experiment described in Problem 1.65 is a crude but convenient way to determine the density of some solids. Describe a similar experiment that would enable you to measure the density of ice. Specifically, what would be the requirements for the liquid used in your experiment?

1.69 A lead sphere of diameter 48.6 cm has a mass of 6.852×10^5 g. Calculate the density of lead.

1.70 Lithium is the least dense metal known (density: 0.53 g/cm^3). What is the volume occupied by 1.20×10^3 g of lithium?

1.71 The medicinal thermometer commonly used in homes can be read ±0.1°F, whereas those in the doctor's office may be accurate to ±0.1°C. In degrees Celsius, express the percent error expected from each of these thermometers in measuring a person's body temperature of 38.9°C.

1.72 Vanillin (used to flavor vanilla ice cream and other foods) is the substance whose aroma the human nose detects in the smallest amount. The threshold limit is 2.0×10^{-11} g per liter of air. If the current price of 50 g of vanillin is $112, determine the cost to supply enough vanillin so that the aroma could be detected in a large aircraft hangar with a volume of 5.0×10^7 ft^3.

1.73 At what temperature does the numerical reading on a Celsius thermometer equal that on a Fahrenheit thermometer?

1.74 Suppose that a new temperature scale has been devised on which the melting point of ethanol

(−117.3°C) and the boiling point of ethanol (78.3°C) are taken as 0°S and 100°S, respectively, where S is the symbol for the new temperature scale. Derive an equation relating a reading on this scale to a reading on the Celsius scale. What would this thermometer read at 25°C?

1.75 A resting adult requires about 240 mL of pure oxygen/min and breathes about 12 times every minute. If inhaled air contains 20% oxygen by volume and exhaled air 16%, what is the volume of air per breath? (Assume that the volume of inhaled air is equal to that of exhaled air.)

1.76 (a) Referring to Problem 1.75, calculate the total volume (in liters) of air an adult breathes in a day. (b) In a city with heavy traffic, the air contains 2.1×10^{-6} L of carbon monoxide (a poisonous gas) per liter. Calculate the average daily intake of carbon monoxide in liters by a person.

1.77 Three different 25.0-g samples of solid pellets are added to 20.0 mL of water in three different measuring cylinders. The results are shown here. Given the densities of the three metals used, identify the cylinder that contains each sample of solid pellets: A(2.9 g/cm^3), B(8.3 g/cm^3), and C(3.3 g/cm^3).

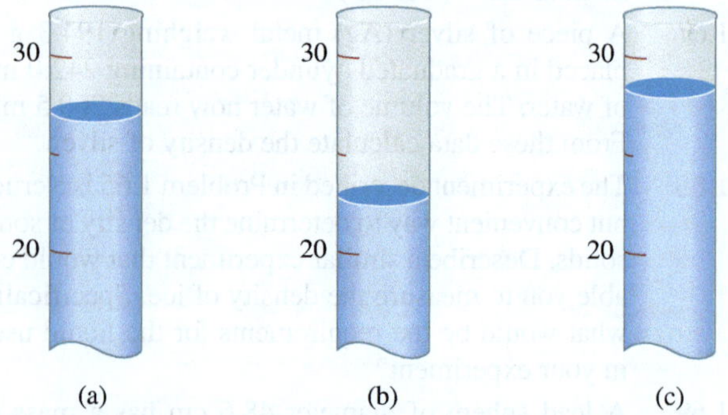

1.78 The circumference of an NBA-approved basketball is 29.6 in. Given that the radius of Earth is about 6400 km, how many basketballs would it take to circle around the equator with the basketballs touching one another? Round off your answer to an integer with three significant figures.

1.79 A student is given a crucible and asked to prove whether it is made of pure platinum. She first weighs the crucible in air and then weighs it suspended in water (density = 0.9986 g/mL). The readings are 860.2 g and 820.2 g, respectively. Based on these measurements and given that the density of platinum is 21.45 g/cm^3, what should her conclusion be? (*Hint:* An object suspended in a fluid is buoyed up by the mass of the fluid displaced by the object. Neglect the buoyance of air.)

1.80 The surface area and average depth of the Pacific Ocean are 1.8×10^8 km^2 and 3.9×10^3 m, respectively. Calculate the volume of water in the ocean in liters.

1.81 The unit "troy ounce" is often used for precious metals such as gold (Au) and platinum (Pt). (1 troy ounce = 31.103 g.) (a) A gold coin weighs 2.41 troy ounces. Calculate its mass in grams. (b) Is a troy ounce heavier or lighter than an ounce? (1 lb = 16 oz; 1 lb = 453.6 g.)

1.82 Osmium (Os) is the densest element known (density = 22.57 g/cm^3). Calculate the mass in pounds and in kilograms of an Os sphere 15 cm in diameter (about the size of a grapefruit). [The volume of a sphere of radius r is $(4/3)\pi r^3$.]

1.83 Percent error is often expressed as the absolute value of the difference between the true value and the experimental value, divided by the true value:

$$\text{percent error} = \frac{|\text{true value} - \text{experimental value}|}{|\text{true value}|} \times 100\%$$

The vertical lines indicate absolute value. Calculate the percent error for the following measurements: (a) The density of alcohol (ethanol) is found to be 0.802 g/mL. (True value: 0.798 g/mL.) (b) The mass of gold in an earring is analyzed to be 0.837 g. (True value: 0.864 g.)

1.84 The natural abundances of elements in the human body, expressed as percent by mass, are: oxygen (O), 65 percent; carbon (C), 18 percent; hydrogen (H), 10 percent; nitrogen (N), 3 percent; calcium (Ca), 1.6 percent; phosphorus (P), 1.2 percent; all other elements, 1.2 percent. Calculate the mass in grams of each element in the body of a 62-kg person.

1.85 The men's world record for running a mile outdoors (as of 1999) is 3 min 43.13 s. At this rate, how long would it take to run a 1500-m race? (1 mi = 1609 m.)

1.86 Venus, the second closest planet to the sun, has a surface temperature of 7.3×10^2 K. Convert this temperature to °C and °F.

1.87 Chalcopyrite, the principal ore of copper (Cu), contains 34.63% Cu by mass. How many grams of Cu can be obtained from 5.11×10^3 kg of the ore?

1.88 It has been estimated that 8.0×10^4 tons of gold (Au) have been mined. Assume gold costs $948 per ounce. What is the total worth of this quantity of gold?

1.89 A 1.0-mL volume of seawater contains about 4.0×10^{-12} g of gold. The total volume of ocean water is 1.5×10^{21} L. Calculate the total amount of gold (in grams) that is present in seawater, and the worth of the gold in dollars (see Problem 1.88). With so much gold out there, why hasn't someone become rich by mining gold from the ocean?

1.90 Measurements show that 1.0 g of iron (Fe) contains 1.1×10^{22} Fe atoms. How many Fe atoms are in 4.9 g of Fe, which is the total amount of iron in the body of an average adult?

1.91 The thin outer layer of Earth, called the crust, contains only 0.50% of Earth's total mass and yet is the source of almost all the elements (the atmosphere provides elements such as oxygen, nitrogen, and a few other gases). Silicon (Si) is the second most abundant element in Earth's crust (27.2% by mass).

Calculate the mass of silicon in kilograms in Earth's crust. (The mass of Earth is 5.9×10^{21} tons. 1 ton = 2000 lb; 1 lb = 453.6 g.)

1.92 The radius of a copper (Cu) atom is roughly 1.3×10^{-10} m. How many times can you divide evenly a piece of 10-cm copper wire until it is reduced to two separate copper atoms? (Assume there are appropriate tools for this procedure and that copper atoms are lined up in a straight line, in contact with each other. Round off your answer to an integer.)

1.93 One gallon of gasoline in an automobile's engine produces on the average 9.5 kg of carbon dioxide, which is a greenhouse gas, that is, it promotes the warming of Earth's atmosphere. Calculate the annual production of carbon dioxide in kilograms if there are 250 million cars in the United States and each car covers a distance of 5000 mi at a consumption rate of 20 miles per gallon.

1.94 A sheet of aluminum (Al) foil has a total area of 1.000 ft^2 and a mass of 3.636 g. What is the thickness of the foil in millimeters? (Density of Al = 2.699 g/cm^3.)

1.95 Comment on whether each of the following is a homogeneous mixture or a heterogeneous mixture: (a) air in a closed bottle, (b) air over New York City.

1.96 Chlorine is used to disinfect swimming pools. The accepted concentration for this purpose is 1 ppm chlorine, or 1 g of chlorine per million grams of water. Calculate the volume of a chlorine solution (in milliliters) a homeowner should add to her swimming pool if the solution contains 6.0% chlorine by mass and there are 2.0×10^4 gallons of water in the pool. (1 gallon = 3.79 L; density of liquids = 1.0 g/mL.)

1.97 An aluminum cylinder is 10.0 cm in length and has a radius of 0.25 cm. If the mass of a single Al atom is 4.48×10^{-23} g, calculate the number of Al atoms present in the cylinder. The density of aluminum is 2.70 g/cm^3.

1.98 A pycnometer is a device for measuring the density of liquids. It is a glass flask with a close-fitting ground glass stopper having a capillary hole through it. (a) The volume of the pycnometer is determined by using distilled water at 20°C with a known density of 0.99820 g/mL. First, the water is filled to the rim. With the stopper in place, the fine hole allows the excess liquid to escape. The pycnometer is then carefully dried with filter paper. Given that the masses of the empty pycnometer and the same one filled with water are 32.0764 g and 43.1195 g, respectively, calculate the volume of the pycnometer. (b) If the mass of the pycnometer filled with ethanol at 20°C is 40.8051 g, calculate the density of ethanol. (c) Pycnometers can also be used to measure the density of solids. First, small zinc granules weighing 22.8476 g are placed in the pycnometer, which is then filled with water. If the combined mass of the pycnometer plus the zinc granules and water is 62.7728 g, what is the density of zinc?

H. S. Photos/Science Source

1.99 In 1849 a gold prospector in California collected a bag of gold nuggets plus sand. Given that the density of gold and sand are 19.3 g/cm^3 and 2.95 g/cm^3, respectively, and that the density of the mixture is 4.17 g/cm^3, calculate the percent by mass of gold in the mixture.

1.100 The average time it takes for a molecule to diffuse a distance of x cm is given by

$$t = \frac{x^2}{2D}$$

where t is the time in seconds and D is the diffusion coefficient. Given that the diffusion coefficient of glucose is 5.7×10^{-7} cm^2/s, calculate the time it would take for a glucose molecule to diffuse 10 μm, which is roughly the size of a cell.

1.101 A human brain weighs about 1 kg and contains about 10^{11} cells. Assuming that each cell is completely filled with water (density = 1 g/mL), calculate the length of one side of such a cell if it were a cube. If the cells are spread out in a thin layer that is a single cell thick, what is the surface area in square meters?

1.102 (a) Carbon monoxide (CO) is a poisonous gas because it binds very strongly to the oxygen carrier hemoglobin in blood. A concentration of 8.00×10^2 ppm by volume of carbon monoxide is considered lethal to humans. Calculate the volume in liters occupied by carbon monoxide in a room that measures 17.6 m long, 8.80 m wide, and 2.64 m high at this concentration. (b) Prolonged exposure to mercury (Hg) vapor can cause neurological disorders and respiratory problems. For safe air quality control, the concentration of mercury vapor must be under 0.050 mg/m^3. Convert this number to g/L. (c) The general test for type II diabetes is that the blood sugar (glucose) level should be below 120 mg per deciliter (mg/dL). Convert this number to micrograms per milliliter (μg/mL).

1.103 A bank teller is asked to assemble "one-dollar" sets of coins for his clients. Each set is made of three quarters, one nickel, and two dimes. The masses of the coins are: quarter: 5.645 g; nickel: 4.967 g; dime: 2.316 g. What is the maximum number of sets that can be assembled from 33.871 kg of quarters, 10.432 kg of nickels, and 7.990 kg of dimes? What is the total mass (in g) of the assembled sets of coins?

1.104 A graduated cylinder is filled to the 40.00-mL mark with a mineral oil. The masses of the cylinder before and after the addition of the mineral oil are 124.966 g and 159.446 g, respectively. In a separate experiment, a metal ball bearing of mass 18.713 g is placed in the cylinder and the cylinder is again filled to the 40.00-mL mark with the mineral oil. The combined mass of the ball bearing and mineral oil is 50.952 g. Calculate the density and radius of the ball bearing. [The volume of a sphere of radius r is $(4/3)\pi r^3$.]

1.105 A cobalt bar (density = 8.90 g/cm^3) is shown here. What is the mass of this bar to the appropriate number of significant figures?

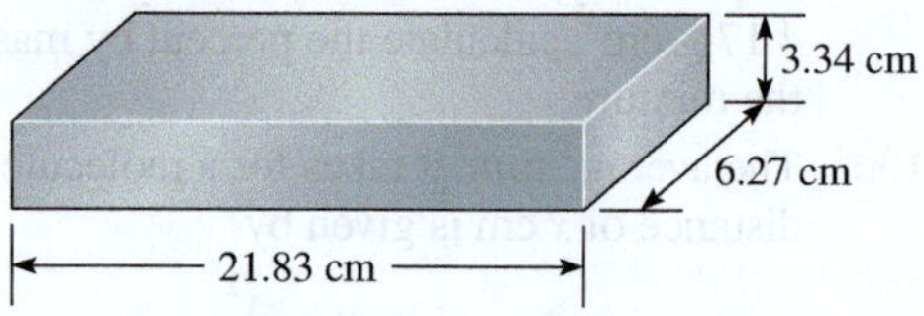

1.106 Bronze is an alloy made of copper (Cu) and tin (Sn) used in applications that require low metal-on-metal friction. Calculate the mass of a bronze cylinder of radius 6.44 cm and length 44.37 cm. The composition of the bronze is 79.42% Cu and 20.58% Sn and the densities of Cu and Sn are 8.94 g/cm^3 and 7.31 g/cm^3, respectively. What assumption should you make in this calculation?

1.107 You are given a liquid. Briefly describe steps you would take to show whether it is a pure substance or a homogeneous mixture.

1.108 A chemist mixes two liquids A and B to form a homogeneous mixture. The densities of the liquids are 2.0514 g/mL for A and 2.6678 g/mL for B. When she drops a small object into the mixture, she finds that the object becomes suspended in the liquid; that is, it neither sinks nor floats. If the mixture is made of 41.37% A and 58.63% B by volume, what is the density of the metal? Can this procedure be used in general to determine the densities of solids? What assumptions must be made in applying this method?

1.109 Tums is a popular remedy for acid indigestion. A typical Tums tablet contains calcium carbonate plus some inert substances. When ingested, it reacts with the gastric juice (hydrochloric acid) in the stomach to give off carbon dioxide gas. When a 1.328-g tablet reacted with 40.00 mL of hydrochloric acid (density: 1.140 g/mL), carbon dioxide gas was given off and the resulting solution weighed 46.699 g. Calculate the number of liters of carbon dioxide gas released if its density is 1.81 g/L.

1.110 A 250-mL glass bottle was filled with 242 mL of water at 20°C and tightly capped. It was then left outdoors overnight, where the average temperature was −5°C. Predict what would happen. The density of water at 20°C is 0.998 g/cm^3 and that of ice at −5°C is 0.916 g/cm^3.

Interpreting, Modeling, & Estimating

1.111 Fusing "nanofibers" with diameters of 100 to 300 nm gives junctures with very small volumes that would potentially allow the study of reactions involving only a few molecules. Estimate the volume in liters of the junction formed between two such fibers with internal diameters of 200 nm. The scale reads 1 μm.

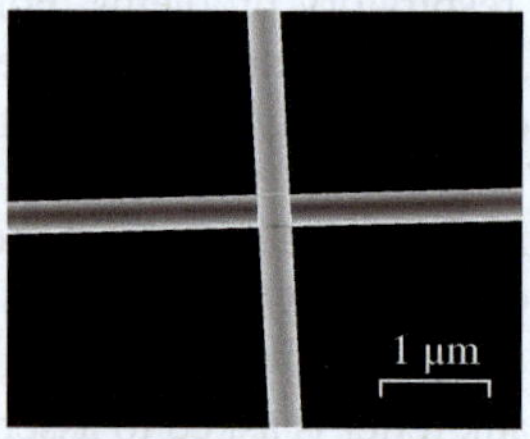

1.112 Estimate the annual consumption of gasoline by passenger cars in the United States.

1.113 Estimate the total amount of ocean water in liters.

1.114 Estimate the volume of blood in an adult in liters.

1.115 Estimate the distance (in miles) covered by an NBA player in a professional basketball game.

1.116 In water conservation, chemists spread a thin film of a certain inert material over the surface of water to cut down on the rate of evaporation of water in reservoirs. This technique was pioneered by Benjamin Franklin three centuries ago. Franklin found that 0.10 mL of oil could spread over the surface of water about 40 m^2 in area. Assuming that the oil forms a *monolayer,* that is, a layer that is only one molecule thick, estimate the length of each oil molecule in nanometers. (1 nm = 1×10^{-9} m.)

Answers to Practice Exercises

1.1 96.5 g. **1.2** 341 g. **1.3** (a) 621.5°F, (b) 78.3°C, (c) −196°C. **1.4** (a) Two, (b) four, (c) three, (d) two, (e) three or two. **1.5** (a) 26.76 L, (b) 4.4 g, (c) 1.6×10^7 dm^2, (d) 0.0756 g/ml, (e) 6.69×10^4 m, (f) 2.0386, (g) 1.71×10^{17}. **1.6** 2.36 lb. **1.7** 1.08×10^5 m^3. **1.8** 0.534 g/cm^3. **1.9** Roughly 0.03 g.

Answers to Review of Concepts & Facts

1.2.1 (c). **1.2.2** Qualitative: (b), (c). Quantitative: (a), (d). **1.3.1** 0.522 cm^3. **1.3.2** 543 K. **1.3.3** (a). **1.4.1** Top ruler = 4.6 cm. Bottom ruler = 4.57 cm. **1.4.2** 10.1 g/cm^3. **1.4.3.** Accurate: no. Precise: yes. **1.5.1** 84.6 kcal. **1.5.2** 1.42 L. **1.5.3** 1.45×10^4 J/g. **1.6.1** 1 ft. **1.7.1** Elements: (b), (d). Compounds: (a), (c). **1.8.1** (a). **1.9.1** Chemical change: (b), (c). Physical change: (d). **1.9.2** Intensive: (a), (b). Extensive: (c).

CHAPTER

2

Atoms, Ions, and Molecules

The nuclear model of the atom with a nucleus and orbiting electrons was devised from the work of Ernest Rutherford.

zoom-zoom/iStock/Getty Images

CHAPTER OUTLINE

Since ancient times humans have pondered the nature of matter. Our modern ideas of the structure of matter began to take shape in the early nineteenth century with Dalton's atomic theory. We now know that all matter is made of atoms, molecules, and ions. All of chemistry is concerned in one way or another with these species.

2.1 The Atomic Theory

Learning Objectives

- Outline Dalton's hypotheses about the nature of matter.
- Differentiate the law of definite proportions, law of multiple proportions, and law of conservation of mass.

In the fifth century B.C. the Greek philosopher Democritus expressed the belief that all matter consists of very small, indivisible particles, which he named *atomos* (meaning "uncuttable or indivisible"). Although Democritus's idea was not accepted by many of his contemporaries (notably Plato and Aristotle), somehow it endured. Experimental evidence from early scientific investigations provided support for the notion of "atomism" and gradually gave rise to the modern definitions of elements and compounds. In 1808 an English scientist and schoolteacher, John Dalton,† formulated a precise definition of the indivisible building blocks of matter that we call atoms.

Dalton's work marked the beginning of the modern era of chemistry. According to Dalton's atomic theory, the hypotheses about the nature of matter can be summarized as follows:

1. Elements are composed of extremely small particles called atoms.
2. All atoms of a given element are identical, having the same size, mass, and chemical properties. The atoms of one element are different from the atoms of all other elements.
3. Compounds are composed of atoms of more than one element. For any given compound, the atoms present are always in the same ratio.
4. A chemical reaction involves only the separation, combination, or rearrangement of atoms; it does not result in the creation or destruction of atoms.

Figure 2.1 is a schematic representation of the last three hypotheses.

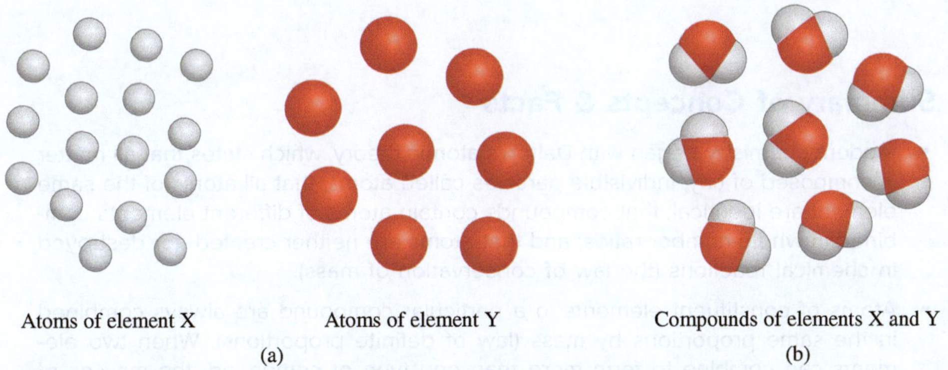

Figure 2.1 *(a) According to Dalton's atomic theory, atoms of the same element are identical, but atoms of one element are different from atoms of other elements. (b) Compound formed from atoms of elements X and Y. In this case, the ratio of the atoms of element X to the atoms of element Y is 2:1. Note that a chemical reaction results only in the rearrangement of atoms, not in their destruction or creation.*

†John Dalton (1766–1844). English chemist, mathematician, and philosopher. In addition to the atomic theory, he also formulated several gas laws and gave the first detailed description of color blindness, which he experienced. Dalton was described as an indifferent experimenter, and singularly wanting in the language and power of illustration. His only recreation was lawn bowling on Thursday afternoons. Perhaps it was the sight of those wooden balls that provided him with the idea of the atomic theory.

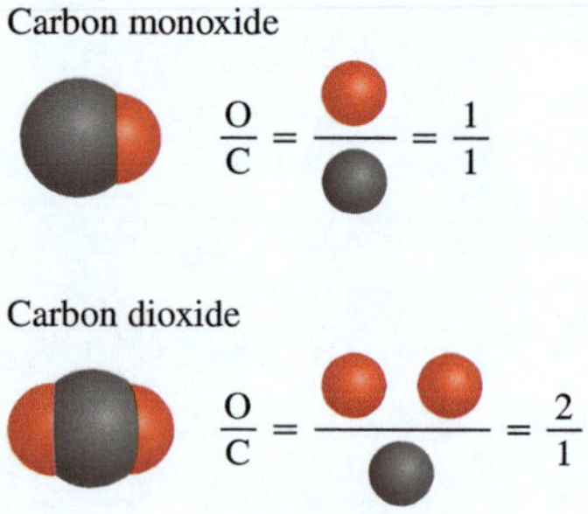

Figure 2.2 *An illustration of the law of multiple proportions.*

Dalton's concept of an atom was far more detailed and specific than Democritus's concept. The second hypothesis states that atoms of one element are different from atoms of all other elements. Dalton made no attempt to describe the structure or composition of atoms—he had no idea what an atom is really like. But he did realize that the different properties shown by elements such as hydrogen and oxygen can be explained by assuming that hydrogen atoms are not the same as oxygen atoms.

The third hypothesis suggests that, to form a certain compound, we need not only atoms of the right kinds of elements, but specific numbers of these atoms as well. This idea is an extension of a law published in 1799 by Joseph Proust,† a French chemist. Proust's ***law of definite proportions*** states that *different samples of the same compound always contain its constituent elements in the same proportion by mass.* Thus, if we were to analyze samples of carbon dioxide gas obtained from different sources, we would find in each sample the same ratio by mass of carbon to oxygen. It stands to reason that if the ratio of the masses of different elements in a given compound is fixed, the ratio of the atoms of these elements in the compound also must be constant.

Dalton's third hypothesis supports another important law, the ***law of multiple proportions.*** According to the law, *if two elements can combine to form more than one compound, the masses of one element that combine with a fixed mass of the other element are in ratios of small whole numbers.* Dalton's theory explains the law of multiple proportions quite simply: Different compounds made up of the same elements differ in the number of atoms of each kind that combine. For example, carbon forms two stable compounds with oxygen—namely, carbon monoxide and carbon dioxide. Modern measurement techniques indicate that one atom of carbon combines with one atom of oxygen in carbon monoxide and with two atoms of oxygen in carbon dioxide. Thus, the ratio of oxygen in carbon monoxide to oxygen in carbon dioxide is 1:2. This result is consistent with the law of multiple proportions (Figure 2.2).

Dalton's fourth hypothesis is another way of stating the ***law of conservation of mass,*** which is that *matter can be neither created nor destroyed.* Because matter is made of atoms that are unchanged in a chemical reaction, it follows that mass must be conserved as well. Dalton's brilliant insight into the nature of matter was the main stimulus for the rapid progress of chemistry during the nineteenth century. (Note that according to Albert Einstein, mass and energy are alternate aspects of a single entity called *mass-energy.* Chemical reactions usually involve a gain or loss of heat and other forms of energy. Thus, when energy is lost in a reaction, for example, mass is also lost. Except for nuclear reactions [see Chapter 19], however, changes of mass in chemical reactions are generally too small to detect. Therefore, for all practical purposes mass is conserved.)

Summary of Concepts & Facts

- Modern chemistry began with Dalton's atomic theory, which states that all matter is composed of tiny, indivisible particles called atoms; that all atoms of the same element are identical; that compounds contain atoms of different elements combined in whole-number ratios; and that atoms are neither created nor destroyed in chemical reactions (the law of conservation of mass).
- Atoms of constituent elements in a particular compound are always combined in the same proportions by mass (law of definite proportions). When two elements can combine to form more than one type of compound, the masses of one element that combine with a fixed mass of the other element are in a ratio of small whole numbers (law of multiple proportions).

†Joseph Louis Proust (1754–1826). French chemist. Proust was the first person to isolate sugar from grapes.

Review of Concepts & Facts

2.1.1 The atoms of elements A (blue) and B (red) form two compounds shown here. Do these compounds obey the law of multiple proportions?

2.1.2 Consider the chemical process shown. Which of the mass laws (mass conservation, definite proportions, and multiple proportions) does this process demonstrate?

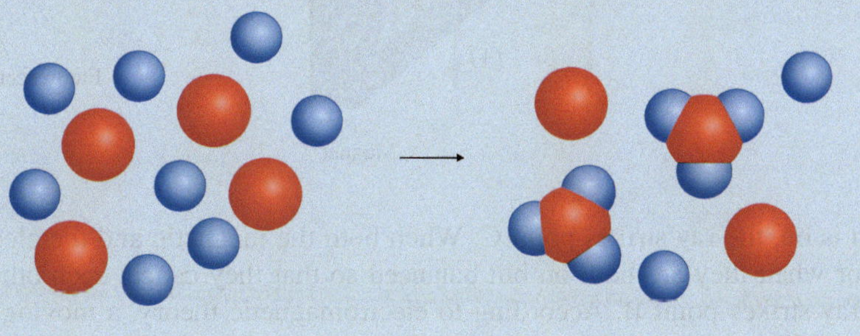

2.2 The Structure of the Atom

Learning Objectives

- Understand the concept of the atom and the nature of an element.
- Assess the importance of experiments conducted by Thomson, Millikan, Röntgen, and Rutherford, and how they influenced our understanding of the nature and structure of atoms.
- Summarize the different types of radiation that radioactive substances can produce.

On the basis of Dalton's atomic theory, we can define an ***atom*** as *the basic unit of an element that can enter into chemical combination.* Dalton imagined an atom that was both extremely small and indivisible. However, a series of investigations that began in the 1850s and extended into the twentieth century clearly demonstrated that atoms actually possess internal structure; that is, they are made up of even smaller particles, which are called *subatomic particles.* This research led to the discovery of three such particles—electrons, protons, and neutrons.

The Electron

In the 1890s, many scientists became caught up in the study of ***radiation,*** *the emission and transmission of energy through space in the form of waves.* Information gained from this research contributed greatly to our understanding of atomic structure. One device used to investigate this phenomenon was a cathode ray tube, the forerunner of the television tube (Figure 2.3). It is a glass tube from which most of the air has been evacuated. When the two metal plates are connected to a high-voltage source, the negatively charged plate, called the *cathode,* emits an invisible ray. The cathode ray is drawn to the positively charged plate, called the *anode,* where it passes through a hole and continues traveling to the other end of the tube. When the ray strikes the specially coated surface, it produces a strong fluorescence, or bright light.

Video: Cathode Ray Tube

In some experiments, two electrically charged plates and a magnet were added to the *outside* of the cathode ray tube (see Figure 2.3). When the magnetic field is on and the electric field is off, the cathode ray strikes point A. When only the electric

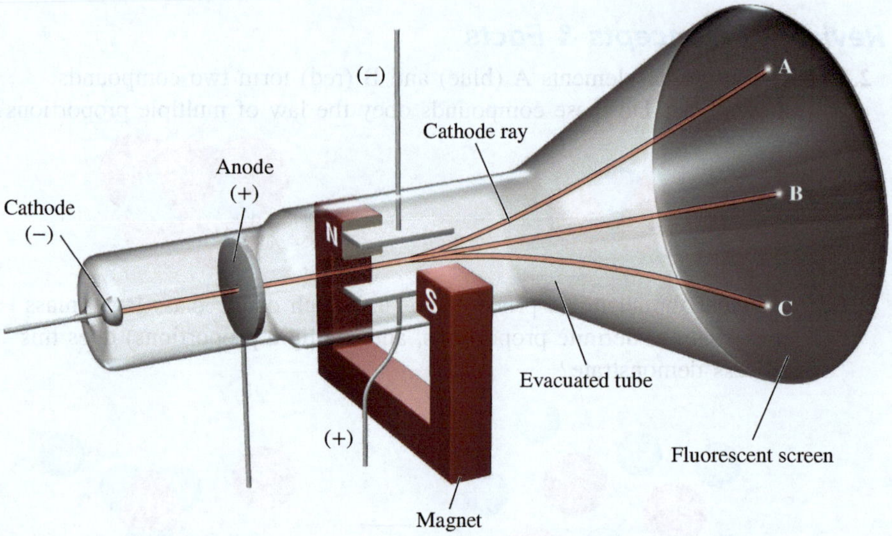

Figure 2.3 *A cathode ray tube with an electric field perpendicular to the direction of the cathode rays and an external magnetic field. The symbols N and S denote the north and south poles of the magnet. The cathode rays will strike the end of the tube at A in the presence of a magnetic field, at C in the presence of an electric field, and at B when there are no external fields present or when the effects of the electric field and magnetic field cancel each other.*

field is on, the ray strikes point C. When both the magnetic and the electric fields are off or when they are both on but balanced so that they cancel each other's influence, the ray strikes point B. According to electromagnetic theory, a moving charged body behaves like a magnet and can interact with electric and magnetic fields through which it passes. Because the cathode ray is attracted by the plate bearing positive charges and repelled by the plate bearing negative charges, it must consist of negatively charged particles. We know these *negatively charged particles* as ***electrons.*** Figure 2.4 shows the effect of a bar magnet on the cathode ray.

Video: Millikan Oil Drop

An English physicist, J. J. Thomson,† used a cathode ray tube and his knowledge of electromagnetic theory to determine the ratio of electric charge to the mass of an individual electron. The number he came up with was -1.76×10^8 C/g, where C stands for *coulomb,* which is the unit of electric charge. Thereafter, in a series of experiments carried out between 1908 and 1917, R. A. Millikan‡ succeeded in measuring the charge of the electron with great precision. His work proved that the charge on each electron was exactly the same. In his experiment, Millikan examined the motion of single tiny drops of oil that picked up static charge from ions in the air. He suspended the charged drops in air by applying an electric field and followed their motions through a microscope (Figure 2.5). Using his knowledge of electrostatics,

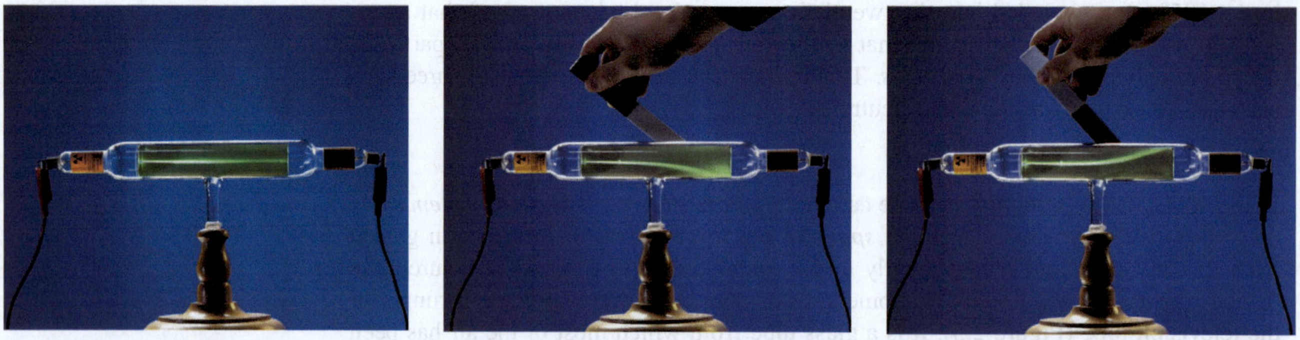

(a) (b) (c)

Figure 2.4 *(a) A cathode ray produced in a discharge tube traveling from the cathode (left) to the anode (right). The ray itself is invisible, but the fluorescence of a zinc sulfide coating on the glass causes it to appear green. (b) The cathode ray is bent downward when a bar magnet is brought toward it. (c) When the polarity of the magnet is reversed, the ray bends in the opposite direction.*
Charles D. Winters/McGraw Hill

†Joseph John Thomson (1856–1940). British physicist who received the Nobel Prize in Physics in 1906 for discovering the electron.

‡Robert Andrews Millikan (1868–1953). American physicist who was awarded the Nobel Prize in Physics in 1923 for determining the charge of the electron.

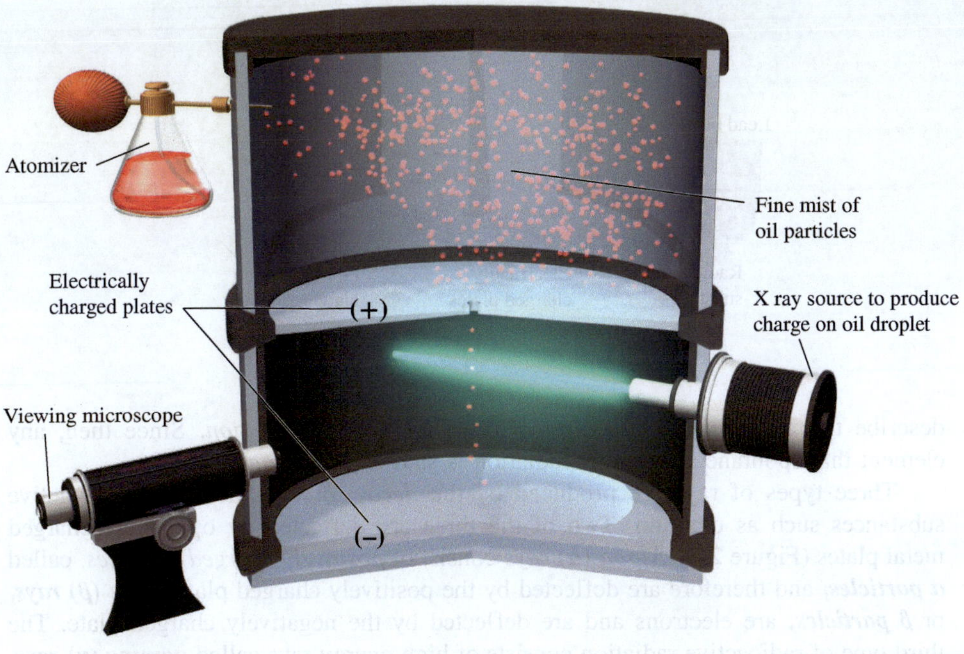

Figure 2.5 *Schematic diagram of Millikan's oil drop experiment.*

Millikan found the charge of an electron to be -1.6022×10^{-19} C. From these data he calculated the mass of an electron:

$$\text{mass of an electron} = \frac{\text{charge}}{\text{charge/mass}}$$
$$= \frac{-1.6022 \times 10^{-19}\ \text{C}}{-1.76 \times 10^{8}\ \text{C/g}}$$
$$= 9.10 \times 10^{-28}\ \text{g}$$

This is an exceedingly small mass.

Radioactivity

In 1895 the German physicist Wilhelm Röntgen[§] noticed that cathode rays caused glass and metals to emit very unusual rays. This highly energetic radiation penetrated matter, darkened covered photographic plates, and caused a variety of substances to fluoresce. Because these rays could not be deflected by a magnet, they could not contain charged particles as cathode rays do. Röntgen called them X rays because their nature was not known.

Not long after Röntgen's discovery, Antoine Becquerel,[†] a professor of physics in Paris, began to study the fluorescent properties of substances. Purely by accident, he found that exposing thickly wrapped photographic plates to a certain uranium compound caused them to darken, even without the stimulation of cathode rays. Like X rays, the rays from the uranium compound were highly energetic and could not be deflected by a magnet, but they differed from X rays because they arose spontaneously. One of Becquerel's students, Marie Curie,[‡] suggested the name ***radioactivity*** to

[§]Wilhelm Konrad Röntgen (1845–1923). German physicist who received the Nobel Prize in Physics in 1901 for the discovery of X rays.

[†]Antoine Henri Becquerel (1852–1908). French physicist who was awarded the Nobel Prize in Physics in 1903 for discovering radioactivity in uranium.

[‡]Marie (Marya Sklodowska) Curie (1867–1934). Polish-born chemist and physicist. In 1903 she and her French husband, Pierre Curie, were awarded the Nobel Prize in Physics for their work on radioactivity. In 1911, she again received the Nobel prize, this time in chemistry, for her work on the radioactive elements radium and polonium. She is one of only four people to have received two Nobel prizes in science. Despite her great contribution to science, her nomination to the French Academy of Sciences in 1911 was rejected by one vote because she was a woman! Her daughter Irene, and son-in-law Frederic Joliot-Curie, shared the Nobel Prize in Chemistry in 1935.

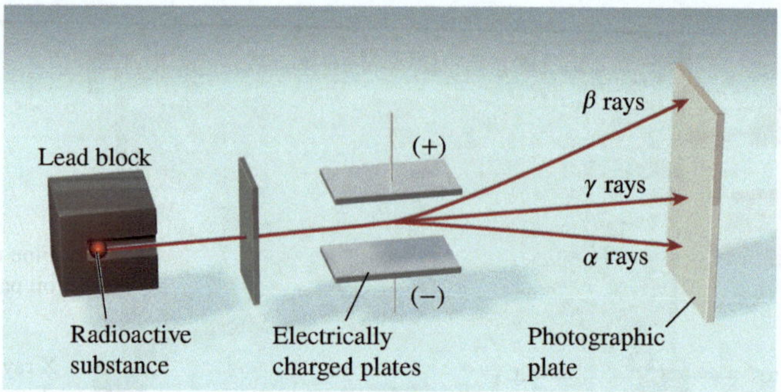

Figure 2.6 *Three types of rays emitted by radioactive elements. β rays consist of negatively charged particles (electrons) and are therefore attracted by the positively charged plate. The opposite holds true for α rays—they are positively charged and are drawn to the negatively charged plate. Because γ rays have no charges, their path is unaffected by an external electric field.*

describe this *spontaneous emission of particles and/or radiation.* Since then, any element that spontaneously emits radiation is said to be *radioactive.*

Video: Alpha, Beta, and Gamma Rays

Three types of rays are produced by the *decay,* or breakdown, of radioactive substances such as uranium. Two of the three are deflected by oppositely charged metal plates (Figure 2.6). ***Alpha (α) rays*** consist of *positively charged particles,* called ***α particles,*** and therefore are deflected by the positively charged plate. ***Beta (β) rays,*** or ***β particles,*** are electrons and are deflected by the negatively charged plate. The third type of radioactive radiation consists of high-energy rays called ***gamma (γ) rays.*** Like X rays, γ rays have no charge and are not affected by an external field.

The Proton and the Nucleus

By the early 1900s, two features of atoms had become clear: They contain electrons, and they are electrically neutral. To maintain electric neutrality, an atom must contain an equal number of positive and negative charges. Therefore, Thomson proposed that an atom could be thought of as a uniform, positive sphere of matter in which electrons are embedded like raisins in a cake (Figure 2.7). This so-called "plum-pudding" model was the accepted theory for a number of years.

Video: α-Particle Scattering

Video: Rutherford's Experiment

In 1910 the New Zealand physicist Ernest Rutherford,[†] who had studied with Thomson at Cambridge University, decided to use α particles to probe the structure of atoms. Together with his associate Hans Geiger[‡] and an undergraduate named Ernest Marsden,[§] Rutherford carried out a series of experiments using very thin foils of gold and other metals as targets for α particles from a radioactive source (Figure 2.8). They observed that the majority of particles penetrated the foil either undeflected or with only a slight deflection. But every now and then an α particle was scattered (or deflected) at a large angle. In some instances, an α particle actually bounced back in the direction from which it had come! This was a most surprising finding, for in Thomson's model the positive charge of the atom was so diffuse that the positive α particles should have passed through the foil with very little deflection. To quote Rutherford's initial reaction when told of this discovery: "It was as incredible as if you had fired a 15-inch shell at a piece of tissue paper and it came back and hit you."

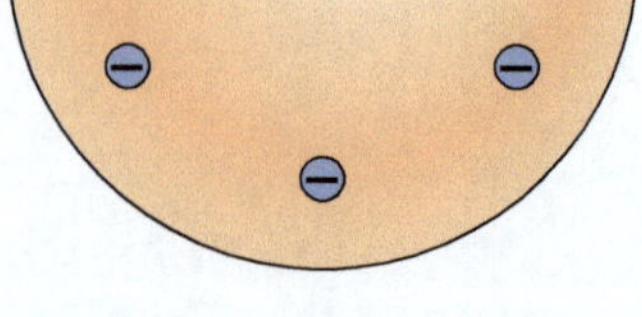

Figure 2.7 *Thomson's model of the atom, sometimes described as the "plum-pudding" model, after a traditional English dessert containing raisins. The electrons are embedded in a uniform, positively charged sphere.*

[†]Ernest Rutherford (1871–1937). New Zealand physicist. Rutherford did most of his work in England (Manchester and Cambridge Universities). He received the Nobel Prize in Chemistry in 1908 for his investigations into the structure of the atomic nucleus. His often-quoted comment to his students was that "all science is either physics or stamp-collecting."

[‡]Johannes Hans Wilhelm Geiger (1882–1945). German physicist. Geiger's work focused on the structure of the atomic nucleus and on radioactivity. He invented a device for measuring radiation that is now commonly called the Geiger counter.

[§]Ernest Marsden (1889–1970). English physicist. It is gratifying to know that at times an undergraduate can assist in winning a Nobel prize. Marsden went on to contribute significantly to the development of science in New Zealand.

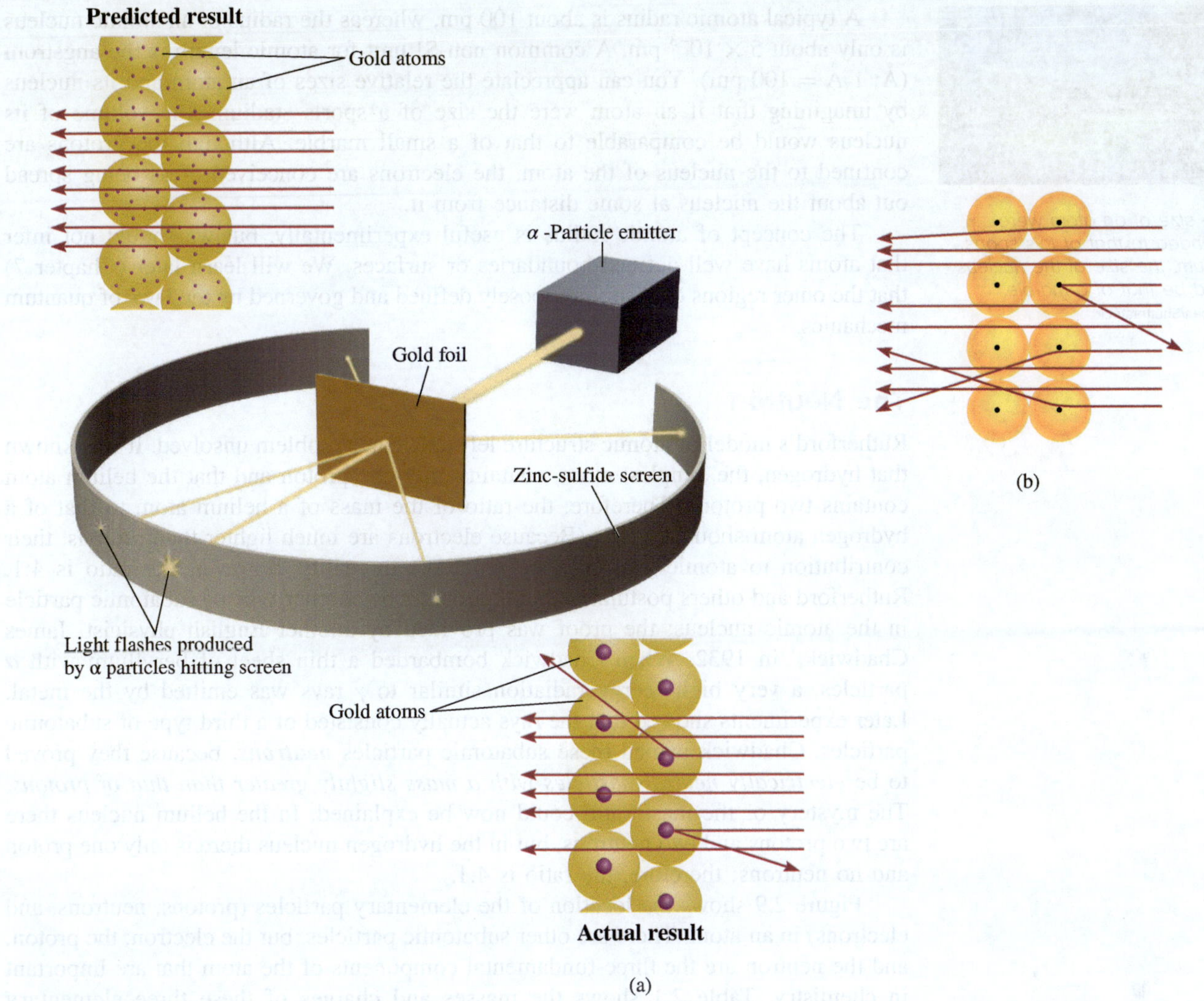

Figure 2.8 *(a) Rutherford's experimental design for measuring the scattering of α particles by a piece of gold foil. Most of the α particles passed through the gold foil with little or no deflection. A few were deflected at wide angles. Occasionally an α particle was turned back. (b) Magnified view of α particles passing through and being deflected by nuclei.*

Rutherford explained the results of the α-scattering experiment by introducing a new model for the atom. According to Rutherford, most of the atom must be empty space. This explains why the majority of α particles passed through the gold foil with little or no deflection. The atom's positive charges, Rutherford proposed, are all concentrated in the ***nucleus,*** which is *a dense central core within the atom.* Whenever an α particle came close to a nucleus in the scattering experiment, it experienced a large repulsive force and therefore a large deflection. Moreover, an α particle traveling directly toward a nucleus would be completely repelled and its direction would be reversed.

The positively charged particles in the nucleus are called ***protons.*** In separate experiments, it was found that each proton carries the same *quantity* of charge as an electron and has a mass of 1.67262×10^{-24} g—about 1840 times the mass of the oppositely charged electron.

At this stage of investigation, scientists perceived the atom as follows: The mass of a nucleus constitutes most of the mass of the entire atom, but the nucleus occupies only about $1/10^{13}$ of the volume of the atom. We express atomic (and molecular) dimensions in terms of the SI unit called the *picometer (pm),* where

$$1 \text{ pm} = 1 \times 10^{-12} \text{ m}$$

If the size of an atom were expanded to that of this sports stadium, the size of the nucleus would be that of a marble.
Kanuman/Shutterstock

A typical atomic radius is about 100 pm, whereas the radius of an atomic nucleus is only about 5×10^{-3} pm. A common non-SI unit for atomic length is the angstrom (Å; 1 Å = 100 pm). You can appreciate the relative sizes of an atom and its nucleus by imagining that if an atom were the size of a sports stadium, the volume of its nucleus would be comparable to that of a small marble. Although the protons are confined to the nucleus of the atom, the electrons are conceived of as being spread out about the nucleus at some distance from it.

The concept of atomic radius is useful experimentally, but we should not infer that atoms have well-defined boundaries or surfaces. We will learn later (Chapter 7) that the outer regions of atoms are loosely defined and governed by the laws of quantum mechanics.

The Neutron

Rutherford's model of atomic structure left one major problem unsolved. It was known that hydrogen, the simplest atom, contains only one proton and that the helium atom contains two protons. Therefore, the ratio of the mass of a helium atom to that of a hydrogen atom should be 2:1. (Because electrons are much lighter than protons, their contribution to atomic mass can be ignored.) In reality, however, the ratio is 4:1. Rutherford and others postulated that there must be another type of subatomic particle in the atomic nucleus; the proof was provided by another English physicist, James Chadwick,† in 1932. When Chadwick bombarded a thin sheet of beryllium with α particles, a very high-energy radiation similar to γ rays was emitted by the metal. Later experiments showed that the rays actually consisted of a third type of subatomic particles. Chadwick named these subatomic particles ***neutrons,*** because they proved to be *electrically neutral particles with a mass slightly greater than that of protons.* The mystery of the mass ratio could now be explained. In the helium nucleus there are two protons and two neutrons, but in the hydrogen nucleus there is only one proton and no neutrons; therefore, the ratio is 4:1.

Figure 2.9 shows the location of the elementary particles (protons, neutrons, and electrons) in an atom. There are other subatomic particles, but the electron, the proton, and the neutron are the three fundamental components of the atom that are important in chemistry. Table 2.1 shows the masses and charges of these three elementary particles.

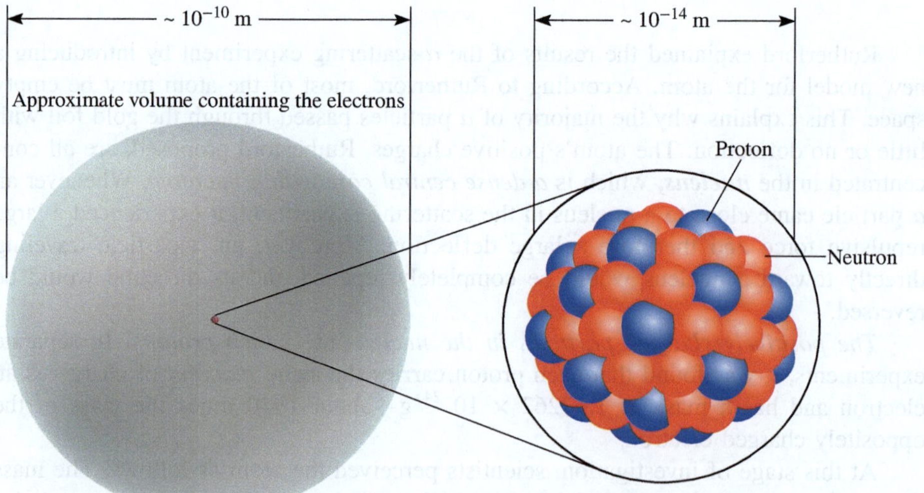

Figure 2.9 *The protons and neutrons of an atom are packed in an extremely small nucleus. Electrons are shown as "clouds" around the nucleus.*

†James Chadwick (1891–1972). British physicist. In 1935 he received the Nobel Prize in Physics for proving the existence of neutrons.

Table 2.1 Mass and Charge of Subatomic Particles

		Charge	
Particle	**Mass (g)**	**Coulomb**	**Charge Unit**
Electron*	9.10938×10^{-28}	-1.6022×10^{-19}	−1
Proton	1.67262×10^{-24}	$+1.6022 \times 10^{-19}$	+1
Neutron	1.67493×10^{-24}	0	0

*More refined measurements have given us a more accurate value of an electron's mass than Millikan's.

Summary of Concepts & Facts

- An atom consists of a very dense central nucleus containing protons and neutrons, with electrons moving about the nucleus at a relatively large distance from it.
- Protons are positively charged, neutrons have no charge, and electrons are negatively charged. Protons and neutrons have roughly the same mass, which is about 1840 times greater than the mass of an electron.

2.3 Atomic Number, Mass Number, and Isotopes

Learning Objectives

- Explain the nature and importance of isotopes.
- Calculate the mass number of an isotope.

All atoms can be identified by the number of protons and neutrons they contain. The ***atomic number (Z)*** is *the number of protons in the nucleus of each atom of an element.* In a neutral atom the number of protons is equal to the number of electrons, so the atomic number also indicates the number of electrons present in the atom. The chemical identity of an atom can be determined solely from its atomic number. For example, the atomic number of fluorine is 9. This means that each fluorine atom has 9 protons and 9 electrons. Or, viewed another way, every atom in the universe that contains 9 protons is correctly named "fluorine."

The ***mass number (A)*** is *the total number of neutrons and protons present in the nucleus of an atom of an element.* Protons and neutrons in the nucleus are collectively known as nucleons. Except for the most common form of hydrogen, which has one proton and no neutrons, all atomic nuclei contain both protons and neutrons. In general, the mass number is given by

$$\begin{aligned}\text{mass number} &= \text{number of protons} + \text{number of neutrons}\\ &= \text{atomic number} + \text{number of neutrons}\end{aligned} \tag{2.1}$$

The number of neutrons in an atom is equal to the difference between the mass number and the atomic number, or $(A - Z)$. For example, if the mass number of a particular boron atom is 12 and the atomic number is 5 (indicating 5 protons in the nucleus), then the number of neutrons is $12 - 5 = 7$. Note that all three quantities (atomic number, number of neutrons, and mass number) must be positive integers, or whole numbers.

Atoms of a given element do not all have the same mass despite what the first hypothesis of Dalton's atomic theory states. Most elements have two or more ***isotopes,*** *atoms that have the same atomic number (Z) but different mass numbers (A).* For example, there are three isotopes of hydrogen. One, simply known as hydrogen, has one proton and no neutrons. The *deuterium* isotope contains one proton and one neutron, and *tritium* has one proton and two neutrons. The accepted way to denote the atomic number and mass number of an atom of an element (X) is as follows:

$$\text{mass number} \rightarrow {}^{A}_{Z}\text{X} \leftarrow \text{atomic number}$$

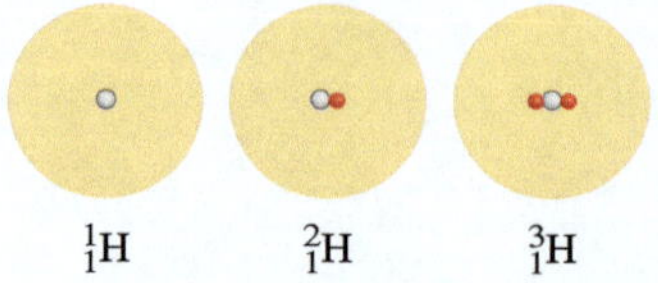

Thus, for the isotopes of hydrogen, we write

$$\underset{\text{hydrogen}}{{}^{1}_{1}\text{H}} \qquad \underset{\text{deuterium}}{{}^{2}_{1}\text{H}} \qquad \underset{\text{tritium}}{{}^{3}_{1}\text{H}}$$

As another example, consider two common isotopes of uranium with mass numbers of 235 and 238, respectively:

$${}^{235}_{92}\text{U} \qquad {}^{238}_{92}\text{U}$$

The first isotope is used in nuclear reactors and atomic bombs, whereas the second isotope lacks the properties necessary for these applications. With the exception of hydrogen, which has different names for each of its isotopes, isotopes of elements are identified by their mass numbers. Thus, the preceding two isotopes are called uranium-235 (pronounced "uranium two thirty-five") and uranium-238 (pronounced "uranium two thirty-eight").

The chemical properties of an element are determined primarily by the protons and electrons in its atoms; neutrons do not take part in chemical changes under normal conditions. Therefore, isotopes of the same element have similar chemistries, forming the same types of compounds and displaying similar reactivities.

Example 2.1 shows how to calculate the number of protons, neutrons, and electrons using atomic numbers and mass numbers.

Example 2.1

Give the number of protons, neutrons, and electrons in each of the following species: (a) ${}^{20}_{11}$Na, (b) ${}^{22}_{11}$Na, (c) ^{17}O, (d) carbon-14.

Strategy Recall that the superscript denotes the mass number (*A*) and the subscript denotes the atomic number (*Z*). Mass number is always greater than atomic number. (The only exception is ${}^{1}_{1}$H, where the mass number is equal to the atomic number.) In a case where no subscript is shown, as in parts (c) and (d), the atomic number can be deduced from the element symbol or name. To determine the number of electrons, remember that because atoms are electrically neutral, the number of electrons is equal to the number of protons.

Solution

(a) The atomic number is 11, so there are 11 protons. The mass number is 20, so the number of neutrons is 20 − 11 = 9. The number of electrons is the same as the number of protons; that is, 11.

(b) The atomic number is the same as that in (a), or 11. The mass number is 22, so the number of neutrons is 22 − 11 = 11. The number of electrons is 11. Note that the species in (a) and (b) are chemically similar isotopes of sodium.

(c) The atomic number of O (oxygen) is 8, so there are 8 protons. The mass number is 17, so there are 17 − 8 = 9 neutrons. There are 8 electrons.

(d) Carbon-14 can also be represented as ^{14}C. The atomic number of carbon is 6, so there are 14 − 6 = 8 neutrons. The number of electrons is 6.

Practice Exercise How many protons, neutrons, and electrons are in the following isotope of copper: ^{63}Cu

Similar problems: 2.15, 2.16.

Summary of Concepts & Facts

- The atomic number of an element is the number of protons in the nucleus of an atom of the element; it determines the identity of an element. The mass number is the sum of the number of protons and the number of neutrons in the nucleus.
- Isotopes are atoms of the same element with the same number of protons but different numbers of neutrons.

Review of Concepts & Facts

2.3.1 What is the atomic number of an element if one of its isotopes has 117 neutrons and a mass number of 195?
2.3.2 How many neutrons are in an atom of ^{114}Cd?
2.3.3 Which of the following two symbols provides more information: ^{17}O or $_{8}O$?

2.4 The Periodic Table

Learning Objective

- Recognize the general organization of the periodic table with respect to metals, metalloids, nonmetals, groups, and periods.

More than half of the elements known today were discovered between 1800 and 1900. During this period, chemists noted that many elements show strong similarities to one another. Recognition of periodic regularities in physical and chemical behavior and the need to organize the large volume of available information about the structure and properties of elemental substances led to the development of the ***periodic table,*** *a chart in which elements having similar chemical and physical properties are grouped together.* Figure 2.10 shows the modern periodic table in which the elements are arranged by atomic number (Z) (shown above the element symbol) in *horizontal rows* called ***periods*** and in *vertical columns* known as ***groups*** or ***families,*** according to similarities in their chemical properties. The 1 through 18 group designation has been recommended by the International Union of Pure and Applied Chemistry (IUPAC) and will be used exclusively in this text. The older numbering system used 1 through 8 followed by either A or B to designate the group (Group 1A, Group 2A, etc.).

The elements can be divided into three categories—metals, nonmetals, and metalloids. A ***metal*** is *a good conductor of heat and electricity,* whereas a ***nonmetal*** is usually *a poor conductor of heat and electricity.* A ***metalloid*** *has properties that are intermediate between those of metals and nonmetals.* Figure 2.10 shows that the majority of known elements are metals; only 19 elements are nonmetals, and 6 are commonly recognized as metalloids. From left to right across any period, the physical and chemical properties of the elements change gradually from metallic to nonmetallic.

Elements are often referred to collectively by their periodic table group number (Group 1, Group 2, and so on). However, for convenience, some element groups have been given special names. *The Group 1 elements (Li, Na, K, Rb, Cs, and Fr)* are called ***alkali metals,*** and *the Group 2 elements (Be, Mg, Ca, Sr, Ba, and Ra)* are called ***alkaline earth metals.*** *Elements in Group 17 (F, Cl, Br, I, and At)* are known as ***halogens,*** and *elements in Group 18 (He, Ne, Ar, Kr, Xe, and Rn)* are called ***noble gases,*** or *rare gases.*

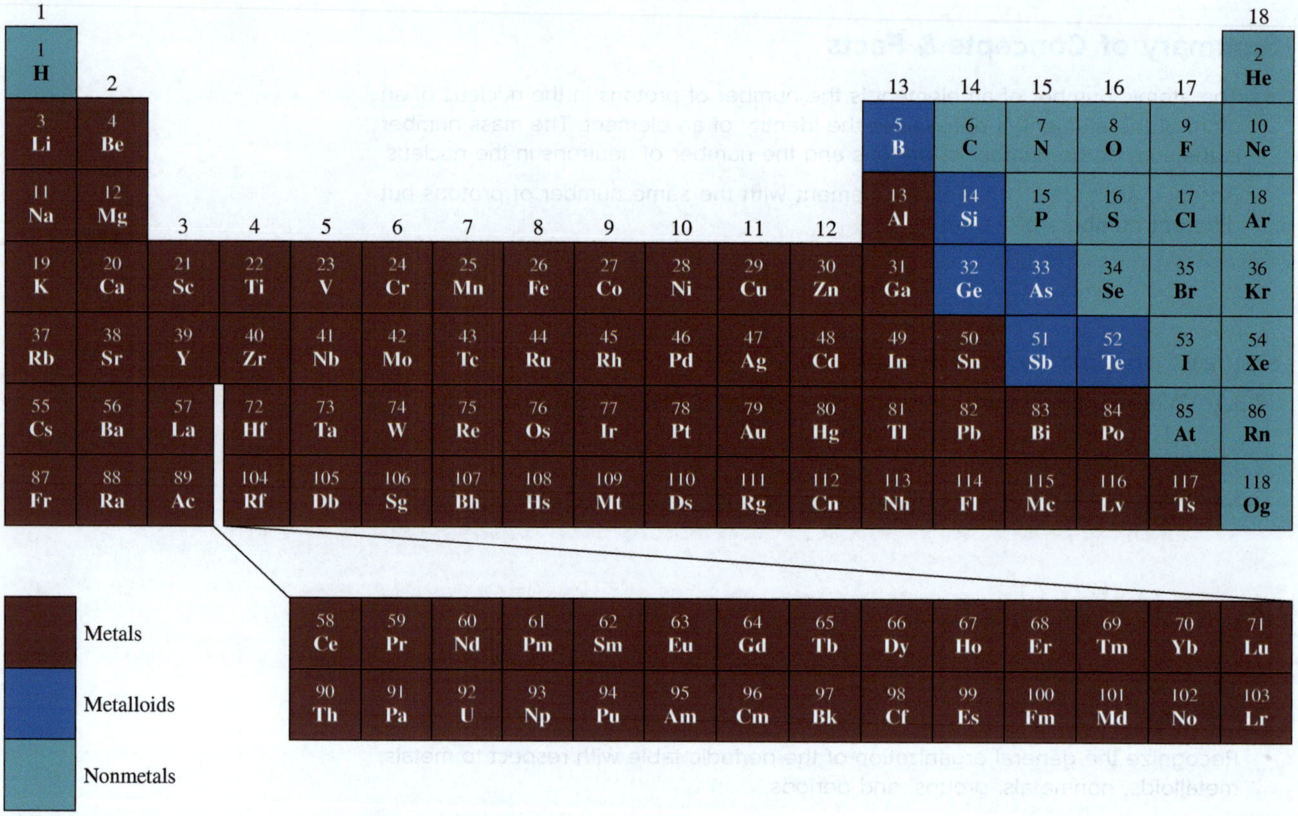

Figure 2.10 *The modern periodic table. The elements are arranged according to the atomic numbers above their symbols. With the exception of hydrogen (H), nonmetals appear at the far right of the table. The two rows of metals beneath the main body of the table are conventionally set apart to keep the table from being too wide. Actually, cerium (Ce) should follow lanthanum (La), and thorium (Th) should come right after actinium (Ac).*

The periodic table is a handy tool that correlates the properties of the elements in a systematic way and helps us to make predictions about chemical behavior. We will take a closer look at this keystone of chemistry in Chapter 8.

The Chemistry in Action essay "Distribution of Elements on Earth and in Living Systems" describes the distribution of the elements on Earth and in the human body.

Summary of Concepts & Facts

- Elements are grouped together in a systematic manner according to their chemical and physical properties on the periodic table. The horizontal rows are known as periods while the vertical columns are groups. The IUPAC system numbers the columns 1 through 18.
- The elements in general can be classified as metals, nonmetals, or metalloids.
- Many groups on the periodic table are known by their historical names: Group 1 (alkali metals), Group 2 (alkaline earth metals), Group 17 (halogens), and Group 18 (noble gases).

Review of Concepts & Facts

2.4.1 In viewing the periodic table, do chemical properties change more markedly across a period or down a group?

2.4.2 Identify the following as a metal, metalloid, or nonmetal: (a) K, (b) Se, (c) Sb, (d) W.

CHEMISTRY *in Action*

Distribution of Elements on Earth and in Living Systems

The majority of elements are naturally occurring. How are these elements distributed on Earth, and which are essential to living systems?

Earth's crust extends from the surface to a depth of about 40 km (about 25 mi). Because of technical difficulties, scientists have not been able to study the inner portions of Earth as easily as the crust. Nevertheless, it is believed that there is a solid core consisting mostly of iron at the center of Earth. Surrounding the core is a layer called the *mantle,* which consists of hot fluid containing iron, carbon, silicon, and sulfur.

Of the 83 elements that are found in nature, 12 make up 99.7 percent of Earth's crust by mass. They are, in decreasing order of natural abundance, oxygen (O), silicon (Si), aluminum (Al), iron (Fe), calcium (Ca), magnesium (Mg), sodium (Na), potassium (K), titanium (Ti), hydrogen (H), phosphorus (P), and manganese (Mn). In discussing the natural abundance of the elements, we should keep in mind that (1) the elements are not evenly distributed throughout Earth's crust, and (2) most elements occur in combined forms. These facts provide the basis for most methods of obtaining pure elements from their compounds, as we will see in later chapters.

The accompanying table lists the essential elements in the human body. Of special interest are the *trace elements,* such as iron (Fe), copper (Cu), zinc (Zn), iodine (I), and cobalt (Co), which together make up about 0.1 percent of the body's mass. These elements are necessary for biological functions such as growth, transport of oxygen for metabolism, and defense against disease. There is a delicate balance in the amounts of these elements in the human body. Too much or too little over an extended period of time can lead to serious illness, development delays, or even death.

Structure of Earth's interior.
Leonello Calvetti/Stocktrek Images/Getty Images

Essential Elements in the Human Body

Element	Percent by Mass*	Element	Percent by Mass*
Oxygen	65	Sodium	0.1
Carbon	18	Magnesium	0.05
Hydrogen	10	Iron	<0.05
Nitrogen	3	Cobalt	<0.05
Calcium	1.6	Copper	<0.05
Phosphorus	1.2	Zinc	<0.05
Potassium	0.2	Iodine	<0.05
Sulfur	0.2	Selenium	<0.01
Chlorine	0.2	Fluorine	<0.01

*Percent by mass *gives the mass of the element in grams present in a 100-g sample.*

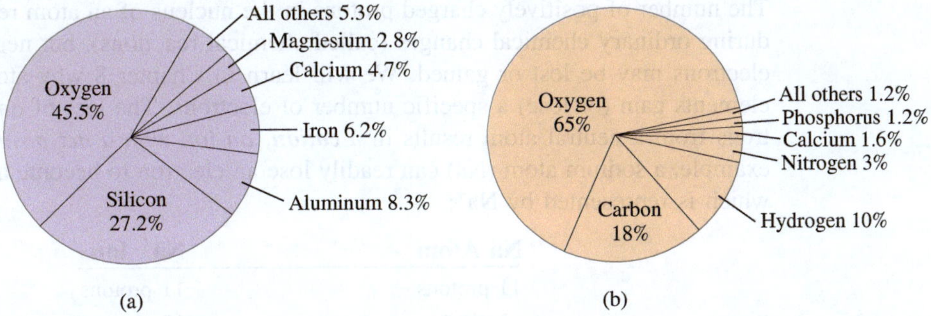

(a) Natural abundance of the elements in percent by mass. For example, oxygen's abundance is 45.5 percent. This means that in a 100-g sample of Earth's crust there are, on the average, 45.5 g of the element oxygen. (b) Abundance of elements in the human body in percent by mass.

2.5 Molecules and Ions

Learning Objective

- Differentiate between molecules and ions.

Of all the elements, only six noble gases in Group 18 of the periodic table (He, Ne, Ar, Kr, Xe, and Rn) exist in nature as single atoms. For this reason, they are called *monatomic* (meaning "one atom") gases. Most matter is composed of molecules or ions formed by atoms.

Molecules

A ***molecule*** *is an aggregate of at least two atoms in a definite arrangement held together by chemical forces* (also called *chemical bonds*). (We will discuss the nature of chemical bonds in Chapters 9 and 10.) A molecule may contain atoms of the same element or atoms of two or more elements joined in a fixed ratio, in accordance with the law of definite proportions stated in Section 2.1. Thus, a molecule is not necessarily a compound, which, by definition, is made up of two or more elements (see Section 1.7). Hydrogen gas, for example, is a pure element, but it consists of molecules made up of two H atoms each. Water, on the other hand, is a molecular compound that contains hydrogen and oxygen in a ratio of two H atoms and one O atom. Like atoms, molecules are electrically neutral.

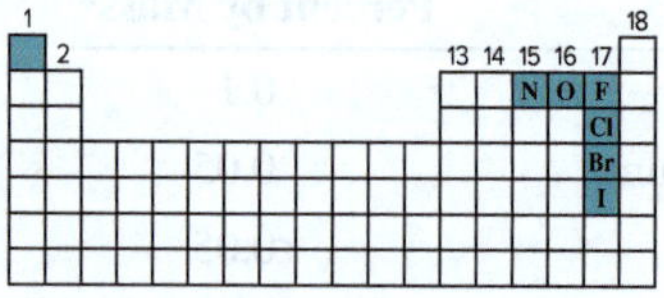

Elements that exist as diatomic molecules.

The hydrogen molecule, symbolized as H_2, is called a ***diatomic molecule,*** because it *contains only two atoms.* Other elements that normally exist as diatomic molecules are nitrogen (N_2) and oxygen (O_2), as well as the Group 17 elements—fluorine (F_2), chlorine (Cl_2), bromine (Br_2), and iodine (I_2). Of course, a diatomic molecule can contain atoms of different elements. Examples are hydrogen chloride (HCl) and carbon monoxide (CO).

The vast majority of molecules contain more than two atoms. They can be atoms of the same element, as in ozone (O_3), which is made up of three atoms of oxygen, or they can be combinations of two or more different elements. *Molecules containing more than two atoms* are called ***polyatomic molecules.*** Like ozone, water (H_2O) and ammonia (NH_3) are polyatomic molecules.

Ions

An ***ion*** *is an atom or a group of atoms that has a net positive or negative charge.* The number of positively charged protons in the nucleus of an atom remains the same during ordinary chemical changes (called chemical reactions), but negatively charged electrons may be lost or gained. We will learn in Chapter 8 why atoms of different elements gain (or lose) a specific number of electrons. The loss of one or more electrons from a neutral atom results in a ***cation,*** *an ion with a net positive charge.* For example, a sodium atom (Na) can readily lose an electron to become a sodium cation, which is represented by Na^+:

Na Atom	**Na^+ Ion**
11 protons	11 protons
11 electrons	10 electrons

On the other hand, an ***anion*** is *an ion whose net charge is negative* due to an increase in the number of electrons. A chlorine atom (Cl), for instance, can gain an electron to become the chloride ion Cl^-:

Cl Atom	**Cl^- Ion**
17 protons	17 protons
17 electrons	18 electrons

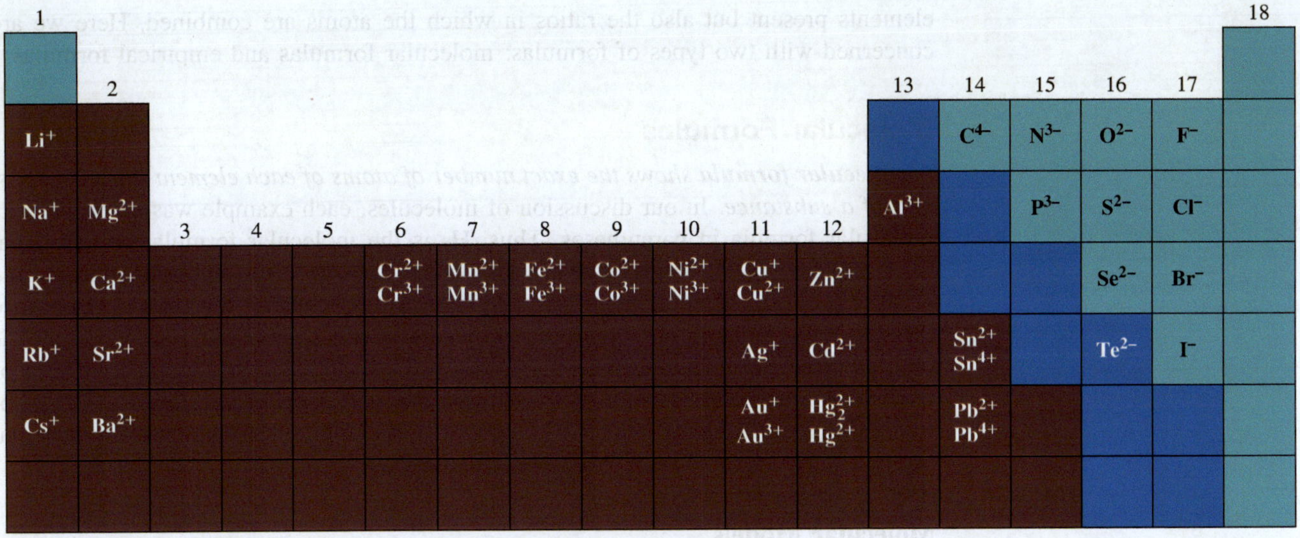

Figure 2.11 *Common monatomic ions arranged according to their positions in the periodic table. Note that the Hg_2^{2+} ion contains two atoms.*

Sodium chloride (NaCl), ordinary table salt, is called an ***ionic compound,*** because it is *formed from cations and anions.*

An atom can lose or gain more than one electron. Examples of ions formed by the loss or gain of more than one electron are Mg^{2+}, Fe^{3+}, S^{2-}, and N^{3-}. These ions, as well as Na^+ and Cl^-, are called ***monatomic ions,*** because they *contain only one atom.* Figure 2.11 shows the charges of a number of monatomic ions. With very few exceptions, metals tend to form cations and nonmetals form anions.

In addition, two or more atoms can combine to form an ion that has a net positive or net negative charge. ***Polyatomic ions*** such as OH^- (hydroxide ion), CN^- (cyanide ion), and NH_4^+ (ammonium ion) are *ions containing more than one atom.*

Summary of Concepts & Facts

- Differentiate between molecules and ions.

Review of Concepts & Facts

2.5.1 What does S_8 signify? How does it differ from 8S?

2.5.2 Determine the number of protons and electrons for the following ions: (a) P^{3-}, (b) Ti^{4+}, (c) Ge^{2+}.

2.5.3 What is the symbol and charge for an ion containing 24 protons and 22 electrons?

2.6 Chemical Formulas

Learning Objectives

- Classify chemical formulas as either molecular or empirical.
- Determine formulas of ionic compounds.

Chemists use ***chemical formulas*** to *express the composition of molecules and ionic compounds in terms of chemical symbols.* By composition we mean not only the elements present but also the ratios in which the atoms are combined. Here we are concerned with two types of formulas: molecular formulas and empirical formulas.

Molecular Formulas

A ***molecular formula*** *shows the exact number of atoms of each element in the smallest unit of a substance.* In our discussion of molecules, each example was given with its molecular formula in parentheses. Thus, H_2 is the molecular formula for hydrogen, O_2 is oxygen, O_3 is ozone, and H_2O is water. The subscript numeral indicates the number of atoms of an element present. There is no subscript for O in H_2O because there is only one atom of oxygen in a molecule of water, and so the number "one" is omitted from the formula. Note that oxygen (O_2) and ozone (O_3) are allotropes of oxygen. An ***allotrope*** is *one of two or more distinct forms of an element.* Two allotropic forms of the element carbon—diamond and graphite—are dramatically different not only in properties but also in their relative cost.

Molecular Models

Molecules are too small for us to observe directly. An effective means of visualizing them is by the use of molecular models. Two standard types of molecular models are currently in use: *ball-and-stick* models and *space-filling* models (Figure 2.12). (See Appendix 7 for the color scheme for atoms.) In ball-and-stick model kits, the atoms are wooden or plastic balls with holes in them. Sticks or springs are used to represent chemical bonds. The angles they form between atoms approximate the bond angles in actual molecules. With the exception of the H atom, the balls are all the same size and each type of atom is represented by a specific color. In space-filling models, atoms are represented by truncated balls held together by snap fasteners, so that the bonds are not visible. The balls are proportional in size to atoms. The first step toward building a molecular model is writing the ***structural formula,*** which *shows how atoms are bonded to one another in a molecule.* For example, it is known that each of the two H atoms is bonded to an O atom in the water molecule. Therefore, the structural formula of water is H—O—H. A line connecting the two atomic symbols represents a chemical bond.

Ball-and-stick models show the three-dimensional arrangement of atoms clearly, and they are fairly easy to construct. However, the balls are not proportional to the size of atoms. Furthermore, the sticks greatly exaggerate the space between atoms in

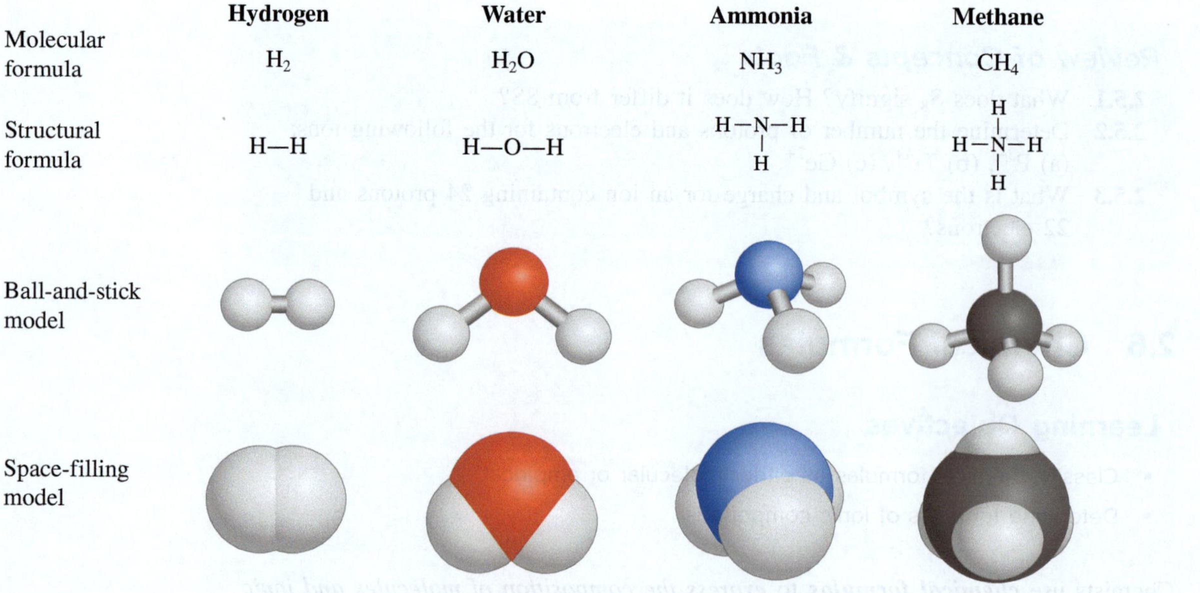

Figure 2.12 *Molecular and structural formulas and molecular models of four common molecules.*

a molecule. Space-filling models are more accurate because they show the variation in atomic size. Their drawbacks are that they are time consuming to put together and they do not show the three-dimensional positions of atoms very well. Molecular modeling software can also be used to create ball-and-stick and space-filling models. We will use both models extensively in this text.

Empirical Formulas

The molecular formula of hydrogen peroxide, a substance used as an antiseptic and as a bleaching agent for textiles and hair, is H_2O_2. This formula indicates that each hydrogen peroxide molecule consists of two hydrogen atoms and two oxygen atoms. The ratio of hydrogen to oxygen atoms in this molecule is 2:2 or 1:1. The empirical formula of hydrogen peroxide is HO. Thus, the ***empirical formula** tells us which elements are present and the simplest whole-number ratio of their atoms.* However, empirical formulas do not necessarily tell us the actual number of atoms in a given molecule. As another example, consider the compound hydrazine (N_2H_4), which is used as a rocket fuel. The empirical formula of hydrazine is NH_2. Although the ratio of nitrogen to hydrogen is 1:2 in both the molecular formula (N_2H_4) and the empirical formula (NH_2), only the molecular formula tells us the actual number of N atoms (two) and H atoms (four) present in a hydrazine molecule.

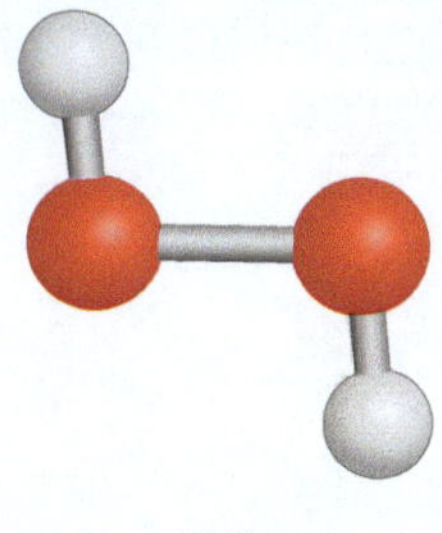

H_2O_2

Empirical formulas are the *simplest* chemical formulas; they are written by reducing the subscripts in the molecular formulas to the smallest possible whole numbers. Molecular formulas are the *true* formulas of molecules. If we know the molecular formula, we also know the empirical formula, but the reverse is not true. Why, then, do chemists bother with empirical formulas? The word *empirical* means "derived from experiment." As we explain in Chapter 3, when chemists analyze an unknown compound, the first step is usually the determination of the compound's empirical formula. With additional information, it is possible to deduce the molecular formula.

For many molecules, the molecular formula and the empirical formula are one and the same. Some examples are water (H_2O), ammonia (NH_3), carbon dioxide (CO_2), and methane (CH_4).

Examples 2.2 and 2.3 deal with writing molecular formulas from molecular models and writing empirical formulas from molecular formulas.

Example 2.2

Write the molecular formula of methylamine, a colorless gas used in the production of pharmaceuticals and pesticides, from its ball-and-stick model, shown here.

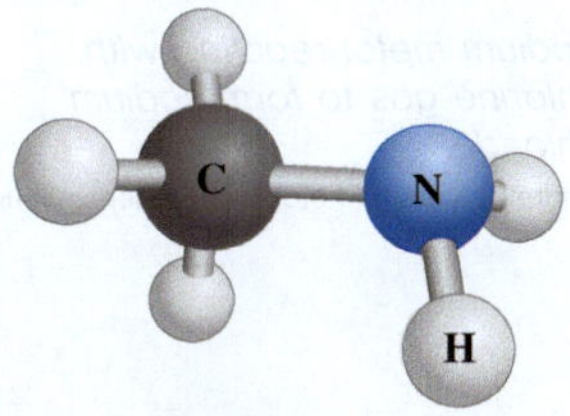

Methylamine

Solution Refer to the labels (also see end papers). There are five H atoms, one C atom, and one N atom. Therefore, the molecular formula is CH_5N. However, the standard way of writing the molecular formula for methylamine is CH_3NH_2 because it shows how the atoms are joined in the molecule.

Practice Exercise Write the molecular formula of chloroform, which is used as a solvent and a cleaning agent. The ball-and-stick model of chloroform is shown here.

Similar problem: 2.49.

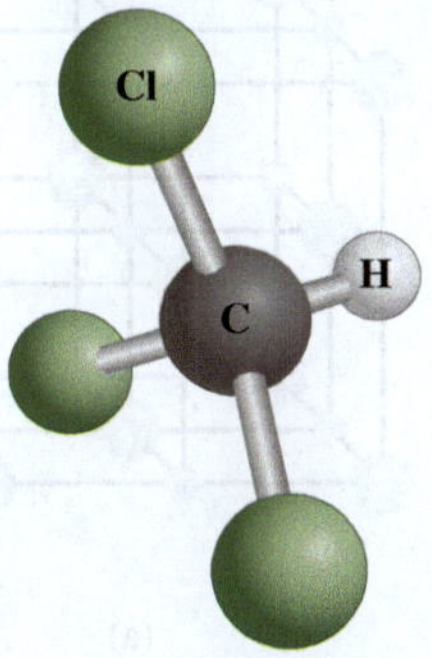

Chloroform

Example 2.3

Write the empirical formulas for the following molecules: (a) diborane (B_2H_6), used in rocket propellants; (b) dimethyl fumarate ($C_8H_{12}O_4$), a substance used to treat psoriasis, a skin disease; and (c) vanillin ($C_8H_8O_3$), a flavoring agent used in foods and beverages.

(Continued)

Strategy Recall that to write the empirical formula, the subscripts in the molecular formula must be converted to the smallest possible whole numbers.

Solution

(a) There are two boron atoms and six hydrogen atoms in diborane. Dividing the subscripts by 2, we obtain the empirical formula BH_3.

(b) In dimethyl fumarate there are 8 carbon atoms, 12 hydrogen atoms, and 4 oxygen atoms. Dividing the subscripts by 4, we obtain the empirical formula C_2H_3O. Note that if we had divided the subscripts by 2, we would have obtained the formula $C_4H_6O_2$. Although the ratio of carbon to hydrogen to oxygen atoms in $C_4H_6O_2$ is the same as that in C_2H_3O (2:3:1), $C_4H_6O_2$ is not the simplest formula because its subscripts are not in the smallest whole-number ratio.

(c) Because the subscripts in $C_8H_8O_3$ are already the smallest possible whole numbers, the empirical formula for vanillin is the same as its molecular formula.

Practice Exercise Write the empirical formula for caffeine ($C_8H_{10}N_4O_2$), a stimulant found in tea and coffee.

Similar problem: 2.47.

Formula of Ionic Compounds

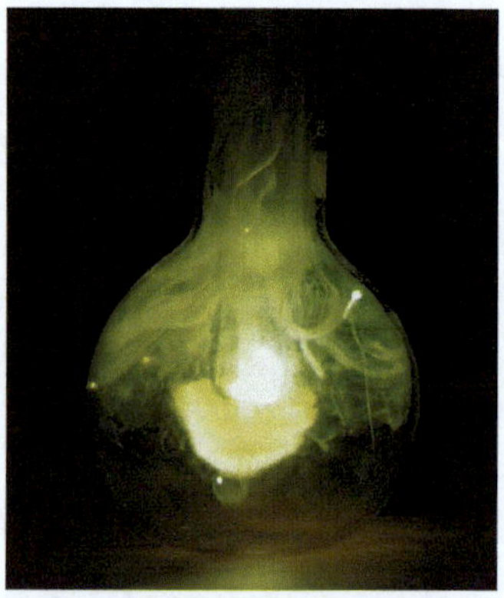

Sodium metal reacting with chlorine gas to form sodium chloride.
Andrew Lambert Photography/Science Source

The formulas of ionic compounds are usually the same as their empirical formulas, because ionic compounds do not consist of discrete molecular units. For example, a solid sample of sodium chloride (NaCl) consists of equal numbers of Na^+ and Cl^- ions arranged in a three-dimensional network called a ***lattice*** (Figure 2.13). In such a compound there is a 1:1 ratio of cations to anions so that the compound is electrically neutral. As you can see in Figure 2.13, no Na^+ ion in NaCl is associated with just one particular Cl^- ion. In fact, each Na^+ ion is equally held by six surrounding Cl^- ions and vice versa. Thus, NaCl is the empirical formula for sodium chloride. In other ionic compounds, the actual structure may be different, but the arrangement of cations and anions is always such that the compounds are all electrically neutral. Note that the charges on the cation and anion are not shown in the formula for an ionic compound.

For ionic compounds to be electrically neutral, the sum of the charges on the cation and anion in each formula unit must be zero. If the charges on the cation and anion are numerically different, we apply the following rule to make the formula electrically neutral: *The subscript of the cation is numerically equal to the charge on the anion, and the subscript of the anion is numerically equal to the charge on the cation.* If the charges are numerically equal, then no subscripts are necessary. This

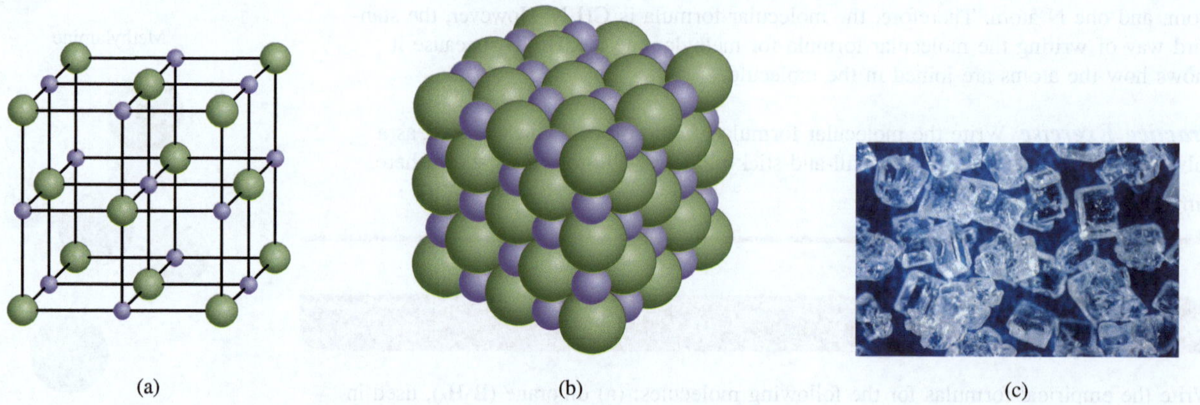

Figure 2.13 *(a) Structure of solid NaCl. (b) In reality, the cations are in contact with the anions. In both (a) and (b), the smaller spheres represent Na^+ ions and the larger spheres, Cl^- ions. (c) Crystals of NaCl.*
(c): oBebee/Shutterstock

rule follows from the fact that because the formulas of ionic compounds are usually empirical formulas, the subscripts must always be reduced to the smallest ratios. Let us consider some examples.

- **Potassium Bromide.** The potassium cation K^+ and the bromine anion Br^- combine to form the ionic compound potassium bromide. The sum of the charges is $+1 + (-1) = 0$, so no subscripts are necessary. The formula is KBr.
- **Zinc Iodide.** The zinc cation Zn^{2+} and the iodine anion I^- combine to form zinc iodide. The sum of the charges of one Zn^{2+} ion and one I^- ion is $+2 + (-1) = +1$. To make the charges add up to zero we multiply the -1 charge of the anion by 2 and add the subscript "2" to the symbol for iodine. Therefore, the formula for zinc iodide is ZnI_2.
- **Aluminum Oxide.** The cation is Al^{3+} and the oxygen anion is O^{2-}. The following diagram helps us determine the subscripts for the compound formed by the cation and the anion:

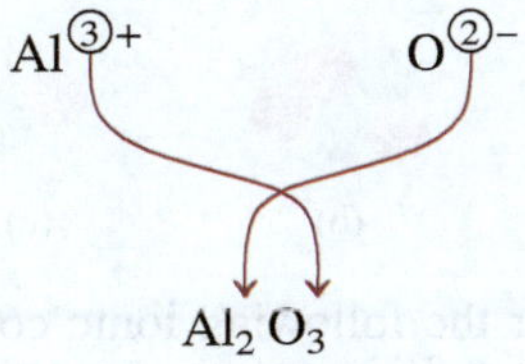

The sum of the charges is $2(+3) + 3(-2) = 0$. Thus, the formula for aluminum oxide is Al_2O_3.

Student Hot Spot

Student data indicate you may struggle with formulas of ionic compounds. Access your eBook for additional Learning Resources on this topic.

Example 2.4

Magnesium nitride is used to prepare Borazon, a very hard compound used in cutting tools and machine parts. Write the formula of magnesium nitride, containing the Mg^{2+} and N^{3-} ions.

Strategy Our guide for writing formulas for ionic compounds is electrical neutrality; that is, the total charge on the cation(s) must be equal to the total charge on the anion(s). Because the charges on the Mg^{2+} and N^{3-} ions are not equal, we know the formula cannot be MgN. Instead, we write the formula as Mg_xN_y, where x and y are subscripts to be determined.

Solution To satisfy electrical neutrality, the following relationship must hold:

$$(+2)x + (-3)y = 0$$

Solving, we obtain $x/y = 3/2$. Setting $x = 3$ and $y = 2$, we write

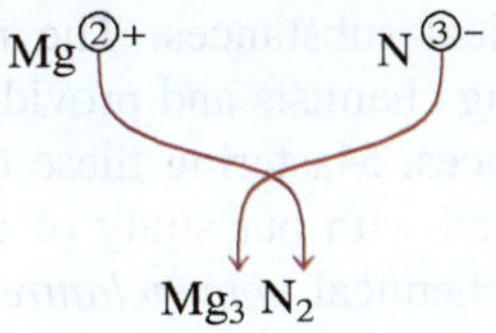

Check The subscripts are reduced to the smallest whole-number ratio of the atoms because the chemical formula of an ionic compound is usually its empirical formula.

Practice Exercise Write the formulas of the following ionic compounds: (a) chromium sulfate (containing the Cr^{3+} and SO_4^{2-} ions) and (b) titanium oxide (containing the Ti^{4+} and O^{2-} ions).

Similar problems: 2.45, 2.46.

When magnesium burns in air, it forms both magnesium oxide and magnesium nitride.

Charles D. Winters/Science Source

Summary of Concepts & Facts

- Chemical formulas combine the symbols for the constituent elements with whole-number subscripts to show the type and number of atoms contained in the smallest unit of a compound.
- The molecular formula conveys the specific number and type of atoms combined in each molecule of a compound. The empirical formula shows the simplest ratios of the atoms combined in a molecule.

Review of Concepts & Facts

2.6.1 Match each of the diagrams shown here with the following ionic compounds: Al_2O_3, LiH, Na_2S, $Mg(NO_3)_2$. (Green spheres represent cations and red spheres represent anions.)

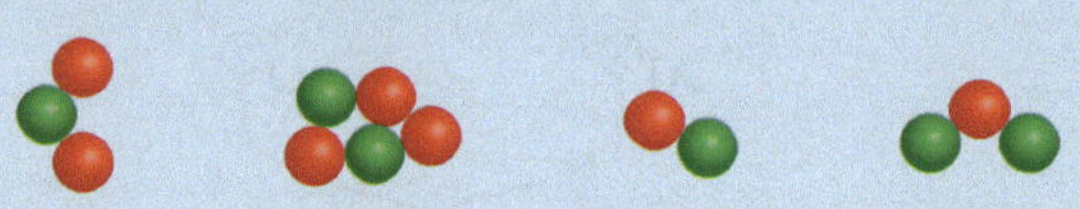

(a) (b) (c) (d)

2.6.2 Write the formulas for the following ionic compounds: (a) calcium iodide (containing the Ca^{2+} and I^- ions), (b) gallium sulfide (containing the Ga^{3+} and S^{2-} ions).

2.7 Naming Compounds

Learning Objectives

- Utilize the rules of nomenclature to name the different types of compounds, including ionic compounds, molecular compounds, acids, bases, oxoacids, and hydrates.
- Identify the common polyatomic ions along with their formulas and charges.

When chemistry was a young science and the number of known compounds was small, it was possible to memorize their names. Many of the names were derived from their physical appearance, properties, origin, or application—for example, milk of magnesia, laughing gas, limestone, caustic soda, lye, washing soda, and baking soda.

Today the number of known compounds is well over 66 million. Fortunately, it is not necessary to memorize their names. Over the years chemists have devised a clear system for naming chemical substances. The rules are accepted worldwide, facilitating communication among chemists and providing a useful way of labeling an overwhelming variety of substances. Mastering these rules now will prove beneficial almost immediately as we proceed with our study of chemistry.

To begin our discussion of chemical *nomenclature,* the naming of chemical compounds, we must first distinguish between inorganic and organic compounds. ***Organic compounds*** *contain carbon, usually in combination with elements such as hydrogen, oxygen, nitrogen, and sulfur.* All other compounds are classified as ***inorganic compounds.*** For convenience, some carbon-containing compounds, such as carbon monoxide (CO), carbon dioxide (CO_2), carbon disulfide (CS_2), compounds containing the cyanide group (CN^-), and carbonate $\left(CO_3^{2-}\right)$ and bicarbonate (HCO_3^-) groups are considered to be inorganic compounds. Section 2.8 gives a brief introduction to organic compounds.

To organize and simplify our venture into naming compounds, we can divide inorganic compounds into four categories: ionic compounds, molecular compounds, acids and bases, and hydrates.

Table 2.2 The "-ide" Nomenclature of Some Common Monatomic Anions According to Their Positions in the Periodic Table

Group 14	Group 15	Group 16	Group 17
C carbide (C^{4-})*	N nitride (N^{3-})	O oxide (O^{2-})	F fluoride (F^-)
Si silicide (Si^{4-})	P phosphide (P^{3-})	S sulfide (S^{2-})	Cl chloride (Cl^-)
		Se selenide (Se^{2-})	Br bromide (Br^-)
		Te telluride (Te^{2-})	I iodide (I^-)

*The word "carbide" is also used for the anion C_2^{2-}.

Ionic Compounds

In Section 2.5 we learned that ionic compounds are made up of cations (positive ions) and anions (negative ions). With the important exception of the ammonium ion, NH_4^+, all cations of interest to us are derived from metal atoms. Metal cations take their names from the elements. For example,

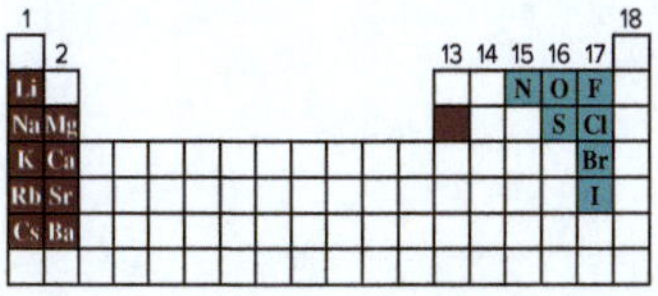

The most reactive metals (orange) and the most reactive nonmetals (blue) combine to form ionic compounds.

Element		Name of Cation	
Na	sodium	Na^+	sodium ion (or sodium cation)
K	potassium	K^+	potassium ion (or potassium cation)
Mg	magnesium	Mg^{2+}	magnesium ion (or magnesium cation)
Al	aluminum	Al^{3+}	aluminum ion (or aluminum cation)

Many ionic compounds are ***binary compounds,*** or *compounds formed from just two elements.* For binary compounds, the first element named is the metal cation, followed by the nonmetallic anion. Thus, NaCl is sodium chloride. The anion is named by taking the first part of the element name (chlorine) and adding "-ide." Potassium bromide (KBr), zinc iodide (ZnI_2), and aluminum oxide (Al_2O_3) are also binary compounds. Table 2.2 shows the "-ide" nomenclature of some common monatomic anions according to their positions in the periodic table.

Video: Formation of an Ionic Compound

The "-ide" ending is also used for certain anion groups containing different elements, such as hydroxide (OH^-) and cyanide (CN^-). Thus, the compounds LiOH and KCN are named lithium hydroxide and potassium cyanide, respectively. These and a number of other such ionic substances are called ***ternary compounds,*** meaning *compounds consisting of three elements.* Table 2.3 lists alphabetically the names of a number of common cations and anions.

Certain metals, especially the *transition metals,* can form more than one type of cation. Take iron as an example. Iron can form two cations: Fe^{2+} and Fe^{3+}. An older nomenclature system that is still in limited use assigns the ending "-ous" to the cation with fewer positive charges and the ending "-ic" to the cation with more positive charges:

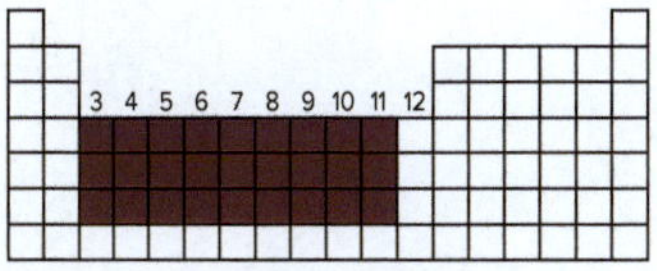

The transition metals are the elements in Groups 3 through 11 (see Figure 2.10).

Fe^{2+} ferrous ion
Fe^{3+} ferric ion

The names of the compounds that these iron ions form with chlorine would thus be

$FeCl_2$ ferrous chloride
$FeCl_3$ ferric chloride

$FeCl_2$ (left) and $FeCl_3$ (right).
Ken Karp/McGraw Hill

This method of naming ions has some distinct limitations. First, the "-ous" and "-ic" suffixes do not provide information regarding the actual charges of the two cations involved. Thus, the ferric ion is Fe^{3+}, but the cation of copper named cupric has the formula Cu^{2+}. In addition, the "-ous" and "-ic" designations provide names for only two different elemental cations.

Some metals can form three or more ions with different positive charges in compounds. Therefore, it has become increasingly common to designate different cations

Table 2.3 Names and Formulas of Some Common Inorganic Cations and Anions

Cation	Anion
aluminum (Al^{3+})	bromide (Br^-)
ammonium (NH_4^+)	carbonate (CO_3^{2-})
barium (Ba^{2+})	chlorate (ClO_3^-)
cadmium (Cd^{2+})	chloride (Cl^-)
calcium (Ca^{2+})	chromate (CrO_4^{2-})
cesium (Cs^+)	cyanide (CN^-)
chromium(III) or chromic (Cr^{3+})	dichromate ($Cr_2O_7^{2-}$)
cobalt(II) or cobaltous (Co^{2+})	dihydrogen phosphate ($H_2PO_4^-$)
copper(I) or cuprous (Cu^+)	fluoride (F^-)
copper(II) or cupric (Cu^{2+})	hydride (H^-)
hydrogen (H^+)	hydrogen carbonate or bicarbonate (HCO_3^-)
iron(II) or ferrous (Fe^{2+})	hydrogen phosphate (HPO_4^{2-})
iron(III) or ferric (Fe^{3+})	hydrogen sulfate or bisulfate (HSO_4^-)
lead(II) or plumbous (Pb^{2+})	hydroxide (OH^-)
lithium (Li^+)	iodide (I^-)
magnesium (Mg^{2+})	nitrate (NO_3^-)
manganese(II) or manganous (Mn^{2+})	nitride (N^{3-})
mercury(I) or mercurous (Hg_2^{2+})*	nitrite (NO_2^-)
mercury(II) or mercuric (Hg^{2+})	oxide (O^{2-})
potassium (K^+)	permanganate (MnO_4^-)
rubidium (Rb^+)	peroxide (O_2^{2-})
silver (Ag^+)	phosphate (PO_4^{3-})
sodium (Na^+)	sulfate (SO_4^{2-})
strontium (Sr^{2+})	sulfide (S^{2-})
tin(II) or stannous (Sn^{2+})	sulfite (SO_3^{2-})
zinc (Zn^{2+})	thiocyanate (SCN^-)

*Mercury(I) exists as a pair as shown.

with Roman numerals. This is called the Stock[†] system. In this system, the Roman numeral I indicates one positive charge on the metal cation, II means two positive charges on the metal cation, and so on. For example, manganese (Mn) atoms can assume several different positive charges:

Mn^{2+}: MnO manganese(II) oxide

Mn^{3+}: Mn_2O_3 manganese(III) oxide

Mn^{4+}: MnO_2 manganese(IV) oxide

These names are pronounced "manganese-two oxide," "manganese-three oxide," and "manganese-four oxide." Using the Stock system, we denote the ferrous ion and the ferric ion as iron(II) and iron(III), respectively; ferrous chloride becomes iron(II) chloride, and ferric chloride is called iron(III) chloride. In keeping with modern practice, we will favor the Stock system of naming compounds in this textbook. Note that some nontransition metals such as tin (Sn) and lead (Pb) can also form more than one type of cation.

Student Hot Spot

Student data indicate you may struggle with naming an ionic compound from its formula. Access your eBook for additional Learning Resources on this topic.

Examples 2.5 and 2.6 illustrate how to name ionic compounds and write formulas for ionic compounds based on the information given in Figure 2.11 and Tables 2.2 and 2.3.

[†]Alfred E. Stock (1876–1946). German chemist. Stock did most of his research in the synthesis and characterization of boron, beryllium, and silicon compounds. He was the first scientist to explore the dangers of mercury poisoning.

Example 2.5

Name the following compounds: (a) $Fe(NO_3)_2$, (b) Na_2HPO_4, (c) $(NH_4)_2SO_3$.

Strategy Our reference for the names of cations and anions is Table 2.3. Keep in mind that if a metal can form cations of different charges (see Figure 2.11), we need to use the Stock system.

Solution

(a) The nitrate ion (NO_3^-) bears one negative charge, so the iron ion must have two positive charges. Because iron forms both Fe^{2+} and Fe^{3+} ions, we need to use the Stock system and call the compound iron(II) nitrate.

(b) The cation is Na^+ and the anion is HPO_4^{2-} (hydrogen phosphate). Because sodium only forms one type of ion (Na^+), there is no need to use sodium(I) in the name. The compound is sodium hydrogen phosphate.

(c) The cation is NH_4^+ (ammonium ion) and the anion is SO_3^{2-} (sulfite ion). The compound is ammonium sulfite.

Practice Exercise Name the following compounds: (a) PbO and (b) $LiClO_3$.

Similar problems: 2.59(b), (e), (f).

Example 2.6

Write chemical formulas for the following compounds: (a) mercury(I) nitrate, (b) cesium oxide, (c) strontium nitride.

Strategy We refer to Table 2.3 for the formulas of cations and anions. Recall that the Roman numerals in the Stock system provide useful information about the charges of the cation.

Solution

(a) The Roman numeral shows that the mercury ion bears a +1 charge. According to Table 2.3, however, the mercury(I) ion is diatomic (that is, Hg_2^{2+}) and the nitrate ion is NO_3^-. Therefore, the formula is $Hg_2(NO_3)_2$.

(b) Each oxide ion bears two negative charges, and each cesium ion bears one positive charge (cesium is in Group 1, as is sodium). Therefore, the formula is Cs_2O.

(c) Each strontium ion (Sr^{2+}) bears two positive charges, and each nitride ion (N^{3-}) bears three negative charges. To make the sum of the charges equal zero, we must adjust the numbers of cations and anions:

$$3(+2) + 2(-3) = 0$$

Thus, the formula is Sr_3N_2.

Practice Exercise Write formulas for the following ionic compounds: (a) rubidium sulfate, (b) barium hydride.

Similar problems: 2.60(a), (b), (d), (h), (i).

Molecular Compounds

Unlike ionic compounds, molecular compounds contain individual molecular units. They are usually composed of nonmetallic elements (see Figure 2.10). Many molecular compounds are binary compounds. Naming binary molecular compounds is similar to naming binary ionic compounds. We place the name of the first element in the

Table 2.4

Greek Prefixes Used in Naming Molecular Compounds

Prefix	Meaning
mono-	1
di-	2
tri-	3
tetra-	4
penta-	5
hexa-	6
hepta-	7
octa-	8
nona-	9
deca-	10

formula first, and the second element is named by adding "-ide" to the root of the element name. Some examples are

HCl	hydrogen chloride
HBr	hydrogen bromide
SiC	silicon carbide

It is quite common for one pair of elements to form several different compounds. In these cases, confusion in naming the compounds is avoided by the use of Greek prefixes to denote the number of atoms of each element present (Table 2.4). Consider the following examples:

CO	carbon monoxide
CO_2	carbon dioxide
SO_2	sulfur dioxide
SO_3	sulfur trioxide
NO_2	nitrogen dioxide
N_2O_4	dinitrogen tetroxide

The following guidelines are helpful in naming compounds with prefixes:

- The prefix "mono-" may be omitted for the first element. For example, PCl_3 is named phosphorus trichloride, not monophosphorus trichloride. Thus, the absence of a prefix for the first element usually means there is only one atom of that element present in the molecule.
- For oxides, the ending "a" in the prefix is sometimes omitted. For example, N_2O_4 may be called dinitrogen tetroxide rather than dinitrogen tetraoxide.

Exceptions to the use of Greek prefixes are molecular compounds containing hydrogen. Traditionally, many of these compounds are called either by their common, nonsystematic names or by names that do not specifically indicate the number of H atoms present:

B_2H_6	diborane
CH_4	methane
SiH_4	silane
NH_3	ammonia
PH_3	phosphine
H_2O	water
H_2S	hydrogen sulfide

Note that even the order of writing the elements in the formulas for hydrogen compounds is irregular. In water and hydrogen sulfide, H is written first, whereas it appears last in the other compounds. But in methane and other binary organic compounds containing carbon and hydrogen, C is written first.

Writing formulas for molecular compounds is usually straightforward. Thus, the name arsenic trifluoride means that there are three F atoms and one As atom in each molecule, and the molecular formula is AsF_3. Note that the order of elements in the formula is the same as in its name.

Student Hot Spot

Student data indicate you may struggle with naming molecular compounds. Access your eBook for additional Learning Resources on this topic.

Example 2.7

Name the following molecular compounds: (a) PBr_5 and (b) As_2O_5.

Strategy We refer to Table 2.4 for the prefixes used in naming molecular compounds.

(Continued)

Solution

(a) Because there are five bromine atoms present, the compound is phosphorus pentabromide.

(b) There are two arsenic atoms and five oxygen atoms present, so the compound is diarsenic pentoxide. Note that the "a" is omitted in "penta."

Practice Exercise Name the following molecular compounds: (a) NF_3, (b) Cl_2O_7.

Similar problems: 2.59(c), (i), (j).

Example 2.8

Write chemical formulas for the following molecular compounds: (a) bromine trifluoride and (b) diboron trioxide.

Strategy We refer to Table 2.4 for the prefixes used in naming molecular compounds.

Solution

(a) Because there are three fluorine atoms and one bromine atom present, the formula is BrF_3.

(b) There are two boron atoms and three oxygen atoms present, so the formula is B_2O_3.

Practice Exercise Write chemical formulas for the following molecular compounds: (a) sulfur tetrafluoride, (b) dinitrogen pentoxide.

Similar problems: 2.61(g), (j).

Figure 2.14 summarizes the steps for naming ionic and binary molecular compounds.

Acids and Bases

Naming Acids

An ***acid*** can be described as *a substance that yields hydrogen ions (H^+) when dissolved in water.* (H^+ is equivalent to one proton, and is often referred to that way.) Formulas for acids contain one or more hydrogen atoms as well as an anionic group. Anions whose names end in "-ide" form acids with a "hydro-" prefix and an "-ic" ending, as shown in Table 2.5. In some cases two different names seem to be assigned to the same chemical formula.

HCl	hydrogen chloride
HCl	hydrochloric acid

When dissolved in water, the HCl molecule is converted to the H^+ and Cl^- ions. The H^+ ion is associated with one or more water molecules, and is usually represented as H_3O^+.

The name assigned to the compound depends on its physical state. In the gaseous or pure liquid state, HCl is a molecular compound called hydrogen chloride. When it is dissolved in water, the molecules break up into H^+ and Cl^- ions; in this state, the substance is called hydrochloric acid.

Oxoacids are acids that *contain hydrogen, oxygen, and another element (the central element).* The formulas of oxoacids are usually written with the H first, followed by the central element and then O. We use the following five common acids as our references in naming oxoacids:

H_2CO_3	carbonic acid
$HClO_3$	chloric acid
HNO_3	nitric acid
H_3PO_4	phosphoric acid
H_2SO_4	sulfuric acid

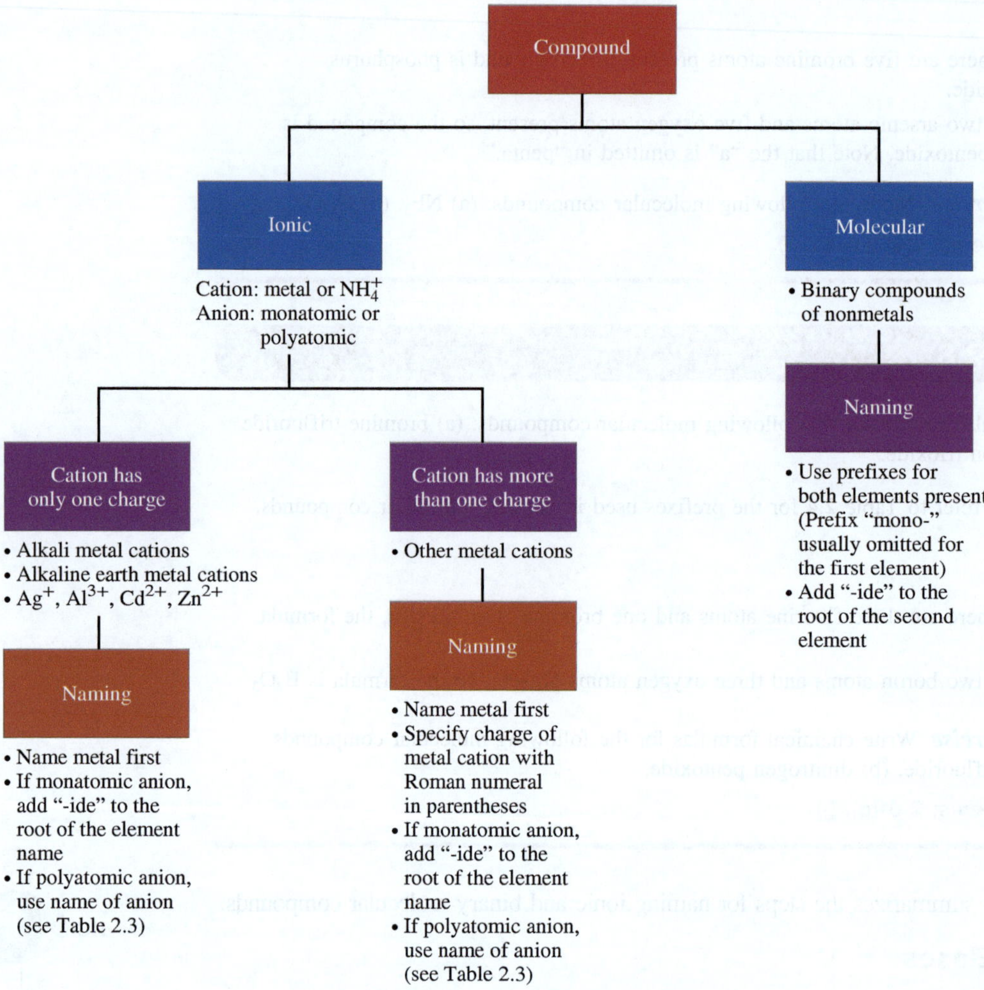

Figure 2.14 *Steps for naming ionic and binary molecular compounds.*

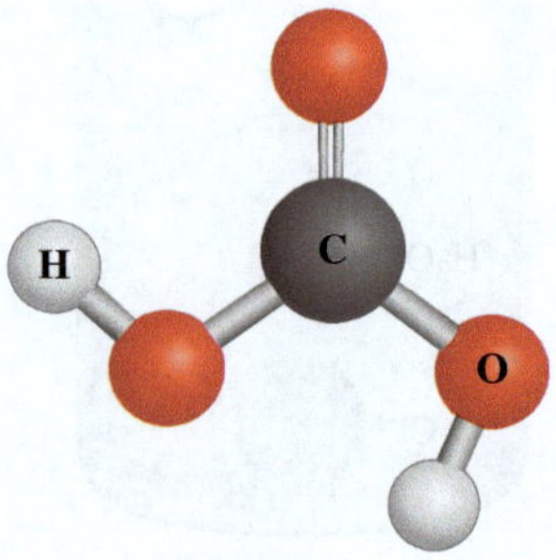

H_2CO_3

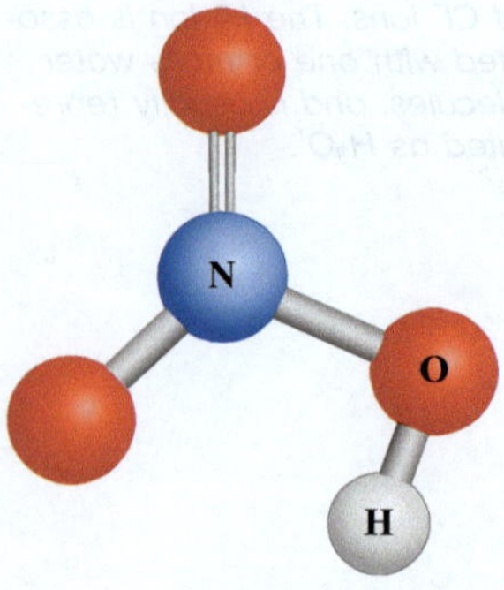

HNO_3

Often two or more oxoacids have the same central atom but a different number of O atoms. Starting with our reference oxoacids whose names all end with "-ic," we use the following rules to name these compounds.

1. Addition of one O atom to the "-ic" acid: The acid is called "per …-ic" acid. Thus, adding an O atom to $HClO_3$ changes chloric acid to perchloric acid, $HClO_4$.
2. Removal of one O atom from the "-ic" acid: The acid is called "-ous" acid. Thus, nitric acid, HNO_3, becomes nitrous acid, HNO_2.
3. Removal of two O atoms from the "-ic" acid: The acid is called "hypo …-ous" acid. Thus, when $HBrO_3$ is converted to $HBrO$, the acid is called hypobromous acid.

Table 2.5 Some Simple Acids

Molecular Compound	Acid	Corresponding Anion
HF (hydrogen fluoride)	HF (hydrofluoric acid)	F^- (fluoride)
HCl (hydrogen chloride)	HCl (hydrochloric acid)	Cl^- (chloride)
HBr (hydrogen bromide)	HBr (hydrobromic acid)	Br^- (bromide)
HI (hydrogen iodide)	HI (hydroiodic acid)	I^- (iodide)
HCN (hydrogen cyanide)	HCN (hydrocyanic acid)	CN^- (cyanide)
H_2S (hydrogen sulfide)	H_2S (hydrosulfuric acid)	S^{2-} (sulfide)

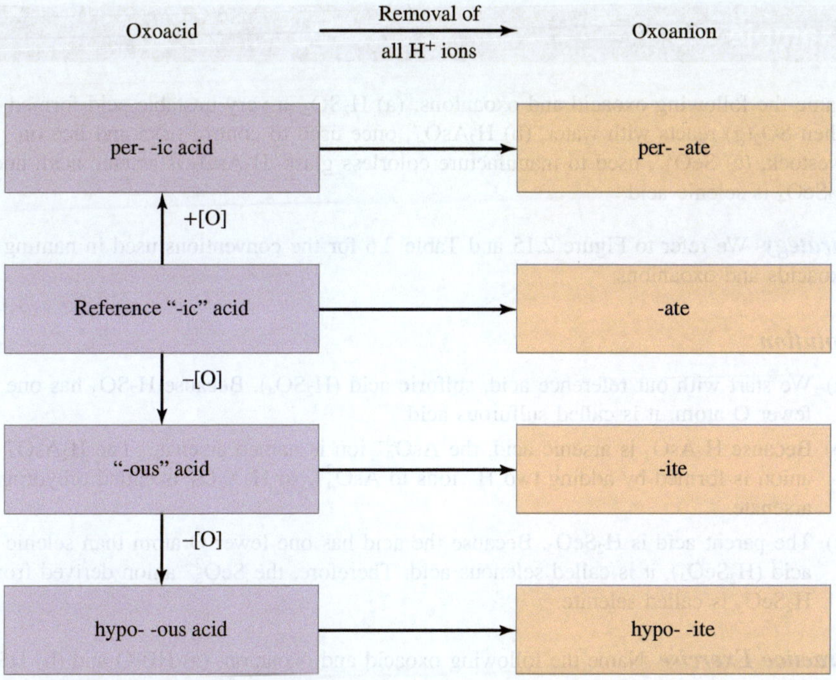

Figure 2.15 *Naming oxoacids and oxoanions.*

The rules for naming ***oxoanions,*** *anions of oxoacids,* are as follows:

1. When all the H ions are removed from the "-ic" acid, the anion's name ends with "-ate." For example, the anion CO_3^{2-} derived from H_2CO_3 is called carbonate.
2. When all the H ions are removed from the "-ous" acid, the anion's name ends with "-ite." Thus, the anion ClO_2^- derived from $HClO_2$ is called chlorite.
3. The names of anions in which one or more but not all the hydrogen ions have been removed must indicate the number of H ions present. For example, consider the anions derived from phosphoric acid:

H_3PO_4	phosphoric acid
$H_2PO_4^-$	dihydrogen phosphate
HPO_4^{2-}	hydrogen phosphate
PO_4^{3-}	phosphate

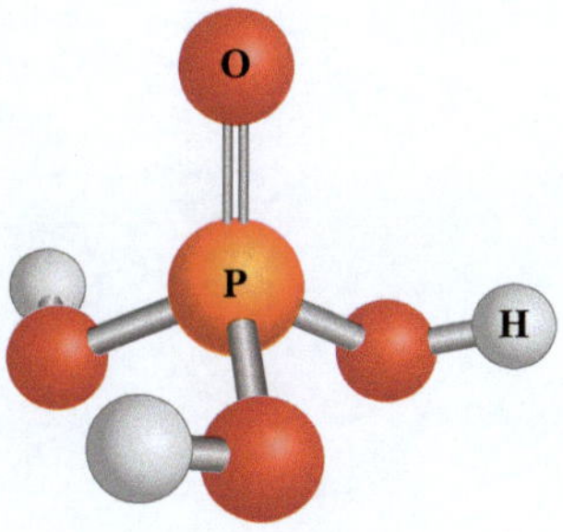

H_3PO_4

Note that we usually omit the prefix "mono-" when there is only one H in the anion. Figure 2.15 summarizes the nomenclature for the oxoacids and oxoanions, and Table 2.6 gives the names of the oxoacids and oxoanions that contain chlorine. Example 2.9 deals with the nomenclature for an oxoacid and an oxoanion.

Table 2.6 Names of Oxoacids and Oxoanions That Contain Chlorine

Acid	Corresponding Anion
$HClO_4$ (perchloric acid)	ClO_4^- (perchlorate)
$HClO_3$ (chloric acid)	ClO_3^- (chlorate)
$HClO_2$ (chlorous acid)	ClO_2^- (chlorite)
$HClO$ (hypochlorous acid)	ClO^- (hypochlorite)

Student Hot Spot

Student data indicate you may struggle with naming oxoacids and compounds with oxoanions. Access your eBook for additional Learning Resources on this topic.

Example 2.9

Name the following oxoacid and oxoanions: (a) H_2SO_3, a very unstable acid formed when $SO_2(g)$ reacts with water, (b) $H_2AsO_4^-$, once used to control ticks and lice on livestock, (c) SeO_3^{2-}, used to manufacture colorless glass. H_3AsO_4 is arsenic acid, and H_2SeO_4 is selenic acid.

Strategy We refer to Figure 2.15 and Table 2.6 for the conventions used in naming oxoacids and oxoanions.

Solution

(a) We start with our reference acid, sulfuric acid (H_2SO_4). Because H_2SO_3 has one fewer O atom, it is called sulfurous acid.

(b) Because H_3AsO_4 is arsenic acid, the AsO_4^{3-} ion is named arsenate. The $H_2AsO_4^-$ anion is formed by adding two H^+ ions to AsO_4^{3-}, so $H_2AsO_4^-$ is called dihydrogen arsenate.

(c) The parent acid is H_2SeO_3. Because the acid has one fewer O atom than selenic acid (H_2SeO_4), it is called selenous acid. Therefore, the SeO_3^{2-} anion derived from H_2SeO_3 is called selenite.

Practice Exercise Name the following oxoacid and oxoanion: (a) HBrO and (b) HSO_4^-.

Similar problems: 2.60(f).

Naming Bases

A ***base*** can be described as *a substance that yields hydroxide ions (OH^-) when dissolved in water.* Some examples are

NaOH	sodium hydroxide
KOH	potassium hydroxide
$Ba(OH)_2$	barium hydroxide

Ammonia (NH_3), a molecular compound in the gaseous or pure liquid state, is also classified as a common base. At first glance this may seem to be an exception to the definition of a base. But note that as long as a substance *yields* hydroxide ions when dissolved in water, it need not contain hydroxide ions in its structure to be considered a base. In fact, when ammonia dissolves in water, NH_3 reacts partially with water to yield NH_4^+ and OH^- ions. Thus, it is properly classified as a base.

Hydrates

Hydrates are *compounds that have a specific number of water molecules attached to them.* For example, in its normal state, each unit of copper(II) sulfate has five water molecules associated with it. The systematic name for this compound is copper(II) sulfate pentahydrate, and its formula is written as $CuSO_4 \cdot 5H_2O$. The water molecules can be driven off by heating. When this occurs, the resulting compound is $CuSO_4$, which is sometimes called *anhydrous* copper(II) sulfate; "anhydrous" means that the compound no longer has water molecules associated with it (Figure 2.16). Some other hydrates are

$BaCl_2 \cdot 2H_2O$	barium chloride dihydrate
$LiCl \cdot H_2O$	lithium chloride monohydrate
$MgSO_4 \cdot 7H_2O$	magnesium sulfate heptahydrate
$Sr(NO_3)_2 \cdot 4H_2O$	strontium nitrate tetrahydrate

Figure 2.16 $CuSO_4 \cdot 5H_2O$ (left) is blue; $CuSO_4$ (right) is white.
Ken Karp/McGraw Hill

Table 2.7 Common and Systematic Names of Some Compounds

Formula	Common Name	Systematic Name
H_2O	Water	Dihydrogen monoxide
NH_3	Ammonia	Trihydrogen nitride
CO_2	Dry ice	Solid carbon dioxide
NaCl	Table salt	Sodium chloride
N_2O	Laughing gas	Dinitrogen monoxide
$CaCO_3$	Marble, chalk, limestone	Calcium carbonate
CaO	Quicklime	Calcium oxide
$Ca(OH)_2$	Slaked lime	Calcium hydroxide
$NaHCO_3$	Baking soda	Sodium hydrogen carbonate
$Na_2CO_3 \cdot 10H_2O$	Washing soda	Sodium carbonate decahydrate
$MgSO_4 \cdot 7H_2O$	Epsom salt	Magnesium sulfate heptahydrate
$Mg(OH)_2$	Milk of magnesia	Magnesium hydroxide
$CaSO_4 \cdot 2H_2O$	Gypsum	Calcium sulfate dihydrate

Familiar Inorganic Compounds

Some compounds are better known by their common names than by their systematic chemical names. Familiar examples are listed in Table 2.7.

Summary of Concepts & Facts

- Chemical compounds are either molecular compounds (in which the smallest units are discrete, individual molecules) or ionic compounds, which are made of cations and anions.
- The names of many inorganic compounds can be deduced from a set of simple rules. The formulas can be written from the names of the compounds.

Review of Concepts & Facts

2.7.1 Why is it that the name for $SeCl_2$, selenium dichloride, contains a prefix, but the name for $SrCl_2$, strontium chloride, does not?

2.7.2 Why is the following question ambiguous: What is the name of HF? What additional information is needed to answer the question?

2.7.3 Name the following compounds: (a) Cs_2SO_3, (b) $Cu(NO_2)_2$.

2.7.4 Write formulas for the following compounds: (a) cobalt(II) hydrogen sulfate, (b) chromium(III) cyanide.

2.7.5 What are the correct names for the following compounds? (a) BrF_5, (b) HIO_4.

2.8 Introduction to Organic Compounds

Learning Objective

- Name simple organic compounds, including alkanes.

The simplest type of organic compounds is the *hydrocarbons,* which contain only carbon and hydrogen atoms. The hydrocarbons are used as fuels for domestic and industrial heating, for generating electricity and powering internal combustion engines, and as starting materials for the chemical industry. One class of hydrocarbons is called the *alkanes.* Table 2.8 shows the names, formulas, and molecular models of the first 10 *straight-chain* alkanes, in which the carbon chains have no branches. The name of an alkane depends on the number of carbon atoms in the molecule and all the names end with "-ane." Starting with C_5H_{12}, we use the Greek prefixes in Table 2.4 to indicate the number of carbon atoms present.

The chemistry of organic compounds is largely determined by the *functional groups,* which consist of one or a few atoms bonded in a specific way. For example,

Table 2.8 The First 10 Straight-Chain Alkanes

Name	Formula	Molecular Model
Methane	CH_4	
Ethane	C_2H_6	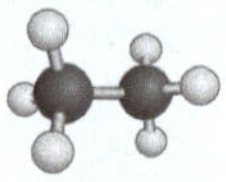
Propane	C_3H_8	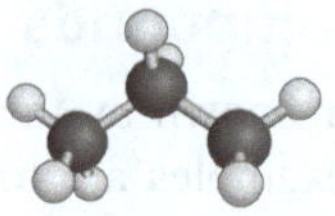
Butane	C_4H_{10}	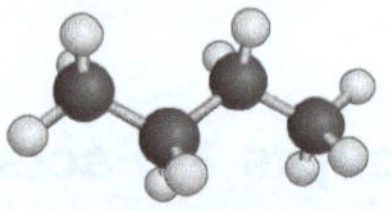
Pentane	C_5H_{12}	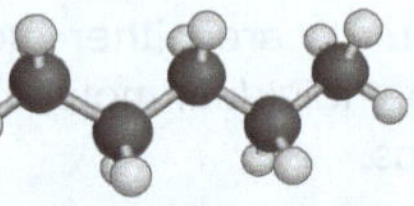
Hexane	C_6H_{14}	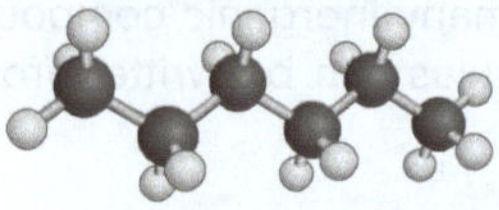
Heptane	C_7H_{16}	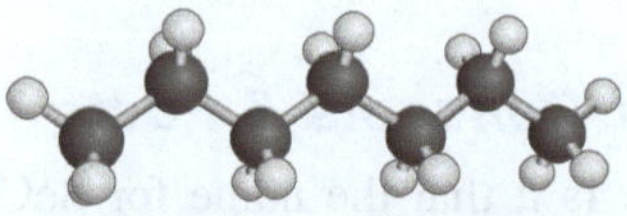
Octane	C_8H_{18}	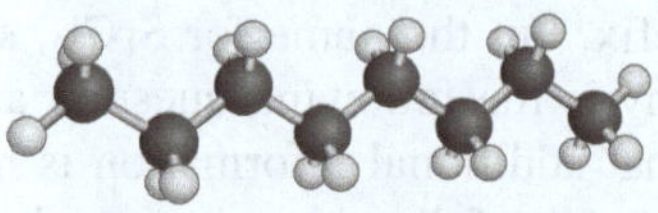
Nonane	C_9H_{20}	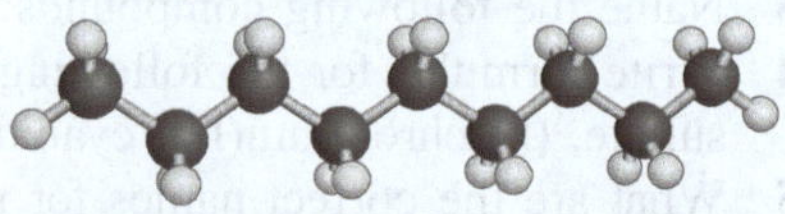
Decane	$C_{10}H_{22}$	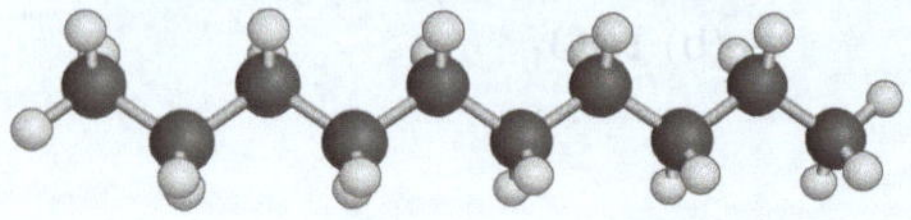

when an H atom in methane is replaced by a hydroxyl group (—OH), an amino group (—NH_2), and a carboxyl group (—COOH), the following molecules are generated:

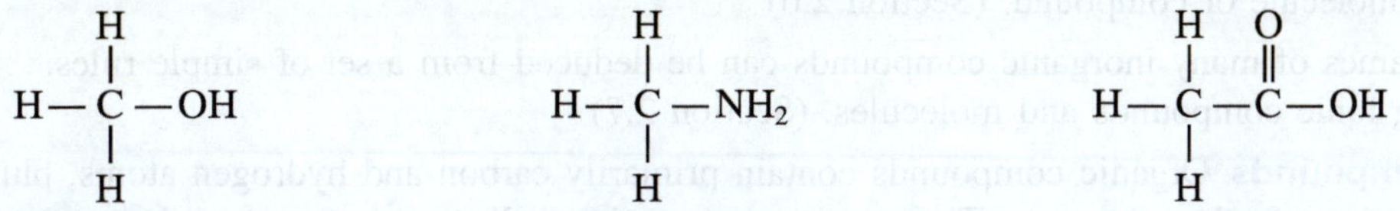

Methanol Methylamine Acetic acid

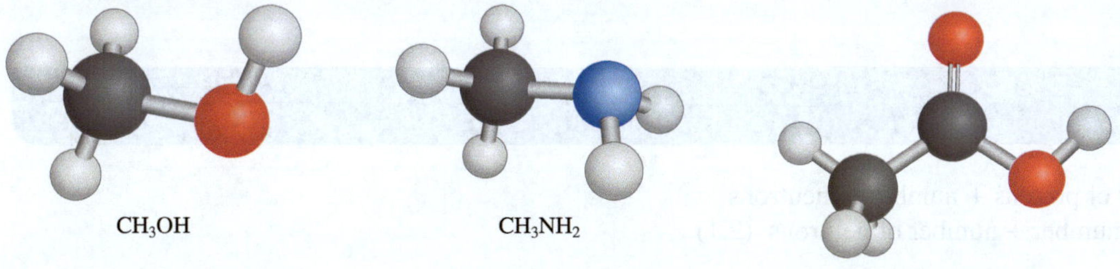

CH_3OH CH_3NH_2 CH_3COOH

The chemical properties of these molecules can be predicted based on the reactivity of the functional groups. Although the nomenclature of the major classes of organic compounds and their properties in terms of the functional groups will not be discussed until Chapter 24, we will frequently use organic compounds as examples to illustrate chemical bonding, acid-base reactions, and other properties throughout the book.

Summary of Concepts & Facts

- Organic compounds contain carbon and elements such as hydrogen, oxygen, and nitrogen. Hydrocarbons are the simplest type of organic compound.

Review of Concepts & Facts

2.8.1 How many different molecules can you generate by replacing one H atom with a hydroxyl group (—OH) in butane (see Table 2.8)?

Chapter Summary

The Atomic Theory The search for the fundamental units of matter began in ancient times. The modern version of atomic theory was laid out by John Dalton, who postulated that elements are made of extremely small particles called atoms and that all atoms of a given element are identical, but they are different from atoms of all other elements. (Section 2.1)

The Structure of the Atom An atom is composed of three elementary particles: proton, electron, and neutron. The proton has a positive charge, the electron has a negative charge, and the neutron has no charge. Protons and neutrons are located in a small region at the center of the atom, called the nucleus, and electrons are spread out about the nucleus at some distance from it. (Section 2.2)

Atomic Number, Mass Number, and Isotopes Atomic number is the number of protons in a nucleus; atoms of different elements have different atomic numbers. Isotopes are atoms of the same element having different numbers of neutrons. Mass number is the sum of the number of protons and neutrons in an atom. Because an atom is electrically neutral, the number of protons is equal to the number of electrons in it. (Section 2.3)

The Periodic Table Elements can be grouped together according to their chemical and physical properties in a chart called the periodic table. The periodic table enables us to classify elements (as metals, metalloids, and nonmetals) and correlate their properties in a systematic way. (Section 2.4)

Molecules and Ions Most matter is composed of molecules or ions. A molecule is a group of at least two atoms held together with chemical bonds. An ion is an atom or group of atoms with an overall positive or negative charge. (Section 2.5)

Chemical Formulas Atoms of most elements interact to form compounds, which are classified as molecules or ionic compounds made of positive (cations) and negative (anions) ions. Chemical formulas tell us the type and number of atoms present in a molecule or compound. (Section 2.6)

Naming Compounds The names of many inorganic compounds can be deduced from a set of simple rules. Specific rules exist for naming ionic compounds and molecules. (Section 2.7)

Introduction to Organic Compounds Organic compounds contain primarily carbon and hydrogen atoms, plus nitrogen, oxygen, sulfur, and atoms of other elements. The parent compounds of all organic compounds are the hydrocarbons of which alkanes are the simplest. (Section 2.8)

Key Equation

$$\begin{aligned}\text{mass number} &= \text{number of protons} + \text{number of neutrons}\\ &= \text{atomic number} + \text{number of neutrons} \quad (2.1)\end{aligned}$$

Key Words

Questions & Problems

Red numbered problems solved in Student Solutions Manual

2.2 The Structure of the Atom

Review Questions

2.1 Define the following terms: (a) α particle, (b) β particle, (c) γ ray, (d) X ray

2.2 Name the types of radiation known to be emitted by radioactive elements.

2.3 Compare the properties of the following: α particles, cathode rays, protons, neutrons, electrons.

2.4 What is meant by the term "fundamental particle"?

2.5 Describe the contributions of the following scientists to our knowledge of atomic structure: J. J. Thomson, R. A. Millikan, Ernest Rutherford, James Chadwick.

2.6 Describe the experimental basis for believing that the nucleus occupies a very small fraction of the volume of the atom.

Problems

2.7 The diameter of a helium atom is about 1×10^2 pm. Suppose that we could line up helium atoms side by side in contact with one another. Approximately how many atoms would it take to make the distance from end to end 1 cm?

2.8 Roughly speaking, the radius of an atom is about 10,000 times greater than that of its nucleus. If an atom were magnified so that the radius of its nucleus became 2.0 cm, about the size of a marble, what would be the radius of the atom in miles? (1 mi = 1609 m.)

2.3 Atomic Number, Mass Number, and Isotopes

Review Questions

2.9 Use the helium-4 isotope to define atomic number and mass number. Why does a knowledge of atomic number enable us to deduce the number of electrons present in an atom?

2.10 Why do all atoms of an element have the same atomic number, although they may have different mass numbers?

2.11 What do we call atoms of the same elements with different mass numbers?

2.12 Explain the meaning of each term in the symbol ${}^{A}_{Z}X$.

Problems

2.13 What is the mass number of an iron atom that has 28 neutrons?

2.14 Calculate the number of neutrons in ${}^{239}Pu$.

2.15 For each of the following species, determine the number of protons and the number of neutrons in the nucleus:

$${}^{3}_{2}He,\ {}^{4}_{2}He,\ {}^{24}_{12}Mg,\ {}^{25}_{12}Mg,\ {}^{48}_{22}Ti,\ {}^{79}_{35}Br,\ {}^{195}_{78}Pt$$

2.16 Indicate the number of protons, neutrons, and electrons in each of the following species:

$${}^{15}_{7}N,\ {}^{33}_{16}S,\ {}^{63}_{29}Cu,\ {}^{84}_{38}Sr,\ {}^{130}_{56}Ba,\ {}^{186}_{74}W,\ {}^{202}_{80}Hg$$

2.17 Write the appropriate symbol for each of the following isotopes: (a) $Z = 11, A = 23$; (b) $Z = 28, A = 64$.

2.18 Write the appropriate symbol for each of the following isotopes: (a) $Z = 74$, $A = 186$; (b) $Z = 80$, $A = 201$.

2.4 The Periodic Table

Review Questions

2.19 What is the periodic table, and what is its significance in the study of chemistry?

2.20 State two differences between a metal and a nonmetal.

2.21 Write the names and symbols for four elements in each of the following categories: (a) nonmetal, (b) metal, (c) metalloid.

2.22 Define, with two examples, the following terms: (a) alkali metals, (b) alkaline earth metals, (c) halogens, (d) noble gases.

Problems

2.23 Elements whose names end with "-ium" are usually metals; sodium is one example. Identify a nonmetal whose name also ends with "-ium."

2.24 Describe the changes in properties (from metals to nonmetals or from nonmetals to metals) as we move (a) down a periodic group and (b) across the periodic table from left to right.

2.25 Consult a handbook of chemical and physical data (ask your instructor where you can locate a copy of the handbook) to find (a) two metals less dense than water, (b) two metals more dense than mercury, (c) the densest known solid metallic element, (d) the densest known solid nonmetallic element.

2.26 Group the following elements in pairs that you would expect to show similar chemical properties: K, F, P, Na, Cl, and N.

2.5 Molecules and Ions

Review Questions

2.27 What is the difference between an atom and a molecule?

2.28 What are allotropes? Give an example. How are allotropes different from isotopes?

2.29 Describe the two commonly used molecular models.

2.30 Give an example of each of the following: (a) a monatomic cation, (b) a monatomic anion, (c) a polyatomic cation, (d) a polyatomic anion.

Problems

2.31 Which of the following diagrams represent diatomic molecules, polyatomic molecules, molecules that are not compounds, molecules that are compounds, or an elemental form of the substance?

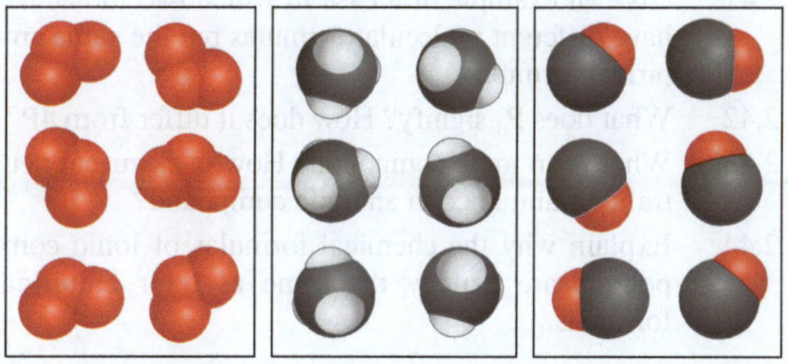

(a) (b) (c)

2.32 Which of the following diagrams represent diatomic molecules, polyatomic molecules, molecules that are not compounds, molecules that are compounds, or an elemental form of the substance?

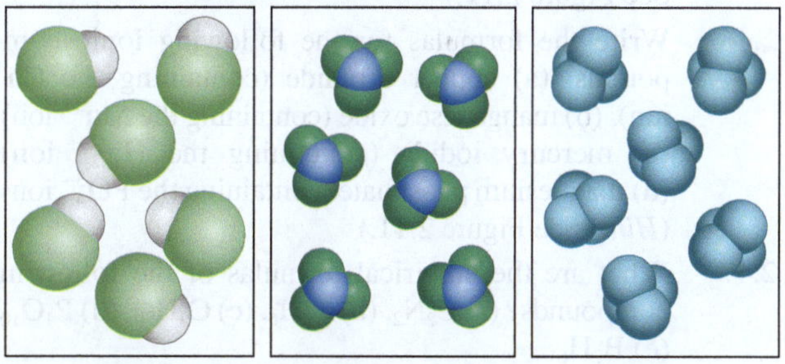

(a) (b) (c)

2.33 Identify the following as elements or compounds: NH_3, N_2, S_8, NO, CO, CO_2, H_2, SO_2.

2.34 Give two examples of each of the following: (a) a diatomic molecule containing atoms of the same element, (b) a diatomic molecule containing atoms of different elements, (c) a polyatomic molecule containing atoms of the same element, (d) a polyatomic molecule containing atoms of different elements.

2.35 Give the number of protons and electrons in each of the following common ions: Na^{+}, Ca^{2+}, Al^{3+}, Fe^{2+}, I^{-}, F^{-}, S^{2-}, O^{2-}, and N^{3-}.

2.36 Give the number of protons and electrons in each of the following common ions: K^{+}, Mg^{2+}, Fe^{3+}, Br^{-}, Mn^{2+}, C^{4-}, Cu^{2+}.

2.37 Pair the following species that contain the same number of electrons: Ar, Sn^{4+}, F^{-}, Fe^{3+}, P^{3-}, V, Ag^{+}, N^{3-}.

2.38 Write the correct symbols for the atoms that contain (a) 25 protons, 25 electrons, and 27 neutrons; (b) 10 protons, 10 electrons, and 12 neutrons; (c) 47 protons, 47 electrons, and 60 neutrons; (d) 53 protons, 53 electrons, and 74 neutrons; (e) 94 protons, 94 electrons, and 145 neutrons.

2.6 Chemical Formulas

Review Questions

2.39 What does a chemical formula represent? What is the ratio of the atoms in the following molecular formulas? (a) NO, (b) NCl_3, (c) N_2O_4, (d) P_4O_6.

2.40 Define molecular formula and empirical formula. What are the similarities and differences between the empirical formula and molecular formula of a compound?

2.41 Give an example of a case in which two molecules have different molecular formulas but the same empirical formula.

2.42 What does P_4 signify? How does it differ from 4P?

2.43 What is an ionic compound? How is electrical neutrality maintained in an ionic compound?

2.44 Explain why the chemical formulas of ionic compounds are usually the same as their empirical formulas.

Problems

2.45 Write the formulas for the following ionic compounds: (a) sodium oxide, (b) iron sulfide (containing the Fe^{2+} ion), (c) cobalt sulfate (containing the Co^{3+} and SO_4^{2-} ions), (d) barium fluoride. (*Hint:* See Figure 2.11.)

2.46 Write the formulas for the following ionic compounds: (a) copper bromide (containing the Cu^+ ion), (b) manganese oxide (containing the Mn^{3+} ion), (c) mercury iodide (containing the Hg_2^{2+} ion), (d) magnesium phosphate (containing the PO_4^{3-} ion). (*Hint:* See Figure 2.11.)

2.47 What are the empirical formulas of the following compounds? (a) C_2N_2, (b) C_6H_6, (c) C_9H_{20}, (d) P_4O_{10}, (e) B_2H_6

2.48 What are the empirical formulas of the following compounds? (a) Al_2Br_6, (b) $Na_2S_2O_4$, (c) N_2O_5, (d) $K_2Cr_2O_7$

2.49 Write the molecular formula of glycine, an amino acid present in proteins. The color codes are black (carbon), blue (nitrogen), red (oxygen), and gray (hydrogen).

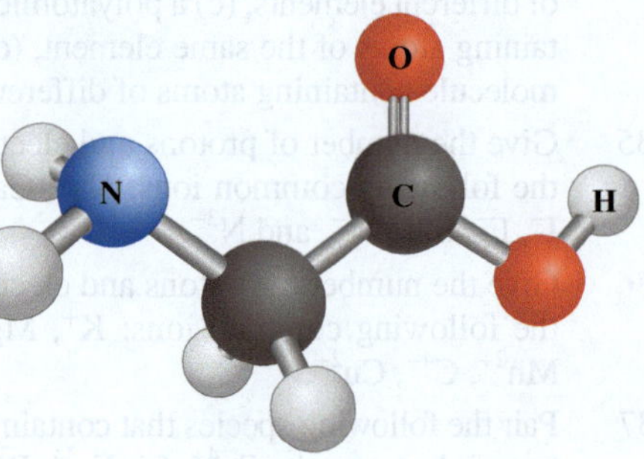

2.50 Write the molecular formula of ethanol. The color codes are black (carbon), red (oxygen), and gray (hydrogen).

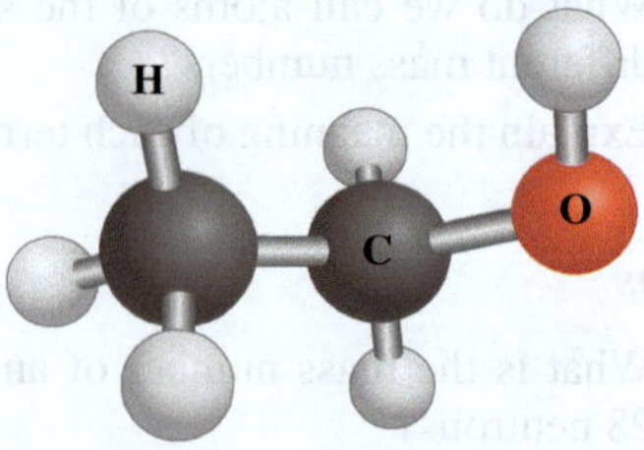

2.51 Which of the following compounds are likely to be ionic? Which are likely to be molecular? $SiCl_4$, LiF, $BaCl_2$, B_2H_6, KCl, C_2H_4

2.52 Which of the following compounds are likely to be ionic? Which are likely to be molecular? CH_4, NaBr, BaF_2, CCl_4, ICl, CsCl, NF_3

2.7 Naming Compounds

Review Questions

2.53 What is the difference between inorganic compounds and organic compounds?

2.54 What are the four major categories of inorganic compounds?

2.55 Give an example each for a binary compound and a ternary compound.

2.56 What is the Stock system? What are its advantages over the older system of naming cations?

2.57 Explain why the formula HCl can represent two different chemical systems.

2.58 Define the following terms: *acids, bases, oxoacids, oxoanions, hydrates.*

Problems

2.59 Name these compounds: (a) Na_2CrO_4, (b) K_2HPO_4, (c) HBr (gas), (d) HBr (in water), (e) Li_2CO_3, (f) $K_2Cr_2O_7$, (g) NH_4NO_2, (h) PF_3, (i) PF_5, (j) P_4O_6, (k) CdI_2, (l) $SrSO_4$, (m) $Al(OH)_3$, (n) $Na_2CO_3 \cdot 10H_2O$.

2.60 Name these compounds: (a) KClO, (b) Ag_2CO_3, (c) $FeCl_2$, (d) $KMnO_4$, (e) $CsClO_3$, (f) HIO, (g) FeO, (h) Fe_2O_3, (i) $TiCl_4$, (j) NaH, (k) Li_3N, (l) Na_2O, (m) Na_2O_2, (n) $FeCl_3 \cdot 6H_2O$.

2.61 Write the formulas for the following compounds: (a) rubidium nitrite, (b) potassium sulfide, (c) sodium hydrogen sulfide, (d) magnesium phosphate, (e) calcium hydrogen phosphate, (f) potassium dihydrogen phosphate, (g) iodine heptafluoride, (h) ammonium sulfate, (i) silver perchlorate, (j) boron trichloride.

2.62 Write the formulas for the following compounds: (a) copper(I) cyanide, (b) strontium chlorite, (c) perbromic acid, (d) hydroiodic acid, (e) disodium ammonium phosphate, (f) lead(II) carbonate, (g) tin(II) fluoride, (h) tetraphosphorus decasulfide, (i) mercury(II) oxide, (j) mercury(I) iodide, (k) selenium hexafluoride.

2.63 Sulfur (S) and fluorine (F) form several different compounds. One of them, SF_6, contains 3.55 g of F for every gram of S. Use the law of multiple proportions to determine n, which represents the number of F atoms in SF_n, given that it contains 2.37 g of F for every gram of S.

2.64 Name the following compounds.

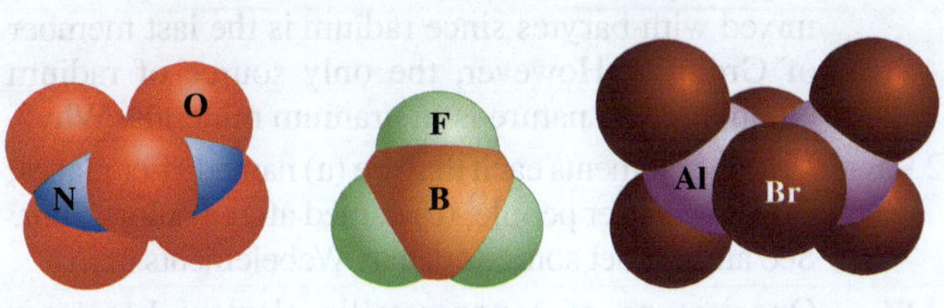

(a) (b) (c)

Additional Problems

2.65 A sample of a uranium compound is found to be losing mass gradually. Explain what is happening to the sample.

2.66 In which one of the following pairs do the two species resemble each other most closely in chemical properties? Explain. (a) $^{1}_{1}H$ and $^{1}_{1}H^{+}$, (b) $^{14}_{7}N$ and $^{14}_{7}N^{3-}$, (c) $^{12}_{6}C$ and $^{13}_{6}C$.

2.67 One isotope of a metallic element has mass number 65 and 35 neutrons in the nucleus. The cation derived from the isotope has 28 electrons. Write the symbol for this cation.

2.68 One isotope of a nonmetallic element has mass number 127 and 74 neutrons in the nucleus. The anion derived from the isotope has 54 electrons. Write the symbol for this anion.

2.69 Determine the molecular and empirical formulas of the compounds shown here. (Black spheres are carbon and gray spheres are hydrogen.)

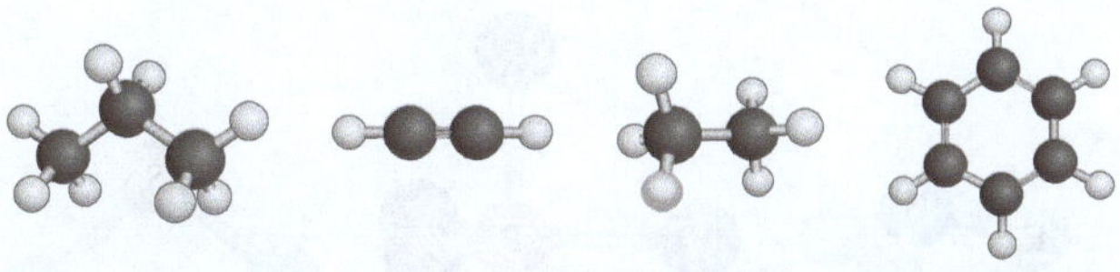

(a) (b) (c) (d)

2.70 What is wrong with or ambiguous about the phrase "four molecules of NaCl"?

2.71 The following phosphorus sulfides are known: P_4S_3, P_4S_7, and P_4S_{10}. Do these compounds obey the law of multiple proportions?

2.72 Which of the following are elements, which are molecules but not compounds, which are compounds but not molecules, and which are both compounds and molecules? (a) SO_2, (b) S_8, (c) Cs, (d) N_2O_5, (e) O, (f) O_2, (g) O_3, (h) CH_4, (i) KBr, (j) S, (k) P_4, (l) LiF

2.73 The following table gives numbers of electrons, protons, and neutrons in atoms or ions of a number of elements. Answer the following: (a) Which of the species are neutral? (b) Which are negatively charged? (c) Which are positively charged? (d) What are the conventional symbols for all the species?

Atom or Ion of Element	**A**	**B**	**C**	**D**	**E**	**F**	**G**
Number of electrons	5	10	18	28	36	5	9
Number of protons	5	7	19	30	35	5	9
Number of neutrons	5	7	20	36	46	6	10

2.74 Identify the elements represented by the following symbols and give the number of protons and neutrons in each case: (a) $^{20}_{10}X$, (b) $^{63}_{29}X$, (c) $^{107}_{47}X$, (d) $^{182}_{74}X$, (e) $^{203}_{84}X$, (f) $^{234}_{94}X$.

2.75 Each of the following pairs of elements will react to form an ionic compound. Write the formulas and name these compounds: (a) barium and oxygen, (b) calcium and phosphorus, (c) aluminum and sulfur, (d) lithium and nitrogen.

2.76 Match the descriptions [(a)–(h)] with each of the following elements: P, Cu, Kr, Sb, Cs, Al, Sr, Cl. (a) A transition metal, (b) a nonmetal that forms a −3 ion, (c) a noble gas, (d) an alkali metal, (e) a metal that forms a +3 ion, (f) a metalloid, (g) an element that exists as a diatomic gas molecule, (h) an alkaline earth metal.

2.77 Explain why anions are always larger than the atoms from which they are derived, whereas cations are always smaller than the atoms from which they are derived. (*Hint:* Consider the electrostatic attraction between protons and electrons.)

2.78 (a) Describe Rutherford's experiment and how it led to the structure of the atom. How was he able to estimate the number of protons in a nucleus from the scattering of the α particles? (b) Consider the ^{23}Na atom. Given that the radius and mass of the nucleus are 3.04×10^{-15} m and 3.82×10^{-23} g, respectively, calculate the density of the nucleus in g/cm^3. The radius of a ^{23}Na atom is 186 pm. Calculate the density of the space occupied by the electrons in the sodium atom. Do your results support Rutherford's model of an atom? [The volume of a sphere of radius r is $(4/3)\pi r^3$.]

2.79 Caffeine, shown here, is a psychoactive stimulant drug. Write the molecular formula and empirical formula of the compound.

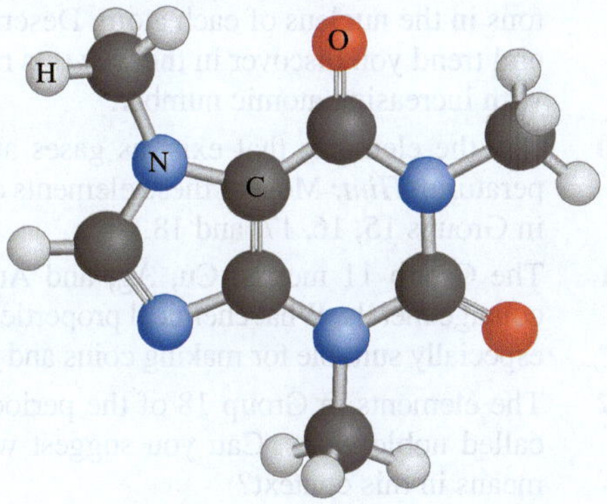

2.80 Acetaminophen, shown here, is the active ingredient in Tylenol. Write the molecular formula and empirical formula of the compound.

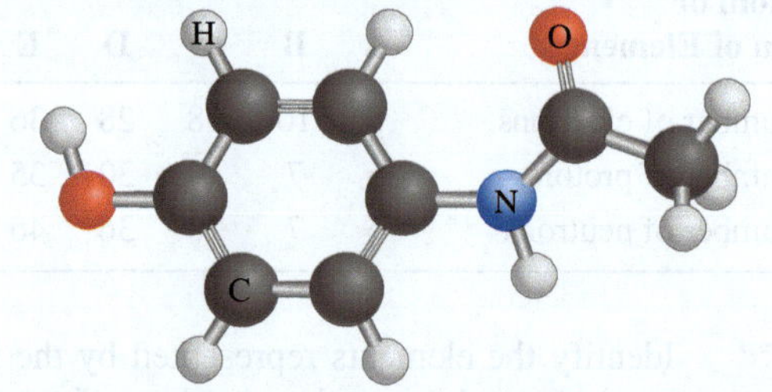

2.81 What is wrong with the chemical formula for each of the following compounds? (a) magnesium iodate, [$Mg(IO_4)_2$], (b) phosphoric acid (H_3PO_3), (c) barium sulfite (BaS), (d) ammonium bicarbonate (NH_3HCO_3)

2.82 What is wrong with the names (in parentheses) for each of the following compounds? (a) $SnCl_4$ (tin chloride), (b) Cu_2O [copper(II) oxide], (c) $Co(NO_3)_2$ (cobalt nitrate), (d) $Na_2Cr_2O_7$ (sodium chromate)

2.83 Fill in the blanks in the following table.

Symbol		$^{54}_{26}Fe^{2+}$			
Protons	5			79	86
Neutrons	6		16	117	136
Electrons	5		18	79	
Net charge			−3		0

2.84 (a) Which elements are most likely to form ionic compounds? (b) Which metallic elements are most likely to form cations with different charges?

2.85 Write the formula of the common ion derived from each of the following: (a) Li, (b) S, (c) I, (d) N, (e) Al, (f) Cs, (g) Mg.

2.86 Which of the following symbols provides more information about the atom: ^{23}Na or $_{11}Na$? Explain.

2.87 Write the chemical formulas and names of binary acids and oxoacids that contain Group 17 elements. Do the same for elements in Groups 13, 14, 15, and 16.

2.88 Of the 118 elements known, only two are liquids at room temperature (25°C). What are they? (*Hint:* One element is a familiar metal and the other element is in Group 17.)

2.89 For the noble gases (the Group 18 elements), $^{4}_{2}He$, $^{20}_{10}Ne$, $^{40}_{18}Ar$, $^{84}_{36}Kr$, and $^{132}_{54}Xe$, (a) determine the number of protons and neutrons in the nucleus of each atom, and (b) determine the ratio of neutrons to protons in the nucleus of each atom. Describe any general trend you discover in the way this ratio changes with increasing atomic number.

2.90 List the elements that exist as gases at room temperature. (*Hint:* Most of these elements can be found in Groups 15, 16, 17, and 18.)

2.91 The Group 11 metals, Cu, Ag, and Au, are called coinage metals. What chemical properties make them especially suitable for making coins and jewelry?

2.92 The elements in Group 18 of the periodic table are called noble gases. Can you suggest what "noble" means in this context?

2.93 The formula for calcium oxide is CaO. What are the formulas for magnesium oxide and strontium oxide?

2.94 A common mineral of barium is barytes, or barium sulfate ($BaSO_4$). Because elements in the same periodic group have similar chemical properties, we might expect to find some radium sulfate ($RaSO_4$) mixed with barytes since radium is the last member of Group 2. However, the only source of radium compounds in nature is in uranium minerals. Why?

2.95 List five elements each that are (a) named after places, (b) named after people, (c) named after a color. (*Hint:* See an Internet source such as Webelements.com.)

2.96 One isotope of a nonmetallic element has mass number 77 and 43 neutrons in the nucleus. The anion derived from the isotope has 36 electrons. Write the symbol for this anion.

2.97 Fluorine reacts with hydrogen (H) and deuterium (D) to form hydrogen fluoride (HF) and deuterium fluoride (DF), where deuterium ($^{2}_{1}H$) is an isotope of hydrogen. Would a given amount of fluorine react with different masses of the two hydrogen isotopes? Does this violate the law of definite proportion? Explain.

2.98 Predict the formula and name of a binary compound formed from the following elements: (a) Na and H, (b) B and O, (c) Na and S, (d) Al and F, (e) F and O, (f) Sr and Cl.

2.99 Identify each of the following elements: (a) a halogen whose anion contains 36 electrons, (b) a radioactive noble gas with 86 protons, (c) a Group 16 element whose anion contains 36 electrons, (d) an alkali metal cation that contains 36 electrons, (e) a Group 14 cation that contains 80 electrons.

2.100 Write the molecular formulas for and names of the following compounds.

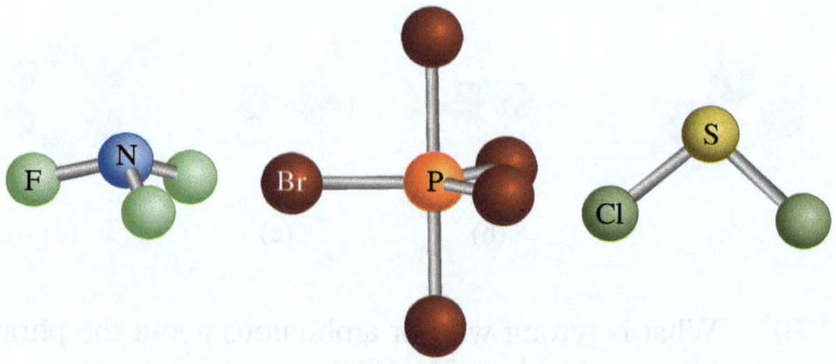

2.101 Show the locations of (a) alkali metals, (b) alkaline earth metals, (c) the halogens, and (d) the noble gases in the following outline of a periodic table. Also draw dividing lines between metals and metalloids and between metalloids and nonmetals.

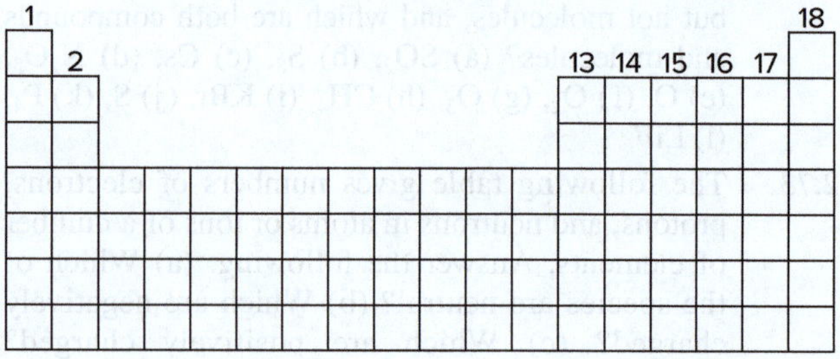

2.102 Fill the blanks in the following table.

Cation	Anion	Formula	Name
			Magnesium bicarbonate
		$SrCl_2$	
Fe^{3+}	NO_2^-		
			Manganese(II) chlorate
		$SnBr_4$	
Co^{2+}	PO_4^{3-}		
Hg_2^{2+}	I^-		
		Cu_2CO_3	
			Lithium nitride
Al^{3+}	S^{2-}		

2.103 Some compounds are better known by their common names than by their systematic chemical names. Give the chemical formulas of the following substances: (a) dry ice, (b) table salt, (c) laughing gas, (d) marble (chalk, limestone), (e) quicklime, (f) slaked lime, (g) baking soda, (h) washing soda, (i) gypsum, (j) milk of magnesia.

2.104 In Section 2.1 it was pointed out that mass and energy are alternate aspects of a single entity called *mass-energy*. The relationship between these two physical quantities is Einstein's famous equation, $E = mc^2$, where E is energy, m is mass, and c is the speed of light. In a combustion experiment, it was found that 12.096 g of hydrogen molecules combined with 96.000 g of oxygen molecules to form water and released 1.715×10^3 kJ of heat. Calculate the corresponding mass change in this process and comment on whether the law of conservation of mass holds for ordinary chemical processes. (*Hint:* The Einstein equation can be used to calculate the change in mass as a result of the change in energy. $1\ J = 1\ kg \cdot m^2/s^2$ and $c = 3.00 \times 10^8$ m/s.)

2.105 Draw all possible structural formulas of the following hydrocarbons: CH_4, C_2H_6, C_3H_8, C_4H_{10}, C_5H_{12}.

2.106 (a) Assuming nuclei are spherical in shape, show that its radius r is proportional to the cube root of mass number (A). (b) In general, the radius of a nucleus is given by $r = r_0A^{1/3}$, where r_0 is a proportionality constant given by 1.2×10^{-15} m. Calculate the volume of the ${}^7_3Li$ nucleus. (c) Given that the radius of a Li atom is 152 pm, calculate the fraction of the atom's volume occupied by the nucleus. Does your result support Rutherford's model of an atom?

2.107 Draw two different structural formulas based on the molecular formula C_2H_6O. Is the fact that you can have more than one compound with the same molecular formula consistent with Dalton's atomic theory?

2.108 Ethane and acetylene are two gaseous hydrocarbons. Chemical analyses show that in one sample of ethane, 2.65 g of carbon are combined with 0.665 g of hydrogen, and in one sample of acetylene, 4.56 g of carbon are combined with 0.383 g of hydrogen. (a) Are these results consistent with the law of multiple proportions? (b) Write reasonable molecular formulas for these compounds.

2.109 A cube made of platinum (Pt) has an edge length of 1.0 cm. (a) Calculate the number of Pt atoms in the cube. (b) Atoms are spherical in shape. Therefore, the Pt atoms in the cube cannot fill all of the available space. If only 74 percent of the space inside the cube is taken up by Pt atoms, calculate the radius in picometers of a Pt atom. The density of Pt is 21.45 g/cm^3 and the mass of a single Pt atom is 3.240×10^{-22} g. [The volume of a sphere of radius r is $(4/3)\pi r^3$.]

2.110 A monatomic ion has a charge of +2. The nucleus of the parent atom has a mass number of 55. If the number of neutrons in the nucleus is 1.2 times that of the number of protons, what is the name and symbol of the element?

2.111 In the following 2 × 2 crossword, each letter must be correct four ways: horizontally, vertically, diagonally, and by itself. When the puzzle is complete, the four spaces will contain the overlapping symbols of 10 elements. Use capital letters for each square. There is only one correct solution.*

1	2
3	4

Horizontal

1–2: Two-letter symbol for a metal used in ancient times

3–4: Two-letter symbol for a metal that burns in air and is found in Group 15

Vertical

1–3: Two-letter symbol for a metalloid

2–4: Two-letter symbol for a metal used in U.S. coins

Single Squares

1: A colorful nonmetal

2: A colorless gaseous nonmetal

3: An element that makes fireworks green

4: An element that has medicinal uses

Diagonal

1–4: Two-letter symbol for an element used in electronics

2–3: Two-letter symbol for a metal used with Zr to make wires for superconducting magnets

2.112 Name the following acids.

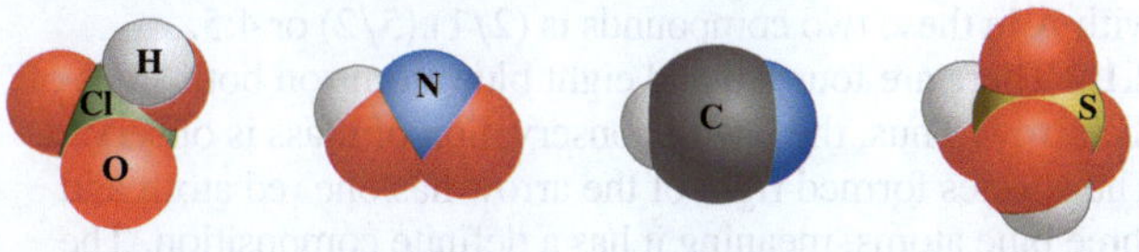

(a) (b) (c) (d)

*Source: S. J. Cyvin of the University of Trondheim (Norway). This puzzle appeared in *Chemical & Engineering News*, December 14, 1987 (p. 86) and in *Chem Matters*, October 1988.

2.113 Calculate the density of the nucleus of a $^{56}_{26}Fe$ atom, given that the nuclear mass is 9.229×10^{-23} g. From your result, comment on the fact that any nucleus containing more than one proton must have neutrons present as well. (*Hint:* See Problem 2.106.)

2.114 Element X reacts with element Y to form an ionic compound containing X^{4+} and Y^{2-} ions. Write a formula for the compound and suggest in which periodic groups these elements are likely to be found. Name a representative compound.

2.115 Methane, ethane, and propane are shown in Table 2.8. Show that the following data are consistent with the law of multiple proportions.

	Mass of Carbon in 1-g Sample	Mass of Hydrogen in 1-g Sample
Methane	0.749 g	0.251 g
Ethane	0.799 g	0.201 g
Propane	0.817 g	0.183 g

Interpreting, Modeling, & Estimating

2.116 In the Rutherford scattering experiment, an α particle is heading directly toward a gold nucleus. The particle will come to a halt when its kinetic energy is completely converted to electrical potential energy. When this happens, how close will the α particle with a kinetic energy of 6.0×10^{-14} J be from the nucleus? [According to Coulomb's law, the electrical potential energy between two charged particles is $E = kQ_1Q_2/r$, where Q_1 and Q_2 are the charges (in coulombs) of the α particle and the gold nucleus, r is the distance of separation in meters, and k is a constant equal to 9.0×10^9 kg · m^3/s^2 · C^2. Joule (J) is the unit of energy where 1 J = 1 kg · m^2/s^2.]

2.117 Estimate the relative sizes of the following species: Li, Li^+, Li^-.

2.118 Compare the atomic size of the following two magnesium isotopes: ^{24}Mg and ^{26}Mg.

2.119 Using visible light, we humans cannot see any object smaller than 2×10^{-5} cm with an unaided eye. Roughly how many This question atoms must be lined up for us to see the atoms?

2.120 If the size of the nucleus of an atom were that of a pea, how far would the electrons be (on average) from the nucleus in meters?

2.121 Sodium and potassium are roughly equal in natural abundance in Earth's crust and most of their compounds are soluble. However, the composition of seawater is much higher in sodium than potassium. Explain.

2.122 One technique proposed for recycling plastic grocery bags is to heat them at 700°C and high pressure to form carbon microspheres that can be used in a number of applications. Electron microscopy shows some representative carbon microspheres obtained in this manner, where the scale is given on the figure. Determine the number of carbon atoms in a typical carbon microsphere.

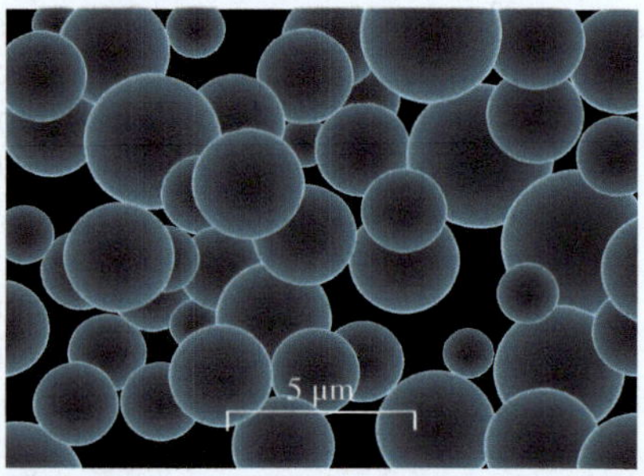

Answers to Practice Exercises

2.1 29 protons, 34 neutrons, and 29 electrons. **2.2** $CHCl_3$. **2.3** $C_4H_5N_2O$. **2.4** (a) $Cr_2(SO_4)_3$. (b) TiO_2. **2.5** (a) Lead(II) oxide. (b) Lithium chlorate. **2.6** (a) Rb_2SO_4. (b) BaH_2. **2.7** (a) Nitrogen trifluoride. (b) Dichlorine heptoxide. **2.8** (a) SF_4. (b) N_2O_5. **2.9** (a) Hypobromous acid. (b) Hydrogen sulfate ion.

Answers to Review of Concepts & Facts

2.1.1 Yes, the ratio of atoms represented by B that combine with A in these two compounds is (2/1):(5/2) or 4:5. **2.1.2** There are four red and eight blue atoms on both sides of the arrow. Thus, the law of conservation of mass is observed. The species formed right of the arrow has one red atom and three blue atoms, meaning it has a definite composition. The law of definite proportions is observed. Only one species is formed in this process, meaning the law of multiple proportions has not been demonstrated. **2.3.1** 78. **2.3.2** 66. **2.3.3** ^{17}O. **2.4.1** Chemical properties change more markedly across a period. **2.4.2** (a) Metal. (b) Nonmetal. (c) Metalloid. (d) Metal. **2.5.1** S_8 signifies one molecule of sulfur that is composed of 8S atoms. 8S represents 8 individual atoms of sulfur. **2.5.2** (a) 15 protons, 18 electrons. (b) 22 protons, 18 electrons. (c) 32 protons, 30 electrons. **2.5.3** Cr^{2+}. **2.6.1** (a) $Mg(NO_3)_2$. (b) Al_2O_3. (c) LiH. (d) Na_2S. **2.6.2** (a) CaI_2. (b) Ga_2S_3. **2.7.1** Se and Cl are both nonmetals, so $SeCl_2$ is a molecule and prefixes are used when naming the compound. Sr is a metal, so $SrCl_2$ is an ionic compound and prefixes are not necessary. **2.7.2** HF can be named as a molecule (hydrogen fluoride) or as an acid (hydrofluoric acid). We need to know if the compound is dissolved in water. **2.7.3** (a) Cesium sulfite. (b) Copper(II) nitrite. **2.7.4** (a) $Co(HSO_4)_2$. (b) $Cr(CN)_3$. **2.7.5** (a) Bromine pentafluoride. (b) Periodic acid. **2.8.1** Two.

CHAPTER

3

Mass Relationships in Chemical Reactions

The reaction between aluminum and ammonium perchlorate fuels the solid rocket boosters to send this rocket into space.
NASA

CHAPTER OUTLINE

In this chapter, we will consider the masses of atoms and molecules and what happens to them when chemical changes occur. Our guide for this discussion will be the law of conservation of mass.

3.1 Atomic Mass

Learning Objectives

- Discuss the nature of the atomic mass scale.
- Determine the average atomic mass of an element from isotopic mass and relative abundance information.

In this chapter, we will use what we have learned about chemical structure and formulas in studying the mass relationships of atoms and molecules. These relationships in turn will help us to explain the composition of compounds and the ways in which composition changes.

The mass of an atom depends on the number of electrons, protons, and neutrons it contains. Knowledge of an atom's mass is important in laboratory work. But atoms are extremely small particles—even the smallest speck of dust that our unaided eyes can detect contains as many as 1×10^{16} atoms! Clearly we cannot weigh a single atom, but it is possible to determine the mass of one atom *relative* to another experimentally. The first step is to assign a value to the mass of one atom of a given element so that it can be used as a standard.

By international agreement, ***atomic mass*** (sometimes called *atomic weight*) is *the mass of the atom in atomic mass units (amu).* One ***atomic mass unit*** is defined as *a mass exactly equal to one-twelfth the mass of one carbon-12 atom.* One atomic mass unit is also known as one Dalton, named after John Dalton. Carbon-12 is the carbon isotope that has six protons and six neutrons. Setting the atomic mass of carbon-12 at 12 amu provides the standard for measuring the atomic mass of the other elements. For example, experiments have shown that, on average, a hydrogen atom is only 8.400 percent as massive as the carbon-12 atom. Thus, if the mass of one carbon-12 atom is exactly 12 amu, the atomic mass of hydrogen must be 0.08400×12 amu or 1.008 amu. Similar calculations show that the atomic mass of oxygen is 16.00 amu and that of iron is 55.85 amu. Thus, although we do not know just how much an average iron atom's mass is, we know that it is approximately 56 times as massive as a hydrogen atom.

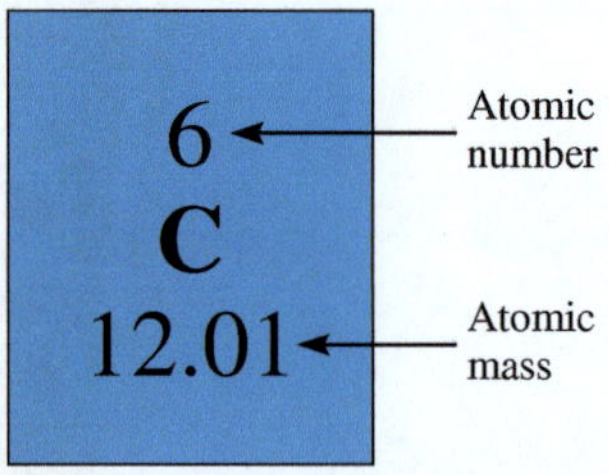

Average Atomic Mass

When you look up the atomic mass of carbon in a periodic table, you will find that its value is not 12.00 amu but 12.01 amu. The reason for the difference is that most naturally occurring elements (including carbon) have more than one isotope. This means that when we measure the atomic mass of an element, we must generally settle for the *average* mass of the naturally occurring mixture of isotopes. For example, the natural abundances of carbon-12 and carbon-13 are 98.90 percent and 1.10 percent, respectively. The atomic mass of carbon-13 has been determined to be 13.00335 amu. Thus, the average atomic mass of carbon can be calculated as follows:

$$\text{average atomic mass of natural carbon} = (0.9890)(12 \text{ amu}) + (0.0110)(13.00335 \text{ amu})$$
$$= 12.01 \text{ amu}$$

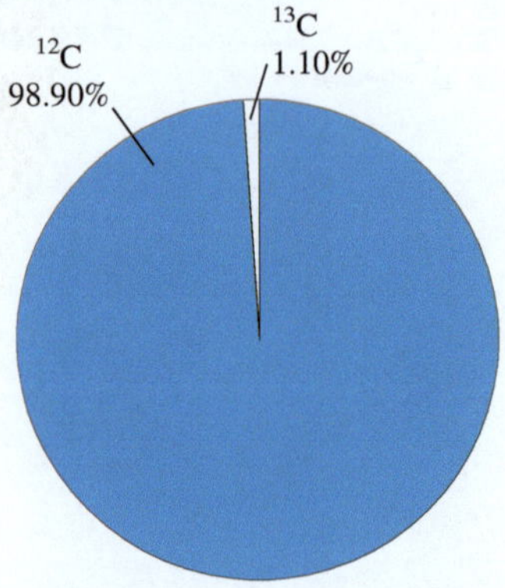

Natural abundances of C-12 and C-13 isotopes.

Note that in calculations involving percentages, we need to convert percentages to fractions. For example, 98.90 percent becomes 98.90/100, or 0.9890. Because there are many more carbon-12 atoms than carbon-13 atoms in naturally occurring carbon, the average atomic mass is much closer to 12 amu than to 13 amu.

It is important to understand that when we say that the atomic mass of carbon is 12.01 amu, we are referring to the *average* value. If carbon atoms could be examined individually, we would find either an atom of atomic mass exactly 12 amu or one of 13.00335 amu, but never one of 12.01 amu. Example 3.1 shows how to calculate the average atomic mass of an element.

Example 3.1

Boron is used in the manufacture of ceramics and polymers such as fiberglass. The atomic masses of its two stable isotopes, $^{10}_{5}B$ (19.80 percent) and $^{11}_{5}B$ (80.20 percent), are 10.0129 amu and 11.0093 amu, respectively. The boron-10 isotope is also important as a neutron-capturing agent in nuclear reactors. Calculate the average atomic mass of boron.

Strategy Each isotope contributes to the average atomic mass based on its relative abundance. Multiplying the mass of an isotope by its fractional abundance (not percent) will give the contribution to the average atomic mass of that particular isotope.

Solution First the percent abundances are converted to fractions: 19.80 percent to 19.80/100 or 0.1980 and 80.20 percent to 80.20/100 or 0.8020. We find the contribution to the average atomic mass for each isotope, and then add the contributions together to obtain the average atomic mass.

$$(0.1980)(10.0129 \text{ amu}) + (0.8020)(11.0093 \text{ amu}) = 10.8129 \text{ amu}$$

Check The average atomic mass should be between the two isotopic masses; therefore, the answer is reasonable. Note that because there are more $^{11}_{5}B$ isotopes than $^{10}_{5}B$ isotopes, the average atomic mass is closer to 11.0093 amu than to 10.0129 amu.

Practice Exercise The atomic masses of the two stable isotopes of copper, $^{63}_{29}Cu$ (69.17 percent) and $^{65}_{29}Cu$ (30.83 percent), are 62.9296 amu and 64.9278 amu, respectively. Calculate the average atomic mass of copper.

Similar problems: 3.5, 3.6.

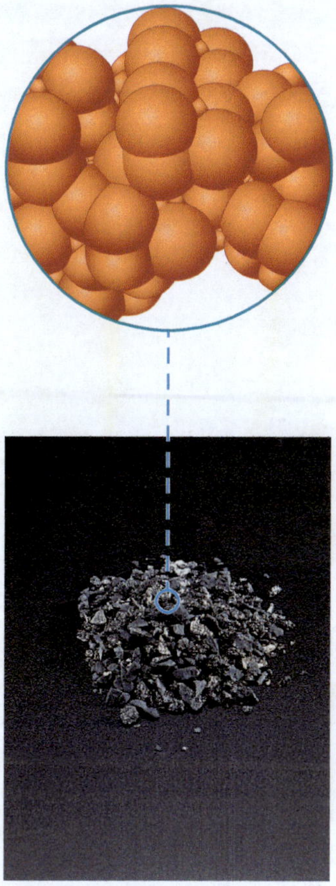

Boron and the solid-state structure of boron.

Stephen Frisch/McGraw Hill

The atomic masses of many elements have been accurately determined to five or six significant figures. However, for our purposes we will normally use atomic masses accurate only to four significant figures (see List of the Elements with Their Symbols and Atomic Masses). For simplicity, we will omit the word "average" when we discuss the atomic masses of the elements.

Summary of Concepts & Facts

- Atomic masses are measured in atomic mass units (amu), a relative unit based on a value of exactly 12 for the C-12 isotope. The atomic mass given for the atoms of a particular element is the average of the naturally occurring isotope distribution of that element.

Review of Concepts & Facts

3.1.1 There are two stable isotopes of iridium: ^{191}Ir (190.96 amu) and ^{193}Ir (192.96 amu). If you were to randomly pick an iridium atom from a large collection of iridium atoms, which isotope are you more likely to select?

3.1.2 The hypothetical element Q occurs as a mixture of 37.50 percent ^{47}Q (47.054 amu) and 62.50 percent ^{51}Q (50.924 amu). Calculate the average atomic mass of Q.

3.2 Avogadro's Number and the Molar Mass of an Element

Learning Objective

- Describe and apply the concept of a mole.

Atomic mass units provide a relative scale for the masses of the elements. But because atoms have such small masses, no usable scale can be devised to weigh them in calibrated units of atomic mass units. In any real situation, we deal with macroscopic samples containing enormous numbers of atoms. Therefore, it is convenient to have a special unit to describe a very large number of atoms. The idea of a unit to denote a particular number of objects is not new. For example, the pair (2 items), the dozen (12 items), and the gross (144 items) are all familiar units. Chemists measure atoms and molecules in moles.

In the SI system the ***mole (mol)*** is *the amount of a substance that contains exactly 6.0221413×10^{23} elementary entities (atoms, molecules, or other particles)*. The previous definition of the mole related it to a specific mass: the amount of substance that contains as many elementary particles as there are atoms in exactly 12 g (or 0.012 kg) of the carbon-12 isotope. The actual number of atoms in 12 g of carbon-12 is determined experimentally, thus, the current definition is an exact number. This number is called ***Avogadro's number*** **(N_A)**, in honor of the Italian scientist Amedeo Avogadro.[†] The currently accepted value is

$$N_A = 6.0221413 \times 10^{23}$$

Generally, we round Avogadro's number to 6.022×10^{23}. Thus, just as 1 dozen oranges contains 12 oranges, 1 mole of hydrogen atoms contains 6.022×10^{23} H atoms. Figure 3.1 shows samples containing 1 mole each of several common elements.

Figure 3.1 *One mole each of several common elements. Carbon (black charcoal powder), sulfur (yellow powder), iron (as nails), copper wires, and mercury (shiny liquid metal).*
Stephen Frisch/McGraw Hill

[†]Lorenzo Romano Amedeo Carlo Avogadro di Quaregua e di Cerreto (1776–1856). Italian mathematical physicist. He practiced law for many years before he became interested in science. His most famous work, now known as Avogadro's law (see Chapter 5), was largely ignored during his lifetime, although it became the basis for determining atomic masses in the late nineteenth century.

The enormity of Avogadro's number is difficult to imagine. For example, spreading 6.022×10^{23} oranges over the entire surface of Earth would produce a layer 9 mi into space! It is also estimated that in the entire observable universe, there are only 1×10^{24} stars, which is just smaller than Avogadro's number! But because atoms (and molecules) are so tiny, we need a huge number to study them in manageable quantities.

We have seen that 1 mole of carbon-12 atoms has a mass of exactly 12 g and contains 6.022×10^{23} atoms. This mass of carbon-12 is its ***molar mass*** **($\mathcal{M}$)**, defined as *the mass (in grams or kilograms) of 1 mole of units* (such as atoms or molecules) *of a substance*. Note that the molar mass of carbon-12 (in grams [g]) is numerically equal to its atomic mass in atomic mass units. Likewise, the atomic mass of sodium (Na) is 22.99 amu and its molar mass is 22.99 g; the atomic mass of phosphorus is 30.97 amu and its molar mass is 30.97 g; and so on. If we know the atomic mass of an element, we also know its molar mass. Note that in many chemical calculations, molar mass is best represented with units of g/mol or kg/mol.

Knowing the molar mass and Avogadro's number, we can calculate the mass of a single atom in grams. For example, we know the molar mass of carbon-12 is 12 g and there are 6.022×10^{23} carbon-12 atoms in 1 mole of the substance; therefore, the mass of one carbon-12 atom is given by

$$\frac{12 \text{ g carbon-12 atoms}}{6.022 \times 10^{23} \text{ carbon-12 atoms}} = 1.993 \times 10^{-23} \text{ g}$$

We can use the preceding result to determine the relationship between atomic mass units and grams. Because the mass of every carbon-12 atom is exactly 12 amu, the number of atomic mass units equivalent to 1 g is

$$\begin{aligned} \frac{\text{amu}}{\text{gram}} &= \frac{12 \text{ amu}}{1 \cancel{\text{carbon-12 atom}}} \times \frac{1 \cancel{\text{carbon-12 atom}}}{1.993 \times 10^{-23} \text{ g}} \\ &= 6.022 \times 10^{23} \text{ amu/g} \end{aligned}$$

Thus,

$$1 \text{ g} = 6.022 \times 10^{23} \text{ amu}$$

and

$$1 \text{ amu} = 1.6605 \times 10^{-24} \text{ g}$$

This example shows that Avogadro's number can be used to convert from the atomic mass units to mass in grams and vice versa.

The notions of Avogadro's number and molar mass enable us to carry out conversions between mass and moles of atoms and between moles and number of atoms (Figure 3.2). We will employ the following conversion factors in the calculations:

$$\frac{1 \text{ mol X}}{\text{molar mass of X}} \quad \text{and} \quad \frac{1 \text{ mol X}}{6.022 \times 10^{23} \text{ X atoms}}$$

where X represents the symbol of an element. Using the proper conversion factors we can convert one quantity to another, as Examples 3.2–3.4 show.

Student Hot Spot

Student data indicate you may struggle with converting mass of an element into moles and atoms. Access your eBook for additional Learning Resources on this topic.

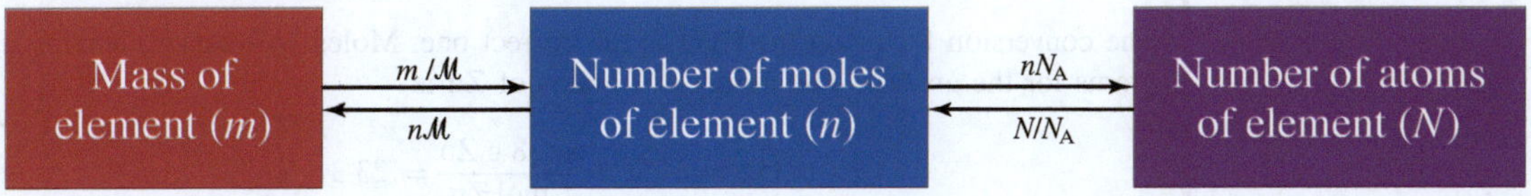

Figure 3.2 *The relationships between mass (m in grams) of an element and number of moles of an element (n) and between number of moles of an element and number of atoms (N) of an element. $\mathcal{M}$ is the molar mass (g/mol) of the element and N_A is Avogadro's number.*

Example 3.2

A scientific research helium balloon.

National Scientific Balloon Facility/Palestine, Texas

Helium (He) is a valuable gas used in industry, low-temperature research, deep-sea diving tanks, and balloons. How many moles of He atoms are in 6.46 g of He?

Strategy We are given grams of helium and asked to solve for moles of helium. What conversion factor do we need to convert between grams and moles? Arrange the appropriate conversion factor so that grams cancel and the unit moles is obtained for your answer.

Solution The conversion factor needed to convert between grams and moles is the molar mass. In the periodic table, we see that the molar mass of He is 4.003 g. This can be expressed as

$$1 \text{ mol He} = 4.003 \text{ g He}$$

From this equality, we can write two conversion factors

$$\frac{1 \text{ mol He}}{4.003 \text{ g He}} \quad \text{and} \quad \frac{4.003 \text{ g He}}{1 \text{ mol He}}$$

The conversion factor on the left is the correct one. Grams will cancel, leaving the unit mol for the answer—that is,

$$6.46 \text{ g He} \times \frac{1 \text{ mol He}}{4.003 \text{ g He}} = 1.61 \text{ mol He}$$

Thus, there are 1.61 moles of He atoms in 6.46 g of He.

Check Because the given mass (6.46 g) is larger than the molar mass of He, we expect to have more than 1 mole of He.

Practice Exercise How many moles of magnesium (Mg) are there in 87.3 g of Mg?

Similar problem: 3.15.

Example 3.3

Zinc.

Mark Dierker/McGraw Hill

Zinc (Zn) is a silvery metal that is used in making brass (with copper) and in plating iron to prevent corrosion. How many grams of Zn are in 0.356 mole of Zn?

Strategy We are trying to solve for grams of zinc. What conversion factor do we need to convert between moles and grams? Arrange the appropriate conversion factor so that moles cancel and the unit grams are obtained for your answer.

Solution The conversion factor needed to convert between moles and grams is the molar mass. In the periodic table, we see the molar mass of Zn is 65.39 g. This can be expressed as

$$1 \text{ mol Zn} = 65.38 \text{ g Zn}$$

From this equality, we can write two conversion factors

$$\frac{1 \text{ mol Zn}}{65.38 \text{ g Zn}} \quad \text{and} \quad \frac{65.38 \text{ g Zn}}{1 \text{ mol Zn}}$$

The conversion factor on the right is the correct one. Moles will cancel, leaving unit of grams for the answer. The number of grams of Zn is

$$0.356 \text{ mol Zn} \times \frac{65.38 \text{ g Zn}}{1 \text{ mol Zn}} = 23.3 \text{ g Zn}$$

Thus, there are 23.3 g of Zn in 0.356 mole of Zn.

(Continued)

Check Does a mass of 23.3 g for 0.356 mole of Zn seem reasonable? What is the mass of 1 mole of Zn?

Practice Exercise Calculate the number of grams of lead (Pb) in 12.4 moles of lead.

Similar problem: 3.16.

Example 3.4

The C_{60} molecule is called buckminsterfullerene because its shape resembles the geodesic domes designed by the visionary architect R. Buckminster Fuller. What is the mass (in grams) of one C_{60} molecule?

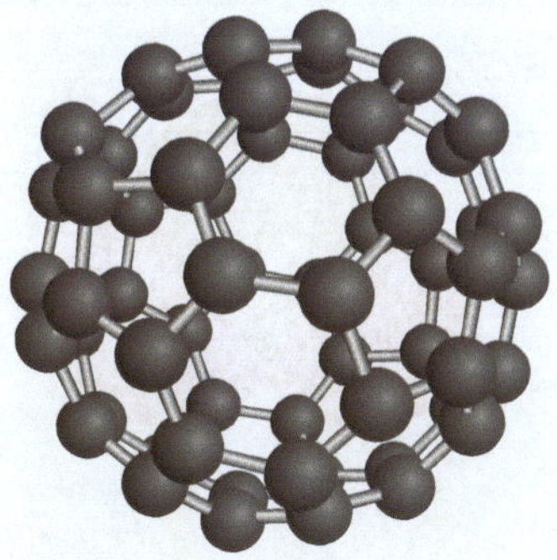

Buckminsterfullerene (C_{60}) or "buckyball."

Strategy The question asks for the mass of one C_{60} molecule. Determine the moles of C atoms in one C_{60} molecule, and then use the molar mass of C to calculate the mass of one molecule in grams.

Solution Because one C_{60} molecule contains 60 C atoms, and 1 mole of C contains 6.022×10^{23} C atoms and has a mass of 12.011 g, we can calculate the mass of one C_{60} molecule as follows:

$$1\ \cancel{C_{60}\ \text{molecule}} \times \frac{60\ \cancel{\text{C atoms}}}{1\ \cancel{C_{60}\ \text{molecule}}} \times \frac{1\ \cancel{\text{mol C}}}{6.022 \times 10^{23}\ \cancel{\text{C atoms}}} \times \frac{12.01\ \text{g}}{1\ \cancel{\text{mol C}}} = 1.197 \times 10^{-21}\ \text{g}$$

Check Because 6.022×10^{23} atoms of C have a mass of 12.01 g, a molecule containing only 60 carbon atoms should have a significantly smaller mass.

Practice Exercise Gold atoms form small clusters containing a fixed number of atoms. What is the mass (in grams) of one Au_{31} cluster?

Similar problems: 3.20, 3.21.

Summary of Concepts & Facts

- A mole is Avogadro's number (6.022×10^{23}) of atoms, molecules, or other particles.

Review of Concepts & Facts

3.2.1 Determine which of the following contains the largest number of atoms: (a) 7.68 g of He, (b) 112 g of Fe, (c) 389 g of Hg.

3.2.2 How many moles of rubidium (Rb) are there in 3.75×10^{24} Rb atoms?

3.2.3 What is the mass in grams of 1.68 moles of vanadium (V)?

3.3 Molecular Mass

Learning Objectives

- Calculate the molecular and molar mass of compounds.
- Interconvert among mass, moles, and number of atoms or molecules.

If we know the atomic masses of the component atoms, we can calculate the mass of a molecule. The ***molecular mass*** (sometimes called *molecular weight*) is *the mass (in atomic mass units) of one molecule.* It is determined by the sum of the atomic masses of each atom in a molecule. For example, the molecular mass of H_2O is

$$2(\text{atomic mass of H}) + \text{atomic mass of O}$$

or

$$2(1.008 \text{ amu}) + 16.00 \text{ amu} = 18.02 \text{ amu}$$

In general, we need to multiply the atomic mass of each element by the number of atoms of that element present in the molecule and sum over all the elements. Example 3.5 illustrates this approach.

SO_2

Example 3.5

Calculate the molecular masses (in atomic mass units) of the following compounds: (a) sulfur dioxide (SO_2), a gas that is responsible for acid rain, and (b) caffeine ($C_8H_{10}N_4O_2$), a stimulant present in tea, coffee, and cola beverages.

Strategy How do atomic masses of different elements combine to give the molecular mass of a compound?

Solution To calculate molecular mass, we need to sum all the atomic masses in the molecule. For each element, we multiply the atomic mass of the element by the number of atoms of that element in the molecule. We find atomic masses in the periodic table.

(a) There are two O atoms and one S atom in SO_2, so that

$$\begin{aligned}\text{molecular mass of } SO_2 &= 32.06 \text{ amu} + 2(16.00 \text{ amu})\\ &= 64.06 \text{ amu}\end{aligned}$$

(b) There are eight C atoms, ten H atoms, four N atoms, and two O atoms in caffeine, so the molecular mass of $C_8H_{10}N_4O_2$ is given by

$$8(12.01 \text{ amu}) + 10(1.008 \text{ amu}) + 4(14.01 \text{ amu}) + 2(16.00 \text{ amu}) = 194.20 \text{ amu}$$

Practice Exercise What is the molecular mass of methanol (CH_3OH)?

Similar problems: 3.23, 3.24.

Student Hot Spot
Student data indicate you may struggle with determining moles from a given mass. Access your eBook to view additional Learning Resources on this topic.

From the molecular mass we can determine the molar mass of a molecule or compound. The molar mass of a compound (in grams) is numerically equal to its molecular mass (in atomic mass units). For example, the molecular mass of water is 18.02 amu, so its molar mass is 18.02 g. Note that 1 mole of water weighs 18.02 g and contains 6.022×10^{23} H_2O *molecules,* just as 1 mole of elemental carbon contains 6.022×10^{23} carbon *atoms.*

As Examples 3.6 and 3.7 show, a knowledge of the molar mass enables us to calculate the numbers of moles and individual atoms in a given quantity of a compound.

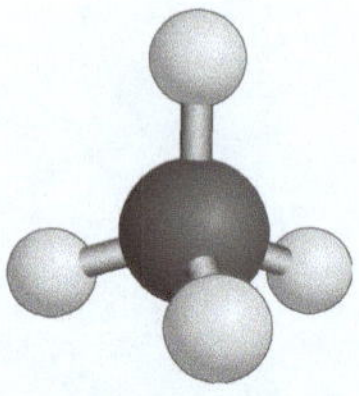

CH_4

Example 3.6

Methane (CH_4) is the principal component of natural gas. How many moles of CH_4 are present in 6.07 g of CH_4?

Strategy We are given grams of CH_4 and asked to solve for moles of CH_4. What conversion factor do we need to convert between grams and moles? Arrange the appropriate conversion factor so that grams cancel and the unit moles are obtained for your answer.

Solution The conversion factor needed to convert between grams and moles is the molar mass. First we need to calculate the molar mass of CH_4, following the procedure in Example 3.5:

$$\begin{aligned}\text{molar mass of } CH_4 &= 12.01 \text{ g} + 4(1.008 \text{ g})\\ &= 16.04 \text{ g}\end{aligned}$$

(Continued)

Because

$$1 \text{ mol } CH_4 = 16.04 \text{ g } CH_4$$

the conversion factor we need should have grams in the denominator so that the unit g will cancel, leaving the unit mol in the numerator:

$$\frac{1 \text{ mol } CH_4}{16.04 \text{ g } CH_4}$$

We now write

$$6.07 \text{ g } \cancel{CH_4} \times \frac{1 \text{ mol } CH_4}{16.04 \text{ g } \cancel{CH_4}} = 0.378 \text{ mol } CH_4$$

Thus, there is 0.378 mole of CH_4 in 6.07 g of CH_4.

Check Should 6.07 g of CH_4 equal less than 1 mole of CH_4? What is the mass of 1 mole of CH_4?

Practice Exercise Calculate the number of moles of chloroform ($CHCl_3$) in 198 g of chloroform.

Similar problem: 3.26.

Methane gas burning on a cooking range.
Steve Allen/Stockbyte/Getty Images

Example 3.7

How many hydrogen atoms are present in 25.6 g of urea [$(NH_2)_2CO$], which is used as a fertilizer, in animal feed, and in the manufacture of polymers? The molar mass of urea is 60.06 g.

Strategy We are asked to solve for atoms of hydrogen in 25.6 g of urea. We cannot convert directly from grams of urea to atoms of hydrogen. How should molar mass and Avogadro's number be used in this calculation? How many moles of H are in 1 mole of urea?

Solution To calculate the number of H atoms, we first must convert grams of urea to moles of urea using the molar mass of urea. This part is similar to Example 3.2. The molecular formula of urea shows there are four moles of H atoms in one mole of urea molecule, so the mole ratio is 4:1. Finally, knowing the number of moles of H atoms, we can calculate the number of H atoms using Avogadro's number. We need two conversion factors: molar mass and Avogadro's number. We can combine these conversions

$$\text{grams of urea} \longrightarrow \text{moles of urea} \longrightarrow \text{moles of H} \longrightarrow \text{atoms of H}$$

into one step:

$$25.6 \text{ g } \cancel{(NH_2)_2CO} \times \frac{1 \text{ mol } \cancel{(NH_2)_2CO}}{60.06 \text{ g } \cancel{(NH_2)_2CO}} \times \frac{4 \text{ } \cancel{\text{mol H}}}{1 \text{ mol } \cancel{(NH_2)_2CO}} \times \frac{6.022 \times 10^{23} \text{ H atoms}}{1 \text{ } \cancel{\text{mol H}}}$$

$$= 1.03 \times 10^{24} \text{ H atoms}$$

Check Does the answer look reasonable? How many atoms of H would 60.06 g of urea contain?

Practice Exercise How many H atoms are in 72.5 g of isopropanol (rubbing alcohol), C_3H_8O?

Similar problems: 3.27, 3.28.

Urea.
Ken Karp/McGraw Hill

Finally, note that for ionic compounds like NaCl and MgO that do not contain discrete molecular units, we use the term *formula mass* instead. The formula unit of NaCl consists of one Na^+ ion and one Cl^- ion. Thus, the formula mass of NaCl is the mass of one formula unit:

$$\text{formula mass of NaCl} = 22.99 \text{ amu} + 35.45 \text{ amu}$$
$$= 58.44 \text{ amu}$$

and its molar mass is 58.44 g. The combined mass of a Na^+ ion and a Cl^- ion is equal to the combined mass of a Na atom and a Cl atom.

Summary of Concepts & Facts

- The molar mass (in grams) of a compound is numerically equal to its mass in atomic mass units (amu) and contains Avogadro's number of molecules (in the case of molecular substances), or simplest formula units (in the case of ionic compounds).

Review of Concepts & Facts

3.3.1 Determine the molecular mass and the molar mass of citric acid, $H_3C_6H_5O_7$.
3.3.2 What is the mass in grams of 0.382 moles of caffeine, $C_8H_{10}O_2N_4$?
3.3.3 How many oxygen atoms are in 124 g of calcium phosphate, $Ca_3(PO_4)_2$ (molar mass = 310.2 g)?

3.4 The Mass Spectrometer

Learning Objective

- Understand how a mass spectrometer provides information about atomic and molecular masses.

The most direct and accurate method for determining atomic and molecular masses is mass spectrometry, depicted in Figure 3.3. In one type of a *mass spectrometer,* a gaseous sample is bombarded by a stream of high-energy electrons. Collisions between the electrons and the gaseous atoms (or molecules) produce positive ions by dislodging an electron from each atom or molecule. These positive ions (of mass m and charge e) are accelerated by two oppositely charged plates as they pass through the

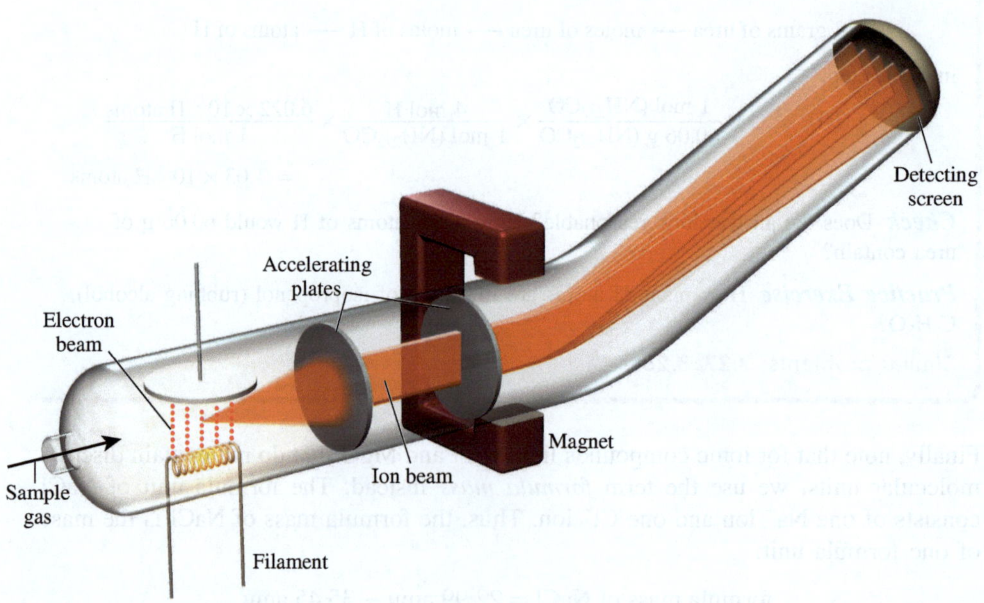

Figure 3.3 *Schematic diagram of one type of mass spectrometer.*

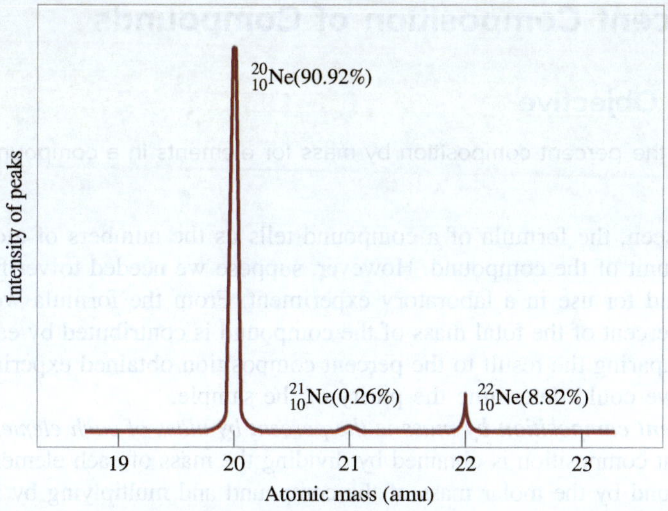

Figure 3.4 *The mass spectrum of the three isotopes of neon.*

plates. The emerging ions are deflected into a circular path by a magnet. The radius of the path depends on the mass-to-charge ratio (that is, m/z). Ions of smaller m/z ratio trace a wider curve than those having a larger m/z ratio, so that ions with equal charges but different masses are separated from one another. The mass of each ion (and hence its parent atom or molecule) is determined from the magnitude of its deflection. Eventually the ions arrive at the detector, which registers a current for each type of ion. The amount of current generated is directly proportional to the number of ions, so it enables us to determine the relative abundance of isotopes.

The first mass spectrometer, developed in the 1920s by the English physicist F. W. Aston,† was crude by today's standards. Nevertheless, it provided indisputable evidence of the existence of isotopes—neon-20 (atomic mass 19.9924 amu and natural abundance 90.92 percent) and neon-22 (atomic mass 21.9914 amu and natural abundance 8.82 percent). When more sophisticated and sensitive mass spectrometers became available, scientists were surprised to discover that neon has a third stable isotope with an atomic mass of 20.9940 amu and natural abundance 0.257 percent (Figure 3.4). This example illustrates how very important experimental accuracy is to a quantitative science like chemistry. Early experiments failed to detect neon-21 because its natural abundance is just 0.257 percent. In other words, only 26 in 10,000 Ne atoms are neon-21. The masses of molecules can be determined in a similar manner by the mass spectrometer. Indeed, it is possible to determine the molar mass of a compound without knowing its chemical formula.

Summary of Concepts & Facts

- Atomic and molecular masses can be accurately determined with a mass spectrometer by determining the relative abundances of isotopes present in a sample.

Review of Concepts & Facts

3.4.1 Explain how the mass spectrometer enables chemists to determine the average atomic mass of chlorine, which has two stable isotopes (^{35}Cl and ^{37}Cl).

†Francis William Aston (1877–1945). English chemist and physicist. He was awarded the Nobel Prize in Chemistry in 1922 for developing the mass spectrometer.

3.5 Percent Composition of Compounds

Learning Objective

- Deduce the percent composition by mass for elements in a compound.

As we have seen, the formula of a compound tells us the numbers of atoms of each element in a unit of the compound. However, suppose we needed to verify the purity of a compound for use in a laboratory experiment. From the formula we could calculate what percent of the total mass of the compound is contributed by each element. Then, by comparing the result to the percent composition obtained experimentally for our sample, we could determine the purity of the sample.

The ***percent composition by mass*** is *the percent by mass of each element in a compound.* Percent composition is obtained by dividing the mass of each element in 1 mole of the compound by the molar mass of the compound and multiplying by 100 percent. Mathematically, the percent composition of an element in a compound is expressed as

$$\text{percent composition of an element} = \frac{n \times \text{molar mass of element}}{\text{molar mass of compound}} \times 100\% \qquad (3.1)$$

where n is the number of moles of the element in 1 mole of the compound. For example, in 1 mole of hydrogen peroxide (H_2O_2) there are 2 moles of H atoms and 2 moles of O atoms. The molar masses of H_2O_2, H, and O are 34.02 g, 1.008 g, and 16.00 g, respectively. Therefore, the percent composition of H_2O_2 is calculated as follows:

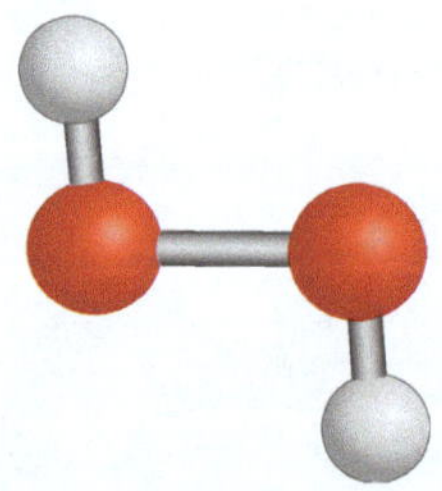

H_2O_2

$$\%\text{H} = \frac{2 \times 1.008 \text{ g H}}{34.02 \text{ g } H_2O_2} \times 100\% = 5.926\%$$

$$\%\text{O} = \frac{2 \times 16.00 \text{ g O}}{34.02 \text{ g } H_2O_2} \times 100\% = 94.06\%$$

Student Hot Spot

Student data indicate you may struggle with percent composition by mass. Access your eBook for additional Learning Resources on this topic.

The sum of the percentages is 5.926% + 94.06% = 99.99%. The small discrepancy from 100 percent is due to the way we rounded off the molar masses of the elements. If we had used the empirical formula HO for the calculation, we would have obtained the same percentages. This is so because both the molecular formula and empirical formula tell us the percent composition by mass of the compound.

Example 3.8

Phosphoric acid (H_3PO_4) is a colorless, syrupy liquid used in detergents, fertilizers, toothpastes, and in carbonated beverages for a "tangy" flavor. Calculate the percent composition by mass of H, P, and O in this compound.

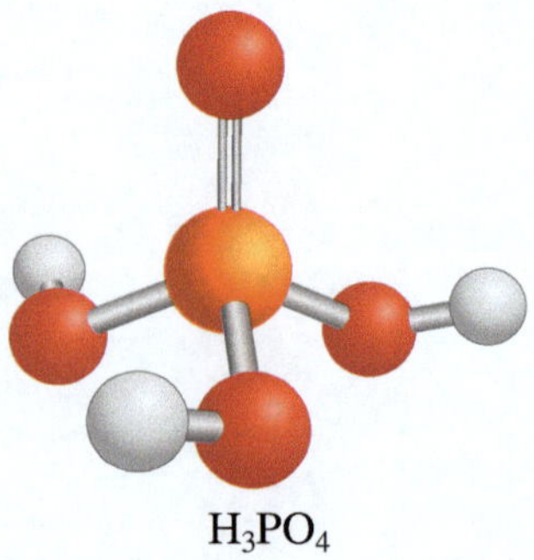

H_3PO_4

Strategy Recall the procedure for calculating a percentage. Assume that we have 1 mole of H_3PO_4. The percent by mass of each element (H, P, and O) is given by the combined molar mass of the atoms of the element in 1 mole of H_3PO_4 divided by the molar mass of H_3PO_4, then multiplied by 100 percent.

Solution The molar mass of H_3PO_4 is 97.99 g. The percent by mass of each of the elements in H_3PO_4 is calculated as follows:

$$\%\text{H} = \frac{3(1.008 \text{ g) H}}{97.99 \text{ g } H_3PO_4} \times 100\% = 3.086\%$$

$$\%\text{P} = \frac{30.97 \text{ g P}}{97.99 \text{ g } H_3PO_4} \times 100\% = 31.61\%$$

$$\%\text{O} = \frac{4(16.00 \text{ g) O}}{97.99 \text{ g } H_3PO_4} \times 100\% = 65.31\%$$

(Continued)

Check Do the percentages add to 100 percent? The sum of the percentages is (3.086% + 31.61% + 65.31%) = 100.01%. The small discrepancy from 100 percent is due to the way we rounded off.

Practice Exercise Calculate the percent composition by mass of each of the elements in sulfuric acid (H_2SO_4).

Similar problem: 3.38.

Student Hot Spot

Student data indicate you may struggle with determining empirical formulas. Access your eBook for additional Learning Resources on this topic.

The procedure used in the example can be reversed if necessary. Given the percent composition by mass of a compound, we can determine the empirical formula of the compound (Figure 3.5). Because we are dealing with percentages and the sum of all the percentages is 100 percent, it is convenient to assume that we started with 100 g of a compound, as Example 3.9 shows.

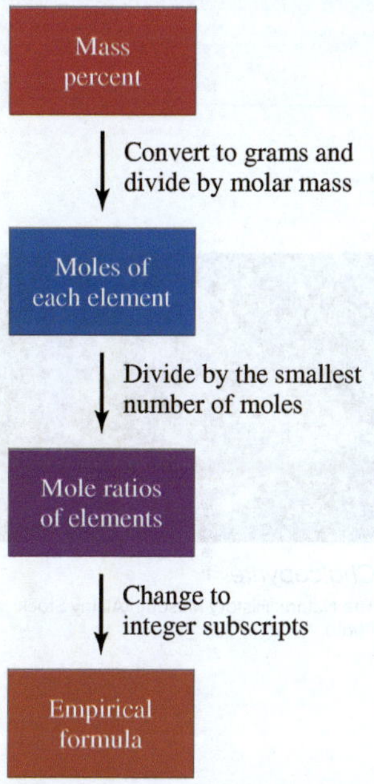

Figure 3.5 *Procedure for calculating the empirical formula of a compound from its percent compositions.*

Example 3.9

Ascorbic acid (vitamin C) cures scurvy. It is composed of 40.92 percent carbon (C), 4.58 percent hydrogen (H), and 54.50 percent oxygen (O) by mass. Determine its empirical formula.

Strategy In a chemical formula, the subscripts represent the ratio of the number of moles of each element that combine to form one mole of the compound. How can we convert from mass percent to moles? If we assume an exactly 100-g sample of the compound, do we know the mass of each element in the compound? How do we then convert from grams to moles?

Solution If we have 100 g of ascorbic acid, then each percentage can be converted directly to grams. In this sample, there will be 40.92 g of C, 4.58 g of H, and 54.50 g of O. Because the subscripts in the formula represent a mole ratio, we need to convert the grams of each element to moles. The conversion factor needed is the molar mass of each element. Let n represent the number of moles of each element so that

$$n_C = 40.92\ \cancel{g\ C} \times \frac{1\ mol\ C}{12.01\ \cancel{g\ C}} = 3.407\ mol\ C$$

$$n_H = 4.58\ \cancel{g\ H} \times \frac{1\ mol\ H}{1.008\ \cancel{g\ H}} = 4.54\ mol\ H$$

$$n_O = 54.50\ \cancel{g\ O} \times \frac{1\ mol\ O}{16.00\ \cancel{g\ O}} = 3.406\ mol\ O$$

Thus, we arrive at the formula $C_{3.407}H_{4.54}O_{3.406}$, which gives the identity and the mole ratios of atoms present. However, chemical formulas are written with whole numbers. Try to convert to whole numbers by dividing all the subscripts by the smallest subscript (3.406):

$$C: \frac{3.407}{3.406} \approx 1 \qquad H: \frac{4.54}{3.406} = 1.33 \qquad O: \frac{3.406}{3.406} = 1$$

where the ≈ sign means "approximately equal to." This gives $CH_{1.33}O$ as the formula for ascorbic acid. Next, we need to convert 1.33, the subscript for H, into an integer. This can be done by a trial-and-error procedure:

$$1.33 \times 1 = 1.33$$
$$1.33 \times 2 = 2.66$$
$$1.33 \times 3 = 3.99 \approx 4$$

Because 1.33 × 3 gives us an integer (4), we multiply all the subscripts by 3 and obtain $C_3H_4O_3$ as the empirical formula for ascorbic acid.

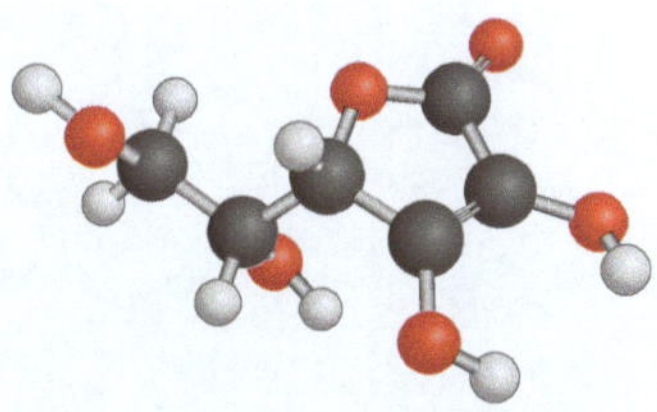

The molecular formula of ascorbic acid is $C_6H_8O_6$.

Check Are the subscripts in $C_3H_4O_3$ reduced to the smallest whole numbers?

(Continued)

Practice Exercise Determine the empirical formula of a compound having the following percent composition by mass: K: 24.75 percent; Mn: 34.77 percent; O: 40.51 percent.

Similar problems: 3.43, 3.44.

Chemists often want to know the actual mass of an element in a certain mass of a compound. For example, in the mining industry, this information will tell scientists about the quality of the ore. Because the percent composition by mass of the elements in the substance can be readily calculated, such a problem can be solved in a rather direct way.

Example 3.10

Chalcopyrite.
The Natural History Museum/Alamy Stock Photo

Chalcopyrite ($CuFeS_2$) is a principal mineral of copper. Calculate the number of kilograms of Cu in 3.71×10^3 kg of chalcopyrite.

Strategy Chalcopyrite is composed of Cu, Fe, and S. The mass due to Cu is based on its percentage by mass in the compound. How do we calculate mass percent of an element?

Solution The molar masses of Cu and $CuFeS_2$ are 63.55 g and 183.51 g, respectively. The mass percent of Cu is therefore

$$\%Cu = \frac{\text{molar mass of Cu}}{\text{molar mass of } CuFeS_2} \times 100\%$$

$$= \frac{63.55 \text{ g}}{183.51 \text{ g}} \times 100\% = 34.63\%$$

To calculate the mass of Cu in a 3.71×10^3 kg sample of $CuFeS_2$, we need to convert the percentage to a fraction (that is, convert 34.63 percent to 34.63/100, or 0.3463) and write

$$\text{mass of Cu in } CuFeS_2 = 0.3463 \times (3.71 \times 10^3 \text{ kg}) = 1.28 \times 10^3 \text{ kg}$$

Check As a ballpark estimate, note that the mass percent of Cu is roughly 33 percent, so that a third of the mass should be Cu—that is, $\frac{1}{3} \times 3.71 \times 10^3 \text{ kg} \approx 1.24 \times 10^3$ kg. This quantity is quite close to the answer.

Practice Exercise Calculate the number of grams of Al in 371 g of Al_2O_3.

Similar problem: 3.41.

Summary of Concepts & Facts

- The percent composition by mass of a compound is the percent by mass of each element present.

Review of Concepts & Facts

3.5.1 Without doing detailed calculations, estimate whether the percent composition by mass of Sr is greater than or smaller than that of O in strontium nitrate [$Sr(NO_3)_2$].

3.5.2 What is the percent composition by mass of each of the elements present in fumaric acid, $C_4H_4O_4$?

3.5.3 Determine the empirical formula of a compound having the following percent composition by mass: P: 43.64 percent; O: 56.36 percent.

3.6 Experimental Determination of Empirical Formulas

Learning Objective

- Solve for the empirical formula of a compound from percent composition or from combustion analysis.

The fact that we can determine the empirical formula of a compound if we know the percent composition enables us to identify compounds experimentally. The procedure is as follows. First, chemical analysis tells us the number of grams of each element present in a given amount of a compound. Then, we convert the quantities in grams to number of moles of each element. Finally, using the method given in Example 3.9, we find the empirical formula of the compound.

As a specific example, let us consider the compound ethanol. When ethanol is burned in an apparatus such as that shown in Figure 3.6, carbon dioxide (CO_2) and water (H_2O) are given off. Because neither carbon nor hydrogen was in the inlet gas, we can conclude that both carbon (C) and hydrogen (H) were present in ethanol and that oxygen (O) may also be present. (Molecular oxygen was added in the combustion process, but some of the oxygen may also have come from the original ethanol sample.)

The masses of CO_2 and of H_2O produced can be determined by measuring the increase in mass of the CO_2 and H_2O absorbers, respectively. Suppose that in one experiment the combustion of 11.5 g of ethanol produced 22.0 g of CO_2 and 13.5 g of H_2O. We can calculate the mass of carbon and hydrogen in the original 11.5-g sample of ethanol as follows:

$$\text{mass of C} = 22.0\ \text{g}\ \cancel{\text{CO}_2} \times \frac{1\ \cancel{\text{mol CO}_2}}{44.01\ \cancel{\text{g CO}_2}} \times \frac{1\ \cancel{\text{mol C}}}{1\ \cancel{\text{mol CO}_2}} \times \frac{12.01\ \text{g C}}{1\ \cancel{\text{mol C}}}$$
$$= 6.00\ \text{g C}$$

$$\text{mass of H} = 13.5\ \text{g}\ \cancel{\text{H}_2\text{O}} \times \frac{1\ \cancel{\text{mol H}_2\text{O}}}{18.02\ \cancel{\text{g H}_2\text{O}}} \times \frac{2\ \cancel{\text{mol H}}}{1\ \cancel{\text{mol H}_2\text{O}}} \times \frac{1.008\ \text{g H}}{1\ \cancel{\text{mol H}}}$$
$$= 1.51\ \text{g H}$$

Thus, 11.5 g of ethanol contains 6.00 g of carbon and 1.51 g of hydrogen. The remainder must be oxygen, whose mass is

$$\begin{aligned}\text{mass of O} &= \text{mass of sample} - (\text{mass of C} + \text{mass of H}) \\ &= 11.5\ \text{g} - (6.00\ \text{g} + 1.51\ \text{g}) \\ &= 4.0\ \text{g}\end{aligned}$$

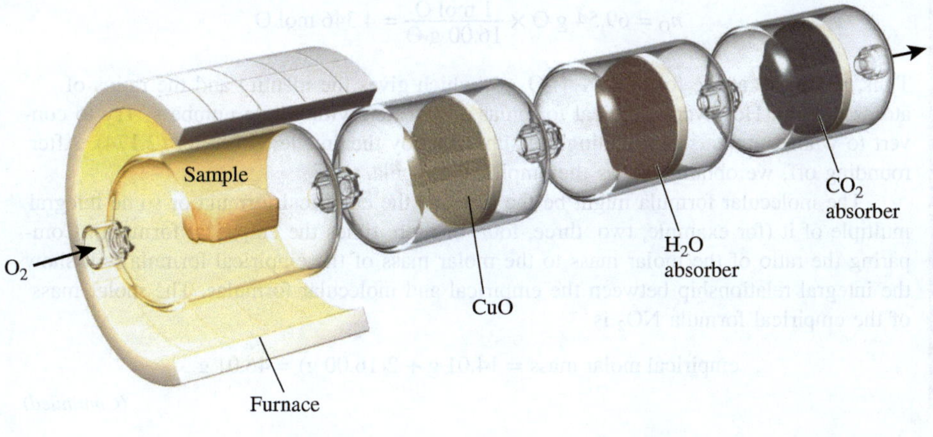

Figure 3.6 *Apparatus for determining the empirical formula of ethanol. The absorbers are substances that can retain water and carbon dioxide, respectively. CuO is used to ensure complete combustion of all carbon to CO_2.*

The number of moles of each element present in 11.5 g of ethanol is

$$\text{moles of C} = 6.00\ \cancel{\text{g C}} \times \frac{1\ \text{mol C}}{12.01\ \cancel{\text{g C}}} = 0.500\ \text{mol C}$$

$$\text{moles of H} = 1.51\ \cancel{\text{g H}} \times \frac{1\ \text{mol H}}{1.008\ \cancel{\text{g H}}} = 1.50\ \text{mol H}$$

$$\text{moles of O} = 4.00\ \cancel{\text{g O}} \times \frac{1\ \text{mol O}}{16.00\ \cancel{\text{g O}}} = 0.250\ \text{mol O}$$

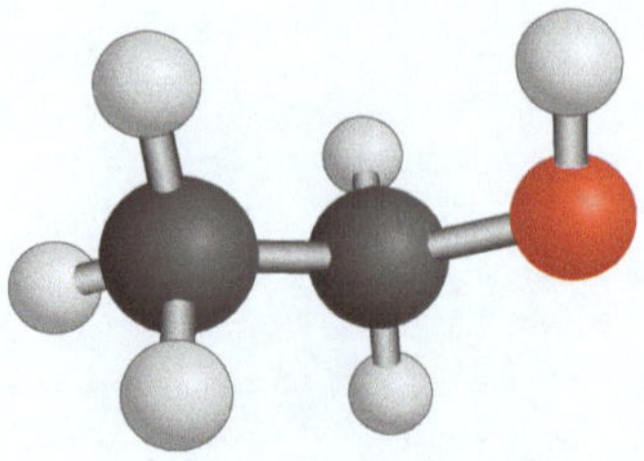

It happens that the molecular formula of ethanol is the same as its empirical formula.

The formula of ethanol is therefore $C_{0.50}H_{1.5}O_{0.25}$ (we round off the number of moles to two significant figures). Because the number of atoms must be an integer, we divide the subscripts by 0.250, the smallest subscript, and obtain for the empirical formula C_2H_6O.

Now we can better understand the word *empirical,* which literally means "based only on observation and measurement." The empirical formula of ethanol is determined from analysis of the compound in terms of its component elements. No knowledge of how the atoms are linked together in the compound is required.

Determination of Molecular Formulas

The formula calculated from percent composition by mass is always the empirical formula because the subscripts in the formula are always reduced to the smallest whole numbers. To calculate the actual, molecular formula we must know the *approximate* molar mass of the compound in addition to its empirical formula. Knowing that the molar mass of a compound must be an integral multiple of the molar mass of its empirical formula, we can use the molar mass to find the molecular formula, as Example 3.11 demonstrates.

Student Hot Spot

Student data indicate you may struggle with combustion analysis and determining molecular formulas. Access your eBook for additional Learning Resources on this topic.

Example 3.11

A sample of a compound contains 30.46 percent nitrogen and 69.54 percent oxygen by mass, as determined by a mass spectrometer. In a separate experiment, the molar mass of the compound is found to be between 90 g and 95 g. Determine the molecular formula and the accurate molar mass of the compound.

Strategy To determine the molecular formula, we first need to determine the empirical formula. Comparing the empirical molar mass to the experimentally determined molar mass will reveal the relationship between the empirical formula and molecular formula.

Solution We start by assuming that there are 100 g of the compound. Then each percentage can be converted directly to grams—that is, 30.46 g of N and 69.54 g of O. Let *n* represent the number of moles of each element so that

$$n_N = 30.46\ \cancel{\text{g N}} \times \frac{1\ \text{mol N}}{14.01\ \cancel{\text{g N}}} = 2.174\ \text{mol N}$$

$$n_O = 69.54\ \cancel{\text{g O}} \times \frac{1\ \text{mol O}}{16.00\ \cancel{\text{g O}}} = 4.346\ \text{mol O}$$

Thus, we arrive at the formula $N_{2.174}O_{4.346}$, which gives the identity and the ratios of atoms present. However, chemical formulas are written with whole numbers. Try to convert to whole numbers by dividing the subscripts by the smaller subscript (2.174). After rounding off, we obtain NO_2 as the empirical formula.

The molecular formula might be the same as the empirical formula or some integral multiple of it (for example, two, three, four, or more times the empirical formula). Comparing the ratio of the molar mass to the molar mass of the empirical formula will show the integral relationship between the empirical and molecular formulas. The molar mass of the empirical formula NO_2 is

$$\text{empirical molar mass} = 14.01\ \text{g} + 2(16.00\ \text{g}) = 46.01\ \text{g}$$

(Continued)

Next, we determine the ratio between the molar mass and the empirical molar mass

$$\frac{\text{molar mass}}{\text{empirical molar mass}} = \frac{90\text{ g}}{46.01\text{ g}} \approx 2$$

The molar mass is twice the empirical molar mass. This means that there are two NO_2 units in each molecule of the compound and the molecular formula is $(NO_2)_2$ or N_2O_4.

The actual molar mass of the compound is two times the empirical molar mass—that is, 2(46.01 g) or 92.02 g—which is between 90 g and 95 g.

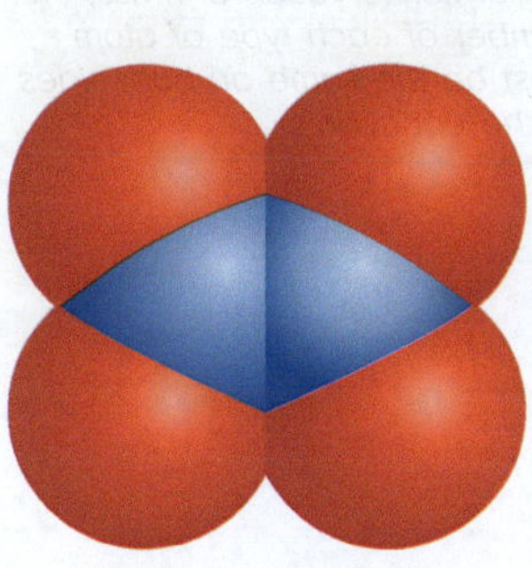

N_2O_4

Check Note that in determining the molecular formula from the empirical formula, we need only know the *approximate* molar mass of the compound. The reason is that the true molar mass is an integral multiple (1 ×, 2 ×, 3 ×, ...) of the empirical molar mass. Therefore, the ratio (molar mass/empirical molar mass) will always be close to an integer.

Practice Exercise A sample of a compound containing boron (B) and hydrogen (H) contains 6.444 g of B and 1.803 g of H. The molar mass of the compound is about 30 g. What is its molecular formula?

Similar problems: 3.48, 3.49, 3.52.

Summary of Concepts & Facts

- If we know the percent composition by mass of a compound, we can deduce the empirical formula of the compound and also the molecular formula of the compound if the approximate molar mass is known.

Review of Concepts & Facts

3.6.1 What is the molecular formula of a compound containing only carbon and hydrogen if combustion of 1.05 g of the compound produces 3.30 g CO_2 and 1.35 g H_2O and its molar mass is about 70 g?

3.7 Chemical Reactions and Chemical Equations

Learning Objective

- Balance chemical equations using stoichiometric coefficients.

Having discussed the masses of atoms and molecules, we turn next to what happens to atoms and molecules in a ***chemical reaction,*** *a process in which a substance (or substances) is changed into one or more new substances.* To communicate with one another about chemical reactions, chemists have devised a standard way to represent them using chemical equations. A ***chemical equation*** *uses chemical symbols to show what happens during a chemical reaction.* In this section, we will learn how to write chemical equations and balance them.

Writing Chemical Equations

Consider what happens when hydrogen gas (H_2) burns in air (which contains oxygen, O_2) to form water (H_2O). This reaction can be represented by the chemical equation

$$H_2 + O_2 \longrightarrow H_2O \tag{3.2}$$

where the "plus" sign means "reacts with" and the arrow means "to yield." Thus, this symbolic expression can be read: "Molecular hydrogen reacts with molecular oxygen

Two hydrogen molecules	+	One oxygen molecule	$\longrightarrow$	Two water molecules
$2H_2$	+	O_2	$\longrightarrow$	$2H_2O$
2 moles H_2	+	1 mole O_2	$\longrightarrow$	2 moles H_2O
$2(2.02\ g) = 4.04\ g\ H_2$	+	$32.00\ g\ O_2$	$\longrightarrow$	$2(18.02\ g) = 36.04\ g\ H_2O$
36.04 g reactants			$\longrightarrow$	36.04 g product

Figure 3.7 *Three ways of representing the combustion of hydrogen. In accordance with the law of conservation of mass, the number of each type of atom must be the same on both sides of the equation.*

to yield water." The reaction is assumed to proceed from left to right as the arrow indicates.

Equation (3.2) is not complete, however, because there are twice as many oxygen atoms on the left side of the arrow (two) as on the right side (one). To conform with the law of conservation of mass, there must be the same number of each type of atom on both sides of the arrow—that is, we must have as many atoms after the reaction ends as we did before it started. We can *balance* Equation (3.2) by placing the appropriate coefficient (2 in this case) in front of H_2 and H_2O:

$$2H_2 + O_2 \longrightarrow 2H_2O$$

This *balanced chemical equation* shows that "two hydrogen molecules can combine or react with one oxygen molecule to form two water molecules" (Figure 3.7). When the coefficient is 1, as in the case of O_2, it is not shown. Because the ratio of the number of molecules is equal to the ratio of the number of moles, the equation can also be read as "2 moles of hydrogen molecules react with 1 mole of oxygen molecules to produce 2 moles of water molecules." We know the mass of a mole of each of these substances, so we can also interpret the equation as "4.04 g of H_2 react with 32.00 g of O_2 to give 36.04 g of H_2O." These three ways of reading the equation are summarized in Figure 3.7.

We refer to H_2 and O_2 in Equation (3.2) as ***reactants,*** which are *the starting materials in a chemical reaction.* Water is the ***product,*** which is *the substance formed as a result of a chemical reaction.* A chemical equation, then, is just the chemist's shorthand description of a reaction. In a chemical equation, the reactants are conventionally written on the left and the products on the right of the arrow:

$$\text{reactants} \longrightarrow \text{products}$$

To provide additional information, chemists often indicate the physical states of the reactants and products by using the letters *g, l,* and *s* to denote gas, liquid, and solid, respectively. For example,

$$2CO(g) + O_2(g) \longrightarrow 2CO_2(g)$$
$$2HgO(s) \longrightarrow 2Hg(l) + O_2(g)$$

To represent what happens when sodium chloride (NaCl) is added to water, we write

$$NaCl(s) \xrightarrow{H_2O} NaCl(aq)$$

where *aq* denotes the aqueous (that is, water) environment. Writing H_2O above the arrow symbolizes the physical process of dissolving a substance in water, although it is sometimes left out for simplicity.

Knowing the states of the reactants and products is especially useful in the laboratory. For example, when potassium bromide (KBr) and silver nitrate ($AgNO_3$) react in an aqueous environment, a solid, silver bromide (AgBr), is formed. This reaction can be represented by the equation

$$KBr(aq) + AgNO_3(aq) \longrightarrow KNO_3(aq) + AgBr(s)$$

If the physical states of reactants and products are not given, an uninformed person might try to bring about the reaction by mixing solid KBr with solid $AgNO_3$. These solids would react very slowly or not at all. Imagining the process on the microscopic level, we can understand that for a product like silver bromide to form, the Ag^+ and Br^- ions would have to come in contact with each other. However, these ions are locked in place in their solid compounds and have little mobility. (Here is an example of how we explain a phenomenon by thinking about what happens at the molecular level, as discussed in Section 1.1.)

Student Hot Spot

Student data indicate you may struggle with interpreting chemical equations. Access your eBook for additional Learning Resources on this topic.

Balancing Chemical Equations

Suppose we want to write an equation to describe a chemical reaction that we have just carried out in the laboratory. How should we go about doing it? Because we know the identities of the reactants, we can write their chemical formulas. The identities of products are more difficult to establish. For simple reactions it is often possible to guess the product(s). For more complicated reactions involving three or more products, chemists may need to perform further tests to establish the presence of specific compounds.

Once we have identified all the reactants and products and have written the correct formulas for them, we assemble them in the conventional sequence—reactants on the left separated by an arrow from products on the right. The equation written at this point is likely to be *unbalanced;* that is, the number of each type of atom on one side of the arrow differs from the number on the other side. In general, we can balance a chemical equation by the following steps:

1. Identify all reactants and products and write their correct formulas on the left side and right side of the equation, respectively.
2. Begin balancing the equation by trying different coefficients to make the number of atoms of each element the same on both sides of the equation. We can change the coefficients (the numbers preceding the formulas) but not the subscripts (the numbers within formulas). Changing the subscripts would change the identity of the substance. For example, $2NO_2$ means "two molecules of nitrogen dioxide," but if we double the subscripts, we have N_2O_4, which is the formula of dinitrogen tetroxide, a completely different compound.
3. First, look for elements that appear only once on each side of the equation with the same number of atoms on each side: The formulas containing these elements must have the same coefficient. Therefore, there is no need to adjust the coefficients of these elements at this point. Next, look for elements that appear only once on each side of the equation but in unequal numbers of atoms. Balance these elements. Finally, balance elements that appear in two or more formulas on the same side of the equation.
4. Check your balanced equation to be sure that you have the same total number of each type of atoms on both sides of the equation arrow.

Let's consider a specific example. In the laboratory, small amounts of oxygen gas can be prepared by heating potassium chlorate ($KClO_3$). The products are oxygen gas (O_2) and potassium chloride (KCl). From this information, we write

$$KClO_3 \longrightarrow KCl + O_2$$

(For simplicity, we omit the physical states of reactants and products.) All three elements (K, Cl, and O) appear only once on each side of the equation, but only for K

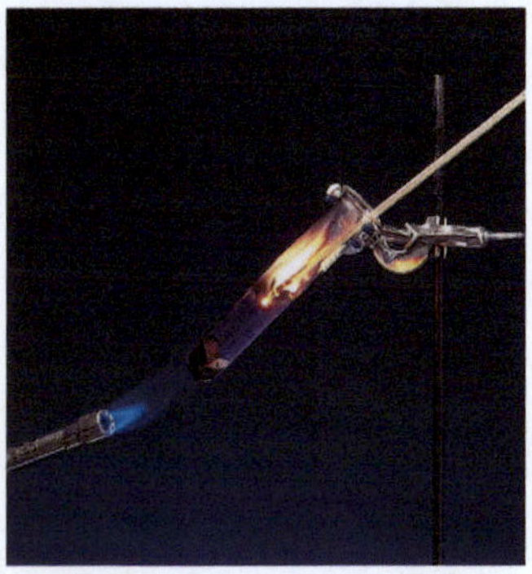

Heating potassium chlorate produces oxygen, which supports the combustion of a wood splint.
Ken Karp/McGraw Hill

and Cl do we have equal numbers of atoms on both sides. Thus, $KClO_3$ and KCl must have the same coefficient. The next step is to make the number of O atoms the same on both sides of the equation. Because there are three O atoms on the left and two O atoms on the right of the equation, we can balance the O atoms by placing a 2 in front of $KClO_3$ and a 3 in front of O_2.

$$2KClO_3 \longrightarrow KCl + 3O_2$$

Finally, we balance the K and Cl atoms by placing a 2 in front of KCl:

$$2KClO_3 \longrightarrow 2KCl + 3O_2 \quad (3.3)$$

As a final check, we can draw up a balance sheet for the reactants and products where the number in parentheses indicates the number of atoms of each element:

Reactants	**Products**
K(2)	K(2)
Cl(2)	Cl(2)
O(6)	O(6)

Note that this equation could also be balanced with coefficients that are multiples of 2 (for $KClO_3$), 2 (for KCl), and 3 (for O_2); for example,

$$4KClO_3 \longrightarrow 4KCl + 6O_2$$

However, it is common practice to use the *simplest* possible set of whole-number coefficients to balance the equation. Equation (3.3) conforms to this convention.

Now let us consider the combustion (that is, burning) of the natural gas component ethane (C_2H_6) in oxygen or air, which yields carbon dioxide (CO_2) and water. The unbalanced equation is

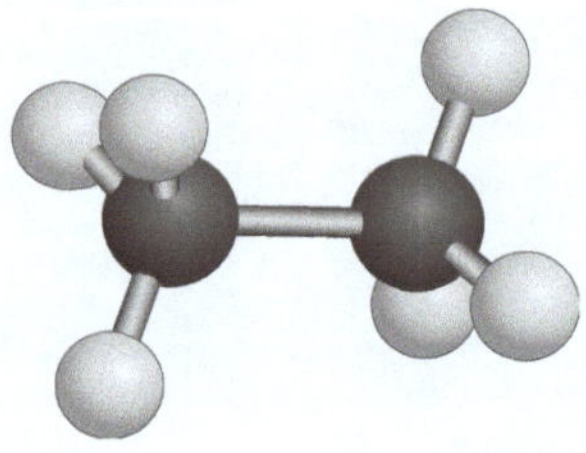

C_2H_6

$$C_2H_6 + O_2 \longrightarrow CO_2 + H_2O$$

We see that the number of atoms is not the same on both sides of the equation for any of the elements (C, H, and O). In addition, C and H appear only once on each side of the equation; O appears in two compounds on the right side (CO_2 and H_2O). To balance the C atoms, we place a 2 in front of CO_2:

$$C_2H_6 + O_2 \longrightarrow 2CO_2 + H_2O$$

To balance the H atoms, we place a 3 in front of H_2O:

$$C_2H_6 + O_2 \longrightarrow 2CO_2 + 3H_2O$$

At this stage, the C and H atoms are balanced, but the O atoms are not because there are seven O atoms on the right-hand side and only two O atoms on the left-hand side of the equation. This inequality of O atoms can be eliminated by writing $\frac{7}{2}$ in front of the O_2 on the left-hand side:

$$C_2H_6 + \tfrac{7}{2}O_2 \longrightarrow 2CO_2 + 3H_2O$$

The "logic" for using $\frac{7}{2}$ as a coefficient is that there were seven oxygen atoms on the right-hand side of the equation, but only a pair of oxygen atoms (O_2) on the left. To balance them we ask how many *pairs* of oxygen atoms are needed to equal seven oxygen atoms. Just as 3.5 pairs of shoes equal seven shoes, $\frac{7}{2}O_2$ molecules equal seven O atoms. As the following tally shows, the equation is now balanced.

Reactants	**Products**
C(2)	C(2)
H(6)	H(6)
O(7)	O(7)

However, we normally prefer to express the coefficients as whole numbers rather than as fractions. Therefore, we multiply the entire equation by 2 to convert $\frac{7}{2}$ to 7:

$$2C_2H_6 + 7O_2 \longrightarrow 4CO_2 + 6H_2O$$

The final tally is

Reactants	Products
C(4)	C(4)
H(12)	H(12)
O(14)	O(14)

Note that the coefficients used in balancing the last equation are the smallest possible set of whole numbers.

In Example 3.12 we will continue to practice our equation-balancing skills.

Example 3.12

When aluminum metal is exposed to air, a protective layer of aluminum oxide (Al_2O_3) forms on its surface. This layer prevents further reaction between aluminum and oxygen, and it is the reason that aluminum beverage cans do not corrode. [In the case of iron, the rust, or iron(III) oxide, that forms is too porous to protect the iron metal underneath, so rusting continues.] Write a balanced equation for the formation of Al_2O_3.

Strategy Remember that the formula of an element or compound cannot be changed when balancing a chemical equation. The equation is balanced by placing the appropriate coefficients in front of the formulas. Follow the procedure for balancing equations described earlier.

Solution The unbalanced equation is

$$Al + O_2 \longrightarrow Al_2O_3$$

In a balanced equation, the number and types of atoms on each side of the equation must be the same. We see that there is one Al atom on the reactants side and there are two Al atoms on the product side. We can balance the Al atoms by placing a coefficient of 2 in front of Al on the reactants side.

$$2Al + O_2 \longrightarrow Al_2O_3$$

There are two O atoms on the reactants side, and three O atoms on the product side of the equation. We can balance the O atoms by placing a coefficient of $\frac{3}{2}$ in front of O_2 on the reactants side.

$$2Al + \tfrac{3}{2}O_2 \longrightarrow Al_2O_3$$

This is a balanced equation. However, equations are normally balanced with the smallest set of *whole*-number coefficients. Multiplying both sides of the equation by 2 gives whole-number coefficients.

$$2(2Al + \tfrac{3}{2}O_2 \longrightarrow Al_2O_3)$$

or

$$4Al + 3O_2 \longrightarrow 2Al_2O_3$$

Check For an equation to be balanced, the number and types of atoms on each side of the equation must be the same. The final tally is

Reactants	Products
Al(4)	Al(4)
O(6)	O(6)

The equation is balanced. Also, the coefficients are reduced to the simplest set of whole numbers.

(Continued)

Practice Exercise Balance the equation representing the reaction between iron(III) oxide (Fe_2O_3) and carbon monoxide (CO) to yield iron (Fe) and carbon dioxide (CO_2).

Similar problems: 3.57, 3.58.

Summary of Concepts & Facts

- Chemical changes, called chemical reactions, are represented by chemical equations. Substances that undergo change—the reactants—are written on the left and the substances formed—the products—appear to the right of the arrow. Chemical equations must be balanced, in accordance with the law of conservation of mass. The number of atoms of each element in the reactants must equal the number in the products.

Review of Concepts & Facts

3.7.1 Which parts of the equation shown here are essential for a balanced equation and which parts are helpful if we want to carry out the reaction in the laboratory?

$$BaH_2(s) + 2H_2O(l) \longrightarrow Ba(OH)_2(aq) + 2H_2(g)$$

3.7.2 What is the balanced equation representing the reaction between potassium thiosulfate ($K_2S_2O_3$) and iodine (I_2) to produce potassium tetrathionate ($K_2S_4O_6$) and potassium iodide (KI)?

3.8 Amounts of Reactants and Products

Learning Objective

- Determine amounts of reactant required or product formed using stoichiometry.

A basic question raised in the chemical laboratory is, "How much product will be formed from specific amounts of starting materials (reactants)?" Or in some cases, we might ask the reverse question, "How much starting material must be used to obtain a specific amount of product?" To interpret a reaction quantitatively, we need to apply our knowledge of molar masses and the mole concept. ***Stoichiometry*** is *the quantitative study of reactants and products in a chemical reaction.*

Whether the units given for reactants (or products) are moles, grams, liters (for gases), or some other units, we use moles to calculate the amount of product formed in a reaction. This approach is called the ***mole method,*** which means simply that *the stoichiometric coefficients in a chemical equation can be interpreted as the number of moles of each substance.* For example, industrially ammonia is synthesized from hydrogen and nitrogen as follows:

$$N_2(g) + 3H_2(g) \longrightarrow 2NH_3(g)$$

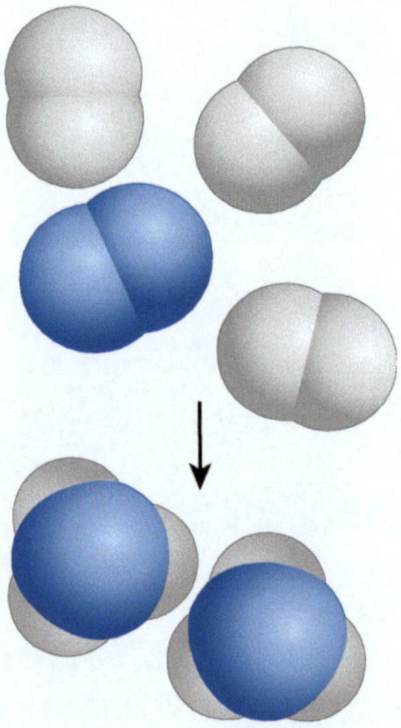

The synthesis of NH_3 from H_2 and N_2.

The stoichiometric coefficients show that one molecule of N_2 reacts with three molecules of H_2 to form two molecules of NH_3. It follows that the relative numbers of moles are the same as the relative number of molecules:

$N_2(g)$	+	$3H_2(g)$	$\longrightarrow$	$2NH_3(g)$
1 molecule		3 molecules		2 molecules
6.022×10^{23} molecules		$3(6.022 \times 10^{23}$ molecules)		$2(6.022 \times 10^{23}$ molecules)
1 mol		3 mol		2 mol

Thus, this equation can also be read as "1 mole of N_2 gas combines with 3 moles of H_2 gas to form 2 moles of NH_3 gas." In stoichiometric calculations, we say that three moles of H_2 are equivalent to two moles of NH_3, that is,

$$3 \text{ mol } H_2 \simeq 2 \text{ mol } NH_3$$

where the symbol $\simeq$ means "stoichiometrically equivalent to" or simply "equivalent to." This relationship enables us to write the conversion factors

$$\frac{3 \text{ mol } H_2}{2 \text{ mol } NH_3} \quad \text{and} \quad \frac{2 \text{ mol } NH_3}{3 \text{ mol } H_2}$$

Similarly, we have 1 mol $N_2 \simeq$ 2 mol NH_3 and 1 mol $N_2 \simeq$ 3 mol H_2.

Let's consider a simple example in which 6.0 moles of H_2 react completely with N_2 to form NH_3. To calculate the amount of NH_3 produced in moles, we use the conversion factor that has H_2 in the denominator and write

$$\text{moles of } NH_3 \text{ produced} = 6.0 \cancel{\text{mol } H_2} \times \frac{2 \text{ mol } NH_3}{3 \cancel{\text{mol } H_2}}$$

$$= 4.0 \text{ mol } NH_3$$

Now suppose 16.0 g of H_2 react completely with N_2 to form NH_3. How many grams of NH_3 will be formed? To do this calculation, we note that the link between H_2 and NH_3 is the mole ratio from the balanced equation. So we need to first convert grams of H_2 to moles of H_2, then to moles of NH_3, and finally to grams of NH_3. The conversion steps are

$$\text{grams of } H_2 \longrightarrow \text{moles of } H_2 \longrightarrow \text{moles of } NH_3 \longrightarrow \text{grams of } NH_3$$

First, we convert 16.0 g of H_2 to number of moles of H_2, using the molar mass of H_2 as the conversion factor.

$$\text{moles of } H_2 = 16.0 \cancel{\text{g } H_2} \times \frac{1 \text{ mol } H_2}{2.016 \cancel{\text{g } H_2}}$$

$$= 7.94 \text{ mol } H_2$$

Next, we calculate the number of moles of NH_3 produced.

$$\text{moles of } NH_3 = 7.94 \cancel{\text{mol } H_2} \times \frac{2 \text{ mol } NH_3}{3 \cancel{\text{mol } H_2}}$$

$$= 5.29 \text{ mol } NH_3$$

Finally, we calculate the mass of NH_3 produced in grams using the molar mass of NH_3 as the conversion factor.

$$\text{grams of } NH_3 = 5.29 \cancel{\text{mol } NH_3} \times \frac{17.03 \text{ g } NH_3}{1 \cancel{\text{mol } NH_3}}$$

$$= 90.1 \text{ g } NH_3$$

These three separate calculations can be combined in a single step as follows:

$$\text{grams of } NH_3 = 16.0 \cancel{\text{g } H_2} \times \frac{1 \cancel{\text{mol } H_2}}{2.016 \cancel{\text{g } H_2}} \times \frac{2 \cancel{\text{mol } NH_3}}{3 \cancel{\text{mol } H_2}} \times \frac{17.03 \text{ g } NH_3}{1 \cancel{\text{mol } NH_3}}$$

$$= 90.1 \text{ g } NH_3$$

Similarly, we can calculate the mass in grams of N_2 consumed in this reaction. The conversion steps are

$$\text{grams of } H_2 \longrightarrow \text{moles of } H_2 \longrightarrow \text{moles of } N_2 \longrightarrow \text{grams of } N_2$$

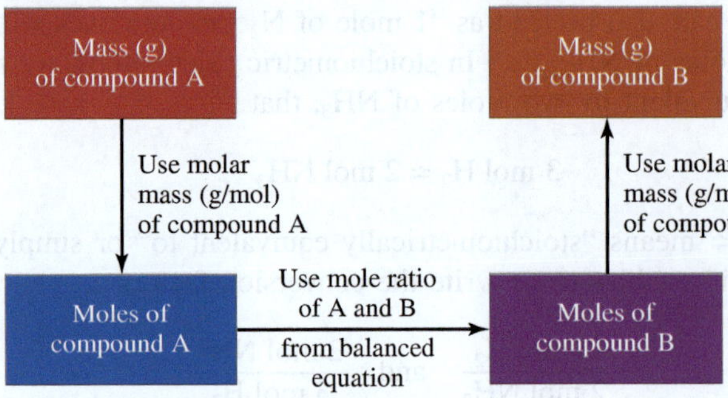

Figure 3.8 *The procedure for calculating the amounts of reactants or products in a reaction using the mole method.*

By using the relationship 1 mol $N_2 \simeq$ 3 mol H_2, we write

$$\text{grams of } N_2 = 16.0 \text{ g } H_2 \times \frac{1 \text{ mol } H_2}{2.016 \text{ g } H_2} \times \frac{1 \text{ mol } N_2}{3 \text{ mol } H_2} \times \frac{28.02 \text{ g } N_2}{1 \text{ mol } N_2}$$
$$= 74.1 \text{ g } N_2$$

The general approach for solving stoichiometry problems is summarized next.

1. Write a balanced equation for the reaction.
2. Convert the given amount of the reactant (in grams or other units) to number of moles.
3. Use the mole ratio from the balanced equation to calculate the number of moles of product formed.
4. Convert the moles of product to grams (or other units) of product.

Student Hot Spot

Student data indicate you may struggle with stoichiometry conversions. Access your eBook for additional Learning Resources on this topic.

Figure 3.8 shows these steps. Sometimes we may be asked to calculate the amount of a reactant needed to form a specific amount of product. In those cases, we can reverse the steps shown in Figure 3.8.

Examples 3.13 and 3.14 illustrate the application of this approach.

Example 3.13

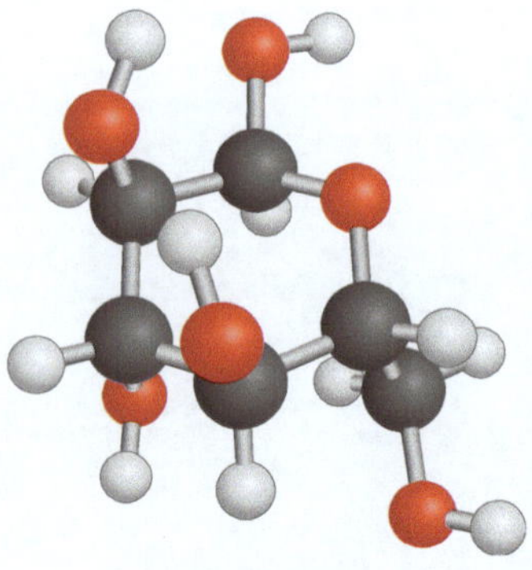

$C_6H_{12}O_6$

The food we eat is degraded, or broken down, in our bodies to provide energy for growth and function. A general overall equation for this very complex process represents the degradation of glucose ($C_6H_{12}O_6$) to carbon dioxide (CO_2) and water (H_2O):

$$C_6H_{12}O_6 + 6O_2 \longrightarrow 6CO_2 + 6H_2O$$

If 856 g of $C_6H_{12}O_6$ is consumed by a person over a certain period, what is the mass of CO_2 produced?

Strategy Looking at the balanced equation, how do we compare the amounts of $C_6H_{12}O_6$ and CO_2? We can compare them based on the *mole ratio* from the balanced equation. Starting with grams of $C_6H_{12}O_6$, how do we convert to moles of $C_6H_{12}O_6$? Once moles of CO_2 are determined using the mole ratio from the balanced equation, how do we convert to grams of CO_2?

Solution We follow the preceding steps and Figure 3.8.

Step 1: The balanced equation is given in the problem.

Step 2: To convert grams of $C_6H_{12}O_6$ to moles of $C_6H_{12}O_6$, we write

$$856 \text{ g } C_6H_{12}O_6 \times \frac{1 \text{ mol } C_6H_{12}O_6}{180.2 \text{ g } C_6H_{12}O_6} = 4.75 \text{ mol } C_6H_{12}O_6$$

Step 3: From the mole ratio, we see that 1 mol $C_6H_{12}O_6 \simeq$ 6 mol CO_2. Therefore, the number of moles of CO_2 formed is

$$4.75 \text{ mol } C_6H_{12}O_6 \times \frac{6 \text{ mol } CO_2}{1 \text{ mol } C_6H_{12}O_6} = 28.5 \text{ mol } CO_2$$

(Continued)

Step 4: Finally, the number of grams of CO_2 formed is given by

$$28.5 \text{ mol } CO_2 \times \frac{44.01 \text{ g } CO_2}{1 \text{ mol } CO_2} = 1.25 \times 10^3 \text{ g } CO_2$$

After some practice, we can combine the conversion steps

$$\text{grams of } C_6H_{12}O_6 \longrightarrow \text{moles of } C_6H_{12}O_6 \longrightarrow \text{moles of } CO_2 \longrightarrow \text{grams of } CO_2$$

into one equation:

$$\text{mass of } CO_2 = 856 \text{ g } C_6H_{12}O_6 \times \frac{1 \text{ mol } C_6H_{12}O_6}{180.2 \text{ g } C_6H_{12}O_6} \times \frac{6 \text{ mol } CO_2}{1 \text{ mol } C_6H_{12}O_6} \times \frac{44.01 \text{ g } CO_2}{1 \text{ mol } CO_2}$$
$$= 1.25 \times 10^3 \text{ g } CO_2$$

Check Does the answer seem reasonable? Should the mass of CO_2 produced be larger than the mass of $C_6H_{12}O_6$ reacted, even though the molar mass of CO_2 is considerably less than the molar mass of $C_6H_{12}O_6$? What is the mole ratio between CO_2 and $C_6H_{12}O_6$?

Practice Exercise Methanol (CH_3OH) burns in air according to the equation

$$2CH_3OH + 3O_2 \longrightarrow 2CO_2 + 4H_2O$$

If 209 g of methanol are used up in a combustion process, what is the mass of H_2O produced?

Similar problem: 3.72.

Example 3.14

All alkali metals react with water to produce hydrogen gas and the corresponding alkali metal hydroxide. A typical reaction is that between lithium and water:

$$2Li(s) + 2H_2O(l) \longrightarrow 2LiOH(aq) + H_2(g)$$

How many grams of Li are needed to produce 9.89 g of H_2?

Strategy The question asks for number of grams of reactant (Li) to form a specific amount of product (H_2). Therefore, we need to reverse the steps shown in Figure 3.8. From the equation we see that 2 mol Li ≏ 1 mol H_2.

Solution The conversion steps are

$$\text{grams of } H_2 \longrightarrow \text{moles of } H_2 \longrightarrow \text{moles of Li} \longrightarrow \text{grams of Li}$$

Combining these steps into one equation, we write

$$9.89 \text{ g } H_2 \times \frac{1 \text{ mol } H_2}{2.016 \text{ g } H_2} \times \frac{2 \text{ mol Li}}{1 \text{ mol } H_2} \times \frac{6.941 \text{ g Li}}{1 \text{ mol Li}} = 68.1 \text{ g Li}$$

Check There are roughly 5 moles of H_2 in 9.89 g H_2, so we need 10 moles of Li. From the approximate molar mass of Li (7 g), does the answer seem reasonable?

Practice Exercise The reaction between nitric oxide (NO) and oxygen to form nitrogen dioxide (NO_2) is a key step in photochemical smog formation:

$$2NO(g) + O_2(g) \longrightarrow 2NO_2(g)$$

How many grams of O_2 are needed to produce 2.21 g of NO_2?

Similar problem: 3.64.

Lithium reacting with water to produce hydrogen gas.

Ken Karp/McGraw Hill

Summary of Concepts & Facts

- Stoichiometry is the quantitative study of products and reactants in chemical reactions. Stoichiometric calculations are best done by expressing both the known and unknown quantities in terms of moles and then converting to other units if necessary.

Review of Concepts & Facts

3.8.1 Which of the following statements is correct for the equation shown here?

$$4NH_3(g) + 5O_2(g) \longrightarrow 4NO(g) + 6H_2O(g)$$

(a) 6 g of H_2O are produced for every 4 g of NH_3 reacted.
(b) 1 mole of NO is produced per mole of NH_3 reacted.
(c) 2 moles of NO are produced for every 3 moles of O_2 reacted.

3.8.2 Silicon reacts with chromium(III) oxide according to the equation

$$3Si(s) + 2Cr_2O_3(s) \longrightarrow 3SiO_2(s) + 4Cr(s)$$

If 59.4 g of silicon is consumed in the reaction, what is the mass of SiO_2 produced?

3.8.3 At high temperatures, magnesium reacts with nitrogen gas according to the equation

$$3Mg(s) + N_2(g) \longrightarrow Mg_3N_2(s)$$

How many grams of magnesium are needed to produce 25.0 g of Mg_3N_2?

3.9 Limiting Reactants

Learning Objectives

- Identify limiting reactants in a chemical reaction.
- Calculate amount of a product based on the limiting reactant in a chemical reaction.

Video: Limiting Reactants

When a chemist carries out a reaction, the reactants are usually not present in exact ***stoichiometric amounts,*** that is, *in the proportions indicated by the balanced equation.* Because the goal of a reaction is to produce the maximum quantity of a useful compound from the starting materials, frequently a large excess of one reactant is supplied to ensure that the more expensive reactant is completely converted to the desired product. Consequently, some reactant will be left over at the end of the reaction. *The reactant used up first in a reaction* is called the ***limiting reactant,*** because the maximum amount of product formed depends on how much of this reactant was originally present. When this reactant is used up, no more product can be formed. ***Excess reactants*** are *the reactants present in quantities greater than necessary to react with the quantity of the limiting reagent.*

The concept of the limiting reactant is analogous to the outfitting of skaters with equipment where each requires two skates. If there are 11 skaters and only 9 pairs of skates, then only 9 of the skaters can have a pair of skates. This leaves two skaters without any skates. The number of skates *limits* the number of skaters that can be provided with a pair of skates, and there is an *excess* of skaters.

Consider the industrial synthesis of methanol (CH_3OH) from carbon monoxide and hydrogen at high temperatures:

$$CO(g) + 2H_2(g) \longrightarrow CH_3OH(g)$$

Student Hot Spot

Student data indicate you may struggle with finding the limiting reactant. Access your eBook for additional Learning Resources on this topic.

Suppose initially we have 4 moles of CO and 6 moles of H_2 (Figure 3.9). One way to determine which of two reactants is the limiting reagent is to calculate the number of moles of CH_3OH obtained based on the initial quantities of CO and H_2. From the preceding definition, we see that only the limiting reactant will yield the *smaller* amount of the product. Starting with 4 moles of CO, we find the number of moles of CH_3OH produced is

$$4 \text{ mol CO} \times \frac{1 \text{ mol } CH_3OH}{1 \text{ mol CO}} = 4 \text{ mol } CH_3OH$$

and starting with 6 moles of H_2, the number of moles of CH_3OH formed is

$$6 \text{ mol } H_2 \times \frac{1 \text{ mol } CH_3OH}{2 \text{ mol } H_2} = 3 \text{ mol } CH_3OH$$

Because H_2 results in a smaller amount of CH_3OH, it must be the limiting reagent. Therefore, CO is the excess reagent.

In stoichiometric calculations involving limiting reactants, the first step is to decide which reactant is the limiting reactant. After the limiting reactant has been identified, the rest of the problem can be solved as outlined in Section 3.8. Example 3.15 illustrates this approach.

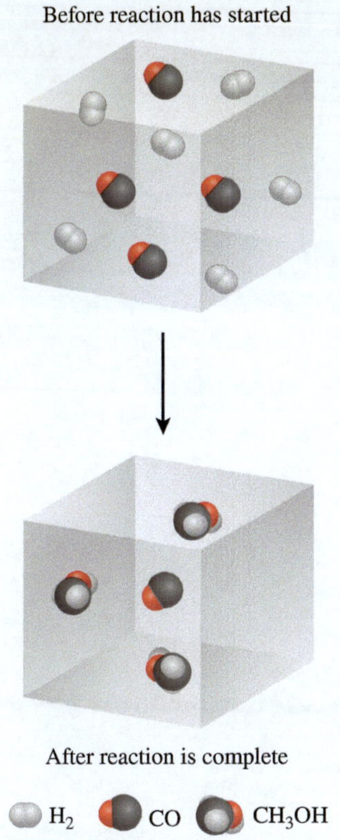

Figure 3.9 *At the start of the reaction, there were six H_2 molecules and four CO molecules. At the end, all the H_2 molecules are gone and only one CO molecule is left. Therefore, H_2 molecule is the limiting reactant and CO is the excess reagent. Each molecule can also be treated as one mole of the substance in this reaction.*

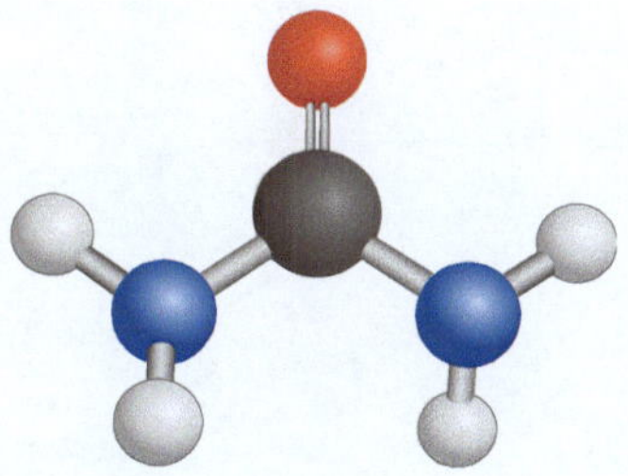

$(NH_2)_2CO$

Example 3.15

The synthesis of urea, $(NH_2)_2CO$, is considered to be the first recognized example of preparing a biological compound from nonbiological reactants, challenging the notion that biological processes involved a "vital force" present only in living systems. Today urea is produced industrially by reacting ammonia with carbon dioxide:

$$2NH_3(g) + CO_2(g) \longrightarrow (NH_2)_2CO(aq) + H_2O(l)$$

In one process, 637.2 g of NH_3 are treated with 1142 g of CO_2. (a) Which of the two reactants is the limiting reactant? (b) Calculate the mass of $(NH_2)_2CO$ formed. (c) How much excess reagent (in grams) is left at the end of the reaction?

(a) Strategy The reactant that produces fewer moles of product is the limiting reactant because it limits the amount of product that can be formed. How do we convert from the amount of reactant to the amount of product? Perform this calculation for each reactant, then compare the moles of product, $(NH_2)_2CO$, formed by the given amounts of NH_3 and CO_2 to determine which reactant is the limiting reagent.

Solution We carry out two separate calculations. First, starting with 637.2 g of NH_3, we calculate the number of moles of $(NH_2)_2CO$ that could be produced if all the NH_3 reacted according to the following conversions.

$$\text{grams of } NH_3 \longrightarrow \text{moles of } NH_3 \longrightarrow \text{moles of } (NH_2)_2CO$$

Combining these conversions in one step, we write

$$\text{moles of } (NH_2)_2CO = 637.2 \text{ g } NH_3 \times \frac{1 \text{ mol } NH_3}{17.03 \text{ g } NH_3} \times \frac{1 \text{ mol } (NH_2)_2CO}{2 \text{ mol } NH_3}$$
$$= 18.71 \text{ mol } (NH_2)_2CO$$

Second, for 1142 g of CO_2, the conversions are

$$\text{grams of } CO_2 \longrightarrow \text{moles of } CO_2 \longrightarrow \text{moles of } (NH_2)_2CO$$

The number of moles of $(NH_2)_2CO$ that could be produced if all the CO_2 reacted is

$$\text{moles of } (NH_2)_2CO = 1142 \text{ g } CO_2 \times \frac{1 \text{ mol } CO_2}{44.01 \text{ g } CO_2} \times \frac{1 \text{ mol } (NH_2)_2CO}{1 \text{ mol } CO_2}$$
$$= 25.95 \text{ mol } (NH_2)_2CO$$

It follows, therefore, that NH_3 must be the limiting reactant because it produces a smaller amount of $(NH_2)_2CO$.

(Continued)

(b) Strategy We determined the moles of $(NH_2)_2CO$ produced in part (a), using NH_3 as the limiting reactant. How do we convert from moles to grams?

Solution The molar mass of $(NH_2)_2CO$ is 60.06 g. We use this as a conversion factor to convert from moles of $(NH_2)_2CO$ to grams of $(NH_2)_2CO$:

$$\text{mass of } (NH_2)_2CO = 18.71 \text{ mol } (NH_2)_2CO \times \frac{60.06 \text{ g } (NH_2)_2CO}{1 \text{ mol } (NH_2)_2CO}$$
$$= 1124 \text{ g } (NH_2)_2CO$$

Check Does your answer seem reasonable? Note that 18.71 moles of product are formed. What is the mass of 1 mole of $(NH_2)_2CO$?

(c) Strategy Working backward, we can determine the amount of CO_2 that reacted to produce 18.71 moles of $(NH_2)_2CO$. The amount of CO_2 left over is the difference between the initial amount and the amount reacted.

Solution Starting with 18.71 moles of $(NH_2)_2CO$, we can determine the mass of CO_2 that reacted using the mole ratio from the balanced equation and the molar mass of CO_2. The conversion steps are

$$\text{grams of } CO_2 \longrightarrow \text{moles of } CO_2 \longrightarrow \text{moles of } (NH_2)_2CO$$

so that

$$\text{mass of } CO_2 \text{ reacted} = 18.71 \text{ mol } (NH_2)_2CO \times \frac{1 \text{ mol } CO_2}{1 \text{ mol } (NH_2)_2CO} \times \frac{44.01 \text{ g } CO_2}{1 \text{ mol } CO_2}$$
$$= 823.4 \text{ g } CO_2$$

The amount of CO_2 remaining (in excess) is the difference between the initial amount (1142 g) and the amount reacted (823.4 g):

$$\text{mass of } CO_2 \text{ remaining} = 1142 \text{ g} - 823.4 \text{ g} = 319 \text{ g}$$

Practice Exercise The reaction between aluminum and iron(III) oxide can generate temperatures approaching 3000°C and is used in welding metals:

$$2Al + Fe_2O_3 \longrightarrow Al_2O_3 + 2Fe$$

In one process, 124 g of Al are reacted with 601 g of Fe_2O_3. (a) Calculate the mass (in grams) of Al_2O_3 formed. (b) How much of the excess reactant is left at the end of the reaction?

Similar problem: 3.86.

Example 3.15 brings out an important point. In practice, chemists usually choose the more expensive chemical as the limiting reactant so that all or most of it will be converted to products in the reaction. In the synthesis of urea, NH_3 is invariably the limiting reactant because it is more expensive than CO_2. At other times, an excess of one reactant is used to help drive the reaction to completion, or to compensate for a side reaction that consumes that reactant. Synthetic chemists often have to calculate the amount of reactant to use based on this need to have one or more components in excess, as Example 3.16 shows.

Example 3.16

The reaction between alcohols and halogen compounds to form ethers is important in organic chemistry, as illustrated here for the reaction between methanol (CH_3OH) and methyl bromide (CH_3Br) to form dimethylether (CH_3OCH_3), which is a useful precursor to other organic compounds and an aerosol propellant.

$$CH_3OH + CH_3Br + LiC_4H_9 \longrightarrow CH_3OCH_3 + LiBr + C_4H_{10}$$

(Continued)

This reaction is carried out in a dry (water-free) organic solvent, and the butyl lithium (LiC_4H_9) serves to remove a hydrogen ion from CH_3OH. Butyl lithium will also react with any residual water in the solvent, so the reaction is typically carried out with 2.5 molar equivalents of that reagent. How many grams of CH_3Br and LiC_4H_9 will be needed to carry out the preceding reaction with 10.0 g of CH_3OH?

Solution We start with the knowledge that CH_3OH and CH_3Br are present in stoichiometric amounts and that LiC_4H_9 is the excess reactant. To calculate the quantities of CH_3Br and LiC_4H_9 needed, we proceed as shown in Example 3.14.

$$\text{grams of } CH_3Br = 10.0 \text{ g } CH_3OH \times \frac{1 \text{ mol } CH_3OH}{32.04 \text{ g } CH_3OH} \times \frac{1 \text{ mol } CH_3Br}{1 \text{ mol } CH_3OH} \times \frac{94.93 \text{ g } CH_3Br}{1 \text{ mol } CH_3Br}$$

$$= 29.6 \text{ g } CH_3Br$$

$$\text{grams of } LiC_4H_9 = 10.0 \text{ g } CH_3OH \times \frac{1 \text{ mol } CH_3OH}{32.04 \text{ g } CH_3OH} \times \frac{2.5 \text{ mol } LiC_4H_9}{1 \text{ mol } CH_3OH} \times \frac{64.05 \text{ g } LiC_4H_9}{1 \text{ mol } LiC_4H_9}$$

$$= 50.0 \text{ g } LiC_4H_9$$

Practice Exercise The reaction between benzoic acid (C_6H_5COOH) and octanol ($C_8H_{17}OH$) to yield octyl benzoate ($C_6H_5COOC_8H_{17}$) and water

$$C_6H_5COOH + C_8H_{17}OH \longrightarrow C_6H_5COOC_8H_{17} + H_2O$$

is carried out with an excess of $C_8H_{17}OH$ to help drive the reaction to completion and maximize the yield of product. If an organic chemist wants to use 1.5 molar equivalents of $C_8H_{17}OH$, how many grams of $C_8H_{17}OH$ would be required to carry out the reaction with 15.7 g of C_6H_5COOH?

Similar problems: 3.141, 3.142.

Summary of Concepts & Facts

- A limiting reactant is the reactant that is present in the smallest stoichiometric amount. It limits the amount of product that can be formed.

Review of Concepts & Facts

3.9.1 Aluminum and bromine vigorously react according to the equation

$$2Al(s) + 3Br_2(l) \longrightarrow 2AlBr_3(s)$$

(a) If 5.00 g of aluminum and 22.2 g of bromine react, what mass of $AlBr_3$ is produced? (b) What mass of the excess reactant remains at the end of the reaction?

3.9.2 Consider the following reaction:

$$2NO(g) + O_2(g) \longrightarrow 2NO_2(g)$$

Starting with the reactants shown in (a), which of the diagrams shown in (b)–(d) best represents the situation in which the limiting reactant has completely reacted?

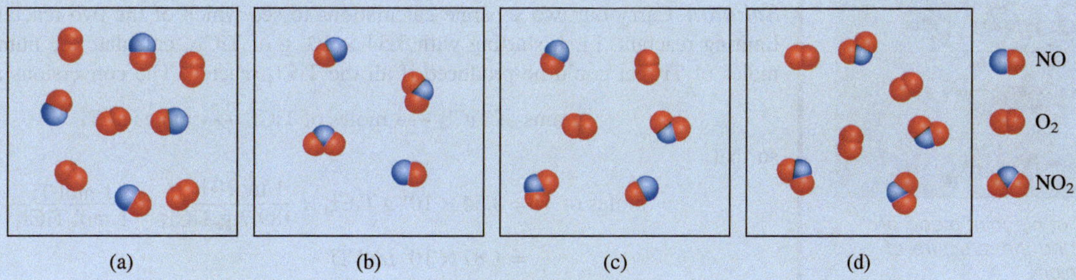

3.10 Reaction Yield and Atom Economy

Learning Objectives

- Calculate theoretical yield, actual yield, and percent yield of a chemical reaction.
- Evaluate the atom economy of a chemical reaction.

The amount of limiting reactant present at the start of a reaction determines the ***theoretical yield*** of the reaction—that is, *the amount of product that would result if all the limiting reactant reacted.* The theoretical yield, then, is the *maximum* obtainable yield, predicted by the balanced equation. In practice, the ***actual yield,*** or *the amount of product actually obtained from a reaction,* is almost always less than the theoretical yield. There are many reasons for the difference between actual and theoretical yields. For instance, many reactions are reversible, and so they do not proceed 100 percent from left to right. Even when a reaction is 100 percent complete, it may be difficult to recover all of the product from the reaction medium (say, from an aqueous solution). Some reactions are complex in the sense that the products formed may react further among themselves or with the reactants to form still other products. These additional reactions will reduce the yield of the first reaction.

To determine how efficient a given reaction is, chemists often figure the ***percent yield,*** which describes *the proportion of the actual yield to the theoretical yield.* It is calculated as follows:

$$\%\text{yield} = \frac{\text{actual yield}}{\text{theoretical yield}} \times 100\% \qquad (3.4)$$

Percent yields may range from a fraction of 1 percent to 100 percent. Chemists strive to maximize the percent yield in a reaction. Factors that can affect the percent yield include temperature and pressure. We will study these effects later.

In Example 3.17 we will calculate the yield of an industrial process.

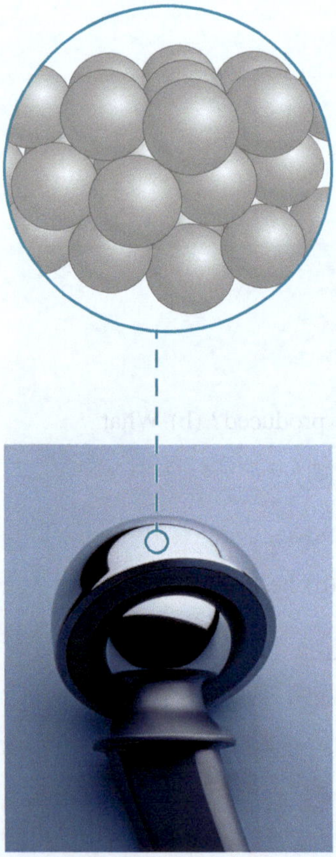

An artificial hip joint made of titanium and the structure of solid titanium.

Super Stock/age fotostock

Example 3.17

Titanium is a strong, lightweight, corrosion-resistant metal that is used in rockets, aircraft, jet engines, and bicycle frames. It is prepared by the reaction of titanium(IV) chloride with molten magnesium between 950°C and 1150°C:

$$TiCl_4(g) + 2Mg(l) \longrightarrow Ti(s) + 2MgCl_2(l)$$

In a certain industrial operation 3.54×10^7 g of $TiCl_4$ are reacted with 1.13×10^7 g of Mg. (a) Calculate the theoretical yield of Ti in grams. (b) Calculate the percent yield if 7.91×10^6 g of Ti are actually obtained.

(a) Strategy Because there are two reactants, this is likely to be a limiting reactant problem. The reactant that produces fewer moles of product is the limiting reactant. How do we convert from amount of reactant to amount of product? Perform this calculation for each reactant, then compare the moles of product, Ti, formed.

Solution Carry out two separate calculations to see which of the two reactants is the limiting reactant. First, starting with 3.54×10^7 g of $TiCl_4$, calculate the number of moles of Ti that could be produced if all the $TiCl_4$ reacted. The conversions are

$$\text{grams of } TiCl_4 \longrightarrow \text{moles of } TiCl_4 \longrightarrow \text{moles of Ti}$$

so that

$$\text{moles of Ti} = 3.54 \times 10^7\ \cancel{\text{g TiCl}_4} \times \frac{1\ \cancel{\text{mol TiCl}_4}}{189.7\ \cancel{\text{g TiCl}_4}} \times \frac{1\ \text{mol Ti}}{1\ \cancel{\text{mol TiCl}_4}}$$

$$= 1.87 \times 10^5\ \text{mol Ti}$$

(Continued)

Next, we calculate the number of moles of Ti formed from 1.13×10^7 g of Mg. The conversion steps are

$$\text{grams of Mg} \longrightarrow \text{moles of Mg} \longrightarrow \text{moles of Ti}$$

and we write

$$\text{moles of Ti} = 1.13 \times 10^7 \, \cancel{\text{g Mg}} \times \frac{1 \, \cancel{\text{mol Mg}}}{24.30 \, \cancel{\text{g Mg}}} \times \frac{1 \text{ mol Ti}}{2 \, \cancel{\text{mol Mg}}}$$

$$= 2.33 \times 10^5 \text{ mol Ti}$$

Therefore, $TiCl_4$ is the limiting reactant because it produces a smaller amount of Ti. The mass of Ti formed is

$$1.87 \times 10^5 \, \cancel{\text{mol Ti}} \times \frac{47.87 \text{ g Ti}}{1 \, \cancel{\text{mol Ti}}} = 8.95 \times 10^6 \text{ g Ti}$$

(b) Strategy The mass of Ti determined in part (a) is the theoretical yield. The amount given in part (b) is the actual yield of the reaction.

Solution The percent yield is given by

$$\%\text{yield} = \frac{\text{actual yield}}{\text{theoretical yield}} \times 100\%$$

$$= \frac{7.91 \times 10^6 \text{ g}}{8.95 \times 10^6 \text{ g}} \times 100\%$$

$$= 88.4\%$$

Check Should the percent yield be less than 100 percent?

Practice Exercise Industrially, vanadium metal, which is used in steel alloys, can be obtained by reacting vanadium (V) oxide with calcium at high temperatures:

$$5Ca + V_2O_5 \longrightarrow 5CaO + 2V$$

In one process, 1.54×10^3 g of V_2O_5 react with 1.96×10^3 g of Ca. (a) Calculate the theoretical yield of V. (b) Calculate the percent yield if 803 g of V are obtained.

Similar problems: 3.89, 3.90.

Industrial processes usually involve huge quantities (thousands to millions of tons) of products. Thus, even a slight improvement in the yield can significantly reduce the cost of production. A case in point is the manufacture of chemical fertilizers, discussed in the Chemistry in Action essay "Chemical Fertilizers."

Atom Economy

Reaction yield is a good measure of the efficiency in converting reactants into a particular product. However, reaction yield does not consider the fate of any other products that may be formed during a reaction. Many reactions do generate multiple products, not all of which are useful or desired. These by-products can lead to waste disposal issues and increased production costs.

A measure of the overall efficiency of converting reactants into useful products is ***atom economy.*** One of the goals in sustainable chemistry (also known as green chemistry) is minimizing the production of waste and undesirable by-products. The atom economy of a reaction allows chemists to determine how well reactant atoms are incorporated into the desired product. The atom economy of a reaction is found by

$$\text{atom economy} = \frac{\text{mass of atoms in desired product(s)}}{\text{mass of atoms in reactants}} \times 100\% \qquad (3.5)$$

CHEMISTRY *in Action*

Chemical Fertilizers

Feeding the world's rapidly increasing population requires that farmers produce ever-larger and healthier crops. Every year they add hundreds of millions of tons of chemical fertilizers to the soil to increase crop quality and yield. In addition to carbon dioxide and water, plants need at least six elements for satisfactory growth. They are N, P, K, Ca, S, and Mg. The preparation and properties of several nitrogen- and phosphorus-containing fertilizers illustrate some of the principles introduced in this chapter.

Nitrogen fertilizers contain nitrate (NO_3^-) salts, ammonium (NH_4^+) salts, and other compounds. Plants can absorb nitrogen in the form of nitrate directly, but ammonium salts and ammonia (NH_3) must first be converted to nitrates by the action of soil bacteria. The principal raw material of nitrogen fertilizers is ammonia, prepared by the reaction between hydrogen and nitrogen:

$$3H_2(g) + N_2(g) \longrightarrow 2NH_3(g)$$

(This reaction will be discussed in detail in Chapters 13 and 14.) In its liquid form, ammonia can be injected directly into the soil.

Alternatively, ammonia can be converted to ammonium nitrate (NH_4NO_3), ammonium sulfate [$(NH_4)_2SO_4$], or ammonium hydrogen phosphate [$(NH_4)_2HPO_4$] in the following acid-base reactions:

$$NH_3(aq) + HNO_3(aq) \longrightarrow NH_4NO_3(aq)$$

$$2NH_3(aq) + H_2SO_4(aq) \longrightarrow (NH_4)_2SO_4(aq)$$

$$2NH_3(aq) + H_3PO_4(aq) \longrightarrow (NH_4)_2HPO_4(aq)$$

Liquid ammonia being applied to the soil before planting.

Glyn Thomas/Alamy Stock Photo

Another method of preparing ammonium sulfate requires two steps:

$$2NH_3(aq) + CO_2(aq) + H_2O(l) \longrightarrow (NH_4)_2CO_3(aq) \quad (1)$$

$$(NH_4)_2CO_3(aq) + CaSO_4(aq) \longrightarrow (NH_4)_2SO_4(aq) + CaCO_3(s) \quad (2)$$

This approach is desirable because the starting materials—carbon dioxide and calcium sulfate—are less costly than sulfuric acid. To increase the yield, ammonia is made the limiting reagent in Reaction (1) and ammonium carbonate is made the limiting reagent in Reaction (2).

The table lists the percent composition by mass of nitrogen in some common fertilizers. The preparation of urea was discussed in Example 3.15.

Percent Composition by Mass of Nitrogen in Five Common Fertilizers

Fertilizer	% N by Mass
NH_3	82.4
NH_4NO_3	35.0
$(NH_4)_2SO_4$	21.2
$(NH_4)_2HPO_4$	21.2
$(NH_2)_2CO$	46.7

Several factors influence the choice of one fertilizer over another: (1) cost of the raw materials needed to prepare the fertilizer; (2) ease of storage, transportation, and utilization; (3) percent composition by mass of the desired element; and (4) suitability of the compound, that is, whether the compound is soluble in water and whether it can be readily taken up by plants. Considering all these factors together, we find that NH_4NO_3 is the most important nitrogen-containing fertilizer in the world, even though ammonia has the highest percentage by mass of nitrogen.

Phosphorus fertilizers are derived from phosphate rock, called *fluorapatite,* $Ca_5(PO_4)_3F$. Fluorapatite is insoluble in water, so it must first be converted to water-soluble calcium dihydrogen phosphate [$Ca(H_2PO_4)_2$]:

$$2Ca_5(PO_4)_3F(s) + 7H_2SO_4(aq) \longrightarrow 3Ca(H_2PO_4)_2(aq) + 7CaSO_4(aq) + 2HF(g)$$

For maximum yield, fluorapatite is made the limiting reagent in this reaction.

The reactions we have discussed for the preparation of fertilizers all appear relatively simple, yet much effort has been expended to improve the yields by changing conditions such as temperature, pressure, and so on. Industrial chemists usually run promising reactions first in the laboratory and then test them in a pilot facility before putting them into mass production.

The atom economy of a reaction may be calculated using molecular mass or molar mass because the mass units will cancel one another. For convenience, mass units are often not shown in a calculation.

Atom economy is a function of a given reaction and can be determined before a reaction is ever conducted because it is not dependent on the mass of desired product recovered. We can illustrate the use of atom economy in the synthesis of phenylalanine ($C_6H_5NH_2$) from nitrobenzene ($C_6H_5NO_2$):

$$\underset{123.11}{4C_6H_5NO_2} + \underset{55.85}{9Fe} + \underset{18.02}{4H_2O} \longrightarrow \underset{93.13}{4C_6H_5NH_2} + 3Fe_3O_4$$

The atom economy of this reaction is

$$\text{atom economy} = \frac{4 \times 93.13}{(4 \times 123.11) + (9 \times 55.85) + (4 \times 18.02)} \times 100\% = 34.91\%$$

Although this reaction has a high experimental reaction yield, its atom economy is low in comparison. We can explain this by examining the actual chemical transformations occurring in the reaction. The $—NO_2$ group in the reactant nitrobenzene ($C_6H_5NO_2$) is converted to a $—NH_2$ group in phenylamine ($C_6H_5NH_2$). The hydrogen atoms from the reactant water replace the oxygen atoms in the nitro group, and those oxygen atoms ultimately are gained by iron to form the undesired product Fe_3O_4. Thus, the low atom economy indicates that few atoms of the reactants Fe and H_2O become part of the phenylamine ($C_6H_5NH_2$).

A more atom economical synthesis of phenylamine is accomplished by the following reaction:

$$\underset{123.11}{C_6H_5NO_2} + \underset{2.02}{3H_2} \longrightarrow \underset{93.13}{C_6H_5NH_2} + 2H_2O$$

The atom economy of this process is

$$\text{atom economy} = \frac{93.13}{123.11 + (3 \times 2.02)} \times 100\% = 72.10\%$$

This reaction has utilized more of the atoms in the reactants to prepare phenylalanine ($C_6H_5NH_2$). And the by-product of the reaction is water, which can be isolated and easily reused making this synthesis more efficient with less waste. From the perspective of sustainable or green chemistry, the second synthesis is preferable. Note, however, that the reaction yield must always be taken into consideration in reality. A synthesis with a slightly lower atom economy but higher reaction yield can often be preferable to one with a higher atom economy.

Example 3.18

The 2006 Presidential Green Chemistry Award was given for an improved synthesis of propylene glycol ($C_3H_8O_2$), a component of airplane de-icing fluids, and an important precursor to a variety of polyester resins. This synthesis used a chromium-based catalyst to convert glycerol ($C_3H_8O_3$) into propylene glycol according to the reaction:

$$C_3H_8O_3 + H_2 \longrightarrow C_3H_8O_2 + H_2O$$

What is the atom economy of this reaction?

Strategy How do we calculate the atom economy of a reaction? To determine the atom economy, the total mass of the atoms in the desired product, propylene glycol, is needed. The total mass of all the atoms in the reactants is the sum of the masses of glycerol and hydrogen.

(Continued)

Solution The required masses of both reactants and the desired product are $C_3H_8O_3$ = 92.1 amu, H_2 = 2.02 amu, and $C_3H_8O_2$ = 76.1 amu. The atom economy is calculated as follows:

$$\text{atom economy} = \frac{\text{total mass of atoms in product}}{\text{total mass of atoms in all reactants}} \times 100\%$$

$$= \frac{76.1}{92.1 + 2.02} \times 100\%$$

$$= 80.9\%$$

Check Because this reaction has two products, propylene glycol and water, not all of the atoms in the reactants are incorporated into the desired product, propylene glycol. Thus, the atom economy of this reaction must be less than 100%.

Practice Exercise A laboratory synthesis of propylene glycol is through oxidation of propylene in the presence of potassium permanganate:

$$3C_3H_6 + 2KMnO_4 + 4H_2O \longrightarrow 3C_3H_8O_2 + 2MnO_2 + 2KOH$$

What is the atom economy of this reaction?

Similar problems: 3.95, 3.96.

Summary of Concepts & Facts

- The amount of product obtained in a reaction (the actual yield) may be less than the maximum possible amount (the theoretical yield). The ratio of the actual yield to the theoretical yield multiplied by 100 percent is the percent yield.
- Atom economy evaluates the overall efficiency of a chemical reaction by determining how many of the atoms in the reactants are found in the desired product(s).

Review of Concepts & Facts

3.10.1 Can the percent yield ever exceed the theoretical yield of a reaction?

3.10.2 Sulfur trioxide (SO_3) is prepared from the oxidation of sulfur dioxide (SO_2) according to the equation

$$2SO_2(g) + O_2(g) \longrightarrow 2SO_3(g)$$

If 16.4 g of SO_2 produce 18.1 g of SO_3, what is the percent yield of the reaction?

3.10.3 What is the atom economy of the reaction to produce molten iron by reduction of iron(III) oxide?

$$Fe_2O_3(s) + 3CO(g) \longrightarrow 2Fe(l) + 3CO_2(g)$$

Chapter Summary

Atomic Mass and Molar Mass The mass of an atom, which is extremely small, is based on the carbon-12 isotope scale. An atom of the carbon-12 isotope is assigned a mass of exactly 12 atomic mass units (amu). To work with the more convenient scale of grams, chemists use the molar mass. The molar mass of carbon-12 is exactly 12 g and contains an Avogadro's number (6.022×10^{23}) of atoms. The molar masses of other elements are also expressed in grams and contain the same number of atoms. The molar mass of a molecule is the sum of the molar masses of its constituent atoms. (Sections 3.1, 3.2, 3.3)

Percent Composition of a Compound The makeup of a compound is most conveniently expressed in terms of its percent composition, which is the percent by mass of each element the compound contains. A knowledge of its chemical formula enables us to calculate the percent composition. Experimental determination of percent composition and the molar mass of a compound enables us to determine its chemical formula. (Sections 3.4, 3.5, 3.6)

Writing Chemical Equations An effective way to represent the outcome of a chemical reaction is to write a chemical equation, which uses chemical formulas to describe what happens. A chemical equation must be balanced so that we have the same number and types of atoms for the reactants, the starting materials, and the products, the substances formed at the end of the reaction. (Section 3.7)

Mass Relationships of a Chemical Reaction A chemical equation enables us to predict the amount of product(s) formed, called the yield, knowing how much reactant(s) was (were) used. This information is of great importance for reactions run on the laboratory or industrial scale. In practice, the actual yield is almost always less than that predicted from the equation because of various complications. The atom economy evaluates the efficiency of a chemical reaction by determining how many of the atoms in the reactants are found in the desired product(s). (Sections 3.8, 3.9, 3.10)

Key Equations

$$\text{percent composition of an element in a compound} = \frac{n \times \text{molar mass of element}}{\text{molar mass of compound}} \times 100\% \quad (3.1)$$

$$\%\text{yield} = \frac{\text{actual yield}}{\text{theoretical yield}} \times 100\% \quad (3.4)$$

$$\text{atom economy} = \frac{\text{mass of atoms in desired product(s)}}{\text{mass of atoms in reactants}} \times 100\% \quad (3.5)$$

Key Words

Actual yield, p. 108
Atom economy, p. 109
Atomic mass, p. 80
Atomic mass unit (amu), p. 80
Avogadro's number (N_A), p. 82
Chemical equation, p. 95
Chemical reaction, p. 95
Excess reactant, p. 104
Limiting reactant, p. 104
Molar mass ($\mathcal{M}$), p. 83
Mole (mol), p. 82
Mole method, p. 100
Molecular mass, p. 85
Percent composition by mass, p. 90
Percent yield, p. 108
Product, p. 96
Reactant, p. 96
Stoichiometric amounts, p. 104
Stoichiometry, p. 100
Theoretical yield, p. 108

Questions & Problems

Red numbered problems solved in Student Solutions Manual

3.1 Atomic Mass

Review Questions

3.1 What is an atomic mass unit? Why is it necessary to introduce such a unit?

3.2 What is the mass (in amu) of a carbon-12 atom? Why is the atomic mass of carbon listed as 12.01 amu in the List of the Elements with Their Symbols and Atomic Masses?

3.3 Explain clearly what is meant by the statement, "The atomic mass of gold is 197.0 amu."

3.4 What information would you need to calculate the average atomic mass of an element?

Problems

3.5 The atomic masses of $^{35}_{17}Cl$ (75.53 percent) and $^{37}_{17}Cl$ (24.47 percent) are 34.968 amu and 36.956 amu, respectively. Calculate the average atomic mass of chlorine. The percentages in parentheses denote the relative abundances.

3.6 The atomic masses of $^{6}_{3}Li$ and $^{7}_{3}Li$ are 6.0151 amu and 7.0160 amu, respectively. Calculate the natural abundances of these two isotopes. The average atomic mass of Li is 6.941 amu.

3.7 What is the mass in grams of 13.2 amu?

3.8 How many amu are there in 8.4 g?

3.2 Avogadro's Number and the Molar Mass of an Element

Review Questions

3.9 Define the term *mole*. What is the unit for mole in calculations? What does the mole have in common with the pair, the dozen, and the gross? What does Avogadro's number represent?

3.10 What is the molar mass of an atom? What are the commonly used units for molar mass?

Problems

3.11 Earth's population is about 7.2 billion. Suppose that every person on Earth participates in a process of counting identical particles at the rate of two particles per second. How many years would it take to count 6.0×10^{23} particles? Assume that there are 365 days in a year.

3.12 The thickness of a piece of paper is 0.0036 in. Suppose a certain book has an Avogadro's number of pages; calculate the thickness of the book in light-years. (*Hint:* See Problem 1.51 for the definition of light-year.)

3.13 How many atoms are there in 5.10 moles of sulfur (S)?

3.14 How many moles of cobalt (Co) atoms are there in 6.00×10^9 (6 billion) Co atoms?

3.15 How many moles of calcium (Ca) atoms are in 77.4 g of Ca?

3.16 How many grams of gold (Au) are there in 15.3 moles of Au?

3.17 What is the mass in grams of a single atom of each of the following elements? (a) Hg, (b) Ne.

3.18 What is the mass in grams of a single atom of each of the following elements? (a) As, (b) Ni.

3.19 What is the mass in grams of 1.00×10^{12} lead (Pb) atoms?

3.20 A modern penny weighs 2.5 g but contains only 0.063 g of copper (Cu). How many copper atoms are present in a modern penny?

3.21 Which of the following has more atoms: 1.10 g of hydrogen atoms or 14.7 g of chromium atoms?

3.22 Which of the following has a greater mass: 2 atoms of lead or 5.1×10^{-23} mole of helium?

3.3 Molecular Mass

Problems

3.23 Calculate the molecular mass or formula mass (in amu) of each of the following substances: (a) CH_4, (b) NO_2, (c) SO_3, (d) C_6H_6, (e) NaI, (f) K_2SO_4, (g) $Ca_3(PO_4)_2$.

3.24 Calculate the molar mass of the following substances: (a) Li_2CO_3, (b) CS_2, (c) $CHCl_3$ (chloroform), (d) $C_6H_8O_6$ (ascorbic acid, or vitamin C), (e) KNO_3, (f) Mg_3N_2.

3.25 Calculate the molar mass of a compound if 0.372 mole of it has a mass of 152 g.

3.26 How many molecules of ethane (C_2H_6) are present in 0.334 g of C_2H_6?

3.27 Calculate the number of C, H, and O atoms in 1.50 g of glucose ($C_6H_{12}O_6$), a sugar.

3.28 Dimethyl sulfoxide [$(CH_3)_2SO$], also called DMSO, is an important solvent that penetrates the skin, enabling it to be used as a topical drug-delivery agent. Calculate the number of C, S, H, and O atoms in 7.14×10^3 g of dimethyl sulfoxide.

3.29 Pheromones are a special type of compound secreted by the females of many insect species to attract the males for mating. One pheromone has the molecular formula $C_{19}H_{38}O$. Normally, the amount of this pheromone secreted by a female insect is about 1.0×10^{-12} g. How many molecules are there in this quantity?

3.30 The density of water is 1.00 g/mL at 4°C. How many water molecules are present in 2.56 mL of water at this temperature?

3.4 The Mass Spectrometer

Review Questions

3.31 Describe the operation of a mass spectrometer.

3.32 Describe how you would determine the isotopic abundance of an element from its mass spectrum.

Problems

3.33 Carbon has two stable isotopes, $^{12}_{6}C$ and $^{13}_{6}C$, and fluorine has only one stable isotope, $^{19}_{9}F$. How many peaks would you observe in the mass spectrum of the positive ion of CF_4^{+}? Assume that the ion does not break up into smaller fragments.

3.34 Hydrogen has two stable isotopes, $^{1}_{1}H$ and $^{2}_{1}H$, and sulfur has four stable isotopes, $^{32}_{16}S$, $^{33}_{16}S$, $^{34}_{16}S$, and $^{36}_{16}S$. How many peaks would you observe in the mass spectrum of the positive ion of hydrogen sulfide, H_2S^{+}? Assume no decomposition of the ion into smaller fragments.

3.5 Percent Composition of Compounds

Review Questions

3.35 Use ammonia (NH_3) to explain what is meant by the percent composition by mass of a compound.

3.36 Describe how the knowledge of the percent composition by mass of an unknown compound can help us identify the compound.

Problems

3.37 Tin (Sn) exists in Earth's crust as SnO_2. Calculate the percent composition by mass of Sn and O in SnO_2.

3.38 For many years chloroform ($CHCl_3$) was used as an inhalation anesthetic in spite of the fact that it is also a toxic substance that may cause severe liver, kidney, and heart damage. Calculate the percent composition by mass of this compound.

3.39 Cinnamic alcohol is used mainly in perfumery, particularly in soaps and cosmetics. Its molecular formula is $C_9H_{10}O$. (a) Calculate the percent composition by mass of C, H, and O in cinnamic alcohol. (b) How many molecules of cinnamic alcohol are contained in a sample of mass 0.469 g?

3.40 All of the substances listed here are fertilizers that contribute nitrogen to the soil. Which of these is the richest source of nitrogen on a mass percentage basis?

(a) Urea, $(NH_2)_2CO$

(b) Ammonium nitrate, NH_4NO_3

(c) Guanidine, $HNC(NH_2)_2$

(d) Ammonia, NH_3

3.41 The formula for rust can be represented by Fe_2O_3. How many moles of Fe are present in 24.6 g of the compound?

3.42 Tin(II) fluoride (SnF_2) is often added to toothpaste as an ingredient to prevent tooth decay. What is the mass of F in grams in 24.6 g of the compound?

3.43 What are the empirical formulas of the compounds with the following compositions? (a) 2.1 percent H, 65.3 percent O, 32.6 percent S; (b) 20.2 percent Al, 79.8 percent Cl.

3.44 What are the empirical formulas of the compounds with the following compositions? (a) 40.1 percent C, 6.6 percent H, 53.3 percent O; (b) 18.4 percent C, 21.5 percent N, 60.1 percent K.

3.45 The anticaking agent added to Morton salt is calcium silicate, $CaSiO_3$. This compound can absorb up to 2.5 times its mass of water and still remain a free-flowing powder. Calculate the percent composition of $CaSiO_3$.

3.6 Experimental Determination of Empirical Formulas

Review Questions

3.46 What does the word *empirical* in empirical formula mean?

3.47 If we know the empirical formula of a compound, what additional information do we need to determine its molecular formula?

Problems

3.48 The empirical formula of a compound is CH. If the molar mass of this compound is about 78 g, what is its molecular formula?

3.49 The molar mass of caffeine is 194.19 g. Is the molecular formula of caffeine $C_4H_5N_2O$ or $C_8H_{10}N_4O_2$?

3.50 Peroxyacyl nitrate (PAN) is one of the components of smog. It is a compound of C, H, N, and O. Determine the percent composition of oxygen and the empirical formula from the following percent composition by mass: 19.8 percent C, 2.50 percent H, 11.6 percent N. What is its molecular formula given that its molar mass is about 120 g?

3.51 Allicin is the compound responsible for the characteristic smell of garlic. An analysis of the compound gives the following percent composition by mass: C: 44.4 percent, H: 6.21 percent, S: 39.5 percent, O: 9.86 percent. Calculate its empirical formula. What is its molecular formula given that its molar mass is about 162 g?

3.52 Monosodium glutamate (MSG), a food-flavor enhancer, has been blamed for "Chinese restaurant syndrome," the symptoms of which are headaches and chest pains. MSG has the following composition by mass: 35.51 percent C, 4.77 percent H, 37.85 percent O, 8.29 percent N, and 13.60 percent Na. What is its molecular formula if its molar mass is about 169 g?

3.7 Chemical Reactions and Chemical Equations

Review Questions

3.53 Use the formation of water from hydrogen and oxygen to explain the following terms: *chemical reaction, reactant, product.*

3.54 What is the difference between a chemical reaction and a chemical equation?

3.55 Why must a chemical equation be balanced? What law is obeyed by a balanced chemical equation?

3.56 Write the symbols used to represent gas, liquid, solid, and the aqueous phase in chemical equations.

Problems

3.57 Balance the following equations using the method outlined in Section 3.7.

(a) $C + O_2 \longrightarrow CO$

(b) $CO + O_2 \longrightarrow CO_2$

(c) $H_2 + Br_2 \longrightarrow HBr$

(d) $K + H_2O \longrightarrow KOH + H_2$

(e) $Mg + O_2 \longrightarrow MgO$

(f) $O_3 \longrightarrow O_2$

(g) $H_2O_2 \longrightarrow H_2O + O_2$

(h) $N_2 + H_2 \longrightarrow NH_3$

(i) $Zn + AgCl \longrightarrow ZnCl_2 + Ag$

(j) $S_8 + O_2 \longrightarrow SO_2$

(k) $NaOH + H_2SO_4 \longrightarrow Na_2SO_4 + H_2O$

(l) $Cl_2 + NaI \longrightarrow NaCl + I_2$

(m) $KOH + H_3PO_4 \longrightarrow K_3PO_4 + H_2O$

(n) $CH_4 + Br_2 \longrightarrow CBr_4 + HBr$

3.58 Balance the following equations using the method outlined in Section 3.7.

(a) $N_2O_5 \longrightarrow N_2O_4 + O_2$

(b) $KNO_3 \longrightarrow KNO_2 + O_2$

(c) $NH_4NO_3 \longrightarrow N_2O + H_2O$

(d) $NH_4NO_2 \longrightarrow N_2 + H_2O$

(e) $NaHCO_3 \longrightarrow Na_2CO_3 + H_2O + CO_2$

(f) $P_4O_{10} + H_2O \longrightarrow H_3PO_4$

(g) $HCl + CaCO_3 \longrightarrow CaCl_2 + H_2O + CO_2$

(h) $Al + H_2SO_4 \longrightarrow Al_2(SO_4)_3 + H_2$

(i) $CO_2 + KOH \longrightarrow K_2CO_3 + H_2O$

(j) $CH_4 + O_2 \longrightarrow CO_2 + H_2O$

(k) $Be_2C + H_2O \longrightarrow Be(OH)_2 + CH_4$

(l) $Cu + HNO_3 \longrightarrow Cu(NO_3)_2 + NO + H_2O$

(m) $S + HNO_3 \longrightarrow H_2SO_4 + NO_2 + H_2O$

(n) $NH_3 + CuO \longrightarrow Cu + N_2 + H_2O$

3.8 Amounts of Reactants and Products

Review Questions

3.59 On what law is stoichiometry based? Why is it essential to use balanced equations in solving stoichiometric problems?

3.60 Describe the steps involved in the mole method.

Problems

3.61 Which of the following equations best represents the reaction shown in the diagram?
(a) $8A + 4B \longrightarrow C + D$
(b) $4A + 8B \longrightarrow 4C + 4D$
(c) $2A + B \longrightarrow C + D$
(d) $4A + 2B \longrightarrow 4C + 4D$
(e) $2A + 4B \longrightarrow C + D$

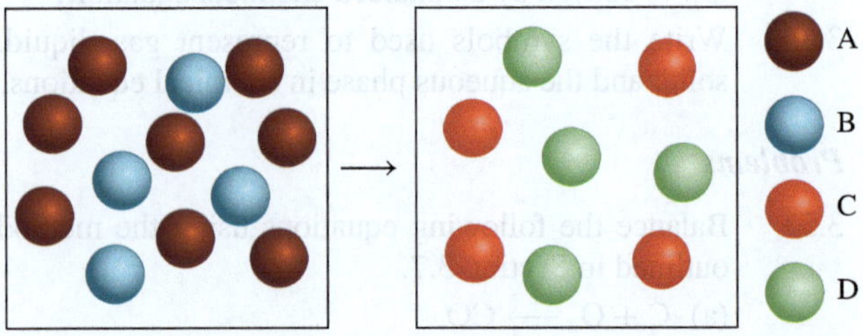

3.62 Which of the following equations best represents the reaction shown in the diagram?
(a) $A + B \longrightarrow C + D$
(b) $6A + 4B \longrightarrow C + D$
(c) $A + 2B \longrightarrow 2C + D$
(d) $3A + 2B \longrightarrow 2C + D$
(e) $3A + 2B \longrightarrow 4C + 2D$

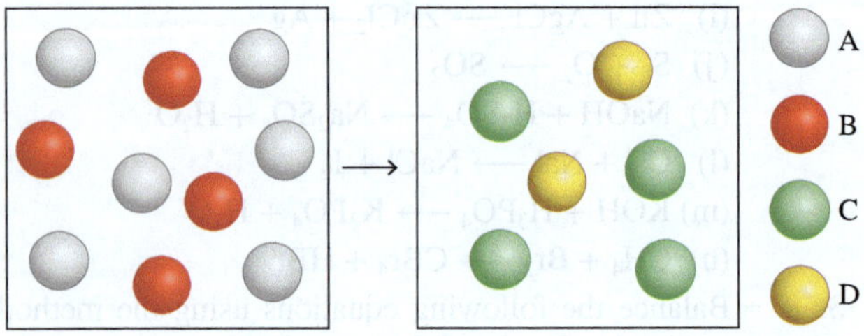

3.63 Consider the combustion of carbon monoxide (CO) in oxygen gas:

$$2CO(g) + O_2(g) \longrightarrow 2CO_2(g)$$

Starting with 3.60 moles of CO, calculate the number of moles of CO_2 produced if there is enough oxygen gas to react with all of the CO.

3.64 Silicon tetrachloride ($SiCl_4$) can be prepared by heating Si in chlorine gas:

$$Si(s) + 2Cl_2(g) \longrightarrow SiCl_4(l)$$

In one reaction, 0.507 mole of $SiCl_4$ is produced. How many moles of molecular chlorine were used in the reaction?

3.65 Ammonia is a principal nitrogen fertilizer. It is prepared by the reaction between hydrogen and nitrogen.

$$3H_2(g) + N_2(g) \longrightarrow 2NH_3(g)$$

In a particular reaction, 6.0 moles of NH_3 were produced. How many moles of H_2 and how many moles of N_2 were reacted to produce this amount of NH_3?

3.66 Certain race cars use methanol (CH_3OH, also called wood alcohol) as a fuel. The combustion of methanol occurs according to the following equation:

$$2CH_3OH(l) + 3O_2(g) \longrightarrow 2CO_2(g) + 4H_2O(l)$$

In a particular reaction, 9.8 moles of CH_3OH are reacted with an excess of O_2. Calculate the number of moles of H_2O formed.

3.67 Calculate the mass in grams of iodine (I_2) that will react completely with 20.4 g of aluminum (Al) to form aluminum iodide (AlI_3).

3.68 How many grams of sulfur (S) are needed to react completely with 246 g of mercury (Hg) to form HgS?

3.69 The annual production of sulfur dioxide from burning coal and fossil fuels, auto exhaust, and other sources is about 26 million tons. The equation for the reaction is

$$S(s) + O_2(g) \longrightarrow SO_2(g)$$

How much sulfur (in tons), present in the original materials, would result in that quantity of SO_2?

3.70 When baking soda (sodium bicarbonate or sodium hydrogen carbonate, $NaHCO_3$) is heated, it releases carbon dioxide gas, which is responsible for the rising of cookies, donuts, and bread. (a) Write a balanced equation for the decomposition of the compound (one of the products is Na_2CO_3). (b) Calculate the mass of $NaHCO_3$ required to produce 20.5 g of CO_2.

3.71 If chlorine bleach is mixed with other cleaning products containing ammonia, the toxic gas $NCl_3(g)$ can form according to the equation

$$3NaClO(aq) + NH_3(aq) \longrightarrow 3NaOH(aq) + NCl_3(g)$$

When 2.94 g of NH_3 reacts with an excess of NaClO according to the preceding reaction, how many grams of NCl_3 are formed?

3.72 Fermentation is a complex chemical process of wine making in which glucose is converted into ethanol and carbon dioxide:

$$\underset{\text{glucose}}{C_6H_{12}O_6} \longrightarrow \underset{\text{ethanol}}{2C_2H_5OH} + 2CO_2$$

Starting with 500.4 g of glucose, what is the maximum amount of ethanol in grams and in liters that can be obtained by this process? (Density of ethanol = 0.789 g/mL.)

3.73 Each copper(II) sulfate unit is associated with five water molecules in crystalline copper(II) sulfate pentahydrate ($CuSO_4 \cdot 5H_2O$). When this compound is heated in air above 100°C, it loses the water molecules and also its blue color:

$$CuSO_4 \cdot 5H_2O \longrightarrow CuSO_4 + 5H_2O$$

If 9.60 g of $CuSO_4$ are left after heating 15.01 g of the blue compound, calculate the number of moles of H_2O originally present in the compound.

3.74 For many years the recovery of gold—that is, the separation of gold from other materials—involved the use of potassium cyanide:

$$4Au + 8KCN + O_2 + 2H_2O \longrightarrow 4KAu(CN)_2 + 4KOH$$

What is the minimum amount of KCN in moles needed to extract 29.0 g (about an ounce) of gold?

3.75 Limestone ($CaCO_3$) is decomposed by heating to quicklime (CaO) and carbon dioxide. Calculate how many grams of quicklime can be produced from 1.0 kg of limestone.

3.76 Nitrous oxide (N_2O) is also called "laughing gas." It can be prepared by the thermal decomposition of ammonium nitrate (NH_4NO_3). The other product is H_2O. (a) Write a balanced equation for this reaction. (b) How many grams of N_2O are formed if 0.46 mole of NH_4NO_3 is used in the reaction?

3.77 The fertilizer ammonium sulfate [$(NH_4)_2SO_4$] is prepared by the reaction between ammonia (NH_3) and sulfuric acid:

$$2NH_3(g) + H_2SO_4(aq) \longrightarrow (NH_4)_2SO_4(aq)$$

How many kilograms of NH_3 are needed to produce 1.00×10^5 kg of $(NH_4)_2SO_4$?

3.78 A common laboratory preparation of oxygen gas is the thermal decomposition of potassium chlorate ($KClO_3$). Assuming complete decomposition, calculate the number of grams of O_2 gas that can be obtained from 46.0 g of $KClO_3$. (The products are KCl and O_2.)

3.9 Limiting Reactants

Review Questions

3.79 Define limiting reactant and excess reactant. What is the significance of the limiting reactant in predicting the amount of the product obtained in a reaction? Can there be a limiting reactant if only one reactant is present?

3.80 Give an everyday example that illustrates the limiting reactant concept.

Problems

3.81 Consider the reaction

$$2A + B \longrightarrow C$$

(a) In the diagram here that represents the reaction, which reactant, A or B, is the limiting reactant?
(b) Assuming complete reaction, draw a molecular-model representation of the amounts of reactants and products left after the reaction. The atomic arrangement in C is ABA.

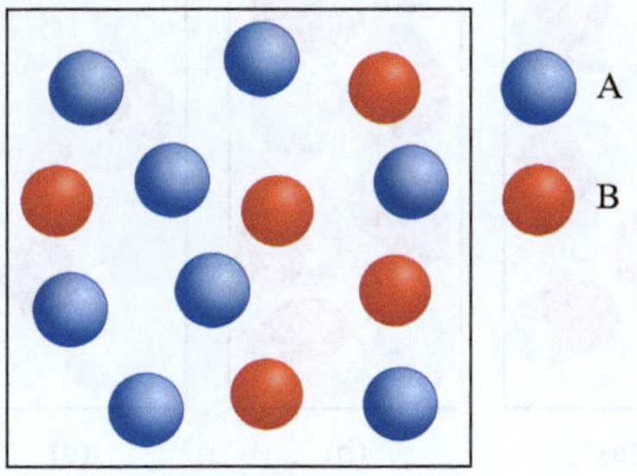

3.82 Consider the reaction

$$N_2 + 3H_2 \longrightarrow 2NH_3$$

Assuming each model represents 1 mole of the substance, show the number of moles of the product and the excess reactant left after the complete reaction.

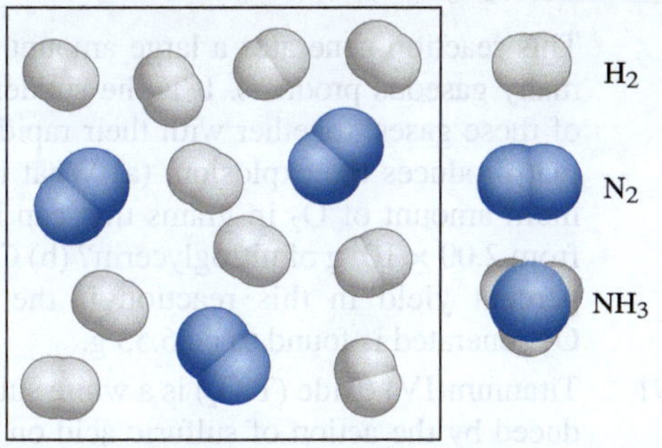

3.83 Nitric oxide (NO) reacts with oxygen gas to form nitrogen dioxide (NO_2), a dark brown gas:

$$2NO(g) + O_2(g) \longrightarrow 2NO_2(g)$$

In one experiment 0.886 mole of NO is mixed with 0.503 mole of O_2. Calculate which of the two reactants is the limiting reactant. Calculate also the number of moles of NO_2 produced.

3.84 Ammonia and sulfuric acid react to form ammonium sulfate. (a) Write an equation for the reaction. (b) Determine the starting mass (in grams) of each reactant if 20.3 g of ammonium sulfate are produced and 5.89 g of sulfuric acid remain unreacted.

3.85 Propane (C_3H_8) is a component of natural gas and is used in domestic cooking and heating. (a) Balance the following equation representing the combustion of propane in air.

$$C_3H_8 + O_2 \longrightarrow CO_2 + H_2O$$

(b) How many grams of carbon dioxide can be produced by burning 3.65 moles of propane? Assume that oxygen is the excess reactant in this reaction.

3.86 Consider the reaction

$$MnO_2 + 4HCl \longrightarrow MnCl_2 + Cl_2 + 2H_2O$$

If 0.86 mole of MnO_2 and 48.2 g of HCl react, which reactant will be used up first? How many grams of Cl_2 will be produced?

3.10 Reaction Yield and Atom Economy

Review Questions

3.87 Why is the theoretical yield of a reaction determined only by the amount of the limiting reactant?

3.88 Why is the actual yield of a reaction almost always smaller than the theoretical yield?

Problems

3.89 Hydrogen fluoride is used in the manufacture of Freons (which destroy ozone in the stratosphere) and in the production of aluminum metal. It is prepared by the reaction

$$CaF_2 + H_2SO_4 \longrightarrow CaSO_4 + 2HF$$

In one process, 6.00 kg of CaF_2 are treated with an excess of H_2SO_4 and yield 2.86 kg of HF. Calculate the percent yield of HF.

3.90 Nitroglycerin ($C_3H_5N_3O_9$) is a powerful explosive. Its decomposition may be represented by

$$4C_3H_5N_3O_9 \longrightarrow 6N_2 + 12CO_2 + 10H_2O + O_2$$

This reaction generates a large amount of heat and many gaseous products. It is the sudden formation of these gases, together with their rapid expansion, that produces the explosion. (a) What is the maximum amount of O_2 in grams that can be obtained from 2.00×10^2 g of nitroglycerin? (b) Calculate the percent yield in this reaction if the amount of O_2 generated is found to be 6.55 g.

3.91 Titanium(IV) oxide (TiO_2) is a white substance produced by the action of sulfuric acid on the mineral ilmenite ($FeTiO_3$):

$$FeTiO_3 + H_2SO_4 \longrightarrow TiO_2 + FeSO_4 + H_2O$$

Its opaque and nontoxic properties make it suitable as a pigment in plastics and paints. In one process, 8.00×10^3 kg of $FeTiO_3$ yielded 3.67×10^3 kg of TiO_2. What is the percent yield of the reaction?

3.92 Ethylene (C_2H_4), an important industrial organic chemical, can be prepared by heating hexane (C_6H_{14}) at 800°C:

$$C_6H_{14} \longrightarrow C_2H_4 + \text{other products}$$

If the yield of ethylene production is 42.5 percent, what mass of hexane must be reacted to produce 481 g of ethylene?

3.93 When heated, lithium reacts with nitrogen to form lithium nitride:

$$6Li(s) + N_2(g) \longrightarrow 2Li_3N(s)$$

What is the theoretical yield of Li_3N in grams when 12.3 g of Li are heated with 33.6 g of N_2? If the actual yield of Li_3N is 5.89 g, what is the percent yield of the reaction?

3.94 Disulfide dichloride (S_2Cl_2) is used in the vulcanization of rubber, a process that prevents the slippage of rubber molecules past one another when stretched. It is prepared by heating sulfur in an atmosphere of chlorine:

$$S_8(l) + 4Cl_2(g) \longrightarrow 4S_2Cl_2(l)$$

What is the theoretical yield of S_2Cl_2 in grams when 4.06 g of S_8 are heated with 6.24 g of Cl_2? If the actual yield of S_2Cl_2 is 6.55 g, what is the percent yield?

3.95 The important industrial compound maleic anhydride ($C_4H_2O_3$) is prepared by the reaction of benzene (C_6H_6) with oxygen:

$$2C_6H_6 + 9O_2 \longrightarrow 2C_4H_2O_3 + 4CO_2 + 4H_2O$$

What is the atom economy of this reaction?

3.96 In the Cativa process, acetic acid is produced by the reaction of methanol with carbon monoxide:

$$CH_3OH + CO \longrightarrow CH_3CO_2H$$

What is the atom economy of the Cativa process?

Additional Problems

3.97 Gallium is an important element in the production of semiconductors. The average atomic mass of ${}^{69}_{31}Ga$ (68.9256 amu) and ${}^{71}_{31}Ga$ (70.9247 amu) is 69.72 amu. Calculate the natural abundances of the gallium isotopes.

3.98 Rubidium is used in "atomic clocks" and other precise electronic equipment. The average atomic mass of ${}^{85}_{37}Rb$ (84.912 amu) and ${}^{87}_{37}Rb$ (86.909 amu) is 85.47 amu. Calculate the natural abundances of the rubidium isotopes.

3.99 This diagram represents the products (CO_2 and H_2O) formed after the combustion of a hydrocarbon (a compound containing only C and H atoms). Write an equation for the reaction. (*Hint:* The molar mass of the hydrocarbon is about 30 g.)

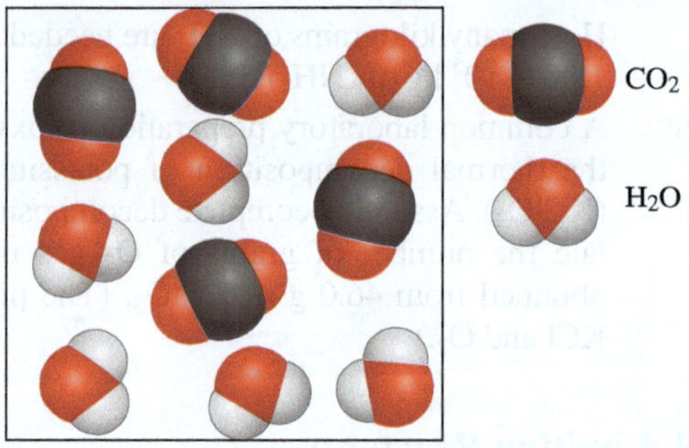

3.100 Consider the reaction of hydrogen gas with oxygen gas:

$$2H_2(g) + O_2(g) \longrightarrow 2H_2O(g)$$

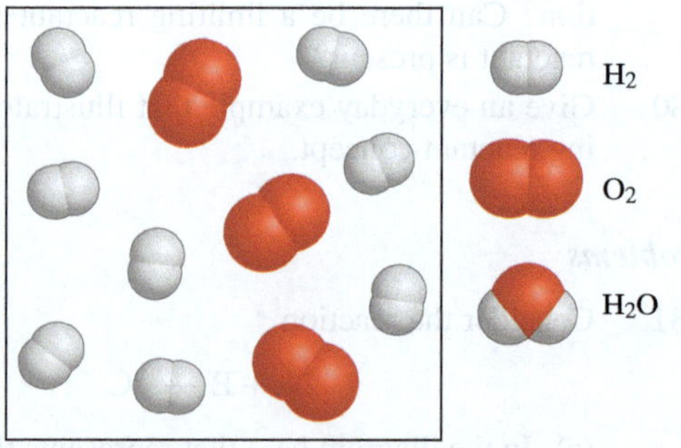

Assuming complete reaction, which of the diagrams (a)–(d) represents the amounts of reactants and products left after the reaction?

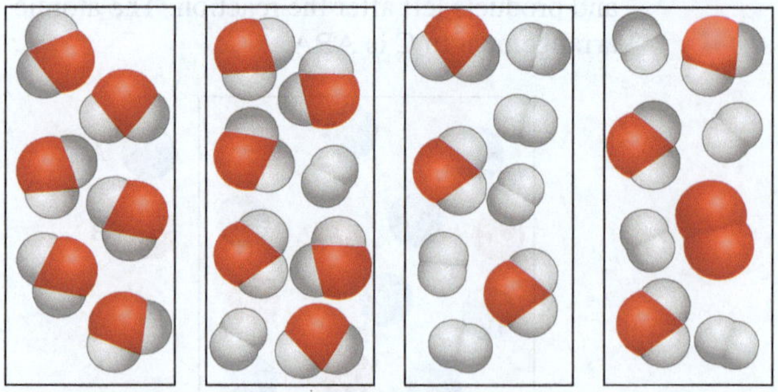

(a) (b) (c) (d)

3.101 Ethylene reacts with hydrogen chloride to form ethyl chloride:

$$C_2H_4(g) + HCl(g) \longrightarrow C_2H_5Cl(g)$$

Calculate the mass of ethyl chloride formed if 4.66 g of ethylene react with an 89.4 percent yield.

3.102 Write balanced equations for the following reactions described in words.

(a) Pentane burns in oxygen to form carbon dioxide and water.

(b) Sodium bicarbonate reacts with hydrochloric acid to form carbon dioxide, sodium chloride, and water.

(c) When heated in an atmosphere of nitrogen, lithium forms lithium nitride.

(d) Phosphorus trichloride reacts with water to form phosphorus acid and hydrogen chloride.

(e) Copper(II) oxide heated with ammonia will form copper, nitrogen gas, and water.

3.103 Industrially, nitric acid is produced by the Ostwald process represented by the following equations:

$$4NH_3(g) + 5O_2(g) \longrightarrow 4NO(g) + 6H_2O(l)$$
$$2NO(g) + O_2(g) \longrightarrow 2NO_2(g)$$
$$2NO_2(g) + H_2O(l) \longrightarrow HNO_3(aq) + HNO_2(aq)$$

What mass of NH_3 (in grams) must be used to produce 1.00 ton of HNO_3 by the above procedure, assuming an 80 percent yield in each step? (1 ton = 2000 lb; 1 lb = 453.6 g.)

3.104 A sample of a compound of Cl and O reacts with an excess of H_2 to give 0.233 g of HCl and 0.403 g of H_2O. Determine the empirical formula of the compound.

3.105 How many grams of H_2O will be produced from the complete combustion of 26.7 g of butane (C_4H_{10})?

3.106 A 26.2-g sample of oxalic acid hydrate ($H_2C_2O_4 \cdot 2H_2O$) is heated in an oven until all the water is driven off. How much of the anhydrous acid is left?

3.107 The atomic mass of element X is 33.42 amu. A 27.22-g sample of X combines with 84.10 g of another element Y to form a compound XY. Calculate the atomic mass of Y.

3.108 How many moles of O are needed to combine with 0.212 mole of C to form (a) CO and (b) CO_2?

3.109 A research chemist used a mass spectrometer to study the two isotopes of an element. Over time, she recorded a number of mass spectra of these isotopes. On analysis, she noticed that the ratio of the taller peak (the more abundant isotope) to the shorter peak (the less abundant isotope) gradually increased with time. Assuming that the mass spectrometer was functioning normally, what do you think was causing this change?

3.110 The aluminum sulfate hydrate [$Al_2(SO_4)_3 \cdot xH_2O$] contains 8.10 percent Al by mass. Calculate x—that is, the number of water molecules associated with each $Al_2(SO_4)_3$ unit.

3.111 The explosive nitroglycerin ($C_3H_5N_3O_9$) has also been used as a drug to treat heart patients to relieve pain (angina pectoris). We now know that nitroglycerin produces nitric oxide (NO), which causes muscles to relax and allows the arteries to dilate. If each nitroglycerin molecule releases one NO per atom of N, calculate the mass percent of NO available from nitroglycerin.

3.112 The carat is the unit of mass used by jewelers. One carat is exactly 200 mg. How many carbon atoms are present in a 24-carat diamond?

3.113 An iron bar weighed 664 g. After the bar had been standing in moist air for a month, exactly one-eighth of the iron turned to rust (Fe_2O_3). Calculate the final mass of the iron bar and rust.

3.114 A certain metal oxide has the formula MO, where M denotes the metal. A 39.46-g sample of the compound is strongly heated in an atmosphere of hydrogen to remove oxygen as water molecules. At the end, 31.70 g of the metal is left over. If O has an atomic mass of 16.00 amu, calculate the atomic mass of M and identify the element.

3.115 An impure sample of zinc (Zn) is treated with an excess of sulfuric acid (H_2SO_4) to form zinc sulfate ($ZnSO_4$) and molecular hydrogen (H_2). (a) Write a balanced equation for the reaction. (b) If 0.0764 g of H_2 is obtained from 3.86 g of the sample, calculate the percent purity of the sample. (c) What assumptions must you make in (b)?

3.116 One of the reactions that occurs in a blast furnace, where iron ore is converted to cast iron, is

$$Fe_2O_3 + 3CO \longrightarrow 2Fe + 3CO_2$$

Suppose that 1.64×10^3 kg of Fe are obtained from a 2.62×10^3 kg sample of Fe_2O_3. Assuming that the reaction goes to completion, what is the percent purity of Fe_2O_3 in the original sample?

3.117 Carbon dioxide (CO_2) is the gas that is mainly responsible for global warming (the greenhouse effect). The burning of fossil fuels is a major cause of the increased concentration of CO_2 in the atmosphere. Carbon dioxide is also the end product of metabolism (see Example 3.13). Using glucose as an example of food, calculate the annual human production of CO_2 in grams, assuming that each person consumes 5.0×10^2 g of glucose per day. The world's population is 7.2 billion, and there are 365 days in a year.

3.118 Carbohydrates are compounds containing carbon, hydrogen, and oxygen in which the hydrogen to oxygen ratio is 2:1. A certain carbohydrate contains 40.0 percent carbon by mass. Calculate the empirical and molecular formulas of the compound if the approximate molar mass is 178 g.

3.119 Which of the following has the greater mass: 0.72 g of O_2 or 0.0011 mole of chlorophyll ($C_{55}H_{72}MgN_4O_5$)?

3.120 Analysis of a metal chloride XCl_3 shows that it contains 67.2 percent Cl by mass. Calculate the molar mass of X and identify the element.

3.121 Acetylsalicylic acid ($C_9H_8O_4$), also known as aspirin, is prepared by the reaction of salicylic acid ($C_7H_6O_3$) and acetic anhydride ($C_4H_6O_3$):

$$C_7H_6O_3 + C_4H_6O_3 \longrightarrow C_9H_8O_4 + C_2H_4O_2$$

What is the atom economy of this reaction?

3.122 The compound hydrazine (N_2H_4) has been used as rocket fuel for the Apollo missions. It can be prepared by the reaction of ammonia (NH_3) with sodium hypochlorite (NaOCl).

$$2NH_3(aq) + NaOCl(aq) \longrightarrow N_2H_4(aq) + NaCl(aq) + H_2O(l)$$

What is the atom economy of this preparation of hydrazine?

3.123 Hemoglobin ($C_{2952}H_{4664}N_{812}O_{832}S_8Fe_4$) is the oxygen carrier in blood. (a) Calculate its molar mass. (b) An average adult has about 5.0 L of blood. Every milliliter of blood has approximately 5.0×10^9 erythrocytes, or red blood cells, and every red blood cell has about 2.8×10^8 hemoglobin molecules. Calculate the mass of hemoglobin molecules in grams in an average adult.

3.124 Myoglobin stores oxygen for metabolic processes in muscle. Chemical analysis shows that it contains 0.34 percent Fe by mass. What is the molar mass of myoglobin? (There is one Fe atom per molecule.)

3.125 Calculate the number of cations and anions in each of the following compounds: (a) 0.764 g of CsI, (b) 72.8 g of $K_2Cr_2O_7$, (c) 6.54 g of $Hg_2(NO_3)_2$.

3.126 A mixture of NaBr and Na_2SO_4 contains 29.96 percent Na by mass. Calculate the percent by mass of each compound in the mixture.

3.127 Consider the reaction $3A + 2B \longrightarrow 3C$. A student mixed 4.0 moles of A with 4.0 moles of B and obtained 2.8 moles of C. What is the percent yield of the reaction?

3.128 Balance the equation shown here in molecular models.

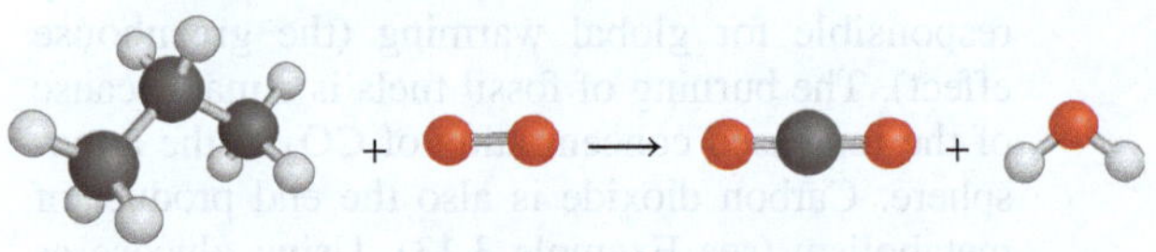

3.129 Aspirin or acetyl salicylic acid is synthesized by reacting salicylic acid with acetic anhydride:

$$\underset{\text{salicylic acid}}{C_7H_6O_3} + \underset{\text{acetic anhydride}}{C_4H_6O_3} \longrightarrow \underset{\text{aspirin}}{C_9H_8O_4} + \underset{\text{acetic acid}}{C_2H_4O_2}$$

(a) How much salicylic acid is required to produce 0.400 g of aspirin (about the content in a tablet), assuming acetic anhydride is present in excess? (b) Calculate the amount of salicylic acid needed if only 74.9 percent of salicylic acid is converted to aspirin. (c) In one experiment, 9.26 g of salicylic acid is reacted with 8.54 g of acetic anhydride. Calculate the theoretical yield of aspirin and the percent yield if only 10.9 g of aspirin is produced.

3.130 Calculate the percent composition by mass of all the elements in calcium phosphate [$Ca_3(PO_4)_2$], a major component of bone.

3.131 Lysine, an essential amino acid in the human body, contains C, H, O, and N. In one experiment, the complete combustion of 2.175 g of lysine gave 3.94 g CO_2 and 1.89 g H_2O. In a separate experiment, 1.873 g of lysine gave 0.436 g NH_3. (a) Calculate the empirical formula of lysine. (b) The approximate molar mass of lysine is 150 g. What is the molecular formula of the compound?

3.132 Does 1 g of hydrogen molecules contain as many H atoms as 1 g of hydrogen atoms?

3.133 Avogadro's number has sometimes been described as a conversion factor between atomic mass units and grams. Use the fluorine atom (19.00 amu) as an example to show the relation between the atomic mass unit and the gram.

3.134 The natural abundances of the two stable isotopes of hydrogen (hydrogen and deuterium) are ${}^1_1H$: 99.985 percent and ${}^2_1H$: 0.015 percent. Assume that water exists as either H_2O or D_2O. Calculate the number of D_2O molecules in exactly 400 mL of water. (Density = 1.00 g/mL.)

3.135 A compound containing only C, H, and Cl was examined in a mass spectrometer. The highest mass peak seen corresponds to an ion mass of 52 amu. The most abundant mass peak seen corresponds to an ion mass of 50 amu and is about three times as intense as the peak at 52 amu. Deduce a reasonable molecular formula for the compound and explain the positions and intensities of the mass peaks mentioned. (*Hint:* Chlorine is the only element that has isotopes in comparable abundances: ${}^{35}_{17}Cl$: 75.5 percent; ${}^{35}_{17}Cl$: 24.5 percent. For H, use ${}^1_1H$; for C, use ${}^{12}_{1}C$.)

3.136 In the formation of carbon monoxide, CO, it is found that 2.445 g of carbon combine with 3.257 g of oxygen. What is the atomic mass of oxygen if the atomic mass of carbon is 12.01 amu?

3.137 What mole ratio of molecular chlorine (Cl_2) to molecular oxygen (O_2) would result from the breakup of the compound Cl_2O_7 into its constituent elements?

3.138 Which of the following substances contains the greatest mass of chlorine? (a) 5.0 g Cl_2, (b) 60.0 g $NaClO_3$, (c) 0.10 mol KCl, (d) 30.0 g $MgCl_2$, (e) 0.50 mol Cl_2.

3.139 A compound made up of C, H, and Cl contains 55.0 percent Cl by mass. If 9.00 g of the compound contain 4.19×10^{23} H atoms, what is the empirical formula of the compound?

3.140 Platinum forms two different compounds with chlorine. One contains 26.7 percent Cl by mass, and the other contains 42.1 percent Cl by mass. Determine the empirical formulas of the two compounds.

3.141 The following reaction is stoichiometric as written

$$C_4H_9Cl + NaOC_2H_5 \longrightarrow C_4H_8 + C_2H_5OH + NaCl$$

but it is often carried out with an excess of $NaOC_2H_5$ to react with any water present in the reaction mixture that might reduce the yield. If the reaction shown was carried out with 6.83 g of C_4H_9Cl, how many grams of $NaOC_2H_5$ would be needed to have a 50 percent molar excess of that reactant?

3.142 Compounds containing ruthenium(II) and bipyridine, $C_{10}H_8N_2$, have received considerable interest because of their role in systems that convert solar energy to electricity. The compound $[Ru(C_{10}H_8N_2)_3]Cl_2$ is synthesized by reacting $RuCl_3 \cdot 3H_2O(s)$ with three molar equivalents of $C_{10}H_8N_2(s)$, along with an excess of triethylamine, $N(C_2H_5)_3(l)$, to convert ruthenium(III) to ruthenium(II). The density of triethylamine is 0.73 g/mL, and typically eight molar equivalents are used in the synthesis. (a) Assuming that you start with 6.5 g of $RuCl_3 \cdot 3H_2O$, how many grams of $C_{10}H_8N_2$ and what volume of $N(C_2H_5)_3$ should be used in the reaction? (b) Given that the yield of this reaction is 91 percent, how many grams of $[Ru(C_{10}H_8N_2)_3]Cl_2$ will be obtained?

3.143 Heating 2.40 g of the oxide of metal X (molar mass of X = 55.9 g/mol) in carbon monoxide (CO) yields the pure metal and carbon dioxide. The mass of the metal product is 1.68 g. From the data given, show that the simplest formula of the oxide is X_2O_3 and write a balanced equation for the reaction.

3.144 A compound X contains 63.3 percent manganese (Mn) and 36.7 percent O by mass. When X is heated, oxygen gas is evolved and a new compound Y containing 72.0 percent Mn and 28.0 percent O is formed. (a) Determine the empirical formulas of X and Y. (b) Write a balanced equation for the conversion of X to Y.

3.145 The formula of a hydrate of barium chloride is $BaCl_2 \cdot xH_2O$. If 1.936 g of the compound gives 1.864 g of anhydrous $BaSO_4$ upon treatment with sulfuric acid, calculate the value of *x*.

3.146 It is estimated that the day Mt. St. Helens erupted (May 18, 1980), about 4.0×10^5 tons of SO_2 were released into the atmosphere. If all the SO_2 was eventually converted to sulfuric acid, how many tons of H_2SO_4 were produced?

3.147 Cysteine, shown here, is one of the 20 amino acids found in proteins in humans. Write the molecular formula and calculate its percent composition by mass.

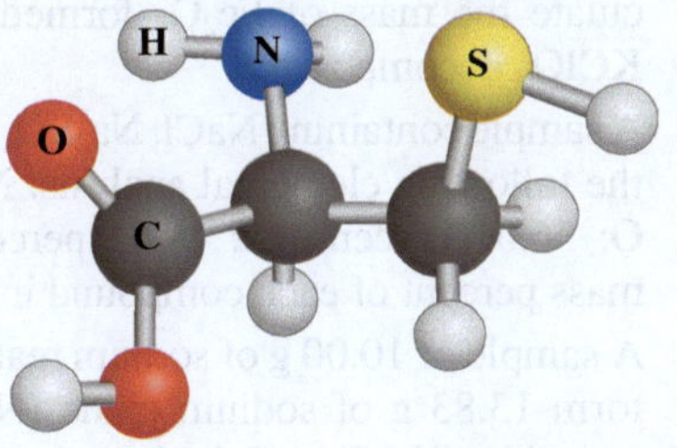

3.148 Isoflurane, shown here, is a common inhalation anesthetic. Write its molecular formula and calculate its percent composition by mass.

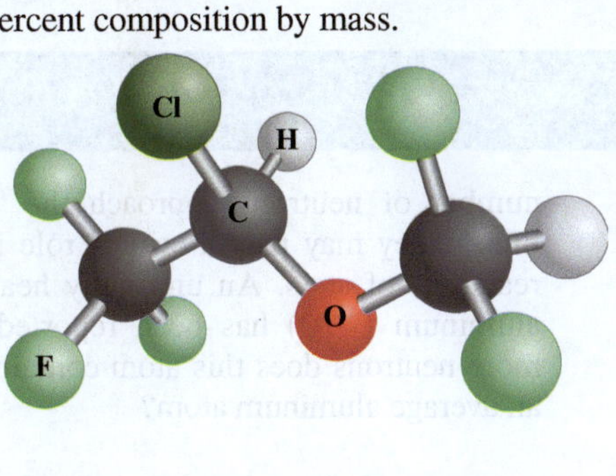

3.149 A mixture of $CuSO_4 \cdot 5H_2O$ and $MgSO_4 \cdot 7H_2O$ is heated until all the water is lost. If 5.020 g of the mixture give 2.988 g of the anhydrous salts, what is the percent by mass of $CuSO_4 \cdot 5H_2O$ in the mixture?

3.150 When 0.273 g of Mg is heated strongly in a nitrogen (N_2) atmosphere, a chemical reaction occurs. The product of the reaction weighs 0.378 g. Calculate the empirical formula of the compound containing Mg and N. Name the compound.

3.151 A mixture of methane (CH_4) and ethane (C_2H_6) of mass 13.43 g is completely burned in oxygen. If the total mass of CO_2 and H_2O produced is 64.84 g, calculate the fraction of CH_4 in the mixture.

3.152 Leaded gasoline contains an additive to prevent engine "knocking." On analysis, the additive compound is found to contain carbon, hydrogen, and lead (Pb) (hence, "leaded gasoline"). When 51.36 g of this compound are burned in an apparatus such as that shown in Figure 3.6, 55.90 g of CO_2 and 28.61 g of H_2O are produced. Determine the empirical formula of the gasoline additive.

3.153 Because of its detrimental effect on the environment, the lead compound described in Problem 3.148 has been replaced by methyl *tert*-butyl ether (a compound of C, H, and O) to enhance the performance of gasoline. (This compound is also being phased out because of its contamination of drinking water.) When 12.1 g of the compound are burned in an apparatus like the one shown in Figure 3.6, 30.2 g of CO_2 and 14.8 g of H_2O are formed. What is the empirical formula of the compound?

3.154 Suppose you are given a cube made of magnesium (Mg) metal of edge length 1.0 cm. (a) Calculate the number of Mg atoms in the cube. (b) Atoms are spherical in shape. Therefore, the Mg atoms in the cube cannot fill all of the available space. If only 74 percent of the space inside the cube is taken up by Mg atoms, calculate the radius in picometers of a Mg atom. (The density of Mg is 1.74 g/cm^3 and the volume of a sphere of radius *r* is $\frac{4}{3}\pi r^3$.)

3.155 A certain sample of coal contains 1.6 percent sulfur by mass. When the coal is burned, the sulfur is converted to sulfur dioxide. To prevent air pollution, this sulfur dioxide is treated with calcium oxide (CaO) to form calcium sulfite ($CaSO_3$). Calculate the daily mass (in kilograms) of CaO needed by a power plant that uses 6.60×10^6 kg of coal per day.

3.156 Air is a mixture of many gases. However, in calculating its "molar mass" we need consider only the three major components: nitrogen, oxygen, and argon. Given that 1 mole of air at sea level is made up of 78.08 percent nitrogen, 20.95 percent oxygen, and 0.97 percent argon, what is the molar mass of air?

3.157 (a) Determine the mass of calcium metal that contains the same number of moles as 89.6 g of zinc metal. (b) Calculate the number of moles of molecular fluorine that has the same mass as 36.9 moles of argon. (c) What is the mass of sulfuric acid that contains 0.56 mole of oxygen atoms? (d) Determine the number of moles of phosphoric acid that contains 2.12 g of hydrogen atoms.

3.158 A major industrial use of hydrochloric acid is in metal pickling. This process involves the removal of metal oxide layers from metal surfaces to prepare them for coating. (a) Write an equation between iron(III) oxide, which represents the rust layer over iron, and HCl to form iron(III) chloride and water. (b) If 1.22 moles of Fe_2O_3 and 289.2 g of HCl react, how many grams of $FeCl_3$ will be produced?

3.159 Octane (C_8H_{18}) is a component of gasoline. Complete combustion of octane yields H_2O and CO_2. Incomplete combustion produces H_2O and CO, which not only reduces the efficiency of the engine using the fuel but is also toxic. In a certain test run, 1.000 gal of octane is burned in an engine. The total mass of CO, CO_2, and H_2O produced is 11.53 kg. Calculate the efficiency of the process; that is, calculate the fraction of octane converted to CO_2. The density of octane is 2.650 kg/gal.

3.160 Industrially, hydrogen gas can be prepared by reacting propane gas (C_3H_8) with steam at about 400°C. The products are carbon monoxide (CO) and hydrogen gas (H_2). (a) Write a balanced equation for the reaction. (b) How many kilograms of H_2 can be obtained from 2.84×10^3 kg of propane?

3.161 In a natural product synthesis, a chemist prepares a complex biological molecule entirely from nonbiological starting materials. The target molecules are often known to have some promise as therapeutic agents, and the organic reactions that are developed along the way benefit all chemists. The overall synthesis, however, requires many steps, so it is important to have the best possible percent yields at each step. What is the overall percent yield for such a synthesis that has 24 steps with an 80 percent yield at each step?

3.162 What is wrong or ambiguous with each of the statements here?

(a) NH_4NO_2 is the limiting reactant in the reaction

$$NH_4NO_2(s) \longrightarrow N_2(g) + 2H_2O(l)$$

(b) The limiting reactants for the reaction shown here are NH_3 and NaCl.

$$NH_3(aq) + NaCl(aq) + H_2CO_3(aq) \longrightarrow NaHCO_3(aq) + NH_4Cl(aq)$$

3.163 (a) For molecules having small molecular masses, mass spectrometry can be used to identify their formulas. To illustrate this point, identify the molecule that most likely accounts for the observation of a peak in a mass spectrum at 16 amu, 17 amu, 18 amu, and 64 amu. (b) Note that there are (among others) two likely molecules that would give rise to a peak at 44 amu, namely, C_3H_8 and CO_2. In such cases, a chemist might try to look for other peaks generated when some of the molecules break apart in the spectrometer. For example, if a chemist sees a peak at 44 amu and also one at 15 amu, which molecule is producing the 44-amu peak? Why? (c) Using the following precise atomic masses—^{1}H (1.00797 amu), ^{12}C (12.00000 amu), and ^{16}O (15.99491 amu)—how precisely must the masses of C_3H_8 and CO_2 be measured to distinguish between them?

3.164 Potash is any potassium mineral that is used for its potassium content. Most of the potash produced in the United States goes into fertilizer. The major sources of potash are potassium chloride (KCl) and potassium sulfate (K_2SO_4). Potash production is often reported as the potassium oxide (K_2O) equivalent or the amount of K_2O that could be made from a given mineral. (a) If KCl costs \$0.55 per kg, for what price (dollar per kg) must K_2SO_4 be sold to supply the same amount of potassium on a per dollar basis? (b) What mass (in kg) of K_2O contains the same number of moles of K atoms as 1.00 kg of KCl?

3.165 A 21.496-g sample of magnesium is burned in air to form magnesium oxide and magnesium nitride. When the products are treated with water, 2.813 g of gaseous ammonia are generated. Calculate the amounts of magnesium nitride and magnesium oxide formed.

3.166 A certain metal M forms a bromide containing 53.79 percent Br by mass. What is the chemical formula of the compound?

3.167 A sample of iron weighing 15.0 g was heated with potassium chlorate ($KClO_3$) in an evacuated container. The oxygen generated from the decomposition of $KClO_3$ converted some of the Fe to Fe_2O_3. If the combined mass of Fe and Fe_2O_3 was 17.9 g, calculate the mass of Fe_2O_3 formed and the mass of $KClO_3$ decomposed.

3.168 A sample containing NaCl, Na_2SO_4, and $NaNO_3$ gives the following elemental analysis: Na: 32.08 percent; O: 36.01 percent; Cl: 19.51 percent. Calculate the mass percent of each compound in the sample.

3.169 A sample of 10.00 g of sodium reacts with oxygen to form 13.83 g of sodium oxide (Na_2O) and sodium peroxide (Na_2O_2). Calculate the percent composition of the mixture.

Interpreting, Modeling, & Estimating

3.170 While most isotopes of light elements such as oxygen and phosphorus contain relatively equal numbers of protons and neutrons, recent results indicate that a new class of isotopes called neutron-rich isotopes can be prepared. These neutron-rich isotopes push the limits of nuclear stability as the large number of neutrons approach the "neutron drip line." They may play a critical role in the nuclear reactions of stars. An unusually heavy isotope of aluminum ($^{43}_{13}$Al) has been reported. How many more neutrons does this atom contain compared to an average aluminum atom?

3.171 Without doing any detailed calculations, arrange the following substances in the increasing order of number of moles: 20.0 g Cl, 35.0 g Br, and 94.0 g I.

3.172 Without doing any detailed calculations, estimate which element has the highest percent composition by mass in each of the following compounds:
(a) $Hg(NO_3)_2$
(b) NF_3
(c) $K_2Cr_2O_7$
(d) $C_{2952}H_{4664}N_{812}O_{832}S_8Fe_4$

3.173 Consider the reaction

$$6Li(s) + N_2(g) \longrightarrow 2Li_3N(s)$$

Without doing any detailed calculations, choose one of the following combinations in which nitrogen is the limiting reactant:
(a) 44 g Li and 38 g N_2
(b) 1380 g Li and 842 g N_2
(c) 1.1 g Li and 0.81 g N_2

3.174 Estimate how high in miles you can stack up an Avogadro's number of oranges covering the entire Earth.

3.175 The following is a crude but effective method for estimating the *order of magnitude* of Avogadro's number using stearic acid ($C_{18}H_{36}O_2$) shown here. When stearic acid is added to water, its molecules collect at the surface and form a monolayer; that is, the layer is only one molecule thick. The cross-sectional area of each stearic acid molecule has been measured to be 0.21 nm^2. In one experiment it is found that 1.4×10^{-4} g of stearic acid is needed to form a monolayer over water in a dish of diameter 20 cm. Based on these measurements, what is Avogadro's number?

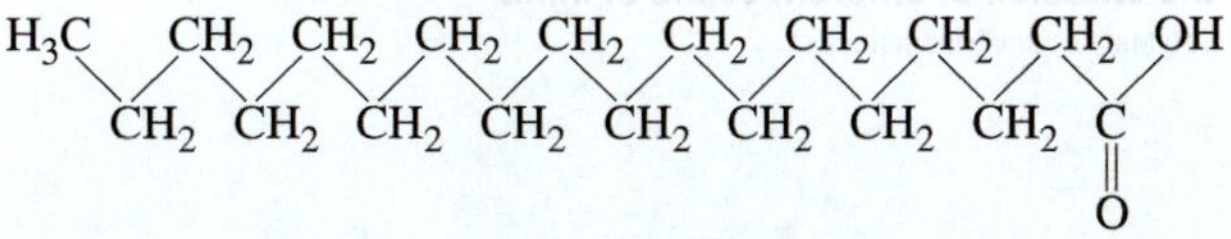

Answers to Practice Exercises

3.1 63.55 amu. **3.2** 3.59 moles. **3.3** 2.57×10^3 g. **3.4** 1.0×10^{-20} g. **3.5** 32.04 amu. **3.6** 1.66 moles. **3.7** 5.81×10^{24} H atoms. **3.8** H: 2.055%; S: 32.69%; O: 65.25%. **3.9** $KMnO_4$ (potassium permanganate). **3.10** 196 g. **3.11** B_2H_6. **3.12** $Fe_2O_3 + 3CO \rightarrow 2Fe + 3CO_2$. **3.13** 235 g. **3.14** 0.769 g. **3.15** (a) 234 g, (b) 234 g. **3.16** 25.1 g. **3.17** (a) 863 g, (b) 93.0%. **3.18** 44.38%.

Answers to Review of Concepts & Facts

3.1.1 ^{193}Ir. **3.1.2** 49.473 amu. **3.2.1** (b). **3.2.2** 6.23 moles. **3.2.3** 85.6 g. **3.3.1** Molecular mass = 192.12 amu, molar mass = 192.12 g. **3.3.2** 74.2 g. **3.3.3** 1.93×10^{24} O atoms. **3.4.1** When isotopes of the two chlorine atoms arrive at the detector of a mass spectrometer, a current is registered for each type of ion. The amount of current generated is directly proportional to the number of ions, so it enables us to determine the relative abundance of each isotope. A weighted average of the masses of the two isotopes based on relative abundance gives the average mass of chlorine. **3.5.1** The percent composition by mass of Sr is smaller than that of O. You need only to compare the relative masses of one Sr atom and six O atoms. **3.5.2** C: 41.39%; H: 3.47%; O: 55.14%. **3.5.3** P_2O_5. **3.6.1** C_5H_{10}. **3.7.1** Essential part: The number of each type of atom on both sides of the reaction arrow. Helpful part: The physical states of the reactants and products. **3.7.2** $2K_2S_2O_3 + I_2 \rightarrow K_2S_4O_6 + 2KI$. **3.8.1** (b). **3.8.2** 127 g. **3.8.3** 18.1 g. **3.9.1** (a) 24.7 g, (b) 2.50 g Al. **3.9.2** Diagram (d) shows that NO is the limiting reactant. **3.10.1** No. **3.10.2** 88.3%. **3.10.3** 45.83%.

CHAPTER

4

Reactions in Aqueous Solutions

By varying the size and concentration of quantum dots in solution, chemists can tune the emission of different colors of light.
Leo Matyushkin/Shutterstock

CHAPTER OUTLINE

Many chemical reactions and virtually all biological processes take place in water. In this chapter, we will discuss three major categories of reactions that occur in aqueous solutions: precipitation reactions, acid-base reactions, and redox reactions. In later chapters, we will study the structural characteristics and properties of water—the so-called *universal solvent*—and its solutions.

4.1 General Properties of Aqueous Solutions

Learning Objective

- Evaluate substances in aqueous solutions as nonelectrolytes, weak electrolytes, or strong electrolytes.

A ***solution*** is *a homogeneous mixture of two or more substances.* The ***solute*** is *the substance present in a smaller amount,* and the ***solvent*** is *the substance present in a larger amount.* A solution may be gaseous (such as air), solid (such as an alloy), or liquid (seawater, for example). In this section, we will discuss only ***aqueous solutions,*** in which *the solute initially is a liquid or a solid and the solvent is water.*

Electrolytic Properties

All solutes that dissolve in water fit into one of two categories: electrolytes and nonelectrolytes. An ***electrolyte*** is *a substance that, when dissolved in water, results in a solution that can conduct electricity.* A ***nonelectrolyte*** *does not conduct electricity when dissolved in water.* Figure 4.1 shows an easy and straightforward method of distinguishing between electrolytes and nonelectrolytes. A pair of inert electrodes (copper or platinum) is immersed in a beaker of water. To light the bulb, electric current must flow from one electrode to the other, thus completing the circuit. Pure water is a very poor conductor of electricity (though tap water does conduct electricity because it contains many dissolved ions). However, if we add a small amount of sodium chloride (NaCl), the bulb will glow as soon as the salt dissolves in the water. Solid NaCl, an ionic compound, breaks up into Na^+ and Cl^- ions when it dissolves in water. The Na^+ ions are attracted to the negative electrode and the Cl^- ions to the positive electrode. The movement of these ions allows the solution to conduct electricity. Because the NaCl solution conducts electricity, we say that NaCl is an electrolyte. Pure water contains very few ions, so it cannot conduct electricity.

Comparing the lightbulb's brightness for the same molar amounts of dissolved substances helps us distinguish between strong and weak electrolytes. A characteristic of strong electrolytes is that the solute is assumed to be 100 percent dissociated into

Video:
Strong Electrolytes, Weak Electrolytes, and Nonelectrolytes

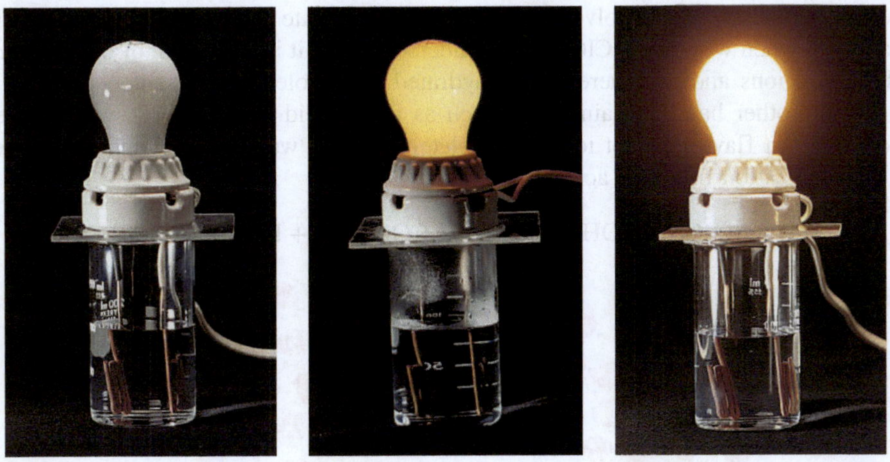

(a) (b) (c)

Figure 4.1 *An arrangement for distinguishing between electrolytes and nonelectrolytes. A solution's ability to conduct electricity depends on the number of ions it contains. (a) A nonelectrolyte solution does not contain ions, and the lightbulb is not lit. (b) A weak electrolyte solution contains a small number of ions, and the lightbulb is dimly lit. (c) A strong electrolyte solution contains a large number of ions, and the lightbulb is brightly lit. The molar amounts of the dissolved solutes are equal in all three cases.*

(a, b, c): Stephen Frisch/McGraw Hill

Table 4.1 Classification of Solutes in Aqueous Solution

Strong Electrolyte	Weak Electrolyte	Nonelectrolyte
HCl	CH_3COOH	$(NH_2)_2CO$ (urea)
HNO_3	HF	CH_3OH (methanol)
$HClO_4$	HNO_2	C_2H_5OH (ethanol)
$H_2SO_4^*$	NH_3	$C_6H_{12}O_6$ (glucose)
NaOH	$H_2O^\dagger$	$C_{12}H_{22}O_{11}$ (sucrose)
$Ba(OH)_2$		
Ionic compounds		

*H_2SO_4 has two ionizable H^+ ions, but only one of the H^+ ions is totally ionized.
†Pure water is an extremely weak electrolyte as will be shown in Chapter 15.

ions in solution. (By *dissociation* we mean the breaking up of the compound into cations and anions.) Thus, we can represent sodium chloride dissolving in water as

$$NaCl(s) \xrightarrow{H_2O} Na^+(aq) + Cl^-(aq)$$

This equation says that all sodium chloride that enters the solution ends up as Na^+ and Cl^- ions; there are no undissociated NaCl units in solution.

Table 4.1 lists examples of strong electrolytes, weak electrolytes, and nonelectrolytes. Ionic compounds, such as sodium chloride, potassium iodide (KI), and calcium nitrate [$Ca(NO_3)_2$], are strong electrolytes. It is interesting to note that human body fluids contain many strong and weak electrolytes.

Water is a very effective solvent for ionic compounds. Although water is an electrically neutral molecule, it has a positive region (the H atoms) and a negative region (the O atom), or positive and negative "poles"; for this reason, it is a *polar* solvent. When an ionic compound such as sodium chloride dissolves in water, the three-dimensional lattice of ions in the solid is broken apart. The Na^+ and Cl^- ions are separated from each other and undergo ***hydration,*** *the process in which an ion is surrounded by water molecules.* Each Na^+ ion is surrounded by a number of water molecules orienting their negative poles toward the cation. Similarly, each Cl^- ion is surrounded by water molecules with their positive poles oriented toward the anion (Figure 4.2). Hydration helps to stabilize ions in solution and prevents cations from recombining with anions.

Video: Hydration

Acids and bases are also electrolytes. Some acids, including hydrochloric acid (HCl) and nitric acid (HNO_3), are strong electrolytes. These acids are assumed to ionize completely in water; for example, when hydrogen chloride gas dissolves in water, it produces hydrated H^+ and Cl^- ions:

$$HCl(g) \xrightarrow{H_2O} H^+(aq) + Cl^-(aq)$$

In other words, *all* the dissolved HCl molecules separate into hydrated H^+ and Cl^- ions. Thus, when we write HCl(*aq*), it is understood that it is a solution of only $H^+(aq)$ and $Cl^-(aq)$ ions and that there are no hydrated HCl molecules present.

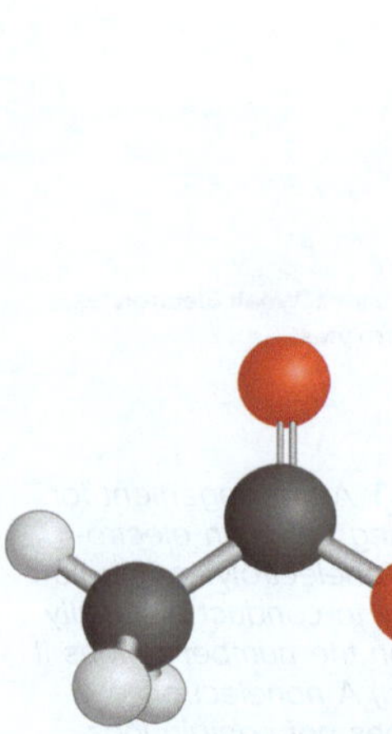

CH_3COOH

On the other hand, certain acids, such as acetic acid (CH_3COOH), which gives vinegar its tart flavor, do not ionize completely and are weak electrolytes. We represent the ionization of acetic acid as

$$CH_3COOH(aq) \rightleftharpoons CH_3COO^-(aq) + H^+(aq)$$

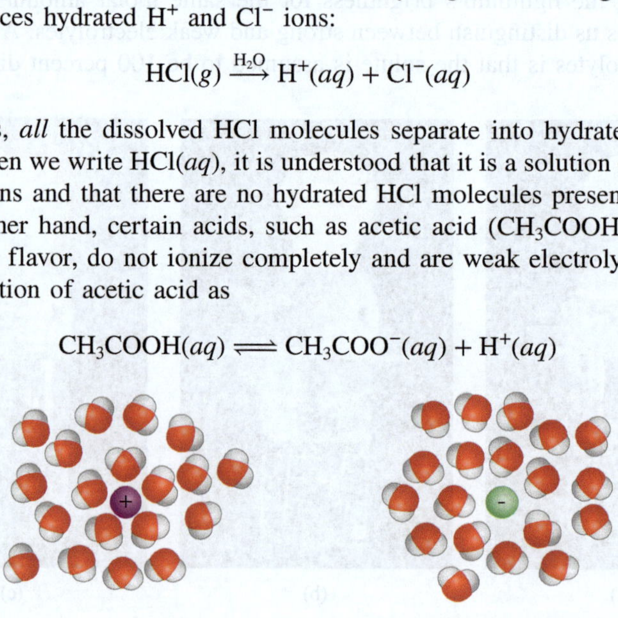

Figure 4.2 *Hydration of Na^+ and Cl^- ions.*

where CH_3COO^- is called the acetate ion. We use the term *ionization* to describe the separation of acids and bases into ions. By writing the formula of acetic acid as CH_3COOH, we indicate that the ionizable proton is in the COOH group.

The ionization of acetic acid is written with a double arrow to show that it is a ***reversible reaction,*** meaning *the reaction can occur in both directions.* Initially, a number of CH_3COOH molecules break up into CH_3COO^- and H^+ ions. As time goes on, some of the CH_3COO^- and H^+ ions recombine into CH_3COOH molecules. Eventually, a state is reached in which the acid molecules ionize as fast as the ions recombine. Such a chemical state, in which no net change can be observed (although activity is continuous on the molecular level), is called *chemical equilibrium.* Acetic acid, then, is a weak electrolyte because its ionization in water is incomplete. By contrast, in a hydrochloric acid solution the H^+ and Cl^- ions have no tendency to recombine and form molecular HCl. We use a single arrow to represent complete ionizations.

Summary of Concepts & Facts

- Aqueous solutions are electrically conducting if the solutes are electrolytes. If the solutes are nonelectrolytes, the solutions do not conduct electricity.

Review of Concepts & Facts

4.1.1 The diagrams (a)–(c) show three compounds, (a) AB_2, (b) AC_2, and (c) AD_2, dissolved in water. Which is the strongest electrolyte and which is the weakest? (For simplicity, water molecules are not shown.)

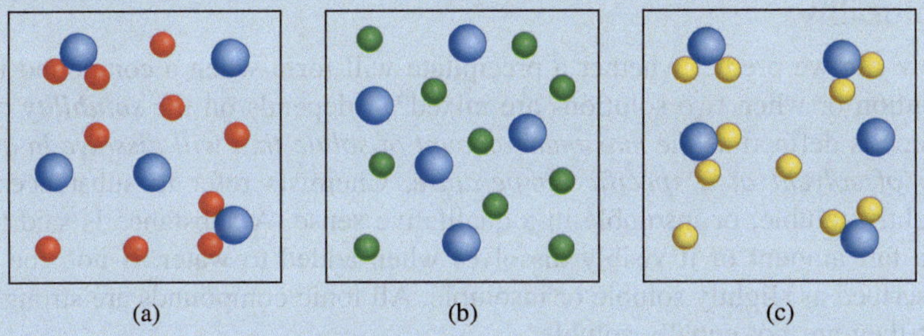

4.1.2 Predict whether the following compounds are a strong electrolyte, weak electrolyte, or nonelectrolyte: (a) KBr, (b) NaI, (c) C_3H_7OH.

4.2 Precipitation Reactions

Learning Objectives

- Apply solubility guidelines toward determining whether a reaction will produce a precipitate.
- Compose the molecular, ionic, and net ionic equations for a reaction.
- Identify the spectator ions in a reaction.

One common type of reaction that occurs in aqueous solution is the ***precipitation reaction,*** which *results in the formation of an insoluble product, known as a precipitate.* A ***precipitate*** is *an insoluble solid that separates from the solution.* Precipitation reactions usually involve ionic compounds. For example, when an aqueous solution of lead(II) nitrate [$Pb(NO_3)_2$] is added to an aqueous solution of potassium iodide (KI), a yellow precipitate of lead(II) iodide (PbI_2) is formed:

$$Pb(NO_3)_2(aq) + 2KI(aq) \longrightarrow PbI_2(s) + 2KNO_3(aq)$$

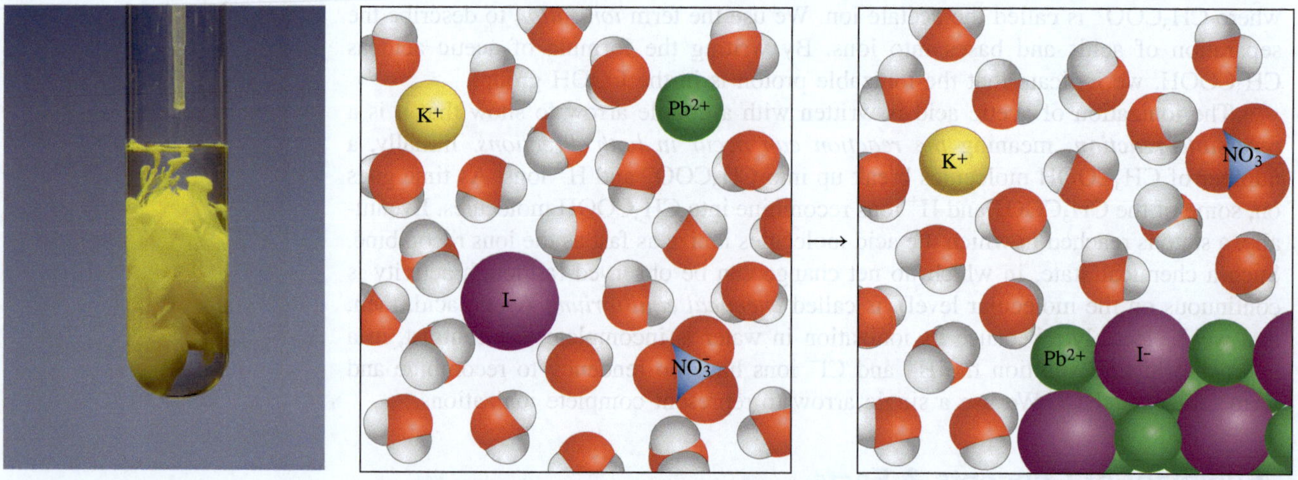

Figure 4.3 *Formation of yellow* PbI_2 *precipitate as a solution of* $Pb(NO_3)_2$ *is added to a solution of KI.*
(left): Charles D. Winters/McGraw Hill

Potassium nitrate remains in solution. Figure 4.3 shows this reaction in progress.

The preceding reaction is an example of a ***metathesis reaction*** (also called a double-displacement reaction), *a reaction that involves the exchange of parts between two compounds.* (In this case, the cations in the two compounds exchange anions, so Pb^{2+} ends up with I^- as PbI_2 and K^+ end up with NO_3^- as KNO_3.) As we will see, the precipitation reactions discussed in this chapter are examples of metathesis reactions.

Solubility

How can we predict whether a precipitate will form when a compound is added to a solution or when two solutions are mixed? It depends on the ***solubility*** of the solute, which is defined as *the maximum amount of solute that will dissolve in a given quantity of solvent at a specific temperature.* Chemists refer to substances as soluble, slightly soluble, or insoluble in a qualitative sense. A substance is said to be soluble if a fair amount of it visibly dissolves when added to water. If not, the substance is described as slightly soluble or insoluble. All ionic compounds are strong electrolytes, but they are not equally soluble.

Table 4.2 classifies a number of common ionic compounds as soluble or insoluble. Keep in mind, however, that even insoluble compounds dissolve to a certain extent. Figure 4.4 shows several precipitates.

Table 4.2 Solubility Rules for Common Ionic Compounds in Water at 25°C

Soluble Compounds	**Insoluble Exceptions**
Compounds containing alkali metal ions (Li^+, Na^+, K^+, Rb^+, Cs^+) and the ammonium ion (NH_4^+)	
Nitrates (NO_3^-), acetates (CH_3COO^-), bicarbonates (HCO_3^-), chlorates (ClO_3^-), and perchlorates (ClO_4^-)	
Halides (Cl^-, Br^-, I^-)	Halides of Ag^+, Hg_2^{2+}, and Pb^{2+}
Sulfates (SO_4^{2-})	Sulfates of Ag^+, Ca^{2+}, Sr^{2+}, Ba^{2+}, Hg_2^{2+}, and Pb^{2+}
Insoluble Compounds	**Soluble Exceptions**
Carbonates (CO_3^{2-}), phosphates (PO_4^{3-}), chromates (CrO_4^{2-}), sulfides (S^{2-})	Compounds containing alkali metal ions and the ammonium ion
Hydroxides (OH^-)	Compounds containing alkali metal ions and the Ba^{2+} ion

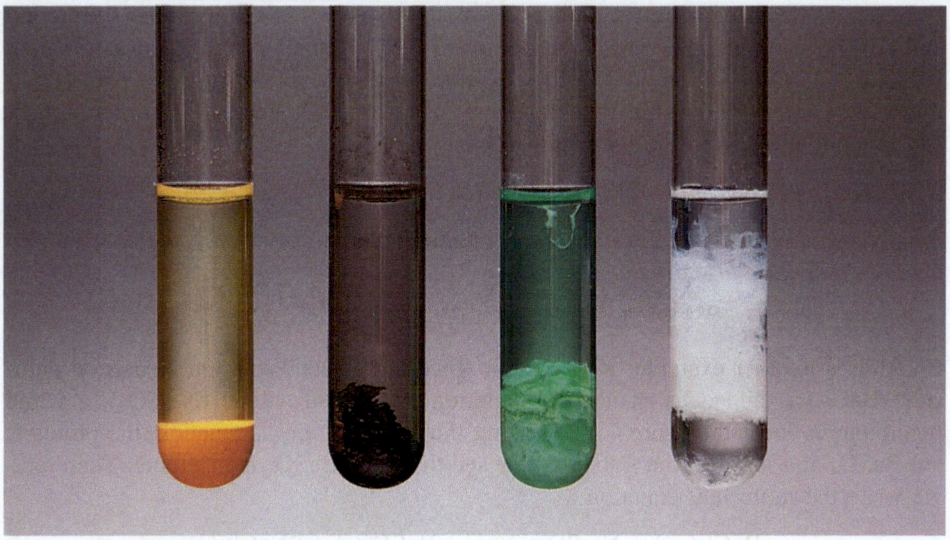

Figure 4.4 *Appearance of several precipitates. From left to right: CdS, PbS, Ni(OH)$_2$, and Al(OH)$_3$.*
Ken Karp/McGraw Hill

Example 4.1 applies the solubility rules from Table 4.2.

Example 4.1

Classify the following ionic compounds as soluble or insoluble: (a) silver sulfate (Ag_2SO_4), (b) calcium carbonate ($CaCO_3$), (c) sodium phosphate (Na_3PO_4).

Strategy Although it is not necessary to memorize the solubilities of compounds, you should keep in mind the following useful rules: All ionic compounds containing alkali metal cations; the ammonium ion; and the nitrate, bicarbonate, and chlorate ions are soluble. For other compounds, we need to refer to Table 4.2.

Solution

(a) According to Table 4.2, Ag_2SO_4 is insoluble.

(b) This is a carbonate and Ca is a Group 2 metal. Therefore, $CaCO_3$ is insoluble.

(c) Sodium is an alkali metal (Group 1) so Na_3PO_4 is soluble.

Practice Exercise Classify the following ionic compounds as soluble or insoluble: (a) CuS, (b) $Ca(OH)_2$, (c) $Zn(NO_3)_2$.

Similar problems: 4.19, 4.20.

Molecular Equations, Ionic Equations, and Net Ionic Equations

The earlier equation describing the precipitation of lead(II) iodide is called a ***molecular equation*** because *the formulas of the compounds are written as though all species existed as molecules or whole units.* A molecular equation is useful because it identifies the reagents [that is, lead(II) nitrate and potassium iodide]. If we wanted to bring about this reaction in the laboratory, we would use the molecular equation. However, a molecular equation does not describe in detail what actually is happening in solution.

As pointed out earlier, when ionic compounds dissolve in water, they break apart into their component cations and anions. To be more realistic, the equations should show the dissociation of dissolved ionic compounds into ions. Therefore, returning to the reaction between potassium iodide and lead(II) nitrate, we would write

$$Pb^{2+}(aq) + 2NO_3^-(aq) + 2K^+(aq) + 2I^-(aq) \longrightarrow PbI_2(s) + 2K^+(aq) + 2NO_3^-(aq)$$

The preceding equation is an example of an ***ionic equation,*** which *shows dissolved species as free, hydrated ions.* To see whether a precipitate might form from this solution, we first combine the cation and anion from different compounds; that is,

Figure 4.5 *Formation of $BaSO_4$ precipitate.*
Charles D. Winters/McGraw Hill

PbI_2 and KNO_3. Referring to Table 4.2, we see that PbI_2 is an insoluble compound and KNO_3 is soluble. Therefore, the dissolved KNO_3 remains in solution as separate K^+ and NO_3^- ions, which are called ***spectator ions,*** or *ions that are not involved in the overall reaction.* Because spectator ions appear on both sides of an equation, they can be eliminated from the ionic equation

$$Pb^{2+}(aq) + \cancel{2NO_3^-(aq)} + \cancel{2K^+(aq)} + 2I^-(aq) \longrightarrow PbI_2(s) + \cancel{2K^+(aq)} + \cancel{2NO_3^-(aq)}$$

Finally, we end up with the ***net ionic equation,*** which *shows only the species that actually take part in the reaction:*

$$Pb^{2+}(aq) + 2I^-(aq) \longrightarrow PbI_2(s)$$

Looking at another example, we find that when an aqueous solution of barium chloride ($BaCl_2$) is added to an aqueous solution of sodium sulfate (Na_2SO_4), a white precipitate is formed (Figure 4.5). Treating this as a metathesis reaction, the products are $BaSO_4$ and NaCl. From Table 4.2 we see that only $BaSO_4$ is insoluble. Therefore, we write the molecular equation as

$$BaCl_2(aq) + Na_2SO_4(aq) \longrightarrow BaSO_4(s) + 2NaCl(aq)$$

The ionic equation for the reaction is

$$Ba^{2+}(aq) + 2Cl^-(aq) + 2Na^+(aq) + SO_4^{2-}(aq) \longrightarrow BaSO_4(s) + 2Na^+(aq) + 2Cl^-(aq)$$

Canceling the spectator ions (Na^+ and Cl^-) on both sides of the equation gives us the net ionic equation

$$Ba^{2+}(aq) + SO_4^{2-}(aq) \longrightarrow BaSO_4(s)$$

The following four steps summarize the procedure for writing ionic and net ionic equations.

1. Write a balanced molecular equation for the reaction, using the correct formulas for the reactant and product ionic compounds. Refer to Table 4.2 to decide which of the products is insoluble and therefore will appear as a precipitate.
2. Write the ionic equation for the reaction. The compound that does not appear as the precipitate should be shown as free ions.
3. Identify and cancel the spectator ions on both sides of the equation. Write the net ionic equation for the reaction.
4. Check that the charges and number of atoms balance in the net ionic equation.

These steps are applied in Example 4.2.

Student Hot Spot

Student data indicate you may struggle with precipitation reactions. Access your eBook for additional Learning Resources on this topic.

Example 4.2

Predict what happens when a potassium phosphate (K_3PO_4) solution is mixed with a calcium nitrate [$Ca(NO_3)_2$] solution. Write a net ionic equation for the reaction.

Strategy From the given information, it is useful to first write the unbalanced equation

$$K_3PO_4(aq) + Ca(NO_3)_2(aq) \longrightarrow ?$$

What happens when ionic compounds dissolve in water? What ions are formed from the dissociation of K_3PO_4 and $Ca(NO_3)_2$? What happens when the cations encounter the anions in solution?

Solution In solution, K_3PO_4 dissociates into K^+ and PO_4^{3-} ions and $Ca(NO_3)_2$ dissociates into Ca^{2+} and NO_3^- ions. According to Table 4.2, calcium ions (Ca^{2+}) and phosphate ions (PO_4^{3-}) will form an insoluble compound, calcium phosphate [$Ca_3(PO_4)_2$], while the other product, KNO_3, is soluble and remains in solution. Therefore, this is a precipitation reaction. We follow the stepwise procedure just outlined.

(Continued)

CHEMISTRY *in Action*

An Undesirable Precipitation Reaction

Limestone ($CaCO_3$) and dolomite ($CaCO_3 \cdot MgCO_3$), which are widespread on Earth's surface, often enter the water supply. According to Table 4.2, calcium carbonate is insoluble in water. However, in the presence of dissolved carbon dioxide (from the atmosphere), calcium carbonate is converted to soluble calcium bicarbonate [$Ca(HCO_3)_2$]:

$$CaCO_3(s) + CO_2(aq) + H_2O(l) \longrightarrow Ca^{2+}(aq) + 2HCO_3^-(aq)$$

where HCO_3^- is the bicarbonate ion.

Water containing Ca^{2+} and/or Mg^{2+} ions is called *hard water,* and water that is mostly free of these ions is called *soft water.* Hard water is unsuitable for some household and industrial uses.

When water containing Ca^{2+} and HCO_3^- ions is heated or boiled, the solution reaction is reversed to produce the $CaCO_3$ precipitate

$$Ca^{2+}(aq) + 2HCO_3^-(aq) \longrightarrow CaCO_3(s) + CO_2(aq) + H_2O(l)$$

and gaseous carbon dioxide is driven off:

$$CO_2(aq) \longrightarrow CO_2(g)$$

Solid calcium carbonate formed in this way is the main component of the scale that accumulates in boilers, water heaters, pipes, and tea kettles. A thick layer of scale reduces heat transfer and decreases the efficiency and durability of boilers, pipes, and appliances. In household hot-water pipes it can restrict or totally block the flow of water. A simple method used by plumbers to remove scale deposits is to introduce a small amount of hydrochloric acid, which reacts with (and therefore dissolves) $CaCO_3$:

$$CaCO_3(s) + 2HCl(aq) \longrightarrow CaCl_2(aq) + H_2O(l) + CO(g)$$

In this way, $CaCO_3$ is converted to soluble $CaCl_2$.

Boiler scale almost fills this hot-water pipe. The deposits consist mostly of $CaCO_3$ with some $MgCO_3$.
Paradorn Kotan/Shutterstock

Step 1: The balanced molecular equation for this reaction is

$$2K_3PO_4(aq) + 3Ca(NO_3)_2(aq) \longrightarrow Ca_3(PO_4)_2(s) + 6KNO_3(aq)$$

Step 2: To write the ionic equation, the soluble compounds are shown as dissociated ions:

$$6K^+(aq) + 2PO_4^{3-}(aq) + 3Ca^{2+}(aq) + 6NO_3^-(aq) \longrightarrow 6K^+(aq) + 6NO_3^-(aq) + Ca_3(PO_4)_2(s)$$

Step 3: Canceling the spectator ions (K^+ and NO_3^-) on each side of the equation, we obtain the net ionic equation:

$$3Ca^{2+}(aq) + 2PO_4^{3-}(aq) \longrightarrow Ca_3(PO_4)_2(s)$$

Step 4: Note that because we balanced the molecular equation first, the net ionic equation is balanced as to the number of atoms on each side, and the number of positive (+6) and negative (−6) charges on the left-hand side is the same.

Practice Exercise Predict the precipitate produced by mixing an $Al(NO_3)_3$ solution with a NaOH solution. Write the net ionic equation for the reaction.

Similar problems: 4.21, 4.22.

A precipitate formed by the reaction between $K_3PO_4(aq)$ and $Ca(NO_3)_2(aq)$.
Ken Karp/McGraw Hill

The Chemistry in Action essay, "An Undesirable Precipitation Reaction," discusses some practical problems associated with precipitation reactions.

Summary of Concepts & Facts

- From general rules about the solubilities of ionic compounds, we can predict whether a precipitate will form in a reaction.

Review of Concepts & Facts

4.2.1 Which of the diagrams (a)–(c) accurately describes the reaction between $Ca(NO_3)_2(aq)$ and $Na_2CO_3(aq)$? For simplicity, only the Ca^{2+} (green) and CO_3^{2-} (gray) ions are shown.

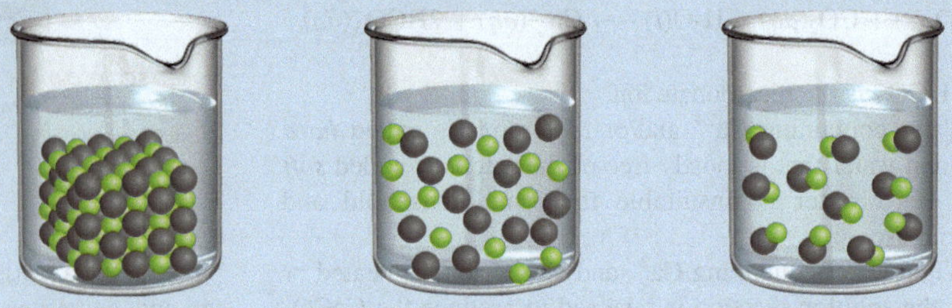

(a) (b) (c)

4.2.2 Classify each of the following compounds as soluble or insoluble in water: (a) $Fe(NO_3)_3$, (b) AgBr, (c) NH_4Cl.

4.2.3 Write the net ionic equation for the precipitation reaction between $Cu(NO_3)_2$ and $(NH_4)_2S$.

4.3 Acid-Base Reactions

Learning Objectives

- Compare the definitions of Brønsted acids and bases.
- Classify acids and bases as weak or strong.
- Categorize an acid as monoprotic, diprotic, or triprotic.
- Summarize a neutralization reaction between an acid and a base.

Acids and bases are as familiar as aspirin and milk of magnesia although many people do not know their chemical names—acetylsalicylic acid (aspirin) and magnesium hydroxide (milk of magnesia). In addition to being the basis of many medicinal and household products, acid-base chemistry is important in industrial processes and essential in sustaining biological systems. Before we can discuss acid-base reactions, we need to know more about acids and bases themselves.

General Properties of Acids and Bases

In Section 2.7 we defined acids as substances that ionize in water to produce H^+ ions and bases as substances that ionize in water to produce OH^- ions. These definitions were formulated in the late nineteenth century by the Swedish chemist Svante Arrhenius[†] to classify substances whose properties in aqueous solutions were well known.

[†]Svante August Arrhenius (1859–1927). Swedish chemist. Arrhenius made important contributions in the study of chemical kinetics and electrolyte solutions. He also speculated that life had come to Earth from other planets, a theory now known as *panspermia.* Arrhenius was awarded the Nobel Prize in Chemistry in 1903.

Acids

- Acids have a sour taste; for example, vinegar owes its sourness to acetic acid, and lemons and other citrus fruits contain citric acid.
- Acids cause color changes in plant dyes; for example, they change the color of litmus from blue to red.
- Acids react with certain metals, such as zinc, magnesium, and iron, to produce hydrogen gas. A typical reaction is that between hydrochloric acid and magnesium:

$$2HCl(aq) + Mg(s) \longrightarrow MgCl_2(aq) + H_2(g)$$

- Acids react with carbonates and bicarbonates, such as Na_2CO_3, $CaCO_3$, and $NaHCO_3$, to produce carbon dioxide gas (Figure 4.6). For example,

$$2HCl(aq) + CaCO_3(s) \longrightarrow CaCl_2(aq) + H_2O(l) + CO_2(g)$$
$$HCl(aq) + NaHCO_3(s) \longrightarrow NaCl(aq) + H_2O(l) + CO_2(g)$$

- Aqueous acid solutions conduct electricity.

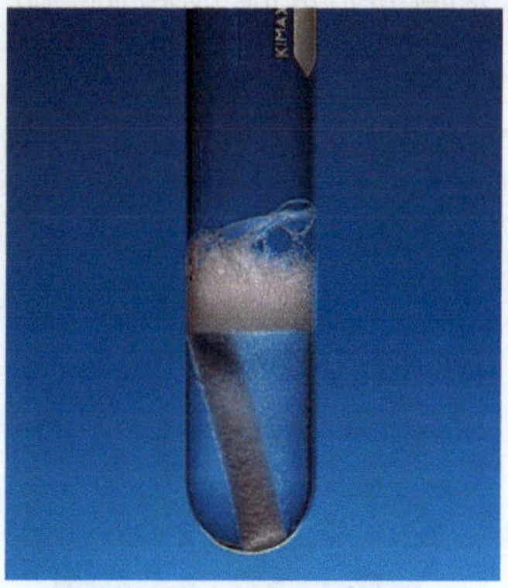

Figure 4.6 *A piece of blackboard chalk, which is mostly $CaCO_3$, reacts with hydrochloric acid.*
Ken Karp/McGraw Hill

Bases

- Bases have a bitter taste.
- Bases feel slippery; for example, soaps, which contain bases, exhibit this property.
- Bases cause color changes in plant dyes; for example, they change the color of litmus from red to blue.
- Aqueous base solutions conduct electricity.

Brønsted Acids and Bases

Arrhenius's definitions of acids and bases are limited in that they apply only to aqueous solutions. Broader definitions were proposed by the Danish chemist Johannes Brønsted‡ in 1932; a ***Brønsted acid*** is *a proton donor*, and a ***Brønsted base*** is *a proton acceptor*. Note that Brønsted's definitions do not require acids and bases to be in aqueous solution.

Hydrochloric acid is a Brønsted acid because it donates a proton, symbolized by $H^+(aq)$, in water:

$$HCl(aq) \longrightarrow H^+(aq) + Cl^-(aq)$$

Note that the H^+ ion is a hydrogen atom that has lost its electron; that is, it is just a bare proton. The size of a proton is about 10^{-15} m, compared to a diameter of 10^{-10} m for an average atom or ion. Such an exceedingly small charged particle cannot exist as a separate entity in aqueous solution owing to its strong attraction for the negative pole (the O atom) in H_2O. Consequently, the proton exists in the hydrated form, as shown in Figure 4.7. Therefore, the ionization of hydrochloric acid should be written as

$$HCl(aq) + H_2O(l) \longrightarrow H_3O^+(aq) + Cl^-(aq)$$

The *hydrated proton*, $H_3O^+(aq)$, is called the ***hydronium ion***. This equation shows a reaction in which a Brønsted acid (HCl) donates a proton to a Brønsted base (H_2O).

‡Johannes Nicolaus Brønsted (1879–1947). Danish chemist. In addition to his theory of acids and bases, Brønsted worked on thermodynamics and the separation of mercury isotopes. In some texts, Brønsted acids and bases are called Brønsted-Lowry acids and bases. Thomas Martin Lowry (1874–1936). English chemist. Brønsted and Lowry developed essentially the same acid-base theory independently in 1923.

Figure 4.7 *Ionization of HCl in water to form the hydronium ion and the chloride ion.*

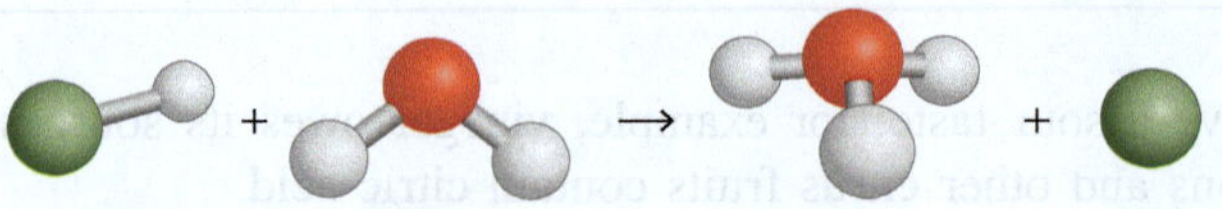

$$HCl + H_2O \longrightarrow H_3O^+ + Cl^-$$

Electrostatic potential map of the H_3O^+ ion. In the rainbow color spectrum representation, the most electron-rich region is red and the most electron-poor region is blue.

Student Hot Spot

Student data indicate you may struggle with acid ionization. Access your eBook for additional Learning Resources on this topic.

Experiments show that the hydronium ion is further hydrated so that the proton may have several water molecules associated with it. Because the acidic properties of the proton are unaffected by the degree of hydration, in this text we will generally use $H^+(aq)$ to represent the hydrated proton. This notation is for convenience, but H_3O^+ is closer to reality. Keep in mind that both notations represent the same species in aqueous solution.

Acids commonly used in the laboratory include hydrochloric acid (HCl), nitric acid (HNO_3), acetic acid (CH_3COOH), sulfuric acid (H_2SO_4), and phosphoric acid (H_3PO_4). The first three are ***monoprotic acids;*** that is, *each unit of the acid yields one hydrogen ion upon ionization:*

$$HCl(aq) \longrightarrow H^+(aq) + Cl^-(aq)$$
$$HNO_3(aq) \longrightarrow H^+(aq) + NO_3^-(aq)$$
$$CH_3COOH(aq) \rightleftharpoons CH_3COO^-(aq) + H^+(aq)$$

As mentioned earlier, because the ionization of acetic acid is incomplete (note the double arrows), it is a weak electrolyte. For this reason it is called a weak acid (see Table 4.1). On the other hand, HCl and HNO_3 are strong acids because they are strong electrolytes, so they are completely ionized in solution (note the use of single arrows).

Sulfuric acid (H_2SO_4) is a ***diprotic acid,*** because *each unit of the acid gives up two H^+ ions,* in two separate steps:

$$H_2SO_4(aq) \longrightarrow H^+(aq) + HSO_4^-(aq)$$
$$HSO_4^-(aq) \rightleftharpoons H^+(aq) + SO_4^{2-}(aq)$$

H_2SO_4 is a strong electrolyte or strong acid (the first step of ionization is complete), but HSO_4^- is a weak acid or weak electrolyte, and we need a double arrow to represent its incomplete ionization.

Triprotic acids, which *yield three H^+ ions,* are relatively few in number. The best known triprotic acid is phosphoric acid, whose ionizations are

$$H_3PO_4(aq) \rightleftharpoons H^+(aq) + H_2PO_4^-(aq)$$
$$H_2PO_4^-(aq) \rightleftharpoons H^+(aq) + HPO_4^{2-}(aq)$$
$$HPO_4^{2-}(aq) \rightleftharpoons H^+(aq) + PO_4^{3-}(aq)$$

All three species (H_3PO_4, $H_2PO_4^-$, and HPO_4^{2-}) in this case are weak acids, and we use the double arrows to represent each ionization step. Anions such as $H_2PO_4^-$ and HPO_4^{2-} are found in aqueous solutions of phosphates such as NaH_2PO_4 and Na_2HPO_4. Table 4.3 lists several common strong and weak acids.

Table 4.1 shows that sodium hydroxide (NaOH) and barium hydroxide [$Ba(OH)_2$] are strong electrolytes. This means that they are completely ionized in solution:

$$NaOH(s) \xrightarrow{H_2O} Na^+(aq) + OH^-(aq)$$
$$Ba(OH)_2(s) \xrightarrow{H_2O} Ba^{2+}(aq) + 2OH^-(aq)$$

The OH^- ion can accept a proton as follows:

$$H^+(aq) + OH^-(aq) \longrightarrow H_2O(l)$$

Table 4.3

Some Common Strong and Weak Acids

Strong Acids	
Hydrochloric acid	HCl
Hydrobromic acid	HBr
Hydroiodic acid	HI
Nitric acid	HNO_3
Sulfuric acid	H_2SO_4
Perchloric acid	$HClO_4$
Weak Acids	
Hydrofluoric acid	HF
Nitrous acid	HNO_2
Phosphoric acid	H_3PO_4
Acetic acid	CH_3COOH

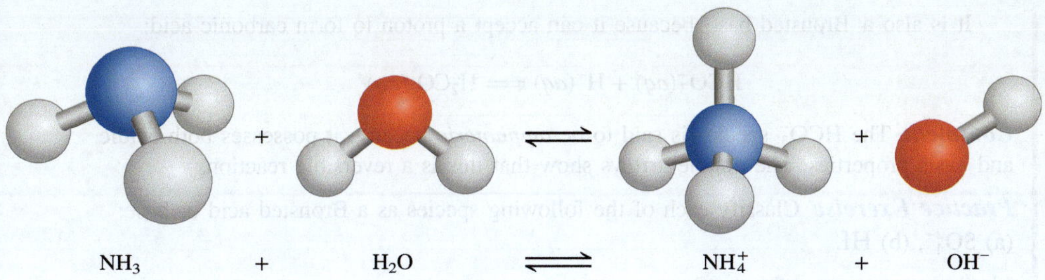

Figure 4.8 *Ionization of ammonia in water to form the ammonium ion and the hydroxide ion.*

Thus, OH^- is a Brønsted base.

Ammonia (NH_3) is classified as a Brønsted base because it can accept a H^+ ion (Figure 4.8):

$$NH_3(aq) + H_2O(l) \rightleftharpoons NH_4^+(aq) + OH^-(aq)$$

Ammonia is a weak electrolyte (and therefore a weak base) because only a small fraction of dissolved NH_3 molecules react with water to form NH_4^+ and OH^- ions.

The most commonly used strong base in the laboratory is sodium hydroxide. It is cheap and soluble. (In fact, all of the alkali metal hydroxides are soluble.) The most commonly used weak base is aqueous ammonia solution, which is sometimes erroneously called ammonium hydroxide. There is no evidence that the species NH_4OH actually exists, other than the NH_4^+ and OH^- ions in solution. All of the Group 2 elements form hydroxides of the type $M(OH)_2$, where M denotes an alkaline earth metal. Of these hydroxides, only $Ba(OH)_2$ is soluble. Magnesium and calcium hydroxides are used in medicine and industry. Hydroxides of other metals, such as $Al(OH)_3$ and $Zn(OH)_2$, are insoluble and are not used as bases.

Example 4.3 classifies substances as Brønsted acids or Brønsted bases.

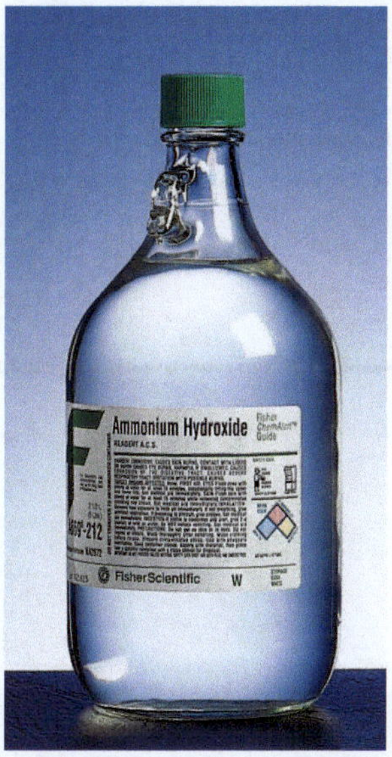

Note that this bottle of aqueous ammonia is erroneously labeled.
Ken Karp/McGraw Hill

Student Hot Spot
Student data indicate you may struggle with base ionization. Access your eBook for additional Learning Resources on this topic.

Example 4.3

Classify each of the following species in aqueous solution as a Brønsted acid or base: (a) HBr, (b) NO_2^-, (c) HCO_3^-.

Strategy What are the characteristics of a Brønsted acid? Does it contain at least an H atom? With the exception of ammonia, most Brønsted bases that you will encounter at this stage are anions.

Solution

(a) We know that HCl is an acid. Because Br and Cl are both halogens (Group 17), we expect HBr, like HCl, to ionize in water as follows:

$$HBr(aq) \longrightarrow H^+(aq) + Br^-(aq)$$

Therefore HBr is a Brønsted acid.

(b) In solution the nitrite ion can accept a proton from water to form nitrous acid:

$$NO_2^-(aq) + H^+(aq) \longrightarrow HNO_2(aq)$$

This property makes NO_2^- a Brønsted base.

(c) The bicarbonate ion is a Brønsted acid because it ionizes in solution as follows:

$$HCO_3^-(aq) \rightleftharpoons H^+(aq) + CO_3^{2-}(aq)$$

(Continued)

It is also a Brønsted base because it can accept a proton to form carbonic acid:

$$HCO_3^-(aq) + H^+(aq) \rightleftharpoons H_2CO_3(aq)$$

Comment The HCO_3^- species is said to be *amphoteric* because it possesses both acidic and basic properties. The double arrows show that this is a reversible reaction.

Practice Exercise Classify each of the following species as a Brønsted acid or base: (a) SO_4^{2-}, (b) HI.

Similar problems: 4.31, 4.32.

Acid-Base Neutralization

Video: Neutralization Reactions

A ***neutralization reaction*** is *a reaction between an acid and a base.* Generally, aqueous acid-base reactions produce water and a ***salt,*** which is *an ionic compound made up of a cation from the base and an anion from the acid:*

$$\text{acid} + \text{base} \longrightarrow \text{salt} + \text{water}$$

The substance we know as table salt, NaCl, is a product of the acid-base reaction

$$HCl(aq) + NaOH(aq) \longrightarrow NaCl(aq) + H_2O(l)$$

However, because both the acid and the base are strong electrolytes, they are completely ionized in solution. The ionic equation is

$$H^+(aq) + Cl^-(aq) + Na^+(aq) + OH^-(aq) \longrightarrow Na^+(aq) + Cl^-(aq) + H_2O(l)$$

Therefore, the reaction can be represented by the net ionic equation

$$H^+(aq) + OH^-(aq) \longrightarrow H_2O(l)$$

Both Na^+ and Cl^- are spectator ions.

If we had started the preceding reaction with equal molar amounts of the acid and the base, at the end of the reaction we would have only a salt and no leftover acid or base. This is a characteristic of acid-base neutralization reactions.

A reaction between a weak acid such as hydrocyanic acid (HCN) and a strong base is

$$HCN(aq) + NaOH(aq) \longrightarrow NaCN(aq) + H_2O(l)$$

Because HCN is a weak acid, it does not ionize appreciably in solution. Thus, the ionic equation is written as

$$HCN(aq) + Na^+(aq) + OH^-(aq) \longrightarrow Na^+(aq) + CN^-(aq) + H_2O(l)$$

and the net ionic equation is

$$HCN(aq) + OH^-(aq) \longrightarrow CN^-(aq) + H_2O(l)$$

Note that only Na^+ is a spectator ion; OH^- and CN^- are not.

The following are also examples of acid-base neutralization reactions, represented by molecular equations:

$$HF(aq) + KOH(aq) \longrightarrow KF(aq) + H_2O(l)$$
$$H_2SO_4(aq) + 2NaOH(aq) \longrightarrow Na_2SO_4(aq) + 2H_2O(l)$$
$$HNO_3(aq) + NH_3(aq) \longrightarrow NH_4NO_3(aq)$$

The last equation looks different because it does not show water as a product. However, if we express $NH_3(aq)$ as $NH_4^+(aq)$ and $OH^-(aq)$, as discussed earlier, then the equation becomes

$$HNO_3(aq) + NH_4^+(aq) + OH^-(aq) \longrightarrow NH_4NO_3(aq) + H_2O(l)$$

Student Hot Spot

Student data indicate you may struggle with predicting the products of a weak acid–strong base reaction. Access your eBook for additional Learning Resources on this topic.

Example 4.4

Write molecular, ionic, and net ionic equations for each of the following acid-base reactions:

(a) hydrobromic acid(*aq*) + barium hydroxide(*aq*) $\longrightarrow$

(b) sulfuric acid(*aq*) + potassium hydroxide(*aq*) $\longrightarrow$

Strategy The first step is to identify the acids and bases as strong or weak. We see that HBr is a strong acid and H_2SO_4 is a strong acid for the first step ionization and a weak acid for the second step ionization. Both $Ba(OH)_2$ and KOH are strong bases.

Solution

(a) Molecular equation:

$$2HBr(aq) + Ba(OH)_2(aq) \longrightarrow BaBr_2(aq) + 2H_2O(l)$$

Ionic equation:

$$2H^+(aq) + 2Br^-(aq) + Ba^{2+}(aq) + 2OH^-(aq) \longrightarrow Ba^{2+}(aq) + 2Br^-(aq) + 2H_2O(l)$$

Net ionic equation:

$$2H^+(aq) + 2OH^-(aq) \longrightarrow 2H_2O(l)$$

or

$$H^+(aq) + OH^-(aq) \longrightarrow H_2O(l)$$

Both Ba^{2+} and Br^- are spectator ions.

(b) Molecular equation:

$$H_2SO_4(aq) + 2KOH(aq) \longrightarrow K_2SO_4(aq) + 2H_2O(l)$$

Ionic equation:

$$H^+(aq) + HSO_4^-(aq) + 2K^+(aq) + 2OH^-(aq) \longrightarrow 2K^+(aq) + SO_4^{2-}(aq) + 2H_2O(l)$$

Net ionic equation:

$$H^+(aq) + HSO_4^-(aq) + 2OH^-(aq) \longrightarrow SO_4^{2-}(aq) + 2H_2O(l)$$

Note that because HSO_4^- is a weak acid and does not ionize appreciably in water, the only spectator ion is K^+.

Practice Exercise Write a molecular equation, an ionic equation, and a net ionic equation for the reaction between aqueous solutions of phosphoric acid and sodium hydroxide.

Similar problem: 4.33(b).

Acid-Base Reactions Leading to Gas Formation

Certain salts like carbonates (containing the CO_3^{2-} ion), bicarbonates (containing the HCO_3^- ion), sulfites (containing the SO_3^{2-} ion), and sulfides (containing the S^{2-} ion) react with acids to form gaseous products. For example, the molecular equation for the reaction between sodium carbonate (Na_2CO_3) and $HCl(aq)$ (see Figure 4.6) is

$$Na_2CO_3(aq) + 2HCl(aq) \longrightarrow 2NaCl(aq) + H_2CO_3(aq)$$

Carbonic acid is unstable and if present in solution in sufficient concentrations decomposes as follows:

$$H_2CO_3(aq) \longrightarrow H_2O(l) + CO_2(g)$$

Similar reactions involving other mentioned salts are

$$NaHCO_3(aq) + HCl(aq) \longrightarrow NaCl(aq) + H_2O(l) + CO_2(g)$$
$$Na_2SO_3(aq) + 2HCl(aq) \longrightarrow 2NaCl(aq) + H_2O(l) + SO_2(g)$$
$$K_2S(aq) + 2HCl(aq) \longrightarrow 2KCl(aq) + H_2S(g)$$

Summary of Concepts & Facts

- Arrhenius acids ionize in water to give H^+ ions, and Arrhenius bases ionize in water to give OH^- ions. Brønsted acids donate protons, and Brønsted bases accept protons.
- The reaction of an acid and a base is called neutralization.

Review of Concepts & Facts

4.3.1 Which of the diagrams (a)–(c) best represents a weak acid? Which represents a very weak acid? Which represents a strong acid? The proton exists in water as the hydronium ion. All acids are monoprotic. (For simplicity, water molecules are not shown.)

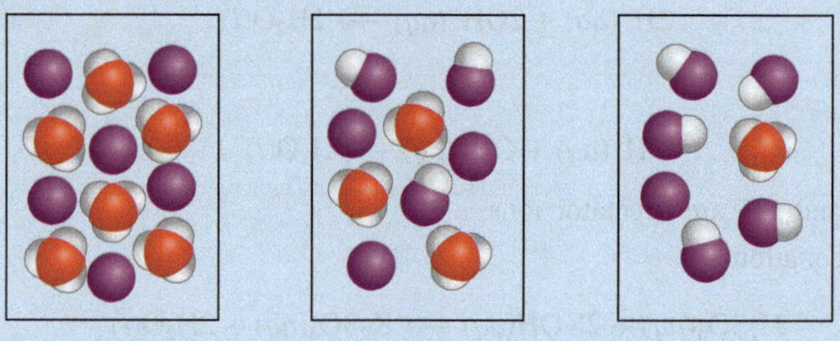

(a) (b) (c)

4.3.2 Identify the Brønsted acid and Brønsted base in the following equation:

$$HBr(aq) + NH_3(aq) \longrightarrow NH_4^+(aq) + Br^-(aq)$$

4.3.3 Write the net ionic equation for the following acid-base reaction:

$$HNO_3(aq) + KOH(aq) \longrightarrow$$

4.4 Oxidation-Reduction Reactions

Learning Objectives

- Identify the various components of an oxidation-reduction reaction, including reducing/oxidizing agents and half-reactions.
- Apply oxidation number rules toward determining the oxidation number of each element in a compound or polyatomic ion.
- Differentiate the types of reactions that may be oxidation-reduction reactions, including combustion, decomposition, displacement, and disproportionation.
- Utilize the activity series to determine whether a metal displacement reaction occurs.

Video: Oxidation-Reduction Reactions

Whereas acid-base reactions can be characterized as proton-transfer processes, the class of reactions called ***oxidation-reduction,*** or ***redox, reactions*** are considered *electron-transfer reactions.* Oxidation-reduction reactions are very much a part of the world around us. They range from the burning of fossil fuels to the action of household bleach. Additionally, most metallic and nonmetallic elements are obtained from their ores by the process of oxidation or reduction.

Many important redox reactions take place in water, but not all redox reactions occur in aqueous solution. We begin our discussion with a reaction in which two elements combine to form a compound. Consider the formation of magnesium oxide (MgO) from magnesium and oxygen (Figure 4.9):

$$2Mg(s) + O_2(g) \longrightarrow 2MgO(s)$$

Magnesium oxide (MgO) is an ionic compound made up of Mg^{2+} and O^{2-} ions. In this reaction, two Mg atoms give up or transfer four electrons to two O atoms (in O_2). For convenience, we can think of this process as two separate steps, one involving the loss of four electrons by the two Mg atoms and the other being the gain of four electrons by an O_2 molecule:

$$2Mg \longrightarrow 2Mg^{2+} + 4e^-$$

$$O_2 + 4e^- \longrightarrow 2O^{2-}$$

Each of these steps is called a ***half-reaction,*** which *explicitly shows the electrons involved in a redox reaction.* In an oxidation half-reaction, electrons appear as a product; in a reduction half-reaction, electrons appear as a reactant. The sum of the half-reactions gives the overall reaction:

$$2Mg + O_2 + 4e^- \longrightarrow 2Mg^{2+} + 2O^{2-} + 4e^-$$

or, if we cancel the electrons that appear on both sides of the equation,

$$2Mg + O_2 \longrightarrow 2Mg^{2+} + 2O^{2-}$$

Finally, the Mg^{2+} and O^{2-} ions combine to form MgO:

$$2Mg^{2+} + 2O^{2-} \longrightarrow 2MgO$$

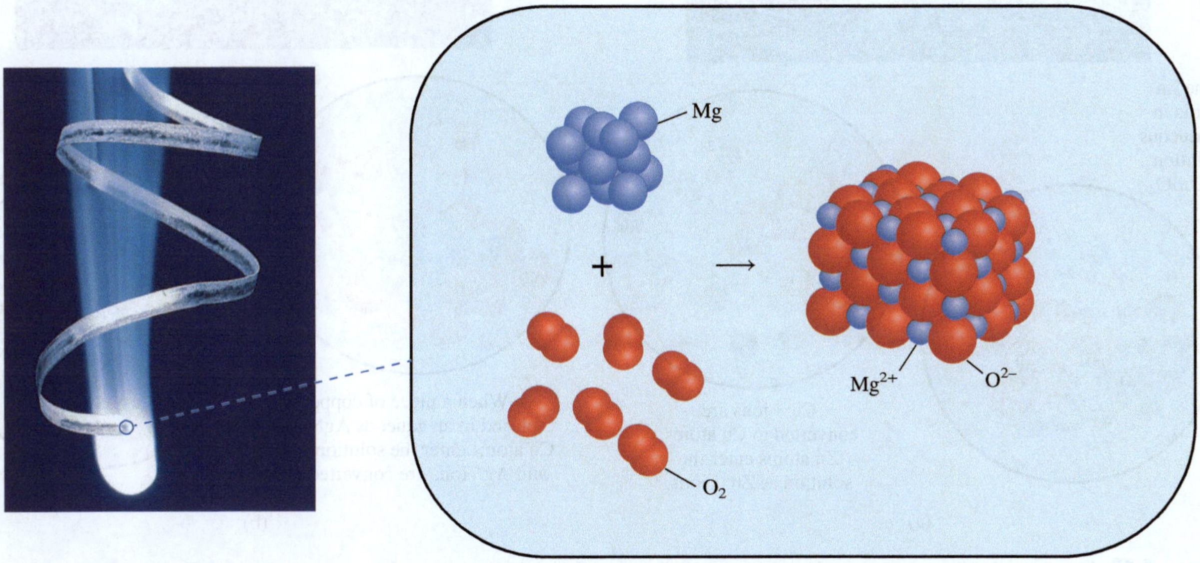

Figure 4.9 *Magnesium burns in oxygen to form magnesium oxide.*
(left): Ken Karp/McGraw Hill

The term ***oxidation reaction*** refers to *the half-reaction that involves loss of electrons.* Chemists originally used "oxidation" to denote the combination of elements with oxygen. However, it now has a broader meaning that includes reactions not involving oxygen. A ***reduction reaction*** is *a half-reaction that involves gain of electrons.* A useful mnemonic for redox is OIL RIG: oxidation is loss (of electrons) and reduction is gain (of electrons).

In the formation of magnesium oxide, magnesium is oxidized. It is said to act as a ***reducing agent*** because it *donates electrons* to oxygen and causes oxygen to be reduced. Oxygen is reduced and acts as an ***oxidizing agent,*** because it *accepts electrons* from magnesium, causing magnesium to be oxidized. Note that the extent of oxidation in a redox reaction must be equal to the extent of reduction; that is, the number of electrons lost by a reducing agent must be equal to the number of electrons gained by an oxidizing agent. Further, oxidizing agents are always reduced, and reducing agents are always oxidized.

The occurrence of electron transfer is more apparent in some redox reactions than others. When metallic zinc is added to a solution containing copper(II) sulfate ($CuSO_4$), zinc reduces Cu^{2+} by donating two electrons to it:

$$Zn(s) + CuSO_4(aq) \longrightarrow ZnSO_4(aq) + Cu(s)$$

In the process, the solution loses the blue color that characterizes the presence of hydrated Cu^{2+} ions (Figure 4.10):

$$Zn(s) + Cu^{2+}(aq) \longrightarrow Zn^{2+}(aq) + Cu(s)$$

Figure 4.10 *Metal displacement reactions in solution. (a) First beaker: A zinc strip is placed in a blue $CuSO_4$ solution. Immediately Cu^{2+} ions are reduced to metallic Cu in the form of a dark layer. Second beaker: In time, most of the Cu^{2+} ions are reduced and the solution becomes colorless. (b) First beaker: A piece of Cu wire is placed in a colorless $AgNO_3$ solution. Ag^+ ions are reduced to metallic Ag. Second beaker: As time progresses, most of the Ag^+ ions are reduced and the solution acquires the characteristic blue color due to the presence of hydrated Cu^{2+} ions.*

The oxidation and reduction half-reactions are

$$\mathrm{Zn} \longrightarrow \mathrm{Zn}^{2+} + 2e^-$$
$$\mathrm{Cu}^{2+} + 2e^- \longrightarrow \mathrm{Cu}$$

Similarly, metallic copper reduces silver ions in a solution of silver nitrate ($AgNO_3$):

Video: Reaction of Cu with $AgNO_3$

$$\mathrm{Cu}(s) + 2\mathrm{AgNO_3}(aq) \longrightarrow \mathrm{Cu(NO_3)_2}(aq) + 2\mathrm{Ag}(s)$$

or

$$\mathrm{Cu}(s) + 2\mathrm{Ag}^+(aq) \longrightarrow \mathrm{Cu}^{2+}(aq) + 2\mathrm{Ag}(s)$$

Oxidation Number

The definitions of oxidation and reduction in terms of loss and gain of electrons apply to the formation of ionic compounds such as MgO and the reduction of Cu^{2+} ions by Zn. However, these definitions do not accurately characterize the formation of hydrogen chloride (HCl) and sulfur dioxide (SO_2):

$$\mathrm{H_2}(g) + \mathrm{Cl_2}(g) \longrightarrow 2\mathrm{HCl}(g)$$
$$\mathrm{S}(s) + \mathrm{O_2}(g) \longrightarrow \mathrm{SO_2}(g)$$

Because HCl and SO_2 are not ionic but molecular compounds, no electrons are actually transferred in the formation of these compounds, as they are in the case of MgO. Nevertheless, chemists find it convenient to treat these reactions as redox reactions, because experimental measurements show that there is a partial transfer of electrons (from H to Cl in HCl and from S to O in SO_2).

To keep track of electrons in redox reactions, it is useful to assign oxidation numbers to the reactants and products. An atom's ***oxidation number,*** also called ***oxidation state,*** signifies *the number of charges the atom would have in a molecule (or an ionic compound) if electrons were transferred completely.* For example, we can rewrite the previous equations for the formation of HCl and SO_2 as follows:

$$\overset{0}{\mathrm{H_2}}(g) + \overset{0}{\mathrm{Cl_2}}(g) \longrightarrow 2\overset{+1}{\mathrm{H}}\overset{-1}{\mathrm{Cl}}(g)$$
$$\overset{0}{\mathrm{S}}(s) + \overset{0}{\mathrm{O_2}}(g) \longrightarrow \overset{+4}{\mathrm{S}}\overset{-2}{\mathrm{O_2}}(g)$$

The numbers above the element symbols are the oxidation numbers. In both of the reactions shown, there is no charge on the atoms in the reactant molecules. Thus, their oxidation number is zero. For the product molecules, however, it is assumed that complete electron transfer has taken place and that atoms have gained or lost electrons. The oxidation numbers reflect the number of electrons "transferred."

Oxidation numbers enable us to identify elements that are oxidized and reduced at a glance. The elements that show an increase in oxidation number—hydrogen and sulfur in the preceding examples—are oxidized. Chlorine and oxygen are reduced, so their oxidation numbers show a decrease from their initial values. Note that the sum of the oxidation numbers of H and Cl in HCl (+1 and −1) is zero. Likewise, if we add the oxidation numbers of S (+4) and two atoms of O [2 × (−2)], the total is zero. The reason is that the HCl and SO_2 molecules are neutral, so the charges must cancel.

We use the following rules to assign oxidation numbers.

1. In free elements (that is, in the uncombined state), each atom has an oxidation number of zero. Thus, each atom in H_2, Br_2, Na, Be, K, O_2, and P_4 has the same oxidation number: zero.
2. For ions composed of only one atom (that is, monatomic ions), the oxidation number is equal to the charge on the ion. Thus, Li^+ ion has an oxidation number

of +1; Br^{2+} ion, +2; Fe^{3+} ion, +3; I^- ion, −1; O^{2-} ion, −2; and so on. All alkali metals have an oxidation number of +1 and all alkaline earth metals have an oxidation number of +2 in their compounds. Aluminum has an oxidation number of +3 in all its compounds.

3. The oxidation number of oxygen in most compounds (for example, MgO and H_2O) is −2, but in hydrogen peroxide (H_2O_2) and peroxide ion (O_2^{2-}), it is −1.
4. The oxidation number of hydrogen is +1, except when it is bonded to metals in binary compounds. In these cases (for example, LiH, NaH, CaH_2), its oxidation number is −1.
5. Fluorine has an oxidation number of −1 in *all* its compounds. Other halogens (Cl, Br, and I) have negative oxidation numbers when they occur as halide ions in their compounds. When combined with oxygen—for example in oxoacids and oxoanions (see Section 2.7)—they have positive oxidation numbers.
6. In a neutral molecule, the sum of the oxidation numbers of all the atoms must be zero. In a polyatomic ion, the sum of oxidation numbers of all the elements in the ion must be equal to the net charge of the ion. For example, in the ammonium ion, NH_4^+, the oxidation number of N is −3 and that of H is +1. Thus, the sum of the oxidation numbers is $-3 + 4(+1) = +1$, which is equal to the net charge of the ion.
7. Oxidation numbers do not have to be integers. For example, the oxidation number of O in the superoxide ion, O_2^-, is $-\frac{1}{2}$. We apply the preceding rules to assign oxidation numbers in Example 4.5.

Student Hot Spot

Student data indicate you may struggle with determining oxidation numbers. Access your eBook for additional Learning Resources on this topic.

Example 4.5

Assign oxidation numbers to all the elements in the following compounds and ion: (a) Li_2O, (b) HNO_3, (c) $Cr_2O_7^{2-}$.

Strategy In general, we follow the rules just listed for assigning oxidation numbers. Remember that all alkali metals have an oxidation number of +1, and in most cases, hydrogen has an oxidation number of +1 and oxygen has an oxidation number of −2 in their compounds.

Solution

(a) By rule 2 we see that lithium has an oxidation number of +1 (Li^+) and oxygen's oxidation number is −2 (O^{2-}).

(b) This is the formula for nitric acid, which yields a H^+ ion and a NO_3^- ion in solution. From rule 4 we see that H has an oxidation number of +1. Thus the other group (the nitrate ion) must have a net oxidation number of −1. Oxygen has an oxidation number of −2, and if we use x to represent the oxidation number of nitrogen, then the nitrate ion can be written as

$$[N^{(x)}O_3^{(2-)}]^-$$

or

$$x = +5$$

(c) From rule 6 we see that the sum of the oxidation numbers in the dichromate ion $Cr_2O_7^{2-}$ must be −2. We know that the oxidation number of O is −2, so all that remains is to determine the oxidation number of Cr, which we call y. The dichromate ion can be written as

$$[Cr_2^{(y)}O_7^{(2-)}]^{2-}$$

or

$$y = +6$$

(Continued)

Check In each case, does the sum of the oxidation numbers of all the atoms equal the net charge on the species?

Practice Exercise Assign oxidation numbers to all the elements in the following compound and ion: (a) PF_3, (b) MnO_4^-.

Similar problems: 4.47, 4.49.

Figure 4.11 shows the known oxidation numbers of the familiar elements, arranged according to their positions in the periodic table. We can summarize the content of this figure as follows:

- Metallic elements have only positive oxidation numbers, whereas nonmetallic elements may have either positive or negative oxidation numbers.
- The highest oxidation number an element in Groups 1 and 2 can have is its group number. For Groups 13 through 17, the highest oxidation number is equal to the group number minus 10. For example, the heavier halogens Cl, Br, and I are in Group 17, so their highest possible oxidation number is +7 (17–10).
- The transition metals (Groups 3–11) usually have several possible oxidation numbers.

1	2	3	4	5	6	7	8	9	10	11	12	13	14	15	16	17	18
1 **H** +1 −1																	2 **He**
3 **Li** +1	4 **Be** +2											5 **B** +3	6 **C** +4 +2 −4	7 **N** +5 +4 +3 +2 +1 −3	8 **O** +2 $-\frac{1}{2}$ −1 −2	9 **F** −1	10 **Ne**
11 **Na** +1	12 **Mg** +2											13 **Al** +3	14 **Si** +4 −4	15 **P** +5 +3 −3	16 **S** +6 +4 +2 −2	17 **Cl** +7 +6 +5 +4 +3 +1 −1	18 **Ar**
19 **K** +1	20 **Ca** +2	21 **Sc** +3	22 **Ti** +4 +3 +2	23 **V** +5 +4 +3 +2	24 **Cr** +6 +5 +4 +3 +2	25 **Mn** +7 +6 +4 +3 +2	26 **Fe** +3 +2	27 **Co** +3 +2	28 **Ni** +2	29 **Cu** +2 +1	30 **Zn** +2	31 **Ga** +3	32 **Ge** +4 −4	33 **As** +5 +3 −3	34 **Se** +6 +4 −2	35 **Br** +5 +3 +1 −1	36 **Kr** +4 +2
37 **Rb** +1	38 **Sr** +2	39 **Y** +3	40 **Zr** +4	41 **Nb** +5 +4	42 **Mo** +6 +4 +3	43 **Tc** +7 +6 +4	44 **Ru** +8 +6 +4 +3	45 **Rh** +4 +3 +2	46 **Pd** +4 +2	47 **Ag** +1	48 **Cd** +2	49 **In** +3	50 **Sn** +4 +2	51 **Sb** +5 +3 −3	52 **Te** +6 +4 −2	53 **I** +7 +5 +1 −1	54 **Xe** +6 +4 +2
55 **Cs** +1	56 **Ba** +2	57 **La** +3	72 **Hf** +4	73 **Ta** +5	74 **W** +6 +4	75 **Re** +7 +6 +4	76 **Os** +8 +4	77 **Ir** +4 +3	78 **Pt** +4 +2	79 **Au** +3 +1	80 **Hg** +2 +1	81 **Tl** +3 +1	82 **Pb** +4 +2	83 **Bi** +5 +3	84 **Po** +2	85 **At** −1	86 **Rn**

Figure 4.11 *The oxidation numbers of elements in their compounds. The more common oxidation numbers are in color.*

(a) (b) (c)

Figure 4.12 *Some simple combination redox reactions. (a) Sulfur burning in air to form sulfur dioxide. (b) Sodium burning in chlorine to form sodium chloride. (c) Aluminum reacting with bromine to form aluminum bromide.*

(a): Charles D. Winters/McGraw Hill; (b): Andrew Lambert Photography/Science Source; (c): Charles D. Winters/Science Source

Types of Redox Reactions

Among the most common oxidation-reduction reactions are combination, decomposition, combustion, and displacement reactions. A more involved type is called disproportionation reactions, which will also be discussed in this section.

Combination Reactions

A ***combination reaction*** is *a reaction in which two or more substances combine to form a single product.* Figure 4.12 shows some combination reactions. For example,

$$\overset{0}{\mathrm{S}}(s) + \overset{0}{\mathrm{O}}_2(g) \longrightarrow \overset{+4\,-2}{\mathrm{SO}_2}(g)$$

$$2\overset{0}{\mathrm{Al}}(s) + 3\overset{0}{\mathrm{Br}}(l) \longrightarrow 2\overset{+3\,-1}{\mathrm{AlBr}_3}(s)$$

Decomposition Reactions

Decomposition reactions are the opposite of combination reactions. Specifically, a ***decomposition reaction*** is *the breakdown of a compound into two or more components* (Figure 4.13). For example,

$$2\overset{+2\,-2}{\mathrm{HgO}}(s) \longrightarrow 2\overset{0}{\mathrm{Hg}}(l) + \overset{0}{\mathrm{O}}_2(g)$$

$$2\overset{+5\,-2}{\mathrm{KClO}_3}(s) \longrightarrow 2\mathrm{K}\overset{-1}{\mathrm{Cl}}(s) + 3\overset{0}{\mathrm{O}}_2(g)$$

$$2\overset{+1\,-1}{\mathrm{NaH}}(s) \longrightarrow 2\overset{0}{\mathrm{Na}}(s) + \overset{0}{\mathrm{H}}_2(g)$$

We show oxidation numbers only for elements that are oxidized or reduced.

Combustion Reactions

A ***combustion reaction*** is *a reaction in which a substance reacts with oxygen, usually with the release of heat and light to produce a flame.* All combustion reactions are redox processes. The reactions between magnesium and sulfur with oxygen described earlier are combustion reactions. Another example is the burning of propane (C_3H_8), a component of natural gas that is used for domestic heating and cooking:

$$\mathrm{C_3H_8}(g) + 5\mathrm{O_2}(g) \longrightarrow 3\mathrm{CO_2}(g) + 4\mathrm{H_2O}(l)$$

Assigning an oxidation number to C atoms in organic compounds is more involved. Here, we focus only on the oxidation number of O atoms, which changes from 0 to −2.

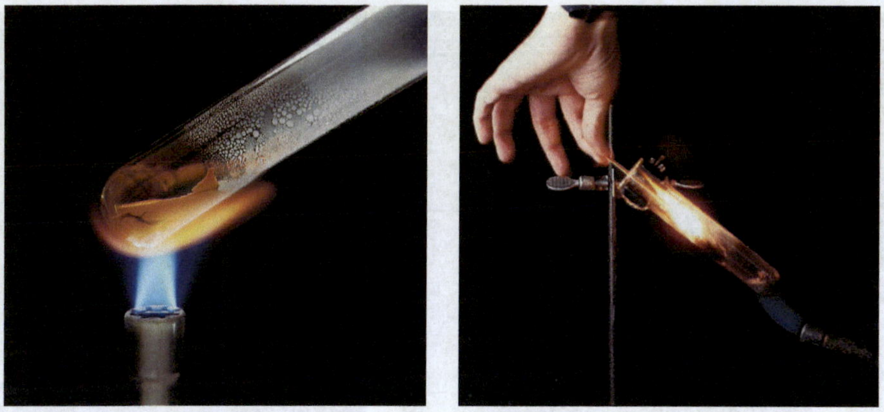

(a) (b)

Figure 4.13 *(a) On heating, mercury(II) oxide (HgO) decomposes to form mercury and oxygen. (b) Heating potassium chlorate ($KClO_3$) produces oxygen, which supports the combustion of the wood splint.*

(a): Charles D. Winters/McGraw Hill; (b): Ken Karp/McGraw Hill

Displacement Reactions

In a ***displacement reaction,*** *an ion (or atom) in a compound is replaced by an ion (or atom) of another element.* Most displacement reactions fit into one of three subcategories: hydrogen displacement, metal displacement, or halogen displacement.

1. Hydrogen Displacement. All alkali metals and some alkaline earth metals (Ca, Sr, and Ba), which are the most reactive of the metallic elements, will displace hydrogen from cold water (Figure 4.14):

$$2\overset{0}{\text{Na}}(s) + 2\overset{+1}{\text{H}}_2\text{O}(l) \longrightarrow 2\overset{+1}{\text{Na}}\text{O}\overset{+1}{\text{H}}(aq) + \overset{0}{\text{H}}_2(g)$$

$$\overset{0}{\text{Ca}}(s) + 2\overset{+1}{\text{H}}_2\text{O}(l) \longrightarrow \overset{+2}{\text{Ca}}(\text{O}\overset{+1}{\text{H}})_2 + \overset{0}{\text{H}}_2(g)$$

Many metals, including those that do not react with water, are capable of displacing hydrogen from acids. For example, zinc (Zn) and magnesium (Mg) do not react with cold water but do react with hydrochloric acid, as follows:

$$\overset{0}{\text{Zn}}(s) + 2\overset{+1}{\text{H}}\text{Cl}(aq) \longrightarrow \overset{+2}{\text{Zn}}\text{Cl}_2(aq) + \overset{0}{\text{H}}_2(g)$$

$$\overset{0}{\text{Mg}}(s) + 2\overset{+1}{\text{H}}\text{Cl}(aq) \longrightarrow \overset{+2}{\text{Mg}}\text{Cl}_2(aq) + \overset{0}{\text{H}}_2(g)$$

Figure 4.15 shows the reactions between hydrochloric acid (HCl) and iron (Fe), zinc (Zn), and magnesium (Mg). These reactions are used to prepare hydrogen gas in the laboratory.

2. Metal Displacement. A metal in a compound can be displaced by another metal in the elemental state. We have already seen examples of zinc replacing copper ions and copper replacing silver ions (see Figure 4.10). Reversing the roles of the metals would result in no reaction. Thus, copper metal will not displace zinc ions from zinc sulfate, and silver metal will not displace copper ions from copper nitrate.

An easy way to predict whether a metal or hydrogen displacement reaction will actually occur is to refer to an ***activity series*** (sometimes called the *electrochemical series*), shown in Figure 4.16. An activity series is *a convenient summary of the results of many possible displacement reactions* similar to the ones already discussed. According to this series, any metal above hydrogen will displace it from water or from an acid, but metals below hydrogen will not react with either water or an acid. In fact, any metal listed in the series will react with any metal (in a compound) below it. For example, Zn is above Cu, so zinc metal will displace copper ions from copper sulfate.

Metal displacement reactions find many applications in metallurgical processes, the goal of which is to separate pure metals from their ores. For example, vanadium is obtained by treating vanadium(V) oxide with metallic calcium:

$$\text{V}_2\text{O}_5(s) + 5\text{Ca}(l) \longrightarrow 2\text{V}(l) + 5\text{CaO}(s)$$

(a)

(b)

Figure 4.14 *Reactions of (a) sodium (Na) and (b) calcium (Ca) with cold water. Note that the reaction is more vigorous with Na than with Ca.*

(a and b): Ken Karp/McGraw Hill

(a) (b) (c)

Figure 4.15 *Reactions of (a) iron (Fe), (b) zinc (Zn), and (c) magnesium (Mg) with hydrochloric acid to form hydrogen gas and the metal chlorides ($FeCl_2$, $ZnCl_2$, $MgCl_2$). The reactivity of these metals is reflected in the rate of hydrogen gas evolution, which is slowest for the least reactive metal, Fe, and fastest for the most reactive metal, Mg.*

(a, b, c): Stephen Frisch/McGraw Hill

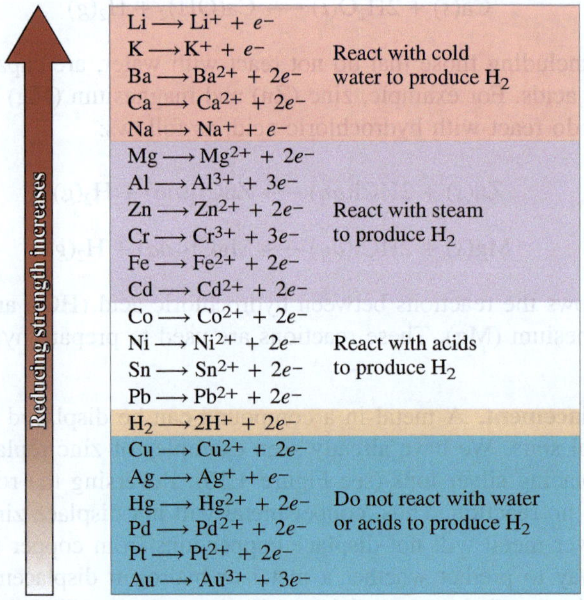

Figure 4.16 *The activity series for metals. The metals are arranged according to their ability to displace hydrogen from an acid or water. Li (lithium) is the most reactive metal, and Au (gold) is the least reactive.*

Similarly, titanium is obtained from titanium(IV) chloride according to the reaction

$$TiCl_4(g) + 2Mg(l) \longrightarrow Ti(s) + 2MgCl_2(l)$$

In each case, the metal that acts as the reducing agent lies above the metal that is reduced (that is, Ca is above V and Mg is above Ti) in the activity series. We will see more examples of this type of reaction in Chapter 18.

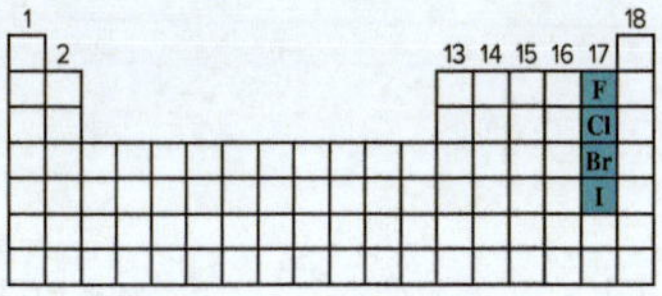

The halogens.

3. Halogen Displacement. Another activity series summarizes the halogens' behavior in halogen displacement reactions:

$$F_2 > Cl_2 > Br_2 > I_2$$

The power of these elements as oxidizing agents decreases as we move down Group 17 from fluorine to iodine, so molecular fluorine can replace chloride, bromide, and iodide ions in solution. In fact, molecular fluorine is so reactive that it also attacks water; thus these reactions cannot be carried out in aqueous solutions. On the other hand, molecular chlorine can displace bromide and iodide ions in aqueous solution. The displacement equations are

$$\overset{0}{Cl_2}(g) + 2K\overset{-1}{Br}(aq) \longrightarrow 2K\overset{-1}{Cl}(aq) + \overset{0}{Br_2}(l)$$

$$\overset{0}{Cl_2}(g) + 2Na\overset{-1}{I}(aq) \longrightarrow 2Na\overset{-1}{Cl}(aq) + \overset{0}{I_2}(s)$$

The ionic equations are

$$\overset{0}{Cl_2}(g) + 2\overset{-1}{Br^-}(aq) \longrightarrow 2\overset{-1}{Cl^-}(aq) + \overset{0}{Br_2}(l)$$

$$\overset{0}{Cl_2}(g) + 2\overset{-1}{I^-}(aq) \longrightarrow 2\overset{-1}{Cl^-}(aq) + \overset{0}{I_2}(s)$$

Molecular bromine, in turn, can displace iodide ion in solution:

$$\overset{0}{Br_2}(l) + 2\overset{-1}{I^-}(aq) \longrightarrow 2\overset{-1}{Br^-}(aq) + \overset{0}{I_2}(s)$$

Reversing the roles of the halogens produces no reaction. Thus, bromine cannot displace chloride ions, and iodine cannot displace bromide and chloride ions.

The halogen displacement reactions have a direct industrial application. The halogens as a group are the most reactive of the nonmetallic elements. They are all strong oxidizing agents. As a result, they are found in nature in the combined state (with metals) as halides and never as free elements. Of these four elements, chlorine is by far the most important industrial chemical. In 2010 the amount of chlorine produced in the United States was about 25 billion pounds, making chlorine the tenth-ranking industrial chemical. The annual production of bromine is only one-hundredth that of chlorine, while the amounts of fluorine and iodine produced are even less.

Recovering the halogens from their halides requires an oxidation process, which is represented by

$$2X^- \longrightarrow X_2 + 2e^-$$

where X denotes a halogen element. Seawater and natural brine (for example, underground water in contact with salt deposits) are rich sources of Cl^-, Br^-, and I^- ions. Minerals such as fluorite (CaF_2) and cryolite (Na_3AlF_6) are used to prepare fluorine. Because fluorine is the strongest oxidizing agent known, there is no way to convert F^- ions to F_2 by chemical means. The only way to carry out the oxidation is by electrolytic means, the details of which will be discussed in Chapter 18. Industrially, chlorine, like fluorine, is produced electrolytically.

Bromine is prepared industrially by oxidizing Br^- ions with chlorine, which is a strong enough oxidizing agent to oxidize Br^- ions but not water:

$$2Br^-(aq) \longrightarrow Br_2(l) + 2e^-$$

One of the richest sources of Br^- ions is the Dead Sea—about 4000 parts per million (ppm) by mass of all dissolved substances in the Dead Sea is Br. Following the oxidation of Br^- ions, bromine is removed from the solution by blowing air over the solution, and the air-bromine mixture is then cooled to condense the bromine (Figure 4.17).

Figure 4.17 *The condensation of liquid bromine.*
Stephen Frisch/McGraw Hill

Iodine is also prepared from seawater and natural brine by the oxidation of I^- ions with chlorine. Because Br^- and I^- ions are invariably present in the same source, they are both oxidized by chlorine. However, it is relatively easy to separate Br_2 from I_2, because iodine is a solid that is sparingly soluble in water. The air-blowing procedure will remove most of the bromine formed but will not affect the iodine present.

Disproportionation Reaction

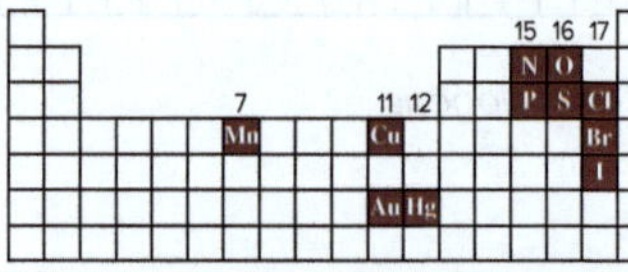

Elements that are most likely to undergo disproportionation reactions.

In this special type of redox reaction, called a ***disproportionation reaction,*** *an element in one oxidation state is simultaneously oxidized and reduced.* One reactant in a disproportionation reaction *always* contains an element that can have at least three oxidation states. The element itself is in an intermediate oxidation state, that is, both higher and lower oxidation states exist for that element in the products. The decomposition of hydrogen peroxide is an example of a disproportionation reaction:

$$2H_2\overset{-1}{O}_2(aq) \longrightarrow 2H_2\overset{-2}{O}(l) + \overset{0}{O}_2(g)$$

Here the oxidation number of oxygen in the reactant (−1) both increases to zero in O_2 and decreases to −2 in H_2O while the oxidation number of hydrogen remains unchanged at +1. Another example is the reaction between molecular chlorine and NaOH solution:

$$\overset{0}{Cl}_2(g) + 2OH^-(aq) \longrightarrow \overset{+1}{Cl}O^-(aq) + \overset{-1}{Cl^-}(aq) + H_2O(l)$$

This reaction describes the formation of household bleaching agents, for it is the hypochlorite ion (ClO^-) that oxidizes the color-bearing substances in stains, converting them to colorless compounds.

Finally, it is interesting to compare redox reactions and acid-base reactions. They are analogous in that acid-base reactions involve the transfer of protons while redox reactions involve the transfer of electrons. However, while acid-base reactions are quite easy to recognize (because they always involve an acid and a base), there is no simple procedure for identifying a redox process. The only sure way is to compare the oxidation numbers of all the elements in the reactants and products. Any change in oxidation number *guarantees* that the reaction is redox in nature.

The classification of different types of redox reactions is illustrated in Example 4.6.

Example 4.6

Classify the following redox reactions and indicate changes in the oxidation numbers of the elements:

(a) $2N_2O(g) \longrightarrow 2N_2(g) + O_2(g)$

(b) $6Li(s) + N_2(g) \longrightarrow 2Li_3N(s)$

(c) $Ni(s) + Pb(NO_3)_2(aq) \longrightarrow Pb(s) + Ni(NO_3)_2(aq)$

(d) $2NO_2(g) + H_2O(l) \longrightarrow HNO_2(aq) + HNO_3(aq)$

Strategy Review the definitions of combination reactions, decomposition reactions, displacement reactions, and disproportionation reactions.

Solution

(a) This is a decomposition reaction because one reactant is converted to two different products. The oxidation number of N changes from +1 to 0, while that of O changes from −2 to 0.

(b) This is a combination reaction (two reactants form a single product). The oxidation number of Li changes from 0 to +1 while that of N changes from 0 to −3.

(c) This is a metal displacement reaction. The Ni metal replaces (reduces) the Pb^{2+} ion. The oxidation number of Ni increases from 0 to +2 while that of Pb decreases from +2 to 0.

(d) The oxidation number of N is +4 in NO_2 and it is +3 in HNO_2 and +5 in HNO_3. Because the oxidation number of the *same* element both increases and decreases, this is a disproportionation reaction.

(Continued)

CHEMISTRY *in Action*

Breathalyzer

Every year in the United States about 25,000 people are killed and 500,000 more are injured as a result of drunk driving. In spite of efforts to educate the public about the dangers of driving while intoxicated and stiffer penalties for drunk driving offenses, law enforcement agencies still have to devote a great deal of work to removing drunk drivers from America's roads.

The police often use a device called a breathalyzer to test drivers suspected of being drunk. The chemical basis of this device is a redox reaction. A sample of the driver's breath is drawn into the breathalyzer, where it is treated with an acidic solution of potassium dichromate. The alcohol (ethanol) in the breath is converted to acetic acid as shown in the following equation:

$$\underset{\text{ethanol}}{3CH_3CH_2OH} + \underset{\substack{\text{potassium}\\\text{dichromate}\\\text{(orange yellow)}}}{2K_2Cr_2O_7} + \underset{\substack{\text{sulfuric}\\\text{acid}}}{8H_2SO_4} \longrightarrow$$

$$\underset{\text{acetic acid}}{3CH_3COOH} + \underset{\substack{\text{chromium(III)}\\\text{sulfate (green)}}}{2Cr_2(SO_4)_3} + \underset{\substack{\text{potassium}\\\text{sulfate}}}{2K_2SO_4} + 11H_2O$$

In this reaction, the ethanol is oxidized to acetic acid and the chromium(VI) in the orange-yellow dichromate ion is reduced to the green chromium(III) ion (see Figure 4.22). The driver's blood alcohol level can be determined readily by measuring the degree of this color change (read from a calibrated meter on the instrument). The current legal limit of blood alcohol content is 0.08 percent by mass. Anything higher constitutes intoxication.

A driver being tested for blood alcohol content with a handheld breathalyzer.
Piotr290/iStock/Getty Images

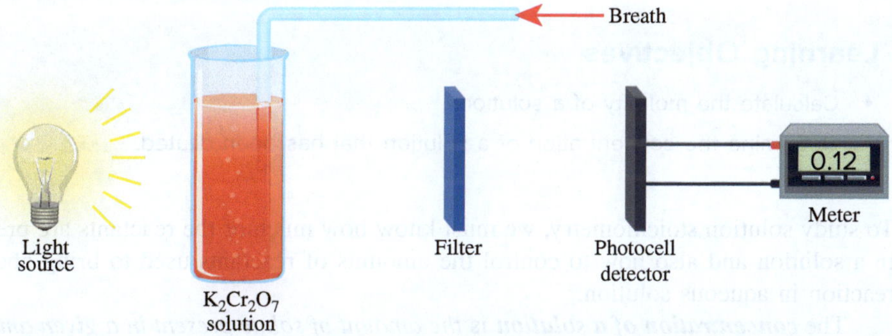

Schematic diagram of a breathalyzer. The alcohol in the driver's breath is reacted with a potassium dichromate solution. The change in the absorption of light due to the formation of chromium(III) sulfate is registered by the detector and shown on a meter, which directly displays the alcohol content in blood. The filter selects only one wavelength of light for measurement.

Practice Exercise Identify the following redox reactions by type:

(a) $Fe + H_2SO_4 \longrightarrow FeSO_4 + H_2$

(b) $S + 3F_2 \longrightarrow SF_6$

(c) $2CuCl \longrightarrow Cu + CuCl_2$

(d) $2Ag + PtCl_2 \longrightarrow 2AgCl + Pt$

Similar problems: 4.55, 4.56.

The Chemistry in Action essay, "Breathalyzer," describes how law enforcement makes use of a redox reaction to apprehend drunk drivers.

Summary of Concepts & Facts

- In redox reactions, oxidation and reduction always occur simultaneously. Oxidation is characterized by the loss of electrons, reduction by the gain of electrons.
- Oxidation numbers help us keep track of charge distribution and are assigned to all atoms in a compound or ion according to specific rules. Oxidation can be defined as an increase in oxidation number; reduction can be defined as a decrease in oxidation number.
- Many redox reactions can be subclassified as combination, decomposition, combustion, displacement, or disproportionation reactions.

Review of Concepts & Facts

4.4.1 Which of the following combination reactions is not a redox reaction?
(a) $2Mg(s) + O_2(g) \longrightarrow 2MgO(s)$
(b) $H_2(g) + F_2(g) \longrightarrow 2HF(g)$
(c) $NH_3(g) + HCl(g) \longrightarrow NH_4Cl(s)$
(d) $2Na(s) + S(s) \longrightarrow Na_2S(s)$

4.4.2 Determine the oxidation number of the underlined atoms in the following compounds and ions: (a) $\underline{Mn}O_2$, (b) $\underline{N}_2F_4$, (c) $\underline{Cr}O_2^-$.

4.5 Concentration of Solutions

Learning Objectives

- Calculate the molarity of a solution.
- Determine the concentration of a solution that has been diluted.

To study solution stoichiometry, we must know how much of the reactants are present in a solution and also how to control the amounts of reactants used to bring about a reaction in aqueous solution.

The ***concentration of a solution*** is *the amount of solute present in a given amount of solvent or a given amount of solution.* (For this discussion, we will assume the solute is a liquid or a solid and the solvent is a liquid.) The concentration of a solution can be expressed in many different ways, as we will see in Chapter 12. Here we will consider one of the most commonly used units in chemistry, ***molarity (M),*** or ***molar concentration,*** which is *the number of moles of solute per liter of solution.* Molarity is defined as

$$\text{molarity} = \frac{\text{moles of solute}}{\text{liters of solution}} \quad (4.1)$$

Equation (4.1) can also be expressed algebraically as

$$M = \frac{n}{V} \quad (4.2)$$

where n denotes the number of moles of solute and V is the volume of the solution in liters.

A 1.46 molar glucose ($C_6H_{12}O_6$) solution, written as 1.46 M $C_6H_{12}O_6$, contains 1.46 moles of the solute ($C_6H_{12}O_6$) in 1 L of the solution. Of course, we do not always work with solution volumes of 1 L. Thus, a 500-mL solution containing 0.730 moles of $C_6H_{12}O_6$ also has a concentration of 1.46 *M*

$$\text{molarity} = \frac{0.730 \text{ mol } C_6H_{12}O_6}{500 \cancel{\text{mL soln}}} \times \frac{1000 \cancel{\text{mL soln}}}{1 \text{ L soln}} = 1.46\,M\,C_6H_{12}O_6$$

Note that concentration, like density, is an intensive property, so its value does not depend on how much of the solution is present.

It is important to keep in mind that molarity refers only to the amount of solute originally dissolved in water and does not take into account any subsequent processes, such as the dissociation of a salt or the ionization of an acid. Consider what happens when a sample of potassium chloride (KCl) is dissolved in enough water to make a 1 *M* solution:

$$KCl(s) \xrightarrow{H_2O} K^+(aq) + Cl^-(aq)$$

Because KCl is a strong electrolyte, it undergoes complete dissociation in solution. Thus, a 1 *M* KCl solution contains 1 mole of K^+ ions and 1 mole of Cl^- ions, and no KCl units are present. The concentrations of the ions can be expressed as $[K^+] = 1\,M$ and $[Cl^-] = 1\,M$, where the square brackets [ ] indicate that the concentration is expressed in molarity. Similarly, in a 1 *M* barium nitrate $[Ba(NO_3)_2]$ solution

$$Ba(NO_3)_2(s) \xrightarrow{H_2O} Ba^{2+}(aq) + 2NO_3^-(aq)$$

we have $[Ba^{2+}] = 1\,M$ and $[NO_3^-] = 2\,M$ and no $Ba(NO_3)_2$ units at all.

The procedure for preparing a solution of known molarity is as follows. First, the solute is accurately weighed and transferred to a volumetric flask through a funnel (Figure 4.18). Next, water is added to the flask, which is carefully swirled to dissolve the solid. After *all* the solid has dissolved, more water is added slowly to bring the level of solution exactly to the volume mark. Knowing the volume of the solution in the flask and the quantity of compound (the number of moles) dissolved, we can calculate the molarity of the solution using Equation (4.1). Note that this procedure does not require knowing the amount of water added, as long as the volume of the final solution is known.

Examples 4.7 and 4.8 illustrate the applications of Equations (4.1) and (4.2).

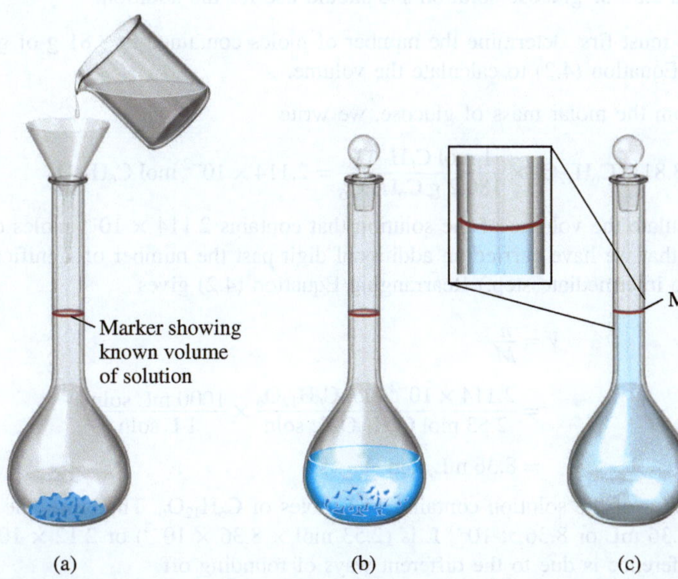

(a) (b) (c)

Figure 4.18 *Preparing a solution of known molarity. (a) A known amount of a solid solute is transferred into the volumetric flask; then water is added through a funnel. (b) The solid is slowly dissolved by gently swirling the flask. (c) After the solid has completely dissolved, more water is added to bring the level of solution to the mark. Knowing the volume of the solution and the amount of solute dissolved in it, we can calculate the molarity of the prepared solution.*

A $K_2Cr_2O_7$ solution.
Ken Karp/McGraw Hill

Example 4.7

How many grams of potassium dichromate ($K_2Cr_2O_7$) are required to prepare a 250.-mL solution whose concentration is 2.16 *M*?

Strategy How many moles of $K_2Cr_2O_7$ does a 1-L (or 1000 mL) 2.16 *M* $K_2Cr_2O_7$ solution contain? A 250.-mL solution? How would you convert moles to grams?

Solution The first step is to determine the number of moles of $K_2Cr_2O_7$ in 250. mL or 0.250 L of a 2.16 *M* solution. Rearranging Equation (4.1) gives

$$\text{moles of solute} = \text{molarity} \times \text{L soln}$$

Thus,

$$\text{moles of } K_2Cr_2O_7 = \frac{2.16 \text{ mol } K_2Cr_2O_7}{1 \text{ }\cancel{\text{L soln}}} \times 0.250 \text{ }\cancel{\text{L soln}}$$
$$= 0.540 \text{ mol } K_2Cr_2O_7$$

The molar mass of $K_2Cr_2O_7$ is 294.2 g, so we write

$$\text{grams of } K_2Cr_2O_7 \text{ needed} = 0.540 \text{ }\cancel{\text{mol } K_2Cr_2O_7} \times \frac{294.2 \text{ g } K_2Cr_2O_7}{1 \text{ }\cancel{\text{mol } K_2Cr_2O_7}}$$
$$= 159 \text{ g } K_2Cr_2O_7$$

Check As a ballpark estimate, the mass should be given by [molarity (mol/L) × volume (L) × molar mass (g/mol)] or [2 mol/L × 0.25 L × 300 g/mol] = 150 g. So the answer is reasonable.

Practice Exercise What is the molarity of an 85.0-mL ethanol (C_2H_5OH) solution containing 1.77 g of ethanol?

Similar problems: 4.65, 4.68.

Example 4.8

A chemist needs to add 3.81 g of glucose to a reaction mixture. Calculate the volume in milliliters of a 2.53 *M* glucose solution she should use for the addition.

Strategy We must first determine the number of moles contained in 3.81 g of glucose and then use Equation (4.2) to calculate the volume.

Solution From the molar mass of glucose, we write

$$3.81 \text{ }\cancel{\text{g } C_6H_{12}O_6} \times \frac{1 \text{ mol } C_6H_{12}O_6}{180.2 \text{ }\cancel{\text{g } C_6H_{12}O_6}} = 2.114 \times 10^{-2} \text{ mol } C_6H_{12}O_6$$

Next, we calculate the volume of the solution that contains 2.114×10^{-2} moles of the solute. (Note that we have carried an additional digit past the number of significant figures for the intermediate step.) Rearranging Equation (4.2) gives

$$V = \frac{n}{M}$$
$$= \frac{2.114 \times 10^{-2} \text{ mol } C_6H_{12}O_6}{2.53 \text{ mol } C_6H_{12}O_6/\text{L soln}} \times \frac{1000 \text{ mL soln}}{1 \text{ L soln}}$$
$$= 8.36 \text{ mL soln}$$

Check One liter of the solution contains 2.53 moles of $C_6H_{12}O_6$. Therefore, the number of moles in 8.36 mL or 8.36×10^{-3} L is ($2.53 \text{ mol} \times 8.36 \times 10^{-3}$) or 2.12×10^{-2} mol. The small difference is due to the different ways of rounding off.

Practice Exercise What volume (in milliliters) of a 0.315 *M* NaOH solution contains 6.22 g of NaOH?

Similar problem: 4.67.

Dilution of Solutions

Concentrated solutions are often stored in the laboratory stockroom for use as needed. Frequently we dilute these "stock" solutions before working with them. ***Dilution*** *is the procedure for preparing a less concentrated solution from a more concentrated one.*

Suppose that we want to prepare 1 L of a 0.400 *M* $KMnO_4$ solution from a solution of 1.00 *M* $KMnO_4$. For this purpose, we need 0.400 moles of $KMnO_4$. Because there is 1.00 mole of $KMnO_4$ in 1 L of a 1.00 *M* $KMnO_4$ solution, there is 0.400 moles of $KMnO_4$ in 0.400 L of the same solution:

$$\frac{1.00 \text{ mol}}{1 \text{ L soln}} = \frac{0.400 \text{ mol}}{0.400 \text{ L soln}}$$

Two $KMnO_4$ solutions of different concentrations.
Ken Karp/McGraw Hill

Therefore, we must withdraw 400 mL from the 1.00 *M* $KMnO_4$ solution and dilute it to 1000 mL by adding water (in a 1-L volumetric flask). This method gives us 1 L of the desired solution of 0.400 *M* $KMnO_4$.

In carrying out a dilution process, it is useful to remember that adding more solvent to a given amount of the stock solution changes (decreases) the concentration of the solution without changing the number of moles of solute present in the solution (Figure 4.19). In other words,

$$\text{moles of solute before dilution} = \text{moles of solute after dilution}$$

Molarity is defined as moles of solute in 1 L of solution, so the number of moles of solute is given by [see Equation (4.2)]

$$\underbrace{\frac{\text{moles of solute}}{\text{liters of soln}}}_{M} \times \underbrace{\text{volume of soln (in liters)}}_{V} = \underbrace{\text{moles of solute}}_{n}$$

or

$$MV = n$$

Because all the solute comes from the original stock solution, we can conclude that *n* remains the same; that is,

$$\underset{\substack{\text{moles of solute}\\ \text{before dilution}}}{M_iV_i} = \underset{\substack{\text{moles of solute}\\ \text{after dilution}}}{M_fV_f} \qquad (4.3)$$

where M_i and M_f are the initial and final concentrations of the solution in molarity and V_i and V_f are the initial and final volumes of the solution, respectively. Of course, the units of V_i and V_f must be the same (mL or L) for the calculation to work. To check the reasonableness of your results, be sure that $M_i > M_f$ and $V_f > V_i$.

We apply Equation (4.3) in Example 4.9.

Student Hot Spot
Student data indicate you may struggle with dilution calculations. Access your eBook for additional Learning Resources on this topic.

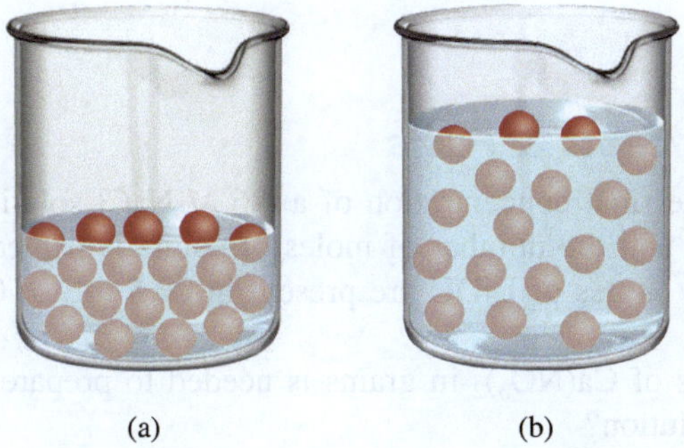

Figure 4.19 *The dilution of a more concentrated solution (a) to a less concentrated one (b) does not change the total number of solute particles (18).*

Example 4.9

Describe how you would prepare 5.00×10^2 mL of a 1.75 *M* H_2SO_4 solution, starting with an 8.61 *M* stock solution of H_2SO_4.

Strategy Because the concentration of the final solution is less than that of the original one, this is a dilution process. Keep in mind that in dilution, the concentration of the solution decreases but the number of moles of the solute remains the same.

Solution We prepare for the calculation by tabulating our data:

$$M_i = 8.61\ M \qquad M_f = 1.75\ M$$
$$V_i = ? \qquad V_f = 5.00 \times 10^2\ \text{mL}$$

Substituting in Equation (4.3),

$$(8.61\ M)(V_i) = (1.75\ M)(5.00 \times 10^2\ \text{mL})$$
$$V_i = \frac{(1.75\ M)(5.00 \times 10^2\ \text{mL})}{8.61\ M}$$
$$= 102\ \text{mL}$$

Thus, we must dilute 102 mL of the 8.61 *M* H_2SO_4 solution with sufficient water to give a final volume of 5.00×10^2 mL in a 500-mL volumetric flask to obtain the desired concentration.

Check The initial volume is less than the final volume, so the answer is reasonable.

Practice Exercise How would you prepare 2.00×10^2 mL of a 0.866 *M* NaOH solution, starting with a 5.07 *M* stock solution?

Similar problems: 4.75, 4.76.

Now that we have discussed the concentration and dilution of solutions, we can examine the quantitative aspects of reactions in aqueous solution or *solution stoichiometry.* Sections 4.6, and 4.7 focus on two techniques for studying solution stoichiometry: gravimetric analysis and titration. These techniques are important tools of ***quantitative analysis,*** which is *the determination of the amount or concentration of a substance in a sample.*

Summary of Concepts & Facts

- The concentration of a solution is the amount of solute present in a given amount of solution. Molarity expresses concentration as the number of moles of solute in 1 L of solution.
- Adding a solvent to a solution, a process known as dilution, decreases the concentration (molarity) of the solution without changing the total number of moles of solute present in the solution.

Review of Concepts & Facts

4.5.1 What is the final concentration of a 0.6 *M* NaCl solution if its volume is doubled and the number of moles of solute is tripled?

4.5.2 How many moles of LiOH are present in 27.5 mL of 0.250 *M* LiOH solution?

4.5.3 What mass of $Ca(NO_3)_2$ in grams is needed to prepare 375 mL of a 1.25 *M* solution?

4.6 Gravimetric Analysis

Learning Objective

- Solve gravimetric analysis problems using solution stoichiometry techniques.

Gravimetric analysis *is an analytical technique based on the measurement of mass.* One type of gravimetric analysis experiment involves the formation, isolation, and mass determination of a precipitate. Generally, this procedure is applied to ionic compounds. First, a sample substance of unknown composition is dissolved in water and allowed to react with another substance to form a precipitate. Then the precipitate is filtered off, dried, and weighed. Knowing the mass and chemical formula of the precipitate formed, we can calculate the mass of a particular chemical component (that is, the anion or cation) of the original sample. Finally, from the mass of the component and the mass of the original sample, we can determine the percent composition by mass of the component in the original compound.

A reaction that is often studied in gravimetric analysis, because the reactants can be obtained in pure form, is

$$AgNO_3(aq) + NaCl(aq) \longrightarrow NaNO_3(aq) + AgCl(s)$$

The net ionic equation is

$$Ag^+(aq) + Cl^-(aq) \longrightarrow AgCl(s)$$

The precipitate is silver chloride (see Table 4.2). As an example, let us say that we wanted to determine *experimentally* the percent by mass of Cl in NaCl. First, we would accurately weigh out a sample of NaCl and dissolve it in water. Next, we would add enough $AgNO_3$ solution to the NaCl solution to cause the precipitation of all the Cl^- ions present in solution as AgCl. In this procedure, NaCl is the limiting reactant and $AgNO_3$ the excess reactant. The AgCl precipitate is separated from the solution by filtration, dried, and weighed. From the measured mass of AgCl, we can calculate the mass of Cl using the percent by mass of Cl in AgCl. Because this same amount of Cl was present in the original NaCl sample, we can calculate the percent by mass of Cl in NaCl. Figure 4.20 shows how this procedure is performed.

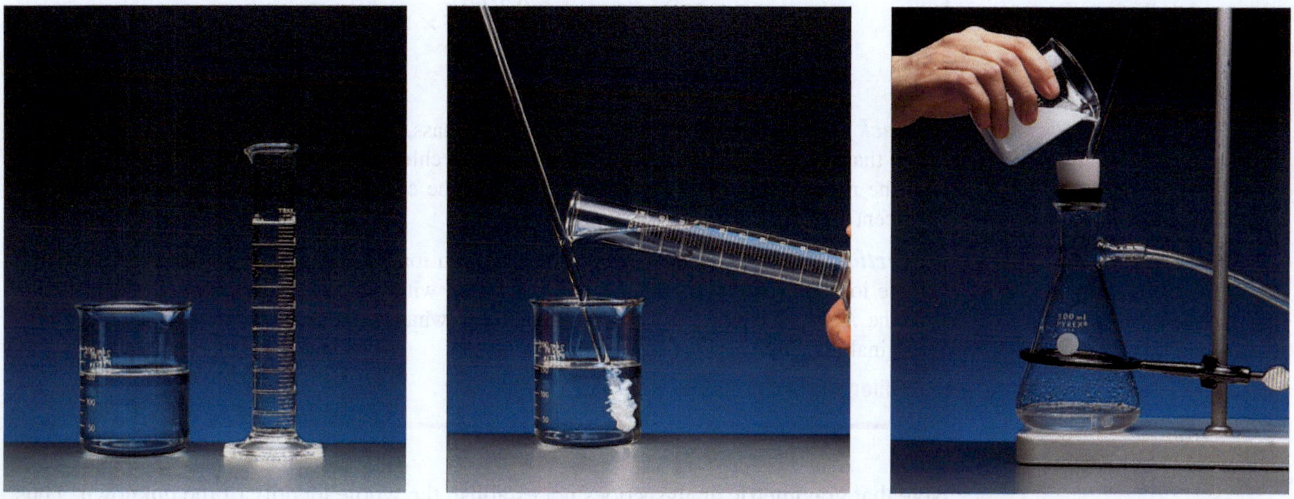

(a) (b) (c)

Figure 4.20 *Basic steps for gravimetric analysis. (a) A solution containing a known amount of NaCl in a beaker. (b) The precipitation of AgCl upon the addition of $AgNO_3$ solution from a measuring cylinder. In this reaction, $AgNO_3$ is the excess reactant and NaCl is the limiting reactant. (c) The solution containing the AgCl precipitate is filtered through a preweighed sintered-disk crucible, which allows the liquid (but not the precipitate) to pass through. The crucible is then removed from the apparatus, dried in an oven, and weighed again. The difference between this mass and that of the empty crucible gives the mass of the AgCl precipitate.*

(a, b, c): Ken Karp/McGraw Hill

Gravimetric analysis is a highly accurate technique because the mass of a sample can be measured accurately. However, this procedure is applicable only to reactions that go to completion, or have nearly 100 percent yield. Thus, if AgCl were slightly soluble instead of being insoluble, it would not be possible to remove all the Cl^- ions from the NaCl solution and the subsequent calculation would be in error.

Example 4.10 shows the calculations involved in a gravimetric experiment.

Example 4.10

A 0.5662-g sample of an ionic compound containing chloride ions and an unknown metal is dissolved in water and treated with an excess of $AgNO_3$. If 1.0882 g of AgCl precipitate forms, what is the percent by mass of Cl in the original compound?

Strategy We are asked to calculate the percent by mass of Cl in the unknown sample, which is

$$\%\text{Cl} = \frac{\text{mass of Cl}}{0.5662 \text{ g sample}} \times 100\%$$

The only source of Cl^- ions is the original compound. These chloride ions eventually end up in the AgCl precipitate. Can we calculate the mass of the Cl^- ions if we know the percent by mass of Cl in AgCl?

Solution The molar masses of Cl and AgCl are 35.45 g and 143.4 g, respectively. Therefore, the percent by mass of Cl in AgCl is given by

$$\begin{aligned}\%\text{Cl} &= \frac{35.45 \text{ g Cl}}{143.4 \text{ g AgCl}} \times 100\% \\ &= 24.72\%\end{aligned}$$

Next, we calculate the mass of Cl in 1.0882 g of AgCl. To do so we convert 24.72 percent to 0.2472 and write

$$\begin{aligned}\text{mass of Cl} &= 0.2472 \times 1.0882 \text{ g} \\ &= 0.2690 \text{ g}\end{aligned}$$

Because the original compound also contained this amount of Cl^- ions, the percent by mass of Cl in the compound is

$$\begin{aligned}\%\text{Cl} &= \frac{0.2690 \text{ g}}{0.5662 \text{ g}} \times 100\% \\ &= 47.51\%\end{aligned}$$

Check AgCl is about 25 percent chloride by mass, so the roughly 1 g of AgCl precipitate that formed corresponds to about 0.25 g of chloride, which is a little less than half of the mass of the original sample. Therefore, the calculated percent chloride of 47.51 percent is reasonable.

Practice Exercise A sample of 0.3220 g of an ionic compound containing the bromide ion (Br^-) is dissolved in water and treated with an excess of $AgNO_3$. If the mass of the AgBr precipitate that forms is 0.6964 g, what is the percent by mass of Br in the original compound?

Similar problem: 4.82.

Note that gravimetric analysis does not establish the whole identity of the unknown. Thus, in Example 4.10 we still do not know what the cation is. However, knowing the percent by mass of Cl greatly helps us to narrow the possibilities. Because no two compounds containing the same anion (or cation) have the same percent composition by mass, comparison of the percent by mass obtained from gravimetric analysis with that calculated from a series of known compounds would reveal the identity of the unknown.

Summary of Concepts & Facts

- Gravimetric analysis is a technique for determining the identity of a compound and/or the concentration of a solution by measuring mass. Gravimetric experiments often involve precipitation reactions.

Review of Concepts & Facts

4.6.1 Calculate the mass of AgBr formed if a solution containing 6.00 g of KBr is treated with an excess of $AgNO_3$.

4.7 Titrations

Learning Objectives

- Explain the basis of a titration.
- Solve acid-base and redox titration problems.

Quantitative studies of many reactions conducted in solution are conveniently carried out using a technique known as titration. In a ***titration,*** *a solution of accurately known concentration,* called a ***standard solution,*** *is added gradually to another solution of unknown concentration, until the chemical reaction between the two solutions is complete.* If we know the volumes of the standard and unknown solutions used in the titration, along with the concentration of the standard solution, we can calculate the concentration of the unknown solution.

Acid-Base Titrations

Acid-base neutralization reactions are commonly studied using titrations. Sodium hydroxide is one of the bases commonly used in the laboratory. However, it is difficult to obtain solid sodium hydroxide in a pure form because it has a tendency to absorb water from air, and its solution reacts with carbon dioxide. For these reasons, a solution of sodium hydroxide must be *standardized* before it can be used in accurate analytical work. We can standardize the sodium hydroxide solution by titrating it against an acid solution of accurately known concentration. The acid often chosen for this task is a monoprotic, weak acid called potassium hydrogen phthalate (KHP), for which the molecular formula is $KHC_8H_4O_4$ (molar mass = 204.2 g). KHP is a white, soluble solid that is commercially available in highly pure form. The reaction between KHP and sodium hydroxide is

$$KHC_8H_4O_4(aq) + NaOH(aq) \longrightarrow KNaC_8H_4O_4(aq) + H_2O(l)$$

and the net ionic equation is

$$HC_8H_4O_4^-(aq) + OH^-(aq) \longrightarrow C_8H_4O_4^{2-}(aq) + H_2O(l)$$

The procedure for the titration is shown in Figure 4.21. First, a known amount of KHP is transferred to an Erlenmeyer flask and some distilled water is added to make up a solution. Next, NaOH solution is carefully added to the KHP solution from a buret until we reach the ***equivalence point,*** that is, *the point at which the acid has completely reacted with or been neutralized by the base.* The equivalence point is

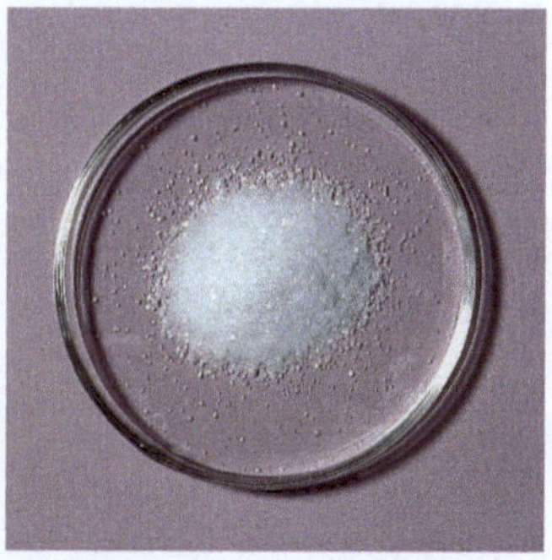

Potassium hydrogen phthalate (KHP).
Ken Karp/McGraw Hill

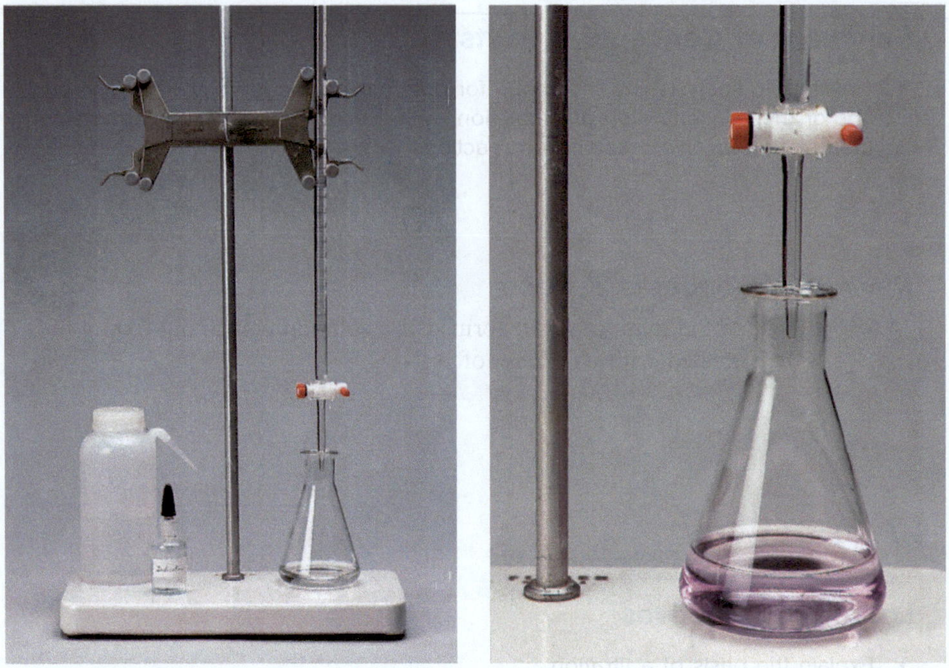

(a) (b)

Figure 4.21 *(a) Apparatus for acid-base titration. A NaOH solution is added from the buret to a KHP solution in an Erlenmeyer flask. (b) A reddish-pink color appears when the equivalence point is reached. The color here has been intensified for visual display.*

(a and b): Ken Karp/McGraw Hill

usually signaled by a sharp change in the color of an indicator in the acid solution. In acid-base titrations, ***indicators*** are *substances that have distinctly different colors in acidic and basic media.* One commonly used indicator is phenolphthalein, which is colorless in acidic and neutral solutions but reddish-pink in basic solutions. At the equivalence point, all the KHP present has been neutralized by the added NaOH and the solution is still colorless. However, if we add just one more drop of NaOH solution from the buret, the solution will immediately turn pink because the solution is now basic. Example 4.11 illustrates such a titration.

Example 4.11

In a titration experiment, a student finds that 23.48 mL of a NaOH solution are needed to neutralize 0.5468 g of KHP. What is the concentration (in molarity) of the NaOH solution?

Strategy We want to determine the molarity of the NaOH solution. What is the definition of molarity?

$$\underset{\text{want to calculate}}{\text{molarity of NaOH}} = \frac{\overset{\text{need to find}}{\text{mol NaOH}}}{\underset{\text{given}}{\text{L soln}}}$$

The volume of NaOH solution is given in the problem. Therefore, we need to find the number of moles of NaOH to solve for molarity. From the preceding equation for the reaction between KHP and NaOH shown in the text, we see that 1 mole of KHP neutralizes 1 mole of NaOH. How many moles of KHP are contained in 0.5468 g of KHP?

Solution First, we calculate the number of moles of KHP consumed in the titration:

$$\text{moles of KHP} = 0.5468\ \text{g KHP} \times \frac{1\ \text{mol KHP}}{204.2\ \text{g KHP}}$$
$$= 2.678 \times 10^{-3}\ \text{mol KHP}$$

(Continued)

Because 1 mol KHP ≏ 1 mol NaOH, there must be 2.678×10^{-3} mol of NaOH in 23.48 mL of NaOH solution. Finally, we calculate the number of moles of NaOH in 1 L of the solution or the molarity as follows:

$$\text{molarity of NaOH soln} = \frac{2.678 \times 10^{-3}\ \text{mol NaOH}}{23.48\ \cancel{\text{mL soln}}} \times \frac{1000\ \cancel{\text{mL soln}}}{1\ \text{L soln}}$$
$$= 0.1141\ \text{mol NaOH}/1\ \text{L soln} = 0.1141\ M$$

Practice Exercise How many grams of KHP are needed to neutralize 18.64 mL of a 0.1004 *M* NaOH solution?

Similar problems: 4.91, 4.92.

The neutralization reaction between NaOH and KHP is one of the simplest types of acid-base neutralization known. Suppose, though, that instead of KHP, we wanted to use a diprotic acid such as H_2SO_4 for the titration. The reaction is represented by

$$2\text{NaOH}(aq) + \text{H}_2\text{SO}_4(aq) \longrightarrow \text{Na}_2\text{SO}_4(aq) + 2\text{H}_2\text{O}(l)$$

Because 2 mol NaOH ≏ 1 mol H_2SO_4, we need twice as much NaOH to react completely with a H_2SO_4 solution of the *same* molar concentration and volume as a monoprotic acid like HCl. On the other hand, we would need twice the amount of HCl to neutralize a $Ba(OH)_2$ solution compared to a NaOH solution having the same concentration and volume because 1 mole of $Ba(OH)_2$ yields 2 moles of OH^- ions:

$$2\text{HCl}(aq) + \text{Ba(OH)}_2(aq) \longrightarrow \text{BaCl}_2(aq) + 2\text{H}_2\text{O}(l)$$

In calculations involving acid-base titrations, regardless of the acid or base that takes place in the reaction, keep in mind that the total number of moles of H^+ ions that have reacted at the equivalence point must be equal to the total number of moles of OH^- ions that have reacted.

Example 4.12 shows the titration of a NaOH solution with a diprotic acid.

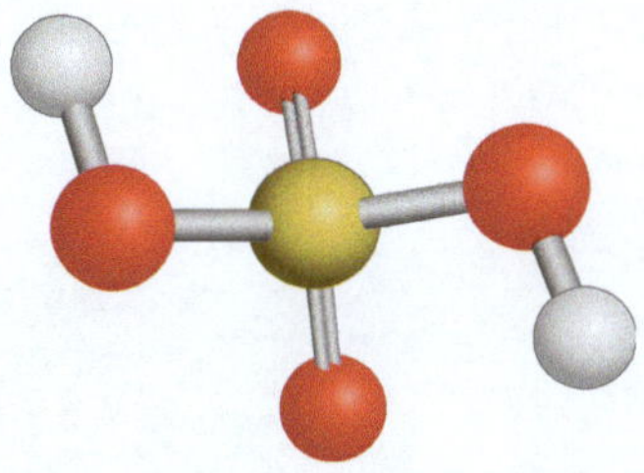

H_2SO_4 has two ionizable protons.

Student Hot Spot

Student data indicate you may struggle with titration calculations. Access your eBook for additional Learning Resources on this topic.

Example 4.12

The sodium hydroxide solution standardized in Example 4.11 is used to titrate 25.00 mL of a sulfuric acid solution. The titration requires 43.79 mL of the 0.1172 *M* NaOH solution to completely neutralize the acid. What is the concentration of the H_2SO_4 solution?

Strategy We want to calculate the concentration of the H_2SO_4 solution. Starting with the volume of NaOH solution required to neutralize the acid, we calculate the moles of NaOH.

$$\underset{\text{measured}}{\text{L soln}} \times \frac{\text{mol NaOH}}{\underset{\text{given}}{\text{L soln}}} = \overset{\text{want to find}}{\text{mol NaOH}}$$

From the equation for the neutralization reaction just shown, we see that 2 moles of NaOH neutralize 1 mole of H_2SO_4. How many moles of NaOH are contained in 43.79 mL of a 0.1172 *M* NaOH solution? How many moles of H_2SO_4 would this quantity of NaOH neutralize? What would be the concentration of the H_2SO_4 solution?

Solution First, we calculate the number of moles of NaOH contained in 43.79 mL of solution:

$$43.79\ \text{mL} \times \frac{1\ \text{L soln}}{1000\ \text{mL soln}} \times \frac{0.1172\ \text{mol NaOH}}{\text{L soln}} = 5.132 \times 10^{-3}\ \text{mol NaOH}$$

(Continued)

From the stoichiometry we see that 1 mol $H_2SO_4 \simeq$ 2 mol NaOH. Therefore, the number of moles of H_2SO_4 reacted must be

$$5.132 \times 10^{-3} \text{ mol NaOH} \times \frac{1 \text{ mol } H_2SO_4}{2 \text{ mol NaOH}} = 2.566 \times 10^{-3} \text{ mol } H_2SO_4$$

From the definition of molarity [see Equation (4.1)], we have

$$\text{molarity} = \frac{\text{moles of solute}}{\text{liters of soln}}$$

So the molarity of the H_2SO_4 solution is

$$\frac{2.566 \times 10^{-3} \text{ mol } H_2SO_4}{25.00 \text{ mL} \times (1 \text{ L}/1000 \text{ mL})} = 0.1026\, M\, H_2SO_4$$

Practice Exercise If 60.02 mL of 0.427 *M* KOH solution are required to neutralize 10.1 mL of H_2SO_4 solution, what is the concentration of the H_2SO_4 solution in molarity?

Similar problem: 4.93(b), (c).

Redox Titrations

As mentioned earlier, redox reactions involve the transfer of electrons, and acid-base reactions involve the transfer of protons. Just as an acid can be titrated against a base, we can titrate an oxidizing agent against a reducing agent, using a similar procedure. We can, for example, carefully add a solution containing an oxidizing agent to a solution containing a reducing agent. The *equivalence point* is reached when the reducing agent is completely oxidized by the oxidizing agent.

Like acid-base titrations, redox titrations normally require an indicator that clearly changes color. In the presence of large amounts of reducing agent, the color of the indicator is characteristic of its reduced form. The indicator assumes the color of its oxidized form when it is present in an oxidizing medium. At or near the equivalence point, a sharp change in the indicator's color will occur as it changes from one form to the other, so the equivalence point can be readily identified.

Two common oxidizing agents are potassium permanganate ($KMnO_4$) and potassium dichromate ($K_2Cr_2O_7$). As Figure 4.22 shows, the colors of the permanganate and dichromate anions are distinctly different from those of the reduced species:

$$\underset{\text{purple}}{MnO_4^-} \longrightarrow \underset{\text{light pink}}{Mn^{2+}}$$

$$\underset{\text{orange yellow}}{Cr_2O_7^{2-}} \longrightarrow \underset{\text{green}}{Cr^{3+}}$$

Figure 4.22 *Left to right: Solutions containing the MnO_4^-, Mn^{2+}, $Cr_2O_7^{2-}$, and Cr^{3+} ions.*
Ken Karp/McGraw Hill

Thus, these oxidizing agents can themselves be used as an *internal* indicator in a redox titration because they have distinctly different colors in the oxidized and reduced forms.

Redox titrations require the same type of calculations (based on the mole method) as acid-base neutralizations. The difference is that the equations and the stoichiometry tend to be more complex for redox reactions. Example 4.13 describes a redox titration.

Example 4.13

A 16.42-mL volume of 0.1327 *M* $KMnO_4$ solution is needed to oxidize 25.00 mL of a $FeSO_4$ solution in an acidic medium. What is the concentration of the $FeSO_4$ solution in molarity? The net ionic equation is

$$5Fe^{2+} + MnO_4^- + 8H^+ \longrightarrow Mn^{2+} + 5Fe^{3+} + 4H_2O$$

Strategy We want to calculate the molarity of the $FeSO_4$ solution. From the definition of molarity

need to find

$$\text{molarity of } FeSO_4 = \frac{\text{mol } FeSO_4}{\text{L soln}}$$

want to calculate given

The volume of the $FeSO_4$ solution is given in the problem. Therefore, we need to find the number of moles of $FeSO_4$ to solve for the molarity. From the net ionic equation, what is the stoichiometric equivalence between Fe^{2+} and MnO_4^-? How many moles of $KMnO_4$ are contained in 16.42 mL of 0.1327 *M* $KMnO_4$ solution?

Solution The number of moles of $KMnO_4$ in 16.42 mL of the solution is

$$\text{moles of } KMnO_4 = \frac{0.1327 \text{ mol } KMnO_4}{1000 \text{ mL soln}} \times 16.42 \text{ mL}$$

$$= 2.179 \times 10^{-3} \text{ mol } KMnO_4$$

From the net ionic equation, we see that 5 mol $Fe^{2+} \simeq 1$ mol MnO_4^-. Therefore, the number of moles of $FeSO_4$ oxidized is

$$\text{moles } FeSO_4 = 2.179 \times 10^{-3} \text{ mol } KMnO_4 \times \frac{5 \text{ mol } FeSO_4}{1 \text{ mol } KMnO_4}$$

$$= 1.090 \times 10^{-2} \text{ mol } FeSO_4$$

The concentration of the $FeSO_4$ solution in moles of $FeSO_4$ per liter of solution is

$$\text{molarity of } FeSO_4 = \frac{\text{mol } FeSO_4}{\text{L soln}}$$

$$= \frac{1.090 \times 10^{-2} \text{ mol } FeSO_4}{25.00 \text{ mL soln}} \times \frac{1000 \text{ mL soln}}{1 \text{ L soln}}$$

$$= 0.4360\ M$$

Practice Exercise How many milliliters of a 0.206 *M* HI solution are needed to reduce 22.5 mL of a 0.374 *M* $KMnO_4$ solution according to the following equation?

$$10HI + 2KMnO_4 + 3H_2SO_4 \longrightarrow 5I_2 + 2MnSO_4 + K_2SO_4 + 8H_2O$$

Similar problems: 4.95, 4.96.

Addition of a $KMnO_4$ solution from a buret to an acidic $FeSO_4$ solution.

Ken Karp/McGraw Hill

The Chemistry in Action essay, "Metal from the Sea," describes an industrial process that involves the types of reactions discussed in this chapter.

CHEMISTRY *in Action*

Metal from the Sea

Magnesium is a valuable, lightweight metal used as a structural material as well as in alloys, in batteries, and in chemical synthesis. Although magnesium is plentiful in Earth's crust, it is cheaper to "mine" the metal from seawater. Magnesium forms the second most abundant cation in the sea (after sodium); there are about 1.3 g of magnesium in a kilogram of seawater. The process for obtaining magnesium from seawater employs all three types of reactions discussed in this chapter: precipitation, acid-base, and redox reactions.

In the first stage in the recovery of magnesium, limestone ($CaCO_3$) is heated at high temperatures to produce quicklime, or calcium oxide (CaO):

$$CaCO_3(s) \longrightarrow CaO(s) + CO_2(g)$$

When calcium oxide is treated with seawater, it forms calcium hydroxide [$Ca(OH)_2$], which is slightly soluble and ionizes to give Ca^{2+} and OH^- ions:

$$CaO(s) + H_2O(l) \longrightarrow Ca^{2+}(aq) + 2OH^-(aq)$$

The surplus hydroxide ions cause the much less soluble magnesium hydroxide to precipitate:

$$Mg^{2+}(aq) + 2OH^-(aq) \longrightarrow Mg(OH)_2(s)$$

The solid magnesium hydroxide is filtered and reacted with hydrochloric acid to form magnesium chloride ($MgCl_2$):

$$Mg(OH)_2(s) + 2HCl(aq) \longrightarrow MgCl_2(aq) + H_2O(l)$$

After the water is evaporated, the solid magnesium chloride is melted in a steel cell. The molten magnesium chloride contains both Mg^{2+} and Cl^- ions. In a process called *electrolysis,* an electric current is passed through the cell to reduce the Mg^{2+} ions and oxidize the Cl^- ions. The half-reactions are

$$Mg^{2+} + 2e^- \longrightarrow Mg$$
$$2Cl^- \longrightarrow Cl_2 + 2e^-$$

The overall reaction is

$$MgCl_2(l) \longrightarrow Mg(l) + Cl_2(g)$$

This is how magnesium metal is produced. The chlorine gas generated can be converted to hydrochloric acid and recycled through the process.

Magnesium hydroxide was precipitated from processed seawater in settling ponds at the Dow Chemical Company once operated in Freeport, Texas.
Courtesy of Dow Chemical Company

Summary of Concepts & Facts

- In an acid-base titration, a solution of known concentration (say, a base) is added gradually to a solution of unknown concentration (say, an acid) with the goal of determining the unknown concentration. The point at which the reaction in the titration is complete, as shown by the change in the indicator's color, is called the equivalence point.
- Redox titrations are similar to acid-base titrations. The point at which the oxidation-reduction reaction is complete is also called the equivalence point.

Review of Concepts & Facts

4.7.1 A NaOH solution is initially mixed with an acid solution shown in (a). Which of the diagrams (b)–(d) corresponds to one of the following acids: HCl, H_2SO_4, H_3PO_4? Color codes: blue spheres (OH^- ions); red spheres (acid molecules); green spheres (anions of the acids). Assume all the acid-base neutralization reactions go to completion.

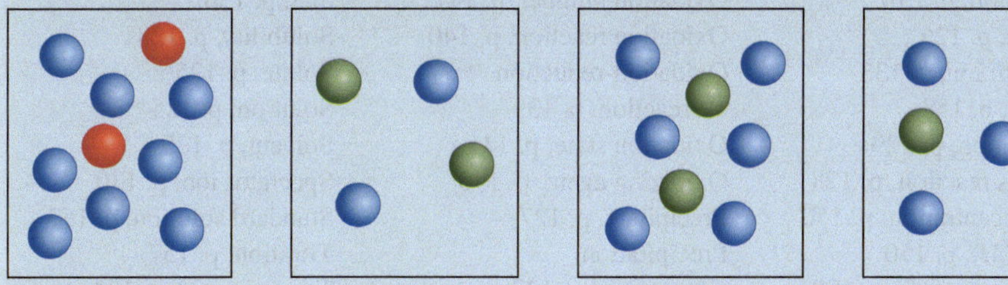

(a) (b) (c) (d)

4.7.2 What volume of a 0.215 *M* HNO_3 solution is required to neutralize 25.0 mL of 0.370 *M* NaOH?

4.7.3 If a solution of a reducing agent is titrated with a solution of an oxidizing agent, and the initial concentrations of the two solutions are the same, does that mean the equivalence point will be reached when an equal volume of oxidizing has been added? Explain.

4.7.4 The concentration of a $KMnO_4$ solution can be determined by redox titration with oxalate ion, $C_2O_4^{2-}$, according to the net ionic equation

$$2MnO_4^- + 5C_2O_4^{2-} + 16H^+ \longrightarrow 2Mn^{2+} + 10CO_2 + 8H_2O$$

What is the concentration of a $KMnO_4$ solution if 20.24 mL of 0.2558 *M* $C_2O_4^{2-}$ solution is needed to reduce 25.00 mL of $KMnO_4$ solution?

Chapter Summary

Reactions in Aqueous Solution Many chemical and almost all biological reactions occur in the aqueous medium. Substances (solutes) that dissolve in water (solvent) can be divided into two categories: electrolytes and nonelectrolytes, depending on their ability to conduct electricity. (Section 4.1)

Three Major Types of Reactions In a precipitation reaction, the product, an insoluble substance, separates from solution. Acid-base reactions involve the transfer of a proton (H^+) from an acid to a base. In an oxidation-reduction reaction or redox reaction, electrons are transferred from a reducing agent to an oxidizing agent. These three types of reactions represent the majority of reactions in chemical and biological systems. (Sections 4.2, 4.3, and 4.4)

Solution Stoichiometry Quantitative studies of reactions in solution require that we know the concentration of the solution, which is usually represented by the molarity unit. Each concentration unit has its advantages and limitations. These studies include gravimetric analysis, which involves the measurement of mass, and titrations in which the unknown concentration of a solution is determined by reaction with a solution of known concentration. (Sections 4.5, 4.6, and 4.7)

Key Equations

$\text{molarity} = \dfrac{\text{moles of solute}}{\text{liters of solution}}$ (4.1)	Calculating molarity
$M = \dfrac{n}{V}$ (4.2)	Calculating molarity
$M_iV_i = M_fV_f$ (4.3)	Dilution of solution

Key Words

Activity series, p. 145
Aqueous solution, p. 125
Brønsted acid, p. 133
Brønsted base, p. 133
Combination reaction, p. 144
Combustion reaction, p. 144
Concentration of a solution, p. 150
Decomposition reaction, p. 144
Dilution, p. 153
Diprotic acid, p. 134
Displacement reaction, p. 145
Disproportionation reaction, p. 148
Electrolyte, p. 125
Equivalence point, p. 157
Gravimetric analysis, p. 155
Half-reaction, p. 139
Hydration, p. 126
Hydronium ion, p. 133
Indicators, p. 158
Ionic equation, p. 129
Metathesis reaction, p. 128
Molar concentration, p. 150
Molarity (*M*), p. 150
Molecular equation, p. 129
Monoprotic acid, p. 134
Net ionic equation, p. 130
Neutralization reaction, p. 136
Nonelectrolyte, p. 125
Oxidation number, p. 141
Oxidation reaction, p. 140
Oxidation-reduction reaction, p. 139
Oxidation state, p. 141
Oxidizing agent, p. 140
Precipitate, p. 127
Precipitation reaction, p. 127
Quantitative analysis, p. 154
Redox reaction, p. 139
Reducing agent, p. 140
Reduction reaction, p. 140
Reversible reaction, p. 127
Salt, p. 136
Solubility, p. 128
Solute, p. 125
Solution, p. 125
Solvent, p. 125
Spectator ion, p. 130
Standard solution, p. 157
Titration, p. 157
Triprotic acid, p. 134

Questions & Problems

Red numbered problems solved in Student Solutions Manual

4.1 General Properties of Aqueous Solutions

Review Questions

4.1 Define *solute, solvent,* and *solution* by describing the process of dissolving a solid in a liquid.

4.2 What is the difference between a nonelectrolyte and an electrolyte? Between a weak electrolyte and a strong electrolyte?

4.3 Describe hydration. What properties of water enable its molecules to interact with ions in solution?

4.4 What is the difference between the following symbols in chemical equations: $\longrightarrow$ and $\rightleftharpoons$?

4.5 Water is an extremely weak electrolyte and therefore cannot conduct electricity. Why are we often cautioned not to operate electrical appliances when our hands are wet?

4.6 Sodium sulfate (Na_2SO_4) is a strong electrolyte. What species are present in $Na_2SO_4(aq)$?

Problems

4.7 The aqueous solutions of three compounds are shown. Identify each compound as a nonelectrolyte, a weak electrolyte, and a strong electrolyte.

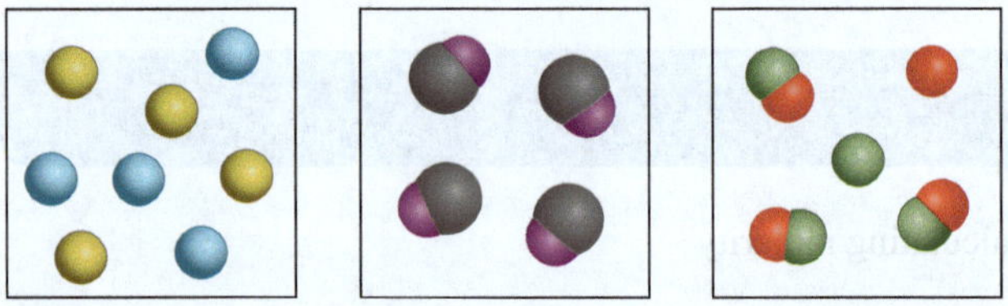

(a) (b) (c)

4.8 Which diagram (a)–(c) best represents the hydration of NaCl when dissolved in water? The Cl^- ion is larger in size than the Na^+ ion.

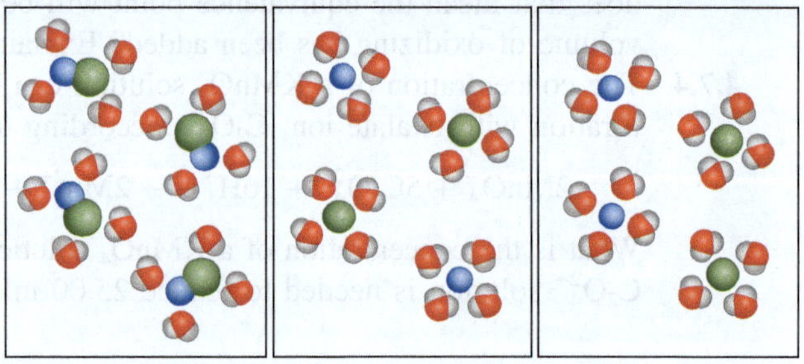

(a) (b) (c)

4.9 Identify each of the following substances as a strong electrolyte, weak electrolyte, or nonelectrolyte: (a) H_2O, (b) KCl, (c) HNO_3, (d) CH_3COOH, (e) $C_{12}H_{22}O_{11}$.

4.10 Identify each of the following substances as a strong electrolyte, weak electrolyte, or nonelectrolyte: (a) $Ba(NO_3)_2$, (b) Ne, (c) NH_3, (d) NaOH.

4.11 The passage of electricity through an electrolyte solution is caused by the movement of (a) electrons only, (b) cations only, (c) anions only, (d) both cations and anions.

4.12 Predict and explain which of the following systems are electrically conducting: (a) solid NaCl, (b) molten NaCl, (c) an aqueous solution of NaCl.

4.13 You are given a water-soluble compound X. Describe how you would determine whether it is an electrolyte or a nonelectrolyte. If it is an electrolyte, how would you determine whether it is strong or weak?

4.14 Explain why a solution of HCl in benzene does not conduct electricity but in water it does.

4.2 Precipitation Reactions

Review Questions

4.15 What is the difference between an ionic equation and a molecular equation?

4.16 What is the advantage of writing net ionic equations?

Problems

4.17 Two aqueous solutions of $AgNO_3$ and NaCl are mixed. Which of the diagrams (a)–(d) best represents the mixture? For simplicity, water molecules are not shown. (Color codes: Ag^+ = gray, Cl^- = green, Na^+ = purple, NO_3^- = blue.)

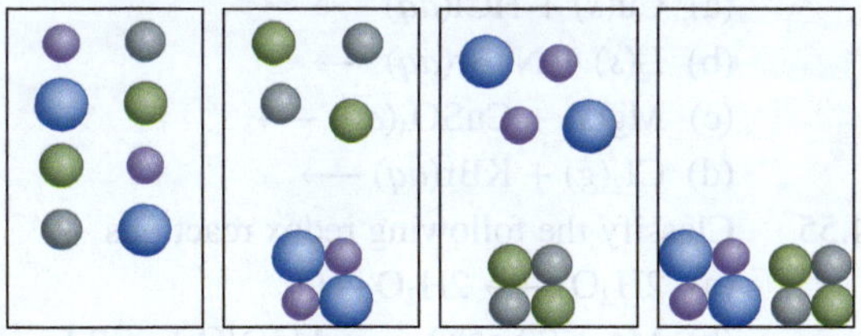

4.18 Two aqueous solutions of KOH and $MgCl_2$ are mixed. Which of the diagrams (a)–(d) best represents the mixture? For simplicity, water molecules are not shown. (Color codes: K^+ = yellow, OH^- = red, Mg^{2+} = purple, Cl^- = green.)

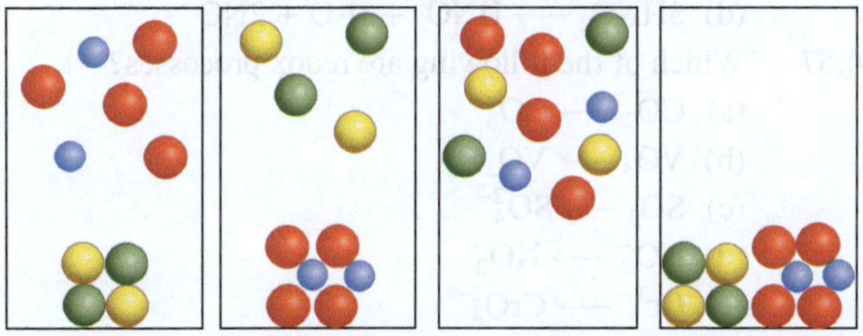

4.19 Characterize the following compounds as soluble or insoluble in water: (a) $Ca_3(PO_4)_2$, (b) $Mn(OH)_2$, (c) $AgClO_3$, (d) K_2S.

4.20 Characterize the following compounds as soluble or insoluble in water: (a) $CaCO_3$, (b) $ZnSO_4$, (c) $Hg(NO_3)_2$, (d) $HgSO_4$, (e) NH_4ClO_4.

4.21 Write ionic and net ionic equations for the following reactions.

(a) $AgNO_3(aq) + Na_2SO_4(aq) \longrightarrow$

(b) $BaCl_2(aq) + ZnSO_4(aq) \longrightarrow$

(c) $(NH_4)_2CO_3(aq) + CaCl_2(aq) \longrightarrow$

4.22 Write ionic and net ionic equations for the following reactions.

(a) $Na_2S(aq) + ZnCl_2(aq) \longrightarrow$

(b) $K_3PO_4(aq) + 3Sr(NO_3)_2(aq) \longrightarrow$

(c) $Mg(NO_3)_2(aq) + 2NaOH(aq) \longrightarrow$

4.23 Which of the following processes will likely result in a precipitation reaction: (a) mixing a $NaNO_3$ solution with a $CuSO_4$ solution, or (b) mixing a $BaCl_2$ solution with a K_2SO_4 solution? Write a net ionic equation for the precipitation reaction.

4.24 With reference to Table 4.2, suggest one method by which you might separate (a) K^+ from Ag^+, (b) Ba^{2+} from Pb^{2+}, (c) NH_4^+ from Ca^{2+}, (d) Ba^{2+} from Cu^{2+}. All cations are assumed to be in aqueous solution, and the common anion is the nitrate ion.

4.3 Acid-Base Reactions

Review Questions

4.25 List the general properties of acids and bases.

4.26 Give Arrhenius's and Brønsted's definitions of an acid and a base. Why are Brønsted's definitions more useful in describing acid-base properties?

4.27 Give an example of a monoprotic acid, a diprotic acid, and a triprotic acid.

4.28 What are the characteristics of an acid-base neutralization reaction?

4.29 What factors qualify a compound as a salt? Specify which of the following compounds are salts: CH_4, NaF, NaOH, CaO, $BaSO_4$, HNO_3, NH_3, KBr.

4.30 Identify the following as a weak or strong acid or base: (a) NH_3, (b) H_3PO_4, (c) LiOH, (d) HCOOH (formic acid), (e) H_2SO_4, (f) HF, (g) $Ba(OH)_2$.

Problems

4.31 Identify each of the following species as a Brønsted acid, base, or both: (a) HI, (b) CH_3COO^-, (c) $H_2PO_4^-$, (d) HSO_4^-.

4.32 Identify each of the following species as a Brønsted acid, base, or both: (a) PO_4^{3-}, (b) ClO_2^-, (c) NH_4^+, (d) HCO_3^-.

4.33 Balance the following equations and write the corresponding ionic and net ionic equations (if appropriate).

(a) $HBr(aq) + NH_3(aq) \longrightarrow$

(b) $Ba(OH)_2(aq) + H_3PO_4(aq) \longrightarrow$

(c) $HClO_4(aq) + Mg(OH)_2(s) \longrightarrow$

4.34 Balance the following equations and write the corresponding ionic and net ionic equations (if appropriate).

(a) $CH_3COOH(aq) + KOH(aq) \longrightarrow$

(b) $H_2CO_3(aq) + NaOH(aq) \longrightarrow$

(c) $HNO_3(aq) + Ba(OH)_2(aq) \longrightarrow$

4.4 Oxidation-Reduction Reactions

Review Questions

4.35 Give an example of a combination redox reaction, a decomposition redox reaction, and a displacement redox reaction.

4.36 True or false: All combustion reactions are redox reactions. Explain your answer.

4.37 What is an oxidation number? How is it used to identify redox reactions? Explain why, except for ionic compounds, oxidation number does not have any physical significance.

4.38 (a) Without referring to Figure 4.11, give the oxidation numbers of the alkali and alkaline earth metals in their compounds. (b) Give the highest oxidation numbers that the Groups 13 through 17 elements can have.

4.39 How is the activity series organized? How is it used in the study of redox reactions?

4.40 Use the following reaction to define *redox reaction, half-reaction, oxidizing agent,* and *reducing agent.*

$$4Na(s) + O_2(g) \longrightarrow 2Na_2O(s)$$

4.41 Is it possible to have a reaction in which oxidation occurs and reduction does not? Explain.

4.42 What is the requirement for an element to undergo disproportionation reactions? Name five common elements that are likely to take part in such reactions.

Problems

4.43 For the complete redox reactions given here, (i) break down each reaction into its half-reactions; (ii) identify the oxidizing agent; (iii) identify the reducing agent.

(a) $2Sr + O_2 \longrightarrow 2SrO$
(b) $2Li + H_2 \longrightarrow 2LiH$
(c) $2Cs + Br_2 \longrightarrow 2CsBr$
(d) $3Mg + N_2 \longrightarrow Mg_3N_2$

4.44 For the complete redox reactions given here, write the half-reactions and identify the oxidizing and reducing agents.

(a) $4Fe + 3O_2 \longrightarrow 2Fe_2O_3$
(b) $Cl_2 + 2NaBr \longrightarrow 2NaCl + Br_2$
(c) $Si + 2F_2 \longrightarrow SiF_4$
(d) $H_2 + Cl_2 \longrightarrow 2HCl$

4.45 Arrange the following species in order of increasing oxidation number of the sulfur atom: (a) H_2S, (b) S_8, (c) H_2SO_4, (d) S^{2-}, (e) HS^-, (f) SO_2, (g) SO_3.

4.46 Phosphorus forms many oxoacids. Indicate the oxidation number of phosphorus in each of the following acids: (a) HPO_3, (b) H_3PO_2, (c) H_3PO_3, (d) H_3PO_4, (e) $H_4P_2O_7$, (f) $H_5P_3O_{10}$.

4.47 Give the oxidation number of the underlined atoms in the following molecules and ions: (a) $\underline{Cl}F$, (b) $\underline{I}F_7$, (c) $\underline{C}H_4$, (d) $\underline{C}_2H_2$, (e) $\underline{C}_2H_4$, (f) $K_2\underline{Cr}O_4$, (g) $K_2\underline{Cr}_2O_7$, (h) $K\underline{Mn}O_4$, (i) $NaH\underline{C}O_3$, (j) $\underline{Li}_2$, (k) $Na\underline{I}O_3$, (l) $K\underline{O}_2$, (m) $\underline{P}F_6^-$, (n) $K\underline{Au}Cl_4$.

4.48 Give the oxidation number for the following species: H_2, Se_8, P_4, O, U, As_4, B_{12}.

4.49 Give oxidation number for the underlined atoms in the following molecules and ions: (a) $\underline{Cs}_2O$, (b) $Ca\underline{I}_2$, (c) $\underline{Al}_2O_3$, (d) $H_3\underline{As}O_3$, (e) $\underline{Ti}O_2$, (f) $\underline{Mo}O_4^{2-}$, (g) $\underline{Pt}Cl_4^{2-}$, (h) $\underline{Pt}Cl_6^{2-}$, (i) $\underline{Sn}F_2$, (j) $\underline{Cl}F_3$, (k) $\underline{Sb}F_6^-$.

4.50 Give the oxidation number of the underlined atoms in the following molecules and ions: (a) $Mg_3\underline{N}_2$, (b) $Cs\underline{O}_2$, (c) $Ca\underline{C}_2$, (d) $\underline{C}O_3^{2-}$, (e) $\underline{C}_2O_4^{2-}$, (f) $Zn\underline{O}_2^{2-}$, (g) $Na\underline{B}H_4$, (h) $\underline{W}O_4^{2-}$.

4.51 Nitric acid is a strong oxidizing agent. State which of the following species is *least* likely to be produced when nitric acid reacts with a strong reducing agent such as zinc metal, and explain why: N_2O, NO, NO_2, N_2O_4, N_2O_5, NH_4^+.

4.52 Which of the following metals can react with water? (a) Au, (b) Li, (c) Hg, (d) Ca, (e) Pt.

4.53 On the basis of oxidation number considerations, one of the following oxides would not react with molecular oxygen: NO, N_2O, SO_2, SO_3, P_4O_6. Which one is it? Why?

4.54 Predict the outcome of the reactions represented by the following equations by using the activity series, and balance the equations.

(a) $Cu(s) + HCl(aq) \longrightarrow$
(b) $I_2(s) + NaBr(aq) \longrightarrow$
(c) $Mg(s) + CuSO_4(aq) \longrightarrow$
(d) $Cl_2(g) + KBr(aq) \longrightarrow$

4.55 Classify the following redox reactions.

(a) $2H_2O_2 \longrightarrow 2H_2O + O_2$
(b) $Mg + 2AgNO_3 \longrightarrow Mg(NO_3)_2 + 2Ag$
(c) $NH_4NO_2 \longrightarrow N_2 + 2H_2O$
(d) $H_2 + Br_2 \longrightarrow 2HBr$

4.56 Classify the following redox reactions.

(a) $P_4 + 10Cl_2 \longrightarrow 4PCl_5$
(b) $2NO \longrightarrow N_2 + O_2$
(c) $Cl_2 + 2KI \longrightarrow 2KCl + I_2$
(d) $3HNO_2 \longrightarrow HNO_3 + H_2O + 2NO$

4.57 Which of the following are redox processes?

(a) $CO_2 \longrightarrow CO_3^{2-}$
(b) $VO_3 \longrightarrow VO_2$
(c) $SO_3 \longrightarrow SO_4^{2-}$
(d) $NO_2^- \longrightarrow NO_3^-$
(e) $Cr^{3+} \longrightarrow CrO_4^{2-}$

4.58 Of the following, which is most likely to be the strongest oxidizing agent: O_2, O_2^+, O_2^-, O_2^{2-}?

4.5 Concentration of Solutions

Review Questions

4.59 Write the equation for calculating molarity. Why is molarity a convenient concentration unit in chemistry?

4.60 Describe the steps involved in preparing a solution of known molar concentration using a volumetric flask.

4.61 Describe the basic steps involved in diluting a solution of known concentration.

4.62 Write the equation that enables us to calculate the concentration of a diluted solution. Give units for all the terms.

Problems

4.63 Calculate the mass of KI in grams required to prepare 5.00×10^2 mL of a 2.80 *M* solution.

4.64 Describe how you would prepare 250 mL of a 0.707 *M* $NaNO_3$ solution.

4.65 How many moles of $MgCl_2$ are present in 60.0 mL of 0.100 *M* $MgCl_2$ solution?

4.66 How many grams of KOH are present in 35.0 mL of a 5.50 *M* solution?

4.67 Calculate the molarity of each of the following solutions: (a) 29.0 g of ethanol (C_2H_5OH) in 545 mL of solution, (b) 15.4 g of sucrose ($C_{12}H_{22}O_{11}$) in 74.0 mL of solution, (c) 9.00 g of sodium chloride (NaCl) in 86.4 mL of solution.

4.68 Calculate the molarity of each of the following solutions: (a) 6.57 g of methanol (CH_3OH) in 1.50×10^2 mL of solution, (b) 10.4 g of calcium chloride ($CaCl_2$) in 2.20×10^2 mL of solution, (c) 7.82 g of naphthalene ($C_{10}H_8$) in 85.2 mL of benzene solution.

4.69 Calculate the volume in milliliters of a solution required to provide the following: (a) 2.14 g of sodium chloride from a 0.270 *M* solution, (b) 4.30 g of ethanol from a 1.50 *M* solution, (c) 0.85 g of acetic acid (CH_3COOH) from a 0.30 *M* solution.

4.70 Determine how many grams of each of the following solutes would be needed to make 2.50×10^2 mL of a 0.100 *M* solution: (a) cesium iodide (CsI), (b) sulfuric acid (H_2SO_4), (c) sodium carbonate (Na_2CO_3), (d) potassium dichromate ($K_2Cr_2O_7$), (e) potassium permanganate ($KMnO_4$).

4.71 What volume of 0.416 *M* $Mg(NO_3)_2$ should be added to 255 mL of 0.102 *M* KNO_3 to produce a solution with a concentration of 0.278 *M* NO_3^- ions? Assume volumes are additive.

4.72 Barium hydroxide, often used to titrate weak organic acids, is obtained as the octahydrate, $Ba(OH)_2 \cdot 8H_2O$. What mass of $Ba(OH)_2 \cdot 8H_2O$ would be required to make 500.0 mL of a solution that is 0.1500 *M* in hydroxide ions?

4.73 Describe how to prepare 1.00 L of 0.646 *M* HCl solution, starting with a 2.00 *M* HCl solution.

4.74 Water is added to 25.0 mL of a 0.866 *M* KNO_3 solution until the volume of the solution is exactly 500 mL. What is the concentration of the final solution?

4.75 How would you prepare 60.0 mL of 0.200 *M* HNO_3 from a stock solution of 4.00 *M* HNO_3?

4.76 You have 505 mL of a 0.125 *M* HCl solution and you want to dilute it to exactly 0.100 *M*. How much water should you add? Assume volumes are additive.

4.77 A 35.2-mL, 1.66 *M* $KMnO_4$ solution is mixed with 16.7 mL of 0.892 *M* $KMnO_4$ solution. Calculate the concentration of the final solution.

4.78 A 46.2-mL, 0.568 *M* calcium nitrate [$Ca(NO_3)_2$] solution is mixed with 80.5 mL of 1.396 *M* calcium nitrate solution. Calculate the concentration of the final solution.

4.6 Gravimetric Analysis

Review Questions

4.79 Describe the basic steps involved in gravimetric analysis. How does this procedure help us determine the identity of a compound or the purity of a compound if its formula is known?

4.80 Distilled water must be used in the gravimetric analysis of chlorides. Why?

Problems

4.81 If 30.0 mL of 0.150 *M* $CaCl_2$ is added to 15.0 mL of 0.100 *M* $AgNO_3$, what is the mass in grams of AgCl precipitate?

4.82 A sample of 0.6760 g of an unknown compound containing barium ions (Ba^{2+}) is dissolved in water and treated with an excess of Na_2SO_4. If the mass of the $BaSO_4$ precipitate formed is 0.4105 g, what is the percent by mass of Ba in the original unknown compound?

4.83 How many grams of NaCl are required to precipitate most of the Ag^+ ions from 2.50×10^2 mL of 0.0113 *M* $AgNO_3$ solution? Write the net ionic equation for the reaction.

4.84 The concentration of sulfate in water can be determined by adding a solution of barium chloride to precipitate the sulfate ion. Write the net ionic equation for this reaction. Treating a 145-mL sample of water with excess $BaCl_2(aq)$ precipitated 0.330 g of $BaSO_4$. Determine the concentration of sulfate in the original water sample.

4.7 Titrations

Review

4.85 Describe the basic steps involved in an acid-base titration. Why is this technique of great practical value?

4.86 How does an acid-base indicator work?

4.87 A student carried out two titrations using a NaOH solution of unknown concentration in the buret. In one titration she weighed out 0.2458 g of KHP (see Section 4.7) and transferred it to an Erlenmeyer flask. She then added 20.00 mL of distilled water to dissolve the acid. In the other titration she weighed out 0.2507 g of KHP but added 40.00 mL of distilled water to dissolve the acid. Assuming no experimental error, would she obtain the same result for the concentration of the NaOH solution?

4.88 Would the volume of a 0.10 *M* NaOH solution needed to titrate 25.0 mL of a 0.10 *M* HNO_2 (a weak acid) solution be different from that needed to titrate 25.0 mL of a 0.10 *M* HCl (a strong acid) solution?

4.89 What are the similarities and differences between acid-base titrations and redox titrations?

4.90 Explain why potassium permanganate ($KMnO_4$) and potassium dichromate ($K_2Cr_2O_7$) can serve as internal indicators in redox titrations.

Problems

4.91 A quantity of 18.68 mL of a KOH solution is needed to neutralize 0.4218 g of KHP. What is the concentration (in molarity) of the KOH solution?

4.92 Calculate the concentration (in molarity) of a NaOH solution if 25.0 mL of the solution are needed to neutralize 17.4 mL of a 0.312 *M* HCl solution.

4.93 Calculate the volume in milliliters of a 1.420 *M* NaOH solution required to titrate the following solutions.
(a) 25.0 mL of a 2.430 *M* HCl solution
(b) 25.0 mL of a 4.500 *M* H_2SO_4 solution
(c) 25.0 mL of a 1.500 *M* H_3PO_4 solution

4.94 What volume of a 0.500 *M* HCl solution is needed to neutralize each of the following?
(a) 10.0 mL of a 0.300 *M* NaOH solution
(b) 10.0 mL of a 0.200 *M* $Ba(OH)_2$ solution

4.95 Iron(II) can be oxidized by an acidic $K_2Cr_2O_7$ solution according to the net ionic equation

$$Cr_2O_7^{2-} + 6Fe^{2+} + 14H^+ \longrightarrow 2Cr^{3+} + 6Fe^{3+} + 7H_2O$$

If it takes 26.0 mL of 0.0250 *M* $K_2Cr_2O_7$ to titrate 25.0 mL of a solution containing Fe^{2+}, what is the molar concentration of Fe^{2+}?

4.96 The SO_2 present in air is mainly responsible for the acid rain phenomenon. Its concentration can be determined by titrating against a standard permanganate solution as follows:

$$5SO_2 + 2MnO_4^- + 2H_2O \longrightarrow 5SO_4^{2-} + 2Mn^{2+} + 4H^+$$

Calculate the number of grams of SO_2 in a sample of air if 7.37 mL of 0.00800 *M* $KMnO_4$ solution are required for the titration.

4.97 A sample of iron ore (containing only Fe^{2+} ions) weighing 0.2792 g was dissolved in dilute acid solution, and all the Fe(II) was converted to Fe(III) ions. The solution required 23.30 mL of 0.0194 *M* $K_2Cr_2O_7$ for titration. Calculate the percent by mass of iron in the ore. (*Hint:* See Problem 4.95 for the balanced equation.)

4.98 The concentration of a hydrogen peroxide solution can be conveniently determined by titration against a standardized potassium permanganate solution in an acidic medium according to the equation

$$2MnO_4^- + 5H_2O_2 + 6H^+ \longrightarrow 5O_2 + 2Mn^{2+} + 8H_2O$$

If 36.44 mL of a 0.01652 *M* $KMnO_4$ solution are required to oxidize 25.00 mL of a H_2O_2 solution, calculate the molarity of the H_2O_2 solution.

4.99 Oxalic acid ($H_2C_2O_4$) is present in many plants and vegetables. If 24.0 mL of 0.0100 *M* $KMnO_4$ solution is needed to titrate 1.00 g of a sample of $H_2C_2O_4$ to the equivalence point, what is the percent by mass of $H_2C_2O_4$ in the sample? The net ionic equation is

$$2MnO_4^- + 16H^+ + 5C_2O_4^{2-} \longrightarrow 2Mn^{2+} + 10CO_2 + 8H_2O$$

4.100 A 15.0-mL sample of an oxalic acid solution requires 25.2 mL of 0.149 *M* NaOH for neutralization. Calculate the volume of a 0.122 *M* $KMnO_4$ solution needed to react with a second 15.0-mL sample of the oxalic acid solution. (*Hint:* Oxalic acid is a diprotic acid. See Problem 4.99 for redox equation.)

4.101 Iodate ion, IO_3^-, oxidizes SO_3^{2-} in acidic solution. The half-reaction for the oxidation is

$$SO_3^{2-} + H_2O \longrightarrow SO_4^{2-} + 2H^+ + 2e^-$$

A 100.0-mL sample of solution containing 1.390 g of KIO_3 reacts with 32.5 mL of 0.500 *M* Na_2SO_3. What is the final oxidation state of the iodine after the reaction has occurred?

4.102 Calcium oxalate (CaC_2O_4), the main component of kidney stones, is insoluble in water. For this reason it can be used to determine the amount of Ca^{2+} ions in fluids such as blood. The calcium oxalate isolated from blood is dissolved in acid and titrated against a standardized $KMnO_4$ solution, as shown in Problem 4.99. In one test it is found that the calcium oxalate isolated from a 10.0-mL sample of blood requires 24.2 mL of 9.56×10^{-4} *M* $KMnO_4$ for titration. Calculate the number of milligrams of calcium per milliliter of blood.

Additional Problems

4.103 Classify the following reactions according to the types discussed in the chapter.
(a) $Cl_2 + 2OH^- \longrightarrow Cl^- + ClO^- + H_2O$
(b) $Ca^{2+} + CO_3^{2-} \longrightarrow CaCO_3$
(c) $NH_3 + H^+ \longrightarrow NH_4^+$
(d) $2CCl_4 + CrO_4^{2-} \longrightarrow 2COCl_2 + CrO_2Cl_2 + 2Cl^-$
(e) $Ca + F_2 \longrightarrow CaF_2$
(f) $2Li + H_2 \longrightarrow 2LiH$
(g) $Ba(NO_3)_2 + Na_2SO_4 \longrightarrow 2NaNO_3 + BaSO_4$
(h) $CuO + H_2 \longrightarrow Cu + H_2O$
(i) $Zn + 2HCl \longrightarrow ZnCl_2 + H_2$
(j) $2FeCl_2 + Cl_2 \longrightarrow 2FeCl_3$
(k) $LiOH + HNO_3 \longrightarrow LiNO_3 + H_2O$

4.104 Oxygen (O_2) and carbon dioxide (CO_2) are colorless and odorless gases. Suggest two chemical tests that would enable you to distinguish between these two gases.

4.105 Which of the following aqueous solutions would you expect to be the best conductor of electricity at 25°C? Explain your answer.
(a) 0.20 *M* NaCl
(b) 0.60 *M* CH_3COOH
(c) 0.25 *M* HCl
(d) 0.20 *M* $Mg(NO_3)_2$

4.106 A 5.00×10^2 mL sample of 2.00 *M* HCl solution is treated with 4.47 g of magnesium. Calculate the concentration of the acid solution after all the metal has reacted. Assume that the volume remains unchanged.

4.107 Shown are two aqueous solutions containing various ions. The volume of each solution is 200 mL. (a) Calculate the mass of the precipitate (in grams) after the solutions are mixed. (b) What are the concentrations (in *M*) of the ions in the final solution? Treat each sphere as 0.100 mol. Assume the volumes are additive.

4.108 Shown are two aqueous solutions containing various ions. The volume of each solution is 200 mL. (a) Calculate the mass of the precipitate (in grams) after the solutions are mixed. (b) What are the concentrations (in *M*) of the ions in the final solution? Treat each sphere as 0.100 mol. Assume the volumes are additive.

4.109 Calculate the volume of a 0.156 *M* $CuSO_4$ solution that would react with 7.89 g of zinc.

4.110 Sodium carbonate (Na_2CO_3) is available in very pure form and can be used to standardize acid solutions. What is the molarity of a HCl solution if 28.3 mL of the solution are required to react with 0.256 g of Na_2CO_3?

4.111 A 3.664-g sample of a monoprotic acid was dissolved in water. It took 20.27 mL of a 0.1578 *M* NaOH solution to neutralize the acid. Calculate the molar mass of the acid.

4.112 Acetic acid (CH_3COOH) is an important ingredient of vinegar. A sample of 50.0 mL of a commercial vinegar is titrated against a 1.00 *M* NaOH solution. What is the concentration (in *M*) of acetic acid present in the vinegar if 5.75 mL of the base are needed for the titration?

4.113 A 15.00-mL solution of potassium nitrate (KNO_3) was diluted to 125.0 mL, and 25.00 mL of this solution were then diluted to 1.000×10^3 mL. The concentration of the final solution is 0.00383 *M*. Calculate the concentration of the original solution.

4.114 When a 2.50-g zinc strip was placed in a $AgNO_3$ solution, silver metal formed on the surface of the strip. After some time had passed, the strip was removed from the solution, dried, and weighed. If the mass of the strip was 3.37 g, calculate the mass of Ag and Zn metals present.

4.115 Calculate the mass of the precipitate formed when 2.27 L of 0.0820 *M* $Ba(OH)_2$ are mixed with 3.06 L of 0.0664 *M* Na_2SO_4.

4.116 Calculate the concentration of the acid (or base) remaining in solution when 10.7 mL of 0.211 *M* HNO_3 are added to 16.3 mL of 0.258 *M* NaOH.

4.117 (a) Describe a preparation for magnesium hydroxide [$Mg(OH)_2$] and predict its solubility. (b) Milk of magnesia contains mostly $Mg(OH)_2$ and is effective in treating acid (mostly hydrochloric acid) indigestion. Calculate the volume of a 0.035 *M* HCl solution (a typical acid concentration in an upset stomach) needed to react with two spoonfuls (approximately 10 mL) of milk of magnesia [at 0.080 g $Mg(OH)_2$/mL].

4.118 A 1.00-g sample of a metal X (that is known to form X^{2+} ions) was added to 0.100 L of 0.500 *M* H_2SO_4. After all the metal had reacted, the remaining acid required 0.0334 L of 0.500 *M* NaOH solution for neutralization. Calculate the molar mass of the metal and identify the element.

4.119 Carbon dioxide in air can be removed by an aqueous metal hydroxide solution such as LiOH and $Ba(OH)_2$. (a) Write equations for the reactions. (Carbon dioxide reacts with water to form carbonic acid.) (b) Calculate the mass of CO_2 that can be removed by 5.00×10^{-2} mL of a 0.800 *M* LiOH and a 0.800 *M* $Ba(OH)_2$ solution. (c) Which solution would you choose for use in a space capsule and which for use in a submarine?

4.120 The molecular formula of malonic acid is $C_3H_4O_4$. If a solution containing 0.762 g of the acid requires 12.44 mL of 1.174 *M* NaOH for neutralization, how many ionizable H atoms are present in the molecule?

4.121 A quantitative definition of solubility is the maximum number of grams of a solute that will dissolve in a given volume of water at a particular temperature. Describe an experiment that would enable you to determine the solubility of a soluble compound.

4.122 A 60.0-mL 0.513 *M* glucose ($C_6H_{12}O_6$) solution is mixed with 120.0 mL of 2.33 *M* glucose solution. What is the concentration of the final solution? Assume the volumes are additive.

4.123 An ionic compound X is only slightly soluble in water. What test would you employ to show that the compound does indeed dissolve in water to a certain extent?

4.124 A student is given an unknown that is either iron(II) sulfate or iron(III) sulfate. Suggest a chemical procedure for determining its identity. (Both iron compounds are water soluble.)

4.125 You are given a colorless liquid. Describe three chemical tests you would perform on the liquid to show that it is water.

4.126 Using the apparatus shown in Figure 4.1, a student found that a sulfuric acid solution caused the lightbulb to glow brightly. However, after the addition of a certain amount of a barium hydroxide [$Ba(OH)_2$]

solution, the light began to dim even though $Ba(OH)_2$ is also a strong electrolyte. Explain.

4.127 The molar mass of a certain metal carbonate, MCO_3, can be determined by adding an excess of HCl acid to react with all the carbonate and then "back titrating" the remaining acid with a NaOH solution. (a) Write an equation for these reactions. (b) In a certain experiment, 20.00 mL of 0.0800 *M* HCl were added to a 0.1022-g sample of MCO_3. The excess HCl required 5.64 mL of 0.1000 *M* NaOH for neutralization. Calculate the molar mass of the carbonate and identify M.

4.128 A 5.012-g sample of an iron chloride hydrate was dried in an oven. The mass of the anhydrous compound was 3.195 g. The compound was then dissolved in water and reacted with an excess of $AgNO_3$. The AgCl precipitate formed weighed 7.225 g. What is the formula of the original compound?

4.129 You are given a soluble compound of unknown molecular formula. (a) Describe three tests that would show that the compound is an acid. (b) Once you have established that the compound is an acid, describe how you would determine its molar mass using a NaOH solution of known concentration. (Assume the acid is monoprotic.) (c) How would you find out whether the acid is weak or strong? You are provided with a sample of NaCl and an apparatus like that shown in Figure 4.1 for comparison.

4.130 You are given two colorless solutions, one containing NaCl and the other sucrose ($C_{12}H_{22}O_{11}$). Suggest a chemical and a physical test that would allow you to distinguish between these two solutions.

4.131 The concentration of lead ions (Pb^{2+}) in a sample of polluted water that also contains nitrate ions (NO_3^-) is determined by adding solid sodium sulfate (Na_2SO_4) to exactly 500 mL of the water. (a) Write the molecular and net ionic equations for the reaction. (b) Calculate the molar concentration of Pb^{2+} if 0.00450 g of Na_2SO_4 was needed for the complete precipitation of Pb^{2+} ions as $PbSO_4$.

4.132 Hydrochloric acid is not an oxidizing agent in the sense that sulfuric acid and nitric acid are. Explain why the chloride ion is not a strong oxidizing agent like SO_4^{2-} and NO_3^-.

4.133 Explain how you would prepare potassium iodide (KI) by means of (a) an acid-base reaction and (b) a reaction between an acid and a carbonate compound.

4.134 Sodium reacts with water to yield hydrogen gas. Why is this reaction not used in the laboratory preparation of hydrogen?

4.135 Describe how you would prepare the following compounds: (a) $Mg(OH)_2$, (b) AgI, (c) $Ba_3(PO_4)_2$.

4.136 Someone spilled concentrated sulfuric acid on the floor of a chemistry laboratory. To neutralize the acid, would it be preferable to pour concentrated sodium hydroxide solution or spray solid sodium bicarbonate over the acid? Explain your choice and the chemical basis for the action.

4.137 Describe in each case how you would separate the cations or anions in an aqueous solution of (a) $NaNO_3$ and $Ba(NO_3)_2$, (b) $Mg(NO_3)_2$ and KNO_3, (c) KBr and KNO_3, (d) K_3PO_4 and KNO_3, (e) Na_2CO_3 and $NaNO_3$.

4.138 The following are common household compounds: table salt (NaCl), table sugar (sucrose), vinegar (contains acetic acid), baking soda ($NaHCO_3$), washing soda ($Na_2CO_3 \cdot 10H_2O$), boric acid (H_3BO_3, used in eyewash), epsom salt ($MgSO_4 \cdot 7H_2O$), sodium hydroxide (used in drain openers), ammonia, milk of magnesia [$Mg(OH)_2$], and calcium carbonate. Based on what you have learned in this chapter, describe test(s) that would enable you to identify each of these compounds.

4.139 Sulfites (compounds containing the SO_3^{2-} ions) are used as preservatives in dried fruit and vegetables and in winemaking. In an experiment to test the presence of sulfite in fruit, a student first soaked several dried apricots in water overnight and then filtered the solution to remove all solid particles. She then treated the solution with hydrogen peroxide (H_2O_2) to oxidize the sulfite ions to sulfate ions. Finally, the sulfate ions were precipitated by treating the solution with a few drops of barium chloride ($BaCl_2$) solution. Write a balanced equation for each of the preceding steps.

4.140 A 0.8870-g sample of a mixture of NaCl and KCl is dissolved in water, and the solution is then treated with an excess of $AgNO_3$ to yield 1.913 g of AgCl. Calculate the percent by mass of each compound in the mixture.

4.141 Based on oxidation number consideration, explain why carbon monoxide (CO) is flammable but carbon dioxide (CO_2) is not.

4.142 Which of the diagrams (a)–(c) corresponds to the reaction between AgOH(*s*) and HNO_3(*aq*)? Write a balanced equation for the reaction. The gray spheres represent the Ag^+ ions and the blue spheres represent the NO_3^- ions.

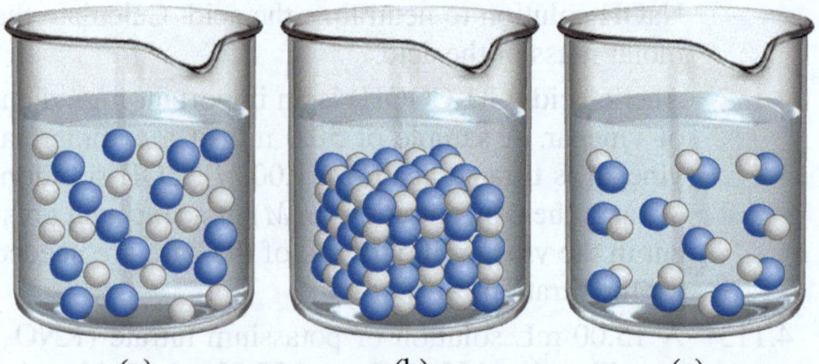

4.143 Chlorine forms a number of oxides with the following oxidation numbers: +1, +3, +4, +6, and +7. Write a formula for each of these compounds.

4.144 A useful application of oxalic acid is the removal of rust (Fe_2O_3) from, say, bathtub rings according to the reaction

$$Fe_2O_3(s) + 6H_2C_2O_4(aq) \longrightarrow 2Fe(C_2O_4)_3^{3-}(aq) + 3H_2O + 6H^+(aq)$$

Calculate the number of grams of rust that can be removed by 5.00×10^2 mL of a 0.100 *M* solution of oxalic acid.

4.145 Acetylsalicylic acid ($C_9H_8O_4$) is a monoprotic acid commonly known as "aspirin." A typical aspirin tablet, however, contains only a small amount of the acid. In an experiment to determine its composition, an aspirin tablet was crushed and dissolved in water. It took 12.25 mL of 0.1466 *M* NaOH to neutralize the solution. Calculate the number of grains of aspirin in the tablet. (One grain = 0.0648 g.)

4.146 A 0.9157-g mixture of $CaBr_2$ and NaBr is dissolved in water, and $AgNO_3$ is added to the solution to form AgBr precipitate. If the mass of the precipitate is 1.6930 g, what is the percent by mass of NaBr in the original mixture?

4.147 Hydrogen halides (HF, HCl, HBr, HI) are highly reactive compounds that have many industrial and laboratory uses. (a) In the laboratory, HF and HCl can be generated by reacting CaF_2 and NaCl with concentrated sulfuric acid. Write appropriate equations for the reactions. (*Hint:* These are not redox reactions.) (b) Why is it that HBr and HI cannot be prepared similarly—that is, by reacting NaBr and NaI with concentrated sulfuric acid? (*Hint:* H_2SO_4 is a stronger oxidizing agent than both Br_2 and I_2.) (c) HBr can be prepared by reacting phosphorus tribromide (PBr_3) with water. Write an equation for this reaction.

4.148 A 325-mL sample of solution contains 25.3 g of $CaCl_2$. (a) Calculate the molar concentration of Cl^- in this solution. (b) How many grams of Cl^- are in 0.100 L of this solution?

4.149 Phosphoric acid (H_3PO_4) is an important industrial chemical used in fertilizers, in detergents, and in the food industry. It is produced by two different methods. In the *electric furnace method,* elemental phosphorus (P_4) is burned in air to form P_4O_{10}, which is then reacted with water to give H_3PO_4. In the *wet process,* the mineral phosphate rock fluorapatite [$Ca_5(PO_4)_3F$] is reacted with sulfuric acid to give H_3PO_4 (and HF and $CaSO_4$). Write equations for these processes and classify each step as precipitation, acid-base, or redox reaction.

4.150 Ammonium nitrate (NH_4NO_3) is one of the most important nitrogen-containing fertilizers. Its purity can be analyzed by titrating a solution of NH_4NO_3 with a standard NaOH solution. In one experiment a 0.2041-g sample of industrially prepared NH_4NO_3 required 24.42 mL of 0.1023 *M* NaOH for neutralization.

(a) Write a net ionic equation for the reaction.

(b) What is the percent purity of the sample?

4.151 Is the following reaction a redox reaction? Explain.

$$3O_2(g) \longrightarrow 2O_3(g)$$

4.152 What is the oxidation number of O in HFO?

4.153 Use molecular models like those in Figures 4.7 and 4.8 to represent the following acid-base reactions.

(a) $OH^- + H_3O^+ \longrightarrow 2H_2O$

(b) $NH_4^+ + NH_2^- \longrightarrow 2NH_3$

Identify the Brønsted acid and base in each case.

4.154 The alcohol content in a 10.0-g sample of blood from a driver required 4.23 mL of 0.07654 *M* $K_2Cr_2O_7$ for titration. Should the police prosecute the individual for drunken driving? (*Hint:* See the Chemistry in Action essay in Section 4.4.)

4.155 On standing, a concentrated nitric acid gradually turns yellow. Explain. (*Hint:* Nitric acid slowly decomposes. Nitrogen dioxide is a colored gas.)

4.156 Describe the laboratory preparation for the following gases: (a) hydrogen, (b) oxygen, (c) carbon dioxide, (d) nitrogen. Indicate the physical states of the reactants and products in each case. [*Hint:* Nitrogen can be obtained by heating ammonium nitrite (NH_4NO_2).]

4.157 Referring to Figure 4.18, explain why one must first dissolve the solid completely before making up the solution to the correct volume.

4.158 Can the following decomposition reaction be characterized as an acid-base reaction? Explain.

$$NH_4Cl(s) \longrightarrow NH_3(g) + HCl(g)$$

4.159 Give a chemical explanation for each of the following: (a) When calcium metal is added to a sulfuric acid solution, hydrogen gas is generated. After a few minutes, the reaction slows down and eventually stops even though none of the reactants is used up. (b) In the activity series, aluminum is above hydrogen, yet the metal appears to be unreactive toward steam and hydrochloric acid. (c) Sodium and potassium lie above copper in the activity series. In your explanation, discuss why Cu^{2+} ions in a $CuSO_4$ solution are not converted to metallic copper upon the addition of these metals. (d) A metal M reacts slowly with steam. There is no visible change when it is placed in a pale green iron(II) sulfate solution. Where should we place M in the activity series? (e) Before aluminum metal was obtained by electrolysis, it was produced by reducing its chloride ($AlCl_3$) with an active metal. What metals would you use to produce aluminum in that way?

4.160 The recommended procedure for preparing a very dilute solution is not to weigh out a very small mass or measure a very small volume of a stock solution. Instead, it is done by a series of dilutions. A sample of 0.8214 g of $KMnO_4$ was dissolved in water and made up to the volume in a 500-mL volumetric flask. A 2.000-mL sample of this solution was transferred to a 1000-mL volumetric flask and diluted to the mark with water. Next, 10.00 mL of the diluted solution were transferred to a 250-mL flask and diluted to the mark with water. (a) Calculate the concentration (in molarity) of the final solution.

(b) Calculate the mass of $KMnO_4$ needed to directly prepare the final solution.

4.161 The following "cycle of copper" experiment is performed in some general chemistry laboratories. The series of reactions starts with copper and ends with metallic copper. The steps are as follows: (1) A piece of copper wire of known mass is allowed to react with concentrated nitric acid [the products are copper(II) nitrate, nitrogen dioxide, and water]. (2) The copper(II) nitrate is treated with a sodium hydroxide solution to form copper(II) hydroxide precipitate. (3) On heating, copper(II) hydroxide decomposes to yield copper(II) oxide. (4) The copper(II) oxide is reacted with concentrated sulfuric acid to yield copper(II) sulfate. (5) Copper(II) sulfate is treated with an excess of zinc metal to form metallic copper. (6) The remaining zinc metal is removed by treatment with hydrochloric acid, and metallic copper is filtered, dried, and weighed. (a) Write a balanced equation for each step and classify the reactions. (b) Assuming that a student started with 65.6 g of copper, calculate the theoretical yield at each step. (c) Considering the nature of the steps, comment on why it is possible to recover most of the copper used at the start.

4.162 A quantity of 25.0 mL of a solution containing both Fe^{2+} and Fe^{3+} ions is titrated with 23.0 mL of 0.0200 *M* $KMnO_4$ (in dilute sulfuric acid). As a result, all of the Fe^{2+} ions are oxidized to Fe^{3+} ions. Next, the solution is treated with Zn metal to convert all of the Fe^{3+} ions to Fe^{2+} ions. Finally, the solution containing only the Fe^{2+} ions requires 40.0 mL of the same $KMnO_4$ solution for oxidation to Fe^{3+}. Calculate the molar concentrations of Fe^{2+} and Fe^{3+} in the original solution. The net ionic equation is

$$MnO_4^- + 5Fe^{2+} + 8H^+ \longrightarrow Mn^{2+} + 5Fe^{3+} + 4H_2O$$

4.163 Use the periodic table framework provided to show the names and positions of two metals that can displace hydrogen from (a) cold water, (b) steam, and (c) acid. Also show two metals that can react neither with water nor acid.

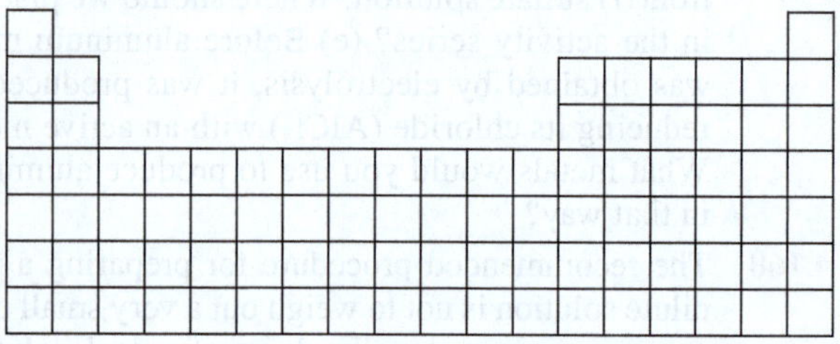

4.164 Referring to the Chemistry in Action essay in Section 4.7, answer the following questions: (a) Identify the precipitation, acid-base, and redox processes. (b) Instead of calcium oxide, why don't we simply add sodium hydroxide to seawater to precipitate magnesium hydroxide? (c) Sometimes a mineral called dolomite (a mixture of $CaCO_3$ and $MgCO_3$) is substituted for limestone to bring about the precipitation of magnesium hydroxide. What is the advantage of using dolomite?

4.165 A 22.02-mL solution containing 1.615 g $Mg(NO_3)_2$ is mixed with a 28.64-mL solution containing 1.073 g NaOH. Calculate the concentrations of the ions remaining in solution after the reaction is complete. Assume volumes are additive.

4.166 Chemical tests of four metals A, B, C, and D show the following results.

(a) Only B and C react with 0.5 *M* HCl to give H_2 gas.

(b) When B is added to a solution containing the ions of the other metals, metallic A, C, and D are formed.

(c) A reacts with 6 *M* HNO_3 but D does not.

Arrange the metals in the increasing order as reducing agents. Suggest four metals that fit these descriptions.

4.167 The antibiotic gramicidin A can transport Na^+ ions into a certain cell at the rate of 5.0×10^7 Na^+ ions s^{-1}. Calculate the time in seconds to transport enough Na^+ ions to increase its concentration by 8.0×10^{-3} *M* in a cell whose intracellular volume is 2.0×10^{-10} mL.

4.168 Shown are two aqueous solutions containing various ions. The volume of each solution is 600 mL. (a) Write a net ionic equation for the reaction after the solutions are mixed. (b) Calculate the mass of the precipitates formed and the concentrations of the ions in the mixed solution. Treat each sphere as 0.0500 mol.

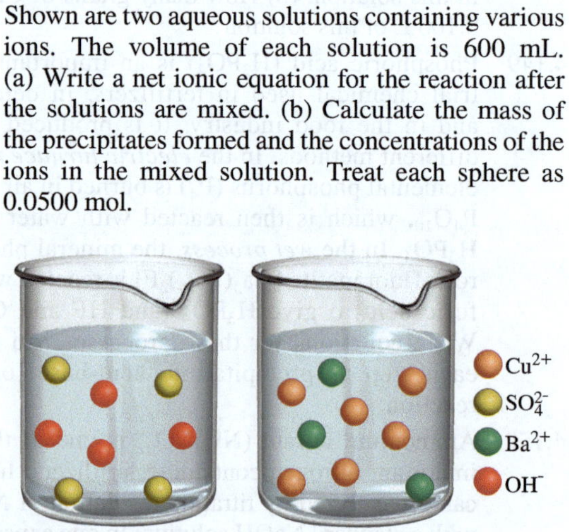

Interpreting, Modeling, & Estimating

4.169 Many proteins contain metal ions for structural and/or redox functions. Which of the following metals fit into one or both categories: Ca, Cu, Fe, Mg, Mn, Ni, Zn?

4.170 The fastest way to introduce therapeutic agents into the bloodstream is by direct delivery into a vein [intravenous (IV) therapy]. A clinical researcher wishes to establish an initial concentration of 6×10^{-4} mmol/L in the bloodstream of an adult male participating in a trial study of a new drug. The drug serum is prepared in the hospital's pharmacy at a concentration of 1.2×10^{-3} mol/L. How much of the serum should be introduced intravenously to achieve the desired initial blood concentration of the drug?

4.171 Public water supplies are often "fluoridated" by the addition of compounds such as NaF, H_2SiF_6, and Na_2SiF_6. It is well established that fluoride helps prevent tooth decay; however, care must be taken not to exceed safe levels of fluoride, which can stain or etch tooth enamel (dental fluorosis). A safe and effective concentration of fluoride in drinking water is generally considered to be around 1 mg/L. How much fluoride would a person consume by drinking fluoridated water in 1 year? What would be the equivalent mass as sodium fluoride?

4.172 Potassium superoxide (KO_2), a useful source of oxygen employed in breathing equipment, reacts with water to form potassium hydroxide, hydrogen peroxide, and oxygen. Furthermore, potassium superoxide also reacts with carbon dioxide to form potassium carbonate and oxygen. (a) Write equations for these two reactions and comment on the effectiveness of potassium superoxide in this application. (b) Focusing only on the reaction between KO_2 and CO_2, estimate the amount of KO_2 needed to sustain a worker in a polluted environment for 30 minutes. See Problem 1.75 for useful information.

Sandro Di Carlo
Darsa/PhotoAlto/Getty Images

4.173 Muriatic acid, a commercial-grade hydrochloric acid used for cleaning masonry surfaces, is typically around 10% HCl by mass and has a density of 1.2 g/cm^3. A 0.5-in layer of boiler scale has accumulated on a 6.0-ft section of hot water pipe with an internal diameter of 2.0 in (see the Chemistry in Action essay in Section 4.7). What is the minimum volume of muriatic acid in gallons that would be needed to remove the boiler scale?

4.174 Because acid-base and precipitation reactions discussed in this chapter all involve ionic species, their progress can be monitored by measuring the electrical conductance of the solution. Match the following reactions with diagrams (a)–(d). The electrical conductance is shown in arbitrary units.

(1) A 1.0 *M* KOH solution is added to 1.0 L of 1.0 *M* CH_3COOH.

(2) A 1.0 *M* NaOH solution is added to 1.0 L of 1.0 *M* HCl.

(3) A 1.0 *M* $BaCl_2$ solution is added to 1.0 L of 1.0 *M* K_2SO_4.

(4) A 1.0 *M* NaCl solution is added to 1.0 L of 1.0 *M* $AgNO_3$.

(5) A 1.0 *M* CH_3COOH solution is added to 1.0 L of 1.0 *M* NH_3.

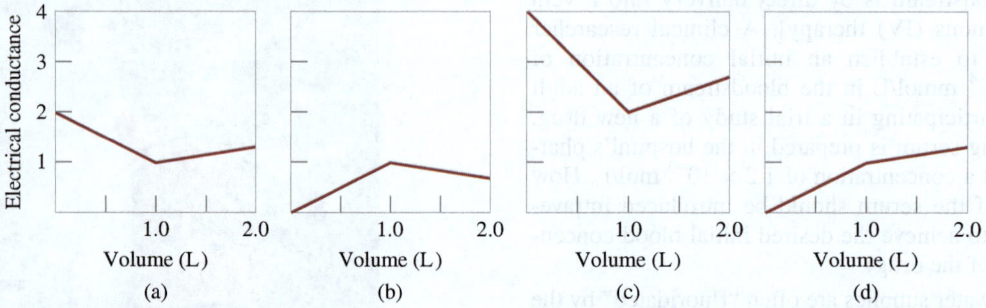

Answers to Practice Exercises

4.1 (a) Insoluble. (b) Insoluble. (c) Soluble. **4.2** $Al^{3+}(aq) + 3OH^-(aq) \longrightarrow Al(OH)_3(s)$. **4.3** (a) Brønsted base. (b) Brønsted acid. **4.4** Molecular equation: $H_3PO_4(aq) + 3NaOH(aq) \longrightarrow Na_3PO_4(aq) + 3H_2O(l)$; ionic equation: $H_3PO_4(aq) + 3Na^+(aq) + 3OH^-(aq) \longrightarrow 3Na^+(aq) + PO_4^{3-}(aq) + 3H_2O(l)$; net ionic equation: $H_3PO_4(aq) + 3OH^-(aq) \longrightarrow PO_4^{3-}(aq) + 3H_2O(l)$. **4.5** (a) P: +3, F: −1; (b) Mn: +7, O: −2. **4.6** (a) Hydrogen displacement reaction, (b) combination reaction, (c) disproportionation reaction, (d) metal displacement reaction. **4.7** 0.452 *M*. **4.8** 494 mL. **4.9** Dilute 34.2 mL of the stock solution to 200 mL. **4.10** 92.02%. **4.11** 0.3822 g. **4.12** 1.27 *M*. **4.13** 204 mL.

Answers to Review of Concepts & Facts

4.1.1 Strongest (b), weakest (c). **4.1.2** (a) Strong electrolyte. (b) Strong electrolyte. (c) Nonelectrolyte. **4.2.1** (a). **4.2.2** (a) Soluble. (b) Insoluble. (c) Soluble. **4.2.3** $Cu^{2+}(aq) + S^{2-}(aq) \longrightarrow CuS(s)$. **4.3.1** Weak acid (b), very weak acid (c), strong acid (a). **4.3.2** Brønsted acid: HBr, Brønsted base: NH_3. **4.3.3** $H^+(aq) + OH^-(aq) \longrightarrow H_2O(l)$. **4.4.1** (c). **4.4.2** (a) +4. (b) +2. (c) +3. **4.5.1** 0.9 *M*. **4.5.2** 6.88×10^{-3} mol. **4.5.3** 76.9 g. **4.6.1** 9.47 g. **4.7.1** HCl (c), H_2SO_4 (d), H_3PO_4 (b). **4.7.2** 43.0 mL. **4.7.3** Not necessarily. The equivalence point will be reached when an equal volume of oxidizing agent is added only if the reducing agent and the oxidizing agent have the same coefficients in the balanced chemical equation for the redox reaction. **4.7.4** 0.1264 *M*.

CHAPTER

5

Gases

A nebula is an interstellar cloud of space dust, hydrogen, helium, and other ionized gases. While these gases are more dense than the surrounding space, most nebulae have densities considerably less dense than any vacuum created on Earth.

STScI/NASA

CHAPTER OUTLINE

Under certain conditions of pressure and temperature, most substances can exist in any one of three states of matter: solid, liquid, or gas. Water, for example, can be solid ice, liquid water, steam, or water vapor. The physical properties of a substance often depend on its state.

Gases, the subject of this chapter, are simpler than liquids and solids in many ways. Molecular motion in gases is totally random, and the forces of attraction between gas molecules are so small that each molecule moves freely and essentially independently of other molecules. Subjected to changes in temperature and pressure, it is easier to predict the behavior of gases. The laws that govern this behavior have played an important role in the development of the atomic theory of matter and the kinetic molecular theory of gases.

5.1 Substances That Exist as Gases

Learning Objective

- Summarize the characteristics of gases that differentiate them from solids and liquids.

We live at the bottom of an ocean of air whose composition by volume is roughly 78 percent N_2, 21 percent O_2, and 1 percent other gases, including CO_2. Today, the chemistry of this vital mixture of gases has become a source of great interest because of the detrimental effects of environmental pollution. The chemistry of the atmosphere and polluting gases is discussed in Chapter 20. Here we will focus generally on the behavior of substances that exist as gases under normal atmospheric conditions, which are defined as 25°C and 1 atmosphere (atm) pressure; in contrast, a vapor is the gaseous form of any substance that is a liquid or solid at normal temperatures and pressures.

Figure 5.1 shows the elements that are gases under normal atmospheric conditions. Note that hydrogen, nitrogen, oxygen, fluorine, and chlorine exist as gaseous diatomic molecules: H_2, N_2, O_2, F_2, and Cl_2. An allotrope of oxygen, ozone (O_3), is also a gas at room temperature. All the elements in Group 18, the noble gases, are monatomic gases: He, Ne, Ar, Kr, Xe, and Rn.

Ionic compounds do not exist as gases at 25°C and 1 atm, because cations and anions in an ionic solid are held together by very strong electrostatic forces—that is,

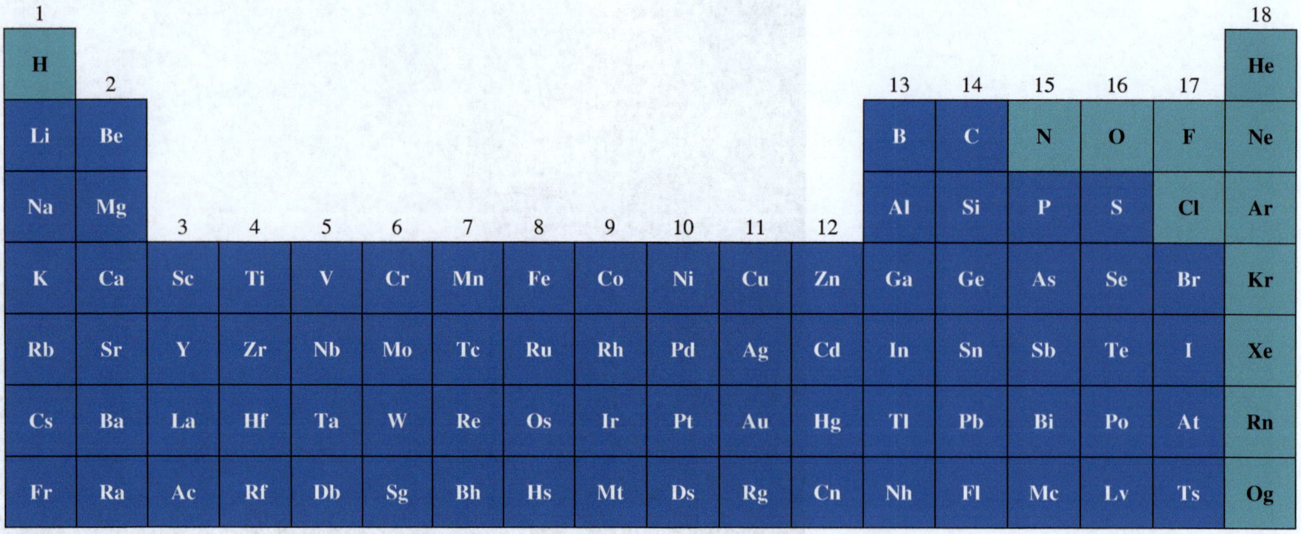

Figure 5.1 *Elements that exist as gases at 25°C and 1 atm. The noble gases (the Group 18 elements) are monatomic species; the other elements exist as diatomic molecules. Ozone (O_3) is also a gas.*

Table 5.1 Some Substances Found as Gases at 1 atm and 25°C

Elements	
H_2 (molecular hydrogen)	N_2 (molecular nitrogen)
O_2 (molecular oxygen)	O_3 (ozone)
F_2 (molecular fluorine)	Cl_2 (molecular chlorine)
He (helium)	Ne (neon)
Ar (argon)	Kr (krypton)
Xe (xenon)	Rn (radon)
Compounds	
HF (hydrogen fluoride)	HCl (hydrogen chloride)
HBr (hydrogen bromide)	HI (hydrogen iodide)
CO (carbon monoxide)	CO_2 (carbon dioxide)
CH_4 (methane)	C_2H_2 (acetylene)
NH_3 (ammonia)	NO (nitric oxide)
NO_2 (nitrogen dioxide)	N_2O (nitrous oxide)
SO_2 (sulfur dioxide)	SF_6 (sulfur hexafluoride)
H_2S (hydrogen sulfide)	HCN (hydrogen cyanide)*

*The boiling point of HCN is 26°C, but it is close enough to qualify as a gas at ordinary atmospheric conditions.

forces between positive and negative charges. To overcome these attractions we must apply a large amount of energy, which in practice means strongly heating the solid. Under normal conditions, all we can do is melt the solid; for example, NaCl melts at the rather high temperature of 801°C. To boil it, we would have to raise the temperature to well above 1000°C.

The behavior of molecular compounds is more varied. Some—for example, CO, CO_2, HCl, NH_3, and CH_4 (methane)—are gases, but the majority of molecular compounds are liquids or solids at room temperature. However, on heating they are converted to gases much more easily than ionic compounds. In other words, molecular compounds usually boil at much lower temperatures than ionic compounds do. There is no simple rule to help us determine whether a certain molecular compound is a gas under normal atmospheric conditions. To make such a determination we need to understand the nature and magnitude of the attractive forces among the molecules, called *intermolecular forces* (discussed in Chapter 11). In general, the stronger these attractions, the less likely a compound can exist as a gas at ordinary temperatures.

Of the gases listed in Table 5.1, only O_2 is essential for our survival. Hydrogen sulfide (H_2S) and hydrogen cyanide (HCN) are deadly poisons. Several others, such as CO, NO_2, O_3, and SO_2, are somewhat less toxic. The gases He, Ne, and Ar are chemically inert; that is, they do not react with any other substance. Most gases are colorless. Exceptions are F_2, Cl_2, and NO_2. The dark brown color of NO_2 is sometimes visible in polluted air. All gases have the following physical characteristics:

NO_2 gas.
Ken Karp/McGraw Hill

- Gases assume the volume and shape of their containers.
- Gases are the most compressible of the states of matter.
- Gases will mix evenly and completely when confined to the same container.
- Gases have much lower densities than liquids and solids.

Summary of Concepts & Facts

- At 25°C and 1 atm, a number of elements and molecular compounds exist as gases. Ionic compounds are solids rather than gases under atmospheric conditions.

5.2 Pressure of a Gas

Learning Objectives

- Understand the definition and the common units of pressure.
- Interconvert units of pressure.

Gases exert pressure on any surface or container with which they come in contact because gas particles are constantly in motion. We humans have adapted so well physiologically to the pressure of the air around us that we are usually unaware of it, perhaps as fish are not conscious of the water's pressure on them.

It is easy to demonstrate atmospheric pressure. One everyday example is the ability to drink a liquid through a straw. Sucking air out of the straw reduces the pressure inside the straw. The greater atmospheric pressure on the liquid pushes it up into the straw to replace the air that has been sucked out.

SI Units of Pressure

Pressure is one of the most readily measurable properties of a gas. To understand how we measure the pressure of a gas, it is helpful to know how the units of measurement are derived. We begin with velocity and acceleration.

Velocity is defined as the change in distance with elapsed time; that is,

$$\text{velocity} = \frac{\text{distance moved}}{\text{elapsed time}}$$

The SI unit for velocity is m/s, although we also use cm/s.

Acceleration is the change in velocity with time, or

$$\text{acceleration} = \frac{\text{change in velocity}}{\text{elapsed time}}$$

Acceleration is measured in m/s^2 (or cm/s^2).

The second law of motion, formulated by Sir Isaac Newton† in the late seventeenth century, defines another *force,* from which the units of pressure are derived. According to this law,

$$\text{force} = \text{mass} \times \text{acceleration}$$

In this context, the *SI unit of force* is the ***newton (N),*** where

$$1\text{ N} = 1\text{ kg m/s}^2$$

Finally, we define ***pressure*** as *force applied per unit area.*

$$\text{pressure} = \frac{\text{force}}{\text{area}}$$

The SI unit of pressure is the ***pascal (Pa),***‡ defined as *one newton per square meter.*

$$1\text{ Pa} = 1\text{ N/m}^2$$

Another unit based on the pascal is the bar:

$$1\text{ bar} = 10^5\text{ Pa} = 0.1\text{ MPa}$$

†Sir Isaac Newton (1642–1726). English mathematician, physicist, and astronomer. Newton is regarded by many as one of the two greatest physicists the world has known (the other is Albert Einstein). There was hardly a branch of physics to which Newton did not make a significant contribution. His book *Principia,* published in 1687, marks a milestone in the history of science.

‡Blaise Pascal (1623–1662). French mathematician and physicist. Pascal's work ranged widely in mathematics and physics, but his specialty was in the area of hydrodynamics (the study of the motion of fluids). He also invented a calculating machine.

Atmospheric Pressure

The atoms and molecules of the gases in the atmosphere, like those of all other matter, are subject to Earth's gravitational pull. As a consequence, the atmosphere is much denser near the surface of Earth than at high altitudes. (The air outside the pressurized cabin of an airplane at 9 km is too thin to breathe.) In fact, the density of air decreases very rapidly with increasing distance from Earth. Measurements show that about 50 percent of the atmosphere lies within 6.4 km of Earth's surface, 90 percent within 16 km, and 99 percent within 32 km. Not surprisingly, the denser the air is, the greater the pressure it exerts. The force experienced by any area exposed to Earth's atmosphere is equal to *the weight of the column of air above it.* ***Atmospheric pressure*** is *the pressure exerted by Earth's atmosphere* (Figure 5.2). The actual value of atmospheric pressure depends on location, temperature, and weather conditions.

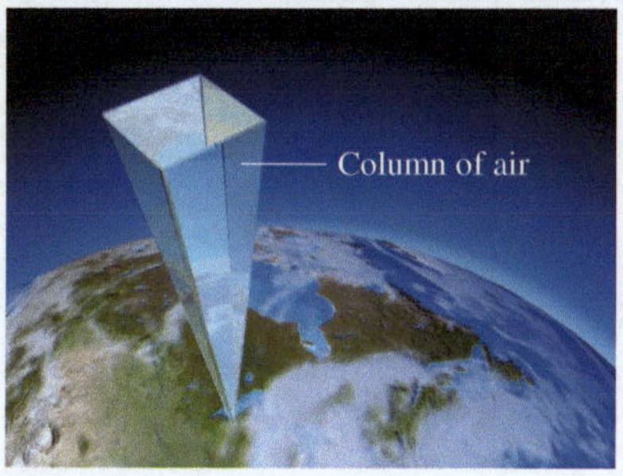

Figure 5.2 *A column of air extending from sea level to the upper atmosphere.*

Does atmospheric pressure act only downward, as you might infer from its definition? Imagine what would happen, then, if you were to hold a piece of paper tight (with both hands) above your head. You might expect the paper to bend due to the pressure of air acting on it, but this does not happen. The reason is that air, like water, is a fluid. The pressure exerted on an object in a fluid comes from all directions—downward and upward, as well as from the left and from the right. At the molecular level, air pressure results from collisions between the air molecules and any surface with which they come in contact. The magnitude of pressure depends on how often and how strongly the molecules impact the surface. It turns out that there are just as many molecules hitting the paper from the top as there are from underneath, so the paper stays flat.

How is atmospheric pressure measured? The ***barometer*** is probably the most familiar *instrument for measuring atmospheric pressure.* A simple barometer consists of a long glass tube, closed at one end and filled with mercury. If the tube is carefully inverted in a dish of mercury so that no air enters the tube, some mercury will flow out of the tube into the dish, creating a vacuum at the top (Figure 5.3). The weight of the mercury remaining in the tube is supported by atmospheric pressure acting on the surface of the mercury in the dish.

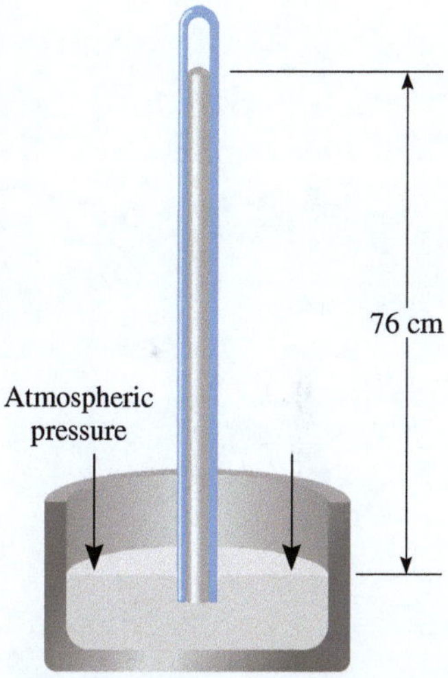

Figure 5.3 *A barometer for measuring atmospheric pressure. Above the mercury in the tube is a vacuum. The column of mercury is supported by the atmospheric pressure.*

Standard atmospheric pressure (1 atm) is equal to *the pressure that supports a column of mercury exactly 760 mm (or 76 cm) high at 0°C at sea level.* In other words, the standard atmosphere equals a pressure of 760 mmHg, where mmHg represents the pressure exerted by a column of mercury 1 mm high. The mmHg unit is also called the *torr,* after the Italian scientist Evangelista Torricelli,† who invented the barometer. Thus,

$$1 \text{ torr} = 1 \text{ mmHg}$$

and

$$1 \text{ atm} = 760 \text{ mmHg (exactly)}$$

The relation between atmospheres and pascals (see Appendix 1) is

$$\begin{aligned} 1 \text{ atm} &= 101{,}325 \text{ Pa} \\ &= 1.01325 \times 10^5 \text{ Pa} \end{aligned}$$

and because 1000 Pa = 1 kPa (kilopascal)

$$1 \text{ atm} = 1.01325 \times 10^2 \text{ kPa}$$

The standard pressure of a gas can be expressed in units of atm, bar, or pascals. In this text, the standard atmospheric pressure specifically refers to a pressure of 1 atm.

Examples 5.1 and 5.2 show the conversion from mmHg to atm and kPa.

†Evangelista Torricelli (1608–1674). Italian mathematician. Torricelli was supposedly the first person to recognize the existence of atmospheric pressure.

Example 5.1

The pressure outside a jet plane flying at high altitude falls considerably below standard atmospheric pressure. Therefore, the air inside the cabin must be pressurized to protect the passengers. What is the pressure in atmospheres in the cabin if the barometer reading is 688 mmHg?

Strategy Because 1 atm = 760 mmHg, the following conversion factor is needed to obtain the pressure in atmospheres:

$$\frac{1 \text{ atm}}{760 \text{ mmHg}}$$

Solution The pressure in the cabin is given by

$$\text{pressure} = 688 \text{ mmHg} \times \frac{1 \text{ atm}}{760 \text{ mmHg}}$$

$$= 0.905 \text{ atm}$$

Practice Exercise Convert 749 mmHg to atmospheres.

Similar problem: 5.13.

Example 5.2

Hurricane Irma in 2017 was one of the most destructive hurricanes in recent years, affecting the Caribbean, Cuba, U.S. Virgin Islands, and Florida, in particular. The lowest pressure recorded for Hurricane Irma was 691 mmHg. What was the pressure in kPa?

Strategy Here we are asked to convert mmHg to kPa. Because

$$1 \text{ atm} = 1.01325 \times 10^5 \text{ Pa} = 760 \text{ mmHg}$$

the conversion factor we need is

$$\frac{1.01325 \times 10^5 \text{ Pa}}{760 \text{ mmHg}}$$

Solution The pressure in kPa is

$$\text{pressure} = 691 \text{ mmHg} \times \frac{1.01325 \times 10^5 \text{ Pa}}{760 \text{ mmHg}}$$

$$= 9.21 \times 10^4 \text{ Pa}$$

$$= 92.1 \text{ kPa}$$

Practice Exercise Convert 295 mmHg to kilopascals.

Similar problem: 5.14.

A ***manometer*** is *a device used to measure the pressure of gases other than the atmosphere.* The principle of operation of a manometer is similar to that of a barometer. There are two types of manometers, shown in Figure 5.4. The *closed-tube manometer* is normally used to measure pressures below atmospheric pressure [Figure 5.4(a)], whereas the *open-tube manometer* is better suited for measuring pressures equal to or greater than atmospheric pressure [Figure 5.4(b)].

Nearly all barometers and many manometers use mercury as the working fluid, despite the fact that it is a toxic substance with a harmful vapor. The reason is that mercury has a very high density (13.6 g/mL) compared with most other liquids. Because the height of the liquid in a column is inversely proportional to the liquid's density, this property enables the construction of manageably small barometers and manometers.

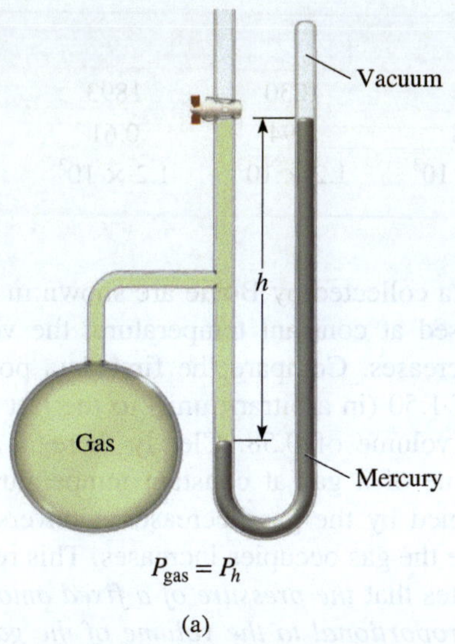

Figure 5.4 *Two types of manometers used to measure gas pressures. (a) Gas pressure may be less or greater than atmospheric pressure. (b) Gas pressure is greater than atmospheric pressure.*

Summary of Concepts & Facts

- Gases exert pressure because their molecules move freely and collide with any surface with which they make contact. Units of gas pressure include millimeters of mercury (mmHg), torr, pascals, and atmospheres. One atmosphere equals 760 mmHg, or 760 torr.

Review of Concepts & Facts

5.2.1 Express 1184 torr in units of mmHg, atm, and kPa.

5.2.2 Rank the following pressures from lowest to highest: (a) 736 mmHg, (b) 0.928 atm, (c) 728 torr, (d) 1.12×10^5 Pa.

5.2.3 Would it be easier to drink water with a straw on top or at the foot of Mt. Everest?

5.3 The Gas Laws

Learning Objective

- Appraise Boyle's, Charles's, and Avogadro's laws and how they may be combined to interconvert measurements of pressure, volume, and temperature.

The gas laws we will study in this chapter are the product of countless experiments on the physical properties of gases that were carried out over several centuries. Each of these generalizations regarding the macroscopic behavior of gaseous substances represents a milestone in the history of science. Together they have played a major role in the development of many ideas in chemistry.

The Pressure-Volume Relationship: Boyle's Law

In the seventeenth century, Robert Boyle† studied the behavior of gases systematically and quantitatively. In one series of studies, Boyle investigated the pressure-volume

Video: The Gas Laws

†Robert Boyle (1627–1691). British chemist and natural philosopher. Although Boyle is commonly associated with the gas law that bears his name, he made many other significant contributions in chemistry and physics. Despite the fact that Boyle was often at odds with scientists of his generation, his book *The Skeptical Chymist* (1661) influenced generations of chemists.

Table 5.2 Typical Pressure-Volume Relationship Obtained by Boyle

P (mmHg)	724	869	951	998	1230	1893	2250
V (arbitrary units)	1.50	1.33	1.22	1.18	0.94	0.61	0.58
PV	1.09×10^3	1.16×10^3	1.16×10^3	1.18×10^3	1.2×10^3	1.2×10^3	1.3×10^3

relationship of a gas sample. Typical data collected by Boyle are shown in Table 5.2. Note that as the pressure (P) is increased at constant temperature, the volume (V) occupied by a given amount of gas decreases. Compare the first data point with a pressure of 724 mmHg and a volume of 1.50 (in arbitrary unit) to the last data point with a pressure of 2250 mmHg and a volume of 0.58. Clearly there is an inverse relationship between pressure and volume of a gas at constant temperature. As the pressure is increased, the volume occupied by the gas decreases. Conversely, if the applied pressure is decreased, the volume the gas occupies increases. This relationship is now known as ***Boyle's law,*** which states that *the pressure of a fixed amount of gas at a constant temperature is inversely proportional to the volume of the gas.*

The apparatus used by Boyle in this experiment was very simple (Figure 5.5). In Figure 5.5(a), the pressure exerted on the gas is equal to atmospheric pressure and the volume of the gas is 100 mL. (Note that the tube is open at the top and is therefore exposed to atmospheric pressure.) In Figure 5.5(b), more mercury has been added to double the pressure on the gas, and the gas volume decreases to 50 mL. Tripling the pressure on the gas decreases its volume to a third of the original value [Figure 5.5(c)].

We can write a mathematical expression showing the inverse relationship between pressure and volume:

$$P \propto \frac{1}{V}$$

where the symbol $\propto$ means "proportional to." We can change $\propto$ to an equals sign and write

$$P = k_1 \times \frac{1}{V} \tag{5.1a}$$

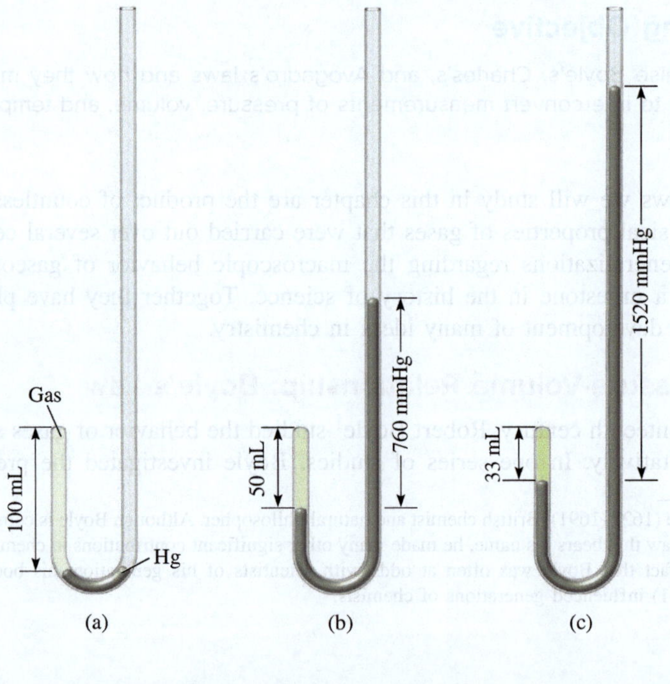

Figure 5.5 *Apparatus for studying the relationship between pressure and volume of a gas. (a) The levels of mercury are equal and the pressure of the gas is equal to the atmospheric pressure (760 mmHg). The gas volume is 100 mL. (b) Doubling the pressure by adding more mercury reduces the gas volume to 50 mL. (c) Tripling the pressure decreases the gas volume to one-third of the original value. The temperature and amount of gas are kept constant.*

where k_1 is a constant called the *proportionality constant.* Equation (5.1a) is the mathematical expression of Boyle's law. We can rearrange Equation (5.1a) and obtain

$$PV = k_1 \tag{5.1b}$$

This form of Boyle's law says that the product of the pressure and volume of a gas at constant temperature and amount of gas is a constant. The top diagram in Figure 5.6

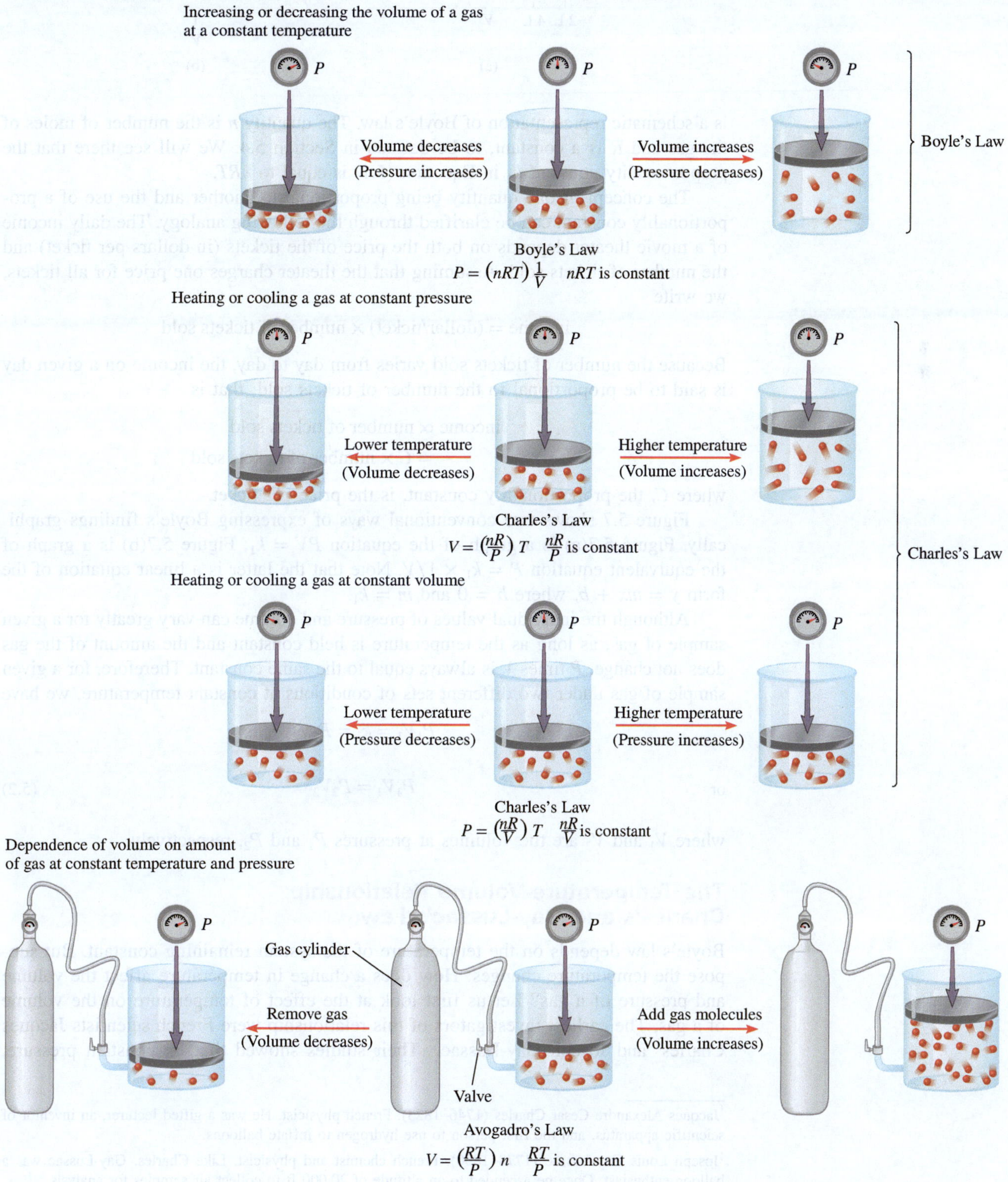

Figure 5.6 *Schematic illustrations of Boyle's law, Charles's law, and Avogadro's law.*

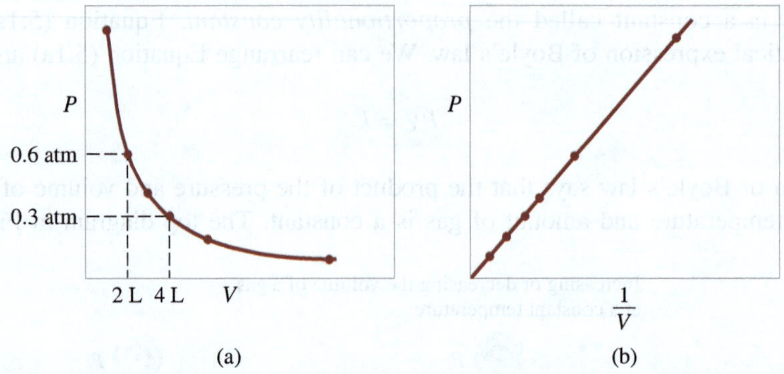

Figure 5.7 *Graphs showing the variation of the volume of a gas with the pressure exerted on the gas, at constant temperature. (a) P versus V. Note that the volume of the gas doubles as the pressure is halved. (b) P versus 1/V. The slope of the line is equal to k_1.*

is a schematic representation of Boyle's law. The quantity n is the number of moles of the gas and R is a constant, to be defined in Section 5.4. We will see there that the proportionality constant k_1 in Equation (5.1) is equal to nRT.

The concept of one quantity being proportional to another and the use of a proportionality constant can be clarified through the following analogy. The daily income of a movie theater depends on both the price of the tickets (in dollars per ticket) and the number of tickets sold. Assuming that the theater charges one price for all tickets, we write

$$\text{income} = (\text{dollar/ticket}) \times \text{number of tickets sold}$$

Because the number of tickets sold varies from day to day, the income on a given day is said to be proportional to the number of tickets sold; that is

$$\text{income} \propto \text{number of tickets sold}$$
$$= C \times \text{number of tickets sold}$$

where C, the proportionality constant, is the price per ticket.

Figure 5.7 shows two conventional ways of expressing Boyle's findings graphically. Figure 5.7(a) is a graph of the equation $PV = k_1$; Figure 5.7(b) is a graph of the equivalent equation $P = k_1 \times 1/V$. Note that the latter is a linear equation of the form $y = mx + b$, where $b = 0$ and $m = k_1$.

Although the individual values of pressure and volume can vary greatly for a given sample of gas, as long as the temperature is held constant and the amount of the gas does not change, P times V is always equal to the same constant. Therefore, for a given sample of gas under two different sets of conditions at constant temperature, we have

$$P_1V_1 = k_1 = P_2V_2$$

or

$$P_1V_1 = P_2V_2 \tag{5.2}$$

where V_1 and V_2 are the volumes at pressures P_1 and P_2, respectively.

The Temperature-Volume Relationship: Charles's and Gay-Lussac's Law

Boyle's law depends on the temperature of the system remaining constant. But suppose the temperature changes: How does a change in temperature affect the volume and pressure of a gas? Let us first look at the effect of temperature on the volume of a gas. The earliest investigators of this relationship were French scientists Jacques Charles† and Joseph Gay-Lussac.‡ Their studies showed that, at constant pressure,

†Jacques Alexandre Cesar Charles (1746–1823). French physicist. He was a gifted lecturer, an inventor of scientific apparatus, and the first person to use hydrogen to inflate balloons.

‡Joseph Louis Gay-Lussac (1778–1850). French chemist and physicist. Like Charles, Gay-Lussac was a balloon enthusiast. Once he ascended to an altitude of 20,000 ft to collect air samples for analysis.

the volume of a gas sample increases when heated and decreases when cooled (Figure 5.8). The quantitative relations involved in changes in gas temperature and volume turn out to be remarkably consistent. For example, we observe an interesting phenomenon when we study the temperature-volume relationship at various pressures. At any given pressure, the plot of volume versus temperature yields a straight line. By extending the line to zero volume, we find the intercept on the temperature axis to be −273.15°C. At any other pressure, we obtain a different straight line for the volume-temperature plot, but we get the *same* zero-volume temperature intercept at −273.15°C (Figure 5.9). (In practice, we can measure the volume of a gas over only a limited temperature range, because all gases condense at low temperatures to form liquids.)

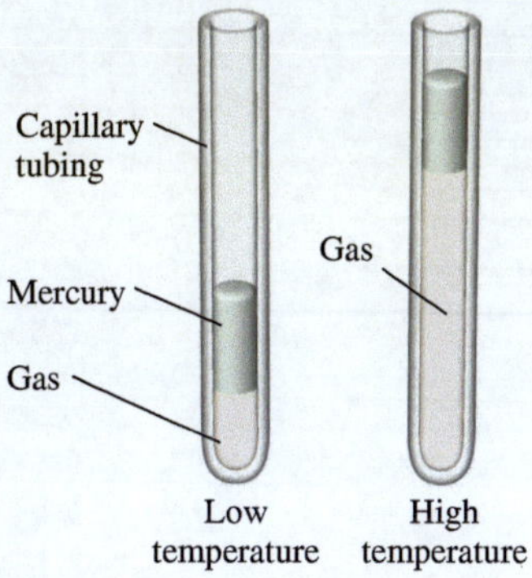

Figure 5.8 *Variation of the volume of a gas sample with temperature, at constant pressure. The pressure exerted on the gas is the sum of the atmospheric pressure and the pressure due to the weight of the mercury.*

In 1848 Lord Kelvin† realized the significance of this phenomenon. He identified −273.15°C as ***absolute zero,*** *theoretically the lowest attainable temperature.* Then he set up an ***absolute temperature scale,*** now called the ***Kelvin temperature scale,*** with *absolute zero as the starting point* (see Section 1.3). On the Kelvin scale, one kelvin (K) is equal *in magnitude* to one degree Celsius. The only difference between the absolute temperature scale and the Celsius scale is that the zero position is shifted. Important points on the two scales match up as follows:

	Kelvin Scale	Celsius Scale
Absolute zero	0 K	−273.15°C
Freezing point of water	273.15 K	0°C
Boiling point of water	373.15 K	100°C

The conversion between °C and K is given in Section 1.3. In most calculations we will use 273 instead of 273.15 as the term relating K and °C. By convention, we use T to denote absolute (kelvin) temperature and t to indicate temperature on the Celsius scale.

The dependence of the volume of a gas on temperature is given by

$$V \propto T$$
$$V = k_2 T$$

or

$$\frac{V}{T} = k_2 \qquad (5.3)$$

where k_2 is the proportionality constant. Equation (5.3) is known as ***Charles's and Gay-Lussac's law,*** or simply ***Charles's law,*** which states that *the volume of a fixed*

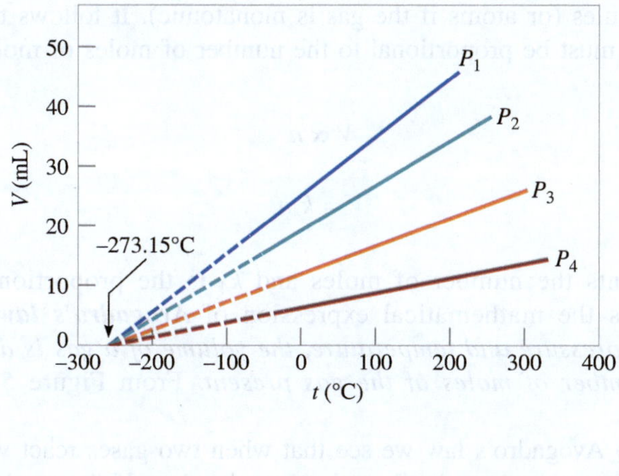

Figure 5.9 *Variation of the volume of a gas sample with temperature, at constant pressure. Each line represents the variation at a certain pressure. The pressures increase from P_1 to P_4. All gases ultimately condense (become liquids) if they are cooled to sufficiently low temperatures; the solid portions of the lines represent the temperature region above the condensation point. When these lines are extrapolated, or extended (the dashed portions), they all intersect at the point representing zero volume and a temperature of −273.15°C.*

†William Thomson, Lord Kelvin (1824–1907). Scottish mathematician and physicist. Kelvin did important work in many branches of physics.

amount of gas maintained at constant pressure is directly proportional to the absolute temperature of the gas. Charles's law is also illustrated in Figure 5.6. We see that the proportionality constant k_2 in Equation (5.3) is equal to nR/P.

Just as we did for pressure-volume relationships at constant temperature, we can compare two sets of volume-temperature conditions for a given sample of gas at constant pressure. From Equation (5.3) we can write

$$\frac{V_1}{T_1} = k_2 = \frac{V_2}{T_2}$$

or

$$\frac{V_1}{T_1} = \frac{V_2}{T_2} \quad (5.4)$$

where V_1 and V_2 are the volumes of the gas at temperatures T_1 and T_2 (both in kelvins), respectively.

Another form of Charles's law shows that at constant amount of gas and volume, the pressure of a gas is proportional to temperature; that is

$$P \propto T$$

$$P = k_3T$$

or

$$\frac{P}{T} = k_3 \quad (5.5)$$

From Figure 5.6 we see that $k_3 = nR/V$. Starting with Equation (5.5), we have

$$\frac{P_1}{T_1} = k_3 = \frac{P_2}{T_2}$$

or

$$\frac{P_1}{T_1} = \frac{P_2}{T_2} \quad (5.6)$$

where P_1 and P_2 are the pressures of the gas at temperatures T_1 and T_2, respectively.

The Volume-Amount Relationship: Avogadro's Law

The work of the Italian scientist Amedeo Avogadro complemented the studies of Boyle, Charles, and Gay-Lussac. In 1811 he published a hypothesis stating that at the same temperature and pressure, equal volumes of different gases contain the same number of molecules (or atoms if the gas is monatomic). It follows that the volume of any given gas must be proportional to the number of moles of molecules present; that is,

$$V \propto n$$

$$V = k_4n \quad (5.7)$$

where n represents the number of moles and k_4 is the proportionality constant. Equation (5.7) is the mathematical expression of ***Avogadro's law,*** which states that *at constant pressure and temperature, the volume of a gas is directly proportional to the number of moles of the gas present.* From Figure 5.6 we see that $k_4 = RT/P$.

According to Avogadro's law we see that when two gases react with each other, their reacting volumes have a simple ratio to each other. If the product is a gas, its volume is related to the volume of the reactants by a simple ratio (a fact demonstrated

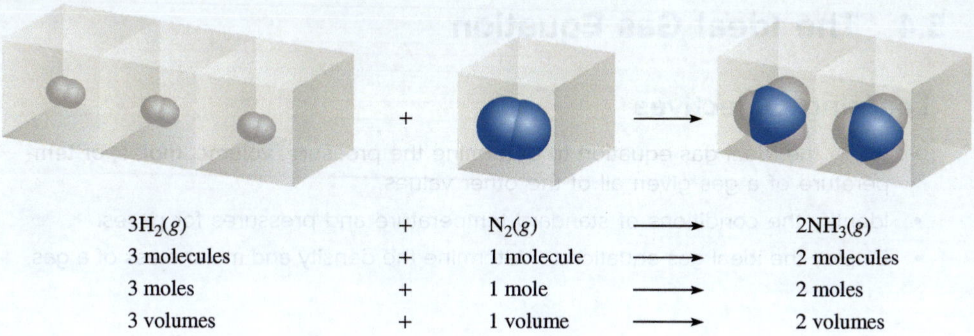

Figure 5.10 *Volume relationship of gases in a chemical reaction. The ratio of the volumes of molecular hydrogen to molecular nitrogen is 3:1, and that of ammonia (the product) to molecular hydrogen and molecular nitrogen combined (the reactants) is 2:4, or 1:2.*

earlier by Gay-Lussac). For example, consider the synthesis of ammonia from molecular hydrogen and molecular nitrogen:

$$\underset{3\text{ mol}}{3H_2(g)} + \underset{1\text{ mol}}{N_2(g)} \longrightarrow \underset{2\text{ mol}}{2NH_3(g)}$$

Because, at the same temperature and pressure, the volumes of gases are directly proportional to the number of moles of the gases present, we can now write

$$\underset{3\text{ volumes}}{3H_2(g)} + \underset{1\text{ volume}}{N_2(g)} \longrightarrow \underset{2\text{ volumes}}{2NH_3(g)}$$

The volume ratio of molecular hydrogen to molecular nitrogen is 3:1, and that of ammonia (the product) to the sum of the volumes of molecular hydrogen and molecular nitrogen (the reactants) is 2:4 or 1:2 (Figure 5.10).

Worked examples illustrating the gas laws are presented in Section 5.4.

Summary of Concepts & Facts

- The pressure-volume relationships of ideal gases are governed by Boyle's law: Volume is inversely proportional to pressure (at constant T and n).
- The temperature-volume relationships of ideal gases are described by Charles's and Gay-Lussac's law: Volume is directly proportional to temperature (at constant P and n).
- Absolute zero (−273.15°C) is the lowest theoretically attainable temperature. The Kelvin temperature scale takes 0 K as absolute zero. In all gas law calculations, temperature must be expressed in kelvins.
- The amount-volume relationships of ideal gases are described by Avogadro's law: Equal volumes of gases contain equal numbers of molecules (at the same T and P).

Review of Concepts & Facts

5.3.1 A gas occupies a volume of 2.50 L at 375 mmHg. What is the volume of the gas if the pressure is increased to 725 mmHg at constant temperature?

5.3.2 Compare the changes in volume when the temperature of a gas is increased at constant pressure from (a) 200 K to 400 K, (b) 200°C to 400°C.

5.3.3 What volume of ClF_3 will be produced when 75.0 mL of F_2 reacts with 25.0 mL of Cl_2 according to the following balanced equation?

$$Cl_2(g) + 3F_2(g) \longrightarrow 2ClF_3(g)$$

Assume the pressure and temperature are held constant.

5.4 The Ideal Gas Equation

Learning Objectives

- Apply the ideal gas equation to determine the pressure, volume, moles, or temperature of a gas given all of the other values.
- Identify the conditions of standard temperature and pressures for gases.
- Employ the ideal gas equation to determine the density and molar mass of a gas.

Let us summarize the gas laws we have discussed so far:

$$\text{Boyle's law: } V \propto \frac{1}{P} \quad \text{(at constant } n \text{ and } T)$$

$$\text{Charles's law: } V \propto T \quad \text{(at constant } n \text{ and } P)$$

$$\text{Avogadro's law: } V \propto n \quad \text{(at constant } P \text{ and } T)$$

We can combine all three expressions to form a single master equation for the behavior of gases:

$$V \propto \frac{nT}{P}$$

$$V = R\frac{nT}{P}$$

or

$$PV = nRT \tag{5.8}$$

where ***R,** the proportionality constant,* is called the ***gas constant.*** Equation (5.8), which is called the ***ideal gas equation,*** *describes the relationship among the four variables P, V, T, and n.* An ***ideal gas*** *is a hypothetical gas whose pressure-volume-temperature behavior can be completely accounted for by the ideal gas equation.* The molecules of an ideal gas do not attract or repel one another, and their volume is negligible compared with the volume of the container. Although there is no such thing in nature as an ideal gas, the ideal gas approximation works rather well for most reasonable temperature and pressure ranges. Thus, we can safely use the ideal gas equation to solve many gas problems.

Before we can apply the ideal gas equation to a real system, we must evaluate the gas constant R. At 0°C (273.15 K) and 1 atm pressure, many real gases behave like an ideal gas. Experiments show that under these conditions, 1 mole of an ideal gas occupies 22.414 L, which is somewhat greater than the volume of a basketball, as shown in Figure 5.11. *The conditions 0°C and 1 atm* are commonly called ***standard temperature and pressure,*** often abbreviated ***STP.*** Note that since 1982, IUPAC defines STP as 0°C and 10^5 Pa. For convenience purposes, we will use 0°C and 1 atm as STP in this text. From Equation (5.8) we can write

$$R = \frac{PV}{nT}$$

$$= \frac{(1 \text{ atm})(22.414 \text{ L})}{(1 \text{ mol})(273.15 \text{ K})}$$

$$= 0.082057 \frac{\text{L} \cdot \text{atm}}{\text{K} \cdot \text{mol}}$$

$$= 0.082057 \text{ L} \cdot \text{atm/K} \cdot \text{mol}$$

The dots between L and atm and between K and mol remind us that both L and atm are in the numerator and both K and mol are in the denominator. For most calculations,

Figure 5.11 *A comparison of the molar volume at STP (which is approximately 22.4 L) with a basketball.*
Ken Karp/McGraw Hill

we will use 22.41 L for the molar volume of a gas at STP. Additional values of *R* are found in Appendix 1; you should always use the appropriate version of *R* that facilitates the cancellation of units in a calculation.

Example 5.3 shows that if we know the quantity, volume, and temperature of a gas, we can calculate its pressure using the ideal gas equation. Unless otherwise stated, we assume that the temperatures given in °C in calculations are exact so that they do not affect the number of significant figures.

Student Hot Spot

Student data indicate you may struggle with the ideal gas law. Access your eBook for additional Learning Resources on this topic.

Example 5.3

Sulfur hexafluoride (SF_6) is a colorless and odorless gas. Due to its lack of chemical reactivity, it is used as an insulator in electronic equipment. Calculate the pressure (in atm) exerted by 1.82 moles of the gas in a steel vessel of volume 5.43 L at 69.5°C.

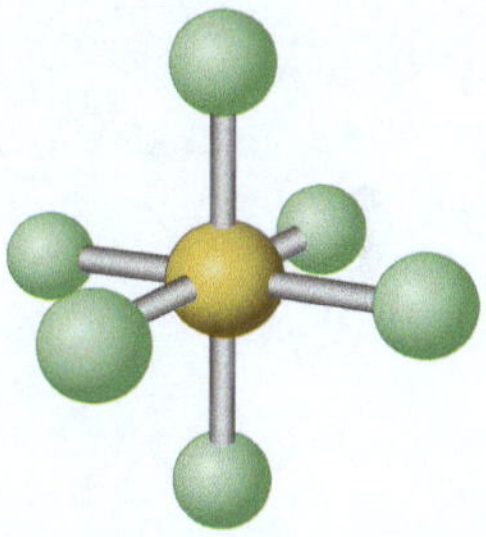

SF_6

Strategy The problem gives the amount of the gas and its volume and temperature. Is the gas undergoing a change in any of its properties? What equation should we use to solve for the pressure? What temperature unit should we use?

Solution Because no changes in gas properties occur, we can use the ideal gas equation to calculate the pressure. Rearranging Equation (5.8), we write

$$P = \frac{nRT}{V}$$

$$= \frac{(1.82 \text{ mol})(0.082057 \text{ L} \cdot \text{atm/K} \cdot \text{mol})(69.5 + 273) \text{ K}}{5.43 \text{ L}}$$

$$= 9.42 \text{ atm}$$

Practice Exercise Calculate the volume (in liters) occupied by 2.12 moles of nitric oxide (NO) at 6.54 atm and 76°C.

Similar problem: 5.32.

By using the fact that the molar volume of a gas occupies 22.41 L at STP, we can calculate the volume of a gas at STP without using the ideal gas equation, as shown in Example 5.4.

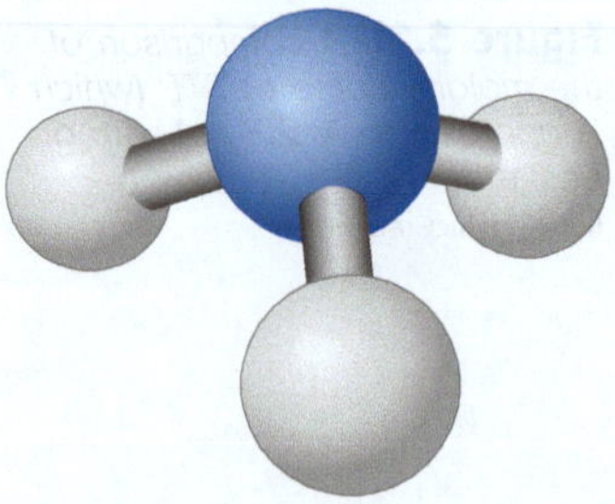

NH_3

Industrial ammonia refrigeration system.
Courtesy of Industrial Refrigeration Service, Inc

Example 5.4

Ammonia gas is used as a refrigerant in food processing and storage industries. Calculate the volume (in liters) occupied by 7.40 g of NH_3 at STP.

Strategy What is the volume of 1 mole of an ideal gas at STP? How many moles are there in 7.40 g of NH_3?

Solution Recognizing that 1 mole of an ideal gas occupies 22.41 L at STP and using the molar mass of NH_3 (17.03 g), we write the sequence of conversions as

$$\text{grams of NH}_3 \longrightarrow \text{moles of NH}_3 \longrightarrow \text{liters of NH}_3 \text{ at STP}$$

so the volume of NH_3 is given by

$$V = 7.40\ \text{g NH}_3 \times \frac{1\ \text{mol NH}_3}{17.03\ \text{g NH}_3} \times \frac{22.41\ \text{L}}{1\ \text{mol NH}_3}$$

$$= 9.74\ \text{L}$$

It is often true in chemistry, particularly in gas law calculations, that a problem can be solved in more than one way. Here the problem can also be solved by first converting 7.40 g of NH_3 to number of moles of NH_3, and then applying the ideal gas equation ($V = nRT/P$). Try it.

Check Because 7.40 g of NH_3 is smaller than its molar mass, its volume at STP should be smaller than 22.41 L. Therefore, the answer is reasonable.

Practice Exercise What is the volume (in liters) occupied by 49.8 g of HCl at STP?

Similar problem: 5.40.

The ideal gas equation is useful for problems that do not involve changes in P, V, T, and n for a gas sample. Thus, if we know any three of the variables we can calculate the fourth one using the equation. At times, however, we need to deal with changes in pressure, volume, and temperature, or even in the amount of gas. When conditions change, we must employ a modified form of the ideal gas equation that takes into account the initial and final conditions. We derive the modified equation as follows. From Equation (5.8),

$$R = \frac{P_1V_1}{n_1T_1} \text{ (before change)} \quad \text{and} \quad R = \frac{P_2V_2}{n_2T_2} \text{ (after change)}$$

Therefore,

$$\frac{P_1V_1}{n_1T_1} = \frac{P_2V_2}{n_2T_2} \tag{5.9}$$

It is interesting to note that all the gas laws discussed in Section 5.3 can be derived from Equation (5.9). If $n_1 = n_2$, as is usually the case because the amount of gas normally does not change, the equation then becomes

$$\frac{P_1V_1}{T_1} = \frac{P_2V_2}{T_2} \tag{5.10}$$

Applications of Equation (5.9) are shown in Examples 5.5 through 5.7.

Example 5.5

An inflated helium balloon with a volume of 0.55 L at sea level (1.0 atm) is allowed to rise to a height of 6.5 km, where the pressure is about 0.40 atm. Assuming that the temperature remains constant, what is the final volume of the balloon?

A scientific research helium balloon.
National Scientific Balloon Facility/Palestine, Texas

Strategy The amount of gas inside the balloon and its temperature remain constant, but both the pressure and the volume change. What gas law do you need?

Solution We start with Equation (5.9)

$$\frac{P_1V_1}{n_1T_1} = \frac{P_2V_2}{n_2T_2}$$

Because $n_1 = n_2$ and $T_1 = T_2$,

$$P_1V_1 = P_2V_2$$

which is Boyle's law [see Equation (5.2)]. The given information is tabulated:

Initial Conditions	Final Conditions
$P_1 = 1.0$ atm	$P_2 = 0.40$ atm
$V_1 = 0.55$ L	$V_2 = ?$

Therefore,

$$V_2 = V_1 \times \frac{P_1}{P_2}$$

$$= 0.55\ \text{L} \times \frac{1.0\ \text{atm}}{0.40\ \text{atm}}$$

$$= 1.4\ \text{L}$$

Check When pressure applied on the balloon is reduced (at constant temperature), the helium gas expands and the balloon's volume increases. The final volume is greater than the initial volume, so the answer is reasonable.

Practice Exercise A sample of chlorine gas occupies a volume of 946 mL at a pressure of 726 mmHg. Calculate the pressure of the gas (in mmHg) if the volume is reduced at constant temperature to 154 mL.

Similar problem: 5.19.

Example 5.6

Argon is an inert gas used in lightbulbs to retard the vaporization of the tungsten filament. A certain lightbulb containing argon at 1.20 atm and 18°C is heated to 85°C at constant volume. Calculate its final pressure (in atm).

Electric lightbulbs are usually filled with argon.
Buena Vista Images/Digital Vision/Getty Images

Strategy The temperature and pressure of argon change but the amount and volume of gas remain the same. What equation would you use to solve for the final pressure? What temperature unit should you use?

Solution Because $n_1 = n_2$ and $V_1 = V_2$, Equation (5.9) becomes

$$\frac{P_1}{T_1} = \frac{P_2}{T_2}$$

which is Charles's law [see Equation (5.6)]. Next we write

Initial Conditions	Final Conditions
$P_1 = 1.20$ atm	$P_2 = ?$
$T_1 = (18 + 273)\ \text{K} = 291\ \text{K}$	$T_2 = (85 + 273)\ \text{K} = 358\ \text{K}$

(Continued)

The final pressure is given by

$$P_2 = P_1 \times \frac{T_2}{T_1}$$

$$= 1.20 \text{ atm} \times \frac{358 \text{ K}}{291 \text{ K}}$$

$$= 1.48 \text{ atm}$$

Check At constant volume, the pressure of a given amount of gas is directly proportional to its absolute temperature. Therefore, the increase in pressure is reasonable.

Practice Exercise A sample of oxygen gas initially at 0.97 atm is cooled from 21°C to −68°C at constant volume. What is its final pressure (in atm)?

Similar problem: 5.36.

Example 5.7

A small bubble rises from the bottom of a lake, where the temperature and pressure are 8°C and 6.4 atm, to the water's surface, where the temperature is 25°C and the pressure is 1.0 atm. Calculate the final volume (in mL) of the bubble if its initial volume was 2.1 mL.

Strategy In solving this kind of problem where a lot of information is provided, it is often helpful to organize the information. The given information is tabulated here:

Initial Conditions	Final Conditions
$P_1 = 6.4$ atm	$P_2 = 1.0$ atm
$V_1 = 2.1$ mL	$V_2 = ?$
$T_1 = (8 + 273)$ K = 281 K	$T_2 = (25 + 273)$ K = 298 K

What temperature unit should be used in the calculation?

Solution According to Equation (5.9)

$$\frac{P_1V_1}{n_1T_1} = \frac{P_2V_2}{n_2T_2}$$

We assume that the amount of air in the bubble remains constant, that is, $n_1 = n_2$ so that

$$\frac{P_1V_1}{T_1} = \frac{P_2V_2}{T_2}$$

which is Equation (5.10). Rearranging Equation (5.10) gives

$$V_2 = V_1 \times \frac{P_1}{P_2} \times \frac{T_2}{T_1}$$

$$= 2.1 \text{ mL} \times \frac{6.4 \text{ atm}}{1.0 \text{ atm}} \times \frac{298 \text{ K}}{281 \text{ K}}$$

$$= 14 \text{ mL}$$

Check We can use any appropriate units for volume (or pressure) as long as we use the same units on both sides of the equation. We see that the final volume involves multiplying the initial volume by a ratio of pressures (P_1/P_2) and a ratio of temperatures (T_1/T_2). Recall that volume is inversely proportional to pressure, and volume is directly proportional to temperature. Because the pressure decreases and temperature increases as the bubble rises, we expect the bubble's volume to increase. In fact, here the change in pressure plays a greater role in the volume change.

Practice Exercise A gas initially at 4.0 L, 1.2 atm, and 66°C undergoes a change so that its final volume and temperature are 1.7 L and 42°C. What is its final pressure? Assume the number of moles remains unchanged.

Similar problem: 5.35.

Density Calculations

If we rearrange the ideal gas equation, we can calculate the density of a gas as

$$\frac{n}{V} = \frac{P}{RT}$$

The number of moles of the gas, *n*, is given by

$$n = \frac{m}{\mathscr{M}}$$

where *m* is the mass of the gas in grams and $\mathscr{M}$ is its molar mass. Therefore

$$\frac{m}{\mathscr{M}V} = \frac{P}{RT}$$

Because density, *d*, is mass per unit volume, we can write

$$d = \frac{m}{V} = \frac{P\mathscr{M}}{RT} \quad (5.11)$$

Unlike molecules in condensed matter (that is, in liquids and solids), gaseous molecules are separated by distances that are large compared with their size. Consequently, the density of gases is very low under atmospheric conditions. For this reason, gas densities are usually expressed in grams per liter (g/L) rather than grams per milliliter (g/mL), as Example 5.8 shows.

Example 5.8

Calculate the density of carbon dioxide (CO_2) in grams per liter (g/L) at 0.990 atm and 55°C.

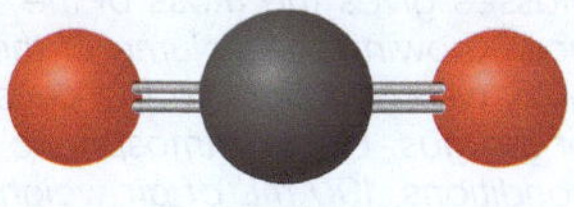

CO_2

Strategy We need Equation (5.11) to calculate gas density. Is sufficient information provided in the problem? What temperature unit should be used?

Solution To use Equation (5.11), we convert temperature to kelvins ($T = 273 + 55 = 328$ K) and use 44.01 g for the molar mass of CO_2; that is,

$$d = \frac{P\mathscr{M}}{RT}$$

$$= \frac{(0.990 \text{ atm})(44.01 \text{ g/mol})}{(0.082057 \text{ L} \cdot \text{atm/K} \cdot \text{mol})(328 \text{ K})} = 1.62 \text{ g/L}$$

Alternatively, we can solve for the density by writing

$$\text{density} \ \frac{\text{mass}}{\text{volume}}$$

Assuming that we have 1 mole of CO_2, the mass is 44.01 g. The volume of the gas can be obtained from the ideal gas equation

$$V = \frac{nRT}{P}$$

$$= \frac{(1 \text{ mol})(0.082057 \text{ L} \cdot \text{atm/K} \cdot \text{mol})(328 \text{ K})}{0.990 \text{ atm}}$$

$$= 27.2 \text{ L}$$

Therefore, the density of CO_2 is given by

$$d = \frac{44.01 \text{ g}}{27.2 \text{ L}} = 1.62 \text{ g/L}$$

(Continued)

Comment In units of grams per milliliter, the gas density is 1.62×10^{-3} g/mL, which is a very small number. In comparison, the density of water is 1.0 g/mL and that of gold is 19.3 g/cm^3.

Practice Exercise What is the density (in g/L) of uranium hexafluoride (UF_6) at 779 mmHg and 62°C?

Similar problem: 5.48.

The Molar Mass of a Gaseous Substance

From what we have seen so far, you may have the impression that the molar mass of a substance is found by examining its formula and summing the molar masses of its component atoms. However, this procedure works only if the actual formula of the substance is known. In practice, chemists often deal with substances of unknown or only partially defined composition. If the unknown substance is gaseous, its molar mass can nevertheless be found thanks to the ideal gas equation. All that is needed is an experimentally determined density value (or mass and volume data) for the gas at a known temperature and pressure. By rearranging Equation (5.11) we get

$$\mathcal{M} = \frac{dRT}{P} \tag{5.12}$$

Figure 5.12 *An apparatus for measuring the density of a gas. A bulb of known volume is filled with the gas under study at a certain temperature and pressure. First the bulb is weighed, and then it is emptied (evacuated) and weighed again. The difference in masses gives the mass of the gas. Knowing the volume of the bulb, we can calculate the density of the gas. Under atmospheric conditions, 100 mL of air weigh about 0.12 g, an easily measured quantity.*

In a typical experiment, a bulb of known volume is filled with the gaseous substance under study. The temperature and pressure of the gas sample are recorded, and the total mass of the bulb plus gas sample is determined (Figure 5.12). The bulb is then evacuated (emptied) and weighed again. The difference in mass is the mass of the gas. The density of the gas is equal to its mass divided by the volume of the bulb. Once we know the density of a gas, we can calculate the molar mass of the substance using Equation (5.12).

The mass spectrometer has become the dominant instrument for determining molar mass, but the determination of molar mass by the density method is still useful, as illustrated by Example 5.9.

Example 5.9

A chemist has synthesized a greenish-yellow gaseous compound of chlorine and oxygen and finds that its density is 7.71 g/L at 36°C and 2.88 atm. Calculate the molar mass of the compound and determine its molecular formula.

Strategy Because Equations (5.11) and (5.12) are rearrangements of each other, we can calculate the molar mass of a gas if we know its density, temperature, and pressure. The molecular formula of the compound must be consistent with its molar mass. What temperature unit should we use?

Solution From Equation (5.12)

$$\begin{aligned}\mathcal{M} &= \frac{dRT}{P}\\ &= \frac{(7.71\ \text{g/L})(0.082057\ \text{L}\cdot\text{atm/K}\cdot\text{mol})(36+273)\ \text{K}}{2.88\ \text{atm}}\\ &= 67.9\ \text{g/mol}\end{aligned}$$

Alternatively, we can solve for the molar mass by writing

$$\text{molar mass of compound} = \frac{\text{mass of compound}}{\text{moles of compound}}$$

(Continued)

From the given density we know there are 7.71 g of the gas in 1 L. The number of moles of the gas in this volume can be obtained from the ideal gas equation

$$n = \frac{PV}{RT}$$

$$= \frac{(2.88 \text{ atm})(1.00 \text{ L})}{(0.082057 \text{ L} \cdot \text{atm/K} \cdot \text{mol})(309 \text{ K})}$$

$$= 0.1135 \text{ mol}$$

Therefore, the molar mass is given by

$$\mathscr{M} = \frac{\text{mass}}{\text{number of moles}} = \frac{7.71 \text{ g}}{0.1135 \text{ mol}} = 67.9 \text{ g/mol}$$

We can determine the molecular formula of the compound by trial and error, using only the knowledge of the molar masses of chlorine (35.45 g) and oxygen (16.00 g). We know that a compound containing one Cl atom and one O atom would have a molar mass of 51.45 g, which is too low, while the molar mass of a compound made up of two Cl atoms and one O atom is 86.90 g, which is too high. Thus, the compound must contain one Cl atom and two O atoms and have the formula ClO_2, which has a molar mass of 67.45 g.

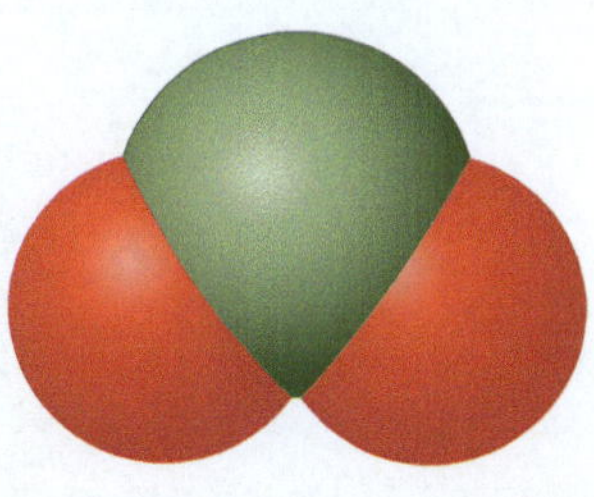

ClO_2

Practice Exercise The density of a gaseous organic compound is 3.38 g/L at 40°C and 1.97 atm. What is its molar mass?

Similar problems: 5.43, 5.47.

Because Equation (5.12) is derived from the ideal gas equation, we can also calculate the molar mass of a gaseous substance using the ideal gas equation, as shown in Example 5.10.

Example 5.10

Chemical analysis of a gaseous compound showed that it contained 33.0 percent silicon (Si) and 67.0 percent fluorine (F) by mass. At 35°C, 0.210 L of the compound exerted a pressure of 1.70 atm. If the mass of 0.210 L of the compound was 2.38 g, calculate the molecular formula of the compound.

Strategy This problem can be divided into two parts. First, it asks for the empirical formula of the compound from the percent by mass of Si and F. Second, the information provided enables us to calculate the molar mass of the compound and hence determine its molecular formula. What is the relationship between empirical molar mass and molar mass calculated from the molecular formula?

Solution We follow the procedure in Example 3.9 (Section 3.5) to calculate the empirical formula by assuming that we have 100 g of the compound, so the percentages are converted to grams. The number of moles of Si and F are given by

$$n_{Si} = 33.0 \text{ g Si} \times \frac{1 \text{ mol Si}}{28.09 \text{ g Si}} = 1.17 \text{ mol Si}$$

$$n_F = 67.0 \text{ g F} \times \frac{1 \text{ mol F}}{19.00 \text{ g F}} = 3.53 \text{ mol F}$$

Therefore, the empirical formula is $Si_{1.17}F_{3.53}$, or, dividing by the smaller subscript (1.17), we obtain SiF_3.

To calculate the molar mass of the compound, we need first to calculate the number of moles contained in 2.38 g of the compound. From the ideal gas equation

$$n = \frac{PV}{RT}$$

$$= \frac{(1.70 \text{ atm})(0.210 \text{ L})}{(0.082057 \text{ L} \cdot \text{atm/K} \cdot \text{mol})(308 \text{ K})} = 0.0141 \text{ mol}$$

(Continued)

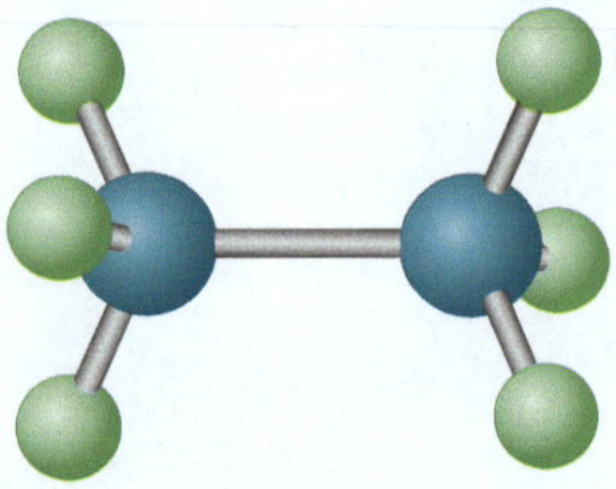

Si_2F_6

Because there are 2.38 g in 0.0141 mole of the compound, the mass in 1 mole, or the molar mass, is given by

$$\mathscr{M} = \frac{2.38\ \text{g}}{0.0141\ \text{mol}} = 169\ \text{g/mol}$$

The molar mass of the empirical formula SiF_3 is 85.09 g. Recall that the ratio (molar mass/empirical molar mass) is always an integer ($169/85.09 \approx 2$). Therefore, the molecular formula of the compound must be $(SiF_3)_2$ or Si_2F_6.

Practice Exercise A gaseous compound is 78.14 percent boron and 21.86 percent hydrogen. At 27°C, 74.3 mL of the gas exerted a pressure of 1.12 atm. If the mass of the gas was 0.0934 g, what is its molecular formula?

Similar problem: 5.49.

Summary of Concepts & Facts

- The ideal gas equation, $PV = nRT$, combines the laws of Boyle, Charles, and Avogadro. This equation describes the behavior of an ideal gas.
- The ideal gas equation can also be used to calculate the density or molar mass of a gas.

Review of Concepts & Facts

5.4.1 The volume of a gas is 52.0 mL at 15.0°C and 2.75 atm. At what temperature will the gas have a pressure of 3.50 atm and a volume of 65.0 mL?

5.4.2 Assuming ideal behavior, which of the following gases will have the greatest volume at STP: (a) 0.82 mole of He, (b) 24 g of N_2, or (c) 5.0×10^{23} molecules of Cl_2? Which gas will have the greatest density?

5.4.3 What is the molar mass of an unknown compound if a 194-mL sample has a mass of 0.477 g at 0.976 atm and 100°C?

5.5 Gas Stoichiometry

Learning Objective

- Manipulate the ideal gas equation to calculate stoichiometric quantities.

In Chapter 3 we used relationships between amounts (in moles) and masses (in grams) of reactants and products to solve stoichiometry problems. When the reactants and/or products are gases, we can also use the relationships between amounts (moles, n) and volume (V) to solve such problems (Figure 5.13). Examples 5.11, 5.12, and 5.13 show how the gas laws are used in these calculations.

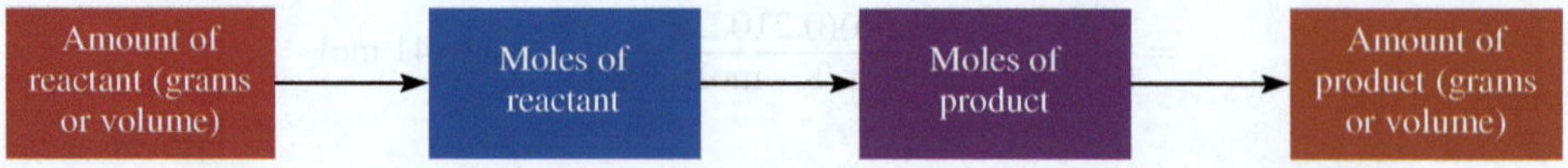

Figure 5.13 *Stoichiometric calculations involving gases.*

Example 5.11

The combustion of acetylene with pure oxygen produces a very high-temperature flame used for welding and cutting metals. Calculate the volume of O_2 (in liters) required for the complete combustion of 7.64 L of acetylene (C_2H_2) measured at the same temperature and pressure.

$$2C_2H_2(g) + 5O_2(g) \longrightarrow 4CO_2(g) + 2H_2O(l)$$

Strategy Note that the temperature and pressure of O_2 and C_2H_2 are the same. Which gas law do we need to relate the volume of the gases to the moles of gases?

Solution According to Avogadro's law, at the same temperature and pressure, the number of moles of gases is directly related to volume. From the equation, we have 5 mol $O_2 \simeq$ 2 mol C_2H_2; therefore, we can also write 5 L $O_2 \simeq$ 2 L C_2H_2. The volume of O_2 that will react with 7.64 L C_2H_2 is given by

$$\text{volume of } O_2 = 7.64\ \text{L}\ \cancel{C_2H_2} \times \frac{5\ \text{L}\ O_2}{2\ \cancel{\text{L}\ C_2H_2}}$$

$$= 19.1\ \text{L}$$

Practice Exercise Assuming no change in temperature and pressure, calculate the volume of O_2 (in liters) required for the complete combustion of 14.9 L of butane (C_4H_{10}).

$$2C_4H_{10}(g) + 13O_2(g) \longrightarrow 8CO_2(g) + 10H_2O(l)$$

Similar problem: 5.26.

The reaction of calcium carbide (CaC_2) with water produces acetylene (C_2H_2), a flammable gas.
Charles D. Winters/McGraw Hill

Example 5.12

Sodium azide (NaN_3) is used in some automobile air bags. The impact of a collision triggers the decomposition of NaN_3 as follows:

$$2NaN_3(s) \longrightarrow 2Na(s) + 3N_2(g)$$

The nitrogen gas produced quickly inflates the bag between the driver and the windshield and dashboard. Calculate the volume of N_2 generated at 80°C and 823 mmHg by the decomposition of 60.0 g of NaN_3.

Strategy From the balanced equation, we see that 2 mol $NaN_3 \simeq$ 3 mol N_2 so the conversion factor between NaN_3 and N_2 is

$$\frac{3\ \text{mol}\ N_2}{2\ \text{mol}\ NaN_3}$$

Because the mass of NaN_3 is given, we can calculate the number of moles of NaN_3 and hence the number of moles of N_2 produced. Finally, we can calculate the volume of N_2 using the ideal gas equation.

Solution First, we calculate the number of moles of N_2 produced by 60.0 g NaN_3 using the following sequence of conversions

$$\text{grams of } NaN_3 \longrightarrow \text{moles of } NaN_3 \longrightarrow \text{moles of } N_2$$

so that

$$\text{moles of } N_2 = 60.0\ \cancel{\text{g}\ NaN_3} \times \frac{1\ \cancel{\text{mol}\ NaN_3}}{65.02\ \cancel{\text{g}\ NaN_3}} \times \frac{3\ \text{mol}\ N_2}{2\ \cancel{\text{mol}\ NaN_3}}$$

$$= 1.38\ \text{mol}\ N_2$$

(Continued)

An air bag can protect the driver in an automobile collision.
Therry/iStock/Getty Images

The volume of 1.38 moles of N_2 can be obtained by using the ideal gas equation

$$V = \frac{nRT}{P} = \frac{(1.38\ \text{mol})(0.082057\ \text{L}\cdot\text{atm/K}\cdot\text{mol})(80 + 273\ \text{K})}{(823/760)\ \text{atm}}$$

$$= 36.9\ \text{L}$$

Practice Exercise The equation for the metabolic breakdown of glucose ($C_6H_{12}O_6$) is the same as the equation for the combustion of glucose in air; that is

$$C_6H_{12}O_6(s) + 6O_2(g) \longrightarrow 6CO_2(g) + 6H_2O(l)$$

Calculate the volume of CO_2 produced at 37°C and 1.00 atm when 5.60 g of glucose is used up in the reaction.

Similar problem: 5.62.

Example 5.13

The air in submerged submarines and space vehicles needs to be purified continuously.
StockTrek/SuperStock

Aqueous lithium hydroxide solution is used to purify air in spacecraft and submarines because it absorbs carbon dioxide, which is an end product of metabolism, according to the equation

$$2LiOH(aq) + CO_2(g) \longrightarrow Li_2CO_3(aq) + H_2O(l)$$

The pressure of carbon dioxide inside the cabin of a submarine having a volume of 2.4×10^5 L is 7.9×10^{-3} atm at 312 K. A solution of lithium hydroxide (LiOH) of negligible volume is introduced into the cabin. Eventually the pressure of CO_2 falls to 1.2×10^{-4} atm. How many grams of lithium carbonate are formed by this process?

Strategy How do we calculate the number of moles of CO_2 reacted from the drop in CO_2 pressure? From the ideal gas equation, we write

$$n = P \times \left(\frac{V}{RT}\right)$$

At constant T and V, the change in pressure of CO_2, ΔP, corresponds to the change in the number of moles of CO_2, Δn. Thus,

$$\Delta n = \Delta P \times \left(\frac{V}{RT}\right)$$

What is the conversion factor between CO_2 and Li_2CO_3?

Solution The drop in CO_2 pressure is (7.9×10^{-3} atm) − (1.2×10^{-4} atm) or 7.8×10^{-3} atm. Therefore, the number of moles of CO_2 reacted is given by

$$\Delta n = 7.8 \times 10^{-3}\ \text{atm} \times \frac{2.4 \times 10^5\ \text{L}}{(0.082057\ \text{L}\cdot\text{atm/K}\cdot\text{mol})(312\ \text{K})}$$

$$= 73\ \text{mol}$$

From the chemical equation, we see that 1 mol $CO_2 \simeq$ 1 mol Li_2CO_3, so the amount of Li_2CO_3 formed is also 73 moles. Then, with the molar mass of Li_2CO_3 (73.89 g), we calculate its mass.

$$\text{mass of } Li_2CO_3 \text{ formed} = 73\ \text{mol}\ \cancel{Li_2CO_3} \times \frac{73.88\ \text{g}\ Li_2CO_3}{1\ \text{mol}\ \cancel{Li_2CO_3}}$$

$$= 5.4 \times 10^3\ \text{g}\ Li_2CO_3$$

Practice Exercise A 2.14-L sample of hydrogen chloride (HCl) gas at 2.61 atm and 28°C is completely dissolved in 668 mL of water to form hydrochloric acid solution. Calculate the molarity of the acid solution. Assume no change in volume.

Similar problem: 5.100.

Summary of Concepts & Facts

- The ideal gas equation can be used to study the stoichiometry of reactions involving gases.

Review of Concepts & Facts

5.5.1 Zinc metal reacts with hydrochloric acid according to the following equation:

$$Zn(s) + 2HCl(aq) \longrightarrow ZnCl_2(aq) + H_2(g)$$

What volume of hydrogen gas is produced at STP when 16.4 g of zinc is used in the reaction?

5.5.2 Alkanes (C_nH_{2n+2}) are discussed in Section 2.8. For which alkanes, if any, does the number of moles of gas remain constant as the gas-phase combustion reaction

$$\text{alkane}(g) + \text{oxygen}(g) \longrightarrow \text{carbon dioxide}(g) + \text{water vapor}(g)$$

proceeds from reactants to products?

5.6 Dalton's Law of Partial Pressures

Learning Objective

- Determine mole fractions and partial pressures for gases in a mixture.

Thus far we have concentrated on the behavior of pure gaseous substances, but experimental studies very often involve mixtures of gases. For example, for a study of air pollution, we may be interested in the pressure-volume-temperature relationship of a sample of air, which contains several gases. In this case, and all cases involving mixtures of gases, the total gas pressure is related to ***partial pressures,*** that is, *the pressures of individual gas components in the mixture.* In 1801 Dalton formulated a law, now known as ***Dalton's law of partial pressures,*** which states that *the total pressure of a mixture of gases is just the sum of the pressures that each gas would exert if it were present alone.* Figure 5.14 illustrates Dalton's law.

Consider a case in which two gases, A and B, are in a container of volume *V*. The pressure exerted by gas A, according to the ideal gas equation, is

$$P_A = \frac{n_A RT}{V}$$

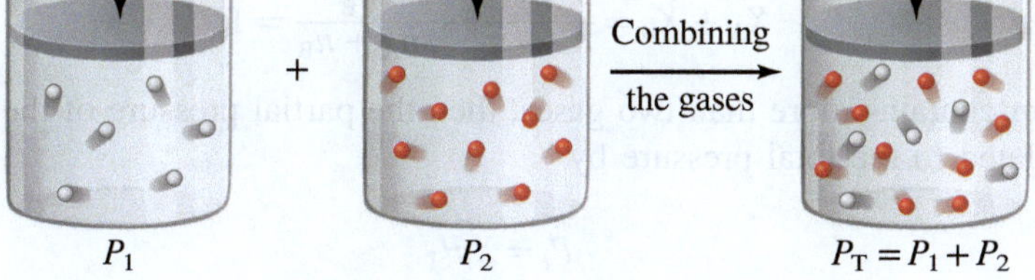

Figure 5.14 *Schematic illustration of Dalton's law of partial pressures.*

where n_A is the number of moles of A present. Similarly, the pressure exerted by gas B is

$$P_B = \frac{n_B RT}{V}$$

In a mixture of gases A and B, the total pressure P_T is the result of the collisions of both types of molecules, A and B, with the walls of the container. Thus, according to Dalton's law,

$$\begin{aligned} P_T &= P_A + P_B \\ &= \frac{n_A RT}{V} + \frac{n_B RT}{V} \\ &= \frac{RT}{V}(n_A + n_B) \\ &= \frac{n_T RT}{V} \end{aligned}$$

where n_T, the total number of moles of gases present, is given by $n = n_A + n_B$, and P_A and P_B are the partial pressures of gases A and B, respectively. For a mixture of gases, then, P_T depends only on the total number of moles of gas present, not on the nature of the gas molecules.

In general, the total pressure of a mixture of gases is given by

$$P_T = P_1 + P_2 + P_3 + \ldots$$

where P_1, P_2, P_3, . . . are the partial pressures of components 1, 2, 3, ... To see how each partial pressure is related to the total pressure, consider again the case of a mixture of two gases, A and B. Dividing P_A by P_T, we obtain

$$\begin{aligned} \frac{P_A}{P_T} &= \frac{n_A RT/V}{(n_A + n_B)RT/V} \\ &= \frac{n_A}{n_A + n_B} \\ &= X_A \end{aligned}$$

where X_A is called the mole fraction of A. The ***mole fraction*** is *the ratio of the number of moles of one component to the number of moles of all components present.* In general, the mole fraction of component i in a mixture is given by

$$X_i = \frac{n_i}{n_T} \tag{5.13}$$

where n_i and n_T are the number of moles of component i and the total number of moles present, respectively. The mole fraction is a dimensionless quantity and is always less than 1. We can now express the partial pressure of A as

$$P_A = X_A P_T$$

Similarly,

$$P_B = X_B P_T$$

Note that the sum of the mole fractions for a mixture of gases must equal 1. If only two components are present, then

$$X_A + X_B = \frac{n_A}{n_A + n_B} + \frac{n_B}{n_A + n_B} = 1$$

If a system contains more than two gases, then the partial pressure of the ith component is related to the total pressure by

$$P_i = X_i P_T \tag{5.14}$$

How are partial pressures determined? A manometer can measure only the total pressure of a gaseous mixture. To obtain the partial pressures, we need to know the mole fractions of the components, which would involve elaborate chemical analyses. The most direct method of measuring partial pressures is using a mass spectrometer. The relative intensities of the peaks in a mass spectrum are directly proportional to the amounts, and hence to the mole fractions, of the gases present.

From mole fractions and total pressure, we can calculate the partial pressures of individual components, as Example 5.14 shows. A direct application of Dalton's law of partial pressures to scuba diving is discussed in this section's Chemistry in Action essay, "Scuba Diving and the Gas Laws."

Student Hot Spot

Student data indicate you may struggle with mole fractions. Access your eBook for additional Learning Resources on this topic.

Example 5.14

A mixture of gases contains 4.46 moles of neon (Ne), 0.74 mole of argon (Ar), and 2.15 moles of xenon (Xe). Calculate the partial pressures of the gases if the total pressure is 2.00 atm at a certain temperature.

Strategy What is the relationship between the partial pressure of a gas and the total gas pressure? How do we calculate the mole fraction of a gas?

Solution According to Equation (5.14), the partial pressure of Ne (P_{Ne}) is equal to the product of its mole fraction (X_{Ne}) and the total pressure (P_T).

$$\underset{\text{want to calculate}}{P_{Ne}} = \overset{\text{need to find}}{X_{Ne}}\underset{\text{given}}{P_T}$$

Using Equation (5.13), we calculate the mole fraction of Ne as follows:

$$X_{Ne} = \frac{n_{Ne}}{n_{Ne} + n_{Ar} + n_{Xe}} = \frac{4.46 \text{ mol}}{4.46 \text{ mol} + 0.74 \text{ mol} + 2.15 \text{ mol}} = 0.607$$

Therefore,

$$P_{Ne} = X_{Ne}P_T = 0.607 \times 2.00 \text{ atm} = 1.21 \text{ atm}$$

Similarly,

$$P_{Ar} = X_{Ar}P_T = 0.10 \times 2.00 \text{ atm} = 0.20 \text{ atm}$$

and

$$P_{Xe} = X_{Xe}P_T = 0.293 \times 2.00 \text{ atm} = 0.586 \text{ atm}$$

Check Make sure that the sum of the partial pressures is equal to the given total pressure; that is, (1.21 + 0.20 + 0.586) atm = 2.00 atm.

Practice Exercise A sample of natural gas contains 8.24 moles of methane (CH_4), 0.421 mole of ethane (C_2H_6), and 0.116 mole of propane (C_3H_8). If the total pressure of the gases is 1.37 atm, what are the partial pressures of the gases?

Similar problem: 5.67.

CHEMISTRY *in Action*

Scuba Diving and the Gas Laws

Scuba diving is an exhilarating sport, and, thanks in part to the gas laws, it is also a safe activity for trained individuals who are in good health. ("Scuba" is an acronym for self-contained underwater breathing apparatus.) Two applications of the gas laws to this popular pastime are the development of guidelines for returning safely to the surface after a dive and the determination of the proper mix of gases to prevent a potentially fatal condition during a dive.

A typical dive might be 40 to 65 ft, but dives to 90 ft are not uncommon. Because seawater has a slightly higher density than fresh water—about 1.03 g/mL, compared with 1.00 g/mL—the pressure exerted by a column of 33 ft of seawater is equivalent to 1 atm pressure. Pressure increases with increasing depth, so at a depth of 66 ft the pressure of the water will be 2 atm, and so on.

What would happen if a diver rose to the surface from a depth of, say, 20 ft rather quickly without breathing? The total decrease in pressure for this change in depth would be (20 ft/33 ft) × 1 atm, or 0.6 atm. When the diver reached the surface, the volume of air trapped in the lungs would have increased by a factor of (1 + 0.6) atm/1 atm, or 1.6 times. This sudden expansion of air can fatally rupture the membranes of the lungs. Another serious possibility is that an *air embolism* might develop. As air expands in the lungs, it is forced into tiny blood vessels called capillaries. The presence of air bubbles in these vessels can block normal blood flow to the brain. As a result, the diver might lose consciousness before reaching the surface. The only cure for an air embolism is recompression. For this painful process, the victim is placed in a chamber filled with compressed air. Here bubbles in the blood are slowly squeezed down to harmless size over the course of several hours to a day. To avoid these unpleasant complications, divers know they must ascend slowly, pausing at certain points to give their body time to adjust to the falling pressure.

Our second example is a direct application of Dalton's law. Oxygen gas is essential for our survival, so it is hard to believe that an excess of oxygen could be harmful. Nevertheless, the toxicity of too much oxygen is well established. For example, newborn infants placed in oxygen tents often sustain damage to the retinal tissue, which can cause partial or total blindness.

Our bodies function best when oxygen gas has a partial pressure of about 0.20 atm, as it does in the air we breathe. The oxygen partial pressure is given by

$$P_{O_2} = X_{O_2}P_T = \frac{n_{O_2}}{n_{O_2} + n_{N_2}} P_T$$

where P_T is the total pressure. However, because volume is directly proportional to the number of moles of gas present (at constant temperature and pressure), we can now write

$$P_{O_2} = \frac{V_{O_2}}{V_{O_2} + V_{N_2}} P_T$$

Video: Collecting a Gas over Water

An important application of Dalton's law of partial pressures involves calculating the amount of a gas collected over water. Gases that are commonly used in the laboratory are generally obtained from pressurized gas cylinders, but if there is an occasional need for a small amount of a certain gas, it may be more convenient to prepare it chemically. For example, when potassium chlorate ($KClO_3$) is heated, it decomposes to KCl and O_2; that is,

$$2KClO_3(s) \longrightarrow 2KCl(s) + 3O_2(g)$$

The oxygen gas can be collected over water, as shown in Figure 5.15. Initially, the inverted bottle is completely filled with water. As oxygen gas is generated, the gas bubbles rise to the top and displace water from the bottle. This method of collecting a gas is based on the assumptions that the gas does not react with water and that it is not appreciably soluble in it. These assumptions are valid for oxygen gas, but not for gases such as NH_3, which dissolves readily in water. The oxygen gas collected in this way is not pure, however, because water vapor is also present in the bottle. The total gas pressure is equal to the sum of the pressures exerted by the oxygen gas and the water vapor:

$$P_T = P_{O_2} + P_{H_2O}$$

Thus, the composition of air is 20 percent oxygen gas and 80 percent nitrogen gas by volume. When a diver is submerged, the pressure of the water on the diver is greater than atmospheric pressure. The air pressure inside the body cavities (for example, lungs, sinuses) must be the same as the pressure of the surrounding water; otherwise they would collapse. A special valve automatically adjusts the pressure of the air breathed from a scuba tank to ensure that the air pressure equals the water pressure at all times. For example, at a depth where the total pressure is 2.0 atm, the oxygen content in air should be reduced to 10 percent by volume to maintain the same partial pressure of 0.20 atm; that is,

$$P_{O_2} = 0.20 \text{ atm} = \frac{V_{O_2}}{V_{O_2} + V_{N_2}} \times 2.0 \text{ atm}$$

$$\frac{V_{O_2}}{V_{O_2} + V_{N_2}} = \frac{0.20 \text{ atm}}{2.0 \text{ atm}} = 0.10 \text{ or } 10\%$$

Scuba divers.

Purestock/SuperStock

Although nitrogen gas may seem to be the obvious choice to mix with oxygen gas, there is a serious problem with it. When the partial pressure of nitrogen gas exceeds 1 atm, enough of the gas dissolves in the blood to cause a condition known as *nitrogen narcosis*. The effects on the diver resemble those associated with alcohol intoxication. Divers suffering from nitrogen narcosis have been known to do strange things, such as dancing on the seafloor and chasing sharks. For this reason, helium is often used to dilute oxygen gas. An inert gas, helium is much less soluble in blood than nitrogen and produces no narcotic effects.

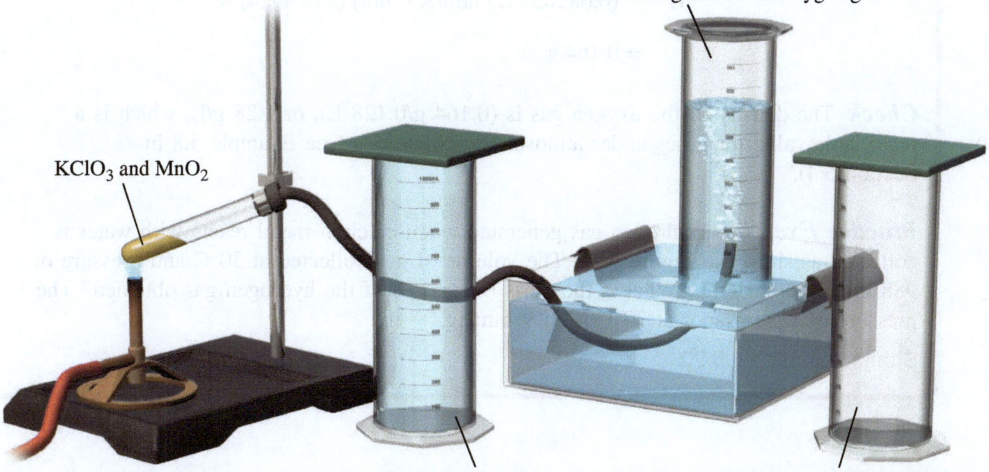

Figure 5.15 *An apparatus for collecting gas over water. The oxygen generated by heating potassium chlorate ($KClO_3$) in the presence of a small amount of manganese dioxide (MnO_2), which speeds up the reaction, is bubbled through water and collected in a bottle as shown. Water originally present in the bottle is pushed into the trough by the oxygen gas.*

Table 5.3

Pressure of Water Vapor at Various Temperatures

Temperature (°C)	Water Vapor Pressure (mmHg)
0	4.58
5	6.54
10	9.21
15	12.79
20	17.55
25	23.78
30	31.86
35	42.22
40	55.39
45	71.97
50	92.65
55	118.22
60	149.61
65	187.83
70	234.03
75	289.49
80	355.63
85	434.04
90	526.41
95	634.61
100	760.00

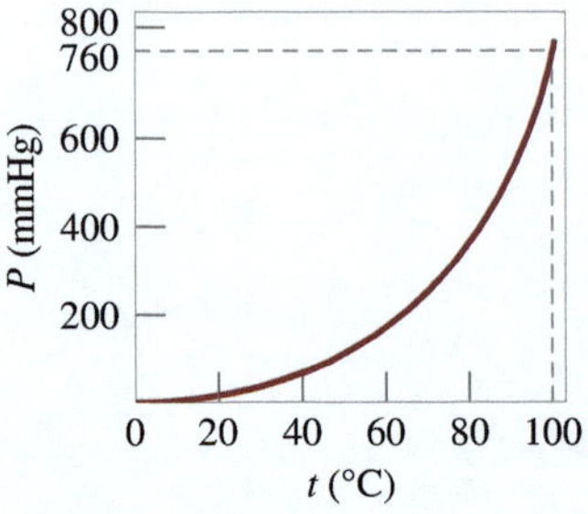

Figure 5.16 *The pressure of water vapor as a function of temperature. Note that at the boiling point of water (100°C) the pressure is 760 mmHg, which is exactly equal to 1 atm.*

Consequently, we must allow for the pressure caused by the presence of water vapor when we calculate the amount of O_2 generated. Table 5.3 shows the pressure of water vapor at various temperatures. These data are plotted in Figure 5.16.

Example 5.15 shows how to use Dalton's law to calculate the amount of a gas collected over water.

Example 5.15

Oxygen gas generated by the decomposition of potassium chlorate is collected as shown in Figure 5.15. The volume of oxygen collected at 24°C and atmospheric pressure of 762.0 mmHg is 128 mL. Calculate the mass (in grams) of oxygen gas obtained. The pressure of the water vapor at 24°C is 22.4 mmHg.

Strategy To solve for the mass of O_2 generated, we must first calculate the partial pressure of O_2 in the mixture. What gas law do we need? How do we convert pressure of O_2 gas to mass of O_2 in grams?

Solution From Dalton's law of partial pressures we know that

$$P_T = P_{O_2} + P_{H_2O}$$

Therefore,

$$\begin{aligned} P_{O_2} &= P_T - P_{H_2O} \\ &= 762.0 \text{ mmHg} - 22.4 \text{ mmHg} \\ &= 739.6 \text{ mmHg} \end{aligned}$$

From the ideal gas equation we write

$$PV = mRT = \frac{n}{\mathcal{M}}RT$$

where m and $\mathcal{M}$ are the mass of O_2 collected and the molar mass of O_2, respectively. Rearranging the equation we obtain

$$\begin{aligned} m = \frac{PV\mathcal{M}}{RT} &= \frac{(739.6/760)\text{atm}(0.128 \text{ L})(32.00 \text{ g/mol})}{(0.082057 \text{ L} \cdot \text{atm/K} \cdot \text{mol})\ (273 + 24)\text{ K}} \\ &= 0.164 \text{ g} \end{aligned}$$

Check The density of the oxygen gas is (0.164 g/0.128 L), or 1.28 g/L, which is a reasonable value for gases under atmospheric conditions (see Example 5.8 in Section 5.4).

Practice Exercise Hydrogen gas generated when calcium metal reacts with water is collected as shown in Figure 5.15. The volume of gas collected at 30°C and pressure of 988 mmHg is 641 mL. What is the mass (in grams) of the hydrogen gas obtained? The pressure of water vapor at 30°C is 31.9 mmHg.

Similar problem: 5.72.

Summary of Concepts & Facts

- Dalton's law of partial pressures states that each gas in a mixture of gases exerts the same pressure as it would if it were alone and occupied the same volume.

Review of Concepts & Facts

5.6.1 Each of the color spheres represents a different gas molecule. Calculate the partial pressures of the gases if the total pressure is 2.6 atm.

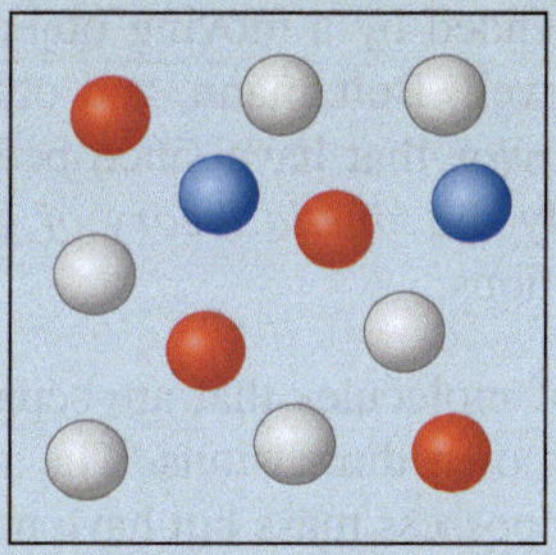

5.6.2 What is the mole fraction of Xe in a mixture of 0.234 moles of Ar, 0.176 moles of Ne, and 0.209 moles of Xe?

5.6.3 Calculate the partial pressure of fluorine in a gas mixture containing 1.45 moles of chlorine, 1.38 moles of fluorine, and 4.23 moles of He. The total pressure of the mixture is 3.59 atm.

5.7 The Kinetic Molecular Theory of Gases

Learning Objectives

- Interpret the basic assumptions of the kinetic molecular theory.
- Compare effusion and diffusion.

The gas laws help us to predict the behavior of gases, but they do not explain what happens at the molecular level to cause the changes we observe in the macroscopic world. For example, why does a gas expand on heating?

In the nineteenth century, a number of physicists, notably Ludwig Boltzmann† and James Clerk Maxwell,‡ found that the physical properties of gases can be explained in terms of the motion of individual molecules. This molecular movement is a form of *energy,* which we define as the capacity to do work or to produce change. In mechanics, *work* is defined as force times distance. Because energy can be measured as work, we can write

$$\begin{aligned} \text{energy} &= \text{work done} \\ &= \text{force} \times \text{distance} \end{aligned}$$

The ***joule (J)***§ is *the SI unit of energy.*

$$\begin{aligned} 1\ \text{J} &= 1\ \text{kg} \cdot \text{m}^2/\text{s}^2 \\ &= 1\ \text{N m} \end{aligned}$$

†Ludwig Eduard Boltzmann (1844–1906). Austrian physicist. Although Boltzmann was one of the greatest theoretical physicists of all time, his work was not recognized by other scientists in his own lifetime. Suffering from poor health and great depression, he committed suicide in 1906.

‡James Clerk Maxwell (1831–1879). Scottish physicist. Maxwell was one of the great theoretical physicists of the nineteenth century; his work covered many areas in physics, including kinetic theory of gases, thermodynamics, and electricity and magnetism.

§James Prescott Joule (1818–1889). English physicist. As a young man, Joule was tutored by John Dalton. He is most famous for determining the mechanical equivalent of heat, the conversion between mechanical energy and thermal energy.

Alternatively, energy can be expressed in kilojoules (kJ).

$$1 \text{ kJ} = 1000 \text{ J}$$

As we will see in Chapter 6, there are many different kinds of energy. ***Kinetic energy (E_k)*** is the type of energy expended by a moving object or *energy of motion*.

The findings of Maxwell, Boltzmann, and others resulted in *a number of generalizations about gas behavior* that have since been known as the ***kinetic molecular theory of gases,*** or simply the *kinetic theory of gases*. Central to the kinetic theory are the following assumptions:

1. A gas is composed of molecules that are separated from each other by distances far greater than their own dimensions. The molecules can be considered to be "points"; that is, they possess mass but have negligible volume. It should be noted that the kinetic theory of gases treats molecules as hard spheres with no internal structure. Because particles in the gas phase are separated by large distances, gases can be compressed easily to occupy less volume.
2. Gas molecules are in constant motion in random directions, and they frequently collide with one another. Collisions among molecules are perfectly elastic. In other words, energy can be transferred from one molecule to another as a result of a collision. Nevertheless, the total energy of all the molecules in a system remains the same.
3. Gas molecules exert neither attractive nor repulsive forces on one another.
4. The average kinetic energy of the molecules is proportional to the temperature of the gas in kelvins. Any two gases at the same temperature will have the same average kinetic energy. The average kinetic energy of a molecule is given by

$$\overline{E_k} = \tfrac{1}{2}m\overline{u^2}$$

where m is the mass of the molecule and u is its speed. The horizontal bar denotes an average value. The quantity $\overline{u^2}$ is called mean square speed; it is the average of the square of the speeds of all the molecules:

$$\overline{u^2} = \frac{u_1^2 + u_2^2 + \cdots + u_N^2}{N}$$

where N is the number of molecules. Assumption 4 enables us to write

$$\overline{E_k} \propto T$$

$$\tfrac{1}{2}m\overline{u^2} \propto T$$

Hence,

$$\overline{E_k} = \tfrac{1}{2}m\overline{u^2} = CT \tag{5.15}$$

where C is the proportionality constant and T is the absolute temperature.

According to the kinetic molecular theory, gas pressure is the result of collisions between molecules and the walls of their container. It depends on the frequency of collision per unit area and on how "hard" the molecules strike the wall. The theory also provides a molecular interpretation of temperature. According to Equation (5.15), the absolute temperature of a gas is a measure of the average kinetic energy of the molecules. In other words, the absolute temperature is an indication of the random motion of the molecules—the higher the temperature, the more energetic the molecules. Because it is related to the temperature of the gas sample, random molecular motion is sometimes referred to as thermal motion.

Application to the Gas Laws

Although the kinetic theory of gases is based on a rather simple model, the mathematical details involved are very complex. However, on a qualitative basis, it is possible to use the theory to account for the general properties of substances in the gaseous state. The following examples illustrate the range of its utility.

- **Compressibility of Gases.** Because molecules in the gas phase are separated by large distances (assumption 1), gases can be compressed easily to occupy less volume.
- **Boyle's Law.** The pressure exerted by a gas results from the impact of its molecules on the walls of the container. The collision rate, or the number of molecular collisions with the walls per second, is proportional to the number density (i.e., number of molecules per unit volume) of the gas. Decreasing the volume of a given amount of gas increases its number density and hence its collision rate. For this reason, the pressure of a gas is inversely proportional to the volume it occupies; as volume decreases, pressure increases and vice versa.
- **Charles's Law.** Because the average kinetic energy of gas molecules is proportional to the sample's absolute temperature (assumption 4), raising the temperature increases the average kinetic energy. Consequently, molecules will collide with the walls of the container more frequently and with greater impact if the gas is heated, and thus the pressure increases. The volume of gas will expand until the gas pressure is balanced by the constant external pressure (see Figure 5.8).
- **Avogadro's Law.** We have shown that the pressure of a gas is directly proportional to both the density and the temperature of the gas. Because the mass of the gas is directly proportional to the number of moles (n) of the gas, we can represent density by n/V. Therefore,

  $$P \propto \frac{n}{V}T$$

  For two gases, 1 and 2, we write

  $$P_1 \propto \frac{n_1T_1}{V_1} = C\frac{n_1T_1}{V_1} \quad P_2 \propto \frac{n_2T_2}{V_2} = C\frac{n_2T_2}{V_2}$$

  where C is the proportionality constant. Thus, for two gases under the same conditions of pressure, volume, and temperature (i.e., when $P_1 = P_2$, $T_1 = T_2$, and $V_1 = V_2$), it follows that $n_1 = n_2$, which is a mathematical expression of Avogadro's law. Another way of stating Avogadro's law is that at the same pressure and temperature, equal volumes of gases, whether they are the same or different gases, contain equal numbers of molecules.
- **Dalton's Law of Partial Pressures.** If molecules do not attract or repel one another (assumption 3), then the pressure exerted by one type of molecule is unaffected by the presence of another gas. Consequently, the total pressure is given by the sum of individual gas pressures.

Distribution of Molecular Speeds

The kinetic theory of gases enables us to investigate molecular motion in more detail. Suppose we have a large number of gas molecules, say, 1 mole, in a container. As long as we hold the temperature constant, the average kinetic energy and the mean-square speed will remain unchanged as time passes. As you might expect, the motion

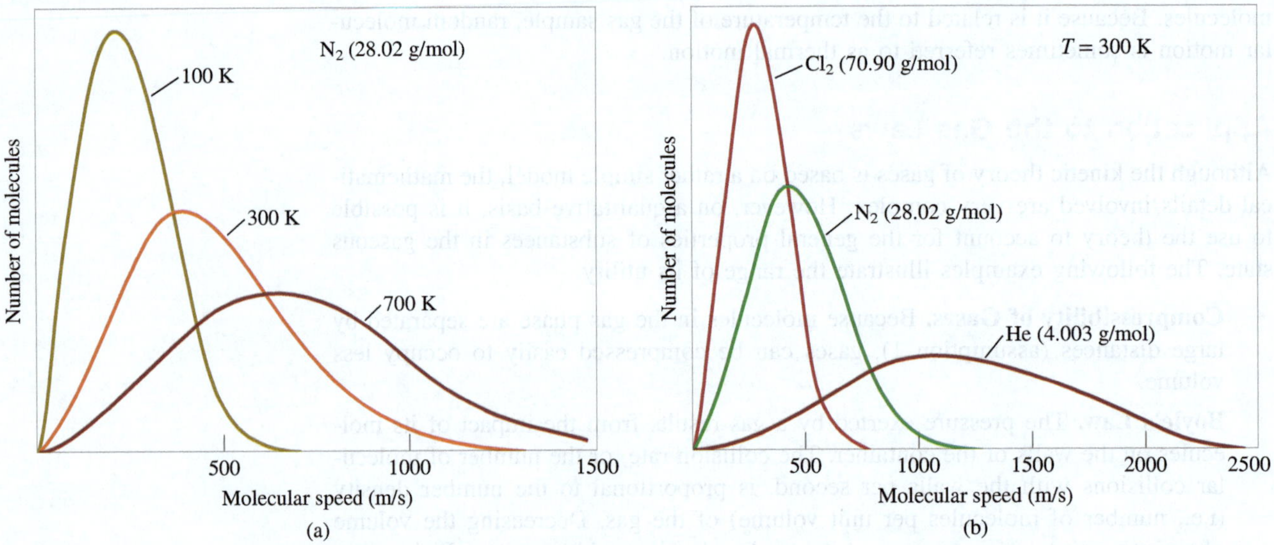

Figure 5.17 *(a) The distribution of speeds for nitrogen gas at three different temperatures. At the higher temperatures, more molecules are moving at faster speeds. (b) The distribution of speeds for three gases at 300 K. At a given temperature, the lighter molecules are moving faster, on the average.*

of the molecules is totally random and unpredictable. At a given instant, how many molecules are moving at a particular speed? To answer this question Maxwell analyzed the behavior of gas molecules at different temperatures.

Figure 5.17(a) shows typical *Maxwell speed distribution curves* for nitrogen gas at three different temperatures. At a given temperature, the distribution curve tells us the number of molecules moving at a certain speed. The peak of each curve represents *the most probable speed,* that is, the speed of the largest number of molecules. Note that the most probable speed increases as temperature increases (the peak shifts toward the right). Furthermore, the curve also begins to flatten out with increasing temperature, indicating that larger numbers of molecules are moving at greater speed. Figure 5.17(b) shows the speed distributions of three gases at the *same* temperature. The difference in the curves can be explained by noting that lighter molecules move faster, on average, than heavier ones.

The distribution of molecular speeds can be demonstrated with the apparatus shown in Figure 5.18. A beam of atoms (or molecules) exits from an oven at a known temperature and passes through a pinhole (to collimate the beam). Two circular plates mounted on the same shaft are rotated by a motor. The first plate is called the

Figure 5.18 *(a) Apparatus for studying molecular speed distribution at a certain temperature. The vacuum pump causes the molecules to travel from left to right as shown. (b) The spread of the deposit on the detector gives the range of molecular speeds, and the density of the deposit is proportional to the fraction of molecules moving at different speeds.*

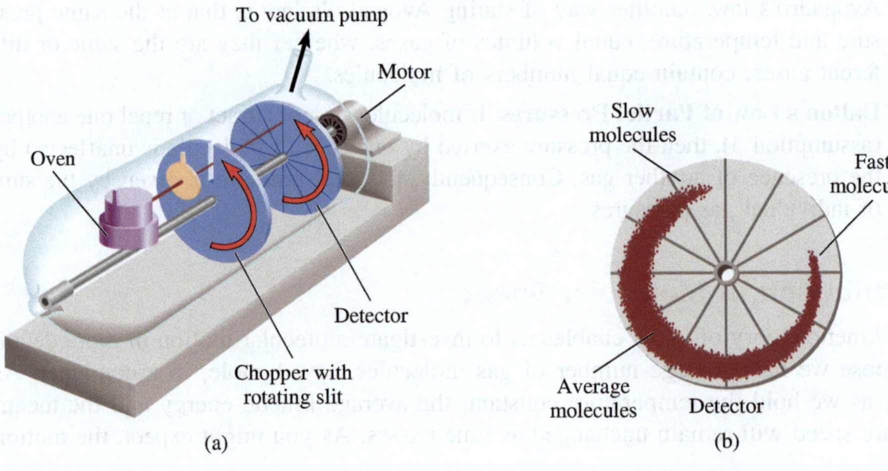

"chopper" and the second is the detector. The purpose of the chopper is to allow small bursts of atoms (or molecules) to pass through it whenever the slit is aligned with the beam. Within each burst, the faster-moving molecules will reach the detector earlier than the slower-moving ones. Eventually, a layer of deposit will accumulate on the detector. Because the two plates are rotating at the same speed, molecules in the next burst will hit the detector plate at approximately the same place as molecules from the previous burst having the same speed. In time, the molecular deposition will become visible. The density of the deposition indicates the distribution of molecular speeds at that particular temperature.

Root-Mean-Square Speed

How fast does a molecule move, on the average, at any temperature T? One way to estimate molecular speed is to calculate the ***root-mean-square (rms) speed*** (u_{rms}), which is *an average molecular speed.* One of the results of the kinetic theory of gases is that the total kinetic energy of a mole of any gas equals $\frac{3}{2}RT$. Earlier we saw that the average kinetic energy of one molecule is $\frac{1}{2}m\overline{u^2}$ and so we can write

$$\overline{E_k} = \tfrac{3}{2}RT$$

$$N_A\left(\tfrac{1}{2}m\overline{u^2}\right) = \tfrac{3}{2}RT$$

where N_A is Avogadro's number and m is the mass of a single molecule. Because $N_Am = \mathcal{M}$, the previous equation can be rearranged to give

$$\overline{u^2} = \frac{3RT}{\mathcal{M}}$$

Taking the square root of both sides gives

$$\sqrt{\overline{u^2}} = u_{rms} = \sqrt{\frac{3RT}{\mathcal{M}}} \quad (5.16)$$

Equation (5.16) shows that the root-mean-square speed of a gas increases with the square root of its temperature (in kelvins). Because $\mathcal{M}$ appears in the denominator, it follows that the heavier the gas, the more slowly its molecules move. If we substitute 8.314 J/K · mol for R (see Appendix 1) and convert the molar mass to kg/mol, then u_{rms} will be calculated in meters per second (m/s). This procedure is illustrated in Example 5.16.

Example 5.16

Calculate the root-mean-square speeds of helium atoms and nitrogen molecules in m/s at 25°C.

Strategy To calculate the root-mean-square speed we need Equation (5.16). What units should we use for R and $\mathcal{M}$ so that u_{rms} will be expressed in m/s?

Solution To calculate u_{rms}, the units of R should be 8.314 J/K · mol and, because 1 J = 1 kg · m^2/s^2, the molar mass must be in kg/mol. The molar mass of He is 4.003 g/mol, or 4.003×10^{-3} kg/mol. From Equation (5.16),

$$u_{rms} = \sqrt{\frac{3RT}{\mathcal{M}}}$$

$$= \sqrt{\frac{3(8.314\ \text{J/K}\cdot\text{mol})(298\ \text{K})}{4.003 \times 10^{-3}\ \text{kg/mol}}}$$

$$= \sqrt{1.86 \times 10^6\ \text{J/kg}}$$

(Continued)

Using the conversion factor 1 J = 1 kg m^2/s^2, we get

$$u_{rms} = \sqrt{1.86 \times 10^6 \text{ kg m}^2/\text{kg} \cdot \text{s}^2}$$
$$= \sqrt{1.86 \times 10^6 \text{ m}^2/\text{s}^2}$$
$$= 1.36 \times 10^3 \text{ m/s}$$

The procedure is the same for N_2, the molar mass of which is 28.02 g/mol, or 2.802 × 10^{-2} kg/mol, so that we write

$$u_{rms} = \sqrt{\frac{3(8.314 \text{ J/K} \cdot \text{mole})(298 \text{ K})}{2.802 \times 10^{-2} \text{ kg/mol}}}$$
$$= \sqrt{2.65 \times 10^5 \text{ m}^2/\text{s}^2}$$
$$= 515 \text{ m/s}$$

Check Because He is a lighter gas, we expect it to move faster, on average, than N_2. A quick way to check the answers is to note that the ratio of the two u_{rms} values ($1.36 \times 10^3/515 \approx 2.6$) should be equal to the square root of the ratios of the molar masses of N_2 to He, that is, $\sqrt{28/4} \approx 2.6$.

Practice Exercise Calculate the root-mean-square speed of molecular chlorine in m/s at 20°C.

Similar problems: 5.81, 5.82.

Jupiter. The interior of this massive planet consists mainly of hydrogen.
NASA-JPL

The calculation in Example 5.16 has an interesting relationship to the composition of Earth's atmosphere. Unlike Jupiter, Earth does not have appreciable amounts of hydrogen or helium in its atmosphere. Why is this the case? A smaller planet than Jupiter, Earth has a weaker gravitational attraction for these lighter molecules. A fairly straightforward calculation shows that to escape Earth's gravitational field, a molecule must possess an escape velocity equal to or greater than 1.1 × 10^4 m/s. Because the average speed of helium is considerably greater than that of molecular nitrogen or molecular oxygen, more helium atoms escape from Earth's atmosphere into outer space. Consequently, only a trace amount of helium is present in our atmosphere. On the other hand, Jupiter, with a mass about 320 times greater than that of Earth, retains both heavy and light gases in its atmosphere.

The Chemistry in Action essay "Super Cold Atoms" describes a fascinating phenomenon involving gases at extremely low temperatures.

Gas Diffusion and Effusion

We will now discuss two phenomena based on gaseous motion.

Gas Diffusion

Video: Diffusion of Gases

Figure 5.19 *The path traveled by a single gas molecule. Each change in direction represents a collision with another molecule.*

A direct demonstration of gaseous random motion is provided by ***diffusion,*** *the gradual mixing of molecules of one gas with molecules of another by virtue of their kinetic properties.* Diffusion always proceeds from a region of high concentration to one where the concentration is lower. Despite the fact that molecular speeds are very great, the diffusion process takes a relatively long time to complete. For example, when a bottle of concentrated ammonia solution is opened at one end of a lab bench, it takes some time before a person at the other end of the bench can smell it. The reason is that a molecule experiences numerous collisions while moving from one end of the bench to the other, as shown in Figure 5.19. Thus, diffusion of gases always happens gradually, and not instantly as molecular speeds seem to suggest. Furthermore, because the root-mean-square speed of a light gas is greater than that of a heavier gas (see Example 5.16 earlier in this section), a lighter gas will diffuse

CHEMISTRY *in Action*

Super Cold Atoms

What happens to a gas when cooled to nearly absolute zero? More than 85 years ago, Albert Einstein, extending work by the Indian physicist Satyendra Nath Bose, predicted that at extremely low temperatures gaseous atoms of certain elements would "merge" or "condense" to form a single quantum entity and a new form of matter. Unlike ordinary gases, liquids, and solids, this supercooled substance, which was named the *Bose-Einstein condensate* (*BEC*), would contain no individual atoms because the original atoms would overlap one another, leaving no space in between.

Einstein's hypothesis inspired an international effort to produce the BEC. But, as sometimes happens in science, the necessary technology was not available until fairly recently, and so early investigations were fruitless. Lasers, which use a process based on another of Einstein's ideas, were not designed specifically for BEC research, but they became a critical tool for this work.

Finally, in 1995, physicists found the evidence they had sought for so long. A team at the University of Colorado was the first to report success. They created a BEC by cooling a sample of gaseous rubidium (Rb) atoms to about 5.0×10^{-8} K using a technique called "laser cooling," a process in which a laser light is directed at a beam of atoms, hitting them head on and dramatically slowing them down. The Rb atoms were further cooled in an "optical molasses" produced by the intersection of six lasers. The slowest, coolest atoms were trapped in a magnetic field while the faster-moving, "hotter" atoms escaped, thereby removing more energy from the gas. Under these conditions, the kinetic energy of the trapped atoms was virtually zero, which accounts for the extremely low temperature of the gas. At this point the Rb atoms formed the condensate, just as Einstein had predicted. Although this BEC was invisible to the naked eye (it measured only 5×10^{-3} cm across), the scientists were able to capture its image on a computer screen by focusing another laser beam on it. The laser caused the BEC to break up after about 15 seconds, but that was long enough to record its existence.

The figure shows the Maxwell velocity distribution† of the Rb atoms at this temperature. The colors indicate the number of atoms having velocity specified by the two horizontal axes. The blue and white portions represent atoms that have merged to form the BEC.

Within weeks of the Colorado team's discovery, a group of scientists at Rice University, using similar techniques, succeeded in producing a BEC with lithium atoms and in 1998 scientists at the Massachusetts Institute of Technology were able to produce a BEC with hydrogen atoms. Since then, many advances have been made in understanding the properties of the BEC in general, and experiments are being extended to molecular systems. It is expected that studies of the BEC will shed light on atomic properties that are still not fully understood (see Chapter 7) and on the mechanism of superconductivity (see the Chemistry in Action essay "High-Temperature Superconductors" in Chapter 11). An additional benefit might be the development of better lasers. Other applications will depend on further study of the BEC itself. Nevertheless, the discovery of a new form of matter has to be one of the foremost scientific achievements of the twentieth century.

†Velocity distribution differs from speed distribution in that velocity has both magnitude and direction. Thus, velocity can have both positive and negative values but speed can have only zero or positive values.

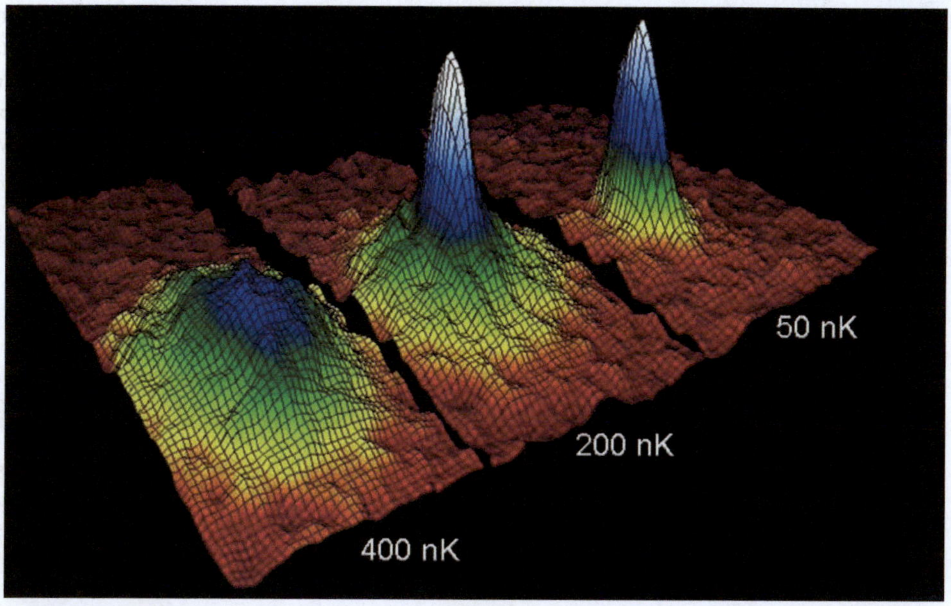

Maxwell velocity distribution of Rb atoms at three different temperatures during the formation of Bose-Einstein condensate. In each case, the velocity increases from the center (zero) outward along the two axes. The red color represents the lower number of Rb atoms and the white color the highest.
Eric Cornell

Figure 5.20 *A demonstration of gas diffusion. NH_3 gas (from a bottle containing aqueous ammonia) combines with HCl gas (from a bottle containing hydrochloric acid) to form solid NH_4Cl. Because NH_3 is lighter and therefore diffuses faster, solid NH_4Cl first appears nearer the HCl bottle (on the right).*
Ken Karp/McGraw Hill

through a certain space more quickly than will a heavier gas. Figure 5.20 illustrates gaseous diffusion.

In 1832 the Scottish chemist Thomas Graham† found that *under the same conditions of temperature and pressure, rates of diffusion for gases are inversely proportional to the square roots of their molar masses.* This statement, now known as ***Graham's law of diffusion,*** is expressed mathematically as

$$\frac{r_1}{r_2} = \sqrt{\frac{\mathscr{M}_2}{\mathscr{M}_1}} \tag{5.17}$$

Student Hot Spot
Student data indicate you may struggle with rates of diffusion and molar masses. Access your eBook for additional Learning Resources on this topic.

where r_1 and r_2 are the diffusion rates of gases 1 and 2, and $\mathscr{M}_1$ and $\mathscr{M}_2$ are their molar masses, respectively.

Gas Effusion

Whereas diffusion is a process by which one gas gradually mixes with another, ***effusion*** *is the process by which a gas under pressure escapes from one compartment of a container to another by passing through a small opening.* Figure 5.21 shows the effusion of a gas into a vacuum. Although effusion differs from diffusion in nature, the rate of effusion of a gas has the same form as Graham's law of diffusion [see Equation (5.17)]. A helium-filled rubber balloon deflates faster than an air-filled one because the rate of effusion through the pores of the rubber is faster for the lighter helium atoms than for the air molecules. Industrially, gas effusion is used to separate uranium isotopes in the forms of gaseous $^{235}UF_6$ and $^{238}UF_6$. By subjecting the gases to many stages of effusion, scientists were able to obtain highly enriched ^{235}U isotope, which was used in the construction of atomic bombs during World War II.

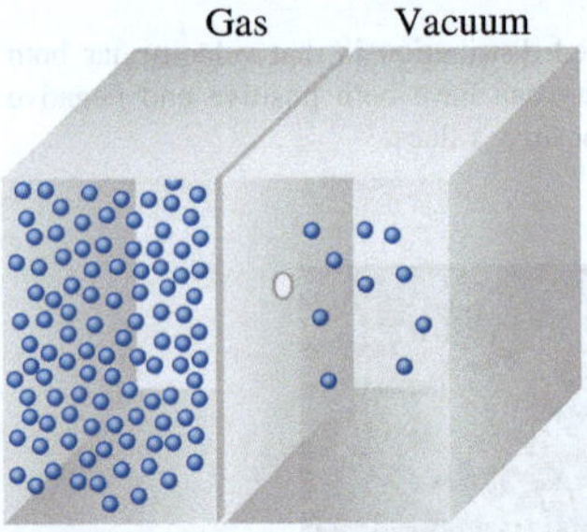

Figure 5.21 *Gas effusion. Gas molecules move from a high-pressure region (left) to a low-pressure one through a pinhole.*

Example 5.17 shows an application of Graham's law.

Example 5.17

A flammable gas made up only of carbon and hydrogen is found to effuse through a porous barrier in 1.50 min. Under the same conditions of temperature and pressure, it takes an equal volume of bromine vapor 4.73 min to effuse through the same barrier. Calculate the molar mass of the unknown gas, and suggest what this gas might be.

(Continued)

†Thomas Graham (1805–1869). Scottish chemist. Graham did important work on osmosis and characterized a number of phosphoric acids.

Strategy The rate of effusion is the number of molecules passing through a porous barrier in a given time. The longer the time it takes, the slower is the rate. Therefore, the rate is *inversely* proportional to the time required for effusion. Equation (5.17) can now be written as $r_1/r_2 = t_2/t_1 = \sqrt{\mathscr{M}_2/\mathscr{M}_1}$, where t_1 and t_2 are the times for effusion for gases 1 and 2, respectively.

Solution From the molar mass of Br_2, we write

$$\frac{1.50 \text{ min}}{4.73 \text{ min}} = \sqrt{\frac{\mathscr{M}}{159.8 \text{ g/mol}}}$$

where $\mathscr{M}$ is the molar mass of the unknown gas. Solving for $\mathscr{M}$, we obtain

$$\mathscr{M} = \left(\frac{1.50 \text{ min}}{4.73 \text{ min}}\right)^2 \times 159.8 \text{ g/mol}$$

$$= 16.1 \text{ g/mol}$$

Because the molar mass of carbon is 12.01 g and that of hydrogen is 1.008 g, the gas is methane (CH_4).

Practice Exercise It takes 192 s for an unknown gas to effuse through a porous wall and 84 s for the same volume of N_2 gas to effuse at the same temperature and pressure. What is the molar mass of the unknown gas?

Similar problems: 5.87, 5.88.

Summary of Concepts & Facts

- The kinetic molecular theory, a mathematical way of describing the behavior of gas molecules, is based on the following assumptions: Gas molecules are separated by distances far greater than their own dimensions, they possess mass but have negligible volume, they are in constant motion, and they frequently collide with one another. The molecules neither attract nor repel one another.
- A Maxwell speed distribution curve shows how many gas molecules are moving at various speeds at a given temperature. As temperature increases, more molecules move at greater speeds.
- In diffusion, two gases gradually mix with each other. In effusion, gas molecules move through a small opening under pressure. Both processes are governed by the same mathematical law—Graham's law of diffusion and effusion.

Review of Concepts & Facts

5.7.1 If one mole each of He and Cl_2 gases are compared at STP, which of the following quantities will be equal to each other?
(a) Root-mean-square speed
(b) effusion rate
(c) average kinetic energy
(d) volume

5.7.2 Which gas effuses faster, Ne or Xe, and how much faster does it effuse?

5.8 Deviation from Ideal Behavior

Learning Objectives

- Explain the factors that cause gases to deviate from ideal gas behavior.
- Demonstrate the use of the van der Waals equation to determine the pressure of a real gas.

The gas laws and the kinetic molecular theory assume that molecules in the gaseous state do not exert any force, either attractive or repulsive, on one another. The other assumption is that the volume of the molecules is negligibly small compared with that of the container. A gas that satisfies these two conditions is said to exhibit *ideal behavior.*

Although we can assume that real gases behave like an ideal gas, we cannot expect them to do so under all conditions. For example, without intermolecular forces, gases could not condense to form liquids. The important question is: Under what conditions will gases most likely exhibit nonideal behavior?

Figure 5.22 shows PV/RT plotted against P for three real gases and an ideal gas at a given temperature. This graph provides a test of ideal gas behavior. According to the ideal gas equation (for 1 mole of gas), PV/RT equals 1, regardless of the actual gas pressure. (When $n = 1$, $PV = nRT$ becomes $PV = RT$, or $PV/RT = 1$.) For real gases, this is true only at moderately low pressures (≤ 5 atm); significant deviations occur as pressure increases. Attractive forces operate among molecules at relatively short distances. At atmospheric pressure, the molecules in a gas are far apart and the attractive forces are negligible. At high pressures, the density of the gas increases; the molecules are much closer to one another. Intermolecular forces can then be significant enough to affect the motion of the molecules, and the gas will not behave ideally.

Another way to observe the nonideal behavior of gases is to lower the temperature. Cooling a gas decreases the molecules' average kinetic energy, which in a sense deprives molecules of the drive they need to break from their mutual attraction.

To study real gases accurately, then, we need to modify the ideal gas equation, taking into account intermolecular forces and finite molecular volumes. Such an analysis was first made by the Dutch physicist J. D. van der Waals† in 1873. Besides being mathematically simple, his treatment provides us with an interpretation of real gas behavior at the molecular level.

Consider the approach of a particular molecule toward the wall of a container (Figure 5.23). The intermolecular attractions exerted by its neighbors tend to soften the impact made by this molecule against the wall. The overall effect is a lower gas

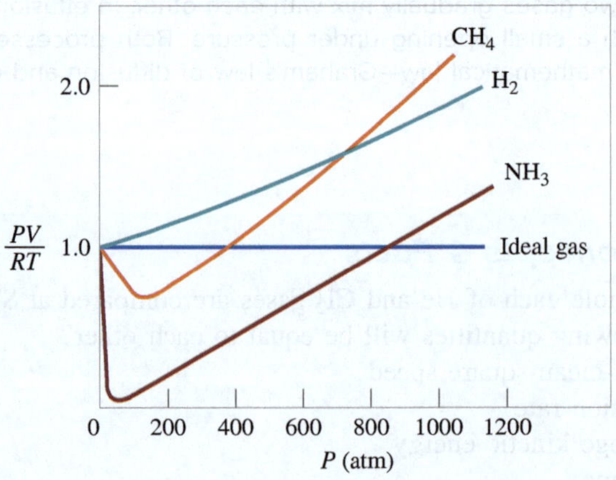

Figure 5.22 *Plot of PV/RT versus P of 1 mole of a gas at 0°C. For 1 mole of an ideal gas, PV/RT is equal to 1, no matter what the pressure of the gas is. For real gases, we observe various deviations from ideality at high pressures. At very low pressures, all gases exhibit ideal behavior; that is, their PV/RT values all converge to 1 as P approaches zero.*

†Johannes Diderck van der Waals (1837–1923). Dutch physicist. Van der Waals received the Nobel Prize in Physics in 1910 for his work on the properties of gases and liquids.

pressure than we would expect for an ideal gas. Van der Waals suggested that the pressure exerted by an ideal gas, P_{ideal}, is related to the experimentally measured pressure, P_{real}, by the equation

$$P_{ideal} = \underset{\substack{\uparrow \\ \text{observed} \\ \text{pressure}}}{P_{real}} + \underset{\substack{\uparrow \\ \text{correction} \\ \text{term}}}{\frac{an^2}{V^2}}$$

where a is a constant and n and V are the number of moles and volume of the container, respectively. The correction term for pressure (an^2/V^2) can be understood as follows: The intermolecular interaction that gives rise to nonideal behavior depends on how frequently any two molecules approach each other closely. The frequency of such "encounters" increases with the square of the number of molecules per unit volume (n^2/V^2), because the probability of finding each of the two molecules in a particular region is proportional to n/V. Thus, a is just a proportionality constant.

Another correction concerns the volume occupied by the gas molecules. In the ideal gas equation, V represents the volume of the container. However, each molecule does occupy a finite, although small, intrinsic volume, so the effective volume of the gas becomes $(V - nb)$, where n is the number of moles of the gas and b is a constant. The term nb represents the volume occupied by n moles of the gas.

Having taken into account the corrections for pressure and volume, we can rewrite the ideal gas equation as follows:

$$\underbrace{\left(P + \frac{an^2}{V^2}\right)}_{\text{corrected pressure}}\underbrace{(V - nb)}_{\text{corrected volume}} = nRT \quad (5.18)$$

Equation (5.18), *relating P, V, T, and n for a nonideal gas,* is known as the ***van der Waals equation.*** The van der Waals constants a and b are selected to give the best possible agreement between Equation (5.18) and observed behavior of a particular gas.

Table 5.4 lists the values of a and b for a number of gases. The value of a indicates how strongly molecules of a given type of gas attract one another. We see that helium atoms have the weakest attraction for one another, because helium has the smallest a value. There is also a rough correlation between molecular size and b. Generally, the larger the molecule (or atom), the greater b is, but the relationship between b and molecular (or atomic) size is not a simple one.

Example 5.18 compares the pressure of a gas calculated using the ideal gas equation and the van der Waals equation.

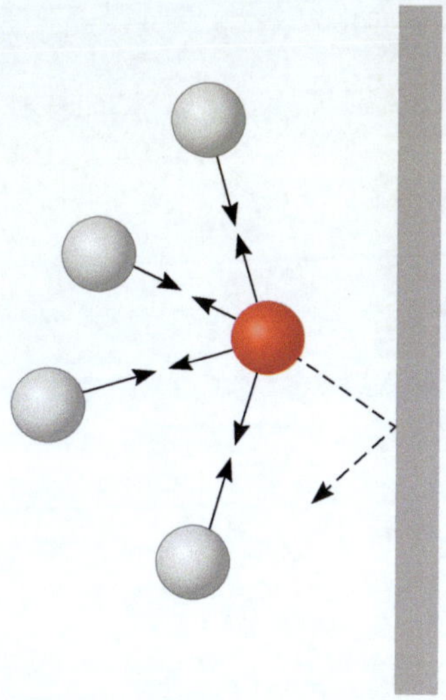

Figure 5.23 *Effect of intermolecular forces on the pressure exerted by a gas. The speed of a molecule that is moving toward the container wall (red sphere) is reduced by the attractive forces exerted by its neighbors (gray spheres). Consequently, the impact this molecule makes with the wall is not as great as it would be if no intermolecular forces were present. In general, the measured gas pressure is lower than the pressure the gas would exert if it behaved ideally.*

Student Hot Spot

Student data indicate you may struggle with the van der Waals equation. Access your eBook for additional Learning Resources on this topic.

Table 5.4

van der Waals Constants of Some Common Gases

Gas	$a\left(\frac{atm \cdot L^2}{mol^2}\right)$	$b\left(\frac{L}{mol}\right)$
He	0.034	0.0237
Ne	0.211	0.0171
Ar	1.34	0.0322
Kr	2.32	0.0398
Xe	4.19	0.0266
H_2	0.244	0.0266
N_2	1.39	0.0391
O_2	1.36	0.0318
Cl_2	6.49	0.0562
CO_2	3.59	0.0427
CH_4	2.25	0.0428
CCl_4	20.4	0.138
NH_3	4.17	0.0371
H_2O	5.46	0.0305

Example 5.18

Given that 3.50 moles of NH_3 occupy 5.20 L at 47°C, calculate the pressure of the gas (in atm) using (a) the ideal gas equation and (b) the van der Waals equation.

Strategy To calculate the pressure of NH_3 using the ideal gas equation, we proceed as in Example 5.3 (Section 5.4). What corrections are made to the pressure and volume terms in the van der Waals equation?

(Continued)

Solution

(a) We have the following data:

$$V = 5.20\ \text{L}$$
$$T = (47 + 273)\ \text{K} = 320.\ \text{K}$$
$$n = 3.50\ \text{mol}$$
$$R = 0.082057\ \text{L} \cdot \text{atm/K} \cdot \text{mol}$$

Substituting these values in the ideal gas equation, we write

$$P = \frac{nRT}{V}$$
$$= \frac{(3.50\ \text{mol})(0.082057\ \text{L} \cdot \text{atm/K} \cdot \text{mol})(320.\ \text{K})}{5.20\ \text{L}}$$
$$= 17.7\ \text{atm}$$

(b) We need Equation (5.18). It is convenient to first calculate the correction terms in Equation (5.18) separately. From Table 5.4, we have

$$a = 4.17\ \text{atm} \cdot \text{L}^2/\text{mol}^2$$
$$b = 0.0371\ \text{L/mol}$$

so that the correction terms for pressure and volume are

$$\frac{an^2}{V^2} = \frac{(4.17\ \text{atm} \cdot \text{L}^2/\text{mol}^2)(3.50\ \text{mol})^2}{(5.20\ \text{L})^2} = 1.89\ \text{atm}$$

$$nb = (3.50\ \text{mol})(0.0371\ \text{L/mol}) = 0.130\ \text{L}$$

Finally, substituting these values in the van der Waals equation, we have

$$(P + 1.89\ \text{atm})(5.20\ \text{L} - 0.130\ \text{L}) = (3.50\ \text{mol})(0.082057\ \text{L} \cdot \text{atm/K} \cdot \text{mol})(320.\ \text{K})$$
$$P = 16.2\ \text{atm}$$

Check Based on your understanding of nonideal gas behavior, is it reasonable that the pressure calculated using the van der Waals equation should be smaller than that using the ideal gas equation? Why?

Practice Exercise Using the data shown in Table 5.4, calculate the pressure exerted by 4.37 moles of molecular chlorine confined in a volume of 2.45 L at 38°C. Compare the pressure with that calculated using the ideal gas equation.

Similar problem: 5.93.

Summary of Concepts & Facts

- The van der Waals equation is a modification of the ideal gas equation that takes into account the nonideal behavior of real gases. It corrects for the fact that real gas molecules do exert forces on each other and that they do have volume. The van der Waals constants are determined experimentally for each gas.

Review of Concepts & Facts

5.8.1 What pressure and temperature conditions cause the most deviation from ideal gas behavior?

5.8.2 Using the van der Waals equation, determine the pressure exerted by 2.25 moles of CH_4 in a 5.25-L vessel at 25°C.

Chapter Summary

Properties of Gases Gases assume the volume and shape of their containers, they are easily compressible, they mix evenly and completely, and they have much lower densities than liquids and solids. (Section 5.1)

Gas Pressures Pressure is one of the most readily measurable properties of a gas. A barometer measures atmospheric pressure, and a manometer measures the pressure of a gas in the laboratory. (Section 5.2)

The Gas Laws Over the years, a number of laws have been developed to explain the physical behavior of gases. These laws show the relationships among the pressure, temperature, volume, and amount of a gas. (Section 5.3)

The Ideal Gas Equation The molecules of an ideal gas possess no volume and exert no forces on one another. At low pressures and high temperatures, most gases can be assumed to behave ideally; their physical behavior is described by the ideal gas equation. (Sections 5.4, 5.5, and 5.6)

Kinetic Molecular Theory of Gases Macroscopic properties like pressure and temperature of a gas can be related to the kinetic motion of molecules. The kinetic molecular theory of gases assumes that the molecules are ideal, the number of molecules is very large, and their motions are totally random. Both gas diffusion and gas effusion demonstrate random molecular motion and are governed by the same mathematical laws. (Section 5.7)

Nonideal Behavior of Gases To account for the behavior of real gases, the ideal gas equation is modified to include the finite volume of molecules and the attractive forces among them. (Section 5.8)

Key Equations

$P_1V_1 = P_2V_2$ (5.2)	Boyle's law. For calculating pressure or volume changes.
$\frac{V_1}{T_1} = \frac{V_2}{T_2}$ (5.4)	Charles's law. For calculating temperature or volume changes.
$\frac{P_1}{T_1} = \frac{P_2}{T_2}$ (5.6)	Charles's law. For calculating temperature or pressure changes.
$V = k_4 n$ (5.7)	Avogadro's law. Constant P and T.
$PV = nRT$ (5.8)	Ideal gas equation.
$\frac{P_1V_1}{n_1T_1} = \frac{P_2V_2}{n_2T_2}$ (5.9)	For calculating changes in pressure, temperature, volume, or amount of gas.
$\frac{P_1V_1}{T_1} = \frac{P_2V_2}{T_2}$ (5.10)	For calculating changes in pressure, temperature, or volume when n is constant.
$d = \frac{m}{V} = \frac{P\mathcal{M}}{RT}$ (5.11)	For calculating density or molar mass.
$X_i = \frac{n_i}{n_T}$ (5.13)	Definition of mole fraction.
$P_i = X_i P_T$ (5.14)	Dalton's law of partial pressures. For calculating partial pressures.
$E_k = \frac{1}{2}m\overline{u^2} = CT$ (5.15)	Relating the average kinetic energy of a gas to its absolute temperature.
$u_{rms} = \sqrt{\frac{3RT}{\mathcal{M}}}$ (5.16)	For calculating the root-mean-square speed of gas molecules.
$\frac{r_1}{r_2} = \sqrt{\frac{\mathcal{M}_2}{\mathcal{M}_1}}$ (5.17)	Graham's law of diffusion and effusion.
$\left(P + \frac{an^2}{V^2}\right)(V - nb) = nRT$ (5.18)	Van der Waals equation. For calculating the pressure of a nonideal gas.

Key Words

Absolute temperature scale, p. 185
Absolute zero, p. 185
Atmospheric pressure, p. 179
Avogadro's law, p. 186
Barometer, p. 179
Boyle's law, p. 182
Charles's and Gay-Lussac's law, p. 185
Charles's law, p. 185
Dalton's law of partial pressures, p. 199
Diffusion, p. 210
Effusion, p. 212
Gas constant (R), p. 188
Graham's law of diffusion, p. 212
Ideal gas, p. 188
Ideal gas equation, p. 188
Joule (J), p. 205
Kelvin temperature scale, p. 185
Kinetic energy (E_K), p. 206
Kinetic molecular theory of gases, p. 206
Manometer, p. 180
Mole fraction, p. 200
Newton (N), p. 178
Partial pressure, p. 199
Pascal (Pa), p. 178
Pressure, p. 178
Root-mean-square (rms) speed (u_{rms}), p. 209
Standard atmospheric pressure (1 atm), p. 179
Standard temperature and pressure (STP), p. 188
van der Waals equation, p. 215

Questions & Problems

Red numbered problems solved in Student Solutions Manual

5.1 Substances That Exist as Gases

Review Questions

5.1 Name five elements and five compounds that exist as gases at room temperature.

5.2 List the physical characteristics of gases.

5.2 Pressure of a Gas

Review Questions

5.3 Define pressure and give the common units for pressure.

5.4 When you are in a plane flying at high altitudes, your ears often experience pain. This discomfort can be temporarily relieved by yawning or swallowing some water. Explain.

5.5 Why is mercury a more suitable substance to use in a barometer than water?

5.6 Explain why the height of mercury in a barometer is independent of the cross-sectional area of the tube. Would the barometer still work if the tubing were tilted at an angle, say 15 degrees (see Figure 5.3)?

5.7 Explain how a unit of length (mmHg) can be used as a unit for pressure.

5.8 Describe what would happen to the column of mercury in these manometers when the stopcock is opened.

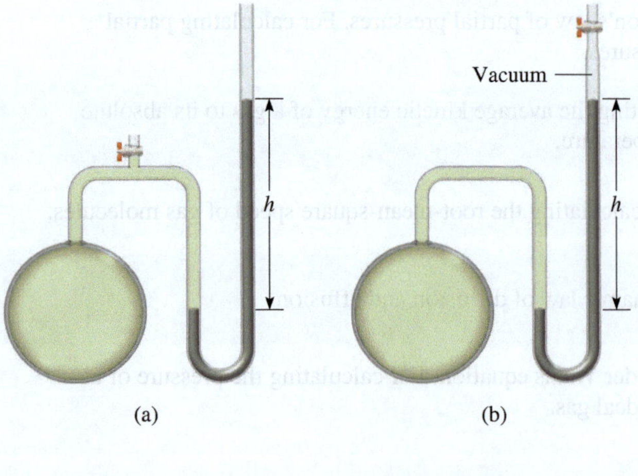

(a) (b)

5.9 What is the difference between a gas and a vapor? At 25°C, which of the following substances in the gas phase should be properly called a gas and which should be called a vapor: molecular nitrogen (N_2), mercury?

5.10 If the maximum distance that water may be brought up a well by a suction pump is 34 ft (10.3 m), how is it possible to obtain water and oil from hundreds of feet below the surface of Earth?

5.11 Why is it that if the barometer reading falls in one part of the world, it must rise somewhere else?

5.12 Why do astronauts have to wear protective suits when they are on the surface of the moon?

Problems

5.13 Convert 562 mmHg to atm.

5.14 The atmospheric pressure at the summit of Denali (formerly known as Mt. McKinley) is 606 mmHg on a certain day. What is the pressure in atm and in kPa?

5.3 The Gas Laws

Review Questions

5.15 State the following gas laws in words and also in the form of an equation: Boyle's law, Charles's law, Avogadro's law. In each case, indicate the conditions under which the law is applicable, and give the units for each quantity in the equation.

5.16 A certain amount of gas is contained in a closed mercury manometer as shown here. Assuming no other parameters change, would h increase, decrease, or remain the same if (a) the amount of the gas was increased; (b) the molar mass of the gas was doubled; (c) the temperature of the gas was increased; (d) the atmospheric pressure in the room was increased; (e) the mercury in the tube was replaced with a less dense fluid; (f) some gas was added to the vacuum at the top of the right-side tube; (g) a hole was drilled in the top of the right-side tube?

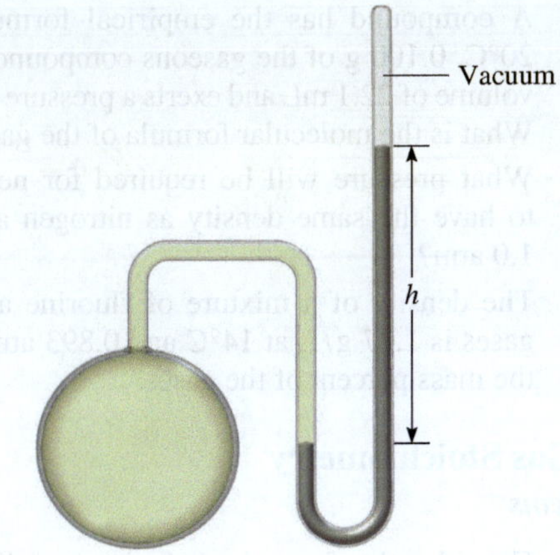

Problems

5.17 A gaseous sample of a substance is cooled at constant pressure. Which of the diagrams (a)–(d) best represents the situation if the final temperature is (i) above the boiling point of the substance, and (ii) below the boiling point but above the freezing point of the substance?

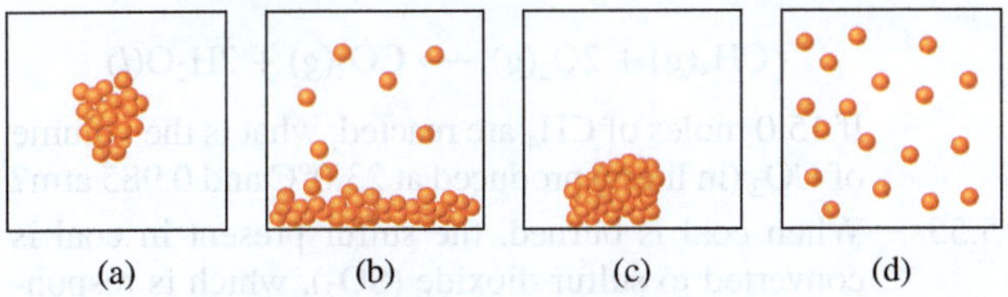

5.18 Consider the following gaseous sample in a cylinder fitted with a movable piston. Initially there are n moles of the gas at temperature T, pressure P, and volume V.

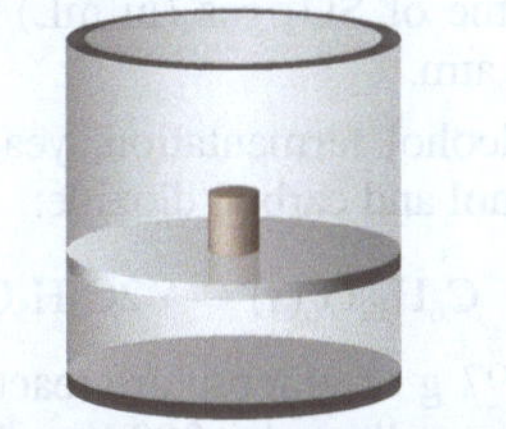

Choose the cylinder (a)–(c) that correctly represents the gas after each of the following changes: (1) The pressure on the piston is tripled at constant n and T. (2) The temperature is doubled at constant n and P. (3) n moles of another gas are added at constant T and P. (4) T is halved and pressure on the piston is reduced to a quarter of its original value.

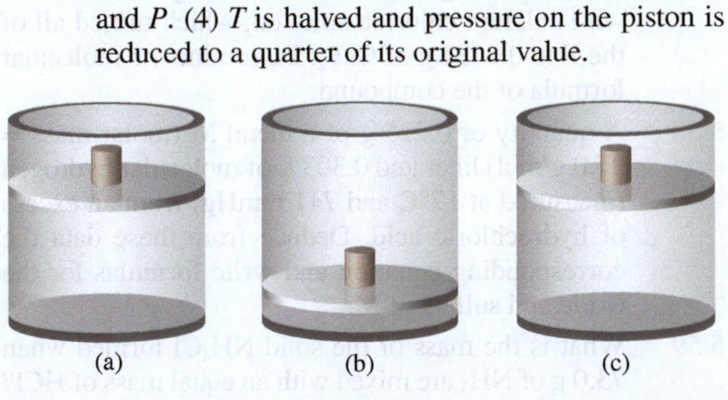

5.19 A gas occupying a volume of 725 mL at a pressure of 0.970 atm is allowed to expand at constant temperature until its pressure reaches 0.541 atm. What is its final volume?

5.20 At 46°C a sample of ammonia gas exerts a pressure of 5.3 atm. What is the pressure when the volume of the gas is reduced to one-tenth (0.10) of the original value at the same temperature?

5.21 The volume of a gas is 5.80 L, measured at 1.00 atm. What is the pressure of the gas in mmHg if the volume is changed to 9.65 L? (The temperature remains constant.)

5.22 A sample of air occupies 3.8 L when the pressure is 1.2 atm. (a) What volume does it occupy at 6.6 atm? (b) What pressure is required to compress it to 0.075 L? (The temperature is kept constant.)

5.23 A 36.4-L volume of methane gas is heated from 25°C to 88°C at constant pressure. What is the final volume of the gas?

5.24 Under constant-pressure conditions a sample of hydrogen gas initially at 88°C and 9.6 L is cooled until its final volume is 3.4 L. What is its final temperature?

5.25 Ammonia burns in oxygen gas to form nitric oxide (NO) and water vapor. How many volumes of NO are obtained from one volume of ammonia at the same temperature and pressure?

5.26 Molecular chlorine and molecular fluorine combine to form a gaseous product. Under the same conditions of temperature and pressure it is found that one volume of Cl_2 reacts with three volumes of F_2 to yield two volumes of the product. What is the formula of the product?

5.4 The Ideal Gas Equation

Review Questions

5.27 List the characteristics of an ideal gas. Write the ideal gas equation and also state it in words. Give the units for each term in the equation.

5.28 Use Equation (5.9) to derive all the gas laws.

5.29 What are standard temperature and pressure (STP)? What is the significance of STP in relation to the volume of 1 mole of an ideal gas?

5.30 Why is the density of a gas much lower than that of a liquid or solid under atmospheric conditions? What units are normally used to express the density of gases?

Problems

5.31 A sample of nitrogen gas kept in a container of volume 2.3 L and at a temperature of 32°C exerts a pressure of 4.7 atm. Calculate the number of moles of gas present.

5.32 Given that 6.9 moles of carbon monoxide gas are present in a container of volume 30.4 L, what is the pressure of the gas (in atm) if the temperature is 62°C?

5.33 What volume will 5.6 moles of sulfur hexafluoride (SF_6) gas occupy if the temperature and pressure of the gas are 128°C and 9.4 atm?

5.34 A certain amount of gas at 25°C and at a pressure of 0.800 atm is contained in a glass vessel. Suppose that the vessel can withstand a pressure of 2.00 atm. How high can you raise the temperature of the gas without bursting the vessel?

5.35 A gas-filled balloon having a volume of 2.50 L at 1.2 atm and 25°C is allowed to rise to the stratosphere (about 30 km above the surface of Earth), where the temperature and pressure are −23°C and 3.00×10^{-3} atm, respectively. Calculate the final volume of the balloon.

5.36 The temperature of 2.5 L of a gas initially at STP is raised to 250°C at constant volume. Calculate the final pressure of the gas in atm.

5.37 The pressure of 6.0 L of an ideal gas in a flexible container is decreased to one-third of its original pressure, and its absolute temperature is decreased by one-half. What is the final volume of the gas?

5.38 A gas evolved during the fermentation of glucose (wine making) has a volume of 0.78 L at 20.1°C and 1.00 atm. What was the volume of this gas at the fermentation temperature of 36.5°C and 1.00 atm pressure?

5.39 An ideal gas originally at 0.85 atm and 66°C was allowed to expand until its final volume, pressure, and temperature were 94 mL, 0.60 atm, and 45°C, respectively. What was its initial volume?

5.40 Calculate its volume (in liters) of 88.4 g of CO_2 at STP.

5.41 A gas at 772 mmHg and 35.0°C occupies a volume of 6.85 L. Calculate its volume at STP.

5.42 Dry ice is solid carbon dioxide. A 0.050-g sample of dry ice is placed in an evacuated 4.6-L vessel at 30°C. Calculate the pressure inside the vessel after all the dry ice has been converted to CO_2 gas.

5.43 At STP, 0.280 L of a gas weighs 0.400 g. Calculate the molar mass of the gas.

5.44 At 741 torr and 44°C, 7.10 g of a gas occupy a volume of 5.40 L. What is the molar mass of the gas?

5.45 Ozone molecules in the stratosphere absorb much of the harmful radiation from the sun. Typically, the temperature and pressure of ozone in the stratosphere are 250 K and 1.0×10^{-3} atm, respectively. How many ozone molecules are present in 1.0 L of air under these conditions?

5.46 Assuming that air contains 78% N_2, 21% O_2, and 1% Ar, all by volume, how many molecules of each type of gas are present in 1.0 L of air at STP?

5.47 A 2.10-L vessel contains 4.65 g of a gas at 1.00 atm and 27.0°C. (a) Calculate the density of the gas in grams per liter. (b) What is the molar mass of the gas?

5.48 Calculate the density of hydrogen bromide (HBr) gas in grams per liter at 733 mmHg and 46°C.

5.49 A certain anesthetic contains 64.9% C, 13.5% H, and 21.6% O by mass. At 120°C and 750 mmHg, 1.00 L of the gaseous compound weighs 2.30 g. What is the molecular formula of the compound?

5.50 A compound has the empirical formula SF_4. At 20°C, 0.100 g of the gaseous compound occupies a volume of 22.1 mL and exerts a pressure of 1.02 atm. What is the molecular formula of the gas?

5.51 What pressure will be required for neon at 30°C to have the same density as nitrogen at 20°C and 1.0 atm?

5.52 The density of a mixture of fluorine and chlorine gases is 1.77 g/L at 14°C and 0.893 atm. Calculate the mass percent of the gases.

5.5 Gas Stoichiometry

Problems

5.53 Consider the formation of nitrogen dioxide from nitric oxide and oxygen:

$$2NO(g) + O_2(g) \longrightarrow 2NO_2(g)$$

If 9.0 L of NO are reacted with excess O_2 at STP, what is the volume in liters of the NO_2 produced?

5.54 Methane, the principal component of natural gas, is used for heating and cooking. The combustion process is

$$CH_4(g) + 2O_2(g) \longrightarrow CO_2(g) + 2H_2O(l)$$

If 15.0 moles of CH_4 are reacted, what is the volume of CO_2 (in liters) produced at 23.0°C and 0.985 atm?

5.55 When coal is burned, the sulfur present in coal is converted to sulfur dioxide (SO_2), which is responsible for the acid rain phenomenon.

$$S(s) + O_2(g) \longrightarrow SO_2(g)$$

If 2.54 kg of S are reacted with oxygen, calculate the volume of SO_2 gas (in mL) formed at 30.5°C and 1.12 atm.

5.56 In alcohol fermentation, yeast converts glucose to ethanol and carbon dioxide:

$$C_6H_{12}O_6(s) \longrightarrow 2C_2H_5OH(l) + 2CO_2(g)$$

If 5.97 g of glucose are reacted and 1.44 L of CO_2 gas are collected at 293 K and 0.984 atm, what is the percent yield of the reaction?

5.57 A compound of P and F was analyzed as follows: Heating 0.2324 g of the compound in a 378-cm^3 container turned all of it to gas, which had a pressure of 97.3 mmHg at 77°C. Then the gas was mixed with calcium chloride solution, which turned all of the F to 0.2631 g of CaF_2. Determine the molecular formula of the compound.

5.58 A quantity of 0.225 g of a metal M (molar mass = 27.0 g/mol) liberated 0.303 L of molecular hydrogen (measured at 17°C and 741 mmHg) from an excess of hydrochloric acid. Deduce from these data the corresponding equation and write formulas for the oxide and sulfate of M.

5.59 What is the mass of the solid NH_4Cl formed when 73.0 g of NH_3 are mixed with an equal mass of HCl? What is the volume of the gas remaining, measured at 14.0°C and 752 mmHg? What gas is it?

5.60 Dissolving 3.00 g of an impure sample of calcium carbonate in hydrochloric acid produced 0.656 L of carbon dioxide (measured at 20.0°C and 792 mmHg). Calculate the percent by mass of calcium carbonate in the sample. State any assumptions.

5.61 Calculate the mass in grams of hydrogen chloride produced when 5.6 L of molecular hydrogen measured at STP react with an excess of molecular chlorine gas.

5.62 Ethanol (C_2H_5OH) burns in air:

$$C_2H_5OH(l) + O_2(g) \longrightarrow CO_2(g) + H_2O(l)$$

Balance the equation and determine the volume of air in liters at 35.0°C and 790 mmHg required to burn 227 g of ethanol. Assume that air is 21.0% O_2 by volume.

5.63 (a) What volumes (in liters) of ammonia and oxygen must react to form 12.8 L of nitric oxide according to the equation at the same temperature and pressure?

$$4NH_3(g) + 5O_2(g) \longrightarrow 4NO(g) + 6H_2O(g)$$

(b) What volumes (in liters) of propane and water vapor must react to form 8.96 L of hydrogen according to the equation at the same temperature and pressure?

$$C_3H_8(g) + 3H_2O(g) \longrightarrow 3CO(g) + 7H_2(g)$$

5.64 A 4.00-g sample of FeS containing nonsulfide impurities reacted with HCl to give 896 mL of H_2S at 14°C and 782 mmHg. Calculate mass percent purity of the sample.

5.6 Dalton's Law of Partial Pressures

Review Questions

5.65 State Dalton's law of partial pressures and explain what mole fraction is. Does mole fraction have units?

5.66 A sample of air contains only nitrogen and oxygen gases whose partial pressures are 0.80 atm and 0.20 atm, respectively. Calculate the total pressure and the mole fractions of the gases.

Problems

5.67 A mixture of gases contains 0.31 mol CH_4, 0.25 mol C_2H_6, and 0.29 mol C_3H_8. The total pressure is 1.50 atm. Calculate the partial pressures of the gases.

5.68 A 2.5-L flask at 15°C contains a mixture of N_2, He, and Ne at partial pressures of 0.32 atm for N_2, 0.15 atm for He, and 0.42 atm for Ne. (a) Calculate the total pressure of the mixture. (b) Calculate the volume in liters at STP occupied by He and Ne if the N_2 is removed selectively.

5.69 Dry air near sea level has the following composition by volume: N_2, 78.08%; O_2, 20.94%; Ar, 0.93%; CO_2, 0.05%. The atmospheric pressure is 1.00 atm. Calculate (a) the partial pressure of each gas in atm and (b) the concentration of each gas in moles per liter at 0°C. (*Hint:* Because volume is proportional to the number of moles present, mole fractions of gases can be expressed as ratios of volumes at the same temperature and pressure.)

5.70 A mixture of helium and neon gases is collected over water at 28.0°C and 745 mmHg. If the partial pressure of helium is 368 mmHg, what is the partial pressure of neon? (Vapor pressure of water at 28°C = 28.3 mmHg.)

5.71 A piece of sodium metal reacts completely with water as follows:

$$2Na(s) + 2H_2O(l) \longrightarrow 2NaOH(aq) + H_2(g)$$

The hydrogen gas generated is collected over water at 25.0°C. The volume of the gas is 246 mL measured at 1.00 atm. Calculate the number of grams of sodium used in the reaction. (Vapor pressure of water at 25°C = 0.0313 atm.)

5.72 A sample of zinc metal reacts completely with an excess of hydrochloric acid:

$$Zn(s) + 2HCl(aq) \longrightarrow ZnCl_2(aq) + H_2(g)$$

The hydrogen gas produced is collected over water at 25.0°C using an arrangement similar to that shown in Figure 5.15. The volume of the gas is 7.80 L, and the pressure is 0.980 atm. Calculate the amount of zinc metal in grams consumed in the reaction. (Vapor pressure of water at 25°C = 23.8 mmHg.)

5.73 Helium is mixed with oxygen gas for deep-sea divers. Calculate the percent by volume of oxygen gas in the mixture if the diver has to submerge to a depth where the total pressure is 4.2 atm. The partial pressure of oxygen is maintained at 0.20 atm at this depth.

5.74 A sample of ammonia (NH_3) gas is completely decomposed to nitrogen and hydrogen gases over heated iron wool. If the total pressure is 866 mmHg, calculate the partial pressures of N_2 and H_2.

5.75 Consider the three gas containers (i)–(iii). All of them have the same volume and are at the same temperature. (a) Which container has the smallest mole fraction of gas A (blue sphere)? (b) Which container has the highest partial pressure of gas B (green sphere)?

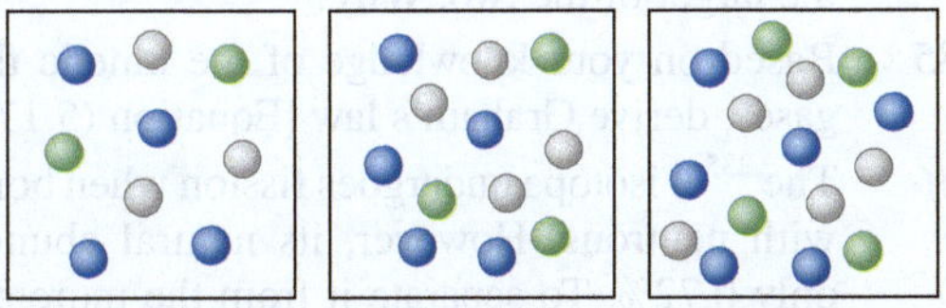

(i) (ii) (iii)

5.76 The volume of the box on the right is twice that of the box on the left. The boxes contain helium atoms (red) and hydrogen molecules (gray) at the same temperature. (a) Which box has a higher total pressure? (b) Which box has a lower partial pressure of helium?

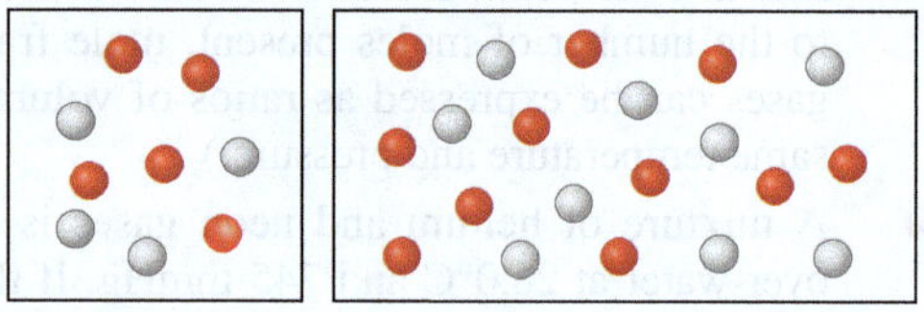

5.7 The Kinetic Molecular Theory of Gases

Review Questions

5.77 What are the basic assumptions of the kinetic molecular theory of gases? How does the kinetic molecular theory explain Boyle's law, Charles's law, Avogadro's law, and Dalton's law of partial pressures?

5.78 What does the Maxwell speed distribution curve tell us? Does Maxwell's theory work for a sample of 200 molecules? Explain.

5.79 Which of the following statements is correct? (a) Heat is produced by the collision of gas molecules against one another. (b) When a gas is heated, the molecules collide with one another more often.

5.80 What is the difference between gas diffusion and effusion? State Graham's law and define the terms in Equation (5.17).

Problems

5.81 Compare the root-mean-square speeds of O_2 and UF_6 at 65°C.

5.82 The temperature in the stratosphere is −23°C. Calculate the root-mean-square speeds of N_2, O_2, and O_3 molecules in this region.

5.83 The average distance traveled by a molecule between successive collisions is called *mean free path*. For a given amount of a gas, how does the mean free path of a gas depend on (a) density, (b) temperature at constant volume, (c) pressure at constant temperature, (d) volume at constant temperature, and (e) size of the atoms?

5.84 At a certain temperature the speeds of six gaseous molecules in a container are 2.0 m/s, 2.2 m/s, 2.6 m/s, 2.7 m/s, 3.3 m/s, and 3.5 m/s. Calculate the root-mean-square speed and the average speed of the molecules. These two average values are close to each other, but the root-mean-square value is always the larger of the two. Why?

5.85 Based on your knowledge of the kinetic theory of gases, derive Graham's law [Equation (5.17)].

5.86 The ^{235}U isotope undergoes fission when bombarded with neutrons. However, its natural abundance is only 0.72%. To separate it from the more abundant ^{238}U isotope, uranium is first converted to UF_6, which is easily vaporized above room temperature. The mixture of the $^{235}UF_6$ and $^{238}UF_6$ gases is then subjected to many stages of effusion. Calculate the separation factor, that is, the enrichment of ^{235}U relative to ^{238}U after one stage of effusion.

5.87 A gas evolved from the fermentation of glucose is found to effuse through a porous barrier in 15.0 min. Under the same conditions of temperature and pressure, it takes an equal volume of N_2 12.0 min to effuse through the same barrier. Calculate the molar mass of the gas and suggest what the gas might be.

5.88 Nickel forms a gaseous compound of the formula $Ni(CO)_x$. What is the value of x given the fact that under the same conditions of temperature and pressure, methane (CH_4) effuses 3.3 times faster than the compound?

5.8 Deviation from Ideal Behavior

Review Questions

5.89 Cite two pieces of evidence to show that gases do not behave ideally under all conditions.

5.90 Under what set of conditions would a gas be expected to behave most ideally: (a) high temperature and low pressure, (b) high temperature and high pressure, (c) low temperature and high pressure, or (d) low temperature and low pressure?

5.91 Shown are plots of PV/RT against P for one mole of a nonideal gas at two different temperatures. Which curve is at the higher temperature?

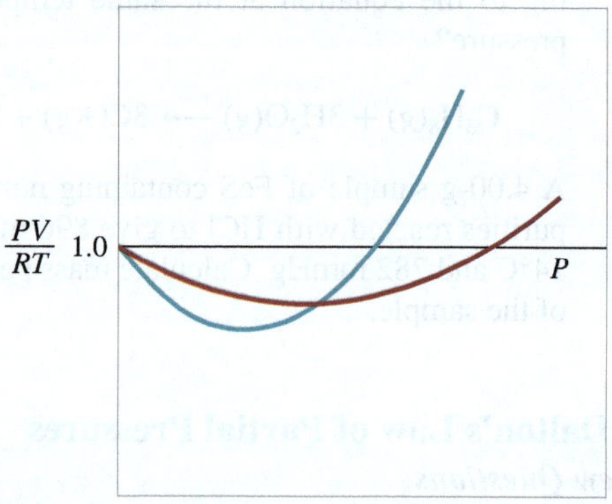

5.92 (a) A real gas is introduced into a flask of volume V. Is the corrected volume of the gas greater or less than V? (b) Ammonia has a larger a value than neon (see Table 5.4). What can you conclude about the relative strength of the attractive forces between molecules of ammonia and between atoms of neon?

Problems

5.93 Using the data shown in Table 5.4, calculate the pressure exerted by 2.50 moles of CO_2 confined in a volume of 5.00 L at 450 K. Compare the pressure with that predicted by the ideal gas equation.

5.94 At 27°C, 10.0 moles of a gas in a 1.50-L container exert a pressure of 130 atm. Is this an ideal gas?

Additional Problems

5.95 Discuss the following phenomena in terms of the gas laws: (a) the pressure increase in an automobile tire on a hot day; (b) the "popping" of a paper bag; (c) the expansion of a weather balloon as it rises in the air; (d) the loud noise heard when a lightbulb shatters.

5.96 Under the same conditions of temperature and pressure, which of the following gases would behave most ideally: Ne, N_2, or CH_4? Explain.

5.97 Nitroglycerin, an explosive compound, decomposes according to the equation

$$4C_3H_5(NO_3)_3(s) \longrightarrow 12CO_2(g) + 10H_2O(g) + 6N_2(g) + O_2(g)$$

Calculate the total volume of gases when collected at 1.2 atm and 25°C from 2.6×10^2 g of nitroglycerin. What are the partial pressures of the gases under these conditions?

5.98 The empirical formula of a compound is CH. At 200°C, 0.145 g of this compound occupies 97.2 mL at a pressure of 0.74 atm. What is the molecular formula of the compound?

5.99 When ammonium nitrite (NH_4NO_2) is heated, it decomposes to give nitrogen gas. This property is used to inflate some tennis balls. (a) Write a balanced equation for the reaction. (b) Calculate the quantity (in grams) of NH_4NO_2 needed to inflate a tennis ball to a volume of 86.2 mL at 1.20 atm and 22°C.

5.100 The percent by mass of bicarbonate (HCO_3^-) in a certain Alka-Seltzer product is 32.5%. Calculate the volume of CO_2 generated (in mL) at 37°C and 1.00 atm when a person ingests a 3.29-g tablet. (*Hint:* The reaction is between HCO_3^- and HCl acid in the stomach.)

5.101 The boiling point of liquid nitrogen is −196°C. On the basis of this information alone, do you think nitrogen is an ideal gas?

5.102 In the metallurgical process of refining nickel, the metal is first combined with carbon monoxide to form tetracarbonylnickel, which is a gas at 43°C:

$$Ni(s) + 4CO(g) \longrightarrow Ni(CO)_4(g)$$

This reaction separates nickel from other solid impurities. (a) Starting with 86.4 g of Ni, calculate the pressure of $Ni(CO)_4$ in a container of volume 4.00 L. (Assume the previous reaction goes to completion.) (b) At temperatures above 43°C, the pressure of the gas is observed to increase much more rapidly than predicted by the ideal gas equation. Explain.

5.103 The partial pressure of carbon dioxide varies with seasons. Would you expect the partial pressure in the Northern Hemisphere to be higher in the summer or winter? Explain.

5.104 A healthy adult exhales about 5.0×10^2 mL of a gaseous mixture with each breath. Calculate the number of molecules present in this volume at 37°C and 1.1 atm. List the major components of this gaseous mixture.

5.105 Sodium bicarbonate ($NaHCO_3$) is called baking soda because when heated, it releases carbon dioxide gas, which is responsible for the rising of cookies, doughnuts, and bread. (a) Calculate the volume (in liters) of CO_2 produced by heating 5.0 g of $NaHCO_3$ at 180°C and 1.3 atm. (b) Ammonium bicarbonate (NH_4HCO_3) has also been used for the same purpose. Suggest one advantage and one disadvantage of using NH_4HCO_3 instead of $NaHCO_3$ for baking.

5.106 A barometer having a cross-sectional area of 1.00 cm^2 at sea level measures a pressure of 76.0 cm of mercury. The pressure exerted by this column of mercury is equal to the pressure exerted by all the air on 1 cm^2 of Earth's surface. Given that the density of mercury is 13.6 g/mL and the average radius of Earth is 6371 km, calculate the total mass of Earth's atmosphere in kilograms. (*Hint:* The surface area of a sphere is $4\pi r^2$, where r is the radius of the sphere.)

5.107 Some commercial drain cleaners contain a mixture of sodium hydroxide and aluminum powder. When the mixture is poured down a clogged drain, the following reaction occurs:

$$2NaOH(aq) + 2Al(s) + 6H_2O(l) \longrightarrow 2NaAl(OH)_2(aq) + 3H_2(g)$$

The heat generated in this reaction helps melt away obstructions such as grease, and the hydrogen gas released stirs up the solids clogging the drain. Calculate the volume of H_2 formed at 23°C and 1.00 atm if 3.12 g of Al are treated with an excess of NaOH.

5.108 The volume of a sample of pure HCl gas was 189 mL at 25°C and 108 mmHg. It was completely dissolved in about 60 mL of water and titrated with an NaOH solution; 15.7 mL of the NaOH solution were required to neutralize the HCl. Calculate the molarity of the NaOH solution.

5.109 Propane (C_3H_8) burns in oxygen to produce carbon dioxide gas and water vapor. (a) Write a balanced equation for this reaction. (b) Calculate the number of liters of carbon dioxide measured at STP that could be produced from 7.45 g of propane.

5.110 Consider this apparatus. Calculate the partial pressures of helium and neon after the stopcock is open. The temperature remains constant at 16°C.

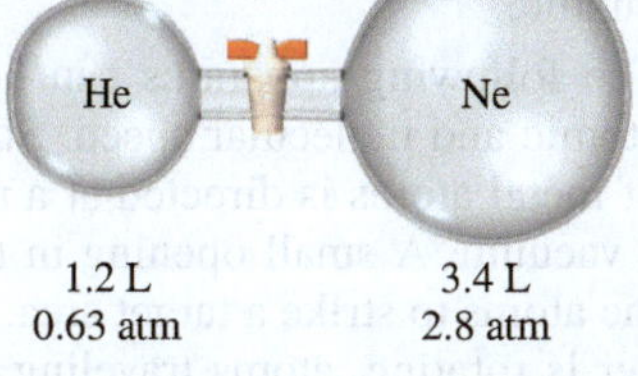

5.111 Nitric oxide (NO) reacts with molecular oxygen as follows:

$$C_2H_5OH(l) + O_2(g) \longrightarrow CO_2(g)H_2O(l)$$

Initially NO and O_2 are separated as shown here. When the valve is opened, the reaction quickly goes to completion. Determine what gases remain at the end and calculate their partial pressures. Assume that the temperature remains constant at 25°C.

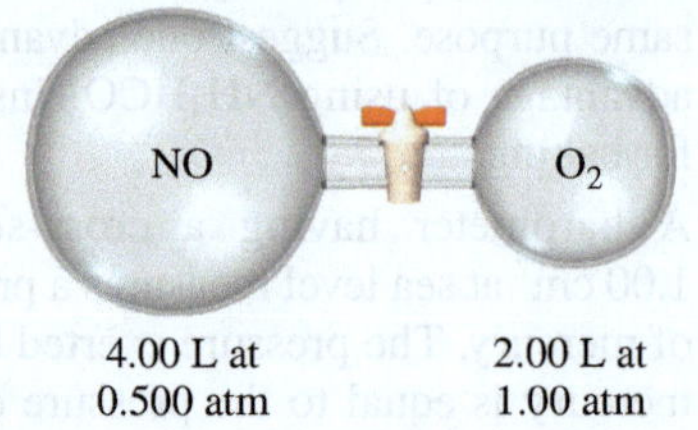

5.112 Consider this apparatus. When a small amount of water is introduced into the flask by squeezing the bulb of the medicine dropper, water is squirted upward out of the long glass tubing. Explain this observation. (*Hint:* Hydrogen chloride gas is soluble in water.)

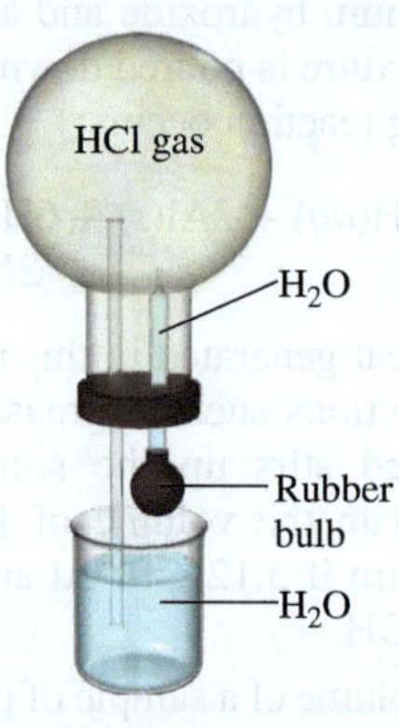

5.113 Describe how you would measure, by either chemical or physical means, the partial pressures of a mixture of gases of the following composition: CO_2 and H_2, (b) He and N_2.

5.114 A certain hydrate has the formula $MgSO_4 \cdot xH_2O$. A quantity of 54.2 g of the compound is heated in an oven to drive off the water. If the steam generated exerts a pressure of 24.8 atm in a 2.00-L container at 120°C, calculate x.

5.115 A mixture of Na_2CO_3 and $MgCO_3$ of mass 7.63 g is reacted with an excess of hydrochloric acid. The CO_2 gas generated occupies a volume of 1.67 L at 1.24 atm and 26°C. From these data, calculate the percent composition by mass of Na_2CO_3 in the mixture.

5.116 The following apparatus can be used to measure atomic and molecular speed. Suppose that a beam of metal atoms is directed at a rotating cylinder in a vacuum. A small opening in the cylinder allows the atoms to strike a target area. Because the cylinder is rotating, atoms traveling at different speeds will strike the target at different positions. In time, a layer of the metal will deposit on the target area, and the variation in its thickness is found to correspond to Maxwell's speed distribution. In one experiment it is found that at 850°C some bismuth (Bi) atoms struck the target at a point 2.80 cm from the spot directly opposite the slit. The diameter of the cylinder is 15.0 cm and it is rotating at 130 revolutions per second. (a) Calculate the speed (m/s) at which the target is moving. (*Hint:* The circumference of a circle is given by $2\pi r$, where r is the radius.) (b) Calculate the time (in seconds) it takes for the target to travel 2.80 cm. (c) Determine the speed of the Bi atoms. Compare your result in (c) with the u_{rms} of Bi at 850°C. Comment on the difference.

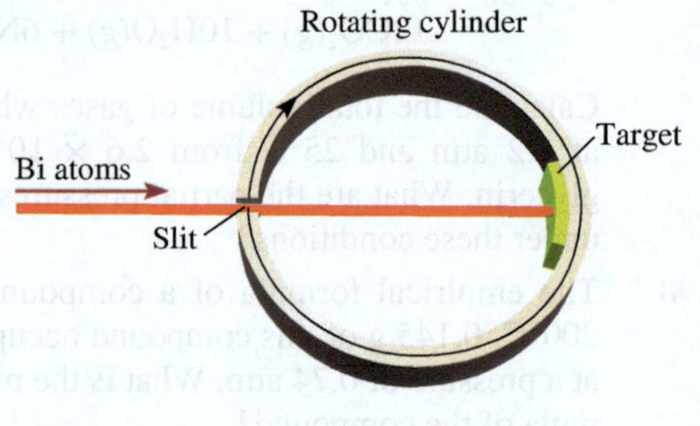

5.117 If 10.00 g of water are introduced into an evacuated flask of volume 2.500 L at 65°C, calculate the mass of water vaporized. (*Hint:* Assume that the volume of the remaining liquid water is negligible; the vapor pressure of water at 65°C is 187.5 mmHg.)

5.118 Commercially, compressed oxygen is sold in metal cylinders. If a 120-L cylinder is filled with oxygen to a pressure of 132 atm at 22°C, what is the mass (in grams) of O_2 present? How many liters of O_2 gas at 1.00 atm and 22°C could the cylinder produce? (Assume ideal behavior.)

5.119 The shells of hard-boiled eggs sometimes crack due to the rapid thermal expansion of the shells at high temperatures. Suggest another reason why the shells may crack.

5.120 Ethylene gas (C_2H_4) is emitted by fruits and is known to be responsible for their ripening. Based on this information, explain why a bunch of bananas ripens faster in a closed paper bag than in a bowl.

5.121 About 8.0×10^6 tons of urea $[(NH_2)_2CO]$ are used annually as a fertilizer. The urea is prepared at 200°C and under high-pressure conditions from carbon dioxide and ammonia (the products are urea and steam). Calculate the volume of ammonia (in liters) measured at 150 atm needed to prepare 1.0 ton of urea.

5.122 Some ballpoint pens have a small hole in the main body of the pen. What is the purpose of this hole?

5.123 The gas laws are vitally important to scuba divers. The pressure exerted by 33 ft of seawater is equivalent to 1 atm pressure. (a) A diver ascends quickly to the surface of the water from a depth of 36 ft without exhaling gas from his lungs. By what factor will the volume of his lungs increase by the time he reaches the surface? Assume that the temperature is constant. (b) The partial pressure of oxygen in air is about 0.20 atm. (Air is 20% oxygen by volume.) In deep-sea diving, the composition of air the diver breathes must be changed to maintain this partial pressure. What must the oxygen content (in percent by volume) be when the total pressure exerted on the diver is 4.0 atm? (At constant temperature and pressure, the volume of a gas is directly proportional to the number of moles of gases.) (*Hint:* See the Chemistry in Action essay "Scuba Diving and the Gas Laws" in Section 5.6.)

5.124 Nitrous oxide (N_2O) can be obtained by the thermal decomposition of ammonium nitrate (NH_4NO_3). (a) Write a balanced equation for the reaction. (b) In a certain experiment, a student obtains 0.340 L of the gas at 718 mmHg and 24°C. If the gas weighs 0.580 g, calculate the value of the gas constant.

5.125 Two vessels are labeled A and B. Vessel A contains NH_3 gas at 70°C, and vessel B contains Ne gas at the same temperature. If the average kinetic energy of NH_3 is 7.1×10^{-21} J/molecule, calculate the root-mean-square speed of Ne atoms.

5.126 Which of the following molecules has the largest *a* value: CH_4, F_2, C_6H_6, Ne?

5.127 The following procedure is a simple though somewhat crude way to measure the molar mass of a gas. A liquid of mass 0.0184 g is introduced into a syringe like the one shown here by injection through the rubber tip using a hypodermic needle. The syringe is then transferred to a temperature bath heated to 45°C, and the liquid vaporizes. The final volume of the vapor (measured by the outward movement of the plunger) is 5.58 mL and the atmospheric pressure is 760 mmHg. Given that the compound's empirical formula is CH_2, determine the molar mass of the compound.

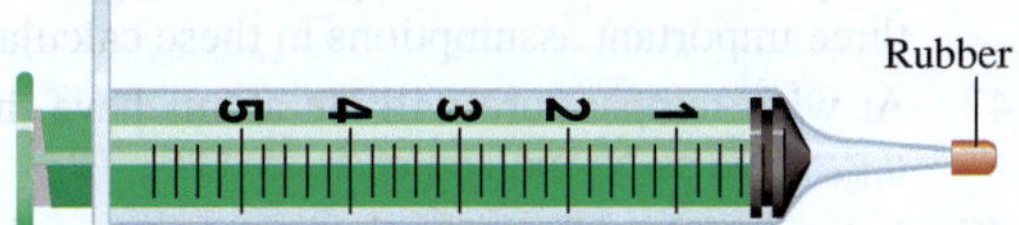

5.128 In 1995 a person suffocated as they walked by an abandoned mine in England. At that moment there was a sharp drop in atmospheric pressure due to a change in the weather. Suggest what might have caused the person's death.

5.129 Acidic oxides such as carbon dioxide react with basic oxides like calcium oxide (CaO) and barium oxide (BaO) to form salts (metal carbonates). (a) Write equations representing these two reactions. (b) A student placed a mixture of BaO and CaO of combined mass 4.88 g in a 1.46-L flask containing carbon dioxide gas at 35°C and 746 mmHg. After the reactions were complete, she found that the CO_2 pressure had dropped to 252 mmHg. Calculate the percent composition by mass of the mixture. Assume volumes of the solids are negligible.

5.130 Identify the Maxwell speed distribution curves shown here with the following gases: Br_2, CH_4, N_2, SO_3.

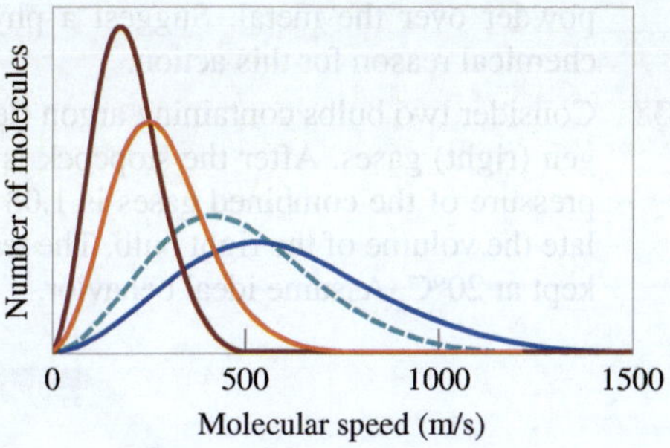

5.131 The running engine of an automobile produces carbon monoxide (CO), a toxic gas, at the rate of about 188 g CO per hour. A car is left idling in a poorly ventilated garage that is 6.0 m long, 4.0 m wide, and 2.2 m high at 20°C. (a) Calculate the rate of CO production in moles per minute. (b) How long would it take to build up a lethal concentration of CO of 1000 ppmv (parts per million by volume)?

5.132 Interstellar space contains mostly hydrogen atoms at a concentration of about 1 atom/cm^3. (a) Calculate the pressure of the H atoms. (b) Calculate the volume (in liters) that contains 1.0 g of H atoms. The temperature is 3 K.

5.133 Atop Mt. Everest, the atmospheric pressure is 210 mmHg and the air density is 0.426 kg/m^3. (a) Calculate the air temperature, given that the molar mass of air is 29.0 g/mol. (b) Assuming no change in air composition, calculate the percent decrease in oxygen gas from sea level to the top of Mt. Everest.

5.134 Relative humidity is defined as the ratio (expressed as a percentage) of the partial pressure of water vapor in the air to the equilibrium vapor pressure (see Table 5.3) at a given temperature. On a certain summer day in North Carolina the partial pressure of water vapor in the air is 3.9×10^3 Pa at 30°C. Calculate the relative humidity.

5.135 Under the same conditions of temperature and pressure, why does 1 L of moist air weigh less than 1 L of dry air? In weather forecasts, an oncoming low-pressure front usually means imminent rainfall. Explain.

5.136 Air entering the lungs ends up in tiny sacs called alveoli. It is from the alveoli that oxygen diffuses into the blood. The average radius of the alveoli is 0.0050 cm and the air inside contains 14% oxygen. Assuming that the pressure in the alveoli is 1.0 atm and the temperature is 37°C, calculate the number of oxygen molecules in one of the alveoli. (*Hint:* The volume of a sphere of radius r is $\frac{4}{3}\pi r^3$.)

5.137 A student breaks a thermometer and spills most of the mercury (Hg) onto the floor of a laboratory that measures 15.2 m long, 6.6 m wide, and 2.4 m high. (a) Calculate the mass of mercury vapor (in grams) in the room at 20°C. The vapor pressure of mercury at 20°C is 1.7×10^{-6} atm. (b) Does the concentration of mercury vapor exceed the air quality regulation of 0.050 mg Hg/m^3 of air? (c) One way to treat small quantities of spilled mercury is to spray sulfur powder over the metal. Suggest a physical and a chemical reason for this action.

5.138 Consider two bulbs containing argon (left) and oxygen (right) gases. After the stopcock is opened, the pressure of the combined gases is 1.08 atm. Calculate the volume of the right bulb. The temperature is kept at 20°C. Assume ideal behavior.

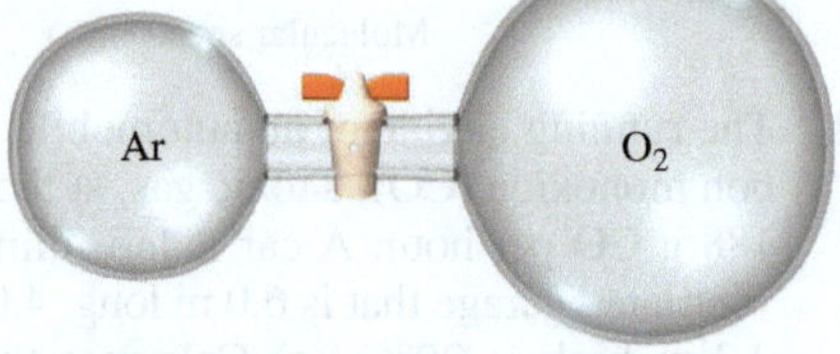

$n = 0.227$ mol
$V = 3.60$ L

$n = 0.144$ mol
$V = ?$

5.139 Nitrogen dioxide (NO_2) cannot be obtained in a pure form in the gas phase because it exists as a mixture of NO_2 and N_2O_4. At 25°C and 0.98 atm, the density of this gas mixture is 2.7 g/L. What is the partial pressure of each gas?

5.140 The Chemistry in Action essay "Super Cold Atoms" in Section 5.7 describes the cooling of rubidium vapor to 5.0×10^{-8} K. Calculate the root-mean-square speed and average kinetic energy of a Rb atom at this temperature.

5.141 Lithium hydride reacts with water as follows:

$$C_2H_5OH(l) + O_2(g) \longrightarrow CO_2(g) + H_2O(l)$$

During World War II, U.S. pilots carried LiH tablets. In the event of a crash landing at sea, the LiH would react with the seawater and fill their life belts and lifeboats with hydrogen gas. How many grams of LiH are needed to fill a 4.1-L life belt at 0.97 atm and 12°C?

5.142 The atmosphere on Mars is composed mainly of carbon dioxide. The surface temperature is 220 K and the atmospheric pressure is about 6.0 mmHg. Taking these values as Martian "STP," calculate the molar volume in liters of an ideal gas on Mars.

5.143 The atmosphere on Venus is composed of 96.5% CO_2, 3.5% N_2, and 0.015% SO_2 by volume. Its standard atmospheric pressure is 9.0×10^6 Pa. Calculate the partial pressures of the gases in pascals.

5.144 A student tries to determine the volume of a bulb like the one shown in Figure 5.12. These are her results: Mass of the bulb filled with dry air at 23°C and 744 mmHg = 91.6843 g; mass of evacuated bulb = 91.4715 g. Assume the composition of air is 78% N_2, 21% O_2, and 1% argon. What is the volume (in milliliters) of the bulb? (*Hint:* First calculate the average molar mass of air, as shown in Problem 3.156.)

5.145 Apply your knowledge of the kinetic theory of gases to the following situations. (a) Two flasks of volumes V_1 and V_1 and V_2 ($V_2 > V_1$) contain the same number of helium atoms at the same temperature. (i) Compare the root-mean-square (rms) speeds and average kinetic energies of the helium (He) atoms in the flasks. (ii) Compare the frequency and the force with which the He atoms collide with the walls of their containers. (b) Equal numbers of He atoms are placed in two flasks of the same volume at temperatures T_1 and T_2 ($T_2 > T_1$). (i) Compare the rms speeds of the atoms in the two flasks. (ii) Compare the frequency and the force with which the He atoms collide with the walls of their containers. (c) Equal numbers of He and neon (Ne) atoms are placed in two flasks of the same volume, and the temperature of both gases is 74°C. Comment on the validity of the following statements: (i) The rms speed of He is equal to that of Ne. (ii) The average kinetic energies of the two gases are equal. (iii) The rms speed of each He atom is 1.47×10^3 m/s.

5.146 It has been said that every breath we take, on average, contains molecules that were once exhaled by Wolfgang Amadeus Mozart (1756–1791). The following calculations demonstrate the validity of this statement. (a) Calculate the total number of molecules in the atmosphere. (*Hint:* Use the result in Problem 5.106 and 29.0 g/mol as the molar mass of air.) (b) Assuming the volume of every breath (inhale or exhale) is 500 mL, calculate the number of molecules exhaled in each breath at 37°C, which is the body temperature. (c) If Mozart's life span was exactly 35 years, what is the number of molecules he exhaled in that period? (Given that an average person breathes 12 times per minute.) (d) Calculate the fraction of molecules in the atmosphere that was exhaled by Mozart. How many of Mozart's molecules do we breathe in with every inhalation of air? Round off your answer to one significant figure. (e) List three important assumptions in these calculations.

5.147 At what temperature will He atoms have the same u_{rms} value as N_2 molecules at 25°C?

5.148 Estimate the distance (in nanometers) between molecules of water vapor at 100°C and 1.0 atm. Assume ideal behavior. Repeat the calculation for liquid water at 100°C, given that the density of water is 0.96 g/cm^3 at that temperature. Comment on your results. (Assume water molecule to be a sphere with a diameter of 0.3 nm.) (*Hint:* First calculate the number density of water molecules. Next, convert the number density to linear density, that is, number of molecules in one direction.)

5.149 Which of the noble gases would not behave ideally under any circumstance? Why?

5.150 A relation known as the barometric formula is useful for estimating the change in atmospheric pressure with altitude. The formula is given by $P = P_0e^{-g\mathcal{M}h/RT}$, where P and P_0 are the pressures at height h and sea level, respectively; g is the acceleration due to gravity (9.8 m/s^2); $\mathcal{M}$ is the average molar mass of air (29.0 g/mol); and R is the gas constant. Calculate the atmospheric pressure in atm at a height of 5.0 km, assuming the temperature is constant at 5°C and $P_0 = 1.0$ atm.

5.151 A 5.72-g sample of graphite was heated with 68.4 g of O_2 in a 8.00-L flask. The reaction that took place was

$$C(graphite) + O_2(g) \longrightarrow CO_2(g)$$

After the reaction was complete, the temperature in the flask was 182°C. What was the total pressure inside the flask?

5.152 An equimolar mixture of H_2 and D_2 effuses through an orifice (small hole) at a certain temperature. Calculate the composition (in mole fractions) of the gases that pass through the orifice. The molar mass of D_2 is 2.014 g/mol.

5.153 A mixture of calcium carbonate ($CaCO_3$) and magnesium carbonate ($MgCO_3$) of mass 6.26 g reacts completely with hydrochloric acid (HCl) to generate 1.73 L of CO_2 at 48°C and 1.12 atm. Calculate the mass percentages of $CaCO_3$ and $MgCO_3$ in the mixture.

5.154 A 6.11-g sample of a Cu-Zn alloy reacts with HCl acid to produce hydrogen gas. If the hydrogen gas has a volume of 1.26 L at 22°C and 728 mmHg, what is the percent of Zn in the alloy? (*Hint:* Cu does not react with HCl.)

5.155 A stockroom supervisor measured the contents of a partially filled 25.0-gallon acetone drum on a day when the temperature was 18.0°C and atmospheric pressure was 750 mmHg, and found that 15.4 gallons of the solvent remained. After tightly sealing the drum, an assistant dropped the drum while carrying it upstairs to the organic laboratory. The drum was dented and its internal volume was decreased to 20.4 gallons. What is the total pressure inside the drum after the accident? The vapor pressure of acetone at 18.0°C is 400 mmHg. (*Hint:* At the time the drum was sealed, the pressure inside the drum, which is equal to the sum of the pressures of air and acetone, was equal to the atmospheric pressure.)

5.156 In 2.00 min, 29.7 mL of He effuse through a small hole. Under the same conditions of pressure and temperature, 10.0 mL of a mixture of CO and CO_2 effuse through the hole in the same amount of time. Calculate the percent composition by volume of the mixture.

5.157 Referring to Figure 5.22, explain the following: (a) Why do the curves dip below the horizontal line labeled ideal gas at low pressures and then why do they arise above the horizontal line at high pressures? (b) Why do the curves all converge to 1 at very low pressures? (c) Each curve intercepts the horizontal line labeled ideal gas. Does it mean that at that point the gas behaves ideally?

5.158 A mixture of methane (CH_4) and ethane (C_2H_6) is stored in a container at 294 mmHg. The gases are burned in air to form CO_2 and H_2O. If the pressure of CO_2 is 356 mmHg measured at the same temperature and volume as the original mixture, calculate the mole fractions of the gases.

5.159 Use the kinetic theory of gases to explain why hot air rises.

5.160 One way to gain a physical understanding of b in the van der Waals equation is to calculate the "excluded volume." Assume that the distance of closest approach between two similar atoms is the sum of their radii ($2r$). (a) Calculate the volume around each atom into which the center of another atom cannot penetrate. (b) From your result in (a), calculate the excluded volume for 1 mole of the atoms, which is the constant b. How does this volume compare with the sum of the volumes of 1 mole of the atoms?

5.161 Use the van der Waals constants in Table 5.4 to estimate the radius of argon in picometers. (*Hint:* See Problem 5.160.)

5.162 Identify the gas whose root-mean-square speed is 2.82 times that of hydrogen iodide (HI) at the same temperature.

5.163 A 5.00-mole sample of NH_3 gas is kept in a 1.92-L container at 300 K. If the van der Waals equation is assumed to give the correct answer for the pressure of the gas, calculate the percent error made in using the ideal gas equation to calculate the pressure.

5.164 The root-mean-square speed of a certain gaseous oxide is 493 m/s at 20°C. What is the molecular formula of the compound?

5.165 Referring to Figure 5.17, we see that the maximum of each speed distribution plot is called the most probable speed (u_{mp}) because it is the speed possessed by the largest number of molecules. It is given by $u_{mp} = \sqrt{2RT/\mathcal{M}}$. (a) Compare u_{mp} with u_{rms} for nitrogen at 25°C. (b) The following diagram shows the Maxwell speed distribution curves for an ideal gas at two different temperatures T_1 and T_2. Calculate the value of T_2.

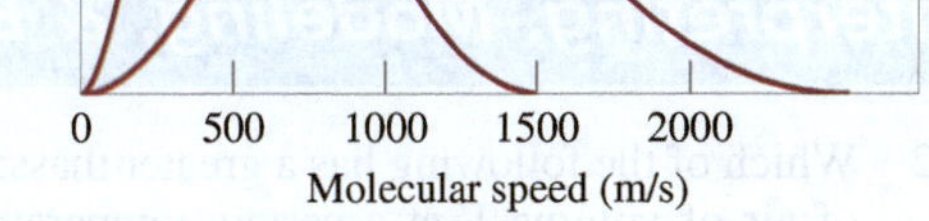

5.166 A gaseous reaction takes place at constant volume and constant pressure in this cylinder. Which of the following equations best describes the reaction? The initial temperature (T_1) is twice that of the final temperature (T_2).

(a) $A + B \longrightarrow C$
(b) $AB \longrightarrow C + D$
(c) $A + B \longrightarrow C + D$
(d) $A + B \longrightarrow 2C + D$

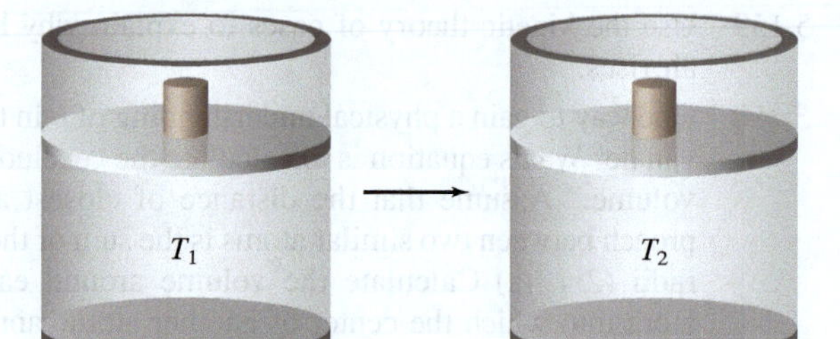

5.167 A gaseous hydrocarbon (containing C and H atoms) in a container of volume 20.2 L at 350 K and 6.63 atm reacts with an excess of oxygen to form 205.1 g of CO_2 and 168.0 g of H_2O. What is the molecular formula of the hydrocarbon?

5.168 Three flasks (a)–(c) contain gases A (red) and B (blue). (i) If the pressure in (a) is 4.0 atm, what are the pressures in (b) and (c)? (ii) Calculate the total pressure and partial pressure of each gas after the valves are opened. The volumes of (a) and (c) are 4.0 L each and that of (b) is 2.0 L. The temperature is the same throughout.

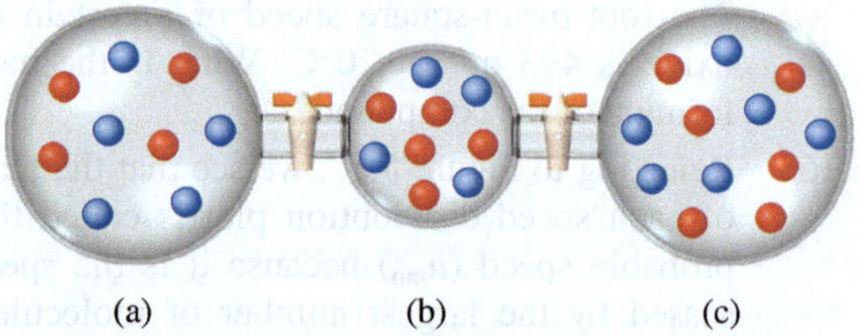

5.169 (a) Show that the pressure exerted by a fluid P (in pascals) is given by $P = hdg$, where h is the column of the fluid in meters, d is the density in kg/m^3, and g is the acceleration due to gravity (9.81 m/s^2). (*Hint:* See Appendix 1.) (b) The volume of an air bubble that starts at the bottom of a lake at 5.24°C increases by a factor of 6 as it rises to the surface of water where the temperature is 18.73°C and the air pressure is 0.973 atm. The density of the lake water is 1.02 g/cm^3. Use the equation in (a) to determine the depth of the lake in meters.

5.170 A student first measured the total pressure of a mixture of gases methane (CH_4), ethane (C_2H_6), and propane (C_3H_8) at a certain temperature, which turned out to be 4.50 atm. She then recorded the mass spectra of the gases shown here. Calculate the partial pressures of the gases.

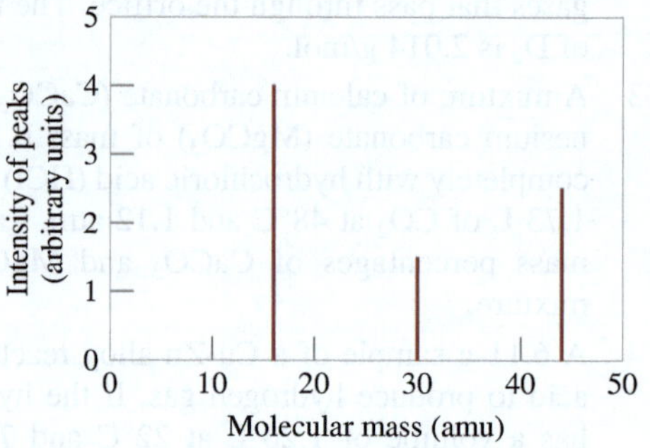

5.171 In 2012, Felix Baumgartner jumped from a balloon roughly 24 mi above Earth, breaking the record for the highest skydive. He reached speeds of more than 700 miles per hour and became the first skydiver to exceed the speed of sound during free fall. The helium-filled plastic balloon used to carry Baumgartner to the edge of space was designed to expand to 8.5 × 10^8 L to accommodate the low pressures at the altitude required to break the record. (a) Calculate the mass of helium in the balloon from the conditions at the time of the jump (8.5 × 10^8 L, −67.8°C, 0.027 mmHg). (b) Determine the volume of the helium in the balloon just before it was released, assuming a pressure of 1.0 atm and a temperature of 23°C.

Interpreting, Modeling, & Estimating

5.172 Which of the following has a greater mass: a sample of air of volume V at a certain temperature T and pressure P or a sample of air plus water vapor having the same volume and at the same temperature and pressure?

5.173 A flask with a volume of 14.5 L contains 1.25 moles of helium gas. Estimate the average distance between He atoms in nanometers.

5.174 Hyperbaric oxygen therapy (HBOT) is very effective in treating burns, crush injuries that impede blood flow, and tissue-damaging infections, as well as carbon monoxide poisoning. However, it has generated some controversy in its application to other maladies (for example, autism, multiple sclerosis). A typical oxygen hyperbaric chamber is shown here. HBOT can be administered using pressure up to 6 atmospheres, but lower pressures are more common. (a) If this chamber was pressurized to 3.0 atm with pure oxygen, how many moles of O_2 would be contained in an empty chamber? (b) Given that a full tank of oxygen contains about 2500 moles of the gas, how many times could the chamber be filled with a single tank of oxygen?

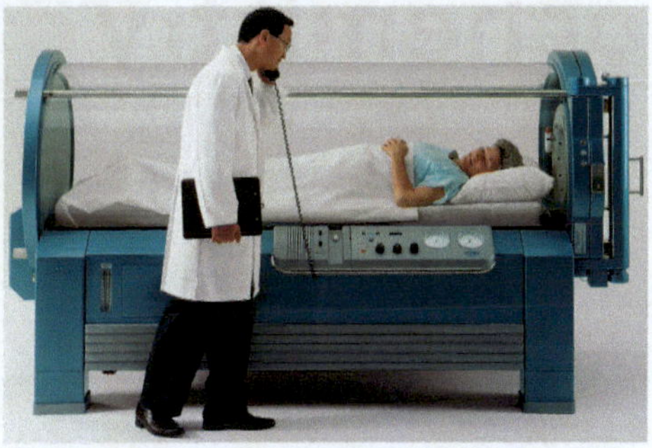

Sechrist Industries

5.175 (a) Fluorescent lightbulbs contain a small amount of mercury, giving a mercury vapor pressure of around 1×10^{-5} atm. When excited electrically, the Hg atoms emit UV light, which excites the phosphor coating of the inner tube, which then emits visible (white) light. Estimate the mass of Hg vapor present in the type of long, thin fluorescent tubes used in offices. (b) Ordinary tungsten incandescent lightbulbs used in households are filled with argon gas at about 0.5 atm to retard the sublimation of the tungsten filament. Estimate the number of moles of Ar in a typical lightbulb.

5.176 (a) Estimate the volume of air at 1.0 atm and 22°C needed to fill a bicycle tire to a pressure of 5.0 atm at the same temperature. (Note that the 5.0 atm is the gauge pressure, which is the difference between the pressure in the tire and atmospheric pressure.) (b) The tire is pumped by filling the cylinder of a hand pump with air at 1.0 atm and then, by compressing the gas in the cylinder, adding all the air in the pump to the air in the tire. If the volume of the pump is 33 percent of the tire's volume, what is the gauge pressure in the tire after three full strokes of the pump?

5.177 On October 15, 2009, a homemade helium balloon was released, and for a while authorities were led to believe that a 6-year-old boy had been carried away in the balloon. (The incident was later revealed to be a hoax.) The balloon traveled more than 50 mi and reached a height of 7000 ft. The shape and span of the balloon are shown in the figure. How much weight could this balloon lift? (A helium balloon can lift a mass equal to the difference in the mass of air and the mass of helium that would be contained in the balloon.) Could it actually lift a 6-year-old boy?

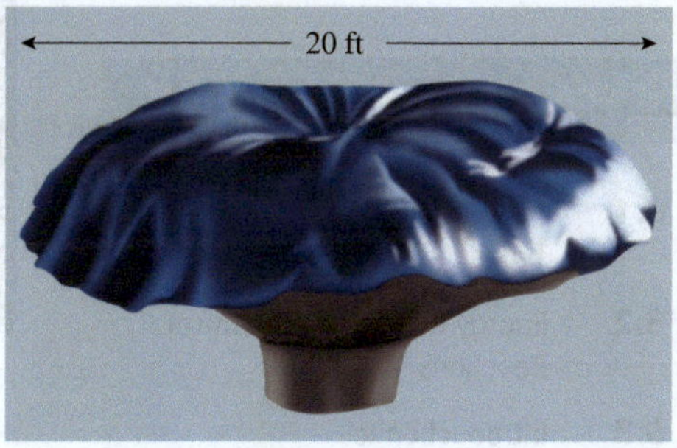

Answers to Practice Exercises

5.1 0.986 atm. **5.2** 39.3 kPa. **5.3** 9.29 L. **5.4** 30.6 L. **5.5** 4.46×10^3 mmHg. **5.6** 0.68 atm. **5.7** 2.6 atm. **5.8** 13.1 g/L. **5.9** 44.1 g/mol. **5.10** B_2H_6. **5.11** 96.9 L. **5.12** 4.75 L. **5.13** 0.338 *M*. **5.14** CH_4: 1.29 atm; C_2H_6: 0.0657 atm; C_3H_8: 0.0181 atm. **5.15** 0.0653 g. **5.16** 321 m/s. **5.17** 146 g/mol. **5.18** 30.0 atm; 45.5 atm using the ideal gas equation.

Answers to Review of Concepts & Facts

5.2.1 1184 mmHg, 1.558 atm, 157.9 kPa. **5.2.2** (b) < (c) < (a) < (d). **5.2.3** It would be easier to drink water with a straw at the foot of Mt. Everest because the atmospheric pressure is greater there, which helps to push the water up the straw. **5.3.1** 1.29 L. **5.3.2** (a) Volume doubles. (b) Volume increases 1.4 times. **5.3.3** 50.0 mL. **5.4.1** 458 K or 185°C. **5.4.2** Greatest volume, (b); greatest density, (c). **5.4.3** 77.1 g/mol. **5.5.1** 11.3 L. **5.5.2** Only for the combustion of methane, $CH_4(g) + 2O_2(g) \longrightarrow CO_2(l) + 2H_2O(g)$. **5.6.1** Blue sphere: 0.43 atm, green sphere: 1.3 atm, red sphere: 0.87 atm. **5.6.2** $X_{Ne} = 0.338$. **5.6.3** 0.702 atm. **5.7.1** (c) and (d). **5.7.2** Ne effuses faster. It effuses 2.551 times faster. **5.8.1** High pressure and low temperature. **5.8.2** 10.3 atm.

CHAPTER

6 Thermochemistry

Biofuels, fuels derived from renewable sources such as corn and soybeans, represent a viable alternative to the common fossil fuels that are nonrenewable.
ThamKC/iStock/Getty Images

CHAPTER OUTLINE

Every chemical reaction obeys two fundamental laws: the law of conservation of mass and the law of conservation of energy. We discussed the mass relationship between reactants and products in Chapter 3; here we will look at the energy changes that accompany chemical reactions.

6.1 The Nature of Energy and Types of Energy

Learning Objective

- Distinguish between kinetic energy and potential, and identify specific examples of each.

"Energy" is a much-used term that represents a rather abstract concept. For instance, when we feel tired, we might say we haven't any *energy;* and we read about the need to find alternatives to nonrenewable *energy* sources. Unlike matter, energy is known and recognized by its effects. It cannot be seen, touched, smelled, or weighed.

Energy is usually defined as *the capacity to do work.* In Chapter 5 we defined work as "force × distance," but we will soon see that there are other kinds of work. All forms of energy are capable of doing work (i.e., of exerting a force over a distance), but not all of them are equally relevant to chemistry. The energy contained in tidal waves, for example, can be harnessed to perform useful work, but the relationship between tidal waves and chemistry is minimal. Chemists define ***work*** as *directed energy change resulting from a process.*

Forms of Energy

All energy is classified as either kinetic or potential. Kinetic energy—the energy produced by a moving object—is one form of energy that is of particular interest to chemists. We first introduced kinetic energy in Section 5.7. Other forms of kinetic energy of interest to chemists include radiant energy and thermal energy.

Radiant energy, or *solar energy, comes from the sun* and is Earth's primary energy source. Solar energy heats the atmosphere and Earth's surface, stimulates the growth of vegetation through the process known as photosynthesis, and influences global climate patterns.

Thermal energy is *the energy associated with the random motion of atoms and molecules.* In general, thermal energy can be calculated from temperature measurements. The more vigorous the motion of the atoms and molecules in a sample of matter, the hotter the sample is, and the greater its thermal energy. However, we need to distinguish carefully between thermal energy and temperature. A cup of coffee at 70°C has a higher temperature than a bathtub filled with warm water at 40°C, but much more thermal energy is stored in the bathtub water because it has a much larger volume and greater mass than the coffee and therefore more water molecules and more molecular motion. Recall from Section 1.9 that temperature is an intensive property (it does not depend on the amount of matter being considered) but thermal energy is an extensive property (it does depend on the amount of matter being considered).

As the water falls over the dam, its potential energy is converted to kinetic energy. Use of this energy to generate electricity is called hydroelectric power.
Jacques Jangoux/Science Source

In contrast to kinetic energy, ***potential energy*** is *energy due to an object's position.* For instance, because of its altitude, a rock at the top of a cliff has more potential energy and will make a bigger splash if it falls into the water below than a similar rock located partway down the cliff. To a chemist, the most important examples of potential energy are chemical energy and electrostatic energy.

Chemical energy is *stored within the structural units of chemical substances;* its quantity is determined by the type and arrangement of constituent atoms. When substances participate in chemical reactions, chemical energy is released, stored, or converted to other forms of energy.

Electrostatic energy is *potential energy due to the interaction of charged particles.* Opposite charges attract each other and the electrostatic energy is negative. Like charges repel each other and the electrostatic energy is positive. The magnitude of the electrostatic energy increases with increasing charges on the particles. The magnitude of the electrostatic energy is directly proportional to the product of the charges and indirectly proportional to the distance between the two particles.

All forms of energy can be converted (at least in principle) from one form to another. We feel warm when we stand in sunlight because radiant energy is converted to thermal energy on our skin. When we exercise, chemical energy stored in the body is used to produce kinetic energy. When a ball starts to roll downhill, its potential energy is converted to kinetic energy. You can undoubtedly think of many other examples. Although energy can assume many different forms that are interconvertible, scientists have concluded that energy can be neither destroyed nor created. When one form of energy disappears, some other form of energy (of equal magnitude) must appear, and vice versa. This principle is summarized by the ***law of conservation of energy:*** *The total quantity of energy in the universe is assumed constant.*

Summary of Concepts & Facts

- Energy is the capacity to do work.
- There are two forms of energy, kinetic and potential. There are many types of kinetic and potential energy.
- Kinetic energy and potential energy are interconvertible as the law of conservation of energy states that the total amount of energy in the universe is constant.

6.2 Energy Changes in Chemical Reactions

Learning Objectives

- Discriminate between the system and surroundings for a given experiment.
- Classify a system as being open, closed, or isolated.
- Categorize a process as endothermic or exothermic.

Often the energy changes that take place during chemical reactions are of as much practical interest as the mass relationships we discussed in Chapter 3. For example, combustion reactions involving fuels such as natural gas and oil are carried out in daily life more for the thermal energy they release than for their products, which are water and carbon dioxide.

Almost all chemical reactions absorb or produce (release) energy, generally in the form of heat. It is important to understand the distinction between thermal energy and heat. ***Heat*** is *the transfer of thermal energy between two bodies that are at different temperatures.* Thus, we often speak of the "heat flow" from a hot object to a cold one. Although the term "heat" by itself implies the transfer of energy, we customarily talk of "heat absorbed" or "heat released" when describing the energy changes that occur during a process. ***Thermochemistry*** is *the study of heat change in chemical reactions.*

To analyze energy changes associated with chemical reactions we must first define the ***system,*** or *the specific part of the universe that is of interest to us.* For chemists, systems usually include substances involved in chemical and physical changes. For example, in an acid-base neutralization experiment, the system may be a beaker containing 50 mL of HCl to which 50 mL of NaOH is added. The ***surroundings*** are *the rest of the universe outside the system.* When heat is absorbed or released during a process, energy is conserved according to the law of conservation of energy (see Section 6.1), but it is transferred between a system and its surroundings.

There are three types of systems. An ***open system*** *can exchange mass and energy, usually in the form of heat with its surroundings.* For example, an open system may

Video: Heat Flow

This infrared photo shows where energy (heat) leaks through the house. The more red the color, the more energy is lost to the outside.
Ivan Smuk/Alamy Stock Photo

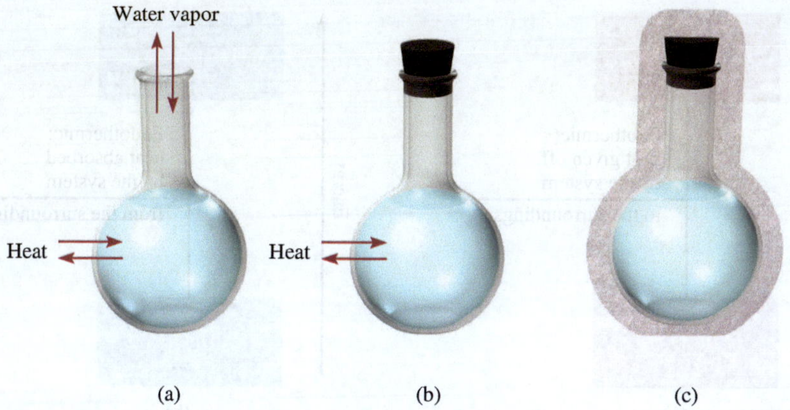

Figure 6.1 *Three systems represented by water in a flask: (a) an open system, which allows the exchange of both energy and mass with surroundings; (b) a closed system, which allows the exchange of energy but not mass; and (c) an isolated system, which allows neither energy nor mass to be exchanged (here the flask is enclosed by a vacuum jacket).*

consist of a quantity of water in an open container, as shown in Figure 6.1(a). If we close the flask, as in Figure 6.1(b), so that no water vapor can escape from or condense into the container, we create a ***closed system,*** which *allows the transfer of energy (heat) but not mass.* By placing the water in a totally insulated container, we can construct an ***isolated system,*** which *does not allow the transfer of either mass or energy,* as shown in Figure 6.1(c).

The combustion of hydrogen gas in oxygen is one of many chemical reactions that release considerable quantities of energy (Figure 6.2):

$$2H_2(g) + O_2(g) \longrightarrow 2H_2O(l) + \text{energy}$$

In this case, we label the reacting mixture (hydrogen, oxygen, and water molecules) the *system* and the rest of the universe the *surroundings.* Because energy cannot be created or destroyed, any energy lost by the system must be gained by the surroundings. Thus, the heat generated by the combustion process is transferred from the system to its surroundings. This reaction is an example of an ***exothermic process,*** which is *any process that gives off heat*—that is, *transfers thermal energy to the surroundings.* Figure 6.3(a) shows the energy change for the combustion of hydrogen gas.

Now consider another reaction, the decomposition of mercury(II) oxide (HgO) at high temperatures:

$$\text{energy} + 2HgO(s) \longrightarrow 2Hg(l) + O_2(g)$$

This reaction is an ***endothermic process,*** *in which heat has to be supplied to the system* (that is, to HgO) *by the surroundings* [Figure 6.3(b)].

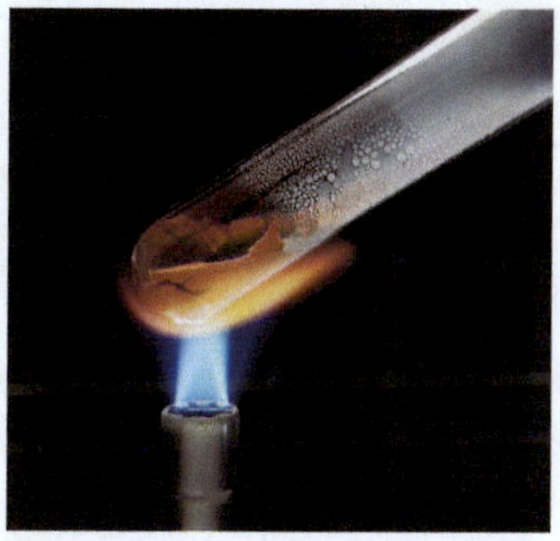

On heating, HgO decomposes to give Hg and O_2

Charles D. Winters/McGraw Hill

Figure 6.2 *The* Hindenburg *disaster. The* Hindenburg, *a German airship filled with hydrogen gas, was destroyed in a spectacular fire at Lakehurst, New Jersey, in 1937.*

akg-images/Newscom

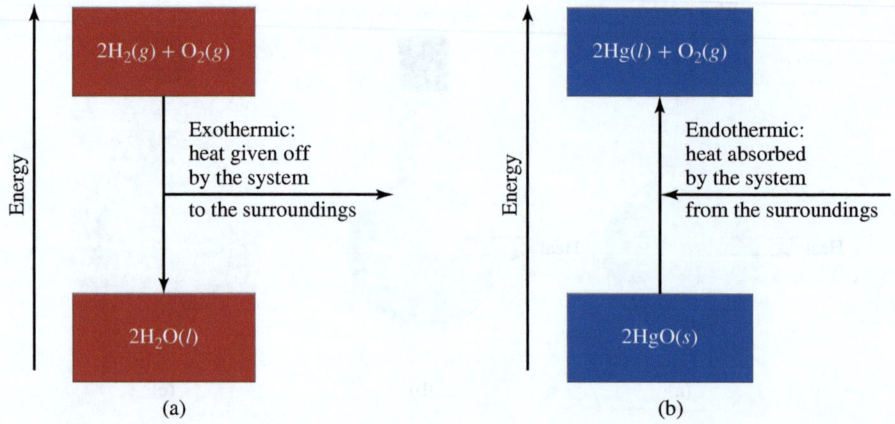

Figure 6.3 *(a) An exothermic process. (b) An endothermic process. Parts (a) and (b) are not drawn to the same scale; that is, the heat released in the formation of H_2O from H_2 and O_2 is not equal to the heat absorbed in the decomposition of HgO.*

From Figure 6.3 you can see that in exothermic reactions, the total energy of the products is less than the total energy of the reactants. The difference is the heat supplied by the system to the surroundings. Just the opposite happens in endothermic reactions. Here, the difference between the energy of the products and the energy of the reactants is equal to the heat supplied to the system by the surroundings.

Summary of Concepts & Facts

- A process that gives off heat to the surroundings is exothermic; a process that absorbs heat from the surroundings is endothermic.

Review of Concepts & Facts

6.2.1 Classify each of the following as an open system, a closed system, or an isolated system.
(a) Milk kept in a closed thermo flask
(b) A student reading in her dorm room
(c) Air inside a tennis ball

6.2.2 Determine if the following processes are endothermic or exothermic.
(a) ZnS(*s*) decomposes to Zn(*s*) and S(*s*).
(b) Propane gas burns in the presence of oxygen in a grill.
(c) Ice cubes melt in a glass.

6.3 Introduction to Thermodynamics

Learning Objectives

- Explain what a state function is and be able to identify state functions and nonstate functions.
- Summarize the first law of thermodynamics.

Thermochemistry is part of a broader subject called ***thermodynamics,*** which is *the scientific study of the interconversion of heat and other kinds of energy.* The laws of thermodynamics provide useful guidelines for understanding the energetics and directions of processes. In this section we will concentrate on the first law of thermodynamics, which is particularly relevant to the study of thermochemistry. We will continue our discussion of thermodynamics in Chapter 17.

In thermodynamics, we study changes in the ***state of a system,*** which is defined by *the values of all relevant macroscopic properties, for example, composition, energy, temperature, pressure, and volume.* Energy, pressure, volume, and temperature are said to be ***state functions***—*properties that are determined by the state of the system, regardless of how that condition was achieved.* In other words, when the state of a system changes, the magnitude of change in any state function depends only on the initial and final states of the system and not on how the change is accomplished or the pathway taken during the change.

The state of a given amount of a gas is specified by its volume, pressure, and temperature. Consider a gas at 2 atm, 300 K, and 1 L (the initial state). Suppose a process is carried out at constant temperature such that the gas pressure decreases to 1 atm. According to Boyle's law, its volume must increase to 2 L. The final state then corresponds to 1 atm, 300 K, and 2 L. The change in volume (ΔV) is

$$\begin{aligned}\Delta V &= V_f - V_i \\ &= 2\text{ L} - 1\text{ L} \\ &= 1\text{ L}\end{aligned}$$

where V_i and V_f denote the initial and final volume, respectively. No matter how we arrive at the final state (for example, the pressure of the gas can be increased first and then decreased to 1 atm), the change in volume is always 1 L. Thus, the volume of a gas is a state function. In a similar manner, we can show that pressure and temperature are also state functions.

Energy is another state function. Using potential energy as an example, we find that the net increase in gravitational potential energy when we go from the same starting point to the top of a mountain is always the same, regardless of how we get there (Figure 6.4).

Student Hot Spot

Student data indicate you may struggle with state functions. Access your eBook for additional Learning Resources on this topic.

The First Law of Thermodynamics

The ***first law of thermodynamics,*** which is based on the law of conservation of energy, states that *energy can be converted from one form to another, but cannot be created or destroyed.*[†] How do we know this is so? It would be impossible to prove the validity of the first law of thermodynamics if we had to determine the total energy content of the universe. Even determining the total energy content of 1 g of iron, say, would be extremely difficult. Fortunately, we can test the validity of the first law by measuring only the *change* in the internal energy of a system between its *initial state* and its *final state* in a process. The change in internal energy ΔU is given by

$$\Delta U = U_f - U_i$$

where U_i and U_f are the internal energies of the system in the initial and final states, respectively.

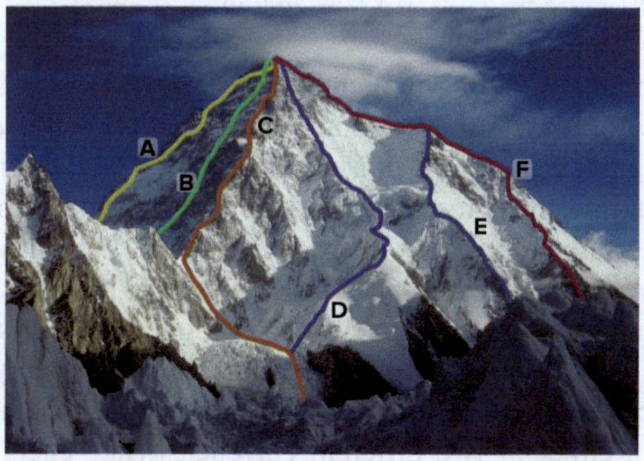

Figure 6.4 *The gain in gravitational potential energy that occurs when a person climbs from the base to the top of a mountain is independent of the path taken.*

Uwe Steffens/Ullstein Bild/Getty Images

[†]See Section 2.1 for a discussion of mass and energy relationship in chemical reactions.

The internal energy of a system has two components: kinetic energy and potential energy. The kinetic energy component consists of various types of molecular motion and the movement of electrons within molecules. Potential energy is determined by the attractive interactions between electrons and nuclei and by repulsive interactions between electrons and between nuclei in individual molecules, as well as by interaction between molecules. It is impossible to measure all these contributions accurately, so we cannot calculate the total energy of a system with any certainty. Changes in energy, on the other hand, can be determined experimentally.

Sulfur burning in air to form SO_2.
Andrew Lambert Photography/Science Source

Consider the reaction between 1 mole of sulfur and 1 mole of oxygen gas to produce 1 mole of sulfur dioxide:

$$S(s) + O_2(g) \longrightarrow SO_2(g)$$

In this case, our system is composed of the reactant molecules S and O_2 (the initial state) and the product molecules SO_2 (the final state). We do not know the internal energy content of either the reactant molecules or the product molecules, but we can accurately measure the *change* in energy content, ΔU, given by

$$\Delta U = U(\text{product}) - U(\text{reactants})$$
$$= \text{energy content of 1 mol } SO_2(g) - \text{energy content of } [1 \text{ mol } S(s) + 1 \text{ mol } O_2(g)]$$

We find that this reaction gives off heat. Therefore, the energy of the product is less than that of the reactants, and ΔU is negative.

Interpreting the release of heat in this reaction to mean that some of the chemical energy contained in the molecules has been converted to thermal energy, we conclude that the transfer of energy from the system to the surroundings does not change the total energy of the universe. That is, the sum of the energy changes must be zero:

$$\Delta U_{\text{sys}} + \Delta U_{\text{surr}} = 0$$

or

$$\Delta U_{\text{sys}} = -\Delta U_{\text{surr}}$$

where the subscripts "sys" and "surr" denote system and surroundings, respectively. Thus, if one system undergoes an energy change ΔU_{sys}, the rest of the universe, or the surroundings, must undergo a change in energy that is equal in magnitude but opposite in sign ($-\Delta U_{\text{surr}}$); energy gained in one place must have been lost somewhere else. Furthermore, because energy can be changed from one form to another, the energy lost by one system can be gained by another system in a different form. For example, the energy lost by burning oil in a power plant may ultimately turn up in our homes as electrical energy, heat, light, and so on.

In chemistry, we are normally interested in the energy changes associated with the system (which may be a flask containing reactants and products), not with its surroundings. Therefore, a more useful form of the first law is

$$\Delta U = q + w \tag{6.1}$$

(We drop the subscript "sys" for simplicity.) Equation (6.1) says that the change in the internal energy, ΔU, of a system is the sum of the heat exchange q between the system and the surroundings and the work done w on (or by) the system. We use lowercase letters (such as q and w) to represent thermodynamic quantities that are not state functions. The sign conventions for q and w are as follows: q is positive for an endothermic process and negative for an exothermic process and w is positive for work done on the system by the surroundings and negative for work done by the system on the surroundings. We can think of the first law of thermodynamics as an energy balance sheet, much like a money balance sheet kept in a bank that does currency exchange. You can withdraw or deposit money in either of two different currencies (like energy change due to heat exchange and work done). However, the value of your bank account depends only on the net amount of money left in it after these transactions, not on which currency you used.

Table 6.1 Sign Conventions for Work and Heat

Process	Sign
Work done by the system on the surroundings	−
Work done on the system by the surroundings	+
Heat absorbed by the system from the surroundings (endothermic process)	+
Heat absorbed by the surroundings from the system (exothermic process)	−

Equation (6.1) may seem abstract, but it is actually quite logical. If a system loses heat to the surroundings or does work on the surroundings, we would expect its internal energy to decrease because those are energy-depleting processes. For this reason, both q and w are negative. Conversely, if heat is added to the system or if work is done on the system, then the internal energy of the system would increase. In this case, both q and w are positive. Table 6.1 summarizes the sign conventions for q and w.

Work and Heat

We will now look at the nature of work and heat in more detail.

Work

We have seen that work can be defined as force F multiplied by distance d:

$$w = F \times d \tag{6.2}$$

In thermodynamics, work has a broader meaning that includes mechanical work (for example, a crane lifting a steel beam), electrical work (a battery supplying electrons to light the bulb of a flashlight), and surface work (blowing up a soap bubble). In this section we will concentrate on mechanical work; in Chapter 18 we will discuss the nature of electrical work.

One way to illustrate mechanical work is to study the expansion or compression of a gas. Many chemical and biological processes involve gas volume changes. Breathing and exhaling air involves the expansion and contraction of the tiny sacs called alveoli in the lungs. Another example is the internal combustion engine of an automobile. The successive expansion and compression of the cylinders due to the combustion of the gasoline-air mixture provide power to the vehicle. Figure 6.5 shows a gas in a cylinder fitted with a weightless, frictionless movable piston at a certain temperature, pressure, and volume. As it expands, the gas pushes the piston upward against a constant opposing external atmospheric pressure P. The work done by the gas on the surroundings is

$$w = -P\Delta V \tag{6.3}$$

where ΔV, the change in volume, is given by $V_f - V_i$. The minus sign in Equation (6.3) takes care of the sign convention for w. For gas expansion (work done *by* the

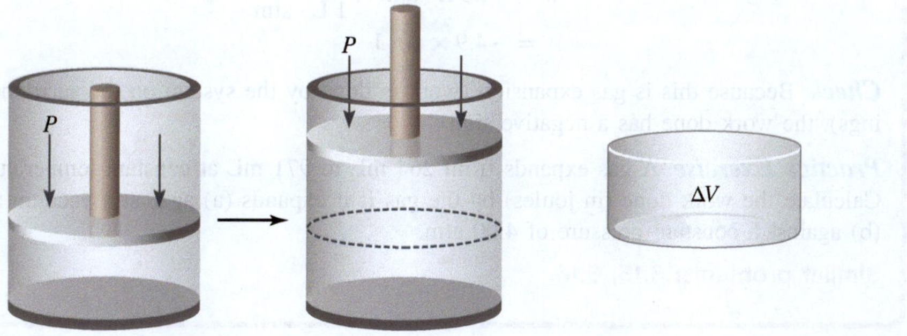

Figure 6.5 *The expansion of a gas against a constant external pressure (such as atmospheric pressure). The gas is in a cylinder fitted with a weightless movable piston. The work done is given by $-P\Delta V$. Because $\Delta V > 0$, the work done is a negative quantity.*

system), $\Delta V > 0$, so $-P\Delta V$ is a negative quantity. For gas compression (work done *on* the system), $\Delta V > 0$, and $-P\Delta V$ is a positive quantity.

Equation (6.3) derives from the fact that pressure × volume can be expressed as (force/area) × volume; that is,

$$P \times V = \underbrace{\frac{F}{d^2}}_{\text{pressure}} \times \underbrace{d^3}_{\text{volume}} = F \times d = w$$

where F is the opposing force and d has the dimension of length, d^2 has the dimensions of area, and d^3 has the dimensions of volume. Thus, the product of pressure and volume is equal to force times distance, or work. You can see that for a given increase in volume (that is, for a certain value of ΔV), the work done depends on the magnitude of the external, opposing pressure P. If P is zero (that is, if the gas is expanding against a vacuum), the work done must also be zero. If P is some positive, nonzero value, then the work done is given by $-P\Delta V$.

According to Equation (6.3), the units for work done by or on a gas are liter atmospheres. To express the work done in the more familiar unit of joules, we use the conversion factor (see Appendix 1).

$$1 \text{ L} \cdot \text{atm} = 101.3 \text{ J}$$

Example 6.1

A certain gas expands in volume from 2.0 L to 6.0 L at constant temperature. Calculate the work done by the gas if it expands (a) against a vacuum and (b) against a constant pressure of 1.2 atm.

Strategy Organize the data from the problem:

Initial volume = 2.0 L $\Delta V = (6.0 - 2.0)\text{ L} = 4.0\text{ L}$ *Final volume* = 6.0 L

(a) $P = 0$, (b) $P = 1.2$ atm.

The work done in gas expansion is equal to the product of the external, opposing pressure and the change in volume. What is the conversion factor between L · atm and J?

Solution

(a) Because the external pressure is zero, no work is done in the expansion.

$$\begin{aligned} w &= -P\Delta V \\ &= -(0)(6.0 - 2.0)\text{ L} \\ &= 0 \end{aligned}$$

(b) The external, opposing pressure is 1.2 atm, so

$$\begin{aligned} w &= -P\Delta V \\ &= -(1.2\text{ atm})(6.0 - 2.0)\text{ L} \\ &= -4.8\text{ L} \cdot \text{atm} \end{aligned}$$

To convert the answer to joules, we write

$$\begin{aligned} w &= -4.8\text{ L} \cdot \text{atm} \times \frac{101.3\text{ J}}{1\text{ L} \cdot \text{atm}} \\ &= -4.9 \times 10^2\text{ J} \end{aligned}$$

Check Because this is gas expansion (work is done by the system on the surroundings), the work done has a negative sign.

Practice Exercise A gas expands from 264 mL to 971 mL at constant temperature. Calculate the work done (in joules) by the gas if it expands (a) against a vacuum and (b) against a constant pressure of 4.00 atm.

Similar problems: 6.15, 6.16.

Example 6.1 shows that work is not a state function. Although the initial and final states are the same in (a) and (b), the amount of work done is different because the external, opposing pressures are different. We *cannot* write $\Delta w = w_f - w_i$ for a change. Work done depends not only on the initial state and final state, but also on how the process is carried out, that is, on the path.

Heat

The other component of internal energy is heat, *q*. Like work, heat is not a state function. For example, it takes 4184 J of energy to raise the temperature of 100 g of water from 20°C to 30°C. This energy can be gained (a) directly as heat energy from a Bunsen burner, without doing any work on the water; (b) by doing work on the water without adding heat energy (for example, by stirring the water with a magnetic stir bar); or (c) by some combination of the procedures described in (a) and (b). This simple illustration shows that heat associated with a given process, like work, depends on how the process is carried out. It is important to note that regardless of which procedure is taken, the change in internal energy of the system, ΔU, depends on the sum of $(q + w)$. If changing the path from the initial state to the final state increases the value of *q*, then it will decrease the value of *w* by the same amount and vice versa, so that ΔU remains unchanged.

In summary, heat and work are not state functions because they are not properties of a system. They manifest themselves only during a process (during a change). Thus, their values depend on the path of the process and vary accordingly.

Example 6.2

The work done when a gas is compressed in a cylinder like that shown in Figure 6.5 is 462 J. During this process, there is a heat transfer of 128 J from the gas to the surroundings. Calculate the energy change for this process.

Strategy Compression is work done on the gas, so what is the sign for *w*? Heat is released by the gas to the surroundings. Is this an endothermic or exothermic process? What is the sign for *q*?

Solution To calculate the energy change of the gas, we need Equation (6.1). Work of compression is positive and because heat is released by the gas, *q* is negative. Therefore, we have

$$\begin{aligned}\Delta U &= q + w \\ &= -128\text{ J} + 462\text{ J} \\ &= 334\text{ J}\end{aligned}$$

As a result, the energy of the gas increases by 334 J.

Practice Exercise A gas expands and does *P-V* work on the surroundings equal to 279 J. At the same time, it absorbs 216 J of heat from the surroundings. What is the change in energy of the system?

Similar problems: 6.17, 6.18.

Summary of Concepts & Facts

- The state of a system is defined by properties such as composition, volume, temperature, and pressure. These properties are called state functions.
- The change in a state function for a system depends only on the initial and final states of the system, and not on the path by which the change is accomplished. Energy is a state function; work and heat are not.

CHEMISTRY *in Action*

Making Snow and Inflating a Bicycle Tire

Many phenomena in everyday life can be explained by the first law of thermodynamics. Here we will discuss two examples of interest to lovers of the outdoors.

Making Snow

If you are an avid downhill skier, you have probably skied on artificial snow. How is this stuff made in quantities large enough to meet the needs of skiers on snowless days? The secret of snowmaking is in the equation $\Delta U = q + w$. A snowmaking machine contains a mixture of compressed air and water vapor at about 20 atm. Because of the large difference in pressure between the tank and the outside atmosphere, when the mixture is sprayed into the atmosphere it expands so rapidly that, as a good approximation, no heat exchange occurs between the system (air and water) and its surroundings; that is, $q = 0$. (In thermodynamics, such a process is called an *adiabatic process.*) Thus, we write

$$\Delta U = q + w = w$$

Because the system does work on the surroundings, w is a negative quantity, and there is a decrease in the system's energy.

Kinetic energy is part of the total energy of the system. In Section 5.7 we saw that the average kinetic energy of a gas is directly proportional to the absolute temperature [Equation (5.15)]. It follows, therefore, that the change in energy ΔU is given by

$$\Delta U = C\Delta T$$

where C is the proportionality constant. Because ΔU is negative, ΔT must also be negative, and it is this cooling effect (or the decrease in the kinetic energy of the water molecules) that is responsible for the formation of snow. Although we need only water to form snow, the presence of air, which also cools on expansion, helps to lower the temperature of the water vapor.

Inflating a Bicycle Tire

If you have ever pumped air into a bicycle tire, you probably noticed a warming effect at the valve stem. This phenomenon, too, can be explained by the first law of thermodynamics. The action of the pump compresses the air inside the pump and the tire. The process is rapid enough to be treated as approximately adiabatic, so that $q = 0$ and $\Delta U = w$. Because work is done on the gas in this case (it is being compressed), w is positive, and there is an increase in energy. Hence, the temperature of the system increases also, according to the equation

$$\Delta U = C\Delta T$$

A snowmaking machine in operation.
Sergey Podkolzin/Alamy Stock Photo

Review of Concepts & Facts

6.3.1 Two ideal gases at the same temperature and pressure are placed in two equal-volume containers. One container has a fixed volume, while the other is a cylinder fitted with a weightless movable piston like that shown in Figure 6.5. Initially, the gas pressures are equal to the external atmospheric pressure. The gases are then heated with a Bunsen burner. What are the signs of q and w for the gases under these conditions?

6.3.2 Calculate the work done when a gas at a pressure of 2.4 atm and constant temperature expands in volume from 1.0 L to 2.2 L.

6.3.3 What is the change in internal energy of a gas when the work done to compress the gas is 58 J, and 79 J of heat is released to the surroundings?

6.4 Enthalpy of Chemical Reactions

Learning Objectives

- Discuss enthalpy and enthalpy changes.
- Calculate the enthalpy change of a reaction and illustrate how it depends on the stoichiometry of the reactants and products.

Our next step is to see how the first law of thermodynamics can be applied to processes carried out under different conditions. Specifically, we will consider two situations most commonly encountered in the laboratory: one in which the volume of the system is kept constant and one in which the pressure applied on the system is kept constant.

If a chemical reaction is run at constant volume, then $\Delta V = 0$ and no P-V work (that is, work of gas expansion or gas compression) will result from this change ($-P\Delta V = 0$). From Equation (6.1) it follows that

$$\begin{aligned} \Delta U &= q - P\Delta V \\ &= q_v \end{aligned} \quad (6.4)$$

We add the subscript "v" to remind us that this is a constant-volume process. This equality may seem strange at first, for we showed earlier that q is not a state function. The process is carried out under constant-volume conditions, however, so that the heat change can have only a specific value, which is equal to ΔU.

Enthalpy

Constant-volume conditions are often inconvenient and sometimes impossible to achieve. Most reactions occur under conditions of constant pressure (usually atmospheric pressure). If such a reaction results in a net increase in the number of moles of a gas, then the system does work on the surroundings (expansion). This result follows from the fact that for the gas formed to enter the atmosphere, it must push the surrounding air back. Conversely, if more gas molecules are consumed than are produced, work is done on the system by the surroundings (compression). Finally, no work is done if there is no net change in the number of moles of gases from reactants to products.

In general, for a constant-pressure process we write

$$\begin{aligned} \Delta U &= q + w \\ &= q_p - P\Delta V \end{aligned}$$

or

$$q_p = \Delta U + P\Delta V \quad (6.5)$$

where the subscript "p" denotes constant-pressure condition.

We now introduce a new thermodynamic function of a system called ***enthalpy (H)***, which is defined by the equation

$$H = U + PV \quad (6.6)$$

where U is the internal energy of the system and P and V are the pressure and volume of the system, respectively. Because U and PV have energy units, enthalpy also has energy units. Furthermore, U, P, and V are all state functions; that is, the changes in $(U + PV)$ depend only on the initial and final states. It follows, therefore, that the change in H, or ΔH, also depends only on the initial and final states. Thus, H is a state function.

For any process, the change in enthalpy according to Equation (6.6) is given by

$$\Delta H = \Delta U + \Delta(PV) \quad (6.7)$$

If the pressure is held constant, then

$$\Delta H = \Delta U + P\Delta V \tag{6.8}$$

Comparing Equation (6.8) with Equation (6.5), we see that for a constant-pressure process, $q_p = \Delta H$. Again, although q is not a state function, the heat change at constant pressure is equal to ΔH because the "path" is defined and therefore it can have only a specific value.

We now have two quantities—ΔU and ΔH—that can be associated with a reaction. If the reaction occurs under constant-volume conditions, then the heat change, q_v, is equal to ΔU. On the other hand, when the reaction is carried out at constant pressure, the heat change, q_p, is equal to ΔH. In Section 6.5, we will discuss ways to measure heat changes at constant volume and constant pressure.

Enthalpy of Reactions

Because most reactions are constant-pressure processes, we can equate the heat change in these cases to the change in enthalpy. For any reaction of the type

$$\text{reactants} \longrightarrow \text{products}$$

we define the change in enthalpy, called the ***enthalpy of reaction,*** ΔH_{rxn}, *as the difference between the enthalpies of the products and the enthalpies of the reactants:*

$$\Delta H = H(\text{products}) - H(\text{reactants}) \tag{6.9}$$

The enthalpy of reaction can be positive or negative, depending on the process. For an endothermic process (heat absorbed by the system from the surroundings), ΔH is positive (that is, $\Delta H > 0$). For an exothermic process (heat released by the system to the surroundings), ΔH is negative (that is, $\Delta H < 0$).

An analogy for enthalpy change is a change in the balance in your bank account. Suppose your initial balance is \$100. After a transaction (deposit or withdrawal), the change in your bank balance, ΔX, is given by

$$\Delta X = X_{\text{final}} - X_{\text{initial}}$$

where X represents the bank balance. If you deposit \$80 into your account, then $\Delta X = \$180 - \$100 = \$80$. This corresponds to an endothermic reaction. (The balance increases and so does the enthalpy of the system.) On the other hand, a withdrawal of \$60 means $\Delta X = \$40 - \$100 = -\$60$. The negative sign of ΔX means your balance has decreased. Similarly, a negative value of ΔH reflects a decrease in enthalpy of the system as a result of an exothermic process. The difference between this analogy and Equation (6.9) is that while you always know your exact bank balance, there is no way to know the enthalpies of individual products and reactants. In practice, we can only measure the *difference* in their values. This analogy assumes that you will not overdraw your account. The enthalpy of a substance cannot be negative.

Now let us apply the idea of enthalpy changes to two common processes, the first involving a physical change, the second a chemical change.

Thermochemical Equations

At 0°C and a pressure of 1 atm, ice melts to form liquid water. Measurements show that for every mole of ice converted to liquid water under these conditions, 6.01 kilojoules (kJ) of heat energy are absorbed by the system (ice). Because the pressure is constant, the heat change is equal to the enthalpy change, ΔH. Furthermore, this is an endothermic process, as expected for the energy-absorbing change of melting ice [Figure 6.6(a)]. Therefore, ΔH is a positive quantity. The equation for this physical change is

$$H_2O(s) \longrightarrow H_2O(l) \qquad \Delta H = 6.01 \text{ kJ/mol}$$

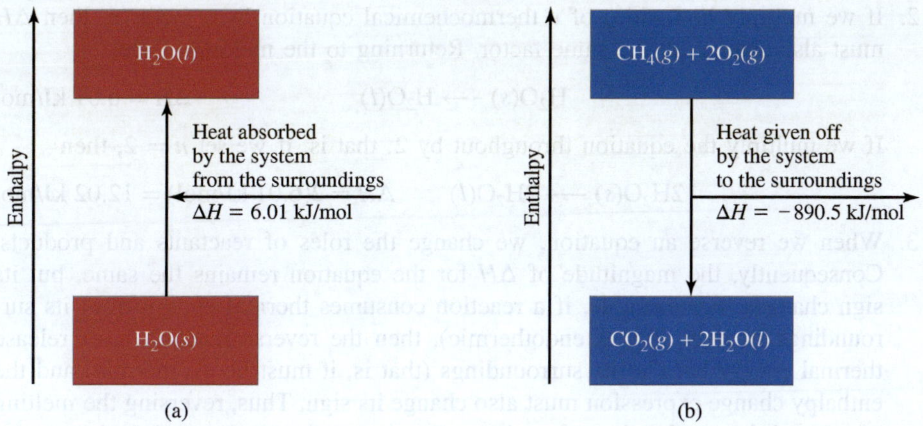

Figure 6.6 *(a) Melting 1 mole of ice at 0°C (an endothermic process) results in an enthalpy increase in the system of 6.01 kJ. (b) Burning 1 mole of methane in oxygen gas (an exothermic process) results in an enthalpy decrease in the system of 890.5 kJ. Parts (a) and (b) are not drawn to the same scale.*

The "per mole" in the unit for ΔH means that this is the enthalpy change *per mole of the reaction (or process) as it is written*—that is, when 1 mole of ice is converted to 1 mole of liquid water. For simplicity, we use "per mole" instead of "per mole of reaction" for ΔH in thermochemical equations.

As another example, consider the combustion of methane (CH_4), the principal component of natural gas:

$$CH_4(g) + 2O_2(g) \longrightarrow CO_2(g) + 2H_2O(l) \qquad \Delta H = -890.5 \text{ kJ/mol}$$

From experience we know that burning natural gas releases heat to the surroundings, so it is an exothermic process. Under a constant-pressure condition this heat change is equal to enthalpy change and ΔH must have a negative sign [Figure 6.6(b)]. Again, the per mole of reaction unit for ΔH means that when 1 mole of CH_4 reacts with 2 moles of O_2 to produce 1 mole of CO_2 and 2 moles of liquid H_2O, 890.5 kJ of heat energy are released to the surroundings. It is important to keep in mind that the ΔH value does not refer to a particular reactant or product. It simply means that the quoted ΔH value refers to all the reacting species in molar quantities. Thus, the following conversion factors can be created:

$$\frac{-890.5 \text{ kJ}}{1 \text{ mol } CH_4} \qquad \frac{-890.5 \text{ kJ}}{2 \text{ mol } O_2} \qquad \frac{-890.5 \text{ kJ}}{1 \text{ mol } CO_2} \qquad \frac{-890.5 \text{ kJ}}{2 \text{ mol } H_2O}$$

Expressing ΔH in units of kJ/mol (rather than just kJ) conforms to the standard convention; its merit will become apparent when we continue our study of thermodynamics in Chapter 17.

The equations for the melting of ice and the combustion of methane are examples of ***thermochemical equations,*** which *show the enthalpy changes as well as the mass relationships.* It is essential to specify a balanced equation when quoting the enthalpy change of a reaction. The following guidelines are helpful in writing and interpreting thermochemical equations.

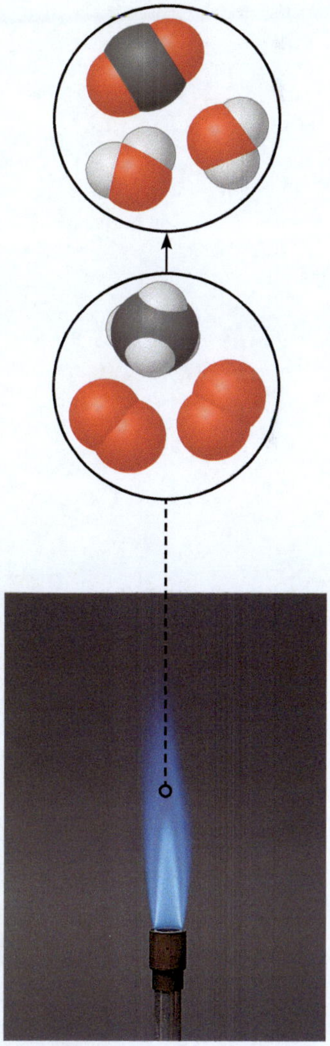

Methane gas burning from a Bunsen burner.
Stephen Frisch/McGraw Hill

1. When writing thermochemical equations, we must always specify the physical states of all reactants and products, because they help determine the actual enthalpy changes. For example, in the equation for the combustion of methane, if we show water vapor rather than liquid water as a product,

$$CH_4(g) + 2O_2(g) \longrightarrow CO_2(g) + 2H_2O(g) \qquad \Delta H = -802.5 \text{ kJ/mol}$$

the enthalpy change is −802.5 kJ rather than −890.5 kJ because 88.0 kJ are needed to convert 2 moles of liquid water to water vapor; that is,

$$2H_2O(l) \longrightarrow 2H_2O(g) \qquad \Delta H = 88.0 \text{ kJ/mol}$$

2. If we multiply both sides of a thermochemical equation by a factor n, then ΔH must also change by the same factor. Returning to the melting of ice,

$$H_2O(s) \longrightarrow H_2O(l) \qquad \Delta H = 6.01 \text{ kJ/mol}$$

If we multiply the equation throughout by 2; that is, if we set $n = 2$, then

$$2H_2O(s) \longrightarrow 2H_2O(l) \qquad \Delta H = 2(6.01 \text{ kJ/mol}) = 12.02 \text{ kJ/mol}$$

3. When we reverse an equation, we change the roles of reactants and products. Consequently, the magnitude of ΔH for the equation remains the same, but its sign changes. For example, if a reaction consumes thermal energy from its surroundings (that is, if it is endothermic), then the reverse reaction must release thermal energy back to its surroundings (that is, it must be exothermic) and the enthalpy change expression must also change its sign. Thus, reversing the melting of ice and the combustion of methane, the thermochemical equations become

$$H_2O(l) \longrightarrow H_2O(s) \qquad \Delta H = -6.01 \text{ kJ/mol}$$
$$CO_2(g) + 2H_2O(l) \longrightarrow CH_4(g) + 2O_2(g) \qquad \Delta H = 890.5 \text{ kJ/mol}$$

and what was an endothermic process becomes exothermic, and vice versa.

Example 6.3

Given the thermochemical equation

$$2SO_2(g) + O_2(g) \longrightarrow 2SO_3(g) \qquad \Delta H = -197.8 \text{ kJ/mol}$$

calculate the heat evolved when 87.9 g of SO_2 (molar mass = 64.07 g/mol) is converted to SO_3.

Strategy The thermochemical equation shows that for every 2 moles of SO_2 reacted, 198.2 kJ of heat are given off (note the negative sign). Therefore, the conversion factor is

$$\frac{-197.8 \text{ kJ}}{2 \text{ mol } SO_2}$$

How many moles of SO_2 are in 87.9 g of SO_2? What is the conversion factor between grams and moles?

Solution We need to first calculate the number of moles of SO_2 in 87.9 g of the compound and then find the number of kilojoules produced from the exothermic reaction. The sequence of conversions is as follows:

$$\text{grams of } SO_2 \longrightarrow \text{moles of } SO_2 \longrightarrow \text{kilojoules of heat generated}$$

Therefore, the enthalpy change for this reaction is given by

$$\Delta H = 87.9 \text{ g } SO_2 \times \frac{1 \text{ mol } SO_2}{64.07 \text{ g } SO_2} \times \frac{-197.8 \text{ kJ}}{2 \text{ mol } SO_2} = -136 \text{ kJ}$$

and the heat released to the surroundings is 136 kJ. Keep in mind that the ΔH for a reaction can be positive or negative, but the heat released or absorbed is always a positive quantity. The word "released" or "absorbed" gives the direction of heat transfer, so no sign is needed.

Check Because 87.9 g is less than twice the molar mass of SO_2 (2×64.07 g) as shown in the preceding thermochemical equation, we expect the heat released to be smaller than 198.2 kJ.

Practice Exercise Calculate the heat evolved when 266 g of white phosphorus (SO_2) burns in air according to the equation

$$P_4(s) + 5O_2(g) \longrightarrow P_4O_{10}(s) \qquad \Delta H = -2984 \text{ kJ/mol}$$

Similar problem: 6.26.

A Comparison of ΔH and ΔU

What is the relationship between ΔH and ΔU for a process? To find out, let us consider the reaction between sodium metal and water:

$$2\text{Na}(s) + 2\text{H}_2\text{O}(l) \longrightarrow 2\text{NaOH}(aq) + \text{H}_2(g) \qquad \Delta H = -367.6 \text{ kJ/mol}$$

This thermochemical equation says that when two moles of sodium react with an excess of water, 367.5 kJ of heat are given off. Note that one of the products is hydrogen gas, which must push back air to enter the atmosphere. Consequently, some of the energy produced by the reaction is used to do work of pushing back a volume of air (ΔV) against atmospheric pressure (P) (Figure 6.7). To calculate the change in internal energy, we rearrange Equation (6.8) as follows:

$$\Delta U = \Delta H - P\Delta V$$

Sodium reacting with water to form hydrogen gas.
Charles D. Winters/McGraw Hill

If we assume the temperature to be 25°C and ignore the small change in the volume of the solution, we can show that the volume of 1 mole of H_2 gas at 1.0 atm and 298 K is 24.5 L, so that $-P\Delta V = -24.5 \text{ L} \cdot \text{atm}$ or -2.5 kJ (recall that $1 \text{ L} \cdot \text{atm} = 101.3 \text{ J}$). Finally,

$$\begin{aligned}\Delta U &= -367.6 \text{ kJ/mol} - 2.5 \text{ kJ/mol}\\ &= -370.1 \text{ kJ/mol}\end{aligned}$$

This calculation shows that ΔU and ΔH are approximately the same. The reason ΔH is smaller than ΔU in magnitude is that some of the internal energy released is used to do gas expansion work, so less heat is evolved. For reactions that do not involve gases, ΔV is usually very small and so ΔU is practically the same as ΔH.

Another way to calculate the internal energy change of a gaseous reaction is to assume ideal gas behavior and constant temperature. In this case,

$$\begin{aligned}\Delta U &= \Delta H - \Delta(PV)\\ &= \Delta H - \Delta(nRT)\\ &= \Delta H - RT\Delta n \end{aligned} \qquad (6.10)$$

where Δn is defined as

$$\Delta n = \text{number of moles of product gases} - \text{number of moles of reactant gases}$$

and R is the gas constant $8.314 \text{ J/K} \cdot \text{mol}$. Note that the temperature must be expressed in kelvins. This equation therefore means that if the number of moles of product(s) and reactant(s) are equal, as in the reaction

$$\text{H}_2(g) + \text{Cl}_2(g) \longrightarrow 2\text{HCl}(g)$$

the difference between ΔU and ΔH is negligible because $\Delta n = 0$.

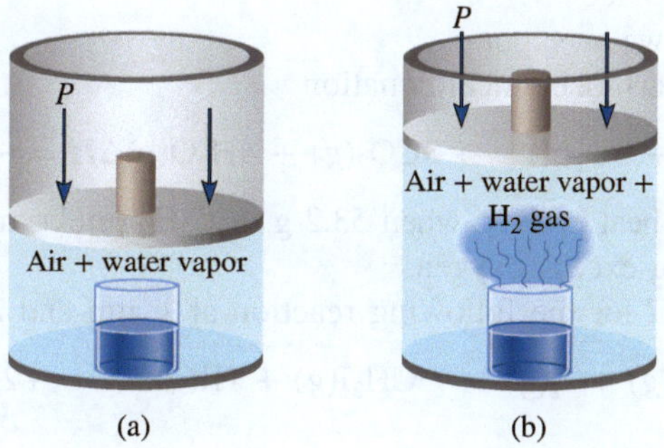

Figure 6.7 *(a) A beaker of water inside a cylinder fitted with a movable piston. The pressure inside is equal to the atmospheric pressure. (b) As the sodium metal reacts with water, the hydrogen gas generated pushes the piston upward (doing work on the surroundings) until the pressure inside is again equal to that of outside.*

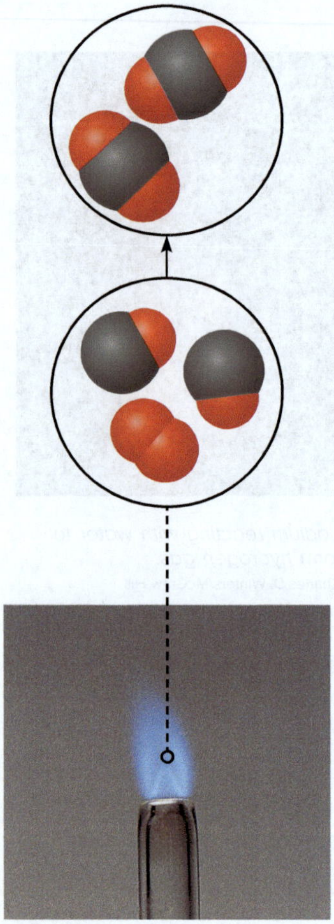

Carbon monoxide burns in air to form carbon dioxide.
Stephen Frisch/McGraw Hill

Example 6.4

Calculate the change in internal energy when 2 moles of CO are converted to 2 moles of CO_2 at 1 atm and 25°C:

$$2CO(g) + O_2(g) \longrightarrow 2CO_2(g) \qquad \Delta H = -566.0 \text{ kJ/mol}$$

Strategy We are given the enthalpy change, ΔH, for the reaction and are asked to calculate the change in internal energy, ΔU. Therefore, we need Equation (6.10). What is the change in the number of moles of gases? ΔH is given in kilojoules, so what units should we use for R?

Solution From the chemical equation we see that 3 moles of gases are converted to 2 moles of gases so that

$$\begin{aligned}\Delta n &= \text{number of moles of product gas} - \text{number of moles of reactant gases}\\ &= 2 - 3\\ &= -1\end{aligned}$$

Using 8.314 J/K · mol for R and $T = 298$ K in Equation (6.10), we write

$$\begin{aligned}\Delta U &= \Delta H - RT\Delta n\\ &= -566.0 \text{ kJ/mol} - (8.314 \text{ J/K} \cdot \text{mol})\left(\frac{1 \text{ kJ}}{1000 \text{ J}}\right)(298 \text{ K})(-1)\\ &= -563.5 \text{ kJ/mol}\end{aligned}$$

Check Knowing that the reacting gaseous system undergoes a compression (3 moles to 2 moles), is it reasonable to have $\Delta H > \Delta U$ in magnitude?

Practice Exercise What is ΔU for the formation of 1 mole of CO at 1 atm and 25°C?

$$C(\text{graphite}) + \frac{1}{2}O_2(g) \longrightarrow CO(g) \qquad \Delta H = -110.5 \text{ kJ/mol}$$

Similar problem: 6.27.

Summary of Concepts & Facts

- Enthalpy is a state function. A change in enthalpy ΔH is equal to $\Delta U + P\Delta V$ for a constant-pressure process.
- The change in enthalpy (ΔH, usually given in kilojoules) is a measure of the heat of reaction (or any other process) at constant pressure.

Review of Concepts & Facts

6.4.1 Which of the constant-pressure processes shown here has the smallest difference between ΔU and ΔH?
(a) water ⟶ water vapor
(b) water ⟶ ice
(c) ice ⟶ water vapor

6.4.2 Given the thermochemical equation

$$C_3H_8(g) + 5O_2(g) \longrightarrow 3CO_2(g) + 4H_2O(g) \Delta H = -2219.9 \text{ kJ/mol}$$

calculate the heat evolved when 53.2 g of C_3H_8 (molar mass = 44.10 g/mol) are burned in excess oxygen.

6.4.3 Calculate ΔU for the following reaction at 1 atm and 25°C:

$$CH_4(g) + I_2(g) \longrightarrow CH_3I(g) + HI(g) \Delta H = +23.0 \text{ kJ/mol}$$

6.5 Calorimetry

Learning Objectives

- Perform calorimetric calculations involving specific heat or heat capacity.
- Describe constant-pressure calorimetry.

In the laboratory, heat changes in physical and chemical processes are measured with a *calorimeter,* a closed container designed specifically for this purpose. Our discussion of ***calorimetry,*** *the measurement of heat changes,* will depend on an understanding of specific heat and heat capacity, so let us consider them first.

Specific Heat and Heat Capacity

The ***specific heat*** **(*c*)** of a substance is *the amount of heat required to raise the temperature of one gram of the substance by one degree Celsius.* It has the units J/g · °C. The ***heat capacity*** **(*C*)** of a substance is *the amount of heat required to raise the temperature of a given quantity of the substance by one degree Celsius.* Its units are J/°C. Specific heat is an intensive property, whereas heat capacity is an extensive property. The relationship between the heat capacity and specific heat of a substance is

$$C = mc \tag{6.11}$$

where *m* is the mass of the substance in grams. For example, the specific heat of water is 4.184 J/g · °C, and the heat capacity of 60.0 g of water is

$$(60.0 \text{ g})(4.184 \text{ J/g} \cdot {}^\circ\text{C}) = 251 \text{ J/}{}^\circ\text{C}$$

Table 6.2 shows the specific heat of some common substances.

If we know the specific heat and the amount of a substance, then the change in the sample's temperature (Δt) will tell us the amount of heat (q) that has been absorbed or released in a particular process. The equations for calculating the heat change are given by

$$q = mc\Delta t \tag{6.12}$$

$$q = C\Delta t \tag{6.13}$$

where Δt is the temperature change:

$$\Delta t = t_{\text{final}} - t_{\text{initial}}$$

The sign convention for q is the same as that for enthalpy change; q is positive for endothermic processes and negative for exothermic processes.

Table 6.2

The Specific Heats of Some Common Substances

Substance	Specific Heat (J/g · °C)
Al	0.897
Au	0.129
C (graphite)	0.709
C (diamond)	0.51
Cu	0.385
Fe	0.449
Hg	0.139
Pb	0.13
H_2O	4.184
C_2H_5OH (ethanol)	2.438

Student Hot Spot

Student data indicate you may struggle with specific heat. Access your eBook for additional Learning Resources on this topic.

Example 6.5

A 466-g sample of water is heated from 8.50°C to 74.60°C. Calculate the amount of heat absorbed (in kilojoules) by the water.

Strategy We know the quantity of water and the specific heat of water. With this information and the temperature rise, we can calculate the amount of heat absorbed (q).

(Continued)

Solution Using Equation (6.12), we write

$$\begin{aligned} q &= mc\Delta t \\ &= (466\text{ g})(4.184\text{ J/g}\cdot{}^\circ\text{C})(74.60^\circ\text{C} - 8.50^\circ\text{C}) \\ &= 1.29\times10^5\text{ J}\times\frac{1\text{ kJ}}{1000\text{ J}} \\ &= 129\text{ kJ} \end{aligned}$$

Check The units g and °C cancel, and we are left with the desired unit kJ. Because heat is absorbed by the water from the surroundings, it has a positive sign.

Practice Exercise An iron bar of mass 869 g cools from 94°C to 5°C. Calculate the heat released (in kilojoules) by the metal.

Similar problem: 6.33.

Constant-Volume Calorimetry

Heat of combustion is usually measured by placing a known mass of a compound in a steel container called a *constant-volume bomb calorimeter,* which is filled with oxygen at about 30 atm of pressure. "Constant volume" refers to the volume of the container, which does not change during the reaction. Note that the container remains intact after the measurement. The term "bomb calorimeter" connotes the explosive nature of the reaction (on a small scale) in the presence of excess oxygen gas. The closed bomb is immersed in a known amount of water, as shown in Figure 6.8. The sample is ignited electrically, and the heat produced by the combustion reaction can be calculated accurately by recording the rise in temperature of the water. The heat given off by the sample is absorbed by the water and the bomb. The special design of the calorimeter enables us to assume that no heat (or mass) is lost to the surroundings during the time it takes to make measurements. Therefore, we can call the bomb and the water in which it is submerged an isolated system. Because no heat enters or leaves the system throughout the process, the heat change of the system (q_{system}) must be zero and we can write

$$\begin{aligned} q_{\text{system}} &= q_{\text{cal}} + q_{\text{rxn}} \\ &= 0 \end{aligned} \tag{6.14}$$

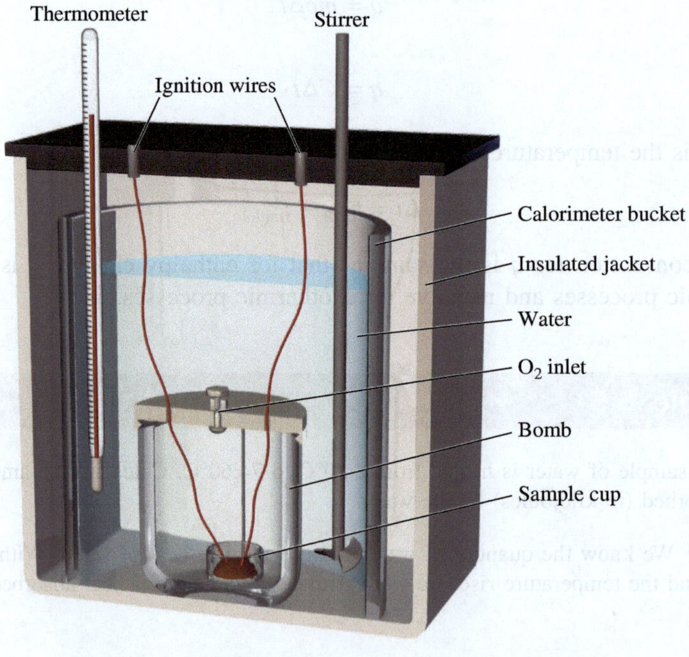

Figure 6.8 *A constant-volume bomb calorimeter. The calorimeter is filled with oxygen gas before it is placed in the bucket. The sample is ignited electrically, and the heat produced by the reaction can be accurately determined by measuring the temperature increase in the known amount of surrounding water.*

where q_{cal} and q_{rxn} are the heat changes for the calorimeter and the reaction, respectively. Thus,

$$q_{rxn} = -q_{cal} \tag{6.15}$$

To calculate q_{cal}, we need to know the heat capacity of the calorimeter (C_{cal}) (note that C_{cal} comprises both the bomb and the surrounding water) and the temperature rise, that is,

$$q_{cal} = C_{cal}\Delta t \tag{6.16}$$

The quantity C_{cal} is calibrated by burning a substance with an accurately known heat of combustion. For example, it is known that the combustion of 1 g of benzoic acid (C_6H_5COOH) releases 26.42 kJ of heat. Note that although the combustion reaction is exothermic, q_{cal} is a positive quantity because it represents the heat absorbed by the calorimeter. If the temperature rise is 4.673°C, then the heat capacity of the calorimeter is given by

$$C_{cal} = \frac{q_{cal}}{\Delta t}$$
$$= \frac{26.42 \text{ kJ}}{4.673°\text{C}} = 5.654 \text{ kJ/°C}$$

Once C_{cal} has been determined, the calorimeter can be used to measure the heat of combustion of other substances.

Note that because reactions in a bomb calorimeter occur under constant-volume rather than constant-pressure conditions, the heat changes correspond to ΔU, *not* the enthalpy change ΔH (see Section 6.4). Equation (6.10) can be used to correct the measured heat changes so that they correspond to ΔH values, but the corrections usually are quite small so we will not concern ourselves with the details here. Finally, it is interesting to note that the energy contents of food and fuel (usually expressed in calories where 1 cal = 4.184 J) are measured with constant-volume calorimeters.

Example 6.6

A quantity of 1.435 g of naphthalene ($C_{10}H_8$), a pungent-smelling substance used in moth repellents, was burned in a constant-volume bomb calorimeter. Consequently, the temperature of the water rose from 20.28°C to 25.95°C. If the heat capacity of the bomb plus water was 10.17 kJ/°C, calculate the heat of combustion of naphthalene on a molar basis; that is, find the molar heat of combustion.

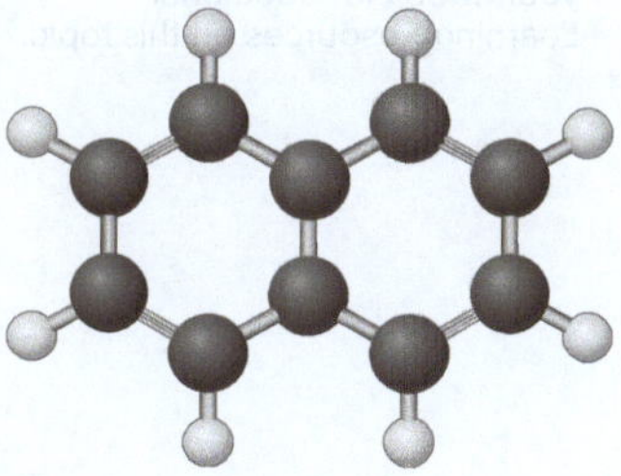

$C_{10}H_8$

Strategy Knowing the heat capacity and the temperature rise, how do we calculate the heat absorbed by the calorimeter? What is the heat generated by the combustion of 1.435 g of naphthalene? What is the conversion factor between grams and moles of naphthalene?

Solution The heat absorbed by the bomb and water is equal to the product of the heat capacity and the temperature change. From Equation (6.16), assuming no heat is lost to the surroundings, we write

$$q_{cal} = C_{cal}\Delta t$$
$$= (10.17 \text{ kJ/°C})(25.95°\text{C} - 20.28°\text{C})$$
$$= 57.66 \text{ kJ}$$

(Continued)

Because $q_{sys} = q_{cal} + q_{rxn} = 0$, $q_{cal} = -q_{rxn}$. The heat change of the reaction is −57.66 kJ. This is the heat released by the combustion of 1.435 g of $C_{10}H_8$; therefore, we can write the conversion factor as

$$\frac{-57.66\ \text{kJ}}{1.435\ \text{g}\ C_{10}H_8}$$

The molar mass of naphthalene is 128.2 g, so the heat of combustion of 1 mole of naphthalene is

$$\text{molar heat of combustion} = \frac{-57.66\ \text{kJ}}{1.435\ \text{g}\ C_{10}H_8} \times \frac{128.2\ \text{g}\ C_{10}H_8}{1\ \text{mol}\ C_{10}H_8}$$

$$= -5.151 \times 10^3\ \text{kJ/mol}$$

Check Knowing that the combustion reaction is exothermic and that the molar mass of naphthalene is much greater than 1.4 g, is the answer reasonable? Under the reaction conditions, can the heat change (−57.66 kJ) be equated to the enthalpy change of the reaction?

Practice Exercise A quantity of 1.922 g of methanol (CH_3OH) was burned in a constant-volume bomb calorimeter. Consequently, the temperature of the water rose by 4.20°C. If the heat capacity of the bomb plus water was 10.4 kJ/°C, calculate the molar heat of combustion of methanol.

Similar problem: 6.37.

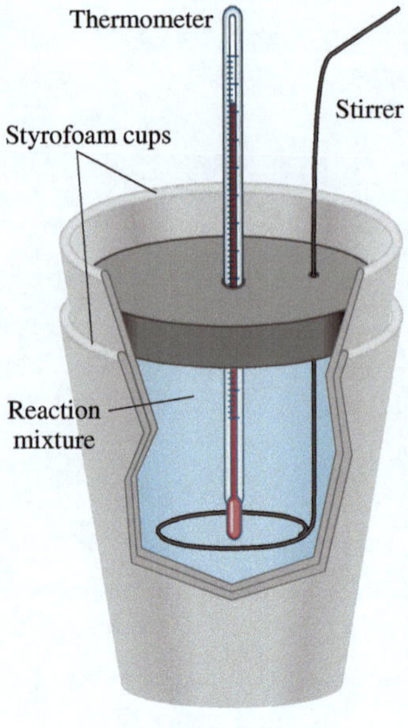

Figure 6.9 *A constant-pressure calorimeter made of two Styrofoam coffee cups. The outer cup helps to insulate the reacting mixture from the surroundings. Two solutions of known volume containing the reactants at the same temperature are carefully mixed in the calorimeter. The heat produced or absorbed by the reaction can be determined by measuring the temperature change.*

Student Hot Spot

Student data indicate you may struggle with constant-pressure calorimetry. Access your eBook for additional Learning Resources on this topic.

Constant-Pressure Calorimetry

A simpler device than the constant-volume calorimeter is the constant-pressure calorimeter, which is used to determine the heat changes for noncombustion reactions. A crude constant-pressure calorimeter can be constructed from two Styrofoam coffee cups, as shown in Figure 6.9. This device measures the heat effects of a variety of reactions, such as acid-base neutralization, as well as the heat of solution and heat of dilution. Because the pressure is constant, the heat change for the process (q_{rxn}) is equal to the enthalpy change (ΔH). As in the case of a constant-volume calorimeter, we treat the calorimeter as an isolated system. Furthermore, we neglect the small heat capacity of the coffee cups in our calculations. Table 6.3 lists some reactions that have been studied with the constant-pressure calorimeter.

Table 6.3 Heats of Some Typical Reactions Measured at Constant Pressure

Type of Reaction	Example	ΔH (kJ/mol)
Heat of neutralization	$HCl(aq) + NaOH(aq) \longrightarrow NaCl(aq) + H_2O(l)$	−58.8
Heat of ionization	$H_2O(l) \longrightarrow H^+(aq) + OH^-(aq)$	58.8
Heat of fusion	$H_2O(s) \longrightarrow H_2O(l)$	6.01
Heat of vaporization	$H_2O(l) \longrightarrow H_2O(g)$	44.0*
Heat of reaction	$MgCl_2(s) + 2Na(l) \longrightarrow 2NaCl(s) + Mg(s)$	−181.1

*Measured at 25°C. At 100°C, the value is 40.79 kJ.

CHEMISTRY *in Action*

White Fat Cells, Brown Fat Cells, and a Potential Cure for Obesity

The food we eat is broken down, or metabolized, in stages by a group of complex biological molecules called enzymes. Most of the energy released at each stage is captured for function and growth. One interesting aspect of metabolism is that the overall change in energy is the same as it is in combustion. For example, the total enthalpy change for the conversion of glucose ($C_6H_{12}O_6$) to carbon dioxide and water is the same whether we burn the substance in air or digest it in the body:

$$C_6H_{12}O_6(s) + 6CO_2(g) \longrightarrow 6CO_2(g) + 6H_2O(l)$$
$$\Delta H = -2803 \text{ kJ/mol}$$

The energy content of food is generally measured in calories. The *calorie* (*cal*) is a non-SI unit of energy that is equivalent to 4.184 J:

$$1 \text{ cal} = 4.184 \text{ J}$$

In the context of nutrition, however, the calorie we speak of (sometimes called a "big calorie") is actually equal to a kilocalorie; that is,

$$1 \text{ Cal} = 1000 \text{ cal} = 4184 \text{ J}$$

The bomb calorimeter described in Figure 6.8 is ideally suited for measuring the energy content, or "fuel values," of foods (see table).

Fuel Values of Foods

Substance	$\Delta H_{\text{combustion}}$ (kJ/g)
Apple	−2
Beef	−8
Beer	−1.5
Bread	−11
Butter	−34
Cheese	−18
Eggs	−6
Milk	−3
Potatoes	−3

The excess energy from food is stored in the body in the form of fats. Fats are a group of organic compounds (triesters of glycerol and fatty acids) that are soluble in organic solvents but insoluble in water. There are two types of fat cells called the white fat cells (WFC) and brown fat cells (BFC). The WFC are designed to store energy for use in time of need for body function. They accumulate under the skin and around internal organs and they cushion and insulate the body. Obese people have a high content of WFC in their bodies. BFC, on the other hand, contain a high concentration of mitochondria, which are specialized subunits within a cell. The main role of BFC is to burn fat molecules and generate heat. Its name is derived from the fact that mitochondria contain iron, giving the tissue a reddish-brown color. In general, women have more BFC than men.

We lose our brown fat as we age, but several studies carried out in 2009 showed that adults possess metabolically active BFC. In one experiment, PET/CT (positron emission tomography and computerized tomography) scans of 24 men exposed to cold and room temperature show that the chilly temperature activates the BFC as they burn off fat molecules to generate heat. Furthermore, it was found that lean people have more active BFC than obese people.

Mice have the same type of fat cells as humans. In 2013 it was demonstrated by genetically labeling the fat cells of mice that WFC could be converted into BFC by exposure to cold (8°C) for one week. Unfortunately, BFC were converted back to WFC a few weeks after the mice were returned to normal room temperature. A separate study suggested that a different type of BFC can be formed from WFC by exercise.

Obesity is a major health hazard in the United States. Treatments for obesity so far are focused on diet to lower the amount of energy consumed, or exercise to increase the amount of energy the body needs. Most current antiobesity drugs work on the diet half of treatment. If scientists can find a way to convert WFC to BFC by biological means, and signs are encouraging, drugs will one day be developed that would fight obesity based on energy expenditure rather than appetite. And one can accomplish this goal without having to exercise in a cold environment.

Example 6.7

A lead (Pb) pellet having a mass of 26.47 g at 89.98°C was placed in a constant-pressure calorimeter of negligible heat capacity containing 100.0 mL of water. The water temperature rose from 22.50°C to 23.05°C. What is the specific heat of the lead pellet?

(Continued)

Strategy We know the masses of water and the lead pellet as well as the initial and final temperatures. Assuming no heat is lost to the surroundings, we can equate the heat lost by the lead pellet to the heat gained by the water. Knowing the specific heat of water, we can then calculate the specific heat of lead.

Solution Treating the calorimeter as an isolated system (no heat lost to the surroundings), we write

$$q_{Pb} + q_{H_2O} = 0$$

or

$$q_{Pb} = -q_{H_2O}$$

The heat gained by the water is given by

$$q_{H_2O} = ms\Delta t$$

where m and s are the mass and specific heat, and $\Delta t = t_{final} - t_{initial}$. Therefore,

$$\begin{aligned} q_{H_2O} &= (100.0\ g)(4.184\ J/g \cdot °C)(23.05°C - 22.50°C) \\ &= 230.1\ J \end{aligned}$$

Because the heat lost by the lead pellet is equal to the heat gained by the water, $q_{Pb} = -280.3$ J. Solving for the specific heat of Pb, we write

$$\begin{aligned} q_{Pb} &= ms\Delta t \\ -230.1\ J &= (26.47\ g)(s)(23.05°C - 89.98°C) \\ s &= 0.130\ J/g \cdot °C \end{aligned}$$

Practice Exercise A 30.14-g stainless steel ball bearing at 117.82°C is placed in a constant-pressure calorimeter containing 120.0 mL of water at 18.44°C. If the specific heat of the ball bearing is 0.474 J/g · °C, calculate the final temperature of the water. Assume the calorimeter to have negligible heat capacity.

Similar problem: 6.88.

Example 6.8

A quantity of 1.00×10^2 mL of 0.500 M HCl was mixed with 1.00×10^2 mL of 0.500 M NaOH in a constant-pressure calorimeter of negligible heat capacity. The initial temperature of the HCl and NaOH solutions was the same, 22.50°C, and the final temperature of the mixed solution was 25.86°C. Calculate the heat change for the neutralization reaction on a molar basis:

$$NaOH(aq) + HCl(aq) \longrightarrow NaCl(aq) + H_2O(l)$$

Assume that the densities and specific heats of the solutions are the same as for water (1.00 g/mL and 4.184 J/g · °C, respectively).

Strategy Because the temperature rose, the neutralization reaction is exothermic. How do we calculate the heat absorbed by the combined solution? What is the heat of the reaction? What is the conversion factor for expressing the heat of reaction on a molar basis?

Solution Assuming no heat is lost to the surroundings, $q_{sys} = q_{soln} + q_{rxn} = 0$, so $q_{rxn} = -q_{soln}$, where q_{soln} is the heat absorbed by the combined solution. Because the density of the solution is 1.00 g/mL, the mass of a 100-mL solution is 100 g. Thus,

$$\begin{aligned} q_{soln} &= ms\Delta t \\ &= (1.00 \times 10^2\ g + 1.00 \times 10^2\ g)(4.184\ J/g \cdot °C)(25.86°C - 22.50°C) \\ &= 2.81 \times 10^3\ J \\ &= 2.81\ kJ \end{aligned}$$

Because $q_{rxn} = -q_{soln}$, $q_{rxn} = -2.81$ kJ.

(Continued)

From the molarities given, the number of moles of both HCl and NaOH in 1.00×10^2 mL solution is

$$\frac{0.500 \text{ mol}}{1 \text{ L}} \times 0.100 \text{ L} = 0.0500 \text{ mol}$$

Therefore, the heat of neutralization when 1.00 mole of HCl reacts with 1.00 mole of NaOH is

$$\text{heat of neutralization} = \frac{-2.81 \text{ kJ}}{0.0500 \text{ mol}} = -56.2 \text{ kJ/mol}$$

Check Is the sign consistent with the nature of the reaction? Under the reaction condition, can the heat change be equated to the enthalpy change?

Practice Exercise A quantity of 4.00×10^2 mL of 0.600 *M* HNO_3 is mixed with 4.00×10^2 mL of 0.300 *M* $Ba(OH_2)$ in a constant-pressure calorimeter of negligible heat capacity. The initial temperature of both solutions is the same at 18.46°C. What is the final temperature of the solution? (Use the result in Example 6.8 for your calculation.)

Similar problem: 6.38.

Summary of Concepts & Facts

- Constant-volume and constant-pressure calorimeters are used to measure heat changes that occur in physical and chemical processes.

Review of Concepts & Facts

6.5.1 A 1-g sample of Al and a 1-g sample of Fe are heated from 40°C to 100°C. Which metal has absorbed a greater amount of heat?

6.5.2 A 1.252-g sample of cyclohexanol ($C_6H_{12}O$) was burned in a constant-volume bomb calorimeter where the temperature of the water increased from 27.00°C to 31.11°C. If the heat capacity of the bomb plus water is 11.3 kJ/°C, what is the molar heat of combustion of cyclohexanol?

6.5.3 A 100.0-g sample of an unknown metal at 125°C is added to 500.0 g of water at 27.5°C. The final temperature of the mixture is 30.0°C. What is the specific heat capacity of the metal?

6.6 Standard Enthalpy of Formation and Reaction

Learning Objectives

- Employ standard heats of formation of substances to determine the enthalpy change of a reaction.
- Apply Hess's law to determine the enthalpy change of a reaction.

So far we have learned that we can determine the enthalpy change that accompanies a reaction by measuring the heat absorbed or released (at constant pressure). From Equation (6.9) we see that ΔH can also be calculated if we know the actual enthalpies of all reactants and products. However, as mentioned earlier, there is no way to measure the *absolute* value of the enthalpy of a substance. Only values *relative* to an arbitrary reference can be determined. This problem is similar to the one geographers face in expressing the elevations of specific mountains or valleys. Rather than trying to devise some type of "absolute" elevation scale (perhaps based on distance from the center of Earth?), by common agreement all geographic heights and depths are expressed relative

Graphite (top) and diamond (bottom).

(top): Ken Karp/McGraw Hill; (bottom): JewelryStock/Alamy Stock Photo

to sea level, an arbitrary reference with a defined elevation of "zero" meters or feet. Similarly, chemists have agreed on an arbitrary reference point for enthalpy.

The "sea level" reference point for all enthalpy expressions is called the ***standard enthalpy of formation*** **(ΔH_f°).** Substances are said to be in the ***standard state*** at 1 atm,† hence the term "standard enthalpy." The superscript "°" represents standard-state conditions (1 atm), and the subscript "f" stands for formation. By convention, *the standard enthalpy of formation of any element in its most stable form is zero.* Take the element oxygen as an example. Molecular oxygen (O_2) is more stable than the other allotropic form of oxygen, ozone (O_3), at 1 atm and 25°C. Thus, we can write $\Delta H_f^\circ(O_2) = 0$, but $\Delta H_f^\circ(O_3) = 142.7$ kJ/mol. Similarly, graphite is a more stable allotropic form of carbon than diamond at 1 atm and 25°C, so we have ΔH_f°(C, graphite) = 0 and ΔH_f°(C, diamond) = 1.90 kJ/mol. Based on this reference for elements, we can now define the standard enthalpy of formation of a compound as *the heat change that results when 1 mole of the compound is formed from its elements at a pressure of 1 atm.* Table 6.4 lists the standard enthalpies of formation for a number of elements and compounds. (For a more complete list of ΔH_f° values, see Appendix 2.) Note that although the standard state does not specify a temperature, we will always use ΔH_f° values measured at 25°C for our discussion because most of the thermodynamic data are collected at this temperature.

Table 6.4 Standard Enthalpies of Formation of Some Inorganic Substances at 25°C

Substance	ΔH_f°(kJ/mol)	Substance	ΔH_f°(kJ/mol)
Ag(*s*)	0	$H_2O_2(l)$	−187.8
AgCl(*s*)	−127.0	Hg(*l*)	0
Al(*s*)	0	$I_2(s)$	0
$Al_2O_3(s)$	−1675.7	HI(*g*)	26.5
$Br_2(l)$	0	Mg(*s*)	0
HBr(*g*)	−36.2	MgO(*s*)	−601.6
C(graphite)	0	$MgCO_3(s)$	−1095.8
C(diamond)	1.9	$N_2(g)$	0
CO(*g*)	−110.5	$NH_3(g)$	−45.9
$CO_2(g)$	−393.5	NO(*g*)	91.3
Ca(*s*)	0	$NO_2(g)$	33.2
CaO(*s*)	−634.9	$N_2O(g)$	81.6
$CaCO_3(s)$	−1207.6	$N_2O_4(g)$	11.1
$Cl_2(g)$	0	O(*g*)	294.2
HCl(*g*)	−92.3	$O_2(g)$	0
Cu(*s*)	0	$O_3(g)$	142.7
CuO(*s*)	−157.3	S(rhombic)	0
$F_2(g)$	0	S(monoclinic)	0.3
HF(*g*)	−273.3	$SO_2(g)$	−296.8
H(*g*)	218.0	$SO_3(g)$	−395.7
$H_2(g)$	0	$H_2S(g)$	−20.6
$H_2O(g)$	−241.8	Zn(*s*)	0
$H_2O(l)$	−285.8	ZnO(*s*)	−350.5

†In thermodynamics, the standard pressure is defined as 10^5 Pa, where 10^5 Pa = 1 bar = 0.987 atm. Because 1 bar differs from 1 atm by only 1.3 percent, we will continue to use 1 atm as the standard pressure. Note that the normal melting point and boiling point of a substance are defined in terms of 1 atm.

The importance of the standard enthalpies of formation is that once we know their values, we can readily calculate the ***standard enthalpy of reaction,*** ΔH°_{rxn}, defined as *the enthalpy of a reaction carried out at 1 atm.* For example, consider the hypothetical reaction

$$aA + bB \longrightarrow cC + dD$$

where *a, b, c,* and *d* are stoichiometric coefficients. For this reaction, ΔH°_{rxn} is given by

$$\Delta H^\circ_{rxn} = [c\Delta H^\circ_f(C) + d\Delta H^\circ_f(D)] - [a\Delta H^\circ_f(A) + b\Delta H^\circ_f(B)] \quad (6.17)$$

We can generalize Equation (6.17) as

$$\Delta H^\circ_{rxn} = \Sigma n\Delta H^\circ_f(\text{products}) - \Sigma m\Delta H^\circ_f(\text{reactants}) \quad (6.18)$$

where *m* and *n* denote the stoichiometric coefficients for the reactants and products, and Σ (sigma) means "the sum of." Note that in calculations, the stoichiometric coefficients are just numbers without units.

To use Equation (6.18) to calculate ΔH°_{rxn}, we must know the ΔH°_f values of the compounds that take part in the reaction. These values can be determined by applying the direct method or the indirect method (Hess's law).

The Direct Method

This method of measuring ΔH°_f works for compounds that can be readily synthesized from their elements. Suppose we want to know the enthalpy of formation of carbon dioxide. We must measure the enthalpy of the reaction when carbon (graphite) and molecular oxygen in their standard states are converted to carbon dioxide in its standard state:

$$C(\text{graphite}) + O_2(g) \longrightarrow CO_2(g) \qquad \Delta H^\circ_{rxn} = -393.5 \text{ kJ/mol}$$

We know from experience that this combustion easily goes to completion. Thus, from Equation (6.18) we can write

$$\Delta H^\circ_{rxn} = \Delta H^\circ_f(CO_2, g) - [\Delta H^\circ_f(C, \text{graphite}) + \Delta H^\circ_f(O_2, g)]$$
$$= -393.5 \text{ kJ/mol}$$

Because both graphite and O_2 are stable allotropic forms of the elements, it follows that ΔH°_f(C, graphite) and $\Delta H^\circ_f(O_2, g)$ are zero. Therefore,

$$\Delta H^\circ_{rxn} = \Delta H^\circ_f(CO_2, g) = -393.5 \text{ kJ/mol}$$

or

$$\Delta H^\circ_f(CO_2, g) = -393.5 \text{ kJ/mol}$$

Note that arbitrarily assigning zero ΔH°_f for each element in its most stable form at the standard state does not affect our calculations in any way. Remember, in thermochemistry we are interested only in enthalpy *changes* because they can be determined experimentally, whereas the absolute enthalpy values cannot. The choice of a zero "reference level" for enthalpy makes calculations easier to handle. Again referring to the terrestrial altitude analogy, we find that Mt. Everest is 8708 ft higher than Mt. McKinley (now officially known as Denali). This difference in altitude is unaffected by the decision to set sea level at 0 ft or at 1000 ft.

Other compounds that can be studied by the direct method are SF_6, P_4O_{10}, and CS_2. The equations representing their syntheses are

$$S(\text{rhombic}) + 3F_2(g) \longrightarrow SF_6(g)$$
$$P_4(\text{white}) + 5O_2(g) \longrightarrow P_4O_{10}(s)$$
$$C(\text{graphite}) + 2S(\text{rhombic}) \longrightarrow CS_2(l)$$

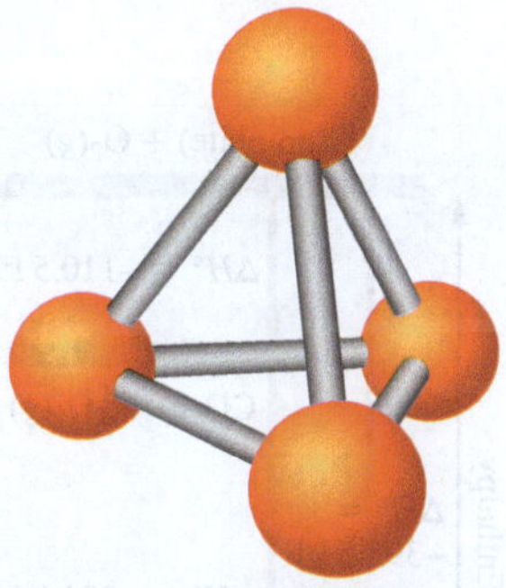

P_4

White phosphorus burns in air to form P_4O_{10}.
Charles D. Winters/McGraw Hill

Student Hot Spot

Student data indicate you may struggle with standard enthalpy of formation. Access your eBook for additional Learning Resources on this topic.

Note that S(rhombic) and P(white) are the most stable allotropes of sulfur and phosphorus, respectively, at 1 atm and 25°C, so their ΔH_f° values are zero.

The Indirect Method

Many compounds cannot be directly synthesized from their elements. In some cases, the reaction proceeds too slowly, or side reactions produce substances other than the desired compound. In these cases, ΔH_f° can be determined by an indirect approach, which is based on Hess's law of heat summation, or simply Hess's law, named after the Swiss-Russian chemist Germain Hess.† ***Hess's law*** can be stated as follows: *When reactants are converted to products, the change in enthalpy is the same whether the reaction takes place in one step or in a series of steps.* In other words, if we can break down the reaction of interest into a series of reactions for which ΔH_{rxn}° can be measured, we can calculate ΔH_{rxn}° for the overall reaction. Hess's law is based on the fact that because H is a state function, ΔH depends only on the initial and final state (that is, only on the nature of reactants and products). The enthalpy change would be the same whether the overall reaction takes place in one step or many steps.

An analogy for Hess's law is as follows: Suppose you go from the first floor to the sixth floor of a building by elevator. The gain in your gravitational potential energy (which corresponds to the enthalpy change for the overall process) is the same whether you go directly there or stop at each floor on your way up (breaking the trip into a series of steps).

Let's say we are interested in the standard enthalpy of formation of carbon monoxide (CO). We might represent the reaction as

$$\text{C(graphite)} + \tfrac{1}{2}\text{O}_2(g) \longrightarrow \text{CO}(g)$$

However, burning graphite also produces some carbon dioxide (CO_2), so we cannot measure the enthalpy change for CO directly as shown. Instead, we must employ an indirect route, based on Hess's law. It is possible to carry out the following two separate reactions, which do go to completion:

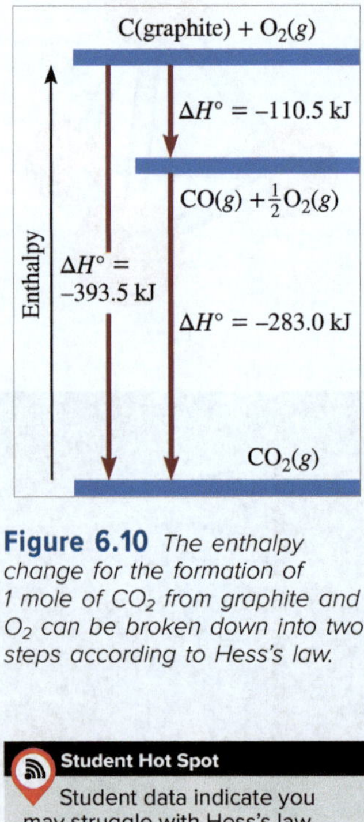

Figure 6.10 *The enthalpy change for the formation of 1 mole of CO_2 from graphite and O_2 can be broken down into two steps according to Hess's law.*

(a) $$\text{C(graphite)} + \text{O}_2(g) \longrightarrow \text{CO}_2(g) \qquad \Delta H_{rxn}^\circ = -393.5 \text{ kJ/mol}$$

(b) $$\text{CO}(g) + \tfrac{1}{2}\text{O}_2(g) \longrightarrow \text{CO}_2(g) \qquad \Delta H_{rxn}^\circ = -283.0 \text{ kJ/mol}$$

First, we reverse Equation (b) to get

(c) $$\text{CO}_2(g) \longrightarrow \text{CO}(g) + \tfrac{1}{2}\text{O}_2(g) \qquad \Delta H_{rxn}^\circ = +283.0 \text{ kJ/mol}$$

Because chemical equations can be added and subtracted just like algebraic equations, we carry out the operation (a) + (c) and obtain

(a) $$\text{C(graphite)} + \text{O}_2(g) \longrightarrow \text{CO}_2(g) \qquad \Delta H_{rxn}^\circ = -393.5 \text{ kJ/mol}$$

(c) $$\text{CO}_2(g) \longrightarrow \text{CO}(g) + \tfrac{1}{2}\text{O}_2(g) \qquad \Delta H_{rxn}^\circ = +283.0 \text{ kJ/mol}$$

(d) $$\text{C(graphite)} + \tfrac{1}{2}\text{O}_2(g) \longrightarrow \text{CO}(g) \qquad \Delta H_{rxn}^\circ = -110.5 \text{ kJ/mol}$$

Thus, $\Delta H_f^\circ(\text{CO}) = -110.5$ kJ/mol. Looking back, we see that the overall reaction is the formation of CO_2 [Equation (a)], which can be broken down into two parts [Equations (d) and (b)]. Figure 6.10 shows the overall scheme of our procedure.

The general rule in applying Hess's law is to arrange a series of chemical equations (corresponding to a series of steps) in such a way that, when added together, all species will cancel except for the reactants and products that appear in the overall reaction. This means that we want the elements on the left and the compound of interest on the right of the arrow. Further, we often need to multiply some or all of the equations representing the individual steps by the appropriate coefficients.

Student Hot Spot

Student data indicate you may struggle with Hess's law. Access your eBook for additional Learning Resources on this topic.

†Germain Henri Hess (1802–1850). Swiss-Russian chemist. Hess was born in Switzerland but spent most of his life in Russia. For formulating Hess's law, he is called the father of thermochemistry.

CHEMISTRY *in Action*

How a Bombardier Beetle Defends Itself

Survival techniques of insects and small animals in a fiercely competitive environment take many forms. For example, chameleons have developed the ability to change color to match their surroundings, and the butterfly *Limenitis* has evolved into a form that mimics the poisonous and unpleasant-tasting monarch butterfly (*Danaus*). A less passive defense mechanism is employed by bombardier beetles (*Brachinus*), which repel predators with a "chemical spray."

The bombardier beetle has a pair of glands at the tip of its abdomen. Each gland consists of two compartments. The inner compartment contains an aqueous solution of hydroquinone and hydrogen peroxide, and the outer compartment holds a mixture of enzymes. (Enzymes are biological molecules that can speed up a reaction.) When threatened, the beetle squeezes some fluid from the inner compartment into the outer compartment, where, in the presence of the enzymes, an exothermic reaction takes place:

A bombardier beetle discharging a chemical spray.
Maria Eisner

(a) $\underset{\text{hydroquinone}}{C_6H_4(OH)_2(aq)} + H_2O_2(aq) \longrightarrow \underset{\text{quinone}}{C_6H_4O_2(aq)} + 2H_2O(l)$

To estimate the heat of reaction, let us consider the following steps:

(b) $C_6H_4(OH)_2(aq) \longrightarrow C_6H_4O_2(aq) + H_2(g)$ $\quad \Delta H° = 177\text{ kJ/mol}$

(c) $H_2O_2(aq) \longrightarrow H_2O(l) + \frac{1}{2}O_2(g)$ $\quad \Delta H° = -94.6\text{ kJ/mol}$

(d) $H_2(g) + \frac{1}{2}O_2(g) \longrightarrow H_2O(l)$ $\quad \Delta H° = -286\text{ kJ/mol}$

Recalling Hess's law, we find that the heat of reaction for (a) is simply the *sum* of those for (b), (c), and (d).

Therefore, we write

$$\begin{aligned}\Delta H°_a &= \Delta H°_b + \Delta H°_c + \Delta H°_d \\ &= (177 - 94.6 - 286)\text{ kJ/mol} \\ &= -204\text{ kJ/mol}\end{aligned}$$

The large amount of heat generated is sufficient to bring the mixture to its boiling point. By rotating the tip of its abdomen, the beetle can quickly discharge the vapor in the form of a fine mist toward an unsuspecting predator. In addition to the thermal effect, the quinones also act as a repellent to other insects and animals. One bombardier beetle carries enough reagents to produce 20 to 30 discharges in quick succession, each with an audible detonation.

Example 6.9

Calculate the standard enthalpy of formation of acetylene (C_2H_2) from its elements:

$$2C(\text{graphite}) + H_2(g) \longrightarrow C_2H_2(g)$$

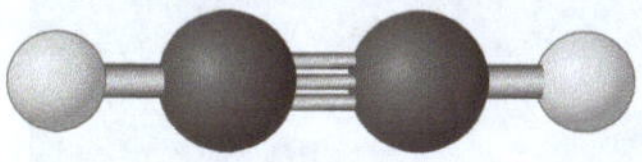

C_2H_2

The equations for each step and the corresponding enthalpy changes are

(a) $C(\text{graphite}) + O_2(g) \longrightarrow CO_2(g)$ $\quad \Delta H°_{rxn} = -393.5\text{ kJ/mol}$

(b) $H_2(g) + \frac{1}{2}O_2(g) \longrightarrow H_2O(l)$ $\quad \Delta H°_{rxn} = -285.8\text{ kJ/mol}$

(c) $2C_2H_2(g) + 5O_2(g) \longrightarrow 4CO_2(g) + 2H_2O(l)$ $\quad \Delta H°_{rxn} = -2600.4\text{ kJ/mol}$

Strategy Our goal here is to calculate the enthalpy change for the formation of C_2H_2 from its elements C and H_2. The reaction does not occur directly, however, so we must use an indirect route using the information given by Equations (a), (b), and (c).

(Continued)

An oxyacetylene torch has a high flame temperature (3000°C) and is used to weld metals.
terekhov igor/Shutterstock

Solution Looking at the synthesis of C_2H_2, we need 2 moles of graphite as reactant. So we multiply Equation (a) by 2 to get

(d) $$2C(\text{graphite}) + 2O_2(g) \longrightarrow 2CO_2(g) \qquad \Delta H^\circ_{rxn} = 2(-393.5\text{ kJ/mol}) = -787.0\text{ kJ/mol}$$

Next, we need 1 mole of H_2 as a reactant and this is provided by Equation (b). Last, we need 1 mole of C_2H_2 as a product. Equation (c) has 2 moles of C_2H_2 as a reactant so we need to reverse the equation and divide it by 2:

(e) $$2CO_2(g) + H_2O(l) \longrightarrow C_2H_2(g) + \tfrac{5}{2}O_2(g) \qquad \Delta H^\circ_{rxn} = \tfrac{1}{2}(2600.4\text{ kJ/mol}) = 1300.2\text{ kJ/mol}$$

Adding Equations (d), (b), and (e) together, we get

$$\begin{array}{lr} 2C(\text{graphite}) + 2O_2(g) \longrightarrow 2CO_2(g) & \Delta H^\circ_{rxn} = -787.0\text{ kJ/mol} \\ H_2(g) + \tfrac{1}{2}O_2(g) \longrightarrow H_2O(l) & \Delta H^\circ_{rxn} = -285.8\text{ kJ/mol} \\ 2CO_2(g) + H_2O(l) \longrightarrow C_2H_2(g) + \tfrac{5}{2}O_2(g) & \Delta H^\circ_{rxn} = 1300.2\text{ kJ/mol} \\ \hline 2C(\text{graphite}) + H_2(g) \longrightarrow C_2H_2(g) & \Delta H^\circ_{rxn} = 227.4\text{ kJ/mol} \end{array}$$

Therefore, $\Delta H^\circ_f = \Delta H^\circ_{rxn} = 227.4$ kJ/mol. The ΔH°_f value means that when 1 mole of C_2H_2 is synthesized from 2 moles of C(graphite) and 1 mole of H_2, 227.4 kJ of heat are absorbed by the reacting system from the surroundings. Thus, this is an endothermic process.

Practice Exercise Calculate the standard enthalpy of formation of carbon disulfide (CS_2) from its elements, given that

$$\begin{array}{lr} C(\text{graphite}) + O_2(g) \longrightarrow CO_2(g) & \Delta H^\circ_{rxn} = -393.5\text{ kJ/mol} \\ S(\text{rhombic}) + O_2(g) \longrightarrow SO_2(g) & \Delta H^\circ_{rxn} = -296.8\text{ kJ/mol} \\ CS_2(l) + 3O_2(g) \longrightarrow CO_2(g) + 2SO_2(g) & \Delta H^\circ_{rxn} = -1076.1\text{ kJ/mol} \end{array}$$

Similar problems: 6.62, 6.63.

We can calculate the enthalpy of reactions from the values of ΔH°_f, as shown in Example 6.10.

Example 6.10

The thermite reaction involves aluminum and iron(III) oxide

$$2Al(s) + Fe_2O_3(s) \longrightarrow Al_2O_3(s) + 2Fe(l)$$

This reaction is highly exothermic and the liquid iron formed is used to weld metals. Calculate the heat released in kilojoules per gram of Al reacted with Fe_2O_3. The ΔH°_f for Fe(l) is 12.40 kJ/mol.

The molten iron formed in a thermite reaction is run down into a mold between the ends of two railroad rails. On cooling, the rails are welded together.
Courtesy of Orgo-Thermit, Inc

Strategy The enthalpy of a reaction is the difference between the sum of the enthalpies of the products and the sum of the enthalpies of the reactants. The enthalpy of each species (reactant or product) is given by its stoichiometric coefficient times the standard enthalpy of formation of the species.

Solution Using the given ΔH°_f value for Fe(l) and other ΔH°_f values in Appendix 2 and Equation (6.18), we write

$$\begin{aligned} \Delta H^\circ_{rxn} &= [\Delta H^\circ_f(Al_2O_3) + 2\Delta H^\circ_f(Fe)] - [2\Delta H^\circ_f(Al) + \Delta H^\circ_f(Fe_2O_3)] \\ &= [(-1675.7\text{ kJ/mol}) + 2(12.40\text{ kJ/mol})] - [2(0) + (-824.2\text{ kJ/mol})] \\ &= -826.7\text{ kJ/mol} \end{aligned}$$

(Continued)

This is the amount of heat released for 2 moles of Al reacted. We use the ratio

$$\frac{-826.7 \text{ kJ}}{2 \text{ mol Al}}$$

to convert to kJ/g Al. The molar mass of Al is 26.98 g, so

$$\text{heat released per gram of Al} = \frac{-826.7 \text{ kJ}}{2 \text{ mol Al}} \times \frac{1 \text{ mol Al}}{26.98 \text{ g Al}}$$
$$= -15.32 \text{ kJ/g}$$

Check Is the negative sign consistent with the exothermic nature of the reaction? As a quick check, we see that 2 moles of Al weigh about 54 g and give off about 826 kJ of heat when reacted with Fe_2O_3. Therefore, the heat given off per gram of Al reacted is approximately −830 kJ/54 g or −15.4 kJ/g.

Practice Exercise Benzene (C_6H_6) burns in air to produce carbon dioxide and liquid water. Calculate the heat released (in kilojoules) per gram of the compound reacted with oxygen. The standard enthalpy of formation of benzene is 49.1 kJ/mol.

Similar problems: 6.54, 6.57.

Summary of Concepts & Facts

- The standard enthalpy of a reaction can be calculated from the standard enthalpies of formation of reactants and products.
- Hess's law states that the overall enthalpy change in a reaction is equal to the sum of enthalpy changes for individual steps in the overall reaction.

Review of Concepts & Facts

6.6.1 Which of the following does not have $\Delta H_f^\circ = 0$ at 25°C?

$N_2(g)$ $Cu(s)$ $Kr(g)$ $Hg(s)$ $H_2(g)$ $I_2(s)$

6.6.2 Explain why reactions involving reactant compounds with positive ΔH_f° values are generally more exothermic than those with negative ΔH_f° values.

6.6.3 Using data from Appendix 2, calculate ΔH_{rxn}° for

$$N_2O(g) + 3H_2(g) \longrightarrow N_2H_4(l) + H_2O(l)$$

6.6.4 Given the following information

$$Pb(s) + PbO_2(s) + 2H_2SO_4(l) \longrightarrow 2PbSO_4(s) + 2H_2O(l) \quad \Delta H_{rxn}^\circ = -506.2 \text{ kJ/mol}$$
$$SO_3(g) + H_2O(l) \longrightarrow H_2SO_4(l) \quad \Delta H_{rxn}^\circ = -132.5 \text{ kJ/mol}$$

determine ΔH_{rxn}° for $Pb(s) + PbO_2(s) + 2SO_3(g) \longrightarrow 2PbSO_4(s)$.

6.7 Heat of Solution and Dilution

Learning Objective

- Appraise the physical processes of heat of solution and heat of dilution.

Although we have focused so far on the thermal energy effects resulting from chemical reactions, many physical processes, such as the melting of ice or the condensation of a vapor, also involve the absorption or release of heat. Enthalpy changes occur as well when a solute dissolves in a solvent or when a solution is diluted. Let us look at these two related physical processes, involving heat of solution and heat of dilution.

Heat of Solution

In the vast majority of cases, dissolving a solute in a solvent produces measurable heat change. At constant pressure, the heat change is equal to the enthalpy change. The ***heat of solution,*** or ***enthalpy of solution,*** ΔH_{soln}, is *the heat generated or absorbed when a certain amount of solute dissolves in a certain amount of solvent.* The quantity ΔH_{soln} represents the difference between the enthalpy of the final solution and the enthalpies of its original components (that is, solute and solvent) before they are mixed. Thus,

$$\Delta H_{soln} = H_{soln} - H_{components} \quad (6.19)$$

Neither H_{soln} nor $H_{components}$ can be measured, but their difference, ΔH_{soln}, can be readily determined in a constant-pressure calorimeter. Like other enthalpy changes, ΔH_{soln} is positive for endothermic (heat-absorbing) processes and negative for exothermic (heat-generating) processes.

Consider the heat of solution of a process in which an ionic compound is the solute and water is the solvent. For example, what happens when solid NaCl dissolves in water? In solid NaCl, the Na^+ and Cl^- ions are held together by strong positive-negative (electrostatic) forces, but when a small crystal of NaCl dissolves in water, the three-dimensional network of ions breaks into its individual units. (The structure of solid NaCl is shown in Figure 2.13.) The separated Na^+ and Cl^- ions are stabilized in solution by their interaction with water molecules (see Figure 4.2). These ions are said to be *hydrated.* In this case, water plays a role similar to that of a good electrical insulator. Water molecules shield the ions (Na^+ and Cl^-) from each other and effectively reduce the electrostatic attraction that held them together in the solid state. The heat of solution is defined by the following process:

$$\text{NaCl}(s) \xrightarrow{H_2O} \text{Na}^+(aq) + \text{Cl}^-(aq) \qquad \Delta H_{soln} = ?$$

Dissolving an ionic compound such as NaCl in water involves complex interactions among the solute and solvent species. However, for the sake of analysis we can imagine that the solution process takes place in two separate steps, illustrated in Figure 6.11. First, the Na^+ and Cl^- ions in the solid crystal are separated from each other and converted to the gaseous state:

$$\text{energy} + \text{NaCl}(s) \longrightarrow \text{Na}^+(g) + \text{Cl}^-(g)$$

The energy required to completely separate one mole of a solid ionic compound into gaseous ions is called ***lattice energy (U).*** The word "lattice" describes arrangement in space of isolated points (occupied by ions) in a regular pattern. Lattice energy is a positive quantity. Beware that lattice energy and internal energy share the same symbol. The lattice energy of NaCl is 788 kJ/mol. In other words, we would need to supply 788 kJ of energy to break 1 mole of solid NaCl into 1 mole of Na^+ ions and 1 mole of Cl^- ions.

Next, the "gaseous" Na^+ and Cl^- ions enter the water and become hydrated:

$$\text{Na}^+(g) + \text{Cl}^-(g) \xrightarrow{H_2O} \text{Na}^+(aq) + \text{Cl}^-(aq) + \text{energy}$$

The enthalpy change associated with the hydration process is called the ***heat of hydration,*** ΔH_{hydr} (heat of hydration is a negative quantity for cations and anions). Applying Hess's law, it is possible to consider ΔH_{soln} as the sum of two related quantities, lattice energy (U) and heat of hydration (ΔH_{hydr}), as shown in Figure 6.11:

$$\Delta H_{soln} = U + \Delta H_{hydr} \quad (6.20)$$

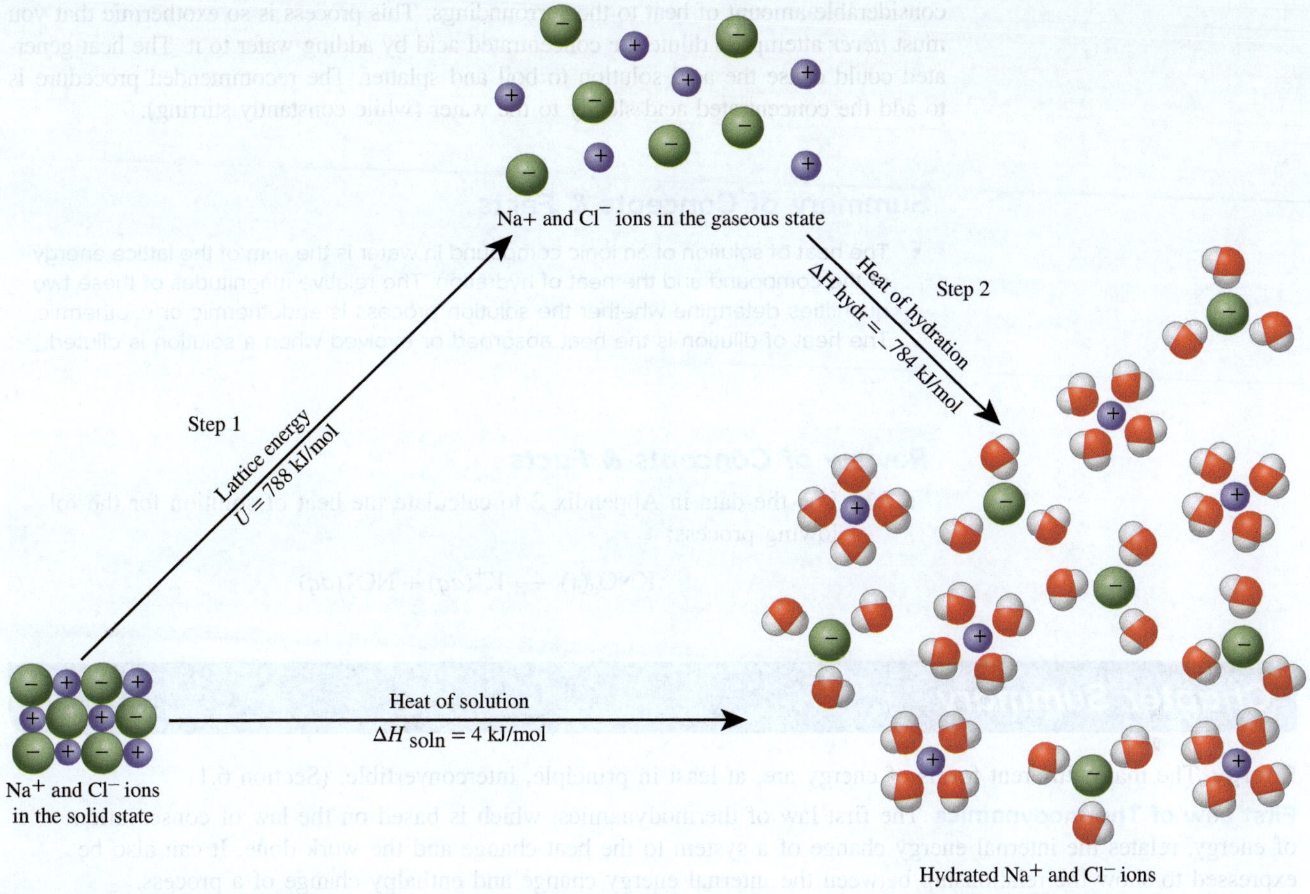

Figure 6.11 *The solution process for NaCl. The process can be considered to occur in two separate steps: (1) separation of ions from the crystal state to the gaseous state and (2) hydration of the gaseous ions. The heat of solution is equal to the energy changes for these two steps,* $\Delta H_{soln} = U + \Delta H_{hydr}$.

Therefore,

$$NaCl(s) \longrightarrow Na^+(g) + Cl^-(g) \qquad U = 788 \text{ kJ/mol}$$
$$Na^+(g) + Cl^-(g) \xrightarrow{H_2O} Na^+(aq) + Cl^-(aq) \qquad \Delta H_{hydr} = -784 \text{ kJ/mol}$$
$$NaCl(s) \xrightarrow{H_2O} Na^+(aq) + Cl^-(aq) \qquad \Delta H_{soln} = 4 \text{ kJ/mol}$$

Thus, when 1 mole of NaCl dissolves in water, 4 kJ of heat will be absorbed from the surroundings. We would observe this effect by noting that the beaker containing the solution becomes slightly colder. Table 6.5 lists the ΔH_{soln} of several ionic compounds. Depending on the nature of the cation and anion involved, ΔH_{soln} for an ionic compound may be either negative (exothermic) or positive (endothermic).

Table 6.5

Heats of Solution of Some Ionic Compounds

Compound	ΔH_{soln} (kJ/mol)
LiCl	−37.1
$CaCl_2$	−82.8
NaCl	4.0
KCl	17.2
NH_4Cl	15.2
NH_4NO_3	26.6

Heat of Dilution

When a previously prepared solution is *diluted,* that is, when more solvent is added to lower the overall concentration of the solute, additional heat is usually given off or absorbed. The ***heat of dilution*** is *the heat change associated with the dilution process.* If a certain solution process is endothermic and the solution is subsequently diluted, *more* heat will be absorbed by the same solution from the surroundings. The converse holds true for an exothermic solution process—more heat will be liberated if additional solvent is added to dilute the solution. Therefore, always be cautious when working on a dilution procedure in the laboratory. Because of its highly exothermic heat of dilution, concentrated sulfuric acid (H_2SO_4) poses a particularly hazardous problem if its concentration must be reduced by mixing it with additional water. Concentrated H_2SO_4 is composed of 98 percent acid and 2 percent water by mass. Diluting it with water releases a considerable amount of heat to the surroundings. This process is so exothermic that you must *never* attempt to dilute the concentrated acid by adding water to it. The heat generated could cause the acid solution to boil and splatter. The recommended procedure is to add the concentrated acid slowly to the water (while constantly stirring).

Summary of Concepts & Facts

- The heat of solution of an ionic compound in water is the sum of the lattice energy of the compound and the heat of hydration. The relative magnitudes of these two quantities determine whether the solution process is endothermic or exothermic. The heat of dilution is the heat absorbed or evolved when a solution is diluted.

Review of Concepts & Facts

6.7.1 Use the data in Appendix 2 to calculate the heat of solution for the following process:

$$KNO_3(s) \longrightarrow K^+(aq) + NO_3^-(aq)$$

Chapter Summary

Energy The many different forms of energy are, at least in principle, interconvertible. (Section 6.1)

First Law of Thermodynamics The first law of thermodynamics, which is based on the law of conservation of energy, relates the internal energy change of a system to the heat change and the work done. It can also be expressed to show the relationship between the internal energy change and enthalpy change of a process. (Sections 6.2, 6.3, and 6.4)

Thermochemistry Most chemical reactions involve the absorption or release of heat. At constant pressure, the heat change is equal to the enthalpy change. The heat change is measured by a calorimeter. Constant-pressure and constant-volume calorimeters are devices for measuring heat changes under the stated conditions. (Section 6.5)

Standard Enthalpy of Reaction Standard enthalpy of reaction is the enthalpy change when the reaction is carried out at 1 atm pressure. It can be calculated from the standard enthalpies of formation of reactants and products. Hess's law enables us to measure the standard enthalpy of formation of a compound in an indirect way. (Sections 6.6 and 6.7)

Key Equations

$\Delta U = q + w$ (6.1)	Mathematical statement of the first law of thermodynamics.
$w = -P\Delta V$ (6.3)	Calculating work done in gas expansion or gas compression.
$H = U + PV$ (6.6)	Definition of enthalpy.
$\Delta H = \Delta U + P\Delta V$ (6.8)	Calculating enthalpy (or energy) change for a constant-pressure process.
$C = mc$ (6.11)	Definition of heat capacity.
$q = mc\Delta t$ (6.12)	Calculating heat change in terms of specific heat.
$q = C\Delta t$ (6.13)	Calculating heat change in terms of heat capacity.
$\Delta H^\circ_{rxn} = \Sigma n\Delta H^\circ_f(\text{products}) - \Sigma m\Delta H^\circ_f(\text{reactants})$ (6.18)	Calculating standard enthalpy of reaction.
$\Delta H_{soln} = U + \Delta H_{hydr}$ (6.20)	Lattice energy and hydration contributions to heat of solution.

Key Words

Calorimetry, p. 247	Exothermic process, p. 233	Lattice energy (U), p. 260	Standard state, p. 254
Chemical energy, p. 231	First law of thermodynamics, p. 235	Law of conservation of energy, p. 232	State function, p. 235
Closed system, p. 233	Heat, p. 232	Open system, p. 232	State of a system, p. 235
Endothermic process, p. 233	Heat capacity (C), p. 247	Potential energy, p. 231	Surroundings, p. 232
Electrostatic energy, p. 231	Heat of dilution, p. 261	Radiant energy, p. 231	System, p. 232
Energy, p. 231	Heat of hydration (ΔH_{hydr}), p. 260	Specific heat (c), p. 247	Thermal energy, p. 231
Enthalpy (H), p. 241	Heat of solution (ΔH_{soln}), p. 260	Standard enthalpy of formation (ΔH_f°), p. 254	Thermochemical equation, p. 243
Enthalpy of reaction (ΔH_{rxn}), p. 242	Hess's law, p. 256	Standard enthalpy of reaction (ΔH_{rxn}°), p. 255	Thermochemistry, p. 232
Enthalpy of solution (ΔH_{soln}), p. 260	Isolated system, p. 233		Thermodynamics, p. 234
			Work, p. 231

Questions & Problems

Red numbered problems solved in Student Solutions Manual

6.1 The Nature of Energy and Types of Energy

Review Questions

6.1 Define these terms: *system, surroundings, open system, closed system, isolated system, thermal energy, chemical energy, potential energy, kinetic energy, law of conservation of energy.*

6.2 What is heat? How does heat differ from thermal energy? Under what condition is heat transferred from one system to another?

6.3 What are the units for energy commonly employed in chemistry?

6.4 A truck initially traveling at 60 km per hour is brought to a complete stop at a traffic light. Does this change violate the law of conservation of energy? Explain.

6.5 These are various forms of energy: chemical, heat, light, mechanical, and electrical. Suggest ways of interconverting these forms of energy.

6.6 Describe the interconversions of forms of energy occurring in these processes: (a) You throw a softball up into the air and catch it. (b) You switch on a flashlight. (c) You ride the ski lift to the top of the hill and then ski down. (d) You strike a match and let it burn down.

6.2 Energy Changes in Chemical Reactions

Review Questions

6.7 Define these terms: *thermochemistry, exothermic process, endothermic process.*

6.8 Stoichiometry is based on the law of conservation of mass. On what law is thermochemistry based?

6.9 Describe two exothermic processes and two endothermic processes.

6.10 Decomposition reactions are usually endothermic, whereas combination reactions are usually exothermic. Give a qualitative explanation for these trends.

6.3 Introduction to Thermodynamics

Review Questions

6.11 On what law is the first law of thermodynamics based? Explain the sign conventions in the equation $\Delta U = q + w$.

6.12 Explain what is meant by a state function. Give two examples of quantities that are state functions and two that are not.

6.13 The internal energy of an ideal gas depends only on its temperature. Do a first-law analysis of this process. A sample of an ideal gas is allowed to expand at constant temperature against atmospheric pressure. (a) Does the gas do work on its surroundings? (b) Is there heat exchange between the system and the surroundings? If so, in which direction? (c) What is ΔU for the gas for this process?

6.14 Consider these changes:

(a) $Hg(l) \longrightarrow Hg(g)$

(b) $3O_2(g) \longrightarrow 2O_3(g)$

(c) $CuSO_4 \cdot 5H_2O(s) \longrightarrow CuSO_4(s) + 5H_2O(g)$

(d) $H_2(g) + F_2(g) \longrightarrow 2HF(g)$

At constant pressure, in which of the reactions is work done by the system on the surroundings? By the surroundings on the system? In which of them is no work done?

Problems

6.15 A sample of nitrogen gas expands in volume from 1.6 L to 5.4 L at constant temperature. Calculate the work done in joules if the gas expands (a) against a vacuum, (b) against a constant pressure of 0.80 atm, and (c) against a constant pressure of 3.7 atm.

6.16 A gas expands in volume from 26.7 mL to 89.3 mL at constant temperature. Calculate the work done (in joules) if the gas expands (a) against a vacuum, (b) against a constant pressure of 1.5 atm, and (c) against a constant pressure of 2.8 atm.

6.17 A gas expands and does *P-V* work on the surroundings equal to 325 J. At the same time, it absorbs 127 J of heat from the surroundings. Calculate the change in energy of the gas.

6.18 The work done to compress a gas is 74 J. As a result, 26 J of heat is given off to the surroundings. Calculate the change in energy of the gas.

6.19 Calculate the work done when 50.0 g of tin dissolves in excess acid at 1.00 atm and 25°C:

$$\mathrm{Sn}(s) + 2\mathrm{H}^+(aq) \longrightarrow \mathrm{Sn}^{2+}(aq) + \mathrm{H}_2(g)$$

Assume ideal gas behavior.

6.20 Calculate the work done in joules when 1.0 mole of water vaporizes at 1.0 atm and 100°C. Assume that the volume of liquid water is negligible compared with that of steam at 100°C, and ideal gas behavior.

6.4 Enthalpy of Chemical Reactions

Review Questions

6.21 Define these terms: *enthalpy, enthalpy of reaction.* Under what condition is the heat of a reaction equal to the enthalpy change of the same reaction?

6.22 In writing thermochemical equations, why is it important to indicate the physical state (that is, gaseous, liquid, solid, or aqueous) of each substance?

6.23 Explain the meaning of this thermochemical equation:

$$4\mathrm{NH}_3(g) + 5\mathrm{O}_2(g) \longrightarrow 4\mathrm{NO}(g) + 6\mathrm{H}_2\mathrm{O}(g) \qquad \Delta H = -902 \text{ kJ/mol}$$

6.24 Consider this reaction:

$$2\mathrm{CH}_3\mathrm{OH}(l) + 3\mathrm{O}_2(g) \longrightarrow 4\mathrm{H}_2\mathrm{O}(l) + 2\mathrm{CO}_2(g) \qquad \Delta H = -1451.8 \text{ kJ/mol}$$

What is the value of ΔH if (a) the equation is multiplied throughout by 2, (b) the direction of the reaction is reversed so that the products become the reactants and vice versa, (c) water vapor instead of liquid water is formed as the product?

Problems

6.25 The first step in the industrial recovery of zinc from the zinc sulfide ore is roasting, that is, the conversion of ZnS to ZnO by heating:

$$2\mathrm{ZnS}(s) + 3\mathrm{O}_2(g) \longrightarrow 2\mathrm{ZnO}(s) + 2\mathrm{SO}_2(g) \qquad \Delta H = -883 \text{ kJ/mol}$$

Calculate the heat evolved (in kJ) per gram of ZnS roasted.

6.26 Determine the amount of heat (in kJ) given off when 1.26×10^4 g of NO_2 are produced according to the equation

$$2\mathrm{NO}(g) + \mathrm{O}_2(g) \longrightarrow 2\mathrm{NO}_2(g) \qquad \Delta H = -116.2 \text{ kJ/mol}$$

6.27 Consider the reaction

$$2\mathrm{H}_2\mathrm{O}(g) \longrightarrow 2\mathrm{H}_2(g) + \mathrm{O}_2(g) \qquad \Delta H = 483.6 \text{ kJ/mol}$$

If 2.0 moles of $H_2O(g)$ are converted to $H_2(g)$ and $O_2(g)$ against a pressure of 1.0 atm at 125°C, what is ΔU for this reaction?

6.28 Consider the reaction

$$\mathrm{H}_2(g) + \mathrm{Cl}_2(g) \longrightarrow 2\mathrm{HCl}(g) \qquad \Delta H = -184.6 \text{ kJ/mol}$$

If 3 moles of H_2 react with 3 moles of Cl_2 to form HCl, calculate the work done (in joules) against a pressure of 1.0 atm at 25°C. What is ΔU for this reaction? Assume the reaction goes to completion.

6.5 Calorimetry

Review Questions

6.29 What is the difference between specific heat and heat capacity? What are the units for these two quantities? Which is the intensive property and which is the extensive property?

6.30 Define calorimetry and describe two commonly used calorimeters. In a calorimetric measurement, why is it important that we know the heat capacity of the calorimeter? How is this value determined?

Problems

6.31 Consider the following data:

Metal	Al	Cu
Mass (g)	10	30
Specific heat (J/g · °C)	0.897	0.385
Temperature (°C)	40	60

When these two metals are placed in contact, which of the following will take place?

(a) Heat will flow from Al to Cu because Al has a larger specific heat.

(b) Heat will flow from Cu to Al because Cu has a larger mass.

(c) Heat will flow from Cu to Al because Cu has a larger heat capacity.

(d) Heat will flow from Cu to Al because Cu is at a higher temperature.

(e) No heat will flow in either direction.

6.32 A piece of silver of mass 362 g has a heat capacity of 85.7 J/°C. What is the specific heat of silver?

6.33 A 6.22-kg piece of copper metal is heated from 20.5°C to 324.3°C. Calculate the heat absorbed (in kJ) by the metal.

6.34 Calculate the amount of heat liberated (in kJ) from 366 g of mercury when it cools from 77.0°C to 12.0°C.

6.35 A sheet of gold weighing 10.0 g and at a temperature of 18.0°C is placed flat on a sheet of iron weighing 20.0 g and at a temperature of 55.6°C. What is the final temperature of the combined metals? Assume that no heat is lost to the surroundings. (*Hint:* The heat gained by the gold must be equal to the heat lost by the iron. The specific heats of the metals are given in Table 6.2.)

6.36 To a sample of water at 23.4°C in a constant-pressure calorimeter of negligible heat capacity is added a 12.1-g piece of aluminum whose temperature is 81.7°C. If the final temperature of water is 24.9°C, calculate the mass of the water in the calorimeter. (*Hint:* See Table 6.2.)

6.37 A 0.1375-g sample of solid magnesium is burned in a constant-volume bomb calorimeter that has a heat capacity of 3024 J/°C. The temperature increases by 1.126°C. Calculate the heat given off by the burning Mg, in kJ/g and in kJ/mol.

6.38 A quantity of 85.0 mL of 0.900 *M* HCl is mixed with 85.0 mL of 0.900 *M* KOH in a constant-pressure calorimeter that has a heat capacity of 325 J/°C. If the initial temperatures of both solutions are the same at 18.24°C, what is the final temperature of the mixed solution? The heat of neutralization is −56.2 kJ/mol. Assume the density and specific heat of the solutions are the same as those for water.

6.6 Standard Enthalpy of Formation and Reaction

Review Questions

6.39 What is meant by the standard-state condition?

6.40 How are the standard enthalpies of an element and a compound determined?

6.41 What is meant by the standard enthalpy of a reaction?

6.42 Write the equation for calculating the enthalpy of a reaction. Define all the terms.

6.43 State Hess's law. Explain, with one example, the usefulness of this law in thermochemistry.

6.44 Describe how chemists use Hess's law to determine the ΔH_f° of a compound by measuring its heat (enthalpy) of combustion.

Problems

6.45 Which of the following standard enthalpy of formation values is not zero at 25°C? $Na(s)$, $Ne(g)$, $CH_4(g)$, $S_8(s)$, $Hg(l)$, $H(g)$.

6.46 The ΔH_f° values of the two allotropes of oxygen, O_2 and O_3, are 0 and 142.7 kJ/mol, respectively, at 25°C. Which is the more stable form at this temperature?

6.47 Which is the more negative quantity at 25°C: ΔH_f° for $H_2O(l)$ or ΔH_f° for $H_2O(g)$?

6.48 Predict the value of ΔH_f° (greater than, less than, or equal to zero) for these elements at 25°C: (a) $Br_2(g)$; $Br_2(l)$. (b) $I_2(g)$; $I_2(s)$.

6.49 In general, compounds with negative ΔH_f° values are more stable than those with positive ΔH_f° values. $H_2O_2(l)$ has a negative ΔH_f° (see Table 6.4). Why, then, does $H_2O_2(l)$ have a tendency to decompose to $H_2O(l)$ and $O_2(g)$?

6.50 Suggest ways (with appropriate equations) that would enable you to measure the ΔH_f° values of $Ag_2O(s)$ and $CaCl_2(s)$ from their elements. No calculations are necessary.

6.51 Calculate the heat of decomposition for this process at constant pressure and 25°C:

$$CaCO_3(s) \longrightarrow CaO(s) + CO_2(g)$$

(Look up the standard enthalpy of formation of the reactant and products in Table 6.4.)

6.52 The standard enthalpies of formation of ions in aqueous solutions are obtained by arbitrarily assigning a value of zero to H^+ ions; that is, $\Delta H_f^\circ[H^+(aq)] = 0$.

(a) For the following reaction

$$HCl(g) \xrightarrow{H_2O} H^+(aq) + Cl^-(aq) \qquad \Delta H^\circ = -74.9 \text{ kJ/mol}$$

calculate ΔH_f° for the Cl^- ions.

(b) Given that ΔH_f° for OH^- ions is −230.0 kJ/mol, calculate the enthalpy of neutralization when 1 mole of a strong monoprotic acid (such as HCl) is titrated by 1 mole of a strong base (such as KOH) at 25°C.

6.53 Calculate the heats of combustion for the following reactions from the standard enthalpies of formation listed in Appendix 2:

(a) $2H_2(g) + O_2(g) \longrightarrow 2H_2O(l)$

(b) $2C_2H_2(g) + 5O_2(g) \longrightarrow 4CO_2(g) + 2H_2O(l)$

6.54 Calculate the heats of combustion for the following reactions from the standard enthalpies of formation listed in Appendix 2:

(a) $C_2H_4(g) + 3O_2(g) \longrightarrow 2CO_2(g) + 2H_2O(l)$

(b) $2H_2S(g) + 3O_2(g) \longrightarrow 2H_2O(l) + 2SO_2(g)$

6.55 Methanol, ethanol, and *n*-propanol are three common alcohols. When 1.00 g of each of these alcohols is burned in air, heat is liberated as shown by the following data: (a) methanol (CH_3OH), −22.6 kJ; (b) ethanol (C_2H_5OH), −29.7 kJ; (c) *n*-propanol (C_3H_7OH), −33.4 kJ. Calculate the heats of combustion of these alcohols in kJ/mol.

6.56 The standard enthalpy change for the following reaction is 436 kJ/mol:

$$H_2(g) \longrightarrow H(g) + H(g)$$

Calculate the standard enthalpy of formation of atomic hydrogen (H).

6.57 From the standard enthalpies of formation, calculate ΔH_{rxn}° for the reaction

$$C_6H_{12}(l) + 9O_2(g) \longrightarrow 6CO_2(g) + 6H_2O(l)$$

For $C_6H_{12}(l)$, $\Delta H_f^\circ = -151.9$ kJ/mol.

6.58 Pentaborane-9, B_5H_9, is a colorless, highly reactive liquid that will burst into flame when exposed to oxygen. The reaction is

$$2B_5H_9(l) + 12O_2(g) \longrightarrow 5B_2O_3(s) + 9H_2O(l)$$

Calculate the kilojoules of heat released per gram of the compound reacted with oxygen. The standard enthalpy of formation of B_5H_9 is 73.2 kJ/mol.

6.59 Determine the amount of heat (in kJ) given off when 1.26×10^4 g of ammonia are produced according to the equation

$$N_2(g) + 3H_2(g) \longrightarrow 2NH_3(g) \quad \Delta H^\circ_{rxn} = -91.8 \text{ kJ/mol}$$

Assume that the reaction takes place under standard-state conditions at 25°C.

6.60 At 850°C, $CaCO_3$ undergoes substantial decomposition to yield CaO and CO_2. Assuming that the ΔH°_f values of the reactant and products are the same at 850°C as they are at 25°C, calculate the enthalpy change (in kJ) if 66.8 g of CO_2 are produced in one reaction.

6.61 From these data,

$$S(\text{rhombic}) + O_2(g) \longrightarrow SO_2(g) \quad \Delta H^\circ_{rxn} = -296.8 \text{ kJ/mol}$$

$$S(\text{monoclinic}) + O_2(g) \longrightarrow SO_2(g) \quad \Delta H^\circ_{rxn} = -297.1 \text{ kJ/mol}$$

calculate the enthalpy change for the transformation

$$S(\text{rhombic}) \longrightarrow S(\text{monoclinic})$$

(Monoclinic and rhombic are different allotropic forms of elemental sulfur.)

6.62 From the following data,

$$C(\text{graphite}) + O_2(g) \longrightarrow CO_2(g) \quad \Delta H^\circ_{rxn} = -393.5 \text{ kJ/mol}$$

$$H_2(g) + \tfrac{1}{2}O_2(g) \longrightarrow H_2O(l) \quad \Delta H^\circ_{rxn} = -285.8 \text{ kJ/mol}$$

$$2C_2H_6(g) + 7O_2(g) \longrightarrow 4CO_2(g) + 6H_2O(l) \quad \Delta H^\circ_{rxn} = -3120.8 \text{ kJ/mol}$$

calculate the enthalpy change for the reaction

$$2C(\text{graphite}) + 3H_2(g) \longrightarrow C_2H_6(g)$$

6.63 From the following heats of combustion,

$$CH_3OH(l) + \tfrac{3}{2}O_2(g) \longrightarrow CO_2(g) + 2H_2O(l) \quad \Delta H^\circ_{rxn} = -725.9 \text{ kJ/mol}$$

$$C(\text{graphite}) + O_2(g) \longrightarrow CO_2(g) \quad \Delta H^\circ_{rxn} = -393.5 \text{ kJ/mol}$$

$$H_2(g) + \tfrac{1}{2}O_2(g) \longrightarrow H_2O(l) \quad \Delta H^\circ_{rxn} = -285.8 \text{ kJ/mol}$$

calculate the enthalpy of formation of methanol (CH_3OH) from its elements:

$$C(\text{graphite}) + 2H_2(g) + \tfrac{1}{2}O_2(g) \longrightarrow CH_3OH(l)$$

6.64 Calculate the standard enthalpy change for the reaction

$$2Al(s) + Fe_2O_3(s) \longrightarrow 2Fe(s) + Al_2O_3(s)$$

given that

$$2Al(s) + \tfrac{3}{2}O_2(g) \longrightarrow Al_2O_3(s) \quad \Delta H^\circ_{rxn} = -1675.7 \text{ kJ/mol}$$

$$2Fe(s) + \tfrac{3}{2}O_2(g) \longrightarrow Fe_2O_3(s) \quad \Delta H^\circ_{rxn} = -824.2 \text{ kJ/mol}$$

6.7 Heat of Solution and Dilution

Review Questions

6.65 Define the following terms: *enthalpy of solution, heat of hydration, lattice energy, heat of dilution.*

6.66 Why is the lattice energy of a solid always a positive quantity? Why is the hydration of ions always a negative quantity?

6.67 Consider two ionic compounds A and B. A has a larger lattice energy than B. Which of the two compounds is more stable?

6.68 Mg^{2+} is a smaller cation than Na^+ and also carries more positive charge. Which of the two species has a larger hydration energy (in kJ/mol)? Explain.

6.69 Consider the dissolution of an ionic compound such as potassium fluoride in water. Break the process into the following steps: separation of the cations and anions in the vapor phase and the hydration of the ions in the aqueous medium. Discuss the energy changes associated with each step. How does the heat of solution of KF depend on the relative magnitudes of these two quantities? On what law is the relationship based?

6.70 Why is it dangerous to add water to a concentrated acid such as sulfuric acid in a dilution process?

Additional Problems

6.71 Which of the following does not have $\Delta H^\circ_f = 0$ at 25°C?

He(g) Fe(s) Cl(g) $S_8(s)$ $O_2(g)$ $Br_2(l)$

6.72 Calculate the expansion work done when 3.70 moles of ethanol are converted to vapor at its boiling point (78.3°C) and 1.0 atm.

6.73 The convention of arbitrarily assigning a zero enthalpy value for the most stable form of each element in the standard state at 25°C is a convenient way of dealing with enthalpies of reactions. Explain why this convention cannot be applied to nuclear reactions.

6.74 Given the thermochemical equations:

$$Br_2(l) + F_2(g) \longrightarrow 2BrF(g) \quad \Delta H^\circ = -188 \text{ kJ/mol}$$

$$Br_2(l) + 3F_2(g) \longrightarrow 2BrF_3(g) \quad \Delta H^\circ = -768 \text{ kJ/mol}$$

calculate the ΔH°_{rxn} for the reaction

$$BrF(g) + F_2(g) \longrightarrow BrF_3(g)$$

6.75 The standard enthalpy change $\Delta H°$ for the thermal decomposition of silver nitrate according to the following equation is +78.67 kJ:

$$AgNO_3(s) \longrightarrow AgNO_2(s) + \frac{1}{2}O_2(g)$$

The standard enthalpy of formation of $AgNO_3(s)$ is −124.4 kJ/mol. Calculate the standard enthalpy of formation of $AgNO_2(s)$.

6.76 Hydrazine, N_2H_4, decomposes according to the following reaction:

$$3N_2H_4(l) \longrightarrow 4NH_3(g) + N_2(g)$$

(a) Given that the standard enthalpy of formation of hydrazine is 50.6 kJ/mol, calculate $\Delta H°$ for its decomposition. (b) Both hydrazine and ammonia burn in oxygen to produce $H_2O(l)$ and $N_2(g)$. Write balanced equations for each of these processes and calculate $\Delta H°$ for each of them. On a mass basis (per kg), would hydrazine or ammonia be the better fuel?

6.77 A quantity of 2.00×10^2 mL of 0.862 *M* HCl is mixed with an equal volume of 0.431 *M* $Ba(OH)_2$ in a constant-pressure calorimeter of negligible heat capacity. The initial temperature of the HCl and $Ba(OH)_2$ solutions is the same at 20.48°C, For the process

$$H^+(aq) + OH^-(aq) \longrightarrow H_2O(l)$$

the heat of neutralization is −55.8 kJ/mol. What is the final temperature of the mixed solution?

6.78 A 3.53-g sample of ammonium nitrate (NH_4NO_3) was added to 80.0 mL of water in a constant-pressure calorimeter of negligible heat capacity. As a result, the temperature of the water decreased from 21.6°C to 18.1°C. Calculate the heat of solution (ΔH_{soln}) of ammonium nitrate.

6.79 Consider the reaction

$$N_2(g) + 3H_2(g) \longrightarrow 2NH_3(g) \qquad \Delta H°_{rxn} = -91.8 \text{ kJ/mol}$$

If 2.0 moles of N_2 react with 6.0 moles of H_2 to form NH_3, calculate the work done (in joules) against a pressure of 1.0 atm at 25°C. What is ΔU for this reaction? Assume the reaction goes to completion.

6.80 Calculate the heat released when 2.00 L of $Cl_2(g)$ with a density of 1.88 g/L react with an excess of sodium metal at 25°C, and 1 atm to form sodium chloride.

6.81 Photosynthesis produces glucose, $C_6H_{12}O_6$, and oxygen from carbon dioxide and water:

$$6CO_2 + 6H_2O \longrightarrow C_6H_{12}O_6 + 6O_2$$

(a) How would you determine experimentally the $\Delta H°_{rxn}$ value for this reaction? (b) Solar radiation produces about 7.0×10^{14} kg glucose a year on Earth. What is the corresponding $\Delta H°$ change?

6.82 A 2.10-mole sample of crystalline acetic acid, initially at 17.0°C, is allowed to melt at 17.0°C and is then heated to 118.1°C (its normal boiling point) at 1.00 atm. The sample is allowed to vaporize at 118.1°C and is then rapidly quenched to 17.0°C, so that it recrystallizes. Calculate $\Delta H°$ for the total process as described.

6.83 Calculate the work done in joules by the reaction

$$2Na(s) + 2H_2O(l) \longrightarrow 2NaOH(aq) + H_2(g)$$

when 0.34 g of Na reacts with water to form hydrogen gas at 0°C and 1.0 atm.

6.84 You are given the following data:

$$H_2(g) \longrightarrow 2H(g) \qquad \Delta H° = 436 \text{ kJ/mol}$$
$$Br_2(g) \longrightarrow 2Br(g) \qquad \Delta H° = 194 \text{ kJ/mol}$$
$$H_2(g) + Br_2(g) \longrightarrow 2HBr(g) \qquad \Delta H° = -72.4 \text{ kJ/mol}$$

Calculate $\Delta H°$ for the reaction

$$H(g) + Br(g) \longrightarrow HBr(g)$$

6.85 A gaseous mixture consists of 28.4 mole percent of hydrogen and 71.6 mole percent of methane. A 15.6-L gas sample, measured at 19.4°C and 2.23 atm, is burned in air. Calculate the heat released.

6.86 When 2.740 g of Ba reacts with O_2 at 298 K and 1 atm to form BaO, 11.14 kJ of heat are released. What is $\Delta H_f°$ for BaO?

6.87 Methanol (CH_3OH) is an organic solvent and is also used as a fuel in some automobile engines. From the following data, calculate the standard enthalpy of formation of methanol:

$$2CH_3OH(l) + 3O_2(g) \longrightarrow 2CO_2(g) + 4H_2O(l) \qquad \Delta H°_{rxn} = -1451.8 \text{ kJ/mol}$$

6.88 A 44.0-g sample of an unknown metal at 99.0°C was placed in a constant-pressure calorimeter containing 80.0 g of water at 24.0°C. The final temperature of the system was found to be 28.4°C. Calculate the specific heat of the metal. (The heat capacity of the calorimeter is 12.4 J/°C.)

6.89 Using the data in Appendix 2, calculate the enthalpy change for the gaseous reaction shown here. (*Hint:* First determine the limiting reagent.)

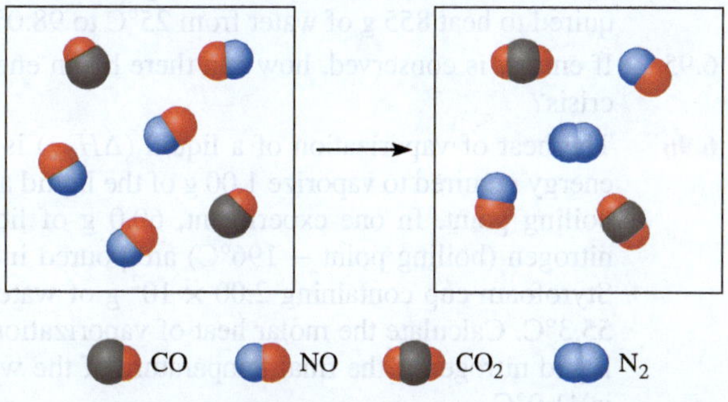

6.90 Producer gas (carbon monoxide) is prepared by passing air over red-hot coke:

$$C(s) + \frac{1}{2}O_2(g) \longrightarrow CO(g)$$

Water gas (mixture of carbon monoxide and hydrogen) is prepared by passing steam over red-hot coke:

$$C(s) + H_2O(g) \longrightarrow CO(g) + H_2(g)$$

For many years, both producer gas and water gas were used as fuels in industry and for domestic cooking. The large-scale preparation of these gases was carried out alternately, that is, first producer gas, then water gas, and so on. Using thermochemical reasoning, explain why this procedure was chosen.

6.91 Compare the heat produced by the complete combustion of 1 mole of methane (CH_4) with a mole of water gas (0.50 mole H_2 and 0.50 mole CO) under the same conditions. On the basis of your answer, would you prefer methane over water gas as a fuel? Can you suggest two other reasons why methane is preferable to water gas as a fuel?

6.92 The so-called hydrogen economy is based on hydrogen produced from water using solar energy. The gas may be burned as a fuel:

$$2H_2(g) + O_2(g) \longrightarrow 2H_2O(l)$$

A primary advantage of hydrogen as a fuel is that it is nonpolluting. A major disadvantage is that it is a gas and therefore is harder to store than liquids or solids. Calculate the volume of hydrogen gas at 25°C, and 1.00 atm required to produce an amount of energy equivalent to that produced by the combustion of a gallon of octane (C_8H_{18}). The density of octane is 2.66 kg/gal, and its standard enthalpy of formation is −249.9 kJ/mol.

6.93 Ethanol (C_2H_5OH) and gasoline (assumed to be all octane) (C_8H_{18}) are both used as automobile fuel. If gasoline is selling for \$4.50/gal, what would the price of ethanol have to be to provide the same amount of heat per dollar? The density and ΔH_f° of octane are 0.7025 g/mL and −249.9 kJ/mol and of ethanol are 0.7894 g/mL and −277.6 kJ/mol, respectively. 1 gal = 3.785 L.

6.94 The combustion of what volume of ethane (C_2H_6), measured at 23.0°C and 752 mmHg, would be required to heat 855 g of water from 25°C to 98.0°C?

6.95 If energy is conserved, how can there be an energy crisis?

6.96 The heat of vaporization of a liquid (ΔH_{vap}) is the energy required to vaporize 1.00 g of the liquid at its boiling point. In one experiment, 60.0 g of liquid nitrogen (boiling point − 196°C) are poured into a Styrofoam cup containing 2.00×10^2 g of water at 55.3°C. Calculate the molar heat of vaporization of liquid nitrogen if the final temperature of the water is 41.0°C.

6.97 Explain the cooling effect experienced when ethanol is rubbed on your skin, given that

$$C_2H_5OH(l) \longrightarrow C_2H_5OH(g) \quad \Delta H^\circ = 38.2 \text{ kJ/mol}$$

6.98 For which of the following reactions does $\Delta H^\circ_{rxn} = \Delta H_f^\circ$?
(a) $H_2(g) + S(\text{rhombic}) \longrightarrow H_2S(g)$
(b) $C(\text{diamond}) + O_2(g) \longrightarrow CO_2(g)$
(c) $H_2(g) + CuO(s) \longrightarrow H_2O(l) + Cu(s)$
(d) $O(g) + O_2(g) \longrightarrow O_3(g)$

6.99 Calculate the work done (in joules) when 1.0 mole of water is frozen at 0°C and 1.0 atm. The volumes of 1 mole of water and ice at 0°C are 0.0180 L and 0.0196 L, respectively.

6.100 A quantity of 0.020 mole of a gas initially at 0.050 L and 20°C undergoes a constant-temperature expansion until its volume is 0.50 L. Calculate the work done (in joules) by the gas if it expands (a) against a vacuum and (b) against a constant pressure of 0.20 atm. (c) If the gas in (b) is allowed to expand unchecked until its pressure is equal to the external pressure, what would its final volume be before it stopped expanding, and what would be the work done?

6.101 Calculate the standard enthalpy of formation for diamond, given that

$$C(\text{graphite}) + O_2(g) \longrightarrow CO_2(g) \quad \Delta H^\circ = -393.5 \text{ kJ/mol}$$

$$C(\text{diamond}) + O_2(g) \longrightarrow CO_2(g) \quad \Delta H^\circ = -395.4 \text{ kJ/mol}$$

6.102 (a) For most efficient use, refrigerator freezer compartments should be fully packed with food. What is the thermochemical basis for this recommendation? (b) Starting at the same temperature, tea and coffee remain hot longer in a thermal flask than chicken noodle soup. Explain.

6.103 Calculate the standard enthalpy change for the fermentation process. (See Problem 3.72.)

6.104 Portable hot packs are available for skiers and people engaged in other outdoor activities in a cold climate. The air-permeable paper packet contains a mixture of powdered iron, sodium chloride, and other components, all moistened by a little water. The exothermic reaction that produces the heat is a very common one—the rusting of iron:

$$4Fe(s) + 3O_2(g) \longrightarrow 2Fe_2O_3(s)$$

When the outside plastic envelope is removed, O_2 molecules penetrate the paper, causing the reaction to begin. A typical packet contains 250 g of iron to warm your hands or feet for up to 4 hours. How much heat (in kJ) is produced by this reaction? (*Hint:* See Appendix 2 for ΔH_f° values.)

6.105 A person ate 0.50 pound of cheese (an energy intake of 4000 kJ). Suppose that none of the energy was stored in his body. What mass (in grams) of water would he need to perspire to maintain his original temperature? (It takes 44.0 kJ to vaporize 1 mole of water.)

6.106 The total volume of the Pacific Ocean is estimated to be 7.2×10^8 km^3. A medium-sized atomic bomb produces 1.0×10^{15} J of energy upon explosion. Calculate the number of atomic bombs needed to release enough energy to raise the temperature of the water in the Pacific Ocean by 1°C.

6.107 A 19.2-g quantity of dry ice (solid carbon dioxide) is allowed to sublime (evaporate) in an apparatus like the one shown in Figure 6.5. Calculate the expansion work done against a constant external pressure of 0.995 atm and at a constant temperature of 22°C. Assume that the initial volume of dry ice is negligible and that CO_2 behaves like an ideal gas.

6.108 The enthalpy of combustion of benzoic acid (C_6H_5COOH) is commonly used as the standard for calibrating constant-volume bomb calorimeters; its value has been accurately determined to be −3226.7 kJ/mol. When 1.9862 g of benzoic acid are burned in a calorimeter, the temperature rises from 21.84°C to 25.67°C. What is the heat capacity of the bomb? (Assume that the quantity of water surrounding the bomb is exactly 2000 g.)

6.109 The combustion of a 25.0-g gaseous mixture of H_2 and CH_4 releases 2354 kJ of heat. Calculate the amounts of the gases in grams.

6.110 Calcium oxide (CaO) is used to remove sulfur dioxide generated by coal-burning power stations:

$$2CaO(s) + 2SO_2(g) + O_2(g) \longrightarrow 2CaSO_4(s)$$

Calculate the enthalpy change for this process if 6.6×10^5 g of SO_2 are removed by this process every day.

6.111 Glauber's salt, sodium sulfate decahydrate ($Na_2SO_4 \cdot 10H_2O$), undergoes a phase transition (that is, melting or freezing) at a convenient temperature of about 32°C:

$$Na_2SO_4 \cdot 10H_2O(s) \longrightarrow Na_2SO_4 \cdot 10H_2O(l) \qquad \Delta H° = 74.4 \text{ kJ/mol}$$

As a result, this compound is used to regulate the temperature in homes. It is placed in plastic bags in the ceiling of a room. During the day, the endothermic melting process absorbs heat from the surroundings, cooling the room. At night, it gives off heat as it freezes. Calculate the mass of Glauber's salt in kilograms needed to lower the temperature of air in a room by 8.2°C at 1.0 atm. The dimensions of the room are 2.80 m × 10.6 m × 17.2 m, the specific heat of air is 1.2 J/g · °C, and the molar mass of air may be taken as 29.0 g/mol.

6.112 A balloon 16 m in diameter is inflated with helium at 18°C. (a) Calculate the mass of He in the balloon, assuming ideal behavior. (b) Calculate the work done (in joules) during the inflation process if the atmospheric pressure is 98.7 kPa.

6.113 Acetylene (C_2H_2) can be hydrogenated (reacting with hydrogen) first to ethylene (C_2H_4) and then to ethane (C_2H_6). Starting with 1 mole of C_2H_2, label the diagram shown here analogous to Figure 6.10. Use the data in Appendix 2.

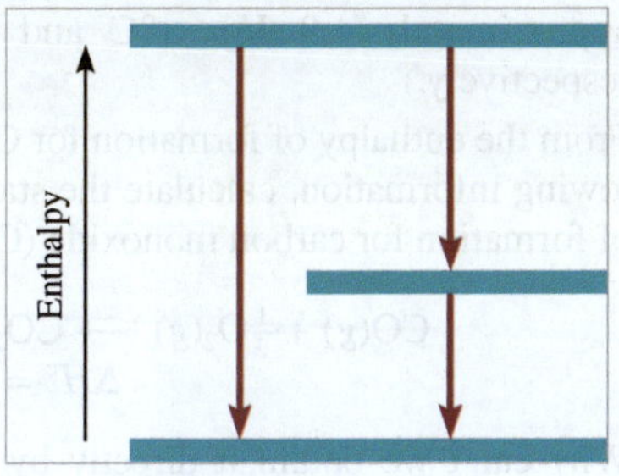

6.114 Calculate the $\Delta H°$ for the reaction

$$Fe^{3+}(aq) + 3OH^-(aq) \longrightarrow Fe(OH)_3(s)$$

6.115 An excess of zinc metal is added to 50.0 mL of a 0.100 M $AgNO_3$ solution in a constant-pressure calorimeter like the one pictured in Figure 6.9. As a result of the reaction

$$Zn(s) + 2Ag^+(aq) \longrightarrow Zn^{2+}(aq) + 2Ag(s)$$

the temperature rises from 19.25°C to 22.17°C. If the heat capacity of the calorimeter is 98.6 J/°C, calculate the enthalpy change for the abovementioned reaction on a molar basis. Assume that the density and specific heat of the solution are the same as those for water, and ignore the specific heats of the metals.

6.116 (a) A person drinks four glasses of cold water (3.0°C) every day. The volume of each glass is 2.5×10^2 mL. How much heat (in kJ) does the body have to supply to raise the temperature of the water to 37°C, the body temperature? (b) How much heat would your body lose if you were to ingest (3.0°C) every day? The volume of each glass is 8.0×10^2 g of snow at 0°C to quench thirst. (The amount of heat necessary to melt snow is 6.01 kJ/mol.)

6.117 A driver's manual states that the stopping distance quadruples as the speed doubles; that is, if it takes 30 ft to stop a car moving at 25 mph then it would take 120 ft to stop a car moving at 50 mph. Justify this statement by using mechanics and the first law of thermodynamics. [Assume that when a car is stopped, its kinetic energy $\left(\frac{1}{2}mu^2\right)$ is totally converted to heat.]

6.118 At 25°C, the standard enthalpy of formation of $HF(aq)$ is given by −320.1 kJ/mol; of $OH^-(aq)$, it is −230.0 kJ/mol; of $F^-(aq)$, it is −332.6 kJ/mol; and of $H_2O(l)$, it is −285.8 kJ/mol.

(a) Calculate the standard enthalpy of neutralization of $HF(aq)$:

$$HF(aq) + OH^-(aq) \longrightarrow F^-(aq) + H_2O(l)$$

(b) Using the value of −55.8 kJ as the standard enthalpy change for the reaction

$$H^+(aq) + OH^-(aq) \longrightarrow H_2O(l)$$

calculate the standard enthalpy change for the reaction

$$HF(aq) \longrightarrow H^+(aq) + F^-(aq)$$

6.119 Why are cold, damp air and hot, humid air more uncomfortable than dry air at the same temperatures? (The specific heats of water vapor and air are approximately 1.9 J/g · °C and 1.0 J/g · °C, respectively.)

6.120 From the enthalpy of formation for CO_2 and the following information, calculate the standard enthalpy of formation for carbon monoxide (CO).

$$CO(g) + \frac{1}{2}O_2(g) \longrightarrow CO_2(g)$$
$$\Delta H° = -283.0 \text{ kJ/mol}$$

Why can't we obtain it directly by measuring the enthalpy of the following reaction?

$$C(\text{graphite}) + \frac{1}{2}O_2(g) \longrightarrow CO(g)$$

6.121 A 46-kg person drinks 500 g of milk, which has a "caloric" value of approximately 3.0 kJ/g. If only 17% of the energy in milk is converted to mechanical work, how high (in meters) can the person climb based on this energy intake? [*Hint:* The work done in ascending is given by *mgh,* where *m* is the mass (in kilograms), *g* the gravitational acceleration (9.8 m/s^2), and *h* the height (in meters).]

6.122 The height of Niagara Falls on the American side is 51 m. (a) Calculate the potential energy of 1.0 g of water at the top of the falls relative to the ground level. (b) What is the speed of the falling water if all of the potential energy is converted to kinetic energy? (c) What would be the increase in temperature of the water if all the kinetic energy were converted to heat? (See Problem 6.121 for suggestions.)

6.123 In the nineteenth century two scientists named Dulong and Petit noticed that for a solid element, the product of its molar mass and its specific heat is approximately 25 J/°C. This observation, now called Dulong and Petit's law, was used to estimate the specific heat of metals. Verify the law for the metals listed in Table 6.2. The law does not apply to one of the metals. Which one is it? Why?

6.124 Determine the standard enthalpy of formation of ethanol (C_2H_5OH) from its standard enthalpy of combustion (−1367.4 kJ/mol).

6.125 Acetylene (C_2H_2) and benzene (C_6H_6) have the same empirical formula. In fact, benzene can be made from acetylene as follows:

$$3C_2H_2(g) \longrightarrow C_6H_6(l)$$

The enthalpies of combustion for C_2H_2 and C_6H_6 are −1300.2 kJ/mol and −3266.9 kJ/mol, respectively. Calculate the standard enthalpies of formation of C_2H_2 and C_6H_6 and hence the enthalpy change for the formation of C_6H_6 from C_2H_2.

6.126 Ice at 0°C is placed in a Styrofoam cup containing 361 g of a soft drink at 23°C. The specific heat of the drink is about the same as that of water. Some ice remains after the ice and soft drink reach an equilibrium temperature of 0°C. Determine the mass of ice that has melted. Ignore the heat capacity of the cup. (*Hint:* It takes 334 J to melt 1 g of ice at 0°C.)

6.127 After a dinner party, the host performed the following trick. First, he blew out one of the burning candles. He then quickly brought a lighted match to about 1 in above the wick. To everyone's surprise, the candle was relighted. Explain how the host was able to accomplish the task without touching the wick.

Sofia Fernandez

6.128 How much heat is required to decompose 89.7 g of NH_4Cl? (*Hint:* You may use the enthalpy of formation values at 25°C, for the calculation.)

6.129 A gas company in Massachusetts charges \$1.30 for 15 ft^3 of natural gas (CH_4) measured at 20°C and 1.0 atm. Calculate the cost of heating 200 mL of water (enough to make a cup of coffee or tea) from 20°C to 100°C. Assume that only 50% of the heat generated by the combustion is used to heat the water; the rest of the heat is lost to the surroundings.

6.130 Calculate the internal energy of a Goodyear blimp filled with helium gas at 1.2×10^5 Pa. The volume of the blimp is 5.5×10^3 m^3. If all the energy were used to heat 10.0 tons of copper at 21°C, calculate the final temperature of the metal. (*Hint:* See Section 5.7 for help in calculating the internal energy of a gas. 1 ton = 9.072×10^5 g.)

6.131 Decomposition reactions are usually endothermic, whereas combination reactions are usually exothermic. Give a qualitative explanation for these trends.

6.132 Acetylene (C_2H_2) can be made by reacting calcium carbide (CaC_2) with water. (a) Write an equation for the reaction. (b) What is the maximum amount of heat (in joules) that can be obtained from the combustion of acetylene, starting with 74.6 g of CaC_2?

6.133 The average temperature in deserts is high during the day but quite cool at night, whereas that in regions along the coastline is more moderate. Explain.

6.134 When 1.034 g of naphthalene ($C_{10}H_8$) are burned in a constant-volume bomb calorimeter at 298 K, 41.56 kJ of heat are evolved. Calculate ΔU and ΔH for the reaction on a molar basis.

6.135 From a thermochemical point of view, explain why a carbon dioxide fire extinguisher or water should not be used on a magnesium fire.

6.136 Calculate the ΔU for the following reaction at 298 K:

$$2H_2(g) + O_2(g) \longrightarrow 2H_2O(l)$$

6.137 Lime is a term that includes calcium oxide (CaO, also called quicklime) and calcium hydroxide [$Ca(OH)_2$, also called slaked lime]. It is used in the steel industry to remove acidic impurities, in air-pollution control to remove acidic oxides such as SO_2, and in water treatment. Quicklime is made industrially by heating limestone ($CaCO_3$) above 2000°C:

$$CaCO_3(s) \longrightarrow CaO(s) + CO_2(g) \quad \Delta H° = 179.2 \text{ kJ/mol}$$

Slaked lime is produced by treating quicklime with water:

$$CaO(s) + H_2O(l) \longrightarrow Ca(OH)_2(s) \quad \Delta H° = -64.5 \text{ kJ/mol}$$

The exothermic reaction of quicklime with water and the rather small specific heats of both quicklime (0.946 J/g · °C) and slaked lime (1.20 J/g · °C) make it hazardous to store and transport lime in vessels made of wood. Wooden sailing ships carrying lime would occasionally catch fire when water leaked into the hold. (a) If a 500-g sample of water reacts with an equimolar amount of CaO (both at an initial temperature of 25°C), what is the final temperature of the product, $Ca(OH)_2$? Assume that the product absorbs all of the heat released in the reaction. (b) Given that the standard enthalpies of formation of CaO and H_2O are −634.9 kJ/mol and −285.8 kJ/mol, respectively, calculate the standard enthalpy of formation of $Ca(OH)_2$.

6.138 A 4.117-g impure sample of glucose ($C_6H_{12}O_6$) was burned in a constant-volume calorimeter having a heat capacity of 19.65 kJ/°C. If the rise in temperature is 3.134°C, calculate the percent by mass of the glucose in the sample. Assume that the impurities are unaffected by the combustion process. See Appendix 2 for thermodynamic data.

6.139 Construct a table with the headings q, w, ΔU, and ΔH. For each of the following processes, deduce whether each of the quantities listed is positive (+), negative (−), or zero (0). (a) Freezing of benzene. (b) Compression of an ideal gas at constant temperature. (c) Reaction of sodium with water. (d) Boiling liquid ammonia. (e) Heating a gas at constant volume. (f) Melting of ice.

6.140 The combustion of 0.4196 g of a hydrocarbon releases 17.55 kJ of heat. The masses of the products are CO_2 = 1.419 g and H_2O = 0.290 g. (a) What is the empirical formula of the compound? (b) If the approximate molar mass of the compound is 76 g, calculate its standard enthalpy of formation.

6.141 Metabolic activity in the human body releases approximately 1.0×10^4 kJ of heat per day. Assuming the body is 50 kg of water, how much would the body temperature rise if it were an isolated system? How much water must the body eliminate as perspiration to maintain the normal body temperature (98.6°F)? Comment on your results. The heat of vaporization of water may be taken as 2.41 kJ/g.

6.142 Give an example for each of the following situations: (a) Adding heat to a system raises its temperature, (b) adding heat to a system does not change (raise) its temperature, and (c) a system's temperature is changed even though no heat is added or removed from it.

6.143 From the following data, calculate the heat of solution for KI:

	NaCl	NaI	KCl	KI
Lattice energy (kJ/mol)	788	686	699	632
Heat of solution (kJ/mol)	4.0	−5.1	17.2	?

6.144 Starting at A, an ideal gas undergoes a cyclic process involving expansion and compression, as shown here. Calculate the total work done. Does your result support the notion that work is not a state function?

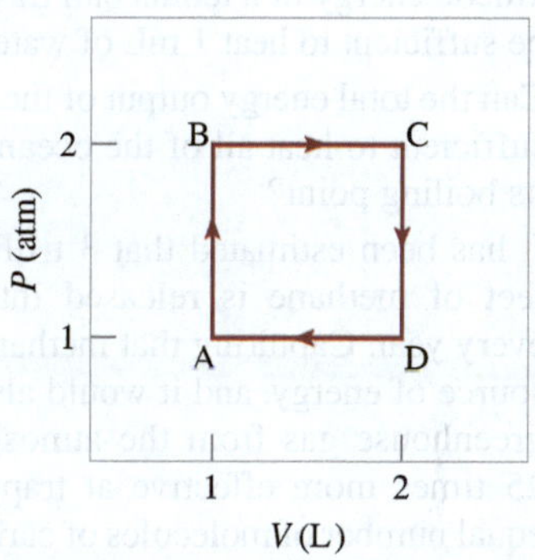

6.145 For reactions in condensed phases (liquids and solids), the difference between ΔH and ΔU is usually quite small. This statement holds for reactions carried out under atmospheric conditions. For certain geochemical processes, however, the external pressure may be so great that ΔH and ΔU can differ by a significant amount. A well-known example is the slow conversion of graphite to diamond under Earth's surface. Calculate ($\Delta H - \Delta U$) for the conversion of 1 mole of graphite to 1 mole of diamond at a pressure of 50,000 atm. The densities of graphite and diamond are 2.25 g/cm³ and 3.52 g/cm³, respectively.

6.146 The diagrams (a)–(d) represent various physical and chemical processes: (a) $2A(g) \longrightarrow A_2(g)$; (b) $MX(s) \longrightarrow M^+(aq) + X^-(aq)$; (c) $AB(g) + C(g) \longrightarrow AC(g) + B(g)$; (d) $B(l) \longrightarrow B(g)$. Predict whether the situations shown are endothermic or exothermic. Explain why in some cases no clear conclusions can be made.

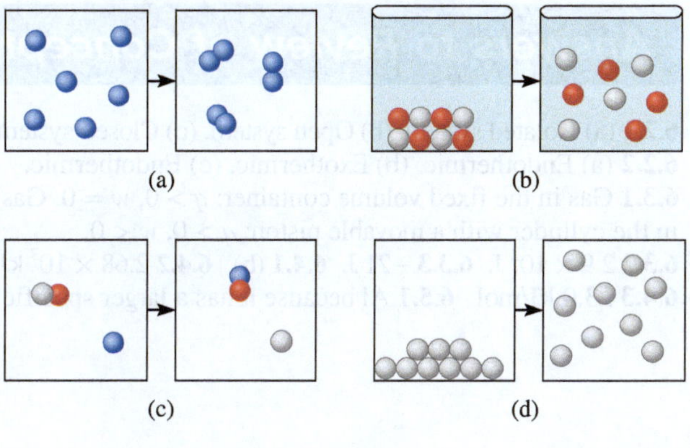

6.147 A 20.3-g sample of an unknown metal and a 28.5-g sample of copper, both at 80.6°C, are added to 100 g of water at 11.2°C in a constant-pressure calorimeter of negligible heat capacity. If the final temperature of the metals and water is 13.7°C, determine the specific heat of the unknown metal.

Interpreting, Modeling, & Estimating

6.148 For most biological processes, $\Delta H \approx \Delta U$. Explain.

6.149 Estimate the potential energy expended by an average adult male in going from the ground to the top floor of the Empire State Building using the staircase.

6.150 The fastest serve in tennis is about 150 mph. Can the kinetic energy of a tennis ball traveling at this speed be sufficient to heat 1 mL of water by 30°C?

6.151 Can the total energy output of the sun in 1 second be sufficient to heat all of the ocean water on Earth to its boiling point?

6.152 It has been estimated that 3 trillion standard cubic feet of methane is released into the atmosphere every year. Capturing that methane would provide a source of energy, and it would also remove a potent greenhouse gas from the atmosphere (methane is 25 times more effective at trapping heat than an equal number of molecules of carbon dioxide). Standard cubic feet is measured at 60°F and 1 atm. Determine the amount of energy that could be obtained by combustion of the methane that escapes each year.

6.153 Biomass plants generate electricity from waste material such as wood chips. Some of these plants convert the feedstock to ethanol (C_2H_5OH) for later use as a fuel. (a) How many grams of ethanol can be produced from 1.0 ton of wood chips, if 85% of the carbon is converted to C_2H_5OH? (b) How much energy would be released by burning the ethanol obtained from 1.0 ton of wood chips? (*Hint:* Treat the wood chips as cellulose.)

6.154 Suppose an automobile carried hydrogen gas in its fuel tank instead of gasoline. At what pressure would the hydrogen gas need to be kept for the tank to contain an equivalent amount of chemical energy as a tank of gasoline?

6.155 A press release announcing a new fuel-cell car to the public stated that hydrogen is "relatively cheap" and "some stations in California sell hydrogen for $5 a kilogram. A kg has the same energy as a gallon of gasoline, so it's like paying $5 a gallon. But you go two to three times as far on the hydrogen." Analyze this claim.

6.156 We hear a lot about how the burning of hydrocarbons produces the greenhouse gas CO_2, but what about the effect of increasing energy consumption on the amount of oxygen in the atmosphere required to sustain life? The figure shows past and projected world energy consumption. (a) How many moles of oxygen would be required to generate the additional energy expenditure for the next decade? (b) What would be the resulting decrease in atmospheric oxygen?

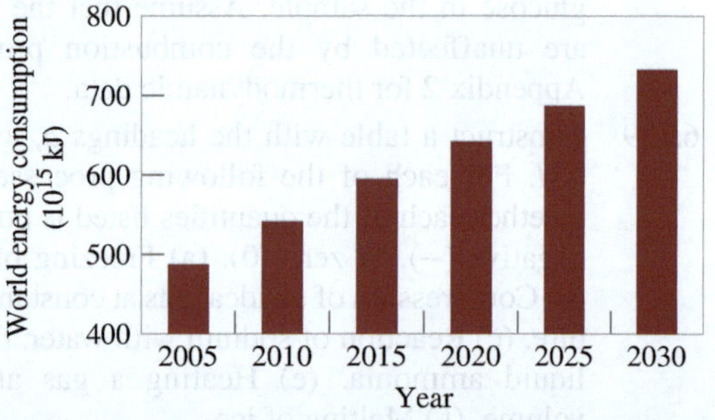

Answers to Practice Exercises

6.1 (a) 0. (b) −286 J. **6.2** −63 J. **6.3** -6.41×10^3 kJ. **6.4** −111.7 kJ/mol. **6.5** −34.3 kJ. **6.6** −728 kJ/mol. **6.7** 21.19°C. **6.8** 22.49°C. **6.9** 89.0 kJ/mol. **6.10** −41.8 kJ/g.

Answers to Review of Concepts & Facts

6.2.1 (a) Isolated system. (b) Open system. (c) Closed system. **6.2.2** (a) Endothermic. (b) Exothermic. (c) Endothermic. **6.3.1** Gas in the fixed volume container: $q > 0$, $w = 0$. Gas in the cylinder with a movable piston: $q > 0$, $w < 0$. **6.3.2** 2.9×10^2 J. **6.3.3** −21 J. **6.4.1** (b). **6.4.2** 2.68×10^3 kJ. **6.4.3** 23.0 kJ/mol. **6.5.1** Al because it has a larger specific heat. **6.5.2** 3720 kJ/mol. **6.5.3** 0.55 J/g · °C. **6.6.1** Hg(*s*). **6.6.2** The negative sign before ΔH_f° for reactants means that reactants with positive ΔH_f° values will likely result in a negative ΔH_{rxn}° (exothermic reaction). Review Equation (6.18) in the text. **6.6.3** −316.8 kJ/mol. **6.6.4** −771.2 kJ/mol. **6.7.1** 34.9 kJ/mol.

CHAPTER

7

Quantum Theory and the Electronic Structure of Atoms

Imaging atomic surfaces illustrates one of the fundamental principles of quantum mechanics where precise measurements are on the atomic scale are not possible.
Andrew Dunn/Alamy Stock Photo

CHAPTER OUTLINE

Quantum theory enables us to predict and understand the critical role that electrons play in chemistry. In one sense, studying atoms amounts to asking the following questions:

1. How many electrons are present in a particular atom?
2. What energies do individual electrons possess?
3. Where in the atom can electrons be found?

The answers to these questions have a direct relationship to the behavior of all substances in chemical reactions, and the story of the search for answers provides a fascinating backdrop for our discussion.

7.1 From Classical Physics to Quantum Theory

Learning Objectives

- Summarize the properties of waves including wavelength, frequency, and amplitude.
- Interconvert wavelength and frequency knowing the speed of light constant.
- Categorize examples of the electromagnetic spectrum in regard to wavelength and type of radiation.
- Evaluate the basis of quantum theory according to Planck's work.

Early attempts by nineteenth-century physicists to understand atoms and molecules met with only limited success. By assuming that molecules behave like rebounding balls, physicists were able to predict and explain some macroscopic phenomena, such as the pressure exerted by a gas. However, this model did not account for the stability of molecules; that is, it could not explain the forces that hold atoms together. It took a long time to realize—and an even longer time to accept—that the properties of atoms and molecules are *not* governed by the same physical laws as larger objects.

The new era in physics started in 1900 with a young German physicist named Max Planck.† While analyzing the data on radiation emitted by solids heated to various temperatures, Planck discovered that atoms and molecules emit energy only in certain discrete quantities, or *quanta.* Physicists had always assumed that energy is continuous and that any amount of energy could be released in a radiation process. Planck's *quantum theory* turned physics upside down. Indeed, the flurry of research that ensued altered our concept of nature forever.

Properties of Waves

To understand Planck's quantum theory, we must first know something about the nature of waves. A ***wave*** can be thought of as *a vibrating disturbance by which energy is transmitted.* The fundamental properties of a wave are illustrated by a familiar type—water waves (Figure 7.1). The regular variation of the peaks and troughs enable us to sense the propagation of the waves.

Figure 7.1 *Ocean water waves.*
Joslin Stevens/Shutterstock

Waves are characterized by their length and height and by the number of waves that pass through a certain point in 1 second (Figure 7.2). ***Wavelength*** λ (lambda) is *the distance between identical points on successive waves.* The ***frequency*** ν (nu) is *the number of waves that pass through a particular point in 1 second.* ***Amplitude*** is *the vertical distance from the midline of a wave to the peak or trough. The point at which the amplitude of a wave is zero* is known as a ***node.*** Figure 7.2(b) shows two waves that have the same amplitude but different wavelengths and frequencies.

†Max Karl Ernst Ludwig Planck (1858–1947). German physicist. Planck received the Nobel Prize in Physics in 1918 for his quantum theory. He also made significant contributions in thermodynamics and other areas of physics.

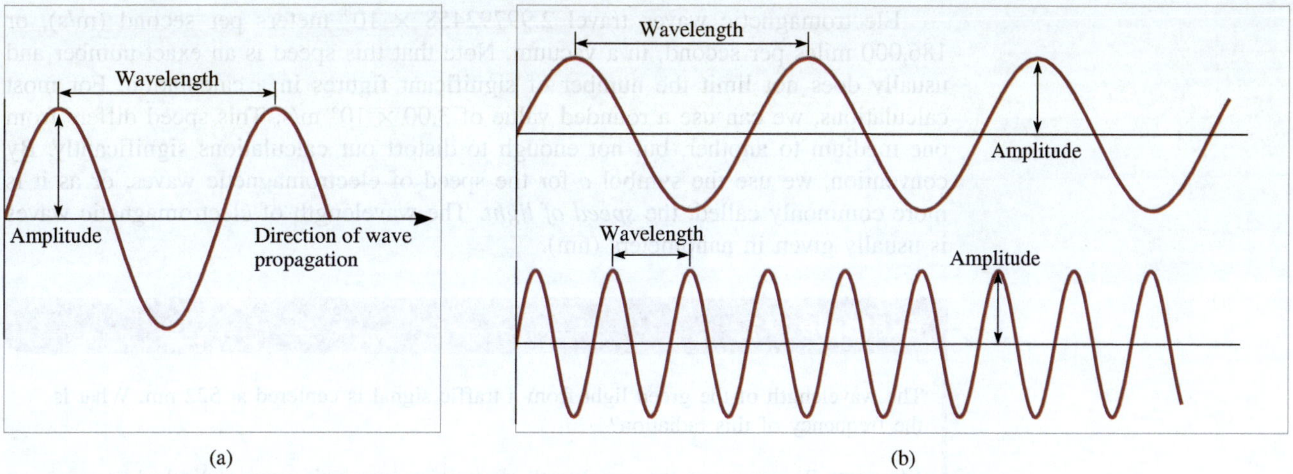

Figure 7.2 *(a) Wavelength and amplitude. (b) Two waves having different wavelengths and frequencies. The wavelength of the top wave is three times that of the lower wave, but its frequency is only one-third that of the lower wave. Both waves have the same speed and amplitude.*

Another important property of waves is their speed, which depends on the type of wave and the nature of the medium through which the wave is traveling (for example, air, water, or a vacuum). The speed (u) of a wave is the product of its wavelength and its frequency; that is:

$$u = \lambda\nu \qquad (7.1)$$

The inherent "sensibility" of Equation (7.1) becomes apparent if we analyze the physical dimensions involved in the three terms. The wavelength (λ) expresses the length of a wave, or distance/wave. The frequency (ν) indicates the number of these waves that pass any reference point per unit of time, or waves/time. Thus, the product of these terms results in dimensions of distance/time, which is speed:

$$\frac{\text{distance}}{\text{time}} = \frac{\text{distance}}{\cancel{\text{wave}}} \times \frac{\cancel{\text{waves}}}{\text{time}}$$

Wavelength is usually expressed in units of meters, centimeters, or nanometers, and frequency is measured in hertz (Hz), where

$$1 \text{ Hz} = 1 \text{ cycle/s}$$

The word "cycle" may be left out and the frequency expressed as, for example, 25/s or 25 s^{-1} (read as "25 per second").

Electromagnetic Radiation

There are many kinds of waves, such as water waves, sound waves, and light waves. In 1873 James Clerk Maxwell proposed that visible light consists of electromagnetic waves. According to Maxwell's theory, an ***electromagnetic wave*** *has an electric field component and a magnetic field component.* These two components have the same wavelength and frequency, and hence the same speed, but they travel in mutually perpendicular planes (Figure 7.3). The significance of Maxwell's theory is that it provides a mathematical description of the general behavior of light. In particular, his model accurately describes how energy in the form of radiation can be propagated through space as vibrating electric and magnetic fields. ***Electromagnetic radiation*** *is the emission and transmission of energy in the form of electromagnetic waves.*

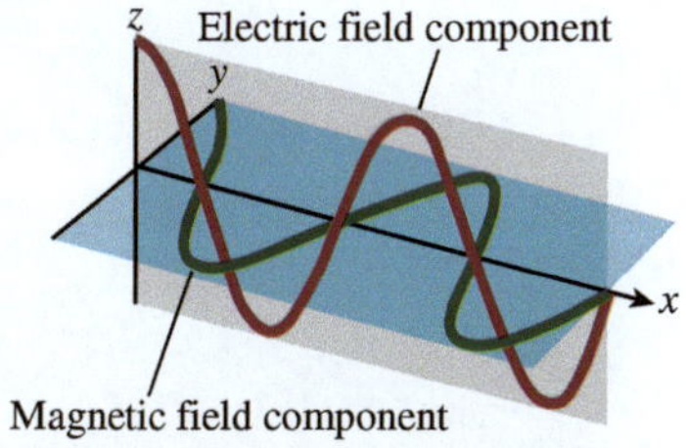

Figure 7.3 *The electric field and magnetic field components of an electromagnetic wave. These two components have the same wavelength, frequency, and amplitude, but they oscillate in two mutually perpendicular planes.*

Electromagnetic waves travel 2.99792458 × 10^8 meters per second (m/s), or 186,000 miles per second, in a vacuum. Note that this speed is an exact number and usually does not limit the number of significant figures in a calculation. For most calculations, we can use a rounded value of 3.00 × 10^8 m/s. This speed differs from one medium to another, but not enough to distort our calculations significantly. By convention, we use the symbol *c* for the speed of electromagnetic waves, or as it is more commonly called, the *speed of light.* The wavelength of electromagnetic waves is usually given in nanometers (nm).

Example 7.1

The wavelength of the green light from a traffic signal is centered at 522 nm. What is the frequency of this radiation?

Strategy We are given the wavelength of an electromagnetic wave and asked to calculate its frequency. Rearranging Equation (7.1) and replacing *u* with *c* (the speed of light) gives

$$\nu = \frac{c}{\lambda}$$

Solution Because the speed of light is given in meters per second, it is convenient to first convert wavelength to meters. Recall that 1 nm = 1 × 10^{-9} m (see Table 1.2). We write

$$\lambda = 522\ \cancel{\text{nm}} \times \frac{1 \times 10^{-9}\ \text{m}}{1\ \cancel{\text{nm}}} = 522 \times 10^{-9}\ \text{m}$$

$$= 5.22 \times 10^{-7}\ \text{m}$$

Substituting in the wavelength and the speed of light (3.00 × 10^8 m/s), the frequency is

$$\nu = \frac{3.00 \times 10^8\ \text{m/s}}{5.22 \times 10^{-7}\ \text{m}}$$

$$= 5.75 \times 10^{14}/\text{s, or } 5.75 \times 10^{14}\ \text{Hz}$$

Check The answer shows that 5.75 × 10^{14} waves pass a fixed point every second. This very high frequency is in accordance with the very high speed of light.

Practice Exercise What is the wavelength (in meters) of an electromagnetic wave whose frequency is 3.64 × 10^7 Hz?

Similar problem: 7.7.

Figure 7.4 shows various types of electromagnetic radiation, which differ from one another in wavelength and frequency. The long radio waves are emitted by large antennas, such as those used by broadcasting stations. The shorter, visible light waves are produced by the motions of electrons within atoms and molecules. The shortest waves, which also have the highest frequency, are associated with γ (gamma) rays, which result from changes within the nucleus of the atom (see Chapter 2). As we will see shortly, the higher the frequency, the more energetic the radiation. Thus, ultraviolet radiation, X rays, and γ rays are high-energy radiation.

The Double Slit Experiment

An important property of electromagnetic waves is their ability to undergo interference with one another. This phenomenon was convincingly demonstrated by the double slit experiment wherein light is shown through two small openings called slits and a repeating pattern of light and dark lines is produced as seen in Figure 7.5. This pattern is known as an interference pattern.

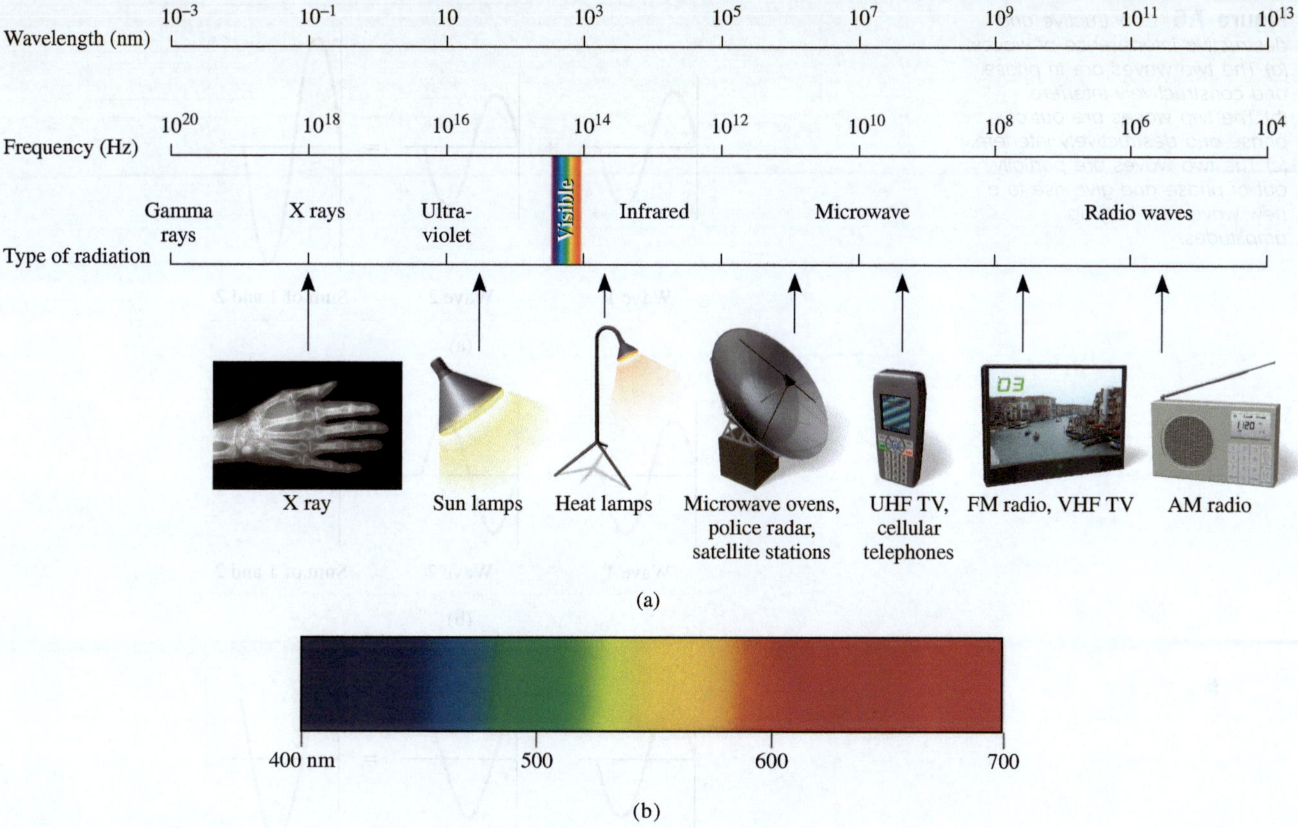

Figure 7.4 *(a) Types of electromagnetic radiation. Gamma rays have the shortest wavelength and highest frequency; radio waves have the longest wavelength and the lowest frequency. Each type of radiation is spread over a specific range of wavelengths (and frequencies). (b) Visible light ranges from a wavelength of 400 nm (violet) to 700 nm (red).*

a: (X ray): Ted Kinsman/Science Source

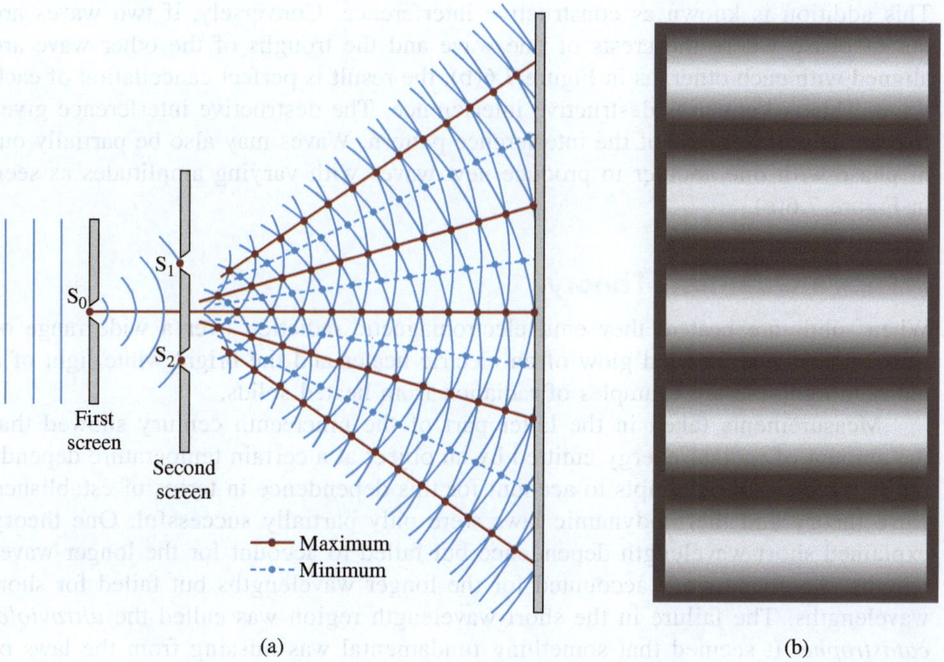

Figure 7.5 *The double slit experiment. (a) The red lines are a result of maximum interactions of waves. The blue lines are a result of the cancellation of waves interacting. (b) The resulting interference pattern with alternating light and dark regions.*

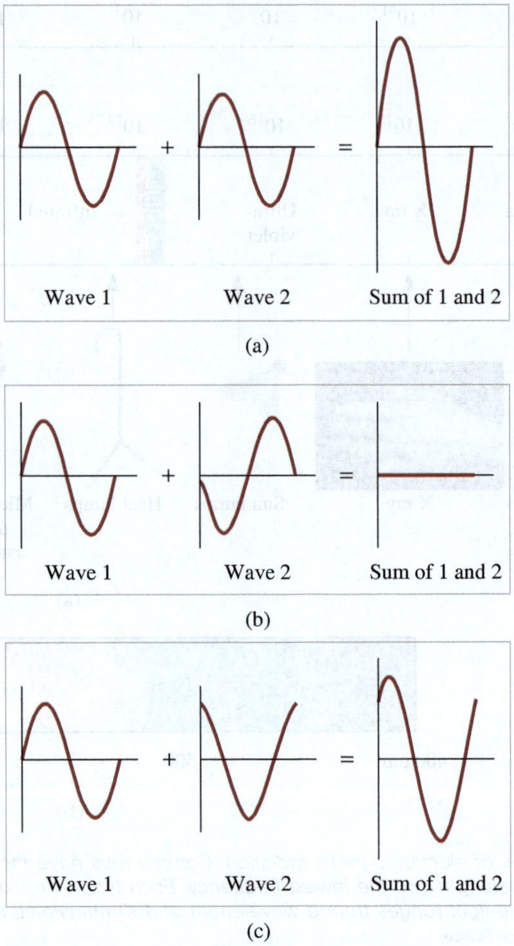

Figure 7.6 *Constructive and destructive interference of waves. (a) The two waves are in phase and constructively interfere. (b) The two waves are out of phase and destructively interfere. (c) The two waves are partially out of phase and give rise to a new wave with varying amplitudes.*

Interference patterns are the result of the interactions of two waves. If two waves are in phase—that is, the crests and the troughs of the waves are identical—the resulting light pattern is due to the addition of the two waves as in Figure 7.6(a). This addition is known as constructive interference. Conversely, if two waves are out of phase where the crests of one wave and the troughs of the other wave are aligned with each other, as in Figure 7.6(b), the result is perfect cancellation of each wave. This is known as destructive interference. The destructive interference gives rise to the dark regions of the interference pattern. Waves may also be partially out of phase with one another to produce new waves with varying amplitudes as seen in Figure 7.6(c).

Planck's Quantum Theory

When solids are heated, they emit electromagnetic radiation over a wide range of wavelengths. The dull red glow of an electric heater and the bright white light of a tungsten lightbulb are examples of radiation from heated solids.

Measurements taken in the latter part of the nineteenth century showed that the amount of radiant energy emitted by an object at a certain temperature depends on its wavelength. Attempts to account for this dependence in terms of established wave theory and thermodynamic laws were only partially successful. One theory explained short-wavelength dependence but failed to account for the longer wavelengths. Another theory accounted for the longer wavelengths but failed for short wavelengths. The failure in the short-wavelength region was called the *ultraviolet catastrophe*. It seemed that something fundamental was missing from the laws of classical physics.

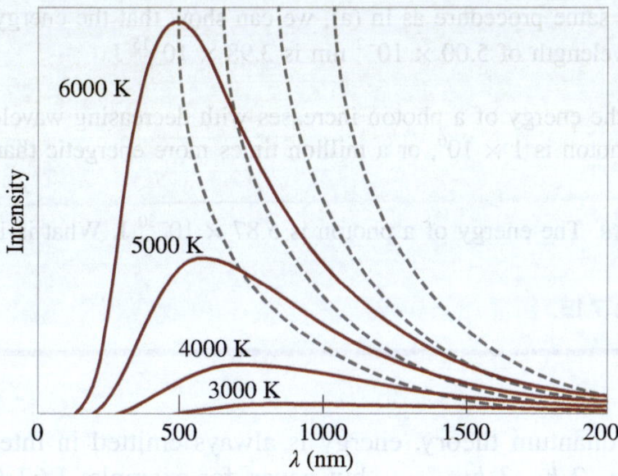

Planck solved the problem with an assumption that departed drastically from accepted concepts. Classical physics assumed that atoms and molecules could emit (or absorb) any arbitrary amount of radiant energy. Planck said that atoms and molecules could emit (or absorb) energy only in discrete quantities, like small packages or bundles. Planck gave the name ***quantum*** to *the smallest quantity of energy that can be emitted (or absorbed) in the form of electromagnetic radiation.* The energy E of a single quantum of energy is given by

$$E = h\nu \qquad (7.2)$$

where h is called *Planck's constant* and ν is the frequency of radiation. The value of Planck's constant is 6.63×10^{-34} J · s. The extremely small magnitude of Planck's constant is particularly important as it highlights the tiny scale of the quantum world. Its presence in an equation indicates the calculation of something on the quantum scale. Because $\nu = c/\lambda$, Equation (7.2) can also be expressed as

$$E = h\frac{c}{\lambda} \qquad (7.3)$$

Example 7.2

Calculate the energy (in joules) of (a) a photon with a wavelength of 5.00×10^4 nm (infrared region) and (b) a photon with a wavelength of 5.00×10^{-2} nm (X-ray region).

Strategy In both (a) and (b) we are given the wavelength of a photon and asked to calculate its energy. We need to use Equation (7.3) to calculate the energy. Planck's constant is given in the text and also in Appendix 6.

Solution

(a) From Equation (7.3),

$$E = h\frac{c}{\lambda}$$

$$= \frac{(6.63 \times 10^{-34}\ \text{J} \cdot \text{s})(3.00 \times 10^8\ \text{m/s})}{(5.00 \times 10^4\ \text{nm})\dfrac{1 \times 10^{-9}\ \text{m}}{1\ \text{nm}}}$$

$$= 3.98 \times 10^{-21}\ \text{J}$$

This is the energy of a single photon with a 5.00×10^4 nm wavelength.

(Continued)

(b) Following the same procedure as in (a), we can show that the energy of the photon that has a wavelength of 5.00×10^{-2} nm is 3.98×10^{-15} J.

Check Because the energy of a photon increases with decreasing wavelength, we see that an "X-ray" photon is 1×10^6, or a million times more energetic than an "infrared" photon.

Practice Exercise The energy of a photon is 5.87×10^{-20} J. What is its wavelength (in nanometers)?

Similar problem: 7.15.

According to quantum theory, energy is always emitted in integral multiples of $h\nu$; for example, $h\nu$, $2\ h\nu$, $3\ h\nu$, . . . , but never, for example, $1.67\ h\nu$ or $4.98\ h\nu$. At the time Planck presented his theory, he could not explain why energies should be fixed or quantized in this manner. Starting with this hypothesis, however, he had no trouble correlating the experimental data for emission by solids over the *entire* range of wavelengths; they all supported the quantum theory.

The idea that energy should be quantized or "bundled" may seem strange, but the concept of quantization has many analogies. For example, an electric charge is also quantized; there can be only whole-number multiples of *e*, the charge of one electron. Matter itself is quantized, for the numbers of electrons, protons, and neutrons and the numbers of atoms in a sample of matter must also be integers. The U.S. money system is based on a "quantum" of value called a penny. Even processes in living systems involve quantized phenomena. The eggs laid by hens are quantized, and a pregnant cat gives birth to an integral number of kittens, not to one-half or three-quarters of a kitten.

Summary of Concepts & Facts

- The quantum theory developed by Planck successfully explains the emission of radiation by heated solids. The quantum theory states that radiant energy is emitted by atoms and molecules in small discrete amounts (quanta), rather than over a continuous range. This behavior is governed by the relationship $E = h\nu$, where E is the energy of the radiation, h is Planck's constant, and ν is the frequency of the radiation. Energy is always emitted in whole-number multiples of $h\nu$ ($1\ h\nu$, $2\ h\nu$, $3\ h\nu$, . . .).

Review of Concepts & Facts

7.1.1 Which of the waves (a)–(c) has (i) the highest frequency, (ii) the longest wavelength, (iii) the greatest amplitude?

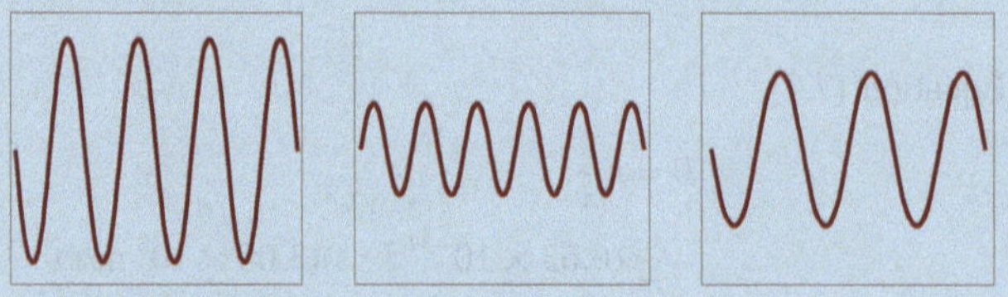

(a) (b) (c)

7.1.2 Why is radiation only in the UV but not the visible or infrared region responsible for suntanning?

7.2 The Photoelectric Effect

Learning Objective

- Describe the photoelectric effect and the use of the Planck equation in explaining the phenomenon.

In 1905, only five years after Planck presented his quantum theory, Albert Einstein† used the theory to solve another mystery in physics, the ***photoelectric effect,*** a phenomenon in which *electrons are ejected from the surface of certain metals exposed to light of at least a certain minimum frequency,* called *the threshold frequency* (Figure 7.7). The number of electrons ejected was proportional to the intensity (or brightness) of the light, but the energies of the ejected electrons were not. Below the threshold frequency no electrons were ejected no matter how intense the light.

The photoelectric effect could not be explained by the wave theory of light. Einstein, however, made an extraordinary assumption. He suggested that a beam of light is really a stream of particles. These *particles of light* are now called ***photons.*** Using Planck's quantum theory of radiation as a starting point, Einstein deduced that each photon must possess energy E, given by the equation

$$E = h\nu$$

where ν is the frequency of light. This equation has the same form as Equation (7.2) because, as we will see shortly, electromagnetic radiation is emitted as well as absorbed in the form of photons.

Electrons are held in a metal by attractive forces, and so removing them from the metal requires light of a sufficiently high frequency (which corresponds to sufficiently high energy; this frequency is the *threshold frequency*) to break them free (Figure 7.8). Shining a beam of light onto a metal surface can be thought of as shooting a beam of particles—photons—at the metal atoms. If the frequency of photons is such that $h\nu$ is exactly equal to *the energy that binds the electrons in the metal,* known as the ***work function (Φ),*** then the light will have just enough energy to knock the electrons loose. If we use light of a higher frequency, then not only will the electrons be knocked loose, but they will also acquire some kinetic energy (Figure 7.9). This situation is summarized by the equation

$$h\nu = E_k + \Phi \qquad (7.4)$$

where E_k is the kinetic energy of the ejected electron and Φ is the work function, which is a measure of how strongly the electrons are held in the metal. Rewriting Equation (7.4) as

$$E_k = h\nu - \Phi$$

shows that the more energetic the photon (that is, the higher the frequency), the greater the kinetic energy of the ejected electron.

†Albert Einstein (1879–1955). German-born American physicist. Regarded by many as one of the two greatest physicists the world has known (the other is Isaac Newton). The three papers (on special relativity, Brownian motion, and the photoelectric effect) that he published in 1905 while employed as a technical assistant in the Swiss patent office in Berne have profoundly influenced the development of physics. He received the Nobel Prize in Physics in 1921 for his explanation of the photoelectric effect.

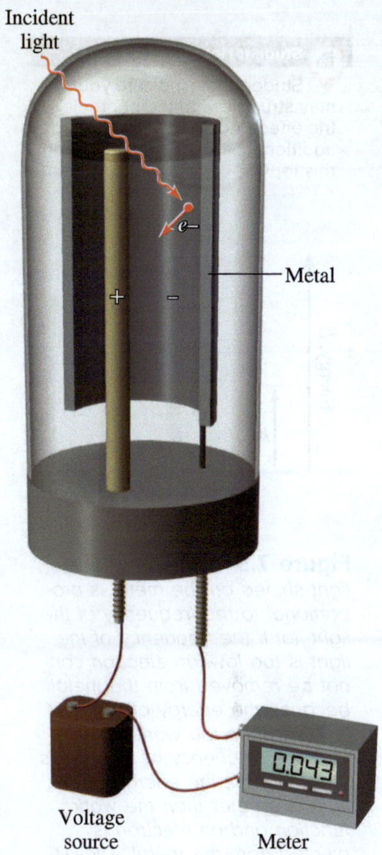

Figure 7.7 *An apparatus for studying the photoelectric effect. Light of a certain frequency falls on a clean metal surface. Ejected electrons are attracted toward the positive electrode. The flow of electrons is registered by a detecting meter. Light meters used in cameras are based on the photoelectric effect.*

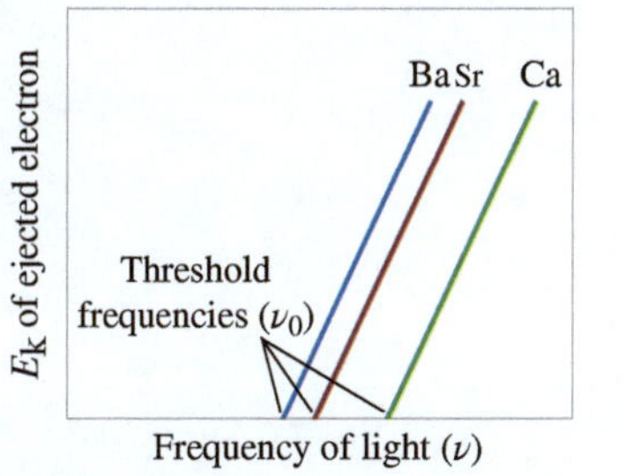

Figure 7.8 *In the photoelectric effect, no electrons are ejected when the frequency of light is below the threshold frequency of a metal. When the frequency of light is above the threshold frequency, electrons are ejected with kinetic energy that increases with increasing frequency of light.*

Student Hot Spot

Student data indicate you may struggle with the photoelectric effect. Access your eBook for additional Learning Resources on this topic.

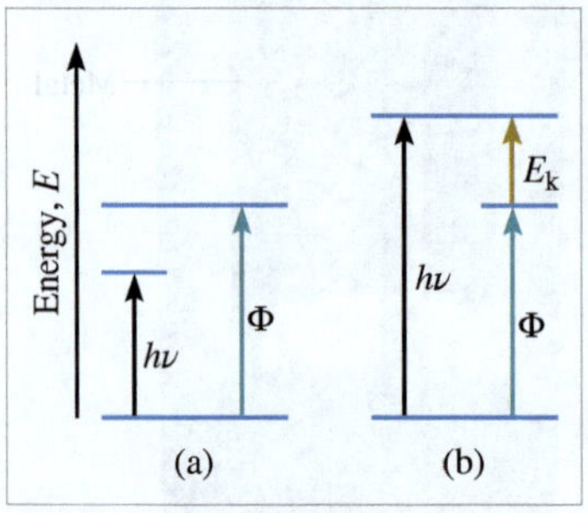

Figure 7.9 *The energy of the light shined on the metal is proportional to the frequency of the light. (a) If the frequency of the light is too low, an electron cannot be removed from the metal because the energy of the light is lower than the work function. (b) If the frequency of the light is high enough, the energy of the light is greater than the work function and an electron is ejected from the metal. The difference between the energy of the light and the work function is the kinetic energy of the ejected electron.*

Now consider two beams of light having the same frequency (which is greater than the threshold frequency) but different intensities. The more intense beam of light consists of a larger number of photons; consequently, it ejects more electrons from the metal's surface than the weaker beam of light. Thus, the more intense the light, the greater the number of electrons emitted by the target metal; the higher the frequency of the light, the greater the kinetic energy of the ejected electrons.

Example 7.3

The work function of cesium metal is 3.42×10^{-19} J. (a) Calculate the minimum frequency of light required to release electrons from the metal. (b) Calculate the kinetic energy of the ejected electron if light of frequency $1.00 \times 10^{15}\ s^{-1}$ is used for irradiating the metal.

Strategy (a) The relationship between the work function of an element and the frequency of light is given by Equation (7.4). The minimum frequency of light needed to dislodge an electron is the point where the kinetic energy of the ejected electron is zero. (b) Knowing both the work function and the frequency of light, we can solve for the kinetic energy of the ejected electron.

Solution

(a) Setting $E_k = 0$ in Equation (7.4), we write

$$h\nu = \Phi$$

Thus,

$$\nu = \frac{\Phi}{h} = \frac{3.42 \times 10^{-19}\ \text{J}}{6.63 \times 10^{-34}\ \text{J} \cdot \text{s}}$$

$$= 5.16 \times 10^{14}\ \text{s}^{-1}$$

(b) Rearranging Equation (7.4) gives

$$\begin{aligned} E_k &= h\nu - \Phi \\ &= (6.63 \times 10^{-34}\ \text{J} \cdot \text{s})(1.00 \times 10^{15}\ \text{s}^{-1}) - 3.42 \times 10^{-19}\ \text{J} \\ &= 3.21 \times 10^{-19}\ \text{J} \end{aligned}$$

Check The kinetic energy of the ejected electron (3.21×10^{-19} J) is smaller than the energy of the photon (6.63×10^{-19} J). Therefore, the answer is reasonable.

Practice Exercise The work function of titanium metal is 6.93×10^{-19} J. Calculate the kinetic energy of the ejected electrons if light of frequency $2.50 \times 10^{15}\ s^{-1}$ is used to irradiate the metal.

Similar problems: 7.21, 7.22.

Einstein's theory of light posed a dilemma for scientists. On the one hand, it explains the photoelectric effect satisfactorily. On the other hand, the particle theory of light is not consistent with the known wave behavior of light. The only way to resolve the dilemma is to accept the idea that light possesses *both* particlelike and wavelike properties. Depending on the experiment, light behaves either as a wave or as a stream of particles. This concept, called particle-wave duality, was totally alien to the way physicists had thought about matter and radiation, and it took a long time for them to accept it. We will see in Section 7.4 that a dual nature (particles and waves) is not unique to light but is also characteristic of all matter, including electrons.

Summary of Concepts & Facts

- Using quantum theory, Einstein solved another mystery of physics—the photoelectric effect. Einstein proposed that light can behave like a stream of particles (photons).

Review of Concepts & Facts

7.2.1 A photon has a frequency of 7.25×10^{14} s^{-1}. Determine the energy (in joules) of this photon.

7.2.2 What is the wavelength (in nm) of a photon with energy 6.69×10^{-19} J/photon?

7.2.3 A clean metal surface is irradiated with light of three different wavelengths λ_1, λ_2, and λ_3. The kinetic energies of the ejected electrons are as follows: λ_1: 2.9×10^{-20} J; λ_2: approximately zero; λ_3: 4.2×10^{-19} J. Which light has the shortest wavelength and which has the longest wavelength?

7.3 Bohr's Theory of the Hydrogen Atom

Learning Objective

- Generalize the Bohr model of the hydrogen atom and how it explains atomic emission spectra in hydrogen.

Einstein's work paved the way for the solution of yet another nineteenth-century "mystery" in physics: the emission spectra of atoms.

Emission Spectra

Since the seventeenth century, when Newton showed that sunlight is composed of various color components that can be recombined to produce white light, chemists and physicists have studied the characteristics of ***emission spectra,*** that is, *either continuous or line spectra of radiation emitted by substances.* The emission spectrum of a substance can be seen by energizing a sample of material either with thermal energy or with some other form of energy (such as a high-voltage electrical discharge). A "red-hot" or "white-hot" iron bar freshly removed from a high-temperature source produces a characteristic glow. This visible glow is the portion of its emission spectrum that is sensed by the eye. The warmth of the same iron bar represents another portion of its emission spectrum—the infrared region. A feature common to the emission spectra of the sun and of a heated solid is that both are continuous; that is, all wavelengths of visible light are represented in the spectra (see the visible region in Figure 7.4).

The emission spectra of atoms in the gas phase, on the other hand, do not show a continuous spread of wavelengths from red to violet; rather, the atoms produce bright lines in different parts of the visible spectrum. These ***line spectra*** are *the light emission only at specific wavelengths.* Figure 7.10 is a schematic diagram of a discharge tube that is used to study emission spectra, and Figure 7.11 shows the color emitted by hydrogen atoms in a discharge tube.

Every element has a unique emission spectrum. The characteristic lines in atomic spectra can be used in chemical analysis to identify unknown atoms, much as fingerprints are used to identify people. When the lines of the emission spectrum of a known element exactly match the lines of the emission spectrum of an unknown sample, the identity of the sample is established. Although the utility of this procedure was recognized some time ago in chemical analysis, the origin of these lines was unknown until early in the twentieth century. Figure 7.12 shows the emission spectra of several elements.

Video: Line Spectra

When a high voltage is applied between the forks, some of the sodium ions in the pickle are converted to sodium atoms in an excited state. These atoms emit the characteristic yellow light as they relax to the ground state.
Stephen Frisch/McGraw Hill

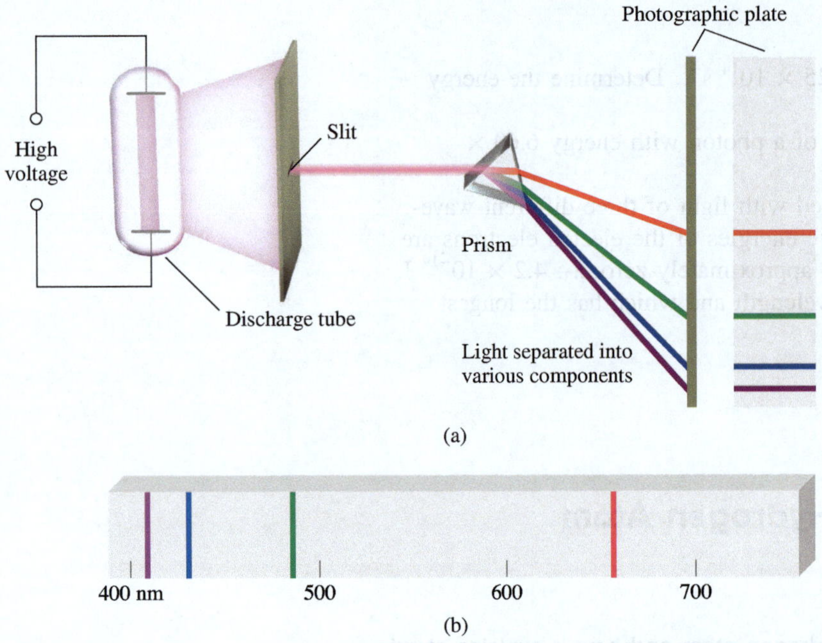

Figure 7.10 *(a) An experimental arrangement for studying the emission spectra of atoms and molecules. The gas under study is in a discharge tube containing two electrodes. As electrons flow from the negative electrode to the positive electrode, they collide with the gas. This collision process eventually leads to the emission of light by the atoms (or molecules). The emitted light is separated into its components by a prism. Each component color is focused at a definite position, according to its wavelength, and forms a colored image of the slit on the photographic plate. The colored images are called spectral lines. (b) The line emission spectrum of hydrogen atoms.*

Figure 7.11 *Color emitted by hydrogen atoms in a discharge tube. The color observed results from the combination of the colors emitted in the visible spectrum.*

SPL/Science Source

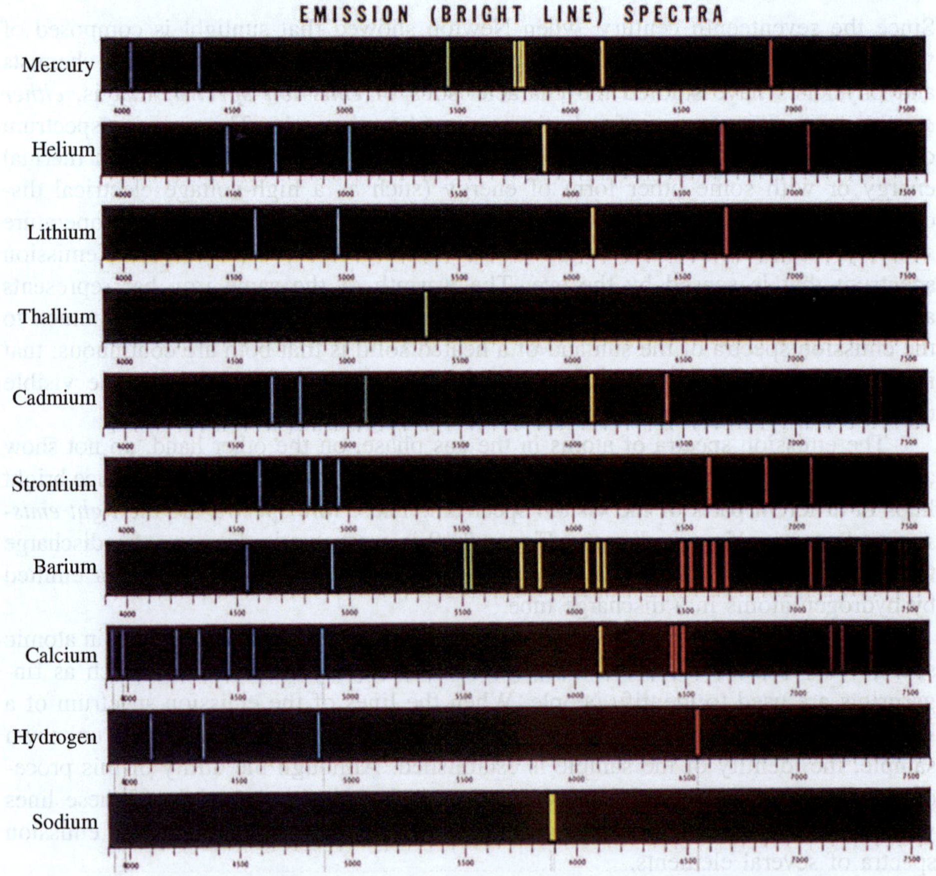

Figure 7.12 *The emission spectra of various elements. The units for the spectral lines are angstroms (Å), where 1 Å = 1×10^{-10} m.*

Courtesy of Wabash Instrument Corporation

Emission Spectrum of the Hydrogen Atom

▶▶▶ Video: Atomic Line Spectra

In 1913, not too long after Planck's and Einstein's discoveries, a theoretical explanation of the emission spectrum of the hydrogen atom was presented by the Danish physicist Niels Bohr.† Bohr's treatment is very complex and is no longer considered to be correct in all its details. Thus, we will concentrate only on his important assumptions and final results, which account for the spectral lines.

When Bohr first tackled this problem, physicists already knew that the atom contains electrons and protons. They thought of an atom as an entity in which electrons whirled around the nucleus in circular orbits at high velocities. This was an appealing model because it resembled the motions of the planets around the sun. In the hydrogen atom, it was believed that the electrostatic attraction between the positive "solar" proton and the negative "planetary" electron pulls the electron inward and that this force is balanced exactly by the outward acceleration due to the circular motion of the electron.

According to the laws of classical physics, however, an electron moving in an orbit of a hydrogen atom would experience an acceleration toward the nucleus by radiating away energy in the form of electromagnetic waves. Thus, such an electron would quickly spiral into the nucleus and annihilate itself with the proton. To explain why this does not happen, Bohr postulated that the electron is allowed to occupy only certain orbits of specific energies. In other words, the energies of the electron are quantized. An electron in any of the allowed orbits will not spiral into the nucleus and therefore will not radiate energy. Bohr attributed the emission of radiation by an energized hydrogen atom to the electron dropping from a higher-energy allowed orbit to a lower one and emitting a quantum of energy (a photon) in the form of light (Figure 7.13). Bohr showed that the energies that an electron in a hydrogen atom can occupy are given by

$$E_n = -R_H\left(\frac{1}{n^2}\right) \tag{7.5}$$

where R_H, the Rydberg‡ constant for the hydrogen atom, has the value 2.18×10^{-18} J. The number n is an integer called the principal quantum number; it has the values $n = 1, 2, 3, \ldots$.

The negative sign in Equation (7.5) is an arbitrary convention, signifying that the energy of the electron in the atom is *lower* than the energy of a *free electron,* which is an electron that is infinitely far from the nucleus. The energy of a free electron is arbitrarily assigned a value of zero. Mathematically, this corresponds to setting n equal to infinity in Equation (7.5), so that $E_\infty = 0$. As the electron gets closer to the nucleus (as n decreases), E_n becomes larger in absolute value, but also more negative. The most negative value, then, is reached when $n = 1$, which corresponds to the most stable energy state. We call this the ***ground state,*** or the ***ground level,*** which refers to *the lowest energy state of a system* (which is an atom in our discussion). The stability of the electron diminishes for $n = 2, 3, \ldots$. Each of these levels is called an ***excited state,*** or ***excited level,*** which is *higher in energy than the ground state.* A hydrogen electron for which n is greater than 1 is said to be in an excited state. The radius of each circular orbit in Bohr's model depends on n^2. Thus, as n increases from 1 to 2 to 3, the orbit radius increases very rapidly. The higher the excited state, the farther away the electron is from the nucleus (and the less tightly it is held by the nucleus).

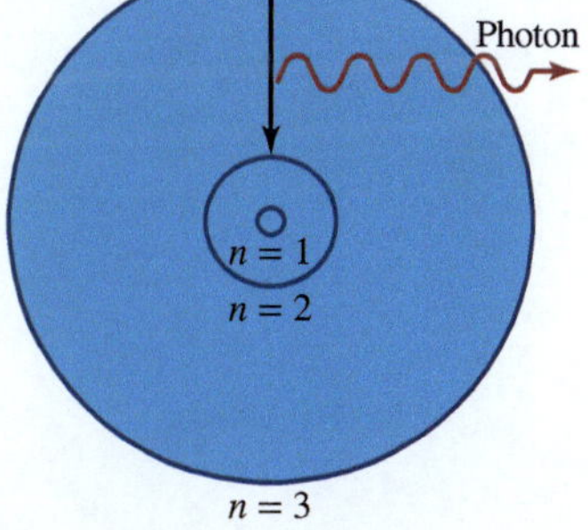

Figure 7.13 *The emission process in an excited hydrogen atom, according to Bohr's theory. An electron originally in a higher-energy orbit (n = 3) falls back to a lower-energy orbit (n = 2). As a result, a photon with energy hν is given off. The value of hν is equal to the difference in energies of the two orbits occupied by the electron in the emission process. For simplicity, only three orbits are shown.*

†Niels Henrik David Bohr (1885–1962). Danish physicist. One of the founders of modern physics, he received the Nobel Prize in Physics in 1922 for his theory explaining the spectrum of the hydrogen atom.

‡Johannes Robert Rydberg (1854–1919). Swedish physicist. Rydberg's major contribution to physics was his study of the line spectra of many elements.

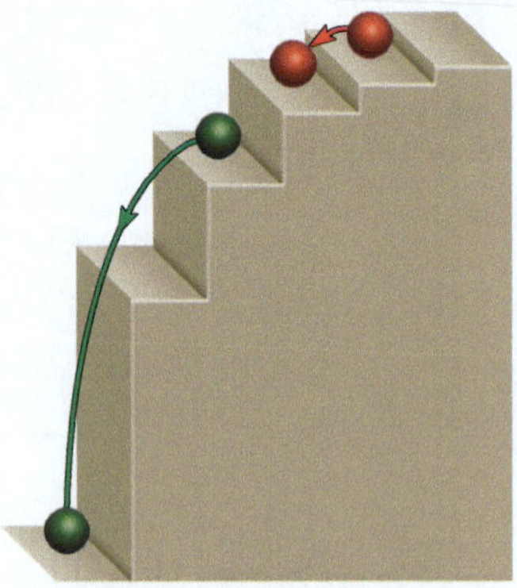

Figure 7.14 *A mechanical analogy for the emission processes. The ball can rest on any step but not between steps.*

Bohr's theory enables us to explain the line spectrum of the hydrogen atom. Radiant energy absorbed by the atom causes the electron to move from a lower-energy state (characterized by a smaller n value) to a higher-energy state (characterized by a larger n value). Conversely, radiant energy (in the form of a photon) is emitted when the electron moves from a higher-energy state to a lower-energy state. The quantized movement of the electron from one energy state to another is analogous to the movement of a tennis ball either up or down a set of stairs (Figure 7.14). The ball can be on any of several steps but never between steps. The journey from a lower step to a higher one is an energy-requiring process, whereas movement from a higher step to a lower step is an energy-releasing process. The quantity of energy involved in either type of change is determined by the distance between the beginning and ending steps. Similarly, the amount of energy needed to move an electron in the Bohr atom depends on the difference in energy levels between the initial and final states.

To apply Equation (7.5) to the emission process in a hydrogen atom, let us suppose that the electron is initially in an excited state characterized by the principal quantum number n_i. During emission, the electron drops to a lower energy state characterized by the principal quantum number n_f (the subscripts i and f denote the initial and final states, respectively). This lower energy state may be either a less excited state or the ground state. The difference between the energies of the initial and final states is

$$\Delta E = E_f - E_i$$

From Equation (7.5),

$$E_f = -R_H\left(\frac{1}{n_f^2}\right)$$

and

$$E_i = -R_H\left(\frac{1}{n_i^2}\right)$$

Therefore,

$$\Delta E = \left(\frac{-R_H}{n_f^2}\right) - \left(\frac{-R_H}{n_i^2}\right)$$

$$= R_H\left(\frac{1}{n_i^2} - \frac{1}{n_f^2}\right)$$

Because this transition results in the emission of a photon of frequency ν and energy $h\nu$, we can write

$$\Delta E = h\nu = R_H\left(\frac{1}{n_i^2} - \frac{1}{n_f^2}\right) \quad (7.6)$$

When a photon is emitted, $n_i > n_f$. Consequently the term in parentheses is negative and ΔE is negative (energy is lost to the surroundings). When energy is absorbed, $n_i < n_f$ and the term in parentheses is positive, so ΔE is positive. Each spectral line in the emission spectrum corresponds to a particular transition in a hydrogen atom. When we study a large number of hydrogen atoms, we observe all possible transitions and hence the corresponding spectral lines. The brightness of a spectral line depends on how many photons of the same wavelength are emitted.

The emission spectrum of hydrogen includes a wide range of wavelengths from the infrared to the ultraviolet. Table 7.1 lists the series of transitions in the hydrogen spectrum; they are named after their discoverers. The Balmer series was particularly easy to study because a number of its lines fall in the visible range.

Figure 7.13 shows a single transition. However, it is more informative to express transitions as shown in Figure 7.15. Each horizontal line represents an allowed energy level for the electron in a hydrogen atom. The energy levels are labeled with their principal quantum numbers.

Table 7.1 The Various Series in Atomic Hydrogen Emission Spectrum

Series	n_f	n_i	Spectrum Region
Lyman	1	2, 3, 4, . . .	Ultraviolet
Balmer	2	3, 4, 5, . . .	Visible and ultraviolet
Paschen	3	4, 5, 6, . . .	Infrared
Brackett	4	5, 6, 7, . . .	Infrared
Pfund	5	6, 7, 8, . . .	Infrared
Humphreys	6	7, 8, 9, . . .	Infrared

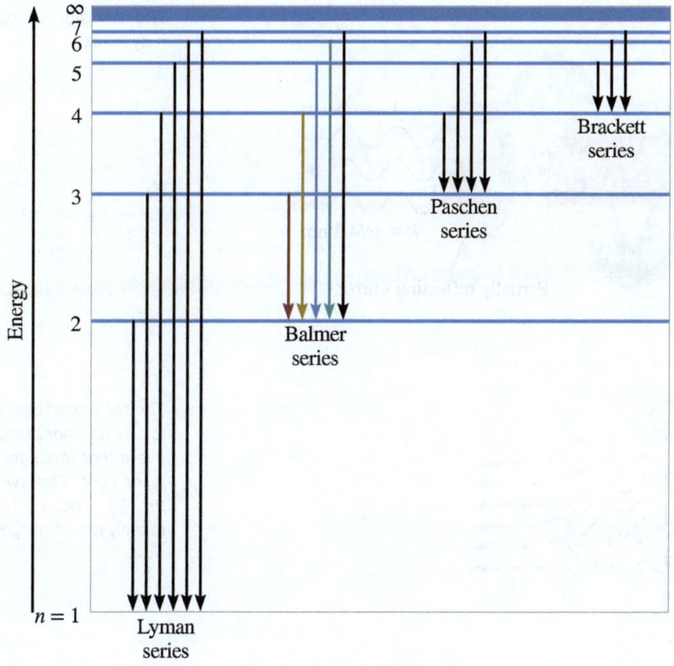

Figure 7.15 *The energy levels in the hydrogen atom and the various emission series. Each energy level corresponds to the energy associated with an allowed energy state for an orbit, as postulated by Bohr and shown in Figure 7.13. The emission lines are labeled according to the scheme in Table 7.1.*

Example 7.4 illustrates the use of Equation (7.6).

Student Hot Spot

Student data indicate you may struggle with calculating energy levels in the Bohr hydrogen atom. Access your eBook for additional Learning Resources on this topic.

Example 7.4

What is the wavelength of a photon (in nanometers) emitted during a transition from the $n_i = 5$ state to the $n_f = 2$ state in the hydrogen atom?

Strategy We are given the initial and final states in the emission process. We can calculate the energy of the emitted photon using Equation (7.6). Then from Equations (7.2) and (7.1) we can solve for the wavelength of the photon. The value of Rydberg's constant is given in the text.

Solution From Equation (7.6) we write

$$\Delta E = R_H\left(\frac{1}{n_i^2} - \frac{1}{n_f^2}\right)$$

$$= 2.18 \times 10^{-18}\ \text{J}\left(\frac{1}{5^2} - \frac{1}{2^2}\right)$$

$$= -4.58 \times 10^{-19}\ \text{J}$$

(Continued)

CHEMISTRY *in Action*

Laser—The Splendid Light

Laser is an acronym for "light amplification by stimulated emission of radiation." It is a special type of emission that involves either atoms or molecules. Since the discovery of laser in 1960, it has been used in numerous systems designed to operate in the gas, liquid, and solid states. These systems emit radiation with wavelengths ranging from infrared through visible and ultraviolet. The advent of laser has truly revolutionized science, medicine, and technology.

Ruby laser was the first known laser. Ruby is a deep-red mineral containing corundum, Al_2O_3, in which some of the Al^{3+} ions have been replaced by Cr^{3+} ions. A flashlamp is used to excite the chromium atoms to a higher energy level. The excited

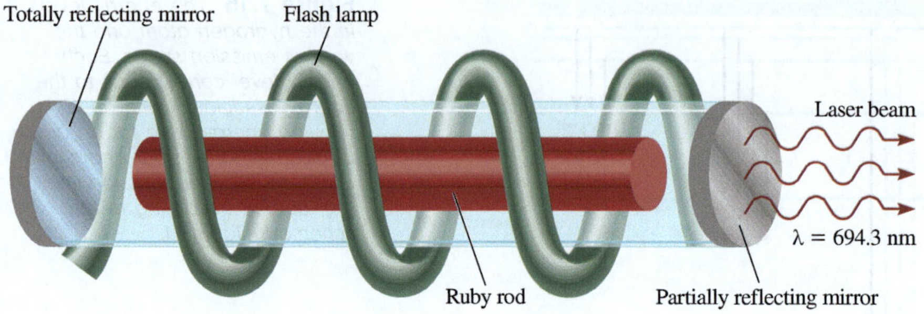

The emission of laser light from a ruby laser.

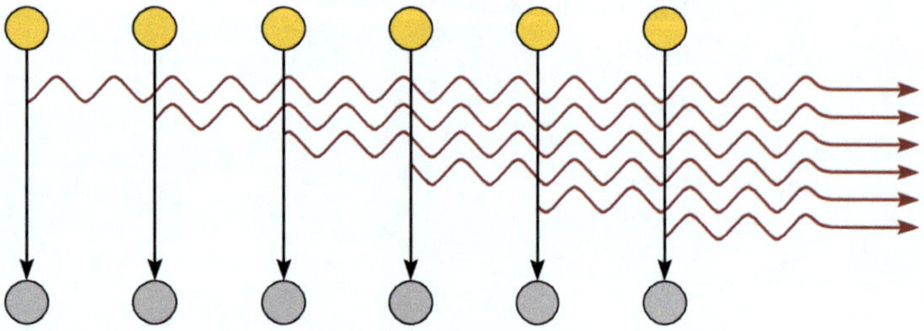

The stimulated emission of one photon by another photon in a cascade event that leads to the emission of laser light. The synchronization of the light waves produces an intensely penetrating laser beam.

The negative sign indicates that this is energy associated with an emission process. To calculate the wavelength, we will omit the minus sign for ΔE because the wavelength of the photon must be positive. Because $\Delta E = h\nu$ or $\nu = \Delta E/h$, we can calculate the wavelength of the photon by writing

$$\begin{aligned}\lambda &= \frac{c}{\nu}\\ &= \frac{ch}{\Delta E}\\ &= \frac{(3.00 \times 10^8 \text{ m/s})(6.63 \times 10^{-34} \text{ J}\cdot\text{s})}{4.58 \times 10^{-19} \text{ J}}\\ &= 4.34 \times 10^{-7} \text{ m}\\ &= 4.34 \times 10^{-7} \text{ m}\left(\frac{1 \text{ nm}}{1 \times 10^{-9} \text{ m}}\right) = 434 \text{ nm}\end{aligned}$$

Check The wavelength is in the visible region of the electromagnetic region (see Figure 7.4). This is consistent with the fact that because $n_f = 2$, this transition gives rise to a spectral line in the Balmer series (see Figure 7.15).

(Continued)

atoms are unstable, so at a given instant some of them will return to the ground state by emitting a photon in the red region of the spectrum. The photon bounces back and forth many times between mirrors at opposite ends of the laser tube. This photon can stimulate the emission of photons of exactly the same wavelength from other excited chromium atoms; these photons in turn can stimulate the emission of more photons, and so on. Because the light waves are *in phase*—that is, their maxima and minima coincide—the photons enhance one another, increasing their power with each passage between the mirrors. One of the mirrors is only partially reflecting, so that when the light reaches a certain intensity it emerges from the mirror as a laser beam. Depending on the mode of operation, the laser light may be emitted in pulses (as in the ruby laser case) or in continuous waves.

Laser light is characterized by three properties: It is intense, it has precisely known wavelength and hence energy, and it is coherent. By *coherent* we mean that the light waves are all in phase. The applications of lasers are quite numerous. Their high intensity and ease of focus make them suitable for doing eye surgery, for drilling holes in metals and welding, and for carrying out nuclear fusion. The fact that they are highly directional and have precisely known wavelengths makes them very useful for telecommunications. Lasers are also used in isotope separation, in holography (three-dimensional photography), in compact disc players, and in supermarket scanners. Lasers have played an important role in the spectroscopic investigation of molecular properties and of many chemical and biological processes.

State-of-the-art lasers used in the research laboratory of Dr. A. H. Zewail at the California Institute of Technology.
Courtesy of Professor Ahmed Zewail/Arthur Amos Noyes Laboratory of Chemical Physics/California Institute of Technology

Practice Exercise What is the wavelength (in nanometers) of a photon emitted during a transition from $n_i = 6$ to $n_f = 4$ state in the H atom?

Similar problems: 7.31, 7.32.

The Chemistry in Action essay "Laser—The Splendid Light" discusses a special type of atomic emission—lasers.

Summary of Concepts & Facts

- The line spectrum of hydrogen, yet another mystery to nineteenth-century physicists, was also explained by applying the quantum theory. Bohr developed a model of the hydrogen atom in which the energy of its single electron is quantized—limited to certain energy values determined by an integer, the principal quantum number.
- An electron in its most stable energy state is said to be in the ground state, and an electron at an energy level higher than its most stable state is said to be in an

excited state. In the Bohr model, an electron emits a photon when it drops from a higher-energy state (an excited state) to a lower-energy state (the ground state or another, less excited state). The release of specific amounts of energy in the form of photons accounts for the lines in the hydrogen emission spectrum.

Review of Concepts & Facts

7.3.1 Which transition in the hydrogen atom would emit light of a shorter wavelength: (a) $n_i = 5 \longrightarrow n_f = 3$ or (b) $n_i = 4 \longrightarrow n_f = 2$?

7.3.2 What is the energy of an electron in the $n = 4$ state in a hydrogen atom?

7.3.3 What is the wavelength (in nm) of a photon emitted when an electron in a hydrogen atom transitions from $n_i = 7$ to $n_f = 2$?

7.4 The Dual Nature of the Electron

Learning Objective

- Assess the importance of the de Broglie wave equation in regard to the dual nature of the electron.

Physicists were both mystified and intrigued by Bohr's theory. They questioned why the energies of the hydrogen electron are quantized. Or, phrasing the question in a more concrete way, Why is the electron in a Bohr atom restricted to orbiting the nucleus at certain fixed distances? For a decade no one, not even Bohr himself, had a logical explanation. In 1924 Louis de Broglie† provided a solution to this puzzle. De Broglie reasoned that if light waves can behave like a stream of particles (photons), then perhaps particles such as electrons can possess wave properties.

According to de Broglie, an electron bound to the nucleus behaves like a *standing wave.* Standing waves can be generated by plucking, say, a guitar string (Figure 7.16). The waves are described as standing, or stationary, because they do not travel along the string. Some points on the string, called *nodes,* do not move at all; that is, *the amplitude of the wave at these points is zero.* There is a node at each end, and there may be nodes between the ends. The greater the frequency of vibration, the shorter the wavelength of the standing wave and the greater the number of nodes. As Figure 7.16 shows, there can be only certain wavelengths in any of the allowed motions of the string.

De Broglie argued that if an electron does behave like a standing wave in the hydrogen atom, the length of the wave must fit the circumference of the orbit exactly (Figure 7.17). Otherwise the wave would partially cancel itself on each successive orbit. Eventually the amplitude of the wave would be reduced to zero, and the wave would not exist.

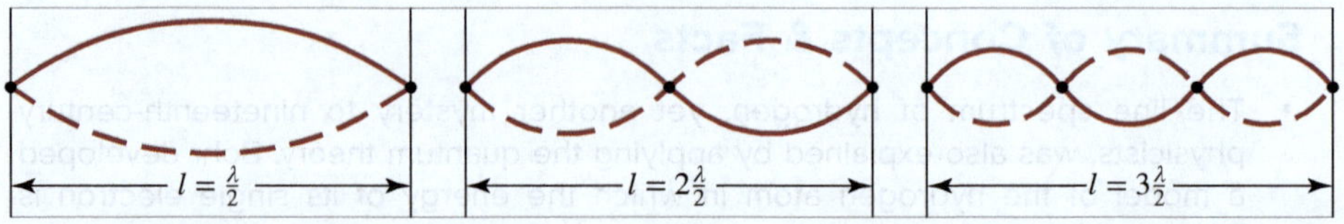

Figure 7.16 *The standing waves generated by plucking a guitar string. Each dot represents a node. The length of the string (l) must be equal to a whole number times one-half the wavelength (λ/2).*

†Louis Victor Pierre Raymond Duc de Broglie (1892–1987). French physicist. Member of an old and noble family in France, he held the title of a prince. In his doctoral dissertation, he proposed that matter and radiation have the properties of both wave and particle. For this work, de Broglie was awarded the Nobel Prize in Physics in 1929.

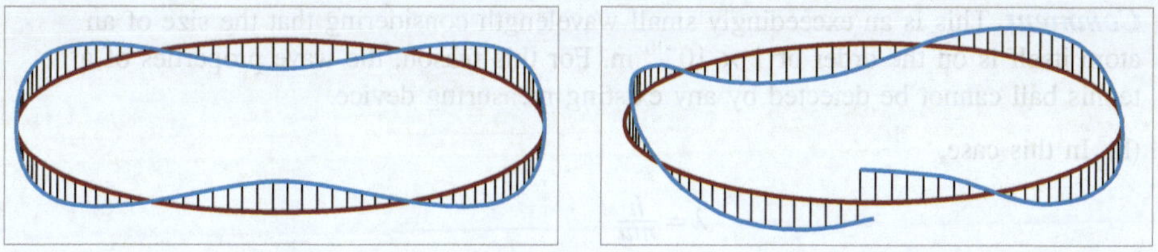

(a) (b)

Figure 7.17 *(a) The circumference of the orbit is equal to an integral number of wavelengths. This is an allowed orbit. (b) The circumference of the orbit is not equal to an integral number of wavelengths. As a result, the electron wave does not close in on itself. This is a nonallowed orbit.*

The relation between the circumference of an allowed orbit ($2\pi r$) and the wavelength (λ) of the electron is given by

$$2\pi r = n\lambda \quad (7.7)$$

where r is the radius of the orbit, λ is the wavelength of the electron wave, and $n = 1, 2, 3, \ldots$. Because n is an integer, it follows that r can have only certain values as n increases from 1 to 2 to 3 and so on. And because the energy of the electron depends on the size of the orbit (or the value of r), its value must be quantized.

De Broglie's reasoning led to the conclusion that waves can behave like particles and particles can exhibit wavelike properties. De Broglie deduced that the particle and wave properties are related by the expression

$$\lambda = \frac{h}{mu} \quad (7.8)$$

where λ, m, and u are the wavelengths (in m) associated with a moving particle, its mass (in kg), and its velocity (in m/s), respectively. Equation (7.8) implies that a particle in motion can be treated as a wave, and a wave can exhibit the properties of a particle. Note that the left side of Equation (7.8) involves the wavelike property of wavelength, whereas the right side makes references to mass, a distinctly particlelike property. Any wavelength calculated using Equation (7.8) is commonly known as a de Broglie wavelength. Also note the presence of Planck's constant in the equation. As noted in Section 7.1, this is an indication of a quantum-scale quantity.

Student Hot Spot

Student data indicate you may struggle with the de Broglie wavelength. Access your eBook for additional Learning Resources on this topic.

Example 7.5

Calculate the wavelength of the "particle" in the following two cases: (a) The fastest serve in tennis is about 150 miles per hour, or 68 m/s. Calculate the wavelength associated with a 6.0×10^{-2} kg tennis ball traveling at this speed. (b) Calculate the wavelength associated with an electron (9.1094×10^{-31} kg) moving at 68 m/s.

Strategy We are given the mass and the speed of the particle in (a) and (b) and asked to calculate the wavelength, so we need Equation (7.8). Note that because the units of Planck's constants are J·s, m and u must be in kg and m/s (1 J = 1 kg · m^2/s^2), respectively.

Solution

(a) Using Equation (7.8) we write

$$\begin{aligned}\lambda &= \frac{h}{mu} \\ &= \frac{6.63 \times 10^{-34}\ \text{J}\cdot\text{s}}{(6.0 \times 10^{-2}\ \text{kg}) \times 68\ \text{m/s}} \\ &= 1.6 \times 10^{-34}\ \text{m}\end{aligned}$$

(Continued)

Comment This is an exceedingly small wavelength considering that the size of an atom itself is on the order of 1×10^{-10} m. For this reason, the wave properties of a tennis ball cannot be detected by any existing measuring device.

(b) In this case,

$$\lambda = \frac{h}{mu}$$

$$= \frac{6.63 \times 10^{-34}\ \text{J} \cdot \text{s}}{(9.1094 \times 10^{-31}\ \text{kg}) \times 68\ \text{m/s}}$$

$$= 1.1 \times 10^{-5}\ \text{m}$$

Comment This wavelength (1.1×10^{-5} m or 1.1×10^{4} nm) is in the infrared region. This calculation shows that only electrons (and other submicroscopic particles) have measurable wavelengths.

Practice Exercise Calculate the wavelength (in nanometers) of a H atom (mass = 1.674×10^{-27} kg) moving at 7.00×10^{2} cm/s.

Similar problems: 7.40, 7.41.

Example 7.5 shows that although de Broglie's equation can be applied to diverse systems, the wave properties become observable only for submicroscopic objects.

Shortly after de Broglie introduced his equation, Clinton Davisson[†] and Lester Germer[‡] in the United States and G. P. Thomson[§] in England demonstrated that electrons do indeed possess wavelike properties. By directing a beam of electrons through a thin piece of gold foil, Thomson obtained a set of concentric rings on a screen, similar to the pattern observed when X rays (which are waves) were used. Figure 7.18 shows such a pattern for aluminum.

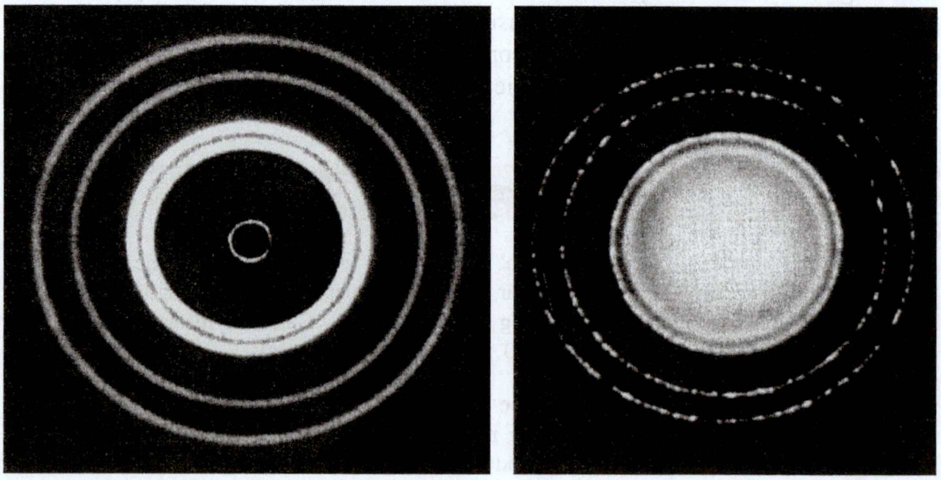

(a) (b)

Figure 7.18 *(a) X-ray diffraction pattern of aluminum foil. (b) Electron diffraction of aluminum foil. The similarity of these two patterns shows that electrons can behave like X rays and display wave properties.*

(a and b): Education Development Center, Inc.

[†]Clinton Joseph Davisson (1881–1958). American physicist. He and G. P. Thomson shared the Nobel Prize in Physics in 1937 for demonstrating wave properties of electrons.

[‡]Lester Halbert Germer (1896–1972). American physicist. Discoverer (with Davisson) of the wave properties of electrons.

[§]George Paget Thomson (1892–1975). English physicist. Son of J. J. Thomson, he received the Nobel Prize in Physics in 1937, along with Clinton Davisson, for demonstrating wave properties of electrons.

CHEMISTRY *in Action*

Electron Microscopy

The electron microscope is an extremely valuable application of the wavelike properties of electrons because it produces images of objects that cannot be seen with the naked eye or with light microscopes. According to the laws of optics, it is impossible to form an image of an object that is smaller than half the wavelength of the light used for the observation. Because the range of visible light wavelengths starts at around 400 nm, or 4×10^{-5} cm, we cannot see anything smaller than 2×10^{-5} cm. In principle, we can see objects on the atomic and molecular scale by using X rays, whose wavelengths range from about 0.01 nm to 10 nm. However, X rays cannot be focused, so they do not produce well-formed images. Electrons, on the other hand, are charged particles, which can be focused in the same way as the image on a TV screen is focused, that is, by applying an electric field or a magnetic field. According to Equation (7.8), the wavelength of an electron is inversely proportional to its velocity. By accelerating electrons to very high velocities, we can obtain wavelengths as short as 0.004 nm.

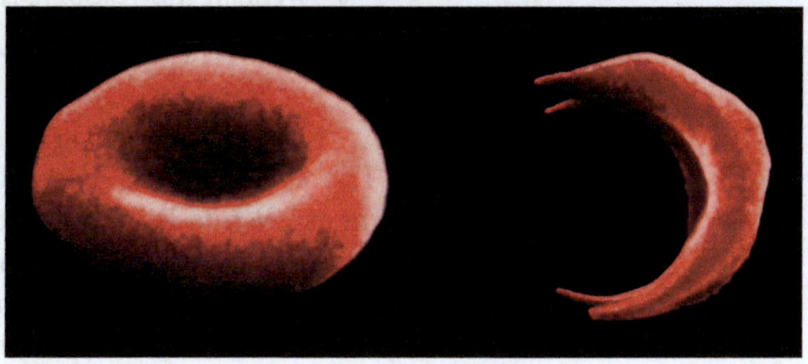

An electron micrograph showing a normal red blood cell and a sickled red blood cell from the same person.
Mary Martin/Science Source

A different type of electron microscope, called the *scanning tunneling microscope* (*STM*), makes use of another quantum mechanical property of the electron to produce an image of the atoms on the surface of a sample. Because of its extremely small mass, an electron is able to move or "tunnel" through an energy barrier (instead of going over it). The STM consists of a tungsten metal needle with a very fine point, the source of the tunneling electrons. A voltage is maintained between the needle and the surface of the sample to induce electrons to tunnel through space to the sample. As the needle moves over the sample, at a distance of a few atomic diameters from the surface, the tunneling current is measured. This current decreases with increasing distance from the sample. By using a feedback loop, the vertical position of the tip can be adjusted to a constant distance from the surface. The extent of these adjustments, which profile the sample, is recorded and displayed as a three-dimensional false-colored image.

Both the electron microscope and the STM are among the most powerful tools in chemical and biological research.

The wave properties of matter give rise to experimental techniques such as electron microscopy and neutron diffraction, which can be used to study the structure and dynamics of liquids and solids. The Chemistry in Action essay "Electron Microscopy" describes electron microscopy.

Summary of Concepts & Facts

- De Broglie extended Einstein's wave-particle description of light to all matter in motion. The wavelength of a moving particle of mass *m* and velocity *u* is given by the de Broglie equation $\lambda = h/mu$.

Review of Concepts & Facts

7.4.1 Which quantity in Equation (7.8) is responsible for the fact that macroscopic objects do not show observable wave properties?

7.4.2 What is the wavelength (in nm) of a neutron (1.67493×10^{-27} kg) traveling at 5.0 m/s?

7.5 Quantum Mechanics

Learning Objectives

- Explain the relevance of the Heisenberg uncertainty principle.
- Appraise how electron density defines the shape of atomic orbitals.

The spectacular success of Bohr's theory was followed by a series of disappointments. While Bohr's approach accounted for the observed emission spectra of hydrogen and hydrogen-like species such as He^+ and Li^{2+}, it could not account for the emission spectra of atoms containing more than one electron, such as atoms of helium and lithium. Further, it did not explain why extra lines appear in the hydrogen emission spectrum when a magnetic field is applied. Another problem arose with the discovery that electrons are wavelike: How can the "position" of a wave be specified? We cannot define the precise location of a wave because a wave extends in space.

The Heisenberg Uncertainty Principle

The dual nature of electrons was particularly troublesome because of the electron's exceedingly small mass. To describe the problem of trying to locate a subatomic particle that behaves like a wave, Werner Heisenberg† formulated what is now known as the ***Heisenberg uncertainty principle:*** *It is impossible to know simultaneously both the momentum p* (defined as mass times velocity) *and the position of a particle with certainty.* Stated mathematically,

$$\Delta x \Delta p \geq \frac{h}{4\pi} \tag{7.9}$$

where Δx and Δp are the uncertainties in measuring the position and momentum of the particle, respectively. To calculate momentum (p) using Equation (7.9), m must be in kg and u must be in m/s. The $\geq$ signs have the following meaning: If the measured uncertainties of position and momentum are large (say, in a crude experiment), their product can be substantially greater than $h/4\pi$ (hence, the $>$ sign). The significance of Equation (7.9) is that even in the most favorable conditions for measuring position and momentum, the product of the uncertainties can never be less than $h/4\pi$ (hence, the $=$ sign). Thus, making measurement of the momentum of a particle *more* precise (that is, making Δp a *small* quantity) means that the position must become correspondingly *less* precise (that is, Δx will become *larger*). Similarly, if the position of the particle is known *more* precisely, its momentum measurement must become less precise. The Heisenberg uncertainty principle imposes a fundamental limit on what we are able to know about the nature of matter on the quantum scale.

Applying the Heisenberg uncertainty principle to the hydrogen atom, we see that in reality, the electron does not orbit the nucleus in a well-defined path, as Bohr thought. Thus, it is not appropriate to imagine the electron circling the nucleus in well-defined orbitals. If it did, we could determine precisely both the position of the electron (from its location on a particular orbit) and its momentum (from its kinetic energy) at the same time, a violation of the uncertainty principle.

†Werner Karl Heisenberg (1901–1976). German physicist. One of the founders of modern quantum theory, Heisenberg received the Nobel Prize in Physics in 1932.

Example 7.6

(a) An electron is moving at a speed of 8.0×10^6 m/s. If the uncertainty in measuring the speed is 1.0 percent of the speed, calculate the uncertainty in the electron's position. The mass of the electron is 9.1094×10^{-31} kg. (b) A baseball of mass 0.15 kg thrown at 100 mph has a momentum of 6.7 kg · m/s. If the uncertainty in measuring this momentum is 1.0×10^{-7} of the momentum, calculate the uncertainty in the baseball's position.

Strategy To calculate the *minimum* uncertainty in both (a) and (b), we use an equal sign in Equation (7.9).

Solution

(a) The uncertainty in the electron's speed u is

$$\begin{aligned}\Delta u &= 0.010 \times 8.0 \times 10^6 \text{ m/s}\\ &= 8.0 \times 10^4 \text{ m/s}\end{aligned}$$

Momentum (p) is $p = mu$, so that

$$\begin{aligned}\Delta p &= m\Delta u\\ &= 9.1094 \times 10^{-31} \text{ kg} \times 8.0 \times 10^4 \text{ m/s}\\ &= 7.3 \times 10^{-26} \text{ kg} \cdot \text{m/s}\end{aligned}$$

From Equation (7.9), the uncertainty in the electron's position is

$$\begin{aligned}\Delta x &= \frac{h}{4\pi\Delta p}\\ &= \frac{6.63 \times 10^{-34} \text{ J} \cdot \text{s}}{4\pi(7.3 \times 10^{-26} \text{ kg} \cdot \text{m/s})}\\ &= 7.24 \times 10^{-10} \text{ m}\end{aligned}$$

This uncertainty corresponds to about 4 atomic diameters.

(b) The uncertainty in the position of the baseball is

$$\begin{aligned}\Delta x &= \frac{h}{4\pi\Delta p}\\ &= \frac{6.63 \times 10^{-34} \text{ J} \cdot \text{s}}{4\pi \times 1.0 \times 10^{-7} \times 6.7 \text{ kg} \cdot \text{m/s}}\\ &= 7.9 \times 10^{-29} \text{ m}\end{aligned}$$

This is such a small number as to be of no consequence; that is, there is practically no uncertainty in determining the position of the baseball in the macroscopic world.

Practice Exercise Estimate the uncertainty in the speed of an oxygen molecule if its position is known to be ±3 nm. The mass of an oxygen molecule is 5.31×10^{-26} kg.

Similar problems: 7.128, 7.146.

The Schrödinger Equation

To be sure, Bohr made a significant contribution to our understanding of atoms, and his suggestion that the energy of an electron in an atom is quantized remains unchallenged. But his theory did not provide a complete description of electronic behavior in atoms. In 1926 the Austrian physicist Erwin Schrödinger,† using a complicated mathematical technique, formulated an equation that describes the behavior and energies of submicroscopic particles in general, an equation analogous to Newton's laws of motion for macroscopic objects. The *Schrödinger equation* requires advanced

†Erwin Schrödinger (1887–1961). Austrian physicist. Schrödinger formulated wave mechanics, which laid the foundation for modern quantum theory. He received the Nobel Prize in Physics in 1933.

calculus to solve, and we will not discuss it here. It is important to know, however, that the equation incorporates both particle behavior, in terms of mass *m,* and wave behavior, in terms of a *wave function* Ψ (psi), which depends on the location in space of the system (such as an electron in an atom).

The wave function itself has no direct physical meaning. However, the probability of finding the electron in a certain region in space is proportional to the square of the wave function, Ψ^2. The idea of relating Ψ^2 to probability stemmed from a wave theory analogy. According to wave theory, the intensity of light is proportional to the square of the amplitude of the wave, or Ψ^2. The most likely place to find a photon is where the intensity is greatest, that is, where the value of Ψ^2 is greatest. A similar argument associates Ψ^2 with the likelihood of finding an electron in regions surrounding the nucleus.

Schrödinger's equation began a new era in physics and chemistry, for it launched a new field, *quantum mechanics* (also called *wave mechanics*). This equation cannot be derived from other principles as it is a fundamental postulate of quantum mechanics. We now refer to the developments in quantum theory from 1913—the time Bohr presented his analysis for the hydrogen atom—to 1926 as "old quantum theory."

The Quantum Mechanical Description of the Hydrogen Atom

The Schrödinger equation specifies the possible energy states the electron can occupy in a hydrogen atom and identifies the corresponding wave functions (Ψ). These energy states and wave functions are characterized by a set of quantum numbers (to be discussed shortly), with which we can construct a comprehensive model of the hydrogen atom.

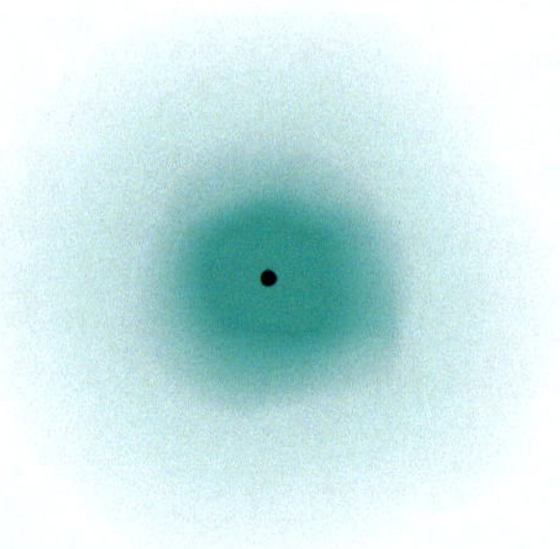

Figure 7.19 *A representation of the electron density distribution surrounding the nucleus in the hydrogen atom. It shows a high probability of finding the electron closer to the nucleus.*

Although quantum mechanics tells us that we cannot pinpoint an electron in an atom, it does define the region where the electron might be at a given time. The concept of ***electron density*** *gives the probability that an electron will be found in a particular region of an atom.* The square of the wave function, Ψ^2, defines the distribution of electron density in three-dimensional space around the nucleus. Regions of high electron density represent a high probability of locating the electron, whereas the opposite holds for regions of low electron density (Figure 7.19).

To distinguish the quantum mechanical description of an atom from Bohr's model, we speak of an atomic orbital, rather than an orbit. An ***atomic orbital*** can be thought of as *the wave function of an electron in an atom.* When we say that an electron is in a certain orbital, we mean that the distribution of the electron density or the probability of locating the electron in space is described by the square of the wave function associated with that orbital. An atomic orbital, therefore, has a characteristic energy, as well as a characteristic distribution of electron density.

The Schrödinger equation works nicely for the simple hydrogen atom with its one proton and one electron, but it turns out that it cannot be solved exactly for any atom containing more than one electron! Fortunately, chemists and physicists have learned to get around this kind of difficulty by approximation. For example, although the behavior of electrons in ***many-electron atoms*** (i.e., *atoms containing two or more electrons*) is not the same as in the hydrogen atom, we assume that the difference is probably not too great. Thus, we can use the energies and wave functions obtained from the hydrogen atom as good approximations of the behavior of electrons in more complex atoms. In fact, this approach provides fairly reliable descriptions of electronic behavior in many-electron atoms.

Summary of Concepts & Facts

- The Heisenberg uncertainty principle sets the limits for measurement of quantum mechanical systems.

- The Schrödinger equation describes the motions and energies of submicroscopic particles. This equation launched quantum mechanics and a new era in physics.
- The Schrödinger equation tells us the possible energy states of the electron in a hydrogen atom and the probability of its location in a particular region surrounding the nucleus. These results can be applied with reasonable accuracy to many-electron atoms.
- An atomic orbital is a function (Ψ) that defines the distribution of electron density (Ψ^2) in space. Orbitals are represented by electron density diagrams or boundary surface diagrams.

Review of Concepts & Facts

7.5.1 What is the difference between Ψ and Ψ^2 for the electron in a hydrogen atom?

7.5.2 A proton is moving at a speed of 6.0×10^6 m/s. If the uncertainty in measuring the speed is 1.0 percent of the speed, calculate the uncertainty in the proton's position. The mass of the proton is 1.67262×10^{-27} kg.

7.6 Quantum Numbers

Learning Objectives

- Recognize the meaning of the four quantum numbers required to describe an electron in an atom.
- Use quantum number rules to determine allowed values for each quantum number.

In quantum mechanics, three ***quantum numbers*** are required to *describe the distribution of electrons in hydrogen and other atoms.* These numbers are derived from the mathematical solution of the Schrödinger equation for the hydrogen atom. They are called the *principal quantum number,* the *angular momentum quantum number,* and the *magnetic quantum number.* These quantum numbers will be used to describe atomic orbitals and to label electrons that reside in them. A fourth quantum number—the *spin quantum number*—describes the behavior of a specific electron and completes the description of electrons in atoms.

The Principal Quantum Number (n)

The principal quantum number (n) can have integral values 1, 2, 3, and so forth; it corresponds to the quantum number in Equation (7.5). [Note that Equation (7.5) holds only for the hydrogen atom.] In a hydrogen atom, the value of n determines the energy of an orbital. As we will see shortly, this is not the case for a many-electron atom. The principal quantum number also relates to the average distance of the electron from the nucleus in a particular orbital. The larger n is, the greater the average distance of an electron in the orbital from the nucleus and therefore the larger the orbital.

The Angular Momentum Quantum Number (ℓ)

The angular momentum quantum number (ℓ) tells us the "shape" of the orbitals (see Section 7.7). The values of ℓ depend on the value of the principal quantum number, n. For a given value of n, ℓ has possible integral values from 0 to $(n - 1)$. If $n = 1$, there is only one possible value of ℓ; that is, $\ell = n - 1 = 1 - 1 = 0$. If $n = 2$, there are two values of ℓ, given by 0 and 1. If $n = 3$, there are three values of ℓ, given by 0, 1, and 2. The value of ℓ is generally designated by the letters *s*, *p*, *d*, . . . as follows:

ℓ	0	1	2	3	4	5
Name of orbital	*s*	*p*	*d*	*f*	*g*	*h*

Thus, if $\ell = 0$, we have an *s* orbital; if $\ell = 1$, we have a *p* orbital; and so on.

The unusual sequence of letters (*s*, *p*, and *d*) has a historical origin. Physicists who studied atomic emission spectra tried to correlate the observed spectral lines with the particular energy states involved in the transitions. They noted that some of the lines were *s*harp; some were rather spread out, or *d*iffuse; and some were very strong and hence referred to as *p*rincipal lines. Subsequently, the initial letters of each adjective were assigned to those energy states. However, after the letter *d* and starting with the letter *f* (for *f*undamental), the orbital designations follow alphabetical order though *j* is omitted.

A collection of orbitals with the same value of *n* is frequently called a shell. One or more orbitals with the same *n* and ℓ values are referred to as a subshell. For example, the shell with $n = 2$ is composed of two subshells, $\ell = 0$ and 1 (the allowed values for $n = 2$). These subshells are called the 2*s* and 2*p* subshells, where 2 denotes the value of *n*, and *s* and *p* denote the values of ℓ.

The Magnetic Quantum Number (m_ℓ)

The magnetic quantum number (m_ℓ) describes the orientation of the orbital in space (to be discussed in Section 7.7). Within a subshell, the value of m_ℓ depends on the value of the angular momentum quantum number, ℓ. For a certain value of ℓ, there are $(2\ell + 1)$ integral values of m_ℓ as follows:

$$-\ell, (-\ell + 1), \ldots 0, \ldots (+\ell - 1), +\ell$$

If $\ell = 0$, then $m_\ell = 0$. If $\ell = 1$, then there are $[(2 \times 1) + 1]$, or three values of m_ℓ, namely, −1, 0, and 1. If $\ell = 2$, there are $[(2 \times 2) + 1]$, or five values of m_ℓ, namely, −2, −1, 0, 1, and 2. The number of m_ℓ values indicates the number of orbitals in a subshell with a particular ℓ value. The various combinations of ℓ and m_ℓ represent different quantum states of an electron but they have the same energy and are referred to as ***degenerate orbitals***.

Student Hot Spot

Student data indicate you may struggle with describing energy levels with quantum numbers. Access your eBook for additional Learning Resources on this topic.

To conclude our discussion of these three quantum numbers, let us consider a situation in which $n = 2$ and $\ell = 1$. The values of *n* and ℓ indicate that we have a 2*p* subshell, and in this subshell we have *three* 2*p* orbitals (because there are three values of m_ℓ, given by −1, 0, and 1).

Example 7.7

What are the allowed values for m_ℓ when $n = 4$ and $\ell = 1$?

Strategy What are the relationships among *n*, ℓ, and m_ℓ?

Solution The values of m_ℓ depend only on the value of ℓ, not on the value of *n*. The values of m_ℓ can vary from $-\ell$ to ℓ. Therefore, m_ℓ can be −1, 0, or 1.

Check The values of *n* and ℓ are fixed, but m_ℓ can have any one of the three values, which correspond to the three *p* orbitals.

Practice Exercise What are the allowed values for m_ℓ when $n = 5$ and $\ell = 2$?

Similar problem: 7.56.

The Electron Spin Quantum Number (m_s)

Experiments on the emission spectra of hydrogen and sodium atoms indicated that lines in the emission spectra could be split by the application of an external magnetic field. The only way physicists could explain these results was to assume that electrons act like tiny magnets. If electrons are thought of as spinning on their own axes, as Earth does, their magnetic properties can be accounted for. According to electromagnetic theory, a spinning charge generates a magnetic field, and it is this motion that causes an electron to behave like a magnet. Figure 7.20 shows the two possible spinning motions of an electron, one clockwise and the other counterclockwise. To take the electron spin into account, it is necessary to introduce a fourth quantum number, called the electron spin quantum number (m_s), which has a value of $+\frac{1}{2}$ or $-\frac{1}{2}$.

Conclusive proof of electron spin was provided by Otto Stern† and Walther Gerlach‡ in 1924. Figure 7.21 shows the basic experimental arrangement. In their experiment, Stern and Gerlach used silver atoms, which contain just one unpaired electron. To illustrate the principle, we can assume that hydrogen atoms are used in the study. A beam of gaseous atoms generated in a hot furnace passes through a nonhomogeneous magnetic field. The interaction between an electron and the magnetic field causes the atom to be deflected from its straight-line path. Because the spinning motion is completely random, the electrons in half of the atoms will be spinning in one direction, and those atoms will be deflected in one way; the electrons in the other half of the atoms will be spinning in the opposite direction, and those atoms will be deflected in the other direction. Thus, two spots of equal intensity are observed on the detecting screen.

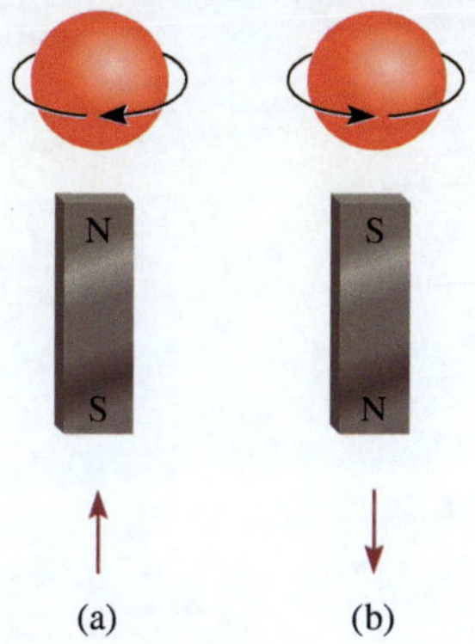

Figure 7.20 *The (a) clockwise and (b) counterclockwise spins of an electron. The magnetic fields generated by these two spinning motions are analogous to those from the two magnets. The upward and downward arrows are used to denote the direction of spin.*

Summary of Concepts & Facts

- Four quantum numbers characterize each electron in an atom: the principal quantum number n identifies the main energy level, or shell, of the orbital; the angular momentum quantum number ℓ indicates the shape of the orbital; the magnetic quantum number m_ℓ specifies the orientation of the orbital in space; and the electron spin quantum number m_s indicates the direction of the electron's spin on its own axis.

Review of Concepts & Facts

7.6.1 What are the allowed values of ℓ when $n = 3$?

7.6.2 Give the four quantum numbers for each of the two electrons in a 6*s* orbital.

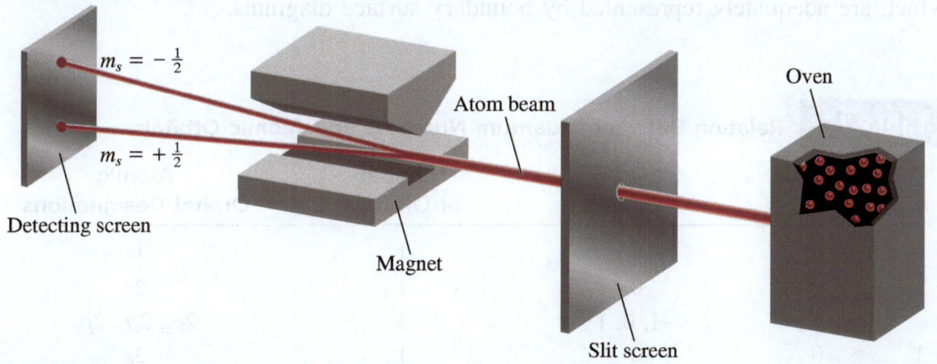

Figure 7.21 *Experimental arrangement for demonstrating the spinning motion of electrons. A beam of atoms is directed through a magnetic field. For example, when a hydrogen atom with a single electron passes through the field, it is deflected in one direction or the other, depending on the direction of the spin. In a stream consisting of many atoms, there will be equal distributions of the two kinds of spins, so that two spots of equal intensity are detected on the screen.*

†Otto Stern (1888–1969). German physicist. He made important contributions to the study of magnetic properties of atoms and the kinetic theory of gases. Stern was awarded the Nobel Prize in Physics in 1943.

‡Walther Gerlach (1889–1979). German physicist. Gerlach's main area of research was in quantum theory.

7.7 Atomic Orbitals

Learning Objectives

- Distinguish the basis of atomic orbitals and how they correspond to quantum numbers.
- Compare the arrangement of atomic orbitals by energy levels in multi-electron atoms.

Table 7.2 shows the relation between quantum numbers and atomic orbitals. We see that when $\ell = 0$, $(2\ell + 1) = 1$, and there is only one value of m_ℓ, thus we have an *s* orbital. When $\ell = 1$, $(2\ell + 1) = 3$, so there are three values of m_ℓ or three *p* orbitals, labeled p_x, p_y, and p_z. When $\ell = 2$, $(2\ell + 1) = 5$, and there are five values of m_ℓ, and the corresponding five *d* orbitals are labeled with more elaborate subscripts. In the following sections we will consider the *s*, *p*, and *d* orbitals separately.

One of the important questions we ask when studying the properties of atomic orbitals is, What are the shapes of the orbitals? Strictly speaking, an orbital does not have a well-defined shape because the wave function characterizing the orbital extends from the nucleus to infinity. In that sense, it is difficult to say what an orbital looks like. On the other hand, it is certainly convenient to think of orbitals as having specific shapes, particularly in discussing the formation of chemical bonds between atoms, as we will do in Chapters 9 and 10.

s Orbitals

Although in principle an electron can be found anywhere, we know that most of the time it is quite close to the nucleus. Figure 7.22(a) shows the distribution of electron density in a hydrogen 1*s* orbital moving outward from the nucleus. As you can see, the electron density falls off rapidly as the distance from the nucleus increases. Roughly speaking, there is about a 90 percent probability of finding the electron within a sphere of radius 100 pm (1 pm $= 1 \times 10^{-12}$ m) surrounding the nucleus. Thus, we can represent the 1*s* orbital by drawing a ***boundary surface diagram*** that *encloses about 90 percent of the total electron density in an orbital,* as shown in Figure 7.22(b). A 1*s* orbital represented in this manner is merely a sphere.

Figure 7.23 shows boundary surface diagrams for the 1*s*, 2*s*, and 3*s* hydrogen atomic orbitals. All *s* orbitals are spherical in shape but differ in size, which increases as the principal quantum number increases. Although the details of electron density variation within each boundary surface are lost, there is no serious disadvantage. For us the most important features of atomic orbitals are their shapes and *relative* sizes, which are adequately represented by boundary surface diagrams.

Table 7.2 Relation Between Quantum Numbers and Atomic Orbitals

n	ℓ	m_ℓ	Number of Orbitals	Atomic Orbital Designations
1	0	0	1	1*s*
2	0	0	1	2*s*
	1	−1, 0, 1	3	$2p_x$, $2p_y$, $2p_z$
3	0	0	1	3*s*
	1	−1, 0, 1	3	$3p_x$, $3p_y$, $3p_z$
	2	−2, −1, 0, 1, 2	5	$3d_{xy}$, $3d_{yz}$, $3d_{xz}$, $3d_{x^2-y^2}$, $3d_{z^2}$
⋮	⋮	⋮	⋮	⋮

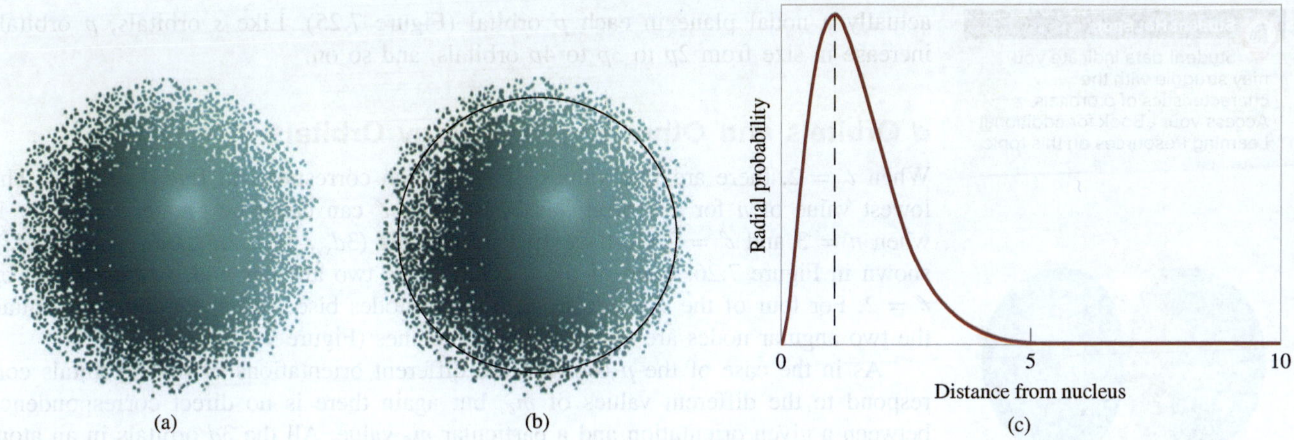

Figure 7.22 *(a) Plot of electron density in the hydrogen 1s orbital as a function of the distance from the nucleus. The electron density falls off rapidly as the distance from the nucleus increases. (b) Boundary surface diagram of the hydrogen 1s orbital. (c) A more realistic way of viewing electron density distribution is to divide the 1s orbital into successive spherical thin shells. A plot of the probability of finding the electron in each shell, called radial probability, as a function of distance shows a maximum at 52.9 pm from the nucleus. Interestingly, this is equal to the radius of the innermost orbit in the Bohr model.*

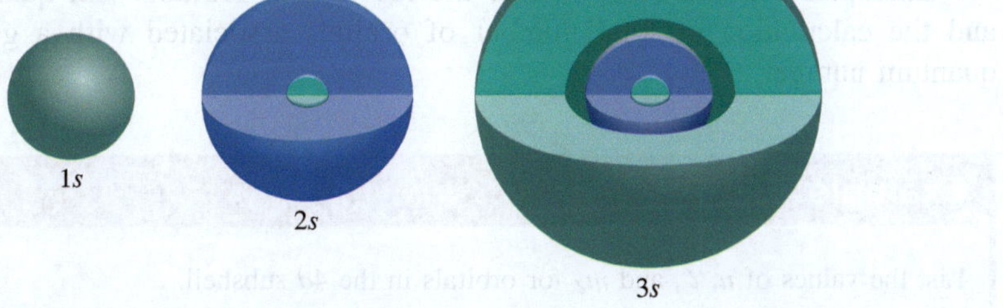

Figure 7.23 *Boundary surface diagrams of the hydrogen 1s, 2s, and 3s orbitals. Each sphere contains about 90 percent of the total electron density. All s orbitals are spherical. Roughly speaking, the size of an orbital is proportional to n^2, where n is the principal quantum number.*

p Orbitals

It should be clear that the *p* orbitals start with the principal quantum number $n = 2$. If $n = 1$, then the angular momentum quantum number ℓ can assume only the value of zero; therefore, there is only a 1*s* orbital. As we saw earlier, when $\ell = 1$, the magnetic quantum number m_ℓ can have values of −1, 0, 1. Starting with $n = 2$ and $\ell = 1$, we therefore have three 2*p* orbitals: $2p_x$, $2p_y$, and $2p_z$ (Figure 7.24). The letter subscripts indicate the axes along which the orbitals are oriented. These three *p* orbitals are identical in size, shape, and energy; they differ from one another only in orientation. Note, however, that there is no simple relation between the values of m_ℓ and the *x*, *y*, and *z* directions. For our purpose, you need only remember that because there are three possible values of m_ℓ, there are three *p* orbitals with different orientations.

The boundary surface diagrams of *p* orbitals in Figure 7.24 show that each *p* orbital can be thought of as two lobes on opposite sides of the nucleus. There is

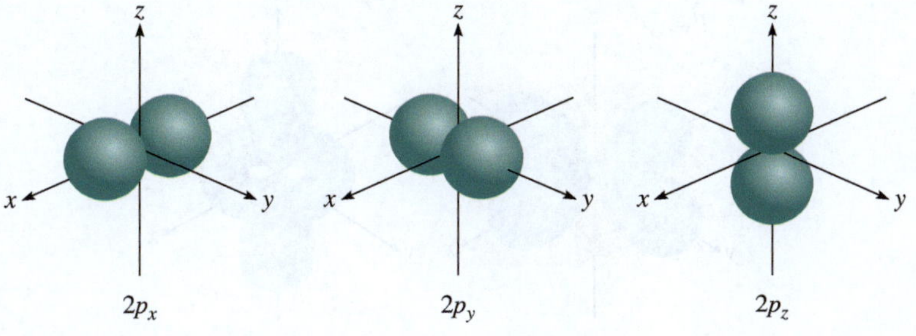

Figure 7.24 *The boundary surface diagrams of the three 2p orbitals. These orbitals are identical in shape and energy, but their orientations are different. The p orbitals of higher principal quantum numbers have similar shapes.*

Student Hot Spot

Student data indicate you may struggle with the characteristics of *p* orbitals. Access your eBook for additional Learning Resources on this topic.

actually a nodal plane in each *p* orbital (Figure 7.25). Like *s* orbitals, *p* orbitals increase in size from 2*p* to 3*p* to 4*p* orbitals, and so on.

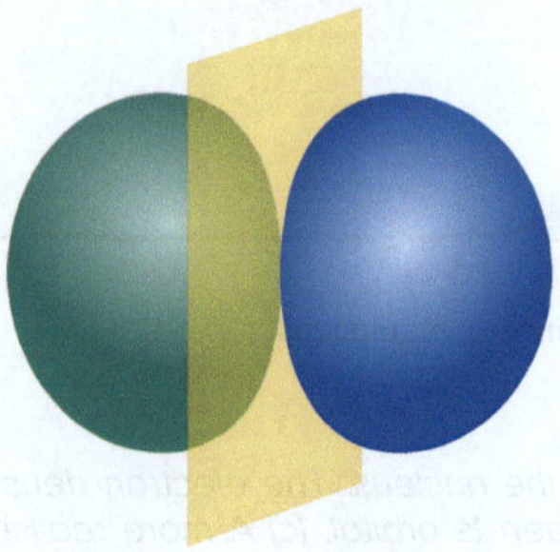

Figure 7.25 *A p orbital has a nodal plane.*

Student Hot Spot

Student data indicate you may struggle with the characteristics of *d* orbitals. Access your eBook for additional Learning Resources on this topic.

d Orbitals and Other Higher-Energy Orbitals

When $\ell = 2$, there are five values of m_ℓ, which correspond to five *d* orbitals. The lowest value of *n* for a *d* orbital is 3. Because ℓ can never be greater than $n - 1$, when $n = 3$ and $\ell = 2$ we have five 3*d* orbitals ($3d_{xy}$, $3d_{yz}$, $3d_{xz}$, $3d_{x^2-y^2}$, and $3d_{z^2}$), shown in Figure 7.26. Each of the *d* orbitals has two angular nodes as predicted by $\ell = 2$. For four of the *d* orbitals, the angular nodes bisect axes; for the $3d_{z^2}$ orbital, the two angular nodes are cones instead of planes (Figure 7.27).

As in the case of the *p* orbitals, the different orientations of the *d* orbitals correspond to the different values of m_ℓ, but again there is no direct correspondence between a given orientation and a particular m_ℓ value. All the 3*d* orbitals in an atom are identical in energy. The *d* orbitals for which *n* is greater than 3 (4*d*, 5*d*, . . .) have similar shapes.

Orbitals having higher energy than *d* orbitals are labeled *f*, *g*, *h*, and so on. The *f* orbitals are important in accounting for the behavior of elements with atomic numbers greater than 57, but their shapes are difficult to represent. In general chemistry, we are not concerned with orbitals having ℓ values greater than 3 (the *g* orbitals and beyond).

Examples 7.8 and 7.9 illustrate the labeling of orbitals with quantum numbers and the calculation of total number of orbitals associated with a given principal quantum number.

Example 7.8

List the values of *n*, ℓ, and m_ℓ for orbitals in the 4*d* subshell.

Strategy What are the relationships among *n*, ℓ, and m_ℓ? What do "4" and "*d*" represent in 4*d?*

(Continued)

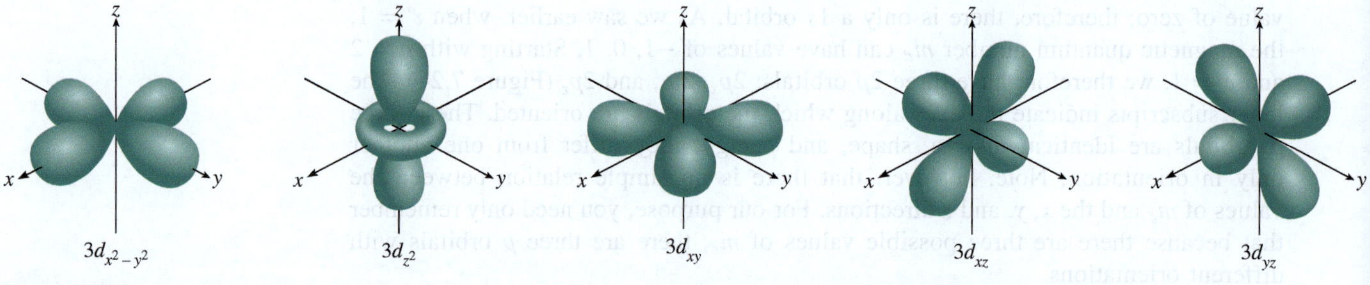

Figure 7.26 *Boundary surface diagrams of the five 3d orbitals. Although the $3d_{z^2}$ orbital looks different, it is equivalent to the other four orbitals in all other respects. The d orbitals of higher principal quantum numbers have similar shapes.*

Figure 7.27 *For the $d_{x^2-y^2}$ orbital, the angular nodes are planes that bisect the x and y axes. For the d_{z^2} orbital, the angular nodes are cones instead of planes.*

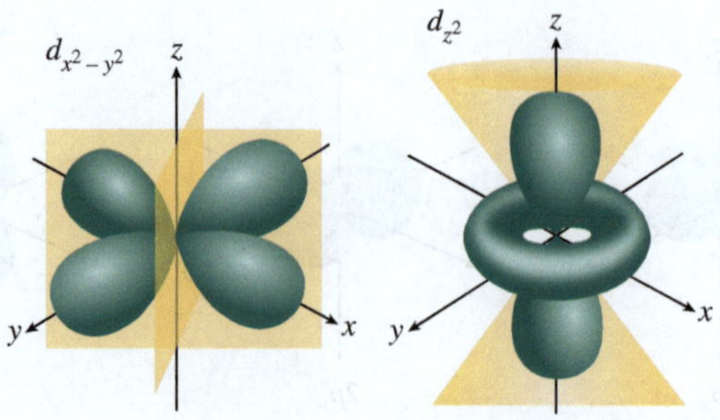

Solution As we saw earlier, the number given in the designation of the subshell is the principal quantum number, so in this case $n = 4$. The letter designates the type of orbital. Because we are dealing with *d* orbitals, $\ell = 2$. The values of m_ℓ can vary from $-\ell$ to ℓ. Therefore, m_ℓ can be $-2, -1, 0, 1$, or 2.

Check The values of n and ℓ are fixed for 4*d*, but m_ℓ can have any one of the five values, which correspond to the five *d* orbitals.

Practice Exercise Give the values of the quantum numbers associated with the orbitals in the 3*p* subshell.

Similar problem: 7.53.

Example 7.9

What is the total number of orbitals associated with the principal quantum number $n = 3$?

Strategy To calculate the total number of orbitals for a given n value, we need to first write the possible values of ℓ. We then determine how many m_ℓ values are associated with each value of ℓ. The total number of orbitals is equal to the sum of all the m_ℓ values.

Solution For $n = 3$, the possible values of ℓ are 0, 1, and 2. Thus, there is one 3*s* orbital ($n = 3$, $\ell = 0$, and $m_\ell = 0$); there are three 3*p* orbitals ($n = 3$, $\ell = 1$, and $m_\ell = -1, 0, 1$); there are five 3*d* orbitals ($n = 3$, $\ell = 2$, and $m_\ell = -2, -1, 0, 1, 2$). The total number of orbitals is $1 + 3 + 5 = 9$.

Check The total number of orbitals for a given value of n is n^2. So here we have $3^2 = 9$. Can you prove the validity of this relationship?

Practice Exercise What is the total number of orbitals associated with the principal quantum number $n = 4$?

Similar problem: 7.56.

The Energies of Orbitals

Now that we have some understanding of the shapes and sizes of atomic orbitals, we are ready to inquire into their relative energies and look at how energy levels affect the actual arrangement of electrons in atoms.

According to Equation (7.5), the energy of an electron in a hydrogen atom is determined solely by its principal quantum number. Thus, the energies of hydrogen orbitals increase as follows (Figure 7.28):

$$1s < 2s = 2p < 3s = 3p = 3d < 4s = 4p = 4d = 4f < \ldots$$

Although the electron density distributions are different in the 2*s* and 2*p* orbitals, hydrogen's electron has the same energy whether it is in the 2*s* orbital or a 2*p* orbital. The 1*s* orbital in a hydrogen atom corresponds to the most stable condition, the ground state. An electron residing in this orbital is most strongly held by the nucleus because it is closest to the nucleus. An electron in the 2*s*, 2*p*, or higher orbitals in a hydrogen atom is in an excited state.

The energy picture is more complex for many-electron atoms than for hydrogen. The energy of an electron in such an atom depends on its angular momentum quantum number as well as on its principal quantum number (Figure 7.29). For many-electron atoms, the 3*d* energy level is very close to the 4*s* energy level. The total energy of an atom, however, depends not only on the sum of the orbital energies but also on the energy of repulsion between the electrons in these orbitals (each orbital can

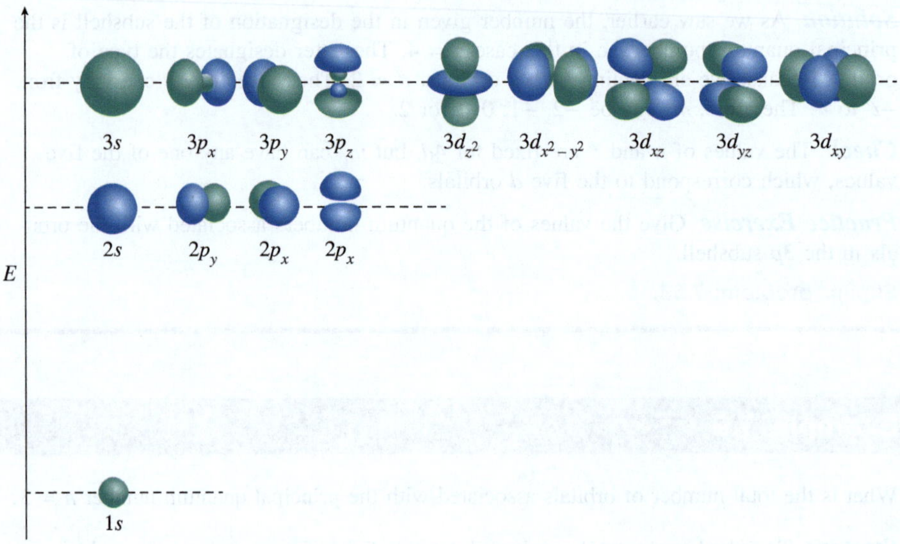

Figure 7.28 *Orbital energy levels in the hydrogen atom. Each short horizontal line represents one orbital. Orbitals with the same principal quantum number (n) all have the same energy.*

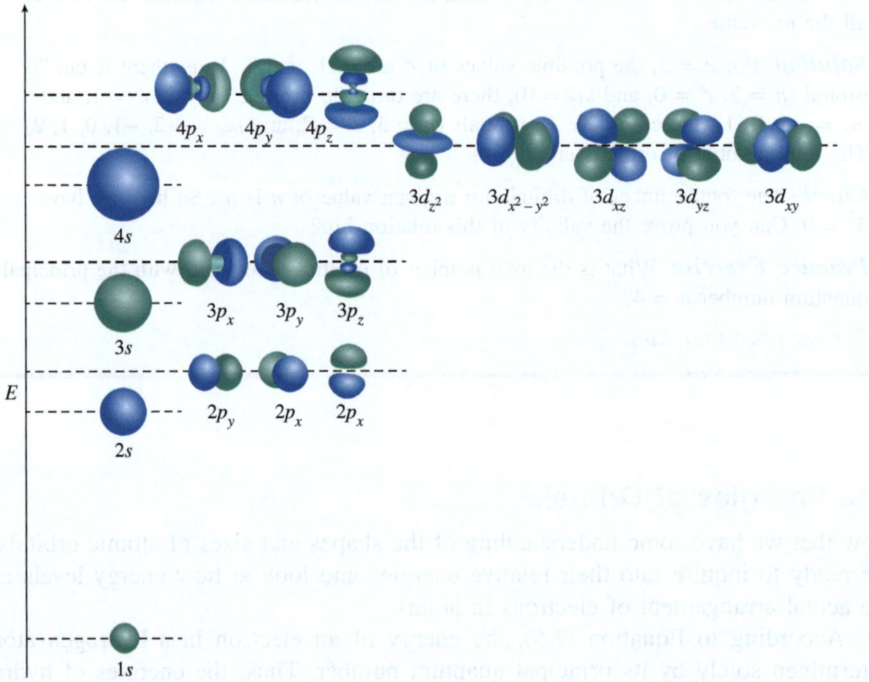

Figure 7.29 *Orbital energy levels in a many-electron atom. Note that the energy level depends on both n and ℓ values.*

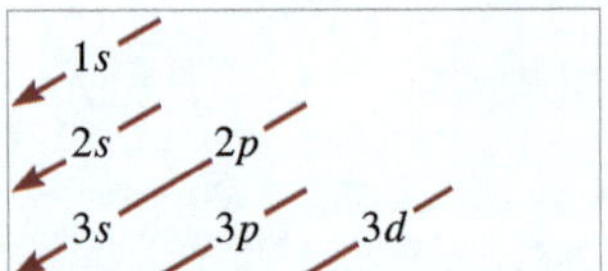

Figure 7.30 *The order in which atomic subshells are filled in a many-electron atom. Start with the 1s orbital and move downward, following the direction of the arrows. Thus, the order goes as follows: 1s < 2s < 2p < 3s < 3p < 4s < 3d < . . . .*

accommodate up to two electrons, as we will see in Section 7.8). It turns out that the total energy of an atom is lower when the 4*s* subshell is filled before a 3*d* subshell. Figure 7.30 depicts the order in which atomic orbitals are filled in a many-electron atom. We will consider specific examples in Section 7.8.

Summary of Concepts & Facts

- The single *s* orbital for each energy level is spherical and centered on the nucleus. The three *p* orbitals present at $n = 2$ and higher; each has two lobes, and the pairs of lobes are arranged at right angles to one another. Starting with $n = 3$, there are five *d* orbitals, with more complex shapes and orientations.

- The energy of the electron in a hydrogen atom is determined solely by its principal quantum number. In many-electron atoms, the principal quantum number and the angular momentum quantum number together determine the energy of an electron.

Review of Concepts & Facts

7.7.1 Why is it not possible to have a 2*d* orbital, but a 3*d* orbital is allowed?
7.7.2 What are the n, ℓ, and m_ℓ values for orbitals in the 4*p* subshell?
7.7.3 How many orbitals are there in the 5*f* subshell?

7.8 Electron Configuration

Learning Objective

- Apply the Pauli exclusion principle and Hund's rule in determining electron configurations and drawing electron orbital diagrams.

The four quantum numbers n, ℓ, m_ℓ, and m_s enable us to label completely an electron in any orbital in any atom. In a sense, we can regard the set of four quantum numbers as the "address" of an electron in an atom, somewhat in the same way that a street address, city, state, and postal ZIP code specify the address of an individual. For example, the four quantum numbers for a 2*s* orbital electron are $n = 2$, $\ell = 0$, $m_\ell = 0$, and $m_s = +\frac{1}{2}$ or $-\frac{1}{2}$. It is inconvenient to write out all the individual quantum numbers, and so we use the simplified notation (n, ℓ, m_ℓ, m_s). For the preceding example, the quantum numbers are either $(2, 0, 0, +\frac{1}{2})$ or $(2, 0, 0, -\frac{1}{2})$. The value of m_s has no effect on the energy, size, shape, or orientation of an orbital, but it determines how electrons are arranged in an orbital.

Example 7.10 shows how quantum numbers of an electron in an orbital are assigned.

Example 7.10

Write the four quantum numbers for an electron in a 3*p* orbital.

Strategy What do the "3" and "*p*" designate in 3*p*? How many orbitals (values of m_ℓ) are there in a 3*p* subshell? What are the possible values of electron spin quantum number?

Solution To start with, we know that the principal quantum number n is 3 and the angular momentum quantum number ℓ must be 1 (because we are dealing with a *p* orbital).

For $\ell = 1$, there are three values of m_ℓ given by −1, 0, and 1. Because the electron spin quantum number m_s can be either $+\frac{1}{2}$ or $-\frac{1}{2}$, we conclude that there are six possible ways to designate the electron using the (n, ℓ, m_ℓ, m_s) notation:

$$(3, 1, -1, +\tfrac{1}{2}) \qquad (3, 1, -1, -\tfrac{1}{2})$$
$$(3, 1, 0, +\tfrac{1}{2}) \qquad (3, 1, 0, -\tfrac{1}{2})$$
$$(3, 1, 1, +\tfrac{1}{2}) \qquad (3, 1, 1, -\tfrac{1}{2})$$

Check In these six designations we see that the values of n and ℓ are constant, but the values of m_ℓ and m_s can vary.

Practice Exercise Write the four quantum numbers for an electron in a 4*d* orbital.

Similar problem: 7.54.

Video: Electron Configurations

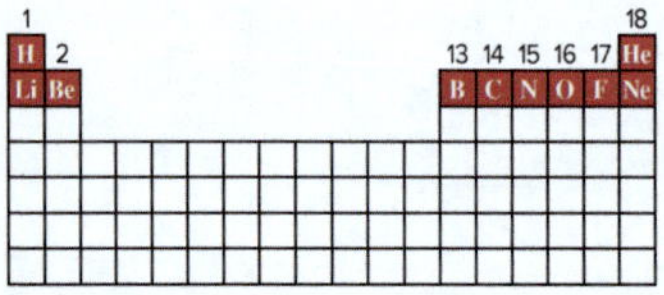

The hydrogen atom is a particularly simple system because it contains only one electron. The electron may reside in the 1*s* orbital (the ground state), or it may be found in some higher-energy orbital (an excited state). For many-electron atoms, however, we must know the ***electron configuration*** of the atom, that is, *how the electrons are distributed among the various atomic orbitals,* to understand electronic behavior. We will use the first 10 elements (hydrogen to neon) to illustrate the rules for writing electron configurations for atoms in the *ground state.* (Section 7.9 will describe how these rules can be applied to the remainder of the elements in the periodic table.) For this discussion, recall that the number of electrons in an atom is equal to its atomic number *Z.*

From Figure 7.28 we know that the electron in a ground-state hydrogen atom must be in the 1*s* orbital, so its electron configuration is $1s^1$:

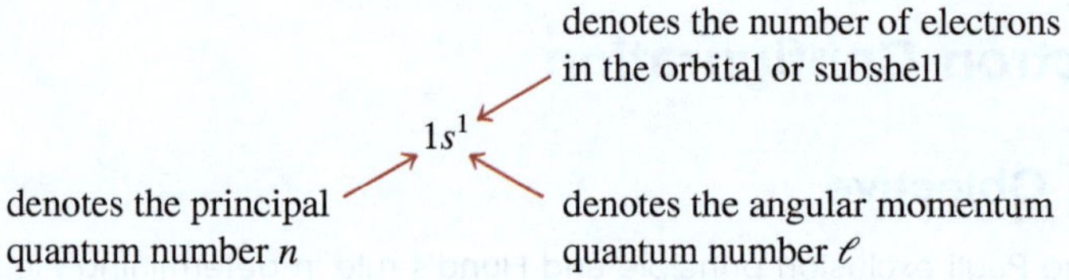

The electron configuration can also be represented by an *orbital diagram* that shows the spin of the electron (see Figure 7.20):

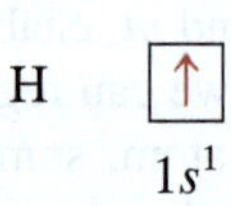

The upward arrow denotes one of the two possible spinning motions of the electron. (Alternatively, we could have represented the electron with a downward arrow.) Remember that the direction of electron spin has no effect on the energy of the electron. The box represents an atomic orbital.

The Pauli Exclusion Principle

For many-electron atoms we use the ***Pauli*† *exclusion principle*** to determine electron configurations. This principle states that *no two electrons in an atom can have the same set of four quantum numbers.* If two electrons in an atom should have the same n, ℓ, and m_ℓ values (that is, these two electrons are in the *same* atomic orbital), then they must have different values of m_s. In other words, only two electrons may occupy the same atomic orbital, and these electrons must have opposite spins. Consider the helium atom, which has two electrons. The three possible ways of placing two electrons in the 1*s* orbital are as follows:

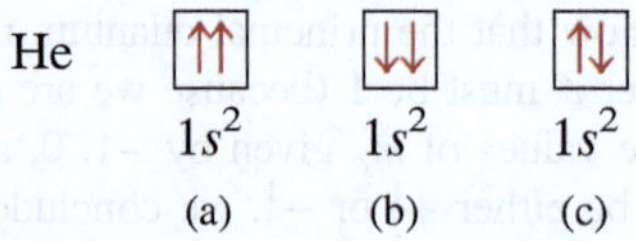

Diagrams (a) and (b) are ruled out by the Pauli exclusion principle. In (a), both electrons have the same upward spin and would have the quantum numbers $(1, 0, 0, +\frac{1}{2})$; in (b), both electrons have downward spins and would have the quantum numbers $(1, 0, 0, -\frac{1}{2})$. Only the configuration in (c) is physically acceptable, because one electron has the quantum numbers $(1, 0, 0, +\frac{1}{2})$ and the other has $(1, 0, 0, -\frac{1}{2})$.

†Wolfgang Pauli (1900–1958). Austrian physicist. One of the founders of quantum mechanics, Pauli was awarded the Nobel Prize in Physics in 1945.

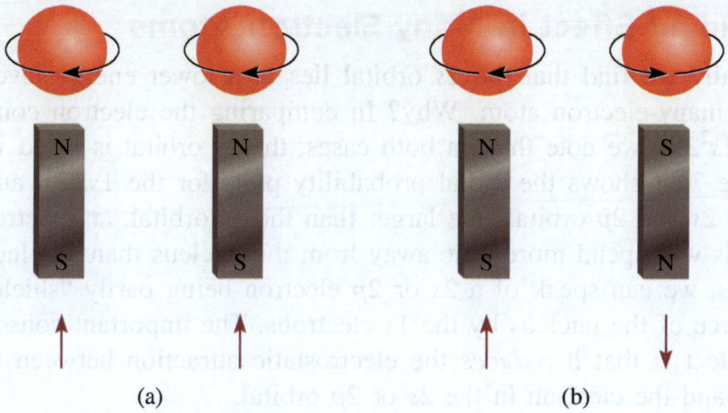

Figure 7.31 *The (a) parallel and (b) antiparallel spins of two electrons. In (a) the two magnetic fields reinforce each other. In (b) the two magnetic fields cancel each other.*

Electrons that have opposite spins in a given orbital are said to be paired. Thus, the helium atom has the following configuration:

He [↑↓]
$1s^2$

Note that $1s^2$ is read "one *s* two," not "one *s* squared."

Diamagnetism and Paramagnetism

The Pauli exclusion principle is one of the fundamental principles of quantum mechanics. It can be tested by a simple observation. If the two electrons in the 1*s* orbital of a helium atom had the same, or parallel, spins (↑↑ or ↓↓), their net magnetic fields would reinforce each other [Figure 7.31(a)]. Such an arrangement would make the helium gas paramagnetic. ***Paramagnetic*** substances are those that *contain net unpaired spins and are attracted by a magnet.* On the other hand, if the electron spins are paired, or antiparallel to each other (↑↓ or ↓↑), the magnetic effects cancel out [Figure 7.31(b)]. ***Diamagnetic*** substances *do not contain net unpaired spins and are slightly repelled by a magnet.*

Measurements of magnetic properties provide the most direct evidence for specific electron configurations of elements. Advances in instrument design during the last 30 years or so enable us to determine the number of unpaired electrons in an atom (Figure 7.32). By experiment we find that the helium atom in its ground state has no net magnetic field. Therefore, the two electrons in the 1*s* orbital must be paired in accord with the Pauli exclusion principle and the helium gas is diamagnetic. A useful rule to keep in mind is that any atom with an *odd* number of electrons will always contain one or more unpaired spins because we need an even number of electrons for complete pairing. On the other hand, atoms containing an even number of electrons may or may not contain unpaired spins. We will see the reason for this behavior shortly.

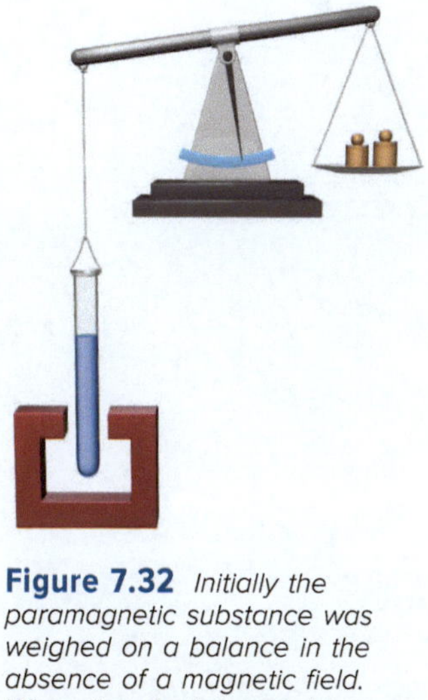

Figure 7.32 *Initially the paramagnetic substance was weighed on a balance in the absence of a magnetic field. When the electromagnet is turned on, the balance is offset because the sample tube is drawn into the magnetic field. Knowing the concentration and the additional mass needed to reestablish balance, it is possible to calculate the number of unpaired electrons in the sample.*

As another example, consider the lithium atom ($Z = 3$), which has three electrons. The third electron cannot go into the 1*s* orbital because it would inevitably have the same set of four quantum numbers as one of the first two electrons. Therefore, this electron "enters" the next (energetically) higher orbital, which is the 2*s* orbital (see Figure 7.29). The electron configuration of lithium is $1s^22s^1$, and its orbital diagram is

Li [↑↓] [↑]
$1s^2$ $2s^1$

The lithium atom contains one unpaired electron and the lithium metal is therefore paramagnetic.

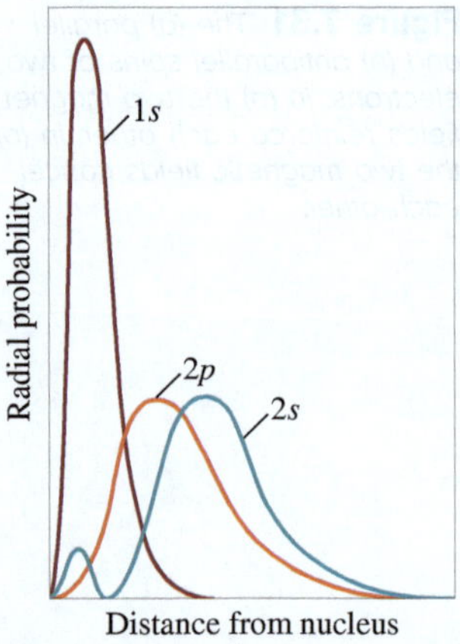

Figure 7.33 *Radial probability plots (see Figure 7.22) for the 1s, 2s, and 2p orbitals. The 1s electrons effectively shield both the 2s and 2p electrons from the nucleus. The 2s orbital is more penetrating than the 2p orbital.*

The Shielding Effect in Many-Electron Atoms

Experimentally we find that the 2*s* orbital lies at a lower energy level than the 2*p* orbital in a many-electron atom. Why? In comparing the electron configurations of $1s^22s^1$ and $1s^22p^1$, we note that, in both cases, the 1*s* orbital is filled with two electrons. Figure 7.33 shows the radial probability plots for the 1*s*, 2*s*, and 2*p* orbitals. Because the 2*s* and 2*p* orbitals are larger than the 1*s* orbital, an electron in either of these orbitals will spend more time away from the nucleus than an electron in the 1*s* orbital. Thus, we can speak of a 2*s* or 2*p* electron being partly "shielded" from the attractive force of the nucleus by the 1*s* electrons. The important consequence of the shielding effect is that it *reduces* the electrostatic attraction between the protons in the nucleus and the electron in the 2*s* or 2*p* orbital.

The manner in which the electron density varies as we move from the nucleus outward depends on the type of orbital. Although a 2*s* electron spends most of its time (on average) slightly farther from the nucleus than a 2*p* electron, the electron density near the nucleus is actually greater for the 2*s* electron (see the small maximum for the 2*s* orbital in Figure 7.33). For this reason, the 2*s* orbital is said to be more "penetrating" than the 2*p* orbital. Therefore, a 2*s* electron is less shielded by the 1*s* electrons and is more strongly held by the nucleus. In fact, for the same principal quantum number *n*, the penetrating power decreases as the angular momentum quantum number ℓ increases, or

$$s > p > d > f > \ldots$$

Because the stability of an electron is determined by the strength of its attraction to the nucleus, it follows that a 2*s* electron will be lower in energy than a 2*p* electron. To put it another way, less energy is required to remove a 2*p* electron than a 2*s* electron, because a 2*p* electron is not held quite as strongly by the nucleus. The hydrogen atom has only one electron and, therefore, is without such a shielding effect.

Continuing our discussion of atoms of the first 10 elements, we go next to beryllium ($Z = 4$). The ground-state electron configuration of beryllium is $1s^22s^2$, or

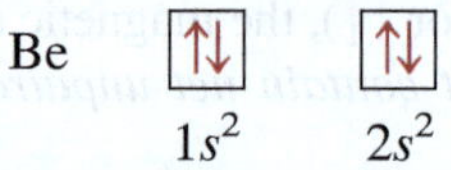

Beryllium is diamagnetic, as we would expect.

The electron configuration of boron ($Z = 5$) is $1s^22s^22p^1$, or

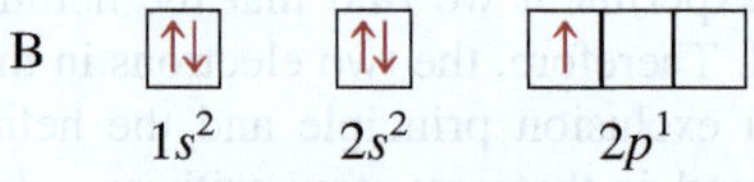

Note that the unpaired electron can be in the $2p_x$, $2p_y$, or $2p_z$ orbital. The choice is completely arbitrary because the three *p* orbitals are equivalent in energy. As the diagram shows, boron is paramagnetic.

Hund's Rule

The electron configuration of carbon ($Z = 6$) is $1s^22s^22p^2$. The following are different ways of distributing two electrons among three *p* orbitals:

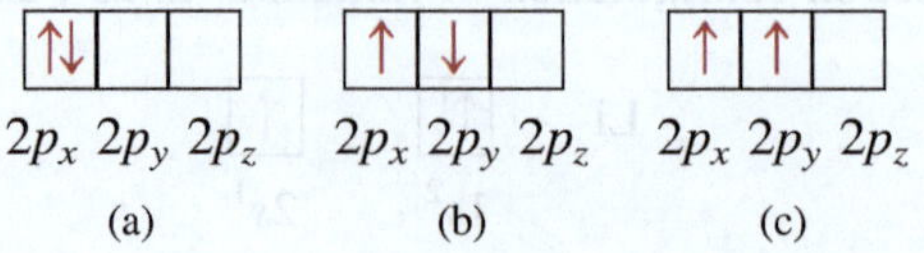

None of the three arrangements violates the Pauli exclusion principle, so we must determine which one will give the greatest stability. The answer is provided by

Hund's rule, † which states that *the most stable arrangement of electrons in subshells is the one with the greatest number of parallel spins.* The arrangement shown in (c) satisfies this condition. In both (a) and (b) the two spins cancel each other. Thus, the orbital diagram for carbon is

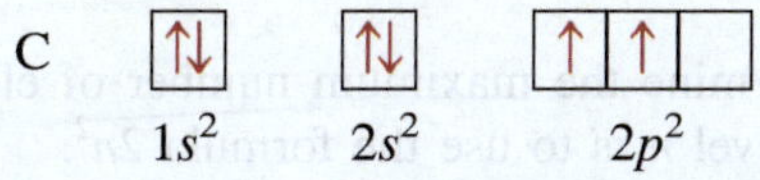

Qualitatively, we can understand why (c) is preferred to (a). In (a), the two electrons are in the same $2p_x$ orbital, and their proximity results in a greater mutual repulsion than when they occupy two separate orbitals, say $2p_x$ and $2p_y$. The choice of (c) over (b) is more subtle but can be justified on theoretical grounds. The necessity for electrons with parallel spins to occupy different degenerate orbitals is due to *spin correlation*, a *quantum mechanical effect that makes electrons electrons with parallel spins have a tendency to be far apart from one another*. The fact that carbon atoms contain two unpaired electrons is thus in accord with Hund's rule.

The electron configuration of nitrogen ($Z = 7$) is $1s^22s^22p^3$:

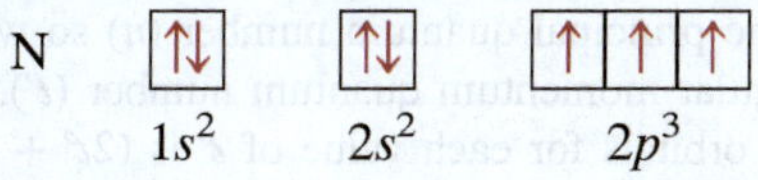

Again, Hund's rule dictates that all three $2p$ electrons have spins parallel to one another; the nitrogen atom contains three unpaired electrons.

The electron configuration of oxygen ($Z = 8$) is $1s^22s^22p^4$. An oxygen atom has two unpaired electrons:

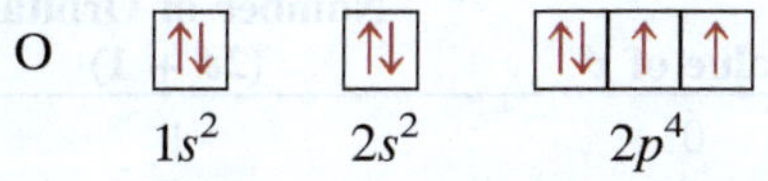

The electron configuration of fluorine ($Z = 9$) is $1s^22s^22p^5$. The nine electrons are arranged as follows:

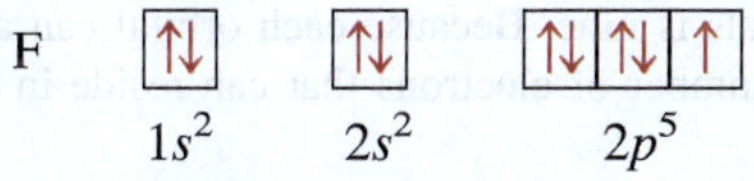

The fluorine atom has one unpaired electron.

In neon ($Z = 10$), the $2p$ subshell is completely filled. The electron configuration of neon is $1s^22s^22p^6$, and *all* the electrons are paired, as follows:

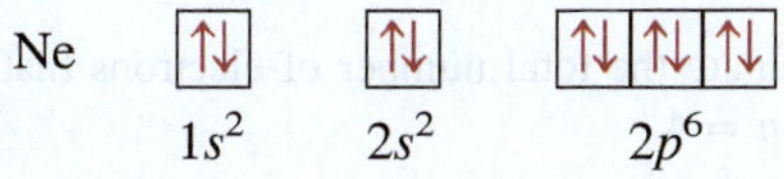

The neon gas should be diamagnetic, and experimental observation bears out this prediction.

General Rules for Assigning Electrons to Atomic Orbitals

Based on the preceding examples we can formulate some general rules for determining the maximum number of electrons that can be assigned to the various subshells and orbitals for a given value of n:

1. Each shell or principal level of quantum number n contains n subshells. For example, if $n = 2$, then there are two subshells (two values of ℓ) of angular momentum quantum numbers 0 and 1.

† Frederick Hund (1896–1997). German physicist. Hund's work was mainly in quantum mechanics. He also helped to develop the molecular orbital theory of chemical bonding.

2. Each subshell of quantum number ℓ contains $(2\ell + 1)$ orbitals. For example, if $\ell = 1$, then there are three p orbitals.
3. No more than two electrons can be placed in each orbital. Therefore, the maximum number of electrons is simply twice the number of orbitals that are employed.
4. A quick way to determine the maximum number of electrons that an atom can have in a principal level n is to use the formula $2n^2$.

Student Hot Spot

Student data indicate you may struggle with understanding electron configurations. Access your eBook for additional Learning Resources on this topic.

Examples 7.11 and 7.12 illustrate the procedure for calculating the number of electrons in orbitals and labeling electrons with the four quantum numbers.

Example 7.11

What is the maximum number of electrons that can be present in the principal level for which $n = 3$?

Strategy We are given the principal quantum number (n) so we can determine all the possible values of the angular momentum quantum number (ℓ). The preceding rule shows that the number of orbitals for each value of ℓ is $(2\ell + 1)$. Thus, we can determine the total number of orbitals. How many electrons can each orbital accommodate?

Solution When $n = 3$, $\ell = 0$, 1, and 2. The number of orbitals for each value of ℓ is given by

Value of ℓ	Number of Orbitals $(2\ell + 1)$
0	1
1	3
2	5

The total number of orbitals is nine. Because each orbital can accommodate two electrons, the maximum number of electrons that can reside in the orbitals is 2×9, or 18.

Check If we use the formula (n^2) in Example 7.9, we find that the total number of orbitals is 3^2 and the total number of electrons is $2(3^2)$ or 18. In general, the number of electrons in a given principal energy level n is $2n^2$.

Practice Exercise Calculate the total number of electrons that can be present in the principal level for which $n = 4$.

Similar problems: 7.64, 7.65.

Example 7.12

An oxygen atom has a total of eight electrons. Write the four quantum numbers for each of the eight electrons in the ground state.

Strategy We start with $n = 1$ and proceed to fill orbitals in the order shown in Figure 7.30. For each value of n we determine the possible values of ℓ. For each value of ℓ, we assign the possible values of m_ℓ. We can place electrons in the orbitals according to the Pauli exclusion principle and Hund's rule.

(Continued)

Solution We start with $n = 1$, so $\ell = 0$, a subshell corresponding to the 1*s* orbital. This orbital can accommodate a total of two electrons. Next, $n = 2$, and ℓ may be either 0 or 1. The $\ell = 0$ subshell contains one 2*s* orbital, which can accommodate two electrons. The remaining four electrons are placed in the $\ell = 1$ subshell, which contains three 2*p* orbitals. The orbital diagram is

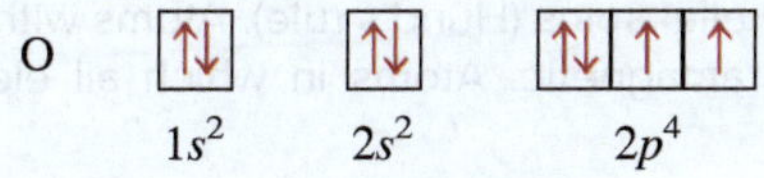

The results are summarized in the following table:

Electron	n	ℓ	m_ℓ	m_s	Orbital
1	1	0	0	$+\frac{1}{2}$	1*s*
2	1	0	0	$-\frac{1}{2}$	
3	2	0	0	$+\frac{1}{2}$	2*s*
4	2	0	0	$-\frac{1}{2}$	
5	2	1	−1	$+\frac{1}{2}$	$2p_x$, $2p_y$, $2p_z$
6	2	1	0	$-\frac{1}{2}$	
7	2	1	1	$+\frac{1}{2}$	
8	2	1	1	$-\frac{1}{2}$	

Of course, the placement of the eighth electron in the orbital labeled $m_\ell = 1$ is completely arbitrary. It would be equally correct to assign it to $m_\ell = 0$ or $m_\ell = -1$.

Practice Exercise Write a complete set of quantum numbers for each of the electrons in boron (B).

Similar problem: 7.91.

At this point let's summarize what our examination of the first 10 elements has revealed about ground-state electron configurations and the properties of electrons in atoms:

1. No two electrons in the same atom can have the same four quantum numbers. This is the Pauli exclusion principle.
2. Each orbital can be occupied by a maximum of two electrons. They must have opposite spins, or different electron spin quantum numbers.
3. The most stable arrangement of electrons in a subshell is the one that has the greatest number of parallel spins. This is Hund's rule.
4. Atoms in which one or more electrons are unpaired are paramagnetic. Atoms in which all the electron spins are paired are diamagnetic.
5. In a hydrogen atom, the energy of the electron depends only on its principal quantum number n. In a many-electron atom, the energy of an electron depends on both n and its angular momentum quantum number ℓ.
6. In a many-electron atom the subshells are filled in the order shown in Figure 7.29.
7. For electrons of the same principal quantum number, their penetrating power, or proximity to the nucleus, decreases in the order $s > p > d > f$. This means that, for example, more energy is required to separate a 3*s* electron from a many-electron atom than is required to remove a 3*p* electron.

Summary of Concepts & Facts

- No two electrons in the same atom can have the same four quantum numbers (the Pauli exclusion principle).
- The most stable arrangement of electrons in a subshell is the one that has the greatest number of parallel spins (Hund's rule). Atoms with one or more unpaired electron spins are paramagnetic. Atoms in which all electrons are paired are diamagnetic.

Review of Concepts & Facts

7.8.1 The ground-state electron configuration of an atom is $1s^22s^22p^63s^23p^3$. Which of the four quantum numbers would be the same for the three $3p$ electrons?

7.8.2 Determine the maximum number of electrons that can be found in the $5d$ subshell.

7.8.3 Identify the error in each of the following sets of quantum numbers (n, ℓ, m_ℓ, m_s): (a) $(2, 0, -1, +\frac{1}{2})$; (b) $(3, 3, 1, -\frac{1}{2})$; (c) 2, 1, 1, 0.

7.9 The Building-Up Principle

Learning Objectives

- Construct electron configurations of atoms using the Aufbau principle.
- Demonstrate the use of the periodic table to write electron configurations of atoms.

Here we will extend the rules used in writing electron configurations for the first 10 elements to the rest of the elements. This process is based on the Aufbau principle. The ***Aufbau principle*** dictates that *as protons are added one by one to the nucleus to build up the elements, electrons are similarly added to the atomic orbitals.* Through this process (the German word *Aufbau* means "building up") we gain a detailed knowledge of the ground-state electron configurations of the elements. As we will see later, knowledge of electron configurations helps us to understand and predict the properties of the elements; it also explains why the periodic table works so well.

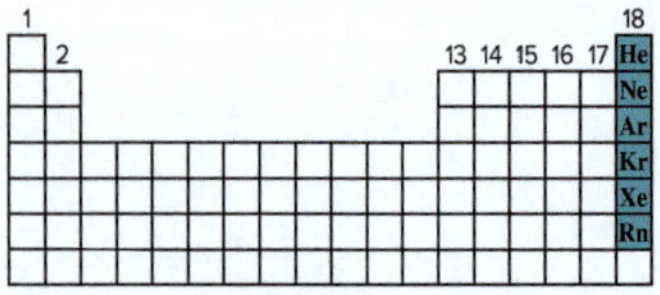

The noble gases.

Table 7.3 gives the ground-state electron configurations of elements from H ($Z = 1$) through the named elements up to Og ($Z = 118$). The electron configurations of all elements except hydrogen and helium are represented by a ***noble gas core,*** which *shows in brackets the noble gas element that most nearly precedes the element being considered,* followed by the symbol for the highest filled subshells in the outermost shells. Notice that the electron configurations of the highest filled subshells in the outermost shells for the elements sodium ($Z = 11$) through argon ($Z = 18$) follow a pattern similar to those of lithium ($Z = 3$) through neon ($Z = 10$).

As mentioned in Section 7.7, the $4s$ subshell is filled before the $3d$ subshell in a many-electron atom (see Figure 7.30). Thus, the electron configuration of potassium ($Z = 19$) is $1s^22s^22p^63s^23p^64s^1$. Because $1s^22s^22p^63s^23p^6$ is the electron configuration of argon, we can simplify the electron configuration of potassium by writing $[Ar]4s^1$, where [Ar] denotes the "argon core." Similarly, we can write the electron configuration

Table 7.3 The Ground-State Electron Configurations of the Elements*

Atomic Number	Symbol	Electron Configuration	Atomic Number	Symbol	Electron Configuration	Atomic Number	Symbol	Electron Configuration
1	H	$1s^1$	41	Nb	$[Kr]5s^1 4d^4$	81	Tl	$[Xe]6s^2 4f^{14} 5d^{10} 6p^1$
2	He	$1s^2$	42	Mo	$[Kr]5s^1 4d^5$	82	Pb	$[Xe]6s^2 4f^{14} 5d^{10} 6p^2$
3	Li	$[He]2s^1$	43	Tc	$[Kr]5s^2 4d^5$	83	Bi	$[Xe]6s^2 4f^{14} 5d^{10} 6p^3$
4	Be	$[He]2s^2$	44	Ru	$[Kr]5s^1 4d^7$	84	Po	$[Xe]6s^2 4f^{14} 5d^{10} 6p^4$
5	B	$[He]2s^2 2p^1$	45	Rh	$[Kr]5s^1 4d^8$	85	At	$[Xe]6s^2 4f^{14} 5d^{10} 6p^5$
6	C	$[He]2s^2 2p^2$	46	Pd	$[Kr]4d^{10}$	86	Rn	$[Xe]6s^2 4f^{14} 5d^{10} 6p^6$
7	N	$[He]2s^2 2p^3$	47	Ag	$[Kr]5s^1 4d^{10}$	87	Fr	$[Rn]7s^1$
8	O	$[He]2s^2 2p^4$	48	Cd	$[Kr]5s^2 4d^{10}$	88	Ra	$[Rn]7s^2$
9	F	$[He]2s^2 2p^5$	49	In	$[Kr]5s^2 4d^{10} 5p^1$	89	Ac	$[Rn]7s^2 6d^1$
10	Ne	$[He]2s^2 2p^6$	50	Sn	$[Kr]5s^2 4d^{10} 5p^2$	90	Th	$[Rn]7s^2 6d^2$
11	Na	$[Ne]3s^1$	51	Sb	$[Kr]5s^2 4d^{10} 5p^3$	91	Pa	$[Rn]7s^2 5f^2 6d^1$
12	Mg	$[Ne]3s^2$	52	Te	$[Kr]5s^2 4d^{10} 5p^4$	92	U	$[Rn]7s^2 5f^3 6d^1$
13	Al	$[Ne]3s^2 3p^1$	53	I	$[Kr]5s^2 4d^{10} 5p^5$	93	Np	$[Rn]7s^2 5f^4 6d^1$
14	Si	$[Ne]3s^2 3p^2$	54	Xe	$[Kr]5s^2 4d^{10} 5p^6$	94	Pu	$[Rn]7s^2 5f^6$
15	P	$[Ne]3s^2 3p^3$	55	Cs	$[Xe]6s^1$	95	Am	$[Rn]7s^2 5f^7$
16	S	$[Ne]3s^2 3p^4$	56	Ba	$[Xe]6s^2$	96	Cm	$[Rn]7s^2 5f^7 6d^1$
17	Cl	$[Ne]3s^2 3p^5$	57	La	$[Xe]6s^2 5d^1$	97	Bk	$[Rn]7s^2 5f^9$
18	Ar	$[Ne]3s^2 3p^6$	58	Ce	$[Xe]6s^2 4f^1 5d^1$	98	Cf	$[Rn]7s^2 5f^{10}$
19	K	$[Ar]4s^1$	59	Pr	$[Xe]6s^2 4f^3$	99	Es	$[Rn]7s^2 5f^{11}$
20	Ca	$[Ar]4s^2$	60	Nd	$[Xe]6s^2 4f^4$	100	Fm	$[Rn]7s^2 5f^{12}$
21	Sc	$[Ar]4s^2 3d^1$	61	Pm	$[Xe]6s^2 4f^5$	101	Md	$[Rn]7s^2 5f^{13}$
22	Ti	$[Ar]4s^2 3d^2$	62	Sm	$[Xe]6s^2 4f^6$	102	No	$[Rn]7s^2 5f^{14}$
23	V	$[Ar]4s^2 3d^3$	63	Eu	$[Xe]6s^2 4f^7$	103	Lr	$[Rn]7s^2 5f^{14} 6d^1$
24	Cr	$[Ar]4s^1 3d^5$	64	Gd	$[Xe]6s^2 4f^7 5d^1$	104	Rf	$[Rn]7s^2 5f^{14} 6d^2$
25	Mn	$[Ar]4s^2 3d^5$	65	Tb	$[Xe]6s^2 4f^9$	105	Db	$[Rn]7s^2 5f^{14} 6d^3$
26	Fe	$[Ar]4s^2 3d^6$	66	Dy	$[Xe]6s^2 4f^{10}$	106	Sg	$[Rn]7s^2 5f^{14} 6d^4$
27	Co	$[Ar]4s^2 3d^7$	67	Ho	$[Xe]6s^2 4f^{11}$	107	Bh	$[Rn]7s^2 5f^{14} 6d^5$
28	Ni	$[Ar]4s^2 3d^8$	68	Er	$[Xe]6s^2 4f^{12}$	108	Hs	$[Rn]7s^2 5f^{14} 6d^6$
29	Cu	$[Ar]4s^1 3d^{10}$	69	Tm	$[Xe]6s^2 4f^{13}$	109	Mt	$[Rn]7s^2 5f^{14} 6d^7$
30	Zn	$[Ar]4s^2 3d^{10}$	70	Yb	$[Xe]6s^2 4f^{14}$	110	Ds	$[Rn]7s^2 5f^{14} 6d^8$
31	Ga	$[Ar]4s^2 3d^{10} 4p^1$	71	Lu	$[Xe]6s^2 4f^{14} 5d^1$	111	Rg	$[Rn]7s^2 5f^{14} 6d^9$
32	Ge	$[Ar]4s^2 3d^{10} 4p^2$	72	Hf	$[Xe]6s^2 4f^{14} 5d^2$	112	Cn	$[Rn]7s^2 5f^{14} 6d^{10}$
33	As	$[Ar]4s^2 3d^{10} 4p^3$	73	Ta	$[Xe]6s^2 4f^{14} 5d^3$	113	Nh	$[Rn]7s^2 5f^{14} 6d^{10} 7p^1$
34	Se	$[Ar]4s^2 3d^{10} 4p^4$	74	W	$[Xe]6s^2 4f^{14} 5d^4$	114	Fl	$[Rn]7s^2 5f^{14} 6d^{10} 7p^2$
35	Br	$[Ar]4s^2 3d^{10} 4p^5$	75	Re	$[Xe]6s^2 4f^{14} 5d^5$	115	Mc	$[Rn]7s^2 5f^{14} 6d^{10} 7p^3$
36	Kr	$[Ar]4s^2 3d^{10} 4p^6$	76	Os	$[Xe]6s^2 4f^{14} 5d^6$	116	Lv	$[Rn]7s^2 5f^{14} 6d^{10} 7p^4$
37	Rb	$[Kr]5s^1$	77	Ir	$[Xe]6s^2 4f^{14} 5d^7$	117	Ts	$[Rn]7s^2 5f^{14} 6d^{10} 7p^5$
38	Sr	$[Kr]5s^2$	78	Pt	$[Xe]6s^1 4f^{14} 5d^9$	118	Og	$[Rn]7s^2 5f^{14} 6d^{10} 7p^6$
39	Y	$[Kr]5s^2 4d^1$	79	Au	$[Xe]6s^1 4f^{14} 5d^{10}$			
40	Zr	$[Kr]5s^2 4d^2$	80	Hg	$[Xe]6s^2 4f^{14} 5d^{10}$			

*The symbol [He] is called the helium core and represents $1s^2$. [Ne] is called the neon core and represents $1s^2 2s^2 2p^6$. [Ar] is called the argon core and represents $[Ne]3s^2 3p^6$. [Kr] is called the krypton core and represents $[Ar]4s^2 3d^{10} 4p^6$. [Xe] is called the xenon core and represents $[Kr]5s^2 4d^{10} 5p^6$. [Rn] is called the radon core and represents $[Xe]6s^2 4f^{14} 5d^{10} 6p^6$.

CHEMISTRY *in Action*

Quantum Dots

We normally consider the color of a chemical substance to be an intensive property (see Section 1.9), because the color does not depend on the amount of that substance that is being considered. As we are learning in this chapter, however, the "normal" behavior of matter is much harder to define as we enter the quantum world of the very small.

Quantum dots are tiny pieces of matter, typically on the order of a few nanometers in diameter, composed of a metal or a semiconductor (see Section 21.3 for a description of semiconductors). By confining the electrons to such a small volume, the allowed energies of these electrons are quantized. Therefore, if quantum dots are excited to higher energy, only certain wavelengths of light are emitted when the electrons go back to their ground states, just as in the case of the emission spectra of atoms. But unlike atoms, the energy of light emitted from a quantum dot can be "tuned" by varying the size of the quantum

Emission from dispersed solutions of CdSe quantum dots arranged from left to right in order of increasing diameter (2 nm to 7 nm).

Prof. Dr. Horst Weller, University of Hamburg, Institute of Physical Chemistry

of calcium (Z = 20) as [Ar]$4s^2$. The placement of the outermost electron in the 4*s* orbital (rather than in the 3*d* orbital) of potassium is strongly supported by experimental evidence. The following comparison also suggests that this is the correct configuration. The chemistry of potassium is very similar to that of lithium and sodium, the first two alkali metals. The outermost electron of both lithium and sodium is in an *s* orbital (there is no ambiguity in assigning their electron configurations); therefore, we expect the last electron in potassium to occupy the 4*s* rather than the 3*d* orbital.

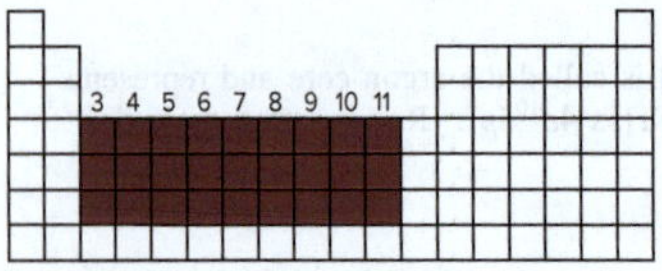

The transition metals.

The elements from scandium (Z = 21) to copper (Z = 29) are transition metals. ***Transition metals*** either *have incompletely filled d subshells or readily give rise to cations that have incompletely filled d subshells.* Consider the first transition metal series, from scandium through copper. In this series additional electrons are placed in the 3*d* orbitals, according to Hund's rule. However, there are two irregularities. The electron configuration of chromium (Z = 24) is [Ar]$4s^1 3d^5$ and not [Ar]$4s^2 3d^4$, as we might expect. A similar break in the pattern is observed for copper, whose electron configuration is [Ar]$4s^1 3d^{10}$ rather than [Ar]$4s^2 3d^9$. The reason for these irregularities is that a slightly greater stability is associated with the half-filled ($3d^5$) and completely dot because that changes the volume within which the electrons are confined. This phenomenon is due to the wavelike behavior of electrons and is analogous to changing the pitch (frequency) of the sound made by plucking a guitar string (see Figure 7.16) by pressing against the neck of the instrument, effectively shortening the string. The ability to regulate the energy of light emitted by a quantum dot is quite remarkable, enabling one to generate the visible spectrum using a single chemical substance by simply varying the diameter of the quantum dots over a range of a few nanometers.

Besides illustrating the quantum behavior of matter and enabling that behavior to be studied on the nanometer scale (as opposed to on a picometer scale at the atomic level), quantum dots offer great promise for yielding important applications in the fields of technology and medicine. Like bulk semiconducting materials, quantum dots can be made to function as LEDs (light-emitting diodes), but unlike these bulk materials, quantum dots emit light symmetrically in very narrow ranges. By combining three quantum dots that emit light of appropriate colors, it is possible to create devices that produce white light at much lower energy costs than required for incandescent bulbs or even fluorescent bulbs, which carry an additional environmental concern because they contain mercury. Quantum dots can also be used to label biological tissue. Besides offering the advantage of greater stability over traditional biological dyes, the surface of quantum dots can be chemically modified to target certain cells such as cancer cells. In addition to enabling tumors to be imaged, these modified quantum dots have the added potential to act therapeutically, either by being incorporated into the more permeable cancer cells and destroying those cells, or by attaching a known antitumor agent to the quantum dot. Other potential applications for quantum dots include quantum computing and photovoltaic cells for harvesting solar energy.

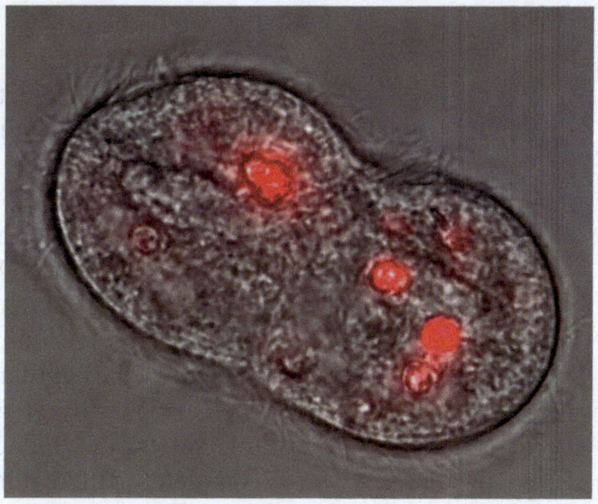

A light micrograph showing the fluorescence of quantum dots in a protozoan, used to study the movement of nanoparticles through the food chain and their potential for accumulation by way of bioconcentration.

The National Institute of Standards and Technology

filled ($3d^{10}$) subshells. Electrons in the same subshell (in this case, the *d* orbitals) have equal energy but different spatial distributions. Consequently, their shielding of one another is relatively small, and the electrons are more strongly attracted by the nucleus when they have the $3d^5$ configuration. According to Hund's rule, the orbital diagram for Cr is

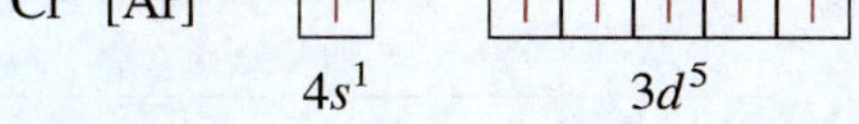

Thus, Cr has a total of six unpaired electrons. The orbital diagram for copper is

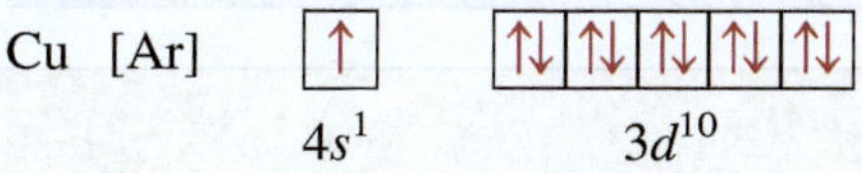

Again, extra stability is gained in this case by having the 3*d* subshell completely filled. In general, half-filled and completely filled subshells have extra stability.

For elements Zn ($Z = 30$) through Kr ($Z = 36$), the 4*s* and 4*p* subshells fill in a straightforward manner. With rubidium ($Z = 37$), electrons begin to enter the $n = 5$ energy level.

The electron configurations in the second transition metal series [yttrium ($Z = 39$) to silver ($Z = 47$)] are also irregular, but we will not be concerned with the details here.

The sixth period of the periodic table begins with cesium ($Z = 55$) and barium ($Z = 56$), whose electron configurations are $[Xe]6s^1$ and $[Xe]6s^2$, respectively. Next we come to lanthanum ($Z = 57$). From Figure 7.30 we would expect that after filling the 6*s* orbital we would place the additional electrons in 4*f* orbitals. In reality, the energies of the 5*d* and 4*f* orbitals are very close; in fact, for lanthanum 4*f* is slightly higher in energy than 5*d*. Thus, lanthanum's electron configuration is $[Xe]6s^25d^1$ and not $[Xe]6s^24f^1$.

Following lanthanum are the 14 elements known as the ***lanthanoids,*** or ***rare earth series*** [cerium ($Z = 58$) to lutetium ($Z = 71$)]. This series was formerly known as the lanthanides. The rare earth metals *have incompletely filled 4f subshells or readily give rise to cations that have incompletely filled 4f subshells.* In this series, the added electrons are placed in 4*f* orbitals. After the 4*f* subshell is completely filled, the next electron enters the 5*d* subshell of lutetium. Note that the electron configuration of gadolinium ($Z = 64$) is $[Xe]6s^24f^75d^1$ rather than $[Xe]6s^24f^8$. Like chromium, gadolinium gains extra stability by having a half-filled subshell ($4f^7$).

The third transition metal series, including lanthanum and hafnium ($Z = 72$) and extending through gold ($Z = 79$), is characterized by the filling of the 5*d* subshell. With Hg ($Z = 80$), both the 6*s* and 5*d* orbitals are now filled. The 6*p* subshell is filled next, which takes us to radon ($Z = 86$).

The *last row of elements* is the ***actinoid series,*** which starts at thorium ($Z = 90$). This series was previously referred to as the actinides. *Most of these elements are not found in nature but have been synthesized.*

With few exceptions, you should be able to write the electron configuration of any element, using Figure 7.30 as a guide. Elements that require particular care are the transition metals, the lanthanoids, and the actinoids. As we noted earlier, at larger values of the principal quantum number *n*, the order of subshell filling may reverse from one element to the next. Figure 7.34 groups the elements according to the type of subshell in which the outermost electrons are placed.

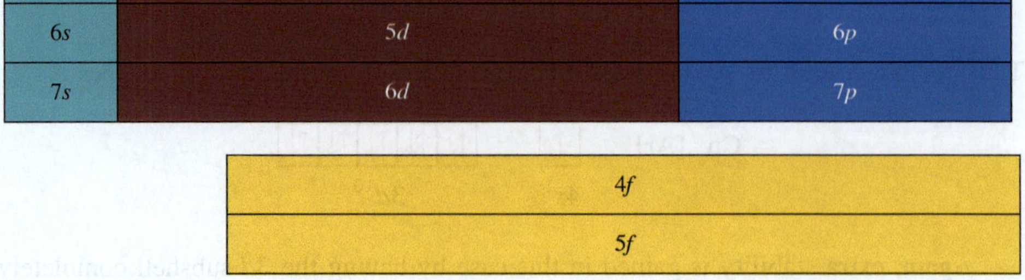

Figure 7.34 *Classification of groups of elements in the periodic table according to the type of subshell being filled with electrons.*

Example 7.13

Write the ground-state electron configurations for (a) sulfur (S) and (b) palladium (Pd), which is diamagnetic.

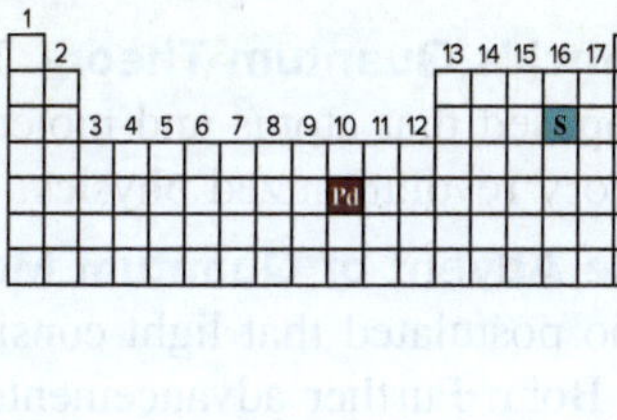

(a) Strategy How many electrons are in the S ($Z = 16$) atom? We start with $n = 1$ and proceed to fill orbitals in the order shown in Figure 7.30. For each value of ℓ, we assign the possible values of m_ℓ. We can place electrons in the orbitals according to the Pauli exclusion principle and Hund's rule and then write the electron configuration. The task is simplified if we use the noble gas core preceding S for the inner electrons.

Solution Sulfur has 16 electrons. The noble gas core in this case is [Ne]. (Ne is the noble gas in the period preceding sulfur.) [Ne] represents $1s^22s^22p^6$. This leaves us 6 electrons to fill the 3*s* subshell and partially fill the 3*p* subshell. Thus, the electron configuration of S is $1s^22s^22p^63s^23p^4$ or $[Ne]3s^23p^4$.

(b) Strategy We use the same approach as that in (a). What does it mean to say that Pd is a diamagnetic element?

Solution Palladium has 46 electrons. The noble gas core in this case is [Kr]. (Kr is the noble gas in the period preceding palladium.) [Kr] represents

$$1s^22s^22p^63s^23p^64s^23d^{10}4p^6$$

The remaining 10 electrons are distributed among the 4*d* and 5*s* orbitals. The three choices are (1) $4d^{10}$, (2) $4d^95s^1$, and (3) $4d^85s^2$. Because palladium is diamagnetic, all the electrons are paired and its electron configuration must be

$$1s^22s^22p^63s^23p^64s^23d^{10}4p^64d^{10}$$

or simply $[Kr]4d^{10}$. The configurations in (2) and (3) both represent paramagnetic elements.

Check To confirm the answer, write the orbital diagrams for (1), (2), and (3).

Practice Exercise Write the ground-state electron configuration for phosphorus (P).

Similar problems: 7.87, 7.88.

Summary of Concepts & Facts

- The Aufbau principle provides the guideline for building up the elements. The periodic table classifies the elements according to their atomic numbers and thus also by the electronic configurations of their atoms.

Review of Concepts & Facts

7.9.1 What element is represented by the following ground-state electron configuration: $1s^22s^22p^63s^23p^5$?

7.9.2 Identify the element that has the following ground-state electron configuration: $[Ar]4s^23d^6$.

7.9.3 Write the electron configuration for an atom of Zr.

Chapter Summary

Planck's Quantum Theory To explain the dependence of radiation emitted by objects on wavelength, Planck proposed that atoms and molecules could emit (or absorb) energy in discrete quantities called quanta. Planck's theory revolutionized physics. (Section 7.1)

The Advent of Quantum Mechanics Planck's work led to the explanation of the photoelectric effect by Einstein, who postulated that light consists of particles called photons, and the emission spectrum of the hydrogen atom by Bohr. Further advancements to quantum theory were made by de Broglie, who demonstrated that an electron possesses both particle and wave properties, and Heisenberg, who derived an inherent limitation to measuring submicroscopic systems. These developments culminated in the Schrödinger equation, which describes the behavior and energy of electrons, atoms, and molecules. (Sections 7.2, 7.3, 7.4, and 7.5)

The Hydrogen Atom The solution to the Schrödinger equation for the hydrogen atom shows quantized energies for the electron and a set of wave functions called atomic orbitals. The atomic orbitals are labeled with specific quantum numbers; the orbitals tell us the regions in which an electron can be located. The results obtained for hydrogen, with minor modifications, can be applied to more complex atoms. (Sections 7.6 and 7.7)

The Building-Up Principle The periodic table can be constructed by increasing atomic number and adding electrons in a stepwise fashion. Specific guidelines (the Pauli exclusion principle and Hund's rule) help us write ground-state electron configurations of the elements, which tell us how electrons are distributed among the atomic orbitals. (Sections 7.8 and 7.9)

Key Equations

$u = \lambda\nu$ (7.1)	Relating speed of a wave to its wavelength and frequency.
$E = h\nu$ (7.2)	Relating energy of a quantum (and of a photon) to the frequency.
$E = h\frac{c}{\lambda}$ (7.3)	Relating energy of a quantum (and of a photon) to the wavelength.
$h\nu = E_k + \Phi$ (7.4)	The photoelectric effect.
$E_n = -R_H\left(\frac{1}{n^2}\right)$ (7.5)	Energy of an electron in the *n*th state in a hydrogen atom.
$\Delta E = h\nu = R_H\left(\frac{1}{n_i^2} - \frac{1}{n_f^2}\right)$ (7.6)	Energy of a photon absorbed or emitted as the electron undergoes a transition from the n_i level to the n_f level.
$\lambda = \frac{h}{mu}$ (7.8)	Relating wavelength of a particle to its mass *m* and velocity *u*.
$\Delta x\Delta p \geq \frac{h}{4\pi}$ (7.9)	Calculating the uncertainty in the position or in the momentum of a particle.

Key Words

Actinoid series, p. 316
Amplitude, p. 274
Atomic orbital, p. 296
Aufbau principle, p. 312
Boundary surface diagram, p. 300
Degenerate orbitals, p. 298
Diamagnetic, p. 307
Electromagnetic radiation, p. 275
Electromagnetic wave, p. 275
Electron configuration, p. 306
Electron density, p. 296
Emission spectra, p. 283
Excited state (or level), p. 285
Frequency (ν), p. 274
Ground state (or level), p. 285
Heisenberg uncertainty principle, p. 294
Hund's rule, p. 309
Lanthanide (rare earth) series, p. 316
Line spectra, p. 283
Many-electron atoms, p. 296
Noble gas core, p. 312
Node, p. 274

Questions & Problems

Red numbered problems solved in Student Solutions Manual

7.1 From Classical Physics to Quantum Theory

Review Questions

7.1 What is a wave? Explain the following terms associated with waves: *wavelength, frequency, amplitude.*

7.2 What are the units for wavelength and frequency of electromagnetic waves? What is the speed of light in meters per second and miles per hour?

7.3 List the types of electromagnetic radiation, starting with the radiation having the longest wavelength and ending with the radiation having the shortest wavelength.

7.4 Give the high and low wavelength values that define the visible region of the electromagnetic spectrum.

7.5 Briefly explain Planck's quantum theory and explain what a quantum is. What are the units for Planck's constant?

7.6 Give two everyday examples that illustrate the concept of quantization.

Problems

7.7 (a) What is the wavelength (in nm) of light having a frequency of 8.6×10^{13} Hz? (b) What is the frequency (in Hz) of light having a wavelength of 566 nm?

7.8 (a) What is the frequency of light having a wavelength of 456 nm? (b) What is the wavelength (in nm) of radiation having a frequency of 2.45×10^9 Hz? (This is the type of radiation used in microwave ovens.)

7.9 The average distance between Mars and Earth is about 1.3×10^8 miles. How long would it take TV pictures transmitted from the *Viking* space vehicle on the Martian surface to reach Earth? (1 mile = 1.61 km)

7.10 How many minutes would it take a radio wave to travel from the planet Venus to Earth? (Average distance from Venus to Earth is 28 million miles.)

7.11 The SI unit of time is the second, which is defined as 9,192,631,770 cycles of radiation associated with a certain emission process in the cesium atom. Calculate the wavelength of this radiation (to three significant figures). In which region of the electromagnetic spectrum is this wavelength found?

7.12 The SI unit of length is the meter, which is defined as the length equal to 1,650,763.73 wavelengths of the light emitted by a particular energy transition in krypton atoms. Calculate the frequency of the light to three significant figures.

7.2 The Photoelectric Effect

Review Questions

7.13 What are photons? What role did Einstein's explanation of the photoelectric effect play in the development of the particle-wave interpretation of the nature of electromagnetic radiation?

7.14 Consider the plots shown here for the photoelectric effect of two different metals A (blue line) and B (red line). (a) Which metal has a greater work function? (b) What does the slope of the lines tell us?

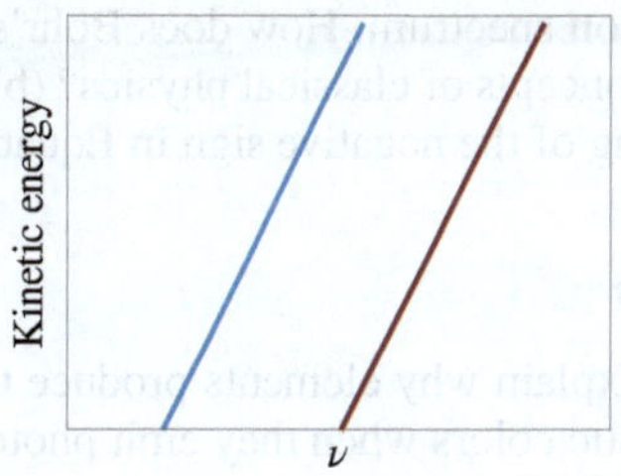

Problems

7.15 A photon has a wavelength of 624 nm. Calculate the energy of the photon in joules.

7.16 The blue color of the sky results from the scattering of sunlight by air molecules. The blue light has a frequency of about 7.5×10^{14} Hz. (a) Calculate the wavelength, in nanometers, associated with this radiation, and (b) calculate the energy, in joules, of a single photon associated with this frequency.

7.17 A photon has a frequency of 6.0×10^4 Hz. (a) Convert this frequency into wavelength (nm). Does this frequency fall in the visible region? (b) Calculate the energy (in joules) of this photon. (c) Calculate the energy (in joules) of 1 mole of photons all with this frequency.

7.18 What is the wavelength, in nm, of radiation that has an energy content of 1.0×10^3 kJ/mol? In which region of the electromagnetic spectrum is this radiation found?

7.19 When copper is bombarded with high-energy electrons, X rays are emitted. Calculate the energy (in joules) associated with the photons if the wavelength of the X rays is 0.154 nm.

7.20 A particular form of electromagnetic radiation has a frequency of 8.11×10^{14} Hz. (a) What is its wavelength in nanometers? In meters? (b) To what region of the electromagnetic spectrum would you assign it? (c) What is the energy (in joules) of one quantum of this radiation?

7.21 The work function of potassium is 3.68×10^{-19} J. (a) What is the minimum frequency of light needed to eject electrons from the metal? (b) Calculate the kinetic energy of the ejected electrons when light of frequency equal to 8.62×10^{14} s^{-1} is used for irradiation.

7.22 When light of frequency equal to 2.11×10^{15} s^{-1} shines on the surface of gold metal, the kinetic energy of ejected electrons is found to be 5.83×10^{-19} J. What is the work function of gold?

7.3 Bohr's Theory of the Hydrogen Atom

Review Questions

7.23 (a) What is an energy level? Explain the difference between ground state and excited state. (b) What are emission spectra? How do line spectra differ from continuous spectra?

7.24 (a) Briefly describe Bohr's theory of the hydrogen atom and how it explains the appearance of an emission spectrum. How does Bohr's theory differ from concepts of classical physics? (b) Explain the meaning of the negative sign in Equation (7.5).

Problems

7.25 Explain why elements produce their own characteristic colors when they emit photons.

7.26 Some copper compounds emit green light when they are heated in a flame. How would you determine whether the light is of one wavelength or a mixture of two or more wavelengths?

7.27 Is it possible for a fluorescent material to emit radiation in the ultraviolet region after absorbing visible light? Explain your answer.

7.28 Explain how astronomers are able to tell which elements are present in distant stars by analyzing the electromagnetic radiation emitted by the stars.

7.29 Consider the following energy levels of a hypothetical atom:

E_4 ——— -1.0×10^{-19} J
E_3 ——— -5.0×10^{-19} J
E_2 ——— -10×10^{-19} J
E_1 ——— -15×10^{-19} J

(a) What is the wavelength of the photon needed to excite an electron from E_1 to E_4? (b) What is the energy (in joules) a photon must have to excite an electron from E_2 to E_3? (c) When an electron drops from the E_3 level to the E_1 level, the atom is said to undergo emission. Calculate the wavelength of the photon emitted in this process.

7.30 The first line of the Balmer series occurs at a wavelength of 656.3 nm. What is the energy difference between the two energy levels involved in the emission that results in this spectral line?

7.31 Calculate the wavelength (in nm) of a photon emitted by a hydrogen atom when its electron drops from the $n = 5$ state to the $n = 3$ state.

7.32 Calculate the frequency (Hz) and wavelength (nm) of the emitted photon when an electron drops from the $n = 4$ to the $n = 2$ level in a hydrogen atom.

7.33 Careful spectral analysis shows that the familiar yellow light of sodium lamps (such as street lamps) is made up of photons of two wavelengths, 589.0 nm and 589.6 nm. What is the difference in energy (in joules) between photons with these wavelengths?

7.34 An electron in the hydrogen atom makes a transition from an energy state of principal quantum numbers n_i to the $n = 2$ state. If the photon emitted has a wavelength of 434 nm, what is the value of n_i?

7.4 The Dual Nature of the Electron

Review Questions

7.35 Explain the statement, Matter and radiation have a "dual nature."

7.36 How does de Broglie's hypothesis account for the fact that the energies of the electron in a hydrogen atom are quantized?

7.37 Why is Equation (7.8) meaningful only for submicroscopic particles, such as electrons and atoms, and not for macroscopic objects?

7.38 (a) If a H atom and a He atom are traveling at the same speed, what will be the relative wavelengths of the two atoms? (b) If a H atom and a He atom have the same kinetic energy, what will be the relative wavelengths of the two atoms?

Problems

7.39 Thermal neutrons move at speeds comparable to those of air molecules at room temperature. These neutrons are most effective in initiating a nuclear chain reaction among ^{235}U isotopes. Calculate the wavelength (in nm) associated with a beam of neutrons moving at 7.00×10^2 m/s. (Mass of a neutron = 1.675×10^{-27} kg.)

7.40 Protons can be accelerated to speeds near that of light in particle accelerators. Estimate the wavelength (in nm) of such a proton moving at 2.90×10^8 m/s. (Mass of a proton = 1.673×10^{-27} kg.)

7.41 What is the de Broglie wavelength, in centimeters, of a 12.4-g hummingbird flying at 1.20×10^2 mph? (1 mile = 1.61 km)

7.42 What is the de Broglie wavelength (in nm) associated with a 2.5-g Ping-Pong ball traveling 35 mph?

7.5 Quantum Mechanics

Review Questions

7.43 What are the inadequacies of Bohr's theory?

7.44 What is the Heisenberg uncertainty principle? What is the Schrödinger equation?

7.45 What is the physical significance of the wave function?

7.46 How is the concept of electron density used to describe the position of an electron in the quantum mechanical treatment of an atom?

7.6 Quantum Numbers

Review Questions

7.47 Describe the four quantum numbers used to characterize an electron in an atom.

7.48 Which quantum number defines a shell? Which quantum numbers define a subshell?

7.49 Which of the following orbitals do not exist: 1*p*, 2*s*, 2*d*, 3*p*, 3*d*, 3*f*, 4*g*?

7.50 Which of the four quantum numbers (n, ℓ, m_ℓ, m_s) determine (a) the energy of an electron in a hydrogen atom and in a many-electron atom, (b) the size of an orbital, (c) the shape of an orbital, (d) the orientation of an orbital in space?

Problems

7.51 An electron in a certain atom is in the $n = 2$ quantum level. List the possible values of ℓ and m_ℓ that it can have.

7.52 An electron in an atom is in the $n = 3$ quantum level. List the possible values of ℓ and m_ℓ that it can have.

7.53 Give the values of the quantum numbers associated with the following orbitals: (a) 2*p*, (b) 3*s*, (c) 5*d*.

7.54 Give the values of the four quantum numbers of an electron in the following orbitals: (a) 3*s*, (b) 4*p*, (c) 3*d*.

7.55 List all the possible subshells and orbitals associated with the principal quantum number n, if $n = 5$.

7.56 List all the possible subshells and orbitals associated with the principal quantum number n, if $n = 6$.

7.7 Atomic Orbitals

Review Questions

7.57 What is an atomic orbital? How does an atomic orbital differ from an orbit?

7.58 Describe the shapes of *s*, *p*, and *d* orbitals. How are these orbitals related to the quantum numbers n, ℓ, and m_ℓ?

7.59 List the hydrogen orbitals in increasing order of energy.

7.60 Why is a boundary surface diagram useful in representing an atomic orbital?

Problems

7.61 Discuss the similarities and differences between a 1*s* and a 2*s* orbital.

7.62 What is the difference between a $2p_x$ and a $2p_y$ orbital?

7.63 Calculate the total number of electrons that can occupy (a) one *s* orbital, (b) three *p* orbitals, (c) five *d* orbitals, (d) seven *f* orbitals.

7.64 What is the total number of electrons that can be held in all orbitals having the same principal quantum number n?

7.65 Determine the maximum number of electrons that can be found in each of the following subshells: 3*s*, 3*d*, 4*p*, 4*f*, 5*f*.

7.66 Indicate the total number of (a) *p* electrons in N ($Z = 7$), (b) *s* electrons in Si ($Z = 14$), and (c) 3*d* electrons in S ($Z = 16$).

7.67 Make a chart of all allowable orbitals in the first four principal energy levels of the hydrogen atom. Designate each by type (for example, *s*, *p*) and indicate how many orbitals of each type there are.

7.68 Why do the 3*s*, 3*p*, and 3*d* orbitals have the same energy in a hydrogen atom but different energies in a many-electron atom?

7.69 For each of the following pairs of hydrogen orbitals, indicate which is higher in energy: (a) 1*s*, 2*s*; (b) 2*p*, 3*p*; (c) $3d_{xy}$, $3d_{yz}$; (d) 3*s*, 3*d*; (e) 4*f*, 5*s*.

7.70 Which orbital in each of the following pairs is lower in energy in a many-electron atom? (a) 2*s*, 2*p*; (b) 3*p*, 3*d*; (c) 3*s*, 4*s*; (d) 4*d*, 5*f*.

7.8 Electron Configuration

Review Questions

7.71 What is electron configuration? Describe the roles that the Pauli exclusion principle and Hund's rule play in writing the electron configuration of elements.

7.72 Explain the meaning of the symbol $4d^6$.

7.73 Explain the meaning of *diamagnetic* and *paramagnetic*. Give an example of an element that is diamagnetic and one that is paramagnetic. What does it mean when we say that electrons are paired?

7.74 What is meant by the term "shielding of electrons" in an atom? Using the Li atom as an example, describe the effect of shielding on the energy of electrons in an atom.

Problems

7.75 Indicate which of the following sets of quantum numbers in an atom are unacceptable and explain why: (a) $(1, 0, \frac{1}{2}, \frac{1}{2})$, (b) $(3, 0, 0, +\frac{1}{2})$, (c) $(2, 2, 1, +\frac{1}{2})$, (d) $(4, 3, -2, +\frac{1}{2})$, (e) $(3, 2, 1, 1)$.

7.76 The ground-state electron configurations listed here are incorrect. Explain what mistakes have been made in each and write the correct electron configurations.

Al: $1s^22s^22p^43s^23p^3$

B: $1s^22s^22p^5$

F: $1s^22s^22p^6$

7.77 The atomic number of an element is 73. Is this element diamagnetic or paramagnetic?

7.78 Indicate the number of unpaired electrons present in each of the following atoms: B, Ne, P, Sc, Mn, Se, Kr, Fe, Cd, I, Pb.

7.9 The Building-Up Principle

Review Questions

7.79 State the Aufbau principle and explain the role it plays in classifying the elements in the periodic table.

7.80 Describe the characteristics of the following groups of elements: transition metals, lanthanides, actinides.

7.81 What is the noble gas core? How does it simplify the writing of electron configurations?

7.82 What are the group and period of the element osmium?

7.83 Define the following terms and give an example of each: *transition metals, lanthanoids, actinoids.*

7.84 Explain why the ground-state electron configurations of Cr and Cu are different from what we might expect.

7.85 Explain what is meant by a noble gas core. Write the electron configuration of a xenon core.

7.86 Comment on the correctness of the following statement: The probability of finding two electrons with the same four quantum numbers in an atom is zero.

Problems

7.87 Use the Aufbau principle to obtain the ground-state electron configuration of selenium.

7.88 Use the Aufbau principle to obtain the ground-state electron configuration of technetium.

7.89 Write the ground-state electron configurations for the following elements: B, V, Ni, As, I, Au.

7.90 Write the ground-state electron configurations for the following elements: Ge, Fe, Zn, Ni, W, Tl.

7.91 The electron configuration of a neutral atom is $1s^22s^22p^63s^2$. Write a complete set of quantum numbers for each of the electrons. Name the element.

7.92 Which of the following species has the most unpaired electrons: S^+, S, or S^-? Explain how you arrive at your answer.

Additional Problems

7.93 A sample tube consisted of atomic hydrogens in their ground state. A student illuminated the atoms with monochromatic light, that is, light of a single wavelength. If only two spectral emission lines in the visible region are observed, what is the wavelength(s) of the incident radiation?

7.94 A laser produces a beam of light with a wavelength of 532 nm. If the power output is 25.0 mW, how many photons does the laser emit per second? (1 W = 1 J/s)

7.95 When a compound containing cesium ion is heated in a Bunsen burner flame, photons with an energy of 4.30×10^{-19} J are emitted. What color is the cesium flame?

7.96 Discuss the current view of the correctness of the following statements. (a) The electron in the hydrogen atom is in an orbit that never brings it closer than 100 pm to the nucleus. (b) Atomic absorption spectra result from transitions of electrons from lower to higher energy levels. (c) A many-electron atom behaves somewhat like a solar system that has a number of planets.

7.97 What is the basis for thinking that atoms are spherical in shape even though the atomic orbitals *p*, *d*, . . . have distinctly nonspherical shapes?

7.98 What is the maximum number of electrons in an atom that can have the following quantum numbers? Specify the orbitals in which the electrons would be found. (a) $n = 2$, $m_s = +\frac{1}{2}$; (b) $n = 4$, $m_\ell = +1$; (c) $n = 3$, $\ell = 2$; (d) $n = 2$, $\ell = 0$, $m_s = -\frac{1}{2}$; (e) $n = 4$, $\ell = 3$, $m_\ell = -2$.

7.99 Identify the following individuals and their contributions to the development of quantum theory: Bohr, de Broglie, Einstein, Planck, Heisenberg, Schrödinger.

7.100 What properties of electrons are used in the operation of an electron microscope?

7.101 In a photoelectric experiment a student uses a light source whose frequency is greater than that needed to eject electrons from a certain metal. However, after continuously shining the light on the same area of the metal for a long period of time the student notices that the maximum kinetic energy of ejected electrons begins to decrease, even though the frequency of the light is held constant. How would you account for this behavior?

7.102 A certain pitcher's fastballs have been clocked at about 100 mph. (a) Calculate the wavelength of a 0.141-kg baseball (in nm) at this speed. (b) What is the wavelength of a hydrogen atom at the same speed? (1 mile = 1609 m)

7.103 A student carried out a photoelectric experiment by shining visible light on a clean piece of cesium metal. The table here shows the kinetic energies (E_k) of the ejected electrons as a function of wavelengths (λ). Determine graphically the work function and the Planck constant.

λ (nm)	405	435.8	480	520	577.7
E_k (J)	2.360×10^{-19}	2.029×10^{-19}	1.643×10^{-19}	1.417×10^{-19}	1.067×10^{-19}

7.104 (a) What is the lowest possible value of the principal quantum number (n) when the angular momentum quantum number (ℓ) is 1? (b) What are the possible values of the angular momentum quantum number (ℓ) when the magnetic quantum number (m_ℓ) is 0, given than $n \leq 4$?

7.105 Considering only the ground-state electron configuration, are there more diamagnetic or paramagnetic elements? Explain.

7.106 A ruby laser produces radiation of wavelength 633 nm in pulses whose duration is 1.00×10^{-9} s. (a) If the laser produces 0.376 J of energy per pulse, how many photons are produced in each pulse? (b) Calculate the power (in watts) delivered by the laser per pulse. (1 W = 1 J/s)

7.107 A 368-g sample of water absorbs infrared radiation at 1.06×10^4 nm from a carbon dioxide laser. Suppose all the absorbed radiation is converted to heat. Calculate the number of photons at this wavelength required to raise the temperature of the water by 5.00°C.

7.108 Photodissociation of water

$$H_2O(l) + h\nu \longrightarrow H_2(g) + \frac{1}{2}O_2(g)$$

has been suggested as a source of hydrogen. The ΔH°_{rxn} for the reaction, calculated from thermochemical data, is 285.8 kJ per mole of water decomposed. Calculate the maximum wavelength (in nm) that would provide the necessary energy. In principle, is it feasible to use sunlight as a source of energy for this process?

7.109 Spectral lines of the Lyman and Balmer series do not overlap. Verify this statement by calculating the longest wavelength associated with the Lyman series and the shortest wavelength associated with the Balmer series (in nm).

7.110 An atom moving at its root-mean-square speed at 20°C has a wavelength of 3.28×10^{-11} m. Identify the atom.

7.111 Certain sunglasses have small crystals of silver chloride (AgCl) incorporated in the lenses. When the lenses are exposed to light of the appropriate wavelength, the following reaction occurs:

$$AgCl \longrightarrow Ag + Cl$$

The Ag atoms formed produce a uniform gray color that reduces the glare. If ΔH for the preceding reaction is 248 kJ/mol, calculate the maximum wavelength of light that can induce this process.

7.112 The He^+ ion contains only one electron and is therefore a hydrogen-like ion. Calculate the wavelengths, in increasing order, of the first four transitions in the Balmer series of the He^+ ion. Compare these wavelengths with the same transitions in a H atom. Comment on the differences. (The Rydberg constant for He^+ is 8.72×10^{-18} J.)

7.113 Ozone (O_3) in the stratosphere absorbs the harmful radiation from the sun by undergoing decomposition: $O_3 \longrightarrow O + O_2$. (a) Referring to Table 6.4, calculate the ΔH° for this process. (b) Calculate the maximum wavelength of photons (in nm) that possesses this energy to cause the decomposition of ozone photochemically.

7.114 The retina of a human eye can detect light when radiant energy incident on it is at least 4.0×10^{-17} J. For light of 600-nm wavelength, how many photons does this correspond to?

7.115 A helium atom and a xenon atom have the same kinetic energy. Calculate the ratio of the de Broglie wavelength of the helium atom to that of the xenon atom.

7.116 A laser is used in treating retina detachment. The wavelength of the laser beam is 514 nm and the power is 1.6 W. If the laser is turned on for 0.060 s during surgery, calculate the number of photons emitted by the laser. (1 W = 1 J/s.)

7.117 An electron in an excited state in a hydrogen atom can return to the ground state in two different ways: (a) via a direct transition in which a photon of wavelength λ_1 is emitted and (b) via an intermediate excited state reached by the emission of a photon of wavelength λ_2. This intermediate excited state then decays to the ground state by emitting another photon of wavelength λ_3. Derive an equation that relates λ_1 to λ_2 and λ_3.

7.118 A photoelectric experiment was performed by separately shining a laser at 450 nm (blue light) and a laser at 560 nm (yellow light) on a clean metal surface and measuring the number and kinetic energy of the ejected electrons. Which light would generate more electrons? Which light would eject electrons with greater kinetic energy? Assume that the same amount of energy is delivered to the metal surface by each laser and that the frequencies of the laser lights exceed the threshold frequency.

7.119 Draw the shapes (boundary surfaces) of the following orbitals: (a) $2p_y$, (b) $3d_{z^2}$, (c) $3d_{x^2-y^2}$. (Show coordinate axes in your sketches.)

7.120 The electron configurations described in this chapter all refer to gaseous atoms in their ground states. An atom may absorb a quantum of energy and promote one of its electrons to a higher-energy orbital. When this happens, we say that the atom is in an excited state. The electron configurations of some excited atoms are given. Identify these atoms and write their ground-state configurations:

(a) $1s^12s^1$
(b) $1s^22s^22p^23d^1$
(c) $1s^22s^22p^64s^1$
(d) $[Ar]4s^13d^{10}4p^4$
(e) $[Ne]3s^23p^43d^1$

7.121 Draw orbital diagrams for atoms with the following electron configurations:

(a) $1s^22s^22p^5$
(b) $1s^22s^22p^63s^23p^3$
(c) $1s^22s^22p^63s^23p^64s^23d^7$

7.122 If Rutherford and his coworkers had used electrons instead of alpha particles to probe the structure of the nucleus as described in Section 2.2, what might they have discovered?

7.123 Scientists have found interstellar hydrogen atoms with quantum number n in the hundreds. Calculate the wavelength of light emitted when a hydrogen atom undergoes a transition from $n = 236$ to $n = 235$. In what region of the electromagnetic spectrum does this wavelength fall?

7.124 Calculate the wavelength of a helium atom whose speed is equal to the root-mean-square speed at 20°C.

7.125 Ionization energy is the minimum energy required to remove an electron from an atom. It is usually expressed in units of kJ/mol, that is, the energy in kilojoules required to remove one mole of electrons from one mole of atoms. (a) Calculate the ionization energy for the hydrogen atom. (b) Repeat the calculation, assuming in this second case that the electrons are removed from the $n = 2$ state.

7.126 An electron in a hydrogen atom is excited from the ground state to the $n = 4$ state. Comment on the correctness of the following statements (true or false).

(a) $n = 4$ is the first excited state.
(b) It takes more energy to ionize (remove) the electron from $n = 4$ than from the ground state.
(c) The electron is farther from the nucleus (on average) in $n = 4$ than in the ground state.
(d) The wavelength of light emitted when the electron drops from $n = 4$ to $n = 1$ is longer than that from $n = 4$ to $n = 2$.
(e) The wavelength the atom absorbs in going from $n = 1$ to $n = 4$ is the same as that emitted as it goes from $n = 4$ to $n = 1$.

7.127 The ionization energy of a certain element is 412 kJ/mol (see Problem 7.125). However, when the atoms of this element are in the first excited state, the ionization energy is only 126 kJ/mol. Based on this information, calculate the wavelength of light emitted in a transition from the first excited state to the ground state.

7.128 Alveoli are the tiny sacs of air in the lungs (see Problem 5.136) whose average diameter is 5.0×10^{-5} m. Consider an oxygen molecule (5.3×10^{-26} kg) trapped within a sac. Calculate the uncertainty in the velocity of the oxygen molecule. (*Hint:* The maximum uncertainty in the position of the molecule is given by the diameter of the sac.)

7.129 How many photons at 660 nm must be absorbed to melt 5.0×10^2 g of ice? On average, how many H_2O molecules does one photon convert from ice to water? (*Hint:* It takes 334 J to melt 1 g of ice at 0°C.)

7.130 Shown are portions of orbital diagrams representing the ground-state electron configurations of certain elements. Which of them violate the Pauli exclusion principle? Hund's rule?

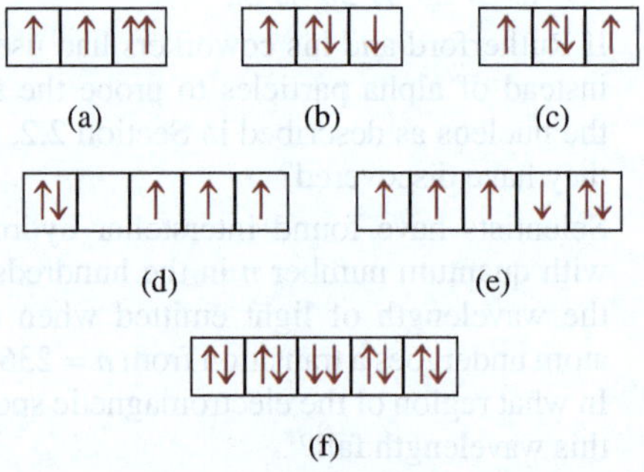

7.131 The UV light that is responsible for tanning the skin falls in the 320- to 400-nm region. Calculate the total energy (in joules) absorbed by a person exposed to this radiation for 2.0 h, given that there are 2.0×10^{16} photons hitting Earth's surface per square centimeter per second over a 80-nm (320 nm to 400 nm) range and that the exposed body area is 0.45 m^2. Assume that only half of the radiation is absorbed and the other half is reflected by the body. (*Hint:* Use an average wavelength of 360 nm in calculating the energy of a photon.)

7.132 The sun is surrounded by a white circle of gaseous material called the corona, which becomes visible during a total eclipse of the sun. The temperature of the corona is in the millions of degrees Celsius, which is high enough to break up molecules and remove some or all of the electrons from atoms. One way astronomers have been able to estimate the temperature of the corona is by studying the emission lines of ions of certain elements. For example, the emission spectrum of Fe^{14+} ions has been recorded and analyzed. Knowing that it takes 3.5×10^4 kJ/mol to convert Fe^{13+} to Fe^{14+}, estimate the temperature of the sun's corona. (*Hint:* The average kinetic energy of one mole of a gas is $\frac{3}{2}RT$.)

7.133 In 1996 physicists created an anti-atom of hydrogen. In such an atom, which is the antimatter equivalent of an ordinary atom, the electrical charges of all the component particles are reversed. Thus, the nucleus of an anti-atom is made of an anti-proton, which has the same mass as a proton but bears a negative charge, while the electron is replaced by an anti-electron (also called positron) with the same mass as an electron, but bearing a positive charge. Would you expect the energy levels, emission spectra, and atomic orbitals of an antihydrogen atom to be different from those of a hydrogen atom? What would happen if an anti-atom of hydrogen collided with a hydrogen atom?

7.134 Use Equation (5.16) to calculate the de Broglie wavelength of a N_2 molecule at 300 K.

7.135 When an electron makes a transition between energy levels of a hydrogen atom, there are no restrictions on the initial and final values of the principal quantum number n. However, there is a quantum mechanical rule that restricts the initial and final values of the orbital angular momentum ℓ. This is the *selection rule,* which states that $\Delta\ell = \pm 1$; that is, in a transition, the value of ℓ can only increase or decrease by one. According to this rule, which of the following transitions are allowed: (a) $2s \longrightarrow 1s$, (b) $3p \longrightarrow 1s$, (c) $3d \longrightarrow 4f$, (d) $4d \longrightarrow 3s$? In view of this selection rule, explain why it is possible to observe the various emission series shown in Figure 7.15.

7.136 In an electron microscope, electrons are accelerated by passing them through a voltage difference. The kinetic energy thus acquired by the electrons is equal to the voltage times the charge on the electron. Thus, a voltage difference of 1 V imparts a kinetic energy of 1.602×10^{-19} C × V or 1.602×10^{-19} J. Calculate the wavelength associated with electrons accelerated by 5.00×10^3 V.

7.137 A microwave oven operating at 1.22×10^8 nm is used to heat 150 mL of water (roughly the volume of a tea cup) from 20°C to 100°C. Calculate the number of photons needed if 92.0% of microwave energy is converted to the thermal energy of water.

7.138 The radioactive Co-60 isotope is used in nuclear medicine to treat certain types of cancer. Calculate the wavelength and frequency of an emitted gamma photon having the energy of 1.29×10^{11} J/mol.

7.139 (a) An electron in the ground state of the hydrogen atom moves at an average speed of 5×10^6 m/s. If the speed is known to an uncertainty of 1%, what is the uncertainty in knowing its position? Given that the radius of the hydrogen atom in the ground state is 5.29×10^{-11} m, comment on your result. The mass of an electron is 9.1094×10^{-31} kg. (b) A 3.2-g Ping-Pong ball moving at 50 mph has a momentum of 0.073 kg · m/s. If the uncertainty in measuring the momentum is 1.0×10^{-7} of the momentum, calculate the uncertainty in the Ping-Pong ball's position.

7.140 One wavelength in the hydrogen emission spectrum is 1280 nm. What are the initial and final states of the transition responsible for this emission?

7.141 Owls have good night vision because their eyes can detect a light intensity as low as 5.0×10^{-13} W/m^2. Calculate the number of photons per second that an owl's eye can detect if its pupil has a diameter of 9.0 mm and the light has a wavelength of 500 nm. (1 W = 1 J/s.)

7.142 For hydrogen-like ions, that is, ions containing only one electron, Equation (7.5) is modified as follows: $E_n = -R_H Z^2(1/n^2)$, where Z is the atomic number of the parent atom. The figure here represents the emission spectrum of such a hydrogen-like ion in the gas phase. All the lines result from the electronic transitions from the excited states to the $n = 2$ state. (a) What electronic transitions correspond to lines B and C? (b) If the wavelength of line C is 27.1 nm, calculate the wavelengths of lines A and B. (c) Calculate the energy needed to remove the electron from the ion in the $n = 4$ state. (d) What is the physical significance of the continuum?

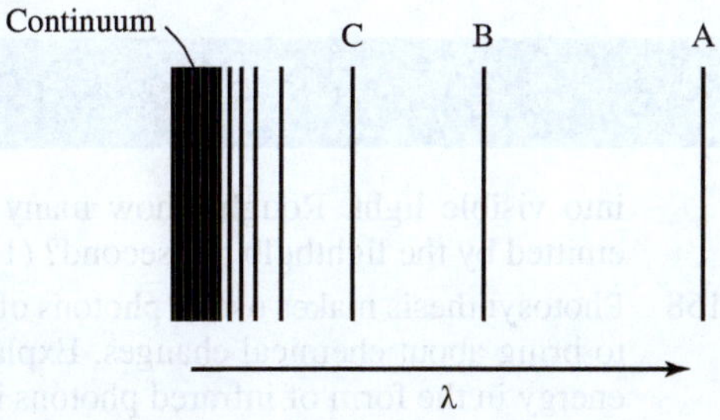

7.143 When two atoms collide, some of their kinetic energy may be converted into electronic energy in one or both atoms. If the average kinetic energy is about equal to the energy for some allowed electronic transition, an appreciable number of atoms can absorb enough energy through an inelastic collision to be raised to an excited electronic state. (a) Calculate the average kinetic energy per atom in a gas sample at 298 K. (b) Calculate the energy difference between the $n = 1$ and $n = 2$ levels in hydrogen. (c) At what temperature is it possible to excite a hydrogen atom from the $n = 1$ level to $n = 2$ level by collision? [The average kinetic energy of 1 mole of an ideal gas is $\frac{3}{2}RT$.]

7.144 Calculate the energies needed to remove an electron from the $n = 1$ state and the $n = 5$ state in the Li^{2+} ion. What is the wavelength (in nm) of the emitted photon in a transition from $n = 5$ to $n = 1$? The Rydberg constant for hydrogen-like ions is $(2.18 \times 10^{-18}\text{ J})Z^2$, where Z is the atomic number.

7.145 The de Broglie wavelength of an accelerating proton in the Large Hadron Collider is 2.5×10^{-14} m. What is the kinetic energy (in joules) of the proton?

7.146 The minimum uncertainty in the position of a certain moving particle is equal to its de Broglie wavelength. If the speed of the particle is 1.2×10^5 m/s, what is the minimum uncertainty in its speed?

7.147 According to Einstein's special theory of relativity, the mass of a moving particle, m_{moving}, is related to its mass at rest, m_{rest}, by the equation

$$m_{\text{moving}} = \frac{m_{\text{rest}}}{\sqrt{1 - \left(\frac{u}{c}\right)^2}}$$

where u and c are the speeds of the particle and light, respectively. (a) In particle accelerators, protons, electrons, and other charged particles are often accelerated to speeds close to the speed of light. Calculate the wavelength (in nm) of a proton moving at 50.0% the speed of light. The mass of a proton is 1.673×10^{-27} kg. (b) Calculate the mass of a 6.0×10^{-2} kg tennis ball moving at 63 m/s. Comment on your results.

7.148 The mathematical equation for studying the photoelectric effect is

$$h\nu = \Phi + \tfrac{1}{2}m_e u^2$$

where ν is the frequency of light shining on the metal, Φ is the work function, and m_e and u are the mass and speed of the ejected electron. In an experiment, a student found that a maximum wavelength of 351 nm is needed to just dislodge electrons from a zinc metal surface. Calculate the speed (in m/s) of an ejected electron when she employed light with a wavelength of 313 nm.

7.149 In the beginning of the twentieth century, some scientists thought that a nucleus may contain both electrons and protons. Use the Heisenberg uncertainty principle to show that an electron cannot be confined within a nucleus. Repeat the calculation for a proton. Comment on your results. Assume the radius of a nucleus to be 1.0×10^{-15} m. The masses of an electron and a proton are 9.109×10^{-31} kg and 1.673×10^{-27} kg, respectively. (*Hint:* Treat the diameter of the nucleus as the uncertainty in position.)

7.150 Blackbody radiation is the term used to describe the dependence of the radiation energy emitted by an object on wavelength at a certain temperature. Planck proposed the quantum theory to account for this dependence. Shown in the figure is a plot of the radiation energy emitted by our sun versus wavelength. This curve is characteristic of the temperature at the surface of the sun. At a higher temperature, the curve has a similar shape but the maximum will shift to a shorter wavelength. What does this curve reveal about two consequences of great biological significance on Earth?

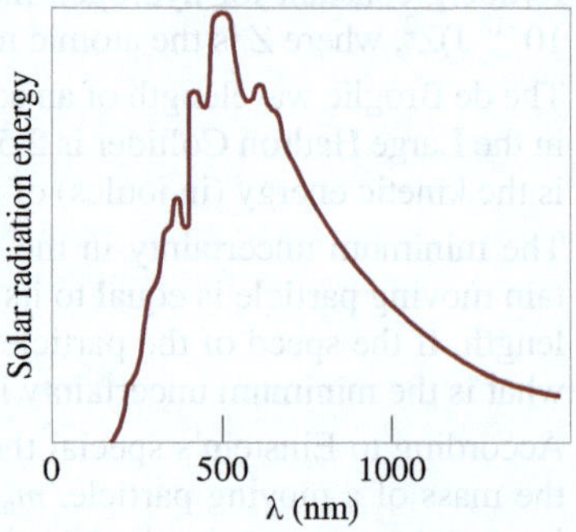

7.151 All molecules undergo vibrational motions. Quantum mechanical treatment shows that the vibrational energy, E_{vib}, of a diatomic molecule like HCl is given by

$$E_{vib} = \left(n + \frac{1}{2}\right)h\nu$$

where n is a quantum number given by $n = 0, 1, 2, 3, \ldots$ and ν is the fundamental frequency of vibration. (a) Sketch the first three vibrational energy levels for HCl. (b) Calculate the energy required to excite a HCl molecule from the ground level to the first excited level. The fundamental frequency of vibration for HCl is $8.66 \times 10^{13}\ s^{-1}$. (c) The fact that the lowest vibrational energy in the ground level is not zero but equal to $\frac{1}{2}h\nu$ means that molecules will vibrate at all temperatures, including the absolute zero. Use the Heisenberg uncertainty principle to justify this prediction. (*Hint:* Consider a nonvibrating molecule and predict the uncertainty in the momentum and hence the uncertainty in the position.)

7.152 The wave function for the 2s orbital in the hydrogen atom is

$$\Psi_{2s} = \frac{1}{\sqrt{2a_0^3}}\left(1 - \frac{\rho}{2}\right)e^{-\rho/2}$$

where a_0 is the value of the radius of the first Bohr orbit, equal to 0.529 nm, ρ is $Z(r/a_0)$, and r is the distance from the nucleus in meters. Calculate the location of the node of the 2s wave function from the nucleus.

7.153 A student placed a large unwrapped chocolate bar in a microwave oven without a rotating glass plate. After turning the oven on for less than a minute, she noticed there were evenly spaced dents (due to melting) about 6 cm apart. Based on her observations, calculate the speed of light given that the microwave frequency is 2.45 GHz. (*Hint:* The energy of a wave is proportional to the square of its amplitude.)

7.154 The wave properties of matter can generally be ignored for macroscopic objects such as tennis balls; however, wave properties have been measured at the fringe of detection for some very large molecules. For example, wave patterns were detected for $C_{60}(C_{12}F_{25})_8$ molecules moving at a velocity of 63 m/s. (a) Calculate the wavelength of a $C_{60}(C_{12}F_{25})_8$ molecule moving at this velocity. (b) How does the wavelength compare to the size of the molecule given that its diameter is roughly 3000 pm?

Interpreting, Modeling, & Estimating

7.155 Atoms of an element have only two accessible excited states. In an emission experiment, however, three spectral lines were observed. Explain. Write an equation relating the shortest wavelength to the other two wavelengths.

7.156 According to Wien's law, the wavelength of maximum intensity in blackbody radiation, λ_{max}, is given by

$$\lambda_{max} = \frac{b}{T}$$

where b is a constant (2.898×10^6 nm · K) and T is the temperature of the radiating body in kelvins. (a) Estimate the temperature at the surface of the sun. (b) How are astronomers able to determine the temperature of stars in general? (See Problem 7.150 for a definition of blackbody radiation.)

7.157 Only a fraction of the electrical energy supplied to an incandescent-tungsten lightbulb is converted to visible light. The rest of the energy shows up as infrared radiation (that is, heat). A 60-W lightbulb converts about 15.0% of the energy supplied to it into visible light. Roughly how many photons are emitted by the lightbulb per second? (1 W = 1 J/s.)

7.158 Photosynthesis makes use of photons of visible light to bring about chemical changes. Explain why heat energy in the form of infrared photons is ineffective for photosynthesis. (*Hint:* Typical chemical bond energies are 200 kJ/mol or greater.)

7.159 A typical red laser pointer has a power of 5 mW. How long would it take a red laser pointer to emit the same number of photons emitted by a 1-W blue laser in 1 s? (1 W = 1 J/s.)

7.160 Referring to the Chemistry in Action essay "Quantum Dots" in Section 7.9, estimate the wavelength of light that would be emitted by a cadmium selenide (CdSe) quantum dot with a diameter of 10 nm. Would the emitted light be visible to the human eye? The diameter and emission wavelength for a series of quantum dots are given here.

Diameter (nm)	2.2	2.5	3.3	4.2	4.9	6.3
Wavelength (nm)	462	503	528	560	583	626

Answers to Practice Exercises

7.1 8.24 m. **7.2** 3.39×10^3 nm. **7.3** 9.65×10^{-19} J. **7.4** 2.63×10^3 nm. **7.5** 56.6 nm. **7.6** 0.2 m/s. **7.7** $m_\ell = -2, -1, 0, 1, 2$. **7.8** $n = 3$, $\ell = 1$, $m_\ell = -1, 0, 1$. **7.9** For $n = 4$, the possible values of ℓ are 0, 1, 2, and 3. Thus, there is one 4*s* orbital ($n = 4$, $\ell = 0$, and $m_\ell = 0$); there are three 4*p* orbitals ($n = 4$, $\ell = 1$, and $m_\ell = -1, 0, 1$); there are five 4*d* orbitals ($n = 4$, $\ell = 2$, and $m_\ell = -2, -1, 0, 1, 2$); there are seven 4*f* orbitals ($n = 4$, $\ell = 3$, and $m_\ell = -3, -2, -1, 0, 1, 2, 3$). The total number of orbitals is $1 + 3 + 5 + 7 = 16$. **7.10** $(4, 2, -2, +\frac{1}{2})$, $(4, 2, -1, +\frac{1}{2})$, $(4, 2, 0, +\frac{1}{2})$, $(4, 2, 1, +\frac{1}{2})$, $(4, 2, 2, +\frac{1}{2})$, $(4, 2, -2, -\frac{1}{2})$, $(4, 2, -1, -\frac{1}{2})$, $(4, 2, 0, -\frac{1}{2})$, $(4, 2, 1, -\frac{1}{2})$, $(4, 2, 2, -\frac{1}{2})$. **7.11** 32 **12** $(1, 0, 0, +\frac{1}{2})$, $(1, 0, 0, -\frac{1}{2})$, $(2, 0, 0, +\frac{1}{2})$, $(2, 0, 0, -\frac{1}{2})$, $(2, 1, -1, -\frac{1}{2})$. There are five other acceptable ways to write the quantum numbers for the last electron (in the 2*p* orbital). **7.13** [Ne]$3s^23p^3$.

Answers to Review of Concepts & Facts

7.1.1 Highest frequency, (b). Longest wavelength, (c). Greatest amplitude, (a). **7.1.2** The wavelengths of visible and infrared radiation are not short enough (and hence not energetic enough) to affect the dark pigment-producing melanocyte cells beneath the skin. **7.2.1** 4.80×10^{-19} J. **7.2.2** 297 nm. **7.2.3** Shortest wavelength: λ_3. Longest wavelength: λ_2. **7.3.1** (b). **7.3.2** -1.36×10^{-19} J. **7.3.3** 397 nm. **7.4.1** The Planck constant, *h*. Because *h* is such a small number, only atomic and molecular systems that have extremely small masses will exhibit measurable wave properties. **7.4.2** 79 nm. **7.5.1** ψ is the wave function of the electron, and ψ^2 represents the probability of finding the electron in a particular region of space around the nucleus. **7.5.2** 5.3×10^{13} m. **7.6.1** When $n = 3$, the allowed ℓ values are 0, 1, and 2. **7.6.2** $(6, 0, 0, +\frac{1}{2})$ and $(6, 0, 0, -\frac{1}{2})$. **7.7.1** When $n = 2$, the allowed ℓ values are 0 and 1, corresponding to *s* and *p* orbitals. When $n = 3$, the allowed ℓ values are 0, 1, and 2, corresponding to *s*, *p*, and *d* orbitals. **7.7.2** $n = 4$, $\ell = 1$, $m_\ell = -1, 0$, or 1. **7.7.3** 7. **7.8.1** n, ℓ, and m_s. **7.8.2** 10. **7.8.3** (a) When $\ell = 0$, m_ℓ must equal 0. (b) When $n = 3$, ℓ can at most be up to $n - 1$. (c) m_s but be either $+\frac{1}{2}$ or $-\frac{1}{2}$. **7.9.1** Cl. **7.9.2** Fe. **7.9.3** [Kr]$5s^24d^2$.

CHAPTER

8

Periodic Relationships Among the Elements

The recurring or "periodic" trends in the properties of elements are most commonly illustrated in tabular form.
welcomia/Shutterstock

CHAPTER OUTLINE

Many of the chemical properties of the elements can be understood in terms of their electron configurations. Because electrons fill atomic orbitals in a fairly regular fashion, it is not surprising that elements with similar electron configurations, such as sodium and potassium, behave similarly in many respects and that, in general, the properties of the elements exhibit observable trends. Chemists in the nineteenth century recognized periodic trends in the physical and chemical properties of the elements, long before quantum theory came onto the scene. Although these chemists were not aware of the existence of electrons and protons, their efforts to systematize the chemistry of the elements were remarkably successful. Their main sources of information were the atomic masses of the elements and other known physical and chemical properties.

8.1 Development of the Periodic Table

Learning Objective

- Explain how elements are arranged in the periodic table.

In the nineteenth century, when chemists had only a vague idea of atoms and molecules and did not know of the existence of electrons and protons, they devised the periodic table using their knowledge of atomic masses. Accurate measurements of the atomic masses of many elements had already been made. Arranging elements according to their atomic masses in a periodic table seemed logical to those chemists, who felt that chemical behavior should somehow be related to atomic mass.

In 1864 the English chemist John Newlands[†] noticed that when the elements were arranged in order of atomic mass, every eighth element had similar properties. Newlands referred to this peculiar relationship as the *law of octaves*. However, this "law" turned out to be inadequate for elements beyond calcium, and Newlands's work was not accepted by the scientific community.

In 1869 the Russian chemist Dmitri Mendeleev[‡] and the German chemist Lothar Meyer[§] independently proposed a much more extensive tabulation of the elements based on the regular, periodic recurrence of properties. Mendeleev's classification system was a great improvement over Newlands's for two reasons: It grouped the elements more accurately, according to their properties; and, equally important, it made possible the prediction of the properties of several elements that had not yet been discovered. For example, Mendeleev proposed the existence of an unknown element that he called eka-aluminum and predicted a number of its properties. (*Eka* is a Sanskrit word meaning "first"; thus eka-aluminum would be the first element under aluminum in the same group.) When gallium was discovered four years later, its properties matched the predicted properties of eka-aluminum remarkably well:

	Eka-Aluminum (Ea)	Gallium (Ga)
Atomic mass	68 amu	69.9 amu
Melting point	Low	29.78°C
Density	5.9 g/cm^3	5.94 g/cm^3
Formula of oxide	Ea_2O_3	Ga_2O_3

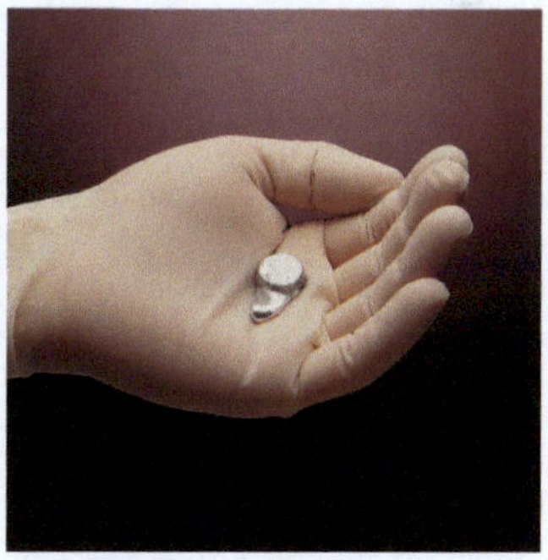

Gallium melts in a person's hand (body temperature is about 37°C).
Stephen Frisch/McGraw Hill

[†]John Alexander Reina Newlands (1838–1898). English chemist. Newlands's work was a step in the right direction in the classification of the elements. Unfortunately, because of its shortcomings, he was subjected to much criticism, and even ridicule. At one meeting he was asked if he had ever examined the elements according to the order of their initial letters! Nevertheless, in 1887 Newlands was honored by the Royal Society of London for his contribution.

[‡]Dmitri Ivanovich Mendeleev (1836–1907). Russian chemist. His work on the periodic classification of elements is regarded by many as the most significant achievement in chemistry in the nineteenth century.

[§]Julius Lothar Meyer (1830–1895). German chemist. In addition to his contribution to the periodic table, Meyer also discovered the chemical affinity of hemoglobin for oxygen.

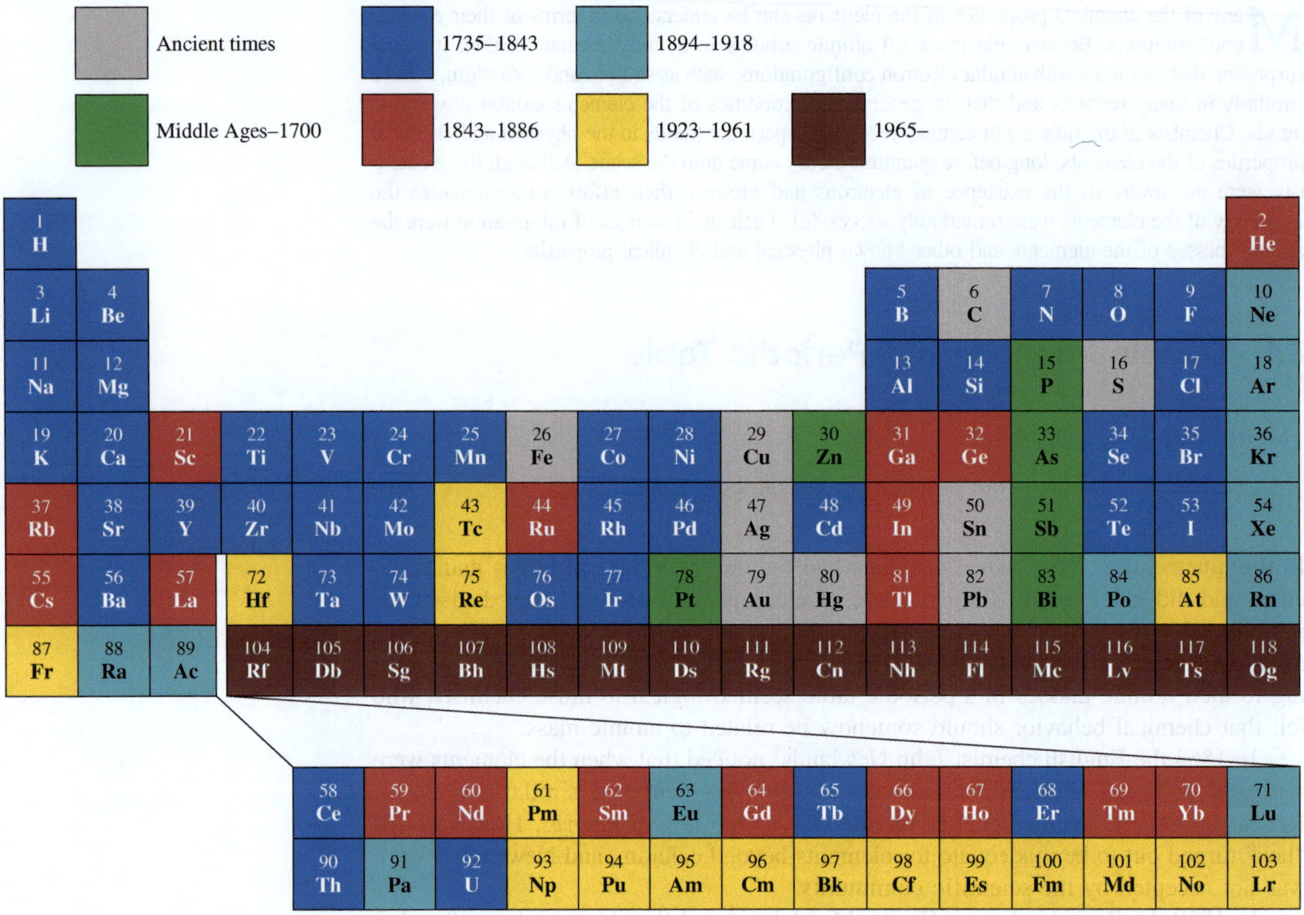

Figure 8.1 *A chronological chart of the discovery of the elements. To date, 118 elements have been identified.*

Mendeleev's periodic table included 66 known elements. By 1900, some 30 more had been added to the list, filling in some of the empty spaces. Figure 8.1 charts the discovery of the elements chronologically.

Although this periodic table was a celebrated success, the early versions had some glaring inconsistencies. For example, the atomic mass of argon (39.95 amu) is greater than that of potassium (39.10 amu). If elements were arranged solely according to increasing atomic mass, argon would appear in the position occupied by potassium in our modern periodic table (see the Periodic Table in the Appendix). But no chemist would place argon, an inert gas, in the same group as lithium and sodium, two very reactive metals. This and other discrepancies suggested that some fundamental property other than atomic mass must be the basis of periodicity. This property turned out to be associated with atomic number, a concept unknown to Mendeleev and his contemporaries.

Using data from α-particle scattering experiments (see Section 2.2), Rutherford estimated the number of positive charges in the nucleus of a few elements, but the significance of these numbers was overlooked for several more years. In 1913 a young English physicist, Henry Moseley,† discovered a correlation between what he called *atomic number* and the frequency of X rays generated by bombarding an element with high-energy electrons. Moseley noticed that the frequencies of X rays emitted from the elements could be correlated by the equation

$$\sqrt{\nu} = a(Z - b) \tag{8.1}$$

†Henry Gwyn-Jeffreys Moseley (1887–1915). English physicist. Moseley discovered the relationship between X-ray spectra and atomic number. A lieutenant in the Royal Engineers, he was killed in action at the age of 28 during the British campaign in Gallipoli, Turkey.

where v is the frequency of the emitted X rays and a and b are constants that are the same for all the elements. Thus, from the square root of the measured frequency of the X rays emitted, we can determine the atomic number of the element.

With a few exceptions, Moseley found that atomic number increases in the same order as atomic mass. For example, calcium is the twentieth element in order of increasing atomic mass, and it has an atomic number of 20. The discrepancies that had puzzled earlier scientists now made sense. The atomic number of argon is 18 and that of potassium is 19, so potassium should follow argon in the periodic table.

A modern periodic table usually shows the atomic number along with the element symbol. As you already know, the atomic number also indicates the number of electrons in the atoms of an element. Electron configurations of elements help to explain the recurrence of physical and chemical properties. The importance and usefulness of the periodic table lie in the fact that we can use our understanding of the general properties and trends within a group or a period to predict with considerable accuracy the properties of any element, even though that element may be unfamiliar to us.

Summary of Concepts & Facts

- Nineteenth-century chemists developed the periodic table by arranging elements in the increasing order of their atomic masses. Discrepancies in early versions of the periodic table were resolved by arranging the elements in order of their atomic numbers.

8.2 Periodic Classification of the Elements

Learning Objectives

- Predict the characteristics of elements based on their position in the periodic table.
- Determine the electron configuration of an element based on its position in the periodic table.
- Describe the significance of valence electrons.
- Predict the electron configurations of cations and anions for main group and transition metals.
- Define isoelectronic species.

Figure 8.2 shows the periodic table together with the outermost ground-state electron configurations of the elements. (The electron configurations of the elements are also given in Table 7.3.) Starting with hydrogen, we see that subshells are filled in the order shown in Figure 7.30. According to the type of subshell being filled, the elements can be divided into categories—the representative elements, the noble gases, the transition elements (or transition metals), the lanthanides, and the actinides. The ***representative elements*** (also called *main group elements*) are *the elements in Group 1, Group 2, and Groups 13 through 17, all of which have incompletely filled s or p subshells of the highest principal quantum number*. With the exception of helium, the *noble gases* (the Group 18 elements) all have a completely filled p subshell. (The electron configurations are $1s^2$ for helium and ns^2np^6 for the other noble gases, where n is the principal quantum number for the outermost shell.)

The transition metals are the elements in Groups 3 through 11, which have incompletely filled d subshells, or readily produce cations with incompletely filled d subshells. (These metals are sometimes referred to as the d-block transition elements.) The Group 12 elements, Zn, Cd, and Hg, are neither representative elements nor

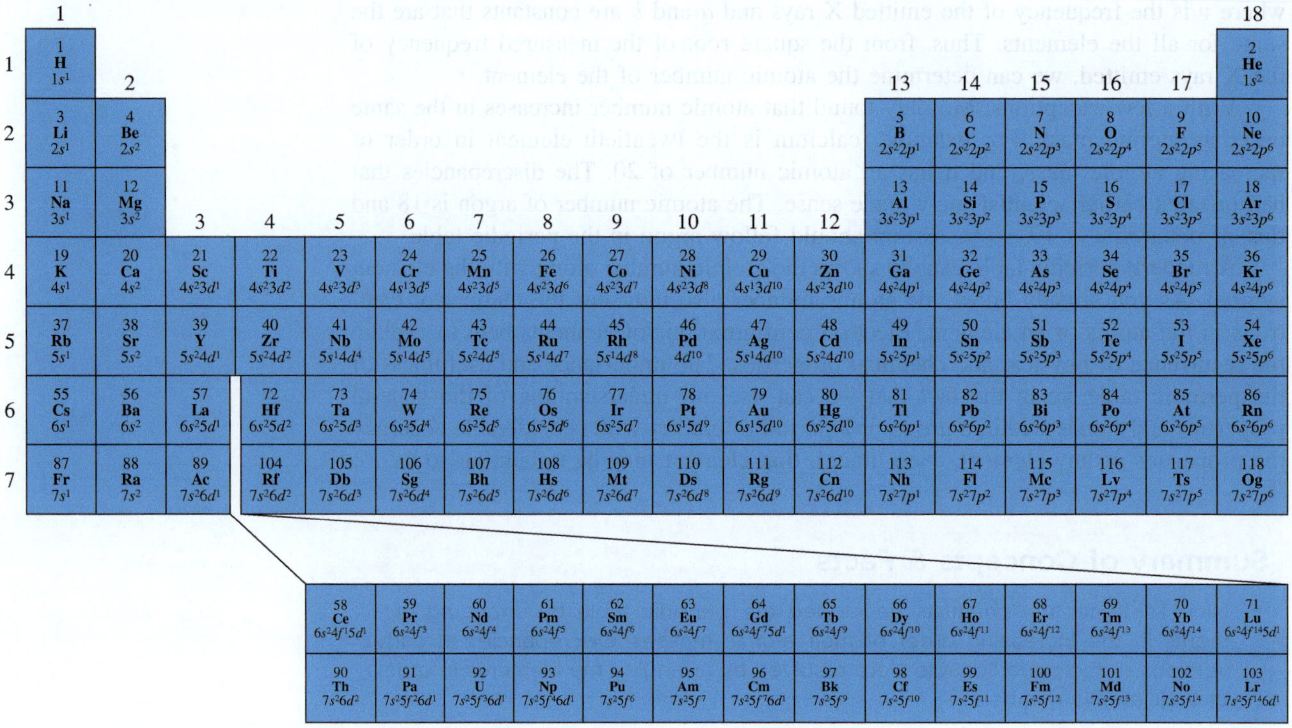

Figure 8.2 *The ground-state electron configurations of the elements. For simplicity, only the configurations of the outer electrons are shown.*

transition metals. There is no special name for this group of metals. The lanthanoids and actinoids are sometimes called *f*-block transition elements because they have incompletely filled *f* subshells. Figure 8.3 distinguishes the groups of elements discussed here.

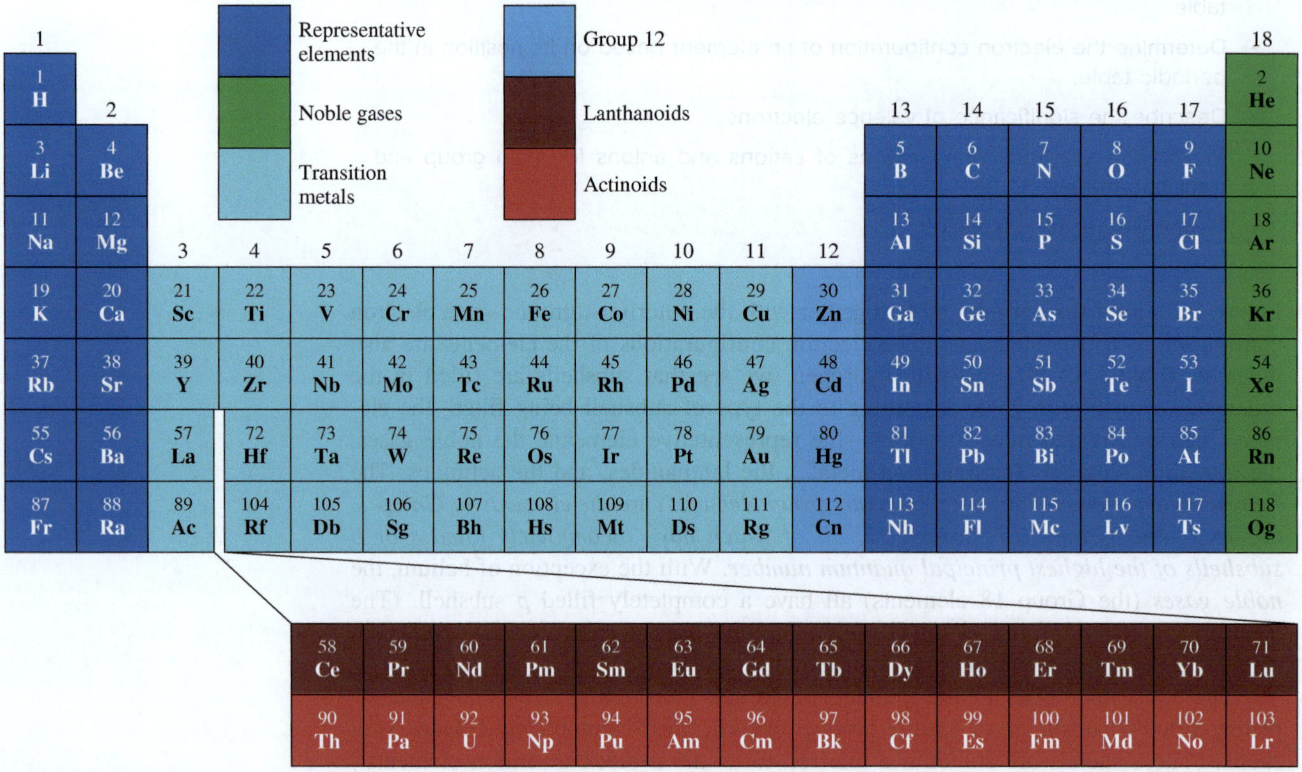

Figure 8.3 *Classification of the elements. Note that the Group 12 elements are often classified as transition metals even though they do not exhibit the characteristics of the transition metals.*

The chemical reactivity of the elements is largely determined by their ***valence electrons,*** which are *the outermost electrons*. For the representative elements, the valence electrons are those in the highest occupied n shell. A convenient method to determine the number of valence electrons for elements in Groups 13 through 18 is to subtract 10 from the group number. *All nonvalence electrons in an atom* are referred to as ***core electrons.*** Looking at the electron configurations of the representative elements once again, a clear pattern emerges: All the elements in a given group have the same number and type of valence electrons. The similarity of the valence electron configurations is what makes the elements in the same group resemble one another in chemical behavior. Thus, for instance, the alkali metals (the Group 1 elements) all have the valence electron configuration of ns^1 (Table 8.1) and they all tend to lose one electron to form the unipositive cations. Similarly, the alkaline earth metals (the Group 2 elements) all have the valence electron configuration of ns^2, and they all tend to lose two electrons to form the dipositive cations. We must be careful, however, in predicting element properties for Groups 13 through 16 based solely on "group membership." For example, the elements in Group 14 all have the same valence electron configuration ns^2np^2, but there is a notable variation in chemical properties among the elements: Carbon is a nonmetal, silicon and germanium are metalloids, and tin and lead are metals.

As a group, the noble gases behave very similarly. Helium and neon are chemically inert, and there are few examples of compounds formed by the other noble gases. This lack of chemical reactivity is due to the completely filled ns and np subshells, a condition that often correlates with great stability. Although the valence electron configuration of the transition metals is not always the same within a group and there is no regular pattern in the change of the electron configuration from one metal to the next in the same period, all transition metals share many characteristics that set them apart from other elements. The reason is that these metals all have an incompletely filled d subshell. Likewise, the lanthanoid (and the actinoid) elements resemble one another because they have incompletely filled f subshells.

Table 8.1

Electron Configurations of Group 1 and Group 2 Elements

Group 1		Group 2	
Li	$[He]2s^1$	Be	$[He]2s^2$
Na	$[Ne]3s^1$	Mg	$[Ne]3s^2$
K	$[Ar]4s^1$	Ca	$[Ar]4s^2$
Rb	$[Kr]5s^1$	Sr	$[Kr]5s^2$
Cs	$[Xe]6s^1$	Ba	$[Xe]6s^2$
Fr	$[Rn]7s^1$	Ra	$[Rn]7s^2$

Example 8.1

An atom of a certain element has 15 electrons. Without consulting a periodic table, answer the following questions: (a) What is the ground-state electron configuration of the element? (b) How should the element be classified? (c) Is the element diamagnetic or paramagnetic?

Strategy (a) We refer to the building-up principle discussed in Section 7.9 and start writing the electron configuration with principal quantum number $n = 1$ and continuing upward until all the electrons are accounted for. (b) What are the electron configuration characteristics of representative elements? Transition elements? Noble gases? (c) Examine the pairing scheme of the electrons in the outermost shell. What determines whether an element is diamagnetic or paramagnetic?

Solution

(a) We know that for $n = 1$ we have a $1s$ orbital (two electrons); for $n = 2$ we have a $2s$ orbital (two electrons) and three $2p$ orbitals (six electrons); for $n = 3$ we have a $3s$ orbital (two electrons). The number of electrons left is $15 - 12 = 3$ and these three electrons are placed in the $3p$ orbitals. The electron configuration is $1s^22s^22p^63s^23p^3$.

(b) Because the $3p$ subshell is not completely filled, this is a representative element. Based on the information given, we cannot say whether it is a metal, a nonmetal, or a metalloid.

(c) According to Hund's rule, the three electrons in the $3p$ orbitals have parallel spins (three unpaired electrons). Therefore, the element is paramagnetic.

(Continued)

Check For (b), note that a transition metal possesses an incompletely filled *d* subshell and a noble gas has a completely filled outer shell. For (c), recall that if the atoms of an element contain an odd number of electrons, then the element must be paramagnetic.

Practice Exercise An atom of a certain element has 20 electrons. (a) Write the ground-state electron configuration of the element, (b) classify the element, and (c) determine whether the element is diamagnetic or paramagnetic.

Similar problem: 8.20.

Representing Free Elements in Chemical Equations

Having classified the elements according to their ground-state electron configurations, we can now look at the way chemists represent metals, metalloids, and nonmetals as free elements in chemical equations. Because metals do not exist in discrete molecular units, we always use their empirical formulas in chemical equations. The empirical formulas are the same as the symbols that represent the elements. For example, the empirical formula for iron is Fe, the same as the symbol for the element.

For nonmetals there is no single rule. Carbon, for example, exists as an extensive three-dimensional network of atoms, and so we use its empirical formula (C) to represent elemental carbon in chemical equations. But hydrogen, nitrogen, oxygen, and the halogens exist as diatomic molecules, and so we use their molecular formulas (H_2, N_2, O_2, F_2, Cl_2, Br_2, I_2) in equations. The stable form of phosphorus is molecular (P_4), and so we use P_4. For sulfur, chemists often use the empirical formula (S) in chemical equations, rather than S_8, which is the stable form. Thus, instead of writing the equation for the combustion of sulfur as

$$S_8(s) + 8O_2(g) \longrightarrow 8SO_2(g)$$

we usually write

$$S(s) + O_2(g) \longrightarrow SO_2(g)$$

Note that these two equations for the combustion of sulfur have identical stoichiometry. This correspondence should not be surprising, because both equations describe the same chemical system. In both cases, a number of sulfur atoms react with twice as many oxygen atoms.

All the noble gases are monatomic species; thus we use their symbols: He, Ne, Ar, Kr, Xe, and Rn. The metalloids, like the metals, all have complex three-dimensional networks, and we represent them, too, with their empirical formulas, that is, their symbols: B, Si, Ge, and so on.

Electron Configurations of Cations and Anions

Because many ionic compounds are made up of monatomic anions and cations, it is helpful to know how to write the electron configurations of these ionic species. Just as for neutral atoms, we use the Pauli exclusion principle and Hund's rule in writing the ground-state electron configurations of cations and anions. We will group the ions in two categories for discussion.

Ions Derived from Representative Elements

Ions formed from atoms of most representative elements have the noble-gas outer-electron configuration of ns^2np^6. In the formation of a cation from the atom of a representative element, one or more electrons are removed from the highest occupied *n* shell. The electron configurations of some atoms and their corresponding cations are as follows:

Na: $[Ne]3s^1$	Na^+: [Ne]
Ca: $[Ar]4s^2$	Ca^{2+}: [Ar]
Al: $[Ne]3s^23p^1$	Al^{3+}: [Ne]

Note that each ion has a stable noble gas configuration.

For the heavier elements in the *p* block, two common cations are possible due to the stability of a filled valence *s* orbital. Consider the elements tin ($Z = 50$) and thallium ($Z = 81$). The electron configuration of tin is $[Kr]5s^24d^{10}5p^2$ and can lose the four valence electrons ($5s^2$ and $5p^2$) to form Sn^{4+} (with a configuration of $[Kr]4d^{10}$) or it can lose just the $5p^2$ electrons to form Sn^{2+} (with a configuration of $[Kr]5s^24d^{10}$). Likewise, thallium can form both Tl^{3+} ($[Xe]4f^{14}5d^{10}$) and Tl^+ ($[Xe]6s^24f^{14}5d^{10}$) ions. The stabilizing presence of the valence pair of *s* electrons (known as an inert pair) in Sn^{2+} and Tl^+ is a result of relativistic effects, which are corrections to the Schrödinger equation to account for special relativity when inner electrons' velocities approach the speed of light. Such relativistic effects are present for heavier atoms in Groups 13 through 16. *The tendency for heavier Group 13 to 16 atoms to form ions with charges two less than their normal valency* is termed the ***inert pair effect.***

In the formation of an anion, one or more electrons are added to the highest partially filled *n* shell. Consider the following examples:

H: $1s^1$	H^-: $1s^2$ or [He]
F: $1s^22s^22p^5$	F^-: $1s^22s^22p^6$ or [Ne]
O: $1s^22s^22p^4$	O^{2-}: $1s^22s^22p^6$ or [Ne]
N: $1s^22s^22p^3$	N^{3-}: $1s^22s^22p^6$ or [Ne]

All of these anions also have stable noble gas configurations. Notice that F^-, Na^+, and Ne (and Al^{3+}, O^{2-}, and N^{3-}) have the same electron configuration. They are said to be ***isoelectronic*** because they *have the same number of electrons, and hence the same ground-state electron configuration.* Thus, H^- and He are also isoelectronic.

Cations Derived from Transition Metals

In Section 7.9 we saw that in the first-row transition metals (Sc to Cu), the 4*s* orbital is always filled before the 3*d* orbitals. Consider manganese, whose electron configuration is $[Ar]4s^23d^5$. When the Mn^{2+} ion is formed, we might expect the two electrons to be removed from the 3*d* orbitals to yield $[Ar]4s^23d^3$. In fact, the electron configuration of Mn^{2+} is $[Ar]3d^5$! The reason is that the electron–electron and electron–nucleus interactions in a neutral atom can be quite different from those in its ion. Thus, whereas the 4*s* orbital is always filled before the 3*d* orbital in Mn, electrons are removed from the 4*s* orbital in forming Mn^{2+} because the 3*d* orbital is more stable than the 4*s* orbital in transition metal ions. Therefore, when a cation is formed from an atom of a transition metal, electrons are always removed first from the *ns* orbital and then from the $(n - 1)d$ orbitals. Keep in mind that most transition metals can form more than one cation and that frequently the cations are not isoelectronic with the preceding noble gases.

Summary of Concepts & Facts

- Electron configuration determines the properties of an element. The modern periodic table classifies the elements according to their atomic numbers, and thus also by their electron configurations. The configuration of the valence electrons directly affects the properties of the atoms of the representative elements.

Review of Concepts & Facts

8.2.1 Identify the elements that fit the following descriptions: (a) an alkaline earth metal ion that is isoelectronic with Kr; (b) an anion with a −3 charge that is isoelectronic with K^+; (c) an ion with a +2 charge that is isoelectronic with Co^{3+}.

8.2.2 What is the ground-state electron configuration for Cr^{3+}?

8.2.3 Identify the ion with a +3 charge that has the electron configuration $[Ne]3s^2$.

8.3 Periodic Variation in Physical Properties

Learning Objectives

- Define effective nuclear charge.
- Predict the trends in atomic and ionic radii among elements.

As we have seen, the electron configurations of the elements show a periodic variation with increasing atomic number. Consequently, there are also periodic variations in physical and chemical behavior. In this section and Sections 8.4 and 8.5, we will examine some physical properties of elements that are in the same group or period and additional properties that influence the chemical behavior of the elements. First, let's look at the concept of effective nuclear charge, which has a direct bearing on many atomic properties.

Effective Nuclear Charge

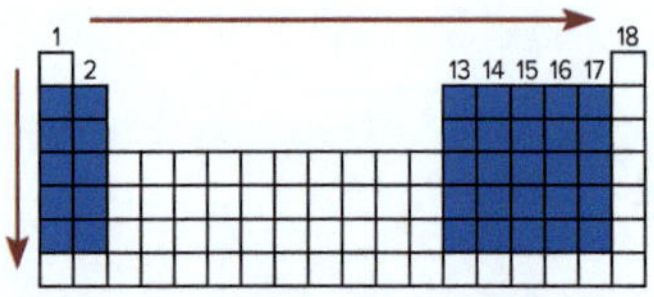

The increase in effective nuclear charge from left to right across a period and from top to bottom in a group for representative elements.

In Chapter 7 we discussed the shielding effect that electrons close to the nucleus have on outer-shell electrons in many-electron atoms. The presence of other electrons in an atom reduces the electrostatic attraction between a given electron and the positively charged protons in the nucleus. The ***effective nuclear charge*** **(Z_{eff})** is *the nuclear charge felt by an electron when both the actual nuclear charge (Z) and the repulsive effects (shielding) of the other electrons are taken into account.* In general, Z_{eff} is given by

$$Z_{eff} = Z - \sigma \tag{8.2}$$

where σ (sigma) is called the *shielding constant* (also called the *screening constant*). The shielding constant is greater than zero but smaller than Z.

One way to illustrate how electrons in an atom shield one another is to consider the amounts of energy required to remove the two electrons from a helium atom. Experiments show that it takes 3.94×10^{-18} J to remove the first electron and 8.72×10^{-18} J to remove the second electron. There is no shielding once the first electron is removed, so the second electron feels the full effect of the +2 nuclear charge.

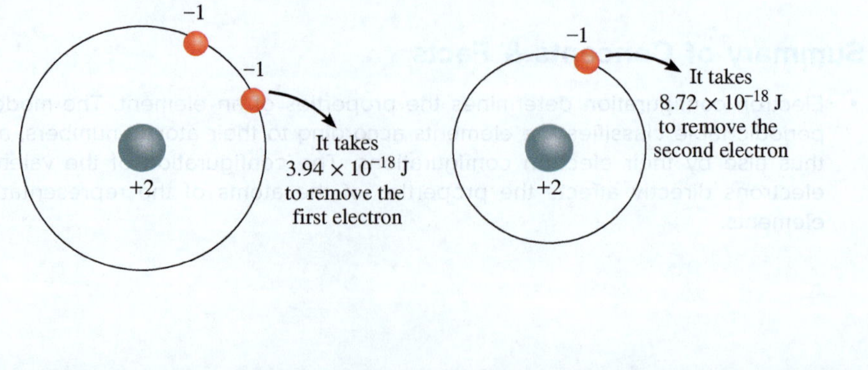

Because the core electrons are, on average, closer to the nucleus than valence electrons, core electrons shield valence electrons much more than valence electrons shield one another. Consider the second-period elements from Li to Ne. Moving from left to right, we find the number of core electrons ($1s^2$) remains constant while the nuclear charge increases. However, because the added electron is a valence electron and valence electrons do not shield each other well, the net effect of moving across the period is a greater effective nuclear charge felt by the valence electrons, as shown here.

	Li	Be	B	C	N	O	F	Ne
Z	3	4	5	6	7	8	9	10
Z_{eff}	1.28	1.91	2.42	3.14	3.83	4.45	5.10	5.76

The effective nuclear charge also increases as we go down a particular periodic group. The attractive force between the nucleus and a particular electron is directly proportional to the effective nuclear charge and inversely proportional to the square of the distance of separation. However, because the valence electrons are now added to increasingly large shells as n increases, the electrostatic attraction between the nucleus and the valence electrons actually decreases.

Video:
Atomic and Ionic Radius

Student Hot Spot

Student data indicate you may struggle with effective nuclear charge. Access your eBook for additional Learning Resources on this topic.

	Li	Na	K	Rb	Cs
Z	3	11	19	37	55
Z_{eff}	1.28	2.51	3.50	4.99	6.36

Atomic Radius

A number of physical properties, including density, melting point, and boiling point, are related to the sizes of atoms, but atomic size is difficult to define. As we saw in Chapter 7, the electron density in an atom extends far beyond the nucleus, but we normally think of atomic size as the volume containing about 90 percent of the total electron density around the nucleus. When we must be even more specific, we define the size of an atom in terms of its ***atomic radius,*** which is *one-half the distance between the two nuclei in two adjacent metal atoms or in a diatomic molecule.*

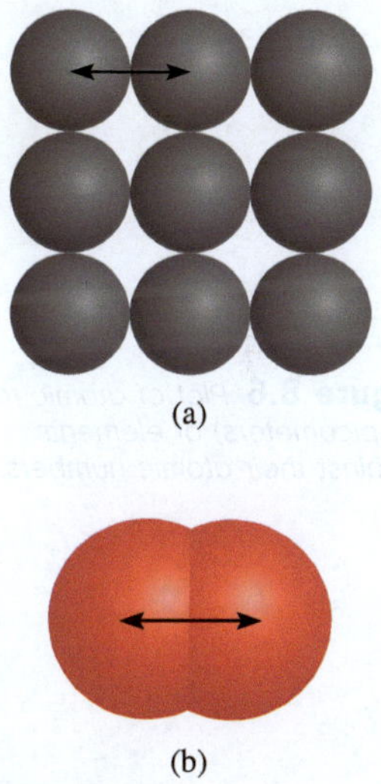

Figure 8.4 *(a) In metals such as polonium, the atomic radius is defined as one-half the distance between the centers of two adjacent atoms. (b) For elements that exist as diatomic molecules, such as iodine, the radius of the atom is defined as one-half the distance between the centers of the atoms in the molecule.*

For atoms linked together to form an extensive three-dimensional network, atomic radius is simply one-half the distance between the nuclei in two neighboring atoms [Figure 8.4(a)]. For elements that exist as simple diatomic molecules, the atomic radius is one-half the distance between the nuclei of the two atoms in a particular molecule [Figure 8.4(b)].

Figure 8.5 shows the atomic radius of many elements according to their positions in the periodic table, and Figure 8.6 plots the atomic radii of these elements against their atomic numbers. Periodic trends are clearly evident. Consider the second-period elements. Because the effective nuclear charge increases from left to right, the added valence electron at each step is more strongly attracted by the nucleus than the one before. Therefore, we expect and indeed find the atomic radius decreases from Li to Ne. Within a group we find that atomic radius increases with atomic number. For the alkali metals in Group 1, the valence electron resides in the ns orbital. Because orbital size increases with the increasing principal quantum number n, the size of the atomic radius increases even though the effective nuclear charge also increases from Li to Cs.

Student Hot Spot

Student data indicate you may struggle with atomic radii. Access your eBook for additional Learning Resources on this topic.

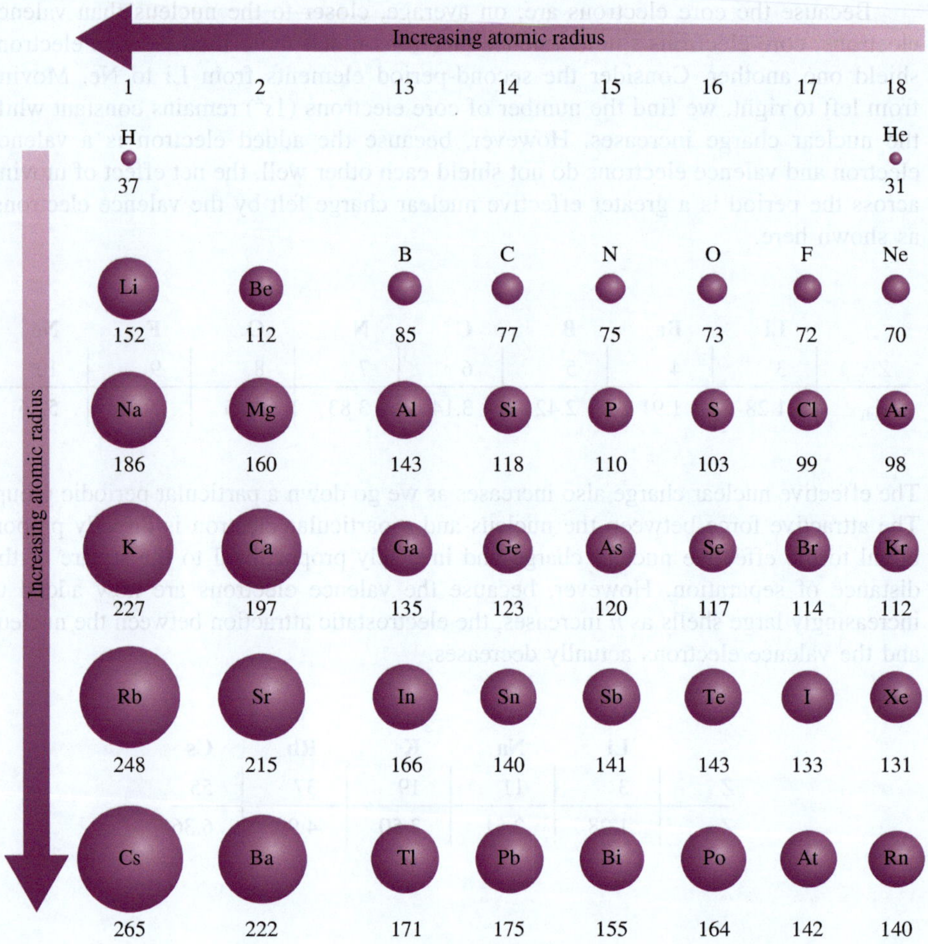

Figure 8.5 *Atomic radii (in picometers) of representative elements according to their positions in the periodic table. Note that there is no general agreement on the size of atomic radii. We focus only on the trends in atomic radii, not on their precise values.*

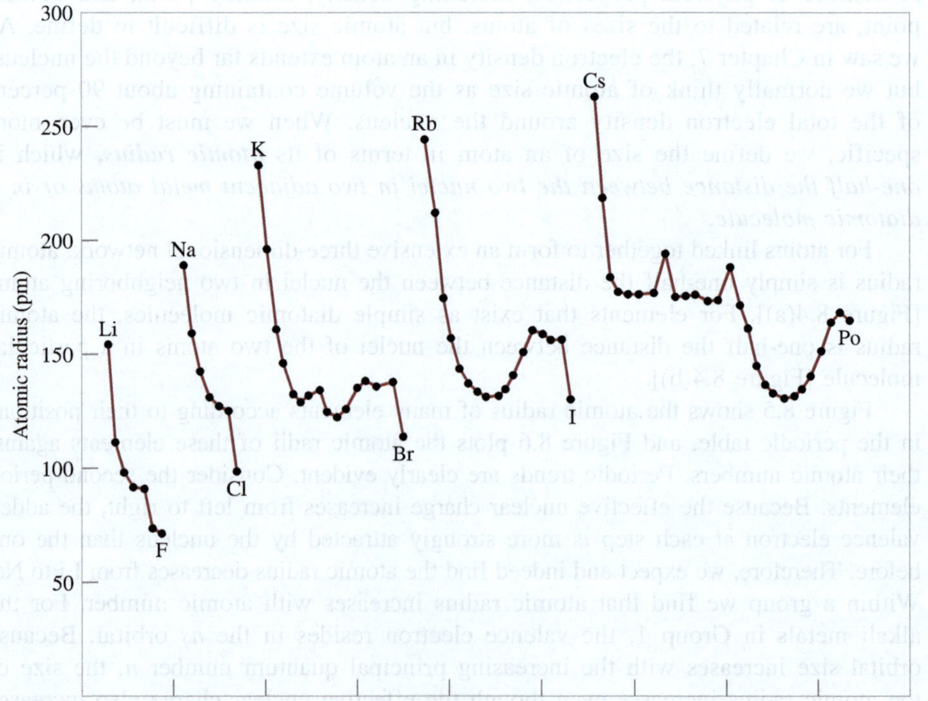

Figure 8.6 *Plot of atomic radii (in picometers) of elements against their atomic numbers.*

Example 8.2

Referring to a periodic table, arrange the following atoms in order of increasing atomic radius: P, Si, N.

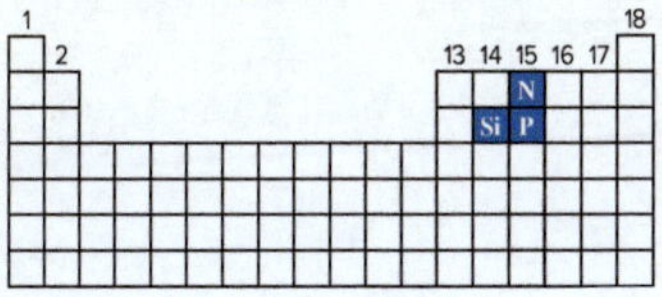

Strategy What are the trends in atomic radii in a periodic group and in a particular period? Which of the preceding elements are in the same group? In the same period?

Solution From Figure 8.1 we see that N and P are in the same group (Group 15). Therefore, the radius of N is smaller than that of P (atomic radius increases as we go down a group). Both Si and P are in the third period, and Si is to the left of P. Therefore, the radius of P is smaller than that of Si (atomic radius decreases as we move from left to right across a period). Thus, the order of increasing radius is N < P < Si.

Practice Exercise Arrange the following atoms in order of decreasing radius: C, Li, Be.

Similar problems: 8.37, 8.38.

Ionic Radius

Ionic radius is *the radius of a cation or an anion.* It can be measured by X-ray diffraction (see Chapter 11). Ionic radius affects the physical and chemical properties of an ionic compound. For example, the three-dimensional structure of an ionic compound depends on the relative sizes of its cations and anions.

When a neutral atom is converted to an ion, we expect a change in size. If the atom forms an anion, its size (or radius) increases, because the nuclear charge remains the same but the repulsion resulting from the additional electron(s) enlarges the domain of the electron cloud. On the other hand, removing one or more electrons from an atom reduces electron–electron repulsion but the nuclear charge remains the same, so the electron cloud shrinks, and the cation is smaller than the atom. Figure 8.7 shows the changes in size that result when alkali metals are converted to cations and halogens are converted to anions; Figure 8.8 shows the changes in size that occur when a lithium atom reacts with a fluorine atom to form a LiF unit.

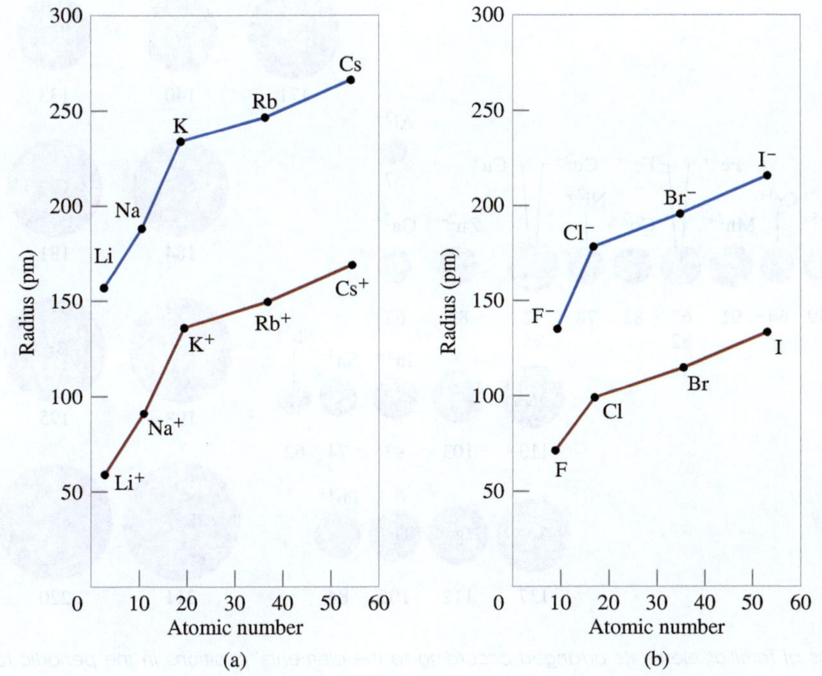

Figure 8.7 *Comparison of atomic radii with ionic radii. (a) Alkali metals and alkali metal cations. (b) Halogens and halide ions.*

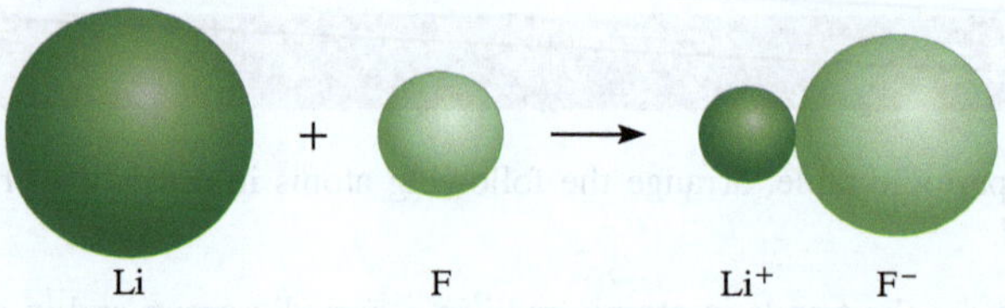

Figure 8.8 *Changes in the sizes of Li and F when they react to form LiF.*

Figure 8.9 shows the radii of ions derived from the familiar elements, arranged according to the elements' positions in the periodic table. We can see parallel trends between atomic radii and ionic radii. For example, from top to bottom both the atomic radius and the ionic radius increase within a group. For ions derived from elements in different groups, a size comparison is meaningful only if the ions are isoelectronic. If we examine isoelectronic ions, we find that cations are smaller than anions. For example, Na^+ is smaller than F^-. Both ions have the same number of electrons, but Na ($Z = 11$) has more protons than F ($Z = 9$). The larger effective nuclear charge of Na^+ results in a smaller radius.

Focusing on isoelectronic cations, we see that the radii of *tripositive ions* (ions that bear three positive charges) are smaller than those of *dipositive ions* (ions that bear two positive charges), which in turn are smaller than *unipositive ions* (ions that bear one positive charge). This trend is nicely illustrated by the sizes of three isoelectronic ions in the third period: Al^{3+}, Mg^{2+}, and Na^+ (see Figure 8.9). The Al^{3+} ion has the same number of electrons as Mg^{2+}, but it has one more proton. Thus, the electron cloud in Al^{3+} is pulled inward more than that in Mg^{2+}. The smaller radius of Mg^{2+} compared with that of Na^+ can be similarly explained. Turning to isoelectronic anions, we find that the radius increases as we go from ions with uninegative charge (−) to those with dinegative charge (2−), and so on. Thus, the oxide ion is larger than the fluoride ion because oxygen has one fewer proton than fluorine; the electron cloud is spread out more in O^{2-}.

Student Hot Spot

Student data indicate you may struggle with ionic radii. Access your eBook for additional Learning Resources on this topic.

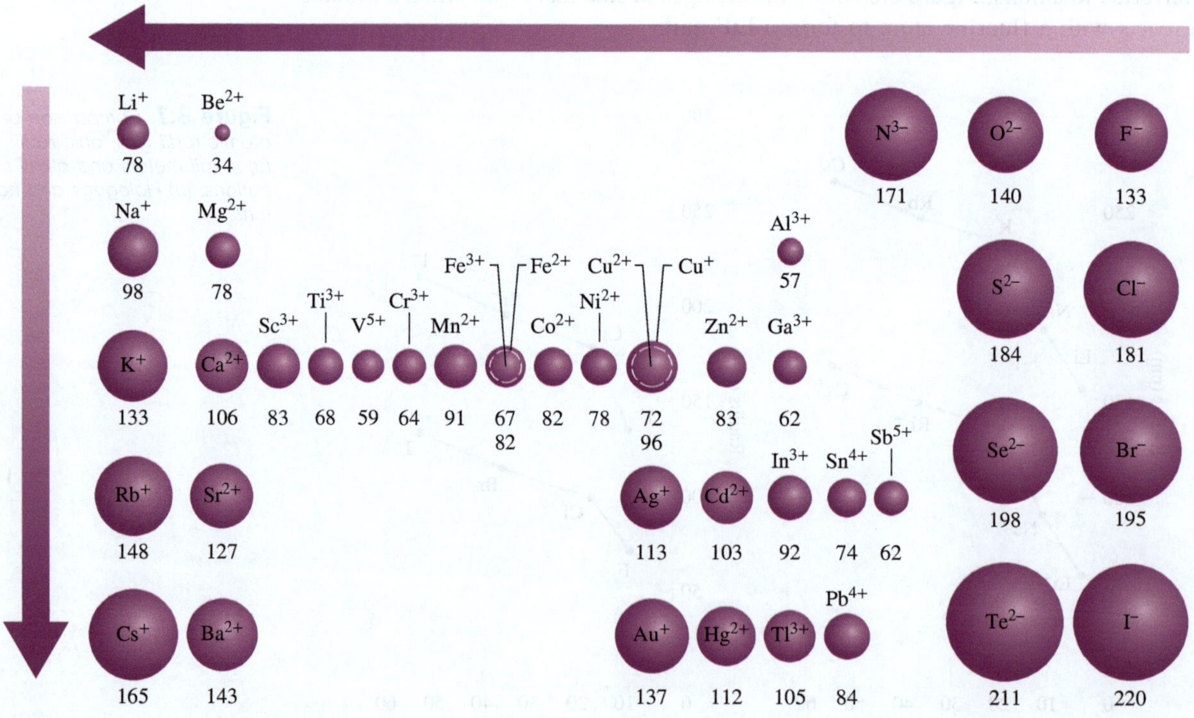

Figure 8.9 *The radii (in picometers) of ions of familiar elements arranged according to the elements' positions in the periodic table.*

Example 8.3

For each of the following pairs, indicate which one of the two species is larger: (a) N^{3-} or F^-; (b) Mg^{2+} or Ca^{2+}; (c) Fe^{2+} or Fe^{3+}.

Strategy In comparing ionic radii, it is useful to classify the ions into three categories: (1) isoelectronic ions, (2) ions that carry the same charges and are generated from atoms of the same periodic group, and (3) ions that carry different charges but are generated from the same atom. In case (1), ions that carry a greater negative charge are always larger; in case (2), ions from atoms having a greater atomic number are always larger; in case (3), ions having a smaller positive charge are always larger.

Solution

(a) N^{3-} and F^- are isoelectronic anions, both containing 10 electrons. Because N^{3-} has only seven protons and F^- has nine, the smaller attraction exerted by the nucleus on the electrons results in a larger N^{3-} ion.

(b) Both Mg and Ca belong to Group 2 (the alkaline earth metals). Thus, Ca^{2+} ion is larger than Mg^{2+} because Ca's valence electrons are in a larger shell ($n = 4$) than are Mg's valence electrons ($n = 3$).

(c) Both ions have the same nuclear charge, but Fe^{2+} has one more electron (24 electrons compared to 23 electrons for Fe^{3+}) and hence greater electron–electron repulsion. The radius of Fe^{2+} is larger.

Practice Exercise Select the smaller ion in each of the following pairs: (a) K^+, Li^+; (b) Au^+, Au^{3+}; (c) P^{3-}, N^{3-}.

Similar problems: 8.43, 8.45.

Variation of Physical Properties Across a Period and Within a Group

From left to right across a period there is a transition from metals to metalloids to nonmetals. Consider the third-period elements from sodium to argon (Figure 8.10). Sodium, the first element in the third period, is a very reactive metal, whereas chlorine, the second-to-last element of that period, is a very reactive nonmetal. In between, the elements show a gradual transition from metallic properties to nonmetallic properties. Sodium, magnesium, and aluminum all have extensive three-dimensional atomic networks, which are held together by forces characteristic of the metallic state. Silicon is a metalloid; it has a giant three-dimensional structure in which the Si atoms are held together very strongly. Starting with phosphorus, the elements exist in simple, discrete molecular units (P_4, S_8, Cl_2, and Ar) that have low melting points and boiling points.

Within a periodic group the physical properties vary more predictably, especially if the elements are in the same physical state. For example, the melting points of argon and xenon are −189.2°C and −111.9°C, respectively. We can estimate the melting point of the intermediate element krypton by taking the average of these two values as follows:

$$\text{melting point of Kr} = \frac{[(-189.2^\circ\text{C}) + (-111.9^\circ\text{C})]}{2} = -150.6^\circ\text{C}$$

This value is quite close to the actual melting point of −156.6°C.

The Chemistry in Action essay "The Third Liquid Element?" illustrates one interesting application of periodic group properties.

Figure 8.10 *The third-period elements. The photograph of argon, which is a colorless, odorless gas, shows the color emitted by the gas from a discharge tube.*

(Na, Cl_2, Ar): Ken Karp/McGraw Hill; (Mg): David J. Green/Alamy Stock Photo; (Al): Kerstin Waurick/E+/Getty Images; (Si): Charles D. Winters/Timeframe Photography/McGraw Hill; (P_4): Al Fenn/The LIFE Images Collection/Getty Images; (S_8): Tom Grundy/Shutterstock

Summary of Concepts & Facts

- Atomic radius varies periodically with the arrangement of the elements in the periodic table. It decreases from left to right and increases from top to bottom.
- Periodic variations in the physical properties of the elements reflect differences in atomic structure. The metallic character of elements decreases across a period from metals through the metalloids to nonmetals and increases from top to bottom within a particular group of representative elements.

Review of Concepts & Facts

8.3.1 Compare the size of each pair of atoms listed here: (a) Be, Ba; (b) Al, S; (c) ^{12}C, ^{13}C.

8.3.2 Arrange the following species in order of increasing size: S^{2-}, Ca^{2+}, Cl^-, K^+, Ar.

(Continued)

CHEMISTRY *in Action*

The Third Liquid Element?

Of the 118 known elements, 11 are gases under atmospheric conditions. Six of these are the Group 18 elements (the noble gases He, Ne, Ar, Kr, Xe, and Rn), and the other five are hydrogen (H_2), nitrogen (N_2), oxygen (O_2), fluorine (F_2), and chlorine (Cl_2). Curiously, only two elements are liquids at 25°C: mercury (Hg) and bromine (Br_2).

We do not know the properties of all the known elements because some of them have never been prepared in quantities large enough for investigation. In these cases, we must rely on periodic trends to predict their properties. What are the chances, then, of discovering a third liquid element?

Let us look at francium (Fr), the last member of Group 1, to see if it might be a liquid at 25°C. All of francium's isotopes are radioactive. The most stable isotope is francium-223, which has a half-life of 21 minutes. (*Half-life* is the time it takes for one-half of the nuclei in any given amount of a radioactive substance to disintegrate.) This short half-life means that only very small traces of francium could possibly exist on Earth. And although it is feasible to prepare francium in the laboratory, no weighable quantity of the element has been prepared or isolated. Thus, we know very little about francium's physical and chemical properties. Yet we can use the group periodic trends to predict some of those properties.

Take francium's melting point as an example. The plot shows how the melting points of the alkali metals vary with atomic number. From lithium to sodium, the melting point drops 81.4°; from sodium to potassium, 34.6°; from potassium to rubidium, 24°; from rubidium to cesium, 11°. On the basis of this trend, we can predict that the change from cesium to francium would be about 5°. If so, the melting point of francium would be about 23°C, which would make it a liquid under atmospheric conditions.

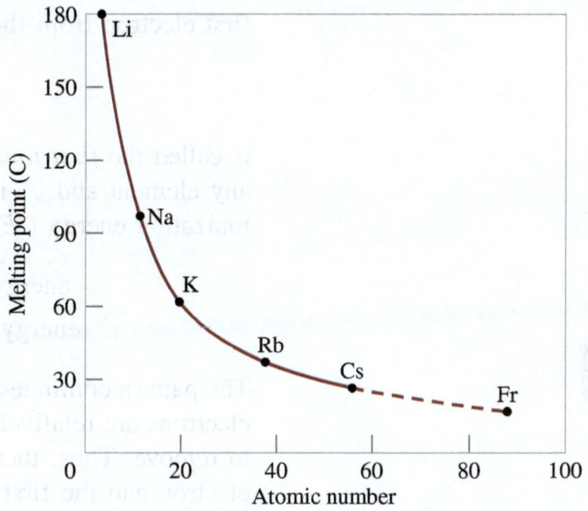

A plot of the melting points of the alkali metals versus their atomic numbers. By extrapolation, the melting point of francium should be 23°C.

8.3.3 Identify the spheres shown here with each of the following: S^{2-}, Mg^{2+}, F^-, Na^+.

8.4 Ionization Energy

Learning Objective

- Predict the trends in ionization energy among elements.

Not only is there a correlation between electron configuration and physical properties, but a close correlation also exists between electron configuration (a microscopic property) and chemical behavior (a macroscopic property). As we will see throughout this book, the chemical properties of any atom are determined by the configuration of the atom's valence electrons. The stability of these outermost electrons is reflected directly in the atom's ionization energies. ***Ionization energy (IE)*** is *the minimum energy (in kJ/mol) required to remove an electron from a gaseous atom in its ground state*. In other words, ionization energy is the amount of energy in kilojoules needed to strip 1 mole of electrons from 1 mole of gaseous atoms. Gaseous atoms are specified in this definition because an atom in the gas phase is virtually uninfluenced by its neighbors and so there are no intermolecular forces (that is, forces between molecules) to take into account when measuring ionization energy.

The magnitude of ionization energy is a measure of how "tightly" the electron is held in the atom. The higher the ionization energy, the more difficult it is to remove the electron. For a many-electron atom, the amount of energy required to remove the first electron from the atom in its ground state,

$$\text{energy} + X(g) \longrightarrow X^+(g) + e^- \quad (8.3)$$

is called the *first ionization energy* (IE_1). In Equation (8.3), X represents an atom of any element and e^- is an electron. The second ionization energy (IE_2) and the third ionization energy (IE_3) are shown in the following equations:

$$\text{energy} + X^+(g) \longrightarrow X^{2+}(g) + e^- \quad \text{second ionization}$$
$$\text{energy} + X^{2+}(g) \longrightarrow X^{3+}(g) + e^- \quad \text{third ionization}$$

The pattern continues for the removal of subsequent electrons. Note that while valence electrons are relatively easy to remove from the atom, core electrons are much harder to remove. Thus, there is a large jump in ionization energy between the last valence electron and the first core electron.

When an electron is removed from an atom, the repulsion among the remaining electrons decreases. Because the nuclear charge remains constant, more energy is needed to remove another electron from the positively charged ion. Thus, ionization energies always increase in the following order:

$$IE_1 < IE_2 < IE_3 < \ldots$$

Table 8.2 lists the ionization energies of the first 20 elements. Ionization is always an endothermic process. By convention, energy absorbed by atoms (or ions) in the ionization process has a positive value. Thus, ionization energies are all positive quantities. Figure 8.11 shows the variation of the first ionization energy with atomic number. The plot clearly exhibits the periodicity in the stability of the most loosely held electron. Note that, apart from small irregularities, the first ionization energies of elements in a period increase with increasing atomic number. This trend is due to the increase in effective nuclear charge from left to right (as in the case of atomic radii variation). A larger effective nuclear charge means a more tightly held valence electron, and hence a higher first ionization energy. A notable feature of Figure 8.11 is the peaks, which correspond to the noble gases. We tend to associate full valence-shell electron configurations with an inherent degree of chemical stability. The high ionization energies of the noble gases, stemming from their large effective nuclear charge, comprise one of the reasons for this stability. In fact, helium ($1s^2$) has the highest first ionization energy of all the elements.

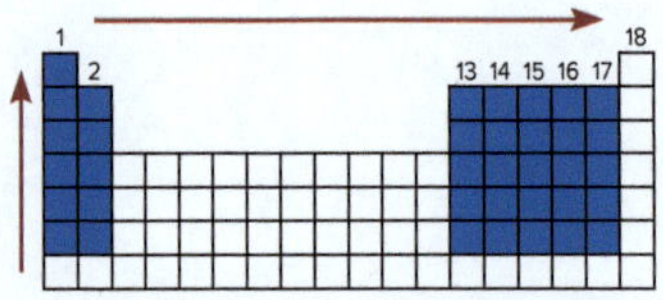

The increase in first ionization energy from left to right across a period and from bottom to top in a group for representative elements.

At the bottom of the graph in Figure 8.11 are the Group 1 elements (the alkali metals), which have the lowest first ionization energies. Each of these metals has one valence electron (the outermost electron configuration is ns^1), which is effectively shielded by the completely filled inner shells. Consequently, it is energetically easy to remove an electron from the atom of an alkali metal to form a unipositive ion (Li^+, Na^+, K^+, ...). Significantly, the electron configurations of these cations are isoelectronic with those noble gases just preceding them in the periodic table.

The Group 2 elements (the alkaline earth metals) have higher first ionization energies than the alkali metals do. The alkaline earth metals have two valence electrons

Table 8.2 The Ionization Energies (kJ/mol) of the First 20 Elements

Z	Element	First	Second	Third	Fourth	Fifth	Sixth
1	H	1,312					
2	He	2,373	5,251				
3	Li	520	7,300	11,815			
4	Be	899	1,757	14,850	21,005		
5	B	801	2,430	3,660	25,000	32,820	
6	C	1,086	2,350	4,620	6,220	38,000	47,261
7	N	1,400	2,860	4,580	7,500	9,400	53,000
8	O	1,314	3,390	5,300	7,470	11,000	13,000
9	F	1,680	3,370	6,050	8,400	11,000	15,200
10	Ne	2,080	3,950	6,120	9,370	12,200	15,000
11	Na	495.9	4,560	6,900	9,540	13,400	16,600
12	Mg	738.1	1,450	7,730	10,500	13,600	18,000
13	Al	577.9	1,820	2,750	11,600	14,800	18,400
14	Si	786.3	1,580	3,230	4,360	16,000	20,000
15	P	1,012	1,904	2,910	4,960	6,240	21,000
16	S	999.5	2,250	3,360	4,660	6,990	8,500
17	Cl	1,251	2,297	3,820	5,160	6,540	9,300
18	Ar	1,521	2,666	3,900	5,770	7,240	8,800
19	K	418.7	3,052	4,410	5,900	8,000	9,600
20	Ca	589.5	1,145	4,900	6,500	8,100	11,000

(the outermost electron configuration is ns^2). Because these two *s* electrons do not shield each other well, the effective nuclear charge for an alkaline earth metal atom is larger than that for the preceding alkali metal. Most alkaline earth compounds contain dipositive ions (Mg^{2+}, Ca^{2+}, Sr^{2+}, Ba^{2+}). The Be^{2+} ion is isoelectronic with Li^+ and with He, Mg^{2+} is isoelectronic with Na^+ and with Ne, and so on.

As Figure 8.11 shows, metals have relatively low ionization energies compared to nonmetals. The ionization energies of the metalloids generally fall between those of metals and nonmetals. The difference in ionization energies suggests why metals

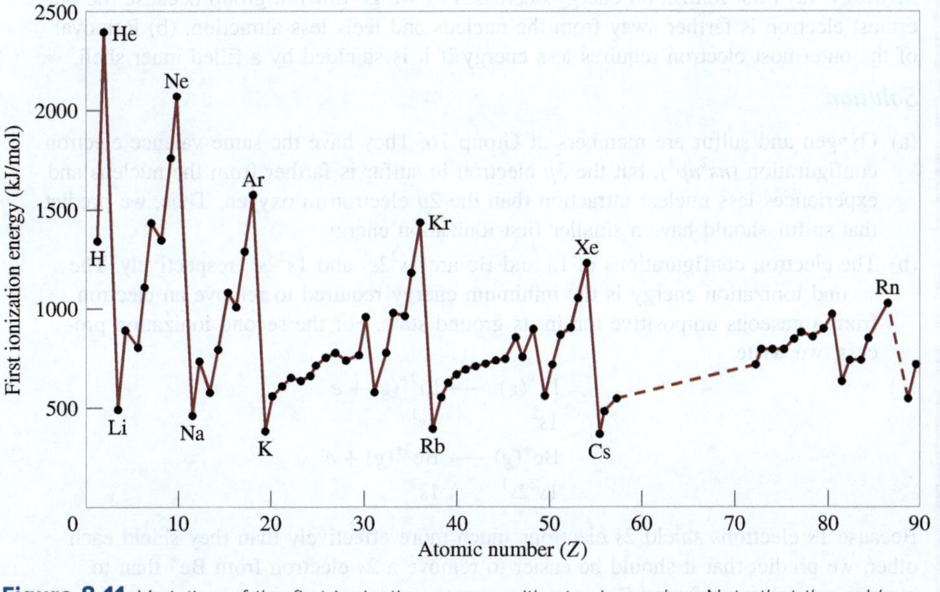

Figure 8.11 *Variation of the first ionization energy with atomic number. Note that the noble gases have high ionization energies, whereas the alkali metals and alkaline earth metals have low ionization energies.*

always form cations and nonmetals form anions in ionic compounds. (The only important nonmetallic cation is the ammonium ion, NH_4^+.) For a given group, ionization energy decreases with increasing atomic number (that is, as we move down the group). Elements in the same group have similar outer electron configurations. However, as the principal quantum number *n* increases, so does the average distance of a valence electron from the nucleus. A greater separation between the electron and the nucleus means a weaker attraction, so that it becomes easier to remove the first electron as we go from element to element down a group even though the effective nuclear charge also increases in the same direction. Thus, the metallic character of the elements within a group increases from top to bottom. This trend is particularly noticeable for elements in Groups 13 to 17. For example, in Group 14, carbon is a nonmetal, silicon and germanium are metalloids, and tin and lead are metals.

Although the general trend in the periodic table is for first ionization energies to increase from left to right, some irregularities do exist. The first exception occurs between Group 2 and 13 elements in the same period (for example, between Be and B and between Mg and Al). The Group 13 elements have lower first ionization energies than Group 2 elements because they all have a single electron in the outermost *p* subshell (ns^2np^1), which is well shielded by the inner electrons and the ns^2 electrons. Therefore, less energy is needed to remove a single *p* electron than to remove an *s* electron from the same principal energy level. The second irregularity occurs between Groups 15 and 16 (for example, between N and O and between P and S). In the Group 15 elements (ns^2np^3), the *p* electrons are in three separate orbitals according to Hund's rule. In Group 16 (ns^2np^4), the additional electron must be paired with one of the three *p* electrons. The proximity of two electrons in the same orbital results in greater electrostatic repulsion, which makes it easier to ionize an atom of the Group 16 element, even though the nuclear charge has increased by one unit. Thus, the ionization energies for Group 16 elements are lower than those for Group 15 elements in the same period.

Student Hot Spot

Student data indicate you may struggle with ionization energy. Access your eBook for additional Learning Resources on this topic.

Example 8.4 compares the ionization energies of some elements.

Example 8.4

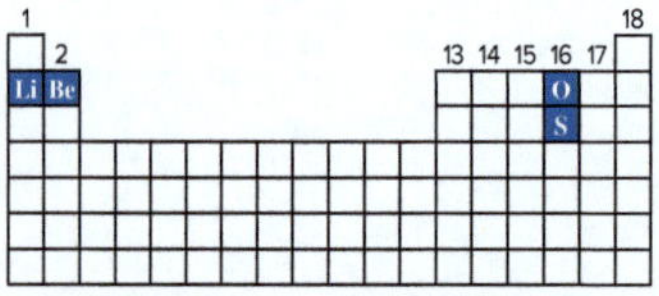

(a) Which atom should have a lower first ionization energy: oxygen or sulfur?
(b) Which atom should have a higher second ionization energy: lithium or beryllium?

Strategy (a) First ionization energy decreases as we go down a group because the outermost electron is farther away from the nucleus and feels less attraction. (b) Removal of the outermost electron requires less energy if it is shielded by a filled inner shell.

Solution

(a) Oxygen and sulfur are members of Group 16. They have the same valence electron configuration (ns^2np^4), but the 3*p* electron in sulfur is farther from the nucleus and experiences less nuclear attraction than the 2*p* electron in oxygen. Thus, we predict that sulfur should have a smaller first ionization energy.

(b) The electron configurations of Li and Be are $1s^22s^1$ and $1s^22s^2$, respectively. The second ionization energy is the minimum energy required to remove an electron from a gaseous unipositive ion in its ground state. For the second ionization process, we write

$$Li^+(g) \longrightarrow Li^{2+}(g) + e^-$$
$$1s^2 \qquad\qquad 1s^1$$
$$Be^+(g) \longrightarrow Be^{2+}(g) + e^-$$
$$1s^22s^1 \qquad\qquad 1s^2$$

Because 1*s* electrons shield 2*s* electrons much more effectively than they shield each other, we predict that it should be easier to remove a 2*s* electron from Be^+ than to remove a 1*s* electron from Li^+.

(Continued)

Check Compare your result with the data shown in Table 8.2. In (a), is your prediction consistent with the fact that the metallic character of the elements increases as we move down a periodic group? In (b), does your prediction account for the fact that alkali metals form +1 ions while alkaline earth metals form +2 ions?

Practice Exercise (a) Which of the following atoms should have a larger first ionization energy: N or P? (b) Which of the following atoms should have a smaller second ionization energy: Na or Mg?

Similar problem: 8.55.

Summary of Concepts & Facts

- Ionization energy is a measure of the tendency of an atom to resist the loss of an electron. The higher the ionization energy, the stronger the attraction between the nucleus and an electron.
- Metals usually have low ionization energies.

Review of Concepts & Facts

8.4.1 Arrange the following atoms in order of increasing first ionization energies: Al, Rb, F.

8.4.2 Label the plots shown here for the first, second, and third ionization energies for Mg, Al, and K.

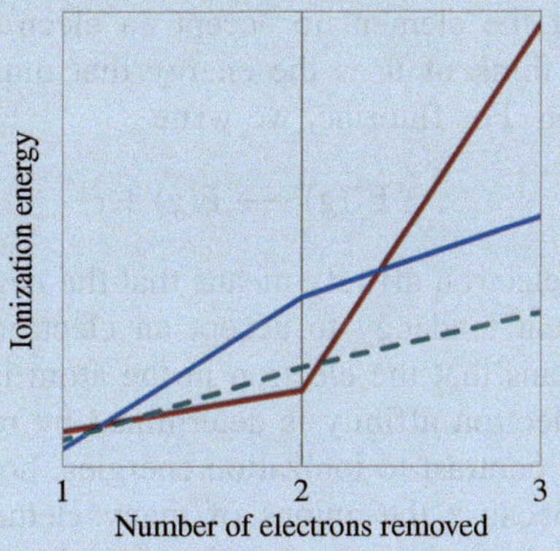

8.5 Electron Affinity

Learning Objective

- Predict the trends in electron affinity among elements.

Another property that greatly influences the chemical behavior of atoms is their ability to accept one or more electrons. This property is called ***electron affinity (EA),*** which is *the negative of the energy change that occurs when an electron is accepted by an atom in the gaseous state to form an anion.*

$$X(g) + e^- \longrightarrow X^-(g) \qquad (8.4)$$

Consider the process in which a gaseous fluorine atom accepts an electron:

$$F(g) + e^- \longrightarrow F^-(g) \qquad \Delta H = -328 \text{ kJ/mol}$$

Table 8.3 **Electron Affinities (kJ/mol) of Some Representative Elements and the Noble Gases***

1	2	13	14	15	16	17	18
H							He
73							< 0
Li	Be	B	C	N	O	F	Ne
60	≤ 0	27	122	0	141	328	< 0
Na	Mg	Al	Si	P	S	Cl	Ar
53	≤ 0	44	134	72	200	349	< 0
K	Ca	Ga	Ge	As	Se	Br	Kr
48	2.4	29	118	77	195	325	< 0
Rb	Sr	In	Sn	Sb	Te	I	Xe
47	4.7	29	121	101	190	295	< 0
Cs	Ba	Tl	Pb	Bi	Po	At	Rn
45	14	30	110	110	?	?	< 0

*The electron affinities of the noble gases Be and Mg have not been determined experimentally, but are believed to be close to zero or negative.

The electron affinity of fluorine is therefore assigned a value of +328 kJ/mol. Electron affinity is positive if the reaction is exothermic and negative if the reaction is endothermic. The more positive the electron affinity is of an element, the greater is the affinity of an atom of the element to accept an electron. Another way of viewing electron affinity is to think of it as the energy that must be supplied to remove an electron from the anion. For fluorine, we write

$$F^-(g) \longrightarrow F(g) + e^- \qquad \Delta H = +328 \text{ kJ/mol}$$

Thus, a large positive electron affinity means that the negative ion is very stable (that is, the atom has a great tendency to accept an electron), just as a high ionization energy of an atom means that the electron in the atom is very stable.

Experimentally, electron affinity is determined by removing the additional electron from an anion. In contrast to ionization energies, however, electron affinities are difficult to measure because the anions of many elements are unstable. Table 8.3 shows the electron affinities of some representative elements and the noble gases, and Figure 8.12 plots the electron affinities of the first 56 elements versus atomic number. The overall trend is an increase in the tendency to accept electrons (electron affinity values become more positive) from left to right across a period. The electron affinities of metals are generally lower than those of nonmetals. The values vary little within a given group. The halogens (Group 17) have the highest electron affinity values.

There is a general correlation between electron affinity and effective nuclear charge, which also increases from left to right in a given period (see Section 8.3). However, as in the case of ionization energies, there are some irregularities. For example, the electron affinity of a Group 2 element is lower than that for the corresponding Group 1 element, and the electron affinity of a Group 15 element is lower than that for the corresponding Group 14 element. These exceptions are due to the valence electron configurations of the elements involved. An electron added to a Group 2 element must end up in a higher-energy *np* orbital, where it is effectively shielded by the ns^2 electrons and therefore experiences a weaker attraction to the nucleus. Therefore, it has a lower electron affinity than the corresponding Group 1 element. Likewise, it is harder to add an electron to a Group 15 element (ns^2np^3) than to the corresponding Group 14 element (ns^2np^2) because the electron added to the Group 15 element must be placed in an *np* orbital that already contains an electron and will therefore experience a greater electrostatic repulsion. Finally, in spite of the

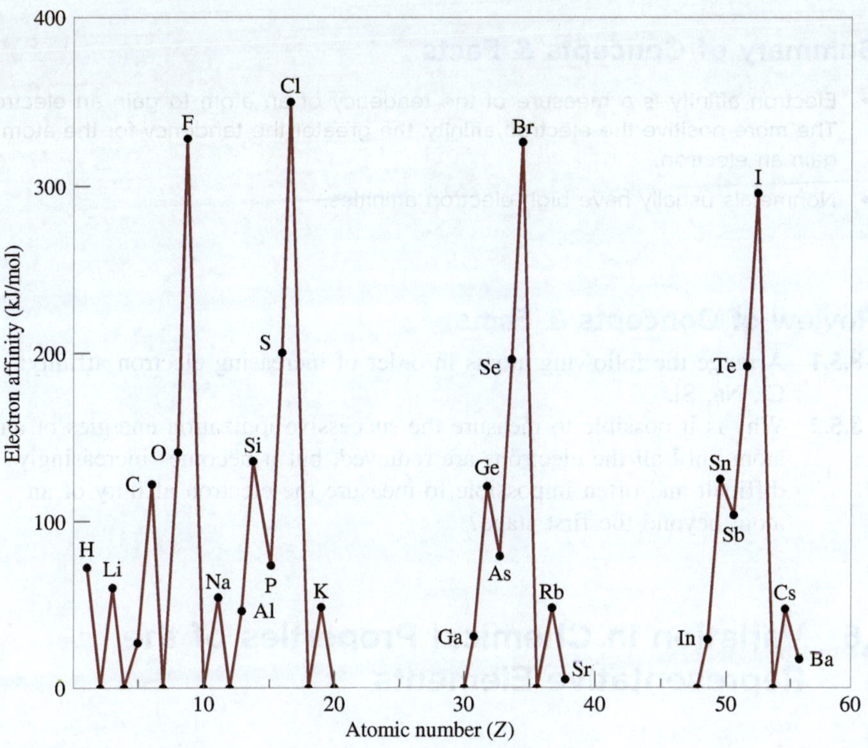

Figure 8.12 *A plot of electron affinity against atomic number from hydrogen to barium.*

fact that noble gases have high effective nuclear charge, they have extremely low electron affinities (zero or negative values). The reason is that an electron added to an atom with an ns^2np^6 configuration has to enter an $(n + 1)s$ orbital, where it is well shielded by the core electrons and will only be very weakly attracted by the nucleus. This analysis also explains why species with complete valence shells tend to be chemically stable.

Example 8.5 shows why the alkaline earth metals do not have a great tendency to accept electrons.

Example 8.5

Why are the electron affinities of the alkaline earth metals, shown in Table 8.3, either negative or small positive values?

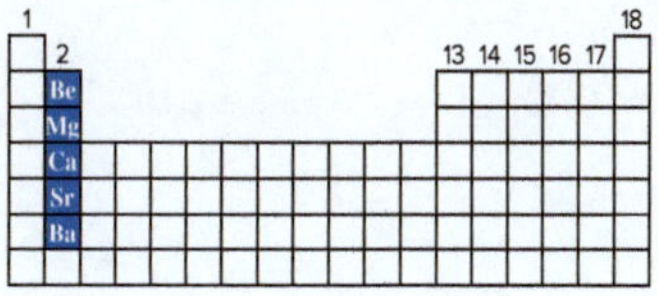

Strategy What are the electron configurations of alkaline earth metals? Would the added electron to such an atom be held strongly by the nucleus?

Solution The valence electron configuration of the alkaline earth metals is ns^2, where n is the highest principal quantum number. For the process

$$\underset{ns^2}{M(g)} + e^- \longrightarrow \underset{ns^2np^1}{M^-(g)}$$

where M denotes a member of the Group 2 family, the extra electron must enter the np subshell, which is effectively shielded by the two ns electrons (the ns electrons are more penetrating than the np electrons) and the inner electrons. Consequently, alkaline earth metals have little tendency to pick up an extra electron.

Practice Exercise Is it likely that Ar will form the anion Ar^-?

Similar problem: 8.63.

Summary of Concepts & Facts

- Electron affinity is a measure of the tendency of an atom to gain an electron. The more positive the electron affinity, the greater the tendency for the atom to gain an electron.
- Nonmetals usually have high electron affinities.

Review of Concepts & Facts

8.5.1 Arrange the following atoms in order of increasing electron affinity: Cl, Na, Si.

8.5.2 Why is it possible to measure the successive ionization energies of an atom until all the electrons are removed, but it becomes increasingly difficult and often impossible to measure the electron affinity of an atom beyond the first stage?

8.6 Variation in Chemical Properties of the Representative Elements

Learning Objective

- Distinguish among the variations in chemical properties and reactivities of main group elements.

Ionization energy and electron affinity help chemists understand the types of reactions that elements undergo and the nature of the elements' compounds. On a conceptual level, these two measures are related in a simple way: Ionization energy measures the attraction of an atom for its own electrons, whereas electron affinity expresses the attraction of an atom for an additional electron from some other source. Together they give us insight into the general attraction of an atom for electrons. With these concepts we can survey the chemical behavior of the elements systematically, paying particular attention to the relationship between their chemical properties and their electron configurations.

We have seen that the metallic character of the elements *decreases* from left to right across a period and *increases* from top to bottom within a group. On the basis of these trends and the knowledge that metals usually have low ionization energies while nonmetals usually have high electron affinities, we can frequently predict the outcome of a reaction involving some of these elements.

General Trends in Chemical Properties

Before we study the elements in individual groups, let us look at some overall trends. We have said that elements in the same group resemble one another in chemical behavior because they have similar valence electron configurations. This statement, although correct in the general sense, must be applied with caution. Chemists have long known that the first member of each group (the element in the second period from lithium to fluorine) differs from the rest of the members of the same group. Lithium, for example, exhibits many, but not all, of the properties characteristic of the alkali metals. Similarly, beryllium is a somewhat atypical member of Group 2, and so on. The difference can be attributed to the unusually small size of the first element in each group (see Figure 8.5).

Another trend in the chemical behavior of the representative elements is the diagonal relationship. ***Diagonal relationships*** are *similarities between pairs of* *elements in different groups and periods of the periodic table.* Specifically, the first three members of the second period (Li, Be, and B) exhibit many similarities to those elements located diagonally below them in the periodic table (Figure 8.13). The reason for this phenomenon is the closeness of the charge densities of their cations. (*Charge density* is the charge of an ion divided by its volume.) Cations with comparable charge densities react similarly with anions and therefore form the same type of compounds. Thus, the chemistry of lithium resembles that of magnesium in some ways; the same holds for beryllium and aluminum and for boron and silicon. Each of these pairs is said to exhibit a diagonal relationship. We will see a number of examples of this relationship later.

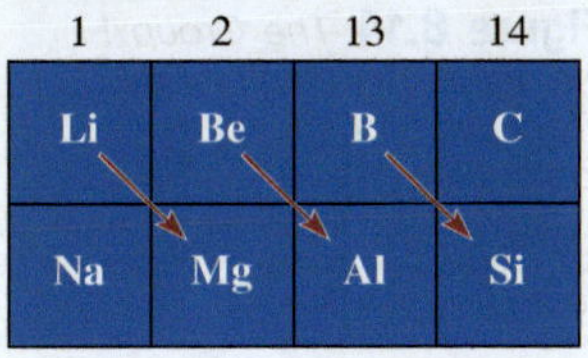

Figure 8.13 *Diagonal relationships in the periodic table.*

Bear in mind that a comparison of the properties of elements in the same group is most valid if we are dealing with elements of the same type with respect to their metallic character. This guideline applies to the elements in Groups 1 and 2, which are all metals, and to the elements in Groups 17 and 18, which are all nonmetals. In Groups 13 through 16, where the elements change either from nonmetals to metals or from nonmetals to metalloids, it is natural to expect greater variation in chemical properties even though the members of the same group have similar outer electron configurations.

Now let us take a closer look at the chemical properties of the representative elements and the noble gases. (We will consider the chemistry of the transition metals in Chapter 23.)

Hydrogen ($1s^1$)

There is no totally suitable position for hydrogen in the periodic table. Traditionally hydrogen is shown in Group 1, but it really could be a class by itself. Like the alkali metals, it has a single *s* valence electron and forms a unipositive ion (H^+), which is hydrated in solution. On the other hand, hydrogen also forms the hydride ion (H^-) in ionic compounds such as NaH and CaH_2. In this respect, hydrogen resembles the halogens, all of which form uninegative ions (F^-, Cl^-, Br^-, and I^-) in ionic compounds. Ionic hydrides react with water to produce hydrogen gas and the corresponding metal hydroxides:

$$2NaH(s) + 2H_2O(l) \longrightarrow 2NaOH(aq) + H_2(g)$$
$$CaH_2(s) + 2H_2O(l) \longrightarrow Ca(OH)_2(s) + 2H_2(g)$$

Of course, the most important compound of hydrogen is water, which forms when hydrogen burns in air:

$$2H_2(g) + O_2(g) \longrightarrow 2H_2O(l)$$

Group 1 Elements (ns^1, $n \geq 2$)

Figure 8.14 shows the Group 1 elements, the alkali metals. All of these elements have low ionization energies and therefore a great tendency to lose the single valence electron. In fact, in the vast majority of their compounds, they are unipositive ions. These metals are so reactive that they are never found in the pure state in nature. They react with water to produce hydrogen gas and the corresponding metal hydroxide:

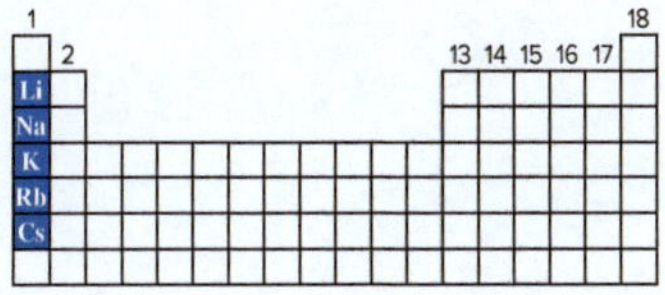

$$2M(s) + 2H_2O(l) \longrightarrow 2MOH(aq) + H_2(g)$$

where M denotes an alkali metal. When exposed to air, they gradually lose their shiny appearance as they combine with oxygen gas to form oxides. Lithium forms lithium oxide (containing the O^{2-} ion):

$$4Li(s) + O_2(g) \longrightarrow 2Li_2O(s)$$

The other alkali metals all form oxides and *peroxides* (containing the O_2^{2-} ion). For example,

$$2Na(s) + O_2(g) \longrightarrow Na_2O_2(s)$$

Figure 8.14 *The Group 1 elements: the alkali metals. Francium (not shown) is radioactive.*

(Li, Na): Ken Karp/McGraw Hill; (K): Al Fenn/The LIFE Images Collection/Getty Images; (Rb, Cs): Stephen Frisch/McGraw Hill

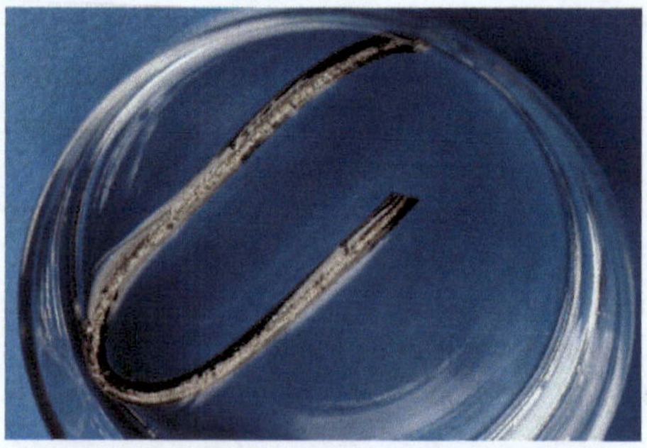

Lithium (Li)

Sodium (Na)

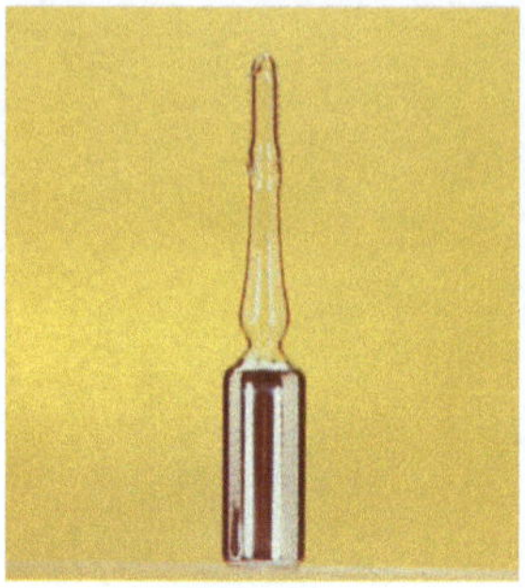

Potassium (K)

Rubidium (Rb)

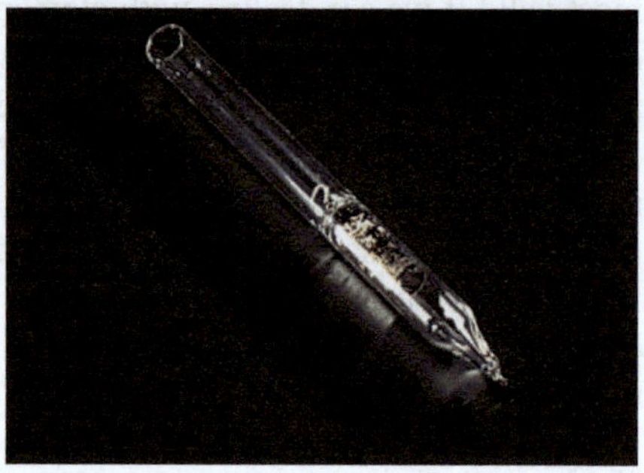

Cesium (Cs)

Potassium, rubidium, and cesium also form *superoxides* (containing the O_2^- ion):

$$K(s) + O_2(g) \longrightarrow KO_2(s)$$

The reason that different types of oxides are formed when alkali metals react with oxygen has to do with the stability of the oxides in the solid state. Because these oxides are all ionic compounds, their stability depends on how strongly the cations and anions attract one another. Lithium tends to form predominantly lithium oxide because this compound is more stable than lithium peroxide. The formation of other alkali metal oxides can be explained similarly.

Group 2 Elements (ns^2, $n \geq 2$)

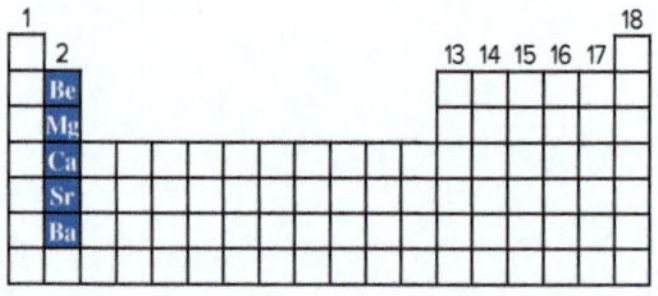

Figure 8.15 shows the Group 2 elements. As a group, the alkaline earth metals are somewhat less reactive than the alkali metals. Both the first and the second ionization energies decrease from beryllium to barium. Thus, the tendency is to form M^{2+} ions (where M denotes an alkaline earth metal atom), and hence the metallic character increases from top to bottom. Most beryllium compounds (BeH_2 and beryllium halides, such as $BeCl_2$) and some magnesium compounds (MgH_2, for example) are molecular rather than ionic in nature.

The reactivities of alkaline earth metals with water vary quite markedly. Beryllium does not react with water; magnesium reacts slowly with steam; calcium, strontium, and barium are reactive enough to attack cold water:

$$Ba(s) + 2H_2O(l) \longrightarrow Ba(OH)_2(aq) + H_2(g)$$

The reactivities of the alkaline earth metals toward oxygen also increase from Be to Ba. Beryllium and magnesium form oxides (BeO and MgO) only at elevated temperatures, whereas CaO, SrO, and BaO form at room temperature.

Magnesium reacts with acids in aqueous solution, liberating hydrogen gas:

$$Mg(s) + 2H^+(aq) \longrightarrow Mg^{2+}(aq) + H_2(g)$$

Calcium, strontium, and barium also react with aqueous acid solutions to produce hydrogen gas. However, because these metals also attack water, two different reactions will occur simultaneously.

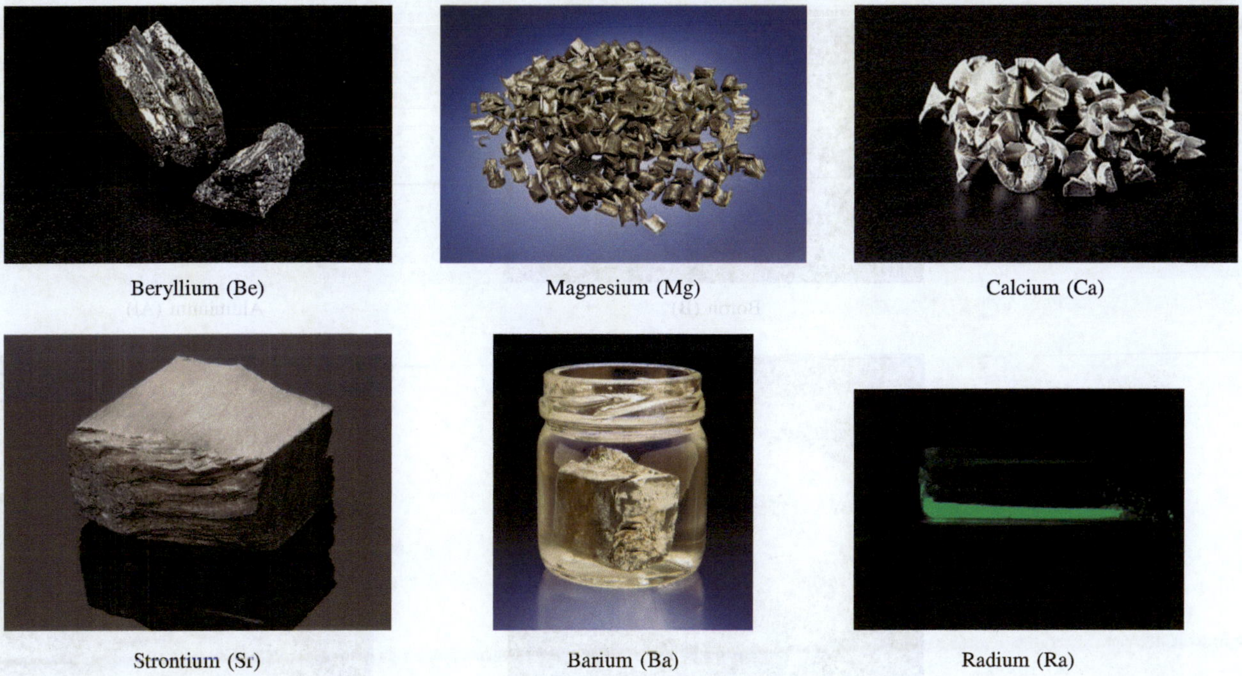

Figure 8.15 *The Group 2 elements: the alkaline earth metals.*
(Be, Ca): Stephen Frisch/McGraw Hill; (Mg, Sr, Ba): Charles D. Winters/McGraw Hill; (Ra): Fred Bayer

The chemical properties of calcium and strontium provide an interesting example of periodic group similarity. Strontium-90, a radioactive isotope, is a major product of an atomic bomb explosion. If an atomic bomb is exploded in the atmosphere, the strontium-90 formed will eventually settle on land and water, and it will reach our bodies via a relatively short food chain. For example, if cows eat contaminated grass and drink contaminated water, they will pass along strontium-90 in their milk. Because calcium and strontium are chemically similar, Sr^{2+} ions can replace Ca^{2+} ions in our bones. Constant exposure of the body to the high-energy radiation emitted by the strontium-90 isotopes can lead to anemia, leukemia, and other chronic illnesses.

Group 13 Elements (ns^2np^1, $n \geq 2$)

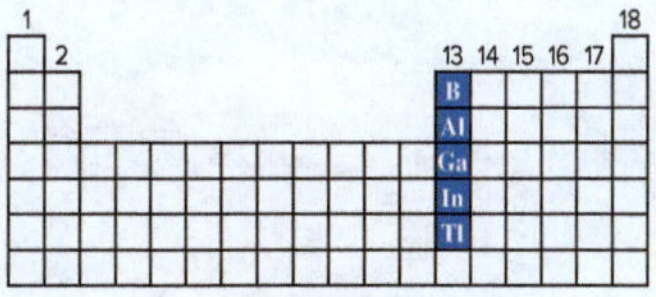

The first member of Group 13, boron, is a metalloid; the rest are metals (Figure 8.16). Boron does not form binary ionic compounds and is unreactive toward oxygen gas and water. The next element, aluminum, readily forms aluminum oxide when exposed to air:

$$4Al(s) + 3O_2(g) \longrightarrow 2Al_2O_3(s)$$

Aluminum that has a protective coating of aluminum oxide is less reactive than elemental aluminum. Aluminum forms only tripositive ions. It reacts with hydrochloric acid as follows:

$$2Al(s) + 6H^+(aq) \longrightarrow 2Al^{3+}(aq) + 3H_2(g)$$

The other Group 13 metallic elements form both unipositive and tripositive ions. Moving down the group, we find that the unipositive ion becomes more stable than the tripositive ion.

The metallic elements in Group 13 also form many molecular compounds. For example, aluminum reacts with hydrogen to form AlH_3, which resembles BeH_2 in its properties. (Here is an example of the diagonal relationship.) Thus, from left to right across the periodic table, we are seeing a gradual shift from metallic to nonmetallic character in the representative elements.

Figure 8.16 *The Group 13 elements. The low melting point of gallium (29.8°C) causes it to melt when held in hand.*
(Ga, B, In): Stephen Frisch/McGraw Hill; (Al): Kerstin Waurick/E+/Getty Images

Group 14 Elements (ns^2np^2, $n \geq 2$)

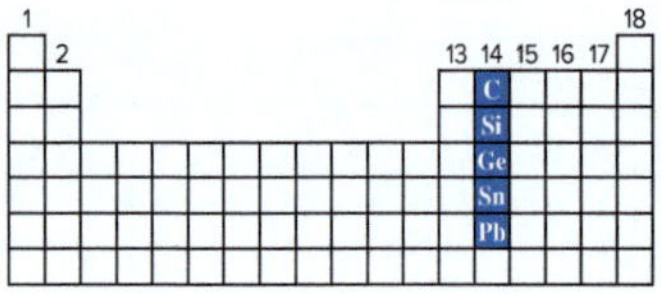

The first member of Group 14, carbon, is a nonmetal, and the next two members, silicon and germanium, are metalloids (Figure 8.17). The metallic elements of this group, tin and lead, do not react with water, but they do react with acids (hydrochloric acid, for example) to liberate hydrogen gas:

$$Sn(s) + 2H^+(aq) \longrightarrow Sn^{2+}(aq) + H_2(g)$$

$$Pb(s) + 2H^+(aq) \longrightarrow Pb^{2+}(aq) + H_2(g)$$

Figure 8.17 *The Group 14 elements.*
(graphite, Ge, Pb): Ken Karp/McGraw Hill; (Diamond): JewelryStock/Alamy Stock Photo; (Si, Sn): Charles D. Winters/Timeframe Photography/McGraw Hill

The Group 14 elements form compounds in both the +2 and +4 oxidation states (recall the inert pair effect, Section 8.2). For carbon and silicon, the +4 oxidation state is the more stable one. For example, CO_2 is more stable than CO; and SiO_2 is a stable compound, but SiO does not exist under normal conditions. As we move down the group, however, the trend in stability is reversed. In tin compounds the +4 oxidation state is only slightly more stable than the +2 oxidation state. In lead compounds the +2 oxidation state is unquestionably the more stable one. The outer electron configuration of lead is $6s^2 6p^2$, and lead tends to lose only the 6*p* electrons (to form Pb^{2+}) rather than both the 6*p* and 6*s* electrons (to form Pb^{4+}).

Group 15 Elements (ns^2np^3, $n \geq 2$)

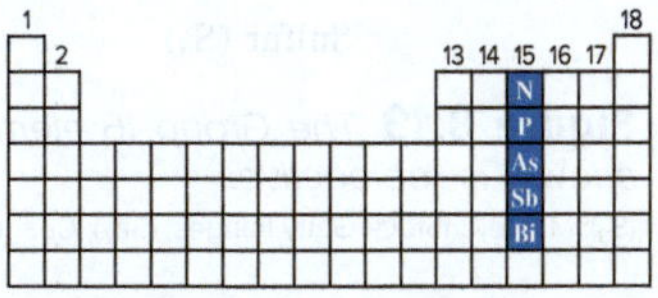

In Group 15, nitrogen and phosphorus are nonmetals, arsenic and antimony are metalloids, and bismuth is a metal (Figure 8.18). Thus, we expect a greater variation in properties within the group.

Elemental nitrogen is a diatomic gas (N_2). It forms a number of oxides (NO, N_2O, NO_2, N_2O_4, and N_2O_5), of which only N_2O_5 is a solid; the others are gases. Nitrogen has a tendency to accept three electrons to form the nitride ion, N^{3-} (thus achieving the electron configuration $1s^22s^22p^6$, which is isoelectronic with neon). Most metallic nitrides (Li_3N and Mg_3N_2, for example) are ionic compounds. Phosphorus exists as P_4 molecules. It forms two solid oxides with the formulas P_4O_6 and P_4O_{10}. The important oxoacids HNO_3 and H_3PO_4 are formed when the following oxides react with water:

$$N_2O_5(s) + H_2O(l) \longrightarrow 2HNO_3(aq)$$
$$P_4O_{10}(s) + 6H_2O(l) \longrightarrow 4H_3PO_4(aq)$$

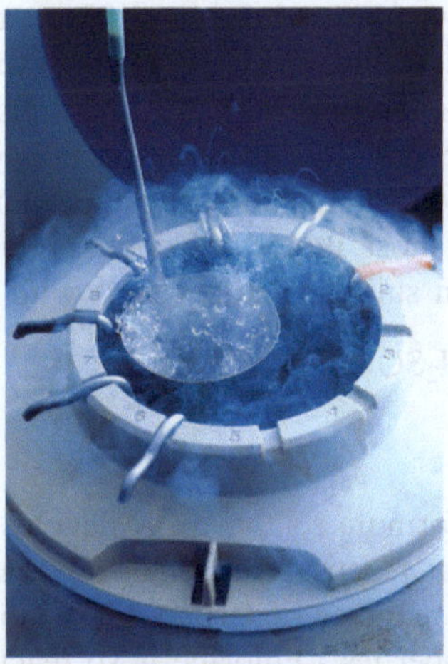

Liquid nitrogen (N_2)

White and red phosphorus (P)

Arsenic (As)

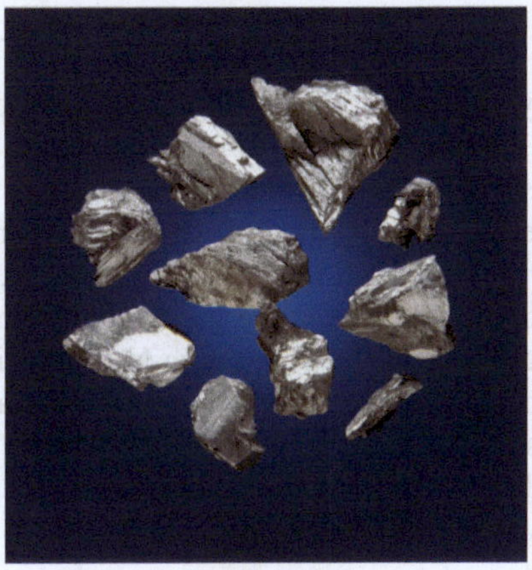

Antimony (Sb)

Bismuth (Bi)

Figure 8.18 *The Group 15 elements. Molecular nitrogen is a colorless, odorless gas.*

(N_2): Science Photo Library/Alamy Stock Photo; (P): Al Fenn/The LIFE Images Collection/Getty Images; (As, Bi): Stephen Frisch/McGraw Hill; (Sb): Charles D. Winters/Timeframe Photography/McGraw Hill

Sulfur (S_8) Selenium (Se_8) Tellurium (Te)

Figure 8.19 *The Group 16 elements sulfur, selenium, and tellurium. Molecular oxygen is a colorless, odorless gas. Polonium (not shown) is radioactive.*

(S_8): MarcelC/iStock/Getty Images; (Se_8): Charles D. Winters/Timeframe Photography/McGraw Hill; (Te): Ken Karp/McGraw Hill

Arsenic, antimony, and bismuth have extensive three-dimensional structures. Bismuth is a far less reactive metal than those in the preceding groups.

Group 16 Elements (ns^2np^4, $n \geq 2$)

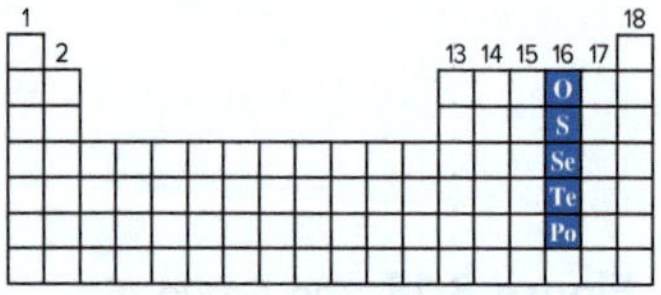

The first three members of Group 16 (oxygen, sulfur, and selenium) are nonmetals, tellurium is a metalloid, and polonium is a metal (Figure 8.19). Oxygen is a diatomic gas; elemental sulfur and selenium have the molecular formulas S_8 and Se_8, respectively; tellurium and polonium have more extensive three-dimensional structures. (Polonium, the last member, is a radioactive element that is difficult to study in the laboratory.) Oxygen has a tendency to accept two electrons to form the oxide ion (O^{2-}) in many ionic compounds. Sulfur, selenium, and tellurium also form dinegative anions (S^{2-}, Se^{2-}, and Te^{2-}). The elements in this group (especially oxygen) form a large number of molecular compounds with nonmetals. The important compounds of sulfur are SO_2, SO_3, and H_2S. The most important commercial sulfur compound is sulfuric acid, which is formed when sulfur trioxide reacts with water:

$$SO_3(g) + H_2O(l) \longrightarrow H_2SO_4(aq)$$

Group 17 Elements (ns^2np^5, $n \geq 2$)

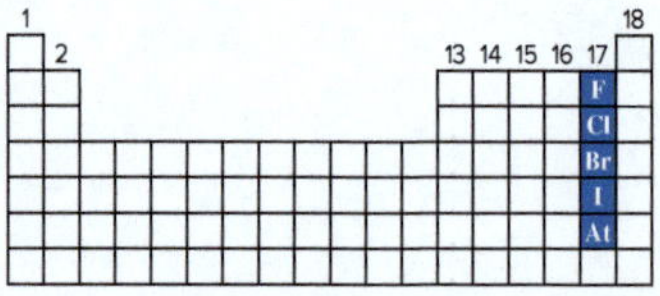

All the halogens are nonmetals with the general formula X_2, where X denotes a halogen element (Figure 8.20). Because of their great reactivity, the halogens are never found in the elemental form in nature. (The last member of Group 17, astatine, is a radioactive element. Little is known about its properties.) Fluorine is so reactive that it attacks water to generate oxygen:

$$2F_2(g) + 2H_2O(l) \longrightarrow 4HF(aq) + O_2(g)$$

Actually the reaction between molecular fluorine and water is quite complex; the products formed depend on reaction conditions. The reaction shown in Figure 8.20 is one of several possible chemical changes.

The halogens have high ionization energies and large positive electron affinities. Anions derived from the halogens (F^-, Cl^-, Br^-, and I^-) are called *halides*. They are isoelectronic with the noble gases immediately to their right in the periodic table. For example, F^- is isoelectronic with Ne, Cl^- with Ar, and so on. The vast majority of the alkali metal halides and alkaline earth metal halides are ionic compounds. The halogens also form many molecular compounds among themselves (such as ICl and BrF_3) and with nonmetallic elements in other groups (such as NF_3, PCl_5, and SF_6). The halogens react with hydrogen to form hydrogen halides:

$$H_2(g) + X_2(g) \longrightarrow 2HX(g)$$

Figure 8.20 *The Group 17 elements chlorine, bromine, and iodine. Fluorine is a greenish-yellow gas that attacks ordinary glassware. Astatine is radioactive.*
SPL/Science Source

When this reaction involves fluorine, it is explosive, but it becomes less and less violent as we substitute chlorine, bromine, and iodine. The hydrogen halides dissolve in water to form hydrohalic acids. Hydrofluoric acid (HF) is a weak acid (that is, it is a weak electrolyte), but the other hydrohalic acids (HCl, HBr, and HI) are all strong acids (strong electrolytes).

Group 18 Elements (ns^2np^6, $n \geq 2$)

All noble gases exist as monatomic species (Figure 8.21). Their atoms have completely filled outer *ns* and *np* subshells, which give them great stability. (Helium is $1s^2$.) The Group 18 ionization energies are among the highest of all elements, and these gases have no tendency to accept extra electrons. For years these elements were called inert gases, and rightly so. Until 1963 no one had been able to prepare a compound containing any of these elements. The British chemist Neil Bartlett[†] shattered chemists'

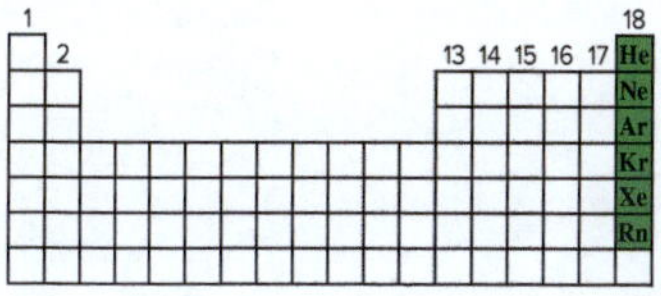

[†]Neil Bartlett (1932–2008). English chemist. Bartlett's work was mainly in the preparation and study of compounds with unusual oxidation states and in solid-state chemistry.

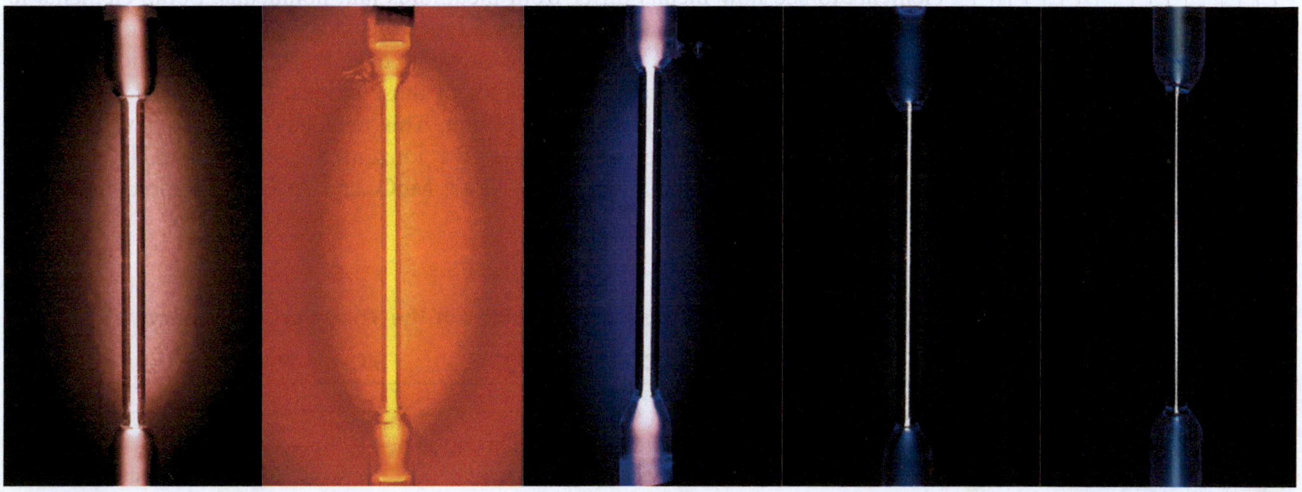

Figure 8.21 *All noble gases are colorless and odorless. These pictures show the colors emitted by the gases from a discharge tube.*
(all): Ken Karp/McGraw Hill

In 2013 astronomers reported finding the emission of HAr^+ in the Crab Nebula, making it the first molecular noble gas species to be detected in space.
ESA, J. Hester and A. Loll (Arizona State University)/NASA

long-held views of these elements when he exposed xenon to platinum hexafluoride, a strong oxidizing agent, and brought about the following reaction:

$$Xe(g) + 2PtF_6(g) \longrightarrow XeF^+Pt_2F_{11}^-(s)$$

Since then, a number of xenon compounds (XeF_4, XeO_3, XeO_4, $XeOF_4$) and a few krypton compounds (KrF_2, for example) have been prepared (Figure 8.22). In 2000, chemists prepared a compound containing argon (HArF), which is stable only below 40 K. Despite the immense interest in the chemistry of the noble gases, however, their compounds do not have any major commercial applications, and they are not involved in natural biological processes. No compounds of helium and neon are known.

Comparison of Group 1 and Group 11 Elements

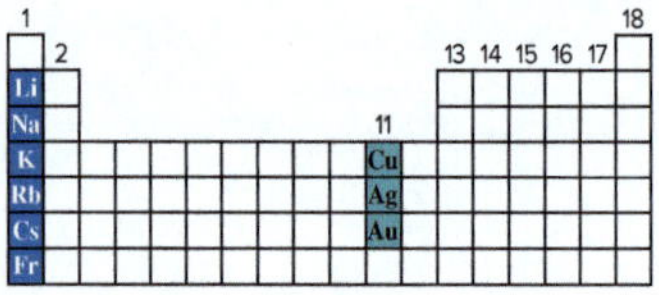

When we compare the Group 1 elements (alkali metals) and the Group 11 elements (copper, silver, and gold), we arrive at an interesting conclusion. Although the metals in these two groups have similar outer electron configurations, with one electron in the outermost *s* orbital, their chemical properties are quite different.

The first ionization energies of Cu, Ag, and Au are 745 kJ/mol, 731 kJ/mol, and 890 kJ/mol, respectively. Because these values are considerably larger than those of the alkali metals (see Table 8.2), the Group 11 elements are much less reactive. The higher ionization energies of the Group 11 elements result from incomplete shielding of the nucleus by the inner *d* electrons (compared with the more effective shielding of the completely filled noble gas cores). Consequently the outer *s* electrons of these elements are more strongly attracted by the nucleus. In fact, copper, silver, and gold are so unreactive that they are usually found in the uncombined state in nature. The inertness and rarity of these metals make them valuable in the manufacture of coins and in jewelry. For this reason, these metals are also called "coinage metals." The difference in chemical properties between the Group 2 elements (the alkaline earth metals) and the Group 12 metals (zinc, cadmium, and mercury) can be explained in a similar way.

Properties of Oxides Across a Period

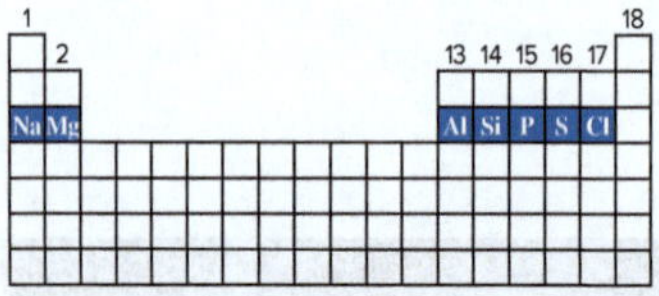

One way to compare the properties of the representative elements across a period is to examine the properties of a series of similar compounds. Because oxygen combines with almost all elements, we will compare the properties of oxides of the third-period elements to see how metals differ from metalloids and nonmetals. Some elements in the third period (P, S, and Cl) form several types of oxides, but for simplicity we will consider only those oxides in which the elements have the highest oxidation number. Table 8.4 lists a few general characteristics of these oxides. We observed earlier that oxygen has a tendency to form the oxide ion. This tendency is greatly favored when oxygen combines with metals that have low ionization energies, namely, those in Groups 1 and 2, plus aluminum. Thus, Na_2O, MgO, and Al_2O_3 are ionic compounds,

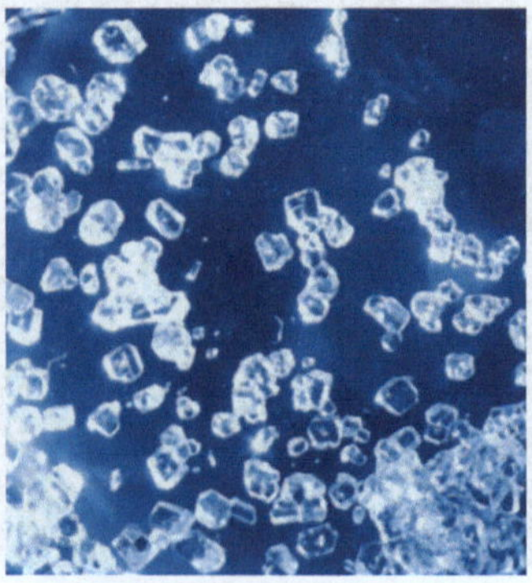

Figure 8.22 *Crystals of xenon tetrafluoride (XeF_4).*
Ken Karp/McGraw Hill

Table 8.4 Some Properties of Oxides of the Third-Period Elements

	Na_2O	MgO	Al_2O_3	SiO_2	P_4O_{10}	SO_3	Cl_2O_7
Type of compound	← Ionic →			← Molecular →			
Structure	← Extensive three-dimensional →				← Discrete molecular units →		
Melting point (°C)	1275	2800	2045	1610	580	16.8	−91.5
Boiling point (°C)	?	3600	2980	2230	?	44.8	82
Acid-base nature	Basic	Basic	Amphoteric	← Acidic →			

CHEMISTRY *in Action*

Discovery of the Noble Gases

In the late 1800s John William Strutt, Third Baron of Rayleigh, who was a professor of physics at the Cavendish Laboratory in Cambridge, England, accurately determined the atomic masses of a number of elements, but he obtained a puzzling result with nitrogen. One of his methods of preparing nitrogen was by the thermal decomposition of ammonia:

$$2NH_3(g) \longrightarrow N_2(g) + 3H_2(g)$$

Another method was to start with air and remove from it oxygen, carbon dioxide, and water vapor. Invariably, the nitrogen from air was a little denser (by about 0.5 percent) than the nitrogen from ammonia.

Lord Rayleigh's work caught the attention of Sir William Ramsay, a professor of chemistry at the University College, London. In 1898 Ramsay passed nitrogen, which he had obtained from air by Rayleigh's procedure, over red-hot magnesium to convert it to magnesium nitride:

$$3Mg(s) + N_2(g) \longrightarrow Mg_3N_2(s)$$

After all of the nitrogen had reacted with magnesium, Ramsay was left with an unknown gas that would not combine with anything.

With the help of Sir William Crookes, the inventor of the discharge tube, Ramsay and Lord Rayleigh found that the emission spectrum of the gas did not match any of the known elements. The gas was a new element! They determined its atomic mass to be 39.95 amu and called it *argon,* which means "the lazy one" in Greek.

Once argon had been discovered, other noble gases were quickly identified. Also in 1898 Ramsay isolated helium from uranium ores. From the atomic masses of helium and argon, their lack of chemical reactivity, and what was then known about the periodic table, Ramsay was convinced that there were other unreactive gases and that they were all members of one periodic group. He and his student Morris Travers set out to find the unknown gases. They used a refrigeration machine to first produce liquid air. Applying a technique called *fractional distillation,* they then allowed the liquid air to warm up gradually and collected components that boiled off at different temperatures. In this manner, they analyzed and identified three new elements—neon, krypton, and xenon—in only three months. Three new elements in three months is a record that may never be broken!

The discovery of the noble gases helped to complete the periodic table. Their atomic masses suggested that these elements should be placed to the right of the halogens. The apparent discrepancy with the position of argon was resolved by Moseley, as discussed in the chapter.

Finally, the last member of the noble gases, radon, was discovered by the German chemist Frederick Dorn in 1900. A radioactive element and the heaviest elemental gas known, radon's discovery not only completed the Group 18 elements, but also advanced our understanding about the nature of radioactive decay and transmutation of elements.

Lord Rayleigh and Ramsay both won Nobel Prizes in 1904 for the discovery of argon. Lord Rayleigh received the prize in physics and Ramsay's award was in chemistry.

Sir William Ramsay (1852–1916).
Library of Congress [LC-USZ62-99748]

as indicated by their high melting points and boiling points. They have extensive three-dimensional structures in which each cation is surrounded by a specific number of anions, and vice versa. As the ionization energies of the elements increase from left to right, so does the molecular nature of the oxides that are formed. Silicon is a metalloid; its oxide (SiO_2) also has a huge three-dimensional network, although no ions are present. The oxides of phosphorus, sulfur, and chlorine are molecular compounds composed of small discrete units. The weak attractions among these molecules result in relatively low melting points and boiling points.

Most oxides can be classified as acidic or basic depending on whether they produce acids or bases when dissolved in water or react as acids or bases in certain processes. Some oxides are ***amphoteric,*** which means that they *display both acidic and basic properties*. The first two oxides of the third period, Na_2O and MgO, are basic oxides. For example, Na_2O reacts with water to form the base sodium hydroxide:

$$Na_2O(s) + H_2O(l) \longrightarrow 2NaOH(aq)$$

Magnesium oxide is quite insoluble; it does not react with water to any appreciable extent. However, it does react with acids in a manner that resembles an acid-base reaction:

$$MgO(s) + 2HCl(aq) \longrightarrow MgCl_2(aq) + H_2O(l)$$

Note that the products of this reaction are a salt ($MgCl_2$) and water, the usual products of an acid-base neutralization.

Aluminum oxide is even less soluble than magnesium oxide; it too does not react with water. However, it shows basic properties by reacting with acids:

$$Al_2O_3(s) + 6HCl(aq) \longrightarrow 2AlCl_3(aq) + 3H_2O(l)$$

It also exhibits acidic properties by reacting with bases:

$$Al_2O_3(s) + 2NaOH(aq) + 3H_2O(l) \longrightarrow 2NaAl(OH)_4(aq)$$

Thus, Al_2O_3 is classified as an amphoteric oxide because it has properties of both acids and bases. This acid-base neutralization produces a salt but no water. Other amphoteric oxides are ZnO, BeO, and Bi_2O_3.

Silicon dioxide is insoluble and does not react with water. It has acidic properties, however, because it reacts with very concentrated bases:

$$SiO_2(s) + 2NaOH(aq) \longrightarrow Na_2SiO_3(aq) + H_2O(l)$$

For this reason, concentrated aqueous, strong bases such as NaOH(*aq*) should not be stored in Pyrex glassware, which is made of SiO_2.

The remaining third-period oxides are acidic. They react with water to form phosphoric acid (H_3PO_4), sulfuric acid (H_2SO_4), and perchloric acid ($HClO_4$):

$$P_4O_{10}(s) + 6H_2O(l) \longrightarrow 4H_3PO_4(aq)$$
$$SO_3(g) + H_2O(l) \longrightarrow H_2SO_4(aq)$$
$$Cl_2O_7(l) + H_2O(l) \longrightarrow 2HClO_4(aq)$$

Certain oxides such as CO and NO are neutral; that is, they do not react with water to produce an acidic or basic solution. In general, oxides containing nonmetallic elements are not basic.

This brief examination of oxides of the third-period elements shows that as the metallic character of the elements decreases from left to right across the period, their oxides change from basic to amphoteric to acidic. Metallic oxides are usually basic, and most oxides of nonmetals are acidic. The intermediate properties of the oxides (as shown by the amphoteric oxides) are exhibited by elements whose positions are intermediate within the period. Note also that because the metallic character of the elements increases from top to bottom within a group of representative elements, we would expect oxides of elements with higher atomic numbers to be more basic than the lighter elements. This is indeed the case.

Example 8.6

Classify the following oxides as acidic, basic, or amphoteric: (a) Rb_2O, (b) BeO, (c) As_2O_5.

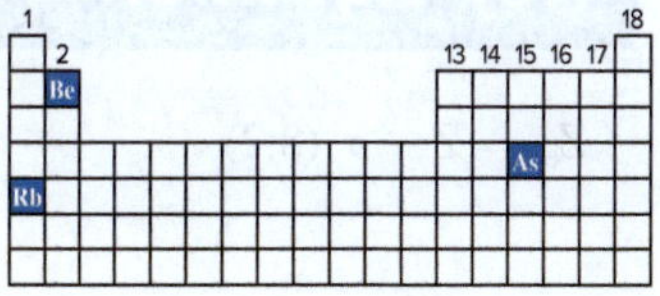

Strategy What type of elements forms acidic oxides? Basic oxides? Amphoteric oxides?

Solution

(a) Because rubidium is an alkali metal, we would expect Rb_2O to be a basic oxide.

(b) Beryllium is an alkaline earth metal. However, because it is the first member of Group 2, we expect that it may differ somewhat from the other members of the group. In the text, we saw that Al_2O_3 is amphoteric. Because beryllium and aluminum exhibit a diagonal relationship, BeO may resemble Al_2O_3 in properties. It turns out that BeO is also an amphoteric oxide.

(c) Because arsenic is a nonmetal, we expect As_2O_5 to be an acidic oxide.

Practice Exercise Classify the following oxides as acidic, basic, or amphoteric: (a) ZnO, (b) P_4O_{10}, (c) CaO.

Similar problem: 8.72.

Summary of Concepts & Facts

- Noble gases are very stable because their outer *ns* and *np* subshells are completely filled. The metals among the representative elements (in Groups 1, 2, and 13) tend to lose electrons until their cations become isoelectronic with the noble gases that precede them in the periodic table. The nonmetals in Groups 15, 16, and 17 tend to accept electrons until their anions become isoelectronic with the noble gases that follow them in the periodic table.

Review of Concepts & Facts

8.6.1 An oxide of an element was determined to be basic. Which of the following could be that element: (a) Ba, (b) Al, or (c) Sb?

Chapter Summary

Development of the Periodic Table In the nineteenth century, chemists noticed a regular, periodic recurrence of chemical and physical properties of elements. In particular, the periodic table drawn up by Mendeleev grouped the elements accurately and was able to predict the properties of several elements that had not yet been discovered. (Section 8.1)

Periodic Classification of the Elements Elements are grouped according to their outer-shell electron configurations, which account for their similar chemical behavior. Special names are assigned to these various groups. (Section 8.2)

Periodic Variation in Properties Overall, physical properties such as atomic and ionic radii of the elements vary in a regular and periodic fashion. Similar variation is also noted in their chemical properties. Chemical properties of special importance are ionization energy, which measures the tendency of an atom of an element to lose an electron, and electron affinity, which measures the tendency of an atom to accept an electron. Ionization energy and electron affinity form the basis for understanding chemical bond formation. (Sections 8.3, 8.4, 8.5, and 8.6)

Key Equation

$Z_{eff} = Z - \sigma$ (8.2)	Definition of effective nuclear charge.

Key Words

Amphoteric oxide, p. 360
Atomic radius, p. 337
Core electrons, p. 333
Diagonal relationship, p. 350
Effective nuclear charge (Z_{eff}), p. 336
Electron affinity (*EA*), p. 347
Inert pair effect, p. 335
Ionic radius, p. 339
Ionization energy (*IE*), p. 344
Isoelectronic, p. 335
Representative elements, p. 331
Valence electrons, p. 333

Questions & Problems

Red numbered problems solved in Student Solutions Manual

8.1 Development of the Periodic Table

Review Questions

8.1 Briefly describe the significance of Mendeleev's periodic table.

8.2 What is Moseley's contribution to the modern periodic table?

8.3 Describe the general layout of a modern periodic table.

8.4 What is the most important relationship among elements in the same group in the periodic table?

8.2 Periodic Classification of the Elements

Review Questions

8.5 Which of the following elements are metals, nonmetals, or metalloids? As, Xe, Fe, Li, B, Cl, Ba, P, I, Si.

8.6 Compare the physical and chemical properties of metals and nonmetals.

8.7 Draw a rough sketch of a periodic table (no details are required). Indicate regions where metals, nonmetals, and metalloids are located.

8.8 What is a representative element? Give names and symbols of four representative elements.

8.9 Without referring to a periodic table, write the name and give the symbol for an element in each of the following groups: 1, 2, 13, 14, 15, 16, 17, 18, transition metals.

8.10 Indicate whether the following elements exist as atomic species, molecular species, or extensive three-dimensional structures in their most stable states at 25°C and 1 atm and write the molecular or empirical formula for each one: phosphorus, iodine, magnesium, neon, carbon, sulfur, cesium, and oxygen.

8.11 You are given a dark shiny solid and asked to determine whether it is iodine or a metallic element. Suggest a nondestructive test that would enable you to arrive at the correct answer.

8.12 What are valence electrons? For representative elements, the number of valence electrons of an element is equal to its group number. Show that this is true for the following elements: Al, Sr, K, Br, P, S, C.

8.13 Write the outer electron configurations for the (a) alkali metals, (b) alkaline earth metals, (c) halogens, (d) noble gases.

8.14 Use the first-row transition metals (Sc to Cu) as an example to illustrate the characteristics of the electron configurations of transition metals.

8.15 The electron configurations of ions derived from representative elements follow a common pattern. What is the pattern, and how does it relate to the stability of these ions?

8.16 What do we mean when we say that two ions or an atom and an ion are isoelectronic?

8.17 What is wrong with the statement "The atoms of element X are isoelectronic with the atoms of element Y"?

8.18 Give three examples of first-row transition metal (Sc to Cu) ions whose electron configurations are represented by the argon core.

Problems

8.19 In the periodic table, the element hydrogen is sometimes grouped with the alkali metals (as in this book) and sometimes with the halogens. Explain why hydrogen can resemble the Group 1 and the Group 17 elements.

8.20 A neutral atom of a certain element has 17 electrons. Without consulting a periodic table, (a) write the ground-state electron configuration of the element, (b) classify the element, (c) determine whether this element is diamagnetic or paramagnetic.

8.21 Group the following electron configurations in pairs that would represent similar chemical properties of their atoms:
(a) $1s^22s^22p^63s^2$
(b) $1s^22s^22p^3$
(c) $1s^22s^22p^63s^23p^64s^23d^{10}4p^6$
(d) $1s^22s^2$
(e) $1s^22s^22p^6$
(f) $1s^22s^22p^63s^23p^3$

8.22 Group the following electron configurations in pairs that would represent similar chemical properties of their atoms:
(a) $1s^22s^22p^5$
(b) $1s^22s^1$
(c) $1s^22s^22p^6$
(d) $1s^22s^22p^63s^23p^5$
(e) $1s^22s^22p^63s^23p^64s^1$
(f) $1s^22s^22p^63s^23p^64s^23d^{10}4p^6$

8.23 Without referring to a periodic table, write the electron configuration of elements with the following atomic numbers: (a) 9, (b) 20, (c) 26, (d) 33. Classify the elements.

8.24 Specify the group of the periodic table in which each of the following elements is found: (a) $[Ne]3s^1$, (b) $[Ne]3s^23p^3$, (c) $[Ne]3s^23p^6$, (d) $[Ar]4s^23d^8$.

8.25 A M^{2+} ion derived from a metal in the first transition metal series has four electrons in the 3*d* subshell. What element might M be?

8.26 A metal ion with a net +3 charge has five electrons in the 3*d* subshell. Identify the metal.

8.27 Write the ground-state electron configurations of the following ions: (a) Li^+, (b) H^-, (c) N^{3-}, (d) F^-, (e) S^{2-}, (f) Al^{3+}, (g) Se^{2-}, (h) Br^-, (i) Rb^+, (j) Sr^{2+}, (k) Sn^{2+}, (l) Te^{2-}, (m) Ba^{2+}, (n) Pb^{2+}, (o) In^{3+}, (p) Tl^+, (q) Tl^{3+}.

8.28 Write the ground-state electron configurations of the following ions, which play important roles in biochemical processes in the body: (a) Na^+, (b) Mg^{2+}, (c) Cl^-, (d) K^+, (e) Ca^{2+}, (f) Fe^{2+}, (g) Cu^{2+}, (h) Zn^{2+}.

8.29 Write the ground-state electron configurations of the following transition metal ions: (a) Sc^{3+}, (b) Ti^{4+}, (c) V^{5+}, (d) Cr^{3+}, (e) Mn^{2+}, (f) Fe^{2+}, (g) Fe^{3+}, (h) Co^{2+}, (i) Ni^{2+}, (j) Cu^+, (k) Cu^{2+}, (l) Ag^+, (m) Au^+, (n) Au^{3+}, (o) Pt^{2+}.

8.30 Name the ions with +3 charges that have the following electron configurations: (a) $[Ar]3d^3$, (b) [Ar], (c) $[Kr]4d^6$, (d) $[Xe]4f^{14}5d^6$.

8.31 Which of the following species are isoelectronic with each other? C, Cl^-, Mn^{2+}, B^-, Ar, Zn, Fe^{3+}, Ge^{2+}

8.32 Group the species that are isoelectronic: Be^{2+}, F^-, Fe^{2+}, N^{3-}, He, S^{2-}, Co^{3+}, Ar.

8.3 Periodic Variation in Physical Properties

Review Questions

8.33 Define atomic radius. Does the size of an atom have a precise meaning?

8.34 How does atomic radius change (a) from left to right across a period and (b) from top to bottom in a group?

8.35 Define ionic radius. How does the size of an atom change when it is converted to (a) an anion and (b) a cation?

8.36 Explain why, for isoelectronic ions, the anions are larger than the cations.

Problems

8.37 On the basis of their positions in the periodic table, select the atom with the larger atomic radius in each of the following pairs: (a) Na, Cs, (b) Be, Ba, (c) N, Sb, (d) F, Br, (e) Ne, Xe.

8.38 Arrange the following atoms in order of decreasing atomic radius: Na, Al, P, Cl, Mg.

8.39 Which is the largest atom in Group 14?

8.40 Which is the smallest atom in Group 17?

8.41 Why is the radius of the lithium atom considerably larger than the radius of the hydrogen atom?

8.42 Use the second period of the periodic table as an example to show that the size of atoms decreases as we move from left to right. Explain the trend.

8.43 Indicate which one of the two species in each of the following pairs is smaller: (a) Cl or Cl^-, (b) Na or Na^+, (c) O^{2-} or S^{2-}, (d) Mg^{2+} or Al^{3+}, (e) Au^+ or Au^{3+}.

8.44 List the following ions in order of increasing ionic radius: N^{3-}, Na^+, F^-, Mg^{2+}, O^{2-}.

8.45 Explain which of the following cations is larger, and why: Cu^+ or Cu^{2+}.

8.46 Explain which of the following anions is larger, and why: Se^{2-} or Te^{2-}.

8.47 Give the physical states (gas, liquid, or solid) of the representative elements in the fourth period (K, Ca, Ga, Ge, As, Se, Br) at 1 atm and 25°C.

8.48 Both H^- and He contain two 1*s* electrons. Which species is larger? Explain your choice.

8.4 Ionization Energy

Review Questions

8.49 Define ionization energy. Ionization energy measurements are usually made when atoms are in the gaseous state. Why? Why is the second ionization energy always greater than the first ionization energy for any element?

8.50 Sketch the outline of the periodic table and show group and period trends in the first ionization energy of the elements. What types of elements have the highest ionization energies and what types the lowest ionization energies?

Problems

8.51 Arrange the following in order of increasing first ionization energy: Na, Cl, Al, S, and Cs.

8.52 Arrange the following in order of increasing first ionization energy: F, K, P, Ca, and Ne.

8.53 Use the third period of the periodic table as an example to illustrate the change in first ionization energies of the elements as we move from left to right. Explain the trend.

8.54 In general, ionization energy increases from left to right across a given period. Aluminum, however, has a lower ionization energy than magnesium. Explain.

8.55 The first and second ionization energies of K are 419 kJ/mol and 3052 kJ/mol, and those of Ca are 590 kJ/mol and 1145 kJ/mol, respectively. Compare their values and comment on the differences.

8.56 Two atoms have the electron configurations $1s^22s^22p^6$ and $1s^22s^22p^63s^1$. The first ionization energy of one is 2080 kJ/mol, and that of the other is 496 kJ/mol. Match each ionization energy with one of the given electron configurations. Justify your choice.

8.57 A hydrogenlike ion contains only one electron. The energies of the electron in a hydrogenlike ion are given by

$$E_n = -(2.18 \times 10^{-18}\ \text{J})\, Z^2\left(\frac{1}{n^2}\right)$$

where n is the principal quantum number and Z is the atomic number of the element. Calculate the ionization energy (in kJ/mol) of the He^+ ion.

8.58 Plasma is a state of matter consisting of positive gaseous ions and electrons. In the plasma state, a mercury atom could be stripped of its 80 electrons and therefore would exist as Hg^{80+}. Use the equation in Problem 8.57 to calculate the energy required for the last ionization step, that is,

$$Hg^{79+}(g) \longrightarrow Hg^{80+}(g) + e^-$$

8.5 Electron Affinity

Review Questions

8.59 (a) Define electron affinity. (b) Electron affinity measurements are made with gaseous atoms. Why? (c) Ionization energy is always a positive quantity, whereas electron affinity may be either positive or negative. Explain.

8.60 Explain the trends in electron affinity from aluminum to chlorine (see Table 8.3).

Problems

8.61 Arrange the elements in each of the following groups in increasing order of the most positive electron affinity: (a) Li, Na, K; (b) F, Cl, Br, I; (c) O, Si, P, Ca, Ba.

8.62 Specify which of the following elements you would expect to have the greatest electron affinity and which would have the least: He, K, Co, S, Cl.

8.63 Considering their electron affinities, do you think it is possible for the alkali metals to form an anion like M^-, where M represents an alkali metal?

8.64 Explain why alkali metals have a greater affinity for electrons than alkaline earth metals.

8.6 Variation in Chemical Properties of the Representative Elements

Review Questions

8.65 What is meant by the diagonal relationship? Name two pairs of elements that show this relationship.

8.66 Which elements are more likely to form acidic oxides? Basic oxides? Amphoteric oxides?

Problems

8.67 Use the alkali metals and alkaline earth metals as examples to show how we can predict the chemical properties of elements simply from their electron configurations.

8.68 Based on your knowledge of the chemistry of the alkali metals, predict some of the chemical properties of francium, the last member of the group.

8.69 As a group, the noble gases are very stable chemically (only Kr and Xe are known to form compounds). Use the concepts of shielding and the effective nuclear charge to explain why the noble gases tend to neither give up electrons nor accept additional electrons.

8.70 Why are Group 11 elements more stable than Group 1 elements even though they seem to have the same outer electron configuration, ns^1, where n is the principal quantum number of the outermost shell?

8.71 How do the chemical properties of oxides change from left to right across a period? From top to bottom within a particular group?

8.72 Write balanced equations for the reactions between each of the following oxides and water: (a) Li_2O, (b) CaO, (c) SO_3.

8.73 Write formulas for and name the binary hydrogen compounds of the second-period elements (Li to F). Describe how the physical and chemical properties of these compounds change from left to right across the period.

8.74 Which oxide is more basic, MgO or BaO? Why?

Additional Problems

8.75 State whether each of the following properties of the representative elements generally increases or decreases (a) from left to right across a period and

(b) from top to bottom within a group: metallic character, atomic size, ionization energy, acidity of oxides.

8.76 With reference to the periodic table, name (a) a halogen element in the fourth period, (b) an element similar to phosphorus in chemical properties, (c) the most reactive metal in the fifth period, (d) an element that has an atomic number smaller than 20 and is similar to strontium.

8.77 Write equations representing the following processes:
(a) The electron affinity of S^-
(b) The third ionization energy of titanium
(c) The electron affinity of Mg^{2+}
(d) The ionization energy of O^{2-}

8.78 List all the common ions of representative elements and transition metals that are isoelectronic with Ar.

8.79 Write the empirical (or molecular) formulas of compounds that the elements in the third period (sodium to chlorine) should form with (a) molecular oxygen and (b) molecular chlorine. In each case indicate whether you would expect the compound to be ionic or molecular in character.

8.80 Element M is a shiny and highly reactive metal (melting point 63°C), and element X is a highly reactive nonmetal (melting point −7.2°C). They react to form a compound with the empirical formula MX, a colorless, brittle white solid that melts at 734°C. When dissolved in water or when in the molten state, the substance conducts electricity. When chlorine gas is bubbled through an aqueous solution containing MX, a reddish-brown liquid appears and Cl^- ions are formed. From these observations, identify M and X. (You may need to consult a handbook of chemistry for the melting-point values.)

8.81 Match each of the elements on the right with its description on the left:

(a) A dark-red liquid	Calcium (Ca)
(b) A colorless gas that burns in oxygen gas	Gold (Au)
(c) A reactive metal that attacks water	Hydrogen (H_2)
(d) A shiny metal that is used in jewelry	Argon (Ar)
(e) An inert gas	Bromine (Br_2)

8.82 Arrange the following species in isoelectronic pairs: O^+, Ar, S^{2-}, Ne, Zn, Cs^+, N^{3-}, As^{3+}, N, Xe.

8.83 In which of the following are the species written in decreasing order by size of radius? (a) Be, Mg, Ba; (b) N^{3-}, O^{2-}, F^-; (c) Tl^{3+}, Tl^{2+}, Tl^+.

8.84 Which of the following properties show a clear periodic variation: (a) first ionization energy, (b) molar mass of the elements, (c) number of isotopes of an element, (d) atomic radius.

8.85 When carbon dioxide is bubbled through a clear calcium hydroxide solution, the solution appears milky. Write an equation for the reaction and explain how this reaction illustrates that CO_2 is an acidic oxide.

8.86 You are given four substances: a fuming red liquid, a dark metallic-looking solid, a pale-yellow gas, and a yellow-green gas that attacks glass. You are told that these substances are the first four members of Group 17, the halogens. Name each one.

8.87 Calculate the change in energy for the following processes:
(a) $Na(g) + Cl(g) \longrightarrow Na^+(g) + Cl^-(g)$
(b) $Ca(g) + 2Br(g) \longrightarrow Ca^{2+}(g) + 2Br^-(g)$

8.88 Calculate the change in energy for the following processes:
(a) $Mg(g) + 2F(g) \longrightarrow Mg^{2+}(g) + 2F^-(g)$
(b) $2Al(g) + 3O(g) \longrightarrow 2Al^{3+}(g) + 3O^{2-}(g)$
The electron affinity of O^- is −844 kJ/mol.

8.89 For each pair of elements listed, give three properties that show their chemical similarity: (a) sodium and potassium, (b) chlorine and bromine.

8.90 Name the element that forms compounds, under appropriate conditions, with every other element in the periodic table except He, Ne, and Ar.

8.91 Explain why the first electron affinity of sulfur is 200 kJ/mol but the second electron affinity is −649 kJ/mol.

8.92 The H^- ion and the He atom have two 1*s* electrons each. Which of the two species is larger? Explain.

8.93 Predict the products of the following oxides with water: Na_2O, BaO, CO_2, N_2O_5, P_4O_{10}, SO_3. Write an equation for each of the reactions. Specify whether the oxides are acidic, basic, or amphoteric.

8.94 Write the formulas and names of the oxides of the second-period elements (Li to N). Identify the oxides as acidic, basic, or amphoteric.

8.95 State whether each of the following elements is a gas, a liquid, or a solid under atmospheric conditions. Also state whether it exists in the elemental form as atoms, as molecules, or as a three-dimensional network: Mg, Cl, Si, Kr, O, I, Hg, Br.

8.96 What factors account for the unique nature of hydrogen?

8.97 The air in a human-operated spacecraft or submarine needs to be purified of exhaled carbon dioxide. Write equations for the reactions between carbon dioxide and (a) lithium oxide (Li_2O), (b) sodium peroxide (Na_2O_2), and (c) potassium superoxide (KO_2).

8.98 The formula for calculating the energies of an electron in a hydrogenlike ion is given in Problem 8.57. This equation cannot be applied to many-electron atoms. One way to modify it for the more complex atoms is to replace Z with $(Z - \sigma)$, where Z is the atomic number and σ is a positive dimensionless quantity called the shielding constant. Consider the helium atom as an example. The physical significance of σ is that it represents the extent of shielding that the two 1*s* electrons exert on each other. Thus, the quantity $(Z - \sigma)$ is appropriately called the "effective nuclear charge." Calculate the value of σ if the first ionization energy of helium is 3.94×10^{-18} J per atom. (Ignore the minus sign in the given equation in your calculation.)

8.99 Why do noble gases have negative electron affinity values?

8.100 The atomic radius of K is 227 pm and that of K^+ is 133 pm. Calculate the percent decrease in volume that occurs when $K(g)$ is converted to $K^+(g)$. [The volume of a sphere is $(\frac{4}{3})\pi r^3$, where r is the radius of the sphere.]

8.101 The atomic radius of F is 72 pm and that of F^- is 133 pm. Calculate the percent increase in volume that occurs when $F(g)$ is converted to $F^-(g)$. (See Problem 8.100 for the volume of a sphere.)

8.102 A technique called photoelectron spectroscopy is used to measure the ionization energy of atoms. A sample is irradiated with UV light, and electrons are ejected from the valence shell. The kinetic energies of the ejected electrons are measured. Because the energy of the UV photon and the kinetic energy of the ejected electron are known, we can write

$$h\nu = IE + \frac{1}{2}mu^2$$

where ν is the frequency of the UV light, and m and u are the mass and velocity of the electron, respectively. In one experiment the kinetic energy of the ejected electron from potassium is found to be 5.34×10^{-19} J using a UV source of wavelength 162 nm. Calculate the ionization energy of potassium. How can you be sure that this ionization energy corresponds to the electron in the valence shell (that is, the most loosely held electron)?

8.103 Referring to the Chemistry in Action essay "Discovery of the Noble Gases" in Section 8.6, answer the following questions: (a) Why did it take so long to discover the first noble gas (argon) on Earth? (b) Once argon had been discovered, why did it take relatively little time to discover the rest of the noble gases? (c) Why was helium not isolated by the fractional distillation of liquid air?

8.104 The energy needed for the following process is 1.96×10^4 kJ/mol:

$$Li(g) \longrightarrow Li^{3+}(g) + 3e^-$$

If the first ionization energy of lithium is 520 kJ/mol, calculate the second ionization energy of lithium, that is, the energy required for the process

$$Li(g) \longrightarrow Li^{3+}(g) + 3e^-$$

(*Hint:* You need the equation in Problem 8.57.)

8.105 An element X reacts with hydrogen gas at 200°C to form compound Y. When Y is heated to a higher temperature, it decomposes to the element X and hydrogen gas in the ratio of 559 mL of H_2 (measured at STP) for 1.00 g of X reacted. X also combines with chlorine to form a compound Z, which contains 63.89 percent by mass of chlorine. Deduce the identity of X.

8.106 A student is given samples of three elements, X, Y, and Z, which could be an alkali metal, a member of Group 14, and a member of Group 15. She makes the following observations: Element X has a metallic luster and conducts electricity. It reacts slowly with hydrochloric acid to produce hydrogen gas. Element Y is a light-yellow solid that does not conduct electricity. Element Z has a metallic luster and conducts electricity. When exposed to air, it slowly forms a white powder. A solution of the white powder in water is basic. What can you conclude about the elements from these observations?

8.107 Identify the ions whose orbital diagrams for the valence electrons are shown. The charges of the ions are (a) 1+, (b) 3+, (c) 4+, (d) 2+.

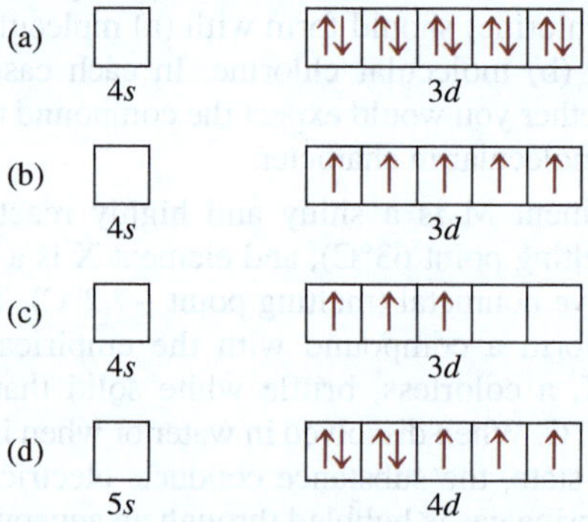

8.108 What is the electron affinity of the Na^+ ion?

8.109 The ionization energies of sodium (in kJ/mol), starting with the first and ending with the eleventh, are 495.9, 4560, 6900, 9540, 13,400, 16,600, 20,120, 25,490, 28,930, 141,360, 170,000. Plot the log of ionization energy (y-axis) versus the number of ionization (x-axis); for example, log 495.9 is plotted versus 1 (labeled IE_1, the first ionization energy), log 4560 is plotted versus 2 (labeled IE_2, the second ionization energy), and so on. (a) Label IE_1 through IE_{11} with the electrons in orbitals such as 1s, 2s, 2p, and 3s. (b) What can you deduce about electron shells from the breaks in the curve?

8.110 Experimentally, the electron affinity of an element can be determined by using a laser light to ionize the anion of the element in the gas phase:

$$X^-(g) + h\nu \longrightarrow X(g) + e^-$$

Referring to Table 8.3, calculate the photon wavelength (in nanometers) corresponding to the electron affinity for chlorine. In what region of the electromagnetic spectrum does this wavelength fall?

8.111 Explain, in terms of their electron configurations, why Fe^{2+} is more easily oxidized to Fe^{3+} than Mn^{2+} is to Mn^{3+}.

8.112 The standard enthalpy of atomization of an element is the energy required to convert one mole of an element in its most stable form at 25°C to one mole of monatomic gas. Given that the standard enthalpy of atomization for sodium is 108.4 kJ/mol, calculate the energy in kilojoules required to convert one mole of sodium metal at 25°C to one mole of gaseous Na^+ ions.

8.113 Write the formulas and names of the hydrides of the following second-period elements: Li, C, N, O, F. Predict their reactions with water.

8.114 Based on knowledge of the electronic configuration of titanium, state which of the following compounds of titanium is unlikely to exist: K_3TiF_6, $K_2Ti_2O_5$, $TiCl_3$, K_2TiO_4, K_2TiF_6.

8.115 Name an element in Group 1 or Group 2 that is an important constituent of each of the following substances: (a) remedy for acid indigestion, (b) coolant in nuclear reactors, (c) Epsom salt, (d) baking powder, (e) gunpowder, (f) a light alloy, (g) fertilizer that also neutralizes acid rain, (h) cement, (i) grit for icy roads. You may need to ask your instructor about some of the items.

8.116 In halogen displacement reactions a halogen element can be generated by oxidizing its anions with a halogen element that lies above it in the periodic table. This means that there is no way to prepare elemental fluorine, because it is the first member of Group 17. Indeed, for years the only way to prepare elemental fluorine was to oxidize F^- ions by electrolytic means. Then, in 1986, a chemist reported that by reacting potassium hexafluoromanganate(IV) (K_2MnF_6) with antimony pentafluoride (SbF_5) at 150°C, he had generated elemental fluorine. Balance the following equation representing the reaction:

$$K_2MnF_6 + SbF_5 \longrightarrow KSbF_6 + MnF_3 + F_2$$

8.117 Write a balanced equation for the preparation of (a) molecular oxygen, (b) ammonia, (c) carbon dioxide, (d) molecular hydrogen, (e) calcium oxide. Indicate the physical state of the reactants and products in each equation.

8.118 Write chemical formulas for oxides of nitrogen with the following oxidation numbers: +1, +2, +3, +4, +5. (*Hint:* There are *two* oxides of nitrogen with +4 oxidation number.)

8.119 Most transition metal ions are colored. For example, a solution of $CuSO_4$ is blue. How would you show that the blue color is due to the hydrated Cu^{2+} ions and not the SO_4^{2-} ions?

8.120 In general, atomic radius and ionization energy have opposite periodic trends. Why?

8.121 Explain why the electron affinity of nitrogen is approximately zero, while the elements on either side, carbon and oxygen, have substantial positive electron affinities.

8.122 Consider the halogens chlorine, bromine, and iodine. The melting point and boiling point of chlorine are −101.0°C and −34.6°C while those of iodine are 113.5°C and 184.4°C, respectively. Thus, chlorine is a gas and iodine is a solid under room conditions. Estimate the melting point and boiling point of bromine. Compare your values with those from a handbook of chemistry.

8.123 Write a balanced equation that predicts the reaction of rubidium (Rb) with (a) $H_2O(l)$, (b) $Cl_2(g)$, (c) $H_2(g)$.

8.124 The successive *IE* of the first four electrons of a representative element are 738.1 kJ/mol, 1450 kJ/mol, 7730 kJ/mol, and 10,500 kJ/mol. Characterize the element according to the periodic group.

8.125 Little is known of the chemistry of astatine, the last member of Group 17. Describe the physical characteristics that you would expect this halogen to have. Predict the products of the reaction between sodium astatide (NaAt) and sulfuric acid. (*Hint:* Sulfuric acid is an oxidizing agent.)

8.126 As discussed in the chapter, the atomic mass of argon is greater than that of potassium. This observation created a problem in the early development of the periodic table because it meant that argon should be placed after potassium. (a) How was this difficulty resolved? (b) From the following data, calculate the average atomic masses of argon and potassium: Ar-36 (35.9675 amu; 0.337%), Ar-38 (37.9627 amu; 0.063%), Ar-40 (39.9624 amu; 99.60%); K-39 (38.9637 amu; 93.258%), K-40 (39.9640 amu; 0.0117%), K-41 (40.9618 amu; 6.730%).

8.127 Calculate the maximum wavelength of light (in nanometers) required to ionize a single sodium atom.

8.128 Predict the atomic number and ground-state electron configuration of the next member of the alkali metals after francium.

8.129 Why do elements that have high ionization energies also have more positive electron affinities? Which group of elements would be an exception to this generalization?

8.130 The first four ionization energies of an element are approximately 579 kJ/mol, 1980 kJ/mol, 2963 kJ/mol, and 6180 kJ/mol. To which periodic group does this element belong?

8.131 Some chemists think that helium should properly be called "helon." Why? What does the ending in helium (*-ium*) suggest?

8.132 (a) The formula of the simplest hydrocarbon is CH_4 (methane). Predict the formulas of the simplest compounds formed between hydrogen and the following elements: silicon, germanium, tin, and lead. (b) Sodium hydride (NaH) is an ionic compound. Would you expect rubidium hydride (RbH) to be more or less ionic than NaH? (c) Predict the reaction between radium (Ra) and water. (d) When exposed to air, aluminum forms a tenacious oxide (Al_2O_3) coating that protects the metal from corrosion. Which metal in Group 2 would you expect to exhibit similar properties? Why?

8.133 Give equations to show that molecular hydrogen can act both as a reducing agent and an oxidizing agent.

8.134 Both Mg^{2+} and Ca^{2+} are important biological ions. One of their functions is to bind to the phosphate groups of ATP molecules or amino acids of proteins. For Group 2A metals in general, the tendency for binding to the anions increases in the order $Ba^{2+} < Sr^{2+} < Ca^{2+} < Mg^{2+}$. Explain the trend.

8.135 Match each of the elements on the right with its description on the left:

(a) A pale yellow gas that reacts with water.	Nitrogen (N_2)
(b) A soft metal that reacts with water to produce hydrogen.	Boron (B)
(c) A metalloid that is hard and has a high melting point.	Aluminum (Al)
(d) A colorless, odorless gas.	Fluorine (F_2)
(e) A metal that is more reactive than iron, but does not corrode in air.	Sodium (Na)

8.136 Write an account on the importance of the periodic table. Pay particular attention to the significance of the position of an element in the table and how the position relates to the chemical and physical properties of the element.

8.137 On the same graph, plot the effective nuclear charge (see Section 8.3) and atomic radius (see Figure 8.5) versus atomic number for the second-period elements Li to Ne. Comment on the trends.

8.138 One allotropic form of an element X is a colorless crystalline solid. The reaction of X with an excess amount of oxygen produces a colorless gas. This gas dissolves in water to yield an acidic solution. Choose one of the following elements that matches X: (a) sulfur, (b) phosphorus, (c) carbon, (d) boron, (e) silicon.

8.139 When magnesium metal is burned in air, it forms two products A and B. A reacts with water to form a basic solution. B reacts with water to form a similar solution as that of A plus a gas with a pungent odor. Identify A and B and write equations for the reactions. (*Hint:* See Chemistry in Action essay "Discovery of the Noble Gases" in Section 8.6.)

8.140 The ionization energy of a certain element is 412 kJ/mol. When the atoms of this element are in the first excited state, however, the ionization energy is only 126 kJ/mol. Based on this information, calculate the wavelength of light emitted in a transition from the first excited state to the ground state.

8.141 Use your knowledge of thermochemistry to calculate the ΔH for the following processes: (a) $Cl^-(g) \longrightarrow Cl^+(g) + 2e^-$, (b) $K^+(g) + 2e^- \longrightarrow K^-(g)$.

8.142 Referring to Table 8.2, explain why the first ionization energy of helium is less than twice the ionization energy of hydrogen, but the second ionization energy of helium is greater than twice the ionization energy of hydrogen. [*Hint:* According to Coulomb's law, the energy between two charges Q_1 and Q_2 separated by distance r is proportional to (Q_1Q_2/r).]

8.143 As mentioned in the Chemistry in Action essay "Chemical Fertilizers" in Section 3.10, ammonium nitrate (NH_4NO_3) is the most important nitrogen-containing fertilizer in the world. Describe how you would prepare this compound, given only air and water as the starting materials. You may have any device at your disposal for this task.

8.144 One way to estimate the effective charge (Z_{eff}) of a many-electron atom is to use the equation $IE_1 = (1312\ \text{kJ/mol})(Z_{eff}^2/n^2)$, where IE_1 is the first ionization energy and n is the principal quantum number of the shell in which the electron resides. Use this equation to calculate the effective charges of Li, Na, and K. Also calculate Z_{eff}/n for each metal. Comment on your results.

8.145 To prevent the formation of oxides, peroxides, and superoxides, alkali metals are sometimes stored in an inert atmosphere. Which of the following gases should not be used for lithium: Ne, Ar, N_2, Kr? Explain. (*Hint:* As mentioned in the chapter, Li and Mg exhibit a diagonal relationship. Compare the common compounds of these two elements.)

8.146 Describe the biological role of the elements in the human body shown in the following periodic table. (You may need to do research at websites such as www.webelements.com.)

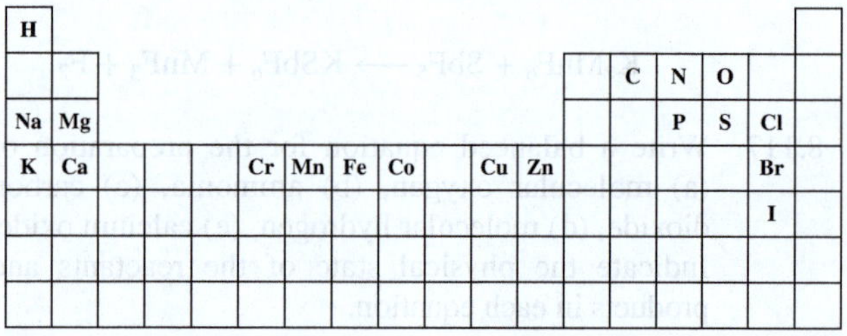

8.147 Recent theoretical calculations suggest that astatine may be a monoatomic metal rather than a diatomic molecule like the other halogens. (a) Rationalize this prediction based on astatine's position in the periodic table. (b) The energy required to remove an electron from one At atom was determined by laser ionization to be 9.3175 eV. Given that 1 eV = 1.602×10^{-19} J, calculate the first ionization energy of astatine in kJ/mol. (c) Comment on whether or not the following first ionization energies support your answer to part (a): Pb, 715.6 kJ/mol; Bi, 702.9 kJ/mol; Po, 811.8 kJ/mol; Rn, 1037 kJ/mol.

Interpreting, Modeling, & Estimating

8.148 Consider the first 18 elements from hydrogen to argon. Would you expect the atoms of half of them to be diamagnetic and half of them to be paramagnetic? Explain.

8.149 Compare the work function for cesium (206 kJ/mol) with its first ionization energy (376 kJ/mol). Explain the difference.

8.150 The only confirmed compound of radon is radon difluoride, RnF_2. One reason that it is difficult to study the chemistry of radon is that all isotopes of radon are radioactive so it is dangerous to handle the substance. Can you suggest another reason why there are so few known radon compounds? (*Hint:* Radioactive decays are exothermic processes.)

8.151 Arsenic (As) is not an essential element for the human body. (a) Based on its position in the periodic table, suggest a reason for its toxicity. (b) When arsenic enters a person's body, it quickly shows up in the follicle of thc growing hair. This action has enabled detectives to solve many murder mysteries by analyzing a victim's hair. Where else might one look for the accumulation of the element if arsenic poisoning is suspected?

8.152 The boiling points of neon and krypton are −245.9°C and −152.9°C, respectively. Using these data, estimate the boiling point of argon.

8.153 Using the following boiling-point data, estimate the boiling point of francium:

Metal	Li	Na	K	Rb	Cs	Fr
B.p. (°C)	1347	882.9	774	688	678.4	?

8.154 The energy gap between the 6*s* and 5*d* levels in gold is 4.32×10^{-19} J. Based on this information, predict the perceived color of gold vapor. (*Hint:* You need to be familiar with the notion of complementary color; see Figure 23.15.)

8.155 Calculate the volume of 1 mole of K atoms (scc Figure 8.5) and compare the result by using the density of K (0.856 g/cm^3). Account for the difference.

Answers to Practice Exercises

8.1 (a) $1s^22s^22p^63s^23p^64s^2$. (b) It is a representative element. (c) Diamagnetic. **8.2** Li > Be > C. **8.3** (a) Li^+. (b) Au^{3+}. (c) N^{3-}. **8.4** (a) N. (b) Mg. **8.5** No. **8.6** (a) Amphoteric. (b) Acidic. (c) Basic.

Answers to Review of Concepts & Facts

8.2.1 (a) Strontium. (b) Phosphorus. (c) Iron. **8.2.2** [Ar]$3d^3$. **8.2.3** P^{3+}. **8.3.1** (a) Ba > Be. (b) Al > S. (c) Same size. Number of neutrons has no effect on atomic radius. **8.3.2** $Ca^{2+} < K^+ < Ar < Cl^- < S^{2-}$. **8.3.3** In decreasing order of the sphere sizes: $S^{2-} > F^- > Na^+ > Mg^{2+}$. **8.4.1** Rb < Al < F. **8.4.2** Blue curve: K; green curve: Al; red curve: Mg. **8.5.1** Na < Si < Cl. **8.5.2** Electrons can be removed from atoms successively because the cations formed are stable. (The remaining electrons are held more tightly by the nucleus.) On the other hand, adding electrons to an atom results in an increasing electrostatic repulsion in the anions, leading to instability. For this reason, it is difficult and often impossible to carry electron affinity measurements beyond the first stage in most cases. **8.6.1** (a).

CHAPTER

9

Compounds and Bonding

Electrostatic potential maps indicate regions of high and low electron density in molecules.
Jason Overby

CHAPTER OUTLINE

Why do atoms of different elements react? What are the forces that hold atoms together in molecules and ionic compounds? What shapes do they assume? These are some of the questions addressed in this chapter and in Chapter 10. We begin by looking at the two types of bonds—ionic and covalent—and the forces that stabilize them.

9.1 Lewis Dot Symbols

Learning Objective

- Determine Lewis dot symbols of elements.

The development of the periodic table and concept of electron configuration gave chemists a rationale for molecule and compound formation. This explanation, formulated by Gilbert Lewis,† is that atoms combine to achieve a more stable electron configuration. Maximum stability results when an atom is isoelectronic with a noble gas.

When atoms interact to form a chemical bond, only their outer regions are in contact. For this reason, when we study chemical bonding, we are concerned primarily with the valence electrons of the atoms. To keep track of valence electrons in a chemical reaction, and to make sure that the total number of electrons does not change, chemists use a system of dots devised by Lewis called Lewis dot symbols. A ***Lewis dot symbol*** *consists of the symbol of an element and one dot for each valence electron in an atom of the element.* Figure 9.1 shows the Lewis dot symbols for the representative elements and the noble gases. Note that, except for helium, the number of valence electrons each atom has is the same as the group number of the element. For example, Li is a Group 1 element and has one dot for one valence electron; Be, a Group 2 element, has two valence electrons (two dots); B, a Group 13 element, has three valence electrons (three dots; recall the convenient method for determining the number of valence electrons, group number minus 10). Elements in the same group have similar outer electron configurations and hence similar

Student Hot Spot

Student data indicate you may struggle with Lewis dot symbols. Access your eBook for additional Learning Resources on this topic.

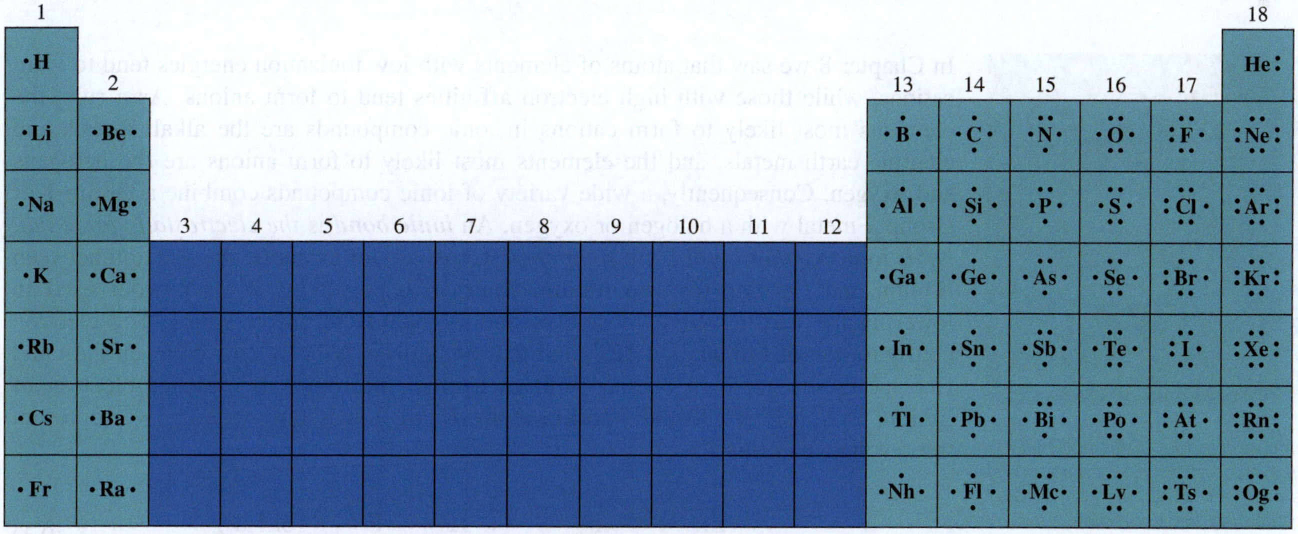

Figure 9.1 *Lewis dot symbols for the representative elements and the noble gases. The number of unpaired dots corresponds to the number of bonds an atom of the element can form in a molecular compound without expanding the octet (see Section 9.4).*

†Gilbert Newton Lewis (1875–1946). American chemist. Lewis made many important contributions in the areas of chemical bonding, thermodynamics, acids and bases, and spectroscopy. Despite the significance of Lewis's work, he was never awarded a Nobel Prize.

Lewis dot symbols. The transition metals, lanthanides, and actinides all have incompletely filled inner shells, and in general we cannot write simple Lewis dot symbols for them.

In this chapter, we will learn to use electron configurations and the periodic table to predict the type of bond atoms will form, as well as the number of bonds an atom of a particular element can form and the stability of the product.

Summary of Concepts & Facts

- A Lewis dot symbol shows the number of valence electrons possessed by an atom of a given element. Lewis dot symbols are useful mainly for the representative elements.

Review of Concepts & Facts

9.1.1 What is the maximum number of dots that can be drawn around the atom of a representative element?

9.1.2 Using only the periodic table, determine the Lewis dot symbol for a selenium (Se) atom.

9.1.3 To what group on the periodic table does element X belong if the Lewis dot symbol is $\cdot\ddot{\text{X}}\cdot$?

9.2 The Ionic Bond

Learning Objective

- Summarize ionic bonding and provide examples of compounds containing ionic bonds.

Lithium fluoride. Industrially, LiF (like most other ionic compounds) is obtained by purifying minerals containing the compound.
Ken Karp/McGraw Hill

In Chapter 8 we saw that atoms of elements with low ionization energies tend to form cations, while those with high electron affinities tend to form anions. As a rule, the elements most likely to form cations in ionic compounds are the alkali metals and alkaline earth metals, and the elements most likely to form anions are the halogens and oxygen. Consequently, a wide variety of ionic compounds combine a Group 1 or Group 2 metal with a halogen or oxygen. An ***ionic bond*** is *the electrostatic force that holds ions together in an ionic compound.* Consider, for example, the reaction between lithium and fluorine to form lithium fluoride, a poisonous white powder used in lowering the melting point of solders and in manufacturing ceramics. The electron configuration of lithium is $1s^22s^1$, and that of fluorine is $1s^22s^22p^5$. When lithium and fluorine atoms come in contact with each other, the outer $2s^1$ valence electron of lithium is transferred to the fluorine atom. Using Lewis dot symbols, we represent the reaction like this:

$$\begin{array}{ccccc} \cdot\text{Li} & + & :\ddot{\underset{\cdot\cdot}{\text{F}}}\cdot & \longrightarrow & \text{Li}^+ \quad :\ddot{\underset{\cdot\cdot}{\text{F}}}:^- \quad \text{(or LiF)} \\ 1s^22s^1 & & 1s^22s^22p^5 & & 1s^2 \quad 1s^22s^22p^6 \end{array} \tag{9.1}$$

For convenience, imagine that this reaction occurs in separate steps—first the ionization of Li:

$$\cdot\text{Li} \longrightarrow \text{Li}^+ + e^-$$

and then the acceptance of an electron by F:

$$:\ddot{\underset{\cdot\cdot}{F}}\cdot + e^- \longrightarrow\ :\ddot{\underset{\cdot\cdot}{F}}:^-$$

Next, imagine the two separate ions joining to form a LiF unit:

$$Li^+ + :\ddot{\underset{\cdot\cdot}{F}}:^- \longrightarrow Li^+ :\ddot{\underset{\cdot\cdot}{F}}:^-$$

Note that the sum of these three equations is

$$Li + :\ddot{\underset{\cdot\cdot}{F}}\cdot \longrightarrow Li^+ :\ddot{\underset{\cdot\cdot}{F}}:^-$$

which is the same as Equation (9.1). The ionic bond in LiF is the electrostatic attraction between the positively charged lithium ion and the negatively charged fluoride ion. The compound itself is electrically neutral. Remember that we normally write the empirical formulas of ionic compounds without showing the charges. The + and − are shown here to emphasize the transfer of electrons.

Many other common reactions lead to the formation of ionic bonds. For instance, calcium burns in oxygen to form calcium oxide:

$$2Ca(s) + O_2(g) \longrightarrow 2CaO(s)$$

▶▶▶ Video: Reactions of Magnesium and Oxygen

Assuming that the diatomic O_2 molecule first splits into separate oxygen atoms (we will look at the energetics of this step later), we can represent the reaction with Lewis symbols:

$$\underset{[Ar]4s^2}{\cdot Ca\cdot} + \underset{1s^22s^22p^4}{\cdot\ddot{\underset{\cdot\cdot}{O}}\cdot} \longrightarrow \underset{[Ar]}{Ca^{2+}}\quad \underset{[Ne]}{:\ddot{\underset{\cdot\cdot}{O}}:^{2-}}$$

There is a transfer of two electrons from the calcium atom to the oxygen atom. Note that the resulting calcium ion (Ca^{2+}) has the argon electron configuration, the oxide ion (O^{2-}) is isoelectronic with neon, and the compound (CaO) is electrically neutral.

In many cases, the cation and the anion in a compound do not carry the same charges. For instance, when lithium burns in air to form lithium oxide (Li_2O), the balanced equation is

$$4Li(s) + O_2(g) \longrightarrow 2Li_2O(s)$$

Using Lewis dot symbols, we write

$$\underset{1s^22s^1}{2\ \cdot Li} + \underset{1s^22s^22p^4}{\cdot\ddot{\underset{\cdot\cdot}{O}}\cdot} \longrightarrow \underset{[He]}{2Li^+}\quad \underset{[Ne]}{:\ddot{\underset{\cdot\cdot}{O}}:^{2-}}\ \text{(or } Li_2O)$$

In this process, the oxygen atom receives two electrons (one from each of the two lithium atoms) to form the oxide ion. The Li^+ ion is isoelectronic with helium.

When magnesium reacts with nitrogen at elevated temperatures, a white solid compound, magnesium nitride (Mg_3N_2), forms:

$$3Mg(s) + N_2(g) \longrightarrow Mg_3N_2(s)$$

or

$$\underset{[Ne]3s^2}{3\ \cdot Mg\cdot} + \underset{1s^22s^22p^3}{2\ \cdot\dot{\underset{\cdot}{N}}\cdot} \longrightarrow \underset{[Ne]}{3Mg^{2+}}\quad \underset{[Ne]}{2\ :\ddot{\underset{\cdot\cdot}{N}}:^{3-}}\ \text{(or } Mg_3N_2)$$

The reaction involves the transfer of six electrons (two from each Mg atom) to two nitrogen atoms. The resulting magnesium ion (Mg^{2+}) and the nitride ion (N^{3-}) are both isoelectronic with neon. Because there are three +2 ions and two −3 ions, the charges balance and the compound is electrically neutral.

In Example 9.1, we apply the Lewis dot symbols to study the formation of an ionic compound.

The mineral corundum (Al_2O_3).
mirecca/iStock/Getty Images

Example 9.1

Aluminum oxide, obtained from the mineral corundum, is used primarily for the production of aluminum metal. Use Lewis dot symbols to show the formation of aluminum oxide (Al_2O_3).

Strategy We use electroneutrality as our guide in writing formulas for ionic compounds; that is, the total positive charges on the cations must be equal to the total negative charges on the anions.

Solution According to Figure 9.1, the Lewis dot symbols of Al and O are

$$\cdot\dot{\text{Al}}\cdot \qquad \cdot\ddot{\underset{\cdot\cdot}{\text{O}}}\cdot$$

Because aluminum tends to form the cation (Al^{3+}) and oxygen the anion (O^{2-}) in ionic compounds, the transfer of electrons is from Al to O. There are three valence electrons in each Al atom; each O atom needs two electrons to form the O^{2-} ion, which is isoelectronic with neon. Thus, the simplest neutralizing ratio of Al^{3+} to O^{2-} is 2:3; two Al^{3+} ions have a total charge of +6, and three O^{2-} ions have a total charge of −6. So the empirical formula of aluminum oxide is Al_2O_3, and the reaction is

$$2\ \cdot\dot{\text{Al}}\cdot \ +\ 3\ \cdot\ddot{\underset{\cdot\cdot}{\text{O}}}\cdot \ \longrightarrow\ 2\text{Al}^{3+}\quad 3\ :\ddot{\underset{\cdot\cdot}{\text{O}}}:^{2-}\quad (\text{or } Al_2O_3)$$

$$[\text{Ne}]3s^23p^1 \qquad 1s^22s^22p^4 \qquad [\text{Ne}] \qquad [\text{Ne}]$$

Check Make sure that the number of valence electrons (24) is the same on both sides of the equation. Are the subscripts in Al_2O_3 reduced to the smallest possible whole numbers?

Practice Exercise Use Lewis dot symbols to represent the formation of barium hydride.

Similar problems: 9.17, 9.18.

Summary of Concepts & Facts

- The elements most likely to form ionic compounds have low ionization energies (such as the alkali metals and the alkaline earth metals, which form cations) or high electron affinities (such as the halogens and oxygen, which form anions).

Review of Concepts & Facts

9.2.1 Use Lewis dot symbols to represent the formation of rubidium sulfide.

9.3 Lattice Energy of Ionic Compounds

Learning Objectives

- Solve for lattice energies using a Born-Haber cycle.
- Rank lattice energies of ionic compounds.

We can predict which elements are likely to form ionic compounds based on ionization energy and electron affinity, but how do we evaluate the stability of an ionic compound? Ionization energy and electron affinity are defined for processes occurring in the gas phase, but at 1 atm and 25°C all ionic compounds are solids. The solid state is a very different environment because each cation in a solid is surrounded by a specific number of anions, and vice versa. Thus, the overall stability of a solid ionic compound depends on the interactions of all these ions and not merely on the interaction of a single cation with a single anion. A quantitative measure of the stability of any ionic solid is its *lattice energy (U),* defined as *the energy required to completely separate one mole of a solid ionic compound into gaseous ions* (see Section 6.7). Note that lattice energy and internal energy share the same symbol (U).

Lattice Energy

Table 9.1 lists the lattice energies and the melting points of several common ionic compounds. There is a rough correlation between lattice energy and melting point. The larger the lattice energy, the more stable the solid and the more tightly held the ions. It takes more energy to melt such a solid, and so the solid has a higher melting point than one with a smaller lattice energy. Note that $MgCl_2$, Na_2O, MgO, CaO, and ScN have unusually high lattice energies. The first of these ionic compounds has a doubly charged cation (Mg^{2+}) and the second a doubly charged anion (O^{2-}); in the third and fourth compounds there is an interaction between two doubly charged species (Mg^{2+} and O^{2-}). The final compound has strong interactions between two triple-charged ions (Sc^{3+} and N^{3-}). The coulombic attractions between two triply charged

Table 9.1 Lattice Energies and Melting Points of Some Alkali Metal and Alkaline Earth Metal Halides and Oxides

Compound	Lattice Energy (kJ/mol)	Melting Point (°C)
LiF	1017	845
LiCl	860	610
LiBr	787	550
LiI	732	450
NaCl	787	801
NaBr	736	750
NaI	686	662
KCl	699	772
KBr	689	735
KI	632	680
$MgCl_2$	2527	714
Na_2O	2570	Sub*
MgO	3890	2800
CaO	3414	2580
ScN	7547	>3000

*Na_2O sublimes at 1275°C.

species, two doubly charged species, or between a doubly charged ion and a singly charged ion, are much stronger than those between singly charged anions and cations.

The Born-Haber Cycle for Determining Lattice Energies

Lattice energy cannot be measured directly. However, if we know the structure and composition of an ionic compound, we can calculate the compound's lattice energy by using ***Coulomb's*[†] *law,*** which states that *the potential energy (E) between two ions is directly proportional to the product of their charges and inversely proportional to the distance of separation between them.* For a single Li^+ ion and a single F^- ion separated by distance r, the potential energy of the system is given by

$$E \propto \frac{Q_{Li^+}Q_{F^-}}{r}$$

$$= k\frac{Q_{Li^+}Q_{F^-}}{r} \qquad (9.2)$$

Student Hot Spot

Student data indicate you may struggle with lattice energy and ionic bond strength. Access your eBook for additional Learning Resources on this topic.

where Q_{Li^+} and Q_{F^-} are the charges on the Li^+ and F^- ions and k is the proportionality constant. Because Q_{Li^+} is positive and Q_{F^-} is negative, E is a negative quantity, and the formation of an ionic bond from Li^+ and F^- is an exothermic process. Consequently, energy must be supplied to reverse the process (in other words, the lattice energy of LiF is positive), and so a bonded pair of Li^+ and F^- ions is more stable than separate Li^+ and F^- ions.

We can also determine lattice energy indirectly, by assuming that the formation of an ionic compound takes place in a series of steps. This procedure, known as the ***Born-Haber cycle,*** *relates lattice energies of ionic compounds to ionization energies, electron affinities, and other atomic and molecular properties.* It is based on Hess's law (see Section 6.6). Developed by Max Born[‡] and Fritz Haber,[§] the Born-Haber cycle defines the various steps that precede the formation of an ionic solid. We will illustrate its use to find the lattice energy of lithium fluoride.

Consider the reaction between lithium and fluorine:

$$Li(s) + \tfrac{1}{2}F_2(g) \longrightarrow LiF(s)$$

The standard enthalpy change for this reaction is −594.1 kJ/mol. (Because the reactants and product are in their standard states, that is, at 1 atm, the enthalpy change is also the standard enthalpy of formation for LiF.) Keeping in mind that the sum of enthalpy changes for the steps is equal to the enthalpy change for the overall reaction (−594.1 kJ/mol), we can trace the formation of LiF from its elements through five separate steps. The process may not occur exactly this way, but this pathway enables us to analyze the energy changes of ionic compound formation, with the application of Hess's law.

1. Convert solid lithium to lithium vapor (the direct conversion of a solid to a gas is called sublimation):

$$Li(s) \longrightarrow Li(g) \qquad \Delta H_1^\circ = 155.2 \text{ kJ/mol}$$

[†]Charles Augustin de Coulomb (1736–1806). French physicist. Coulomb did research in electricity and magnetism and applied Newton's inverse square law to electricity. He also invented a torsion balance.

[‡]Max Born (1882–1970). German physicist. Born was one of the founders of modern physics. His work covered a wide range of topics. He received the Nobel Prize in Physics in 1954 for his interpretation of the wave function for particles.

[§]Fritz Haber (1868–1934). German chemist. Haber's process for synthesizing ammonia from atmospheric nitrogen kept Germany supplied with nitrates for explosives during World War I. He also did work on gas warfare. In 1918 Haber received the Nobel Prize in Chemistry.

The energy of sublimation for lithium is 155.2 kJ/mol.

2. Dissociate $\frac{1}{2}$ mole of F_2 gas into separate gaseous F atoms:

$$\tfrac{1}{2}F_2(g) \longrightarrow F(g) \qquad \Delta H_2^\circ = 75.3 \text{ kJ/mol}$$

The energy needed to break the bonds in 1 mole of F_2 molecules is 150.6 kJ. Here we are breaking the bonds in 0.5 mole of F_2, so the enthalpy change is 150.6/2, or 75.3, kJ.

3. Ionize 1 mole of gaseous Li atoms (see Table 8.2):

$$Li(g) \longrightarrow Li^+(g) + e^- \qquad \Delta H_3^\circ = 520 \text{ kJ/mol}$$

This process corresponds to the first ionization of lithium.

4. Add 1 mole of electrons to 1 mole of gaseous F atoms. As discussed in Section 8.5, the energy change for this process is just the opposite of electron affinity (see Table 8.3):

$$F(g) + e^- \longrightarrow F^-(g) \qquad \Delta H_4^\circ = -328 \text{ kJ/mol}$$

5. Combine 1 mole of gaseous Li^+ and 1 mole of F^- to form 1 mole of solid LiF:

$$Li^+(g) + F^-(g) \longrightarrow LiF(s) \qquad \Delta H_5^\circ = ?$$

The reverse of step 5,

$$\text{energy} + LiF(s) \longrightarrow Li^+(g) + F^-(g)$$

defines the lattice energy of LiF. Thus, the lattice energy must have the same magnitude as ΔH_5° but an opposite sign. Although we cannot determine ΔH_5° directly, we can calculate its value by the following procedure.

$$\begin{array}{lrcll}
1. & Li(s) & \longrightarrow & Li(g) & \Delta H_1^\circ = 155.2 \text{ kJ/mol} \\
2. & \tfrac{1}{2}F_2(g) & \longrightarrow & F(g) & \Delta H_2^\circ = 75.3 \text{ kJ/mol} \\
3. & Li(g) & \longrightarrow & Li^+(g) + e^- & \Delta H_3^\circ = 520 \text{ kJ/mol} \\
4. & F(g) + e^- & \longrightarrow & F^-(g) & \Delta H_4^\circ = -328 \text{ kJ/mol} \\
5. & Li^+(g) + F^-(g) & \longrightarrow & LiF(s) & \Delta H_5^\circ = ? \\
\hline
 & Li(s) + \tfrac{1}{2}F_2(g) & \longrightarrow & LiF(s) & \Delta H_{\text{overall}}^\circ = -594.1 \text{ kJ/mol}
\end{array}$$

According to Hess's law, we can write

$$\Delta H_{\text{overall}}^\circ = \Delta H_1^\circ + \Delta H_2^\circ + \Delta H_3^\circ + \Delta H_4^\circ + \Delta H_5^\circ$$

or

$$-594.1 \text{ kJ/mol} = 155.2 \text{ kJ/mol} + 75.3 \text{ kJ/mol} + 520 \text{ kJ/mol} - 328 \text{ kJ/mol} + \Delta H_5^\circ$$

Hence,

$$\Delta H_5^\circ = -1017 \text{ kJ/mol}$$

and the lattice energy of LiF is +1017 kJ/mol.

Figure 9.2 summarizes the Born-Haber cycle for LiF. Steps 1, 2, and 3 all require the input of energy. On the other hand, steps 4 and 5 release energy. Because ΔH_5° is a large negative quantity, the lattice energy of LiF is a large positive quantity, which accounts for the stability of solid LiF. The greater the lattice energy, the more stable the ionic compound. Keep in mind that lattice energy is *always* a positive quantity because the separation of ions in a solid into ions in the gas phase is, by Coulomb's law, an endothermic process.

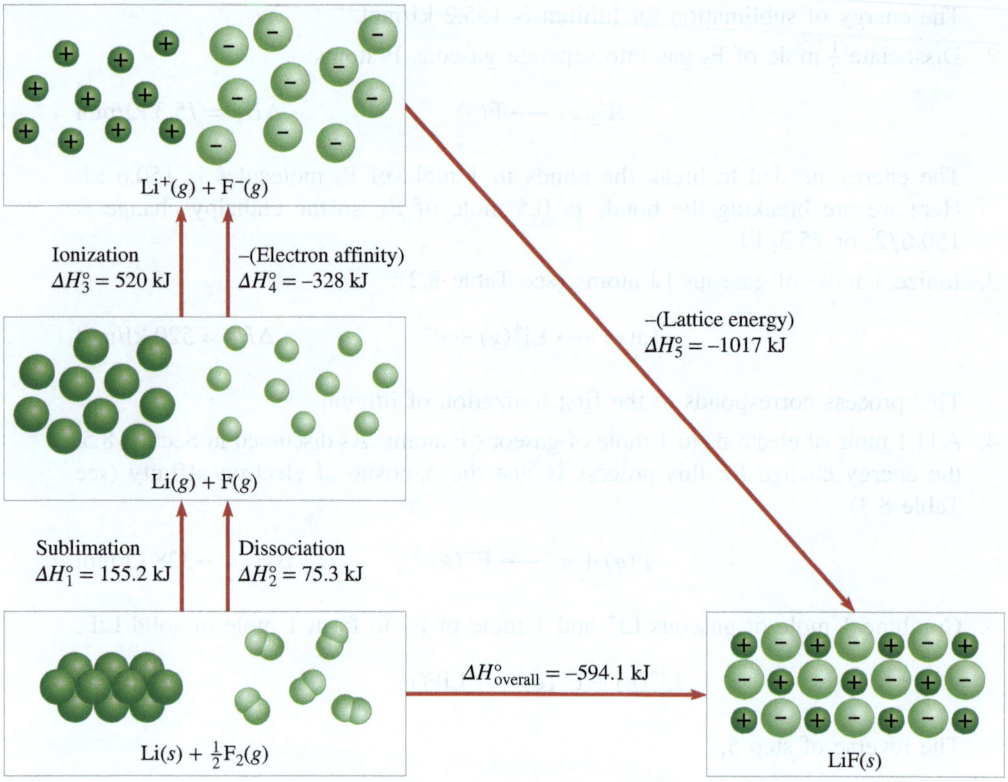

Figure 9.2 *The Born-Haber cycle for the formation of 1 mole of solid LiF.*

Lattice Energy and the Formulas of Ionic Compounds

Because lattice energy is a measure of the stability of ionic compounds, its value can help us explain the formulas of these compounds. Consider magnesium chloride as an example. We have seen that the ionization energy of an element increases rapidly as successive electrons are removed from its atom. For example, the first ionization energy of magnesium is 738 kJ/mol, and the second ionization energy is 1450 kJ/mol, almost twice the first. We might ask why, from the standpoint of energy, magnesium does not prefer to form unipositive ions in its compounds. Why doesn't magnesium chloride have the formula MgCl (containing the Mg^+ ion) rather than $MgCl_2$ (containing the Mg^{2+} ion)? Admittedly, the Mg^{2+} ion has the noble gas configuration [Ne], which represents stability because of its completely filled shells. But the stability gained through the filled shells does not, in fact, outweigh the energy input needed to remove an electron from the Mg^+ ion. The reason the formula is $MgCl_2$ lies in the extra stability gained by the formation of solid magnesium chloride. The lattice energy of $MgCl_2$ is 2527 kJ/mol, which is more than enough to compensate for the energy needed to remove the first two electrons from a Mg atom (738 kJ/mol + 1450 kJ/mol = 2188 kJ/mol).

What about sodium chloride? Why is the formula for sodium chloride NaCl and not $NaCl_2$ (containing the Na^{2+} ion)? Although Na^{2+} does not have the noble gas electron configuration, we might expect the compound to be $NaCl_2$ because Na^{2+} has a higher charge and therefore the hypothetical $NaCl_2$ should have a greater lattice energy. Again, the answer lies in the balance between energy input (that is, ionization energies) and the stability gained from the formation of the solid. The sum of the first two ionization energies of sodium is

$$496 \text{ kJ/mol} + 4560 \text{ kJ/mol} = 5056 \text{ kJ/mol}$$

The compound $NaCl_2$ does not exist, but if we assume a value of 2527 kJ/mol as its lattice energy (same as that for $MgCl_2$), we see that the energy yield would be far too small to compensate for the energy required to produce the Na^{2+} ion.

CHEMISTRY *in Action*

Sodium Chloride—A Common and Important Ionic Compound

We are all familiar with sodium chloride as table salt. It is a typical ionic compound, a brittle solid with a high melting point (801°C) that conducts electricity in the molten state and in aqueous solution. The structure of solid NaCl is shown in Figure 2.13.

One source of sodium chloride is rock salt, which is found in subterranean deposits often hundreds of meters thick. It is also obtained from seawater or brine (a concentrated NaCl solution) by solar evaporation. Sodium chloride also occurs in nature as the mineral *halite*.

Sodium chloride is used more often than any other material in the manufacture of inorganic chemicals. World consumption of this substance is about 200 million tons per year. The major use of sodium chloride is in the production of other essential inorganic chemicals such as chlorine gas, sodium hydroxide, sodium metal, hydrogen gas, and sodium carbonate. It is also used to melt ice and snow on highways and roads. However, because sodium chloride is harmful to plant life and promotes corrosion of cars, its use for this purpose is of considerable environmental concern.

Solar evaporation process for obtaining sodium chloride.
Alfred Pasieka/Science Photo Library/Alamy Stock Photo

Underground rock salt mining.
Eckehard Schulz/AP Images

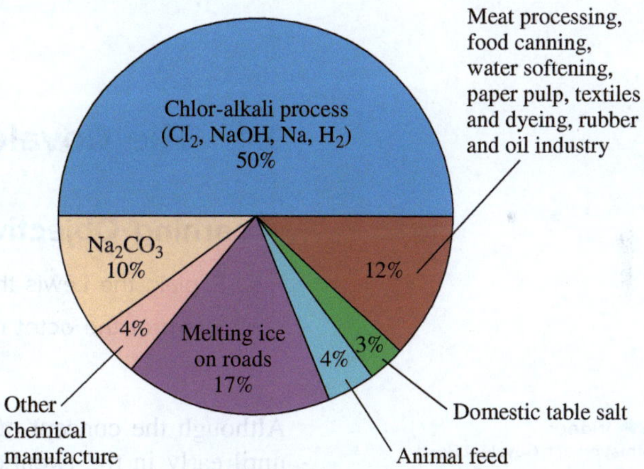

Uses of sodium chloride.

What has been said about the cations applies also to the anions. In Section 8.5 we observed that the electron affinity of oxygen is 141 kJ/mol, meaning that the following process releases energy (and is therefore favorable):

$$O(g) + e^- \longrightarrow O^-(g)$$

As we would expect, adding another electron to the O^- ion

$$O^-(g) + e^- \longrightarrow O^{2-}(g)$$

would be unfavorable in the gas phase because of the increase in electrostatic repulsion. Indeed, the electron affinity of O^- is negative (−844 kJ/mol). Yet compounds containing the oxide ion (O^{2-}) do exist and are very stable, whereas compounds containing the O^- ion are not known. Again, the high lattice energy resulting from the O^{2-} ions in compounds such as Na_2O and MgO far outweighs the energy needed to produce the O^{2-} ion.

Summary of Concepts & Facts

- An ionic bond is the product of the electrostatic forces of attraction between positive and negative ions. An ionic compound consists of a large network of ions in which positive and negative charges are balanced. The structure of a solid ionic compound maximizes the net attractive forces among the ions.
- Lattice energy is a measure of the stability of an ionic solid. It can be calculated by means of the Born-Haber cycle, which is based on Hess's law.

Review of Concepts & Facts

9.3.1 Which of the following compounds has a larger lattice energy, LiCl or CsBr?

9.3.2 Arrange the compounds NaF, MgO, and AlN in order of increasing lattice energy.

9.4 The Covalent Bond

Learning Objectives

- Explain the Lewis theory of bonding.
- Interpret the octet rule as it pertains to Lewis structures of compounds.

Video: Formation of a Covalent Bond

Although the concept of molecules goes back to the seventeenth century, it was not until early in the twentieth century that chemists began to understand how and why molecules form. The first major breakthrough was Gilbert Lewis's suggestion that a chemical bond involves electron sharing by atoms. He depicted the formation of a chemical bond in H_2 as

$$H\cdot + \cdot H \longrightarrow H:H$$

This type of electron pairing is an example of a ***covalent bond,*** *a bond in which two electrons are shared by two atoms.* ***Covalent compounds*** *are compounds that contain only covalent bonds.* For the sake of simplicity, the shared pair of electrons is often represented by a single line. Thus, the covalent bond in the hydrogen molecule can be written as H—H. In a covalent bond, each electron in a shared pair is attracted to the nuclei of both atoms. This attraction holds the two atoms in H_2 together and is responsible for the formation of covalent bonds in other molecules.

Covalent bonding between many-electron atoms involves only the valence electrons. Consider the fluorine molecule, F_2. The electron configuration of F is $1s^22s^22p^5$. The 1*s* electrons are low in energy and stay near the nucleus most of the time. For this reason they do not participate in bond formation. Thus, each F atom has seven valence electrons (the 2*s* and 2*p* electrons). According to Figure 9.1, there is only one unpaired electron on F, so the formation of the F_2 molecule can be represented as follows:

$$:\ddot{\underset{..}{F}}\cdot + \cdot\ddot{\underset{..}{F}}: \longrightarrow :\ddot{\underset{..}{F}}:\ddot{\underset{..}{F}}: \quad \text{or} \quad :\ddot{\underset{..}{F}}—\ddot{\underset{..}{F}}:$$

Note that only two valence electrons participate in the formation of F_2. The other, nonbonding electrons, are called ***lone pairs***—*pairs of valence electrons that are not involved in covalent bond formation.* Thus, each F in F_2 has three lone pairs of electrons:

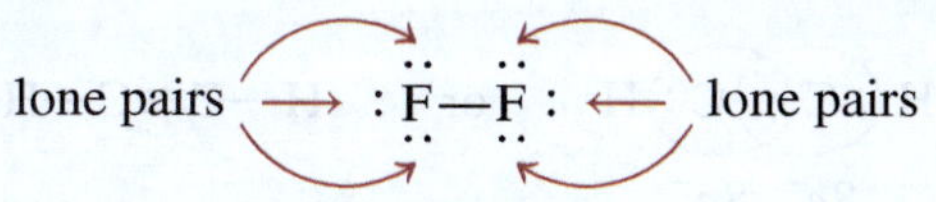

The structures we use to represent covalent compounds, such as H_2 and F_2, are called Lewis structures. A ***Lewis structure*** *is a representation of covalent bonding in which shared electron pairs are shown either as lines or as pairs of dots between two atoms, and lone pairs are shown as pairs of dots on individual atoms.* Only valence electrons are shown in a Lewis structure.

Let us consider the Lewis structure of the water molecule. Figure 9.1 shows the Lewis dot symbol for oxygen with two unpaired dots or two unpaired electrons, so we expect that O might form two covalent bonds. Because hydrogen has only one electron, it can form only one covalent bond. Thus, the Lewis structure for water is

$$H:\ddot{\underset{..}{O}}:H \quad \text{or} \quad H—\ddot{\underset{..}{O}}—H$$

In this case, the O atom has two lone pairs. The hydrogen atom has no lone pairs because its only electron is used to form a covalent bond.

In the F_2 and H_2O molecules, the F and O atoms achieve a noble gas configuration by sharing electrons:

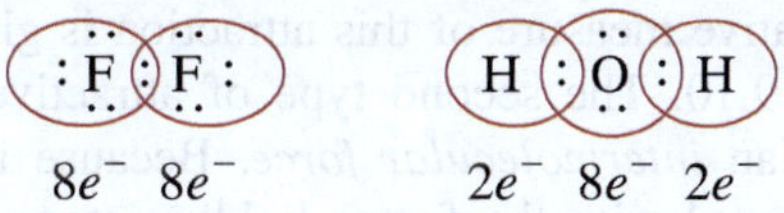

The formation of these molecules illustrates the ***octet rule,*** formulated by Lewis: *An atom other than hydrogen tends to form bonds until it is surrounded by eight valence electrons.* In other words, a covalent bond forms when there are not enough electrons for each individual atom to have a complete octet. By sharing electrons in a covalent bond, the individual atoms can complete their octets. The requirement for hydrogen is that it attain the electron configuration of helium, or a total of two electrons.

The octet rule works mainly for elements in the second period of the periodic table. These elements have only 2*s* and 2*p* subshells, which can hold a total of eight electrons. When an atom of one of these elements forms a covalent compound, it can attain the noble gas electron configuration [Ne] by sharing electrons with other atoms in the same compound. Later, we will discuss a number of important exceptions to the octet rule that give us further insight into the nature of chemical bonding.

Atoms can form different types of covalent bonds. In a ***single bond,*** *two atoms are held together by one electron pair.* Many compounds are held together by ***multiple bonds,*** that is, bonds formed when *two atoms share two or more pairs of electrons.* If *two atoms share two pairs of electrons,* the covalent bond is called a ***double bond.*** Double bonds are found in molecules of carbon dioxide (CO_2) and ethylene (C_2H_4):

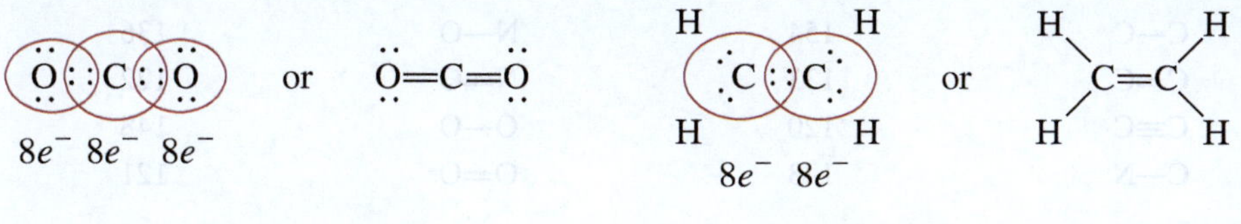

A ***triple bond*** arises when *two atoms share three pairs of electrons,* as in the nitrogen molecule (N_2):

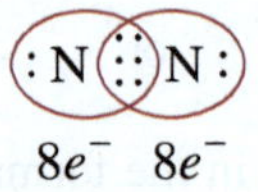

or :N≡N:

The acetylene molecule (C_2H_2) also contains a triple bond, in this case between two carbon atoms:

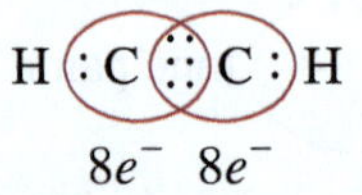

or H—C≡C—H

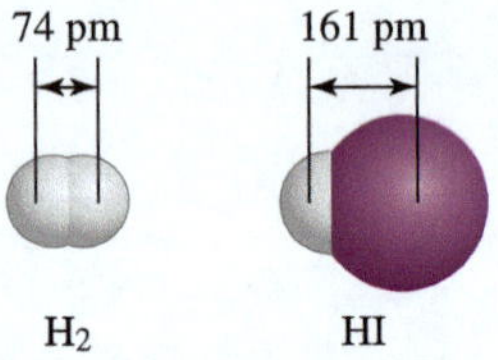

Figure 9.3 *Bond length (in pm) in H_2 and HI.*

Note that in ethylene and acetylene all the valence electrons are used in bonding; there are no lone pairs on the carbon atoms. In fact, with the exception of carbon monoxide, stable molecules containing carbon do not have lone pairs on the carbon atoms.

Multiple bonds are shorter than single covalent bonds. ***Bond length*** is *the distance between the nuclei of two covalently bonded atoms in a molecule* (Figure 9.3). Table 9.2 shows some experimentally determined bond lengths. For a given pair of atoms, such as carbon and nitrogen, triple bonds are shorter than double bonds, which, in turn, are shorter than single bonds. The shorter multiple bonds are also more stable than single bonds, as we will see later.

Comparison of the Properties of Covalent and Ionic Compounds

Video: Ionic and Covalent Bonding

Video: Ionic vs. Covalent Bonding

Ionic and covalent compounds differ markedly in their general physical properties because of differences in the nature of their bonds. There are two types of attractive forces in covalent compounds. The first type is the force that holds the atoms together in a molecule. A quantitative measure of this attraction is given by bond enthalpy, to be discussed in Section 9.10. The second type of attractive force operates *between* molecules and is called an *intermolecular force.* Because intermolecular forces are usually quite weak compared with the forces holding atoms together within a molecule, molecules of a covalent compound are not held together tightly. Consequently covalent compounds are usually gases, liquids, or low-melting solids. Further, if intermolecular forces are weak, it is relatively easy to break up aggregates of molecules to form liquids from solids (corresponding to low melting points) and gases from liquids (corresponding to low boiling points).

On the other hand, the electrostatic forces holding ions together in an ionic compound are usually very strong, so ionic compounds are solids at room temperature and

Table 9.2 Average Bond Lengths of Some Common Single, Double, and Triple Bonds

Bond Type	Bond Length (pm)	Bond Type	Bond Length (pm)
C—H	107	C=N	138
O—H	96	C≡N	116
C—O	143	N—N	147
C=O	121	N=N	124
C≡O	113	N≡N	110
C—C	154	N—O	136
C=C	133	N=O	122
C≡C	120	O—O	148
C—N	143	O=O	121

Table 9.3 Comparison of Some General Properties of an Ionic Compound and a Covalent Compound

Property	NaCl	CCl_4
Appearance	White solid	Colorless liquid
Melting point (°C)	802.02	−22.8
Molar heat of fusion* (kJ/mol)	28.16	2.56
Boiling point (°C)	1465	76.7
Molar heat of vaporization* (kJ/mol)	600	29.82
Density (g/cm^3)	2.17	1.59
Solubility in water	High	Very low
Electrical conductivity		
Solid	Poor	Poor
Liquid	Good	Poor

*Molar heat of fusion and molar heat of vaporization are the amounts of heat needed to melt 1 mole of the solid and to vaporize 1 mole of the liquid, respectively.

have high melting points. Many ionic compounds are soluble in water, and the resulting aqueous solutions conduct electricity, because the compounds are strong electrolytes. Most covalent compounds are insoluble in water, or if they do dissolve, their aqueous solutions generally do not conduct electricity because the compounds are nonelectrolytes. Molten ionic compounds conduct electricity because they contain mobile cations and anions; liquid or molten covalent compounds do not conduct electricity because no ions are present. Table 9.3 compares some of the general properties of a typical ionic compound, sodium chloride, with those of a covalent compound, carbon tetrachloride (CCl_4).

Summary of Concepts & Facts

- In a covalent bond, two electrons (one pair) are shared by two atoms. In multiple covalent bonds, two or three pairs of electrons are shared by two atoms. Some covalently bonded atoms also have lone pairs, that is, pairs of valence electrons that are not involved in bonding. The arrangement of bonding electrons and lone pairs in a molecule is represented by a Lewis structure.

Review of Concepts & Facts

9.4.1 Why is it not possible for hydrogen to form double or triple bonds in covalent compounds?

9.4.2 Arrange the bonds N═N, N≡N, and N—N in order of decreasing bond length.

9.5 Electronegativity

Learning Objective

- Define electronegativity and appraise its role in predicting bond polarity.

A covalent bond, as we have said, is the sharing of an electron pair by two atoms. In a molecule like H_2, in which the atoms are identical, we expect the electrons to be equally shared—that is, the electrons spend the same amount of time in the vicinity

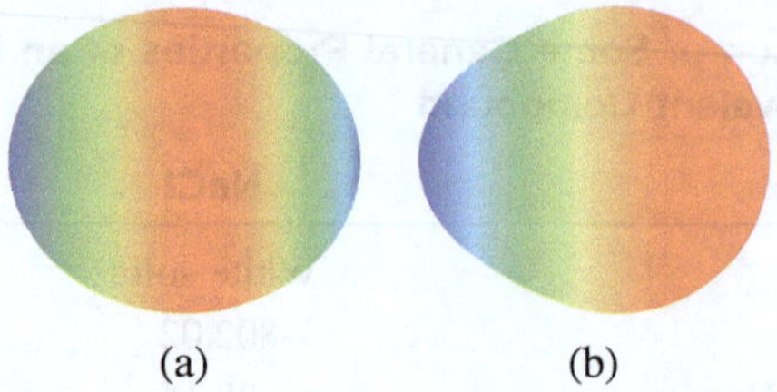

Figure 9.4 *(a) Electrostatic potential map of the H_2 molecule. (b) Electrostatic potential map of the HF molecule. The distribution varies according to the colors of the rainbow. The most electron-rich region is red; the most electron-poor region is blue.*

of each atom. We can see this in Figure 9.4(a) by visualizing the distribution of electrons in the H_2 molecule using an electrostatic potential map. These models show regions where electrons have a high probability density (red) and regions where there is a low probability density of electrons (blue).

However, in the covalently bonded HF molecule, the H and F atoms do not share the bonding electrons equally because H and F are different atoms:

$$\mathrm{H{-}\ddot{\underset{\cdot\cdot}{F}}:}$$

The bond in HF is called a ***polar covalent bond,*** or simply a *polar bond,* because *the electrons spend more time in the vicinity of one atom than the other.* Experimental evidence indicates that in the HF molecule the electrons spend more time near the F atom. We can think of this unequal sharing of electrons as a partial electron transfer or a shift in electron density, as it is more commonly described, from H to F [Figure 9.4(b)]. This "unequal sharing" of the bonding electron pair results in a relatively greater electron density near the fluorine atom and a correspondingly lower electron density near hydrogen. The HF bond and other polar bonds can be thought of as being intermediate between a (nonpolar) covalent bond, in which the sharing of electrons is exactly equal, and an ionic bond, in which the *transfer of the electron(s) is nearly complete.*

A property that helps us distinguish a nonpolar covalent bond from a polar covalent bond is ***electronegativity,*** *the ability of an atom to attract toward itself the electrons in a chemical bond.* Elements with high electronegativity have a greater tendency to attract electrons than do elements with low electronegativity. As we might expect, electronegativity is related to electron affinity and ionization energy. Thus, an atom such as fluorine, which has a high electron affinity (tends to pick up electrons easily) and a high ionization energy (does not lose electrons easily), has a high electronegativity. On the other hand, sodium has a low electron affinity, a low ionization energy, and a low electronegativity.

Electronegativity is a relative concept, meaning that an element's electronegativity can be measured only in relation to the electronegativity of other elements. Linus Pauling† devised a method for calculating *relative* electronegativities of most elements. These values are shown in Figure 9.5. Electronegativity values have no units. A careful examination of this chart reveals trends and relationships among electronegativity values of different elements. In general, electronegativity increases from left to right across a period in the periodic table, as the metallic character of the elements decreases. Within each group, electronegativity decreases with increasing atomic number and increasing metallic character. Note that the transition metals do not follow these trends. The most electronegative elements—the halogens, oxygen, nitrogen, and sulfur—are found in the upper right-hand corner of the periodic table, and the least electronegative elements (the alkali and alkaline earth metals) are clustered near the lower left-hand corner. These trends are readily apparent on a graph, as shown in Figure 9.6.

†Linus Carl Pauling (1901–1994). American chemist. Regarded by many as the most influential chemist of the twentieth century, Pauling did research in a remarkably broad range of subjects, from chemical physics to molecular biology. Pauling received the Nobel Prize in Chemistry in 1954 for his work on protein structure, and the Nobel Peace Prize in 1962. He is the only person to be the sole recipient of two Nobel prizes.

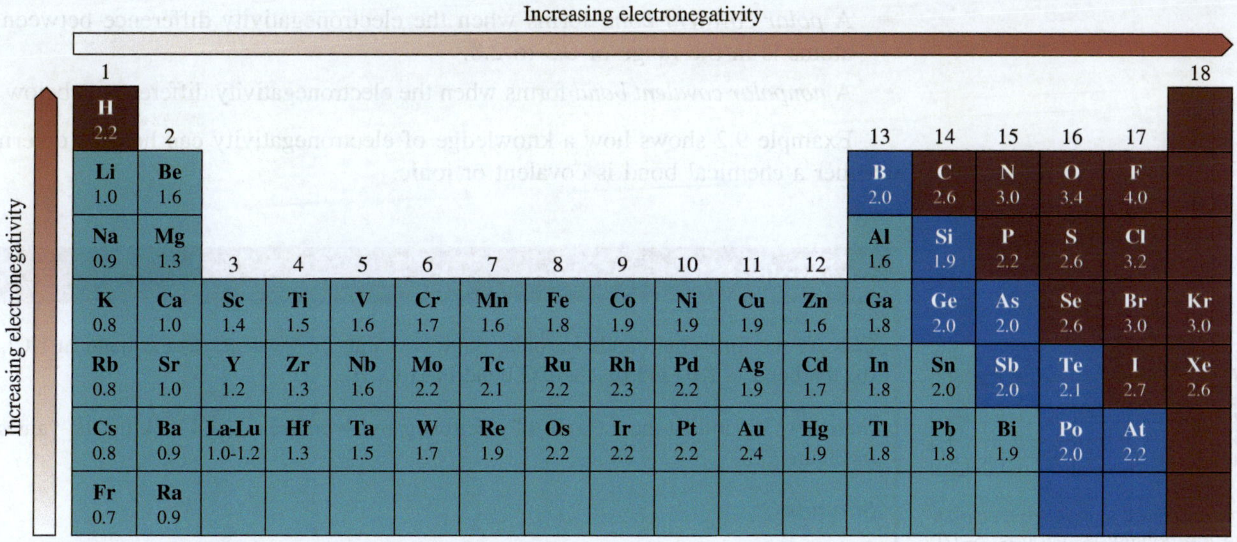

Figure 9.5 *The electronegativities of common elements.*

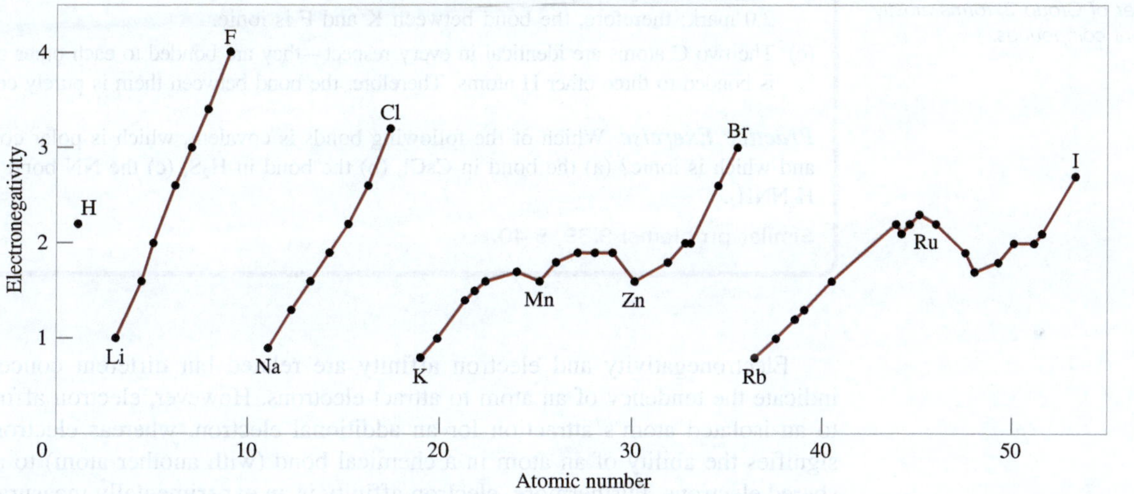

Figure 9.6 *Variation of electronegativity with atomic number. The halogens have the highest electronegativities, and the alkali metals the lowest.*

Atoms of elements with widely different electronegativities tend to form ionic bonds (such as those that exist in NaCl and CaO compounds) with each other because the atom of the less electronegative element gives up its electron(s) to the atom of the more electronegative element. An ionic bond generally joins an atom of a metallic element and an atom of a nonmetallic element. Atoms of elements with comparable electronegativities tend to form polar covalent bonds with each other because the shift in electron density is usually small. Most covalent bonds involve atoms of nonmetallic elements. Only atoms of the same element, which have the same electronegativity, can be joined by a pure covalent bond. These trends and characteristics are what we would expect, given our knowledge of ionization energies and electron affinities.

There is no sharp distinction between a polar bond and an ionic bond, but the following general rules are helpful as a rough guide.

- An *ionic bond* forms when the electronegativity difference between the two bonding atoms is 2.0 or more. This rule applies to most but not all ionic compounds.

- A *polar covalent bond* forms when the electronegativity difference between the atoms is in the range of 0.3 to 2.0.
- A *nonpolar covalent bond* forms when the electronegativity difference is below 0.3.

Example 9.2 shows how a knowledge of electronegativity can help us determine whether a chemical bond is covalent or ionic.

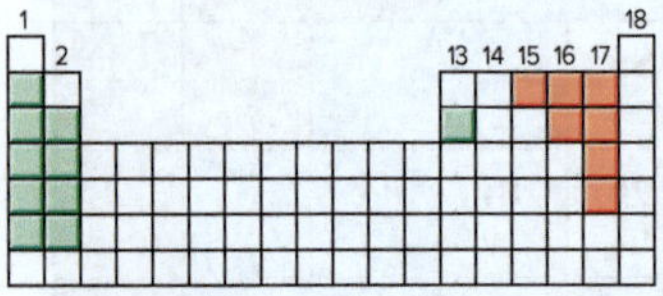

The most electronegative elements are the nonmetals (Groups 15–17) and the least electronegative elements are the alkali and alkaline earth metals (Groups 1–2) and aluminum. Beryllium, the first member of Group 2, forms mostly covalent compounds.

Example 9.2

Classify the following bonds as ionic, polar covalent, or covalent: (a) the bond in HCl, (b) the bond in KF, and (c) the CC bond in H_3CCH_3.

Strategy We follow the 2.0 rule of electronegativity difference and look up the values in Figure 9.5.

Solution

(a) The electronegativity difference between H and Cl is 1.0, which is appreciable but not large enough (by the 2.0 rule) to qualify HCl as an ionic compound. Therefore, the bond between H and Cl is polar covalent.

(b) The electronegativity difference between K and F is 3.2, which is well above the 2.0 mark; therefore, the bond between K and F is ionic.

(c) The two C atoms are identical in every respect—they are bonded to each other and each is bonded to three other H atoms. Therefore, the bond between them is purely covalent.

Practice Exercise Which of the following bonds is covalent, which is polar covalent, and which is ionic? (a) the bond in CsCl, (b) the bond in H_2S, (c) the NN bond in H_2NNH_2.

Similar problems: 9.39, 9.40.

Electronegativity and electron affinity are related but different concepts. Both indicate the tendency of an atom to attract electrons. However, electron affinity refers to an isolated atom's attraction for an additional electron, whereas electronegativity signifies the ability of an atom in a chemical bond (with another atom) to attract the shared electrons. Furthermore, electron affinity is an experimentally measurable quantity, whereas electronegativity is an estimated number that cannot be measured.

Sometimes chemists use the quantity *percent ionic character* to better describe the nature of a bond. A purely ionic bond would have 100 percent ionic character, although no such bond is known, whereas a nonpolar or purely covalent bond has 0 percent ionic character. As Figure 9.7 shows, there is a correlation between the

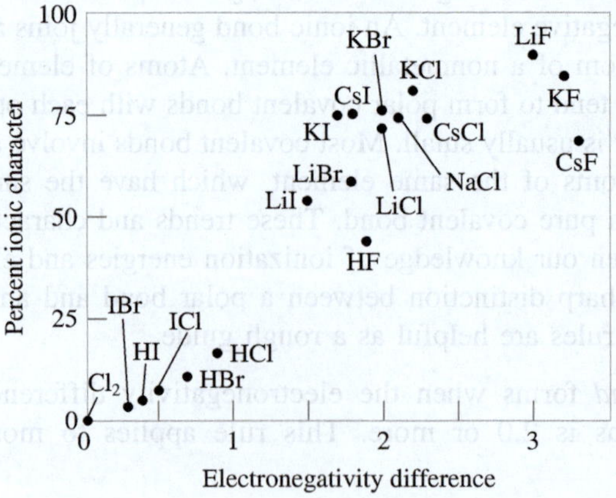

Figure 9.7 *Relation between percent ionic character and electronegativity difference.*

percent ionic character of a bond and the electronegativity difference between the bonding atoms.

Electronegativity and Oxidation Number

In Chapter 4 we introduced the rules for assigning oxidation numbers of elements in their compounds. The concept of electronegativity is the basis for these rules. In essence, oxidation number refers to the number of charges an atom would have if electrons were transferred completely to the more electronegative of the bonded atoms in a molecule.

Consider the NH_3 molecule, in which the N atom forms three single bonds with the H atoms. Because N is more electronegative than H, electron density will be shifted from H to N. If the transfer were complete, each H would donate an electron to N, which would have a total charge of −3 while each H would have a charge of +1. Thus, we assign an oxidation number of −3 to N and an oxidation number of +1 to H in NH_3.

Oxygen usually has an oxidation number of −2 in its compounds, except in hydrogen peroxide (H_2O_2), whose Lewis structure is

H—O—O—H

A bond between identical atoms makes no contribution to the oxidation number of those atoms because the electron pair of that bond is *equally* shared. Because H has an oxidation number of +1, each O atom has an oxidation number of −1.

Can you see now why fluorine always has an oxidation number of −1? It is the most electronegative element known, and it *always* forms a single bond in its compounds. Therefore, it would bear a −1 charge if electron transfer were complete.

Summary of Concepts & Facts

- Electronegativity is a measure of an atom's ability to attract electrons in a chemical bond.

Review of Concepts & Facts

9.5.1 Rank the bonds B—H, C—Cl, and P—P in order of increasing bond polarity.

9.5.2 Classify the following bonds as covalent, polar covalent, or ionic: (a) the bonds in NH_3, (b) the C—C bond in C_2H_6, (c) the bond in NaF.

9.5.3 Identify the electrostatic potential maps shown here with HCl and LiH. In both diagrams, the H atom is on the left.

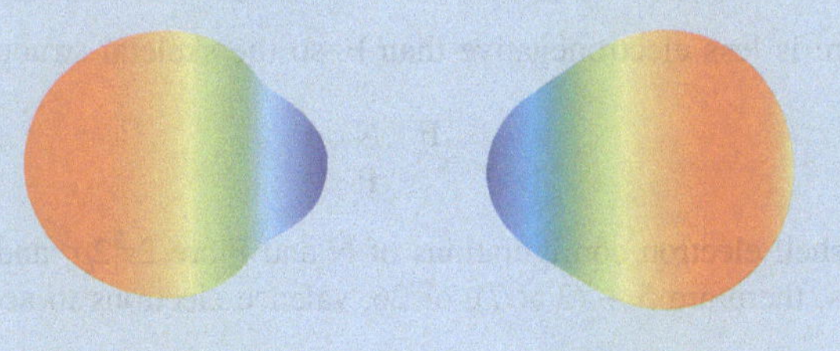

9.6 Writing Lewis Structures

Learning Objective

- Utilize the guidelines for writing Lewis structures of compounds and polyatomic ions.

Although the octet rule and Lewis structures do not present a complete picture of covalent bonding, they do help to explain the bonding scheme in many compounds and account for the properties and reactions of molecules. For this reason, you should practice writing Lewis structures of compounds. The basic steps are as follows:

1. Write the skeletal structure of the compound, using chemical symbols and placing bonded atoms next to one another. For simple compounds, this task is fairly easy. For more complex compounds, we must either be given the information or make an intelligent guess about it. In general, the least electronegative atom occupies the central position. Hydrogen and fluorine usually occupy the terminal (end) positions in the Lewis structure.
2. Count the total number of valence electrons present, referring, if necessary, to Figure 9.1. For polyatomic anions, add the number of negative charges to that total. (For example, for the CO_3^{2-} ion we add two electrons because the 2− charge indicates that there are two more electrons than are provided by the atoms.) For polyatomic cations, we subtract the number of positive charges from this total. (Thus, for NH_4^+ we subtract one electron because the 1+ charge indicates a loss of one electron from the group of atoms.)
3. Draw a single covalent bond between the central atom and each of the surrounding atoms. Complete the octets of the atoms bonded to the central atom. (Remember that the valence shell of a hydrogen atom is complete with only two electrons; hydrogen is said to follow the "duet rule.") Electrons belonging to the central or surrounding atoms must be shown as lone pairs if they are not involved in bonding. The total number of electrons to be used is that determined in step 2.
4. After completing steps 1 through 3, if the central atom has fewer than eight electrons, try adding double or triple bonds between the surrounding atoms and the central atom, using lone pairs from the surrounding atoms to complete the octet of the central atom.

Student Hot Spot

Student data indicate you may struggle with writing Lewis structures. Access your eBook for additional Learning Resources on this topic.

Examples 9.3, 9.4, and 9.5 illustrate the four-step procedure for writing Lewis structures of compounds and an ion.

Example 9.3

Write the Lewis structure for nitrogen trifluoride (NF_3) in which all three F atoms are bonded to the N atom.

Solution We follow the preceding procedure for writing Lewis structures.

Step 1: The N atom is less electronegative than F, so the skeletal structure of NF_3 is

F N F
F

Step 2: The outer-shell electron configurations of N and F are $2s^22p^3$ and $2s^22p^5$, respectively. Thus, there are $5 + (3 \times 7)$, or 26, valence electrons to account for in NF_3.

NF_3 is a colorless, odorless, unreactive gas.

(Continued)

Step 3: We draw a single covalent bond between N and each F, and complete the octets for the F atoms. We place the remaining two electrons on N:

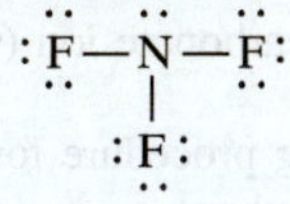

Because this structure satisfies the octet rule for all the atoms, step 4 is not required.

Check Count the valence electrons in NF_3 (in bonds and in lone pairs). The result is 26, the same as the total number of valence electrons on three F atoms ($3 \times 7 = 21$) and one N atom (5).

Practice Exercise Write the Lewis structure for carbon disulfide (CS_2).

Similar problem: 9.45.

Example 9.4

Write the Lewis structure for nitric acid (HNO_3) in which the three O atoms are bonded to the central N atom and the ionizable H atom is bonded to one of the O atoms.

Solution We follow the procedure already outlined for writing Lewis structures.

Step 1: The skeletal structure of HNO_3 is

O N O H
O

Step 2: The outer-shell electron configurations of N, O, and H are $2s^2 2p^3$, $2s^2 2p^4$, and $1s^1$, respectively. Thus, there are $5 + (3 \times 6) + 1$, or 24, valence electrons to account for in HNO_3.

Step 3: We draw a single covalent bond between N and each of the three O atoms and between one O atom and the H atom. Then we fill in electrons to comply with the octet rule for the O atoms:

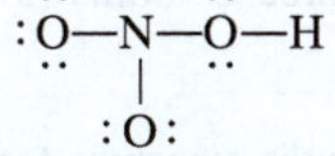

Step 4: We see that this structure satisfies the octet rule for all the O atoms but not for the N atom. The N atom has only six electrons. Therefore, we move a lone pair from one of the end O atoms to form another bond with N. Now the octet rule is also satisfied for the N atom:

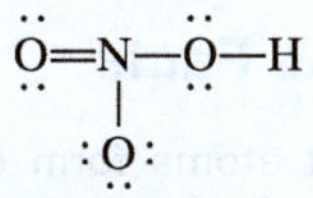

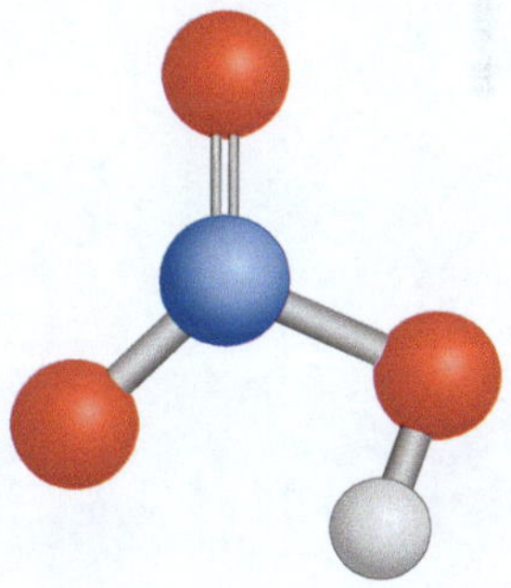

HNO_3 is a strong electrolyte.

Check Make sure that all the atoms (except H) satisfy the octet rule. Count the valence electrons in HNO_3 (in bonds and in lone pairs). The result is 24, the same as the total number of valence electrons on three O atoms ($3 \times 6 = 18$), one N atom (5), and one H atom (1).

Practice Exercise Write the Lewis structure for formic acid (HCOOH).

Similar problem: 9.45.

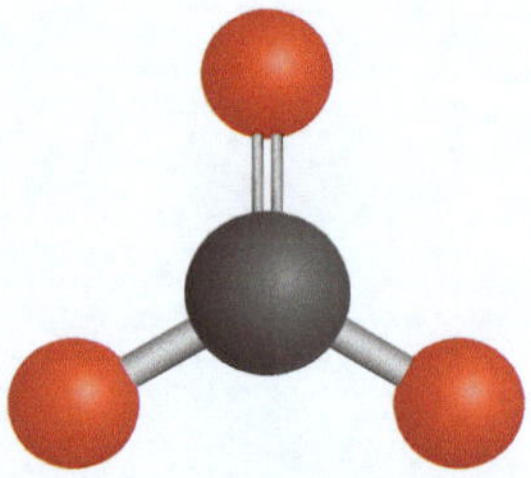

CO_3^{2-}

Example 9.5

Write the Lewis structure for the carbonate ion (CO_3^{2-}).

Solution We follow the preceding procedure for writing Lewis structures and note that this is an anion with two negative charges.

Step 1: We can deduce the skeletal structure of the carbonate ion by recognizing that C is less electronegative than O. Therefore, it is most likely to occupy a central position as follows:

O
O C O

Step 2: The outer-shell electron configurations of C and O are $2s^22p^2$ and $2s^22p^4$, respectively, and the ion itself has two negative charges. Thus, the total number of electrons is $4 + (3 \times 6) + 2$, or 24.

Step 3: We draw a single covalent bond between C and each O and comply with the octet rule for the O atoms:

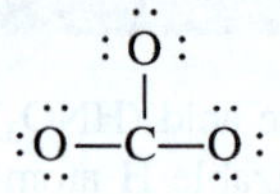

This structure shows all 24 electrons.

Step 4: Although the octet rule is satisfied for the O atoms, it is not for the C atom. Therefore, we move a lone pair from one of the O atoms to form another bond with C. Now the octet rule is also satisfied for the C atom. We use the brackets to indicate that the −2 charge is on the whole ion.

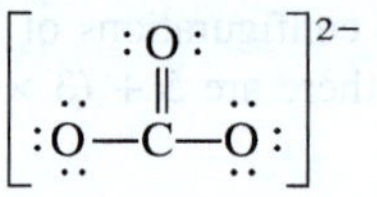

Check Make sure that all the atoms satisfy the octet rule. Count the valence electrons in CO_3^{2-} (in chemical bonds and in lone pairs). The result is 24, the same as the total number of valence electrons on three O atoms ($3 \times 6 = 18$), one C atom (4), and two negative charges (2).

Practice Exercise Write the Lewis structure for the nitrite ion (NO_2^-).

Similar problem: 9.44.

Summary of Concepts & Facts

- The octet rule predicts that atoms form enough covalent bonds to surround themselves with eight electrons each.

Review of Concepts & Facts

9.6.1 Write the Lewis structure for PCl_3.
9.6.2 Write the Lewis structure for CN^-.

9.6.3 The molecular model shown here represents guanine, a component of a DNA molecule. Only the connections between the atoms are shown in this model. Draw a complete Lewis structure of the molecule, showing all the multiple bonds and lone pairs.

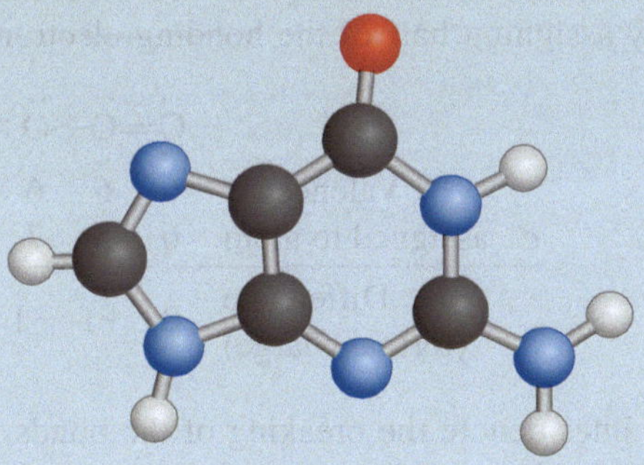

9.7 Formal Charge and Lewis Structure

Learning Objectives

- Calculate formal charges on atoms in a Lewis structure.
- Employ formal charges to identify the most likely structure of a compound when multiple Lewis structures are possible.

By comparing the number of electrons in an isolated atom with the number of electrons that are associated with the same atom in a Lewis structure, we can determine the distribution of electrons in the molecule and draw the most plausible Lewis structure. The bookkeeping procedure is as follows: In an isolated atom, the number of electrons associated with the atom is simply the number of valence electrons. (As usual, we need not be concerned with the inner electrons.) In a molecule, electrons associated with the atom are the nonbonding electrons plus the electrons in the bonding pair(s) between the atom and other atom(s). However, because electrons are shared in a bond, we must divide the electrons in a bonding pair equally between the atoms forming the bond. An atom's ***formal charge*** is *the electrical charge difference between the valence electrons in an isolated atom and the number of electrons assigned to that atom in a Lewis structure.*

To assign the number of electrons on an atom in a Lewis structure, we proceed as follows:

- All the atom's nonbonding electrons are assigned to the atom.
- We break the bond(s) between the atom and other atom(s) and assign half of the bonding electrons to the atom.

Let us illustrate the concept of formal charge using the ozone molecule (O_3). Proceeding by steps, as we did in Examples 9.3 and 9.4, we draw the skeletal structure of O_3 and then add bonds and electrons to satisfy the octet rule for the two end atoms:

$$:\ddot{\underset{\cdot\cdot}{O}}—\underset{\cdot\cdot}{O}—\ddot{\underset{\cdot\cdot}{O}}:$$

You can see that although all available electrons are used, the octet rule is not satisfied for the central atom. To remedy this, we convert a lone pair on one of

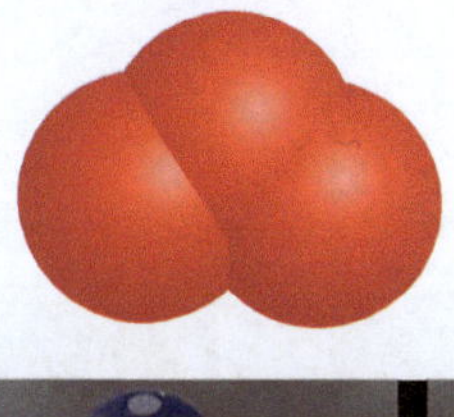

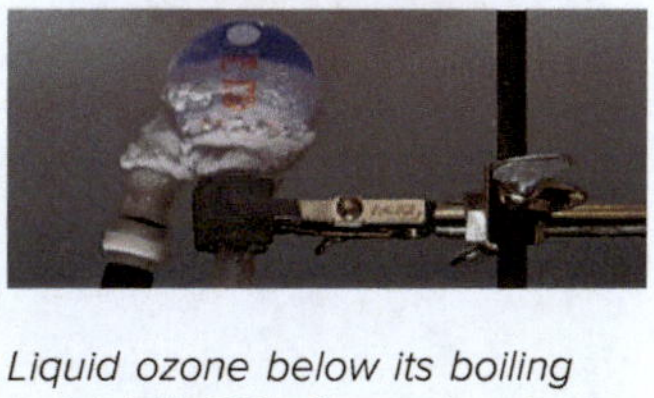

Liquid ozone below its boiling point (−111.3°C). Ozone is a toxic, light blue gas with a pungent odor.
Ken Karp/McGraw Hill

the end atoms to a second bond between that end atom and the central atom, as follows:

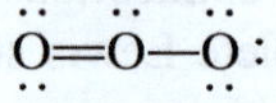

The formal charge on each atom in O_3 can now be calculated according to the following scheme by assigning half of the bonding electrons to each atom:

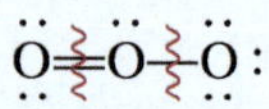

Valence e^-	6	6	6
e^- assigned to atom	6	5	7
Difference (formal charge)	0	+1	−1

where the wavy red lines denote the breaking of the bonds. Note that the breaking of a single bond results in the transfer of an electron, the breaking of a double bond results in a transfer of two electrons to each of the bonding atoms, and so on. Thus, the formal charges of the atoms in O_3 are

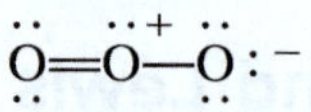

For single positive and negative charges, we normally omit the numeral 1.

When you write formal charges, these rules are helpful:

1. For molecules, the sum of the charges must add up to zero because molecules are electrically neutral species. (This rule applies, for example, to the O_3 molecule.)
2. For cations, the sum of formal charges must equal the positive charge. For anions, the sum of formal charges must equal the negative charge.

Student Hot Spot

Student data indicate you may struggle with formal charges. Access your eBook for additional Learning Resources on this topic.

Note that formal charges help us keep track of valence electrons and gain a qualitative picture of charge distribution in a molecule. We should not interpret formal charges as actual, complete transfer of electrons. In the O_3 molecule, for example, experimental studies do show that the central O atom bears a partial positive charge while the end O atoms bear a partial negative charge, but there is no evidence that there is a complete transfer of electrons from one atom to another. In determining formal charges, ask yourself if the atom in the molecule (or ion) has more electrons than its valence electrons (thus corresponding to a negative formal charge) or does the atom have fewer electrons than its valence electrons (and thus giving rise to a positive formal charge).

Example 9.6

Write formal charges for the carbonate ion.

Strategy The Lewis structure for the carbonate ion was developed in Example 9.5:

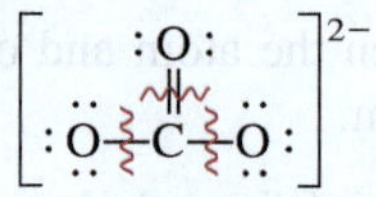

The formal charges on the atoms can be calculated using the given procedure.

Solution We subtract the number of nonbonding electrons and half of the bonding electrons from the valence electrons of each atom.

The C atom: The C atom has four valence electrons and there are no nonbonding electrons on the atom in the Lewis structure. The breaking of the double bond and two single bonds results in the transfer of four electrons to the C atom. Therefore, the formal charge is 4 − 4 = 0.

(Continued)

The O atom in C═O: The O atom has six valence electrons and there are four nonbonding electrons on the atom. The breaking of the double bond results in the transfer of two electrons to the O atom. Here the formal charge is $6 - 4 - 2 = 0$.

The O atom in C—O: This atom has six nonbonding electrons and the breaking of the single bond transfers another electron to it. Therefore, the formal charge is $6 - 6 - 1 = -1$.

Thus, the Lewis structure for CO_3^{2-} with formal charges is

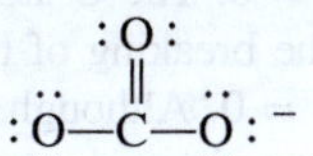

Check Note that the sum of the formal charges is −2, the same as the charge on the carbonate ion.

Practice Exercise Write formal charges for the nitrite ion (NO_2^-).

Similar problem: 9.52.

Sometimes there is more than one acceptable Lewis structure for a given species. In such cases, we can often select the most plausible Lewis structure by using formal charges and the following guidelines:

- For molecules, a Lewis structure in which there are no formal charges is preferable to one in which formal charges are present.
- Lewis structures with large formal charges (+2, +3, and/or −2, −3, and so on) are less plausible than those with small formal charges.
- Among Lewis structures having similar distributions of formal charges, the most plausible structure is the one in which negative formal charges are placed on the more electronegative atoms.

Example 9.7 shows how formal charges facilitate the choice of the correct Lewis structure for a molecule.

Example 9.7

Formaldehyde (CH_2O), a liquid with a disagreeable odor, traditionally has been used to preserve laboratory specimens. Draw the most likely Lewis structure for the compound.

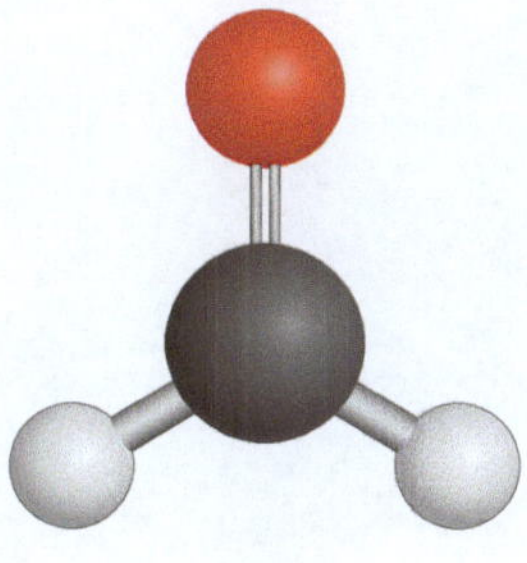

CH_2O

Strategy A plausible Lewis structure should satisfy the octet rule for all the elements, except H, and have the formal charges (if any) distributed according to electronegativity guidelines.

Solution The two possible skeletal structures are

H C O H	H, H C O
(a)	(b)

First we draw the Lewis structures for each of these possibilities:

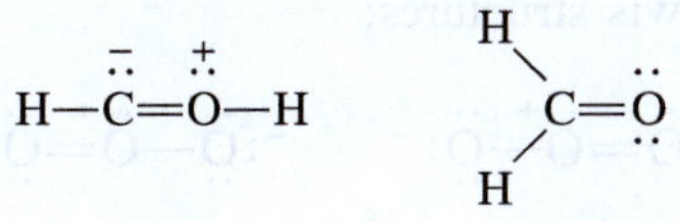

(a) (b)

(Continued)

To show the formal charges, we follow the procedure given in Example 9.6. In (a) the C atom has a total of five electrons (one lone pair plus three electrons from the breaking of a single and a double bond). Because C has four valence electrons, the formal charge on the atom is 4 − 5 = −1. The O atom has a total of five electrons (one lone pair and three electrons from the breaking of a single and a double bond). Because O has six valence electrons, the formal charge on the atom is 6 − 5 = +1. In (b) the C atom has a total of four electrons from the breaking of two single bonds and a double bond, so its formal charge is 4 − 4 = 0. The O atom has a total of six electrons (two lone pairs and two electrons from the breaking of the double bond). Therefore, the formal charge on the atom is 6 − 6 = 0. Although both structures satisfy the octet rule, (b) is the more likely structure because it carries no formal charges.

Check In each case make sure that the total number of valence electrons is 12. Can you suggest two other reasons why (a) is less plausible?

Practice Exercise Draw the most reasonable Lewis structure of a molecule that contains a N atom, a C atom, and a H atom.

Similar problem: 9.47.

Summary of Concepts & Facts

- When one atom in a covalently bonded pair donates two electrons to the bond, the Lewis structure can include the formal charge on each atom as a means of keeping track of the valence electrons.

Review of Concepts & Facts

9.7.1 Write the formal charges on H, C, and N, respectively, in HCN.

9.7.2 Write the formal charges on O and Cl, respectively, in OCl^-.

9.7.3 Consider three possible atomic arrangements for cyanamide (CH_2N_2): (a) H_2CNN, (b) H_2NCN, (c) HNNCH. Using formal charges as a guide, determine which is the most plausible arrangement.

9.8 The Concept of Resonance

Learning Objectives

- Employ formal charges to identify the most likely structure of a compound when multiple Lewis structures are possible.
- Evaluate the concept of resonance and draw resonance structures of a given compound or polyatomic ion.

Our drawing of the Lewis structure for ozone (O_3) satisfied the octet rule for the central atom because we placed a double bond between it and one of the two end O atoms. In fact, we can put the double bond at either end of the molecule, as shown by these two equivalent Lewis structures:

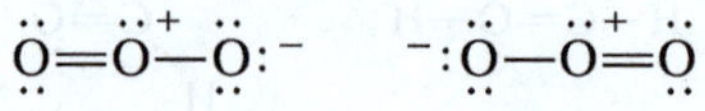

However, neither one of these two Lewis structures accounts for the known bond lengths in O_3.

We would expect the O—O bond in O_3 to be longer than the O═O bond because double bonds are known to be shorter than single bonds. Yet experimental evidence shows that both oxygen-to-oxygen bonds are equal in length (128 pm). We resolve this discrepancy by using *both* Lewis structures to represent the ozone molecule:

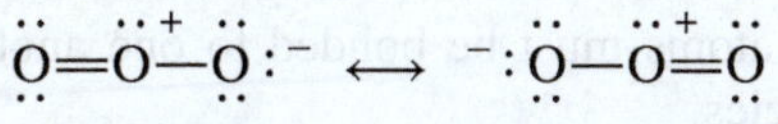

Each of these structures is called a resonance structure. A ***resonance structure,*** then, is *one of two or more Lewis structures for a single molecule that cannot be represented accurately by only one Lewis structure.* The double-headed arrow indicates that the structures shown are resonance structures.

Video: Resonance

The term ***resonance*** itself means *the use of two or more Lewis structures to represent a particular molecule.* Like the medieval European traveler to Africa who described a rhinoceros as a cross between a griffin and a unicorn, two familiar but imaginary animals, we describe ozone, a real molecule, in terms of two familiar but nonexistent structures.

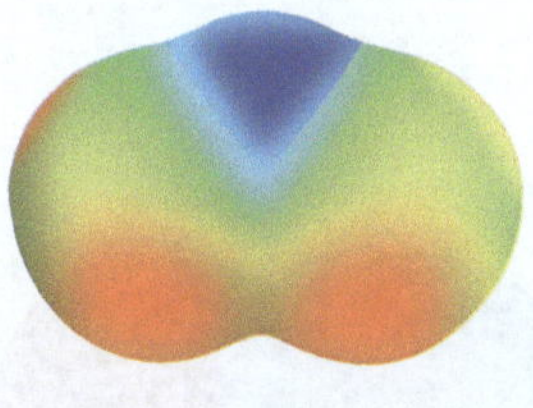

Electrostatic potential map of O_3. The electron density is evenly distributed between the two end O atoms.

A common misconception about resonance is the notion that a molecule such as ozone somehow shifts quickly back and forth from one resonance structure to the other. Keep in mind that *neither* resonance structure adequately represents the actual molecule, which has its own unique, stable structure. "Resonance" is a human invention, designed to address the limitations in these simple bonding models. To extend the animal analogy, a rhinoceros is a distinct creature, not some oscillation between a mythical griffin and unicorn!

The carbonate ion provides another example of resonance:

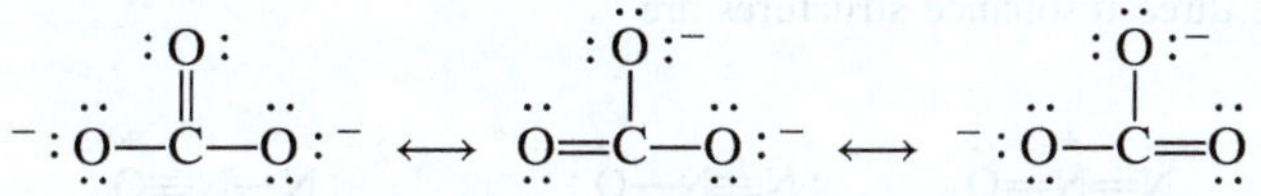

According to experimental evidence, all carbon-to-oxygen bonds in CO_3^{2-} are equivalent. Therefore, the properties of the carbonate ion are best explained by considering its resonance structures together.

The concept of resonance applies equally well to organic systems. A good example is the benzene molecule (C_6H_6):

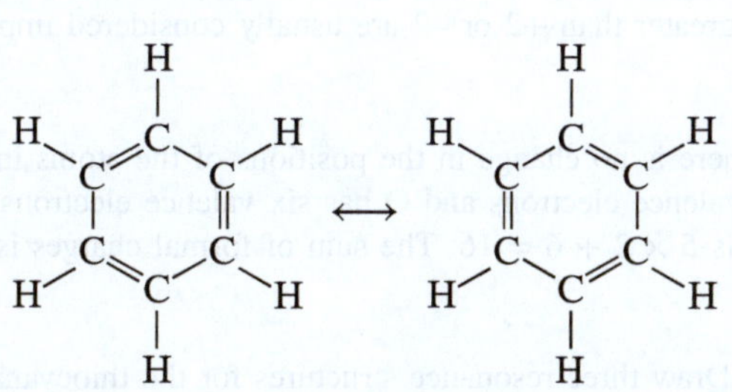

If one of these resonance structures corresponded to the actual structure of benzene, there would be two different bond lengths between adjacent C atoms, one characteristic of the single bond and the other of the double bond. In fact, the distance between all adjacent C atoms in benzene is 140 pm, which is shorter than a C—C bond (154 pm) and longer than a C═C bond (133 pm).

A simpler way of drawing the structure of the benzene molecule and other compounds containing the "benzene ring" is to show only the skeleton and not the carbon and hydrogen atoms. By this convention the resonance structures are represented by

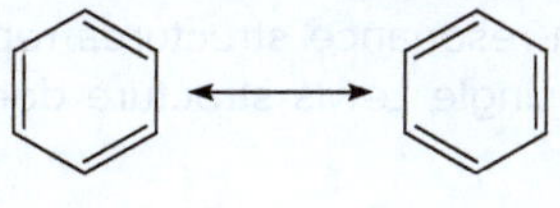

Note that the C atoms at the corners of the hexagon and the H atoms are all omitted, although they are understood to exist. Only the bonds between the C atoms are shown.

Student Hot Spot

Student data indicate you may struggle with resonance structures. Access your eBook for additional Learning Resources on this topic.

Remember this important rule for drawing resonance structures: The positions of electrons, but not those of atoms, can be rearranged in different resonance structures. In other words, the same atoms must be bonded to one another in all the resonance structures for a given species.

So far, the resonance structures shown in the examples all contribute equally to the real structure of the molecules and ion. This is not always the case, as we will see in Example 9.8.

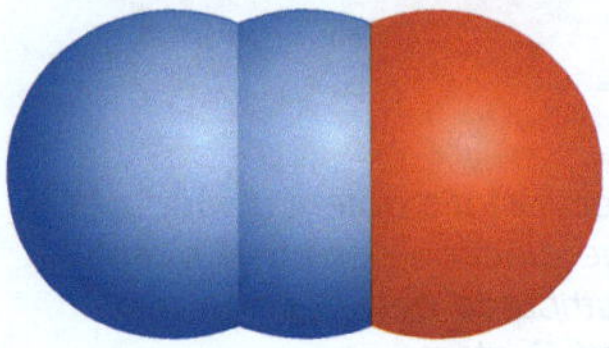

Example 9.8

Draw three resonance structures for the molecule nitrous oxide, N_2O (the atomic arrangement is NNO). Indicate formal charges. Rank the structures in their relative importance to the overall properties of the molecule.

Strategy The skeletal structure for N_2O is

N N O

We follow the procedure used for drawing Lewis structures and calculating formal charges in Examples 9.5 and 9.6.

Solution The three resonance structures are

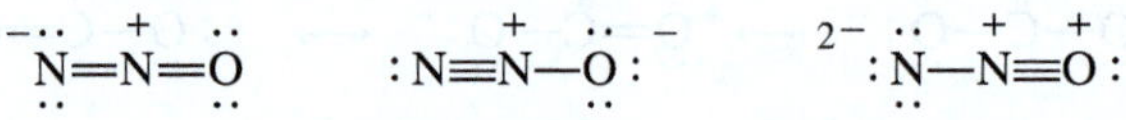

(a) (b) (c)

We see that all three structures show formal charges. Structure (b) is the most important one because the negative charge is on the more electronegative oxygen atom. Structure (c) is the least important one because it has a larger separation of formal charges. Also, the positive charge is on the more electronegative oxygen atom. Resonance structures with formal charges greater than +2 or –2 are usually considered implausible and often can be discarded.

Check Make sure there is no change in the positions of the atoms in the structures. Because N has five valence electrons and O has six valence electrons, the total number of valence electrons is $5 \times 2 + 6 = 16$. The sum of formal charges is zero in each structure.

Practice Exercise Draw three resonance structures for the thiocyanate ion, SCN^-. Rank the structures in decreasing order of importance.

Similar problems: 9.55, 9.56.

Summary of Concepts & Facts

- For some molecules or polyatomic ions, two or more Lewis structures based on the same skeletal structure satisfy the octet rule and appear chemically reasonable. Taken together, such resonance structures represent the molecule or ion more accurately than any single Lewis structure does.

Review of Concepts & Facts

9.8.1 Draw two resonance structures for SO_2.

9.8.2 The molecular model shown here represents acetamide, which is used as an organic solvent. Only the connections between the atoms are shown in this model. Draw two resonance structures for the molecule, showing the positions of multiple bonds and formal charges.

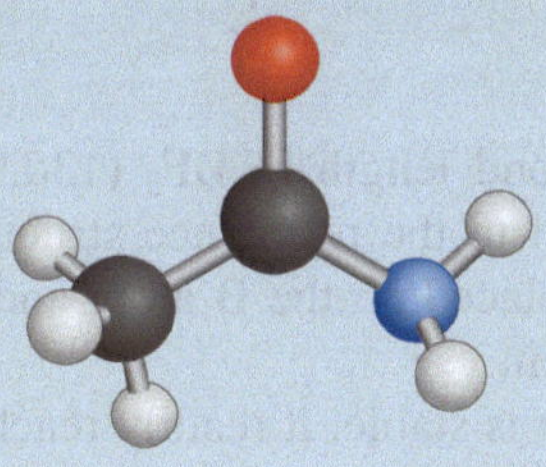

9.9 Exceptions to the Octet Rule

Learning Objectives

- Write Lewis structures for species that do not obey the octet rule.
- Draw Lewis structures with and without expanded octets for species where both are possible.

As mentioned earlier, the octet rule applies mainly to the second-period elements. Exceptions to the octet rule fall into three categories characterized by an incomplete octet, an odd number of electrons, or more than eight valence electrons around the central atom.

The Incomplete Octet

In some compounds, the number of electrons surrounding the central atom in a stable molecule is fewer than eight. Consider, for example, beryllium, which is a Group 2 (and a second-period) element. The electron configuration of beryllium is $1s^2 2s^2$; it has two valence electrons in the 2s orbital. In the gas phase, beryllium hydride (BeH_2) exists as discrete molecules. The Lewis structure of BeH_2 is

$$\text{H—Be—H}$$

Beryllium, unlike the other Group 2 elements, forms mostly covalent compounds of which BeH_2 is an example.

As you can see, only four electrons surround the Be atom, and there is no way to satisfy the octet rule for beryllium in this molecule.

Elements in Group 13, particularly boron and aluminum, also tend to form compounds in which they are surrounded by fewer than eight electrons. Take boron as an example. Because its electron configuration is $1s^2 2s^2 2p^1$, it has a total of three valence electrons. Boron reacts with the halogens to form a class of compounds having the general formula BX_3, where X is a halogen atom. Thus, in boron trifluoride there are only six electrons around the boron atom:

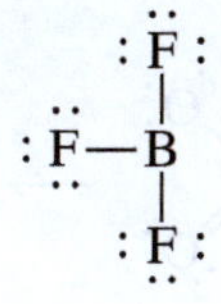

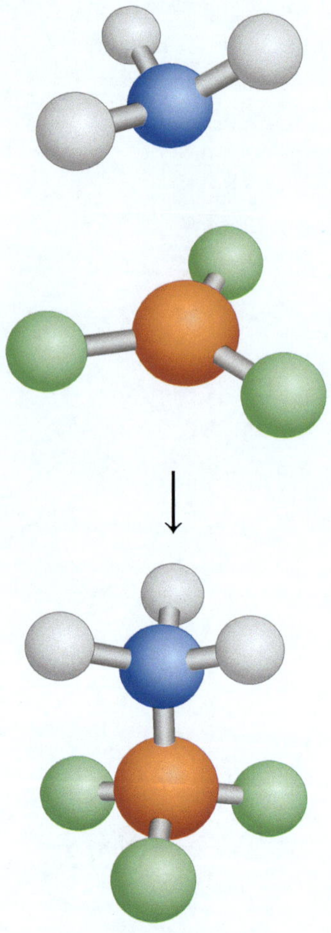

$NH_3 + BF_3 \longrightarrow H_3N—BF_3$

The following resonance structures all contain a double bond between B and F and satisfy the octet rule for boron:

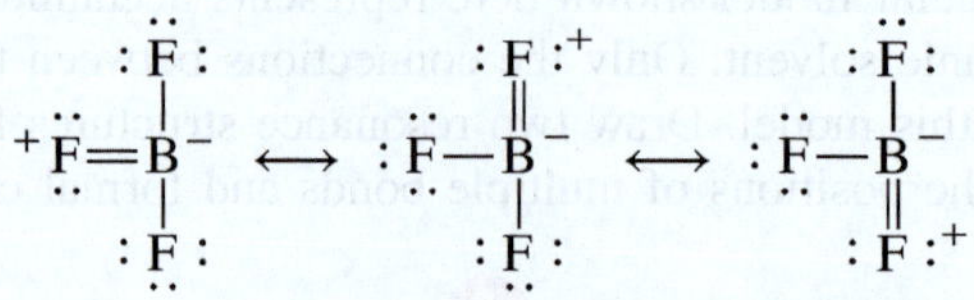

The fact that the B—F bond length in BF_3 (130.9 pm) is shorter than a single bond (137.3 pm) lends support to the resonance structures even though in each case the negative formal charge is placed on the B atom and the positive formal charge on the more electronegative F atom.

Although boron trifluoride is stable, it readily reacts with ammonia. This reaction is better represented by using the Lewis structure in which boron has only six valence electrons around it:

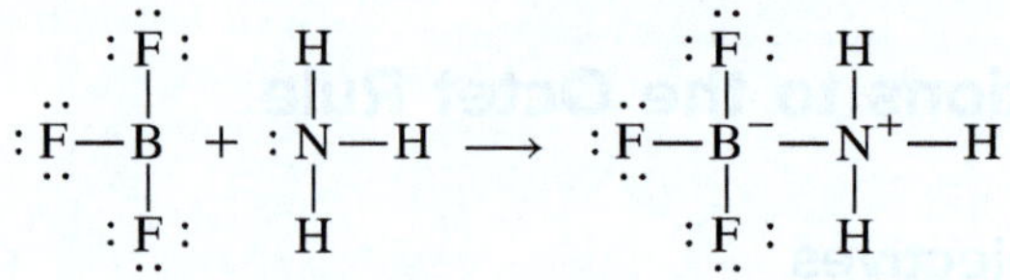

It seems that the properties of BF_3 are best explained by all four resonance structures.

The B—N bond in the abovementioned compound is different from the covalent bonds discussed so far in the sense that both electrons are contributed by the N atom. This type of bond is called a ***coordinate covalent bond*** (also referred to as a *dative bond*), defined as *a covalent bond in which one of the atoms donates both electrons.* Although the properties of a coordinate covalent bond do not differ from those of a normal covalent bond (because all electrons are alike no matter what their source), the distinction is useful for keeping track of valence electrons and assigning formal charges.

Odd-Electron Molecules

Some molecules contain an *odd* number of electrons. Among them are nitric oxide (NO) and nitrogen dioxide (NO_2):

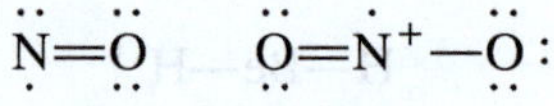

Because we need an even number of electrons for complete pairing (to reach eight), the octet rule clearly cannot be satisfied for all the atoms in any of these molecules.

Odd-electron molecules are sometimes called *radicals*. Many radicals are highly reactive. The reason is that there is a tendency for the unpaired electron to form a covalent bond with an unpaired electron on another molecule. For example, when two nitrogen dioxide molecules collide, they form dinitrogen tetroxide in which the octet rule is satisfied for both the N and O atoms:

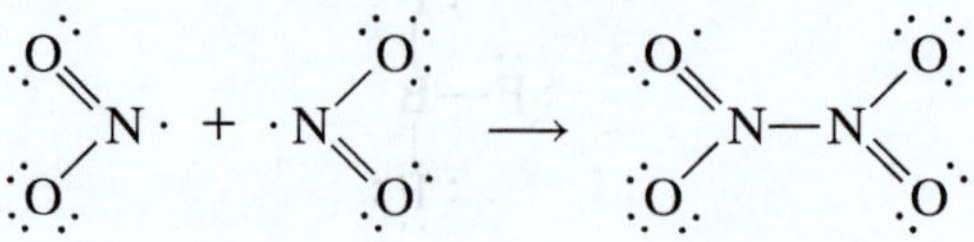

The Expanded Octet

Atoms of the second-period elements cannot have more than eight valence electrons around the central atom, but atoms of elements in and beyond the third period of the periodic table form some compounds in which more than eight electrons surround the central atom. In addition to the 3*s* and 3*p* orbitals, elements in the third period also have 3*d* orbitals that can be used in bonding. These orbitals enable an atom to form an *expanded octet*. Compounds with expanded octets are also commonly called *hypervalent*. One compound in which there is an expanded octet is sulfur hexafluoride, a very stable compound. The electron configuration of sulfur is $[Ne]3s^23p^4$. In SF_6, each of sulfur's six valence electrons forms a covalent bond with a fluorine atom, so there are 12 electrons around the central sulfur atom:

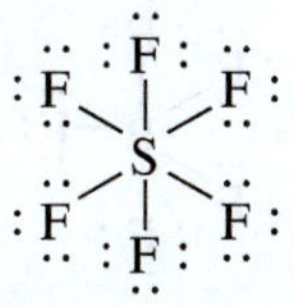

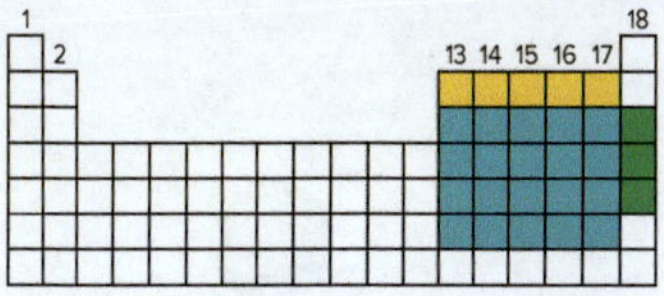

Yellow: second-period elements cannot have an expanded octet. Blue: third-period elements and beyond can have an expanded octet. Green: the noble gases usually only have an expanded octet.

In Chapter 10 we will see that these 12 electrons, or six bonding pairs, are accommodated in six orbitals that originate from the one 3*s*, the three 3*p*, and two of the five 3*d* orbitals. Sulfur also forms many compounds in which it obeys the octet rule. In sulfur dichloride, for instance, S is surrounded by only eight electrons:

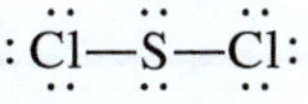

Examples 9.9, 9.10, 9.11 concern compounds that do not obey the octet rule.

Example 9.9

At high temperatures aluminum iodide (Al_2I_6) dissociates into AlI_3 molecules. Draw the Lewis structure for AlI_3.

Strategy We follow the procedures used in Examples 9.5 and 9.6 to draw the Lewis structure and calculate formal charges.

Solution The outer-shell electron configurations of Al and I are $3s^23p^1$ and $5s^25p^5$, respectively. The total number of valence electrons is $3 + 3 \times 7$ or 24. Because Al is less electronegative than I, it occupies a central position and forms three bonds with the I atoms:

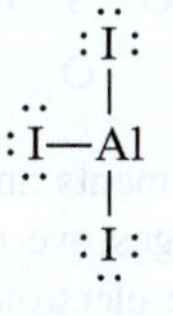

Note that there are no formal charges on the Al and I atoms.

Check Although the octet rule is satisfied for the I atoms, there are only six valence electrons around the Al atom. Thus, AlI_3 is an example of the incomplete octet.

Practice Exercise Draw the Lewis structure for BeF_2.

Similar problem: 9.68.

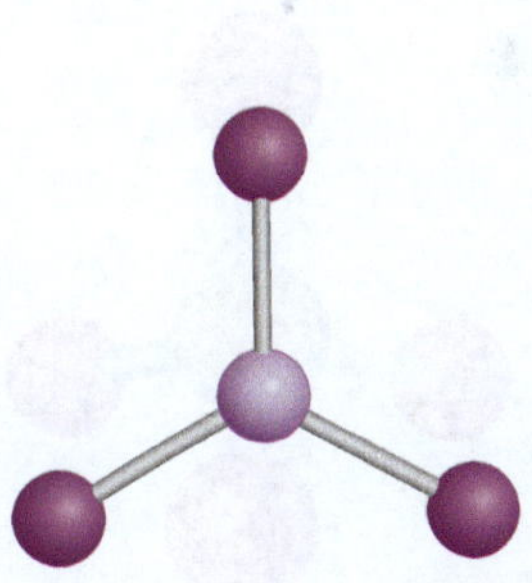

AlI_3 has a tendency to dimerize or combine two units to form Al_2I_6.

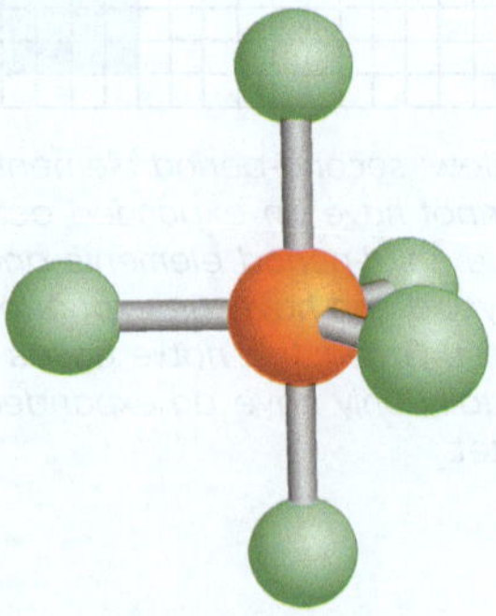

PF_5 is a reactive gaseous compound.

Example 9.10

Draw the Lewis structure for phosphorus pentafluoride (PF_5), in which all five F atoms are bonded to the central P atom.

Strategy Note that P is a third-period element. We follow the procedures given in Examples 9.5 and 9.6 to draw the Lewis structure and calculate formal charges.

Solution The outer-shell electron configurations for P and F are $3s^23p^3$ and $2s^22p^5$, respectively, and so the total number of valence electrons is 5 + (5 × 7), or 40. Phosphorus, like sulfur, is a third-period element, and therefore it can have an expanded octet. The Lewis structure of PF_5 is

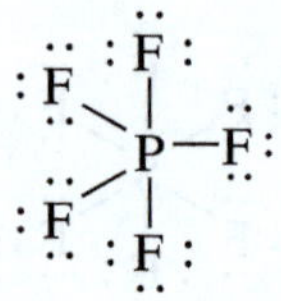

Note that there are no formal charges on the P and F atoms.

Check Although the octet rule is satisfied for the F atoms, there are 10 valence electrons around the P atom, giving it an expanded octet.

Practice Exercise Draw the Lewis structure for arsenic pentafluoride (AsF_5).

Similar problem: 9.70.

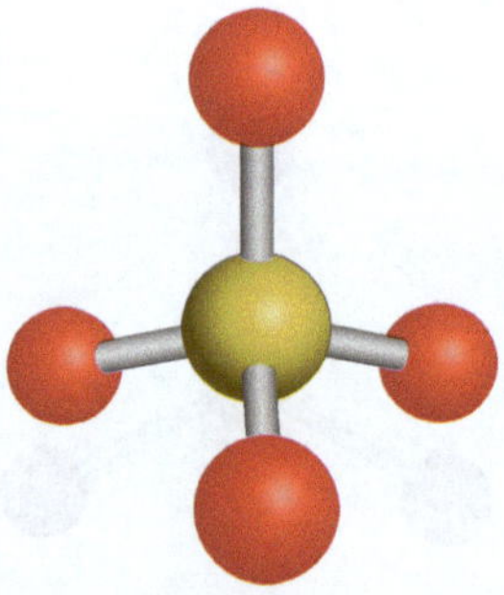

SO_4^{2-}

Example 9.11

Draw a Lewis structure for the sulfate ion (SO_4^{2-}) in which all four O atoms are bonded to the central S atom.

Strategy Note that S is a third-period element. We follow the procedures given in Examples 9.5 and 9.6 to draw the Lewis structure and calculate formal charges.

Solution The outer-shell electron configurations of S and O are $3s^23p^4$ and $2s^22p^4$, respectively.

Step 1: The skeletal structure of (SO_4^{2-}) is

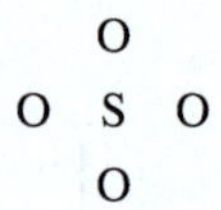

Step 2: Both O and S are Group 16 elements and so have six valence electrons each. Including the two negative charges, we must therefore account for a total of 6 + (4 × 6) + 2, or 32, valence electrons in SO_4^{2-}.

Step 3: We draw a single covalent bond between all the bonding atoms:

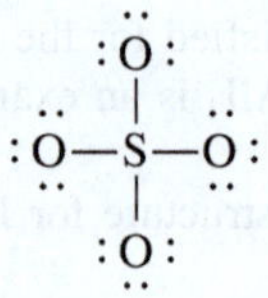

(Continued)

Next we show formal charges on the S and O atoms:

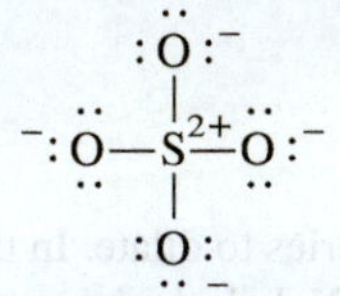

Note that we can eliminate some of the formal charges for SO_4^{2-} by expanding the S atom's octet as follows (note that this structure is only one of the six equivalent structures for SO_4^{2-}):

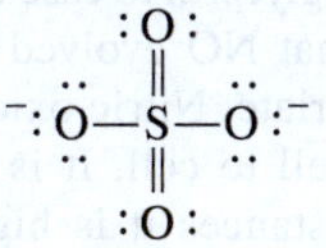

The question of which of these two structures is more important, that is, the one in which the S atom obeys the octet rule but bears more formal charges or the one in which the S atom expands its octet, has been the subject of some debate among chemists. In many cases, only elaborate quantum mechanical calculations can provide a clearer answer. At this stage of learning, you should realize that both representations are valid Lewis structures and you should be able to draw both types of structures. One helpful rule is that in trying to minimize formal charges by expanding the central atom's octet, only add enough double bonds to make the formal charge on the central atom zero. Thus, the following structure would give formal charges on S(−2) and O(0) that are inconsistent with the electronegativities of these elements and should therefore not be included to represent the SO_4^{2-} ion.

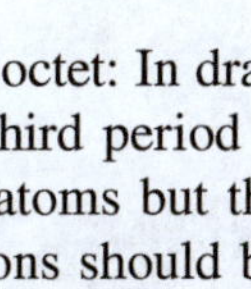

Practice Exercise Draw reasonable Lewis structures of sulfuric acid (H_2SO_4).

Similar problem: 9.91.

A final note about the expanded octet: In drawing Lewis structures of compounds containing a central atom from the third period and beyond, sometimes we find that the octet rule is satisfied for all the atoms but there are still valence electrons left to place. In such cases, the extra electrons should be placed as lone pairs on the central atom. Example 9.12 shows this approach.

Example 9.12

Draw a Lewis structure of the noble gas compound xenon tetrafluoride (XeF_4) in which all F atoms are bonded to the central Xe atom.

Strategy Note that Xe is a fifth-period element. We follow the procedures in Examples 9.5 and 9.6 for drawing the Lewis structure and calculating formal charges.

Solution

Step 1: The skeletal structure of XeF_4 is

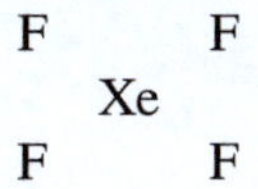

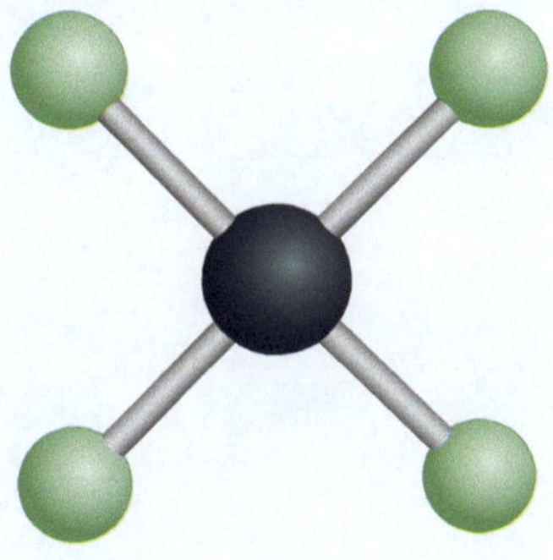

XeF_4

(Continued)

CHEMISTRY *in Action*

Just Say NO

Nitric oxide (NO), the simplest nitrogen oxide, is an odd-electron molecule, and therefore it is paramagnetic. A colorless gas (boiling point: −152°C), NO can be prepared in the laboratory by reacting sodium nitrite ($NaNO_2$) with a reducing agent such as Fe^{2+} in an acidic medium.

$$NO_2^-(aq) + Fe^{2+}(aq) + 2H^+(aq) \longrightarrow NO(g) + Fe^{3+}(aq) + H_2O(l)$$

Environmental sources of nitric oxide include the burning of fossil fuels containing nitrogen compounds and the reaction between nitrogen and oxygen inside the automobile engine at high temperatures:

$$N_2(g) + O_2(g) \longrightarrow 2NO(g)$$

Lightning also contributes to the atmospheric concentration of NO. Exposed to air, nitric oxide quickly forms brown nitrogen dioxide gas:

$$2NO(g) + O_2(g) \longrightarrow 2NO_2(g)$$

Nitrogen dioxide is a major component of smog.

About 40 years ago scientists studying muscle relaxation discovered that our bodies produce nitric oxide for use as a neurotransmitter. (A *neurotransmitter* is a small molecule that serves to facilitate cell-to-cell communications.) Since then, it has been detected in at least a dozen cell types in various parts of the body. Cells in the brain, the liver, the pancreas, the gastrointestinal tract, and the blood vessels can synthesize nitric oxide. This molecule also functions as a cellular toxin to kill harmful bacteria. And that's not all: In 1996 it was reported that NO binds to hemoglobin, the oxygen-carrying protein in the blood. No doubt it helps to regulate blood pressure.

The discovery of the biological role of nitric oxide has shed light on how nitroglycerin ($C_3H_5N_3O_9$) works as a drug. For many years, nitroglycerin tablets have been prescribed for heart patients to relieve pain (*angina pectoris*) caused by a brief interference in the flow of blood to the heart, although how it worked was not understood. We now know that nitroglycerin produces nitric oxide, which causes muscles to relax and allows the arteries to dilate. In this respect, it is interesting to note that Alfred Nobel, the inventor of dynamite (a mixture of nitroglycerin and clay that stabilizes the explosive before use), who established the prizes bearing his name, had heart trouble. But he refused his doctor's recommendation to ingest a small amount of nitroglycerin to ease the pain.

That NO evolved as a messenger molecule is entirely appropriate. Nitric oxide is small and so can diffuse quickly from cell to cell. It is a stable molecule, but under certain circumstances it is highly reactive, which accounts for its protective function. The enzyme that brings about muscle relaxation contains iron for which nitric oxide has a high affinity. It is the binding of NO to the iron that activates the enzyme. Nevertheless, in the cell, where biological effectors are typically very large molecules, the pervasive effects of one of the smallest known molecules are unprecedented. Because of the immense importance of NO in many biological roles, it was named "Molecule of the Year" in 1992 by *Science* magazine.

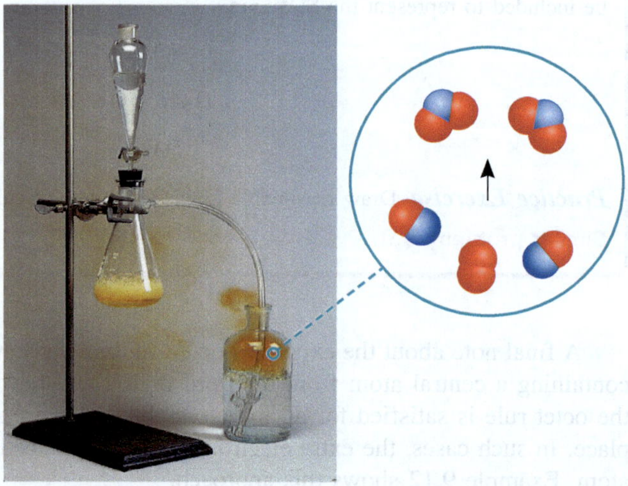

Colorless nitric oxide gas is produced by the action of Fe^{2+} on an acidic sodium nitrite solution. The gas is bubbled through water and immediately reacts with oxygen to form the brown NO_2 gas when exposed to air.
Ken Karp/McGraw Hill

Step 2: The outer-shell electron configurations of Xe and F are $5s^25p^6$ and $2s^22p^5$, respectively, and so the total number of valence electrons is $8 + (4 \times 7)$ or 36.

Step 3: We draw a single covalent bond between all the bonding atoms. The octet rule is satisfied for the F atoms, each of which has three lone pairs. The sum of the lone pair electrons on the four F atoms (4×6) and the four bonding pairs (4×2) is 32. Therefore, the remaining four electrons are shown as two lone pairs on the Xe atom:

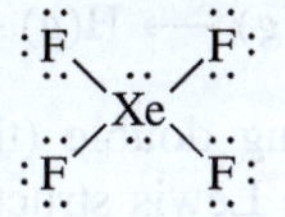

We see that the Xe atom has an expanded octet. There are no formal charges on the Xe and F atoms.

Practice Exercise Write the Lewis structure of sulfur tetrafluoride (SF_4).

Similar problem: 9.69.

Summary of Concepts & Facts

- There are exceptions to the octet rule, particularly for covalent beryllium compounds, elements in Group 13, odd-electron molecules, and elements in the third period and beyond in the periodic table.

Review of Concepts & Facts

9.9.1 Both boron and aluminum tend to form compounds in which they are surrounded with fewer than eight electrons. However, aluminum is able to form compounds and polyatomic ions where it is surrounded by more than eight electrons (e.g., AlF_6^-). Why is it possible for aluminum, but not boron, to expand the octet?

9.9.2 Draw a Lewis structure for IF_5.

9.9.3 Draw a Lewis structure for ClO.

9.10 Bond Enthalpy

Learning Objective

- Demonstrate the use of bond enthalpies to estimate the enthalpy change in a reaction.

A measure of the stability of a molecule is its ***bond enthalpy,*** which is *the enthalpy change required to break a particular bond in 1 mole of gaseous molecules.* (Remember that it takes energy to break a bond so that energy is released when a bond is formed.) The experimentally determined bond enthalpy of the diatomic hydrogen molecule, for example, is

$$H_2(g) \longrightarrow H(g) + H(g) \qquad \Delta H° = 436 \text{ kJ/mol}$$

This equation tells us that breaking the covalent bonds in 1 mole of gaseous H_2 molecules requires 436 kJ of energy. For the less stable chlorine molecule,

$$Cl_2(g) \longrightarrow Cl(g) + Cl(g) \qquad \Delta H^\circ = 243 \text{ kJ/mol}$$

Bond enthalpies can also be directly measured for diatomic molecules containing unlike elements, such as HCl,

$$HCl(g) \longrightarrow H(g) + Cl(g) \qquad \Delta H^\circ = 431 \text{ kJ/mol}$$

as well as for molecules containing double (the Lewis structure of O_2 contains a double bond) and triple bonds (the Lewis structure of N_2 contains a triple bond):

$$O_2(g) \longrightarrow O(g) + O(g) \qquad \Delta H^\circ = 498 \text{ kJ/mol}$$

$$N_2(g) \longrightarrow N(g) + N(g) \qquad \Delta H^\circ = 945 \text{ kJ/mol}$$

Measuring the strength of covalent bonds in polyatomic molecules is more complicated. For example, measurements show that the energy needed to break the first O—H bond in H_2O is different from that needed to break the second bond:

$$H_2O(g) \longrightarrow H(g) + OH(g) \qquad \Delta H^\circ = 502 \text{ kJ/mol}$$

$$OH(g) \longrightarrow H(g) + O(g) \qquad \Delta H^\circ = 427 \text{ kJ/mol}$$

In each case, an O—H bond is broken, but the first step is more endothermic than the second. The difference between the two ΔH° values suggests that the second O—H bond itself has undergone change, because of the changes in the chemical environment.

Now we can understand why the bond enthalpy of the same O—H bond in two different molecules such as methanol (CH_3OH) and water (H_2O) will not be the same: Their environments are different. Thus, for polyatomic molecules, we speak of the *average* bond enthalpy of a particular bond. For example, we can measure the energy of the O—H bond in 10 different polyatomic molecules and obtain the average O—H bond enthalpy by dividing the sum of the bond enthalpies by 10. Table 9.4 lists the average bond enthalpies of a number of diatomic and polyatomic molecules. As stated earlier, triple bonds are stronger than double bonds, which, in turn, are stronger than single bonds.

Use of Bond Enthalpies in Thermochemistry

A comparison of the thermochemical changes that take place during a number of reactions (Chapter 6) reveals a strikingly wide variation in the enthalpies of different reactions. For example, the combustion of hydrogen gas in oxygen gas is fairly exothermic:

$$H_2(g) + \tfrac{1}{2}O_2 \longrightarrow H_2O(l) \qquad \Delta H^\circ = -285.8 \text{ kJ/mol}$$

On the other hand, the formation of glucose ($C_6H_{12}O_6$) from water and carbon dioxide, best achieved by photosynthesis, is highly endothermic:

$$6CO_2(g) + 6H_2O(l) \longrightarrow C_6H_{12}O_6(s) + 6O_2(g) \qquad \Delta H^\circ = 2803 \text{ kJ/mol}$$

We can account for such variations by looking at the stability of individual reactant and product molecules. After all, most chemical reactions involve the making and breaking of bonds. Therefore, knowing the bond enthalpies and hence the stability of molecules tells us something about the thermochemical nature of reactions that molecules undergo.

Table 9.4 Bond Enthalpies

Bond	Bond Enthalpy (kJ/mol)	Bond	Bond Enthalpy (kJ/mol)
H—H	436	C—I	253
H—N	391	C—P	263
H—O	463	C—S	259
H—S	354	C═S	477
H—P	326	N—N	163
H—F	570	N═N	418
H—Cl	431	N≡N	945
H—Br	366	N—O	201
H—I	298	N═O	607
C—H	413	O—O	146
C—C	348	O═O	498
C═C	614	O—P	589
C≡C	839	O═S	523
C—N	293	P—P	197
C═N	615	P═P	489
C≡N	891	S—S	266
C—O	358	S═S	418
C═O*	745	F—F	159
C≡O	1077	Cl—Cl	243
C—F	485	Cl—F	261
C—Cl	328	Br—Br	194
C—Br	276	I—I	152

*The C═O bond enthalpy in CO_2 is 799 kJ/mol.

In many cases, it is possible to predict the approximate enthalpy of reaction by using the average bond enthalpies. Because energy is always required to break chemical bonds and chemical bond formation is always accompanied by a release of energy, we can estimate the enthalpy of a reaction by counting the total number of bonds broken and formed in the reaction and recording all the corresponding energy changes. The enthalpy of reaction in the *gas phase* is given by

$$\begin{aligned}\Delta H^\circ &= \Sigma\text{BE(reactants)} - \Sigma\text{BE(products)} \\ &= \text{total energy input} - \text{total energy released}\end{aligned} \quad (9.3)$$

where BE stands for average bond enthalpy and Σ is the summation sign. As written, Equation (9.3) takes care of the sign convention for ΔH°. Thus, if the total energy input is greater than the total energy released, ΔH° is positive and the reaction is endothermic. On the other hand, if more energy is released than absorbed, ΔH° is negative and the reaction is exothermic (Figure 9.8). If reactants and products are all diatomic molecules, then Equation (9.3) will yield accurate results because the bond enthalpies of diatomic molecules are accurately known. If some or all of the reactants and products are polyatomic molecules, Equation (9.3) will yield only approximate results because the bond enthalpies used will be averages.

For diatomic molecules, Equation (9.3) is equivalent to Equation (6.18), so the results obtained from these two equations should correspond, as Example 9.13 illustrates.

Figure 9.8 *Bond enthalpy changes in (a) an endothermic reaction and (b) an exothermic reaction.*

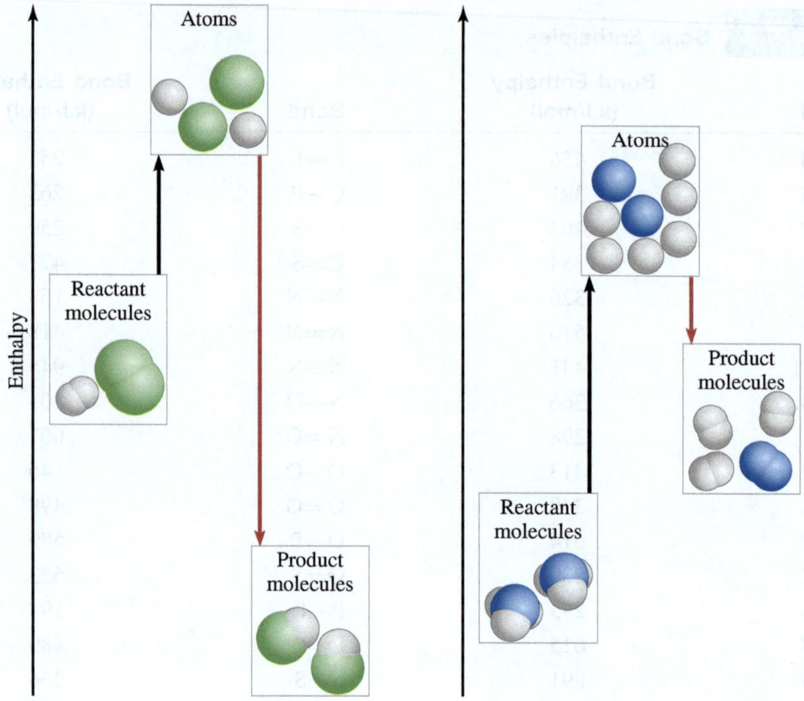

$$H_2(g) + F_2(g) \longrightarrow 2HF(g)$$
$$\Delta H° = -546.6 \text{ kJ/mol}$$
(a)

$$2NH_3(g) \longrightarrow N_2(g) + 3H_2(g)$$
$$\Delta H° = 91.8 \text{ kJ/mol}$$
(b)

Example 9.13

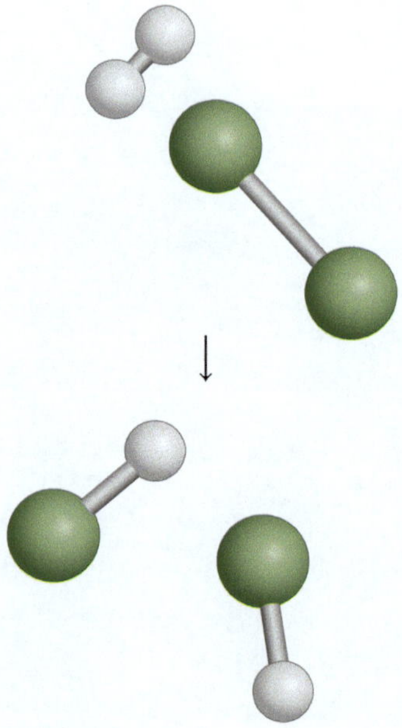

Use Equation (9.3) to calculate the enthalpy of reaction for the process

$$H_2(g) + Cl_2(g) \longrightarrow 2HCl(g)$$

Compare your result with that obtained using Equation (6.18).

Strategy Keep in mind that bond breaking is an energy-absorbing (endothermic) process and bond making is an energy-releasing (exothermic) process. Therefore, the overall energy change is the difference between these two opposing processes, as described by Equation (9.3).

Solution We start by counting the number of bonds broken and the number of bonds formed and the corresponding energy changes. This is best done by creating a table:

Type of bonds broken	*Number of bonds broken*	*Bond enthalpy (kJ/mol)*	*Energy change (kJ/mol)*
H—H (H_2)	1	436	436
Cl—Cl (Cl_2)	1	243	243

Type of bonds formed	*Number of bonds formed*	*Bond enthalpy (kJ/mol)*	*Energy change (kJ/mol)*
H—Cl (HCl)	2	431	862

Next, we obtain the total energy input and total energy released:

$$\text{total energy input} = 436 \text{ kJ/mol} + 243 \text{ kJ/mol} = 679 \text{ kJ/mol}$$
$$\text{total energy released} = 862 \text{ kJ/mol}$$

(Continued)

Using Equation (9.3), we write

$$\Delta H° = 679 \text{ kJ/mol} - 862 \text{ kJ/mol} = -183 \text{ kJ/mol}$$

Alternatively, we can use Equation (6.18) and the data in Appendix 2 to calculate the enthalpy of reaction:

$$\begin{aligned}\Delta H° &= 2\Delta H_f°(\text{HCl}) - [\Delta H_f°(\text{H}_2) + \Delta H_f°(\text{Cl}_2)]\\ &= (2)(-92.3 \text{ kJ/mol}) - 0 - 0\\ &= -184.6 \text{ kJ/mol}\end{aligned}$$

Check Because the reactants and products are all diatomic molecules, we expect the results of Equations (9.3) and (6.18) to be the same. The small discrepancy here is due to different ways of rounding off.

Practice Exercise Calculate the enthalpy of the reaction

$$H_2(g) + F_2(g) \longrightarrow 2HF(g)$$

using (a) Equation (9.3) and (b) Equation (6.18).

Similar problem: 9.110.

Example 9.14 uses Equation (9.3) to estimate the enthalpy of a reaction involving a polyatomic molecule.

Example 9.14

Estimate the enthalpy change for the combustion of hydrogen gas:

$$2H_2(g) + O_2(g) \longrightarrow 2H_2O(g)$$

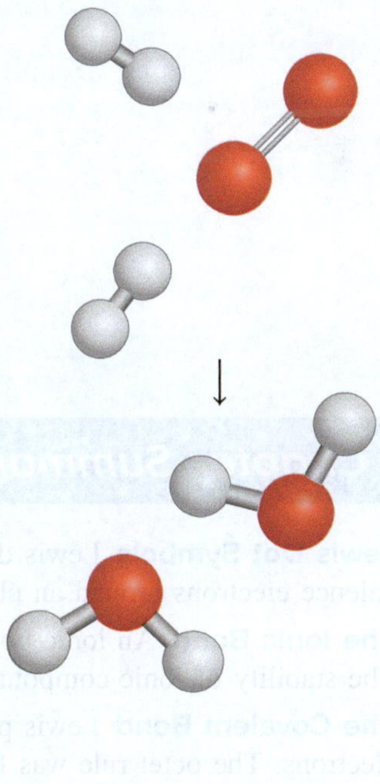

Strategy We basically follow the same procedure as that in Example 9.13. Note, however, that H_2O is a polyatomic molecule, and so we need to use the average bond enthalpy value for the O—H bond.

Solution We construct the following table:

Type of bonds broken	*Number of bonds broken*	*Bond enthalpy (kJ/mol)*	*Energy change (kJ/mol)*
H—H (H_2)	2	436	872
O=O (O_2)	1	498	498

Type of bonds formed	*Number of bonds formed*	*Bond enthalpy (kJ/mol)*	*Energy change (kJ/mol)*
O—H (H_2O)	4	463	1852

Next, we obtain the total energy input and total energy released:

$$\text{total energy input} = 872 \text{ kJ/mol} + 498 \text{ kJ/mol} = 1370 \text{ kJ/mol}$$

$$\text{total energy released} = 1852 \text{ kJ/mol}$$

Using Equation (9.3), we write

$$\Delta H° = 1370 \text{ kJ/mol} - 1852 \text{ kJ/mol} = -482 \text{ kJ/mol}$$

(Continued)

This result is only an estimate because the bond enthalpy of O—H is an average quantity. Alternatively, we can use Equation (6.18) and the data in Appendix 2 to calculate the enthalpy of reaction:

$$\begin{aligned}\Delta H^\circ &= 2\Delta H_f^\circ(H_2O) - [2\Delta H_f^\circ(H_2) + \Delta H_f^\circ(O_2)] \\ &= 2(-241.8 \text{ kJ/mol}) - 0 - 0 \\ &= -483.6 \text{ kJ/mol}\end{aligned}$$

Check Note that the estimated value based on average bond enthalpies is marginally close to the value calculated using ΔH_f° data. In general, Equation (9.3) works best for reactions that are either quite endothermic or quite exothermic, that is, reactions for which $\Delta H^\circ_{rxn} > 100$ kJ/mol or for which $\Delta H^\circ_{rxn} < -100$ kJ/mol.

Practice Exercise For the reaction

$$H_2(g) + C_2H_4(g) \longrightarrow C_2H_6(g)$$

(a) Estimate the enthalpy of reaction, using the bond enthalpy values in Table 9.4.

(b) Calculate the enthalpy of reaction, using standard enthalpies of formation. (ΔH_f° for H_2, C_2H_4, and C_2H_6 are 0, 52.4 kJ/mol, and −84.0 kJ/mol, respectively.)

Similar problem: 9.76.

Summary of Concepts & Facts

- The strength of a covalent bond is measured in terms of its bond enthalpy. Bond enthalpies can be used to estimate the enthalpy of reactions.

Review of Concepts & Facts

9.10.1 Why does ΔH°_{rxn} calculated using bond enthalpies not always agree with that calculated using ΔH_f° values?

9.10.2 Using bond enthalpy values, predict the enthalpy of reaction for the reaction of fluorine and chlorine to produce chlorine trifluoride:

$$3F_2(g) + Cl_2(g) \longrightarrow 2ClF_3(g)$$

Chapter Summary

Lewis Dot Symbols Lewis dot symbols depict an atom or ion of main group elements with dots representing the valence electrons around an element symbol. (Section 9.1)

The Ionic Bond An ionic bond is the electrostatic force that holds the cations and anions in an ionic compound. The stability of ionic compounds is determined by their lattice energies. (Sections 9.2 and 9.3)

The Covalent Bond Lewis postulated the formation of a covalent bond in which atoms share one or more pairs of electrons. The octet rule was formulated to predict the correctness of Lewis structures. This rule says that an atom other than hydrogen tends to form bonds until it is surrounded by eight valence electrons. (Section 9.4)

Electronegativity This property allows us to distinguish a nonpolar covalent bond from a polar covalent bond. Electronegativity is a measure of the ability of an atom to attract electrons to itself in a molecule. (Section 9.5)

Characteristics of Lewis Structures In addition to covalent bonds, a Lewis structure also shows lone pairs, which are pairs of electrons not involved in bonding, on atoms and formal charges, which are the result of bookkeeping of electrons used in bonding. A resonance structure is one of two or more Lewis structures for a single molecule that cannot be described fully with only one Lewis structure. (Sections 9.6, 9.7, and 9.8)

Exceptions to the Octet Rule The octet rule applies mainly to the second-period elements. The three categories of exceptions to the octet rule are the incomplete octet, in which an atom in a molecule has fewer than eight valence electrons, the odd-electron molecules, which have an odd number of valence electrons, and the expanded octet, in which an atom has more than eight valence electrons. These exceptions can be explained by more refined theories of chemical bonding. (Section 9.9)

Bond Enthalpy From a knowledge of the strengths of covalent bonds or bond enthalpies, it is possible to estimate the enthalpy change of a reaction. (Section 9.10)

Key Equation

$\Delta H^\circ = \Sigma BE(\text{reactants}) - \Sigma BE(\text{products})$ (9.3)	Calculating enthalpy change of a reaction from bond enthalpies.

Key Words

Bond enthalpy, p. 403
Bond length, p. 382
Born-Haber cycle, p. 376
Coordinate covalent bond, p. 398
Coulomb's law, p. 376
Covalent bond, p. 380
Covalent compound, p. 380
Double bond, p. 381
Electronegativity, p. 384
Formal charge, p. 391
Ionic bond, p. 372
Lewis dot symbol, p. 371
Lewis structure, p. 381
Lone pair, p. 381
Multiple bond, p. 381
Octet rule, p. 381
Polar covalent bond, p. 384
Resonance, p. 395
Resonance structure, p. 395
Single bond, p. 381
Triple bond, p. 382

Questions & Problems

Red numbered problems solved in Student Solutions Manual

9.1 Lewis Dot Symbols

Review Questions

9.1 What is a Lewis dot symbol? To what elements does the symbol mainly apply?

9.2 Use the second member of each group from Group 1 to Group 17 to show that the number of valence electrons on an atom of the element is the same as its group number.

9.3 Without referring to Figure 9.1, write Lewis dot symbols for atoms of the following elements: (a) Be, (b) K, (c) Ca, (d) Ga, (e) O, (f) Br, (g) N, (h) I, (i) As, (j) F.

9.4 Write Lewis dot symbols for the following ions: (a) Li^+, (b) Cl^-, (c) S^{2-}, (d) Sr^{2+}, (e) N^{3-}.

9.5 Write Lewis dot symbols for the following atoms and ions: (a) I, (b) I^-, (c) S, (d) S^{2-}, (e) P, (f) P^{3-}, (g) Na, (h) Na^+, (i) Mg, (j) Mg^{2+}, (k) Al, (l) Al^{3+}, (m) Pb, (n) Pb^{2+}.

9.2 The Ionic Bond

Review Questions

9.6 Explain what an ionic bond is.

9.7 Explain how ionization energy and electron affinity determine whether atoms of elements will combine to form ionic compounds.

9.8 Name five metals and five nonmetals that are very likely to form ionic compounds. Write formulas for compounds that might result from the combination of these metals and nonmetals. Name these compounds.

9.9 Name one ionic compound that contains only nonmetallic elements.

9.10 Name one ionic compound that contains a polyatomic cation and a polyatomic anion (see Table 2.3).

9.11 Explain why ions with charges greater than 3 are seldom found in ionic compounds.

9.12 The term "molar mass" was introduced in Chapter 3. What is the advantage of using the term "molar mass" when we discuss ionic compounds?

9.13 In which of the following states would NaCl be electrically conducting: (a) solid, (b) molten (that is, melted), (c) dissolved in water? Explain your answers.

9.14 Beryllium forms a compound with chlorine that has the empirical formula $BeCl_2$. How would you determine whether it is an ionic compound? (The compound is not soluble in water.)

Problems

9.15 An ionic bond is formed between a cation A^+ and an anion B^-. How would the energy of the ionic bond [see Equation (9.2)] be affected by the following changes? (a) doubling the radius of A^+, (b) tripling the charge on A^+, (c) doubling the charges on A^+ and B^-, (d) decreasing the radii of A^+ and B^- to half their original values.

9.16 Give the empirical formulas and names of the compounds formed from the following pairs of ions: (a) Rb^+ and I^-, (b) Cs^+ and SO_4^{2-}, (c) Sr^{2+} and N^{3-}, (d) Al^{3+} and S^{2-}.

9.17 Use Lewis dot symbols to show the transfer of electrons between the following atoms to form cations and anions: (a) Na and F, (b) K and S, (c) Ba and O, (d) Al and N.

9.18 Write the Lewis dot symbols of the reactants and products in the following reactions. (First balance the equations.)

(a) $Sr + Se \longrightarrow SrSe$

(b) $Ca + H_2 \longrightarrow CaH_2$

(c) $Li + N_2 \longrightarrow Li_3N$

(d) $Al + S \longrightarrow Al_2S_3$

9.19 For each of the following pairs of elements, state whether the binary compound they form is likely to be ionic or covalent. Write the empirical formula and name of the compound: (a) I and Cl, (b) Mg and F.

9.20 For each of the following pairs of elements, state whether the binary compound they form is likely to be ionic or covalent. Write the empirical formula and name of the compound: (a) B and F, (b) K and Br.

9.3 Lattice Energy of Ionic Compounds

Review Questions

9.21 What is lattice energy and what role does it play in the stability of ionic compounds?

9.22 Explain how the lattice energy of an ionic compound such as KCl can be determined using the Born-Haber cycle. On what law is this procedure based?

9.23 Specify which compound in the following pairs of ionic compounds has the higher lattice energy: (a) KCl or MgO, (b) LiF or LiBr, (c) Mg_3N_2 or NaCl. Explain your choice.

9.24 Compare the stability (in the solid state) of the following pairs of compounds: (a) LiF and LiF_2 (containing the Li^{2+} ion), (b) Cs_2O and CsO (containing the O^- ion), (c) $CaBr_2$ and $CaBr_3$ (containing the Ca^{3+} ion).

Problems

9.25 Use the Born-Haber cycle outlined in Section 9.3 for LiF to calculate the lattice energy of NaCl. [The heat of sublimation of Na is 108 kJ/mol and ΔH_f° (NaCl) = −411 kJ/mol. Energy needed to dissociate $\frac{1}{2}$ mole of Cl_2 into Cl atoms = 121.4 kJ.]

9.26 Calculate the lattice energy of calcium chloride given that the heat of sublimation of Ca is 121 kJ/mol and $\Delta H_f^\circ(CaCl_2) = -795$ kJ/mol. (See Tables 8.2 and 8.3 for other data.)

9.4 The Covalent Bond

Review Questions

9.27 What is Lewis's contribution to our understanding of the covalent bond?

9.28 Use an example to illustrate each of the following terms: *lone pairs, Lewis structure, the octet rule, bond length.*

9.29 What is the difference between a Lewis dot symbol and a Lewis structure?

9.30 How many lone pairs are on the underlined atoms in these compounds? H<u>Br</u>, H_2<u>S</u>, <u>C</u>H_4

9.31 Compare single, double, and triple bonds in a molecule, and give an example of each. For the same bonding atoms, how does the bond length change from single bond to triple bond?

9.32 Compare the properties of ionic compounds and covalent compounds.

9.5 Electronegativity

Review Questions

9.33 Define electronegativity, and explain the difference between electronegativity and electron affinity. Describe in general how the electronegativities of the elements change according to position in the periodic table.

9.34 What is a polar covalent bond? Name two compounds that contain one or more polar covalent bonds.

Problems

9.35 List the following bonds in order of increasing ionic character: the lithium-to-fluorine bond in LiF, the potassium-to-oxygen bond in K_2O, the nitrogen-to-nitrogen bond in N_2, the sulfur-to-oxygen bond in SO_2, the chlorine-to-fluorine bond in ClF_3.

9.36 Arrange the following bonds in order of increasing ionic character: carbon to hydrogen, fluorine to hydrogen, bromine to hydrogen, sodium to chlorine, potassium to fluorine, lithium to chlorine.

9.37 Four atoms are arbitrarily labeled D, E, F, and G. Their electronegativities are as follows: D = 3.8, E = 3.3, F = 2.8, and G = 1.3. If the atoms of these elements form the molecules DE, DG, EG, and DF, how would you arrange these molecules in order of increasing covalent bond character?

9.38 List the following bonds in order of increasing ionic character: cesium to fluorine, chlorine to chlorine, bromine to chlorine, silicon to carbon.

9.39 Classify the following bonds as ionic, polar covalent, or covalent, and give your reasons: (a) the CC bond in H_3CCH_3, (b) the KI bond in KI, (c) the NB bond in H_3NBCl_3, (d) the CF bond in CF_4.

9.40 Classify the following bonds as ionic, polar covalent, or covalent, and give your reasons: (a) the SiSi bond in $Cl_3SiSiCl_3$, (b) the SiCl bond in $Cl_3SiSiCl_3$, (c) the CaF bond in CaF_2, (d) the NH bond in NH_3.

9.6 Writing Lewis Structures

Review Questions

9.41 Summarize the essential features of the Lewis octet rule.

9.42 The octet rule applies mainly to the second-period elements. Explain.

Problems

9.43 Write Lewis structures for the following molecules and ions: (a) NCl_3, (b) OCS, (c) H_2O_2, (d) CH_3COO^-, (e) CN^-, (f) $CH_3CH_2NH_3^+$.

9.44 Write Lewis structures for the following molecules and ions: (a) OF_2, (b) N_2F_2, (c) Si_2H_6, (d) OH^-, (e) CH_2ClCOO^-, (f) $CH_3NH_3^+$.

9.45 Write Lewis structures for the following molecules: (a) ICl, (b) PH_3, (c) P_4 (each P is bonded to three other P atoms), (d) H_2S, (e) N_2H_4, (f) $HClO_3$, (g) $COBr_2$ (C is bonded to O and Br atoms).

9.46 Write Lewis structures for the following molecules: (a) BrF_3, (b) H_2Te, (c) NH_2OH, (d) $POCl_3$ (P is bonded to O and Cl atoms), (e) CH_3CH_2F, (f) NF_3, (g) CH_3NH_2.

9.47 The following Lewis structures for (a) HCN, (b) C_2H_2, (c) SnO_2, (d) BF_3, (e) HOF, (f) HCOF, (g) NF_3 are incorrect. Explain what is wrong with each one and give a correct structure for the molecule. (Relative positions of atoms are shown correctly.)

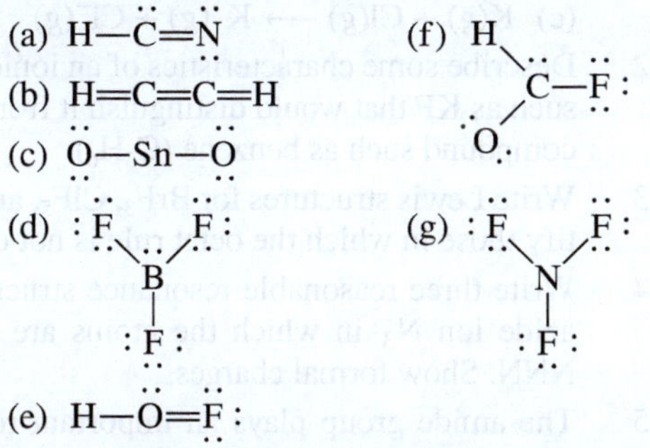

9.48 The skeletal structure of acetic acid shown here is correct, but some of the bonds are wrong. (a) Identify the incorrect bonds and explain what is wrong with them. (b) Write the correct Lewis structure for acetic acid.

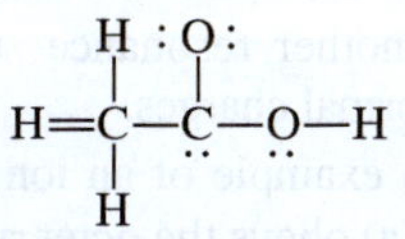

9.7 Formal Charge and Lewis Structure

Review Questions

9.49 Explain the concept of formal charge.

9.50 Do formal charges represent actual separation of charges?

Problems

9.51 Write Lewis structures for the following ions: (a) NO_2^+, (b) S_2^{2-}, (c) BrF_2^+, (d) SCN^-. Show formal charges.

9.52 Write Lewis structures for the following ions: (a) O_2^{2-}, (b) C_2^{2-}, (c) NO^+, (d) NH_4^+. Show formal charges.

9.8 The Concept of Resonance

Review Questions

9.53 Define *bond length, resonance,* and *resonance structure.* What are the rules for writing resonance structures?

9.54 Is it possible to "trap" a resonance structure of a compound for study? Explain.

Problems

9.55 Write Lewis structures for the following species, including all resonance forms, and show formal charges: (a) HCO_2^-, (b) $CH_2NO_2^-$. Relative positions of the atoms are as follows:

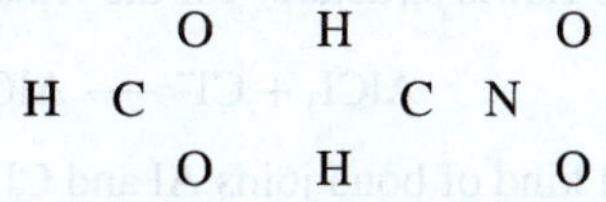

9.56 Draw three resonance structures for the chlorate ion, ClO_3^-. Show formal charges.

9.57 Write three resonance structures for hydrazoic acid, HN_3. The atomic arrangement is HNNN. Show formal charges.

9.58 Draw two resonance structures for diazomethane, CH_2N_2. Show formal charges. The skeletal structure of the molecule is

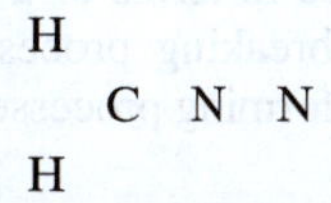

9.59 Draw three resonance structures for the molecule N_2O_3 (atomic arrangement is $ONNO_2$). Show formal charges.

9.60 Draw three reasonable resonance structures for the OCN^- ion. Show formal charges.

9.9 Exceptions to the Octet Rule

Review Questions

9.61 Why does the octet rule not hold for many compounds containing elements in the third period of the periodic table and beyond?

9.62 Give three examples of compounds that do not satisfy the octet rule. Write a Lewis structure for each.

9.63 Because fluorine has seven valence electrons ($2s^22p^5$), seven covalent bonds in principle could form around the atom. Such a compound might be FH_7 or FCl_7. These compounds have never been prepared. Why?

9.64 What is a coordinate covalent bond? Is it different from a normal covalent bond?

Problems

9.65 The AlI_3 molecule has an incomplete octet around Al. Draw three resonance structures of the molecule in which the octet rule is satisfied for both the Al and the I atoms. Show formal charges.

9.66 In the vapor phase, beryllium chloride consists of discrete $BeCl_2$ molecules. Is the octet rule satisfied for Be in this compound? If not, can you form an octet around Be by drawing another resonance structure? How plausible is this structure?

9.67 Of the noble gases, only Kr, Xe, and Rn are known to form a few compounds with O and/or F. Write Lewis structures for the following molecules: (a) XeF_2, (b) XeF_4, (c) XeF_6, (d) $XeOF_4$, (e) XeO_2F_2. In each case Xe is the central atom.

9.68 Write a Lewis structure for $SbCl_5$. Does this molecule obey the octet rule?

9.69 Write Lewis structures for SeF_4 and SeF_6. Is the octet rule satisfied for Se?

9.70 Write Lewis structures for the reaction

$$AlCl_3 + Cl^- \longrightarrow AlCl_4^-$$

What kind of bond joins Al and Cl in the product?

9.10 Bond Enthalpy

Review Questions

9.71 What is bond enthalpy? Bond enthalpies of polyatomic molecules are average values, whereas those of diatomic molecules can be accurately determined. Why?

9.72 Explain why the bond enthalpy of a molecule is usually defined in terms of a gas-phase reaction. Why are bond-breaking processes always endothermic and bond-forming processes always exothermic?

Problems

9.73 From the following data, calculate the average bond enthalpy for the N—H bond:

$$NH_3(g) \longrightarrow NH_2(g) + H(g) \quad \Delta H° = 434 \text{ kJ/mol}$$
$$NH_2(g) \longrightarrow NH(g) + H(g) \quad \Delta H° = 380 \text{ kJ/mol}$$
$$NH(g) \longrightarrow N(g) + H(g) \quad \Delta H° = 359 \text{ kJ/mol}$$

9.74 For the reaction

$$O(g) + O_2(g) \longrightarrow O_3(g) \quad \Delta H° = -106.5 \text{ kJ/mol}$$

Calculate the average bond enthalpy in O_3.

9.75 The bond enthalpy of $F_2(g)$ is 159 kJ/mol. Calculate ΔH_f° for F(*g*).

9.76 For the reaction

$$2C_2H_6(g) + 7O_2(g) \longrightarrow 4CO_2(g) + 6H_2O(g)$$

(a) Estimate the enthalpy of reaction from the average bond enthalpies in Table 9.4.

(b) Calculate the enthalpy of reaction from the standard enthalpies of formation (see Appendix 2) of the reactant and product molecules, and compare the result with your answer for part (a).

Additional Problems

9.77 Classify the following substances as ionic compounds or covalent compounds containing discrete molecules: CH_4, KF, CO, $SiCl_4$, $BaCl_2$.

9.78 Which of the following are ionic compounds? Which are covalent compounds? RbCl, PF_5, BrF_3, KO_2, CI_4

9.79 Match each of the following energy changes with one of the processes given: ionization energy, electron affinity, bond enthalpy, and standard enthalpy of formation.

(a) $F(g) + e^- \longrightarrow F^-(g)$

(b) $F_2(g) \longrightarrow 2F(g)$

(c) $Na(g) \longrightarrow Na^+(g) + e^-$

(d) $Na(s) + \frac{1}{2}F_2(g) \longrightarrow NaF(s)$

9.80 The formulas for the fluorides of the third-period elements are NaF, MgF_2, AlF_3, SiF_4, PF_5, SF_6, and ClF_3. Classify these compounds as covalent or ionic.

9.81 Use ionization energy (see Table 8.2) and electron affinity values (see Table 8.3) to calculate the energy change (in kJ/mol) for the following reactions:

(a) $Li(g) + I(g) \longrightarrow Li^+(g) + I^-(g)$

(b) $Na(g) + F(g) \longrightarrow Na^+(g) + F^-(g)$

(c) $K(g) + Cl(g) \longrightarrow K^+(g) + Cl^-(g)$

9.82 Describe some characteristics of an ionic compound such as KF that would distinguish it from a covalent compound such as benzene (C_6H_6).

9.83 Write Lewis structures for BrF_3, ClF_5, and IF_7. Identify those in which the octet rule is not obeyed.

9.84 Write three reasonable resonance structures for the azide ion N_3^- in which the atoms are arranged as NNN. Show formal charges.

9.85 The amide group plays an important role in determining the structure of proteins:

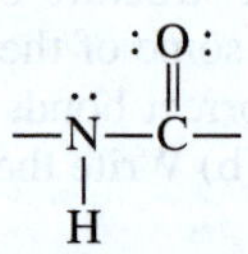

Draw another resonance structure for this group. Show formal charges.

9.86 Give an example of an ion or molecule containing Al that (a) obeys the octet rule, (b) has an expanded octet, and (c) has an incomplete octet.

9.87 Draw four reasonable resonance structures for the PO_3F^{2-} ion. The central P atom is bonded to the three O atoms and to the F atom. Show formal charges.

9.88 Attempts to prepare the compounds listed here as stable species under atmospheric conditions have failed. Suggest possible reasons for the failure. CF_2, LiO_2, $CsCl_2$, PI_5

9.89 Draw reasonable resonance structures for the following ions: (a) HSO_4^-, (b) PO_4^{3-}, (c) HSO_3^-, (d) SO_3^{2-}. (*Hint:* See comment on Example 9.11.)

9.90 Are the following statements true or false? (a) Formal charges represent actual separation of charges. (b) ΔH°_{rxn} can be estimated from the bond enthalpies of reactants and products. (c) All second-period elements obey the octet rule in their compounds. (d) The resonance structures of a molecule can be separated from one another.

9.91 A rule for drawing plausible Lewis structures is that the central atom is invariably less electronegative than the surrounding atoms. Explain why this is so. Why does this rule not apply to compounds like H_2O and NH_3?

9.92 Using the following information and the fact that the average C—H bond enthalpy is 413 kJ/mol, estimate the standard enthalpy of formation of methane (CH_4).

$$C(s) \longrightarrow C(g) \qquad \Delta H^\circ_{rxn} = 717 \text{ kJ/mol}$$
$$2H_2(g) \longrightarrow 4H(g) \qquad \Delta H^\circ_{rxn} = 872 \text{ kJ/mol}$$

9.93 Based on energy considerations, which of the following reactions will occur more readily?
(a) $Cl(g) + CH_4(g) \longrightarrow CH_3Cl(g) + H(g)$
(b) $Cl(g) + CH_4(g) \longrightarrow CH_3(g) + HCl(g)$
(*Hint:* Refer to Table 9.4, and assume that the average bond enthalpy of the C—Cl bond is 328 kJ/mol.)

9.94 Which of the following molecules has the shortest nitrogen-to-nitrogen bond: N_2H_4, N_2O, N_2, N_2O_4? Explain.

9.95 Most organic acids can be represented as RCOOH, where COOH is the carboxyl group and R is the rest of the molecule. (For example, R is CH_3 in acetic acid, CH_3COOH.) (a) Draw a Lewis structure for the carboxyl group. (b) Upon ionization, the carboxyl group is converted to the carboxylate group, COO^-. Draw resonance structures for the carboxylate group.

9.96 Which of the following species are isoelectronic: NH_4^+, C_6H_6, CO, CH_4, N_2, $B_3N_3H_6$?

9.97 The following species have been detected in interstellar space: (a) CH, (b) OH, (c) C_2, (d) HNC, (e) HCO. Draw Lewis structures for these species and indicate whether they are diamagnetic or paramagnetic.

9.98 The amide ion, NH_2^-, is a Brønsted base. Represent the reaction between the amide ion and water.

9.99 Draw Lewis structures for the following organic molecules: (a) tetrafluoroethylene (C_2F_4), (b) propane (C_3H_8), (c) butadiene ($CH_2CHCHCH_2$), (d) propyne (CH_3CCH), (e) benzoic acid (C_6H_5COOH). (To draw C_6H_5COOH, replace a H atom in benzene with a COOH group.)

9.100 The triiodide ion (I_3^-) in which the I atoms are arranged in a straight line is stable, but the corresponding F_3^- ion does not exist. Explain.

9.101 Compare the bond enthalpy of F_2 with the energy change for the following process:

$$F_2(g) \longrightarrow F^+(g) + F^-(g)$$

Which is the preferred dissociation for F_2, energetically speaking?

9.102 Methyl isocyanate (CH_3NCO) is used to make certain pesticides. In December 1984, water leaked into a tank containing this substance at a chemical plant, producing a toxic cloud that killed thousands of people in Bhopal, India. Draw Lewis structures for CH_3NCO, showing formal charges.

9.103 The chlorine nitrate molecule ($ClONO_2$) is believed to be involved in the destruction of ozone in the Antarctic stratosphere. Draw a plausible Lewis structure for this molecule.

9.104 Several resonance structures for the molecule CO_2 are shown. Explain why some of them are likely to be of little importance in describing the bonding in this molecule.

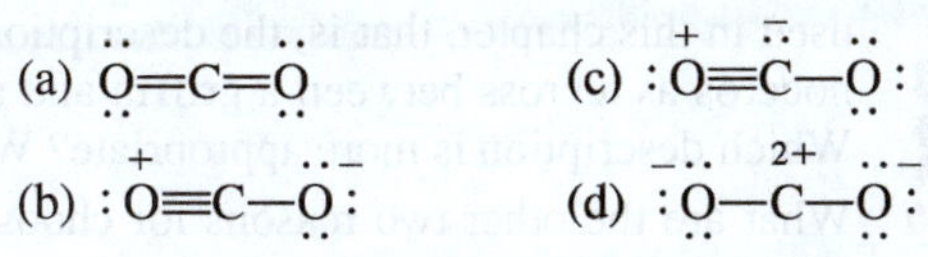

9.105 For each of the following organic molecules draw a Lewis structure in which the carbon atoms are bonded to each other by single bonds: (a) C_2H_6, (b) C_4H_{10}, (c) C_5H_{12}. For (b) and (c), show only structures in which each C atom is bonded to no more than two other C atoms.

9.106 Draw Lewis structures for the following chlorofluorocarbons (CFCs), which are partly responsible for the depletion of ozone in the stratosphere: (a) $CFCl_3$, (b) CF_2Cl_2, (c) CHF_2Cl, (d) CF_3CHF_2.

9.107 Draw Lewis structures for the following organic molecules. In each there is one C═C bond, and the rest of the carbon atoms are joined by C—C bonds. C_2H_3F, C_3H_6, C_4H_8

9.108 Calculate ΔH° for the reaction

$$H_2(g) + I_2(g) \longrightarrow 2HI(g)$$

using (a) Equation (9.3) and (b) Equation (6.18), given that ΔH°_f for $I_2(g)$ is 62.4 kJ/mol.

9.109 Draw Lewis structures for the following organic molecules: (a) methanol (CH_3OH); (b) ethanol (CH_3CH_2OH); (c) tetraethyllead [$Pb(CH_2CH_3)_4$], which was used in "leaded gasoline"; (d) methylamine (CH_3NH_2), which is used in tanning; (e) mustard gas ($ClCH_2CH_2SCH_2CH_2Cl$), a poisonous gas used in World War I; (f) urea [$(NH_2)_2CO$], a fertilizer; and (g) glycine (NH_2CH_2COOH), an amino acid.

9.110 Write Lewis structures for the following four isoelectronic species: (a) CO, (b) NO^+, (c) CN^-, (d) N_2. Show formal charges.

9.111 Oxygen forms three types of ionic compounds in which the anions are oxide (O^{2-}), peroxide (O_2^{2-}), and superoxide (O_2^-). Draw Lewis structures of these ions.

9.112 Comment on the correctness of the statement, "All compounds containing a noble gas atom violate the octet rule."

9.113 Write three resonance structures for (a) the cyanate ion (NCO^-) and (b) the isocyanate ion (CNO^-). In each case, rank the resonance structures in order of increasing importance.

9.114 (a) From the following data calculate the bond enthalpy of the F_2^- ion.

$$F_2(g) \longrightarrow 2F(g) \qquad \Delta H^\circ_{rxn} = 159 \text{ kJ/mol}$$
$$F^-(g) \longrightarrow F(g) + e^- \qquad \Delta H^\circ_{rxn} = 328 \text{ kJ/mol}$$
$$F_2^-(g) \rightarrow F_2(g) + e^- \qquad \Delta H^\circ_{rxn} = 290 \text{ kJ/mol}$$

(b) Explain the difference between the bond enthalpies of F_2 and F_2^-.

9.115 The resonance concept is sometimes described by analogy to a mule, which is a cross between a horse and a donkey. Compare this analogy with the one used in this chapter, that is, the description of a rhinoceros as a cross between a griffin and a unicorn. Which description is more appropriate? Why?

9.116 What are the other two reasons for choosing (b) in Example 9.7?

9.117 In the Chemistry in Action essay "Just Say NO," nitric oxide is said to be one of about ten of the smallest stable molecules known. Based on what you have learned in the course so far, write all the diatomic molecules you know, give their names, and show their Lewis structures.

9.118 The N—O bond distance in nitric oxide is 115 pm, which is intermediate between a triple bond (106 pm) and a double bond (120 pm). (a) Draw two resonance structures for NO and comment on their relative importance. (b) Is it possible to draw a resonance structure having a triple bond between the atoms?

9.119 Write the formulas of the binary hydride for the second-period elements LiH to HF. Comment on the change from ionic to covalent character of these compounds. Note that beryllium behaves differently from the rest of the Group 2 metals (see Section 8.6).

9.120 Hydrazine borane, $NH_2NH_2BH_3$, has been proposed as a hydrogen storage material. When reacted with lithium hydride (LiH), hydrogen gas is released:

$$NH_2NH_2BH_3 + LiH \longrightarrow LiNH_2NHBH_3 + H_2$$

Write Lewis structures for $NH_2NH_2BH_3$ and $NH_2NHBH_3^-$ and assign all formal charges.

9.121 Although nitrogen dioxide (NO_2) is a stable compound, there is a tendency for two such molecules to combine to form dinitrogen tetroxide (N_2O_4). Why? Draw four resonance structures of N_2O_4, showing formal charges.

9.122 Another possible skeletal structure for the CO_3^{2-} (carbonate) ion besides the one presented in Example 9.5 is O C O O. Why would we not use this structure to represent CO_3^{2-}?

9.123 Draw a Lewis structure for nitrogen pentoxide (N_2O_5) in which each N is bonded to three O atoms.

9.124 In the gas phase, aluminum chloride exists as a dimer (a unit of two) with the formula Al_2Cl_6. Its skeletal structure is given by

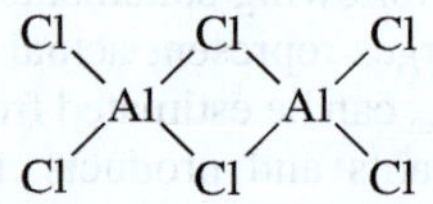

Complete the Lewis structure and indicate the coordinate covalent bonds in the molecule.

9.125 The hydroxyl radical (OH) plays an important role in atmospheric chemistry. It is highly reactive and has a tendency to combine with a H atom from other compounds, causing them to break up. Thus, OH is sometimes called a "detergent" radical because it helps to clean up the atmosphere. (a) Write the Lewis structure for the radical. (b) Refer to Table 9.4 and explain why the radical has a high affinity for H atoms. (c) Estimate the enthalpy change for the reaction

$$OH(g) + CH_4(g) \longrightarrow CH_3(g) + H_2O(g)$$

(d) The radical is generated when sunlight hits water vapor. Calculate the maximum wavelength (in nanometers) required to break an O—H bond in H_2O.

9.126 Experiments show that it takes 1656 kJ/mol to break all the bonds in methane (CH_4) and 4006 kJ/mol to break all the bonds in propane (C_3H_8). Based on these data, calculate the average bond enthalpy of the C—C bond.

9.127 Calculate ΔH°_{rxn} at 25°C of the reaction between carbon monoxide and hydrogen shown here using both bond enthalpy and ΔH°_f values.

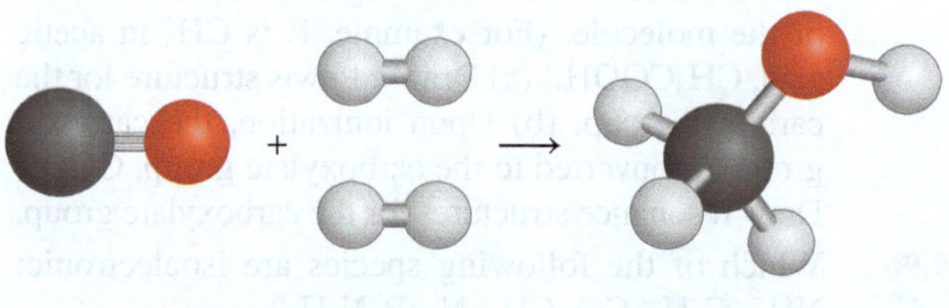

9.128 Calculate ΔH°_{rxn} at 25°C of the reaction between ethylene and chlorine shown here using both bond enthalpy and ΔH°_f values. (ΔH°_f for $C_2H_4Cl_2$ is −132 kJ/mol.)

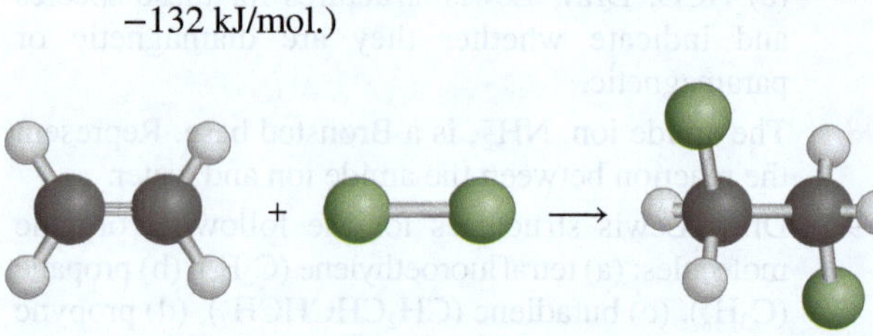

9.129 Draw three resonance structures of sulfur dioxide (SO_2). Indicate the most plausible structure(s).

9.130 Vinyl chloride (C_2H_3Cl) differs from ethylene (C_2H_4) in that one of the H atoms is replaced with a Cl atom. Vinyl chloride is used to prepare poly(vinyl chloride), which is an important polymer used in pipes. (a) Draw the Lewis structure of vinyl chloride. (b) The repeating unit in poly(vinyl chloride) is —CH_2—CHCl—. Draw a portion of the molecule showing three such repeating units. (c) Calculate the enthalpy change when 1.0×10^3 kg of vinyl chloride forms poly(vinyl chloride).

9.131 In 1998 scientists using a special type of electron microscope were able to measure the force needed to break a *single* chemical bond. If 2.0×10^{29} N was needed to break a C—Si bond, estimate the bond enthalpy in kJ/mol. Assume that the bond had to be stretched by a distance of 2 Å (2×10^{-10} m) before it is broken.

9.132 The American chemist Robert S. Mulliken suggested a different definition for the electronegativity (EN) of an element, given by

$$EN = \frac{IE + EA}{2}$$

where IE is the first ionization energy and EA the electron affinity of the element. Calculate the electronegativities of O, F, and Cl using this equation. Compare the electronegativities of these elements on the Mulliken and Pauling scale. (To convert to the Pauling scale, divide each EN value by 230 kJ/mol.)

9.133 Among the common inhaled anesthetics are:

halothane: $CF_3CHClBr$
enflurane: $CHFClCF_2OCHF_2$
isoflurane: $CF_3CHClOCHF_2$
methoxyflurane: $CHCl_2CF_2OCH_3$

Draw Lewis structures of these molecules.

9.134 A student in your class claims that magnesium oxide actually consists of Mg^+ and O^- ions, not Mg^{2+} and O^{2-} ions. Suggest some experiments one could do to show that your classmate is wrong.

9.135 Shown here is a skeletal structure of borazine ($B_3N_3H_6$). Draw two resonance structures of the molecule, showing all the bonds and formal charges. Compare its properties with the isoelectronic molecule benzene.

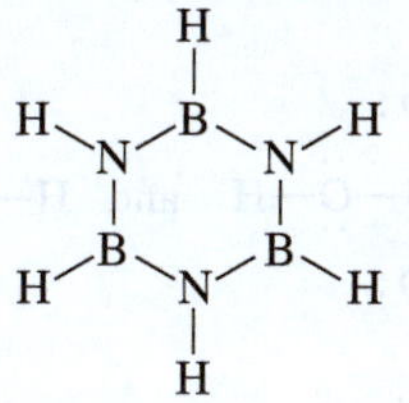

9.136 Calculate the wavelength of light needed to carry out the reaction

$$H_2 \longrightarrow H^+ + H^-$$

9.137 Sulfuric acid (H_2SO_4), the most important industrial chemical in the world, is prepared by oxidizing sulfur to sulfur dioxide and then to sulfur trioxide. Although sulfur trioxide reacts with water to form sulfuric acid, it forms a mist of fine droplets of H_2SO_4 with water vapor that is hard to condense. Instead, sulfur trioxide is first dissolved in 98 percent sulfuric acid to form oleum ($H_2S_2O_7$). On treatment with water, concentrated sulfuric acid can be generated. Write equations for all the steps and draw Lewis structures of oleum based on the discussion in Example 9.11.

9.138 From the lattice energy of KCl in Table 9.1 and the ionization energy of K and electron affinity of Cl in Tables 8.2 and 8.3, calculate the $\Delta H°$ for the reaction

$$K(g) + Cl(g) \longrightarrow KCl(s)$$

9.139 The species H_3^+ is the simplest polyatomic ion. The geometry of the ion is that of an equilateral triangle. (a) Draw three resonance structures to represent the ion. (b) Given the following information

$$2H + H^+ \longrightarrow H_3^+ \qquad \Delta H° = -849 \text{ kJ/mol}$$

$$\text{and} \qquad H_2 \longrightarrow 2H \qquad \Delta H° = 436.4 \text{ kJ/mol}$$

calculate $\Delta H°$ for the reaction

$$H^+ + H_2 \longrightarrow H_3^+$$

9.140 The bond enthalpy of the C—N bond in the amide group of proteins (see Problem 9.85) can be treated as an average of C—N and C═N bonds. Calculate the maximum wavelength of light needed to break the bond.

9.141 In 1999 an unusual cation containing only nitrogen (N_5^+) was prepared. Draw three resonance structures of the ion, showing formal charges. (*Hint:* The N atoms are joined in a linear fashion.)

9.142 Nitroglycerin, one of the most commonly used explosives, has the structure

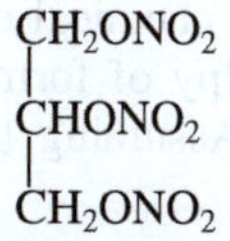

The decomposition reaction is

$$4C_3H_5N_3O_9(l) \longrightarrow 12CO_2(g) + 10H_2O(g) + 6N_2(g) + O_2(g)$$

The explosive action is the result of the heat released and the large increase in gaseous volume. (a) Calculate the $\Delta H°$ for the decomposition of one mole of nitroglycerin using both standard enthalpy of formation values and bond enthalpies. Assume that the two O atoms in the NO_2 groups are attached to N with one single bond and one double bond. (b) Calculate the combined volume of the gases at STP. (c) Assuming an initial explosion temperature of 3000 K, estimate the pressure exerted by the gases using the result from (b). (The standard enthalpy of formation of nitroglycerin is −371.1 kJ/mol.)

9.143 Give a brief description of the medical uses of the following ionic compounds: $AgNO_3$, $BaSO_4$, $CaSO_4$, KI, Li_2CO_3, $Mg(OH)_2$, $MgSO_4$, $NaHCO_3$,

Na_2CO_3, NaF, TiO_2, ZnO. You would need to do a Web search of some of these compounds.

9.144 Use Table 9.4 to estimate the bond enthalpy of the C—C, N—N, and O—O bonds in C_2H_6, N_2H_4, and H_2O_2, respectively. What effect do lone pairs on adjacent atoms have on the strength of the particular bonds?

9.145 The isolated O^{2-} ion is unstable so it is not possible to measure the electron affinity of the O^- ion directly. Show how you can calculate its value by using the lattice energy of MgO and the Born-Haber cycle. [Useful information: $Mg(s) \longrightarrow Mg(g)\ \Delta H° = 148$ kJ/mol.]

9.146 When irradiated with light of wavelength 471.7 nm, the chlorine molecule dissociates into chlorine atoms. One Cl atom is formed in its ground electronic state while the other is in an excited state that is 10.5 kJ/mol above the ground state. What is the bond enthalpy of the Cl_2 molecule?

9.147 Recall from Chapter 8 that the product of the reaction between Xe(*g*) and $PtF_6(g)$ was originally thought to be an ionic compound composed of Xe^+ cations and PtF_6^- anions (see Figure 8.22). This prediction was based on the theoretical enthalpy of formation of $XePtF_6$ calculated using a Born-Haber cycle. (a) The lattice energy for $XePtF_6$ was estimated to be 460 kJ/mol. Explain whether or not this value is consistent with the lattice energies in Table 9.1. (b) Calculate $\Delta H_f°$ for $XePtF_6$ given IE_1 for Xe(*g*) is 1170 kJ/mol and EA_1 for $PtF_6(g)$ is 770 kJ/mol. Comment on the expected stability of $XePtF_6$ based on your calculation.

Interpreting, Modeling, & Estimating

9.148 The reaction between fluorine (F_2) with ethane (C_2H_6) produces predominantly CF_4 rather than C_2F_6 molecules. Explain.

9.149 A new allotrope of oxygen, O_4, has been reported. The exact structure of O_4 is unknown, but the simplest possible structure would be a four-member ring consisting of oxygen-oxygen single bonds. The report speculated that the O_4 molecule might be useful as a fuel "because it packs a lot of oxygen in a small space, so it might be even more energy-dense than the liquefied ordinary oxygen used in rocket fuel." (a) Draw a Lewis structure for O_4 and write a balanced chemical equation for the reaction between ethane, $C_2H_6(g)$, and $O_4(g)$ to give carbon dioxide and water vapor. (b) Estimate $\Delta H°$ for the reaction. (c) Write a chemical equation illustrating the standard enthalpy of formation of $O_4(g)$ and estimate $\Delta H_f°$. (d) Assuming the oxygen allotropes are in excess, which will release more energy when reacted with ethane (or any other fuel): $O_2(g)$ or $O_4(g)$? Explain using your answers to parts (a)–(c).

9.150 Because bond formation is exothermic, when two gas-phase atoms come together to form a diatomic molecule it is necessary for a third atom or molecule to absorb the energy that is released. Otherwise the molecule will undergo dissociation. If two atoms of hydrogen combine to form $H_2(g)$, what would be the increase in velocity of a third hydrogen atom that absorbs the energy released from this process?

9.151 Estimate $\Delta H_f°$ for sodium astatide (NaAt) according to the equation

$$Na(s) + \tfrac{1}{2}At_2(s) \longrightarrow NaAt(s)$$

The information in Problem 8.147 may be useful.

Answers to Practice Exercises

9.1 $Ba\cdot + 2\cdot H \longrightarrow Ba^{2+}\ 2H{:}^-$ (or BaH_2)
$[Xe]6s^2$ $1s^1$ [Xe] [He]

9.2 (a) Ionic. (b) Polar covalent. (c) Covalent.

9.3 S=C=S **9.4**

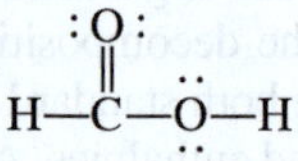

9.5 [O=N—O:]⁻

9.6 O=N—O:⁻ **9.7** H—C≡N:

9.8 S=C=N⁻ ⟷ :S—C≡N: ⟷ :S≡C—N:²⁻

The first structure is the most important; the last structure is the least important.

9.9 :F—Be—F: **9.10**

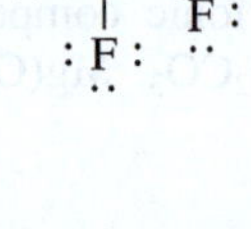

9.11

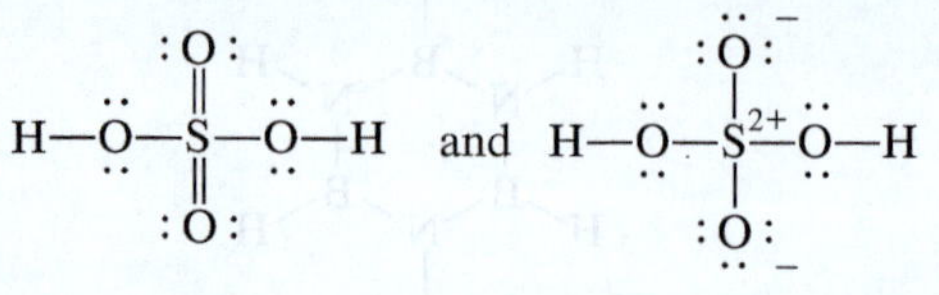

9.12

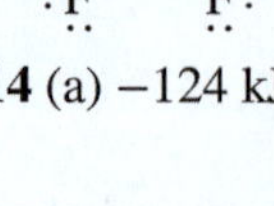

9.13 (a) −545 kJ/mol. (b) −546.6 kJ/mol.

9.14 (a) −124 kJ/mol. (b) −136.4 kJ/mol.

Answers to Review of Concepts & Facts

9.1.1 7. **9.1.2** · Se · **9.1.3** 5A.

9.2.1 2 ·Rb + ·S· ⟶ 2 Rb^+ :S:$^{2-}$

[Kr]$5s^1$ [Ar]$3s^1 3p^4$ [Kr] [Ar]

(or Rb_2S). **9.3.1** LiCl. **9.3.2** NaF < MgO < AlN. **9.4.1** Hydrogen can only use the 1*s* orbital for bonding, so it only forms single bonds; *p* orbitals (or higher orbitals) are required to form double and triple bonds. **9.4.2** N—N > N=N > N≡N. **9.5.1** P—P < B—H < C—Cl. **9.5.2** (a) Polar covalent. (b) Covalent. (c) Ionic. **9.5.3** Left: LiH. Right: HCl.

9.6.1 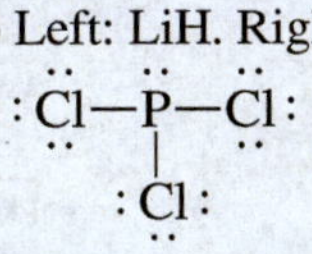 **9.6.2** [:C≡N:]$^-$

9.6.3 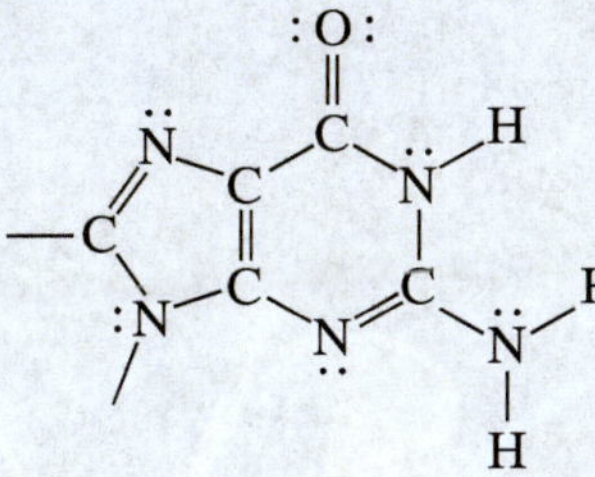

9.7.1 H: 0, C: 0, N: 0. **9.7.2** O: –2, Cl: +1. **9.7.3** (b).

9.8.1 O=S—O: ⟷ :O—S=O

9.8.2 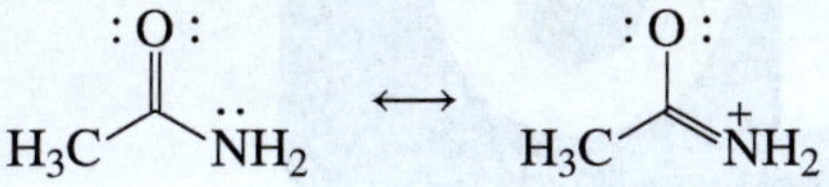

9.9.1 Aluminum is in the third period meaning it has 3*d* orbitals that can be used in bonding.

9.9.2 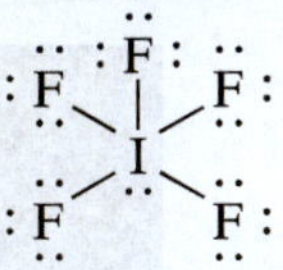 **9.9.3** :Cl—O·

9.10.1 In many cases, average bond enthalpies are used in the calculation. **9.10.2** –846 kJ/mol.

CHAPTER 10

Structure and Bonding Theories

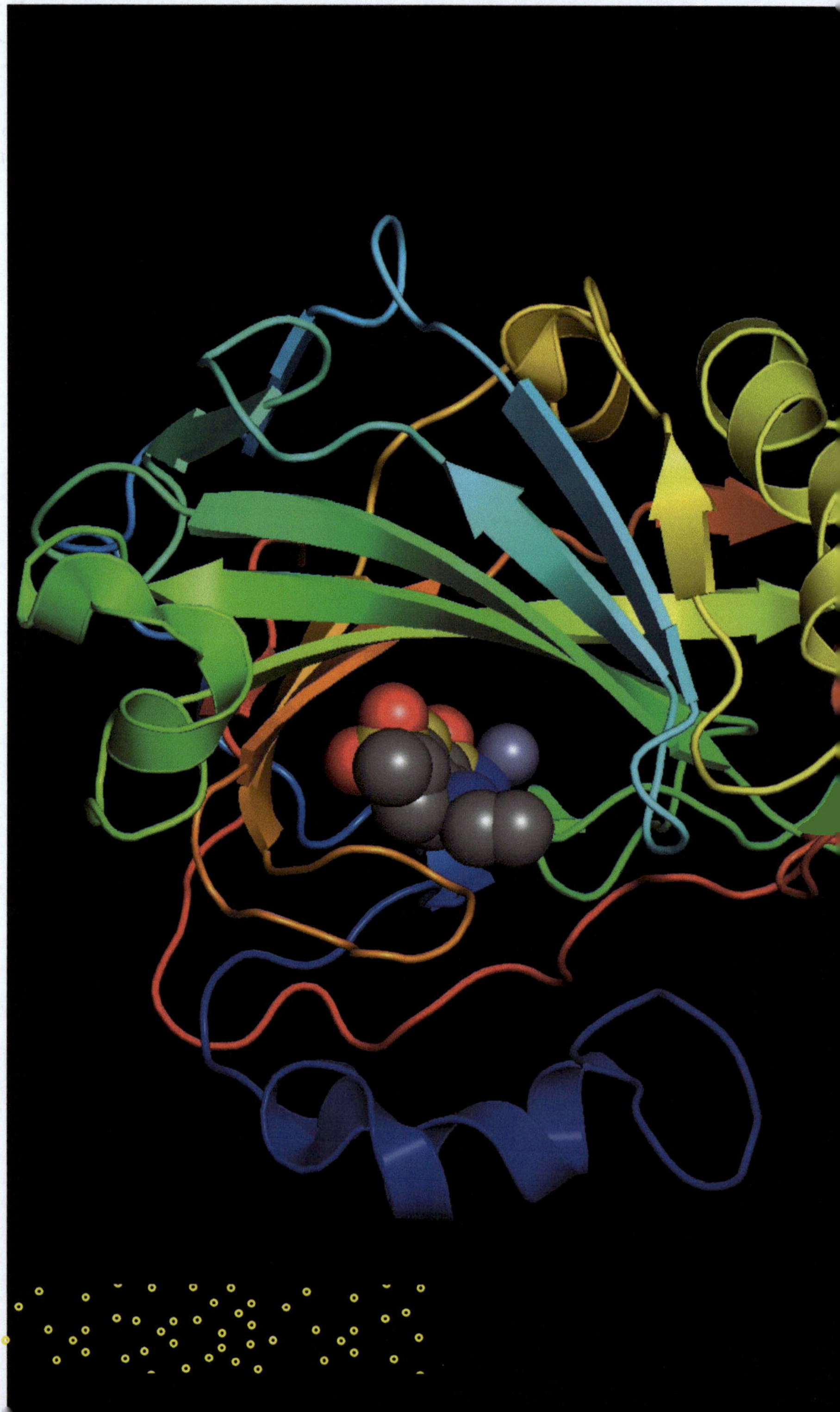

The drug dorzolamide bound to the active site of carbonic anhydrase. Dorzolamide is the first drug to market that resulted from structure-based drug design.
Courtesy of Jason Overby

CHAPTER OUTLINE

In Chapter 9, we discussed bonding in terms of the Lewis theory. Here we will study the shape, or geometry, of molecules. Geometry has an important influence on the physical and chemical properties of molecules, such as density, melting point, boiling point, and reactivity. We will see that we can predict the shapes of molecules with considerable accuracy using a simple method based on Lewis structures.

The Lewis theory of chemical bonding, although useful and easy to apply, does not explain how and why bonds form. A proper understanding of bonding comes from quantum mechanics. Therefore, in the second part of the chapter we will apply quantum mechanics to the study of the geometry and stability of molecules.

10.1 Molecular Geometry

Learning Objectives

- Apply the valence-shell electron-pair repulsion (VSEPR) model to predict the shape of a molecule or polyatomic ion.
- Predict deviations from ideal bond angles in structures based on the presence of lone pairs on a central atom.

Molecular geometry is the three-dimensional arrangement of atoms in a molecule. A molecule's geometry affects its physical and chemical properties, such as melting point, boiling point, density, and the types of reactions it undergoes. In general, bond lengths and bond angles must be determined by experiment. However, there is a simple procedure that enables us to predict with considerable success the overall geometry of a molecule or ion if we know the number of electrons (which we often refer to as electron domains; electron domains are either a pair of electrons or a bond, regardless if the bond is single, double, or triple) surrounding a central atom in its Lewis structure. The term "central atom" means an atom that is not a terminal atom in a polyatomic molecule. The basis of this approach is the assumption that electron pairs in the valence shell of an atom repel one another. The ***valence shell*** is *the outermost electron-occupied shell of an atom; it holds the electrons that are usually involved in bonding.*

In a covalent bond, a pair of electrons, often called the *bonding pair,* is responsible for holding two atoms together. However, in a polyatomic molecule, where there are two or more bonds between the central atom and the surrounding atoms, the repulsion between electrons in different bonding pairs causes them to remain as far apart as possible. The geometry that the molecule ultimately assumes (as defined by the positions of all the atoms) minimizes the repulsion. This approach to the study of molecular geometry is called the ***valence-shell electron-pair repulsion (VSEPR) model,*** because *it accounts for the geometric arrangements of electron pairs around a central atom in terms of the electrostatic repulsion between electron pairs.*

Video: VSEPR

Video: VSEPR Theory

Two general rules govern the use of the VSEPR model:

1. As far as electron-pair repulsion is concerned, double bonds and triple bonds can be treated like single bonds. This approximation is good for qualitative purposes. However, you should realize that in reality multiple bonds are "larger" than single bonds; that is, because there are two or three bonds between two atoms, the electron density occupies more space.
2. If a molecule has two or more resonance structures, we can apply the VSEPR model to any one of them. Formal charges are usually not shown.

With this model in mind, we can predict the geometry of molecules (and ions) in a systematic way. For this purpose, it is convenient to divide molecules into two categories, according to whether or not the central atom has lone pairs.

Molecules in Which the Central Atom Has No Lone Pairs

For simplicity we will consider molecules that contain atoms of only two elements, A and B, of which A is the central atom. These molecules have the general formula AB_x, where x is an integer 2, 3, ... (If $x = 1$, we have the diatomic molecule AB, which is linear by definition.) In the vast majority of cases, x is between 2 and 6.

Table 10.1 shows five possible arrangements of electron pairs around the central atom A. As a result of mutual repulsion, the electron pairs stay as far from one another as possible. Note that the table shows arrangements of the electron pairs but not the positions of the atoms that surround the central atom. Molecules in which the central atom has no lone pairs have one of these five arrangements of bonding pairs. Using Table 10.1 as a reference, let us take a close look at the geometry of molecules with the formulas AB_2, AB_3, AB_4, AB_5, and AB_6.

AB_2: Beryllium Chloride ($BeCl_2$)

The Lewis structure of beryllium chloride in the gaseous state is

$$:\ddot{\text{Cl}}—\text{Be}—\ddot{\text{Cl}}:$$

Because the bonding pairs repel each other, they must be at opposite ends of a straight line for them to be as far apart as possible. Thus, the Cl—Be—Cl angle is predicted to be 180°, and the molecule is linear (see Table 10.1). The "ball-and-stick" model of $BeCl_2$ is

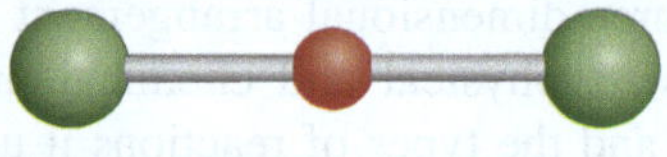

Table 10.1 Electron-Pair Geometry and Molecular Geometry

Number of Electron Pairs	Arrangement of Electron Pairs*	Electron-Pair Geometry	Molecular Geometry*
2	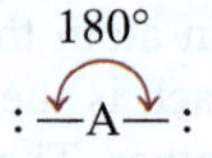 Linear	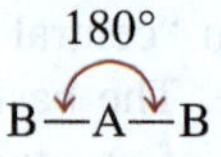 Linear	
3	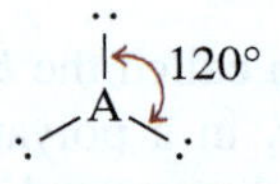 Trigonal planar	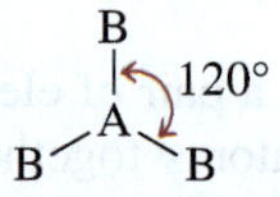 Trigonal planar	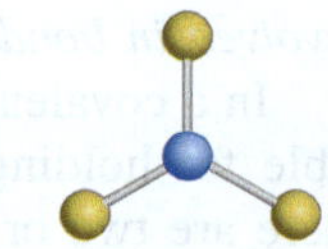
4	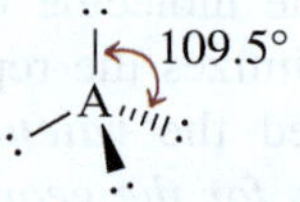 Tetrahedral	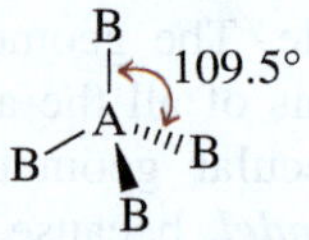 Tetrahedral	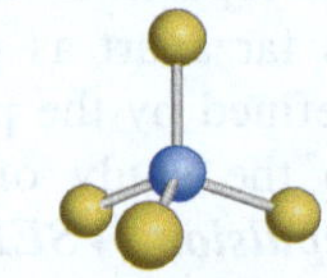
5	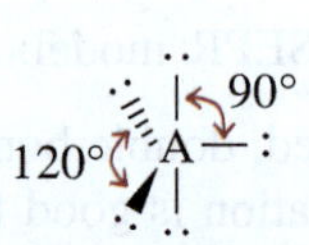 Trigonal bipyramidal	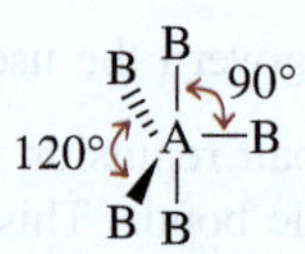 Trigonal bipyramidal	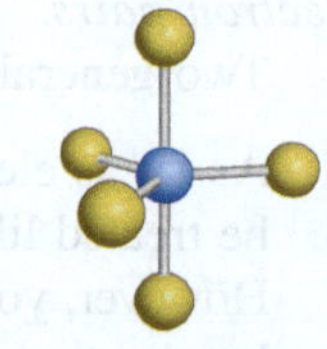
6	 Octahedral	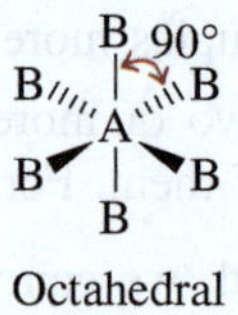 Octahedral	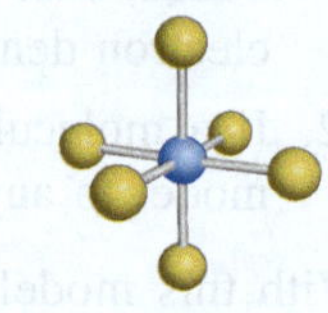

*Bonds coming out of the page are represented as solid wedges. Bonds going into the page are represented as dashed wedges. Bonds in the plane of the page are represented as solid lines.

AB_3: Boron Trifluoride (BF_3)

Boron trifluoride contains three covalent bonds, or bonding pairs.

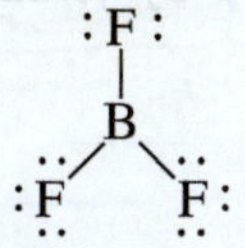

In the most stable arrangement, the three B—F bonds point to the corners of an equilateral triangle with B in the center of the triangle. According to Table 10.1, the geometry of BF_3 is trigonal planar because all four atoms lie in the same plane and the three end atoms form an equilateral triangle:

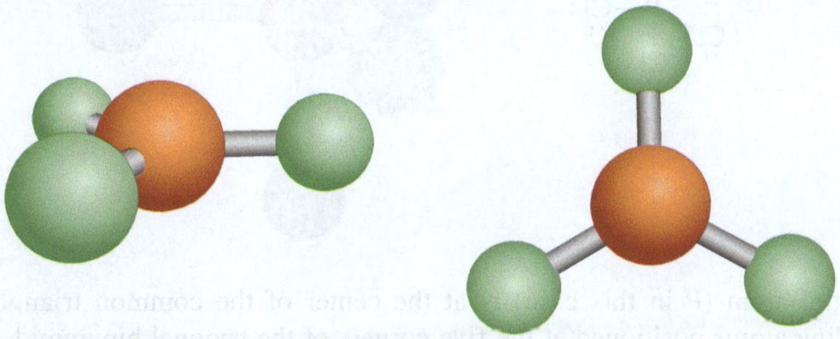

Thus, each of the three F—B—F angles is 120°.

AB_4: Methane (CH_4)

The Lewis structure of methane is

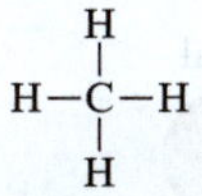

Because there are four bonding pairs, the geometry of CH_4 is tetrahedral (see Table 10.1). A *tetrahedron* has four sides (the prefix *tetra-* means "four"), or faces, all of which are equilateral triangles. In a tetrahedral molecule, the central atom (C in this case) is located at the center of the tetrahedron and the other four atoms are at the corners. The bond angles are all 109.5°. While Lewis structures by themselves do not imply any geometry, it is possible to represent the three-dimensional structure using different bond styles. A solid line is used to show a bond in a plane, a wedge shows a bond in front of the plane, and a dashed wedge shows a bond behind the plane.

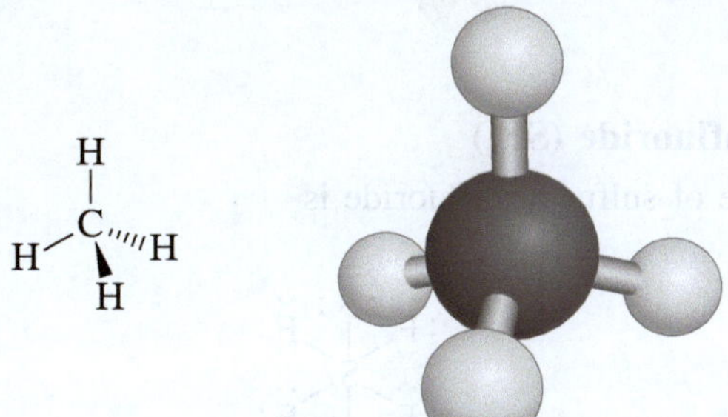

AB_5: Phosphorus Pentachloride (PCl_5)

The Lewis structure of phosphorus pentachloride (in the gas phase) is

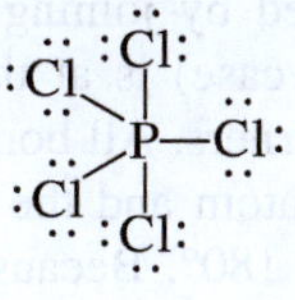

The only way to minimize the repulsive forces among the five bonding pairs is to arrange the P—Cl bonds in the form of a trigonal bipyramid (see Table 10.1). A trigonal bipyramid can be generated by joining two tetrahedrons along a common triangular base:

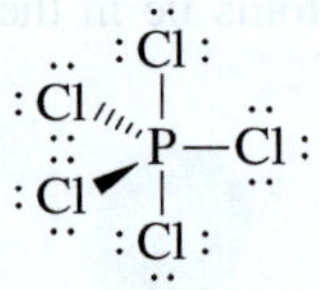

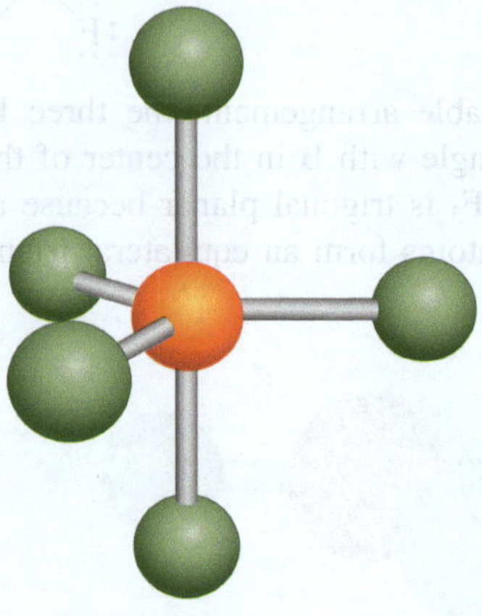

The central atom (P in this case) is at the center of the common triangle with the surrounding atoms positioned at the five corners of the trigonal bipyramid. The atoms that are above and below the triangular plane are said to occupy *axial* positions, and those that are in the triangular plane are said to occupy *equatorial* positions. The angle between any two equatorial bonds is 120°; that between an axial bond and an equatorial bond is 90°, and that between the two axial bonds is 180°.

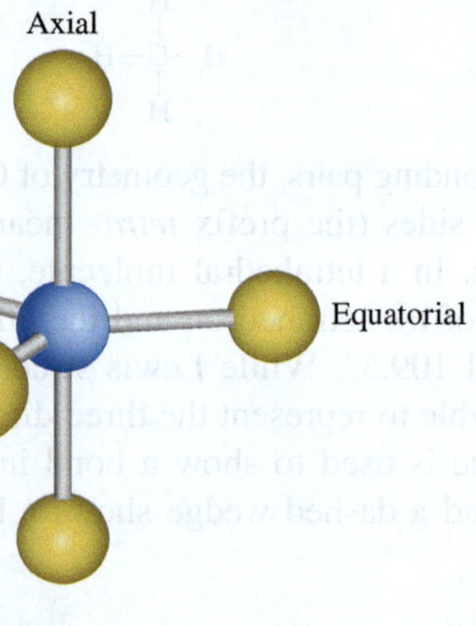

AB_6: Sulfur Hexafluoride (SF_6)

The Lewis structure of sulfur hexafluoride is

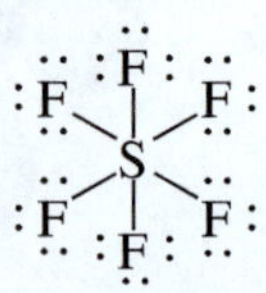

The most stable arrangement of the six S—F bonding pairs is in the shape of an octahedron, as shown in Table 10.1. An octahedron has eight sides (the prefix *octa-* means "eight"). It can be generated by joining two square pyramids on a common base. The central atom (S in this case) is at the center of the square base and the surrounding atoms are at the six corners. All bond angles are 90° except the one made by the bonds between the central atom and the pairs of atoms that are diametrically opposite each other. That angle is 180°. Because the six bonds are equivalent in an octahedral molecule, we cannot use the terms *axial* and *equatorial* as in a trigonal bipyramidal molecule.

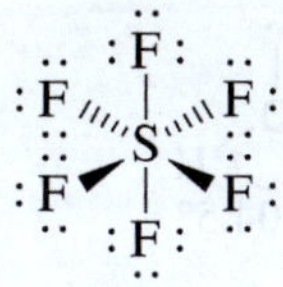

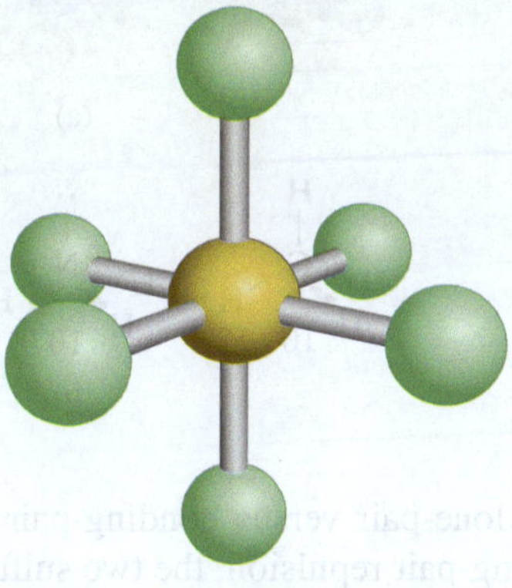

Molecules in Which the Central Atom Has One or More Lone Pairs

Determining the geometry of a molecule is more complicated if the central atom has both lone pairs and bonding pairs. In such molecules there are three types of repulsive forces—those between bonding pairs, those between lone pairs, and those between a bonding pair and a lone pair. In general, according to the VSEPR model, the repulsive forces decrease in the following order:

lone-pair vs. lone-pair repulsion > lone-pair vs. bonding-pair repulsion > bonding-pair vs. bonding-pair repulsion

Electrons in a bond are held by the attractive forces exerted by the nuclei of the two bonded atoms. These electrons have less "spatial distribution" than lone pairs; that is, they take up less space than lone-pair electrons, which are associated with only one particular atom. Because lone-pair electrons in a molecule occupy more space, they experience greater repulsion from neighboring lone pairs and bonding pairs. To keep track of the total number of bonding pairs and lone pairs, we designate molecules with lone pairs as AB_xE_y, where A is the central atom, B is a surrounding atom, and E is a lone pair on A. Both x and y are integers; $x = 2, 3, \ldots$, and $y = 1, 2, \ldots$. Thus, the values of x and y indicate the number of surrounding atoms and number of lone pairs on the central atom, respectively. For $x = 1$, we have a diatomic molecule, which by definition has a linear geometry. The simplest such molecule would be a triatomic molecule with one lone pair on the central atom and the formula is AB_2E.

As the following examples show, in most cases the presence of lone pairs on the central atom makes it difficult to predict the bond angles accurately.

AB_2E: Sulfur Dioxide (SO_2)

The Lewis structure of sulfur dioxide is

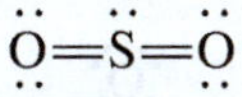

Because VSEPR treats double bonds as though they were single (i.e., double and triple bonds count as one electron domain), the SO_2 molecule can be viewed as consisting of three electron pairs on the central S atom. Of these, two are bonding pairs and one is a lone pair. In Table 10.1 we see that the overall arrangement of three electron pairs is trigonal planar. But because one of the electron pairs is a lone pair, the SO_2 molecule has a "bent" shape.

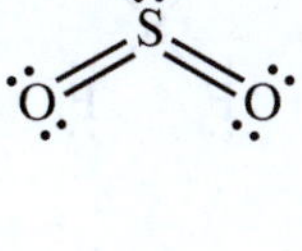

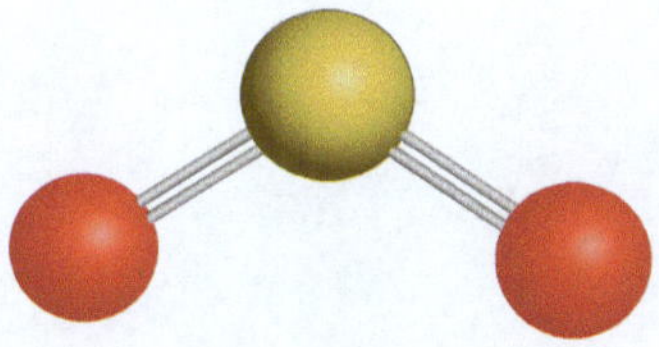

SO_2

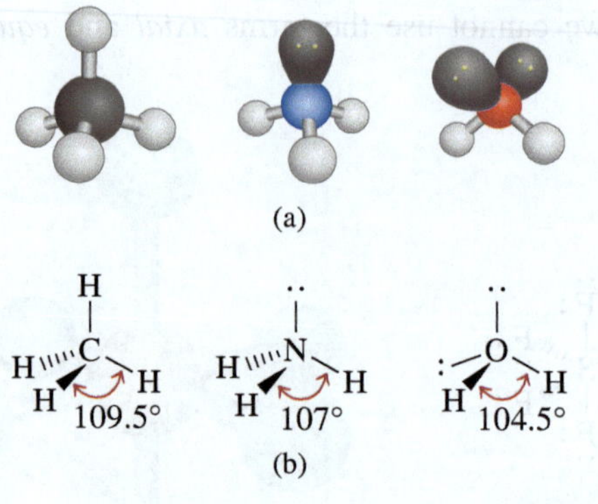

Figure 10.1 *(a) The relative sizes of bonding pairs and lone pairs in CH_4, NH_3, and H_2O. (b) The bond angles in CH_4, NH_3, and H_2O. Note that the solid wedges represent electron pairs (bonding pairs or lone pairs) coming out above the plane of the paper, the dashed wedges represent electron pairs going back behind the plane of the paper, and the solid lines represent electron pairs in the plane of the paper.*

Because the lone-pair versus bonding-pair repulsion is greater than the bonding-pair versus bonding-pair repulsion, the two sulfur-to-oxygen bonds are pushed together slightly and the O—S—O angle is less than 120°.

AB_3E: Ammonia (NH_3)

The ammonia molecule contains three bonding pairs and one lone pair:

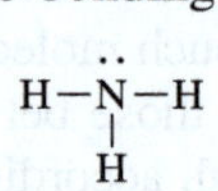

As Table 10.1 shows, the overall arrangement of four electron pairs is tetrahedral. But in NH_3 one of the electron pairs is a lone pair, so the geometry of NH_3 is trigonal pyramidal (so-called because it looks like a pyramid, with the N atom at the apex). Because the lone pair repels the bonding pairs more strongly, the three N—H bonding pairs are pushed closer together:

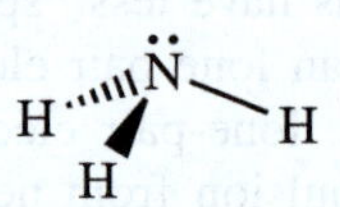

Thus, the H—N—H angle in ammonia is smaller than the ideal tetrahedral angle of 109.5° (Figure 10.1).

AB_2E_2: Water (H_2O)

A water molecule contains two bonding pairs and two lone pairs:

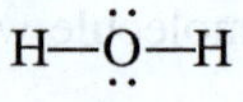

The overall arrangement of the four electron pairs in water is tetrahedral, the same as in ammonia. However, unlike ammonia, water has two lone pairs on the central O atom. These lone pairs tend to be as far from each other as possible. Consequently, the two O—H bonding pairs are pushed toward each other, and we predict an even greater deviation from the tetrahedral angle than in NH_3. As Figure 10.1 shows, the H—O—H angle is 104.5°. The geometry of H_2O is bent:

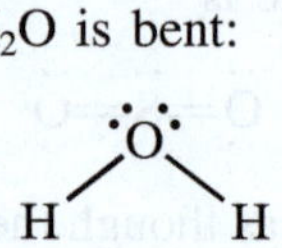

AB_4E: Sulfur Tetrafluoride (SF_4)

The Lewis structure of SF_4 is

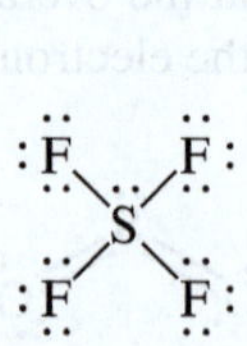

The central sulfur atom has five electron pairs whose arrangement, according to Table 10.1, is trigonal bipyramidal. In the SF_4 molecule, however, one of the electron pairs is a lone pair, so the molecule must have one of the following geometries:

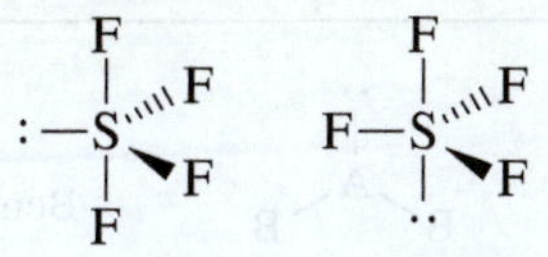

(a) (b)

SF_4

In (a) the lone pair occupies an equatorial position, and in (b) it occupies an axial position. The axial position has three neighboring pairs at 90° and one at 180°, while the equatorial position has two neighboring pairs at 90° and two more at 120°. The repulsion is smaller for (a), and indeed (a) is the structure observed experimentally. This shape is sometimes described as a distorted tetrahedron (or more commonly as seesaw if you turn the structure 90° to the right to view it). The angle between the axial F atoms and S is 173°, and that between the equatorial F atoms and S is 102°.

Table 10.2 shows the geometries of simple molecules in which the central atom has one or more lone pairs, including some that we have not discussed.

Geometry of Molecules with More Than One Central Atom

So far we have discussed the geometry of molecules having only one central atom. The overall geometry of molecules with more than one central atom is difficult to define in most cases. Often we can only describe the shape around each of the central atoms. For example, consider methanol, CH_3OH, whose Lewis structure is

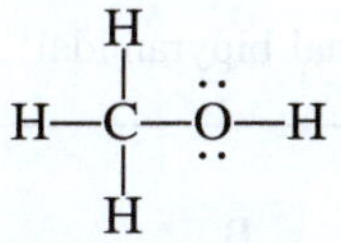

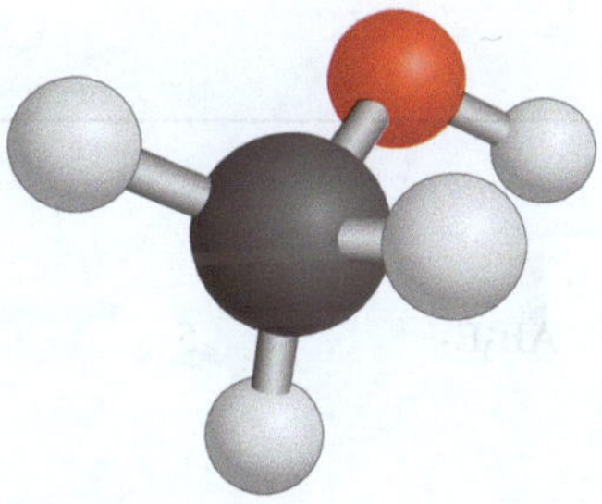

Figure 10.2 *The geometry of CH_3OH.*

The two central (nonterminal) atoms in methanol are C and O. We can say that the three C—H and the C—O bonding pairs are tetrahedrally arranged about the C atom. The H—C—H and O—C—H bond angles are approximately 109°. The O atom here is like the one in water in that it has two lone pairs and two bonding pairs. Therefore, the H—O—C portion of the molecule is bent, and the angle H—O—C is approximately equal to 105° (Figure 10.2).

Guidelines for Applying the VSEPR Model

Having studied the geometries of molecules in two categories (central atoms with and without lone pairs), let us consider some rules for applying the VSEPR model to all types of molecules:

1. Write the Lewis structure of the molecule, considering only the electron pairs around the central atom (that is, the atom that is bonded to more than one other atom).
2. Count the number of electron pairs around the central atom (bonding pairs and lone pairs). Treat double and triple bonds as though they were single bonds. Refer to Table 10.1 to predict the overall arrangement of the electron pairs.
3. Use Tables 10.1 and 10.2 to predict the geometry of the molecule.
4. In predicting bond angles, note that a lone pair repels another lone pair or a bonding pair more strongly than a bonding pair repels another bonding pair. Remember that in general there is no easy way to predict bond angles accurately when the central atom possesses one or more lone pairs.

Table 10.2 **Electron-Pair Geometry and Molecular Geometry of Molecules and Ions**

Class of Molecule	Total Number of Electron Pairs	Number of Bonding Pairs	Number of Lone Pairs	Arrangement of Electron Pairs	Geometry of Molecule or Ion	Examples
AB_2E	3	2	1	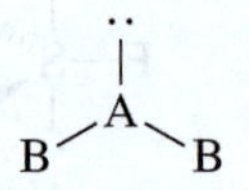 Trigonal Planar	Bent (or V-shaped)	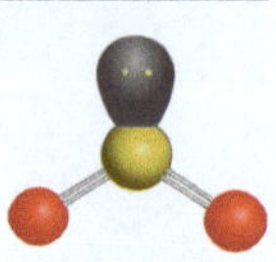 SO_2
AB_3E	4	3	1	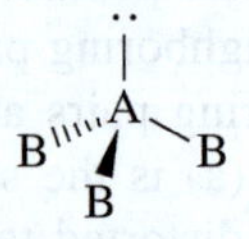 Tetrahedral	Trigonal pyramidal	 NH_3
AB_2E_2	4	2	2	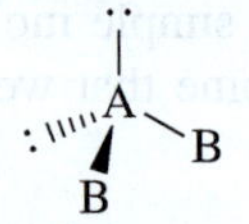 Tetrahedral	Bent (or V-shaped)	 H_2O
AB_4E	5	4	1	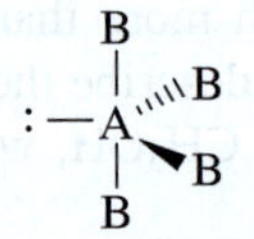 Trigonal bipyramidal	Seesaw (or distorted tetrahedron)	 SF_4
AB_3E_2	5	3	2	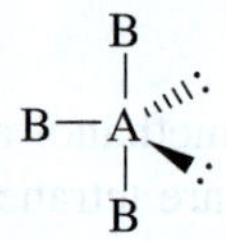 Trigonal bipyramidal	T-shaped	 ClF_3
AB_2E_3	5	2	3	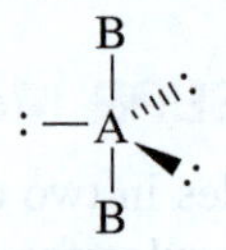 Trigonal bipyramidal	Linear	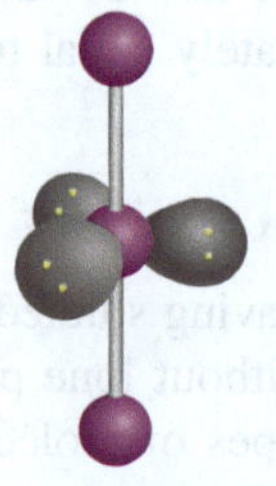 I_3^-
AB_5E	6	5	1	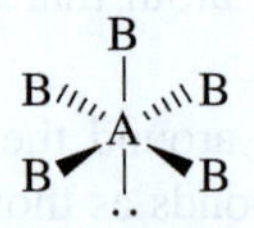 Octahedral	Square pyramidal	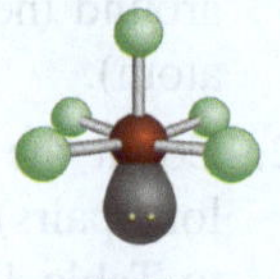 BrF_5
AB_4E_2	6	4	2	 Octahedral	Square planar	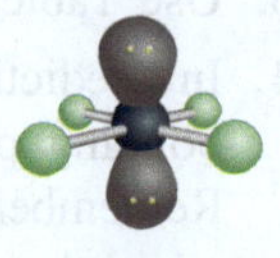 XeF_4

The VSEPR model generates reliable predictions of the geometries of a variety of molecular structures. Chemists use the VSEPR approach because of its simplicity. Although there are some theoretical concerns about whether "electron-pair repulsion" actually determines molecular shapes, the assumption that it does leads to useful (and generally reliable) predictions. We need not ask more of any model at this stage in our study of chemistry. Example 10.1 illustrates the application of VSEPR.

Student Hot Spot

Student data indicate you may struggle with VSEPR. Access your eBook for additional Learning Resources on this topic.

Example 10.1

Use the VSEPR model to predict the geometry of the following molecules and ions: (a) AsH_3, (b) OF_2, (c) $AlCl_4^-$, (d) I_3^-, (e) C_2H_4.

Strategy The sequence of steps in determining molecular geometry is as follows:

draw Lewis structure ⟶ find arrangement of electron pairs ⟶ find arrangement of bonding pairs ⟶ determine geometry based on bonding pairs

Solution

(a) The Lewis structure of AsH_3 is

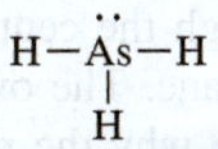

There are four electron pairs around the central atom; therefore, the electron pair arrangement is tetrahedral (see Table 10.1). Recall that the geometry of a molecule is determined only by the arrangement of atoms (in this case the As and H atoms). Thus, removing the lone pair leaves us with three bonding pairs and a trigonal pyramidal geometry, like NH_3. We cannot predict the H—As—H angle accurately, but we know that it is less than 109.5° because the repulsion of the bonding electron pairs in the As—H bonds by the lone pair on As is greater than the repulsion between the bonding pairs.

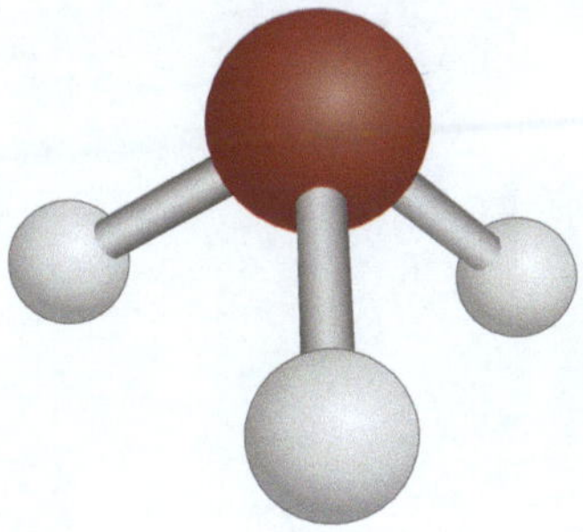

AsH_3

(b) The Lewis structure of OF_2 is

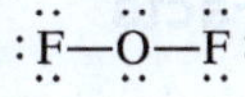

There are four electron pairs around the central atom; therefore, the electron pair arrangement is tetrahedral (see Table 10.1). Recall that the geometry of a molecule is determined only by the arrangement of atoms (in this case the O and F atoms). Thus, removing the two lone pairs leaves us with two bonding pairs and a bent geometry, like H_2O. We cannot predict the F—O—F angle accurately, but we know that it must be less than 109.5° because the repulsion of the bonding electron pairs in the O—F bonds by the lone pairs on O is greater than the repulsion between the bonding pairs.

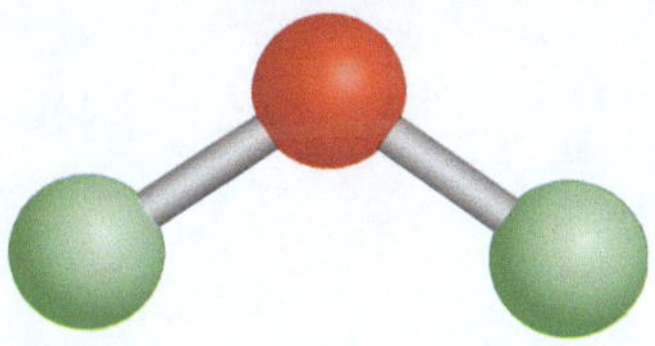

OF_2

(c) The Lewis structure of $AlCl_4^-$ is

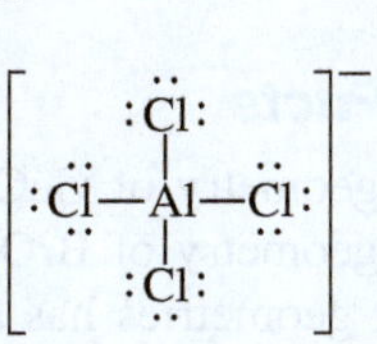

There are four electron pairs around the central atom; therefore, the electron pair arrangement is tetrahedral. Because there are no lone pairs present, the arrangement of the bonding pairs is the same as the electron pair arrangement. Therefore, $AlCl_4^-$ has a tetrahedral geometry and the Cl—Al—Cl angles are all 109.5°.

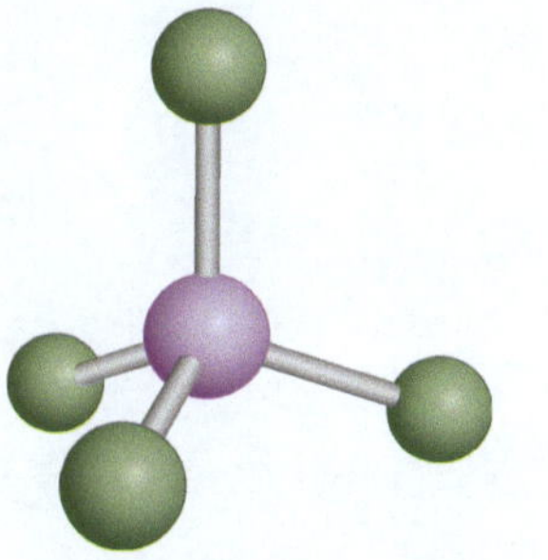

$AlCl_4^-$

(d) The Lewis structure of I_3^- is

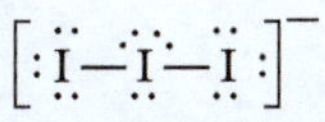

There are five electron pairs around the central I atom; therefore, the electron pair arrangement is trigonal bipyramidal. Of the five electron pairs, three are lone pairs

I_3^-

(Continued)

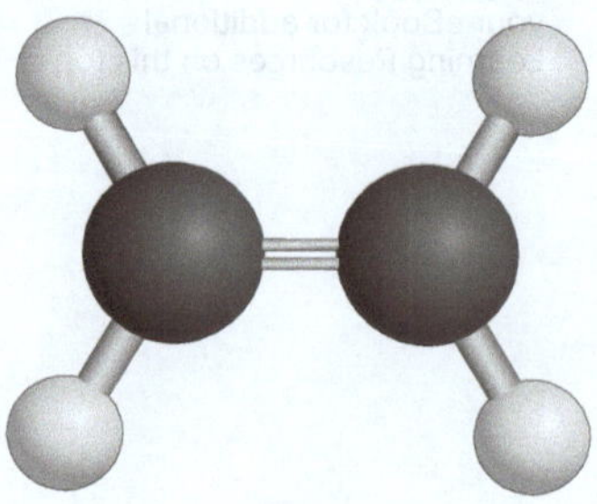

C_2H_4

and two are bonding pairs. Recall that the lone pairs preferentially occupy the equatorial positions in a trigonal bipyramid (see Table 10.2). Thus, removing the lone pairs leaves us with a linear geometry for I_3^-; that is, all three I atoms lie in a straight line.

(e) The Lewis structure of C_2H_4 is

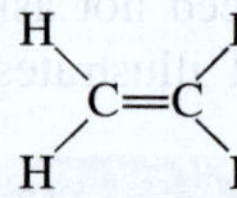

The C=C bond is treated as though it were a single bond (one electron domain) in the VSEPR model. Because there are three electron pairs around each C atom and there are no lone pairs present, the arrangement around each C atom has a trigonal planar shape like BF_3, discussed earlier. Thus, the predicted bond angles in C_2H_4 are all 120°.

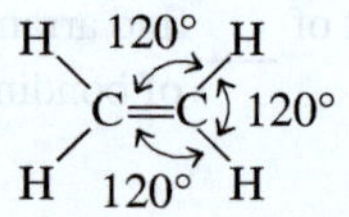

Comment (1) The I_3^- ion is one of the few structures for which the bond angle (180°) can be predicted accurately even though the central atom contains lone pairs. (2) In C_2H_4, all six atoms lie in the same plane. The overall planar geometry is not predicted by the VSEPR model, but we will see why the molecule prefers to be planar later. In reality, the angles are close, but not equal, to 120° because the bonds are not all equivalent.

Practice Exercise Use the VSEPR model to predict the geometry of (a) $SiBr_4$, (b) CS_2, (c) NO_3^-.

Similar problems: 10.7, 10.8, 10.9.

Summary of Concepts & Facts

- The VSEPR model for predicting molecular geometry is based on the assumption that valence-shell electron pairs repel one another and tend to stay as far apart as possible.
- According to the VSEPR model, molecular geometry can be predicted from the number of bonding electron pairs and lone pairs. Lone pairs repel other pairs more forcefully than bonding pairs do and thus distort bond angles from the ideal geometry.

Review of Concepts & Facts

10.1.1 What is the molecular geometry of $GeCl_4$?

10.1.2 What is the molecular geometry of BrO_3^-?

10.1.3 Which of the following geometries has a greater stability for tin(IV) hydride (SnH_4)?

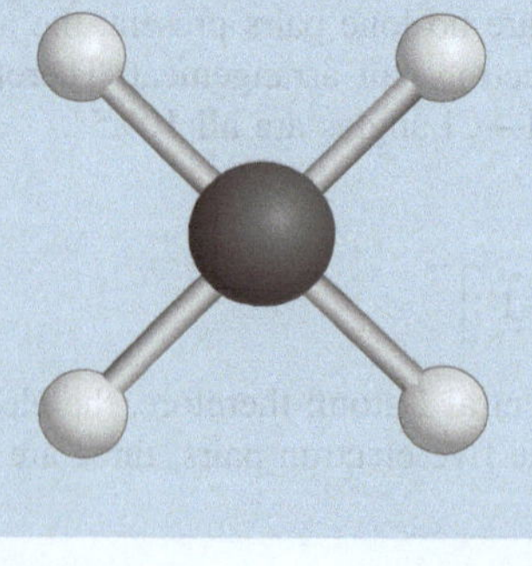

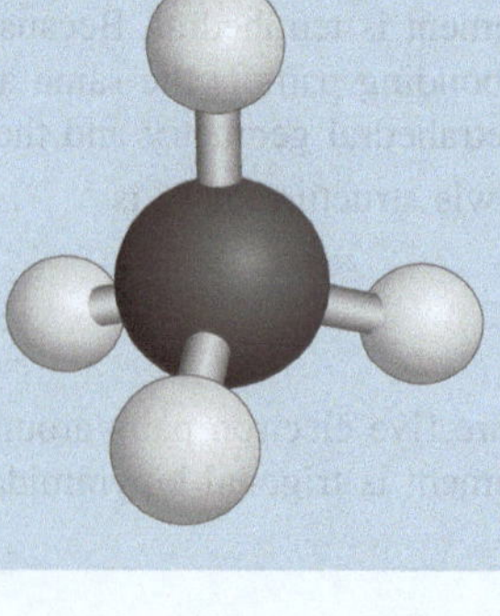

10.2 Dipole Moments

Learning Objective

- Assess whether or not a molecule will have a dipole moment.

In Section 9.5 we learned that hydrogen fluoride is a covalent compound with a polar bond. There is a shift of electron density from H to F because the F atom is more electronegative than the H atom (see Figure 9.4). The shift of electron density is symbolized by placing a crossed arrow (⟼) above the Lewis structure to indicate the direction of the shift. For example,

$$\overset{\longmapsto}{\mathrm{H{-}\ddot{\underset{..}{F}}:}}$$

The consequent charge separation can be represented as

$$\overset{\delta+}{\mathrm{H}}{-}\overset{\delta-}{\mathrm{\ddot{\underset{..}{F}}:}}$$

where δ (delta) denotes a partial charge. This separation of charges can be confirmed in an electric field (Figure 10.3). When the field is turned on, HF molecules orient their negative ends toward the positive plate and their positive ends toward the negative plate. This alignment of molecules can be detected experimentally.

A quantitative measure of the polarity of a bond is its ***dipole moment (μ),*** which is *the product of the charge Q and the distance r between the charges:*

$$\mu = Q \times r \qquad (10.1)$$

To maintain electrical neutrality, the charges on both ends of an electrically neutral diatomic molecule must be equal in magnitude and opposite in sign. However, in Equation (10.1), Q refers only to the magnitude of the charge and not to its sign, so μ is always positive. In a diatomic molecule like HF, the charge Q is equal to δ^+ and δ^-. Dipole moments are usually expressed in debye units (D), named for Peter Debye.†
The conversion factor is

$$1\ \mathrm{D} = 3.336 \times 10^{-30}\ \mathrm{C\ m}$$

where C is coulomb and m is meter.

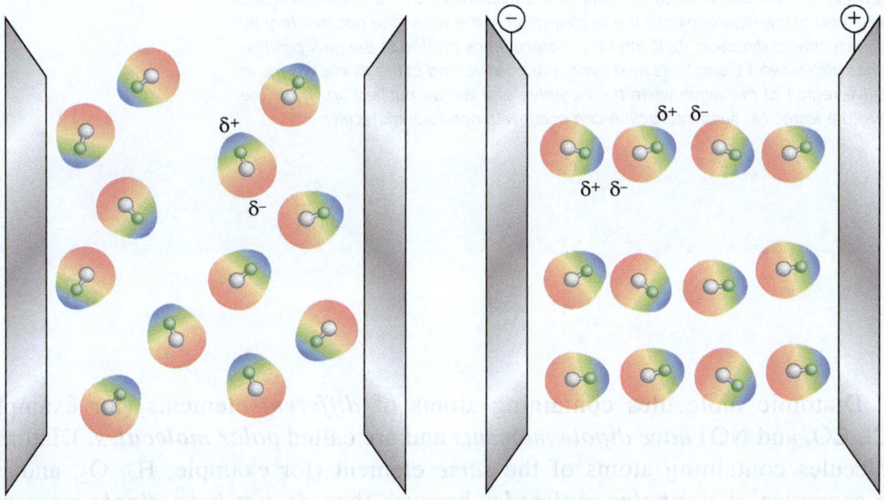

Figure 10.3 *Behavior of polar molecules (a) in the absence of an external electric field and (b) when the electric field is turned on. Nonpolar molecules are not affected by an electric field.*

†Peter Joseph William Debye (1884–1966). American chemist and physicist of Dutch origin. Debye made many significant contributions in the study of molecular structure, polymer chemistry, X-ray analysis, and electrolyte solution. He was awarded the Nobel Prize in Chemistry in 1936.

CHEMISTRY *in Action*

Microwave Ovens—Dipole Moments at Work

In the last 40 years, the microwave oven has become a ubiquitous appliance. Microwave technology enables us to thaw and cook food much more rapidly than conventional appliances do. How do microwaves heat food so quickly?

In Chapter 7 we saw that microwaves are a form of electromagnetic radiation (see Figure 7.3). Microwaves are generated by a magnetron, which was invented during World War II when radar technology was being developed. The magnetron is a hollow cylinder encased in a horseshoe-shaped magnet. In the center of the cylinder is a cathode rod. The walls of the cylinder act as an anode. When heated, the cathode emits electrons that travel toward the anode. The magnetic field forces the electrons to move in a circular path. This motion of charged particles generates microwaves, which are adjusted to a frequency of 2.45 GHz (2.45×10^9 Hz) for cooking. A "waveguide" directs the microwaves into the cooking compartment. Rotating fan blades reflect the microwaves to all parts of the oven.

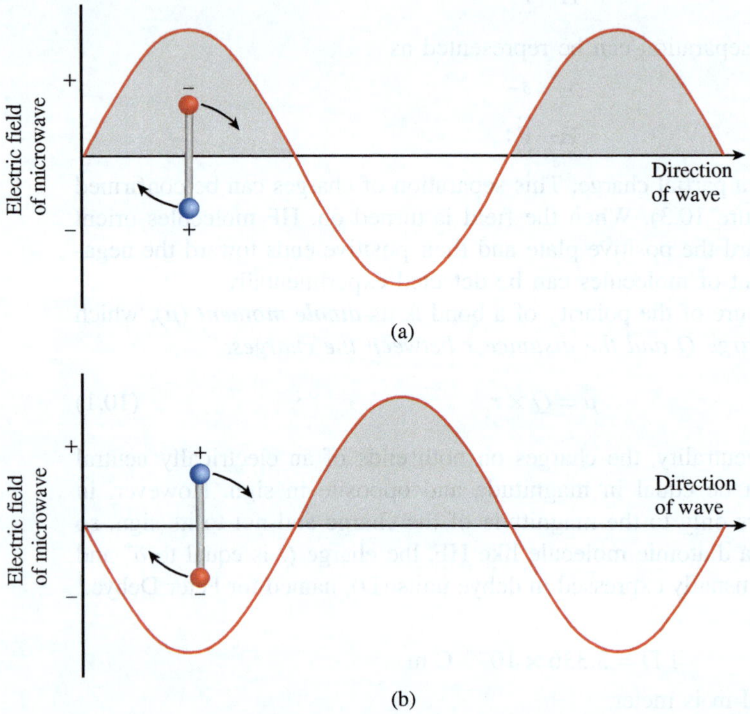

Interaction between the electric field component of the microwave and a polar molecule. (a) The negative end of the dipole follows the propagation of the wave (the positive region) and rotates in a clockwise direction. (b) If, after the molecule has rotated to the new position the radiation has also moved along to its next cycle, the positive end of the dipole will move into the negative region of the wave while the negative end will be pushed up. Thus, the molecule will rotate faster. No such interaction can occur with nonpolar molecules.

Video: Polarity of Molecules

Video: Influence of Shape on Polarity

Diatomic molecules containing atoms of *different* elements (for example, HCl, CO, and NO) *have dipole moments* and are called ***polar molecules.*** Diatomic molecules containing atoms of the *same* element (for example, H_2, O_2, and F_2) are examples of ***nonpolar molecules*** because they *do not have dipole moments.* For a molecule made up of three or more atoms, both the polarity of the bonds and the molecular geometry determine whether there is a dipole moment. Even if polar bonds are present, the molecule will not necessarily have a dipole moment.

The cooking action in a microwave oven results from the interaction between the electric field component of the radiation with the polar molecules—mostly water—in food. All molecules rotate at room temperature. If the frequency of the radiation and that of the molecular rotation are equal, energy can be transferred from the microwave to the polar molecule. As a result, the molecule will rotate faster. This is what happens in a gas. In the condensed state (for example, in food), a molecule cannot execute the free rotation. Nevertheless, it still experiences a torque (a force that causes rotation) that tends to align its dipole moment with the oscillating field of the microwave. Consequently, there is friction between the molecules, which appears as heat in the food.

The reason that a microwave oven can cook food so fast is that the radiation is not absorbed by nonpolar molecules and can, therefore, reach different parts of food at the same time. (Depending on the amount of water present, microwaves can penetrate food to a depth of several inches.) In a conventional oven, heat can affect the center of foods only by conduction (that is, by transfer of heat from hot air molecules to cooler molecules in food in a layer-by-layer fashion), which is a very slow process.

The following points are relevant to the operation of a microwave oven. Plastics and Pyrex glassware do not contain polar molecules and are therefore not affected by microwave radiation. (Styrofoam and certain plastics cannot be used in microwaves because they melt from the heat of the food.) Metals, however, reflect microwaves, thereby shielding the food and possibly returning enough energy to the microwave emitter to overload it. Because microwaves can induce a current in the metal, this action can lead to sparks jumping between the container and the bottom or walls of the oven. Finally, although water molecules in ice are locked in position and therefore cannot rotate, we routinely thaw food in a microwave oven. The reason is that at room temperature a thin film of liquid water quickly forms on the surface of frozen food and the mobile molecules in that film can absorb the radiation to start the thawing process.

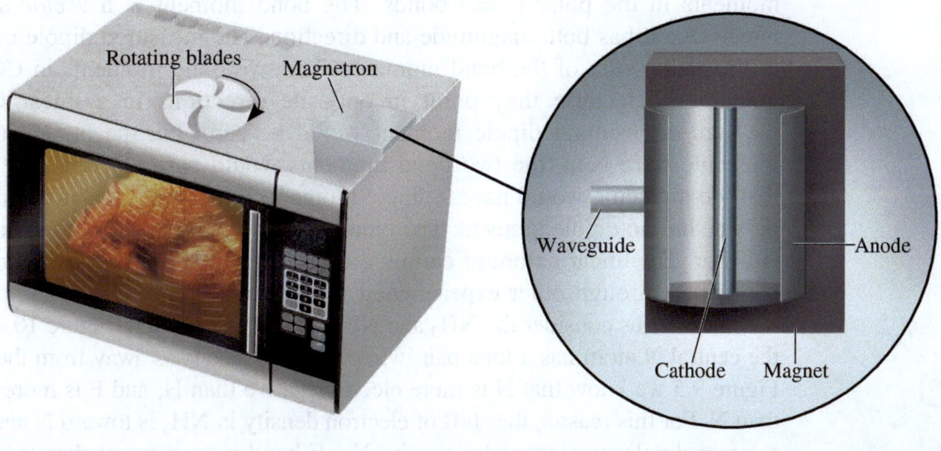

A microwave oven. The microwaves generated by the magnetron are reflected to all parts of the oven by the rotating fan blades.

Carbon dioxide (CO_2), for example, is a triatomic molecule, so its geometry is either linear or bent:

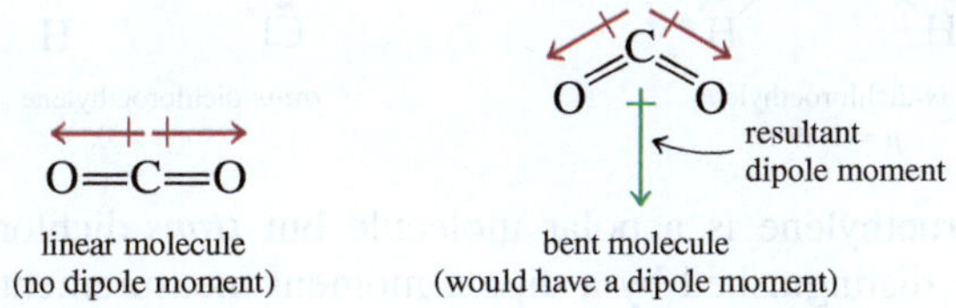

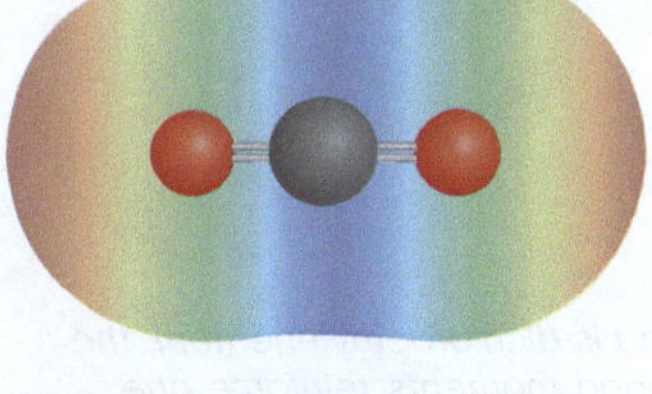

Each carbon-to-oxygen bond is polar, with the electron density shifted toward the more electronegative oxygen atom. However, the linear geometry of the molecule results in the cancellation of the two bond moments.

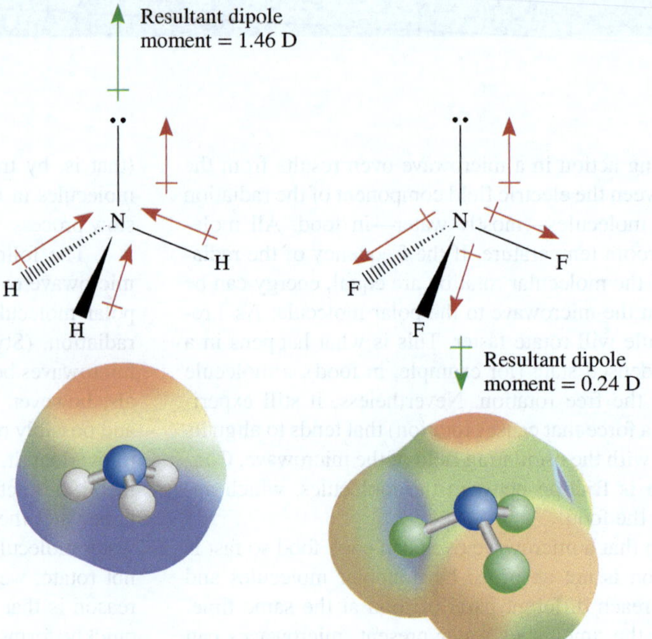

Figure 10.4 *Bond moments and resultant dipole moments in NH_3 and NF_3. The electrostatic potential maps show the electron density distributions in these molecules.*

The arrows show the shift of electron density from the less electronegative carbon atom to the more electronegative oxygen atom. In each case, the dipole moment of the entire molecule is made up of two *bond moments,* that is, individual dipole moments in the polar C═O bonds. The bond moment is a *vector quantity,* which means that it has both magnitude and direction. The measured dipole moment is equal to the vector sum of the bond moments. The two bond moments in CO_2 are equal in magnitude. Because they point in opposite directions in a linear CO_2 molecule, the sum or resultant dipole moment would be zero. On the other hand, if the CO_2 molecule were bent, the two bond moments would partially reinforce each other, so that the molecule would have a dipole moment. Experimentally it is found that carbon dioxide has no dipole moment. Therefore, we conclude that the carbon dioxide molecule is linear. The linear nature of carbon dioxide predicted by the VSEPR model has been confirmed through other experimental measurements.

Next let us consider the NH_3 and NF_3 molecules shown in Figure 10.4. In both cases, the central N atom has a lone pair, whose charge density is away from the N atom. From Figure 9.5 we know that N is more electronegative than H, and F is more electronegative than N. For this reason, the shift of electron density in NH_3 is toward N and so contributes a larger dipole moment, whereas the N—F bond moments are directed away from the N atom and so together they offset the contribution of the lone pair to the dipole moment. Thus, the resultant dipole moment in NH_3 is larger than that in NF_3.

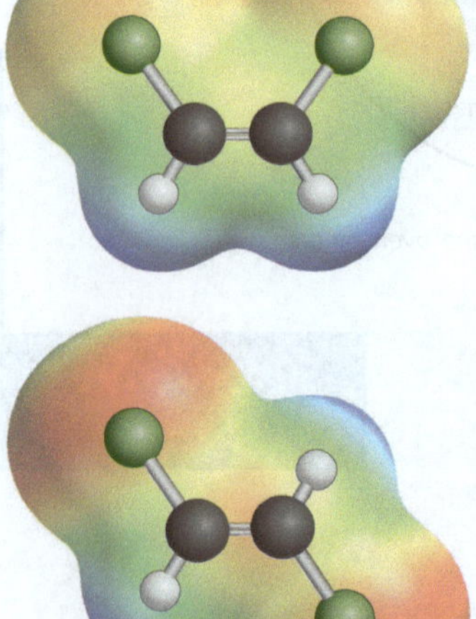

In cis-dichloroethylene (top), the bond moments reinforce one another and the molecule is polar. The opposite holds for trans-dichloroethylene and the molecule is nonpolar.

Dipole moments can be used to distinguish between molecules that have the same formula but different structures. For example, the following molecules both exist; they have the same molecular formula ($C_2H_2Cl_2$), the same number and type of bonds, but different molecular structures:

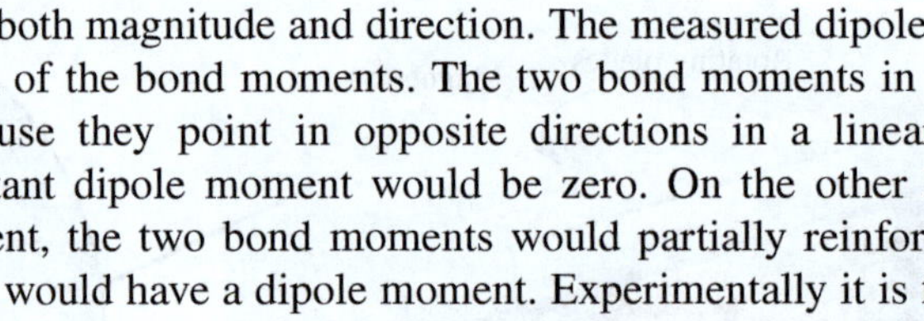

cis-dichloroethylene
$\mu = 1.89$ D

trans-dichloroethylene
$\mu = 0$

Because *cis*-dichloroethylene is a polar molecule but *trans*-dichloroethylene is not, they can readily be distinguished by a dipole moment measurement. Additionally, as

Table 10.3 Dipole Moments of Some Polar Molecules

Molecule	Geometry	Dipole Moment (D)
HF	Linear	1.92
HCl	Linear	1.08
HBr	Linear	0.78
HI	Linear	0.38
H_2O	Bent	1.87
H_2S	Bent	1.10
NH_3	Trigonal pyramidal	1.46
SO_2	Bent	1.60

we will see in Chapter 11, the strength of intermolecular forces is partially determined by whether molecules possess a dipole moment. Table 10.3 lists the dipole moments of several polar molecules.

Example 10.2 shows how we can predict whether a molecule possesses a dipole moment if we know its molecular geometry.

Example 10.2

Predict whether each of the following molecules has a dipole moment: (a) BrCl, (b) BF_3 (trigonal planar), (c) CH_2Cl_2 (tetrahedral).

Strategy Keep in mind that the dipole moment of a molecule depends on both the difference in electronegativities of the elements present and its geometry. A molecule can have polar bonds (if the bonded atoms have different electronegativities), but it may not possess a dipole moment if it has a highly symmetrical geometry.

Solution

(a) Because bromine chloride is diatomic, it has a linear geometry. Chlorine is more electronegative than bromine (see Figure 9.5), so BrCl is polar with chlorine at the negative end:

$$\overset{\longleftrightarrow}{\text{Br—Cl}}$$

Thus, the molecule does have a dipole moment. In fact, all diatomic molecules containing different elements possess a dipole moment.

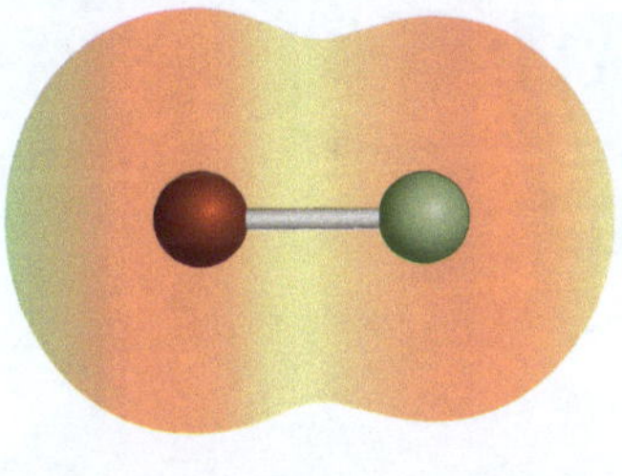

Electrostatic potential map of BrCl shows that the electron density is shifted toward the Cl atom.

(b) Because fluorine is more electronegative than boron, each B—F bond in BF_3 (boron trifluoride) is polar and the three bond moments are equal. However, the symmetry of a trigonal planar shape means that the three bond moments exactly cancel one another:

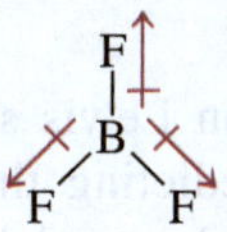

An analogy is an object that is pulled in the directions shown by the three bond moments. If the forces are equal, the object will not move. Consequently, BF_3 has no dipole moment; it is a nonpolar molecule.

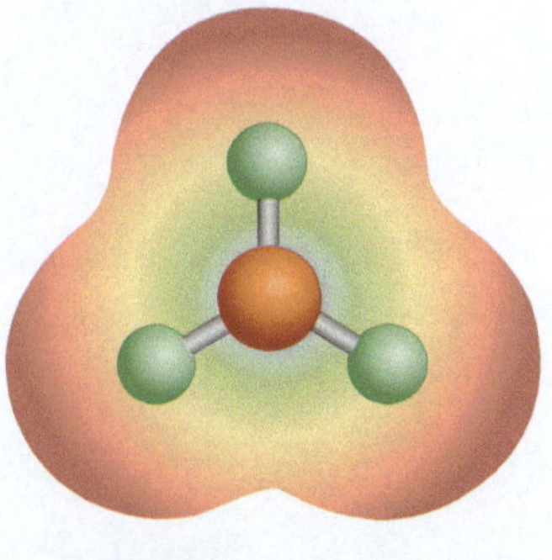

Electrostatic potential map shows that the electron density is symmetrically distributed in the BF_3 molecule.

(c) The Lewis structure of CH_2Cl_2 (methylene chloride) is

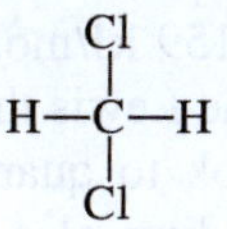

(Continued)

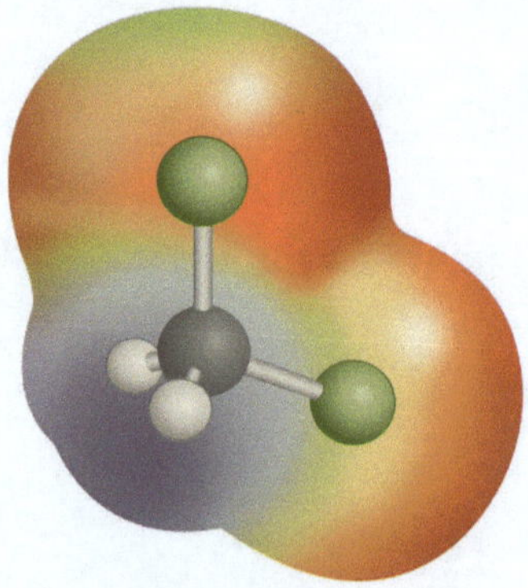

Electrostatic potential map of CH_2Cl_2. The electron density is shifted toward the electronegative Cl atoms.

This molecule is similar to CH_4 in that it has an overall tetrahedral shape. However, because not all the bonds are identical, there are three different bond angles: H—C—H, H—C—Cl, and Cl—C—Cl. These bond angles are close to, but not equal to, 109.5°. Because chlorine is more electronegative than carbon, which is more electronegative than hydrogen, the bond moments do not cancel and the molecule possesses a dipole moment:

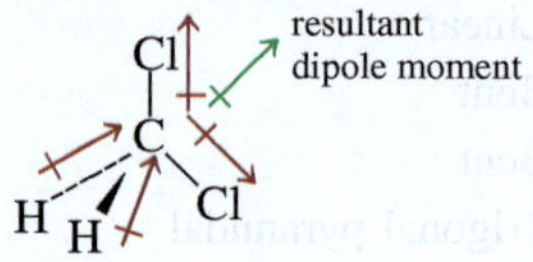

Thus, CH_2Cl_2 is a polar molecule.

Practice Exercise Does the $AlCl_3$ molecule have a dipole moment?
Similar problems: 10.21, 10.22, 10.23.

Summary of Concepts & Facts

- Dipole moment is a measure of the charge separation in molecules containing atoms of different electronegativities. The dipole moment of a molecule is the resultant of whatever bond moments are present. Information about molecular geometry can be obtained from dipole moment measurements.

Review of Concepts & Facts

10.2.1 Predict whether PF_5 has a dipole moment.

10.2.2 The molecule CF_4 does not have a dipole moment whereas SF_4 does have a dipole moment. Explain.

10.2.3 Carbon dioxide has a linear geometry and is nonpolar. Yet we know that the molecule executes bending and stretching motions that create a dipole moment. How would you reconcile these conflicting descriptions about CO_2?

10.3 Valence Bond Theory

Learning Objective

- Employ valence bond theory to describe the bonding in covalent molecules.

The VSEPR model, based largely on Lewis structures, provides a relatively simple and straightforward method for predicting the geometry of molecules. But as we noted earlier, the Lewis theory of chemical bonding does not clearly explain why chemical bonds exist. Relating the formation of a covalent bond to the pairing of electrons was a step in the right direction, but it did not go far enough. For example, the Lewis theory describes the single bond between the H atoms in H_2 and that between the F atoms in F_2 in essentially the same way—as the pairing of two electrons. Yet these two molecules have quite different bond enthalpies and bond lengths (436 kJ/mol and 74 pm for H_2 and 159 kJ/mol and 142 pm for F_2). These and many other facts cannot be explained by the Lewis theory. For a more complete explanation of chemical bond formation we look to quantum mechanics. In fact, the quantum mechanical study of chemical bonding also provides a means for understanding molecular geometry.

At present, two quantum mechanical theories are used to describe covalent bond formation and the electronic structure of molecules. *Valence bond (VB) theory* assumes that the electrons in a molecule occupy atomic orbitals of the individual atoms. It enables us to retain a picture of individual atoms taking part in the bond formation. The second theory, called *molecular orbital (MO) theory* (Section 10.6), assumes the formation of molecular orbitals from the atomic orbitals. Neither theory perfectly explains all aspects of bonding, but each has contributed something to our understanding of many observed molecular properties.

Let us start our discussion of valence bond theory by considering the formation of a H_2 molecule from two H atoms. The Lewis theory describes the H—H bond in terms of the pairing of the two electrons on the H atoms. In the framework of valence bond theory, the covalent H—H bond is formed by the *overlap* of the two 1*s* orbitals in the H atoms. By overlap, we mean that the two orbitals share a common region in space.

What happens to two H atoms as they move toward each other and form a bond? Initially, when the two atoms are far apart, there is no interaction. We say that the potential energy of this system (i.e., the two H atoms) is zero. Recall that an object has potential energy by virtue of its position. As the atoms approach each other, each electron is attracted by the nucleus of the other atom; at the same time, the electrons repel each other, as do the nuclei. While the atoms are still separated, attraction is stronger than repulsion, so that the potential energy of the system *decreases* (i.e., it becomes negative) as the atoms approach each other (Figure 10.5). This trend continues until the potential energy reaches a minimum value. At this point, when the system has the lowest potential energy, it is most stable. This condition corresponds to substantial overlap of the 1*s* orbitals and the formation of a stable H_2 molecule. If the distance between nuclei were to decrease further, the potential energy would rise steeply and finally become positive as a result of the increased electron-electron and nuclear-nuclear repulsions. In accordance with the law of conservation of energy, the decrease in potential energy as a result of H_2 formation must be accompanied by a release of energy. Experiments show that as a H_2 molecule is formed from two H atoms, heat is given off. The converse is also true. To break a H—H bond, energy must be supplied to the molecule. Figure 10.6 is another way of viewing the formation of a H_2 molecule.

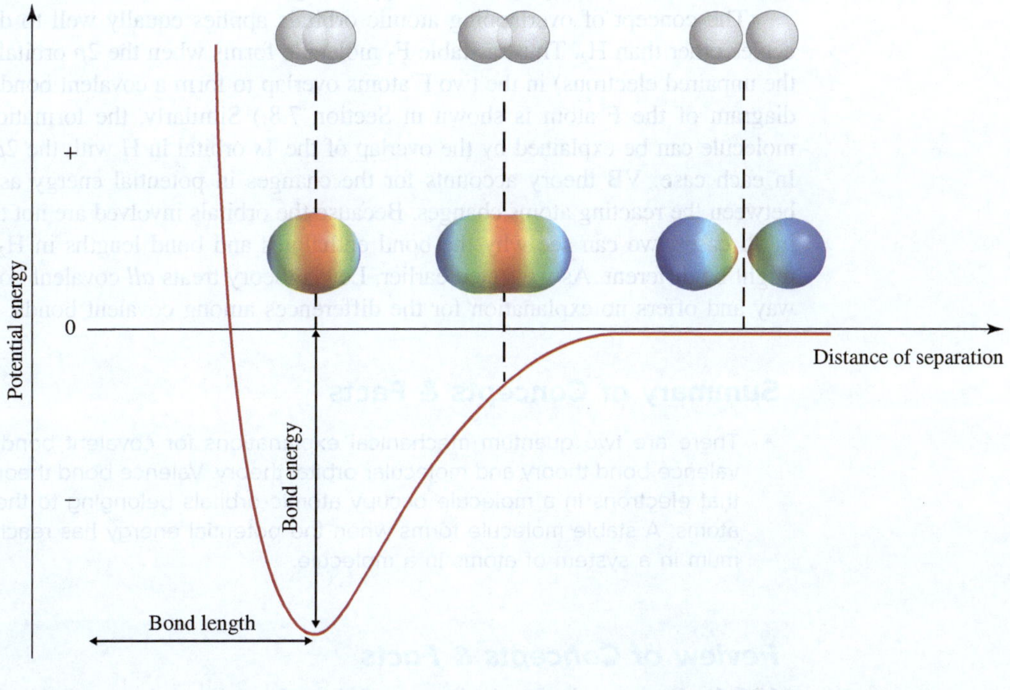

Figure 10.5 *Change in potential energy of two H atoms with their distance of separation. At the point of minimum potential energy, the H_2 molecule is in its most stable state and the bond length is 74 pm. The spheres represent the 1s orbitals.*

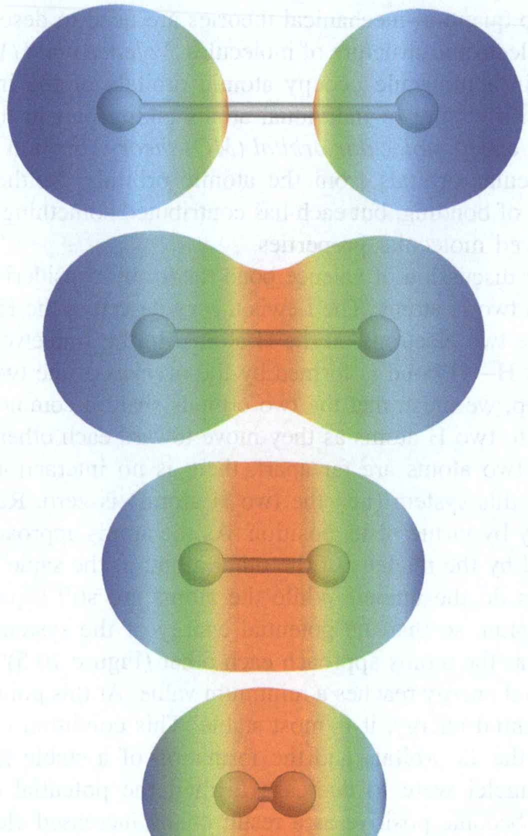

Figure 10.6 *Top to bottom: As two H atoms approach each other, their 1s orbitals begin to interact and each electron begins to feel the attraction of the other proton. Gradually, the electron density builds up in the region between the two nuclei (red color). Eventually, a stable H_2 molecule is formed with an internuclear distance of 74 pm.*

Thus, valence bond theory gives a clearer picture of chemical bond formation than the Lewis theory does. Valence bond theory states that a stable molecule forms from reacting atoms when the potential energy of the system has decreased to a minimum; the Lewis theory ignores energy changes in chemical bond formation.

The concept of overlapping atomic orbitals applies equally well to diatomic molecules other than H_2. Thus, a stable F_2 molecule forms when the 2*p* orbitals (containing the unpaired electrons) in the two F atoms overlap to form a covalent bond. (The orbital diagram of the F atom is shown in Section 7.8.) Similarly, the formation of the HF molecule can be explained by the overlap of the 1*s* orbital in H with the 2*p* orbital in F. In each case, VB theory accounts for the changes in potential energy as the distance between the reacting atoms changes. Because the orbitals involved are not the same kind in all cases, we can see why the bond enthalpies and bond lengths in H_2, F_2, and HF might be different. As we stated earlier, Lewis theory treats *all* covalent bonds the same way and offers no explanation for the differences among covalent bonds.

Summary of Concepts & Facts

- There are two quantum mechanical explanations for covalent bond formation: valence bond theory and molecular orbital theory. Valence bond theory assumes that electrons in a molecule occupy atomic orbitals belonging to the individual atoms. A stable molecule forms when the potential energy has reached a minimum in a system of atoms in a molecule.

Review of Concepts & Facts

10.3.1 Compare the Lewis theory and the valence bond theory of chemical bonding.

10.4 Hybridization of Atomic Orbitals

Learning Objective

- Determine the hybridization of an atom in a molecule or polyatomic ion.

The concept of atomic orbital overlap should apply also to polyatomic molecules. However, a satisfactory bonding scheme must account for molecular geometry. We will discuss three examples of the valence bond (VB) treatment of bonding in polyatomic molecules.

sp^3 Hybridization

Consider the CH_4 molecule. Focusing only on the valence electrons, we can represent the orbital diagram of C as shown in the figure below. Because the carbon atom has two unpaired electrons (one in each of the two 2*p* orbitals), it can form only two bonds with hydrogen in its ground state. Although the species CH_2 is known, it is very unstable. To account for the four C—H bonds in methane, we can try to promote (that is, energetically excite) an electron from the 2*s* orbital to the 2*p* orbital. Now there are four unpaired electrons on C that could form four C—H bonds. However, the geometry is wrong, because three of the H—C—H bond angles would have to be 90° (remember that the three 2*p* orbitals on carbon are mutually perpendicular), and yet *all* H—C—H angles are 109.5°.

To explain the bonding in methane, VB theory uses hypothetical ***hybrid orbitals,*** which are *atomic orbitals obtained when two or more nonequivalent orbitals of the same atom combine in preparation for covalent bond formation.* ***Hybridization*** is the term applied to *the mixing of atomic orbitals in an atom (usually a central atom) to generate a set of hybrid orbitals.* We can generate four equivalent hybrid orbitals for carbon by mixing the 2*s* orbital and the three 2*p* orbitals:

Video: Hybridization

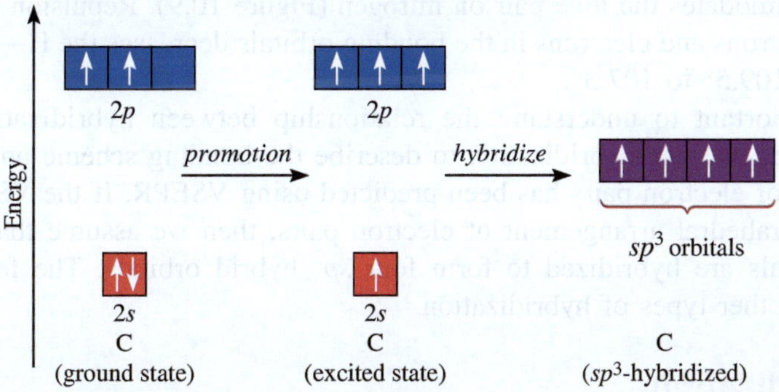

Because the new orbitals are formed from one *s* and three *p* orbitals, they are called sp^3 hybrid orbitals (pronounced "*s*-*p* three"). Figure 10.7 shows the shape and orientations of the sp^3 orbitals. These four hybrid orbitals are directed toward the four corners of a regular tetrahedron. Figure 10.8 shows the formation of four covalent bonds between the carbon sp^3 hybrid orbitals and the hydrogen 1*s* orbitals in CH_4. Thus, CH_4

Video: Molecular Shape and Orbital Hybridization

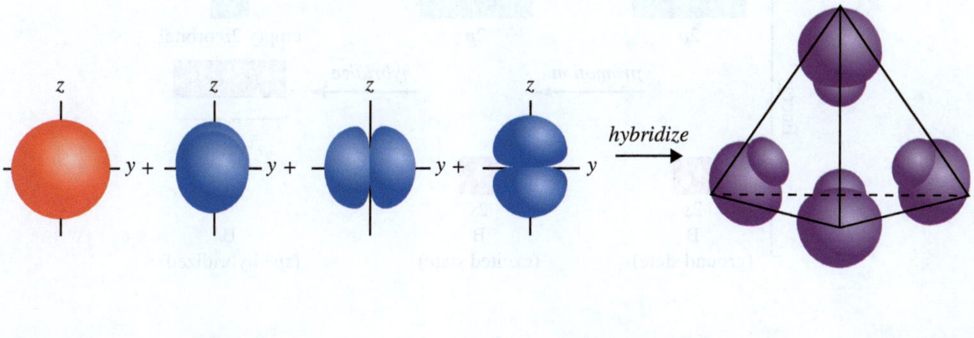

Figure 10.7 *Formation of four sp^3 hybrid orbitals from one 2s and three 2p orbitals. The sp^3 orbitals point to the corners of a tetrahedron.*

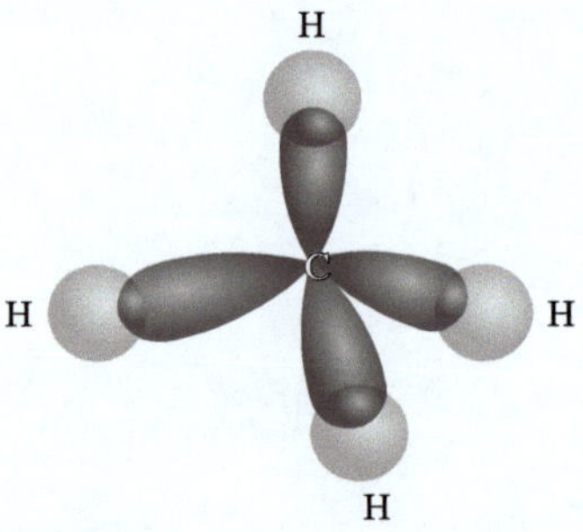

Figure 10.8 *Formation of four bonds between the carbon sp^3 hybrid orbitals and the hydrogen 1s orbitals in CH_4. The smaller lobes are not shown.*

has a tetrahedral shape, and all the H—C—H angles are 109.5°. Note that although energy is required to bring about hybridization, this input is more than compensated for by the energy released upon the formation of C—H bonds. (Recall that bond formation is an exothermic process.)

The following analogy is useful for understanding hybridization. Suppose that we have a beaker of a red solution and three beakers of blue solutions and that the volume of each is 50 mL. The red solution corresponds to one 2*s* orbital, the blue solutions represent three 2*p* orbitals, and the four equal volumes symbolize four separate orbitals. By mixing the solutions we obtain 200 mL of a purple solution, which can be divided into four 50-mL portions (that is, the hybridization process generates four sp^3 orbitals). Just as the purple color is made up of the red and blue components of the original solutions, the sp^3 hybrid orbitals possess both *s* and *p* orbital characteristics.

Another example of sp^3 hybridization is ammonia (NH_3). Table 10.1 shows that the arrangement of four electron pairs is tetrahedral, so that the bonding in NH_3 can be explained by assuming that N, like C in CH_4, is sp^3-hybridized. The ground-state electron configuration of N is $1s^22s^22p^3$, so that the orbital diagram for the sp^3-hybridized N atom is

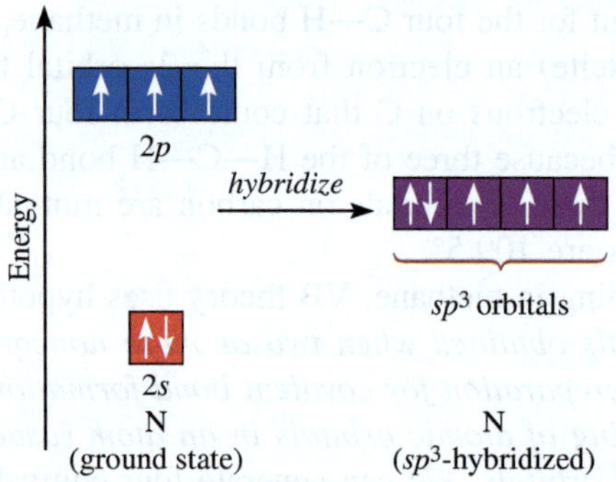

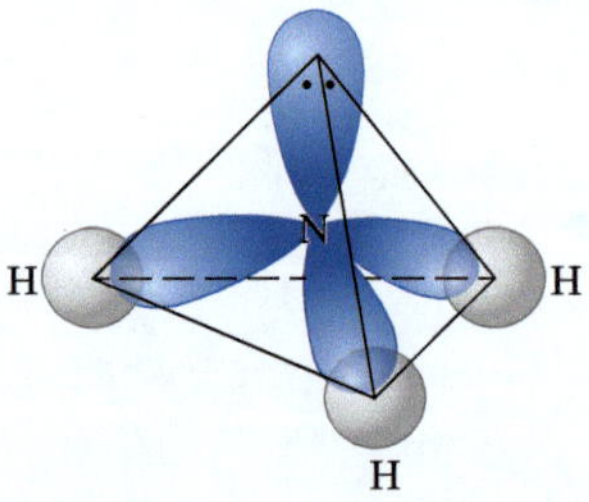

Figure 10.9 *The sp^3-hybridized N atom in NH_3. Three sp^3 hybrid orbitals form bonds with the H atoms. The fourth is occupied by nitrogen's lone pair.*

Three of the four hybrid orbitals form covalent N—H bonds, and the fourth hybrid orbital accommodates the lone pair on nitrogen (Figure 10.9). Repulsion between the lone-pair electrons and electrons in the bonding orbitals decreases the H—N—H bond angles from 109.5° to 107.3°.

It is important to understand the relationship between hybridization and the VSEPR model. We use hybridization to describe the bonding scheme only when the arrangement of electron pairs has been predicted using VSEPR. If the VSEPR model predicts a tetrahedral arrangement of electron pairs, then we assume that one *s* and three *p* orbitals are hybridized to form four sp^3 hybrid orbitals. The following are examples of other types of hybridization.

sp^2 Hybridization

Next we will look at the BF_3 (boron trifluoride) molecule, known to have a planar geometry. Considering only the valence electrons, the orbital diagram of B is shown in the figure below. First, we promote a 2*s* electron to an empty 2*p* orbital. Mixing the 2*s* orbital with the two 2*p* orbitals generates three sp^2 hybrid orbitals (pronounced "*s*-*p* two").

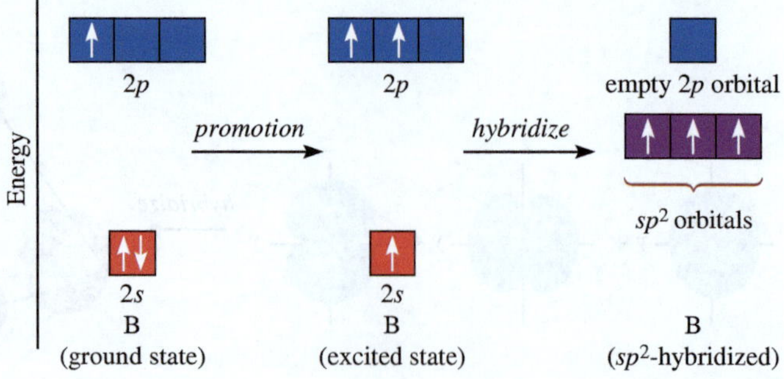

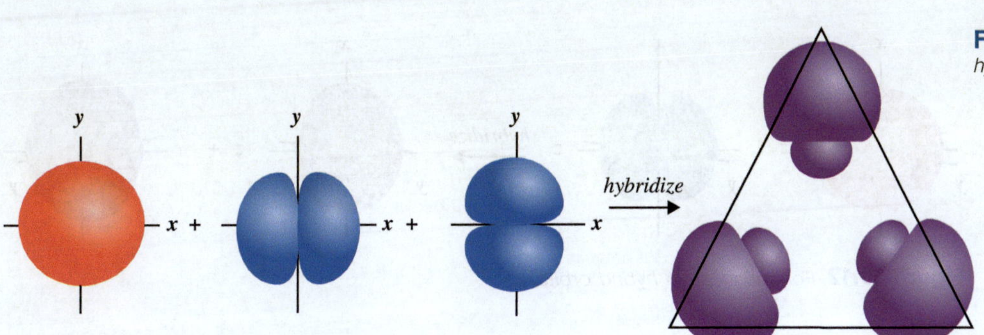

Figure 10.10 *Formation of sp^2 hybrid orbitals.*

These three sp^2 orbitals lie in the same plane, and the angle between any two of them is 120° (Figure 10.10). Each of the B—F bonds is formed by the overlap of a boron sp^2 hybrid orbital and a fluorine 2p orbital (Figure 10.11). The BF_3 molecule is planar with all the F—B—F angles equal to 120°. This result conforms to experimental findings and also to VSEPR predictions.

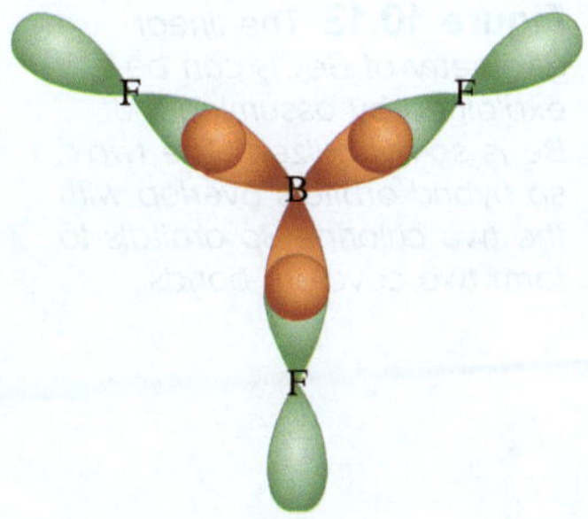

Figure 10.11 *The sp^2 hybrid orbitals of boron overlap with the 2p orbitals of fluorine. The BF_3 molecule is planar, and all the F—B—F angles are 120°.*

sp Hybridization

The beryllium chloride ($BeCl_2$) molecule is predicted to be linear by VSEPR. The orbital diagram for the valence electrons in Be is shown in the figure below. We know that in its ground state, Be does not form covalent bonds with Cl because its electrons are paired in the 2s orbital. So we turn to hybridization for an explanation of Be's bonding behavior. First, we promote a 2s electron to a 2p orbital. Now there are two Be orbitals available for bonding, the 2s and 2p. However, if two Cl atoms were to combine with Be in this excited state, one Cl atom would share a 2s electron and the other Cl would share a 2p electron, making two nonequivalent Cl—Be—Cl bonds. This scheme contradicts experimental evidence. In the actual $BeCl_2$ molecule, the two Be—Cl bonds are identical in every respect. Thus, the 2s and 2p orbitals must be mixed, or hybridized, to form two equivalent sp hybrid orbitals:

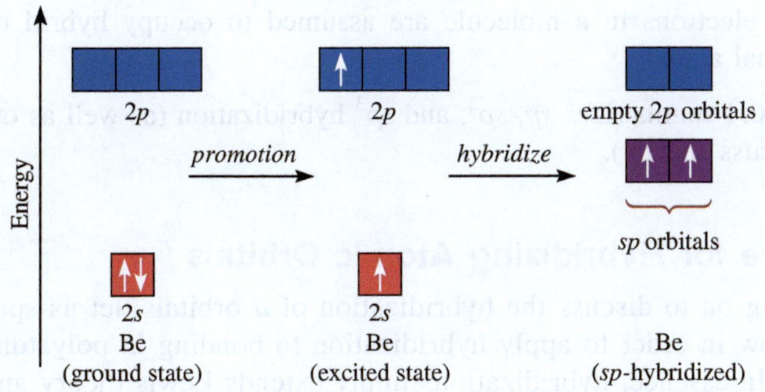

Figure 10.12 shows the shape and orientation of the sp orbitals. These two hybrid orbitals lie on the same line, the x-axis, so that the angle between them is 180°. Each of the Be—Cl bonds is then formed by the overlap of a Be sp hybrid orbital and a Cl 3p orbital, and the resulting $BeCl_2$ molecule has a linear geometry (Figure 10.13).

You may have noticed an interesting connection between hybridization and the octet rule. Regardless of the type of hybridization, an atom starting with one s and three p orbitals would still possess four orbitals, enough to accommodate a total of

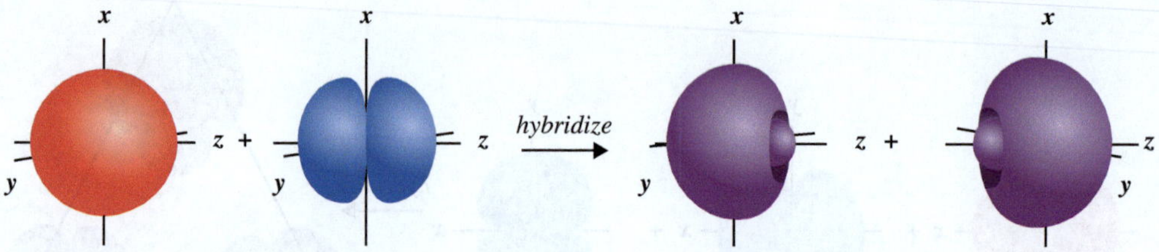

Figure 10.12 *Formation of sp hybrid orbitals.*

Figure 10.13 *The linear geometry of $BeCl_2$ can be explained by assuming that Be is sp-hybridized. The two sp hybrid orbitals overlap with the two chlorine 3p orbitals to form two covalent bonds.*

eight electrons in a compound. For elements in the second period of the periodic table, eight is the maximum number of electrons that an atom of any of these elements can accommodate in the valence shell. This is the reason that the octet rule is usually obeyed by the second-period elements.

The situation is different for an atom of a third-period element. If we use only the 3*s* and 3*p* orbitals of the atom to form hybrid orbitals in a molecule, then the octet rule applies. However, in some molecules the same atom may use one or more 3*d* orbitals, in addition to the 3*s* and 3*p* orbitals, to form hybrid orbitals. In these cases, the octet rule does not hold. We will see specific examples of the participation of the 3*d* orbital in hybridization shortly.

To summarize our discussion of hybridization, we note that

- The concept of hybridization is not applied to isolated atoms. It is a theoretical model used only to explain covalent bonding.
- Hybridization is the mixing of at least two nonequivalent atomic orbitals, for example, *s* and *p* orbitals. Therefore, a hybrid orbital is not a pure atomic orbital. Hybrid orbitals and pure atomic orbitals have very different shapes.
- The number of hybrid orbitals generated is equal to the number of pure atomic orbitals that participate in the hybridization process.
- Hybridization requires an input of energy; however, the system more than recovers this energy during bond formation.
- Covalent bonds in polyatomic molecules and ions are formed by the overlap of hybrid orbitals, or of hybrid orbitals with unhybridized ones. Therefore, the hybridization bonding scheme is still within the framework of valence bond theory; electrons in a molecule are assumed to occupy hybrid orbitals of the individual atoms.

Table 10.4 summarizes *sp*, sp^2, and sp^3 hybridization (as well as other types that we will discuss shortly).

Procedure for Hybridizing Atomic Orbitals

Before going on to discuss the hybridization of *d* orbitals, let us specify what we need to know in order to apply hybridization to bonding in polyatomic molecules in general. In essence, hybridization simply extends Lewis theory and the VSEPR model. To assign a suitable state of hybridization to the central atom in a molecule, we must have some idea about the geometry of the molecule. The steps are as follows:

1. Draw the Lewis structure of the molecule.
2. Predict the overall arrangement of the electron pairs (both bonding pairs and lone pairs) using the VSEPR model (see Table 10.1).
3. Deduce the hybridization of the central atom by matching the arrangement of the electron pairs with those of the hybrid orbitals shown in Table 10.4.

Student Hot Spot

Student data indicate you may struggle with hybridization. Access your eBook for additional Learning Resources on this topic.

Table 10.4 Important Hybrid Orbitals and Their Shapes

Pure Atomic Orbitals of the Central Atom	Hybridization of the Central Atom	Number of Hybrid Orbitals	Shape of Hybrid Orbitals	Examples
s, p	*sp*	2	 Linear	$BeCl_2$
s, p, p	sp^2	3	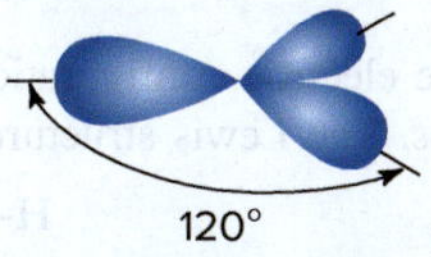 Trigonal planar	BF_3
s, p, p, p	sp^3	4	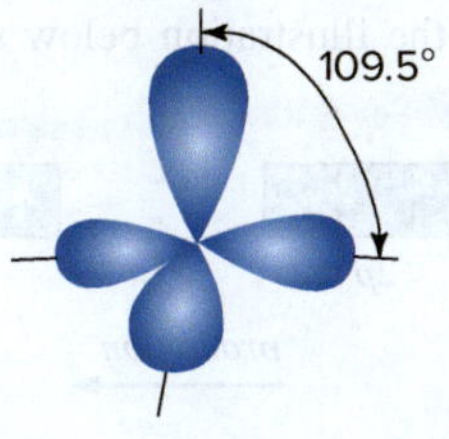 Tetrahedral	CH_4, NH_4^+
s, p, p, p, d	sp^3d	5	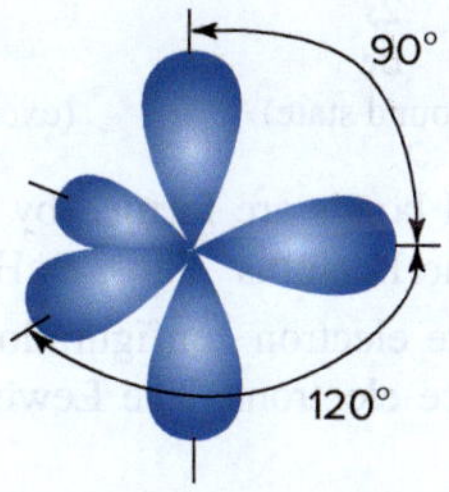 Trigonal bipyramidal	PCl_5
s, p, p, p, d, d	sp^3d^2	6	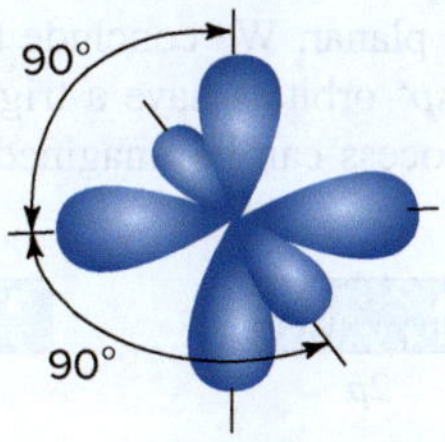 Octahedral	SF_6

Example 10.3 illustrates this procedure.

Example 10.3

Determine the hybridization state of the central (underlined) atom in each of the following molecules: (a) <u>Be</u>H_2, (b) <u>Al</u>I_3, (c) <u>P</u>F_3. Describe the hybridization process and determine the molecular geometry in each case.

Strategy The steps for determining the hybridization of the central atom in a molecule are

draw Lewis structure of the molecule ⟶ use VSEPR to determine the electron pair arrangement surrounding the central atom (see Table 10.1) ⟶ use Table 10.4 to determine the hybridization state of the central atom

Solution

BeH_2

(a) The ground-state electron configuration of Be is $1s^22s^2$ and the Be atom has two valence electrons. The Lewis structure of BeH_2 is

H—Be—H

There are two bonding pairs around Be; therefore, the electron pair arrangement is linear. We conclude that Be uses *sp* hybrid orbitals in bonding with H, because *sp* orbitals have a linear arrangement (see Table 10.4). The hybridization process can be imagined as the illustration below shows.

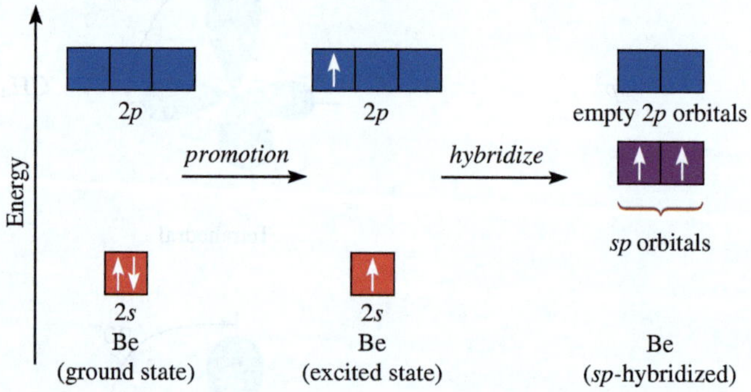

The two Be—H bonds are formed by the overlap of the Be *sp* orbitals with the 1*s* orbitals of the H atoms. Thus, BeH_2 is a linear molecule.

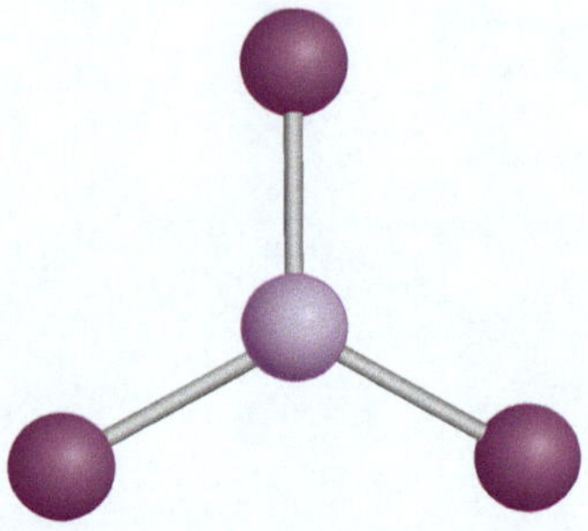

AlI_3

(b) The ground-state electron configuration of Al is $[Ne]3s^23p^1$. Therefore, the Al atom has three valence electrons. The Lewis structure of AlI_3 is

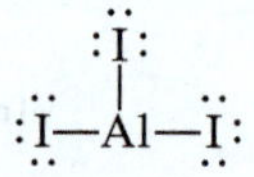

There are three pairs of electrons around Al; therefore, the electron pair arrangement is trigonal planar. We conclude that Al uses sp^2 hybrid orbitals in bonding with I because sp^2 orbitals have a trigonal planar arrangement (see Table 10.4). The hybridization process can be imagined as the illustration below shows.

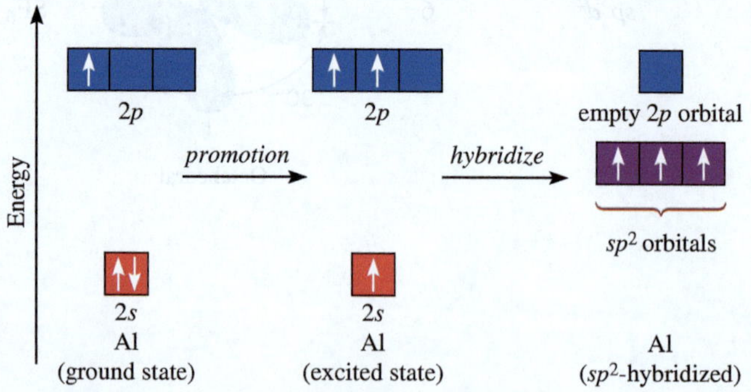

The sp^2 hybrid orbitals overlap with the 5*p* orbitals of I to form three covalent Al—I bonds. We predict that the AlI_3 molecule is trigonal planar and all the I—Al—I angles are 120°.

(Continued)

(c) The ground-state electron configuration of P is $[Ne]3s^23p^3$. Therefore, P atom has five valence electrons. The Lewis structure of PF_3 is

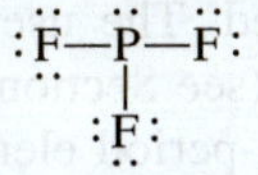

There are four pairs of electrons around P; therefore, the electron pair arrangement is tetrahedral. We conclude that P uses sp^3 hybrid orbitals in bonding to F, because sp^3 orbitals have a tetrahedral arrangement (see Table 10.4). The hybridization process can be imagined to take place as shown below.

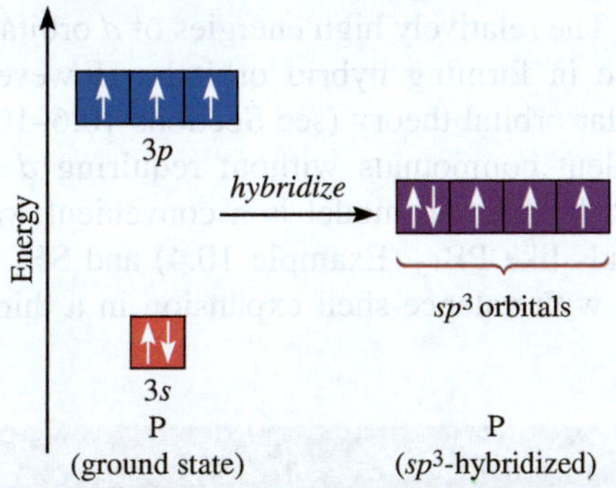

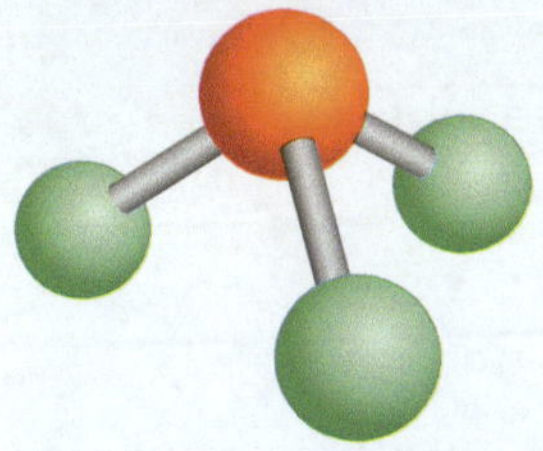

PF_3

As in the case of NH_3, one of the sp^3 hybrid orbitals is used to accommodate the lone pair on P. The other three sp^3 hybrid orbitals form covalent P—F bonds with the 2*p* orbitals of F. We predict the geometry of the molecule to be trigonal pyramidal; the F—P—F angle should be somewhat less than 109.5°.

Practice Exercise Determine the hybridization state of the underlined atoms in the following compounds: (a) <u>Si</u>Br_4, (b) <u>B</u>Cl_3.

Similar problems: 10.31, 10.33.

Hybridization of *s*, *p*, and *d* Orbitals

We have seen that hybridization neatly explains bonding that involves *s* and *p* orbitals. For elements in the third period and beyond, however, we cannot always account for molecular geometry by assuming that only *s* and *p* orbitals hybridize. To understand the formation of molecules with trigonal bipyramidal and octahedral geometries, for instance, we must include *d* orbitals in the hybridization concept.

Consider the SF_6 molecule as an example. In Section 10.1 we saw that this molecule has octahedral geometry, which is also the arrangement of the six electron pairs. Table 10.4 shows that the S atom is sp^3d^2-hybridized in SF_6. The ground-state electron configuration of S is $[Ne]3s^23p^4$. Focusing only on the valence electrons, we have the orbital diagram shown in the figure below. Because the 3*d* level is quite close in energy to the 3*s* and 3*p* levels, we can promote 3*s* and 3*p* electrons to two of the 3*d* orbitals.

Mixing the 3*s*, three 3*p*, and two 3*d* orbitals generates six sp^3d^2 hybrid orbitals (pronounced "*s-p* three *d* two"). Notice there are three unused 3*d* orbitals.

SF_6

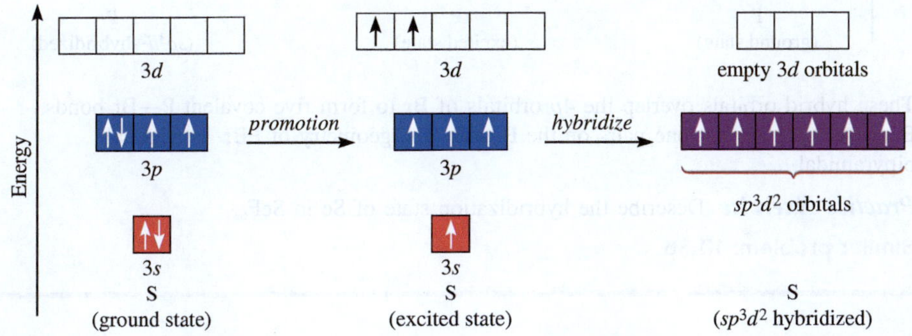

The six S—F bonds are formed by the overlap of the hybrid orbitals of the S atom with the $2p$ orbitals of the F atoms. Because there are 12 electrons around the S atom, the octet rule is violated. The use of d orbitals in addition to s and p orbitals to form an expanded octet (see Section 9.9) is an example of *valence-shell expansion (or hypervalency).* Second-period elements, unlike third-period elements, do not have $2d$ energy levels, so they can never expand their valence shells. (Recall that when $n = 2$, $l = 0$ and 1. Thus, we can only have $2s$ and $2p$ orbitals.) Hence atoms of second-period elements can never be surrounded by more than eight electrons in any of their compounds.

There is controversy surrounding the use of d orbitals to make hybrid orbitals for hypervalent compounds. The relatively high energies of d orbitals in main group atoms should prohibit their use in forming hybrid orbitals. However, a different bonding model known as molecular orbital theory (see Sections 10.6–10.8) can easily describe the bonding in hypervalent compounds without requiring d orbitals. Nevertheless, using d orbitals in the hybridization model is a convenient way to begin describing the bonding in compounds like PBr_5 (Example 10.4) and SF_6.

Example 10.4 deals with valence-shell expansion in a third-period element.

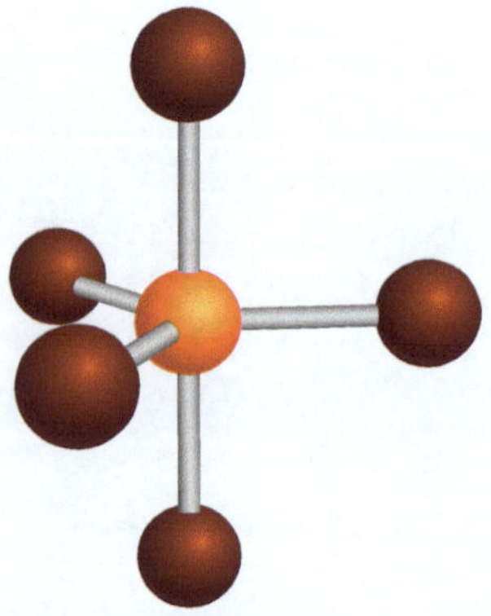

PBr_5

Example 10.4

Describe the hybridization state of phosphorus in phosphorus pentabromide (PBr_5).

Strategy Follow the same procedure shown in Example 10.3.

Solution The ground-state electron configuration of P is $[Ne]3s^23p^3$. Therefore, the P atom has five valence electrons. The Lewis structure of PBr_5 is

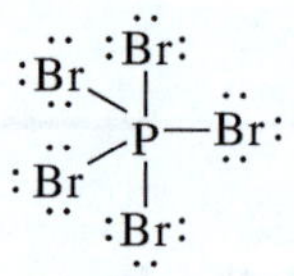

There are five pairs of electrons around P; therefore, the electron pair arrangement is trigonal bipyramidal. We conclude that P uses sp^3d hybrid orbitals in bonding to Br, because sp^3d hybrid orbitals have a trigonal bipyramidal arrangement (see Table 10.4). The hybridization process can be imagined as follows. The orbital diagram of the ground-state P atom is

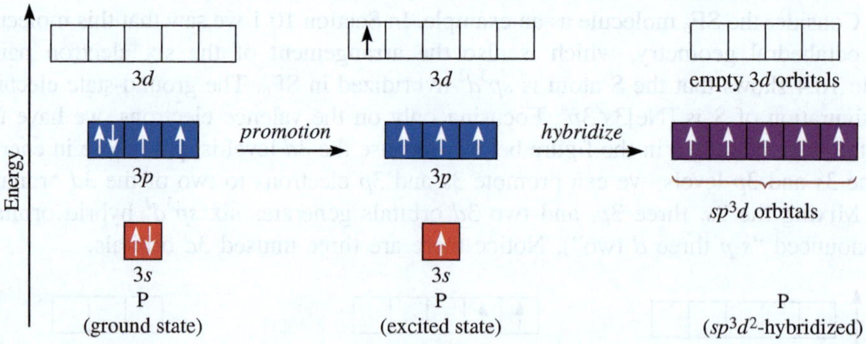

These hybrid orbitals overlap the $4p$ orbitals of Br to form five covalent P—Br bonds. Because there are no lone pairs on the P atom, the geometry of PBr_5 is trigonal bipyramidal.

Practice Exercise Describe the hybridization state of Se in SeF_6.

Similar problem: 10.36.

Summary of Concepts & Facts

- In valence bond theory, hybridized atomic orbitals are formed by the combination and rearrangement of orbitals from the same atom. The hybridized orbitals are all of equal energy and electron density, and the number of hybridized orbitals is equal to the number of pure atomic orbitals that combine.
- Valence-shell expansion can be explained by assuming hybridization of *s*, *p*, and *d* orbitals.
- In *sp* hybridization, the two hybrid orbitals lie in a straight line; in sp^2 hybridization, the three hybrid orbitals are directed toward the corners of an equilateral triangle; in sp^3 hybridization, the four hybrid orbitals are directed toward the corners of a tetrahedron; in sp^3d hybridization, the five hybrid orbitals are directed toward the corners of a trigonal bipyramid; in sp^3d^2 hybridization, the six hybrid orbitals are directed toward the corners of an octahedron.

Review of Concepts & Facts

10.4.1 How many orbitals does a set of sp^3d hybrid orbitals contain?
10.4.2 What is the hybridization of P in PH_4^+?
10.4.3 What is the hybridization of Xe in XeF_4 (see Example 9.12)?

10.5 Hybridization in Molecules Containing Double and Triple Bonds

Learning Objective

- Analyze the bonding in molecules containing multiple bonds.

The concept of hybridization is useful also for molecules with double and triple bonds. Consider the ethylene molecule, C_2H_4, as an example. In Example 10.1 we saw that C_2H_4 contains a carbon-carbon double bond and has planar geometry. Both the geometry and the bonding can be understood if we assume that each carbon atom is sp^2-hybridized. Figure 10.14 shows orbital diagrams of this hybridization process. We assume that only the $2p_x$ and $2p_y$ orbitals combine with the $2s$ orbital, and that the $2p_z$ orbital remains unchanged. Figure 10.15 shows that the $2p_z$ orbital is perpendicular to the plane of the hybrid orbitals. Now how do we account for the bonding of the C atoms? As Figure 10.16(a) shows, each carbon atom uses the three sp^2 hybrid orbitals to form two bonds with the two hydrogen $1s$ orbitals and one bond with the sp^2 hybrid orbital of the adjacent C atom. In addition, the two unhybridized $2p_z$ orbitals of the C atoms form another bond by overlapping sideways [see Figure 10.16(b)].

Video: Sigma and Pi Bonds

A distinction is made between the two types of covalent bonds in C_2H_4. The three bonds formed by each C atom in Figure 10.16(a) are all ***sigma (σ) bonds,*** *covalent*

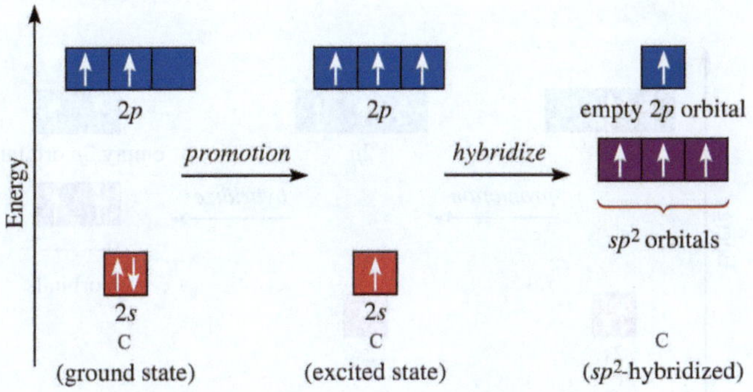

Figure 10.14 *The sp^2 hybridization of a carbon atom. The 2s orbital is mixed with only two 2p orbitals to form three equivalent sp^2 hybrid orbitals. This process leaves an electron in the unhybridized orbital, the $2p_z$ orbital.*

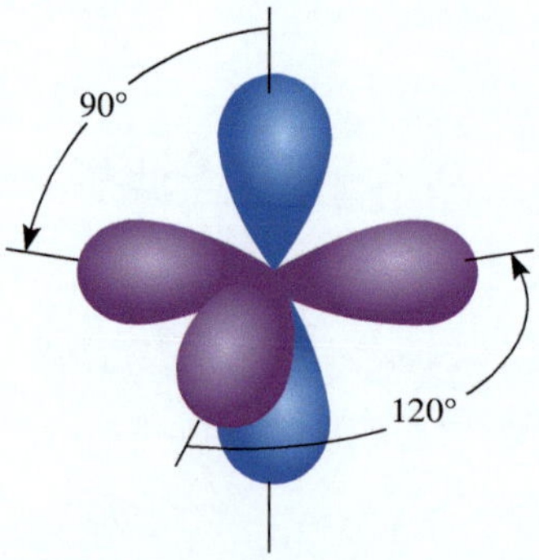

Figure 10.15 *Each carbon atom in the C_2H_4 molecule has three sp^2 hybrid orbitals (purple) and one unhybridized $2p_z$ orbital (blue), which is perpendicular to the plane of the hybrid orbitals.*

bonds formed by orbitals overlapping end-to-end, with the electron density concentrated between the nuclei of the bonding atoms. The second type is called a ***pi*** **(*π*) *bond,*** which is defined as *a covalent bond formed by sideways overlapping orbitals with electron density concentrated above and below the plane of the nuclei of the bonding atoms.* The two C atoms form a pi bond as shown in Figure 10.16(b). It is this pi bond formation that gives ethylene its planar geometry. Figure 10.16(c) shows the orientation of the sigma and pi bonds. Figure 10.17 is yet another way of looking at the planar C_2H_4 molecule and the formation of the pi bond. Although we normally represent the carbon-carbon double bond as C═C (as in a Lewis structure), it is important to keep in mind that the two bonds are different types: One is a sigma bond and the other is a pi bond. In fact, the bond enthalpies of the carbon-carbon pi and sigma bonds are about 270 kJ/mol and 350 kJ/mol, respectively.

The acetylene molecule (C_2H_2) contains a carbon-carbon triple bond. Because the molecule is linear, we can explain its geometry and bonding by assuming that each C atom is *sp*-hybridized by mixing the 2*s* with the $2p_x$ orbital (Figure 10.18). As

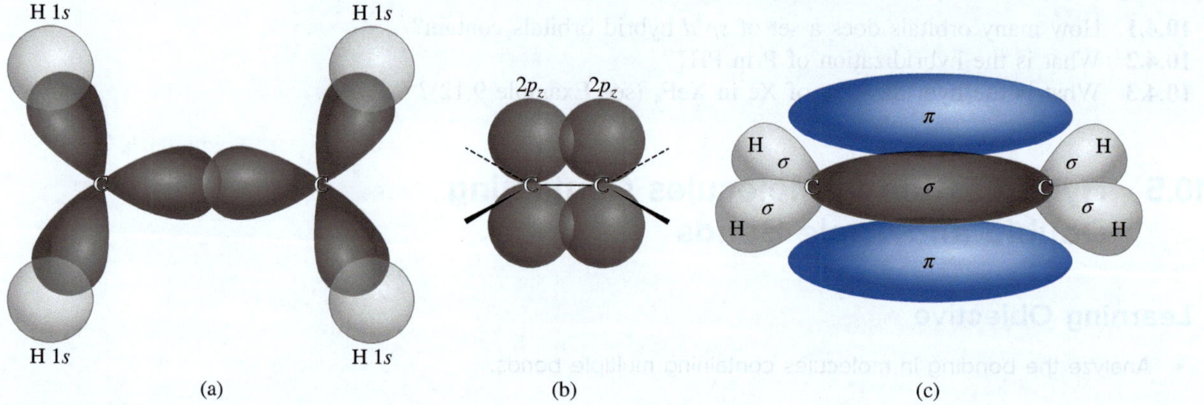

Figure 10.16 *Bonding in ethylene, C_2H_4. (a) Top view of the sigma bonds between carbon atoms and between carbon and hydrogen atoms. All the atoms lie in the same plane, making C_2H_4 a planar molecule. (b) Side view showing how the two $2p_z$ orbitals on the two carbon atoms overlap, leading to the formation of a pi bond. The solid, dashed, and wedged lines show the directions of the sigma bonds. (c) The interactions in (a) and (b) lead to the formation of the sigma bonds and the pi bond in ethylene. Note that the pi bond lies above and below the plane of the molecule.*

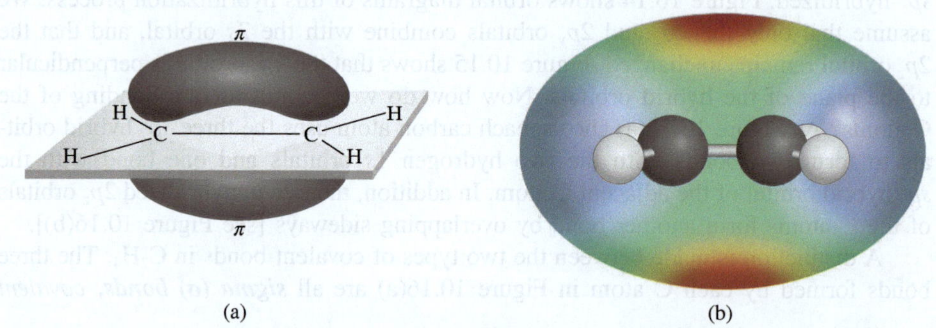

Figure 10.17 *(a) Another view of the pi bond in the C_2H_4 molecule. Note that all six atoms are in the same plane. It is the overlap of the $2p_z$ orbitals that causes the molecule to assume a planar structure. (b) Electrostatic potential map of C_2H_4.*

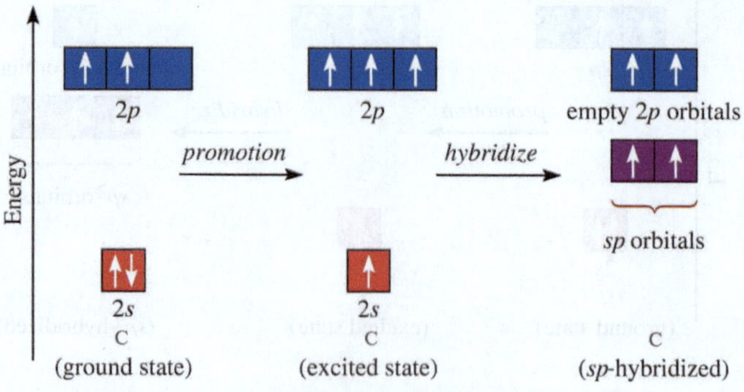

Figure 10.18 *The sp hybridization of a carbon atom. The 2s orbital is mixed with only one 2p orbital to form two sp hybrid orbitals. This process leaves an electron in each of the two unhybridized 2p orbitals, namely, the $2p_y$ and $2p_z$ orbitals.*

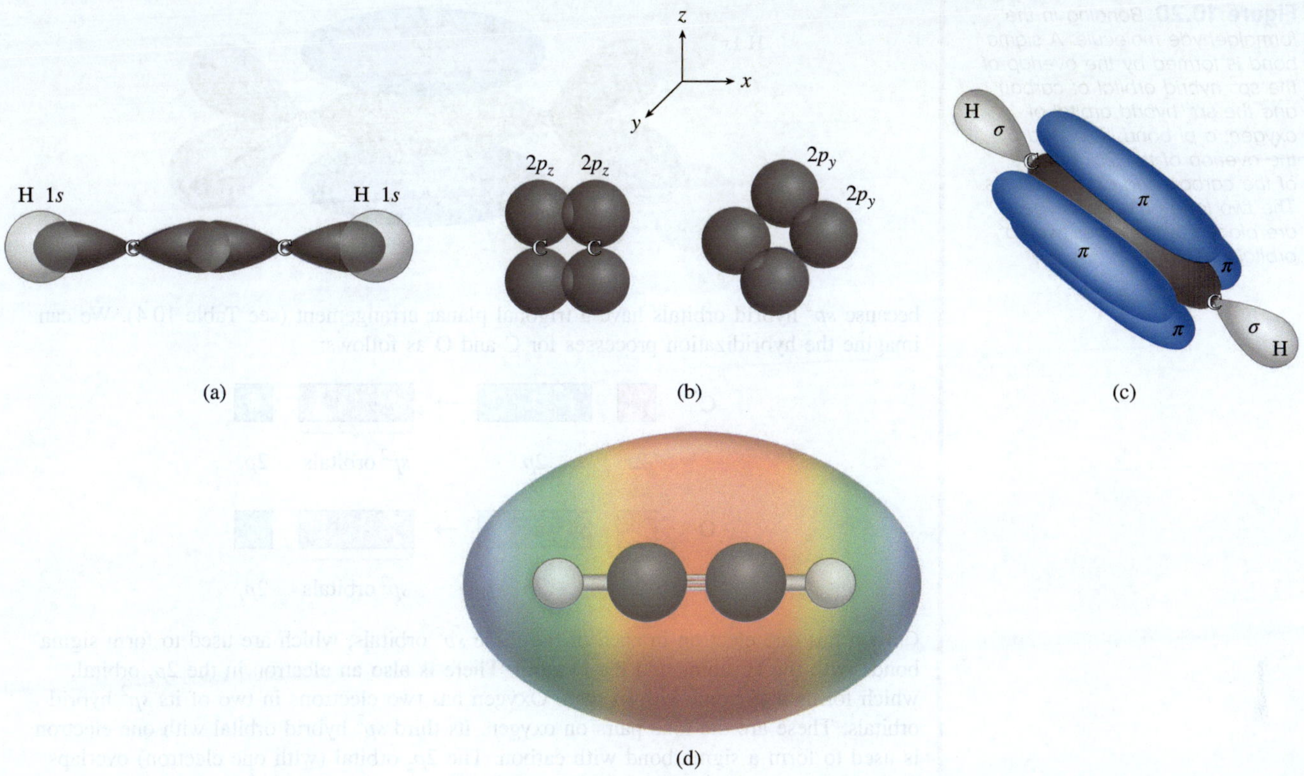

Figure 10.19 *Bonding in acetylene, C_2H_2. (a) Top view showing the overlap of the sp orbitals between the C atoms and the overlap of the sp orbital with the 1s orbital between the C and H atoms. All the atoms lie along a straight line; therefore, acetylene is a linear molecule. (b) Side view showing the overlap of the two $2p_y$ orbitals and of the two $2p_z$ orbitals of the two carbon atoms, which leads to the formation of two pi bonds. (c) Formation of the sigma and pi bonds as a result of the interactions in (a) and (b). (d) Electrostatic potential map of C_2H_2.*

Figure 10.19 shows, the two *sp* hybrid orbitals of each C atom form one sigma bond with a hydrogen 1*s* orbital and another sigma bond with the other C atom. In addition, two pi bonds are formed by the sideways overlap of the unhybridized $2p_y$ and $2p_z$ orbitals. Thus, the C≡C bond is made up of one sigma bond and two pi bonds.

The following rule helps us predict hybridization in molecules containing multiple bonds: If the central atom forms a double bond, it is sp^2-hybridized; if it forms two double bonds or a triple bond, it is *sp*-hybridized. Note that this rule applies only to atoms of the second-period elements. Atoms of third-period elements and beyond that form multiple bonds present a more complicated picture and will not be dealt with here.

Student Hot Spot

Student data indicate you may struggle with hybridization in multiple bonds. Access your eBook for additional Learning Resources on this topic.

Example 10.5

Describe the bonding in the formaldehyde molecule whose Lewis structure is

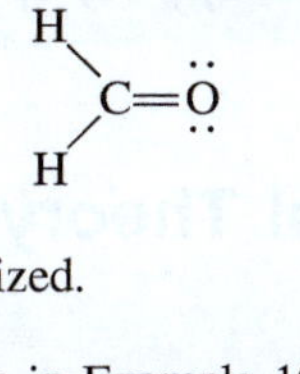

Assume that the O atom is sp^2-hybridized.

Strategy Follow the procedure shown in Example 10.3.

Solution There are three pairs of electrons around the C atom; therefore, the electron pair arrangement is trigonal planar. (Recall that a double bond is treated as a single bond in the VSEPR model.) We conclude that C uses sp^2 hybrid orbitals in bonding,

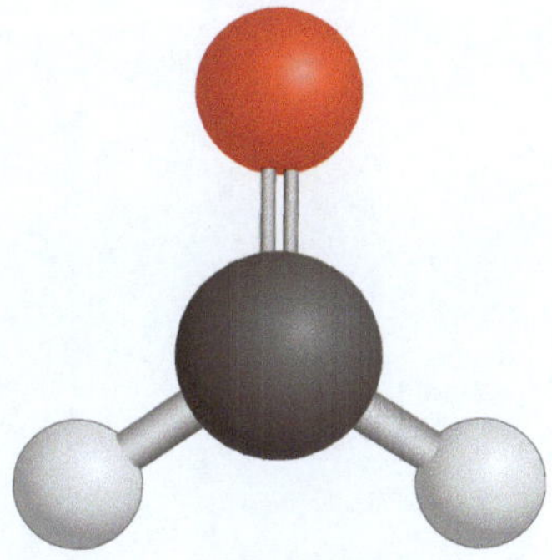

CH_2O

(Continued)

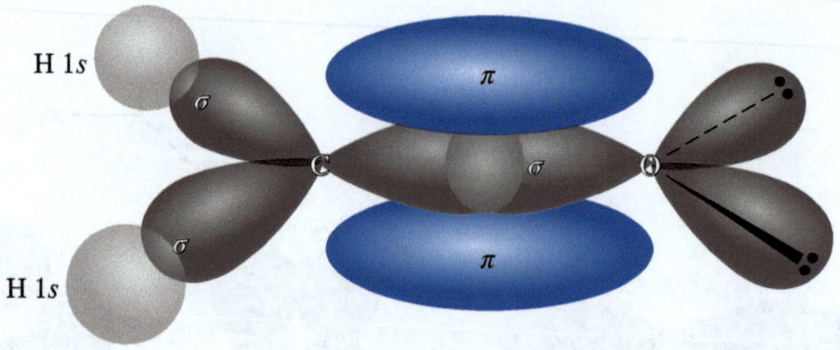

Figure 10.20 *Bonding in the formaldehyde molecule. A sigma bond is formed by the overlap of the sp^2 hybrid orbital of carbon and the sp^2 hybrid orbital of oxygen; a pi bond is formed by the overlap of the $2p_z$ orbitals of the carbon and oxygen atoms. The two lone pairs on oxygen are placed in the other two sp^2 orbitals of oxygen.*

because sp^2 hybrid orbitals have a trigonal planar arrangement (see Table 10.4). We can imagine the hybridization processes for C and O as follows:

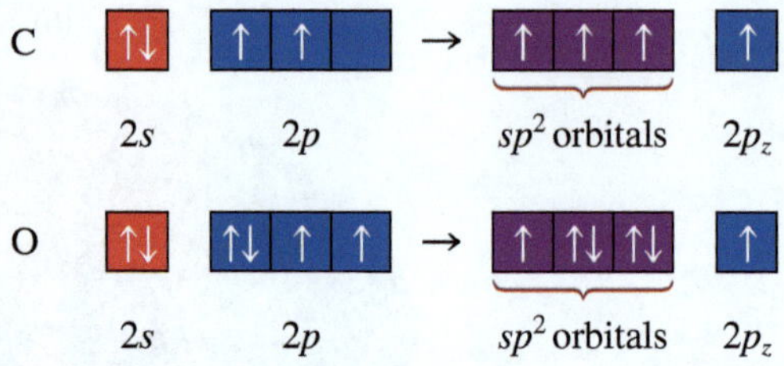

Carbon has one electron in each of the three sp^2 orbitals, which are used to form sigma bonds with the H atoms and the O atom. There is also an electron in the $2p_z$ orbital, which forms a pi bond with oxygen. Oxygen has two electrons in two of its sp^2 hybrid orbitals. These are the lone pairs on oxygen. Its third sp^2 hybrid orbital with one electron is used to form a sigma bond with carbon. The $2p_z$ orbital (with one electron) overlaps with the $2p_z$ orbital of C to form a pi bond (Figure 10.20).

Practice Exercise Describe the bonding in the hydrogen cyanide molecule, HCN. Assume that N is *sp*-hybridized.

Similar problems: 10.40, 10.41, 10.43.

Summary of Concepts & Facts

- In an sp^2-hybridized atom (for example, carbon), the one unhybridized *p* orbital can form a pi bond with another *p* orbital. A carbon-carbon double bond consists of a sigma bond and a pi bond. In an *sp*-hybridized carbon atom, the two unhybridized *p* orbitals can form two pi bonds with two *p* orbitals on another atom (or atoms). A carbon-carbon triple bond consists of one sigma bond and two pi bonds.

Review of Concepts & Facts

10.5.1 How many pi bonds are present in CS_2?

10.5.2 Which of the following pairs of atomic orbitals on adjacent nuclei can overlap to form a sigma bond? A pi bond? Which cannot overlap (no bond)? Consider the *x*-axis to be the internuclear axis. (a) 1*s* and 2*s*, (b) 1*s* and $2p_x$, (c) $2p_y$ and $2p_y$, (d) $3p_y$ and $2p_z$, (e) $2p_x$ and $3p_x$.

10.6 Molecular Orbital Theory

Learning Objectives

- Explain how atomic orbitals overlap to form molecular orbitals according to molecular orbital theory.
- Identify bonding and antibonding molecular orbitals.

Valence bond theory is one of the two quantum mechanical approaches that explain bonding in molecules. It accounts, at least qualitatively, for the stability of the covalent bond in terms of overlapping atomic orbitals. Using the concept of hybridization, valence bond theory can explain molecular geometries predicted by the VSEPR model. However, the assumption that electrons in a molecule occupy atomic orbitals of the individual atoms can only be an approximation, because each bonding electron in a molecule must be in an orbital that is characteristic of the molecule as a whole.

In some cases, valence bond theory cannot satisfactorily account for observed properties of molecules. Consider the oxygen molecule, whose Lewis structure is

$$\overset{\cdot\cdot}{\underset{\cdot\cdot}{O}}=\overset{\cdot\cdot}{\underset{\cdot\cdot}{O}}$$

According to this description, all the electrons in O_2 are paired and oxygen should therefore be diamagnetic. But experiments have shown that the oxygen molecule has two unpaired electrons (Figure 10.21). This finding suggests a fundamental deficiency in valence bond theory, one that justifies searching for an alternative bonding approach that accounts for the properties of O_2 and other molecules that do not match the predictions of valence bond theory.

Magnetic and other properties of molecules are sometimes better explained by another quantum mechanical approach called *molecular orbital (MO) theory.* Molecular orbital theory describes covalent bonds in terms of ***molecular orbitals,*** which *result from interaction of the atomic orbitals of the bonding atoms and are associated with the entire molecule.* The difference between a molecular orbital and an atomic orbital is that an atomic orbital is associated with only one atom.

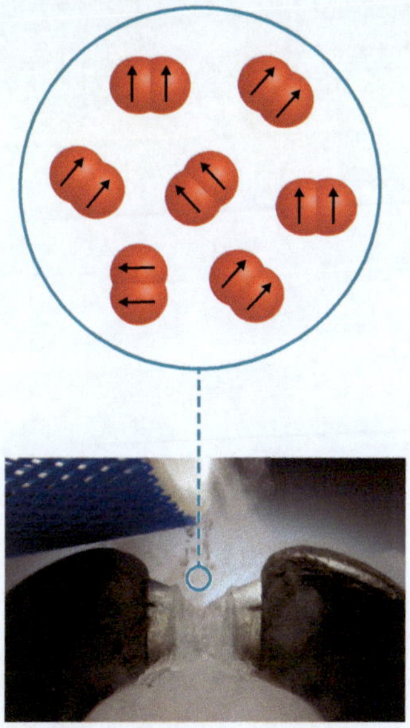

Figure 10.21 *Liquid oxygen caught between the poles of a magnet, because the O_2 molecules are paramagnetic, having two parallel spins.*
Charles D. Winters/Timeframe Photography/McGraw Hill

Bonding and Antibonding Molecular Orbitals

According to MO theory, the overlap of the 1*s* orbitals of two hydrogen atoms leads to the formation of two molecular orbitals: one bonding molecular orbital and one antibonding molecular orbital. A ***bonding molecular orbital*** has *lower energy and greater stability than the atomic orbitals from which it was formed.* An ***antibonding molecular orbital*** has *higher energy and lower stability than the atomic orbitals from which it was formed.* As the names "bonding" and "antibonding" suggest, placing electrons in a bonding molecular orbital yields a stable covalent bond, whereas placing electrons in an antibonding molecular orbital results in an unstable bond.

In the bonding molecular orbital, the electron density is greatest between the nuclei of the bonding atoms. In the antibonding molecular orbital, on the other hand, the electron density decreases to zero between the nuclei. We can understand this distinction if we recall that electrons in orbitals have wave characteristics. A property unique to waves enables waves of the same type to interact in such a way that the resultant wave has either an enhanced amplitude or a diminished amplitude. In the former case, we call the interaction *constructive interference;* in the latter case, it is *destructive interference* (Figure 10.22).

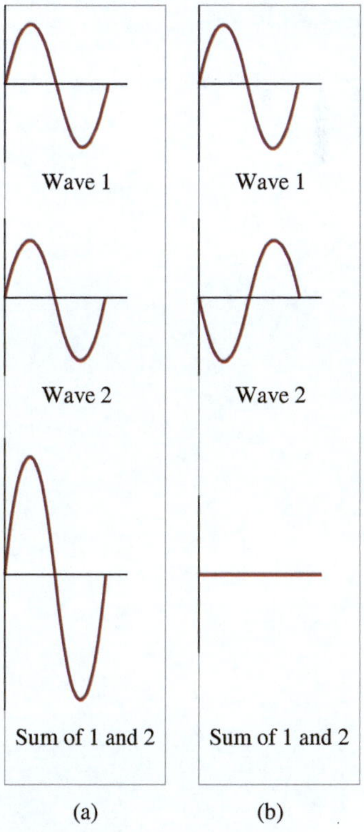

Figure 10.22 *Constructive interference (a) and destructive interference (b) of two waves of the same wavelength and amplitude.*

The formation of bonding molecular orbitals corresponds to constructive interference (the increase in amplitude is analogous to the buildup of electron density between the two nuclei). The formation of antibonding molecular orbitals corresponds to destructive interference (the decrease in amplitude is analogous to the decrease in electron density between the two nuclei). The constructive and destructive interactions between the two 1*s* orbitals in the H_2 molecule, then, lead to the formation of a sigma bonding molecular orbital σ_{1s} and a sigma antibonding molecular orbital σ_{1s}^{*}:

a sigma bonding molecular orbital	a sigma antibonding molecular orbital
↘ σ_{1s} ↖	↘ $\sigma_{1s}^{\star}$ ↖
formed from 1*s* orbitals	formed from 1*s* orbitals

where the star denotes an antibonding molecular orbital.

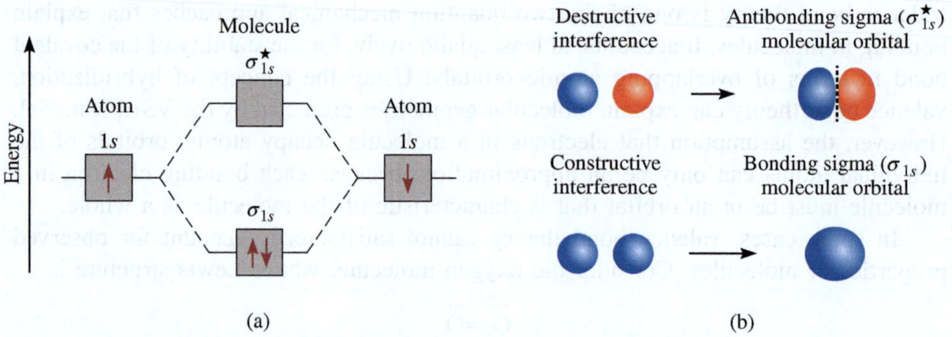

Figure 10.23 *(a) Energy levels of bonding and antibonding molecular orbitals in the H_2 molecule. Note that the two electrons in the σ_{1s} orbital must have opposite spins in accord with the Pauli exclusion principle. Keep in mind that the higher the energy of the molecular orbital, the less stable the electrons in that molecular orbital. (b) Constructive and destructive interferences between the two hydrogen 1s orbitals lead to the formation of a bonding and an antibonding molecular orbital. In the bonding molecular orbital, there is a buildup between the nuclei of electron density, which acts as a negatively charged "glue" to hold the positively charged nuclei together. In the antibonding molecular orbital, there is a nodal plane between the nuclei, where the electron density is zero.*

In a ***sigma molecular orbital*** (bonding or antibonding) *the electron density is concentrated symmetrically around a line between the two nuclei of the bonding atoms.* Two electrons in a sigma molecular orbital form a sigma bond (see Section 10.5). The two electrons in a sigma molecular orbital are paired indicating that the Pauli exclusion principle applies to molecules as well as to atoms. Remember that a single covalent bond (such as H—H or F—F) is almost always a sigma bond.

Figure 10.23 shows the *molecular orbital energy level diagram*—that is, the relative energy levels of the orbitals produced in the formation of the H_2 molecule—and the constructive and destructive interferences between the two 1*s* orbitals. The colors shown for the atomic and molecular orbitals represent the phases of the orbitals as described by quantum mechanics. These colored phases help us visualize the formation of molecular orbitals. Notice that in the antibonding molecular orbital there is a *nodal plane* between the nuclei that signifies zero electron density. The nuclei are repelled by each other's positive charges, rather than held together. Electrons in the antibonding molecular orbital have higher energy (and less stability) than they would have in the isolated atoms. On the other hand, electrons in the bonding molecular orbital have less energy (and hence greater stability) than they would have in the isolated atoms.

Although we have used the hydrogen molecule to illustrate molecular orbital formation, the concept is equally applicable to other molecules. In the H_2 molecule, we consider only the interaction between 1*s* orbitals; with more complex molecules we need to consider additional atomic orbitals as well. Nevertheless, for all *s* orbitals, the process is the same as for 1*s* orbitals. Thus, the interaction between two 2*s* or 3*s* orbitals can be understood in terms of the molecular orbital energy level diagram and the formation of bonding and antibonding molecular orbitals shown in Figure 10.23.

For *p* orbitals, the process is more complex because they can interact with each other in two different ways. For example, two 2*p* orbitals can approach each other end-to-end to produce a sigma bonding and a sigma antibonding molecular orbital, as shown in Figure 10.24(a). Alternatively, the two *p* orbitals can overlap sideways to generate a bonding and an antibonding pi molecular orbital [see Figure 10.24(b)].

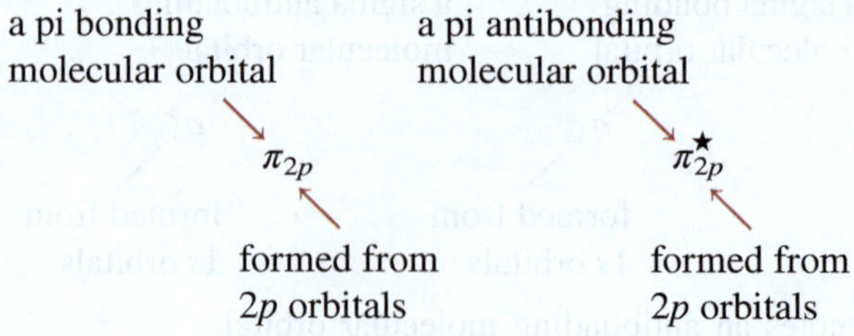

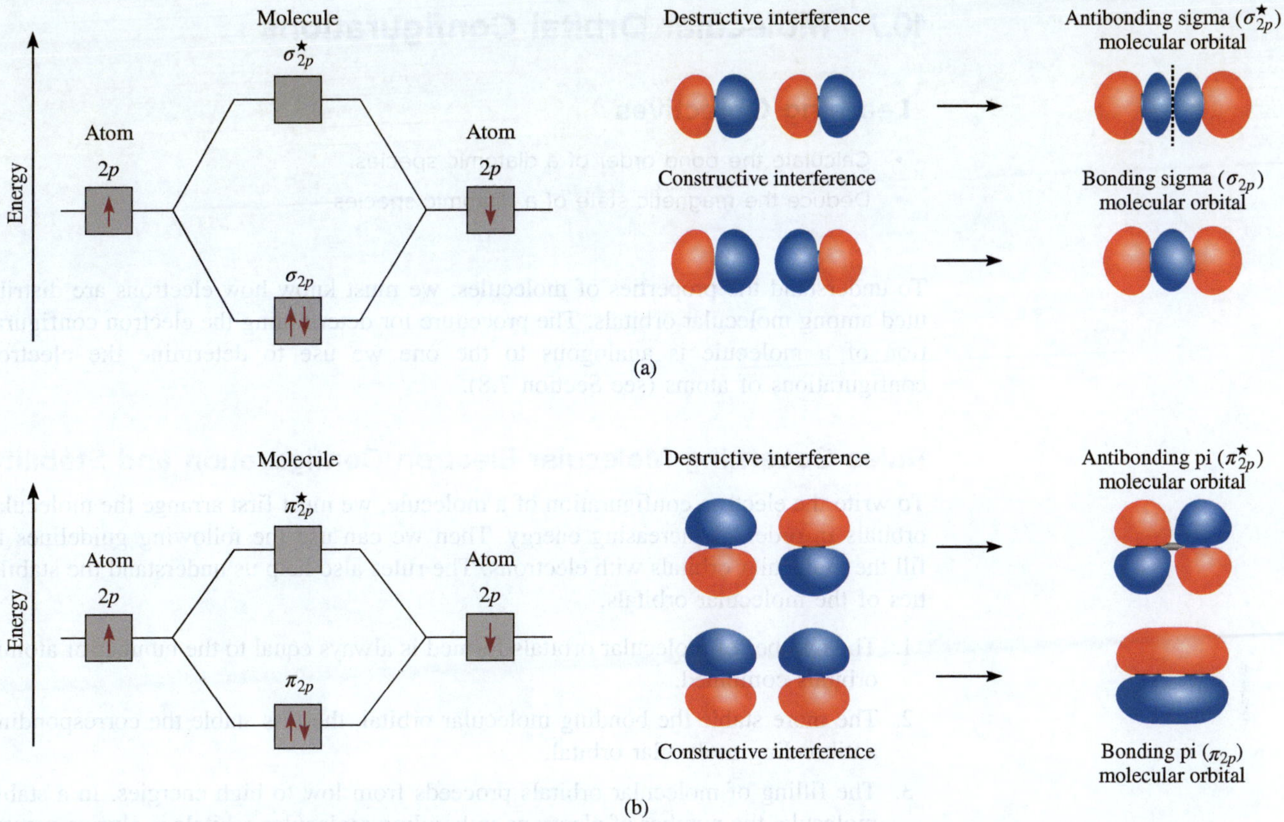

Figure 10.24 *Two possible interactions between two equivalent p orbitals and the corresponding molecular orbitals. (a) When the p orbitals overlap end-to-end, a sigma bonding and a sigma antibonding molecular orbital form. (b) When the p orbitals overlap side-to-side, a pi bonding and a pi antibonding molecular orbital form. Normally, a sigma bonding molecular orbital is more stable than a pi bonding molecular orbital, because side-to-side interaction leads to a smaller overlap of the p orbitals than does end-to-end interaction. We assume that the $2p_x$ orbitals take part in the sigma molecular orbital formation. The $2p_y$ and $2p_z$ orbitals can interact to form only π molecular orbitals. The behavior shown in (b) represents the interaction between the $2p_y$ orbitals or the $2p_z$ orbitals. In both cases, the dashed line represents a nodal plane between the nuclei, where the electron density is zero.*

In a ***pi molecular orbital*** (bonding or antibonding), *the electron density is concentrated above and below a line joining the two nuclei of the bonding atoms.* Two electrons in a pi molecular orbital form a pi bond (see Section 10.5). A double bond is almost always composed of a sigma bond and a pi bond; a triple bond is always a sigma bond plus two pi bonds.

Summary of Concepts & Facts

- Molecular orbital theory describes bonding in terms of the combination and rearrangement of atomic orbitals to form orbitals that are associated with the molecule as a whole.
- Bonding molecular orbitals increase electron density between the nuclei and are lower in energy than individual atomic orbitals. Antibonding molecular orbitals have a region of zero electron density between the nuclei, and an energy level higher than that of the individual atomic orbitals.

Review of Concepts & Facts

10.6.1 One way to account for the fact that an O_2 molecule contains two unpaired electrons is to draw the following Lewis structure:

$$\cdot\ddot{\underset{\cdot\cdot}{O}}-\ddot{\underset{\cdot\cdot}{O}}\cdot$$

Suggest two reasons why this structure is unsatisfactory.

10.7 Molecular Orbital Configurations

Learning Objectives

- Calculate the bond order of a diatomic species.
- Deduce the magnetic state of a diatomic species.

To understand the properties of molecules, we must know how electrons are distributed among molecular orbitals. The procedure for determining the electron configuration of a molecule is analogous to the one we use to determine the electron configurations of atoms (see Section 7.8).

Rules Governing Molecular Electron Configuration and Stability

To write the electron configuration of a molecule, we must first arrange the molecular orbitals in order of increasing energy. Then we can use the following guidelines to fill the molecular orbitals with electrons. The rules also help us understand the stabilities of the molecular orbitals.

1. The number of molecular orbitals formed is always equal to the number of atomic orbitals combined.
2. The more stable the bonding molecular orbital, the less stable the corresponding antibonding molecular orbital.
3. The filling of molecular orbitals proceeds from low to high energies. In a stable molecule, the number of electrons in bonding molecular orbitals is always greater than that in antibonding molecular orbitals because we place electrons first in the lower-energy bonding molecular orbitals.
4. Like an atomic orbital, each molecular orbital can accommodate up to two electrons with opposite spins in accordance with the Pauli exclusion principle.
5. When electrons are added to molecular orbitals of the same energy, the most stable arrangement is predicted by Hund's rule; that is, electrons enter these molecular orbitals with parallel spins.
6. The number of electrons in the molecular orbitals is equal to the sum of all the electrons on the bonding atoms.

Hydrogen and Helium Molecules

Later in this section we will study molecules formed by atoms of the second-period elements. Before we do, it will be instructive to predict the relative stabilities of the simple species H_2^+, H_2, He_2^+, and He_2, using the energy-level diagrams shown in Figure 10.25. The σ_{1s} and σ_{1s}^* orbitals can accommodate a maximum of four electrons. The total number of electrons increases from one for H_2^+ to four for He_2. The Pauli exclusion principle stipulates that each molecular orbital can accommodate a maximum of two electrons with opposite spins. We are concerned only with the ground-state electron configurations in these cases.

To evaluate the stabilities of these species we determine their ***bond order,*** defined as

$$\text{bond order} = \frac{1}{2}\left(\begin{matrix}\text{number of electrons}\\ \text{in bonding MOs}\end{matrix} - \begin{matrix}\text{number of electrons}\\ \text{in antibonding MOs}\end{matrix}\right) \qquad (10.2)$$

The bond order indicates the approximate strength of a bond. Recall that the quantitative measure of the strength of a bond is the bond enthalpy (see Section 9.10). For example, if there are two electrons in the bonding molecular orbital and none in the antibonding molecular orbital, the bond order is one, which means that there is

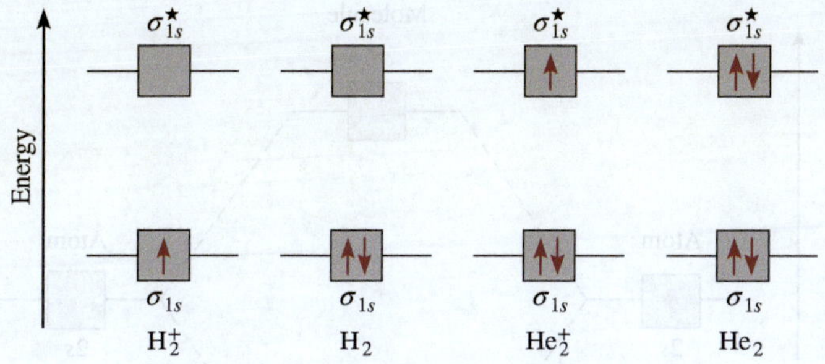

Figure 10.25 *Energy levels of the bonding and antibonding molecular orbitals in H_2^+, H_2, He_2^+, and He_2. In all these species, the molecular orbitals are formed by the interaction of two 1s orbitals.*

one covalent bond and that the molecule is stable. Note that the bond order can be a fraction, but a bond order of zero (or a negative value) means the bond has no stability and the molecule cannot exist. Bond order can be used only qualitatively for purposes of comparison. For example, a bonding sigma molecular orbital with two electrons and a bonding pi molecular orbital with two electrons would each have a bond order of one. Yet, these two bonds must differ in bond strength (and bond length) because of the differences in the extent of atomic orbital overlap.

We are ready now to make predictions about the stability of H_2^+, H_2, He_2^+, and He_2 (see Figure 10.25). The H_2^+ molecular ion has only one electron in the σ_{1s} orbital. Because a covalent bond consists of two electrons in a bonding molecular orbital, H_2^+ has only half of one bond, or a bond order of $\frac{1}{2}$. Thus, we predict that the H_2^+ molecular ion may be a stable species. The electron configuration of H_2^+ is written as $(\sigma_{1s})^1$. The superscript in $(\sigma_{1s})^1$ indicates that there is one electron in the sigma bonding molecular orbital.

The H_2 molecule has two electrons, both of which are in the σ_{1s} orbital. According to our scheme, two electrons equal one full bond; therefore, the H_2 molecule has a bond order of one, or one full covalent bond. The electron configuration of H_2 is $(\sigma_{1s})^2$.

As for the He_2^+ molecular ion, we place the first two electrons in the σ_{1s} orbital and the third electron in the σ_{1s}^* orbital. Because the antibonding molecular orbital is destabilizing, we expect He_2^+ to be less stable than H_2. Roughly speaking, the instability resulting from the electron in the σ_{1s}^* orbital is balanced by one of the σ_{1s} electrons. The bond order is $\frac{1}{2}(2 - 1) = \frac{1}{2}$ and the overall stability of He_2^+ is similar to that of the H_2^+ molecule. The electron configuration of He_2^+ is $(\sigma_{1s})^2(\sigma_{1s}^*)^1$.

In He_2 there would be two electrons in the σ_{1s} orbital and two electrons in the σ_{1s}^* orbital, so the molecule would have a bond order of zero and no net stability. The electron configuration of He_2 would be $(\sigma_{1s})^2(\sigma_{1s}^*)^2$.

To summarize, we can arrange our examples in order of decreasing stability:

$$H_2 > H_2^+, He_2^+ > He_2$$

We know that the hydrogen molecule is a stable species. Our simple molecular orbital method predicts that H_2^+ and He_2^+ also possess some stability, because both have bond orders of $\frac{1}{2}$. Indeed, their existence has been confirmed by experiment. It turns out that H_2^+ is somewhat more stable than He_2^+, because there is only one electron in the hydrogen molecular ion and therefore it has no electron-electron repulsion. Furthermore, H_2^+ also has less nuclear repulsion than He_2^+. Our prediction about He_2 is that it would have no stability, but in 1993 He_2 gas was found to exist. The "molecule" is extremely unstable and has only a transient existence under specially created conditions.

Homonuclear Diatomic Molecules of Second-Period Elements

We are now ready to study the ground-state electron configuration of molecules containing second-period elements. We will consider only the simplest case, that of ***homonuclear diatomic molecules,*** or *diatomic molecules containing atoms of the same elements.*

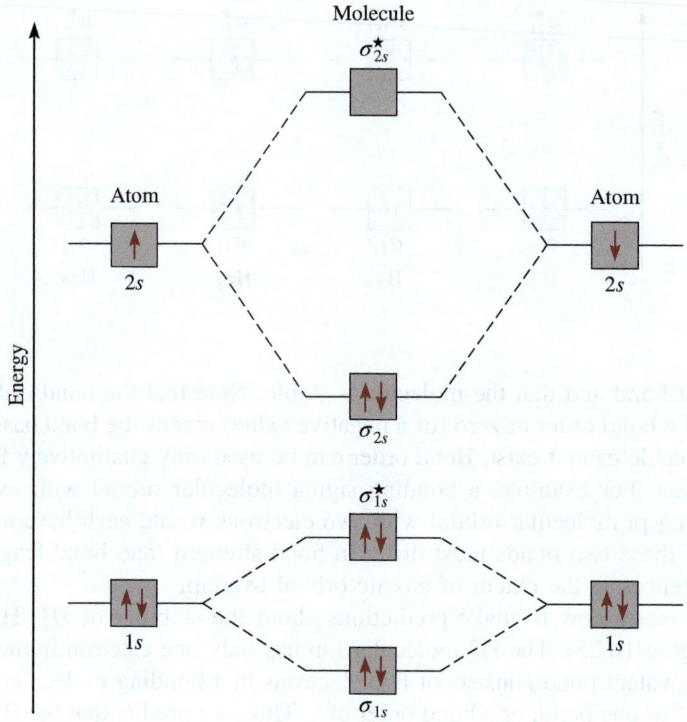

Figure 10.26 *Molecular orbital energy level diagram for the Li_2 molecule. The six electrons in Li_2 (Li's electron configuration $1s^2 2s^1$) are in the σ_{1s}, σ_{1s}^*, and σ_{2s} orbitals. Because there are two electrons each in σ_{1s} and σ_{1s}^* (just as in He_2), there is no net bonding or antibonding effect. Therefore, the single covalent bond in Li_2 is formed by the two electrons in the bonding molecular orbital σ_{2s}. Note that although the antibonding orbital (σ_{1s}^*) has higher energy and is thus less stable than the bonding orbital (σ_{1s}), this antibonding orbital has less energy and greater stability than the σ_{2s} bonding orbital.*

Student Hot Spot

Student data indicate you may struggle with molecular orbital diagrams. Access your eBook for additional Learning Resources on this topic.

Figure 10.26 shows the molecular orbital energy level diagram for the first member of the second period, Li_2. These molecular orbitals are formed by the overlap of 1*s* and 2*s* orbitals. We will use this diagram to build up all the diatomic molecules, as we will see shortly.

The situation is more complex when the bonding also involves *p* orbitals. Two *p* orbitals can form either a sigma bond or a pi bond. Because there are three *p* orbitals for each atom of a second-period element, we know that one sigma and two pi molecular orbitals will result from the constructive interaction. The sigma molecular orbital is formed by the overlap of the $2p_x$ orbitals along the internuclear axis, that is, the *x*-axis. The $2p_y$ and $2p_z$ orbitals are perpendicular to the *x*-axis, and they will overlap sideways to give two pi molecular orbitals. The molecular orbitals are called σ_{2p_x}, π_{2p_y}, and π_{2p_z} orbitals, where the subscripts indicate which atomic orbitals take part in forming the molecular orbitals. As shown in Figure 10.24, overlap of the two *p* orbitals is normally greater in a σ molecular orbital than in a π molecular orbital, so we would expect the former to be lower in energy. However, the energies of molecular orbitals actually increase as follows:

$$\sigma_{1s} < \sigma_{1s}^* < \sigma_{2s} < \sigma_{2s}^* < \pi_{2p_y} = \pi_{2p_z} < \sigma_{2p_x} < \pi_{2p_y}^* = \pi_{2p_z}^* < \sigma_{2p_x}^*$$

The inversion of the σ_{2p_x} orbital and the π_{2p_y} and π_{2p_z} orbitals is due to the interaction between the 2*s* orbital on one atom with the 2*p* orbital on the other. In MO terminology, we say there is mixing between these orbitals. The condition for mixing is that the 2*s* and 2*p* orbitals must be close in energy. This condition is met for the lighter molecules B_2, C_2, and N_2 with the result that the σ_{2p_x} orbital is raised in energy relative to the π_{2p_y} and π_{2p_z} orbitals as already shown. The mixing is less pronounced for O_2 and F_2 so the σ_{2p_x} orbital lies lower in energy than the π_{2p_y} and π_{2p_z} orbitals in these molecules.

With these concepts and Figure 10.27, which shows the order of increasing energies for 2*p* molecular orbitals, we can write the electron configurations and predict the magnetic properties and bond orders of second-period homonuclear diatomic molecules. We will consider a few examples.

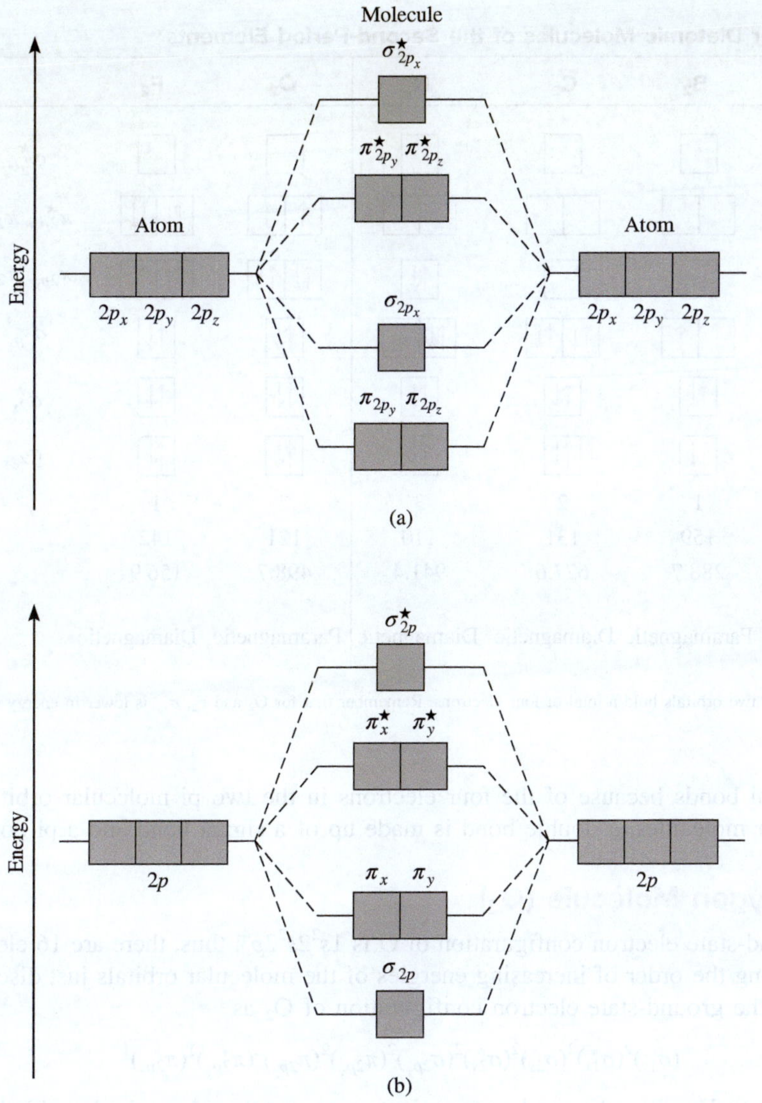

Figure 10.27 *General molecular orbital energy level diagram for the second-period homonuclear diatomic molecules Li_2, Be_2, B_2, C_2, and N_2. For simplicity, the σ_{1s} and σ_{2s} orbitals have been omitted. Note that in these molecules, the σ_{2p_x} orbital is higher in energy than either the π_{2p_y} or the π_{2p_z} orbitals. This means that electrons in the σ_{2p_x} orbitals are less stable than those in π_{2p_y} and π_{2p_z}. This aberration stems from the different interactions between the electrons in the σ_{2p_x} orbital, on one hand, and π_{2p_y} and π_{2p_z} orbitals, on the other hand, with the electrons in the lower-energy σ_s orbitals. For O_2 and F_2, the σ_{2p_x} orbital is lower in energy than π_{2p_y} and π_{2p_z}.*

The Lithium Molecule (Li_2)

The electron configuration of Li is $1s^2 2s^1$, so Li_2 has a total of six electrons. According to Figure 10.26, these electrons are placed (two each) in the σ_{1s}, σ^*_{1s}, and σ_{2s} molecular orbitals. The electrons of σ_{1s} and σ^*_{1s} make no net contribution to the bonding in Li_2. Thus, the electron configuration of the molecular orbitals in Li_2 is $(\sigma_{1s})^2(\sigma^*_{1s})^2(\sigma_{2s})^2$. Since there are two more electrons in the bonding molecular orbitals than in antibonding orbitals, the bond order is 1 [see Equation (10.2)]. We conclude that the Li_2 molecule is stable, and because it has no unpaired electron spins, it should be diamagnetic. Indeed, diamagnetic Li_2 molecules are known to exist in the vapor phase.

The Carbon Molecule (C_2)

The carbon atom has the electron configuration $1s^2 2s^2 2p^2$; thus, there are 12 electrons in the C_2 molecule. Referring to Figures 10.26 and 10.27, we place the last four electrons in the π_{2p_y} and π_{2p_z} orbitals. Therefore, C_2 has the electron configuration

$$(\sigma_{1s})^2(\sigma^*_{1s})^2(\sigma_{2s})^2(\sigma^*_{2s})^2(\pi_{2p_y})^2(\pi_{2p_z})^2$$

Its bond order is 2, and the molecule has no unpaired electrons. Again, diamagnetic C_2 molecules have been detected in the vapor state. Note that the double bonds in C_2

Table 10.5 **Properties of Homonuclear Diatomic Molecules of the Second-Period Elements***

	Li_2	B_2	C_2	N_2	O_2	F_2	
$\sigma^{\star}_{2p_x}$	□	□	□	□	□	□	$\sigma^{\star}_{2p_x}$
$\pi^{\star}_{2p_y}, \pi^{\star}_{2p_z}$	□□	□□	□□	□□	↑ ↑	↑↓ ↑↓	$\pi^{\star}_{2p_y}, \pi^{\star}_{2p_z}$
σ_{2p_x}	□	□	□	↑↓	↑↓ ↑↓	↑↓ ↑↓	π_{2p_y}, π_{2p_z}
π_{2p_y}, π_{2p_z}	□□	↑ ↑	↑↓ ↑↓	↑↓ ↑↓	↑↓	↑↓	σ_{2p_x}
$\sigma^{\star}_{2s}$	□	↑↓	↑↓	↑↓	↑↓	↑↓	$\sigma^{\star}_{2s}$
σ_{2s}	↑↓	↑↓	↑↓	↑↓	↑↓	↑↓	σ_{2s}
Bond order	1	1	2	3	2	1	
Bond length (pm)	267	159	131	110	121	142	
Bond enthalpy (kJ/mol)	104.6	288.7	627.6	941.4	498.7	156.9	
Magnetic properties	Diamagnetic	Paramagnetic	Diamagnetic	Diamagnetic	Paramagnetic	Diamagnetic	

*For simplicity the σ_{1s} and σ^{*}_{1s} orbitals are omitted. These two orbitals hold a total of four electrons. Remember that for O_2 and F_2, σ_{2p_x} is lower in energy than π_{2p_y} and π_{2p_z}.

are both pi bonds because of the four electrons in the two pi molecular orbitals. In most other molecules, a double bond is made up of a sigma bond and a pi bond.

The Oxygen Molecule (O_2)

The ground-state electron configuration of O is $1s^22s^22p^4$; thus, there are 16 electrons in O_2. Using the order of increasing energies of the molecular orbitals just discussed, we write the ground-state electron configuration of O_2 as

$$(\sigma_{1s})^2(\sigma^{*}_{1s})^2(\sigma_{2s})^2(\sigma^{*}_{2s})^2(\sigma_{2p_x})^2(\pi_{2p_y})^2(\pi_{2p_z})^2(\pi^{*}_{2p_y})^1(\pi^{*}_{2p_z})^1$$

According to Hund's rule, the last two electrons enter the $\pi^{*}_{2p_y}$ and $\pi^{*}_{2p_z}$ orbitals with parallel spins. Ignoring the σ_{1s} and σ_{2s} orbitals (because their net effects on bonding are zero), we calculate the bond order of O_2 using Equation (10.2):

$$\text{bond order} = \frac{1}{2}(6 - 2) = 2$$

Therefore, the O_2 molecule has a bond order of 2 and oxygen is paramagnetic, a prediction that corresponds to experimental observations.

Table 10.5 summarizes the general properties of the stable diatomic molecules of the second period.

Example 10.6 shows how MO theory can help predict molecular properties of ions.

Example 10.6

The N_2^+ ion can be prepared by bombarding the N_2 molecule with fast-moving electrons. Predict the following properties of N_2^+: (a) electron configuration, (b) bond order, (c) magnetic properties, and (d) bond length relative to the bond length of N_2 (that is, is it longer or shorter?).

Strategy From Table 10.5 we can deduce the properties of ions generated from the homonuclear molecules. How does the stability of a molecule depend on the number of

(Continued)

electrons in bonding and antibonding molecular orbitals? From what molecular orbital is an electron removed to form the N_2^+ ion from N_2? What properties determine whether a species is diamagnetic or paramagnetic?

Solution From Table 10.5 we can deduce the properties of ions generated from the homonuclear diatomic molecules.

(a) Because N_2^+ has one fewer electron than N_2, its electron configuration is

$$(\sigma_{1s})^2(\sigma_{1s}^\star)^2(\sigma_{2s})^2(\sigma_{2s}^\star)^2(\pi_{2p_y})^2(\pi_{2p_z})^2(\sigma_{2p_x})^1$$

(b) The bond order of N_2^+ is found by using Equation (10.2):

$$\text{bond order} = \frac{1}{2}(9 - 4) = 2.5$$

(c) N_2^+ has one unpaired electron, so it is paramagnetic.

(d) Because the electrons in the bonding molecular orbitals are responsible for holding the atoms together, N_2^+ should have a weaker and, therefore, longer bond than N_2. (In fact, the bond length of N_2^+ is 112 pm, compared with 110 pm for N_2.)

Check Because an electron is removed from a bonding molecular orbital, we expect the bond order to decrease. The N_2^+ ion has an odd number of electrons (13), so it should be paramagnetic.

Practice Exercise Which of the following species has a longer bond length: F_2 or F_2^-?

Similar problems: 10.57, 10.58.

Summary of Concepts & Facts

- We write electron configurations for molecular orbitals as we do for atomic orbitals, filling in electrons in the order of increasing energy levels. The number of molecular orbitals always equals the number of atomic orbitals that were combined. The Pauli exclusion principle and Hund's rule govern the filling of molecular orbitals.
- Molecules are stable if the number of electrons in bonding molecular orbitals is greater than that in antibonding molecular orbitals.

Review of Concepts & Facts

10.7.1 Calculate the bond order of F_2^+.
10.7.2 Determine if N_2^+ is diamagnetic or paramagnetic.
10.7.3 Estimate the bond enthalpy (kJ/mol) of the H_2^+ ion.

10.8 Delocalized Molecular Orbitals

Learning Objective

- Appraise localized and delocalized bonds in molecules.

So far we have discussed chemical bonding only in terms of electron pairs. However, the properties of a molecule cannot always be explained accurately by a single structure. A case in point is the O_3 molecule, discussed in Section 9.8. There we overcame the dilemma by introducing the concept of resonance. In this section, we will tackle the problem in another way—by applying the molecular orbital approach. As in Section 9.8, we will use the benzene molecule and the carbonate ion as examples. Note that in discussing the bonding of polyatomic molecules or ions, it is convenient to

CHEMISTRY *in Action*

Buckyball, Anyone?

In 1985 chemists at Rice University in Texas used a high-powered laser to vaporize graphite in an effort to create unusual molecules believed to exist in interstellar space. Mass spectrometry revealed that one of the products was an unknown species with the formula C_{60}. Because of its size and the fact that it is pure carbon, this molecule has an exotic shape, which the researchers worked out using paper, scissors, and tape. Subsequent spectroscopic and X-ray measurements confirmed that C_{60} is shaped like a hollow sphere with a carbon atom at each of the 60 vertices. Geometrically, buckyball (short for "buckminsterfullerene") is the most symmetrical molecule known. In spite of its unique features, however, its bonding scheme is straightforward. Each carbon is sp^2-hybridized, and there are extensive delocalized molecular orbitals over the entire structure.

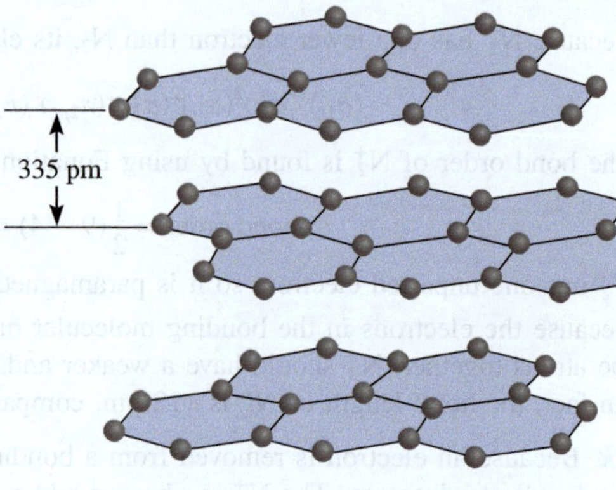

Graphite is made up of layers of six-membered rings of carbon.

The discovery of buckyball generated tremendous interest within the scientific community. Here was a new allotrope of carbon with an intriguing geometry and unknown properties to investigate. Since 1985 chemists have created a whole class of *fullerenes,* with 70, 76, and even larger numbers of carbon atoms. Moreover, buckyball has been found to be a natural component of soot.

Buckyball and its heavier members represent a whole new concept in molecular architecture with far-reaching implications. For example, buckyball has been prepared with a helium atom trapped in its cage. Buckyball also reacts with potassium to give K_3C_{60}, which acts as a superconductor at 18 K. It is also possible to attach transition metals to buckyball. These derivatives show promise as catalysts. Because of its unique shape, buckyball can be used as a lubricant.

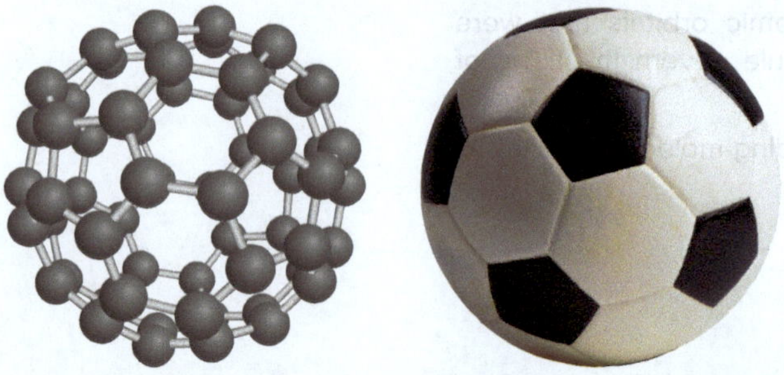

The geometry of a buckyball C_{60} (left) resembles a soccer ball (right). Scientists arrived at this structure by fitting together paper cutouts of enough hexagons and pentagons to accommodate 60 carbon atoms at the points where they intersect.

(ball): Comstock Images/Alamy Stock Photo

One fascinating discovery, made in 1991 by Japanese scientists, was the identification of structural relatives of buckyball. These molecules are hundreds of nanometers long with a tubular shape and an internal cavity about 15 nm in diameter. Dubbed "buckytubes" or "nanotubes" (because of their size), these molecules have two distinctly different structures. One is a single sheet of graphite that is capped at both ends with a kind of truncated buckyball. The other is a scroll-like tube having anywhere from 2 to 30 graphitelike layers. Nanotubes are many times stronger than steel wires of similar dimensions. Numerous potential applications have been proposed for them, including conducting and high-strength materials, hydrogen storage media, molecular sensors, semiconductor devices, and molecular probes. The study of these materials has created a new field called *nanotechnology,* so called because scientists can manipulate materials on a molecular scale to create useful devices.

determine first the hybridization state of the atoms present (a valence bond approach), followed by the formation of appropriate molecular orbitals.

The Benzene Molecule

Benzene (C_6H_6) is a planar hexagonal molecule with carbon atoms situated at the six corners. All carbon-carbon bonds are equal in length and strength, as are all carbon-hydrogen bonds, and the CCC and HCC angles are all 120°. Therefore, each carbon atom is sp^2-hybridized; it forms three sigma bonds with two adjacent carbon atoms

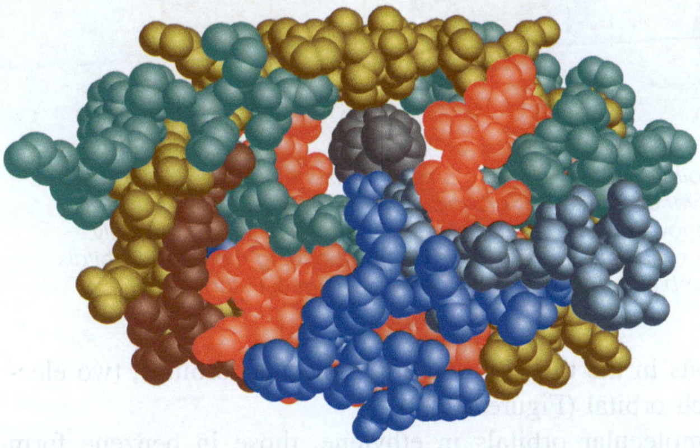

Computer-generated model of the binding of a buckyball derivative to the site of HIV-protease that normally attaches to a protein needed for the reproduction of HIV. The buckyball structure (purple color) fits tightly into the active site, thus preventing the enzyme from carrying out its function.

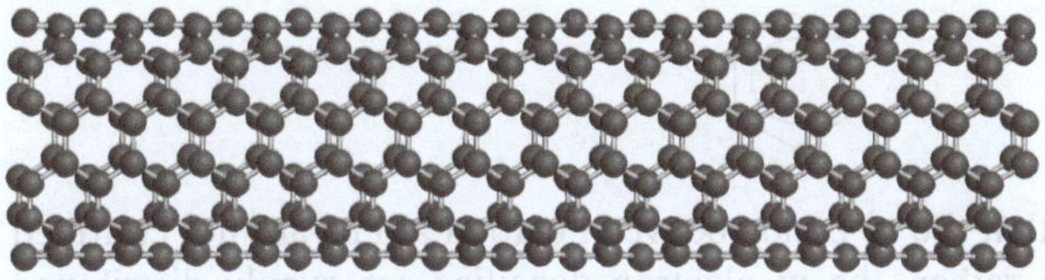

The structure of a buckytube that consists of a single layer of carbon atoms. Note that the truncated buckyball "cap," which has been separated from the rest of the buckytube in this view, has a different structure than the graphite-like cylindrical portion of the tube. Chemists have devised ways to open the cap in order to place other molecules inside the tube.

In the first biological application of buckyball, chemists at the University of California at San Francisco and Santa Barbara made a discovery in 1993 that could help in designing drugs to treat AIDS. The human immunodeficiency virus (HIV) that causes AIDS reproduces by synthesizing a long protein chain, which is cut into smaller segments by an enzyme called HIV-protease. One way to stop AIDS, then, might be to inactivate the enzyme. When the chemists reacted a water-soluble derivative of buckyball with HIV-protease, they found that it binds to the portion of the enzyme that would ordinarily cleave the reproductive protein, thereby preventing the HIV from reproducing. Consequently the virus could no longer infect the human cells they had grown in the laboratory. The buckyball compound itself is not a suitable drug for use against AIDS because of potential side effects and delivery difficulties, but it does provide a model for the development of such drugs.

In a recent development, scientists used a piece of adhesive tape (like Scotch tape) to peel off a flake of carbon from a piece of graphite (as is found in pencils) with the thickness of just one atom. This newfound material, called graphene, is a two-dimensional crystal with unusual electrical and optical properties. It is an excellent heat conductor. Graphene is almost totally transparent yet the carbon atoms are packed so dense that not even helium, the smallest gaseous atom, can pass through it. It seems like many interesting and useful discoveries will come from the study of this unusual substance in the coming years.

A model of graphene showing the honeycomb structure.
Robert Brook/Science Photo Library/Getty Images

and a hydrogen atom (Figure 10.28). This arrangement leaves an unhybridized $2p_z$ orbital on each carbon atom, perpendicular to the plane of the benzene molecule, or *benzene ring,* as it is often called. So far the description resembles the configuration of ethylene (C_2H_4), discussed in Section 10.5, except that in this case there are six unhybridized $2p_z$ orbitals in a cyclic arrangement.

Because of their similar shape and orientation, each $2p_z$ orbital overlaps two others, one on each adjacent carbon atom. According to the rules listed in Section 10.7, the interaction of six $2p_z$ orbitals leads to the formation of six pi molecular orbitals, of which three are bonding and three antibonding. A benzene molecule in the ground

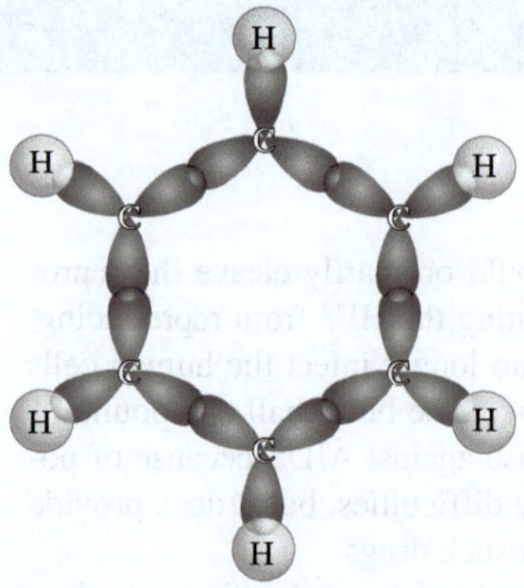

Figure 10.28 *The sigma bond framework in the benzene molecule. Each carbon atom is* sp^2*-hybridized and forms sigma bonds with two adjacent carbon atoms and another sigma bond with a hydrogen atom.*

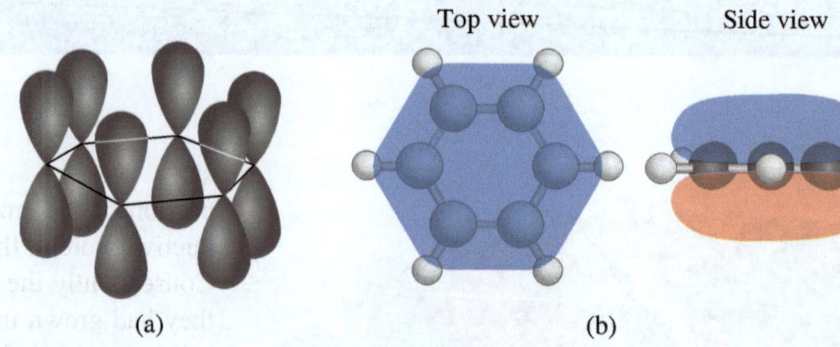

Figure 10.29 *(a) The six* $2p_z$ *orbitals on the carbon atoms in benzene. (b) The delocalized molecular orbital formed by the overlap of the* $2p_z$ *orbitals. The delocalized molecular orbital possesses pi symmetry and lies above and below the plane of the benzene ring. Actually, these* $2p_z$ *orbitals can combine in six different ways to yield three bonding molecular orbitals and three antibonding molecular orbitals. The one shown here is the most stable.*

state therefore has six electrons in the three pi bonding molecular orbitals, two electrons with paired spins in each orbital (Figure 10.29).

Unlike the pi bonding molecular orbitals in ethylene, those in benzene form ***delocalized molecular orbitals,*** which *are not confined between two adjacent bonding atoms, but actually extend over three or more atoms.* Therefore, electrons residing in any of these orbitals are free to move around the benzene ring. For this reason, the structure of benzene is sometimes represented as

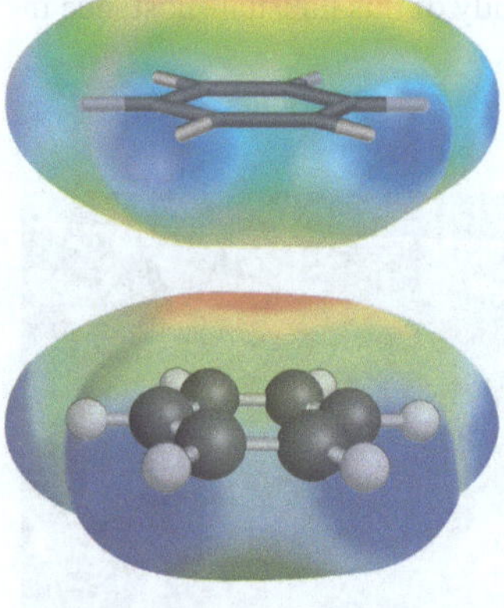

Electrostatic potential map of benzene shows the electron density (red color) above and below the plane of the molecule. For simplicity, only the framework of the molecule is shown.

in which the circle indicates that the pi bonds between carbon atoms are not confined to individual pairs of atoms; rather, the pi electron densities are evenly distributed throughout the benzene molecule. The carbon and hydrogen atoms are not shown in the simplified diagram.

We can now state that each carbon-to-carbon linkage in benzene contains a sigma bond and a "partial" pi bond. The bond order between any two adjacent carbon atoms is therefore between 1 and 2. Thus, molecular orbital theory offers an alternative to the resonance approach, which is based on valence bond theory. (The resonance structures of benzene are shown in Section 9.8.)

The Carbonate Ion

Cyclic compounds like benzene are not the only ones with delocalized molecular orbitals. Let's look at bonding in the carbonate ion (CO_3^{2-}). VSEPR predicts a trigonal planar geometry for the carbonate ion, like that for BF_3. The planar structure of the carbonate ion can be explained by assuming that the carbon atom is sp^2-hybridized. The C atom forms sigma bonds with three O atoms. Thus, the unhybridized $2p_z$ orbital of the C atom can simultaneously overlap the $2p_z$ orbitals of all three O atoms (Figure 10.30). The result is a delocalized molecular orbital that extends over all four nuclei in such a way that the electron densities (and hence the bond orders) in the carbon-to-oxygen bonds are all the same. Molecular orbital theory therefore provides

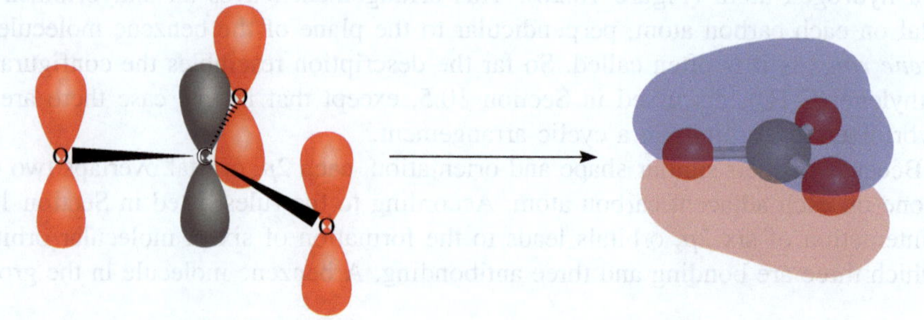

Figure 10.30 *Bonding in the carbonate ion. The carbon atom forms three sigma bonds with the three oxygen atoms. In addition, the* $2p_z$ *orbitals of the carbon and oxygen atoms overlap to form delocalized molecular orbitals, so that there is also a partial pi bond between the carbon atom and each of the three oxygen atoms.*

an acceptable alternative explanation of the properties of the carbonate ion as compared with the resonance structures of the ion shown in Section 9.8.

We should note that molecules with delocalized molecular orbitals are generally more stable than those containing molecular orbitals extending over only two atoms. For example, the benzene molecule, which contains delocalized molecular orbitals, is chemically less reactive (and hence more stable) than molecules containing "localized" C=C bonds, such as ethylene.

Summary of Concepts & Facts

- Delocalized molecular orbitals, in which electrons are free to move around a whole molecule or group of atoms, are formed by electrons in *p* orbitals of adjacent atoms. Delocalized molecular orbitals are an alternative to resonance structures in explaining observed molecular properties.

Review of Concepts & Facts

10.8.1 Describe the bonding in the nitrate ion (NO_3^-) in terms of resonance structures and delocalized molecular orbitals.

Chapter Summary

Molecular Geometry Molecular geometry refers to the three-dimensional arrangement of atoms in a molecule. For relatively small molecules, in which the central atom contains two to six bonds, geometries can be reliably predicted by the valence-shell electron-pair repulsion (VSEPR) model. This model is based on the assumption that chemical bonds and lone pairs tend to remain as far apart as possible to minimize repulsion. (Section 10.1)

Dipole Moments In a diatomic molecule, the difference in the electronegativities of bonding atoms results in a polar bond and a dipole moment. The dipole moment of a molecule made up of three or more atoms depends on both the polarity of the bonds and molecular geometry. Dipole moment measurements can help us distinguish between different possible geometries of a molecule. (Section 10.2)

Valence Bond Theory Valence bond theory is a quantum mechanical description of bonding that assumes electrons in a molecule occupy atomic orbitals of the individual atoms and these atomic orbitals overlap to form a covalent bond. Valence bond theory states that stable molecules form when the potential energy of the system of atoms has decreased to a minimum. Diatomic molecules can be easily explained by this minimization of potential energy with changes in distance between reacting atoms. (Section 10.3)

Hybridization of Atomic Orbitals Hybridization is the quantum mechanical description of chemical bonding. Atomic orbitals are hybridized, or mixed, to form hybrid orbitals. These orbitals then interact with other atomic orbitals to form chemical bonds. Various molecular geometries can be generated by different hybridizations. The hybridization concept accounts for the exception to the octet rule and also explains the formation of double and triple bonds. (Sections 10.4, 10.5)

Molecular Orbital Theory Molecular orbital theory describes bonding in terms of the combination of atomic orbitals to form orbitals that are associated with the molecule as a whole. Molecules are stable if the number of electrons in bonding molecular orbitals is greater than that in antibonding molecular orbitals. We write electron configurations for molecular orbitals as we do for atomic orbitals, using the Pauli exclusion principle and Hund's rule. (Sections 10.6, 10.7, 10.8)

Key Equations

$\mu = Q \times r$ (10.1)	Expressing dipole moment in terms of charge (Q) and distance of separation (r) between charges.

$$\text{bond order} = \frac{1}{2}\left(\begin{matrix}\text{number of electrons} \\ \text{in bonding MOs}\end{matrix} - \begin{matrix}\text{number of electrons} \\ \text{in antibonding MOs}\end{matrix}\right) \quad (10.2)$$

Key Words

Antibonding molecular orbital, p. 449
Bond order, p. 452
Bonding molecular orbital, p. 449
Delocalized molecular orbital, p. 460
Dipole moment (μ), p. 429
Homonuclear diatomic molecule, p. 453
Hybrid orbital, p. 437
Hybridization, p. 437
Molecular orbital, p. 449
Nonpolar molecule, p. 430
Pi (π) bond, p. 446
Pi molecular orbital, p. 451
Polar molecule, p. 430
Sigma (σ) bond, p. 445
Sigma molecular orbital, p. 450
Valence shell, p. 419
Valence-shell electron-pair repulsion (VSEPR) model, p. 419

Questions & Problems

Red numbered problems solved in Student Solutions Manual

10.1 Molecular Geometry

Review Questions

10.1 How is the geometry of a molecule defined and why is the study of molecular geometry important?

10.2 Sketch the shape of a linear triatomic molecule, a trigonal planar molecule containing four atoms, a tetrahedral molecule, a trigonal bipyramidal molecule, and an octahedral molecule. Give the bond angles in each case.

10.3 How many atoms are directly bonded to the central atom in a tetrahedral molecule, a trigonal bipyramidal molecule, and an octahedral molecule?

10.4 Discuss the basic features of the VSEPR model. Explain why the magnitude of repulsion decreases in the following order: lone pair–lone pair > lone pair–bonding pair > bonding pair–bonding pair.

10.5 In the trigonal bipyramidal arrangement, why does a lone pair occupy an equatorial position rather than an axial position?

10.6 The geometry of CH_4 could be square planar, with the four H atoms at the corners of a square and the C atom at the center of the square. Sketch this geometry and compare its stability with that of a tetrahedral CH_4 molecule.

Problems

10.7 Predict the geometries of the following species using the VSEPR method: (a) PCl_3, (b) $CHCl_3$, (c) SiH_4, (d) $TeCl_4$.

10.8 Predict the geometries of the following species: (a) $AlCl_3$, (b) $ZnCl_2$, (c) $ZnCl_4^{2-}$.

10.9 Predict the geometry of the following molecules and ion using the VSEPR model: (a) CBr_4, (b) BCl_3, (c) NF_3, (d) H_2Se, (e) NO_2^-.

10.10 Predict the geometry of the following molecules and ion using the VSEPR model: (a) CH_3I, (b) ClF_3, (c) H_2S, (d) SO_3, (e) SO_4^{2-}.

10.11 Predict the geometry of the following molecules using the VSEPR method: (a) $HgBr_2$, (b) N_2O (arrangement of atoms is NNO), (c) SCN^- (arrangement of atoms is SCN).

10.12 Predict the geometries of the following ions: (a) NH_4^+, (b) NH_2^-, (c) CO_3^{2-}, (d) ICl_2^-, (e) ICl_4^-, (f) AlH_4^-, (g) $SnCl_5^-$, (h) H_3O^+, (i) BeF_4^{2-}.

10.13 Describe the geometry around each of the three central atoms in the CH_3COOH molecule.

10.14 Which of the following species are tetrahedral? $SiCl_4$, SeF_4, XeF_4, CI_4, $CdCl_4^{2-}$

10.2 Dipole Moments

Review Questions

10.15 Define dipole moment. What are the units and symbol for dipole moment?

10.16 What is the relationship between the dipole moment and the bond moment? How is it possible for a molecule to have bond moments and yet be nonpolar?

10.17 Explain why an atom cannot have a permanent dipole moment.

10.18 The bonds in beryllium hydride (BeH_2) molecules are polar, and yet the dipole moment of the molecule is zero. Explain.

Problems

10.19 Referring to Table 10.3, arrange the following molecules in order of increasing dipole moment: H_2O, H_2S, H_2Te, H_2Se.

10.20 The dipole moments of the hydrogen halides decrease from HF to HI (see Table 10.3). Explain this trend.

10.21 List the following molecules in order of increasing dipole moment: H_2O, CBr_4, H_2S, HF, NH_3, CO_2.

10.22 Does the molecule OCS have a higher or lower dipole moment than CS_2?

10.23 Which of the molecules (a) or (b) has a higher dipole moment?

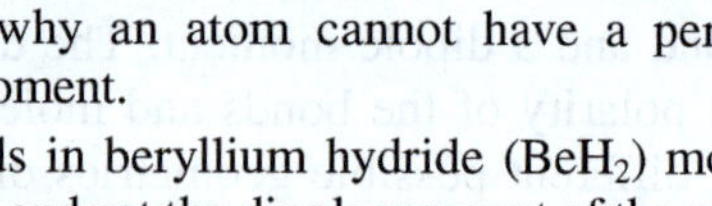

(a) (b)

10.24 Arrange the following compounds in order of increasing dipole moment:

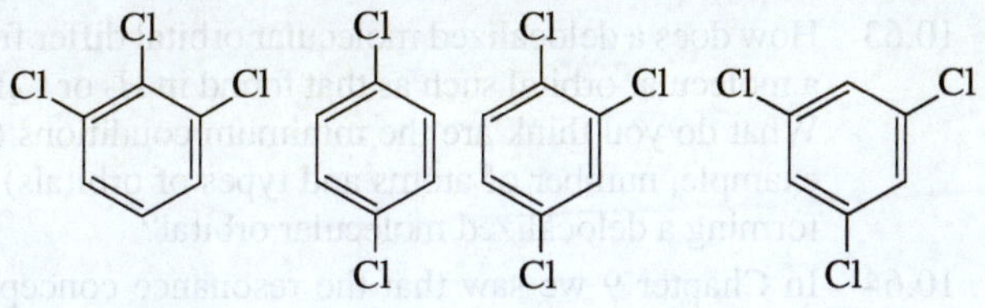

(a) (b) (c) (d)

10.3 Valence Bond Theory

Review Questions

10.25 What is valence bond theory? How does it differ from the Lewis concept of chemical bonding?

10.26 Use valence bond theory to explain the bonding in Cl_2 and HCl. Show how the atomic orbitals overlap when a bond is formed.

10.27 Draw a potential energy curve for the bond formation in F_2.

10.4 Hybridization of Atomic Orbitals

Review Questions

10.28 What is the hybridization of atomic orbitals? Why is it impossible for an isolated atom to exist in the hybridized state?

10.29 How does a hybrid orbital differ from a pure atomic orbital? Can two 2*p* orbitals of an atom hybridize to give two hybridized orbitals?

10.30 What is the angle between the following two hybrid orbitals on the same atom? (a) *sp* and *sp* hybrid orbitals, (b) sp^2 and sp^2 hybrid orbitals, (c) sp^3 and sp^3 hybrid orbitals

Problems

10.31 Describe the bonding scheme of the AsH_3 molecule in terms of hybridization.

10.32 What is the hybridization state of Si in SiH_4 and in $H_3Si—SiH_3$?

10.33 Describe the change in hybridization (if any) of the Al atom in the following reaction:

$$AlCl_3 + Cl^- \longrightarrow AlCl_4^-$$

10.34 Consider the reaction

$$BF_3 + NH_3 \longrightarrow F_3B—NH_3$$

Describe the changes in hybridization (if any) of the B and N atoms as a result of this reaction.

10.35 What hybrid orbitals are used by nitrogen atoms in the following species? (a) NH_3, (b) $H_2N—NH_2$, (c) NO_3^-

10.36 Describe the hybridization of phosphorus in PF_5.

10.37 Give the formula of a cation comprised of iodine and fluorine in which the iodine atom is sp^3d-hybridized.

10.38 Give the formula of an anion comprised of iodine and fluorine in which the iodine atom is sp^3d^2-hybridized.

10.5 Hybridization in Molecules Containing Double and Triple Bonds

Review Question

10.39 How would you distinguish between a sigma bond and a pi bond?

Problems

10.40 What are the hybrid orbitals of the carbon atoms in the following molecules?

(a) $H_3C—CH_3$
(b) $H_3C—CH{=}CH_2$
(c) $CH_3—C{\equiv}C—CH_2OH$
(d) $CH_3CH{=}O$
(e) CH_3COOH

10.41 Specify which hybrid orbitals are used by carbon atoms in the following species: (a) CO, (b) CO_2, (c) CN^-.

10.42 What is the hybridization state of the central N atom in the azide ion, N_3^-? (Arrangement of atoms: NNN.)

10.43 The allene molecule $H_2C{=}C{=}CH_2$ is linear (the three C atoms lie on a straight line). What are the hybridization states of the carbon atoms? Draw diagrams to show the formation of sigma bonds and pi bonds in allene.

10.44 How many pi bonds and sigma bonds are there in the tetracyanoethylene molecule?

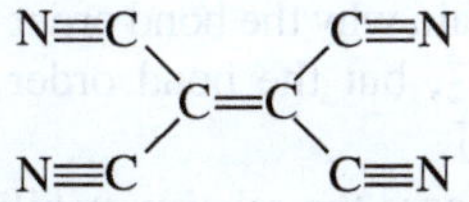

10.45 How many sigma bonds and pi bonds are there in each of the molecules (a), (b), and (c)?

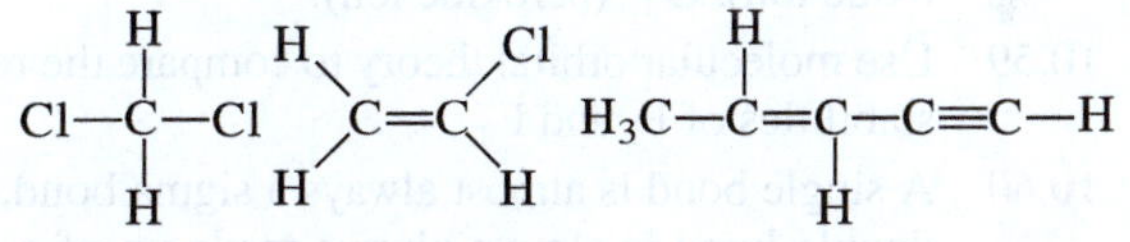

(a) (b) (c)

10.6 Molecular Orbital Theory

Review Questions

10.46 What is molecular orbital theory? How does it differ from valence bond theory?

10.47 Sketch the shapes of the following molecular orbitals: σ_{1s}, σ_{1s}^*, π_{2p}, and π_{2p}^*. How do their energies compare?

10.7 Molecular Orbital Configurations

Review Questions

10.48 Explain the significance of bond order. Can bond order be used for quantitative comparisons of the strengths of chemical bonds?

Problems

10.49 Explain in molecular orbital terms the changes in H—H internuclear distance that occur as the molecular H_2 is ionized first to H_2^+ and then to H_2^{2+}.

10.50 The formation of H_2 from two H atoms is an energetically favorable process. Yet statistically there is less than a 100 percent chance that any two H atoms will undergo the reaction. Apart from energy considerations, how would you account for this observation based on the electron spins in the two H atoms?

10.51 Draw a molecular orbital energy level diagram for each of the following species: H_2, HHe, He_2^+. Compare their relative stabilities in terms of bond orders. (Treat HHe as a diatomic molecule with three electrons.)

10.52 Arrange the following species in order of increasing stability: Li_2, Li_2^+, Li_2^-. Justify your choice with a molecular orbital energy level diagram.

10.53 Use molecular orbital theory to explain why the Be_2 molecule does not exist.

10.54 Which of these species has a longer bond, B_2 or B_2^+? Explain in terms of molecular orbital theory.

10.55 Acetylene (C_2H_2) has a tendency to lose two protons (H^+) and form the carbide ion (C_2^{2-}), which is present in a number of ionic compounds, such as CaC_2 and MgC_2. Describe the bonding scheme in the C_2^{2-} ion in terms of molecular orbital theory. Compare the bond order in C_2^{2-} with that in C_2.

10.56 Compare the Lewis and molecular orbital treatments of the oxygen molecule.

10.57 Explain why the bond order of N_2 is greater than that of N_2^+, but the bond order of O_2 is less than that of O_2^+.

10.58 Compare the relative stability of the following species and indicate their magnetic properties (that is, diamagnetic or paramagnetic): O_2, O_2^+, O_2^- (superoxide ion), O_2^{2-} (peroxide ion).

10.59 Use molecular orbital theory to compare the relative stabilities of F_2 and F_2^+.

10.60 A single bond is almost always a sigma bond, and a double bond is almost always made up of a sigma bond and a pi bond. There are very few exceptions to this rule. Show that the B_2 and C_2 molecules are examples of the exceptions.

10.61 In 2009 the ion N_2^{3-} was isolated. Use a molecular orbital diagram to compare its properties (bond order and magnetism) with the isoelectronic ion O_2^-.

10.62 The following potential energy curve represents the formation of F_2 from two F atoms. Describe the state of bonding at the marked regions.

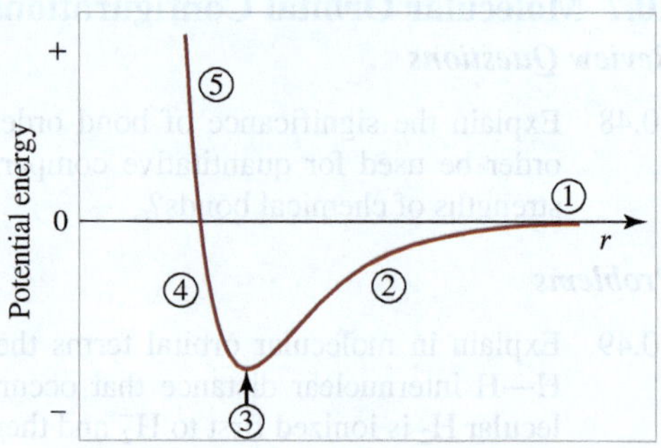

10.8 Delocalized Molecular Orbitals

Review Questions

10.63 How does a delocalized molecular orbital differ from a molecular orbital such as that found in H_2 or C_2H_4? What do you think are the minimum conditions (for example, number of atoms and types of orbitals) for forming a delocalized molecular orbital?

10.64 In Chapter 9 we saw that the resonance concept is useful for dealing with species such as the benzene molecule and the carbonate ion. How does molecular orbital theory deal with these species?

Problems

10.65 Both ethylene (C_2H_4) and benzene (C_6H_6) contain the C=C bond. The reactivity of ethylene is greater than that of benzene. For example, ethylene readily reacts with molecular bromine, whereas benzene is normally quite inert toward molecular bromine and many other compounds. Explain this difference in reactivity.

10.66 Explain why the symbol on the left is a better representation of benzene molecules than that on the right.

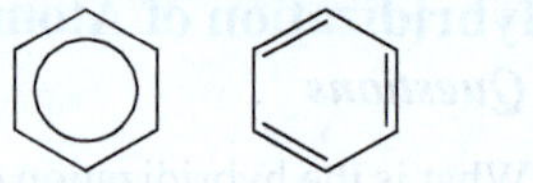

10.67 Determine which of these molecules has a more delocalized orbital and justify your choice.

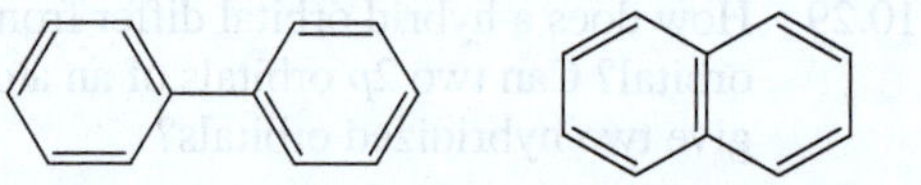

(*Hint:* Both molecules contain two benzene rings. In naphthalene, the two rings are fused together. In biphenyl, the two rings are joined by a single bond, around which the two rings can rotate.)

10.68 Nitryl fluoride (FNO_2) is very reactive chemically. The fluorine and oxygen atoms are bonded to the nitrogen atom. (a) Write a Lewis structure for FNO_2. (b) Indicate the hybridization of the nitrogen atom. (c) Describe the bonding in terms of molecular orbital theory. Where would you expect delocalized molecular orbitals to form?

10.69 Describe the bonding in the nitrate ion NO_3^- in terms of delocalized molecular orbitals.

10.70 What is the state of hybridization of the central O atom in O_3? Describe the bonding in O_3 in terms of delocalized molecular orbitals.

Additional Problems

10.71 Which of the following species is not likely to have a tetrahedral shape? (a) $SiBr_4$, (b) NF_4^+, (c) SF_4, (d) $BeCl_4^{2-}$, (e) BF_4^-, (f) $AlCl_4^-$

10.72 Draw the Lewis structure of mercury(II) bromide. Is this molecule linear or bent? How would you establish its geometry?

10.73 Sketch the bond moments and resultant dipole moments for the following molecules: H_2O, PCl_3, XeF_4, PCl_5, SF_6.

10.74 Although both carbon and silicon are in Group 14, very few Si=Si bonds are known. Account for the instability of silicon-to-silicon double bonds in general. (*Hint:* Compare the atomic radii of C and Si in Figure 8.3. What effect would the larger size have on pi bond formation?)

10.75 Acetaminophen is the active ingredient in Tylenol. (a) Write the molecular formula of the compound. (b) What is the hybridization state of each C, N, and O atom? (c) Describe the geometry about each C, N, and O atom.

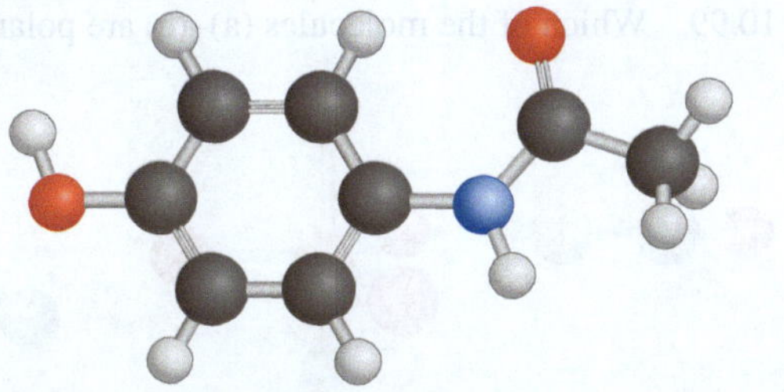

10.76 Caffeine is a stimulant drug present in coffee. (a) Write the molecular formula of the compound. (b) What is the hybridization state of each C, N, and O atom? (c) Describe the geometry about each C, N, and O atom.

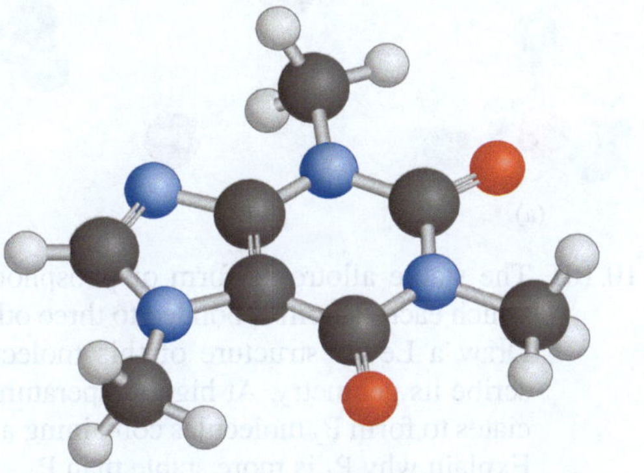

10.77 Predict the geometry of sulfur dichloride (SCl_2) and the hybridization of the sulfur atom.

10.78 Antimony pentafluoride, SbF_5, reacts with XeF_4 and XeF_6 to form ionic compounds, $XeF_3^+SbF_6^-$ and $XeF_5^+SbF_6^-$. Describe the geometries of the cations and anion in these two compounds.

10.79 Draw Lewis structures and give the other information requested for the following molecules: (a) BF_3. Shape: planar or nonplanar? (b) ClO_3^-. Shape: planar or nonplanar? (c) H_2O. Show the direction of the resultant dipole moment. (d) OF_2. Polar or nonpolar molecule? (e) NO_2. Estimate the ONO bond angle.

10.80 Predict the bond angles for the following molecules: (a) $BeCl_2$, (b) BCl_3, (c) CCl_4, (d) CH_3Cl, (e) Hg_2Cl_2 (arrangement of atoms: ClHgHgCl), (f) $SnCl_2$, (g) H_2O_2, (h) SnH_4.

10.81 Briefly compare the VSEPR and hybridization approaches to the study of molecular geometry.

10.82 Describe the hybridization state of arsenic in arsenic pentafluoride (AsF_5).

10.83 Draw Lewis structures and give the other information requested for the following: (a) SO_3. Polar or nonpolar molecule? (b) PF_3. Polar or nonpolar molecule? (c) F_3SiH. Show the direction of the resultant dipole moment. (d) SiH_3^-. Planar or pyramidal shape? (e) Br_2CH_2. Polar or nonpolar molecule?

10.84 Which of the following molecules and ions are linear? ICl_2^-, IF_2^+, OF_2, SnI_2, $CdBr_2$.

10.85 Draw the Lewis structure for the $BeCl_4^{2-}$ ion. Predict its geometry and describe the hybridization state of the Be atom.

10.86 The N_2F_2 molecule can exist in either of the following two forms:

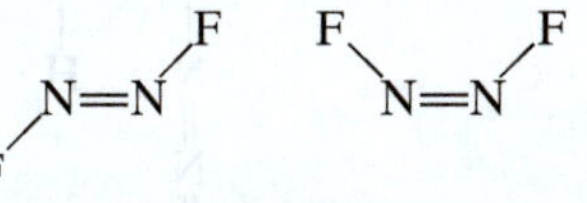

(a) What is the hybridization of N in the molecule?

(b) Which structure has a dipole moment?

10.87 Cyclopropane (C_3H_6) has the shape of a triangle in which a C atom is bonded to two H atoms and two other C atoms at each corner. Cubane (C_8H_8) has the shape of a cube in which a C atom is bonded to one H atom and three other C atoms at each corner. (a) Draw Lewis structures of these molecules. (b) Compare the CCC angles in these molecules with those predicted for an sp^3-hybridized C atom. (c) Would you expect these molecules to be easy to make?

10.88 The compound 1,2-dichloroethane ($C_2H_4Cl_2$) is nonpolar, while *cis*-dichloroethylene ($C_2H_2Cl_2$) has a dipole moment:

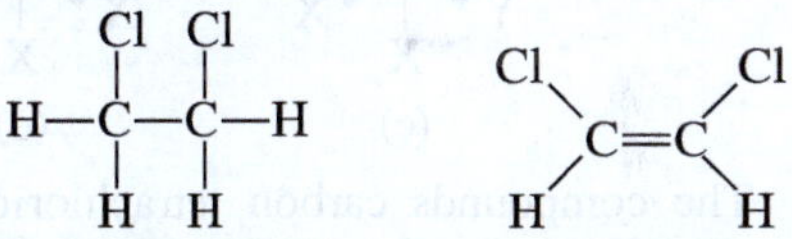

1,2-dichloroethane *cis*-dichloroethylene

The reason for the difference is that groups connected by a single bond can rotate with respect to each other, but no rotation occurs when a double bond connects the groups. On the basis of bonding considerations, explain why rotation occurs in 1,2-dichloroethane but not in *cis*-dichloroethylene.

10.89 Does the following molecule have a dipole moment?

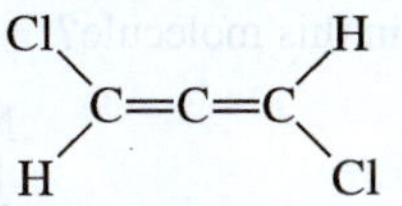

(*Hint:* See the answer to Problem 10.39.)

10.90 So-called greenhouse gases, which contribute to global warming, have a dipole moment or can be bent or distorted into shapes that have a dipole moment. Which of the following gases are greenhouse gases: N_2, O_2, O_3, CO, CO_2, NO_2, N_2O, CH_4, $CFCl_3$?

10.91 The bond angle of SO_2 is very close to 120°, even though there is a lone pair on S. Explain.

10.92 3′-azido-3′-deoxythymidine, shown here, commonly known as AZT, is one of the drugs used to treat acquired immune deficiency syndrome (AIDS). What are the hybridization states of the C and N atoms in this molecule?

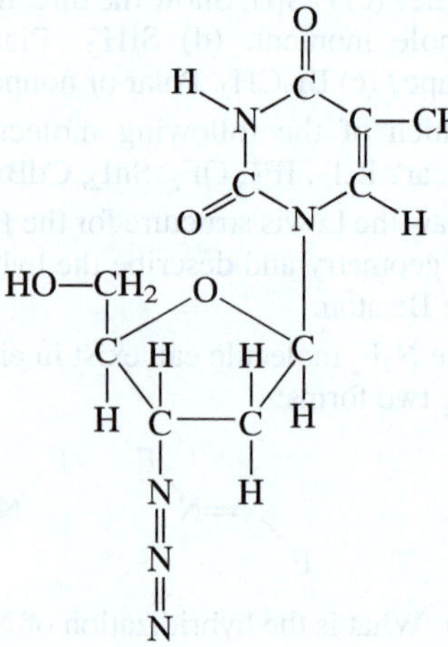

10.93 The following molecules (AX_4Y_2) all have octahedral geometry. Group the molecules that are equivalent to each other.

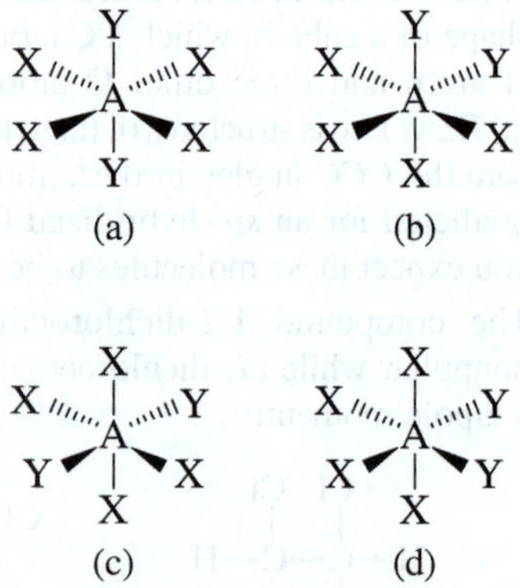

10.94 The compounds carbon tetrachloride (CCl_4) and silicon tetrachloride ($SiCl_4$) are similar in geometry and hybridization. However, CCl_4 does not react with water but $SiCl_4$ does. Explain the difference in their chemical reactivities. (*Hint:* The first step of the reaction is believed to be the addition of a water molecule to the Si atom in $SiCl_4$.)

10.95 Write the ground-state electron configuration for B_2. Is the molecule diamagnetic or paramagnetic?

10.96 What are the hybridization states of the C and N atoms in this molecule?

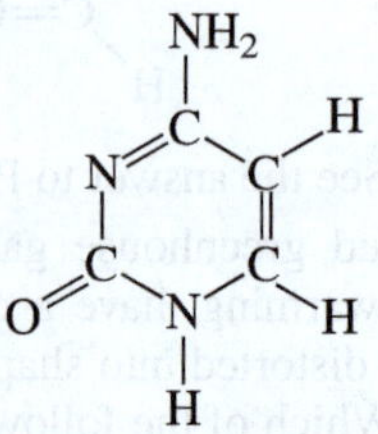

10.97 Use molecular orbital theory to explain the difference between the bond enthalpies of F_2 and F_2^- (see Problem 9.116).

10.98 Referring to the Chemistry in Action essay "Microwave Ovens–Dipole Moments at Work" in Section 10.2, answer the following questions: (a) If you wanted to cook a roast (beef or lamb), would you use a microwave oven or a conventional oven? (b) Radar is a means of locating an object by measuring the time for the echo of a microwave from the object to return to the source and the direction from which it returns. Would radar work if oxygen, nitrogen, and carbon dioxide were polar molecules? (c) In early tests of radar at the English Channel during World War II, the results were inconclusive even though there was no equipment malfunction. Why? (*Hint:* The weather is often foggy in the region.)

10.99 Which of the molecules (a)–(c) are polar?

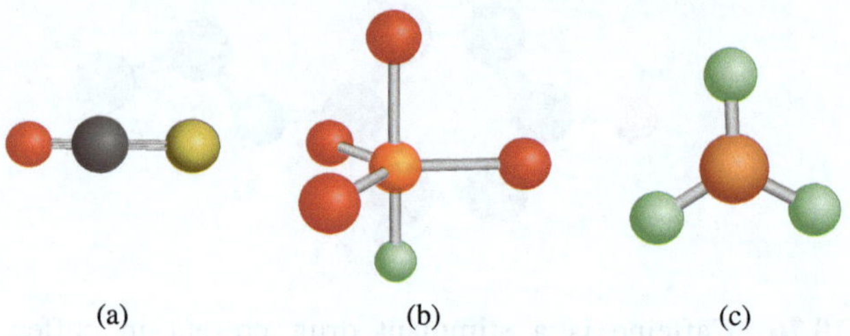

10.100 Which of the molecules (a)–(c) are polar?

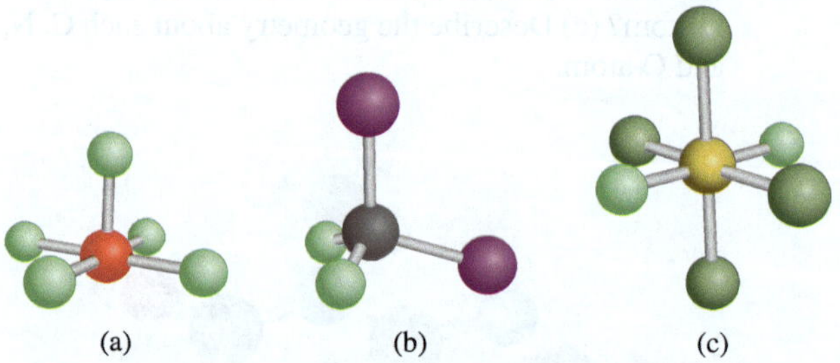

10.101 The stable allotropic form of phosphorus is P_4, in which each P atom is bonded to three other P atoms. Draw a Lewis structure of this molecule and describe its geometry. At high temperatures, P_4 dissociates to form P_2 molecules containing a P=P bond. Explain why P_4 is more stable than P_2.

10.102 Referring to Table 9.4, explain why the bond enthalpy for Cl_2 is greater than that for F_2. (*Hint:* The bond lengths of F_2 and Cl_2 are 142 pm and 199 pm, respectively.)

10.103 Use molecular orbital theory to explain the bonding in the azide ion (N_3^-). (Arrangement of atoms is NNN.)

10.104 The ionic character of the bond in a diatomic molecule can be estimated by the formula

$$\frac{\mu}{ed} \times 100\%$$

where μ is the experimentally measured dipole moment (in C m), e the electronic charge, and d the bond length in meters. (The quantity ed is the hypothetical dipole moment for the case in which the transfer of an electron from the less electronegative to the more electronegative atom is complete.) Given that the dipole moment and bond length of HF are 1.92 D and 91.7 pm, respectively, calculate the percent ionic character of the molecule.

10.105 Draw three Lewis structures for compounds with the formula $C_2H_2F_2$. Indicate which of the compound(s) are polar.

10.106 Greenhouse gases absorb (and trap) outgoing infrared radiation (heat) from Earth and contribute to global warming. The molecule of a greenhouse gas either possesses a permanent dipole moment or has a changing dipole moment during its vibrational motions. Consider three of the vibrational modes of carbon dioxide,

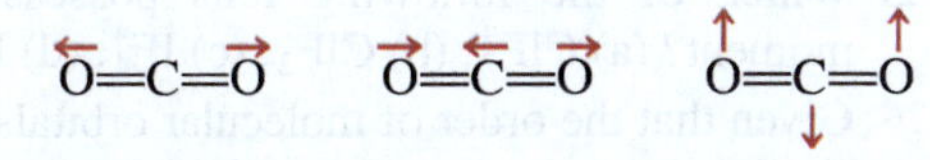

where the arrows indicate the movement of the atoms. (During a complete cycle of vibration, the atoms move toward one extreme position and then reverse their direction to the other extreme position.) Which of the preceding vibrations are responsible for CO_2 to behave as a greenhouse gas? Which of the following molecules can act as a greenhouse gas: N_2, O_2, CO, NO_2, and N_2O?

10.107 Aluminum trichloride ($AlCl_3$) is an electron-deficient molecule. It has a tendency to form a dimer (a molecule made of two $AlCl_3$ units):

$$AlCl_3 + AlCl_3 \longrightarrow Al_2Cl_6$$

(a) Draw a Lewis structure for the dimer. (b) Describe the hybridization state of Al in $AlCl_3$ and Al_2Cl_6. (c) Sketch the geometry of the dimer. (d) Do these molecules possess a dipole moment?

10.108 The molecules *cis*-dichloroethylene and *trans*-dichloroethylene shown in Section 10.2 can be interconverted by heating or irradiation. (a) Starting with *cis*-dichloroethylene, show that rotating the C═C bond by 180° will break only the pi bond but will leave the sigma bond intact. Explain the formation of *trans*-dichloroethylene from this process. (Treat the rotation as two stepwise 90° rotations.) (b) Account for the difference in the bond enthalpies for the pi bond (about 270 kJ/mol) and the sigma bond (about 350 kJ/mol). (c) Calculate the longest wavelength of light needed to bring about this conversion.

10.109 Progesterone is a hormone responsible for female sex characteristics. In the usual shorthand structure, each point where lines meet represents a C atom, and most H atoms are not shown. Draw the complete structure of the molecule, showing all C and H atoms. Indicate which C atoms are sp^2- and sp^3-hybridized.

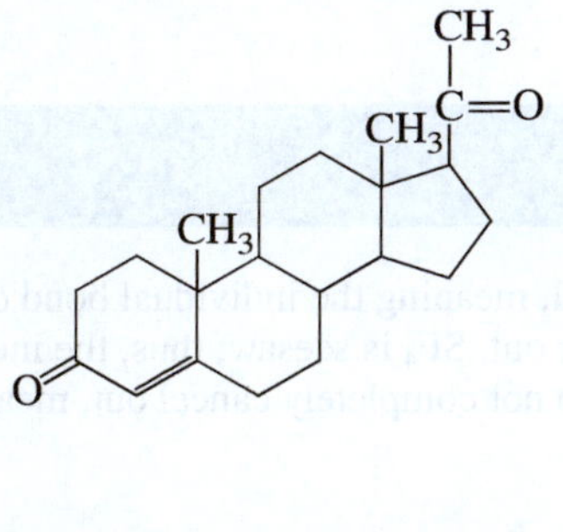

10.110 For each pair listed here, state which one has a higher first ionization energy and explain your choice: (a) H or H_2, (b) N or N_2, (c) O or O_2, (d) F or F_2.

10.111 The molecule benzyne (C_6H_4) is a very reactive species. It resembles benzene in that it has a six-membered ring of carbon atoms. Draw a Lewis structure of the molecule and account for the molecule's high reactivity.

10.112 Assume that the third-period element phosphorus forms a diatomic molecule, P_2, in an analogous way as nitrogen does to form N_2. (a) Write the electronic configuration for P_2. Use $[Ne_2]$ to represent the electron configuration for the first two periods. (b) Calculate its bond order. (c) What are its magnetic properties (diamagnetic or paramagnetic)?

10.113 Consider a N_2 molecule in its first excited electronic state, that is, when an electron in the highest occupied molecular orbital is promoted to the lowest empty molecular orbital. (a) Identify the molecular orbitals involved and sketch a diagram to show the transition. (b) Compare the bond order and bond length of N_2^* with N_2, where the asterisk denotes the excited molecule. (c) Is N_2^* diamagnetic or paramagnetic? (d) When N_2^* loses its excess energy and converts to the ground state N_2, it emits a photon of wavelength 470 nm, which makes up part of the aurora's lights. Calculate the energy difference between these levels.

10.114 As mentioned in the chapter, the Lewis structure for O_2 is

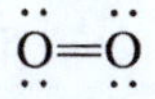

Use the molecular orbital theory to show that the structure actually corresponds to an excited state of the oxygen molecule.

10.115 Referring to Problem 9.143, describe the hybridization state of the N atoms and the overall shape of the ion.

10.116 Describe the geometry and hybridization for the reactants and product in the following reaction:

$$ClF_3 + AsF_5 \longrightarrow [ClF_2^+][AsF_6^-]$$

10.117 Draw the Lewis structure of ketene (C_2H_2O) and describe the hybridization states of the C atoms. The molecule does not contain O—H bonds. On separate diagrams, sketch the formation of sigma and pi bonds.

10.118 TCDD, or 2,3,7,8-tetrachlorodibenzo-*p*-dioxin, is a highly toxic compound:

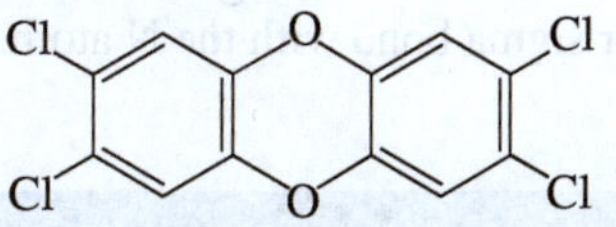

It gained considerable notoriety in 2004 when it was implicated in the murder plot of a Ukrainian politician. (a) Describe its geometry and state whether the molecule has a dipole moment. (b) How many pi bonds and sigma bonds are there in the molecule?

10.119 Write the electron configuration of the cyanide ion (CN^-). Name a stable molecule that is isoelectronic with the ion.

10.120 Carbon monoxide (CO) is a poisonous compound due to its ability to bind strongly to Fe^{2+} in the hemoglobin molecule. The molecular orbitals of CO have the same energy order as those of the N_2 molecule. (a) Draw a Lewis structure of CO and assign formal charges. Explain why CO has a rather small dipole moment of 0.12 D. (b) Compare the bond order of CO with that from molecular orbital theory. (c) Which of the atoms (C or O) is more likely to form bonds with the Fe^{2+} ion in hemoglobin?

10.121 The geometries discussed in this chapter all lend themselves to fairly straightforward elucidation of bond angles. The exception is the tetrahedron, because its bond angles are hard to visualize. Consider the CCl_4 molecule, which has a tetrahedral geometry and is nonpolar. By equating the bond moment of a particular C—Cl bond to the resultant bond moments of the other three C—Cl bonds in opposite directions, show that the bond angles are all equal to 109.5°.

10.122 Carbon suboxide (C_3O_2) is a colorless pungent-smelling gas. Does it possess a dipole moment?

10.123 Which of the following ions possess a dipole moment? (a) ClF_2^+, (b) ClF_2^-, (c) IF_4^+, (d) IF_4^-

10.124 Given that the order of molecular orbitals for NO is similar to that for O_2, arrange the following species in increasing bond orders: NO^{2-}, NO^-, NO, NO^+, NO^{2+}.

Interpreting, Modeling, & Estimating

10.125 Shown here are molecular models of SX_4 for X = F, Cl, and Br. Comment on the trends in the bond angle between the axial S—X bonds in these molecules.

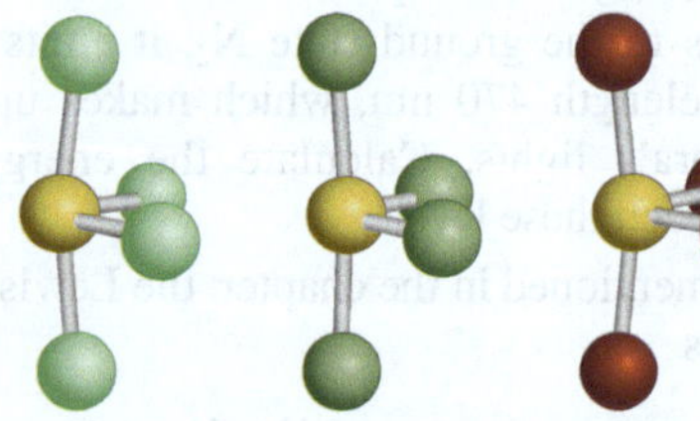

SF_4 SCl_4 SBr_4

10.126 Based on what you have learned from this chapter and Chapter 9, name a diatomic molecule that has the strongest known chemical bond and one with the weakest known chemical bond.

10.127 The stability of benzene is due to the fact that we can draw reasonable resonance structures for the molecule, which is equivalent to saying that there is electron delocalization. *Resonance energy* is a measure of how much more stable benzene is compared to the hypothetical molecule, which can be represented by just a single resonance structure. The enthalpies of hydrogenation (the addition of hydrogen) of cyclohexene (C_6H_{10}) to cyclohexane (C_6H_{12}) and benzene to cyclohexane are as follows:

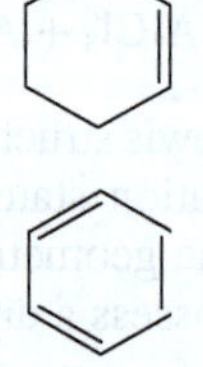

(In these simplified structures, each point where lines meet represents a C atom. There is a H atom attached to a sp^2-hybridized C atom and there are two H atoms attached to a sp^3-hybridized C atom.) Estimate the resonance energy of benzene from these data.

10.128 How many carbon atoms are contained in one square centimeter of graphene? (See the Chemistry in Action essay "Buckyball, Anyone?" in Section 10.8 for a description of graphene.) What would be the mass of a 1-cm^2 section of graphene?

Answers to Practice Exercises

10.1 (a) Tetrahedral, (b) Linear, (c) Trigonal planar. **10.2** No. **10.3** (a) sp^3, (b) sp^2. **10.4** sp^3d^2. **10.5** The C atom is *sp*-hybridized. It forms a sigma bond with the H atom and another sigma bond with the N atom. The two unhybridized *p* orbitals on the C atom are used to form two pi bonds with the N atom. The lone pair on the N atom is placed in the *sp* orbital. **10.6** F_2^-.

Answers to Review of Concepts & Facts

10.1.1 Tetrahedral. **10.1.2** Trigonal pyramidal. **10.1.3** The geometry on the right because the bond angles are larger (109.5° versus 90°). **10.2.1** No dipole moment. **10.2.2** CF_4 is tetrahedral, meaning the individual bond dipoles cancel each other out. SF_4 is seesaw; thus, the individual bond dipoles do not completely cancel out, meaning the molecule has a dipole moment. **10.2.3** At a given moment CO_2 can possess a dipole moment due to some of its vibrational motions. However, at the next instant the dipole moment changes sign because the vibrational motion reverses its direction. Over time (for example, the time it takes to make a dipole moment measurement), the net dipole moment averages to zero and the molecule is nonpolar. **10.3.1** The Lewis theory, which describes the bond formation as the pairing of electrons, fails to account for different bond lengths and bond strength in molecules. Valence bond theory explains chemical bond formation in terms of the overlap of atomic orbitals and can therefore account for different molecular properties. In essence, the Lewis theory is a classical approach to chemical bonding whereas the valence bond theory is a quantum mechanical treatment of chemical bonding. **10.4.1** 5. **10.4.2** sp^3. **10.4.3** sp^3d^2. **10.5.1** 2. **10.5.2** Sigma bonds: (a), (b), (e). Pi bond: (c). No bond: (d). **10.6.1** (1) The structure shows a single bond. The O_2 molecule has a double bond (from bond enthalpy measurements). (2) The structure violates the octet rule. **10.7.1** 2. **10.7.2** Paramagnetic. **10.7.3** H_2^+ has a bond order of ½, so we would expect its bond enthalpy to be about half that of H_2, which (436 kJ/mol)/2 or 218 kJ/mol (see Table 9.4). In reality, its bond enthalpy is 268 kJ/mol. The bond enthalpy is greater than that predicted from bond order because there is less repulsion (only one electron) in the ion. **10.8.1** The resonance structures of the nitrate ion are:

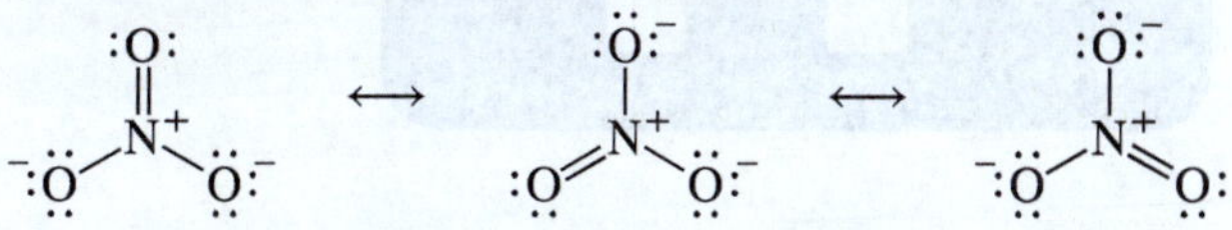

The delocalized molecular orbitals in the nitrate ion are similar to those in the carbonate ion. The N atom is sp^2-hybridized. The $2p_z$ orbital on N overlaps with the $2p_z$ orbitals on the three O atoms to form delocalized molecular orbitals over the four atoms. The resulting nitrogen-to-oxygen bonds are all the same in length and strength.

CHAPTER 11

Intermolecular Forces and Liquids and Solids

Two layers of the two-dimensional solid structure of graphite, an allotrope of carbon. A single sheet of graphite is known as graphene.

Anna Kireieva/Shutterstock

CHAPTER OUTLINE

Although we live immersed in a mixture of gases that make up Earth's atmosphere, we are more familiar with the behavior of liquids and solids because they are more visible. Every day we use water and other liquids for drinking, bathing, cleaning, and cooking, and we handle, sit upon, and wear solids.

Molecular motion is more restricted in liquids than in gases; and in solids the atoms and molecules are packed even more tightly together. In fact, in a solid they are held in well-defined positions and are capable of little free motion relative to one another. In this chapter we will examine the structure of liquids and solids and discuss some of the fundamental properties of these two states of matter. We will also study the nature of transitions among gases, liquids, and solids.

11.1 The Kinetic Molecular Theory of Liquids and Solids

Learning Objective

- Summarize the characteristic properties of gases, liquids, and solids.

In Chapter 5 we used the kinetic molecular theory to explain the behavior of gases in terms of the constant, random motion of gas molecules. In gases, the distances between molecules are so great (compared with their diameters) that at ordinary temperatures and pressures (say, 25°C and 1 atm), there is no appreciable interaction between the molecules. Because there is a great deal of empty space in a gas—that is, space that is not occupied by molecules—gases can be readily compressed. The lack of strong forces between molecules also allows a gas to expand to fill the volume of its container. Furthermore, the large amount of empty space explains why gases have very low densities under normal conditions.

Liquids and solids are quite a different story. The principal difference between the condensed states (liquids and solids) and the gaseous state is the distance between molecules. In a liquid, the molecules are so close together that there is very little empty space. Thus, liquids are much more difficult to compress than gases, and they are also much denser under normal conditions. Molecules in a liquid are held together by one or more types of attractive forces, which will be discussed in Section 11.2. A liquid also has a definite volume, because molecules in a liquid do not break away from the attractive forces. The molecules can, however, move past one another freely, and so a liquid can flow, can be poured, and will assume the shape of its container.

In a solid, molecules are held rigidly in position with virtually no freedom of motion. Many solids are characterized by long-range order; that is, the molecules are arranged in regular configurations in three dimensions. There is even less empty space in a solid than in a liquid. Thus, solids are almost incompressible and possess definite shape and volume. With very few exceptions (water being the most important), the density of the solid form is higher than that of the liquid form for a given substance. It is not uncommon for two states of a substance to coexist. An ice cube (solid) floating in a glass of water (liquid) is a familiar example. Chemists refer to the different states of a substance that are present in a system as phases. A ***phase*** is *a homogeneous part of the system in contact with other parts of the system but separated from them by a well-defined boundary.* Thus, our glass of ice water contains both the solid phase and the liquid phase of water. In this chapter we will use the term "phase" when talking about changes of state involving one substance, as well as systems containing more than one phase of a substance. Table 11.1 summarizes some of the characteristic properties of the three phases of matter.

Summary of Concepts & Facts

- All substances exist in one of three common states: gas, liquid, or solid. The major difference between the condensed state and the gaseous state is the distance separating molecules.

Table 11.1 Characteristic Properties of Gases, Liquids, and Solids

State of Matter	Volume/Shape	Density	Compressibility	Motion of Molecules
Gas	Assumes the volume and shape of its container	Low	Very compressible	Very free motion
Liquid	Has a definite volume but assumes the shape of its container	High	Only slightly compressible	Slide past one another freely
Solid	Has a definite volume and shape	High	Virtually incompressible	Vibrate about fixed positions

11.2 Intermolecular Forces

Learning Objectives

- Differentiate the different types of intermolecular forces: dipole-dipole, hydrogen bonding, dispersion forces, and ion-dipole.
- Evaluate the intermolecular forces present in a given substance.

Intermolecular forces are *attractive forces between molecules.* Intermolecular forces are responsible for the nonideal behavior of gases described in Chapter 5. They exert even more influence in the condensed phases of matter—liquids and solids. As the temperature of a gas drops, the average kinetic energy of its molecules decreases. Eventually, at a sufficiently low temperature, the molecules no longer have enough energy to break away from the attraction of neighboring molecules. At this point, the molecules aggregate to form small drops of liquid. This transition from the gaseous to the liquid phase is known as *condensation.*

In contrast to intermolecular forces, ***intramolecular forces*** *hold atoms together in a molecule.* (Chemical bonding, discussed in Chapters 9 and 10, involves intramolecular forces.) Intramolecular forces stabilize individual molecules, whereas intermolecular forces are primarily responsible for the bulk properties of matter (for example, melting and boiling points).

Generally, intermolecular forces are much weaker than intramolecular forces (Figure 11.1). It usually requires much less energy to evaporate a liquid than to break the bonds in the molecules of the liquid. For example, it takes about 41 kJ of energy to vaporize 1 mole of water at its boiling point, but about 930 kJ of energy are necessary to break the two bonds in 1 mole of water molecules. The boiling points of substances often reflect the strength of the intermolecular forces operating among the molecules. At the boiling point, enough energy must be supplied to overcome the attractive forces among molecules before they can enter the vapor phase. If it takes more energy to separate molecules of substance A than of substance B, because A molecules are held together by stronger intermolecular forces, then the boiling point of A is higher than that of B. The same principle applies also to the melting points of the substances. In general, the melting points of substances increase with the strength of the intermolecular forces.

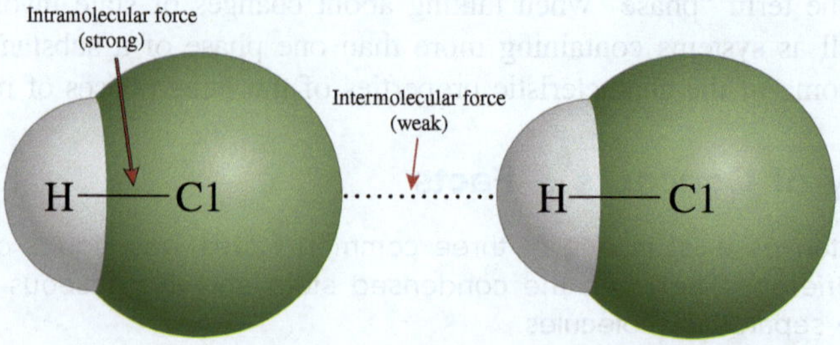

Figure 11.1 *Intramolecular and intermolecular forces.*

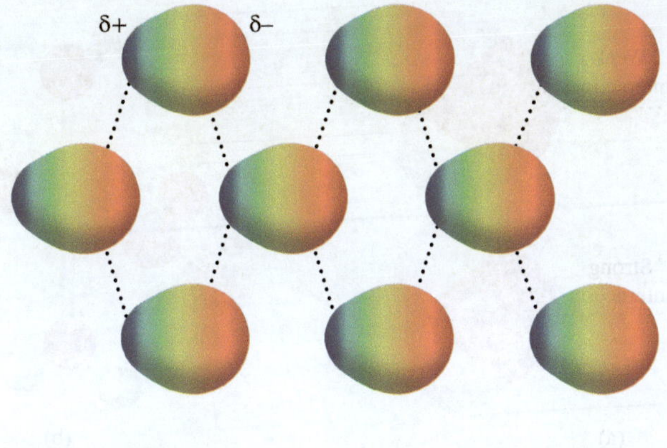

Figure 11.2 *Molecules that have a permanent dipole moment tend to align with opposite polarities in the solid phase for maximum attractive interaction.*

To discuss the properties of condensed matter, we must understand the different types of intermolecular forces. *Dipole-dipole, dipole-induced dipole,* and *dispersion forces* make up what chemists commonly refer to as ***van der Waals forces,*** after the Dutch physicist Johannes van der Waals (see Section 5.8). Ions and dipoles are attracted to one another by electrostatic forces called *ion-dipole forces,* which are *not* van der Waals forces. *Hydrogen bonding* is a particularly strong type of dipole-dipole interaction. Because only a few elements can participate in hydrogen bond formation, it is treated as a separate category. Depending on the phase of a substance, the nature of chemical bonds, and the types of elements present, more than one type of interaction may contribute to the total attraction between molecules, as we will see in this chapter.

Dipole-Dipole Forces

Dipole-dipole forces are *attractive forces between polar molecules,* that is, between molecules that possess dipole moments (see Section 10.2). Their origin is electrostatic, and they can be understood in terms of Coulomb's law. The larger the dipole moment, the greater the force. Figure 11.2 shows the orientation of polar molecules in a solid. In liquids, polar molecules are not held as rigidly as in a solid, but they tend to align in a way that, on average, maximizes the attractive interaction.

Ion-Dipole Forces

Coulomb's law also explains ***ion-dipole forces,*** which *attract an ion (either a cation or an anion) and a polar molecule to each other* (Figure 11.3). The strength of this

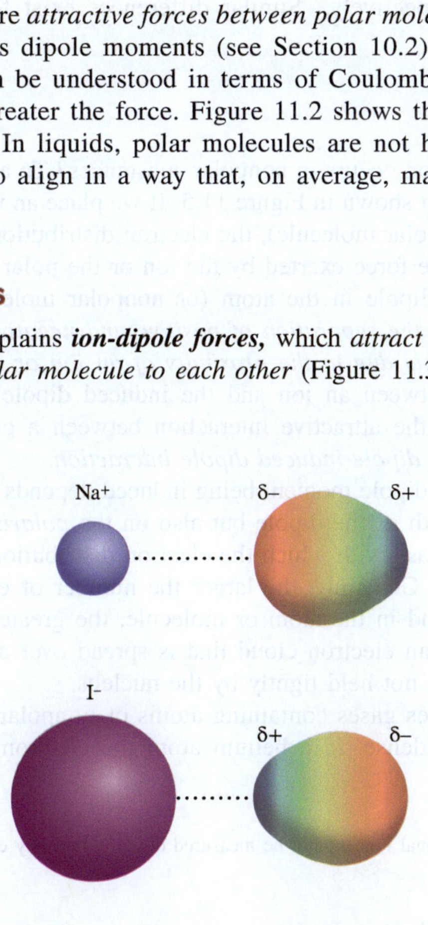

Figure 11.3 *Two types of ion-dipole interaction.*

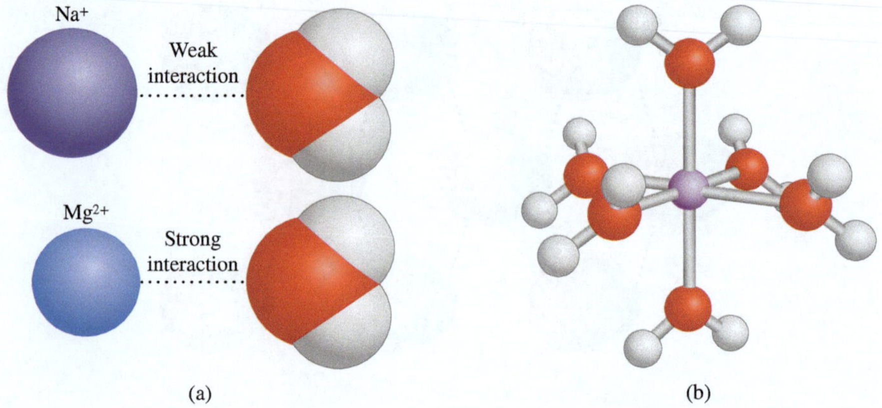

Figure 11.4 *(a) Interaction of a water molecule with a Na^+ ion and a Mg^{2+} ion. (b) In aqueous solutions, metal ions are usually surrounded by six water molecules in an octahedral arrangement.*

interaction depends on the charge and size of the ion and on the magnitude of the dipole moment and size of the molecule. The charges on cations are generally more concentrated, because cations are usually smaller than anions. Therefore, a cation interacts more strongly with dipoles than does an anion having a charge of the same magnitude.

Hydration, discussed in Section 4.1, is one example of ion-dipole interaction. Heat of hydration (see Section 6.7) is the result of the favorable interaction between the cations and anions of an ionic compound with water. Figure 11.4 shows the ion-dipole interaction between the Na^+ and Mg^{2+} ions with a water molecule, which has a large dipole moment (1.87 D). Because the Mg^{2+} ion has a higher charge and a smaller ionic radius (78 pm) than that of the Na^+ ion (98 pm), it interacts more strongly with water molecules. (In reality, each ion is surrounded by six water molecules in solution.) Consequently, the heats of hydration for the Na^+ and Mg^{2+} ions are −405 kJ/mol and −1926 kJ/mol, respectively.† Similar differences exist for anions of different charges and sizes.

Dispersion Forces

What attractive interaction occurs in nonpolar substances? To answer to this question, consider the arrangement shown in Figure 11.5. If we place an ion or a polar molecule near an atom (or a nonpolar molecule), the electron distribution of the atom (or molecule) is distorted by the force exerted by the ion or the polar molecule, resulting in a kind of dipole. The dipole in the atom (or nonpolar molecule) is said to be an ***induced dipole*** because *the separation of positive and negative charges in the atom (or nonpolar molecule) is due to the proximity of an ion or a polar molecule.* The attractive interaction between an ion and the induced dipole is called *ion-induced dipole interaction,* and the attractive interaction between a polar molecule and the induced dipole is called *dipole-induced dipole interaction.*

Figure 11.5 *(a) Spherical charge distribution in a helium atom. (b) Distortion caused by the approach of a cation. (c) Distortion caused by the approach of a dipole.*

The likelihood of a dipole moment being induced depends not only on the charge on the ion or the strength of the dipole but also on the *polarizability* of the atom or molecule—that is, the ease with which the electron distribution in the atom (or molecule) can be distorted. Generally, the larger the number of electrons and the more diffuse the electron cloud in the atom or molecule, the greater its polarizability. By *diffuse cloud* we mean an electron cloud that is spread over an appreciable volume, so that the electrons are not held tightly by the nucleus.

Polarizability enables gases containing atoms or nonpolar molecules (for example, He and N_2) to condense. In a helium atom the electrons are moving at some

†Heats of hydration of individual ions cannot be measured directly, but they can be reliably estimated.

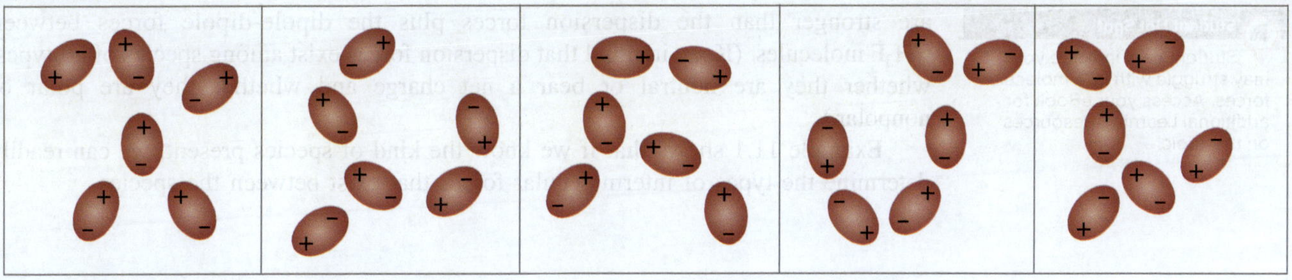

Figure 11.6 *Induced dipoles interacting with each other. Such patterns exist only momentarily; new arrangements are formed in the next instant. This type of interaction is responsible for the condensation of nonpolar gases.*

distance from the nucleus. At any instant it is likely that the atom has a dipole moment created by the specific positions of the electrons. This dipole moment is called an *instantaneous dipole* because it lasts for just a tiny fraction of a second. In the next instant the electrons are in different locations and the atom has a new instantaneous dipole, and so on. Averaged over time (that is, the time it takes to make a dipole moment measurement), however, the atom has no dipole moment because the instantaneous dipoles all cancel one another. In a collection of He atoms, an instantaneous dipole of one He atom can induce a dipole in each of its nearest neighbors (Figure 11.6). At the next moment, a different instantaneous dipole can create temporary dipoles in the surrounding He atoms. The important point is that this kind of interaction produces ***dispersion forces,*** *attractive forces that arise as a result of temporary dipoles induced in atoms or molecules.* At very low temperatures (and reduced atomic speeds), dispersion forces are strong enough to hold He atoms together, causing the gas to condense. The attraction between nonpolar molecules can be explained similarly.

A quantum mechanical interpretation of temporary dipoles was provided by Fritz London† in 1930. London showed that the magnitude of this attractive interaction is directly proportional to the polarizability of the atom or molecule. As we might expect, dispersion forces may be quite weak. This is certainly true for helium, which has a boiling point of only 4.2 K, or −269°C. (Note that helium has only two electrons, which are tightly held in the 1*s* orbital. Therefore, the helium atom has a very low polarizability.)

Dispersion forces, which are also called London forces, usually increase with molar mass because molecules with larger molar mass tend to have more electrons, and dispersion forces increase in strength with the number of electrons. Furthermore, larger molar mass often means a bigger atom whose electron distribution is more easily disturbed because the outer electrons are less tightly held by the nuclei. Table 11.2 compares the melting points of similar substances that consist of nonpolar molecules. As expected, the melting point increases as the number of electrons in the molecule increases. Because these are all nonpolar molecules, the only attractive intermolecular forces present are the dispersion forces.

In many cases, dispersion forces are comparable to or even greater than the dipole-dipole forces between polar molecules. For a dramatic illustration, let us compare the boiling points of CH_3F (−78.4°C) and CCl_4 (76.5°C). Although CH_3F has a dipole moment of 1.8 D, it boils at a much lower temperature than CCl_4, a nonpolar molecule. CCl_4 boils at a higher temperature simply because it contains more electrons. As a result, the dispersion forces between CCl_4 molecules

†Fritz London (1900–1954). German physicist. London was a theoretical physicist whose major work was on superconductivity in liquid helium.

Table 11.2

Melting Points of Similar Nonpolar Compounds

Compound	Melting Point (°C)
CH_4	−182.5
CF_4	−150.0
CCl_4	−23.0
CBr_4	90.0
CI_4	171.0

> **Student Hot Spot**
> Student data indicate you may struggle with intermolecular forces. Access your eBook for additional Learning Resources on this topic.

are stronger than the dispersion forces plus the dipole-dipole forces between CH_3F molecules. (Keep in mind that dispersion forces exist among species of all types, whether they are neutral or bear a net charge and whether they are polar or nonpolar.)

Example 11.1 shows that if we know the kind of species present, we can readily determine the types of intermolecular forces that exist between the species.

Example 11.1

What type(s) of intermolecular forces exist between the following pairs: (a) HBr and H_2S, (b) Cl_2 and CBr_4, (c) I_2 and NO_3^-, (d) NH_3 and C_6H_6?

Strategy Classify the species into three categories: ionic, polar (possessing a dipole moment), and nonpolar. Keep in mind that dispersion forces exist between *all* species.

Solution

(a) Both HBr and H_2S are polar molecules:

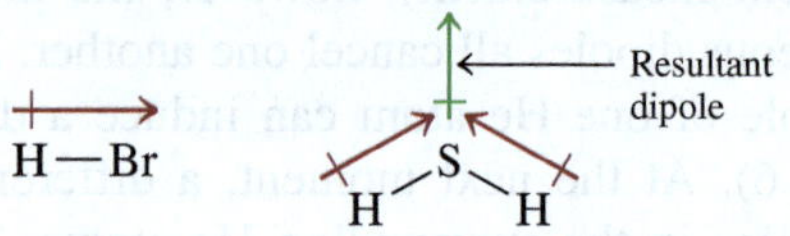

Therefore, the intermolecular forces present are dipole-dipole forces, as well as dispersion forces.

(b) Both Cl_2 and CBr_4 are nonpolar, so there are only dispersion forces between these molecules.

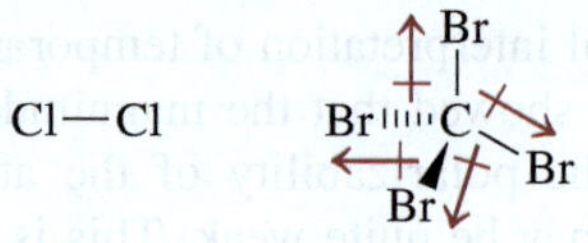

$\mu = 0$ $\mu = 0$

(c) I_2 is a homonuclear diatomic molecule and therefore nonpolar, so the forces between it and the ion NO_3^- are ion-induced dipole forces and dispersion forces.

(d) NH_3 is polar, and C_6H_6 is nonpolar. The forces are dipole-induced dipole forces and dispersion forces.

Practice Exercise Name the type(s) of intermolecular forces that exist(s) between molecules (or basic units) in each of the following species: (a) LiF, (b) CH_4, (c) SO_2.

Similar problem: 11.10.

The Hydrogen Bond

Normally, the boiling points of a series of similar compounds containing elements in the same periodic group increase with increasing molar mass. This increase in boiling point is due to the increase in dispersion forces for molecules with more electrons. Hydrogen compounds of Group 14 follow this trend, as Figure 11.7 shows. The lightest compound, CH_4, has the lowest boiling point, and the heaviest compound, SnH_4, has the highest boiling point. However, hydrogen compounds of the elements in Groups 15, 16, and 17 do not follow this trend. In each of these series, the lightest compound (NH_3, H_2O, and HF) has the highest boiling point, contrary to our expectations based

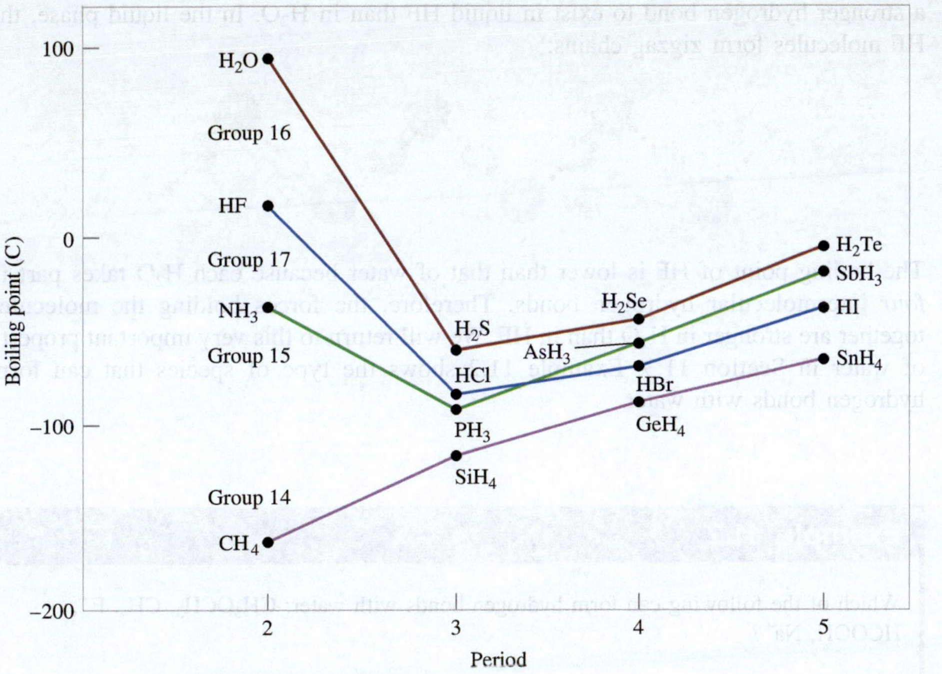

Figure 11.7 *Boiling points of the hydrogen compounds of Groups 14, 15, 16, and 17 elements. Although normally we expect the boiling point to increase as we move down a group, we see that three compounds (NH_3, H_2O, and HF) behave differently. The anomaly can be explained in terms of intermolecular hydrogen bonding.*

on molar mass. This observation must mean there are stronger intermolecular attractions in NH_3, H_2O, and HF, compared to other molecules in the same groups. In fact, this particularly strong type of intermolecular attraction is called the ***hydrogen bond,*** which is *a special type of dipole-dipole interaction between the hydrogen atom in a polar bond, such as N—H, O—H, or F—H, and an electronegative O, N, or F atom.* The interaction is written

$$\text{A—H} \cdots \text{:B} \quad \text{or} \quad \text{A—H} \cdots \text{:A}$$

A and B represent O, N, or F; A—H is one molecule or part of a molecule and B is a part of another molecule; and the dotted line represents the hydrogen bond. The three atoms usually lie in a straight line, but the angle AHB (or AHA) can deviate as much as 30° from linearity. Note that the O, N, and F atoms all possess at least one lone pair that can interact with the hydrogen atom in hydrogen bonding.

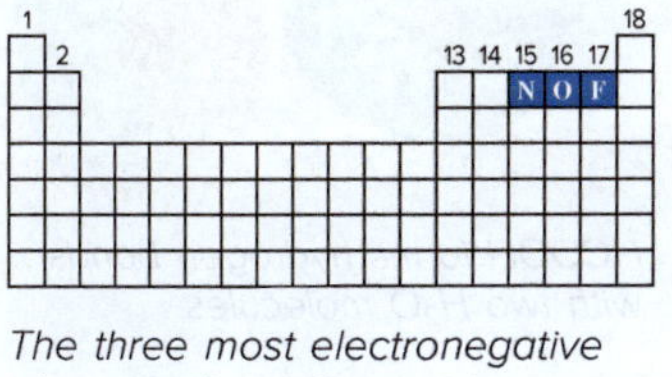

The three most electronegative elements that take part in hydrogen bonding.

The average strength of a hydrogen bond is quite large for a dipole-dipole interaction (up to 40 kJ/mol). Thus, hydrogen bonds have a powerful effect on the structures and properties of many compounds. Figure 11.8 shows several examples of hydrogen bonding.

The strength of a hydrogen bond is determined by the coulombic interaction between the lone-pair electrons of the electronegative atom and the hydrogen nucleus. For example, fluorine is more electronegative than oxygen, and so we would expect

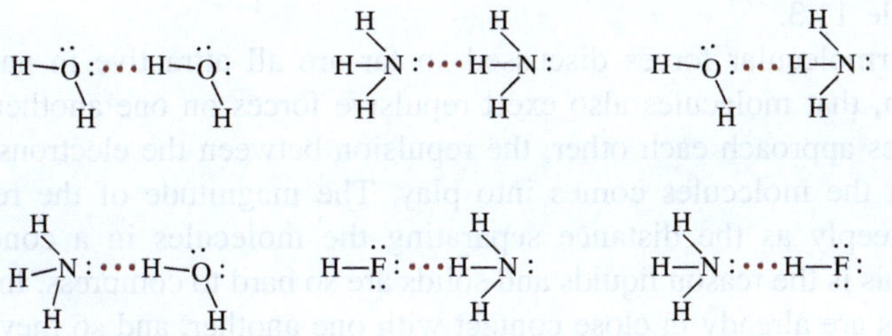

Figure 11.8 *Hydrogen bonding in water, ammonia, and hydrogen fluoride. Solid lines represent covalent bonds, and dotted lines represent hydrogen bonds.*

a stronger hydrogen bond to exist in liquid HF than in H_2O. In the liquid phase, the HF molecules form zigzag chains:

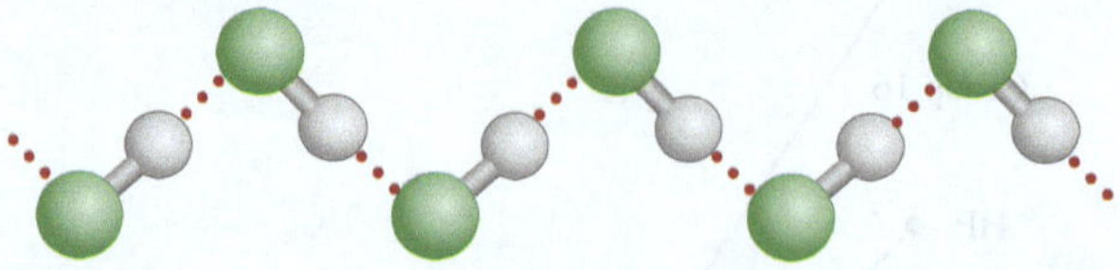

The boiling point of HF is lower than that of water because each H_2O takes part in *four* intermolecular hydrogen bonds. Therefore, the forces holding the molecules together are stronger in H_2O than in HF. We will return to this very important property of water in Section 11.3. Example 11.2 shows the type of species that can form hydrogen bonds with water.

Example 11.2

Which of the following can form hydrogen bonds with water: CH_3OCH_3, CH_4, F^-, HCOOH, Na^+?

Strategy A species can form hydrogen bonds with water if it contains one of the three electronegative elements (F, O, or N) or it has a H atom bonded to one of these three elements.

Solution There are no electronegative elements (F, O, or N) in either CH_4 or Na^+. Therefore, only CH_3OCH_3, F^-, and HCOOH can form hydrogen bonds with water.

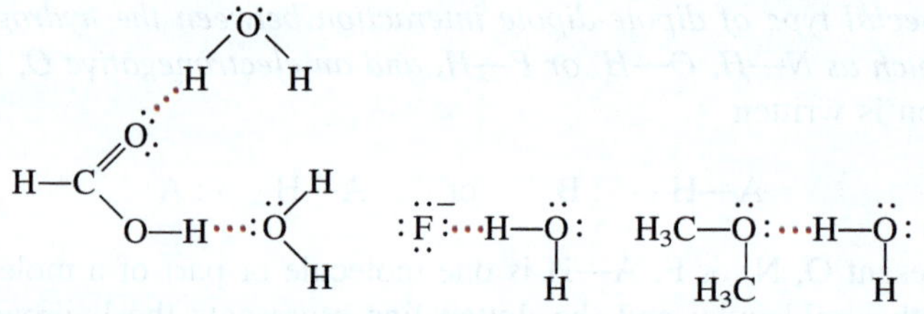

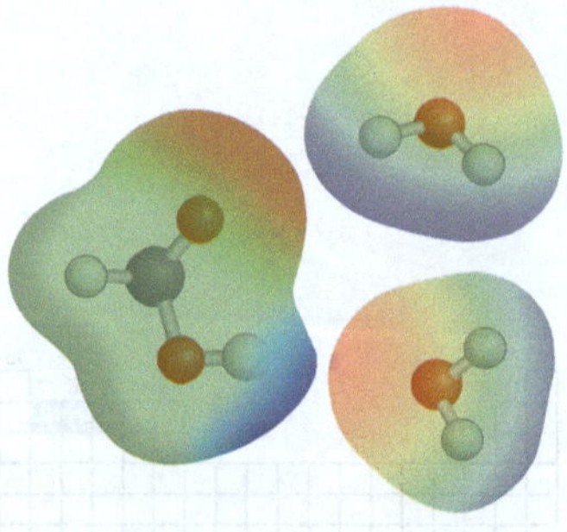

HCOOH forms hydrogen bonds with two H_2O molecules.

Check Note that HCOOH (formic acid) can form hydrogen bonds with water in two different ways.

Practice Exercise Which of the following species are capable of hydrogen bonding among themselves: (a) H_2S, (b) C_6H_6, (c) CH_3OH?

Similar problem: 11.12.

As noted earlier, intermolecular forces are considerably weaker than intramolecular forces. A summary of these intermolecular forces and their typical energies are provided in Table 11.3.

The intermolecular forces discussed so far are all attractive in nature. Keep in mind, though, that molecules also exert repulsive forces on one another. Thus, when two molecules approach each other, the repulsion between the electrons and between the nuclei in the molecules comes into play. The magnitude of the repulsive force rises very steeply as the distance separating the molecules in a condensed phase decreases. This is the reason liquids and solids are so hard to compress. In these phases, the molecules are already in close contact with one another, and so they greatly resist being compressed further.

Table 11.3 Intermolecular Forces Summary

Intermolecular Force	Interaction	Example	Typical Energy (kJ/mol)
Dispersion	All types of atoms and molecules	He, H_2, CH_4, NH_3, HCl	~5*
Dipole-dipole	Polar molecules	HCl	2*
Dipole-induced dipole	Polar molecule and another molecule (polar or nonpolar)	Between HCl and Cl_2 or between HCl and CH_4	<2*
Ion-dipole	Polar molecule and an ion	Metal ion dissolved in H_2O	15
Hydrogen bond	Molecule containing O, N, or F bonded to H and molecule containing O, N, or F (the H atom links two electronegative atoms)	H_2O, NH_3, HF	10–40†

*In the gas phase.
†The hydrogen bond in $[HF_2]^-$ is 165 kJ/mol.

Summary of Concepts & Facts

- Intermolecular forces act between molecules or between molecules and ions. Generally, these attractive forces are much weaker than bonding forces.
- Dipole-dipole forces and ion-dipole forces attract molecules with dipole moments to other polar molecules or ions.
- Dispersion forces are the result of temporary dipole moments induced in ordinarily nonpolar molecules. The extent to which a dipole moment can be induced in a molecule is called its polarizability. The term "van der Waals forces" refers to dipole-dipole, dipole-induced dipole, and dispersion forces.
- Hydrogen bonding is a relatively strong dipole-dipole interaction between a polar bond containing a hydrogen atom and an electronegative O, N, or F atom. Hydrogen bonds between water molecules are particularly strong.

Review of Concepts & Facts

11.2.1 What intermolecular forces exist between molecules of the following substances: (a) HCl, (b) F_2, (c) C_6H_{14}?

11.2.2 Which of the following compounds is most likely to exist as a liquid at room temperature: ethane (C_2H_6), hydrazine (N_2H_4), fluoromethane (CH_3F)?

11.3 Properties of Liquids

Learning Objective

- Explain surface tension and viscosity.

Intermolecular forces give rise to a number of structural features and properties of liquids. In this section we will look at two such phenomena associated with liquids in general: surface tension and viscosity. Then we will discuss the structure and properties of water.

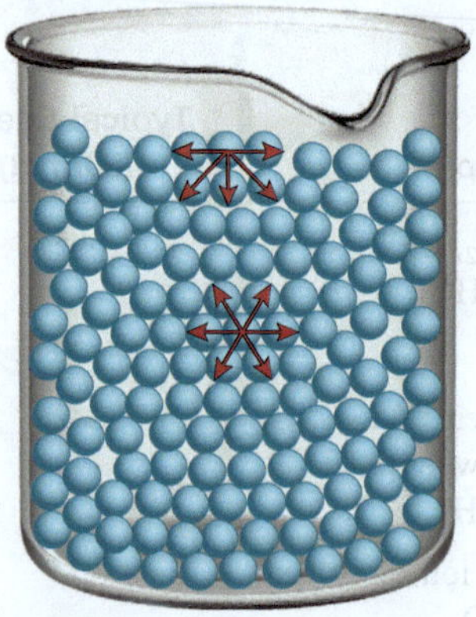

Figure 11.9 *Intermolecular forces acting on a molecule in the surface layer of a liquid and in the interior region of the liquid.*

Figure 11.10 *Water beads on an apple, which has a waxy surface.*
Ken Karp/McGraw Hill

Surface Tension

Molecules within a liquid are pulled in all directions by intermolecular forces; there is no tendency for them to be pulled in any one way. However, molecules at the surface are pulled downward and sideways by other molecules, but not upward away from the surface (Figure 11.9). These intermolecular attractions thus tend to pull the molecules into the liquid and cause the surface to tighten like an elastic film. Because there is little or no attraction between polar water molecules and, say, the nonpolar wax molecules on a freshly waxed car, a drop of water assumes the shape of a small round bead, because a sphere minimizes the surface area of a liquid. The waxy surface of a wet apple also produces this effect (Figure 11.10).

Surface tension enables water to form a thin layer over this swimmer as she breaks the surface of the water.
Courtesy of K.K. McSharry

A measure of the elastic force in the surface of a liquid is surface tension. The ***surface tension*** is *the amount of energy required to stretch or increase the surface of a liquid by a unit area* (for example, by 1 cm^2). Liquids that have strong intermolecular forces also have high surface tensions. Thus, because of hydrogen bonding, water has a considerably greater surface tension than most other liquids.

Another example of surface tension is *capillary action.* Figure 11.11(a) shows water rising spontaneously in a capillary tube. A thin film of water adheres to the wall of the glass tube. The surface tension of water causes this film to contract,

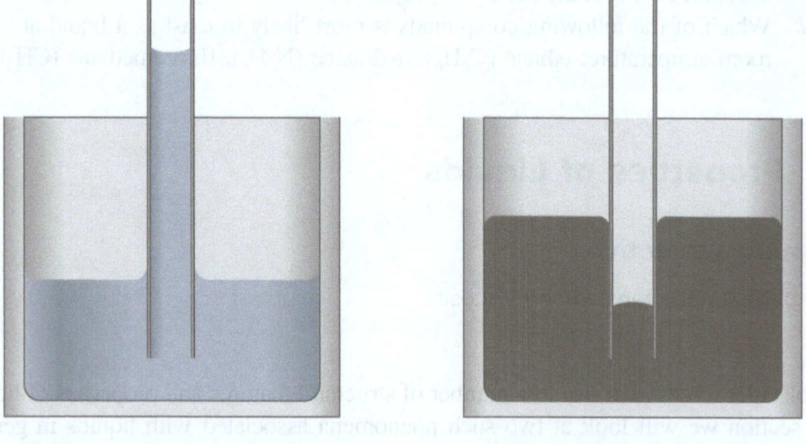

(a) (b)

Figure 11.11 *(a) When adhesion is greater than cohesion, the liquid (for example, water) rises in the capillary tube. (b) When cohesion is greater than adhesion, as it is for mercury, a depression of the liquid in the capillary tube results. Note that the meniscus in the tube of water is concave, or rounded downward, whereas that in the tube of mercury is convex, or rounded upward.*

and as it does, it pulls the water up the tube. Two types of forces bring about capillary action. One is ***cohesion,*** which is *the intermolecular attraction between like molecules* (in this case, the water molecules). The second force, called ***adhesion,*** is *an attraction between unlike molecules,* such as those in water and in the sides of a glass tube. If adhesion is stronger than cohesion, as it is in Figure 11.11(a), the contents of the tube will be pulled upward. This process continues until the adhesive force is balanced by the weight of the water in the tube. This action is by no means universal among liquids, as Figure 11.11(b) shows. In mercury, cohesion is greater than the adhesion between mercury and glass, so that when a capillary tube is dipped in mercury, the result is a depression or lowering, at the mercury level—that is, the height of the liquid in the capillary tube is below the surface of the mercury.

Viscosity

The expression "slow as molasses in January" owes its truth to another physical property of liquids called viscosity. ***Viscosity*** is *a measure of a fluid's resistance to flow.* The greater the viscosity, the more slowly the liquid flows. The viscosity of a liquid usually decreases as temperature increases; thus, hot molasses flows much faster than cold molasses.

Liquids that have strong intermolecular forces have higher viscosities than those that have weak intermolecular forces (Table 11.4). Water has a higher viscosity than many other liquids because of its ability to form hydrogen bonds. Interestingly, the viscosity of glycerol is significantly higher than that of all the other liquids listed in Table 11.4. Glycerol has the structure

$$\begin{array}{l} CH_2\text{—}OH \\ | \\ CH\text{—}OH \\ | \\ CH_2\text{—}OH \end{array}$$

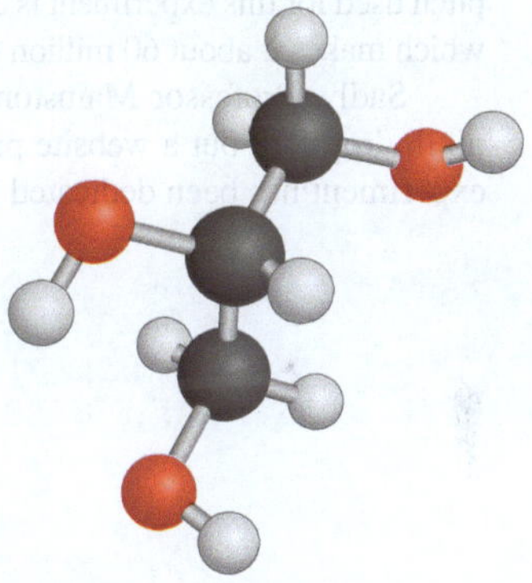

Glycerol is a clear, odorless, syrupy liquid used to make explosives, ink, and lubricants.

Like water, glycerol can form hydrogen bonds. Each glycerol molecule has three groups that can participate in hydrogen bonding with other glycerol molecules. Furthermore, because of their shape, the molecules have a great tendency to become entangled rather than to slip past one another as the molecules of less viscous liquids do. These interactions contribute to its high viscosity.

Table 11.4 Viscosity of Some Common Liquids at 20°C

Liquid	Viscosity ($N \cdot s/m^2$)*
Acetone (C_3H_6O)	3.16×10^{-4}
Benzene (C_6H_6)	6.25×10^{-4}
Blood	4×10^{-3}
Carbon tetrachloride (CCl_4)	9.69×10^{-4}
Diethyl ether ($C_2H_5OC_2H_5$)	2.33×10^{-4}
Ethanol (C_2H_5OH)	1.20×10^{-3}
Glycerol ($C_3H_8O_3$)	1.49
Mercury (Hg)	1.55×10^{-3}
Water (H_2O)	1.01×10^{-3}

*The SI units of viscosity are newton-second per meter squared.

CHEMISTRY *in Action*

A Very Slow Pitch

In 1927, Thomas Parnell started what may be the longest-running laboratory experiment in the history of laboratory experiments. Professor Parnell wanted to show his physics students at the University of Queensland an interesting property of pitch, a derivative of tar. Pitch is viscoelastic, which means that it will break into pieces if struck with enough force, but like a viscous liquid, it also flows slowly. Very slowly!

Parnell heated a sample of pitch to a temperature that allowed it to be poured into a funnel, and the funnel and a receiving beaker were covered with a bell jar and placed on display outside of the lecture hall. And then they waited.

The first drop fell 8 or 9 years after the pitch settled in the funnel and the stem was cut, but no one saw it fall. Drops fell at a rate of roughly one drop per decade, still with no witnesses. A few years after the third drop fell, Professor John Mainstone took over as curator and guardian of the experiment, but he did not get to witness any of the five drops that fell during his 52 years as curator, including several near misses and a webcam failure when the eighth drop fell on November 28, 2000. Based on the rate at which the drops have fallen over the past eight decades, the viscosity of the pitch used for this experiment is estimated to be $2.3 \times 10^8 \ N \cdot s/m^2$, which makes it about 60 million times "slower than molasses."

Sadly, Professor Mainstone died before he could see the ninth drop fall, but a website providing a live video feed to the experiment has been dedicated in his honor.

John Mainstone watching the Pitch Drop Experiment.
Courtesy of The University of Queensland

The Structure and Properties of Water

Water is so common a substance on Earth that we often overlook its unique nature. All life processes involve water. Water is an excellent solvent for many ionic compounds, as well as for other substances capable of forming hydrogen bonds with water. If water did not have the ability to form hydrogen bonds, it would be a gas at room temperature.

As Table 6.2 shows, water has a high specific heat. The reason is that to raise the temperature of water (that is, to increase the average kinetic energy of water molecules), we must first break the many intermolecular hydrogen bonds. Thus, water can absorb a substantial amount of heat while its temperature rises only slightly. The converse is also true: Water can give off much heat with only a slight decrease in its temperature. For this reason, the huge quantities of water that are present in our lakes and oceans can effectively moderate the climate of adjacent land areas by absorbing heat in the summer and giving off heat in the winter, with only small changes in the temperature of the body of water.

The most striking property of water is that its solid form is less dense than its liquid form: Ice floats at the surface of liquid water. The density of almost all other substances is greater in the solid state than in the liquid state (Figure 11.12).

To understand why water is different, we have to examine the electronic structure of the H_2O molecule. As we saw in Chapter 9, there are two pairs of nonbonding electrons, or two lone pairs, on the oxygen atom:

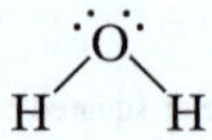

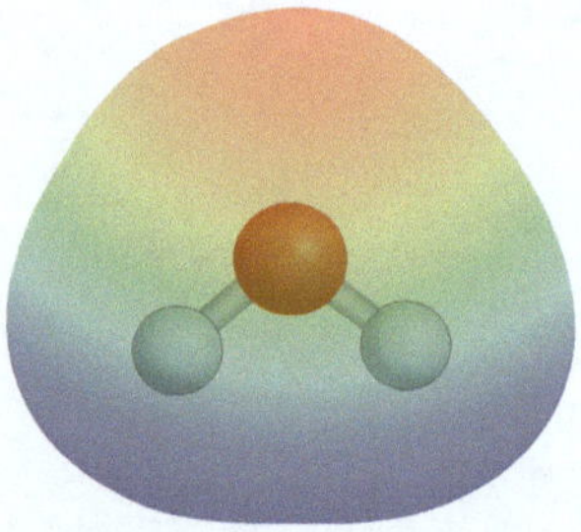

Electrostatic potential map of water.

Figure 11.12 *Left: Ice cubes float on water. Right: Solid benzene sinks to the bottom of liquid benzene.*
Ken Karp/McGraw Hill

Although many compounds can form intermolecular hydrogen bonds, the difference between H_2O and other polar molecules, such as NH_3 and HF, is that each oxygen atom can form *two* hydrogen bonds, the same as the number of lone electron pairs on the oxygen atom. Thus, water molecules are joined together in an extensive three-dimensional network in which each oxygen atom is approximately tetrahedrally bonded to four hydrogen atoms, two by covalent bonds and two by hydrogen bonds. This equality in the number of hydrogen atoms and lone pairs is not characteristic of NH_3 or HF or, for that matter, of any other molecule capable of forming hydrogen bonds. Consequently, these other molecules can form rings or chains, but not three-dimensional structures.

The highly ordered three-dimensional structure of ice (Figure 11.13) prevents the molecules from getting too close to one another. But consider what happens when ice melts. At the melting point, a number of water molecules have enough kinetic energy to break free of the intermolecular hydrogen bonds. These molecules become trapped in the cavities of the three-dimensional structure, which is broken down into smaller clusters. As a result, there are more molecules per unit volume in liquid water than in ice. Thus, because density = mass/volume, the density of water is greater than that

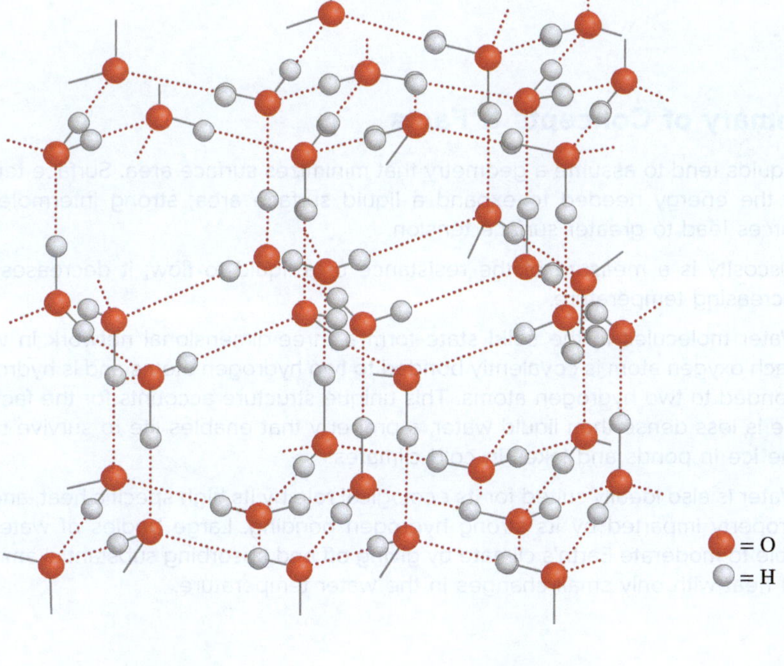

Figure 11.13 *The three-dimensional structure of ice. Each O atom is bonded to four H atoms. The covalent bonds are shown by short solid lines and the weaker hydrogen bonds by long dotted lines between O and H. The empty space in the structure accounts for the low density of ice.*

CHEMISTRY *in Action*

Why Do Lakes Freeze from the Top Down?

The fact that ice is less dense than water has a profound ecological significance. Consider, for example, the temperature changes in the fresh water of a lake in a cold climate. As the temperature of the water near the surface drops, the density of this water increases. The colder water then sinks toward the bottom, while warmer water, which is less dense, rises to the top. This normal convection motion continues until the temperature throughout the water reaches 4°C. Below this temperature, the density of water begins to decrease with decreasing temperature (see Figure 11.13), so that it no longer sinks. On further cooling, the water begins to freeze at the surface. The ice layer formed does not sink because it is less dense than the liquid; it even acts as a thermal insulator for the water below it. Were ice heavier, it would sink to the bottom of the lake and eventually the water would freeze upward. Most living organisms in the body of water could not survive being frozen in ice. Fortunately, lake water does not freeze upward from the bottom. This unusual property of water makes the sport of ice fishing possible.

Ice fishing. The ice layer that forms on the surface of a lake insulates the water beneath and maintains a high enough temperature to sustain aquatic life.
Staffan Widstrand/The Image Bank/Getty Images

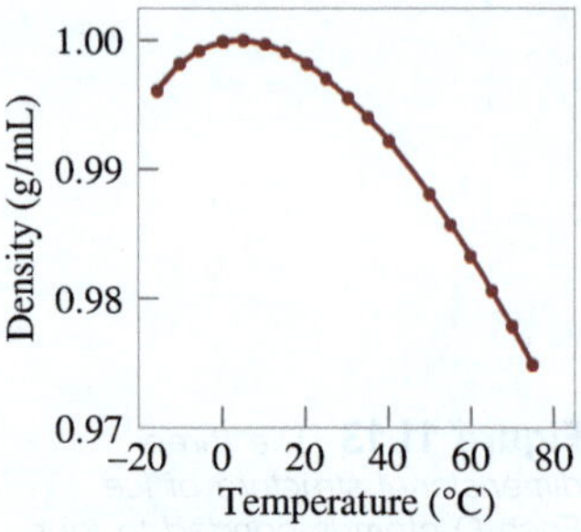

Figure 11.14 *Plot of density versus temperature for liquid water. The maximum density of water is reached at 4°C. The density of ice at 0°C is about 0.92 g/cm^3.*

of ice. With further heating, more water molecules are released from intermolecular hydrogen bonding, so that the density of water tends to increase with rising temperature just above the melting point. Of course, at the same time, water expands as it is being heated so that its density is decreased. These two processes—the trapping of free water molecules in cavities and thermal expansion—act in opposite directions. From 0°C to 4°C, the trapping prevails and water becomes progressively denser. Beyond 4°C, however, thermal expansion predominates and the density of water decreases with increasing temperature (Figure 11.14).

Summary of Concepts & Facts

- Liquids tend to assume a geometry that minimizes surface area. Surface tension is the energy needed to expand a liquid surface area; strong intermolecular forces lead to greater surface tension.
- Viscosity is a measure of the resistance of a liquid to flow; it decreases with increasing temperature.
- Water molecules in the solid state form a three-dimensional network in which each oxygen atom is covalently bonded to two hydrogen atoms and is hydrogen-bonded to two hydrogen atoms. This unique structure accounts for the fact that ice is less dense than liquid water, a property that enables life to survive under the ice in ponds and lakes in cold climates.
- Water is also ideally suited for its ecological role by its high specific heat, another property imparted by its strong hydrogen bonding. Large bodies of water are able to moderate Earth's climate by giving off and absorbing substantial amounts of heat with only small changes in the water temperature.

Review of Concepts & Facts

11.3.1 Why are motorists advised to use more viscous oils for their engines in the summer and less viscous oils in the winter?

11.4 Crystal Structure

Learning Objectives

- Define unit cell, lattice point, and coordination number.
- Distinguish the major types of unit cells in terms of the packing of spheres.
- Determine the atomic radius of an atom given its density and crystal type.

Solids can be divided into two categories: crystalline and amorphous. Ice is a ***crystalline solid,*** which *possesses rigid and long-range order; its atoms, molecules, or ions occupy specific positions.* The arrangement of such particles in a crystalline solid is such that the net attractive intermolecular forces are at their maximum. The forces responsible for the stability of a crystal can be ionic forces, covalent bonds, van der Waals forces, hydrogen bonds, or a combination of these forces. *Amorphous solids* such as glass lack a well-defined arrangement and long-range molecular order. We will discuss them in Section 11.7. In this section, we will concentrate on the structure of crystalline solids.

A ***unit cell*** is *the basic repeating structural unit of a crystalline solid.* Figure 11.15 shows a unit cell and its extension in three dimensions. Each sphere represents an atom, ion, or molecule and is called a *lattice point.* In many crystals, the lattice point does not actually contain such a particle. Rather, there may be several atoms, ions, or molecules identically arranged about each lattice point. For simplicity, however, we can assume that each lattice point is occupied by an atom. This is certainly the case with most metals. Every crystalline solid can be described in terms of one of the seven types of unit cells shown in Figure 11.16. The geometry of the cubic unit cell is particularly simple because all sides and all angles are equal. Any of the unit cells, when repeated in space in all three dimensions, forms the lattice structure characteristic of a crystalline solid.

Video: Cubic Unit Cells and Their Origins

Packing Spheres

We can understand the general geometric requirements for crystal formation by considering the different ways of packing a number of identical spheres (Ping-Pong balls,

Video: Packing Spheres

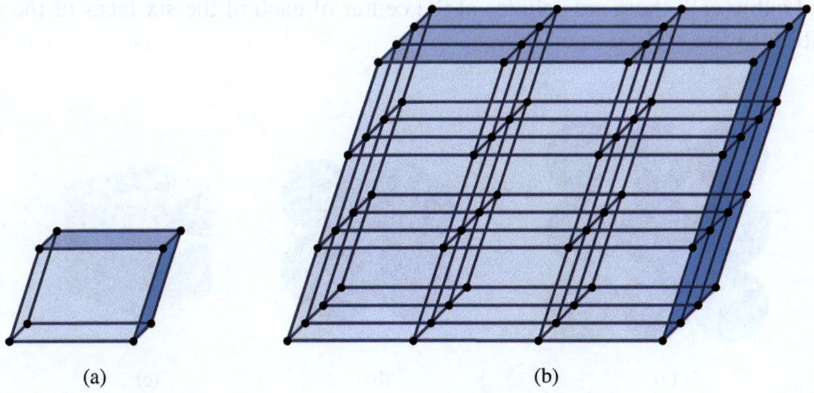

Figure 11.15 *(a) A unit cell and (b) its extension in three dimensions. The black spheres represent either atoms or molecules.*

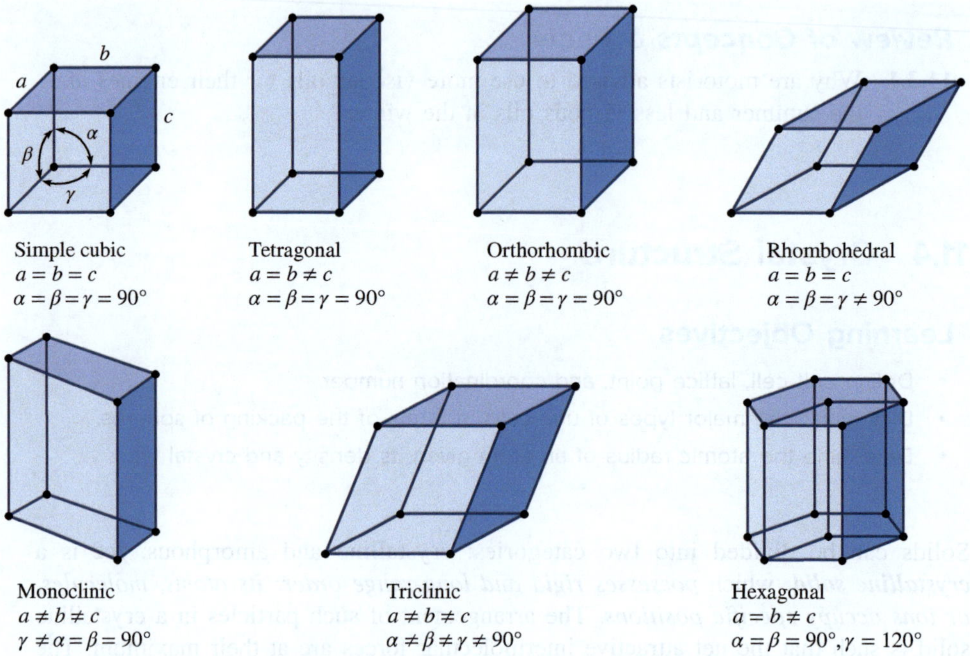

Figure 11.16 *The seven types of unit cells. Angle α is defined by edges b and c, angle β by edges a and c, and angle γ by edges a and b.*

for example) to form an ordered three-dimensional structure. The way the spheres are arranged in layers determines what type of unit cell we have.

In the simplest case, a layer of spheres can be arranged as shown in Figure 11.17(a). The three-dimensional structure can be generated by placing a layer above and below this layer in such a way that spheres in one layer are directly over the spheres in the layer below it. This procedure can be extended to generate many, many layers, as in the case of a crystal. Focusing on the sphere labeled with an "*x*," we see that it is in contact with four spheres in its own layer, one sphere in the layer above, and one sphere in the layer below. Each sphere in this arrangement is said to have a *coordination number* of 6 because it has six immediate neighbors. The ***coordination number*** is defined as *the number of atoms (or ions) surrounding an atom (or ion) in a crystal lattice.* Its value gives us a measure of how tightly the spheres are packed together—the larger the coordination number, the closer the spheres are to each other. The basic, repeating unit in the array of spheres is called a *simple cubic cell* (scc) [see Figure 11.17(b)].

The other types of cubic cells are the *body-centered cubic cell* (bcc) and the *face-centered cubic cell* (fcc) (Figure 11.18). A body-centered cubic arrangement differs from a simple cube in that the second layer of spheres fits into the depressions of the first layer and the third layer into the depressions of the second layer (Figure 11.19). The coordination number of each sphere in this structure is 8 (each sphere is in contact with four spheres in the layer above and four spheres in the layer below). In the face-centered cubic cell, there are spheres at the center of each of the six faces of the cube, in addition to the eight corner spheres.

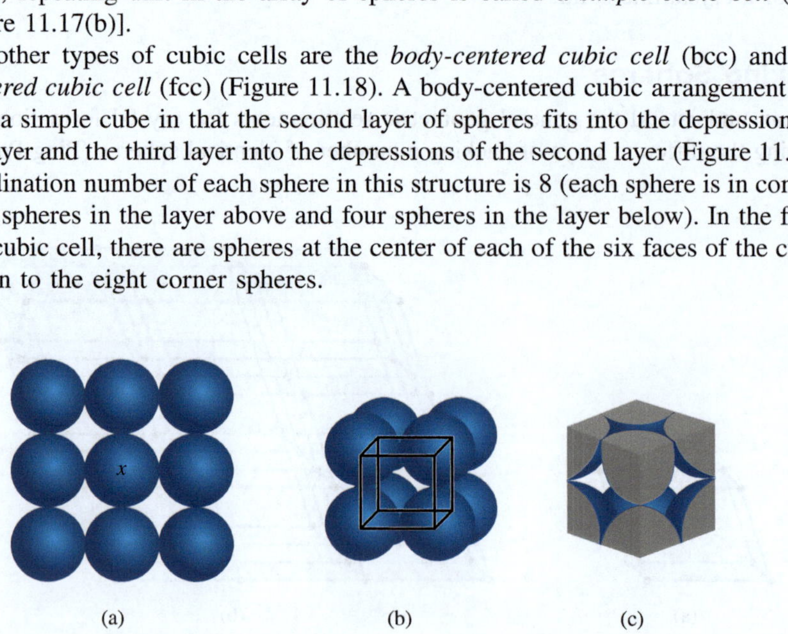

Figure 11.17 *Arrangement of identical spheres in a simple cubic cell. (a) Top view of one layer of spheres. (b) Definition of a simple cubic cell. (c) Because each sphere is shared by eight unit cells and there are eight corners in a cube, there is the equivalent of one complete sphere inside a simple cubic unit cell.*

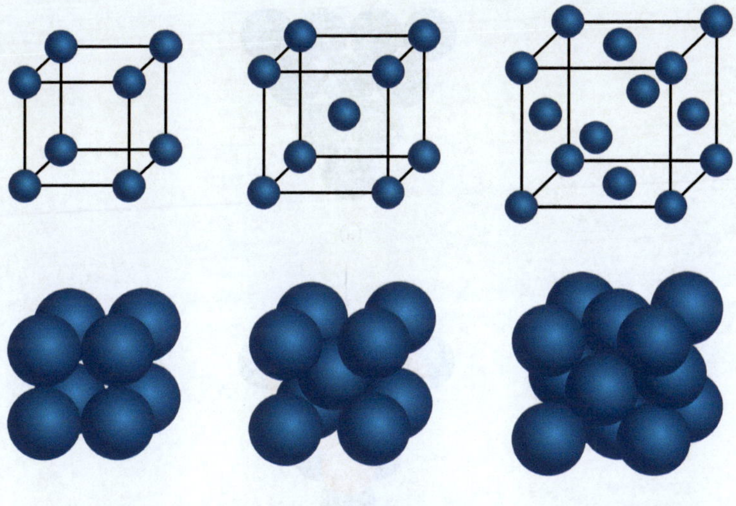

Figure 11.18 *Three types of cubic cells. In reality, the spheres representing atoms, molecules, or ions are in contact with one another in these cubic cells.*

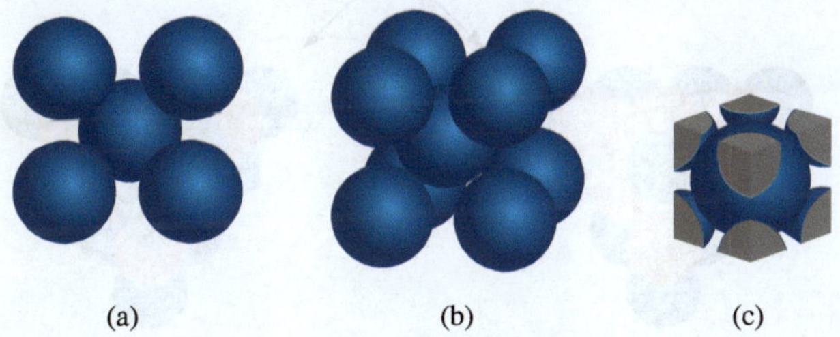

Figure 11.19 *Arrangement of identical spheres in a body-centered cube. (a) Top view. (b) Definition of a body-centered cubic unit cell. (c) There is the equivalent of two complete spheres inside a body-centered cubic unit cell.*

Because every unit cell in a crystalline solid is adjacent to other unit cells, most of a cell's atoms are shared by neighboring cells. For example, in all types of cubic cells, each corner atom belongs to eight unit cells [Figure 11.20(a)]; an edge atom is shared by four unit cells [see Figure 11.20(b)], and a face-centered atom is shared by two unit cells [see Figure 11.20(c)]. Because each corner sphere is shared by eight unit cells and there are eight corners in a cube, there will be the equivalent of only one complete sphere inside a simple cubic unit cell (see Figure 11.18). A body-centered cubic cell contains the equivalent of two complete spheres, one in the center and eight shared corner spheres. A face-centered cubic cell contains four complete spheres—three from the six face-centered atoms and one from the eight shared corner spheres.

Closest Packing

Clearly there is more empty space in the simple cubic and body-centered cubic cells than in the face-centered cubic cell. ***Closest packing,*** *the most efficient arrangement*

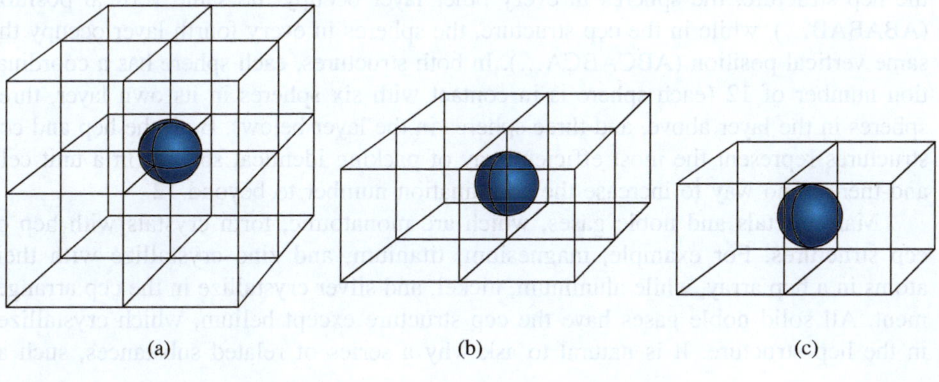

Figure 11.20 *(a) A corner atom in any cell is shared by eight unit cells. (b) An edge atom is shared by four unit cells. (c) A face-centered atom in a cubic cell is shared by two unit cells.*

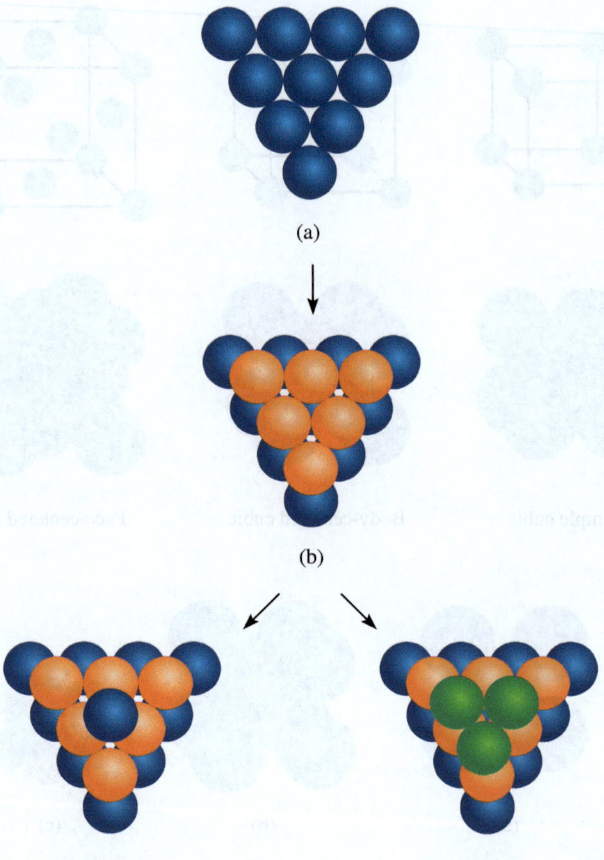

Figure 11.21 *(a) In a close-packed layer, each sphere is in contact with six others. (b) Spheres in the second layer fit into the depressions between the first-layer spheres. (c) In the hexagonal close-packed structure, each third-layer sphere is directly over a first-layer sphere. (d) In the cubic close-packed structure, each third-layer sphere fits into a depression that is directly over a depression in the first layer.*

of spheres, starts with the structure shown in Figure 11.21(a), which we call layer A. Focusing on the only enclosed sphere, we see that it has six immediate neighbors in that layer. In the second layer (which we call layer B), spheres are packed into the depressions between the spheres in the first layer so that all the spheres are as close together as possible [see Figure 11.21(b)].

There are two ways that a third-layer sphere may cover the second layer to achieve closest packing. The spheres may fit into the depressions so that each third-layer sphere is directly over a first-layer sphere [see Figure 11.21(c)]. Because there is no difference between the arrangement of the first and third layers, we also call the third layer, layer A. Alternatively, the third-layer spheres may fit into the depressions that lie directly over the depressions in the first layer [see Figure 11.21(d)]. In this case, we call the third layer, layer C. Figure 11.22 shows the "exploded views" and the structures resulting from these two arrangements. The ABA arrangement is known as the *hexagonal close-packed (hcp) structure,* and the ABC arrangement is the *cubic close-packed (ccp) structure,* which corresponds to the face-centered cube already described. Note that in the hcp structure, the spheres in every other layer occupy the same vertical position (ABABAB…), while in the ccp structure, the spheres in every fourth layer occupy the same vertical position (ABCABCA…). In both structures, each sphere has a coordination number of 12 (each sphere is in contact with six spheres in its own layer, three spheres in the layer above, and three spheres in the layer below). Both the hcp and ccp structures represent the most efficient way of packing identical spheres in a unit cell, and there is no way to increase the coordination number to beyond 12.

Many metals and noble gases, which are monatomic, form crystals with hcp or ccp structures. For example, magnesium, titanium, and zinc crystallize with their atoms in a hcp array, while aluminum, nickel, and silver crystallize in the ccp arrangement. All solid noble gases have the ccp structure except helium, which crystallizes in the hcp structure. It is natural to ask why a series of related substances, such as

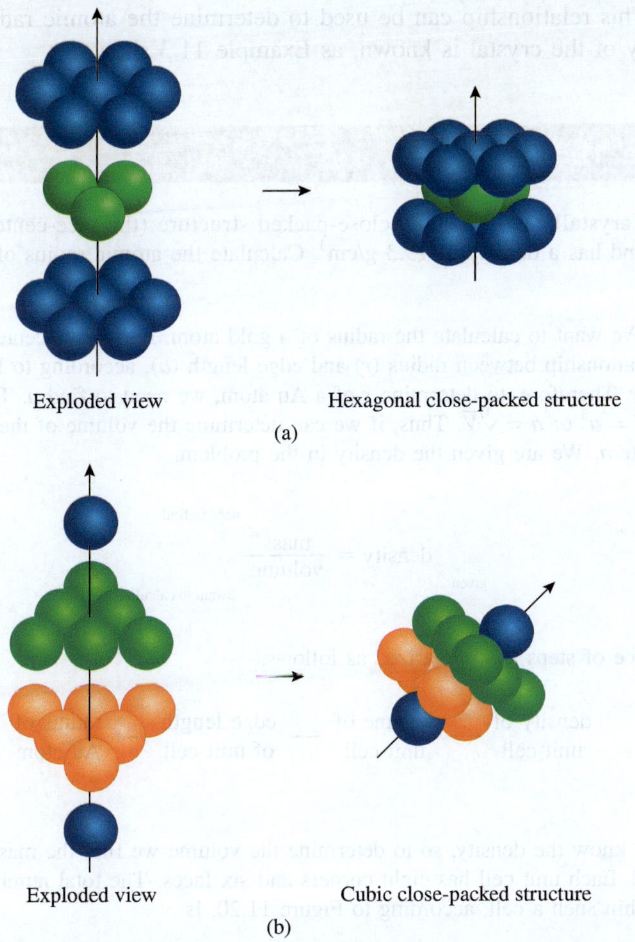

Figure 11.22 *Exploded views of (a) a hexagonal close-packed structure and (b) a cubic close-packed structure. The arrow is tilted to show the face-centered cubic unit cell more clearly. Note that this arrangement is the same as the face-centered unit cell.*

the transition metals or the noble gases, would form different crystal structures. The answer lies in the relative stability of a particular crystal structure, which is governed by intermolecular forces. Thus, magnesium metal has the hcp structure because this arrangement of Mg atoms results in the greatest stability of the solid.

Figure 11.23 summarizes the relationship between the atomic radius r and the edge length a of a simple cubic cell, a body-centered cubic cell, and a face-centered

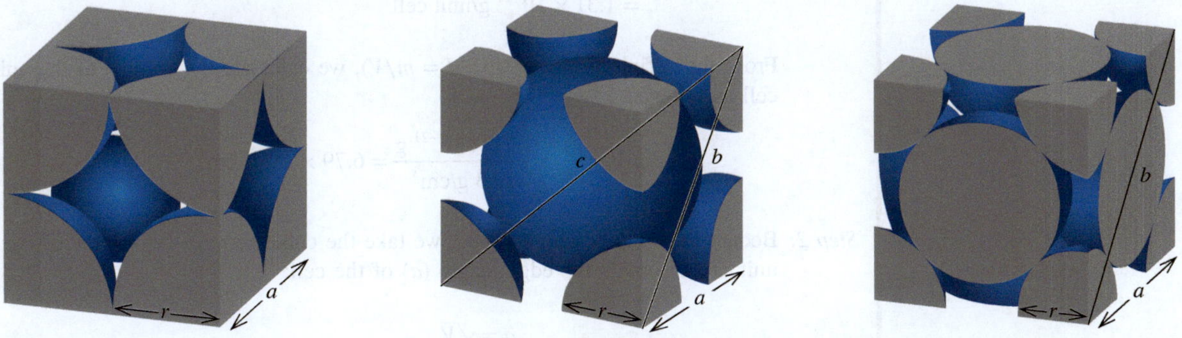

scc

$a = 2r$

bcc

$$b^2 = a^2 + a^2$$
$$c^2 = a^2 + b^2 = 3a^2$$
$$c = \sqrt{3}a = 4r$$
$$a = \frac{4r}{\sqrt{3}}$$

fcc

$$b = 4r$$
$$b^2 = a^2 + a^2$$
$$16r^2 = 2a^2$$
$$a = \sqrt{8}r$$

Figure 11.23 *The relationship between the edge length (a) and radius (r) of atoms in the simple cubic cell, body-centered cubic cell, and face-centered cubic cell.*

cubic cell. This relationship can be used to determine the atomic radius of a sphere if the density of the crystal is known, as Example 11.3 shows.

Example 11.3

Gold (Au) crystallizes in a cubic close-packed structure (the face-centered cubic unit cell) and has a density of 19.3 g/cm^3. Calculate the atomic radius of gold in picometers.

Strategy We want to calculate the radius of a gold atom. For a face-centered cubic unit cell, the relationship between radius (r) and edge length (a), according to Figure 11.23, is $a = \sqrt{8}r$. Therefore, to determine r of a Au atom, we need to find a. The volume of a cube is $V = a^3$ or $a = \sqrt[3]{V}$. Thus, if we can determine the volume of the unit cell, we can calculate a. We are given the density in the problem.

$$\text{density} = \frac{\text{mass}}{\text{volume}}$$

need to find (mass)

given (density)

want to calculate (volume)

The sequence of steps is summarized as follows:

$$\begin{matrix}\text{density of} \\ \text{unit cell}\end{matrix} \longrightarrow \begin{matrix}\text{volume of} \\ \text{unit cell}\end{matrix} \longrightarrow \begin{matrix}\text{edge length} \\ \text{of unit cell}\end{matrix} \longrightarrow \begin{matrix}\text{radius of} \\ \text{Au atom}\end{matrix}$$

Solution

Step 1: We know the density, so to determine the volume we find the mass of the unit cell. Each unit cell has eight corners and six faces. The total number of atoms within such a cell, according to Figure 11.20, is

$$\left(8 \times \frac{1}{8}\right) + \left(6 \times \frac{1}{2}\right) = 4$$

The mass of a unit cell in grams is

$$m = \frac{4\ \cancel{\text{atoms}}}{1\ \text{unit cell}} \times \frac{1\ \cancel{\text{mol}}}{6.022 \times 10^{23}\ \cancel{\text{atoms}}} \times \frac{197.0\ \text{g Au}}{1\ \cancel{\text{mol Au}}}$$

$$= 1.31 \times 10^{-21}\ \text{g/unit cell}$$

From the definition of density ($d = m/V$), we calculate the volume of the unit cell as follows:

$$V = \frac{m}{d} = \frac{1.31 \times 10^{-21}\ \text{g}}{19.3\ \text{g/cm}^3} = 6.79 \times 10^{-23}\ \text{cm}^3$$

Step 2: Because volume is length cubed, we take the cubic root of the volume of the unit cell to obtain the edge length (a) of the cell:

$$a = \sqrt[3]{V}$$

$$= \sqrt[3]{6.79 \times 10^{-23}\ \text{cm}^3}$$

$$= 4.08 \times 10^{-8}\ \text{cm}$$

Step 3: From Figure 11.23 we see that the radius of an Au sphere (r) is related to the edge length by

$$a = \sqrt{8}\, r$$

(Continued)

Therefore,

$$r = \frac{a}{\sqrt{8}} = \frac{4.08 \times 10^{-8}\text{ cm}}{\sqrt{8}}$$
$$= 1.44 \times 10^{-8}\text{ cm}$$
$$= 1.44 \times 10^{-8}\ \cancel{\text{cm}} \times \frac{1 \times 10^{-2}\ \cancel{\text{m}}}{1\ \cancel{\text{cm}}} \times \frac{1\text{ pm}}{1 \times 10^{-12}\ \cancel{\text{m}}}$$
$$= 144\text{ pm}$$

Practice Exercise When silver crystallizes, it forms face-centered cubic cells. The unit cell edge length is 408.7 pm. Calculate the density of silver.

Similar problem: 11.39.

Summary of Concepts & Facts

- Crystalline solids have a regular structure of atoms, ions, or molecules.
- The basic structural unit of a crystalline solid is the unit cell, which is repeated to form a three-dimensional crystal lattice.
- Determine the atomic radius of an atom given its density and crystal type.

Review of Concepts & Facts

11.4.1 Tungsten crystallizes in a body-centered cubic lattice (the W atoms occupy only the lattice points). How many W atoms are present in a unit cell?

11.4.2 Cerium forms face-centered cubic cells when it crystallizes. If the cell edge length is 760 pm, what is the radius of a cerium atom (in pm)?

11.5 X-Ray Diffraction by Crystals

Learning Objective

- Apply the Bragg equation to the diffraction of X rays by solids.

Virtually all we know about crystal structure has been learned from X-ray diffraction studies. ***X-ray diffraction*** refers to *the scattering of X rays by the units of a crystalline solid.* The scattering, or diffraction, patterns produced are used to deduce the arrangement of particles in the solid lattice.

In Section 10.6 we discussed the interference phenomenon associated with waves (see Figure 10.22). Because X rays are one form of electromagnetic radiation, and therefore waves, we would expect them to exhibit such behavior under suitable conditions. In 1912 the German physicist Max von Laue[†] correctly suggested that, because the wavelength of X rays is comparable in magnitude to the distances

[†]Max Theodor Felix von Laue (1879–1960). German physicist von Laue received the Nobel Prize in Physics in 1914 for his discovery of X-ray diffraction.

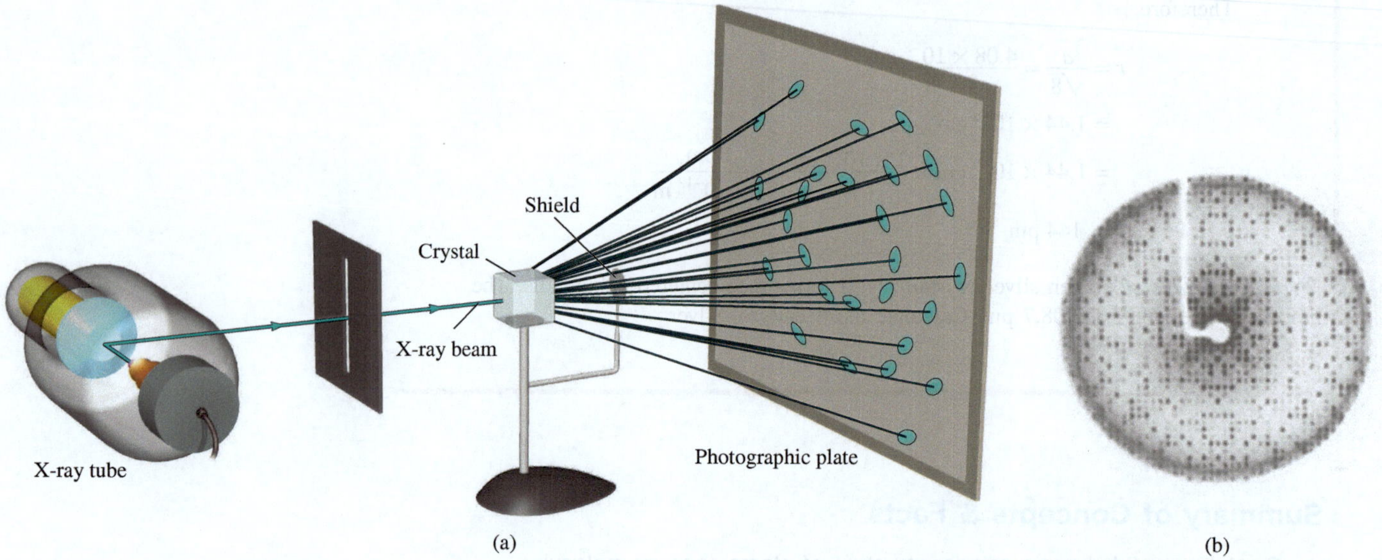

Figure 11.24 *(a) An arrangement for obtaining the X-ray diffraction pattern of a crystal. The shield prevents the strong undiffracted X rays from damaging the photographic plate. (b) X-ray diffraction pattern of crystalline lysozyme, a protein. The white "L" is a shadow of the sample holder and shield.*

between lattice points in a crystal, the lattice should be able to *diffract* X rays. An X-ray diffraction pattern is the result of interference in the waves associated with X rays.

Figure 11.24 shows a typical X-ray diffraction setup. A beam of X rays is directed at a mounted crystal. Atoms in the crystal absorb some of the incoming radiation and then reemit it; the process is called the *scattering of X rays.*

To understand how a diffraction pattern may be generated, consider the scattering of X rays by atoms in two parallel planes (Figure 11.25). Initially, the two incident rays are *in phase* with each other (their maxima and minima occur at the same positions). The upper wave is scattered, or reflected, by an atom in the first layer, while the lower wave is scattered by an atom in the second layer. For these two scattered waves to be in phase again, the extra distance traveled by

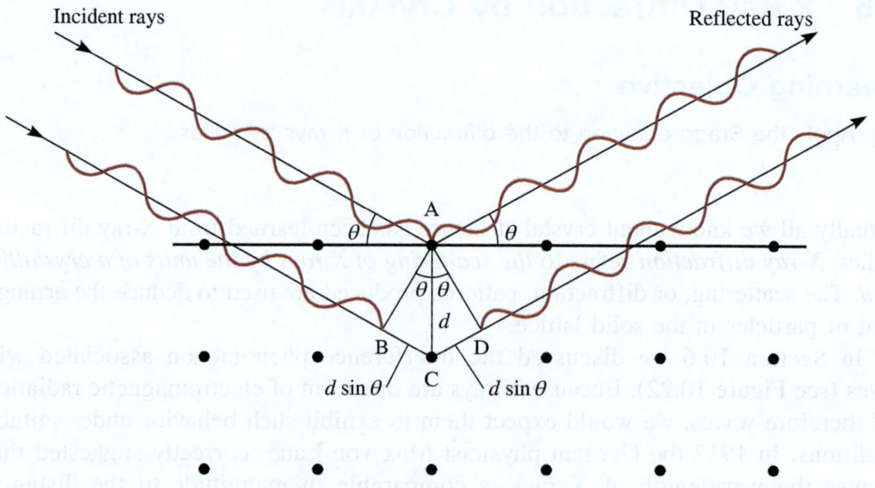

Figure 11.25 *Reflection of X rays from two layers of atoms. The lower wave travels a distance 2d sin θ longer than the upper wave does. For the two waves to be in phase again after reflection, it must be true that 2d sin θ = nλ, where λ is the wavelength of the X ray and n = 1, 2, 3. . . . The sharply defined spots in Figure 11.24 are observed only if the crystal is large enough to consist of hundreds of parallel layers.*

the lower wave must be an integral multiple of the wavelength (λ) of the X ray; that is,

$$BC + CD = 2d \sin \theta = n\lambda \qquad n = 1, 2, 3, \ldots$$

or

$$2d \sin \theta = n\lambda \tag{11.1}$$

where θ is the angle between the X rays and the plane of the crystal and d is the distance between adjacent planes. Equation (11.1) is known as the Bragg equation after William H. Bragg† and Sir William L. Bragg.‡ The reinforced waves (waves that have interacted constructively, see Figure 10.22) produce a dark spot on a photographic film for each value of θ that satisfies the Bragg equation.

Example 11.4 illustrates the use of Equation (11.1).

Example 11.4

X rays of wavelength 0.154 nm strike an aluminum crystal; the rays are reflected at an angle of 19.3°. Assuming that $n = 1$, calculate the spacing between the planes of aluminum atoms (in pm) that is responsible for this angle of reflection. The conversion factor is obtained from 1 nm = 1000 pm.

Strategy This is an application of Equation (11.1).

Solution Converting the wavelength to picometers and using the angle of reflection (19.3°), we write

$$d = \frac{n\lambda}{2 \sin \theta} = \frac{\lambda}{2 \sin \theta}$$

$$= \frac{0.154 \text{ mn} \times \dfrac{1000 \text{ pm}}{1 \text{ nm}}}{2 \sin 19.3°}$$

$$= 177 \text{ pm}$$

Practice Exercise X rays of wavelength 0.154 nm are diffracted from a crystal at an angle of 14.17°. Assuming that $n = 1$, calculate the distance (in pm) between layers in the crystal.

Similar problems: 11.47, 11.48.

The X-ray diffraction technique offers the most accurate method for determining bond lengths and bond angles in molecules in the solid state. Because X rays are scattered by electrons, chemists can construct an electron-density contour map from the diffraction patterns by using a complex mathematical procedure. Basically, an *electron-density contour map* tells us the relative electron densities at various locations in a molecule. The densities reach a maximum near the center of each atom. In this manner, we can determine the positions of the nuclei and hence the geometric parameters of the molecule.

†William Henry Bragg (1862–1942). English physicist. Bragg's work was mainly in X-ray crystallography. He shared the Nobel Prize in Physics with his son Sir William Bragg in 1915.

‡Sir William Lawrence Bragg (1890–1972). English physicist. Bragg formulated the fundamental equation for X-ray diffraction and shared the Nobel Prize in Physics with his father in 1915.

Summary of Concepts & Facts

- X-ray diffraction has provided much of our knowledge about crystal structure. The Bragg equation allows the determination of spacing within a unit cell.

Review of Concepts & Facts

11.5.1 Why can the X-ray diffraction technique not be used to study molecular structure in a liquid?

11.5.2 X rays of wavelength 0.154 nm are diffracted from a crystalline solid at an angle of 11.2°. Assuming that $n = 1$, calculate the distance (in pm) between layers in the crystal?

11.6 Types of Crystals

Learning Objective

- Identify the major characteristics and examples of the major types of crystals: ionic, covalent, molecular, and metallic.

These giant potassium dihydrogen phosphate crystals were grown in the laboratory. The largest one weighs 701 lb!

Courtesy of Lawrence Livermore National Laboratory

The structures and properties of crystals, such as melting point, density, and hardness, are determined by the kinds of forces that hold the particles together. We can classify any crystal as one of four types: ionic, covalent, molecular, or metallic.

Ionic Crystals

Ionic crystals have two important characteristics: (1) They are composed of charged species, and (2) anions and cations are generally quite different in size. Knowing the radii of the ions is helpful in understanding the structure and stability of these compounds. There is no way to measure the radius of an individual ion, but sometimes it is possible to come up with a reasonable estimate. For example, if we know the radius of I^- in KI is about 216 pm, we can determine the radius of K^+ ion in KI, and from that, the radius of Cl^- in KCl, and so on. The ionic radii in Figure 8.9 are average values derived from many different compounds. Let us consider the NaCl crystal, which has a face-centered cubic lattice (see Figure 2.13). Figure 11.26

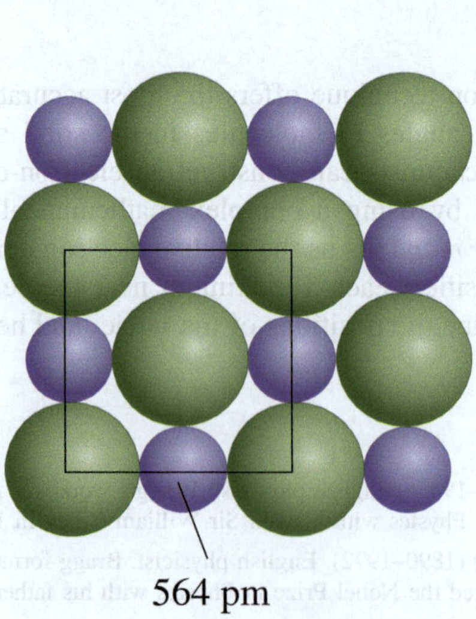

Figure 11.26 *Relation between the radii of Na^+ and Cl^- ions and the unit cell dimensions. Here the cell edge length is equal to twice the sum of the two ionic radii.*

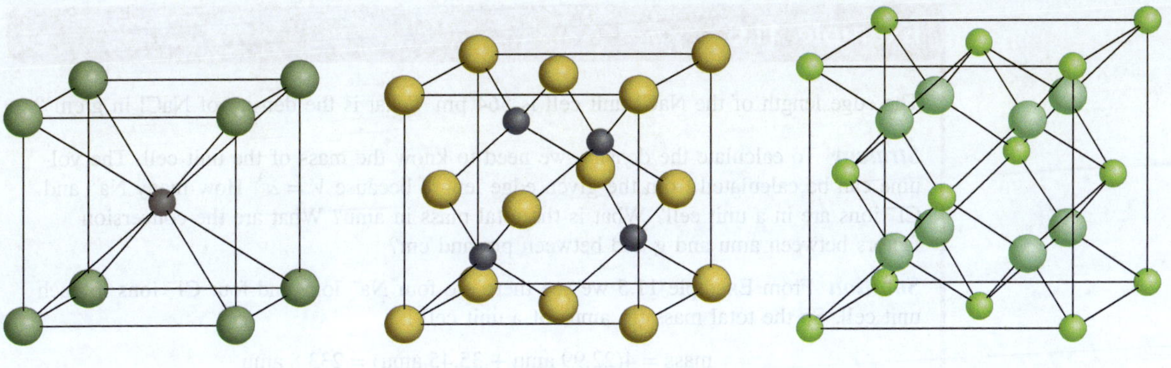

Figure 11.27 *Crystal structures of (a) CsCl, (b) ZnS, and (c) CaF_2. In each case, the cation is the smaller sphere.*

shows that the edge length of the unit cell of NaCl is twice the sum of the ionic radii of Na^+ and Cl^-. Using the values given in Figure 8.9, we calculate the edge length to be 2(95 + 181) pm, or 552 pm. But the edge length shown in Figure 11.26 was determined by X-ray diffraction to be 564 pm. The discrepancy between these two values tells us that the radius of an ion actually varies slightly from one compound to another.

Figure 11.27 shows the crystal structures of three ionic compounds: CsCl, ZnS, and CaF_2. Because Cs^+ is considerably larger than Na^+, CsCl has the simple cubic lattice. ZnS has the *zincblende* structure, which is based on the face-centered cubic lattice. If the S^{2-} ions occupy the lattice points, the Zn^{2+} ions are located one-fourth of the distance along each body diagonal. Other ionic compounds that have the zincblende structure include CuCl, BeS, CdS, and HgS. CaF_2 has the *fluorite* structure. The Ca^{2+} ions occupy the lattice points, and each F^- ion is tetrahedrally surrounded by four Ca^{2+} ions. The compounds SrF_2, BaF_2, $BaCl_2$, and PbF_2 also have the fluorite structure.

Examples 11.5 and 11.6 show how to calculate the number of ions in and the density of a unit cell.

Example 11.5

How many Na^+ and Cl^- ions are in each NaCl unit cell?

Solution NaCl has a structure based on a face-centered cubic lattice. As Figure 2.13 shows, one whole Na^+ ion is at the center of the unit cell, and there are twelve Na^+ ions at the edges. Because each edge Na^+ ion is shared by four unit cells [see Figure 11.20(b)], the total number of Na^+ ions is $1 + \left(12 \times \frac{1}{4}\right) = 4$. Similarly, there are six Cl^- ions at the face centers and eight Cl^- ions at the corners. Each face-centered ion is shared by two unit cells, and each corner ion is shared by eight unit cells [see Figure 11.20(a) and (c)], so the total number of Cl^- ions is $\left(6 \times \frac{1}{2}\right) + \left(8 \times \frac{1}{8}\right) = 4$. Thus, there are four Na^+ ions and four Cl^- ions in each NaCl unit cell. Figure 11.28 shows the portions of the Na^+ and Cl^- ions *within* a unit cell.

Check This result agrees with sodium chloride's empirical formula.

Practice Exercise How many atoms are in a body-centered cube, assuming that all atoms occupy lattice points?

Similar problem: 11.41.

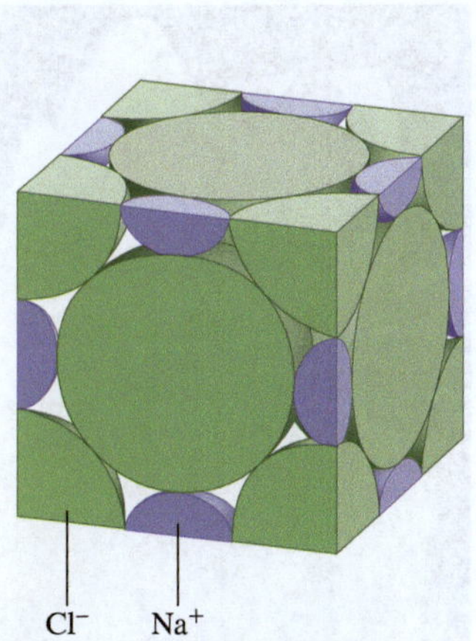

Figure 11.28 *Portions of Na^+ and Cl^- ions within a face-centered cubic unit cell.*

Example 11.6

The edge length of the NaCl unit cell is 564 pm. What is the density of NaCl in g/cm^3?

Strategy To calculate the density, we need to know the mass of the unit cell. The volume can be calculated from the given edge length because $V = a^3$. How many Na^+ and Cl^- ions are in a unit cell? What is the total mass in amu? What are the conversion factors between amu and g and between pm and cm?

Solution From Example 11.5 we see there are four Na^+ ions and four Cl^- ions in each unit cell. So the total mass (in amu) of a unit cell is

$$\text{mass} = 4(22.99 \text{ amu} + 35.45 \text{ amu}) = 233.8 \text{ amu}$$

Converting amu to grams, we write

$$233.8 \text{ amu} \times \frac{1 \text{ g}}{6.022 \times 10^{23} \text{ amu}} = 3.882 \times 10^{-22} \text{ g}$$

The volume of the unit cell is $V = a^3 = (564 \text{ pm})^3$. Converting pm^3 to cm^3, the volume is given by

$$V = (564 \text{ pm})^3 \times \left(\frac{1 \times 10^{-12} \text{ m}}{1 \text{ pm}}\right)^3 \times \left(\frac{1 \text{ cm}}{1 \times 10^{-2} \text{ m}}\right)^3 = 1.794 \times 10^{-22} \text{ cm}^3$$

Finally, from the definition of density,

$$\text{density} = \frac{\text{mass}}{\text{volume}} = \frac{3.882 \times 10^{-22} \text{ g}}{1.794 \times 10^{-22} \text{ cm}^3}$$
$$= 2.16 \text{ g/cm}^3$$

Practice Exercise Copper crystallizes in a face-centered cubic lattice (the Cu atoms are at the lattice points only). If the density of the metal is 8.96 g/cm^3, what is the unit cell edge length in pm?

Similar problem: 11.42.

Most ionic crystals have high melting points, an indication of the strong cohesive forces holding the ions together. A measure of the stability of ionic crystals is the lattice energy (see Section 9.3); the higher the lattice energy, the more stable the compound. These solids do not conduct electricity because the ions are fixed in position. However, in the molten state (that is, when melted) or dissolved in water, the ions are free to move and the resulting liquid is electrically conducting.

Quartz.
Doug Sherman/Geofile

Covalent Crystals

In covalent crystals, atoms are held together in an extensive three-dimensional network entirely by covalent bonds. Well-known examples are the two allotropes of carbon: diamond and graphite (see Figure 8.17). In diamond, each carbon atom is sp^3-hybridized; it is bonded to four other atoms (Figure 11.29). The strong covalent bonds in three dimensions contribute to diamond's unusual hardness (it is the hardest material known) and very high melting point (3550°C). In graphite, carbon atoms are arranged in six-membered rings. The atoms are all sp^2-hybridized; each atom is covalently bonded to three other atoms. The remaining unhybridized 2*p* orbital is used in pi bonding. In fact, each layer of graphite has the kind of delocalized molecular orbital that is present in benzene (see Section 10.8). Because electrons are free to move around in this extensively delocalized molecular orbital, graphite is a good conductor of electricity in directions along the planes of carbon atoms. The layers are held together by weak van der Waals forces. The covalent bonds in graphite account for its hardness; however, because the layers can slide over one another, graphite is slippery to the touch and is effective as a lubricant. It is also used in pencils and laser printer toner.

Another covalent crystal is quartz (SiO_2). The arrangement of silicon atoms in quartz is similar to that of carbon in diamond, but in quartz there is an oxygen atom

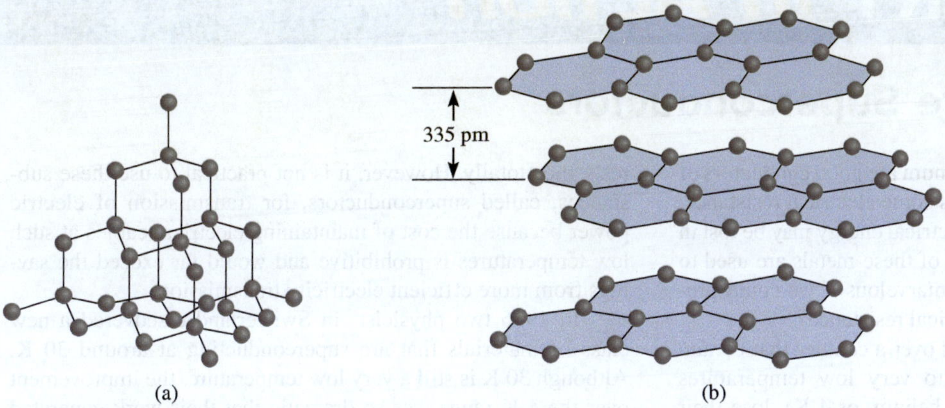

Figure 11.29 *(a) The structure of diamond. Each carbon is tetrahedrally bonded to four other carbon atoms. (b) The structure of graphite. The distance between successive layers is 335 pm.*

between each pair of Si atoms. Because Si and O have different electronegativities, the Si—O bond is polar. Nevertheless, SiO_2 is similar to diamond in many respects, such as hardness and high melting point (1610°C).

Molecular Crystals

In a molecular crystal, the lattice points are occupied by molecules, and the attractive forces between them are van der Waals forces and/or hydrogen bonding. An example of a molecular crystal is solid sulfur dioxide (SO_2), in which the predominant attractive force is a dipole-dipole interaction. Intermolecular hydrogen bonding is mainly responsible for maintaining the three-dimensional lattice of ice (see Figure 11.13). Other examples of molecular crystals are I_2, P_4, and S_8.

In general, except in ice, molecules in molecular crystals are packed together as closely as their size and shape allow. Because van der Waals forces and hydrogen bonding are generally quite weak compared with covalent and ionic bonds, molecular crystals are more easily broken apart than ionic and covalent crystals. Indeed, most molecular crystals melt at temperatures below 100°C.

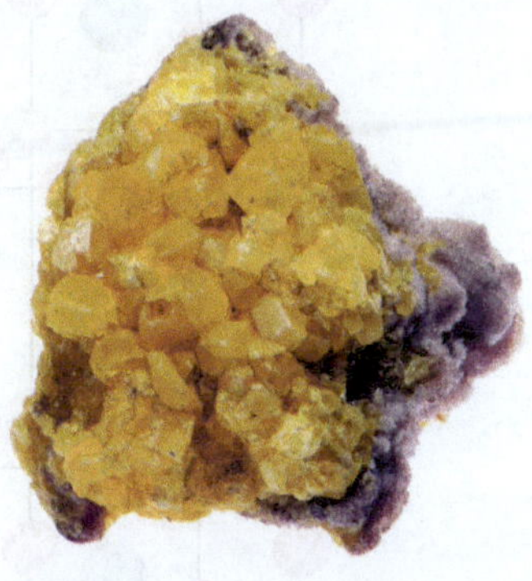

Sulfur.
Branko Jovanovic/Shutterstock

Metallic Crystals

In a sense, the structure of metallic crystals is the simplest because every lattice point in a crystal is occupied by an atom of the same metal. Metallic crystals are generally body-centered cubic, face-centered cubic, or hexagonal close-packed (Figure 11.30). Consequently, metallic elements are usually very dense.

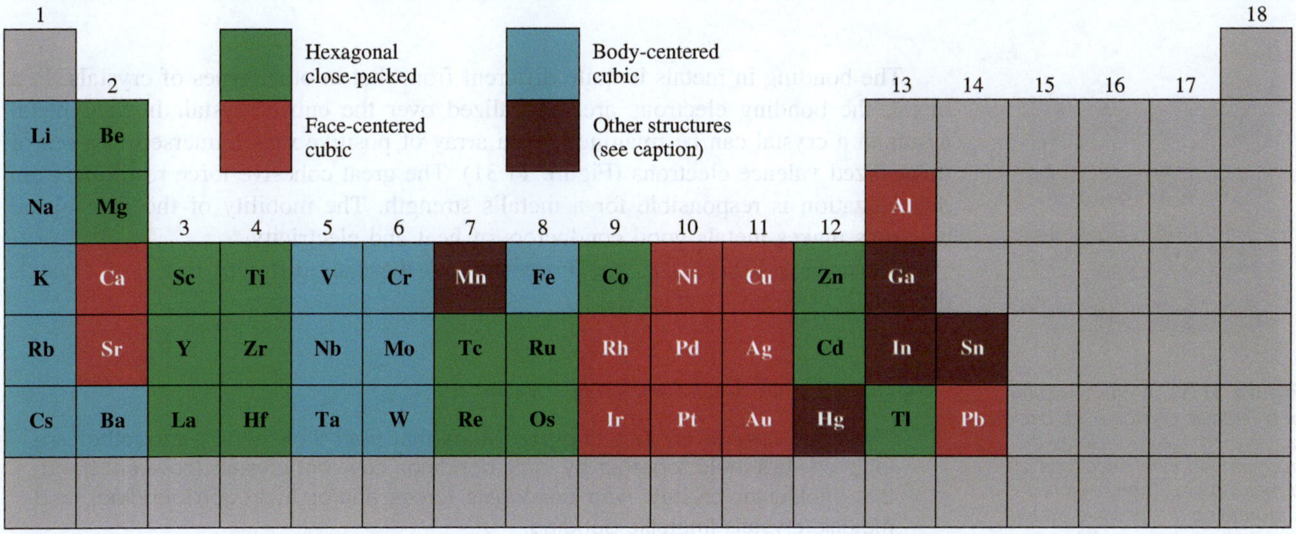

Figure 11.30 *Crystal structures of metals. The metals are shown in their positions in the periodic table. Mn has a cubic structure, Ga an orthorhombic structure, In and Sn a tetragonal structure, and Hg a rhombohedral structure (see Figure 11.16).*

CHEMISTRY *in Action*

High-Temperature Superconductors

Metals such as copper and aluminum are good conductors of electricity, but they do possess some electrical resistance. In fact, up to about 20 percent of electrical energy may be lost in the form of heat when cables made of these metals are used to transmit electricity. Wouldn't it be marvelous if we could produce cables that possessed no electrical resistance?

Actually it has been known for over a century that certain metals and alloys, when cooled to very low temperatures (around the boiling point of liquid helium, or 4 K), lose their resistance totally. However, it is not practical to use these substances, called superconductors, for transmission of electric power because the cost of maintaining electrical cables at such low temperatures is prohibitive and would far exceed the savings from more efficient electricity transmission.

In 1986 two physicists in Switzerland discovered a new class of materials that are superconducting at around 30 K. Although 30 K is still a very low temperature, the improvement over the 4-K range was so dramatic that their work generated immense interest and triggered a flurry of research activity. Within months, scientists synthesized compounds that are

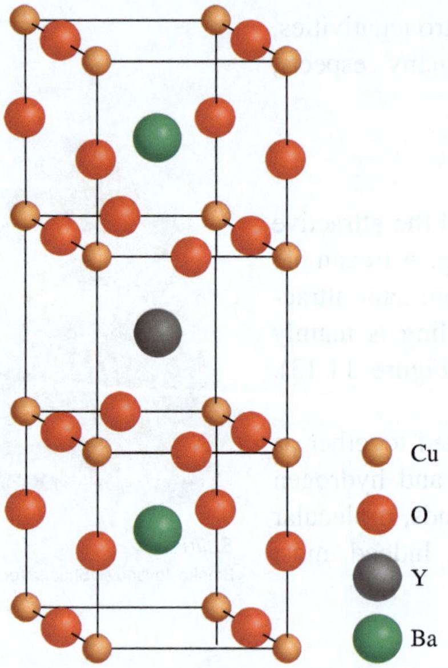

Crystal structure of $YBa_2Cu_3O_x$ (x = 6 or 7). Because some of the O atom sites are vacant, the formula is not constant.

The levitation of a magnet above a high-temperature superconductor immersed in liquid nitrogen.
ktsimage/iStockphoto/Getty Images

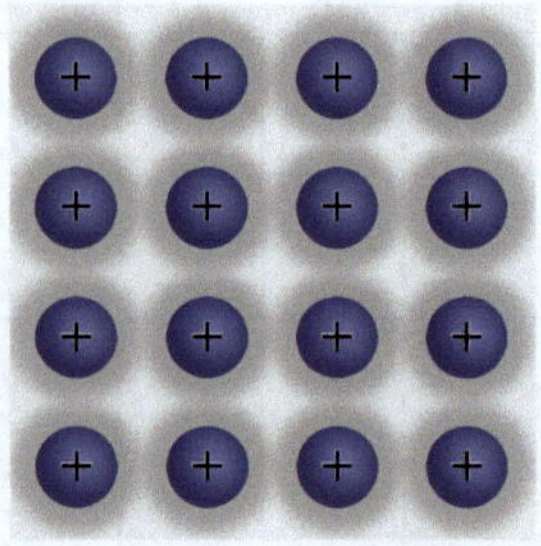

Figure 11.31 *A cross section of a metallic crystal. Each circled positive charge represents the nucleus and inner electrons of a metal atom. The gray area surrounding the positive metal ions indicates the mobile sea of valence electrons.*

The bonding in metals is quite different from that in other types of crystals. In a metal, the bonding electrons are delocalized over the entire crystal. In fact, metal atoms in a crystal can be imagined as an array of positive ions immersed in a sea of delocalized valence electrons (Figure 11.31). The great cohesive force resulting from delocalization is responsible for a metal's strength. The mobility of the delocalized electrons makes metals good conductors of heat and electricity.

Table 11.5 summarizes the properties of the four different types of crystals discussed.

Summary of Concepts & Facts

- The four types of crystals and the forces that hold their particles together are ionic crystals (held together by ionic bonding), covalent crystals (covalent bonding), molecular crystals (van der Waals forces and/or hydrogen bonding), and metallic crystals (metallic bonding).

superconducting around 95 K, which is well above the boiling point of liquid nitrogen (77 K). The crystal structure of one of these compounds shows a mixed oxide of yttrium, barium, and copper with the formula $YBa_2Cu_3O_x$ (where $x = 6$ or 7). The photo shows a magnet being levitated above such a superconductor, which is immersed in liquid nitrogen.

Despite the initial excitement, this class of high-temperature superconductors has not fully lived up to its promise. After more than 30 years of intense research and development, scientists still puzzle over how and why these compounds superconduct. It has also proved difficult to make wires of these compounds, and other technical problems have limited their large-scale commercial applications thus far.

In another encouraging development, in 2001 scientists in Japan discovered that magnesium diboride (MgB_2) becomes superconducting at about 40 K. Although liquid neon (b.p. 27 K) must be used as coolant instead of liquid nitrogen, it is still much cheaper than using liquid helium. Magnesium diboride has several advantages as a high-temperature superconductor. First, it is an inexpensive compound (about $2 per gram) so large quantities are available for testing. Second, the mechanism of superconductivity in MgB_2 is similar to the well-understood metal alloy superconductors at 4 K. Third, it is much easier to fabricate this compound, that is, to make it into wires or thin films. With further research effort, it is hoped that someday soon different types of high-temperature superconductors will be used to build supercomputers, whose speeds are limited by how fast electric current flows, more powerful particle accelerators, efficient devices for nuclear fusion, and more accurate magnetic resonance imaging (MRI) machines for medical use. The progress in high-temperature superconductors is just warming up!

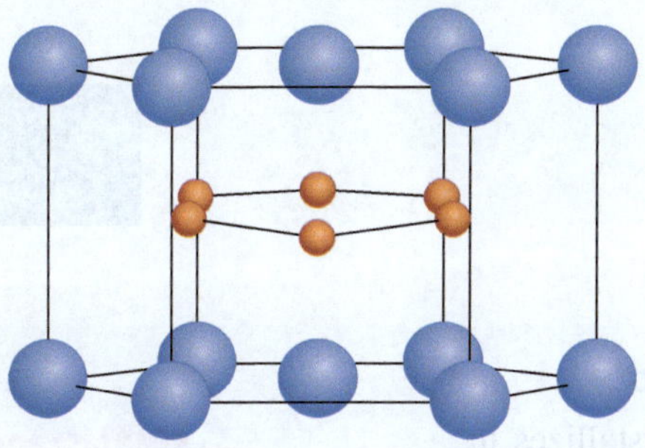

Crystal structure of MgB_2. The Mg atoms (blue) form a hexagonal layer, while the B atoms (gold) form a graphitelike honeycomb layer.

An experimental levitation train that operates on superconducting material at the temperature of liquid helium.
Imaginechina/AP Images

Table 11.5 Types of Crystals and General Properties

Type of Crystal	Force(s) Holding the Units Together	General Properties	Examples
Ionic	Electrostatic attraction	Hard, brittle, high melting point, poor conductor of heat and electricity	NaCl, LiF, MgO, $CaCO_3$
Covalent	Covalent bond	Hard, high melting point, poor conductor of heat and electricity	C (diamond),† SiO_2 (quartz)
Molecular*	Dispersion forces, dipole-dipole forces, hydrogen bonds	Soft, low melting point, poor conductor of heat and electricity	Ar, CO_2, I_2, H_2O, $C_{12}H_{22}O_{11}$ (sucrose)
Metallic	Metallic bond	Soft to hard, low to high melting point, good conductor of heat and electricity	All metallic elements; for example, Na, Mg, Fe, Cu

*Included in this category are crystals made up of individual atoms.
†Diamond is a good thermal conductor.

CHEMISTRY *in Action*

And All for the Want of a Button

In June 1812, Napoleon's mighty army, some 600,000 strong, marched into Russia. By early December, however, his forces were reduced to fewer than 10,000 men. An intriguing theory for Napoleon's defeat has to do with the tin buttons on his soldiers' coats! Tin has two allotropic forms called α (gray tin) and β (white tin). White tin, which has a cubic structure and a shiny metallic appearance, is stable at room temperature and above. Below 13°C, it slowly changes into gray tin. The random growth of the microcrystals of gray tin, which has a tetragonal structure, weakens the metal and makes it crumble. Thus, in the severe Russian winter, the soldiers were probably more busy holding their coats together with their hands than carrying weapons.

Actually, the so-called "tin disease" has been known for centuries. In the unheated cathedrals of medieval Europe, organ pipes made of tin were found to crumble as a result of the allotropic transition from white tin to gray tin. It is puzzling, therefore, that Napoleon, a great believer in keeping his troops fit for battle, would permit the use of tin for buttons. The tin story, if true, could be paraphrased in the Old English Nursery Rhyme: "And all for the want of a button."

Is Napoleon trying to instruct his soldiers how to keep their coats tight?
Antoine-Jean Gros (French painter 1771-1835)/Photos.com/360/Getty Images

Review of Concepts & Facts

11.6.1 Lithium fluoride (LiF) crystallizes in a face-centered cubic cell (same as NaCl). What is the density of LiF if the edge length of the unit cell is 402 pm?

11.6.2 Shown here is a zinc oxide unit cell. How many Zn^{2+} and O^{2-} ions are in the unit cell? What is the formula of the compound?

11.7 Amorphous Solids

Learning Objective

- Describe the nature of an amorphous solid.

Solids are most stable in crystalline form. However, if a solid is formed rapidly (for example, when a liquid is cooled quickly), its atoms or molecules do not have time to align themselves and may become locked in positions other than those of a regular crystal. The resulting solid is said to be *amorphous*. ***Amorphous solids,*** such as glass, *lack a regular three-dimensional arrangement of atoms.* In this section, we will discuss briefly the properties of glass.

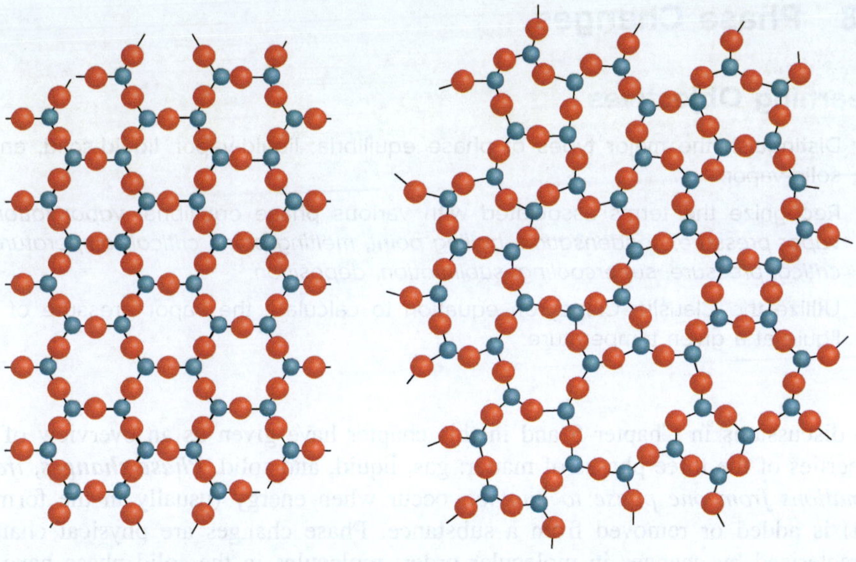

Figure 11.32 *Two-dimensional representation of (a) crystalline quartz and (b) noncrystalline quartz glass. The small spheres represent silicon. In reality, the structure of quartz is three-dimensional. Each Si atom is tetrahedrally bonded to four O atoms.*

Glass is one of civilization's most valuable and versatile materials. It is also one of the oldest—glass articles date back as far as 1000 B.C. ***Glass*** commonly refers to *an optically transparent fusion product of inorganic materials that has cooled to a rigid state without crystallizing.* By fusion product we mean that the glass is formed by mixing molten silicon dioxide (SiO_2), its chief component, with compounds such as sodium oxide (Na_2O), boron oxide (B_2O_3), and certain transition metal oxides for color and other properties. In some respects glass behaves more like a liquid than a solid. X-ray diffraction studies show that glass lacks long-range periodic order.

There are about 800 different types of glass in common use today. Figure 11.32 shows two-dimensional schematic representations of crystalline quartz and amorphous quartz glass. Table 11.6 shows the composition and properties of quartz, Pyrex, and soda-lime glass.

The color of glass is due largely to the presence of metal ions (as oxides). For example, green glass contains iron(III) oxide, Fe_2O_3, or copper(II) oxide, CuO; yellow glass contains uranium(IV) oxide, UO_2; blue glass contains cobalt(II) and copper(II) oxides, CoO and CuO; and red glass contains small particles of gold and copper. Note that most of the ions mentioned here are derived from the transition metals.

Summary of Concepts & Facts

- Solids without a regular structure are amorphous. Glass is an example of an amorphous solid.

Table 11.6 **Composition and Properties of Three Types of Glass**

Name	Composition	Properties and Uses
Pure quartz glass	100% SiO_2	Low thermal expansion, transparent to wide range of wavelengths. Used in optical research.
Pyrex glass	SiO_2, 60–80% B_2O_3, 10–25% Al_2O_3, small amount	Low thermal expansion; transparent to visible and infrared, but not to UV radiation. Used mainly in laboratory and household cooking glassware.
Soda-lime glass	SiO_2, 75% Na_2O, 15% CaO, 10%	Easily attacked by chemicals and sensitive to thermal shocks. Transmits visible light, but absorbs UV radiation. Used mainly in windows and bottles.

11.8 Phase Changes

Learning Objectives

- Distinguish the major types of phase equilibria: liquid-vapor, liquid-solid, and solid-vapor.
- Recognize the terms associated with various phase equilibria: *vaporization, vapor pressure, condensation, boiling point, melting point, critical temperature, critical pressure, supercooling, sublimation, deposition.*
- Utilize the Clausius-Clapeyron equation to calculate the vapor pressure of a liquid at a given temperature.

The discussions in Chapter 5 and in this chapter have given us an overview of the properties of the three phases of matter: gas, liquid, and solid. ***Phase changes,*** *transformations from one phase to another,* occur when energy (usually in the form of heat) is added or removed from a substance. Phase changes are physical changes characterized by changes in molecular order; molecules in the solid phase have the greatest order, and those in the gas phase have the greatest randomness. Keeping in mind the relationship between energy change and the increase or decrease in molecular order will help us understand the nature of these physical changes.

Liquid-Vapor Equilibrium

Molecules in a liquid are not fixed in a rigid lattice. Although they lack the total freedom of gaseous molecules, these molecules are in constant motion. Because liquids are denser than gases, the collision rate among molecules is much higher in the liquid phase than in the gas phase. When the molecules in a liquid have sufficient energy to escape from the surface a phase change occurs. ***Evaporation,*** or ***vaporization,*** is *the process in which a liquid is transformed into a gas.*

How does evaporation depend on temperature? Figure 11.33 shows the kinetic energy distribution of molecules in a liquid at two different temperatures. As we can see, the higher the temperature, the greater the kinetic energy, and hence more molecules leave the liquid.

Vapor Pressure

When a liquid evaporates, its gaseous molecules exert a vapor pressure (recall the difference between a gas and a vapor in Section 5.1). Consider the apparatus shown in Figure 11.34. Before the evaporation process starts, the mercury levels in the U-shaped manometer tube are equal. As soon as some molecules leave the

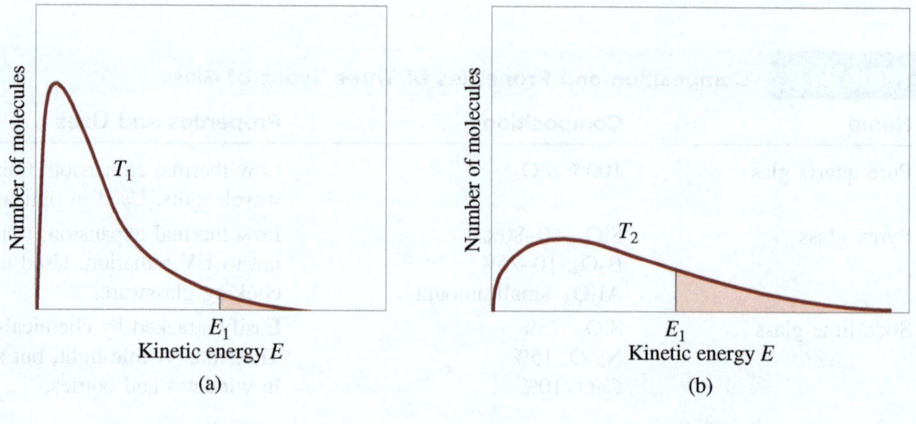

Figure 11.33 *Kinetic energy distribution curves for molecules in a liquid (a) at a temperature T_1 and (b) at a higher temperature T_2. Note that at the higher temperature the curve flattens out. The shaded areas represent the number of molecules possessing kinetic energy equal to or greater than a certain kinetic energy E_1. The higher the temperature, the greater the number of molecules with high kinetic energy.*

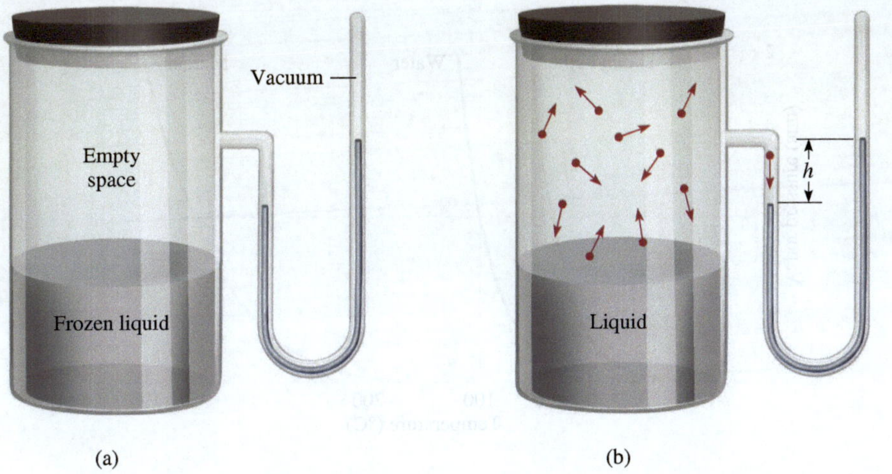

Figure 11.34 *Apparatus for measuring the vapor pressure of a liquid. (a) Initially the liquid is frozen so there are no molecules in the vapor phase. (b) On heating, a liquid phase is formed and vaporization begins. At equilibrium, the number of molecules leaving the liquid is equal to the number of molecules returning to the liquid. The difference in the mercury levels (h) gives the equilibrium vapor pressure of the liquid at the specified temperature.*

liquid, a vapor phase is established. The vapor pressure is measurable only when a fair amount of vapor is present. The process of evaporation does not continue indefinitely, however. Eventually, the mercury levels stabilize and no further changes are seen.

What happens at the molecular level during evaporation? In the beginning, the traffic is only one way: Molecules are moving from the liquid to the empty space. Soon the molecules in the space above the liquid establish a vapor phase. As the concentration of molecules in the vapor phase increases, some molecules *condense;* that is, they return to the liquid phase. ***Condensation,*** *the change from the gas phase to the liquid phase,* occurs because a molecule strikes the liquid surface and becomes trapped by intermolecular forces in the liquid.

The rate of evaporation is constant at any given temperature, and the rate of condensation increases with the increasing concentration of molecules in the vapor phase. A state of ***dynamic equilibrium,*** in which *the rate of a forward process is exactly balanced by the rate of the reverse process,* is reached when the rates of condensation and evaporation become equal (Figure 11.35). The ***equilibrium vapor pressure*** is *the vapor pressure measured when a dynamic equilibrium exists between condensation and evaporation.* We often use the simpler term "vapor pressure" when we talk about the equilibrium vapor pressure of a liquid. This practice is acceptable as long as we know the meaning of the abbreviated term.

Video: Equilibrium Vapor Pressure

It is important to note that the equilibrium vapor pressure is the *maximum* vapor pressure of a liquid at a given temperature and that it is constant at a constant temperature. (It is independent of the amount of liquid as long as there is some liquid present.) From the foregoing discussion we expect the vapor pressure of a liquid to increase with temperature. Plots of vapor pressure versus temperature for three different liquids in Figure 11.36 confirm this expectation.

Student Hot Spot
Student data indicate you may struggle with vapor pressure. Access your eBook for additional Learning Resources on this topic.

Molar Heat of Vaporization and Boiling Point

A measure of the strength of intermolecular forces in a liquid is the ***molar heat of vaporization*** (ΔH_{vap}), defined as *the energy* (usually in kilojoules) *required to vaporize 1 mole of a liquid.* The molar heat of vaporization is directly related to the strength of intermolecular forces that exist in the liquid. If the intermolecular attraction is strong, it takes a lot of energy to free the molecules from the liquid phase and the molar heat of vaporization will be high. Such liquids will also have a low vapor pressure.

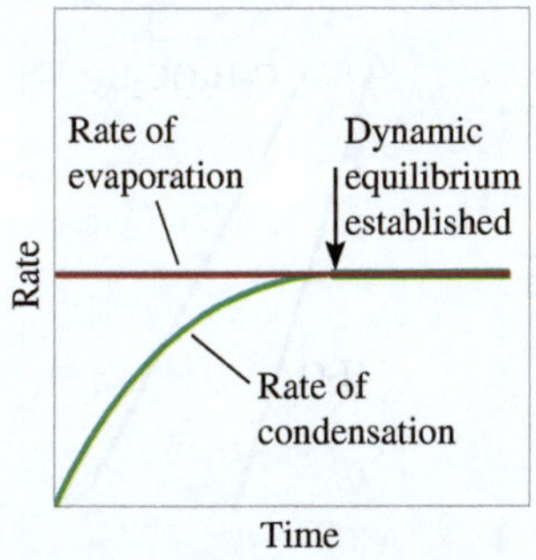

Figure 11.35 *Comparison of the rates of evaporation and condensation as the system approaches equilibrium at constant temperature.*

The previous discussion predicts that the equilibrium vapor pressure (P) of a liquid should increase with increasing temperature, as shown in Figure 11.36. Analysis of this behavior reveals that the quantitative relationship between the

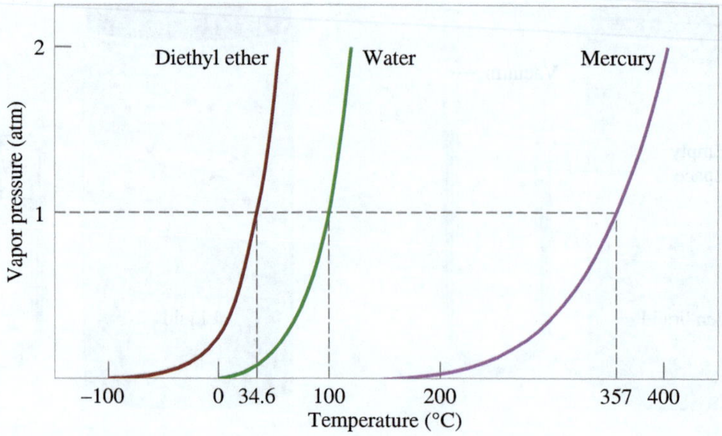

Figure 11.36 *The increase in vapor pressure with temperature for three liquids. The normal boiling points of the liquids (at 1 atm) are shown on the horizontal axis. The strong metallic bonding in mercury results in a much lower vapor pressure of the liquid at room temperature.*

vapor pressure P of a liquid and the absolute temperature T is given by the Clausius[†]-Clapeyron[‡] equation

$$\ln P = -\frac{\Delta H_{vap}}{RT} + C \tag{11.2}$$

where ln is the natural logarithm, R is the gas constant (8.314 J/K · mol), and C is a constant. The Clausius-Clapeyron equation has the form of the linear equation $y = mx + b$:

$$\ln P = \left(-\frac{\Delta H_{vap}}{R}\right)\left(\frac{1}{T}\right) + C$$

$$\updownarrow \qquad \updownarrow \qquad \updownarrow \quad \updownarrow$$

$$y = \qquad m \qquad x \; + \; b$$

By measuring the vapor pressure of a liquid at different temperatures (see Figure 11.36) and plotting ln P versus $1/T$, we determine the slope, which is equal to $-\Delta H_{vap}/R$. (ΔH_{vap} is assumed to be independent of temperature.) This method is used to determine heats of vaporization (Table 11.7). Figure 11.37 shows plots of ln P versus $1/T$ for water and diethyl ether. Note that the straight line for water has a steeper slope because water has a larger ΔH_{vap}.

If we know the values of ΔH_{vap} and P of a liquid at one temperature, we can use the Clausius-Clapeyron equation to calculate the vapor pressure of the liquid at a

[†]Rudolf Julius Emanuel Clausius (1822–1888). German physicist. Clausius's work was mainly in electricity, kinetic theory of gases, and thermodynamics.

[‡]Benoit Paul Emile Clapeyron (1799–1864). French engineer. Clapeyron made contributions to the thermodynamic aspects of steam engines.

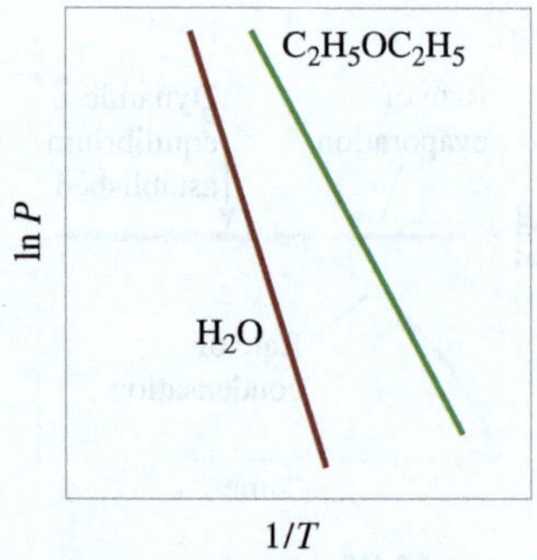

Figure 11.37 *Plots of ln P versus 1/T for water and diethyl ether. The slope in each case is equal to $-\Delta H_{vap}/R$.*

Table 11.7 Molar Heats of Vaporization for Selected Liquids

Substance	Boiling Point* (°C)	ΔH_{vap} (kJ/mol)
Argon (Ar)	−186	6.4
Benzene (C_6H_6)	80.1	30.7
Diethyl ether ($C_2H_5OC_2H_5$)	34.4	26.5
Ethanol ($C_2H_5OH_2$)	78.2	38.6
Mercury (Hg)	357	59.1
Methane (CH_4)	−161.5	8.2
Water (H_2O)	100	40.65

*Measured at 1 atm.

different temperature. At temperatures T_1 and T_2, the vapor pressures are P_1 and P_2. From Equation (11.2) we can write

$$\ln P_1 = -\frac{\Delta H_{vap}}{RT_1} + C \quad (11.3)$$

$$\ln P_2 = -\frac{\Delta H_{vap}}{RT_2} + C \quad (11.4)$$

Subtracting Equation (11.4) from Equation (11.3) we obtain

$$\ln P_1 - \ln P_2 = -\frac{\Delta H_{vap}}{RT_1} - \left(-\frac{\Delta H_{vap}}{RT_2}\right) = \frac{\Delta H_{vap}}{R}\left(\frac{1}{T_2} - \frac{1}{T_1}\right)$$

Hence,

$$\ln \frac{P_1}{P_2} = \frac{\Delta H_{vap}}{R}\left(\frac{1}{T_2} - \frac{1}{T_1}\right)$$

or

$$\ln \frac{P_1}{P_2} = \frac{\Delta H_{vap}}{R}\left(\frac{T_1 - T_2}{T_1 T_2}\right) \quad (11.5)$$

Example 11.7 illustrates the use of Equation (11.5).

Example 11.7

Diethyl ether is a volatile, highly flammable organic liquid that is used mainly as a solvent. The vapor pressure of diethyl ether is 401 mmHg at 18°C. Calculate its vapor pressure at 32°C.

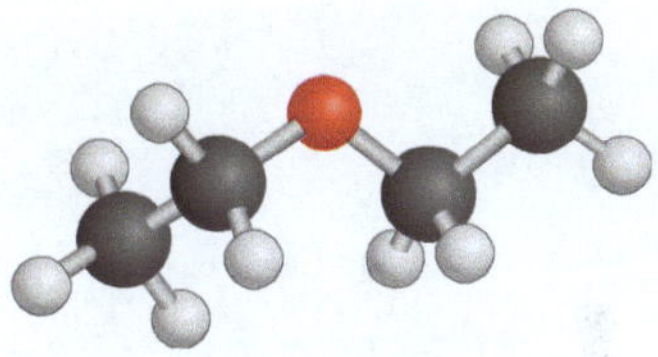

$C_2H_5OC_2H_5$

Strategy We are given the vapor pressure of diethyl ether at one temperature and asked to find the pressure at another temperature. Therefore, we need Equation (11.5).

Solution Table 11.7 tells us that ΔH_{vap} = 26.5 kJ/mol. The data are

$$P_1 = 401 \text{ mmHg} \qquad P_2 = ?$$
$$T_1 = 18°C = 291 \text{ K} \qquad T_2 = 32°C = 305 \text{ K}$$

From Equation (11.5) we have

$$\ln \frac{401}{P_2} = \frac{26{,}500 \text{ J/mol}}{8.314 \text{ J/K} \cdot \text{mol}}\left[\frac{291 \text{ K} - 305 \text{ K}}{(291 \text{ K})(305 \text{ K})}\right] = -0.503$$

Taking the antilog of both sides (see Appendix 5), we obtain

$$\frac{401}{P_2} = e^{-0.503} = 0.605$$

Hence,

$$P_2 = 663 \text{ mmHg}$$

Check We expect the vapor pressure to be greater at the higher temperature. Therefore, the answer is reasonable.

Practice Exercise The vapor pressure of ethanol is 100 mmHg at 34.9°C. What is its vapor pressure at 63.5°C? (ΔH_{vap} for ethanol is 38.6 kJ/mol.)

Similar problem: 11.84.

A practical way to demonstrate the molar heat of vaporization is by rubbing an alcohol such as ethanol (C_2H_5OH) or isopropanol (C_3H_7OH), or rubbing alcohol, on your hands. These alcohols have a lower ΔH_{vap} than water, so that the heat from your hands is enough to increase the kinetic energy of the alcohol molecules and evaporate them. As a result of the loss of heat, your hands feel cool. This process is similar to perspiration, which is one of the means by which the human body maintains a constant temperature. Because of the strong intermolecular hydrogen bonding that exists in water, a considerable amount of energy is needed to vaporize the water in perspiration from the body's surface. This energy is supplied by the heat generated in various metabolic processes.

You have already seen that the vapor pressure of a liquid increases with temperature. Every liquid has a temperature at which it begins to boil. The ***boiling point*** is *the temperature at which the vapor pressure of a liquid is equal to the external pressure.* The *normal* boiling point of a liquid is the temperature at which it boils when the external pressure is 1 atm.

At the boiling point, bubbles form within the liquid. When a bubble forms, the liquid originally occupying that space is pushed aside, and the level of the liquid in the container is forced to rise. The pressure exerted *on* the bubble is largely atmospheric pressure, plus some *hydrostatic pressure* (that is, pressure due to the presence of liquid). The pressure *inside* the bubble is due solely to the vapor pressure of the liquid. When the vapor pressure becomes equal to the external pressure, the bubble rises to the surface of the liquid and bursts. If the vapor pressure in the bubble were lower than the external pressure, the bubble would collapse before it could rise. We can thus conclude that the boiling point of a liquid depends on the external pressure. (We usually ignore the small contribution due to the hydrostatic pressure.) For example, at 1 atm, water boils at 100°C, but if the pressure is reduced to 0.5 atm, water boils at only 82°C.

Because the boiling point is defined in terms of the vapor pressure of the liquid, we expect the boiling point to be related to the molar heat of vaporization: The higher ΔH_{vap}, the higher the boiling point. The data in Table 11.7 roughly confirm our prediction. Ultimately, both the boiling point and ΔH_{vap} are determined by the strength of intermolecular forces. For example, argon (Ar) and methane (CH_4), which have weak dispersion forces, have low boiling points and small molar heats of vaporization. Diethyl ether ($C_2H_5OC_2H_5$) has a dipole moment, and the dipole-dipole forces account for its moderately high boiling point and ΔH_{vap}. Both ethanol (C_2H_5OH) and water have strong hydrogen bonding, which accounts for their high boiling points and large ΔH_{vap} values. Strong metallic bonding causes mercury to have the highest boiling point and ΔH_{vap} of this group of liquids. Interestingly, the boiling point of benzene, which is nonpolar, is comparable to that of ethanol. Benzene has a high polarizability due to the distribution of its electrons in the delocalized pi molecular orbitals, and the dispersion forces among benzene molecules can be as strong as or even stronger than dipole-dipole forces and/or hydrogen bonds.

Student Hot Spot

Student data indicate you may struggle with boiling point. Access your eBook for additional Learning Resources on this topic.

Critical Temperature and Pressure

The opposite of evaporation is condensation. In principle, a gas can be made to liquefy by either one of two techniques. By cooling a sample of gas we decrease the kinetic energy of its molecules, so that eventually molecules aggregate to form small drops of liquid. Alternatively, we can apply pressure to the gas. Compression reduces the average distance between molecules so that they are held together by mutual attraction. Industrial liquefaction processes combine these two methods.

Every substance has a ***critical temperature*** (T_c), *above which its gas phase cannot be made to liquefy, no matter how great the applied pressure.* This is also *the highest temperature at which a substance can exist as a liquid.* Putting it another way, above the critical temperature there is no fundamental distinction between a liquid and a gas—we simply have a fluid. ***Critical pressure*** (P_c) is *the minimum pressure that must*

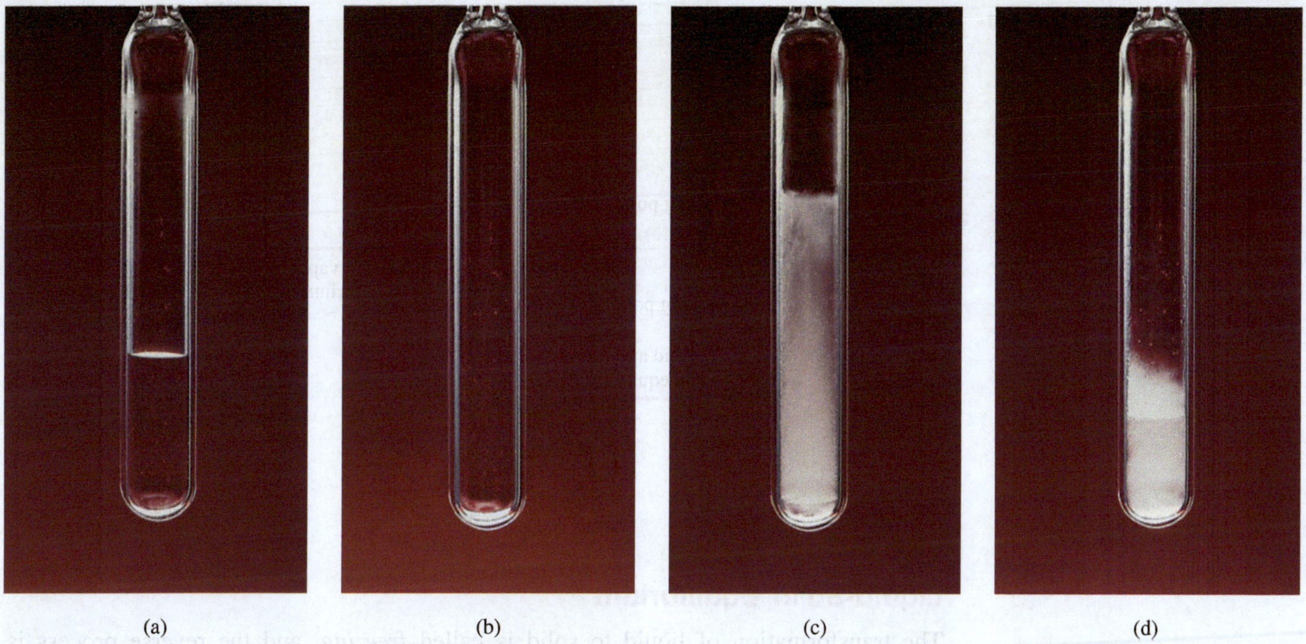

Figure 11.38 *The critical phenomenon of sulfur hexafluoride. (a) Below the critical temperature the clear liquid phase is visible. (b) Above the critical temperature the liquid phase has disappeared. (c) The substance is cooled just below its critical temperature. The fog represents the condensation of vapor. (d) Finally, the liquid phase reappears.*
Ken Karp/McGraw Hill

be applied to bring about liquefaction at the critical temperature. The existence of the critical temperature can be qualitatively explained as follows: Intermolecular attraction is a finite quantity for any given substance and it is independent of temperature. Below T_c, this force is sufficiently strong to hold the molecules together (under some appropriate pressure) in a liquid. Above T_c, molecular motion becomes so energetic that the molecules can break away from this attraction. Figure 11.38 shows what happens when sulfur hexafluoride is heated above its critical temperature (45.5°C) and then cooled down to below 45.5°C.

Table 11.8 lists the critical temperatures and critical pressures of a number of common substances. The critical temperature of a substance reflects the strength of its intermolecular forces. Benzene, ethanol, mercury, and water, which have strong intermolecular forces, also have high critical temperatures compared with the other substances listed in the table.

Table 11.8 **Critical Temperatures and Critical Pressures of Selected Substances**

Substance	T_c (°C)	P_c (atm)
Ammonia (NH_3)	132.4	112.1
Argon (Ar)	−122.5	48.0
Benzene (C_6H_6)	288.9	48.4
Carbon dioxide (CO_2)	31.0	73.0
Diethyl ether ($C_2H_5OC_2H_5$)	193.7	35.9
Ethanol (C_2H_5OH)	242	61.7
Mercury (Hg)	1491	1648
Methane (CH_4)	−82.6	45.6
Molecular hydrogen (H_2)	−249.2	12.7
Molecular nitrogen (N_2)	−147.0	33.5
Molecular oxygen (O_2)	−118.6	49.8
Sulfur hexafluoride (SF_6)	45.6	37.2
Water (H_2O)	374.4	217.7

Figure 11.39 *A typical heating curve, from the solid phase through the liquid phase to the gas phase of a substance. Because ΔH_{fus} is smaller than ΔH_{vap}, a substance melts in less time than it takes to boil. This explains why AB is shorter than CD. The steepness of the solid, liquid, and vapor heating lines is determined by the specific heat of the substance in each state.*

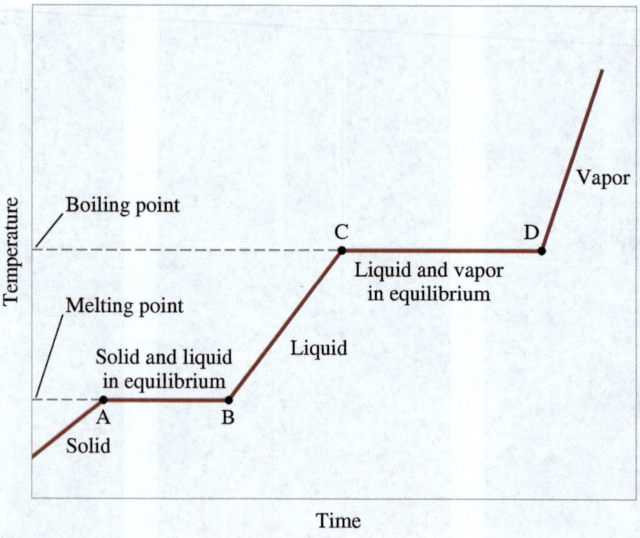

Liquid-Solid Equilibrium

The transformation of liquid to solid is called *freezing,* and the reverse process is called *melting,* or *fusion.* Fusion refers to the process of melting; thus, a "fuse" breaks an electrical circuit when a metallic strip melts due to the heat generated by excessively high electrical current. The ***melting point*** of a solid or the ***freezing point*** of a liquid is *the temperature at which solid and liquid phases coexist in equilibrium.* The *normal* melting (or freezing) point of a substance is the temperature at which a substance melts (or freezes) at 1 atm pressure. We generally omit the word "normal" when the pressure is at 1 atm.

The most familiar liquid-solid equilibrium is that of water and ice. At 0°C and 1 atm, the dynamic equilibrium is represented by

$$\text{ice} \rightleftharpoons \text{water}$$

A practical illustration of this dynamic equilibrium is provided by a glass of ice water. As the ice cubes melt to form water, some of the water between ice cubes may freeze, thus joining the cubes together. This is not a true dynamic equilibrium, however, because the glass is not kept at 0°C; thus, all the ice cubes will eventually melt away.

Figure 11.39 shows how the temperature of a substance changes as it absorbs heat from its surroundings. We see that as a solid is heated, its temperature increases until it reaches its melting point. At this temperature, the average kinetic energy of the molecules has become sufficiently large to begin overcoming the intermolecular forces that hold the molecules together in the solid state. A transition from the solid to liquid phase begins in which the absorption of heat is used to break apart more and more of the molecules in the solid. It is important to note that during this transition ($A \longrightarrow B$) the average kinetic energy of the molecules does not change, so the temperature stays constant. Once the substance has completely melted, further absorption of heat increases its temperature until the boiling point is reached ($B \longrightarrow C$). Here, the transition from the liquid to the gaseous phase occurs ($C \longrightarrow D$) in which the absorbed heat is used to break the intermolecular forces holding the molecules in the liquid phase so the temperature again remains constant. Once this transition has been completed, the temperature of the gas increases on further heating.

Molar heat of fusion (ΔH_{fus}) is *the energy* (usually in kilojoules) *required to melt 1 mole of a solid.* Table 11.9 shows the molar heats of fusion for the substances listed in Table 11.7. A comparison of the data in the two tables shows that for each substance ΔH_{fus} is smaller than ΔH_{vap}. This is consistent with the fact that molecules in a liquid are still fairly closely packed together, so that some energy is needed to bring about

Table 11.9 Molar Heats of Fusion for Selected Substances

Substance	Melting Point* (°C)	ΔH_{fus} (kJ/mol)
Argon (Ar)	−189.3	1.18
Benzene (C_6H_6)	5.5	9.87
Diethyl ether ($C_2H_5OC_2H_5$)	−116.2	7.19
Ethanol (C_2H_5OH)	−114.1	4.93
Mercury (Hg)	−39	2.30
Methane (CH_4)	−182.5	0.94
Water (H_2O)	0	6.01

*Measured at 1 atm.

the rearrangement from solid to liquid. On the other hand, when a liquid evaporates, its molecules become completely separated from one another and considerably more energy is required to overcome the attractive force.

As we would expect, *cooling* a substance has the opposite effect of heating it. If we remove heat from a gas sample at a steady rate, its temperature decreases. As the liquid is being formed, heat is given off by the system, because its potential energy is decreasing. For this reason, the temperature of the system remains constant over the condensation period (D $\longrightarrow$ C). After all the vapor has condensed, the temperature of the liquid begins to drop. Continued cooling of the liquid finally leads to freezing (B $\longrightarrow$ A).

The phenomenon known as ***supercooling*** refers to the situation in which *a liquid can be temporarily cooled to below its freezing point.* Supercooling occurs when heat is removed from a liquid so rapidly that the molecules literally have no time to assume the ordered structure of a solid. A supercooled liquid is unstable; gentle stirring or the addition to it of a small "seed" crystal of the same substance will cause it to solidify quickly.

Solid-Vapor Equilibrium

Solids, too, undergo evaporation and, therefore, possess a vapor pressure. Consider the following dynamic equilibrium:

$$\text{solid} \rightleftharpoons \text{vapor}$$

Sublimation is *the process in which molecules go directly from the solid into the vapor phase.* ***Deposition*** is the reverse process, that is, *molecules make the transition from vapor to solid directly.* Naphthalene, which is the substance used to make mothballs, has a fairly high (equilibrium) vapor pressure for a solid (1 mmHg at 53°C); thus, its pungent vapor quickly permeates an enclosed space. Iodine also sublimes. Above room temperature, the violet color of iodine vapor is easily visible in a closed container.

Solid iodine in equilibrium with its vapor.
Ken Karp/McGraw Hill

Because molecules are more tightly held in a solid, the vapor pressure of a solid is generally much less than that of the corresponding liquid. ***Molar heat of sublimation*** (**ΔH_{sub}**) of a substance is *the energy* (usually in kilojoules) *required to sublime 1 mole of a solid.* It is equal to the sum of the molar heats of fusion and vaporization:

$$\Delta H_{sub} = \Delta H_{fus} + \Delta H_{vap} \tag{11.6}$$

Equation (11.6) is an illustration of Hess's law (see Section 6.6). The enthalpy, or heat change, for the overall process is the same whether the substance changes directly from the solid to the vapor form or from the solid to the liquid

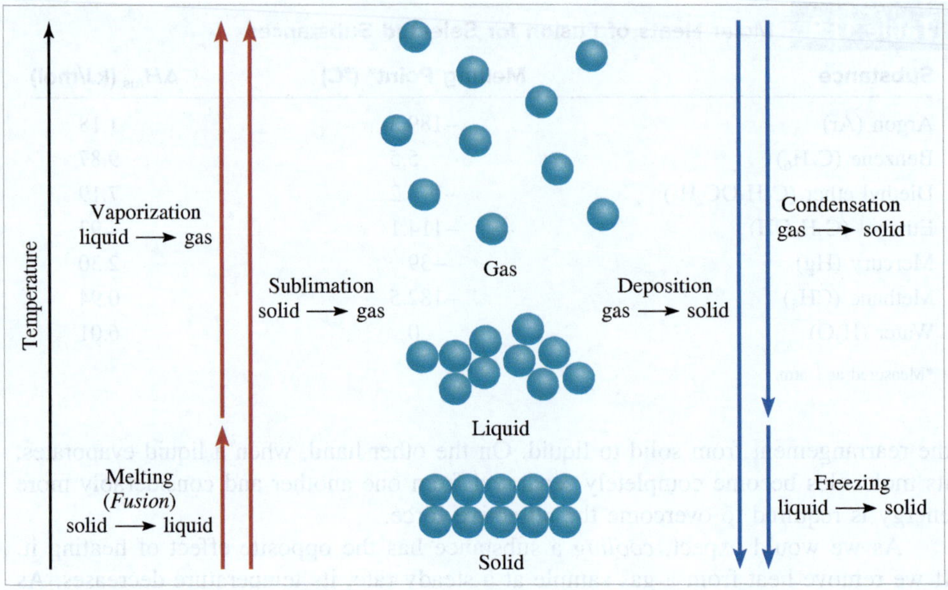

Figure 11.40 *The various phase changes that a substance can undergo.*

and then to the vapor. Note that Equation (11.6) holds only if all the phase changes occur at the *same* temperature. If not, the equation can be used only as an approximation.

Figure 11.40 summarizes the types of phase changes discussed in this section. When a substance is heated, its temperature will rise and eventually it will undergo a phase transition. To calculate the total energy change for such a process we must include all of the steps, shown in Example 11.8.

> **Student Hot Spot**
> Student data indicate you may struggle with enthalpy changes for phase changes. Access your eBook for additional Learning Resources on this topic.

Example 11.8

Calculate the amount of energy (in kilojoules) needed to heat 346 g of liquid water from 0°C to 182°C. Assume that the specific heat of water is 4.184 J/g · °C over the entire liquid range and that the specific heat of steam is 1.99 J/g · °C.

Strategy The heat change (q) at each stage is given by $q = ms\Delta t$ (see Section 6.5), where m is the mass of water, s is the specific heat, and Δt is the temperature change. If there is a phase change, such as vaporization, then q is given by $n\Delta H_{vap}$, where n is the number of moles of water.

Solution The calculation can be broken down in three steps.

Step 1: Heating water from 0°C to 100°C
Using Equation (6.12) we write

$$\begin{aligned} q_1 &= ms\Delta t \\ &= (346\ \text{g})(4.184\ \text{J/g}\cdot{}^\circ\text{C})(100^\circ\text{C} - 0^\circ\text{C}) \\ &= 1.45 \times 10^5\ \text{J} \\ &= 145\ \text{kJ} \end{aligned}$$

Step 2: Evaporating 346 g of water at 100°C (a phase change)
In Table 11.7 we see ΔH_{vap} = 40.65 kJ/mol for water, so

$$\begin{aligned} q_2 &= 346\ \text{g H}_2\text{O} \times \frac{1\ \text{mol H}_2\text{O}}{18.02\ \text{g H}_2\text{O}} \times \frac{40.65\ \text{kJ}}{1\ \text{mol H}_2\text{O}} \\ &= 781\ \text{kJ} \end{aligned}$$

(Continued)

Step 3: Heating steam from 100°C to 182°C

$$q_3 = ms\Delta t$$
$$= (346\ \text{g})(1.99\ \text{J/g} \cdot {}^\circ\text{C})(182^\circ\text{C} - 100^\circ\text{C})$$
$$= 5.65 \times 10^4\ \text{J}$$
$$= 56.5\ \text{kJ}$$

The overall energy required is given by

$$q_{\text{overall}} = q_1 + q_2 + q_3$$
$$= 145\ \text{kJ} + 781\ \text{kJ} + 56.5\ \text{kJ}$$
$$= 983\ \text{kJ}$$

Check All the qs have a positive sign, which is consistent with the fact that heat is absorbed to raise the temperature from 0°C to 182°C. Also, as expected, much more heat is absorbed during the phase transition.

Practice Exercise Calculate the heat released when 68.0 g of steam at 124°C is converted to water at 45°C.

Similar problem: 11.76.

Summary of Concepts & Facts

- A liquid in a closed vessel eventually establishes a dynamic equilibrium between evaporation and condensation. The vapor pressure over the liquid under these conditions is the equilibrium vapor pressure, which is often referred to simply as "vapor pressure."
- At the boiling point, the vapor pressure of a liquid equals the external pressure. The molar heat of vaporization of a liquid is the energy required to vaporize 1 mole of the liquid. It can be determined by measuring the vapor pressure of the liquid as a function of temperature and using the Clausius-Clapeyron equation. The molar heat of fusion of a solid is the energy required to melt 1 mole of the solid.
- For every substance there is a temperature, called the critical temperature, above which its gas phase cannot be made to liquefy.

Review of Concepts & Facts

11.8.1 A student studies the ln P versus $1/T$ plots for two organic liquids: methanol (CH_3OH) and dimethyl ether (CH_3OCH_3), such as those shown in Figure 11.37. The slopes are -2.32×10^3 K and -4.50×10^3 K, respectively. How should she assign the ΔH_{vap} values to these two compounds?

11.8.2 The vapor pressure of hexafluorobenzene is 92.47 mmHg at 27°C. If ΔH_{vap} of hexafluorobenzene is 35.5 kJ/mol, calculate the vapor pressure at 47°C.

11.9 Phase Diagrams

Learning Objective

- Employ phase diagrams to identify the triple point and to determine the phase of a substance at a given temperature and pressure.

CHEMISTRY *in Action*

Hard-Boiling an Egg on a Mountaintop, Pressure Cookers, and Ice Skating

Phase equilibria are affected by external pressure. Depending on atmospheric conditions, the boiling point and freezing point of water may deviate appreciably from 100°C and 0°C, respectively, as we see in the following discussion.

Hard-Boiling an Egg on a Mountaintop

Suppose you have just scaled Pike's Peak in Colorado. To help regain your strength following the strenuous work, you decide to hard-boil an egg and eat it. To your surprise, water seems to boil more quickly than usual, but after 10 min in boiling water, the egg is still not cooked. A little knowledge of phase equilibria could have saved you the disappointment of cracking open an uncooked egg (especially if it is the only egg you brought with you). The summit of Pike's Peak is 14,115 ft above sea level. At this altitude, the atmospheric pressure is only about 0.6 atm. From Figure 11.42(b), we see that the boiling point of water decreases with decreasing pressure, so at the lower pressure water will boil at about 86°C. However, it is not the boiling action but the amount of heat delivered to the egg that does the actual cooking, and the amount of heat delivered is proportional to the temperature of the water. For this reason, it would take considerably longer, perhaps 30 min, to hard-boil your egg.

Pressure Cookers

The effect of pressure on boiling point also explains why pressure cookers save time in the kitchen. A pressure cooker is a sealed container that allows steam to escape only when it exceeds a certain pressure. The pressure above the water in the cooker is the sum of the atmospheric pressure and the pressure of the steam. Consequently, the water in the pressure cooker will boil at a higher temperature than 100°C and the food in it will be hotter and cook faster.

Ice Skating

Let us now turn to the ice-water equilibrium. The negative slope of the solid-liquid curve means that the melting point of ice decreases with increasing external pressure, as shown in Figure 11.42(b). This phenomenon helps to make ice skating possible. Because skates have very thin runners, a 130-lb person can exert a pressure equivalent to 500 atm on the ice. (Remember that pressure is defined as force per unit area.) Consequently, at a temperature lower than 0°C, the ice under the skates melts and the film of water formed under the runner facilitates the movement of the skater over ice. Calculations show that the melting point of ice decreases by 7.4×10^{-3} °C when the pressure increases by 1 atm. Thus, when the pressure exerted on the ice by the skater is 500 atm, the melting point falls to $-(500 \times 7.4 \times 10^{-3})$, or −3.7°C. Actually, it turns out that friction between the blades and the ice is the major cause for melting the ice. This explains why it is possible to skate outdoors even when the temperature drops below −20°C.

The pressure exerted by the skater on ice lowers its melting point, and the film of water formed under the blades acts as a lubricant between the skate and the ice.
Atsushi Tomura/Getty Images

Video: Phase Diagrams and the States of Matter

The overall relationships among the solid, liquid, and vapor phases are best represented in a single graph known as a phase diagram. A ***phase diagram*** *summarizes the conditions at which a substance exists as a solid, liquid, or gas.* A phase diagram plots where phase changes occur based on changes in pressure and temperature.

A typical phase diagram (Figure 11.41) is divided into three regions, each of which represents a pure phase. The line separating any two regions indicates conditions under which these two phases can exist in equilibrium. These curves are sometimes known as ***phase boundaries.*** For example, the phase boundary between the liquid and vapor

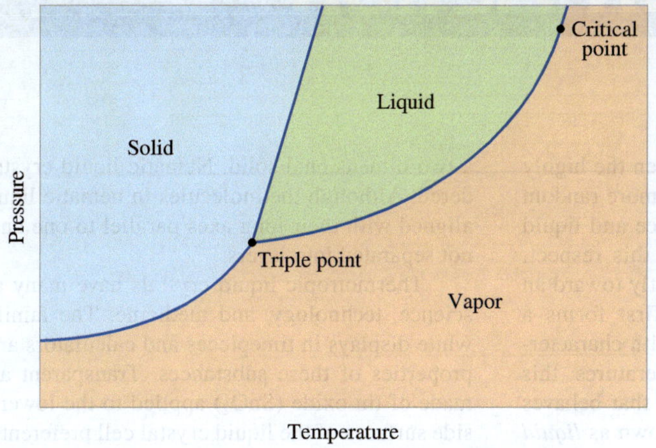

Figure 11.41 *A typical phase diagram. Phase diagrams show which phase of a substance is stable at a particular temperature and pressure. Each curve is a phase boundary The point at which all three phases can exist in equilibrium is the triple point. The critical point is the end of the phase boundary of the liquid and vapor equilibrium and provides the critical temperature and critical pressure of a substance.*

phases shows the variation of vapor pressure with temperature. The other two-phase boundaries similarly indicate conditions for equilibrium between solid and liquid and between solid and vapor. The point at which all three curves meet is called the ***triple point,*** which is *the point at which the vapor, liquid, and solid states of a substance are in equilibrium*. The triple point also represents the lowest temperature at which a substance can exist as a liquid. Where the liquid-vapor phase boundary stops is the ***critical point*** and represents the highest pressure (critical pressure, P_c) and temperature (critical temperature, T_c) at which the substance can exist as a liquid (see Section 11.8).

Phase diagrams enable us to predict changes in the melting point and boiling point of a substance as a result of changes in the external pressure; we can also anticipate directions of phase transitions brought about by changes in temperature and pressure. In this section we will briefly discuss the phase diagrams of water and carbon dioxide.

Water

Figure 11.42(a) shows the phase diagram of water. The normal melting point and boiling point of water at 1 atm are 0°C and 100°C, respectively. What would happen if melting and boiling were carried out at some other pressure? Figure 11.42(b) shows that increasing the pressure above 1 atm will raise the boiling point and lower the melting point. A decrease in pressure will lower the boiling point and raise the melting point. For water, the triple point is at 0.01°C and 0.006 atm. At the critical point of water (not shown on the phase diagram), the critical temperature and pressure are 374.4°C and 219.5 atm, respectively.

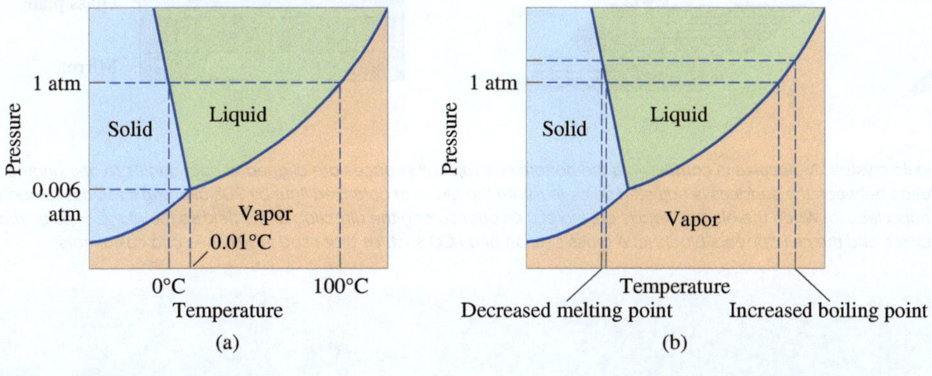

Figure 11.42 *(a) The phase diagram of water. Each solid line between two phases specifies the conditions of pressure and temperature under which the two phases can exist in equilibrium. The point at which all three phases can exist in equilibrium (0.006 atm and 0.01°C) is called the triple point. (b) This phase diagram tells us that increasing the pressure on ice lowers its melting point and that increasing the pressure of liquid water raises its boiling point.*

CHEMISTRY *in Action*

Liquid Crystals

Ordinarily, there is a sharp distinction between the highly ordered state of a crystalline solid and the more random molecular arrangement of liquids. Crystalline ice and liquid water, for example, differ from each other in this respect. One class of substances, however, tends so greatly toward an ordered arrangement that a melting crystal first forms a milky liquid, called the *paracrystalline state,* with characteristically crystalline properties. At higher temperatures, this milky fluid changes sharply into a clear liquid that behaves like an ordinary liquid. Such substances are known as *liquid crystals.*

Molecules that exhibit liquid crystallinity are usually long and rodlike. An important class of liquid crystals is called thermotropic liquid crystals, which form when the solid is heated. The two common structures of thermotropic liquid crystals are nematic and smectic. In smectic liquid crystals, the long axes of the molecules are perpendicular to the plane of the layers. The layers are free to slide over one another so that the substance has the mechanical properties of a two-dimensional solid. Nematic liquid crystals are less ordered. Although the molecules in nematic liquid crystals are aligned with their long axes parallel to one another, they are not separated into layers.

Thermotropic liquid crystals have many applications in science, technology, and medicine. The familiar black-and-white displays in timepieces and calculators are based on the properties of these substances. Transparent aligning agents made of tin oxide (SnO_2) applied to the lower and upper inside surfaces of the liquid crystal cell preferentially orient the molecules in the nematic phase by 90° relative to each other. In this way, the molecules become "twisted" through the liquid crystal phase. When properly adjusted, this twist rotates the plane of polarization by 90° and allows the light to pass through the two polarizers (arranged at 90° to each other). When an electric field is applied, the nematic molecules experience a torque (a torsion or rotation) that forces them to align along the direction of the field. Now the incident polarized light cannot pass through the top polarizer. In

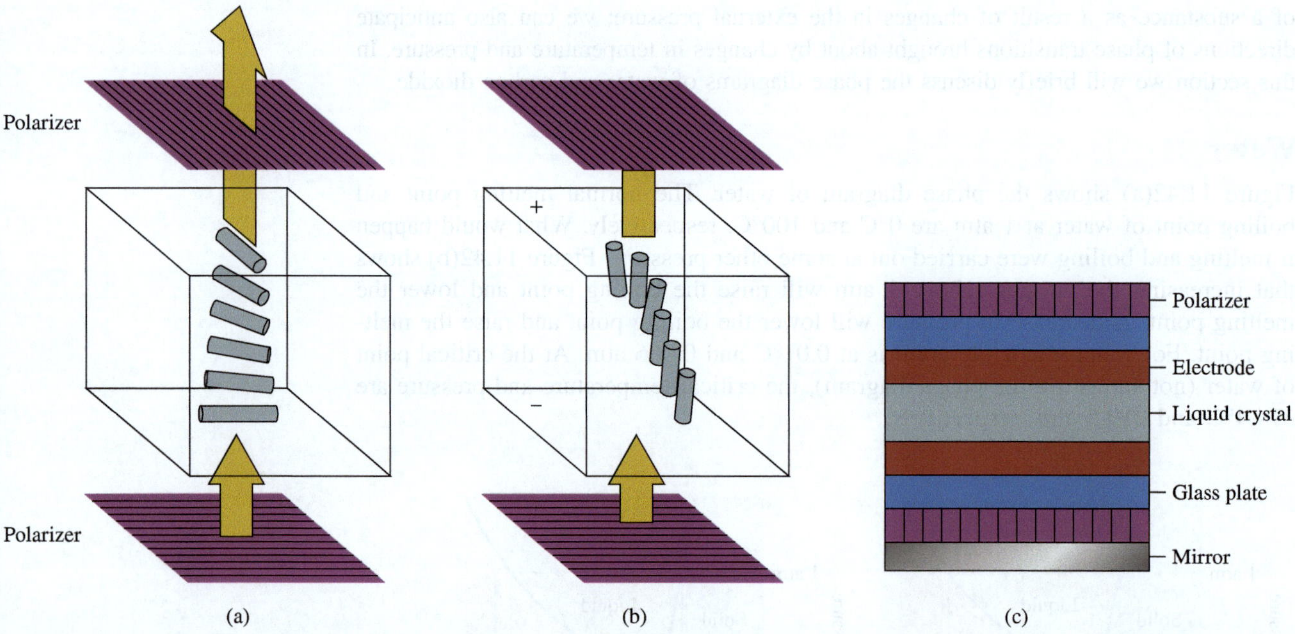

A liquid crystal display (LCD) using nematic liquid crystals. Molecules in contact with the bottom and top cell surfaces are aligned at right angles to one another. (a) The extent of twist in the molecular orientation between the surfaces is adjusted so as to rotate the plane of polarized light by 90°, allowing it to pass through the top polarizer. Consequently, the cell appears clear. (b) When the electric field is on, molecules orient along the direction of the field so the plane of polarized light can no longer pass through the top polarizer, and the cell appears black. (c) A cross section of a LCD such as that used in watches and calculators.

watches and calculators, a mirror is placed under the bottom polarizer. In the absence of an electric field, the reflected light goes through both polarizers and the cell looks clear from the top. When the electric field is turned on, the incident light from the top cannot pass through the bottom polarizer to reach the reflector and the cell becomes dark. Typically a few volts are applied across a nematic layer about 10 μm thick (1 μm = 10^{-6} m). The response time for molecules to align and relax when the electric field is turned on and off is in the ms range (1 ms = 10^{-3} s).

Another type of thermotropic liquid crystals is called cholesteric liquid crystals. The color of cholesteric liquid crystals changes with temperature and therefore they are suitable for use as sensitive thermometers. In metallurgy, for example, they are used to detect metal stress, heat sources, and conduction paths. Medically, the temperature of the body at specific sites can be determined with the aid of liquid crystals. This technique has become an important diagnostic tool in treating infection and tumor growth (for example, breast tumors). Because localized infections and tumors increase metabolic rate and hence temperature in the affected tissues, a thin film of liquid crystal can help a physician see whether an infection or tumor is present by responding to a temperature difference with a change of color.

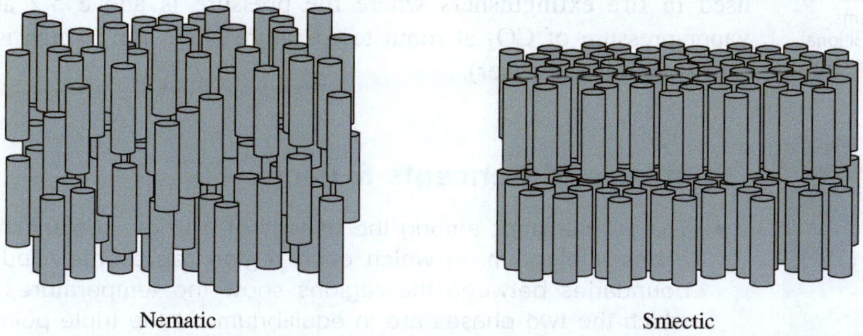

The alignment of molecules in two types of liquid crystals. Nematic liquid crystals behave like a one-dimensional solid and smectic liquid crystals behave like a two-dimensional solid.

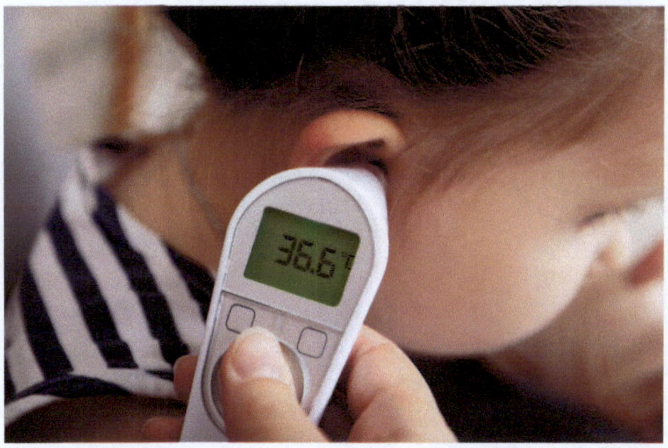

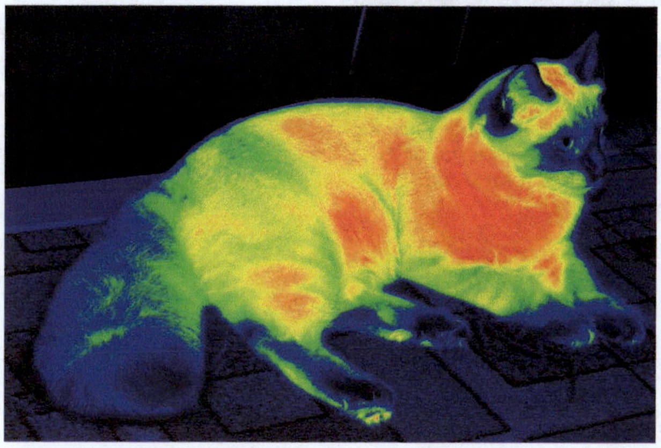

A liquid crystal thermogram. The red color represents the highest temperature and the blue color the lowest temperature.
(left): AndreyPopov/iStock/Getty Images; (right): PandaWild/Shutterstock

Note that the solid-liquid phase boundary for water has a negative slope. The negative slope for the solid-liquid phase boundary is due to the fact that the molar volume of ice is greater than that of liquid water, hence water is denser than ice. An increase in pressure favors the liquid phase in this case.

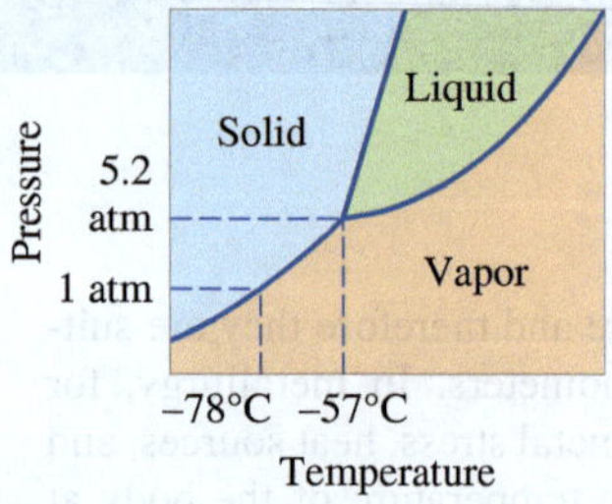

Figure 11.43 *The phase diagram of carbon dioxide. Note that the solid-liquid boundary line has a positive slope. The liquid phase is not stable below 5.2 atm, so that only the solid and vapor phases can exist under atmospheric conditions.*

Carbon Dioxide

The phase diagram of carbon dioxide (Figure 11.43) is generally similar to that of water, with one important exception—the slope of the curve between solid and liquid is positive. In fact, this holds true for almost all other substances. Water behaves differently because ice is *less* dense than liquid water. The triple point of carbon dioxide is at 5.2 atm and −57°C.

An interesting observation can be made about the phase diagram in Figure 11.43. As you can see, the entire liquid phase lies well above atmospheric pressure; therefore, it is impossible for solid carbon dioxide to melt at 1 atm. Instead, when solid CO_2 is heated to −78°C at 1 atm, it sublimes. In fact, solid carbon dioxide is called dry ice because it looks like ice and *does not melt* (Figure 11.44). Because of this property, dry ice is useful as a refrigerant. Liquid CO_2 is often used in fire extinguishers where the pressure is above 5.2 atm (the equilibrium vapor pressure of CO_2 at room temperature is 67 atm, which is the pressure inside a CO_2 fire extinguisher).

Student Hot Spot

Student data indicate you may struggle with phase diagrams. Access your eBook for additional Learning Resources on this topic.

Figure 11.44 *Under atmospheric conditions, solid carbon dioxide does not melt; it can only sublime. The cold carbon dioxide gas causes nearby water vapor to condense and form a fog.*
Ken Karp/McGraw Hill

Summary of Concepts & Facts

- The relationships among the phases of a single substance are illustrated by a phase diagram, in which each region represents a pure phase and the boundaries between the regions show the temperatures and pressures at which the two phases are in equilibrium. At the triple point, all three phases are in equilibrium.

Review of Concepts & Facts

11.9.1 Which phase diagram (a)–(c) corresponds to a substance that will sublime rather than melt as it is heated at 1 atm?

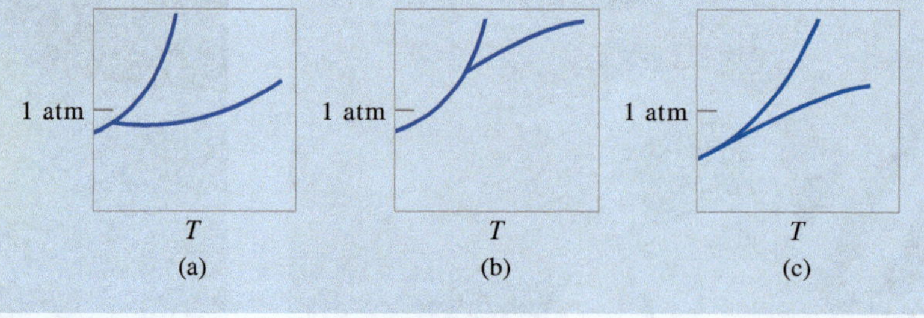

Chapter Summary

Intermolecular Forces Intermolecular forces, which are responsible for the nonideal behavior of gases, also account for the existence of the condensed states of matter—liquids and solids. They exist between polar molecules, between ions and polar molecules, and between nonpolar molecules. A special type of intermolecular force, called the hydrogen bond, describes the interaction between the hydrogen atom in a polar bond and an electronegative atom such as O, N, or F. (Section 11.2)

The Liquid State Liquids tend to assume the shapes of their containers. The surface tension of a liquid is the energy required to increase its surface area. It manifests itself in capillary action, which is responsible for the rise (or depression) of a liquid in a narrow tubing. Viscosity is a measure of a liquid's resistance to flow. It always decreases with increasing temperature. The structure of water is unique in that its solid state (ice) is less dense than its liquid state. (Section 11.3)

The Crystalline State A crystalline solid possesses rigid and long-range order. Different crystal structures can be generated by packing identical spheres in three dimensions. X-ray diffraction is a technique that helps determine the dimensions of a crystal structure. This technique is based on the scattering of X rays by atoms or molecules in a crystal. Some solids can also exist in the amorphous form, which lacks orderly three-dimensional arrangement. A well-known example of an amorphous solid is glass. (Sections 11.4, 11.5, 11.7)

Bonding in Solids Atoms, molecules, or ions are held in a solid by different types of bonding. Electrostatic forces are responsible for ionic solids, intermolecular forces are responsible for molecular solids, covalent bonds are responsible for covalent solids, and a special type of interaction, which involves electrons being delocalized over the entire crystal, accounts for the existence of metals. Intermolecular forces help us understand the structure and physical properties such as density, melting point, and electrical conductivity. (Section 11.6)

Phase Transitions The states of matter can be interconverted by heating or cooling. Two phases are in equilibrium at the transition temperature such as boiling or freezing. Solids can also be directly converted to vapor by sublimation. Above a certain temperature, called the critical temperature, the gas of a substance cannot be made to liquefy. The pressure-temperature relationships of solid, liquid, and vapor phases are best represented by a phase diagram. Phase diagrams help us understand the conditions under which a phase is stable and changes in pressure and/or temperature needed to bring about a phase transition. (Sections 11.8, 11.9)

Key Equations

$2d \sin \theta = n\lambda$ (11.1)	Bragg equation for calculating the distance between planes of atoms in a crystal lattice.
$\ln P = -\frac{\Delta H_{vap}}{RT} + C$ (11.2)	Clausius-Clapeyron equation for determining ΔH_{vap} of a liquid.
$\ln \frac{P_1}{P_2} = \frac{\Delta H_{vap}}{R}\left(\frac{T_1 - T_2}{T_1 T_2}\right)$ (11.5)	For calculating ΔH_{vap}, vapor pressure, or boiling point of a liquid.
$\Delta H_{sub} = \Delta H_{fus} + \Delta H_{vap}$ (11.6)	Application of Hess's law.

Key Words

Adhesion, p. 481
Amorphous solid, p. 500
Boiling point, p. 506
Closest packing, p. 487
Cohesion, p. 481
Condensation, p. 503
Coordination number, p. 486
Critical point, p. 513
Critical pressure (P_c), p. 506
Critical temperature (T_c), p. 506
Crystalline solid, p. 485
Deposition, p. 509
Dipole-dipole forces, p. 473
Dispersion forces, p. 475
Dynamic equilibrium, p. 503
Equilibrium vapor pressure, p. 503
Evaporation, p. 502
Freezing point, p. 508
Glass, p. 501
Hydrogen bond, p. 477
Induced dipole, p. 474
Intermolecular forces, p. 472
Intramolecular forces, p. 472
Ion-dipole forces, p. 473
Melting point, p. 508
Molar heat of fusion (ΔH_{fus}), p. 508
Molar heat of sublimation (ΔH_{sub}), p. 509
Molar heat of vaporization (ΔH_{vap}), p. 503
Phase, p. 471
Phase boundary, p. 512
Phase change, p. 502
Phase diagram, p. 512
Sublimation, p. 509
Supercooling, p. 509
Surface tension, p. 480
Triple point, p. 513
Unit cell, p. 485
van der Waals forces, p. 473
Vaporization, p. 502
Viscosity, p. 481
X-ray diffraction, p. 491

Questions & Problems

Red numbered problems solved in Student Solutions Manual

11.2 Intermolecular Forces

Review Questions

11.1 Give an example for each type of intermolecular force: (a) dipole-dipole interaction, (b) dipole-induced dipole interaction, (c) ion-dipole interaction, (d) dispersion forces, (e) van der Waals forces.

11.2 Explain the term *polarizability*. What kind of molecules tend to have high polarizabilities? What is the relationship between polarizability and intermolecular forces?

11.3 Explain the difference between a temporary dipole moment and the permanent dipole moment.

11.4 Give some evidence that all atoms and molecules exert attractive forces on one another.

11.5 What physical properties should you consider in comparing the strength of intermolecular forces in solids and in liquids?

11.6 Which elements can take part in hydrogen bonding? Why is hydrogen unique in this kind of interaction?

Problems

11.7 The compounds Br_2 and ICl have the same number of electrons, yet Br_2 melts at −7.2°C and ICl melts at 27.2°C. Explain.

11.8 If you lived in Alaska, which of the following natural gases would you keep in an outdoor storage tank in winter: methane (CH_4), propane (C_3H_8), or butane (C_4H_{10})? Explain your answer.

11.9 The binary hydrogen compounds of the Group 14 elements and their boiling points are CH_4, −162°C; SiH_4, −112°C; GeH_4, −88°C; and SnH_4, −52°C. Explain the increase in boiling points from CH_4 to SnH_4.

11.10 List the types of intermolecular forces that exist between molecules (or basic units) in each of the following species: (a) benzene (C_6H_6), (b) CH_3Cl, (c) PF_3, (d) NaCl, (e) CS_2.

11.11 Ammonia is both a donor and an acceptor of hydrogen in hydrogen-bond formation. Draw a diagram showing the hydrogen bonding of an ammonia molecule with two other ammonia molecules.

11.12 Which of the following species are capable of hydrogen-bonding among themselves? (a) C_2H_6, (b) HI, (c) KF, (d) BeH_2, (e) CH_3COOH

11.13 Arrange the following in order of increasing boiling point: RbF, CO_2, CH_3OH, CH_3Br. Explain your reasoning.

11.14 Diethyl ether has a boiling point of 34.5°C, and 1-butanol has a boiling point of 117°C:

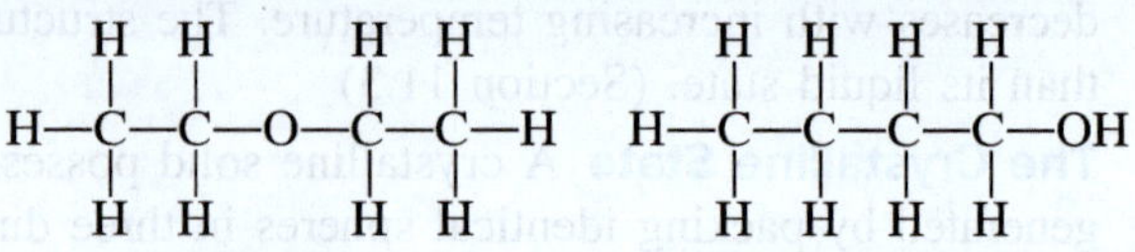

diethyl ether 1-butanol

Both of these compounds have the same numbers and types of atoms. Explain the difference in their boiling points.

11.15 Which member of each of the following pairs of substances would you expect to have a higher boiling point? (a) O_2 and Cl_2, (b) SO_2 and CO_2, (c) HF and HI

11.16 Which substance in each of the following pairs would you expect to have the higher boiling point? (a) Ne or Xe, (b) CO_2 or CS_2, (c) CH_4 or Cl_2, (d) F_2 or LiF, (e) NH_3 or PH_3. Explain your answer.

11.17 Explain in terms of intermolecular forces why (a) NH_3 has a higher boiling point than CH_4, and (b) KCl has a higher melting point than I_2.

11.18 What kind of attractive forces must be overcome in order to (a) melt ice, (b) boil molecular bromine, (c) melt solid iodine, and (d) dissociate F_2 into F atoms?

11.19 The following compounds have the same molecular formulas (C_4H_{10}). Which one would you expect to have a higher boiling point?

(a) (b)

11.20 Explain the difference in the melting points of the following compounds:

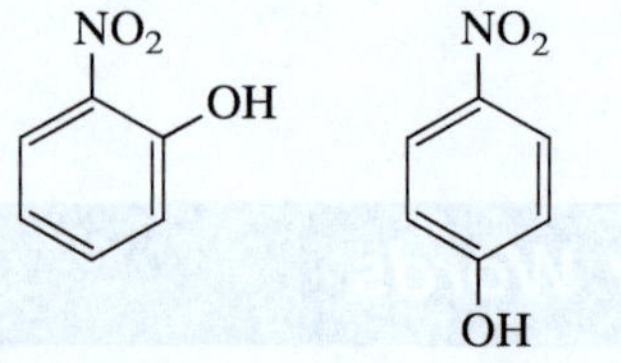

m.p. 45°C m.p. 115°C

(*Hint:* Only one of the two can form intramolecular hydrogen bonds.)

11.3 Properties of Liquids

Review Questions

11.21 Explain why liquids, unlike gases, are virtually incompressible.

11.22 What is surface tension? What is the relationship between intermolecular forces and surface tension? How does surface tension change with temperature?

11.23 Despite the fact that stainless steel is much denser than water, a stainless-steel razor blade can be made to float on water. Why?

11.24 Use water and mercury as examples to explain adhesion and cohesion.

11.25 A glass can be filled slightly above the rim with water. Explain why the water does not overflow.

11.26 Draw diagrams showing the capillary action of (a) water and (b) mercury in three tubes of different radii.

11.27 What is viscosity? What is the relationship between intermolecular forces and viscosity?

11.28 Why does the viscosity of a liquid decrease with increasing temperature?

11.29 Why is ice less dense than water?

11.30 Outdoor water pipes have to be drained or insulated in winter in a cold climate. Why?

Problems

11.31 Predict which of the following liquids has greater surface tension: ethanol (C_2H_5OH) or dimethyl ether (CH_3OCH_3).

11.32 Predict the viscosity of ethylene glycol relative to that of ethanol and glycerol (see Table 11.4).

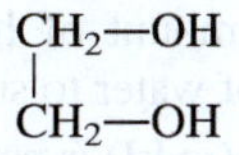

ethylene glycol

11.4 Crystal Structure

Review Questions

11.33 Define the following terms: *crystalline solid, lattice point, unit cell, coordination number, closest packing*.

11.34 Describe the geometries of the following cubic cells: simple cubic, body-centered cubic, face-centered cubic. Which of these structures would give the highest density for the same type of atoms? Which the lowest?

11.35 Classify the solid states in terms of crystal types of the elements in the third period of the periodic table. Predict the trends in their melting points and boiling points.

11.36 The melting points of the oxides of the third-period elements are given in parentheses: Na_2O (1275°C), MgO (2800°C), Al_2O_3 (2045°C), SiO_2 (1610°C), P_4O_{10} (580°C), SO_3 (16.8°C), Cl_2O_7 (−91.5°C). Classify these solids in terms of crystal types.

Problems

11.37 What is the coordination number of each sphere in (a) a simple cubic cell, (b) a body-centered cubic cell, and (c) a face-centered cubic cell? Assume the spheres are all the same.

11.38 Calculate the number of spheres that would be found within a simple cubic, a body-centered cubic, and a face-centered cubic cell. Assume that the spheres are the same.

11.39 Metallic iron crystallizes in a cubic lattice. The unit cell edge length is 287 pm. The density of iron is 7.87 g/cm^3. How many iron atoms are within a unit cell?

11.40 Barium metal crystallizes in a body-centered cubic lattice (the Ba atoms are at the lattice points only). The unit cell edge length is 502 pm, and the density of the metal is 3.50 g/cm^3. Using this information, calculate Avogadro's number. [*Hint:* First calculate the volume (in cm^3) occupied by 1 mole of Ba atoms in the unit cells. Next calculate the volume (in cm^3) occupied by one Ba atom in the unit cell. Assume that 68% of the unit cell is occupied by Ba atoms.]

11.41 Vanadium crystallizes in a body-centered cubic lattice (the V atoms occupy only the lattice points). How many V atoms are present in a unit cell?

11.42 Europium crystallizes in a body-centered cubic lattice (the Eu atoms occupy only the lattice points). The density of Eu is 5.26 g/cm^3. Calculate the unit cell edge length in pm.

11.43 Crystalline silicon has a cubic structure. The unit cell edge length is 543 pm. The density of the solid is 2.33 g/cm^3. Calculate the number of Si atoms in one unit cell.

11.44 A face-centered cubic cell contains 8 X atoms at the corners of the cell and 6 Y atoms at the faces. What is the empirical formula of the solid?

11.5 X-Ray Diffraction by Crystals

Review Questions

11.45 Define X-ray diffraction. What are the typical wavelengths (in nanometers) of X rays (see Figure 7.4)?

11.46 Write the Bragg equation. Define every term and describe how this equation can be used to measure interatomic distances.

Problems

11.47 When X rays of wavelength 0.090 nm are diffracted by a metallic crystal, the angle of first-order diffraction ($n = 1$) is measured to be 15°. What is the distance (in pm) between the layers of atoms responsible for the diffraction?

11.48 The distance between layers in a NaCl crystal is 282 pm. X rays are diffracted from these layers at an angle of 23.0°. Assuming that $n = 1$, calculate the wavelength of the X rays in nm.

11.6 Types of Crystals

Review Questions

11.49 Describe and give examples of the following types of crystals: (a) ionic crystals, (b) covalent crystals, (c) molecular crystals, (d) metallic crystals.

11.50 Why are metals good conductors of heat and electricity? Why does the ability of a metal to conduct electricity decrease with increasing temperature?

Problems

11.51 A solid is hard, brittle, and electrically nonconducting. Its melt (the liquid form of the substance) and an aqueous solution containing the substance conduct electricity. Classify the solid.

11.52 A solid is soft and has a low melting point (below 100°C). The solid, its melt, and an aqueous solution containing the substance are all nonconductors of electricity. Classify the solid.

11.53 A solid is very hard and has a high melting point. Neither the solid nor its melt conducts electricity. Classify the solid.

11.54 Which of the following are molecular solids and which are covalent solids? Se_8, HBr, Si, CO_2, C, P_4O_6, SiH_4

11.55 Classify the solid state of the following substances as ionic crystals, covalent crystals, molecular crystals, or metallic crystals: (a) CO_2, (b) B_{12}, (c) S_8, (d) KBr, (e) Mg, (f) SiO_2, (g) LiCl, (h) Cr.

11.56 Explain why diamond is harder than graphite. Why is graphite an electrical conductor but diamond is not?

11.7 Amorphous Solids

Review Questions

11.57 What is an amorphous solid? How does it differ from crystalline solid?

11.58 Define glass. What is the chief component of glass? Name three types of glass.

11.8 Phase Changes

Review Questions

11.59 What is a phase change? Name all possible changes that can occur among the vapor, liquid, and solid phases of a substance.

11.60 What is the equilibrium vapor pressure of a liquid? How is it measured and how does it change with temperature?

11.61 Use any one of the phase changes to explain what is meant by dynamic equilibrium.

11.62 Define the following terms: (a) *molar heat of vaporization*, (b) *molar heat of fusion*, (c) *molar heat of sublimation*. What are their units?

11.63 How is the molar heat of sublimation related to the molar heats of vaporization and fusion? On what law are these relationships based?

11.64 What can we learn about the intermolecular forces in a liquid from the molar heat of vaporization?

11.65 The greater the molar heat of vaporization of a liquid, the greater its vapor pressure. True or false?

11.66 Define boiling point. How does the boiling point of a liquid depend on external pressure? Referring to Table 5.3, what is the boiling point of water when the external pressure is 187.5 mmHg?

11.67 As a liquid is heated at constant pressure, its temperature rises. This trend continues until the boiling point of the liquid is reached. No further rise in temperature of the liquid can be induced by heating. Explain.

11.68 What is critical temperature? What is the significance of critical temperature in liquefaction of gases?

11.69 What is the relationship between intermolecular forces in a liquid and the liquid's boiling point and critical temperature? Why is the critical temperature of water greater than that of most other substances?

11.70 How do the boiling points and melting points of water and carbon tetrachloride vary with pressure? Explain any difference in behavior of these two substances.

11.71 Why is solid carbon dioxide called dry ice?

11.72 Wet clothes dry more quickly on a hot, dry day than on a hot, humid day. Explain.

11.73 Which of the following phase transitions gives off more heat: (a) 1 mole of steam to 1 mole of water at 100°C, or (b) 1 mole of water to 1 mole of ice at 0°C?

11.74 A beaker of water is heated to boiling by a Bunsen burner. Would adding another burner raise the boiling point of water? Explain.

Problems

11.75 Calculate the amount of heat (in kJ) required to convert 74.6 g of water to steam at 100°C.

11.76 How much heat (in kJ) is needed to convert 866 g of ice at −10°C to steam at 126°C? (The specific heats of ice and steam are 2.03 J/g · °C and 1.99 J/g · °C, respectively.)

11.77 How is the rate of evaporation of a liquid affected by (a) temperature, (b) the surface area of a liquid exposed to air, (c) intermolecular forces?

11.78 The molar heats of fusion and sublimation of molecular iodine are 15.27 kJ/mol and 62.30 kJ/mol, respectively. Estimate the molar heat of vaporization of liquid iodine.

11.79 The following compounds, listed with their boiling points, are liquid at −10°C: butane, −0.5°C; ethanol, 78.2°C; toluene, 110.6°C. At −10°C, which of these liquids would you expect to have the highest vapor pressure? Which the lowest? Explain.

11.80 Freeze-dried coffee is prepared by freezing brewed coffee and then removing the ice component with a vacuum pump. Describe the phase changes taking place during these processes.

11.81 A student hangs wet clothes outdoors on a winter day when the temperature is −15°C. After a few hours, the clothes are found to be fairly dry. Describe the phase changes in this drying process.

11.82 Steam at 100°C causes more serious burns than water at 100°C. Why?

11.83 Vapor pressure measurements at several different temperatures are shown here for mercury. Determine graphically the molar heat of vaporization for mercury.

t (°C)	200	250	300	320	340
P (mmHg)	17.3	74.4	246.8	376.3	557.9

11.84 The vapor pressure of benzene, C_6H_6, is 40.1 mmHg at 7.6°C. What is its vapor pressure at 60.6°C? The molar heat of vaporization of benzene is 30.7 kJ/mol.

11.85 The vapor pressure of liquid X is lower than that of liquid Y at 20°C, but higher at 60°C. What can you deduce about the relative magnitude of the molar heats of vaporization of X and Y?

11.86 Explain why splashing a small amount of liquid nitrogen (b.p. 77 K) is not as harmful as splashing boiling water on your skin.

11.9 Phase Diagrams

Review Questions

11.87 What is a phase diagram? What useful information can be obtained from the study of a phase diagram?

11.88 Explain how water's phase diagram differs from those of most substances. What property of water causes the difference?

Problems

11.89 The phase diagram of sulfur is shown. (a) How many triple points are there? (b) Monoclinic and rhombic are two allotropes of sulfur. Which is more stable under atmospheric conditions? (c) Describe what happens when sulfur at 1 atm is heated from 80°C to 200°C.

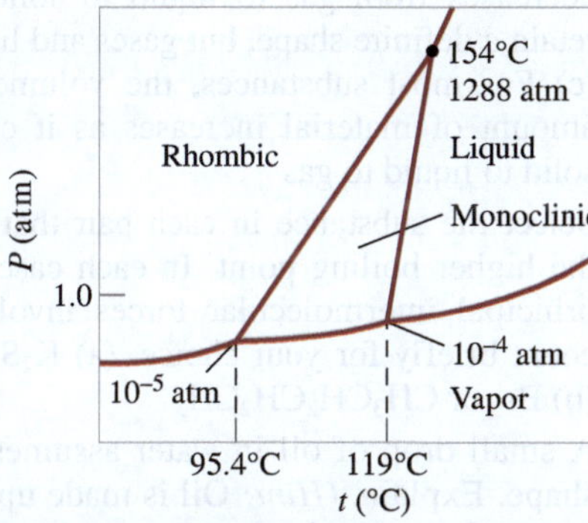

11.90 A length of wire is placed on top of a block of ice. The ends of the wire extend over the edges of the ice, and a heavy weight is attached to each end. It is found that the ice under the wire gradually melts, so that the wire slowly moves through the ice block. At the same time, the water above the wire refreezes. Explain the phase changes that accompany this phenomenon.

11.91 The boiling point and freezing point of sulfur dioxide are −10°C and −72.7°C (at 1 atm), respectively. The triple point is −75.5°C and 1.65×10^{-3} atm, and its critical point is at 157°C and 78 atm. On the basis of this information, draw a rough sketch of the phase diagram of SO_2.

11.92 A phase diagram of water is shown. Label the regions. Predict what would happen as a result of the following changes: (a) Starting at A, we raise the temperature at constant pressure. (b) Starting at C, we lower the temperature at constant pressure. (c) Starting at B, we lower the pressure at constant temperature.

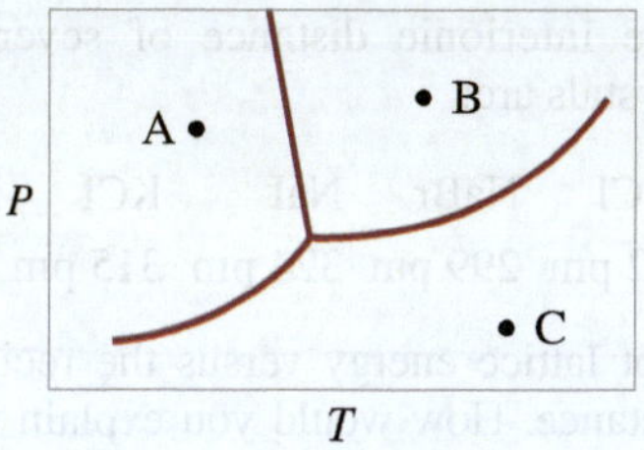

Additional Problems

11.93 Name the kinds of attractive forces that must be overcome to (a) boil liquid ammonia, (b) melt solid phosphorus (P_4), (c) dissolve CsI in liquid HF, (d) melt potassium metal.

11.94 Which of the following properties indicates very strong intermolecular forces in a liquid? (a) Very low surface tension, (b) very low critical temperature, (c) very low boiling point, (d) very low vapor pressure

11.95 At −35°C, liquid HI has a higher vapor pressure than liquid HF. Explain.

11.96 Based on the following properties of elemental boron, classify it as one of the crystalline solids discussed in Section 11.6: high melting point (2300°C), poor conductor of heat and electricity, insoluble in water, very hard substance.

11.97 Referring to Figure 11.43, determine the stable phase of CO_2 at (a) 4 atm and −60°C, and (b) 0.5 atm and −20°C.

11.98 Classify the unit cell of molecular iodine.

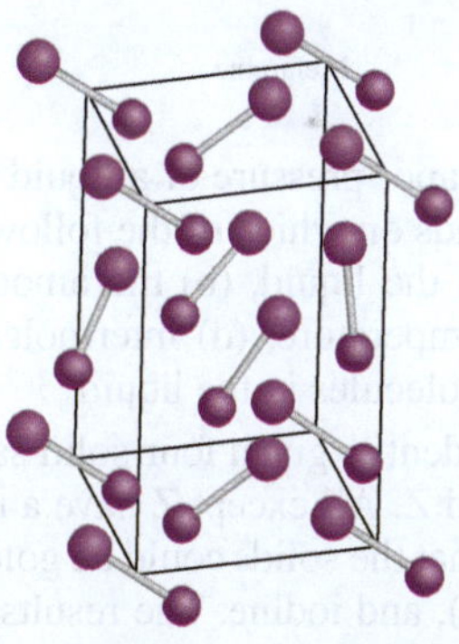

11.99 A CO_2 fire extinguisher is located on the outside of a building in Massachusetts. During the winter months, one can hear a sloshing sound when the extinguisher is gently shaken. In the summertime there is often no sound when it is shaken. Explain. Assume that the extinguisher has no leaks and that it has not been used.

11.100 What is the vapor pressure of mercury at its normal boiling point (357°C)?

11.101 A flask of water is connected to a powerful vacuum pump. When the pump is turned on, the water begins to boil. After a few minutes, the same water begins to freeze. Eventually, the ice disappears. Explain what happens at each step.

11.102 The liquid-vapor boundary line in the phase diagram of any substance always stops abruptly at a certain point. Why?

11.103 The interionic distance of several alkali halide crystals are:

NaCl	NaBr	NaI	KCl	KBr	KI
282 pm	299 pm	324 pm	315 pm	330 pm	353 pm

Plot lattice energy versus the reciprocal interionic distance. How would you explain the plot in terms of the dependence of lattice energy on distance of separation between ions? What law governs this interaction? (For lattice energies, see Table 9.1.)

11.104 Which has a greater density, crystalline SiO_2 or amorphous SiO_2? Why?

11.105 In 2009, thousands of babies in China became ill from drinking contaminated milk. To falsely boost the milk's protein content, melamine ($C_3H_6N_6$) was added to diluted milk because of its high nitrogen composition. Unfortunately, melamine forms a precipitate by hydrogen bonding with cyanuric acid ($C_3H_3N_3O_3$), another contaminant present. The resulting stonelike particles caused severe kidney damage in many babies. Draw the hydrogen-bonded complex formed from these two molecules.

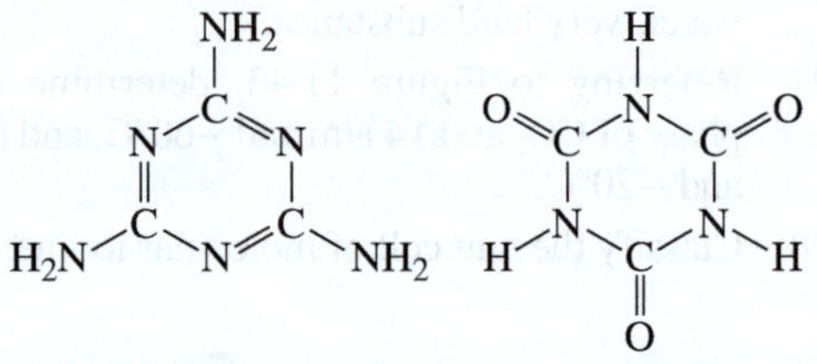

Melamine Cyanuric acid

11.106 The vapor pressure of a liquid in a closed container depends on which of the following? (a) The volume above the liquid, (b) the amount of liquid present, (c) temperature, (d) intermolecular forces between the molecules in the liquid

11.107 A student is given four solid samples labeled W, X, Y, and Z. All except Z have a metallic luster. She is told that the solids could be gold, lead sulfide, quartz (SiO_2), and iodine. The results of her investigations are as follows: (a) W is a good electrical conductor; X, Y, and Z are poor electrical conductors. (b) When the solids are hit with a hammer, W flattens out, X shatters into many pieces, Y is smashed into a powder, and Z is cracked. (c) When the solids are heated with a Bunsen burner, Y melts with some sublimation, but X, W, and Z do not melt. (d) In treatment with 6 *M* HNO_3, X dissolves; there is no effect on W, Y, or Z. On the basis of these test results, identify the solids.

11.108 Which of the following statements are false? (a) Dipole-dipole interactions between molecules are greatest if the molecules possess only temporary dipole moments. (b) All compounds containing hydrogen atoms can participate in hydrogen-bond formation. (c) Dispersion forces exist between all atoms, molecules, and ions. (d) The extent of ion-induced dipole interaction depends only on the charge on the ion.

11.109 Note the kettle of boiling water on a stove. Identify the phases in regions A and B.

NaturaLight/Alamy Stock Photo

11.110 The south pole of Mars is covered with dry ice, which partly sublimes during the summer. The CO_2 vapor re-condenses in the winter when the temperature drops to 150 K. Given that the heat of sublimation of CO_2 is 25.9 kJ/mol, calculate the atmospheric pressure on the surface of Mars. [*Hint:* Use Figure 11.43 to determine the normal sublimation temperature of dry ice and Equation (11.5), which also applies to sublimations.]

11.111 The properties of gases, liquids, and solids differ in a number of respects. How would you use the kinetic molecular theory (see Section 5.7) to explain the following observations? (a) Ease of compressibility decreases from gas to liquid to solid. (b) Solids retain a definite shape, but gases and liquids do not. (c) For most substances, the volume of a given amount of material increases as it changes from solid to liquid to gas.

11.112 Select the substance in each pair that should have the higher boiling point. In each case identify the principal intermolecular forces involved and account briefly for your choice. (a) K_2S or $(CH_3)_3N$, (b) Br_2 or $CH_3CH_2CH_2CH_3$

11.113 A small drop of oil in water assumes a spherical shape. Explain. (*Hint:* Oil is made up of nonpolar molecules, which tend to avoid contact with water.)

11.114 Under the same conditions of temperature and density, which of the following gases would you expect to behave less ideally: CH_4, SO_2? Explain.

11.115 The fluorides of the second-period elements and their melting points are: LiF, 845°C; BeF_2, 800°C; BF_3, −126.7°C; CF_4, −184°C; NF_3, −206.6°C; OF_2, −223.8°C; F_2, −219.6°C. Classify the type(s) of intermolecular forces present in each compound.

11.116 The standard enthalpy of formation of gaseous molecular iodine is 62.4 kJ/mol. Use this information to calculate the molar heat of sublimation of molecular iodine at 25°C.

11.117 The following graph shows approximate plots of ln *P* versus $1/T$ for three compounds: methanol (CH_3OH), methyl chloride (CH_3Cl), and propane (C_3H_8), where *P* is the vapor pressure. Match the lines with these compounds.

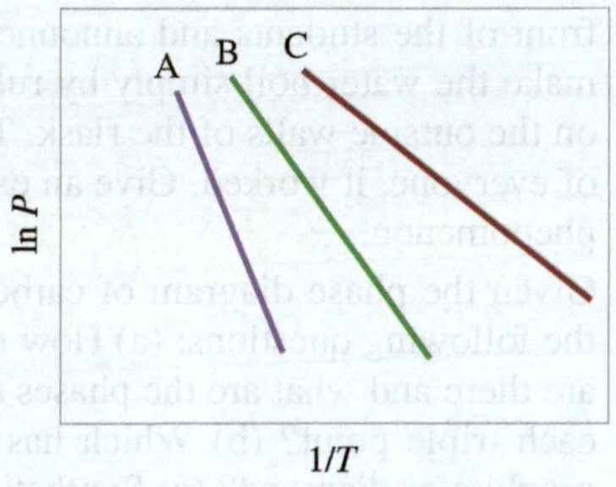

11.118 Determine the final state and its temperature when 150.0 kJ of heat are added to 50.0 g of water at 20°C. The specific heat of steam is 1.99 J/g · C.

11.119 The distance between Li^+ and Cl^- is 257 pm in solid LiCl and 203 pm in a LiCl unit in the gas phase. Explain the difference in the bond lengths.

11.120 Heat of hydration, that is, the heat change that occurs when ions become hydrated in solution, is largely due to ion-dipole interactions. The heats of hydration for the alkali metal ions are Li^+, −520 kJ/mol; Na^+, −405 kJ/mol; K^+, −321 kJ/mol. Account for the trend in these values.

11.121 If water were a linear molecule, (a) would it still be polar, and (b) would the water molecules still be able to form hydrogen bonds with one another?

11.122 Calculate the $\Delta H°$ for the following processes at 25°C: (a) $Br_2(l) \longrightarrow Br_2(g)$ and (b) $Br_2(g) \longrightarrow 2Br_2(g)$. Comment on the relative magnitudes of these $\Delta H°$ values in terms of the forces involved in each case. [*Hint:* See Table 9.4, and given that $\Delta H_f°[Br_2(g)] = 30.9$ kJ/mol.]

11.123 Gaseous or highly volatile liquid anesthetics are often preferred in surgical procedures because once inhaled, these vapors can quickly enter the bloodstream through the alveoli and then enter the brain. Shown here are several common gaseous anesthetics with their boiling points. Based on intermolecular force considerations, explain the advantages of using these anesthetics. (*Hint:* The brain barrier is made of membranes that have a nonpolar interior region.)

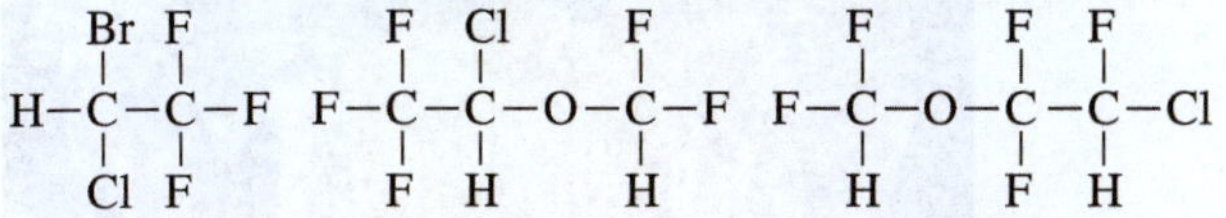

Halothane 50°C | Isoflurane 48.5°C | Enflurane 56.5°C

11.124 A beaker of water is placed in a closed container. Predict the effect on the vapor pressure of the water when (a) its temperature is lowered, (b) the volume of the container is doubled, (c) more water is added to the beaker.

11.125 The phase diagram of helium is shown. Helium is the only known substance that has two different liquid phases called helium-I and helium-II. (a) What is the maximum temperature at which helium-II can exist? (b) What is the minimum pressure at which solid helium can exist? (c) What is the normal boiling point of helium-I? (d) Can solid helium sublime? (e) How many triple points are there?

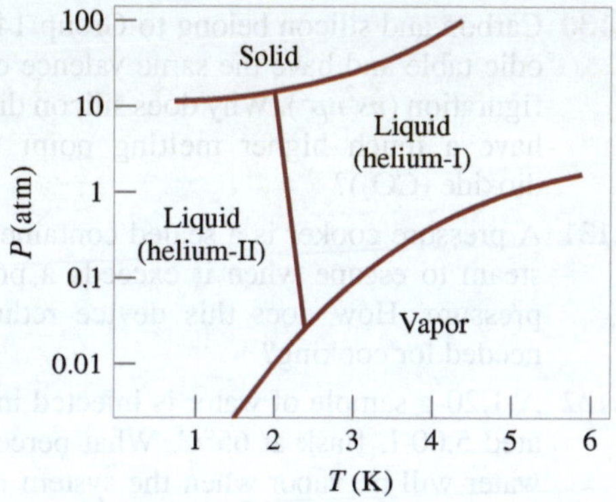

11.126 Referring to Figure 11.27, determine the number of each type of ion within the unit cells.

11.127 Ozone (O_3) is a strong oxidizing agent that can oxidize all the common metals except gold and platinum. A convenient test for ozone is based on its action on mercury. When exposed to ozone, mercury becomes dull looking and sticks to glass tubing (instead of flowing freely through it). Write a balanced equation for the reaction. What property of mercury is altered by its interaction with ozone?

11.128 A sample of limestone ($CaCO_3$) is heated in a closed vessel until it is partially decomposed. Write an equation for the reaction and state how many phases are present.

11.129 Silicon used in computer chips must have an impurity level below 10^{-9} (that is, fewer than one impurity atom for every 10^9 Si atoms). Silicon is prepared by the reduction of quartz (SiO_2) with coke (a form of carbon made by the destructive distillation of coal) at about 2000°C:

$$SiO_2(s) + 2C(s) \longrightarrow Si(l) + 2CO(g)$$

Next, solid silicon is separated from other solid impurities by treatment with hydrogen chloride at 350°C to form gaseous trichlorosilane ($SiCl_3H$):

$$Si(s) + 3HCl(g) \longrightarrow SiCl_3H(g) + H_2(g)$$

Finally, ultrapure Si can be obtained by reversing the previous reaction at 1000°C:

$$SiCl_3H(g) + H_2(g) \longrightarrow Si(s) + 3HCl(g)$$

(a) Trichlorosilane has a vapor pressure of 0.258 atm at −2°C. What is its normal boiling point? Is trichlorosilane's boiling point consistent with the type of intermolecular forces that exist among its molecules? (The molar heat of vaporization of trichlorosilane is 28.8 kJ/mol.) (b) What types of crystals do Si and SiO_2 form? (c) Silicon has a diamond crystal structure (see Figure 11.29). Each cubic unit cell (edge length a = 543 pm) contains eight Si atoms. If there are 1.0×10^{13} boron atoms per cubic centimeter in a sample of pure silicon, how many Si atoms are there for every B atom in the sample? Does this sample satisfy the 10^{-9} purity requirement for the electronic grade silicon?

11.130 Carbon and silicon belong to Group 14 of the periodic table and have the same valence electron configuration (ns^2np^2). Why does silicon dioxide (SiO_2) have a much higher melting point than carbon dioxide (CO_2)?

11.131 A pressure cooker is a sealed container that allows steam to escape when it exceeds a predetermined pressure. How does this device reduce the time needed for cooking?

11.132 A 1.20-g sample of water is injected into an evacuated 5.00-L flask at 65°C. What percentage of the water will be vapor when the system reaches equilibrium? Assume ideal behavior of water vapor and that the volume of liquid water is negligible. The vapor pressure of water at 65°C is 187.5 mmHg.

11.133 What are the advantages of cooking the vegetable broccoli with steam instead of boiling it in water?

11.134 A quantitative measure of how efficiently spheres pack into unit cells is called *packing efficiency,* which is the percentage of the cell space occupied by the spheres. Calculate the packing efficiencies of a simple cubic cell, a body-centered cubic cell, and a face-centered cubic cell. (*Hint:* Refer to Figure 11.23 and use the relationship that the volume of a sphere is $\frac{4}{3}\pi r^3$, where r is the radius of the sphere.)

11.135 Provide an explanation for each of the following phenomena: (a) Solid argon (m.p. −183.3°C; b.p. −185.8°C) can be prepared by immersing a flask containing argon gas in liquid nitrogen (b.p. −195.8°C) until it liquefies and then connecting the flask to a vacuum pump. (b) The melting point of cyclohexane (C_6H_{12}) increases with increasing pressure exerted on the solid cyclohexane. (c) Certain high-altitude clouds contain water droplets at −10°C. (d) When a piece of dry ice is added to a beaker of water, fog forms above the water.

Ken Karp/McGraw Hill

11.136 Argon crystallizes in the face-centered cubic arrangement at 40 K. Given that the atomic radius of argon is 191 pm, calculate the density of solid argon.

11.137 A chemistry instructor performed the following mystery demonstration. Just before the students arrived in class, she heated some water to boiling in an Erlenmeyer flask. She then removed the flask from the flame and closed the flask with a rubber stopper. After the class commenced, she held the flask in front of the students and announced that she could make the water boil simply by rubbing an ice cube on the outside walls of the flask. To the amazement of everyone, it worked. Give an explanation for this phenomenon.

11.138 Given the phase diagram of carbon shown, answer the following questions: (a) How many triple points are there and what are the phases that can coexist at each triple point? (b) Which has a higher density, graphite or diamond? (c) Synthetic diamond can be made from graphite. Using the phase diagram, how would you go about making diamond?

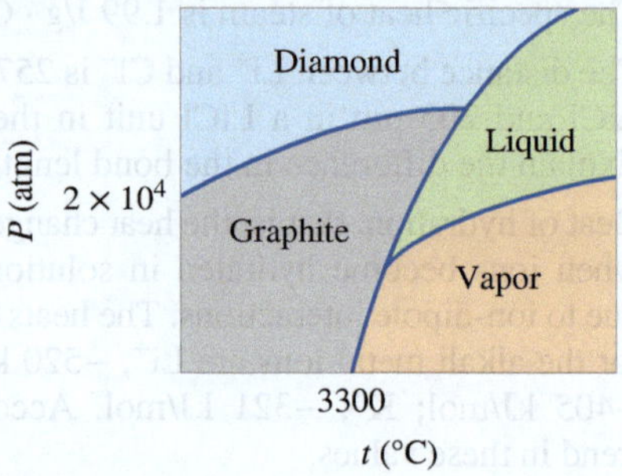

11.139 Swimming coaches sometimes suggest that a drop of alcohol (ethanol) placed in an ear plugged with water "draws out the water." Explain this action from a molecular point of view.

11.140 Use the concept of intermolecular forces to explain why the far end of a walking cane rises when one raises the handle.

11.141 Why do citrus growers spray their trees with water to protect them from freezing?

Tony Ranze/AFP/Getty Images

11.142 What is the origin of dark spots on the inner glass walls of an old tungsten lightbulb? What is the purpose of filling these lightbulbs with argon gas?

11.143 The compound dichlorodifluoromethane (CCl_2F_2) has a normal boiling point of −30°C, a critical temperature of 112°C, and a corresponding critical pressure of 40 atm. If the gas is compressed to 18 atm at 20°C, will the gas condense? Your answer should be based on a graphical interpretation.

11.144 A student heated a beaker of cold water (on a tripod) with a Bunsen burner. When the gas is ignited, she noticed that there was water condensed on the outside of the beaker. Explain what happened.

11.145 Sketch the cooling curves of water from about 110°C to about −10°C. How would you also show the formation of supercooled liquid below 0°C which then freezes to ice? The pressure is at 1 atm throughout the process. The curves need not be drawn quantitatively.

11.146 Iron crystallizes in a body-centered cubic lattice. The cell length as determined by X-ray diffraction is 286.7 pm. Given that the density of iron is 7.874 g/cm^3, calculate Avogadro's number.

11.147 The boiling point of methanol is 64.5°C and the standard enthalpy of formation of methanol vapor is −201.0 kJ/mol. Calculate the vapor pressure of methanol (in mmHg) at 25°C. (*Hint:* See Appendix 2 for other thermodynamic data of methanol.)

11.148 An alkali metal in the form of a cube of edge length 0.171 cm is vaporized in a 0.843-L container at 1235 K. The vapor pressure is 19.2 mmHg. Identify the metal by calculating the atomic radius in picometers and the density. (*Hint:* You need to consult Figures 8.5, 11.23, 11.30, and a chemistry handbook.)

11.149 A closed vessel of volume 9.6 L contains 2.0 g of water. Calculate the temperature (in °C) at which only half of the water remains in the liquid phase. (See Table 5.3 for vapor pressures of water at different temperatures.)

11.150 A sample of water shows the following behavior as it is heated at a constant rate:

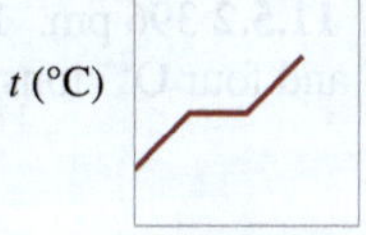

If twice the mass of water has the same amount of heat transferred to it, which of the graphs (a)–(d) best describes the temperature variation? Note that the scales for all the graphs are the same.

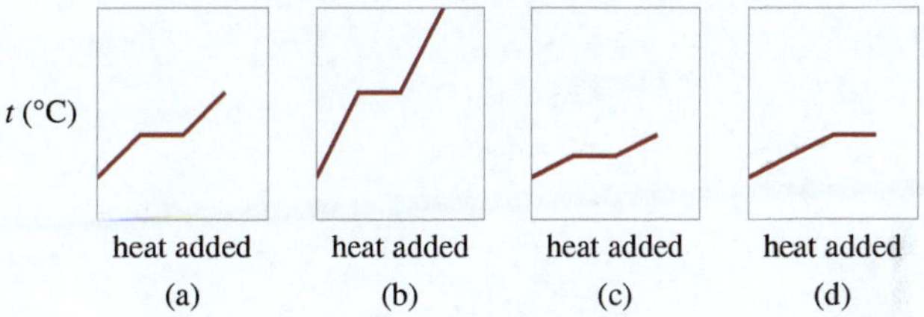

(Used with permission from the *Journal of Chemical Education,* Vol. 79, No. 7, 2002, pp. 889–895; © 2002, Division of Chemical Education, Inc.)

11.151 The electrical conductance of copper metal decreases with temperature, but that of a $CuSO_4$ solution increases with temperature. Explain.

11.152 Assuming ideal behavior, calculate the density of gaseous HF at its normal boiling point (19.5°C). The experimentally measured density under the same conditions is 3.10 g/L. Account for the discrepancy between your calculated value and the experimental result.

Interpreting, Modeling, & Estimating

11.153 Both calcium and strontium crystallize in face-centered cubic unit cells. Which metal has a greater density?

11.154 Is the vapor pressure of a liquid more sensitive to changes in temperature if ΔH_{vap} is small or large?

11.155 Estimate the molar heat of vaporization of a liquid whose vapor pressure doubles when the temperature is raised from 85°C to 95°C.

11.156 On a summer day the temperature and (relative) humidity were 95°F and 65 percent, respectively, in Florida. What would be the volume of water in a typical student dormitory room if all of the water vapor were condensed to liquid?

11.157 Without the aid of instruments, give two examples of evidence that solids exhibit vapor pressure.

Answers to Practice Exercises

11.1 (a) Ionic and dispersion forces, (b) Dispersion forces, (c) Dipole-dipole and dispersion forces. **11.2** Only (c). **11.3** 10.50 g/cm^3. **11.4** 315 pm. **11.5** Two. **11.6** 361 pm. **11.7** 360 mmHg. **11.8** 173 kJ.

Answers to Review of Concepts & Facts

11.2.1 (a) Dispersion and dipole-dipole forces. (b) Dispersion forces. (c) Dispersion forces. **11.2.2** Hydrazine. **11.4.1** 2. **11.4.2** 270 pm. **11.5.1** There is no regular or long-range order in a liquid. **11.5.2** 396 pm. **11.6.1** 2.65 g/cm^3. **11.6.2** Four Zn^{2+} and four O^{2-} ions. ZnO. **11.8.1** According to Equation (11.2), the slope of the curve is given by $-\Delta H_{vap}/R$. CH_3OH has a higher ΔH_{vap} (due to hydrogen bonding), so the steeper curve should be labeled CH_3OH. The results are CH_3OH: $\Delta H_{vap} = 37.4$ kJ/mol; CH_3OCH_3: $\Delta H_{vap} = 19.3$ kJ/mol. **11.8.2** 225.1 mmHg. **11.9.1** (b).

CHAPTER

12

Physical Properties of Solutions

A ferrofluid is a colloid containing nanoparticle ferromagnets suspended in a fluid. Ferrofluids become magnetized in the presence of strong magnetic fields.
MADDRAT/Shutterstock

CHAPTER OUTLINE

Most chemical reactions take place, not between pure solids, liquids, or gases, but among ions and molecules dissolved in water or other solvents. In Chapters 5 and 11 we looked at the properties of gases, liquids, and solids. In this chapter we examine the properties of solutions, concentrating mainly on the role of intermolecular forces in solubility and other physical properties of solution.

12.1 Types of Solutions

Learning Objective

- Describe saturated, unsaturated, and supersaturated as they pertain to solutions.

In Section 4.1 we noted that a solution is a homogeneous mixture of two or more substances. Because this definition places no restriction on the nature of the substances involved, we can distinguish six types of solutions, depending on the original states (solid, liquid, or gas) of the solution components. Table 12.1 gives examples of each type.

Our focus in this chapter will be on solutions involving at least one liquid component—that is, gas-liquid, liquid-liquid, and solid-liquid solutions. And, perhaps not too surprisingly, the liquid solvent in most of the solutions we will study is water.

Chemists also characterize solutions by their capacity to dissolve a solute. A ***saturated solution*** *contains the maximum amount of a solute that will dissolve in a given solvent at a specific temperature.* An ***unsaturated solution*** *contains less solute than it has the capacity to dissolve.* A third type, a ***supersaturated solution,*** *contains more solute than is present in a saturated solution.* Supersaturated solutions are not very stable. In time, some of the solute will come out of a supersaturated solution as crystals. ***Crystallization*** is *the process in which dissolved solute comes out of solution and forms crystals* (Figure 12.1). Note that both precipitation and crystallization describe the

Table 12.1 Types of Solutions

Component 1	Component 2	State of Resulting Solution	Examples
Gas	Gas	Gas	Air
Gas	Liquid	Liquid	Soda water (CO_2 in water)
Gas	Solid	Solid	H_2 gas in palladium
Liquid	Liquid	Liquid	Ethanol in water
Solid	Liquid	Liquid	NaCl in water
Solid	Solid	Solid	Brass (Cu/Zn), solder (Sn/Pb)

Figure 12.1 *In a supersaturated sodium acetate solution (left), sodium acetate crystals rapidly form when a small seed crystal is added.*
Ken Karp/McGraw Hill

separation of excess solid substance from a supersaturated solution. However, solids formed by the two processes differ in appearance. We normally think of precipitates as being made up of small particles, whereas crystals may be large and well formed.

Summary of Concepts & Facts

- Solutions are homogeneous mixtures of two or more substances, which may be solids, liquids, or gases.

12.2 A Molecular View of the Solution Process

Learning Objective

- Summarize the three types of interactions that determine the extent to which a solute dissolves in solution.

The intermolecular attractions that hold molecules together in liquids and solids also play a central role in the formation of solutions. When one substance (the solute) dissolves in another (the solvent), particles of the solute disperse throughout the solvent. The solute particles occupy positions that are normally taken by solvent molecules. The ease with which a solute particle replaces a solvent molecule depends on the relative strengths of three types of interactions:

- Solvent-solvent interaction
- Solute-solute interaction
- Solvent-solute interaction

For simplicity, we can imagine the solution process taking place in three distinct steps (Figure 12.2). Step 1 is the separation of solvent molecules, and step 2 entails the separation of solute molecules. These steps require energy input to break attractive

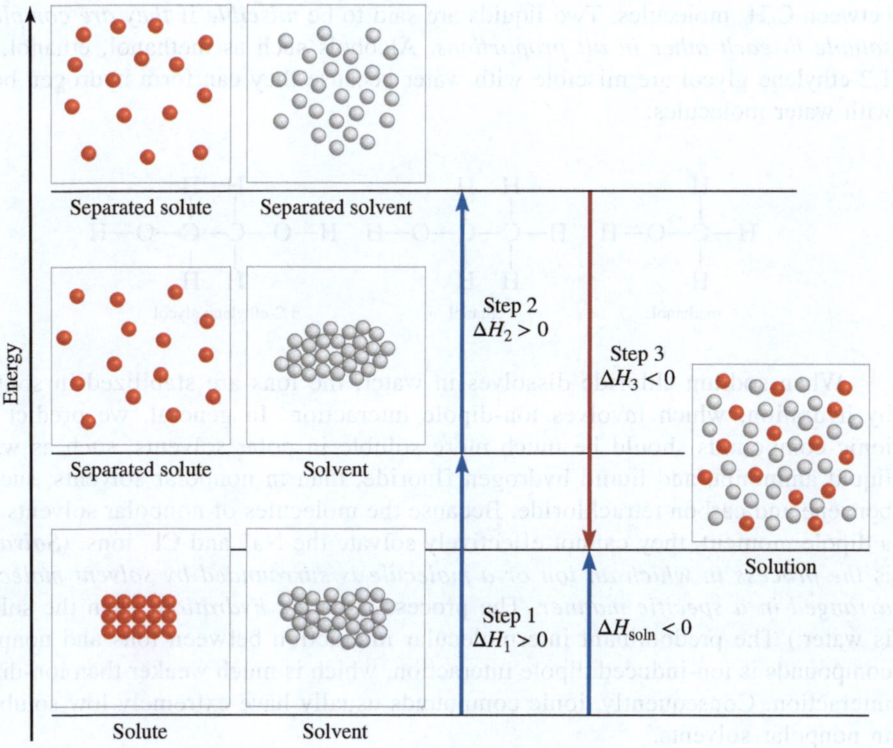

Figure 12.2 *A molecular view of the solution process portrayed as taking place in three steps: First the solvent and solute molecules are separated (steps 1 and 2). Then the solvent and solute molecules mix (step 3).*

intermolecular forces; therefore, they are endothermic. In step 3 the solvent and solute molecules mix. This process can be exothermic or endothermic. The heat of solution ΔH_{soln} is given by

$$\Delta H_{soln} = \Delta H_1 + \Delta H_2 + \Delta H_3$$

If the solute-solvent attraction is stronger than the solvent-solvent attraction and solute-solute attraction, the solution process is favorable, or exothermic ($\Delta H_{soln} < 0$). If the solute-solvent interaction is weaker than the solvent-solvent and solute-solute interactions, then the solution process is endothermic ($\Delta H_{soln} > 0$).

You may wonder why a solute dissolves in a solvent at all if the attraction for its own molecules is stronger than the solute-solvent attraction. The solution process, like all physical and chemical processes, is governed by two factors. One is energy, which determines whether a solution process is exothermic or endothermic. The second factor is an inherent tendency toward disorder in all natural events. In much the same way that a deck of new playing cards becomes mixed up after it has been shuffled a few times, when solute and solvent molecules mix to form a solution, there is an increase in randomness, or disorder. In the pure state, the solvent and solute possess a fair degree of order, characterized by the more or less regular arrangement of atoms, molecules, or ions in three-dimensional space. Much of this order is destroyed when the solute dissolves in the solvent (see Figure 12.2). Therefore, the solution process is accompanied by an increase in disorder. It is the increase in disorder of the system that favors the solubility of any substance, even if the solution process is endothermic.

Video: Dissolution of an Ionic and a Covalent Compound

Solubility is a measure of how much solute will dissolve in a solvent at a specific temperature. The saying "like dissolves like" is helpful in predicting the solubility of a substance in a given solvent. What this expression means is that two substances with intermolecular forces of similar type and magnitude are likely to be soluble in each other. For example, both carbon tetrachloride (CCl_4) and benzene (C_6H_6) are nonpolar liquids. The only intermolecular forces present in these substances are dispersion forces (see Section 11.2). When these two liquids are mixed, they readily dissolve in each other, because the attraction between CCl_4 and C_6H_6 molecules is comparable in magnitude to the forces between CCl_4 molecules and between C_6H_6 molecules. Two liquids are said to be ***miscible*** if *they are completely soluble in each other in all proportions.* Alcohols such as methanol, ethanol, and 1,2-ethylene glycol are miscible with water because they can form hydrogen bonds with water molecules:

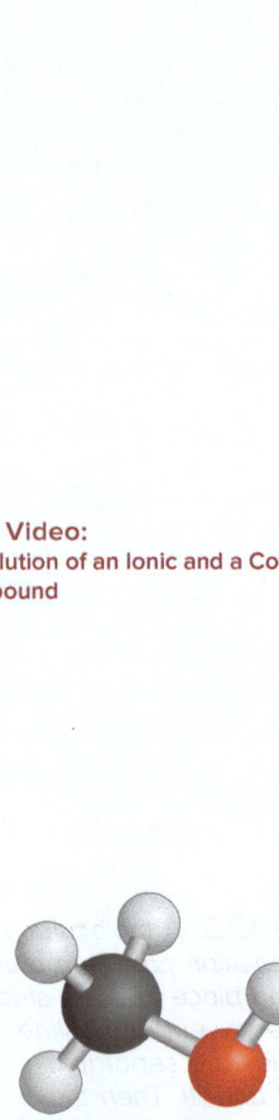

CH_3OH

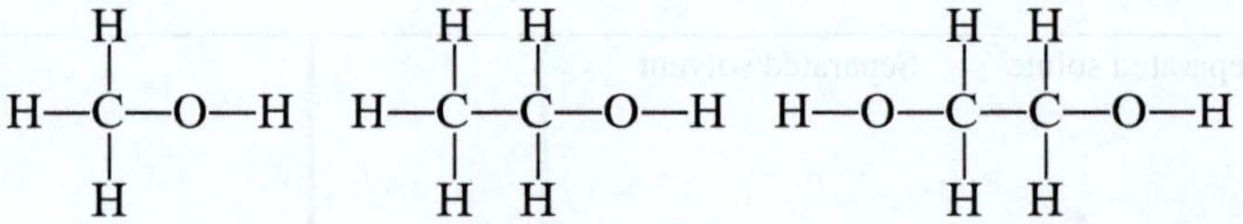

methanol   ethanol   1,2-ethylene glycol

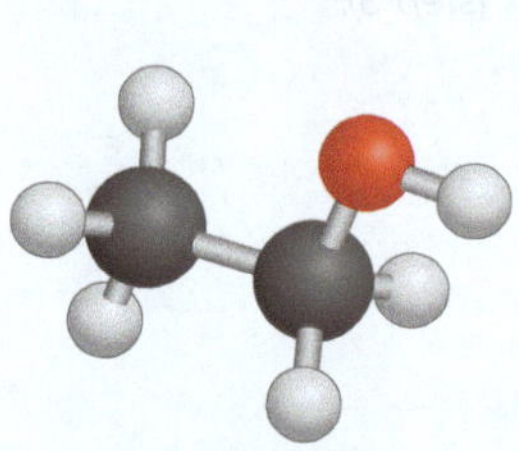

C_2H_5OH

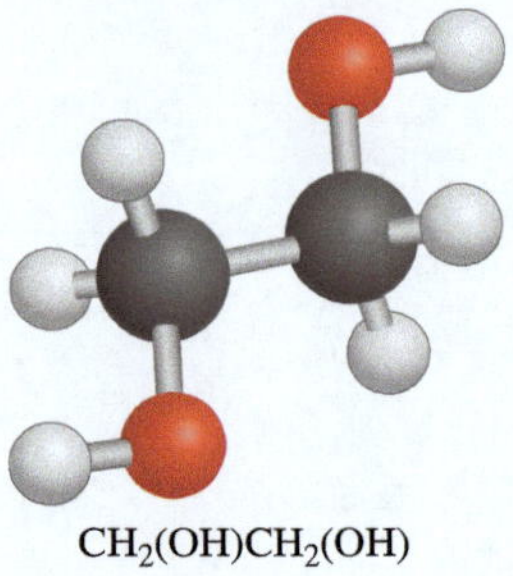

$CH_2(OH)CH_2(OH)$

When sodium chloride dissolves in water, the ions are stabilized in solution by hydration, which involves ion-dipole interaction. In general, we predict that ionic compounds should be much more soluble in polar solvents, such as water, liquid ammonia, and liquid hydrogen fluoride, than in nonpolar solvents, such as benzene and carbon tetrachloride. Because the molecules of nonpolar solvents lack a dipole moment, they cannot effectively solvate the Na^+ and Cl^- ions. (***Solvation*** is *the process in which an ion or a molecule is surrounded by solvent molecules arranged in a specific manner.* The process is called *hydration* when the solvent is water.) The predominant intermolecular interaction between ions and nonpolar compounds is ion-induced dipole interaction, which is much weaker than ion-dipole interaction. Consequently, ionic compounds usually have extremely low solubility in nonpolar solvents.

Example 12.1 illustrates how to predict solubility based on a knowledge of the intermolecular forces in the solute and the solvent.

Example 12.1

Predict the relative solubilities in the following cases: (a) bromine (Br_2) in benzene (C_6H_6, $\mu = 0$ D) and in water ($\mu = 1.87$ D), (b) KCl in carbon tetrachloride (CCl_4, $\mu = 0$ D) and in liquid ammonia (NH_3, $\mu = 1.46$ D), (c) formaldehyde (CH_2O) in carbon disulfide (CS_2, $\mu = 0$ D) and in water.

Strategy In predicting solubility, remember the saying: Like dissolves like. A nonpolar solute will dissolve in a nonpolar solvent; ionic compounds will generally dissolve in polar solvents due to favorable ion-dipole interaction; solutes that can form hydrogen bonds with the solvent will have high solubility in the solvent.

Solution

(a) Br_2 is a nonpolar molecule and therefore should be more soluble in C_6H_6, which is also nonpolar, than in water. The only intermolecular forces between Br_2 and C_6H_6 are dispersion forces.

(b) KCl is an ionic compound. For it to dissolve, the individual K^+ and Cl^- ions must be stabilized by ion-dipole interaction. Because CCl_4 has no dipole moment, KCl should be more soluble in liquid NH_3, a polar molecule with a large dipole moment.

(c) Because CH_2O is a polar molecule and CS_2 (a linear molecule) is nonpolar,

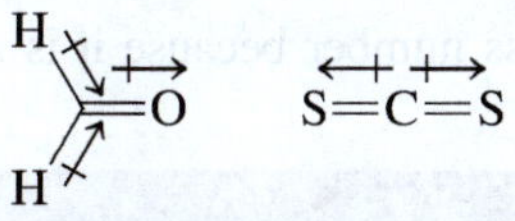

$\mu > 0$ $\mu = 0$

the forces between molecules of CH_2O and CS_2 are dipole-induced dipole and dispersion. On the other hand, CH_2O can form hydrogen bonds with water, so it should be more soluble in that solvent.

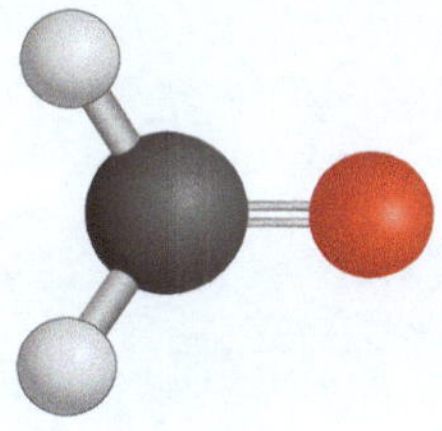

CH_2O

Practice Exercise Is iodine (I_2) more soluble in water or in carbon disulfide (CS_2)?

Similar problem: 12.11.

Summary of Concepts & Facts

- The ease of dissolving a solute in a solvent is governed by intermolecular forces. Energy and the disorder that results when molecules of the solute and solvent mix to form a solution are the forces driving the solution process.

Review of Concepts & Facts

12.2.1 Which of the following would you expect to be more soluble in benzene than in water: C_4H_{10}, HBr, KNO_3, P_4?

12.2.2 What is the strongest type of intermolecular force between the solute and the solvent in each solution? (a) $CH_3OCH_3(g)$ in $H_2O(l)$, (b) Ne(g) in $H_2O(l)$.

12.3 Concentration Units

Learning Objective

- Apply the various concentration units to express the concentration of a given solution or to interchange concentration units.

Quantitative study of a solution requires knowing its *concentration,* that is, the amount of solute present in a given amount of solution. Chemists use several different concentration units, each of which has advantages as well as limitations. Let us examine the four most common units of concentration: percent by mass, mole fraction, molarity, and molality.

Types of Concentration Units

Percent by Mass

The ***percent by mass*** (also called *percent by weight* or *weight percent*) is *the ratio of the mass of a solute to the mass of the solution, multiplied by 100 percent:*

$$\text{percent by mass} = \frac{\text{mass of solute}}{\text{mass of solute} + \text{mass of solvent}} \times 100\%$$

or

$$\text{percent by mass} = \frac{\text{mass of solute}}{\text{mass of soln}} \times 100\% \tag{12.1}$$

The percent by mass is a unitless number because it is a ratio of two similar quantities.

Example 12.2

A sample of 0.892 g of potassium chloride (KCl) is dissolved in 54.6 g of water. What is the percent by mass of KCl in the solution?

Strategy We are given the mass of a solute dissolved in a certain amount of solvent. Therefore, we can calculate the mass percent of KCl using Equation (12.1).

Solution We write

$$\begin{aligned}\text{percent by mass of KCl} &= \frac{\text{mass of solute}}{\text{mass of soln}} \times 100\% \\ &= \frac{0.892\text{ g}}{0.892\text{ g} + 54.6\text{ g}} \times 100\% \\ &= 1.61\%\end{aligned}$$

Practice Exercise A sample of 6.44 g of naphthalene ($C_{10}H_8$) is dissolved in 80.1 g of benzene (C_6H_6). Calculate the percent by mass of naphthalene in this solution.

Similar problem: 12.15.

Mole Fraction (X)

The mole fraction was introduced in Section 5.6. The mole fraction of a component of a solution, say, component A, is written X_A and is defined as

$$\text{mole fraction of component A} = X_A = \frac{\text{moles of A}}{\text{sum of moles of all components}}$$

The mole fraction is also unitless, because it too is a ratio of two similar quantities.

Molarity (M)

In Section 4.5 molarity was defined as the number of moles of solute in 1 L of solution; that is,

$$\text{molarity} = \frac{\text{moles of solute}}{\text{liters of soln}}$$

Thus, the units of molarity are mol/L.

Molality (m)

Molality is *the number of moles of solute dissolved in 1 kg (1000 g) of solvent*—that is,

$$\text{molality} = \frac{\text{moles of solute}}{\text{mass of solvent (kg)}} \tag{12.2}$$

For example, to prepare a 1 molal, or 1 m, sodium sulfate (Na_2SO_4) aqueous solution, we need to dissolve 1 mole (142.0 g) of the substance in 1000 g (1 kg) of water. Depending on the nature of the solute-solvent interaction, the final volume of the solution will be either greater or less than 1000 mL. It is also possible, though very unlikely, that the final volume could be equal to 1000 mL.

Example 12.3 shows how to calculate the molality of a solution.

Example 12.3

Calculate the molality of a sulfuric acid solution containing 24.4 g of sulfuric acid in 198 g of water. The molar mass of sulfuric acid is 98.09 g.

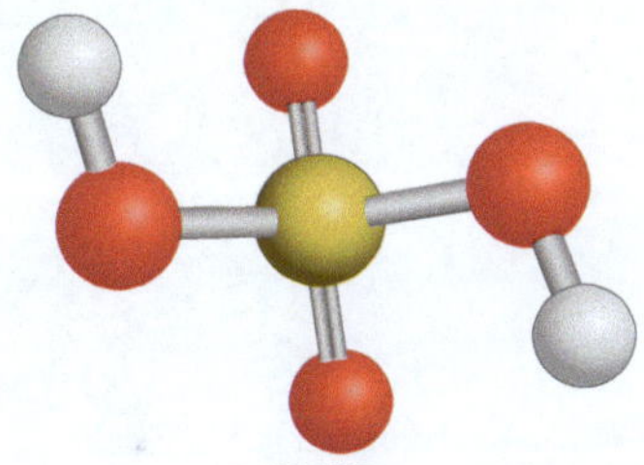

H_2SO_4

Strategy To calculate the molality of a solution, we need to know the number of moles of solute and the mass of the solvent in kilograms.

Solution The definition of molality (m) is

$$m = \frac{\text{moles of solute}}{\text{mass of solvent (kg)}}$$

First, we find the number of moles of sulfuric acid in 24.4 g of the acid, using its molar mass as the conversion factor.

$$\text{moles of } H_2SO_4 = 24.4 \text{ g } H_2SO_4 \times \frac{1 \text{ mol } H_2SO_4}{98.09 \text{ g } H_2SO_4}$$
$$= 0.249 \text{ mol } H_2SO_4$$

The mass of water is 198 g, or 0.198 kg. Therefore,

$$m = \frac{0.249 \text{ mol } H_2SO_4}{0.198 \text{ kg } H_2O}$$
$$= 1.26\, m$$

Practice Exercise What is the molality of a solution containing 7.78 g of urea [$(NH_2)_2CO$] in 203 g of water?

Similar problem: 12.17.

Comparison of Concentration Units

The choice of a concentration unit is based on the purpose of the experiment. For instance, the mole fraction is not used to express the concentrations of solutions for titrations and gravimetric analyses, but it is appropriate for calculating partial pressures of gases (see Section 5.6) and for dealing with vapor pressures of solutions (to be discussed later in this chapter).

The advantage of molarity is that it is generally easier to measure the volume of a solution, using precisely calibrated volumetric flasks, than to weigh the solvent, as we saw in Section 4.5. For this reason, molarity is often preferred over molality. On the other hand, molality is independent of temperature, because the concentration is expressed in number of moles of solute and mass of solvent. The volume of a solution typically increases with increasing temperature, so that a solution that is 1.0 *M* at 25°C may become 0.97 *M* at 45°C because of the increase in volume on warming. This concentration dependence on temperature can significantly affect the accuracy of an experiment. Therefore, it is sometimes preferable to use molality instead of molarity.

Percent by mass is similar to molality in that it is independent of temperature. Furthermore, because it is defined in terms of ratio of mass of solute to mass of solution, we do not need to know the molar mass of the solute to calculate the percent by mass.

Sometimes it is desirable to convert one concentration unit of a solution to another; for example, the same solution may be employed for different experiments that require different concentration units for calculations. Suppose we want to express the concentration of a 0.396 *m* glucose ($C_6H_{12}O_6$) solution in molarity. We know there is 0.396 mole of glucose in 1000 g of the solvent and we need to determine the volume of this solution to calculate molarity. First, we calculate the mass of the solution from the molar mass of glucose:

$$\left(0.396\ \cancel{\text{mol } C_6H_{12}O_6} \times \frac{180.2\ \text{g}}{1\ \cancel{\text{mol } C_6H_{12}O_6}}\right) + 1000\ \text{g } H_2O = 1071\ \text{g}$$

The next step is to experimentally determine the density of the solution, which is found to be 1.16 g/mL. We can now calculate the volume of the solution in liters by writing

$$\begin{aligned}\text{volume} &= \frac{\text{mass}}{\text{density}}\\ &= \frac{1071\ \text{g}}{1.16\ \text{g/mL}} \times \frac{1\ \text{L}}{1000\ \text{mL}}\\ &= 0.923\ \text{L}\end{aligned}$$

Finally, the molarity of the solution is given by

$$\begin{aligned}\text{molarity} &= \frac{\text{moles of solute}}{\text{liters of soln}}\\ &= \frac{0.396\ \text{mol}}{0.923\ \text{L}}\\ &= 0.429\ \text{mol/L} = 0.429\ M\end{aligned}$$

As you can see, the density of the solution serves as a conversion factor between molality and molarity.

Examples 12.4 and 12.5 show concentration unit conversions.

Example 12.4

The density of a 2.45 *M* aqueous solution of methanol (CH_3OH) is 0.976 g/mL. What is the molality of the solution? The molar mass of methanol is 32.04 g.

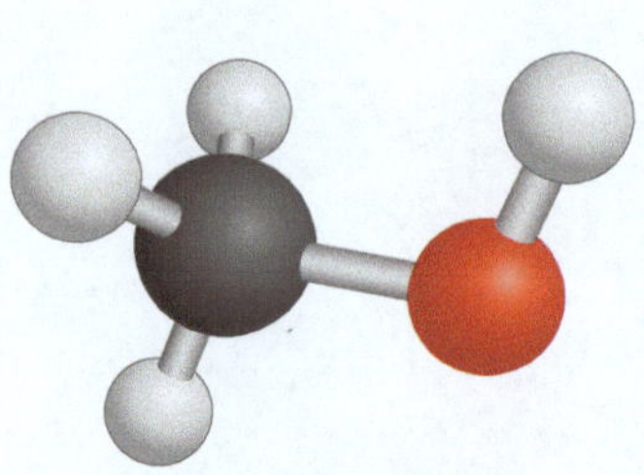

CH_3OH

Strategy To calculate the molality, we need to know the number of moles of methanol and the mass of solvent in kilograms. We assume 1 L of solution, so the number of moles of methanol is 2.45 mol.

$$m = \frac{\text{moles of solute}}{\text{mass of solvent (kg)}}$$

given (moles of solute); want to calculate (m); need to find (mass of solvent)

(Continued)

Solution Our first step is to calculate the mass of water in 1 L of the solution, using density as a conversion factor. The total mass of 1 L of a 2.45 *M* solution of methanol is

$$1 \text{ L soln} \times \frac{1000 \text{ mL soln}}{1 \text{ L soln}} \times \frac{0.976 \text{ g}}{1 \text{ mL soln}} = 976 \text{ g}$$

Because this solution contains 2.45 moles of methanol, the amount of water (solvent) in the solution is

$$\begin{aligned}\text{mass of } H_2O &= \text{mass of soln} - \text{mass of solute}\\ &= 976 \text{ g} - \left(2.45 \text{ mol } CH_3OH \times \frac{32.04 \text{ g } CH_3OH}{1 \text{ mol } CH_3OH}\right)\\ &= 898 \text{ g}\end{aligned}$$

The molality of the solution can be calculated by converting 898 g to 0.898 kg:

$$\begin{aligned}\text{molality} &= \frac{2.45 \text{ mol } CH_3OH}{0.898 \text{ kg } H_2O}\\ &= 2.73\ m\end{aligned}$$

Practice Exercise Calculate the molality of a 5.86 *M* ethanol (C_2H_5OH) solution whose density is 0.927 g/mL.

Similar problems: 12.18(a), 12.19.

Example 12.5

Calculate the molality of a 35.4 percent (by mass) aqueous solution of phosphoric acid (H_3PO_4). The molar mass of phosphoric acid is 97.99 g.

Strategy In solving this type of problem, it is convenient to assume that we start with 100.0 g of the solution. If the mass of phosphoric acid is 35.4 percent, or 35.4 g, the percent by mass and mass of water must be 100.0% − 35.4% = 64.6% and 64.6 g.

Solution From the known molar mass of phosphoric acid, we can calculate the molality in two steps, as shown in Example 12.3. First we calculate the number of moles of phosphoric acid in 35.4 g of the acid:

$$\begin{aligned}\text{moles of } H_3PO_4 &= 35.4 \text{ g } H_3PO_4 \times \frac{1 \text{ mol } H_3PO_4}{97.99 \text{ g } H_3PO_4}\\ &= 0.361 \text{ mol } H_3PO_4\end{aligned}$$

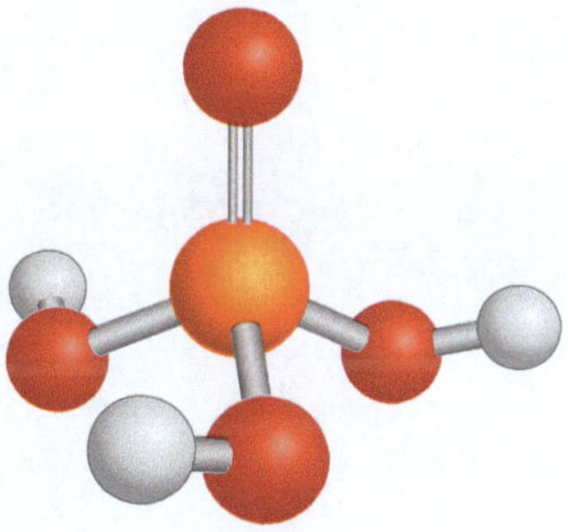

H_3PO_4

The mass of water is 64.6 g, or 0.0646 kg. Therefore, the molality is given by

$$\begin{aligned}\text{molality} &= \frac{0.361 \text{ mol } H_3PO_4}{0.0646 \text{ kg } H_2O}\\ &= 5.59\ m\end{aligned}$$

Practice Exercise Calculate the molality of a 44.6 percent (by mass) aqueous solution of sodium chloride.

Similar problem: 12.18(b).

Summary of Concepts & Facts

- The concentration of a solution can be expressed as percent by mass, mole fraction, molarity, and molality. The choice of units depends on the circumstances.

Review of Concepts & Facts

12.3.1 A solution is prepared at 20°C and its concentration is expressed in three different units: percent by mass, molality, and molarity. The solution is then heated to 88°C. Which of the concentration units will change (increase or decrease)?

12.3.2 Determine the percent composition by mass of LiCl in a solution prepared by dissolving 2.34 g of LiCl in 67.5 mL of water.

12.3.3 At 20°C, an aqueous solution is 24.0 percent HCl by mass and has a density of 1.119 g/mL. Calculate the molarity and the molality of this solution.

12.4 The Effect of Temperature on Solubility

Learning Objective

- Appraise the effect of temperature on solubility.

Recall that solubility is defined as the maximum amount of a solute that will dissolve in a given quantity of solvent *at a specific temperature.* Temperature affects the solubility of most substances. In this section we will consider the effects of temperature on the solubility of solids and gases.

Solid Solubility and Temperature

Figure 12.3 shows the temperature dependence of the solubility of some ionic compounds in water. In most but certainly not all cases, the solubility of a solid substance increases with temperature. However, there is no clear correlation between the sign of ΔH_{soln} and the variation of solubility with temperature. For example, the solution process of $CaCl_2$ is exothermic, and that of NH_4NO_3 is endothermic. But the solubility of both compounds increases with increasing temperature. In general, the effect of temperature on solubility is best determined experimentally.

Fractional Crystallization

The dependence of the solubility of a solid on temperature varies considerably, as Figure 12.3 shows. The solubility of $NaNO_3$, for example, increases sharply with

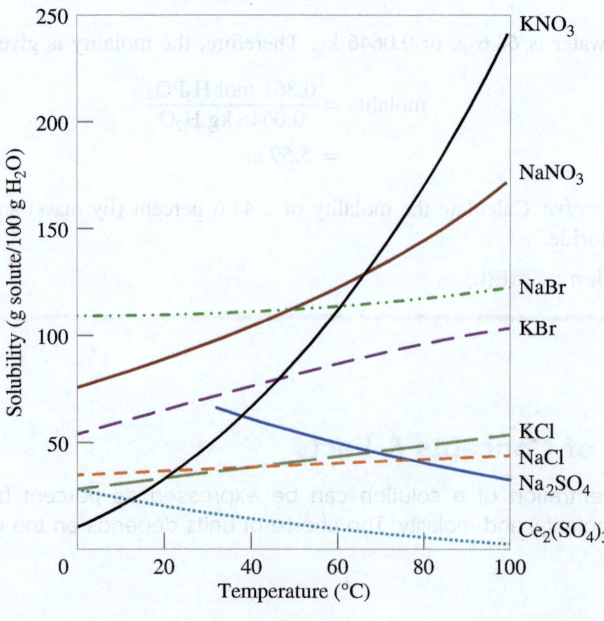

Figure 12.3 *Temperature dependence of the solubility of some ionic compounds in water.*

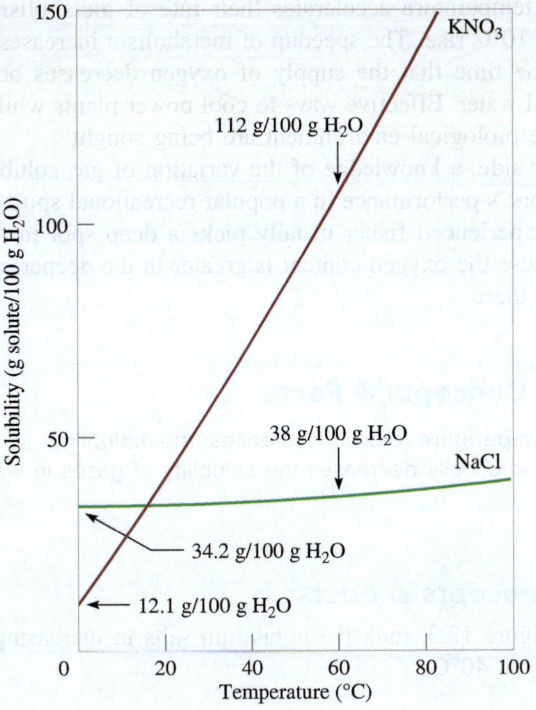

Figure 12.4 *The solubilities of KNO_3 and NaCl at 0°C and 60°C. The difference in temperature dependence enables us to isolate one of these compounds from a solution containing both of them, through fractional crystallization.*

temperature, while that of NaCl changes very little. This wide variation provides a means of obtaining pure substances from mixtures. ***Fractional crystallization*** is *the separation of a mixture of substances into pure components on the basis of their differing solubilities.*

Suppose we have a sample of 90 g of KNO_3 that is contaminated with 10 g of NaCl. To purify the KNO_3 sample, we dissolve the mixture in 100 mL of water at 60°C and then gradually cool the solution to 0°C. At this temperature, the solubilities of KNO_3 and NaCl are 12.1 g/100 g H_2O and 34.2 g/100 g H_2O, respectively. Thus, (90 − 12) g, or 78 g, of KNO_3 will crystallize out of the solution, but all of the NaCl will remain dissolved (Figure 12.4). In this manner, we can obtain about 90 percent of the original amount of KNO_3 in pure form. The KNO_3 crystals can be separated from the solution by filtration.

Many of the solid inorganic and organic compounds that are used in the laboratory were purified by fractional crystallization. Generally, the method works best if the compound to be purified has a steep solubility curve, that is, if it is considerably more soluble at high temperatures than at low temperatures. Otherwise, much of it will remain dissolved as the solution is cooled. Fractional crystallization also works well if the amount of impurity in the solution is relatively small.

Gas Solubility and Temperature

The solubility of gases in water usually decreases with increasing temperature (Figure 12.5). When water is heated in a beaker, you can see bubbles of air forming on the side of the glass before the water boils. As the temperature rises, the dissolved air molecules begin to "boil out" of the solution long before the water itself boils.

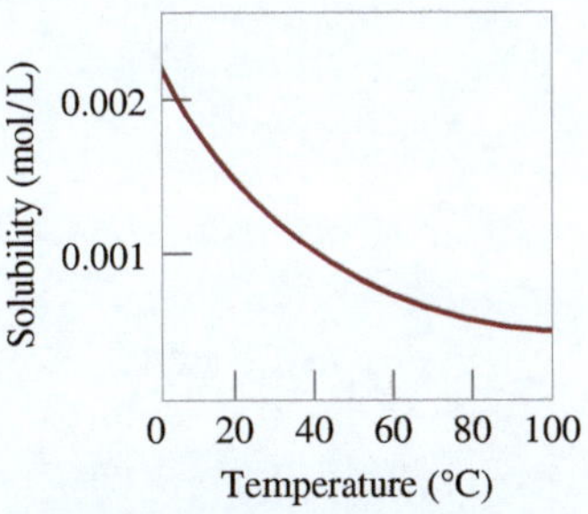

Figure 12.5 *Dependence on temperature of the solubility of O_2 gas in water. Note that the solubility decreases as temperature increases. The pressure of the gas over the solution is 1 atm.*

The reduced solubility of molecular oxygen in hot water has a direct bearing on *thermal pollution*—that is, the heating of the environment (usually waterways) to temperatures that are harmful to its living inhabitants. It is estimated that every year in the United States some 100,000 billion gallons of water are used for industrial cooling, mostly in electric power and nuclear power production. This process heats the water, which is then returned to the rivers and lakes from which it was taken. Ecologists have become increasingly concerned about the effect of thermal pollution on aquatic life. Fish, like all other cold-blooded animals, have much more difficulty coping with rapid temperature fluctuation in the environment than humans do. An increase in water temperature accelerates their rate of metabolism, which generally doubles with each 10°C rise. The speedup of metabolism increases the fish's need for oxygen at the same time that the supply of oxygen decreases because of its lower solubility in heated water. Effective ways to cool power plants while doing only minimal damage to the biological environment are being sought.

On the lighter side, a knowledge of the variation of gas solubility with temperature can improve one's performance in a popular recreational sport—fishing. On a hot summer day, an experienced fisher usually picks a deep spot in the river or lake to cast the bait. Because the oxygen content is greater in the deeper, cooler region, most fish will be found there.

Summary of Concepts & Facts

- Increasing temperature usually increases the solubility of solid and liquid substances and usually decreases the solubility of gases in water.

Review of Concepts & Facts

12.4.1 Using Figure 12.3, rank the potassium salts in increasing order of solubility at 40°C.

12.5 The Effect of Pressure on the Solubility of Gases

Learning Objective

- Utilize Henry's law to determine the solubility of a gas in solution.

For all practical purposes, external pressure has no influence on the solubilities of liquids and solids, but it does greatly affect the solubility of gases. The quantitative relationship between gas solubility and pressure is given by ***Henry's*[†] *law,*** which states that *the solubility of a gas in a liquid is proportional to the pressure of the gas over the solution:*

$$c \propto P$$

$$c = kP \qquad (12.3)$$

Here c is the molar concentration (mol/L) of the dissolved gas; P is the pressure (in atm) of the gas over the solution at equilibrium; and, for a given gas, k is a constant that depends only on temperature. Every gas has a different k value at a given temperature. The constant k has the units mol/L · atm. You can see that when the pressure of the gas is 1 atm, c is *numerically* equal to k. If several gases are present, P is the partial pressure.

Henry's law can be understood qualitatively in terms of the kinetic molecular theory. The amount of gas that will dissolve in a solvent depends on how frequently the gas molecules collide with the liquid surface and become trapped by the condensed phase. Suppose we have a gas in dynamic equilibrium with a solution [Figure 12.6(a)]. At every instant, the number of gas molecules entering the solution is equal to the number of dissolved molecules moving into the gas phase. If the partial pressure of the gas is increased [Figure 12.6(b)], more molecules dissolve in the liquid because more molecules are striking the surface of the liquid. This process continues until the

[†]William Henry (1775–1836). English chemist. Henry's major contribution to science was his discovery of the law describing the solubility of gases, which now bears his name.

Figure 12.6 *A molecular interpretation of Henry's law. When the partial pressure of the gas over the solution increases from (a) to (b), the concentration of the dissolved gas also increases according to Equation (12.3).*

concentration of the solution is again such that the number of molecules leaving the solution per second equals the number entering the solution. Because of the higher concentration of molecules in both the gas and solution phases, this number is greater in (b) than in (a), where the partial pressure is lower.

A practical demonstration of Henry's law is the effervescence of a soft drink when the cap of the bottle is removed. Before the beverage bottle is sealed, it is pressurized with a mixture of air and CO_2 saturated with water vapor. Because of the high partial pressure of CO_2 in the pressurizing gas mixture, the amount dissolved in the soft drink is many times the amount that would dissolve under normal atmospheric conditions. When the cap is removed, the pressurized gases escape, eventually the pressure in the bottle falls to atmospheric pressure, and the amount of CO_2 remaining in the beverage is determined only by the normal atmospheric partial pressure of CO_2, 0.0003 atm. The excess dissolved CO_2 comes out of solution, causing the effervescence.

Example 12.6 applies Henry's law to nitrogen gas.

The effervescence of a soft drink. The bottle was shaken before being opened to dramatize the escape of CO_2.

Ken Karp/McGraw Hill

Example 12.6

The solubility of nitrogen gas at 25°C and 1 atm is 6.8×10^{-4} mol/L. What is the concentration (in molarity) of nitrogen dissolved in water under atmospheric conditions? The partial pressure of nitrogen gas in the atmosphere is 0.78 atm.

Strategy The given solubility enables us to calculate Henry's law constant (k), which can then be used to determine the concentration of the solution.

Solution The first step is to calculate the quantity k in Equation (12.3):

$$c = kP$$
$$6.8 \times 10^{-4} \text{ mol/L} = k\,(1 \text{ atm})$$
$$k = 6.8 \times 10^{-4} \text{ mol/L} \cdot \text{atm}$$

Therefore, the solubility of nitrogen gas in water is

$$\begin{aligned} c &= (6.8 \times 10^{-4} \text{ mol/L} \cdot \text{atm})(0.78 \text{ atm}) \\ &= 5.3 \times 10^{-4} \text{ mol/L} \\ &= 5.3 \times 10^{-4}\ M \end{aligned}$$

The decrease in solubility is the result of lowering the pressure from 1 atm to 0.78 atm.

Check The ratio of the concentrations [$(5.3 \times 10^{-4}\ M/6.8 \times 10^{-4}\ M) = 0.78$] should be equal to the ratio of the pressures (0.78 atm/1.0 atm = 0.78).

Practice Exercise Calculate the molar concentration of oxygen in water at 25°C for a partial pressure of 0.22 atm. The Henry's law constant for oxygen is 1.3×10^{-3} mol/L · atm.

Similar problem: 12.37.

CHEMISTRY *in Action*

The Killer Lake

Disaster struck swiftly and without warning. On August 21, 1986, Lake Nyos in Cameroon, a small nation on the west coast of Africa, suddenly belched a dense cloud of carbon dioxide. Speeding down a river valley, the cloud asphyxiated over 1700 people and many livestock.

How did this tragedy happen? Lake Nyos is stratified into layers that do not mix. A boundary separates the freshwater at the surface from the deeper, denser solution containing dissolved minerals and gases, including CO_2. The CO_2 gas comes from springs of carbonated groundwater that percolate upward into the bottom of the volcanically formed lake. Given the high water pressure at the bottom of the lake, the concentration of CO_2 gradually accumulated to a dangerously high level, in accordance with Henry's law. What triggered the release of CO_2 is not known for certain. It is believed that an earthquake, landslide, or even strong winds may have upset the delicate balance within the lake, creating waves that overturned the water layers. When the deep water rose, dissolved CO_2 came out of solution, just as a soft drink fizzes when the bottle is uncapped. Being heavier than air, the CO_2 traveled close to the ground and literally smothered an entire village 15 miles away.

Now, more than 35 years after the incident, scientists are concerned that the CO_2 concentration at the bottom of Lake Nyos is again reaching saturation level. To prevent a recurrence of the earlier tragedy, an attempt has been made to pump up the deep water, thus releasing the dissolved CO_2. In addition to being costly, this approach is controversial because it might disturb the waters near the bottom of the lake, leading to an uncontrollable release of CO_2 to the surface. In the meantime, a natural time bomb is ticking away.

Deep waters in Lake Nyos are pumped to the surface to remove dissolved CO_2 gas.

Bill Evans, U.S. Geological Survey

Most gases obey Henry's law, but there are some important exceptions. For example, if the dissolved gas *reacts* with water, higher solubilities can result. The solubility of ammonia is much higher than expected because of the reaction

$$NH_3 + H_2O \rightleftharpoons NH_4^+ + OH^-$$

Carbon dioxide also reacts with water, as follows:

$$CO_2 + H_2O \rightleftharpoons H_2CO_3$$

Another interesting example is the dissolution of molecular oxygen in blood. Normally, oxygen gas is only sparingly soluble in water (see Practice Exercise in Example 12.6). However, its solubility in blood is dramatically greater because of the high content of hemoglobin (Hb) molecules. Each hemoglobin molecule can bind up to four oxygen molecules, which are eventually delivered to the tissues for use in metabolism:

$$Hb + 4O_2 \rightleftharpoons Hb(O_2)_4$$

It is this process that accounts for the high solubility of molecular oxygen in blood.

The Chemistry in Action essay "The Killer Lake" explains a natural disaster with Henry's law.

Summary of Concepts & Facts

- According to Henry's law, the solubility of a gas in a liquid is directly proportional to the partial pressure of the gas over the solution.

Review of Concepts & Facts

12.5.1 Which of the following gases has the greatest Henry's law constant in water at 25°C: CH_4, Ne, HCl, H_2?

12.5.2 Calculate the molar concentration of carbon dioxide in water at 25°C for a partial pressure of 0.0025 atm. The Henry's law constant for carbon dioxide is 3.1×10^{-2} mol/L · atm.

12.6 Colligative Properties of Nonelectrolyte Solutions

Learning Objective

- Perform calculations involving various colligative properties of nonelectrolyte solutions, including vapor-pressure lowering, boiling-point elevation, freezing-point depression, and osmotic pressure.

Colligative properties (or collective properties) are *properties that depend only on the number of solute particles in solution and not on the nature of the solute particles.* These properties are bound together by a common origin—they all depend on the number of solute particles present, regardless of whether they are atoms, ions, or molecules. The colligative properties are vapor-pressure lowering, boiling-point elevation, freezing-point depression, and osmotic pressure. For our discussion of colligative properties of nonelectrolyte solutions it is important to keep in mind that we are talking about relatively dilute solutions, that is, solutions whose concentrations are $\leq 0.2\ M$.

Vapor-Pressure Lowering

If a solute is ***nonvolatile*** (that is, *it does not have a measurable vapor pressure*), the vapor pressure of its solution is always less than that of the pure solvent. Thus, the relationship between solution vapor pressure and solvent vapor pressure depends on the concentration of the solute in the solution. This relationship is expressed by ***Raoult's*****[†]** ***law,*** which states that *the vapor pressure of a solvent over a solution, P_1, is given by the vapor pressure of the pure solvent, P_1°, times the mole fraction of the solvent in the solution, X_1:*

$$P_1 = X_1 P_1^\circ \tag{12.4}$$

In a solution containing only one solute, $X_1 = 1 - X_2$, where X_2 is the mole fraction of the solute. Equation (12.4) can therefore be rewritten as

$$P_1 = (1 - X_2)P_1^\circ$$

or

$$P_1 = P_1^\circ - X_2 P_1^\circ$$

so that

$$P_1^\circ - P_1 = \Delta P = X_2 P_1^\circ \tag{12.5}$$

[†]François Marie Raoult (1830–1901). French chemist. Raoult's work was mainly in solution properties and electrochemistry.

Student Hot Spot

Student data indicate you may struggle with Raoult's law. Access your eBook for additional Learning Resources on this topic.

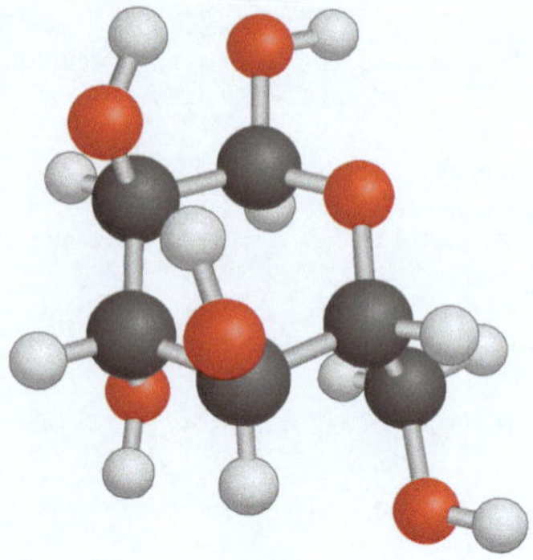

$C_6H_{12}O_6$

We see that the *decrease* in vapor pressure, ΔP, is directly proportional to the solute concentration (measured in mole fraction).

Example 12.7 illustrates the use of Raoult's law [Equation (12.5)].

Example 12.7

Calculate the vapor pressure of a solution made by dissolving 218 g of glucose (molar mass = 180.2 g/mol) in 460 mL of water at 30°C. What is the vapor-pressure lowering? The vapor pressure of pure water at 30°C is given in Table 5.3. Assume the density of the solvent is 1.00 g/mL.

Strategy We need Raoult's law [Equation (12.4)] to determine the vapor pressure of a solution. Note that glucose is a nonvolatile solute.

Solution The vapor pressure of a solution (P_1) is

$$\underset{\text{want to calculate}}{P_1} = \overset{\text{need to find}}{X_1}\underset{\text{given}}{P_1^\circ}$$

First we calculate the number of moles of glucose and water in the solution:

$$n_1(\text{water}) = 460 \text{ mL} \times \frac{1.00 \text{ g}}{1 \text{ mL}} \times \frac{1 \text{ mol}}{18.02 \text{ g}} = 25.5 \text{ mol}$$

$$n_2(\text{glucose}) = 218 \text{ g} \times \frac{1 \text{ mol}}{180.2 \text{ g}} = 1.21 \text{ mol}$$

The mole fraction of water, X_1, is given by

$$X_1 = \frac{n_1}{n_1 + n_2}$$

$$= \frac{25.5 \text{ mol}}{25.5 \text{ mol} + 1.21 \text{ mol}} = 0.955$$

From Table 5.3, we find the vapor pressure of water at 30°C to be 31.82 mmHg. Therefore, the vapor pressure of the glucose solution is

$$P_1 = 0.955 \times 31.82 \text{ mmHg}$$

$$= 30.4 \text{ mmHg}$$

Finally, the vapor-pressure lowering (ΔP) is (31.82 − 30.4) mmHg, or 1.4 mmHg.

Check We can also calculate the vapor pressure lowering by using Equation (12.5). Because the mole fraction of glucose is (1 − 0.955), or 0.045, the vapor pressure lowering is given by (0.045)(31.82 mmHg) or 1.4 mmHg.

Practice Exercise Calculate the vapor pressure of a solution made by dissolving 82.4 g of urea (molar mass = 60.06 g/mol) in 212 mL of water at 35°C. What is the vapor-pressure lowering?

Similar problems: 12.49, 12.50.

Why is the vapor pressure of a solution less than that of the pure solvent? As was mentioned in Section 12.2, one driving force in physical and chemical processes is an increase in disorder—the greater the disorder, the more favorable the process. Vaporization increases the disorder of a system because molecules in a vapor have less order than those in a liquid. Because a solution is more disordered than a pure solvent, the difference in disorder between a solution and a vapor is less than that between a pure solvent and a vapor. Thus, solvent molecules have less of a tendency to leave a solution than to leave the pure solvent to become vapor, and the vapor pressure of a solution is less than that of the solvent.

If both components of a solution are ***volatile*** (that is, *have measurable vapor pressure*), the vapor pressure of the solution is the sum of the individual partial pressures. Raoult's law holds equally well in this case:

$$P_A = X_A P_A^\circ$$
$$P_B = X_B P_B^\circ$$

where P_A and P_B are the partial pressures over the solution for components A and B, P_A° and P_B° are the vapor pressures of the pure substances, and X_A and X_B are their mole fractions. The total pressure is given by Dalton's law of partial pressure (see Section 5.6):

$$P_T = P_A + P_B$$

or

$$P_T = X_A P_A^\circ + X_B P_B^\circ$$

For example, benzene and toluene are volatile components that have similar structures and therefore similar intermolecular forces:

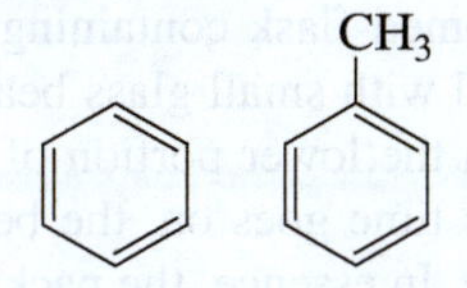

benzene toluene

In a solution of benzene and toluene, the vapor pressure of each component obeys Raoult's law. Figure 12.7 shows the dependence of the total vapor pressure (P_T) in a benzene-toluene solution on the composition of the solution. Note that we need only express the composition of the solution in terms of the mole fraction of one component. For every value of $X_{benzene}$, the mole fraction of toluene, $X_{toluene}$, is given by $(1 - X_{benzene})$. The benzene-toluene solution is one of the few examples of an ***ideal solution,*** which is *any solution that obeys Raoult's law.* One characteristic of an ideal solution is that the heat of solution, ΔH_{soln}, is zero.

Most solutions do not behave ideally in this respect. Designating two volatile substances as A and B, we can consider the following two cases:

Case 1: If the intermolecular forces between A and B molecules are weaker than those between A molecules and between B molecules, then there is a greater tendency for these molecules to leave the solution than in the case of an ideal solution. Consequently, the vapor pressure of the solution is greater than the sum of the vapor pressures as predicted by Raoult's law for the same concentration. This behavior gives rise to the *positive deviation* [Figure 12.8(a)]. In this case, the heat of solution is positive (that is, mixing is an endothermic process).

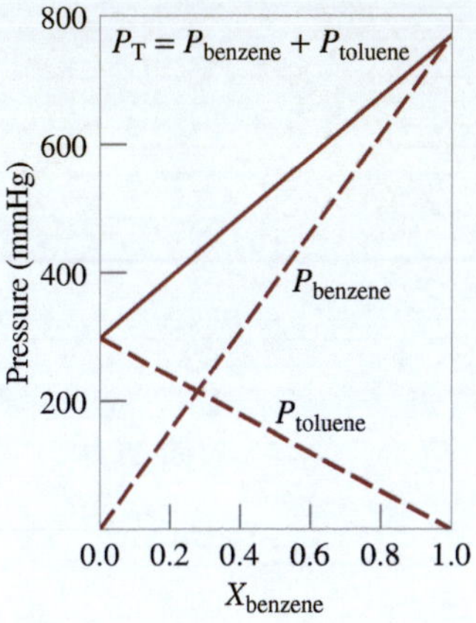

Figure 12.7 *The dependence of the partial pressures of benzene and toluene on their mole fractions in a benzene-toluene solution ($X_{toluene} = 1 - X_{benzene}$) at 80°C. This solution is said to be ideal because the vapor pressures obey Raoult's law.*

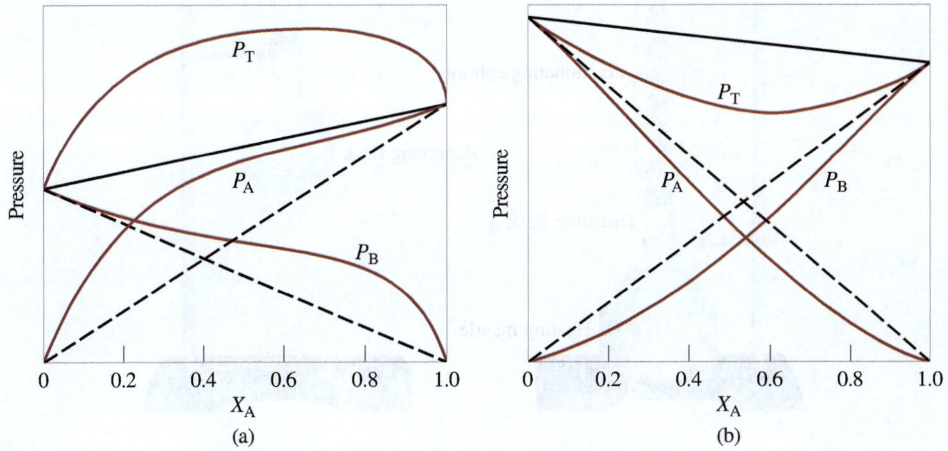

Figure 12.8 *Nonideal solutions. (a) Positive deviation occurs when P_T is greater than that predicted by Raoult's law (the solid black line). (b) Negative deviation. Here, P_T is less than that predicted by Raoult's law (the solid black line).*

Case 2: If A molecules attract B molecules more strongly than they do their own kind, the vapor pressure of the solution is less than the sum of the vapor pressures as predicted by Raoult's law. Here we have a *negative deviation* [Figure 12.8(b)]. In this case, the heat of solution is negative (that is, mixing is an exothermic process).

Fractional Distillation

Solution vapor pressure has a direct bearing on ***fractional distillation,*** *a procedure for separating liquid components of a solution based on their different boiling points.* Fractional distillation is somewhat analogous to fractional crystallization. Suppose we want to separate a *binary system* (a system with two components), say, benzene-toluene. Both benzene and toluene are relatively volatile, yet their boiling points are appreciably different (80.1°C and 110.6°C, respectively). When we boil a solution containing these two substances, the vapor formed is somewhat richer in the more volatile component, benzene. If the vapor is condensed in a separate container and that liquid is boiled again, a still higher concentration of benzene will be obtained in the vapor phase. By repeating this process many times, it is possible to separate benzene completely from toluene.

In practice, chemists use an apparatus like that shown in Figure 12.9 to separate volatile liquids. The round-bottomed flask containing the benzene-toluene solution is fitted with a long column packed with small glass beads. When the solution boils, the vapor condenses on the beads in the lower portion of the column, and the liquid falls back into the distilling flask. As time goes on, the beads gradually heat up, allowing the vapor to move upward slowly. In essence, the packing material causes the benzene-toluene mixture to be subjected continuously to numerous vaporization-condensation steps. At each step the composition of the vapor in the column will be richer in the more volatile, or lower boiling-point, component (in this case, benzene). The vapor that rises to the top of the column is essentially pure benzene, which is then condensed and collected in a receiving flask.

Fractional distillation is as important in industry as it is in the laboratory. The petroleum industry employs fractional distillation on a large scale to separate the components of crude oil. More will be said of this process in Chapter 24.

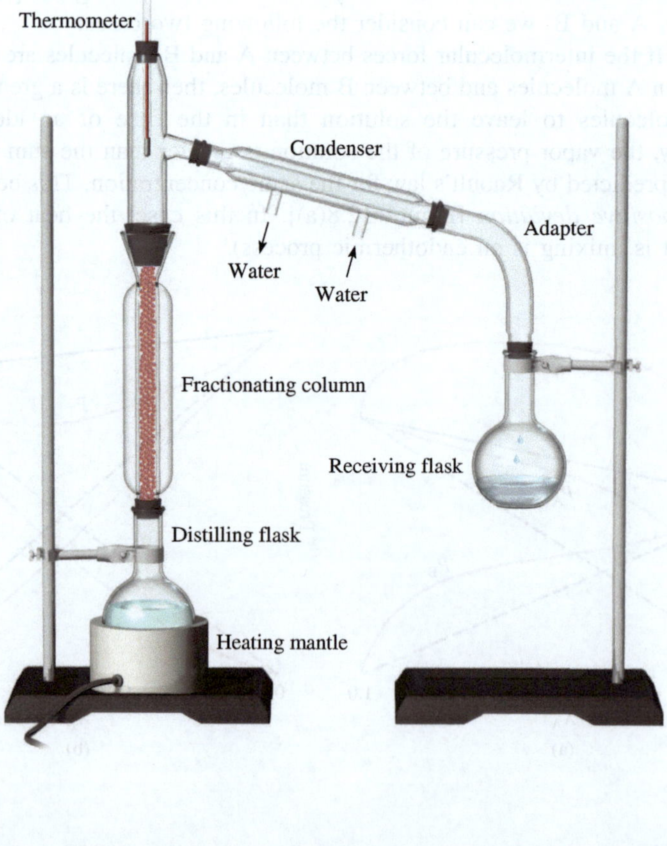

Figure 12.9 *An apparatus for small-scale fractional distillation. The fractionating column is packed with tiny glass beads. The longer the fractionating column, the more complete the separation of the volatile liquids.*

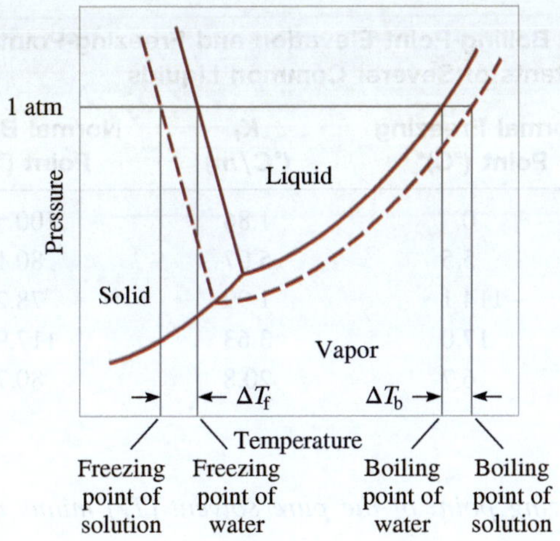

Figure 12.10 *Phase diagram illustrating the boiling-point elevation and freezing-point depression of aqueous solutions. The dashed curves pertain to the solution, and the solid curves to the pure solvent. As you can see, the boiling point of the solution is higher than that of water, and the freezing point of the solution is lower than that of water.*

Boiling-Point Elevation

The boiling point of a solution is the temperature at which its vapor pressure equals the external atmospheric pressure (see Section 11.8). Because the presence of a nonvolatile solute lowers the vapor pressure of a solution, it must also affect the boiling point of the solution. Figure 12.10 shows the phase diagram of water and the changes that occur in an aqueous solution. Because at any temperature the vapor pressure of the solution is lower than that of the pure solvent regardless of temperature, the liquid-vapor curve for the solution lies below that for the pure solvent. Consequently, the dashed solution curve intersects the horizontal line that marks $P = 1$ atm at a *higher* temperature than the normal boiling point of the pure solvent. This graphical analysis shows that the boiling point of the solution is higher than that of water. The ***boiling-point elevation*** **(ΔT_b)** is defined as *the boiling point of the solution (T_b) minus the boiling point of the pure solvent (T_b°):*

$$\Delta T_b = T_b - T_b^\circ$$

Because $T_b > T_b^\circ$, ΔT_b is a positive quantity.

The value of ΔT_b is proportional to the vapor-pressure lowering, and so it is also proportional to the concentration (molality) of the solution. That is,

$$\Delta T_b \propto m$$

$$\Delta T_b = K_b m \qquad (12.6)$$

where m is the molality of the solution and K_b is the *molal boiling-point elevation constant.* In calculating the new boiling point, we add ΔT_b to the normal boiling point of the solvent. The units of K_b are °C/m. It is important to understand the choice of concentration unit here. We are dealing with a system (the solution) whose temperature is *not* constant, so we cannot express the concentration units in molarity because molarity changes with temperature.

Table 12.2 lists values of K_b for several common solvents. Using the boiling-point elevation constant for water and Equation (12.6), you can see that if the molality of an aqueous solution is 1.00 m, the boiling point will be 100.51°C.

Freezing-Point Depression

A nonscientist may remain forever unaware of the boiling-point elevation phenomenon, but a careful observer living in a cold climate is familiar with freezing-point depression. Ice on frozen roads and sidewalks melts when sprinkled with salts such as NaCl or $CaCl_2$. This method of thawing succeeds because it depresses the freezing point of water.

Figure 12.10 shows that lowering the vapor pressure of the solution shifts the solid-liquid curve to the left. Consequently, this line intersects the horizontal line at a temperature *lower* than the freezing point of water. The ***freezing-point depression*** **(ΔT_f)**

Deicing of airplanes is based on freezing-point depression.
Meinzahn/iStockphoto/Getty Images

Table 12.2 **Molal Boiling-Point Elevation and Freezing-Point Depression Constants of Several Common Liquids**

Solvent	Normal Freezing Point (°C)*	K_f (°C/m)	Normal Boiling Point (°C)*	K_b (°C/m)
Water	0	1.86	100	0.51
Benzene	5.5	5.07	80.1	2.64
Ethanol	−114.1	1.99	78.2	1.23
Acetic acid	17.0	3.63	117.9	2.93
Cyclohexane	6.7	20.8	80.7	2.92

*Measured at 1 atm.

is defined as *the freezing point of the pure solvent* (T_f°) *minus the freezing point of the solution* (T_f):

$$\Delta T_f = T_f^\circ - T_f$$

Because $T_f^\circ > T_f$, ΔT_f is a positive quantity. Again, ΔT_f is proportional to the concentration of the solution:

$$\Delta T_f \propto m$$

$$\Delta T_f = K_f m \tag{12.7}$$

where m is the concentration of the solute in molality units, and K_f is the *molal freezing-point depression constant* (see Table 12.2). To determine the new freezing point, we subtract ΔT_f from the normal freezing point of the solvent. Like K_b, K_f has the units °C/m.

A qualitative explanation of the freezing-point depression phenomenon is as follows: Freezing involves a transition from the disordered state to the ordered state. For this to happen, energy must be removed from the system. Because a solution has greater disorder than the solvent, more energy needs to be removed from it to create order than in the case of a pure solvent. Therefore, the solution has a lower freezing point than its solvent. Note that when a solution freezes, the solid that separates is the pure solvent component.

Student Hot Spot

Student data indicate you may struggle with freezing-point depression. Access your eBook for additional Learning Resources on this topic.

For boiling-point elevation to occur, the solute must be nonvolatile, but no such restriction applies to freezing-point depression. For example, methanol (CH_3OH), a fairly volatile liquid that boils at only 65°C, has sometimes been used as an antifreeze in automobile radiators.

A practical application of the freezing-point depression is described in Example 12.8.

In cold climate regions, antifreeze must be used in car radiators in winter.

Ken Karp/McGraw Hill

Example 12.8

Ethylene glycol (EG), $CH_2(OH)CH_2(OH)$, is a common automobile antifreeze. It is water soluble and fairly nonvolatile (b.p. 197°C). Calculate the freezing point of a solution containing 651 g of this substance in 2505 g of water. Would you keep this substance in your car radiator during the summer? The molar mass of ethylene glycol is 62.01 g.

Strategy This question asks for the depression in freezing point of the solution.

$$\Delta T_f = K_f m$$

(ΔT_f: want to calculate; K_f: constant; m: need to find)

The information given enables us to calculate the molality of the solution and we refer to Table 12.2 for the K_f of water.

(Continued)

Solution To solve for the molality of the solution, we need to know the number of moles of EG and the mass of the solvent in kilograms. We find the molar mass of EG, and convert the mass of the solvent to 2.505 kg, and calculate the molality as follows:

$$651\ \text{g EG} \times \frac{1\ \text{mol EG}}{62.07\ \text{g EG}} = 10.5\ \text{mol EG}$$

$$m = \frac{\text{moles of solute}}{\text{mass of solvent (kg)}}$$

$$= \frac{10.5\ \text{mol EG}}{2.505\ \text{kg H}_2\text{O}} = 4.19\ \text{mol EG/kg H}_2\text{O}$$

$$= 4.19\ m$$

From Equation (12.7) and Table 12.2 we write

$$\Delta T_f = K_f m$$

$$= (1.86°\text{C}/m)(4.19\ m)$$

$$= 7.79°\text{C}$$

Because pure water freezes at 0°C, the solution will freeze at (0 − 7.79)°C or −7.79°C. We can calculate boiling-point elevation in the same way as follows:

$$\Delta T_b = K_b m$$

$$= (0.51°\text{C}/m)(4.19\ m)$$

$$= 2.1°\text{C}$$

Because the solution will boil at (100 + 2.1)°C, or 102.1°C, it would be preferable to leave the antifreeze in your car radiator in summer to prevent the solution from boiling.

Practice Exercise Calculate the boiling point and freezing point of a solution containing 478 g of ethylene glycol in 3202 g of water.

Similar problems: 12.56, 12.59.

Osmotic Pressure

Video: Osmosis

Many chemical and biological processes depend on ***osmosis***, *the selective passage of solvent molecules through a porous membrane from a dilute solution to a more concentrated one.* Figure 12.11 illustrates this phenomenon. The left compartment of the apparatus contains pure solvent; the right compartment contains a solution. The two compartments are separated by a ***semipermeable membrane***, which *allows the passage of solvent molecules but blocks the passage of solute molecules.* At the start, the water levels in the two tubes are equal [see Figure 12.11(a)]. After some time, the level in the right tube begins to rise and continues to go up until equilibrium is reached, that is, until no further change can be observed. The ***osmotic pressure*** (π)

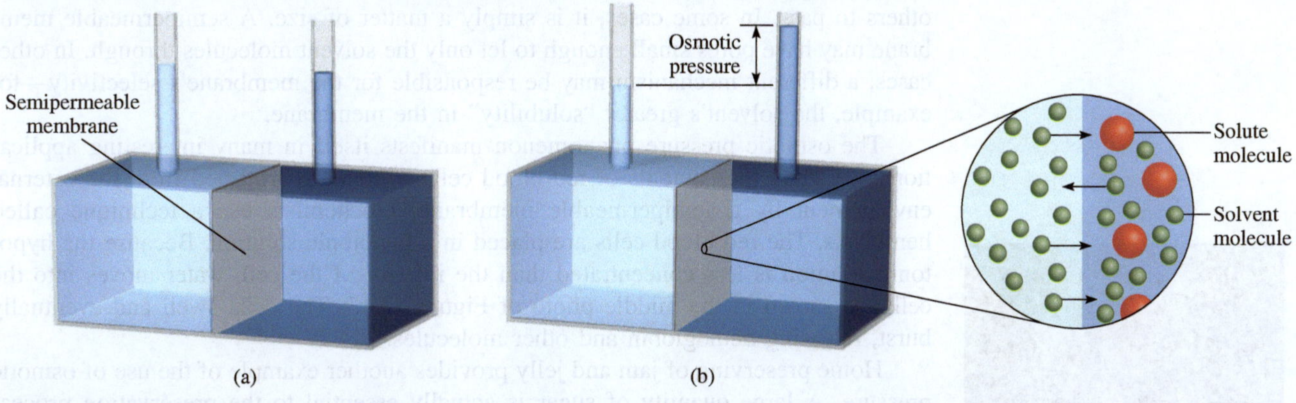

Figure 12.11 *Osmotic pressure. (a) The levels of the pure solvent (left) and of the solution (right) are equal at the start. (b) During osmosis, the level on the solution side rises as a result of the net flow of solvent from left to right. The osmotic pressure is equal to the hydrostatic pressure exerted by the column of fluid in the right tube at equilibrium. Basically the same effect occurs when the pure solvent is replaced by a more dilute solution than that on the right.*

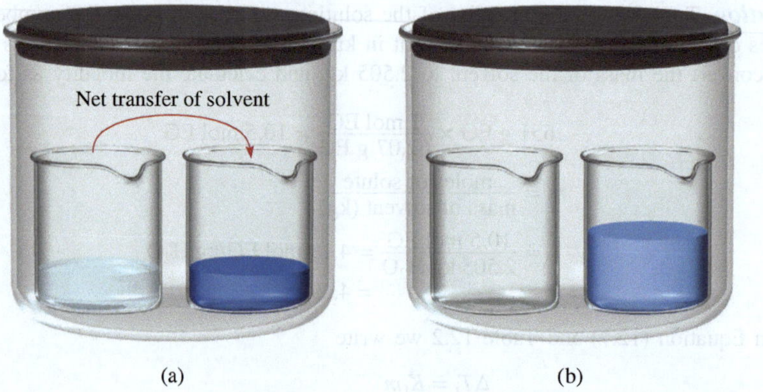

Figure 12.12 *(a) Unequal vapor pressures inside the container lead to a net transfer of water from the left beaker (which contains pure water) to the right one (which contains a solution). (b) At equilibrium, all the water in the left beaker has been transferred to the right beaker. This driving force for solvent transfer is analogous to the osmotic phenomenon that is shown in Figure 12.11.*

of a solution is *the pressure required to stop osmosis.* As shown in Figure 12.11(b), this pressure can be measured directly from the difference in the final fluid levels.

What causes water to move spontaneously from left to right in this case? The situation depicted in Figure 12.12 helps us understand the driving force behind osmosis. Because the vapor pressure of pure water is higher than the vapor pressure of the solution, there is a net transfer of water from the left beaker to the right one. Given enough time, the transfer will continue until no more water remains in the left beaker. A similar driving force causes water to move from the pure solvent into the solution during osmosis.

The osmotic pressure of a solution is given by

$$\pi = MRT \tag{12.8}$$

where M is the molarity of solution, R is the gas constant (0.0821 L · atm/K · mol), and T is the absolute temperature. The osmotic pressure, π, is expressed in atm. Because osmotic pressure measurements are carried out at constant temperature, we express the concentration in terms of the more convenient units of molarity rather than molality.

Like boiling-point elevation and freezing-point depression, osmotic pressure is directly proportional to the concentration of solution. This is what we would expect, because all colligative properties depend only on the number of solute particles in solution. If two solutions are of equal concentration and, hence, have the same osmotic pressure, they are said to be *isotonic.* If two solutions are of unequal osmotic pressures, the more concentrated solution is said to be *hypertonic* and the more dilute solution is described as *hypotonic* (Figure 12.13).

Although osmosis is a common and well-studied phenomenon, relatively little is known about how the semipermeable membrane stops some molecules yet allows others to pass. In some cases, it is simply a matter of size. A semipermeable membrane may have pores small enough to let only the solvent molecules through. In other cases, a different mechanism may be responsible for the membrane's selectivity—for example, the solvent's greater "solubility" in the membrane.

The osmotic pressure phenomenon manifests itself in many interesting applications. To study the contents of red blood cells, which are protected from the external environment by a semipermeable membrane, biochemists use a technique called hemolysis. The red blood cells are placed in a hypotonic solution. Because the hypotonic solution is less concentrated than the interior of the cell, water moves into the cells, as shown in the middle photo of Figure 12.13. The cells swell and eventually burst, releasing hemoglobin and other molecules.

California redwoods.
Bas Vermolen/Moment Open/Getty Images

Home preserving of jam and jelly provides another example of the use of osmotic pressure. A large quantity of sugar is actually essential to the preservation process because the sugar helps to kill bacteria that may cause botulism. As the last photo of Figure 12.13 shows, when a bacterial cell is in a hypertonic (high-concentration) sugar solution, the intracellular water tends to move out of the bacterial cell to the more concentrated solution by osmosis. This process, known as *crenation,* causes the

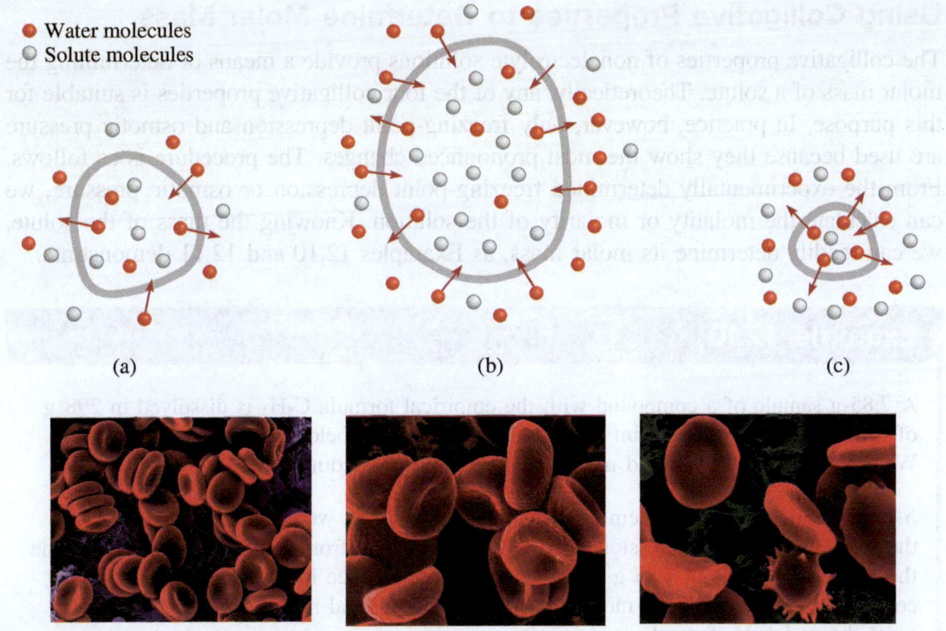

Figure 12.13 *From left to right: a red blood cell in an isotonic solution (cell remains unchanged), in a hypotonic solution (cell swells), and in a hypertonic solution (cell shrinks).*
Dennis Kunkel Microscopy/Science Source

cell to shrink and, eventually, to cease functioning. The natural acidity of fruits also inhibits bacteria growth.

Osmotic pressure also is the major mechanism for transporting water upward in plants. Because leaves constantly lose water to the air, in a process called *transpiration,* the solute concentrations in leaf fluids increase. Water is pulled up through the trunk, branches, and stems of trees by osmotic pressure. Up to 10 to 15 atm pressure is necessary to transport water to the leaves at the tops of California's redwoods, which reach about 120 m in height. (The capillary action discussed in Section 11.3 is responsible for the rise of water only up to a few centimeters.)

Example 12.9 shows that an osmotic pressure measurement can be used to find the concentration of a solution.

Example 12.9

The average osmotic pressure of seawater, measured in the kind of apparatus shown in Figure 12.11, is about 30.0 atm at 25°C. Calculate the molar concentration of an aqueous solution of sucrose ($C_{12}H_{22}O_{11}$) that is isotonic with seawater.

Strategy When we say the sucrose solution is isotonic with seawater, what can we conclude about the osmotic pressures of these two solutions?

Solution A solution of sucrose that is isotonic with seawater must have the same osmotic pressure, 30.0 atm. Using Equation (12.8).

$$\pi = MRT$$

$$M = \frac{\pi}{RT} = \frac{30.0 \text{ atm}}{(0.0821 \text{ L} \cdot \text{atm/K} \cdot \text{mol})(298 \text{ K})}$$

$$= 1.23 \text{ mol/L}$$

$$= 1.23\ M$$

Practice Exercise What is the osmotic pressure (in atm) of a 0.884 M urea solution at 16°C?

Similar problem: 12.63.

Using Colligative Properties to Determine Molar Mass

The colligative properties of nonelectrolyte solutions provide a means of determining the molar mass of a solute. Theoretically, any of the four colligative properties is suitable for this purpose. In practice, however, only freezing-point depression and osmotic pressure are used because they show the most pronounced changes. The procedure is as follows. From the experimentally determined freezing-point depression or osmotic pressure, we can calculate the molality or molarity of the solution. Knowing the mass of the solute, we can readily determine its molar mass, as Examples 12.10 and 12.11 demonstrate.

Example 12.10

A 7.85-g sample of a compound with the empirical formula C_5H_4 is dissolved in 298 g of benzene. The freezing point of the solution is 1.05°C below that of pure benzene. What are the molar mass and molecular formula of this compound?

Strategy Solving this problem requires three steps. First, we calculate the molality of the solution from the depression in freezing point. Next, from the molality we determine the number of moles in 7.85 g of the compound and hence its molar mass. Finally, comparing the experimental molar mass with the empirical molar mass enables us to write the molecular formula.

Solution The sequence of conversions for calculating the molar mass of the compound is

$$\begin{array}{c}\text{freezing–point}\\ \text{depression}\end{array} \longrightarrow \text{molality} \longrightarrow \begin{array}{c}\text{number of}\\ \text{moles}\end{array} \longrightarrow \text{molar mass}$$

Our first step is to calculate the molality of the solution. From Equation (12.7) and Table 12.2 we write

$$\text{molality} = \frac{\Delta T_f}{K_f} = \frac{1.05°\text{C}}{5.07°\text{C}/m} = 0.207\ m$$

Because there is 0.205 mole of the solute in 1 kg of solvent, the number of moles of solute in 301 g, or 0.301 kg, of solvent is

$$0.298\ \text{kg} \times \frac{0.207\ \text{mol}}{1\ \text{kg}} = 0.0617\ \text{mol}$$

Thus, the molar mass of the solute is

$$\text{molar mass} = \frac{\text{grams of compound}}{\text{moles of compound}}$$

$$= \frac{7.85\ \text{g}}{0.0617\ \text{mol}} = 127\ \text{g/mol}$$

Now we can determine the ratio

$$\frac{\text{molar mass}}{\text{empirical molar mass}} = \frac{127\ \text{g/mol}}{64\ \text{g/mol}} \approx 2$$

$C_{10}H_8$

Therefore, the molecular formula is $(C_5H_4)_2$ or $C_{10}H_8$ (naphthalene).

Practice Exercise A solution of 0.85 g of an organic compound in 100.0 g of benzene has a freezing point of 5.17°C. What are the molality of the solution and the molar mass of the solute?

Similar problem: 12.57.

Example 12.11

A solution is prepared by dissolving 35.0 g of hemoglobin (Hb) in enough water to make up 1 L in volume. If the osmotic pressure of the solution is found to be 10.0 mmHg at 25°C, calculate the molar mass of hemoglobin.

(Continued)

Strategy We are asked to calculate the molar mass of Hb. The steps are similar to those outlined in Example 12.10. From the osmotic pressure of the solution, we calculate the molarity of the solution. Then, from the molarity, we determine the number of moles in 35.0 g of Hb and hence its molar mass. What units should we use for π and temperature?

Solution The sequence of conversions is as follows:

osmotic pressure ⟶ molarity ⟶ number of moles ⟶ molar mass

First we calculate the molarity using Equation (12.8)

$$\pi = MRT$$

$$M = \frac{\pi}{RT}$$

$$= \frac{10.0\ \text{mmHg} \times \dfrac{1\ \text{atm}}{760\ \text{mmHg}}}{(0.0821\ \text{L} \cdot \text{atm/K} \cdot \text{mol})(298\ \text{K})}$$

$$= 5.38 \times 10^{-4}\ \text{M}$$

The volume of the solution is 1 L, so it must contain 5.38×10^{-4} mol of Hb. We use this quantity to calculate the molar mass:

$$\text{moles of Hb} = \frac{\text{mass of Hb}}{\text{molar mass of Hb}}$$

$$\text{molar mass of Hb} = \frac{\text{mass of Hb}}{\text{moles of Hb}}$$

$$= \frac{35.0\ \text{g}}{5.38 \times 10^{-4}\ \text{mol}}$$

$$= 6.51 \times 10^{4}\ \text{g/mol}$$

Practice Exercise A 202-mL benzene solution containing 2.47 g of an organic polymer has an osmotic pressure of 8.63 mmHg at 21°C. Calculate the molar mass of the polymer.

Similar problems: 12.64, 12.66.

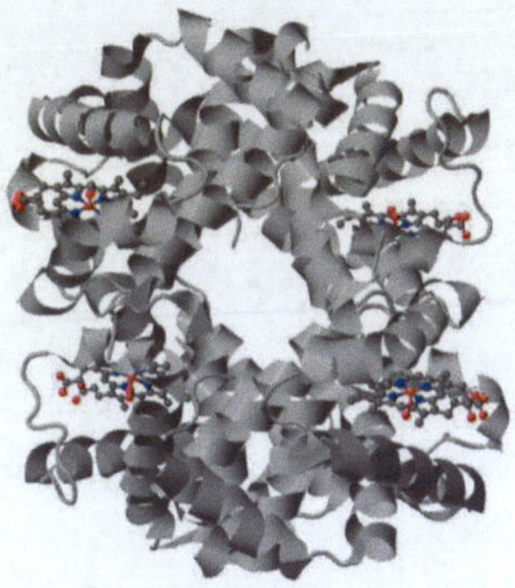

Ribbon representation of the structure of hemoglobin. There are four heme groups in each molecule that bind oxygen molecules.

A pressure of 10.0 mmHg, as in Example 12.11, can be measured easily and accurately. For this reason, osmotic pressure measurements are very useful for determining the molar masses of large molecules, such as proteins. To see how much more practical the osmotic pressure technique is than freezing-point depression would be, let us estimate the change in freezing point of the same hemoglobin solution. If an aqueous solution is quite dilute, we can assume that molarity is roughly equal to molality. (Molarity would be equal to molality if the density of the aqueous solution were 1 g/mL.) Hence, from Equation (12.7) we write

$$\Delta T_f = (1.86°\text{C}/m)(5.38 \times 10^{-4}\ m)$$

$$= 1.00 \times 10^{-3}°\text{C}$$

The freezing-point depression of one-thousandth of a degree is too small a temperature change to measure accurately. For this reason, the freezing-point depression technique is more suitable for determining the molar mass of smaller and more soluble molecules, those having molar masses of 500 g or less, because the freezing-point depressions of their solutions are much greater.

Summary of Concepts & Facts

- Raoult's law states that the partial pressure of a substance A over a solution is equal to the mole fraction (X_A) of A times the vapor pressure (P_A°) of pure A. An ideal solution obeys Raoult's law over the entire range of concentration. In practice, very few solutions exhibit ideal behavior.
- Vapor-pressure lowering, boiling-point elevation, freezing-point depression, and osmotic pressure are colligative properties of solutions; that is, they depend only on the number of solute particles that are present and not on their nature.

Review of Concepts & Facts

12.6.1 A solution contains equal molar amounts of liquids A and B. The vapor pressures of pure A and B are 120 mmHg and 180 mmHg, respectively, at a certain temperature. If the vapor pressure of the solution is 164 mmHg, what can you deduce about the intermolecular forces between A and B molecules compared to the intermolecular forces between A molecules and between B molecules?

12.6.2 Sketch a phase diagram like that shown in Figure 12.10 for nonaqueous solutions such as naphthalene dissolved in benzene. Would the freezing-point depression and boiling-point elevation still apply in this case?

12.6.3 What does it mean when we say that the osmotic pressure of a sample of seawater is 25 atm at a certain temperature?

12.6.4 Calculate the boiling point and freezing point of a solution prepared by dissolving 741 g of glucose ($C_6H_{12}O_6$) in 1.50 kg of water. (For water, $K_b = 0.51°C/m$ and $K_f = 1.86°C/m$.)

12.7 Colligative Properties of Electrolyte Solutions

Learning Objective

- Employ van't Hoff factors to determine the colligative properties of electrolyte solutions.

The study of colligative properties of electrolytes requires a slightly different approach than the one used for the colligative properties of nonelectrolytes. The reason is that electrolytes dissociate into ions in solution, and so one unit of an electrolyte compound separates into two or more particles when it dissolves. (Remember, it is the total number of solute particles that determines the colligative properties of a solution.) For example, each unit of NaCl dissociates into two ions—Na^+ and Cl^-. Thus, the colligative properties of a 0.1 *m* NaCl solution should be twice as great as those of a 0.1 *m* solution containing a nonelectrolyte, such as sucrose. Similarly, we would expect a 0.1 *m* $CaCl_2$ solution to depress the freezing point by three times as much as a 0.1 *m* sucrose solution because each $CaCl_2$ produces three ions. To account for this effect we define a quantity called the ***van't Hoff*[†] *factor (i),*** given by

$$i = \frac{\text{actual number of particles in soln after dissociation}}{\text{number of formula units initially dissolved in soln}} \quad (12.9)$$

Thus, *i* should be 1 for all nonelectrolytes. For strong electrolytes such as NaCl and KNO_3, *i* should be 2 as every unit of NaCl and KNO_3 that dissociates yields two ions. For strong electrolytes such as Na_2SO_4 and $CaCl_2$, *i* should be 3 as every unit of Na_2SO_4 and $CaCl_2$ that dissociates produces three ions. Consequently, the equations for colligative properties must be modified as

$$\Delta T_b = iK_b m \quad (12.10)$$

$$\Delta T_f = iK_f m \quad (12.11)$$

$$\pi = iMRT \quad (12.12)$$

[†]Jacobus Hendricus van't Hoff (1852–1911). Dutch chemist. One of the most prominent chemists of his time, van't Hoff did significant work in thermodynamics, molecular structure and optical activity, and solution chemistry. In 1901 he received the first Nobel Prize in Chemistry.

Table 12.3 The van't Hoff Factor of 0.0500 *M* Electrolyte Solutions at 25°C

Electrolyte	*i*(Measured)	*i*(Calculated)
Sucrose*	1.0	1.0
HCl	1.9	2.0
NaCl	1.9	2.0
$MgSO_4$	1.3	2.0
$MgCl_2$	2.7	3.0
$FeCl_3$	3.4	4.0

*Sucrose is a nonelectrolyte. It is listed here for comparison only.

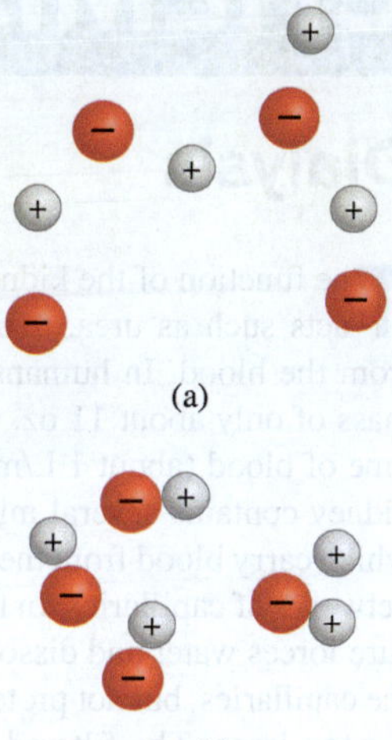

Figure 12.14 *(a) Free ions and (b) ion pairs in solution. Such an ion pair bears no net charge and therefore cannot conduct electricity in solution.*

In reality, the colligative properties of electrolyte solutions are usually smaller than anticipated because at higher concentrations, electrostatic forces come into play and bring about the formation of ion pairs. An ***ion pair*** is made up of *one or more cations and one or more anions held together by electrostatic forces.* The presence of an ion pair reduces the number of particles in solution, causing a reduction in the colligative properties (Figure 12.14). Electrolytes containing multicharged ions such as Mg^{2+}, Al^{3+}, SO_4^{2-}, and PO_4^{3-} have a greater tendency to form ion pairs than electrolytes such as NaCl and KNO_3, which are made up of singly charged ions.

Table 12.3 shows the experimentally measured values of *i* and those calculated assuming complete dissociation. As you can see, the agreement is close but not perfect, indicating that the extent of ion-pair formation in these solutions at that concentration is appreciable.

Example 12.12

The osmotic pressure of a 0.010 *M* potassium iodide (KI) solution at 25°C is 0.465 atm. Calculate the van't Hoff factor for KI at this concentration.

Strategy Note that KI is a strong electrolyte, so we expect it to dissociate completely in solution. If so, its osmotic pressure would be

$$2(0.010\ M)(0.0821\ \text{L} \cdot \text{atm/K} \cdot \text{mol})(298\ \text{K}) = 0.489\ \text{atm}$$

However, the measured osmotic pressure is only 0.465 atm. The smaller than predicted osmotic pressure means that there is ion-pair formation, which reduces the number of solute particles (K^+ and I^- ions) in solution.

Solution From Equation (12.12) we have

$$i = \frac{\pi}{MRT}$$

$$= \frac{0.465\ \text{atm}}{(0.010\ M)(0.0821\ \text{L} \cdot \text{atm/K} \cdot \text{mol})(298\ \text{K})}$$

$$= 1.90$$

Practice Exercise The freezing-point depression of a 0.100 *m* $MgSO_4$ solution is 0.225°C. Calculate the van't Hoff factor of $MgSO_4$ at this concentration.

Similar problem: 12.77.

The Chemistry in Action essay "Dialysis" describes dialysis, a medical procedure through which a person's blood is cleansed of toxins.

Summary of Concepts & Facts

- In electrolyte solutions, the interaction between ions leads to the formation of ion pairs. The van't Hoff factor provides a measure of the extent of dissociation of electrolytes in solution.

CHEMISTRY *in Action*

Dialysis

The function of the kidneys is to filter metabolic waste products such as urea, toxins, excess mineral salts, and water from the blood. In humans, the two kidneys have a combined mass of only about 11 oz. Despite their small size, a large volume of blood (about 1 L/min) flows through the kidneys. Each kidney contains several million filtering units called nephrons, which carry blood from the renal artery to the glomeruli, minute networks of capillaries. In the glomeruli, the normal blood pressure forces water and dissolved substances through the pores of the capillaries, but not proteins and red blood cells, because they are too large. The filtered fluid contains many substances the blood cannot lose in large quantities: salts, sugars, amino acids, and water. Most of the substances are reabsorbed into the blood by a process called active transport (movement of a substance from a low-concentration region to a higher concentration one). Much of the water also returns to the blood by osmosis. Eventually, the excess water, mineral salts, and waste substances are discharged from the body as urine.

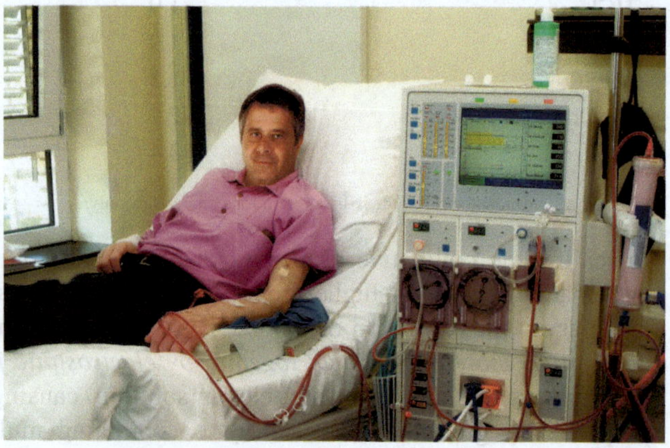

A patient undergoing dialysis.
Gopixa/Shutterstock

Kidney malfunction can be life threatening. When the kidneys fail, the blood must be purified by a dialysis procedure in which waste products are removed from the body by artificial means. In *hemodialysis*, which is one type of treatment, blood is removed from the body and pumped into a machine (called the dialyzer) that filters the toxic substances out of the blood and then returns the purified blood to the patient's bloodstream. By a simple surgical procedure, blood is pumped out of the patient's artery, through a dialyzer, and returned to the vein. Inside the dialyzer, an artificial porous membrane separates the blood from a fluid (the dialysate) that contains many of the blood's components at similar concentrations so that there is no net passage of them between the blood and the dialysate through the membrane. The size of the membrane pores is such that only small waste products can pass through them into the dialysate to be washed away. Again, proteins, cells, and other important blood components are prevented by their size from passing through the membrane and remain in the purified (dialyzed) blood. Typically a hemodialysis procedure lasts 4 to 6 hours. In many cases, one weekly treatment is sufficient to purify the blood.

Review of Concepts & Facts

12.7.1 Indicate which compound in each of the following groups has a greater tendency to form ion pairs in water: (a) NaCl or Na_2SO_4, (b) $MgCl_2$ or $MgSO_4$, (c) LiBr or KBr.

12.7.2 The osmotic pressure of blood is about 7.4 atm. What is the approximate concentration of a saline solution (NaCl solution) a physician should use for intravenous injection? Use 37°C for physiological temperature.

12.8 Colloids

Learning Objective

- Explain colloids and provide examples.

The solutions discussed so far are true homogeneous mixtures. Now consider what happens if we add fine sand to a beaker of water and stir. The sand particles are suspended at first but then gradually settle to the bottom. This is an example of a

Table 12.4 Types of Colloids

Dispersing Medium	Dispersed Phase	Name	Example
Gas	Liquid	Aerosol	Fog, mist
Gas	Solid	Aerosol	Smoke
Liquid	Gas	Foam	Whipped cream
Liquid	Liquid	Emulsion	Mayonnaise
Liquid	Solid	Sol	Milk of magnesia
Solid	Gas	Foam	Plastic foams
Solid	Liquid	Gel	Jelly, butter
Solid	Solid	Solid sol	Certain alloys (steel), opal

heterogeneous mixture. Between these two extremes is an intermediate state called a colloidal suspension, or simply, a colloid. A ***colloid*** is *a dispersion of particles of one substance (the dispersed phase) throughout a dispersing medium made of another substance.* Colloidal particles are much larger than the normal solute molecules; they range from 1×10^3 pm to 1×10^6 pm. Also, a colloidal suspension lacks the homogeneity of an ordinary solution. The dispersed phase and the dispersing medium can be gases, liquids, solids, or a combination of different phases, as shown in Table 12.4.

A number of colloids are familiar to us. An *aerosol* consists of liquid droplets or solid particles dispersed in a gas. Examples are fog and smoke. Mayonnaise, which is made by breaking oil into small droplets in water, is an example of *emulsion,* which consists of liquid droplets dispersed in another liquid. Milk of magnesia is an example of *sol,* a suspension of solid particles in a liquid.

One way to distinguish a solution from a colloid is by the *Tyndall*† *effect.* When a beam of light passes through a colloid, it is scattered by the dispersed phase (Figure 12.15). No such scattering is observed with ordinary solutions because the solute molecules are too small to interact with visible light. Another demonstration of the Tyndall effect is the scattering of sunlight by dust or smoke in the air (Figure 12.16).

Figure 12.15 *Three beams of white light, passing through a colloid of sulfur particles in water, change to orange, pink, and bluish-green. The colors produced depend on the size of the particles and also on the position of the viewer. The smaller the dispersed particles, the shorter (and bluer) the wavelengths.*
Paul Weller/Jane Bixby Weller

Hydrophilic and Hydrophobic Colloids

Among the most important colloids are those in which the dispersing medium is water. Such colloids are divided into two categories called ***hydrophilic,*** or *water-loving,* and ***hydrophobic,*** or *water-fearing.* Hydrophilic colloids are usually solutions containing extremely large molecules such as proteins. In the aqueous phase, a protein like hemoglobin folds in such a way that the hydrophilic parts of the molecule, the parts that can interact favorably with water molecules by ion-dipole forces or hydrogen-bond formation, are on the outside surface (Figure 12.17).

A hydrophobic colloid normally would not be stable in water, and the particles would clump together, like droplets of oil in water merging to form a film of oil at water's surface. They can be stabilized, however, by *adsorption* of ions on their surface (Figure 12.18). (Adsorption refers to adherence onto a surface. It differs from absorption in that the latter means passage to the interior of the medium.) These adsorbed ions can interact with water, thus stabilizing the colloid. At the same time, the electrostatic repulsion between the particles prevents them from clumping together. Soil particles in rivers and streams are hydrophobic particles stabilized in this way. When the freshwater enters the sea, the charges on the particles are neutralized by the high-salt medium, and the particles clump together to form the silt that is seen at the mouth of the river.

Figure 12.16 *Sunlight scattered by dust particles in the air.*
Bob Van Den Berg/Moment Open/Getty Images

†John Tyndall (1820–1893). Irish physicist. Tyndall did important work in magnetism, and explained glacier motion.

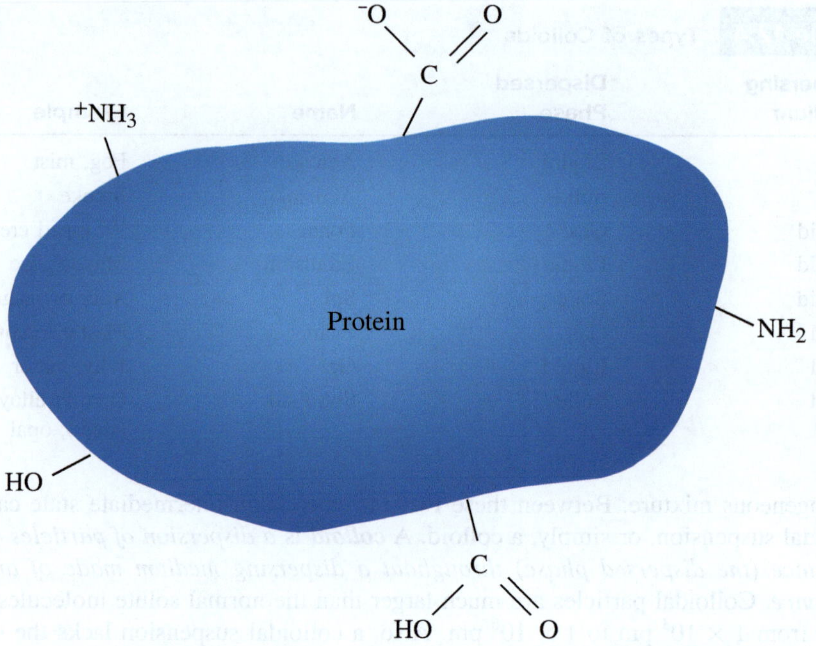

Figure 12.17 *Hydrophilic groups on the surface of a large molecule such as protein stabilizes the molecule in water. Note that all these groups can form hydrogen bonds with water.*

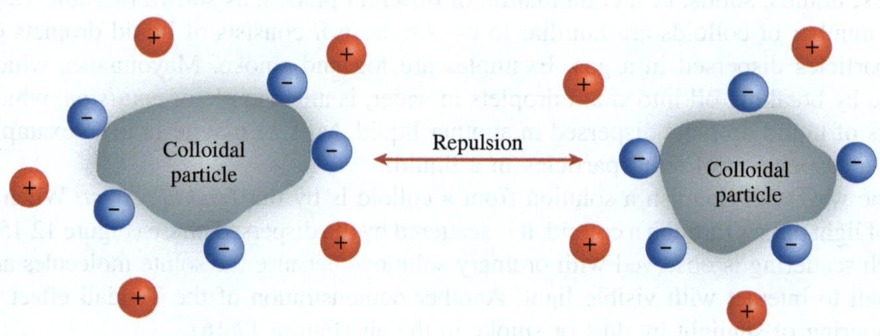

Figure 12.18 *Diagram showing the stabilization of hydrophobic colloids. Negative ions are adsorbed onto the surface and the repulsion between like charges prevents the clumping of the particles.*

Another way hydrophobic colloids can be stabilized is by the presence of other hydrophilic groups on their surfaces. Consider sodium stearate, a soap molecule that has a polar head and a long hydrocarbon tail that is nonpolar (Figure 12.19). The cleansing action of soap is the result of the dual nature of the hydrophobic tail and the hydrophilic end group. The hydrocarbon tail is readily soluble in oily substances, which are also nonpolar, while the ionic —COO^- group remains outside the oily surface. When enough soap molecules have surrounded an oil droplet, as shown in Figure 12.20, the entire system becomes solubilized in water because the exterior portion is now largely hydrophilic. This is how greasy substances are removed by the action of soap.

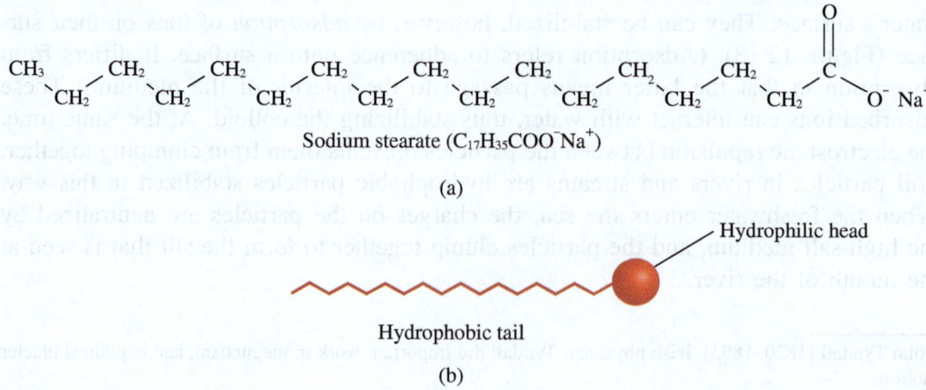

Figure 12.19 *(a) A sodium stearate molecule. (b) The simplified representation of the molecule that shows a hydrophilic head and a hydrophobic tail.*

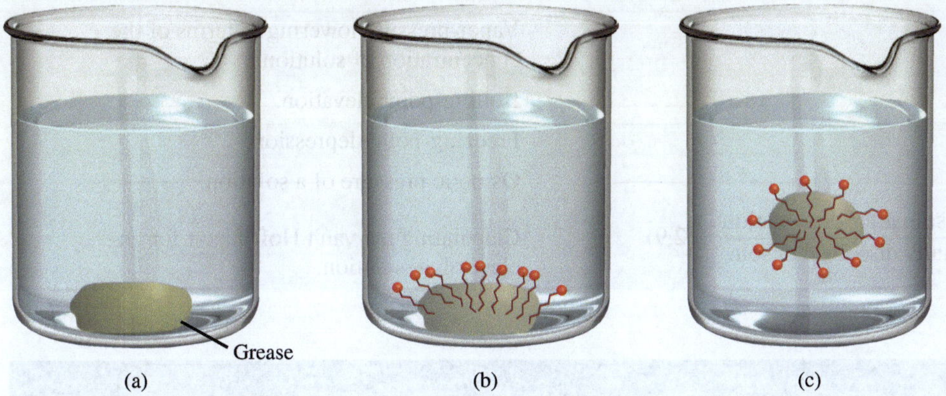

Figure 12.20 *The cleansing action of soap. (a) Grease (oily substance) is not soluble in water. (b) When soap is added to water, the nonpolar tails of soap molecules dissolve in grease. (c) Finally, the grease is removed in the form of an emulsion. Note that each oily droplet now has an ionic exterior that is hydrophilic.*

Summary of Concepts & Facts

- A colloid is a dispersion of particles (about 1×10^3 pm to 1×10^6 pm) of one substance in another substance. A colloid is distinguished from a regular solution by the Tyndall effect, which is the scattering of visible light by colloidal particles. Colloids in water are classified as hydrophilic colloids and hydrophobic colloids.

Chapter Summary

Solutions There are many types of solutions; the most common is the liquid solution in which the solvent is a liquid and the solute is a solid or a liquid. Molecules that possess similar types of intermolecular forces readily mix with each other. Solubility is a quantitative measure of the amount of a solute dissolved in a solvent at a specific temperature. (Sections 12.1, 12.2, and 12.3)

Effect of Temperature and Pressure on Solubility Temperature generally has a marked influence on the solubility of a substance. Pressure can affect the solubility of a gas in a liquid but has little effect if the solute is a solid or liquid. (Sections 12.4 and 12.5)

Colligative Properties The presence of a solute affects the vapor pressure, boiling point, and freezing point of a solvent. In addition, when a solution is separated from the solvent by a semipermeable membrane, osmosis, the passage of solvent molecules from the solvent to the solution, occurs. Equations have been derived that relate the extent of the changes in these properties to the concentration of the solution. (Sections 12.6 and 12.7)

Colloids A dispersion of particles of one substance throughout a dispersing medium made of another substance is a colloid. The particles of a colloid are larger than molecules of solute in a solution. A key factor in a colloid is the lack of homogeneity compared to that of a regular solution. (Section 12.8)

Key Equations

$\text{percent by mass} = \dfrac{\text{mass of solute}}{\text{mass of soln}} \times 100\%$ (12.1)	Calculating the percent by mass of a solution.
$\text{molality } (m) = \dfrac{\text{moles of solute}}{\text{mass of solvent (kg)}}$ (12.2)	Calculating the molality of a solution.
$c = kP$ (12.3)	Henry's law for calculating solubility of gases.
$P_1 = X_1 P_1^\circ$ (12.4)	Raoult's law relating the vapor pressure of a liquid to its vapor pressure in a solution.

$\Delta P = X_2 P_1^\circ$ (12.5)	Vapor-pressure lowering in terms of the concentration of solution.
$\Delta T_b = K_b m$ (12.6)	Boiling-point elevation.
$\Delta T_f = K_f m$ (12.7)	Freezing-point depression.
$\pi = MRT$ (12.8)	Osmotic pressure of a solution.
$i = \dfrac{\text{actual number of particles in soln after dissociation}}{\text{number of formula units initially dissolved in soln}}$ (12.9)	Calculating the van't Hoff factor for an electrolyte solution.

Key Words

Boiling-point elevation (ΔT_b), p. 545
Colligative properties, p. 541
Colloid, p. 555
Crystallization, p. 528
Fractional crystallization, p. 537
Fractional distillation, p. 544
Freezing-point depression (ΔT_f), p. 545
Henry's law, p. 538
Hydrophilic, p. 555
Hydrophobic, p. 555
Ideal solution, p. 543
Ion pair, p. 553
Miscible, p. 530
Molality, p. 533
Nonvolatile, p. 541
Osmosis, p. 547
Osmotic pressure (π), p. 547
Percent by mass, p. 532
Raoult's law, p. 541
Saturated solution, p. 528
Semipermeable membrane, p. 547
Solvation, p. 530
Supersaturated solution, p. 528
Unsaturated solution, p. 528
van't Hoff factor (i), p. 552
Volatile, p. 543

Questions & Problems

Red numbered problems solved in Student Solutions Manual

12.1 Types of Solutions

Review Questions

12.1 Distinguish between an unsaturated solution, a saturated solution, and a supersaturated solution.

12.2 From which type of solution listed in Question 12.1 does crystallization or precipitation occur? How does a crystal differ from a precipitate?

12.2 A Molecular View of the Solution Process

Review Questions

12.3 Briefly describe the solution process at the molecular level. Use the dissolution of a solid in a liquid as an example.

12.4 Basing your answer on intermolecular force considerations, explain what "like dissolves like" means.

12.5 What is solvation? What factors influence the extent to which solvation occurs? Give two examples of solvation; include one that involves ion-dipole interaction and one in which dispersion forces come into play.

12.6 As you know, some solution processes are endothermic and others are exothermic. Provide a molecular interpretation for the difference.

12.7 Explain why the solution process usually leads to an increase in disorder.

12.8 Describe the factors that affect the solubility of a solid in a liquid. What does it mean to say that two liquids are miscible?

Problems

12.9 Why is naphthalene ($C_{10}H_8$) more soluble than CsF in benzene?

12.10 Explain why ethanol (C_2H_5OH) is not soluble in cyclohexane (C_6H_{12}).

12.11 Arrange the following compounds in order of increasing solubility in water: O_2, LiCl, Br_2, methanol (CH_3OH).

12.12 Explain the variations in solubility in water of the alcohols listed here:

Compound	Solubility in Water (g/100 g) at 20°C
CH_3OH	∞
CH_3CH_2OH	∞
$CH_3CH_2CH_2OH$	∞
$CH_3CH_2CH_2CH_2OH$	9
$CH_3CH_2CH_2CH_2CH_2OH$	2.7

(*Note:* ∞ means that the alcohol and water are completely miscible in all proportions.)

12.3 Concentration Units

Review Questions

12.13 Define the following concentration terms and give their units: *percent by mass, mole fraction, molarity, molality*. Compare their advantages and disadvantages.

12.14 Outline the steps required for conversion between molarity, molality, and percent by mass.

Problems

12.15 Calculate the percent by mass of the solute in each of the following aqueous solutions: (a) 5.50 g of NaBr in 78.2 g of solution, (b) 31.0 g of KCl in 152 g of water, (c) 4.5 g of toluene in 29 g of benzene.

12.16 Calculate the amount of water (in grams) that must be added to (a) 5.00 g of urea $(NH_2)_2CO$ in the preparation of a 16.2 percent by mass solution, and (b) 26.2 g of $MgCl_2$ in the preparation of a 1.5 percent by mass solution.

12.17 Calculate the molality of each of the following solutions: (a) 14.3 g of sucrose $(C_{12}H_{22}O_{11})$ in 676 g of water, (b) 7.20 moles of ethylene glycol $(C_2H_6O_2)$ in 3546 g of water.

12.18 Calculate the molality of each of the following aqueous solutions: (a) 2.50 *M* NaCl solution (density of solution = 1.08 g/mL), (b) 48.2 percent by mass KBr solution.

12.19 Calculate the molalities of the following aqueous solutions: (a) 1.22 *M* sugar $(C_{12}H_{22}O_{11})$ solution (density of solution = 1.12 g/mL), (b) 0.87 *M* NaOH solution (density of solution = 1.04 g/mL), (c) 5.24 *M* $NaHCO_3$ solution (density of solution = 1.19 g/mL).

12.20 For dilute aqueous solutions in which the density of the solution is roughly equal to that of the pure solvent, the molarity of the solution is equal to its molality. Show that this statement is correct for a 0.010 *M* aqueous urea $(NH_2)_2CO$ solution.

12.21 The alcohol content of hard liquor is normally given in terms of the "proof," which is defined as twice the percentage by volume of ethanol (C_2H_5OH) present. Calculate the number of grams of alcohol present in 1.00 L of 75-proof gin. The density of ethanol is 0.798 g/mL.

12.22 The concentrated sulfuric acid we use in the laboratory is 98.0 percent H_2SO_4 by mass. Calculate the molality and molarity of the acid solution. The density of the solution is 1.83 g/mL.

12.23 Calculate the molarity and the molality of an NH_3 solution made up of 30.0 g of NH_3 in 70.0 g of water. The density of the solution is 0.982 g/mL.

12.24 The density of an aqueous solution containing 10.0 percent of ethanol (C_2H_5OH) by mass is 0.984 g/mL. (a) Calculate the molality of this solution. (b) Calculate its molarity. (c) What volume of the solution would contain 0.125 mole of ethanol?

12.4 The Effect of Temperature on Solubility

Review Questions

12.25 How do the solubilities of most ionic compounds in water change with temperature? With pressure?

12.26 Describe the fractional crystallization process and its application.

Problems

12.27 A 3.20-g sample of a salt dissolves in 9.10 g of water to give a saturated solution at 25°C. What is the solubility (in g salt/100 g of H_2O) of the salt?

12.28 The solubility of KNO_3 is 155 g per 100 g of water at 75°C and 38.0 g at 25°C. What mass (in grams) of KNO_3 will crystallize out of solution if exactly 100 g of its saturated solution at 75°C is cooled to 25°C?

12.29 A 50-g sample of impure $KClO_3$ (solubility = 7.1 g per 100 g H_2O at 20°C) is contaminated with 10 percent of KCl (solubility = 25.5 g per 100 g of H_2O at 20°C). Calculate the minimum quantity of 20°C water needed to dissolve all the KCl from the sample. How much $KClO_3$ will be left after this treatment? (Assume that the solubilities are unaffected by the presence of the other compound.)

12.5 The Effect of Pressure on the Solubility of Gases

Review Questions

12.30 Discuss the factors that influence the solubility of a gas in a liquid.

12.31 What is thermal pollution? Why is it harmful to aquatic life?

12.32 What is Henry's law? Define each term in the equation, and give its units. How would you account for the law in terms of the kinetic molecular theory of gases? Give two exceptions to Henry's law.

12.33 A student is observing two beakers of water. One beaker is heated to 30°C, and the other is heated to 100°C. In each case, bubbles form in the water. Are these bubbles of the same origin? Explain.

12.34 A man bought a goldfish in a pet shop. Upon returning home, he put the goldfish in a bowl of recently boiled water that had been cooled quickly. A few minutes later the fish was found dead. Explain what happened to the fish.

Problems

12.35 A beaker of water is initially saturated with dissolved air. Explain what happens when He gas at 1 atm is bubbled through the solution for a long time.

12.36 A miner working 260 m below sea level opened a carbonated soft drink during a lunch break. To his surprise, the soft drink tasted rather "flat." Shortly afterward, the miner took an elevator to the surface. During the trip up, he could not stop belching. Why?

12.37 The solubility of CO_2 in water at 25°C and 1 atm is 0.034 mol/L. What is its solubility under atmospheric conditions? (The partial pressure of CO_2 in air is 0.0003 atm.) Assume that CO_2 obeys Henry's law.

12.38 The solubility of N_2 in blood at 37°C and at a partial pressure of 0.80 atm is 5.6×10^{-4} mol/L. A deep-sea diver breathes compressed air with the partial pressure of N_2 equal to 4.0 atm. Assume that the total volume of blood in the body is 5.0 L. Calculate the amount of N_2 gas released (in liters at 37°C and 1 atm) when the diver returns to the surface of the water, where the partial pressure of N_2 is 0.80 atm.

12.6 Colligative Properties of Nonelectrolyte Solutions

Review Questions

12.39 What are colligative properties? What is the meaning of the word "colligative" in this context?

12.40 Write the equation representing Raoult's law, and express it in words.

12.41 Use a solution of benzene in toluene to explain what is meant by an ideal solution.

12.42 Write the equations relating boiling-point elevation and freezing-point depression to the concentration of the solution. Define all the terms, and give their units.

12.43 How is vapor-pressure lowering related to a rise in the boiling point of a solution?

12.44 Use a phase diagram to show the difference in freezing points and boiling points between an aqueous urea solution and pure water.

12.45 What is osmosis? What is a semipermeable membrane?

12.46 Write the equation relating osmotic pressure to the concentration of a solution. Define all the terms and specify their units.

12.47 Explain why molality is used for boiling-point elevation and freezing-point depression calculations and molarity is used in osmotic pressure calculations.

12.48 Describe how you would use freezing-point depression and osmotic pressure measurements to determine the molar mass of a compound. Why are boiling-point elevation and vapor-pressure lowering normally not used for this purpose?

Problems

12.49 A solution is prepared by dissolving 396 g of sucrose ($C_{12}H_{22}O_{11}$) in 624 g of water. What is the vapor pressure of this solution at 30°C? (The vapor pressure of water is 31.8 mmHg at 30°C.)

12.50 How many grams of sucrose ($C_{12}H_{22}O_{11}$) must be added to 552 g of water to give a solution with a vapor pressure 2.0 mmHg less than that of pure water at 20°C? (The vapor pressure of water at 20°C is 17.5 mmHg.)

12.51 The vapor pressure of benzene is 100.0 mmHg at 26.1°C. Calculate the vapor pressure of a solution containing 24.6 g of camphor ($C_{10}H_{16}O$) dissolved in 98.5 g of benzene. (Camphor is a low-volatility solid.)

12.52 The vapor pressures of ethanol (C_2H_5OH) and 1-propanol (C_3H_7OH) at 35°C are 100 mmHg and 37.6 mmHg, respectively. Assume ideal behavior and calculate the partial pressures of ethanol and 1-propanol at 35°C over a solution of ethanol in 1-propanol, in which the mole fraction of ethanol is 0.300.

12.53 The vapor pressure of ethanol (C_2H_5OH) at 20°C is 44 mmHg, and the vapor pressure of methanol (CH_3OH) at the same temperature is 94 mmHg. A mixture of 30.0 g of methanol and 45.0 g of ethanol is prepared (and can be assumed to behave as an ideal solution). (a) Calculate the vapor pressure of methanol and ethanol above this solution at 20°C. (b) Calculate the mole fraction of methanol and ethanol in the vapor above this solution at 20°C. (c) Suggest a method for separating the two components of the solution.

12.54 How many grams of urea [$(NH_2)_2CO$] must be added to 450 g of water to give a solution with a vapor pressure 2.50 mmHg less than that of pure water at 30°C? (The vapor pressure of water at 30°C is 31.8 mmHg.)

12.55 What are the boiling point and freezing point of a 2.47 *m* solution of naphthalene in benzene? (The boiling point and freezing point of benzene are 80.1°C and 5.5°C, respectively.)

12.56 An aqueous solution contains the amino acid glycine (NH_2CH_2COOH). Assuming that the acid does not ionize in water, calculate the molality of the solution if it freezes at −1.1°C.

12.57 Pheromones are compounds secreted by the females of many insect species to attract males. One of these compounds contains 80.78 percent C, 13.56 percent H, and 5.66 percent O. A solution of 1.00 g of this pheromone in 8.50 g of benzene freezes at 3.37°C. What are the molecular formula and molar mass of the compound? (The normal freezing point of pure benzene is 5.50°C.)

12.58 The elemental analysis of an organic solid extracted from gum arabic (a gummy substance used in adhesives, inks, and pharmaceuticals) showed that it contained 40.0 percent C, 6.7 percent H, and 53.3 percent O. A solution of 0.650 g of the solid in 27.8 g of the solvent diphenyl gave a freezing-point depression of 1.56°C. Calculate the molar mass and molecular formula of the solid. (K_f for diphenyl is 8.00°C/*m*.)

12.59 How many liters of the antifreeze ethylene glycol [$CH_2(OH)CH_2(OH)$] would you add to a car radiator containing 6.50 L of water if the coldest winter temperature in your area is −20°C? Calculate the boiling point of this water-ethylene glycol mixture. (The density of ethylene glycol is 1.11 g/mL.)

12.60 A solution is prepared by condensing 4.00 L of a gas, measured at 27°C and 748 mmHg pressure, into 58.0 g of benzene. Calculate the freezing point of this solution.

12.61 The molar mass of benzoic acid (C_6H_5COOH) determined by measuring the freezing-point depression in benzene is twice what we would expect for the molecular formula, $C_7H_6O_2$. Explain this apparent anomaly.

12.62 A solution of 2.50 g of a compound having the empirical formula C_6H_5P in 25.0 g of benzene is observed to freeze at 4.3°C. Calculate the molar mass of the solute and its molecular formula.

12.63 What is the osmotic pressure (in atm) of a 1.36 *M* aqueous solution of urea [$(NH_2)_2CO$] at 22.0°C?

12.64 A solution containing 0.8330 g of a polymer of unknown structure in 170.0 mL of an organic solvent was found to have an osmotic pressure of 5.20 mmHg at 25°C. Determine the molar mass of the polymer.

12.65 A quantity of 7.480 g of an organic compound is dissolved in water to make 300.0 mL of solution. The solution has an osmotic pressure of 1.43 atm at 27°C. The analysis of this compound shows that it contains 41.8 percent C, 4.7 percent H, 37.3 percent O, and 16.3 percent N. Calculate the molecular formula of the compound.

12.66 A solution of 6.85 g of a carbohydrate in 100.0 g of water has a density of 1.024 g/mL and an osmotic pressure of 4.61 atm at 20.0°C. Calculate the molar mass of the carbohydrate.

12.7 Colligative Properties of Electrolyte Solutions

Review Questions

12.67 What are ion pairs? What effect does ion-pair formation have on the colligative properties of a solution? How does the ease of ion-pair formation depend on (a) charges on the ions, (b) size of the ions, (c) nature of the solvent (polar versus nonpolar), (d) concentration?

12.68 What is the van't Hoff factor? What information does it provide?

Problems

12.69 Which of the following aqueous solutions has (a) the higher boiling point, (b) the higher freezing point, and (c) the lower vapor pressure: 0.35 *m* $CaCl_2$ or 0.90 *m* urea? Explain. Assume $CaCl_2$ to undergo complete dissociation.

12.70 Consider two aqueous solutions, one of sucrose ($C_{12}H_{22}O_{11}$) and the other of nitric acid (HNO_3). Both solutions freeze at −1.5°C. What other properties do these solutions have in common?

12.71 Arrange the following solutions in order of decreasing freezing point: 0.10 *m* Na_3PO_4, 0.35 *m* NaCl, 0.20 *m* $MgCl_2$, 0.15 *m* $C_6H_{12}O_6$, 0.15 *m* CH_3COOH.

12.72 Arrange the following aqueous solutions in order of decreasing freezing point, and explain your reasoning: 0.50 *m* HCl, 0.50 *m* glucose, 0.50 *m* acetic acid.

12.73 What are the normal freezing points and boiling points of the following solutions? (a) 21.2 g NaCl in 135 mL of water, (b) 15.4 g of urea in 66.7 mL of water

12.74 At 25°C the vapor pressure of pure water is 23.76 mmHg and that of seawater is 22.98 mmHg. Assuming that seawater contains only NaCl, estimate its molal concentration.

12.75 Both NaCl and $CaCl_2$ are used to melt ice on roads and sidewalks in winter. What advantages do these substances have over sucrose or urea in lowering the freezing point of water?

12.76 A 0.86 percent by mass solution of NaCl is called "physiological saline" because its osmotic pressure is equal to that of the solution in blood cells. Calculate the osmotic pressure of this solution at normal body temperature (37°C). Note that the density of the saline solution is 1.005 g/mL.

12.77 The osmotic pressure of 0.010 *M* solutions of $CaCl_2$ and urea at 25°C are 0.605 atm and 0.245 atm, respectively. Calculate the van't Hoff factor for the $CaCl_2$ solution.

12.78 Calculate the osmotic pressure of a 0.0500 *M* $MgSO_4$ solution at 25°C. (*Hint:* See Table 12.3.)

12.8 Colloids

Review Questions

12.79 What are colloids? Referring to Table 12.4, why is there no colloid in which both the dispersed phase and the dispersing medium are gases?

12.80 Describe how hydrophilic and hydrophobic colloids are stabilized in water.

Additional Problems

12.81 Aqueous solutions A and B both contain urea at different concentrations. On standing while exposed to air, the vapor pressure of A remains constant while that of B gradually decreases. (a) Which solution has a higher boiling point? (b) Eventually the two solutions have the same vapor pressure. Explain.

12.82 Water and methanol are miscible with each other but they are immiscible with octane (C_8H_{18}). Which of the following shows the correct picture when equal volumes of these three liquids are mixed in a test tube at 20°C? Assume volumes to be additive. (The densities of the liquids are methanol: 0.792 g/mL; octane: 0.703 g/mL; water: 0.998 g/mL.)

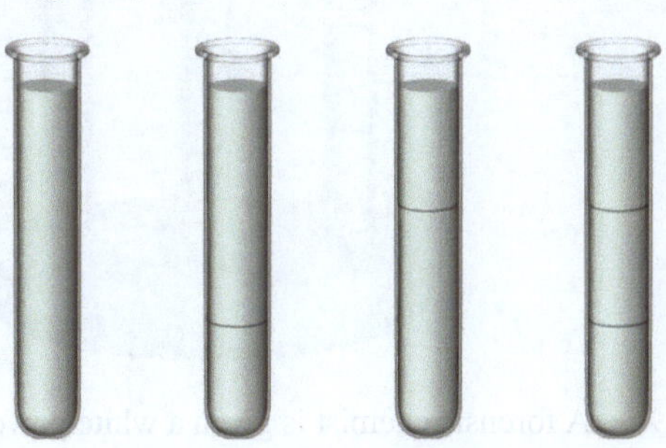

(a) (b) (c) (d)

12.83 Lysozyme is an enzyme that cleaves bacterial cell walls. A sample of lysozyme extracted from egg white has a molar mass of 13,930 g. A quantity of 0.100 g of this enzyme is dissolved in 150 g of water at 25°C. Calculate the vapor-pressure lowering, the depression in freezing point, the elevation in boiling point, and the osmotic pressure of this solution. (The vapor pressure of water at 25°C is 23.76 mmHg.)

12.84 Solutions A and B have osmotic pressures of 2.4 atm and 4.6 atm, respectively, at a certain temperature. What is the osmotic pressure of a solution prepared by mixing equal volumes of A and B at the same temperature?

12.85 A cucumber placed in concentrated brine (salt water) shrivels into a pickle. Explain.

12.86 Two liquids A and B have vapor pressures of 76 mmHg and 132 mmHg, respectively, at 25°C. What is the total vapor pressure of the ideal solution made up of (a) 1.00 mole of A and 1.00 mole of B, and (b) 2.00 moles of A and 5.00 moles of B?

12.87 Calculate the van't Hoff factor of Na_3PO_4 in a 0.40 *m* solution whose freezing point is −2.6°C.

12.88 A 262-mL sample of a sugar solution containing 1.22 g of the sugar has an osmotic pressure of 30.3 mmHg at 35°C. What is the molar mass of the sugar?

12.89 An aqueous solution of a 0.10 *M* monoprotic acid HA has an osmotic pressure of 3.22 atm at 25°C. What is the percent ionization of the acid at this concentration?

12.90 Calculate the mass of naphthalene ($C_{10}H_8$) that must be added to 250 g of benzene (C_6H_6) to give a solution with a freezing point 2.00°C below that of pure benzene.

12.91 Consider the three mercury manometers shown. One of them has 1 mL of water on top of the mercury, another has 1 mL of a 1 *m* urea solution on top of the mercury, and the third one has 1 mL of a 1 *m* NaCl solution placed on top of the mercury. Which of these solutions is in the tube labeled X, which is in Y, and which is in Z?

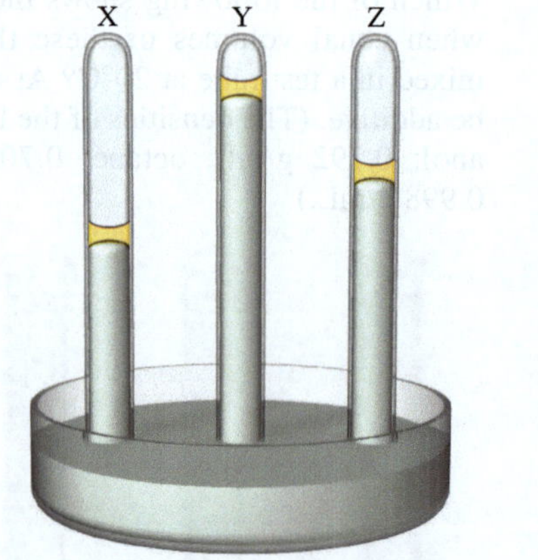

12.92 A forensic chemist is given a white powder for analysis. She dissolves 0.50 g of the substance in 8.0 g of benzene. The solution freezes at 3.9°C. Can the chemist conclude that the compound is cocaine ($C_{17}H_{21}NO_4$)? What assumptions are made in the analysis?

12.93 "Time-release" drugs have the advantage of releasing the drug to the body at a constant rate so that the drug concentration at any time is not too high as to have harmful side effects or too low as to be ineffective. A schematic diagram of a pill that works on this basis is shown. Explain how it works.

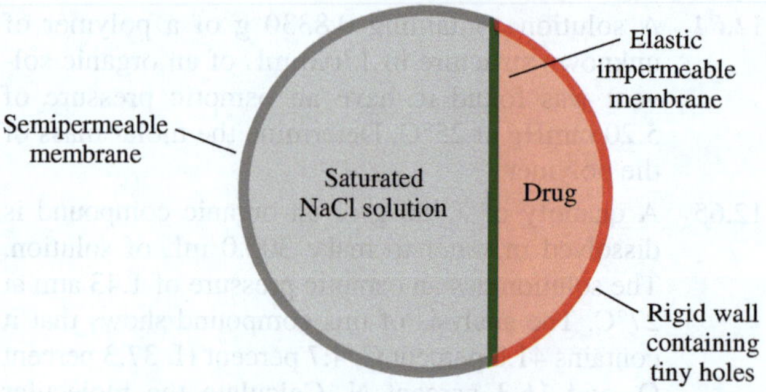

12.94 A solution of 1.00 g of anhydrous aluminum chloride, $AlCl_3$, in 50.0 g of water freezes at −1.11°C. Does the molar mass determined from this freezing point agree with that calculated from the formula? Why?

12.95 Desalination is a process of removing dissolved salts from seawater. (a) Briefly describe how you would apply distillation and freezing for this purpose. (b) Desalination can also be accomplished by *reverse osmosis*, which uses high pressure to force water from a more concentrated solution to a less concentrated one. Assuming a sample of seawater is 0.50 *M* in NaCl, calculate the minimum pressure that needs to be applied for reverse osmosis at 25°C. What is the main advantage of reverse osmosis over distillation and freezing?

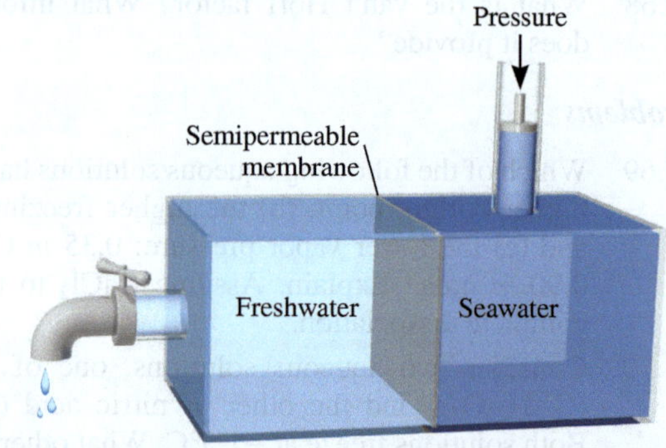

12.96 Fish breathe the dissolved air in water through their gills. Assuming the partial pressures of oxygen and nitrogen in air to be 0.20 atm and 0.80 atm, respectively, calculate the mole fractions of oxygen and nitrogen in water at 298 K. Comment on your results. See Example 12.6 for Henry's law constants.

12.97 A protein has been isolated as a salt with the formula $Na_{20}P$ (this notation means that there are 20 Na^+ ions associated with a negatively charged protein P^{20-}). The osmotic pressure of a 10.0-mL solution containing 0.225 g of the protein is 0.257 atm at 25.0°C. (a) Calculate the molar mass of the protein from these data. (b) Calculate the actual molar mass of the protein.

12.98 A nonvolatile organic compound Z was used to make up two solutions. Solution A contains 5.00 g of Z dissolved in 100 g of water, and solution B contains 2.31 g of Z dissolved in 100 g of benzene. Solution A has a vapor pressure of 754.5 mmHg at the normal boiling point of water, and solution B has the same vapor pressure at the normal boiling point of benzene. Calculate the molar mass of Z in solutions A and B and account for the difference.

12.99 Hydrogen peroxide with a concentration of 3.0 percent (3.0 g of H_2O_2 in 100 mL of solution) is sold in drugstores for use as an antiseptic. For a 10.0-mL 3.0 percent H_2O_2 solution, calculate (a) the oxygen gas produced (in liters) at STP when the compound undergoes complete decomposition, and (b) the ratio of the volume of O_2 collected to the initial volume of the H_2O_2 solution.

12.100 State which of the alcohols listed in Problem 12.12 you would expect to be the best solvent for each of the following substances, and explain why: (a) I_2, (b) KBr, (c) $CH_3CH_2CH_2CH_2CH_3$.

12.101 Before a carbonated beverage bottle is sealed, it is pressurized with a mixture of air and carbon dioxide. (a) Explain the effervescence that occurs when the cap of the bottle is removed. (b) What causes the fog to form near the mouth of the bottle right after the cap is removed?

12.102 Iodine (I_2) is only sparingly soluble in water (left photo). Yet upon the addition of iodide ions (for example, from KI), iodine is converted to the triiodide ion, which readily dissolves (right photo):

$$I_2(s) + I^-(aq) \rightleftharpoons I_3^-(aq)$$

Describe the change in solubility of I_2 in terms of the change in intermolecular forces.

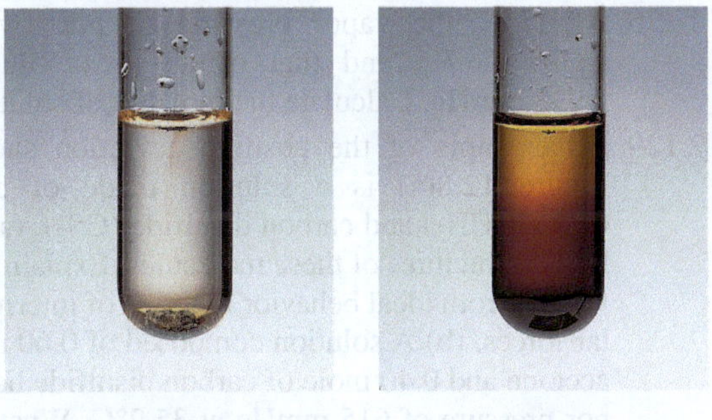

Ken Karp/McGraw Hill

12.103 Two beakers, one containing a 50-mL aqueous 1.0 *M* glucose solution and the other a 50-mL aqueous 2.0 *M* glucose solution, are placed under a tightly sealed bell jar at room temperature. What are the volumes in these two beakers at equilibrium?

12.104 In the apparatus shown, what will happen if the membrane is (a) permeable to both water and the Na^+ and Cl^- ions, (b) permeable to water and Na^+ ions but not to Cl^- ions, (c) permeable to water but not to Na^+ and Cl^- ions?

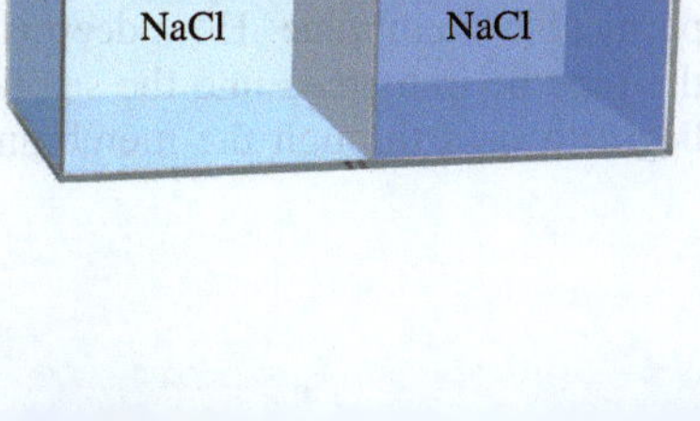

12.105 Explain why it is essential that fluids used in intravenous injections have approximately the same osmotic pressure as blood.

12.106 Concentrated hydrochloric acid is usually available at a concentration of 37.7 percent by mass. What is its molar concentration? (The density of the solution is 1.19 g/mL.)

12.107 Explain each of the following statements: (a) The boiling point of seawater is higher than that of pure water. (b) Carbon dioxide escapes from the solution when the cap is removed from a carbonated soft-drink bottle. (c) Molal and molar concentrations of dilute aqueous solutions are approximately equal. (d) In discussing the colligative properties of a solution (other than osmotic pressure), it is preferable to express the concentration in units of molality rather than in molarity. (e) Methanol (b.p. 65°C) is useful as an antifreeze, but it should be removed from the car radiator during the summer season.

12.108 A mixture of NaCl and sucrose ($C_{12}H_{22}O_{11}$) of combined mass 10.2 g is dissolved in enough water to make up a 250 mL solution. The osmotic pressure of the solution is 7.32 atm at 23°C. Calculate the mass percent of NaCl in the mixture.

12.109 A 0.050 *M* hydrofluoric acid (HF) solution is 11 percent ionized at 25°C. Calculate the osmotic pressure of the solution.

12.110 Shown here is a plot of vapor pressures of two liquids A and B at different concentrations at a certain temperature. Which of the following statements are false? (a) The solutions exhibit negative deviation from Raoult's law. (b) A and B molecules attract each other more weakly than they do their own kind. (c) ΔH_{soln} is positive. (d) At $X_A = 0.20$, the solution has a higher boiling point than liquid B and a lower boiling point than liquid A.

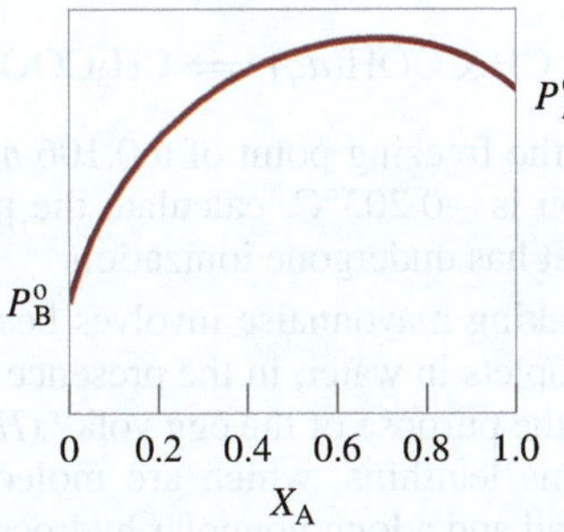

12.111 A 1.32-g sample of a mixture of cyclohexane (C_6H_{12}) and naphthalene ($C_{10}H_8$) is dissolved in 18.9 g of benzene (C_6H_6). The freezing point of the solution is 2.2°C. Calculate the mass percent of the mixture. (See Table 12.2 for constants.)

12.112 How does each of the following affect the solubility of an ionic compound? (a) lattice energy, (b) solvent (polar versus nonpolar), (c) enthalpies of hydration of cation and anion

12.113 A solution contains two volatile liquids A and B. Complete the following table, in which the symbol ↔ indicates attractive intermolecular forces.

Attractive Forces	Deviation from Raoult's Law	ΔH_{soln}
$A \leftrightarrow A, B \leftrightarrow B > A \leftrightarrow B$		
	Negative	
		Zero

12.114 The concentration of commercially available concentrated sulfuric acid is 98.0 percent by mass, or 18 *M*. Calculate the density and the molality of the solution.

12.115 The concentration of commercially available concentrated nitric acid is 70.0 percent by mass, or 15.9 *M*. Calculate the density and the molality of the solution.

12.116 A mixture of ethanol and 1-propanol behaves ideally at 36°C and is in equilibrium with its vapor. If the mole fraction of ethanol in the solution is 0.62, calculate its mole fraction in the vapor phase at this temperature. (The vapor pressures of pure ethanol and 1-propanol at 36°C are 108 mmHg and 40.0 mmHg, respectively.)

12.117 For ideal solutions, the volumes are additive. This means that if 5 mL of A and 5 mL of B form an ideal solution, the volume of the solution is 10 mL. Provide a molecular interpretation for this observation. When 500 mL of ethanol (C_2H_5OH) are mixed with 500 mL of water, the final volume is less than 1000 mL. Why?

12.118 Ammonia (NH_3) is very soluble in water, but nitrogen trichloride (NCl_3) is not. Explain.

12.119 Aluminum sulfate [$Al_2(SO_4)_3$] is sometimes used in municipal water treatment plants to remove undesirable particles. Explain how this process works.

12.120 Acetic acid is a weak acid that ionizes in solution as follows:

$$CH_3COOH(aq) \rightleftharpoons CH_3COO^-(aq) + H^+(aq)$$

If the freezing point of a 0.106 *m* CH_3COOH solution is −0.203°C, calculate the percent of the acid that has undergone ionization.

12.121 Making mayonnaise involves beating oil into small droplets in water, in the presence of egg yolk. What is the purpose of the egg yolk? (*Hint:* Egg yolk contains lecithins, which are molecules with a polar head and a long nonpolar hydrocarbon tail.)

12.122 Acetic acid is a polar molecule and can form hydrogen bonds with water molecules. Therefore, it has a high solubility in water. Yet acetic acid is also soluble in benzene (C_6H_6), a nonpolar solvent that lacks the ability to form hydrogen bonds. A solution of 3.8 g of CH_3COOH in 80 g C_6H_6 has a freezing point of 3.5°C. Calculate the molar mass of the solute and suggest what its structure might be. (*Hint:* Acetic acid molecules can form hydrogen bonds between themselves.)

12.123 A 2.6-L sample of water contains 192 μg of lead. Does this concentration of lead exceed the safety limit of 0.050 ppm of lead per liter of drinking water? [*Hint:* 1 μg = 1×10^{-6} g. Parts per million (ppm) is defined as (mass of component/mass of solution) $\times 10^6$.]

12.124 Certain fishes in the Antarctic Ocean swim in water at about −2°C. (a) To prevent their blood from freezing, what must be the concentration (in molality) of the blood? Is this a reasonable physiological concentration? (b) In recent years scientists have discovered a special type of protein in these fishes' blood which, although present in quite low concentrations (≤ 0.001 *m*), has the ability to prevent the blood from freezing. Suggest a mechanism for its action.

12.125 As we know, if a soft drink can is shaken and then opened, the drink escapes violently. However, if after shaking the can we tap it several times with a metal spoon, no such "explosion" of the drink occurs. Why?

12.126 Why are ice cubes (for example, those you see in the trays in the freezer of a refrigerator) cloudy inside?

12.127 Two beakers are placed in a closed container. Beaker A initially contains 0.15 mole of naphthalene ($C_{10}H_8$) in 100 g of benzene (C_6H_6) and beaker B initially contains 31 g of an unknown compound dissolved in 100 g of benzene. At equilibrium, beaker A is found to have lost 7.0 g of benzene. Assuming ideal behavior, calculate the molar mass of the unknown compound. State any assumptions made.

12.128 At 27°C, the vapor pressure of pure water is 23.76 mmHg and that of an urea solution is 22.98 mmHg. Calculate the molality of solution.

12.129 An example of the positive deviation shown in Figure 12.8(a) is a solution made of acetone (CH_3COCH_3) and carbon disulfide (CS_2). (a) Draw Lewis structures of these molecules. Explain the deviation from ideal behavior in terms of intermolecular forces. (b) A solution composed of 0.60 mole of acetone and 0.40 mole of carbon disulfide has a vapor pressure of 615 mmHg at 35.2°C. What would be the vapor pressure if the solution behaved ideally? The vapor pressure of the pure solvents at the same temperature are acetone: 349 mmHg; carbon disulfide: 501 mmHg. (c) Predict the sign of ΔH_{soln}.

12.130 Liquids A (molar mass 100 g/mol) and B (molar mass 110 g/mol) form an ideal solution. At 55°C, A has a vapor pressure of 95 mmHg and B has a vapor pressure of 42 mmHg. A solution is prepared by mixing equal masses of A and B. (a) Calculate the mole fraction of each component in the solution. (b) Calculate the partial pressures of A and B over the solution at 55°C. (c) Suppose that some of the vapor described in (b) is condensed to a liquid in a separate container. Calculate the mole fraction of each component in this liquid and the vapor pressure of each component above this liquid at 55°C.

12.131 A very long pipe is capped at one end with a semipermeable membrane. How deep (in meters) must the pipe be immersed into the sea for freshwater to begin to pass through the membrane? Assume the water to be at 20°C and treat it as a 0.70 *M* NaCl solution. The density of seawater is 1.03 g/cm^3 and the acceleration due to gravity is 9.81 m/s^2.

12.132 Two beakers, 1 and 2, containing 50 mL of 0.10 *M* urea and 50 mL of 0.20 *M* urea, respectively, are placed under a tightly sealed container (see Figure 12.12) at 298 K. Calculate the mole fraction of urea in the solutions at equilibrium. Assume ideal behavior.

12.133 A mixture of liquids A and B exhibits ideal behavior. At 84°C, the total vapor pressure of a solution containing 1.2 moles of A and 2.3 moles of B is 331 mmHg. Upon the addition of another mole of B to the solution, the vapor pressure increases to 347 mmHg. Calculate the vapor pressures of pure A and B at 84°C.

12.134 Use Henry's law and the ideal gas equation to prove the statement that the volume of a gas that dissolves in a given amount of solvent is *independent* of the pressure of the gas. (*Hint:* Henry's law can be modified as $n = kP$, where n is the number of moles of the gas dissolved in the solvent.)

12.135 (a) Derive the equation relating the molality (m) of a solution to its molarity (M),

$$m = \frac{M}{d - \dfrac{M\mathcal{M}}{1000}}$$

where d is the density of the solution (g/mL) and $\mathcal{M}$ is the molar mass of the solute (g/mol). (*Hint:* Start by expressing the solvent in kilograms in terms of the difference between the mass of the solution and the mass of the solute.) (b) Show that, for dilute aqueous solutions, m is approximately equal to M.

12.136 At 298 K, the osmotic pressure of a glucose solution is 10.50 atm. Calculate the freezing point of the solution. The density of the solution is 1.16 g/mL.

12.137 A student carried out the following procedure to measure the pressure of carbon dioxide in a soft drink bottle. First, she weighed the bottle (853.5 g). Next, she carefully removed the cap to let the CO_2 gas escape. She then reweighed the bottle with the cap (851.3 g). Finally, she measured the volume of the soft drink (452.4 mL). Given that Henry's law constant for CO_2 in water at 25°C is 3.4×10^{-2} mol/L · atm, calculate the pressure of CO_2 in the original bottle. Why is this pressure only an estimate of the true value?

12.138 Valinomycin is an antibiotic. It functions by binding K^+ ions and transporting them across the membrane into cells to offset the ionic balance. The molecule is represented here by its *skeletal* structure in which the end of each straight line corresponds to a carbon atom (unless a N or an O atom is shown at the end of the line). There are as many H atoms attached to each C atom as necessary to give each C atom a total of four bonds. Use the "like dissolves like" guideline to explain its function. (*Hint:* The —CH_3 groups at the two ends of the Y shape are nonpolar.)

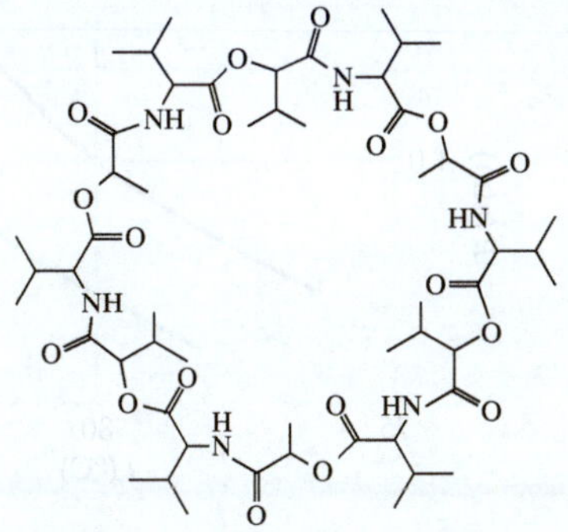

12.139 Often the determination of the molar mass of a compound by osmotic pressure measurement is carried out at several different concentrations to get a more reliable average value. From the following data for the osmotic pressure of poly(methyl methacrylate) in toluene at 25°C, determine graphically the molar mass of the polymer. [*Hint:* Rearrange Equation (12.8) so that π is expressed in terms of c, which is the number of grams of the solute per liter of solution.]

π (atm)	8.40×10^{-4}	1.72×10^{-3}	2.52×10^{-3}	3.23×10^{-3}	7.75×10^{-3}
c (g/L)	8.10	12.31	15.00	18.17	28.05

12.140 Here is an after-dinner trick. With guests still sitting at the table, the host provided each of them with a glass of water containing an ice cube floating on top and a piece of string about 2 to 3 in. in length. He then asked them to find a way to lift the ice cube without touching it by hand or using any other objects such as a spoon or fork. Explain how this task can be accomplished. (*Hint:* The table had not been cleared so the salt and pepper shakers were still there.)

Interpreting, Modeling, & Estimating

12.141 The molecule drawn here has shown promise as an agent for cleaning up oil spills in water. Instead of dispersing the oil into water as soap molecules would do (see Figures 12.19 and 12.20), these molecules bind with the oil to form a gel, which can be easily separated from the body of water. Suggest an explanation for the ability of this compound to remove oil from water.

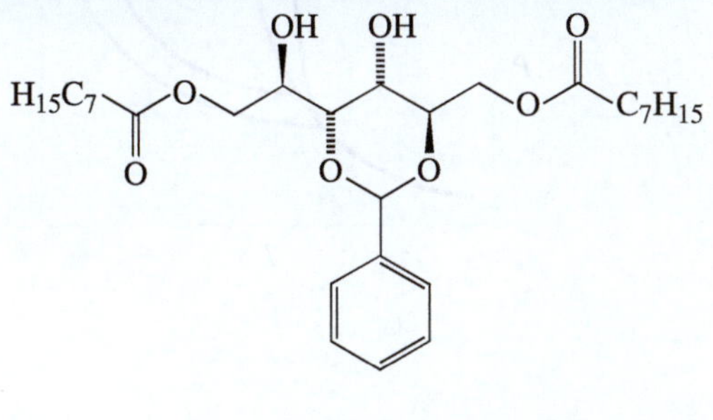

12.142 The Henry's law constant of oxygen in water at 25°C is 1.3×10^{-3} mol/L · atm. Calculate the molarity of oxygen in water under 1 atmosphere of air. Comment on the prospect for our survival without hemoglobin molecules. (Recall from previous problems that the total volume of blood in an adult human is about 5 L.)

12.143 The diagram shows the vapor pressure curves for pure benzene and a solution of a nonvolatile solute in benzene. Estimate the molality of the benzene solution.

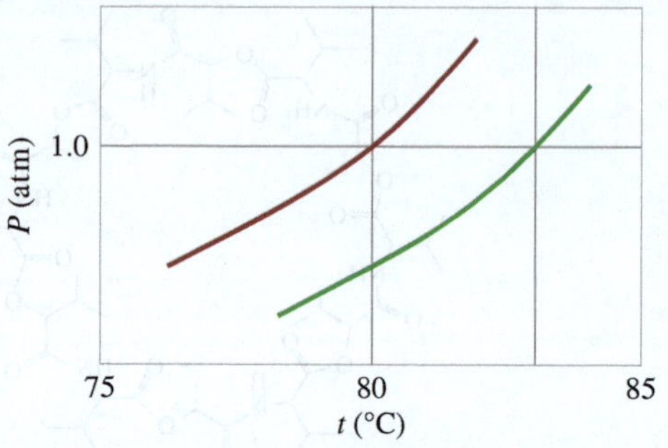

12.144 A common misconception is that adding salt to the water used to cook spaghetti will decrease the cooking time, presumably because it increases the boiling point of the water. Calculate the boiling point of a typical salted water solution used to cook spaghetti. Do you think this increase in temperature will make much difference in the cooking time for spaghetti?

12.145 Estimate the volume of the oil droplet that would be formed by the compound sodium stearate shown in Figure 12.19.

12.146 The diagram here shows the vapor pressure curves of two liquids A and B and a solution of the two liquids. Given that A is more volatile than B, match the curves with the pure liquids and the solution.

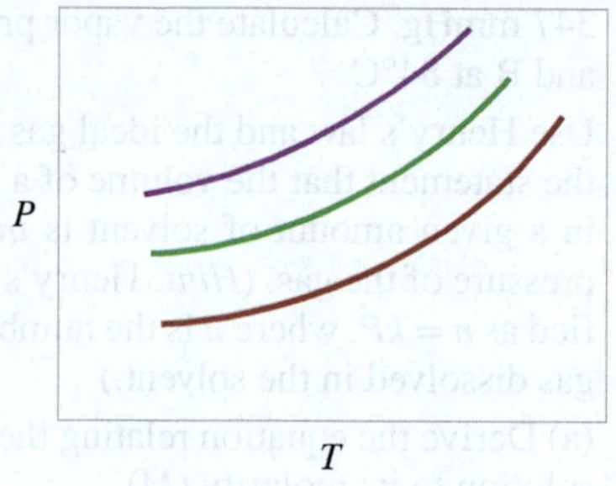

Answers to Practice Exercises

12.1 Carbon disulfide. **12.2** 7.44 percent. **12.3** 0.638 *m*. **12.4** 8.92 *m*. **12.5** 13.8 *m*. **12.6** 2.9×10^{-4} *M*. **12.7** 37.8 mmHg; 4.4 mmHg. **12.8** T_b: 101.2°C; T_f: −4.48°C. **12.9** 21.0 atm. **12.10** 0.065 *m* and 1.3×10^2 g/mol. **12.11** 2.60×10^4 g. **12.12** 1.21.

Answers to Review of Concepts & Facts

12.2.1 C_4H_{10} and P_4. **12.2.2** (a) Hydrogen bonding. (b) Dipole-induced dipole. **12.3.1** Molarity (it decreases because the volume of the solution increases on heating). **12.3.2** 3.35%. **12.3.3** 7.37 *M* and 8.65 *m*. **12.4.1** $KCl < KNO_3 < KBr$. **12.5.1** HCl because it is much more soluble in water. **12.5.2** 7.8×10^{-5} mol/L. **12.6.1** The vapor pressure of an ideal solution would be 150 mmHg. If the vapor pressure of the solution is 164 mmHg, it means that the intermolecular forces between A and B molecules are weaker than the intermolecular forces between A molecules and between B molecules. **12.6.2**

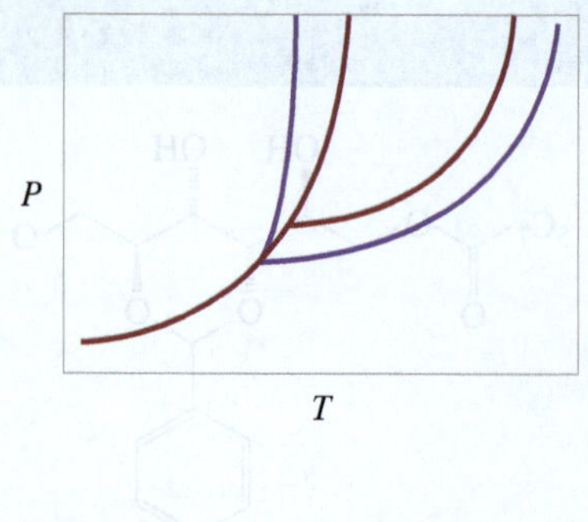

Yes, freezing-point depression and boiling-point elevation would still apply. **12.6.3** When the seawater is placed in an apparatus like that shown in Figure 12.11, it exerts a pressure of 25 atm. **12.6.4** b.p. = 101.4°C; f.p. = −5.1°C. **12.7.1** (a) Na_2SO_4. (b) $MgSO_4$. (c) LiBr. **12.7.2** Assume $i = 2$ for NaCl. The concentration of the saline solution should be about 0.15 *M*.

CHAPTER

13

Chemical Kinetics

Polar stratospheric clouds form at low temperatures in the atmosphere above Antarctica and are implicated in the catalytic depletion of ozone.

Per-Andre Hoffmann/Image Professionals GmbH/ Alamy Stock Photo

CHAPTER OUTLINE

In previous chapters, we studied basic definitions in chemistry, and we examined the properties of gases, liquids, solids, and solutions. We have discussed molecular properties and looked at several types of reactions in some detail. In this chapter and in subsequent chapters, we will look more closely at the relationships and the laws that govern chemical reactions.

How can we predict whether or not a reaction will take place? Once started, how fast does the reaction proceed? How far will the reaction go before it stops? The laws of thermodynamics (to be discussed in Chapter 17) help us answer the first question. Chemical kinetics, the subject of this chapter, provides answers to the question about the speed of a reaction. The last question is one of many answered by the study of chemical equilibrium, which we will consider in Chapters 14, 15, and 16.

13.1 The Rate of a Reaction

Learning Objectives

- Evaluate the average rate of a reaction given appropriate data.
- Infer the instantaneous rate of a reaction given a graph.
- Use the stoichiometry of a reaction to express the rate of a reaction in terms of a reactant or product.

Chemical kinetics is *the area of chemistry concerned with the speeds, or rates, at which a chemical reaction occurs*. The word "kinetic" suggests movement or change; in Chapter 5 we defined kinetic energy as the energy available because of the motion of an object. Here kinetics refers to the rate of a reaction, or the ***reaction rate,*** which is *the change in the concentration of a reactant or a product with time* (*M*/s).

There are many reasons for studying the rate of a reaction. To begin, there is intrinsic curiosity about why reactions have such vastly different rates. Some processes, such as the initial steps in vision and photosynthesis and nuclear chain reactions, take place on a time scale as short as 10^{-12} s to 10^{-6} s. Others, like the curing of cement and the conversion of graphite to diamond, take years or millions of years to complete. On a practical level, knowledge of reaction rates is useful in drug design, in pollution control, and in food processing. Industrial chemists often place more emphasis on speeding up the rate of a reaction than on maximizing its yield.

We know that any reaction can be represented by the general equation

$$\text{reactants} \longrightarrow \text{products}$$

This equation tells us that during the course of a reaction, reactants are consumed while products are formed. As a result, we can follow the progress of a reaction by monitoring either the decrease in concentration of the reactants or the increase in concentration of the products.

Figure 13.1 shows the progress of a simple reaction in which A molecules are converted to B molecules:

$$\text{A} \longrightarrow \text{B}$$

The decrease in the number of A molecules and the increase in the number of B molecules with time are shown in Figure 13.2. In general, it is more convenient to express the reaction rate in terms of the change in concentration with time. Thus, for the reaction A $\longrightarrow$ B we can express the rate as

$$\text{rate} = -\frac{\Delta[\text{A}]}{\Delta t} \quad \text{or} \quad \text{rate} = \frac{\Delta[\text{B}]}{\Delta t}$$

where Δ[A] and Δ[B] are the changes in concentration (molarity) over a time period Δt (recall that Δ denotes the difference between the final and initial states). Because

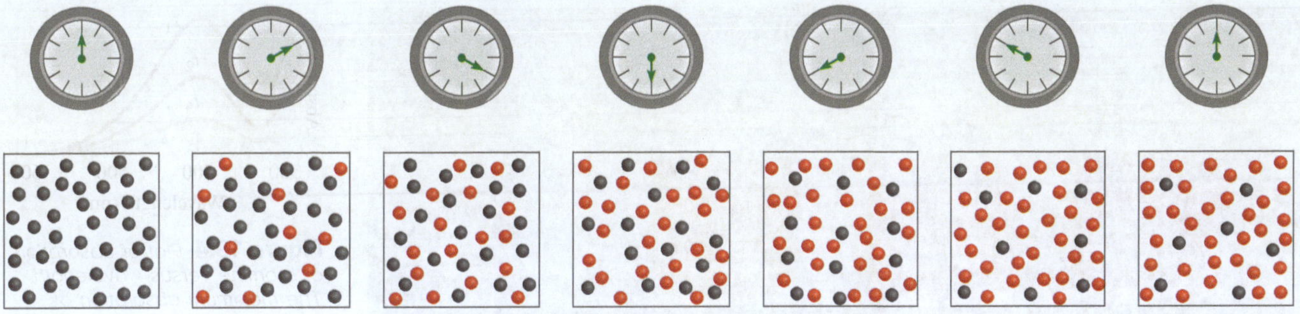

Figure 13.1 *The progress of reaction A ⟶ B at 10-s intervals over a period of 60 s. Initially, only A molecules (gray spheres) are present. As time progresses, B molecules (red spheres) are formed.*

the concentration of A *decreases* during the time interval, Δ[A] is a negative quantity. The rate of a reaction is a positive quantity, so a minus sign is needed in the rate expression to make the rate positive. On the other hand, the rate of product formation does not require a minus sign because Δ[B] is a positive quantity (the concentration of B *increases* with time). These rates are *average rates* because they are averaged over a certain time period Δt.

Our next step is to see how the rate of a reaction is obtained experimentally. By definition, we know that to determine the rate of a reaction we have to monitor the concentration of the reactant (or product) as a function of time. For reactions in solution, the concentration of a species can often be measured by spectroscopic means. If ions are involved, the change in concentration can also be detected by an electrical conductance measurement. Reactions involving gases are most conveniently followed by pressure measurements. We will consider two specific reactions for which different methods are used to measure the reaction rates.

Reaction of Molecular Bromine and Formic Acid

In aqueous solutions, molecular bromine reacts with formic acid (HCOOH) as follows:

$$Br_2(aq) + HCOOH(aq) \longrightarrow 2Br^-(aq) + 2H^+(aq) + CO_2(g)$$

Molecular bromine is reddish-brown in color. All the other species in the reaction are colorless. As the reaction progresses, the concentration of Br_2 steadily decreases and its color fades (Figure 13.3). This loss of color and hence concentration can be monitored easily with a spectrometer, which registers the amount of visible light absorbed by bromine (Figure 13.4).

Measuring the change (decrease) in bromine concentration at some initial time and then at some final time enables us to determine the average rate of the reaction during that interval:

$$\text{average rate} = -\frac{\Delta[Br_2]}{\Delta t}$$

$$= -\frac{[Br_2]_{final} - [Br_2]_{initial}}{t_{final} - t_{initial}}$$

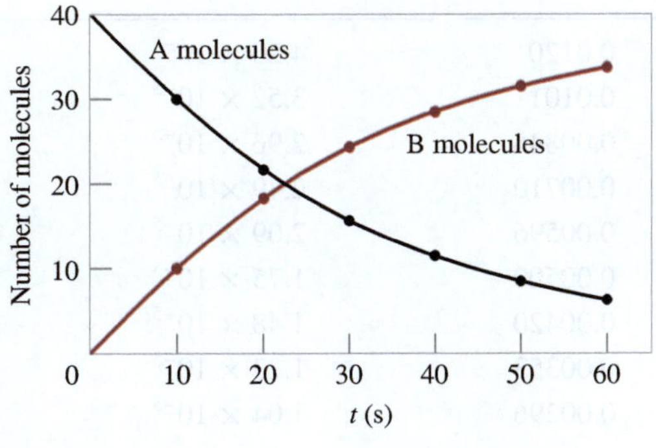

Figure 13.2 *The rate of reaction A ⟶ B, represented as the decrease of A molecules with time and as the increase of B molecules with time.*

Figure 13.3 *From left to right: The decrease in bromine concentration as time elapses shows up as a loss of color.*
Ken Karp/McGraw Hill

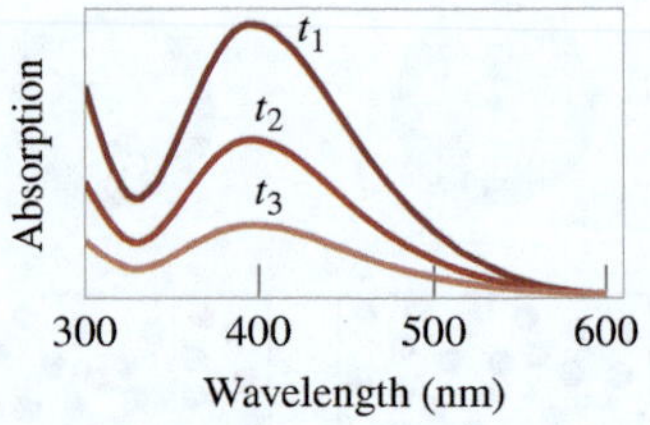

Figure 13.4 *Plot of absorption of bromine versus wavelength. The maximum absorption of visible light by bromine occurs at 393 nm. As the reaction progresses (t_1 to t_3), the absorption, which is proportional to [Br_2], decreases.*

Using the data provided in Table 13.1 we can calculate the average rate over the first 50-s time interval as follows:

$$\text{average rate} = -\frac{(0.0101 - 0.0120)\ M}{50.0\ \text{s}} = 3.80 \times 10^{-5}\ M/\text{s}$$

If we had chosen the first 100 s as our time interval, the average rate would then be given by

$$\text{average rate} = -\frac{(0.00846 - 0.0120)\ M}{100.0\ \text{s}} = 3.54 \times 10^{-5}\ M/\text{s}$$

These calculations demonstrate that the average rate of the reaction depends on the time interval we choose.

By calculating the average reaction rate over shorter and shorter intervals, we can obtain the rate for a specific instant in time, which gives us the *instantaneous rate* of the reaction at that time. Figure 13.5 shows the plot of [Br_2] versus time, based on the data shown in Table 13.1. Graphically, the instantaneous rate at 100 s after the start of the reaction, say, is given by the slope of the tangent to the curve at that instant. The instantaneous rate at any other time can be determined in a similar manner. Note that the instantaneous rate determined in this way will always have the same value for the same concentrations of reactants, as long as the temperature is kept constant. We do not need to be concerned with what time interval to use. Unless otherwise stated, we will refer to the instantaneous rate at a specific time merely as "the rate" at that time.

Table 13.1 **Rates of the Reaction Between Molecular Bromine and Formic Acid at 25°C**

Time (s)	[Br_2] (M)	Rate (M/s)	$k = \frac{\text{rate}}{[Br_2]}$ (s^{-1})
0.0	0.0120	4.20×10^{-5}	3.50×10^{-3}
50.0	0.0101	3.52×10^{-5}	3.49×10^{-3}
100.0	0.00846	2.96×10^{-5}	3.50×10^{-3}
150.0	0.00710	2.49×10^{-5}	3.51×10^{-3}
200.0	0.00596	2.09×10^{-5}	3.51×10^{-3}
250.0	0.00500	1.75×10^{-5}	3.50×10^{-3}
300.0	0.00420	1.48×10^{-5}	3.52×10^{-3}
350.0	0.00353	1.23×10^{-5}	3.48×10^{-3}
400.0	0.00296	1.04×10^{-5}	3.51×10^{-3}

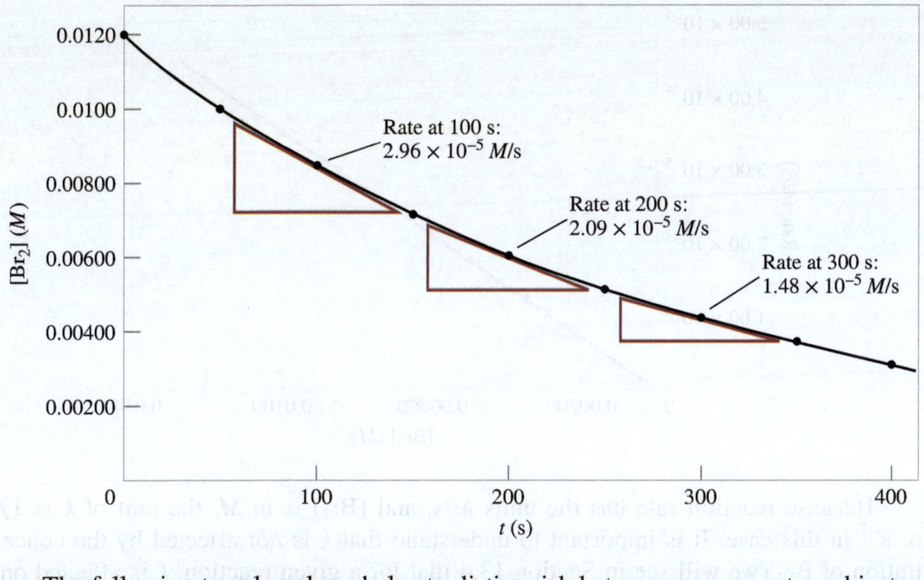

Figure 13.5 *The instantaneous rates of the reaction between molecular bromine and formic acid at t = 100 s, 200 s, and 300 s are given by the slopes of the tangents at these times.*

The following travel analogy helps to distinguish between average rate and instantaneous rate. The distance by car from San Francisco to Los Angeles is 512 mi along a certain route. If it takes a person 11.4 h to go from one city to the other, the average speed is 512 mi/11.4 h or 44.9 mph. But if the car is traveling at 55.3 mph 3 h and 26 min after departure, then the instantaneous speed of the car is 55.3 mph at that time. In other words, instantaneous speed is the speed that you would read from the speedometer. Note that the speed of the car in our example can increase or decrease during the trip, but the instantaneous rate of a reaction always decreases with time.

The rate of the bromine-formic acid reaction also depends on the concentration of formic acid. However, by adding a large excess of formic acid to the reaction mixture we can ensure that the concentration of formic acid remains virtually constant throughout the course of the reaction. Under this condition, the change in the amount of formic acid present in solution has no effect on the measured rate.

Let's consider the effect that the bromine concentration has on the rate of reaction. Look at the data in Table 13.1. Compare the concentration of Br_2 and the reaction rate at $t = 50$ s and $t = 250$ s. At $t = 50$ s, the bromine concentration is 0.0101 M and the rate of reaction is 3.52×10^{-5} M/s. At $t = 250$ s, the bromine concentration is 0.00500 M and the rate of reaction is 1.75×10^{-5} M/s. The concentration at $t = 50$ s is double the concentration at $t = 250$ s (0.0101 M versus 0.00500 M), and the rate of reaction at $t = 50$ s is double the rate at $t = 250$ s (3.52×10^{-5} M/s versus 1.75×10^{-5} M/s). We see that as the concentration of bromine is doubled, the rate of reaction also doubles.

$$\frac{[Br_2]_{50\text{ s}}}{[Br_2]_{250\text{ s}}} \approx 2 \quad \text{and} \quad \frac{\text{rate at 50 s}}{\text{rate at 250 s}} = \frac{3.52 \times 10^{-5}\ M/\text{s}}{1.75 \times 10^{-5}\ M/\text{s}} \approx 2$$

Thus, the rate is directly proportional to the Br_2 concentration, that is

$$\begin{aligned} \text{rate} &\propto [Br_2] \\ &= k[Br_2] \end{aligned}$$

where the term k is known as the ***rate constant,*** *a constant of proportionality between the reaction rate and the concentration of reactant*. This direct proportionality between Br_2 concentration and rate is also supported by plotting the data.

Figure 13.6 is a plot of the rate versus Br_2 concentration. The fact that this graph is a straight line shows that the rate is directly proportional to the concentration; the higher the concentration, the higher the rate. Rearranging the last equation gives

$$k = \frac{\text{rate}}{[Br_2]}$$

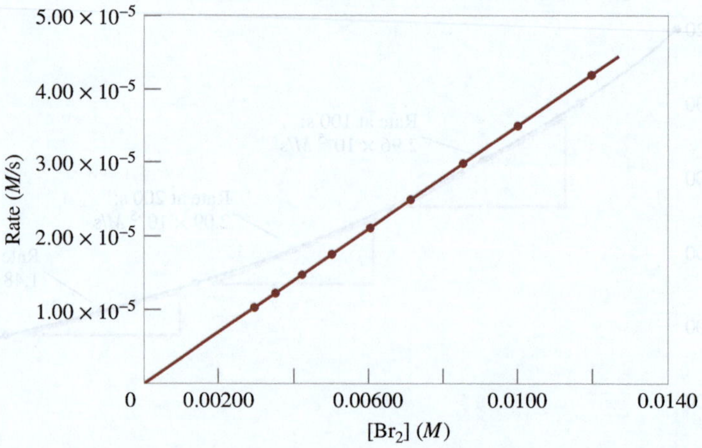

Figure 13.6 *Plot of rate versus molecular bromine concentration for the reaction between molecular bromine and formic acid. The straight-line relationship shows that the rate of reaction is directly proportional to the molecular bromine concentration.*

Because reaction rate has the units *M*/s, and $[Br_2]$ is in *M*, the unit of *k* is 1/s, or s^{-1} in this case. It is important to understand that *k* is *not* affected by the concentration of Br_2 (we will see in Section 13.4 that for a given reaction, *k* is affected only by a change in temperature). To be sure, the rate is greater at a higher concentration and smaller at a lower concentration of Br_2, but the *ratio* of rate/$[Br_2]$ remains the same provided the temperature does not change.

From Table 13.1 we can calculate the rate constant for the reaction. Taking the data for $t = 50$ s, we write

$$k = \frac{\text{rate}}{[Br_2]}$$
$$= \frac{3.52 \times 10^{-5}\ M/\text{s}}{0.0101\ M} = 3.49 \times 10^{-3}\ \text{s}^{-1}$$

We can use the data for any *t* to calculate *k*. The slight variations in the values of *k* listed in Table 13.1 are due to experimental deviations in rate measurements.

Decomposition of Hydrogen Peroxide

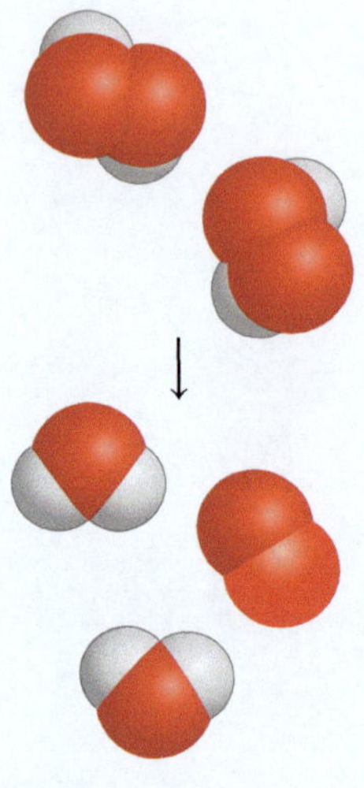

$2H_2O_2 \longrightarrow 2H_2O + O_2$

If one of the products or reactants is a gas, we can use a manometer to find the reaction rate. Consider the decomposition of hydrogen peroxide at 20°C:

$$2H_2O_2(aq) \longrightarrow 2H_2O(l) + O_2(g)$$

In this case, the rate of decomposition can be determined by monitoring the rate of oxygen evolution with a manometer (Figure 13.7). The oxygen pressure can be readily converted to concentration by using the ideal gas equation

$$PV = nRT$$

or

$$P = \frac{n}{V}RT = [O_2]RT$$

where n/V gives the molarity of oxygen gas. Rearranging the equation, we get

$$[O_2] = \frac{1}{RT}P$$

The reaction rate, which is given by the rate of oxygen production, can now be written as

$$\text{rate} = \frac{\Delta[O_2]}{\Delta t} = \frac{1}{RT}\frac{\Delta P}{\Delta t}$$

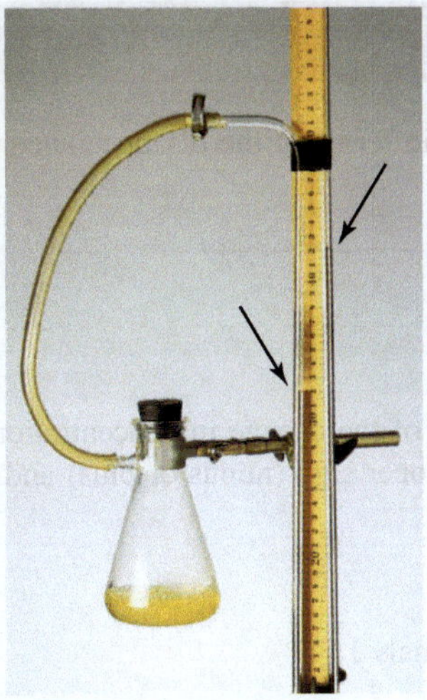

Figure 13.7 *The rate of hydrogen peroxide decomposition can be measured with a manometer, which shows the increase in the oxygen gas pressure with time. The arrows show the mercury levels in the U tube.*
Ken Karp/McGraw Hill

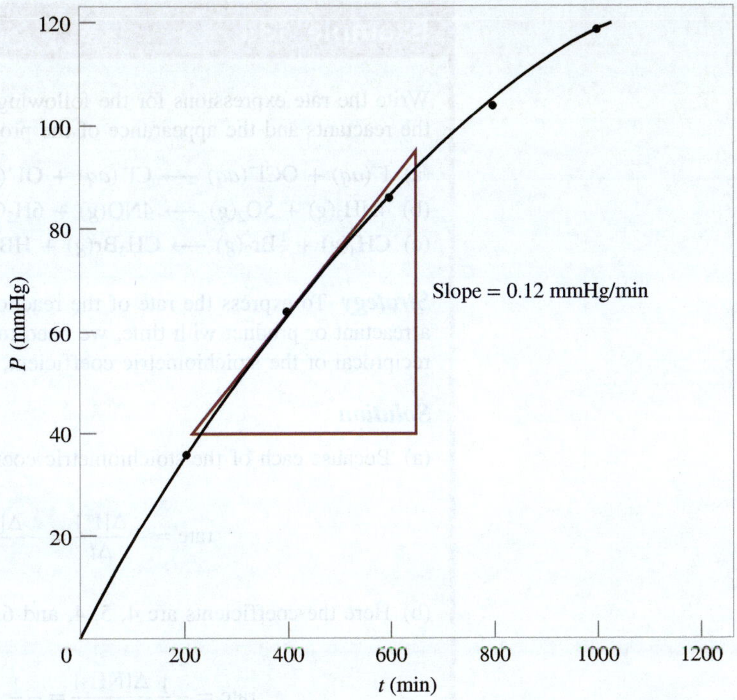

Figure 13.8 *The instantaneous rate for the decomposition of hydrogen peroxide at 400 min is given by the slope of the tangent multiplied by 1/RT.*

Figure 13.8 shows the increase in oxygen pressure with time and the determination of an instantaneous rate at 400 min. To express the rate in the normal units of *M*/s, we convert mmHg/min to atm/s, then multiply the slope of the tangent ($\Delta P/\Delta t$) by $1/RT$, as shown in the previous equation.

Reaction Rates and Stoichiometry

We have seen that for stoichiometrically simple reactions of the type A $\longrightarrow$ B, the rate can be expressed in terms of either the decrease in reactant concentration with time, $-\Delta[\text{A}]/\Delta t$, or the increase in product concentration with time, $\Delta[\text{B}]/\Delta t$. For more complex reactions, we must be careful in writing the rate expressions. Consider, for example, the reaction

$$2\text{A} \longrightarrow \text{B}$$

Two moles of A disappear for each mole of B that forms; that is, the rate at which B forms is one-half the rate at which A disappears. Thus, the rate can be expressed as

$$\text{rate} = -\frac{1}{2}\frac{\Delta[\text{A}]}{\Delta t} \quad \text{or} \quad \text{rate} = \frac{\Delta[\text{B}]}{\Delta t}$$

In general, for the reaction

$$a\text{A} + b\text{B} \longrightarrow c\text{C} + d\text{D}$$

the rate is given by

$$\text{rate} = -\frac{1}{a}\frac{\Delta[\text{A}]}{\Delta t} = -\frac{1}{b}\frac{\Delta[\text{B}]}{\Delta t} = \frac{1}{c}\frac{\Delta[\text{C}]}{\Delta t} = \frac{1}{d}\frac{\Delta[\text{D}]}{\Delta t}$$

Examples 13.1 and 13.2 show writing the reaction rate expressions and calculating rates of product formation and reactant disappearance.

Example 13.1

Write the rate expressions for the following reactions in terms of the disappearance of the reactants and the appearance of the products.

(a) $I^-(aq) + OCl^-(aq) \longrightarrow Cl^-(aq) + OI^-(aq)$
(b) $4NH_3(g) + 5O_2(g) \longrightarrow 4NO(g) + 6H_2O(g)$
(c) $CH_4(g) + \frac{1}{2}Br_2(g) \longrightarrow CH_3Br(g) + HBr(g)$

Strategy To express the rate of the reaction in terms of the change in concentration of a reactant or product with time, we need to use the proper sign (minus or plus) and the reciprocal of the stoichiometric coefficient.

Solution

(a) Because each of the stoichiometric coefficients equals 1,

$$\text{rate} = -\frac{\Delta[I^-]}{\Delta t} = -\frac{\Delta[OCl^-]}{\Delta t} = \frac{\Delta[Cl^-]}{\Delta t} = \frac{\Delta[OI^-]}{\Delta t}$$

(b) Here the coefficients are 4, 5, 4, and 6, so

$$\text{rate} = -\frac{1}{4}\frac{\Delta[NH_3]}{\Delta t} = -\frac{1}{5}\frac{\Delta[O_2]}{\Delta t} = \frac{1}{4}\frac{\Delta[NO]}{\Delta t} = \frac{1}{6}\frac{\Delta[H_2O]}{\Delta t}$$

(c) The coefficient for Br_2 is $\frac{1}{2}$, so to express the rate in terms of the disappearance of dibromine, we multiply by the reciprocal of the coefficient $\left[\left(\frac{1}{2}\right)^{-1} = 2\right]$, giving

$$\text{rate} = -\frac{\Delta[CH_4]}{\Delta t} = -2\frac{\Delta[Br_2]}{\Delta t} = \frac{\Delta[CH_3Br]}{\Delta t} = \frac{\Delta[HBr]}{\Delta t}$$

Practice Exercise Write the rate expressions for the following reaction:

$$CH_4(g) + 2O_2(g) \longrightarrow CO_2(g) + 2H_2O(g)$$

Similar problems: 13.5, 13.6.

Example 13.2

Consider the reaction

$$4NO_2(g) + O_2 \longrightarrow 2N_2O_5(g)$$

Suppose that, at a particular moment during the reaction, molecular oxygen is reacting at the rate of 0.024 *M*/s. (a) At what rate is N_2O_5 being formed? (b) At what rate is NO_2 reacting?

Strategy To calculate the rate of formation of N_2O_5 and disappearance of NO_2, we need to express the rate of the reaction in terms of the stoichiometric coefficients as in Example 13.1:

$$\text{rate} = -\frac{1}{4}\frac{\Delta[NO_2]}{\Delta t} = -\frac{\Delta[O_2]}{\Delta t} = \frac{1}{2}\frac{\Delta[N_2O_5]}{\Delta t}$$

We are given

$$\frac{\Delta[O_2]}{\Delta t} = -0.024\ M/\text{s}$$

where the minus sign shows that the concentration of O_2 is decreasing with time.

(Continued)

Solution

(a) From the preceding rate expression we have

$$-\frac{\Delta[O_2]}{\Delta t} = \frac{1}{2}\frac{\Delta[N_2O_5]}{\Delta t}$$

Therefore,

$$\frac{\Delta[N_2O_5]}{\Delta t} = -2(-0.024\ M/s) = 0.048\ M/s$$

(b) Here we have

$$\frac{1}{4}\frac{\Delta[NO_2]}{\Delta t} = -\frac{\Delta[O_2]}{\Delta t}$$

so

$$\frac{\Delta[NO_2]}{\Delta t} = 4(-0.024\ M/s) = -0.096\ M/s$$

Practice Exercise Consider the reaction

$$4PH_3(g) \longrightarrow P_4(g) + 6H_2(g)$$

Suppose that, at a particular moment during the reaction, molecular hydrogen is being formed at the rate of 0.078 *M*/s. (a) At what rate is P_4 being formed? (b) At what rate is PH_3 reacting?

Similar problems: 13.7, 13.8.

Summary of Concepts & Facts

- The rate of a chemical reaction is the change in the concentration of reactants or products over time. The rate is not constant but varies continuously as concentrations change.

Review of Concepts & Facts

13.1.1 Write a balanced equation for a gas-phase reaction whose rate is given by

$$\text{rate} = -\frac{1}{2}\frac{\Delta[NOCl]}{\Delta t} = \frac{1}{2}\frac{\Delta[NO]}{\Delta t} = \frac{\Delta[Cl_2]}{\Delta t}$$

13.1.2 Write the rate expression for the following reaction.

$$2A + B \longrightarrow C + 2D$$

13.1.3 For the same reaction,

$$2A + B \longrightarrow C + 2D$$

what is the rate of change in the concentration of B if the concentration of A is changing at a rate of 0.070 *M*/s?

13.2 Rate Laws

Learning Objectives

- Summarize reaction order and provide examples of a zeroth-, first-, and second-order reaction rate law.
- Write the rate law of a reaction given experimental data.
- Express the units of the rate constant k for a reaction.

So far we have learned that the rate of a reaction is proportional to the concentration of reactants and that the proportionality constant k is called the rate constant. The ***rate law*** *expresses the relationship of the rate of a reaction to the rate constant and the concentrations of the reactants raised to some powers.* For the general reaction

$$aA + bB \longrightarrow cC + dD$$

the rate law takes the form

$$\text{rate} = k[A]^x[B]^y \qquad (13.1)$$

where x and y are numbers that must be determined experimentally. Note that, in general, x and y are *not* equal to the stoichiometric coefficients a and b. When we know the values of x, y, and k, we can use Equation (13.1) to calculate the rate of the reaction, given the concentrations of A and B.

Student Hot Spot

Student data indicate you may struggle with rate laws. Access your eBook for additional Learning Resources on this topic.

The exponents x and y specify the relationships between the concentrations of reactants A and B and the reaction rate. Added together, they give us the overall ***reaction order,*** defined as *the sum of the powers to which all reactant concentrations appearing in the rate law are raised.* For Equation (13.1) the overall reaction order is $x + y$. Alternatively, we can say that the reaction is xth order in A, yth order in B, and $(x + y)$ th order overall.

To see how to determine the rate law of a reaction, let us consider the reaction between fluorine and chlorine dioxide:

$$F_2(g) + 2ClO_2(g) \longrightarrow 2FClO_2(g)$$

$F_2 + 2ClO_2 \longrightarrow 2FClO_2$

One way to study the effect of reactant concentration on reaction rate is to determine how the *initial rate* depends on the starting concentrations. It is preferable to measure the initial rates because as the reaction proceeds, the concentrations of the reactants decrease and it may become difficult to measure the changes accurately. Also, there may be a reverse reaction of the type

$$\text{products} \longrightarrow \text{reactants}$$

which would introduce error into the rate measurement. Both of these complications are virtually absent during the early stages of the reaction.

Table 13.2 shows three rate measurements for the formation of $FClO_2$. Looking at entries 1 and 3, we see that as we double $[F_2]$ while holding $[ClO_2]$ constant, the reaction rate doubles. Thus, the rate is directly proportional to $[F_2]$. Similarly, the data in entries 1 and 2 show that as we quadruple $[ClO_2]$ at constant $[F_2]$, the rate increases by four times, so that the rate is also directly proportional to $[ClO_2]$. We can summarize our observations by writing the rate law as

$$\text{rate} = k[F_2][ClO_2]$$

Because both $[F_2]$ and $[ClO_2]$ are raised to the first power, the reaction is first order in F_2, first order in ClO_2, and (1 + 1) or second order overall. Note that $[ClO_2]$ is

Table 13.2 Rate Data for the Reaction Between F_2 and ClO_2

$[F_2]$ (*M*)	$[ClO_2]$ (*M*)	Initial Rate (*M*/s)
1. 0.10	0.010	1.2×10^{-3}
2. 0.10	0.040	4.8×10^{-3}
3. 0.20	0.010	2.4×10^{-3}

raised to the power of 1 whereas its stoichiometric coefficient in the overall equation is 2. The equality of reaction order (first) and stoichiometric coefficient (1) for F_2 is coincidental in this case.

From the reactant concentrations and the initial rate, we can also calculate the rate constant. Using the first entry of data in Table 13.2, we can write

$$k = \frac{\text{rate}}{[F_2][ClO_2]} = \frac{1.2 \times 10^{-3}\ M/\text{s}}{(0.10\ M)(0.010\ M)} = 1.2/M \cdot \text{s}$$

Reaction order enables us to understand how the reaction depends on reactant concentrations. Suppose, for example, that for the general reaction $a\text{A} + b\text{B} \longrightarrow c\text{C} + d\text{D}$ we have $x = 1$ and $y = 2$. The rate law for the reaction is [see Equation (13.1)]

$$\text{rate} = k[\text{A}][\text{B}]^2$$

This reaction is first order in A, second order in B, and third order overall ($1 + 2 = 3$). Let us assume that initially $[\text{A}] = 1.0\ M$ and $[\text{B}] = 1.0\ M$. The rate law tells us that if we double the concentration of A from 1.0 *M* to 2.0 *M* at constant [B], we also double the reaction rate:

for $[\text{A}] = 1.0\ M$ $\quad \text{rate}_1 = k(1.0\ M)(1.0\ M)^2 = k(1.0\ M^3)$

for $[\text{A}] = 2.0\ M$ $\quad \text{rate}_2 = k(2.0\ M)(1.0\ M)^2 = k(2.0\ M^3)$

Hence, $\quad \text{rate}_2 = 2(\text{rate}_1)$

On the other hand, if we double the concentration of B from 1.0 *M* to 2.0 *M* at constant $[\text{A}] = 1\ M$, the rate will increase by a factor of 4 because of the power 2 in the exponent:

for $[\text{B}] = 1.0\ M$ $\quad \text{rate}_1 = k(1.0\ M)(1.0\ M)^2 = k(1.0\ M^3)$

for $[\text{B}] = 2.0\ M$ $\quad \text{rate}_2 = k(1.0\ M)(2.0\ M)^2 = k(4.0\ M^3)$

Hence, $\quad \text{rate}_2 = 4(\text{rate}_1)$

If, for a certain reaction, $x = 0$ and $y = 1$, then the rate law is

$$\text{rate} = k[\text{A}]^0[\text{B}] = k[\text{B}]$$

This reaction is zero order in A, first order in B, and first order overall. The exponent zero tells us that the rate of this reaction is *independent* of the concentration of A.

Table 13.3 Units of the Rate Constant *k*

Overall Reaction Order	Sample Rate Law	Units of *k*
0	rate $= k$	$M \cdot s^{-1}$
1	rate $= k[A]$	s^{-1}
2	rate $= k[A]^2$ or	$M^{-1} \cdot s^{-1}$
	rate $= k[A][B]$	
3	rate $= k[A]^2[B]$ or	$M^{-2} \cdot s^{-1}$
	rate $= k[A][B]^2$	

To emphasize that rate is independent of the concentration of A, zero order does not mean that the rate is zero. Note that reaction order can also be a fraction.

The following points summarize our discussion of the rate law:

1. Rate laws are always determined experimentally. From the concentrations of reactants and the initial reaction rates we can determine the reaction order and then the rate constant of the reaction.
2. Reaction order is always defined in terms of reactant (not product) concentrations.
3. The order of a reactant is not related to the stoichiometric coefficient of the reactant in the overall balanced equation.

Student Hot Spot

Student data indicate you may struggle with determining reaction orders. Access your eBook for additional Learning Resources on this topic.

It is important to consider the units of the rate constant for a particular reaction. Because the units of reaction rate must always be $M \cdot s^{-1}$, the rate constant units depend on the overall order of a reaction. Table 13.3 summarizes the units of the rate constant for various overall reaction orders.

Example 13.3 illustrates the procedure for determining the rate law of a reaction.

Example 13.3

The reaction of nitric oxide with hydrogen at 1280°C is

$$2NO(g) + 2H_2(g) \longrightarrow N_2(g) + 2H_2O(g)$$

From the following data collected at this temperature, determine (a) the rate law, (b) the rate constant, and (c) the rate of the reaction when $[NO] = 12.0 \times 10^{-3}\ M$ and $[H_2] = 6.0 \times 10^{-3}\ M$.

Experiment	[NO] (M)	[H_2] (M)	Initial Rate (M/s)
1	5.0×10^{-3}	2.0×10^{-3}	1.3×10^{-5}
2	10.0×10^{-3}	2.0×10^{-3}	5.0×10^{-5}
3	10.0×10^{-3}	4.0×10^{-3}	10.0×10^{-5}

$2NO + 2H_2 \longrightarrow N_2 + 2H_2O$

Strategy We are given a set of concentration and reaction rate data and asked to determine the rate law and the rate constant. We assume that the rate law takes the form

$$\text{rate} = k[NO]^x[H_2]^y$$

How do we use the data to determine x and y? Once the orders of the reactants are known, we can calculate k from any set of rate and concentrations. Finally, the rate law enables us to calculate the rate at any concentrations of NO and H_2.

Solution

(a) Experiments 1 and 2 show that when we double the concentration of NO at constant concentration of H_2, the rate quadruples. Taking the ratio of the rates from these two experiments

$$\frac{\text{rate}_2}{\text{rate}_1} = \frac{5.0 \times 10^{-5}\ M/s}{1.3 \times 10^{-5}\ M/s} \approx 4 = \frac{k(10.0 \times 10^{-3}\ M)^x\,(2.0 \times 10^{-3}\ M)^y}{k(5.0 \times 10^{-3}\ M)^x\,(2.0 \times 10^{-3}\ M)^y}$$

(Continued)

Therefore,

$$\frac{(10.0 \times 10^{-3}\ M)^x}{(5.0 \times 10^{-3}\ M)^x} = 2^x = 4$$

or $x = 2$; that is, the reaction is second order in NO. Experiments 2 and 3 indicate that doubling $[H_2]$ at constant [NO] doubles the rate. Here we write the ratio as

$$\frac{\text{rate}_3}{\text{rate}_4} = \frac{10.0 \times 10^{-5}\ M/\text{s}}{5.0 \times 10^{-5}\ M/\text{s}} = 2 = \frac{k(10.0 \times 10^{-3}\ M)^x\,(4.0 \times 10^{-3}\ M)^y}{k(10.0 \times 10^{-3}\ M)^x\,(2.0 \times 10^{-3}\ M)^y}$$

Therefore,

$$\frac{(4.0 \times 10^{-3}\ M)^y}{(2.0 \times 10^{-3}\ M)^y} = 2^y = 2$$

or $y = 1$; that is, the reaction is first order in H_2. Hence the rate law is given by

$$\text{rate} = k[\text{NO}]^2[\text{H}_2]$$

which shows that it is a (2 + 1) or third-order reaction overall.

(b) The rate constant k can be calculated using the data from any one of the experiments. Rearranging the rate law, we get

$$k = \frac{\text{rate}}{[\text{NO}]^2[\text{H}_2]}$$

The data from experiment 2 give us

$$k = \frac{5.0 \times 10^{-5}\ M/\text{s}}{(10.0 \times 10^{-3}\ M)^2(2.0 \times 10^{-3}\ M)}$$
$$= 2.5 \times 10^2/M^2 \cdot \text{s}$$

(c) Using the known rate constant and concentrations of NO and H_2, we write

$$\text{rate} = (2.5 \times 10^2/M^2 \cdot \text{s})(12.0 \times 10^{-3}\ M)^2(6.0 \times 10^{-3}\ M)$$
$$= 2.2 \times 10^{-4}\ M/\text{s}$$

Comment Note that the reaction is first order in H_2, whereas the stoichiometric coefficient for H_2 in the balanced equation is 2. The order of a reactant is not related to the stoichiometric coefficient of the reactant in the overall balanced equation.

Practice Exercise The reaction of peroxydisulfate ion ($S_2O_8^{2-}$) with iodide ion (I^-) is

$$S_2O_8^{2-}(aq) + 3I^-(aq) \longrightarrow 2SO_4^{2-}(aq) + I_3^-(aq)$$

From the following data collected at a certain temperature, determine the rate law and calculate the rate constant.

Experiment	$[S_2O_8^{2-}]$ (M)	$[I^-]$ (M)	Initial Rate (M/s)
1	0.080	0.034	2.2×10^{-4}
2	0.080	0.017	1.1×10^{-4}
3	0.16	0.017	2.2×10^{-4}

Similar problem: 13.15.

Summary of Concepts & Facts

- The rate law expresses the relationship of the rate of a reaction to the rate constant and the concentrations of the reactants raised to appropriate powers. The rate constant k for a given reaction changes only with temperature.

(Continued)

- Reaction order is the power to which the concentration of a given reactant is raised in the rate law. Overall reaction order is the sum of the powers to which reactant concentrations are raised in the rate law. The rate law and the reaction order cannot be determined from the stoichiometry of the overall equation for a reaction; they must be determined by experiment. For a zero-order reaction, the reaction rate is equal to the rate constant.

Review of Concepts & Facts

13.2.1 For the reaction $A + 2B \longrightarrow C + 2D$, use the following data collected at a certain temperature to determine the rate law and the rate constant.

Experiment	A (*M*)	B (*M*)	Initial Rate (*M*/s)
1	0.24	0.020	3.3×10^{-3}
2	0.72	0.020	9.9×10^{-3}
3	0.24	0.040	3.3×10^{-3}

13.2.2 The relative rates of the reaction $2A + B \longrightarrow$ products shown in the diagrams (a)–(c) are 1:2:4. The red spheres represent A molecules and the green spheres represent B molecules. Write a rate law for this reaction.

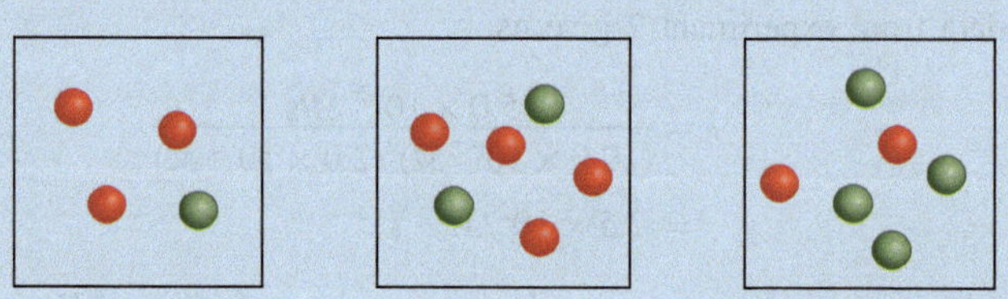
(a) (b) (c)

13.3 The Relation Between Reactant Concentration and Time

Learning Objectives

- Apply the integrated rate law for a reaction to find the concentration of a reactant.
- Compute the half-life ($t_{\frac{1}{2}}$) of a reaction.
- Calculate the amount of reactant remaining after a certain period of time given the half-life and order of a reaction.

Rate law expressions enable us to calculate the rate of a reaction from the rate constant and reactant concentrations. The rate laws can also be used to determine the concentrations of reactants at any time during the course of a reaction. We will illustrate this application by first considering two of the most common rate laws—those applying to reactions that are first order overall and those applying to reactions that are second order overall.

First-Order Reactions

A ***first-order reaction*** is *a reaction whose rate depends on the reactant concentration raised to the first power*. In a first-order reaction of the type

$$A \longrightarrow \text{product}$$

the rate is

$$\text{rate} = -\frac{\Delta[\text{A}]}{\Delta t}$$

From the rate law we also know that

$$\text{rate} = k[\text{A}]$$

To obtain the units of k for this rate law, we write

$$k = \frac{\text{rate}}{[\text{A}]} = \frac{M/\text{s}}{M} = 1/\text{s} \quad \text{or} \quad \text{s}^{-1}$$

Combining the first two equations for the rate we get

$$-\frac{\Delta[\text{A}]}{\Delta t} = k[\text{A}] \qquad (13.2)$$

Using calculus, we can show from Equation (13.2) that

$$\ln \frac{[\text{A}]_t}{[\text{A}]_0} = -kt \qquad (13.3)$$

where ln is the natural logarithm, and $[\text{A}]_0$ and $[\text{A}]_t$ are the concentrations of A at times $t = 0$ and $t = t$, respectively. It should be understood that $t = 0$ need not correspond to the beginning of the experiment; it can be any time when we choose to start monitoring the change in the concentration of A.

Equation (13.3) can be rearranged as follows:

$$\ln [\text{A}]_t = -kt + \ln [\text{A}]_0 \qquad (13.4)$$

Equation (13.4) has the form of the linear equation $y = mx + b$, in which m is the slope of the line that is the graph of the equation

$$\begin{array}{ccccc} \ln [\text{A}]_t & = & (-k)(t) & + & \ln [\text{A}]_0 \\ \updownarrow & & \updownarrow \quad \updownarrow & & \updownarrow \\ y & = & m \quad x & + & b \end{array}$$

Consider Figure 13.9. As we would expect during the course of a reaction, the concentration of the reactant A decreases with time [Figure 13.9(a)]. For a first-order reaction, if we plot $\ln [\text{A}]_t$ versus time (y versus x), we obtain a straight line with a slope equal to $-k$ and a y intercept equal to $\ln [\text{A}]_0$ [Figure 13.9(b)]. Thus, we can calculate the rate constant from the slope of this plot.

There are many first-order reactions. An example is the decomposition of ethane (C_2H_6) into highly reactive fragments called methyl radicals (CH_3):

$$C_2H_6 \longrightarrow 2CH_3$$

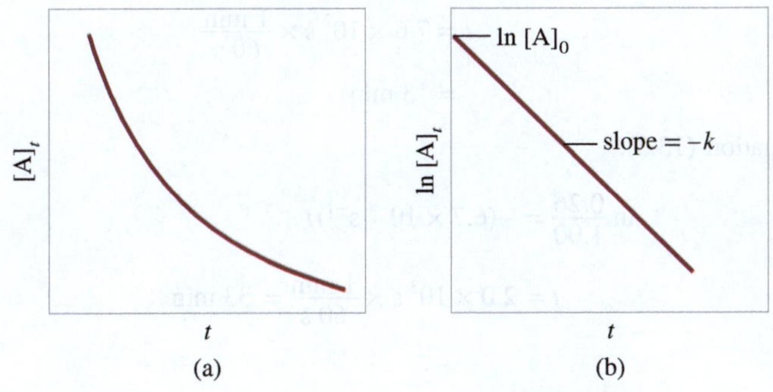

Figure 13.9 *First-order reaction characteristics: (a) the exponential decrease of reactant concentration with time; (b) a plot of ln $[A]_t$ versus t. The slope of the line is equal to $-k$.*

Student Hot Spot

Student data indicate you may struggle with the first-order integrated rate law. Access your eBook to view additional Learning Resources on this topic.

The decomposition of N_2O_5 is also a first-order reaction

$$2N_2O_5(g) \longrightarrow 4NO_2(g) + O_2(g)$$

In Example 13.4 we apply Equation (13.3) to an organic reaction.

Example 13.4

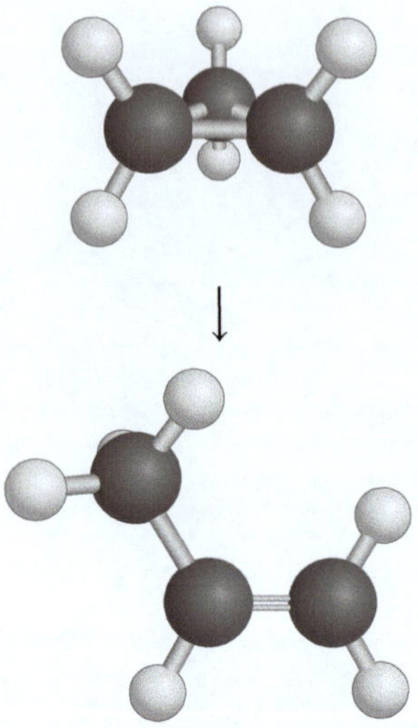

The conversion of cyclopropane to propene in the gas phase is a first-order reaction with a rate constant of $6.7 \times 10^{-4}\ s^{-1}$ at 500°C.

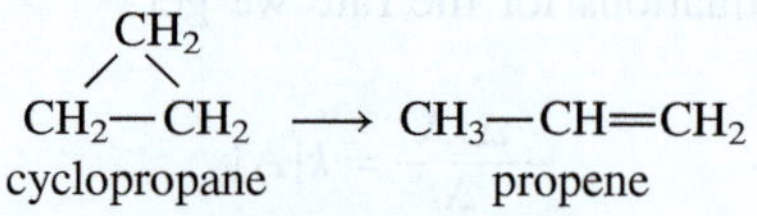

(a) If the initial concentration of cyclopropane was 0.25 *M*, what is the concentration after 8.8 min? (b) How long (in minutes) will it take for the concentration of cyclopropane to decrease from 0.25 *M* to 0.15 *M*? (c) How long (in minutes) will it take to convert 74 percent of the starting material?

Strategy The relationship between the concentrations of a reactant at different times in a first-order reaction is given by Equation (13.3) or (13.4). In (a) we are given $[A]_0 = 0.25\ M$ and asked for $[A]_t$ after 8.8 min. In (b) we are asked to calculate the time it takes for cyclopropane to decrease in concentration from 0.25 *M* to 0.15 *M*. No concentration values are given for (c). However, if initially we have 100 percent of the compound and 74 percent has reacted, then what is left must be (100% − 74%), or 26%. Thus, the ratio of the percentages will be equal to the ratio of the actual concentrations; that is, $[A]_t/[A]_0 = 26\%/100\%$, or 0.26/1.00.

Solution

(a) In applying Equation (13.4), we note that because *k* is given in units of s^{-1}, we must first convert 8.8 min to seconds:

$$8.8 \text{ min} \times \frac{60 \text{ s}}{1 \text{ min}} = 528 \text{ s}$$

We write

$$\begin{aligned}\ln [A]_t &= -kt + \ln [A]_0 \\ &= -(6.7 \times 10^{-4}\ s^{-1})(528\ s) + \ln (0.25) \\ &= -1.74\end{aligned}$$

Hence, $\quad [A]_t = e^{-1.74} = 0.18\ M$

Note that in the ln $[A]_0$ term, $[A]_0$ is expressed as a dimensionless quantity (0.25) because we cannot take the logarithm of units.

(b) Using Equation (13.3),

$$\ln \frac{0.15\ M}{0.25\ M} = -(6.7 \times 10^{-4}\ s^{-1})t$$

$$t = 7.6 \times 10^2 \text{ s} \times \frac{1 \text{ min}}{60 \text{ s}}$$

$$= 13 \text{ min}$$

(c) From Equation (13.3),

$$\ln \frac{0.26}{1.00} = -(6.7 \times 10^{-4}\ s^{-1})t$$

$$t = 2.0 \times 10^3 \text{ s} \times \frac{1 \text{ min}}{60 \text{ s}} = 33 \text{ min}$$

(Continued)

Practice Exercise The reaction $2A \longrightarrow B$ is first order in A with a rate constant of $2.8 \times 10^{-2}\ s^{-1}$ at 80°C. How long (in seconds) will it take for A to decrease from 0.88 *M* to 0.14 *M*?

Similar problem: 13.94.

Now let us determine graphically the order and rate constant of the decomposition of dinitrogen pentoxide in carbon tetrachloride (CCl_4) solvent at 45°C:

$$2N_2O_5(CCl_4) \longrightarrow 4NO_2(g) + O_2(g)$$

The following table shows the variation of N_2O_5 concentration with time, and the corresponding ln $[N_2O_5]$ values.

t (s)	$[N_2O_5]$ (*M*)	ln $[N_2O_5]$
0	0.91	−0.094
300	0.75	−0.29
600	0.64	−0.45
1200	0.44	−0.82
3000	0.16	−1.83

Applying Equation (13.4) we plot ln $[N_2O_5]$ versus *t*, as shown in Figure 13.10. The fact that the points lie on a straight line shows that the rate law is first order. Next, we determine the rate constant from the slope. We select two points far apart on the line and subtract their *y* and *x* values as follows:

$$\text{slope } (m) = \frac{\Delta y}{\Delta x}$$

$$= \frac{-1.50 - (-0.34)}{(2430 - 400)\ s}$$

$$= -5.7 \times 10^{-4}\ s^{-1}$$

Because $m = -k$, we get $k = 5.7 \times 10^{-4}\ s^{-1}$.

For gas-phase reactions, we can replace the concentration terms in Equation (13.3) with the pressures of the gaseous reactant. Consider the first-order reaction

$$A(g) \longrightarrow \text{product}$$

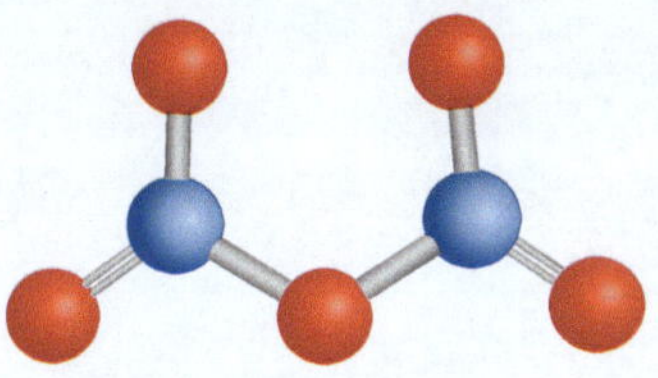

N_2O_5

N_2O_5 decomposes to give NO_2 (brown color).

Ken Karp/McGraw Hill

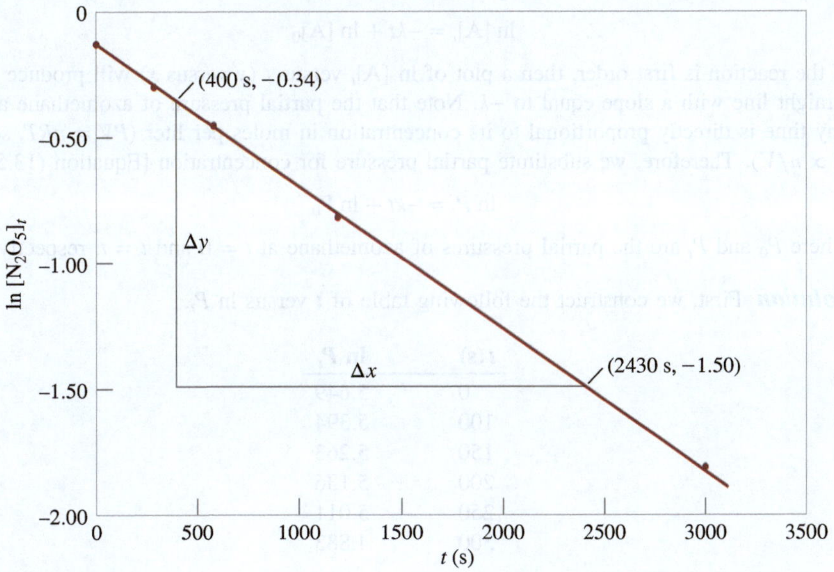

Figure 13.10 *Plot of ln $[N_2O_5]_t$ versus time. The rate constant can be determined from the slope of the straight line.*

Using the ideal gas equation we write

$$PV = n_A RT$$

or

$$\frac{n_A}{V} = [A] = \frac{P}{RT}$$

Substituting $[A] = P/RT$ in Equation (13.3), we get

$$\ln \frac{[A]_t}{[A]_0} = \ln \frac{P_t/RT}{P_0/RT} = \ln \frac{P_t}{P_0} = -kt$$

The equation corresponding to Equation (13.4) now becomes

$$\ln P_t = -kt + \ln P_0 \qquad (13.5)$$

Example 13.5 shows the use of pressure measurements to study the kinetics of a first-order reaction.

Example 13.5

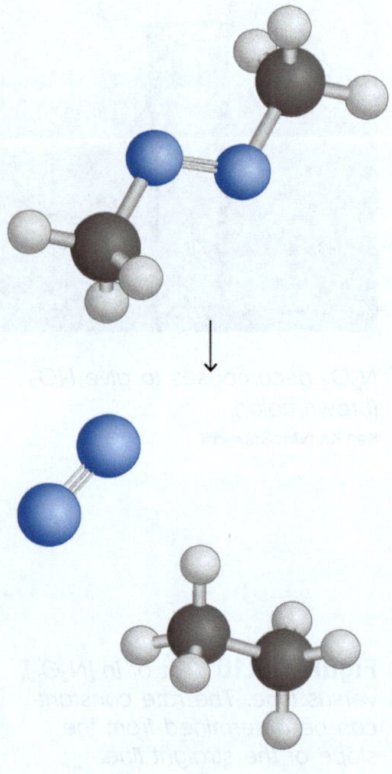

$C_2H_6N_2 \longrightarrow N_2 + C_2H_6$

The rate of decomposition of azomethane ($C_2H_6N_2$) is studied by monitoring the partial pressure of the reactant as a function of time:

$$CH_3{-}N{=}N{-}CH_3(g) \longrightarrow N_2(g) + C_2H_6(g)$$

The data obtained at 300°C are shown in the following table.

Time (s)	Partial Pressure of Azomethane (mmHg)
0	284
100	220
150	193
200	170
250	150
300	132

Are these values consistent with first-order kinetics? If so, determine the rate constant.

Strategy To test for first-order kinetics, we consider the integrated first-order rate law that has a linear form, which is Equation (13.4):

$$\ln [A]_t = -kt + \ln [A]_0$$

If the reaction is first order, then a plot of $\ln [A]_t$ versus t (y versus x) will produce a straight line with a slope equal to $-k$. Note that the partial pressure of azomethane at any time is directly proportional to its concentration in moles per liter ($PV = nRT$, so $P \propto n/V$). Therefore, we substitute partial pressure for concentration [Equation (13.5)]:

$$\ln P_t = -kt + \ln P_0$$

where P_0 and P_t are the partial pressures of azomethane at $t = 0$ and $t = t$, respectively.

Solution First, we construct the following table of t versus $\ln P_t$.

t (s)	$\ln P_t$
0	5.649
100	5.394
150	5.263
200	5.136
250	5.011
300	4.883

(Continued)

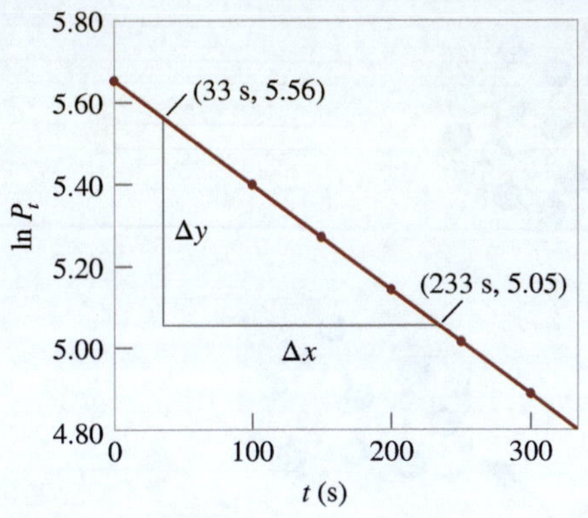

Figure 13.11 *Plot of ln P_t versus time for the decomposition of azomethane.*

Figure 13.11, which is based on the data given in the table, shows that a plot of ln P_t versus t yields a straight line, so the reaction is indeed first order. The slope of the line is given by

$$\text{slope} = \frac{5.05 - 5.56}{(233 - 33)\ \text{s}} = -2.55 \times 10^{-3}\ \text{s}^{-1}$$

According to Equation (13.4), the slope is equal to $-k$, so $k = 2.55 \times 10^{-3}\ \text{s}^{-1}$.

Practice Exercise Ethyl iodide (C_2H_5I) decomposes at a certain temperature in the gas phase as follows:

$$C_2H_5I(g) \longrightarrow C_2H_4(g) + HI(g)$$

From the following data determine the order of the reaction and the rate constant.

Time (min)	$[C_2H_5I]$ (M)
0	0.36
15	0.30
30	0.25
48	0.19
75	0.13

Similar problems: 13.19, 13.20.

Reaction Half-Life

As a reaction proceeds, the concentration of the reactant(s) decreases. Another measure of the rate of a reaction, relating concentration to time, is the ***half-life,*** $t_{\frac{1}{2}}$, which is *the time required for the concentration of a reactant to decrease to half of its initial concentration.* We can obtain an expression for $t_{\frac{1}{2}}$ for a first-order reaction as follows. Equation (13.3) can be rearranged to give

$$t = \frac{1}{k} \ln \frac{[A]_0}{[A]_t}$$

By the definition of half-life, when $t = t_{\frac{1}{2}}$, $[A]_t = [A]_0/2$, so

$$t_{\frac{1}{2}} = \frac{1}{k} \ln \frac{[A]_0}{[A]_0/2}$$

or

$$t_{\frac{1}{2}} = \frac{1}{k} \ln 2 = \frac{0.693}{k} \qquad (13.6)$$

Equation (13.6) tells us that the half-life of a first-order reaction is *independent* of the initial concentration of the reactant. Thus, it takes the same time for the concentration of the reactant to decrease from 1.0 M to 0.50 M, say, as it does for a decrease in concentration from 0.10 M to 0.050 M (Figure 13.12). Measuring the half-life of a reaction is one way to determine the rate constant of a first-order reaction.

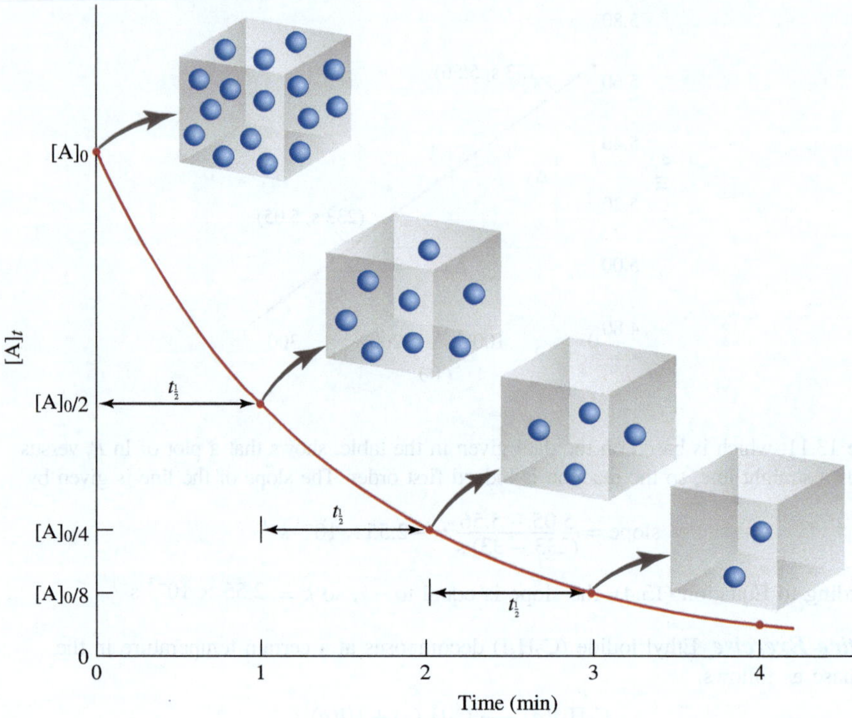

Figure 13.12 *A plot of $[A]_t$ versus time for the first-order reaction $A \longrightarrow$ products. The half-life of the reaction is 1 min. After the elapse of each half-life, the concentration of A is halved.*

The following analogy may be helpful for understanding Equation (13.6). If a college student takes 4 yr to graduate, the half-life of his or her stay at the college is 2 yr. Thus, half-life is not affected by how many other students are present. Similarly, the half-life of a first-order reaction is concentration independent.

The usefulness of $t_{\frac{1}{2}}$ is that it gives us a measure of the magnitude of the rate constant—the shorter the half-life, the larger the k. Consider, for example, two radioactive isotopes used in nuclear medicine: ^{24}Na ($t_{\frac{1}{2}}$ = 14.7 h) and ^{60}Co ($t_{\frac{1}{2}}$ = 5.3 yr). It is obvious that the ^{24}Na isotope decays faster because it has a shorter half-life. If we started with 1 mole each of the isotopes, most of the ^{24}Na would be gone in a week while the ^{60}Co sample would be mostly intact.

In Example 13.6 we calculate the half-life of a first-order reaction.

Student Hot Spot

Student data indicate you may struggle with half-life equations. Access your eBook to view additional Learning Resources on this topic.

Example 13.6

The decomposition of ethane (C_2H_6) to methyl radicals is a first-order reaction with a rate constant of $5.36 \times 10^{-4}\ s^{-1}$ at 700°C:

$$C_2H_6(g) \longrightarrow 2CH_3(g)$$

Calculate the half-life of the reaction in minutes.

Strategy To calculate the half-life of a first-order reaction, we use Equation (13.6). A conversion is needed to express the half-life in minutes.

Solution For a first-order reaction, we only need the rate constant to calculate the half-life of the reaction. From Equation (13.6)

$$t_{\frac{1}{2}} = \frac{0.693}{k}$$

$$= \frac{0.693}{5.36 \times 10^{-4}\ s^{-1}}$$

$$= 1.29 \times 10^3\ s \times \frac{1\ min}{60\ s}$$

$$= 21.5\ min$$

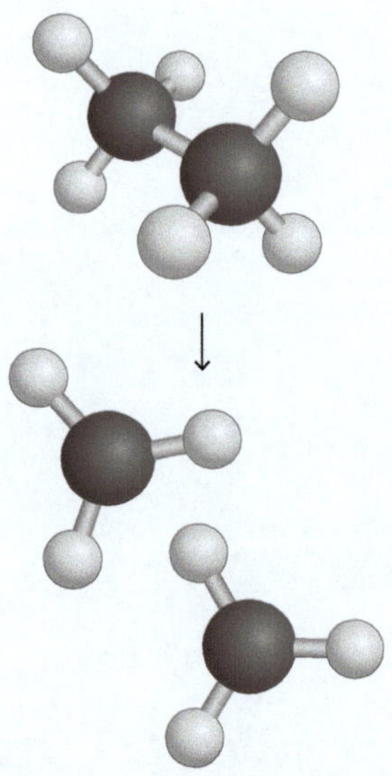

$C_2H_6 \longrightarrow 2CH_3$

(Continued)

Practice Exercise Calculate the half-life of the decomposition of N_2O_5, plotted in Figure 13.10.

Similar problem: 13.26.

Second-Order Reactions

A ***second-order reaction*** *is a reaction whose rate depends on the concentration of one reactant raised to the second power or on the concentrations of two different reactants, each raised to the first power*. The simpler type involves only one kind of reactant molecule:

$$A \longrightarrow \text{product}$$

where

$$\text{rate} = -\frac{\Delta[A]}{\Delta t}$$

From the rate law,

$$\text{rate} = k[A]^2$$

As before, we can determine the units of k by writing

$$k = \frac{\text{rate}}{[A]^2} = \frac{M/\text{s}}{M^2} = 1/M \cdot \text{s}$$

Another type of second-order reaction is

$$A + B \longrightarrow \text{product}$$

and the rate law is given by

$$\text{rate} = k[A][B]$$

The reaction is first order in A and first order in B, so it has an overall reaction order of 2.

Using calculus, we can obtain the following expressions for "A ⟶ product" second-order reactions:

$$\frac{1}{[A]_t} = kt + \frac{1}{[A]_0} \qquad (13.7)$$

Student Hot Spot

Student data indicate you may struggle with the second-order integrated rate law. Access your eBook to view additional Learning Resources on this topic.

Equation (13.7) has the form of a linear equation. As Figure 13.13 shows, a plot of $1/[A]_t$ versus t gives a straight line with slope = k and y intercept = $1/[A]_0$.

Pseudo-First-Order Reactions

The other type of second-order reaction

$$A + B \longrightarrow \text{product}$$

and the corresponding rate law

$$\text{rate} = k[A][B]$$

are actually more common than the $k[A]^2$ second-order kinetics already shown. However, it is considerably more difficult to treat mathematically. While it is possible to solve the integrated form of the rate law, a common approach is to measure the second-order reaction rates under *pseudo-first-order kinetics* conditions.

If this reaction is carried out under the conditions where one of the reactants is in large excess over the other, then the concentration of the excess reactant will not

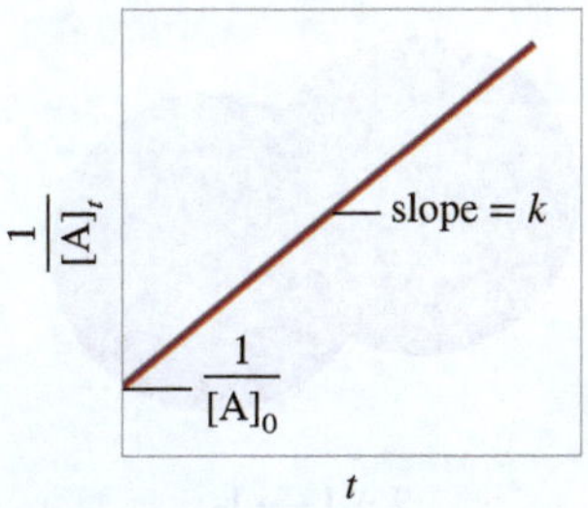

Figure 13.13 *A plot of $1/[A]_t$ versus t for the second-order reaction A ⟶ products. The slope of the line is equal to k.*

change appreciably over the course of the reaction. For example, if [B] ≫ [A], then [B] will be essentially constant and we have

$$\text{rate} = k[\text{A}][\text{B}] = k_{\text{obs}}[\text{A}]$$

Note that the rate law now has the appearance of a first-order reaction. The rate constant k_{obs}, called the pseudo-first-order rate constant, is given by $k_{\text{obs}} = k[\text{B}]$, where the subscript "obs" denotes observed and k is the second-order rate constant. If we measure k_{obs} for many different initial concentrations of B, then a plot of k_{obs} versus [B] will yield a straight line with a slope equal to k.

Previously, we saw that the reaction between bromine and formic acid can be treated as a first-order reaction because formic acid is present in excess (see Section 13.1). Another well-studied example is the hydrolysis (meaning reaction with water) of ethyl acetate to yield acetic acid and ethanol:

$$CH_3COOC_2H_5 + H_2O \longrightarrow CH_3COOH + C_2H_5OH$$

Because the concentration of water, the solvent, is about 55.5 M[†] compared to 1 M or less for ethyl acetate, [H_2O] can be treated as a constant so the rate is given by

$$\text{rate} = k[CH_3COOC_2H_5][H_2O] = k_{\text{obs}}[CH_3COOC_2H_5]$$

where $k_{\text{obs}} = k[H_2O]$.

Reaction Half-Life

We can obtain an equation for the half-life of a second-order reaction of the type A ⟶ product by setting $[\text{A}]_t = [\text{A}]_0/2$ in Equation (13.7)

$$\frac{1}{[\text{A}]_0/2} = kt_{\frac{1}{2}} + \frac{1}{[\text{A}]_0}$$

Solving for $t_{\frac{1}{2}}$ we obtain

$$t_{\frac{1}{2}} = \frac{1}{k[\text{A}]_0} \qquad (13.8)$$

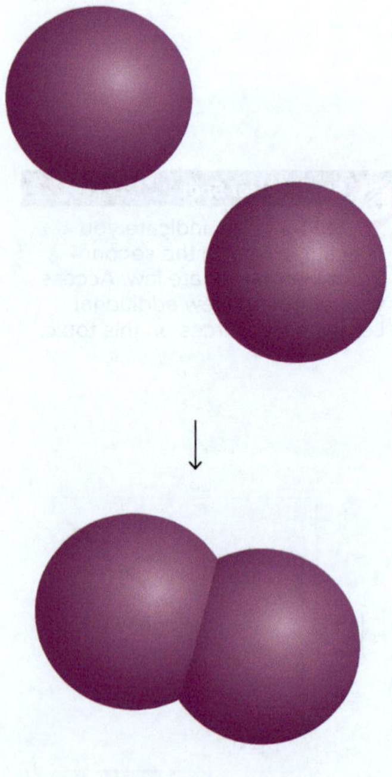
$I + I \longrightarrow I_2$

Note that the half-life of a second-order reaction is inversely proportional to the initial reactant concentration. This result makes sense because the half-life should be shorter in the early stage of the reaction when more reactant molecules are present to collide with each other. Measuring the half-lives at different initial concentrations is one way to distinguish between a first-order and a second-order reaction.

The kinetic analysis of a second-order reaction is shown in Example 13.7.

Example 13.7

Iodine atoms combine to form molecular iodine in the gas phase

$$\text{I}(g) + \text{I}(g) \longrightarrow \text{I}_2(g)$$

This reaction follows second-order kinetics and has the high rate constant $7.0 \times 10^9/M \cdot s$ at 23°C. (a) If the initial concentration of I was 0.086 M, calculate the concentration after 2.0 min. (b) Calculate the half-life of the reaction if the initial concentration of I is 0.60 M and if it is 0.42 M.

(Continued)

[†]In 1 L of a relatively dilute solution, the mass of water is approximately 1000 g so there are 1000 g/(18.02 g/mol) or 55.5 mole of water. Thus, the concentration of water is 55.5 M.

Strategy (a) The relationship between the concentrations of a reactant at different times is given by the integrated rate law. Because this is a second-order reaction, we use Equation (13.7). (b) We are asked to calculate the half-life. The half-life for a second-order reaction is given by Equation (13.8).

Solution

(a) To calculate the concentration of a species at a later time of a second-order reaction, we need the initial concentration and the rate constant. Applying Equation (13.7)

$$\frac{1}{[A]_t} = kt + \frac{1}{[A]_0}$$

$$= (7.0 \times 10^9/M \cdot s)(2.0 \text{ min} \times)\frac{60 \text{ s}}{1 \text{ min}} + \frac{1}{0.086\, M}$$

where $[A]_t$ is the concentration at $t = 2.0$ min. Solving the equation, we get

$$[A]_t = 1.2 \times 10^{-12}\, M$$

This is such a low concentration that it is virtually undetectable. The very large rate constant for the reaction means that practically all the I atoms combine after only 2.0 min of reaction time.

(b) We need Equation (13.8) for this part.

For $[I]_0 = 0.60\, M$

$$t_{\frac{1}{2}} = \frac{1}{k[A]_0}$$

$$= \frac{1}{(7.0 \times 10^9/M \cdot s)\,(0.60\, M)}$$

$$= 2.4 \times 10^{-10} \text{ s}$$

For $[I]_0 = 0.42\, M$

$$t_{\frac{1}{2}} = \frac{1}{(7.0 \times 10^9/M \cdot s)(0.42\, M)}$$

$$= 3.4 \times 10^{-10} \text{ s}$$

Check These results confirm that the half-life of a second-order reaction, unlike that of a first-order reaction, is not a constant but depends on the initial concentration of the reactant(s). Does it make sense that a larger initial concentration should have a shorter half-life?

Practice Exercise The reaction $2A \longrightarrow B$ is second order with a rate constant of $51/M \cdot$ min at 24°C. (a) Starting with $[A]_0 = 0.0092\, M$, how long will it take for $[A]_t = 3.7 \times 10^{-3}\, M$? (b) Calculate the half-life of the reaction.

Similar problems: 13.27, 13.28.

Zero-Order Reactions

First- and second-order reactions are the most common reaction types. Reactions whose order is zero are rare. For the zero-order reaction

$$A \longrightarrow \text{product}$$

the rate law is given by

$$\text{rate} = k[A]^0$$

$$= k$$

Thus, the rate of a zero-order reaction is a *constant,* independent of reactant concentration. (Recall that any number raised to the power zero is equal to one.) Using calculus, we can show that

$$[A]_t = -kt + [A]_0 \qquad (13.9)$$

CHEMISTRY *in Action*

Radiocarbon Dating

How do scientists determine the ages of artifacts from archaeological excavations? If someone tried to sell you a manuscript supposedly dating from 1000 B.C., how could you be certain of its authenticity? Is a mummy found in an Egyptian pyramid *really* 3000 years old? Is the so-called Shroud of Turin truly the burial cloth of Jesus Christ? The answers to these and other similar questions can usually be found by applying chemical kinetics and the *radiocarbon dating technique*.

Earth's atmosphere is constantly being bombarded by cosmic rays of extremely high penetrating power. These rays, which originate in outer space, consist of electrons, neutrons, and atomic nuclei. One of the important reactions between the atmosphere and cosmic rays is the capture of neutrons by atmospheric nitrogen (nitrogen-14 isotope) to produce the radioactive carbon-14 isotope and hydrogen. The unstable carbon atoms eventually form $^{14}CO_2$, which mixes with the ordinary carbon dioxide ($^{12}CO_2$) in the air. As the carbon-14 isotope decays, it emits β particles (electrons). The rate of decay (as measured by the number of electrons emitted per second) obeys first-order kinetics. It is customary in the study of radioactive decay to write the rate law as

$$\text{rate} = kN$$

where k is the first-order rate constant and N the number of ^{14}C nuclei present. The half-life of the decay, $t_{\frac{1}{2}}$, is 5.73×10^3 yr, so that from Equation (13.6) we write

$$k = \frac{0.693}{5.73 \times 10^3 \text{ yr}} = 1.21 \times 10^{-4} \text{ yr}^{-1}$$

The Shroud of Turin. For generations there has been controversy about whether the Shroud, a piece of linen bearing the image of a man, was the burial cloth of Jesus Christ.
Perseomed/iStock/Getty Images

The carbon-14 isotopes enter the biosphere when carbon dioxide is taken up in plant photosynthesis. Plants are eaten by animals, which exhale carbon-14 in CO_2. Eventually, carbon-14 participates in many aspects of the carbon cycle. The ^{14}C lost by radioactive decay is constantly replenished by the production of new isotopes in the atmosphere. In this decay-replenishment process, a dynamic equilibrium is established whereby the ratio of

Equation (13.9) has the form of a linear equation. As Figure 13.14 shows, a plot of $[A]_t$ versus t gives a straight line with slope $= -k$ and y intercept $= [A]_0$. To calculate the half-life of a zero-order reaction, we set $[A]_t = [A]_0/2$ in Equation (13.9) and obtain

$$t_{\frac{1}{2}} = \frac{[A]_0}{2k} \qquad (13.10)$$

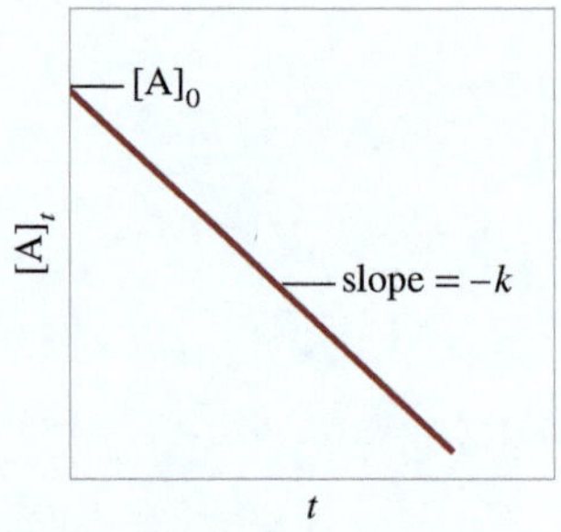

Figure 13.14 *A plot of $[A]_t$ versus t for a zero-order reaction. The slope of the line is equal to −k.*

Many of the known zero-order reactions take place on a metal surface. An example is the decomposition of nitrous oxide (N_2O) to nitrogen and oxygen in the presence of platinum (Pt):

$$2N_2O(g) \longrightarrow 2N_2(g) + O_2(g)$$

When all the binding sites on Pt are occupied, the rate becomes constant regardless of the amount of N_2O present in the gas phase. As we will see in Section 13.6, another well-studied zero-order reaction occurs in enzyme catalysis.

^{14}C to ^{12}C remains constant in living matter. But when an individual plant or an animal dies, the carbon-14 isotope in it is no longer replenished, so the ratio decreases as ^{14}C decays. This same change occurs when carbon atoms are trapped in coal, petroleum, or wood preserved underground, and, of course, in Egyptian mummies. After a number of years, there are proportionately fewer ^{14}C nuclei in, say, a mummy than in a living person.

In 1955, Willard F. Libby[†] suggested that this fact could be used to estimate the length of time the carbon-14 isotope in a particular specimen has been decaying without replenishment. Rearranging Equation (13.3), we can write

$$\ln \frac{N_0}{N_t} = kt$$

where N_0 and N_t are the number of ^{14}C nuclei present at $t = 0$ and $t = t$, respectively. Because the rate of decay is directly proportional to the number of ^{14}C nuclei present, the preceding equation can be rewritten as

$$t = \frac{1}{k} \ln \frac{N_0}{N_t}$$

$$= \frac{1}{1.21 \times 10^{-4}\ \text{yr}^{-1}} \ln \frac{\text{decay rate at } t = 0}{\text{decay rate at } t = 1}$$

$$= \frac{1}{1.21 \times 10^{-4}\ \text{yr}^{-1}} \ln \frac{\text{decay rate of fresh sample}}{\text{decay rate of old sample}}$$

[†]Willard Frank Libby (1908–1980). American chemist. Libby received the Nobel Prize in Chemistry in 1960 for his work on radiocarbon dating.

Knowing k and the decay rates for the fresh sample and the old sample, we can calculate t, which is the age of the old sample. This ingenious technique is based on a remarkably simple idea. Its success depends on how accurately we can measure the rate of decay. In fresh samples, the ratio $^{14}C/^{12}C$ is about $1/10^{12}$, so the equipment used to monitor the radioactive decay must be very sensitive. Precision is more difficult with older samples because they contain even fewer ^{14}C nuclei. Nevertheless, radiocarbon dating has become an extremely valuable tool for estimating the age of archaeological artifacts, paintings, and other objects dating back 1000 to 50,000 years.

A well-publicized application of radiocarbon dating was the determination of the age of the Shroud of Turin. In 1988 three laboratories in Europe and the United States, working on samples of less than 50 mg of the Shroud, showed by carbon-14 dating that the Shroud dates from between A.D. 1260 and A.D. 1390. These findings seem to indicate that the Shroud could not have been the burial cloth of Christ. The most recent data analysis in 2020 points to a similar timeframe of the thirteenth or fourteenth century; however, some research contends the finding was invalid because the dating analysis was based on contaminants introduced by repairs to the Shroud in later years. It seems the controversy will continue for some time, and further testing on the Shroud is warranted.

Third- and higher-order reactions are quite complex; they are not presented in this book. Table 13.4 summarizes the kinetics of zero-order, first-order, and second-order reactions. The Chemistry in Action essay "Radiocarbon Dating" in Section 13.3 describes the application of chemical kinetics to estimating the ages of objects.

Table 13.4 Summary of the Kinetics of Zero-Order, First-Order, and Second-Order Reactions

Order	Rate Law	Concentration-Time Equation	Half-Life
0	Rate $= k$	$[A]_t = -kt + [A]_0$	$\frac{[A]_0}{2k}$
1	Rate $= k[A]$	$\ln \frac{[A]_t}{[A]_0} = -kt$	$\frac{0.693}{k}$
2[†]	Rate $= k[A]^2$	$\frac{1}{[A]_t} = kt + \frac{1}{[A]_0}$	$\frac{1}{k[A]_0}$

[†]A ⟶ product.

Summary of Concepts & Facts

- Integrated rate laws for zero-, first-, or second-order reaction can be used to find the concentration of a reactant at any point in time in a given reaction.
- The half-life of a reaction (the time it takes for the concentration of a reactant to decrease by one-half) can be used to determine the rate constant of a first-order reaction.

Review of Concepts & Facts

13.3.1 Consider the first-order reaction A $\longrightarrow$ B in which A molecules (blue spheres) are converted to B molecules (orange spheres). (a) What are the half-life and rate constant for the reaction? (b) How many molecules of A and B are present at $t = 20$ s and $t = 30$ s?

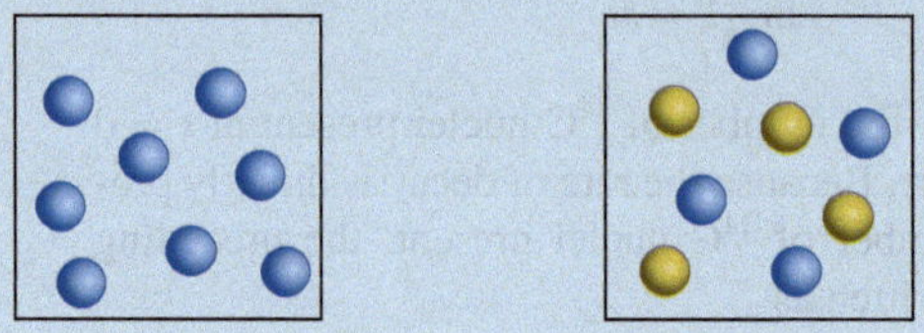

13.3.2 Consider the reaction A $\longrightarrow$ products. The half-life of the reaction depends on the initial concentration of A. Which of the following statements is inconsistent with the given information? (a) The half-life of the reaction decreases as the initial concentration increases. (b) A plot of ln $[A]_t$ versus t yields a straight line. (c) Doubling the concentration of A quadruples the rate.

13.3.3 Consider the first-order reaction A $\longrightarrow$ products. The initial concentration of A is 1.56 M and after 48.0 min, the concentration of A is 0.869 M. What is the half-life of the reaction?

13.3.4 What is the initial concentration of a reactant in a first-order reaction (A $\longrightarrow$ products) if, after 2.0 min, the concentration of reactant is 0.075 M? The half-life for the reaction is 45 s.

13.4 Activation Energy and Temperature Dependence of Rate Constants

Learning Objectives

- Appraise the main factors that affect the rate of a reaction.
- Define effective collision and activation energy.
- Produce a rate constant or the activation energy of a reaction using the Arrhenius equation.

With very few exceptions, reaction rates increase with increasing temperature. For example, the time required to hard-boil an egg in water is much shorter if the "reaction" is carried out at 100°C (about 10 min) than at 80°C (about 30 min). Conversely, an effective way to preserve foods is to store them at subzero temperatures, thereby slowing the rate of bacterial decay.

Table 13.5 **Kinetic Data for $C_2H_5Br + OH^-$ at Various Temperatures**

Experiment	$[C_2H_5Br]$	$[OH^-]$	*T* (K)	Rate (*M*/s)	*k* ($M^{-1}\ s^{-1}$)
1	0.200	0.200	298	3.52×10^{-6}	8.80×10^{-5}
2	0.200	0.200	303	6.40×10^{-6}	1.60×10^{-4}
3	0.200	0.200	308	1.12×10^{-5}	2.80×10^{-4}
4	0.200	0.200	313	2.00×10^{-5}	5.00×10^{-4}
5	0.200	0.200	318	3.40×10^{-5}	8.50×10^{-4}

Table 13.5 shows kinetic data for the reaction of bromoethane with hydroxide ions in aqueous solution:

$$C_2H_5Br(aq) + OH^-(aq) \longrightarrow C_2H_5OH(aq) + Br^-(aq)$$

To determine the effect of temperature on reactions, the concentrations and rates for various temperatures were measured. The rate constant *k* for each temperature is calculated from the rate law (this reaction is second-order overall and first order in each reactant) and then plotted (Figure 13.15), showing that *k* increases exponentially with increasing *T*.

To explain this behavior, we must ask how reactions get started in the first place.

The Collision Theory of Chemical Kinetics

The kinetic molecular theory of gases (Section 5.7) postulates that gas molecules frequently collide with one another. Therefore, it seems logical to assume—and it is generally true—that chemical reactions occur as a result of collisions between reacting molecules. In terms of the *collision theory* of chemical kinetics, then, we expect the rate of a reaction to be directly proportional to the number of molecular collisions per second, or to the frequency of molecular collisions:

$$\text{rate} \propto \frac{\text{number of collisions}}{\text{s}}$$

This simple relationship explains the dependence of reaction rate on concentration.

Consider the reaction of A molecules with B molecules to form some product. Suppose that each product molecule is formed by the direct combination of an A

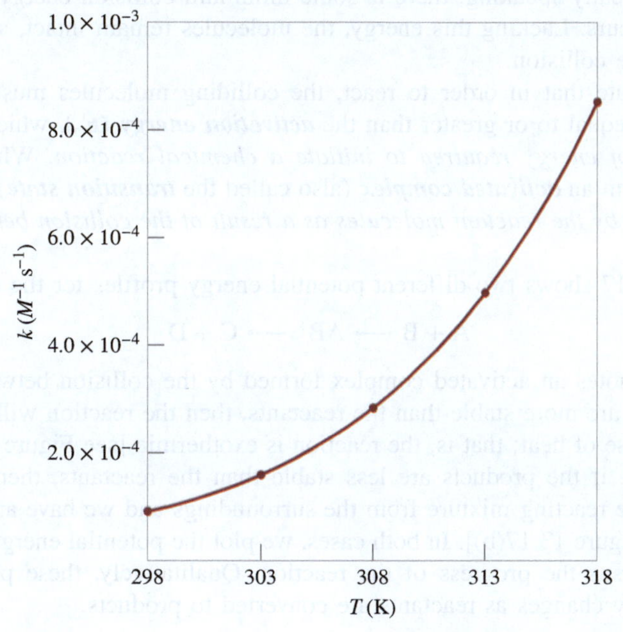

Figure 13.15 *Dependence of rate constant on temperature. The rate constants of most reactions increase with increasing temperature.*

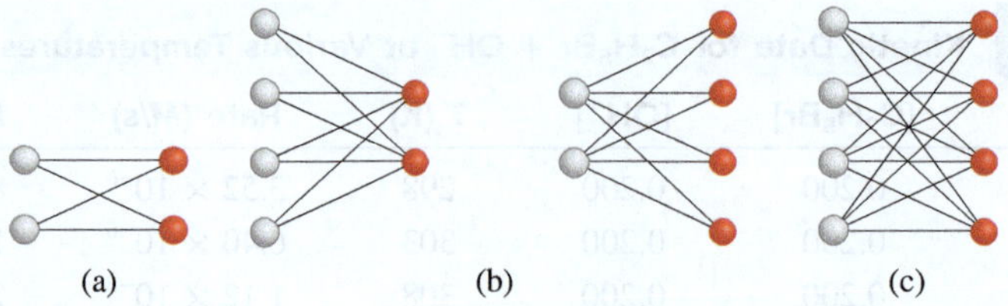

Figure 13.16 *Dependence of number of collisions on concentration. We consider here only A-B collisions, which can lead to formation of products. (a) There are four possible collisions among two A and two B molecules. (b) Doubling the number of either type of molecule (but not both) increases the number of collisions to eight. (c) Doubling both the A and B molecules increases the number of collisions to sixteen. In each case, the collision between a red sphere and a gray sphere can only be counted once.*

molecule and a B molecule. If we doubled the concentration of A, then the number of A-B collisions would also double, because there would be twice as many A molecules that could collide with B molecules in any given volume (Figure 13.16). Consequently, the rate would increase by a factor of 2. Similarly, doubling the concentration of B molecules would increase the rate twofold. Thus, we can express the rate law as

$$\text{rate} = k[\text{A}][\text{B}]$$

The reaction is first order in both A and B and obeys second-order kinetics.

The collision theory is intuitively appealing, but the relationship between rate and molecular collisions is more complicated than you might expect. The implication of the collision theory is that a reaction always occurs when an A and a B molecule collide. However, not all collisions lead to reactions. Calculations based on the kinetic molecular theory show that, at ordinary pressures (say, 1 atm) and temperatures (say, 298 K), there are about 1×10^{27} binary collisions (collisions between two molecules) in 1 mL of volume every second in the gas phase. Even more collisions per second occur in liquids. If every binary collision led to a product, then most reactions would be complete almost instantaneously. In practice, we find that the rates of reactions differ greatly. This means that, in many cases, collisions alone do not guarantee that a reaction will take place.

Any molecule in motion possesses kinetic energy; the faster it moves, the greater the kinetic energy. But a fast-moving molecule will not break up into fragments on its own. To react, it must collide with another molecule. When molecules collide, part of their kinetic energy is converted to vibrational energy. If the initial kinetic energies are large, then the colliding molecules will vibrate so strongly as to break some of the chemical bonds. This bond fracture is the first step toward product formation. If the initial kinetic energies are small, the molecules will merely bounce off each other intact. Energetically speaking, there is some minimum collision energy below which no reaction occurs. Lacking this energy, the molecules remain intact, and no change results from the collision.

Video: Activation Energy

We postulate that in order to react, the colliding molecules must have a total kinetic energy equal to or greater than the ***activation energy (E_a),*** which is *the minimum amount of energy required to initiate a chemical reaction.* When molecules collide they form an ***activated complex*** (also called the ***transition state***), *a temporary species formed by the reactant molecules as a result of the collision before they form the product.*

Figure 13.17 shows two different potential energy profiles for the reaction

$$\text{A} + \text{B} \longrightarrow \text{AB}^{\ddagger} \longrightarrow \text{C} + \text{D}$$

where $AB^{\ddagger}$ denotes an activated complex formed by the collision between A and B. If the products are more stable than the reactants, then the reaction will be accompanied by a release of heat; that is, the reaction is exothermic [see Figure 13.17(a)]. On the other hand, if the products are less stable than the reactants, then heat will be absorbed by the reacting mixture from the surroundings and we have an endothermic reaction [see Figure 13.17(b)]. In both cases, we plot the potential energy of the reacting system versus the progress of the reaction. Qualitatively, these plots show the potential energy changes as reactants are converted to products.

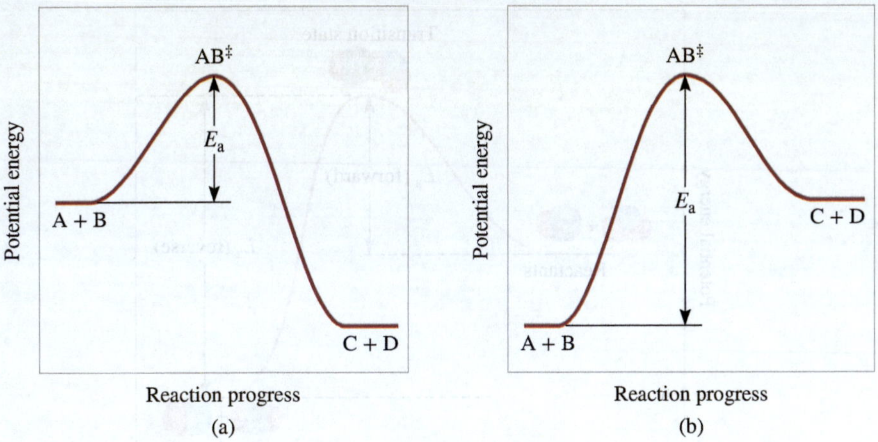

Figure 13.17 *Potential energy profiles for (a) exothermic and (b) endothermic reactions. These plots show the change in potential energy as reactants A and B are converted to products C and D. The activated complex ($AB^{\ddagger}$) is a highly unstable species with a high potential energy. The activation energy is defined for the forward reaction in both (a) and (b). Note that the products C and D are more stable than the reactants in (a) and less stable than those in (b).*

We can think of activation energy as a barrier that prevents less energetic molecules from reacting. Because the number of reactant molecules in an ordinary reaction is very large, the speeds, and hence also the kinetic energies of the molecules, vary greatly. Normally, only a small fraction of the colliding molecules—the fastest-moving ones—have enough kinetic energy to exceed the activation energy. These molecules can therefore take part in the reaction. The increase in the rate (or the rate constant) with temperature can now be explained: The speeds of the molecules obey the Maxwell distributions shown in Figure 5.17. Compare the speed distributions at two different temperatures. Because more high-energy molecules are present at the higher temperature, the rate of product formation is also greater at the higher temperature.

Overcoming the activation energy is only one requirement for a collision to result in a reaction. For more complex reactions involving molecules, we must also consider the orientation of how molecules collide relative to each other. The reaction between carbon monoxide (CO) and nitrogen dioxide (NO_2) to form carbon dioxide (CO_2) and nitric oxide (NO) illustrates this point:

$$CO(g) + NO_2(g) \longrightarrow CO_2(g) + NO(g)$$

This reaction is most favorable when the reacting molecules approach each other according to that shown in Figure 13.18(a). Otherwise, few or no products are formed [see Figure 13.18(b) and (c)]. Figure 13.19 shows a plot of the potential energy as the reaction progresses. The difference in energy between the transition state and the reactants is the activation energy for the forward reaction. We can see here that the reverse of the reaction also has an activation energy that can be found on the potential energy profile.

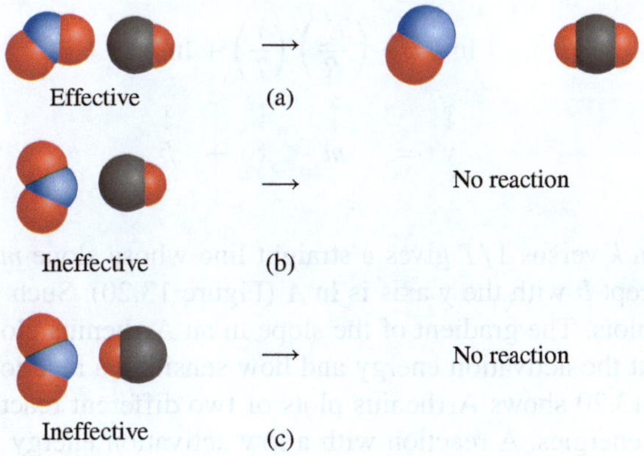

Figure 13.18 *The orientations of the molecules shown in (a) are effective and will likely lead to the formation of products. The orientations shown in (b) and (c) are ineffective, and no products will be formed.*

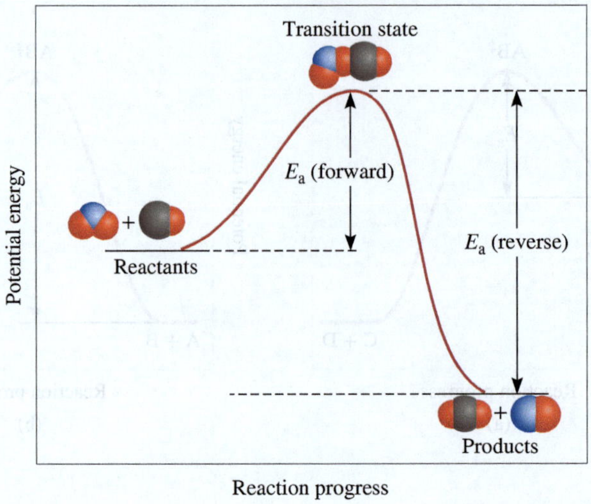

Figure 13.19 *Potential energy profile for the reaction of CO with NO_2.*

The Arrhenius Equation

The dependence of the rate constant of a reaction on temperature can be expressed by the following equation, known as the *Arrhenius equation:*

$$k = Ae^{-E_a/RT} \tag{13.11}$$

where E_a is the activation energy of the reaction (in kJ/mol), R the gas constant (8.314 J/K · mol), T the absolute temperature, and e the base of the natural logarithm scale (see Appendix 5). The quantity A represents the collision frequency and is called the frequency factor. It can be treated as a constant for a given reacting system over a fairly wide temperature range. Equation (13.11) shows that the rate constant is directly proportional to A and, therefore, to the collision frequency. In addition, because of the minus sign associated with the exponent E_a/RT, the rate constant decreases with increasing activation energy and increases with increasing temperature. This equation can be expressed in a more useful form by taking the natural logarithm of both sides:

$$\ln k = \ln Ae^{-E_a/RT}$$

$$\ln k = \ln A - \frac{E_a}{RT} \tag{13.12}$$

Equation (13.12) can be rearranged to a linear equation:

$$\begin{array}{ccccccc} \ln k & = & -\left(\dfrac{E_a}{R}\right) & \left(\dfrac{1}{T}\right) & + & \ln A \\ \updownarrow & & \updownarrow & \updownarrow & & \updownarrow \\ y & = & m & x & + & b \end{array} \tag{13.13}$$

Thus, a plot of ln k versus $1/T$ gives a straight line whose slope m is equal to $-E_a/R$ and whose intercept b with the y axis is ln A (Figure 13.20). Such graphs are referred to as Arrhenius plots. The gradient of the slope in an Arrhenius plot provides valuable information about the activation energy and how sensitive a reaction is to temperature changes. Figure 13.20 shows Arrhenius plots of two different reactions with very different activation energies. A reaction with a low activation energy (15 kJ/mol or less)

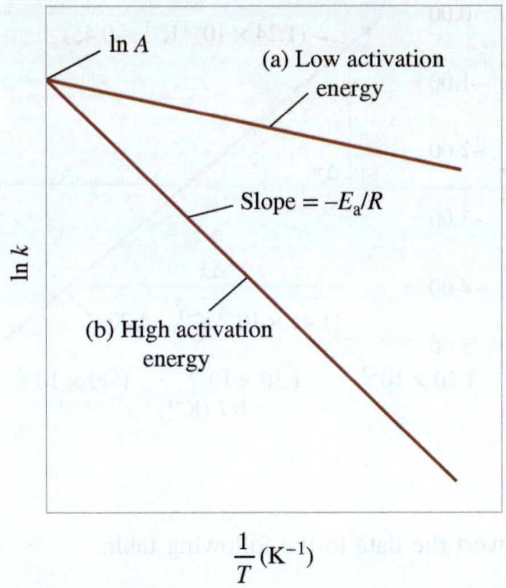

Figure 13.20 *Arrhenius plots of ln k versus 1/T. The activation energy is equal to the slope and ln A is the y-intercept. (a) A reaction with a low activation energy (approximately 10 kJ/mol) has an Arrhenius plot with a not very steep slope. (b) A reaction with a high activation energy (≥60 kJ/mol) has an Arrhenius plot with a very steep slope.*

has an Arrhenius plot with a not very steep slope; thus, the rates change only slightly with temperature as k changes slowly with temperature. A reaction with a high activation energy (60 kJ/mol or higher) has an Arrhenius plot with a very steep slope. This indicates that the reaction rate depends strongly on the temperature as k increases rapidly with temperature.

Example 13.8 demonstrates a graphical method for determining the activation energy of a reaction.

Example 13.8

The rate constants for the decomposition of acetaldehyde

$$CH_3CHO(g) \longrightarrow CH_4(g) + CO(g)$$

were measured at five different temperatures. The data are shown in the accompanying table. Plot ln k versus $1/T$, and determine the activation energy (in kJ/mol) for the reaction. Note that the reaction is "$\frac{3}{2}$" order in CH_3CHO, so k has the units of $1/M^{\frac{1}{2}} \cdot$ s.

k (1/$M^{1/2}$ · s)	T (K)
0.011	700
0.035	730
0.105	760
0.343	790
0.789	810

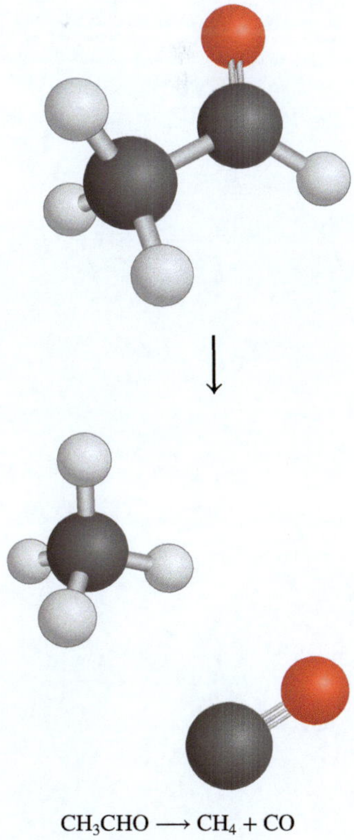

$CH_3CHO \longrightarrow CH_4 + CO$

Strategy Consider the Arrhenius equation written as a linear equation

$$\ln k = \left(-\frac{E_a}{R}\right)\left(\frac{1}{T}\right) + \ln A$$

A plot of ln k versus $1/T$ (y versus x) will produce a straight line with a slope equal to $-E_a/R$. Thus, the activation energy can be determined from the slope of the plot.

(Continued)

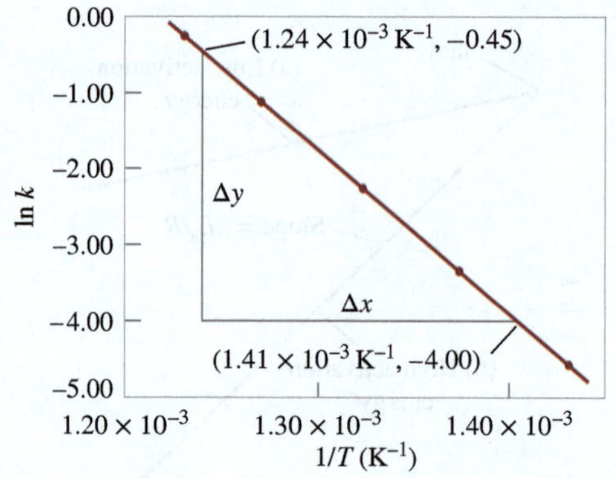

Figure 13.21 *Plot of ln k versus 1/T. The slope of the line is equal to $-E_a/R$.*

Solution First we convert the data to the following table.

ln k	$1/T$ (K^{-1})
−4.51	1.43×10^{-3}
−3.35	1.37×10^{-3}
−2.254	1.32×10^{-3}
−1.070	1.27×10^{-3}
−0.237	1.23×10^{-3}

A plot of these data yields the graph in Figure 13.21. The slope of the line is calculated from two pairs of coordinates:

$$\text{slope} = \frac{-4.00 - (-0.45)}{(1.41 - 1.24) \times 10^{-3}\ \text{K}^{-1}} = -2.09 \times 10^4\ \text{K}$$

From the linear form of Equation (13.13)

$$\text{slope} = -\frac{E_a}{R} = -2.09 \times 10^4\ \text{K}$$

$$E_a = (8.314\ \text{J/K} \cdot \text{mol})(2.09 \times 10^4\ \text{K})$$
$$= 1.74 \times 10^5\ \text{J/mol}$$
$$= 1.74 \times 10^2\ \text{kJ/mol}$$

Check It is important to note that although the rate constant itself has the units $1/M^{\frac{1}{2}} \cdot \text{s}$, the quantity ln k has no units (we cannot take the logarithm of a unit).

Practice Exercise The second-order rate constant for the decomposition of nitrous oxide (N_2O) into nitrogen molecule and oxygen atom has been measured at different temperatures:

$k(1/M \cdot s)$	t(°C)
1.87×10^{-3}	600
0.0113	650
0.0569	700
0.244	750

Determine graphically the activation energy for the reaction.

Similar problem: 13.40.

An equation relating the rate constants k_1 and k_2 at temperatures T_1 and T_2 can be used to calculate the activation energy or to find the rate constant at another temperature if the activation energy is known. To derive such an equation we start with Equation (13.12):

$$\ln k_1 = \ln A - \frac{E_a}{RT_1}$$

$$\ln k_2 = \ln A - \frac{E_a}{RT_2}$$

Subtracting $\ln k_2$ from $\ln k_1$ gives

$$\ln k_1 - \ln k_2 = \frac{E_a}{R}\left(\frac{1}{T_2} - \frac{1}{T_1}\right)$$

$$\ln \frac{k_1}{k_2} = \frac{E_a}{R}\left(\frac{1}{T_2} - \frac{1}{T_1}\right)$$

$$\ln \frac{k_1}{k_2} = \frac{E_a}{R}\left(\frac{T_1 - T_2}{T_1T_2}\right) \quad (13.14)$$

Student Hot Spot

Student data indicate you may struggle with determining activation energy. Access your eBook for additional Learning Resources on this topic.

Example 13.9 illustrates the use of the equation we have just derived.

Example 13.9

The rate constant of a first-order reaction is $3.46 \times 10^{-2}\ s^{-1}$ at 298 K. What is the rate constant at 350 K if the activation energy for the reaction is 50.2 kJ/mol?

Strategy A modified form of the Arrhenius equation relates two rate constants at two different temperatures [see Equation (13.14)]. Make sure the units of R and E_a are consistent.

Solution The data are

$$k_1 = 3.46 \times 10^{-2}\ s^{-1} \qquad k_2 = ?$$

$$T_1 = 298\ K \qquad T_2 = 350\ K$$

Substituting in Equation (13.14),

$$\ln \frac{3.46 \times 10^{-2}\ s^{-1}}{k_2} = \frac{50.2 \times 10^3\ J/mol}{8.314\ J/K \cdot mol}\left[\frac{298\ K - 350\ K}{(298\ K)(350\ K)}\right]$$

We convert E_a to units of J/mol to match the units of R. Solving the equation gives

$$\ln \frac{3.46 \times 10^{-2}\ s^{-1}}{k_2} = -3.01$$

$$\frac{3.46 \times 10^{-2}\ s^{-1}}{k_2} = e^{-3.01} = 0.0493$$

$$k_2 = 0.702\ s^{-1}$$

Check The rate constant is expected to be greater at a higher temperature. Therefore, the answer is reasonable.

Practice Exercise The first-order rate constant for the reaction of methyl chloride (CH_3Cl) with water to produce methanol (CH_3OH) and hydrochloric acid (HCl) is $3.32 \times 10^{-10}\ s^{-1}$ at 25°C. Calculate the rate constant at 40°C if the activation energy is 116 kJ/mol.

Similar problem: 13.42.

Video: Orientation of Collision

For simple reactions (for example, reactions between atoms), we can equate the frequency factor (A) in the Arrhenius equation with the frequency of collision between the reacting species. For more complex reactions, we must also consider the "orientation factor," that is, how reacting molecules are oriented relative to each other. As we saw earlier in this section, the reaction between carbon monoxide (CO) and nitrogen dioxide (NO_2) to form carbon dioxide (CO_2) and nitric oxide (NO) highlights this point [see Figure 13.18(a)]. The quantitative treatment of orientation factor is to modify Equation (13.11) as follows:

$$k = pAe^{-E_a/RT} \tag{13.15}$$

where p is the orientation factor. The orientation factor is a unitless quantity; its value ranges from 1 for reactions involving atoms such as $I + I \longrightarrow I_2$ to 10^{-6} or smaller for reactions involving molecules.

Summary of Concepts & Facts

- In terms of collision theory, a reaction occurs when molecules collide with sufficient energy, called the activation energy, to break the bonds and initiate the reaction. The rate constant and the activation energy are related by the Arrhenius equation.

Review of Concepts & Facts

13.4.1 What is the activation energy of a particular reaction if the rate constant for the reaction is $2.7 \times 10^{-2}\ s^{-1}$ at 25°C and $6.2 \times 10^{-2}\ s^{-1}$ at 115°C?

13.4.2 (a) What can you deduce about the magnitude of the activation energy of a reaction if its rate constant changes appreciably with a small change in temperature? (b) If a reaction occurs every time two reacting molecules collide, what can you say about the orientation factor and the activation energy of the reaction?

13.5 Reaction Mechanisms

Learning Objectives

- Define reaction mechanism.
- Generate the rate law of a reaction given its rate-determining step in a reaction mechanism.

As we mentioned earlier, an overall balanced chemical equation does not tell us much about how a reaction actually takes place. In many cases, it merely represents the sum of several ***elementary steps,*** or *elementary reactions, a series of simple reactions that represent the progress of the overall reaction at the molecular level.* The term for *the sequence of elementary steps that leads to product formation* is ***reaction mechanism.*** The reaction mechanism is comparable to the route of travel followed during a trip; the overall chemical equation specifies only the origin and destination.

As an example of a reaction mechanism, let us consider the reaction between nitric oxide and oxygen:

$$2NO(g) + O_2(g) \longrightarrow 2NO_2(g)$$

We know that the products are not formed directly from the collision of two NO molecules with an O_2 molecule because N_2O_2 is detected during the course of the reaction. Let us assume that the reaction actually takes place via two elementary steps as follows:

$$2NO(g) \longrightarrow N_2O_2(g)$$

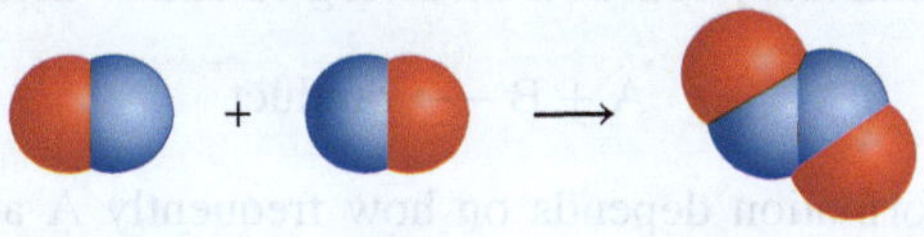

$$N_2O_2(g) + O_2(g) \longrightarrow 2NO_2(g)$$

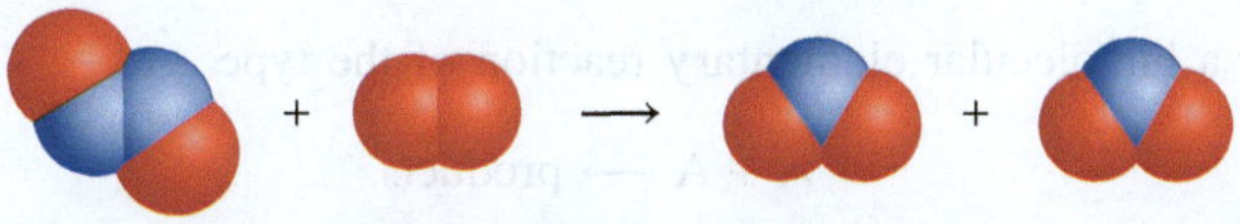

In the first elementary step, two NO molecules collide to form a N_2O_2 molecule. This event is followed by the reaction between N_2O_2 and O_2 to give two molecules of NO_2. The net chemical equation, which represents the overall change, is given by the sum of the elementary steps:

$$\begin{array}{ll} \text{Step 1:} & NO + NO \longrightarrow N_2O_2 \\ \text{Step 2:} & N_2O_2 + O_2 \longrightarrow 2NO_2 \\ \text{Overall reaction:} & 2NO + \cancel{N_2O_2} + O_2 \longrightarrow \cancel{N_2O_2} + 2NO_2 \end{array}$$

Species such as N_2O_2 are called ***intermediates*** because they *appear in the mechanism of the reaction (that is, the elementary steps) but not in the overall balanced equation.* Keep in mind that an intermediate is always formed in an early elementary step and consumed in a later elementary step.

The ***molecularity of a reaction*** is *the number of molecules reacting in an elementary step*. These molecules may be of the same or different types. Each of the elementary steps discussed earlier is called a ***bimolecular reaction,*** *an elementary step that involves two molecules*. An example of a ***unimolecular reaction,*** *an elementary step in which only one reacting molecule participates*, is the conversion of cyclopropane to propene discussed in Example 13.4. Very few ***termolecular reactions,*** *reactions that involve the participation of three molecules in one elementary step,* are known, because the simultaneous encounter of three molecules is a far less likely event than a bimolecular collision.

Rate Laws and Elementary Steps

Knowing the elementary steps of a reaction enables us to deduce the rate law. Suppose we have the following elementary reaction:

$$A \longrightarrow \text{products}$$

Because there is only one molecule present, this is a unimolecular reaction. It follows that the larger the number of A molecules present, the faster the rate of product formation. Thus, the rate of a unimolecular reaction is directly proportional to the concentration of A, or is first order in A:

$$\text{rate} = k[A]$$

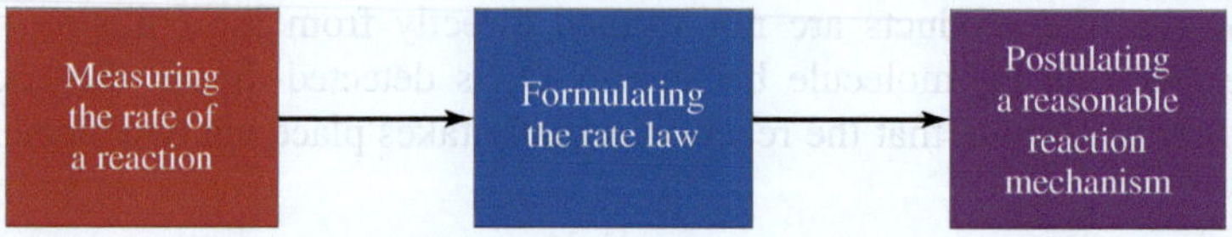

Figure 13.22 *Sequence of steps in the study of a reaction mechanism.*

For a bimolecular elementary reaction involving A and B molecules

$$A + B \longrightarrow \text{product}$$

the rate of product formation depends on how frequently A and B collide, which in turn depends on the concentrations of A and B. Thus, we can express the rate as

$$\text{rate} = k[A][B]$$

Similarly, for a bimolecular elementary reaction of the type

$$A + A \longrightarrow \text{products}$$

or

$$2A \longrightarrow \text{products}$$

the rate becomes

$$\text{rate} = k[A]^2$$

The preceding examples show that the reaction order for each reactant in an elementary reaction is equal to its stoichiometric coefficient in the chemical equation for that step. The rate law for a reaction can be written directly from the coefficients of an elementary step. But, in general, we cannot tell by merely looking at the overall balanced equation whether the reaction occurs as shown or in a series of steps. This determination is made in the laboratory.

When we study a reaction that has more than one elementary step, the rate law for the overall process is given by the ***rate-determining step,*** which is *the slowest step in the sequence of steps leading to product formation.*

An analogy for the rate-determining step is the flow of traffic along a narrow road. Assuming the cars cannot pass one another on the road, the rate at which the cars travel is governed by the slowest-moving car.

Experimental studies of reaction mechanisms begin with the collection of data (rate measurements). Next, we analyze the data to determine the rate constant and order of the reaction, and we write the rate law. Finally, we suggest a plausible mechanism for the reaction in terms of elementary steps (Figure 13.22). The elementary steps must satisfy two requirements:

- The sum of the elementary steps must give the overall balanced equation for the reaction.
- The rate-determining step should predict the same rate law as is determined experimentally.

Remember that for a proposed reaction scheme, we must be able to detect the presence of any intermediate(s) formed in one or more elementary steps.

The decomposition of hydrogen peroxide and the formation of hydrogen iodide from molecular hydrogen and molecular iodine illustrate the elucidation of reaction mechanisms by experimental studies.

Figure 13.23 *The decomposition of hydrogen peroxide is catalyzed by the iodide ion. A few drops of liquid soap have been added to the solution to dramatize the evolution of oxygen gas. (Some of the iodide ions are oxidized to molecular iodine, which then reacts with iodide ions to form the brown triiodide I_3^- ion.)*
Ken Karp/McGraw Hill

Hydrogen Peroxide Decomposition

The decomposition of hydrogen peroxide is facilitated by iodide ions (Figure 13.23). The overall reaction is

$$2H_2O_2(aq) \longrightarrow 2H_2O(l) + O_2(g)$$

By experiment, the rate law is found to be

$$\text{rate} = k[H_2O_2][I^-]$$

Thus, the reaction is first order with respect to both H_2O_2 and I^-.

You can see that H_2O_2 decomposition does not occur in a single elementary step corresponding to the overall balanced equation. If it did, the reaction would be second order in H_2O_2 (as a result of the collision of two H_2O_2 molecules). What's more, the I^- ion, which is not even part of the overall equation, appears in the rate law expression. How can we reconcile these facts? First, we can account for the observed rate law by assuming that the reaction takes place in two separate elementary steps, each of which is bimolecular:

$$\text{Step 1:}\quad H_2O_2 + I^- \xrightarrow{k_1} H_2O + IO^-$$
$$\text{Step 2:}\quad H_2O_2 + IO^- \xrightarrow{k_2} H_2O + O_2 + I^-$$

If we further assume that step 1 is the rate-determining step, then the rate of the reaction can be determined from the first step alone:

$$\text{rate} = k_1[H_2O_2][I^-]$$

where $k_1 = k$. Note that the IO^- ion is an intermediate because it does not appear in the overall balanced equation. Although the I^- ion also does not appear in the overall equation, I^- differs from IO^- in that the former is present at the start of the reaction and at its completion. The function of I^- is to speed up the reaction—that is, it is a *catalyst*. We will discuss catalysis in Section 13.6. Figure 13.24 shows the potential energy profile for a reaction like the decomposition of H_2O_2. We see that the first step, which is rate determining, has a larger activation energy than the second step. The intermediate, although stable enough to be observed, reacts quickly to form the products.

Example 13.10 concerns the mechanistic study of a relatively simple reaction.

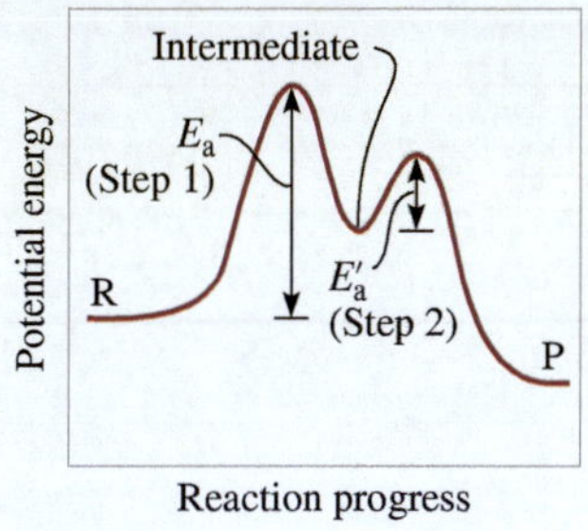

Figure 13.24 *Potential energy profile for a two-step reaction in which the first step is rate-determining. R and P represent reactants and products, respectively.*

Student Hot Spot

Student data indicate you may struggle with reaction mechanisms. Access your eBook for additional Learning Resources on this topic.

Example 13.10

The gas-phase decomposition of nitrous oxide (N_2O) is believed to occur via two elementary steps:

$$\text{Step 1:}\quad N_2O \xrightarrow{k_1} N_2 + O$$
$$\text{Step 2:}\quad N_2O + O \xrightarrow{k_2} N_2 + O_2$$

Experimentally the rate law is found to be rate $= k[N_2O]$. (a) Write the equation for the overall reaction. (b) Identify the intermediate. (c) What can you say about the relative rates of steps 1 and 2?

Strategy (a) Because the overall reaction can be broken down into elementary steps, knowing the elementary steps would enable us to write the overall reaction. (b) What are the characteristics of an intermediate? Does it appear in the overall reaction? (c) What determines which elementary step is rate determining? How does a knowledge of the rate-determining step help us write the rate law of a reaction?

Solution

(a) Adding the equations for steps 1 and 2 gives the overall reaction

$$2N_2O \longrightarrow 2N_2 + O_2$$

(b) Because the O atom is produced in the first elementary step and it does not appear in the overall balanced equation, it is an intermediate.

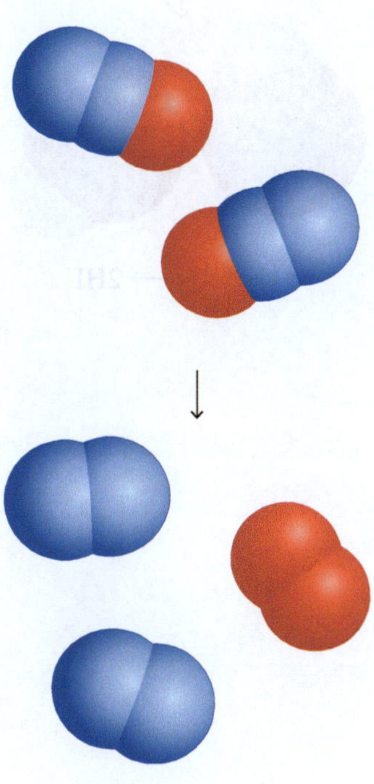

$2N_2O \longrightarrow 2N_2 + O_2$

(Continued)

(c) If we assume that step 1 is the rate-determining step, then the rate of the overall reaction is given by

$$\text{rate} = k_1[N_2O]$$

and $k = k_1$.

Check There are two criteria that must be met for a proposed reaction mechanism to be plausible: (1) The individual steps (elementary steps) must sum to the corrected overall reaction. (2) The rate-determining step (the slow step) must have the same rate law as the experimentally determined rate law.

Practice Exercise The reaction between NO_2 and CO to produce NO and CO_2 is believed to occur via two steps:

$$\text{Step 1:} \quad NO_2 + NO_2 \longrightarrow NO + NO_3$$
$$\text{Step 2:} \quad NO_3 + CO \longrightarrow NO_2 + CO_2$$

The experimental rate law is rate $= k[NO_2]^2$. (a) Write the equation for the overall reaction. (b) Identify the intermediate. (c) What can you say about the relative rates of steps 1 and 2?

Similar problem: 13.55.

Mechanisms with an Equilibrium Step

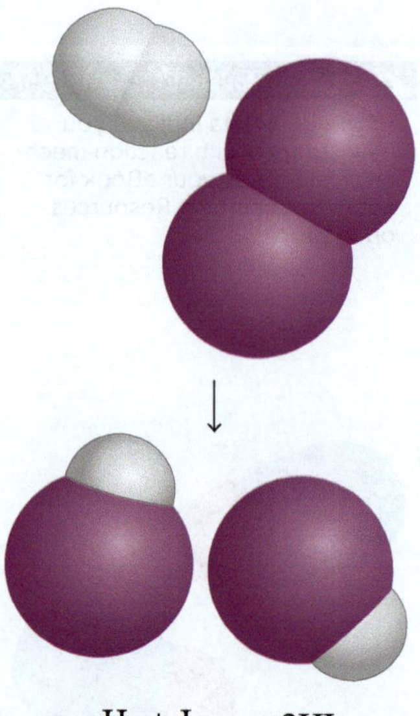

$H_2 + I_2 \longrightarrow 2HI$

A common reaction mechanism is one that involves at least two elementary steps, the first of which is very rapid in both the forward and reverse directions compared with the second step. An example is the reaction between molecular hydrogen and molecular iodine to produce hydrogen iodide:

$$H_2(g) + I_2(g) \longrightarrow 2HI(g)$$

Experimentally, the rate law is found to be

$$\text{rate} = k[H_2][I_2]$$

For many years it was thought that the reaction occurred just as written; that is, it is a bimolecular reaction involving a hydrogen molecule and an iodine molecule. However, in the 1960s chemists found that the actual mechanism is more complicated. A two-step mechanism was proposed:

$$\text{Step 1:} \quad I_2 \underset{k_{-1}}{\overset{k_1}{\rightleftharpoons}} 2I$$

$$\text{Step 2:} \quad H_2 + 2I \xrightarrow{k_2} 2HI$$

where k_1, k_{-1}, and k_2 are the rate constants for the reactions. The I atoms are the intermediate in this reaction.

When the reaction begins, there are very few I atoms present. But as I_2 dissociates, the concentration of I_2 decreases while that of I increases. Therefore, the forward rate of step 1 decreases and the reverse rate increases. Soon the two rates become equal, and a chemical equilibrium is established (chemical equilibrium will be discussed in Chapter 14). Because the elementary reactions in step 1 are much faster than the one in step 2, equilibrium is reached before any significant reaction with hydrogen occurs, and it persists throughout the reaction.

In the equilibrium condition of step 1 the forward rate is equal to the reverse rate; that is,

$$k_1[I_2] = k_{-1}[I]^2$$

or

$$[I]^2 = \frac{k_1}{k_{-1}}[I_2]$$

The rate of the reaction is given by the slow, rate-determining step, which is step 2:

$$\text{rate} = k_2[H_2][I]^2$$

Substituting the expression for $[I]^2$ into this rate law, we obtain

$$\text{rate} = \frac{k_1 k_2}{k_{-1}}[H_2][I_2]$$
$$= k[H_2][I_2]$$

where $k = k_1k_2/k_{-1}$. As you can see, this two-step mechanism also gives the correct rate law for the reaction. This agreement along with the observation of intermediate I atoms provides strong evidence that the mechanism is correct.

Example 13.11 examines a reaction with an equilibrium step in the mechanism.

Example 13.11

The overall reaction of NO gas with bromine vapor is

$$2NO(g) + Br_2(g) \longrightarrow 2NOBr(g)$$

Is the following mechanism possible for this reaction? The experimentally determined rate law is rate = $k[NO]^2[Br_2]$.

$$\text{Step 1:} \quad NO(g) + Br_2(g) \underset{k_{-1}}{\overset{k_1}{\rightleftharpoons}} NOBr_2(g) \quad \text{(fast)}$$

$$\text{Step 2:} \quad NOBr_2(g) + NO(g) \xrightarrow{k_2} 2NOBr(g) \quad \text{(slow)}$$

Strategy To determine if the mechanism is plausible, we must compare the rate-determining step to the experimentally determined rate law. The rate-determining step, the slow step, has the intermediate $NOBr_2$ and there should be only reactants from the overall reaction in the rate law. We must solve for the $NOBr_2$ concentration in terms of reactant concentrations.

Solution The rate law based on the rate-determining step is

$$\text{rate} = k[NOBr_2][NO]$$

At equilibrium, the forward and reverse reactions of the equilibrium step are occurring at the same rate allowing us to set the rates equal to each other. We then rearrange this equation to solve for the concentration of $NOBr_2$.

$$k_1[NO][Br_2] = k_{-1}[NOBr_2]$$
$$[NOBr_2] = \frac{k_1[NO][Br_2]}{k_{-1}}$$

Substituting this intermediate expression into the original rate law gives

$$\text{rate} = k_2\frac{k_1[NO][Br_2]}{k_{-1}}[NO] = k[NO]^2[Br_2] \quad \text{where } k = \frac{k_2k_1}{k_{-1}}$$

Check Removing the intermediate by redefining it in terms of reactant concentrations produces a rate law that matches the experimentally determined rate. Therefore, the mechanism is plausible.

Practice Exercise For the reaction $2NO(g) + O_2(g) \longrightarrow 2NO_2(g)$ where the experimentally determined rate law is rate = $k[NO]^2[O_2]$, show that the following mechanism is plausible.

$$\text{Step 1:} \quad NO(g) + NO(g) \underset{k_{-1}}{\overset{k_1}{\rightleftharpoons}} N_2O_2(g) \quad \text{(fast)}$$

$$\text{Step 2:} \quad N_2O_2(g) + O_2(g) \xrightarrow{k_2} 2NO_2(g) \quad \text{(slow)}$$

Similar problem: 13.57.

Finally, we note that not all reactions have a single rate-determining step. A reaction may have two or more comparably slow steps. The kinetic analysis of such reactions is generally more involved.

Experimental Support for Reaction Mechanisms

How can we find out whether the proposed mechanism for a particular reaction is correct? In the case of hydrogen peroxide decomposition we might try to detect the presence of the IO^- ions by spectroscopic means. Evidence of their presence would support the reaction scheme. Similarly, for the hydrogen iodide reaction, detection of iodine atoms would lend support to the two-step mechanism. For example, I_2 dissociates into atoms when it is irradiated with visible light. Thus, we might predict that the formation of HI from H_2 and I_2 would speed up as the intensity of light is increased because that should increase the concentration of I atoms. Indeed, this is just what is observed.

In another case, chemists wanted to know which C—O bond is broken in the reaction between methyl acetate and water to better understand the reaction mechanism

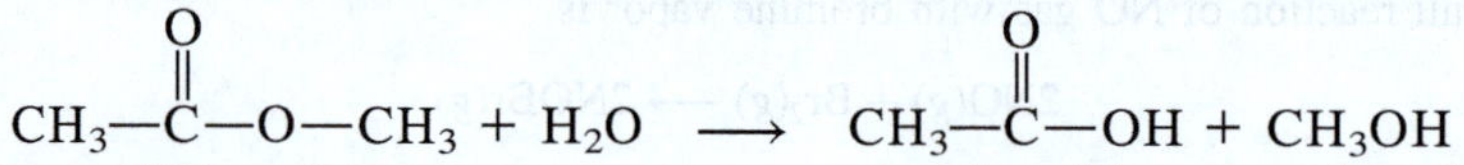

methyl acetate — acetic acid — methanol

The two possibilities are

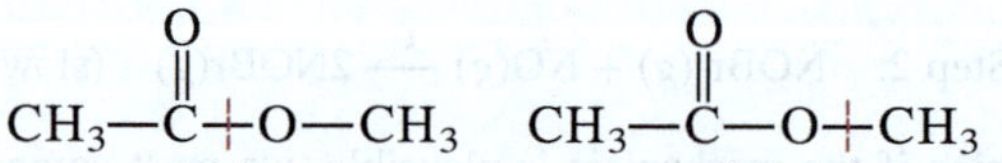

(a) (b)

To distinguish between schemes (a) and (b), chemists used water containing the oxygen-18 isotope instead of ordinary water (which contains the oxygen-16 isotope). When the oxygen-18 water was used, only the acetic acid formed contained oxygen-18:

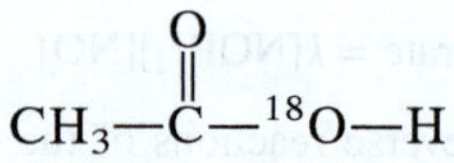

Thus, the reaction must have occurred via bond-breaking scheme (a), because the product formed via scheme (b) would retain both of its original oxygen atoms.

Another example is photosynthesis, the process by which green plants produce glucose from carbon dioxide and water:

$$6CO_2 + 6H_2O \longrightarrow C_6H_{12}O_6 + 6O_2$$

A question that arose early in studies of photosynthesis was whether the molecular oxygen was derived from water, from carbon dioxide, or from both. By using water containing the oxygen-18 isotope, it was demonstrated that the evolved oxygen came from water, and none came from carbon dioxide, because the O_2 contained only the ^{18}O isotopes. This result supported the mechanism in which water molecules are "split" by light:

$$2H_2O + h\nu \longrightarrow O_2 + 4H^+ + 4e^-$$

where $h\nu$ represents the energy of a photon. The protons and electrons are used to drive energetically unfavorable reactions that are necessary for plant growth and function.

These examples give some idea of how inventive chemists must be in studying reaction mechanisms. For complex reactions, however, it is virtually impossible to prove the uniqueness of any particular mechanism.

Summary of Concepts & Facts

- The overall balanced equation for a reaction may be the sum of a series of simple reactions, called elementary steps. The complete series of elementary steps for a reaction is the reaction mechanism.
- If one step in a reaction mechanism is much slower than all other steps, it is the rate-determining step.

Review of Concepts & Facts

13.5.1 The rate law for the reaction $H_2 + 2IBr \longrightarrow I_2 + 2HBr$ is rate $= k[H_2][IBr]$. Given that HI is an intermediate, write a two-step mechanism for the reaction.

13.5.2 For the reaction between NO and O_2, the following mechanism has been suggested:

$$NO(g) + NO(g) \rightleftharpoons N_2O_2(g) \quad \textit{fast}$$
$$N_2O_2(g) + O_2(g) \longrightarrow 2NO_2(g) \quad \textit{slow}$$

What is the rate law for this reaction?

13.6 Catalysis

Learning Objectives

- Summarize catalysts and reaction intermediates.
- Differentiate the major types of catalysts: heterogeneous, homogeneous, and enzymatic.

Video: Catalysis

For the decomposition of hydrogen peroxide, we saw that the reaction rate depends on the concentration of iodide ions even though I^- does not appear in the overall equation. We noted that I^- acts as a catalyst for that reaction. A ***catalyst*** is *a substance that increases the rate of a reaction by lowering the activation energy.* It does so by providing an alternative reaction pathway. The catalyst may react to form an intermediate with the reactant, but it is regenerated in a subsequent step so it is not consumed in the reaction.

In the laboratory preparation of molecular oxygen, a sample of potassium chlorate is heated, as shown in Figure 4.13(b). The reaction is

$$2KClO_3(s) \longrightarrow 2KCl(s) + 3O_2(g)$$

However, this thermal decomposition process is very slow in the absence of a catalyst. The rate of decomposition can be increased dramatically by adding a small amount of the catalyst manganese(IV) dioxide (MnO_2), a black powdery substance. All of the MnO_2 can be recovered at the end of the reaction, just as all the I^- ions remain following H_2O_2 decomposition.

A catalyst speeds up a reaction by providing a set of elementary steps with more favorable kinetics than those that exist in its absence. From Equation (13.11) we know that the rate constant k (and hence the rate) of a reaction depends on the frequency factor A and the activation energy E_a—the larger A or the smaller E_a, the greater the rate. In many cases, a catalyst increases the rate by lowering the activation energy for the reaction.

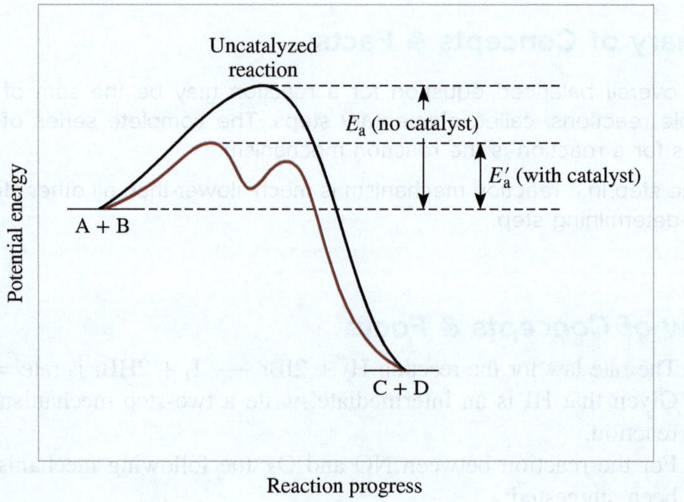

Figure 13.25 *Comparison of the activation energy barriers of an uncatalyzed reaction and the same reaction with a catalyst. The catalyst lowers the energy barrier but does not affect the actual energies of the reactants or products. Although the reactants and products are the same in both cases, the reaction mechanisms and rate laws are different in (a) and (b).*

Let us assume that the following reaction has a certain rate constant k and an activation energy E_a.

$$A + B \xrightarrow{k} C + D$$

In the presence of a catalyst, however, the rate constant is k_c (called the *catalytic rate constant*):

$$A + B \xrightarrow{k_c} C + D$$

By the definition of a catalyst,

$$\text{rate}_{\text{catalyzed}} > \text{rate}_{\text{uncatalyzed}}$$

Figure 13.25 shows the potential energy profiles for both reactions. Note that the total energies of the reactants (A and B) and those of the products (C and D) are unaffected by the catalyst; the only difference between the two is a lowering of the overall activation energy. Because the activation energy for the reverse reaction is also lowered, a catalyst enhances the rates of the reverse and forward reactions equally.

As noted earlier, a catalyst changes the mechanism of a reaction by providing a lower energy pathway. As seen in Figure 13.25, the catalyzed reaction proceeds through a two-step pathway whereas the uncatalyzed reaction has a single transition state. Table 13.6 lists the activation energies for some catalyzed and uncatalyzed reactions.

Table 13.6 Activation Energies of Catalyzed and Uncatalyzed Reactions

Reaction	Uncatalyzed E_a (kJ/mol)	Catalyzed E'_a (kJ/mol)	Catalyst
$2HI(g) \longrightarrow H_2(g) + I_2(g)$	+184	+105	gold
$2N_2O(g) \longrightarrow 2N_2(g) + O_2(g)$	+245	+121	gold
$2NH_3(g) \longrightarrow N_2(g) + 3H_2(g)$	+350	+162	tungsten
$2HI(g) \longrightarrow H_2(g) + I_2(g)$	+184	+121	gold
		+59	platinum
$2H_2O_2(aq) \longrightarrow 2H_2O(l) + O_2(aq)$	+75	+49	platinum
		+23	enzyme

There are three general types of catalysis, depending on the nature of the rate-increasing substance: heterogeneous catalysis, homogeneous catalysis, and enzyme catalysis.

Heterogeneous Catalysis

In *heterogeneous catalysis,* the reactants and the catalyst are in different phases. Usually the catalyst is a solid and the reactants are either gases or liquids. Heterogeneous catalysis is by far the most important type of catalysis in industrial chemistry, especially in the synthesis of many key chemicals. Here we describe three specific examples of heterogeneous catalysis that account for millions of tons of chemicals produced annually on an industrial scale.

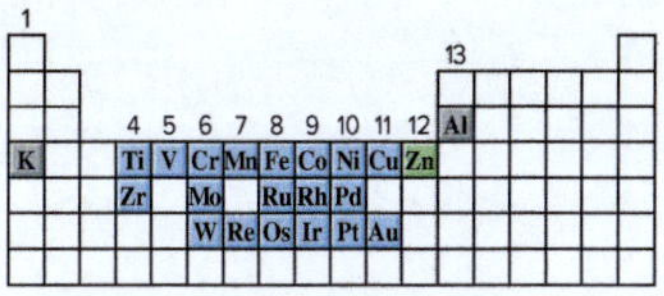

Metals and compounds of metals that are most frequently used in heterogeneous catalysis.

The Haber Synthesis of Ammonia

Ammonia is an extremely valuable inorganic substance used in the fertilizer industry, the manufacture of explosives, and many other applications. Around the turn of the twentieth century, many chemists strove to synthesize ammonia from nitrogen and hydrogen. The supply of atmospheric nitrogen is virtually inexhaustible, and hydrogen gas can be produced readily by passing steam over heated coal:

$$H_2O(g) + C(s) \longrightarrow CO(g) + H_2(g)$$

Hydrogen is also a by-product of petroleum refining.

The formation of NH_3 from N_2 and H_2 is exothermic:

$$N_2(g) + 3H_2(g) \longrightarrow 2NH_3(g) \qquad \Delta H^\circ = -92.6 \text{ kJ/mol}$$

But the reaction rate is extremely slow at room temperature. To be practical on a large scale, a reaction must occur at an appreciable rate *and* it must have a high yield of the desired product. Raising the temperature does accelerate the reaction, but at the same time it promotes the decomposition of NH_3 molecules into N_2 and H_2, thus lowering the yield of NH_3.

In 1905, after testing literally hundreds of compounds at various temperatures and pressures, Fritz Haber discovered that iron plus a few percent of oxides of potassium and aluminum catalyze the reaction of hydrogen with nitrogen to yield ammonia at about 500°C. This procedure is known as the *Haber process.*

In heterogeneous catalysis, the surface of the solid catalyst is usually the site of the reaction. The initial step in the Haber process involves the dissociation of N_2 and H_2 on the metal surface (Figure 13.26). Although the dissociated species are not truly free atoms because they are bonded to the metal surface, they are highly reactive. The two reactant molecules behave very differently on the catalyst surface. Studies show that H_2 dissociates into atomic hydrogen at temperatures as low as −196°C (the boiling point of liquid nitrogen). Nitrogen molecules, on the other hand, dissociate at about 500°C. The highly reactive N and H atoms combine rapidly at high temperatures to produce the desired NH_3 molecules:

$$N + 3H \longrightarrow NH_3$$

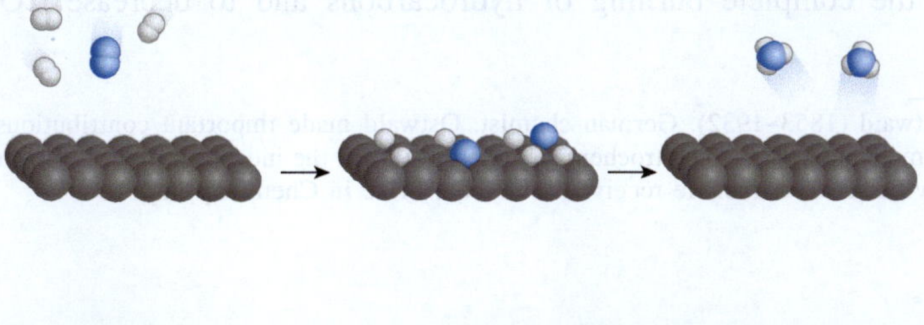

Figure 13.26 *The catalytic action in the synthesis of ammonia. First the H_2 and N_2 molecules bind to the surface of the catalyst. This interaction weakens the covalent bonds within the molecules and eventually causes the molecules to dissociate. The highly reactive H and N atoms combine to form NH_3 molecules, which then leave the surface.*

Figure 13.27 *Platinum-rhodium catalyst used in the Ostwald process.*
Reuters/Alamy Stock Photo

The Manufacture of Nitric Acid

Nitric acid is one of the most important inorganic acids. It is used in the production of fertilizers, dyes, drugs, and explosives. The major industrial method of producing nitric acid is the *Ostwald*† *process.* The starting materials, ammonia and molecular oxygen, are heated in the presence of a platinum-rhodium catalyst (Figure 13.27) to about 800°C:

$$4NH_3(g) + 5O_2(g) \longrightarrow 4NO(g) + 6H_2O(g)$$

The nitric oxide readily oxidizes (without catalysis) to nitrogen dioxide:

$$2NO(g) + O_2(g) \longrightarrow 2NO_2(g)$$

When dissolved in water, NO_2 forms both nitrous acid and nitric acid:

$$2NO_2(g) + H_2O(l) \longrightarrow HNO_2(aq) + HNO_3(aq)$$

On heating, nitrous acid is converted to nitric acid as follows:

$$3HNO_2(aq) \longrightarrow HNO_3(aq) + H_2O(l) + 2NO(g)$$

The NO generated can be recycled to produce NO_2 in the second step.

Catalytic Converters

At high temperatures inside a running car's engine, nitrogen and oxygen gases react to form nitric oxide:

$$N_2(g) + O_2(g) \longrightarrow 2NO(g)$$

When released into the atmosphere, NO rapidly combines with O_2 to form NO_2. Nitrogen dioxide and other gases emitted by an automobile, such as carbon monoxide (CO) and various unburned hydrocarbons, make automobile exhaust a major source of air pollution.

Most new cars are equipped with catalytic converters (Figure 13.28). An efficient catalytic converter serves two purposes: It oxidizes CO and unburned hydrocarbons to CO_2 and H_2O, and it reduces NO and NO_2 to N_2 and O_2. Hot exhaust gases into which air has been injected are passed through the first chamber of one converter to accelerate the complete burning of hydrocarbons and to decrease CO emission.

†Wilhelm Ostwald (1853–1932). German chemist. Ostwald made important contributions to chemical kinetics, thermodynamics, and electrochemistry. He developed the industrial process for preparing nitric acid that now bears his name. He received the Nobel Prize in Chemistry in 1909.

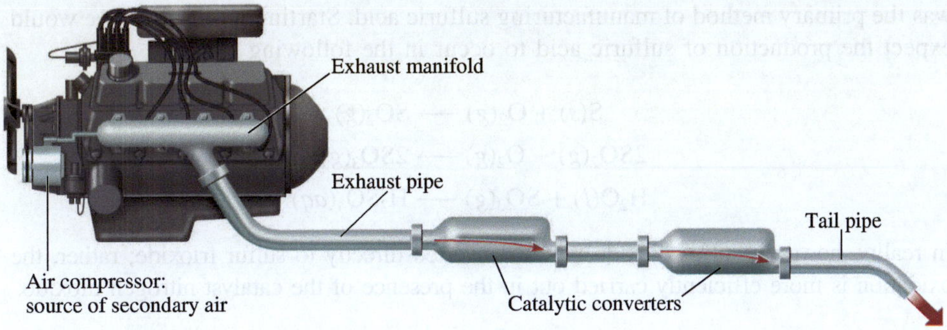

Figure 13.28 *A two-stage catalytic converter for an automobile.*

(A cross section of the catalytic converter is shown in Figure 13.29.) However, because high temperatures increase NO production, a second chamber containing a different catalyst (a transition metal or a transition metal oxide such as CuO or Cr_2O_3) and operating at a lower temperature is required to dissociate NO into N_2 and O_2 before the exhaust is discharged through the tailpipe.

Homogeneous Catalysis

In *homogeneous catalysis* the reactants and catalyst are dispersed in a single phase, usually liquid. Acid and base catalyses are the most important types of homogeneous catalysis in liquid solution. For example, the reaction of ethyl acetate with water to form acetic acid and ethanol normally occurs too slowly to be measured.

$$\underset{\text{ethyl acetate}}{CH_3-\overset{\overset{\displaystyle O}{\|}}{C}-O-C_2H_5} + H_2O \longrightarrow \underset{\text{acetic acid}}{CH_3-\overset{\overset{\displaystyle O}{\|}}{C}-OH} + \underset{\text{ethanol}}{C_2H_5OH}$$

In the absence of the catalyst, the rate law is given by

$$\text{rate} = k[CH_3COOC_2H_5]$$

However, the reaction can be catalyzed by an acid. In the presence of hydrochloric acid, the rate is faster and the rate law is given by

$$\text{rate} = k_c[CH_3COOC_2H_5][H^+]$$

Note that because $k_c > k$, the rate is determined solely by the catalyzed portion of the reaction.

Homogeneous catalysis can also take place in the gas phase. A well-known example of catalyzed gas-phase reactions is the lead chamber process, which for many years

Figure 13.29 *A cross-sectional view of a catalytic converter. The beads contain platinum, palladium, and rhodium, which catalyze the conversion of CO and hydrocarbons to carbon dioxide and water.*
Dorling Kindersley/Universal Images Group/Newscom

was the primary method of manufacturing sulfuric acid. Starting with sulfur, we would expect the production of sulfuric acid to occur in the following steps:

$$S(s) + O_2(g) \longrightarrow SO_2(g)$$
$$2SO_2(g) + O_2(g) \longrightarrow 2SO_3(g)$$
$$H_2O(l) + SO_3(g) \longrightarrow H_2SO_4(aq)$$

In reality, however, sulfur dioxide is not converted directly to sulfur trioxide; rather, the oxidation is more efficiently carried out in the presence of the catalyst nitrogen dioxide:

$$2SO_2(g) + 2NO_2(g) \longrightarrow 2SO_3(g) + 2NO(g)$$
$$2NO(g) + O_2(g) \longrightarrow 2NO_2(g)$$

Overall reaction: $2SO_2(g) + O_2(g) \longrightarrow 2SO_3(g)$

Note that there is no net loss of NO_2 in the overall reaction, so that NO_2 meets the criteria for a catalyst.

In recent years, chemists have devoted much effort to developing a class of transition metal compounds to serve as homogeneous catalysts. These compounds are soluble in various organic solvents and therefore can catalyze reactions in the same phase as the dissolved reactants. Many of the processes they catalyze are organic. For example, a red-violet compound of rhodium, $[(C_6H_5)_3P]_3RhCl$, catalyzes the conversion of a carbon-carbon double bond to a carbon-carbon single bond as follows:

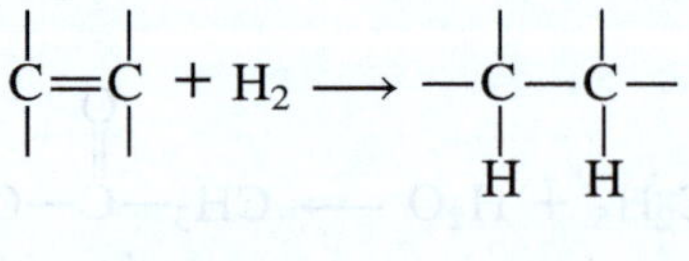

Homogeneous catalysis has several advantages over heterogeneous catalysis. For one thing, the reactions can often be carried out under atmospheric conditions, thus reducing production costs and minimizing the decomposition of products at high temperatures. In addition, homogeneous catalysts can be designed to function selectively for a particular type of reaction, and homogeneous catalysts cost less than the precious metals (for example, platinum and gold) used in heterogeneous catalysis.

Enzyme Catalysis

Of all the intricate processes that have evolved in living systems, none is more striking or more essential than enzyme catalysis. ***Enzymes*** are *biological catalysts*. Enzymes can increase the rate of biochemical reactions by factors ranging from 10^6 to 10^{18} as most biochemical reactions do not occur at appreciable rates without enzymes to sustain life. A selection of enzymes and their rate enhancements are given in Table 13.7.

Another key feature of enzymes is that they act only on certain molecules, called *substrates* (that is, reactants), while leaving the rest of the system unaffected. It has

Table 13.7 Catalytic Effect of Various Enzymes

Enzyme	Reaction Half-Life Without Enzyme	Uncatalyzed Rate Constant, k_{un} (s^{-1})	Catalyzed Rate Constant, k_{cat} (s^{-1})	Enzymatic Rate Enhancement (k_{cat}/k_{un})
Carbonic anhydrase	5 s	0.13	1×10^6	7.7×10^6
Chorismate mutase	7.4 h	2.6×10^{-5}	50	1.9×10^6
Ketosteroid isomerase	7 wk	1.7×10^{-7}	66,000	3.9×10^{11}
Carboxypeptidase A	7.3 y	3.0×10^{-9}	578	1.9×10^{11}
Staphylococcal nuclease	130,000 y	1.7×10^{-13}	95	5.6×10^{14}

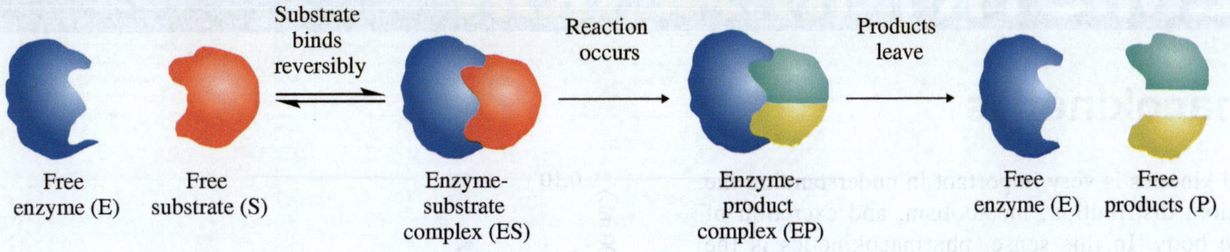

$$E + S \rightleftharpoons ES \longrightarrow EP \longrightarrow E + P$$

Figure 13.30 *The lock-and-key model of an enzyme's specificity for substrate molecules.*

been estimated that an average living *cell* may contain some 3000 different enzymes, each of them catalyzing a specific reaction in which a substrate is converted into the appropriate products. Enzyme catalysis is usually homogeneous because the substrate and enzyme are present in aqueous solution.

An enzyme is typically a large protein molecule that contains one or more *active sites* where interactions with substrates take place. These sites are structurally compatible with specific substrate molecules, in much the same way as a key fits a particular lock. In fact, the notion of a rigid enzyme structure that binds only to molecules whose shape exactly matches that of the active site was the basis of an early theory of enzyme catalysis, the so-called lock-and-key theory developed by the German chemist Emil Fischer† in 1894 (Figure 13.30). Fischer's hypothesis accounts for the specificity of enzymes, but it contradicts research evidence that a single enzyme binds to substrates of different sizes and shapes. Chemists now know that an enzyme molecule (or at least its active site) has a fair amount of structural flexibility and can modify its shape to accommodate more than one type of substrate. Figure 13.31 shows a molecular model of an enzyme in action.

The mathematical treatment of enzyme kinetics is quite complex, even when we know the basic steps involved in the reaction. A simplified scheme is given by the following elementary steps:

$$E + S \underset{k_{-1}}{\overset{k_1}{\rightleftharpoons}} ES$$

$$ES \xrightarrow{k_2} E + P$$

where E, S, and P represent enzyme, substrate, and product, and ES is the enzyme-substrate intermediate. It is often assumed that the formation of ES and its decomposition back to enzyme and substrate molecules occur rapidly and that the rate-determining step is the formation of product. (This is similar to the formation of HI discussed in Section 13.5.)

†Emil Fischer (1852–1919). German chemist. Regarded by many as the greatest organic chemist of the nineteenth century, Fischer made many significant contributions in the synthesis of sugars and other important molecules. He was awarded the Nobel Prize in Chemistry in 1902.

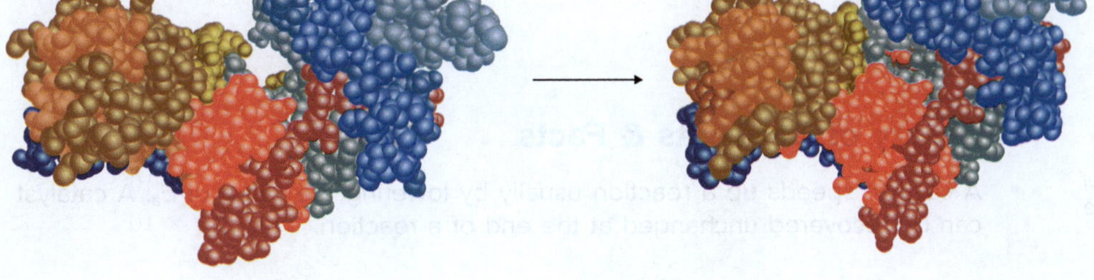

Figure 13.31 *Left to right: The binding of a glucose molecule (red) to hexokinase (an enzyme in the metabolic pathway). Note how the region at the active site closes around glucose after binding. Frequently, the geometries of both the substrate and the active site are altered to fit each other.*

CHEMISTRY *in Action*

Pharmacokinetics

Chemical kinetics is very important in understanding the absorption, distribution, metabolism, and excretion of drugs in the body. In this sense, pharmacokinetics is the study of what the body does to a drug (as opposed to pharmacodynamics, which is the study of what a drug does to the body). Knowledge of the rates of drug absorption and distribution in the body is essential to achieving and maintaining proper dosages as well as understanding the mechanisms of action.

Drug concentrations are typically measured in blood plasma or urine at various times to give a drug concentration versus time plot. As the drug is absorbed into the bloodstream, it is distributed to the various tissues and organs in the body and simultaneously eliminated by a combination of excretion and metabolism (biotransformation), all of which occur at different rates depending on the drug. The sum of all these processes is the mechanism of drug delivery and distribution. Because the drug must be distributed between different organs and cross between aqueous (blood and urine) and lipid (fat) tissue, and because many biological processes involve enzymes, kinetic behavior that is zero order in the drug is much more common in pharmacokinetics than it is in homogeneous solution reaction kinetics. For example, the decomposition of ethanol by the enzyme alcohol dehydrogenase is zero order in ethanol.

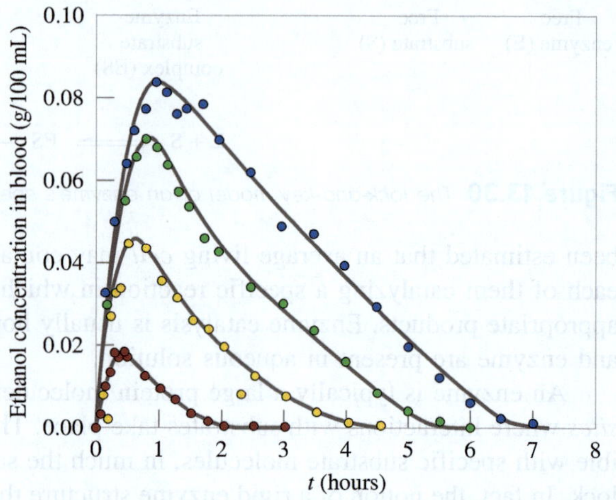

Concentration of ethanol in blood versus time after oral administration of various doses: red (14 g), yellow (28 g), green (42 g), blue (56 g).

Usually the absorption of a drug is more rapid than the elimination, giving a steeper rise in concentration compared to the more gradual decline. The *minimum effective concentration* (MEC) is the minimum concentration required for the drug to provide the desired therapeutic effect. The *minimum toxic concentration* (MTC) is the drug concentration at which

In general, the rate of such a reaction is given by the equation

$$\text{rate} = \frac{\Delta[\text{P}]}{\Delta t} = k_2[\text{ES}]$$

The concentration of the ES intermediate is itself proportional to the amount of the substrate present, and a plot of the rate versus the concentration of substrate typically yields a curve like that shown in Figure 13.32. Initially the rate rises rapidly with increasing substrate concentration. However, above a certain concentration, all the active sites are occupied, and the reaction becomes zero order in the substrate. In other words, the rate remains the same even though the substrate concentration increases. At and beyond this point, the rate of formation of product depends only on how fast the ES intermediate breaks down, not on the number of substrate molecules present.

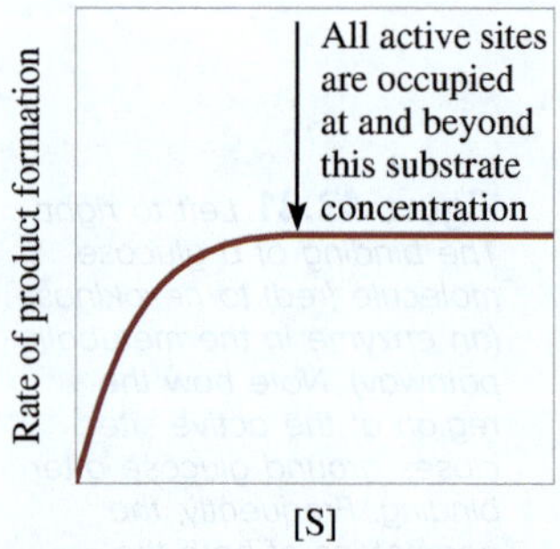

Figure 13.32 *Plot of the rate of product formation versus substrate concentration in an enzyme-catalyzed reaction.*

Summary of Concepts & Facts

- A catalyst speeds up a reaction usually by lowering the value of E_a. A catalyst can be recovered unchanged at the end of a reaction.

the drug becomes toxic or other undesired side effects outweigh the benefit of the drug. Taken together, the MEC and MTC define a *therapeutic index,* and one of the goals of pharmacokinetics is to determine a dosing regimen that keeps the drug concentration within the therapeutic index (that is, above the MEC but below the MTC). For example, most antibiotics such as amoxicillin have a fairly wide therapeutic index, but anticoagulant (blood thinner) medications such as Coumadin have a narrow therapeutic index. Determination of the correct dosage is based on the kinetics of the drug's delivery as well as the rate of disappearance due to decomposition, biotransformation, and excretion. Often this dosage will depend on the body weight of the person, because blood volume will be roughly proportional to body weight and the drug concentration will depend on the volume of distribution (blood in this case) as well as the amount of drug administered. Doctors and nurses refer to dosage charts in references such as the *Physician's Desk Reference* (PDR), which are based on pharmacokinetic studies.

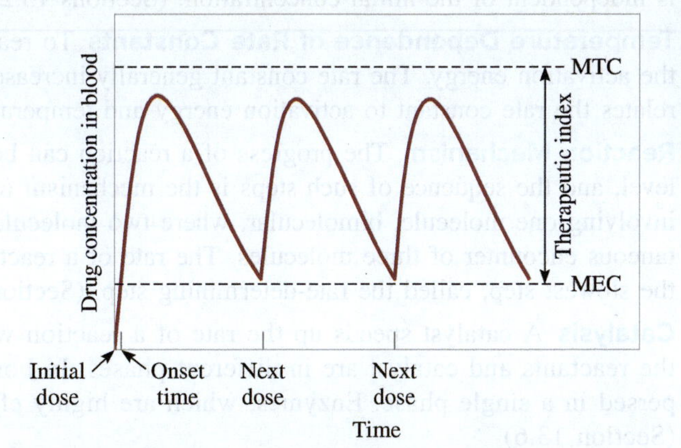

Drug concentration in blood as a function of time. The concentration of the drug increases after the drug is administered, and then decreases as the drug is metabolized and excreted. This process is repeated when the next dose is administered, giving the plot its characteristic "sawtooth" shape.

The *onset time* is the time required after the drug is administered for the concentration to reach the MEC and enter the therapeutic range. Sometimes a doctor will prescribe a higher first dose of the drug (*loading dose*) to reduce the onset time. After that, the drug must be administered at intervals to keep the concentration within the therapeutic index, giving a characteristic "sawtooth" profile of drug concentration versus time. The dosage is continued for as long as the drug is needed; however, physiological adaptations to the drug may require an adjustment in the regimen. For some drugs (for example, certain steroids), the drug dosage is tapered off rather than abruptly stopped to avoid shock to the system.

- In heterogeneous catalysis, which is of great industrial importance, the catalyst is a solid and the reactants are gases or liquids. In homogeneous catalysis, the catalyst and the reactants are in the same phase. Enzymes are catalysts in living systems.

Review of Concepts & Facts

13.6.1 Which of the following is false regarding catalysis? (a) E_a is lower for a catalyzed reaction. (b) ΔH°_{rxn} is lower for a catalyzed reaction. (c) A catalyzed reaction has a different mechanism.

Chapter Summary

Rate of a Reaction The rate of a reaction measures how fast a reactant is consumed or how fast a product is formed. The rate is expressed as a ratio of the change in concentration to elapsed time. (Section 13.1)

Rate Laws Experimental measurement of the rate leads to the rate law for the reaction, which expresses the rate in terms of the rate constant and the concentrations of the reactants. The dependence of rate on concentrations gives the order of a reaction. A reaction can be described as zero order if the rate does not depend on the concentration of the reactant, or first order if it depends on the reactant raised to the first power. Higher orders and fractional orders are also known. An important characteristic of reaction rates is the time required for the concentration of a reactant to decrease to half of its initial concentration, called the half-life. For first-order reactions, the half-life is independent of the initial concentration. (Sections 13.2, 13.3)

Temperature Dependence of Rate Constants To react, molecules must possess energy equal to or greater than the activation energy. The rate constant generally increases with increasing temperature. The Arrhenius equation relates the rate constant to activation energy and temperature. (Section 13.4)

Reaction Mechanism The progress of a reaction can be broken into a series of elementary steps at the molecular level, and the sequence of such steps is the mechanism of the reaction. Elementary steps can be unimolecular, involving one molecule; bimolecular, where two molecules react; or in rare cases, termolecular, involving the simultaneous encounter of three molecules. The rate of a reaction having more than one elementary step is governed by the slowest step, called the rate-determining step. (Section 13.5)

Catalysis A catalyst speeds up the rate of a reaction without itself being consumed. In heterogeneous catalysis, the reactants and catalyst are in different phases. In homogeneous catalysis, the reactants and catalyst are dispersed in a single phase. Enzymes, which are highly efficient catalysts, play a central role in all living systems. (Section 13.6)

Key Equations

$\text{rate} = k[A]^x[B]^y$ (13.1)	Rate law expressions. The sum $(x + y)$ gives the overall order of the reaction.
$\ln \frac{[A]_t}{[A]_0} = -kt$ (13.3)	Relationship between concentration and time for a first-order reaction.
$\ln [A]_t = -kt + \ln [A]_0$ (13.4)	Equation for the graphical determination of k for a first-order reaction.
$t_{\frac{1}{2}} = \frac{0.693}{k}$ (13.6)	Half-life for a first-order reaction.
$\frac{1}{[A]_t} = kt + \frac{1}{[A]_0}$ (13.7)	Relationship between concentration and time for a second-order reaction.
$k = Ae^{-E_a/RT}$ (13.11)	The Arrhenius equation expressing the dependence of the rate constant on activation energy and temperature.
$\ln k = \left(-\frac{E_a}{R}\right)\left(\frac{1}{T}\right) + \ln A$ (13.13)	Equation for the graphical determination of activation energy.
$\ln \frac{k_1}{k_2} = \frac{E_a}{R}\left(\frac{T_1 - T_2}{T_1T_2}\right)$ (13.14)	Relationships of rate constants at two different temperatures.

Key Words

Activated complex, p. 594
Activation energy (E_a), p. 594
Bimolecular reaction, p. 601
Catalyst, p. 607
Chemical kinetics, p. 568
Elementary steps, p. 600
Enzyme, p. 612
First-order reaction, p. 580
Half-life ($t_{\frac{1}{2}}$), p. 585
Intermediate, p. 601
Molecularity of a reaction, p. 601
Rate constant (k), p. 571
Rate-determining step, p. 602
Rate law, p. 576
Reaction mechanism, p. 600
Reaction order, p. 576
Reaction rate, p. 568
Second-order reaction, p. 587
Termolecular reaction, p. 601
Transition state, p. 594
Unimolecular reaction, p. 601

Questions & Problems

Red numbered problems solved in Student Solutions Manual

13.1 The Rate of a Reaction

Review Questions

13.1 What is meant by the rate of a chemical reaction? What are the units of the rate of a reaction?

13.2 Distinguish between average rate and instantaneous rate. Which of the two rates gives us an unambiguous measurement of reaction rate? Why?

13.3 What are the advantages of measuring the initial rate of a reaction?

13.4 Can you suggest two reactions that are very slow (take days or longer to complete) and two reactions that are very fast (reactions that are over in minutes or seconds)?

Problems

13.5 Write the reaction rate expressions for the following reactions in terms of the disappearance of the reactants and the appearance of products:
(a) $H_2(g) + I_2(g) \longrightarrow 2HI(g)$
(b) $5Br^-(aq) + BrO_3^-(aq) + 6H^+(aq) \longrightarrow 3Br_2(aq) + 3H_2O(l)$

13.6 Write the reaction rate expressions for the following reactions in terms of the disappearance of the reactants and the appearance of products:
(a) $2H_2(g) + O_2(g) \longrightarrow 2H_2O(g)$
(b) $4NH_3(g) + 5O_2(g) \longrightarrow 4NO(g) + 6H_2O(g)$

13.7 Consider the reaction

$$2NO(g) + O_2(g) \longrightarrow 2NO_2(g)$$

Suppose that at a particular moment during the reaction nitric oxide (NO) is reacting at the rate of 0.066 *M*/s. (a) At what rate is NO_2 being formed? (b) At what rate is molecular oxygen reacting?

13.8 Consider the reaction

$$N_2(g) + 3H_2(g) \longrightarrow 2NH_3(g)$$

Suppose that at a particular moment during the reaction molecular hydrogen is reacting at the rate of 0.074 *M*/s. (a) At what rate is ammonia being formed? (b) At what rate is molecular nitrogen reacting?

13.2 Rate Laws

Review Questions

13.9 Explain what is meant by the rate law of a reaction.

13.10 What are the units for the rate constants of zero-order, first-order, and second-order reactions?

13.11 Consider the zero-order reaction: A $\longrightarrow$ product. (a) Write the rate law for the reaction. (b) What are the units for the rate constant? (c) Plot the rate of the reaction versus [A].

13.12 On which of the following properties does the rate constant of a reaction depend: (a) reactant concentrations, (b) nature of reactants, or (c) temperature?

Problems

13.13 The rate law for the reaction

$$NH_4^+(aq) + NO_2^-(aq) \longrightarrow N_2(g) + 2H_2O(l)$$

is given by rate = $k[NH_4^+][NO_2^-]$. At 25°C, the rate constant is $3.0 \times 10^{-4}/M \cdot s$. Calculate the rate of the reaction at this temperature if $[NH_4^+] = 0.26\ M$ and $[NO_2^-] = 0.080\ M$.

13.14 Use the data in Table 13.2 to calculate the rate of the reaction at the time when $[F_2] = 0.010\ M$ and $[ClO_2] = 0.020\ M$.

13.15 Consider the reaction

$$A + B \longrightarrow \text{products}$$

From the following data obtained at a certain temperature, determine the order of the reaction and calculate the rate constant.

[A] (*M*)	[B] (*M*)	Rate (*M*/s)
1.50	1.50	3.20×10^{-1}
1.50	2.50	3.20×10^{-1}
3.00	1.50	6.40×10^{-1}

13.16 Consider the reaction

$$X + Y \longrightarrow Z$$

From the following data, obtained at 360 K, (a) determine the order of the reaction, and (b) determine the initial rate of disappearance of X when the concentration of X is 0.30 *M* and that of Y is 0.40 *M*.

Initial Rate of Disappearance of X (*M*/s)	[X] (*M*)	[Y] (*M*)
0.053	0.10	0.50
0.127	0.20	0.30
1.02	0.40	0.60
0.254	0.20	0.60
0.509	0.40	0.30

13.17 Determine the overall orders of the reactions to which the following rate laws apply: (a) rate = $k[NO_2]^2$, (b) rate = k, (c) rate = $k[H_2][Br_2]^{\frac{1}{2}}$, (d) rate = $k[NO]^2[O_2]$.

13.18 Consider the reaction

$$A \longrightarrow B$$

The rate of the reaction is 1.6×10^{-2} *M*/s when the concentration of A is 0.35 *M*. Calculate the rate constant if the reaction is (a) first order in A, and (b) second order in A.

13.19 Cyclobutane decomposes to ethylene according to the equation

$$C_4H_8(g) \longrightarrow 2C_2H_4(g)$$

Determine the order of the reaction and the rate constant based on the following pressures, which were recorded when the reaction was carried out at 430°C in a constant-volume vessel.

Time (s)	$P_{C_4H_8}$ (mmHg)
0	400
2,000	316
4,000	248
6,000	196
8,000	155
10,000	122

13.20 The following gas-phase reaction was studied at 290°C by observing the change in pressure as a function of time in a constant-volume vessel:

$$ClCO_2CCl_3(g) \longrightarrow 2COCl_2(g)$$

Determine the order of the reaction and the rate constant based on the following data:

Time (s)	*P* (mmHg)
0	15.76
181	18.88
513	22.79
1164	27.08

where *P* is the total pressure.

13.3 The Relation Between Reactant Concentration and Time

Review Questions

13.21 Write an equation relating the concentration of a reactant A at $t = 0$ to that at $t = t$ for a first-order reaction. Define all the terms and give their units. Do the same for a second-order reaction.

13.22 Define half-life. Write the equation relating the half-life of a first-order reaction to the rate constant.

13.23 Write the equations relating the half-life of a second-order reaction to the rate constant. How does it differ from the equation for a first-order reaction?

13.24 For a first-order reaction, how long will it take for the concentration of reactant to fall to one-eighth its original value? Express your answer in terms of the half-life ($t_{\frac{1}{2}}$) and in terms of the rate constant *k*.

Problems

13.25 What is the half-life of a compound if 75 percent of a given sample of the compound decomposes in 60 min? Assume first-order kinetics.

13.26 The thermal decomposition of phosphine (PH_3) into phosphorus and molecular hydrogen is a first-order reaction:

$$4PH_3(g) \longrightarrow P_4(g) + 6H_2(g)$$

The half-life of the reaction is 35.0 s at 680°C. Calculate (a) the first-order rate constant for the reaction, and (b) the time required for 95 percent of the phosphine to decompose.

13.27 The rate constant for the second-order reaction

$$2NOBr(g) \longrightarrow 2NO(g) + Br_2(g)$$

is $0.80/M \cdot s$ at 10°C. (a) Starting with a concentration of 0.086 *M*, calculate the concentration of NOBr after 22 s. (b) Calculate the half-lives when $[NOBr]_0 = 0.072\ M$ and $[NOBr]_0 = 0.054\ M$.

13.28 The rate constant for the second-order reaction

$$2NO_2(g) \longrightarrow 2NO(g) + O_2(g)$$

is $0.54/M \cdot s$ at 300°C. How long (in seconds) would it take for the concentration of NO_2 to decrease from 0.62 *M* to 0.28 *M*?

13.29 Consider the first-order reaction $A \longrightarrow B$ shown here. (a) What is the rate constant of the reaction? (b) How many A (yellow) and B (blue) molecules are present at $t = 20$ s and 30 s?

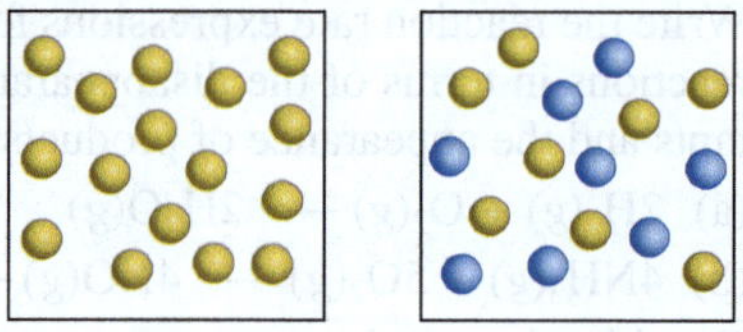

$t = 0$ s $t = 10$ s

13.30 The reaction $X \longrightarrow Y$ shown here follows first-order kinetics. Initially different amounts of X molecules are placed in three equal-volume containers at the same temperature. (a) What are the relative rates of the reaction in these three containers? (b) How would the relative rates be affected if the volume of each container were doubled? (c) What are the relative half-lives of the reactions in (i) to (iii)?

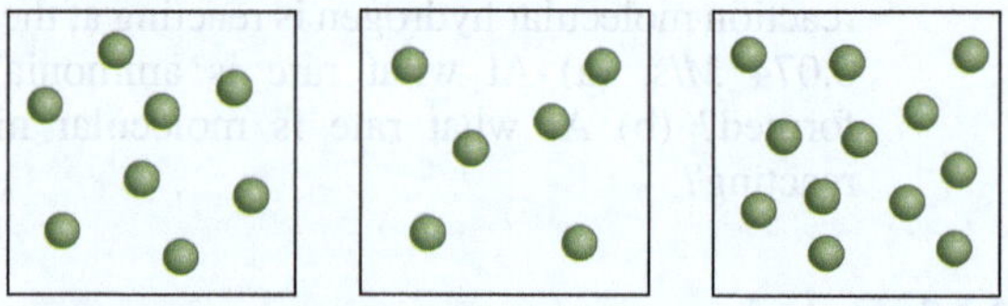

(i) (ii) (iii)

13.4 Activation Energy and Temperature Dependence of Rate Constants

Review Questions

13.31 Define activation energy. What role does activation energy play in chemical kinetics?

13.32 Write the Arrhenius equation and define all terms.

13.33 Use the Arrhenius equation to show why the rate constant of a reaction (a) decreases with increasing activation energy and (b) increases with increasing temperature.

13.34 The burning of methane in oxygen is a highly exothermic reaction. Yet a mixture of methane and oxygen gas can be kept indefinitely without any apparent change. Explain.

13.35 Sketch a potential energy versus reaction progress plot for the following reactions:
(a) $S(s) + O_2(g) \longrightarrow SO_2(g)$ $\Delta H° = -296$ kJ/mol
(b) $Cl_2(g) \longrightarrow Cl(g) + Cl(g)$ $\Delta H° = 243$ kJ/mol

13.36 The reaction $H + H_2 \longrightarrow H_2 + H$ has been studied for many years. Sketch a potential energy versus reaction progress diagram for this reaction.

Problems

13.37 (1) Diagram (a) shows the plots of ln k versus $1/T$ for two first-order reactions, where k is the rate constant and T is the absolute temperature. Which reaction has a greater activation energy? (2) Diagram (b) shows the plots for a first-order reaction at two different temperatures. Which plot corresponds to a higher temperature?

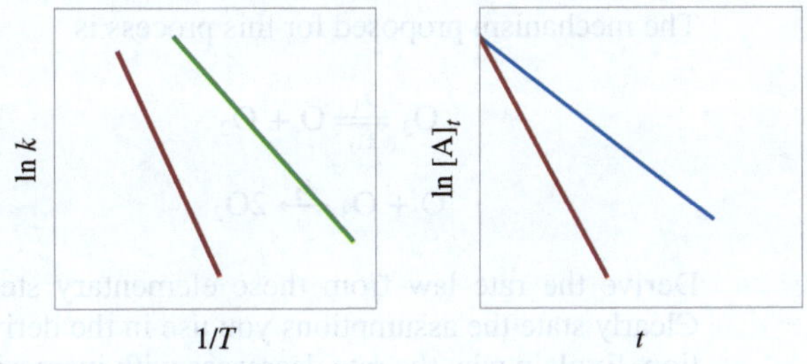

(a) (b)

13.38 Given the same reactant concentrations, the reaction

$$CO(g) + Cl_2(g) \longrightarrow COCl_2(g)$$

at 250°C is 1.50×10^3 times as fast as the same reaction at 150°C. Calculate the activation energy for this reaction. Assume that the frequency factor is constant.

13.39 Some reactions are described as parallel in that the reactant simultaneously forms different products with different rate constants. An example is

$$A \xrightarrow{k_1} B$$

and

$$A \xrightarrow{k_2} C$$

The activation energies are 45.3 kJ/mol for k_1 and 69.8 kJ/mol for k_2. If the rate constants are equal at 320 K, at what temperature will $k_1/k_2 = 2.00$?

13.40 Variation of the rate constant with temperature for the first-order reaction

$$2N_2O_5(g) \longrightarrow 2N_2O_4(g) + O_2(g)$$

is given in the following table. Determine graphically the activation energy for the reaction.

***T* (K)**	***k* (s^{-1})**
298	1.74×10^{-5}
308	6.61×10^{-5}
318	2.51×10^{-4}
328	7.59×10^{-4}
338	2.40×10^{-3}

13.41 For the reaction

$$NO(g) + O_3(g) \longrightarrow NO_2(g) + O_2(g)$$

the frequency factor A is 8.7×10^{12} s^{-1} and the activation energy is 63 kJ/mol. What is the rate constant for the reaction at 75°C?

13.42 The rate constant of a first-order reaction is 4.60×10^{-4} s^{-1} at 350°C. If the activation energy is 104 kJ/mol, calculate the temperature at which its rate constant is 8.80×10^{-4} s^{-1}.

13.43 The rate constants of some reactions double with every 10-degree rise in temperature. Assume that a reaction takes place at 295 K and 305 K. What must the activation energy be for the rate constant to double as described?

13.44 Consider the first-order reaction

$$CH_3NCl(g) \longrightarrow CH_3CN(g)$$

Given that the frequency factor and activation energy for the reaction are 3.98×10^{13} s^{-1} and 161 kJ/mol, respectively, calculate the rate constant at 600°C.

13.45 Consider the second-order reaction

$$NO(g) + Cl_2(g) \longrightarrow NOCl(g) + Cl(g)$$

Given that the frequency factor and activation energy for the reaction are $4.0 \times 10^9/M \cdot s$ and 85 kJ/mol, respectively, calculate the rate constant at 500°C.

13.46 The rate at which tree crickets chirp is 2.0×10^2 per minute at 27°C but only 39.6 per minute at 5°C. From these data, calculate the "activation energy" for the chirping process. (*Hint:* The ratio of rates is equal to the ratio of rate constants.)

13.47 The diagram here describes the initial state of the reaction $A_2 + B_2 \longrightarrow 2AB$.

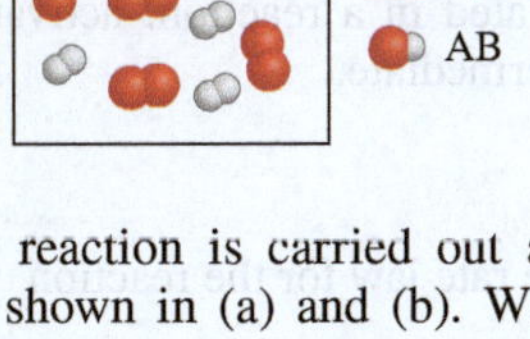

Suppose the reaction is carried out at two temperatures as shown in (a) and (b). Which picture represents the result at the higher temperature? (The reaction proceeds for the same amount of time at both temperatures.)

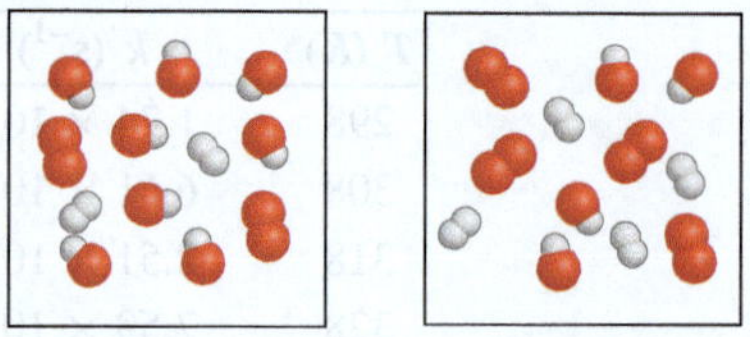

(a) (b)

13.5 Reaction Mechanisms

Review Questions

13.48 What do we mean by the mechanism of a reaction? What is an elementary step? What is the molecularity of a reaction?

13.49 Classify each of the following elementary steps as unimolecular, bimolecular, or termolecular.

(a)

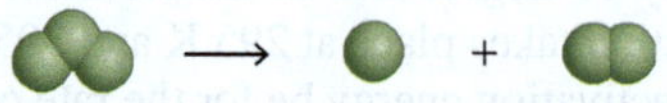

(b)

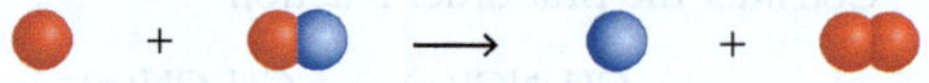

(c)

13.50 Reactions can be classified as unimolecular, bimolecular, and so on. Why are there no zero-molecular reactions? Explain why termolecular reactions are rare.

13.51 Determine the molecularity and write the rate law for each of the following elementary steps:

(a) $X \longrightarrow$ products

(b) $X + Y \longrightarrow$ products

(c) $X + Y + Z \longrightarrow$ products

(d) $X + X \longrightarrow$ products

(e) $X + 2Y \longrightarrow$ products

13.52 What is the rate-determining step of a reaction? Give an everyday analogy to illustrate the meaning of "rate determining."

13.53 The equation for the combustion of ethane (C_2H_6) is

$$2C_2H_6(g) + 7O_2(g) \longrightarrow 4CO_2(g) + 6H_2O(l)$$

Explain why it is unlikely that this equation also represents the elementary step for the reaction.

13.54 Specify which of the following species cannot be isolated in a reaction: activated complex, product, intermediate.

Problems

13.55 The rate law for the reaction

$$2NO(g) + Cl_2(g) \longrightarrow 2NOCl_2(g)$$

is given by rate $= k[NO][Cl_2]$. (a) What is the order of the reaction? (b) A mechanism involving the following steps has been proposed for the reaction:

$$NO(g) + Cl_2(g) \longrightarrow NOCl_2(g)$$
$$NOCl_2(g) + NO(g) \longrightarrow 2NOCl(g)$$

If this mechanism is correct, what does it imply about the relative rates of these two steps?

13.56 For the reaction $X_2 + Y + Z \longrightarrow XY + XZ$ it is found that doubling the concentration of X_2 doubles the reaction rate, tripling the concentration of Y triples the rate, and doubling the concentration of Z has no effect. (a) What is the rate law for this reaction? (b) Why is it that the change in the concentration of Z has no effect on the rate? (c) Suggest a mechanism for the reaction that is consistent with the rate law.

13.57 The rate law for the decomposition of ozone to molecular oxygen

$$2O_3(g) \longrightarrow 3O_2(g)$$

is

$$\text{rate} = k\frac{[O_3]^2}{[O_2]}$$

The mechanism proposed for this process is

$$O_3 \underset{k_{-1}}{\overset{k_1}{\rightleftharpoons}} O + O_2$$

$$O + O_3 \xrightarrow{k_2} 2O_2$$

Derive the rate law from these elementary steps. Clearly state the assumptions you use in the derivation. Explain why the rate decreases with increasing O_2 concentration.

13.58 The rate law for the reaction

$$2H_2(g) + 2NO(g) \longrightarrow N_2(g) + 2H_2O(g)$$

is rate $= k[H_2][NO]^2$. Which of the following mechanisms can be ruled out on the basis of the observed rate expression?

Mechanism I

$$H_2 + NO \longrightarrow H_2O + N \quad \text{(slow)}$$
$$N + NO \longrightarrow N_2 + O \quad \text{(fast)}$$
$$O + H_2 \longrightarrow H_2O \quad \text{(fast)}$$

Mechanism II

$$H_2 + 2NO \longrightarrow N_2O + H_2O \quad \text{(slow)}$$
$$N_2O + H_2 \longrightarrow N_2 + H_2O \quad \text{(fast)}$$

Mechanism III

$$2NO \rightleftharpoons N_2O_2 \quad \text{(fast equilibrium)}$$
$$N_2O_2 + H_2 \longrightarrow N_2O + H_2O \quad \text{(slow)}$$
$$N_2O + H_2 \longrightarrow N_2 + H_2O \quad \text{(fast)}$$

13.6 Catalysis

Review Questions

13.59 How does a catalyst increase the rate of a reaction?

13.60 What are the characteristics of a catalyst?

13.61 A certain reaction is known to proceed slowly at room temperature. Is it possible to make the reaction proceed at a faster rate without changing the temperature?

13.62 Distinguish between homogeneous catalysis and heterogeneous catalysis. Describe three important industrial processes that utilize heterogeneous catalysis.

13.63 Are enzyme-catalyzed reactions examples of homogeneous or heterogeneous catalysis? Explain.

13.64 The concentrations of enzymes in cells are usually quite small. What is the biological significance of this fact?

Problems

13.65 The diagram shown here represents a two-step mechanism. (a) Write the equation for each step and the overall reaction. (b) Identify the intermediate and catalyst. The color codes are A = blue and B = red.

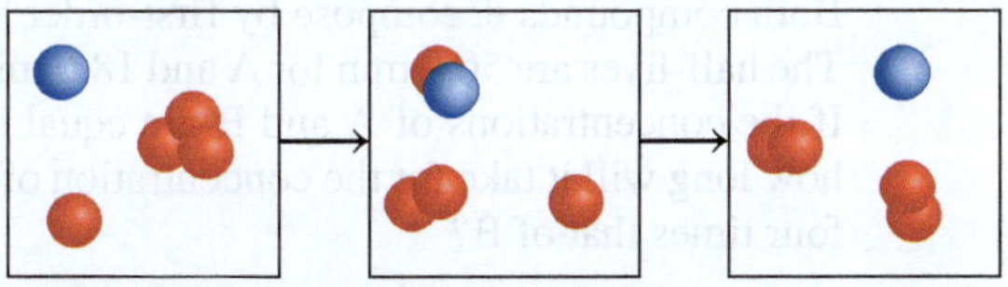

13.66 Consider the following mechanism for the enzyme-catalyzed reaction:

$$E + S \underset{k_{-1}}{\overset{k_1}{\rightleftharpoons}} ES \quad \text{(fast equilibrium)}$$

$$ES \xrightarrow{k_2} E + P \quad \text{(slow)}$$

Derive an expression for the rate law of the reaction in terms of the concentrations of E and S. (*Hint:* To solve for [ES], make use of the fact that, at equilibrium, the rate of forward reaction is equal to the rate of the reverse reaction.)

Additional Problems

13.67 The following diagrams represent the progress of the reaction A ⟶ B, where the red spheres represent A molecules and the blue spheres represent B molecules. Calculate the rate constant of the reaction.

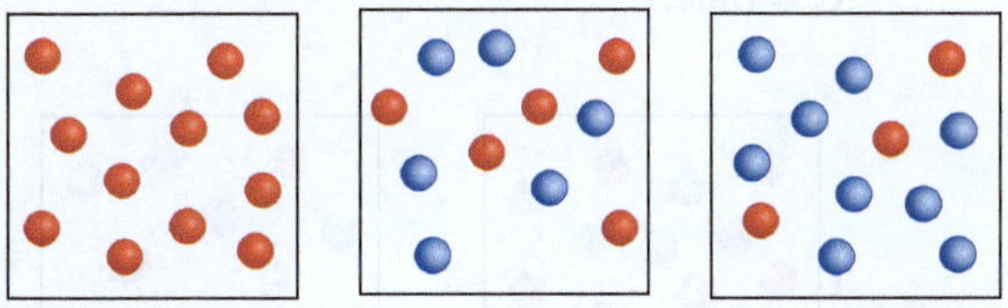

$t = 0$ s $\quad$ $t = 20$ s $\quad$ $t = 40$ s

13.68 The following diagrams show the progress of the reaction $2A \longrightarrow A_2$. Determine whether the reaction is first order or second order and calculate the rate constant.

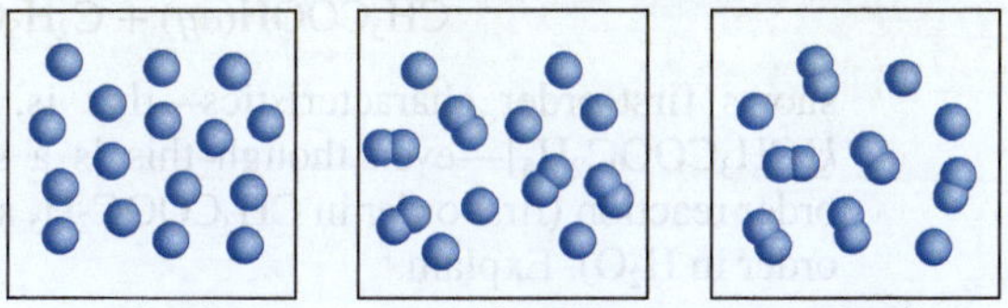

$t = 0$ min $\quad$ $t = 15$ min $\quad$ $t = 30$ min

13.69 Suggest experimental means by which the rates of the following reactions could be followed:

(a) $CaCO_3(s) \longrightarrow CaO(s) + CO_2(g)$

(b) $Cl_2(g) + 2Br^-(aq) \longrightarrow Br_2(aq) + 2Cl^-(aq)$

(c) $C_2H_6(g) \longrightarrow C_2H_4(g) + H_2(g)$

(d) $C_2H_5I(g) + H_2O(l) \longrightarrow C_2H_5OH(aq) + H^+(aq) + I^-(aq)$

13.70 List four factors that influence the rate of a reaction.

13.71 "The rate constant for the reaction

$$NO_2(g) + CO(g) \longrightarrow NO(g) + CO_2(g)$$

is $1.64 \times 10^{-6}/M \cdot s$." What is incomplete about this statement?

13.72 In a certain industrial process involving a heterogeneous catalyst, the volume of the catalyst (in the shape of a sphere) is 10.0 cm^3. Calculate the surface area of the catalyst. If the sphere is broken down into eight spheres, each having a volume of 1.25 cm^3, what is the total surface area of the spheres? Which of the two geometric configurations of the catalyst is more effective? (The surface area of a sphere is $4\pi r^2$, where r is the radius of the sphere.) Based on your analysis here, explain why it is sometimes dangerous to work in grain elevators.

13.73 Use the data in Example 13.5 to determine graphically the half-life of the reaction.

13.74 The following data were collected for the reaction between hydrogen and nitric oxide at 700°C:

$$2H_2(g) + 2NO(g) \longrightarrow 2H_2O(g) + N_2(g)$$

Experiment	$[H_2]$	[NO]	Initial Rate (M/s)
1	0.010	0.025	2.4×10^{-6}
2	0.0050	0.025	1.2×10^{-6}
3	0.010	0.0125	0.60×10^{-6}

(a) Determine the order of the reaction. (b) Calculate the rate constant. (c) Suggest a plausible mechanism that is consistent with the rate law. (*Hint:* Assume that the oxygen atom is the intermediate.)

13.75 When methyl phosphate is heated in acid solution, it reacts with water:

$$CH_3OPO_3H_2 + H_2O \longrightarrow CH_3OH + H_3PO_4$$

If the reaction is carried out in water enriched with ^{18}O, the oxygen-18 isotope is found in the phosphoric acid product but not in the methanol. What does this tell us about the mechanism of the reaction?

13.76 The rate of the reaction

$$CH_3COOC_2H_5(aq) + H_2O(l) \longrightarrow CH_3COOH(aq) + C_2H_5OH(aq)$$

shows first-order characteristics—that is, rate = $k[CH_3COOC_2H_5]$—even though this is a second-order reaction (first order in $CH_3COOC_2H_5$ and first order in H_2O). Explain.

13.77 Which of the following equations best describes the diagram shown: (a) A ⟶ B, (b) A ⟶ 3B, (c) 3A ⟶ B?

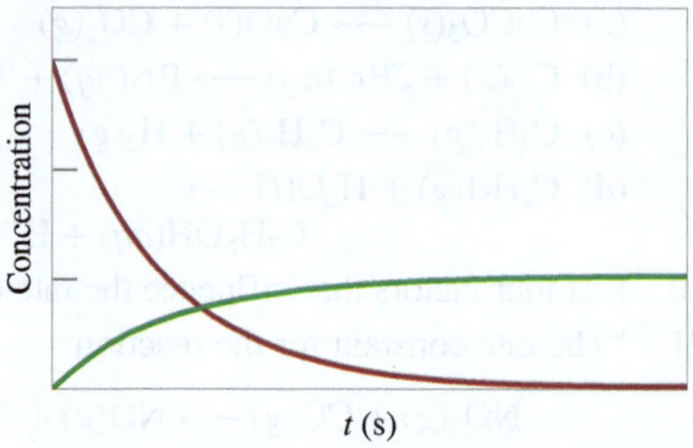

13.78 The reaction 2A + 3B ⟶ C is first order with respect to A and B. When the initial concentrations are $[A] = 1.6 \times 10^{-2}\ M$ and $[B] = 2.4 \times 10^{-3}\ M$, the rate is $4.1 \times 10^{-4}\ M/s$. Calculate the rate constant of the reaction.

13.79 The bromination of acetone is acid-catalyzed:

$$CH_3COCH_3 + Br_2 \xrightarrow[\text{catalyst}]{H^+} CH_3COCH_2Br + H^+ + Br^-$$

The rate of disappearance of bromine was measured for several different concentrations of acetone, bromine, and H^+ ions at a certain temperature:

	$[CH_3COCH_3]$	$[Br_2]$	$[H^+]$	Rate of Disappearance of Br_2 (M/s)
(1)	0.30	0.050	0.050	5.7×10^{-5}
(2)	0.30	0.10	0.050	5.7×10^{-5}
(3)	0.30	0.050	0.20	1.2×10^{-4}
(4)	0.40	0.050	0.20	3.1×10^{-4}
(5)	0.40	0.050	0.050	7.6×10^{-5}

(a) What is the rate law for the reaction? (b) Determine the rate constant. (c) The following mechanism has been proposed for the reaction:

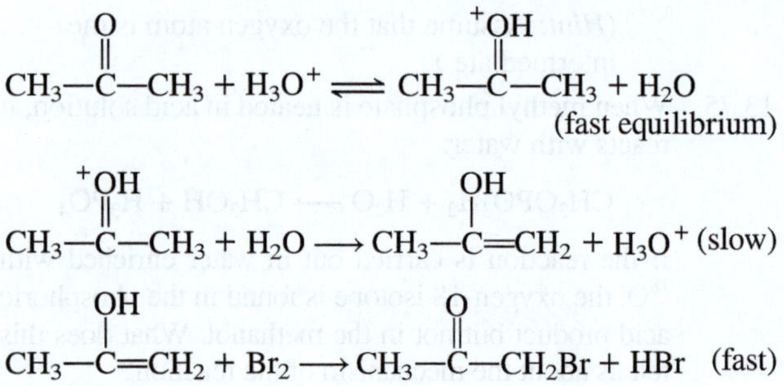

Show that the rate law deduced from the mechanism is consistent with that shown in (a).

13.80 The decomposition of N_2O to N_2 and O_2 is a first-order reaction. At 730°C the half-life of the reaction is 3.58×10^3 min. If the initial pressure of N_2O is 2.10 atm at 730°C, calculate the total gas pressure after one half-life. Assume that the volume remains constant.

13.81 The reaction $S_2O_8^{2-} + 2I^- \longrightarrow 2SO_4^{2-} + I_2$ proceeds slowly in aqueous solution, but it can be catalyzed by the Fe^{3+} ion. Given that Fe^{3+} can oxidize I^- and Fe^{2+} can reduce $S_2O_8^{2-}$, write a plausible two-step mechanism for this reaction. Explain why the uncatalyzed reaction is slow.

13.82 What are the units of the rate constant for a third-order reaction?

13.83 The integrated rate law for the zero-order reaction A ⟶ B is $[A]_t = [A]_0 - kt$. (a) Sketch the following plots: (i) rate versus $[A]_t$ and (ii) $[A]_t$ versus t. (b) Derive an expression for the half-life of the reaction. (c) Calculate the time in half-lives when the integrated rate law is no longer valid, that is, when $[A]_t = 0$.

13.84 A flask contains a mixture of compounds A and B. Both compounds decompose by first-order kinetics. The half-lives are 50.0 min for A and 18.0 min for B. If the concentrations of A and B are equal initially, how long will it take for the concentration of A to be four times that of B?

13.85 Shown here are plots of concentration of reactant versus time for two first-order reactions at the same temperature. In each case, determine which reaction has a greater rate constant.

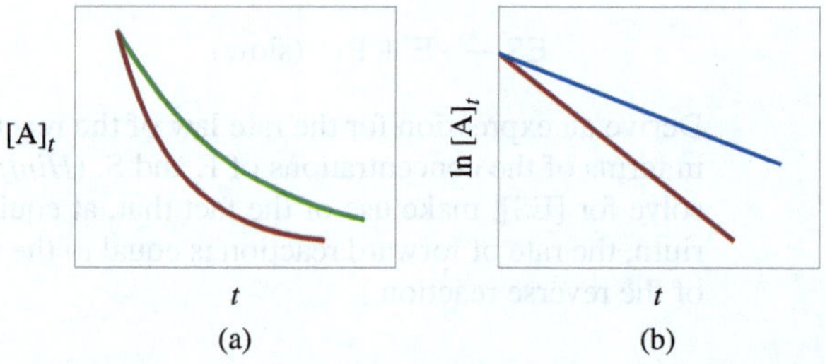

13.86 The diagrams here represent the reaction A + B ⟶ C carried out under different initial concentrations of A and B. Determine the rate law of the reaction. (The color codes are A = red, B = purple, C = blue.)

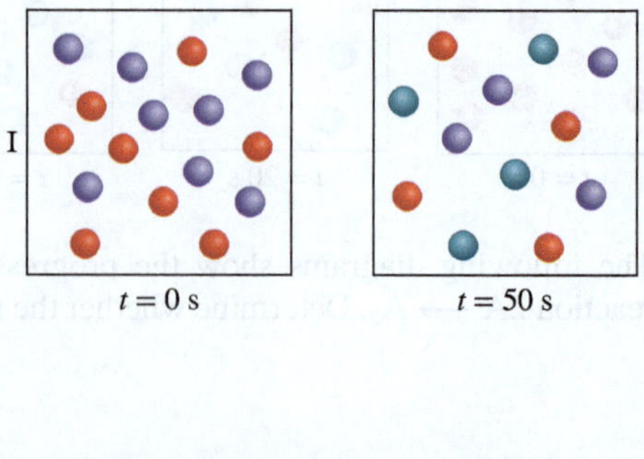

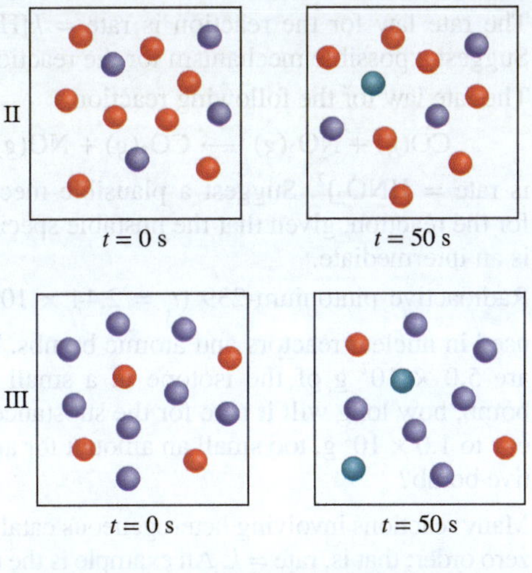

13.87 Referring to Example 13.5, explain how you would measure the partial pressure of azomethane experimentally as a function of time.

13.88 The rate law for the reaction $2NO_2(g) \longrightarrow N_2O_4(g)$ is rate $= k[NO_2]^2$. Which of the following changes will change the value of k? (a) The pressure of NO_2 is doubled. (b) The reaction is run in an organic solvent. (c) The volume of the container is doubled. (d) The temperature is decreased. (e) A catalyst is added to the container.

13.89 The reaction of G_2 with E_2 to form 2EG is exothermic, and the reaction of G_2 with X_2 to form 2XG is endothermic. The activation energy of the exothermic reaction is greater than that of the endothermic reaction. Sketch the potential energy profile diagrams for these two reactions on the same graph.

13.90 In the nuclear industry, workers use a rule of thumb that the radioactivity from any sample will be relatively harmless after 10 half-lives. Calculate the fraction of a radioactive sample that remains after this time period. (*Hint:* Radioactive decays obey first-order kinetics.)

13.91 Briefly comment on the effect of a catalyst on each of the following: (a) activation energy, (b) reaction mechanism, (c) enthalpy of reaction, (d) rate of forward step, (e) rate of reverse step.

13.92 When 6 g of granulated Zn is added to a solution of 2 M HCl in a beaker at room temperature, hydrogen gas is generated. For each of the following changes (at constant volume of the acid) state whether the rate of hydrogen gas evolution will be increased, decreased, or unchanged: (a) 6 g of powdered Zn is used; (b) 4 g of granulated Zn is used; (c) 2 M acetic acid is used instead of 2 M HCl; (d) temperature is raised to 40°C.

13.93 Strictly speaking, the rate law derived for the reaction in Problem 13.74 applies only to certain concentrations of H_2. The general rate law for the reaction takes the form

$$\text{rate} = \frac{k_1[NO]^2[H_2]}{1 + k_2[H_2]}$$

where k_1 and k_2 are constants. Derive rate law expressions under the conditions of very high and very low hydrogen concentrations. Does the result from Problem 13.74 agree with one of the rate expressions here?

13.94 A certain first-order reaction is 35.5 percent complete in 4.90 min at 25°C. What is its rate constant?

13.95 The decomposition of dinitrogen pentoxide has been studied in carbon tetrachloride solvent (CCl_4) at a certain temperature:

$$2N_2O_5 \longrightarrow 4NO_2 + O_2$$

$[N_2O_5]$	Initial Rate (M/s)
0.92	0.95×10^{-5}
1.23	1.20×10^{-5}
1.79	1.93×10^{-5}
2.00	2.10×10^{-5}
2.21	2.26×10^{-5}

Determine graphically the rate law for the reaction and calculate the rate constant.

13.96 The thermal decomposition of N_2O_5 obeys first-order kinetics. At 45°C, a plot of ln $[N_2O_5]$ versus t gives a slope of -6.18×10^{-4} min^{-1}. What is the half-life of the reaction?

13.97 When a mixture of methane and bromine is exposed to visible light, the following reaction occurs slowly:

$$CH_4(g) + Br_2(g) \longrightarrow CH_3Br(g) + HBr(g)$$

Suggest a reasonable mechanism for this reaction. (*Hint:* Bromine vapor is deep red; methane is colorless.)

13.98 The rate of the reaction between H_2 and I_2 to form HI (discussed in Section 13.5) increases with the intensity of visible light. (a) Explain why this fact supports the two-step mechanism given. (The color of I_2 vapor is shown in Section 13.5.) (b) Explain why the visible light has no effect on the formation of H atoms.

13.99 The carbon-14 decay rate of a sample obtained from a young tree is 0.260 disintegration per second per gram of the sample. Another wood sample prepared from an object recovered at an archaeological excavation gives a decay rate of 0.186 disintegration per second per gram of the sample. What is the age of the object? (*Hint:* See the Chemistry in Action essay "Radiocarbon Dating" in Section 13.3.)

13.100 Consider the following elementary step:

$$X + 2Y \longrightarrow XY_2$$

(a) Write a rate law for this reaction. (b) If the initial rate of formation of XY_2 is 3.8×10^{-3} M/s and the initial concentrations of X and Y are 0.26 M and 0.88 M, what is the rate constant of the reaction?

13.101 In recent years ozone in the stratosphere has been depleted at an alarmingly fast rate by chlorofluorocarbons

(CFCs). A CFC molecule such as $CFCl_3$ is first decomposed by UV radiation:

$$CFCl_3 \longrightarrow CFCl_2 + Cl$$

The chlorine radical then reacts with ozone as follows:

$$Cl + O_3 \longrightarrow ClO + O_2$$
$$ClO + O \longrightarrow Cl + O_2$$

The O atom is from the photochemical decomposition of O_2 molecules. (a) Write the overall reaction for the last two steps. (b) What are the roles of Cl and ClO? (c) Why is the fluorine radical not important in this mechanism? (d) One suggestion to reduce the concentration of chlorine radicals is to add hydrocarbons such as ethane (C_2H_6) to the stratosphere. How will this work? (e) Draw potential energy versus reaction progress diagrams for the uncatalyzed and catalyzed (by Cl) destruction of ozone: $O_3 + O \longrightarrow 2O_2$ Use the thermodynamic data in Appendix 2 to determine whether the reaction is exothermic or endothermic.

13.102 Chlorine oxide (ClO), which plays an important role in the depletion of ozone (see Problem 13.101), decays rapidly at room temperature according to the equation

$$2ClO(g) \longrightarrow Cl_2(g) + O_2(g)$$

From the following data, determine the reaction order and calculate the rate constant of the reaction.

Time (s)	[ClO] (*M*)
0.12×10^{-3}	8.49×10^{-6}
0.96×10^{-3}	7.10×10^{-6}
2.24×10^{-3}	5.79×10^{-6}
3.20×10^{-3}	5.20×10^{-6}
4.00×10^{-3}	4.77×10^{-6}

13.103 A compound X undergoes two *simultaneous* first-order reactions as follows: $X \longrightarrow Y$ with rate constant k_1 and $X \longrightarrow Z$ with rate constant k_2. The ratio of k_1/k_2 at 40°C is 8.0. What is the ratio at 300°C? Assume that the frequency factors of the two reactions are the same.

13.104 Consider a car fitted with a catalytic converter. The first 5 minutes or so after it is started are the most polluting. Why?

13.105 The following scheme in which A is converted to B, which is then converted to C is known as a consecutive reaction.

$$A \longrightarrow B \longrightarrow C$$

Assuming that both steps are first order, sketch on the same graph the variations of [A], [B], and [C] with time.

13.106 Hydrogen and iodine monochloride react as follows:

$$H_2(g) + 2ICl(g) \longrightarrow 2HCl(g) + I_2(g)$$

The rate law for the reaction is rate $= k[H_2][ICl]$. Suggest a possible mechanism for the reaction.

13.107 The rate law for the following reaction

$$CO(g) + NO_2(g) \longrightarrow CO_2(g) + NO(g)$$

is rate $= k[NO_2]^2$. Suggest a plausible mechanism for the reaction, given that the unstable species NO_3 is an intermediate.

13.108 Radioactive plutonium-239 ($t_{\frac{1}{2}} = 2.44 \times 10^5$ yr) is used in nuclear reactors and atomic bombs. If there are 5.0×10^2 g of the isotope in a small atomic bomb, how long will it take for the substance to decay to 1.0×10^2 g, too small an amount for an effective bomb?

13.109 Many reactions involving heterogeneous catalysts are zero order; that is, rate $= k$. An example is the decomposition of phosphine (PH_3) over tungsten (W):

$$4PH_3(g) \longrightarrow P_4(g) + 6H_2(g)$$

It is found that the reaction is independent of $[PH_3]$ as long as phosphine's pressure is sufficiently high (≥ 1 atm). Explain.

13.110 Thallium(I) is oxidized by cerium(IV) as follows:

$$Tl^+ + 2Ce^{4+} \longrightarrow Tl^{3+} + 2Ce^{3+}$$

The elementary steps, in the presence of Mn(II), are as follows:

$$Ce^{4+} + Mn^{2+} \longrightarrow Ce^{3+} + Mn^{3+}$$
$$Ce^{4+} + Mn^{3+} \longrightarrow Ce^{3+} + Mn^{4+}$$
$$Tl^+ + Mn^{4+} \longrightarrow Tl^{3+} + Mn^{2+}$$

(a) Identify the catalyst, intermediates, and the rate-determining step if the rate law is rate $= k[Ce^{4+}][Mn^{2+}]$. (b) Explain why the reaction is slow without the catalyst. (c) Classify the type of catalysis (homogeneous or heterogeneous).

13.111 Sucrose ($C_{12}H_{22}O_{11}$), commonly called table sugar, undergoes hydrolysis (reaction with water) to produce fructose ($C_6H_{12}O_6$) and glucose ($C_6H_{12}O_6$):

$$C_{12}H_{22}O_{11} + H_2O \longrightarrow \underset{\text{fructose}}{C_6H_{12}O_6} + \underset{\text{glucose}}{C_6H_{12}O_6}$$

This reaction is of considerable importance in the candy industry. First, fructose is sweeter than sucrose. Second, a mixture of fructose and glucose, called *invert sugar,* does not crystallize, so the candy containing this sugar would be chewy rather than brittle as candy containing sucrose crystals would be. (a) From the following data determine the order of the reaction. (b) How long does it take to hydrolyze 95 percent of sucrose? (c) Explain why the rate law does not include $[H_2O]$ even though water is a reactant.

Time (min)	$[C_{12}H_{22}O_{11}]$
0	0.500
60.0	0.400
96.4	0.350
157.5	0.280

13.112 The first-order rate constant for the decomposition of dimethyl ether

$$(CH_3)_2O(g) \longrightarrow CH_4(g) + H_2(g) + CO(g)$$

is $3.2 \times 10^{-4}\ s^{-1}$ at 450°C. The reaction is carried out in a constant-volume flask. Initially only dimethyl ether is present and the pressure is 0.350 atm. What is the pressure of the system after 8.0 min? Assume ideal behavior.

13.113 At 25°C, the rate constant for the ozone-depleting reaction $O(g) + O_3(g) \longrightarrow 2O_2(g)$ is 7.9×10^{-15} cm^3/molecule · s. Express the rate constant in units of $1/M \cdot s$.

13.114 Consider the following elementary steps for a consecutive reaction:

$$A \xrightarrow{k_1} B \xrightarrow{k_2} C$$

(a) Write an expression for the rate of change of B. (b) Derive an expression for the concentration of B under steady-state conditions; that is, when B is decomposing to C at the same rate as it is formed from A.

13.115 Ethanol is a toxic substance that, when consumed in excess, can impair respiratory and cardiac functions by interference with the neurotransmitters of the nervous system. In the human body, ethanol is metabolized by the enzyme alcohol dehydrogenase to acetaldehyde, which causes "hangovers." (a) Based on your knowledge of enzyme kinetics, explain why binge drinking (that is, consuming too much alcohol too fast) can prove fatal. (b) Methanol is even more toxic than ethanol. It is also metabolized by alcohol dehydrogenase, and the product, formaldehyde, can cause blindness or death. An antidote to methanol poisoning is ethanol. Explain how this procedure works.

13.116 Strontium-90, a radioactive isotope, is a major product of an atomic bomb explosion. It has a half-life of 28.1 yr. (a) Calculate the first-order rate constant for the nuclear decay. (b) Calculate the fraction of ^{90}Sr that remains after 10 half-lives. (c) Calculate the number of years required for 99.0 percent of ^{90}Sr to disappear.

13.117 Consider the potential energy profiles for the following three reactions (from left to right). (1) Rank the rates (slowest to fastest) of the reactions. (2) Calculate ΔH for each reaction and determine which reaction(s) are exothermic and which reaction(s) are endothermic. Assume the reactions have roughly the same frequency factors.

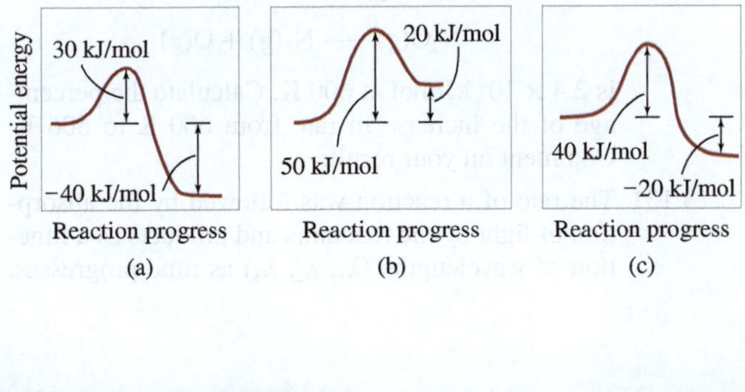

13.118 Consider the following potential energy profile for the $A \longrightarrow D$ reaction. (a) How many elementary steps are there? (b) How many intermediates are formed? (c) Which step is rate-determining? (d) Is the overall reaction exothermic or endothermic?

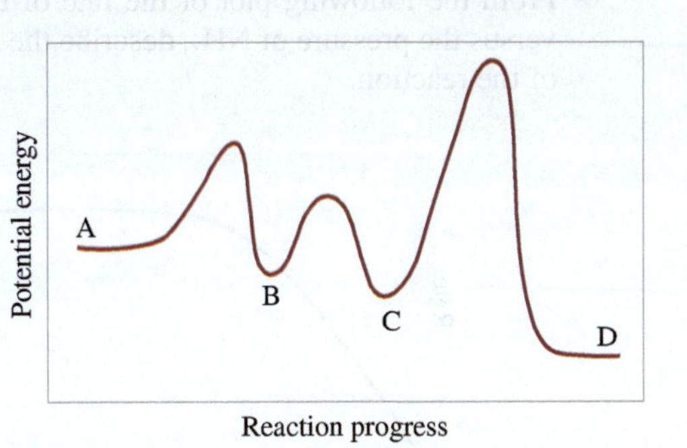

13.119 A factory that specializes in the refinement of transition metals such as titanium was on fire. The firefighters were advised not to douse the fire with water. Why?

13.120 The activation energy for the decomposition of hydrogen peroxide

$$2H_2O_2(aq) \longrightarrow 2H_2O_2(l) + O_2(g)$$

is 42 kJ/mol, whereas when the reaction is catalyzed by the enzyme catalase, it is 7.0 kJ/mol. Calculate the temperature that would cause the uncatalyzed reaction to proceed as rapidly as the enzyme-catalyzed decomposition at 20°C. Assume the frequency factor A to be the same in both cases.

13.121 The *activity* of a radioactive sample is the number of nuclear disintegrations per second, which is equal to the first-order rate constant times the number of radioactive nuclei present. The fundamental unit of radioactivity is the *curie* (Ci), where 1 Ci corresponds to exactly 3.70×10^{10} disintegrations per second. This decay rate is equivalent to that of 1 g of radium-226. Calculate the rate constant and half-life for the radium decay. Starting with 1.0 g of the radium sample, what is the activity after 500 yr? The molar mass of Ra-226 is 226.03 g/mol.

13.122 To carry out metabolism, oxygen is taken up by hemoglobin (Hb) to form oxyhemoglobin (HbO_2) according to the simplified equation

$$Hb(aq) + O_2(aq) \xrightarrow{k} HbO_2(aq)$$

where the second-order rate constant is $2.1 \times 10^6/M \cdot s$ at 37°C. (The reaction is first order in Hb and O_2.) For an average adult, the concentrations of Hb and O_2 in the blood at the lungs are $8.0 \times 10^{-6}\ M$ and $1.5 \times 10^{-6}\ M$, respectively. (a) Calculate the rate of formation of HbO_2. (b) Calculate the rate of consumption of O_2. (c) The rate of formation of HbO_2 increases to $1.4 \times 10^{-4}\ M/s$ during exercise to meet the demand of increased metabolism rate. Assuming the Hb concentration to remain the same, what must be the oxygen concentration to sustain this rate of HbO_2 formation?

13.123 At a certain elevated temperature, ammonia decomposes on the surface of tungsten metal as follows:

$$2NH_3 \longrightarrow N_2 + 3H_2$$

From the following plot of the rate of the reaction versus the pressure of NH_3, describe the mechanism of the reaction.

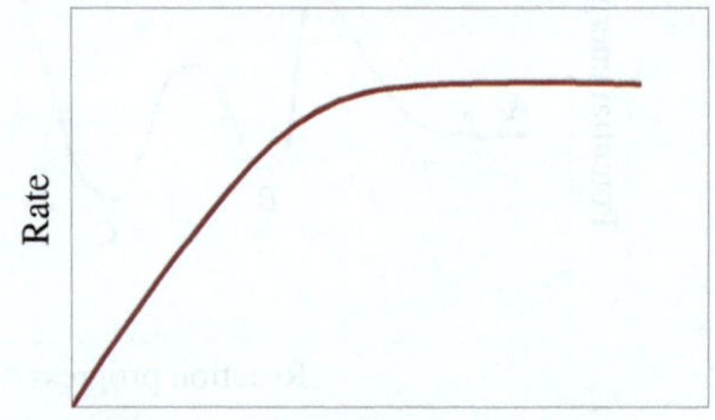

13.124 The following expression shows the dependence of the half-life of a reaction ($t_{\frac{1}{2}}$) on the initial reactant concentration $[A]_0$:

$$t_{\frac{1}{2}} \propto \frac{1}{[A]_0^{n-1}}$$

where n is the order of the reaction. Verify this dependence for zero-, first-, and second-order reactions.

13.125 Polyethylene is used in many items, including water pipes, bottles, electrical insulation, toys, and mailer envelopes. It is a *polymer,* a molecule with a very high molar mass made by joining many ethylene molecules together. (Ethylene is the basic unit, or monomer for polyethylene.) The initiation step is

$$R_2 \xrightarrow{k_1} 2R\cdot \qquad \text{initiation}$$

The R · species (called a radical) reacts with an ethylene molecule (M) to generate another radical

$$R\cdot + M \longrightarrow M_1\cdot$$

Reaction of $M_1\cdot$ with another monomer leads to the growth or propagation of the polymer chain

$$M_1\cdot + M \xrightarrow{k_p} M_2\cdot \qquad \text{propagation}$$

This step can be repeated with hundreds of monomer units. The propagation terminates when two radicals combine

$$M'\cdot + M''\cdot \xrightarrow{k_t} M' - M'' \qquad \text{termination}$$

The initiator frequently used in the polymerization of ethylene is benzoyl peroxide $[(C_6H_5COO)_2]$:

$$[(C_6H_5COO)_2] \longrightarrow 2C_6H_5COO\cdot$$

This is a first-order reaction. The half-life of benzoyl peroxide at 100°C is 19.8 min. (a) Calculate the rate constant (in min^{-1}) of the reaction. (b) If the half-life of benzoyl peroxide is 7.30 h, or 438 min, at 70°C, what is the activation energy (in kJ/mol) for the decomposition of benzoyl peroxide? (c) Write the rate laws for the elementary steps in the polymerization process, and identify the reactant, product, and intermediates. (d) What condition would favor the growth of long, high-molar-mass polyethylenes?

13.126 The rate constant for the gaseous reaction

$$H_2(g) + I_2(g) \longrightarrow 2HI(g)$$

is $2.42 \times 10^{-2}/M \cdot s$ at 400°C. Initially an equimolar sample of H_2 and I_2 is placed in a vessel at 400°C and the total pressure is 1658 mmHg. (a) What is the initial rate (M/min) of formation of HI? (b) What are the rate of formation of HI and the concentration of HI (in molarity) after 10.0 min?

13.127 A protein molecule, P, of molar mass $\mathcal{M}$ dimerizes when it is allowed to stand in solution at room temperature. A plausible mechanism is that the protein molecule is first denatured (that is, loses its activity due to a change in overall structure) before it dimerizes:

$$P \xrightarrow{k} P^* \text{ (denatured)} \qquad \text{slow}$$
$$2P^* \longrightarrow P_2 \qquad \text{fast}$$

where the asterisk denotes a denatured protein molecule. Derive an expression for the average molar mass (of P and P_2), $\overline{\mathcal{M}}$, in terms of the initial protein concentration $[P]_0$ and the concentration at time t, $[P]_t$, and $\mathcal{M}$. Describe how you would determine k from molar mass measurements.

13.128 When the concentration of A in the reaction $A \longrightarrow B$ was changed from 1.20 M to 0.60 M, the half-life increased from 2.0 min to 4.0 min at 25°C. Calculate the order of the reaction and the rate constant. (*Hint:* Use the equation in Problem 13.124.)

13.129 At a certain elevated temperature, ammonia decomposes on the surface of tungsten metal as follows:

$$NH_3 \longrightarrow \frac{1}{2}N_2 + \frac{3}{2}H_2$$

The kinetic data are expressed as the variation of the half-life with the initial pressure of NH_3:

P (mmHg)	264	130	59	16
$t_{\frac{1}{2}}$ (s)	456	228	102	60

(a) Determine the order of the reaction. (b) How does the order depend on the initial pressure? (c) How does the mechanism of the reaction vary with pressure? (*Hint:* You need to use the equation in Problem 13.124 and plot log $t_{\frac{1}{2}}$ versus log P.)

13.130 The activation energy for the reaction

$$N_2O(g) \longrightarrow N_2(g) + O(g)$$

is 2.4×10^2 kJ/mol at 600 K. Calculate the percentage of the increase in rate from 600 K to 606 K. Comment on your results.

13.131 The rate of a reaction was followed by the absorption of light by the reactants and products as a function of wavelengths (λ_1, λ_2, λ_3) as time progresses.

Which of the following mechanisms is consistent with the experimental data?

(a) $A \longrightarrow B, A \longrightarrow C$

(b) $A \longrightarrow B + C$

(c) $A \longrightarrow B, B \longrightarrow C + D$

(d) $A \longrightarrow B, B \longrightarrow C$

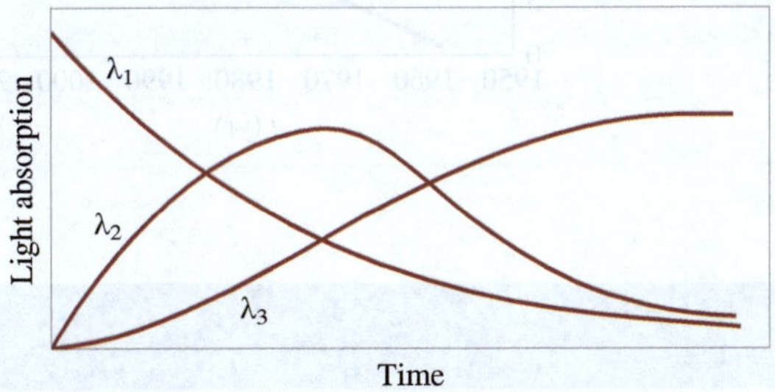

13.132 A gas mixture containing CH_3 fragments, C_2H_6 molecules, and an inert gas (He) was prepared at 600 K with a total pressure of 5.42 atm. The elementary reaction

$$CH_3 + C_2H_6 \longrightarrow CH_4 + C_2H_5$$

has a second-order rate constant of $3.0 \times 10^4/M \cdot s$. Given that the mole fractions of CH_3 and C_2H_6 are 0.00093 and 0.00077, respectively, calculate the initial rate of the reaction at this temperature.

13.133 To prevent brain damage, a drastic medical procedure is to lower the body temperature of someone who has suffered cardiac arrest. What is the physiochemical basis for this treatment?

13.134 The activation energy (E_a) for the reaction

$$2N_2O(g) \longrightarrow 2N_2(g) + O_2(g) \quad \Delta H^\circ = -164 \text{ kJ/mol}$$

is 240 kJ/mol. What is E_a for the reverse reaction?

13.135 The rate constants for the first-order decomposition of an organic compound in solution are measured at several temperatures:

k (s^{-1})	0.00492	0.0216	0.0950	0.326	1.15
T (K)	278	288	298	308	318

Determine graphically the activation energy and frequency factor for the reaction.

13.136 Assume that the formation of nitrogen dioxide:

$$2NO(g) + O_2(g) \longrightarrow 2NO_2(g)$$

is an elementary reaction. (a) Write the rate law for this reaction. (b) A sample of air at a certain temperature is contaminated with 2.0 ppm of NO by volume. Under these conditions, can the rate law be simplified? If so, write the simplified rate law. (c) Under the condition described in (b), the half-life of the reaction has been estimated to be 6.4×10^3 min. What would be the half-life if the initial concentration of NO were 10 ppm?

Interpreting, Modeling, & Estimating

13.137 An instructor performed a lecture demonstration of the thermite reaction (see Example 6.10). He mixed aluminum with iron(III) oxide in a metal bucket placed on a block of ice. After the extremely exothermic reaction started, there was an enormous bang, which was not characteristic of thermite reactions. Give a plausible chemical explanation for the unexpected sound effect. The bucket was open to air.

13.138 Account for the variation of the rate of an enzyme-catalyzed reaction versus temperature shown here. What is the approximate temperature that corresponds to the maximum rate in the human body?

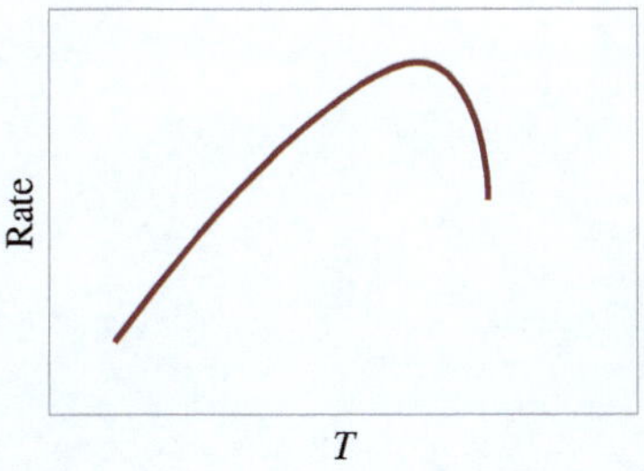

13.139 Is the rate constant (k) of a reaction more sensitive to changes in temperature if E_a is small or large?

13.140 Shown here is a plot of $[A]_t$ versus t for the reaction $A \longrightarrow$ product. (a) Determine the order and the rate constant of the reaction. (b) Estimate the initial rate and the rate at 30 s.

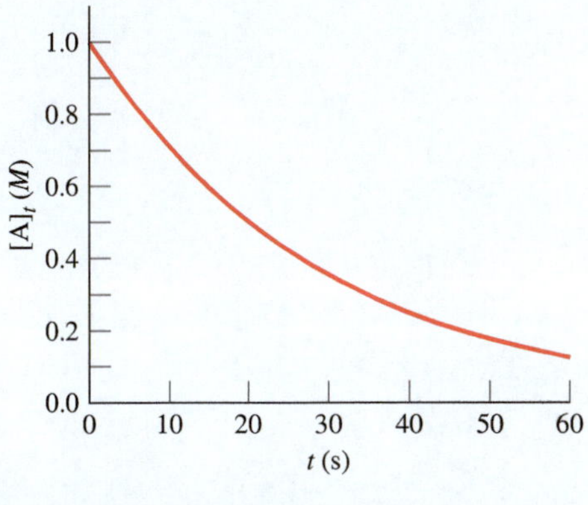

13.141 What are the shortest and longest times (in years) that can be estimated by carbon-14 dating?

13.142 In addition to chemical and biological systems, kinetic treatments can sometimes be applied to behavioral and social processes such as the evolution of technology. For example, in 1965, Gordon Moore, a co-founder of Intel, described a trend that the number of transistors on an integrated circuit (N) roughly doubles every 1.5 yr. Now referred to as *Moore's law,* this trend has persisted for the past several decades. A plot of ln N versus year is shown here. (a) Determine the rate constant for the growth in the number of transistors on an integrated circuit. (b) Based on the rate constant, how long does it take for N to double? (c) If *Moore's law* continues until the end of the century, what will be the number of transistors on an integrated circuit in the year 2100? Comment on your result.

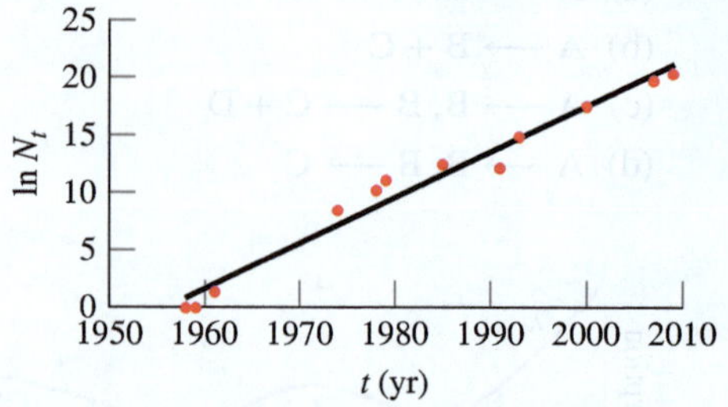

Answers to Practice Exercises

13.1 rate $= -\dfrac{\Delta[CH_4]}{\Delta t} = -\dfrac{1}{2}\dfrac{\Delta[O_2]}{\Delta t} = \dfrac{\Delta[CO_2]}{\Delta t} = \dfrac{1}{2}\dfrac{\Delta[H_2O]}{\Delta t}$. **13.2** (a) 0.013 M/s; (b) −0.052 M/s. **13.3** rate $= k[S_2O_8^{2-}][I^-]$; $k = 8.1 \times 10^{-2}/M \cdot s$. **13.4** 66 s. **13.5** First order. $1.4 \times 10^{-2}\ min^{-1}$. **13.6** 1.2×10^3 s. **13.7** (a) 3.2 min; (b) 2.1 min. **13.8** 240 kJ/mol. **13.9** $3.13 \times 10^{-9}\ s^{-1}$. **13.10** (a) $NO_2 + CO \longrightarrow NO + CO_2$; (b) NO_3; (c) The first step is rate-determining. **13.11** Because at equilibrium the forward and reverse processes are occurring at the same rate, we can set their rates equal to each other and solve for the intermediate concentration. $k_1[NO]^2 = k_{-1}[N_2O_2]$. Substituting the solution into the original rate law (rate $= k[N_2O_2][O_2]$) gives rate $= k_2\dfrac{k_1[NO]^2}{k_{-1}}[O_2] = k[NO]^2[O_2]$, where $k = \dfrac{k_2k_1}{k_{-1}}$.

Answers to Review of Concepts & Facts

13.1.1 $2NOCl(g) \longrightarrow 2NO(g) + Cl_2(g)$. **13.1.2** rate $= -\dfrac{1}{2}\dfrac{\Delta A}{\Delta t} = -\dfrac{\Delta B}{\Delta t} = \dfrac{\Delta C}{\Delta t} = \dfrac{1}{2}\dfrac{\Delta D}{\Delta t}$. **13.1.3** 0.035 M/s. **13.2.1** rate $= k[A]$. **13.2.2** rate $= k[A][B]^2$. **13.3.1** (a) $t_{\frac{1}{2}} = 10$ s, $k = 0.069\ s^{-1}$. (b) At $t = 20$ s; two A and six B molecules. At $t = 30$ s; one A and seven B molecules. **13.3.2** (b). **13.3.3** $t_{\frac{1}{2}} = 56.9$ min. **13.3.4** 0.48 M. **13.4.1** 8.9 kJ/mol. **13.4.2** (a) The reaction has a large E_a. (b) The reaction has a small E_a and the orientation factor is approximately 1.

13.5.1
$H_2 + IBr \longrightarrow HI + HBr$ (show)
$HI + IBr \longrightarrow I_2 + HBr$ (last)
$H_2 + 2IBr \longrightarrow I_2 + 2HBr$

13.5.2 rate $= k[NO]^2[O_2]$ **13.6.1** (b).

CHAPTER

14

Chemical Equilibrium

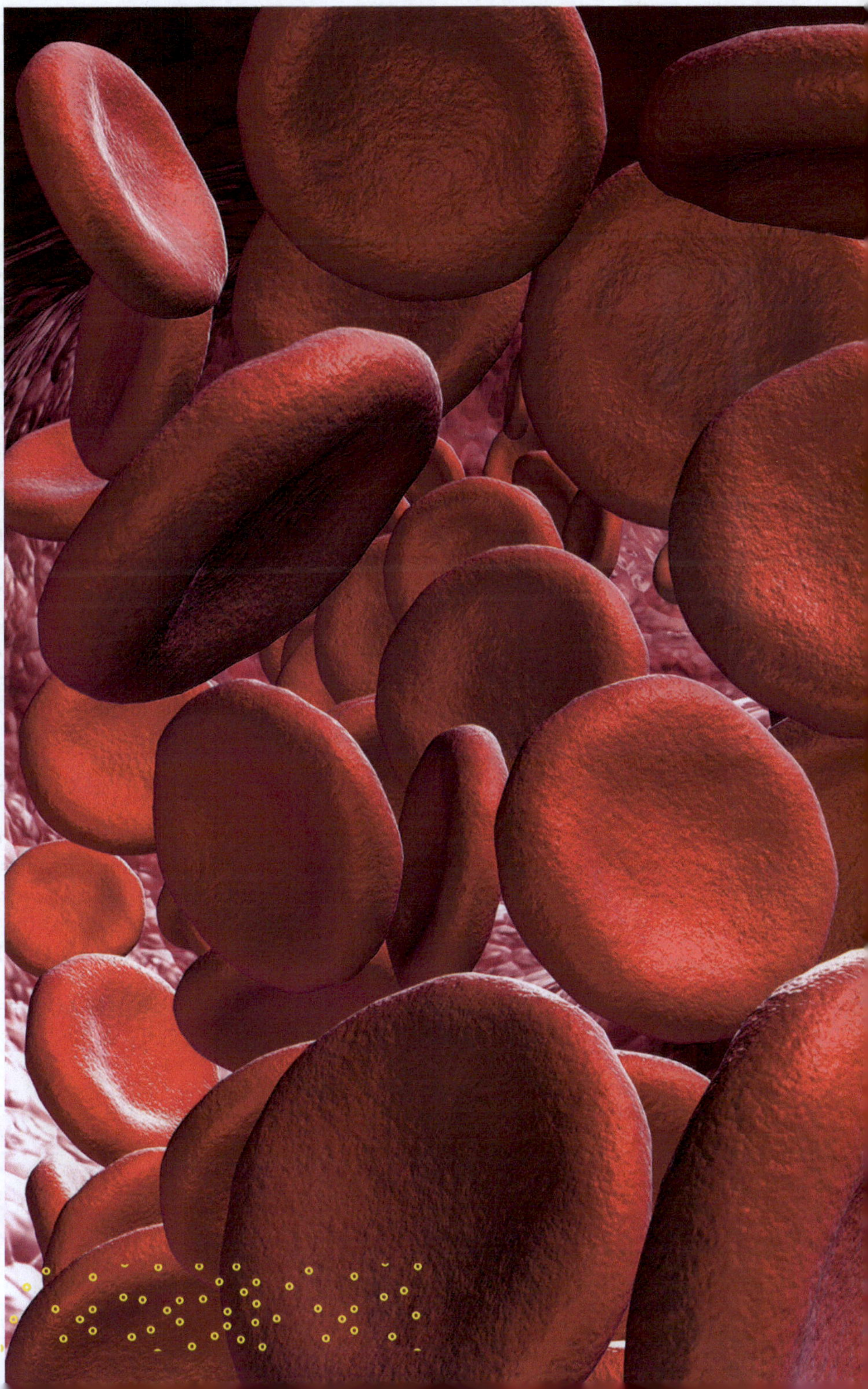

Hemoglobin in red blood cells transports oxygen to establish a dynamic equilibrium between oxygen and carbon dioxide in humans.

yodiyim/Shutterstock

CHAPTER OUTLINE

Equilibrium is a state in which there are no observable changes as time goes by. When a chemical reaction has reached the equilibrium state, the concentrations of reactants and products remain constant over time, and there are no visible changes in the system. However, there is much activity at the molecular level because reactant molecules continue to form product molecules while product molecules react to yield reactant molecules. This dynamic equilibrium situation is the subject of this chapter. Here we will discuss different types of equilibrium reactions, the meaning of the equilibrium constant and its relationship to the rate constant, and factors that can disrupt a system at equilibrium.

14.1 The Concept of Equilibrium and the Equilibrium Constant

Learning Objectives

- Summarize equilibrium as it applies to chemical and physical processes.
- Write the equilibrium constant expression for a given reaction.
- Use the equilibrium constant to predict the relative amounts of reactants and products at equilibrium.

Video: Chemical Equilibrium

Liquid water in equilibrium with its vapor in a closed system at room temperature.
Ken Karp/McGraw Hill

Few chemical reactions proceed in only one direction. Most are reversible, at least to some extent. At the start of a reversible process, the reaction proceeds toward the formation of products. As soon as some product molecules are formed, the reverse process begins to take place and reactant molecules are formed from product molecules. ***Chemical equilibrium*** is achieved when *the rates of the forward and reverse reactions are equal and the concentrations of the reactants and products remain constant.*

Chemical equilibrium is a dynamic process. As such, it can be likened to the movement of skiers at a busy ski resort, where the number of skiers carried up the mountain on the chair lift is equal to the number coming down the slopes. Although there is a constant transfer of skiers, the number of people at the top and the number at the bottom of the slope do not change.

Note that chemical equilibrium involves different substances as reactants and products. Equilibrium between two phases of the same substance is called ***physical equilibrium*** because *the changes that occur are physical processes*. The vaporization of water in a closed container at a given temperature is an example of physical equilibrium. In this instance, the number of H_2O molecules leaving and the number returning to the liquid phase are equal:

$$H_2O(l) \rightleftharpoons H_2O(g)$$

(Recall from Chapter 4 that the double arrow means that the reaction is reversible.)

The study of physical equilibrium yields useful information, such as the equilibrium vapor pressure (see Section 11.8). However, chemists are particularly interested in chemical equilibrium processes, such as the reversible reaction involving nitrogen dioxide (NO_2) and dinitrogen tetroxide (N_2O_4) (Figure 14.1). The progress of the reaction

$$N_2O_4(g) \rightleftharpoons 2NO_2(g)$$

Figure 14.1 *A reversible reaction between N_2O_4 and NO_2 molecules.*

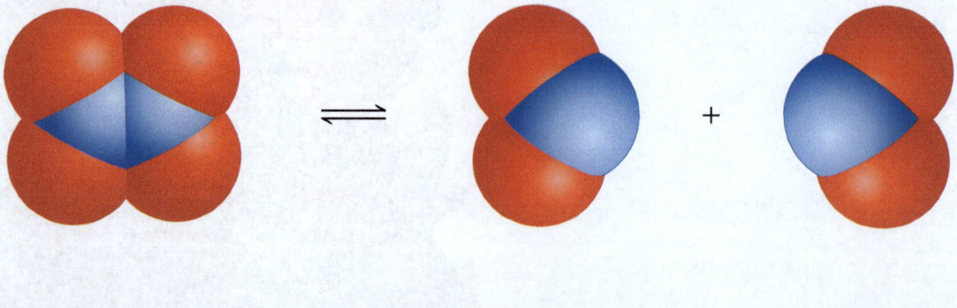

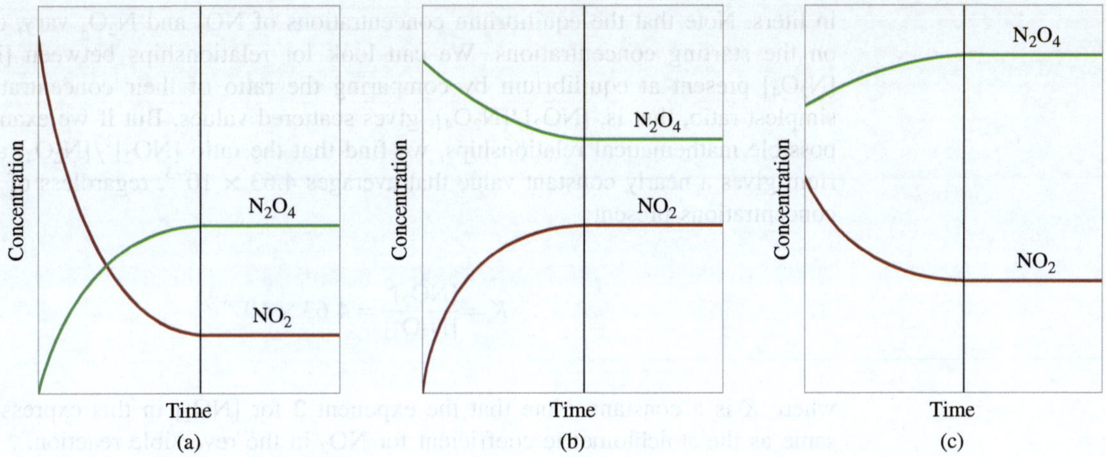

Figure 14.2 *Change in the concentrations of NO_2 and N_2O_4 with time, in three situations. (a) Initially only NO_2 is present. (b) Initially only N_2O_4 is present. (c) Initially a mixture of NO_2 and N_2O_4 is present. In each case, equilibrium is established to the right of the vertical line.*

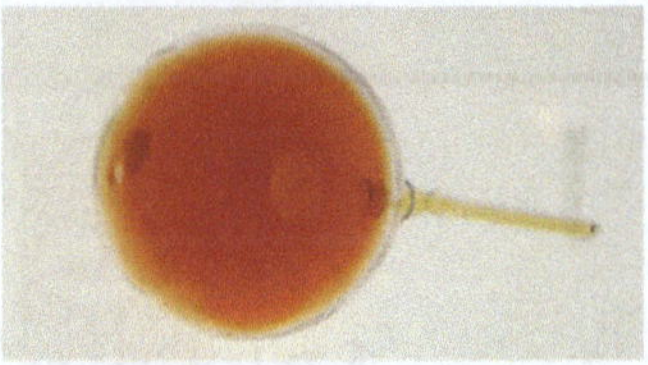

NO_2 and N_2O_4 gases at equilibrium.
Ken Karp/McGraw Hill

can be monitored easily because N_2O_4 is a colorless gas, whereas NO_2 has a dark brown color that makes it sometimes visible in polluted air. Suppose that N_2O_4 is injected into an evacuated flask. Some brown color appears immediately, indicating the formation of NO_2 molecules. The color intensifies as the dissociation of N_2O_4 continues until eventually equilibrium is reached. Beyond that point, no further change in color is evident because the concentrations of both N_2O_4 and NO_2 remain constant. We can also bring about an equilibrium state by starting with pure NO_2. As some of the NO_2 molecules combine to form N_2O_4, the color fades. Yet another way to create an equilibrium state is to start with a mixture of NO_2 and N_2O_4 and monitor the system until the color stops changing. These studies demonstrate that the preceding reaction is indeed reversible because a pure component (N_2O_4 or NO_2) reacts to give the other gas. The important thing to keep in mind is that at equilibrium, the conversions of N_2O_4 to NO_2 and of NO_2 to N_2O_4 are still going on. We do not see a color change because the two rates are equal—the removal of NO_2 molecules takes place as fast as the production of NO_2 molecules, and N_2O_4 molecules are formed as quickly as they dissociate. Figure 14.2 summarizes these three situations.

The Equilibrium Constant

Table 14.1 shows some experimental data for the reaction just described at 25°C. The gas concentrations are expressed in molarity, which can be calculated from the number of moles of gases present initially and at equilibrium and the volume of the flask

Table 14.1 The NO_2–N_2O_4 System at 25°C

Initial Concentrations (*M*)		Equilibrium Concentrations (*M*)		Ratio of Concentrations at Equilibrium	
$[NO_2]$	$[N_2O_4]$	$[NO_2]$	$[N_2O_4]$	$\frac{[NO_2]}{[N_2O_4]}$	$\frac{[NO_2]^2}{[N_2O_4]}$
0.000	0.670	0.0547	0.643	0.0851	4.65×10^{-3}
0.0500	0.446	0.0457	0.448	0.102	4.66×10^{-3}
0.0300	0.500	0.0475	0.491	0.0967	4.60×10^{-3}
0.0400	0.600	0.0523	0.594	0.0880	4.60×10^{-3}
0.200	0.000	0.0204	0.0898	0.227	4.63×10^{-3}

in liters. Note that the equilibrium concentrations of NO_2 and N_2O_4 vary, depending on the starting concentrations. We can look for relationships between $[NO_2]$ and $[N_2O_4]$ present at equilibrium by comparing the ratio of their concentrations. The simplest ratio, that is, $[NO_2]/[N_2O_4]$, gives scattered values. But if we examine other possible mathematical relationships, we find that the ratio $[NO_2]^2/[N_2O_4]$ at equilibrium gives a nearly constant value that averages 4.63×10^{-3}, regardless of the initial concentrations present:

$$K = \frac{[NO_2]^2}{[N_2O_4]} = 4.63 \times 10^{-3} \tag{14.1}$$

where K is a constant. Note that the exponent 2 for $[NO_2]$ in this expression is the same as the stoichiometric coefficient for NO_2 in the reversible reaction.

We can generalize this phenomenon with the following reaction at equilibrium:

$$a\text{A} + b\text{B} \rightleftharpoons c\text{C} + d\text{D}$$

where a, b, c, and d are the stoichiometric coefficients for the reacting species A, B, C, and D. For the reaction at a particular temperature

$$K = \frac{[\text{C}]^c[\text{D}]^d}{[\text{A}]^a[\text{B}]^b} \tag{14.2}$$

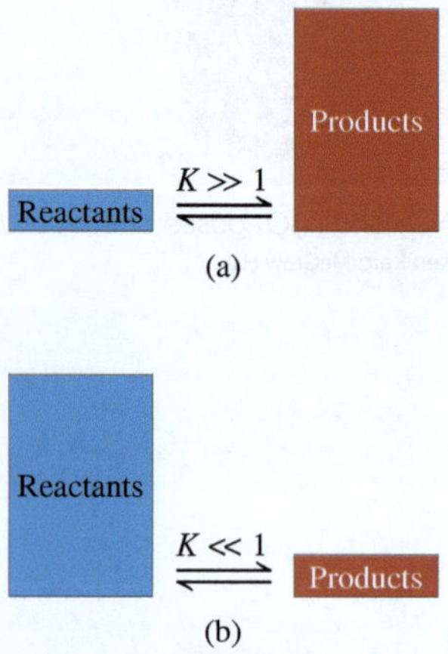

Figure 14.3 *(a) At equilibrium, there are more products than reactants, and the equilibrium is said to lie to the right. (b) In the opposite situation, when there are more reactants than products, the equilibrium is said to lie to the left.*

where K is the ***equilibrium constant.*** Equation (14.2) was formulated by two Norwegian chemists, Cato Guldberg[†] and Peter Waage,[‡] in 1864. It is the mathematical expression of their ***law of mass action,*** which holds that *for a reversible reaction at equilibrium and a constant temperature, a certain ratio of reactant and product concentrations has a constant value, K (the equilibrium constant).* Note that although the concentrations may vary, as long as a given reaction is at equilibrium and the temperature does not change, according to the law of mass action, the value of K remains constant. The validity of Equation (14.2) and the law of mass action has been established by studying many reversible reactions.

The equilibrium constant, then, is defined by a *quotient,* the numerator of which is obtained by multiplying together the equilibrium concentrations of the *products,* each raised to a power equal to its stoichiometric coefficient in the balanced equation. Applying the same procedure to the equilibrium concentrations of *reactants* gives the denominator. The magnitude of the equilibrium constant tells us whether an equilibrium reaction favors the products or reactants. If K is much greater than 1 (that is, $K \gg 1$), the equilibrium will lie to the right and favor the products. Conversely, if the equilibrium constant is much smaller than 1 (that is, $K \ll 1$), the equilibrium will lie to the left and favor the reactants (Figure 14.3). In this context, any number greater than 10 is considered to be much greater than 1, and any number less than 0.1 is much less than 1.

Student Hot Spot

Student data indicate you may struggle with understanding the equilibrium constant. Access your eBook for additional Learning Resources on this topic.

Although the use of the words "reactants" and "products" may seem confusing because any substance serving as a reactant in the forward reaction also is a product of the reverse reaction, it is in keeping with the convention of referring to substances on the left of the equilibrium arrows as "reactants" and those on the right as "products."

[†]Cato Maximilian Guldberg (1836–1902). Norwegian chemist and mathematician. Guldberg's research was mainly in thermodynamics.

[‡]Peter Waage (1833–1900). Norwegian chemist. Like that of his coworker, Guldberg, Waage's research was primarily in thermodynamics.

Summary of Concepts & Facts

- Dynamic equilibria between phases are called physical equilibria. Chemical equilibrium is a reversible process in which the rates of the forward and reverse reactions are equal and the concentrations of reactants and products do not change with time.
- For the general chemical reaction $aA + bB \rightleftharpoons cC + dD$, the concentrations of reactants and products at equilibrium (in moles per liter) are related by the equilibrium constant expression.

Review of Concepts & Facts

14.1.1 Consider the equilibrium $X \rightleftharpoons Y$, where the forward reaction rate constant is greater than the reverse reaction rate constant. Which of the following is true about the equilibrium constant: (a) $K_c > 1$, (b) $K_c < 1$, (c) $K_c = 1$?

14.1.2 The equilibrium constant K_c for a particular process is 3.2×10^{16}. Does this process have predominantly reactants or products present at equilibrium?

14.2 Writing Equilibrium Constant Expressions

Learning Objectives

- Differentiate between heterogeneous and homogeneous equilibria.
- Convert between K_c and K_P for a reaction involving gaseous species.
- Determine the value of the equilibrium constant given equilibrium concentrations of reactants and products.

The concept of equilibrium constants is extremely important in chemistry. As you will soon see, equilibrium constants are the key to solving a wide variety of stoichiometry problems involving equilibrium systems. For example, an industrial chemist who wants to maximize the yield of sulfuric acid, say, must have a clear understanding of the equilibrium constants for all the steps in the process, starting from the oxidation of sulfur and ending with the formation of the final product. A physician specializing in clinical cases of acid-base imbalance needs to know the equilibrium constants of weak acids and bases. And knowledge of equilibrium constants of pertinent gas-phase reactions will help an atmospheric chemist better understand the process of ozone destruction in the stratosphere.

To use equilibrium constants, we must express them in terms of the reactant and product concentrations. Our only guide is the law of mass action [Equation (14.2)], which is the general formula for finding equilibrium concentrations. However, because the concentrations of the reactants and products can be expressed in different units and because the reacting species are not always in the same phase, there may be more than one way to express the equilibrium constant for the *same* reaction. To begin with, we will consider reactions in which the reactants and products are in the same phase.

Homogeneous Equilibria

The term ***homogeneous equilibrium*** applies to reactions in which *all reacting species are in the same phase*. An example of homogeneous gas-phase equilibrium is the dissociation of N_2O_4. The equilibrium constant, as given in Equation (14.1), is

$$K_c = \frac{[NO_2]^2}{[N_2O_4]}$$

Note that the subscript in K_c indicates that the concentrations of the reacting species are expressed in molarity or moles per liter. The concentrations of reactants and products in gaseous reactions can also be expressed in terms of their partial pressures. From Equation (5.8) we see that at constant temperature the pressure P of a gas is directly related to the concentration in mol/L of the gas; that is, $P = (n/V)RT$. Thus, for the equilibrium process

$$N_2O_4(g) \rightleftharpoons 2NO_2(g)$$

we can write

$$K_P = \frac{P_{NO_2}^2}{P_{N_2O_4}} \quad (14.3)$$

where P_{NO_2} and $P_{N_2O_4}$ are the equilibrium partial pressures (in atm) of NO_2 and N_2O_4, respectively. The subscript in K_P tells us that equilibrium concentrations are expressed in terms of pressure.

In general, K_c is not equal to K_P, because the partial pressures of reactants and products are not equal to their concentrations expressed in moles per liter. A simple relationship between K_P and K_c can be derived as follows. Let us consider the following equilibrium in the gas phase:

$$aA(g) \rightleftharpoons bB(g)$$

where a and b are stoichiometric coefficients. The equilibrium constant K_c is given by

$$K_c = \frac{[B]^b}{[A]^a}$$

and the expression for K_P is
$$K_P = \frac{P_B^b}{P_A^a}$$

where P_A and P_B are the partial pressures of A and B. Assuming ideal gas behavior,

$$P_AV = n_ART$$

$$P_A = \frac{n_ART}{V}$$

where V is the volume of the container in liters. Also

$$P_BV = n_BRT$$

$$P_B = \frac{n_BRT}{V}$$

Substituting these relations into the expression for K_P, we obtain

$$K_P = \frac{\left(\frac{n_BRT}{V}\right)^b}{\left(\frac{n_ART}{V}\right)^a} = \frac{\left(\frac{n_B}{V}\right)^b}{\left(\frac{n_A}{V}\right)^a}(RT)^{b-a}$$

Now both n_A/V and n_B/V have units of mol/L and can be replaced by [A] and [B], so that

$$K_P = \frac{[B]^b}{[A]^a}(RT)^{\Delta n} = K_c(RT)^{\Delta n} \quad (14.4)$$

where

$$\Delta n = b - a$$
$$= \text{moles of gaseous products} - \text{moles of gaseous reactants}$$

Because pressures are usually expressed in atm, the gas constant R is given by 0.0821 L · atm/K · mol (note that other pressure units can be used but the value of R must be adjusted appropriately), and we can write the relationship between K_P and K_c as

$$K_P = K_c(0.0821T)^{\Delta n} \quad (14.5)$$

Student Hot Spot

Student data indicate you may struggle with converting between K_c and K_P. Access your eBook for additional Learning Resources on this topic.

In general, $K_P \neq K_c$ except in the special case in which $\Delta n = 0$ as in the equilibrium mixture of molecular hydrogen, molecular bromine, and hydrogen bromide:

$$H_2(g) + Br_2(g) \rightleftharpoons 2HBr(g)$$

In this case, Equation (14.5) can be written as

$$K_P = K_c(0.0821T)^0 = K_c$$

As another example of homogeneous equilibrium, let us consider the ionization of acetic acid (CH_3COOH) in water:

$$CH_3COOH(aq) + H_2O(l) \rightleftharpoons CH_3COO^-(aq) + H_3O^+(aq)$$

The equilibrium constant is

$$K'_c = \frac{[CH_3COO^-][H_3O^+]}{[CH_3COOH][H_2O]}$$

(We use the prime for K_c here to distinguish it from the final form of equilibrium constant to be derived later.) In 1 L, or 1000 g, of water, there are 1000 g/(18.02 g/mol), or 55.5 moles, of water. Therefore, the "concentration" of water, or $[H_2O]$, is 55.5 mol/L, or 55.5 M (see Section 13.3). This is a large quantity compared to the concentrations of other species in solution (usually 1 M or smaller), and we can assume that it does not change appreciably during the course of a reaction. Thus, we can treat $[H_2O]$ as a constant and rewrite the equilibrium constant as

$$K_c = \frac{[CH_3COO^-][H_3O^+]}{[CH_3COOH]}$$

where

$$K_c = K'_c[H_2O]$$

Equilibrium Constant and Units

Note that it is general practice not to include units for the equilibrium constant. In thermodynamics, the equilibrium constant is defined in terms of *activities* rather than concentrations. For an ideal system, the activity of a substance is the ratio of its concentration (or partial pressure) to a standard value, which is 1 *M* (or 1 atm). This procedure eliminates all units but does not alter the numerical parts of the concentration or pressure. Consequently, *K* has no units. We will extend this practice to acid-base equilibria and solubility equilibria in Chapters 15 and 16.

For nonideal systems, the activities are not exactly numerically equal to concentrations. In some cases, the differences can be appreciable. Unless otherwise noted, we will treat all systems as ideal.

Examples 14.1, 14.2, 14.3, 14.4 illustrate the procedure for writing equilibrium constant expressions and calculating equilibrium constants and equilibrium concentrations.

Example 14.1

Write expressions for K_c, and K_P if applicable, for the following reversible reactions at equilibrium:

(a) $HF(aq) + H_2O(l) \rightleftharpoons H_3O^+(aq) + F^-(aq)$

(b) $2NO(g) + O_2(g) \rightleftharpoons 2NO_2(g)$

(c) $CH_3COOH(aq) + C_2H_5OH(aq) \rightleftharpoons CH_3COOC_2H_5(aq) + H_2O(l)$

Strategy Keep in mind the following facts: (1) The K_P expression applies only to gaseous reactions, and (2) the concentration of solvent (usually water) does not appear in the equilibrium constant expression.

Solution

(a) Because there are no gases present, K_P does not apply and we have only K_c:

$$K'_c = \frac{[H_3O^+][F^-]}{[HF][H_2O]}$$

HF is a weak acid, so that the amount of water consumed in acid ionizations is negligible compared with the total amount of water present as solvent. Thus, we can rewrite the equilibrium constant as

$$K_c = \frac{[H_3O^+][F^-]}{[HF]}$$

(b) $$K_c = \frac{[NO_2]^2}{[NO]^2[O_2]} \qquad K_P = \frac{P^2_{NO_2}}{P^2_{NO}P_{O_2}}$$

(c) The equilibrium constant K'_c is given by

$$K'_c = \frac{[CH_3COOC_2H_5][H_2O]}{[CH_3COOH][C_2H_5OH]}$$

Because the water produced in the reaction is negligible compared with the water solvent, the concentration of water does not change. Thus, we can write the new equilibrium constant as

$$K_c = \frac{[CH_3COOC_2H_5]}{[CH_3COOH][C_2H_5OH]}$$

Practice Exercise Write K_c and K_P for the decomposition of dinitrogen pentoxide:

$$2N_2O_5(g) \rightleftharpoons 4NO_2(g) + O_2(g)$$

Similar problems: 14.7, 14.8.

Example 14.2

The following equilibrium process has been studied at 230°C:

$$2NO(g) + O_2(g) \rightleftharpoons 2NO_2(g)$$

In one experiment, the concentrations of the reacting species at equilibrium are found to be [NO] = 0.0542 *M*, $[O_2]$ = 0.127 *M*, and $[NO_2]$ = 15.5 *M*. Calculate the equilibrium constant (K_c) of the reaction at this temperature.

Strategy The concentrations given are equilibrium concentrations. They have units of mol/L, so we can calculate the equilibrium constant (K_c) using the law of mass action [Equation (14.2)].

Solution The equilibrium constant is given by

$$K_c = \frac{[NO_2]^2}{[NO]^2[O_2]}$$

Substituting the concentrations, we find that

$$K_c = \frac{(15.5)^2}{(0.0542)^2(0.127)} = 6.44 \times 10^5$$

Check Note that K_c is given without units. Also, the large magnitude of K_c is consistent with the high product (NO_2) concentration relative to the concentrations of the reactants (NO and O_2).

Practice Exercise Carbonyl chloride ($COCl_2$), also called phosgene, was used in World War I as a poisonous gas. The equilibrium concentrations for the reaction between carbon monoxide and molecular chlorine to form carbonyl chloride

$$CO(g) + Cl_2(g) \rightleftharpoons COCl_2(g)$$

at 74°C are [CO] = 1.2×10^{-2} *M*, $[Cl_2]$ = 0.054 *M*, and $[COCl_2]$ = 0.14 *M*. Calculate the equilibrium constant (K_c).

Similar problem: 14.16.

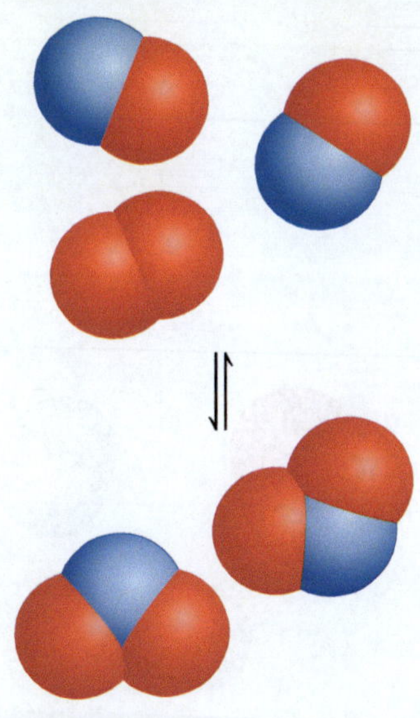

$2NO + O_2 \rightleftharpoons 2NO_2$

Example 14.3

The equilibrium constant K_P for the decomposition of phosphorus pentachloride to phosphorus trichloride and molecular chlorine

$$PCl_5(g) \rightleftharpoons PCl_3(g) + Cl_2(g)$$

is found to be 1.05 at 250°C. If the equilibrium partial pressures of PCl_5 and PCl_3 are 0.875 atm and 0.463 atm, respectively, what is the equilibrium partial pressure of Cl_2 at 250°C?

Strategy The concentrations of the reacting gases are given in atm, so we can express the equilibrium constant in K_P. From the known K_P value and the equilibrium pressures of PCl_3 and PCl_5, we can solve for P_{Cl_2}.

Solution First, we write K_P in terms of the partial pressures of the reacting species

$$K_P = \frac{P_{PCl_3}P_{Cl_2}}{P_{PCl_5}}$$

Knowing the partial pressures, we write

$$1.05 = \frac{(0.463)(P_{Cl_2})}{(0.875)}$$

or

$$P_{Cl_2} = \frac{(1.05)(0.875)}{(0.463)} = 1.98 \text{ atm}$$

(Continued)

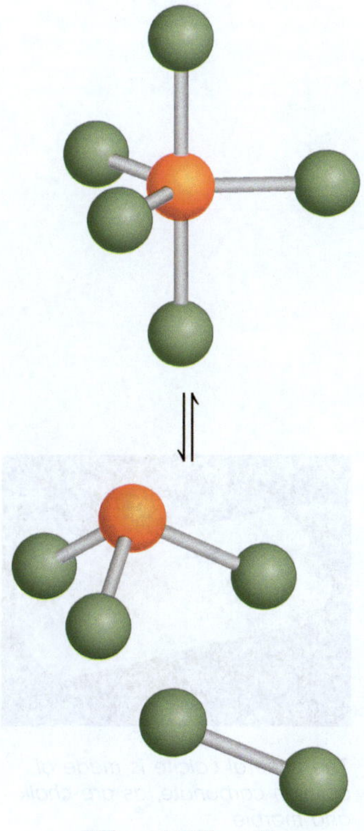

$PCl_5 \rightleftharpoons PCl_3 + Cl_2$

Check Note that we have added atm as the unit for P_{Cl_2}.

Practice Exercise The equilibrium constant K_P for the reaction

$$2NO_2(g) \rightleftharpoons 2NO(g) + O_2(g)$$

is 158 at 1000 K. Calculate P_{O_2} if $P_{NO_2} = 0.400$ atm and $P_{NO} = 0.270$ atm.

Similar problem: 14.19.

Example 14.4

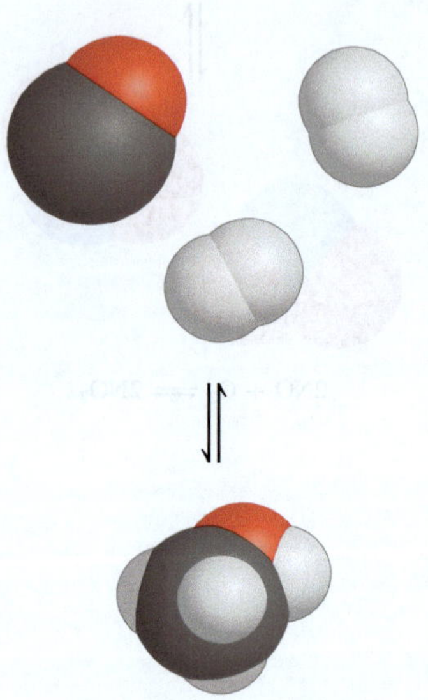

$CO + 2H_2 \rightleftharpoons CH_3OH$

Methanol (CH_3OH) is manufactured industrially by the reaction

$$CO(g) + 2H_2(g) \rightleftharpoons CH_3OH(g)$$

The equilibrium constant (K_c) for the reaction is 10.5 at 220°C. What is the value of K_P at this temperature?

Strategy The relationship between K_c and K_P is given by Equation (14.5). What is the change in the number of moles of gases from reactants to product? Recall that

$$\Delta n = \text{moles of gaseous products} - \text{moles of gaseous reactants}$$

What unit of temperature should we use?

Solution The relationship between K_c and K_P is

$$K_P = K_c(0.0821T)^{\Delta n}$$

Because $T = 273 + 220 = 493$ K and $\Delta n = 1 - 3 = -2$, we have

$$K_P = (10.5)(0.0821 \times 493)^{-2}$$
$$= 6.41 \times 10^{-3}$$

Check Note that K_P, like K_c, is a dimensionless quantity. This example shows that we can get a quite different value for the equilibrium constant for the same reaction, depending on whether we express the concentrations in moles per liter or in atmospheres.

Practice Exercise For the reaction

$$N_2(g) + 3H_2(g) \rightleftharpoons 2NH_3(g)$$

K_P is 4.3×10^{-4} at 375°C. Calculate K_c for the reaction.

Similar problems: 14.17, 14.18.

Heterogeneous Equilibria

As you might expect, a ***heterogeneous equilibrium*** results from *a reversible reaction involving reactants and products that are in different phases*. For example, when calcium carbonate is heated in a closed vessel, the following equilibrium is attained:

$$CaCO_3(s) \rightleftharpoons CaO(s) + CO_2(g)$$

The mineral calcite is made of calcium carbonate, as are chalk and marble.

Collection Varin-Visage/Science Source

The two solids and one gas constitute three separate phases. At equilibrium, we might write the equilibrium constant as

$$K'_c = \frac{[CaO][CO_2]}{[CaCO_3]} \tag{14.6}$$

(Again, the prime for K_c here is to distinguish it from the final form of equilibrium constant to be derived shortly.) However, the "concentration" of a solid, like its

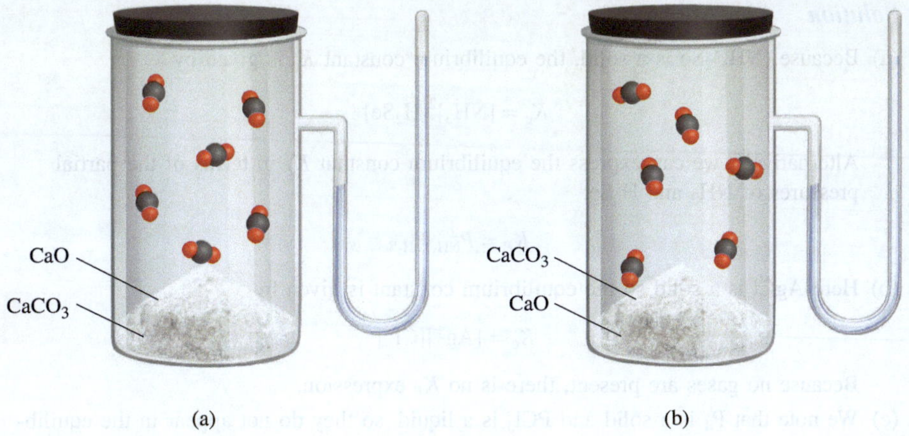

Figure 14.4 *In (a) and (b) the equilibrium pressure of CO_2 is the same at the same temperature, despite the presence of different amounts of $CaCO_3$ and CaO.*

density, is an intensive property and does not depend on how much of the substance is present. For example, the "molar concentration" of copper (density: 8.96 g/cm^3) at 20°C is the same, whether we have 1 gram or 1 ton of the metal:

$$[\text{Cu}] = \frac{8.96\,\text{g}}{1\,\text{cm}^3} \times \frac{1\,\text{mol}}{63.55\,\text{g}} = 0.141\,\text{mol/cm}^3 = 141\,\text{mol/L}$$

For this reason, the terms $[CaCO_3]$ and [CaO] are themselves constants and can be combined with the equilibrium constant. We can simplify Equation (14.6) by writing

$$\frac{[\text{CaCO}_3]}{[\text{CaO}]}K_c' = K_c = [\text{CO}_2] \quad (14.7)$$

where K_c, the "new" equilibrium constant, is conveniently expressed in terms of a single concentration, that of CO_2. Note that the value of K_c does not depend on how much $CaCO_3$ and CaO are present, as long as some of each is present at equilibrium (Figure 14.4).

The situation becomes simpler if we replace concentrations with activities. In thermodynamics, the activity of a pure solid is 1. Thus, the concentration terms for $CaCO_3$ and CaO are both unity, and from the preceding equilibrium equation we can immediately write $K_c = [CO_2]$. Similarly, the activity of a pure liquid is also 1. Thus, if a reactant or a product is a liquid, we can omit it in the equilibrium constant expression.

Alternatively, we can express the equilibrium constant as

$$K_P = P_{\text{CO}_2} \quad (14.8)$$

The equilibrium constant in this case is numerically equal to the pressure of CO_2 gas, an easily measurable quantity.

Example 14.5

Write the equilibrium constant expression K_c, and K_P if applicable, for each of the following heterogeneous systems:

(a) $(NH_4)_2Se(s) \rightleftharpoons 2NH_3(g) + H_2Se(g)$

(b) $AgCl(s) \rightleftharpoons Ag^+(aq) + Cl^-(aq)$

(c) $P_4(s) + 6Cl_2(g) \rightleftharpoons 4PCl_3(l)$

Strategy We omit any pure solids or pure liquids in the equilibrium constant expression because their activities are unity.

(Continued)

Solution

(a) Because $(NH_4)_2Se$ is a solid, the equilibrium constant K_c is given by

$$K_c = [NH_3]^2[H_2Se]$$

Alternatively, we can express the equilibrium constant K_P in terms of the partial pressures of NH_3 and H_2Se:

$$K_P = P_{NH_3}^2 P_{H_2Se}$$

(b) Here AgCl is a solid so the equilibrium constant is given by

$$K_c = [Ag^+][Cl^-]$$

Because no gases are present, there is no K_P expression.

(c) We note that P_4 is a solid and PCl_3 is a liquid, so they do not appear in the equilibrium constant expression. Thus, K_c is given by

$$K_c = \frac{1}{[Cl_2]^6}$$

Alternatively, we can express the equilibrium constant in terms of the pressure of Cl_2:

$$K_P = \frac{1}{P_{Cl_2}^6}$$

Practice Exercise Write equilibrium constant expressions for K_c and K_P for the formation of nickel tetracarbonyl, which is used to separate nickel from other impurities:

$$Ni(s) + 4CO(g) \rightleftharpoons Ni(CO)_4(g)$$

Similar problem: 14.8.

Example 14.6

Consider the following heterogeneous equilibrium:

$$CaCO_3(s) \rightleftharpoons CaO(s) + CO_2(g)$$

At 800°C, the pressure of CO_2 is 0.236 atm. Calculate (a) K_P and (b) K_c for the reaction at this temperature.

Strategy Remember that pure solids do not appear in the equilibrium constant expression. The relationship between K_P and K_c is given by Equation (14.5).

Solution

(a) Using Equation (14.8) we write

$$K_P = P_{CO_2} = 0.236$$

(b) From Equation (14.5), we know

$$K_P = K_c(0.0821T)^{\Delta n}$$

In this case, $T = 800 + 273 = 1073$ K and $\Delta n = 1$, so we substitute these values in the equation and obtain

$$0.236 = K_c(0.0821 \times 1073)$$

$$K_c = 2.68 \times 10^{-3}$$

Practice Exercise Consider the following equilibrium at 395 K:

$$NH_4HS(s) \rightleftharpoons NH_3(g) + H_2S(g)$$

The partial pressure of each gas is 0.265 atm. Calculate K_P and K_c for the reaction.

Similar problem: 14.22.

Multiple Equilibria

The reactions we have considered so far are all relatively simple. A more complicated situation is one in which the product molecules in one equilibrium system are involved in a second equilibrium process:

$$A + B \rightleftharpoons C + D$$
$$C + D \rightleftharpoons E + F$$

The products formed in the first reaction, C and D, react further to form products E and F. At equilibrium we can write two separate equilibrium constants:

$$K'_c = \frac{[C][D]}{[A][B]}$$

and

$$K''_c = \frac{[E][F]}{[C][D]}$$

The overall reaction is given by the sum of the two reactions

$$\begin{array}{lll} & A + B \rightleftharpoons C + D & K'_c \\ & C + D \rightleftharpoons E + F & K''_c \\ \hline \text{Overall reaction:} & A + B \rightleftharpoons E + F & K_c \end{array}$$

and the equilibrium constant K_c for the overall reaction is

$$K_c = \frac{[E][F]}{[A][B]}$$

We obtain the same expression if we take the product of the expressions for K'_c and K''_c:

$$K'_c K''_c = \frac{[C][D]}{[A][B]} \times \frac{[E][F]}{[C][D]} = \frac{[E][F]}{[A][B]}$$

Therefore,

$$K_c = K'_c K''_c \qquad (14.9)$$

We can now make an important statement about multiple equilibria: *If a reaction can be expressed as the sum of two or more reactions, the equilibrium constant for the overall reaction is given by the product of the equilibrium constants of the individual reactions.*

Among the many known examples of multiple equilibria is the ionization of diprotic acids in aqueous solution. The following equilibrium constants have been determined for carbonic acid (H_2CO_3) at 25°C:

$$H_2CO_3(aq) \rightleftharpoons H^+(aq) + HCO_3^-(aq) \qquad K'_c = \frac{[H^+][HCO_3^-]}{[H_2CO_3]} = 4.2 \times 10^{-7}$$

$$HCO_3^-(aq) \rightleftharpoons H^+(aq) + CO_3^{2-}(aq) \qquad K''_c = \frac{[H^+][CO_3^{2-}]}{[HCO_3^-]} = 4.8 \times 10^{-11}$$

The overall reaction is the sum of these two reactions

$$H_2CO_3(aq) \rightleftharpoons 2H^+(aq) + CO_3^{2-}(aq)$$

and the corresponding equilibrium constant is given by

$$K_c = \frac{[H^+]^2[CO_3^{2-}]}{[H_2CO_3]}$$

Using Equation (14.9) we arrive at

$$\begin{aligned} K_c &= K'_c K''_c \\ &= (4.2 \times 10^{-7})(4.8 \times 10^{-11}) \\ &= 2.0 \times 10^{-17} \end{aligned}$$

The Form of K and the Equilibrium Equation

Before concluding this section, let us look at two important rules for writing equilibrium constants:

1. When the equation for a reversible reaction is written in the opposite direction, the equilibrium constant becomes the reciprocal of the original equilibrium constant. Thus, if we write the NO_2–N_2O_4 equilibrium as

$$N_2O_4(g) \rightleftharpoons 2NO_2(g)$$

then at 25°C,

$$K_c = \frac{[NO_2]^2}{[N_2O_4]} = 4.63 \times 10^{-3}$$

However, we can represent the equilibrium equally well as

$$2NO_2(g) \rightleftharpoons N_2O_4(g)$$

and the equilibrium constant is now given by

$$K_c' = \frac{[N_2O_4]}{[NO_2]^2} = \frac{1}{K_c} = \frac{1}{4.63 \times 10^{-3}} = 216$$

You can see that $K_c = 1/K_c'$ or $K_cK_c' = 1.00$. Either K_c or K_c' is a valid equilibrium constant, but it is meaningless to say that the equilibrium constant for the NO_2–N_2O_4 system is 4.63×10^{-3}, or 216, unless we also specify how the equilibrium equation is written.

2. The value of K also depends on how the equilibrium equation is balanced. Consider the following ways of describing the same equilibrium:

$$\tfrac{1}{2}N_2O_4(g) \rightleftharpoons NO_2(g) \qquad K_c' = \frac{[NO_2]}{[N_2O_4]^{\frac{1}{2}}}$$

$$N_2O_4(g) \rightleftharpoons 2NO_2(g) \qquad K_c = \frac{[NO_2]^2}{[N_2O_4]}$$

Looking at the exponents we see that $K_c' = \sqrt{K_c}$. In Table 14.1 we find $K_c = 4.63 \times 10^{-3}$; therefore $K_c' = 0.0680$. According to the law of mass action, each concentration term in the equilibrium constant expression is raised to a power equal to its stoichiometric coefficient. Thus, if you double a chemical equation throughout, the corresponding equilibrium constant will be the square of the original value; if you triple the equation, the equilibrium constant will be the cube of the original value, and so on. The NO_2–N_2O_4 example illustrates once again the need to write the chemical equation when quoting the numerical value of an equilibrium constant.

Student Hot Spot

Student data indicate you may struggle with manipulating equilibrium constants. Access your eBook for additional Learning Resources on this topic.

Example 14.7 deals with the relationship between the equilibrium constants for differently balanced equations describing the same reaction.

Example 14.7

The reaction for the production of ammonia can be written in a number of ways:

(a) $N_2(g) + 3H_2(g) \rightleftharpoons 2NH_3(g)$

(b) $\tfrac{1}{2}N_2(g) + \tfrac{3}{2}H_2(g) \rightleftharpoons NH_3(g)$

(c) $\tfrac{1}{3}N_2(g) + H_2(g) \rightleftharpoons \tfrac{2}{3}NH_3(g)$

(Continued)

Write the equilibrium constant expression for each formulation. (Express the concentrations of the reacting species in mol/L.)
(d) How are the equilibrium constants related to one another?

Strategy We are given three different expressions for the same reacting system. Remember that the equilibrium constant expression depends on how the equation is balanced, that is, on the stoichiometric coefficients used in the equation.

Solution

(a) $$K_a = \frac{[NH_3]^2}{[N_2][H_2]^3}$$

(b) $$K_b = \frac{[NH_3]}{[N_2]^{\frac{1}{2}}[H_2]^{\frac{3}{2}}}$$

(c) $$K_c = \frac{[NH_3]^{\frac{2}{3}}}{[N_2]^{\frac{1}{3}}[H_2]}$$

(d) $$K_a = K_b^2$$
$$K_a = K_c^3$$
$$K_b^2 = K_c^3 \quad \text{or} \quad K_b = K_c^{\frac{3}{2}}$$

Practice Exercise Write the equilibrium expression (K_c) for each of the following reactions and show how they are related to each other: (a) $3O_2(g) \rightleftharpoons 2O_3(g)$, (b) $O_2(g) \rightleftharpoons \frac{2}{3}O_3(g)$.

Similar problem: 14.20.

Summary of Guidelines for Writing Equilibrium Constant Expressions

1. The concentrations of the reacting species in the condensed phase are expressed in mol/L; in the gaseous phase, the concentrations can be expressed in mol/L or in atm. K_c is related to K_P by a simple equation [Equation (14.5)].
2. The concentrations of pure solids, pure liquids (in heterogeneous equilibria), and solvents (in homogeneous equilibria) do not appear in the equilibrium constant expressions.
3. The equilibrium constant (K_c or K_P) is a dimensionless quantity.
4. In quoting a value for the equilibrium constant, we must specify the balanced equation and the temperature.
5. If a reaction can be expressed as the sum of two or more reactions, the equilibrium constant for the overall reaction is given by the product of the equilibrium constants of the individual reactions.

Summary of Concepts & Facts

- A chemical equilibrium process in which all reactants and products are in the same phase is homogeneous. If the reactants and products are not all in the same phase, the equilibrium is heterogeneous. The concentrations of pure solids, pure liquids, and solvents are constant and do not appear in the equilibrium constant expression of a reaction.
- If a reaction can be expressed as the sum of two or more reactions, the equilibrium constant for the overall reaction is given by the product of the equilibrium constants of the individual reactions.
- The value of K depends on how the chemical equation is balanced, and the equilibrium constant for the reverse of a particular reaction is the reciprocal of the equilibrium constant of that reaction.

Review of Concepts & Facts

14.2.1 For which of the following reactions is K_c equal to K_P?
(a) $4NH_3(g) + 5O_2(g) \rightleftharpoons 4NO(g) + 6H_2O(g)$
(b) $2H_2O_2(aq) \rightleftharpoons 2H_2O(l) + O_2(g)$
(c) $PCl_3(g) + 3NH_3(g) \rightleftharpoons 3HCl(g) + P(NH_2)_3(g)$

14.2.2 You are given the equilibrium constant for the reaction

$$N_2(g) + O_2(g) \rightleftharpoons 2NO(g)$$

Suppose you want to calculate the equilibrium constant for the reaction

$$N_2(g) + 2O_2(g) \rightleftharpoons 2NO_2(g)$$

What additional equilibrium constant value (for another reaction) would you need for this calculation? Assume all the equilibria are studied at the same temperature.

14.2.3 From the following equilibrium constant expression, write a balanced chemical equation for the gas-phase reaction.

$$K_c = \frac{[NH_3]^2[H_2O]^4}{[NO_2]^2[H_2]^7}$$

14.2.4 Write the equilibrium constant expression for the following reaction:

$$A(s) + B(g) \rightleftharpoons C(s)$$

14.3 The Relationship Between Chemical Kinetics and Chemical Equilibrium

Learning Objective

- Infer the relationship between rate constants and the equilibrium constant.

We have seen that K, defined in Equation (14.2), is constant at a given temperature regardless of variations in individual equilibrium concentrations (review Table 14.1). We can find out why this is so and at the same time gain insight into the equilibrium process by considering the kinetics of chemical reactions.

Let us suppose that the following reversible reaction occurs via a mechanism consisting of a single *elementary step* in both the forward and reverse directions:

$$A + 2B \underset{k_r}{\overset{k_f}{\rightleftharpoons}} AB_2$$

The forward rate is given by

$$\text{rate}_f = k_f[A][B]^2$$

and the reverse rate is given by

$$\text{rate}_r = k_r[AB_2]$$

where k_f and k_r are the rate constants for the forward and reverse directions. At equilibrium, when no net changes occur, the two rates must be equal:

$$\text{rate}_f = \text{rate}_r$$

or

$$k_f[A][B]^2 = k_r[AB_2]$$

$$\frac{k_f}{k_r} = \frac{[AB_2]}{[A][B]^2}$$

Because both k_f and k_r are constants at a given temperature, their ratio is also a constant, which is equal to the equilibrium constant K_c.

$$\frac{k_f}{k_r} = K_c = \frac{[AB_2]}{[A][B]^2}$$

So K_c is always a constant regardless of the equilibrium concentrations of the reacting species because it is always equal to k_f/k_r, the quotient of two quantities that are themselves constant at a given temperature. Because rate constants are temperature-dependent [see Equation (13.11)], it follows that the equilibrium constant must also change with temperature.

Now suppose the same reaction has a mechanism with more than one elementary step. Suppose it occurs via a two-step mechanism as follows:

$$\textit{Step 1:} \quad 2B \underset{k'_r}{\overset{k'_f}{\rightleftharpoons}} B_2$$

$$\textit{Step 2:} \quad A + B_2 \underset{k''_r}{\overset{k''_f}{\rightleftharpoons}} AB_2$$

$$\textit{Overall reaction:} \quad A + 2B \rightleftharpoons AB_2$$

This is an example of multiple equilibria, discussed in Section 14.2. We write the expressions for the equilibrium constants:

$$K' = \frac{k'_f}{k'_r} = \frac{[B_2]}{[B]^2} \tag{14.10}$$

$$K'' = \frac{k''_f}{k''_r} = \frac{[AB_2]}{[A][B_2]} \tag{14.11}$$

Multiplying Equation (14.10) by Equation (14.11), we get

$$K'K'' = \frac{[B_2][AB_2]}{[B]^2[A][B_2]} = \frac{[AB_2]}{[A][B]^2}$$

For the overall reaction, we can write

$$K_c = \frac{[AB_2]}{[A][B]^2} = K'K''$$

Because both K' and K'' are constants, K_c is also a constant. This result lets us generalize our treatment of the reaction

$$aA + bB \rightleftharpoons cC + dD$$

Regardless of whether this reaction occurs via a single-step or a multistep mechanism, we can write the equilibrium constant expression according to the law of mass action shown in Equation (14.2):

$$K = \frac{[C]^c[D]^d}{[A]^a[B]^b}$$

In summary, we see that in terms of chemical kinetics, the equilibrium constant of a reaction can be expressed as a ratio of the rate constants of the forward and reverse reactions. This analysis explains why the equilibrium constant is a constant and why its value changes with temperature.

Summary of Concepts & Facts

- The equilibrium constant is the ratio of the rate constant for the forward reaction to that for the reverse reaction.

Review of Concepts & Facts

14.3.1 The equilibrium constant (K_c) for reaction $A \rightleftharpoons B + C$ is 4.8×10^{-2} at 80°C. If the forward rate constant is $3.2 \times 10^2\ s^{-1}$, calculate the reverse rate constant.

14.4 What Does the Equilibrium Constant Tell Us?

Learning Objectives

- Appraise the equilibrium constant as it relates to the reaction quotient (*Q*).
- Predict the direction of a reaction given initial concentrations of reactants and products and the value of the equilibrium constant.
- Calculate the equilibrium concentration of reactants or products given initial concentrations.
- Construct an equilibrium table for a reaction and use it to determine equilibrium, initial or final concentrations of a reactant or product.

We have seen that the equilibrium constant for a given reaction can be calculated from known equilibrium concentrations. Once we know the value of the equilibrium constant, we can use Equation (14.2) to calculate unknown equilibrium concentrations—remembering, of course, that the equilibrium constant has a constant value only if the temperature does not change. In general, the equilibrium constant helps us to predict the direction in which a reaction mixture will proceed to achieve equilibrium and to calculate the concentrations of reactants and products once equilibrium has been reached. These uses of the equilibrium constant will be explored in this section.

Predicting the Direction of a Reaction

The equilibrium constant K_c for the formation of hydrogen iodide from molecular hydrogen and molecular iodine in the gas phase

$$H_2(g) + I_2(g) \rightleftharpoons 2HI(g)$$

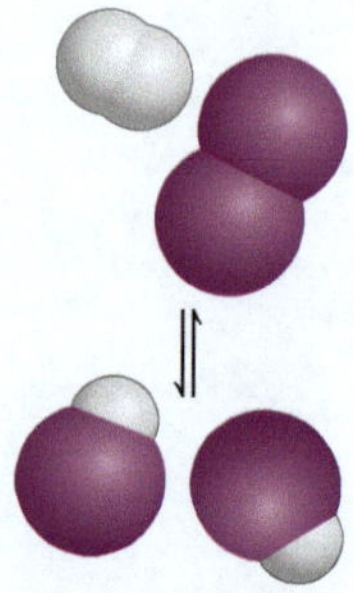

$H_2 + I_2 \rightleftharpoons 2HI$

is 54.3 at 430°C. Suppose that in a certain experiment we place 0.243 mole of H_2, 0.146 mole of I_2, and 1.98 moles of HI all in a 1.00-L container at 430°C. Will there be a net reaction to form more H_2 and I_2 or more HI? Inserting the starting concentrations in the equilibrium constant expression, we write

$$\frac{[HI]_0^2}{[H_2]_0[I_2]_0} = \frac{(1.98)^2}{(0.243)(0.146)} = 111$$

where the subscript 0 indicates initial concentrations (before equilibrium is reached). Because the quotient $[HI]_0^2/[H_2]_0[I_2]_0$ is greater than K_c, this system is not at equilibrium.

For reactions that have not reached equilibrium, such as the formation of HI considered earlier, we obtain the ***reaction quotient*** ***(Q_c)***, instead of the equilibrium constant *by substituting the initial concentrations into the equilibrium constant expression.* To determine the direction in which the net reaction will proceed to achieve equilibrium, we compare the values of Q_c and K_c. The three possible cases are as follows:

- $Q_c < K_c$ The ratio of initial concentrations of products to reactants is too small. To reach equilibrium, reactants must be converted to products. The system proceeds from left to right (consuming reactants, forming products) to reach equilibrium.
- $Q_c = K_c$ The initial concentrations are equilibrium concentrations. The system is at equilibrium.
- $Q_c > K_c$ The ratio of initial concentrations of products to reactants is too large. To reach equilibrium, products must be converted to reactants. The system proceeds from right to left (consuming products, forming reactants) to reach equilibrium.

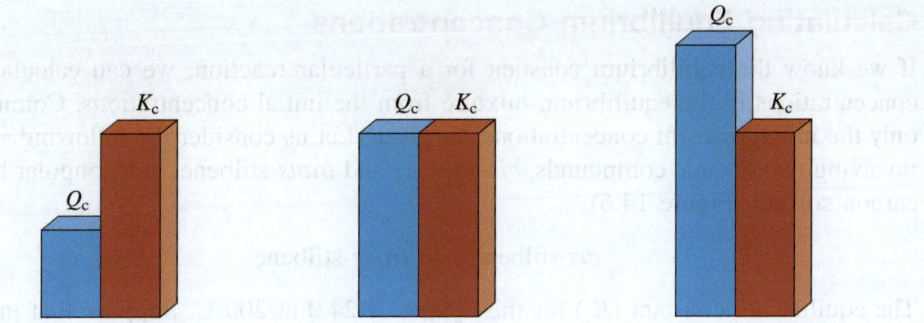

Figure 14.5 *The direction of a reversible reaction to reach equilibrium depends on the relative magnitudes of Q_c and K_c. Note that K_c is a constant at a given temperature, but Q_c varies according to the relative amounts of reactants and products present.*

Figure 14.5 shows a comparison of K_c with Q_c.

Example 14.8 shows how the value of Q_c can help us determine the direction of net reaction toward equilibrium.

Student Hot Spot

Student data indicate you may struggle with reaction quotients. Access your eBook for additional Learning Resources on this topic.

Example 14.8

At the start of a reaction, there are 0.249 mol N_2, 3.21×10^{-2} mol H_2, and 6.42×10^{-4} mol NH_3 in a 3.50-L reaction vessel at 375°C. If the equilibrium constant (K_c) for the reaction

$$N_2(g) + 3H_2(g) \rightleftharpoons 2NH_3(g)$$

is 1.2 at this temperature, decide whether the system is at equilibrium. If it is not, predict which way the net reaction will proceed.

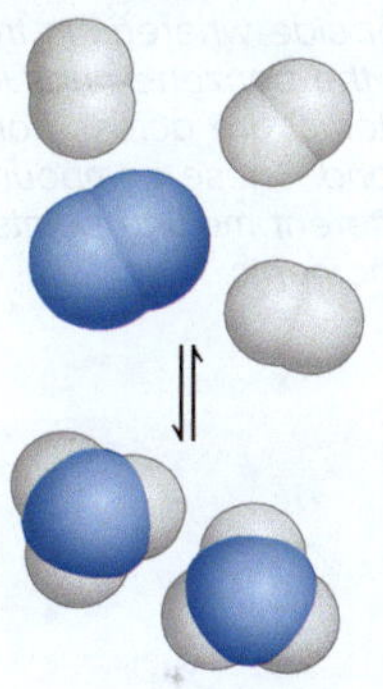

Strategy We are given the initial amounts of the gases (in moles) in a vessel of known volume (in liters), so we can calculate their molar concentrations and hence the reaction quotient (Q_c). How does a comparison of Q_c with K_c enable us to determine if the system is at equilibrium or, if not, in which direction will the net reaction proceed to reach equilibrium?

Solution The initial concentrations of the reacting species are

$$[N_2]_0 = \frac{0.249 \text{ mol}}{3.50 \text{ L}} = 0.0711\ M$$

$$[H_2]_0 = \frac{3.21 \times 10^{-2} \text{ mol}}{3.50 \text{ L}} = 9.17 \times 10^{-3}\ M$$

$$[NH_3]_0 = \frac{6.42 \times 10^{-4} \text{ mol}}{3.50 \text{ L}} = 1.83 \times 10^{-4}\ M$$

Next we write

$$Q_c = \frac{[NH_3]_0^2}{[N_2]_0[H_2]_0^3} = \frac{(1.83 \times 10^{-4})^2}{(0.0711)(9.17 \times 10^{-3})^3} = 0.611$$

Because Q_c is smaller than K_c (1.2), the system is not at equilibrium. The net result will be an increase in the concentration of NH_3 and a decrease in the concentrations of N_2 and H_2. That is, the net reaction will proceed from left to right until equilibrium is reached.

Practice Exercise The equilibrium constant (K_c) for the formation of nitrosyl chloride, an orange-yellow compound, from nitric oxide and molecular chlorine

$$2NO(g) + Cl_2(g) \rightleftharpoons 2NOCl(g)$$

is 6.5×10^4 at 35°C. In a certain experiment, 2.0×10^{-2} mole of NO, 8.3×10^{-3} mole of Cl_2, and 6.8 moles of NOCl are mixed in a 2.0-L flask. In which direction will the system proceed to reach equilibrium?

Similar problems: 14.39, 14.40.

Calculating Equilibrium Concentrations

If we know the equilibrium constant for a particular reaction, we can calculate the concentrations in the equilibrium mixture from the initial concentrations. Commonly, only the initial reactant concentrations are given. Let us consider the following system involving two organic compounds, *cis*-stilbene and *trans*-stilbene, in a nonpolar hydrocarbon solvent (Figure 14.6):

$$cis\text{-stilbene} \rightleftharpoons trans\text{-stilbene}$$

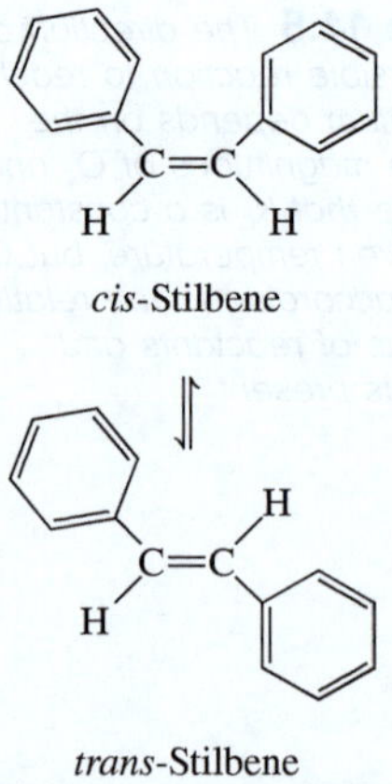

Figure 14.6 *The equilibrium between cis-stilbene and trans-stilbene. Note that both molecules have the same molecular formula ($C_{14}H_{12}$) and also the same type of bonds. However, in cis-stilbene, the benzene rings are on one side of the C═C bond and the H atoms are on the other side whereas in trans-stilbene the benzene rings (and the H atoms) are across from the C═C bond. These compounds have different melting points and dipole moments.*

The equilibrium constant (K_c) for this system is 24.0 at 200°C. Suppose that initially only *cis*-stilbene is present at a concentration of 0.850 mol/L. How do we calculate the concentrations of *cis*- and *trans*-stilbene at equilibrium? From the stoichiometry of the reaction we see that for every mole of *cis*-stilbene converted, 1 mole of *trans*-stilbene is formed. Let x be the equilibrium concentration of *trans*-stilbene in mol/L; therefore, the equilibrium concentration of *cis*-stilbene must be $(0.850 - x)$ mol/L. It is useful to summarize the changes in concentration as follows:

	cis-stilbene	⇌	*trans*-stilbene
Initial (*M*):	0.850		0
Change (*M*):	$-x$		$+x$
Equilibrium (*M*):	$(0.850 - x)$		x

A positive (+) change represents an increase and a negative (−) change a decrease in concentration at equilibrium. This procedure of solving equilibrium concentrations is sometimes referred to as the ICE method, where the acronym stands for Initial, Change, and Equilibrium. Next, we set up the equilibrium constant expression

$$K_c = \frac{[trans\text{-stilbene}]}{[cis\text{-stilbene}]}$$

$$24.0 = \frac{x}{0.850 - x}$$

$$x = 0.816\ M$$

Having solved for x, we calculate the equilibrium concentrations of *cis*-stilbene and *trans*-stilbene as follows:

$$[cis\text{-stilbene}] = (0.850 - 0.816)\ M = 0.034\ M$$

$$[trans\text{-stilbene}] = 0.816\ M$$

To check the results we could use the equilibrium concentrations to calculate K_c.

We summarize our approach to solving equilibrium constant problems as follows:

1. Express the equilibrium concentrations of all species in terms of the initial concentrations and a single unknown x, which represents the change in concentration.
2. Write the equilibrium constant expression in terms of the equilibrium concentrations. Knowing the value of the equilibrium constant, solve for x.
3. Having solved for x, calculate the equilibrium concentrations of all species.

Examples 14.9 and 14.10 illustrate the application of this three-step procedure.

Student Hot Spot

Student data indicate you may struggle with determining equilibrium concentrations. Access your eBook for additional Learning Resources on this topic.

Example 14.9

A mixture of 0.500 mol H_2 and 0.500 mol I_2 was placed in a 1.00-L stainless-steel flask at 430°C. The equilibrium constant K_c for the reaction $H_2(g) + I_2(g) \rightleftharpoons 2HI(g)$ is 54.3 at this temperature. Calculate the concentrations of H_2, I_2, and HI at equilibrium.

(Continued)

Strategy We are given the initial amounts of the gases (in moles) in a vessel of known volume (in liters), so we can calculate their molar concentrations. Because initially no HI was present, the system could not be at equilibrium. Therefore, some H_2 would react with the same amount of I_2 (why?) to form HI until equilibrium was established.

Solution We follow the preceding procedure to calculate the equilibrium concentrations.

Step 1: The stoichiometry of the reaction is 1 mol H_2 reacting with 1 mol I_2 to yield 2 mol HI. Let x be the depletion in concentration (mol/L) of H_2 and I_2 at equilibrium. It follows that the equilibrium concentration of HI must be $2x$. We summarize the changes in concentrations as follows:

	H_2	+	I_2	$\rightleftharpoons$	2HI
Initial (M):	0.500		0.500		0.000
Change (M):	$-x$		$-x$		$+2x$
Equilibrium (M):	$(0.500 - x)$		$(0.500 - x)$		$2x$

Step 2: The equilibrium constant is given by

$$K_c = \frac{[HI]^2}{[H_2][I_2]}$$

Substituting, we get

$$54.3 = \frac{(2x)^2}{(0.500 - x)(0.500 - x)}$$

Taking the square root of both sides, we get

$$7.37 = \frac{2x}{0.500 - x}$$

$$x = 0.393\ M$$

Step 3: At equilibrium, the concentrations are

$$[H_2] = (0.500 - 0.393)\ M = 0.107\ M$$

$$[I_2] = (0.500 - 0.393)\ M = 0.107\ M$$

$$[HI] = 2 \times 0.393\ M = 0.786\ M$$

Check You can check your answers by calculating K_c using the equilibrium concentrations. Remember that K_c is a constant for a particular reaction at a given temperature.

Practice Exercise Consider the reaction in Example 14.9. Starting with a concentration of 0.040 M for HI, calculate the concentrations of HI, H_2, and I_2 at equilibrium.

Similar problem: 14.48.

Example 14.10

For the same reaction and temperature as in Example 14.9, suppose that the initial concentrations of H_2, I_2, and HI are 0.00623 M, 0.00414 M, and 0.0224 M, respectively. Calculate the concentrations of these species at equilibrium.

Strategy From the initial concentrations we can calculate the reaction quotient (Q_c) to see if the system is at equilibrium or, if not, in which direction the net reaction will proceed to reach equilibrium. A comparison of Q_c with K_c also enables us to determine if there will be a depletion in H_2 and I_2 or HI as equilibrium is established.

Solution First we calculate Q_c as follows:

$$Q_c = \frac{[HI]_0^2}{[H_2]_0[I_2]_0} = \frac{(0.0224)^2}{(0.00623)(0.00414)} = 19.5$$

Because Q_c (19.5) is smaller than K_c (54.3), we conclude that the net reaction will proceed from left to right until equilibrium is reached (see Figure 14.4); that is, there will be a depletion of H_2 and I_2 and a gain in HI.

(Continued)

Step 1: Let x be the depletion in concentration (mol/L) of H_2 and I_2 at equilibrium. From the stoichiometry of the reaction it follows that the increase in concentration for HI must be $2x$. Next we write

	H_2	+	I_2	$\rightleftharpoons$	$2HI$
Initial (M):	0.00623		0.00414		0.0224
Change (M):	$-x$		$-x$		$+2x$
Equilibrium (M):	$(0.00623 - x)$		$(0.00414 - x)$		$(0.0224 + 2x)$

Step 2: The equilibrium constant is

$$K_c = \frac{[HI]^2}{[H_2][I_2]}$$

Substituting, we get

$$54.3 = \frac{(0.0224 + 2x)^2}{(0.00623 - x)(0.00414 - x)}$$

It is not possible to solve this equation by the square root shortcut, as the starting concentrations $[H_2]$ and $[I_2]$ are unequal. Instead, we must first carry out the multiplications

$$54.3(2.58 \times 10^{-5} - 0.0104x + x^2) = 5.02 \times 10^{-4} + 0.0896x + 4x^2$$

Collecting terms, we get

$$50.3x^2 - 0.654x + 8.98 \times 10^{-4} = 0$$

This is a quadratic equation of the form $ax^2 + bx + c = 0$. The solution for a quadratic equation (see Appendix 5) is

$$x = \frac{-b \pm \sqrt{b^2 - 4ac}}{2a}$$

Here we have $a = 50.3$, $b = -0.654$, and $c = 8.98 \times 10^{-4}$, so that

$$x = \frac{0.654 \pm \sqrt{(-0.654)^2 - 4(50.3)(8.98 \times 10^{-4})}}{2 \times 50.3}$$

$$x = 0.0114\,M \quad \text{or} \quad x = 0.00156\,M$$

The first solution is physically impossible because the amounts of H_2 and I_2 reacted would be more than those originally present. The second solution gives the correct answer. Note that in solving quadratic equations of this type, one answer is always physically impossible, so choosing a value for x is easy.

Step 3: At equilibrium, the concentrations are

$$[H_2] = (0.00623 - 0.00156)\,M = 0.00467\,M$$

$$[I_2] = (0.00414 - 0.00156)\,M = 0.00258\,M$$

$$[HI] = (0.0224 + 2 \times 0.00156)\,M = 0.0255\,M$$

Check You can check the answers by calculating K_c using the equilibrium concentrations. Remember that K_c is a constant for a particular reaction at a given temperature.

Practice Exercise At 1280°C the equilibrium constant (K_c) for the reaction

$$Br_2(g) \rightleftharpoons 2Br(g)$$

is 1.1×10^{-3}. If the initial concentrations are $[Br_2] = 6.3 \times 10^{-2}\,M$ and $[Br] = 1.2 \times 10^{-2}\,M$, calculate the concentrations of these species at equilibrium.

Similar problem: 14.90.

Examples 14.9 and 14.10 show that we can calculate the concentrations of all the reacting species at equilibrium if we know the equilibrium constant and the initial concentrations. This information is valuable if we need to estimate the yield of a reaction. For example, if the reaction between H_2 and I_2 to form HI were to go to completion, the number of moles of HI formed in Example 14.9 would be 2 × 0.500 mol, or 1.00 mol. However, because of the equilibrium process, the actual amount of HI formed can be no more than 2 × 0.393 mol, or 0.786 mol, a 78.6 percent yield.

Summary of Concepts & Facts

- The reaction quotient Q has the same form as the equilibrium constant, but it applies to a reaction that may not be at equilibrium. If $Q > K$, the reaction will proceed from right to left to achieve equilibrium. If $Q < K$, the reaction will proceed from left to right to achieve equilibrium.

Review of Concepts & Facts

Use the following information to answer questions 14.4.1 and 14.4.2: K_c for the reaction

$$PCl_3(g) + Cl_2(g) \rightleftharpoons PCl_5(g)$$

is 10.0 at a certain temperature.

14.4.1 At a certain point during the reaction, there are 0.50 mole each of PCl_3 and Cl_2 and 0.25 mole of PCl_5 present in a 5.00-L container. Determine whether this system is at equilibrium. If it is not, predict which way the net reaction will proceed.

14.4.2 What is the equilibrium concentration of PCl_3 if 1.00 mole each of PCl_3 and Cl_2 are placed into a 5.00-L container?

14.4.3 The equilibrium constant (K_c) for the $A_2 + B_2 \rightleftharpoons 2AB$ reaction is 3 at a certain temperature. Which of the diagrams shown here corresponds to the reaction at equilibrium? For those mixtures that are not at equilibrium, will the net reaction move in the forward or reverse direction to reach equilibrium?

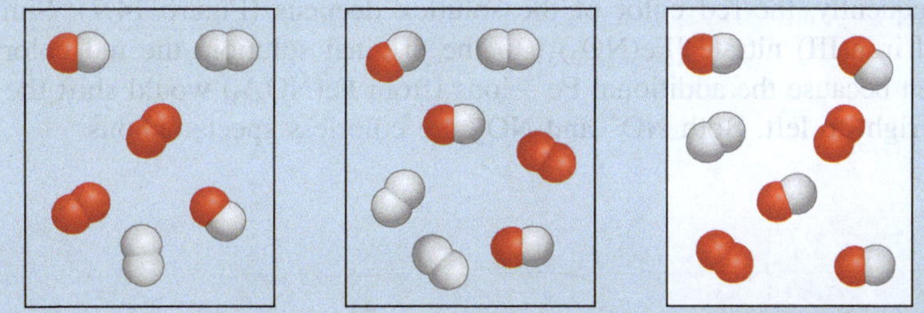

(a) (b) (c)

14.5 Factors That Affect Chemical Equilibrium

Learning Objectives

- Describe Le Châtelier's principle.
- Apply Le Châtelier's principle toward determining the shift of a reaction at equilibrium given a change in one of the following: removal or addition of reactant or product, change in volume or pressure, temperature change, or addition of a catalyst.

Chemical equilibrium represents a balance between forward and reverse reactions. In most cases, this balance is quite delicate. Changes in experimental conditions may disturb the balance and shift the equilibrium position so that more or less of the desired product is formed. When we say that an equilibrium position shifts to the right, for example, we mean that the net reaction is now from left to right. Variables that can be controlled experimentally are concentration, pressure, volume, and temperature. Here we will examine how each of these variables affects a reacting system at equilibrium. In addition, we will examine the effect of a catalyst on equilibrium.

Le Châtelier's Principle

Video: Le Châtelier's Principle

There is a general rule that helps us to predict the direction in which an equilibrium reaction will move when a change in concentration, pressure, volume, or temperature occurs. The rule, known as ***Le Châtelier's*† *principle,*** states that *if an external stress is applied to a system at equilibrium, the system adjusts in such a way that the stress is partially offset as the system reaches a new equilibrium position.* The word "stress" here means a change in concentration, pressure, volume, or temperature that removes the system from the equilibrium state. We will use Le Châtelier's principle to assess the effects of such changes.

Changes in Concentration

Iron(III) thiocyanate [$Fe(SCN)_3$] dissolves readily in water to give a red solution. The red color is due to the presence of hydrated $FeSCN^{2+}$ ion. The equilibrium between undissociated $FeSCN^{2+}$ and the Fe^{3+} and SCN^- ions is given by

$$\underset{\text{red}}{FeSCN^{2+}(aq)} \rightleftharpoons \underset{\text{pale yellow}}{Fe^{3+}(aq)} + \underset{\text{colorless}}{SCN^-(aq)}$$

What happens if we add some sodium thiocyanate (NaSCN) to this solution? In this case, the stress applied to the equilibrium system is an increase in the concentration of SCN^- (from the dissociation of NaSCN). To offset this stress, some Fe^{3+} ions react with the added SCN^- ions, and the equilibrium shifts from right to left:

$$FeSCN^{2+}(aq) \longleftarrow Fe^{3+}(aq) + SCN^-(aq)$$

Consequently, the red color of the solution deepens (Figure 14.7). Similarly, if we added iron(III) nitrate [$Fe(NO_3)_3$] to the original solution, the red color would also deepen because the additional Fe^{3+} ions [from $Fe(NO_3)_3$] would shift the equilibrium from right to left. Both NO^+ and NO_3^- are colorless spectator ions.

(a) (b) (c) (d)

Figure 14.7 *Effect of concentration change on the position of equilibrium. (a) An aqueous $Fe(SCN)_3$ solution. The color of the solution is due to both the red $FeSCN^{2+}$ and the yellow Fe^{3+} ions. (b) After the addition of some NaSCN to the solution in (a), the equilibrium shifts to the left. (c) After the addition of some $Fe(NO_3)_3$ to the solution in (a), the equilibrium shifts to the left. (d) After the addition of some $H_2C_2O_4$ to the solution in (a), the equilibrium shifts to the right. The yellow color is due to the $Fe(C_2O_4)_3^{3-}$ ions.*
Ken Karp/McGraw Hill

†Henri Louis Le Châtelier (1850–1936). French chemist. Le Châtelier did work on metallurgy, cements, glasses, fuels, and explosives. He was also noted for his skills in industrial management.

Now suppose we add some oxalic acid ($H_2C_2O_4$) to the original solution. Oxalic acid ionizes in water to form the oxalate ion, $C_2O_4^{2-}$, which binds strongly to the Fe^{3+} ions. The formation of the stable yellow ion $Fe(C_2O_4)_3^{3-}$ removes free Fe^{3+} ions in solution. Consequently, more $FeSCN^{2+}$ units dissociate and the equilibrium shifts from left to right:

$$FeSCN^{2+}(aq) \longrightarrow Fe^{3+}(aq) + SCN^-(aq)$$

The red solution will turn yellow due to the formation of $Fe(C_2O_4)_3^{3-}$ ions.

This experiment demonstrates that all reactants and products are present in the reacting system at equilibrium. Second, increasing the concentrations of the products (Fe^{3+} or SCN^-) shifts the equilibrium to the left, and decreasing the concentration of the product Fe^{3+} shifts the equilibrium to the right. These results are just as predicted by Le Châtelier's principle. Remember that Le Châtelier's principle summarizes the observed behavior of equilibrium systems; therefore, it is incorrect to say that a given equilibrium shift occurs "because of" Le Châtelier's principle.

The effect of a change in concentration on the equilibrium position is shown in Example 14.11.

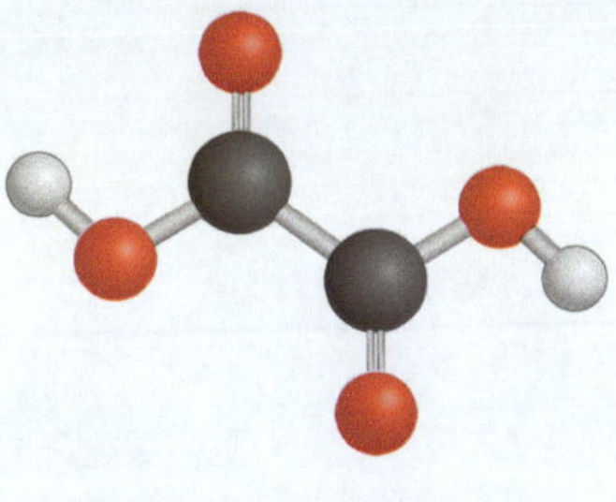

Oxalic acid is sometimes used to remove bathtub rings that consist of rust, or Fe_2O_3.

Student Hot Spot

Student data indicate you may struggle with how concentration changes affect equilibria. Access your eBook for additional Learning Resources on this topic.

Example 14.11

At 720°C, the equilibrium constant K_c for the reaction

$$N_2(g) + 3H_2(g) \rightleftharpoons 2NH_3(g)$$

is 2.37×10^{-3}. In a certain experiment, the equilibrium concentrations are $[N_2] = 0.683\ M$, $[H_2] = 8.80\ M$, and $[NH_3] = 1.05\ M$. Suppose some NH_3 is added to the mixture so that its concentration is increased to $3.65\ M$. (a) Use Le Châtelier's principle to predict the shift in direction of the net reaction to reach a new equilibrium. (b) Confirm your prediction by calculating the reaction quotient Q_c and comparing its value with K_c.

Strategy (a) What is the stress applied to the system? How does the system adjust to offset the stress? (b) At the instant when some NH_3 is added, the system is no longer at equilibrium. How do we calculate the Q_c for the reaction at this point? How does a comparison of Q_c with K_c tell us the direction of the net reaction to reach equilibrium?

Solution

(a) The stress applied to the system is the addition of NH_3. To offset this stress, some NH_3 reacts to produce N_2 and H_2 until a new equilibrium is established. The net reaction therefore shifts from right to left; that is,

$$N_2(g) + 3H_2(g) \longleftarrow 2NH_3(g)$$

(b) At the instant when some of the NH_3 is added, the system is no longer at equilibrium. The reaction quotient is given by

$$Q_c = \frac{[NH_3]_0^2}{[N_2]_0[H_2]_0^3} = \frac{(3.65)^2}{(0.683)(8.80)^3} = 2.86 \times 10^{-2}$$

Because this value is greater than 2.37×10^{-3}, the net reaction shifts from right to left until Q_c equals K_c.

Figure 14.8 shows qualitatively the changes in concentrations of the reacting species.

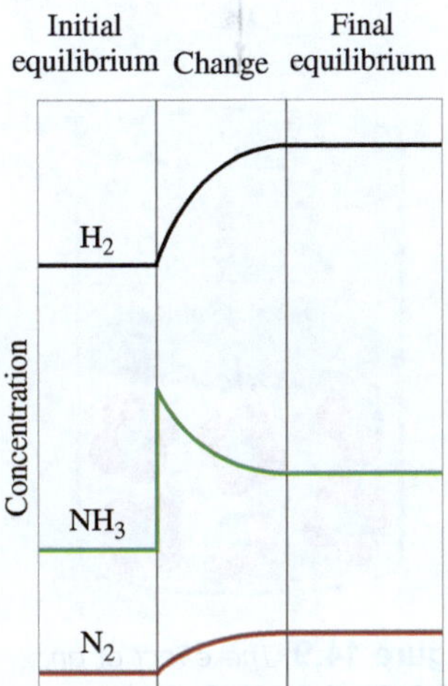

Figure 14.8 *Changes in concentration of H_2, N_2, and NH_3 after the addition of NH_3 to the equilibrium mixture. When the new equilibrium is established, all the concentrations are changed but K_c remains the same because temperature remains constant.*

Practice Exercise At 430°C, the equilibrium constant (K_P) for the reaction

$$2NO(g) + O_2(g) \rightleftharpoons 2NO_2(g)$$

(Continued)

is 1.5×10^5. In one experiment, the initial pressures of NO, O_2, and NO_2 are 2.1×10^{-3} atm, 1.1×10^{-2} atm, and 0.14 atm, respectively. Calculate Q_P and predict the direction that the net reaction will shift to reach equilibrium.

Similar problem: 14.46.

Changes in Volume and Pressure

Changes in pressure ordinarily do not affect the concentrations of reacting species in condensed phases (say, in an aqueous solution) because liquids and solids are virtually incompressible. On the other hand, concentrations of gases are greatly affected by changes in pressure. Let us look again at Equation (5.8):

$$PV = nRT$$
$$P = \left(\frac{n}{V}\right)RT$$

Note that P and V are related to each other inversely: The greater the pressure, the smaller the volume, and vice versa. Note, too, that the term (n/V) is the concentration of the gas in mol/L, and it varies directly with pressure.

Suppose that the equilibrium system

$$N_2O_4(g) \rightleftharpoons 2NO_2(g)$$

is in a cylinder fitted with a movable piston. What happens if we increase the pressure on the gases by pushing down on the piston at constant temperature? Because the volume decreases, the concentration (n/V) of both NO_2 and N_2O_4 increases. Note that the concentration of NO_2 is squared in the equilibrium constant expression, so the increase in pressure increases the numerator more than the denominator. The system is no longer at equilibrium and we write

$$Q_c = \frac{[NO_2]_0^2}{[N_2O_4]_0}$$

Thus, $Q_c > K_c$ and the net reaction will shift to the left until $Q_c = K_c$ (Figure 14.9). Conversely, a decrease in pressure (increase in volume) would result in $Q_c < K_c$, and the net reaction would shift to the right until $Q_c = K_c$. (This conclusion is also predicted by Le Châtelier's principle.)

In general, an increase in pressure (decrease in volume) favors the net reaction that decreases the total number of moles of gases (the reverse reaction, in this case), and a decrease in pressure (increase in volume) favors the net reaction that increases the total number of moles of gases (here, the forward reaction). For reactions in which there is no change in the number of moles of gases, a pressure (or volume) change has no effect on the position of equilibrium.

It is possible to change the pressure of a system without changing its volume. Suppose the NO_2–N_2O_4 system is contained in a stainless-steel vessel whose volume is constant. We can increase the total pressure in the vessel by adding an inert gas (helium, for example) to the equilibrium system. Adding helium to the equilibrium mixture at constant volume increases the total gas pressure and decreases the mole fractions of both NO_2 and N_2O_4; but the partial pressure of each gas, given by the product of its mole fraction and total pressure (see Section 5.6), does not change. Thus, the presence of an inert gas in such a case does not affect the equilibrium.

Example 14.12 illustrates the effect of a pressure change on the equilibrium position.

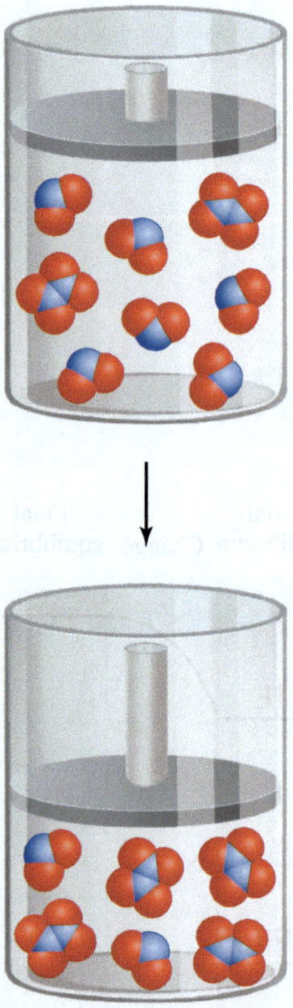

Figure 14.9 *The effect of an increase in pressure on the $N_2O_4(g) \rightleftharpoons 2NO_2(g)$ equilibrium.*

Example 14.12

Consider the following equilibrium systems:

(a) $2PbS(s) + 3O_2(g) \rightleftharpoons 2PbO(s) + 2SO_2(g)$

(b) $PCl_5(g) \rightleftharpoons PCl_3(g) + Cl_2(g)$

(c) $H_2(g) + CO_2(g) \rightleftharpoons H_2O(g) + CO(g)$

(Continued)

Predict the direction of the net reaction in each case as a result of increasing the pressure (decreasing the volume) on the system at constant temperature.

Strategy A change in pressure can affect only the volume of a gas, but not that of a solid because solids (and liquids) are much less compressible. The stress applied is an increase in pressure. According to Le Châtelier's principle, the system will adjust to partially offset this stress. In other words, the system will adjust to decrease the pressure. This can be achieved by shifting to the side of the equation that has fewer moles of gas. Recall that pressure is directly proportional to moles of gas: $PV = nRT$ so $P \propto n$.

Solution

(a) Consider only the gaseous molecules. In the balanced equation, there are 3 moles of gaseous reactants and 2 moles of gaseous products. Therefore, the net reaction will shift toward the products (to the right) when the pressure is increased.

(b) The number of moles of products is 2 and that of reactants is 1; therefore, the net reaction will shift to the left, toward the reactant.

(c) The number of moles of products is equal to the number of moles of reactants, so a change in pressure has no effect on the equilibrium.

Check In each case, the prediction is consistent with Le Châtelier's principle.

Practice Exercise Consider the equilibrium reaction involving nitrosyl chloride, nitric oxide, and molecular chlorine:

$$2NOCl(g) \rightleftharpoons 2NO(g) + Cl_2(g)$$

Predict the direction of the net reaction as a result of decreasing the pressure (increasing the volume) on the system at constant temperature.

Similar problem: 14.56.

Changes in Temperature

A change in concentration, pressure, or volume may alter the equilibrium position, that is, the relative amounts of reactants and products, but it does not change the value of the equilibrium constant. Only a change in temperature can alter the equilibrium constant. To see why, let us consider the reaction

$$N_2O_4(g) \rightleftharpoons 2NO_2(g)$$

The forward reaction is endothermic (absorbs heat, $\Delta H° > 0$):

$$\text{heat} + N_2O_4(g) \longrightarrow 2NO_2(g) \qquad \Delta H° = 58.0 \text{ kJ/mol}$$

so the reverse reaction is exothermic (releases heat, $\Delta H° < 0$):

$$2NO_2(g) \longrightarrow N_2O_4(g) + \text{heat} \qquad \Delta H° = -58.0 \text{ kJ/mol}$$

At equilibrium at a certain temperature, the heat effect is zero because there is no net reaction. If we treat heat as though it were a chemical reagent, then a rise in temperature "adds" heat to the system and a drop in temperature "removes" heat from the system. As with a change in any other parameter (concentration, pressure, or volume), the system shifts to reduce the effect of the change. Therefore, a temperature increase favors the endothermic direction of the reaction (from left to right of the equilibrium equation), which decreases $[N_2O_4]$ and increases $[NO_2]$. A temperature decrease favors the exothermic direction of the reaction (from right to left of the equilibrium equation), which decreases $[NO_2]$ and increases $[N_2O_4]$. Consequently, the equilibrium constant, given by

$$K_c = \frac{[NO_2]^2}{[N_2O_4]}$$

increases when the system is heated and decreases when the system is cooled (Figure 14.10).

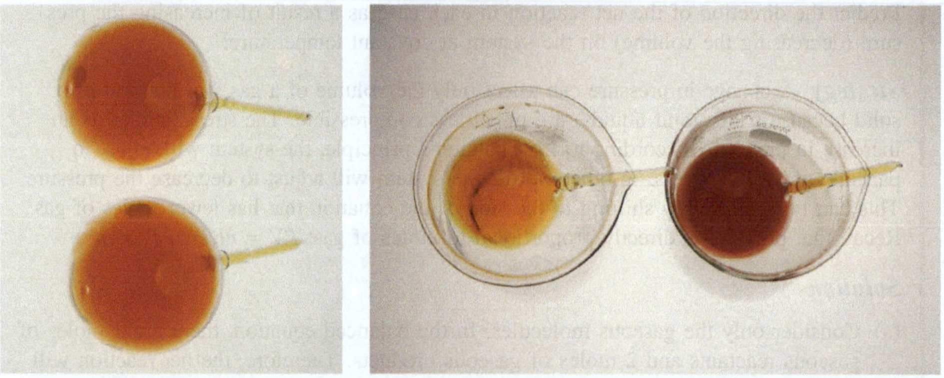

(a) (b)

Figure 14.10 *(a) Two bulbs containing a mixture of NO_2 and N_2O_4 gases at equilibrium. (b) When one bulb is immersed in ice water (left), its color becomes lighter, indicating the formation of colorless N_2O_4 gas. When the other bulb is immersed in hot water, its color darkens, indicating an increase in NO_2.*
(both): Ken Karp/McGraw Hill

As another example, consider the equilibrium between the following ions:

$$\underset{\text{blue}}{CoCl_4^{2-}} + 6H_2O \rightleftharpoons \underset{\text{pink}}{Co(H_2O)_6^{2+}} + 4Cl^-$$

The formation of $CoCl_4^{2-}$ is endothermic. On heating, the equilibrium shifts to the left and the solution turns blue. Cooling favors the exothermic reaction [the formation of $Co(H_2O)_6^{2+}$] and the solution turns pink (Figure 14.11).

In summary, *a temperature increase favors an endothermic reaction, and a temperature decrease favors an exothermic reaction.*

The Effect of a Catalyst

We know that a catalyst enhances the rate of a reaction by lowering the reaction's activation energy (see Section 13.6). However, as Figure 13.25 shows, a catalyst lowers the activation energy of the forward reaction and the reverse reaction to the same extent. We can therefore conclude that the presence of a catalyst does not alter the equilibrium constant, nor does it shift the position of an equilibrium system. Adding a catalyst to a reaction mixture that is not at equilibrium will simply cause the mixture to reach equilibrium sooner. The same equilibrium mixture could be obtained without the catalyst, but we might have to wait much longer for it to happen.

Summary of Factors That May Affect the Equilibrium Position

We have considered four ways to affect a reacting system at equilibrium. It is important to remember that, of the four, *only a change in temperature changes the value of the equilibrium constant.* Changes in concentration, pressure, and volume can alter the

Figure 14.11 *(Left) Heating favors the formation of the blue $CoCl_4^{2-}$ ion. (Right) Cooling favors the formation of the pink $Co(H_2O)_6^{2+}$ ion.*
Ken Karp/McGraw Hill

CHEMISTRY *in Action*

Life at High Altitudes and Hemoglobin Production

In the human body, countless chemical equilibria must be maintained to ensure physiological well-being. If environmental conditions change, the body must adapt to keep functioning. The consequences of a sudden change in altitude dramatize this fact. Flying from San Francisco, which is at sea level, to Mexico City, where the elevation is 2.3 km (1.4 mi), or scaling a 3-km mountain in two days can cause headache, nausea, extreme fatigue, and other discomforts. These conditions are all symptoms of hypoxia, a deficiency in the amount of oxygen reaching body tissues. In serious cases, the victim may slip into a coma and die if not treated quickly. And yet a person living at a high altitude for weeks or months gradually recovers from altitude sickness and adjusts to the low oxygen content in the atmosphere, so that he or she can function normally.

The combination of oxygen with the hemoglobin (Hb) molecule, which carries oxygen through the blood, is a complex reaction, but for our purposes it can be represented by a simplified equation:

$$\text{Hb}(aq) + \text{O}_2(aq) \rightleftharpoons \text{HbO}_2(aq)$$

where HbO_2 is oxyhemoglobin, the hemoglobin-oxygen complex that actually transports oxygen to tissues. The equilibrium constant is

$$K_c = \frac{[\text{HbO}_2]}{[\text{Hb}][\text{O}_2]}$$

At an altitude of 3 km the partial pressure of oxygen is only about 0.14 atm, compared with 0.2 atm at sea level. According to Le Châtelier's principle, a decrease in oxygen concentration will shift the equilibrium shown in the equation above from right to left. This change depletes the supply of oxyhemoglobin, causing hypoxia. Given enough time, the body copes with this problem by producing more hemoglobin molecules. The equilibrium will then gradually shift back toward the formation of oxyhemoglobin. It takes two to three weeks for the increase in hemoglobin production to meet the body's basic needs adequately. A return to full capacity may require several years to occur. Studies show that long-time residents of high-altitude areas have high hemoglobin levels in their blood—sometimes as much as 50 percent more than individuals living at sea level!

Mountaineers need weeks or even months to become acclimatized before scaling summits such as Mount Everest.
Daniel Prudek/iStock/Getty Images

equilibrium concentrations of the reacting mixture, but they cannot change the equilibrium constant as long as the temperature does not change. A catalyst will speed up the process, but it has no effect on the equilibrium constant or on the equilibrium concentrations of the reacting species. Two processes that illustrate the effects of changed conditions on equilibrium processes are discussed in the Chemistry in Action essays "Life at High Altitudes and Hemoglobin Production" and "The Haber Process."

The effects of temperature, concentration, and pressure change, as well as the addition of an inert gas, on an equilibrium system are treated in Example 14.13.

Example 14.13

Consider the following equilibrium process between dinitrogen tetrafluoride (N_2F_4) and nitrogen difluoride (NF_2):

$$\text{N}_2\text{F}_4(g) \rightleftharpoons 2\text{NF}_2(g) \qquad \Delta H° = 38.5 \text{ kJ/mol}$$

(Continued)

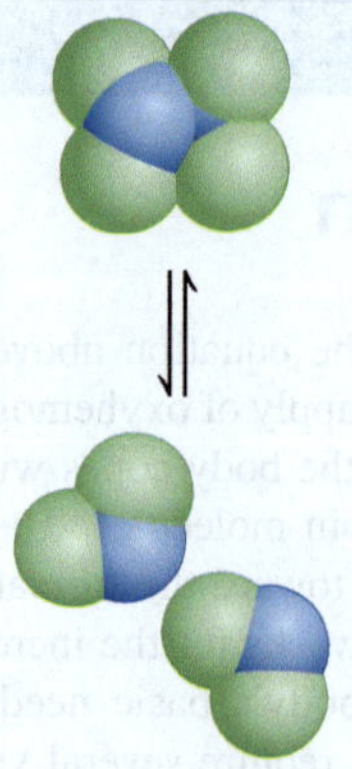

$N_2F_4 \rightleftharpoons 2NF_2$

Predict the changes in the equilibrium if (a) the reacting mixture is heated at constant volume; (b) some N_2F_4 gas is removed from the reacting mixture at constant temperature and volume; (c) the pressure on the reacting mixture is decreased at constant temperature; and (d) a catalyst is added to the reacting mixture.

Strategy (a) What does the sign of $\Delta H°$ indicate about the heat change (endothermic or exothermic) for the forward reaction? (b) Would the removal of some N_2F_4 increase or decrease the Q_c of the reaction? (c) How would the decrease in pressure change the volume of the system? (d) What is the function of a catalyst? How does it affect a reacting system not at equilibrium? at equilibrium?

Solution

(a) The stress applied is the heat added to the system. Note that the $N_2F_4 \longrightarrow 2NF_2$ reaction is an endothermic process ($\Delta H° > 0$), which absorbs heat from the surroundings. Therefore, we can think of heat as a reactant

$$\text{heat} + N_2F_4(g) \rightleftharpoons 2NF_2(g)$$

The system will adjust to remove some of the added heat by undergoing a decomposition reaction (from left to right). The equilibrium constant

$$K_c = \frac{[NF_2]^2}{[N_2F_4]}$$

will therefore increase with increasing temperature because the concentration of NF_2 has increased and that of N_2F_4 has decreased. Recall that the equilibrium constant is a constant only at a particular temperature. If the temperature is changed, then the equilibrium constant will also change.

(b) The stress here is the removal of N_2F_4 gas. The system will shift to replace some of the N_2F_4 removed. Therefore, the system shifts from right to left until equilibrium is reestablished. As a result, some NF_2 combines to form N_2F_4.

Comment The equilibrium constant remains unchanged in this case because temperature is held constant. It might seem that K_c should change because NF_2 combines to produce N_2F_4. Remember, however, that initially some N_2F_4 was removed. The system adjusts to replace only some of the N_2F_4 that was removed, so that overall the amount of N_2F_4 has decreased. In fact, by the time the equilibrium is reestablished, the amounts of both NF_2 and N_2F_4 have decreased. Looking at the equilibrium constant expression, we see that dividing a smaller numerator by a smaller denominator gives the same value of K_c.

(c) The stress applied is a decrease in pressure (which is accompanied by an increase in gas volume). The system will adjust to remove the stress by increasing the pressure. Recall that pressure is directly proportional to the number of moles of a gas. In the balanced equation we see that the formation of NF_2 from N_2F_4 will increase the total number of moles of gases and hence the pressure. Therefore, the system will shift from left to right to reestablish equilibrium. The equilibrium constant will remain unchanged because temperature is held constant.

(d) The function of a catalyst is to increase the rate of a reaction. If a catalyst is added to a reacting system not at equilibrium, the system will reach equilibrium faster than if left undisturbed. If a system is already at equilibrium, as in this case, the addition of a catalyst will not affect either the concentrations of NF_2 and N_2F_4 or the equilibrium constant.

Practice Exercise Consider the equilibrium between molecular oxygen and ozone:

$$3O_2(g) \rightleftharpoons 2O_3(g) \qquad \Delta H° = 284 \text{ kJ/mol}$$

What would be the effect of (a) increasing the pressure on the system by decreasing the volume, (b) adding O_2 to the system at constant volume, (c) decreasing the temperature, and (d) adding a catalyst?

Similar problems: 14.57, 14.58.

CHEMISTRY *in Action*

The Haber Process

Knowing the factors that affect chemical equilibrium has great practical value for industrial applications, such as the synthesis of ammonia. The Haber process for synthesizing ammonia from molecular hydrogen and nitrogen uses a heterogeneous catalyst to speed up the reaction (see Section 13.6). Let us look at the equilibrium reaction for ammonia synthesis to determine whether there are factors that could be manipulated to enhance the yield.

Suppose, as a prominent industrial chemist at the turn of the twentieth century, you are asked to design an efficient procedure for synthesizing ammonia from hydrogen and nitrogen. Your main objective is to obtain a high yield of the product while keeping the production costs down. Your first step is to take a careful look at the balanced equation for the production of ammonia:

$$N_2(g) + 3H_2(g) \rightleftharpoons 2NH_3(g) \qquad \Delta H^\circ = -92.6 \text{ kJ/mol}$$

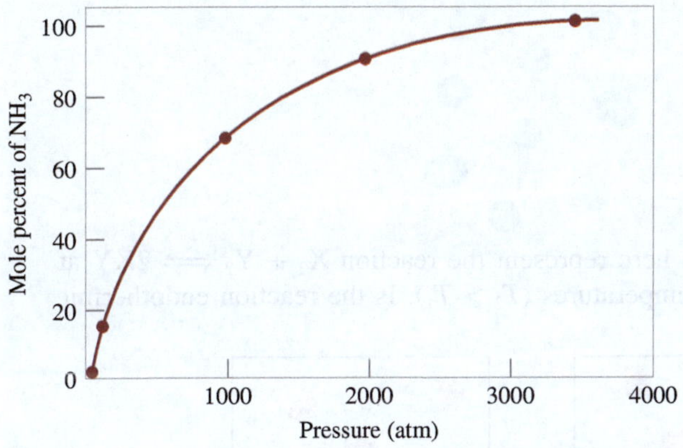

Mole percent of NH_3 as a function of the total pressure of the gases at 425°C.

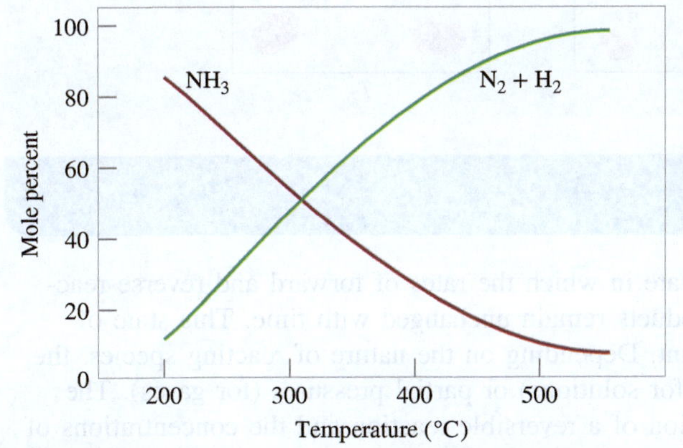

The composition (mole percent) of $H_2 + N_2$ and NH_3 at equilibrium (for a certain starting mixture) as a function of temperature.

Two ideas strike you: *First,* because 1 mole of N_2 reacts with 3 moles of H_2 to produce 2 moles of NH_3, a higher yield of NH_3 can be obtained at equilibrium if the reaction is carried out under high pressures. This is indeed the case, as shown by the plot of mole percent of NH_3 versus the total pressure of the reacting system. *Second,* the exothermic nature of the forward reaction tells you that the equilibrium constant for the reaction will decrease with increasing temperature. Thus, for maximum yield of NH_3, the reaction should be run at the lowest possible temperature. The second graph shows that the yield of ammonia increases with decreasing temperature. A low-temperature operation (say, 220 K or −53°C) is desirable in other respects, too. The boiling point of NH_3 is −33.5°C, so as it formed it would quickly condense to a liquid, which could be conveniently removed from the reacting system. (Both H_2 and N_2 are still gases at this temperature.) Consequently, the net reaction would shift from left to right, just as desired.

On paper, then, these are your conclusions. Let us compare your recommendations with the actual conditions in an industrial plant. Typically, the operating pressures are between 500 atm and 1000 atm, so you are right to advocate high pressure. Furthermore, in the industrial process the NH_3 never reaches its equilibrium value but is constantly removed from the reaction mixture in a continuous process operation. This design makes sense, too, as you had anticipated. The only discrepancy is that the operation is usually carried out at about 500°C. This high-temperature operation is costly and the yield of NH_3 is low. The justification for this choice is that the *rate* of NH_3 production increases with increasing temperature. Commercially, faster production of NH_3 is preferable even if it means a lower yield and higher operating cost. For this reason, the combination of high-pressure, high-temperature conditions, and the proper catalyst is the most efficient way to produce ammonia on a large scale.

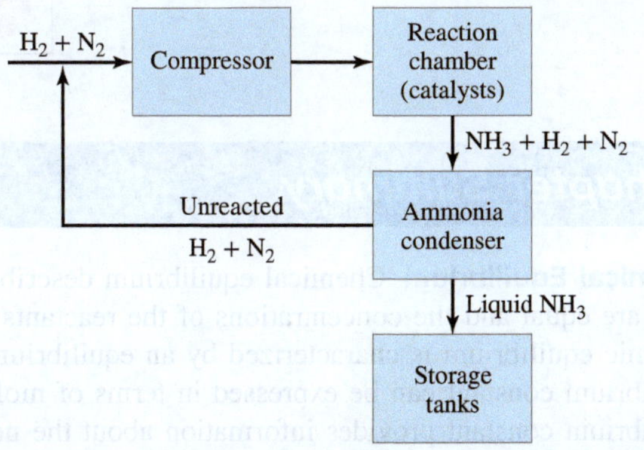

Schematic diagram of the Haber process for ammonia synthesis. The heat generated from the reaction is used to heat the incoming gases.

Summary of Concepts & Facts

- Le Châtelier's principle states that if an external stress is applied to a system at chemical equilibrium, the system will adjust to partially offset the stress.
- Only a change in temperature changes the value of the equilibrium constant for a particular reaction. Changes in concentration, pressure, or volume may change the equilibrium concentrations of reactants and products. The addition of a catalyst hastens the attainment of equilibrium but does not affect the equilibrium concentrations of reactants and products.

Review of Concepts & Facts

14.5.1 For the equilibrium system

$$2SO_2(g) + O_2(g) \rightleftharpoons 2SO_3 \quad \Delta H° = -198 \text{ kJ/mol}$$

determine in which direction the equilibrium will shift in response to each change: (a) increase in temperature, (b) decrease in pressure, (c) addition of $O_2(g)$, (d) addition of a catalyst.

14.5.2 The diagram here shows the gaseous reaction $2A \rightleftharpoons A_2$ at equilibrium. If the pressure is decreased by increasing the volume at constant temperature, how would the concentrations of A and A_2 change when a new equilibrium is established?

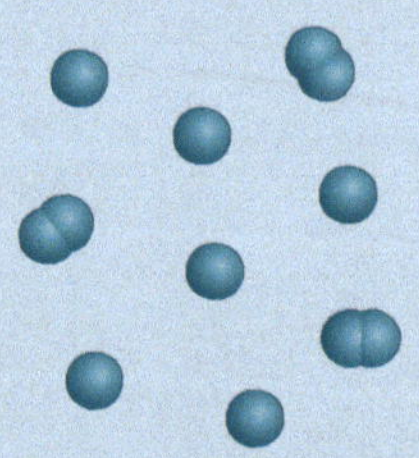

14.5.3 The diagrams shown here represent the reaction $X_2 + Y_2 \rightleftharpoons 2XY$ at equilibrium at two temperatures ($T_2 > T_1$). Is the reaction endothermic or exothermic?

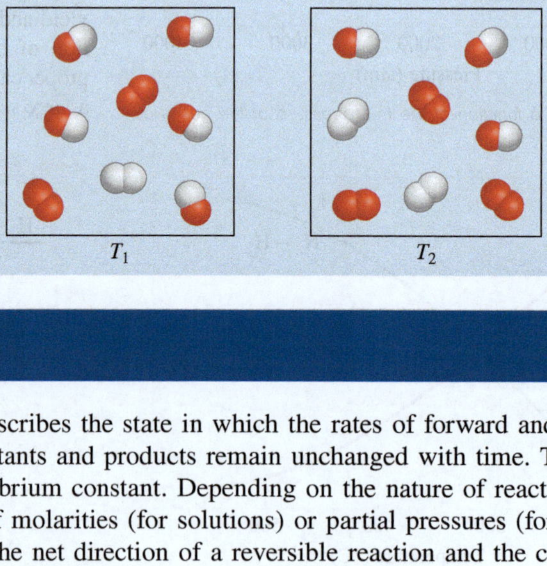

Chapter Summary

Chemical Equilibrium Chemical equilibrium describes the state in which the rates of forward and reverse reactions are equal and the concentrations of the reactants and products remain unchanged with time. This state of dynamic equilibrium is characterized by an equilibrium constant. Depending on the nature of reacting species, the equilibrium constant can be expressed in terms of molarities (for solutions) or partial pressures (for gases). The equilibrium constant provides information about the net direction of a reversible reaction and the concentrations of the equilibrium mixture. (Sections 14.1, 14.2, 14.3, and 14.4)

Factors That Affect Chemical Equilibrium Changes in concentration can affect the position of an equilibrium state—that is, the relative amounts of reactants and products. Changes in pressure and volume may have the same effect for gaseous systems at equilibrium. Only a change in temperature can alter the value of the equilibrium constant. A catalyst can establish the equilibrium state faster by speeding the forward and reverse reactions, but it can change neither the equilibrium position nor the equilibrium constant. (Section 14.5)

Key Equations

$K = \frac{[C]^c[D]^d}{[A]^a[B]^b}$ (14.2) Law of mass action. General expression of equilibrium constant.

$K_P = K_c(0.0821T)^{\Delta n}$ (14.5) Relationship between K_P and K_c.

$K_c = K_c' K_c''$ (14.9) The equilibrium constant for the overall reaction is given by the product of the equilibrium constants for the individual reactions.

Key Words

Chemical equilibrium, p. 630
Equilibrium constant (K), p. 632
Heterogeneous equilibrium, p. 638
Homogeneous equilibrium, p. 634
Law of mass action, p. 632
Le Châtelier's principle, p. 652
Physical equilibrium, p. 630
Reaction quotient (Q_c), p. 646

Questions & Problems

Red numbered problems solved in Student Solutions Manual

14.1 The Concept of Equilibrium and the Equilibrium Constant

Review Questions

14.1 Define equilibrium. Give two examples of a dynamic equilibrium.

14.2 Explain the difference between physical equilibrium and chemical equilibrium. Give two examples of each.

14.3 What is the law of mass action?

14.4 Briefly describe the importance of equilibrium in the study of chemical reactions.

14.2 Writing Equilibrium Constant Expressions

Review Questions

14.5 Define homogeneous equilibrium and heterogeneous equilibrium. Give two examples of each.

14.6 What do the symbols K_c and K_P represent?

14.7 Write the expressions for the equilibrium constants K_P of the following thermal decomposition reactions:
(a) $2NaHCO_3(s) \rightleftharpoons Na_2CO_3(s) + CO_2(g) + H_2O(g)$
(b) $2CaSO_4(s) \rightleftharpoons 2CaO(s) + 2SO_2(g) + O_2(g)$

14.8 Write equilibrium constant expressions for K_c, and for K_P if applicable, for the following processes:
(a) $2CO_2(g) \rightleftharpoons 2CO(g) + O_2(g)$
(b) $3O_2(g) \rightleftharpoons 2O_3(g)$
(c) $CO(g) + Cl_2(g) \rightleftharpoons COCl_2(g)$
(d) $H_2O(g) + C(s) \rightleftharpoons CO(g) + H_2(g)$
(e) $HCOOH(aq) \rightleftharpoons H^+(aq) + HCOO^-(aq)$
(f) $2HgO(s) \rightleftharpoons 2Hg(l) + O_2(g)$

14.9 Write the equilibrium constant expressions for K_c, and K_P if applicable, for the following reactions:
(a) $2NO_2(g) + 7H_2(g) \rightleftharpoons 2NH_3(g) + 4H_2O(l)$
(b) $2ZnS(s) + 3O_2(g) \rightleftharpoons 2ZnO(s) + 2SO_2(g)$
(c) $C(s) + CO_2(g) \rightleftharpoons 2CO(g)$
(d) $C_6H_5COOH(aq) \rightleftharpoons C_6H_5COO^-(aq) + H^+(aq)$

14.10 Write the equation relating K_c to K_P, and define all the terms.

14.11 What is the rule for writing the equilibrium constant for the overall reaction involving two or more reactions?

14.12 Give an example of a multiple equilibria reaction.

Problems

14.13 The equilibrium constant for the reaction A $\rightleftharpoons$ B is $K_c = 10$ at a certain temperature. (1) Starting with only reactant A, which of the diagrams shown here best represents the system at equilibrium? (2) Which of the diagrams best represents the system at equilibrium if $K_c = 0.10$? Explain why you can calculate K_c in each case without knowing the volume of the container. The gray spheres represent the A molecules and the green spheres represent the B molecules.

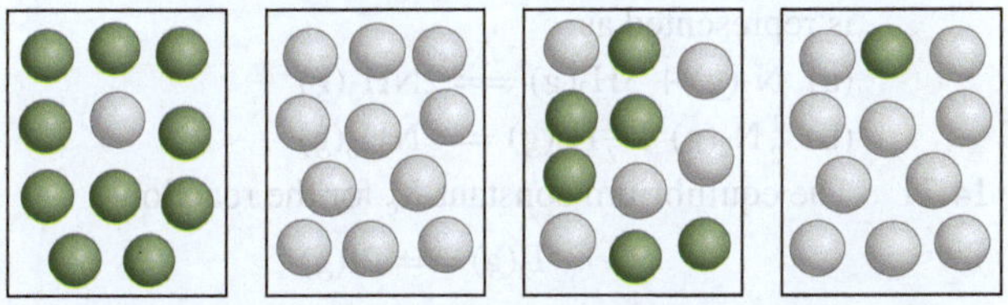

(a) (b) (c) (d)

14.14 The following diagrams represent the equilibrium state for three different reactions of the type A + X $\rightleftharpoons$ AX (X = B, C, or D):

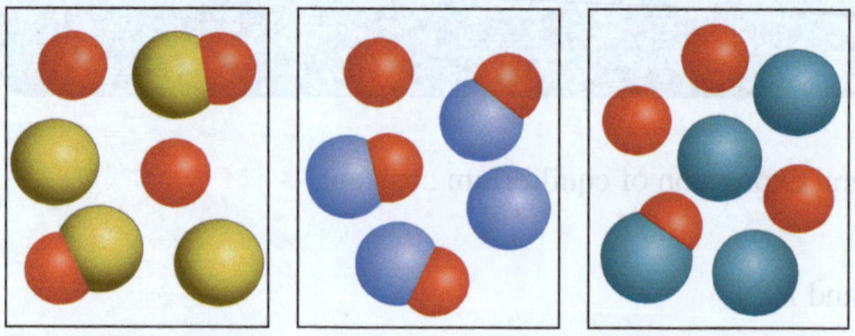

$A + B \rightleftharpoons AB$ $A + C \rightleftharpoons AC$ $A + D \rightleftharpoons AD$

(a) Which reaction has the largest equilibrium constant?

(b) Which reaction has the smallest equilibrium constant?

14.15 The equilibrium constant (K_c) for the reaction

$$2HCl(g) \rightleftharpoons H_2(g) + Cl_2(g)$$

is 4.17×10^{-34} at 25°C. What is the equilibrium constant for the reaction

$$H_2(g) + Cl_2(g) \rightleftharpoons 2HCl(g)$$

at the same temperature?

14.16 Consider the following equilibrium process at 700°C:

$$2H_2(g) + S_2(g) \rightleftharpoons 2H_2S(g)$$

Analysis shows that there are 2.50 moles of H_2, 1.35×10^{-5} mole of S_2, and 8.70 moles of H_2S present in a 12.0-L flask. Calculate the equilibrium constant K_c for the reaction.

14.17 What is K_P at 1273°C for the reaction

$$2CO(g) + O_2(g) \rightleftharpoons 2CO_2(g)$$

if K_c is 2.24×10^{22} at the same temperature?

14.18 The equilibrium constant K_P for the reaction

$$2SO_3(g) \rightleftharpoons 2SO_2(g) + O_2(g)$$

is 1.8×10^{-5} at 350°C. What is K_c for this reaction?

14.19 Consider the following reaction:

$$N_2(g) + O_2(g) \rightleftharpoons 2NO(g)$$

If the equilibrium partial pressures of N_2, O_2, and NO are 0.15 atm, 0.33 atm, and 0.050 atm, respectively, at 2200°C, what is K_P?

14.20 A reaction vessel contains NH_3, N_2, and H_2 at equilibrium at a certain temperature. The equilibrium concentrations are $[NH_3] = 0.25\ M$, $[N_2] = 0.11\ M$, and $[H_2] = 1.91\ M$. Calculate the equilibrium constant K_c for the synthesis of ammonia if the reaction is represented as

(a) $N_2(g) + 3H_2(g) \rightleftharpoons 2NH_3(g)$

(b) $\frac{1}{2}N_2(g) + \frac{3}{2}H_2(g) \rightleftharpoons NH_3(g)$

14.21 The equilibrium constant K_c for the reaction

$$I_2(g) \rightleftharpoons 2I(g)$$

is 3.8×10^{-5} at 727°C. Calculate K_c and K_P for the equilibrium

$$2I(g) \rightleftharpoons I_2(g)$$

at the same temperature.

14.22 At equilibrium, the pressure of the reacting mixture

$$CaCO_3(s) \rightleftharpoons CaO(s) + CO_2(g)$$

is 0.105 atm at 350°C. Calculate K_P and K_c for this reaction.

14.23 The equilibrium constant K_P for the reaction

$$PCl_5(g) \rightleftharpoons PCl_3(g) + Cl_2(g)$$

is 1.05 at 250°C. The reaction starts with a mixture of PCl_5, PCl_3, and Cl_2 at pressures 0.177 atm, 0.223 atm, and 0.111 atm, respectively, at 250°C. When the mixture comes to equilibrium at that temperature, which pressures will have decreased and which will have increased? Explain why.

14.24 Ammonium carbamate, $NH_4CO_2NH_2$, decomposes as follows:

$$NH_4CO_2NH_2(s) \rightleftharpoons 2NH_3(g) + CO_2(g)$$

Starting with only the solid, it is found that at 40°C the total gas pressure (NH_3 and CO_2) is 0.363 atm. Calculate the equilibrium constant K_P.

14.25 Consider the following reaction at 1600°C.

$$Br_2(g) \rightleftharpoons 2Br(g)$$

When 1.05 moles of Br_2 are put in a 0.980-L flask, 1.20 percent of the Br_2 undergoes dissociation. Calculate the equilibrium constant K_c for the reaction.

14.26 Pure phosgene gas ($COCl_2$), 3.00×10^{-2} mol, was placed in a 1.50-L container. It was heated to 800 K, and at equilibrium the pressure of CO was found to be 0.497 atm. Calculate the equilibrium constant K_P for the reaction

$$CO(g) + Cl_2(g) \rightleftharpoons COCl_2(g)$$

14.27 Consider the equilibrium

$$2NOBr(g) \rightleftharpoons 2NO(g) + Br_2(g)$$

If nitrosyl bromide, NOBr, is 34 percent dissociated at 25°C and the total pressure is 0.25 atm, calculate K_P and K_c for the dissociation at this temperature.

14.28 A 2.50-mole quantity of NOCl was initially in a 1.50-L reaction chamber at 400°C. After equilibrium was established, it was found that 28.0 percent of the NOCl had dissociated:

$$2NOCl(g) \rightleftharpoons 2NO(g) + Cl_2(g)$$

Calculate the equilibrium constant K_c for the reaction.

14.29 The following equilibrium constants have been determined for hydrosulfuric acid at 25°C:

$$H_2S(aq) \rightleftharpoons H^+(aq) + HS^-(aq) \qquad K_c' = 9.5 \times 10^{-8}$$

$$HS^-(aq) \rightleftharpoons H^+(aq) + S^{2-}(aq) \qquad K_c'' = 1.0 \times 10^{-19}$$

Calculate the equilibrium constant for the following reaction at the same temperature:

$$H_2S(aq) \rightleftharpoons 2H^+(aq) + S^{2-}(aq)$$

14.30 The following equilibrium constants have been determined for oxalic acid at 25°C:

$$H_2C_2O_4(aq) \rightleftharpoons H^+(aq) + HC_2O_4^-(aq) \quad K_c' = 6.5 \times 10^{-2}$$

$$HC_2O_4^-(aq) \rightleftharpoons H^+(aq) + C_2O_4^{2-}(aq) \quad K_c'' = 6.1 \times 10^{-5}$$

Calculate the equilibrium constant for the following reaction at the same temperature:

$$H_2C_2O_4(aq) \rightleftharpoons 2H^+(aq) + C_2O_4^{2-}(aq)$$

14.31 The following equilibrium constants were determined at 1123 K:

$$C(s) + CO_2(g) \rightleftharpoons 2CO(g) \quad K_P' = 1.3 \times 10^{14}$$

$$CO(g) + Cl_2(g) \rightleftharpoons COCl_2(g) \quad K_P'' = 6.0 \times 10^{-3}$$

Write the equilibrium constant expression K_P, and calculate the equilibrium constant at 1123 K for

$$C(s) + CO_2(g) + 2Cl_2(g) \rightleftharpoons 2COCl_2(g)$$

14.32 At a certain temperature the following reactions have the constants shown:

$$S(s) + O_2(g) \rightleftharpoons SO_2(g) \quad K_c' = 4.2 \times 10^{52}$$

$$2S(s) + 3O_2(g) \rightleftharpoons 2SO_3(g) \quad K_c'' = 9.8 \times 10^{128}$$

Calculate the equilibrium constant K_c for the following reaction at that temperature:

$$2SO_2(g) + O_2(g) \rightleftharpoons 2SO_3(g)$$

14.3 The Relationship Between Chemical Kinetics and Chemical Equilibrium

Review Questions

14.33 Based on rate constant considerations, explain why the equilibrium constant depends on temperature.

14.34 Explain why reactions with large equilibrium constants, such as the formation of rust (Fe_2O_3), may have very slow rates.

Problems

14.35 Water is a very weak electrolyte that undergoes the following ionization (called autoionization):

$$H_2O(l) \underset{k_{-1}}{\overset{k_1}{\rightleftharpoons}} H^+(aq) + OH^-(aq)$$

(a) If $k_1 = 2.4 \times 10^{-5}\ s^{-1}$ and $k_{-1} = 1.3 \times 10^{11}/M \cdot s$, calculate the equilibrium constant K where $K = [H^+][OH^-]/[H_2O]$.

(b) Calculate the product $[H^+][OH^-]$ and $[H^+]$ and $[OH^-]$.

14.36 Consider the following reaction, which takes place in a single elementary step:

$$2A + B \underset{k_{-1}}{\overset{k_1}{\rightleftharpoons}} A_2B$$

If the equilibrium constant K_c is 12.6 at a certain temperature and if $k_r = 5.1 \times 10^{-2}\ s^{-1}$, calculate the value of k_f.

14.4 What Does the Equilibrium Constant Tell Us?

Review Questions

14.37 Define reaction quotient. How does it differ from equilibrium constant?

14.38 Outline the steps for calculating the concentrations of reacting species in an equilibrium reaction.

Problems

14.39 The equilibrium constant K_P for the reaction

$$2SO_2(g) + O_2(g) \rightleftharpoons 2SO_3(g)$$

is 5.60×10^4 at 350°C. The initial pressures of SO_2 and O_2 in a mixture are 0.350 atm and 0.762 atm, respectively, at 350°C. When the mixture equilibrates, is the total pressure less than or greater than the sum of the initial pressures (1.112 atm)?

14.40 For the synthesis of ammonia

$$N_2(g) + 3H_2(g) \rightleftharpoons 2NH_3(g)$$

the equilibrium constant K_c at 375°C is 1.2. Starting with $[H_2]_0 = 0.76\ M$, $[N_2]_0 = 0.60\ M$, and $[NH_3]_0 = 0.48\ M$, which gases will have increased in concentration and which will have decreased in concentration when the mixture comes to equilibrium?

14.41 For the reaction

$$H_2(g) + CO_2(g) \rightleftharpoons H_2O(g) + CO(g)$$

at 700°C, $K_c = 0.534$. Calculate the number of moles of H_2 that are present at equilibrium if a mixture of 0.300 mole of CO and 0.300 mole of H_2O is heated to 700°C in a 10.0-L container.

14.42 At 1000 K, a sample of pure NO_2 gas decomposes:

$$2NO_2(g) \rightleftharpoons 2NO(g) + O_2(g)$$

The equilibrium constant K_P is 158. Analysis shows that the partial pressure of O_2 is 0.25 atm at equilibrium. Calculate the pressure of NO and NO_2 in the mixture.

14.43 The equilibrium constant K_c for the reaction

$$H_2(g) + Br_2(g) \rightleftharpoons 2HBr(g)$$

is 2.18×10^6 at 730°C. Starting with 3.20 moles of HBr in a 12.0-L reaction vessel, calculate the concentrations of H_2, Br_2, and HBr at equilibrium.

14.44 The dissociation of molecular iodine into iodine atoms is represented as

$$I_2(g) \rightleftharpoons 2I(g)$$

At 1000 K, the equilibrium constant K_c for the reaction is 3.80×10^{-5}. Suppose you start with 0.0456 mole of I_2 in a 2.30-L flask at 1000 K. What are the concentrations of the gases at equilibrium?

14.45 The equilibrium constant K_c for the decomposition of phosgene, $COCl_2$, is 4.63×10^{-3} at 527°C:

$$COCl_2(g) \rightleftharpoons CO(g) + Cl_2(g)$$

Calculate the equilibrium partial pressure of all the components, starting with pure phosgene at 0.760 atm.

14.46 Consider the following equilibrium process at 686°C:

$$CO_2(g) + H_2(g) \rightleftharpoons CO(g) + H_2O(g)$$

The equilibrium concentrations of the reacting species are [CO] = 0.050 *M*, $[H_2]$ = 0.045 *M*, $[CO_2]$ = 0.086 *M*, and $[H_2O]$ = 0.040 *M*. (a) Calculate K_c for the reaction at 686°C. (b) If we add CO_2 to increase its concentration to 0.50 mol/L, what will the concentrations of all the gases be when equilibrium is reestablished?

14.47 Consider the heterogeneous equilibrium process:

$$C(s) + CO_2(g) \rightleftharpoons 2CO(g)$$

At 700°C, the total pressure of the system is found to be 4.50 atm. If the equilibrium constant K_P is 1.52, calculate the equilibrium partial pressures of CO_2 and CO.

14.48 The equilibrium constant K_c for the reaction

$$H_2(g) + CO_2(g) \rightleftharpoons H_2O(g) + CO(g)$$

is 4.2 at 1650°C. Initially 0.80 mol H_2 and 0.80 mol CO_2 are injected into a 5.0-L flask. Calculate the concentration of each species at equilibrium.

14.5 Factors That Affect Chemical Equilibrium

Review Questions

14.49 Explain Le Châtelier's principle. How can this principle help us maximize the yields of reactions?

14.50 Use Le Châtelier's principle to explain why the equilibrium vapor pressure of a liquid increases with increasing temperature.

14.51 List four factors that can shift the position of an equilibrium. Only one of these factors can alter the value of the equilibrium constant. Which one is it?

14.52 Does the addition of a catalyst have any effects on the position of an equilibrium?

Problems

14.53 Consider the following equilibrium system involving SO_2, Cl_2, and SO_2Cl_2 (sulfuryl dichloride):

$$SO_2(g) + Cl_2(g) \rightleftharpoons SO_2Cl_2(g)$$

Predict how the equilibrium position would change if (a) Cl_2 gas were added to the system; (b) SO_2Cl_2 were removed from the system; (c) SO_2 were removed from the system. The temperature remains constant.

14.54 Heating solid sodium bicarbonate in a closed vessel establishes the following equilibrium:

$$2NaHCO_3(s) \rightleftharpoons NaCO_3(s) + H_2O(g) + CO_2(g)$$

What would happen to the equilibrium position if (a) some of the CO_2 were removed from the system; (b) some solid Na_2CO_3 were added to the system; (c) some of the solid $NaHCO_2$ were removed from the system? The temperature remains constant.

14.55 Consider the following equilibrium systems:

(a) $A \rightleftharpoons 2B$ $\Delta H° = 20.0$ kJ/mol

(b) $A + B \rightleftharpoons C$ $\Delta H° = -5.4$ kJ/mol

(c) $A \rightleftharpoons B$ $\Delta H° = 0.0$ kJ/mol

Predict the change in the equilibrium constant K_c that would occur in each case if the temperature of the reacting system were raised.

14.56 What effect does an increase in pressure have on each of the following systems at equilibrium? The temperature is kept constant and, in each case, the reactants are in a cylinder fitted with a movable piston.

(a) $A(s) \rightleftharpoons 2B(s)$

(b) $2A(l) \rightleftharpoons B(l)$

(c) $A(s) \rightleftharpoons B(g)$

(d) $A(g) \rightleftharpoons B(g)$

(e) $A(g) \rightleftharpoons 2B(g)$

14.57 Consider the equilibrium

$$2I(g) \rightleftharpoons I_2(g)$$

What would be the effect on the position of equilibrium of (a) increasing the total pressure on the system by decreasing its volume; (b) adding gaseous I_2 to the reaction mixture; and (c) decreasing the temperature at constant volume?

14.58 Consider the following equilibrium process:

$$PCl_5(g) \rightleftharpoons PCl_3(g) + Cl_2(g) \quad \Delta H° = 87.9 \text{ kJ/mol}$$

Predict the direction of the shift in equilibrium when (a) the temperature is raised; (b) more chlorine gas is added to the reaction mixture; (c) some PCl_3 is removed from the mixture; (d) the pressure on the gases is increased; (e) a catalyst is added to the reaction mixture.

14.59 Consider the reaction

$$2SO_2(g) + O_2(g) \rightleftharpoons 2SO_3(g)$$
$$\Delta H° = -197.8 \text{ kJ/mol}$$

Comment on the changes in the concentrations of SO_2, O_2, and SO_3 at equilibrium if we were to (a) increase the temperature; (b) increase the pressure; (c) increase SO_2; (d) add a catalyst; (e) add helium at constant volume.

14.60 In the uncatalyzed reaction

$$N_2O_4(g) \rightleftharpoons 2NO_2(g)$$

the pressures of the gases at equilibrium are $P_{N_2O_4}$ = 0.377 and P_{NO_2} = 1.56 atm at 100°C. What would happen to these pressures if a catalyst were added to the mixture?

14.61 Consider the gas-phase reaction

$$2CO(g) + O_2(g) \rightleftharpoons 2CO_2(g)$$

Predict the shift in the equilibrium position when helium gas is added to the equilibrium mixture at (a) constant pressure and (b) constant volume.

14.62 Consider this equilibrium reaction in a closed container:

$$CaCO_3(s) \rightleftharpoons CaO(s) + CO_2(g)$$

What will happen if the following occurs? (a) The volume is increased. (b) Some CaO is added to the mixture. (c) Some $CaCO_3$ is removed. (d) Some CO_2 is added to the mixture. (e) A few drops of a NaOH solution are added to the mixture. (f) A few drops of a HCl solution are added to the mixture (ignore the reaction between CO_2 and water). (g) Temperature is increased.

Additional Problems

14.63 Consider the statement: "The equilibrium constant of a reacting mixture of solid NH_4Cl and gaseous NH_3 and HCl is 0.316." List three important pieces of information that are missing from this statement.

14.64 Pure nitrosyl chloride (NOCl) gas was heated to 240°C in a 1.00-L container. At equilibrium the total pressure was 1.00 atm and the NOCl pressure was 0.64 atm.

$$2NOCl(g) \rightleftharpoons 2NO(g) + Cl_2(g)$$

(a) Calculate the partial pressures of NO and Cl_2 in the system.

(b) Calculate the equilibrium constant K_P.

14.65 Determine the initial and equilibrium concentrations of HI if the initial concentrations of H_2 and I_2 are both 0.16 *M* and their equilibrium concentrations are both 0.072 *M* at 430°C. The equilibrium constant (K_c) for the reaction $H_2(g) + I_2(g) \rightleftharpoons 2HI(g)$ is 54.2 at 430°C.

14.66 Diagram (a) shows the reaction $A_2(g) + B_2(g) \rightleftharpoons 2AB(g)$ at equilibrium at a certain temperature, where the blue spheres represent A and the yellow spheres represent B. If each sphere represents 0.020 mole and the volume of the container is 1.0 L, calculate the concentration of each species when the reaction in (b) reaches equilibrium.

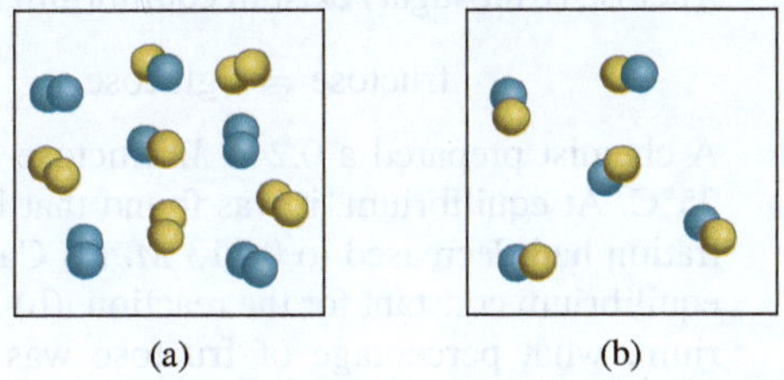

14.67 The equilibrium constant (K_P) for the formation of the air pollutant nitric oxide (NO) in an automobile engine at 530°C is 2.9×10^{-11}:

$$N_2(g) + O_2(g) \rightleftharpoons 2NO(g)$$

(a) Calculate the partial pressure of NO under these conditions if the partial pressures of nitrogen and oxygen are 3.0 atm and 0.012 atm, respectively. (b) Repeat the calculation for atmospheric conditions where the partial pressures of nitrogen and oxygen are 0.78 atm and 0.21 atm and the temperature is 25°C. (The K_P for the reaction is 4.0×10^{31} at this temperature.) (c) Is the formation of NO endothermic or exothermic? (d) What natural phenomenon promotes the formation of NO? Why?

14.68 Baking soda (sodium bicarbonate) undergoes thermal decomposition as follows:

$$2NaHCO_3(s) \rightleftharpoons Na_2CO_3(s) + CO_2(g) + H_2O(g)$$

Would we obtain more CO_2 and H_2O by adding extra baking soda to the reaction mixture in (a) a closed vessel or (b) an open vessel?

14.69 Consider the following reaction at equilibrium:

$$A(g) \rightleftharpoons 2B(g)$$

From the data shown here, calculate the equilibrium constant (both K_P and K_c) at each temperature. Is the reaction endothermic or exothermic?

Temperature (°C)	[A] (*M*)	[B] (*M*)
200	0.0125	0.843
300	0.171	0.764
400	0.250	0.724

14.70 The equilibrium constant K_P for the reaction

$$2H_2O(g) \rightleftharpoons 2H_2(g) + O_2(g)$$

is 2×10^{-42} at 25°C. (a) What is K_c for the reaction at the same temperature? (b) The very small value of K_P (and K_c) indicates that the reaction overwhelmingly favors the formation of water molecules. Explain why, despite this fact, a mixture of hydrogen and oxygen gases can be kept at room temperature without any change.

14.71 Consider the following reacting system:

$$2NO(g) + Cl_2(g) \rightleftharpoons 2NOCl(g)$$

What combination of temperature and pressure (high or low) would maximize the yield of nitrosyl chloride (NOCl)? (*Hint:* You will also need to consult Appendix 2.)

14.72 At a certain temperature and a total pressure of 1.2 atm, the partial pressures of an equilibrium mixture

$$2A(g) \rightleftharpoons B(g)$$

are $P_A = 0.60$ atm and $P_B = 0.60$ atm. (a) Calculate the K_P for the reaction at this temperature. (b) If the total pressure were increased to 1.5 atm, what would be the partial pressures of A and B at equilibrium?

14.73 The decomposition of ammonium hydrogen sulfide

$$NH_4HS(s) \rightleftharpoons NH_3(g) + H_2S(g)$$

is an endothermic process. A 6.1589-g sample of the solid is placed in an evacuated 4.000-L vessel at exactly 24°C. After equilibrium has been established, the total pressure inside is 0.709 atm. Some solid NH_4HS remains in the vessel. (a) What is the K_P for the reaction? (b) What percentage of the solid has decomposed? (c) If the volume of the vessel were doubled at constant temperature, what would happen to the amount of solid in the vessel?

14.74 Consider the reaction

$$2NO(g) + O_2(g) \rightleftharpoons 2NO_2(g)$$

At 430°C, an equilibrium mixture consists of 0.020 mole of O_2, 0.040 mole of NO, and 0.096 mole of NO_2. Calculate K_P for the reaction, given that the total pressure is 0.20 atm.

14.75 When heated, ammonium carbamate decomposes as follows:

$$NH_4CO_2NH_2(s) \rightleftharpoons 2NH_3(g) + CO_2(g)$$

At a certain temperature the equilibrium pressure of the system is 0.318 atm. Calculate K_P for the reaction.

14.76 A mixture of 0.47 mole of H_2 and 3.59 moles of HCl is heated to 2800°C. Calculate the equilibrium partial pressures of H_2, Cl_2, and HCl if the total pressure is 2.00 atm. For the reaction

$$H_2(g) + Cl_2(g) \rightleftharpoons 2HCl(g)$$

K_P is 193 at 2800°C.

14.77 When heated at high temperatures, iodine vapor dissociates as follows:

$$I_2(g) \rightleftharpoons 2I(g)$$

In one experiment, a chemist finds that when 0.054 mole of I_2 was placed in a flask of volume 0.48 L at 587 K, the degree of dissociation (that is, the fraction of I_2 dissociated) was 0.0252. Calculate K_c and K_P for the reaction at this temperature.

14.78 One mole of N_2 and three moles of H_2 are placed in a flask at 375°C. Calculate the total pressure of the system at equilibrium if the mole fraction of NH_3 is 0.21. The K_P for the reaction is 4.31×10^{-4}.

14.79 At 1130°C the equilibrium constant (K_c) for the reaction

$$2H_2S(g) \rightleftharpoons 2H_2(g) + S_2(g)$$

is 2.25×10^{-4}. If $[H_2S] = 4.84 \times 10^{-3}$ *M* and $[H_2] = 1.50 \times 10^{-3}$ *M*, calculate $[S_2]$.

14.80 A quantity of 6.75 g of SO_2Cl_2 was placed in a 2.00-L flask. At 648 K, there is 0.0345 mole of SO_2 present. Calculate K_c for the reaction

$$SO_2Cl_2(g) \rightleftharpoons SO_2(g) + Cl_2(g)$$

14.81 The formation of SO_3 from SO_2 and O_2 is an intermediate step in the manufacture of sulfuric acid, and it is also responsible for the acid rain phenomenon. The equilibrium constant K_P for the reaction

$$2SO_2(g) + O_2(g) \rightleftharpoons 2SO_3(g)$$

is 0.13 at 830°C. In one experiment 2.00 mol SO_2 and 2.00 mol O_2 were initially present in a flask. What must the total pressure at equilibrium be to have an 80.0 percent yield of SO_3?

14.82 Consider the dissociation of iodine:

$$I_2(g) \rightleftharpoons 2I(g)$$

A 1.00-g sample of I_2 is heated to 1200°C in a 500-mL flask. At equilibrium the total pressure is 1.51 atm. Calculate K_P for the reaction. [*Hint:* Use the result in Problem 14.117(a). The degree of dissociation α can be obtained by first calculating the ratio of observed pressure over calculated pressure, assuming no dissociation.]

14.83 Eggshells are composed mostly of calcium carbonate ($CaCO_3$) formed by the reaction

$$Ca^{2+}(aq) + CO_3^{2-}(aq) \rightleftharpoons CaCO_3(s)$$

The carbonate ions are supplied by carbon dioxide produced as a result of metabolism. Explain why eggshells are thinner in the summer when the rate of panting by chickens is greater. Suggest a remedy for this situation.

14.84 The equilibrium constant K_P for the following reaction is 4.31×10^{-4} at 375°C:

$$N_2(g) + 3H_2(g) \rightleftharpoons 2NH_3(g)$$

In a certain experiment a student starts with 0.862 atm of N_2 and 0.373 atm of H_2 in a constant-volume vessel at 375°C. Calculate the partial pressures of all species when equilibrium is reached.

14.85 A quantity of 0.20 mole of carbon dioxide was heated to a certain temperature with an excess of graphite in a closed container until the following equilibrium was reached:

$$C(s) + CO_2(g) \rightleftharpoons 2CO(g)$$

Under these conditions, the average molar mass of the gases was 35 g/mol. (a) Calculate the mole fractions of CO and CO_2. (b) What is K_P if the total pressure is 11 atm? (*Hint:* The average molar mass is the sum of the products of the mole fraction of each gas and its molar mass.)

14.86 When dissolved in water, glucose (corn sugar) and fructose (fruit sugar) exist in equilibrium as follows:

$$\text{fructose} \rightleftharpoons \text{glucose}$$

A chemist prepared a 0.244 *M* fructose solution at 25°C. At equilibrium, it was found that its concentration had decreased to 0.113 *M*. (a) Calculate the equilibrium constant for the reaction. (b) At equilibrium, what percentage of fructose was converted to glucose?

14.87 At room temperature, solid iodine is in equilibrium with its vapor through sublimation and deposition (see Section 11.8). Describe how you would use radioactive iodine, in either solid or vapor form, to show that there is a dynamic equilibrium between these two phases.

14.88 At 1024°C, the pressure of oxygen gas from the decomposition of copper(II) oxide (CuO) is 0.49 atm:

$$4CuO(s) \rightleftharpoons 2Cu_2O(s) + O_2(g)$$

(a) What is K_P for the reaction? (b) Calculate the fraction of CuO that will decompose if 0.16 mole of it is placed in a 2.0-L flask at 1024°C. (c) What would the fraction be if a 1.0 mole sample of CuO were used? (d) What is the smallest amount of CuO (in moles) that would establish the equilibrium?

14.89 A mixture containing 3.9 moles of NO and 0.88 mole of CO_2 was allowed to react in a flask at a certain temperature according to the equation

$$NO(g) + CO_2(g) \rightleftharpoons NO_2(g) + CO(g)$$

At equilibrium, 0.11 mole of CO_2 was present. Calculate the equilibrium constant K_c of this reaction.

14.90 The equilibrium constant K_c for the reaction

$$H_2(g) + I_2(g) \rightleftharpoons 2HI(g)$$

is 54.3 at 430°C. At the start of the reaction there are 0.714 mole of H_2, 0.984 mole of I_2, and 0.886 mole of HI in a 2.40-L reaction chamber. Calculate the concentrations of the gases at equilibrium.

14.91 When heated, a gaseous compound A dissociates as follows:

$$A(g) \rightleftharpoons B(g) + C(g)$$

In an experiment, A was heated at a certain temperature until its equilibrium pressure reached $0.14P$, where P is the total pressure. Calculate the equilibrium constant K_P of this reaction.

14.92 When a gas was heated under atmospheric conditions, its color deepened. Heating above 150°C caused the color to fade, and at 550°C the color was barely detectable. However, at 550°C, the color was partially restored by increasing the pressure of the system. Which of the following best fits this description? Justify your choice. (a) A mixture of hydrogen and bromine, (b) pure bromine, (c) a mixture of nitrogen dioxide and dinitrogen tetroxide. (*Hint:* Bromine has a reddish color and nitrogen dioxide is a brown gas. The other gases are colorless.)

14.93 In this chapter we learned that a catalyst has no effect on the position of an equilibrium because it speeds up both the forward and reverse rates to the same extent. To test this statement, consider a situation in which an equilibrium of the type

$$2A(g) \rightleftharpoons B(g)$$

is established inside a cylinder fitted with a weightless piston. The piston is attached by a string to the cover of a box containing a catalyst. When the piston moves upward (expanding against atmospheric pressure), the cover is lifted and the catalyst is exposed to the gases. When the piston moves downward, the box is closed. Assume that the catalyst speeds up the forward reaction ($2A \longrightarrow B$) but does not affect the reverse process ($B \longrightarrow 2A$). Suppose the catalyst is suddenly exposed to the equilibrium system as shown here. Describe what would happen subsequently. How does this "thought" experiment convince you that no such catalyst can exist?

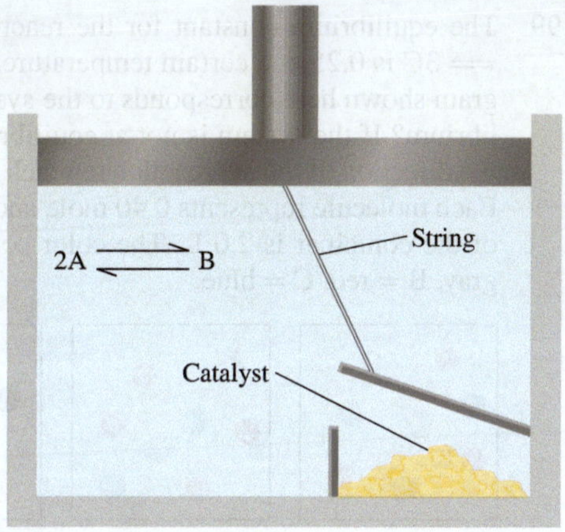

14.94 The equilibrium constant K_c for the following reaction is 1.2 at 375°C.

$$N_2(g) + 3H_2(g) \rightleftharpoons 2NH_3(g)$$

(a) What is the value of K_P for this reaction?
(b) What is the value of the equilibrium constant K_c for $2NH_3(g) \rightleftharpoons N_2(g) + 3H_2(g)$?
(c) What is the value of K_c for $\frac{1}{2}N_2(g) + \frac{3}{2}H_2(g) \rightleftharpoons NH_3(g)$?
(d) What are the values of K_P for the reactions described in (b) and (c)?

14.95 A sealed glass bulb contains a mixture of NO_2 and N_2O_4 gases. Describe what happens to the following properties of the gases when the bulb is heated from 20°C to 40°C: (a) color, (b) pressure, (c) average molar mass, (d) degree of dissociation (from N_2O_4 to NO_2), (e) density. Assume that volume remains constant. (*Hint:* NO_2 is a brown gas; N_2O_4 is colorless.)

14.96 At 20°C, the vapor pressure of water is 0.0231 atm. Calculate K_P and K_c for the process

$$H_2O(l) \rightleftharpoons H_2O(g)$$

14.97 Industrially, sodium metal is obtained by electrolyzing molten sodium chloride. The reaction at the cathode is $Na^+ + e^- \longrightarrow Na$. We might expect that potassium metal would also be prepared by electrolyzing molten potassium chloride. However, potassium metal is soluble in molten potassium chloride and therefore is hard to recover. Furthermore, potassium vaporizes readily at the operating temperature, creating hazardous conditions. Instead, potassium is prepared by the distillation of molten potassium chloride in the presence of sodium vapor at 892°C:

$$Na(g) + KCl(l) \rightleftharpoons NaCl(l) + K(g)$$

In view of the fact that potassium is a stronger reducing agent than sodium, explain why this approach works. (The boiling points of sodium and potassium are 892°C and 770°C, respectively.)

14.98 In the gas phase, nitrogen dioxide is actually a mixture of nitrogen dioxide (NO_2) and dinitrogen tetroxide (N_2O_4). If the density of such a mixture is 2.3 g/L at 74°C and 1.3 atm, calculate the partial pressures of the gases and K_P for the dissociation of N_2O_4.

14.99 The equilibrium constant for the reaction $A + 2B \rightleftharpoons 3C$ is 0.25 at a certain temperature. Which diagram shown here corresponds to the system at equilibrium? If the system is not at equilibrium, predict the direction of the net reaction to reach equilibrium. Each molecule represents 0.40 mole and the volume of the container is 2.0 L. The color codes are A = gray, B = red, C = blue.

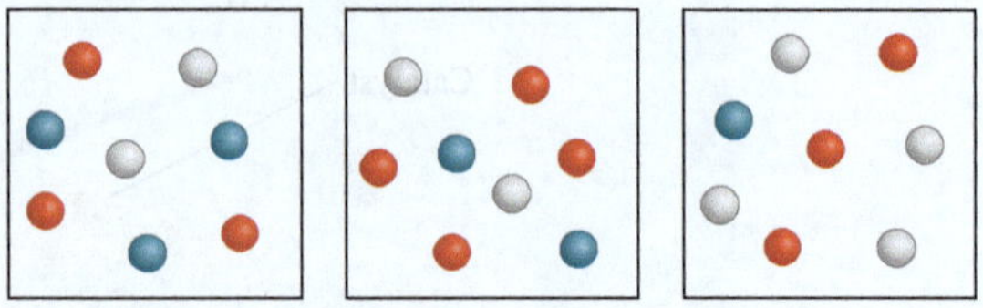

(a) (b) (c)

14.100 The equilibrium constant for the reaction $4X + Y \rightleftharpoons 3Z$ is 33.3 at a certain temperature. Which diagram shown here corresponds to the system at equilibrium? If the system is not at equilibrium, predict the direction of the net reaction to reach equilibrium. Each molecule represents 0.20 mole and the volume of the container is 1.0 L. The color codes are X = blue, Y = gray, Z = red.

(a) (b) (c)

14.101 About 75 percent of hydrogen for industrial use is produced by the *steam-reforming* process. This process is carried out in two stages called primary and secondary reforming. In the primary stage, a mixture of steam and methane at about 30 atm is heated over a nickel catalyst at 800°C to give hydrogen and carbon monoxide:

$$CH_4(g) + H_2O(g) \rightleftharpoons CO(g) + 3H_2(g) \qquad \Delta H° = 260 \text{ kJ/mol}$$

The secondary stage is carried out at about 1000°C, in the presence of air, to convert the remaining methane to hydrogen:

$$CH_4(g) + \tfrac{1}{2}O_2(g) \rightleftharpoons CO(g) + 2H_2(g) \qquad \Delta H° = 35.7 \text{ kJ/mol}$$

(a) What conditions of temperature and pressure would favor the formation of products in both the primary and secondary stage? (b) The equilibrium constant K_c for the primary stage is 18 at 800°C. (i) Calculate K_P for the reaction. (ii) If the partial pressures of methane and steam were both 15 atm at the start, what are the pressures of all the gases at equilibrium?

14.102 Photosynthesis can be represented by

$$6CO_2(g) + 6H_2O(l) \rightleftharpoons C_6H_{12}O_6(s) + 6O_2(g) \qquad \Delta H° = 2801 \text{ kJ/mol}$$

Explain how the equilibrium would be affected by the following changes: (a) Partial pressure of CO_2 is increased. (b) O_2 is removed from the mixture. (c) $C_6H_{12}O_6$ (glucose) is removed from the mixture. (d) More water is added. (e) A catalyst is added. (f) Temperature is decreased.

14.103 Consider the decomposition of ammonium chloride at a certain temperature:

$$NH_4Cl(s) \rightleftharpoons NH_3(g) + HCl(g)$$

Calculate the equilibrium constant K_P if the total pressure is 2.2 atm at that temperature.

14.104 At 25°C, the equilibrium partial pressures of NO_2 and N_2O_4 are 0.15 atm and 0.20 atm, respectively. If the volume is doubled at constant temperature, calculate the partial pressures of the gases when a new equilibrium is established.

14.105 In 1899 the German chemist Ludwig Mond developed a process for purifying nickel by converting it to the volatile nickel tetracarbonyl [$Ni(CO)_4$] (b.p. = 42.2°C):

$$Ni(s) + 4CO(g) \rightleftharpoons Ni(CO)_4(g)$$

(a) Describe how you can separate nickel and its solid impurities. (b) How would you recover nickel? [ΔH_f° for $Ni(CO)_4$ is −602.9 kJ/mol.]

14.106 Consider the equilibrium reaction described in Problem 14.23. A quantity of 2.50 g of PCl_5 is placed in an evacuated 0.500-L flask and heated to 250°C. (a) Calculate the pressure of PCl_5, assuming it does not dissociate. (b) Calculate the partial pressure of PCl_5 at equilibrium. (c) What is the total pressure at equilibrium? (d) What is the degree of dissociation of PCl_5? (The degree of dissociation is given by the fraction of PCl_5 that has undergone dissociation.)

14.107 Consider the equilibrium system $3A \rightleftharpoons B$. Sketch the changes in the concentrations of A and B over time for the following situations: (a) Initially only A is present. (b) Initially only B is present. (c) Initially both A and B are present (with A in higher concentration). In each case, assume that the concentration of B is higher than that of A at equilibrium.

14.108 The vapor pressure of mercury is 0.0020 mmHg at 26°C. (a) Calculate K_c and K_P for the process $Hg(l) \rightleftharpoons Hg(g)$. (b) A chemist breaks a thermometer and spills mercury onto the floor of a laboratory measuring 6.1 m long, 5.3 m wide, and 3.1 m high. Calculate the mass of mercury (in grams) vaporized at equilibrium and the concentration of mercury vapor in mg/m^3. Does this concentration exceed the safety limit of 0.05 mg/m^3? (Ignore the volume of furniture and other objects in the laboratory.)

14.109 At 25°C, a mixture of NO_2 and N_2O_4 gases are in equilibrium in a cylinder fitted with a movable piston. The concentrations are $[NO_2] = 0.0475\ M$ and $[N_2O_4] = 0.487\ M$. The volume of the gas mixture is halved by pushing down on the piston at constant temperature. Calculate the concentrations of the gases when equilibrium is reestablished. Will the color become darker or lighter after the change? [*Hint:* K_c for the dissociation of N_2O_4 to NO_2 is 4.63×10^{-3}. $N_2O_4(g)$ is colorless and $NO_2(g)$ has a brown color.]

14.110 A student placed a few ice cubes in a drinking glass with water. A few minutes later she noticed that some of the ice cubes were fused together. Explain what happened.

14.111 Consider the potential energy diagrams for two types of reactions $A \rightleftharpoons B$. In each case, answer the following questions for the system at equilibrium. (a) How would a catalyst affect the forward and reverse rates of the reaction? (b) How would a catalyst affect the energies of the reactant and product? (c) How would an increase in temperature affect the equilibrium constant? (d) If the only effect of a catalyst is to lower the activation energies for the forward and reverse reactions, show that the equilibrium constant remains unchanged if a catalyst is added to the reacting mixture.

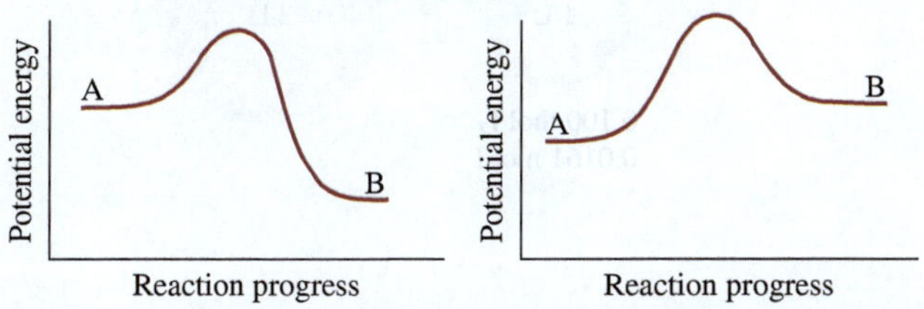

14.112 The equilibrium constant K_c for the reaction $2NH_3(g) \rightleftharpoons N_2(g) + 3H_2(g)$ is 0.83 at 375°C. A 14.6-g sample of ammonia is placed in a 4.00-L flask and heated to 375°C. Calculate the concentrations of all the gases when equilibrium is reached.

14.113 A quantity of 1.0 mole of N_2O_4 was introduced into an evacuated vessel and allowed to attain equilibrium at a certain temperature

$$N_2O_4(g) \rightleftharpoons 2NO_2(g)$$

The average molar mass of the reacting mixture was 70.6 g/mol. (a) Calculate the mole fractions of the gases. (b) Calculate K_P for the reaction if the total pressure was 1.2 atm. (c) What would be the mole fractions if the pressure were increased to 4.0 atm by reducing the volume at the same temperature?

14.114 The equilibrium constant (K_P) for the reaction

$$C(s) + CO_2(g) \rightleftharpoons 2CO(g)$$

is 1.9 at 727°C. What total pressure must be applied to the reacting system to obtain 0.012 mole of CO_2 and 0.025 mole of CO?

14.115 The forward and reverse rate constants for the reaction $A(g) + B(g) \rightleftharpoons C(g)$ are $3.6 \times 10^{-3}/M \cdot s$ and $8.7 \times 10^{-4}\ s^{-1}$, respectively, at 323 K. Calculate the equilibrium pressures of all the species starting at $P_A = 1.6$ atm and $P_B = 0.44$ atm.

14.116 The equilibrium constant (K_P) for the reaction $PCl_3(g) + Cl_2(g) \rightleftharpoons PCl_5(g)$ is 2.93 at 127°C. Initially there were 2.00 moles of PCl_3 and 1.00 mole of Cl_2 present. Calculate the partial pressures of the gases at equilibrium if the total pressure is 2.00 atm.

14.117 Consider the reaction between NO_2 and N_2O_4 in a closed container:

$$N_2O_4(g) \rightleftharpoons 2NO_2(g)$$

Initially, 1 mole of N_2O_4 is present. At equilibrium, α mole of N_2O_4 has dissociated to form NO_2. (a) Derive an expression for K_P in terms of α and P, the total pressure. (b) How does the expression in (a) help you predict the shift in equilibrium due to an increase in P? Does your prediction agree with Le Châtelier's principle?

14.118 The dependence of the equilibrium constant of a reaction on temperature is given by the van't Hoff equation:

$$\ln K = -\frac{\Delta H°}{RT} + C$$

where C is a constant. The following table gives the equilibrium constant (K_P) for the reaction at various temperatures:

$$2NO(g) + O_2(g) \rightleftharpoons 2NO_2(g)$$

K_P	138	5.12	0.436	0.0626	0.0130
T(K)	600	700	800	900	1000

Determine graphically the $\Delta H°$ for the reaction.

14.119 (a) Use the van't Hoff equation in Problem 14.118 to derive the following expression, which relates the equilibrium constants at two different temperatures:

$$\ln\frac{K_1}{K_2} = \frac{\Delta H°}{R}\left(\frac{1}{T_2} - \frac{1}{T_1}\right)$$

How does this equation support the prediction based on Le Châtelier's principle about the shift in equilibrium with temperature? (b) The vapor pressures of water are 31.82 mmHg at 30°C and 92.51 mmHg at 50°C. Calculate the molar heat of vaporization of water.

14.120 The K_P for the reaction

$$SO_2Cl_2(g) \rightleftharpoons SO_2(g) + Cl_2(g)$$

is 2.05 at 648 K. A sample of SO_2Cl_2 is placed in a container and heated to 648 K while the total pressure is kept constant at 9.00 atm. Calculate the partial pressures of the gases at equilibrium.

14.121 The "boat" form and "chair" form of cyclohexane (C_6H_{12}) interconverts as shown here:

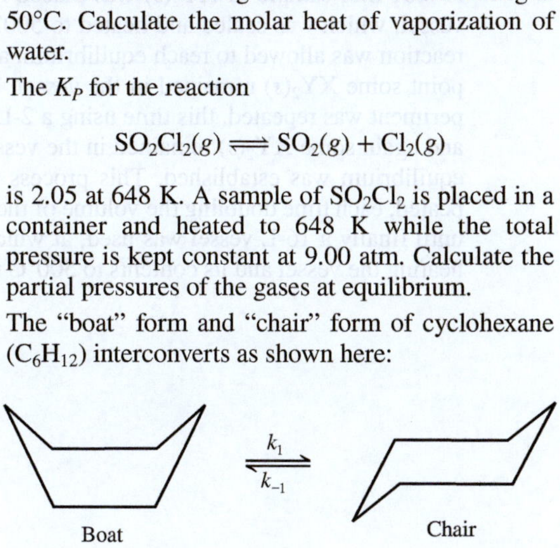

In this representation, the H atoms are omitted and a C atom is assumed to be at each intersection of two lines (bonds). The conversion is first order in each direction. The activation energy for the chair → boat conversion is 41 kJ/mol. If the frequency factor is $1.0 \times 10^{12}\ s^{-1}$, what is k_1 at 298 K? The equilibrium constant K_c for the reaction is 9.83×10^3 at 298 K.

14.122 Consider the following reaction at a certain temperature:

$$A_2 + B_2 \rightleftharpoons 2AB$$

The mixing of 1 mole of A_2 with 3 moles of B_2 gives rise to x mole of AB at equilibrium. The addition of 2 more moles of A_2 produces another x mole of AB. What is the equilibrium constant for the reaction?

14.123 Iodine is sparingly soluble in water but much more so in carbon tetrachloride (CCl_4). The equilibrium constant, also called the partition coefficient, for the distribution of I_2 between these two phases

$$I_2(aq) \rightleftharpoons I_2(CCl_4)$$

is 83 at 20°C. (a) A student adds 0.030 L of CCl_4 to 0.200 L of an aqueous solution containing 0.032 g I_2. The mixture is shaken and the two phases are then allowed to separate. Calculate the fraction of I_2 remaining in the aqueous phase. (b) The student now repeats the extraction of I_2 with another 0.030 L of CCl_4. Calculate the fraction of the I_2 from the original solution that remains in the aqueous phase. (c) Compare the result in (b) with a single extraction using 0.060 L of CCl_4. Comment on the difference.

14.124 Consider the following equilibrium system:

$$N_2O_4(g) \rightleftharpoons 2NO_2(g) \quad \Delta H° = 58.0\ \text{kJ/mol}$$

(a) If the volume of the reacting system is changed at constant temperature, describe what a plot of P versus $1/V$ would look like for the system. (*Hint:* See Figure 5.7.) (b) If the temperatures of the reacting system is changed at constant pressure, describe what a plot of V versus T would look like for the system. (*Hint:* See Figure 5.9.)

14.125 At 1200°C, the equilibrium constant (K_c) for the reaction $I_2(g) \rightleftharpoons 2I(g)$ is 2.59×10^{-3}. Calculate the concentrations of I_2 and I after the stopcock is opened and the system reestablishes equilibrium at the same temperature.

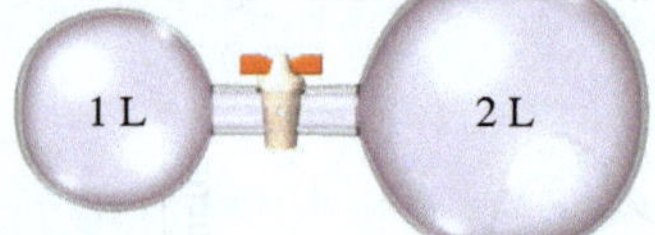

Interpreting, Modeling, & Estimating

14.126 Estimate the vapor pressure of water at 60°C (see Problem 14.119).

14.127 A compound $XY_2(s)$ decomposes to form X(g) and Y(g) according to the following chemical equation:

$$XY_2(s) \longrightarrow X(g) + 2Y(g)$$

A 0.01-mol sample of $XY_2(s)$ was placed in a 1-L vessel, which was sealed and heated to 500°C. The reaction was allowed to reach equilibrium, at which point some $XY_2(s)$ remained in the vessel. The experiment was repeated, this time using a 2-L vessel, and again some $XY_2(s)$ remained in the vessel after equilibrium was established. This process was repeated, each time doubling the volume of the vessel, until finally a 16-L vessel was used, at which point heating the vessel and its contents to 500°C resulted in decomposition of the entire 0.01 mole of $XY_2(s)$ according to the above reaction. Estimate K_c and K_P for the reaction at 500°C.

14.128 Using the simplified chemical equilibrium given in the Chemistry in Action essay "Life at High Altitudes and Hemoglobin Production" in Section 14.5, by how much would the concentration of hemoglobin, Hb, in a person's blood need to increase if she moved to an altitude of 2 km above sea level, to give the same concentration of HbO_2 as when she was living at sea level?

14.129 The equilibrium constant (K_P) for the reaction

$$I_2(g) \longrightarrow 2I(g)$$

is 1.8×10 at 872 K and 0.048 at 1173 K. From these data, estimate the bond enthalpy of I_2. (*Hint:* See van't Hoff's equation in Problem 14.119.)

Answers to Practice Exercises

14.1 $K_c = \frac{[NO_2]^4[O_2]}{[N_2O_5]^2}$; $K_P = \frac{P_{NO_2}^4 P_{O_2}}{P_{N_2O_5}^2}$.
14.2 2.2×10^2. **14.3** 347 atm. **14.4** 1.2.
14.5 $K_c = \frac{[Ni(CO)_4]}{[CO]^4}$; $K_P = \frac{P_{Ni(CO)_4}}{P_{CO}^4}$.
14.6 $K_P = 0.0702$; $K_c = 6.68 \times 10^{-5}$.
14.7 (a) $K_a = \frac{[O_3]^2}{[O_2]^3}$, (b) $K_b = \frac{[O_3]^{\frac{2}{3}}}{[O_2]}$; $K_a = K_b^3$.
14.8 From right to left. **14.9** [HI] = 0.031 *M*, $[H_2] = 4.3 \times 10^{-3}$ *M*, $[I_2] = 4.3 \times 10^{-3}$ *M*. **14.10** $[Br_2] = 0.065$ *M*, $[Br] = 8.4 \times 10^{-3}$ *M*. **14.11** $Q_P = 4.0 \times 10^5$; the net reaction will shift from right to left. **14.12** Left to right. **14.13** The equilibrium will shift from (a) left to right, (b) left to right, and (c) right to left. (d) A catalyst has no effect on the equilibrium.

Answers to Review of Concepts & Facts

14.1.1 (a) $K_c > 1$. **14.1.2** Products. **14.2.1** (c). **14.2.2** $2NO(g) + O_2(g) \rightleftharpoons 2NO_2(g)$. **14.2.3** $2NO_2(g) + 7H_2(g) \rightleftharpoons 2NH_3(g) + 4H_2O(g)$. **14.2.4** $K_c = \frac{1}{[B]}$. **14.3.1** $k_1 = 6.7 \times 10^3$ *M*/s. **14.4.1** No. Right toward products. **14.4.2** 0.10 *M*. **14.4.3** (a) Right toward products. (b) At equilibrium. (c) Left toward reactants. **14.5.1** (a) Left toward reactants. (b) Left toward reactants. (c) Right toward products. (d) No effect. **14.5.2** A will increases and A_2 will decrease. **14.5.3** The reaction is exothermic.

CHAPTER

15

Acids and Bases

Mono Lake, California, contains extremely high levels of dissolved salts making the water very alkaline and inhospitable to most marine life.

stevemendenhall/RooM/Getty Images

CHAPTER OUTLINE

Some of the most important processes in chemical and biological systems are acid-base reactions in aqueous solutions. In this first of two chapters on the properties of acids and bases, we will study the definitions of acids and bases, the pH scale, the ionization of weak acids and weak bases, and the relationship between acid strength and molecular structure. We will also look at oxides that can act as acids and bases.

15.1 Brønsted Acids and Bases

Learning Objective

- Explain the nature of a Brønsted acid and base.

In Chapter 4 we defined a Brønsted acid as a substance capable of donating a proton, and a Brønsted base as a substance that can accept a proton. These definitions are generally suitable for a discussion of the properties and reactions of acids and bases.

An extension of the Brønsted definition of acids and bases is the concept of the ***conjugate acid-base pair,*** which can be defined as *an acid and its conjugate base or a base and its conjugate acid.* The conjugate base of a Brønsted acid is the species that remains when one proton has been removed from the acid. Conversely, a conjugate acid results from the addition of a proton to a Brønsted base.

The proton is always associated with water molecules in aqueous solution. The H_3O^+ ion is the simplest formula of a hydrated proton.

Every Brønsted acid has a conjugate base, and every Brønsted base has a conjugate acid. For example, the chloride ion (Cl^-) is the conjugate base formed from the acid HCl, and H_3O^+ (hydronium ion) is the conjugate acid of the base H_2O.

$$HCl + H_2O \longrightarrow H_3O^+ + Cl^-$$

Similarly, the ionization of acetic acid can be represented as

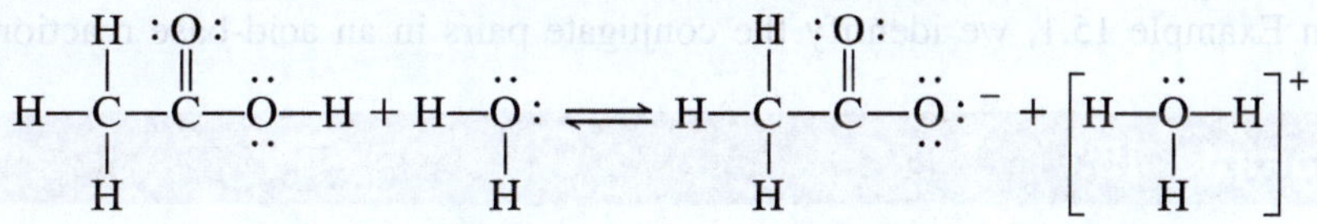

$$\underset{\text{acid}_1}{CH_3COOH(aq)} + \underset{\text{base}_2}{H_2O(l)} \rightleftharpoons \underset{\text{base}_1}{CH_3COO^-(aq)} + \underset{\text{acid}_2}{H_3O^+(aq)}$$

The subscripts 1 and 2 designate the two conjugate acid-base pairs. Thus, the acetate ion (CH_3COO^-) is the conjugate base of CH_3COOH. Both the ionization of HCl (see Section 4.3) and the ionization of CH_3COOH are examples of Brønsted acid-base reactions.

The Brønsted definition also enables us to classify ammonia as a base because of its ability to accept a proton:

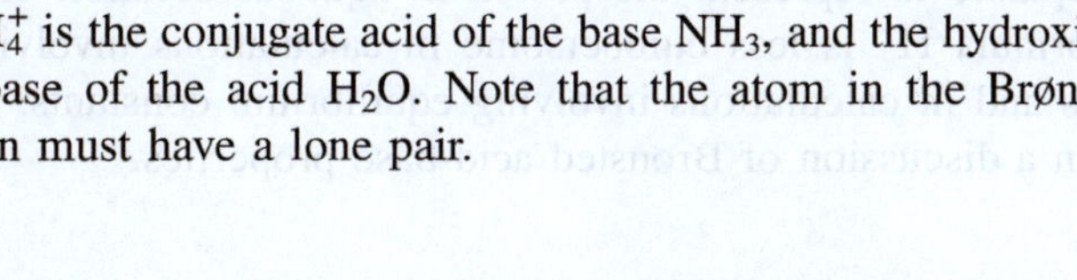

$$\underset{\text{base}_1}{NH_3(aq)} + \underset{\text{acid}_2}{H_2O(l)} \rightleftharpoons \underset{\text{acid}_1}{NH_4^+(aq)} + \underset{\text{base}_2}{OH^-(aq)}$$

In this case, NH_4^+ is the conjugate acid of the base NH_3, and the hydroxide ion OH^- is the conjugate base of the acid H_2O. Note that the atom in the Brønsted base that accepts a H^+ ion must have a lone pair.

A more accurate way to illustrate what actually occurs in a Brønsted acid-base reaction shows the movement of electrons using Lewis structures with curved arrows.

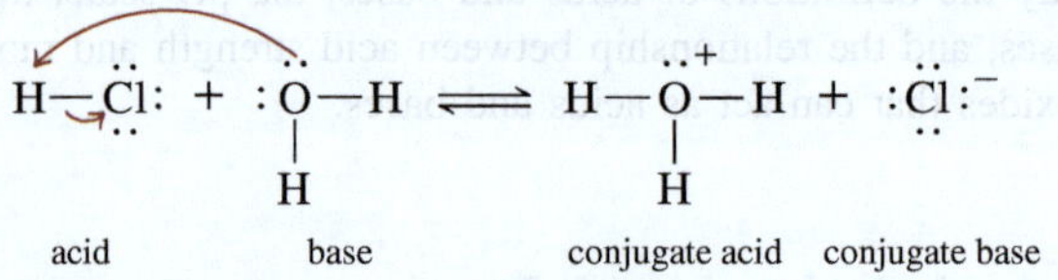

A curved arrow indicates the movement of an electron pair. For example, a shared pair of electrons between two atoms may move to one atom and thus become a lone pair, or a lone pair on one atom may move to make a new bond. The left-to-right curved arrow illustrated above shows the movement of a bonding pair of electrons in an H–Cl bond to the Cl, thereby forming a Cl^- ion while breaking the H–Cl bond. The right-to-left arrow shows the use of a lone pair on the O atom in H_2O to form a new bond to the H in HCl, thus making H_3O^+. The use of curved arrows shows the movement of electrons to break and form bonds as opposed to using the terms "donation" and "acceptance" of protons. The result of both methods produces the same result as HCl and H_2O react to form H_3O^+ and Cl^-.

The reaction of NH_3 with H_2O to generate NH_4^+ and OH^- can also be illustrated using curved arrows.

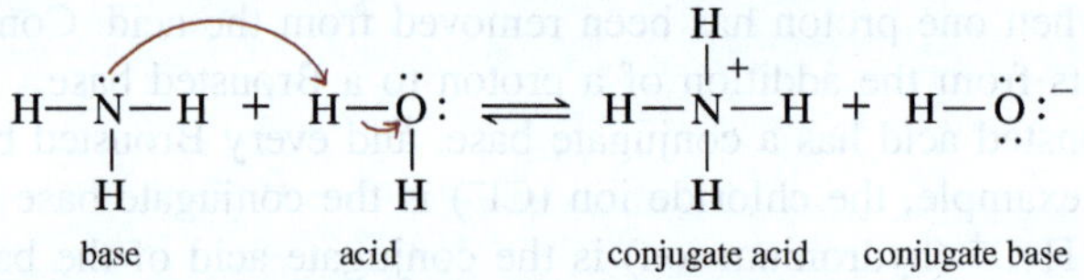

Here the left-to-right arrow shows the lone pair on the N in NH_3 moving to form a new bond with one of the H atoms in H_2O, thereby producing NH_4^+. The smaller left-to-right arrow shows the simultaneous movement of the bonding electron pair in H_2O moving to become a lone pair on the O atom thus, making the OH^- ion.

Student Hot Spot

Student data indicate you may struggle with conjugate acid-base pairs. Access your eBook for additional Learning Resources on this topic.

In Example 15.1, we identify the conjugate pairs in an acid-base reaction.

Example 15.1

Identify the conjugate acid-base pairs in the reaction between ammonia and hydrofluoric acid in aqueous solution

$$NH_3(aq) + HF(aq) \rightleftharpoons NH_4^+(aq) + F^-(aq)$$

Strategy Remember that a conjugate base always has one fewer H atom and one more negative charge (or one fewer positive charge) than the formula of the corresponding acid.

Solution NH_3 has one fewer H atom and one fewer positive charge than NH_4^+. F^- has one fewer H atom and one more negative charge than HF. Therefore, the conjugate acid-base pairs are (1) NH_4^+ and NH_3 and (2) HF and F^-.

Practice Exercise Identify the conjugate acid-base pairs for the reaction

$$CN^- + H_2O \rightleftharpoons HCN + OH^-$$

Similar problem: 15.5.

It is acceptable to represent the proton in aqueous solution either as H^+ or as H_3O^+. The formula H^+ is less cumbersome in calculations involving hydrogen ion concentrations and in calculations involving equilibrium constants, whereas H_3O^+ is more useful in a discussion of Brønsted acid-base properties.

Summary of Concepts & Facts

- Brønsted acids donate protons, and Brønsted bases accept protons. These are the definitions that normally underlie the use of the terms "acid" and "base."
- Brønsted acid-base reactions can be described by the movement of electron pairs to break and form bonds or lone pairs. A curved arrow indicates where an electron pair moves from and where it moves to.

Review of Concepts & Facts

15.1.1 Which of the following does not constitute a conjugate acid-base pair: (a) HNO_2–NO_2^-, (b) H_2CO_3–CO_3^{2-}, (c) $CH_3NH_3^+$–CH_3NH_2?

15.1.2 Write the formulas of the conjugate acid and conjugate base for each of the following substances: (a) HSO_4^-, (b) HPO_4^{2-}.

15.2 The Acid-Base Properties of Water

Learning Objectives

- Define the autoionization of water.
- Utilize the ion-product constant (K_w) to calculate $[H_3O]^+$ or $[OH]^-$ in aqueous solutions.

Water, as we know, is a unique solvent. One of its special properties is its ability to act either as an acid or as a base. Water functions as a base in reactions with acids such as HCl and CH_3COOH, and it functions as an acid in reactions with bases such as NH_3. Water is a very weak electrolyte and therefore a poor conductor of electricity, but it does undergo ionization to a small extent:

$$H_2O(l) \rightleftharpoons H^+(aq) + OH^-(aq)$$

This reaction is sometimes called the *autoionization* of water. To describe the acid-base properties of water in the Brønsted framework, we express its autoionization as follows (also shown in Figure 15.1):

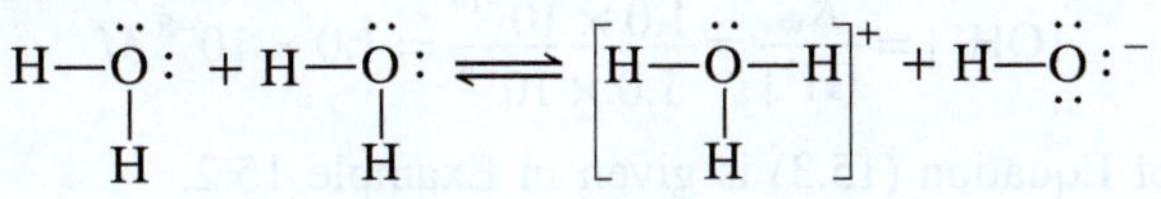

or

$$\underset{\text{acid}_1}{H_2O} + \underset{\text{base}_2}{H_2O} \rightleftharpoons \underset{\text{acid}_2}{H_3O^+} + \underset{\text{base}_1}{OH^-} \qquad (15.1)$$

The acid-base conjugate pairs are (1) H_2O (acid) and OH^- (base) and (2) H_3O^+ (acid) and H_2O (base).

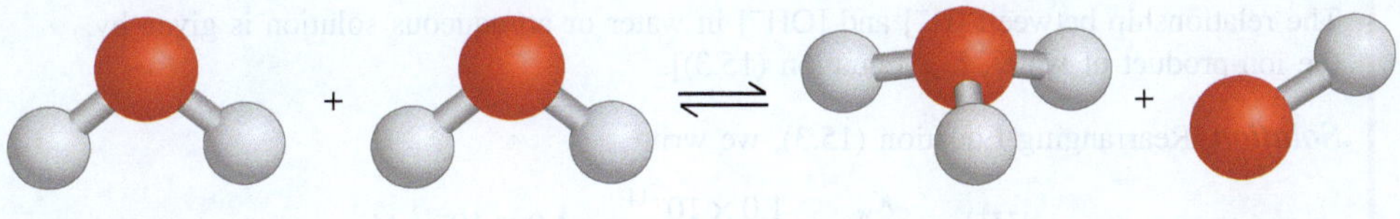

Figure 15.1 *Reaction between two water molecules to form hydronium and hydroxide ions.*

Using curved arrows and Lewis structures, the reaction can be illustrated as

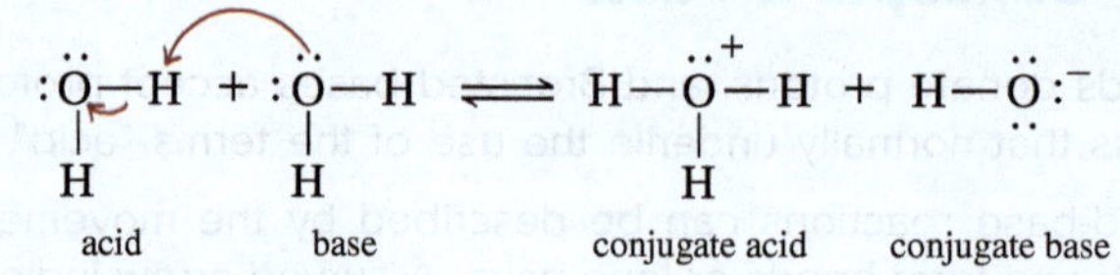

The Ion Product of Water

In the study of acid-base reactions, the hydrogen ion concentration is key; its value indicates the acidity or basicity of the solution. Because only a very small fraction of water molecules are ionized, the concentration of water, $[H_2O]$, remains virtually unchanged (recall that in pure water, $[H_2O] = 55.5\ M$). Therefore, the equilibrium constant for the autoionization of water, according to Equation (15.1), is

$$K_c = [H_3O^+][OH^-]$$

Because we use $H^+(aq)$ and $H_3O^+(aq)$ interchangeably to represent the hydrated proton, the equilibrium constant can also be expressed as

$$K_c = [H^+][OH^-]$$

To indicate that the equilibrium constant refers to the autoionization of water, we replace K_c with K_w:

$$K_w = [H_3O^+][OH^-] = [H^+][OH^-] \qquad (15.2)$$

where K_w is called the ***ion-product constant,*** which is *the product of the molar concentrations of H^+ and OH^- ions at a particular temperature.*

In pure water at 25°C, the concentrations of H^+ and OH^- ions are equal and found to be $[H^+] = 1.0 \times 10^{-7}\ M$ and $[OH^-] = 1.0 \times 10^{-7}\ M$. Thus, from Equation (15.2), at 25°C

$$K_W = (1.0 \times 10^{-7})(1.0 \times 10^{-7}) = 1.0 \times 10^{-14}$$

Whether we have pure water or an aqueous solution of dissolved species, the following relation *always* holds at 25°C:

$$K_w = [H^+][OH^-] = 1.0 \times 10^{-14} \qquad (15.3)$$

Whenever $[H^+] = [OH^-]$, the aqueous solution is said to be neutral. In an acidic solution there is an excess of H^+ ions and $[H^+] > [OH^-]$. In a basic solution there is an excess of hydroxide ions, so $[H^+] < [OH^-]$. In practice we can change the concentration of either H^+ or OH^- ions in solution, but we cannot vary both of them independently. If we adjust the solution so that $[H^+] = 1.0 \times 10^{-6}\ M$, the OH^- concentration *must* change to

$$[OH^-] = \frac{K_W}{[H^+]} = \frac{1.0 \times 10^{-14}}{1.0 \times 10^{-6}} = 1.0 \times 10^{-8}\ M$$

An application of Equation (15.3) is given in Example 15.2.

Example 15.2

The concentration of OH^- ions in a certain household ammonia cleaning solution is $0.0025\ M$. Calculate the concentration of H^+ ions.

Strategy We are given the concentration of the OH^- ions and asked to calculate $[H^+]$. The relationship between $[H^+]$ and $[OH^-]$ in water or an aqueous solution is given by the ion-product of water, K_w [Equation (15.3)].

Solution Rearranging Equation (15.3), we write

$$[H^+] = \frac{K_w}{[OH^-]} = \frac{1.0 \times 10^{-14}}{0.0025} = 4.0 \times 10^{-12}\ M$$

(Continued)

Check Because $[H^+] < [OH^-]$, the solution is basic, as we would expect from the earlier discussion of the reaction of ammonia with water.

Practice Exercise Calculate the concentration of OH^- ions in a HCl solution whose hydrogen ion concentration is 1.3 *M*.

Similar problems: 15.15, 15.16.

Summary of Concepts & Facts

- Water undergoes ionization to produce $[H_3O]^+$ and $[OH]^-$ ions. This is known as autoionization, the reaction of water with itself in the Brønsted definition of acids and bases.
- The equilibrium constant of the autoionization of water is the ion-product constant, K_w. In acidic solutions, $[H_3O]^+ > [OH]^-$ while in basic solutions $[OH]^- > [H_3O]^+$.

Review of Concepts & Facts

15.2.1 If the H^+ ion concentration in an aqueous solution is 0.0010 *M*, why is it not possible for the OH^- ion concentration to be 1.0×10^{-10} *M*?

15.2.2 Determine $[H^+]$ in a solution where $[OH^-] = 2.3 \times 10^{-5}$ *M*.

15.3 pH—A Measure of Acidity

Learning Objectives

- Utilize the pH scale to classify a solution as being acidic, basic, or neutral.
- Use pH, pOH, the concentration of protons, or the concentration of hydroxide ions to conduct calculations asking for one of these terms.

Because the concentrations of H^+ and OH^- ions in aqueous solutions are frequently very small numbers and therefore inconvenient to work with, Soren Sorensen† in 1909 proposed a more practical measure called pH. The ***pH*** of a solution is defined as *the negative logarithm of the hydrogen ion concentration (in mol/L):*

$$\mathrm{pH} = -\log [H_3O^+] \quad \text{or} \quad \mathrm{pH} = -\log [H^+] \qquad (15.4)$$

Keep in mind that Equation (15.4) is simply a definition designed to give us convenient numbers to work with. The negative logarithm gives us a positive number for pH, which otherwise would be negative due to the small value of $[H^+]$. The pH of concentrated acid solutions can be negative; for example, the pH of 2.0 *M* HCl is –0.30. Furthermore, the term $[H^+]$ in Equation (15.4) pertains only to the *numerical part* of the expression for hydrogen ion concentration, for we cannot take the logarithm of units. Thus, like the equilibrium constant, the pH of a solution is a dimensionless quantity.

Because pH is simply a way to express hydrogen ion concentration, acidic and basic solutions at 25°C can be distinguished by their pH values, as follows:

Acidic solutions:	$[H^+] > 1.0 \times 10^{-7}$ *M*, pH < 7.00
Basic solutions:	$[H^+] < 1.0 \times 10^{-7}$ *M*, pH > 7.00
Neutral solutions:	$[H^+] = 1.0 \times 10^{-7}$ *M*, pH = 7.00

Notice that pH increases as $[H^+]$ decreases.

†Soren Peer Lauritz Sorensen (1868–1939). Danish biochemist. Sorensen originally wrote the symbol as p_H and called p the "hydrogen ion exponent" (*Wasserstoffionexponent*); it is the initial letter of *Potenz* (German), *puissance* (French), and *power* (English). It is now customary to write the symbol as pH.

Sometimes we may be given the pH value of a solution and asked to calculate the H^+ ion concentration. In that case, we need to take the antilog of Equation (15.4) as follows:

$$[H_3O^+] = 10^{-pH} \quad \text{or} \quad [H^+] = 10^{-pH} \tag{15.5}$$

Be aware that the definition of pH just shown, and indeed all the calculations involving solution concentrations (expressed either as molarity or molality) discussed in previous chapters, are subject to error because we have implicitly assumed ideal behavior. In reality, ion-pair formation and other types of intermolecular interactions may affect the actual concentrations of species in solution. The situation is analogous to the relationships between ideal gas behavior and the behavior of real gases discussed in Chapter 5. Depending on temperature, volume, and amount and type of gas present, the measured gas pressure may differ from that calculated using the ideal gas equation. Similarly, the actual or "effective" concentration of a solute may not be what we think it is, knowing the amount of substance originally dissolved in solution. Just as we have the van der Waals and other equations to reconcile discrepancies between the ideal gas and nonideal gas behavior, we can account for nonideal behavior in solution.

One way is to replace the concentration term with *activity,* which is the effective concentration. Strictly speaking, then, the pH of solution should be defined as

$$\text{pH} = -\log a_{H^+} \tag{15.6}$$

where a_{H^+} is the activity of the H^+ ion. As mentioned in Section 14.2, for an ideal solution activity is numerically equal to concentration. For real solutions, activity usually differs from concentration, sometimes appreciably. Knowing the solute concentration, there are reliable ways based on thermodynamics for estimating its activity, but the details are beyond the scope of this text. Keep in mind, therefore, that, except for dilute solutions, the measured pH is usually not the same as that calculated from Equation (15.4) because the concentration of the H^+ ion in molarity is not numerically equal to its activity value. Although we will continue to use concentration in our discussion, it is important to know that this approach will give us only an approximation of the chemical processes that actually take place in the solution phase.

In the laboratory, the pH of a solution is measured with a pH meter (Figure 15.2). Table 15.1 lists the pH of a number of common fluids. As you can see, the pH of body fluids varies greatly, depending on location and function. The low pH (high acidity) of gastric juices facilitates digestion whereas a higher pH of blood is necessary for the transport of oxygen. These pH-dependent actions will be illustrated in Chemistry in Action essays in this chapter and Chapter 16.

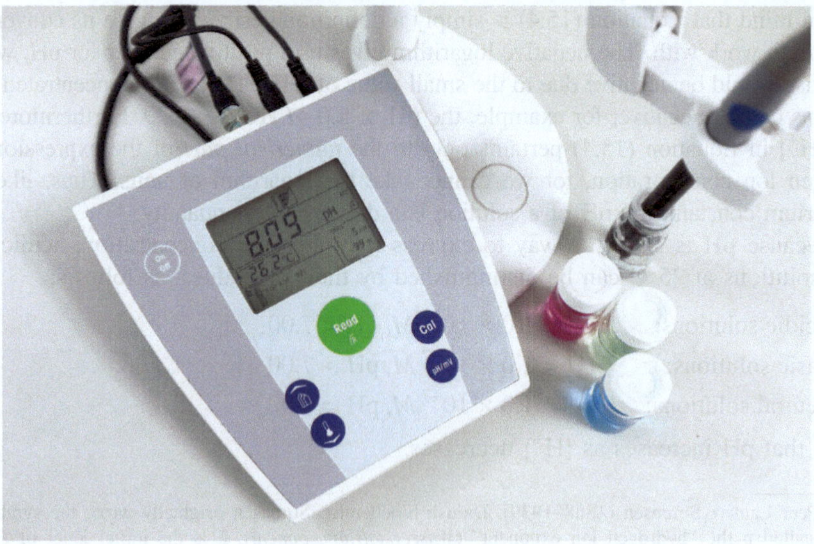

Figure 15.2 *A pH meter is commonly used in the laboratory to determine the pH of a solution. Although many pH meters have scales marked with values from 1 to 14, pH values can, in fact, be less than 1 and greater than 14.*
photongpix/iStock/Getty Images

A pOH scale analogous to the pH scale can be devised using the negative logarithm of the hydroxide ion concentration of a solution. Thus, we define pOH as

$$pOH = -\log [OH^-] \tag{15.7}$$

If we are given the pOH value of a solution and asked to calculate the OH^- ion concentration, we can take the antilog of Equation (15.7) as follows

$$[OH^-] = 10^{-pOH} \tag{15.8}$$

Now consider again the ion-product constant for water at 25°C:

$$[H^+][OH^-] = K_w = 1.0 \times 10^{-14}$$

Taking the negative logarithm of both sides, we obtain

$$-(\log [H^+] + \log [OH^-]) = -\log (1.0 \times 10^{-14})$$
$$-\log [H^+] - \log [OH^-] = 14.00$$

From the definitions of pH and pOH we obtain

$$pH + pOH = 14.00 \tag{15.9}$$

Equation (15.9) provides us with another way to express the relationship between the H^+ ion concentration and the OH^- ion concentration.

Examples 15.3, 15.4, and 15.5 illustrate calculations involving pH.

Table 15.1

The pH of Some Common Fluids

Sample	pH Value
Gastric juice in the stomach	1.0–2.0
Lemon juice	2.4
Vinegar	3.0
Grapefruit juice	3.2
Orange juice	3.5
Urine	4.8–7.5
Water exposed to air*	5.5
Saliva	6.4–6.9
Milk	6.5
Pure water	7.0
Blood	7.35–7.45
Tears	7.4
Milk of magnesia	10.6
Household ammonia	11.5

*Water exposed to air for a long period of time absorbs atmospheric CO_2 to form carbonic acid, H_2CO_3.

Example 15.3

The concentration of H^+ ions in a bottle of table wine was 3.2×10^{-4} *M* right after the cork was removed. Only half of the wine was consumed. The other half, after it had been standing open to the air for a month, was found to have a hydrogen ion concentration equal to 1.0×10^{-3} *M*. Calculate the pH of the wine on these two occasions.

Strategy We are given the H^+ ion concentration and asked to calculate the pH of the solution. What is the definition of pH?

Solution According to Equation (15.4), pH = $-\log [H^+]$. When the bottle was first opened, $[H^+] = 3.2 \times 10^{-4}$ *M*, which we substitute in Equation (15.4)

$$pH = -\log [H^+]$$
$$= -\log (3.2 \times 10^{-4}) = 3.49$$

On the second occasion, $[H^+] = 1.0 \times 10^{-3}$ *M*, so that

$$pH = -\log (1.0 \times 10^{-3}) = 3.00$$

Comment In each case, the pH has only two significant figures. The two digits to the right of the decimal in 3.49 tell us that there are two significant figures in the original number.

The increase in hydrogen ion concentration (or decrease in pH) is largely the result of the conversion of some of the alcohol (ethanol) to acetic acid, a reaction that takes place in the presence of molecular oxygen.

Practice Exercise Nitric acid (HNO_3) is used in the production of fertilizer, dyes, drugs, and explosives. Calculate the pH of a HNO_3 solution having a hydrogen ion concentration of 0.76 *M*.

Similar problems: 15.17, 15.18.

CHEMISTRY *in Action*

Antacids and the pH Balance in Your Stomach

An average adult produces between 2 and 3 L of gastric juice daily. Gastric juice is a thin, acidic digestive fluid secreted by glands in the mucous membrane that lines the stomach. It contains, among other substances, hydrochloric acid. The pH of gastric juice is about 1.5, which corresponds to a hydrochloric acid concentration of 0.03 *M*—a concentration strong enough to dissolve zinc metal! What is the purpose of this highly acidic medium? Where do the H^+ ions come from? What happens when there is an excess of H^+ ions present in the stomach?

A simplified diagram of the stomach is shown here. The inside lining is made up of parietal cells, which are fused together to form tight junctions. The interiors of the cells are protected from the surroundings by cell membranes. These membranes allow water and neutral molecules to pass in and out of the stomach, but they usually block the movement of ions such as H^+, Na^+, K^+, and Cl^-. The H^+ ions come from carbonic acid (H_2CO_3) formed as a result of the hydration of CO_2, an end product of metabolism:

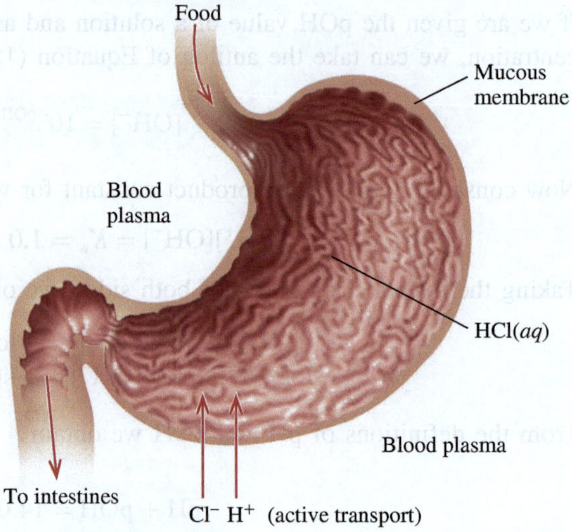

A simplified diagram of the human stomach.

$$CO_2(g) + H_2O(l) \rightleftharpoons H_2CO_3(aq)$$
$$H_2CO_2(aq) \rightleftharpoons H^+(aq) + HCO_3^-(aq)$$

These reactions take place in the blood plasma bathing the cells in the mucosa. By a process known as *active transport*, H^+ ions move across the membrane into the stomach interior. (Active transport processes are aided by enzymes.) To maintain electrical balance, an equal number of Cl^- ions also move from the blood plasma into the stomach. Once in the stomach, most of these ions are prevented from diffusing back into the blood plasma by cell membranes.

The purpose of the highly acidic medium within the stomach is to digest food and to activate certain digestive enzymes. Eating stimulates H^+ ion secretion. A small fraction of these ions normally are reabsorbed by the mucosa, causing many tiny hemorrhages. About half a million cells are shed by the lining every minute, and a healthy stomach is completely relined every three days or so. However, if the acid content is excessively high, the constant influx of H^+ ions through the membrane back to the blood plasma can cause muscle contraction, pain, swelling, inflammation, and bleeding.

One way to temporarily reduce the H^+ ion concentration in the stomach is to take an antacid. The major function of antacids is to neutralize excess HCl in gastric juice. The table "Some Common Commercial Antacid Preparations" lists the active ingredients of some popular antacids. The reactions by which these antacids neutralize stomach acid are as follows:

$$NaHCO_3(aq) + HCl(aq) \longrightarrow NaCl(aq) + H_2O(l) + CO_2(g)$$
$$CaCO_3(s) + 2HCl(aq) \longrightarrow CaCl_2(aq) + H_2O(l) + CO_2(g)$$
$$MgCO_3(s) + 2HCl(aq) \longrightarrow MgCl_2(aq) + H_2O(l) + CO_2(g)$$
$$Mg(OH)_2(s) + 2HCl(aq) \longrightarrow MgCl_2(aq) + 2H_2O(l)$$
$$Al(OH)_2NaCO_3(s) + 4HCl(aq) \longrightarrow AlCl_3(aq) + NaCl(aq) + 3H_2O(l) + CO_2(g)$$

When an Alka-Seltzer tablet dissolves in water, the bicarbonate ions in it react with the acid component in the tablet to produce carbon dioxide gas.
Ken Karp/McGraw Hill

Some Common Commercial Antacid Preparations

Commercial Name	Active Ingredients
Alka-2	Calcium carbonate
Alka-Seltzer	Aspirin, sodium bicarbonate, citric acid
Bufferin	Aspirin, magnesium carbonate, aluminum glycinate
Buffered aspirin	Aspirin, magnesium carbonate, aluminum hydroxide-glycine
Milk of magnesia	Magnesium hydroxide
Rolaids	Dihydroxy aluminum sodium carbonate
Tums	Calcium carbonate

The CO_2 released by most of these reactions increases gas pressure in the stomach, causing the person to belch. The fizzing that takes place when an Alka-Seltzer tablet dissolves in water is caused by carbon dioxide, which is released by the reaction between citric acid and sodium bicarbonate:

$$\underset{\text{citric acid}}{C_4H_7O_5(COOH)(aq)} + NaHCO_3(aq) \longrightarrow \underset{\text{sodium citrate}}{C_4H_7O_5COONa(aq)} + H_2O(l) + CO_2(g)$$

This action helps to disperse the ingredients and even enhances the taste of the solution.

The mucosa of the stomach is also damaged by the action of aspirin, the chemical name of which is acetylsalicylic acid. Aspirin is itself a moderately weak acid:

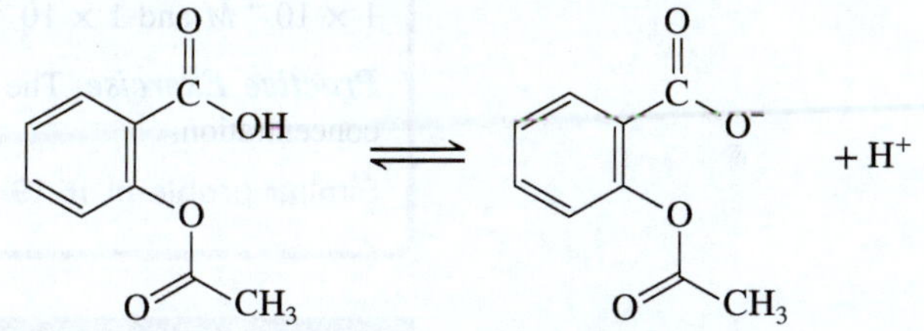

acetylsalicylic acid   acetylsalicylate ion

In the presence of the high concentration of H^+ ions in the stomach, this acid remains largely nonionized. A relatively nonpolar molecule, acetylsalicylic acid has the ability to penetrate membrane barriers that are also made up of nonpolar molecules. However, inside the membrane are many small water pockets, and when an acetylsalicylic acid molecule enters such a pocket, it ionizes into H^+ and acetylsalicylate ions. These ionic species become trapped in the interior regions of the membrane. The continued buildup of ions in this fashion weakens the structure of the membrane and eventually causes bleeding. Approximately 2 mL of blood are usually lost for every aspirin tablet taken, an amount not generally considered harmful. However, the action of aspirin can result in severe bleeding in some individuals. It is interesting to note that the presence of alcohol makes acetylsalicylic acid even more soluble in the membrane, and so further promotes the bleeding.

Example 15.4

The pH of rainwater collected in a certain region of the northeastern United States on a particular day was 4.82. Calculate the H^+ ion concentration of the rainwater.

Strategy Here we are given the pH of a solution and asked to calculate $[H^+]$. Because pH is defined as $pH = -\log [H^+]$, we can solve for $[H^+]$ by taking the antilog of the pH; that is, $[H^+] = 10^{-pH}$, as shown in Equation (15.5).

Solution From Equation (15.4)

$$pH = -\log [H^+] = 4.82$$

Therefore,

$$\log [H^+] = -4.82$$

To calculate $[H^+]$, we need to take the antilog of −4.82:

$$[H^+] = 10^{-4.82} = 1.5 \times 10^{-5}\ M$$

Check Because the pH is between 4 and 5, we can expect $[H^+]$ to be between $1 \times 10^{-4}\ M$ and $1 \times 10^{-5}\ M$. Therefore, the answer is reasonable.

Practice Exercise The pH of a certain orange juice is 3.33. Calculate the $[H^+]$ ion concentration.

Similar problem: 15.19.

Example 15.5

In a NaOH solution $[OH^-]$ is $2.9 \times 10^{-4}\ M$. Calculate the pH of the solution.

Strategy Solving this problem takes two steps. First, we need to calculate pOH using Equation (15.7). Next, we use Equation (15.9) to calculate the pH of the solution.

Solution We use Equation (15.7):

$$\begin{aligned} pOH &= -\log [OH^-] \\ &= -\log (2.9 \times 10^{-4}) \\ &= 3.54 \end{aligned}$$

Now we use Equation (15.9):

$$\begin{aligned} pH + pOH &= 14.00 \\ pH &= 14.00 - pOH \\ &= 14.00 - 3.54 = 10.46 \end{aligned}$$

Alternatively, we can use the ion-product constant of water, $K_w = [H^+][OH^-]$, to calculate $[H^+]$, and then we can calculate the pH from the $[H^+]$. Try it.

Check The answer shows that the solution is basic (pH > 7), which is consistent with a NaOH solution.

Practice Exercise The OH^- ion concentration of a blood sample is $2.5 \times 10^{-7}\ M$. What is the pH of the blood?

Similar problem: 15.18.

Summary of Concepts & Facts

- The acidity of an aqueous solution is expressed as its pH, which is defined as the negative logarithm of the hydrogen ion concentration (in mol/L).
- At 25°C, an acidic solution has pH < 7, a basic solution has pH > 7, and a neutral solution has pH = 7.

Review of Concepts & Facts

15.3.1 What is the pH of a solution in which $[H^+] = 5.8 \times 10^{-7}\ M$? What is the pOH of this solution?

15.3.2 Determine $[H^+]$ in a solution in which pOH = 10.93.

15.3.3 What is the pOH of a solution in which $[H^+] = 9.1 \times 10^{-11}\ M$?

15.3.4 Which is more acidic: a solution where $[H^+] = 2.5 \times 10^{-3}\ M$ or a solution where pOH = 11.6?

15.4 Strength of Acids and Bases

Learning Objectives

- Calculate the pH of a strong acid or strong base solution.
- Appraise the value of the equilibrium constant in an acid-base reaction using relative acid strengths.

***Strong acids** are strong electrolytes that, for practical purposes, are assumed to ionize completely in water* (Figure 15.3). Most of the strong acids are inorganic acids: hydrochloric acid (HCl), nitric acid (HNO_3), perchloric acid ($HClO_4$), and sulfuric acid (H_2SO_4):

$$HCl(aq) + H_2O(l) \longrightarrow H_3O^+(aq) + Cl^-(aq)$$
$$HNO_3(aq) + H_2O(l) \longrightarrow H_3O^+(aq) + NO_3^-(aq)$$
$$HClO_4(aq) + H_2O(l) \longrightarrow H_3O^+(aq) + ClO_4^-(aq)$$
$$H_2SO_4(aq) + H_2O(l) \longrightarrow H_3O^+(aq) + HSO_4^-(aq)$$

Video: The Dissociation of Strong and Weak Acids

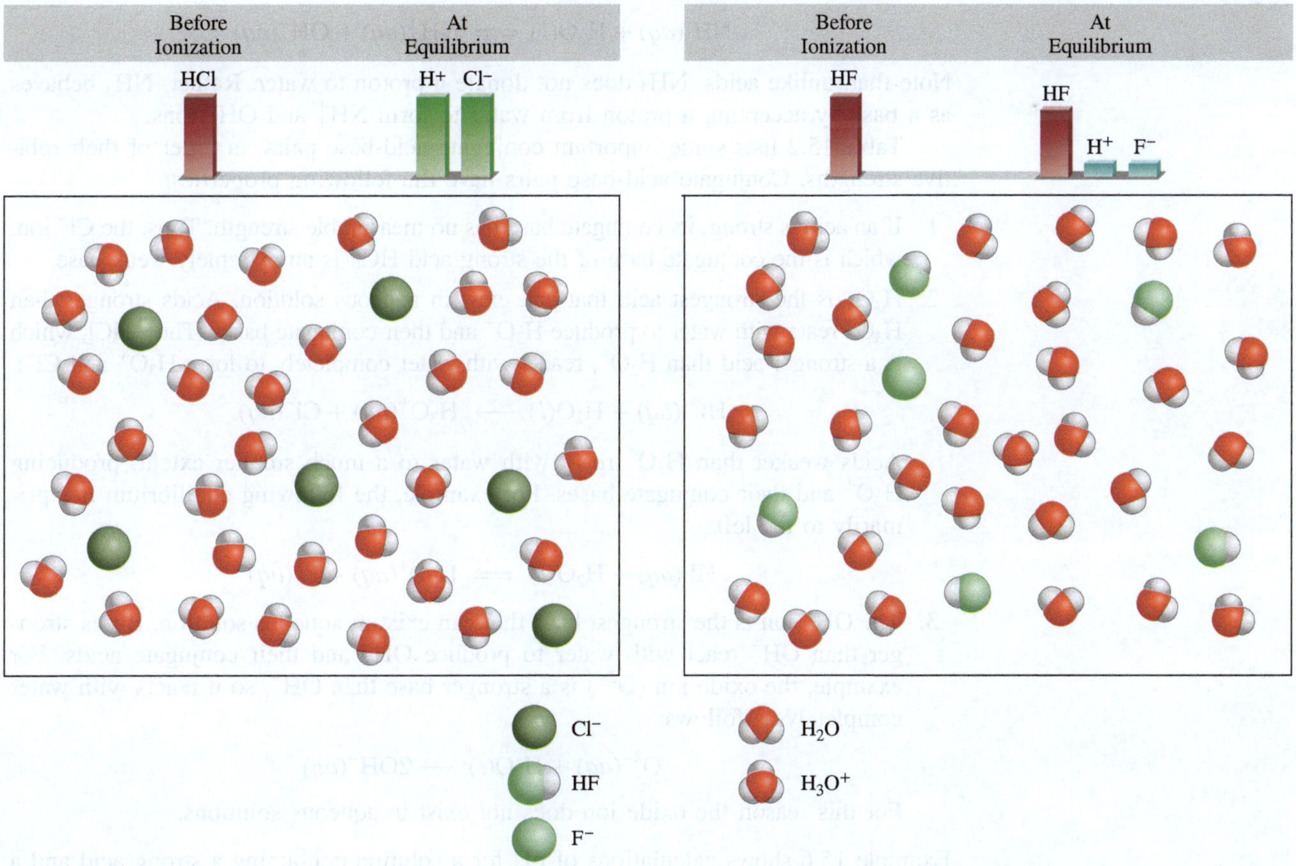

Figure 15.3 *The extent of ionization of a strong acid such as HCl (left) and a weak acid such as HF (right). Initially, there were 6 HCl and 6 HF molecules present. The strong acid is assumed to be completely ionized in solution. The proton exists in solution as the hydronium ion (H_3O^+).*

Zn reacts more vigorously with a strong acid like HCl (left) than with a weak acid like CH_3COOH (right) of the same concentration because there are more H^+ ions in the former solution.
Stephen Frisch/McGraw Hill

Note that H_2SO_4 is a diprotic acid; we show only the first stage of ionization here. At equilibrium, solutions of strong acids will not contain any nonionized acid molecules.

Most acids are ***weak acids,*** which *ionize only to a limited extent in water.* At equilibrium, aqueous solutions of weak acids contain a mixture of nonionized acid molecules, H_3O^+ ions, and the conjugate base. Examples of weak acids are hydrofluoric acid (HF), acetic acid (CH_3COOH), and the ammonium ion (NH_4^+). The limited ionization of weak acids is related to the equilibrium constant for ionization, which we will study in the next section.

Like strong acids, ***strong bases*** *are strong electrolytes that ionize completely in water.* Hydroxides of alkali metals and certain alkaline earth metals are strong bases. [All alkali metal hydroxides are soluble. Of the alkaline earth hydroxides, $Be(OH)_2$ and $Mg(OH)_2$ are insoluble; $Ca(OH)_2$ and $Sr(OH)_2$ are slightly soluble; and $Ba(OH)_2$ is soluble.] Some examples of strong bases are

$$NaOH(s) \xrightarrow{H_2O} Na^+(aq) + OH^-(aq)$$
$$KOH(s) \xrightarrow{H_2O} K^+(aq) + OH^-(aq)$$
$$Ba(OH)_2(s) \xrightarrow{H_2O} Ba^{2+}(aq) + 2OH^-(aq)$$

Strictly speaking, these metal hydroxides are not Brønsted bases because they cannot accept a proton. However, the hydroxide ion (OH^-) formed when they ionize *is* a Brønsted base because it can accept a proton:

$$H_3O^+(aq) + OH^-(aq) \longrightarrow 2H_2O(l)$$

Thus, when we call NaOH or any other metal hydroxide a base, we are actually referring to the OH^- species derived from the hydroxide.

Weak bases, like weak acids, *are weak electrolytes.* Ammonia is a weak base. It ionizes to a very limited extent in water:

$$NH_3(aq) + H_2O(l) \rightleftharpoons NH_4^+(aq) + OH^-(aq)$$

Note that, unlike acids, NH_3 does not donate a proton to water. Rather, NH_3 behaves as a base by accepting a proton from water to form NH_4^+ and OH^- ions.

Table 15.2 lists some important conjugate acid-base pairs, in order of their relative strengths. Conjugate acid-base pairs have the following properties:

1. If an acid is strong, its conjugate base has no measurable strength. Thus, the Cl^- ion, which is the conjugate base of the strong acid HCl, is an extremely weak base.
2. H_3O^+ is the strongest acid that can exist in aqueous solution. Acids stronger than H_3O^+ react with water to produce H_3O^+ and their conjugate bases. Thus, HCl, which is a stronger acid than H_3O^+, reacts with water completely to form H_3O^+ and Cl^-:

   $$HCl(aq) + H_2O(l) \longrightarrow H_3O^+(aq) + Cl^-(aq)$$

   Acids weaker than H_3O^+ react with water to a much smaller extent, producing H_3O^+ and their conjugate bases. For example, the following equilibrium lies primarily to the left:

   $$HF(aq) + H_2O(l) \rightleftharpoons H_3O^+(aq) + F^-(aq)$$

3. The OH^- ion is the strongest base that can exist in aqueous solution. Bases stronger than OH^- react with water to produce OH^- and their conjugate acids. For example, the oxide ion (O^{2-}) is a stronger base than OH^-, so it reacts with water completely as follows:

   $$O^{2-}(aq) + H_2O(l) \longrightarrow 2OH^-(aq)$$

   For this reason the oxide ion does not exist in aqueous solutions.

Example 15.6 shows calculations of pH for a solution containing a strong acid and a solution of a strong base.

Table 15.2 Relative Strengths of Conjugate Acid-Base Pairs

	Acid	Conjugate Base
Strong acids	$HClO_4$ (perchloric acid)	ClO_4^- (perchlorate ion)
	HI (hydroiodic acid)	I^- (iodide ion)
	HBr (hydrobromic acid)	Br^- (bromide ion)
	HCl (hydrochloric acid)	Cl^- (chloride ion)
	H_2SO_4 (sulfuric acid)	HSO_4^- (hydrogen sulfate ion)
	HNO_3 (nitric acid)	NO_3^- (nitrate ion)
	H_3O^+ (hydronium ion)	H_2O (water)
Weak acids	HSO_4^- (hydrogen sulfate ion)	SO_4^{2-} (sulfate ion)
	HF (hydrofluoric acid)	F^- (fluoride ion)
	HNO_2 (nitrous acid)	NO_2^- (nitrite ion)
	HCOOH (formic acid)	$HCOO^-$ (formate ion)
	CH_3COOH (acetic acid)	CH_3COO^- (acetate ion)
	NH_4^+ (ammonium ion)	NH_3 (ammonia)
	HCN (hydrocyanic acid)	CN^- (cyanide ion)
	H_2O (water)	OH^- (hydroxide ion)
	NH_3 (ammonia)	NH_2^- (amide ion)

Acid strength increases (upward); Base strength increases (downward)

Example 15.6

Calculate the pH of (a) a 1.0×10^{-3} *M* HCl solution and (b) a 0.020 *M* $Ba(OH)_2$ solution.

Strategy Keep in mind that HCl is a strong acid and $Ba(OH)_2$ is a strong base. Thus, these species are completely ionized and no HCl or $Ba(OH)_2$ will be left in solution.

Solution

(a) The ionization of HCl is

$$HCl(aq) \longrightarrow H^+(aq) + Cl^-(aq)$$

The concentrations of all the species (HCl, H^+, and Cl^-) before and after ionization can be represented using the ICE method for solving equilibrium concentrations as follows:

	$HCl(aq)$	$\longrightarrow$	$H^+(aq)$	+	$Cl^-(aq)$
Initial (*M*):	1.0×10^{-3}		0.0		0.0
Change (*M*):	-1.0×10^{-3}		$+1.0 \times 10^{-3}$		$+1.0 \times 10^{-3}$
Final (*M*):	0.0		1.0×10^{-3}		1.0×10^{-3}

A positive (+) change represents an increase and a negative (−) change indicates a decrease in concentration. Thus,

$$[H^+] = 1.0 \times 10^{-3}\ M$$
$$pH = -\log(1.0 \times 10^{-3}) = 3.00$$

(b) $Ba(OH)_2$ is a strong base; each $Ba(OH)_2$ unit produces two OH^- ions:

$$Ba(OH)_2(aq) \rightleftharpoons Ba^{2+}(aq) + 2OH^-(aq)$$

The changes in the concentrations of all the species can be represented as follows:

	$Ba(OH)_2(aq)$	$\longrightarrow$	$Ba^{2+}(aq)$	+	$2OH^-(aq)$
Initial (*M*):	0.020		0.00		0.00
Change (*M*):	−0.020		+0.020		+2(0.020)
Final (*M*):	0.00		0.020		0.040

(Continued)

Thus,

$$[OH^-] = 0.040\ M$$
$$pOH = -\log 0.040 = 1.40$$

Therefore, from Equation (15.8),

$$pH = 14.00 - pOH$$
$$= 14.00 - 1.40$$
$$= 12.60$$

Check Note that in both (a) and (b) we have neglected the contribution of the autoionization of water to $[H^+]$ and $[H^-]$ because $1.0 \times 10^{-7}\ M$ is so small compared with $1.0 \times 10^{-3}\ M$ and $0.040\ M$.

Practice Exercise Calculate the pH of a $1.8 \times 10^{-2}\ M$ $Ba(OH)_2$ solution.

Similar problem: 15.18.

If we know the relative strengths of two acids, we can predict the position of equilibrium between one of the acids and the conjugate base of the other, as illustrated in Example 15.7.

Example 15.7

Predict the direction of the following reaction in aqueous solution:

$$HNO_2(aq) + CN^-(aq) \rightleftharpoons HCN(aq) + NO_2^-(aq)$$

Strategy The problem is to determine whether, at equilibrium, the reaction will be shifted to the right, favoring HCN and NO_2^-, or to the left, favoring HNO_2 and CN^-. Which of the two is the stronger acid and hence a stronger proton donor: HNO_2 or HCN? Which of the two is the stronger base and hence a stronger proton acceptor: CN^- or NO_2^-? Remember that the stronger the acid, the weaker its conjugate base.

Solution In Table 15.2 we see that HNO_2 is a stronger acid than HCN. Thus, CN^- is a stronger base than NO_2^-. The net reaction will proceed from left to right as written because HNO_2 is a better proton donor than HCN (and CN^- is a better proton acceptor than NO_2^-).

Practice Exercise Predict whether the equilibrium constant for the following reaction is greater than or smaller than 1:

$$CH_3COOH(aq) + HCOO^-(aq) \rightleftharpoons CH_3COO^-(aq) + HCOOH(aq)$$

Similar problem: 15.37.

Summary of Concepts & Facts

- In aqueous solution, the following are classified as strong acids: $HClO_4$, HI, HBr, HCl, H_2SO_4 (first stage of ionization), and HNO_3. Strong bases in aqueous solution include hydroxides of alkali metals and of alkaline earth metals (except beryllium).

Review of Concepts & Facts

15.4.1 What is the concentration of $Ca(OH)_2$ in a solution if the pH is 7.92?

15.4.2 Predict whether the equilibrium constant for the following reaction is greater than or smaller than 1:

$$HF(aq) + NH_3(aq) \rightleftharpoons F^-(aq) + NH_4^+(aq)$$

15.4.3 List in order of decreasing concentration all the ionic and molecular species in the following acid solutions: (i) HNO_3, (ii) HF.

15.4.4 List in order of decreasing concentration all the ionic and molecular species in the following base solutions: (i) NH_3, (ii) KOH.

15.5 Weak Acids and Acid Ionization Constants

Learning Objectives

- Apply the acid-dissociation constant (K_a) in equilibrium calculations.
- Determine the percent ionization of a weak acid.

As we have seen, there are relatively few strong acids. The vast majority of acids are weak acids. Consider a weak monoprotic acid, HA. Its ionization in water is represented by

$$HA(aq) + H_2O(l) \rightleftharpoons H_3O^+(aq) + A^-(aq)$$

or simply

$$HA(aq) \rightleftharpoons H^+(aq) + A^-(aq)$$

The equilibrium expression for this ionization is

$$K_a = \frac{[H_3O^+][A^-]}{[HA]} \quad \text{or} \quad K_a = \frac{[H^+][A^-]}{[HA]} \tag{15.10}$$

where K_a, the ***acid ionization constant,*** is the *equilibrium constant for the ionization of an acid*. At a given temperature, the strength of the acid HA is measured quantitatively by the magnitude of K_a. The larger K_a, the stronger the acid—that is, the greater the concentration of H^+ ions at equilibrium due to its ionization. Keep in mind, however, that only weak acids have K_a values associated with them.

Video: Acid Ionization

Table 15.3 lists a number of weak acids and their K_a values at 25°C in order of decreasing acid strength. Although all these acids are weak, within the group there is great variation in their strengths. For example, K_a for HF (6.3×10^{-4}) is about 1 million times that for HCN (6.2×10^{-10}).

Generally, we can calculate the hydrogen ion concentration or pH of an acid solution at equilibrium, given the initial concentration of the acid and its K_a value. Alternatively, if we know the pH of a weak acid solution and its initial concentration, we can determine its K_a. The basic approach for solving these problems, which deal with equilibrium concentrations, is the same one outlined in Chapter 14. However, because acid ionization represents a major category of chemical equilibrium in aqueous solution, we will develop a systematic procedure for solving this type of problem that will also help us to understand the chemistry involved.

Suppose we are asked to calculate the pH of a 0.50 *M* HF solution at 25°C. The ionization of HF is given by

$$HF(aq) \rightleftharpoons H^+(aq) + F^-(aq)$$

From Table 15.3 we write

$$K_a = \frac{[H^+][F^-]}{[HF]} = 6.3 \times 10^{-4}$$

The first step is to identify all the species present in solution that may affect its pH. Because weak acids ionize to a small extent, at equilibrium the major species present are nonionized HF and some H^+ and F^- ions. Another major species is H_2O, but its very small K_w (1.0×10^{-14}) means that water is not a significant contributor to the H^+ ion concentration. Therefore, unless otherwise stated, we will always ignore the H^+ ions produced by the autoionization of water. Note that we need not be concerned with the OH^- ions that are also present in solution. The OH^- concentration can be determined from Equation (15.3) after we have calculated $[H^+]$.

Table 15.3 Ionization Constants of Some Weak Acids and Their Conjugate Bases at 25°C

Name of Acid	Formula	Structure	K_a	Conjugate Base	K_b†
Hydrofluoric acid	HF	H—F	6.3×10^{-4}	F^-	1.6×10^{-11}
Nitrous acid	HNO_2	O=N—O—H	5.6×10^{-4}	NO_2^-	1.8×10^{-11}
Acetylsalicylic acid (aspirin)	$C_9H_8O_4$	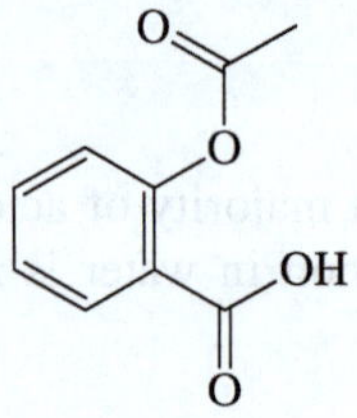	3.0×10^{-4}	$C_9H_7O_4^-$	3.3×10^{-11}
Formic acid	HCOOH	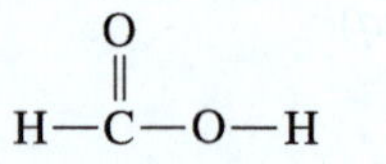	1.8×10^{-4}	$HCOO^-$	5.6×10^{-11}
Ascorbic acid*	$C_6H_8O_6$	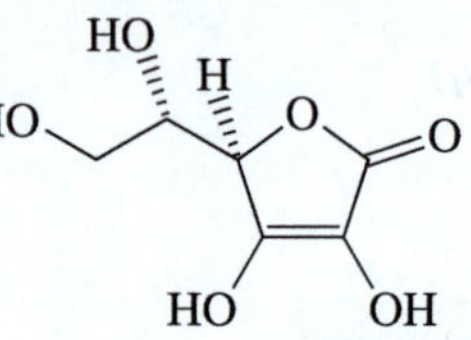	9.1×10^{-5}	$C_6H_7O_6^-$	1.1×10^{-10}
Benzoic acid	C_6H_5COOH	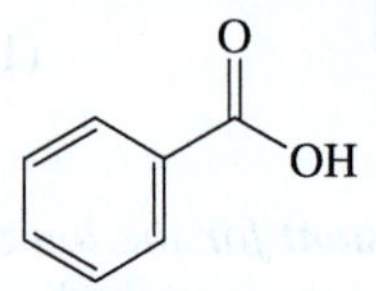	6.3×10^{-5}	$C_6H_5COO^-$	1.6×10^{-10}
Acetic acid	CH_3COOH	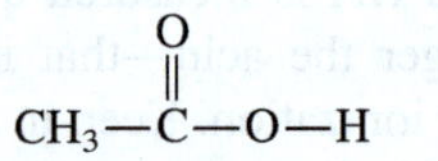	1.8×10^{-5}	CH_3COO^-	5.6×10^{-10}
Hydrocyanic acid	HCN	H—C≡N	6.2×10^{-10}	CN^-	1.6×10^{-5}
Phenol	C_6H_5OH	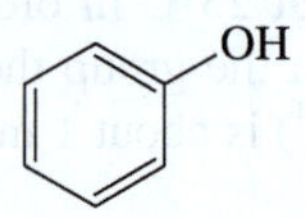	1.0×10^{-10}	$C_6H_5O^-$	1.0×10^{-4}

*For ascorbic acid it is the upper left hydroxyl group that is associated with this ionization constant.
†The base ionization constant K_b is discussed in Section 15.6.

We can summarize the changes in the concentrations of HF, H^+, and F^- according to the steps shown in Section 14.4 as follows:

	$HF(aq)$	$\rightleftharpoons$	$H^+(aq)$	+	$F^-(aq)$
Initial (*M*):	0.50		0.00		0.00
Change (*M*):	$-x$		$+x$		$+x$
Equilibrium (*M*):	$0.50 - x$		x		x

The equilibrium concentrations of HF, H^+, and F^-, expressed in terms of the unknown *x*, are substituted into the ionization constant expression to give

$$K_a = \frac{(x)(x)}{0.50 - x} = 6.3 \times 10^{-4}$$

Rearranging this expression, we write

$$x^2 + 6.3 \times 10^{-4}\,x - 3.2 \times 10^{-4} = 0$$

This is a quadratic equation, which can be solved using the quadratic formula (see Appendix 5). Or we can try using a shortcut to solve for *x*. Because HF is a weak acid and weak acids ionize only to a slight extent, we reason that *x* must be small compared to 0.50. Therefore, we can make the approximation

$$0.50 - x \approx 0.50$$

Now the ionization constant expression becomes

$$\frac{x^2}{0.50 - x} \approx \frac{x^2}{0.50} = 6.3 \times 10^{-4}$$

Rearranging, we get

$$x^2 = (0.50)(6.3 \times 10^{-4}) = 3.15 \times 10^{-4}$$
$$x = \sqrt{3.15 \times 10^{-4}} = 0.18\ M$$

Thus, we have solved for x without having to use the quadratic equation. At equilibrium, we have

$$[\text{HF}] = (0.50 - 0.018)\ M = 0.48\ M$$
$$[\text{H}^+] = 0.018\ M$$
$$[\text{F}^-] = 0.018\ M$$

and the pH of the solution is

$$\text{pH} = -\log(0.018) = 1.74$$

How good is this approximation? Because K_a values for weak acids are generally known to an accuracy of only ±5 percent, it is reasonable to require x to be less than 5 percent of 0.50, the number from which it is subtracted. In other words, the approximation is valid if the following expression is equal to or less than 5 percent:

$$\frac{0.018\ M}{0.50\ M} \times 100\% = 3.6\%$$

Thus, the approximation we made is acceptable.

Now consider a different situation. If the initial concentration of HF is 0.050 M, and we use the abovementioned procedure to solve for x, we would get $5.6 \times 10^{-3}\ M$. However, the following test shows that this answer is not a valid approximation because it is greater than 5 percent of 0.050 M:

$$\frac{5.6 \times 10^{-3}\ M}{0.050\ M} \times 100\% = 11\%$$

In this case, we can get an accurate value for x by solving the quadratic equation.

The Quadratic Equation

We start by writing the ionization expression in terms of the unknown x:

$$\frac{x^2}{0.050 - x} = 6.3 \times 10^{-4}$$

$$x^2 + 6.3 \times 10^{-4}x - 3.2 \times 10^{-5} = 0$$

This expression fits the quadratic equation $ax^2 + bx + c = 0$. Using the quadratic formula, we write

$$x = \frac{-b \pm \sqrt{b^2 - 4ac}}{2a}$$
$$= \frac{-6.3 \times 10^{-4} \pm \sqrt{(6.3 \times 10^{-4})^2 - 4(1)(-3.2 \times 10^{-5})}}{2(1)}$$
$$= \frac{-6.3 \times 10^{-4} \pm 0.011}{2}$$
$$= 5.2 \times 10^{-3}\ M \quad \text{or} \quad -5.8 \times 10^{-3}\ M$$

The second solution ($x = -5.8 \times 10^{-3}\ M$) is physically impossible because the concentration of ions produced as a result of ionization cannot be negative. Choosing $x = 5.2 \times 10^{-3}\ M$, we can solve for [HF], $[H^+]$, and $[F^-]$ as follows:

$$[\text{HF}] = (0.050 - 5.2 \times 10^{-3})\ M = 0.045\ M$$
$$[\text{H}^+] = 5.2 \times 10^{-3}\ M$$
$$[\text{F}^-] = 5.2 \times 10^{-3}\ M$$

The pH of the solution, then, is

$$\text{pH} = -\log(5.2 \times 10^{-3}) = 2.28$$

In summary, the main steps for solving weak acid ionization problems are:

1. Identify the major species that can affect the pH of the solution. In most cases we can ignore the ionization of water. We omit the hydroxide ion because its concentration is determined by that of the H^+ ion.
2. Express the equilibrium concentrations of these species in terms of the initial concentration of the acid and a single unknown x, which represents the change in concentration.
3. Write the weak acid ionization and express the ionization constant K_a in terms of the equilibrium concentrations of H^+, the conjugate base, and the unionized acid. First solve for x by the approximate method. If the approximate method is not valid, use the quadratic equation to solve for x.
4. Having solved for x, calculate the equilibrium concentrations of all species and/or the pH of the solution.

Student Hot Spot

Student data indicate you may struggle with determining the pH of a weak acid solution. Access your eBook for additional Learning Resources on this topic.

Example 15.8 provides another illustration of this procedure.

Example 15.8

Calculate the pH of a 0.036 M nitrous acid (HNO_2) solution:

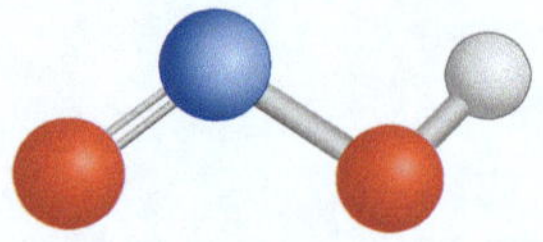

HNO_2

$$\text{HNO}_2(aq) \rightleftharpoons \text{H}^+(aq) + \text{NO}_2^-(aq)$$

Strategy Recall that a weak acid only partially ionizes in water. We are given the initial concentration of a weak acid and asked to calculate the pH of the solution at equilibrium. As in Example 15.6, we ignore the ionization of H_2O so the major source of H^+ ions is the acid. The concentration of OH^- ions is very small as we would expect from an acidic solution so it is present as a minor species.

Solution We follow the procedure already outlined.

Step 1: The species that can affect the pH of the solution are HNO_2, H^+, and the conjugate base NO_2^-. We ignore water's contribution to $[H^+]$.

Step 2: Letting x be the equilibrium concentration of H^+ and NO_2^- ions in mol/L, we summarize:

	$HNO_2(aq)$	$\rightleftharpoons$	$H^+(aq)$	+	$NO_2^-(aq)$
Initial (M):	0.036		0.00		0.00
Change (M):	$-x$		$+x$		$+x$
Equilibrium (M):	$0.036 - x$		x		x

(Continued)

Step 3: From Table 15.3 we write

$$K_a = \frac{[H^+][NO_2^-]}{[HNO_2]}$$

$$5.6 \times 10^{-4} = \frac{x^2}{0.036 - x}$$

Applying the approximation $0.036 - x \approx 0.036$, we obtain

$$5.6 \times 10^{-4} = \frac{x^2}{0.036 - x} \approx \frac{x^2}{0.036}$$

$$x^2 = 2.02 \times 10^{-5}$$

$$x = 4.5 \times 10^{-3}\ M$$

To test the approximation,

$$\frac{4.5 \times 10^{-3}\ M}{0.036\ M} \times 100\% = 13\%$$

Because this is greater than 5 percent, our approximation is not valid and we must solve the quadratic equation, as follows:

$$x^2 + 5.6 \times 10^{-4}\ x - 2.02 \times 10^{-5} = 0$$

$$x = \frac{-5.6 \times 10^{-4} \pm \sqrt{(5.6 \times 10^{-4})^2 - 4(1)(-2.02 \times 10^{-5})}}{2(1)}$$

$$= 4.2 \times 10^{-3}\ M \quad \text{or} \quad -4.8 \times 10^{-3}\ M$$

The second solution is physically impossible, because the concentration of ions produced as a result of ionization cannot be negative. Therefore, the solution is given by the positive root, $x = 4.2 \times 10^{-3}\ M$.

Step 4: At equilibrium

$$[H^+] = 4.2 \times 10^{-3}\ M$$

$$pH = -\log(4.2 \times 10^{-3})$$

$$= 2.38$$

Check Note that the calculated pH indicates that the solution is acidic, which is what we would expect for a weak acid solution. Compare the calculated pH with that of a 0.036 *M* strong acid solution such as HCl to convince yourself of the difference between a strong acid and a weak acid.

Practice Exercise What is the pH of a 0.122 *M* monoprotic acid whose K_a is 5.7×10^{-4}?

Similar problem: 15.43.

Student Hot Spot

Student data indicate you may struggle with calculating the K_a of a weak acid from the pH. Access your eBook for additional Learning Resources on this topic.

One way to determine K_a of an acid is to measure the pH of the acid solution of known concentration at equilibrium. Example 15.9 shows this approach.

Example 15.9

The pH of a 0.10 *M* solution of formic acid (HCOOH) is 2.39. What is the K_a of the acid?

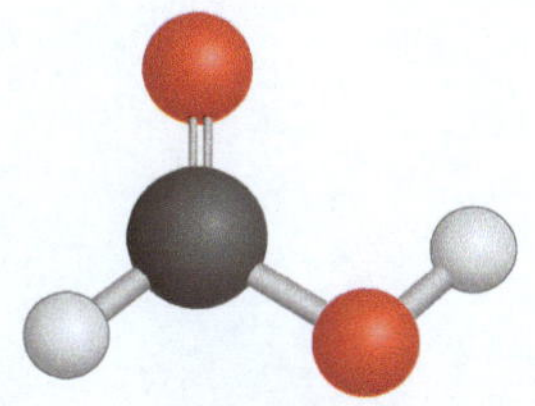

HCOOH

Strategy Formic acid is a weak acid. It only partially ionizes in water. Note that the concentration of formic acid refers to the initial concentration, before ionization has started. The pH of the solution, on the other hand, refers to the equilibrium state. To calculate K_a, then, we need to know the concentrations of all three species: $[H^+]$, $[HCOO^-]$, and $[HCOOH]$ at equilibrium. As usual, we ignore the ionization of water.

(Continued)

Solution We proceed as follows.

Step 1: The major species in solution are HCOOH, H^+, and the conjugate base $HCOO^-$.

Step 2: First we need to calculate the hydrogen ion concentration from the pH value:

$$pH = -\log [H^+]$$
$$2.39 = -\log [H^+]$$

Taking the antilog of both sides, we get

$$[H^+] = 10^{-2.39} = 4.1 \times 10^{-3}\ M$$

Next we summarize the changes:

	$HCOOH(aq)$	$\rightleftharpoons$	$H^+(aq)$	+	$HCOO^-(aq)$
Initial (*M*):	0.10		0.00		0.00
Change (*M*):	-4.1×10^{-3}		$+4.1 \times 10^{-3}$		$+4.1 \times 10^{-3}$
Equilibrium (*M*):	$(0.10 - 4.1 \times 10^{-3})$		4.1×10^{-3}		4.1×10^{-3}

Note that because the pH and hence the H^+ ion concentration is known, it follows that we also know the concentrations of HCOOH and $HCOO^-$ at equilibrium.

Step 3: The ionization constant of formic acid is given by

$$K_a = \frac{[H^+][HCOO^-]}{[HCOOH]} = \frac{(4.1 \times 10^{-3})(4.1 \times 10^{-3})}{(0.10 - 4.1 \times 10^{-3})} = 1.8 \times 10^{-4}$$

Check The K_a value matches the one listed in Table 15.3.

Practice Exercise The pH of a 0.060 *M* weak monoprotic acid is 3.44. Calculate the K_a of the acid.

Similar problem: 15.45.

Percent Ionization

We have seen that the magnitude of K_a indicates the strength of an acid. Another measure of the strength of an acid is its ***percent ionization,*** which is defined as

$$\text{percent ionization} = \frac{\text{ionized acid concentration at equilibrium}}{\text{initial concentration of acid}} \times 100\% \quad (15.11)$$

The stronger the acid, the greater the percent ionization. For a monoprotic acid HA, the concentration of the acid that undergoes ionization is equal to the concentration of the H^+ ions or the concentration of the A^- ions at equilibrium. Therefore, we can write the percent ionization as

$$\text{percent ionization} = \frac{[H^+]}{[HA]_0} \times 100\%$$

where $[H^+]$ is the concentration at equilibrium and $[HA]_0$ is the initial concentration.

Referring to Example 15.8, we see that the percent ionization of a 0.036 *M* HNO_2 solution is

$$\text{percent ionization} = \frac{4.2 \times 10^{-3}\ M}{0.036\ M} \times 100\% = 12\%$$

Thus, only about one of every nine HNO_2 molecules has ionized. This is consistent with the fact that HNO_2 is a weak acid.

The extent to which a weak acid ionizes depends on the initial concentration of the acid. The more dilute the solution, the greater the percentage ionization (Figure 15.4). In qualitative terms, when an acid is diluted, the concentration of the *particles* in the solution is reduced. According to Le Châtelier's principle (see Section 14.5), this reduction in particle concentration (the stress) is counteracted by shifting the reaction to the side with more particles; that is, the equilibrium shifts from the nonionized acid side (one particle) to the side containing H^+ ions and the conjugate base (two particles): $HA \rightleftharpoons H^+ + A^-$. Consequently, the concentration of *particles* increases in the solution.

Student Hot Spot

Student data indicate you may struggle with the relationship between acid strength and percent ionization. Access your eBook for additional Learning Resources on this topic.

The dependence of percent ionization on initial concentration can be illustrated by the HF case discussed earlier.

0.50 *M* HF:

$$\text{percent ionization} = \frac{0.018\ M}{0.50\ M} \times 100\% = 3.6\%$$

0.050 *M* HF:

$$\text{percent ionization} = \frac{4.9 \times 10^{-3}\ M}{0.050\ M} \times 100\% = 9.8\%$$

We see that, as expected, a more dilute HF solution has a greater percent ionization of the acid.

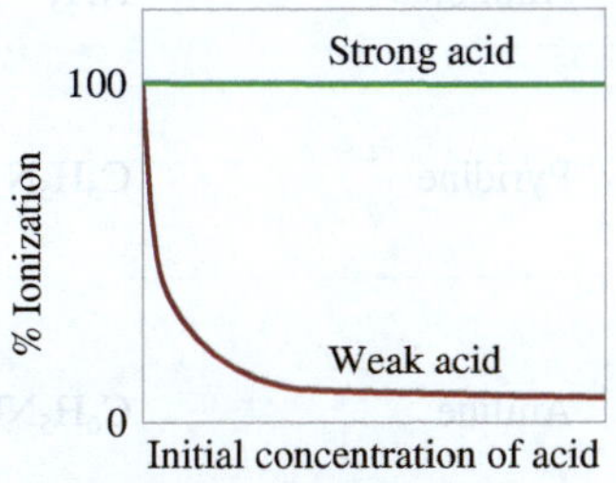

Figure 15.4 *Dependence of percent ionization on initial concentration of acid. Note that at very low concentrations, all acids (weak and strong) are almost completely ionized.*

Summary of Concepts & Facts

- The acid ionization constant K_a increases with acid strength.
- Percent ionization is another measure of the strength of acids. The more dilute a solution of a weak acid, the greater the percent ionization of the acid.

Review of Concepts & Facts

15.5.1 Determine the pH of a 0.065 *M* benzoic acid (C_6H_5COOH) solution.

15.5.2 A 0.15 *M* solution of a weak acid has pH of 5.87. Calculate the K_a of this acid.

15.5.3 The "concentration" of water is 55.5 *M*. Calculate the percent ionization of water.

15.6 Weak Bases and Base Ionization Constants

Learning Objective

- Apply the base-dissociation constant (K_b) in equilibrium calculations.

The ionization of weak bases is treated in the same way as the ionization of weak acids. When ammonia dissolves in water, it undergoes the reaction

$$NH_3(aq) + H_2O(l) \rightleftharpoons NH_4^+(aq) + OH^-(aq)$$

The equilibrium constant is given by

$$K = \frac{[NH_4^+][OH^-]}{[NH_3][H_2O]}$$

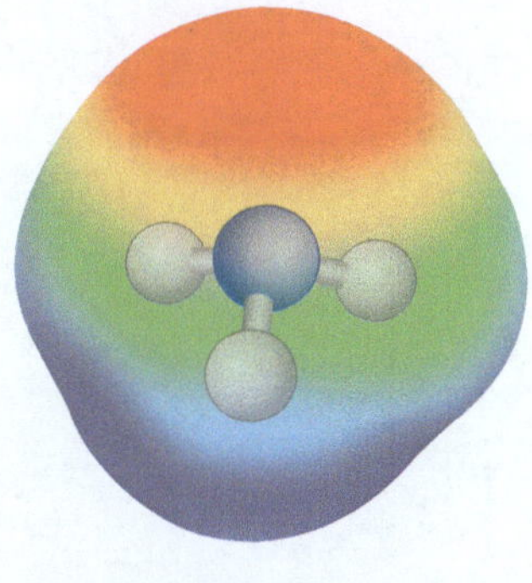

The lone pair (red color) on the N atom accounts for ammonia's basicity.

Table 15.4 Ionization Constants of Some Weak Bases and Their Conjugate Acids at 25°C

Name of Base	Formula	Structure	K_b^*	Conjugate Acid	K_a
Ethylamine	$C_2H_5NH_2$	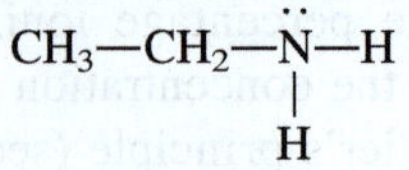	4.5×10^{-4}	$C_2H_5\overset{+}{N}H_3$	2.2×10^{-11}
Methylamine	CH_3NH_2	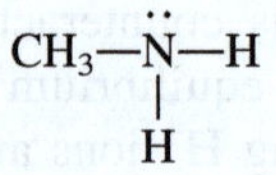	4.6×10^{-4}	$CH_3\overset{+}{N}H_3$	2.2×10^{-11}
Ammonia	NH_3	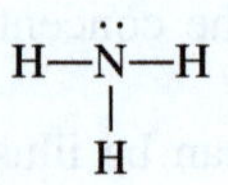	1.8×10^{-5}	NH_4^+	5.6×10^{-10}
Pyridine	C_5H_5N	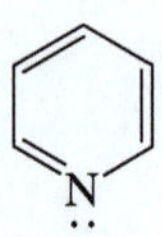	1.7×10^{-9}	$C_5H_5\overset{+}{N}H$	5.9×10^{-6}
Aniline	$C_6H_5NH_2$	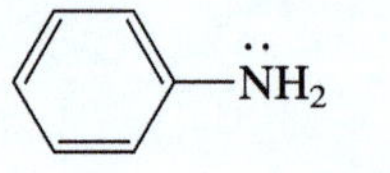	7.4×10^{-10}	$C_6H_5\overset{+}{N}H_3$	1.4×10^{-5}
Urea	$(NH_2)_2CO$	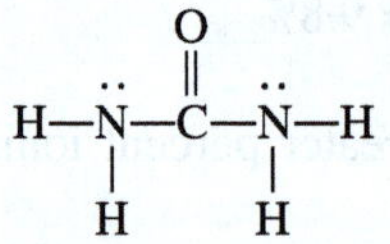	1.3×10^{-14}	$H_2NCO\overset{+}{N}H_3$	0.77

*The nitrogen atom with the lone pair accounts for each compound's basicity. In the case of urea, K_b can be associated with either nitrogen atom.

Video: Base Ionization

Compared with the total concentration of water, very few water molecules are consumed by this reaction, so we can treat $[H_2O]$ as a constant. Thus, we can write the ***base ionization constant (K_b),*** which is *the equilibrium constant for the ionization reaction,* as

$$K_b = K[H_2O] = \frac{[NH_4^+][OH^-]}{[NH_3]}$$

$$= 1.8 \times 10^{-5}$$

Table 15.4 lists a number of common weak bases and their ionization constants. Note that the basicity of all these compounds is attributable to the lone pair of electrons on the nitrogen atom. The ability of the lone pair to accept a H^+ ion makes these substances Brønsted bases.

In solving problems involving weak bases, we follow the same procedure we used for weak acids. The main difference is that we calculate $[OH^-]$ first, rather than $[H^+]$. Example 15.10 shows this approach.

Example 15.10

What is the pH of a 0.40 *M* ammonia solution?

Strategy The procedure here is similar to the one used for a weak acid (see Example 15.8). From the ionization of ammonia, we see that the major species in solution at equilibrium are NH_3, NH_4^+, and OH^-. The hydrogen ion concentration is very small as we would expect from a basic solution, so it is present as a minor species. As before, we ignore the ionization of water.

Solution We proceed according to the following steps.

Step 1: The major species in an ammonia solution are NH_3, NH_4^+, and OH^-. We ignore the very small contribution to OH^- concentration by water.

(Continued)

Step 2: Letting x be the equilibrium concentration of NH_4^+ and OH^- ions in mol/L, we summarize:

	$NH_3(aq) + H_2O(l) \rightleftharpoons$	$NH_4^+(aq)$	+ $OH^-(aq)$
Initial (M):	0.40	0.00	0.00
Change (M):	$-x$	$+x$	$+x$
Equilibrium (M):	$0.40 - x$	x	x

Step 3: Table 15.4 gives us K_b:

$$K_b = \frac{[NH_4^+][OH^-]}{[NH_3]}$$

$$1.8 \times 10^{-5} = \frac{x^2}{0.40 - x}$$

Applying the approximation $0.40 - x \approx 0.40$ we obtain

$$1.8 \times 10^{-5} = \frac{x^2}{0.40 - x} \approx \frac{x^2}{0.40}$$

$$x^2 = 7.2 \times 10^{-6}$$

$$x = 2.7 \times 10^{-3}\ M$$

To test the approximation, we write

$$\frac{2.7 \times 10^{-3}\ M}{0.40\ M} \times 100\% = 0.68\%$$

Therefore, the approximation is valid.

Step 4: At equilibrium, $[OH^-] = 2.7 \times 10^{-3}\ M$. Thus,

$$\text{pOH} = -\log(2.7 \times 10^{-3}) = 2.57$$

$$\text{pH} = 14.00 - 2.57 = 11.43$$

Check Note that the pH calculated is basic, which is what we would expect from a weak base solution. Compare the calculated pH with that of a 0.40 M strong base solution, such as KOH, to convince yourself of the difference between a strong base and a weak base.

Practice Exercise Calculate the pH of a 0.26 M methylamine solution (see Table 15.4).
Similar problem: 15.55.

Summary of Concepts & Facts

- The base ionization constant, K_b, expresses the strength of a base and can be used to determine the pH of a weak base solution.

Review of Concepts & Facts

15.6.1 Determine the pH of a 0.055 M solution of an unknown weak base: ($K_b = 9.1 \times 10^{-8}$).

15.6.2 Consider the following three solutions of equal concentration. Using data in Table 15.4, rank the three solutions from most basic to least basic: (a) aniline, (b) methylamine, (c) urea.

15.7 The Relationship Between the Ionization Constants of Acids and Their Conjugate Bases

Learning Objective

- Summarize the relationship between K_a, K_b, and K_w.

An important relationship between the acid ionization constant and the ionization constant of its conjugate base can be derived as follows, using acetic acid as an example:

$$CH_3COOH(aq) \rightleftharpoons H^+(aq) + CH_3COO^-(aq)$$

$$K_a = \frac{[H^+][CH_3COO^-]}{[CH_3COOH]}$$

The conjugate base, CH_3COO^-, supplied by a sodium acetate (CH_3COONa) solution, reacts with water according to the equation

$$CH_3COO^-(aq) + H_2O(l) \rightleftharpoons CH_3COOH(aq) + OH^-(aq)$$

and we can write the base ionization constant as

$$K_b = \frac{[CH_3COOH][OH^-]}{[CH_3COO^-]}$$

The product of these two ionization constants is given by

$$K_aK_b = \frac{[H^+][CH_3COO^-]}{[CH_3COOH]} \times \frac{[CH_3COOH][OH^-]}{[CH_3COO^-]}$$

$$= [H^+][OH^-]$$

$$= K_w$$

This result may seem strange at first, but if we add the two equations we see that the sum is simply the autoionization of water.

(1)	$CH_3COOH(aq) \rightleftharpoons H^+(aq) + CH_3COO^-(aq)$	K_a
(2)	$CH_3COO^-(aq) + H_2O(l) \rightleftharpoons CH_3COOH(aq) + OH^-(aq)$	K_b
(3)	$H_2O(l) \rightleftharpoons H^+(aq) + OH^-(aq)$	K_w

This example illustrates one of the rules for chemical equilibria: When two reactions are added to give a third reaction, the equilibrium constant for the third reaction is the product of the equilibrium constants for the two added reactions (see Section 14.2). Thus, for any conjugate acid-base pair it is always true that

$$K_aK_b = K_w \tag{15.12}$$

Expressing Equation (15.12) as

$$K_a = \frac{K_w}{K_b} \qquad K_b = \frac{K_w}{K_a}$$

enables us to draw an important conclusion: The stronger the acid (the larger K_a), the weaker its conjugate base (the smaller K_b), and vice versa (see Tables 15.3 and 15.4).

We can use Equation (15.12) to calculate the K_b of the conjugate base (CH_3COO^-) of CH_3COOH as follows. We find the K_a value of CH_3COOH in Table 15.3 and write

$$K_b = \frac{K_w}{K_a}$$
$$= \frac{1.0 \times 10^{-14}}{1.8 \times 10^{-5}}$$
$$= 5.6 \times 10^{-10}$$

Summary of Concepts & Facts

- The product of the ionization constant of an acid and the ionization constant of its conjugate base is equal to the ion-product constant of water.

Review of Concepts & Facts

15.7.1 An unknown organic acid has $K_a = 5.6 \times 10^{-6}$. What is the K_b of the conjugate base?

15.7.2 Consider the following two acids and their ionization constants:

$$HCOOH \quad K_a = 1.8 \times 10^{-4}$$
$$HCN \quad K_a = 6.2 \times 10^{-10}$$

Which conjugate base ($HCOO^-$ or CN^-) is stronger?

15.8 Diprotic and Polyprotic Acids

Learning Objective

- Identify polyprotic acids and evaluate the concentrations of various species in solution given an initial concentration.

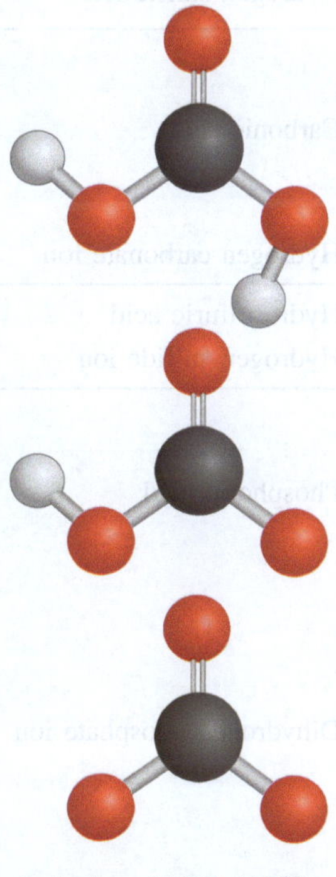

Top to bottom: H_2CO_3, HCO_3^-, and CO_3^{2-}.

The treatment of diprotic and polyprotic acids is more involved than that of monoprotic acids because these substances may yield more than one hydrogen ion per molecule. These acids ionize in a stepwise manner; that is, they lose one proton at a time. An ionization constant expression can be written for each ionization stage. Consequently, two or more equilibrium constant expressions must often be used to calculate the concentrations of species in the acid solution. For example, for carbonic acid, H_2CO_3, we write

$$H_2CO_3(aq) \rightleftharpoons H^+(aq) + HCO_3^-(aq) \qquad K_{a_1} = \frac{[H^+][HCO_3^-]}{[H_2CO_3]}$$

$$HCO_3^-(aq) \rightleftharpoons H^+(aq) + CO_3^{2-}(aq) \qquad K_{a_2} = \frac{[H^+][CO_3^{2-}]}{[HCO_3^-]}$$

Note that the conjugate base in the first ionization stage becomes the acid in the second ionization stage.

Table 15.5 shows the ionization constants of several diprotic acids and one polyprotic acid. For a given acid, the first ionization constant is much larger than the second ionization constant, and so on. This trend is reasonable because it is easier to remove a H^+ ion from a neutral molecule than to remove another H^+ ion from a negatively charged ion derived from the molecule.

Student Hot Spot

Student data indicate you may struggle with equilibrium concentrations of polyprotic acids. Access your eBook to view additional Learning Resources on this topic.

Table 15.5 **Ionization Constants of Some Diprotic Acids and a Polyprotic Acid and Their Conjugate Bases at 25°C**

Name of Acid	Formula	Structure	K_a	Conjugate Base	K_b
Sulfuric acid	H_2SO_4	H—O—S(=O)(=O)—O—H	very large	HSO_4^-	very small
Hydrogen sulfate ion	HSO_4^-	H—O—S(=O)(=O)—O⁻	1.3×10^{-2}	SO_4^{2-}	7.7×10^{-13}
Oxalic acid	$H_2C_2O_4$	H—O—C(=O)—C(=O)—O—H	5.0×10^{-2}	$HC_2O_4^-$	2.0×10^{-13}
Hydrogen oxalate ion	$HC_2O_4^-$	H—O—C(=O)—C(=O)—O⁻	1.6×10^{-5}	$C_2O_4^{2-}$	$6.3E-10 \times 10^{-10}$
Sulfurous acid*	H_2SO_3	H—O—S(=O)—O—H	1.3×10^{-2}	HSO_3^-	7.7×10^{-13}
Hydrogen sulfite ion	HSO_3^-	H—O—S(=O)—O⁻	6.3×10^{-8}	SO_3^{2-}	1.6×10^{-7}
Carbonic acid	H_2CO_3	H—O—C(=O)—O—H	4.0×10^{-7}	HCO_3^-	2.1×10^{-8}
Hydrogen carbonate ion	HCO_3^-	H—O—C(=O)—O⁻	4.7×10^{-11}	CO_3^{2-}	2.0×10^{-4}
Hydrosulfuric acid	H_2S	H—S—H	7.9×10^{-8}	HS^-	1.3×10^{-7}
Hydrogen sulfide ion†	HS^-	H—S⁻	1×10^{-19}	S^{2-}	1×10^{-5}
Phosphoric acid	H_3PO_4	H—O—P(=O)(—O—H)—O—H	6.9×10^{-3}	$H_2PO_4^-$	1.4×10^{-12}
Dihydrogen phosphate ion	$H_2PO_4^-$	H—O—P(=O)(—O—H)—O⁻	6.2×10^{-8}	HPO_4^{2-}	1.6×10^{-7}
Hydrogen phosphate ion	HPO_4^{2-}	H—O—P(=O)(—O⁻)—O⁻	4.8×10^{-13}	PO_4^{3-}	2.1×10^{-2}

*H_2SO_3 has never been isolated and exists in only minute concentration in aqueous solution of SO_2. The K_a value here refers to the process $SO_2(g) + H_2O(l) \rightleftharpoons H^+(aq) + HSO_3^-(aq)$.

†The ionization constant of HS^- is very low and difficult to measure. The value listed here is only an estimate.

In Example 15.11 we calculate the equilibrium concentrations of all the species of a diprotic acid in aqueous solution.

Example 15.11

Oxalic acid ($H_2C_2O_4$) is a poisonous substance used chiefly as a bleaching and cleansing agent (for example, to remove bathtub rings). Calculate the concentrations of all the species present at equilibrium in a 0.10 *M* solution.

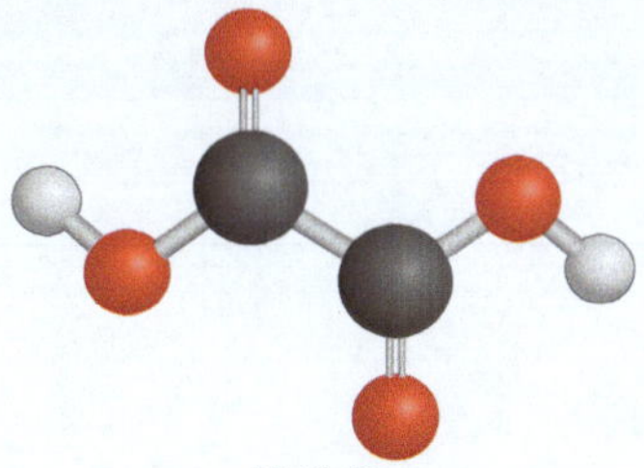

$H_2C_2O_4$

Strategy Determining the equilibrium concentrations of the species of a diprotic acid in aqueous solution is more involved than for a monoprotic acid. We follow the same procedure as that used for a monoprotic acid for each stage, as in Example 15.8. Note that the conjugate base from the first stage of ionization becomes the acid for the second stage of ionization.

Solution We proceed according to the following steps:

Step 1: The major species in solution at this stage are the nonionized acid, H^+ ions, and the conjugate base.

Step 2: Letting x be the equilibrium concentration of H^+ and HC_2O_4 ions in mol/L, we summarize:

	$H_2C_2O_4(aq)$	$\rightleftharpoons$	$H^+(aq)$	+	$HC_2O_4^-(aq)$
Initial (*M*):	0.10		0.00		0.00
Change (*M*):	$-x$		$+x$		$+x$
Equilibrium (*M*):	$0.10 - x$		x		x

Step 3: Table 15.5 gives us

$$K_a = \frac{[H^+][HC_2O_4^-]}{[H_2C_2O_4]}$$

$$5.0 \times 10^{-2} = \frac{x^2}{0.10 - x}$$

Applying the approximation $0.10 - x \approx 0.10$, we obtain

$$5.0 \times 10^{-2} = \frac{x^2}{0.10 - x} \approx \frac{x^2}{0.10}$$

$$x^2 = 5.0 \times 10^{-3}$$

$$x = 7.1 \times 10^{-2}\, M$$

To test the approximation,

$$\frac{7.1 \times 10^{-2}\, M}{0.10\, M} \times 100\% = 71\%$$

Clearly the approximation is not valid. Therefore, we must solve the quadratic equation

$$x^2 + 5.0 \times 10^{-2}\, x - 5.0 \times 10^{-3} = 0$$

The result is $x = 0.50\, M$.

Step 4: When the equilibrium for the first stage of ionization is reached, the concentrations are

$$[H^+] = 0.50\, M$$

$$[HC_2O_4^-] = 0.50\, M$$

$$[H_2C_2O_4] = (0.10 - 0.050)\, M = 0.050\, M$$

(Continued)

Next we consider the second stage of ionization.

Step 1: At this stage, the major species are $HC_2O_4^-$, which acts as the acid in the second stage of ionization, H^+, and the conjugate base $C_2O_4^-$.

Step 2: Letting y be the equilibrium concentration of H^+ and $C_2O_4^-$ ions in mol/L, we summarize:

	$HC_2O_4^-(aq)$	$\rightleftharpoons$	$H^+(aq)$	$+$	$C_2O_4^{2-}(aq)$
Initial (M):	0.050		0.050		0.00
Change (M):	$-y$		$+y$		$+y$
Equilibrium (M):	$0.050 - y$		$0.050 + y$		y

Step 3: Table 15.5 gives us

$$K_a = \frac{[H^+][C_2O_4^{2-}]}{[HC_2O_4^-]}$$

$$1.6 \times 10^{-5} = \frac{(0.050 + y)(y)}{(0.050 - y)}$$

Applying the approximation $0.050 + y \approx 0.050$ and $0.050 - y \approx 0.050$, we obtain

$$\frac{(0.050)(y)}{(0.050)} = y = 1.6 \times 10^{-5}\ M$$

and we test the approximation,

$$\frac{1.6 \times 10^{-5}\ M}{0.050\ M} \times 100\% = 0.032\%$$

The approximation is valid.

Step 4: At equilibrium,

$$[H_2C_2O_4] = 0.050\ M$$
$$[HC_2O_4^-] = (0.050 - 1.6 \times 10^{-5})\ M = 0.050\ M$$
$$[H^+] = (0.050 + 1.6 \times 10^{-5})\ M = 0.050\ M$$
$$[C_2O_4^{2-}] = 1.6 \times 10^{-5}\ M$$
$$[OH^-] = 1.0 \times 10^{-14}/0.050 = 2.0 \times 10^{-13}\ M$$

Practice Exercise Calculate the concentrations of $H_2C_2O_4$, $HC_2O_4^-$, $C_2O_4^{2-}$, and H^+ ions in a 0.20 M oxalic acid solution.

Similar problem: 15.66.

Example 15.11 shows that for diprotic acids, if $K_{a_1} \gg K_{a_2}$, then we can assume that the concentration of H^+ ions is the product of only the first stage of ionization. Furthermore, the concentration of the conjugate base for the second-stage ionization is *numerically* equal to K_{a_2}.

Phosphoric acid (H_3PO_4) is a polyprotic acid with three ionizable hydrogen atoms:

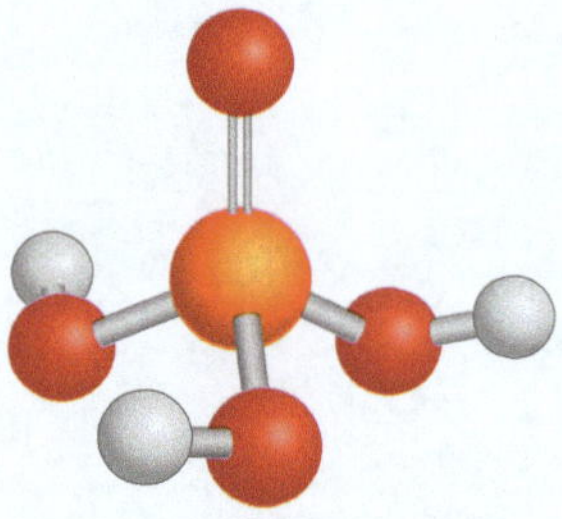

H_3PO_4

$$H_3PO_4(aq) \rightleftharpoons H^+(aq) + H_2PO_4^-(aq) \qquad K_{a_1} = \frac{[H^+][H_2PO_4^-]}{[H_3PO_4]} = 6.3 \times 10^{-3}$$

$$H_2PO_4^-(aq) \rightleftharpoons H^+(aq) + HPO_4^{2-}(aq) \qquad K_{a_2} = \frac{[H^+][HPO_4^{2-}]}{[H_2PO_4^-]} = 6.2 \times 10^{-8}$$

$$HPO_4^{2-}(aq) \rightleftharpoons H^+(aq) + PO_4^{3-}(aq) \qquad K_{a_3} = \frac{[H^+][PO_4^{3-}]}{[HPO_4^{2-}]} = 4.8 \times 10^{-13}$$

We see that phosphoric acid is a weak polyprotic acid and that its ionization constants decrease markedly for the second and third stages. Thus, we can predict that, in a solution containing phosphoric acid, the concentration of the nonionized acid is the highest, and the only other species present in significant concentrations are H^+ and $H_2PO_4^-$ ions.

Summary of Concepts & Facts

- Many acids contain more than one ionizable proton. The equilibrium concentrations of species in solution can be determined using methods similar to that for monoprotic acids.

Review of Concepts & Facts

15.8.1 Which of the diagrams (a)–(c) represents a solution of sulfuric acid? Water molecules have been omitted for clarity.

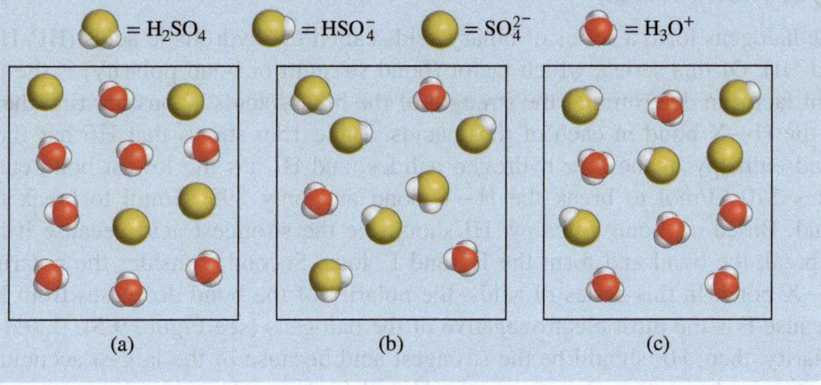

15.9 Molecular Structure and the Strength of Acids

Learning Objective

- Rank acids according to their relative strength based on their molecular structures.

The strength of an acid depends on a number of factors, such as the properties of the solvent, the temperature, and, of course, the molecular structure of the acid. When we compare the strengths of two acids, we can eliminate some variables by considering their properties in the same solvent and at the same temperature and concentration. Then we can focus on the structure of the acids.

Let us consider a certain acid HX. The strength of the acid is measured by its tendency to ionize:

$$HX \longrightarrow H^+ + X^-$$

Two factors influence the extent to which the acid undergoes ionization. One is the strength of the H—X bond—the stronger the bond, the more difficult it is for the HX molecule to break up and hence the *weaker* the acid. The other factor is the polarity of the H—X bond. The difference in the electronegativities between H and X results in a polar bond like

$$\overset{\delta+}{H}—\overset{\delta-}{X}$$

Table 15.6 Bond Enthalpies for Hydrogen Halides and Acid Strengths for Hydrohalic Acids

Bond	Bond Enthalpy (kJ/mol)	Acid Strength
H—F	570	weak
H—Cl	431	strong
H—Br	344	strong
H—I	298	strong

If the bond is highly polarized, that is, if there is a large accumulation of positive and negative charges on the H and X atoms, HX will tend to break up into H^+ and X^- ions. So a high degree of polarity characterizes a *stronger* acid. Later we will consider some examples in which either bond strength or bond polarity plays a prominent role in determining acid strength.

Hydrohalic Acids

The halogens form a series of binary acids called the hydrohalic acids (HF, HCl, HBr, and HI). Of this series, which factor (bond strength or bond polarity) is the predominant factor in determining the strength of the binary acids? Consider first the strength of the H—X bond in each of these acids. Table 15.6 shows that HF has the highest bond enthalpy of the four hydrogen halides, and HI has the lowest bond enthalpy. It takes 570 kJ/mol to break the H—F bond and only 298 kJ/mol to break the H—I bond. Based on bond enthalpy, HI should be the strongest acid because it is easiest to break the bond and form the H^+ and I^- ions. Second, consider the polarity of the H—X bond. In this series of acids, the polarity of the bond decreases from HF to HI because F is the most electronegative of the halogens (see Figure 9.5). Based on bond polarity, then, HF should be the strongest acid because of the largest accumulation of positive and negative charges on the H and F atoms. Thus, we have two competing factors to consider in determining the strength of binary acids. The fact that HI is a strong acid and that HF is a weak acid indicates that bond enthalpy is the predominant factor in determining the acid strength of binary acids. In this series of binary acids, the weaker the bond, the stronger the acid so that the strength of the acids increases as follows:

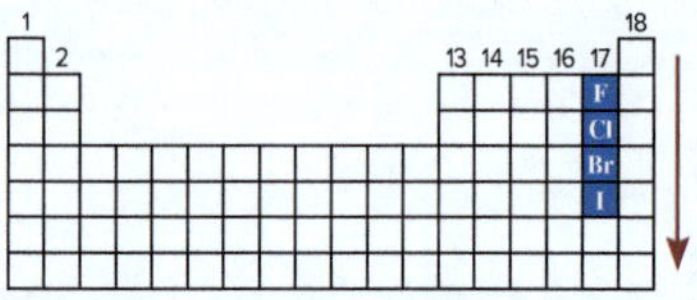

Strength of hydrohalic acids increases from HF to HI.

$$HF \ll HCl < HBr < HI$$

Oxoacids

Now let us consider the oxoacids. Oxoacids, as we learned in Chapter 2, contain hydrogen, oxygen, and one other element Z, which occupies a central position. Figure 15.5 shows the Lewis structures of several common oxoacids. As you can see, these acids

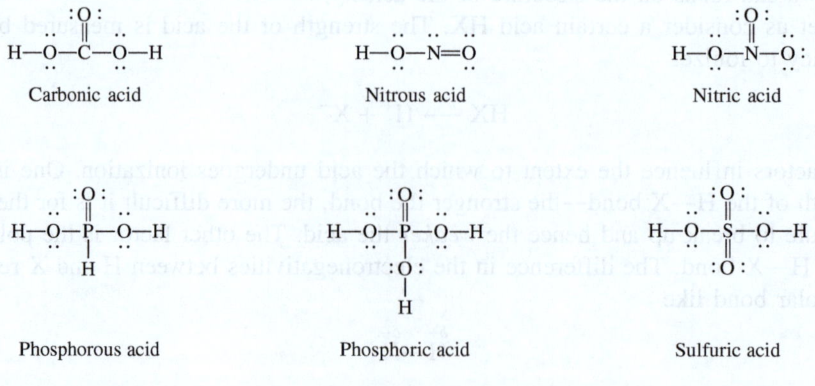

Figure 15.5 *Lewis structures of some common oxoacids. For simplicity, the formal charges have been omitted.*

are characterized by the presence of one or more O—H bonds. The central atom Z might also have other groups attached to it:

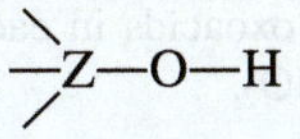

If Z is an electronegative element, or is in a high oxidation state, it will attract electrons, thus making the Z—O bond more covalent and the O—H bond more polar. Consequently, the tendency for the hydrogen to be donated as a H^+ ion increases:

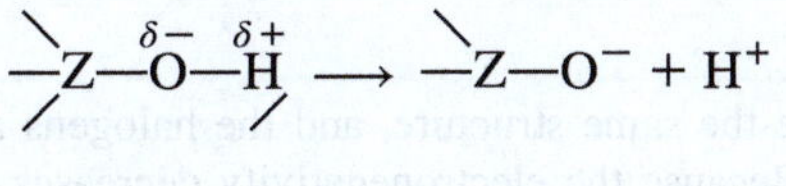

To compare their strengths, it is convenient to divide the oxoacids into two groups.

1. *Oxoacids Having Different Central Atoms That Are from the Same Group of the Periodic Table and That Have the Same Oxidation Number.* Within this group, acid strength increases with increasing electronegativity of the central atom, as $HClO_3$ and $HBrO_3$ illustrate:

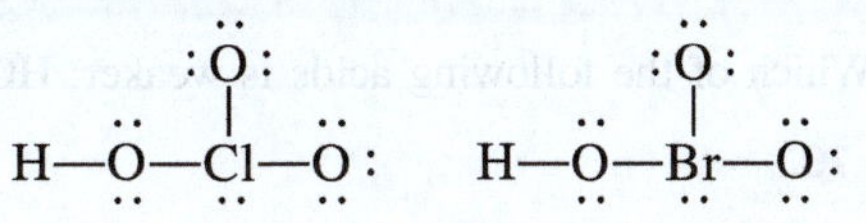

Cl and Br have the same oxidation number, +5. However, because Cl is more electronegative than Br, it attracts the electron pair it shares with oxygen (in the Cl—O—H group) to a greater extent than Br does. Consequently, the O—H bond is more polar in chloric acid than in bromic acid and ionizes more readily. Thus, the relative acid strengths are

$$HClO_3 > HBrO_3$$

2. *Oxoacids Having the Same Central Atom but Different Numbers of Attached Groups.* Within this group, acid strength increases as the oxidation number of the central atom increases. Consider the oxoacids of chlorine shown in Figure 15.6. In this series the ability of chlorine to draw electrons away from the OH group (thus making the O—H bond more polar) increases with the number of electronegative O atoms attached to Cl. Thus, $HClO_4$ is the strongest acid because it has the largest number of O atoms attached to Cl, and the acid strength decreases as follows:

$$HClO_4 > HClO_3 > HClO_2 > HClO$$

Example 15.12 compares the strengths of acids based on their molecular structures.

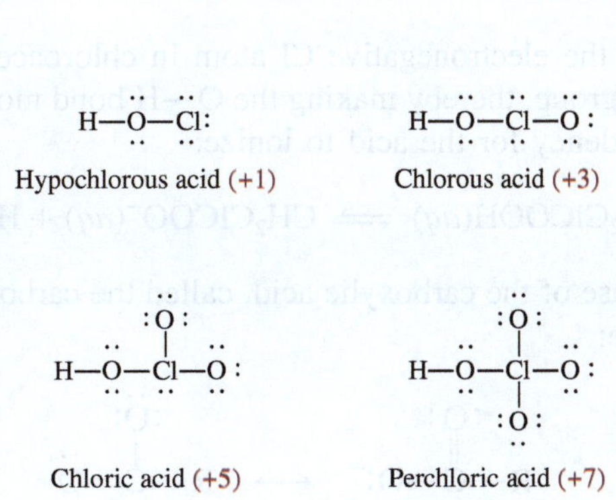

Figure 15.6 *Lewis structures of the oxoacids of chlorine. The oxidation number of the Cl atom is shown in parentheses. For simplicity, the formal charges have been omitted. Note that although hypochlorous acid is written as HClO, the H atom is bonded to the O atom.*

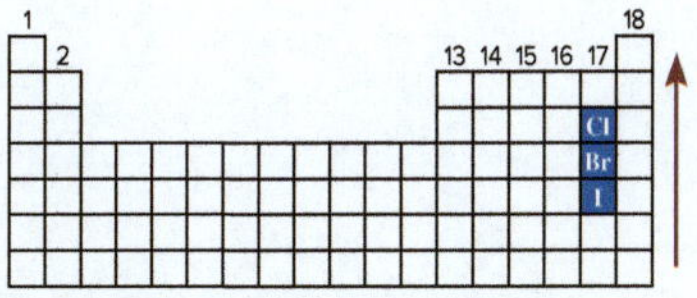

Strength of halogen-containing oxoacids having the same number of O atoms increases from bottom to top.

Student Hot Spot

Student data indicate you may struggle with strengths of oxoacids. Access your eBook for additional Learning Resources on this topic.

Example 15.12

Predict the relative strengths of the oxoacids in each of the following groups: (a) HClO, HBrO, and HIO; (b) HNO_3 and HNO_2.

Strategy Examine the molecular structure. In (a) the three acids have similar structures but differ only in the central atom (Cl, Br, and I). Which central atom is the most electronegative? In (b) the acids have the same central atom (N) but differ in the number of O atoms. What is the oxidation number of N in each of these two acids?

Solution

(a) These acids all have the same structure, and the halogens all have the same oxidation number (+1). Because the electronegativity decreases from Cl to I, the Cl atom attracts the electron pair it shares with the O atom to the greatest extent. Consequently, the bond is the most polar in HClO and least polar in HIO. Thus, the acid strength decreases as follows:

$$HClO > HBrO > HIO$$

(b) The structures of HNO_3 and HNO_2 are shown in Figure 15.5. Because the oxidation number of N is +5 in HNO_3 and +3 in HNO_2, HNO_3 is a stronger acid than HNO_2.

Practice Exercise Which of the following acids is weaker: $HClO_2$ or $HClO_3$?

Similar problem: 15.70.

Carboxylic Acids

So far the discussion has focused on inorganic acids. A group of organic acids that also deserves attention is the carboxylic acids, whose Lewis structures can be represented by

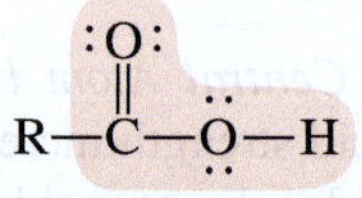

where R is part of the acid molecule and the shaded portion represents the *carboxyl* group, —COOH. The strength of carboxylic acids depends on the nature of the R group. Consider, for example, acetic acid and chloroacetic acid:

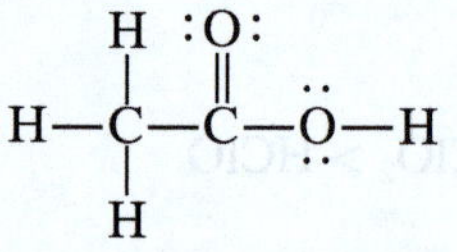

acetic acid ($K_a = 1.8 \times 10^{-5}$)

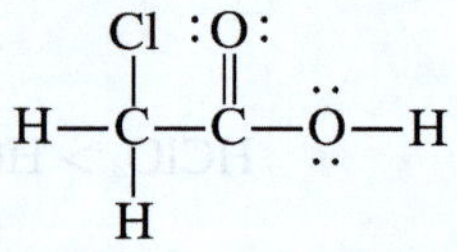

chloroacetic acid ($K_a = 1.5 \times 10^{-3}$)

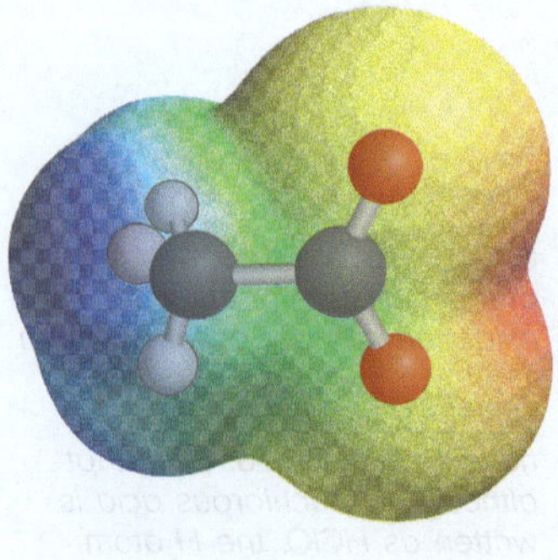

Electrostatic potential map of the acetate ion. The electron density is evenly distributed between the two O atoms.

The presence of the electronegative Cl atom in chloroacetic acid shifts electron density toward the R group, thereby making the O—H bond more polar. Consequently, there is a greater tendency for the acid to ionize:

$$CH_2ClCOOH(aq) \rightleftharpoons CH_2ClCOO^-(aq) + H^+(aq)$$

The conjugate base of the carboxylic acid, called the carboxylate anion ($RCOO^-$), can exhibit resonance:

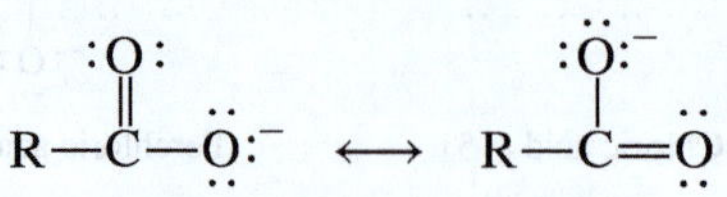

In the language of molecular orbital theory, we attribute the stability of the anion to its ability to spread or delocalize the electron density over several atoms. The greater the extent of electron delocalization, the more stable the anion and the greater the tendency for the acid to undergo ionization. Thus, benzoic acid (C_6H_5COOH, $K_a = 6.3 \times 10^{-5}$) is a stronger acid than acetic acid because the benzene ring (see Section 10.8) facilitates electron delocalization, so that the benzoate anion ($C_6H_5COO^-$) is more stable than the acetate anion (CH_3COO^-).

Summary of Concepts & Facts

- The relative strengths of acids can be explained qualitatively in terms of their molecular structures.

Review of Concepts & Facts

15.9.1 Arrange the following acids in order of increasing strength: $HBrO_4$, $HBrO$, $HBrO_3$.

15.10 Acid-Base Properties of Salts

Learning Objectives

- Solve for the pH of a salt solution.
- Predict the relative pH of a salt solution based upon the acid or base used to form the salt.

As defined in Section 4.3, a salt is an ionic compound formed by the reaction between an acid and a base. Salts are strong electrolytes that completely dissociate into ions in water. The term ***salt hydrolysis*** describes *the reaction of an anion or a cation of a salt, or both, with water*. Salt hydrolysis usually affects the pH of a solution.

Salts That Produce Neutral Solutions

It is generally true that salts containing an alkali metal ion or alkaline earth metal ion (except Be^{2+}) and the conjugate base of a strong acid (for example, Cl^-, Br^-, and NO_3^-) do not undergo hydrolysis to an appreciable extent, and their solutions are assumed to be neutral. For instance, when $NaNO_3$, a salt formed by the reaction of NaOH with HNO_3, dissolves in water, it dissociates completely as follows:

$$NaNO_3(s) \xrightarrow{H_2O} Na^+(aq) + NO_3^-(aq)$$

The hydrated Na^+ ion neither donates nor accepts H^+ ions. The NO_3^- ion is the conjugate base of the strong acid HNO_3, and it has no affinity for H^+ ions. Consequently, a solution containing Na^+ and NO_3^- ions is neutral, with a pH of about 7.

Salts That Produce Basic Solutions

The solution of a salt derived from a strong base and a weak acid is basic. For example, the dissociation of sodium acetate (CH_3COONa) in water is given by

$$CH_3COONa(s) \xrightarrow{H_2O} Na^+(aq) + CH_3COO^-(aq)$$

The hydrated Na^+ ion has no appreciable acidic or basic properties. The acetate ion CH_3COO^-, however, is the conjugate base of the weak acid CH_3COOH and therefore has an affinity for H^+ ions. The hydrolysis reaction is given by

$$CH_3COO^-(aq) + H_2O(l) \rightleftharpoons CH_3COOH(aq) + OH^-(aq)$$

Because this reaction produces OH^- ions, the sodium acetate solution will be basic. The equilibrium constant for this hydrolysis reaction is the same as the base ionization constant expression for CH_3COO^-, so we write (see Section 15.7)

$$K_b = \frac{[CH_3COOH][OH^-]}{[CH_3COO^-]} = 5.6 \times 10^{-10}$$

Because each CH_3COO^- ion that hydrolyzes produces one OH^- ion, the concentration of OH^- at equilibrium is the same as the concentration of CH_3COO^- that hydrolyzed. We can define the *percent hydrolysis* as

$$\% \text{ hydrolysis} = \frac{[CH_3COO^-]_{\text{hydrolyzed}}}{[CH_3COO^-]_{\text{initial}}} \times 100\%$$

$$= \frac{[OH^-]_{\text{equilibrium}}}{[CH_3COO^-]_{\text{initial}}} \times 100\%$$

Student Hot Spot

Student data indicate you may struggle with determining the pH of a salt solution. Access your eBook for additional Learning Resources on this topic.

A calculation based on the hydrolysis of CH_3COONa is illustrated in Example 15.13. In solving salt hydrolysis problems, we follow the same procedure we used for weak acids and weak bases.

Example 15.13

Calculate the pH of a 0.15 *M* solution of sodium acetate (CH_3COONa). What is the percent hydrolysis?

Strategy What is a salt? In solution, CH_3COONa dissociates completely into Na^+ and CH_3COO^- ions. The Na^+ ion, as we saw earlier, does not react with water and has no effect on the pH of the solution. The CH_3COO^- ion is the conjugate base of the weak acid CH_3COOH. Therefore, we expect that it will react to a certain extent with water to produce CH_3COOH and OH^-, and the solution will be basic.

Solution

Step 1: Because we started with a 0.15 *M* sodium acetate solution, the concentrations of the ions are also equal to 0.15 *M* after dissociation:

	$CH_3COONa(aq)$	$\longrightarrow$	$Na^+(aq)$	+	$CH_3COO^-(aq)$
Initial (*M*):	0.15		0		0
Change (*M*):	−0.15		+0.15		+0.15
Equilibrium (*M*):	0		0.15		0.15

Of these ions, only the acetate ion will react with water:

$$CH_3COO^-(aq) + H_2O(l) \rightleftharpoons CH_3COOH(aq) + OH^-(aq)$$

At equilibrium, the major species in solution are CH_3COOH, CH_3COO^-, and OH^-. The concentration of the H^+ ion is very small as we would expect for a basic solution, so it is treated as a minor species. We ignore the ionization of water.

Step 2: Let x be the equilibrium concentration of CH_3COOH and OH^- ions in mol/L; we summarize:

	$CH_3COO^-(aq)$	+ $H_2O(l)$ $\rightleftharpoons$	$CH_3COOH(aq)$	+	$OH^-(aq)$
Initial (*M*):	0.15		0.00		0.00
Change (*M*):	$-x$		$+x$		$+x$
Equilibrium (*M*):	$0.15 - x$		x		x

(Continued)

Step 3: From the preceding discussion and Table 15.3 we write the equilibrium constant of hydrolysis, or the base ionization constant, as

$$K_b = \frac{[CH_3COOH][OH^-]}{[CH_3COO^-]}$$

$$5.6 \times 10^{-10} = \frac{x^2}{0.15 - x}$$

Because K_b is very small and the initial concentration of the base is large, we can apply the approximation $0.15 - x \approx 0.15$:

$$5.6 \times 10^{-10} = \frac{x^2}{0.15 - x} \approx \frac{x^2}{0.15}$$

$$x = 9.2 \times 10^{-6}\ M$$

Step 4: At equilibrium:

$$[OH^-] = 9.2 \times 10^{-6}\ M$$

$$\text{pOH} = -\log(9.2 \times 10^{-6})$$

$$= 5.04$$

$$\text{pH} = 14.00 - 5.04$$

$$= 8.96$$

Thus, the solution is basic, as we would expect. The percent hydrolysis is given by

$$\%\ \text{hydrolysis} = \frac{9.2 \times 10^{-6}\ M}{0.15\ M} \times 100\%$$

$$= 0.0061\%$$

Check The result shows that only a very small amount of the anion undergoes hydrolysis. Note that the calculation of percent hydrolysis takes the same form as the test for the approximation, which is valid in this case.

Practice Exercise Calculate the pH of a 0.24 M sodium formate solution (HCOONa).

Similar problem: 15.81.

Salts That Produce Acidic Solutions

When a salt derived from a strong acid such as HCl and a weak base such as NH_3 dissolves in water, the solution becomes acidic. For example, consider the process

$$NH_4Cl(s) \xrightarrow{H_2O} NH_4^+(aq) + Cl^-(aq)$$

The Cl^- ion, being the conjugate base of a strong acid, has no affinity for H^+ and no tendency to hydrolyze. The ammonium ion NH_4^+ is the weak conjugate acid of the weak base NH_3 and ionizes as follows:

$$NH_4^+(aq) + H_2O(l) \rightleftharpoons NH_3(aq) + H_3O^+(aq)$$

or simply

$$NH_4^+(aq) \rightleftharpoons NH_3(aq) + H^+(aq)$$

Note that this reaction also represents the hydrolysis of the NH_4^+ ion. Because H^+ ions are produced, the pH of the solution decreases. The equilibrium constant (or ionization constant) for this process is given by

$$K_a = \frac{[NH_3][H^+]}{[NH_4^+]} = \frac{K_w}{K_b} = \frac{1.0 \times 10^{-14}}{1.8 \times 10^{-5}} = 5.6 \times 10^{-10}$$

and we can calculate the pH of an ammonium chloride solution following the same procedure used in Example 15.13.

In principle, *all* metal ions react with water to produce an acidic solution. However, because the extent of hydrolysis is most pronounced for the small and highly charged metal cations such as Al^{3+}, Cr^{3+}, Fe^{3+}, Bi^{3+}, and Be^{2+}, we generally neglect

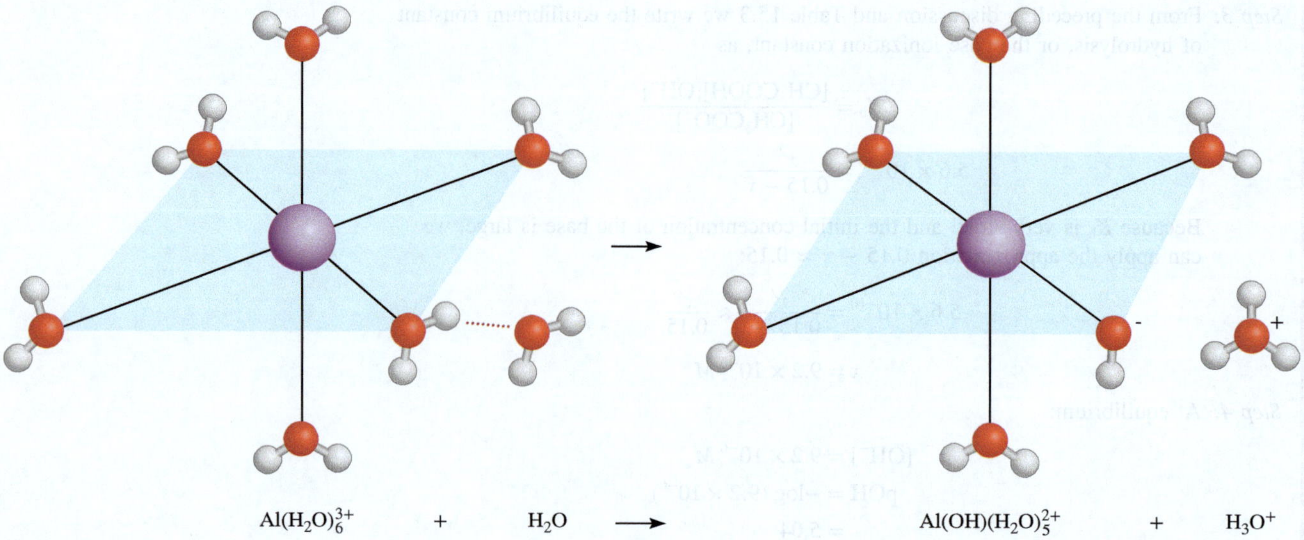

Figure 15.7 *The six H_2O molecules surround the Al^{3+} ion octahedrally. The attraction of the small Al^{3+} ion for the lone pairs on the oxygen atoms is so great that the O—H bonds in a H_2O molecule attached to the metal cation are weakened, allowing the loss of a proton (H^+) to an incoming H_2O molecule. This hydrolysis of the metal cation makes the solution acidic.*

the relatively small interaction of alkali metal ions and most alkaline earth metal ions with water. When aluminum chloride ($AlCl_3$) dissolves in water, the Al^{3+} ions take the hydrated form $Al(H_2O)_6^{3+}$ (Figure 15.7). Let us consider one bond between the metal ion and an oxygen atom from one of the six water molecules in $Al(H_2O)_6^{3+}$:

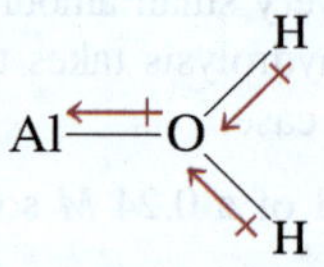

The positively charged Al^{3+} ion draws electron density toward itself, increasing the polarity of the O—H bonds. Consequently, the H atoms have a greater tendency to ionize than those in water molecules not involved in hydration. The resulting ionization process can be written as

$$Al(H_2O)_6^{3+}(aq) + H_2O(l) \rightleftharpoons Al(OH)(H_2O)_5^{2+}(aq) + H_3O^+(aq)$$

or simply

$$Al(H_2O)_6^{3+}(aq) \rightleftharpoons Al(OH)(H_2O)_5^{2+}(aq) + H^+(aq)$$

The equilibrium constant for the metal cation hydrolysis is given by

$$K_a = \frac{[Al(OH)(H_2O)_5^{2+}][H^+]}{[Al(H_2O)_6^{3+}]} = 1.3 \times 10^{-5}$$

Note that $Al(OH)(H_2O)_5^{2+}$ can undergo further ionization:

$$Al(OH)(H_2O)_5^{2+}(aq) \rightleftharpoons Al(OH)_2(H_2O)_4^{+}(aq) + H^+(aq)$$

and so on. However, it is generally sufficient to take into account only the first stage of hydrolysis.

The extent of hydrolysis is greatest for the smallest and most highly charged ions because a "compact" highly charged ion is more effective in polarizing the O—H bond and facilitating ionization. This is why relatively large ions of low charge such as Na^+ and K^+ do not undergo appreciable hydrolysis.

Salts in Which Both the Cation and the Anion Hydrolyze

So far we have considered salts in which only one ion undergoes hydrolysis. For salts derived from a weak acid and a weak base, both the cation and the anion hydrolyze. However, whether a solution containing such a salt is acidic, basic, or neutral depends

Table 15.7 Acid-Base Properties of Salts

Type of Salt	Examples	Ions That Undergo Hydrolysis	pH of Solution
Cation from strong base; anion from strong acid	NaCl, KI, KNO_3, RbBr, $BaCl_2$	None	≈ 7
Cation from strong base; anion from weak acid	CH_3COONa, KNO_2	Anion	> 7
Cation from weak base; anion from strong acid	NH_4Cl, NH_4NO_3	Cation	< 7
Cation from weak base; anion from weak acid	NH_4NO_2, CH_3COONH_4, NH_4CN	Anion and cation	< 7 if $K_b < K_a$ ≈ 7 if $K_b \approx K_a$ > 7 if $K_b > K_a$
Small, highly charged cation; anion from strong acid	$AlCl_3$, $Fe(NO_3)_3$	Hydrated cation	< 7

on the relative strengths of the weak acid and the weak base. Because the mathematics associated with this type of system is rather involved, we will focus on making qualitative predictions about these solutions based on the following guidelines:

- $\mathbf{K_b > K_a}$. If K_b for the anion is greater than K_a for the cation, then the solution must be basic because the anion will hydrolyze to a greater extent than the cation. At equilibrium, there will be more OH^- ions than H^+ ions.
- $\mathbf{K_b < K_a}$. Conversely, if K_b for the anion is smaller than K_a for the cation, the solution will be acidic because cation hydrolysis will be more extensive than anion hydrolysis.
- $\mathbf{K_b \approx K_a}$. If K_a is approximately equal to K_b, the solution will be nearly neutral.

Student Hot Spot

Student data indicate you may struggle with predicting the pH of a salt solution. Access your eBook to view additional Learning Resources on this topic.

Table 15.7 summarizes the behavior in aqueous solution of the salts discussed in this section.

Example 15.14 illustrates how to predict the acid-base properties of salt solutions.

Example 15.14

Predict whether the following solutions will be acidic, basic, or nearly neutral: (a) NH_4I, (b) $NaNO_2$, (c) $FeCl_3$, (d) NH_4F.

Strategy In deciding whether a salt will undergo hydrolysis, ask yourself the following questions: Is the cation a highly charged metal ion or an ammonium ion? Is the anion the conjugate base of a weak acid? If yes to either question, then hydrolysis will occur. In cases where both the cation and the anion react with water, the pH of the solution will depend on the relative magnitudes of K_a for the cation and K_b for the anion (see Table 15.7).

Solution We first break up the salt into its cation and anion components and then examine the possible reaction of each ion with water.

(a) The cation is NH_4^+, which will hydrolyze to produce NH_3 and H^+. The I^- anion is the conjugate base of the strong acid HI. Therefore, I^- will not hydrolyze and the solution is acidic.

(b) The Na^+ cation does not hydrolyze. The NO_2^- is the conjugate base of the weak acid HNO_2 and will hydrolyze to give HNO_2 and OH^-. The solution will be basic.

(c) Fe^{3+} is a small metal ion with a high charge and hydrolyzes to produce H^+ ions. The Cl^- does not hydrolyze. Consequently, the solution will be acidic.

(d) Both the NH_4^+ and F^- ions will hydrolyze. From Tables 15.3 and 15.4 we see that the K_a of NH_4^+ (5.6×10^{-10}) is greater than the K_b for F^- (1.4×10^{-11}). Therefore, the solution will be acidic.

Practice Exercise Predict whether the following solutions will be acidic, basic, or nearly neutral: (a) $LiClO_4$, (b) Na_3PO_4, (c) $Bi(NO_3)_3$, (d) NH_4CN.

Similar problems: 15.77, 15.78.

Finally, we note that some anions can act either as an acid or as a base. For example, the bicarbonate ion (HCO_3^-) can ionize or undergo hydrolysis as follows (see Table 15.5):

$$HCO_3^-(aq) + H_2O(l) \rightleftharpoons H_3O^+(aq) + CO_3^{2-}(aq) \qquad K_a = 5.0 \times 10^{-11}$$

$$HCO_3^-(aq) + H_2O(l) \rightleftharpoons H_2CO_3(aq) + OH^-(aq) \qquad K_b = 2.1 \times 10^{-8}$$

Because $K_b > K_a$, we predict that the hydrolysis reaction will outweigh the ionization process. Thus, a solution of sodium bicarbonate ($NaHCO_3$) will be basic.

Summary of Concepts & Facts

- Most salts are strong electrolytes that dissociate completely into ions in solution. The reaction of these ions with water, called salt hydrolysis, can produce acidic or basic solutions. In salt hydrolysis, the conjugate bases of weak acids yield basic solutions, and the conjugate acids of weak bases yield acidic solutions.
- Small, highly charged metal ions, such as Al^{3+} and Fe^{3+}, hydrolyze to yield acidic solutions.

Review of Concepts & Facts

15.10.1 What is the pH of a 0.74 *M* solution of potassium cyanide (KCN)?

15.10.2 The diagrams shown here represent solutions of three salts NaX (X = A, B, or C). (a) Which X^- has the weakest conjugate acid? (b) Arrange the three X^- anions in order of increasing base strength. The Na^+ ion and water molecules have been omitted for clarity.

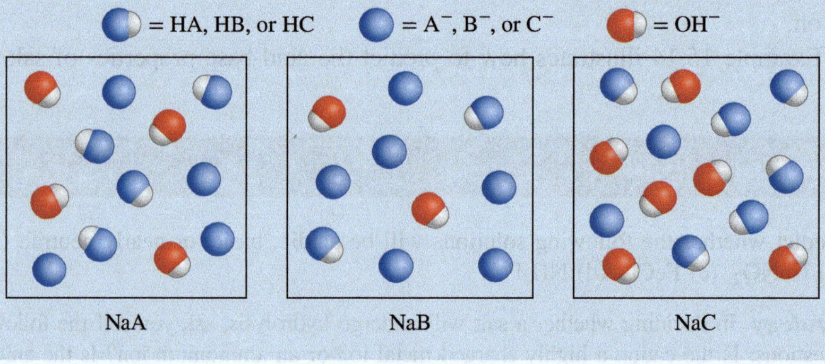

15.11 Acid-Base Properties of Oxides and Hydroxides

Learning Objective

- Classify an oxide as being acidic, basic, or amphoteric.

As we saw in Chapter 8, oxides can be classified as acidic, basic, or amphoteric. Our discussion of acid-base reactions would be incomplete if we did not examine the properties of these compounds.

Figure 15.8 shows the formulas of a number of oxides of the representative elements in their highest oxidation states. Note that all alkali metal oxides and all alkaline earth metal oxides except BeO are basic. Beryllium oxide and several metallic oxides in Groups 13 and 14 are amphoteric. Nonmetallic oxides in which the oxidation number of the representative element is high are acidic (for example, N_2O_5, SO_3, and Cl_2O_7), but those in which the oxidation number of the representative element is low (for example, CO and NO) show no measurable acidic properties. No nonmetallic oxides are known to have basic properties.

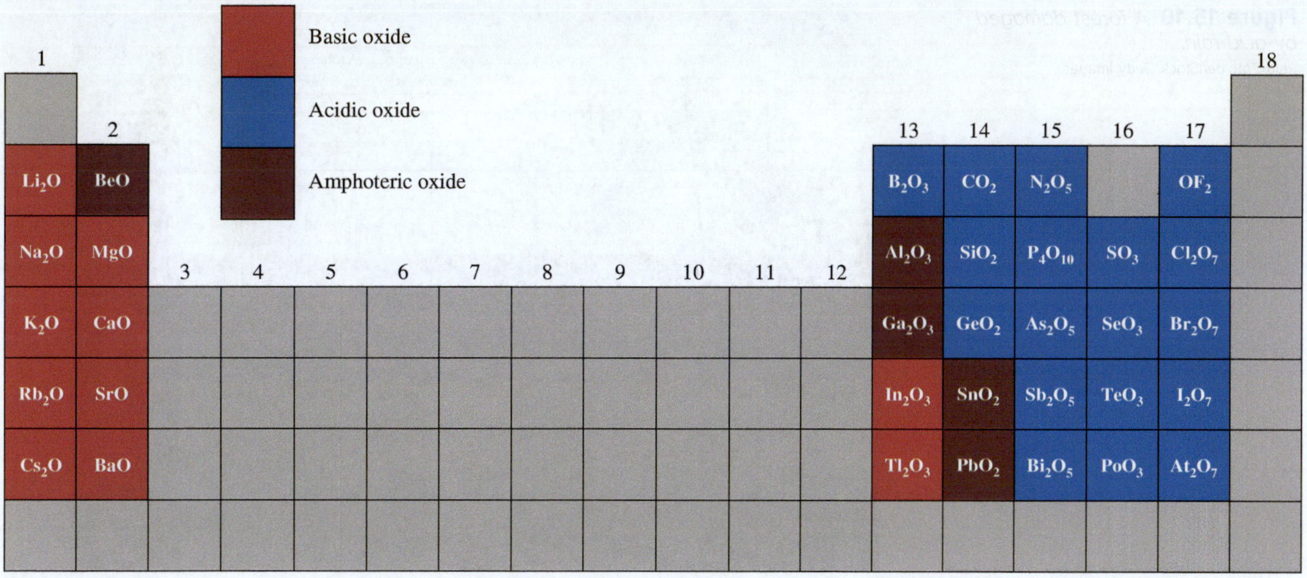

Figure 15.8 *Oxides of the representative elements in their highest oxidation states.*

The basic metallic oxides react with water to form metal hydroxides:

$$Na_2O(s) \xrightarrow{H_2O} 2NaOH(aq)$$
$$BaO(s) \xrightarrow{H_2O} Ba(OH)_2(aq)$$

The reactions between acidic oxides and water are as follows:

$$CO_2(g) + H_2O(l) \rightleftharpoons H_2CO_3(aq)$$
$$SO_3(g) + H_2O(l) \rightleftharpoons H_2SO_4(aq)$$
$$N_2O_5(g) + H_2O(l) \rightleftharpoons 2HNO_3(aq)$$
$$P_4O_{10}(s) + 6H_2O(l) \rightleftharpoons 4H_3PO_4(aq)$$
$$Cl_2O_7(l) + H_2O(l) \rightleftharpoons 2HClO_4(aq)$$

The reaction between CO_2 and H_2O explains why when pure water is exposed to air (which contains CO_2) it gradually reaches a pH of about 5.5 (Figure 15.9). The

Figure 15.9 *(Left) A beaker of water to which a few drops of bromothymol blue indicator have been added. (Right) As dry ice is added to the water, the CO_2 reacts to form carbonic acid, which turns the solution acidic and changes the color from blue to yellow.*
Ken Karp/McGraw Hill

Figure 15.10 *A forest damaged by acid rain.*
RelaxFoto.de/iStock/Getty Images

reaction between SO_3 and H_2O is largely responsible for acid rain (Figure 15.10), which we will study in more detail in Chapter 20.

Reactions between acidic oxides and bases and those between basic oxides and acids resemble normal acid-base reactions in that the products are a salt and water:

$$\underset{\text{acidic oxide}}{CO_2(g)} + \underset{\text{base}}{2NaOH(aq)} \longrightarrow \underset{\text{salt}}{Na_2CO_3(aq)} + \underset{\text{water}}{H_2O(l)}$$

$$\underset{\text{basic oxide}}{BaO(s)} + \underset{\text{acid}}{2HNO_3(aq)} \longrightarrow \underset{\text{salt}}{Ba(NO_3)_2(aq)} + \underset{\text{water}}{H_2O(l)}$$

As Figure 15.8 shows, aluminum oxide (Al_2O_3) is amphoteric. Depending on the reaction conditions, it can behave either as an acidic oxide or as a basic oxide. For example, Al_2O_3 acts as a base with hydrochloric acid to produce a salt ($AlCl_3$) and water:

$$Al_2O_3(s) + 6HCl(aq) \longrightarrow 2AlCl_3(aq) + 3H_2O(l)$$

and acts as an acid with sodium hydroxide:

$$Al_2O_3(s) + 2NaOH(aq) \longrightarrow 2NaAlO_2(aq) + H_2O(l)$$

Some transition metal oxides in which the metal has a high oxidation number act as acidic oxides. The higher the oxidation number of the metal, the more covalent the compound; the lower the oxidation number, the more ionic the compound. Two familiar examples are manganese(VII) oxide (Mn_2O_7) and chromium(VI) oxide (CrO_3), both of which react with water to produce acids:

$$Mn_2O_7(l) + H_2O(l) \longrightarrow \underset{\text{permanganic acid}}{2HMnO_4(aq)}$$

$$CrO_3(s) + H_2O(l) \longrightarrow \underset{\text{chromic acid}}{H_2CrO_4(aq)}$$

Basic and Amphoteric Hydroxides

We have seen that the alkali and alkaline earth metal hydroxides [except $Be(OH)_2$] are basic in properties. The following hydroxides are amphoteric: $Be(OH)_2$, $Al(OH)_3$, $Sn(OH)_2$, $Pb(OH)_2$, $Cr(OH)_3$, $Cu(OH)_2$, $Zn(OH)_2$, and $Cd(OH)_2$. For example, aluminum hydroxide reacts with both acids and bases:

$$Al(OH)_3(s) + 3H^+(aq) \longrightarrow Al^{3+}(aq) + 3H_2O(l)$$

$$Al(OH)_3(s) + OH^-(aq) \rightleftharpoons Al(OH)_4^-(aq)$$

All amphoteric hydroxides are insoluble.

It is interesting that beryllium hydroxide, like aluminum hydroxide, exhibits amphoterism:

$$Be(OH)_2(s) + 2H^+(aq) \longrightarrow Be^{2+}(aq) + 2H_2O(l)$$
$$Be(OH)_2(s) + 2OH^-(aq) \rightleftharpoons Be(OH)_4^{2-}(aq)$$

This is another example of the diagonal relationship between beryllium and aluminum (see Section 8.6).

Summary of Concepts & Facts

- Most oxides can be classified as acidic, basic, or amphoteric. Metal hydroxides are either basic or amphoteric.

Review of Concepts & Facts

15.11.1 Arrange the following oxides in order of increasing basicity: K_2O, Al_2O_3, BaO.

15.12 Lewis Acids and Bases

Learning Objective

- Identify a species as a Lewis acid or a Lewis base.

So far we have discussed acid-base properties in terms of the Brønsted theory. To behave as a Brønsted base, for example, a substance must be able to accept protons. By this definition both the hydroxide ion and ammonia are bases:

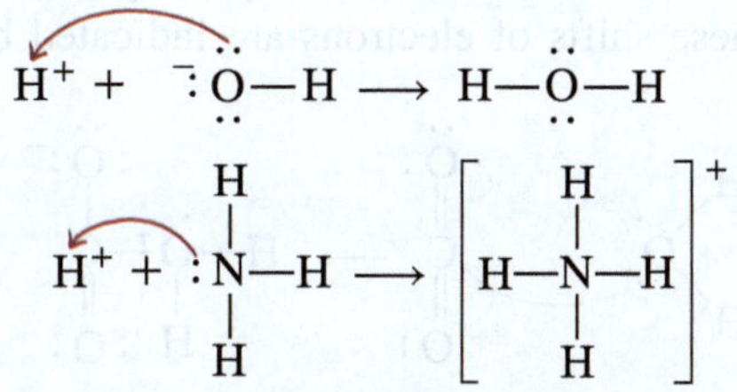

In each case, the atom to which the proton becomes attached possesses at least one unshared pair of electrons. This characteristic property of OH^-, NH_3, and other Brønsted bases suggests a more general definition of acids and bases.

In 1932 the American chemist G. N. Lewis formulated such a definition. He defined what we now call a ***Lewis base*** as *a substance that can donate a pair of electrons*. A ***Lewis acid*** is *a substance that can accept a pair of electrons*. For example, in the protonation of ammonia, NH_3 acts as a Lewis base because it donates a pair of electrons to the proton H^+, which acts as a Lewis acid by accepting the pair of electrons. Lewis acids are either deficient in electrons or the central atoms has a vacant valence orbital. A Lewis acid-base reaction, therefore, is one that involves the donation of a pair of electrons from one species to another. Such a reaction does not produce a salt and water.

The significance of the Lewis concept is that it is more general than other definitions. Lewis acid-base reactions include many reactions that do not involve Brønsted

Figure 15.11 *A Lewis acid-base reaction involving BF_3 and NH_3.*
Ken Karp/McGraw Hill

acids. Consider, for example, the reaction between boron trifluoride (BF_3) and ammonia to form an adduct compound (Figure 15.11):

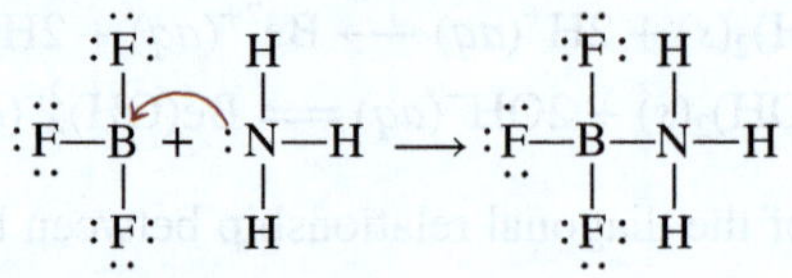

Lewis acid   Lewis base

In Section 10.4 we saw that the B atom in BF_3 is sp^2-hybridized. The vacant, unhybridized $2p_z$ orbital accepts the pair of electrons from NH_3. So BF_3 functions as an acid according to the Lewis definition, even though it does not contain an ionizable proton. Note that a coordinate covalent bond is formed between the B and N atoms, as is the case in all Lewis acid-base reactions.

Another Lewis acid containing boron is boric acid. Boric acid (a weak acid used in eyewash) is an oxoacid with the following structure:

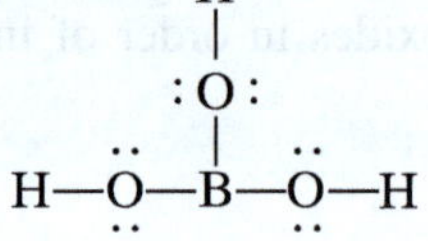

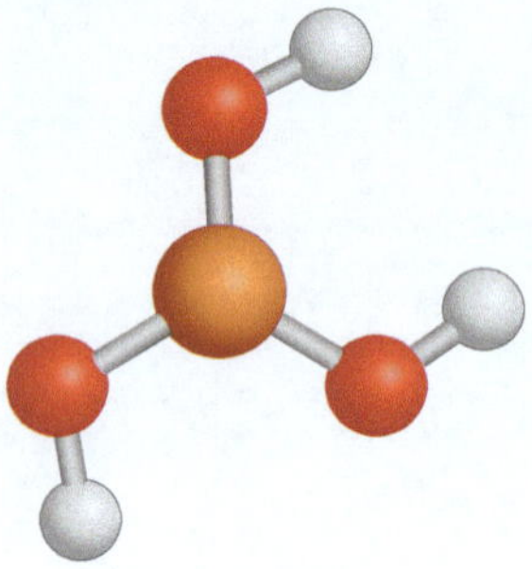

H_3BO_3

Boric acid does not ionize in water to produce a H^+ ion. Its reaction with water is

$$B(OH)_3(aq) + H_2O(l) \rightleftharpoons B(OH)_4^-(aq) + H^+(aq)$$

In this Lewis acid-base reaction, boric acid accepts a pair of electrons from the hydroxide ion that is derived from the H_2O molecule.

The hydration of carbon dioxide to produce carbonic acid

$$CO_2(g) + H_2O(l) \longrightarrow H_2CO_3(aq)$$

can be understood in the Lewis framework as follows: The first step involves donation of a lone pair on the oxygen atom in H_2O to the carbon atom in CO_2. An orbital is vacated on the C atom to accommodate the lone pair by removal of the electron pair in the C—O pi bond. These shifts of electrons are indicated by the curved arrows.

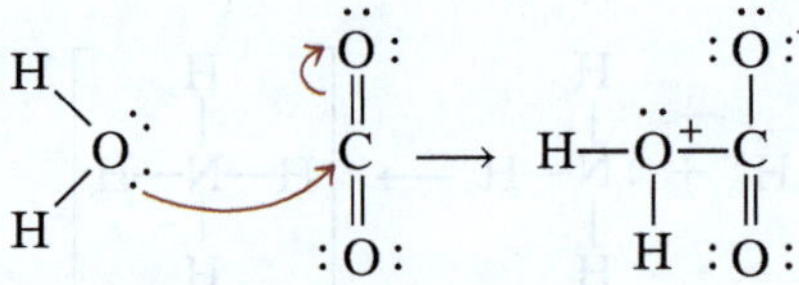

Therefore, H_2O is a Lewis base and CO_2 is a Lewis acid. Next, a proton is transferred onto the O atom bearing a negative charge to form H_2CO_3.

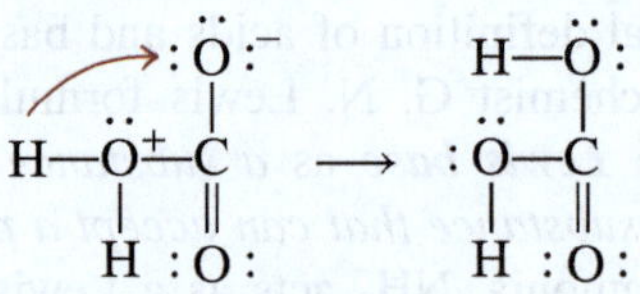

Other examples of Lewis acid-base reactions are

$$\underset{\text{acid}}{Ag^+(aq)} + \underset{\text{base}}{2NH_3(aq)} \rightleftharpoons Ag(NH_3)_2^+(aq)$$

$$\underset{\text{acid}}{Cd^{2+}(aq)} + \underset{\text{base}}{4I^-(aq)} \rightleftharpoons CdI_4^{2-}(aq)$$

$$\underset{\text{acid}}{Ni(s)} + \underset{\text{base}}{4CO(g)} \rightleftharpoons Ni(CO)_4(g)$$

It is important to note that the hydration of metal ions in solution is in itself a Lewis acid-base reaction (see Figure 15.7). Thus, when copper(II) sulfate ($CuSO_4$) dissolves in water, each Cu^{2+} ion is associated with six water molecules as $Cu(H_2O)_6^{2+}$. In this case, the Cu^{2+} ion acts as the acid and the H_2O molecules act as the base.

Although the Lewis definition of acids and bases has greater significance because of its generality, we normally speak of "an acid" and "a base" in terms of the Brønsted definition. The term "Lewis acid" usually is reserved for substances that can accept a pair of electrons but do not contain ionizable hydrogen atoms.

Example 15.15 classifies Lewis acids and Lewis bases.

Student Hot Spot
Student data indicate you may struggle with Lewis acids and bases. Access your eBook for additional Learning Resources on this topic.

Example 15.15

Identify the Lewis acid and Lewis base in each of the following reactions:

(a) $C_2H_5OC_2H_5 + AlCl_3 \rightleftharpoons (C_2H_5)_2OAlCl_3$

(b) $Hg^{2+}(aq) + 4CN^-(aq) \rightleftharpoons Hg(CN)_4^{2-}(aq)$

Strategy In Lewis acid-base reactions, the acid is usually a cation or an electron-deficient molecule, whereas the base is an anion or a molecule containing an atom with lone pairs. (a) Draw the molecular structure for $C_2H_5OC_2H_5$. What is the hybridization state of Al in $AlCl_3$? (b) Which ion is likely to be an electron acceptor? An electron donor?

Solution

(a) The Al is sp^2-hybridized in $AlCl_3$ with an empty $2p_z$ orbital. It is electron deficient, sharing only six electrons. Therefore, the Al atom has a tendency to gain two electrons to complete its octet. This property makes $AlCl_3$ a Lewis acid. On the other hand, the lone pairs on the oxygen atom in $C_2H_5OC_2H_5$ make the compound a Lewis base:

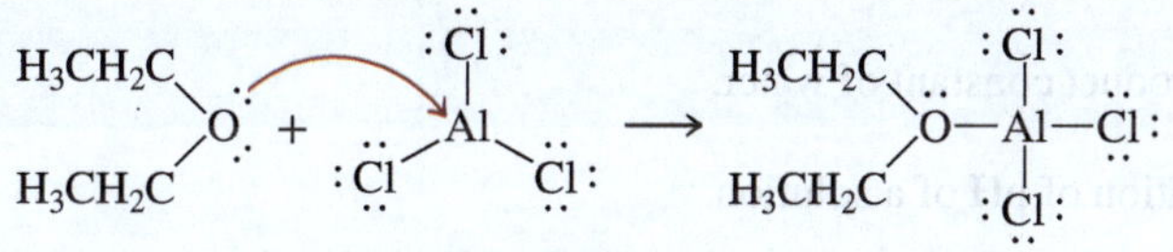

(b) Here the Hg^{2+} ion accepts four pairs of electrons from the CN^- ions. Therefore, Hg^{2+} is the Lewis acid and CN^- is the Lewis base.

Practice Exercise Identify the Lewis acid and Lewis base in the reaction

$$Co^{3+}(aq) + 6NH_3(aq) \rightleftharpoons Co(NH_3)_6^{3+}(aq)$$

Similar problem: 15.94.

Summary of Concepts & Facts

- Lewis acids accept pairs of electrons and Lewis bases donate pairs of electrons. The term "Lewis acid" is generally reserved for substances that can accept electron pairs but do not contain ionizable hydrogen atoms.

Review of Concepts & Facts

15.12.1 Which of the following cannot behave as a Lewis base: (a) NH_3, (b) OF_2, (c) CH_4, (d) OH^-, (e) Fe^{3+}?

Chapter Summary

Brønsted Acids and Bases A Brønsted acid can donate a proton and a Brønsted base can accept a proton. For every Brønsted acid, there exists a conjugate Brønsted base and vice versa. (Section 15.1)

Acid-Base Properties of Water and the pH Scale Water acts both as a Brønsted acid and as a Brønsted base. At 25°C, the concentrations of H^+ and OH^- ions are both at 10^{-7} *M*. The pH scale is established to express the acidity of a solution—the smaller the pH, the higher the H^+ concentration and the greater the acidity. (Sections 15.2 and 15.3)

Acid and Base Ionization Constants Strong acids and strong bases are assumed to ionize completely. Most weak acids and bases ionize to a small extent. The concentrations of the acid, conjugate base, and H^+ ion at equilibrium can be calculated from the acid ionization constant, which is the equilibrium constant for the reaction. (Sections 15.4, 15.5, 15.6, 15.7, and 15.8)

Molecular Structure and Acid Strength The strength of a series of structurally similar acids can be compared using parameters such as bond enthalpy, bond polarity, and oxidation number. (Section 15.9)

Acid-Base Properties of Salts and Oxides Many salts react with water in a process called hydrolysis. From the nature of the cation and anion present in the salt, it is possible to predict the pH of the resulting solution. Most oxides also react with water to produce acidic or basic solutions. (Sections 15.10 and 15.11)

Lewis Acids and Bases A more general definition of acids and bases characterizes an acid as a substance that can accept a pair of electrons and a base as a substance that can donate a pair of electrons. All Brønsted acids and bases are Lewis acids and bases. (Section 15.12)

Key Equations

$K_w = [H^+][OH^-]$ (15.3)	Ion-product constant of water.
$pH = -\log [H^+]$ (15.4)	Definition of pH of a solution.
$[H^+] = 10^{-pH}$ (15.5)	Calculating H^+ ion concentration from pH.
$pOH = -\log [OH^-]$ (15.7)	Definition of pOH of a solution.
$[OH^-] = 10^{-pOH}$ (15.8)	Calculating OH^- ion concentration from pOH.
$pH + pOH = 14.00$ (15.9)	Another form of Equation (15.3).

$$\text{percent ionization} = \frac{\text{ionized acid concentration at equilibrium}}{\text{initial concentration of acid}} \times 100\% \quad (15.11)$$

$K_aK_b = K_w$ (15.12)	Relationship between the acid and base ionization constants of a conjugate acid-base pair.

Key Words

Acid ionization constant (K_a), p. 687
Base ionization constant (K_b), p. 694
Conjugate acid-base pair, p. 673
Ion-product constant, p. 676
Lewis acid, p. 713
Lewis base, p. 713
Percent ionization, p. 692
pH, p. 677
Salt hydrolysis, p. 705
Strong acid, p. 683
Strong base, p. 684
Weak acid, p. 684
Weak base, p. 684

Questions & Problems†

Red numbered problems solved in Student Solutions Manual

15.1 Brønsted Acids and Bases

Review Questions

15.1 Define Brønsted acids and bases. Give an example of a conjugate pair in an acid-base reaction.

15.2 For a species to act as a Brønsted base, an atom in the species must possess a lone pair of electrons. Explain why this is so.

Problems

15.3 Classify each of the following species as a Brønsted acid or base, or both: (a) H_2O, (b) OH^-, (c) H_3O^+, (d) NH_3, (e) NH_4^+, (f) NH_2^-, (g) NO_3^-, (h) CO_3^{2-}, (i) HBr, (j) HCN.

15.4 Write the formulas of the conjugate bases of the following acids: (a) HNO_2, (b) H_2SO_4, (c) H_2S, (d) HCN, (e) HCOOH (formic acid).

15.5 Identify the acid-base conjugate pairs in each of the following reactions:

(a) $CH_3COO^- + HCN \rightleftharpoons CH_3COOH + CN^-$

(b) $HCO_3^- + HCO_3^- \rightleftharpoons H_2CO_3 + CO_3^{2-}$

(c) $H_2PO_4^- + NH_3 \rightleftharpoons HPO_4^{2-} + NH_4^+$

(d) $HClO + CH_3NH_2 \rightleftharpoons CH_3NH_3^+ + ClO^-$

(e) $CO_3^{2-} + H_2O \rightleftharpoons HCO_3^- + OH^-$

15.6 Write the formula for the conjugate acid of each of the following bases: (a) HS^-, (b) HCO_3^-, (c) CO_3^{2-}, (d) $H_2PO_4^-$, (e) HPO_4^{2-}, (f) PO_4^{3-}, (g) HSO_4^-, (h) SO_4^{2-}, (i) SO_3^{2-}.

15.7 Oxalic acid ($H_2C_2O_4$) has the following structure:

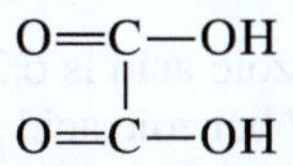

An oxalic acid solution contains the following species in varying concentrations: $H_2C_2O_4$, $HC_2O_4^-$, $C_2O_4^{2-}$, and H^+. (a) Draw Lewis structures of $HC_2O_4^-$ and $C_2O_4^{2-}$. (b) Which of the four species can act only as acids? Which can act only as bases? Which can act as both acids and bases?

15.8 Write the formula for the conjugate base of each of the following acids: (a) $CH_2ClCOOH$, (b) HIO_4, (c) H_3PO_4, (d) $H_2PO_4^-$, (e) HPO_4^{2-}, (f) H_2SO_4, (g) HSO_4^-, (h) HIO_3, (i) HSO_3^-, (j) NH_4^+, (k) H_2S, (l) HS^-, (m) HClO.

15.2 The Acid-Base Properties of Water

Review Questions

15.9 What is the ion-product constant for water?

15.10 Write an equation relating $[H^+]$ and $[OH^-]$ in solution at 25°C.

15.11 The ion-product constant for water is 1.0×10^{-14} at 25°C and 3.8×10^{-14} at 40°C. Is the forward process

$$H_2O(l) \rightleftharpoons H^+(aq) + OH^-(aq)$$

endothermic or exothermic?

15.3 pH—A Measure of Acidity

Review Questions

15.12 Define pH. Why do chemists normally choose to discuss the acidity of a solution in terms of pH rather than hydrogen ion concentration, $[H^+]$?

15.13 The pH of a solution is 6.7. From this statement alone, can you conclude that the solution is acidic? If not, what additional information would you need? Can the pH of a solution be zero or negative? If so, give examples to illustrate these values.

15.14 Define pOH. Write the equation relating pH and pOH.

Problems

15.15 Calculate the concentration of OH^- ions in a 1.4×10^{-3} *M* HCl solution.

15.16 Calculate the concentration of H^+ ions in a 0.62 *M* NaOH solution.

15.17 Calculate the pH of each of the following solutions: (a) 0.0010 *M* HCl, (b) 0.76 *M* KOH.

15.18 Calculate the pH of each of the following solutions: (a) 2.8×10^{-4} *M* $Ba(OH)_2$, (b) 5.2×10^{-4} *M* HNO_3.

15.19 Calculate the hydrogen ion concentration in mol/L for solutions with the following pH values: (a) 2.42, (b) 11.21, (c) 6.96, (d) 15.00.

15.20 Calculate the hydrogen ion concentration in mol/L for each of the following solutions: (a) a solution whose pH is 5.20, (b) a solution whose pH is 16.00, (c) a solution whose hydroxide concentration is 3.7×10^{-9} *M*.

15.21 Complete the following table for a solution:

pH	$[H^+]$	Solution is
< 7		
	$< 1.0 \times 10^{-7}$ *M*	
		Neutral

15.22 Fill in the word *acidic, basic,* or *neutral* for the following solutions:

(a) pOH > 7 solution is

(b) pOH = 7; solution is

(c) pOH < 7; solution is

15.23 The pOH of a strong base solution is 1.88 at 25°C. Calculate the concentration of the base if (a) the base is KOH and (b) the base is $Ba(OH)_2$.

†Unless otherwise stated, the temperature is assumed to be 25°C.

15.24 Calculate the number of moles of KOH in 5.50 mL of a 0.360 *M* KOH solution. What is the pOH of the solution?

15.25 How much NaOH (in grams) is needed to prepare 546 mL of solution with a pH of 10.00?

15.26 A solution is made by dissolving 18.4 g of HCl in 662 mL of water. Calculate the pH of the solution. (Assume that the volume remains constant.)

15.4 Strength of Acids and Bases

Review Questions

15.27 Explain what is meant by the strength of an acid.

15.28 Without referring to the text, write the formulas of four strong acids and four weak acids.

15.29 What is the strongest acid and the strongest base that can exist in water?

15.30 H_2SO_4 is a strong acid, but HSO_4^- is a weak acid. Account for the difference in strength of these two related species.

Problems

15.31 Which of the following diagrams best represents a strong acid, such as HCl, dissolved in water? Which represents a weak acid? Which represents a very weak acid? (The hydrated proton is shown as a hydronium ion. Water molecules are omitted for clarity.)

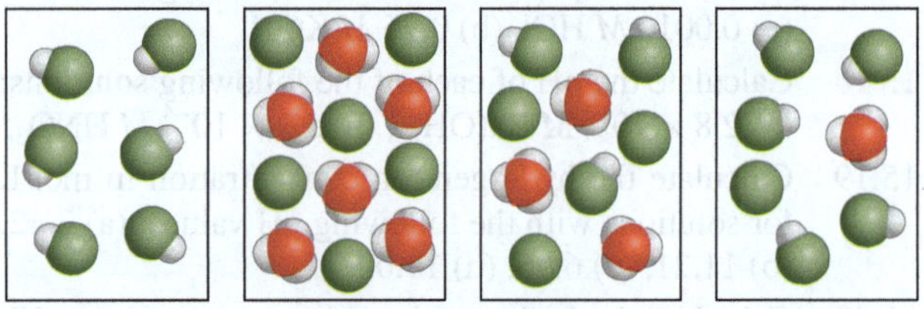

(a) (b) (c) (d)

15.32 (1) Which of the following diagrams represents a solution of a weak diprotic acid? (2) Which diagrams represent chemically implausible situations? (The hydrated proton is shown as a hydronium ion. Water molecules are omitted for clarity.)

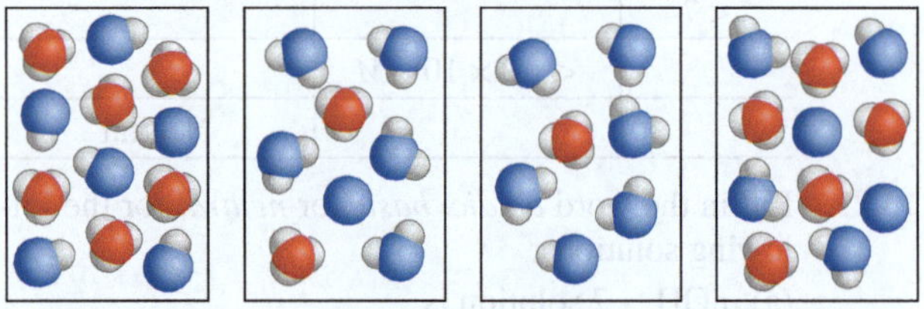

(a) (b) (c) (d)

15.33 Classify each of the following species as a weak or strong acid: (a) HNO_3, (b) HF, (c) H_2SO_4, (d) HSO_4^-, (e) H_2CO_3, (f) HCO_3^-, (g) HCl, (h) HCN, (i) HNO_2.

15.34 Classify each of the following species as a weak or strong base: (a) LiOH, (b) CN^-, (c) H_2O, (d) ClO_4^-, (e) NH_2^-.

15.35 Which of the following statements is/are true for a 0.10 *M* solution of a weak acid HA?
(a) The pH is 1.00.
(b) $[H^+] \gg [A^-]$
(c) $[H^+] = [A^-]$
(d) The pH is less than 1.

15.36 Which of the following statements is/are true regarding a 1.0 *M* solution of a strong acid HA?
(a) $[A^-] > [H^+]$
(b) The pH is 0.00.
(c) $[H^+] = 1.0\ M$
(d) $[HA] = 1.0\ M$

15.37 Predict the direction that predominates in this reaction:

$$F^-(aq) + H_2O(l) \rightleftharpoons HF(aq) + OH^-(aq)$$

15.38 Predict whether the following reaction will proceed from left to right to any measurable extent:

$$CH_3COOH(aq) + Cl^-(aq) \longrightarrow$$

15.5 Weak Acids and Acid Ionization Constants

Review Questions

15.39 What does the ionization constant tell us about the strength of an acid?

15.40 List the factors on which the K_a of a weak acid depends.

15.41 Why do we normally not quote K_a values for strong acids such as HCl and HNO_3? Why is it necessary to specify temperature when giving K_a values?

15.42 Which of the following solutions has the highest pH: (a) 0.40 *M* HCOOH, (b) 0.40 *M* $HClO_4$, (c) 0.40 *M* CH_3COOH?

Problems

15.43 The K_a for benzoic acid is 6.3×10^{-5}. Calculate the pH of a 0.10 *M* benzoic acid solution.

15.44 A 0.0560-g quantity of acetic acid is dissolved in enough water to make 50.0 mL of solution. Calculate the concentrations of H^+, CH_3COO^-, and CH_3COOH at equilibrium. (K_a for acetic acid = 1.8×10^{-5}.)

15.45 The pH of an acid solution is 6.20. Calculate the K_a for the acid. The initial acid concentration is 0.010 *M*.

15.46 What is the original molarity of a solution of formic acid (HCOOH) whose pH is 3.26 at equilibrium?

15.47 Calculate the percent ionization of benzoic acid having the following concentrations: (a) 0.20 *M*, (b) 0.00020 *M*.

15.48 Calculate the percent ionization of hydrofluoric acid at the following concentrations: (a) 0.60 *M*, (b) 0.0046 *M*, (c) 0.00028 *M*. Comment on the trends.

15.49 A 0.040 *M* solution of a monoprotic acid is 14% ionized. Calculate the ionization constant of the acid.

15.50 (a) Calculate the percent ionization of a 0.20 *M* solution of the monoprotic acetylsalicylic acid (aspirin) for which $K_a = 3.0 \times 10^{-4}$. (b) The pH of gastric juice in the stomach of a certain individual is 1.00. After a few aspirin tablets have been swallowed, the concentration of acetylsalicylic acid in the stomach is 0.20 *M*. Calculate the percent ionization of the acid under these conditions. What effect does the nonionized acid have on the membranes lining the stomach? (*Hint:* See the Chemistry in Action essay "Antacids and the pH Balance in Your Stomach" in Section 15.3.)

15.6 Weak Bases and Base Ionization Constants

Review Questions

15.51 Use NH_3 to illustrate what we mean by the strength of a base.

15.52 Which of the following has a higher pH: (a) 0.20 *M* NH_3, (b) 0.20 *M* NaOH?

Problems

15.53 Calculate the pH of a 0.24 *M* solution of a weak base with a K_b of 3.5×10^{-6}.

15.54 The diagrams here represent three different weak base solutions of equal concentration. List the bases in order of increasing K_b value. (Water molecules are omitted for clarity.)

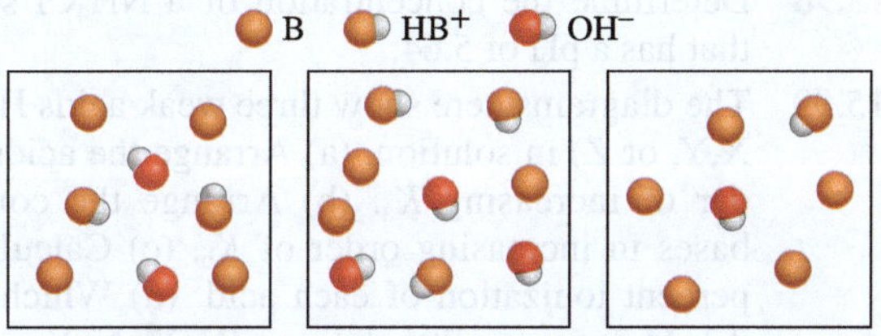

(a) (b) (c)

15.55 Calculate the pH for each of the following solutions: (a) 0.10 *M* NH_3, (b) 0.050 *M* C_5H_5N (pyridine).

15.56 The pH of a 0.30 *M* solution of a weak base is 10.66. What is the K_b of the base?

15.57 What is the original molarity of a solution of ammonia whose pH is 11.22?

15.58 In a 0.080 *M* NH_3 solution, what percent of the NH_3 is present as NH_4^+?

15.7 The Relationship Between the Ionization Constants of Acids and Their Conjugate Bases

Review Questions

15.59 Write the equation relating K_a for a weak acid and K_b for its conjugate base. Use NH_3 and its conjugate acid NH_4^+ to derive the relationship between K_a and K_b.

15.60 From the relationship $K_aK_b = K_w$, what can you deduce about the relative strengths of a weak acid and its conjugate base?

15.8 Diprotic and Polyprotic Acids

Review Questions

15.61 Carbonic acid is a diprotic acid. Explain what that means.

15.62 Write all the species (except water) that are present in a phosphoric acid solution. Indicate which species can act as a Brønsted acid, which as a Brønsted base, and which as both a Brønsted acid and a Brønsted base.

Problems

15.63 The first and second ionization constants of a diprotic acid H_2A are K_{a_1} and K_{a_2} at a certain temperature. Under what conditions will $[A^{2-}] = K_{a_2}$?

15.64 Compare the pH of a 0.040 *M* HCl solution with that of a 0.040 *M* H_2SO_4 solution. (*Hint:* H_2SO_4 is a strong acid; K_a for $HSO_4^- = 1.3 \times 10^{-2}$.)

15.65 What are the concentrations of HSO_4^-, SO_4^{2-}, and H^+ in a 0.20 *M* $KHSO_4$ solution?

15.66 Calculate the concentrations of H^+, HCO_3^-, and CO_3^{2-} in a 0.025 *M* H_2CO_3 solution.

15.9 Molecular Structure and the Strength of Acids

Review Questions

15.67 List four factors that affect the strength of an acid.

15.68 How does the strength of an oxoacid depend on the electronegativity and oxidation number of the central atom?

Problems

15.69 Predict the acid strengths of the following compounds: H_2O, H_2S, and H_2Se.

15.70 Compare the strengths of the following pairs of acids: (a) H_2SO_4 and H_2SeO_4, (b) H_3PO_4 and H_3AsO_4.

15.71 Which of the following is the stronger acid: $CH_2ClCOOH$ or $CHCl_2COOH$? Explain your choice.

15.72 Consider the following compounds:

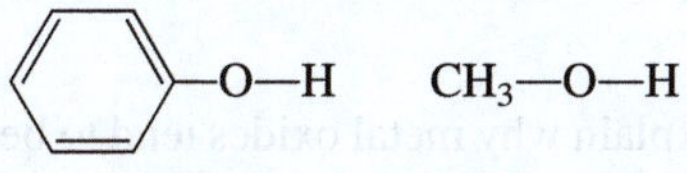

phenol methanol

Experimentally, phenol is found to be a stronger acid than methanol. Explain this difference in terms of the structures of the conjugate bases. (*Hint:* A more stable conjugate base favors ionization. Only one of the conjugate bases can be stabilized by resonance.)

15.10 Acid-Base Properties of Salts

Review Questions

15.73 Define salt hydrolysis. Categorize salts according to how they affect the pH of a solution.

15.74 Explain why small, highly charged metal ions are able to undergo hydrolysis.

15.75 Al^{3+} is not a Brønsted acid but $Al(H_2O)_6^{3+}$ is. Explain.

15.76 Specify which of the following salts will undergo hydrolysis: KF, $NaNO_3$, NH_4NO_2, $MgSO_4$, KCN, C_6H_5COONa, RbI, Na_2CO_3, $CaCl_2$, HCOOK.

Problems

15.77 Predict the pH(> 7, < 7, or ≈ 7) of aqueous solutions containing the following salts: (a) KBr, (b) $Al(NO_3)_3$, (c) $BaCl_2$, (d) $Bi(NO_3)_3$.

15.78 Predict whether the following solutions are acidic, basic, or nearly neutral: (a) NaBr, (b) K_2SO_3, (c) NH_4NO_2, (d) $Cr(NO_3)_3$.

15.79 A certain salt, MX (containing the M^+ and X^- ions), is dissolved in water, and the pH of the resulting solution is 7.0. Can you say anything about the strengths of the acid and the base from which the salt is derived?

15.80 In a certain experiment a student finds that the pHs of 0.10 *M* solutions of three potassium salts KX, KY, and KZ are 7.0, 9.0, and 11.0, respectively. Arrange the acids HX, HY, and HZ in the order of increasing acid strength.

15.81 Calculate the pH of a 0.36 *M* CH_3COONa solution.

15.82 Calculate the pH of a 0.42 *M* NH_4Cl solution.

15.83 Predict the pH(> 7, < 7, ≈ 7) of a $NaHCO_3$ solution.

15.84 Predict whether a solution containing the salt K_2HPO_4 will be acidic, neutral, or basic.

15.11 Acid-Base Properties of Oxides and Hydroxides

Review Questions

15.85 Classify the following oxides as acidic, basic, amphoteric, or neutral: (a) CO_2, (b) K_2O, (c) CaO, (d) N_2O_5, (e) CO, (f) NO, (g) SnO_2, (h) SO_3, (i) Al_2O_3, (j) BaO.

15.86 Write equations for the reactions between (a) CO_2 and NaOH(*aq*), (b) Na_2O and HNO_3(*aq*).

Problems

15.87 Explain why metal oxides tend to be basic if the oxidation number of the metal is low and acidic if the oxidation number of the metal is high. (*Hint:* Metallic compounds in which the oxidation numbers of the metals are low are more ionic than those in which the oxidation numbers of the metals are high.)

15.88 Arrange the oxides in each of the following groups in order of increasing basicity: (a) K_2O, Al_2O_3, BaO, (b) CrO_3, CrO, Cr_2O_3.

15.89 $Zn(OH)_2$ is an amphoteric hydroxide. Write balanced ionic equations to show its reaction with (a) HCl, (b) NaOH [the product is $Zn(OH)_4^{2-}$].

15.90 $Al(OH)_3$ is an insoluble compound. It dissolves in excess NaOH in solution. Write a balanced ionic equation for this reaction. What type of reaction is this?

15.12 Lewis Acids and Bases

Review Questions

15.91 What are the Lewis definitions of an acid and a base? In what way are they more general than the Brønsted definitions?

15.92 In terms of orbitals and electron arrangements, what must be present for a molecule or an ion to act as a Lewis acid (use H^+ and BF_3 as examples)? What must be present for a molecule or ion to act as a Lewis base (use OH^- and NH_3 as examples)?

Problems

15.93 Classify each of the following species as a Lewis acid or a Lewis base: (a) CO_2, (b) H_2O, (c) I^-, (d) SO_2, (e) NH_3, (f) OH^-, (g) H^+, (h) BCl_3.

15.94 Describe the following reaction in terms of the Lewis theory of acids and bases:

$$AlCl_3(s) + Cl^-(aq) \longrightarrow AlCl_4^-(aq)$$

15.95 Which would be considered a stronger Lewis acid: (a) BF_3 or BCl_3, (b) Fe^{2+} or Fe^{3+}? Explain.

15.96 All Brønsted acids are Lewis acids, but the reverse is not true. Give two examples of Lewis acids that are not Brønsted acids.

Additional Problems

15.97 Determine the concentration of a $NaNO_2$ solution that has a pH of 8.22.

15.98 Determine the concentration of a NH_4Cl solution that has a pH of 5.64.

15.99 The diagrams here show three weak acids HA(A = X,Y, or Z) in solution. (a) Arrange the acids in order of increasing K_a. (b) Arrange the conjugate bases in increasing order of K_b. (c) Calculate the percent ionization of each acid. (d) Which of the 0.1 *M* sodium salt solutions (NaX, NaY, or NaZ) has the lowest pH? (The hydrated proton is shown as a hydronium ion. Water molecules are omitted for clarity.)

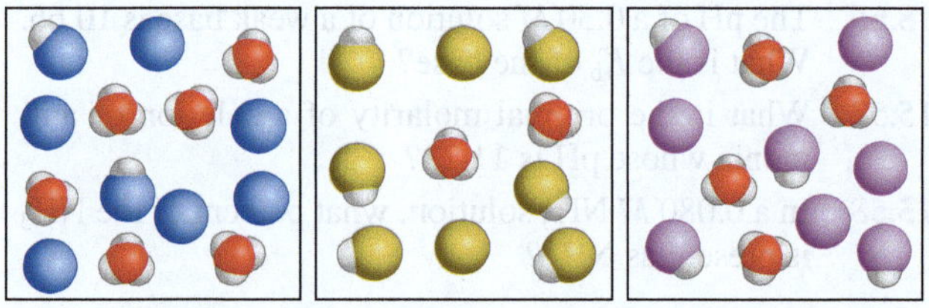

HX HY HZ

15.100 A typical reaction between an antacid and the hydrochloric acid in gastric juice is

$$NaHCO_3(s) + HCl(aq) \rightleftharpoons NaCl(aq) + H_2O(l) + CO_2(g)$$

Calculate the volume (in L) of CO_2 generated from 0.350 g of $NaHCO_3$ and excess gastric juice at 1.00 atm and 37.0°C.

15.101 To which of the following would the addition of an equal volume of 0.60 *M* NaOH lead to a solution having a lower pH: (a) water, (b) 0.30 *M* HCl, (c) 0.70 *M* KOH, (d) 0.40 *M* $NaNO_3$?

15.102 The pH of a 0.0642 *M* solution of a monoprotic acid is 3.86. Is this a strong acid?

15.103 Like water, liquid ammonia undergoes autoionization:

$$NH_3 + NH_3 \rightleftharpoons NH_4^+ + NH_2^-$$

(a) Identify the Brønsted acids and Brønsted bases in this reaction.
(b) What species correspond to H^+ and OH^- and what is the condition for a neutral solution?

15.104 HA and HB are both weak acids although HB is the stronger of the two. Will it take a larger volume of a 0.10 *M* NaOH solution to neutralize 50.0 mL of 0.10 *M* HB than would be needed to neutralize 50.0 mL of 0.10 *M* HA?

15.105 A solution contains a weak monoprotic acid HA and its sodium salt NaA both at 0.1 *M* concentration. Show that $[OH^-] = K_w/K_a$.

15.106 The three common chromium oxides are CrO, Cr_2O_3, and CrO_3. If Cr_2O_3 is amphoteric, what can you say about the acid-base properties of CrO and CrO_3?

15.107 Use the data in Table 15.3 to calculate the equilibrium constant for the following reaction:

$$HCOOH(aq) + OH^-(aq) \rightleftharpoons HCOO^-(aq) + H_2O(l)$$

15.108 Use the data in Table 15.3 to calculate the equilibrium constant for the following reaction:

$$CH_3COOH(aq) + NO_2^-(aq) \rightleftharpoons CH_3COO^-(aq) + HNO_2(aq)$$

15.109 Most of the hydrides of Group 1 and Group 2 metals are ionic (the exceptions are BeH_2 and MgH_2, which are covalent compounds). (a) Describe the reaction between the hydride ion (H^-) and water in terms of a Brønsted acid-base reaction. (b) The same reaction can also be classified as a redox reaction. Identify the oxidizing and reducing agents.

15.110 Calculate the pH of a 0.20 *M* ammonium acetate (CH_3COONH_4) solution.

15.111 Novocaine, used as a local anesthetic by dentists, is a weak base ($K_b = 8.91 \times 10^{-6}$). What is the ratio of the concentration of the base to that of its acid in the blood plasma (pH = 7.40) of a patient?

15.112 Which of the following is the stronger base: NF_3 or NH_3? (*Hint:* F is more electronegative than H.)

15.113 Which of the following is a stronger base: NH_3 or PH_3? (*Hint:* The N—H bond is stronger than the P—H bond.)

15.114 The ion product of D_2O is 1.35×10^{-15} at 25°C. (a) Calculate pD where $pD = -\log[D^+]$. (b) For what values of pD will a solution be acidic in D_2O? (c) Derive a relation between pD and pOD.

15.115 Give an example of the following: (a) a weak acid that contains oxygen atoms, (b) a weak acid that does not contain oxygen atoms, (c) a neutral molecule that acts as a Lewis acid, (d) a neutral molecule that acts as a Lewis base, (e) a weak acid that contains two ionizable H atoms, (f) a conjugate acid-base pair, both of which react with HCl to give carbon dioxide gas.

15.116 What is the pH of 250.0 mL of an aqueous solution containing 0.616 g of the strong acid trifluoromethane sulfonic acid (CF_3SO_3H)?

15.117 (a) Use VSEPR to predict the geometry of the hydronium ion, H_3O^+. (b) The O atom in H_2O has two lone pairs and in principle can accept two H^+ ions. Explain why the species H_4O^{2+} does not exist. What would be its geometry if it did exist?

15.118 HF is a weak acid, but its strength increases with concentration. Explain. (*Hint:* F^- reacts with HF to form HF_2^-. The equilibrium constant for this reaction is 5.2 at 25°C.)

15.119 When chlorine reacts with water, the resulting solution is weakly acidic and reacts with $AgNO_3$ to give a white precipitate. Write balanced equations to represent these reactions. Explain why manufacturers of household bleaches add bases such as NaOH to their products to increase their effectiveness.

15.120 When the concentration of a strong acid is not substantially higher than 1.0×10^{-7} *M*, the ionization of water must be taken into account in the calculation of the solution's pH. (a) Derive an expression for the pH of a strong acid solution, including the contribution to $[H^+]$ from H_2O. (b) Calculate the pH of a 1.0×10^{-7} *M* HCl solution.

15.121 Calculate the pH of a 2.00 *M* NH_4CN solution.

15.122 Calculate the concentrations of all species in a 0.100 *M* H_3PO_4 solution.

15.123 Identify the Lewis acid and Lewis base that lead to the formation of the following species: (a) $AlCl_4^-$, (b) $Cd(CN)_4^{2-}$, (c) HCO_3^-, (d) H_2SO_4.

15.124 Very concentrated NaOH solutions should not be stored in Pyrex glassware. Why? (*Hint:* See Section 11.7.)

15.125 In the vapor phase, acetic acid molecules associate to a certain extent to form dimers:

$$2CH_3COOH(g) \rightleftharpoons (CH_3COOH)_2(g)$$

At 51°C the pressure of a certain acetic acid vapor system is 0.0342 atm in a 360-mL flask. The vapor is condensed and neutralized with 13.8 mL of 0.0568 *M* NaOH. (a) Calculate the degree of dissociation (α) of the dimer under these conditions:

$$(CH_3COOH)_2 \rightleftharpoons 2CH_3COOH$$

(*Hint:* See Problem 14.117 for general procedure.) (b) Calculate the equilibrium constant K_P for the reaction in (a).

15.126 Calculate the concentrations of all the species in a 0.100 *M* Na_2CO_3 solution.

15.127 Henry's law constant for CO_2 at 38°C is 2.28×10^{-3} mol/L · atm. Calculate the pH of a solution of CO_2 at 38°C in equilibrium with the gas at a partial pressure of 3.20 atm.

15.128 Hydrocyanic acid (HCN) is a weak acid and a deadly poisonous compound—in the gaseous form (hydrogen cyanide) it is used in gas chambers. Why is it dangerous to treat sodium cyanide with acids (such as HCl) without proper ventilation?

15.129 How many grams of NaCN would you need to dissolve in enough water to make exactly 250 mL of solution with a pH of 10.00?

15.130 A solution of formic acid (HCOOH) has a pH of 2.53. How many grams of formic acid are there in 100.0 mL of the solution?

15.131 Calculate the pH of a 1-L solution containing 0.150 mole of CH_3COOH and 0.100 mole of HCl.

15.132 A 1.87-g sample of Mg reacts with 80.0 mL of a HCl solution whose pH is −0.544. What is the pH of the solution after all the Mg has reacted? Assume constant volume.

15.133 You are given two beakers, one containing an aqueous solution of strong acid (HA) and the other an aqueous solution of weak acid (HB) of the same concentration. Describe how you would compare the strengths of these two acids by (a) measuring the pH, (b) measuring electrical conductance, (c) studying the rate of hydrogen gas evolution when these solutions are reacted with an active metal such as Mg or Zn.

15.134 Use Le Châtelier's principle to predict the effect of the following changes on the extent of hydrolysis of sodium nitrite ($NaNO_2$) solution: (a) HCl is added, (b) NaOH is added, (c) NaCl is added, (d) the solution is diluted.

15.135 Describe the hydration of SO_2 as a Lewis acid-base reaction. (*Hint:* Refer to the discussion of the hydration of CO_2 in Section 15.12.)

15.136 The disagreeable odor of fish is mainly due to organic compounds (RNH_2) containing an amino group, $—NH_2$, where R is the rest of the molecule. Amines are bases just like ammonia. Explain why putting some lemon juice on fish can greatly reduce the odor.

15.137 A solution of methylamine (CH_3NH_2) has a pH of 10.64. How many grams of methylamine are there in 100.0 mL of the solution?

15.138 A 0.400 *M* formic acid (HCOOH) solution freezes at −0.758°C. Calculate the K_a of the acid at that temperature. (*Hint:* Assume that molarity is equal to molality. Carry your calculations to three significant figures and round off to two for K_a.)

15.139 Both the amide ion (NH_2^-) and the nitride ion (N^{3-}) are stronger bases than the hydroxide ion and hence do not exist in aqueous solutions. (a) Write equations showing the reactions of these ions with water, and identify the Brønsted acid and base in each case. (b) Which of the two is the stronger base?

15.140 The atmospheric sulfur dioxide (SO_2) concentration over a certain region is 0.12 ppm by volume. Calculate the pH of the rainwater due to this pollutant. Assume that the dissolution of SO_2 does not affect its pressure.

15.141 Calcium hypochlorite [$Ca(OCl)_2$] is used as a disinfectant for swimming pools. When dissolved in water it produces hypochlorous acid:

$$Ca(OCl)_2(s) + 2H_2O(l) \rightleftharpoons 2HClO(aq) + Ca(OH)_2(s)$$

which ionizes as follows:

$$HClO(aq) \rightleftharpoons H^+(aq) + ClO^-(aq) \quad K_a = 4.0 \times 10^{-8}$$

As strong oxidizing agents, both HClO and ClO^- can kill bacteria by destroying their cellular components. However, too high a HClO concentration is irritating to the eyes of swimmers and too high a concentration of ClO^- will cause the ions to decompose in sunlight. The recommended pH for pool water is 7.8. Calculate the percent of these species present at this pH.

15.142 Explain the action of smelling salt, which is ammonium carbonate [$(NH_4)_2CO_3$]. (*Hint:* The thin film of aqueous solution that lines the nasal passage is slightly basic.)

15.143 About half of the hydrochloric acid produced annually in the United States (3.0 billion pounds) is used in metal pickling. This process involves the removal of metal oxide layers from metal surfaces to prepare them for coating. (a) Write the overall and net ionic equations for the reaction between iron(III) oxide, which represents the rust layer over iron, and HCl. Identify the Brønsted acid and base. (b) Hydrochloric acid is also used to remove scale (which is mostly $CaCO_3$) from water pipes (see Chemistry in Action essay "An Undesirable Precipitation Reaction" in Section 4.2). Hydrochloric acid reacts with calcium carbonate in two stages; the first stage forms the bicarbonate ion, which then reacts further to form carbon dioxide. Write equations for these two stages and for the overall reaction. (c) Hydrochloric acid is used to recover oil from the ground. It dissolves rocks (often $CaCO_3$) so that the oil can flow more easily. In one process, a 15% (by mass) HCl solution is injected into an oil well to dissolve the rocks. If the density of the acid solution is 1.073 g/mL, what is the pH of the solution?

15.144 Which of the following does not represent a Lewis acid-base reaction?

(a) $H_2O + H^+ \longrightarrow H_3O^+$
(b) $NH_3 + BF_3 \longrightarrow H_3NBF_3$
(c) $PF_3 + F_2 \longrightarrow PF_5$
(d) $Al(OH)_3 + OH^- \longrightarrow Al(OH)_4^-$

15.145 True or false? If false, explain why each statement is wrong. (a) All Lewis acids are Brønsted acids. (b) The conjugate base of an acid always carries a negative charge. (c) The percent ionization of a base increases with its concentration in solution. (d) A solution of barium fluoride is acidic.

15.146 How many milliliters of a strong monoprotic acid solution at pH = 4.12 must be added to 528 mL of the same acid solution at pH = 5.76 to change its pH to 5.34? Assume that the volumes are additive.

15.147 Calculate the pH and percent ionization of a 0.80 *M* HNO_2 solution.

15.148 Consider the two weak acids HX(molar mass = 180 g/mol) and HY(molar mass = 78.0 g/mol). If a solution of 16.9 g/L of HX has the same pH as one containing 9.05 g/L of HY, which is the stronger acid at these concentrations?

15.149 Hemoglobin (Hb) is a blood protein that is responsible for transporting oxygen. It can exist in the protonated form as HbH^+. The binding of oxygen can be represented by the simplified equation

$$HbH^+ + O_2 \rightleftharpoons HbO_2 + H^+$$

(a) What form of hemoglobin is favored in the lungs where oxygen concentration is highest? (b) In body tissues, where the cells release carbon dioxide produced by metabolism, the blood is more acidic due to the formation of carbonic acid. What form of hemoglobin is favored under this condition? (c) When a person hyperventilates, the concentration of CO_2 in his or her blood decreases. How does this action affect the abovementioned equilibrium? Frequently a person who is hyperventilating is advised to breathe into a paper bag. Why does this action help the individual?

15.150 A 1.294-g sample of a metal carbonate (MCO_3) is reacted with 500 mL of a 0.100 *M* HCl solution. The excess HCl acid is then neutralized by 32.80 mL of 0.588 *M* NaOH. Identify M.

15.151 Prove the statement that when the concentration of a weak acid HA decreases by a factor of 10, its percent ionization increases by a factor of $\sqrt{10}$. State any assumptions.

15.152 Calculate the pH of a solution that is 1.00 *M* HCN and 1.00 *M* HF. Compare the concentration (in molarity) of the CN^- ion in this solution with that in a 1.00 *M* HCN solution. Comment on the difference.

15.153 Teeth enamel is hydroxyapatite [$Ca_3(PO_4)_3OH$]. When it dissolves in water (a process called *demineralization*), it dissociates as follows:

$$Ca_5(PO_4)_3OH \longrightarrow 5Ca^{2+} + 3PO_4^{3-} + OH^-$$

The reverse process, called *remineralization,* is the body's natural defense against tooth decay. Acids produced from food remove the OH^- ions and thereby weaken the enamel layer. Most toothpastes contain a fluoride compound such as NaF or SnF_2. What is the function of these compounds in preventing tooth decay?

15.154 Use the van't Hoff equation (see Problem 14.119) and the data in Appendix 2 to calculate the pH of water at its normal boiling point.

15.155 At 28°C and 0.982 atm, gaseous compound HA has a density of 1.16 g/L. A quantity of 2.03 g of this compound is dissolved in water and diluted to exactly 1 L. If the pH of the solution is 5.22 (due to the ionization of HA) at 25°C, calculate the K_a of the acid.

15.156 A 10.0-g sample of white phosphorus was burned in an excess of oxygen. The product was dissolved in enough water to make 500 mL of solution. Calculate the pH of the solution at 25°C.

15.157 Calculate the pH of a 0.20 *M* $NaHCO_3$ solution. (*Hint:* As an approximation, calculate hydrolysis and ionization separately first, followed by partial neutralization.)

15.158 (a) Shown here is a solution containing hydroxide ions and hydronium ions. What is the pH of the solution? (b) How many H_3O^+ ions would you need to draw for every OH^- ion if the pH of the solution is 5.0? The color codes are H_3O^+ (red) and OH^- (blue). Water molecules and counter ions are omitted for clarity.

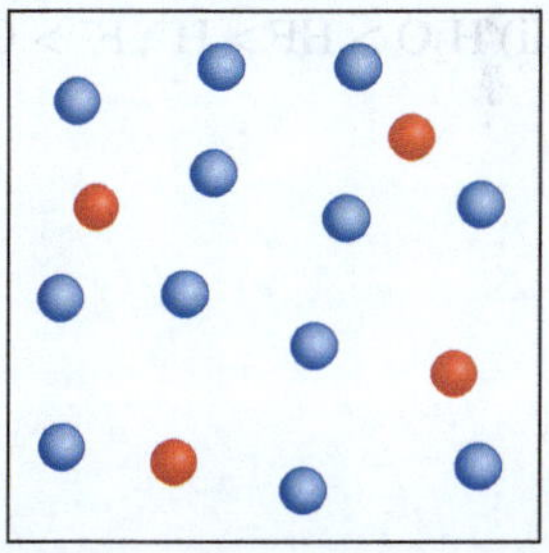

15.159 In this chapter, HCl, HBr, and HI are all listed as strong acids because they are assumed to be ionized completely in water. If, however, we choose a solvent such as acetic acid that is a weaker Brønsted base than water, it is possible to rank the acids in increasing strength as HCl < HBr < HI. (a) Write equations showing proton transfer between the acids and CH_3COOH. Describe how you would compare the strength of the acids in this solvent experimentally. (b) Draw a Lewis structure of the conjugate acid $CH_3COOH_2^+$.

15.160 Use the data in Appendix 2 to calculate the ΔH°_{rxn} for the following reactions: (a) $NaOH(aq) + HCl(aq) \longrightarrow NaCl(aq) + H_2O(l)$; (b) $KOH(aq) + HNO_3(aq) \longrightarrow KNO_3(aq) + H_2O(l)$. Comment on your results.

Interpreting, Modeling, & Estimating

15.161 Malonic acid [$CH_2(COOH)_2$] is a diprotic acid. Compare its two K_a values with that of acetic acid (CH_3COOH) (K_a), and account for the differences in the three K_a values.

15.162 Look up the contents of a Tums tablet. How many tablets are needed to increase the pH of the gastric juice in a person's stomach from 1.2 to 1.5?

15.163 Phosphorous acid, $H_3PO_3(aq)$, is a diprotic acid with $K_{a_1} = 3 \times 10^{-2}$. (a) After examining the K_a values in Table 15.5, estimate K_{a_2} for $H_3PO_3(aq)$ and calculate the pH of a 0.10 *M* solution of $Na_2HPO_3(aq)$. (b) The structure of H_3PO_3 is given in Figure 15.5. Explain why $H_3PO_4(aq)$ is a triprotic acid, but $H_3PO_3(aq)$ is only a diprotic acid.

Answers to Practice Exercises

15.1 (1) H_2O (acid) and OH^- (base); (2) HCN (acid) and CN^- (base). **15.2** 7.7×10^{-15} *M*. **15.3** 0.12. **15.4** 4.7×10^{-4} *M*. **15.5** 7.40. **15.6** 12.56. **15.7** Smaller than 1. **15.8** 2.09. **15.9** 2.2×10^{-6}. **15.10** 12.03. **15.11** $[H_2C_2O_4] = 0.11$ *M*, $[HC_2O_4^-] = 0.086$ *M*, $[C_2O_4^{2-}] = 6.1 \times 10^{-5}$ *M*, $[H^+] = 0.086$ *M*. **15.12** $HClO_2$. **15.13** 8.58. **15.14** (a) pH ≈ 7, (b) pH > 7, (c) pH < 7, (d) pH > 7. **15.15** Lewis acid: Co^{3+}; Lewis base: NH_3.

Answers to Review of Concepts & Facts

15.1.1 (b). **15.1.2** (a) conjugate acid: H_2SO_4; conjugate base: SO_4^{2-}. (b) conjugate acid: $H_2PO_4^-$; conjugate base: PO_4^{3-}. **15.2.1** Because $K_w = 1.0 \times 10^{-14}$, when $[H^+] = 0.0010$ *M*, $[OH^-] = 1.0 \times 10^{-11}$ *M*. **15.2.2** 4.3×10^{-10}. **15.3.1** pH = 6.24. pOH = 7.76. **15.3.2** 8.51×10^{-4}. **15.3.3** 3.96. **15.3.4** The solution with pOH = 11.6 is more acidic. The pH of this solution is 14.0 − 11.6 = 2.4. **15.4.1** 4.2×10^{-7} *M*. **15.4.2** $K > 1$. **15.4.3** (i) $H_2O > H^+$, NO_3^- OH^-. (ii) $H_2O > HF > H^+$, $F^- > OH^-$. **15.4.4** (i) $H_2O > NH_3 > NH_4^+$, $OH^- > H^+$. (ii) $H_2O > K^+$, $OH^- > H^+$. **15.5.1** 2.69. **15.5.2** 1.2×10^{-11}. **15.5.3** % ionization = $\frac{1.0 \times 10^{-7} M}{55.5 M} \times 100\% = 1.8 \times 10^{-7}\%$. **15.6.1** 9.85. **15.6.2** methylamine > aniline > urea. **15.7.1** 1.8×10^{-9}. **15.7.2** CN^-. **15.8.1** (c). **15.9.1** $HBrO < HBrO_3 < HBrO_4$. **15.10.1** 11.31 **15.10.2** (a) C^-. (b) $B^- < A^- < C^-$. **15.11.1** $Al_2O_3 < BaO < K_2O$. **15.12.1** (c) and (e).

CHAPTER

16

Acid-Base Equilibria and Solubility Equilibria

Coral are particularly susceptible to changes in ocean pH as their limestone skeletons have increased solubility in more acidic ocean waters.

Michael McCoy/Science Source

CHAPTER OUTLINE

In this chapter we will continue the study of acid-base reactions with a discussion of buffer action and titrations. We will also look at another type of aqueous equilibrium—that between slightly soluble compounds and their ions in solution.

16.1 The Common Ion Effect

Learning Objectives

- Describe the common ion effect.
- Determine equilibrium concentrations or pH of a solution involving a common ion.

Our discussion of acid-base ionization and salt hydrolysis in Chapter 15 was limited to solutions containing a single solute. In this section, we will consider the acid-base properties of a solution with two dissolved solutes that contain the same ion (cation or anion), called the *common ion*.

The presence of a common ion suppresses the ionization of a weak acid or a weak base. If sodium acetate and acetic acid are dissolved in the same solution, for example, they both dissociate and ionize to produce CH_3COO^- ions:

$$CH_3COONa(s) \xrightarrow{H_2O} CH_3COO^-(aq) + Na^+(aq)$$
$$CH_3COOH(aq) \rightleftharpoons CH_3COO^-(aq) + H^+(aq)$$

CH_3COONa is a strong electrolyte, so it dissociates completely in solution, but CH_3COOH, a weak acid, ionizes only slightly. According to Le Châtelier's principle, the addition of CH_3COO^- ions from CH_3COONa to a solution of CH_3COOH will suppress the ionization of CH_3COOH (that is, shift the equilibrium from right to left), thereby decreasing the hydrogen ion concentration. Thus, a solution containing both CH_3COOH and CH_3COONa will be *less* acidic than a solution containing only CH_3COOH at the same concentration. The shift in equilibrium of the acetic acid ionization is caused by the acetate ions from the salt. CH_3COO^- is the common ion because it is supplied by both CH_3COOH and CH_3COONa.

The ***common ion effect*** is *the shift in equilibrium caused by the addition of a compound having an ion in common with the dissolved substance*. The common ion effect plays an important role in determining the pH of a solution and the solubility of a slightly soluble salt (to be discussed later in this chapter). Here we will study the common ion effect as it relates to the pH of a solution. Keep in mind that despite its distinctive name, the common ion effect is simply a special case of Le Châtelier's principle.

Let us consider the pH of a solution containing a weak acid, HA, and a soluble salt of the weak acid, such as NaA. We start by writing

$$HA(aq) + H_2O(l) \rightleftharpoons H_3O^+(aq) + A^-(aq)$$

or simply

$$HA(aq) \rightleftharpoons H^+(aq) + A^-(aq)$$

The ionization constant K_a is given by

$$K_a = \frac{[H^+][A^-]}{[HA]} \tag{16.1}$$

Rearranging Equation (16.1) gives

$$[H^+] = \frac{K_a[HA]}{[A^-]}$$

Taking the negative logarithm of both sides, we obtain

$$-\log [H^+] = -\log K_a - \log \frac{[HA]}{[A^-]}$$

or

$$-\log [H^+] = -\log K_a + \log \frac{[A^-]}{[HA]}$$

So

$$pH = pK_a + \log \frac{[A^-]}{[HA]} \tag{16.2}$$

where

$$pK_a = -\log K_a \tag{16.3}$$

Equation (16.2) is called the *Henderson-Hasselbalch equation.* A more general form of this expression is

$$pH = pK_a + \log \frac{[\text{conjugate base}]}{[\text{acid}]} \tag{16.4}$$

In our example, HA is the acid and A^- is the conjugate base. Thus, if we know K_a and the concentrations of the acid and the salt of the acid, we can calculate the pH of the solution.

It is important to remember that the Henderson-Hasselbalch equation is derived from the equilibrium constant expression. It is valid regardless of the source of the conjugate base, whether it comes from the acid alone or is supplied by both the acid and its salt.

In problems that involve the common ion effect, we are usually given the starting concentrations of a weak acid HA and its salt, such as NaA. As long as the concentrations of these species are reasonably high (≥ 0.01 *M*), we can neglect the ionization of the acid and the hydrolysis of the salt. This is a valid approximation because HA is a weak acid and the extent of the hydrolysis of the A^- ion is generally very small. Moreover, the presence of A^- (from NaA) further suppresses the ionization of HA and the presence of HA further suppresses the hydrolysis of A^-. Thus, we can use the *starting* concentrations as the equilibrium concentrations in Equation (16.1) or Equation (16.4).

In Example 16.1 we calculate the pH of a solution containing a common ion.

Student Hot Spot

Student data indicate you may struggle with determining the pH of a solution with a common ion. Access your eBook to view additional Learning Resources on this topic.

Example 16.1

(a) Calculate the pH of a 0.20 *M* CH_3COOH solution. (b) What is the pH of a solution containing both 0.20 *M* CH_3COOH and 0.30 *M* CH_3COONa? The K_a of CH_3COOH is 1.8×10^{-5}.

Strategy (a) We calculate $[H^+]$ and hence the pH of the solution by following the procedure in Example 15.8. (b) CH_3COOH is a weak acid ($CH_3COOH \rightleftharpoons CH_3COO^- + H^+$) and CH_3COONa is a soluble salt that is completely dissociated in solution ($CH_3COONa \longrightarrow Na^+ + CH_3COO^-$). The common ion here is the acetate ion, CH_3COO^-. At equilibrium, the major species in solution are CH_3COOH, CH_3COO^-, Na^+, H^+, and H_2O. The Na^+ ion has no acid or base properties and we ignore the ionization of water. Because K_a is an equilibrium constant, its value is the same whether we have just the acid or a mixture of the acid and its salt in solution. Therefore, we can calculate $[H^+]$ at equilibrium and hence pH if we know both $[CH_3COOH]$ and $[CH_3COO^-]$ at equilibrium.

(Continued)

Solution

(a) In this case, the changes are

	$CH_3COOH(aq) \rightleftharpoons$	$H^+(aq)$ +	$CH_3COO^-(aq)$
Initial (M):	0.20	0	0
Change (M):	$-x$	$+x$	$+x$
Equilibrium (M):	$0.20 - x$	x	x

$$K_a = \frac{[H^+][CH_3COO^-]}{[CH_3COOH]}$$

$$1.8 \times 10^{-5} = \frac{x^2}{0.20 - x}$$

Assuming $0.20 - x \approx 0.20$, we obtain

$$1.8 \times 10^{-5} = \frac{x^2}{0.20 - x} \approx \frac{x^2}{0.20}$$

or $x = [H^+] = 1.9 \times 10^{-3}\ M$

Thus, $\mathrm{pH} = -\log(1.9 \times 10^{-3}) = 2.72$

(b) Sodium acetate is a strong electrolyte, so it dissociates completely in solution:

$$CH_3COONa(aq) \longrightarrow \underset{0.30\ M}{Na^+(aq)} + \underset{0.30\ M}{CH_3COO^-(aq)}$$

The initial concentrations, changes, and final concentrations of the species involved in the equilibrium are

	$CH_3COOH(aq) \rightleftharpoons$	$H^+(aq)$ +	$CH_3COO^-(aq)$
Initial (M):	0.20	0	0.30
Change (M):	$-x$	$+x$	$+x$
Equilibrium (M):	$0.20 - x$	x	$0.30 + x$

From Equation (16.1),

$$K_a = \frac{[H^+][CH_3COO^-]}{[CH_3COOH]}$$

$$1.8 \times 10^{-5} = \frac{(x)(0.30 + x)}{0.20 - x}$$

Assuming that $0.30 + x \approx 0.30$ and $0.20 - x \approx 0.20$, we obtain

$$1.8 \times 10^{-5} = \frac{(x)(0.30 + x)}{0.20 - x} \approx \frac{(x)(0.30)}{0.20}$$

or $x = [H^+] = 1.2 \times 10^{-5}\ M$

Thus, $\mathrm{pH} = -\log\,[H^+]$

$$= -\log\,(1.2 \times 10^{-5}) = 4.92$$

Check Comparing the results in (a) and (b), we see that when the common ion (CH_3COO^-) is present, according to Le Châtelier's principle, the equilibrium shifts from right to left. This action decreases the extent of ionization of the weak acid. Consequently, fewer H^+ ions are produced in (b) and the pH of the solution is higher than that in (a). As always, you should check the validity of the assumptions.

Practice Exercise What is the pH of a solution containing 0.30 M HCOOH and 0.52 M HCOOK? Compare your result with the pH of a 0.30 M HCOOH solution.

Similar problem: 16.5.

The common ion effect also operates in a solution containing a weak base, such as NH_3, and a salt of the base, say NH_4Cl. At equilibrium,

$$NH_4^+(aq) \rightleftharpoons NH_3(aq) + H^+(aq)$$

$$K_a = \frac{[NH_3][H^+]}{[NH_4^+]}$$

We can derive the Henderson-Hasselbalch equation for this system as follows. Rearranging the previous equation we obtain

$$[H^+] = \frac{K_a[NH_4^+]}{[NH_3]}$$

Taking the negative logarithm of both sides gives

$$-\log [H^+] = -\log K_a - \log \frac{[NH_4^+]}{[NH_3]}$$

$$-\log [H^+] = -\log K_a + \log \frac{[NH_3]}{[NH_4^+]}$$

or

$$pH = pK_a + \log \frac{[NH_3]}{[NH_4^+]}$$

A solution containing both NH_3 and its salt NH_4Cl is *less* basic than a solution containing only NH_3 at the same concentration. The common ion NH_4^+ suppresses the ionization of NH_3 in the solution containing both the base and the salt.

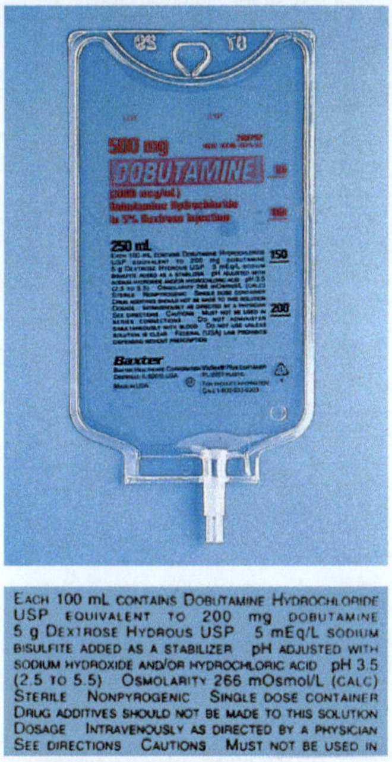

Fluids for intravenous injection must include buffer systems to maintain the proper blood pH.
Ken Karp/McGraw Hill

Summary of Concepts & Facts

- The common ion effect tends to suppress the ionization of a weak acid or a weak base. This action can be explained by Le Châtelier's principle.

Review of Concepts & Facts

16.1.1 Which of the following would cause a decrease in the percent ionization of a solution of hydrofluoric acid (HF) at equilibrium: (a) KF, (b) H_2O, (c) CaF_2, (d) HCl?

16.1.2 What is the pH of a solution containing 0.25 *M* HNO_2 and 0.43 *M* $NaNO_2$?

16.2 Buffer Solutions

Learning Objectives

- Calculate the pH of a buffer using equilibrium concentrations and the Henderson-Hasselbalch equation.
- Summarize how to prepare a buffer with a given pH.
- Select the appropriate weak acid (or weak base) to make a buffer of a given pH.

Video: Buffer Solutions

A ***buffer solution*** is *a solution of (1) a weak acid or a weak base and (2) its salt; both components must be present. The solution has the ability to resist changes in pH upon the addition of small amounts of either acid or base*. Buffers are very important to chemical and biological systems. The pH in the human body varies greatly from one fluid to another; for example, the pH of blood is about 7.4, whereas gastric juice in the stomach has a pH of about 1.5. These pH values, which are crucial for proper enzyme function and the balance of osmotic pressure, are maintained by buffers in most cases.

Video: Properties of Buffers

A buffer solution must contain a relatively large concentration of acid to react with any OH^- ions that are added to it, and it must contain a similar concentration of base to react with any added H^+ ions. Furthermore, the acid and the base components of the buffer must not consume each other in a neutralization reaction. These requirements are satisfied by an acid-base conjugate pair, for example, a weak acid and its conjugate base (supplied by a salt) or a weak base and its conjugate acid (supplied by a salt).

Figure 16.1 *The acid-base indicator bromophenol blue (added to all solutions shown) is used to illustrate buffer action. The indicator's color is blue-purple above pH 4.6 and yellow below pH 3.0. (Left to right) (a) A buffer solution made up of 50 mL of 0.1 M CH_3COOH and 50 mL of 0.1 M CH_3COONa. The solution has a pH of 4.7 and turns the indicator blue-purple. (b) After the addition of 40 mL of 0.1 M HCl solution to the solution in (a), the color remains blue-purple. (c) A 100-mL CH_3COOH solution whose pH is 4.7. (d) After the addition of 6 drops (about 0.3 mL) of 0.1 M HCl solution, the color turns yellow. Without buffer action, the pH of the solution decreases rapidly to less than 3.0 upon the addition of 0.1 M HCl.*

A simple buffer solution can be prepared by adding comparable molar amounts of acetic acid (CH_3COOH) and its salt sodium acetate (CH_3COONa) to water. The equilibrium concentrations of both the acid and the conjugate base (from CH_3COONa) are assumed to be the same as the starting concentrations (see Section 16.1). A solution containing these two substances has the ability to neutralize either added acid or added base. Sodium acetate, a strong electrolyte, dissociates completely in water:

$$CH_3COONa(s) \xrightarrow{H_2O} CH_3COO^-(aq) + Na^+(aq)$$

If an acid is added, the H^+ ions will be consumed by the conjugate base in the buffer, CH_3COO^-, according to the equation

$$CH_3COO^-(aq) + H^+(aq) \longrightarrow CH_3COOH(aq)$$

If a base is added to the buffer system, the OH^- ions will be neutralized by the acid in the buffer:

$$CH_3COOH(aq) + OH^-(aq) \longrightarrow CH_3COO^-(aq) + H_2O(l)$$

As you can see, the two reactions that characterize this buffer system are identical to those for the common ion effect described in Example 16.1. The *buffering capacity,* that is, the effectiveness of the buffer solution, depends on the amount of acid and conjugate base from which the buffer is made. The larger the amount, the greater the buffering capacity.

In general, a buffer system can be represented as salt-acid or conjugate base-acid. Thus, the sodium acetate–acetic acid buffer system just discussed can be written as CH_3COONa/CH_3COOH or simply CH_3COO^-/CH_3COOH. Figure 16.1 shows this buffer system in action.

Example 16.2 distinguishes buffer systems from acid-salt combinations that do not function as buffers.

Example 16.2

Which of the following solutions can be classified as buffer systems? (a) KH_2PO_4/H_3PO_4, (b) $NaClO_4/HClO_4$, (c) C_5H_5N/C_5H_5NHCl (C_5H_5N is pyridine; its K_b is given in Table 15.4). Explain your answer.

Strategy What constitutes a buffer system? Which of the preceding solutions contains a weak acid and its salt (containing the weak conjugate base)? Which of the preceding

(Continued)

solutions contains a weak base and its salt (containing the weak conjugate acid)? Why is the conjugate base of a strong acid not able to neutralize an added acid?

Solution The criteria for a buffer system are that we must have a weak acid and its salt (containing the weak conjugate base) or a weak base and its salt (containing the weak conjugate acid).

(a) H_3PO_4 is a weak acid, and its conjugate base, $H_2PO_4^-$, is a weak base (see Table 15.5). Therefore, this is a buffer system.

(b) Because $HClO_4$ is a strong acid, its conjugate base, ClO_4^-, is an extremely weak base. This means that the ClO_4^- ion will not combine with a H^+ ion in solution to form $HClO_4$. Thus, the system cannot act as a buffer system.

(c) As Table 15.4 shows, C_5H_5N is a weak base and its conjugate acid, $C_5H_5\overset{+}{N}H$ (the cation of the salt C_5H_5NHCl), is a weak acid. Therefore, this is a buffer system.

Practice Exercise Which of the following couples are buffer systems: (a) KF/HF, (b) KBr/HBr, (c) $Na_2CO_3/NaHCO_3$?

Similar problems: 16.9, 16.10.

The effect of a buffer solution on pH is illustrated by Example 16.3.

Example 16.3

(a) Calculate the pH of a buffer system containing 1.0 *M* CH_3COOH and 1.0 *M* CH_3COONa. (b) What is the pH of the buffer system after the addition of 0.10 mole of gaseous HCl to 1.0 L of the solution? Assume that the volume of the solution does not change when the HCl is added.

Strategy (a) The pH of the buffer system before the addition of HCl can be calculated with the procedure described in Example 16.1, because it is similar to the common-ion problem. The K_a of CH_3COOH is 1.8×10^{-5} (see Table 15.3). (b) The reaction describing the buffer action, in this case, is $CH_3COO^- + H^+ \longrightarrow CH_3COOH$.

Solution

(a) We summarize the concentrations of the species at equilibrium as follows:

	$CH_3COOH(aq)$	$\rightleftharpoons$ $H^+(aq)$	+ $CH_3COO^-(aq)$
Initial (*M*):	1.0	0	1.0
Change (*M*):	$-x$	$+x$	$+x$
Equilibrium (*M*):	$1.0 - x$	x	$1.0 + x$

$$K_a = \frac{[H^+][CH_3COO^-]}{[CH_3COOH]}$$

$$1.8 \times 10^{-5} = \frac{(x)(1.0 + x)}{(1.0 - x)}$$

Assuming $1.0 + x \approx 1.0$ and $1.0 - x \approx 1.0$, we obtain

$$1.8 \times 10^{-5} = \frac{(x)(1.0 + x)}{(1.0 - x)} \approx \frac{x(1.0)}{1.0}$$

or $$x = [H^+] = 1.8 \times 10^{-5}\ M$$

Thus, $$pH = -\log(1.8 \times 10^{-5}) = 4.74$$

When the concentrations of the acid and the conjugate base are the same, the pH of the buffer is equal to the pK_a of the acid.

(b) When HCl is added to the solution, the initial changes are

	$HCl(aq)$	$\longrightarrow$ $H^+(aq)$	+ $Cl^-(aq)$
Initial (mol):	0.10	0	0
Change (mol):	−0.10	+0.10	+0.10
Final (mol):	0	0.10	0.10

(Continued)

The Cl^- ion is a spectator ion in solution because it is the conjugate base of a strong acid.

The H^+ ions provided by the strong acid HCl react completely with the conjugate base of the buffer, which is CH_3COO^-. At this point it is more convenient to work with moles rather than molarity. The reason is that in some cases the volume of the solution may change when a substance is added. A change in volume will change the molarity, but not the number of moles. The neutralization reaction is summarized next:

	$CH_3COO^-(aq)$	$+\ H^+(aq)$	$\longrightarrow CH_3COOH(aq)$
Initial (mol):	1.0	0.10	1.0
Change (mol):	−0.10	−0.10	+0.10
Final (mol):	0.90	0	1.1

Finally, to calculate the pH of the buffer after neutralization of the acid, we convert back to molarity by dividing moles by 1.0 L of solution.

	$CH_3COOH(aq)$	$\rightleftharpoons H^+(aq)$	$+\ CH_3COO^-(aq)$
Initial (M):	1.1	0	0.90
Change (M):	$-x$	$+x$	$+x$
Equilibrium (M):	$1.1 - x$	x	$0.90 + x$

$$K_a = \frac{[H^+][CH_3COO^-]}{[CH_3COOH]}$$

$$1.8 \times 10^{-5} = \frac{(x)(0.90 + x)}{1.1 - x}$$

Assuming $0.90 + x \approx 0.90$ and $1.1 - x \approx 1.1$, we obtain

$$1.8 \times 10^{-5} = \frac{(x)(0.90 + x)}{1.1 - x} \approx \frac{x(0.90)}{1.1}$$

or $$x = [H^+] = 2.2 \times 10^{-5}\,M$$

Thus, $$\text{pH} = -\log(2.2 \times 10^{-5}) = 4.66$$

Check The pH decreases by only a small amount upon the addition of HCl. This is consistent with the action of a buffer solution.

Practice Exercise Calculate the pH of the 0.30 M NH_3/0.36 M NH_4Cl buffer system. What is the pH after the addition of 20.0 mL of 0.050 M NaOH to 80.0 mL of the buffer solution?

Similar problem: 16.17.

In the buffer solution examined in Example 16.3, there is a decrease in pH (the solution becomes more acidic) as a result of added HCl. We can also compare the changes in H^+ ion concentration as follows:

$$\text{Before addition of HCl: } [H^+] = 1.8 \times 10^{-5}\,M$$
$$\text{After addition of HCl: } [H^+] = 2.2 \times 10^{-5}\,M$$

Thus, the H^+ ion concentration increases by a factor of

$$\frac{2.2 \times 10^{-5}\,M}{1.8 \times 10^{-5}\,M} = 1.2$$

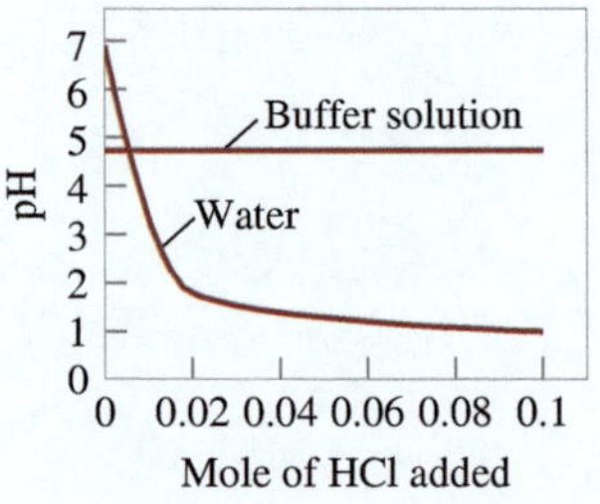

Figure 16.2 *A comparison of the change in pH when 0.10 mol HCl is added to pure water and to an acetate buffer solution as described in Example 16.3.*

To appreciate the effectiveness of the CH_3COONa/CH_3COOH buffer, let us find out what would happen if 0.10 mol HCl were added to 1 L of water, and compare the increase in H^+ ion concentration.

$$\text{Before addition of HCl: } [H^+] = 1.0 \times 10^{-7}\,M$$
$$\text{After addition of HCl: } [H^+] = 0.10\,M$$

As a result of the addition of HCl, the H^+ ion concentration increases by a factor of

$$\frac{0.10\,M}{1.0 \times 10^{-7}\,M} = 1.0 \times 10^{6}$$

amounting to a millionfold increase! This comparison shows that a properly chosen buffer solution can maintain a fairly constant H^+ ion concentration, or pH (Figure 16.2).

Preparing a Buffer Solution with a Specific pH

Now suppose we want to prepare a buffer solution with a specific pH. How do we go about it? Equation (16.4) indicates that if the molar concentrations of the acid and its conjugate base are approximately equal; that is, if [acid] ≈ [conjugate base], then

$$\log \frac{[\text{conjugate base}]}{[\text{acid}]} \approx 0$$

or

$$\text{pH} \approx \text{p}K_a$$

Thus, to prepare a buffer solution, we work backwards. First we choose a weak acid whose $\text{p}K_a$ is close to the desired pH. Next, we substitute the pH and $\text{p}K_a$ values in Equation (16.4) to obtain the ratio [conjugate base]/[acid]. This ratio can then be converted to molar quantities for the preparation of the buffer solution. Example 16.4 shows this approach.

Student Hot Spot

Student data indicate you may struggle with preparing buffers. Access your eBook for additional Learning Resources on this topic.

Example 16.4

Describe how you would prepare a "phosphate buffer" with a pH of about 7.40.

Strategy For a buffer to function effectively, the concentrations of the acid component must be roughly equal to the conjugate base component. According to Equation (16.4), when the desired pH is close to the $\text{p}K_a$ of the acid, that is, when pH ≈ $\text{p}K_a$,

$$\log \frac{[\text{conjugate base}]}{[\text{acid}]} \approx 0$$

or

$$\frac{[\text{conjugate base}]}{[\text{acid}]} \approx 1$$

Solution Because phosphoric acid is a triprotic acid, we write the three stages of ionization as follows: The K_a values are obtained from Table 15.5 and the $\text{p}K_a$ values are found by applying Equation (16.3).

$$H_3PO_4(aq) \rightleftharpoons H^+(aq) + H_2PO_4^-(aq) \quad K_{a_1} = 6.9 \times 10^{-3};\ \text{p}K_{a_1} = 2.16$$
$$H_2PO_4^-(aq) \rightleftharpoons H^+(aq) + HPO_4^{2-}(aq) \quad K_{a_2} = 6.2 \times 10^{-8};\ \text{p}K_{a_2} = 7.21$$
$$HPO_4^{2-}(aq) \rightleftharpoons H^+(aq) + PO_4^{3-}(aq) \quad K_{a_3} = 4.8 \times 10^{-13};\ \text{p}K_{a_3} = 12.32$$

The most suitable of the three buffer systems is $HPO_4^{2-}/H_2PO_4^-$, because the $\text{p}K_a$ of the acid $H_2PO_4^-$ is closest to the desired pH. From the Henderson-Hasselbalch equation we write

$$\text{pH} = \text{p}K_a + \log \frac{[\text{conjugate base}]}{[\text{acid}]}$$

$$7.40 = 7.21 + \log \frac{[HPO_4^{2-}]}{[H_2PO_4^-]}$$

$$\log \frac{[HPO_4^{2-}]}{[H_2PO_4^-]} = 0.19$$

Taking the antilog, we obtain

$$\frac{[HPO_4^{2-}]}{[H_2PO_4^-]} = 10^{0.19} = 1.5$$

Thus, one way to prepare a phosphate buffer with a pH of 7.40 is to dissolve disodium hydrogen phosphate (Na_2HPO_4) and sodium dihydrogen phosphate (NaH_2PO_4) in a mole ratio of 1.5:1.0 in water. For example, we could dissolve 1.5 mole of Na_2HPO_4 and 1.0 mole of NaH_2PO_4 in enough water to make up a 1-L solution.

Practice Exercise How would you prepare a liter of "carbonate buffer" at a pH of 10.10? You are provided with carbonic acid (H_2CO_3), sodium hydrogen carbonate ($NaHCO_3$), and sodium carbonate (Na_2CO_3). See Table 15.5 for K_a values.

Similar problems: 16.19, 16.20.

The Chemistry in Action essay "Maintaining the pH of Blood" illustrates the importance of buffer systems in the human body.

CHEMISTRY *in Action*

Maintaining the pH of Blood

All higher animals need a circulatory system to carry fuel and oxygen for their life processes and to remove wastes. In the human body this vital exchange takes place in the versatile fluid known as blood, of which there are about 5 L (10.6 pints) in an average adult. Blood circulating deep in the tissues carries oxygen and nutrients to keep cells alive, and removes carbon dioxide and other waste materials. Using several buffer systems, nature has provided an extremely efficient method for the delivery of oxygen and the removal of carbon dioxide.

Blood is an enormously complex system, but for our purposes we need look at only two essential components: blood plasma and red blood cells, or *erythrocytes*. Blood plasma contains many compounds, including proteins, metal ions, and inorganic phosphates. The erythrocytes contain hemoglobin molecules, as well as the enzyme *carbonic anhydrase,* which catalyzes both the formation of carbonic acid (H_2CO_3) and its decomposition:

$$CO_2(aq) + H_2O(l) \rightleftharpoons H_2CO_3(aq)$$

The substances inside the erythrocyte are protected from extracellular fluid (blood plasma) by a cell membrane that allows only certain molecules to diffuse through it.

The pH of blood plasma is maintained at about 7.40 by several buffer systems, the most important of which is the HCO_3^-/H_2CO_3 system. In the erythrocyte, where the pH is 7.25, the principal buffer systems are HCO_3^-/H_2CO_3 and hemoglobin. The hemoglobin molecule is a complex protein molecule (molar mass 65,000 g) that contains a number of ionizable protons. As a very rough approximation, we can treat it as a monoprotic acid of the form HHb:

$$HHb(aq) \rightleftharpoons H^+(aq) + Hb^-(aq)$$

where HHb represents the hemoglobin molecule and Hb^- the conjugate base of HHb. Oxyhemoglobin ($HHbO_2$), formed by the combination of oxygen with hemoglobin, is a stronger acid than HHb:

$$HHbO_2(aq) \rightleftharpoons H^+(aq) + HbO_2^-(aq)$$

As the electron micrograph of red blood cells shows, carbon dioxide produced by metabolic processes diffuses into the erythrocyte, where it is rapidly converted to H_2CO_3 by carbonic anhydrase:

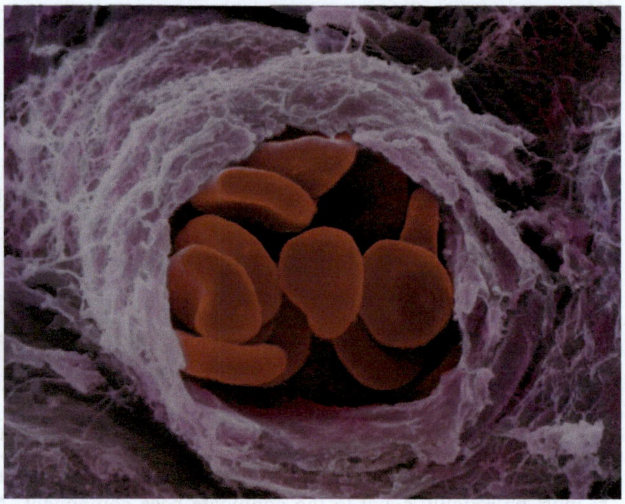

Electron micrograph of red blood cells in a small branch of an artery.
Professors Pietro M. Motta & Silvia Correr/Science Source

$$CO_2(aq) + H_2O(l) \rightleftharpoons H_2CO_3(aq)$$

The ionization of the carbonic acid

$$H_2CO_3(aq) \rightleftharpoons H^+(aq) + HCO_3^-(aq)$$

has two important consequences. First, the bicarbonate ion diffuses out of the erythrocyte and is carried by the blood plasma to the lungs. This is the major mechanism for removing carbon dioxide. Second, the H^+ ions shift the equilibrium in favor of the nonionized oxyhemoglobin molecule:

$$H^+(aq) + HbO_2^-(aq) \rightleftharpoons HHbO_2(aq)$$

Because $HHbO_2$ releases oxygen more readily than its conjugate base (HbO_2^-), the formation of the acid promotes the following reaction from left to right:

$$HHbO_2(aq) \rightleftharpoons HHb(aq) + O_2(aq)$$

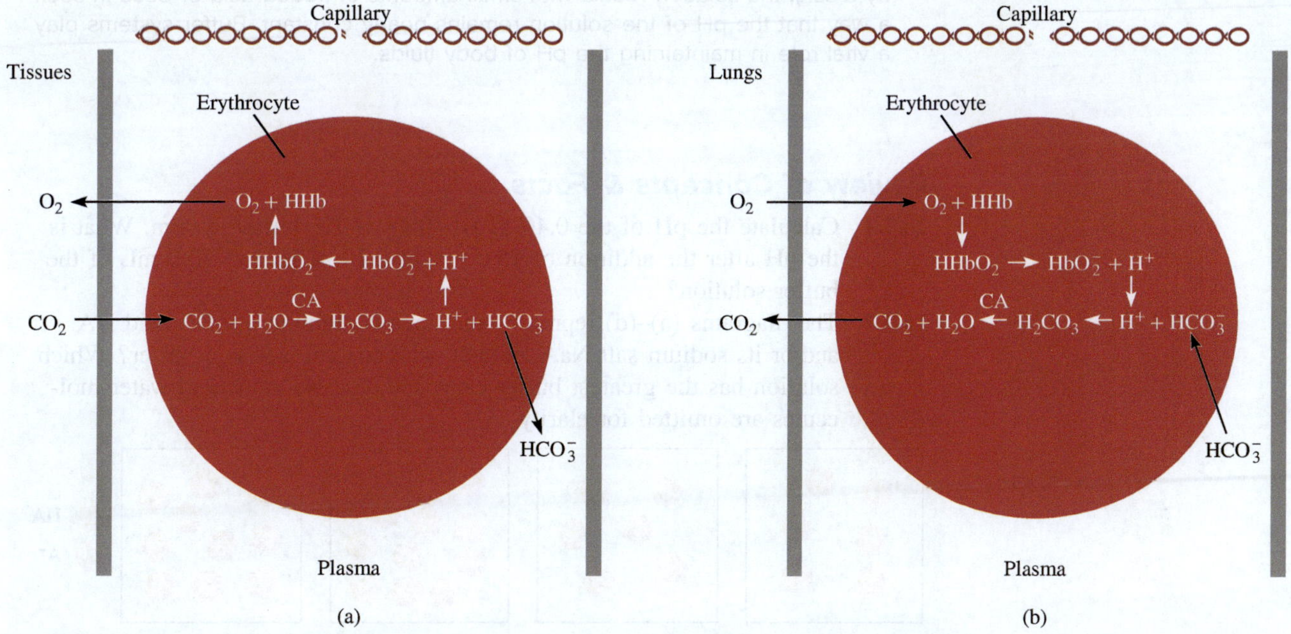

Oxygen–carbon dioxide transport and release by blood. (a) The partial pressure of CO_2 is higher in the metabolizing tissues than in the plasma. Thus, it diffuses into the blood capillaries and then into erythrocytes. There it is converted to carbonic acid by the enzyme carbonic anhydrase (CA). The protons provided by the carbonic acid then combine with the HbO_2^- anions to form $HHbO_2$, which eventually dissociates into HHb and O_2. Because the partial pressure of O_2 is higher in the erythrocytes than in the tissues, oxygen molecules diffuse out of the erythrocytes and then into the tissues. The bicarbonate ions also diffuse out of the erythrocytes and are carried by the plasma to the lungs. (b) In the lungs, the processes are exactly reversed. Oxygen molecules diffuse from the lungs, where they have a higher partial pressure, into the erythrocytes. There they combine with HHb to form $HHbO_2$. The protons provided by $HHbO_2$ combine with the bicarbonate ions diffused into the erythrocytes from the plasma to form carbonic acid. In the presence of carbonic anhydrase, carbonic acid is converted to H_2O and CO_2. The CO_2 then diffuses out of the erythrocytes and into the lungs, where it is exhaled.

The O_2 molecules diffuse out of the erythrocyte and are taken up by other cells to carry out metabolism.

When the venous blood returns to the lungs, the above-mentioned processes are reversed. The bicarbonate ions now diffuse into the erythrocyte, where they react with hemoglobin to form carbonic acid:

$$HHb(aq) + HCO_3^-(aq) \rightleftharpoons Hb^-(aq) + H_2CO_3(aq)$$

Most of the acid is then converted to CO_2 by carbonic anhydrase:

$$H_2CO_3(aq) \rightleftharpoons H_2O(l) + CO_2(aq)$$

The carbon dioxide diffuses to the lungs and is eventually exhaled. The formation of the Hb^- ions (due to the reaction between HHb and HCO_3^-) also favors the uptake of oxygen at the lungs

$$Hb^-(aq) + O_2(aq) \rightleftharpoons HbO_2^-(aq)$$

because Hb^- has a greater affinity for oxygen than HHb.

When the arterial blood flows back to the body tissues, the entire cycle is repeated.

Summary of Concepts & Facts

- A buffer solution is a combination of either a weak acid and its weak conjugate base (supplied by a salt) or a weak base and its weak conjugate acid (supplied by a salt); the solution reacts with small amounts of added acid or base in such a way that the pH of the solution remains nearly constant. Buffer systems play a vital role in maintaining the pH of body fluids.

Review of Concepts & Facts

16.2.1 Calculate the pH of the 0.40 *M* HF/0.48 *M* KF buffer system. What is the pH after the addition of 15.0 mL of 0.10 *M* HCl to 50.0 mL of the buffer solution?

16.2.2 The diagrams (a)–(d) represent solutions containing a weak acid HA and/or its sodium salt NaA. Which solutions can act as a buffer? Which solution has the greatest buffer capacity? The Na^+ ions and water molecules are omitted for clarity.

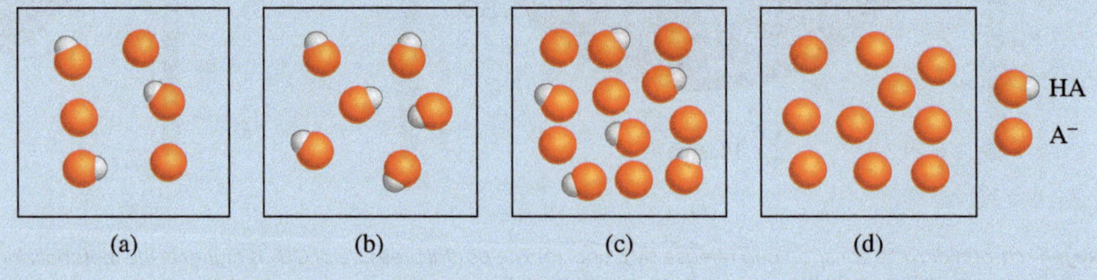

16.3 Acid-Base Titrations

Learning Objectives

- Predict the pH of a titration given the type of acid and base used.
- Evaluate the pH throughout a titration between acids and bases (strong titrated with strong; strong titrated with weak).
- Examine a titration curve to identify what type of titration was conducted (e.g., weak acid titrated with strong base).

Video: Acid-Base Titrations

Having discussed buffer solutions, we can now look in more detail at the quantitative aspects of acid-base titrations (see Section 4.7). We will consider three types of reactions: (1) titrations involving a strong acid and a strong base, (2) titrations involving a weak acid and a strong base, and (3) titrations involving a strong acid and a weak base. Titrations involving a weak acid and a weak base are complicated by the hydrolysis of both the cation and the anion of the salt formed. It is difficult to determine the equivalence point in these cases. Therefore, these titrations will not be dealt with here. Figure 16.3 shows the arrangement for monitoring the pH during the course of a titration.

Strong Acid–Strong Base Titrations

The reaction between a strong acid (say, HCl) and a strong base (say, NaOH) can be represented by

$$NaOH(aq) + HCl(aq) \longrightarrow NaCl(aq) + H_2O(l)$$

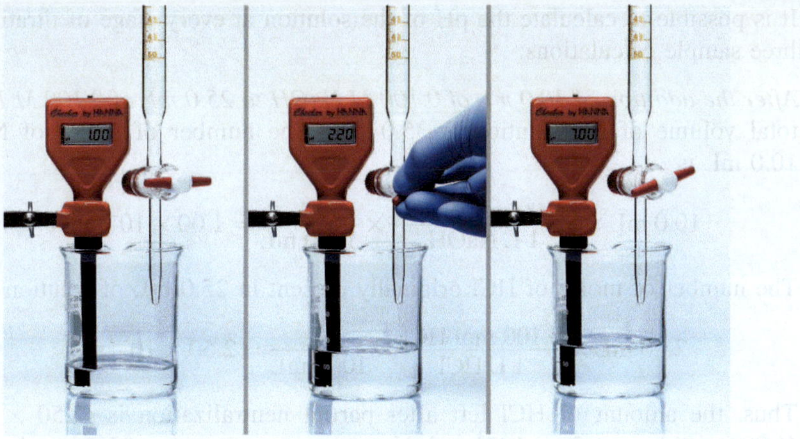

Figure 16.3 *A pH meter is used to monitor an acid-base titration.*
GIPhotoStock Images/Science Source

or in terms of the net ionic equation

$$H^+(aq) + OH^-(aq) \longrightarrow H_2O(l)$$

Consider the addition of a 0.100 M NaOH solution (from a buret) to an Erlenmeyer flask containing 25.0 mL of 0.100 M HCl. For convenience, we will use only three significant figures for volume and concentration and two significant figures for pH. Figure 16.4 shows the pH profile of the titration (also known as the titration curve). Before the addition of NaOH, the pH of the acid is given by −log(0.100), or 1.00. When NaOH is added, the pH of the solution increases slowly at first. Near the equivalence point the pH begins to rise steeply, and at the equivalence point (that is, the point at which equimolar amounts of acid and base have reacted) the curve rises almost vertically. In a strong acid–strong base titration both the hydrogen ion and hydroxide ion concentrations are very small at the equivalence point (approximately 1×10^{-7} M); consequently, the addition of a single drop of the base can cause a large increase in $[OH^-]$ and in the pH of the solution. Beyond the equivalence point, the pH again increases slowly with the addition of NaOH.

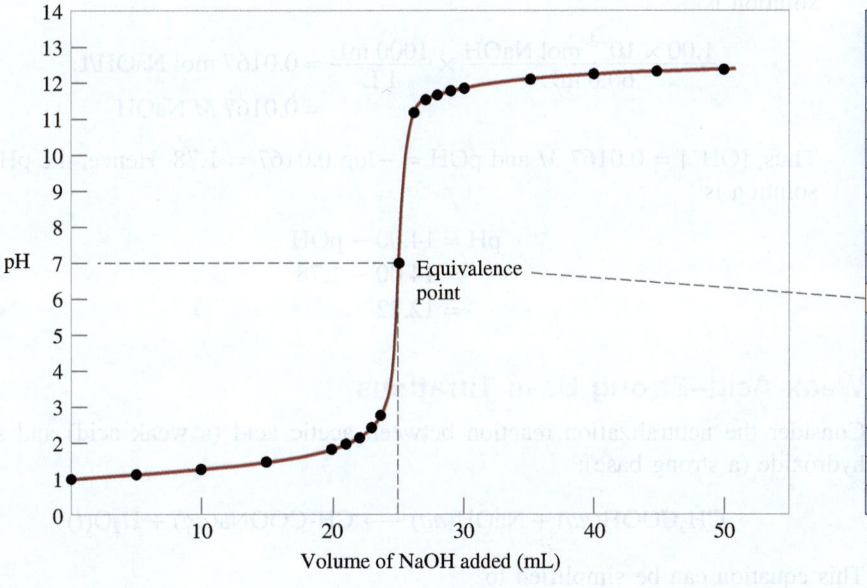

Volume NaOH added (mL)	pH
0.0	1.00
5.0	1.18
10.0	1.37
15.0	1.60
20.0	1.95
22.0	2.20
24.0	2.69
25.0	7.00
26.0	11.29
28.0	11.75
30.0	11.96
35.0	12.22
40.0	12.36
45.0	12.46
50.0	12.52

Figure 16.4 *pH profile of a strong acid–strong base titration. A 0.100 M NaOH solution is added from a buret to 25.0 mL of a 0.100 M HCl solution in an Erlenmeyer flask (see Figure 4.21). This curve is sometimes referred to as a titration curve.*

It is possible to calculate the pH of the solution at every stage of titration. Here are three sample calculations.

1. *After the addition of 10.0 mL of 0.100 M NaOH to 25.0 mL of 0.100 M HCl.* The total volume of the solution is 35.0 mL. The number of moles of NaOH in 10.0 mL is

$$10.0 \text{ mL} \times \frac{0.100 \text{ mol NaOH}}{1 \text{ L NaOH}} \times \frac{1 \text{ L}}{1000 \text{ mL}} = 1.00 \times 10^{-3} \text{ mol}$$

The number of moles of HCl originally present in 25.0 mL of solution is

$$25.0 \text{ mL} \times \frac{0.100 \text{ mol HCl}}{1 \text{ L HCl}} \times \frac{1 \text{ L}}{1000 \text{ mL}} = 2.50 \times 10^{-3} \text{ mol}$$

Thus, the amount of HCl left after partial neutralization is $(2.50 \times 10^{-3}) - (1.00 \times 10^{-3})$, or 1.50×10^{-3} mol. Next, the concentration of H^+ ions in 35.0 mL of solution is found as follows:

$$\frac{1.50 \times 10^{-3} \text{ mol HCl}}{35.0 \text{ mL}} \times \frac{1000 \text{ mL}}{1 \text{ L}} = 0.0429 \text{ mol HCl/L}$$
$$= 0.0429\ M \text{ HCl}$$

Thus, $[H^+] = 0.0429\ M$, and the pH of the solution is

$$\text{pH} = -\log 0.0429 = 1.37$$

2. *After the addition of 25.0 mL of 0.100 M NaOH to 25.0 mL of 0.100 M HCl.* This is a simple calculation because it involves a complete neutralization reaction and the salt (NaCl) does not undergo hydrolysis. At the equivalence point, $[H^+] = [OH^-] = 1.00 \times 10^{-7}\ M$ and the pH of the solution is 7.00.

3. *After the addition of 35.0 mL of 0.100 M NaOH to 25.0 mL of 0.100 M HCl.* The total volume of the solution is now 60.0 mL. The number of moles of NaOH added is

$$35.0 \text{ mL} \times \frac{0.100 \text{ mol NaOH}}{1 \text{ L NaOH}} \times \frac{1 \text{ L}}{1000 \text{ mL}} = 3.50 \times 10^{-3} \text{ mol}$$

The number of moles of HCl in 25.0 mL of solution is 2.50×10^{-3} mol. After complete neutralization of HCl, the number of moles of NaOH left is $(3.50 \times 10^{-3}) - (2.50 \times 10^{-3})$, or 1.00×10^{-3} mol. The concentration of NaOH in 60.0 mL of solution is

$$\frac{1.00 \times 10^{-3} \text{ mol NaOH}}{60.0 \text{ mL}} \times \frac{1000 \text{ mL}}{1 \text{ L}} = 0.0167 \text{ mol NaOH/L}$$
$$= 0.0167\ M \text{ NaOH}$$

Thus, $[OH^-] = 0.0167\ M$ and $\text{pOH} = -\log 0.0167 = 1.78$. Hence, the pH of the solution is

$$\text{pH} = 14.00 - \text{pOH}$$
$$= 14.00 - 1.78$$
$$= 12.22$$

Weak Acid–Strong Base Titrations

Consider the neutralization reaction between acetic acid (a weak acid) and sodium hydroxide (a strong base):

$$CH_3COOH(aq) + NaOH(aq) \longrightarrow CH_3COONa(aq) + H_2O(l)$$

This equation can be simplified to

$$CH_3COOH(aq) + OH^-(aq) \longrightarrow CH_3COO^-(aq) + H_2O(l)$$

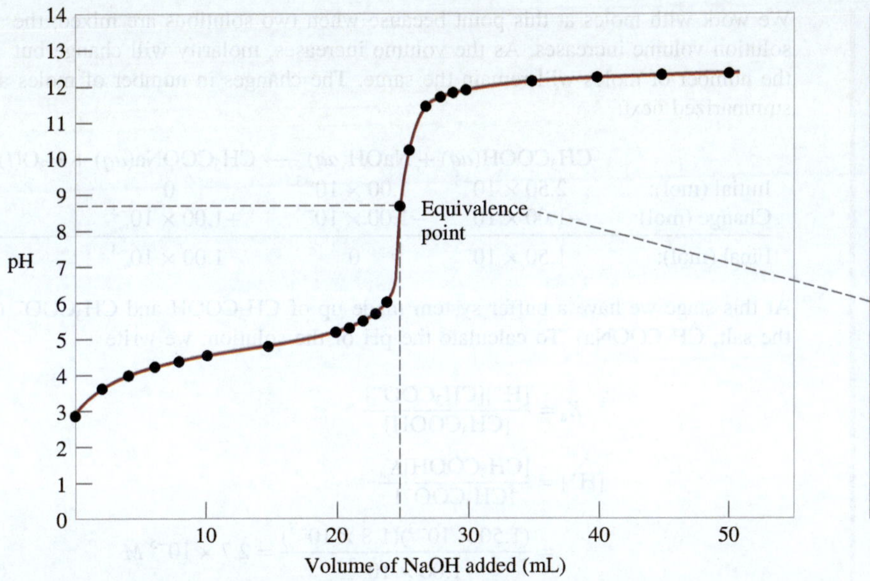

Volume NaOH added (mL)	pH
0.0	2.87
5.0	4.14
10.0	4.57
15.0	4.92
20.0	5.35
22.0	5.61
24.0	6.12
25.0	8.72
26.0	10.29
28.0	11.75
30.0	11.96
35.0	12.22
40.0	12.36
45.0	12.46
50.0	12.52

Figure 16.5 *pH profile of a weak acid–strong base titration. A 0.100 M NaOH solution is added from a buret to 25.0 mL of a 0.100 M CH_3COOH solution in an Erlenmeyer flask. Due to the hydrolysis of the salt formed, the pH at the equivalence point is greater than 7.*

The acetate ion undergoes hydrolysis as follows:

$$CH_3COO^-(aq) + H_2O(l) \rightleftharpoons CH_3COOH(aq) + OH^-(aq)$$

Therefore, at the equivalence point, when we only have sodium acetate present, the pH will be *greater than* 7 as a result of the excess OH^- ions formed (Figure 16.5). Note that this situation is analogous to the hydrolysis of sodium acetate (CH_3COONa) (see Section 15.10).

Example 16.5 deals with the titration of a weak acid with a strong base.

Student Hot Spot

Student data indicate you may struggle with weak acid–strong base titrations. Access your eBook to view additional Learning Resources on this topic.

Example 16.5

Calculate the pH in the titration of 25.0 mL of 0.100 *M* acetic acid by sodium hydroxide after the addition to the acid solution of (a) 10.0 mL of 0.100 *M* NaOH, (b) 25.0 mL of 0.100 *M* NaOH, (c) 35.0 mL of 0.100 *M* NaOH.

Strategy The reaction between CH_3COOH and NaOH is

$$CH_3COOH(aq) + NaOH(aq) \longrightarrow CH_3COONa(aq) + H_2O(l)$$

We see that 1 mol $CH_3COOH \simeq$ 1 mol NaOH. Therefore, at every stage of the titration, we can calculate the number of moles of base reacting with the acid, and the pH of the solution is determined by the excess acid or base left over. At the equivalence point, however, the neutralization is complete and the pH of the solution will depend on the extent of the hydrolysis of the salt formed, which is CH_3COONa.

Solution

(a) The number of moles of NaOH in 10.0 mL is

$$10.0\ \text{mL} \times \frac{0.100\ \text{mol NaOH}}{1\ \text{L NaOH soln}} \times \frac{1\ \text{L}}{1000\ \text{mL}} = 1.00 \times 10^{-3}\ \text{mol}$$

The number of moles of CH_3COOH originally present in 25.0 mL of solution is

$$25.0\ \text{mL} \times \frac{0.100\ \text{mol } CH_3COOH}{1\ \text{L } CH_3COOH\ \text{soln}} \times \frac{1\ \text{L}}{1000\ \text{mL}} = 2.50 \times 10^{-3}\ \text{mol}$$

(Continued)

We work with moles at this point because when two solutions are mixed, the solution volume increases. As the volume increases, molarity will change but the number of moles will remain the same. The changes in number of moles are summarized next:

	$CH_3COOH(aq)$	+ $NaOH(aq)$ ⟶	$CH_3COONa(aq) + H_2O(l)$
Initial (mol):	2.50×10^{-3}	1.00×10^{-3}	0
Change (mol):	-1.00×10^{-3}	-1.00×10^{-3}	$+1.00 \times 10^{-3}$
Final (mol):	1.50×10^{-3}	0	1.00×10^{-3}

At this stage we have a buffer system made up of CH_3COOH and CH_3COO^- (from the salt, CH_3COONa). To calculate the pH of the solution, we write

$$K_a = \frac{[H^+][CH_3COO^-]}{[CH_3COOH]}$$

$$[H^+] = \frac{[CH_3COOH]K_a}{[CH_3COO^-]}$$

$$= \frac{(1.50 \times 10^{-3})(1.8 \times 10^{-5})}{1.00 \times 10^{-3}} = 2.7 \times 10^{-5}\ M$$

Therefore, pH = −log (2.7×10^{-5}) = 4.57.

(b) These quantities (that is, 25.0 mL of 0.100 M NaOH reacting with 25.0 mL of 0.100 M CH_3COOH) correspond to the equivalence point. The number of moles of NaOH in 25.0 mL of the solution is

$$2.50\ \text{mL} \times \frac{0.100\ \text{mol NaOH}}{1\ \text{L NaOH soln}} \times \frac{1\ \text{L}}{1000\ \text{mL}} = 2.50 \times 10^{-3}\ \text{mol}$$

The changes in number of moles are summarized next:

	$CH_3COOH(aq)$	+ $NaOH(aq)$ ⟶	$CH_3COONa(aq) + H_2O(l)$
Initial (mol):	2.50×10^{-3}	2.50×10^{-3}	0
Change (mol):	-2.50×10^{-3}	-2.50×10^{-3}	$+2.50 \times 10^{-3}$
Final (mol):	0	0	2.50×10^{-3}

At the equivalence point, the concentrations of both the acid and the base are zero. The total volume is (25.0 + 25.0) mL or 50.0 mL, so the concentration of the salt is

$$[CH_3COONa] = \frac{2.50 \times 10^{-3}\ \text{mol}}{50.0\ \text{mL}} \times \frac{1000\ \text{mL}}{1\ \text{L}}$$
$$= 0.0500\ \text{mol/L} = 0.0500\ M$$

The next step is to calculate the pH of the solution that results from the hydrolysis of the CH_3COO^- ions. Following the procedure described in Example 15.13 and looking up the base ionization constant (K_b) for CH_3COO^- in Table 15.3, we write

$$K_b = 5.6 \times 10^{-10} = \frac{[CH_3COOH][OH^-]}{[CH_3COO^-]} = \frac{x^2}{0.0500 - x}$$
$$x = [OH^-] = 5.3 \times 10^{-6}\ M,\ \text{pH} = 8.72$$

(c) After the addition of 35.0 mL of NaOH, the solution is well past the equivalence point. The number of moles of NaOH originally present is

$$35.0\ \text{mL} \times \frac{0.100\ \text{mol NaOH}}{1\ \text{L NaOH soln}} \times \frac{1\ \text{L}}{1000\ \text{mL}} = 3.50 \times 10^{-3}\ \text{mol}$$

The changes in number of moles are summarized next:

	$CH_3COOH(aq)$	+ $NaOH(aq)$ ⟶	$CH_3COONa(aq) + H_2O(l)$
Initial (mol):	2.50×10^{-3}	3.50×10^{-3}	0
Change (mol):	-2.50×10^{-3}	-2.50×10^{-3}	$+2.50 \times 10^{-3}$
Final (mol):	0	1.00×10^{-3}	2.50×10^{-3}

(Continued)

At this stage we have two species in solution that are responsible for making the solution basic: OH^- and CH_3COO^- (from CH_3COONa). However, because OH^- is a much stronger base than CH_3COO^-, we can safely neglect the hydrolysis of the CH_3COO^- ions and calculate the pH of the solution using only the concentration of the OH^- ions. The total volume of the combined solutions is (25.0 + 35.0) mL or 60.0 mL, so we calculate OH^- concentration as follows:

$$[OH^-] = \frac{1.00 \times 10^{-3}\ \text{mol}}{60.0\ \text{mL}} \times \frac{1000\ \text{mL}}{1\ \text{L}}$$
$$= 0.0167\ \text{mol/L} = 0.0167\,M$$
$$\text{pOH} = -\log[OH^-] = -\log 0.0167 = 1.78$$
$$\text{pH} = 14.00 - 1.78 = 12.22$$

Practice Exercise Exactly 100 mL of 0.10 *M* nitrous acid (HNO_2) are titrated with a 0.10 *M* NaOH solution. Calculate the pH for (a) the initial solution, (b) the point at which 80 mL of the base has been added, (c) the equivalence point, (d) the point at which 105 mL of the base has been added.

Similar problem: 16.35.

Strong Acid–Weak Base Titrations

Consider the titration of HCl, a strong acid, with NH_3, a weak base:

$$HCl(aq) + NH_3(aq) \longrightarrow NH_4Cl(aq)$$

or simply

$$H^+(aq) + NH_3(aq) \longrightarrow NH_4^+(aq)$$

The pH at the equivalence point is *less than* 7 due to the hydrolysis of the NH_4^+ ion:

$$NH_4^+(aq) + H_2O(l) \rightleftharpoons NH_3(aq) + H_3O^+(aq)$$

or simply

$$NH_4^+(aq) \rightleftharpoons NH_3(aq) + H^+(aq)$$

Because of the volatility of an aqueous ammonia solution, it is more convenient to add hydrochloric acid from a buret to the ammonia solution. Figure 16.6 shows the titration curve for this experiment.

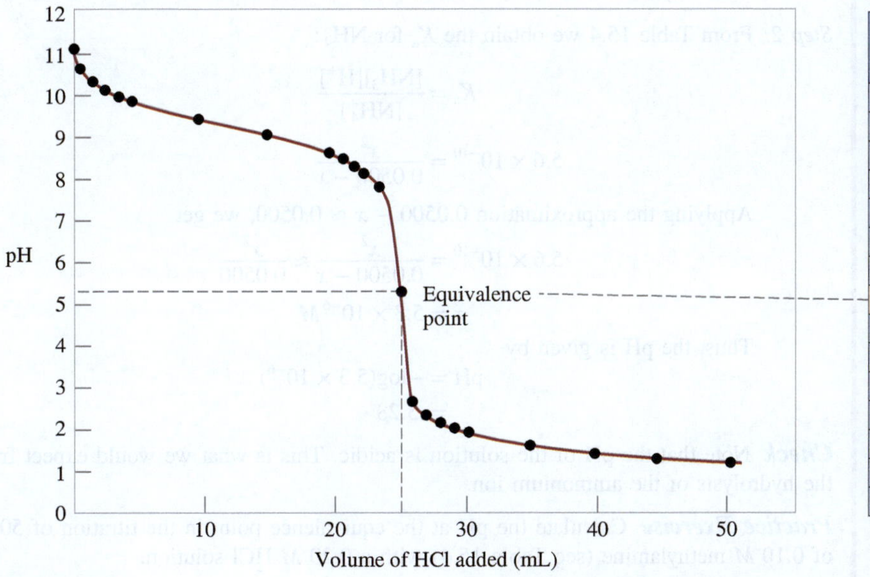

Volume HCl added (mL)	pH
0.0	11.13
5.0	9.86
10.0	9.44
15.0	9.08
20.0	8.66
22.0	8.39
24.0	7.88
25.0	5.28
26.0	2.70
28.0	2.22
30.0	2.00
35.0	1.70
40.0	1.52
45.0	1.40
50.0	1.30

Figure 16.6 *pH profile of a strong acid–weak base titration. A 0.100 M HCl solution is added from a buret to 25.0 mL of a 0.100 M NH_3 solution in an Erlenmeyer flask. As a result of salt hydrolysis, the pH at the equivalence point is lower than 7.*

Example 16.6

Calculate the pH at the equivalence point when 25.0 mL of 0.100 *M* NH_3 is titrated by a 0.100 *M* HCl solution.

Strategy The reaction between NH_3 and HCl is

$$NH_3(aq) + HCl(aq) \longrightarrow NH_4Cl(aq)$$

We see that 1 mol NH_3 ≏ 1 mol HCl. At the equivalence point, the major species in solution are the salt NH_4Cl (dissociated into NH_4^+ and Cl^- ions) and H_2O. First, we determine the concentration of NH_4Cl formed. Then we calculate the pH as a result of the NH_4^+ ion hydrolysis. The Cl^- ion, being the conjugate base of a strong acid HCl, does not react with water. As usual, we ignore the ionization of water.

Solution The number of moles of NH_3 in 25.0 mL of 0.100 *M* solution is

$$25.0 \text{ mL} \times \frac{0.100 \text{ mol } NH_3}{1 \text{ L } NH_3} \times \frac{1 \text{ L}}{1000 \text{ mL}} = 2.50 \times 10^{-3} \text{ mol}$$

At the equivalence point the number of moles of HCl added equals the number of moles of NH_3. The changes in number of moles are summarized as follows:

	$NH_3(aq)$	+	$HCl(aq)$	$\longrightarrow$	$NH_4Cl(aq)$
Initial (mol):	2.50×10^{-3}		2.50×10^{-3}		0
Change (mol):	-2.50×10^{-3}		-2.50×10^{-3}		$+2.50 \times 10^{-3}$
Final (mol):	0		0		2.50×10^{-3}

At the equivalence point, the concentrations of both the acid and the base are zero. The total volume is (25.0 + 25.0) mL, or 50.0 mL, so the concentration of the salt is

$$[NH_4Cl] = \frac{2.50 \times 10^{-3} \text{ mol}}{50.0 \text{ mL}} \times \frac{1000 \text{ mL}}{1 \text{ L}}$$
$$= 0.0500 \text{ mol/L} = 0.0500\, M$$

The pH of the solution at the equivalence point is determined by the hydrolysis of NH_4^+ ions. We follow the procedure in Section 15.10.

Step 1: We represent the hydrolysis of the cation NH_4^+, and let x be the equilibrium concentration of NH_3 and H^+ ions in mol/L:

	$NH_4^+(aq)$	$\rightleftharpoons$	$NH_3(aq)$	+	$H^+(aq)$
Initial (*M*):	0.0500		0.000		0.000
Change (*M*):	$-x$		$+x$		$+x$
Equilibrium (*M*):	$(0.0500 - x)$		x		x

Step 2: From Table 15.4 we obtain the K_a for NH_4^+:

$$K_a = \frac{[NH_3][H^+]}{[NH_4^+]}$$

$$5.6 \times 10^{-10} = \frac{x^2}{0.0500 - x}$$

Applying the approximation $0.0500 - x \approx 0.0500$, we get

$$5.6 \times 10^{-10} = \frac{x^2}{0.0500 - x} \approx \frac{x^2}{0.0500}$$
$$x = 5.3 \times 10^{-6}\, M$$

Thus, the pH is given by

$$pH = -\log(5.3 \times 10^{-6})$$
$$= 5.28$$

Check Note that the pH of the solution is acidic. This is what we would expect from the hydrolysis of the ammonium ion.

Practice Exercise Calculate the pH at the equivalence point in the titration of 50 mL of 0.10 *M* methylamine (see Table 15.4) with a 0.20 *M* HCl solution.

Similar problem: 16.33.

Summary of Concepts & Facts

- The pH at the equivalence point of an acid-base titration depends on hydrolysis of the salt formed in the neutralization reaction. For strong acid–strong base titrations, the pH at the equivalence point is 7; for weak acid–strong base titrations, the pH at the equivalence point is greater than 7; for strong acid–weak base titrations, the pH at the equivalence point is less than 7.

Review of Concepts & Facts

16.3.1 For which of the following titrations will the pH at the equivalence point not be neutral: (a) HNO_2 by NaOH, (b) KOH by $HClO_4$, (c) HCOOH by KOH, (d) CH_3NH_2 by HNO_3?

16.3.2 Calculate the pH at the equivalence point in the titration of 50 mL of 0.15 *M* ammonia with a 0.25 *M* HNO_3 solution.

16.3.3 Calculate the pH in the titration of 50.0 mL of 0.100 *M* hydrofluoric acid (HF) by potassium hydroxide after the addition to the acid solution of 35.0 mL of 0.100 *M* KOH.

16.4 Acid-Base Indicators

Learning Objectives

- Predict the pH of a titration given the type of acid and base used.
- Evaluate the pH throughout a titration between acids and bases (strong titrated with strong; strong titrated with weak).
- Examine a titration curve to identify what type of titration was conducted (e.g., weak acid titrated with strong base).

The equivalence point, as we have seen, is the point at which the number of moles of OH^- ions added to a solution is equal to the number of moles of H^+ ions originally present. To determine the equivalence point in a titration, then, we must know exactly how much volume of a base to add from a buret to an acid in a flask. One way to achieve this goal is to add a few drops of an acid-base indicator to the acid solution at the start of the titration. You will recall from Chapter 4 that an indicator is usually a weak organic acid or base that has distinctly different colors in its nonionized and ionized forms. These two forms are related to the pH of the solution in which the indicator is dissolved. The ***end point*** of a titration *occurs when the indicator changes color*. However, not all indicators change color at the same pH, so the choice of indicator for a particular titration depends on the nature of the acid and base used in the titration (that is, whether they are strong or weak). By choosing the proper indicator for a titration, we can use the end point to determine the equivalence point.

Let us consider a weak monoprotic acid that we will call HIn. To be an effective indicator, HIn and its conjugate base, In^-, must have distinctly different colors. In solution, the acid ionizes to a small extent:

$$HIn(aq) \rightleftharpoons H^+(aq) + In^-(aq)$$

If the indicator is in a sufficiently acidic medium, the equilibrium, according to Le Châtelier's principle, shifts to the left and the predominant color of the indicator is that of the nonionized form (HIn). On the other hand, in a basic medium the equilibrium shifts to the right and the color of the solution will be due mainly to that of the conjugate base (In^-). Roughly speaking, we can use the following concentration ratios to predict the perceived color of the indicator:

$$\frac{[HIn]}{[In^-]} \geq 10 \quad \text{color of acid (HIn) predominates}$$

$$\frac{[HIn]}{[In^-]} \leq 0.1 \quad \text{color of conjugate base (In}^-\text{) predominates}$$

If $[HIn] \approx [In^-]$, then the indicator color is a combination of the colors of HIn and In^-.

The end point of an indicator does not occur at a specific pH; rather, there is a range of pH within which the end point will occur. Typical indicators change color over the pH range given by $pH = pK_a \pm 1$, where K_a is the acid ionization constant of the indicator. In practice, we choose an indicator whose end point lies on the steep part of the titration curve. Because the equivalence point also lies on the steep part of the curve, this choice ensures that the pH at the equivalence point will fall within the range over which the indicator changes color. In Section 4.7 we mentioned that phenolphthalein is a suitable indicator for the titration of NaOH and HCl. Phenolphthalein is colorless in acidic and neutral solutions, but reddish pink in basic solutions. Measurements show that at pH less than 8.3 the indicator is colorless but that it begins to turn reddish pink when the pH exceeds 8.3. As shown in Figure 16.4, the steepness of the pH curve near the equivalence point means that the addition of a very small quantity of NaOH (say, 0.05 mL, which is about the volume of a drop from the buret) brings about a large rise in the pH of the solution. What is important, however, is the fact that the steep portion of the pH profile includes the range over which phenolphthalein changes from colorless to reddish pink. Whenever such a correspondence occurs, the indicator can be used to locate the equivalence point of the titration (Figure 16.7).

Many acid-base indicators are plant pigments. For example, by boiling chopped red cabbage in water we can extract pigments that exhibit many different colors at various pHs (Figure 16.8). Table 16.1 lists a number of indicators commonly used in acid-base titrations. The choice of a particular indicator depends on the strength of the acid and base to be titrated. Example 16.7 illustrates this point.

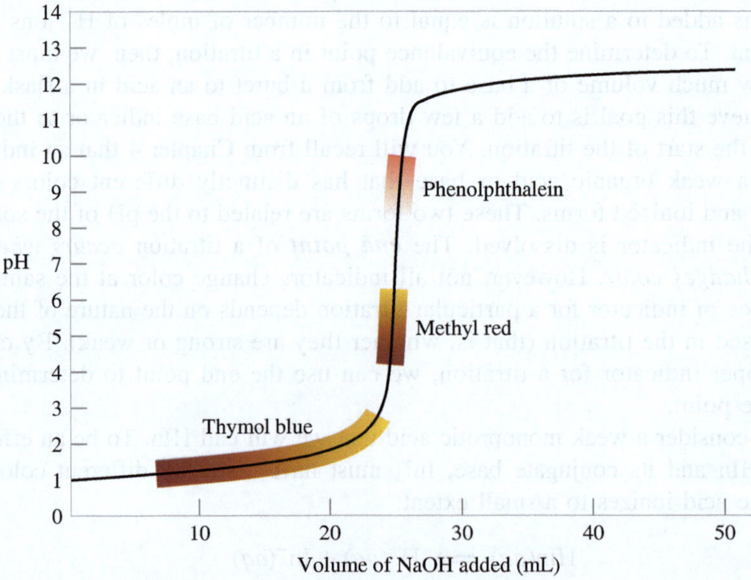

Figure 16.7 *The titration curve of a strong acid with a strong base. Because the regions over which the indicators methyl red and phenolphthalein change color occur along the steep portion of the curve, they can be used to monitor the equivalence point of the titration. Thymol blue, on the other hand, cannot be used for the same purpose because the color change does not match the steep portion of the titration curve (see Table 16.1).*

Figure 16.8 *Solutions containing extracts of red cabbage (obtained by boiling the cabbage in water) produce different colors when treated with an acid and a base. The pH of the solutions increases from left to right.*
Ken Karp/McGraw Hill

Example 16.7

Which indicator(s) listed in Table 16.1 would you use for the acid-base titrations shown in (a) Figure 16.4, (b) Figure 16.5, and (c) Figure 16.6?

Strategy The choice of an indicator for a particular titration is based on the fact that its pH range for color change must overlap the steep portion of the titration curve. Otherwise we cannot use the color change to locate the equivalence point.

Solution

(a) Near the equivalence point, the pH of the solution changes abruptly from 4 to 10. Therefore, all the indicators except thymol blue, bromophenol blue, and methyl orange are suitable for use in the titration.

(b) Here the steep portion covers the pH range between 7 and 10; therefore, the suitable indicators are cresol red and phenolphthalein.

(c) Here the steep portion of the pH curve covers the pH range between 3 and 7; therefore, the suitable indicators are bromophenol blue, methyl orange, methyl red, and chlorophenol blue.

Practice Exercise Referring to Table 16.1, specify which indicator(s) you would use for the following titrations: (a) HBr versus CH_3NH_2, (b) HNO_3 versus NaOH, (c) HNO_2 versus KOH.

Similar problem: 16.43.

Table 16.1 Some Common Acid-Base Indicators

	Color		
Indicator	**In Acid**	**In Base**	**pH Range***
Thymol blue	Red	Yellow	1.2–2.8
Bromophenol blue	Yellow	Bluish purple	3.0–4.6
Methyl orange	Orange	Yellow	3.1–4.4
Methyl red	Red	Yellow	4.2–6.2
Chlorophenol blue	Yellow	Red	4.8–6.4
Bromothymol blue	Yellow	Blue	6.0–7.6
Cresol red	Yellow	Red	7.2–8.8
Phenolphthalein	Colorless	Reddish pink	8.0–10.0

*The pH range is defined as the range over which the indicator changes from the acid color to the base color.

Summary of Concepts & Facts

- Predict the pH of a titration given the type of acid and base used.
- Evaluate the pH throughout a titration between acids and bases (strong titrated with strong; strong titrated with weak).
- Examine a titration curve to identify what type of titration was conducted (e.g., weak acid titrated with strong base).

Review of Concepts & Facts

16.4.1 Under what conditions will the end point of an acid-base titration accurately represent the equivalence point?

16.5 Solubility Equilibria

Learning Objectives

- Predict the pH of a titration given the type of acid and base used.
- Evaluate the pH throughout a titration between acids and bases (strong titrated with strong; strong titrated with weak).
- Examine a titration curve to identify what type of titration was conducted (e.g., weak acid titrated with strong base).

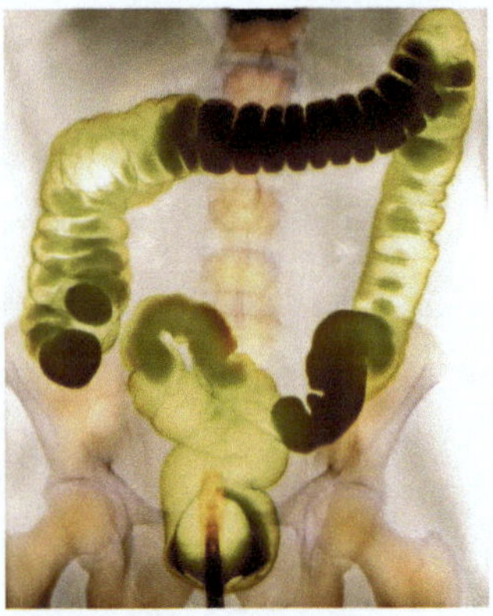

$BaSO_4$ imaging of human large intestine.
MedImage/Science Source

Precipitation reactions are important in industry, medicine, and everyday life. For example, the preparation of many essential industrial chemicals such as sodium carbonate (Na_2CO_3) is based on precipitation reactions. The dissolving of tooth enamel, which is mainly made of hydroxyapatite [$Ca_5(PO_4)_3OH$], in an acidic medium leads to tooth decay. Barium sulfate ($BaSO_4$), an insoluble compound that is opaque to X rays, is used to diagnose ailments of the digestive tract. Stalactites and stalagmites, which consist of calcium carbonate ($CaCO_3$), are produced by a precipitation reaction, and so are many foods, such as fudge.

The general rules for predicting the solubility of ionic compounds in water were introduced in Section 4.2. Although useful, these solubility rules do not enable us to make quantitative predictions about how much of a given ionic compound will dissolve in water. To develop a quantitative approach, we start with what we already know about chemical equilibrium. Unless otherwise stated, in the following discussion the solvent is water and the temperature is 25°C.

Solubility Product

Consider a saturated solution of silver chloride that is in contact with solid silver chloride. This is an example of a heterogeneous equilibrium in which the components are in more than one phase. The solubility equilibrium can be represented as

$$AgCl(s) \rightleftharpoons Ag^+(aq) + Cl^-(aq)$$

Silver chloride is an insoluble salt (see Table 4.2). The small amount of solid AgCl that dissolves in water is assumed to dissociate completely into Ag^+ and Cl^- ions. We know from Chapter 14 that for heterogeneous reactions the concentration of the solid is a constant (recall that the activity of a solid is one). Thus, we can write the equilibrium constant for the dissolution of AgCl (see Example 14.5) as

$$K_{sp} = [Ag^+][Cl^-]$$

where K_{sp} is called the solubility product constant or simply the *solubility product*. In general, the ***solubility product*** of a compound is *the product of the molar concentrations of the constituent ions, each raised to the power of its stoichiometric coefficient in the equilibrium equation.*

Because each AgCl unit contains only one Ag^+ ion and one Cl^- ion, its solubility product expression is particularly simple to write. The following cases are more complex:

- MgF_2:

$$MgF_2(s) \rightleftharpoons Mg^{2+}(aq) + 2F^-(aq) \qquad K_{sp} = [Mg^{2+}][F^-]^2$$

- Ag_2CO_3:

$$Ag_2CO_3(s) \rightleftharpoons 2Ag^+(aq) + CO_3^{2-}(aq) \qquad K_{sp} = [Ag^+]^2[CO_3^{2-}]$$

- $Ca_3(PO_4)_2$:

$$Ca_3(PO_4)_2(s) \rightleftharpoons 3Ca^{2+}(aq) + 2PO_4^{3-}(aq) \qquad K_{sp} = [Ca^{2+}]^3[PO_4^{3-}]^2$$

Table 16.2 lists the solubility products for a number of salts of low solubility. Soluble salts such as NaCl and KNO_3, which have very large K_{sp} values, are not listed in the table for essentially the same reason we did not include K_a values for strong acids in Table 15.3. The value of K_{sp} indicates the solubility of an ionic compound—the smaller the value, the less soluble the compound in water. However, in using K_{sp} values to compare solubilities, you should choose compounds that have similar formulas, such as AgCl and ZnS, or CaF_2 and $Fe(OH)_2$.

A cautionary note: In Section 15.3 we assumed that dissolved substances exhibit ideal behavior for our calculations involving solution concentrations, but this assumption is not always valid. For example, a solution of barium fluoride (BaF_2) may contain both neutral and charged ion pairs, such as BaF_2 and BaF^+, in addition to free Ba^{2+}

Table 16.2 Solubility Products of Some Slightly Soluble Ionic Compounds at 25°C

Compound	K_{sp}	Compound	K_{sp}
Aluminum hydroxide [$Al(OH)_3$]	1.8×10^{-33}	Lead(II) chromate ($PbCrO_4$)	2.0×10^{-14}
Barium carbonate ($BaCO_3$)	2.6×10^{-9}	Lead(II) fluoride (PbF_2)	3.3×10^{-8}
Barium fluoride (BaF_2)	1.8×10^{-6}	Lead(II) iodide (PbI_2)	9.8×10^{-9}
Barium sulfate ($BaSO_4$)	1.1×10^{-10}	Lead(II) sulfide (PbS)	3.4×10^{-28}
Bismuth sulfide (Bi_2S_3)	1.6×10^{-72}	Magnesium carbonate ($MgCO_3$)	6.8×10^{-6}
Cadmium sulfide (CdS)	8.0×10^{-28}	Magnesium hydroxide [$Mg(OH)_2$]	5.6×10^{-12}
Calcium carbonate ($CaCO_3$)	3.4×10^{-9}	Manganese(II) sulfide (MnS)	3.0×10^{-14}
Calcium fluoride (CaF_2)	3.5×10^{-11}	Mercury(I) chloride (Hg_2Cl_2)	1.4×10^{-18}
Calcium hydroxide [$Ca(OH)_2$]	5.0×10^{-6}	Mercury(II) sulfide (HgS)	4.0×10^{-54}
Calcium phosphate [$Ca_3(PO_4)_2$]	2.1×10^{-33}	Nickel(II) sulfide (NiS)	1.4×10^{-24}
Chromium(III) hydroxide [$Cr(OH)_3$]	3.0×10^{-29}	Silver bromide (AgBr)	5.4×10^{-13}
Cobalt(II) sulfide (CoS)	4.0×10^{-21}	Silver carbonate (Ag_2CO_3)	8.5×10^{-12}
Copper(I) bromide (CuBr)	6.3×10^{-9}	Silver chloride (AgCl)	1.8×10^{-10}
Copper(I) iodide (CuI)	1.3×10^{-12}	Silver iodide (AgI)	8.5×10^{-17}
Copper(II) hydroxide [$Cu(OH)_2$]	2.2×10^{-20}	Silver sulfate (Ag_2SO_4)	1.2×10^{-5}
Copper(II) sulfide (CuS)	6.0×10^{-37}	Silver sulfide (Ag_2S)	6.0×10^{-51}
Iron(II) hydroxide [$Fe(OH)_2$]	4.9×10^{-17}	Strontium carbonate ($SrCO_3$)	5.6×10^{-10}
Iron(III) hydroxide [$Fe(OH)_3$]	2.8×10^{-39}	Strontium sulfate ($SrSO_4$)	3.4×10^{-7}
Iron(II) sulfide (FeS)	6.0×10^{-19}	Tin(II) sulfide (SnS)	1.0×10^{-26}
Lead(II) carbonate ($PbCO_3$)	7.4×10^{-14}	Zinc hydroxide [$Zn(OH)_2$]	3.0×10^{-17}
Lead(II) chloride ($PbCl_2$)	1.7×10^{-5}	Zinc sulfide (ZnS)	3.0×10^{-23}

and F^- ions. Furthermore, many anions in the ionic compounds listed in Table 16.2 are conjugate bases of weak acids. Consider copper sulfide (CuS). The S^{2-} ion can hydrolyze as follows:

$$S^{2-}(aq) + H_2O(l) \rightleftharpoons HS^-(aq) + OH^-(aq)$$
$$HS^-(aq) + H_2O(l) \rightleftharpoons H_2S(aq) + OH^-(aq)$$

Highly charged small metal ions such as Al^{3+} and Bi^{3+} will undergo hydrolysis, as discussed in Section 15.10. Both ion-pair formation and salt hydrolysis decrease the concentrations of the ions that appear in the K_{sp} expression, but we need not be concerned with the deviations from ideal behavior here.

For the dissolution of an ionic solid in aqueous solution, any one of the following conditions may exist: (1) The solution is unsaturated, (2) the solution is saturated, or (3) the solution is supersaturated. For concentrations of ions that do not correspond to equilibrium conditions we use the reaction quotient (see Section 14.4), which in this case is called the *ion product (Q)*, to predict whether a precipitate will form. Note that Q has the same form as K_{sp} except that the concentrations of ions are *not* equilibrium concentrations. For example, if we mix a solution containing Ag^+ ions with one containing Cl^- ions, then the ion product is given by

$$Q = [Ag^+]_0[Cl^-]_0$$

The subscript 0 reminds us that these are initial concentrations and do not necessarily correspond to those at equilibrium. The possible relationships between Q and K_{sp} are

$$Q < K_{sp}$$
$$[Ag^+]_0[Cl^-]_0 < 1.8 \times 10^{-10}$$
Unsaturated solution (no precipitation)

$$Q = K_{sp}$$
$$[Ag^+][Cl^-] = 1.8 \times 10^{-10}$$
Saturated solution (no precipitation)

$$Q > K_{sp}$$
$$[Ag^+]_0[Cl^-]_0 > 1.8 \times 10^{-10}$$
Supersaturated solution; AgCl will precipitate out until the product of the ionic concentrations is equal to 1.8×10^{-10}

Molar Solubility and Solubility

There are two other ways to express a substance's solubility: ***molar solubility,*** which is *the number of moles of solute in 1 L of a saturated solution (mol/L)*, and ***solubility,*** which described here is *the number of grams of solute in 1 L of a saturated solution (g/L)*. Note that both these expressions refer to the concentration of saturated solutions at some given temperature (usually 25°C).

Both molar solubility and solubility are convenient to use in the laboratory. We can use them to determine K_{sp} by following the steps outlined in Figure 16.9(a). Example 16.8 illustrates this procedure.

Student Hot Spot

Student data indicate you may struggle with determining K_{sp}. Access your eBook for additional Learning Resources on this topic.

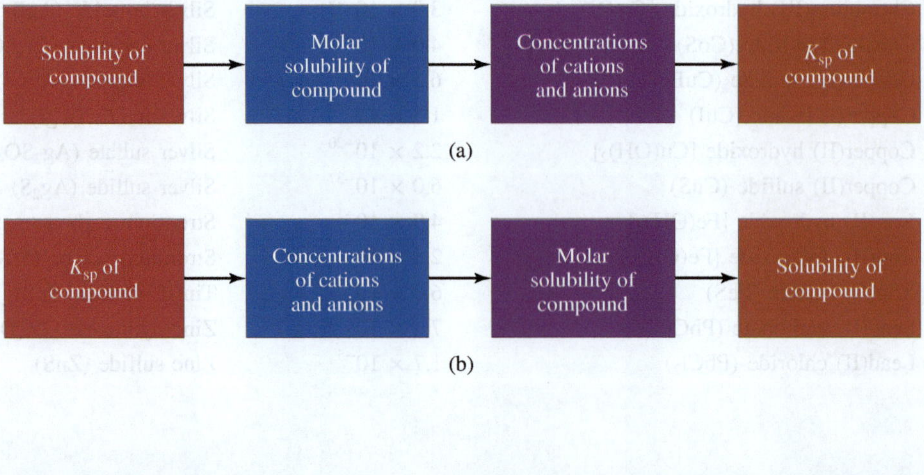

Figure 16.9 *Sequence of steps (a) for calculating K_{sp} from solubility data and (b) for calculating solubility from K_{sp} data.*

Example 16.8

The solubility of calcium sulfate ($CaSO_4$) is found to be 0.67 g/L. Calculate the value of K_{sp} for calcium sulfate.

Strategy We are given the solubility of $CaSO_4$ and asked to calculate its K_{sp}. The sequence of conversion steps, according to Figure 16.9(a), is

$$\begin{array}{c} \text{solubility of} \\ CaSO_4 \text{ in g/L} \end{array} \longrightarrow \begin{array}{c} \text{molar solubility} \\ \text{of } CaSO_4 \end{array} \longrightarrow \begin{array}{c} [Ca^{2+}] \text{ and} \\ [SO_4^{2-}] \end{array} \longrightarrow \begin{array}{c} K_{sp} \text{ of} \\ CaSO_4 \end{array}$$

Solution Consider the dissociation of $CaSO_4$ in water. Let s be the molar solubility (in mol/L) of $CaSO_4$.

	$CaSO_4(s)$ $\rightleftharpoons$	$Ca^{2+}(aq)$ +	$SO_4^{2-}(aq)$
Initial (M):		0	0
Change (M):	$-s$	$+s$	$+s$
Equilibrium (M):		s	s

The solubility product for $CaSO_4$ is

$$K_{sp} = [Ca^{2+}][SO_4^{2-}] = s^2$$

First, we calculate the number of moles of $CaSO_4$ dissolved in 1 L of solution:

$$\frac{0.95 \text{ g } CaSO_4}{1 \text{ L soln}} \times \frac{1 \text{ mol } CaSO_4}{136.2 \text{ g } CaSO_4} = 7.0 \times 10^{-3} \text{ mol/L} = s$$

From the solubility equilibrium we see that for every mole of $CaSO_4$ that dissolves, 1 mole of Ca^{2+} and 1 mole of SO_4^{2-} are produced. Thus, at equilibrium,

$$[Ca^{2+}] = 7.0 \times 10^{-3}\, M \quad \text{and} \quad [SO_4^{2-}] = 7.0 \times 10^{-3}\, M$$

Now we can calculate K_{sp}:

$$\begin{aligned} K_{sp} &= [Ca^{2+}][SO_4^{2-}] \\ &= (7.0 \times 10^{-3})(7.0 \times 10^{-3}) \\ &= 4.9 \times 10^{-5} \end{aligned}$$

Practice Exercise The solubility of lead chromate ($PbCrO_4$) is 4.5×10^{-5} g/L. Calculate the solubility product of this compound.

Similar problem: 16.56.

Calcium sulfate is used as a drying agent and in the manufacture of paints, ceramics, and paper. A hydrated form of calcium sulfate, called plaster of Paris, is used to make casts for broken bones.
Ken Karp/McGraw Hill

Sometimes we are given the value of K_{sp} for a compound and asked to calculate the compound's molar solubility. For example, the K_{sp} of silver bromide (AgBr) is 5.4×10^{-13}. We can calculate its molar solubility by the same procedure as that for acid ionization constants. First we identify the species present at equilibrium. Here we have Ag^+ and Br^- ions. Let s be the molar solubility (in mol/L) of AgBr. Because one unit of AgBr yields one Ag^+ and one Br^- ion, at equilibrium both $[Ag^+]$ and $[Br^-]$ are equal to s. We summarize the changes in concentrations as follows:

	$AgBr(s)$ $\rightleftharpoons$	$Ag^+(aq)$ +	$Br^-(aq)$
Initial (M):		0.00	0.00
Change (M):	$-s$	$+s$	$+s$
Equilibrium (M):		s	s

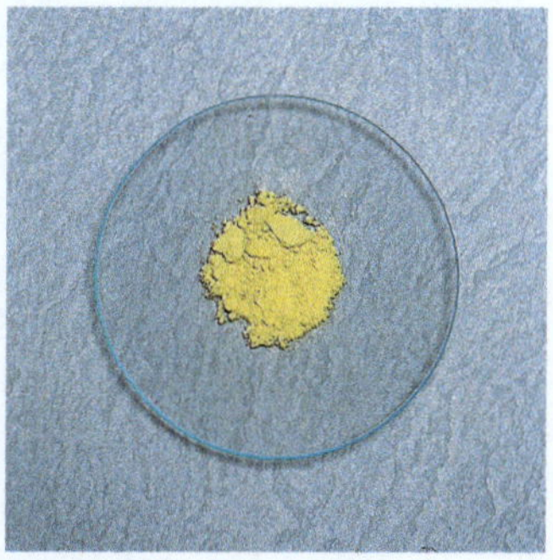

Silver bromide is used in photographic emulsions.
Ken Karp/McGraw Hill

From Table 16.2 we write

$$K_{sp} = [Ag^+][Br^-]$$
$$5.4 \times 10^{-13} = (s)(s)$$
$$s = \sqrt{5.4 \times 10^{-13}} = 7.3 \times 10^{-7}\,M$$

Therefore, at equilibrium

$$[Ag^+] = 7.3 \times 10^{-7}\,M$$
$$[Br^-] = 7.3 \times 10^{-7}\,M$$

Thus, the molar solubility of AgBr also is $7.3 \times 10^{-7}\,M$.

Example 16.9 makes use of this approach.

Student Hot Spot

Student data indicate you may struggle with determining the solubility of a salt. Access your eBook to view additional Learning Resources on this topic.

Copper(II) hydroxide is used as a pesticide and to treat seeds.
Ken Karp/McGraw Hill

Example 16.9

Using the data in Table 16.2, calculate the solubility of copper(II) hydroxide, $Cu(OH)_2$, in g/L.

Strategy We are given the K_{sp} of $Cu(OH)_2$ and asked to calculate its solubility in g/L. The sequence of conversion steps, according to Figure 16.9(b), is

$$\begin{matrix} K_{sp} \text{ of} \\ Cu(OH)_2 \end{matrix} \longrightarrow \begin{matrix} [Cu^{2+}] \text{ and} \\ [OH^-] \end{matrix} \longrightarrow \begin{matrix} \text{molar solubility} \\ \text{of } Cu(OH)_2 \end{matrix} \longrightarrow \begin{matrix} \text{solubility of} \\ Cu(OH)_2 \text{ in g/L} \end{matrix}$$

Solution Consider the dissociation of $Cu(OH)_2$ in water:

	$Cu(OH)_2(s)$	$\rightleftharpoons$	$Cu^{2+}(aq)$	+	$2OH^-(aq)$
Initial (M):			0		0
Change (M):	$-s$		$+s$		$+2s$
Equilibrium (M):			s		$2s$

Note that the molar concentration of OH^- is twice that of Cu^{2+}. The solubility product of $Cu(OH)_2$ is

$$K_{sp} = [Cu^{2+}][OH^-]^2$$
$$= (s)(2s)^2 = 4s^3$$

From the K_{sp} value in Table 16.2, we solve for the molar solubility of $Cu(OH)_2$ as follows:

$$2.2 \times 10^{-20} = 4s^3$$
$$s^3 = \frac{2.2 \times 10^{-20}}{4} = 5.5 \times 10^{-21}$$

Hence, $s = 1.8 \times 10^{-7}\,M$

Finally, from the molar mass of $Cu(OH)_2$ and its molar solubility, we calculate the solubility in g/L:

$$\text{solubility of } Cu(OH)_2 = \frac{1.8 \times 10^{-7} \text{ mol } Cu(OH)_2}{1 \text{ L soln}} \times \frac{97.57 \text{ g } Cu(OH)_2}{1 \text{ mol } Cu(OH)_2}$$
$$= 1.8 \times 10^{-5} \text{ g/L}$$

Practice Exercise Calculate the solubility of silver chloride (AgCl) in g/L.

Similar problem: 16.58.

Table 16.3 Relationship Between K_{sp} and Molar Solubility (s)

Compound	K_{sp} Expression	Cation	Anion	Relation Between K_{sp} and s
AgCl	$[Ag^+][Cl^-]$	s	s	$K_{sp} = s^2;\ s = (K_{sp})^{\frac{1}{2}}$
$BaSO_4$	$[Ba^{2+}][SO_4^{2-}]$	s	s	$K_{sp} = s^2;\ s = (K_{sp})^{\frac{1}{2}}$
Ag_2CO_3	$[Ag^+]^2[CO_3^{2-}]$	$2s$	s	$K_{sp} = 4s^3;\ s = \left(\frac{K_{sp}}{4}\right)^{\frac{1}{3}}$
PbF_2	$[Pb^{2+}][F^-]^2$	s	$2s$	$K_{sp} = 4s^3;\ s = \left(\frac{K_{sp}}{4}\right)^{\frac{1}{3}}$
$Al(OH)_3$	$[Al^{3+}][OH^-]^3$	s	$3s$	$K_{sp} = 27s^4;\ s = \left(\frac{K_{sp}}{27}\right)^{\frac{1}{4}}$
$Ca_3(PO_4)_2$	$[Ca^{2+}]^3[PO_4^{3-}]^2$	$3s$	$2s$	$K_{sp} = 108s^5;\ s = \left(\frac{K_{sp}}{108}\right)^{\frac{1}{5}}$

As Examples 16.8 and 16.9 show, solubility and solubility product are related. If we know one, we can calculate the other, but each quantity provides different information. Table 16.3 shows the relationship between molar solubility and solubility product for a number of ionic compounds.

When carrying out solubility and/or solubility product calculations, keep in mind the following important points:

1. Solubility is the quantity of a substance that dissolves in a certain quantity of water to produce a saturated solution. In solubility equilibria calculations, it is usually expressed as *grams* of solute per liter of solution. Molar solubility is the number of *moles* of solute per liter of solution.
2. Solubility product is an equilibrium constant.
3. Molar solubility, solubility, and solubility product all refer to a *saturated solution.*

Predicting Precipitation Reactions

From a knowledge of the solubility rules (see Section 4.2) and the solubility products listed in Table 16.2, we can predict whether a precipitate will form when we mix two solutions or add a soluble compound to a solution. This ability often has practical value. In industrial and laboratory preparations, we can adjust the concentrations of ions until the ion product exceeds K_{sp} to obtain a given compound (in the form of a precipitate). The ability to predict precipitation reactions is also useful in medicine. For example, kidney stones, which can be extremely painful, consist largely of calcium oxalate, CaC_2O_4 ($K_{sp} = 2.3 \times 10^{-9}$). The normal physiological concentration of calcium ions in blood plasma is about 5 mM (1 mM = 1×10^{-3} M). Oxalate ions ($C_2O_4^{2-}$), derived from oxalic acid present in many vegetables such as rhubarb and spinach, react with the calcium ions to form insoluble calcium oxalate, which can gradually build up in the kidneys. Proper adjustment of a patient's diet can help to reduce precipitate formation. Example 16.10 illustrates the steps involved in predicting precipitation reactions.

A kidney stone.
remik44992/Shutterstock

Example 16.10

Exactly 200 mL of 0.0040 M $BaCl_2$ are mixed with exactly 600 mL of 0.0080 M K_2SO_4. Will a precipitate form?

(Continued)

Strategy Under what condition will an ionic compound precipitate from solution? The ions in solution are Ba^{2+}, Cl^-, K^+, and SO_4^{2-}. According to the solubility rules listed in Table 4.2, the only precipitate that can form is $BaSO_4$. From the information given, we can calculate $[Ba^{2+}]$ and $[SO_4^{2-}]$ because we know the number of moles of the ions in the original solutions and the volume of the combined solution. Next, we calculate the ion product Q ($Q = [Ba^{2+}]_0[SO_4^{2-}]_0$) and compare the value of Q with K_{sp} of $BaSO_4$ to see if a precipitate will form—that is, if the solution is supersaturated.

Solution The number of moles of Ba^{2+} present in the original 200 mL of solution is

$$200 \text{ mL} \times \frac{0.0040 \text{ mol Ba}^{2+}}{1 \text{ L soln}} \times \frac{1 \text{ L}}{1000 \text{ mL}} = 8.0 \times 10^{-4} \text{ mol Ba}^{2+}$$

Assuming the volumes are additive, the total volume after combining the two solutions is 800 mL. The concentration of Ba^{2+} in the 800-mL volume is

$$[\text{Ba}^{2+}] = \frac{8.0 \times 10^{-4} \text{ mol}}{800 \text{ mL}} \times \frac{1000 \text{ mL}}{1 \text{ L soln}}$$
$$= 1.0 \times 10^{-3}\, M$$

The number of moles of SO_4^{2-} in the original 600-mL solution is

$$600 \text{ mL} \times \frac{0.0080 \text{ mol SO}_4^{2-}}{1 \text{ L soln}} \times \frac{1 \text{ L}}{1000 \text{ mL}} = 4.8 \times 10^{-3} \text{ mol SO}_4^{2-}$$

The concentration of SO_4^{2-} in the 800 mL of the combined solution is

$$[\text{SO}_4^{2-}] = \frac{4.8 \times 10^{-3} \text{ mol}}{800 \text{ mL}} \times \frac{1000 \text{ mL}}{1 \text{ L soln}}$$
$$= 6.0 \times 10^{-3}\, M$$

Now we must compare Q and K_{sp}. From Table 16.2,

$$\text{BaSO}_4(s) \rightleftharpoons \text{Ba}^{2+}(aq) + \text{SO}_4^{2-}(aq) \qquad K_{sp} = 1.1 \times 10^{-10}$$

As for Q,

$$Q = [\text{Ba}^{2+}]_0[\text{SO}_4^{2-}]_0 = (1.0 \times 10^{-3})(6.0 \times 10^{-3})$$
$$= 6.0 \times 10^{-6}$$

Therefore,

$$Q > K_{sp}$$

The solution is supersaturated because the value of Q indicates that the concentrations of the ions are too large. Thus, some of the $BaSO_4$ will precipitate out of solution until

$$[\text{Ba}^{2+}][\text{SO}_4^{2-}] = 1.1 \times 10^{-10}$$

Practice Exercise If 2.00 mL of 0.200 M NaOH are added to 1.00 L of 0.100 M $CaCl_2$, will precipitation occur?

Similar problem: 16.61.

Summary of Concepts & Facts

- Predict the pH of a titration given the type of acid and base used.
- Evaluate the pH throughout a titration between acids and bases (strong titrated with strong; strong titrated with weak).
- Examine a titration curve to identify what type of titration was conducted (e.g., weak acid titrated with strong base).

Review of Concepts & Facts

16.5.1 What is the molar solubility of Ag_2CO_3 ($K_{sp} = 8.5 \times 10^{-12}$)?

16.5.2 Will a precipitate form when 50.0 mL of 0.0100 *M* NaF are added to 100.0 mL of 0.0100 *M* $Ba(NO_3)_2$?

16.5.3 The diagrams (a)–(d) represent solutions of AgCl, which may also contain ions such as Na^+ and NO_3^- (not shown) that do not affect the solubility of AgCl. If (a) represents a saturated solution of AgCl, classify the other solutions as unsaturated, saturated, or supersaturated.

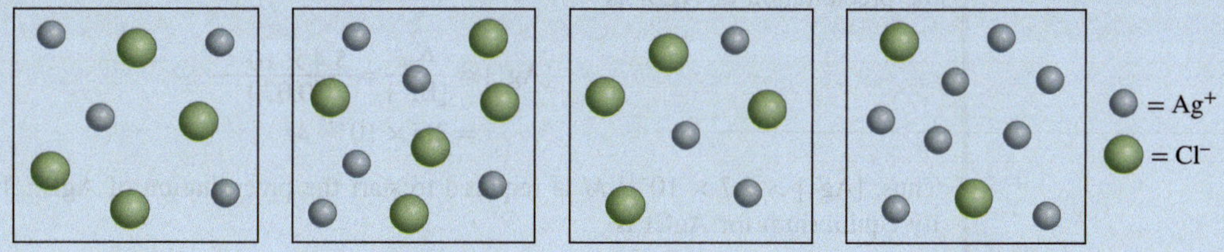

16.6 Separation of Ions by Fractional Precipitation

Learning Objective

- Explain fractional precipitation.

In chemical analysis, it is sometimes desirable to remove one type of ion from solution by precipitation while leaving other ions in solution. For instance, the addition of sulfate ions to a solution containing both potassium and barium ions causes $BaSO_4$ to precipitate out, thereby removing most of the Ba^{2+} ions from the solution. The other "product," K_2SO_4, is soluble and will remain in solution. The $BaSO_4$ precipitate can be separated from the solution by filtration.

Even when *both* products are insoluble, we can still achieve some degree of separation by choosing the proper reagent to bring about precipitation. Consider a solution that contains Cl^-, Br^-, and I^- ions. One way to separate these ions is to convert them to insoluble silver halides. As the K_{sp} values show [AgCl (1.8×10^{-10}), AgBr (5.4×10^{-13}), AgI (8.5×10^{-17})], the solubility of the halides decreases from AgCl to AgI. Thus, when a soluble compound such as silver nitrate is slowly added to this solution, AgI begins to precipitate first, followed by AgBr and then AgCl.

Example 16.11 describes the separation of only two ions (Cl^- and Br^-), but the procedure can be applied to a solution containing more than two different types of ions if precipitates of differing solubility can be formed.

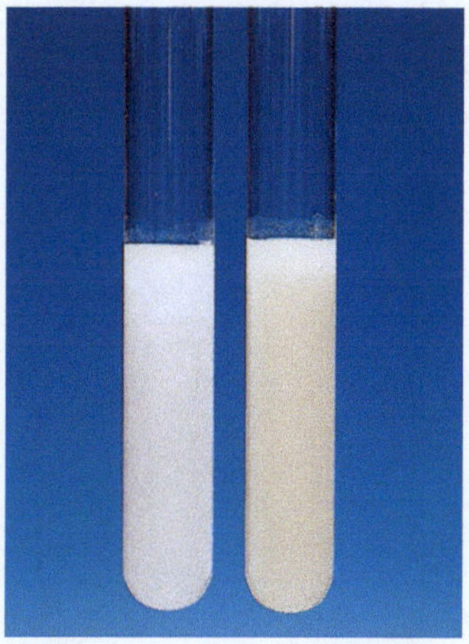

Suspension of AgCl precipitate (left) and AgBr precipitate (right).
Ken Karp/McGraw Hill

Example 16.11

A solution contains 0.020 *M* Cl^- ions and 0.020 *M* Br^- ions. To separate the Cl^- ions from the Br^- ions, solid $AgNO_3$ is slowly added to the solution without changing the volume. What concentration of Ag^+ ions (in mol/L) is needed to precipitate as much AgBr as possible without precipitating AgCl?

Strategy In solution, $AgNO_3$ dissociates into Ag^+ and NO_3^- ions. The Ag^+ ions then combine with the Cl^- and Br^- ions to form AgCl and AgBr precipitates. Because AgBr is less soluble (it has a smaller K_{sp} than that of AgCl), it will precipitate first. Therefore, this is a fractional precipitation problem. Knowing the concentrations of Cl^- and

(Continued)

Br^- ions, we can calculate $[Ag^+]$ from the K_{sp} values. Keep in mind that K_{sp} refers to a saturated solution. To initiate precipitation, $[Ag^+]$ must exceed the concentration in the saturated solution in each case.

Solution The solubility equilibrium for AgBr is

$$AgBr(s) \rightleftharpoons Ag^+(aq) + Br^-(aq) \qquad K_{sp} = [Ag^+][Br^-]$$

Because $[Br^-] = 0.020\ M$, the concentration of Ag^+ that must be exceeded to initiate the precipitation of AgBr is

$$[Ag^+] = \frac{K_{sp}}{[Br^-]} = \frac{5.4 \times 10^{-13}}{0.020}$$
$$= 2.7 \times 10^{-11}\ M$$

Thus, $[Ag^+] > 2.7 \times 10^{-11}\ M$ is required to start the precipitation of AgBr. The solubility equilibrium for AgCl is

$$AgCl(s) \rightleftharpoons Ag^+(aq) + Cl^-(aq) \qquad K_{sp} = [Ag^+][Cl^-]$$

so that

$$[Ag^+] = \frac{K_{sp}}{[Cl^-]} = \frac{1.8 \times 10^{-10}}{0.020}$$
$$= 9.0 \times 10^{-9}\ M$$

Therefore, $[Ag^+] > 9.0 \times 10^{-9}\ M$ is needed to initiate the precipitation of AgCl.

To precipitate the Br^- ions as AgBr without precipitating the Cl^- ions as AgCl, then, $[Ag^+]$ must be greater than $2.7 \times 10^{-11}\ M$ and lower than $9.0 \times 10^{-9}\ M$.

Practice Exercise The solubility products of AgCl and Ag_3PO_4 are 1.8×10^{-10} and 8.9×10^{-17}, respectively. If Ag^+ is added (without changing the volume) to 1.00 L of a solution containing 0.10 mol Cl^- and 0.10 mol PO_4^{3-}, calculate the concentration of Ag^+ ions (in mol/L) required to initiate (a) the precipitation of AgCl and (b) the precipitation of Ag_3PO_4.

Similar problems: 16.63, 16.64.

Example 16.11 raises the question, What is the concentration of Br^- ions remaining in solution just before AgCl begins to precipitate? To answer this question we let $[Ag^+] = 8.0 \times 10^{-9}\ M$. Then

$$[Br^-] = \frac{K_{sp}}{[Ag^+]}$$
$$= \frac{7.7 \times 10^{-13}}{9.0 \times 10^{-9}}$$
$$= 8.6 \times 10^{-5}\ M$$

The percent of Br^- remaining in solution (the *unprecipitated* Br^-) at the critical concentration of Ag^+ is

$$\%\,Br^- = \frac{[Br^-]_{unppt'd}}{[Br^-]_{original}} \times 100\%$$
$$= \frac{8.6 \times 10^{-5}\ M}{0.020\ M} \times 100\%$$
$$= 0.43\%\ \text{unprecipitated}$$

Thus, (100 − 0.43) percent, or 99.57 percent, of Br^- will have precipitated as AgBr just before AgCl begins to precipitate. By this procedure, the Br^- ions can be quantitatively separated from the Cl^- ions.

Summary of Concepts & Facts

- In fractional precipitation, different types of ions can be effectively separated depending on their differing solubility products.

Review of Concepts & Facts

16.6.1 $AgNO_3$ is slowly added to a solution that contains 0.1 *M* each of Br^-, CO_3^{2-}, and SO_4^{2-} ions. Which compound will precipitate first and which compound will precipitate last? (Use the K_{sp} of each compound to calculate $[Ag^+]$ needed to produce a saturated solution.)

16.7 Factors Affecting Solubility

Learning Objectives

- Demonstrate how common ions affect the solubility of salts.
- Assess the relative solubilities of salts in acidic solution.
- Interpret the effect of complex ion formation on solubility of salts in solution.

As we have noted, K_{sp} is an equilibrium constant; precipitation of an ionic compound from solution occurs whenever the product of ion concentrations, each raised to the appropriate power, exceeds K_{sp} for that substance. In a saturated solution of AgCl, for example, the ion product $[Ag^+][Cl^-]$ is, of course, equal to K_{sp}. Furthermore, simple stoichiometry tells us that $[Ag^+] = [Cl^-]$. But this equality does not hold in all situations.

The Common Ion Effect and Solubility

In Section 16.1 we discussed the effect of a common ion on acid and base ionizations. Here we will examine the relationship between the common ion effect and solubility. Suppose we study a solution containing two dissolved substances that share a common ion, say, AgCl and $AgNO_3$. In addition to the dissociation of AgCl, the following process also contributes to the total concentration of the common silver ions in solution:

$$AgNO_3(s) \xrightarrow{H_2O} Ag^+(aq) + NO_3^-(aq)$$

The solubility equilibrium of AgCl is

$$AgCl(s) \rightleftharpoons Ag^+(aq) + Cl^-(aq)$$

If $AgNO_3$ is added to a saturated AgCl solution, the increase in $[Ag^+]$ will make the ion product greater than the solubility product:

$$Q = [Ag^+]_0[Cl^-]_0 > K_{sp}$$

To reestablish equilibrium, some AgCl will precipitate out of the solution, as Le Châtelier's principle would predict, until the ion product is once again equal to K_{sp}. The effect of adding a common ion, then, is a *decrease* in the solubility of the salt (AgCl) in solution. In this case $[Ag^+]$ is no longer equal to $[Cl^-]$ at equilibrium; rather, $[Ag^+] > [Cl^-]$. Note that at a given temperature, only the solubility of a compound is altered (decreased by the common ion effect). Its solubility product (K_{sp}), which is an equilibrium constant, remains the same whether or not other substances are present in the solution.

Student Hot Spot

Student data indicate you may struggle with the effect of a common ion on solubility. Access your eBook for additional Learning Resources on this topic.

Example 16.12 shows the common ion effect on solubility.

Example 16.12

Calculate the solubility of silver chloride (in g/L) in a 6.5×10^{-3} M silver nitrate solution.

Strategy This is a common-ion problem. The common ion here is Ag^+, which is supplied by both AgCl and $AgNO_3$. Remember that the presence of the common ion will affect only the solubility of AgCl (in g/L), but not the K_{sp} value because it is an equilibrium constant.

Solution *Step 1:* The relevant species in solution are Ag^+ ions (from both AgCl and $AgNO_3$) and Cl^- ions. The NO_3^- ions are spectator ions.

Step 2: Because $AgNO_3$ is a soluble strong electrolyte, it dissociates completely:

$$\underset{\phantom{6.5\times10^{-3}M}}{AgNO_3(s)} \xrightarrow{H_2O} \underset{6.5\times 10^{-3}\,M}{Ag^+(aq)} + \underset{6.5\times 10^{-3}\,M}{NO_3^-(aq)}$$

Let s be the molar solubility of AgCl in $AgNO_3$ solution. We summarize the changes in concentrations as follows:

	$AgCl(s)$	$\rightleftharpoons$	$Ag^+(aq)$	+	$Cl^-(aq)$
Initial (M):			6.5×10^{-3}		0.00
Change (M):	$-s$		$+s$		$+s$
Equilibrium (M):			$(6.5 \times 10^{-3} + s)$		s

Step 3:

$$K_{sp} = [Ag^+][Cl^-]$$
$$1.8 \times 10^{-10} = (6.5 \times 10^{-3} + s)(s)$$

Because AgCl is quite insoluble and the presence of Ag^+ ions from $AgNO_3$ further lowers the solubility of AgCl, s must be very small compared with 6.5×10^{-3}. Therefore, applying the approximation $6.5 \times 10^{-3} + s \approx 6.5 \times 10^{-3}$, we obtain

$$1.8 \times 10^{-10} = (6.5 \times 10^{-3})s$$
$$s = 2.8 \times 10^{-8}\,M$$

Step 4: At equilibrium

$$[Ag^+] = (6.5 \times 10^{-3} + 2.8 \times 10^{-8})\,M \approx 6.5 \times 10^{-3}\,M$$
$$[Cl^-] = 2.8 \times 10^{-8}\,M$$

and so our approximation was justified in step 3. Because all the Cl^- ions must come from AgCl, the amount of AgCl dissolved in $AgNO_3$ solution also is 2.8×10^{-8} M. Then, knowing the molar mass of AgCl (143.4 g), we can calculate the solubility of AgCl as follows:

$$\text{solubility of AgCl in AgNO}_3\text{ solution} = \frac{2.8 \times 10^{-8}\text{ mol AgCl}}{1\text{ L soln}} \times \frac{143.4\text{ g AgCl}}{1\text{ mol AgCl}}$$
$$= 4.0 \times 10^{-6}\text{ g/L}$$

Check The solubility of AgCl in pure water is 1.9×10^{-3} g/L (see the Practice Exercise in Example 16.9). Therefore, the lower solubility (4.0×10^{-6} g/L) in the presence of $AgNO_3$ is reasonable. You should also be able to predict the lower solubility using Le Châtelier's principle. Adding Ag^+ ions shifts the equilibrium to the left, thus decreasing the solubility of AgCl.

Practice Exercise Calculate the solubility in g/L of AgBr in (a) pure water and (b) 0.0010 M NaBr.

Similar problem: 16.68.

pH and Solubility

The solubilities of many substances also depend on the pH of the solution. Consider the solubility equilibrium of magnesium hydroxide:

$$\mathrm{Mg(OH)_2}(s) \rightleftharpoons \mathrm{Mg^{2+}}(aq) + \mathrm{2OH^-}(aq)$$

Adding OH^- ions (increasing the pH) shifts the equilibrium from right to left, thereby decreasing the solubility of $Mg(OH)_2$. (This is another example of the common ion effect.) On the other hand, adding H^+ ions (decreasing the pH) shifts the equilibrium from left to right, and the solubility of $Mg(OH)_2$ increases. Thus, insoluble bases tend to dissolve in acidic solutions. Similarly, insoluble acids dissolve in basic solutions.

To explore the quantitative effect of pH on the solubility of $Mg(OH)_2$, let us first calculate the pH of a saturated $Mg(OH)_2$ solution. We write

$$K_{sp} = [\mathrm{Mg^{2+}}][\mathrm{OH^-}]^2 = 1.2 \times 10^{-11}$$

Let s be the molar solubility of $Mg(OH)_2$. Proceeding as in Example 16.9,

$$K_{sp} = (s)(2s)^2 = 4s^3$$
$$4s^3 = 5.6 \times 10^{-12}$$
$$s^3 = 1.4 \times 10^{-12}$$
$$s = 1.1 \times 10^{-4}\,M$$

At equilibrium, therefore,

$$[\mathrm{OH^-}] = 2 \times 1.1 \times 10^{-4}\,M = 2.2 \times 10^{-4}\,M$$
$$\mathrm{pOH} = -\log(2.2 \times 10^{-4}) = 3.66$$
$$\mathrm{pH} = 14.00 - 3.66 = 10.34$$

In a medium with a pH of less than 10.45, the solubility of $Mg(OH)_2$ would increase. This follows from the fact that a lower pH indicates a higher $[H^+]$ and thus a lower $[OH^-]$, as we would expect from $K_w = [H^+][OH^-]$. Consequently, $[Mg^{2+}]$ rises to maintain the equilibrium condition, and more $Mg(OH)_2$ dissolves. The dissolution process and the effect of extra H^+ ions can be summarized as follows:

$$\mathrm{Mg(OH)_2}(s) \rightleftharpoons \mathrm{Mg^{2+}}(aq) + \mathrm{2OH^-}(aq)$$
$$\mathrm{2H^+}(aq) + \mathrm{2OH^-}(aq) \rightleftharpoons \mathrm{2H_2O}(l)$$
$$\text{Overall:} \quad \mathrm{Mg(OH)_2}(s) + \mathrm{2H^+}(aq) \rightleftharpoons \mathrm{Mg^{2+}}(aq) + \mathrm{2H_2O}(l)$$

If the pH of the medium were higher than 10.34, $[OH^-]$ would be higher and the solubility of $Mg(OH)_2$ would decrease because of the common ion (OH^-) effect.

The pH also influences the solubility of salts that contain a basic anion. For example, the solubility equilibrium for BaF_2 is

$$\mathrm{BaF_2}(s) \rightleftharpoons \mathrm{Ba^{2+}}(aq) + \mathrm{2F^-}(aq)$$

and

$$K_{sp} = [\mathrm{Ba^{2+}}][\mathrm{F^-}]^2$$

In an acidic medium, the high $[H^+]$ will shift the following equilibrium from left to right:

$$\mathrm{H^+}(aq) + \mathrm{F^-}(aq) \rightleftharpoons \mathrm{HF}(aq)$$

As $[F^-]$ decreases, $[Ba^{2+}]$ must increase to maintain the equilibrium condition. Thus, more BaF_2 dissolves. The dissolution process and the effect of pH on the solubility of BaF_2 can be summarized as follows:

$$\mathrm{BaF_2}(s) \rightleftharpoons \mathrm{Ba^{2+}}(aq) + \mathrm{2F^-}(aq)$$
$$\mathrm{2H^+}(aq) + \mathrm{2F^-}(aq) \rightleftharpoons \mathrm{2HF}(aq)$$
$$\text{Overall:} \quad \mathrm{BaF_2}(s) + \mathrm{2H^+}(aq) \rightleftharpoons \mathrm{Ba^{2+}}(aq) + \mathrm{2HF}(aq)$$

The solubilities of salts containing anions that do not hydrolyze are unaffected by pH. Examples of such anions are Cl^-, Br^-, and I^-.

Milk of magnesia, which contains $Mg(OH)_2$, is used to treat acid indigestion.

Ken Karp/McGraw Hill

Student Hot Spot

Student data indicate you may struggle with the effect of pH changes on solubility. Access your eBook for additional Learning Resources on this topic.

Examples 16.13 and 16.14 deal with the effect of pH on solubility.

Example 16.13

Which of the following compounds will be more soluble in acidic solution than in water: (a) CuS, (b) AgCl, (c) $PbSO_4$?

Strategy In each case, write the dissociation reaction of the salt into its cation and anion. The cation will not interact with the H^+ ion because they both bear positive charges. The anion will act as a proton acceptor only if it is the conjugate base of a weak acid. How would the removal of the anion affect the solubility of the salt?

Solution

(a) The solubility equilibrium for CuS is

$$CuS(s) \rightleftharpoons Cu^{2+}(aq) + S^{2-}(aq)$$

The sulfide ion is the conjugate base of the weak acid HS^-. Therefore, the S^{2-} ion reacts with the H^+ ion as follows:

$$S^{2-}(aq) + H^+(aq) \longrightarrow HS^-(aq)$$

This reaction removes the S^{2-} ions from solution. According to Le Châtelier's principle, the equilibrium will shift to the right to replace some of the S^{2-} ions that were removed, thereby increasing the solubility of CuS.

(b) The solubility equilibrium is

$$AgCl(s) \rightleftharpoons Ag^+(aq) + Cl^-(aq)$$

Because Cl^- is the conjugate base of a strong acid (HCl), the solubility of AgCl is not affected by an acid solution.

(c) The solubility equilibrium for $PbSO_4$ is

$$PbSO_4(s) \rightleftharpoons Pb^{2+}(aq) + SO_4^{2-}(aq)$$

The sulfate ion is a weak base because it is the conjugate base of the weak acid HSO_4^-. Therefore, the SO_4^{2-} ion reacts with the H^+ ion as follows:

$$SO_4^{2-}(aq) + H^+(aq) \longrightarrow HSO_4^-(aq)$$

This reaction removes the SO_4^{2-} ions from solution. According to Le Châtelier's principle, the equilibrium will shift to the right to replace some of the SO_4^{2-} ions that were removed, thereby increasing the solubility of $PbSO_4$.

Practice Exercise Is the solubility of the following compounds increased in an acidic solution: (a) $Ca(OH)_2$, (b) $Mg_3(PO_4)_2$, (c) $PbBr_2$?

Similar problem: 16.72.

Example 16.14

Calculate the concentration of aqueous ammonia necessary to initiate the precipitation of iron(II) hydroxide from a 0.0030 *M* solution of $FeCl_2$.

Strategy For iron(II) hydroxide to precipitate from solution, the product $[Fe^{2+}][OH^-]^2$ must be greater than its K_{sp}. First, we calculate $[OH^-]$ from the known $[Fe^{2+}]$ and the K_{sp} value listed in Table 16.2. This is the concentration of OH^- in a saturated solution of $Fe(OH)_2$. Next, we calculate the concentration of NH_3 that will supply this concentration of OH^- ions. Finally, any NH_3 concentration greater than the calculated value will initiate the precipitation of $Fe(OH)_2$ because the solution will become supersaturated.

(Continued)

Solution Ammonia reacts with water to produce OH^- ions, which then react with Fe^{2+} to form $Fe(OH)_2$. The equilibria of interest are

$$NH_3(aq) + H_2O(l) \rightleftharpoons NH_4^+(aq) + OH^-(aq)$$
$$Fe^{2+}(aq) + 2OH^-(aq) \rightleftharpoons Fe(OH)_2(s)$$

First we find the OH^- concentration above which $Fe(OH)_2$ begins to precipitate. We write

$$K_{sp} = [Fe^{2+}][OH^-]^2 = 4.9 \times 10^{-17}$$

Because $FeCl_2$ is a strong electrolyte, $[Fe^{2+}] = 0.0030\ M$ and

$$[OH^-]^2 = \frac{4.9 \times 10^{-17}}{0.0030} = 1.6 \times 10^{-14}$$
$$[OH^-] = 1.3 \times 10^{-7}\ M$$

Next, we calculate the concentration of NH_3 that will supply $1.3 \times 10^{-7}\ M$ OH^- ions. Let x be the initial concentration of NH_3 in mol/L. We summarize the changes in concentrations resulting from the ionization of NH_3 as follows.

	$NH_3(aq)$	+	$H_2O(l) \rightleftharpoons$	$NH_4^+(aq)$	+	$OH^-(aq)$
Initial (M):	x			0.00		0.00
Change (M):	-1.3×10^{-7}			$+1.3 \times 10^{-7}$		$+1.3 \times 10^{-7}$
Equilibrium (M):	$(x - 1.3 \times 10^{-7})$			1.3×10^{-7}		1.3×10^{-7}

Substituting the equilibrium concentrations in the expression for the ionization constant (see Table 15.4),

$$K_b = \frac{[NH_4^+][OH^-]}{[NH_3]}$$
$$1.8 \times 10^{-5} = \frac{(1.3 \times 10^{-7})(1.3 \times 10^{-7})}{(x - 1.3 \times 10^{-7})}$$

Solving for x, we obtain

$$x = 1.3 \times 10^{-7}\ M$$

Therefore, the concentration of NH_3 must be slightly greater than $1.3 \times 10^{-7}\ M$ to initiate the precipitation of $Fe(OH)_2$.

Practice Exercise Calculate whether or not a precipitate will form if 2.0 mL of 0.60 M NH_3 are added to 1.0 L of $1.0 \times 10^{-3}\ M$ $ZnSO_4$.

Similar problem: 16.76.

Complex Ion Equilibria and Solubility

Lewis acid-base reactions in which a metal cation combines with a Lewis base result in the formation of complex ions. Thus, we can define a ***complex ion*** as *an ion containing a central metal cation bonded to one or more molecules or ions.* Complex ions are crucial to many chemical and biological processes. Here we will consider the effect of complex ion formation on solubility. In Chapter 23 we will discuss the chemistry of complex ions in more detail.

Transition metals have a particular tendency to form complex ions because they have incompletely filled *d* subshells. This property enables them to act effectively as Lewis acids in reactions with many molecules or ions that serve as electron donors, or as Lewis bases. For example, a solution of cobalt(II) chloride is pink because of the presence of the $Co(H_2O)_6^{2+}$ ions (Figure 16.10). When HCl is added, the solution turns blue as a result of the formation of the complex ion $CoCl_4^{2-}$:

$$Co^{2+}(aq) + 4Cl^-(aq) \rightleftharpoons CoCl_4^{2-}(aq)$$

Figure 16.10 *(Left) An aqueous cobalt(II) chloride solution. The pink color is due to the presence of $Co(H_2O)_6^{2+}$ ions. (Right) After the addition of HCl solution, the solution turns blue because of the formation of the complex $CoCl_4^{2-}$ ions.*
Ken Karp/McGraw Hill

Copper(II) sulfate ($CuSO_4$) dissolves in water to produce a blue solution. The hydrated copper(II) ions are responsible for this color; many other sulfates (Na_2SO_4, for example) are colorless. Adding a *few drops* of concentrated ammonia solution to a $CuSO_4$ solution causes the formation of a light blue precipitate, copper(II) hydroxide:

$$Cu^{2+}(aq) + 2OH^-(aq) \longrightarrow Cu(OH)_2(s)$$

The OH^- ions are supplied by the ammonia solution. If more NH_3 is added, the blue precipitate redissolves to produce a beautiful dark blue solution, this time due to the formation of the complex ion $Cu(NH_3)_4^{2+}$ (Figure 16.11):

$$Cu(OH)_2(s) + 4NH_3(aq) \rightleftharpoons Cu(NH_3)_4^{2+}(aq) + 2OH^-(aq)$$

Thus, the formation of the complex ion $Cu(NH_3)_4^{2+}$ increases the solubility of $Cu(OH)_2$.

A measure of the tendency of a metal ion to form a particular complex ion is given by the ***formation constant K_f*** (also called the *stability constant*), which is *the equilibrium constant for the complex ion formation.* The larger K_f is, the more stable the complex ion is. Table 16.4 lists the formation constants of a number of complex ions.

The formation of the $Cu(NH_3)_4^{2+}$ ion can be expressed as

$$Cu^{2+}(aq) + 4NH_3(aq) \rightleftharpoons Cu(NH_3)_4^{2+}(aq)$$

for which the formation constant is

$$K_f = \frac{[Cu(NH_3)_4^{2+}]}{[Cu^{2+}][NH_3]^4} = 2.1 \times 10^{13}$$

Student Hot Spot

Student data indicate you may struggle with the effect of complex ion formation on solubility. Access your eBook for additional Learning Resources on this topic.

The very large value of K_f in this case indicates that the complex ion is quite stable in solution and accounts for the very low concentration of copper(II) ions at equilibrium.

Figure 16.11 *(Left) An aqueous solution of copper(II) sulfate. (Center) After the addition of a few drops of a concentrated aqueous ammonia solution, a light blue precipitate of $Cu(OH)_2$ is formed. (Right) When more concentrated aqueous ammonia solution is added, the $Cu(OH)_2$ precipitate dissolves to form the dark blue complex ion $Cu(NH_3)_4^{2+}$.*
Ken Karp/McGraw Hill

Table 16.4 Formation Constants of Selected Complex Ions in Water at 25°C

Complex Ion	Equilibrium Expression	Formation Constant (K_f)
$Ag(NH_3)_2^+$	$Ag^+ + 2NH_3 \rightleftharpoons Ag(NH_3)_2^+$	1.1×10^7
$Ag(CN)_2^-$	$Ag^+ + 2CN^- \rightleftharpoons Ag(CN)_2^-$	1.3×10^{21}
$Cu(CN)_4^{2-}$	$Cu^{2+} + 4CN^- \rightleftharpoons Cu(CN)_4^{2-}$	1.0×10^{25}
$Cu(NH_3)_4^{2+}$	$Cu^{2+} + 4NH_3 \rightleftharpoons Cu(NH_3)_4^{2+}$	2.1×10^{13}
$Cd(CN)_4^{2-}$	$Cd^{2+} + 4CN^- \rightleftharpoons Cd(CN)_4^{2-}$	6.0×10^{18}
CdI_4^{2-}	$Cd^{2+} + 4I^- \rightleftharpoons CdI_4^{2-}$	2.6×10^5
$HgCl_4^{2-}$	$Hg^{2+} + 4Cl^- \rightleftharpoons HgCl_4^{2-}$	1.2×10^{15}
HgI_4^{2-}	$Hg^{2+} + 4I^- \rightleftharpoons HgI_4^{2-}$	6.8×10^{29}
$Hg(CN)_4^{2-}$	$Hg^{2+} + 4CN^- \rightleftharpoons Hg(CN)_4^{2-}$	2.5×10^{41}
$Co(NH_3)_6^{3+}$	$Co^{3+} + 6NH_3 \rightleftharpoons Co(NH_3)_6^{3+}$	1.6×10^{35}
$Zn(NH_3)_4^{2+}$	$Zn^{2+} + 4NH_3 \rightleftharpoons Zn(NH_3)_4^{2+}$	4.4×10^9

Example 16.15

A 0.20-mole quantity of $CuSO_4$ is added to a liter of 1.20 M NH_3 solution. What is the concentration of Cu^{2+} ions at equilibrium?

Strategy The addition of $CuSO_4$ to the NH_3 solution results in complex ion formation

$$Cu^{2+}(aq) + 4NH_3(aq) \rightleftharpoons Cu(NH_3)_4^{2+}(aq)$$

From Table 16.4 we see that the formation constant (K_f) for this reaction is very large; therefore, the reaction lies mostly to the right. At equilibrium, the concentration of Cu^{2+} will be very small. As a good approximation, we can assume that essentially all the dissolved Cu^{2+} ions end up as $Cu(NH_3)_4^{2+}$ ions. How many moles of NH_3 will react with 0.20 mole of Cu^{2+}? How many moles of $Cu(NH_3)_4^{2+}$ will be produced? A very small amount of Cu^{2+} will be present at equilibrium. Set up the K_f expression for the preceding equilibrium to solve for $[Cu^{2+}]$.

Solution The amount of NH_3 consumed in forming the complex ion is 4×0.20 mol, or 0.80 mol. (Note that 0.20 mol Cu^{2+} is initially present in solution and four NH_3 molecules are needed to form a complex ion with one Cu^{2+} ion.) The concentration of NH_3 at equilibrium is therefore (1.20 − 0.80) mol/L soln or 0.40 M, and that of $Cu(NH_3)_4^{2+}$ is 0.20 mol/L soln or 0.20 M, the same as the initial concentration of Cu^{2+}. [There is a 1:1 mole ratio between Cu^{2+} and $Cu(NH_3)_4^{2+}$.] Because $Cu(NH_3)_4^{2+}$ does dissociate to a slight extent, we call the concentration of Cu^{2+} at equilibrium x and write

$$K_f = \frac{[Cu(NH_3)_4^{2+}]}{[Cu^{2+}][NH_3]^4}$$

$$2.1 \times 10^{13} = \frac{0.20}{x(0.40)^4}$$

Solving for x and keeping in mind that the volume of the solution is 1 L, we obtain

$$x = [Cu^{2+}] = 3.7 \times 10^{-13}\ M$$

Check The small value of $[Cu^{2+}]$ at equilibrium, compared with 0.20 M, certainly justifies our approximation.

Practice Exercise If 2.50 g of $CuSO_4$ are dissolved in 9.0×10^2 mL of 0.30 M NH_3, what are the concentrations of Cu^{2+}, $Cu(NH_3)_4^{2+}$, and NH_3 at equilibrium?

Similar problem: 16.79.

The effect of complex ion formation generally is to *increase* the solubility of a substance, as Example 16.16 shows.

Example 16.16

Calculate the molar solubility of AgCl in a 1.0 *M* NH_3 solution.

Strategy AgCl is only slightly soluble in water:

$$AgCl(s) \rightleftharpoons Ag^+(aq) + Cl^-(aq)$$

The Ag^+ ions form a complex ion with NH_3 (see Table 16.4):

$$Ag^+(aq) + 2NH_3(aq) \rightleftharpoons Ag(NH_3)_2^+$$

Combining these two equilibria will give the overall equilibrium for the process.

Solution *Step 1:* Initially, the species in solution are Ag^+ and Cl^- ions and NH_3. The reaction between Ag^+ and NH_3 produces the complex ion $Ag(NH_3)_2^+$.

Step 2: The equilibrium reactions are

$$AgCl(s) \rightleftharpoons Ag^+(aq) + Cl^-(aq)$$

$$K_{sp} = [Ag^+][Cl^-] = 1.8 \times 10^{-10}$$

$$Ag^+(aq) + 2NH_3(aq) \rightleftharpoons Ag(NH_3)_2^+(aq)$$

$$K_f = \frac{[Ag(NH_3)_2^+]}{[Ag^+][NH_3]^2} = 1.1 \times 10^7$$

$$\text{Overall:} \quad AgCl(s) + 2NH_3(aq) \rightleftharpoons Ag(NH_3)_2^+(aq) + Cl^-(aq)$$

The equilibrium constant *K* for the overall reaction is the product of the equilibrium constants of the individual reactions (see Section 14.2):

$$K = K_{sp}K_f = \frac{[Ag(NH_3)_2^+][Cl^-]}{[NH_3]^2}$$

$$= (1.8 \times 10^{-10})(1.1 \times 10^7)$$

$$= 2.0 \times 10^{-3}$$

Let *s* be the molar solubility of AgCl (mol/L). We summarize the changes in concentrations that result from formation of the complex ion as follows:

	$AgCl(s)$ +	$2NH_3(aq)$	$\rightleftharpoons$ $Ag(NH_3)_2^+(aq)$ +	$Cl^-(aq)$
Initial (*M*):		1.0	0.0	0.0
Change (*M*):	$-s$	$-2s$	$+s$	$+s$
Equilibrium (*M*):		$(1.0 - 2s)$	s	s

The formation constant for $Ag(NH_3)_2^+$ is quite large, so most of the silver ions exist in the complexed form. In the absence of ammonia we have, at equilibrium, $[Ag^+] = [Cl^-]$. As a result of complex ion formation, however, we can write $[Ag(NH_3)_2^+] = [Cl^-]$.

Step 3:

$$K = \frac{(s)(s)}{(1.0 - 2s)^2}$$

$$2.0 \times 10^{-3} = \frac{s^2}{(1.0 - 2s)^2}$$

Taking the square root of both sides, we obtain

$$0.045 = \frac{s}{1.0 - 2s}$$

$$s = 0.041\ M$$

Step 4: At equilibrium, 0.041 mole of AgCl dissolves in 1 L of 1.0 *M* NH_3 solution.

Check The molar solubility of AgCl in pure water is 1.3×10^{-5} *M*. Thus, the formation of the complex ion $Ag(NH_3)_2^+$ enhances the solubility of AgCl (Figure 16.12).

Practice Exercise Calculate the molar solubility of AgBr in a 1.0 *M* NH_3 solution.

Similar problem: 16.82.

CHEMISTRY *in Action*

How an Eggshell Is Formed

The formation of the shell of a hen's egg is a fascinating example of a natural precipitation process.

An average eggshell weighs about 5 g and is 40 percent calcium. Most of the calcium in an eggshell is laid down within a 16-h period. This means that it is deposited at a rate of about 125 mg per hour. No hen can consume calcium fast enough to meet this demand. Instead, it is supplied by special bony masses in the hen's long bones, which accumulate large reserves of calcium for eggshell formation. [The inorganic calcium component of the bone is calcium phosphate, $Ca_3(PO_4)_2$, an insoluble compound.] If a hen is fed a low-calcium diet, her eggshells become progressively thinner; she might have to mobilize 10 percent of the total amount of calcium in her bones just to lay one egg! When the food supply is consistently low in calcium, egg production eventually stops.

Chicken eggs.
muratart/Shutterstock

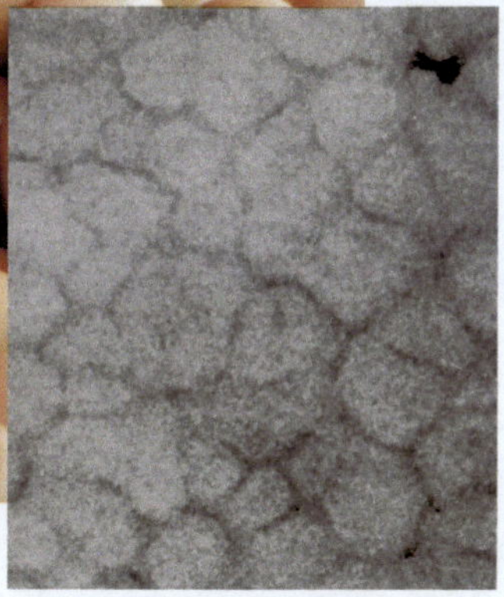

X-ray micrograph of an eggshell, showing columns of calcite.
Courtesy of A. Raymond Terepka

The eggshell is largely composed of calcite, a crystalline form of calcium carbonate ($CaCO_3$). Normally, the raw materials, Ca^{2+} and CO_3^{2-}, are carried by the blood to the shell gland. The calcification process is a precipitation reaction:

$$Ca^{2+}(aq) + CO_3^{2-}(aq) \rightleftharpoons CaCO_3(s)$$

In the blood, free Ca^{2+} ions are in equilibrium with calcium ions bound to proteins. As the free ions are taken up by the shell gland, more are provided by the dissociation of the protein-bound calcium.

The carbonate ions necessary for eggshell formation are a metabolic byproduct. Carbon dioxide produced during metabolism is converted to carbonic acid (H_2CO_3) by the enzyme carbonic anhydrase (CA):

$$CO_2(g) + H_2O(l) \overset{CA}{\rightleftharpoons} H_2CO_3(aq)$$

Carbonic acid ionizes stepwise to produce carbonate ions:

$$H_2CO_3(aq) \rightleftharpoons H^+(aq) + HCO_3^-(aq)$$
$$HCO_3^- \rightleftharpoons H^+(aq) + CO_3^{2-}(aq)$$

Chickens do not perspire and so must pant to cool themselves. Panting expels more CO_2 from the chicken's body than normal respiration does. According to Le Châtelier's principle, panting will shift the CO_2—H_2CO_3 equilibrium shown from right to left, thereby lowering the concentration of the CO_3^{2-} ions in solution and resulting in thin eggshells. One remedy for this problem is to give chickens carbonated water to drink in hot weather. The CO_2 dissolved in the water adds CO_2 to the chicken's body fluids and shifts the CO_2—H_2CO_3 equilibrium to the right.

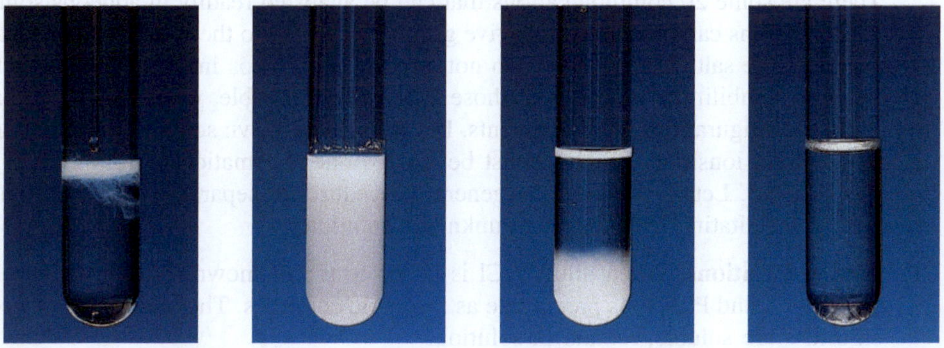

Figure 16.12 *From left to right: Formation of AgCl precipitate when $AgNO_3$ solution is added to NaCl solution. With the addition of NH_3 solution, the AgCl precipitate dissolves as the soluble $Ag(NH_3)_2^+$ forms.*
Ken Karp/McGraw Hill

Finally, we note that there is a class of hydroxides, called *amphoteric hydroxides,* which can react with both acids and bases. Examples are $Al(OH)_3$, $Pb(OH)_2$, $Cr(OH)_3$, $Zn(OH)_2$, and $Cd(OH)_2$. Thus, $Al(OH)_3$ reacts with acids and bases as follows:

$$Al(OH)_3(s) + 3H^+(aq) \longrightarrow Al^{3+}(aq) + 3H_2O(l)$$
$$Al(OH)_3(s) + OH^-(aq) \rightleftharpoons Al(OH)_4^-(aq)$$

The increase in solubility of $Al(OH)_3$ in a basic medium is the result of the formation of the complex ion $Al(OH)_4^-$, in which $Al(OH)_3$ acts as the Lewis acid and OH^- acts as the Lewis base. Other amphoteric hydroxides behave in a similar manner.

Summary of Concepts & Facts

- The presence of a common ion decreases the solubility of a slightly soluble salt.
- The solubility of slightly soluble salts containing basic anions increases as the hydrogen ion concentration increases. The solubility of salts with anions derived from strong acids is unaffected by pH.
- Complex ions are formed in solution by the combination of a metal cation with a Lewis base. The formation constant K_f measures the tendency toward the formation of a specific complex ion. Complex ion formation can increase the solubility of an insoluble substance.

Review of Concepts & Facts

16.7.1 For each pair of solutions, determine the one in which $PbI_2(s)$ will be more soluble: (a) $NaClO_3(aq)$ or $NaI(aq)$, (b) $Pb(NO_3)_2(aq)$ or $Ba(NO_3)_2(aq)$.

16.7.2 Calculate the molar solubility of CaF_2 in 0.0015 *M* NaF.

16.7.3 Which of the following salts will be more soluble in acidic solution than in water: (a) $BaSO_4$, (b) AgI, (c) FeS?

16.7.4 Which compound, when added to water, will increase the solubility of CdS: (a) $LiNO_3$, (b) Na_2SO_4, (c) KCN, (d) $NaClO_3$?

16.8 Solubility and Qualitative Analysis

Learning Objective

- Express the meaning and primary basis of qualitative analysis.

In Section 4.6, we discussed the principle of gravimetric analysis, by which we measure the amount of an ion in an unknown sample. Here we will briefly discuss ***qualitative analysis,*** *the determination of the types of ions present in a solution.* We will focus on the cations.

There are some 20 common cations that can be analyzed readily in aqueous solution. These cations can be divided into five groups according to the solubility products of their insoluble salts (Table 16.5). Do not confuse the groups in Table 16.5, which are based on solubility products, with those in the periodic table, which are based on the electron configurations of the elements. Because an unknown solution may contain from 1 to all 20 ions, any analysis must be carried out systematically from Group 1 through Group 5. Let us consider the general procedure for separating these 20 ions by adding precipitating reagents to an unknown solution.

- **Group 1 Cations.** When dilute HCl is added to the unknown solution, only the Ag^+, Hg_2^{2+}, and Pb^{2+} ions precipitate as insoluble chlorides. The other ions, whose chlorides are soluble, remain in solution.

Table 16.5 Separation of Cations into Groups According to Their Precipitation Reactions with Various Reagents

Group	Cation	Precipitating Reagents	Insoluble Compound	K_{sp}
1	Ag^+	HCl	AgCl	1.8×10^{-10}
	Hg_2^{2+}		Hg_2Cl_2	1.4×10^{-18}
	Pb^{2+}	↓	$PbCl_2$	1.7×10^{-5}
2	Bi^{3+}	H_2S	Bi_2S_3	1.6×10^{-72}
	Cd^{2+}	in acidic	CdS	8.0×10^{-28}
	Cu^{2+}	solutions	CuS	6.0×10^{-37}
	Hg^{2+}		HgS	4.0×10^{-54}
	Sn^{2+}	↓	SnS	1.0×10^{-26}
3	Al^{3+}	H_2S	$Al(OH)_3$	1.8×10^{-33}
	Co^{2+}	in basic	CoS	4.0×10^{-21}
	Cr^{3+}	solutions	$Cr(OH)_3$	3.0×10^{-29}
	Fe^{2+}		FeS	6.0×10^{-19}
	Mn^{2+}		MnS	3.0×10^{-14}
	Ni^{2+}		NiS	1.4×10^{-24}
	Zn^{2+}	↓	ZnS	3.0×10^{-23}
4	Ba^{2+}	Na_2CO_3	$BaCO_3$	2.6×10^{-9}
	Ca^{2+}		$CaCO_3$	3.4×10^{-9}
	Sr^{2+}	↓	$SrCO_3$	5.6×10^{-10}
5	K^+	No precipitating	None	
	Na^+	reagent	None	
	NH_4^+		None	

- **Group 2 Cations.** After the chloride precipitates have been removed by filtration, hydrogen sulfide is reacted with the unknown acidic solution. Under this condition, the concentration of the S^{2-} ion in solution is negligible. Therefore, the precipitation of metal sulfides is best represented as

$$M^{2+}(aq) + H_2S(aq) \rightleftharpoons MS(s) + 2H^+(aq)$$

  Adding acid to the solution shifts this equilibrium to the left so that only the least soluble metal sulfides, that is, those with the smallest K_{sp} values, will precipitate out of solution. These are Bi_2S_3, CdS, CuS, HgS, and SnS (see Table 16.5).

- **Group 3 Cations.** At this stage, sodium hydroxide is added to the solution to make it basic. In a basic solution, the abovementioned equilibrium shifts to the right. Therefore, the more soluble sulfides (CoS, FeS, MnS, NiS, ZnS) now precipitate out of solution. Note that the Al^{3+} and Cr^{3+} ions actually precipitate as the hydroxides $Al(OH)_3$ and $Cr(OH)_3$, rather than as the sulfides, because the hydroxides are less soluble. The solution is then filtered to remove the insoluble sulfides and hydroxides.

- **Group 4 Cations.** After all the Group 1, 2, and 3 cations have been removed from solution, sodium carbonate is added to the basic solution to precipitate Ba^{2+}, Ca^{2+}, and Sr^{2+} ions as $BaCO_3$, $CaCO_3$, and $SrCO_3$. These precipitates, too, are removed from solution by filtration.

- **Group 5 Cations.** At this stage, the only cations possibly remaining in solution are Na^+, K^+, and NH_4^+. The presence of NH_4^+ can be determined by adding sodium hydroxide:

$$NaOH(aq) + NH_4^+(aq) \longrightarrow Na^+(aq) + H_2O(l) + NH_3(g)$$

Figure 16.13 *Left to right: Flame colors of lithium, sodium, potassium, and copper.*
Stephen Frisch/McGraw Hill

The ammonia gas is detected either by noting its characteristic odor or by observing a piece of wet red litmus paper turning blue when placed above (not in contact with) the solution. To confirm the presence of Na^+ and K^+ ions, we often use a flame test, as follows: A piece of platinum wire (chosen because platinum is inert) is moistened with the solution and then held over a Bunsen burner flame. Because NaOH is adding in Group 3 and Na_2CO_3 is added in Group 4, the flame test for Na^+ is carried out using the original solution. Each type of metal ion gives a characteristic color when heated in this manner. For example, the color emitted by Na^+ ions is yellow, that of K^+ ions is violet, and that of Cu^{2+} ions is green (Figure 16.13).

Figure 16.14 summarizes this scheme for separating metal ions.

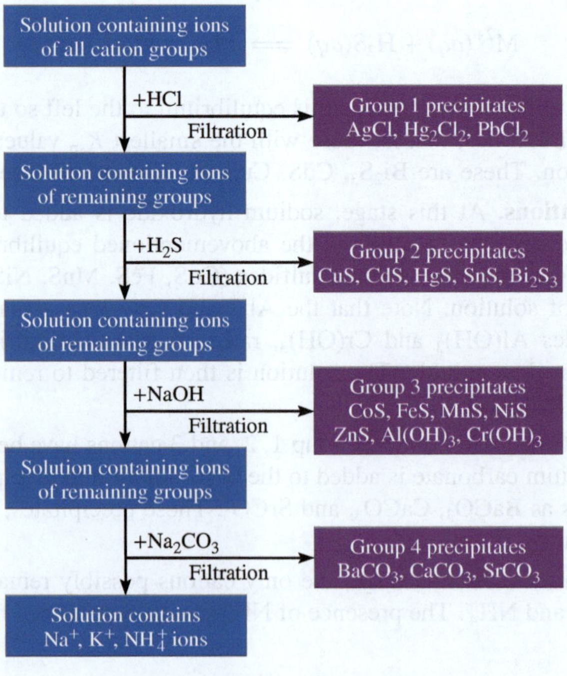

Figure 16.14 *A flow chart for the separation of cations in qualitative analysis.*

Two points regarding qualitative analysis must be mentioned. First, the separation of the cations into groups is made as selective as possible; that is, the anions that are added as reagents must be such that they will precipitate the fewest types of cations. For example, all the cations in Group 1 also form insoluble sulfides. Thus, if H_2S were reacted with the solution at the start, as many as seven different sulfides might precipitate out of solution (Group 1 *and* Group 2 sulfides), an undesirable outcome. Second, the removal of cations at each step must be carried out as completely as possible. For example, if we do not add enough HCl to the unknown solution to remove all the Group 1 cations, they will precipitate with the Group 2 cations as insoluble sulfides, interfering with further chemical analysis and leading to erroneous conclusions.

Summary of Concepts & Facts

- Qualitative analysis is the identification of cations and anions in solution. It is based largely on the principles of solubility equilibria.

Review of Concepts & Facts

16.8.1 An aqueous solution contains both Zn^{2+} and Pb^{2+} ions. Which of the following reagents would allow separation of the ions: (a) Na_2CO_3, (b) HNO_3, (c) HCl, (d) NaOH?

Chapter Summary

Buffer Solutions A buffer solution contains a weak acid and a salt derived from the acid. To maintain a relatively constant pH, the acid and base components of the buffer solution react with added acid or base. Buffer solutions play an important role in many chemical and biological processes. (Sections 16.1 and 16.2)

Acid-Base Titrations The characteristics of an acid-base titration depend on the strength of the acid and base involved. It is possible to calculate the pH during any stage of titration involving strong and/or weak acids and bases. Different indicators are used to determine the end point of a titration. (Sections 16.3 and 16.4)

Solubility Equilibria Another application of the equilibrium concept is the solubility equilibria of sparingly soluble salts, which are expressed as the solubility product. The solubility of such a substance can be affected by the presence of a common cation or anion, or the pH. Complex-ion formation, an example of the Lewis acid-base type reaction, increases the solubility of an insoluble salt. The solubility product principle can be applied to qualitative analysis, which is the identification of ions in solution. (Sections 16.5, 16.6, 16.7, and 16.8)

Key Equations

$pK_a = -\log K_a$ (16.3)	Definition of pK_a.
$pH = pK_a + \log\frac{[\text{conjugate base}]}{[\text{acid}]}$ (16.4)	Henderson-Hasselbalch equation.

Key Words

Buffer solution, p. 729	End point, p. 743	Molar solubility, p. 748	Solubility, p. 748
Common ion effect, p. 726	Formation constant (K_f), p. 760	Qualitative analysis, p. 764	Solubility product (K_{sp}), p. 747
Complex ion, p. 759			

Questions & Problems†

Red numbered problems solved in Student Solutions Manual

16.1 The Common Ion Effect

Review Questions

16.1 Use Le Châtelier's principle to explain how the common ion effect affects the pH of a solution.

16.2 Describe the effect on pH (increase, decrease, or no change) that results from each of the following additions: (a) potassium acetate to an acetic acid solution, (b) ammonium nitrate to an ammonia solution, (c) sodium formate (HCOONa) to a formic acid (HCOOH) solution, (d) potassium chloride to a hydrochloric acid solution, (e) barium iodide to a hydroiodic acid solution.

16.3 Define pK_a for a weak acid. What is the relationship between the value of the pK_a and the strength of the acid? Do the same for a weak base.

16.4 The pK_as of two monoprotic acids HA and HB are 5.9 and 8.1, respectively. Which of the two is the stronger acid?

Problems

16.5 Determine the pH of (a) a 0.40 *M* CH_3COOH solution, (b) a solution that is 0.40 *M* CH_3COOH and 0.20 *M* CH_3COONa.

16.6 Determine the pH of (a) a 0.20 *M* NH_3 solution, (b) a solution that is 0.20 *M* in NH_3 and 0.30 *M* NH_4Cl.

16.2 Buffer Solutions

Review Questions

16.7 What is a buffer solution? What constitutes a buffer solution?

16.8 Which of the following has the greatest buffer capacity: (a) 0.40 *M* CH_3COONa/0.20 *M* CH_3COOH, (b) 0.40 *M* CH_3COONa/0.60 *M* CH_3COOH, (c) 0.30 *M* CH_3COONa/0.60 *M* CH_3COOH?

Problems

16.9 Which of the following solutions can act as a buffer: (a) KCl/HCl, (b) $KHSO_4/H_2SO_4$, (c) Na_2HPO_4/NaH_2PO_4, (d) KNO_2/HNO_2?

16.10 Which of the following solutions can act as a buffer: (a) KCN/HCN, (b) $Na_2SO_4/NaHSO_4$, (c) NH_3/NH_4NO_3, (d) NaI/HI?

16.11 Calculate the pH of the buffer system made up of 0.15 *M* NH_3/0.35 *M* NH_4Cl.

16.12 Calculate the pH of the following two buffer solutions: (a) 2.0 *M* CH_3COONa/2.0 *M* CH_3COOH, (b) 0.20 *M* CH_3COONa/0.20 *M* CH_3COOH. Which is the more effective buffer? Why?

16.13 The pH of a bicarbonate-carbonic acid buffer is 8.00. Calculate the ratio of the concentration of carbonic acid (H_2CO_3) to that of the bicarbonate ion (HCO_3^-).

16.14 What is the pH of the buffer 0.10 *M* Na_2HPO_4/0.15 *M* KH_2PO_4?

16.15 The pH of a sodium acetate–acetic acid buffer is 4.50. Calculate the ratio $[CH_3COO^-]/[CH_3COOH]$.

16.16 The pH of blood plasma is 7.40. Assuming the principal buffer system is HCO_3^-/H_2CO_3, calculate the ratio $[HCO_3^-]/[H_2CO_3]$. Is this buffer more effective against an added acid or an added base?

16.17 Calculate the pH of the 0.20 *M* NH_3/0.20 *M* NH_4Cl buffer. What is the pH of the buffer after the addition of 10.0 mL of 0.10 *M* HCl to 65.0 mL of the buffer?

16.18 Calculate the pH of 1.00 L of the buffer 1.00 *M* CH_3COONa/1.00 *M* CH_3COOH before and after the addition of (a) 0.080 mol NaOH, (b) 0.12 mol HCl. (Assume that there is no change in volume.)

16.19 A diprotic acid, H_2A, has the following ionization constants: $K_{a_1} = 1.1 \times 10^{-3}$ and $K_{a_2} = 2.5 \times 10^{-6}$. To make up a buffer solution of pH 5.80, which combination would you choose: $NaHA/H_2A$ or $Na_2A/NaHA$?

16.20 A student is asked to prepare a buffer solution at pH = 8.60, using one of the following weak acids: HA($K_a = 2.7 \times 10^{-3}$), HB($K_a = 4.4 \times 10^{-6}$), HC($K_a = 2.6 \times 10^{-9}$). Which acid should she choose? Why?

16.21 The diagrams (a)–(d) contain one or more of the compounds H_2A, NaHA, and Na_2A, where H_2A is a weak diprotic acid. (1) Which of the solutions can

†The temperature is assumed to be 25°C for all the problems.

act as buffer solutions? (2) Which solution is the most effective buffer solution? (Water molecules and Na^+ ions have been omitted for clarity.)

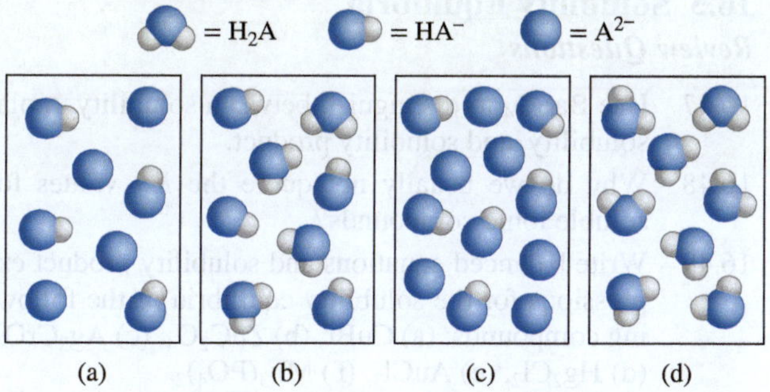

16.22 The diagrams (a)–(d) represent solutions containing a weak acid HA(pK_a = 5.00) and its sodium salt NaA. (1) Calculate the pH of the solutions. (2) What is the pH after the addition of 0.1 mol H^+ ions to solution (a)? (3) What is the pH after the addition of 0.1 mol OH^- ions to solution (d)? Treat each sphere as 0.1 mol.

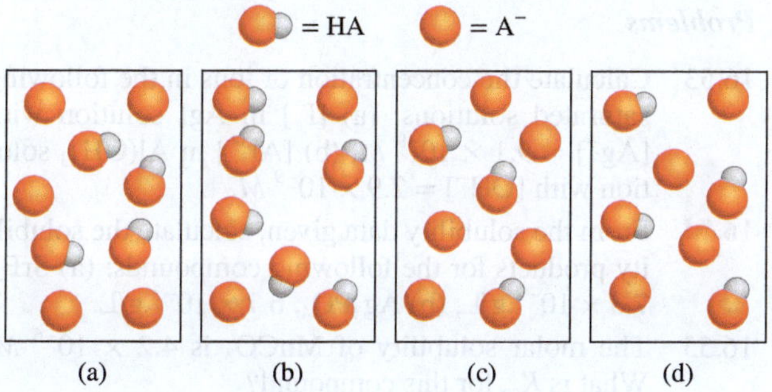

16.23 How much NaOH (in moles) must be added to 1 L of a buffer solution that is 1.8 *M* in acetic acid and 1.2 *M* in sodium acetate to result in a buffer solution of pH 5.22? Assume volume to remain constant.

16.24 How much HCl (in moles) must be added to 1 L of a buffer solution that is 0.84 *M* in ammonia and 0.96 *M* in ammonium chloride to result in a buffer solution of pH 8.56? Assume volume to remain constant.

16.3 Acid-Base Titrations

Review Questions

16.25 Briefly describe what happens in an acid-base titration.

16.26 Sketch titration curves for the following acid-base titrations: (a) HCl versus NaOH, (b) HCl versus CH_3NH_2, (c) CH_3COOH versus NaOH. In each case, the base is added to the acid in an Erlenmeyer flask. Your graphs should show pH on the *y*-axis and volume of base added on the *x*-axis.

Problems

16.27 A 0.2688-g sample of a monoprotic acid neutralizes 16.4 mL of 0.08133 *M* KOH solution. Calculate the molar mass of the acid.

16.28 A 5.00-g quantity of a diprotic acid was dissolved in water and made up to exactly 250 mL. Calculate the molar mass of the acid if 25.0 mL of this solution required 11.1 mL of 1.00 *M* KOH for neutralization. Assume that both protons of the acid were titrated.

16.29 In a titration experiment, 12.5 mL of 0.500 *M* H_2SO_4 neutralize 50.0 mL of NaOH. What is the concentration of the NaOH solution?

16.30 In a titration experiment, 20.4 mL of 0.883 *M* HCOOH neutralize 19.3 mL of $Ba(OH)_2$. What is the concentration of the $Ba(OH)_2$ solution?

16.31 A 0.1276-g sample of an unknown monoprotic acid was dissolved in 25.0 mL of water and titrated with 0.0633 *M* NaOH solution. The volume of base required to bring the solution to the equivalence point was 18.4 mL. (a) Calculate the molar mass of the acid. (b) After 10.0 mL of base had been added during the titration, the pH was determined to be 5.87. What is the K_a of the unknown acid?

16.32 A solution is made by mixing 5.00×10^2 mL of 0.167 *M* NaOH with 5.00×10^2 mL of 0.100 *M* CH_3COOH. Calculate the equilibrium concentrations of H^+, CH_3COOH, CH_3COO^-, OH^-, and Na^+.

16.33 Calculate the pH at the equivalence point for the following titration: 0.20 *M* HCl versus 0.20 *M* methylamine (CH_3NH_2). (See Table 15.4.)

16.34 Calculate the pH at the equivalence point for the following titration: 0.10 *M* HCOOH versus 0.10 *M* NaOH.

16.35 A 25.0-mL solution of 0.100 *M* CH_3COOH is titrated with a 0.200 *M* KOH solution. Calculate the pH after the following additions of the KOH solution: (a) 0.0 mL, (b) 5.0 mL, (c) 10.0 mL, (d) 12.5 mL, (e) 15.0 mL.

16.36 A 10.0-mL solution of 0.300 *M* NH_3 is titrated with a 0.100 *M* HCl solution. Calculate the pH after the following additions of the HCl solution: (a) 0.0 mL, (b) 10.0 mL, (c) 20.0 mL, (d) 30.0 mL, (e) 40.0 mL.

16.37 The diagrams (a)–(d) represent solutions at different stages in the titration of a weak acid HA with NaOH. Identify the solution that corresponds to (1) the initial stage before the addition of NaOH, (2) halfway to the equivalence point, (3) the equivalence point, (4) beyond the equivalence point. Is the pH greater than, less than, or equal to 7 at the equivalence point? (Water molecules and Na^+ ions have been omitted for clarity.)

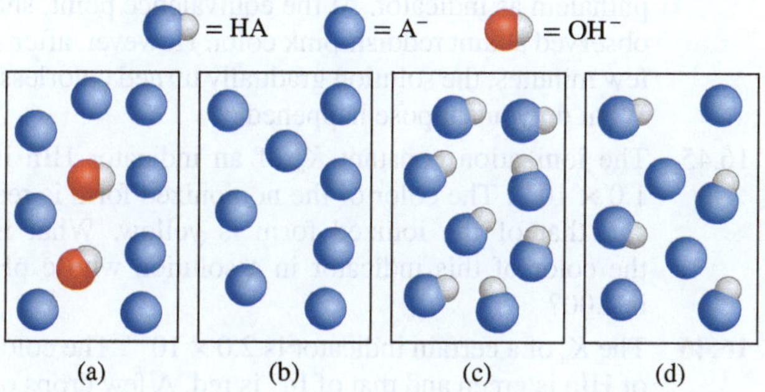

16.38 The diagrams (a)–(d) represent solutions at various stages in the titration of a weak base B (such as NH_3) with HCl. Identify the solution that corresponds to (1) the initial stage before the addition of HCl, (2) halfway to the equivalence point, (3) the equivalence point, (4) beyond the equivalence point. Is the pH greater than, less than, or equal to 7 at the equivalence point? (Water molecules and Cl^- ions have been omitted for clarity.)

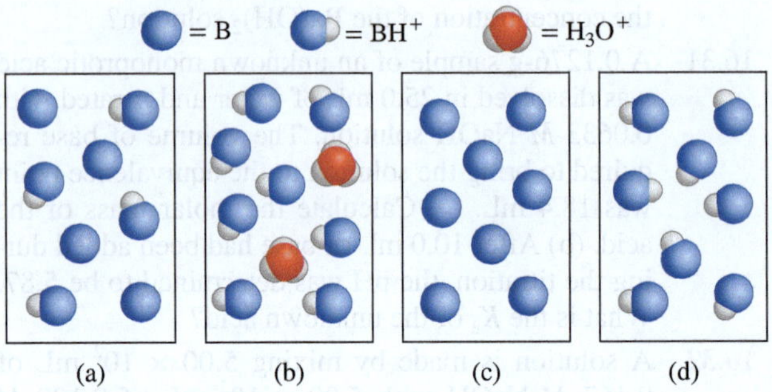

16.39 A 0.054 *M* HNO_2 solution is titrated with a KOH solution. What is $[H^+]$ at half way to the equivalence point?

16.40 A student titrates an unknown monoprotic acid with a NaOH solution from a buret. After the addition of 12.35 mL of NaOH, the pH of the solution read 5.22. The equivalence point is reached at 24.70 mL of NaOH. What is the K_a of the acid?

16.4 Acid-Base Indicators

Review Questions

16.41 Explain how an acid-base indicator works in a titration. What are the criteria for choosing an indicator for a particular acid-base titration?

16.42 The amount of indicator used in an acid-base titration must be small. Why?

Problems

16.43 Referring to Table 16.1, specify which indicator(s) you would use for the following titrations: (a) HCOOH versus NaOH, (b) HCl versus KOH, (c) HNO_3 versus CH_3NH_2.

16.44 A student carried out an acid-base titration by adding NaOH solution from a buret to an Erlenmeyer flask containing HCl solution and using phenolphthalein as indicator. At the equivalence point, she observed a faint reddish-pink color. However, after a few minutes, the solution gradually turned colorless. What do you suppose happened?

16.45 The ionization constant K_a of an indicator HIn is 1.0×10^{-6}. The color of the nonionized form is red and that of the ionized form is yellow. What is the color of this indicator in a solution whose pH is 4.00?

16.46 The K_a of a certain indicator is 2.0×10^{-6}. The color of HIn is green and that of In^- is red. A few drops of the indicator are added to a HCl solution, which is then titrated against a NaOH solution. At what pH will the indicator change color?

16.5 Solubility Equilibria

Review Questions

16.47 Use $BaSO_4$ to distinguish between solubility, molar solubility, and solubility product.

16.48 Why do we usually not quote the K_{sp} values for soluble ionic compounds?

16.49 Write balanced equations and solubility product expressions for the solubility equilibria of the following compounds: (a) CuBr, (b) ZnC_2O_4, (c) Ag_2CrO_4, (d) Hg_2Cl_2, (e) $AuCl_3$, (f) $Mn_3(PO_4)_2$.

16.50 Write the solubility product expression for the ionic compound A_xB_y.

16.51 How can we predict whether a precipitate will form when two solutions are mixed?

16.52 Silver chloride has a larger K_{sp} than silver carbonate (see Table 16.2). Does this mean that AgCl also has a larger molar solubility than Ag_2CO_3?

Problems

16.53 Calculate the concentration of ions in the following saturated solutions: (a) $[I^-]$ in AgI solution with $[Ag^+] = 9.1 \times 10^{-9}$ *M*, (b) $[Al^{3+}]$ in $Al(OH)_3$ solution with $[OH^-] = 2.9 \times 10^{-9}$ *M*.

16.54 From the solubility data given, calculate the solubility products for the following compounds: (a) SrF_2, 7.3×10^{-2} g/L, (b) Ag_3PO_4, 6.7×10^{-3} g/L.

16.55 The molar solubility of $MnCO_3$ is 4.2×10^{-6} *M*. What is K_{sp} for this compound?

16.56 The solubility of an ionic compound MX (molar mass = 346 g) is 4.63×10^{-3} g/L. What is K_{sp} for the compound?

16.57 The solubility of an ionic compound M_2X_3 (molar mass = 288 g) is 3.6×10^{-17} g/L. What is K_{sp} for the compound?

16.58 Using data from Table 16.2, calculate the molar solubility of CaF_2.

16.59 What is the pH of a saturated zinc hydroxide solution?

16.60 The pH of a saturated solution of a metal hydroxide MOH is 9.68. Calculate the K_{sp} for the compound.

16.61 If 20.0 mL of 0.10 *M* $Ba(NO_3)_2$ are added to 50.0 mL of 0.10 *M* Na_2CO_3, will $BaCO_3$ precipitate?

16.62 A volume of 75 mL of 0.060 *M* NaF is mixed with 25 mL of 0.15 *M* $Sr(NO_3)_2$. Calculate the concentrations in the final solution of NO_3^-, Na^+, Sr^{2+}, and F^-. (K_{sp} for $SrF_2 = 4.3 \times 10^{-9}$.)

16.6 Separation of Ions by Fractional Precipitation

Problems

16.63 Solid NaI is slowly added to a solution that is 0.010 *M* in Cu^+ and 0.010 *M* in Ag^+. (a) Which compound will begin to precipitate first? (b) Calculate $[Ag^+]$

when CuI just begins to precipitate. (c) What percent of Ag^+ remains in solution at this point?

16.64 Find the approximate pH range suitable for the separation of Fe^{3+} and Zn^{2+} ions by precipitation of $Fe(OH)_3$ from a solution that is initially 0.010 *M* in both Fe^{3+} and Zn^{2+}. Assume a 99% precipitation of $Fe(OH)_3$.

16.7 Factors Affecting Solubility

Review Questions

16.65 How does the common ion effect influence solubility equilibria? Use Le Châtelier's principle to explain the decrease in solubility of $CaCO_3$ in a Na_2CO_3 solution.

16.66 The molar solubility of AgCl in 6.5×10^{-3} *M* $AgNO_3$ is 2.5×10^{-8} *M*. In deriving K_{sp} from these data, which of the following assumptions are reasonable?
(a) K_{sp} is the same as solubility.
(b) K_{sp} of AgCl is the same in 6.5×10^{-3} *M* $AgNO_3$ as in pure water.
(c) Solubility of AgCl is independent of the concentration of $AgNO_3$.
(d) $[Ag^+]$ in solution does not change significantly upon the addition of AgCl to 6.5×10^{-3} *M* $AgNO_3$.
(e) $[Ag^+]$ in solution after the addition of AgCl to 6.5×10^{-3} *M* $AgNO_3$ is the same as it would be in pure water.

16.67 Explain the formation of complexes in Table 16.4 in terms of Lewis acid-base theory.

16.68 Give an example to illustrate the general effect of complex ion formation on solubility.

Problems

16.69 How many grams of $CaCO_3$ will dissolve in 3.0×10^2 mL of 0.050 *M* $Ca(NO_3)_2$?

16.70 The solubility product of $PbBr_2$ is 6.6×10^{-6}. Determine the molar solubility (a) in pure water, (b) in 0.20 *M* KBr solution, (c) in 0.20 *M* $Pb(NO_3)_2$ solution.

16.71 Calculate the molar solubility of AgCl in a 1.00-L solution containing 10.0 g of dissolved $CaCl_2$.

16.72 Calculate the molar solubility of $BaSO_4$ (a) in water, (b) in a solution containing 1.0 *M* SO_4^{2-} ions.

16.73 Which of the following ionic compounds will be more soluble in acid solution than in water: (a) $BaSO_4$, (b) $PbCl_2$, (c) $Fe(OH)_3$, (d) $CaCO_3$?

16.74 Which of the following will be more soluble in acid solution than in pure water: (a) CuI, (b) Ag_2SO_4, (c) $Zn(OH)_2$, (d) BaC_2O_4, (e) $Ca_3(PO_4)_2$?

16.75 Compare the molar solubility of $Mg(OH)_2$ in water and in a solution buffered at a pH of 9.0.

16.76 Calculate the molar solubility of $Fe(OH)_2$ in a solution buffered at (a) pH 8.00, (b) pH 10.00.

16.77 The solubility product of $Mg(OH)_2$ is 5.6×10^{-12}. What minimum OH^- concentration must be attained (for example, by adding NaOH) to decrease the Mg^{2+} concentration in a solution of $Mg(NO_3)_2$ to less than 1.0×10^{-10} *M*?

16.78 Calculate whether or not a precipitate will form if 2.00 mL of 0.60 *M* NH_3 are added to 1.0 L of 1.0×10^{-3} *M* $FeSO_4$.

16.79 If 2.50 g of $CuSO_4$ are dissolved in 9.0×10^2 mL of 0.30 *M* NH_3, what are the concentrations of Cu^{2+}, $Cu(NH_3)_4^{2+}$, and NH_3 at equilibrium?

16.80 Calculate the concentrations of Cd^{2+}, $Cd(CN)_4^{2-}$, and CN^- at equilibrium when 0.50 g of $Cd(NO_3)_2$ dissolves in 5.0×10^2 mL of 0.50 *M* NaCN.

16.81 If NaOH is added to 0.010 *M* Al^{3+}, which will be the predominant species at equilibrium: $Al(OH)_3$ or $Al(OH)_4^-$? The pH of the solution is 14.00. [K_f for $Al(OH)_4^- = 2.0 \times 10^{33}$.]

16.82 Calculate the molar solubility of AgI in a 1.0 *M* NH_3 solution.

16.83 Both Ag^+ and Zn^{2+} form complex ions with NH_3. Write balanced equations for the reactions. However, $Zn(OH)_2$ is soluble in 6 *M* NaOH, and AgOH is not. Explain.

16.84 Explain, with balanced ionic equations, why (a) CuI_2 dissolves in ammonia solution, (b) AgBr dissolves in NaCN solution, (c) $HgCl_2$ dissolves in KCl solution.

16.8 Solubility and Qualitative Analysis

Review Questions

16.85 Outline the general procedure of qualitative analysis.

16.86 Give two examples of metal ions in each group (1 through 5) in the qualitative analysis scheme.

Problems

16.87 In a Group 1 analysis, a student obtained a precipitate containing both AgCl and $PbCl_2$. Suggest one reagent that would enable her to separate AgCl(*s*) from $PbCl_2$(*s*).

16.88 In a Group 1 analysis, a student adds HCl acid to the unknown solution to make $[Cl^-] = 0.15$ *M*. Some $PbCl_2$ precipitates. Calculate the concentration of Pb^{2+} remaining in solution.

16.89 Both KCl and NH_4Cl are white solids. Suggest one reagent that would enable you to distinguish between these two compounds.

16.90 Describe a simple test that would enable you to distinguish between $AgNO_3$(*s*) and $Cu(NO_3)_2$(*s*).

Additional Problems

16.91 To act as an effective buffer, the concentrations of the acid and the conjugate base should not differ by more than a factor of 10, that is,

$$10 \geq \frac{[\text{conjugate base}]}{[\text{acid}]} \geq 0.1$$

(a) Show that the *buffer range*—that is, the range of the concentration ratio over which the buffer is effective—is given by $pH = pK_a \pm 1$.

(b) Calculate the pH range for the following buffer systems: (i) acetate, (ii) nitrite, (iii) bicarbonate, (iv) phosphate.

16.92 The pK_a of the indicator methyl orange is 3.46. Over what pH range does this indicator change from 90% HIn to 90% In^-?

16.93 The iodide impurity in a 4.50-g sample of a metal nitrate is precipitated as silver iodide. If 5.54 mL of 0.186 *M* $AgNO_3$ solution is needed for the precipitation, calculate the mass percent of iodide in the sample.

16.94 A sodium acetate-acetic acid buffer solution was prepared by adding a 0.020 *M* HCl solution to 500 mL of 0.020 *M* CH_3COONa and then diluting the mixed solution to 1.0 L. Calculate the original volume of the HCl solution needed to prepare a buffer solution of pH 5.00.

16.95 Sketch the titration curve of a weak acid versus a strong base like the one shown in Figure 16.5. On your graph indicate the volume of base used at the equivalence point and also at the half-equivalence point—that is, the point at which half of the acid has been neutralized. Show how you can measure the pH of the solution at the half-equivalence point. Using Equation (16.4), explain how you can determine the pK_a of the acid by this procedure.

16.96 A 200-mL volume of NaOH solution was added to 400 mL of a 2.00 *M* HNO_2 solution. The pH of the mixed solution was 1.50 units greater than that of the original acid solution. Calculate the molarity of the NaOH solution.

16.97 The pK_a of butyric acid (HBut) is 4.7. Calculate K_b for the butyrate ion (But^-).

16.98 A solution is made by mixing 5.00×10^2 mL of 0.167 *M* NaOH with 5.00×10^2 mL 0.100 *M* HCOOH. Calculate the equilibrium concentrations of H^+, HCOOH, $HCOO^-$, OH^-, and Na^+.

16.99 $Cd(OH)_2$ is an insoluble compound. It dissolves in excess NaOH in solution. Write a balanced ionic equation for this reaction. What type of reaction is this?

16.100 A student mixes 50.0 mL of 1.00 *M* $Ba(OH)_2$ with 86.4 mL of 0.494 *M* H_2SO_4. Calculate the mass of $BaSO_4$ formed and the pH of the mixed solution.

16.101 For which of the following reactions is the equilibrium constant called a solubility product?

(a) $Zn(OH)_2(s) + 2OH^-(aq) \rightleftharpoons Zn(OH)_4^{2-}(aq)$

(b) $3Ca^{2+}(aq) + 2PO_4^{3-}(aq) \rightleftharpoons Ca_3(PO_4)_2(s)$

(c) $CaCO_3(s) + 2H^+(aq) \rightleftharpoons Ca^{2+}(aq) + H_2O(l) + CO_2(g)$

(d) $PbI_2(s) \rightleftharpoons Pb^{2+}(aq) + 2I^-(aq)$

16.102 A 2.0-L kettle contains 116 g of boiler scale ($CaCO_3$). How many times would the kettle have to be completely filled with distilled water to remove all of the deposit?

16.103 Equal volumes of 0.12 *M* $AgNO_3$ and 0.14 *M* $ZnCl_2$ solution are mixed. Calculate the equilibrium concentrations of Ag^+, Cl^-, Zn^{2+}, and NO_3^-.

16.104 Calculate the solubility (in g/L) of Ag_2CO_3.

16.105 Find the approximate pH range suitable for separating Mg^{2+} and Zn^{2+} by the precipitation of $Zn(OH)_2$ from a solution that is initially 0.010 *M* in Mg^{2+} and Zn^{2+}.

16.106 A volume of 25.0 mL of 0.100 *M* HCl is titrated against a 0.100 *M* CH_3NH_2 solution added to it from a buret. Calculate the pH values of the solution (a) after 10.0 mL of CH_3NH_2 solution have been added, (b) after 25.0 mL of CH_3NH_2 solution have been added, (c) after 35.0 mL of CH_3NH_2 solution have been added.

16.107 The molar solubility of $Pb(IO_3)_2$ in a 0.10 *M* $NaIO_3$ solution is 2.4×10^{-11} mol/L. What is K_{sp} for $Pb(IO_3)_2$?

16.108 When a KI solution was added to a solution of mercury(II) chloride, a precipitate [mercury(II) iodide] formed. A student plotted the mass of the precipitate versus the volume of the KI solution added and obtained the following graph. Explain the appearance of the graph.

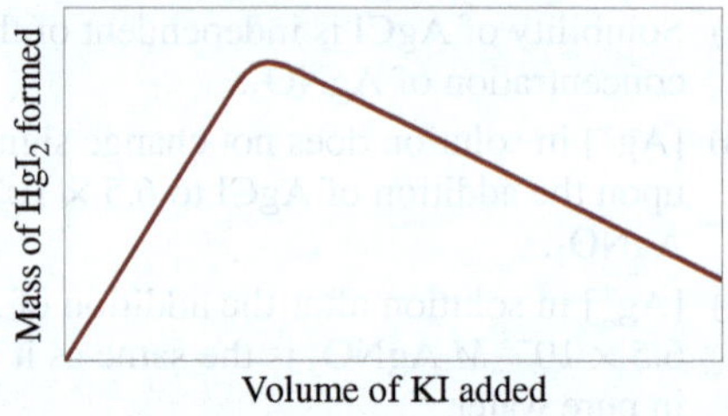

16.109 Barium is a toxic substance that can seriously impair heart function. For an X ray of the gastrointestinal tract, a patient drinks an aqueous suspension of 20 g $BaSO_4$. If this substance were to equilibrate with the 5.0 L of the blood in the patient's body, what would be $[Ba^{2+}]$? For a good estimate, we may assume that the temperature is at 25°C. Why is $Ba(NO_3)_2$ not chosen for this procedure?

16.110 The pK_a of phenolphthalein is 9.10. Over what pH range does this indicator change from 95 percent HIn to 95 percent In^-?

16.111 Solid NaBr is slowly added to a solution that is 0.010 *M* in Cu^+ and 0.010 *M* in Ag^+. (a) Which compound will begin to precipitate first? (b) Calculate $[Ag^+]$ when CuBr just begins to precipitate. (c) What percent of Ag^+ remains in solution at this point?

16.112 Cacodylic acid is $(CH_3)_2AsO_2H$. Its ionization constant is 6.4×10^{-7}. (a) Calculate the pH of 50.0 mL of a 0.10 *M* solution of the acid. (b) Calculate the pH of 25.0 mL of 0.15 *M* $(CH_3)_2AsO_2Na$. (c) Mix the solutions in part (a) and part (b). Calculate the pH of the resulting solution.

16.113 Radiochemical techniques are useful in estimating the solubility product of many compounds. In one experiment, 50.0 mL of a 0.010 *M* $AgNO_3$ solution containing a silver isotope with a radioactivity of 74,025 counts per min per mL were mixed with 100 mL of a 0.030 *M* $NaIO_3$ solution. The mixed solution was diluted to 500 mL and filtered to remove all of the $AgIO_3$ precipitate. The remaining solution was found to have a radioactivity of 44.4 counts per min per mL. What is the K_{sp} of $AgIO_3$?

16.114 The molar mass of a certain metal carbonate, MCO_3, can be determined by adding an excess of HCl acid to react with all the carbonate and then "back titrating" the remaining acid with a NaOH solution. (a) Write equations for these reactions. (b) In a certain experiment, 18.68 mL of 5.653 *M* HCl were added to a 3.542-g sample of MCO_3. The excess HCl required 12.06 mL of 1.789 *M* NaOH for neutralization. Calculate the molar mass of the carbonate and identify M.

16.115 Acid-base reactions usually go to completion. Confirm this statement by calculating the equilibrium constant for each of the following cases: (a) a strong acid reacting with a strong base, (b) a strong acid reacting with a weak base (NH_3), (c) a weak acid (CH_3COOH) reacting with a strong base, (d) a weak acid (CH_3COOH) reacting with a weak base (NH_3). (*Hint:* Strong acids exist as H^+ ions and strong bases exist as OH^- ions in solution. You need to look up K_a, K_b, and K_w.)

16.116 Calculate *x*, which is the number of molecules of water in oxalic acid hydrate, $H_2C_2O_4 \cdot xH_2O$, from the following data: 5.00 g of the compound is made up to exactly 250 mL solution, and 25.0 mL of this solution requires 15.9 mL of 0.500 *M* NaOH solution for neutralization.

16.117 Describe how you would prepare a 1-L 0.20 *M* CH_3COONa/0.20 *M* CH_3COOH buffer system by (a) mixing a solution of CH_3COOH with a solution of CH_3COONa, (b) reacting a solution of CH_3COOH with a solution of NaOH, and (c) reacting a solution of CH_3COONa with a solution of HCl.

16.118 Phenolphthalein is the common indicator for the titration of a strong acid with a strong base. (a) If the pK_a of phenolphthalein is 9.10, what is the ratio of the nonionized form of the indicator (colorless) to the ionized form (reddish pink) at pH 8.00? (b) If 2 drops of 0.060 *M* phenolphthalein are used in a titration involving a 50.0-mL volume, what is the concentration of the ionized form at pH 8.00? (Assume that 1 drop = 0.050 mL.)

16.119 Oil paintings containing lead(II) compounds as constituents of their pigments darken over the years. Suggest a chemical reason for the color change.

16.120 What reagents would you employ to separate the following pairs of ions in solution: (a) Na^+ and Ba^{2+}, (b) K^+ and Pb^{2+}, (c) Zn^{2+} and Hg^{2+}?

16.121 Look up the K_{sp} values for $BaSO_4$ and $SrSO_4$ in Table 16.2. Calculate the concentrations of Ba^{2+}, Sr^{2+}, and SO_4^{2-} in a solution that is saturated with both compounds.

16.122 In principle, amphoteric oxides, such as Al_2O_3 and BeO, can be used to prepare buffer solutions because they possess both acidic and basic properties (see Section 15.11). Explain why these compounds are of little practical use as buffer components.

16.123 $CaSO_4$ ($K_{sp} = 2.4 \times 10^{-5}$) has a larger K_{sp} value than that of Ag_2SO_4 ($K_{sp} = 1.2 \times 10^{-5}$). Does it follow that $CaSO_4$ also has greater solubility (g/L)?

16.124 When lemon juice is squirted into tea, the color becomes lighter. In part, the color change is due to dilution, but the main reason for the change is an acid-base reaction. What is the reaction? (*Hint:* Tea contains "polyphenols," which are weak acids, and lemon juice contains citric acid.)

16.125 How many milliliters of 1.0 *M* NaOH must be added to 200 mL of 0.10 *M* NaH_2PO_4 to make a buffer solution with a pH of 7.50?

16.126 The maximum allowable concentration of Pb^{2+} ions in drinking water is 0.05 ppm (that is, 0.05 g of Pb^{2+} in 1 million g of water). Is this guideline exceeded if an underground water supply is at equilibrium with the mineral anglesite, $PbSO_4$ ($K_{sp} = 1.6 \times 10^{-8}$)?

16.127 One of the most common antibiotics is penicillin G (benzylpenicillinic acid), which has the structure

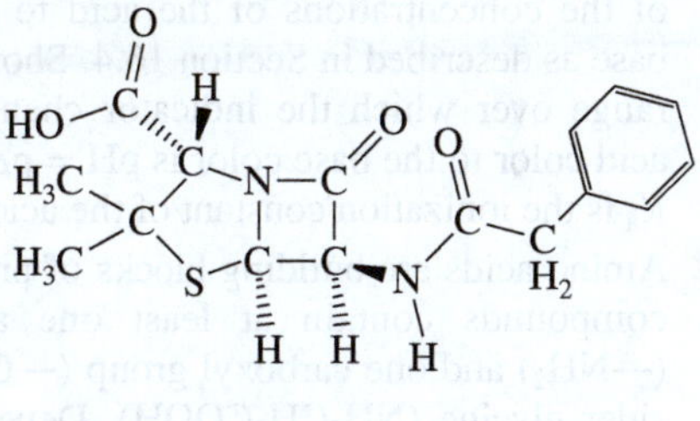

It is a weak monoprotic acid:

$$HP \rightleftharpoons H^+ + P^- \qquad K_a = 1.64 \times 10^{-3}$$

where HP denotes the parent acid and P^- the conjugate base. Penicillin G is produced by growing molds in fermentation tanks at 25°C and a pH range of 4.5 to 5.0. The crude form of this antibiotic is obtained by extracting the fermentation broth with an organic solvent in which the acid is soluble. (a) Identify the acidic hydrogen atom. (b) In one stage of purification, the organic extract of the crude penicillin G is treated with a buffer solution at pH = 6.50. What is the ratio of the conjugate base of penicillin G to the acid at this pH? Would you expect the conjugate base to be more soluble in water than the acid? (c) Penicillin G is not suitable for oral administration, but the sodium salt (NaP) is because it is soluble. Calculate the pH of a 0.12 *M* NaP solution formed when a tablet containing the salt is dissolved in a glass of water.

16.128 Which of the following solutions has the highest $[H^+]$: (a) 0.10 *M* HF, (b) 0.10 *M* HF in 0.10 *M* NaF, (c) 0.10 *M* HF in 0.10 *M* SbF_5? (*Hint:* SbF_5 reacts with F^- to form the complex ion SbF_6^-.)

16.129 Distribution curves show how the fractions of nonionized acid and its conjugate base vary as a function of pH of the medium. Plot distribution curves for CH_3COOH and its conjugate base CH_3COO^- in solution. Your graph should show fraction as the *y*-axis and pH as the *x*-axis. What are the fractions and pH at the point where these two curves intersect?

16.130 Water containing Ca^{2+} and Mg^{2+} ions is called *hard water* and is unsuitable for some household and industrial use because these ions react with soap to form insoluble salts, or curds. One way to remove the Ca^{2+} ions from hard water is by adding washing soda ($Na_2CO_3 \cdot 10H_2O$). (a) The molar solubility of $CaCO_3$ is 3.4×10^{-5} *M*. What is its molar solubility in a 0.050 *M* Na_2CO_3 solution? (b) Why are Mg^{2+} ions not removed by this procedure? (c) The Mg^{2+} ions are removed as $Mg(OH)_2$ by adding slaked lime [$Ca(OH)_2$] to the water to produce a saturated solution. Calculate the pH of a saturated $Ca(OH)_2$ solution. (d) What is the concentration of Mg^{2+} ions at this pH? (e) In general, which ion (Ca^{2+} or Mg^{2+}) would you remove first? Why?

16.131 Consider the ionization of the following acid-base indicator:

$$HIn(aq) \rightleftharpoons H^+(aq) + In^-(aq)$$

The indicator changes color according to the ratios of the concentrations of the acid to its conjugate base as described in Section 16.4. Show that the pH range over which the indicator changes from the acid color to the base color is $pH = pK_a \pm 1$, where K_a is the ionization constant of the acid.

16.132 Amino acids are building blocks of proteins. These compounds contain at least one amino group ($—NH_2$) and one carboxyl group ($—COOH$). Consider glycine (NH_2CH_2COOH). Depending on the pH of the solution, glycine can exist in one of three possible forms:

Fully protonated: $\overset{+}{N}H_3—CH_2—COOH$
Dipolar ion: $\overset{+}{N}H_3—CH_2—COO^-$
Fully ionized: $NH_2—CH_2—COO^-$

Predict the predominant form of glycine at pH 1.0, 7.0, and 12.0. The pK_a of the carboxyl group is 2.3 and that of the ammonium group ($-NH_3^+$) is 9.6.

16.133 (a) Referring to Figure 16.6, describe how you would determine the pK_b of the base. (b) Derive an analogous Henderson-Hasselbalch equation relating pOH to pK_b of a weak base B and its conjugate acid HB^+. Sketch a titration curve showing the variation of the pOH of the base solution versus the volume of a strong acid added from a buret. Describe how you would determine the pK_b from this curve. (*Hint:* $pK_b = -\log K_b$.)

16.134 A 25.0-mL of 0.20 *M* HF solution is titrated with a 0.20 *M* NaOH solution. Calculate the volume of NaOH solution added when the pH of the solution is (a) 2.85, (b) 3.15, (c) 11.89. (Ignore salt hydrolysis.)

16.135 Draw distribution curves for an aqueous carbonic acid solution. Your graph should show fraction of species present as the *y*-axis and pH as the *x*-axis. Note that at any pH, only two of the three species (H_2CO_3, HCO_3^-, and CO_3^{2-}) are present in appreciable concentrations. Use the pK_a values in Table 15.5.

16.136 One way to distinguish a buffer solution with an acid solution is by dilution. (a) Consider a buffer solution made of 0.500 *M* CH_3COOH and 0.500 *M* CH_3COONa. Calculate its pH and the pH after it has been diluted 10-fold. (b) Compare the result in (a) with the pHs of a 0.500 *M* CH_3COOH solution before and after it has been diluted 10-fold.

16.137 Histidine is one of the 20 amino acids found in proteins. Shown here is a fully protonated histidine molecule where the numbers denote the pK_a values of the acidic groups.

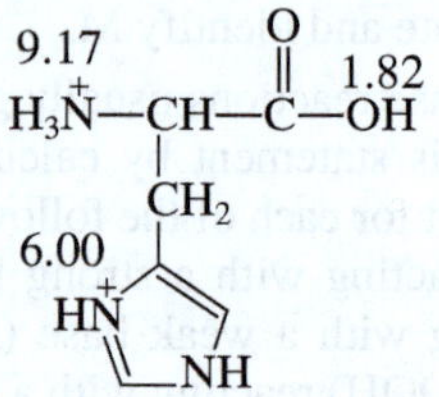

(a) Show stepwise ionization of histidine in solution. (*Hint:* The H^+ ion will first come off from the strongest acid group followed by the next strongest acid group, and so on.) (b) A dipolar ion is one in which the species has an equal number of positive and negative charges. Identify the dipolar ion in (a). (c) The pH at which the dipolar ion predominates is called the isoelectric point, denoted by p*I*. The isoelectric point is the average of the pK_a values leading to and following the formation of the dipolar ion. Calculate the p*I* of histidine. (d) The histidine group plays an important role in buffering blood (see Chemistry in Action essay "Maintaining the pH of Blood" in Section 16.2). Which conjugate acid-base pair shown in (a) is responsible for this action?

16.138 A sample of 0.96 L of HCl at 372 mmHg and 22°C is bubbled into 0.034 L of 0.57 *M* NH_3. What is the pH of the resulting solution? Assume the volume of solution remains constant and that the HCl is totally dissolved in the solution.

16.139 (a) Assuming complete dissociation and no ion-pair formation, calculate the freezing point of a 0.50 *m* NaI solution. (b) What is the freezing point after the addition of sufficient HgI_2, an insoluble compound, to the solution to react with all the free I^- ions in solution? Assume volume to remain constant.

16.140 Calculate the maximum mass (in grams) of each of the following soluble salts that can be added to 200 mL of 0.100 *M* $MgCl_2$ without causing a precipitate to form: (a) Na_2CO_3, (b) $AgNO_3$, (c) KOH. Assume volume to remain constant.

16.141 A 1.0-L saturated silver carbonate solution at 5°C is treated with enough hydrochloric acid to decompose the compound. The carbon dioxide generated is collected in a 19-mL vial and exerts a pressure of 114 mmHg at 25°C. What is the K_{sp} of Ag_2CO_3 at 5°C?

16.142 The two curves shown represent the titration of two weak acids of the same concentration with a strong base such as NaOH. Use three observations to determine which of the two acids is stronger.

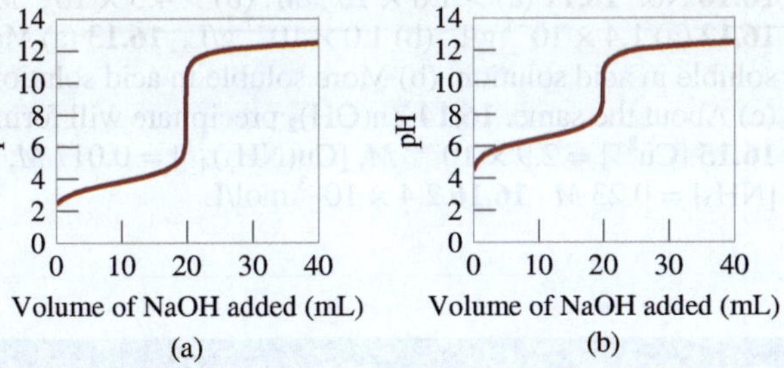

16.143 The two curves shown represent the titration of two weak bases of the same concentration with a strong acid such as HCl. Use three observations to determine which of the two bases is stronger.

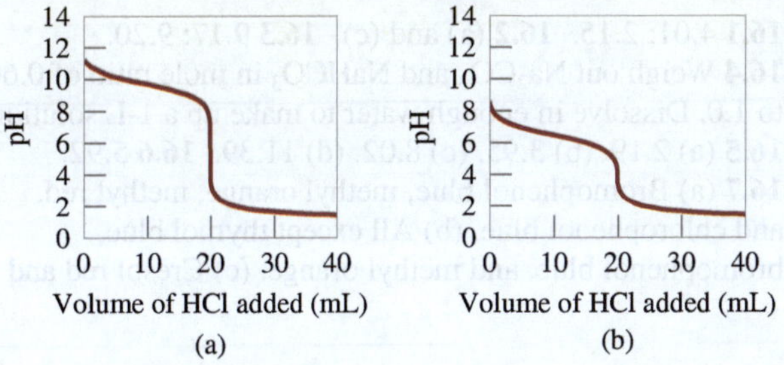

16.144 A 100-mL 0.100 *M* $CuSO_4$ solution is mixed with a 100-mL 0.100 *M* $Ba(OH)_2$ solution. Calculate the concentrations of the ions in the combined solution.

Interpreting, Modeling, & Estimating

16.145 The titration curve shown represents the titration of a weak diprotic acid (H_2A) versus NaOH. Identify the major species present at the marked points and estimate the pK_{a_1} and pK_{a_2} values of the acid.

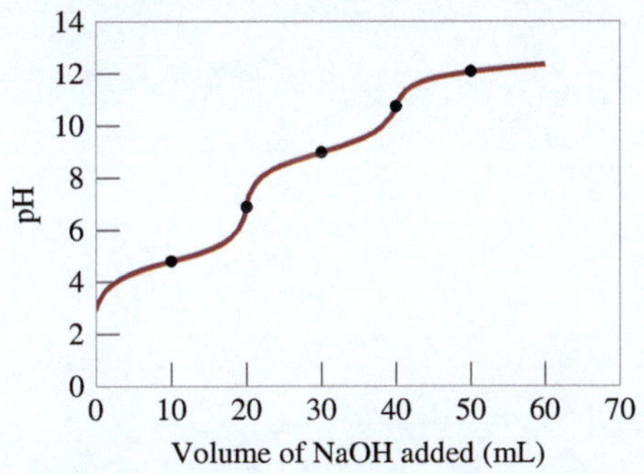

16.146 The titration curve shown represents the titration of a weak dibasic base (for example, a compound that contains two $-NH_2$ groups) versus HCl. Identify the major species present at the marked points and estimate the pK_{a_1} and pK_{a_2} values of the base.

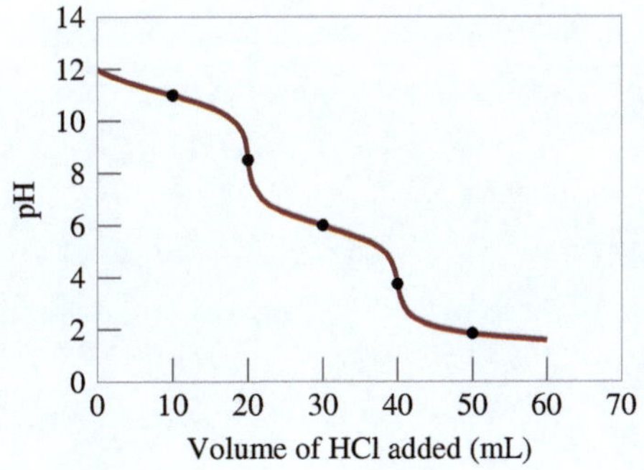

16.147 Use appropriate equations to account for the solubility of the amphoteric aluminum hydroxide [$Al(OH)_3$] at low and high pHs.

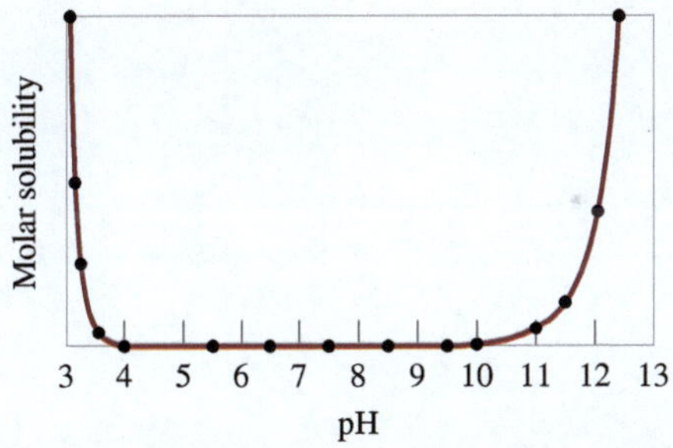

16.148 From Table 16.2 we see that silver bromide (AgBr) has a larger solubility product than iron(II) hydroxide [$Fe(OH)_2$]. Does this mean that AgBr is more soluble than $Fe(OH)_2$?

16.149 Aspirin is a weak acid with $pK_a = 3.5$. What is the ratio of neutral (protonated) aspirin to deprotonated aspirin in the following body fluids: (a) saliva, (b) gastric juices in the stomach, and (c) blood?

Answers to Practice Exercises

16.1 4.01; 2.15. **16.2** (a) and (c). **16.3** 9.17; 9.20. **16.4** Weigh out Na_2CO_3 and $NaHCO_3$ in mole ratio of 0.60 to 1.0. Dissolve in enough water to make up a 1-L solution. **16.5** (a) 2.19. (b) 3.95. (c) 8.02. (d) 11.39. **16.6** 5.92. **16.7** (a) Bromophenol blue, methyl orange, methyl red, and chlorophenol blue. (b) All except thymol blue, bromophenol blue, and methyl orange. (c) Cresol red and phenolphthalein. **16.8** 2.0×10^{-14}. **16.9** 1.9×10^{-3} g/L. **16.10** No. **16.11** (a) $> 1.8 \times 10^{-9}$ *M*. (b) $> 4.5 \times 10^{-5}$ *M*. **16.12** (a) 1.4×10^{-4} g/L. (b) 1.0×10^{-7} g/L. **16.13** (a) More soluble in acid solution. (b) More soluble in acid solution. (c) About the same. **16.14** $Zn(OH)_2$ precipitate will form. **16.15** $[Cu^{2+}] = 2.9 \times 10^{-13}$ *M*, $[Cu(NH_3)_4^{2+}] = 0.017$ *M*, $[NH_3] = 0.23$ *M*. **16.16** 2.4×10^{-3} mol/L.

Answers to Review of Concepts & Facts

16.1.1 (a), (c), and (d). **16.1.2** 3.49. **16.2.1** 3.28; 3.18. **16.2.2** (a) and (c) can act as buffers. (c) has a greater buffer capacity. **16.3.1** (a), (c), and (d). **16.3.2** 5.14. **16.3.3** 3.51. **16.4.1** The end point of an acid-base titration will accurately represent the equivalence point when the pH region over the indicator changes color matches the steep portion of the titration curve. **16.5.1** 1.3×10^{-4} *M*. **16.5.2** No, ($Q < K_{sp}$). **16.5.3** (b) Supersaturated. (c) Unsaturated. (d) Saturated. **16.6.1** AgBr will precipitate first and Ag_2SO_4 will precipitate last. **16.7.1** (a) $NaClO_3(aq)$. (b) $Ba(NO_3)_2(aq)$. **16.7.2** 1.6×10^{-5} *M*. **16.7.3** (a) and (c). **16.7.4** (c). **16.8.1** (c).

CHAPTER

17

Entropy, Gibbs Energy, and Equilibrium

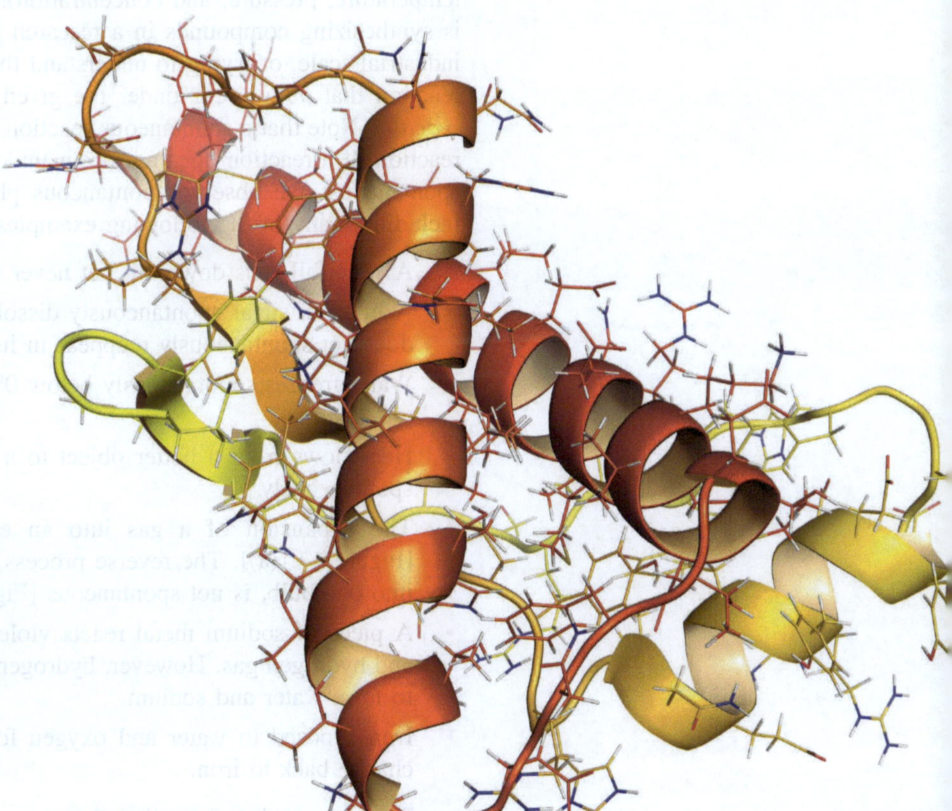

The laws of thermodynamics control many biological processes such as protein folding.
Pieter Van de Vijver/molekuul/123RF

CHAPTER OUTLINE

Thermodynamics is an extensive and far-reaching scientific discipline that deals with the interconversion of heat and other forms of energy. Thermodynamics enables us to use information gained from experiments on a system to draw conclusions about other aspects of the same system without further experimentation. For example, we saw in Chapter 6 that it is possible to calculate the enthalpy of reaction from the standard enthalpies of formation of the reactant and product molecules. This chapter introduces the second and third laws of thermodynamics and the Gibbs-energy function. It also discusses the relationship between Gibbs energy and chemical equilibrium.

17.1 Spontaneous Processes

Learning Objective

- Contrast spontaneous and nonspontaneous processes and cite examples of each.

One of the main objectives in studying thermodynamics, as far as chemists are concerned, is to be able to predict whether or not a reaction will occur when reactants are brought together under a specific set of conditions (for example, at a certain temperature, pressure, and concentration). This knowledge is important whether one is synthesizing compounds in a research laboratory, manufacturing chemicals on an industrial scale, or trying to understand the intricate biological processes in a cell. A reaction that *does* occur under the given set of conditions is called a *spontaneous reaction.* Note that a spontaneous reaction does not necessarily mean an instantaneous reaction. If a reaction does not occur under specified conditions, it is said to be nonspontaneous. We observe spontaneous physical and chemical processes every day, including many of the following examples:

- A waterfall runs downhill, but never up, spontaneously.
- A lump of sugar spontaneously dissolves in a cup of coffee, but dissolved sugar does not spontaneously reappear in its original form.
- Water freezes spontaneously below 0°C, and ice melts spontaneously above 0°C (at 1 atm).
- Heat flows from a hotter object to a colder one, but the reverse never happens spontaneously.
- The expansion of a gas into an evacuated bulb is a spontaneous process [Figure 17.1(a)]. The reverse process, that is, the gathering of all the molecules into one bulb, is not spontaneous [Figure 17.1(b)].
- A piece of sodium metal reacts violently with water to form sodium hydroxide and hydrogen gas. However, hydrogen gas does not react with sodium hydroxide to form water and sodium.
- Iron exposed to water and oxygen forms rust, but rust does not spontaneously change back to iron.

These examples show that processes that occur spontaneously in one direction cannot, under the same conditions, also take place spontaneously in the opposite direction.

If we assume that spontaneous processes occur so as to decrease the energy of a system, we can explain why a ball rolls downhill and why springs in a clock unwind. Similarly, a large number of exothermic reactions are spontaneous. An example is the combustion of methane:

$$CH_4(g) + 2O_2(g) \longrightarrow CO_2(g) + 2H_2O(l) \qquad \Delta H° = -890.5 \text{ kJ/mol}$$

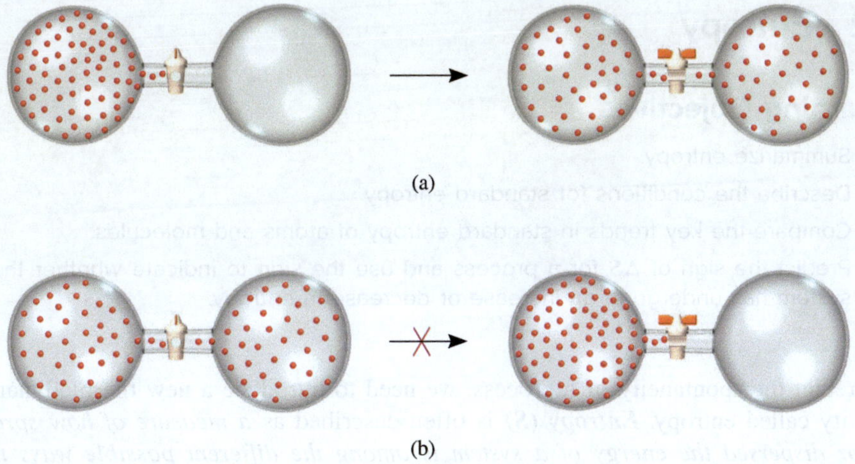

Figure 17.1 *(a) A spontaneous process. After the valve is opened, the molecules distribute evenly between the two bulbs. (b) A nonspontaneous process. After the valve is opened, the molecules preferentially gather in one bulb.*

Another example is the acid-base neutralization reaction:

$$H^+(aq) + OH^-(aq) \longrightarrow H_2O(l) \qquad \Delta H^\circ = -55.8 \text{ kJ/mol}$$

But consider a solid-to-liquid phase transition such as

$$H_2O(s) \longrightarrow H_2O(l) \qquad \Delta H^\circ = 6.01 \text{ kJ/mol}$$

In this case, the assumption that spontaneous processes always decrease a system's energy fails. Experience tells us that ice melts spontaneously above 0°C even though the process is endothermic. Another example that contradicts our assumption is the dissolution of ammonium nitrate in water:

$$NH_4NO_3(s) \xrightarrow{H_2O} NH_4^+(aq) + NO_3^-(aq) \qquad \Delta H^\circ = 25.7 \text{ kJ/mol}$$

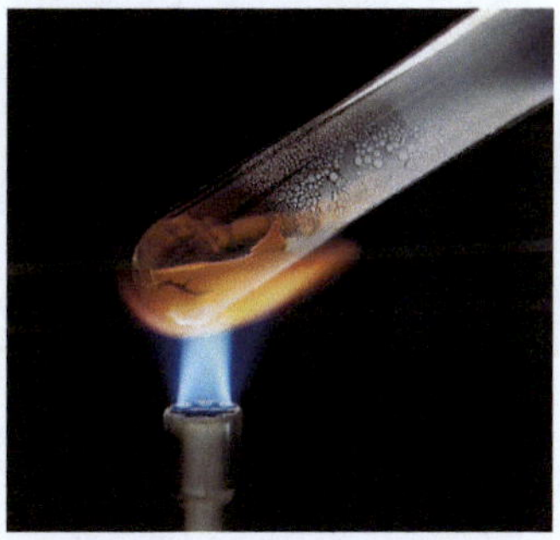

When heated, HgO decomposes to give Hg and O_2.
Ken Karp/McGraw Hill

This process is spontaneous, and yet it is also endothermic. The decomposition of mercury(II) oxide is an endothermic reaction that is nonspontaneous at room temperature, but it becomes spontaneous when the temperature is raised:

$$2HgO(s) \longrightarrow 2Hg(l) + O_2(g) \qquad \Delta H^\circ = 90.8 \text{ kJ/mol}$$

From a study of the examples mentioned and many more cases, we come to the following conclusion: Exothermicity favors the spontaneity of a reaction but does not guarantee it. Just as it is possible for an endothermic reaction to be spontaneous, it is possible for an exothermic reaction to be nonspontaneous. In other words, we cannot decide whether or not a chemical reaction will occur spontaneously solely on the basis of energy changes in the system. To make this kind of prediction we need another thermodynamic quantity, which turns out to be *entropy*.

Summary of Concepts & Facts

- A spontaneous process is one that occurs under a given set of conditions. Spontaneous reactions do not imply how fast the process will occur or how long a process will take. A nonspontaneous process is one that does not occur under a given set of conditions.

17.2 Entropy

Learning Objectives

- Summarize entropy.
- Describe the conditions for standard entropy.
- Compare the key trends in standard entropy of atoms and molecules.
- Predict the sign of ΔS for a process and use the sign to indicate whether the system has undergone an increase or decrease in entropy.

To predict the spontaneity of a process, we need to introduce a new thermodynamic quantity called entropy. ***Entropy (S)*** is often described as *a measure of how spread out or dispersed the energy of a system is among the different possible ways that system can contain energy.* The greater the dispersal, the greater the entropy is. Most processes are accompanied by a change in entropy. A cup of hot water has a certain amount of entropy due to the dispersal of energy among the various energy states of the water molecules (for example, energy states associated with the translational, rotational, and vibrational motions of the water molecules). If left standing on a table, the water loses heat to the cooler surroundings. Consequently, there is an overall increase in entropy because of the dispersal of energy over a great many energy states of the air molecules.

As another example, consider the situation depicted in Figure 17.1. Before the valve is opened, the system possesses a certain amount of entropy. Upon opening the valve, the gas molecules now have access to the combined volume of both bulbs. A larger volume for movement results in a narrowing of the gap between translational energy levels of the molecules. Quantum mechanical analysis shows that the spacing between translational energy levels is inversely proportional to the volume of the container and the mass of the molecules. Consequently, the entropy of the system increases because closely spaced energy levels lead to a greater dispersal among the energy levels. We can visualize this idea with gas particles on energy levels in a box with a changeable volume (Figure 17.2). When the gas flows into the right container, the overall volume of the gas is larger and there are more energy levels available (and the energy levels are closer together on average so more distributions are possible).

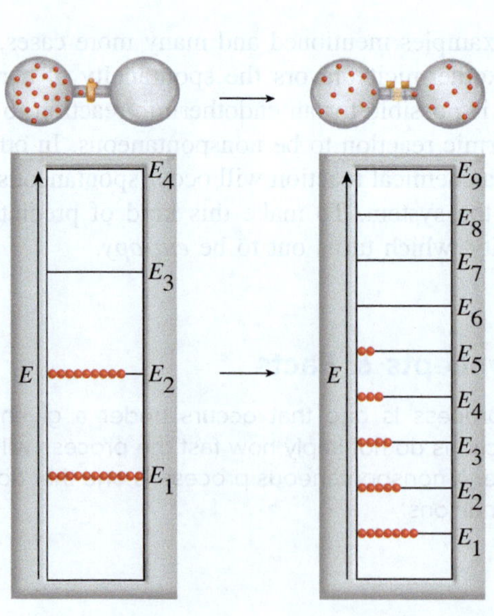

Figure 17.2 *Quantum mechanical analysis shows that the spacing between translational energy levels (E_1, E_2, E_3, etc.) is inversely proportional to the volume of a container. When the volume is increased, more energy levels at the lower energies become available. The total energy is now dispersed to more energy levels. The volumes depicted here are not to scale but illustrate the increase in the number of available energy levels with increasing volume.*

Quantitative Description of Entropy

Before we introduce the second law of thermodynamics, which relates entropy change (increase) to spontaneous processes, it is useful to first provide a proper definition of entropy. To do so let us consider a simple system of four molecules distributed between two equal compartments, as shown in Figure 17.3. Assume that each specific arrangement has an equal probability of occurring and there is no energy difference among the arrangements. There is only one way to arrange all the molecules in the left compartment, four ways to have three molecules in the left compartment and one in the right compartment, and six ways to have two molecules in each of the two compartments. The eleven possible ways of distributing the molecules are called microscopic states or microstates and each set of similar microstates is called a distribution. As you can see, distribution III is the most probable because there are six microstates or six ways to achieve it, and distribution I is the least probable because it has one microstate and therefore there is only one way to achieve it. [There are actually still other possible ways to distribute the four molecules between the two compartments. We can have all four molecules in the right compartment (one way) and three molecules in the right compartment and one molecule in the left compartment (four ways). However, the distributions shown in Figure 17.3 are sufficient for our discussion.]

Based on this analysis, we conclude that the probability of occurrence of a particular distribution (state) depends on the number of ways (microstates) in which the distribution can be achieved. As the number of molecules approaches macroscopic scale, it is not difficult to see that they will be evenly distributed between the two compartments because this distribution has many, many more microstates than all other distributions.

In 1868 Boltzmann showed that the entropy of a system is related to the natural log of the number of microstates (W):

$$S = k_B \ln W \tag{17.1}$$

where k_B is called the Boltzmann constant (1.38×10^{-23} J/K). Thus, the larger the W, the greater is the entropy of the system. Like enthalpy, entropy is a state function (see

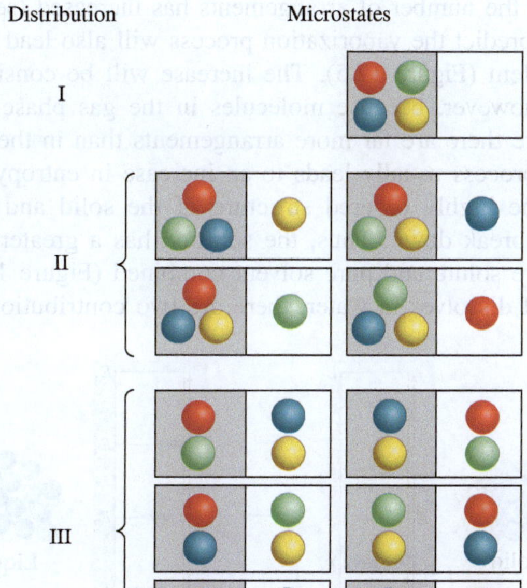

Figure 17.3 *Some possible ways of distributing four molecules between two equal compartments. Distribution I can be achieved in only one way (all four molecules in the left compartment) and has one microstate. Distribution II can be achieved in four ways and has four microstates. Distribution III can be achieved in six ways and has six microstates.*

Engraved on Ludwig Boltzmann's tombstone in Vienna is his famous equation. The "log" stands for "$\log_e$"—the natural logarithm or ln.

Dr. Matthias K. Gobbert/University of Maryland, Baltimore County

Section 6.3). Consider a certain process in a system. The entropy change for the process, ΔS, is

$$\Delta S = S_f - S_i \tag{17.2}$$

where S_i and S_f are the entropies of the system in the initial and final states, respectively. From Equation (17.1) we can write

$$\begin{aligned} \Delta S &= k_B \ln W_f - k_B \ln W_i \\ &= k_B \ln \frac{W_f}{W_i} \end{aligned} \tag{17.3}$$

where W_i and W_f are the corresponding numbers of microstates in the initial and final states. Thus, if $W_f > W_i$, $\Delta S > 0$ and the entropy of the system increases.

Changes in Entropy

Earlier we described the increase in entropy of a system as a result of the increase in the dispersal of energy. There is a connection between the qualitative description of entropy in terms of dispersal of energy and the quantitative definition of entropy in terms of microstates given by Equation (17.1). We conclude the following:

- A system with fewer microstates (smaller W) among which to spread its energy (small dispersal) has a lower entropy.
- A system with more microstates (larger W) among which to spread its energy (large dispersal) has a higher entropy.

Next, we will study several processes that lead to a change in entropy of a system in terms of the change in the number of microstates of the system.

In Figure 17.4, a solid is converted to a liquid at constant temperature. In a solid the positions of the atoms or molecules are confined and the number of possible arrangements is small. Upon melting, these atoms or molecules can occupy many more positions as they move away from the lattice points. Consequently, the number of arrangements increases because there are now many more ways to arrange the particles. Therefore, we predict this "order ⟶ disorder" phase transition to result in an increase in entropy because the number of arrangements has increased (see Figure 17.4).

Similarly, we predict the vaporization process will also lead to an increase in the entropy of the system (Figure 17.5). The increase will be considerably greater than that for melting, however, because molecules in the gas phase occupy much more space, and therefore there are far more arrangements than in the liquid phase.

The solution process usually leads to an increase in entropy. When sucrose dissolves in water, the highly ordered structure of the solid and part of the ordered structure of water break down. Thus, the solution has a greater number of arrangements than the pure solute and pure solvent combined (Figure 17.6). When an ionic solid such as NaCl dissolves in water, there are two contributions to the increase in

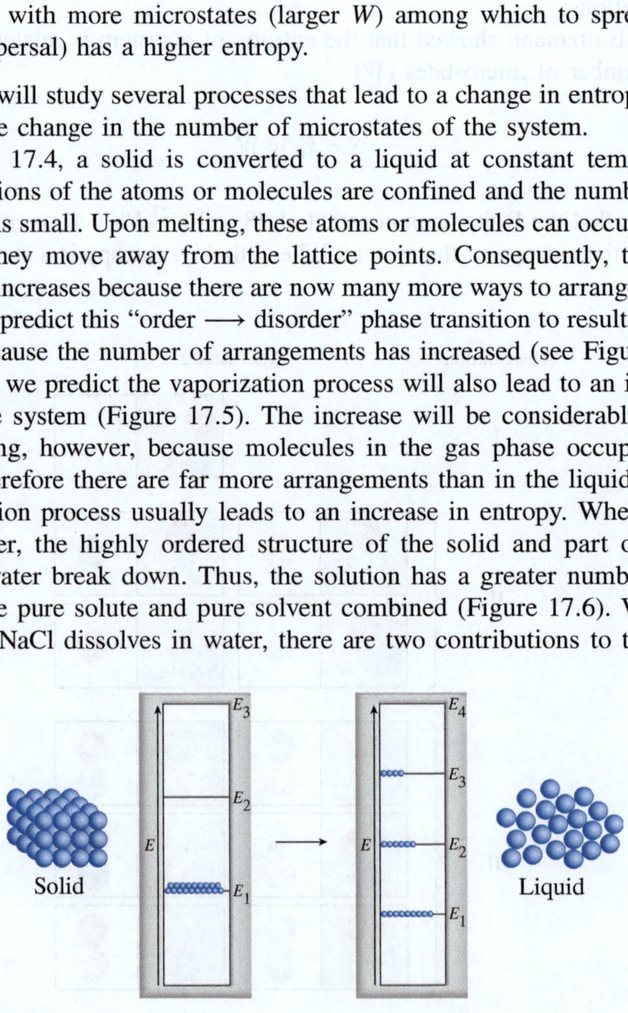

Figure 17.4 *Because of greater mobility, there are more different possible arrangements (W) of atoms or molecules in the liquid phase than there are in the solid phase. The entropy of a substance increases when it is melted (s → l).*

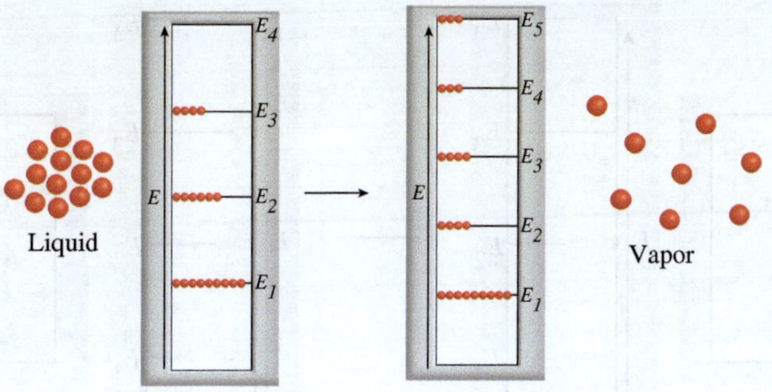

Figure 17.5 There are many more different possible arrangements of molecules in the gas phase than there are in the liquid phase. The entropy of a substance increases when it is vaporized ($l \rightarrow g$).

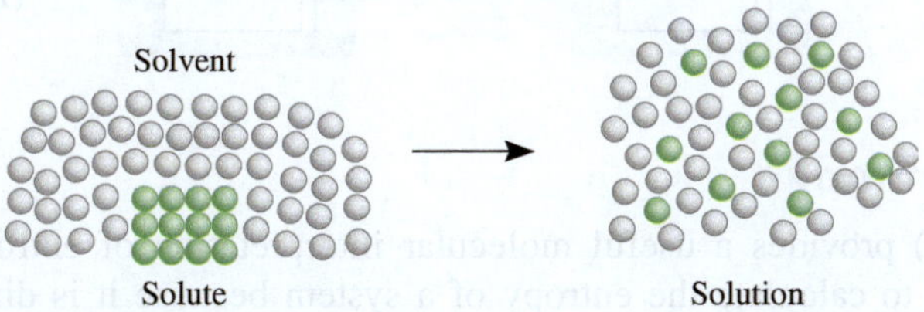

Figure 17.6 The dissolving process in general leads to an increase in entropy of the system because $S_{soln} > S_{solute} + S_{solvent}$.

entropy: the solution process (mixing of solute with solvent) and the dissociation of the compound into ions:

$$NaCl(s) \xrightarrow{H_2O} Na^+(aq) + Cl^-(aq)$$

More particles lead to a greater number of possible arrangements. However, we must also consider hydration, which causes water molecules to become more ordered around the ions. This hydration process decreases entropy because it reduces the number of arrangements of the solvent molecules. For small, highly charged ions such as Al^{3+} and Fe^{3+}, the decrease in entropy due to hydration can outweigh the increase in entropy due to mixing and dissociation so that the entropy change for the overall process can actually be negative. Table 17.1 lists the changes in entropy associated with the spontaneous dissolution of several ionic solids.

Heating also increases the entropy of a system. In addition to translational motion, molecules can also execute rotational motions and vibrational motions (Figure 17.7). As the temperature is increased, the energies associated with all types of molecular motion increase. This increase in energy is distributed or dispersed among the quantized energy levels. Consequently, more arrangements become available at a higher temperature; therefore, the entropy of a system always increases with increasing temperature (Figure 17.8).

Table 17.1 Entropy Changes for the Dissolution of Some Ionic Solids at 25°C

Dissolution Equation	ΔS°_{soln} (J/K · mol)
$NH_4NO_3(s) \longrightarrow NH_4^+(aq) + NO_3^-(aq)$	108.7
$AlCl_3(s) \longrightarrow Al^{3+}(aq) + 3Cl^-(aq)$	−261.5
$FeCl_3(s) \longrightarrow Fe^{3+}(aq) + 3Cl^-(aq)$	−288.7

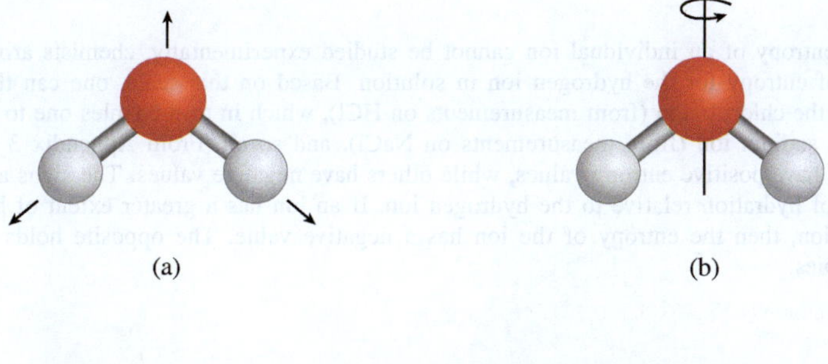

Figure 17.7 (a) A vibrational motion in a water molecule. The atoms are displaced as shown by the arrows and then reverse their directions to complete a cycle of vibration. (b) A rotational motion of a water molecule about an axis through the oxygen atom. The molecule can also vibrate and rotate in other ways.

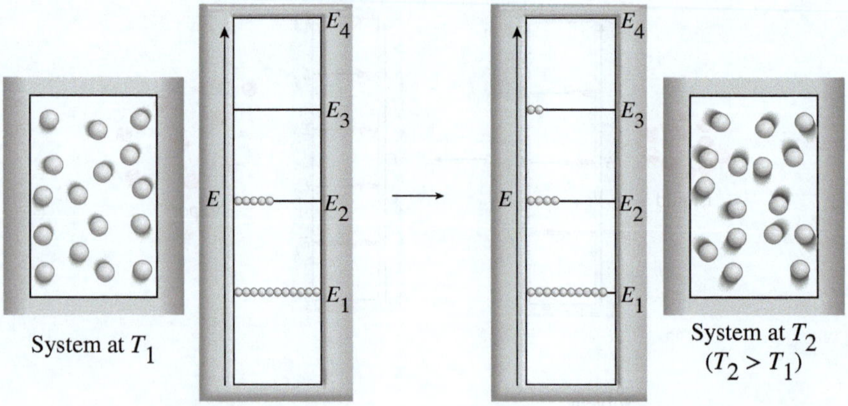

Figure 17.8 *At higher temperatures, molecules have greater kinetic energy, thereby making more energy levels accessible. This increases the number of energy levels within which the energy of the system can be dispersed, causing entropy to increase.*

Standard Entropy

Equation (17.1) provides a useful molecular interpretation of entropy, but is normally not used to calculate the entropy of a system because it is difficult to determine the number of possible arrangements for a macroscopic system containing many molecules. Instead, entropy is obtained by calorimetric methods. In fact, as we will see shortly, it is possible to determine the absolute value of entropy of a substance, called absolute entropy, something we cannot do for energy or enthalpy. *Standard entropy* is the absolute entropy of a substance at 1 atm and usually quoted with its value at 25°C. (Recall that the standard state refers only to 1 atm. The reason for specifying 25°C is that many processes are carried out at room temperature.) Table 17.2 lists standard entropies of a few elements and compounds; Appendix 2 provides a more extensive listing.† The units of entropy are J/K or J/K · mol for 1 mole of the substance. We use joules rather than kilojoules because entropy values are typically quite small. Entropies of elements and compounds are all positive (that is, $S° > 0$). By contrast, the standard enthalpy of formation (ΔH_f°) for elements in their stable form is arbitrarily set equal to zero, and for compounds, it may be positive or negative.

Referring to Table 17.2, we see that the standard entropy of water vapor is greater than that of liquid water. Similarly, bromine vapor has a higher standard entropy than liquid bromine, and iodine vapor has a greater standard entropy than solid iodine. For different substances in the same phase, molecular complexity determines which ones have higher entropies. Both diamond and graphite are solids, but diamond has a more ordered structure and hence a smaller number of energy arrangements (see Figure 11.29). Therefore, diamond has a smaller standard entropy than graphite. Consider the natural gases methane (CH_4) and ethane (C_2H_6). Ethane has a more complex structure and hence more ways to execute molecular motions, which also increase its number of possible arrangements (Figure 17.9). Therefore, ethane has a greater standard entropy than methane. Both helium (He) and neon (Ne) are monatomic gases, which cannot execute rotational or vibrational motions, but neon has a greater standard entropy than helium because its molar mass is greater. Heavier atoms have more closely spaced energy levels so there is a greater distribution of the atoms' energy among the levels (Figure 17.10). Consequently, there are more energy arrangements associated with these atoms.

Table 17.2

Standard Entropy Values ($S°$) for Some Substances at 25°C

Substance	$S°$ (J/K · mol)
$H_2O(l)$	70.0
$H_2O(g)$	188.8
$Br_2(l)$	152.2
$Br_2(g)$	245.5
$I_2(s)$	116.1
$I_2(g)$	260.7
C (diamond)	2.4
C (graphite)	5.7
CH_4 (methane)	186.3
C_2H_6 (ethane)	229.2
He(g)	126.1
Ne(g)	146.2

†Because the entropy of an individual ion cannot be studied experimentally, chemists arbitrarily assign a zero value of entropy for the hydrogen ion in solution. Based on this scale, one can then determine the entropy of the chloride ion (from measurements on HCl), which in turn enables one to determine the entropy of the sodium ion (from measurements on NaCl), and so on. From Appendix 3 you will note that some ions have positive entropy values, while others have negative values. The signs are determined by the extent of hydration relative to the hydrogen ion. If an ion has a greater extent of hydration than the hydrogen ion, then the entropy of the ion has a negative value. The opposite holds for ions with positive entropies.

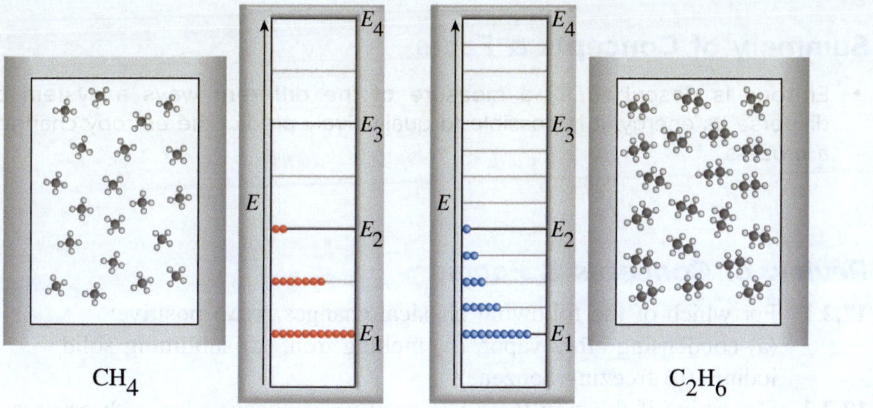

Figure 17.9 *The greater a molecule's complexity, the greater the number of possible ways it can rotate and vibrate. Ethane (C_2H_6) is more complex than methane (CH_4) and exhibits more different kinds of vibrations and rotations. This results in more energy levels within which the energy of the system can be dispersed. The number and spacing of additional energy levels have been simplified to keep the illustration clear.*

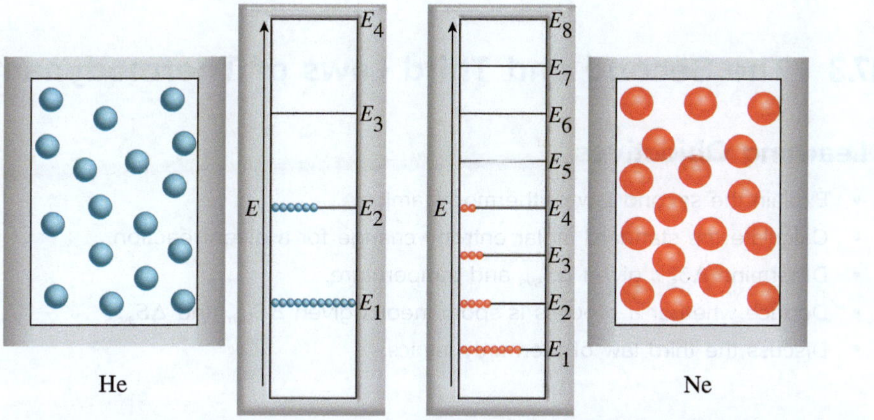

Figure 17.10 *The energy levels for a substance with a larger molar mass are more closely spaced. Ne has roughly four times the molar mass of He and thus has many more energy levels within which the energy of the system can be dispersed. The same amount of energy is being dispersed in each case.*

Example 17.1

Predict whether the entropy change is greater or less than zero for each of the following processes: (a) freezing ethanol, (b) evaporating a beaker of liquid bromine at room temperature, (c) dissolving glucose in water, (d) cooling nitrogen gas from 80°C to 20°C.

Strategy To determine the entropy change in each case, we examine whether the number of microstates of the system increases or decreases. The sign of ΔS will be positive if there is an increase in the number of microstates and negative if the number of microstates decreases.

Solution

(a) Upon freezing, the ethanol molecules are held rigid in position. This phase transition reduces the number of microstates and therefore the entropy decreases; that is, $\Delta S < 0$.

(b) Evaporating bromine increases the number of microstates because the Br_2 molecules can occupy many more positions in nearly empty space. Therefore, $\Delta S > 0$.

(c) Glucose is a nonelectrolyte. The solution process leads to a greater dispersal of matter due to the mixing of glucose and water molecules so we expect $\Delta S > 0$.

(d) The cooling process decreases various molecular motions. This leads to a decrease in microstates and so $\Delta S < 0$.

Practice Exercise How does the entropy of a system change for each of the following processes: (a) condensing water vapor, (b) forming sucrose crystals from a supersaturated solution, (c) heating hydrogen gas from 60°C to 80°C, (d) subliming dry ice?

Similar problem: 17.5.

Bromine is a fuming liquid at room temperature.
Ken Karp/McGraw Hill

Summary of Concepts & Facts

- Entropy is described as a measure of the different ways a system can disperse its energy. It is possible to qualitatively predict the entropy change of a process.

Review of Concepts & Facts

17.2.1 For which of the following physical changes is ΔS positive: (a) condensing ether vapor, (b) melting iron, (c) subliming solid iodine, (d) freezing benzene?

17.2.2 Determine if the sign for ΔS is positive or negative for each process: (a) $H_2O(s) \longrightarrow H_2O(l)$, (b) $C_3H_8(l) \longrightarrow C_3H_8(g)$, (c) $I_2(g) \longrightarrow I_2(s)$.

17.3 The Second and Third Laws of Thermodynamics

Learning Objectives

- Explain the second law of thermodynamics.
- Calculate the standard molar entropy change for a given reaction.
- Determine ΔS_{surr} given ΔS_{sys} and temperature.
- Deduce whether a process is spontaneous given ΔS_{surr} and ΔS_{sys}.
- Discuss the third law of thermodynamics.

The connection between entropy and the spontaneity of a reaction is expressed by the ***second law of thermodynamics:*** *The entropy of the universe increases in a spontaneous process and remains unchanged in an equilibrium process.* Because the universe is made up of the system and the surroundings, the entropy change in the universe (ΔS_{univ}) for any process is the *sum* of the entropy changes in the system (ΔS_{sys}) and in the surroundings (ΔS_{surr}). Mathematically, we can express the second law of thermodynamics as follows:

For a spontaneous process: $$\Delta S_{univ} = \Delta S_{sys} + \Delta S_{surr} > 0 \quad (17.4)$$

For an equilibrium process: $$\Delta S_{univ} = \Delta S_{sys} + \Delta S_{surr} = 0 \quad (17.5)$$

For a spontaneous process, the second law says that ΔS_{univ} must be greater than zero, but it does not place a restriction on either ΔS_{sys} or ΔS_{surr}. Thus, it is possible for either ΔS_{sys} or ΔS_{surr} to be negative, as long as the sum of these two quantities is greater than zero. For an equilibrium process, ΔS_{univ} is zero. In this case, ΔS_{sys} and ΔS_{surr} must be equal in magnitude, but opposite in sign. What if for some hypothetical process we find that ΔS_{univ} is negative? Then this means the process is not spontaneous in the direction described. Rather, it is spontaneous in the *opposite* direction.

Entropy Changes in the System

To calculate ΔS_{univ}, we need to know both ΔS_{sys} and ΔS_{surr}. Let us focus first on ΔS_{sys}. Suppose that the system is represented by the following reaction:

$$aA + bB \longrightarrow cC + dD$$

As is the case for the enthalpy of a reaction [see Equation (6.18)], the ***standard entropy of reaction*** (ΔS°_{rxn}) is given by *the difference in standard entropies between products and reactants:*

$$\Delta S^\circ_{rxn} = [cS^\circ(C) + dS^\circ(D)] - [aS^\circ(A) + bS^\circ(B)] \quad (17.6)$$

or, in general, using Σ to represent summation and m and n for the stoichiometric coefficients in the reaction

$$\Delta S^\circ_{rxn} = \Sigma nS^\circ(\text{products}) - \Sigma mS^\circ(\text{reactants}) \quad (17.7)$$

The standard entropy values of a large number of compounds have been measured in J/K · mol. To calculate ΔS°_{rxn} (which is ΔS_{sys}), we look up their values in Appendix 2 and proceed according to Example 17.2.

Example 17.2

From the standard entropy values in Appendix 2, calculate the standard entropy changes for the following reactions at 25°C.

(a) $CaCO_3(s) \longrightarrow CaO(s) + CO_2(g)$
(b) $N_2(g) + 3H_2(g) \longrightarrow 2NH_3(g)$
(c) $H_2(g) + Cl_2(g) \longrightarrow 2HCl(g)$

Strategy To calculate the standard entropy of a reaction, we look up the standard entropies of reactants and products in Appendix 2 and apply Equation (17.7). As in the calculation of enthalpy of reaction [see Equation (6.18)], the stoichiometric coefficients have no units, so ΔS°_{rxn} is expressed in units of J/K · mol.

Solution

(a) $\Delta S^\circ_{rxn} = [S^\circ(CaO) + S^\circ(CO_2)] - [S^\circ(CaCO_3)]$
$= [(38.1 \text{ J/K} \cdot \text{mol}) + (213.8 \text{ J/K} \cdot \text{mol})] - (91.7 \text{ J/K} \cdot \text{mol})$
$= 160.2 \text{ J/K} \cdot \text{mol}$

Thus, when 1 mole of $CaCO_3$ decomposes to form 1 mole of CaO and 1 mole of gaseous CO_2, there is an increase in entropy equal to 160.2 J/K · mol.

(b) $\Delta S^\circ_{rxn} = [2S^\circ(NH_3)] - [S^\circ(N_2) + 3S^\circ(H_2)]$
$= (2)(192.8 \text{ J/K} \cdot \text{mol}) - [(191.6 \text{ J/K} \cdot \text{mol}) + (3)(130.7 \text{ J/K} \cdot \text{mol})]$
$= -198.1 \text{ J/K} \cdot \text{mol}$

This result shows that when 1 mole of gaseous nitrogen reacts with 3 moles of gaseous hydrogen to form 2 moles of gaseous ammonia, there is a decrease in entropy equal to −198.1 J/K · mol.

(c) $\Delta S^\circ_{rxn} = [2S^\circ(HCl)] - [S^\circ(H_2) + S^\circ(Cl_2)]$
$= (2)(186.9 \text{ J/K} \cdot \text{mol}) - [(130.7 \text{ J/K} \cdot \text{mol}) + (223.1 \text{ J/K} \cdot \text{mol})]$
$= 20.0 \text{ J/K} \cdot \text{mol}$

Thus, the formation of 2 moles of gaseous HCl from 1 mole of gaseous H_2 and 1 mole of gaseous Cl_2 results in a small increase in entropy equal to 20 J/K · mol.

Comment The ΔS°_{rxn} values all apply to the system.

Practice Exercise Calculate the standard entropy change for the following reactions at 25°C:

(a) $2CO(g) + O_2(g) \longrightarrow 2CO_2(g)$
(b) $3O_2(g) \longrightarrow 2O_3(g)$
(c) $2NaHCO_3(s) \longrightarrow Na_2CO_3(s) + H_2O(l) + CO_2(g)$

Similar problems: 17.11 and 17.12.

The results of Example 17.2 are consistent with those observed for many other reactions. Taken together, they support the following general rules:

- If a reaction produces more gas molecules than it consumes [Example 17.2(a)], $\Delta S°$ is positive.
- If the total number of gas molecules diminishes [Example 17.2(b)], $\Delta S°$ is negative.
- If there is no net change in the total number of gas molecules [Example 17.2(c)], then $\Delta S°$ may be positive or negative, but will be relatively small numerically.

These conclusions make sense, given that gases invariably have greater entropy than liquids and solids. For reactions involving only liquids and solids, predicting the sign of $\Delta S°$ is more difficult, but in many such cases an increase in the total number of molecules and/or ions is accompanied by an increase in entropy.

Example 17.3 shows how knowing the nature of reactants and products makes it possible to predict entropy changes.

Example 17.3

Predict whether the entropy change of the system in each of the following reactions is positive or negative.

(a) $2H_2(g) + O_2(g) \longrightarrow 2H_2O(l)$
(b) $NH_4Cl(s) \longrightarrow NH_3(g) + HCl(g)$
(c) $H_2(g) + Br_2(g) \longrightarrow 2HBr(g)$

Strategy We are asked to predict, not calculate, the sign of entropy change in the reactions. The factors that lead to an increase in entropy are (1) a transition from a condensed phase to the vapor phase and (2) a reaction that produces more product molecules than reactant molecules in the same phase. It is also important to compare the relative complexity of the product and reactant molecules. In general, the more complex the molecular structure, the greater the entropy of the compound.

Solution

(a) Two reactant molecules combine to form one product molecule. Even though H_2O is a more complex molecule than either H_2 and O_2, the fact that there is a net decrease of one molecule and gases are converted to liquid ensures that the number of microstates will be diminished and hence $\Delta S°$ is negative.
(b) A solid is converted to two gaseous products. Therefore, $\Delta S°$ is positive.
(c) The same number of molecules is involved in the reactants as in the product. Furthermore, all molecules are diatomic and therefore of similar complexity. As a result, we cannot predict the sign of $\Delta S°$, but we know that the change must be quite small in magnitude.

Practice Exercise Discuss qualitatively the sign of the entropy change expected for each of the following processes:

(a) $I_2(s) \longrightarrow 2I(g)$
(b) $2Zn(s) + O_2(g) \longrightarrow 2ZnO(s)$
(c) $N_2(g) + O_2(g) \longrightarrow 2NO(g)$

Similar problems: 17.13 and 17.14.

Entropy Changes in the Surroundings

Next we see how ΔS_{surr} is calculated. When an exothermic process takes place in the system, the heat transferred to the surroundings enhances motion of the molecules in the surroundings. Consequently, there is an increase in the number of

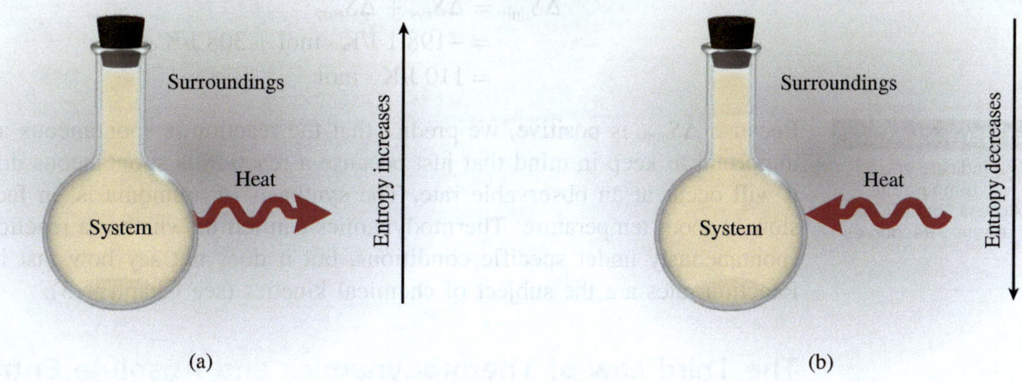

Figure 17.11 *(a) An exothermic process transfers heat from the system to the surroundings and results in an increase in the entropy of the surroundings. (b) An endothermic process absorbs heat from the surroundings and thereby decreases the entropy of the surroundings.*

microstates and the entropy of the surroundings increases. Conversely, an endothermic process in the system absorbs heat from the surroundings and so decreases the entropy of the surroundings because molecular motion decreases (Figure 17.11). For constant-pressure processes the heat change is equal to the enthalpy change of the system, ΔH_{sys}. Therefore, the change in entropy of the surroundings, ΔS_{surr}, is proportional to ΔH_{sys}:

$$\Delta S_{surr} \propto -\Delta H_{sys}$$

The minus sign is used because if the process is exothermic, ΔH_{sys} is negative and ΔS_{surr} is a positive quantity, indicating an increase in entropy. On the other hand, for an endothermic process, ΔH_{sys} is positive and the negative sign ensures that the entropy of the surroundings decreases.

The change in entropy for a given amount of heat absorbed also depends on the temperature. If the temperature of the surroundings is high, the molecules are already quite energetic. Therefore, the absorption of heat from an exothermic process in the system will have relatively little impact on molecular motion and the resulting increase in entropy of the surroundings will be small. However, if the temperature of the surroundings is low, then the addition of the same amount of heat will cause a more drastic increase in molecular motion and hence a larger increase in entropy. By analogy, someone coughing in a crowded restaurant will not disturb too many people, but someone coughing in a library definitely will. From the inverse relationship between ΔS_{surr} and temperature T (in kelvins)—that is, the higher the temperature, the smaller the ΔS_{surr} and vice versa—we can rewrite the abovementioned relationship as

$$\Delta S_{surr} = \frac{-\Delta H_{sys}}{T} \tag{17.8}$$

Let us now apply the procedure for calculating ΔS_{sys} and ΔS_{surr} to the synthesis of ammonia and ask whether the reaction is spontaneous at 25°C:

$$N_2(g) + 3H_2(g) \longrightarrow 2NH_3(g) \qquad \Delta H^\circ_{rxn} = -91.8\ \text{kJ/mol}$$

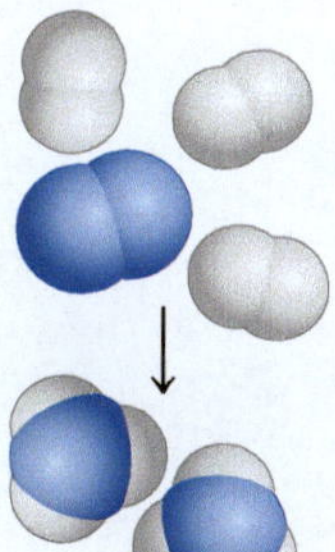

The synthesis of NH_3 from N_2 and H_2.

From Example 17.2(b) we have $\Delta S_{sys} = -198.1$ J/K · mol, and substituting ΔH_{sys} (−91.8 kJ/mol) in Equation (17.8), we obtain

$$\Delta S_{surr} = \frac{-(-91.8 \times 1000)\ \text{J/mol}}{298\ \text{K}} = 308\ \text{J/K} \cdot \text{mol}$$

The change in entropy of the universe is

$$\begin{aligned}\Delta S_{univ} &= \Delta S_{sys} + \Delta S_{surr}\\ &= -198.1 \text{ J/K} \cdot \text{mol} + 308 \text{ J/K} \cdot \text{mol}\\ &= 110 \text{ J/K} \cdot \text{mol}\end{aligned}$$

Student Hot Spot

Student data indicate you may struggle with temperature and entropy. Access your eBook for additional Learning Resources on this topic.

Because ΔS_{univ} is positive, we predict that the reaction is spontaneous at 25°C. It is important to keep in mind that just because a reaction is spontaneous does not mean it will occur at an observable rate. The synthesis of ammonia is, in fact, extremely slow at room temperature. Thermodynamics can tell us whether a reaction will occur spontaneously under specific conditions, but it does not say how fast it will occur. Reaction rates are the subject of chemical kinetics (see Chapter 13).

The Third Law of Thermodynamics and Absolute Entropy

Finally, it is appropriate to consider the third law of thermodynamics briefly in connection with the determination of entropy values. So far we have related entropy to microstates—the greater the number of microstates a system possesses, the larger is the entropy of the system. Consider a perfect crystalline substance at absolute zero (0 K). Under these conditions, molecular motions are kept at a minimum and the number of microstates (W) is one (there is only one way to arrange the atoms or molecules to form a perfect crystal). From Equation (17.1) we write

$$\begin{aligned}S &= k \ln W\\ &= k \ln 1 = 0\end{aligned}$$

According to the ***third law of thermodynamics,*** *the entropy of a perfect crystalline substance is zero at the absolute zero of temperature.* As the temperature increases, the freedom of motion increases and hence also the number of microstates. Thus, the entropy of any substance at a temperature above 0 K is greater than zero. Note also that if the crystal is impure or if it has defects, then its entropy is greater than zero even at 0 K, because it would not be perfectly ordered and the number of microstates would be greater than one.

The important point about the third law of thermodynamics is that it enables us to determine the *absolute* entropies of substances. Starting with the knowledge that the entropy of a pure crystalline substance is zero at absolute zero, we can measure the increase in entropy of the substance when it is heated from 0 K to, say, 298 K. The change in entropy, ΔS, is given by

$$\begin{aligned}\Delta S &= S_f - S_i\\ &= S_f\end{aligned}$$

because S_i is zero. The entropy of the substance at 298 K, then, is given by ΔS or S_f, which is called the absolute entropy because this is the *true* value and not a value derived using some arbitrary reference as in the case of standard enthalpy of formation. Thus, the entropy values quoted so far and those listed in Appendix 2 are all absolute entropies. Because measurements are carried out at 1 atm, we usually refer to absolute entropies as standard entropies. In contrast, we cannot have the absolute energy (U) or enthalpy (H) of a substance because the zero of energy or enthalpy is undefined. Figure 17.12 shows the change (increase) in entropy of a substance with temperature. At absolute zero, it has a zero entropy value (assuming that it is a perfect crystalline substance). As it is heated, its entropy increases gradually because of greater molecular motion. At the melting point, there is a sizable increase in entropy as the liquid state is formed. Further heating increases the entropy of the liquid again due to enhanced molecular motion. At the boiling point there is a large increase in entropy as a result of the liquid-to-vapor transition. Beyond that temperature, the entropy of the gas continues to rise with increasing temperature.

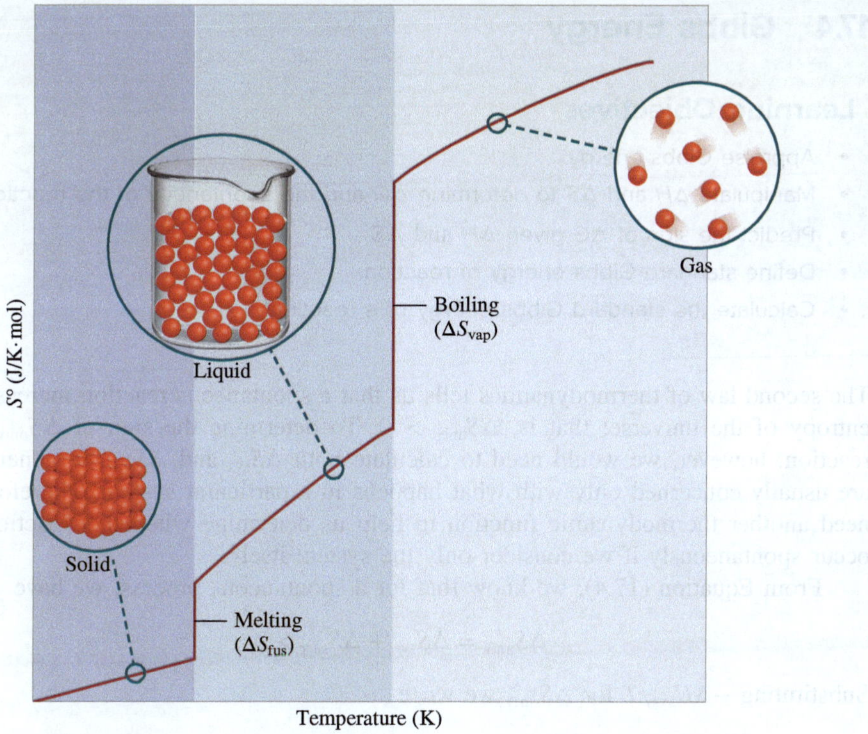

Figure 17.12 *Entropy increase of a substance as the temperature rises from absolute zero.*

Summary of Concepts & Facts

- Any spontaneous process must lead to a net increase in entropy in the universe (second law of thermodynamics).
- The standard entropy of a chemical reaction can be calculated from the absolute entropies of reactants and products.
- The third law of thermodynamics states that the entropy of a perfect crystalline substance is zero at 0 K. This law enables us to measure the absolute entropies of substances.

Review of Concepts & Facts

17.3.1 Calculate the standard entropy change for the following reaction at 25°C:

$$2NO_2(g) \longrightarrow 2NO(g) + O_2(g)$$

17.3.2 Consider the gas-phase reaction of A_2 (blue) and B_2 (orange) to form AB_3.
(a) Write a balanced equation for the reaction.
(b) What is the sign of ΔS for the reaction?

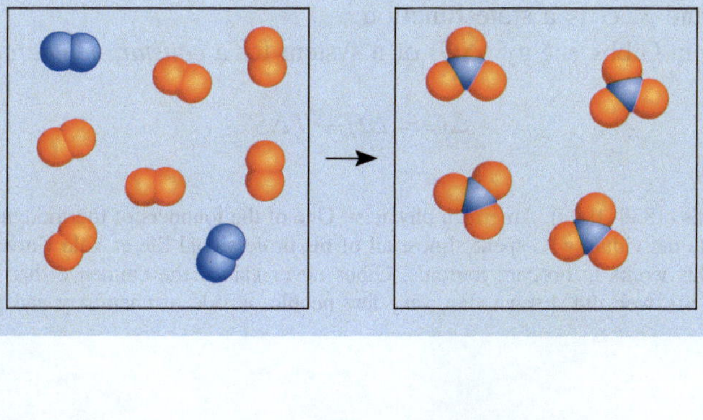

17.4 Gibbs Energy

Learning Objectives

- Appraise Gibbs energy.
- Manipulate ΔH and ΔS to determine ΔG and the spontaneity of the reaction.
- Predict the sign of ΔG given ΔH and ΔS.
- Define standard Gibbs energy of reaction.
- Calculate the standard Gibbs energy of a reaction.

The second law of thermodynamics tells us that a spontaneous reaction increases the entropy of the universe; that is, $\Delta S_{\text{univ}} > 0$. To determine the sign of ΔS_{univ} for a reaction, however, we would need to calculate both ΔS_{sys} and ΔS_{surr}. In general, we are usually concerned only with what happens in a particular system. Therefore, we need another thermodynamic function to help us determine whether a reaction will occur spontaneously if we consider only the system itself.

From Equation (17.4), we know that for a spontaneous process, we have

$$\Delta S_{\text{univ}} = \Delta S_{\text{sys}} + \Delta S_{\text{surr}} > 0$$

Substituting $-\Delta H_{\text{sys}}/T$ for ΔS_{surr}, we write

$$\Delta S_{\text{univ}} = \Delta S_{\text{sys}} - \frac{\Delta H_{\text{sys}}}{T} > 0$$

Multiplying both sides of the equation by T gives

$$T\Delta S_{\text{unix}} = -\Delta H_{\text{sys}} + T\Delta S_{\text{sys}} > 0$$

Now we have a criterion for a spontaneous reaction that is expressed only in terms of the properties of the system (ΔH_{sys} and ΔS_{sys}). For convenience, we can change the preceding equation by multiplying it throughout by -1 and replacing the $>$ sign with $<$:

$$-T\Delta S_{\text{univ}} = \Delta H_{\text{sys}} - T\Delta S_{\text{sys}} < 0$$

This equation says that for a process carried out at constant pressure and temperature T, if the changes in enthalpy and entropy of the system are such that $\Delta H_{\text{sys}} - T\Delta S_{\text{sys}}$ is less than zero, the process must be spontaneous.

In order to express the spontaneity of a reaction more directly, we introduce another thermodynamic function called ***Gibbs***[†] ***energy (G),*** (previously known as ***free energy***):

$$G = H - TS \tag{17.9}$$

All quantities in Equation (17.9) pertain to the system, and T is the temperature of the system. You can see that G has units of energy (both H and TS are in energy units). Like H and S, G is a state function.

The change in Gibbs energy (ΔG) of a system for a *constant-temperature* process is

$$\Delta G = \Delta H - T\Delta S \tag{17.10}$$

[†]Josiah Willard Gibbs (1839–1903). American physicist. One of the founders of thermodynamics, Gibbs was a modest and private individual who spent almost all of his professional life at Yale University. Because he published most of his works in obscure journals, Gibbs never gained the eminence that his contemporary and admirer James Maxwell did. Even today, very few people outside of chemistry and physics have ever heard of Gibbs.

CHEMISTRY *in Action*

The Efficiency of Heat Engines

An engine is a machine that converts energy to work; a *heat engine* is a machine that converts *thermal energy* to work. Heat engines play an essential role in our technological society; they range from automobile engines to the giant steam turbines that run generators to produce electricity. Regardless of the type of heat engine, its efficiency level is of great importance; that is, for a given amount of heat input, how much useful work can we get out of the engine? The second law of thermodynamics helps us answer this question.

The figure shows a simple form of a heat engine. A cylinder fitted with a weightless piston is initially at temperature T_1. Next, the cylinder is heated to a higher temperature T_2. The gas in the cylinder expands and pushes up the piston. Finally the cylinder is cooled down to T_1 and the apparatus returns to its original state. By repeating this cycle, the up-and-down movement of the piston can be made to do mechanical work.

A unique feature of heat engines is that some heat must be given off to the surroundings when they do work. With the piston in the up position, no further work can be done if we do not cool the cylinder back to T_1. The cooling process removes some of the thermal energy that could otherwise be converted to work and thereby places a limit on the efficiency of heat engines.

The figure shows the heat transfer processes in a heat engine. Initially, a certain amount of heat flows from the heat reservoir (at temperature T_h) into the engine. As the engine does work, some of the heat is given off to the surroundings, or the heat sink (at temperature T_c). By definition, the efficiency of a heat engine is

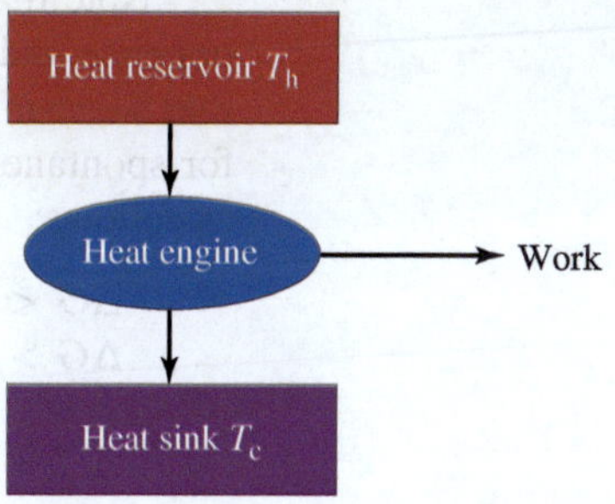

Heat transfers during the operation of a heat engine.

$$\text{efficiency} = \frac{\text{useful work output}}{\text{energy input}} \times 100\%$$

Analysis based on the second law shows that efficiency can also be expressed as

$$\text{efficiency} = \left(1 - \frac{T_c}{T_h}\right) \times 100\%$$

$$= \frac{T_h - T_c}{T_h} \times 100\%$$

Thus, the efficiency of a heat engine is given by the difference in temperatures between the heat reservoir and the heat sink (both in kelvins), divided by the temperature of the heat reservoir. In practice, we can make $(T_h - T_c)$ as large as possible, but because T_c cannot be zero and T_h cannot be infinite, the efficiency of a heat engine can therefore never be 100 percent.

At a power plant, superheated steam at about 560°C (833 K) is used to drive a turbine for electricity generation. The temperature of the heat sink is about 38°C (or 311 K). The efficiency is given by

$$\text{efficiency} = \frac{833\text{ K} - 311\text{ K}}{833\text{ K}} \times 100\%$$

$$= 63\%$$

In practice, because of friction, heat loss, and other complications, the maximum efficiency of a steam turbine is only about 40 percent. Therefore, for every ton of coal used at a power plant, 0.40 ton generates electricity while the rest of it ends up warming the surroundings!

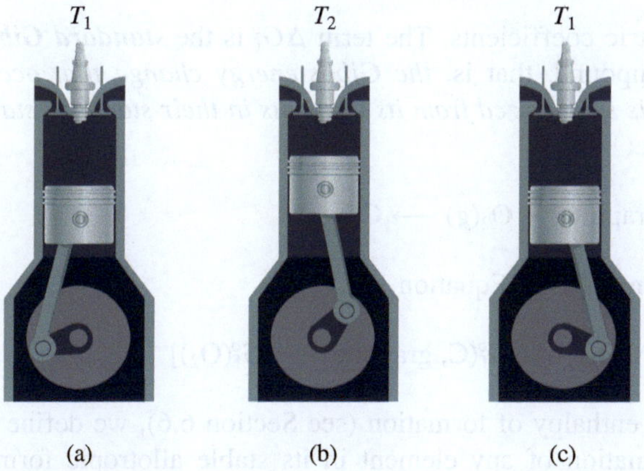

A simple heat engine. (a) The engine is initially at T_1. (b) When heated to T_2, the piston is pushed up due to gas expansion. (c) When cooled to T_1, the piston returns to the original position.

In this context, Gibbs energy is *the energy available to do work*. Thus, if a particular reaction is accompanied by a release of usable energy (that is, if ΔG is negative), this fact alone guarantees that it is spontaneous, and there is no need to worry about what happens to the rest of the universe.

Note that we have merely organized the expression for the entropy change of the universe and equating the Gibbs-energy change of the system (ΔG) with $-T\Delta S_{univ}$, so that we can focus on changes in the system. We can now summarize the conditions for spontaneity and equilibrium at constant temperature and pressure in terms of ΔG as follows:

$\Delta G < 0$	The reaction is spontaneous in the forward direction.
$\Delta G > 0$	The reaction is nonspontaneous. The reaction is spontaneous in the opposite direction.
$\Delta G = 0$	The system is at equilibrium. There is no net change.

Standard Gibbs-Energy Changes

The ***standard Gibbs-energy of reaction*** **(ΔG°_{rxn})** is *the Gibbs-energy change for a reaction when it occurs under standard-state conditions, when reactants in their standard states are converted to products in their standard states*. Table 17.3 summarizes the conventions used by chemists to define the standard states of pure substances as well as solutions. To calculate ΔG°_{rxn} we start with the equation

$$aA + bB \longrightarrow cC + dD$$

The standard Gibbs-energy change for this reaction is given by

$$\Delta G^\circ_{rxn} = [c\Delta G^\circ_f(C) + d\Delta G^\circ_f(D)] - [a\Delta G^\circ_f(A) + b\Delta G^\circ_f(B)] \quad (17.11)$$

or, in general,

$$\Delta G^\circ_{rxn} = \Sigma n\Delta G^\circ_f(\text{products}) - \Sigma m\Delta G^\circ_f(\text{reactants}) \quad (17.12)$$

where m and n are stoichiometric coefficients. The term ΔG°_f is the ***standard Gibbs-energy of formation*** of a compound, that is, *the Gibbs-energy change that occurs when 1 mole of the compound is synthesized from its elements in their standard states*. For the combustion of graphite:

$$C(\text{graphite}) + O_2(g) \longrightarrow CO_2(g)$$

the standard Gibbs-energy change [from Equation (17.12)] is

$$\Delta G^\circ_{rxn} = \Delta G^\circ_f(CO_2) - [\Delta G^\circ_f(C, \text{graphite}) + \Delta G^\circ_f(O_2)]$$

As in the case of the standard enthalpy of formation (see Section 6.6), we define the standard Gibbs energy of formation of any element in its stable allotropic form at 1 atm and 25°C as zero. Thus,

$$\Delta G^\circ_f(C, \text{graphite}) = 0 \quad \text{and} \quad \Delta G^\circ_f(O_2) = 0$$

Therefore, the standard Gibb-energy change for the reaction in this case is equal to the standard Gibbs energy of formation of CO_2:

$$\Delta G^\circ_{rxn} = \Delta G^\circ_f(CO_2)$$

Appendix 2 lists the values of ΔG°_f for a number of compounds.

Table 17.3

Conventions for Standard States

State of Matter	Standard State
Gas	1 atm pressure
Liquid	Pure liquid
Solid	Pure solid
Elements*	$\Delta G^\circ_f = 0$
Solution	1 molar concentration

*The most stable allotropic form at 25°C and 1 atm.

Student Hot Spot

Student data indicate you may struggle with standard Gibbs-energy changes. Access your eBook for additional Learning Resources on this topic.

Calculations of standard Gibbs-energy changes are handled as shown in Example 17.4.

Example 17.4

Calculate the standard Gibbs-energy changes for the following reactions at 25°C.

(a) $CH_4(g) + 2O_2(g) \longrightarrow CO_2(g) + 2H_2O(l)$

(b) $2MgO(s) \longrightarrow 2Mg(s) + O_2(g)$

Strategy To calculate the standard Gibbs-energy change of a reaction, we look up the standard Gibbs energies of formation of reactants and products in Appendix 2 and apply Equation (17.12). Note that all the stoichiometric coefficients have no units so ΔG°_{rxn} is expressed in units of kJ/mol, and ΔG°_f for O_2 is zero because it is the stable allotropic element at 1 atm and 25°C.

(a) According to Equation (17.12), we write

$$\Delta G^\circ_{rxn} = [\Delta G^\circ_f(CO_2) + 2\Delta G^\circ_f(H_2O)] - [\Delta G^\circ_f(CH_4) + 2\Delta G^\circ_f(O_2)]$$

We insert the appropriate values from Appendix 2:

$$\Delta G^\circ_{rxn} = [(-394.4 \text{ kJ/mol}) + (2)(-237.1 \text{ kJ/mol})] - [(-50.8 \text{ kJ/mol}) + (2)(0 \text{ kJ/mol})]$$
$$= -817.8 \text{ kJ/mol}$$

(b) The equation is

$$\Delta G^\circ_{rxn} = [2\Delta G^\circ_f(Mg) + \Delta G^\circ_f(O_2)] - [2\Delta G^\circ_f(MgO)]$$

From data in Appendix 2 we write

$$\Delta G^\circ_{rxn} = [(2)(0 \text{ kJ/mol}) + (0 \text{ kJ/mol})] - [(2)(-569.3 \text{ kJ/mol})]$$
$$= 1139 \text{ kJ/mol}$$

Practice Exercise Calculate the standard Gibbs-energy changes for the following reactions at 25°C:

(a) $H_2(g) + Br_2(l) \longrightarrow 2HBr(g)$

(b) $2C_2H_6(g) + 7O_2(g) \longrightarrow 4CO_2(g) + 6H_2O(l)$

Similar problems: 17.17 and 17.18.

Temperature Dependence of ΔG

To predict the sign of ΔG, according to Equation (17.10), we need to know both ΔH and ΔS. A negative ΔH (an exothermic reaction) and a positive ΔS (a reaction that results in an increase in the microstates of the system) tend to make ΔG negative, although temperature may also influence the *direction* of a spontaneous reaction. The four possible outcomes of this relationship are as follows:

- If both ΔH and ΔS are positive, then ΔG will be negative only when the $T\Delta S$ term is greater in magnitude than ΔH. This condition is met when T is large.
- If ΔH is positive and ΔS is negative, ΔG will always be positive, regardless of temperature.
- If ΔH is negative and ΔS is positive, then ΔG will always be negative regardless of temperature.
- If ΔH is negative and ΔS is negative, then ΔG will be negative only when $T\Delta S$ is smaller in magnitude than ΔH. This condition is met when T is small.

The temperatures that will cause ΔG to be negative for the first and last cases depend on the actual values of ΔH and ΔS of the system. Table 17.4 summarizes the effects of the possibilities just described.

Table 17.4 Factors Affecting the Sign of ΔG in the Relationship $\Delta G = \Delta H - T\Delta S$

ΔH	ΔS	ΔG	Example
+	+	Reaction proceeds spontaneously at high temperatures. At low temperatures, reaction is spontaneous in the reverse direction.	$2HgO(s) \longrightarrow 2Hg(l) + O_2(g)$
+	−	ΔG is always positive. Reaction is spontaneous in the reverse direction at all temperatures.	$3O_2(g) \longrightarrow 2O_3(g)$
−	+	ΔG is always negative. Reaction proceeds spontaneously at all temperatures.	$2H_2O_2(aq) \longrightarrow 2H_2O(l) + O_2(g)$
−	−	Reaction proceeds spontaneously at low temperatures. At high temperatures, the reverse reaction becomes spontaneous.	$NH_3(g) + HCl(g) \longrightarrow NH_4Cl(s)$

Before we apply the change in Gibbs energy to predict reaction spontaneity, it is useful to distinguish between ΔG and $\Delta G°$. Suppose we carry out a reaction in solution with all the reactants in their standard states (that is, all at 1 *M* concentration). As soon as the reaction starts, the standard-state condition no longer exists for the reactants or the products because their concentrations are different from 1 *M*. Under nonstandard-state conditions, we must use the sign of ΔG rather than that of $\Delta G°$ to predict the direction of the reaction. The sign of $\Delta G°$, on the other hand, tells us whether the products or the reactants are favored when the reacting system reaches equilibrium. Thus, a negative value of $\Delta G°$ indicates that the reaction favors product formation whereas a positive value of $\Delta G°$ indicates that there will be more reactants than products at equilibrium.

We will now consider two specific applications of Equation (17.10).

Temperature and Chemical Reactions

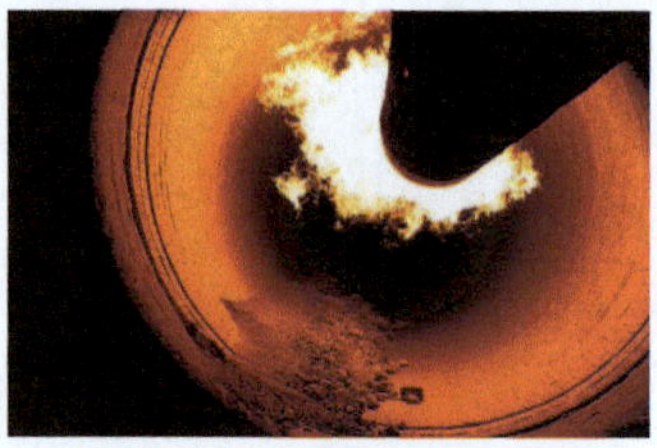

The production of quicklime (CaO) from limestone ($CaCO_3$) in a rotary kiln.
Drpixel/Shutterstock

Calcium oxide (CaO), also called quicklime, is an extremely valuable inorganic substance used in steelmaking, production of calcium metal, the paper industry, water treatment, and pollution control. It is prepared by decomposing limestone ($CaCO_3$) in a kiln at a high temperature:

$$CaCO_3(s) \rightleftharpoons CaO(s) + CO_2(g)$$

The reaction is reversible, and CaO readily combines with CO_2 to form $CaCO_3$. The pressure of CO_2 in equilibrium with $CaCO_3$ and CaO increases with temperature. In the industrial preparation of quicklime, the system is never maintained at equilibrium; rather, CO_2 is constantly removed from the kiln to shift the equilibrium from left to right, promoting the formation of calcium oxide.

The important information for the practical chemist is the temperature at which the decomposition of $CaCO_3$ becomes appreciable (that is, the temperature at which the reaction begins to favor products). We can make a reliable estimate of that temperature as follows. First, we calculate $\Delta H°$ and $\Delta S°$ for the reaction at 25°C, using the data in Appendix 2. To determine $\Delta H°$ we apply Equation (6.18):

$$\begin{aligned}\Delta H° &= [\Delta H_f°(CaO) + \Delta H_f°(CO_2)] - [\Delta H_f°(CaCO_3)]\\ &= [(-635.6 \text{ kJ/mol}) + (-393.5 \text{ kJ/mol})] - (-1206.9 \text{ kJ/mol})\\ &= 177.8 \text{ kJ/mol}\end{aligned}$$

Next, we apply Equation (17.6) to find $\Delta S°$:

$$\begin{aligned}\Delta S° &= [S°(CaO) + S°(CO_2)] - S°(CaCO_3)\\ &= [(39.8 \text{ J/K}\cdot\text{mol}) + (213.6 \text{ J/K}\cdot\text{mol})] - (92.9 \text{ J/K}\cdot\text{mol})\\ &= 160.5 \text{ J/K}\cdot\text{mol}\end{aligned}$$

From Equation (17.10),

$$\Delta G° = \Delta H° - T\Delta S°$$

we obtain

$$\Delta G^\circ = 177.8 \text{ kJ/mol} - (298 \text{ K})(160.5 \text{ J/K} \cdot \text{mol})\left(\frac{1 \text{ kJ}}{1000 \text{ J}}\right)$$
$$= 130.0 \text{ kJ/mol}$$

Because ΔG° is a large positive quantity, we conclude that the reaction is not favored for product formation at 25°C (or 298 K). Indeed, the pressure of CO_2 is so low at room temperature that it cannot be measured. To make ΔG° negative, we first have to find the temperature at which ΔG° is zero; that is,

$$0 = \Delta H^\circ - T\Delta S^\circ$$

or

$$T = \frac{\Delta H^\circ}{\Delta S^\circ}$$
$$= \frac{(177.8 \text{ kJ/mol})(1000 \text{ J/1 kJ})}{160.5 \text{ J/K} \cdot \text{mol}}$$
$$= 1108 \text{ K or } 835°\text{C}$$

At a temperature higher than 835°C, ΔG° becomes negative, indicating that the reaction now favors the formation of CaO and CO_2. For example, at 840°C, or 1113 K,

$$\Delta G^\circ = \Delta H^\circ - T\Delta S^\circ$$
$$= 177.8 \text{ kJ/mol} - (1113 \text{ K})(160.5 \text{ J/K} \cdot \text{mol})\left(\frac{1 \text{ kJ}}{1000 \text{ J}}\right)$$
$$= -0.8 \text{ kJ/mol}$$

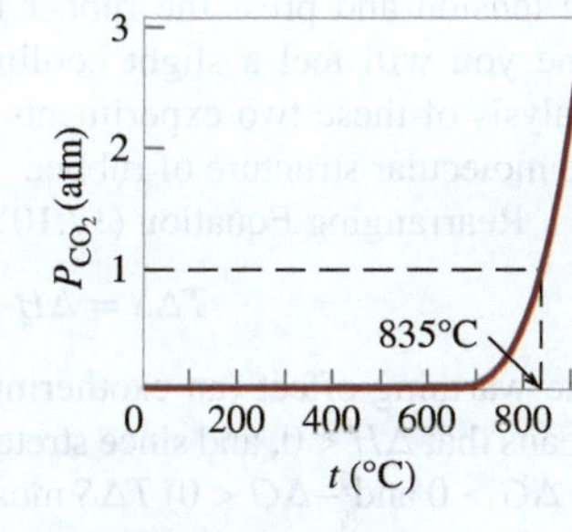

Figure 17.13 *Equilibrium pressure of CO_2 from the decomposition of $CaCO_3$, as a function of temperature. This curve is calculated by assuming that ΔH° and ΔS° of the reaction do not change with temperature.*

Two points are worth making about such a calculation. First, we used the ΔH° and ΔS° values at 25°C to calculate changes that occur at a much higher temperature. Because both ΔH° and ΔS° change with temperature, this approach will not give us an accurate value of ΔG°, but it is good enough for "ballpark" estimates. Second, we should not be misled into thinking that nothing happens below 835°C and that at 835°C $CaCO_3$ suddenly begins to decompose. Far from it. The fact that ΔG° is a positive value at some temperature below 835°C does not mean that no CO_2 is produced, but rather that the pressure of the CO_2 gas formed at that temperature will be below atm (its standard-state value; see Table 17.3). As Figure 17.13 shows, the pressure of CO_2 at first increases very slowly with temperature; it becomes easily measurable above 700°C. The significance of 835°C is that this is the temperature at which the equilibrium pressure of CO_2 reaches 1 atm. Above 835°C, the equilibrium pressure of CO_2 exceeds 1 atm.

Student Hot Spot

Student data indicate you may struggle with the effect of temperature on Gibbs-energy changes. Access your eBook for additional Learning Resources on this topic.

Phase Transitions

At the temperature at which a phase transition occurs (that is, at the melting point or boiling point) the system is at equilibrium ($\Delta G = 0$), so Equation (17.10) becomes

$$\Delta G = \Delta H - T\Delta S$$
$$0 = \Delta H - T\Delta S$$

or

$$\Delta S = \frac{\Delta H}{T}$$

Let us first consider the ice-water equilibrium. For the ice $\longrightarrow$ water transition, ΔH is the molar heat of fusion (see Table 11.9), and T is the melting point. The entropy change is therefore

$$\Delta S_{\text{ice}\rightarrow\text{water}} = \frac{6010 \text{ J/mol}}{273 \text{ K}}$$
$$= 22.0 \text{ J/K} \cdot \text{mol}$$

Thus, when 1 mole of ice melts at 0°C, there is an increase in entropy of 22.0 J/K · mol. The increase in entropy is consistent with the increase in microstates from solid to liquid. Conversely, for the water $\longrightarrow$ ice transition, the decrease in entropy is given by

$$\Delta S_{\text{water}\rightarrow\text{ice}} = \frac{-6010 \text{ J/mol}}{273 \text{ K}}$$
$$= -22.0 \text{ J/K} \cdot \text{mol}$$

CHEMISTRY *in Action*

The Thermodynamics of a Rubber Band

We all know how useful a rubber band can be. But not everyone is aware that a rubber band has some very interesting thermodynamic properties based on its structure.

You can easily perform the following experiments with a rubber band that is at least 0.5 cm wide. Quickly stretch the rubber band and then press it against your lips. You will feel a slight warming effect. Next, reverse the process. Stretch a rubber band and hold it in position for a few seconds. Then quickly release the tension and press the rubber band against your lips. This time you will feel a slight cooling effect. A thermodynamic analysis of these two experiments can tell us something about the molecular structure of rubber.

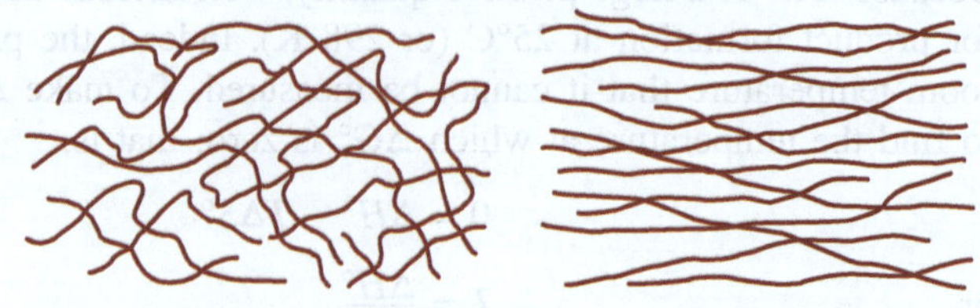

(a) Rubber molecules in their normal state. Note the high degree of entanglement (large number of microstates and a high entropy). (b) Under tension, the molecules line up and the arrangement becomes much more ordered (a small number of microstates and a low entropy).

Rearranging Equation (17.10) ($\Delta G = \Delta H - T\Delta S$) gives

$$T\Delta S = \Delta H - \Delta G$$

The warming effect (an exothermic process) due to stretching means that $\Delta H < 0$, and since stretching is nonspontaneous (that is, $\Delta G > 0$ and $-\Delta G < 0$) $T\Delta S$ must be negative. Because T, the absolute temperature, is always positive, we conclude that ΔS due to stretching must be negative, and therefore rubber in its natural state is more entangled (has more microstates) than when it is under tension.

When the tension is removed, the stretched rubber band spontaneously snaps back to its original shape; that is, ΔG is negative and $-\Delta G$ is positive. The cooling effect means that it is an endothermic process ($\Delta H > 0$), so that $T\Delta S$ is positive. Thus, the entropy of the rubber band increases when it goes from the stretched state to the natural state.

In the laboratory, we normally carry out unidirectional changes, that is, either ice to water or water to ice transition. We can calculate entropy change in each case using the equation $\Delta S = \Delta H/T$ as long as the temperature remains at 0°C. The same procedure can be applied to the water ⟶ steam transition. In this case ΔH is the heat of vaporization and T is the boiling point of water. Example 17.5 examines the phase transitions in benzene.

Example 17.5

Liquid and solid benzene in equilibrium at 5.5°C.
Ken Karp/McGraw Hill

The molar heats of fusion and vaporization of benzene are 10.9 kJ/mol and 31.0 kJ/mol, respectively. Calculate the entropy changes for the solid ⟶ liquid and liquid ⟶ vapor transitions for benzene. At 1 atm pressure, benzene melts at 5.5°C and boils at 80.1°C.

Strategy At the melting point, liquid and solid benzene are at equilibrium, so $\Delta G = 0$. From Equation (17.10) we have $\Delta G = 0 = \Delta H - T\Delta S$ or $\Delta S = \Delta H/T$. To calculate the entropy change for the solid benzene ⟶ liquid benzene transition, we write $\Delta S_{fus} = \Delta H_{fus}/T_f$. Here ΔH_{fus} is positive for an endothermic process, so ΔS_{fus} is also positive, as expected for a solid to liquid transition. The same procedure applies to the liquid benzene ⟶ vapor benzene transition. What temperature unit should be used?

Solution The entropy change for melting 1 mole of benzene at 5.5°C is

$$\begin{aligned}\Delta S_{fus} &= \frac{\Delta H_{fus}}{T_f} \\ &= \frac{(10.9\ \text{kJ/mol})(1000\ \text{J/1 kJ})}{(5.5 + 273)\ \text{K}} \\ &= 39.1\ \text{J/K} \cdot \text{mol}\end{aligned}$$

(Continued)

Similarly, the entropy change for boiling 1 mole of benzene at 80.1°C is

$$\begin{aligned}\Delta S_{\text{vap}} &= \frac{\Delta H_{\text{vap}}}{T_{\text{b}}}\\ &= \frac{(31.0\ \text{kJ/mol})(1000\ \text{J}/1\ \text{kJ})}{(80.1 + 273)\ \text{K}}\\ &= 87.8\ \text{J/K} \cdot \text{mol}\end{aligned}$$

Check Because vaporization creates more microstates than the melting process, $\Delta S_{\text{vap}} > \Delta S_{\text{fus}}$.

Practice Exercise The molar heats of fusion and vaporization of argon are 1.3 kJ/mol and 6.3 kJ/mol, and argon's melting point and boiling point are −190°C and −186°C, respectively. Calculate the entropy changes for fusion and vaporization.

Similar problem: 17.64.

Summary of Concepts & Facts

- Under conditions of constant temperature and pressure, the Gibbs-energy change ΔG is less than zero for a spontaneous process and greater than zero for a nonspontaneous process. For an equilibrium process, $\Delta G = 0$.
- For a chemical or physical process at constant temperature and pressure, $\Delta G = \Delta H - T\Delta S$. This equation can be used to predict the spontaneity of a process.
- The standard Gibbs-energy change for a reaction, $\Delta G°$, can be calculated from the standard Gibbs energies of formation of reactants and products.

Review of Concepts & Facts

17.4.1 Under what circumstances will an endothermic reaction proceed spontaneously?

17.4.2 Explain why, in many reactions in which both the reactants and products are in the solution phase, ΔH often gives a good hint about the spontaneity of a reaction at 298 K.

17.4.3 Calculate ΔG for the following reaction at 25°C:

$$2\text{ICl}(g) \longrightarrow \text{I}_2(g) + \text{Cl}_2(g)$$

if $\Delta H = 26.9$ kJ/mol and $\Delta S = -11.3$ J/K · mol.

17.4.4 Consider the sublimation of iodine (I_2) at 45°C in the closed flask shown. If the enthalpy of sublimation is 62.4 kJ/mol, what is the ΔS for sublimation?

Ken Karp/McGraw Hill

17.5 Gibbs Energy and Chemical Equilibrium

Learning Objective

- Evaluate ΔG and $\Delta G°$ of a reaction at a specified temperature given Q or K.

As mentioned earlier, during the course of a chemical reaction not all the reactants and products will be at their standard states. Under this condition, the relationship between ΔG and $\Delta G°$, which can be derived from thermodynamics (and more particularly, the Gibbs energy due to the mixing of substances), is

$$\Delta G = \Delta G° + RT \ln Q \tag{17.13}$$

where R is the gas constant (8.314 J/K · mol), T is the absolute temperature of the reaction, and Q is the reaction quotient (see Section 14.4). We see that ΔG depends on two quantities: $\Delta G°$ and $RT \ln Q$. For a given reaction at temperature T the value of $\Delta G°$ is fixed but that of $RT \ln Q$ is not, because Q varies according to the composition of the reaction mixture. Let us consider two special cases:

Case 1: A large negative value of $\Delta G°$ will tend to make ΔG also negative. Thus, the net reaction will proceed from left to right until a significant amount of product has been formed. At that point, the $RT \ln Q$ term will become positive enough to match the negative $\Delta G°$ term in magnitude.

Case 2: A large positive $\Delta G°$ term will tend to make ΔG also positive. Thus, the net reaction will proceed from right to left until a significant amount of reactant has been formed. At that point the $RT \ln Q$ term will become negative enough to match the positive $\Delta G°$ term in magnitude.

At equilibrium, by definition, $\Delta G = 0$ and $Q = K$, where K is the equilibrium constant. Thus,

$$0 = \Delta G° + RT \ln K$$

or

$$\Delta G° = -RT \ln K \tag{17.14}$$

In this equation, K_P is used for gases and K_c for reactions in solution. Note that the larger the K is, the more negative $\Delta G°$ is. For chemists, Equation (17.14) is one of the most important equations in thermodynamics because it enables us to find the equilibrium constant of a reaction if we know the change in standard Gibbs energy and vice versa.

It is significant that Equation (17.14) relates the equilibrium constant to the *standard* Gibbs-energy change $\Delta G°$ rather than to the *actual* Gibbs-energy change ΔG. The actual Gibbs-energy change of the system varies as the reaction progresses and becomes zero at equilibrium. On the other hand, $\Delta G°$ is a constant for a particular reaction at a given temperature. Figure 17.14 shows plots of the Gibbs energy of a reacting system versus the extent of the reaction for two types of reactions. As you can see, if $\Delta G° < 0$, the products are favored over reactants at equilibrium. Conversely, if $\Delta G° > 0$, there will be more reactants than products at equilibrium. Table 17.5 summarizes the three possible relations between $\Delta G°$ and K, as predicted by Equation (17.14). Remember this important distinction: It is the sign of ΔG and not that of $\Delta G°$ that determines the direction of reaction spontaneity. The sign of $\Delta G°$ only tells us the relative amounts of products and reactants when equilibrium is reached, not the direction of the net reaction.

For reactions having very large or very small equilibrium constants, it is generally very difficult, if not impossible, to measure the K values by monitoring the

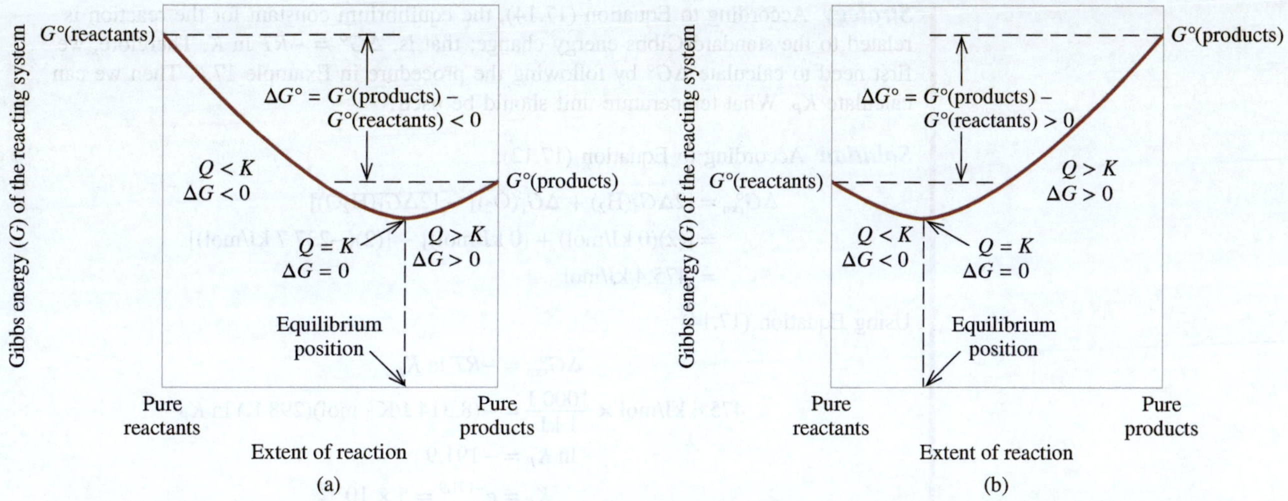

Figure 17.14 *(a) $\Delta G° < 0$. At equilibrium, there is a significant conversion of reactants to products. (b) $\Delta G° > 0$. At equilibrium, reactants are favored over products. In both cases, the net reaction toward equilibrium is from left to right (reactants to products) if $Q < K$ and right to left (products to reactants) if $Q > K$. At equilibrium, $Q = K$.*

Table 17.5 **Relation Between $\Delta G°$ and K as Predicted by the Equation $\Delta G° = -RT \ln K$**

K	$\ln K$	$\Delta G°$	Comments
> 1	Positive	Negative	Products are favored over reactants at equilibrium.
= 1	0	0	Products and reactants are equally favored at equilibrium.
< 1	Negative	Positive	Reactants are favored over products at equilibrium.

concentrations of all the reacting species. Consider, for example, the formation of nitric oxide from molecular nitrogen and molecular oxygen:

$$N_2(g) + O_2(g) \rightleftharpoons 2NO(g)$$

At 25°C, the equilibrium constant K_P is

$$K_P = \frac{P_{NO}^2}{P_{N_2}P_{O_2}} = 4.0 \times 10^{-31}$$

The very small value of K_P means that the concentration of NO at equilibrium will be exceedingly low. In such a case the equilibrium constant is more conveniently obtained from $\Delta G°$. (As we have seen, $\Delta G°$ can be calculated from $\Delta H°$ and $\Delta S°$ or from Gibbs energy of formation data.) On the other hand, the equilibrium constant for the formation of hydrogen iodide from molecular hydrogen and molecular iodine is near unity at room temperature:

$$H_2(g) + I_2(g) \rightleftharpoons 2HI(g)$$

For this reaction, it is easier to measure K_P and then calculate $\Delta G°$ using Equation (17.14) than to measure $\Delta H°$ and $\Delta S°$ and use Equation (17.10).

Examples 17.6, 17.7, and 17.8 illustrate the use of Equations (17.13) and (17.14).

Student Hot Spot

Student data indicate you may struggle with the relationship between Gibbs energy and equilibrium. Access your eBook for additional Learning Resources on this topic.

Example 17.6

Using data listed in Appendix 2, calculate the equilibrium constant (K_P) for the following reaction at 25°C:

$$2H_2O(l) \rightleftharpoons 2H_2(g) + O_2(g)$$

(Continued)

Strategy According to Equation (17.14), the equilibrium constant for the reaction is related to the standard Gibbs-energy change; that is, $\Delta G^\circ = -RT \ln K$. Therefore, we first need to calculate ΔG° by following the procedure in Example 17.4. Then we can calculate K_P. What temperature unit should be used?

Solution According to Equation (17.12),

$$\begin{aligned}\Delta G^\circ_{rxn} &= [2\Delta G^\circ_f(H_2) + \Delta G^\circ_f(O_2)] - [2\Delta G^\circ_f(H_2O)]\\ &= [(2)(0\ \text{kJ/mol}) + (0\ \text{kJ/mol})] - [(2)(-237.7\ \text{kJ/mol})]\\ &= 475.4\ \text{kJ/mol}\end{aligned}$$

Using Equation (17.14)

$$\Delta G^\circ_{rxn} = -RT \ln K_P$$

$$475.4\ \text{kJ/mol} \times \frac{1000\ \text{J}}{1\ \text{kJ}} = -(8.314\ \text{J/K} \cdot \text{mol})(298\ \text{K}) \ln K_P$$

$$\ln K_P = -191.9$$

$$K_P = e^{-191.9} = 5 \times 10^{-84}$$

Comment This extremely small equilibrium constant is consistent with the fact that water does not spontaneously decompose into hydrogen and oxygen gases at 25°C. Thus, a large positive ΔG° favors reactants over products at equilibrium.

Practice Exercise Calculate the equilibrium constant (K_P) for the reaction

$$2O_3(g) \longrightarrow 3O_2(g)$$

at 25°C.

Similar problems: 17.23 and 17.26.

Example 17.7

In Chapter 16 we discussed the solubility product of slightly soluble substances. Using the solubility product of silver chloride at 25°C (1.77×10^{-10}), calculate ΔG° for the process

$$AgCl(s) \rightleftharpoons Ag^+(aq) + Cl^-(aq)$$

Strategy According to Equation (17.14), the equilibrium constant for the reaction is related to standard Gibbs-energy change; that is, $\Delta G^\circ = -RT \ln K$. Because this is a heterogeneous equilibrium, the solubility product (K_{sp}) is the equilibrium constant. We calculate the standard Gibbs-energy change from the K_{sp} value of AgCl. What temperature unit should be used?

Solution The solubility equilibrium for AgCl is

$$AgCl(s) \rightleftharpoons Ag^+(aq) + Cl^-(aq)$$

$$K_{sp} = [Ag^+][Cl^-] = 1.77 \times 10^{-10}$$

Using Equation (17.14) we obtain

$$\begin{aligned}\Delta G^\circ &= -(8.314\ \text{J/K} \cdot \text{mol})(298\ \text{K}) \ln (1.77 \times 10^{-10})\\ &= 5.56 \times 10^4\ \text{J/mol}\\ &= 55.6\ \text{kJ/mol}\end{aligned}$$

Check The large, positive ΔG° indicates that AgCl is slightly soluble and that the equilibrium lies mostly to the left.

Practice Exercise Calculate ΔG° for the following process at 25°C:

$$BaF_2(s) \rightleftharpoons Ba^{2+}(aq) + 2F^-(aq)$$

The K_{sp} of BaF_2 is 1.84×10^{-7}.

Similar problem: 17.25.

Example 17.8

The equilibrium constant (K_P) for the reaction

$$N_2O_4(g) \rightleftharpoons 2NO_2(g)$$

is 0.113 at 298 K, which corresponds to a standard Gibbs-energy change of 5.40 kJ/mol. In a certain experiment, the initial pressures are $P_{NO_2} = 0.122$ atm and $P_{N_2O_4} = 0.453$ atm. Calculate ΔG for the reaction at these pressures and predict the direction of the net reaction toward equilibrium.

Strategy From the information given we see that neither the reactant nor the product is at its standard state of 1 atm. To determine the direction of the net reaction, we need to calculate the Gibbs-energy change under nonstandard-state conditions (ΔG) using Equation (17.13) and the given $\Delta G°$ value. Note that the partial pressures are expressed as dimensionless quantities in the reaction quotient Q_P because they are divided by the standard-state value of 1 atm (see Section 14.2 and Table 17.3).

Solution Equation (17.13) can be written as

$$\begin{aligned}\Delta G &= \Delta G° + RT \ln Q_P \\ &= \Delta G° + RT \ln \frac{P_{NO_2}^2}{P_{N_2O_4}} \\ &= 5.40 \times 10^3 \text{ J/mol} + (8.314 \text{ J/K} \cdot \text{mol})(298 \text{ K}) \times \ln \frac{(0.122)^2}{0.453} \\ &= 5.40 \times 10^3 \text{ J/mol} - 8.46 \times 10^3 \text{ J/mol} \\ &= -3.06 \times 10^3 \text{ J/mol} = -3.06 \text{ kJ/mol}\end{aligned}$$

Because $\Delta G < 0$, the net reaction proceeds from left to right to reach equilibrium.

Check Note that although $\Delta G° > 0$, the reaction can be made to favor product formation initially by having a small concentration (pressure) of the product compared to that of the reactant. Confirm the prediction by showing that $Q_P < K_P$.

Practice Exercise The K_P for the reaction $N_2(g) + 3H_2(g) \rightleftharpoons 2NH_3(g)$ is 4.3×10^{-4} at 375°C. In one experiment, the initial pressures are $P_{H_2} = 0.40$ atm, $P_{N_2} = 0.86$ atm, and $P_{NH_3} = 4.4 \times 10^{-3}$ atm. Calculate ΔG for the reaction and predict the direction of the net reaction.

Similar problems: 17.27 and 17.28.

Summary of Concepts & Facts

- The equilibrium constant of a reaction and the standard Gibbs-energy change of the reaction are related by the equation $\Delta G° = -RT \ln K$.

Review of Concepts & Facts

17.5.1 A reaction has a positive $\Delta H°$ and a negative $\Delta S°$. Is the equilibrium constant (K) for this reaction greater than 1, equal to 1, or less than 1?

17.5.2 For the reaction

$$A(g) + B(g) \rightleftharpoons C(g)$$

$\Delta G° = -33.3$ kJ/mol at 25°C. What is the value of K_P?

17.5.3 For the reaction

$$C(aq) \rightleftharpoons A(aq) + B(aq)$$

$\Delta G° = 1.95$ kJ/mol at 25°C. What is ΔG (in kJ/mol) at 25°C when the concentrations are [A] = [B] = 0.315 M and [C] = 0.405 M?

17.6 Thermodynamics in Living Systems

Learning Objective

- Explain why thermodynamically unfavorable metabolic reactions can occur.

A mechanical analog for coupled reactions. We can make the smaller weight move upward (a nonspontaneous process) by coupling it with the falling of a larger weight.

Many biochemical reactions have a positive $\Delta G°$ value, yet they are essential to the maintenance of life. In living systems, these reactions are coupled to an energetically favorable process, one that has a negative $\Delta G°$ value. The principle of *coupled reactions* is based on a simple concept: We can use a thermodynamically favorable reaction to drive an unfavorable one. Consider an industrial process. Suppose we wish to extract zinc from the ore sphalerite (ZnS). The following reaction will not work because it has a large positive $\Delta G°$ value:

$$ZnS(s) \longrightarrow Zn(s) + S(s) \qquad \Delta G° = 201.3 \text{ kJ/mol}$$

On the other hand, the combustion of sulfur to form sulfur dioxide is favored because of its large negative $\Delta G°$ value:

$$S(s) + O_2(g) \longrightarrow SO_2(g) \qquad \Delta G° = -300.1 \text{ kJ/mol}$$

By coupling the two processes we can bring about the separation of zinc from zinc sulfide. In practice, this means heating ZnS in air so that the tendency of S to form SO_2 will promote the decomposition of ZnS:

$$\begin{array}{ll} ZnS(s) \longrightarrow Zn(s) + S(s) & \Delta G° = 201.3 \text{ kJ/mol} \\ S(s) + O_2(g) \longrightarrow SO_2(g) & \Delta G° = -300.1 \text{ kJ/mol} \\ \hline ZnS(s) + O_2(g) \longrightarrow Zn(s) + SO_2(g) & \Delta G° = -98.8 \text{ kJ/mol} \end{array}$$

Coupled reactions play a crucial role in the survival of living organisms. In biological systems, enzymes facilitate a wide variety of nonspontaneous reactions. For example, in the human body, food molecules, represented by glucose ($C_6H_{12}O_6$), are converted to carbon dioxide and water during metabolism with a substantial release of Gibbs energy:

$$C_6H_{12}O_6(s) + 6O_2(g) \longrightarrow 6CO_2(g) + 6H_2O(l) \qquad \Delta G° = -2878 \text{ kJ/mol}$$

In a living cell, this reaction does not take place in a single step (as burning glucose in a flame would); rather, the glucose molecule is broken down with the aid of enzymes in a series of steps. Much of the Gibbs energy released along the way is used to synthesize adenosine triphosphate (ATP) from adenosine diphosphate (ADP) and phosphoric acid (Figure 17.15):

$$ADP + H_3PO_4 \longrightarrow ATP + H_2O \qquad \Delta G° = +31 \text{ kJ/mol}$$

The function of ATP is to store Gibbs energy until it is needed by cells. Under appropriate conditions, ATP undergoes hydrolysis to give ADP and phosphoric acid, with a release of 31 kJ/mol of Gibbs energy, which can be used to drive energetically unfavorable reactions, such as protein synthesis.

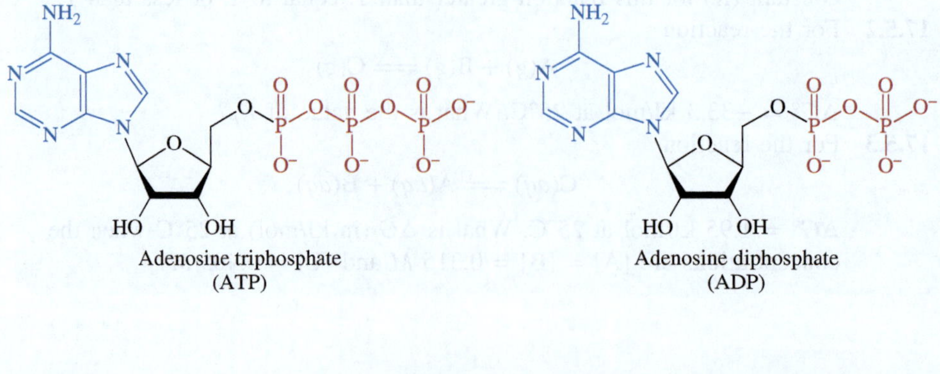

Figure 17.15 *Structure of ATP and ADP in ionized forms. The adenine group is in blue, the ribose group in black, and the phosphate group in red. Note that ADP has one fewer phosphate group than ATP.*

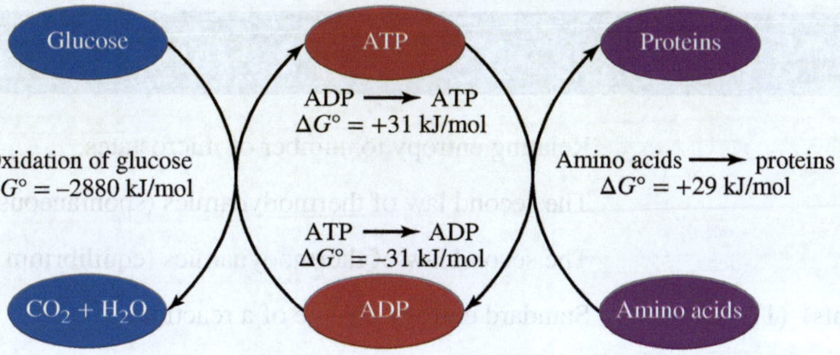

Figure 17.16 *Schematic representation of ATP synthesis and coupled reactions in living systems. The conversion of glucose to carbon dioxide and water during metabolism releases Gibbs energy. The released Gibbs energy is used to convert ADP into ATP. The ATP molecules are then used as an energy source to drive unfavorable reactions, such as protein synthesis from amino acids.*

Proteins are polymers made of amino acids. The stepwise synthesis of a protein molecule involves the joining of individual amino acids. Consider the formation of the dipeptide (a two-amino-acid unit) alanylglycine from alanine and glycine. This reaction represents the first step in the synthesis of a protein molecule:

$$\text{Alanine} + \text{Glycine} \longrightarrow \text{Alanylglycine} \qquad \Delta G° = +29 \text{ kJ/mol}$$

As you can see, this reaction does not favor the formation of product, and so only a little of the dipeptide would be formed at equilibrium. However, with the aid of an enzyme, the reaction is coupled to the hydrolysis of ATP as follows:

$$\text{ATP} + H_2O + \text{Alanine} + \text{Glycine} \longrightarrow \text{ADP} + H_3PO_4 + \text{Alanylglycine}$$

The overall Gibbs-energy change is given by $\Delta G° = -31$ kJ/mol + 29 kJ/mol = −2 kJ/mol, which means that the coupled reaction now favors the formation of product, and an appreciable amount of alanylglycine will be formed under this condition. Figure 17.16 shows the ATP-ADP interconversions that act as energy storage (from metabolism) and Gibbs-energy release (from ATP hydrolysis) to drive essential reactions.

Summary of Concepts & Facts

- Many biological reactions are nonspontaneous. They are driven by the hydrolysis of ATP, for which $\Delta G°$ is negative.

Chapter Summary

Entropy Entropy is the thermodynamic function for predicting the spontaneity of a reaction. On a molecular level, the entropy of a system can in principle be calculated from the number of possible arrangements associated with the system. Entropy is determined by calorimetric methods and standard entropy values are known for many substances. (Sections 17.1 and 17.2)

Second and Third Laws of Thermodynamics The laws of thermodynamics have been successfully applied to the study of chemical and physical processes. The first law of thermodynamics is based on the law of conservation of energy. The second law of thermodynamics deals with natural or spontaneous processes. The function that predicts the spontaneity of a reaction is entropy. The second law states that for a spontaneous process, the change in the entropy of the universe must be positive. The third law enables us to determine absolute entropy values. (Section 17.3)

Gibbs Energy Gibbs energy helps us to determine the spontaneity of a reaction by focusing only on the system. The change in Gibbs energy for a process is made up of two terms: a change in enthalpy and a change in entropy times temperature. At constant temperature and pressure, a decrease in Gibbs energy signals a spontaneous reaction. The change in the standard Gibbs energy can be related to the equilibrium constant of a reaction. (Sections 17.4 and 17.5)

Thermodynamics in Living Systems Many reactions of biological importance are nonspontaneous. By coupling such reactions to those that have a negative Gibbs-energy change with the aid of enzymes, the net reaction can be made to proceed to yield the desired products. (Section 17.6)

Key Equations

$S = k \ln W$ (17.1)	Relating entropy to number of microstates.
$\Delta S_{univ} = \Delta S_{sys} + \Delta S_{surr} > 0$ (17.4)	The second law of thermodynamics (spontaneous process).
$\Delta S_{univ} = \Delta S_{sys} + \Delta S_{surr} = 0$ (17.5)	The second law of thermodynamics (equilibrium process).
$\Delta S^\circ_{rxn} = \Sigma nS^\circ$ (products) $- \Sigma mS^\circ$ (reactants) (17.7)	Standard entropy change of a reaction.
$G = H - TS$ (17.9)	Definition of Gibbs energy.
$\Delta G = \Delta H - T\Delta S$ (17.10)	Gibbs-energy change at constant temperature.
$\Delta G^\circ_{rxn} = \Sigma n\Delta G^\circ_f$ (products) $- \Sigma m\Delta G^\circ_f$ (reactants) (17.12)	Standard Gibbs-energy change of a reaction.
$\Delta G = \Delta G^\circ + RT \ln Q$ (17.13)	Relationship between Gibbs-energy change and standard Gibbs-energy change and reaction quotient.
$\Delta G^\circ = -RT \ln K$ (17.14)	Relationship between standard Gibbs-energy change and the equilibrium constant.

Key Words

Entropy (S), p. 780
Free energy (G), p. 792
Gibbs energy (G), p. 792
Second law of thermodynamics, p. 786
Standard entropy of reaction (ΔS°_{rxn}), p. 787
Standard Gibbs-energy of formation (ΔG°_f), p. 794
Standard Gibbs-energy of reaction (ΔG°_{rxn}), p. 794
Third law of thermodynamics, p. 790

Questions & Problems

Red numbered problems solved in Student Solutions Manual

17.1 Spontaneous Processes

Review Questions

17.1 Explain what is meant by a spontaneous process. Give two examples each of spontaneous and nonspontaneous processes.

17.2 State which of the following processes are spontaneous and which are nonspontaneous: (a) dissolving table salt (NaCl) in hot soup; (b) climbing Mt. Everest; (c) spreading fragrance in a room by removing the cap from a perfume bottle; (d) separating helium and neon from a mixture of the gases.

17.3 Which of the following processes are spontaneous and which are nonspontaneous at a given temperature?

(a) $NaNO_3(s) \xrightarrow{H_2O} NaNO_3(aq)$ saturated soln

(b) $NaNO_3(s) \xrightarrow{H_2O} NaNO_3(aq)$ unsaturated soln

(c) $NaNO_3(s) \xrightarrow{H_2O} NaNO_3(aq)$ supersaturated soln

17.2 Entropy

Review Questions

17.4 Define entropy. What are the units of entropy?

Problems

17.5 How does the entropy of a system change for each of the following processes?

(a) A solid melts.

(b) A liquid freezes.

(c) A liquid boils.

(d) A vapor is converted to a solid.

(e) A vapor condenses to a liquid.

(f) A solid sublimes.

(g) Urea dissolves in water.

17.6 Consider the arrangement in Figure 17.1. Because the volume of the two bulbs is the same, the probability of finding a molecule in either bulb is $\frac{1}{2}$.

Calculate the probability of all the molecules ending up in the same bulb if the number is (a) 2, (b) 100, and (c) 6×10^{23}. Based on your results, explain why the situation shown in Figure 17.1(b) will not be observed for a macroscopic system.

17.3 The Second and Third Laws of Thermodynamics

Review Questions

17.7 State the second law of thermodynamics in words and express it mathematically.

17.8 State the third law of thermodynamics and explain its usefulness in calculating entropy values.

Problems

17.9 For each pair of substances listed here, choose the one having the larger standard entropy value at 25°C. The same molar amount is used in the comparison. Explain the basis for your choice. (a) $Li(s)$ or $Li(l)$; (b) $C_2H_5OH(l)$ or $CH_3OCH_3(l)$ (*Hint:* Which molecule can hydrogen-bond?); (c) $Ar(g)$ or $Xe(g)$; (d) $CO(g)$ or $CO_2(g)$; (e) $O_2(g)$ or $O_3(g)$; (f) $NO_2(g)$ or $N_2O_4(g)$

17.10 Arrange the following substances (1 mole each) in order of increasing entropy at 25°C: (a) $Ne(g)$, (b) $SO_2(g)$, (c) $Na(s)$, (d) $NaCl(s)$, (e) $H_2(g)$. Give the reasons for your arrangement.

17.11 Using the data in Appendix 2, calculate the standard entropy changes for the following reactions at 25°C:
(a) $S(s) + O_2(g) \longrightarrow SO_2(g)$
(b) $MgCO_3(s) \longrightarrow MgO(s) + CO_2(g)$

17.12 Using the data in Appendix 2, calculate the standard entropy changes for the following reactions at 25°C:
(a) $H_2(g) + CuO(s) \longrightarrow Cu(s) + H_2O(g)$
(b) $2Al(s) + 3ZnO(s) \longrightarrow Al_2O_3(s) + 3Zn(s)$
(c) $CH_4(g) + 2O_2(g) \longrightarrow CO_2(g) + 2H_2O(l)$

17.13 Without consulting Appendix 2, predict whether the entropy change is positive or negative for each of the following reactions. Give reasons for your predictions.
(a) $2KClO_4(s) \longrightarrow 2KClO_3(s) + O_2(g)$
(b) $H_2O(g) \longrightarrow H_2O(l)$
(c) $2Na(s) + 2H_2O(l) \longrightarrow 2NaOH(aq) + H_2(g)$
(d) $N_2(g) \longrightarrow 2N(g)$

17.14 State whether the sign of the entropy change expected for each of the following processes will be positive or negative, and explain your predictions.
(a) $PCl_3(l) + Cl_2(g) \longrightarrow PCl_5(s)$
(b) $2H_gO(s) \longrightarrow 2H_g(l) + O_2(g)$
(c) $H_2(g) \longrightarrow 2H(g)$
(d) $U(s) + 3F_2(g) \longrightarrow UF_6(s)$

17.4 Gibbs Energy

Review Questions

17.15 Define Gibbs energy. What are its units?

17.16 Why is it more convenient to predict the direction of a reaction in terms of ΔG_{sys} instead of ΔS_{univ}? Under what conditions can ΔG_{sys} be used to predict the spontaneity of a reaction?

Problems

17.17 Calculate $\Delta G°$ for the following reactions at 25°C:
(a) $N_2(g) + O_2(g) \longrightarrow 2NO(g)$
(b) $H_2O(l) \longrightarrow H_2O(g)$
(c) $2C_2H_2(g) + 5O_2(g) \longrightarrow 4CO_2(g) + 2H_2O(l)$
(*Hint:* Look up the standard Gibbs energies of formation of the reactants and products in Appendix 2.)

17.18 Calculate $\Delta G°$ for the following reactions at 25°C:
(a) $2Mg(s) + O_2(g) \longrightarrow 2MgO(s)$
(b) $2SO_2(g) + O_2(g) \longrightarrow 2SO_3(g)$
(c) $2C_2H_6(g) + 7O_2(g) \longrightarrow 4CO_2(g) + 6H_2O(l)$
See Appendix 2 for thermodynamic data.

17.19 From the values of ΔH and ΔS, predict which of the following reactions would be spontaneous at 25°C: reaction A: $\Delta H = 10.5$ kJ/mol, $\Delta S = 30$ J/K · mol; reaction B: $\Delta H = 1.8$ kJ/mol, $\Delta S = -113$ J/K · mol. If either of the reactions is nonspontaneous at 25°C, at what temperature might it become spontaneous?

17.20 Find the temperatures at which reactions with the following ΔH and ΔS values would become spontaneous: (a) $\Delta H = -126$ kJ/mol, $\Delta S = 84$ J/K · mol; (b) $\Delta H = -11.7$ kJ/mol, $\Delta S = -105$ J/K · mol.

17.5 Gibbs Energy and Chemical Equilibrium

Review Questions

17.21 Explain the difference between ΔG and $\Delta G°$.

17.22 Explain why Equation (17.14) is of great importance in chemistry.

Problems

17.23 Calculate K_P for the following reaction at 25°C:

$$H_2(g) + I_2(g) \rightleftharpoons 2HI(g) \qquad \Delta G° = 3.4 \text{ kJ/mol}$$

17.24 For the autoionization of water at 25°C,

$$H_2O(l) \rightleftharpoons H^+(aq) + OH^-(aq)$$

K_w is 1.0×10^{-14}. What is $\Delta G°$ for the process?

17.25 Consider the following reaction at 25°C:

$$Fe(OH)_2(s) \rightleftharpoons Fe^{2+}(aq) + 2OH^-(aq)$$

Calculate $\Delta G°$ for the reaction. K_{sp} for $Fe(OH)_2$ is 4.9×10^{-17}.

17.26 Calculate $\Delta G°$ and K_P for the following equilibrium reaction at 25°C.

$$2H_2O(g) \rightleftharpoons 2H_2(g) + O_2(g)$$

17.27 (a) Calculate $\Delta G°$ and K_P for the following equilibrium reaction at 25°C. The $\Delta G_f°$ values are 0 for $Cl_2(g)$, −267.8 kJ/mol for $PCl_3(g)$, and −305.0 kJ/mol for $PCl_5(g)$.

$$PCl_5(g) \rightleftharpoons PCl_3(g) + Cl_2(g)$$

(b) Calculate ΔG for the reaction if the partial pressures of the initial mixture are $P_{PCl_5} = 0.0029$ atm, $P_{PCl_3} = 0.27$ atm, and $P_{Cl_2} = 0.40$ atm.

17.28 The equilibrium constant (K_P) for the reaction

$$H_2(g) + CO_2(g) \rightleftharpoons H_2O(g) + CO(g)$$

is 4.40 at 2000 K. (a) Calculate $\Delta G°$ for the reaction. (b) Calculate ΔG for the reaction when the partial pressures are $P_{H_2} = 0.25$ atm, $P_{CO_2} = 0.78$ atm, $P_{H_2O} = 0.66$ atm, and $P_{CO} = 1.20$ atm.

17.29 Consider the decomposition of calcium carbonate:

$$CaCO_3(s) \rightleftharpoons CaO(s) + CO_2(g)$$

Calculate the pressure in atm of CO_2 in an equilibrium process (a) at 25°C and (b) at 800°C. Assume that $\Delta H° = 179.2$ kJ/mol and $\Delta S° = 160.2$ J/K · mol for the temperature range.

17.30 The equilibrium constant K_P for the reaction

$$CO(g) + Cl_2(g) \rightleftharpoons COCl_2(g)$$

is 5.62×10^{35} at 25°C. Calculate $\Delta G_f°$ for $COCl_2$ at 25°C.

17.31 At 25°C, $\Delta G°$ for the process

$$H_2O(l) \rightleftharpoons H_2O(g)$$

is 8.5 kJ/mol. Calculate the vapor pressure of water at this temperature.

17.32 Calculate $\Delta G°$ for the process

$$\text{C(diamond)} \longrightarrow \text{C(graphite)}$$

Is the formation of graphite from diamond favored at 25°C? If so, why is it that diamonds do not become graphite on standing?

17.6 Thermodynamics in Living Systems

Review Questions

17.33 What is a coupled reaction? What is its importance in biological reactions?

17.34 What is the role of ATP in biological reactions?

Problems

17.35 Referring to the metabolic process involving glucose in Section 17.6, calculate the maximum number of moles of ATP that can be synthesized from ADP from the breakdown of one mole of glucose.

17.36 In the metabolism of glucose, the first step is the conversion of glucose to glucose 6-phosphate:

$$\text{glucose} + H_3PO_4 \longrightarrow \text{glucose 6-phosphate} + H_2O \qquad \Delta G° = 13.4 \text{ kJ/mol}$$

Because $\Delta G°$ is positive, this reaction does not favor the formation of products. Show how this reaction can be made to proceed by coupling it with the hydrolysis of ATP. Write an equation for the coupled reaction and estimate the equilibrium constant for the coupled process.

Additional Problems

17.37 Explain the following nursery rhyme in terms of the second law of thermodynamics.

Humpty Dumpty sat on a wall;
Humpty Dumpty had a great fall.
All the King's horses and all the King's men
Couldn't put Humpty together again.

17.38 Calculate ΔG for the reaction

$$H_2O(l) \rightleftharpoons H^+(aq) + OH^-(aq)$$

at 25°C for the following conditions:
(a) $[H^+] = 1.0 \times 10^{-7}$ *M*, $[OH^-] = 1.0 \times 10^{-7}$ *M*
(b) $[H^+] = 1.0 \times 10^{-3}$ *M*, $[OH^-] = 1.0 \times 10^{-4}$ *M*
(c) $[H^+] = 1.0 \times 10^{-12}$ *M*, $[OH^-] = 2.0 \times 10^{-8}$ *M*
(d) $[H^+] = 3.5$ *M*, $[OH^-] = 4.8 \times 10^{-4}$ *M*

17.39 Calculate the $\Delta S°_{soln}$ for the following processes: (a) $NH_4NO_3(s) \longrightarrow NH_4^+(aq) + NO_3^-(aq)$ and (b) $FeCl_3(s) \longrightarrow Fe^{3+}(aq) + 3Cl^-(aq)$. Give a qualitative explanation for the signs.

17.40 The following reaction is spontaneous at a certain temperature *T*. Predict the sign of ΔS_{surr}.

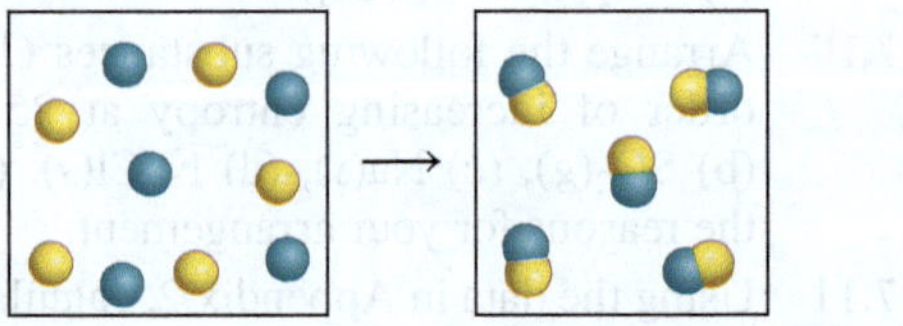

17.41 Which of the following thermodynamic functions are associated only with the first law of thermodynamics: *S*, *U*, *G*, *H*?

17.42 A student placed 1 g of each of three compounds A, B, and C in a container and found that after 1 week no change had occurred. Offer some possible explanations for the fact that no reactions took place. Assume that A, B, and C are totally miscible liquids.

17.43 Use the data in Appendix 2 to calculate the equilibrium constant for the reaction $AgI(s) \rightleftharpoons Ag^+(aq) + I^-(aq)$ at 25°C. Compare your result with the K_{sp} value in Table 16.2.

17.44 Predict the signs of ΔH, ΔS, and ΔG of the system for the following processes at 1 atm: (a) ammonia melts at −60°C, (b) ammonia melts at −77.7°C, (c) ammonia melts at −100°C. (The normal melting point of ammonia is −77.7°C.)

17.45 Consider the following facts: Water freezes spontaneously at −5°C and 1 atm, and ice has a more ordered structure than liquid water. Explain how a spontaneous process can lead to a decrease in entropy.

17.46 Ammonium nitrate (NH_4NO_3) dissolves spontaneously and endothermically in water. What can you deduce about the sign of ΔS for the solution process?

17.47 Calculate the equilibrium pressure of CO_2 due to the decomposition of barium carbonate ($BaCO_3$) at 25°C.

17.48 (a) Trouton's rule states that the ratio of the molar heat of vaporization of a liquid (ΔH_{vap}) to its boiling point in kelvins is approximately 90 J/K · mol. Use the following data to show that this is the case and explain why Trouton's rule holds true.

	t_{bp}(°C)	ΔH_{vap}(kJ/mol)
Benzene	80.1	30.7
Hexane	68.7	28.9
Mercury	357	59.1
Toluene	110.6	33.2

(b) Use the values in Table 11.7 to calculate the same ratio for ethanol and water. Explain why Trouton's rule does not apply to these two substances as well as it does to other liquids.

17.49 Referring to Problem 17.48, explain why the ratio is considerably smaller than 90 J/K · mol for liquid HF.

17.50 Carbon monoxide (CO) and nitric oxide (NO) are polluting gases contained in automobile exhaust. Under suitable conditions, these gases can be made to react to form nitrogen (N_2) and the less harmful carbon dioxide (CO_2). (a) Write an equation for this reaction. (b) Identify the oxidizing and reducing agents. (c) Calculate the K_P for the reaction at 25°C. (d) Under normal atmospheric conditions, the partial pressures are P_{N_2} = 0.80 atm, P_{CO_2} = 3.0×10^{-4} atm, P_{CO} = 5.0×10^{-5} atm, and P_{NO} = 5.0×10^{-7} atm. Calculate Q_P and predict the direction toward which the reaction will proceed. (e) Will raising the temperature favor the formation of N_2 and CO_2?

17.51 For reactions carried out under standard-state conditions, Equation (17.10) takes the form $\Delta G° = \Delta H° - T\Delta S°$. (a) Assuming $\Delta H°$ and $\Delta S°$ are independent of temperature, derive the equation

$$\ln \frac{K_2}{K_1} = \frac{\Delta H°}{R}\left(\frac{T_2 - T_1}{T_1 T_2}\right)$$

where K_1 and K_2 are the equilibrium constants at T_1 and T_2, respectively. (b) Given that at 25°C, K_c is 4.63×10^{-3} for the reaction

$$N_2O_4(g) \rightleftharpoons 2NO_2(g) \qquad \Delta H° = 55.3 \text{ kJ/mol}$$

calculate the equilibrium constant at 65°C.

17.52 Use the thermodynamic data in Appendix 2 to calculate the K_{sp} of AgCl.

17.53 Consider the reaction A ⟶ B + C at 298 K. Given that the forward rate constant (k_f) is 0.46 s^{-1} and the reverse rate constant (k_r) is $1.5 \times 10^{-2}/M \cdot s$, calculate $\Delta G°$ of the reaction.

17.54 The K_{sp} of AgCl is given in Table 16.2. What is its value at 60°C? [*Hint:* You need the result of Problem 17.51(a) and the data in Appendix 2 to calculate $\Delta H°$.]

17.55 Under what conditions does a substance have a standard entropy of zero? Can a substance ever have a negative standard entropy?

17.56 Water gas, a mixture of H_2 and CO, is a fuel made by reacting steam with red-hot coke (a by-product of coal distillation):

$$H_2O(g) + C(s) \rightleftharpoons CO(g) + H_2(g)$$

From the data in Appendix 2, estimate the temperature at which the reaction begins to favor the formation of products.

17.57 Consider the following Brønsted acid-base reaction at 25°C:

$$HF(aq) + Cl^-(aq) \rightleftharpoons HCl(aq) + F^-(aq)$$

(a) Predict whether *K* will be greater or smaller than unity. (b) Does $\Delta S°$ or $\Delta H°$ make a greater contribution to $\Delta G°$? (c) Is $\Delta H°$ likely to be positive or negative?

17.58 Crystallization of sodium acetate from a supersaturated solution occurs spontaneously (see Section 12.1). What can you deduce about the signs of ΔS and ΔH?

17.59 Consider the thermal decomposition of $CaCO_3$:

$$CaCO_3(s) \rightleftharpoons CaO(s) + CO_2(g)$$

The equilibrium vapor pressures of CO_2 are 22.6 mmHg at 700°C and 1829 mmHg at 950°C. Calculate the standard enthalpy of the reaction. [*Hint:* See Problem 17.51(a).]

17.60 A certain reaction is spontaneous at 72°C. If the enthalpy change for the reaction is 19 kJ/mol, what is the *minimum* value of ΔS (in J/K · mol) for the reaction?

17.61 Predict whether the entropy change is positive or negative for each of these reactions:

(a) $Zn(s) + 2HCl(aq) \longrightarrow ZnCl_2(aq) + H_2(g)$

(b) $O(g) + O(g) \longrightarrow O_2(g)$

(c) $NH_4NO_3(s) \longrightarrow N_2O(g) + 2H_2O(g)$

(d) $2H_2O_2(l) \longrightarrow 2H_2O(l) + O_2(g)$

17.62 The reaction $NH_3(g) + HCl(g) \longrightarrow NH_4Cl(s)$ proceeds spontaneously at 25°C even though there is a decrease in the number of microstates of the system (gases are converted to a solid). Explain.

17.63 Use the following data to determine the normal boiling point, in kelvins, of mercury. What assumptions must you make to do the calculation?

Hg(*l*): $\Delta H_f° = 0$ (by definition)
$S° = 75.9$ J/K · mol

Hg(*g*): $\Delta H_f° = 61.4$ kJ/mol
$S° = 175.0$ J/K · mol

17.64 The molar heat of vaporization of ethanol is 38.6 kJ/mol and the boiling point of ethanol is 78.2°C. Calculate ΔS for the vaporization of 0.50 mol ethanol.

17.65 A certain reaction is known to have a $\Delta G°$ value of −122 kJ/mol. Will the reaction necessarily occur if the reactants are mixed together?

17.66 In the Mond process for the purification of nickel, carbon monoxide is reacted with heated nickel to produce $Ni(CO)_4$, which is a gas and can therefore be separated from solid impurities:

$$Ni(s) + 4CO(g) \rightleftharpoons Ni(CO)_4(g)$$

Given that the standard Gibbs energies of formation of $CO(g)$ and $Ni(CO)_4(g)$ are −137.2 kJ/mol and −587.4 kJ/mol, respectively, calculate the equilibrium constant of the reaction at 80°C. Assume that ΔG_f° is temperature independent.

17.67 Calculate ΔG° and K_P for the following processes at 25°C:
(a) $H_2(g) + Br_2(l) \rightleftharpoons 2HBr(g)$
(b) $\frac{1}{2}H_2(g) + \frac{1}{2}Br_2(l) \rightleftharpoons HBr(g)$
Account for the differences in ΔG° and K_P obtained for (a) and (b).

17.68 Calculate the pressure of O_2 (in atm) over a sample of NiO at 25°C if ΔG° = 216 kJ/mol for the reaction

$$NiO(s) \rightleftharpoons Ni(s) + \tfrac{1}{2}O_2(g)$$

17.69 Comment on the statement: "Just talking about entropy increases its value in the universe."

17.70 For a reaction with a negative ΔG° value, which of the following statements is false? (a) The equilibrium constant K is greater than one. (b) The reaction is spontaneous when all the reactants and products are in their standard states. (c) The reaction is always exothermic.

17.71 Consider the reaction

$$N_2(g) + O_2(g) \rightleftharpoons 2NO(g)$$

Given that ΔG° for the reaction at 25°C is 175.2 kJ/mol, (a) calculate the standard Gibbs energy of formation of NO, and (b) calculate K_P of the reaction. (c) One of the starting substances in smog formation is NO. Assuming that the temperature in a running automobile engine is 1100°C, estimate K_P for the above reaction. (d) As farmers know, lightning helps to produce a better crop. Why?

17.72 Heating copper(II) oxide at 400°C does not produce any appreciable amount of Cu:

$$CuO(s) \rightleftharpoons Cu(s) + \tfrac{1}{2}O_2(g) \quad \Delta G^\circ = 129.7 \text{ kJ/mol}$$

However, if this reaction is coupled to the conversion of graphite to carbon monoxide, it becomes spontaneous. Write an equation for the coupled process and calculate the equilibrium constant for the coupled reaction.

17.73 The internal engine of a 1200-kg car is designed to run on octane (C_8H_{18}), whose enthalpy of combustion is 5510 kJ/mol. If the car is moving up a slope, calculate the maximum height (in meters) to which the car can be driven on 1.0 gallon of the fuel. Assume that the engine cylinder temperature is 2200°C and the exit temperature is 760°C, and neglect all forms of friction. The mass of 1 gallon of fuel is 3.1 kg. [*Hint:* See the Chemistry in Action essay "The Efficiency of Heat Engines" in Section 17.4. The work done in moving the car over a vertical distance is mgh, where m is the mass of the car in kg, g the acceleration due to gravity (9.81 m/s^2), and h the height in meters.]

17.74 Consider the decomposition of magnesium carbonate:

$$MgCO_3(s) \rightleftharpoons MgO(s) + CO_2(g)$$

Calculate the temperature at which the decomposition begins to favor products. Assume that both ΔH° and ΔS° are independent of temperature.

17.75 (a) Over the years there have been numerous claims about "perpetual motion machines," machines that will produce useful work with no input of energy. Explain why the first law of thermodynamics prohibits the possibility of such a machine existing. (b) Another kind of machine, sometimes called a "perpetual motion of the second kind," operates as follows. Suppose an ocean liner sails by scooping up water from the ocean and then extracting heat from the water, converting the heat to electric power to run the ship, and dumping the water back into the ocean. This process does not violate the first law of thermodynamics, for no energy is created—energy from the ocean is just converted to electrical energy. Show that the second law of thermodynamics prohibits the existence of such a machine.

17.76 The activity series in Section 4.4 shows that reaction (a) is spontaneous while reaction (b) is nonspontaneous at 25°C:
(a) $Fe(s) + 2H^+ \longrightarrow Fe^{2+}(aq) + H_2(g)$
(b) $Cu(s) + 2H^+ \longrightarrow Cu^{2+}(aq) + H_2(g)$
Use the data in Appendix 2 to calculate the equilibrium constant for these reactions and hence confirm that the activity series is correct.

17.77 The rate constant for the elementary reaction

$$2NO(g) + O_2(g) \longrightarrow 2NO_2(g)$$

is $7.1 \times 10^9/M^2 \cdot s$ at 25°C. What is the rate constant for the reverse reaction at the same temperature?

17.78 The following reaction is the cause of sulfur deposits formed at volcanic sites (see Section 20.4):

$$2H_2S(g) + SO_2(g) \rightleftharpoons 3S(s) + 2H_2O(g)$$

It may also be used to remove SO_2 from powerplant stack gases. (a) Identify the type of redox reaction it is. (b) Calculate the equilibrium constant (K_P) at 25°C and comment on whether this method is feasible for removing SO_2. (c) Would this procedure become more or less effective at a higher temperature?

17.79 Describe two ways that you could measure ΔG° of a reaction.

17.80 The following reaction represents the removal of ozone in the stratosphere:

$$2O_3(g) \rightleftharpoons 3O_2(g)$$

Calculate the equilibrium constant (K_P) for the reaction. In view of the magnitude of the equilibrium constant, explain why this reaction is not considered a major cause of ozone depletion in the absence of man-made pollutants such as the nitrogen oxides and CFCs. Assume the temperature of the stratosphere to be −30°C and ΔG_f° to be temperature independent.

17.81 A 74.6-g ice cube floats in the Arctic Sea. The temperature and pressure of the system and surroundings are at 1 atm and 0°C. Calculate ΔS_{sys}, ΔS_{surr}, and ΔS_{univ} for the melting of the ice cube. What can you conclude about the nature of the process from the value of ΔS_{univ}? (The molar heat of fusion of water is 6.01 kJ/mol.)

17.82 Comment on the feasibility of extracting copper from its ore chalcocite (Cu_2S) by heating:

$$Cu_2S(s) \longrightarrow 2Cu(s) + S(s)$$

Calculate the ΔG° for the overall reaction if the above process is coupled to the conversion of sulfur to sulfur dioxide. Given that $\Delta G_f^\circ(Cu_2S) = -86.2$ kJ/mol.

17.83 Active transport is the process in which a substance is transferred from a region of lower concentration to one of higher concentration. This is a nonspontaneous process and must be coupled to a spontaneous process, such as the hydrolysis of ATP. The concentrations of K^+ ions in the blood plasma and in nerve cells are 15 mM and 400 mM, respectively (1 mM = 1×10^{-3} M). Use Equation (17.13) to calculate ΔG for the process at the physiological temperature of 37°C:

$$K^+(15\ mM) \longrightarrow K^+(400\ mM)$$

In this calculation, the ΔG° term can be set to zero. What is the justification for this step?

17.84 Large quantities of hydrogen are needed for the synthesis of ammonia. One preparation of hydrogen involves the reaction between carbon monoxide and steam at 300°C in the presence of a copper-zinc catalyst:

$$CO(g) + H_2O(g) \rightleftharpoons CO_2(g) + H_2(g)$$

Calculate the equilibrium constant (K_P) for the reaction and the temperature at which the reaction favors the formation of CO and H_2O. Will a larger K_P be attained at the same temperature if a more efficient catalyst is used?

17.85 Shown here are the thermodynamic data for ethanol:

	ΔH_f°(kJ/mol)	S°(J/K · mol)
Liquid	−276.6	160.7
Vapor	−234.8	281.6

Calculate the vapor pressure of ethanol at 25°C. Assume the thermodynamic values are independent of temperature.

17.86 The reaction shown here is spontaneous at a certain temperature T. What is the sign of ΔS_{surr}?

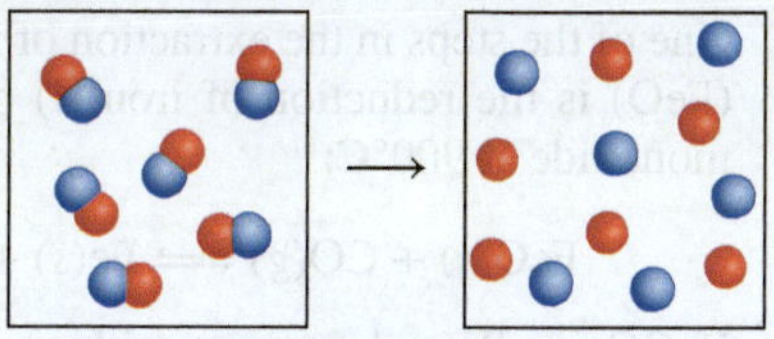

17.87 Consider two carboxylic acids (acids that contain the —COOH group): CH_3COOH (acetic acid, K_a = 1.8×10^{-5}) and $CH_2ClCOOH$ (chloroacetic acid, $K_a = 1.4 \times 10^{-3}$). (a) Calculate ΔG° for the ionization of these acids at 25°C. (b) From the equation $\Delta G^\circ = \Delta H^\circ - T\Delta S^\circ$, we see that the contributions to the ΔG° term are an enthalpy term (ΔH°) and a temperature times entropy term ($T\Delta S^\circ$). These contributions are listed here for the two acids:

	ΔH°(kJ/mol)	$T\Delta S^\circ$(kJ/mol)
CH_3COOH	−0.57	−27.6
$CH_2ClCOOH$	−4.7	−21.1

Which is the dominant term in determining the value of ΔG° (and hence K_a of the acid)? (c) What processes contribute to ΔH°? (Consider the ionization of the acids as a Brønsted acid-base reaction.) (d) Explain why the $T\Delta S^\circ$ term is more negative for CH_3COOH.

17.88 Many hydrocarbons exist as structural isomers, which are compounds that have the same molecular formula but different structures. For example, both butane and isobutane have the same molecular formula of C_4H_{10} (see Problem 11.19). Calculate the mole percent of these molecules in an equilibrium mixture at 25°C, given that the standard Gibbs energy of formation of butane is −15.7 kJ/mol and that of isobutane is −18.0 kJ/mol. Does your result support the notion that straight-chain hydrocarbons (that is, hydrocarbons in which the C atoms are joined along a line) are less stable than branch-chain hydrocarbons?

17.89 Use the thermodynamic data in Appendix 2 to determine the normal boiling point of liquid bromine. Assume the values are independent of temperature.

17.90 In each of the following reactions, there is one species for which the standard entropy value is not listed in Appendix 2. Determine the S° for that species. (a) The ΔS_{rxn}° for the reaction $Na(s) \longrightarrow Na(l)$ is 6.4 J/K · mol. (b) The ΔS_{rxn}° for the reaction $2S(\text{monoclinic}) + Cl_2(g) \longrightarrow S_2Cl_2(g)$ is 43.2 J/K · mol. (c) The ΔS_{rxn}° for the reaction $FeCl_2(s) \longrightarrow Fe^{2+}(aq) + 2Cl^-(aq)$ is −118.3 J/K · mol.

17.91 A rubber band is stretched vertically by attaching a weight to one end and holding the other end by hand. On heating the rubber band with a hot-air blower, it is observed to shrink slightly in length. Give a thermodynamic analysis for this behavior. (*Hint:* See the Chemistry in Action essay "The Thermodynamics of a Rubber Band" in Section 17.4.)

17.92 One of the steps in the extraction of iron from its ore (FeO) is the reduction of iron(II) oxide by carbon monoxide at 900°C:

$$FeO(s) + CO(g) \rightleftharpoons Fe(s) + CO_2(g)$$

If CO is allowed to react with an excess of FeO, calculate the mole fractions of CO and CO_2 at equilibrium. State any assumptions.

17.93 Derive the equation

$$\Delta G = RT \ln(Q/K)$$

where Q is the reaction quotient and describe how you would use it to predict the spontaneity of a reaction.

17.94 The sublimation of carbon dioxide at −78°C is

$$CO_2(s) \longrightarrow CO_2(g) \qquad \Delta H_{sub} = 62.4 \text{ kJ/mol}$$

Calculate ΔS_{sub} when 84.8 g of CO_2 sublimes at this temperature.

17.95 Entropy has sometimes been described as "time's arrow" because it is the property that determines the forward direction of time. Explain.

17.96 Referring to Figure 17.1, we see that the probability of finding all 100 molecules in the same bulb is 8×10^{-31}. Assuming that the age of the universe is 13 billion years, calculate the time in seconds during which this event can be observed.

17.97 A student looked up the ΔG°_f, ΔH°_f, and S° values for CO_2 in Appendix 2. Plugging these values into Equation (17.10), he found that $\Delta G^\circ_f \neq \Delta H^\circ_f - TS^\circ$ at 298 K. What is wrong with his approach?

17.98 Consider the following reaction at 298 K:

$$2H_2(g) + O_2(g) \longrightarrow 2H_2O(l) \qquad \Delta H^\circ = -571.6 \text{ kJ/mol}$$

Calculate ΔS_{sys}, ΔS_{surr}, and ΔS_{univ} for the reaction.

17.99 As an approximation, we can assume that proteins exist either in the native (or physiologically functioning) state and the denatured state:

$$\text{native} \rightleftharpoons \text{denatured}$$

The standard molar enthalpy and entropy of the denaturation of a certain protein are 512 kJ/mol and 1.60 kJ/K · mol, respectively. Comment on the signs and magnitudes of these quantities, and calculate the temperature at which the process favors the denatured state.

17.100 Which of the following are not state functions: S, H, q, w, T?

17.101 Which of the following is not accompanied by an increase in the entropy of the system: (a) mixing of two gases at the same temperature and pressure, (b) mixing of ethanol and water, (c) discharging a battery, (d) expansion of a gas followed by compression to its original temperature, pressure, and volume?

17.102 Hydrogenation reactions (for example, the process of converting C═C bonds to C—C bonds in food industry) are facilitated by the use of a transition metal catalyst, such as Ni or Pt. The initial step is the adsorption, or binding, of hydrogen gas onto the metal surface. Predict the signs of ΔH, ΔS, and ΔG when hydrogen gas is adsorbed onto the surface of Ni metal.

17.103 Give a detailed example of each of the following, with an explanation: (a) a thermodynamically spontaneous process; (b) a process that would violate the first law of thermodynamics; (c) a process that would violate the second law of thermodynamics; (d) an irreversible process; (e) an equilibrium process.

17.104 At 0 K, the entropy of carbon monoxide crystal is not zero but has a value of 4.2 J/K · mol, called the residual entropy. According to the third law of thermodynamics, this means that the crystal does not have a perfect arrangement of the CO molecules. (a) What would be the residual entropy if the arrangement were totally random? (b) Comment on the difference between the result in (a) and 4.2 J/K · mol. [*Hint:* Assume that each CO molecule has two choices for orientation and use Equation (17.1) to calculate the residual entropy.]

17.105 Comment on the correctness of the analogy sometimes used to relate a student's dormitory room becoming untidy to an increase in entropy.

17.106 The standard enthalpy of formation and the standard entropy of gaseous benzene are 82.9 kJ/mol and 269.2 J/K · mol, respectively. Calculate ΔH°, ΔS°, and ΔG° for the process at 25°C.

$$C_6H_6(l) \longrightarrow C_6H_6(g)$$

Comment on your answers.

17.107 In chemistry, the standard state for a solution is 1 M (see Table 17.2). This means that each solute concentration expressed in molarity is divided by 1 M. In biological systems, however, we define the standard state for the H^+ ions to be 1×10^{-7} M because the physiological pH is about 7. Consequently, the change in the standard Gibbs energy according to these two conventions will be different involving uptake or release of H^+ ions, depending on which convention is used. We will therefore replace ΔG° with $\Delta G^{\circ\prime}$, where the prime denotes that it is the standard Gibbs-energy change for a biological process. (a) Consider the reaction

$$A + B \longrightarrow C + xH^+$$

where x is a stoichiometric coefficient. Use Equation (17.13) to derive a relation between ΔG° and $\Delta G^{\circ\prime}$, keeping in mind that ΔG is the same for a process regardless of which convention is used. Repeat the derivation for the reverse process:

$$C + xH^+ \longrightarrow A + B$$

(b) NAD^+ and NADH are the oxidized and reduced forms of nicotinamide adenine dinucleotide, two key compounds in the metabolic pathways. For the oxidation of NADH:

$$NADM + H^+ \longrightarrow NAD^+ + H_2$$

ΔG° is −21.8 kJ/mol at 298 K. Calculate $\Delta G^{\circ\prime}$. Also calculate ΔG using both the chemical and biological conventions when [NADH] = 1.5×10^{-2} M,

$[H^+] = 3.0 \times 10^{-5}\ M$, $[NAD] = 4.6 \times 10^{-3}\ M$, and $P_{H_2} = 0.010$ atm.

17.108 The following diagram shows the variation of the equilibrium constant with temperature for the reaction

$$I_2(g) \rightleftharpoons 2I(g)$$

Calculate $\Delta G°$, $\Delta H°$, and $\Delta S°$ for the reaction at 872 K. (*Hint:* See Problem 17.51.)

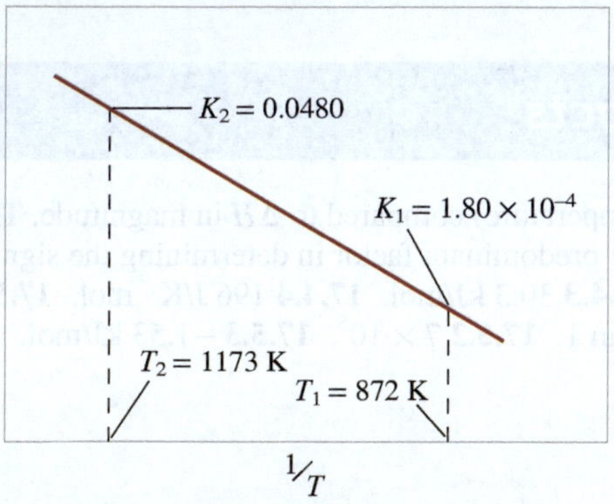

17.109 Consider the gas-phase reaction between A_2 (blue) and B_2 (red) to form AB at 298 K:

$$A_2(g) + B_2(g) \rightleftharpoons 2AB(g) \qquad \Delta G° = -3.4 \text{ kJ/mol}$$

(1) Which of the following reaction mixtures is at equilibrium?

(2) Which of the following reaction mixtures has a negative ΔG value?

(3) Which of the following reaction mixtures has a positive ΔG value?

The partial pressures of the gases in each frame are equal to the number of A_2, B_2, and AB molecules times 0.10 atm. Round your answers to two significant figures.

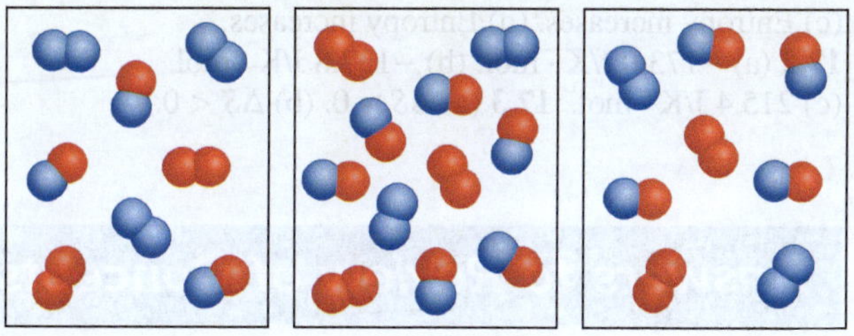

(a) (b) (c)

17.110 The K_P for the reaction

$$N_2 + 3H_2 \rightleftharpoons 2NH_3$$

is 2.4×10^{-3} at 720°C. What is the minimum partial pressure of N_2 required for the reaction to be spontaneous in the forward direction if the partial pressures of H_2 and NH_3 are 1.52 atm and 2.1×10^{-2} atm, respectively?

17.111 The table shown here lists the ion-product constant (K_w) of water at several temperatures. Determine graphically the $\Delta H°$ for the ionization of water.

K_w	0.113×10^{-14}	0.292×10^{-14}	1.008×10^{-14}	2.917×10^{-14}	5.474×10^{-14}
t(°C)	0	10	25	40	50

(*Hint:* See Problem 14.118.)

Interpreting, Modeling, & Estimating

17.112 Draw the missing distributions in Figure 17.2. (*Hint:* See the footnote in Section 17.2.)

17.113 The reaction $NH_3(g) + HCl(g) \longrightarrow NH_4Cl(s)$ is spontaneous at room temperature (see Figure 5.20). Estimate the temperature at which the reaction is no longer spontaneous under standard conditions.

17.114 The boiling point of diethyl ether is 34.6°C. Estimate (a) its molar heat of vaporization and (b) its vapor pressure at 20°C. (*Hint:* See Problems 17.48 and 17.51.)

17.115 Nicotine is the compound in tobacco responsible for addiction to smoking. While most of the nicotine in tobacco exists in the neutral form, roughly 90 percent of the nicotine in the bloodstream is protonated, as represented in the following chemical equation. Estimate $\Delta G°$ for the reaction.

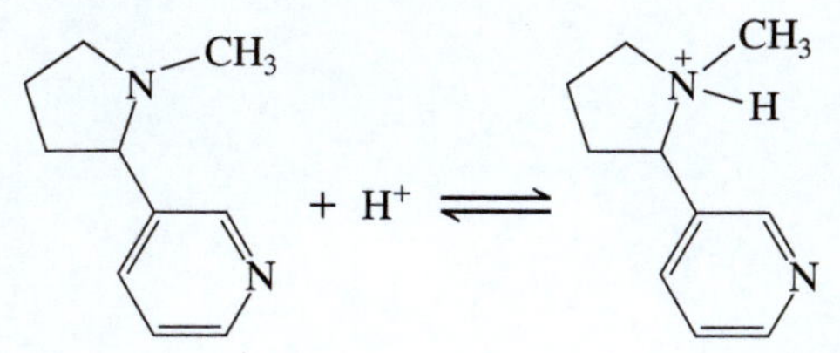

17.116 Estimate ΔS for the process depicted in Figure 17.1(a) if the apparatus contained 20 molecules in the flask on the left in the initial distribution, and each flask contained 10 molecules in the final distribution. *Useful information:* The number of ways to distribute n objects between two bins such that r particles are in one bin is called the number of combinations (C) and is given by the equation

$$C(n, r) = \frac{n!}{r!(n - r)!}$$

where $n!$ ("n factorial") $= 1 \times 2 \times 3 \times \ldots \times n$, and 0! is defined to be 1.

17.117 At what point in the series $H—O_n—H(g)$ (n = 1, 2, 3, . . .) does formation of the compound from the elements $H_2(g)$ and $O_2(g)$ become nonspontaneous?

Answers to Practice Exercises

17.1 (a) Entropy decreases. (b) Entropy decreases. (c) Entropy increases. (d) Entropy increases. **17.2** (a) −173.0 J/K · mol. (b) −137.8 J/K · mol. (c) 215.4 J/K · mol. **17.3** (a) $\Delta S > 0$. (b) $\Delta S < 0$. (c) $\Delta S \approx 0$. **17.4** (a) −107 kJ/mol. (b) −3476 kJ/mol. **17.5** $\Delta S_{fus} = 14$ J/K · mol; $\Delta S_{vap} = 74$ J/K · mol. **17.6** 2×10^{57}. **17.7** 38.4 kJ/mol. **17.8** $\Delta G = -1.0$ kJ/mol; direction is from left to right.

Answers to Review of Concepts & Facts

17.2.1 (b) and (c). **17.2.2** (a) Positive. (b) Positive. (c) Negative. **17.3.1** +146.6 J/K · mol. **17.3.2** (a) $A_2 + 3B_2 \longrightarrow 2AB_3$; (b) $\Delta S < 0$. **17.4.1** ΔS must be positive and $T\Delta S > \Delta H$ in magnitude. **17.4.2** Because S is usually quite small for solution process, the $T\Delta S$ term is small (at room temperature) compared to ΔH in magnitude. Thus, ΔH is the predominant factor in determining the sign of ΔG. **17.4.3** 30.3 kJ/mol. **17.4.4** 196 J/K · mol. **17.5.1** Less than 1. **17.5.2** 7×10^5. **17.5.3** −1.53 kJ/mol.

CHAPTER

18 Electrochemistry

Fuel cells employ chemical reactions to generate electrical energy that can be harnessed for use in automobiles.
Gene Chutka/E+/Getty Images

CHAPTER OUTLINE

One form of energy that has tremendous practical significance is electrical energy. A day without electricity from either the power company or batteries is unimaginable in our technological society. The area of chemistry that deals with the interconversion of electrical energy and chemical energy is electrochemistry.

Electrochemical processes are redox reactions in which the energy released by a spontaneous reaction is converted to electricity or in which electricity is used to drive a nonspontaneous chemical reaction. The latter type is called electrolysis.

This chapter explains the fundamental principles and applications of galvanic cells, the thermodynamics of electrochemical reactions, and the cause and prevention of corrosion by electrochemical means. Some simple electrolytic processes and the quantitative aspects of electrolysis are also discussed.

18.1 Redox Reactions

Learning Objective

- Utilize the half-reaction method to balance oxidation-reduction equations.

Electrochemistry is *the branch of chemistry that deals with the interconversion of electrical energy and chemical energy.* Electrochemical processes are redox (oxidation-reduction) reactions in which the energy released by a spontaneous reaction is converted to electricity or in which electrical energy is used to cause a nonspontaneous reaction to occur. Although redox reactions were discussed in Chapter 4, it is helpful to review some of the basic concepts that will come up again in this chapter.

In redox reactions, electrons are transferred from one substance to another. The reaction between magnesium metal and hydrochloric acid is an example of a redox reaction:

$$\overset{0}{Mg}(s) + 2\overset{+1}{H}Cl(aq) \longrightarrow \overset{+2}{Mg}Cl_2(aq) + \overset{0}{H_2}(g)$$

Recall that the numbers above the elements are the oxidation numbers of the elements. The rules for assigning oxidation numbers are presented in Section 4.4. The loss of electrons by an element during oxidation is marked by an increase in the element's oxidation number. In reduction, there is a decrease in oxidation number resulting from a gain of electrons by an element. In the preceding reaction, Mg metal is oxidized and H^+ ions are reduced; the Cl^- ions are spectator ions.

Balancing Redox Equations

Equations for redox reactions like the preceding one are relatively easy to balance. However, in the laboratory we often encounter more complex redox reactions involving oxoanions such as chromate (CrO_4^{2-}), dichromate ($Cr_2O_7^{2-}$), permanganate (MnO_4^-), nitrate (NO_3^-), and sulfate (SO_4^{2-}). In principle, we can balance any redox equation using the procedure outlined in Section 3.7, but there are some special techniques for handling redox reactions, techniques that also give us insight into electron transfer processes. Here we will discuss one such procedure, called the *ion-electron method.* In this approach, the overall reaction is divided into two half-reactions, one for oxidation and one for reduction. The equations for the two half-reactions are balanced separately and then added together to give the overall balanced equation.

Suppose we are asked to balance the equation showing the oxidation of Fe^{2+} ions to Fe^{3+} ions by dichromate ions ($Cr_2O_7^{2-}$) in an acidic medium. As a result, the $Cr_2O_7^{2-}$ ions are reduced to Cr^{3+} ions. The following steps will help us balance the equation.

Step 1: Write the unbalanced equation for the reaction in ionic form.

$$Fe^{2+} + Cr_2O_7^{2-} \longrightarrow Fe^{3+} + Cr^{3+}$$

Step 2: Separate the equation into two half-reactions.

$$\text{Oxidation:}\quad \overset{+2}{\text{Fe}}{}^{2+} \longrightarrow \overset{+3}{\text{Fe}}{}^{3+}$$

$$\text{Reduction:}\quad \overset{+6}{\text{Cr}}{}_2\text{O}_7^{2-} \longrightarrow \overset{+3}{\text{Cr}}{}^{3+}$$

Step 3: Balance each half-reaction for number and type of atoms and charges. For reactions in an acidic medium, add H_2O to balance the O atoms and H^+ to balance the H atoms.

Oxidation half-reaction: The atoms are already balanced. In an oxidation half-reaction, electrons appear as a product. To balance the charge, we add an electron to the right-hand side of the arrow:

$$Fe^{2+} \longrightarrow Fe^{3+} + e^-$$

Reduction half-reaction: Because the reaction takes place in an acidic medium, we add seven H_2O molecules to the right-hand side of the arrow to balance the O atoms:

$$Cr_2O_7^{2-} \longrightarrow 2Cr^{3+} + 7H_2O$$

To balance the H atoms, we add $14H^+$ ions on the left-hand side:

$$14H^+ + Cr_2O_7^{2-} \longrightarrow 2Cr^{3+} + 7H_2O$$

There are now 12 positive charges on the left-hand side and only six positive charges on the right-hand side. In a reduction half-reaction, electrons appear as a reactant. Therefore, we add six electrons on the left:

$$14H^+ + Cr_2O_7^{2-} + 6e^- \longrightarrow 2Cr^{3+} + 7H_2O$$

Step 4: Add the two half-equations together and balance the final equation by inspection. The electrons on both sides must cancel. If the oxidation and reduction half-reactions contain different numbers of electrons, we need to multiply one or both half-reactions to equalize the number of electrons.

Here we have only one electron for the oxidation half-reaction and six electrons for the reduction half-reaction, so we need to multiply the oxidation half-reaction by 6 and write

$$\begin{array}{c} 6(Fe^{2+} \longrightarrow Fe^{3+} + e^-) \\ 14H^+ + Cr_2O_7^{2-} + 6e^- \longrightarrow 2Cr^{3+} + 7H_2O \\ \hline 6Fe^{2+} + 14H^+ + Cr_2O_7^{2-} + \cancel{6e^-} \longrightarrow 6Fe^{3+} + 2Cr^{3+} + 7H_2O + \cancel{6e^-} \end{array}$$

The electrons on both sides cancel, and we are left with the balanced net ionic equation:

$$6Fe^{2+} + 14H^+ + Cr_2O_7^{2-} \longrightarrow 6Fe^{3+} + 2Cr^{3+} + 7H_2O$$

Step 5: Verify that the equation contains the same type and numbers of atoms and the same charges on both sides of the equation.

A final check shows that the resulting equation is "atomically" and "electrically" balanced.

For reactions in a basic medium, we proceed through step 4 as if the reaction were carried out in an acidic medium. Then, for every H^+ ion we add an equal number of OH^- ions to *both* sides of the equation. Where H^+ and OH^- ions appear on the same side of the equation, we combine the ions to give H_2O. Example 18.1 illustrates this procedure.

Student Hot Spot

Student data indicate you may struggle with balancing redox reactions in acidic solution. Access your eBook for additional Learning Resources on this topic.

Student Hot Spot

Student data indicate you may struggle with balancing redox reactions in basic solution. Access your eBook for additional Learning Resources on this topic.

Example 18.1

Write a balanced ionic equation to represent the oxidation of iodide ion (I^-) by permanganate ion (MnO_4^-) in basic solution to yield molecular iodine (I_2) and manganese(IV) oxide (MnO_2).

Strategy We follow the preceding procedure for balancing redox equations. Note that the reaction takes place in a basic medium.

Solution

Step 1: The unbalanced equation is

$$MnO_4^- + I^- \longrightarrow MnO_2 + I_2$$

Step 2: The two half-reactions are

$$\text{Oxidation:} \quad \overset{-1}{I^-} \longrightarrow \overset{0}{I_2}$$

$$\text{Reduction:} \quad \overset{+7}{MnO_4^-} \longrightarrow \overset{+4}{MnO_2}$$

Step 3: We balance each half-reaction for number and type of atoms and charges. Oxidation half-reaction: We first balance the I atoms:

$$2I^- \longrightarrow I_2$$

To balance charges, we add two electrons to the right-hand side of the equation:

$$2I^- \longrightarrow I_2 + 2e^-$$

Reduction half-reaction: To balance the O atoms, we add two H_2O molecules on the right:

$$MnO_4^- \longrightarrow MnO_2 + 2H_2O$$

To balance the H atoms, we add four H^+ ions on the left:

$$MnO_4^- + 4H^+ \longrightarrow MnO_2 + 2H_2O$$

There are three net positive charges on the left, so we add three electrons to the same side to balance the charges:

$$MnO_4^- + 4H^+ + 3e^- \longrightarrow MnO_2 + 2H_2O$$

Step 4: We now add the oxidation and reduction half-reactions to give the overall reaction. To equalize the number of electrons, we need to multiply the oxidation half-reaction by 3 and the reduction half-reaction by 2 as follows:

$$3(2I^- \longrightarrow I_2 + 2e^-)$$

$$2(MnO_4^- + 4H^+ + 3e^- \longrightarrow MnO_2 + 2H_2O)$$

$$6I^- + 2MnO_4^- + 8H^+ + \cancel{6e^-} \longrightarrow 3I_2 + 2MnO_2 + 4H_2O + \cancel{6e^-}$$

The electrons on both sides cancel, and we are left with the balanced net ionic equation:

$$6I^- + 2MnO_4^- + 8H^+ \longrightarrow 3I_2 + 2MnO_2 + 4H_2O$$

This is the balanced equation in an acidic medium. However, because the reaction is carried out in a basic medium, for every H^+ ion we need to add an equal number of OH^- ions to both sides of the equation:

$$6I^- + 2MnO_4^- + 8H^+ + 8OH^- \longrightarrow 3I_2 + 2MnO_2 + 4H_2O + 8OH^-$$

Finally, combining the H^+ and OH^- ions to form water, we obtain

$$6I^- + 2MnO_4^- + 4H_2O \longrightarrow 3I_2 + 2MnO_2 + 8OH^-$$

Step 5: A final check shows that the equation is balanced in terms of both atoms and charges.

(Continued)

Practice Exercise Balance the following equation for the reaction in an acidic medium by the ion-electron method:

$$Fe^{2+} + MnO_4^- \longrightarrow Fe^{3+} + Mn^{2+}$$

Similar problems: 18.1, 18.2.

Summary of Concepts & Facts

- Redox reactions involve the transfer of electrons. Equations representing redox processes can be balanced using the ion-electron method.
- Determine equilibrium concentrations or pH of a solution involving a common ion.

Review of Concepts & Facts

18.1.1 Determine the number of electrons transferred in the following reaction:

$$2Al(s) + 6H^+(aq) \longrightarrow 2Al^{3+}(aq) + 3H_2(g)$$

18.1.2 For the following reaction in acidic solution, what is the coefficient for NO_2 when the equation is balanced?

$$Sn + NO_3^- \longrightarrow SnO_2 + NO_2$$

18.2 Galvanic Cells

Learning Objective

- Explain key electrochemical terms: *galvanic cell, anode, cathode,* and *salt bridge*.

In Section 4.4 we saw that when a piece of zinc metal is placed in a $CuSO_4$ solution, Zn is oxidized to Zn^{2+} ions while Cu^{2+} ions are reduced to metallic copper (see Figure 4.10):

$$Zn(s) + Cu^{2+}(aq) \longrightarrow Zn^{2+}(aq) + Cu(s)$$

The electrons are transferred directly from the reducing agent (Zn) to the oxidizing agent (Cu^{2+}) in solution. However, if we physically separate the oxidizing agent from the reducing agent, the transfer of electrons can take place via an external conducting medium (a metal wire). As the reaction progresses, it sets up a constant flow of electrons and hence generates electricity (that is, it produces electrical work such as driving an electric motor).

The experimental apparatus for generating electricity through the use of a spontaneous reaction is called a ***galvanic cell*** or *voltaic cell,* after the Italian scientists Luigi Galvani and Alessandro Volta, who constructed early versions of the device. Figure 18.1 shows the essential components of a galvanic cell. A zinc bar is immersed in a $ZnSO_4$ solution, and a copper bar is immersed in a $CuSO_4$ solution. The cell operates on the principle that the oxidation of Zn to Zn^{2+} and the reduction of Cu^{2+} to Cu can be made to take place simultaneously in separate locations with the transfer of electrons between them occurring through an external wire. The zinc and copper bars are called *electrodes.* This particular arrangement of electrodes (Zn and Cu) and solutions ($ZnSO_4$ and $CuSO_4$) is called the Daniell cell. By definition, the ***anode*** in a galvanic cell is *the electrode at which oxidation occurs* and the ***cathode*** is *the electrode at which reduction occurs.* A convenient way to remember which process occurs at which electrode is that alphabetically *a*node precedes *c*athode and *o*xidation precedes *r*eduction. Therefore, the anode is where oxidation occurs and the cathode is where the reduction occurs.

Video: Galvanic Cells

Video: The Cu/Zn Voltaic Cell

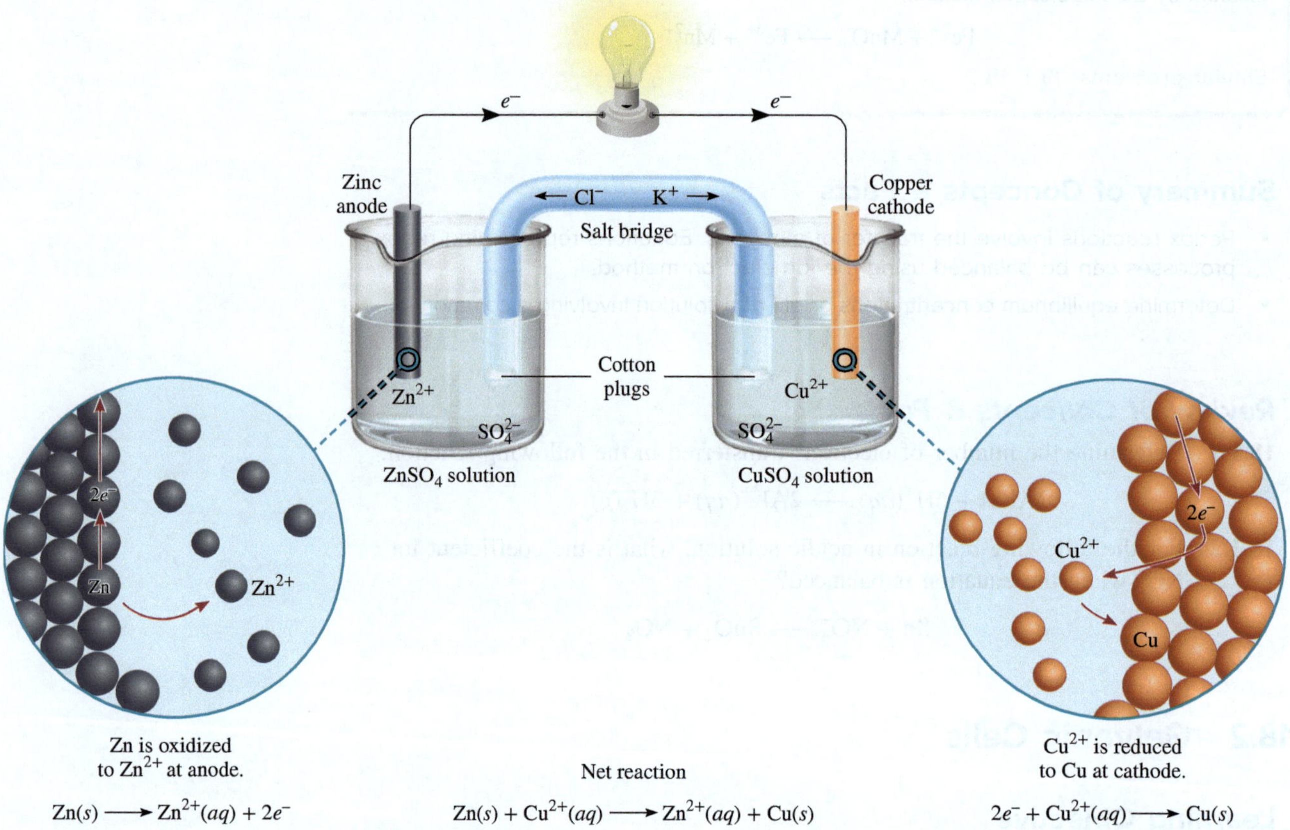

Figure 18.1 *A galvanic cell. The salt bridge (an inverted U tube) containing a KCl solution provides an electrically conducting medium between two solutions. The openings of the U tube are loosely plugged with cotton balls to prevent the KCl solution from flowing into the containers while allowing the anions and cations to move across. The lightbulb is lit as electrons flow externally from the Zn electrode (anode) to the Cu electrode (cathode).*

For the Daniell cell, the ***half-cell reactions,*** that is, *the oxidation and reduction reactions at the electrodes,* are

$$\text{Zn electrode (anode):}\quad Zn(s) \longrightarrow Zn^{2+}(aq) + 2e^-$$
$$\text{Cu electrode (cathode):}\quad Cu^{2+}(aq) + 2e^- \longrightarrow Cu(s)$$

Note that unless the two solutions are separated from each other, the Cu^{2+} ions will react directly with the zinc bar:

$$Cu^{2+}(aq) + Zn(s) \longrightarrow Cu(s) + Zn^{2+}(aq)$$

and no useful electrical work will be obtained.

Video: Operation of a Voltaic Cell

To complete the electrical circuit, the solutions must be connected by a conducting medium through which the cations and anions can move from one electrode compartment to the other. This requirement is satisfied by a *salt bridge,* which, in its simplest form, is an inverted U tube containing an inert electrolyte solution, such as KCl or NH_4NO_3, whose ions will not react with other ions in solution or with the electrodes (see Figure 18.1). During the course of the overall redox reaction, electrons flow externally from the anode (Zn electrode) through the wire to the cathode (Cu electrode). In the solution, the cations (Zn^{2+}, Cu^{2+}, and K^+) move toward the cathode, while the anions (SO_4^{2-} and Cl^-) move toward the anode. Without the salt bridge connecting the two solutions, the buildup of positive charge in the anode compartment (due to the formation of Zn^{2+} ions) and negative charge in the cathode compartment (created when some of the Cu^{2+} ions are reduced to Cu) would quickly prevent the cell from operating.

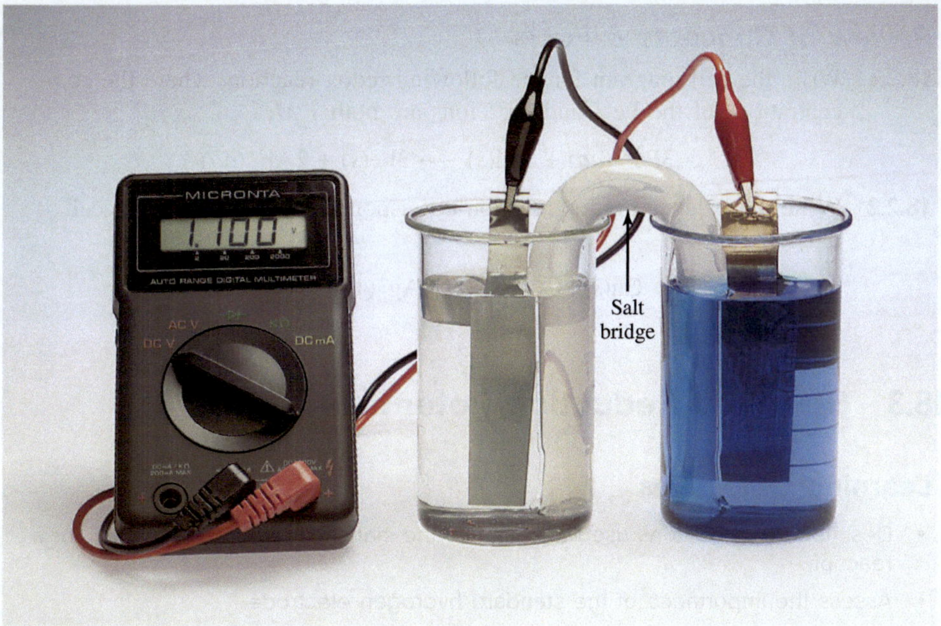

Figure 18.2 *Practical setup of the galvanic cell described in Figure 18.1. Note the U tube (salt bridge) connecting the two beakers. When the concentrations of $ZnSO_4$ and $CuSO_4$ are 1 molar (1 M) at 25°C, the cell voltage is 1.10 V. No current flows between the electrodes during a voltage measurement.*

Gene Chutka/E+/Getty Images

An electric current flows from the anode to the cathode because there is a difference in electrical potential energy between the electrodes. This flow of electric current is analogous to that of water down a waterfall, which occurs because there is a difference in gravitational potential energy, or the flow of gas from a high-pressure region to a low-pressure region. Experimentally, the *difference in electrical potential between the anode and the cathode* is measured by a voltmeter (Figure 18.2). The voltage across the electrodes of a galvanic cell is called the ***cell voltage,*** or *cell potential.* Another common term for cell voltage is the ***electromotive force (emf) (E),*** which, despite its name, is a measure of voltage, not force. We will see that the voltage of a cell depends not only on the nature of the electrodes and the ions, but also on the concentrations of the ions and the temperature at which the cell is operated.

The conventional notation for representing galvanic cells is the *cell diagram.* For the Daniell cell shown in Figure 18.1, if we assume that the concentrations of Zn^{2+} and Cu^{2+} ions are 1 *M*, the cell diagram is

$$Zn(s) \mid Zn^{2+}(1\ M) \parallel Cu^{2+}(1\ M) \mid Cu(s)$$

The single vertical line represents a phase boundary. For example, the zinc electrode is a solid and the Zn^{2+} ions (from $ZnSO_4$) are in solution. Thus, we draw a line between Zn and Zn^{2+} to show the phase boundary. The double vertical lines denote the salt bridge. By convention, the anode is written first, to the left of the double lines, and the other components appear in the order in which we would encounter them in moving from the anode to the cathode.

Student Hot Spot

Student data indicate you may struggle with understanding cell diagrams. Access your eBook for additional Learning Resources on this topic.

Summary of Concepts & Facts

- In a galvanic cell, electricity is produced by a spontaneous chemical reaction. Oxidation and reduction take place separately at the anode and cathode, respectively, and the electrons flow through an external circuit.
- The two parts of a galvanic cell are the half-cells, and the reactions at the electrodes are the half-cell reactions. A salt bridge allows ions to flow between the half-cells.

Review of Concepts & Facts

18.2.1 Write the cell diagram for the following redox reaction, where the concentrations of the Fe^{2+} and Al^{3+} ions are both 1 *M*.

$$3Fe^{2+}(aq) + 2Al(s) \longrightarrow 3Fe(s) + 2Al^{3+}(aq)$$

18.2.2 Write the balanced redox reaction corresponding to the following cell diagram:

$$Cu(s) \mid Cu^{2+}(1\,M) \parallel Ag^{+}(1\,M) \mid Ag(s)$$

18.3 Standard Reduction Potentials

Learning Objectives

- Describe the conditions used to calculate the standard reduction potential for a reaction.
- Assess the importance of the standard hydrogen electrode.

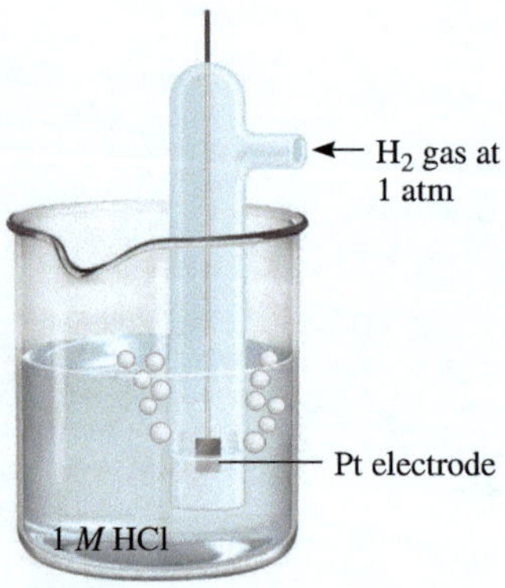

Figure 18.3 *A hydrogen electrode operating under standard-state conditions. Hydrogen gas at 1 atm is bubbled through a 1 M HCl solution. The platinum electrode is part of the hydrogen electrode.*

Student Hot Spot

Student data indicate you may struggle with the standard hydrogen electrode. Access your eBook for additional Learning Resources on this topic.

When the concentrations of the Cu^{2+} and Zn^{2+} ions are both 1.0 *M*, we find that the voltage or emf of the Daniell cell is 1.10 V at 25°C (see Figure 18.2). This voltage must be related directly to the redox reactions, but how? Just as the overall cell reaction can be thought of as the sum of two half-cell reactions, the measured emf of the cell can be treated as the sum of the electrical potentials at the Zn and Cu electrodes. Knowing one of these electrode potentials, we could obtain the other by subtraction (from 1.10 V). It is impossible to measure the potential of just a single electrode, but if we arbitrarily set the potential value of a particular electrode at zero, we can use it to determine the relative potentials of other electrodes. The hydrogen electrode, shown in Figure 18.3, serves as the reference for this purpose. Hydrogen gas is bubbled into a hydrochloric acid solution at 25°C. The platinum electrode has two functions. First, it provides a surface on which the dissociation of hydrogen molecules can take place:

$$H_2 \longrightarrow 2H^+ + 2e^-$$

Second, it serves as an electrical conductor to the external circuit.

Under standard-state conditions (when the pressure of H_2 is 1 atm and the concentration of the HCl solution is 1 *M*; see Table 17.3), the potential for the reduction of H^+ at 25°C is taken to be *exactly* zero:

$$2H^+(1\,M) + 2e^- \longrightarrow H_2(1\text{ atm}) \qquad E° = 0\text{ V}$$

The superscript "°" denotes standard-state conditions, and $E°$ is the ***standard reduction potential,*** or *the voltage associated with a reduction reaction at an electrode when all solutes are 1 M and all gases are at 1 atm.* Thus, the standard reduction potential of the hydrogen electrode is defined as zero. The hydrogen electrode is called the *standard hydrogen electrode (SHE).*

We can use the SHE to measure the potentials of other kinds of electrodes. For example, Figure 18.4(a) shows a galvanic cell with a zinc electrode and a SHE. In this case, the zinc electrode is the anode and the SHE is the cathode. We deduce this fact from the decrease in mass of the zinc electrode during the operation of the cell, which is consistent with the loss of zinc to the solution caused by the oxidation reaction:

$$Zn(s) \longrightarrow Zn^{2+}(aq) + 2e^-$$

The cell diagram is

$$Zn(s) \mid Zn^{2+}(1\,M) \parallel H^+(1\,M) \mid H_2(1\text{ atm}) \mid Pt(s)$$

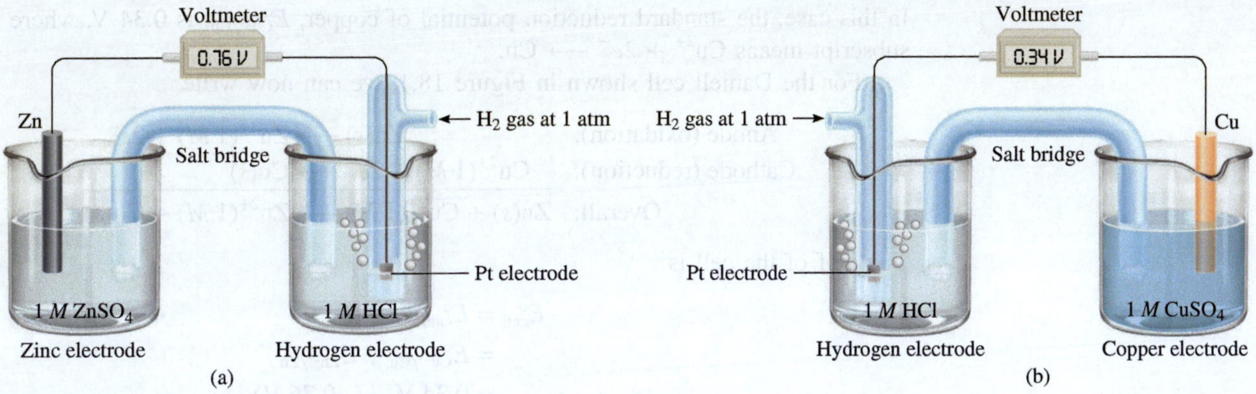

Figure 18.4 *(a) A cell consisting of a zinc electrode and a hydrogen electrode. (b) A cell consisting of a copper electrode and a hydrogen electrode. Both cells are operating under standard-state conditions. Note that in (a) the SHE acts as the cathode, but in (b) it acts as the anode. As mentioned in Figure 18.2, no current flows between the electrodes during a voltage measurement.*

As mentioned earlier, the Pt electrode provides the surface on which the reduction takes place. When all the reactants are in their standard states (that is, H_2 at 1 atm, H^+ and Zn^{2+} ions at 1 *M*), the emf of the cell is 0.76 V at 25°C. We can write the half-cell reactions as follows:

$$\begin{aligned} &\text{Anode (oxidation):} && \text{Zn}(s) \longrightarrow \text{Zn}^{2+}(1\,M) + 2e^- \\ &\text{Cathode (reduction):} && 2\text{H}^+(1\,M) + 2e^- \longrightarrow \text{H}_2(1\text{ atm}) \\ \hline &\text{Overall:} && \text{Zn}(s) + 2\text{H}^+(1\,M) \longrightarrow \text{Zn}^{2+}(1\,M) + \text{H}_2(1\text{ atm}) \end{aligned}$$

By convention, the ***standard emf*** of the cell, E°_{cell}, which is composed of a contribution from the anode and a contribution from the cathode, is given by

$$E^\circ_{\text{cell}} = E^\circ_{\text{cathode}} - E^\circ_{\text{anode}} \quad (18.1)$$

where *both* E°_{cathode} and E°_{anode} are the standard reduction potentials of the electrodes. For the Zn-SHE cell, we write

$$\begin{aligned} E^\circ_{\text{cell}} &= E^\circ_{\text{H}^+/\text{H}_2} - E^\circ_{\text{Zn}^{2+}/\text{Zn}} \\ 0.76\text{ V} &= 0 - E^\circ_{\text{Zn}^{2+}/\text{Zn}} \end{aligned}$$

where the subscript H^+/H_2 means $2H^+ + 2e^- \longrightarrow H_2$ and the subscript Zn^{2+}/Zn means $Zn^{2+} + 2e^- \longrightarrow Zn$. Thus, the standard reduction potential of zinc, $E^\circ_{\text{Zn}^{2+}/\text{Zn}}$, is −0.76 V.

The standard electrode potential of copper can be obtained in a similar fashion, by using a cell with a copper electrode and a SHE [Figure 18.4(b)]. In this case, the copper electrode is the cathode because its mass increases during the operation of the cell, as is consistent with the reduction reaction:

$$\text{Cu}^{2+}(aq) + 2e^- \longrightarrow \text{Cu}(s)$$

The cell diagram is

$$\text{Pt}(s)\,|\,\text{H}_2(1\text{ atm})\,|\,\text{H}^+(1\,M)\,\|\,\text{Cu}^{2+}(1\,M)\,|\,\text{Cu}(s)$$

and the half-cell reactions are

$$\begin{aligned} &\text{Anode (oxidation):} && \text{H}_2(1\text{ atm}) \longrightarrow 2\text{H}^+(1\,M) + 2e^- \\ &\text{Cathode (reduction):} && \text{Cu}^{2+}(1\,M) + 2e^- \longrightarrow \text{Cu}(s) \\ \hline &\text{Overall:} && \text{H}_2(1\text{ atm}) + \text{Cu}^{2+}(1\,M) \longrightarrow 2\text{H}^+(1\,M) + \text{Cu}(s) \end{aligned}$$

Under standard-state conditions and at 25°C, the emf of the cell is 0.34 V, so we write

$$\begin{aligned} E^\circ_{\text{cell}} &= E^\circ_{\text{cathode}} - E^\circ_{\text{anode}} \\ 0.34\text{ V} &= E^\circ_{\text{Cu}^{2+}/\text{Cu}} - E^\circ_{\text{H}^+/\text{H}_2} \\ &= E^\circ_{\text{Cu}^{2+}/\text{Cu}} - 0 \end{aligned}$$

In this case, the standard reduction potential of copper, $E^\circ_{Cu^{2+}/Cu}$, is 0.34 V, where the subscript means $Cu^{2+} + 2e^- \longrightarrow Cu$.

For the Daniell cell shown in Figure 18.1, we can now write

$$\begin{array}{rr} \text{Anode (oxidation):} & Zn(s) \longrightarrow Zn^{2+}(1\,M) + 2e^- \\ \text{Cathode (reduction):} & Cu^{2+}(1\,M) + 2e^- \longrightarrow Cu(s) \\ \hline \text{Overall:} & Zn(s) + Cu^{2+}(1\,M) \longrightarrow Zn^{2+}(1\,M) + Cu(s) \end{array}$$

The emf of the cell is

$$\begin{aligned} E^\circ_{cell} &= E^\circ_{cathode} - E^\circ_{anode} \\ &= E^\circ_{Cu^{2+}/Cu} - E^\circ_{Zn^{2+}/Zn} \\ &= 0.34\text{ V} - (-0.76\text{ V}) \\ &= 1.10\text{ V} \end{aligned}$$

As in the case of ΔG° (see Table 17.3), we can use the sign of E° to predict the extent of a redox reaction. A positive E° means the redox reaction will favor the formation of products at equilibrium. Conversely, a negative E° means that more reactants than products will be formed at equilibrium. We will examine the relationships among E°_{cell}, ΔG°, and K later in this chapter.

Table 18.1 lists standard reduction potentials for a number of half-cell reactions. The activity series in Figure 4.16 is based on data given in Table 18.1. By definition, the SHE has an E° value of 0.00 V. Below the SHE the negative standard reduction potentials increase, and above it the positive standard reduction potentials increase. It is important to know the following points about the table in calculations:

1. The E° values apply to the half-cell reactions as read in the forward (left to right) direction.
2. The more positive E° is, the greater the tendency for the substance to be reduced. For example, the half-cell reaction

   $$F_2(1\text{ atm}) + 2e^- \longrightarrow 2F^-(1\,M) \qquad E^\circ = 2.87\text{ V}$$

   has the highest positive E° value among all the half-cell reactions. Thus, F_2 is the *strongest* oxidizing agent because it has the greatest tendency to be reduced. At the other extreme is the reaction

   $$Li^+(1\,M) + e^- \longrightarrow Li(s) \qquad E^\circ = -3.05\text{ V}$$

   which has the most negative E° value. Thus, Li^+ is the *weakest* oxidizing agent because it is the most difficult species to reduce. Conversely, we say that F^- is the weakest reducing agent and Li metal is the strongest reducing agent. Under standard-state conditions, the oxidizing agents (the species on the left-hand side of the half-reactions in Table 18.1) increase in strength from bottom to top and the reducing agents (the species on the right-hand side of the half-reactions) increase in strength from top to bottom.
3. The half-cell reactions are reversible. Depending on the conditions, any electrode can act either as an anode or as a cathode. Earlier we saw that the SHE is the cathode (H^+ is reduced to H_2) when coupled with zinc in a cell and that it becomes the anode (H_2 is oxidized to H^+) when used in a cell with copper.
4. Under standard-state conditions, any species on the left of a given half-cell reaction will react spontaneously with a species that appears on the right of any half-cell reaction located *below* it in Table 18.1. This principle is sometimes called the *diagonal rule*. In the case of the Daniell cell

   $$Cu^{2+}(1\,M) + 2e^- \longrightarrow Cu(s) \qquad E^\circ = 0.34\text{ V}$$

   $$Zn^{2+}(1\,M) + 2e^- \longrightarrow Zn(s) \qquad E^\circ = -0.76\text{ V}$$

Table 18.1 Standard Reduction Potentials at 25°C*

Half-Reaction	$E°$(V)
$F_2(g) + 2e^- \longrightarrow 2F^-(aq)$	+2.87
$O_3(g) + 2H^+(aq) + 2e^- \longrightarrow O_2(g) + H_2O$	+2.08
$Co^{3+}(aq) + e^- \longrightarrow Co^{2+}(aq)$	+1.92
$H_2O_2(aq) + 2H^+(aq) + 2e^- \longrightarrow 2H_2O$	+1.78
$Ce^{4+}(aq) + e^- \longrightarrow Ce^{3+}(aq)$	+1.72
$PbO_2(s) + 4H^+(aq) + SO_4^{2-}(aq) + 2e^- \longrightarrow PbSO_4(s) + 2H_2O$	+1.69
$Au^+(aq) + e^- \longrightarrow Au(s)$	+1.69
$Ce^{4+}(aq) + e^- \longrightarrow Ce^{3+}(aq)$	+1.61
$MnO_4^-(aq) + 8H^+(aq) + 5e^- \longrightarrow Mn^{2+}(aq) + 4H_2O$	+1.51
$Au^{3+}(aq) + 3e^- \longrightarrow Au(s)$	+1.50
$Cl_2(g) + 2e^- \longrightarrow 2Cl^-(aq)$	+1.36
$Cr_2O_7^{2-}(aq) + 14H^+(aq) + 6e^- \longrightarrow 2Cr^{3+}(aq) + 7H_2O$	+1.36
$MnO_2(s) + 4H^+(aq) + 2e^- \longrightarrow Mn^{2+}(aq) + 2H_2O$	+1.22
$O_2(g) + 4H^+(aq) + 4e^- \longrightarrow 2H_2O$	+1.23
$2IO_3^-(aq) + 12H^+(aq) + 10e^- \longrightarrow I_2(s) + 6H_2O(l)$	+1.20
$Br_2(l) + 2e^- \longrightarrow 2Br^-(aq)$	+1.07
$VO^{2+}(aq) + 2H^+(aq) + e^- \longrightarrow VO_2^+(aq) + H_2O(l)$	+0.99
$HNO_2(aq) + H^+(aq) + e^- \longrightarrow NO(g) + H_2O(l)$	+0.98
$NO_3^-(aq) + 4H^+(aq) + 3e^- \longrightarrow NO(g) + 2H_2O$	+0.96
$2Hg^{2+}(aq) + 2e^- \longrightarrow Hg_2^{2+}(aq)$	+0.92
$Hg_2^{2+}(aq) + 2e^- \longrightarrow 2Hg(l)$	+0.85
$Ag^+(aq) + e^- \longrightarrow Ag(s)$	+0.80
$Fe^{3+}(aq) + e^- \longrightarrow Fe^{2+}(aq)$	+0.77
$O_2(g) + 2H^+(aq) + 2e^- \longrightarrow H_2O_2(aq)$	+0.70
$MnO_4^-(aq) + 2H_2O + 3e^- \longrightarrow MnO_2(s) + 4OH^-(aq)$	+0.60
$I_2(s) + 2e^- \longrightarrow 2I^-(aq)$	+0.54
$Cu^+(aq) + e^- \longrightarrow Cu(s)$	+0.52
$O_2(g) + 2H_2O + 4e^- \longrightarrow 4OH^-(aq)$	+0.40
$Cu^{2+}(aq) + 2e^- \longrightarrow Cu(s)$	+0.34
$AgCl(s) + e^- \longrightarrow Ag(s) + Cl^-(aq)$	+0.22
$HSO_4^-(aq) + 3H^+(aq) + 2e^- \longrightarrow H_2SO_3(aq) + H_2O(l)$	+0.17
$SO_4^{2-}(aq) + 4H^+(aq) + 2e^- \longrightarrow SO_2(g) + 2H_2O$	+0.17
$Cu^{2+}(aq) + e^- \longrightarrow Cu^+(aq)$	+0.15
$Sn^{4+}(aq) + 2e^- \longrightarrow Sn^{2+}(aq)$	+0.15
$2H^+(aq) + 2e^- \longrightarrow H_2(g)$	0.00
$Fe^{3+}(aq) + 3e^- \longrightarrow Fe(s)$	−0.04
$Pb^{2+}(aq) + 2e^- \longrightarrow Pb(s)$	−0.13
$CrO_4^{2-}(aq) + 4H_2O(l) + 3e^- \longrightarrow Cr(OH)_3(s) + 5OH^-(aq)$	−0.13
$Sn^{2+}(aq) + 2e^- \longrightarrow Sn(s)$	−0.14
$N_2(g) + 5H^+(aq) + 4e^- \longrightarrow N_2H_5^+(aq)$	−0.21
$Ni^{2+}(aq) + 2e^- \longrightarrow Ni(s)$	−0.26
$Co^{2+}(aq) + 2e^- \longrightarrow Co(s)$	−0.28
$PbSO_4(s) + 2e^- \longrightarrow Pb(s) + SO_4^{2-}(aq)$	−0.31
$PbSO_4(s) + H^+(aq) + 2e^- \longrightarrow Pb(s) + HSO_4^-(aq)$	−0.36
$Cd^{2+}(aq) + 2e^- \longrightarrow Cd(s)$	−0.40
$Cr^{3+}(aq) + e^- \longrightarrow Cr^{2+}(aq)$	−0.41
$Fe^{2+}(aq) + 2e^- \longrightarrow Fe(s)$	−0.45
$Cr^{3+}(aq) + 3e^- \longrightarrow Cr(s)$	−0.74
$Zn^{2+}(aq) + 2e^- \longrightarrow Zn(s)$	−0.76
$2H_2O + 2e^- \longrightarrow H_2(g) + 2OH^-(aq)$	−0.83
$Cr^{2+}(aq) + e^- \longrightarrow Cr(s)$	−0.91
$N_2(g) + 4H_2O(l) + 4e^- \longrightarrow 4OH^-(aq) + N_2H_4(aq)$	−1.16
$Mn^{2+}(aq) + 2e^- \longrightarrow Mn(s)$	−1.19
$Al^{3+}(aq) + 3e^- \longrightarrow Al(s)$	−1.68
$Be^{2+}(aq) + 2e^- \longrightarrow Be(s)$	−1.85
$Sc^{3+}(aq) + 3e^- \longrightarrow Sc(s)$	−2.08
$Mg^{2+}(aq) + 2e^- \longrightarrow Mg(s)$	−2.37
$Na^+(aq) + e^- \longrightarrow Na(s)$	−2.71
$Ca^{2+}(aq) + 2e^- \longrightarrow Ca(s)$	−2.87
$Sr^{2+}(aq) + 2e^- \longrightarrow Sr(s)$	−2.90
$Ba^{2+}(aq) + 2e^- \longrightarrow Ba(s)$	−2.91
$K^+(aq) + e^- \longrightarrow K(s)$	−2.93
$Li^+(aq) + e^- \longrightarrow Li(s)$	−3.04

↑ Increasing strength as oxidizing agent

↓ Increasing strength as reducing agent

*For all half-reactions the concentration is 1 M for dissolved species and the pressure is 1 bar for gases. These are the standard-state values.

We see that the substance on the left of the first half-cell reaction is Cu^{2+} and the substance on the right in the second half-cell reaction is Zn. The diagonal red line shows that Cu^{2+} is the oxidizing agent and Zn is the reducing agent. Therefore, as we saw earlier, Zn spontaneously reduces Cu^{2+} to form Zn^{2+} and Cu.

5. Changing the stoichiometric coefficients of a half-cell reaction *does not* affect the value of $E°$ because electrode potentials are intensive properties. This means that the value of $E°$ is unaffected by the size of the electrodes or the amount of solutions present. For example,

$$I_2(s) + 2e^- \longrightarrow 2I^-(1\,M) \qquad E° = 0.54\text{ V}$$

but $E°$ does not change if we multiply the half-reaction by 2:

$$2I_2(s) + 4e^- \longrightarrow 4I^-(1\,M) \qquad E° = 0.54\text{ V}$$

6. Like ΔH, ΔG, and ΔS, the sign of $E°$ changes but its magnitude remains the same when we reverse a reaction.

Student Hot Spot

Student data indicate you may struggle with standard reduction potentials. Access your eBook for additional Learning Resources on this topic.

As Examples 18.2 and 18.3 show, Table 18.1 enables us to predict the outcome of redox reactions under standard-state conditions, whether they take place in a galvanic cell, where the reducing agent and oxidizing agent are physically separated from each other, or in a beaker, where the reactants are all mixed together.

Example 18.2

Predict what will happen if molecular bromine (Br_2) is added to a solution containing NaCl and NaI at 25°C. Assume all species are in their standard states.

Strategy To predict what redox reaction(s) will take place, we need to compare the standard reduction potentials of Cl_2, Br_2, and I_2 and apply the diagonal rule.

Solution From Table 18.1, we write the standard reduction potentials as follows:

$$Cl_2(1\text{ atm}) + 2e^- \longrightarrow 2Cl^-(1\,M) \qquad E° = 1.36\text{ V}$$
$$Br_2(l) + 2e^- \longrightarrow 2Br^-(1\,M) \qquad E° = 1.07\text{ V}$$
$$I_2(s) + 2e^- \longrightarrow 2I^-(1\,M) \qquad E° = 0.54\text{ V}$$

Applying the diagonal rule we see that Br_2 will oxidize I^- but will not oxidize Cl^-. Therefore, the only redox reaction that will occur appreciably under standard-state conditions is

$$\begin{aligned} &\text{Oxidation:} && 2I^-(1\,M) \longrightarrow I_2(s) + 2e^- \\ &\text{Reduction:} && Br_2(l) + 2e^- \longrightarrow 2Br^-(1\,M) \\ \hline &\text{Overall:} && 2I^-(1\,M) + Br_2(l) \longrightarrow I_2(s) + 2Br^-(1\,M) \end{aligned}$$

Check We can confirm our conclusion by calculating $E°_{cell}$. Try it. Note that the Na^+ ions are inert and do not enter into the redox reaction.

Practice Exercise Can Sn reduce $Zn^{2+}(aq)$ under standard-state conditions?

Similar problems: 18.14, 18.18.

Example 18.3

A galvanic cell consists of a Mg electrode in a 1.0 *M* $Mg(NO_3)_2$ solution and a Ag electrode in a 1.0 *M* $AgNO_3$ solution. Calculate the standard emf of this cell at 25°C.

Strategy At first it may not be clear how to assign the electrodes in the galvanic cell. From Table 18.1 we write the standard reduction potentials of Ag and Mg and apply the diagonal rule to determine which is the anode and which is the cathode.

(Continued)

Solution The standard reduction potentials are

$$Ag^+(1.0\ M) + e^- \longrightarrow Ag(s) \qquad E° = 0.80\ V$$
$$Mg^{2+}(1.0\ M) + 2e^- \longrightarrow Mg(s) \qquad E° = -2.37\ V$$

Applying the diagonal rule, we see that Ag^+ will oxidize Mg:

$$\text{Anode (oxidation):} \qquad Mg(s) \longrightarrow Mg^{2+}(1.0\ M) + 2e^-$$
$$\text{Cathode (reduction):} \qquad 2Ag^+(1.0\ M) + 2e^- \longrightarrow 2Ag(s)$$
$$\text{Overall:} \qquad Mg(s) + 2Ag^+(1.0\ M) \longrightarrow Mg^{2+}(1.0\ M) + 2Ag(s)$$

Note that to balance the overall equation, we multiplied the reduction of Ag^+ by 2. We can do so because, as an intensive property, $E°$ is not affected by this procedure. We find the emf of the cell by using Equation (18.1) and Table 18.1:

$$\begin{aligned} E°_{cell} &= E°_{cathode} - E°_{anode} \\ &= E°_{Ag^+/Ag} - E°_{Mg^{2+}/Mg} \\ &= 0.80\ V - (-2.37\ V) \\ &= 3.17\ V \end{aligned}$$

Check The positive value of $E°$ shows that the forward reaction is favored.

Practice Exercise What is the standard emf of a galvanic cell made of a Cd electrode in a 1.0 M $Cd(NO_3)_2$ solution and a Cr electrode in a 1.0 M $Cr(NO_3)_3$ solution at 25°C?

Similar problems: 18.11, 18.12.

Summary of Concepts & Facts

- The electromotive force (emf) of a cell is the voltage difference between the two electrodes. In the external circuit, electrons flow from the anode to the cathode in a galvanic cell. In solution, the anions move toward the anode and the cations move toward the cathode.
- Standard reduction potentials show the relative likelihood of half-cell reduction reactions and can be used to predict the products, direction, and spontaneity of redox reactions between various substances.

Review of Concepts & Facts

18.3.1 Which of the following metals will react with (that is, be oxidized by) HNO_3, but not with HCl: Cu, Zn, Ag?

18.3.2 Calculate the standard emf for the following cell:

$$Mn(s) \mid Mn^{2+}(1\ M) \parallel Ag^+(1\ M) \mid Ag(s)$$

18.4 Thermodynamics of Redox Reactions

Learning Objective

- Calculate $E°_{cell}$ of a reaction given the values of ΔG and K and interconvert among these three quantities.

Our next step is to see how $E°_{cell}$ is related to thermodynamic quantities such as $\Delta G°$ and K. In a galvanic cell, chemical energy is converted to electrical energy to do electrical work such as running an electric motor. Electrical energy, in this case, is the product of the emf of the cell and the total electrical charge (in coulombs) that passes through the cell:

$$\begin{aligned}\text{electrical energy} &= \text{coulombs} \times \text{volts}\\ &= \text{joules}\end{aligned}$$

The equality means that

$$1\text{ J} = 1\text{ C} \times 1\text{ V}$$

The total charge is determined by the number of electrons that pass through the cell, so we have

$$\text{total charge} = \text{number of } e^- \times \text{charge of one } e^-$$

In general, it is more convenient to express the total charge in molar quantities. The charge of one mole of electrons is called the ***Faraday constant (F),*** after the English chemist and physicist Michael Faraday,† where

$$\begin{aligned}1\ F &= 6.022 \times 10^{23}\ e^-/\text{mol } e^- \times 1.602 \times 10^{-19}\ \text{C}/e^-\\ &= 9.647 \times 10^4\ \text{C/mol } e^-\end{aligned}$$

Therefore, the total charge can now be expressed as nF, where n is the number of moles of electrons exchanged between the oxidizing agent and reducing agent in the overall redox equation for the electrochemical process. In most calculations, we round the Faraday constant to 96,500 C/mol e^-.

The measured emf (E_{cell}) is the *maximum* voltage the cell can achieve. Therefore, the electrical work done, w_{ele}, which is the maximum work that can be done (w_{max}), is given by the product of the total charge and the emf of the cell:

$$w_{max} = w_{ele} = -nFE_{cell}$$

The negative sign indicates that the electrical work is done by the system (galvanic cell) on the surroundings. The sign convention for electrical work is the same as that for P-V work, discussed in Section 6.3. In Chapter 17 we defined Gibbs energy as the energy available to do work. Specifically, the change in Gibbs energy (ΔG) represents the maximum amount of useful work that can be obtained in a reaction:

$$\Delta G = w_{max} = w_{ele}$$

Therefore, we can write

$$\Delta G = -nFE_{cell} \tag{18.2}$$

For a spontaneous reaction, ΔG is negative. Because both n and F are positive quantities, it follows that E_{cell} must also be positive. For reactions in which reactants and products are in their standard states (1 M or 1 atm), Equation (18.2) becomes

$$\Delta G^\circ = -nFE^\circ_{cell} \tag{18.3}$$

Student Hot Spot

Student data indicate you may struggle with the relationship between ΔG° and E°_{cell}. Access your eBook for additional Learning Resources on this topic.

Now we can relate E°_{cell} to the equilibrium constant (K) of a redox reaction. In Section 17.5 we saw that the standard Gibbs-energy change ΔG° for a reaction is related to its equilibrium constant as follows [see Equation (17.14)]:

$$\Delta G^\circ = -RT \ln K$$

†Michael Faraday (1791–1867). English chemist and physicist. Faraday is regarded by many as the greatest experimental scientist of the nineteenth century. He started as an apprentice to a bookbinder at the age of 13, but became interested in science after reading a book on chemistry. Faraday invented the electric motor and was the first person to demonstrate the principle governing electrical generators. Besides making notable contributions to the fields of electricity and magnetism, Faraday also worked on optical activity, and discovered and named benzene.

Table 18.2 Relationships Among $\Delta G°$, K, and $E°_{cell}$

$\Delta G°$	K	$E°_{cell}$	Reaction Under Standard-State Conditions
Negative	>1	Positive	Favors formation of products.
0	=1	0	Reactants and products are equally favored.
Positive	<1	Negative	Favors formation of reactants.

If we combine Equations (17.14) and (18.3), we obtain

$$-nFE°_{cell} = RT \ln K$$

Solving for $E°_{cell}$

$$E°_{cell} = \frac{RT}{nF} \ln K \quad (18.4)$$

When $T = 298$ K, Equation (18.4) can be simplified by substituting for R and F:

$$E°_{cell} = \frac{(8.314 \text{ J/K} \cdot \text{mol})(298 \text{ K})}{n(96{,}500 \text{ J/V} \cdot \text{mol})} \ln K$$

or

$$E°_{cell} = \frac{0.0257 \text{ V}}{n} \ln K \quad (18.5)$$

Alternatively, Equation (18.5) can be written using the base-10 logarithm of K:

$$E°_{cell} = \frac{0.0592 \text{ V}}{n} \log K \quad (18.6)$$

Thus, if any one of the three quantities $\Delta G°$, K, or $E°_{cell}$ is known, the other two can be calculated using Equation (17.14), Equation (18.3), or Equation (18.4) (Figure 18.5). We summarize the relationships among $\Delta G°$, K, and $E°_{cell}$ and characterize the spontaneity of a redox reaction in Table 18.2. For simplicity, we sometimes omit the subscript "cell" in $E°$ and E.

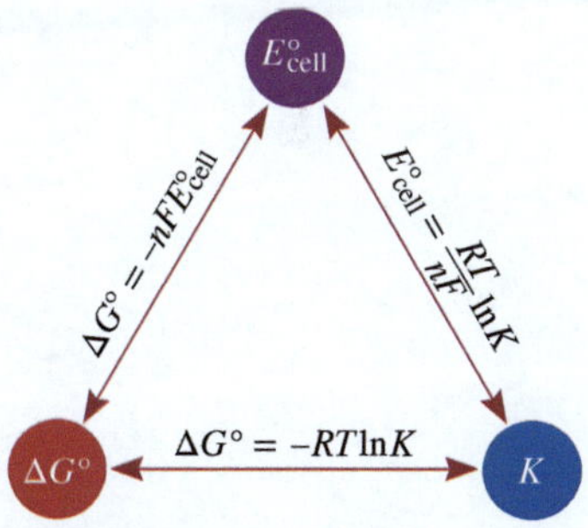

Figure 18.5 *Relationships among $E°$, K, and $\Delta G°$.*

Examples 18.4 and 18.5 apply Equations (18.3) and (18.5).

Example 18.4

Calculate the equilibrium constant for the following reaction at 25°C:

$$Sn(s) + 2Cu^{2+}(aq) \rightleftharpoons Sn^{2+}(aq) + 2Cu^{+}(aq)$$

Strategy The relationship between the equilibrium constant K and the standard emf is given by Equation (18.5): $E°_{cell} = (0.0257 \text{ V}/n)\ln K$. Thus, if we can determine the standard emf, we can calculate the equilibrium constant. We can determine the $E°_{cell}$ of a hypothetical galvanic cell made up of two couples (Sn^{2+}/Sn and Cu^{2+}/Cu^{+}) from the standard reduction potentials in Table 18.1.

Solution The half-cell reactions are

Anode (oxidation): $Sn(s) \longrightarrow Sn^{2+}(aq) + 2e^-$

Cathode (reduction): $2Cu^{2+}(aq) + 2e^- \longrightarrow 2Cu^{+}(aq)$

$$\begin{aligned} E°_{cell} &= E°_{cathode} - E°_{anode} \\ &= E°_{Cu^{2+}/Cu^{+}} - E°_{Sn^{2+}/Sn} \\ &= 0.15 \text{ V} - (-0.14 \text{ V}) \\ &= 0.29 \text{ V} \end{aligned}$$

(Continued)

Equation (18.5) can be written

$$\ln K = \frac{nE^\circ}{0.0257\ \text{V}}$$

In the overall reaction we find $n = 2$. Therefore,

$$\ln K = \frac{(2)(0.29\ \text{V})}{0.0257\ \text{V}} = 22.6$$

$$K = e^{22.6} = 7 \times 10^9$$

Practice Exercise Calculate the equilibrium constant for the following reaction at 25°C:

$$Fe^{2+}(aq) + 2Ag(s) \rightleftharpoons Fe(s) + 2Ag^+(aq)$$

Similar problems: 18.23, 18.24.

Example 18.5

Calculate the standard Gibbs-energy change for the following reaction at 25°C:

$$2Au(s) + 3Ca^{2+}(1.0\ M) \longrightarrow 2Au^{3+}(1.0\ M) + 3Ca(s)$$

Strategy The relationship between the standard Gibbs-energy change and the standard emf of the cell is given by Equation (18.3): $\Delta G^\circ = -nFE^\circ_{\text{cell}}$. Thus, if we can determine E°_{cell}, we can calculate ΔG°. We can determine the E°_{cell} of a hypothetical galvanic cell made up of two couples (Au^{3+}/Au and Ca^{2+}/Ca) from the standard reduction potentials in Table 18.1.

Solution The half-cell reactions are

$$\text{Anode (oxidation):} \quad 2Au(s) \longrightarrow 2Au^{3+}(1.0\ M) + 6e^-$$
$$\text{Cathode (reduction):} \quad 3Ca^{2+}(1.0\ M) + 6e^- \longrightarrow 3Ca(s)$$

$$\begin{aligned} E^\circ_{\text{cell}} &= E^\circ_{\text{cathode}} - E^\circ_{\text{anode}} \\ &= E^\circ_{Ca^{2+}/Ca} - E^\circ_{Au^{3+}/Au} \\ &= -2.87\ \text{V} - 1.50\ \text{V} \\ &= -4.37\ \text{V} \end{aligned}$$

Now we use Equation (18.3):

$$\Delta G^\circ = -nFE^\circ$$

The overall reaction shows that $n = 6$, so

$$\begin{aligned} \Delta G^\circ &= -(6)(96{,}500\ \text{J/V} \cdot \text{mol})(-4.37\ \text{V}) \\ &= 2.53 \times 10^6\ \text{J/mol} \\ &= 2.53 \times 10^3\ \text{kJ/mol} \end{aligned}$$

Check The large positive value of ΔG° tells us that the reaction favors the reactants at equilibrium. The result is consistent with the fact that E° for the galvanic cell is negative.

Practice Exercise Calculate ΔG° for the following reaction at 25°C:

$$2Al^{3+}(aq) + 3Mg(s) \rightleftharpoons 2Al(s) + 3Mg^{2+}(aq)$$

Similar problem: 18.26.

Summary of Concepts & Facts

- The decrease in Gibbs energy of the system in a spontaneous redox reaction is equal to the electrical work done by the system on the surroundings, or $\Delta G = -nFE$.
- The equilibrium constant for a redox reaction can be found from the standard electromotive force of a cell.

Review of Concepts & Facts

18.4.1 Calculate the equilibrium constant for the following reaction at 25°C:

$$Pb^{2+}(aq) + Fe(s) \longrightarrow Pb(s) + Fe^{2+}(aq)$$

18.4.2 Calculate $\Delta G°$ for the following reaction at 25°C:

$$Cl_2(g) + 2F^-(aq) \longrightarrow 2Cl^-(aq) + F_2(g)$$

18.4.3 Compare the ease of measuring the equilibrium constant of a reaction electrochemically with that by chemical means in general [see Equation (17.14)].

18.5 The Effect of Concentration on Cell Emf

Learning Objectives

- Use the Nernst equation to calculate the E_{cell} of a cell under nonstandard conditions.
- Define concentration cell.

So far we have focused on redox reactions in which reactants and products are in their standard states, but standard-state conditions are often difficult, and sometimes impossible, to maintain. However, there is a mathematical relationship between the emf of a galvanic cell and the concentration of reactants and products in a redox reaction under nonstandard-state conditions. This equation is derived next.

The Nernst Equation

Consider a redox reaction of the type

$$a\text{A} + b\text{B} \longrightarrow c\text{C} + d\text{D}$$

From Equation (17.13),

$$\Delta G = \Delta G° + RT \ln Q$$

Because $\Delta G = -nFE$ and $\Delta G° = -nFE°$, the equation can be expressed as

$$-nFE = -nFE° + RT \ln Q$$

Dividing the equation through by $-nF$, we get

$$E = E° - \frac{RT}{nF} \ln Q \qquad (18.7)$$

where Q is the reaction quotient (see Section 14.4). Equation (18.7) is known as the ***Nernst*† *equation.*** At 298 K, Equation (18.7) can be rewritten as

$$E = E^\circ - \frac{0.0257\ \text{V}}{n}\ln Q \qquad (18.8)$$

or, expressing Equation (18.8) using the base-10 logarithm of Q:

$$E = E^\circ - \frac{0.0592\ \text{V}}{n}\log Q \qquad (18.9)$$

During the operation of a galvanic cell, electrons flow from the anode to the cathode, resulting in product formation and a decrease in reactant concentration. Thus, Q increases, which means that E decreases. Eventually, the cell reaches equilibrium. At equilibrium, there is no net transfer of electrons, so $E = 0$ and $Q = K$, where K is the equilibrium constant.

The Nernst equation enables us to calculate E as a function of reactant and product concentrations in a redox reaction. For example, for the Daniell cell in Figure 18.1,

$$\text{Zn}(s) + \text{Cu}^{2+}(aq) \longrightarrow \text{Zn}^{2+}(aq) + \text{Cu}(s)$$

The Nernst equation for this cell at 25°C can be written as

$$E = 1.10\ \text{V} - \frac{0.0257\ \text{V}}{2}\ln\frac{[\text{Zn}^{2+}]}{[\text{Cu}^{2+}]}$$

Student Hot Spot

Student data indicate you may struggle with cell potentials at nonstandard conditions. Access your eBook for additional Learning Resources on this topic.

If the ratio $[\text{Zn}^{2+}]/[\text{Cu}^{2+}]$ is less than 1, $\ln([\text{Zn}^{2+}]/[\text{Cu}^{2+}])$ is a negative number, so that the second term on the right-hand side of the preceding equation is positive. Under this condition E is greater than the standard emf E°. If the ratio is greater than 1, E is smaller than E°.

Example 18.6 illustrates the use of the Nernst equation.

Example 18.6

Predict whether the following reaction would proceed spontaneously as written at 298 K:

$$\text{Co}(s) + \text{Fe}^{2+}(aq) \longrightarrow \text{Co}^{2+}(aq) + \text{Fe}(s)$$

given that $[\text{Co}^{2+}] = 0.15\ M$ and $[\text{Fe}^{2+}] = 0.68\ M$.

Strategy Because the reaction is not run under standard-state conditions (concentrations are not 1 M), we need Nernst's equation [Equation (18.8)] to calculate the emf (E) of a hypothetical galvanic cell and determine the spontaneity of the reaction. The standard emf (E°) can be calculated using the standard reduction potentials in Table 18.1. Remember that solids do not appear in the reaction quotient (Q) term in the Nernst equation. Note that 2 moles of electrons are transferred per mole of reaction, that is, $n = 2$.

Solution The half-cell reactions are

Anode (oxidation): $\text{Co}(s) \longrightarrow \text{Co}^{2+}(aq) + 2e^-$

Cathode (reduction): $\text{Fe}^{2+}(aq) + 2e^- \longrightarrow \text{Fe}(s)$

$$\begin{aligned} E^\circ_{\text{cell}} &= E^\circ_{\text{cathode}} - E^\circ_{\text{anode}} \\ &= E^\circ_{\text{Fe}^{2+}/\text{Fe}} - E^\circ_{\text{Co}^{2+}/\text{Co}} \\ &= -0.45\ \text{V} - (-0.28\ \text{V}) \\ &= -0.017\ \text{V} \end{aligned}$$

(Continued)

†Walther Hermann Nernst (1864–1941). German chemist and physicist. Nernst's work was mainly on electrolyte solution and thermodynamics. He also invented an electric piano. Nernst was awarded the Nobel Prize in Chemistry in 1920 for his contribution to thermodynamics.

From Equation (18.8) we write

$$E = E^\circ - \frac{0.0257\ \text{V}}{n}\ln Q$$
$$= E^\circ - \frac{0.0257\ \text{V}}{n}\ln\frac{[\text{Co}^{2+}]}{\text{Fe}^{2+}}$$
$$= -0.17\ \text{V} - \frac{0.0257\ \text{V}}{2}\ln\frac{0.15}{0.68}$$
$$= -0.17\ \text{V} + 0.019\ \text{V}$$
$$= -0.15\ \text{V}$$

Because E is negative, the reaction is not spontaneous in the direction written.

Practice Exercise Will the following reaction occur spontaneously at 25°C, given that $[Fe^{2+}] = 0.60\ M$ and $[Cd^{2+}] = 0.010\ M$?

$$\text{Cd}(s) + \text{Fe}^{2+}(aq) \longrightarrow \text{Cd}^{2+}(aq) + \text{Fe}(s)$$

Similar problems: 18.31, 18.32.

Now suppose we want to determine at what ratio of $[Co^{2+}]$ to $[Fe^{2+}]$ the reaction in Example 18.6 would become spontaneous. We can use Equation (18.8) as follows:

$$E = E^\circ - \frac{0.0257\ \text{V}}{n}\ln Q$$

We first set E equal to zero, which corresponds to the equilibrium situation ($Q = K$).

$$0 = -0.17\ \text{V} - \frac{0.0257\ \text{V}}{2}\ln\frac{[\text{Co}^{2+}]}{[\text{Fe}^{2+}]}$$
$$\ln\frac{[\text{Co}^{2+}]}{\text{Fe}^{2+}} = -13.2$$
$$\frac{[\text{Co}^{2+}]}{[\text{Fe}^{2+}]} = e^{-13.2} = K$$

or

$$K = 2 \times 10^{-6}$$

Thus, for the reaction to be spontaneous, the ratio $[Co^{2+}]/[Fe^{2+}]$ must be smaller than 2×10^{-6} so that E would become positive.

As Example 18.7 shows, if gases are involved in the cell reaction, their concentrations should be expressed in atm.

Example 18.7

Consider the galvanic cell shown in Figure 18.4(a). In a certain experiment, the emf (E) of the cell is found to be 0.54 V at 25°C. Suppose that $[Zn^{2+}] = 1.0\ M$ and $P_{H_2} = 1.0$ atm. Calculate the molar concentration of H^+.

Strategy The equation that relates standard emf and nonstandard emf is the Nernst equation. The overall cell reaction is

$$\text{Zn}(s) + 2\text{H}^+(?\ M) \longrightarrow \text{Zn}^{2+}(1.0\ M) + \text{H}_2(1.0\ \text{atm})$$

Given the emf of the cell (E), we apply the Nernst equation to solve for $[H^+]$. Note that 2 moles of electrons are transferred per mole of reaction; that is, $n = 2$.

(Continued)

Solution As we saw in Section 18.3, the standard emf ($E°$) for the cell is 0.76 V. The concentrations in Q are divided by their standard-state value of 1 M and pressure is divided by 1 atm. From Equation (18.8) we write

$$E = E° - \frac{0.0257\ \text{V}}{n} \ln Q$$

$$= E° - \frac{0.0257\ \text{V}}{n} \ln \frac{[\text{Zn}^{2+}]P_{\text{H}_2}}{[\text{H}^+]^2}$$

$$0.54\ \text{V} = 0.76\ \text{V} - \frac{0.0257\ \text{V}}{2} \ln \frac{(1.0)(1.0)}{[\text{H}^+]^2}$$

$$-0.22\ \text{V} = -\frac{0.0257\ \text{V}}{2} \ln \frac{1}{[\text{H}^+]^2}$$

$$17.1 = \ln \frac{1}{[\text{H}^+]^2}$$

$$e^{17.1} = \frac{1}{[\text{H}^+]^2}$$

$$[\text{H}^+] = \sqrt{\frac{1}{3 \times 10^7}} = 2 \times 10^{-4}\ M$$

Check The fact that the nonstandard-state emf (E) is given in the problem means that not all the reacting species are in their standard-state concentrations. Thus, because both Zn^{2+} ions and H_2 gas are in their standard states, $[H^+]$ is not 1 M.

Practice Exercise What is the emf of a galvanic cell consisting of a Cd^{2+}/Cd half-cell and a $Pt/H^+/H_2$ half-cell if $[Cd^{2+}] = 0.20\ M$, $[H^+] = 0.16\ M$, and $P_{H_2} = 0.80$ atm?

Similar problem: 18.34.

Example 18.7 shows that a galvanic cell whose cell reaction involves H^+ ions can be used to measure $[H^+]$ or pH. The pH meter described in Section 15.3 is based on this principle. However, the hydrogen electrode (see Figure 18.3) is normally not employed in laboratory work because it is awkward to use. Instead, it is replaced by a *glass electrode,* shown in Figure 18.6. The electrode consists of a very thin glass membrane that is permeable to H^+ ions. A silver wire coated with silver chloride is immersed in a dilute hydrochloric acid solution. When the electrode is placed in a solution whose pH is different from that of the inner solution, the potential difference that develops between the two sides of the membrane can be monitored using a reference electrode. The emf of the cell made up of the glass electrode and the reference electrode is measured with a voltmeter that is calibrated in pH units.

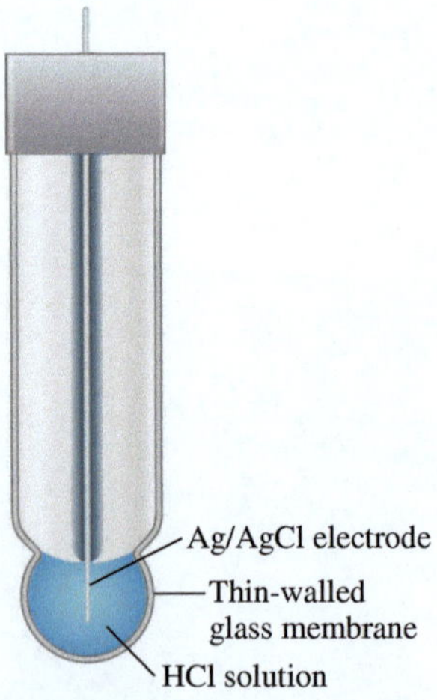

Figure 18.6 *A glass electrode that is used in conjunction with a reference electrode in a pH meter.*

Concentration Cells

Because electrode potential depends on ion concentrations, it is possible to construct a galvanic cell from two half-cells composed of the *same* material but differing in ion concentrations. Such a cell is called a *concentration cell.*

Consider a situation in which zinc electrodes are put into two aqueous solutions of zinc sulfate at 0.10 M and 1.0 M concentrations. The two solutions are connected by a salt bridge, and the electrodes are joined by a piece of wire in an arrangement like that shown in Figure 18.1. According to Le Châtelier's principle, the tendency for the reduction

$$\text{Zn}^{2+}(aq) + 2e^- \longrightarrow \text{Zn}(s)$$

increases with increasing concentration of Zn^{2+} ions. Therefore, reduction should occur in the more concentrated compartment and oxidation should take place on the more dilute side. The cell diagram is

$$\text{Zn}(s)\ |\ \text{Zn}^{2+}(0.10\ M)\ ||\ \text{Zn}^{2+}(1.0\ M)\ |\ \text{Zn}(s)$$

and the half-reactions are

$$\begin{array}{rl} \text{Oxidation:} & Zn(s) \longrightarrow Zn^{2+}(0.10\ M) + 2e^- \\ \text{Reduction:} & Zn^{2+}(1.0\ M) + 2e^- \longrightarrow Zn(s) \\ \hline \text{Overall:} & Zn^{2+}(1.0\ M) \longrightarrow Zn^{2+}(0.10\ M) \end{array}$$

The emf of the cell is

$$E = E° - \frac{0.0257\ \text{V}}{2} \ln \frac{[Zn^{2+}]_{dil}}{[Zn^{2+}]_{conc}}$$

where the subscripts "dil" and "conc" refer to the 0.10 *M* and 1.0 *M* concentrations, respectively. The $E°$ for this cell is zero (the same electrode and the same type of ions are involved), so

$$\begin{aligned} E &= 0 - \frac{0.0257\ \text{V}}{2} \ln \frac{0.10}{1.0} \\ &= 0.0296\ \text{V} \end{aligned}$$

The emf of concentration cells is usually small and decreases continually during the operation of the cell as the concentrations in the two compartments approach each other. When the concentrations of the ions in the two compartments are the same, E becomes zero, and no further change occurs.

A biological cell can be compared to a concentration cell for the purpose of calculating its *membrane potential.* Membrane potential is the electrical potential that exists across the membrane of various kinds of cells, including muscle cells and nerve cells. It is responsible for the propagation of nerve impulses and heartbeat. A membrane potential is established whenever there are unequal concentrations of the same type of ion in the interior and exterior of a cell. For example, the concentrations of K^+ ions in the interior and exterior of a nerve cell are 400 m*M* and 15 m*M*, respectively. Treating the situation as a concentration cell and applying the Nernst equation for just one kind of ion, we can write

$$\begin{aligned} E &= E° - \frac{0.0257\ \text{V}}{1} \ln \frac{[K^+]_{ex}}{[K^+]_{in}} \\ &= -(0.0257\ \text{V}) \ln \frac{15}{400} \\ &= 0.084\ \text{V or } 84\ \text{mV} \end{aligned}$$

where "ex" and "in" denote exterior and interior. Note that we have set $E° = 0$ because the same type of ion is involved. Thus, an electrical potential of 84 mV exists across the membrane due to the unequal concentrations of K^+ ions.

Summary of Concepts & Facts

- The Nernst equation gives the relationship between the cell emf and the concentrations of the reactants and products under nonstandard-state conditions.

Review of Concepts & Facts

18.5.1 Consider the following cell diagram:

$$Mg(s)\ |\ MgSO_4(0.40\ M)\ ||\ NiSO_4(0.60\ M)\ |\ Ni(s)$$

Calculate the cell voltage at 25°C. How does the cell voltage change when (a) $[Mg^{2+}]$ is decreased by a factor of 4 and (b) $[Ni^{2+}]$ is decreased by a factor of 3?

18.5.2 Calculate the cell voltage at 25°C of a concentration cell consisting of an Ag electrode in 0.15 *M* $AgNO_3$ and an Ag electrode in 1.0 *M* $AgNO_3$.

18.6 Batteries and Fuel Cells

Learning Objective

- Distinguish among the different types of common batteries, including dry cell, alkaline, lead storage, lithium, and fuel cells.

A ***battery*** is *a galvanic cell, or a series of combined galvanic cells, that can be used as a source of direct electric current at a constant voltage.* Although the operation of a battery is similar in principle to that of the galvanic cells described in Section 18.2, a battery has the advantage of being completely self-contained and requiring no auxiliary components such as salt bridges. Here we will discuss several types of batteries that are in widespread use.

The Dry Cell Battery

The most common dry cell, that is, a cell without a fluid component, is the *Leclanché cell* used in flashlights and transistor radios. The anode of the cell consists of a zinc can or container that is in contact with manganese dioxide (MnO_2) and an electrolyte. The electrolyte consists of ammonium chloride and zinc chloride in water, to which starch is added to thicken the solution to a pastelike consistency so that it is less likely to leak (Figure 18.7). A carbon rod serves as the cathode, which is immersed in the electrolyte in the center of the cell. The cell reactions are

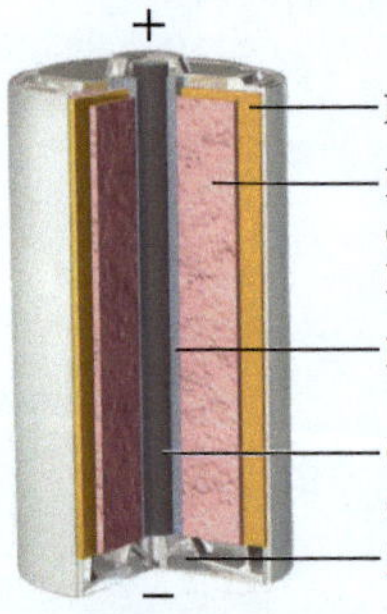

Figure 18.7 *Interior section of a dry cell of the kind used in flashlights and transistor radios. Actually, the cell is not completely dry, as it contains a moist electrolyte paste.*

$$\begin{aligned} \text{Anode:} &\quad Zn(s) \longrightarrow Zn^{2+}(aq) + 2e^- \\ \text{Cathode:} &\quad 2NH_4^+(aq) + 2MnO_2(s) + 2e^- \longrightarrow Mn_2O_3(s) + 2NH_3(aq) + H_2O(l) \\ \hline \text{Overall:} &\quad Zn(s) + 2NH_4^+(aq) + 2MnO_2(s) \longrightarrow Zn^{2+}(aq) + 2NH_3(aq) \\ &\qquad + H_2O(l) + Mn_2O_3(s) \end{aligned}$$

Actually, this equation is an oversimplification of a complex process. The voltage produced by a dry cell is about 1.5 V.

The Mercury Battery

The mercury battery is used extensively in medicine and electronic industries and is more expensive than the common dry cell. Contained in a stainless steel cylinder, the mercury battery consists of a zinc anode (amalgamated with mercury) in contact with a strongly alkaline electrolyte containing zinc oxide and mercury(II) oxide (Figure 18.8). The cell reactions are

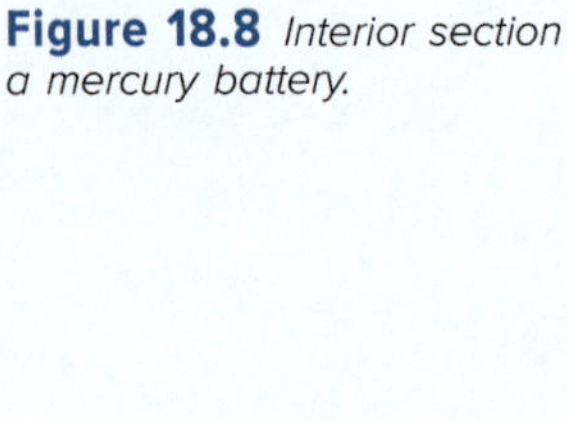

Figure 18.8 *Interior section of a mercury battery.*

$$\begin{aligned} \text{Anode:} &\quad Zn(Hg) + 2OH^-(aq) \longrightarrow ZnO(s) + H_2O(l) + 2e^- \\ \text{Cathode:} &\quad HgO(s) + H_2O(l) + 2e^- \longrightarrow Hg(l) + 2OH^-(aq) \\ \hline \text{Overall:} &\quad Zn(Hg) + HgO(s) \longrightarrow ZnO(s) + Hg(l) \end{aligned}$$

Because there is no change in electrolyte composition during operation—the overall cell reaction involves only solid substances—the mercury battery provides a more constant voltage (1.35 V) than the Leclanché cell. It also has a considerably higher capacity and longer life. These qualities make the mercury battery ideal for use in pacemakers, hearing aids, electric watches, and light meters.

The Lead Storage Battery

The lead storage battery commonly used in automobiles consists of six identical cells joined together in series. Each cell has a lead anode and a cathode made of lead

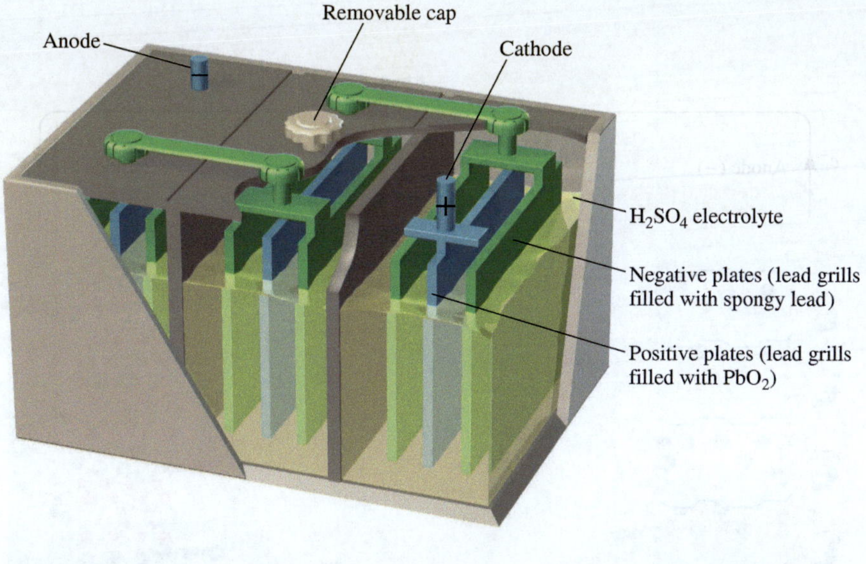

Figure 18.9 *Interior section of a lead storage battery. Under normal operating conditions, the concentration of the sulfuric acid solution is about 38 percent by mass.*

dioxide (PbO_2) packed on a metal plate (Figure 18.9). Both the cathode and the anode are immersed in an aqueous solution of sulfuric acid, which acts as the electrolyte. The cell reactions are

$$\begin{array}{ll} \text{Anode:} & Pb(s) + SO_4^{2-}(aq) \longrightarrow PbSO_4(s) + 2e^- \\ \text{Cathode:} & PbO_2(s) + 4H^+(aq) + SO_4^{2-}(aq) + 2e^- \longrightarrow PbSO_4(s) + 2H_2O(l) \\ \hline \text{Overall:} & Pb(s) + PbO_2(s) + 4H^+(aq) + 2SO_4^{2-}(aq) \longrightarrow 2PbSO_4(s) + 2H_2O(l) \end{array}$$

Under normal operating conditions, each cell produces 2 V; a total of 12 V from the six cells is used to power the ignition circuit of the automobile and its other electrical systems. The lead storage battery can deliver large amounts of current for a short time, such as the time it takes to start up the engine.

Unlike the Leclanché cell and the mercury battery, the lead storage battery is rechargeable. Recharging the battery means reversing the normal electrochemical reaction by applying an external voltage at the cathode and the anode. (This kind of process is called *electrolysis,* see Section 18.8.) The reactions that replenish the original materials are

$$\begin{array}{ll} & PbSO_4(s) + 2e^- \longrightarrow Pb(s) + SO_4^{2-}(aq) \\ & PbSO_4(s) + 2H_2O(l) \longrightarrow PbO_2(s) + 4H^+(aq) + SO_4^{2-}(aq) + 2e^- \\ \hline \text{Overall:} & 2PbSO_4(s) + 2H_2O(l) \longrightarrow Pb(s) + PbO_2(s) + 4H^+(aq) + 2SO_4^{2-}(aq) \end{array}$$

The overall reaction is exactly the opposite of the normal cell reaction.

Two aspects of the operation of a lead storage battery are worth noting. First, because the electrochemical reaction consumes sulfuric acid, the degree to which the battery has been discharged can be checked by measuring the density of the electrolyte with a hydrometer, as is usually done at gas stations. The density of the fluid in a "healthy," fully charged battery should be equal to or greater than 1.2 g/mL. Second, people living in cold climates sometimes have trouble starting their cars because the battery has "gone dead." Thermodynamic calculations show that the emf of many galvanic cells decreases with decreasing temperature. However, for a lead storage battery, the temperature coefficient is about 1.5×10^{-4} V/°C; that is, there is a decrease in voltage of 1.5×10^{-4} V for every degree drop in temperature. Thus, even allowing for a 40°C change in temperature, the decrease in voltage amounts to only 6×10^{-3} V, which is about

$$\frac{6 \times 10^{-3}\text{ V}}{12\text{ V}} \times 100\% = 0.05\%$$

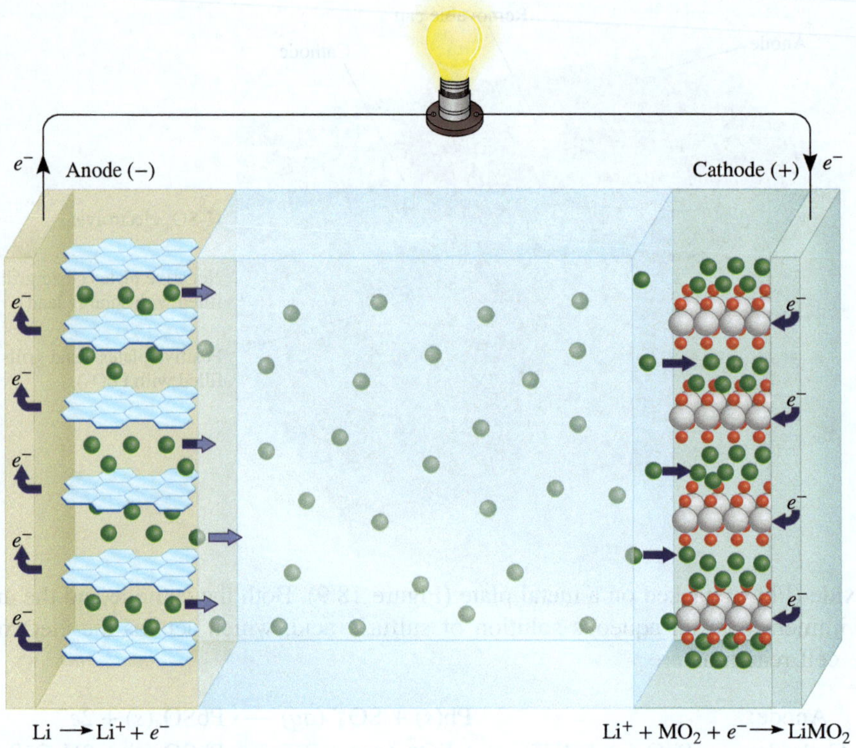

Figure 18.10 *A lithium-ion battery. Lithium atoms (represented by green spheres) are embedded between sheets of graphene (a single layer of graphite), which serve as the anode. A metal oxide material (gray and red spheres) is the cathode. During operation, Li^+ ions migrate through an electrolyte solution from the anode to the cathode while electrons flow externally through the wire from the anode to the cathode to complete the circuit.*

of the operating voltage, an insignificant change. The real cause of a battery's apparent breakdown is an increase in the viscosity of the electrolyte as the temperature decreases. For the battery to function properly, the electrolyte must be fully conducting. However, the ions move much more slowly in a viscous medium, so the resistance of the fluid increases, leading to a decrease in the power output of the battery. If an apparently "dead battery" is warmed to near room temperature on a frigid day, it recovers its ability to deliver normal power.

The Lithium-Ion Battery

Figure 18.10 shows a schematic diagram of a lithium-ion battery. The anode is made of a conducting carbonaceous material, usually graphite, which has tiny spaces in its structure that can hold both Li atoms and Li^+ ions. The cathode is made of a transition metal oxide such as CoO_2, which can also hold Li^+ ions. Because of the high reactivity of the metal, nonaqueous electrolyte (organic solvent plus dissolved salt) must be used. During the discharge of the battery, the half-cell reactions are

$$\text{Anode (oxidation):} \quad Li(s) \longrightarrow Li^+ + e^-$$
$$\text{Cathode (reduction):} \quad Li^+ + CoO_2 + e^- \longrightarrow LiCoO_2(s)$$
$$\text{Overall:} \quad Li(s) + CoO_2 \longrightarrow LiCoO_2(s) \qquad E_{cell} = 3.4\ V$$

The advantage of the battery is that lithium has the most negative standard reduction potential (see Table 18.1) and hence the greatest reducing strength. Furthermore, lithium is the lightest metal so that only 6.941 g of Li (its molar mass) are needed to produce 1 mole of electrons. A lithium-ion battery can be recharged literally hundreds of times without deterioration. These desirable characteristics make it suitable for use in cellular telephones, digital cameras, and laptop computers.

Recent progress in the manufacture of electric and hybrid automobiles, and increasing demand for these vehicles, has in turn created an intense demand for lithium-ion batteries. The batteries used in most highway-rated electric vehicles and some power tools are lithium iron phosphate (LFP) batteries. The design of LFP batteries is functionally the same as that shown in Figure 18.10, except that the cathode is $FePO_4$, and $LiFePO_4$ is formed at the cathode as the battery is discharged. These batteries share many of the advantages of other lithium-ion batteries (low weight, greater tendency for the metal to become oxidized at the anode), but they also have the additional advantage of extremely high chemical and thermal stability. As such, LFP batteries can be recharged many times and withstand very high temperatures without significant decomposition, and they avoid the problems with fires caused by the arrays of conventional lithium-ion batteries used in early prototypes of electric vehicles. Some other advantages of LFP batteries include reduced environmental concerns and a greater ability to retain a charge compared to other batteries. LFP batteries have a somewhat lower energy density than traditional lithium-ion batteries, but that trade-off is considered acceptable in applications that require a more robust battery. Early LFP batteries suffered from poor conductivity, but that problem has been addressed by "doping" the batteries with compounds that improve the conductivity.

The increasing demand for lithium caused by the rapidly growing battery market raises questions about the world supply of this important alkali metal. It is projected that over the next few years the demand for lithium will quickly outpace the supply, which predominantly comes from Chile, Argentina, and China. The discovery in 2010 of a massive lithium deposit in Afghanistan may help to address this growing demand.

Fuel Cells

Fossil fuels are a major source of energy, but conversion of fossil fuel into electrical energy is a highly inefficient process. Consider the combustion of methane:

$$CH_4(g) + 2O_2(g) \longrightarrow CO_2(g) + 2H_2O(l) + \text{energy}$$

To generate electricity, heat produced by the reaction is first used to convert water to steam, which then drives a turbine that drives a generator. An appreciable fraction of the energy released in the form of heat is lost to the surroundings at each step; even the most efficient power plant converts only about 40 percent of the original chemical energy into electricity. Because combustion reactions are redox reactions, it is more desirable to carry them out directly by electrochemical means, thereby greatly increasing the efficiency of power production. This objective can be accomplished by a device known as a ***fuel cell,*** *a galvanic cell that requires a continuous supply of reactants to keep functioning*.

In its simplest form, a hydrogen-oxygen fuel cell consists of an electrolyte solution, such as potassium hydroxide solution, and two inert electrodes. Hydrogen and oxygen gases are bubbled through the anode and cathode compartments (Figure 18.11), where the following reactions take place:

$$\begin{aligned} \text{Anode:} \quad & 2H_2(g) + 4OH^-(aq) \longrightarrow 4H_2O(l) + 4e^- \\ \text{Cathode:} \quad & O_2(g) + 2H_2O(l) + 4e^- \longrightarrow 4OH^-(aq) \\ \hline \text{Overall:} \quad & 2H_2(g) + O_2(g) \longrightarrow 2H_2O(l) \end{aligned}$$

The standard emf of the cell can be calculated as follows, with data from Table 18.1:

$$\begin{aligned} E^\circ_{cell} &= E^\circ_{cathode} - E^\circ_{anode} \\ &= 0.40\text{ V} - (-0.83\text{ V}) \\ &= 1.23\text{ V} \end{aligned}$$

Thus, the cell reaction is spontaneous under standard-state conditions. Note that the reaction is the same as the hydrogen combustion reaction, but the oxidation and reduction are carried out separately at the anode and the cathode. Like platinum in the standard hydrogen electrode, the electrodes have a twofold function. They serve as electrical conductors, and they provide the necessary surfaces for the initial decomposition of the molecules into atomic species prior to electron transfer. They are *electrocatalysts*. Metals such as platinum, nickel, and rhodium are good electrocatalysts.

A car powered by hydrogen fuel cells manufactured by General Motors.
Michael Klinec/Alamy Stock Photo

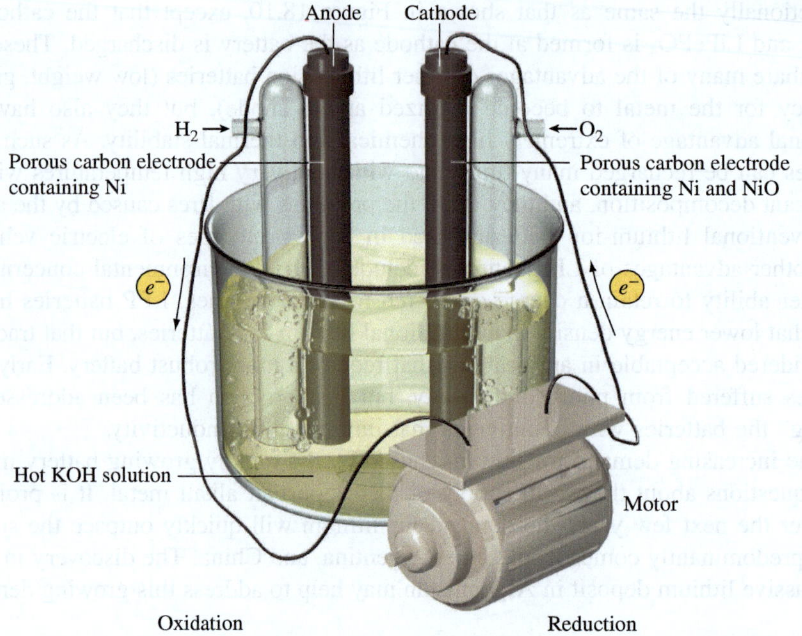

Figure 18.11 *A hydrogen-oxygen fuel cell. The Ni and NiO embedded in the porous carbon electrodes are electrocatalysts.*

In addition to the H_2-O_2 system, a number of other fuel cells have been developed. Among these is the propane-oxygen fuel cell. The half-cell reactions are

$$\begin{aligned}
&\text{Anode:} && C_3H_8(g) + 6H_2O(l) \longrightarrow 3CO_2(g) + 20H^+(aq) + 20e^- \\
&\text{Cathode:} && 5O_2(g) + 20H^+(aq) + 20e^- \longrightarrow 10H_2O(l) \\
&\text{Overall:} && C_3H_8(g) + 5O_2(g) \longrightarrow 3CO_2(g) + 4H_2O(l)
\end{aligned}$$

The overall reaction is identical to the burning of propane in oxygen.

Unlike batteries, fuel cells do not store chemical energy. Reactants must be constantly resupplied, and products must be constantly removed from a fuel cell. In this respect, a fuel cell resembles an engine more than it does a battery. However, the fuel cell does not operate like a heat engine and therefore is not subject to the same kind of thermodynamic limitations in energy conversion (see the Chemistry in Action essay "The Efficiency of Heat Engines" in Section 17.4).

Properly designed fuel cells may be as much as 70 percent efficient, about twice as efficient as an internal combustion engine. In addition, fuel-cell generators are free of the noise, vibration, heat transfer, thermal pollution, and other problems normally associated with conventional power plants. Nevertheless, fuel cells are not yet in widespread use. A major problem lies in the lack of cheap electrocatalysts able to function efficiently for long periods of time without contamination. The most successful application of fuel cells to date has been in space vehicles (Figure 18.12).

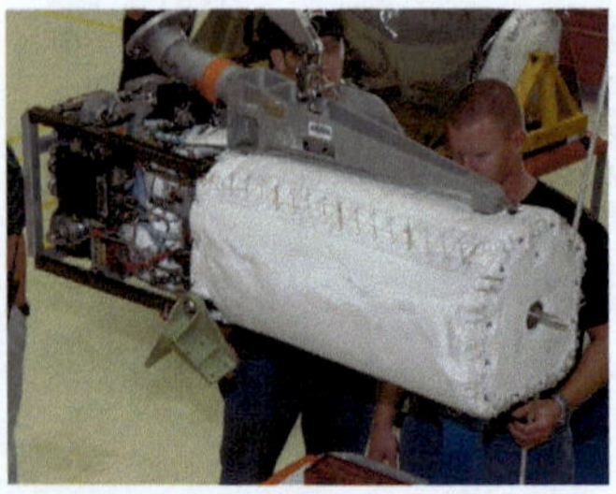

Figure 18.12 *A hydrogen-oxygen fuel cell once used in the space program. The pure water produced by the cell was consumed by the astronauts.*
Kim Shiflett/NASA

Summary of Concepts & Facts

- Batteries, which consist of one or more galvanic cells, are used widely as self-contained power sources. Some of the better-known batteries are the dry cell, such as the Leclanché cell, the mercury battery, and the lead storage battery used in automobiles. Fuel cells produce electrical energy from a continuous supply of reactants.

Review of Concepts & Facts

18.6.1 How many Leclanché cells are contained in a 9-volt battery?

CHEMISTRY *in Action*

Bacteria Power

Usable electricity generated from bacteria? Yes, it is possible. Scientists at the University of Massachusetts at Amherst have discovered an organism known as the *Geobacter* species that do exactly that. The ubiquitous *Geobacter* bacteria normally grow at the bottom of rivers or lakes. They get their energy by oxidizing the decaying organic matter to produce carbon dioxide. The bacteria possess tentacles 10 times the length of their own size to reach the electron acceptors [mostly iron(III) oxide] in the overall anaerobic redox process.

The Massachusetts scientists constructed a bacterial fuel cell using graphite electrodes. The *Geobacter* grow naturally on the surface of the electrode, forming a stable "biofilm." The overall reaction is

$$CH_3COO^- + 2O_2 + H^+ \longrightarrow 2CO_2 + 2H_2O$$

where the acetate ion represents organic matter. The electrons are transferred directly from *Geobacter* to the graphite anode and then flow externally to the graphite cathode. Here, the electron acceptor is oxygen.

So far, the current generated by such a fuel cell is small. With proper development, however, it can someday be used to generate electricity for cooking, lighting, and powering electrical appliances and computers in homes, and in remote sensing devices. This is also a desirable way to clean the environment. Although the end product in the redox process is carbon dioxide, a greenhouse gas, the same product would be formed from the normal decay of the organic wastes.

The oxidizing action of *Geobacter* has another beneficial effect. Tests show that uranium salts can replace iron(III) oxide as the electron acceptor. Thus, by adding acetate ions and the bacteria to the groundwater contaminated with uranium, it is possible to reduce the soluble uranium(VI) salts to the insoluble uranium(IV) salts, which can be readily removed before the water ends up in households and farmlands.

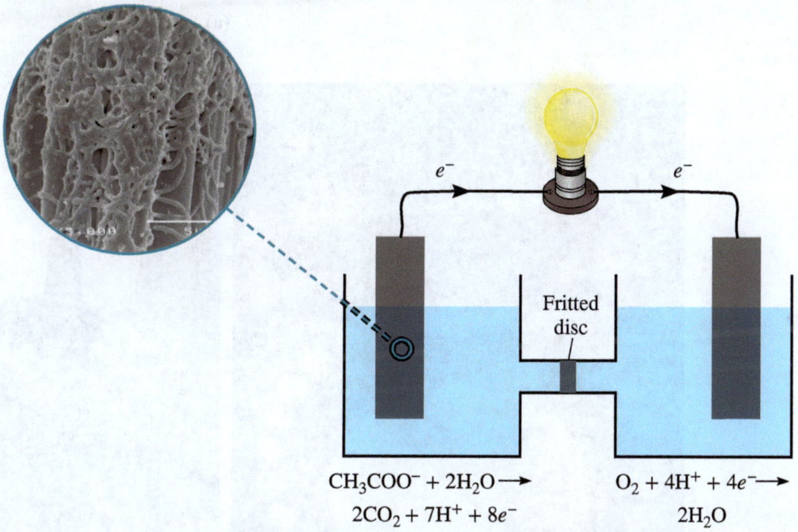

A bacterial fuel cell. The blowup shows the scanning electron micrograph of the bacteria growing on a graphite anode. The fritted disc allows the ions to pass between the compartments.
Derek R. Lovley/Environmental Biotechnology, Univ. of Massachusetts

18.7 Corrosion

Learning Objective

- Discuss the basis of corrosion and galvanization.

Corrosion is the term usually applied to *the deterioration of metals by an electrochemical process.* We see many examples of corrosion around us. Rust on iron, tarnish on silver, and the green patina formed on copper and brass are a few of them (Figure 18.13).

Figure 18.13 *Examples of corrosion: (a) a rusted bridge, (b) a half-tarnished silver dish, and (c) the Statue of Liberty coated with patina before its restoration in 1986.*
(a): Fernando Arias/Moment/Getty Images; (b): Ken Karp/McGraw Hill; (c): Andrew_Howe/Getty Images

Corrosion causes enormous damage to buildings, bridges, ships, and cars. The cost of metallic corrosion to the U.S. economy has been estimated to be well over $275 billion a year! This section discusses some of the fundamental processes that occur in corrosion and methods used to protect metals against it.

By far the most familiar example of corrosion is the formation of rust on iron. Oxygen gas and water must be present for iron to rust. Although the reactions involved are quite complex and not completely understood, the main steps are believed to be as follows. A region of the metal's surface serves as the anode, where oxidation occurs:

$$Fe(s) \longrightarrow Fe^{2+}(aq) + 2e^-$$

The electrons given up by iron reduce atmospheric oxygen to water at the cathode, which is another region of the same metal's surface:

$$O_2(g) + 4H^+(aq) + 4e^- \longrightarrow 2H_2O(l)$$

The overall redox reaction is

$$2Fe(s) + O_2(g) + 4H^+(aq) \longrightarrow 2Fe^{2+}(aq) + 2H_2O(s)$$

With data from Table 18.1, we find the standard emf for this process:

$$\begin{aligned} E^\circ_{cell} &= E^\circ_{cathode} - E^\circ_{anode} \\ &= 1.23\ V - (-0.45\ V) \\ &= 1.68\ V \end{aligned}$$

Note that this reaction occurs in an acidic medium; the H^+ ions are supplied in part by the reaction of atmospheric carbon dioxide with water to form H_2CO_3. The positive standard emf means that the process will favor rust formation.

The Fe^{2+} ions formed at the anode are further oxidized by oxygen:

$$4Fe^{2+}(aq) + O_2(g) + (4 + 2x)H_2O(l) \longrightarrow 2Fe_2O_3 \cdot xH_2O(s) + 8H^+(aq)$$

This hydrated form of iron(III) oxide is known as rust. The amount of water associated with the iron oxide varies, so we represent the formula as $Fe_2O_3 \cdot xH_2O$.

Figure 18.14 shows the mechanism of rust formation. The electrical circuit is completed by the migration of electrons and ions; this is why rusting occurs so rapidly in salt water. In cold climates, salts (NaCl or $CaCl_2$) spread on roadways to melt ice and snow are a major cause of rust formation on automobiles.

Metallic corrosion is not limited to iron. Consider aluminum, a metal used to make many useful things, including airplanes and beverage cans. Aluminum has a much greater tendency to oxidize than iron does; in Table 18.1 we see that Al has a more negative standard reduction potential than Fe. Based on this fact alone, we might expect to see airplanes slowly corrode away in rainstorms, and soda cans transformed into piles of corroded aluminum. These processes do not occur because the layer of insoluble aluminum oxide (Al_2O_3) that forms on its surface when the metal is exposed to air serves to protect the aluminum underneath from further corrosion. The rust that forms on the surface of iron, however, is too porous to protect the underlying metal.

Coinage metals such as copper and silver also corrode, but much more slowly.

$$Cu(s) \longrightarrow Cu^{2+}(aq) + 2e^-$$
$$Ag(s) \longrightarrow Ag^+(aq) + e^-$$

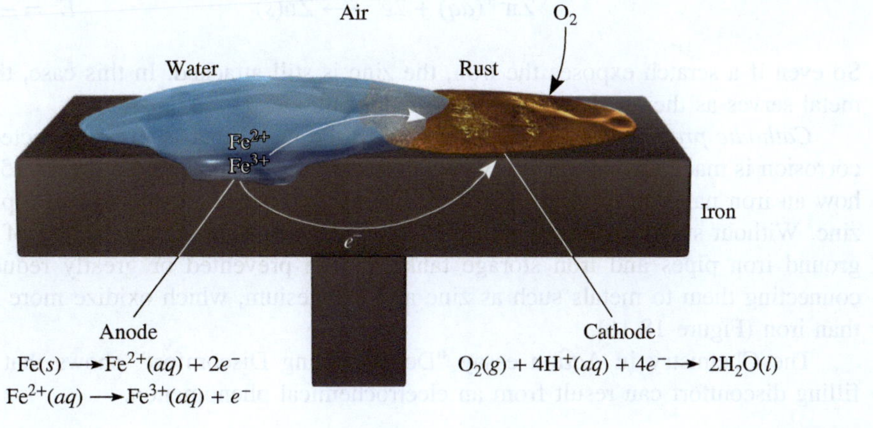

Figure 18.14 *The electrochemical process involved in rust formation. The H^+ ions are supplied by H_2CO_3, which forms when CO_2 dissolves in water.*

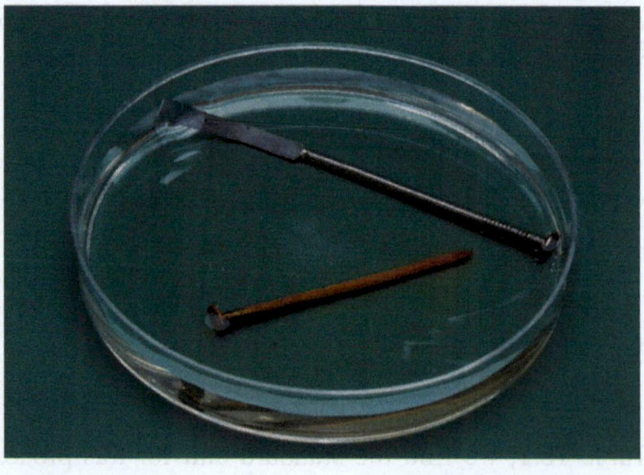

Figure 18.15 *An iron nail that is cathodically protected by a piece of zinc strip does not rust in water, while an iron nail without such protection rusts readily.*
Ken Karp/McGraw Hill

In normal atmospheric exposure, copper forms a layer of copper carbonate ($CuCO_3$), a green substance also called patina, that protects the metal underneath from further corrosion. Likewise, silverware that comes into contact with foodstuffs develops a layer of silver sulfide (Ag_2S).

A number of methods have been devised to protect metals from corrosion. Most of these methods are aimed at preventing rust formation. The most obvious approach is to coat the metal surface with paint. However, if the paint is scratched, pitted, or dented to expose even the smallest area of bare metal, rust will form under the paint layer. The surface of iron metal can be made inactive by a process called *passivation*. A thin oxide layer is formed when the metal is treated with a strong oxidizing agent such as concentrated nitric acid. A solution of sodium chromate is often added to cooling systems and radiators to prevent rust formation.

The tendency for iron to oxidize is greatly reduced when it is alloyed with certain other metals. For example, in stainless steel, an alloy of iron and chromium, a layer of chromium oxide forms that protects the iron from corrosion.

An iron container can be covered with a layer of another metal such as tin or zinc. A "tin" can is made by applying a thin layer of tin over iron. Rust formation is prevented as long as the tin layer remains intact. However, once the surface has been scratched, rusting occurs rapidly. If we look up the standard reduction potentials, according to the diagonal rule, we find that iron acts as the anode and tin as the cathode in the corrosion process:

$$Sn^{2+}(aq) + 2e^- \longrightarrow Sn(s) \qquad E° = -0.14\text{ V}$$
$$Fe^{2+}(aq) + 2e^- \longrightarrow Fe(s) \qquad E° = -0.45\text{ V}$$

The protective process is different for zinc-plated, or *galvanized,* iron. Zinc is more easily oxidized than iron (see Table 18.1):

$$Zn^{2+}(aq) + 2e^- \longrightarrow Zn(s) \qquad E° = -0.76\text{ V}$$

So even if a scratch exposes the iron, the zinc is still attacked. In this case, the zinc metal serves as the anode and the iron is the cathode.

Cathodic protection is a process in which the metal that is to be protected from corrosion is made the cathode in what amounts to a galvanic cell. Figure 18.15 shows how an iron nail can be protected from rusting by connecting the nail to a piece of zinc. Without such protection, an iron nail quickly rusts in water. Rusting of underground iron pipes and iron storage tanks can be prevented or greatly reduced by connecting them to metals such as zinc and magnesium, which oxidize more readily than iron (Figure 18.16).

The Chemistry in Action essay "Dental Filling Discomfort" shows that dental filling discomfort can result from an electrochemical phenomenon.

CHEMISTRY *in Action*

Dental Filling Discomfort

In modern dentistry one of the most commonly used materials to fill decaying teeth is known as *dental amalgam.* (An amalgam is a substance made by combining mercury with another metal or metals.) Dental amalgam actually consists of three solid phases having stoichiometries approximately corresponding to Ag_2Hg_3, Ag_3Sn, and Sn_8Hg. The standard reduction potentials for these solid phases are Hg_2^{2+}/Ag_2Hg_3, 0.85 V; Sn^{2+}/Ag_3Sn, −0.05 V; and Sn^{2+}/Sn_8Hg, −0.13 V.

Anyone who bites a piece of aluminum foil (such as that used for wrapping candies) in such a way that the foil presses against a dental filling will probably experience a momentary sharp pain. In effect, an electrochemical cell has been created in the mouth, with aluminum ($E° = -1.66$ V) as the anode, the filling as the cathode, and saliva as the electrolyte. Contact between the aluminum foil and the filling short-circuits the cell, causing a weak current to flow between the electrodes. This current stimulates the sensitive nerve of the tooth, causing an unpleasant sensation.

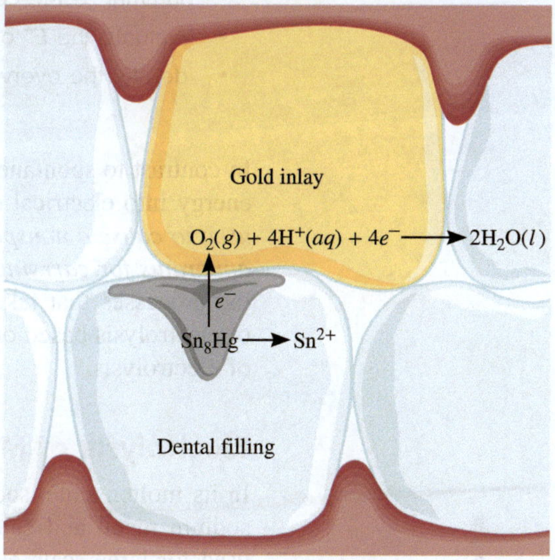

Corrosion of a dental filling brought about by contact with a gold inlay.

Another type of discomfort results when a less electropositive metal touches a dental filling. For example, if a filling makes contact with a gold inlay in a nearby tooth, corrosion of the filling will occur. In this case, the dental filling acts as the anode and the gold inlay as the cathode. Referring to the $E°$ values for the three phases, we see that the Sn_8Hg phase is most likely to corrode. When that happens, release of Sn(II) ions in the mouth produces an unpleasant metallic taste. Prolonged corrosion will eventually result in another visit to the dentist for a replacement filling.

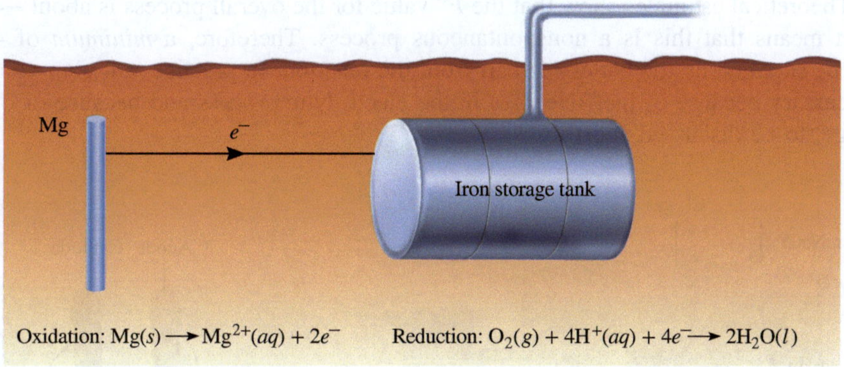

Figure 18.16 *Cathodic protection of an iron storage tank (cathode) by magnesium, a more electropositive metal (anode). Because only the magnesium is depleted in the electrochemical process, it is sometimes called the sacrificial anode.*

Summary of Concepts & Facts

- The corrosion of metals, such as the rusting of iron, is an electrochemical phenomenon.

Review of Concepts & Facts

18.7.1 Which of the following metals can act as a sacrificial anode to protect iron: Sr, Ni, Pb, Co?

18.8 Electrolysis

Learning Objectives

- Summarize the construction and operation of an electrolytic cell.
- Evaluate the $E°$ of a given electrolytic cell.
- Identify the overvoltage in an electrolytic process.

In contrast to spontaneous redox reactions, which result in the conversion of chemical energy into electrical energy, ***electrolysis*** is the process in which *electrical energy is used to cause a nonspontaneous chemical reaction to occur.* An ***electrolytic cell*** is *an apparatus for carrying out electrolysis.* The same principles underlie electrolysis and the processes that take place in galvanic cells. Here we will discuss three examples of electrolysis based on those principles. Then we will look at the quantitative aspects of electrolysis.

Electrolysis of Molten Sodium Chloride

In its molten state, sodium chloride, an ionic compound, can be electrolyzed to form sodium metal and chlorine. Figure 18.17(a) is a diagram of a *Downs cell,* which is used for large-scale electrolysis of NaCl. In molten NaCl, the cations and anions are the Na^+ and Cl^- ions, respectively. Figure 18.17(b) is a simplified diagram showing the reactions that occur at the electrodes. The electrolytic cell contains a pair of electrodes connected to the battery. The battery serves as an "electron pump," driving electrons to the cathode, where reduction occurs, and withdrawing electrons from the anode, where oxidation occurs. The reactions at the electrodes are

$$\begin{aligned} \text{Anode (oxidation):} \quad & 2Cl^-(l) \longrightarrow Cl_2(g) + 2e^- \\ \text{Cathode (reduction):} \quad & 2Na^+(l) + 2e^- \longrightarrow 2Na(l) \\ \hline \text{Overall:} \quad & 2Na^+(l) + 2Cl^-(l) \longrightarrow 2Na(l) + Cl_2(g) \end{aligned}$$

This process is a major source of pure sodium metal and chlorine gas.

Theoretical estimates show that the $E°$ value for the overall process is about −4 V, which means that this is a nonspontaneous process. Therefore, a *minimum* of 4 V must be supplied by the battery to carry out the reaction. In practice, a higher voltage is necessary because of inefficiencies in the electrolytic process and because of overvoltage, to be discussed shortly.

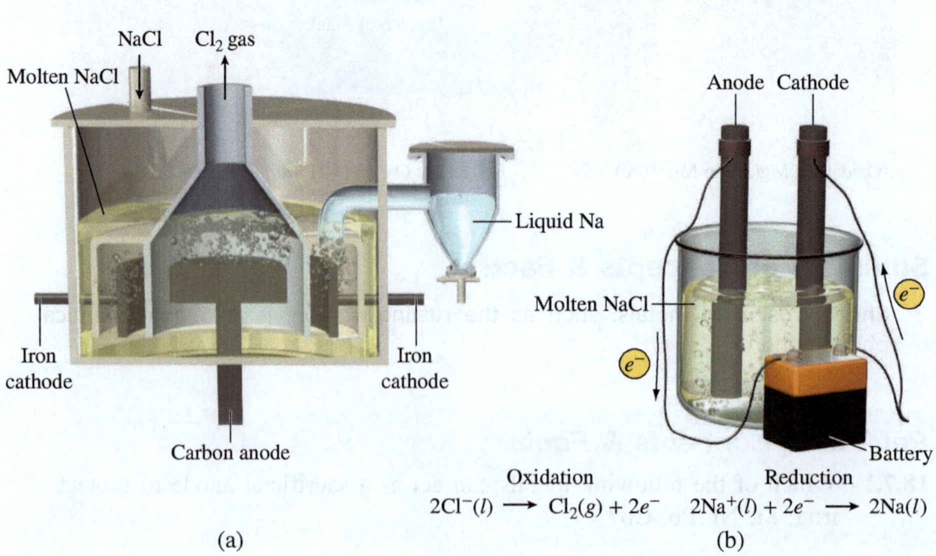

Figure 18.17 *(a) A practical arrangement called a Downs cell for the electrolysis of molten NaCl (m.p. = 801°C). The sodium metal formed at the cathodes is in the liquid state. Since liquid sodium metal is lighter than molten NaCl, the sodium floats to the surface, as shown, and is collected. Chlorine gas forms at the anode and is collected at the top. (b) A simplified diagram showing the electrode reactions during the electrolysis of molten NaCl. The battery is needed to drive the nonspontaneous reactions.*

Electrolysis of Water

Water in a beaker under atmospheric conditions (1 atm and 25°C) will not spontaneously decompose to form hydrogen and oxygen gas because the standard Gibbs-energy change for the reaction is a large positive quantity:

$$2H_2O(l) \longrightarrow 2H_2(g) + O_2(g) \qquad \Delta G° = 571.6 \text{ kJ/mol}$$

However, this reaction can be induced in a cell like the one shown in Figure 18.18. This electrolytic cell consists of a pair of electrodes made of a nonreactive metal, such as platinum, immersed in water. When the electrodes are connected to the battery, nothing happens because there are not enough ions in pure water to carry much of an electric current. (Remember that at 25°C, pure water has only 1×10^{-7} M H^+ ions and 1×10^{-7} M OH^- ions.) On the other hand, the reaction occurs readily in a 0.1 M H_2SO_4 solution because there are a sufficient number of ions to conduct electricity. Immediately, gas bubbles begin to appear at both electrodes.

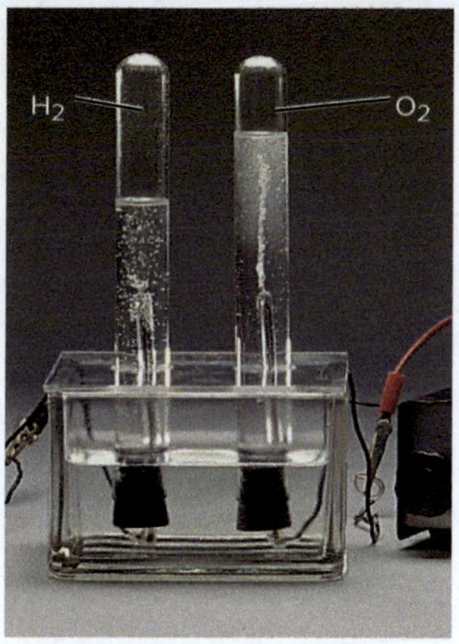

Figure 18.18 *Apparatus for small-scale electrolysis of water. The volume of hydrogen gas generated at the cathode is twice that of oxygen gas generated at the anode.*
Stephen Frisch/McGraw Hill

Figure 18.19 shows the electrode reactions. The process at the anode is

$$2H_2O(l) \longrightarrow O_2(g) + 4H^+(aq) + 4e^-$$

while at the cathode we have

$$2H^+(aq) + 2e^- \longrightarrow H_2(g)$$

The overall reaction is given by

$$\begin{aligned} \text{Anode (oxidation):}\quad & 2H_2O(l) \longrightarrow O_2(g) + 4H^+(aq) + 4e^- \\ \text{Cathode (reduction):}\quad & 2[2H^+(aq) + 2e^- \longrightarrow H_2(g)] \\ \hline \text{Overall:}\quad & 2H_2O(l) \longrightarrow 2H_2(g) + O_2(g) \end{aligned}$$

Note that no net H_2SO_4 is consumed.

Electrolysis of an Aqueous Sodium Chloride Solution

This is the most complicated of the three examples of electrolysis considered here because aqueous sodium chloride solution contains several species that could be oxidized and reduced. The oxidation reactions that might occur at the anode are

$$(1)\quad 2Cl^-(aq) \longrightarrow Cl_2(g) + 2e^-$$
$$(2)\quad 2H_2O(l) \longrightarrow O_2(g) + 4H^+(aq) + 4e^-$$

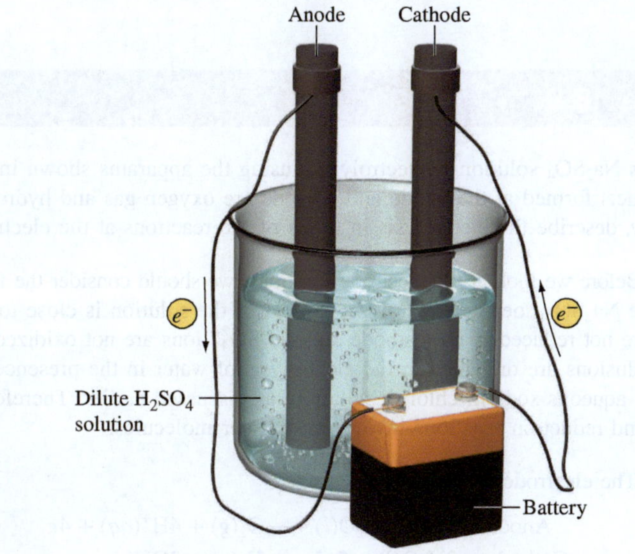

Figure 18.19 *A diagram showing the electrode reactions during the electrolysis of water. Note that the signs of the electrodes are opposite to those of a galvanic cell. In a galvanic cell, the anode is negative because it supplies electrons to the external circuit. In an electrolytic cell, the anode is positive because electrons are withdrawn from it by the battery.*

Referring to Table 18.1, we find

$$Cl_2(g) + 2e^- \longrightarrow 2Cl^-(aq) \qquad E° = 1.36\ V$$

$$O_2(g) + 4H^+(aq) + 4e^- \longrightarrow 2H_2O(l) \qquad E° = 1.23\ V$$

The standard reduction potentials of (1) and (2) are not very different, but the values do suggest that H_2O should be preferentially oxidized at the anode. However, by experiment we find that the gas liberated at the anode is Cl_2, not O_2! In studying electrolytic processes, we sometimes find that the voltage required for a reaction is considerably higher than the electrode potential indicates. The ***overvoltage*** is *the difference between the electrode potential and the actual voltage required to cause electrolysis.* The overvoltage for O_2 formation is quite high. Therefore, under normal operating conditions Cl_2 gas is actually formed at the anode instead of O_2.

The reductions that might occur at the cathode are

$$(3) \quad 2H^+(aq) + 2e^- \longrightarrow H_2(g) \qquad E° = 0.00\ V$$

$$(4) \quad 2H_2O(l) + 2e^- \longrightarrow H_2(g) + 2OH^-(aq) \qquad E° = -0.83\ V$$

$$(5) \quad Na^+(aq) + e^- \longrightarrow Na(s) \qquad E° = -2.71\ V$$

Reaction (5) is ruled out because it has a very negative standard reduction potential. Reaction (3) is preferred over (4) under standard-state conditions. At a pH of 7 (as is the case for a NaCl solution), however, they are equally probable. We generally use (4) to describe the cathode reaction because the concentration of H^+ ions is too low (about 1×10^{-7} M) to make (3) a reasonable choice.

Thus, the half-cell reactions in the electrolysis of aqueous sodium chloride are

$$\text{Anode (oxidation):} \quad 2Cl^-(aq) \longrightarrow Cl_2(g) + 2e^-$$

$$\text{Cathode (reduction):} \quad 2H_2O(l) + 2e^- \longrightarrow H_2(g) + 2OH^-(aq)$$

$$\text{Overall:} \quad 2H_2O(l) + 2Cl^-(aq) \longrightarrow H_2(g) + Cl_2(g) + 2OH^-(aq)$$

As the overall reaction shows, the concentration of the Cl^- ions decreases during electrolysis and that of the OH^- ions increases. Therefore, in addition to H_2 and Cl_2, the useful by-product NaOH can be obtained by evaporating the aqueous solution at the end of the electrolysis.

Student Hot Spot

Student data indicate you may struggle with predicting products of electrolysis. Access your eBook for additional Learning Resources on this topic.

Keep in mind the following from our analysis of electrolysis: Cations are likely to be reduced at the cathode and anions are likely to be oxidized at the anode, and in aqueous solutions water itself may be oxidized and/or reduced. The outcome depends on the nature of other species present.

Example 18.8 deals with the electrolysis of an aqueous solution of sodium sulfate (Na_2SO_4).

Example 18.8

An aqueous Na_2SO_4 solution is electrolyzed, using the apparatus shown in Figure 18.18. If the products formed at the anode and cathode are oxygen gas and hydrogen gas, respectively, describe the electrolysis in terms of the reactions at the electrodes.

Strategy Before we look at the electrode reactions, we should consider the following facts: (1) Because Na_2SO_4 does not hydrolyze, the pH of the solution is close to 7. (2) The Na^+ ions are not reduced at the cathode and the SO_4^{2-} ions are not oxidized at the anode. These conclusions are drawn from the electrolysis of water in the presence of sulfuric acid and in aqueous sodium chloride solution, as discussed earlier. Therefore, both the oxidation and reduction reactions involve only water molecules.

Solution The electrode reactions are

$$\text{Anode:} \quad 2H_2O(l) \longrightarrow O_2(g) + 4H^+(aq) + 4e^-$$

$$\text{Cathode:} \quad 2H_2O(l) + 2e^- \longrightarrow H_2(g) + 2OH^-(aq)$$

(Continued)

The overall reaction, obtained by doubling the cathode reaction coefficients and adding the result to the anode reaction, is

$$6H_2O(l) \longrightarrow 2H_2(g) + O_2(g) + 4H^+(aq) + 4OH^-(aq)$$

If the H^+ and OH^- ions are allowed to mix, then

$$4H^+(aq) + 4OH^-(aq) \longrightarrow 4H_2O(l)$$

and the overall reaction becomes

$$2H_2O(l) \longrightarrow 2H_2(g) + O_2(g)$$

Practice Exercise An aqueous solution of $Mg(NO_3)_2$ is electrolyzed. What are the gaseous products at the anode and cathode?

Similar problem: 18.46.

Electrolysis has many important applications in industry, mainly in the extraction and purification of metals. We will discuss some of these applications in Chapter 21.

Quantitative Aspects of Electrolysis

The quantitative treatment of electrolysis was developed primarily by Faraday. He observed that the mass of product formed (or reactant consumed) at an electrode is proportional to both the amount of electricity transferred at the electrode and the molar mass of the substance in question. For example, in the electrolysis of molten NaCl, the cathode reaction tells us that one Na atom is produced when one Na^+ ion accepts an electron from the electrode. To reduce 1 mole of Na^+ ions, we must supply Avogadro's number (6.02×10^{23}) of electrons to the cathode. On the other hand, the stoichiometry of the anode reaction shows that oxidation of two Cl^- ions yields one chlorine molecule. Therefore, the formation of 1 mole of Cl_2 results in the transfer of 2 moles of electrons from the Cl^- ions to the anode. Similarly, it takes 2 moles of electrons to reduce 1 mole of Mg^{2+} ions and 3 moles of electrons to reduce 1 mole of Al^{3+} ions:

$$Mg^{2+} + 2e^- \longrightarrow Mg$$
$$Al^{3+} + 3e^- \longrightarrow Al$$

In an electrolysis experiment, we generally measure the current (in amperes, A) that passes through an electrolytic cell in a given period of time. The relationship between charge (in coulombs, C) and current is

$$1\ \text{C} = 1\ \text{A} \times 1\ \text{s}$$

that is, a coulomb is the quantity of electrical charge passing any point in the circuit in 1 second when the current is 1 ampere.

Figure 18.20 shows the steps involved in calculating the quantities of substances produced in electrolysis. Let us illustrate the approach by considering molten $CaCl_2$ in an electrolytic cell. Suppose a current of 0.452 A is passed through the cell for 1.50 h. How much product will be formed at the anode and at the cathode? In solving electrolysis problems of this type, the first step is to determine which species will be oxidized at the anode and which species will be reduced at the cathode. Here the choice is straightforward because we only have Ca^{2+} and Cl^- ions in molten $CaCl_2$. Thus, we write the half- and overall cell reactions as

$$\begin{aligned} \text{Anode (oxidation):} \quad & 2Cl^-(l) \longrightarrow Cl_2(g) + 2e^- \\ \text{Cathode (reduction):} \quad & Ca^{2+}(l) + 2e^- \longrightarrow Ca(l) \\ \hline \text{Overall:} \quad & Ca^{2+}(l) + 2Cl^-(l) \longrightarrow Ca(l) + Cl_2(g) \end{aligned}$$

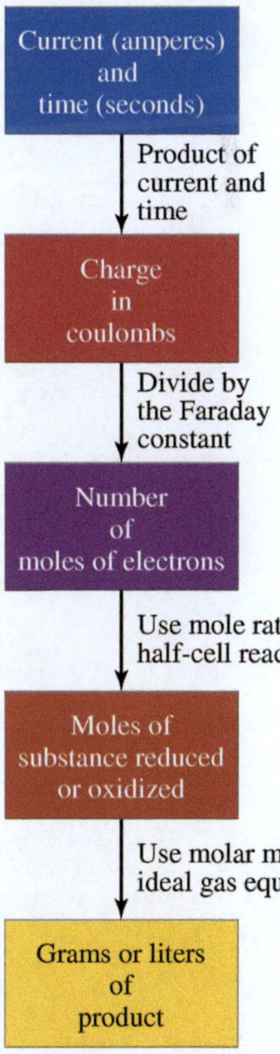

Figure 18.20 *Steps involved in calculating amounts of substances reduced or oxidized in electrolysis.*

The quantities of calcium metal and chlorine gas formed depend on the number of electrons that pass through the electrolytic cell, which in turn depends on current × time, or charge:

$$?C = 0.452 \cancel{A} \times 1.50 \cancel{h} \times \frac{3600 \cancel{s}}{1 \cancel{h}} \times \frac{1\ C}{1 \cancel{A} \cdot \cancel{s}} = 2.44 \times 10^3\ C$$

Because 1 mole e^- = 96,500 C and 2 mole e^- are required to reduce 1 mole of Ca^{2+} ions, the mass of Ca metal formed at the cathode is calculated as follows:

$$?g\ Ca = 2.44 \times 10^3 \cancel{C} \times \frac{1 \cancel{mol\ e^-}}{96{,}500 \cancel{C}} \times \frac{1 \cancel{mol\ Ca}}{2 \cancel{mol\ e^-}} \times \frac{40.08\ g\ Ca}{1 \cancel{mol\ Ca}} = 0.507\ g\ Ca$$

Student Hot Spot

Student data indicate you may struggle with quantitative aspects of electrolysis. Access your eBook to view additional Learning Resources on this topic.

The anode reaction indicates that 1 mole of chlorine is produced per 2 mole e^- of electricity. Hence the mass of chlorine gas formed is

$$?g\ Cl_2 = 2.44 \times 10^3 \cancel{C} \times \frac{1 \cancel{mol\ e^-}}{96{,}500 \cancel{C}} \times \frac{1 \cancel{mol\ Cl_2}}{2 \cancel{mol\ e^-}} \times \frac{70.90\ g\ Cl_2}{1 \cancel{mol\ Cl_2}} = 0.896\ g\ Cl_2$$

Example 18.9 applies this approach to the electrolysis in an aqueous solution.

Example 18.9

A current of 1.26 A is passed through an electrolytic cell containing a dilute sulfuric acid solution for 7.44 h. Write the half-cell reactions and calculate the volume of gases generated at STP.

Strategy Earlier in Section 18.8 we saw that the half-cell reactions for the process are

$$\begin{aligned} \text{Anode (oxidation):} &\quad 2H_2O(l) \longrightarrow O_2(g) + 4H^+(aq) + 4e^- \\ \text{Cathode (reduction):} &\quad 2[2H^+(aq) + 2e^- \longrightarrow H_2(g)] \\ \hline \text{Overall:} &\quad 2H_2O(l) \longrightarrow 2H_2(g) + O_2(g) \end{aligned}$$

According to Figure 18.20, we carry out the following conversion steps to calculate the quantity of O_2 in moles:

$$\text{current} \times \text{time} \longrightarrow \text{coulombs} \longrightarrow \text{moles of } e^- \longrightarrow \text{moles of } O_2$$

Then, using the ideal gas equation we can calculate the volume of O_2 in liters at STP. A similar procedure can be used for H_2.

Solution First we calculate the number of coulombs of electricity that pass through the cell:

$$?C = 1.26 \cancel{A} \times 7.44 \cancel{h} \times \frac{3600 \cancel{s}}{1 \cancel{h}} \times \frac{1\ C}{1 \cancel{A} \cdot \cancel{s}} = 3.37 \times 10^4\ C$$

Next, we convert number of coulombs to number of moles of electrons:

$$3.37 \times 10^4 \cancel{C} \times \frac{1\ mol\ e^-}{96{,}500 \cancel{C}} = 0.349\ mol\ e^-$$

From the oxidation half-reaction we see that 1 mol O_2 ≏ 4 mol e^-. Therefore, the number of moles of O_2 generated is

$$0.349 \cancel{mol\ e^-} \times \frac{1\ mol\ O_2}{4 \cancel{mol\ e^-}} = 0.0873\ mol\ O_2$$

The volume of 0.0873 mol O_2 at STP is given by

$$\begin{aligned} V &= \frac{nRT}{P} \\ &= \frac{(0.0873\ mol)(0.0821\ L \cdot atm/K \cdot mol)(273\ K)}{1\ atm} = 1.96\ L \end{aligned}$$

(Continued)

The procedure for hydrogen is similar. To simplify, we combine the first two steps to calculate the number of moles of H_2 generated:

$$3.37 \times 10^4 \not{C} \times \frac{1 \not{mol\, e^-}}{96{,}500 \not{C}} \times \frac{1 \text{ mol } H_2}{2 \not{mol\, e^-}} = 0.175 \text{ mol } H_2$$

The volume of 0.175 mol H_2 at STP is given by

$$\begin{aligned} V &= \frac{nRT}{P} \\ &= \frac{(0.175 \text{ mol})(0.0821 \text{ L} \cdot \text{atm/K} \cdot \text{mol})(273 \text{ K})}{1 \text{ atm}} \\ &= 3.92 \text{ L} \end{aligned}$$

Check Note that the volume of H_2 is twice that of O_2 (see Figure 18.18), which is what we would expect based on Avogadro's law (at the same temperature and pressure, volume is directly proportional to the number of moles of gases).

Practice Exercise A constant current is passed through an electrolytic cell containing molten $MgCl_2$ for 18 h. If 4.8×10^5 g of Cl_2 are obtained, what is the current in amperes?

Similar problem: 18.51.

Summary of Concepts & Facts

- Electric current from an external source is used to drive a nonspontaneous chemical reaction in an electrolytic cell. The amount of product formed or reactant consumed depends on the quantity of electricity transferred at the electrodes.
- The quantity of electricity carried by 1 mole of electrons is equal to 96,500 C.

Review of Concepts & Facts

18.8.1 What is the minimum voltage needed for the electrolytic process shown in Figure 18.19?

18.8.2 Write the half-cell reactions for the anode and cathode of the following electrolytic cell. Explain why the signs of the anode and cathode are opposite to those in a galvanic cell.

18.8.3 In the electrolysis of molten $CaCl_2$, a current of 1.24 A is passed through the cell for 2.0 h. What is the mass of Ca produced at the cathode?

Chapter Summary

Redox Reactions and Electrochemical Cells Equations representing redox reactions can be balanced using the half-reaction method. These reactions involve the transfer of electrons from a reducing agent to an oxidizing agent. Using separate compartments, such a reaction can be used to generate electrons in an arrangement called a galvanic cell. (Sections 18.1 and 18.2)

Thermodynamics of Galvanic Cells The voltage measured in a galvanic cell can be broken down into the electrode potentials of the anode (where oxidation takes place) and cathode (where reduction takes place). This voltage can be related to the Gibbs-energy change and the equilibrium constant of the redox process. The Nernst equation relates the cell voltage to the cell voltage under standard-state conditions and the concentrations of reacting species. (Sections 18.3, 18.4, and 18.5)

Batteries and Fuel Cells Batteries are electrochemical cells that can supply direct electric current at a constant voltage. There are many different types of batteries used in automobiles, flashlights, and pacemakers. Fuel cells are a special type of electrochemical cell that generates electricity by the oxidation of hydrogen or hydrocarbons. (Section 18.6)

Corrosion Corrosion is a spontaneous redox reaction that results in the formation of rust from iron, silver sulfide from silver, and patina (copper carbonate) from copper. Corrosion causes enormous damage to buildings, structures, ships, and cars. Many methods have been devised to prevent or minimize the effect of corrosion. (Section 18.7)

Electrolysis Electrolysis is the process in which electrical energy is used to cause a nonspontaneous redox reaction to occur. The quantitative relationship between the current supplied and the products formed is provided by Faraday. Electrolysis is the major method for producing active metals and nonmetals and many essential industrial chemicals. (Section 18.8)

Key Equations

$E^\circ_{\text{cell}} = E^\circ_{\text{cathode}} - E^\circ_{\text{anode}}$ (18.1)	Calculating the standard emf of a galvanic cell.
$\Delta G = -nFE_{\text{cell}}$ (18.2)	Relating Gibbs-energy change to the emf of the cell.
$\Delta G^\circ = -nFE^\circ_{\text{cell}}$ (18.3)	Relating the standard Gibbs-energy change to the standard emf of the cell.
$E^\circ_{\text{cell}} = \dfrac{0.0257\text{ V}}{n}\ln K$ (18.5)	Relating the standard emf of the cell to the equilibrium constant.
$E^\circ_{\text{cell}} = \dfrac{0.0592\text{ V}}{n}\log K$ (18.6)	Relating the standard emf of the cell to the equilibrium constant.
$E = E^\circ - \dfrac{0.0257\text{ V}}{n}\ln Q$ (18.8)	Relating the emf of the cell to the concentrations under nonstandard state conditions.
$E = E^\circ - \dfrac{0.0592\text{ V}}{n}\log Q$ (18.9)	Relating the emf of the cell to the concentrations under nonstandard state conditions.

Key Words

Anode, p. 819
Battery, p. 836
Cathode, p. 819
Cell voltage, p. 821
Corrosion, p. 841
Electrochemistry, p. 816
Electrolysis, p. 846
Electrolytic cell, p. 846
Electromotive force (emf) (E), p. 821
Faraday constant (F), p. 828
Fuel cell, p. 839
Galvanic cell, p. 819
Half-cell reaction, p. 820
Nernst equation, p. 832
Overvoltage, p. 848
Standard emf (E°_{cell}), p. 823
Standard reduction potential, p. 822

Questions & Problems

Red numbered problems solved in Student Solutions Manual

18.1 Redox Reactions

Problems

18.1 Balance the following redox equations by the ion-electron method:
(a) $H_2O_2 + Fe^{2+} \longrightarrow Fe^{3+} + H_2O$ (in acidic solution)
(b) $Cu + HNO_3 \longrightarrow Cu^{2+} + NO + H_2O$ (in acidic solution)
(c) $CN^- + MnO_4^- \longrightarrow CNO^- + MnO_2$ (in basic solution)
(d) $Br_2 \longrightarrow BrO_3^- + Br^-$ (in basic solution)
(e) $S_2O_3^{2-} + I_2 \longrightarrow I^- + S_4O_6^{2-}$ (in acidic solution)

18.2 Balance the following redox equations by the ion-electron method:
(a) $Mn^{2+} + H_2O_2 \longrightarrow MnO_2 + H_2O$ (in basic solution)
(b) $Bi(OH)_3 + SnO_2^{2-} \longrightarrow SnO_3^{2-} + Bi$ (in basic solution)
(c) $Cr_2O_7^{2-} + C_2O_4^{2-} \longrightarrow Cr^{3+} + CO_2$ (in acidic solution)
(d) $ClO_3^- + Cl^- \longrightarrow Cl_2 + ClO_2$ (in acidic solution)

18.2 Galvanic Cells

Review Questions

18.3 Define the following terms: *anode, cathode, cell voltage, electromotive force,* and *standard reduction potential.*

18.4 Describe the basic features of a galvanic cell. Why are the two components of the cell separated from each other?

18.5 What is the function of a salt bridge? What kind of electrolyte should be used in a salt bridge?

18.6 What is a cell diagram? Write the cell diagram for a galvanic cell consisting of an Al electrode placed in a 1 *M* $Al(NO_3)_3$ solution and a Ag electrode placed in a 1 *M* $AgNO_3$ solution.

18.7 What is the difference between the half-reactions discussed in redox processes in Chapter 4 and the half-cell reactions discussed in Section 18.2?

18.3 Standard Reduction Potentials

Review Questions

18.8 Define the following terms: *electromotive force* and *standard reduction potential.*

18.9 Use the information in Table 2.1, and calculate the Faraday constant.

18.10 Discuss the spontaneity of an electrochemical reaction in terms of its standard emf (E°_{cell}).

Problems

18.11 Calculate the standard emf of a cell that uses the Mg/Mg^{2+} and Cu/Cu^{2+} half-cell reactions at 25°C. Write the equation for the cell reaction that occurs under standard-state conditions.

18.12 Calculate the standard emf of a cell that uses Ag/Ag^+ and Al/Al^{3+} half-cell reactions. Write the cell reaction that occurs under standard-state conditions.

18.13 Predict whether Fe^{3+} can oxidize I^- to I_2 under standard-state conditions.

18.14 Which of the following reagents can oxidize H_2O to $O_2(g)$ under standard-state conditions: $H^+(aq)$, $Cl^-(aq)$, $Cl_2(g)$, $Cu^{2+}(aq)$, $Pb^{2+}(aq)$, $MnO_4^-(aq)$ (in acid)?

18.15 Consider the following half-reactions:

$$MnO_4^-(aq) + 8H^+(aq) + 5e^- \longrightarrow Mn^{2+}(aq) + 4H_2O(l)$$

$$NO_3^-(aq) + 4H^+(aq) + 3e^- \longrightarrow NO(g) + 2H_2O(l)$$

Predict whether NO_3^- ions will oxidize Mn^{2+} to MnO_4^- under standard-state conditions.

18.16 Predict whether the following reactions would occur spontaneously in aqueous solution at 25°C. Assume that the initial concentrations of dissolved species are all 1.0 *M*.
(a) $Ca(s) + Cd^{2+}(aq) \longrightarrow Ca^{2+}(aq) + Cd(s)$
(b) $2Br^-(aq) + Sn^{2+}(aq) \longrightarrow Br_2(l) + Sn(s)$
(c) $2Ag(s) + Ni^{2+}(aq) \longrightarrow 2Ag^+(aq) + Ni(s)$
(d) $Cu^+(aq) + Fe^{3+}(aq) \longrightarrow Cu^{2+}(aq) + Fe^{2+}(aq)$

18.17 Which species in each pair is a better oxidizing agent under standard-state conditions: (a) Br_2 or Au^{3+}? (b) H_2 or Ag^+? (c) Cd^{2+} or Cr^{3+}? (d) O_2 in acidic media or O_2 in basic media?

18.18 Which species in each pair is a better reducing agent under standard-state conditions: (a) Na or Li? (b) H_2 or I_2? (c) Fe^{2+} or Ag? (d) Br^- or Co^{2+}?

18.19 Consider the electrochemical reaction $Sn^{2+} + X \longrightarrow Sn + X^{2+}$. Given that $E^\circ_{cell} = 0.14$ V, what is the E° for the X^{2+}/X half-reaction?

18.20 The E°_{cell} for the following cell is 1.54 V at 25°C:

$$U(s) \mid U^{3+}(aq) \parallel Ni^{2+}(aq) \mid Ni(s)$$

Calculate the standard reduction potential for the U^{3+}/U half-cell.

18.4 Thermodynamics of Redox Reactions

Review Questions

18.21 Write the equations relating ΔG° and K to the standard emf of a cell. Define all the terms.

18.22 The E° value of one cell reaction is positive and that of another cell reaction is negative. Which cell reaction will proceed toward the formation of more products at equilibrium?

Problems

18.23 What is the equilibrium constant for the following reaction at 25°C?

$$Mg(s) + Zn^{2+}(aq) \rightleftharpoons Mg^{2+}(aq) + Zn(s)$$

18.24 The equilibrium constant for the reaction

$$Sr(s) + Mg^{2+}(aq) \rightleftharpoons Sr^{2+}(aq) + Mg(s)$$

is 3.20×10^{12} at 25°C. Calculate $E°$ for a cell made up of Sr/Sr^{2+} and Mg/Mg^{2+} half-cells.

18.25 Use the standard reduction potentials to find the equilibrium constant for each of the following reactions at 25°C:
(a) $Br_2(l) + 2I^-(aq) \rightleftharpoons 2Br^-(aq) + I_2(s)$
(b) $2Ce^{4+}(aq) + 2Cl^-(aq) \rightleftharpoons Cl_2(g) + 2Ce^{3+}(aq)$
(c) $5Fe^{2+}(aq) + MnO_4^-(aq) + 8H^+(aq) \rightleftharpoons Mn^{2+}(aq) + 4H_2O(l) + 4H_2O(l) + 5Fe^{3+}(aq)$

18.26 Calculate $\Delta G°$; and K_c for the following reactions at 25°C:
(a) $Mg(s) + Pb^{2+}(aq) \rightleftharpoons Mg^{2+}(aq) + Pb(s)$
(b) $Br_2(l) + 2I^-(aq) \rightleftharpoons 2Br^-(aq) + I_2(s)$
(c) $O_2(g) + 4H^+(aq) + 4Fe^{2+}(aq) \rightleftharpoons 2H_2O(l) + 4Fe^{3+}(aq)$
(d) $2Al(s) + 3I_2(s) \rightleftharpoons 2Al^{3+}(aq) + 6I^-(aq)$

18.27 Under standard-state conditions, what spontaneous reaction will occur in aqueous solution among the ions Ce^{4+}, Ce^{3+}, Fe^{3+}, and Fe^{2+}? Calculate $\Delta G°$ and K_c for the reaction.

18.28 Given that $E° = 0.52$ V for the reduction $Cu^+(aq) + e^- \longrightarrow Cu(s)$, calculate $E°$, $\Delta G°$, and K for the following reaction at 25°C:

$$2Cu^+(aq) \longrightarrow Cu^{2+}(aq) + Cu(s)$$

18.5 The Effect of Concentration on Cell Emf

Review Questions

18.29 Write the Nernst equation and explain all the terms.

18.30 Write the Nernst equation for the following processes at some temperature *T*:
(a) $Mg(s) + Sn^{2+}(aq) \longrightarrow Mg^{2+}(aq) + Sn(s)$
(b) $2Cr(s) + 3Pb^{2+}(aq) \longrightarrow 2Cr^{3+}(aq) + 3Pb(s)$

Problems

18.31 What is the potential of a cell made up of Zn/Zn^{2+} and Cu/Cu^{2+} half-cells at 25°C if $[Zn^{2+}] = 0.25\ M$ and $[Cu^{2+}] = 0.15\ M$?

18.32 Calculate $E°$, E, and ΔG for the following cell reactions.
(a) $Mg(s) + Sn^{2+}(aq) \longrightarrow Mg^{2+}(aq) + Sn(s)$
$[Mg^{2+}] = 0.045\ M$, $[Sn^{2+}] = 0.035\ M$
(b) $3Zn(s) + 2Cr^{3+}(aq) \longrightarrow 3Zn^{2+}(aq) + 2Cr(s)$
$[Cr^{3+}] = 0.010\ M$, $[Zn^{2+}] = 0.0085\ M$

18.33 Calculate the standard potential of the cell consisting of the Zn/Zn^{2+} half-cell and the SHE. What will the emf of the cell be if $[Zn^{2+}] = 0.45\ M$, $P_{H_2} = 2.0$ atm, and $[H^+] = 1.8\ M$?

18.34 What is the emf of a cell consisting of a Pb^{2+}/Pb half-cell and a $Pt/H^+/H_2$ half-cell if $[Pb^{2+}] = 0.10\ M$, $[H^+] = 0.050\ M$, and $P_{H_2} = 2.0$ atm?

18.35 Referring to the arrangement in Figure 18.1, calculate the $[Cu^{2+}]/[Zn^{2+}]$ ratio at which the following reaction is spontaneous at 25°C:

$$Cu(s) + Zn^{2+}(aq) \longrightarrow Cu^{2+}(aq) + Zn(s)$$

18.36 Calculate the emf of the following concentration cell:

$$Mg(s) \mid Mg^{2+}(0.24\ M) \parallel Mg^{2+}(0.53\ M) \mid Mg(s)$$

18.6 Batteries and Fuel Cells

Review Questions

18.37 Explain the differences between a primary galvanic cell—one that is not rechargeable—and a storage cell (for example, the lead storage battery), which is rechargeable.

18.38 Discuss the advantages and disadvantages of fuel cells over conventional power plants in producing electricity.

Problems

18.39 The hydrogen-oxygen fuel cell is described in Section 18.6. (a) What volume of $H_2(g)$, stored at 25°C at a pressure of 155 atm, would be needed to run an electric motor drawing a current of 8.5 A for 3.0 h? (b) What volume (liters) of air at 25°C and 1.00 atm will have to pass into the cell per minute to run the motor? Assume that air is 20 percent O_2 by volume and that all the O_2 is consumed in the cell. The other components of air do not affect the fuel-cell reactions. Assume ideal gas behavior.

18.40 Calculate the standard emf of the propane fuel cell discussed in Section 18.6 at 25°C, given that $\Delta G_f°$ for propane is −23.4 kJ/mol.

18.7 Corrosion

Review Questions

18.41 Steel hardware, including nuts and bolts, is often coated with a thin plating of cadmium. Explain the function of the cadmium layer.

18.42 "Galvanized iron" is steel sheet that has been coated with zinc; "tin" cans are made of steel sheet coated with tin. Discuss the functions of these coatings and the electrochemistry of the corrosion reactions that occur if an electrolyte contacts the scratched surface of a galvanized iron sheet or a tin can.

18.43 Tarnished silver contains Ag_2S. The tarnish can be removed by placing silverware in an aluminum pan containing an inert electrolyte solution, such as NaCl. Explain the electrochemical principle for this procedure. [The standard reduction potential for the half-cell reaction $Ag_2S(s) + 2e^- \longrightarrow 2Ag(s) + S^{2-}(aq)$ is −0.71 V.]

18.44 How does the tendency of iron to rust depend on the pH of solution?

18.8 Electrolysis

Review Questions

18.45 What is the difference between a galvanic cell (such as a Daniell cell) and an electrolytic cell?

18.46 Describe the electrolysis of an aqueous solution of KNO_3.

Problems

18.47 The half-reaction at an electrode is

$$Mg^{2+}(\text{molten}) + 2e^- \longrightarrow Mg(s)$$

Calculate the number of grams of magnesium that can be produced by supplying 1.00 *F* to the electrode.

18.48 Consider the electrolysis of molten barium chloride, $BaCl_2$. (a) Write the half-reactions. (b) How many grams of barium metal can be produced by supplying 0.50 A for 30 min?

18.49 Considering only the cost of electricity, would it be cheaper to produce a ton of sodium or a ton of aluminum by electrolysis?

18.50 If the cost of electricity to produce magnesium by the electrolysis of molten magnesium chloride is \$155 per ton of metal, what is the cost (in dollars) of the electricity necessary to produce (a) 10.0 tons of aluminum, (b) 30.0 tons of sodium, (c) 50.0 tons of calcium?

18.51 One of the half-reactions for the electrolysis of water is

$$2H_2O(l) \longrightarrow O_2(g) + 4H^+(aq) + 4e^-$$

If 0.076 L of O_2 is collected at 25°C and 755 mmHg, how many moles of electrons had to pass through the solution?

18.52 How many moles of electrons are required to produce (a) 0.84 L of O_2 at exactly 1 atm and 25°C from aqueous H_2SO_4 solution; (b) 1.50 L of Cl_2 at 750 mmHg and 20°C from molten NaCl; (c) 6.0 g of Sn from molten $SnCl_2$?

18.53 Calculate the amounts of Cu and Br_2 produced in 1.0 h at inert electrodes in a solution of $CuBr_2$ by a current of 4.50 A.

18.54 In the electrolysis of an aqueous $AgNO_3$ solution, 0.67 g of Ag is deposited after a certain period of time. (a) Write the half-reaction for the reduction of Ag^+. (b) What is the probable oxidation half-reaction? (c) Calculate the quantity of electricity used, in coulombs.

18.55 A steady current was passed through molten $CoSO_4$ until 2.35 g of metallic cobalt was produced. Calculate the number of coulombs of electricity used.

18.56 A constant electric current flows for 3.75 h through two electrolytic cells connected in series. One contains a solution of $AgNO_3$ and the second a solution of $CuCl_2$. During this time 2.00 g of silver are deposited in the first cell. (a) How many grams of copper are deposited in the second cell? (b) What is the current flowing, in amperes?

18.57 What is the hourly production rate of chlorine gas (in kg) from an electrolytic cell using aqueous NaCl electrolyte and carrying a current of 1.500×10^3 A? The anode efficiency for the oxidation of Cl^- is 93.0 percent.

18.58 Chromium plating is applied by electrolysis to objects suspended in a dichromate solution, according to the following (unbalanced) half-reaction:

$$Cr_2O_7^{2-}(aq) + e^- + H^+(aq) \longrightarrow Cr(s) + H_2O(l)$$

How long (in hours) would it take to apply a chromium plating 1.0×10^{-2} mm thick to a car bumper with a surface area of 0.25 m^2 in an electrolytic cell carrying a current of 25.0 A? (The density of chromium is 7.19 g/cm^3.)

18.59 The passage of a current of 0.750 A for 25.0 min deposited 0.369 g of copper from a $CuSO_4$ solution. From this information, calculate the molar mass of copper.

18.60 A quantity of 0.300 g of copper was deposited from a $CuSO_4$ solution by passing a current of 3.00 A through the solution for 304 s. Calculate the value of the Faraday constant.

18.61 In a certain electrolysis experiment, 1.44 g of Ag were deposited in one cell (containing an aqueous $AgNO_3$ solution), while 0.120 g of an unknown metal X was deposited in another cell (containing an aqueous XCl_3 solution) in series with the $AgNO_3$ cell. Calculate the molar mass of X.

18.62 One of the half-reactions for the electrolysis of water is

$$2H^+(aq) + 2e^- \longrightarrow H_2(g)$$

If 0.845 L of H_2 is collected at 25°C and 782 mmHg, how many moles of electrons had to pass through the solution?

18.63 A steady current of 10.0 A is passed through three electrolytic cells for 10.0 min. Calculate the mass of the metals formed if the solutions are 0.10 *M* $AgNO_3$, 0.10 *M* $Cu(NO_3)_2$, and 0.10 *M* $Au(NO_3)_3$.

18.64 Industrially, copper metal can be purified electrolytically according to the following arrangement. The anode is made of the impure Cu electrode and the cathode is the pure Cu electrode. The electrodes are immersed in a $CuSO_4$ solution. (a) Write the half-cell reactions at the electrodes. (b) Calculate the mass (in grams) of Cu purified after passing a current of 20 A for 10 h. (c) Explain why impurities such as Zn, Fe, Au, and Ag are not deposited at the electrodes.

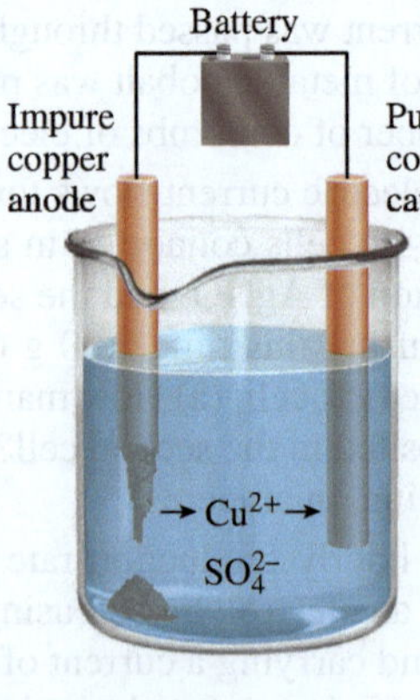

Additional Problems

18.65 A Daniell cell consists of a zinc electrode in 1.00 L of 1.00 *M* $ZnSO_4$ and a Cu electrode in 1.00 L of 1.00 *M* $CuSO_4$ at 25°C. A steady current of 10.0 A is drawn from the cell. Calculate the E_{cell} after 1.00 h. Assume volumes to remain constant.

18.66 A concentration cell is constructed having Cu electrodes in two $CuSO_4$ solutions A and B. At 25°C, the osmotic pressures of the two solutions are 48.9 atm and 4.89 atm, respectively. Calculate the E_{cell}, assuming no ion-pair formation.

18.67 For each of the following redox reactions, (i) write the half-reactions, (ii) write a balanced equation for the whole reaction, (iii) determine in which direction the reaction will proceed spontaneously under standard-state conditions:

(a) $H_2(g) + Ni^{2+}(aq) \longrightarrow H^+(aq) + Ni(s)$

(b) $MnO_4^-(aq) + Cl^-(aq) \longrightarrow Mn^{2+}(aq) + Cl_2(g)$ (in acid solution)

(c) $Cr(s) + Zn^{2+}(aq) \longrightarrow Cr^{3+}(aq) + Zn(s)$

18.68 The oxidation of 25.0 mL of a solution containing Fe^{2+} requires 26.0 mL of 0.0250 *M* $K_2Cr_2O_7$ in acidic solution. Balance the following equation and calculate the molar concentration of Fe^{2+}:

$$Cr_2O_7^{2-} + Fe^{2+} + H^+ \longrightarrow Cr^{3+} + Fe^{3+}$$

18.69 The SO_2 present in air is mainly responsible for the phenomenon of acid rain. The concentration of SO_2 can be determined by titrating against a standard permanganate solution as follows:

$$5SO_2 + 2MnO_4^- + 2H_2O \longrightarrow 5SO_4^{2-} + 2Mn^{2+} + 4H^+$$

Calculate the number of grams of SO_2 in a sample of air if 7.37 mL of 0.00800 *M* $KMnO_4$ solution are required for the titration.

18.70 A sample of iron ore weighing 0.2792 g was dissolved in an excess of a dilute acid solution. All the iron was first converted to Fe(II) ions. The solution then required 23.30 mL of 0.0194 *M* $KMnO_4$ for oxidation to Fe(III) ions. Calculate the percent by mass of iron in the ore.

18.71 The concentration of a hydrogen peroxide solution can be conveniently determined by titration against a standardized potassium permanganate solution in an acidic medium according to the following unbalanced equation:

$$MnO_4^- + H_2O_2 \longrightarrow O_2 + Mn^{2+}$$

(a) Balance the equation. (b) If 36.44 mL of a 0.01652 *M* $KMnO_4$ solution are required to completely oxidize 25.00 mL of a H_2O_2 solution, calculate the molarity of the H_2O_2 solution.

18.72 Oxalic acid ($H_2C_2O_4$) is present in many plants and vegetables. (a) Balance the following equation in acid solution:

$$MnO_4^- + C_2O_4^{2-} \longrightarrow Mn^{2+} + CO_2$$

(b) If a 1.00-g sample of $H_2C_2O_4$ requires 24.0 mL of 0.0100 *M* $KMnO_4$ solution to reach the equivalence point, what is the percent by mass of $H_2C_2O_4$ in the sample?

18.73 Complete the following table. State whether the cell reaction is spontaneous, nonspontaneous, or at equilibrium.

E	Δ*G*	Cell Reaction
> 0		
	> 0	
= 0		

18.74 Calcium oxalate (CaC_2O_4) is insoluble in water. This property has been used to determine the amount of Ca^{2+} ions in blood. The calcium oxalate isolated from blood is dissolved in acid and titrated against a standardized $KMnO_4$ solution as described in Problem 18.72. In one test, it is found that the calcium oxalate isolated from a 10.0-mL sample of blood requires 24.2 mL of 9.56×10^{-4} *M* $KMnO_4$ for titration. Calculate the number of milligrams of calcium per milliliter of blood.

18.75 From the following information, calculate the solubility product of AgBr:

$$Ag^+(aq) + e^- \longrightarrow Ag(s) \qquad E° = 0.80\ V$$
$$AgBr(s) + e^- \longrightarrow Ag(s) + Br^-(aq) \qquad E° = 0.07\ V$$

18.76 Consider a galvanic cell composed of the SHE and a half-cell using the reaction $Ag^+(aq) + e^- \longrightarrow Ag(s)$. (a) Calculate the standard cell potential. (b) What is the spontaneous cell reaction under standard-state conditions? (c) Calculate the cell potential when $[H^+]$ in the hydrogen electrode is changed to (i) 1.0×10^{-2} *M* and (ii) 1.0×10^{-5} *M*, all other reagents being held at standard-state conditions. (d) Based on this cell arrangement, suggest a design for a pH meter.

18.77 A galvanic cell consists of a silver electrode in contact with 346 mL of 0.100 *M* $AgNO_3$ solution and a magnesium electrode in contact with 288 mL of 0.100 *M* $Mg(NO_3)_2$ solution. (a) Calculate *E* for the cell at 25°C. (b) A current is drawn from the cell until 1.20 g of silver have been deposited at the silver electrode. Calculate E for the cell at this stage of operation.

18.78 Explain why chlorine gas can be prepared by electrolyzing an aqueous solution of NaCl but fluorine gas cannot be prepared by electrolyzing an aqueous solution of NaF.

18.79 Calculate the emf of the following concentration cell at 25°C:

$$Cu(s) \mid Cu^{2+}(0.080\ M) \parallel Cu^{2+}(1.2\ M) \mid Cu(s)$$

18.80 The cathode reaction in the Leclanché cell is given by

$$2MnO_2(s) + Zn^{2+}(aq) + 2e^- \longrightarrow ZnMn_2O_4(s)$$

If a Leclanché cell produces a current of 0.0050 A, calculate how many hours this current supply will last if there are initially 4.0 g of MnO_2 present in the cell. Assume that there is an excess of Zn^{2+} ions.

18.81 Suppose you are asked to verify experimentally the electrode reactions shown in Example 18.8. In addition to the apparatus and the solution, you are also given two pieces of litmus paper, one blue and the other red. Describe what steps you would take in this experiment.

18.82 For a number of years it was not clear whether mercury(I) ions existed in solution as Hg^+ or as Hg_2^{2+}. To distinguish between these two possibilities, we could set up the following system:

$$Hg(l) \mid \text{soln A} \parallel \text{soln B} \mid Hg(l)$$

where soln A contained 0.263 g mercury(I) nitrate per liter and soln B contained 2.63 g mercury(I) nitrate per liter. If the measured emf of such a cell is 0.0289 V at 18°C, what can you deduce about the nature of the mercury(I) ions?

18.83 An aqueous KI solution to which a few drops of phenolphthalein have been added is electrolyzed using an apparatus like the one shown here:

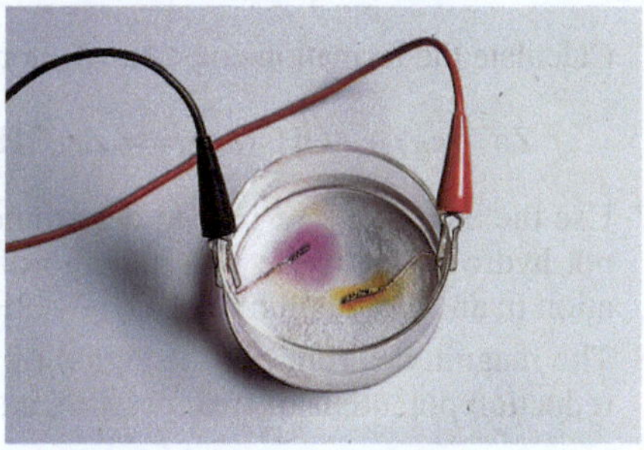

Ken Karp/McGraw Hill

Describe what you would observe at the anode and the cathode. (*Hint:* Molecular iodine is only slightly soluble in water, but in the presence of I^- ions, it forms the brown color of I_3^- ions. See Problem 12.102.)

18.84 A piece of magnesium metal weighing 1.56 g is placed in 100.0 mL of 0.100 M $AgNO_3$ at 25°C. Calculate $[Mg^{2+}]$ and $[Ag^+]$ in solution at equilibrium. What is the mass of the magnesium left? The volume remains constant.

18.85 Describe an experiment that would enable you to determine which is the cathode and which is the anode in a galvanic cell using copper and zinc electrodes.

18.86 An acidified solution was electrolyzed using copper electrodes. A constant current of 1.18 A caused the anode to lose 0.584 g after 1.52×10^3 s. (a) What is the gas produced at the cathode and what is its volume at STP? (b) Given that the charge of an electron is 1.6022×10^{-19} C, calculate Avogadro's number. Assume that copper is oxidized to Cu^{2+} ions.

18.87 In a certain electrolysis experiment involving Al^{3+} ions, 60.2 g of Al is recovered when a current of 0.352 A is used. How many minutes did the electrolysis last?

18.88 Consider the oxidation of ammonia:

$$4NH_3(g) + 3O_2(g) \longrightarrow 2N_2(g) + 6H_2O(l)$$

(a) Calculate the $\Delta G°$ for the reaction. (b) If this reaction were used in a fuel cell, what would the standard cell potential be?

18.89 When an aqueous solution containing gold(III) salt is electrolyzed, metallic gold is deposited at the cathode and oxygen gas is generated at the anode. (a) If 9.26 g of Au is deposited at the cathode, calculate the volume (in liters) of O_2 generated at 23°C and 747 mmHg. (b) What is the current used if the electrolytic process took 2.00 h?

18.90 In an electrolysis experiment, a student passes the same quantity of electricity through two electrolytic cells, one containing a silver salt and the other a gold salt. Over a certain period of time, she finds that 2.64 g of Ag and 1.61 g of Au are deposited at the cathodes. What is the oxidation state of gold in the gold salt?

18.91 People living in cold-climate countries where there is plenty of snow are advised not to heat their garages in the winter. What is the electrochemical basis for this recommendation?

18.92 Given that

$$2Hg^{2+}(aq) + 2e^- \longrightarrow Hg_2^{2+}(aq) \qquad E° = 0.92\ V$$
$$Hg_2^{2+}(aq) + 2e^- \longrightarrow 2Hg(l) \qquad E° = 0.85\ V$$

calculate $\Delta G°$ and K for the following process at 25°C:

$$Hg_2^{2+}(aq) \longrightarrow Hg^{2+}(aq) + Hg(l)$$

(The preceding reaction is an example of a *disproportionation reaction* in which an element in one oxidation state is both oxidized and reduced.)

18.93 A galvanic cell with $E°_{cell} = 0.30$ V can be constructed using an Fe electrode in a 1.0 M $Fe(NO_3)_2$ solution, and either a Sn electrode in a 1.0 M $Sn(NO_3)_2$ solution or a Cr electrode in a 1.0 M $Cr(NO_3)_3$ solution, even though Sn^{2+}/Sn and Cr^{3+}/Cr have different standard reduction potentials. Explain.

18.94 Shown here is a galvanic cell connected to an electrolytic cell. Label the electrodes (anodes and cathodes) and show the movement of electrons along the wires and cations and anions in solution. For simplicity, the salt bridge is not shown for the galvanic cell.

18.95 Fluorine (F_2) is obtained by the electrolysis of liquid hydrogen fluoride (HF) containing potassium fluoride (KF). (a) Write the half-cell reactions and the overall reaction for the process. (b) What is the purpose of KF? (c) Calculate the volume of F_2 (in liters) collected at 24.0°C and 1.2 atm after electrolyzing the solution for 15 h at a current of 502 A.

18.96 A 300-mL solution of NaCl was electrolyzed for 6.00 min. If the pH of the final solution was 12.24, calculate the average current used.

18.97 Industrially, copper is purified by electrolysis. The impure copper acts as the anode, and the cathode is made of pure copper. The electrodes are immersed in a $CuSO_4$ solution. During electrolysis, copper at the anode enters the solution as Cu^{2+} while Cu^{2+} ions are reduced at the cathode. (a) Write half-cell reactions and the overall reaction for the electrolytic process. (b) Suppose the anode was contaminated with Zn and Ag. Explain what happens to these impurities during electrolysis. (c) How many hours will it take to obtain 1.00 kg of Cu at a current of 18.9 A?

18.98 An aqueous solution of a platinum salt is electrolyzed at a current of 2.50 A for 2.00 h. As a result, 9.09 g of metallic Pt are formed at the cathode. Calculate the charge on the Pt ions in this solution.

18.99 Consider a galvanic cell consisting of a magnesium electrode in contact with 1.0 *M* $Mg(NO_3)_2$ and a cadmium electrode in contact with 1.0 *M* $Cd(NO_3)_2$. Calculate $E°$ for the cell, and draw a diagram showing the cathode, anode, and direction of electron flow.

18.100 A current of 6.00 A passes through an electrolytic cell containing dilute sulfuric acid for 3.40 h. If the volume of O_2 gas generated at the anode is 4.26 L (at STP), calculate the charge (in coulombs) on an electron.

18.101 Gold will not dissolve in either concentrated nitric acid or concentrated hydrochloric acid. However, the metal does dissolve in a mixture of the acids (one part HNO_3 and three parts HCl by volume), called *aqua regia*. (a) Write a balanced equation for this reaction. (*Hint:* Among the products are $HAuCl_4$ and NO_2.) (b) What is the function of HCl?

18.102 Explain why most useful galvanic cells give voltages of no more than 1.5 to 2.5 V. What are the prospects for developing practical galvanic cells with voltages of 5 V or more?

18.103 The table here shows the standard reduction potentials of several half-reactions:

Half-Reactions	$E°$ (V)
$A^{2+} + 2e^- \longrightarrow A$	−1.46
$B_2 + 2e^- \longrightarrow 2B^-$	0.33
$C^{3+} + 3e^- \longrightarrow C$	1.13
$D^+ + e^- \longrightarrow D$	−0.87

(a) Which is the strongest oxidizing agent and which is the strongest reducing agent? (b) Which substances can be oxidized by B_2? (c) Which substances can be reduced by B^-? (d) Write the overall equation for a cell that delivers a voltage of 2.59 V under standard-state conditions.

18.104 Consider a concentration cell made of the following two compartments: Cl_2(0.20 atm) | Cl^-(1.0 *M*) and Cl_2(2.0 atm) | Cl^-(1.0 *M*). Platinum is used as the inert electrodes. Draw a cell diagram for the cell and calculate the emf of the cell at 25°C.

18.105 A silver rod and a SHE are dipped into a saturated aqueous solution of silver oxalate, $Ag_2C_2O_4$, at 25°C. The measured potential difference between the rod and the SHE is 0.589 V, the rod being positive. Calculate the solubility product constant for silver oxalate.

18.106 Zinc is an amphoteric metal; that is, it reacts with both acids and bases. The standard reduction potential is −1.36 V for the reaction

$$Zn(OH)_4^{2-}(aq) + 2e^- \longrightarrow Zn(s) + 4OH^-(aq)$$

Calculate the formation constant (K_f) for the reaction

$$Zn^{2+}(aq) + 4OH^-(aq) \rightleftharpoons Zn(OH)_4^{2-}(aq)$$

18.107 Use the data in Table 18.1 to determine whether or not hydrogen peroxide will undergo disproportionation in an acid medium: $2H_2O_2 \longrightarrow 2H_2O + O_2$.

18.108 The magnitudes (but *not* the signs) of the standard reduction potentials of two metals X and Y are

$$Y^{2+} + 2e^- \longrightarrow Y \qquad |E°| = 0.34\ V$$
$$X^{2+} + 2e^- \longrightarrow X \qquad |E°| = 0.25\ V$$

where the | | notation denotes that only the magnitude (but not the sign) of the $E°$ value is shown. When the half-cells of X and Y are connected, electrons flow from X to Y. When X is connected to a SHE, electrons flow from X to SHE. (a) Are the $E°$ values of the half-reactions positive or negative? (b) What is the standard emf of a cell made up of X and Y?

18.109 A galvanic cell is constructed as follows. One half-cell consists of a platinum wire immersed in a solution containing 1.0 *M* Sn^{2+} and 1.0 *M* Sn^{4+}; the other half-cell has a thallium rod immersed in a solution of 1.0 *M* Tl^{+}. (a) Write the half-cell reactions and the overall reaction. (b) What is the equilibrium constant at 25°C? (c) What is the cell voltage if the Tl^{+} concentration is increased 10-fold? ($E^\circ_{Tl^+/Tl} = -0.34$ V.)

18.110 Given the standard reduction potentials for Au^{3+} and Au^{+} in Table 18.1, answer the following questions. (a) Why does gold not tarnish in air? (b) Will the following disproportionation occur spontaneously?

$$3Au^+(aq) \longrightarrow Au^{3+}(aq) + 2Au(s)$$

(c) Predict the reaction between gold and fluorine gas.

18.111 The ingestion of a very small quantity of mercury is not considered too harmful. Would this statement still hold if the gastric juice in your stomach were mostly nitric acid instead of hydrochloric acid?

18.112 When 25.0 mL of a solution containing both Fe^{2+} and Fe^{3+} ions is titrated with 23.0 mL of 0.0200 *M* $KMnO_4$ (in dilute sulfuric acid), all of the Fe^{2+} ions are oxidized to Fe^{3+} ions. Next, the solution is treated with Zn metal to convert all of the Fe^{3+} ions to Fe^{2+} ions. Finally, 40.0 mL of the same $KMnO_4$ solution are added to the solution to oxidize the Fe^{2+} ions to Fe^{3+}. Calculate the molar concentrations of Fe^{2+} and Fe^{3+} in the original solution.

18.113 Consider the Daniell cell in Figure 18.1. When viewed externally, the anode appears negative and the cathode positive (electrons are flowing from the anode to the cathode). Yet in solution anions are moving toward the anode, which means that it must appear positive to the anions. Because the anode cannot simultaneously be negative and positive, give an explanation for this apparently contradictory situation.

18.114 Use the data in Table 18.1 to show that the decomposition of H_2O_2 (a disproportionation reaction) is spontaneous at 25°C:

$$2H_2O_2(aq) \longrightarrow 2H_2O(l) + O_2(g)$$

18.115 Consider two electrolytic cells A and B. Cell A contains a 0.20 *M* $CoSO_4$ solution and platinum electrodes. Cell B differs from cell A only in that cobalt metals are used as electrodes. In each case, a current of 0.20 A is passed through the cell for 1.0 h. (a) Write equations for the half-cell and overall cell reactions for these cells. (b) Calculate the products formed (in grams) at the anode and cathode in each case.

18.116 A galvanic cell consists of a Mg electrode in a 1 *M* $Mg(NO_3)_2$ solution and another metal electrode X in a 1 *M* $X(NO_3)_2$ solution. Listed here are the E°_{cell} values of four such galvanic cells. In each case, identify X from Table 18.1. (a) E°_{cell} = 2.11 V, (b) E°_{cell} = 2.24 V, (c) E°_{cell} = 1.61 V, (d) E°_{cell} = 1.92 V.

18.117 The concentration of sulfuric acid in the lead-storage battery of an automobile over a period of time has decreased from 38.0 percent by mass (density = 1.29 g/mL) to 26.0 percent by mass (1.19 g/mL). Assume the volume of the acid remains constant at 724 mL. (a) Calculate the total charge in coulombs supplied by the battery. (b) How long (in hours) will it take to recharge the battery back to the original sulfuric acid concentration using a current of 22.4 amperes?

18.118 Consider a Daniell cell operating under nonstandard-state conditions. Suppose that the cell's reaction is multiplied by 2. What effect does this have on each of the following quantities in the Nernst equation: (a) *E*, (b) E°, (c) *Q*, (d) ln *Q*, (e) *n*?

18.119 An electrolysis cell was constructed similar to the one shown in Figure 18.18, except 0.1 *M* $MgCl_2(aq)$ was used as the electrolyte solution. Under these conditions, a clear gas was formed at one electrode and a very pale green gas was formed at the other electrode in roughly equal volumes. (a) What gases are formed at these electrodes? (b) Write balanced half-reactions for each electrode. Account for any deviation from the normally expected results.

18.120 Comment on whether F_2 will become a stronger oxidizing agent with increasing H^+ concentration.

18.121 In recent years there has been much interest in electric cars. List some advantages and disadvantages of electric cars compared to automobiles with internal combustion engines.

18.122 Calculate the pressure of H_2 (in atm) required to maintain equilibrium with respect to the following reaction at 25°C:

$$Pb(s) + 2H^+(aq) \rightleftharpoons Pb^{2+}(aq) + H_2(g)$$

Given that $[Pb^{2+}]$ = 0.035 *M* and the solution is buffered at pH 1.60.

18.123 A piece of magnesium ribbon and a copper wire are partially immersed in a 0.1 *M* HCl solution in a beaker. The metals are joined externally by another piece of metal wire. Bubbles are seen to evolve at both the Mg and Cu surfaces. (a) Write equations representing the reactions occurring at the metals. (b) What visual evidence would you seek to show that Cu is not oxidized to Cu^{2+}? (c) At some stage, NaOH solution is added to the beaker to neutralize the HCl acid. Upon further addition of NaOH, a white precipitate forms. What is it?

18.124 The zinc-air battery shows much promise for electric cars because it is lightweight and rechargeable:

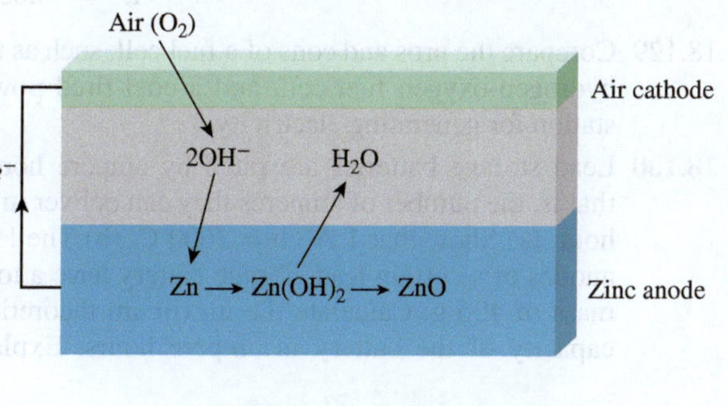

The net transformation is $Zn(s) + \frac{1}{2}O_2(g) \longrightarrow ZnO(s)$. (a) Write the half-reactions at the zinc-air electrodes and calculate the standard emf of the battery at 25°C. (b) Calculate the emf under actual operating conditions when the partial pressure of oxygen is 0.21 atm. (c) What is the energy density (measured as the energy in kilojoules that can be obtained from 1 kg of the metal) of the zinc electrode? (d) If a current of 2.1×10^5 A is to be drawn from a zinc-air battery system, what volume of air (in liters) would need to be supplied to the battery every second? Assume that the temperature is 25°C and the partial pressure of oxygen is 0.21 atm.

18.125 Calculate $E°$ for the reactions of mercury with (a) 1 *M* HCl and (b) 1 *M* HNO_3. Which acid will oxidize Hg to Hg_2^{2+} under standard-state conditions? Can you identify which test tube shown contains HNO_3 and Hg and which contains HCl and Hg?

Ken Karp/McGraw Hill

18.126 Because all alkali metals react with water, it is not possible to measure the standard reduction potentials of these metals directly as in the case of, say, zinc. An indirect method is to consider the following hypothetical reaction:

$$Li^+(aq) + \tfrac{1}{2}H_2(g) \longrightarrow Li(s) + H^+(aq)$$

Using the appropriate equation presented in this chapter and the thermodynamic data in Appendix 2, calculate $E°$ for $Li^+(aq) + e^- \longrightarrow Li(s)$ at 298 K. Compare your result with that listed in Table 18.1. (See Appendix 6 for the Faraday constant.)

18.127 A galvanic cell using Mg/Mg^{2+} and Cu/Cu^{2+} half-cells operates under standard-state conditions at 25°C and each compartment has a volume of 218 mL. The cell delivers 0.22 A for 31.6 h. (a) How many grams of Cu are deposited? (b) What is the $[Cu^{2+}]$ remaining?

18.128 Given the following standard reduction potentials, calculate the ion-product, K_w, for water at 25°C:

$$2H^+(aq) + 2e^- \longrightarrow H_2(g) \qquad E° = 0.00\ V$$

$$2H_2O(l) + 2e^- \longrightarrow H_2(g) + 2OH^-(aq) \qquad E° = -0.83\ V$$

18.129 Compare the pros and cons of a fuel cell, such as the hydrogen-oxygen fuel cell, and a coal-fired power station for generating electricity.

18.130 Lead storage batteries are rated by ampere hours, that is, the number of amperes they can deliver in an hour. (a) Show that 1 A · h = 3600 C. (b) The lead anodes of a certain lead-storage battery have a total mass of 406 g. Calculate the maximum theoretical capacity of the battery in ampere hours. Explain why in practice we can never extract this much energy from the battery. (*Hint:* Assume all of the lead will be used up in the electrochemical reaction and refer to the electrode reactions in Section 18.6.) (c) Calculate $E°_{cell}$ and $\Delta G°$ for the battery.

18.131 Use Equations (17.10) and (18.3) to calculate the emf values of the Daniell cell at 25°C and 80°C. Comment on your results. What assumptions are used in the derivation? (*Hint:* You need the thermodynamic data in Appendix 2.)

18.132 A construction company is installing an iron culvert (a long cylindrical tube) that is 40.0 m long with a radius of 0.900 m. To prevent corrosion, the culvert must be galvanized. This process is carried out by first passing an iron sheet of appropriate dimensions through an electrolytic cell containing Zn^{2+} ions, using graphite as the anode and the iron sheet as the cathode. If the voltage is 3.26 V, what is the cost of electricity for depositing a layer 0.200 mm thick if the efficiency of the process is 95 percent? The electricity rate is \$0.12 per kilowatt hour (kWh), where 1 W = 1 J/s and the density of Zn is 7.14 g/cm³.

18.133 A 9.00×10^2-mL 0.200 *M* MgI_2 was electrolyzed. As a result, hydrogen gas was generated at the cathode and iodine was formed at the anode. The volume of hydrogen collected at 26°C and 779 mmHg was 1.22×10^3 mL. (a) Calculate the charge in coulombs consumed in the process. (b) How long (in min) did the electrolysis last if a current of 7.55 A was used? (c) A white precipitate was formed in the process. What was it and what was its mass in grams? Assume the volume of the solution was constant.

18.134 Based on the following standard reduction potentials:

$$Fe^{2+}(aq) + 2e^- \longrightarrow Fe(s) \qquad E°_1 = -0.45\ V$$

$$Fe^{3+}(aq) + e^- \longrightarrow Fe^{2+}(aq) \qquad E°_2 = 0.77\ V$$

calculate the standard reduction potential for the half-reaction

$$Fe^{3+}(aq) + 3e^- \longrightarrow Fe(s) \qquad E°_3 = ?$$

18.135 A galvanic cell is constructed by immersing a piece of copper wire in 25.0 mL of a 0.20 *M* $CuSO_4$ solution and a zinc strip in 25.0 mL of a 0.20 *M* $ZnSO_4$ solution. (a) Calculate the emf of the cell at 25°C and predict what would happen if a small amount of concentrated NH_3 solution were added to (i) the $CuSO_4$ solution and (ii) the $ZnSO_4$ solution. Assume that the volume in each compartment remains constant at 25.0 mL. (b) In a separate experiment, 25.0 mL of 3.00 *M* NH_3 are added to the $CuSO_4$ solution. If the emf of the cell is 0.68 V, calculate the formation constant (K_f) of $Cu(NH_3)_4^{2+}$.

18.136 Calculate the equilibrium constant for the following reaction at 298 K:

$$Zn(s) + Cu^{2+}(aq) \longrightarrow Zn^{2+}(aq) + Cu(s)$$

18.137 To remove the tarnish (Ag_2S) on a silver spoon, a student carried out the following steps. First, she placed the spoon in a large pan filled with water so the spoon was totally immersed. Next, she added a few tablespoonful of baking soda (sodium bicarbonate), which readily dissolved. Finally, she placed some aluminum foil at the bottom of the pan in contact with the spoon and then heated the solution to about 80°C. After a few minutes, the spoon was removed and rinsed with cold water. The tarnish was gone and the spoon regained its original shiny appearance. (a) Describe with equations the electrochemical basis for the procedure. (b) Adding NaCl instead of $NaHCO_3$ would also work because both compounds are strong electrolytes. What is the added advantage of using $NaHCO_3$? (*Hint:* Consider the pH of the solution.) (c) What is the purpose of heating the solution? (d) Some commercial tarnish removers contain a fluid (or paste) that is a dilute HCl solution. Rubbing the spoon with the fluid will also remove the tarnish. Name two disadvantages of using this procedure compared to the one described earlier.

18.138 The nitrite ion (NO_2^-) in soil is oxidized to nitrate ion (NO_3^-) by the bacteria *Nitrobacter agilis* in the presence of oxygen. The half-reduction reactions are

$$NO_3^- + 2H^+ + 2e^- \longrightarrow NO_2^- + H_2O \quad E° = 0.42\ V$$
$$O_2 + 4H^+ + 4e^- \longrightarrow 2H_2O \quad E° = 1.23\ V$$

Calculate the yield of ATP synthesis per mole of nitrite oxidized. (*Hint:* See Section 17.6.)

18.139 The diagram here shows an electrolytic cell consisting of a Co electrode in a 2.0 *M* $Co(NO_3)_2$ solution and a Mg electrode in a 2.0 *M* $Mg(NO_3)_2$ solution. (a) Label the anode and cathode and show the half-cell reactions. Also label the signs (+ or −) on the battery terminals. (b) What is the minimum voltage to drive the reaction? (c) After the passage of 10.0 A for 2.00 h the battery is replaced with a voltmeter and the electrolytic cell now becomes a galvanic cell. Calculate E_{cell}. Assume volumes to remain constant at 1.00 L in each compartment.

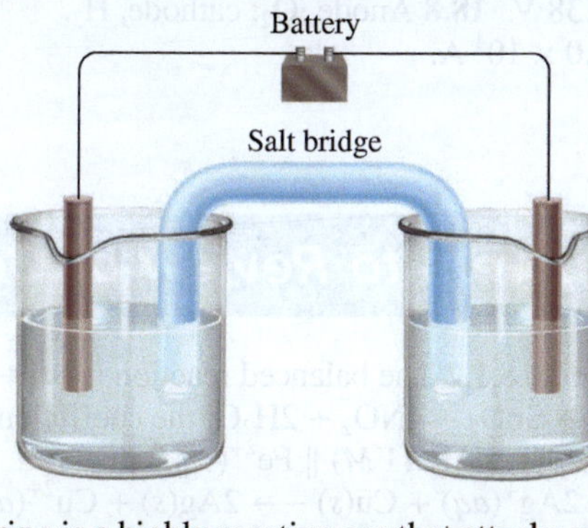

18.140 Fluorine is a highly reactive gas that attacks water to form HF and other products. Follow the procedure in Problem 18.126 to show how you can determine indirectly the standard reduction for fluorine as shown in Table 18.1.

18.141 Show a sketch of a galvanic concentration cell. Each compartment consists of a Co electrode in a $Co(NO_3)_2$ solution. The concentrations in the compartments are 2.0 *M* and 0.10 *M*, respectively. Label the anode and cathode compartments. Show the direction of electron flow. (a) Calculate the E_{cell} at 25°C. (b) What are the concentrations in the compartments when the E_{cell} drops to 0.020 V? Assume volumes to remain constant at 1.00 L in each compartment.

Interpreting, Modeling, & Estimating

18.142 The emf of galvanic cells varies with temperature (either increases or decreases). Starting with Equation (18.3), derive an equation that expresses $E°_{cell}$ in terms of $\Delta H°$ and $\Delta S°$. Predict whether $E°_{cell}$ will increase or decrease if the temperature of a Daniell cell increases. Assume both $\Delta H°$ and $\Delta S°$ to be temperature independent.

18.143 A concentration cell ceases to operate when the concentrations of the two cell compartments are equal. At this stage, is it possible to generate an emf from the cell by adjusting another parameter without changing the concentrations? Explain.

18.144 It has been suggested that a car can be powered from the hydrogen generated by reacting aluminum soda cans with a solution of lye (sodium hydroxide) according to the following reaction:

$$2Al(s) + 2OH^-(aq) + 6H_2O(l) \longrightarrow 2Al(OH)_4^-(aq) + 3H_2(g)$$

How many aluminum soda cans would be required to generate the same amount of chemical energy as contained in one tank of gasoline? Read the Chemistry in Action on aluminum recycling in Section 21.7, and comment on the cost and environmental impact of powering a car with aluminum cans.

18.145 Estimate how long it would take to electroplate a teaspoon with silver from a solution of $AgNO_3$, assuming a constant current of 2 A.

18.146 The potential for a cell based on the standard hydrogen electrode and the half-reaction

$$M^{n+}(aq) + ne^- \longrightarrow M(s)$$

was measured at several concentrations of $M^{n+}(aq)$, giving the following plot. What is the value of *n* in the half-reaction?

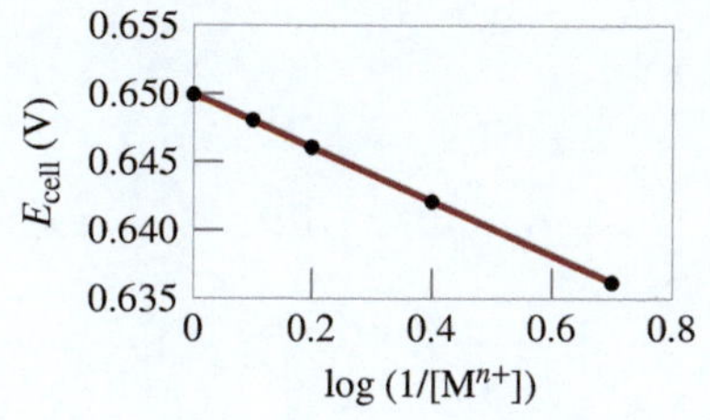

Answers to Practice Exercises

18.1 $5Fe^{2+} + MnO_4^- + 8H^+ \longrightarrow 5Fe^{3+} + Mn^{2+} + 4H_2O$. **18.2** No. **18.3** 0.34 V. **18.4** 1×10^{-42}. **18.5** $\Delta G° = -4.1 \times 10^2$ kJ/mol. **18.6** Yes, $E = +0.01$ V. **18.7** 0.38 V. **18.8** Anode, O_2; cathode, H_2. **18.9** 2.0×10^4 A.

Answers to Review of Concepts & Facts

18.1.1 6. **18.1.2** The balanced reaction is $Sn + 4NO_3^- + 4H^+ \longrightarrow SnO_2 + 4NO_2 + 2H_2O$; the coefficient is 4. **18.2.1** $Al(s) \mid Al^{3+}(1\ M) \parallel Fe^{2+}(1\ M) \mid Fe(s)$. **18.2.2** $2Ag^+(aq) + Cu(s) \longrightarrow 2Ag(s) + Cu^{2+}(aq)$. **18.3.1** Cu, Ag. **18.3.2** 1.99 V. **18.4.1** 6.5×10^{-10}. **18.4.2** 0.049 V. **18.4.3** It is much easier to determine the equilibrium constant electrochemically. It is only necessary to measure the emf of the cell and then use Equations (18.3) and (17.14) to calculate K. In contrast, use of Equation (17.14) alone requires measurements of both $\Delta H°$ and $\Delta S°$ to first determine $\Delta G°$ and then K. Note though that most reactions do not lend themselves to electrochemical measurements. **18.5.1** 2.13 V. (a) 2.14 V. (b) 2.11 V. **18.5.2** 0.049 V. **18.6.1** 9 V/(1.5 V/cell) = 6 cells. **18.7.1** Sr. **18.8.1** 1.23 V. **18.8.2** Anode: $2Cl^- \longrightarrow Cl_2 + 2e^-$. Cathode: $Mg^{2+} + 2e^- \longrightarrow Mg$. In a galvanic cell, the anode is labeled negative because it supplies electrons to the external circuit. In an electrolytic cell, the anode is labeled positive because electrons are withdrawn from it by the battery. The sign of each electrode in the electrolytic cell is the same as the sign of the battery electrode to which it is attached. **18.8.3** 1.9 g.

CHAPTER

19 Nuclear Chemistry

The Large Hadron Collider (LHC) is the largest particle accelerator in the world. By colliding protons moving at nearly the speed of light, scientists hope to create conditions that existed right after the Big Bang.
Xenotar/iStockphoto/Getty Images

CHAPTER OUTLINE

Nuclear chemistry is the study of reactions involving changes in atomic nuclei. This branch of chemistry began with the discovery of natural radioactivity by Antoine Becquerel and grew as a result of subsequent investigations by Pierre and Marie Curie and many others. Nuclear chemistry is very much in the news today. In addition to applications in the manufacture of atomic bombs, hydrogen bombs, and neutron bombs, even the peaceful use of nuclear energy has become controversial, in part because of safety concerns about nuclear power plants and because of problems with radioactive waste disposal. In this chapter, we will study nuclear reactions, the stability of the atomic nucleus, radioactivity, and the effects of radiation on biological systems.

19.1 The Nature of Nuclear Reactions

Learning Objectives

- Distinguish the particles or types of radiation that an unstable nucleus can produce.
- Differentiate between chemical and nuclear reactions.
- Demonstrate use of subatomic particles in balancing nuclear reactions.

With the exception of hydrogen (^1_1H), all nuclei contain two kinds of fundamental particles, called *protons* and *neutrons.* Some nuclei are unstable; they emit particles and/or electromagnetic radiation spontaneously (see Section 2.2). The name for this phenomenon is *radioactivity.* All elements having an atomic number greater than 83 are radioactive. For example, the isotope of polonium, polonium-210 ($^{210}_{84}Po$), decays spontaneously to $^{206}_{82}Pb$ by emitting an α particle.

Another type of radioactivity, known as ***nuclear transmutation,*** *results from the bombardment of nuclei by neutrons, protons, or other nuclei.* An example of a nuclear transmutation is the conversion of atmospheric $^{14}_{7}N$ to $^{14}_{6}C$ and ^1_1H, which results when the nitrogen isotope captures a neutron (from the sun). In some cases, heavier elements are synthesized from lighter elements. This type of transmutation occurs naturally in outer space, but it can also be achieved artificially, as we will see in Section 19.4.

Radioactive decay and nuclear transmutation are *nuclear reactions,* which differ significantly from ordinary chemical reactions. Table 19.1 summarizes the differences.

Balancing Nuclear Equations

To discuss nuclear reactions in any depth, we need to understand how to write and balance the equations. Writing a nuclear equation differs somewhat from writing

Table 19.1 Comparison of Chemical Reactions and Nuclear Reactions

	Chemical Reactions	Nuclear Reactions
Process	Atoms are rearranged by breaking and forming chemical bonds	Elements (or isotopes of the same element) are converted into other elements
Particles involved	Only valence electrons in atomic or molecular orbitals are involved	Protons, neutrons, electrons, and other nuclear particles may be involved
Energy involved	Absorption or release of small amounts of energy (typically kJ/mol)	Absorption or release of tremendous amounts of energy (typically GJ/mol or higher)
Reaction rate	Rates affected by temperature, pressure, concentration, and catalysts	Rates not affected by temperature, pressure, concentration, or catalysts

Table 19.2 Nuclear Particles and Nuclear Radiation

Type	Particle*	Mass Number	Charge
Proton	p, ${}^{1}_{1}\text{p}$	1	+1
Neutron	n, ${}^{1}_{0}\text{n}$	1	0
Electron	e, ${}^{0}_{-1}e$	0	−1
Alpha particle	α, ${}^{4}_{2}\alpha$	4	+2
Beta particle	β, ${}^{0}_{-1}\beta$	0	−1
Gamma ray	γ	0	0
Positron	β^{+}, ${}^{0}_{+1}\beta$	0	+1

*Alternative symbols for the various particles are also given.

equations for chemical reactions. In addition to writing the symbols for various chemical elements, we must also explicitly indicate protons, neutrons, and electrons. In fact, we must show the numbers of protons and neutrons present in *every* species in such an equation.

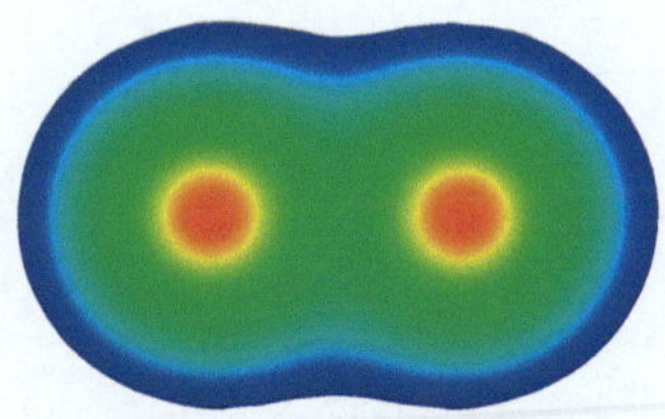

A positron is the antiparticle of the electron. In 2007 physicists prepared dipositronium (Ps_2), which contains only electrons and positrons. The diagram here shows the central nuclear positions containing positrons (red) surrounded by electrons (green). The Ps_2 species exists for less than a nanosecond before the electron and positron annihilate each other with the emission of γ rays.

A summary of the elementary particles commonly involved in nuclear reactions is provided in Table 19.2. We have already discussed the three fundamental particles of an atom—the proton, the neutron, and the electron—in Section 2.2.

In accordance with the notation used in Section 2.3, the superscript in each case denotes the mass number (the total number of neutrons and protons present) and the subscript is the atomic number (the number of protons). Thus, the "atomic number" of a proton is 1, because there is one proton present, and the "mass number" is also 1, because there is one proton but no neutrons present. On the other hand, the "mass number" of a neutron is 1, but its "atomic number" is zero, because there are no protons present. For the electron, the "mass number" is zero (there are neither protons nor neutrons present), but the "atomic number" is −1, because the electron possesses a unit negative charge.

In Section 2.2, we also introduced the three common types of radiation. The α particle has two protons and two neutrons, so its atomic number is 2 and its mass number is 4; the α particle is a helium-4 nucleus. A β particle has the same physical characteristics as an electron; however, we use the symbol ${}^{0}_{-1}e$ to represent an electron in or from an atomic orbital. The symbol ${}^{0}_{-1}\beta$ represents an electron that comes from a nucleus (in a decay process in which a neutron is converted to a proton and an electron) and not from an atomic orbital. A γ ray is a high-energy photon (see Figure 7.4) and thus has no mass or charge.

Antimatter, matter with mass equal to normal matter but with an opposite charge, was predicted in the 1920s and finally discovered in 1932 with the detection of the positron, the first antiparticle. The ***positron*** *has the same mass as the electron, but bears a + 1 charge.* Because of its relationship with the electron, we use the β symbol but designate the positive charge with a +. While we will not discuss them here, there are many other forms of antimatter, including antiprotons and antineutrons.

In balancing any nuclear equation, we observe the following rules:

- The total number of protons plus neutrons in the products and in the reactants must be the same (conservation of mass number).
- The total number of nuclear charges in the products and in the reactants must be the same (conservation of atomic number).

We can summarize these rules in the following manner:

$$^{\text{Total mass}}_{\text{Total charge}}\text{Reactants} = {}^{\text{Total mass}}_{\text{Total charge}}\text{Products}$$

If we know the atomic numbers and mass numbers of all the species but one in a nuclear equation, we can identify the unknown species by applying these rules, as shown in Example 19.1, which illustrates how to balance nuclear decay equations.

Student Hot Spot

Student data indicate you may struggle with balancing nuclear equations. Access your eBook for additional Learning Resources on this topic.

Example 19.1

Balance the following nuclear equations (that is, identify the product X):

(a) $^{212}_{84}\text{Po} \longrightarrow {}^{208}_{82}\text{Pb} + \text{X}$

(b) $^{137}_{55}\text{Cs} \longrightarrow {}^{137}_{56}\text{Ba} + \text{X}$

Strategy In balancing nuclear equations, note that the sum of atomic numbers and that of mass numbers must match on both sides of the equation.

Solution

(a) The mass number and atomic number are 212 and 84, respectively, on the left-hand side and 208 and 82, respectively, on the right-hand side. Thus, X must have a mass number of 4 and an atomic number of 2, which means that it is an α particle. The balanced equation is

$$^{212}_{84}\text{Po} \longrightarrow {}^{208}_{82}\text{Pb} + {}^{4}_{2}\alpha$$

(b) In this case, the mass number is the same on both sides of the equation, but the atomic number of the product is 1 more than that of the reactant. Thus, X must have a mass number of 0 and an atomic number of −1, which means that it is a β particle. The only way this change can come about is to have a neutron in the Cs nucleus transformed into a proton and an electron (which is why we use the ${}^{0}_{-1}\beta$ symbol); that is, ${}^{1}_{0}\text{n} \longrightarrow {}^{1}_{1}\text{p} + {}^{0}_{-1}\beta$ (note that this process does not alter the mass number). Thus, the balanced equation is

$$^{137}_{55}\text{Cs} \longrightarrow {}^{137}_{56}\text{Ba} + {}^{0}_{-1}\beta$$

Check Note that the equations in (a) and (b) are balanced for nuclear particles but not for electrical charges. To balance the charges, we would need to add two electrons on the right-hand side of (a) and express barium as a cation (Ba^{+}) in (b).

Practice Exercise Identify X in the following nuclear equation:

$$^{78}_{33}\text{As} \longrightarrow {}^{0}_{-1}\beta + \text{X}$$

Similar problems: 19.7, 19.8.

Summary of Concepts & Facts

- Radioactive nuclei emit α particles, β particles, positrons, or γ rays. The equation for a nuclear reaction includes the particles emitted, and both the mass numbers and the atomic numbers must balance.

Review of Concepts & Facts

19.1.1 Write a nuclear equation depicting the formation of a positron from a proton.

19.1.2 Identify X in the following nuclear equation:

$$^{235}_{92}\text{U} + {}^{1}_{0}\text{n} \longrightarrow {}^{144}_{55}\text{Cs} + 3{}^{1}_{0}\text{n} + \text{X}$$

19.1.3 $^{55}_{25}\text{Mn}$ is prepared by addition of an electron to what nuclear species?

19.2 Nuclear Stability

Learning Objectives

- Summarize the rules for nuclear stability and use them to predict whether a particular nucleus is stable.
- Explain nuclear binding energy.
- Calculate the nuclear binding energy of a nucleus.

The nucleus occupies a very small portion of the total volume of an atom, but it contains most of the atom's mass because both the protons and the neutrons reside there. In studying the stability of the atomic nucleus, it is helpful to know something about its density, because it tells us how tightly the particles are packed together. As a sample calculation, let us assume that a nucleus has a radius of 5×10^{-3} pm and a mass of 1×10^{-22} g. These figures correspond roughly to a nucleus containing 30 protons and 30 neutrons. Density is mass/volume, and we can calculate the volume from the known radius (the volume of a sphere is $\frac{4}{3}\pi r^3$, where r is the radius of the sphere). First, we convert the pm units to cm. Then we calculate the density in g/cm^3:

$$r = 5 \times 10^{-3} \text{ pm} \times \frac{1 \times 10^{-12} \text{ m}}{1 \text{ pm}} \times \frac{100 \text{ cm}}{1 \text{ m}} = 5 \times 10^{-13} \text{ cm}$$

$$\text{density} = \frac{\text{mass}}{\text{volume}} = \frac{1 \times 10^{-22} \text{ g}}{\frac{4}{3}\pi r^3} = \frac{1 \times 10^{-22} \text{ g}}{\frac{4}{3}\pi (5 \times 10^{-13} \text{ cm})^3}$$

$$= 2 \times 10^{14} \text{ g/cm}^3$$

This is an exceedingly high density. The highest density known for an element is 22.6 g/cm^3, for osmium (Os). Thus, the average atomic nucleus is roughly 9×10^{12} (or 9 trillion) times more dense than the densest element known!

The enormously high density of the nucleus prompts us to wonder what holds the particles together so tightly. From electrostatic interaction (Coulomb's law), we know that like charges repel and opposite charges attract one another. We would thus expect the protons to repel one another strongly, particularly when we consider how close they must be to each other. This indeed is so. However, in addition to the repulsion, there are also short-range attractions between proton and proton, proton and neutron, and neutron and neutron. This short-range force is known as the ***strong nuclear force***. The strong nuclear force is 137 times as strong as the repulsive force but operates only over distances of femtometers (10^{-15} m). The stability of any nucleus is determined by the difference between electrostatic repulsion and the strong nuclear force. If repulsion outweighs attraction, the nucleus is unstable and disintegrates, emitting particles and/or radiation. If attractive forces prevail, the nucleus is stable.

The principal factor that determines whether a nucleus is stable is the *neutron-to-proton ratio (n/p)*. For stable atoms of elements having low atomic number, the n/p value is close to 1. As the atomic number increases, the neutron-to-proton ratios of the stable nuclei become greater than 1. This deviation at higher atomic numbers arises because a larger number of neutrons is needed to counteract the strong repulsion among the protons and stabilize the nucleus. The following rules are useful in predicting nuclear stability:

1. Nuclei that contain 2, 8, 20, 50, 82, or 126 protons or neutrons are generally more stable than nuclei that do not possess these numbers. For example, there are 10 stable isotopes of tin (Sn) with the atomic number 50 and only 2 stable isotopes of antimony (Sb) with the atomic number 51. The numbers 2, 8, 20, 50, 82, and 126 are called *magic numbers*. The significance of these numbers for nuclear

Table 19.3 Number of Stable Isotopes with Even and Odd Numbers of Protons and Neutrons

Protons	Neutrons	Number of Stable Isotopes
Odd	Odd	4
Odd	Even	50
Even	Odd	53
Even	Even	164

stability is similar to the numbers of electrons associated with the very stable noble gases (that is, 2, 10, 18, 36, 54, and 86 electrons).

2. Nuclei with even numbers of both protons and neutrons are generally more stable than those with odd numbers of these particles (Table 19.3).
3. All isotopes of the elements with atomic numbers higher than 83 are radioactive. All isotopes of technetium (Tc, $Z = 43$) and promethium (Pm, $Z = 61$) are radioactive.

Figure 19.1 shows a plot of the number of neutrons versus the number of protons in various isotopes. The stable nuclei are located in an area of the graph known as the *belt of stability.* Most radioactive nuclei lie outside this belt. Above the stability belt, the nuclei have higher neutron-to-proton ratios than those within the belt (for the same number of protons). To lower this ratio (and hence move down toward the belt of stability), these nuclei undergo the following process, called *beta decay* (or *β-particle emission*):

Student Hot Spot

Student data indicate you may struggle with understanding the belt of stability. Access your eBook for additional Learning Resources on this topic.

$$^{1}_{0}\text{n} \longrightarrow {}^{1}_{1}\text{p} + {}^{0}_{-1}\beta$$

Beta decay leads to an increase in the number of protons in the nucleus and a simultaneous decrease in the number of neutrons. Some examples are

$$^{14}_{6}\text{C} \longrightarrow {}^{14}_{7}\text{N} + {}^{0}_{-1}\beta$$
$$^{40}_{19}\text{K} \longrightarrow {}^{40}_{20}\text{Ca} + {}^{0}_{-1}\beta$$
$$^{97}_{40}\text{Zr} \longrightarrow {}^{97}_{41}\text{Nb} + {}^{0}_{-1}\beta$$

As noted, heavy nuclei ($Z > 83$) are naturally radioactive due to the instability of the nucleus. These all lie above the belt of stability and are too heavy to be stable. Such nuclei undergo ***alpha decay,*** which decreases both the number of protons and neutrons. An example of alpha decay is

$$^{226}_{88}\text{Ra} \longrightarrow {}^{222}_{86}\text{Rn} + {}^{4}_{2}\alpha$$

Below the stability belt, the nuclei have lower neutron-to-proton ratios than those in the belt (for the same number of protons). To increase this ratio (and hence move up toward the belt of stability), these nuclei either do ***positron emission***

$$^{1}_{1}\text{p} \longrightarrow {}^{1}_{0}\text{n} + {}^{0}_{+1}\beta$$

or undergo ***electron capture.*** An example of positron emission is

$$^{38}_{19}\text{K} \longrightarrow {}^{38}_{18}\text{Ar} + {}^{0}_{+1}\beta$$

Student Hot Spot

Student data indicate you may struggle with positron emission. Access your eBook for additional Learning Resources on this topic.

Electron capture is the capture of an electron—usually a 1*s* electron—by the nucleus. We use ${}^{0}_{-1}e$ rather than ${}^{0}_{-1}\beta$ to show the particle in a nuclear equation because the electron came from an atomic orbital and not from the nucleus. The captured electron combines with a proton to form a neutron so that the atomic number decreases by

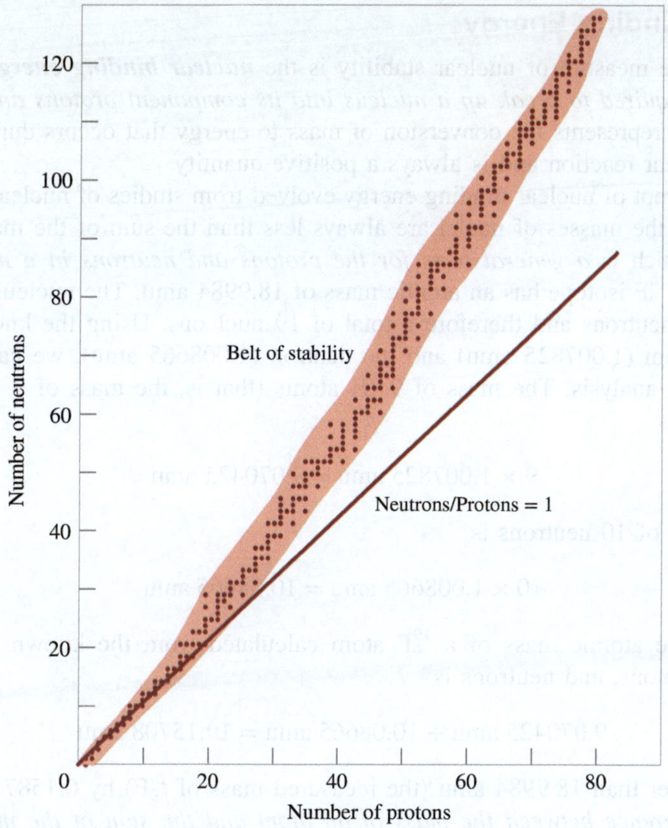

Figure 19.1 *Plot of neutrons versus protons for various stable isotopes, represented by dots. The straight line represents the points at which the neutron-to-proton ratio equals 1. The shaded area represents the belt of stability.*

one while the mass number remains the same. The emission of X rays (and another nearly massless particle known as a neutrino) accompanies electron capture as electrons from higher energy orbitals fill the orbital. This process has the same net effect as positron emission:

Student Hot Spot

Student data indicate you may struggle with electron capture. Access your eBook for additional Learning Resources on this topic.

$$^{37}_{18}Ar + ^{0}_{-1}e \longrightarrow ^{37}_{17}Cl$$

$$^{55}_{26}Fe + ^{0}_{-1}e \longrightarrow ^{55}_{25}Mn$$

The various types of nuclear decay are summarized in Table 19.4. One additional mode of decay is ***gamma emission,*** whereby gamma rays are emitted from an excited nucleus. Such emission of gamma radiation is analogous to an atom in an excited state emitting photons when the atom returns to the ground state. Many nuclear decays result in an excited state of a nucleus so gamma emission often occurs with other types of decay (particularly, β decay). An example of gamma emission is

$$^{99}_{43}Tc \longrightarrow ^{99}_{44}Ru + ^{0}_{-1}\beta + \gamma$$

Table 19.4 Summary of Radioactive Decay Modes

Mode	Emission	Net Decay Process
Alpha decay	α particle	α particle expelled from nucleus
Beta decay	β particle	$^{1}_{0}n \longrightarrow ^{1}_{1}p + ^{0}_{-1}\beta$
Positron emission	β^+ particle	$^{1}_{1}p \longrightarrow ^{1}_{0}n + ^{0}_{+1}\beta$
Electron capture	X rays	$^{0}_{-1}e + ^{1}_{1}p \longrightarrow ^{1}_{0}n$
Gamma emission	γ ray	excited nucleus $\longrightarrow$ stable nucleus

Nuclear Binding Energy

A quantitative measure of nuclear stability is the ***nuclear binding energy,*** which is *the energy required to break up a nucleus into its component protons and neutrons.* This quantity represents the conversion of mass to energy that occurs during an exothermic nuclear reaction and is always a positive quantity.

The concept of nuclear binding energy evolved from studies of nuclear properties showing that the masses of nuclei are always less than the sum of the masses of the ***nucleons,*** which is *a general term for the protons and neutrons in a nucleus.* For example, the $^{19}_{9}F$ isotope has an atomic mass of 18.9984 amu. The nucleus has 9 protons and 10 neutrons and therefore a total of 19 nucleons. Using the known masses of the $^{1}_{1}H$ atom (1.007825 amu) and the neutron (1.008665 amu), we can carry out the following analysis. The mass of 9 $^{1}_{1}H$ atoms (that is, the mass of 9 protons and 9 electrons) is

$$9 \times 1.007825 \text{ amu} = 9.070425 \text{ amu}$$

and the mass of 10 neutrons is

$$10 \times 1.008665 \text{ amu} = 10.08665 \text{ amu}$$

Therefore, the atomic mass of a $^{19}_{9}F$ atom calculated from the known numbers of electrons, protons, and neutrons is

$$9.070425 \text{ amu} + 10.08665 \text{ amu} = 19.15708 \text{ amu}$$

which is larger than 18.9984 amu (the measured mass of $^{19}_{9}F$) by 0.1587 amu.

The difference between the mass of an atom and the sum of the masses of its protons, neutrons, and electrons is called the ***mass defect.*** Relativity theory tells us that the loss in mass shows up as energy (heat) given off to the surroundings. Thus, the formation of $^{19}_{9}F$ is exothermic. Einstein's *mass-energy equivalence relationship* states that

$$E = mc^2 \qquad (19.1)$$

where E is energy, m is mass, and c is the speed of light. We can calculate the amount of energy released by writing

$$\Delta E = (\Delta m)c^2 \qquad (19.2)$$

where ΔE and Δm are defined as follows:

$$\Delta E = \text{energy of product} - \text{energy of reactants}$$
$$\Delta m = \text{mass of product} - \text{mass of reactants}$$

Thus, the change in mass is given by

$$\Delta m = 18.9984 \text{ amu} - 19.15708 \text{ amu}$$
$$= -0.1587 \text{ amu}$$

Because $^{19}_{9}F$ has a mass that is less than the mass calculated from the number of electrons and nucleons present, Δm is a negative quantity. Consequently, ΔE is also a negative quantity; that is, energy is released to the surroundings as a result of the formation of the fluorine-19 nucleus. So we calculate ΔE as follows:

$$\Delta E = (-0.1587 \text{ amu})(3.00 \times 10^8 \text{ m/s})^2$$
$$= -1.43 \times 10^{16} \text{ amu} \cdot \text{m}^2/\text{s}^2$$

With the conversion factors

$$1\ \text{kg} = 6.022 \times 10^{26}\ \text{amu}$$
$$1\ \text{J} = 1\ \text{kg} \cdot \text{m}^2/\text{s}^2$$

we obtain

$$\Delta E = \left(-1.43 \times 10^{16}\ \frac{\text{amu} \cdot \text{m}^2}{\text{s}^2}\right) \times \left(\frac{1.00\ \text{kg}}{6.022 \times 10^{26}\ \text{amu}}\right) \times \left(\frac{1\ \text{J}}{1\ \text{kg} \cdot \text{m}^2/\text{s}^2}\right)$$
$$= -2.37 \times 10^{-11}\ \text{J}$$

This is the amount of energy released when one fluorine-19 nucleus is formed from 9 protons and 10 neutrons. The nuclear binding energy of the nucleus is 2.37×10^{-11} J, which is the amount of energy needed to decompose the nucleus into separate protons and neutrons. In the formation of 1 mole of fluorine nuclei, for instance, the energy released is

$$\Delta E = (-2.37 \times 10^{-11}\ \text{J})(6.022 \times 10^{23}/\text{mol})$$
$$= -1.43 \times 10^{13}\ \text{J/mol}$$
$$= -1.43 \times 10^{10}\ \text{kJ/mol}$$

The nuclear binding energy, therefore, is 1.43×10^{10} kJ for 1 mole of fluorine-19 nuclei, which is a tremendously large quantity when we consider that the enthalpies of ordinary chemical reactions are of the order of only 200 kJ. The procedure we have followed can be used to calculate the nuclear binding energy of any nucleus.

As we have noted, nuclear binding energy is an indication of the stability of a nucleus. However, in comparing the stability of any two nuclei we must account for the fact that they have different numbers of nucleons. For this reason it is more meaningful to use the *nuclear binding energy per nucleon,* defined as

$$\text{nuclear binding energy per nucleon} = \frac{\text{nuclear binding energy}}{\text{number of nucleons}}$$

For the fluorine-19 nucleus,

$$\text{nuclear binding energy per nucleon} = \frac{2.37 \times 10^{-11}\ \text{J}}{19\ \text{nucleons}}$$
$$= 1.25 \times 10^{-12}\ \text{J/nucleon}$$

The nuclear binding energy per nucleon enables us to compare the stability of all nuclei on a common basis. Figure 19.2 shows the variation of nuclear binding energy

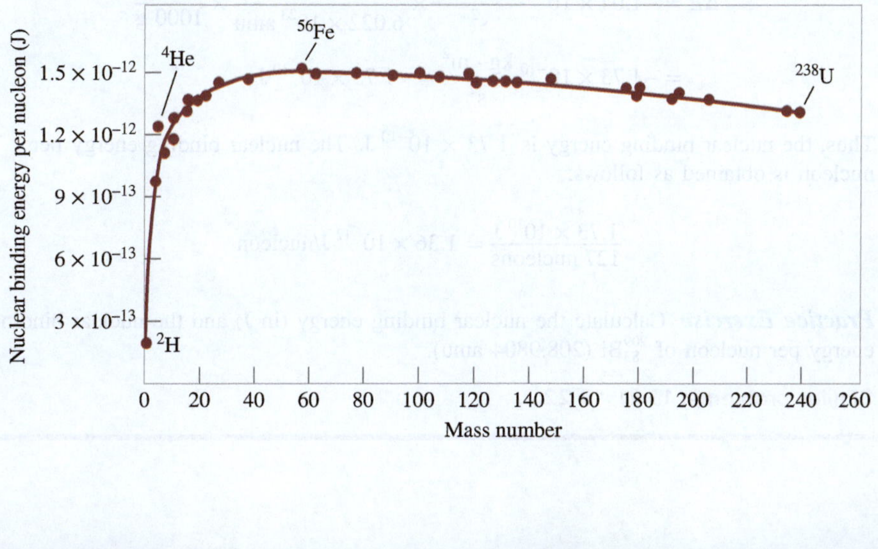

Figure 19.2 *Plot of nuclear binding energy per nucleon versus mass number.*

Student Hot Spot

Student data indicate you may struggle with nuclear binding energy. Access your eBook to view additional Learning Resources on this topic.

per nucleon plotted against mass number. As you can see, the curve rises rather steeply. The highest binding energies per nucleon belong to elements with intermediate mass numbers—between 40 and 100—and are greatest for elements in the iron, cobalt, and nickel region (the Group 8–10 elements) of the periodic table. This means that the *net* attractive forces among the particles (protons and neutrons) are greatest for the nuclei of these elements.

Nuclear binding energy and nuclear binding energy per nucleon are calculated for an iodine nucleus in Example 19.2.

Example 19.2

The atomic mass of $^{127}_{53}I$ is 126.9004 amu. Calculate the nuclear binding energy of this nucleus and the corresponding nuclear binding energy per nucleon.

Strategy To calculate the nuclear binding energy, we first determine the difference between the mass of the nucleus and the mass of all the protons and neutrons, which gives us the mass defect. Next, we apply Equation (19.2) [$\Delta E = (\Delta m)c^2$].

Solution There are 53 protons and 74 neutrons in the iodine nucleus. The neutron-to-proton ratio is 1.4, which places iodine-127 in the belt of stability. The mass of 53 ^1_1H atom is

$$53 \times 1.007825 \text{ amu} = 53.41473 \text{ amu}$$

and the mass of 74 neutrons is

$$74 \times 1.008665 \text{ amu} = 74.64121 \text{ amu}$$

Therefore, the predicted mass for $^{127}_{53}I$ is 53.41473 + 74.64121 = 128.05594 amu, and the mass defect is

$$\begin{aligned}\Delta m &= 126.9004 \text{ amu} - 128.05594 \text{ amu}\\ &= -1.1555 \text{ amu}\end{aligned}$$

The energy released is

$$\begin{aligned}\Delta E &= (\Delta m)c^2\\ &= (-1.1555 \text{ amu})(3.00 \times 10^8 \text{ m/s})^2\\ &= -1.04 \times 10^{17} \text{ amu} \cdot \text{m}^2/\text{s}^2\end{aligned}$$

Let's convert to a more familiar energy unit of joules. Recall that 1 J = 1 kg · m^2/s^2. Therefore, we need to convert amu to kg:

$$\begin{aligned}\Delta E &= -1.04 \times 10^{17} \frac{\cancel{\text{amu}} \cdot \text{m}^2}{\text{s}^2} \times \frac{1.00 \cancel{\text{g}}}{6.022 \times 10^{23} \cancel{\text{amu}}} \times \frac{1 \text{ kg}}{1000 \cancel{\text{g}}}\\ &= -1.73 \times 10^{-10} \frac{\text{kg} \cdot \text{m}^2}{\text{s}^2} = -1.73 \times 10^{-10} \text{ J}\end{aligned}$$

Thus, the nuclear binding energy is 1.73×10^{-10} J. The nuclear binding energy per nucleon is obtained as follows:

$$\frac{1.73 \times 10^{10} \text{ J}}{127 \text{ nucleons}} = 1.36 \times 10^{-12} \text{ J/nucleon}$$

Practice Exercise Calculate the nuclear binding energy (in J) and the nuclear binding energy per nucleon of $^{209}_{83}Bi$ (208.9804 amu).

Similar problems: 19.21, 19.22.

Summary of Concepts & Facts

- For stable nuclei of low atomic number, the neutron-to-proton ratio is close to 1. For heavier stable nuclei, the ratio becomes greater than 1. All nuclei with 84 or more protons are unstable and radioactive. Nuclei with even atomic numbers tend to have a greater number of stable isotopes than those with odd atomic numbers.
- Nuclear binding energy is a quantitative measure of nuclear stability. Nuclear binding energy can be calculated from a knowledge of the mass defect of the nucleus.

Review of Concepts & Facts

19.2.1 The following isotopes are unstable. Use Figure 19.1 to predict whether they will undergo beta decay or positron emission. (a) ^{13}B. (b) ^{188}Au. Write a nuclear equation for each case.

19.2.2 The atomic mass of $^{238}_{92}U$ is 238.0507847 amu. Calculate the nuclear binding energy per nucleon.

19.2.3 What is the change in mass (in kg) for the following reaction?

$$CH_4(g) + 2O_2(g) \longrightarrow CO_2(g) + 2H_2O(l) \qquad \Delta H° = -890.5 \text{ kJ/mol}$$

19.3 Natural Radioactivity

Learning Objective

- Use the half-life of a radioactive decay in calculations.

Nuclei outside the belt of stability, as well as nuclei with more than 83 protons, tend to be unstable. The spontaneous emission by unstable nuclei of particles or electromagnetic radiation, or both, is known as radioactivity. The main types of radiation are α particles (or doubly charged helium nuclei, He^{2+}); β particles (or electrons); γ rays, which are very-short-wavelength (0.1 nm to 10^{-4} nm) electromagnetic waves; positron emission; and electron capture.

The disintegration of a radioactive nucleus is often the beginning of a ***radioactive decay series,*** which is *a sequence of nuclear reactions that ultimately result in the formation of a stable isotope*. Table 19.5 shows the decay series of naturally occurring uranium-238, which involves 14 steps. This decay scheme, known as the *uranium decay series,* also shows the half-lives of all the intermediate products.

Video: Radioactive Decay

It is important to be able to balance the nuclear reaction for each of the steps in a radioactive decay series. For example, the first step in the uranium decay series is the decay of uranium-238 to thorium-234, with the emission of an α particle. Hence, the reaction is

$$^{238}_{92}U \longrightarrow \, ^{234}_{90}Th + \, ^{4}_{2}\alpha$$

The next step is represented by

$$^{234}_{90}Th \longrightarrow \, ^{234}_{91}Pa + \, ^{0}_{-1}\beta$$

and so on. In a discussion of radioactive decay steps, the beginning radioactive isotope is called the *parent* and the product, the *daughter*.

Table 19.5 The Uranium Decay Series*

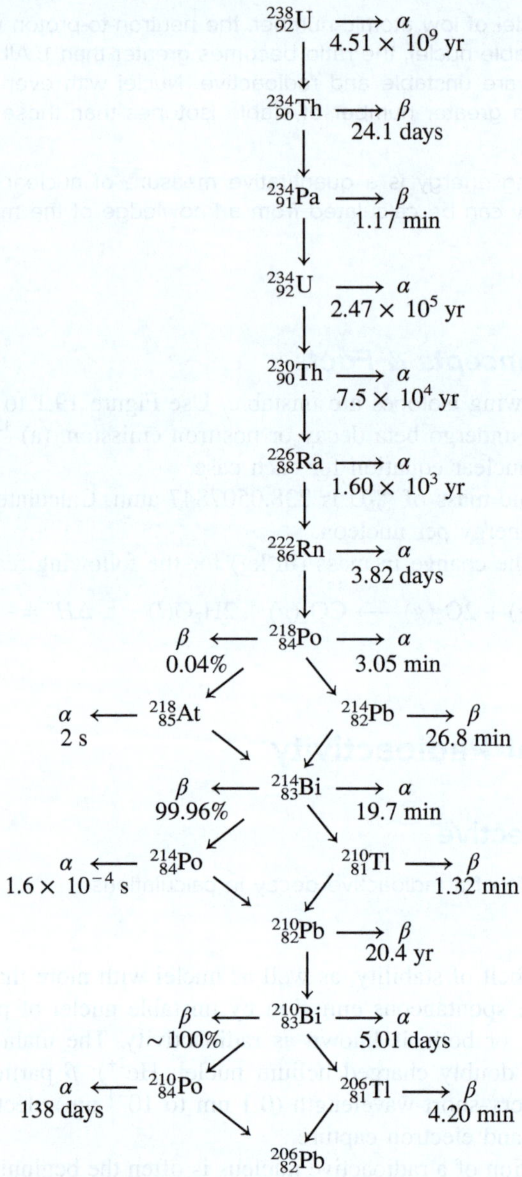

*The times denote the half-lives.

Kinetics of Radioactive Decay

All radioactive decays obey first-order kinetics. Therefore, the rate of radioactive decay at any time t is given by

$$\text{rate of decay at time } t = \lambda N \tag{19.3}$$

where λ is the first-order rate constant and N is the number of radioactive nuclei present at time t. (We use λ instead of k for rate constant in accord with the notation used by nuclear scientists.) According to Equation (13.3), the number of radioactive nuclei at time zero (N_0) and time t (N_t) is

$$\ln \frac{N_t}{N_0} = -\lambda t$$

and the corresponding half-life of the reaction is given by Equation (13.5):

$$t_{\frac{1}{2}} = \frac{0.693}{\lambda}$$

The half-lives (hence the rate constants) of radioactive isotopes vary greatly from nucleus to nucleus. For example, looking at Table 19.3, we find two extreme cases:

$$^{238}_{92}\text{U} \longrightarrow {}^{234}_{90}\text{Th} + {}^{4}_{2}\alpha \qquad t_{\frac{1}{2}} = 4.51 \times 10^9 \text{ yr}$$

$$^{214}_{84}\text{Po} \longrightarrow {}^{210}_{82}\text{Pb} + {}^{4}_{2}\alpha \qquad t_{\frac{1}{2}} = 1.6 \times 10^{-4} \text{ s}$$

We do not have to wait 4.51×10^9 yr to make a half-life measurement of uranium-238. Its value can be calculated from the rate constant using Equation (13.5). The ratio of these two rate constants after conversion to the same time unit is about 1×10^{21}, an enormously large number. Furthermore, the rate constants are unaffected by changes in environmental conditions such as temperature and pressure. These highly unusual features are not seen in ordinary chemical reactions (see Table 19.1).

Dating Based on Radioactive Decay

The half-lives of radioactive isotopes have been used as "atomic clocks" to determine the ages of certain objects. Some examples of dating by radioactive decay measurements will be described here.

Radiocarbon Dating

The carbon-14 isotope is produced when atmospheric nitrogen is bombarded by cosmic rays:

$$^{14}_{7}\text{N} + {}^{1}_{0}\text{n} \longrightarrow {}^{14}_{6}\text{C} + {}^{1}_{1}\text{H}$$

The radioactive carbon-14 isotope decays according to the equation

$$^{14}_{6}\text{C} \longrightarrow {}^{14}_{7}\text{N} + {}^{0}_{-1}\beta$$

The preceding decay series is the basis of the radiocarbon dating technique described in the Chemistry in Action essay "Radiocarbon Dating" in Section 13.3. In 2009, scientists in Sweden applied the technique to settle a controversy regarding heart muscle regeneration. The long-held view was that the heart cannot produce new muscle cells so people die with the same heart they were born with. After a heart attack, part of the cardiac muscle is lost and the heart heals principally through the formation of scar tissues. Carbon-14 concentration in the atmosphere remained relatively stable until 1955, when aboveground nuclear bomb tests caused a sharp increase until a test ban treaty was signed in 1963, after which it gradually decreased. The newly generated carbon-14 isotopes from nuclear blasts were incorporated into plants and animals. Through diets these isotopes finally appeared in the DNA of new cells in our bodies and stay unchanged for the life of the cell. It was found that for people born before 1955 the concentration of carbon-14 in their heart muscle cells exceeded the atmospheric concentration of carbon-14 at the time of their birth. This finding suggests that they continued to build new heart muscle cells after 1955 when atmospheric carbon-14 concentration increased. Those born during or after the atmospheric testing had a lower concentration due to the steady decline in atmospheric carbon-14 after the tests were halted. Because the level of carbon-14 in the atmosphere falls each year, the amount of carbon-14 in the DNA can serve to indicate the cell's birth date. The results show that about 1 percent of heart muscle cells are replaced every year at age 25, and the rate gradually falls to less than half a percent per year by age 75. Knowing that heart muscle cells *do* regenerate opens up possibilities of regulating

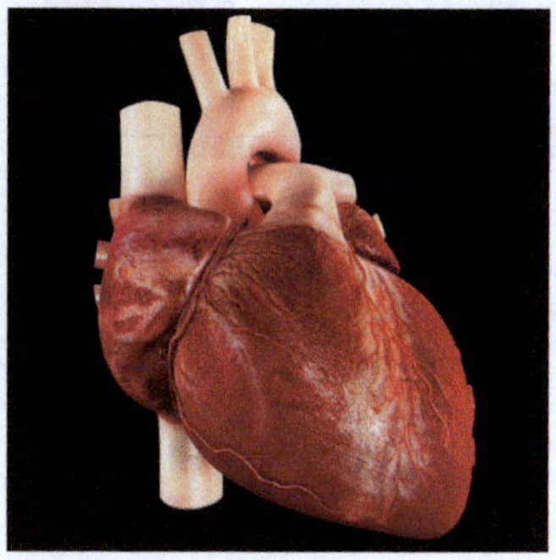

A human heart.
Medi-Mation Ltd./Science Source

that process. It is hoped that someday drugs will be developed to increase heart muscle cells in heart attack patients.

Dating Using Uranium-238 Isotopes

Because some of the intermediate products in the uranium decay series have very long half-lives (see Table 19.3), this series is particularly suitable for estimating the age of rocks in the earth and of extraterrestrial objects. The half-life for the first step ($^{238}_{92}U$ to $^{234}_{90}Th$) is 4.51×10^9 yr. This is about 20,000 times the second largest value (that is, 2.47×10^5 yr), which is the half-life for $^{234}_{92}U$ to $^{230}_{90}Th$. Therefore, as a good approximation, we can assume that the half-life for the overall process (that is, from $^{238}_{92}U$ to $^{206}_{82}Pb$) is governed solely by the first step (that is, the first step is the rate-determining step):

$$^{238}_{92}U \longrightarrow {}^{206}_{82}Pb + 8\,^{4}_{2}\alpha + 6\,^{0}_{-1}\beta \qquad t_{\frac{1}{2}} = 4.51 \times 10^9 \text{ yr}$$

In naturally occurring uranium minerals we should and do find some lead-206 isotopes formed by radioactive decay. Assuming that no lead was present when the mineral was formed and that the mineral has not undergone chemical changes that would allow the lead-206 isotope to be separated from the parent uranium-238, it is possible to estimate the age of the rocks from the mass ratio of $^{206}_{82}Pb$ to $^{238}_{92}U$. The previous equation tells us that for every mole, or 238 g, of uranium that undergoes complete decay, 1 mole, or 206 g, of lead is formed. If only half a mole of uranium-238 has undergone decay, the mass ratio $^{206}Pb/^{238}U$ becomes

$$\frac{206 \text{ g}/2}{238 \text{ g}/2} = 0.866$$

and the process would have taken a half-life of 4.51×10^9 yr to complete (Figure 19.3). Ratios lower than 0.866 mean that the rocks are less than 4.51×10^9 yr old, and higher ratios suggest a greater age. Interestingly, studies based on the uranium series as well as other decay series put the age of the oldest rocks and, therefore, probably the age of Earth itself at 4.5×10^9, or 4.5 billion, years.

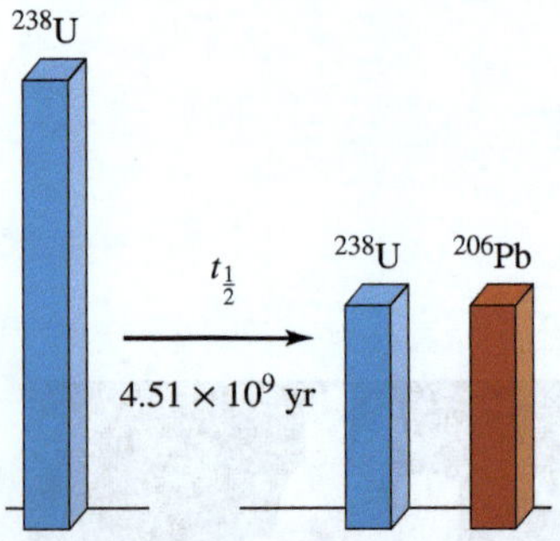

Figure 19.3 *After one half-life, half of the original uranium-238 is converted to lead-206.*

Dating Using Potassium-40 Isotopes

This is one of the most important techniques in geochemistry. The radioactive potassium-40 isotope decays by several different modes, but the relevant one as far as dating is concerned is that of electron capture:

$$^{40}_{19}K + {}^{0}_{-1}e \longrightarrow {}^{40}_{18}Ar \qquad t_{\frac{1}{2}} = 1.2 \times 10^9 \text{ yr}$$

The accumulation of gaseous argon-40 is used to gauge the age of a specimen. When a potassium-40 atom in a mineral decays, argon-40 is trapped in the lattice of the mineral and can escape only if the material is melted. Melting, therefore, is the procedure for analyzing a mineral sample in the laboratory. The amount of argon-40 present can be conveniently measured with a mass spectrometer (see Section 3.4). Knowing the ratio of argon-40 to potassium-40 in the mineral and the half-life of decay makes it possible to establish the ages of rocks ranging from millions to billions of years old.

Summary of Concepts & Facts

- Uranium-238 is the parent of a natural radioactive decay series that can be used to determine the ages of rocks.
- All radioactive decays are first-order rate processes.

Review of Concepts & Facts

19.3.1 What mass of $^{42}_{19}K$ remains after 100 hours if the original sample contained 256 g of $^{42}_{19}K$? ($t_{\frac{1}{2}}$ of $^{42}_{19}K$ = 12.5 hours.)

19.3.2 Iron-59 (red spheres) decays to cobalt (blue spheres) via beta decay with a half-life of 45.1 d. (a) Write a balanced nuclear equation for the process. (b) From the following diagram, determine how many half-lives have elapsed.

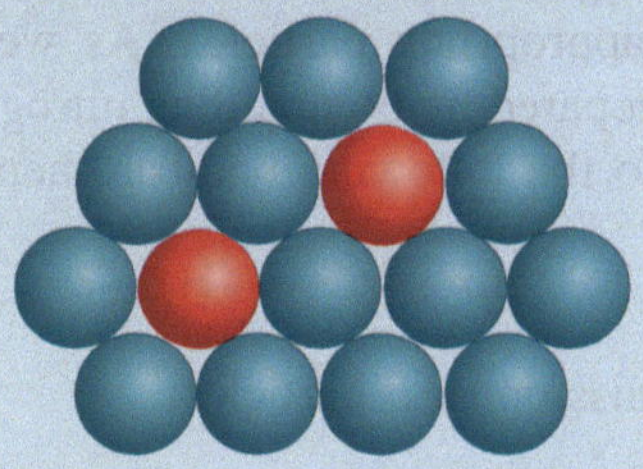

19.4 Nuclear Transmutation

Learning Objective

- Produce the balanced nuclear reaction for a nuclear transmutation.

The scope of nuclear chemistry would be rather narrow if study were limited to natural radioactive elements. An experiment performed by Rutherford in 1919, however, suggested the possibility of producing radioactivity artificially. When he bombarded a sample of nitrogen with α particles, the following reaction took place:

$$^{14}_{7}N + ^{4}_{2}\alpha \longrightarrow ^{17}_{8}O + ^{1}_{1}p$$

An oxygen-17 isotope was produced with the emission of a proton. This reaction demonstrated for the first time the feasibility of converting one element into another, by the process of nuclear transmutation. Nuclear transmutation differs from radioactive decay in that the former is brought about by the collision of two particles.

The preceding reaction can be abbreviated as $^{14}_{7}N(\alpha, p)^{17}_{8}O$. Note that in the parentheses the bombarding particle is written first, followed by the ejected particle. Example 19.3 illustrates the use of this notation to represent nuclear transmutations.

Student Hot Spot

Student data indicate you may struggle with nuclear transmutation. Access your eBook for additional Learning Resources on this topic.

Example 19.3

Write the balanced equation for the nuclear reaction $^{56}_{26}Fe(d, \alpha)^{54}_{25}Mn$, where d represents the deuterium nucleus, $^{2}_{1}H$.

Strategy To write the balanced nuclear equation, remember that the first isotope $^{56}_{26}Fe$ is the reactant and the second isotope $^{54}_{25}Fe$ is the product. The first symbol in parentheses (d) is the bombarding particle and the second symbol in parentheses (α) is the particle emitted as a result of nuclear transmutation.

Solution The abbreviation tells us that when iron-56 is bombarded with a deuterium nucleus, it produces the manganese-54 nucleus plus an α particle. Thus, the equation for this reaction is

$$^{56}_{26}Fe + ^{2}_{1}H \longrightarrow ^{4}_{2}\alpha + ^{54}_{25}Mn$$

(Continued)

Check Make sure that the sum of mass numbers and the sum of atomic numbers are the same on both sides of the equation.

Practice Exercise Write a balanced equation for $^{106}_{46}\text{Pd}(\alpha, \text{p})^{109}_{47}\text{Ag}$.

Similar problems: 19.37, 19.38.

Although light elements are generally not radioactive, they can be made so by bombarding their nuclei with appropriate particles. As we saw earlier, the radioactive carbon-14 isotope can be prepared by bombarding nitrogen-14 with neutrons. Tritium, $^{3}_{1}\text{H}$, is prepared according to the following bombardment:

$$^{6}_{3}\text{Li} + ^{1}_{0}\text{n} \longrightarrow ^{3}_{1}\text{H} + ^{4}_{2}\alpha$$

Tritium decays with the emission of β particles:

$$^{3}_{1}\text{H} \longrightarrow ^{3}_{2}\text{He} + ^{\ 0}_{-1}\beta \qquad t_{\frac{1}{2}} = 12.5 \text{ yr}$$

Particle Accelerator

Many synthetic isotopes are prepared by using neutrons as projectiles. This approach is particularly convenient because neutrons carry no charges and therefore are not repelled by the targets—the nuclei. In contrast, when the projectiles are positively charged particles (for example, protons or α particles), they must have considerable kinetic energy to overcome the electrostatic repulsion between themselves and the target nuclei. The synthesis of phosphorus from aluminum is one example:

$$^{27}_{13}\text{Al} + ^{4}_{2}\alpha \longrightarrow ^{30}_{15}\text{P} + ^{1}_{0}\text{n}$$

A *particle accelerator* uses electric and magnetic fields to increase the kinetic energy of charged species so that a reaction will occur (Figure 19.4). Alternating the polarity (that is, + and −) on specially constructed plates causes the particles to accelerate along a spiral path. When they have the energy necessary to initiate the desired nuclear reaction, they are guided out of the accelerator into a collision with a target substance.

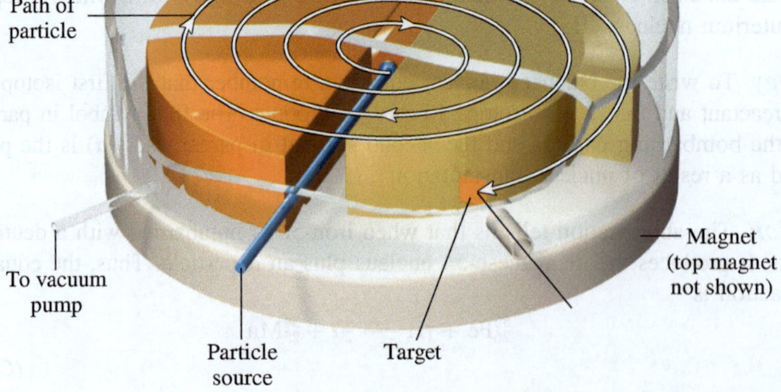

Figure 19.4 *Schematic diagram of a cyclotron particle accelerator. The particle (an ion) to be accelerated starts at the center and is forced to move in a spiral path through the influence of electric and magnetic fields until it emerges at a high velocity. The magnetic fields are perpendicular to the plane of the dees (so-called because of their shape), which are hollow and serve as electrodes.*

Figure 19.5 *A section of a particle accelerator.*
FAYOLLE/SIPA/Newscom

Various designs have been developed for particle accelerators, one of which accelerates particles along a linear path of about 3 km (Figure 19.5). It is now possible to accelerate particles to a speed well above 90 percent of the speed of light. (According to Einstein's theory of relativity, it is impossible for a particle to move *at* the speed of light. The only exception is the photon, which has a zero rest mass.) The extremely energetic particles produced in accelerators are employed by physicists to smash atomic nuclei to fragments. Studying the debris from such disintegrations provides valuable information about nuclear structure and binding forces.

The world's largest accelerator is the Large Hadron Collider (LHC), located 300 ft below ground along the border between France and Switzerland. It is housed inside a circular tunnel 17 mi around. In 2012, scientists at the LHC announced the discovery of the Higgs boson. It takes about 1 trillion proton–proton collisions to produce one Higgs boson event. Though the subatomic particle was predicted in 1964, it took almost 50 years to prove its existence.

The Transuranium Elements

Particle accelerators made it possible to synthesize the so-called ***transuranium elements,*** *elements with atomic numbers greater than 92.* Neptunium ($Z = 93$) was first prepared in 1940. Since then, 25 other transuranium elements have been synthesized. All isotopes of these elements are radioactive. Table 19.6 lists the transuranium elements that have been reported and some of the reactions through which they were formed.

Island of Stability

The latest synthesis of element 117 in 2010 filled the periodic table up to already-created element 118. Nuclear scientists believe that some heavier elements may occupy an "island of stability" in which atoms have longer half-lives (Figure 19.6). Analogous to the electronic structure of atoms, atomic nuclei can be thought of as concentric shells of protons and neutrons. The most stable nuclei occur when the outermost shells are filled. Some theories predict this will happen with 184 neutrons and 114, 120, or 126 protons, the presumed center of the island of stability. Despite considerable technical difficulties, scientists are hopeful that elements 119 and 120 and beyond will someday be synthesized. Isotopes of these elements may have half-lives of seconds, days, or even years. (In contrast, super-heavy elements made so far have half-lives of fractions of a second.) They are likely to find applications in industry and medicine. The intriguing question for scientists is this: Does the periodic table come to an end, and if so, where does it end?

Table 19.6 The Transuranium Elements

Atomic Number	Name	Symbol	Preparation
93	Neptunium	Np	${}^{238}_{92}\text{U} + {}^{1}_{0}\text{n} \longrightarrow {}^{239}_{93}\text{Np} + {}^{0}_{-1}\beta$
94	Plutonium	Pu	${}^{239}_{93}\text{Np} \longrightarrow {}^{239}_{94}\text{Pu} + {}^{0}_{-1}\beta$
95	Americium	Am	${}^{239}_{94}\text{Pu} + {}^{1}_{0}\text{n} \longrightarrow {}^{240}_{95}\text{Am} + {}^{0}_{-1}\beta$
96	Curium	Cm	${}^{239}_{94}\text{Pu} + {}^{4}_{2}\alpha \longrightarrow {}^{242}_{96}\text{Cm} + {}^{1}_{0}\text{n}$
97	Berkelium	Bk	${}^{241}_{95}\text{Am} + {}^{4}_{2}\alpha \longrightarrow {}^{243}_{97}\text{Bk} + 2{}^{1}_{0}\text{n}$
98	Californium	Cf	${}^{242}_{96}\text{Cm} + {}^{4}_{2}\alpha \longrightarrow {}^{245}_{98}\text{Cf} + {}^{1}_{0}\text{n}$
99	Einsteinium	Es	${}^{238}_{92}\text{U} + 15{}^{1}_{0}\text{n} \longrightarrow {}^{253}_{99}\text{Es} + 7{}^{0}_{-1}\beta$
100	Fermium	Fm	${}^{238}_{92}\text{U} + 17{}^{1}_{0}\text{n} \longrightarrow {}^{255}_{100}\text{Fm} + 8{}^{0}_{-1}\beta$
101	Mendelevium	Md	${}^{253}_{99}\text{Es} + {}^{4}_{2}\alpha \longrightarrow {}^{256}_{101}\text{Md} + {}^{1}_{0}\text{n}$
102	Nobelium	No	${}^{246}_{96}\text{Cm} + {}^{12}_{6}\text{C} \longrightarrow {}^{254}_{102}\text{No} + 4{}^{1}_{0}\text{n}$
103	Lawrencium	Lr	${}^{252}_{98}\text{Cf} + {}^{10}_{5}\text{B} \longrightarrow {}^{257}_{103}\text{Lr} + 5{}^{1}_{0}\text{n}$
104	Rutherfordium	Rf	${}^{249}_{98}\text{Cf} + {}^{12}_{6}\text{C} \longrightarrow {}^{257}_{104}\text{Rf} + 4{}^{1}_{0}\text{n}$
105	Dubnium	Db	${}^{249}_{98}\text{Cf} + {}^{15}_{7}\text{N} \longrightarrow {}^{260}_{105}\text{Db} + 4{}^{1}_{0}\text{n}$
106	Seaborgium	Sg	${}^{249}_{98}\text{Cf} + {}^{18}_{8}\text{O} \longrightarrow {}^{263}_{106}\text{Sg} + 4{}^{1}_{0}\text{n}$
107	Bohrium	Bh	${}^{209}_{83}\text{Bi} + {}^{54}_{24}\text{Cr} \longrightarrow {}^{262}_{107}\text{Bh} + {}^{1}_{0}\text{n}$
108	Hassium	Hs	${}^{208}_{82}\text{Pb} + {}^{58}_{26}\text{Fe} \longrightarrow {}^{265}_{108}\text{Hs} + {}^{1}_{0}\text{n}$
109	Meitnerium	Mt	${}^{209}_{83}\text{Bi} + {}^{58}_{26}\text{Fe} \longrightarrow {}^{266}_{109}\text{Mt} + {}^{1}_{0}\text{n}$
110	Darmstadtium	Ds	${}^{208}_{82}\text{Pb} + {}^{62}_{28}\text{Ni} \longrightarrow {}^{269}_{110}\text{Ds} + {}^{1}_{0}\text{n}$
111	Roentgenium	Rg	${}^{209}_{83}\text{Bi} + {}^{64}_{28}\text{Ni} \longrightarrow {}^{272}_{111}\text{Rg} + {}^{1}_{0}\text{n}$
112	Copernicium	Cn	${}^{208}_{82}\text{Pb} + {}^{70}_{30}\text{Zn} \longrightarrow {}^{277}_{112}\text{Cn} + {}^{1}_{0}\text{n}$
113	Nihonium	Nh	${}^{228}_{115}\text{Mc} \longrightarrow {}^{284}_{113}\text{Nh} + {}^{4}_{2}\alpha$
114	Flerovium	Fl	${}^{244}_{94}\text{Pu} + {}^{48}_{20}\text{Ca} \longrightarrow {}^{289}_{114}\text{Fl} + 3{}^{1}_{0}\text{n}$
115	Moscovium	Mc	${}^{243}_{95}\text{Am} + {}^{48}_{20}\text{Ca} \longrightarrow {}^{288}_{115}\text{Mc} + 3{}^{1}_{0}\text{n}$
116	Livermorium	Lv	${}^{248}_{96}\text{Cm} + {}^{48}_{20}\text{Ca} \longrightarrow {}^{293}_{116}\text{Lv} + 3{}^{1}_{0}\text{n}$
117	Tennessine	Ts	${}^{249}_{97}\text{Bk} + {}^{48}_{20}\text{Ca} \longrightarrow {}^{297}_{117}\text{Ts} + 4{}^{1}_{0}\text{n}$
118	Oganesson	Og	${}^{249}_{98}\text{Cf} + {}^{48}_{20}\text{Ca} \longrightarrow {}^{294}_{118}\text{Og} + 3{}^{1}_{0}\text{n}$

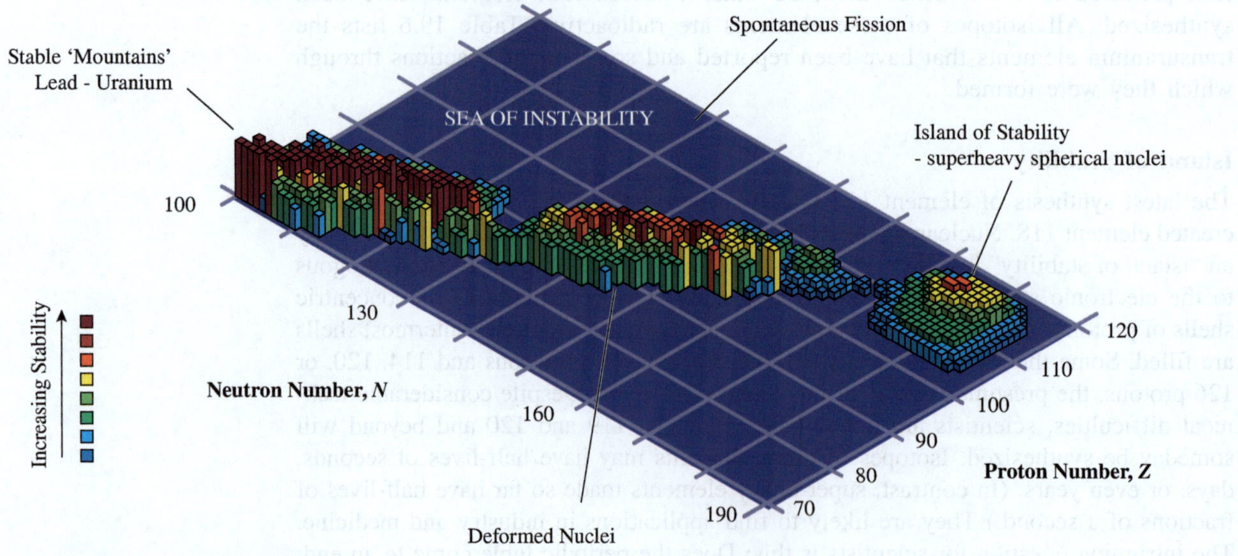

Figure 19.6 *Island of stability.*

Summary of Concepts & Facts

- Artificial radioactive elements are created by bombarding other elements with accelerated neutrons, protons, or α particles.

Review of Concepts & Facts

19.4.1 Write the correct abbreviated form of the equation

$$^{27}_{13}Al + ^{4}_{2}\alpha \longrightarrow ^{30}_{15}P + ^{1}_{0}n$$

19.4.2 The element oganesson (Og) was first created in 2006 in Dubna, Russia. The identification of this nucleus was verified by creating one of its daughter nuclei according to the following nuclear reaction: $^{245}_{96}Cm$ ($^{48}_{20}Ca$, 2n) X. Determine the product X and write the balanced equation for this nuclear reaction.

19.5 Nuclear Fission

Learning Objectives

- Discuss the process of nuclear fission.
- Define critical mass and nuclear chain reaction.
- Generalize the basis of a nuclear reactor.

Nuclear fission is the process in which *a heavy nucleus (mass number > 200) divides to form smaller nuclei of intermediate mass and one or more neutrons.* Because the heavy nucleus is less stable than its products (see Figure 19.2), this process releases a large amount of energy.

Video: Nuclear Fission

The first nuclear fission reaction to be studied was that of uranium-235 bombarded with slow neutrons, whose speed is comparable to that of air molecules at room temperature. Under these conditions, uranium-235 undergoes fission, as shown in Figure 19.7. Actually, this reaction is very complex: More than 30 different elements have been found among the fission products (Figure 19.8). A representative reaction is

$$^{235}_{92}U + ^{1}_{0}n \longrightarrow ^{90}_{38}Sr + ^{143}_{54}Xe + 3^{1}_{0}n$$

Although many heavy nuclei can be made to undergo fission, only the fission of naturally occurring uranium-235 and of the artificial isotope plutonium-239 has any

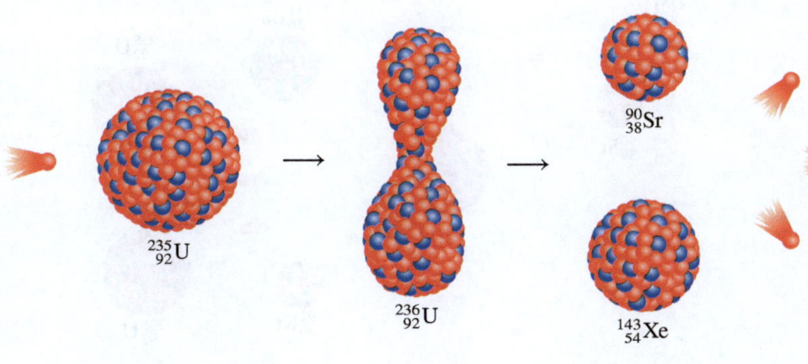

Figure 19.7 *Nuclear fission of ^{235}U. When a ^{235}U nucleus captures a neutron (red sphere), it undergoes fission to yield two smaller nuclei. On the average, 2.4 neutrons are emitted for every ^{235}U nucleus that divides.*

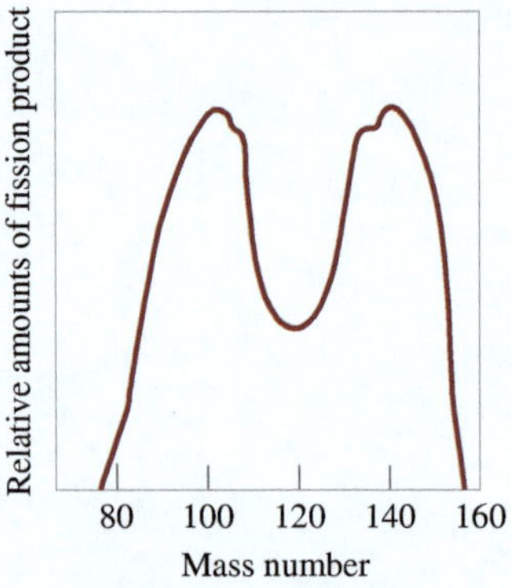

Figure 19.8 *Relative yields of the products resulting from the fission of ^{235}U as a function of mass number.*

practical importance. Table 19.7 shows the nuclear binding energies of uranium-235 and its fission products. As the table shows, the binding energy per nucleon for uranium-235 is less than the sum of the binding energies for strontium-90 and xenon-143. Therefore, when a uranium-235 nucleus is split into two smaller nuclei, a certain amount of energy is released. Let us estimate the magnitude of this energy. The difference between the binding energies of the reactants and products is $(1.23 \times 10^{-10} + 1.92 \times 10^{-10})$ J − (2.82×10^{-10}) J, or 3.3×10^{-11} J per uranium-235 nucleus. For 1 mole of uranium-235, the energy released would be $(3.3 \times 10^{-11}) \times (6.02 \times 10^{23})$, or 2.0×10^{13} J. This is an extremely exothermic reaction, considering that the heat of combustion of 1 ton of coal is only about 5×10^{7} J.

The significant feature of uranium-235 fission is not just the enormous amount of energy released, but the fact that more neutrons are produced than are originally captured in the process. This property makes possible a ***nuclear chain reaction,*** which is *a self-sustaining sequence of nuclear fission reactions.* The neutrons generated during the initial stages of fission can induce fission in other uranium-235 nuclei, which in turn produces more neutrons, and so on. In less than a second, the reaction can become uncontrollable, liberating a tremendous amount of heat to the surroundings.

For a chain reaction to occur, enough uranium-235 must be present in the sample to capture the neutrons. Otherwise, many of the neutrons will escape from the sample and the chain reaction will not occur. In this situation the mass of the sample is said to be *subcritical.* Figure 19.9 shows what happens when the amount of the fissionable material is equal to or greater than the ***critical mass,*** *the minimum mass of fissionable material required to generate a self-sustaining nuclear chain reaction.* In this case, most of the neutrons will be captured by uranium-235 nuclei, and a chain reaction will occur.

Table 19.7

Nuclear Binding Energies of ^{235}U and Its Fission Products

	Nuclear Binding Energy
^{235}U	2.82×10^{-10} J
^{90}Sr	1.23×10^{-10} J
^{143}Xe	1.92×10^{-10} J

The Atomic Bomb

The first application of nuclear fission was in the development of the atomic bomb. How is such a bomb made and detonated? The crucial factor in the bomb's design is the determination of the critical mass for the bomb. A small atomic bomb is equivalent to 20,000 tons of TNT (trinitrotoluene). Because 1 ton of TNT releases about

Figure 19.9 *If a critical mass is present, many of the neutrons emitted during the fission process will be captured by other ^{235}U nuclei and a chain reaction will occur.*

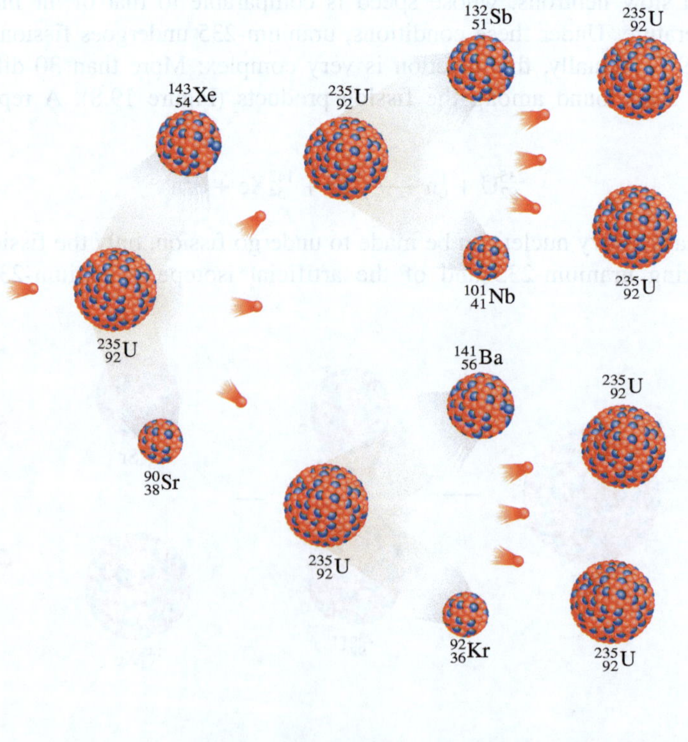

4×10^9 J of energy, 20,000 tons would produce 8×10^{13} J. Earlier we saw that 1 mole, or 235 g, of uranium-235 liberates 2.0×10^{13} J of energy when it undergoes fission. Thus, the mass of the isotope present in a small bomb must be at least

$$235\text{ g} \times \frac{8 \times 10^{13}\text{ J}}{2.0 \times 10^{13}\text{ J}} \approx 1\text{ kg}$$

For obvious reasons, an atomic bomb is never assembled with the critical mass already present. Instead, the critical mass is formed by using a conventional explosive, such as TNT, to force the fissionable sections together, as shown in Figure 19.10. Neutrons from a source at the center of the device trigger the nuclear chain reaction. Uranium-235 was the fissionable material in the bomb dropped on Hiroshima, Japan, on August 6, 1945. Plutonium-239 was used in the bomb exploded over Nagasaki three days later. The fission reactions generated were similar in these two cases, as was the extent of the destruction.

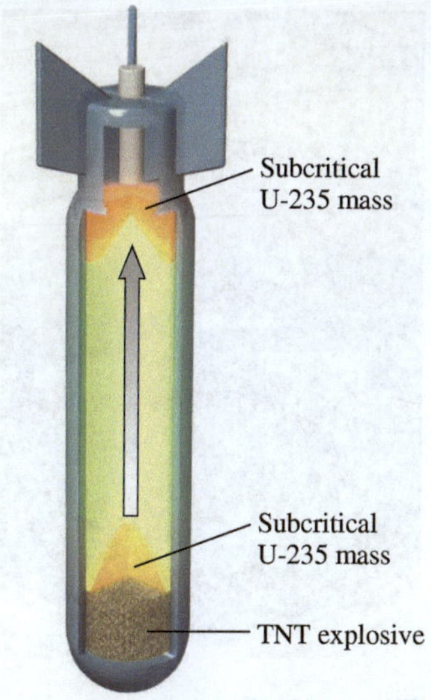

Figure 19.10 *Schematic diagram of an atomic bomb. The TNT explosives are set off first. The explosion forces the sections of fissionable material together to form an amount considerably larger than the critical mass.*

Nuclear Reactors

A peaceful but controversial application of nuclear fission is the generation of electricity using heat from a controlled chain reaction in a nuclear reactor. Currently, nuclear reactors provide about 20 percent of the electrical energy in the United States. This is a small but by no means negligible contribution to the nation's energy production. Several different types of nuclear reactors are in operation; we will briefly discuss the main features of three of them, along with their advantages and disadvantages.

Light Water Reactors

Most of the nuclear reactors in the United States are *light water reactors.* Figure 19.11 is a schematic diagram of such a reactor, and Figure 19.12 shows the refueling process in the core of a nuclear reactor.

An important aspect of the fission process is the speed of the neutrons. Slow neutrons split uranium-235 nuclei more efficiently than do fast ones. Because fission

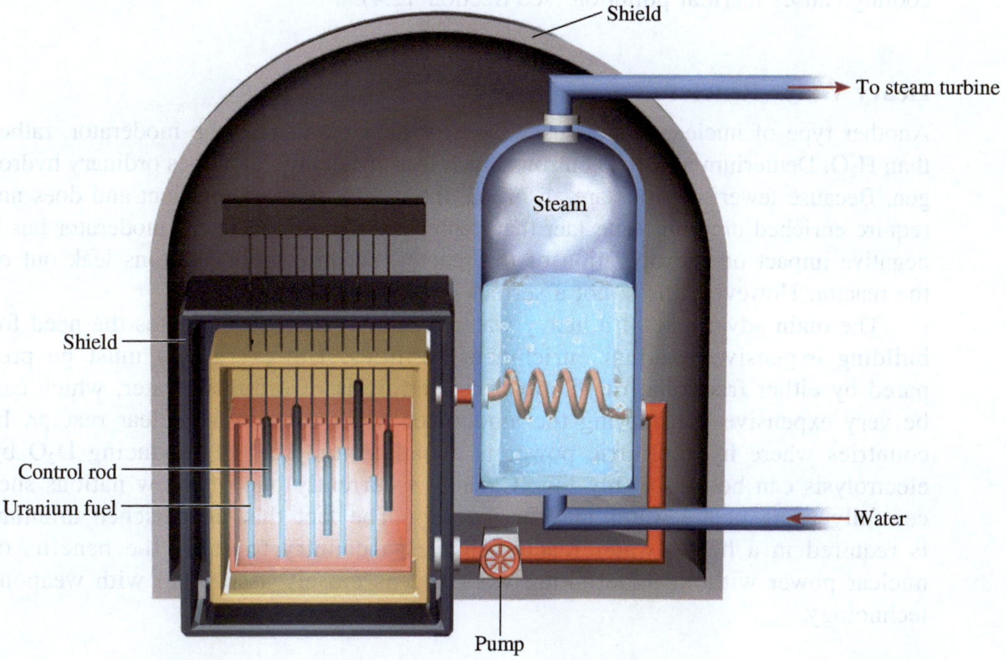

Figure 19.11 *Schematic diagram of a nuclear fission reactor. The fission process is controlled by cadmium or boron rods. The heat generated by the process is used to produce steam for the generation of electricity via a heat exchange system.*

Figure 19.12 *Refueling the core of a nuclear reactor.*
Toby Talbot/AP Images

Figure 19.13 *Uranium oxide, U_3O_8.*
Marvin Lazarus/Science Source

reactions are highly exothermic, the neutrons produced usually move at high velocities. For greater efficiency they must be slowed down before they can be used to induce nuclear disintegration. To accomplish this goal, scientists use ***moderators,*** which are *substances that can reduce the kinetic energy of neutrons.* A good moderator must satisfy several requirements: It should be nontoxic and inexpensive (as very large quantities of it are necessary); and it should resist conversion into a radioactive substance by neutron bombardment. Furthermore, it is advantageous for the moderator to be a fluid so that it can also be used as a coolant. No substance fulfills all these requirements, although water comes closer than many others that have been considered. Nuclear reactors that use light water (H_2O) as a moderator are called light water reactors because ^1_1H is the lightest isotope of the element hydrogen.

The nuclear fuel consists of uranium, usually in the form of its oxide, U_3O_8 (Figure 19.13). Naturally occurring uranium contains about 0.7 percent of the uranium-235 isotope, which is too low a concentration to sustain a small-scale chain reaction. For effective operation of a light water reactor, uranium-235 must be enriched to a concentration of 3 or 4 percent. In principle, the main difference between an atomic bomb and a nuclear reactor is that the chain reaction that takes place in a nuclear reactor is kept under control at all times. The factor limiting the rate of the reaction is the number of neutrons present. This can be controlled by lowering cadmium or boron control rods between the fuel elements. These rods capture neutrons according to the equations

$$^{113}_{48}Cd + ^1_0n \longrightarrow ^{114}_{48}Cd + \gamma$$

$$^{10}_5B + ^1_0n \longrightarrow ^7_3Li + ^4_2\alpha$$

where γ denotes gamma rays. Without the control rods the reactor core would melt from the heat generated and release radioactive materials into the environment.

Nuclear reactors have rather elaborate cooling systems that absorb the heat given off by the nuclear reaction and transfer it outside the reactor core, where it is used to produce enough steam to drive an electric generator. In this respect, a nuclear power plant is similar to a conventional power plant that burns fossil fuel. In both cases, large quantities of cooling water are needed to condense steam for reuse. Thus, most nuclear power plants are built near a river or a lake. Unfortunately this method of cooling causes thermal pollution (see Section 12.4).

Heavy Water Reactors

Another type of nuclear reactor uses D_2O, or heavy water, as the moderator, rather than H_2O. Deuterium absorbs neutrons much less efficiently than does ordinary hydrogen. Because fewer neutrons are absorbed, the reactor is more efficient and does not require enriched uranium. The fact that deuterium is a less efficient moderator has a negative impact on the operation of the reactor, because more neutrons leak out of the reactor. However, this is not a serious disadvantage.

The main advantage of a heavy water reactor is that it eliminates the need for building expensive uranium enrichment facilities. However, D_2O must be prepared by either fractional distillation or electrolysis of ordinary water, which can be very expensive considering the amount of water used in a nuclear reactor. In countries where hydroelectric power is abundant, the cost of producing D_2O by electrolysis can be reasonably low. Canada is currently one of a few nations successfully using heavy water nuclear reactors. The fact that no enriched uranium is required in a heavy water reactor enables a country to enjoy the benefits of nuclear power without undertaking work that is closely associated with weapons technology.

Breeder Reactors

A ***breeder reactor*** uses uranium fuel, but unlike a conventional nuclear reactor, it *produces more fissionable materials than it uses.*

We know that when uranium-238 is bombarded with fast neutrons, the following reactions take place:

$$^{238}_{92}\text{U} + ^{1}_{0}\text{n} \longrightarrow ^{239}_{92}\text{U}$$

$$^{239}_{92}\text{U} \longrightarrow ^{239}_{93}\text{Np} + ^{0}_{-1}\beta \qquad t_{\frac{1}{2}} = 23.4 \text{ min}$$

$$^{239}_{93}\text{Np} \longrightarrow ^{239}_{94}\text{Pu} + ^{0}_{-1}\beta \qquad t_{\frac{1}{2}} = 2.35 \text{ days}$$

In this manner the nonfissionable uranium-238 is transmuted into the fissionable isotope plutonium-239. Plutonium-239 forms plutonium oxide (Figure 19.14), which can be readily separated from uranium.

Figure 19.14 *The red glow of the radioactive plutonium oxide, PuO_2.*
Los Alamos National Laboratory

In a typical breeder reactor, nuclear fuel containing uranium-235 or plutonium-239 is mixed with uranium-238 so that breeding takes place within the core. For every uranium-235 (or plutonium-239) nucleus undergoing fission, more than one neutron is captured by uranium-238 to generate plutonium-239. Thus, the stockpile of fissionable material can be steadily increased as the starting nuclear fuels are consumed. It takes about 7 to 10 yr to regenerate the sizable amount of material needed to refuel the original reactor and to fuel another reactor of comparable size. This interval is called the *doubling time.*

Another fertile isotope is $^{232}_{90}\text{Th}$. Upon capturing slow neutrons, thorium is transmuted to uranium-233, which, like uranium-235, is a fissionable isotope:

$$^{232}_{90}\text{Th} + ^{1}_{0}\text{n} \longrightarrow ^{233}_{90}\text{Th}$$

$$^{233}_{90}\text{Th} \longrightarrow ^{233}_{91}\text{Pa} + ^{0}_{-1}\beta \qquad t_{\frac{1}{2}} = 22 \text{ min}$$

$$^{233}_{91}\text{Pa} \longrightarrow ^{233}_{92}\text{U} + ^{0}_{-1}\beta \qquad t_{\frac{1}{2}} = 27.4 \text{ days}$$

Uranium-233 ($t_{\frac{1}{2}} = 1.6 \times 10^5$ yr) is stable enough for long-term storage.

Although the amounts of uranium-238 and thorium-232 in Earth's crust are relatively plentiful (4 ppm and 12 ppm by mass, respectively), the development of breeder reactors has been very slow. To date, the United States does not have a single operating breeder reactor, and only a few have been built in other countries, such as France and Russia. One problem is economics; breeder reactors are more expensive to build than conventional reactors. There are also more technical difficulties associated with the construction of such reactors. As a result, the future of breeder reactors, in the United States at least, is rather uncertain.

Hazards of Nuclear Energy

Many people, including environmentalists, regard nuclear fission as a highly undesirable method of energy production. Many fission products such as strontium-90 are dangerous radioactive isotopes with long half-lives. Plutonium-239, used as a nuclear fuel and produced in breeder reactors, is one of the most chemically toxic substances known. It is an alpha emitter with a half-life of 24,400 yr.

Accidents, too, present many dangers. An accident at the Three Mile Island reactor in Pennsylvania in 1979 first brought the potential hazards of nuclear plants to public attention. In this instance, very little radiation escaped the reactor, but the plant remained closed for more than a decade while repairs were made and safety issues addressed. Only a few years later, on April 26, 1986, a reactor at the Chernobyl nuclear plant in Ukraine surged out of control. The fire and explosion that followed released much radioactive material into the environment. People working near the plant died within weeks as a result of the exposure to the intense radiation. Thirty-five

CHEMISTRY *in Action*

Nature's Own Fission Reactor

It all started with a routine analysis in May 1972 at the nuclear fuel processing plant in Pierrelatte, France. A staff member was checking the isotope ratio of U-235 to U-238 in a uranium ore and obtained a puzzling result. It had long been known that the relative natural occurrence of U-235 and U-238 is 0.7202 percent and 99.2798 percent, respectively. In this case, however, the amount of U-235 present was only 0.7171 percent. This may seem like a very small deviation, but the measurements were so precise that this difference was considered highly significant. The ore had come from the Oklo mine in the Gabon Republic, a small country on the west coast of Africa. Subsequent analyses of other samples showed that some contained even less U-235, in some cases as little as 0.44 percent.

The logical explanation for the low percentages of U-235 was that a nuclear fission reaction at the mine must have consumed some of the U-235 isotopes. But how did this happen? There are several conditions under which such a nuclear fission reaction could take place. In the presence of heavy water, for example, a chain reaction is possible with unenriched uranium. Without heavy water, such a fission reaction could still occur if the uranium ore and the moderator were arranged according to some specific geometric constraints at the site of the reaction. Both of the possibilities seem rather farfetched. The most plausible explanation is that the uranium ore originally present in the mine was enriched with U-235 and that a nuclear fission reaction took place with light water, as in a conventional nuclear reactor.

As mentioned earlier, the natural abundance of U-235 is 0.7202 percent, but it has not always been that low. The half-lives of U-235 and U-238 are 700 million and 4.51 billion years, respectively. This means that U-235 must have been *more* abundant in the past, because it has a shorter half-life. In fact, at the time Earth was formed, the natural abundance of U-235 was as high as 17 percent! Because the lowest concentration of U-235 required for the operation of a fission reactor is 1 percent, a nuclear chain reaction could have taken place as recently as 400 million years ago. By analyzing the amounts of radioactive fission products left in the ore, scientists concluded that the Gabon "reactor" operated about 2 billion years ago.

Having an enriched uranium sample is only one of the requirements for starting a controlled chain reaction. There must also have been a sufficient amount of the ore and an appropriate moderator present. It appears that as a result of a geological transformation, uranium ore was continually being washed into the Oklo region to yield concentrated deposits. The moderator needed for the fission process was largely water, present as water of crystallization in the sedimentary ore.

Thus, in a series of extraordinary events, a natural nuclear fission reactor operated at the time when the first life forms appeared on Earth. As is often the case in scientific endeavors, humans are not necessarily the innovators but merely the imitators of nature.

Photo showing the open-pit mining of the Oklo uranium deposit in Gabon revealed more than a dozen zones where nuclear fission had once taken place.

From Meshik, A.P., et.al.: The Workings of An Ancient Nuclear Reactor. Scientific American. November 2005; 293(5). Photo appeared on page 82-83. Photo by François Gauthier-Lafaye.

years later, it is still not known how many people died from radiation-induced cancer cases. Estimates vary between a few thousand and hundreds of thousands. The latest large-scale nuclear plant accident occurred in Fukushima, Japan, on March 11, 2011. A powerful earthquake, followed by a destructive tsunami, severely damaged the nuclear reactors at the plant. The long-term harmful effects of the radiation leakage to the environment are not yet fully assessed, but it is believed to be comparable to that at Chernobyl.

Molten glass is poured over nuclear waste before burial.
DOE/Science Source

In addition to the risk of accidents, the problem of radioactive waste disposal has not been satisfactorily resolved even for safely operated nuclear plants. Many suggestions have been made as to where to store or dispose of nuclear waste, including burial underground, burial beneath the ocean floor, and storage in deep geologic formations. But none of these sites has proved absolutely safe in the long run. Leakage of radioactive wastes into underground water, for example, can endanger nearby communities. The ideal disposal site would seem to be the sun, where a bit more radiation would make little difference, but this kind of operation requires 100 percent reliability in space technology.

In 2012, the Nuclear Regulatory Commission approved the construction of two new nuclear reactors. These permits were the first granted in the United States since the Three Mile Island accident, but because of the hazards, the future of nuclear reactors is clouded. What was once hailed as the ultimate solution to our energy needs in the twenty-first century is now being debated and questioned by both the scientific community and laypeople. It seems likely that the controversy will continue for some time.

Summary of Concepts & Facts

- Nuclear fission is the splitting of a large nucleus into two smaller nuclei and one or more neutrons. When the free neutrons are captured efficiently by other nuclei, a chain reaction can occur.
- Nuclear reactors use the heat from a controlled nuclear fission reaction to produce power. The three important types of reactors are light water reactors, heavy water reactors, and breeder reactors.

Review of Concepts & Facts

19.5.1 Why are boron compounds often added to badly damaged nuclear reactors such as those in Fukushima, Japan?

19.6 Nuclear Fusion

Learning Objective

- Evaluate the process of nuclear fusion.

In contrast to the nuclear fission process, ***nuclear fusion,*** *the combining of small nuclei into larger ones,* is largely exempt from the waste disposal problem.

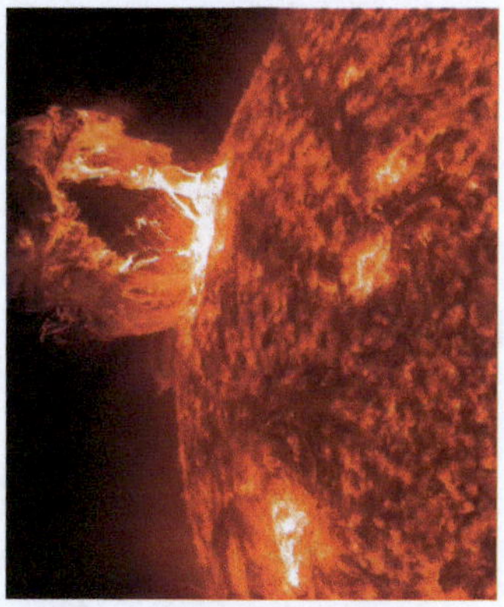

Nuclear fusion keeps the temperature in the interior of the sun at about 15 million °C.
NASA/SDO

Figure 19.2 showed that for the lightest elements, nuclear stability increases with increasing mass number. This behavior suggests that if two light nuclei combine or fuse together to form a larger, more stable nucleus, an appreciable amount of energy will be released in the process. This is the basis for ongoing research into the harnessing of nuclear fusion for the production of energy.

Nuclear fusion occurs constantly in the sun. The sun is made up mostly of hydrogen and helium. In its interior, where temperatures reach about 15 million degrees Celsius, the following fusion reactions are believed to take place:

$$^{1}_{1}\text{H} + ^{2}_{1}\text{H} \longrightarrow ^{3}_{2}\text{He}$$
$$^{3}_{2}\text{He} + ^{3}_{2}\text{He} \longrightarrow ^{4}_{2}\text{He} + 2^{1}_{1}\text{H}$$
$$^{1}_{1}\text{H} + ^{1}_{1}\text{H} \longrightarrow ^{2}_{1}\text{H} + ^{0}_{+1}\beta$$

Because *fusion reactions take place only at very high temperatures,* they are often called ***thermonuclear reactions.***

Fusion Reactors

A major concern in choosing the proper nuclear fusion process for energy production is the temperature necessary to carry out the process. Some promising reactions are

Reaction	Energy Released
$^{2}_{1}\text{H} + ^{2}_{1}\text{H} \longrightarrow ^{3}_{1}\text{H} + ^{1}_{1}\text{H}$	4.9×10^{-13} J
$^{2}_{1}\text{H} + ^{3}_{1}\text{H} \longrightarrow ^{4}_{2}\text{He} + ^{1}_{0}\text{n}$	2.8×10^{-12} J
$^{6}_{3}\text{Li} + ^{2}_{1}\text{H} \longrightarrow 2^{4}_{2}\text{He}$	3.6×10^{-12} J

These reactions take place at extremely high temperatures, on the order of 100 million degrees Celsius, to overcome the repulsive forces between the nuclei. The first reaction is particularly attractive because the world's supply of deuterium is virtually inexhaustible. The total volume of water on Earth is about 1.5×10^{21} L. Because the natural abundance of deuterium is 1.5×10^{-2} percent, the total amount of deuterium present is roughly 4.5×10^{21} g, or 5.0×10^{15} tons. The cost of preparing deuterium is minimal compared with the value of the energy released by the reaction.

In contrast to the fission process, nuclear fusion looks like a very promising energy source, at least "on paper." Although thermal pollution would be a problem, fusion has the following advantages: (1) The fuels are cheap and almost inexhaustible, and (2) the process produces little radioactive waste. If a fusion machine were turned off, it would shut down completely and instantly, without any danger of a meltdown.

If nuclear fusion is so great, why isn't there even one fusion reactor producing energy? Although we command the scientific knowledge to design such a reactor, the technical difficulties have not yet been solved. The basic problem is finding a way to hold the nuclei together long enough, and at the appropriate temperature, for fusion to occur. At temperatures of about 100 million degrees Celsius, molecules cannot exist, and most or all of the atoms are stripped of their electrons. This *state of matter, a gaseous mixture of positive ions and electrons,* is called ***plasma.*** The problem of containing this plasma is a formidable one. What solid container can exist at such temperatures? None, unless the amount of plasma is small; but then the solid surface would immediately cool the sample and quench the fusion reaction. One approach to solving this problem is to use *magnetic confinement.* Because a plasma consists of charged particles moving at high speeds, a magnetic field will exert force on it. As Figure 19.15 shows, the plasma moves through a doughnut-shaped tunnel, confined by a complex magnetic field. Thus, the plasma never comes in contact with the walls of the container.

Another design employs lasers to ignite the nuclear reaction. One approach focuses 192 high-power laser beams on a small fuel pellet containing deuterium and tritium (Figure 19.16). The energy from the lasers heats the pellet to an extremely high temperature, causing it to *implode,* that is, to collapse inward from all sides into

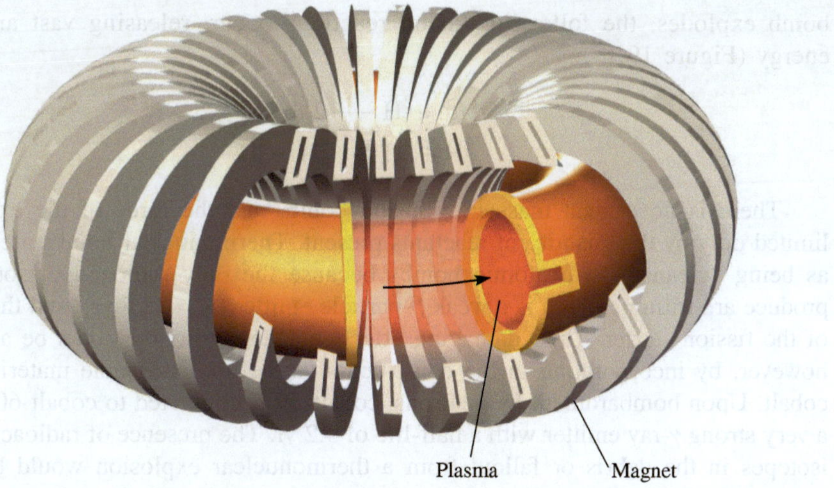

Figure 19.15 *A magnetic plasma confinement design called tokamak.*

Figure 19.16 *The reaction chamber at the National Ignition Facility, where 192 lasers are used to initiate fusion in a small pellet of deuterium and tritium.*
Courtesy of Lawrence Livermore National Laboratory

a tiny volume. Under this condition, the fusion process is initiated. The U.S. National Ignition Facility reported in December 2022 that for the first time the energy produced by fusion in the laboratory (3.15 MJ) exceeded the amount used to induce the process (2.05 MJ). Although many technical difficulties still need to be solved before nuclear fusion can be put to practical use on a large scale, scientists are excited by the first positive result after many years of hard work.

The Hydrogen Bomb

The technical problems inherent in the design of a nuclear fusion reactor do not affect the production of a hydrogen bomb, also called a thermonuclear bomb. In this case, the objective is all power and no control. Hydrogen bombs do not contain gaseous hydrogen or gaseous deuterium; they contain solid lithium deuteride (LiD), which can be packed very tightly. The detonation of a hydrogen bomb occurs in two stages—first a fission reaction and then a fusion reaction. The required temperature for fusion is achieved with an atomic bomb. Immediately after the atomic bomb explodes, the following fusion reactions occur, releasing vast amounts of energy (Figure 19.17):

$$^{6}_{3}\text{Li} + ^{2}_{1}\text{H} \longrightarrow 2^{4}_{2}\alpha$$
$$^{2}_{1}\text{H} + ^{2}_{1}\text{H} \longrightarrow ^{3}_{1}\text{H} + ^{1}_{1}\text{H}$$

Figure 19.17 *Explosion of a thermonuclear bomb.*
Stocktrek/Photodisc/Getty Images

There is no critical mass in a fusion bomb, and the force of the explosion is limited only by the quantity of reactants present. Thermonuclear bombs are described as being "cleaner" than atomic bombs because the only radioactive isotopes they produce are tritium, which is a weak β-particle emitter ($t_{\frac{1}{2}} = 12.5$ yr), and the products of the fission starter. Their damaging effects on the environment can be aggravated, however, by incorporating in the construction some nonfissionable material such as cobalt. Upon bombardment by neutrons, cobalt-59 is converted to cobalt-60, which is a very strong γ-ray emitter with a half-life of 5.2 yr. The presence of radioactive cobalt isotopes in the debris or fallout from a thermonuclear explosion would be fatal to those who survived the initial blast.

Summary of Concepts & Facts

- Nuclear fusion, the type of reaction that occurs in the sun, is the combination of two light nuclei to form one heavy nucleus. Fusion takes place only at very high temperatures, so high that controlled large-scale nuclear fusion has so far not been achieved.

19.7 Uses of Isotopes

Learning Objective

- Identify examples of the uses of isotopes in science and medicine.

Radioactive and stable isotopes alike have many applications in science and medicine. We have previously described the use of isotopes in the study of reaction mechanisms (see Section 13.5) and in dating artifacts (see Chemistry in Action essay "Radiocarbon Dating" in Section 13.3 and Section 19.3). In this section we will discuss a few more examples.

Structural Determination

The formula of the thiosulfate ion is $S_2O_3^{2-}$. For some years chemists were uncertain as to whether the two sulfur atoms occupied equivalent positions in the ion. The thiosulfate ion is prepared by treatment of the sulfite ion with elemental sulfur:

$$\text{SO}_3^{2-}(aq) + \text{S}(s) \longrightarrow \text{S}_2\text{O}_3^{2-}(aq)$$

When thiosulfate is treated with dilute acid, the reaction is reversed. The sulfite ion is reformed and elemental sulfur precipitates:

$$\text{S}_2\text{O}_3^{2-}(aq) \xrightarrow{\text{H}^+} \text{SO}_3^{2-}(aq) + \text{S}(s) \quad (19.4)$$

If this sequence is started with elemental sulfur enriched with the radioactive sulfur-35 isotope, the isotope acts as a "label" for S atoms. All the labels are found in the sulfur precipitate in Equation (19.4); none of them appears in the final sulfite ions. Clearly, then, the two atoms of sulfur in $S_2O_3^{2-}$ are not structurally equivalent, as would be the case if the structure were

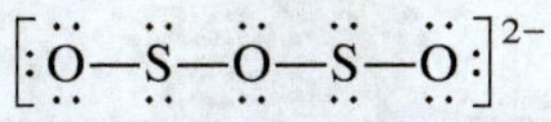

Otherwise, the radioactive isotope would be present in both the elemental sulfur precipitate and the sulfite ion. Based on spectroscopic studies, we now know that the structure of the thiosulfate ion is

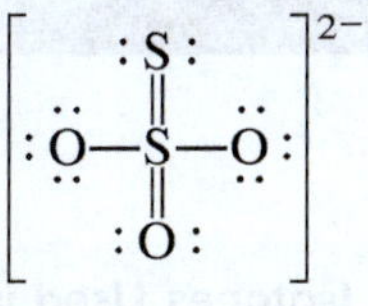

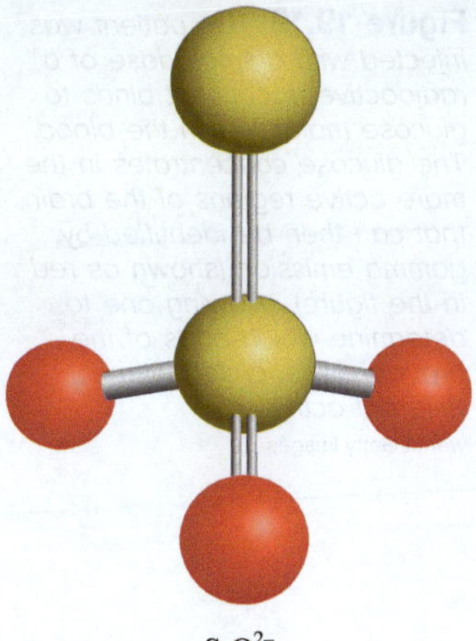

$S_2O_3^{2-}$

Study of Photosynthesis

The study of photosynthesis is also rich with isotope applications. The overall photosynthesis reaction can be represented as

$$6CO_2 + 6H_2O \longrightarrow C_6H_{12}O_6 + 6O_2$$

In Section 13.5 we learned that the ^{18}O isotope was used to determine the source of O_2. The radioactive ^{14}C isotope helped to determine the path of carbon in photosynthesis. Starting with $^{14}CO_2$, it was possible to isolate the intermediate products during photosynthesis and measure the amount of radioactivity of each carbon-containing compound. In this manner, the path from CO_2 through various intermediate compounds to carbohydrate could be clearly charted. *Isotopes, especially radioactive isotopes that are used to trace the path of the atoms of an element in a chemical or biological process,* are called ***tracers.***

Isotopes in Medicine

Tracers are used also for diagnosis in medicine. Sodium-24 (a β emitter with a half-life of 14.8 h) injected into the bloodstream as a salt solution can be monitored to trace the flow of blood and detect possible constrictions or obstructions in the circulatory system. Iodine-131 (a β emitter with a half-life of 8 days) has been used to test the activity of the thyroid gland. A malfunctioning thyroid can be detected by giving the patient a drink of a solution containing a known amount of $Na^{131}I$ and measuring the radioactivity just above the thyroid to see if the iodine is absorbed at the normal rate. Of course, the amounts of radioisotope used in the human body must always be kept small; otherwise, the patient might suffer permanent damage from the high-energy radiation. A radioactive isotope of fluorine, fluorine-18, emits positrons, which are annihilated by electrons, forming γ-rays used to image the brain (Figure 19.18). Table 19.8 shows some of the radioactive isotopes used in medicine.

Technetium, the first artificially prepared element, is one of the most useful elements in nuclear medicine. Although technetium is a transition metal, all its isotopes are radioactive. In the laboratory it is prepared by the nuclear reactions

$$^{98}_{42}Mo + ^{1}_{0}n \longrightarrow ^{99}_{42}Mo$$

$$^{99}_{42}Mo \longrightarrow ^{99m}_{43}Tc + ^{0}_{-1}\beta$$

where the superscript m denotes that the technetium-99 isotope is produced in its excited nuclear state. This isotope has a half-life of about 6 h, decaying by γ radiation

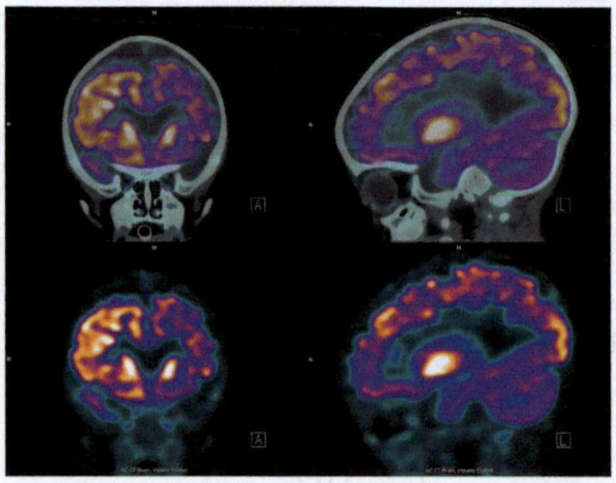

Figure 19.18 *This patient was injected with a small dose of a radioactive tracer that binds to glucose molecules in the blood. The glucose concentrates in the more active regions of the brain that can then be identified by gamma emission (shown as red in the figure), allowing one to determine which parts of the brain are associated with different activities.*
wenht/Getty Images

Table 19.8 **Some Radioactive Isotopes Used in Medicine**

Isotope	Half-Life	Uses
^{18}F	1.8 h	Brain imaging, bone scan
^{24}Na	15 h	Monitoring blood circulation
^{32}P	14.3 d	Location of ocular, brain, and skin tumors
^{43}K	22.4 h	Myocardial scan
^{47}Ca	4.5 d	Study of calcium metabolism
^{51}Cr	27.8 d	Determination of red blood cell volume, spleen imaging, placenta localization
^{60}Co	5.3 yr	Sterilization of medical equipment, cancer treatment
^{99m}Tc	6 h	Imaging of various organs, bones, placenta location
^{125}I	60 d	Study of pancreatic function, thyroid imaging, liver function
^{131}I	8 d	Brain imaging, liver function, thyroid activity

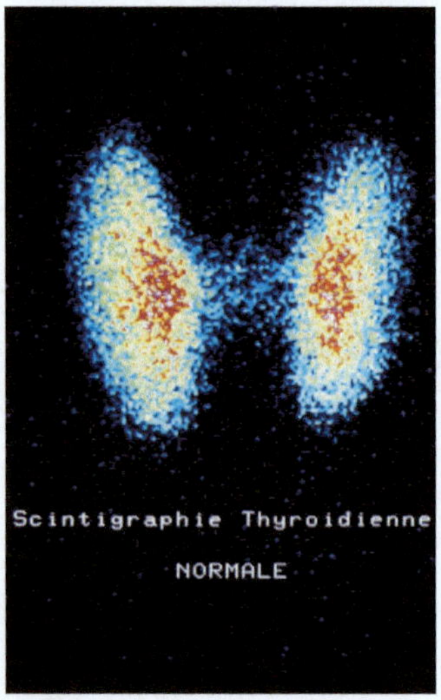

Image of a person's thyroid glands obtained using $^{99m}_{43}$Tc.
BSIP/Newscom

to technetium-99 in its nuclear ground state. Thus, it is a valuable diagnostic tool. The patient either drinks or is injected with a solution containing ^{99m}Tc. By detecting the γ rays emitted by ^{99m}Tc, doctors can obtain images of organs such as the heart, liver, and lungs.

A major advantage of using radioactive isotopes as tracers is that they are easy to detect. Their presence even in very small amounts can be detected by photographic techniques or by devices known as counters. Figure 19.19 is a diagram of a Geiger

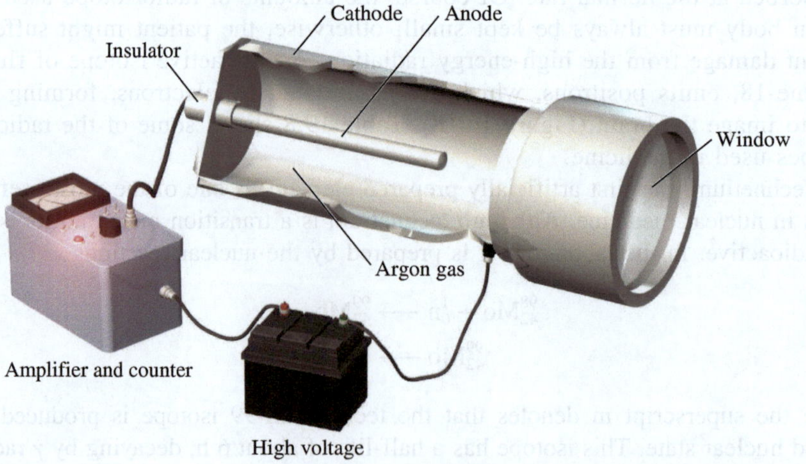

Figure 19.19 *Schematic diagram of a Geiger counter. Radiation (α or β particles, or γ rays) entering through the window ionized the argon gas to generate a small current flow between the electrodes. This current is amplified and is used to flash a light or operate a counter with a clicking sound.*

counter, an instrument widely used in scientific work and medical laboratories to detect radiation.

Summary of Concepts & Facts

- Radioactive isotopes are easy to detect and thus make excellent tracers in chemical reactions and in medical practice.

19.8 Biological Effects of Radiation

Learning Objective

- Explain why high-energy radiation is biologically harmful.

In this section, we will examine briefly the effects of radiation on biological systems. But first let us define quantitative measures of radiation. The fundamental unit of radioactivity is the *curie* (Ci); 1 Ci corresponds to exactly 3.70×10^{10} nuclear disintegrations per second. This decay rate is equivalent to that of 1 g of radium. A *millicurie* (mCi) is one-thousandth of a curie. Thus, 10 mCi of a carbon-14 sample is the quantity that undergoes

$$(10 \times 10^{-3})(3.70 \times 10^{10}) = 3.70 \times 10^{8}$$

disintegrations per second.

The intensity of radiation depends on the number of disintegrations as well as on the energy and type of radiation emitted. One common unit for the absorbed dose of radiation is the *rad* (*r*adiation *a*bsorbed *d*ose), which is the amount of radiation that results in the absorption of 1×10^{-2} J per kilogram of irradiated material. The biological effect of radiation also depends on the part of the body irradiated and the type of radiation. For this reason, the rad is often multiplied by a factor called *RBE* (*r*elative *b*iological *e*ffectiveness). The RBE is approximately 1 for beta and gamma radiation and about 10 for alpha radiation. To measure the biological damage, which depends on dose rate, total dose, and the type of tissue affected, we introduce another term called a *rem* (*r*oentgen *e*quivalent for *m*an), given by

$$\text{number of rems} = (\text{number of rads})\ (\text{RBE}) \tag{19.5}$$

Of the three types of nuclear radiation, alpha particles usually have the least penetrating power. Beta particles are more penetrating than alpha particles, but less so than gamma rays. Gamma rays have very short wavelengths and high energies. Furthermore, because they carry no charge, they cannot be stopped by shielding materials as easily as alpha and beta particles. However, if alpha or beta emitters are ingested, their damaging effects are greatly aggravated because the organs will be constantly subject to damaging radiation at close range. For example, strontium-90, a beta emitter, can replace calcium in bones, where it does the greatest damage.

Table 19.9 lists the average amounts of radiation an American receives every year. It should be pointed out that for short-term exposures to radiation, a dosage of 50–200 rem will cause a decrease in white blood cell counts and other complications,

Table 19.9 Average Yearly Radiation Doses for Americans

Source	Dose (mrem/yr)*
Cosmic rays	20–50
Ground and surroundings	25
Human body†	26
Medical and dental X rays	50–75
Air travel	5
Fallout from weapons tests	5
Nuclear waste	2
Total	133–188

*1 mrem = 1 millirem = 1×10^{-3} rem.

†The radioactivity in the body comes from food and air.

while a dosage of 500 rem or greater may result in death within weeks. Current safety standards permit nuclear workers to be exposed to no more than 5 rem per year and specify a maximum of 0.5 rem of human-made radiation per year for the general public.

The chemical basis of radiation damage is that of ionizing radiation. Radiation of either particles or gamma rays can remove electrons from atoms and molecules in its path, leading to the formation of ions and radicals. ***Radicals*** (also called *free radicals*) are *molecular fragments having one or more unpaired electrons; they are usually short-lived and highly reactive.* For example, when water is irradiated with gamma rays, the following reactions take place:

$$H_2O \xrightarrow{\text{radiation}} H_2O^+ + e^-$$

$$H_2O^+ + H_2O \longrightarrow H_3O^+ + \underset{\text{hydroxyl radical}}{\cdot OH}$$

The electron (in the hydrated form) can subsequently react with water or with a hydrogen ion to form atomic hydrogen, and with oxygen to produce the superoxide ion, O_2^- (a radical):

$$e^- + O_2 \longrightarrow \cdot O_2^-$$

In the tissues the superoxide ions and other free radicals attack cell membranes and a host of organic compounds, such as enzymes and DNA molecules. Organic compounds can themselves be directly ionized and destroyed by high-energy radiation.

It has long been known that exposure to high-energy radiation can induce cancer in humans and other animals. Cancer is characterized by uncontrolled cellular growth. On the other hand, it is also well established that cancer cells can be destroyed by proper radiation treatment. In radiation therapy, a compromise is sought. The radiation to which the patient is exposed must be sufficient to destroy cancer cells without killing too many normal cells and, it is hoped, without inducing another form of cancer.

Radiation damage to living systems is generally classified as *somatic* or *genetic.* Somatic injuries are those that affect the organism during its own lifetime. Sunburn, skin rash, cancer, and cataracts are examples of somatic damage. Genetic damage means inheritable changes or gene mutations. For example, a person whose chromosomes, parts of the cell structure that contain DNA, have been damaged or altered by radiation may pass cellular damage on to offspring.

Student Hot Spot

Student data indicate you may struggle with the biological effects of radiation. Access your eBook for additional Learning Resources on this topic.

CHEMISTRY *in Action*

Food Irradiation

If you eat processed food, you have probably eaten ingredients exposed to radioactive rays. In the United States, up to 10 percent of herbs and spices are irradiated to control mold, zapped with X rays at a dose equal to 60 million chest X rays. Although food irradiation has been used in one way or another for more than 50 years, it faces an uncertain future in this country.

Back in 1953, the U.S. Army started an experimental program of food irradiation so that deployed troops could have fresh food without refrigeration. The procedure is a simple one. Food is exposed to high levels of radiation to kill insects and harmful bacteria. It is then packaged in airtight containers, in which it can be stored for months without deterioration. The radiation sources for most food preservation are cobalt-60 and cesium-137, both of which are γ emitters, although X rays and electron beams can also be used to irradiate food.

The benefits of food irradiation are obvious—it reduces energy demand by eliminating the need for refrigeration, and it prolongs the shelf life of various foods, which is of vital importance for poor countries. Yet there is considerable opposition to this procedure. First, there is a fear that irradiated food may itself become radioactive. No such evidence has been found. A more serious objection is that irradiation can destroy the nutrients such as vitamins and amino acids. Furthermore, the ionizing radiation produces reactive species, such as the hydroxyl radical, which then react with the organic molecules to produce potentially harmful substances. Interestingly, the same effects are produced when food is cooked by heat.

Strawberries irradiated at 200 kilorads (right) are still fresh after 15 days storage at 4°C; those not irradiated are moldy.
Ken Karp/McGraw Hill

Food Irradiation Dosages and Their Effects*

Dosage	Effect
Low dose (up to 100 kilorad)	Inhibits sprouting of potatoes, onions, garlics.
	Inactivates trichinae in pork.
	Kills or prevents insects from reproducing in grains, fruits, and vegetables after harvest.
Medium dose (100 to 1000 kilorads)	Delays spoilage of meat, poultry, and fish by killing spoilage microorganism.
	Reduces salmonella and other foodborne pathogens in meat, fish, and poultry.
	Extends shelf life by delaying mold growth on strawberries and some other fruits.
High dose (1000 to 10,000 kilorads)	Sterilizes meat, poultry, fish, and some other foods.
	Kills microorganisms and insects in spices and seasoning.

*Source: Zurer, Pamela S., "Food Irradiation A Technology at a Turning Point," *Chemical & Engineering News,* vol. 64, issue 18, May 5, 1986, Pages 46–56.

CHEMISTRY *in Action*

Boron Neutron Capture Therapy

Each year more than half a million people in the world contract brain tumors and about 2000 die from the disease. Treatment of a brain tumor is one of the most challenging of cancer cases because of the site of the malignant growth, which makes surgical excision difficult and often impossible. Likewise, conventional radiation therapy using X rays or γ rays from outside the skull is seldom effective.

An ingenious approach to this problem is called boron neutron capture therapy (BNCT). This technique brings together two components, each of which separately has minimal harmful effects on the cells. The first component uses a compound containing a stable boron isotope (^{10}B) that can be concentrated in tumor cells. The second component is a beam of low-energy neutrons. Upon capturing a neutron, the following nuclear reaction takes place:

$$^{10}_{5}\text{B} + ^{1}_{0}\text{n} \longrightarrow ^{7}_{3}\text{Li} + ^{4}_{2}\alpha$$

The recoiling α particle and the lithium nucleus together carry about 3.8×10^{-13} J of energy. Because these high-energy particles are confined to just a few micrometers (about the diameter of a cell), they can preferentially destroy tumor cells without damaging the surrounding tissues. ^{10}B has a large neutron absorption cross section and is therefore particularly suited for this application. Ionizing radiation like X rays requires oxygen to produce reactive hydroxyl and superoxide radicals to enhance their biological effectiveness. However, a rapidly expanding tumor frequently depletes its blood supply and hence also the oxygen content. BNCT does not require oxygen and therefore does not suffer from this limitation. BNCT is currently an active research area involving collaborations of chemists, nuclear physicists, and physicians.

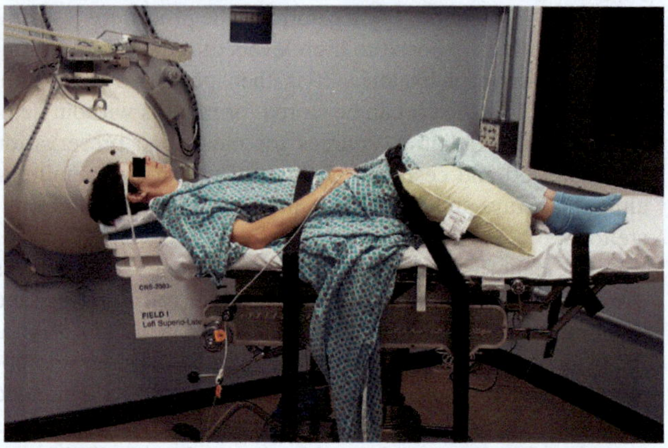

Setup for a lateral BNCT brain irradiation using the fission converter-based epithermal neutron beam at the Massachusetts Institute of Technology with a 12-cm-diameter aperture.

Courtesy of Kent J. Riley

Summary of Concepts & Facts

- High-energy radiation damages living systems by causing ionization and the formation of free radicals.

Chapter Summary

Nuclear Reactions There are many differences between nuclear reactions and ordinary chemical reactions. To balance nuclear equations, various nuclear particles and forms of nuclear radiation are involved. These include electrons, protons, neutrons, and alpha particles. Nuclear equations are balanced by ensuring the conservation of mass number and conservation of atomic number are followed. (Section 19.1)

Nuclear Stability To maintain nuclear stability, the ratio of neutrons to protons must fall within a certain range. A quantitative measure of nuclear stability is nuclear binding energy, which is the energy required to break up a nucleus into its component protons and neutrons. Nuclear binding energy can be calculated from the masses of protons and neutrons and that of the nucleus using Einstein's mass-energy equivalence relationship. (Section 19.2)

Natural Radioactivity and Nuclear Transmutation Unstable nuclei undergo spontaneous decay with the emission of radiation and particles. All nuclear decays obey first-order kinetics. The half-lives of several radioactive nuclei have been used to date objects. Stable nuclei can also be made radioactive by bombardment with elementary particles or atomic nuclei. Many new elements have been created artificially in particle accelerators where such bombardments occur. (Sections 19.3 and 19.4)

Nuclear Fission and Nuclear Fusion Certain nuclei, when bombarded with neutrons, undergo fission to produce smaller nuclei, additional neutrons, and a large amount of energy. When enough nuclei are present to reach critical mass, a nuclear chain reaction, a self-sustaining sequence of nuclear fission reactions, takes place. Nuclear fissions find applications in the construction of atomic bombs and nuclear reactors. Nuclear fusion is the process in which nuclei of light elements are made to fuse at very high temperatures to form a heavier nucleus. Such a process releases even a greater amount of energy than nuclear fission and is used in making hydrogen, or thermonuclear, bombs. (Sections 19.5 and 19.6)

Uses of Isotopes Isotopes, especially radioactive isotopes, are used as tracers to study the mechanisms of chemical and biological reactions and as medical diagnostic tools. (Section 19.7)

Biological Effects of Radiation The penetrating and harmful effects of radiation on biological systems have been thoroughly studied and are well understood. (Section 19.8)

Key Equations

$E = mc^2$ (19.1)	Einstein's mass-energy equivalence relationship
$\Delta E = (\Delta m)c^2$ (19.2)	Relation between mass defect and energy released

Key Words

Alpha decay, p. 868
Breeder reactor, p. 885
Critical mass, p. 882
Electron capture, p. 868
Gamma emission, p. 869
Mass defect, p. 870
Moderators, p. 884
Nuclear binding energy, p. 870
Nuclear chain reaction, p. 882
Nuclear fission, p. 881
Nuclear fusion, p. 887
Nuclear transmutation, p. 864
Nucleon, p. 870
Plasma, p. 888
Positron, p. 865
Positron emission, p. 868
Radical, p. 894
Radioactive decay series, p. 873
Strong nuclear force, p. 867
Thermonuclear reaction, p. 888
Tracer, p. 891
Transuranium elements, p. 879

Questions & Problems

Red numbered problems solved in Student Solutions Manual

19.1 The Nature of Nuclear Reactions

Review Questions

19.1 How do nuclear reactions differ from ordinary chemical reactions?

19.2 What are the steps in balancing nuclear equations?

19.3 What is the difference between ${}^{0}_{-1}e$ and ${}^{0}_{-1}\beta$?

19.4 What is the difference between an electron and a positron?

19.5 Which of the following nuclear decays produces a daughter nucleus with a higher atomic number: (a) γ, (b) ${}^{0}_{+1}\beta$, (c) ${}^{0}_{-1}\beta$, (d) α?

19.6 The table here is a summary of different modes of nuclear decay. Fill in the changes in atomic number (Z), number of neutrons (N), and mass number (A) in each case. Use "+" sign for increase, "−" sign for decrease, and "0" for no change.

Decay Mode	Change in Z	Change in N	Change in A
α decay			
${}^{0}_{-1}\beta$ decay			
${}^{0}_{+1}\beta$ decay			
e^- capture			

Problems

19.7 Complete the following nuclear equations and identify X in each case:
(a) $^{26}_{12}Mg + ^{1}_{1}p \longrightarrow ^{4}_{2}\alpha + X$
(b) $^{59}_{27}Co + ^{2}_{1}H \longrightarrow ^{60}_{27}Co + X$
(c) $^{235}_{92}U + ^{1}_{0}n \longrightarrow ^{94}_{36}Kr + ^{139}_{56}Ba + 3X$
(d) $^{53}_{24}Cr + ^{4}_{2}\alpha \longrightarrow ^{1}_{0}n + X$
(e) $^{20}_{8}O \longrightarrow ^{20}_{9}F + X$

19.8 Complete the following nuclear equations and identify X in each case:
(a) $^{135}_{53}I \longrightarrow ^{135}_{54}Xe + X$
(b) $^{40}_{19}K \longrightarrow ^{0}_{-1}\beta + X$
(c) $^{59}_{27}Co + ^{1}_{0}n \longrightarrow ^{56}_{25}Mn + X$
(d) $^{235}_{92}U + ^{1}_{0}n \longrightarrow ^{99}_{40}Zr + ^{135}_{52}Te + 2X$

19.2 Nuclear Stability

Review Questions

19.9 State the general rules for predicting nuclear stability.

19.10 What is the belt of stability?

19.11 Why is it impossible for the isotope $^{2}_{2}He$ to exist?

19.12 Define nuclear binding energy, mass defect, and nucleon.

19.13 How does Einstein's equation, $E = mc^2$, enable us to calculate nuclear binding energy?

19.14 Why is it preferable to use nuclear binding energy per nucleon for a comparison of the stabilities of different nuclei?

Problems

19.15 The radius of a uranium-235 nucleus is about 7.0×10^{-13} pm. Calculate the density of the nucleus in g/cm^3. (Assume the atomic mass is 235 amu.)

19.16 For each pair of isotopes listed, predict which one is less stable: (a) $^{6}_{3}Li$ or $^{9}_{3}Li$, (b) $^{23}_{11}Na$ or $^{25}_{11}Na$, (c) $^{48}_{20}Ca$ or $^{48}_{21}Sc$.

19.17 For each pair of elements listed, predict which one has more stable isotopes: (a) Co or Ni, (b) F or Se, (c) Ag or Cd.

19.18 In each pair of isotopes shown, indicate which one you would expect to be radioactive: (a) $^{20}_{10}Ne$ and $^{17}_{10}Ne$, (b) $^{40}_{20}Ca$ and $^{45}_{20}Ca$, (c) $^{95}_{42}Mo$ and $^{92}_{43}Tc$, (d) $^{195}_{80}Hg$ and $^{196}_{80}Hg$, (e) $^{209}_{83}Bi$ and $^{242}_{96}Cm$.

19.19 Given that

$$H(g) + H(g) \longrightarrow H_2(g) \qquad \Delta H° = -436.4 \text{ kJ/mol}$$

calculate the change in mass (in kg) per mole of H_2 formed.

19.20 Estimates show that the total energy output of the sun is 5×10^{26} J/s. What is the corresponding mass loss in kg/s of the sun?

19.21 Calculate the nuclear binding energy (in J) and the binding energy per nucleon of the following isotopes: (a) $^{7}_{3}Li$ (7.01600 amu), (b) $^{35}_{17}Cl$ (34.95952 amu).

19.22 Calculate the nuclear binding energy (in J) and the binding energy per nucleon of the following isotopes: (a) $^{4}_{2}He$ (4.0026 amu), (b) $^{184}_{74}W$ (183.9510 amu).

19.3 Natural Radioactivity

Review Questions

19.23 Discuss factors that lead to nuclear decay.

19.24 Outline the principle for dating materials using radioactive isotopes.

Problems

19.25 Fill in the blanks in the following radioactive decay series:
(a) $^{232}Th \xrightarrow{\alpha}$ ____ $\xrightarrow{\beta}$ ____ $\xrightarrow{\beta}$ ^{228}Th
(b) $^{235}U \xrightarrow{\alpha}$ ____ $\xrightarrow{\beta}$ ____ $\xrightarrow{\alpha}$ ^{227}Ac
(c) ____ $\xrightarrow{\alpha}$ $^{233}Pa \xrightarrow{\beta}$ ____ $\xrightarrow{\alpha}$ ____

19.26 A radioactive substance undergoes decay as follows:

Time (days)	Mass (g)
0	500
1	389
2	303
3	236
4	184
5	143
6	112

Calculate the first-order decay constant and the half-life of the reaction.

19.27 The radioactive decay of Tl-206 to Pb-206 has a half-life of 4.20 min. Starting with 5.00×10^{22} atoms of Tl-206, calculate the number of such atoms left after 42.0 min.

19.28 A freshly isolated sample of ^{90}Y was found to have an activity of 9.8×10^5 disintegrations per minute at 1:00 P.M. on December 3, 2020. At 2:15 P.M. on December 17, 2020, its activity was redetermined and found to be 2.6×10^4 disintegrations per minute. Calculate the half-life of ^{90}Y.

19.29 Why do radioactive decay series obey first-order kinetics?

19.30 In the thorium decay series, thorium-232 loses a total of 6 α particles and 4 β particles in a 10-stage process. What is the final isotope produced?

19.31 Strontium-90 is one of the products of the fission of uranium-235. This strontium isotope is radioactive, with a half-life of 28.1 yr. Calculate how long (in yr) it will take for 1.00 g of the isotope to be reduced to 0.200 g by decay.

19.32 Consider the decay series

$$A \longrightarrow B \longrightarrow C \longrightarrow D$$

where A, B, and C are radioactive isotopes with half-lives of 4.50 s, 15.0 days, and 1.00 s, respectively, and D is nonradioactive. Starting with 1.00 mole of A, and none of B, C, or D, calculate the number of moles of A, B, C, and D left after 30 days.

19.33 Two radioactive isotopes X and Y have the same molar amount at $t = 0$. A week later, there are four times as many X as there are Y. If the half-life of X is 2.0 d, calculate the half-life of Y in days.

19.34 Determine the symbol $^{A}_{Z}X$ for the parent nucleus whose α decay produces the same daughter as the $^{0}_{-1}\beta$ decay of $^{220}_{85}At$.

19.4 Nuclear Transmutation

Review Questions

19.35 What is the difference between radioactive decay and nuclear transmutation?

19.36 How is nuclear transmutation achieved in practice?

Problems

19.37 Write balanced nuclear equations for the following reactions and identify X:
(a) X(p, α) $^{12}_{6}C$, (b) $^{27}_{13}Al$ (d, α)X, (c) $^{55}_{25}Mn$ (n, γ)X.

19.38 Write balanced nuclear equations for the following reactions and identify X:
(a) $^{80}_{34}Se$(d, p)X, (b) X(d,2p)$^{9}_{3}Li$, (c) $^{10}_{5}B$(n, α)X.

19.39 Describe how you would prepare astatine-211, starting with bismuth-209.

19.40 A long-cherished dream of alchemists was to produce gold from cheaper and more abundant elements. This dream was finally realized when $^{198}_{80}Hg$ was converted into gold by neutron bombardment. Write a balanced equation for this reaction.

19.5 Nuclear Fission

Review Questions

19.41 Define nuclear fission, nuclear chain reaction, and critical mass.

19.42 Which isotopes can undergo nuclear fission?

19.43 Explain how an atomic bomb works.

19.44 Explain the functions of a moderator and a control rod in a nuclear reactor.

19.45 Discuss the differences between a light water and a heavy water nuclear fission reactor. What are the advantages of a breeder reactor over a conventional nuclear fission reactor?

19.46 No form of energy production is without risk. Make a list of the risks to society involved in fueling and operating a conventional coal-fired electric power plant, and compare them with the risks of fueling and operating a nuclear fission-powered electric plant.

19.6 Nuclear Fusion

Review Questions

19.47 Define nuclear fusion, thermonuclear reaction, and plasma.

19.48 Why do heavy elements such as uranium undergo fission while light elements such as hydrogen and lithium undergo fusion?

19.49 How does a hydrogen bomb work?

19.50 What are the advantages of a fusion reactor over a fission reactor? What are the practical difficulties in operating a large-scale fusion reactor?

19.7 Uses of Isotopes

Problems

19.51 Describe how you would use a radioactive iodine isotope to demonstrate that the following process is in dynamic equilibrium:

$$PbI_2(s) \rightleftharpoons Pb^{2+}(aq) + 2I^-(aq)$$

19.52 Consider the following redox reaction:

$$IO_4^-(aq) + 2I^-(aq) + H_2O(l) \longrightarrow I_2(s) + IO_3^-(aq) + 2OH^-(aq)$$

When KIO_4 is added to a solution containing iodide ions labeled with radioactive iodine-128, all the radioactivity appears in I_2 and none in the IO_3^- ion. What can you deduce about the mechanism for the redox process?

19.53 Explain how you might use a radioactive tracer to show that ions are not completely motionless in crystals.

19.54 Each molecule of hemoglobin, the oxygen carrier in blood, contains four Fe atoms. Explain how you would use the radioactive $^{59}_{26}Fe$($t_{\frac{1}{2}}$ = 46 days) to show that the iron in a certain food is converted into hemoglobin.

Additional Problems

19.55 In the chapter, we saw that the unit curie corresponds to exactly 3.70×10^{10} nuclear disintegration per second for 1 g of radium. Derive this unit given that the half-life of $^{226}_{88}Ra$ is 1.6×10^3 yr.

19.56 Manganese-50 (blue spheres) decays via $_{+1}^{0}\beta$ particle emission with a half-life of 0.282 s. (a) Write a balanced nuclear equation for the process. (b) From the diagram shown here, determine how many half-lives have elapsed. (The green spheres represent the decay product.)

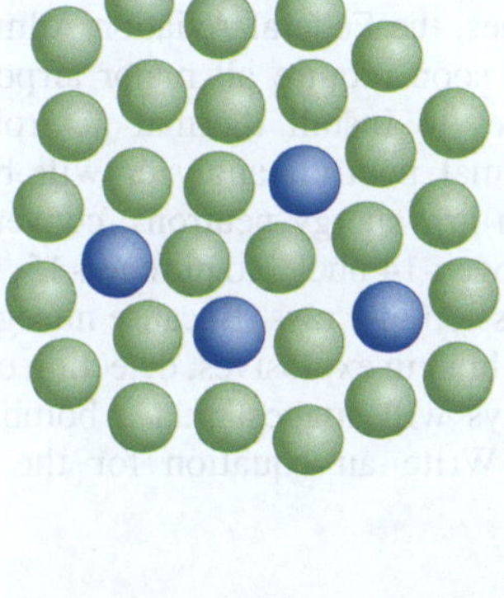

19.57 How does a Geiger counter work?

19.58 Nuclei with an even number of protons and an even number of neutrons are more stable than those with an odd number of protons and/or an odd number of neutrons. What is the significance of the even numbers of protons and neutrons in this case?

19.59 Tritium, ^3H, is radioactive and decays by electron emission. Its half-life is 12.5 yr. In ordinary water the ratio of ^1H to ^3H atoms is 1.0×10^{17} to 1. (a) Write a balanced nuclear equation for tritium decay. (b) How many disintegrations will be observed per minute in a 1.00-kg sample of water?

19.60 (a) What is the activity, in millicuries, of a 0.500-g sample of $^{237}_{93}Np$? (This isotope decays by α-particle emission and has a half-life of 2.20×10^6 yr.) (b) Write a balanced nuclear equation for the decay of $^{237}_{93}Np$.

19.61 The following equations are for nuclear reactions that are known to occur in the explosion of an atomic bomb. Identify X.

(a) $^{235}_{92}U + ^{1}_{0}n \longrightarrow ^{140}_{56}Ba + 3^{1}_{0}n + X$

(b) $^{235}_{92}U + ^{1}_{0}n \longrightarrow ^{144}_{55}Cs + ^{90}_{37}Rb + 2X$

(c) $^{235}_{92}U + ^{1}_{0}n \longrightarrow ^{87}_{35}Br + 3^{1}_{0}n + X$

(d) $^{235}_{92}U + ^{1}_{0}n \longrightarrow ^{160}_{62}Sm + ^{72}_{30}Zn + 4X$

19.62 Calculate the nuclear binding energies, in J/nucleon, for the following species: (a) ^{10}B (10.0129 amu), (b) ^{11}B (11.00931 amu), (c) ^{14}N (14.00307 amu), (d) ^{56}Fe (55.9349 amu).

19.63 Write complete nuclear equations for the following processes: (a) tritium, ^3H, undergoes β decay; (b) ^{242}Pu undergoes α-particle emission; (c) ^{131}I undergoes β decay; (d) ^{251}Cf emits an α particle.

19.64 The nucleus of nitrogen-18 lies above the stability belt. Write an equation for a nuclear reaction by which nitrogen-18 can achieve stability.

19.65 Why is strontium-90 a particularly dangerous isotope for humans?

19.66 How are scientists able to tell the age of a fossil?

19.67 After the Chernobyl accident, people living close to the nuclear reactor site were urged to take large amounts of potassium iodide as a safety precaution. What is the chemical basis for this action?

19.68 Astatine, the last member of Group 17, can be prepared by bombarding bismuth-209 with α particles. (a) Write an equation for the reaction. (b) Represent the equation in the abbreviated form, as discussed in Section 19.4.

19.69 To detect bombs that may be smuggled onto airplanes, the Federal Aviation Administration (FAA) will soon require all major airports in the United States to install thermal neutron analyzers. The thermal neutron analyzer will bombard baggage with low-energy neutrons, converting some of the nitrogen-14 nuclei to nitrogen-15, with simultaneous emission of γ rays. Because nitrogen content is usually high in explosives, detection of a high dosage of γ rays will suggest that a bomb may be present. (a) Write an equation for the nuclear process. (b) Compare this technique with the conventional X-ray detection method.

19.70 Explain why achievement of nuclear fusion in the laboratory requires a temperature of about 100 million degrees Celsius, which is much higher than that in the interior of the sun (15 million degrees Celsius).

19.71 Tritium contains one proton and two neutrons. There is no proton-proton repulsion present in the nucleus. Why, then, is tritium radioactive?

19.72 The carbon-14 decay rate of a sample obtained from a young tree is 0.260 disintegration per second per gram of the sample. Another wood sample prepared from an object recovered at an archaeological excavation gives a decay rate of 0.186 disintegration per second per gram of the sample. What is the age of the object?

19.73 The usefulness of radiocarbon dating is limited to objects no older than 50,000 yr. What percent of the carbon-14, originally present in the sample, remains after this period of time?

19.74 The radioactive potassium-40 isotope decays to argon–40 with a half-life of 1.2×10^9 yr. (a) Write a balanced equation for the reaction. (b) A sample of moon rock is found to contain 18 percent potassium-40 and 82 percent argon by mass. Calculate the age of the rock in years.

19.75 Both barium (Ba) and radium (Ra) are members of Group 2 and are expected to exhibit similar chemical properties. However, Ra is not found in barium ores. Instead, it is found in uranium ores. Explain.

19.76 Nuclear waste disposal is one of the major concerns of the nuclear industry. In choosing a safe and stable environment to store nuclear wastes, consideration must be given to the heat released during nuclear decay. As an example, consider the β decay of ^{90}Sr(89.907738 amu):

$$^{90}_{38}Sr \longrightarrow ^{90}_{39}Y + ^{0}_{-1}\beta \qquad t_{\frac{1}{2}} = 28.1 \text{ yr}$$

The ^{90}Y(89.907152 amu) further decays as follows:

$$^{90}_{39}Y \longrightarrow ^{90}_{40}Zr + ^{0}_{-1}\beta \qquad t_{\frac{1}{2}} = 64 \text{ h}$$

Zirconium-90 (89.904703 amu) is a stable isotope. (a) Use the mass defect to calculate the energy released (in joules) in each of the above two decays. (The mass of the electron is 5.4857×10^{-4} amu.) (b) Starting with one mole of ^{90}Sr, calculate the number of moles of ^{90}Sr that will decay in a year. (c) Calculate the amount of heat released (in kilojoules) corresponding to the number of moles of ^{90}Sr decayed to ^{90}Zr in (b).

19.77 Calculate the energy released (in joules) from the following fusion reaction:

$$^{2}_{1}H + ^{3}_{1}H \longrightarrow ^{4}_{2}He + ^{1}_{0}n$$

The atomic masses are ${}^{2}_{1}H = 2.0140$ amu, ${}^{3}_{1}H = 3.01603$ amu, ${}^{4}_{2}He = 4.00260$ amu, ${}^{1}_{0}n = 1.008665$ amu.

19.78 As a result of being exposed to the radiation released during the Chernobyl nuclear accident, the dose of iodine-131 in a person's body is 7.4 mCi(1 mCi = 1×10^{-3} Ci). Use the relationship rate = λN to calculate the number of atoms of iodine-131 to which this radioactivity corresponds. (The half-life of ${}^{131}I$ is 8.1 d.)

19.79 Referring to the Chemistry in Action essay "Boron Neutron Capture Therapy" (Section 19.8), why is it highly unlikely that irradiated food would become radioactive?

19.80 From the definition of curie, calculate Avogadro's number, given that the molar mass of ${}^{226}Ra$ is 226.03 g/mol and that it decays with a half-life of 1.6×10^{3} yr.

19.81 As of 2011, elements 113 through 118 have all been synthesized. Element 113 (Nh) was formed by the alpha decay of element 115 (Mc); element 114 (Fl) was created by bombarding ${}^{244}Pu$ with ${}^{48}Ca$; element 115 (Mc) was created by bombarding ${}^{243}Am$ with ${}^{48}Ca$; element 116 (Lv) was created by bombarding ${}^{248}Cm$ with ${}^{48}Ca$; element 117 (Ts) was created by bombarding ${}^{249}Bk$ with ${}^{48}Ca$; element 118 (Og) was created by bombarding ${}^{249}Cf$ with ${}^{48}Ca$. Write an equation for each synthesis. Predict the chemical properties of these elements.

19.82 Sources of energy on Earth include fossil fuels, geothermal, gravitational, hydroelectric, nuclear fission, nuclear fusion, solar, and wind. Which of these have a "nuclear origin," either directly or indirectly?

19.83 A person received an anonymous gift of a decorative box, which he placed on his desk. A few months later he became ill and died shortly afterward. After investigation, the cause of his death was linked to the box. The box was airtight and had no toxic chemicals on it. What might have killed the man?

19.84 Identify two of the most abundant radioactive elements that exist on Earth. Explain why they are still present. (You may need to consult a handbook of chemistry.)

19.85 (a) Calculate the energy released when an U-238 isotope decays to Th-234. The atomic masses are given by U-238: 238.0508 amu; Th-234: 234.0436 amu; He-4: 4.0026 amu. (b) The energy released in (a) is transformed into the kinetic energy of the recoiling Th-234 nucleus and the α particle. Which of the two will move away faster? Explain.

19.86 Cobalt-60 is an isotope used in diagnostic medicine and cancer treatment. It decays with γ-ray emission. Calculate the wavelength of the radiation in nanometers if the energy of the γ ray is 2.4×10^{-13} J/photon.

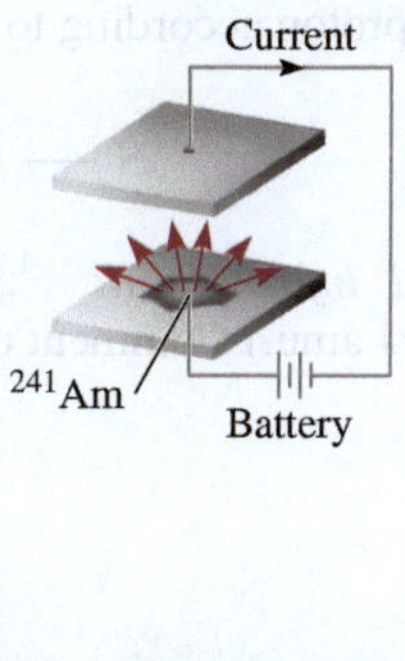

19.87 Americium-241 is used in smoke detectors because it has a long half-life (458 yr) and its emitted α particles are energetic enough to ionize air molecules. Given the schematic diagram of a smoke detector, explain how it works.

19.88 The constituents of wine contain, among others, carbon, hydrogen, and oxygen atoms. A bottle of wine was sealed about 6 yr ago. To confirm its age, which of the isotopes would you choose in a radioactive dating study? The half-lives of the isotopes are ${}^{13}C$: 5730 yr; ${}^{15}O$: 124 s; ${}^{3}H$: 12.5 yr. Assume that the activities of the isotopes were known at the time the bottle was sealed.

19.89 Name two advantages of a nuclear-powered submarine over a conventional submarine.

19.90 In 1997, a scientist at a nuclear research center in Russia placed a thin shell of copper on a sphere of highly enriched uranium-235. Suddenly, there was a huge burst of radiation, which turned the air blue. Three days later, the scientist died of radiation damage. Explain what caused the accident. (*Hint:* Copper is an effective metal for reflecting neutrons.)

19.91 A radioactive isotope of copper decays as follows:

$${}^{64}Cu \longrightarrow {}^{64}Zn + {}^{0}_{-1}\beta \qquad t_{\frac{1}{2}} = 12.8 \text{ h}$$

Starting with 84.0 g of ${}^{64}Cu$, calculate the quantity of ${}^{64}Zn$ produced after 18.4 h.

19.92 A 0.0100-g sample of a radioactive isotope with a half-life of 1.3×10^{9} yr decays at the rate of 2.9×10^{4} dpm. Calculate the molar mass of the isotope.

19.93 In each of the diagrams (a)–(c), identify the isotopes involved and the type of decay process. Use the ${}^{A}_{Z}X$ symbol for each isotope.

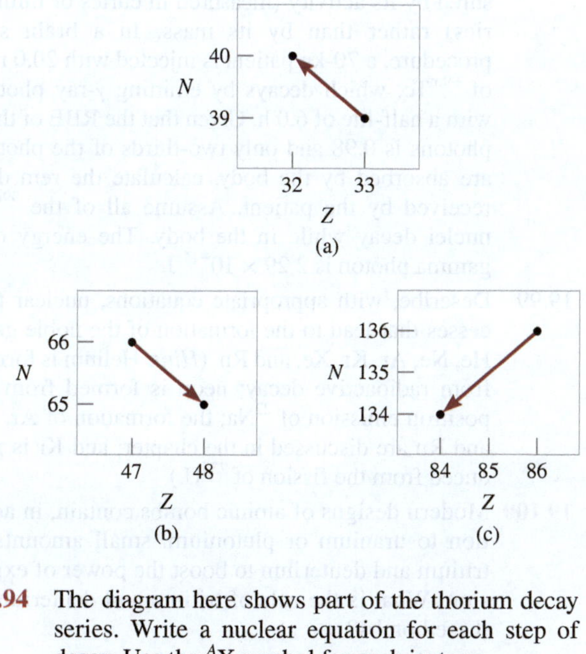

19.94 The diagram here shows part of the thorium decay series. Write a nuclear equation for each step of decay. Use the ${}^{A}_{Z}X$ symbol for each isotope.

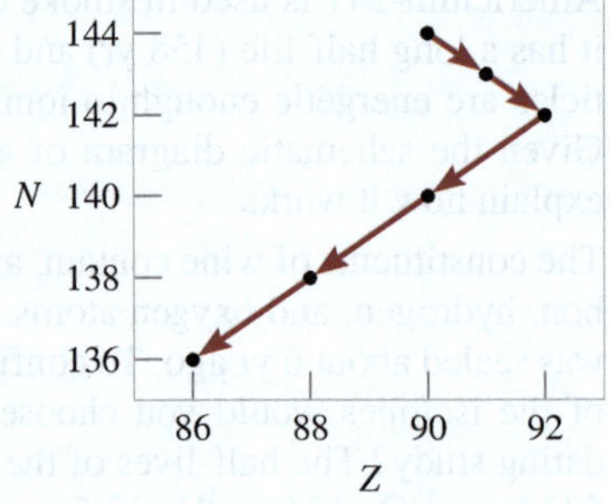

19.95 The half-life of ^{27}Mg is 9.50 min. (a) Initially there were 4.20×10^{12} ^{27}Mg nuclei present. How many ^{27}Mg nuclei are left 30.0 min later? (b) Calculate the ^{27}Mg activities (in Ci) at $t = 0$ and $t = 30.0$ min. (c) What is the probability that any one ^{27}Mg nucleus decays during a 1-s interval? What assumption is made in this calculation?

19.96 The radioactive isotope ^{238}Pu, used in pacemakers, decays by emitting an alpha particle with a half-life of 86 yr. (a) Write an equation for the decay process. (b) The energy of the emitted alpha particle is 9.0×10^{-13} J, which is the energy per decay. Assuming that all the alpha particle energy is used to run the pacemaker, calculate the power output at $t = 0$ and $t = 10$ yr. Initially 1.0 mg of ^{238}Pu was present in the pacemaker. (*Hint:* After 10 yr, the activity of the isotope decreases by 8.0 percent. Power is measured in watts or J/s.)

19.97 (a) Assuming nuclei are spherical in shape, show that its radius (r) is proportional to the cube root of mass number (A). (b) In general, the radius of a nucleus is given by $r = r_0 A^{\frac{1}{3}}$, where r_0, the proportionality constant, is given by 1.2×10^{-15} m. Calculate the volume of the ^{238}U nucleus.

19.98 The quantity of a radioactive material is often measured by its activity (measured in curies or millicuries) rather than by its mass. In a brain scan procedure, a 70-kg patient is injected with 20.0 mCi of ^{99m}Tc, which decays by emitting γ-ray photons with a half-life of 6.0 h. Given that the RBE of these photons is 0.98 and only two-thirds of the photons are absorbed by the body, calculate the rem dose received by the patient. Assume all of the ^{99m}Tc nuclei decay while in the body. The energy of a gamma photon is 2.29×10^{-14} J.

19.99 Describe, with appropriate equations, nuclear processes that lead to the formation of the noble gases He, Ne, Ar, Kr, Xe, and Rn. (*Hint:* Helium is formed from radioactive decay; neon is formed from the positron emission of ^{22}Na; the formation of Ar, Xe, and Rn are discussed in the chapter; and Kr is produced from the fission of ^{235}U.)

19.100 Modern designs of atomic bombs contain, in addition to uranium or plutonium, small amounts of tritium and deuterium to boost the power of explosion. What is the role of tritium and deuterium in these bombs?

19.101 What is the source of heat for volcanic activities on Earth?

19.102 Alpha particles produced from radioactive decays eventually pick up electrons from the surroundings to form helium atoms. Calculate the volume (mL) of He collected at STP when 1.00 g of pure ^{226}Ra is stored in a closed container for 100 yr. (*Hint:* Focusing only on half-lives that are short compared to 100 years and ignoring minor decay schemes in Table 19.3, first show that there are 5 α particles generated per ^{226}Ra decay to ^{206}Pb.)

19.103 In 2006, an ex-KGB agent was murdered in London. Subsequent investigation showed that the cause of death was poisoning with the radioactive isotope ^{210}Po, which was added to his drinks/food. (a) ^{210}Po is prepared by bombarding ^{209}Bi with neutrons. Write an equation for the reaction. (b) Who discovered the element polonium? (*Hint:* See an Internet source such as Webelements.com.) (c) The half-life of ^{210}Po is 138 d. It decays with the emission of an α particle. Write an equation for the decay process. (d) Calculate the energy of an emitted α particle. Assume both the parent and daughter nuclei to have zero kinetic energy. The atomic masses are ^{210}Po (209.98285 amu), ^{206}Pb (205.97444 amu), $^{4}_{2}\alpha$ (4.00150 amu). (e) Ingestion of 1 μg of ^{210}Po could prove fatal. What is the total energy released by this quantity of ^{210}Po?

19.104 An electron and a positron are accelerated to nearly the speed of light before colliding in a particle accelerator. The ensuing collision produces an exotic particle having a mass many times that of a proton. Does the result violate the law of conservation of mass?

19.105 The volume of an atom's nucleus is 1.33×10^{-42} m^3. The nucleus contains 110 neutrons. Identify the atom and write the symbol of the atom as $^{A}_{Z}$X. (*Hint:* See Problem 19.97.)

19.106 In the chapter, we learned to calculate the nuclear binding energy, which pertains to the stability of a particular nucleus. It is also possible to estimate the binding energy of a single nucleon (neutron or proton) to the remainder of the nucleus. (a) From the following nuclear equation and nuclear masses, calculate the binding energy of a single neutron:

$$^{14}_{7}\text{N} \longrightarrow {}^{13}_{7}\text{N} + {}^{1}_{0}\text{n}$$

(*Useful information:* $^{14}_{7}$N: 14.003074 amu; $^{13}_{7}$N: 13.005738 amu; $^{1}_{0}$n: 1.00866 amu.) (b) By a similar procedure, we can calculate the binding energy of a single proton according to the equation

$$^{14}_{7}\text{N} \longrightarrow {}^{13}_{6}\text{C} + {}^{1}_{1}\text{p}$$

(*Useful information:* $^{13}_{6}$C: 13.003355 amu; $^{1}_{1}$p: 1.00794 amu.) Comment on your results.

Interpreting, Modeling, & Estimating

19.107 Which of the following poses a greater health hazard: a radioactive isotope with a short half-life or a radioactive isotope with a long half-life? Assume equal molar amounts and the same type of radiation and comparable energies per particle emitted.

19.108 To start a deuterium-deuterium fusion reaction, it has been estimated that each nucleus needs an initial kinetic energy of about 4×10^{-14} J. What would be the corresponding temperature for the process? Why is this temperature value an overestimate?

19.109 In a science fiction novel a nuclear engineer designed a car powered by deuterium-deuterium fusion. Over a certain period of time, the car consumed 0.2 g of deuterium fuel. How many gallons of gasoline would have to be burned to equal the energy generated by the deuterium fuel? (For useful information about gasoline energy content, see Problem 17.73.)

19.110 The leakage of radioactive materials to the environment when a nuclear reactor core malfunctions is often made worse by explosions at the nuclear plant caused by hydrogen gas, as was the case in Fukushima, Japan, in 2011. Explain what caused the hydrogen explosion. (*Useful information:* The nuclear fuel rods are held in zirconium alloy tubes.)

Answers to Practice Exercises

19.1 $^{78}_{34}\text{Sc}$. **19.2** 2.63×10^{-10} J; 1.26×10^{-12} J/nucleon. **19.3** $^{106}_{46}\text{Pd} + ^{4}_{2}\alpha \longrightarrow ^{110}_{48}\text{Ag} + ^{1}_{1}\text{p}$.

Answers to Review of Concepts & Facts

19.1.1 $^{1}_{1}\text{p} \longrightarrow ^{0}_{+1}\beta + ^{1}_{0}\text{n}$. **19.1.2** $^{89}_{37}\text{Rb}$. **19.1.3** $^{55}_{26}\text{Fe}$. **19.2.1** (a) β emission: $^{13}_{5}\text{B} \longrightarrow ^{13}_{6}\text{C} + ^{0}_{-1}\beta$. (b) Positron emission: $^{188}_{79}\text{Au} \longrightarrow ^{188}_{78}\text{Pt} + ^{0}_{+1}\beta$. **19.2.2** 1.215×10^{-12} J/nucleon. **19.2.3** $\Delta m = -9.9 \times 10^{-12}$ kg. **19.3.1** 1.00 g. **19.3.2** (a) $^{59}_{26}\text{Fe} \longrightarrow ^{59}_{27}\text{Co} + ^{0}_{-1}\beta$. (b) 3. **19.4.1** $^{27}_{13}\text{Al}(\alpha, \text{n})\,^{30}_{15}\text{P}$. **19.4.2** $^{290}_{116}\text{Lv}$; $^{245}_{96}\text{Cm} + ^{48}_{20}\text{Ca} \longrightarrow ^{290}_{116}\text{Lv} + 3^{1}_{0}\text{n}$. **19.5.1** Boron compounds are added to capture neutrons to cool the fuel rods in attempt to prevent meltdown.

CHAPTER

20

Chemistry in the Atmosphere

Lightning causes atmospheric nitrogen and oxygen to form nitric oxide, which is eventually converted to nitrates.
View Stock/age fotostock

CHAPTER OUTLINE

We have studied basic definitions in chemistry, and we have examined the properties of gases, liquids, solids, and solutions. We have discussed chemical bonding and intermolecular forces and seen how chemical kinetics and chemical equilibrium concepts help us understand the nature of chemical reactions. It is appropriate at this stage to apply our knowledge to the study of one extremely important system: the atmosphere. Although Earth's atmosphere is fairly simple in composition, its chemistry is very complex and not fully understood. The chemical processes that take place in our atmosphere are induced by solar radiation, but they are intimately connected to natural events and human activities on Earth's surface.

In this chapter, we will discuss the structure and composition of the atmosphere, together with some of the chemical processes that occur there. In addition, we will take a look at the major sources of air pollution and prospects for controlling them.

20.1 Earth's Atmosphere

Learning Objectives

- Summarize the process of nitrogen fixation.
- Categorize the regions of Earth's atmosphere.

Earth is unique among the planets of our solar system in having an atmosphere that is chemically active and rich in oxygen. Mars, for example, has a much thinner atmosphere that is about 90 percent carbon dioxide. Jupiter, on the other hand, has no solid surface; it is made up of 90 percent hydrogen, 9 percent helium, and 1 percent other substances.

It is generally believed that 3 to 4 billion years ago, Earth's atmosphere consisted mainly of ammonia, methane, and water. There was little, if any, free oxygen present. Ultraviolet (UV) radiation from the sun probably penetrated the atmosphere, rendering the surface of Earth sterile. However, the same UV radiation may have triggered the chemical reactions (perhaps beneath the surface) that eventually led to life on Earth. Primitive organisms used energy from the sun to break down carbon dioxide (produced by volcanic activity) to obtain carbon, which they incorporated in their own cells. The major by-product of this process, called *photosynthesis,* is oxygen. Another important source of oxygen is the *photodecomposition* of water vapor by UV light. Over time, the more reactive gases such as ammonia and methane have largely disappeared, and today our atmosphere consists mainly of oxygen and nitrogen gases. Biological processes determine to a great extent the atmospheric concentrations of these gases, one of which is reactive (oxygen) and the other unreactive (nitrogen).

Table 20.1 shows the composition of dry air at sea level. The total mass of the atmosphere is about 5.3×10^{18} kg. Water is excluded from this table because its concentration in air can vary drastically from location to location.

Figure 20.1 shows the major processes involved in the cycle of nitrogen in nature. Molecular nitrogen, with its triple bond, is a very stable molecule. However, through biological and industrial ***nitrogen fixation,*** *the conversion of molecular nitrogen into nitrogen compounds,* atmospheric nitrogen gas is converted into nitrates and other compounds suitable for assimilation by algae and plants. Another important mechanism for producing nitrates from nitrogen gas is lightning. The steps are

$$N_2(g) + O_2(g) \xrightarrow[\text{energy}]{\text{electrical}} 2NO(g)$$
$$2NO(g) + O_2(g) \longrightarrow 2NO_2(g)$$
$$2NO_2(g) + H_2O(l) \longrightarrow HNO_2(aq) + HNO_3(aq)$$

About 30 million tons of HNO_3 are produced this way annually. Nitric acid is converted to nitrate salts in the soil. These nutrients are taken up by plants, which in turn are ingested by animals. Animals use the nutrients from plants to make proteins and other essential biomolecules. Denitrification reverses nitrogen fixation to complete the cycle.

Table 20.1

Composition of Dry Air at Sea Level

Gas	Composition Gas (% by Volume)
N_2	78.09
O_2	20.95
Ar	0.93
CO_2	0.040
Ne	0.0018
He	0.000524
Kr	0.00011
Xe	0.000006

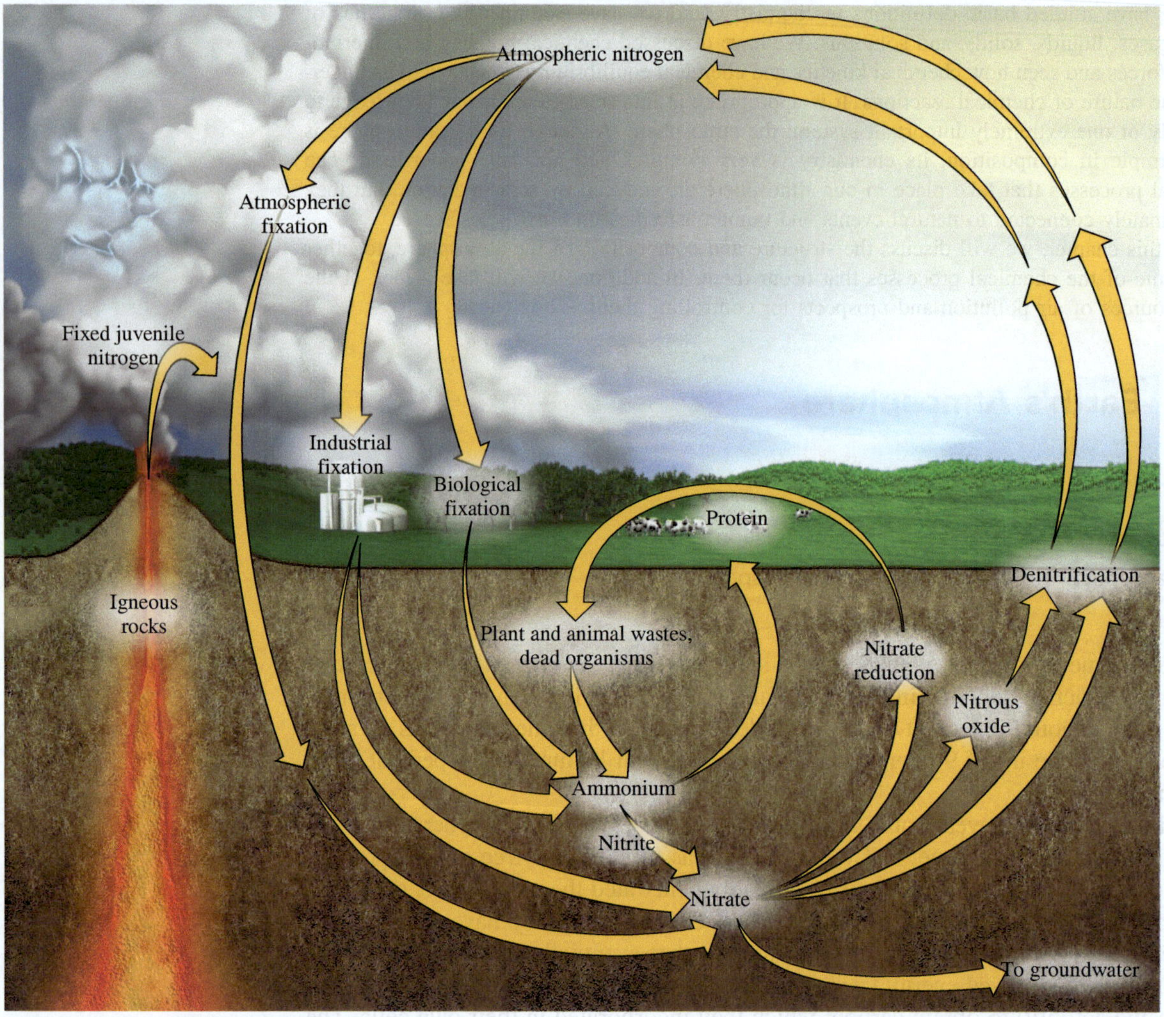

Figure 20.1 *The nitrogen cycle. Although the supply of nitrogen in the atmosphere is virtually inexhaustible, it must be combined with hydrogen or oxygen before it can be assimilated by higher plants, which in turn are consumed by animals. Juvenile nitrogen is nitrogen that has not previously participated in the nitrogen cycle.*

For example, certain anaerobic organisms decompose animal wastes as well as dead plants and animals to produce free molecular nitrogen from nitrates.

The main processes of the global oxygen cycle are shown in Figure 20.2. This cycle is complicated by the fact that oxygen takes so many different chemical forms. Atmospheric oxygen is removed through respiration and various industrial processes (mostly combustion), which produce carbon dioxide. Photosynthesis is the major mechanism by which molecular oxygen is regenerated from carbon dioxide and water.

Scientists divide the atmosphere into several different layers according to temperature variation and composition (Figure 20.3). As far as visible events are concerned, the most active region is the ***troposphere,*** *the layer of the atmosphere that contains about 80 percent of the total mass of air and practically all of the atmosphere's water vapor.* The troposphere is the thinnest layer of the atmosphere (10 km), but it is where all the dramatic events of weather—rain, lightning, hurricanes—occur. Temperature decreases almost linearly with increasing altitude in this region.

Above the troposphere is the ***stratosphere,*** *which consists of nitrogen, oxygen, and ozone.* In the stratosphere, the air temperature *rises* with altitude. This warming effect is the result of exothermic reactions triggered by UV radiation from the sun (to be discussed in Section 20.3). One of the products of this reaction sequence is ozone (O_3), which, as we will see shortly, serves to prevent harmful UV rays from reaching Earth's surface.

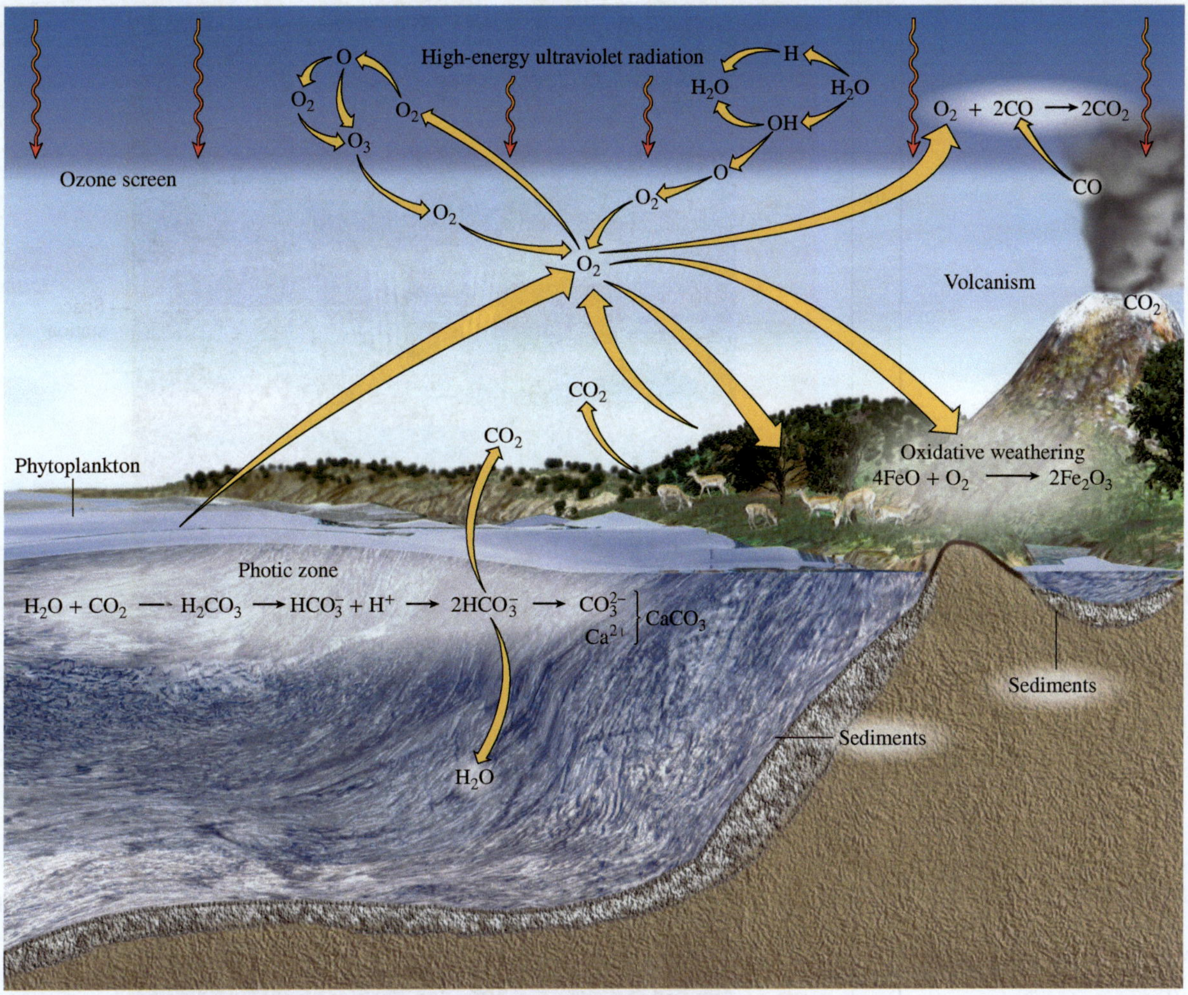

Figure 20.2 *The oxygen cycle. The cycle is complicated because oxygen appears in so many chemical forms and combinations, primarily as molecular oxygen, in water, and in organic and inorganic compounds.*

In the ***mesosphere,*** which is *above the stratosphere,* the concentration of ozone and other gases is low, and the temperature decreases with increasing altitude. The ***thermosphere,*** or ***ionosphere,*** is *the uppermost layer of the atmosphere.* The rise in temperature in this region is the result of the bombardment of molecular oxygen and nitrogen and atomic species by energetic particles, such as electrons and protons, from the sun. Typical reactions are

$$N_2 \longrightarrow 2N \qquad \Delta H^\circ = 945 \text{ kJ/mol}$$
$$N \longrightarrow N^+ + e^- \qquad \Delta H^\circ = 1402 \text{ kJ/mol}$$
$$O_2 \longrightarrow O_2^+ + e^- \qquad \Delta H^\circ = 1176 \text{ kJ/mol}$$

In reverse, these processes liberate the equivalent amount of energy, mostly as heat. Ionized particles are responsible for the reflection of radio waves back toward Earth.

Summary of Concepts & Facts

- Earth's atmosphere is made up mainly of nitrogen and oxygen, plus a number of other trace gases. The chemical processes that go on in the atmosphere are influenced by solar radiation, volcanic eruption, and human activities.
- Nitrogen fixation is the process of converting atmospheric nitrogen into nitrogen compounds that are important for sustaining algae and plants.

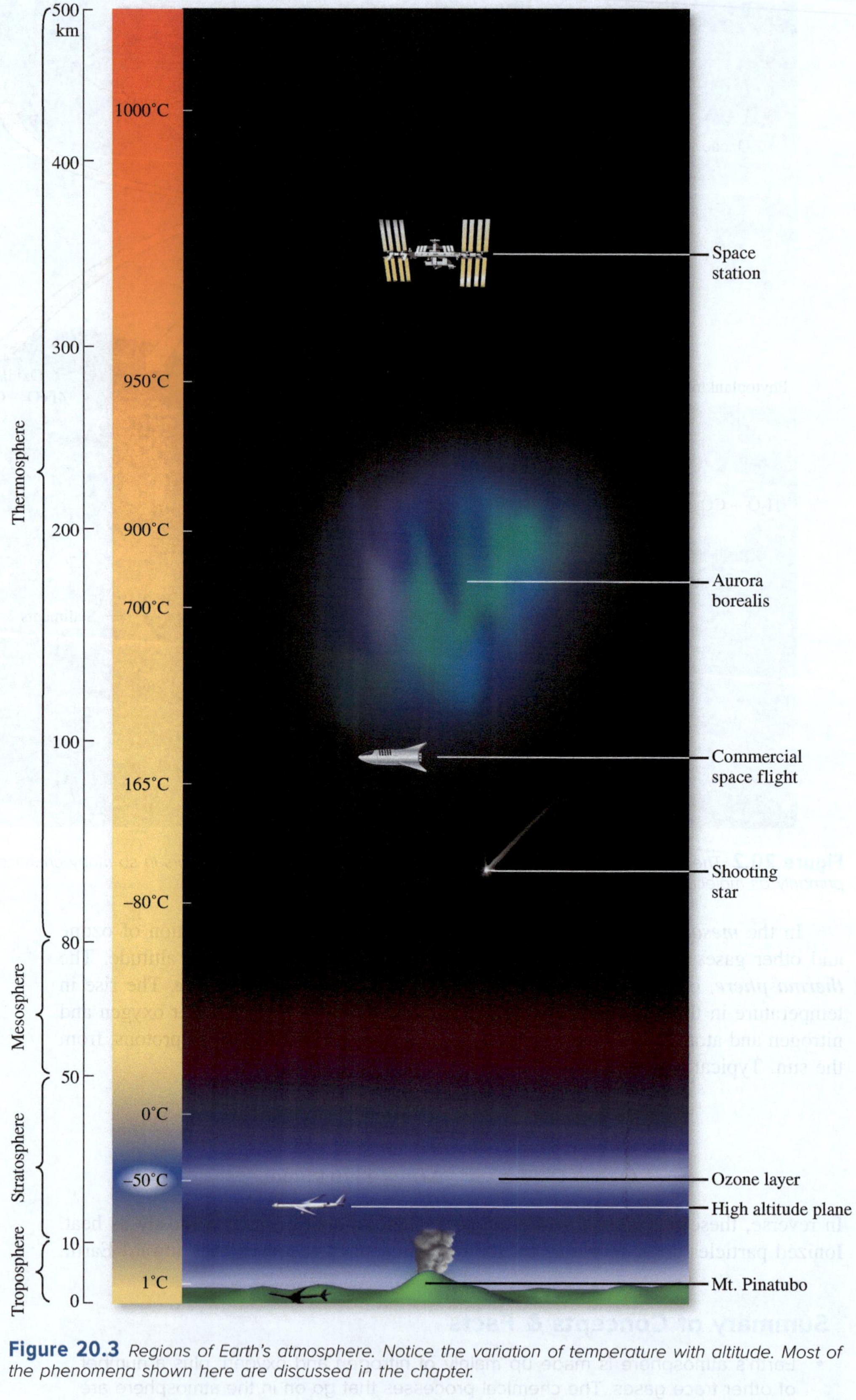

Figure 20.3 *Regions of Earth's atmosphere. Notice the variation of temperature with altitude. Most of the phenomena shown here are discussed in the chapter.*

20.2 Phenomena in the Outer Layers of the Atmosphere

Learning Objective

- Outline the process that gives rise to the aurora borealis and aurora australis.

In this section, we will discuss two dazzling phenomena that occur in the outer regions of the atmosphere. One is a natural event. The other is a curious by-product of human space travel.

Aurora Borealis and Aurora Australis

Violent eruptions on the surface of the sun, called *solar flares,* result in the ejection of myriad electrons and protons into space, where they disrupt radio transmission and provide us with spectacular celestial light shows known as *auroras* (Figure 20.4). These electrons and protons collide with the molecules and atoms in Earth's upper atmosphere, causing them to become ionized and electronically excited. Eventually, the excited molecules and ions return to the ground state with the emission of light. For example, an excited oxygen atom emits photons at wavelengths of 558 nm (green) and between 630 nm and 636 nm (red):

$$O^* \longrightarrow O + h\nu$$

where the asterisk denotes an electronically excited species and $h\nu$ the emitted photon (see Section 7.2). Similarly, the blue and violet colors often observed in auroras result from the transition in the ionized nitrogen molecule:

$$N_2^{+*} \longrightarrow N_2^+ + h\nu$$

The wavelengths for this transition fall between 391 and 470 nm.

The incoming streams of solar protons and electrons are oriented by Earth's magnetic field so that most auroral displays occur in doughnut-shaped zones about 2000 km in diameter centered on the North and South Poles. *Aurora borealis* is the name given to this phenomenon in the Northern Hemisphere. In the Southern Hemisphere, it is called *aurora australis.* Sometimes, the number of solar particles is so immense that auroras are also visible from other locations on Earth.

Figure 20.4 *Aurora borealis, commonly referred to as the northern lights.*
Atli Mar Hafsteinsson/Cultura/Image Source

Example 20.1

The bond enthalpy of O_2 is 498.7 kJ/mol. Calculate the maximum wavelength (nm) of a photon that can cause the dissociation of an O_2 molecule.

Strategy We want to calculate the wavelength of a photon that will break an O═O bond. Therefore, we need the amount of energy in one bond. The bond enthalpy of O_2 is given in units of kJ/mol. The units needed for the energy of one bond are J/molecule. Once we know the energy in one bond, we can calculate the minimum frequency and maximum wavelength needed to dissociate one O_2 molecule. The conversion steps are

$$\text{kJ/mol} \longrightarrow \text{J/molecule} \longrightarrow \text{frequency of photon} \longrightarrow \text{wavelength of photon}$$

Solution First we calculate the energy required to break one O═O bond:

$$\frac{498.7 \times 10^3 \text{ J}}{1 \text{ mol}} \times \frac{1 \text{ mol}}{6.022 \times 10^{23} \text{ molecules}} = 8.281 \times 10^{-19} \frac{\text{J}}{\text{molecule}}$$

The energy of the photon is given by $E = h\nu$ [Equation (7.2)]. Therefore,

$$\nu = \frac{E}{h} = \frac{8.281 \times 10^{-19} \text{ J}}{6.63 \times 10^{-34} \text{ J} \cdot \text{s}}$$
$$= 1.25 \times 10^{-15} \text{ s}^{-1}$$

Finally, we calculate the wavelength of the photon, given by $\lambda = c/\nu$ [see Equation (7.1)], as follows:

$$\lambda = \frac{3.00 \times 10^8 \text{ m/s}}{1.25 \times 10^{15} \text{ s}^{-1}}$$
$$= 2.40 \times 10^{-7} \text{ m} = 240 \text{ nm}$$

Comment In principle, any photon with a wavelength of 240 nm or *shorter* can dissociate an O_2 molecule.

Practice Exercise Calculate the wavelength (in nm) of a photon needed to dissociate an O_3 molecule:

$$O_3 \longrightarrow O + O_2 \qquad \Delta H° = 107.2 \text{ kJ/mol}$$

Similar problem: 20.11.

The Mystery Glow of Space Shuttles

A human-made light show that baffled scientists for several years is produced by space shuttles orbiting Earth. In 1983, astronauts first noticed an eerie orange glow on the outside surface of their spacecraft at an altitude about 300 km above Earth (Figure 20.5). The light, which usually extends about 10 cm away from the protective silica heat tiles and other surface materials, is most pronounced on the parts of the shuttle facing its direction of travel. This fact led scientists to postulate that collision between oxygen atoms in the atmosphere and the fast-moving shuttle somehow produced the orange light. Spectroscopic measurements of the glow, as well as laboratory tests, strongly suggested that nitric oxide (NO) and nitrogen dioxide (NO_2) also played a part. It is believed that oxygen atoms interact with nitric oxide adsorbed on (that is, bound to) the shuttle's surface to form electronically excited nitrogen dioxide:

$$O + NO \longrightarrow NO_2^*$$

As the NO_2^* leaves the shell of the spacecraft, it emits photons at a wavelength of 680 nm (orange).

$$NO_2^* \longrightarrow NO_2 + h\nu$$

Support for this explanation came inadvertently in 1991, when astronauts aboard *Discovery* released various gases, including carbon dioxide, neon, xenon, and nitric oxide,

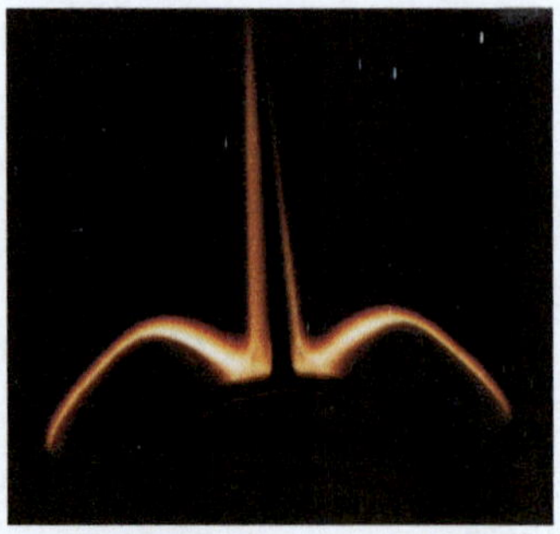

Figure 20.5 *The glowing tail section of the space shuttle viewed from inside the vehicle.*
JSC/NASA

from the cargo bay in the course of an unrelated experiment. Expelled one at a time, these gases scattered onto the surface of the shuttle's tail. The nitric oxide caused the normal shuttle glow to intensify markedly, but the other gases had no effect on it.

What is the source of the nitric oxide on the outside of the spacecraft? Scientists believe that some of it may come from the exhaust gases emitted by the shuttle's rockets and that some of it is present in the surrounding atmosphere. The shuttle glow does not harm the vehicle, but it does interfere with spectroscopic measurements on distant objects made from the spacecraft.

Summary of Concepts & Facts

- In the outer regions of the atmosphere the bombardment of molecules and atoms by solar particles gives rise to auroras. The glow on space shuttles is caused by excitation of molecules adsorbed on the shuttles' surface.

20.3 Depletion of Ozone in the Stratosphere

Learning Objective

- Evaluate the role of ozone in the stratosphere and how chlorofluorocarbons have affected the levels of ozone in the atmosphere.

As mentioned earlier, ozone in the stratosphere prevents UV radiation emitted by the sun from reaching Earth's surface. The formation of ozone in this region begins with the *photodissociation,* the breaking of chemical bonds by radiant energy, of oxygen molecules by solar radiation at wavelengths below 240 nm:

$$O_2 \xrightarrow[< 240\ \text{nm}]{\text{UV}} O + O \qquad (20.1)$$

The highly reactive O atoms combine with oxygen molecules to form ozone as follows:

$$O + O_2 + M \longrightarrow O_3 + M \qquad (20.2)$$

where M is some inert substance such as N_2. The role of M in this exothermic reaction is to absorb some of the excess energy released and prevent the spontaneous decomposition of the O_3 molecule. The energy that is not absorbed by M is given off as heat. (As the M molecules themselves become de-excited, they release more heat to the surroundings.) In addition, ozone itself absorbs UV light between 200 and 300 nm:

$$O_3 \xrightarrow{\text{UV}} O + O_2 \qquad (20.3)$$

The process continues when O and O_2 recombine to form O_3 as shown in Equation (20.2), further warming the stratosphere.

If all the stratospheric ozone were compressed into a single layer at STP on Earth, that layer would be only about 3 mm thick! Although the concentration of ozone in the stratosphere is very low, it is sufficient to filter out (that is, absorb) solar radiation in the 200- to 300-nm range [see Equation (20.3)]. In the stratosphere, it acts as our protective shield against UV radiation, which can induce skin cancer, cause genetic mutations, and destroy crops and other forms of vegetation.

The formation and destruction of ozone by natural processes is a dynamic equilibrium that maintains a constant concentration of ozone in the stratosphere. Since the mid-1970s scientists have been concerned about the harmful effects of certain chlorofluorocarbons (CFCs) on the ozone layer. The CFCs, which are generally known by the trade name Freons, were first synthesized in the 1930s. Some of the common ones are

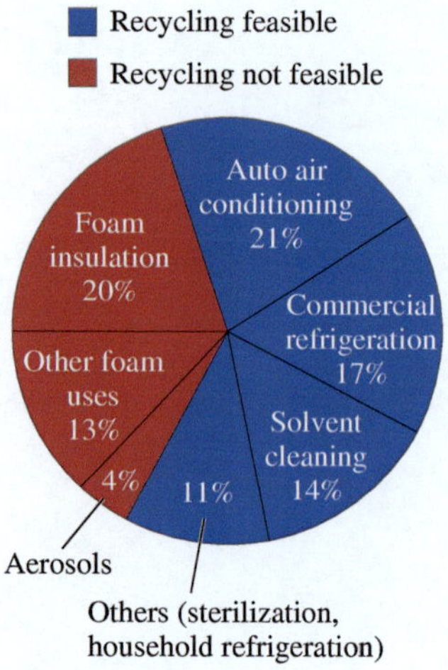

Figure 20.6 *Uses of CFCs. Since 1978, the use of aerosol propellants has been banned in the United States.*

$CFCl_3$ (Freon 11), CF_2Cl_2 (Freon 12), $C_2F_3Cl_3$ (Freon 113), and $C_2F_4Cl_2$ (Freon 114). Because these compounds are readily liquefied, relatively inert, nontoxic, noncombustible, and volatile, they have been used as coolants in refrigerators and air conditioners, in place of highly toxic liquid sulfur dioxide (SO_2) and ammonia (NH_3). Large quantities of CFCs are also used in the manufacture of disposable foam products such as cups and plates, as aerosol propellants in spray cans, and as solvents to clean newly soldered electronic circuit boards (Figure 20.6). In 1977, the peak year of production, nearly 1.5×10^6 tons of CFCs were produced in the United States. Most of the CFCs produced for commercial and industrial use are eventually discharged into the atmosphere.

Because of their relative inertness, the CFCs slowly diffuse unchanged up to the stratosphere (it can take years for CFCs to reach this layer of the atmosphere), where UV radiation of wavelengths between 175 and 220 nm causes them to decompose:

$$CFCl_3 \longrightarrow CFCl_2 + Cl$$
$$CF_2Cl_2 \longrightarrow CF_2Cl + Cl$$

The reactive chlorine atoms then undergo the following reactions:

$$Cl + O_3 \longrightarrow ClO + O_2 \quad (20.4)$$

$$ClO + O \longrightarrow Cl + O_2 \quad (20.5)$$

The overall result [sum of Equations (20.4) and (20.5)] is the net removal of an O_3 molecule from the stratosphere:

$$O_3 + O \longrightarrow 2O_2 \quad (20.6)$$

The oxygen atoms in Equation (20.5) are supplied by the photochemical decomposition of molecular oxygen and ozone described earlier. Note that the Cl atom plays the role of a catalyst in the reaction mechanism scheme represented by Equations (20.4) and (20.5) because it is not used up and therefore can take part in many such reactions. One Cl atom can destroy up to 100,000 O_3 molecules before it is removed by some other reaction. The ClO (chlorine monoxide) species is an intermediate because it is produced in the first elementary step [Equation (20.4)] and consumed in the second step [Equation (20.5)]. The preceding mechanism for the destruction of ozone has been supported by the detection of ClO in the stratosphere in recent years (Figure 20.7). As can be seen, the concentration of O_3 decreases in regions that have high amounts of ClO.

Another group of compounds that can destroy stratospheric ozone are the nitrogen oxides, generally denoted as NO_x. (Examples of NO_x are NO and NO_2.) These compounds come from the exhausts of high-altitude supersonic aircraft and from human and natural

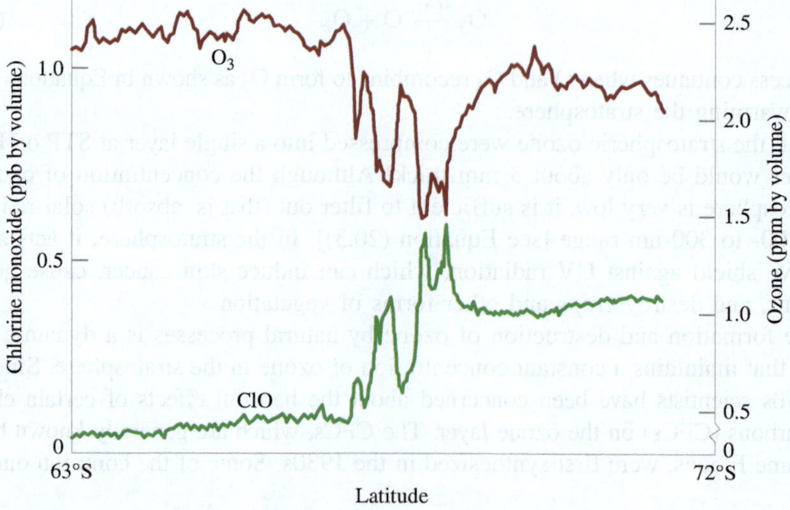

Figure 20.7 *The variations in the concentrations of ClO and O_3 with latitude.*

activities on Earth. Solar radiation decomposes a substantial amount of the other nitrogen oxides to nitric oxide (NO), which participates in the destruction of ozone as follows:

$$O_3 \longrightarrow O_2 + O$$
$$NO + O_3 \longrightarrow NO_2 + O_2$$
$$NO_2 + O \longrightarrow NO + O_2$$
$$\text{Overall:} \quad 2O_3 \longrightarrow 3O_2$$

In this case, NO is the catalyst and NO_2 is the intermediate. Nitrogen dioxide also reacts with chlorine monoxide to form chlorine nitrate:

$$ClO + NO_2 \longrightarrow ClONO_2$$

Chlorine nitrate is relatively stable and serves as a "chlorine reservoir," which plays a role in the depletion of the stratospheric ozone over the North and South Poles.

Polar Ozone Holes

In the mid-1980s, evidence began to accumulate that an "Antarctic ozone hole" developed in late winter, depleting the stratospheric ozone over Antarctica by as much as 50 percent (Figure 20.8). In the stratosphere, a stream of air known as the "polar vortex" circles Antarctica in winter. Air trapped within this vortex becomes extremely cold during the polar night. This condition leads to the formation of ice particles known as polar stratospheric clouds (PSCs) (Figure 20.9). Acting as a heterogeneous catalyst, these PSCs provide a surface for reactions converting HCl (emitted from Earth) and chlorine nitrate to more reactive chlorine molecules:

$$HCl + ClONO_2 \longrightarrow Cl_2 + HNO_3$$

By early spring, the sunlight splits molecular chlorine into chlorine atoms

$$Cl_2 + h\nu \longrightarrow 2Cl$$

which then attack ozone as shown earlier.

The situation is not as severe in the warmer Arctic region, where the vortex does not persist quite as long. Studies have shown that ozone levels in this region have declined between 4 and 8 percent in the past decade. Volcanic eruptions, such as that of Mount Pinatubo in the Philippines in 1991, inject large quantities of dust-sized particles and sulfuric acid aerosols (see Table 12.4) into the atmosphere. These particles can perform the same catalytic function as the ice crystals at the South Pole. As a result, the Arctic hole is expected to grow larger during the next few years.

Recognizing the serious implications of the loss of ozone in the stratosphere, nations throughout the world have acknowledged the need to drastically curtail or totally stop the production of CFCs. In 1978 the United States was one of the few

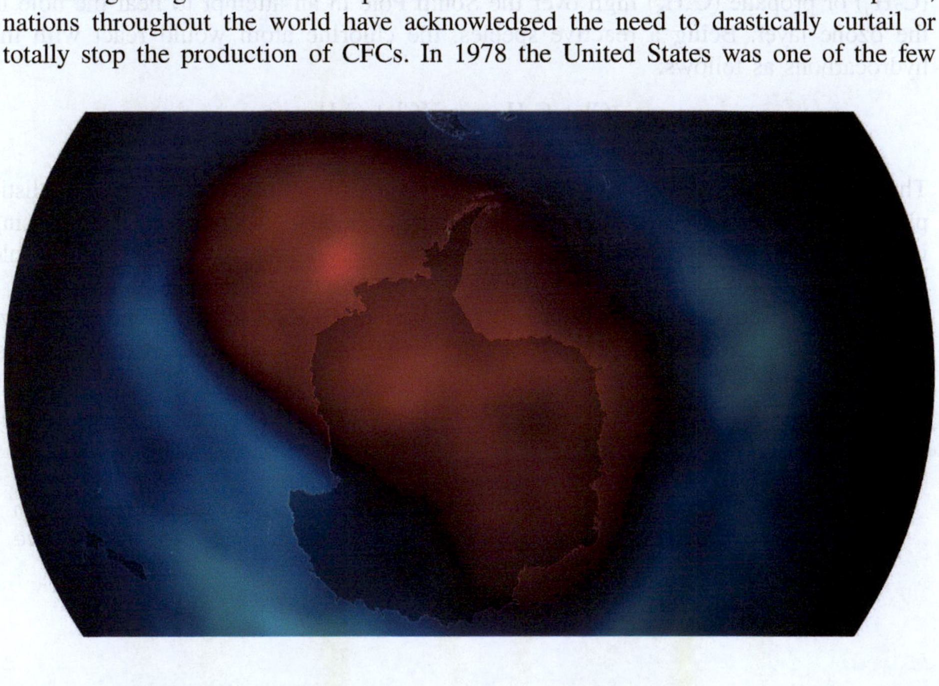

Figure 20.8 *In recent years, scientists have found that the ozone layer in the stratosphere over the South Pole has become thinner. This map, based on data collected over a number of years, shows the depletion of ozone in red.*
NOAA

Figure 20.9 *Polar stratospheric clouds containing ice particles can catalyze the formation of Cl atoms and lead to the destruction of ozone.*
Lamont Poole/NASA

countries to ban the use of CFCs in hair sprays and other aerosols. An international treaty—the Montreal Protocol—was signed by most industrialized nations in 1987, setting targets for cutbacks in CFC production and the complete elimination of these substances by the year 2000. While some progress has been made in this respect, many nations have not been able to abide by the treaty because of the importance of CFCs to their economies. Recycling could play a significant supplementary role in preventing CFCs already in appliances from escaping into the atmosphere. As Figure 20.6 shows, more than half of the CFCs in use are recoverable.

An intense effort is underway to find CFC substitutes that are effective refrigerants but not harmful to the ozone layer. One of the promising candidates is hydrochlorofluorocarbon 134a, or HCFC-134a (CH_2FCF_3). The presence of the hydrogen atoms makes the compound more susceptible to oxidation in the lower atmosphere so that it never reaches the stratosphere. Specifically, it is attacked by the hydroxyl radical (which is formed by a series of complex reactions driven by sunlight) in the troposphere:

$$CH_2FCF_3 + \cdot OH \longrightarrow CHFCF_3 + H_2O$$

The $CHFCF_3$ fragments react with oxygen, eventually decomposing to CO_2, water, and hydrogen fluoride that are removed by rainwater.

Although it is not clear whether the CFCs already released to the atmosphere will eventually result in catastrophic damage to life on Earth, it is conceivable that the depletion of ozone can be slowed by reducing the availability of Cl atoms. Indeed, some chemists have suggested sending a fleet of planes to spray 50,000 tons of ethane (C_2H_6) or propane (C_3H_8) high over the South Pole in an attempt to heal the hole in the ozone layer. Being a reactive species, the chlorine atom would react with the hydrocarbons as follows:

$$Cl + C_2H_6 \longrightarrow HCl + C_2H_5$$
$$Cl + C_3H_8 \longrightarrow HCl + C_3H_7$$

The products of these reactions would not affect the ozone concentration. A less realistic plan is to rejuvenate the ozone layer by producing large quantities of ozone and releasing it into the stratosphere from airplanes. Technically this solution is feasible, but it would be enormously costly and it would require the collaboration of many nations.

Having discussed the chemistry in the outer regions of Earth's atmosphere, we will focus in Sections 20.4, 20.5, 20.6, 20.7, and 20.8 on events closer to us, that is, in the troposphere.

Summary of Concepts & Facts

- Ozone in the stratosphere absorbs harmful UV radiation in the 200- to 300-nm range and protects life underneath. For many years, chlorofluorocarbons have been destroying the ozone layer.

20.4 Volcanoes

Learning Objective

- Identify the effects of volcanic eruptions.

Volcanic eruptions, Earth's most spectacular natural displays of energy, are instrumental in forming large parts of Earth's crust. The upper mantle, immediately under the crust, is nearly molten. A slight increase in heat, such as that generated by the movement of one crustal plate under another, melts the rock. The molten rock, called *magma,* rises to the surface and generates some types of volcanic eruptions (Figure 20.10).

An active volcano emits gases, liquids, and solids. The gases spewed into the atmosphere include primarily N_2, CO_2, HCl, HF, H_2S, and water vapor. It is estimated that volcanoes are the source of about two-thirds of the sulfur in the air. On the slopes of Mount St. Helens, which last erupted in 1980, deposits of elemental sulfur are visible near the eruption site. At high temperatures, the hydrogen sulfide gas given off by a volcano is oxidized by air:

$$2H_2S(g) + 3O_2(g) \longrightarrow 2SO_2(g) + 2H_2O(g)$$

Some of the SO_2 is reduced by more H_2S from the volcano to elemental sulfur and water:

$$2H_2S(g) + SO_2(g) \longrightarrow 3S(s) + 2H_2O(g)$$

The rest of the SO_2 is released into the atmosphere, where it reacts with water to form acid rain (see Section 20.6).

The tremendous force of a volcanic eruption carries a sizable amount of gas into the stratosphere. There SO_2 is oxidized to SO_3, which is eventually converted to sulfuric acid aerosols in a series of complex mechanisms. In addition to destroying ozone in the stratosphere (see Section 20.3), these aerosols can also affect climate. Because the stratosphere is above the atmospheric weather patterns, the aerosol clouds often persist for more than a year. They absorb solar radiation and thereby cause a drop in temperature at Earth's surface. However, this cooling effect is local rather than global, because it depends on the site and frequency of volcanic eruptions.

Summary of Concepts & Facts

- Volcanic eruptions can lead to air pollution, deplete ozone in the stratosphere, and affect climate.

Figure 20.10 *A volcanic eruption on the island of Hawaii.*
J.D. Griggs/U.S. Geological Survey (USGS)

20.5 The Greenhouse Effect

Learning Objective

- Assess the greenhouse effect and how it influences global warming.

Although carbon dioxide is only a trace gas in Earth's atmosphere, with a concentration of about 0.033 percent by volume (see Table 20.1), it plays a critical role in controlling our climate. The so-called ***greenhouse effect*** describes *the trapping of heat near Earth's surface by gases in the atmosphere, particularly carbon dioxide.* The glass roof of a greenhouse transmits visible sunlight and absorbs some of the outgoing infrared (IR) radiation, thereby trapping the heat. Carbon dioxide acts somewhat like a glass roof, except that the temperature rise in the greenhouse is due mainly to the restricted air circulation inside. Calculations show that if the atmosphere did not contain carbon dioxide, Earth would be 30°C cooler! A dramatic illustration of the greenhouse effect is found on Venus where the atmosphere is 97 percent CO_2 and the atmospheric pressure is 9×10^6 Pa (equivalent to 89 atm). The surface temperature of Venus is about 730 K!

Figure 20.11 shows the carbon cycle in our global ecosystem. The transfer of carbon dioxide to and from the atmosphere is an essential part of the carbon cycle.

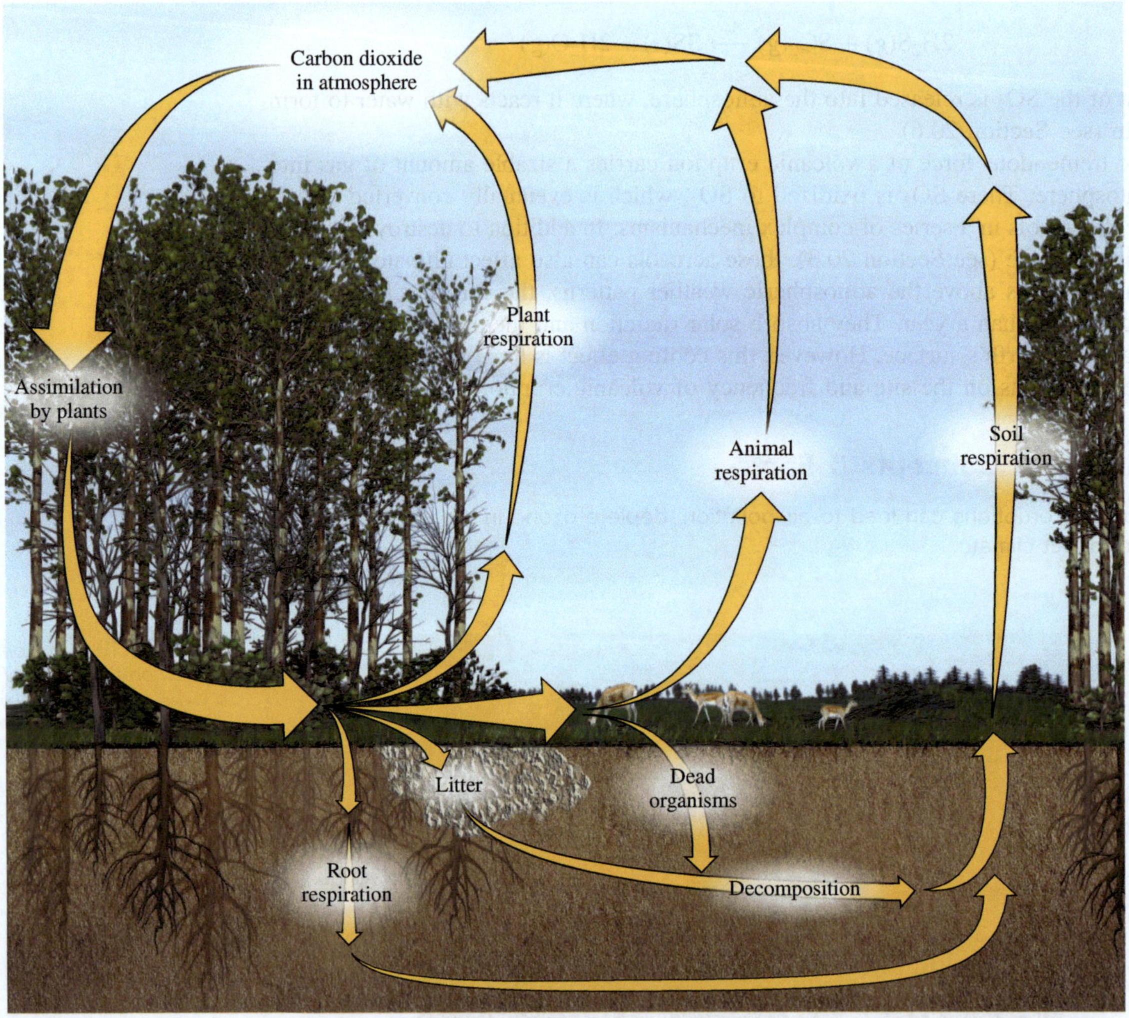

Figure 20.11 *The carbon cycle.*

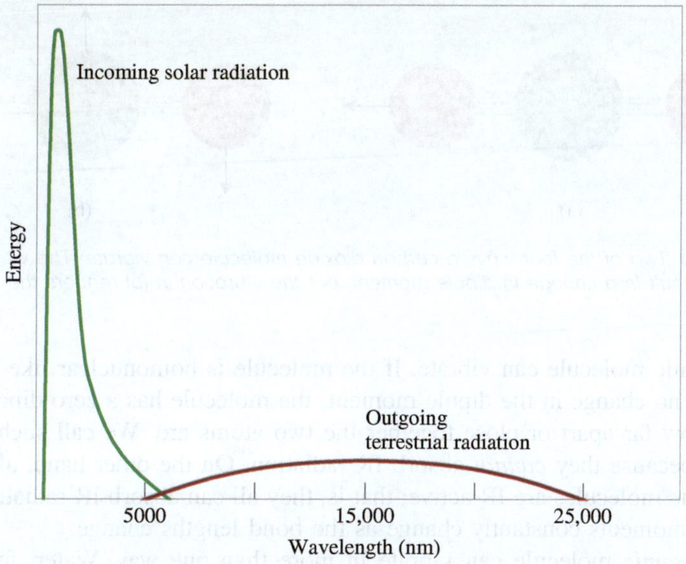

Figure 20.12 The incoming radiation from the sun and the outgoing radiation from Earth's surface.

Carbon dioxide is produced when any form of carbon or a carbon-containing compound is burned in an excess of oxygen. Many carbonates give off CO_2 when heated, and all give off CO_2 when treated with acid:

$$CaCO_3(s) \longrightarrow CaO(s) + CO_2(g)$$
$$CaCO_3(s) + 2HCl(aq) \longrightarrow CaCl_2(aq) + H_2O(l) + CO_2(g)$$

Carbon dioxide is also a by-product of the fermentation of sugar:

$$\underset{\text{glucose}}{C_6H_{12}O_6(aq)} \xrightarrow{\text{yeast}} \underset{\text{ethanol}}{2C_2H_5OH(aq)} + 2CO_2(g)$$

Carbohydrates and other complex carbon-containing molecules are consumed by animals, which respire and release CO_2 as an end product of metabolism:

$$C_6H_{12}O_6(aq) + 6O_2(g) \longrightarrow 6CO_2(g) + 6H_2O(l)$$

As mentioned earlier, another major source of CO_2 is volcanic activity.

Carbon dioxide is removed from the atmosphere by photosynthetic plants and certain microorganisms:

$$6CO_2(g) + 6H_2O(l) \longrightarrow C_6H_{12}O_6(aq) + 6O_2(g)$$

After plants and animals die, the carbon in their tissues is oxidized to CO_2 and returns to the atmosphere. In addition, there is a dynamic equilibrium between atmospheric CO_2 and carbonates in the oceans and lakes.

The solar radiant energy received by Earth is distributed over a band of wavelengths between 100 and 5000 nm, but much of it is concentrated in the 400- to 700-nm range, which is the visible region of the spectrum (Figure 20.12). By contrast, the thermal radiation emitted by Earth's surface is characterized by wavelengths longer than 4000 nm (IR region) because of the much lower average surface temperature compared to that of the sun. The outgoing IR radiation can be absorbed by water and carbon dioxide, but not by nitrogen and oxygen.

All molecules vibrate, even at the lowest temperatures. The energy associated with molecular vibration is quantized, much like the electronic energies of atoms and molecules. To vibrate more energetically, a molecule must absorb a photon of a specific wavelength in the IR region. First, however, its dipole moment *must* change during the course of a vibration. [Recall that the dipole moment of a molecule is the product of the charge and the distance between charges (see Section 10.2).] Figure 20.13 shows

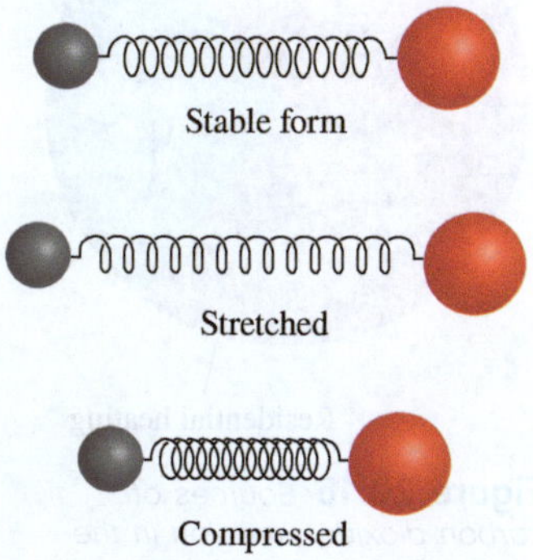

Figure 20.13 Vibrational motion of a diatomic molecule. Chemical bonds can be stretched and compressed like a spring.

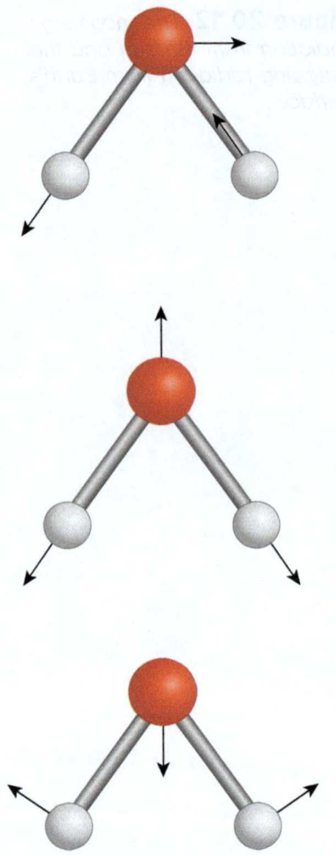

Figure 20.14 *The three different modes of vibration of a water molecule. Each mode of vibration can be imagined by moving the atoms along the arrows and then reversing their directions.*

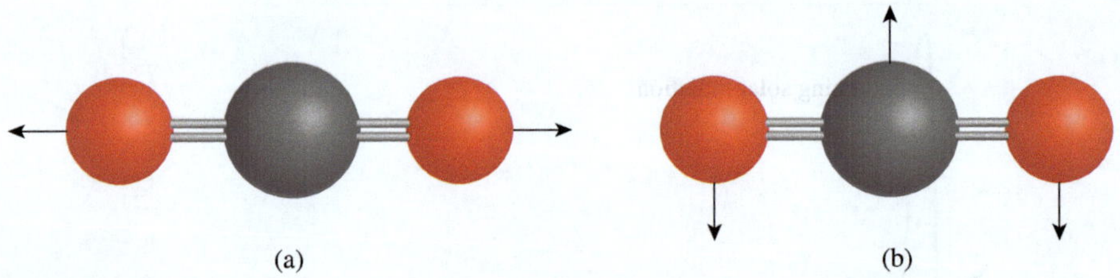

Figure 20.15 *Two of the four ways a carbon dioxide molecule can vibrate. The vibration in (a) does not result in a change in dipole moment, but the vibration in (b) renders the molecule IR-active.*

how a diatomic molecule can vibrate. If the molecule is homonuclear like N_2 and O_2, there can be no change in the dipole moment; the molecule has a zero dipole moment no matter how far apart or close together the two atoms are. We call such molecules IR-inactive because they *cannot* absorb IR radiation. On the other hand, all heteronuclear diatomic molecules are IR-active; that is, they all can absorb IR radiation because their dipole moments constantly change as the bond lengths change.

A *polyatomic* molecule can vibrate in more than one way. Water, for example, can vibrate in three different ways, as shown in Figure 20.14. Because water is a polar molecule, it is easy to see that any of these vibrations results in a change in dipole moment because there is a change in bond length. Therefore, a H_2O molecule is IR-active. Carbon dioxide has a linear geometry and is nonpolar. Figure 20.15 shows two of the four ways a CO_2 molecule can vibrate. One of them [Figure 20.15(a)] symmetrically displaces atoms from the center of gravity and will not create a dipole moment, but the other vibration [Figure 20.15(b)] is IR-active because the dipole moment changes from zero to a maximum value in one direction and then reaches the same maximum value when it changes to the other extreme position.

Upon receiving a photon in the IR region, a molecule of H_2O or CO_2 is promoted to a higher vibrational energy level:

$$H_2O + h\nu \longrightarrow H_2O^*$$
$$CO_2 + h\nu \longrightarrow CO_2^*$$

(the asterisk denotes a vibrationally excited molecule). These energetically excited molecules soon lose their excess energy either by collision with other molecules or by spontaneous emission of radiation. Part of this radiation is emitted to outer space and part returns to Earth's surface.

Although the total amount of water vapor in our atmosphere has not altered noticeably over the years, the concentration of CO_2 has been rising steadily since the turn of the twentieth century as a result of the burning of fossil fuels (petroleum, natural gas, and coal). Figure 20.16 shows the percentages of CO_2 emitted due to human activities in the United States in 1998, and Figure 20.17 shows the variation of carbon dioxide concentration over a period of years, as measured in Hawaii. In the Northern Hemisphere, the seasonal oscillations are caused by removal of carbon dioxide by photosynthesis during the growing season and its buildup during the fall and winter months. Clearly, the trend is toward an increase in CO_2. The current rate of increase is about 1 ppm (1 part CO_2 per million parts air) by volume per year, which is equivalent to 9×10^9 tons of CO_2! Scientists have estimated that by the year 2014 the CO_2 concentration will exceed preindustrial levels by about 40 percent.

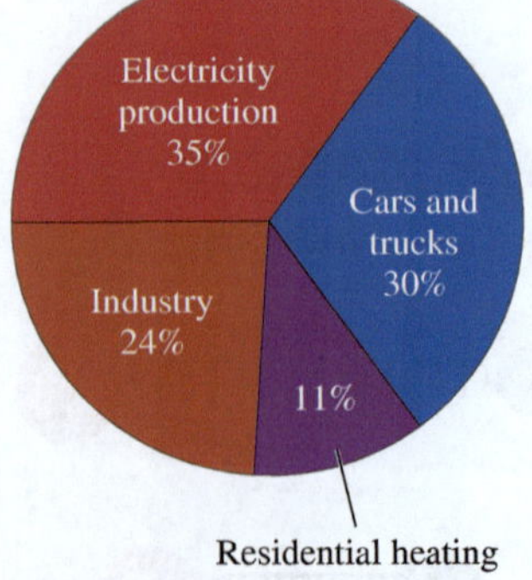

Figure 20.16 *Sources of carbon dioxide emission in the United States. Note that not all of the emitted CO_2 enters the atmosphere. Some of it is taken up by carbon dioxide "sinks," such as the ocean.*

In addition to CO_2 and H_2O, other greenhouse gases, such as the CFCs, CH_4, NO_x, and N_2O, also contribute appreciably to the warming of the atmosphere. Figure 20.18 shows the gradual increase in temperature over the years and Figure 20.19 shows the relative contributions of the greenhouse gases to global warming.

It is predicted by some meteorologists that should the buildup of greenhouse gases continue at its current rate, Earth's average temperature will increase by about 1° to 3°C in this century. Although a temperature increase of a few degrees may seem insignificant, it is actually large enough to disrupt the delicate thermal balance on

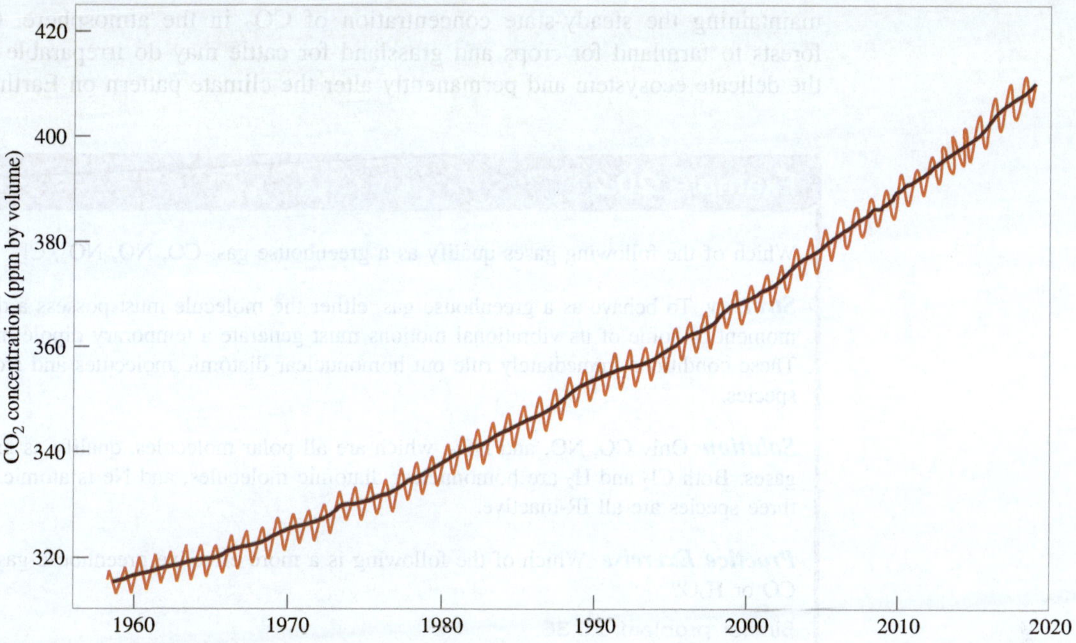

Figure 20.17 *Yearly variation of carbon dioxide concentration at Mauna Loa, Hawaii. The general trend clearly points to an increase of carbon dioxide in the atmosphere. In May 2013, the mean concentration recorded was 400 ppm.*

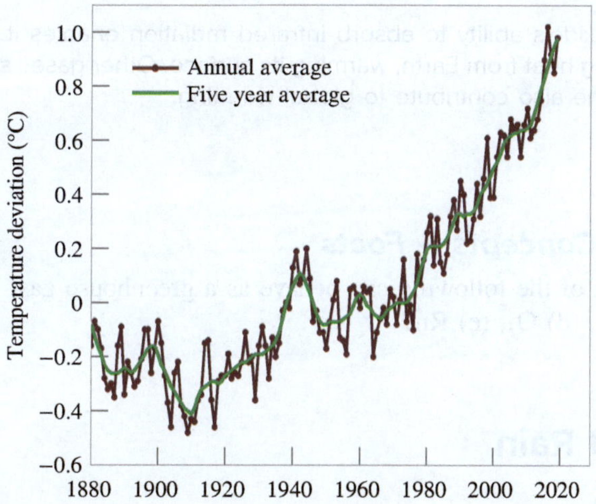

Figure 20.18 *The change in global temperature from 1880 to 2018.*

National Aeronautics and Space Administration (NASA)

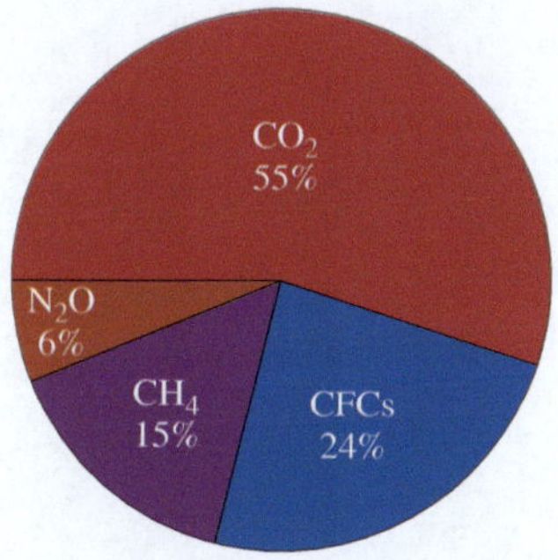

Figure 20.19 *Contribution to global warming by various greenhouse gases. The concentrations of CFCs and methane are much lower than that of carbon dioxide. However, because they can absorb IR radiation much more effectively than CO_2, they make an appreciable contribution to the overall warming effect.*

Earth and could cause glaciers and icecaps to melt. Consequently, the sea level would rise and coastal areas would be flooded.

To combat the greenhouse effect, we must lower carbon dioxide emission. As more nations industrialize across the globe, the production of CO_2 will increase appreciably. Lowering carbon dioxide emissions can be achieved by improving energy efficiency in automobiles and in household heating and lighting, and by developing nonfossil fuel energy sources, such as photovoltaic cells. Nuclear energy is a viable alternative, but its use is highly controversial due to the difficulty of disposing of radioactive waste and the fact that nuclear power stations are more prone to accidents than conventional power stations (see Chapter 19). The proposed phasing out of CFCs, the most potent greenhouse gases, will help to slow the warming trend. The recovery of methane gas generated at landfills and the reduction of natural gas leakages are other steps we could take to control CO_2 emission. Finally, the preservation of the Amazon jungle, tropical forests in Southeast Asia, and other large forests is vital to maintaining the steady-state concentration of CO_2 in the atmosphere. Converting forests to farmland for crops and grassland for cattle may do irreparable damage to the delicate ecosystem and permanently alter the climate pattern on Earth.

Example 20.2

Which of the following gases qualify as a greenhouse gas: CO, NO, NO_2, Cl_2, H_2, Ne?

Strategy To behave as a greenhouse gas, either the molecule must possess a dipole moment or some of its vibrational motions must generate a temporary dipole moment. These conditions immediately rule out homonuclear diatomic molecules and atomic species.

Solution Only CO, NO, and NO_2, which are all polar molecules, qualify as greenhouse gases. Both Cl_2 and H_2 are homonuclear diatomic molecules, and Ne is atomic. These three species are all IR-inactive.

Practice Exercise Which of the following is a more effective greenhouse gas: CO or H_2O?

Similar problem: 20.36.

Summary of Concepts & Facts

- Carbon dioxide's ability to absorb infrared radiation enables it to trap some of the outgoing heat from Earth, warming its surface. Other gases such as the CFCs and methane also contribute to global warming.

Review of Concepts & Facts

20.5.1 Which of the following can behave as a greenhouse gas: (a) CH_4, (b) N_2, (c) O_2, (d) O_3, (e) Rn?

20.6 Acid Rain

Learning Objective

- Recognize how sulfur dioxide causes acid rain.

Scientists have known about acid rain since the late nineteenth century, but it has been a public issue for only about 40 years. Every year acid rain causes hundreds of millions of dollars' worth of damage to stone buildings and statues throughout the world (Figure 20.20). Acid rain is also toxic to vegetation and aquatic life. Many well-documented cases show dramatically how acid rain has destroyed agricultural and forest lands and killed aquatic organisms (see Figure 15.10).

Precipitation in the northeastern United States has an average pH of about 4.3 (Figure 20.21). Because atmospheric CO_2 in equilibrium with rainwater would not be expected to result in a pH less than 5.5, sulfur dioxide (SO_2) and, to a lesser extent, nitrogen oxides from auto emissions are believed to be responsible for the high acidity of rainwater. Acidic oxides, such as SO_2, react with water to give the corresponding acids. There are several sources of atmospheric SO_2. Nature itself contributes much

Figure 20.20 *The effect of acid rain on the marble statue of George Washington in Washington Square, New York City. The photos were taken 50 years apart (1944–1994).*

SO_2 in the form of volcanic eruptions. Also, many metals exist combined with sulfur in nature. Extracting the metals often entails *smelting,* or *roasting,* the ores—that is, heating the metal sulfide in air to form the metal oxide and SO_2. For example,

$$2ZnS(s) + 3O_2(g) \longrightarrow 2ZnO(s) + 2SO_2(g)$$

The metal oxide can be reduced more easily than the sulfide (by a more reactive metal or in some cases by carbon) to the free metal.

Although smelting is a major source of SO_2, the burning of fossil fuels in industry, in power plants, and in homes accounts for most of the SO_2 emitted to the atmosphere. The sulfur content of coal ranges from 0.5 to 5 percent by mass, depending on the source of the coal. The sulfur content of other fossil fuels is similarly variable. Oil from the Middle East, for instance, is low in sulfur, while that from Venezuela has a high sulfur content. To a lesser extent, the nitrogen-containing compounds in oil and coal are converted to nitrogen oxides, which can also acidify rainwater.

All in all, some 50 to 60 million tons of SO_2 are released into the atmosphere each year! In the troposphere, SO_2 is almost all oxidized to H_2SO_4 in the form of aerosol, which ends up in wet precipitation or acid rain. The mechanism for the

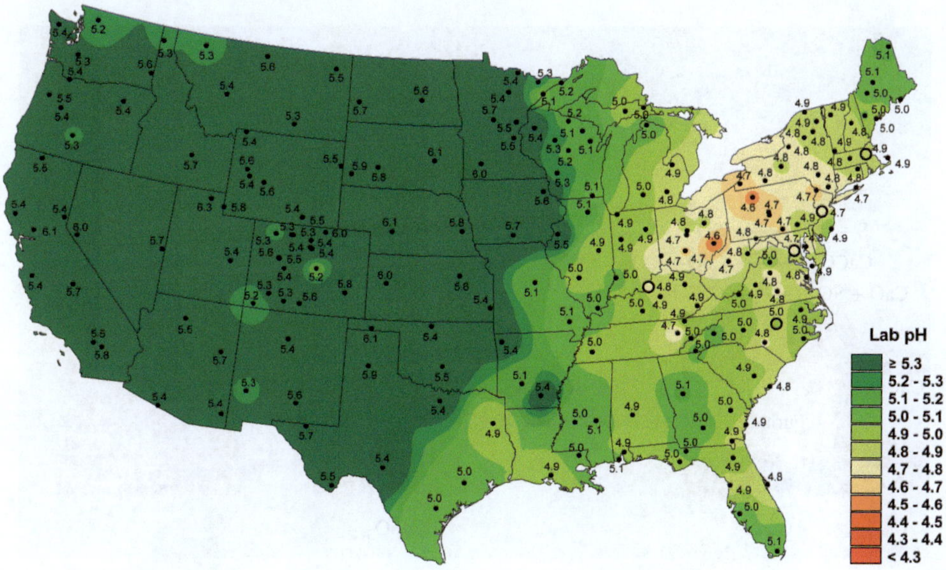

Figure 20.21 *Mean precipitation pH in the United States in 2009. Most SO_2 comes from the midwestern states. Prevailing winds carry the acid droplets formed over the Northeast. Nitrogen oxides also contribute to the acid rain formation.*

conversion of SO_2 to H_2SO_4 is quite complex and not fully understood. The reaction is believed to be initiated by the hydroxyl radical (OH):

$$OH + SO_2 \longrightarrow HOSO_2$$

The $HOSO_2$ radical is further oxidized to SO_3:

$$HOSO_2 + O_2 \longrightarrow HO_2 + SO_3$$

The sulfur trioxide formed would then rapidly react with water to form sulfuric acid:

$$SO_3 + H_2O \longrightarrow H_2SO_4$$

SO_2 can also be oxidized to SO_3 and then converted to H_2SO_4 on particles by heterogeneous catalysis. Eventually, the acid rain can corrode limestone and marble ($CaCO_3$). A typical reaction is

$$CaCO_3(s) + H_2SO_4(aq) \longrightarrow CaSO_4(s) + H_2O(l) + CO_2(g)$$

Sulfur dioxide can also attack calcium carbonate directly:

$$2CaCO_3(s) + 2SO_2(g) + O_2(g) \longrightarrow 2CaSO_4(s) + 2CO_2(g)$$

There are two ways to minimize the effects of SO_2 pollution. The most direct approach is to remove sulfur from fossil fuels before combustion, but this is technologically difficult to accomplish. A cheaper but less efficient way is to remove SO_2 as it is formed. For example, in one process powdered limestone is injected into the power plant boiler or furnace along with coal (Figure 20.22). At high temperatures the following decomposition occurs:

$$\underset{\text{limestone}}{CaCO_3(s)} \longrightarrow \underset{\text{quicklime}}{CaO(s)} + CO_2(g)$$

The quicklime reacts with SO_2 to form calcium sulfite and some calcium sulfate:

$$CaO(s) + SO_2(g) \longrightarrow CaSO_3(s)$$

$$2CaO(s) + 2SO_2(g) + O_2(g) \longrightarrow 2CaSO_4(s)$$

To remove any remaining SO_2, an aqueous suspension of quicklime is injected into a purification chamber prior to the gases' escape through the smokestack. Quicklime is also added to lakes and soils in a process called *liming* to reduce their acidity (Figure 20.23).

Installing a sulfuric acid plant near a metal ore refining site is also an effective way to cut SO_2 emission because the SO_2 produced by roasting metal sulfides can be captured for use in the synthesis of sulfuric acid. This is a very sensible way to turn what is a pollutant in one process into a starting material for another process!

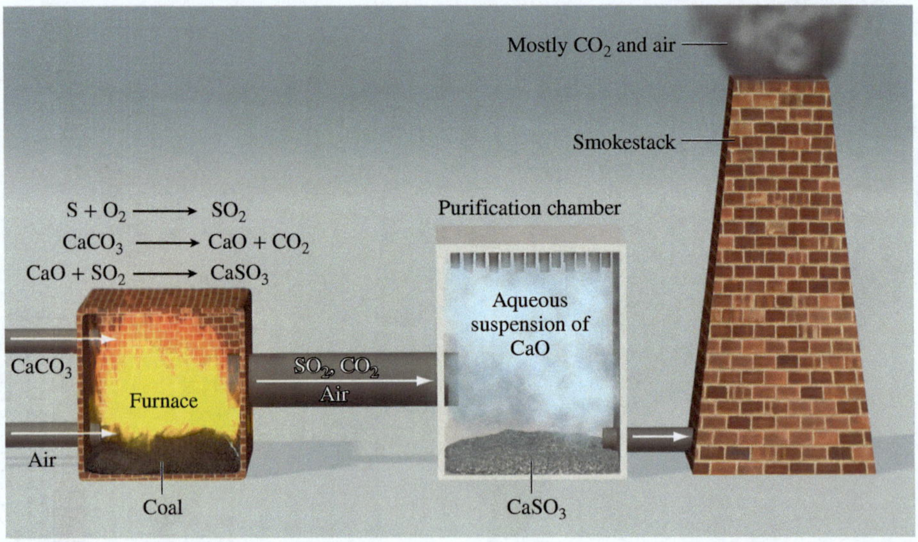

Figure 20.22 *Common procedure for removing SO_2 from burning fossil fuel. Powdered limestone decomposes into CaO, which reacts with SO_2 to form $CaSO_3$. The remaining SO_2 is reacted with an aqueous suspension of CaO to form $CaSO_3$.*

Figure 20.23 *Spreading calcium oxide (CaO) over acidified soil. This process is called liming.*
blickwinkel/McPHOTO/BilderBox/Alamy Stock Photo

Summary of Concepts & Facts

- Sulfur dioxide, and to a lesser extent nitrogen oxides, generated mainly from the burning of fossil fuels and from the roasting of metal sulfides, causes acid rain.

20.7 Photochemical Smog

Learning Objective

- Express how photochemical smog forms.

The word "smog" was coined to describe the combination of smoke and fog that shrouded London during the 1950s. The primary cause of this noxious cloud was sulfur dioxide. Today, however, we are more familiar with ***photochemical smog,*** which *is formed by the reactions of automobile exhaust in the presence of sunlight.*

The heavy use of automobiles is the cause of photochemical smog formation.
Digital Vision Ltd./Superstock

Automobile exhaust consists mainly of NO, CO, and various unburned hydrocarbons. These gases are called *primary pollutants* because they set in motion a series of photochemical reactions that produce *secondary pollutants.* It is the secondary pollutants—chiefly NO_2 and O_3—that are responsible for the buildup of smog.

Nitric oxide is the product of the reaction between atmospheric nitrogen and oxygen at high temperatures inside an automobile engine:

$$N_2(g) + O_2(g) \longrightarrow 2NO(g)$$

Once released into the atmosphere, nitric oxide is oxidized to nitrogen dioxide:

$$2NO(g) + O_2(g) \longrightarrow 2NO_2(g)$$

Sunlight causes the photochemical decomposition of NO_2 (at a wavelength shorter than 400 nm) into NO and O:

$$NO_2(g) + h\nu \longrightarrow NO(g) + O(g)$$

Atomic oxygen is a highly reactive species that can initiate a number of important reactions, one of which is the formation of ozone:

$$O(g) + O_2(g) + M \longrightarrow O_3(g) + M$$

where M is some inert substance such as N_2. Ozone attacks the C═C linkage in rubber:

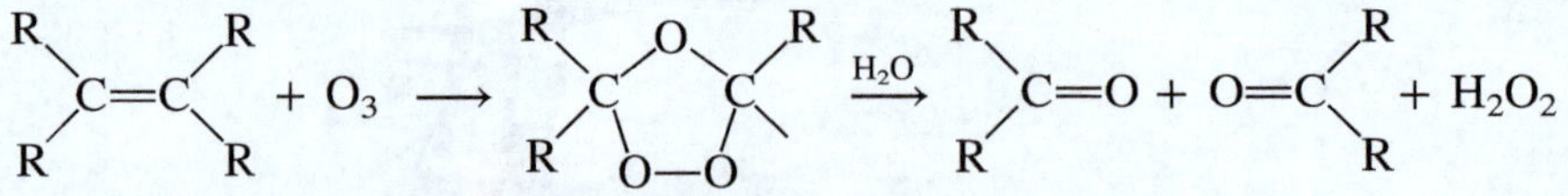

where R represents groups of C and H atoms. In smog-ridden areas, this reaction can cause automobile tires to crack. Similar reactions are also damaging to lung tissues and other biological substances.

Ozone can be formed also by a series of very complex reactions involving unburned hydrocarbons, nitrogen oxides, and oxygen. One of the products of these reactions is peroxyacetyl nitrate (PAN):

$$CH_3{-}\overset{\displaystyle O}{\overset{\|}{C}}{-}O{-}O{-}NO_2$$

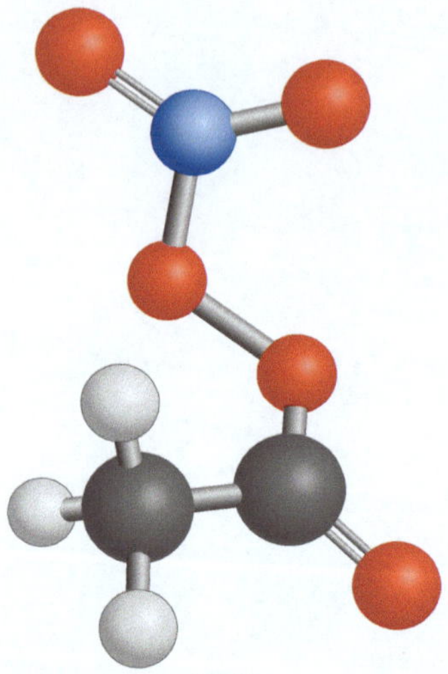

PAN

PAN is a powerful lachrymator, or tear producer, and causes breathing difficulties.

Figure 20.24 shows typical variations with time of primary and secondary pollutants. Initially, the concentration of NO_2 is quite low. As soon as solar radiation penetrates the atmosphere, more NO_2 is formed from NO and O_2. Note that the concentration of ozone remains fairly constant at a low level in the early morning hours. As the concentration of unburned hydrocarbons and aldehydes increases in the air, the concentrations of NO_2 and O_3 also rise rapidly. The actual amounts, of course, depend on the location, traffic, and weather conditions, but their presence is always accompanied by haze (Figure 20.25). The oxidation of hydrocarbons produces various organic intermediates, such as alcohols and carboxylic acids, which are all less volatile than the hydrocarbons themselves. These substances eventually condense into small droplets of liquid. The dispersion of these droplets in air, called *aerosol,* scatters sunlight and reduces visibility. This interaction also makes the air look hazy.

The haze over the Smoky Mountains is caused by aerosols produced by the oxidation of hydrocarbons emitted by pine trees.

Christian Bauer/Sodapix AG, Switzerland/ Fotosearch/Glow Images

As the mechanism of photochemical smog formation has become better understood, major efforts have been made to reduce the buildup of primary pollutants. Most automobiles now are equipped with catalytic converters designed to oxidize CO and unburned hydrocarbons to CO_2 and H_2O and to reduce NO and NO_2 to N_2 and O_2 (see Section 13.6). More efficient automobile engines and better public transportation systems would also help to decrease air pollution in urban areas. A recent technological innovation to combat photochemical smog is to coat automobile radiators and air conditioner compressors with a platinum catalyst. So equipped, a running car can

Figure 20.24 *Typical variations with time in concentration of air pollutants on a smoggy day.*

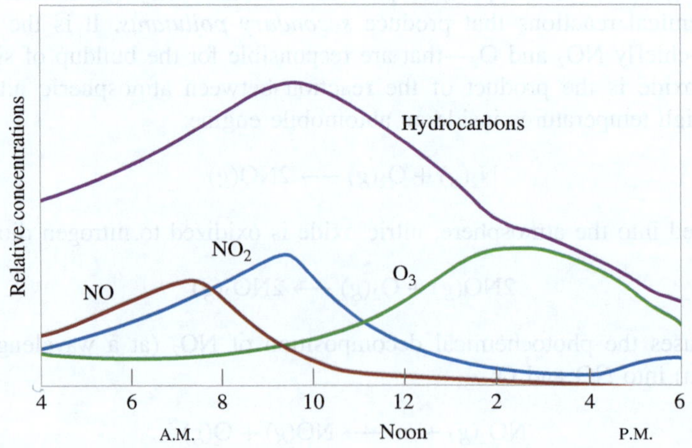

Figure 20.25 *A smoggy day in Beijing.*
narvikk/E+/Getty Images

purify the air that flows under the hood by converting ozone and carbon monoxide to oxygen and carbon dioxide:

$$O_3(g) + CO(g) \xrightarrow{Pt} O_2(g) + CO_2(g)$$

In a city like Los Angeles, where the number of miles driven in one day equals nearly 300 million, this approach would significantly improve the air quality and reduce the "high-ozone level" warnings frequently issued to its residents. In fact, a drive on the freeway would help to clean up the air!

Summary of Concepts & Facts

- Photochemical smog is formed by the photochemical reaction of automobile exhaust in the presence of sunlight. It is a complex reaction involving nitrogen oxides, ozone, and hydrocarbons.

20.8 Indoor Pollution

Learning Objective

- Describe the major indoor air pollutants.

Difficult as it is to avoid air pollution outdoors, it is no easier to avoid indoor pollution. The air quality in homes and in the workplace is affected by human activities, by construction materials, and by other factors in our immediate environment. The common indoor pollutants are radon, carbon monoxide and carbon dioxide, and formaldehyde.

The Risk from Radon

In a highly publicized case in the mid-1980s, an employee reporting for work at the Limerick Nuclear Power Plant in Pennsylvania set off the plant's radiation monitor. Astonishingly, the source of his contamination turned out not to be the plant, but radon in his home!

A lot has been said and written about the potential dangers of radon as an air pollutant. Just what is radon? Where does it come from? And how does it affect our health?

Radon is a member of Group 18 (the noble gases). It is an intermediate product of the radioactive decay of uranium-238. (The uranium decay series is discussed in Chapter 19.) All isotopes of radon are radioactive, but radon-222 is the most hazardous because it has the longest half-life—3.8 days. Radon, which accounts for slightly over half the background radioactivity on Earth, is generated mostly from the phosphate minerals of uranium (Figure 20.26).

Since the 1970s, high levels of radon have been detected in homes built on reclaimed land above uranium mill tailing deposits. The colorless, odorless, and tasteless radon gas enters a building through tiny cracks in the basement floor (Figure 20.27). It is slightly soluble in water, so it can be spread in different media. Radon-222 is an α-emitter. When it decays, it produces radioactive polonium-214 and polonium-218, which can build up to high levels in an enclosed space. These solid radioactive particles can adhere to airborne dust and smoke, which are inhaled into the lungs and deposited in the respiratory tract. Over a long period of time, the α particles emitted by polonium and its decay products, which are also radioactive, can cause lung cancer. After cigarette smoking, radon is the leading cause of lung cancer in the United States. It is responsible for perhaps 20,000 deaths per year.

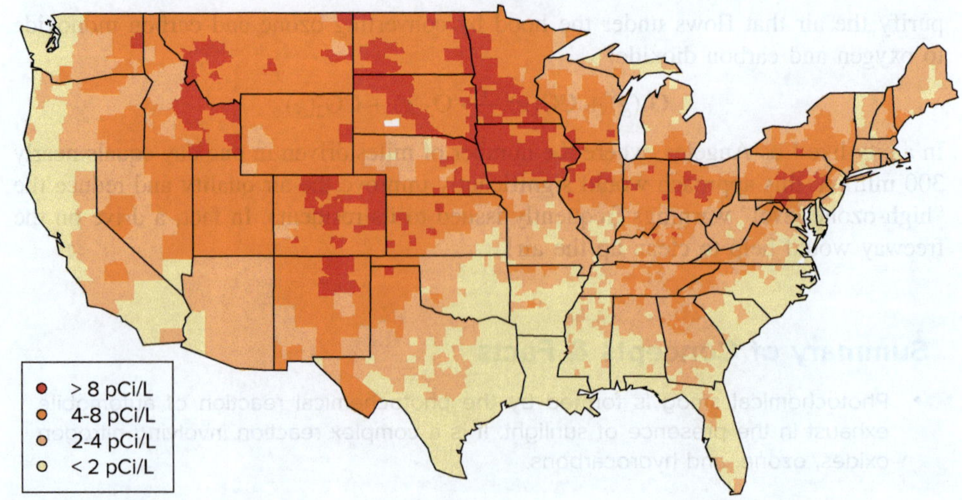

Figure 20.26 *Map of radon emission in the United States measured in picocuries per liter of air.*

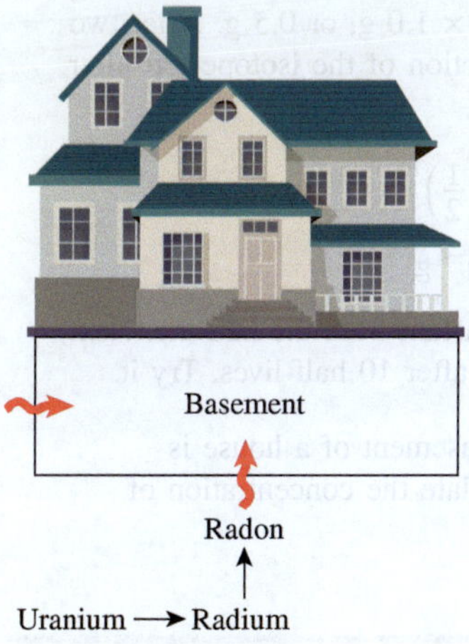

Figure 20.27 *Radon usually enters houses through the foundation or basement walls.*

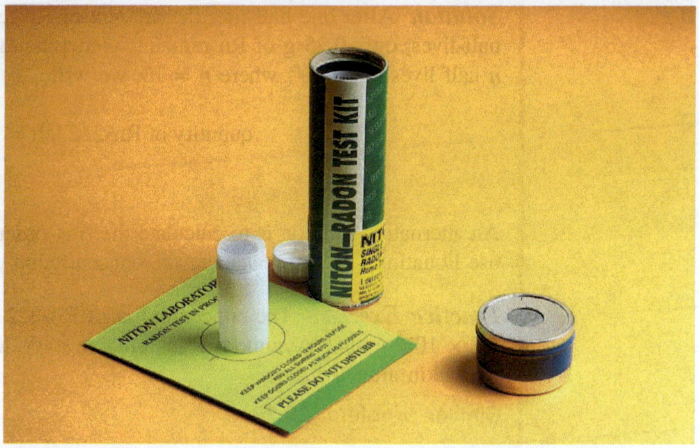

Figure 20.28 *Home radon detectors: Long-term track etch (left) and short-term charcoal canister (right).*
Ken Karp/McGraw Hill

What can be done to combat radon pollution indoors? The first step is to measure the radon level in the basement with a reliable test kit. Short-term and long-term kits are available (Figure 20.28). The short-term tests use activated charcoal (that is, heat-treated charcoal) to collect the decay products of radon over a period of several days. The container is sent to a laboratory where a technician measures the radioactivity (γ rays) from radon-decay products lead-214 and bismuth-214. Knowing the length of exposure, the lab technician back-calculates to determine radon concentration. The long-term test kits use a piece of special polymer film on which an α particle will leave a "track." After several months' exposure, the film is etched with a sodium hydroxide solution and the number of tracks counted. Knowing the length of exposure enables the technician to calculate the radon concentration. If the radon level is unacceptably high, then the house must be regularly ventilated. This precaution is particularly important in recently built houses, which are well insulated. A more effective way to prevent radon pollution is to reroute the gas before it gets into the house, for example, by installing a ventilation duct to draw air from beneath the basement floor to the outside.

Currently there is considerable controversy regarding the health effects of radon. The first detailed studies of the effects of radon on human health were carried out in the 1950s when it was recognized that uranium miners suffered from an abnormally high incidence of lung cancer. Some scientists have challenged the validity of these studies because the miners were also smokers. It seems quite likely that there is a synergistic effect between radon and smoking on the development of lung cancer. Radon-decay products will adhere not only to tobacco tar deposits in the lungs, but also to the solid particles in cigarette smoke, which can be inhaled by smokers and nonsmokers. More systematic studies are needed to evaluate the environmental impact of radon. In the meantime, the Environmental Protection Agency (EPA) has recommended remedial action where the radioactivity level due to radon exceeds 4 picocuries (pCi) per liter of air. (A curie corresponds to 3.70×10^{10} disintegrations of radioactive nuclei per second; a picocurie is a trillionth of a curie, or 3.70×10^{-2} disintegrations per second.)

Example 20.3

The half-life of Rn-222 is 3.8 days. Starting with 1.0 g of Rn-222, how much will be left after 10 half-lives? Recall that radioactive decays obey first-order kinetics.

Strategy All radioactive decays obey first-order kinetics. Therefore, its half-life is independent of the initial concentration.

(Continued)

Solution After one half-life, the amount of Rn left is 0.5×1.0 g, or 0.5 g. After two half-lives, only 0.25 g of Rn remains. Generalizing the fraction of the isotope left after n half lives as $(1/2)^n$, where $n = 10$, we write

$$\text{quantity of Rn-222 left} = 1.0\text{ g} \times \left(\frac{1}{2}\right)^{10}$$
$$= 9.8 \times 10^{-4}\text{ g}$$

An alternative solution is to calculate the first-order rate constant from the half-life. Next, use Equation (13.3) to calculate the concentration of radon after 10 half-lives. Try it.

Practice Exercise The concentration of Rn-222 in the basement of a house is 1.8×10^{-6} mol/L. Assume the air remains static and calculate the concentration of the radon after 2.4 days.

Similar problem: 20.73.

Carbon Dioxide and Carbon Monoxide

Both carbon dioxide (CO_2) and carbon monoxide (CO) are products of combustion. In the presence of an abundant supply of oxygen, CO_2 is formed; in a limited supply of oxygen, both CO and CO_2 are formed. The indoor sources of these gases are gas cooking ranges, woodstoves, space heaters, tobacco smoke, human respiration, and exhaust fumes from cars (in garages). Carbon dioxide is not a toxic gas, but it does have an asphyxiating effect (see Chemistry in Action essay "The Killer Lake" in Section 12.5). In airtight buildings, the concentration of CO_2 can reach as high as 2000 ppm by volume (compared with 300 ppm outdoors). Workers exposed to high concentrations of CO_2 in skyscrapers and other sealed environments become fatigued more easily and have difficulty concentrating. Adequate ventilation is the solution to CO_2 pollution.

Like CO_2, CO is a colorless and odorless gas, but it differs from CO_2 in that it is highly poisonous. The toxicity of CO lies in its unusual ability to bind very strongly to hemoglobin, the oxygen carrier in blood. Both O_2 and CO bind to the Fe(II) ion in hemoglobin, but the affinity of hemoglobin for CO is about 200 times greater than that for O_2 (see Chapter 25). Hemoglobin molecules with tightly bound CO (called carboxyhemoglobin) cannot carry the oxygen needed for metabolic processes. A small amount of CO intake can cause drowsiness and headache; death may result when about half the hemoglobin molecules are complexed with CO. The best first-aid response to CO poisoning is to remove the victim immediately to an area with a plentiful oxygen supply or to give mouth-to-mouth resuscitation.

Formaldehyde

CH_2O

Formaldehyde (CH_2O) is a rather disagreeable-smelling liquid used as a preservative for laboratory specimens. Industrially, formaldehyde resins are used as bonding agents in building and furniture construction materials such as plywood and particle board. In addition, urea-formaldehyde insulation foams are used to fill wall cavities. The resins and foams slowly break down to release free formaldehyde, especially under acid and humid conditions. Low concentrations of formaldehyde in the air can cause drowsiness, nausea, headaches, and other respiratory ailments. Laboratory tests show that breathing high concentrations of formaldehyde can induce cancers in animals, and it is now also classified as a human carcinogen. The safe standard of formaldehyde in indoor air has been set at 0.1 ppm by volume.

Because formaldehyde is a reducing agent, devices have been constructed to remove it by means of a redox reaction. Indoor air is circulated through an air purifier containing an oxidant such as $Al_2O_3/KMnO_4$, which converts formaldehyde to the less harmful and less volatile formic acid (HCOOH). Proper ventilation is the best way to remove formaldehyde. However, care should be taken not to remove the air from a room too quickly without replenishment, because a reduced pressure would cause the formaldehyde resins to decompose faster, resulting in the release of *more* formaldehyde.

Summary of Concepts & Facts

- Indoor air pollution is caused by radon, a radioactive gas formed during uranium decay; carbon monoxide and carbon dioxide, products of combustion; and formaldehyde, a volatile organic substance released from resins used in construction materials.

Chapter Summary

Earth's Atmosphere Earth's atmosphere is made up mainly of nitrogen and oxygen, plus a number of other trace gases. Molecular nitrogen in the atmosphere is incorporated into other compounds via nitrogen fixation. The regions of the atmosphere are the troposphere, the stratosphere, the mesosphere, the thermosphere, and the ionosphere. In the outer regions of the atmosphere, the bombardment of molecules and atoms by solar particles gives rise to the *aurora borealis* in the Northern Hemisphere and the *aurora australis* in the Southern Hemisphere. (Sections 20.1 and 20.2)

Ozone Ozone in the stratosphere absorbs harmful UV radiation in the 200- to 300-nm range and protects life underneath. For many years, chlorofluorocarbons have been destroying the ozone layer. (Section 20.3)

Volcanoes Volcanic eruptions can lead to air pollution, deplete ozone in the stratosphere, and affect climate worldwide. (Section 20.4)

Greenhouse Effect Carbon dioxide can absorb infrared radiation thereby trapping some of the outgoing heat from Earth. This warms the surface in a phenomenon known as the greenhouse effect. Other gases such as the CFCs and methane also contribute to the greenhouse effect. Global warming is the result of the enhanced greenhouse effect caused by human activities. (Section 20.5)

Acid Rain Sulfur dioxide, and to a lesser extent nitrogen oxides, generated mostly from the burning of fossil fuels and from the roasting of metal sulfides, is a direct cause of acid rain. (Section 20.6)

Smog Photochemical smog is formed by chemical reactions from automobile exhaust with sunlight. It is a complex series of reactions involving nitrogen oxides, ozone, and hydrocarbons. (Section 20.7)

Indoor Pollution Indoor air pollution is mainly attributed to radon, a radioactive gas formed during uranium decay; carbon monoxide and carbon dioxide, products of combustion; and formaldehyde, a volatile organic substance released from resins used in construction materials. (Section 20.8)

Key Words

Greenhouse effect, p. 916
Ionosphere, p. 907
Mesosphere, p. 907
Nitrogen fixation, p. 905
Photochemical smog, p. 923
Stratosphere, p. 906
Thermosphere, p. 907
Troposphere, p. 906

Questions & Problems

Red numbered problems solved in Student Solutions Manual

20.1 Earth's Atmosphere

Review Questions

20.1 Describe the regions of Earth's atmosphere.

20.2 Briefly outline the main processes of the nitrogen and oxygen cycles.

20.3 Explain why, for maximum performance, supersonic airplanes need to fly at a high altitude (in the stratosphere).

20.4 Jupiter's atmosphere consists mainly of hydrogen (90 percent) and helium (9 percent). How does this mixture of gases contrast with the composition of Earth's atmosphere? Why does the composition differ?

Problems

20.5 Referring to Table 20.1, calculate the mole fraction of CO_2 and its concentration in parts per million by volume.

20.6 Calculate the partial pressure of CO_2 (in atm) in dry air when the atmospheric pressure is 754 mmHg.

20.7 Describe the processes that result in the warming of the stratosphere.

20.8 Calculate the total mass (in kilograms) of nitrogen, oxygen, and carbon dioxide gases in the atmosphere. (*Hint:* See Problem 5.106 and Table 20.1. Use 29.0 g/mol for the molar mass of air.)

20.2 Phenomena in the Outer Layers of the Atmosphere

Review Questions

20.9 What process gives rise to aurora borealis and aurora australis?

20.10 Why can astronauts not release oxygen atoms to test the mechanism of shuttle glow?

Problems

20.11 The highly reactive OH radical (a species with an unpaired electron) is believed to be involved in some atmospheric processes. Table 9.4 lists the bond enthalpy for the oxygen-to-hydrogen bond in OH as 460 kJ/mol. What is the longest wavelength (in nm) of radiation that can bring about the reaction

$$OH(g) \longrightarrow O(g) + H(g)$$

20.12 The green color observed in aurora borealis is produced by the emission of a photon by an electronically excited oxygen atom at 558 nm. Calculate the energy difference between the two levels involved in the emission process.

20.3 Depletion of Ozone in the Stratosphere

Review Questions

20.13 Briefly describe the absorption of solar radiation in the stratosphere by O_2 and O_3 molecules.

20.14 Explain the processes that have a warming effect on the stratosphere.

20.15 List the properties of CFCs, and name four major uses of these compounds.

20.16 How do CFCs and nitrogen oxides destroy ozone in the stratosphere?

20.17 What causes the polar ozone holes?

20.18 How do volcanic eruptions contribute to ozone destruction?

20.19 Describe ways to curb the destruction of ozone in the stratosphere.

20.20 Discuss the effectiveness of some of the CFC substitutes.

Problems

20.21 Given that the quantity of ozone in the stratosphere is equivalent to a 3.0-mm-thick layer of ozone on Earth at STP, calculate the number of ozone molecules in the stratosphere and their mass in kilograms. (*Hint:* The radius of Earth is 6371 km and the surface area of a sphere is $4\pi r^2$, where r is the radius.)

20.22 Referring to the answer in Problem 20.21, and assuming that the level of ozone in the stratosphere has already fallen 6.0 percent, calculate the number of kilograms of ozone that would have to be manufactured on a daily basis so that we could restore the ozone to the original level in 100 yr. If ozone is made according to the process $3O_2(g) \longrightarrow 2O_3(g)$, how many kilojoules of energy would be required?

20.23 Both Freon-11 and Freon-12 are made by the reaction of carbon tetrachloride (CCl_4) with hydrogen fluoride. Write equations for these reactions.

20.24 Why are CFCs not decomposed by UV radiation in the troposphere?

20.25 The average bond enthalpies of the C—Cl and C—F bonds are 340 kJ/mol and 485 kJ/mol, respectively. Based on this information, explain why the C—Cl bond in a CFC molecule is preferentially broken by solar radiation at 250 nm.

20.26 Like CFCs, certain bromine-containing compounds such as CF_3Br can also participate in the destruction of ozone by a similar mechanism starting with the Br atom:

$$CF_3Br \longrightarrow CF_3 + Br$$

Given that the average C—Br bond enthalpy is 276 kJ/mol, estimate the longest wavelength required to break this bond. Will this compound be decomposed in the troposphere only or in both the troposphere and stratosphere?

20.27 Draw Lewis structures for chlorine nitrate ($ClONO_2$) and chlorine monoxide (ClO).

20.28 Draw Lewis structures for HCFC-123 (CF_3CHCl_2) and CF_3CFH_2.

20.4 Volcanoes

Review Questions

20.29 What are the effects of volcanic eruptions on climate?

20.30 Classify the reaction between H_2S and SO_2 that leads to the formation of sulfur at the site of a volcanic eruption.

20.5 The Greenhouse Effect

Review Questions

20.31 What is the greenhouse effect? What is the criterion for classifying a gas as a greenhouse gas?

20.32 Why is more emphasis placed on the role of carbon dioxide in the greenhouse effect than on that of water?

20.33 Describe three human activities that generate carbon dioxide. List two major mechanisms for the uptake of carbon dioxide.

20.34 Deforestation contributes to the greenhouse effect in two ways. What are they?

20.35 How does an increase in world population enhance the greenhouse effect?

20.36 Is ozone a greenhouse gas? If so, sketch three ways an ozone molecule can vibrate.

20.37 What effects do CFCs and their substitutes have on Earth's temperature?

20.38 Why are CFCs more effective greenhouse gases than methane and carbon dioxide?

Problems

20.39 The annual production of zinc sulfide (ZnS) is 4.0×10^4 tons. Estimate the number of tons of SO_2 produced by roasting it to extract zinc metal.

20.40 Calcium oxide or quicklime (CaO) is used in steelmaking, cement manufacture, and pollution control. It is prepared by the thermal decomposition of calcium carbonate:

$$CaCO_3(s) \longrightarrow CaO(s) + CO_2(g)$$

Calculate the yearly release of CO_2 (in kilograms) to the atmosphere if the annual production of CaO in the United States is 1.7×10^{10} kg.

20.41 The molar heat capacity of a diatomic molecule is 29.1 J/K · mol. Assuming the atmosphere contains only nitrogen gas and there is no heat loss, calculate the total heat intake (in kilojoules) if the atmosphere warms up by 3°C during the next 50 yr. Given that there are 1.8×10^{20} moles of diatomic molecules present, how many kilograms of ice (at the North and South Poles) will this quantity of heat melt at 0°C? (The molar heat of fusion of ice is 6.01 kJ/mol.)

20.42 As mentioned in the chapter, spraying the stratosphere with hydrocarbons such as ethane and propane should eliminate Cl atoms. What is the drawback of this procedure if used on a large scale for an extended period of time?

20.6 Acid Rain

Review Questions

20.43 Name the gas that is largely responsible for the acid rain phenomenon.

20.44 List three detrimental effects of acid rain.

20.45 Briefly discuss two industrial processes that lead to acid rain.

20.46 Discuss ways to curb acid rain.

20.47 Water and sulfur dioxide are both polar molecules and their geometry is similar. Why is SO_2 not considered a major greenhouse gas?

20.48 Describe the removal of SO_2 by CaO (to form $CaSO_3$) in terms of a Lewis acid-base reaction.

Problems

20.49 An electric power station annually burns 3.1×10^7 kg of coal containing 2.4 percent sulfur by mass. Calculate the volume of SO_2 emitted at STP.

20.50 The concentration of SO_2 in the troposphere over a certain region is 0.16 ppm by volume. The gas dissolves in rainwater as follows:

$$SO_2(g) + H_2O(l) \rightleftharpoons H^+(aq) + HSO_3^-(aq)$$

Given that the equilibrium constant for the preceding reaction is 1.3×10^{-2}, calculate the pH of the rainwater. Assume that the reaction does not affect the partial pressure of SO_2.

20.7 Photochemical Smog

Review Questions

20.51 What is photochemical smog? List the factors that favor the formation of photochemical smog.

20.52 What are primary and secondary pollutants?

20.53 Identify the gas that is responsible for the brown color of photochemical smog.

20.54 The safety limits of ozone and carbon monoxide are 120 ppb by volume and 9 ppm by volume, respectively. Why does ozone have a lower limit?

20.55 Suggest ways to minimize the formation of photochemical smog.

20.56 In which region of the atmosphere is ozone beneficial? In which region is it detrimental?

Problems

20.57 Assume that the formation of nitrogen dioxide:

$$2NO(g) + O_2(g) \longrightarrow 2NO_2(g)$$

is an elementary reaction. (a) Write the rate law for this reaction. (b) A sample of air at a certain temperature is contaminated with 2.0 ppm of NO by volume. Under these conditions, can the rate law be simplified? If so, write the simplified rate law. (c) Under the conditions described in (b), the half-life of the reaction has been estimated to be 6.4×10^3 min. What would the half-life be if the initial concentration of NO were 10 ppm?

20.58 The gas-phase decomposition of peroxyacetyl nitrate (PAN) obeys first-order kinetics:

$$CH_3COOONO_2 \longrightarrow CH_3COOO + NO_2$$

with a rate constant of $4.9 \times 10^{-4}\ s^{-1}$. Calculate the rate of decomposition in *M*/s if the concentration of PAN is 0.55 ppm by volume. Assume STP conditions.

20.59 On a smoggy day in a certain city the ozone concentration was 0.42 ppm by volume. Calculate the partial pressure of ozone (in atm) and the number of ozone molecules per liter of air if the temperature and pressure were 20.0°C and 748 mmHg, respectively.

20.60 Which of the following settings is the most suitable for photochemical smog formation: (a) Gobi desert at noon in June, (b) New York City at 1 p.m. in July, (c) Boston at noon in January? Explain your choice.

20.8 Indoor Pollution

Review Questions

20.61 List the major indoor pollutants and their sources.

20.62 What is the best way to deal with indoor pollution?

20.63 Why is it dangerous to idle a car's engine in a poorly ventilated place, such as the garage?

20.64 Describe the properties that make radon an indoor pollutant. Would radon be more hazardous if ^{222}Rn had a longer half-life?

Problems

20.65 A concentration of 8.00×10^2 ppm by volume of CO is considered lethal to humans. Calculate the minimum mass of CO in grams that would become a lethal concentration in a closed room 17.6 m long, 8.80 m wide, and 2.64 m high. The temperature and pressure are 20.0°C and 756 mmHg, respectively.

20.66 A volume of 5.0 L of polluted air at 18.0°C and 747 mmHg is passed through lime water [an aqueous suspension of $Ca(OH)_2$], so that all the carbon dioxide present is precipitated as $CaCO_3$. If the mass of the $CaCO_3$ precipitate is 0.026 g, calculate the percentage by volume of CO_2 in the air sample.

Additional Problems

20.67 Briefly describe the harmful effects of the following substances: O_3, SO_2, NO_2, CO, $CH_3COOONO_2$ (PAN), Rn.

20.68 The equilibrium constant (K_P) for the reaction

$$N_2(g) + O_2(g) \rightleftharpoons 2NO(g)$$

is 4.0×10^{-31} at 25°C and 2.6×10^{-6} at 1100°C, the temperature of a running car's engine. Is this an endothermic or exothermic reaction?

20.69 As stated in the chapter, carbon monoxide has a much higher affinity for hemoglobin than oxygen does. (a) Write the equilibrium constant expression (K_c) for the following process:

$$CO(g) + HbO_2(aq) \rightleftharpoons O_2(g) + HbCO(aq)$$

where HbO_2 and HbCO are oxygenated hemoglobin and carboxyhemoglobin, respectively. (b) The composition of a breath of air inhaled by a person smoking a cigarette is 1.9×10^{-6} mol/L CO and 8.6×10^{-3} mol/L O_2. Calculate the ratio of [HbCO] to [HbO_2], given that K_c is 212 at 37°C.

20.70 Instead of monitoring carbon dioxide, suggest another gas that scientists could study to substantiate the fact that CO_2 concentration is steadily increasing in the atmosphere.

20.71 In 1991 it was discovered that nitrous oxide (N_2O) is produced in the synthesis of nylon. This compound, which is released into the atmosphere, contributes *both* to the depletion of ozone in the stratosphere and to the greenhouse effect. (a) Write equations representing the reactions between N_2O and oxygen atoms in the stratosphere to produce nitric oxide (NO), which is then oxidized by ozone to form nitrogen dioxide. (b) Is N_2O a more effective greenhouse gas than carbon dioxide? Explain. (c) One of the intermediates in nylon manufacture is adipic acid [$HOOC(CH_2)_4COOH$]. About 2.2×10^9 kg of adipic acid are consumed every year. It is estimated that for every mole of adipic acid produced, 1 mole of N_2O is generated. What is the maximum number of moles of O_3 that can be destroyed as a result of this process per year?

20.72 A glass of water initially at pH 7.0 is exposed to dry air at sea level at 20°C. Calculate the pH of the water when equilibrium is reached between atmospheric CO_2 and CO_2 dissolved in the water, given that Henry's law constant for CO_2 at 20°C is 0.032 mol/L · atm. (*Hint:* Assume no loss of water due to evaporation and use Table 20.1 to calculate the partial pressure of CO_2. Your answer should correspond roughly to the pH of rainwater.)

20.73 A 14-m by 10-m by 3.0-m basement had a high radon content. On the day the basement was sealed off from its surroundings so that no exchange of air could take place, the partial pressure of ^{222}Rn was 1.2×10^{-6} mmHg. Calculate the number of ^{222}Rn isotopes ($t_{\frac{1}{2}}$ = 3.8 d) at the beginning and end of 31 days. Assume STP conditions.

20.74 Ozone in the troposphere is formed by the following steps:

$$NO_2 \longrightarrow NO + O \quad (1)$$

$$O + O_2 \longrightarrow O_3 \quad (2)$$

The first step is initiated by the absorption of visible light (NO_2 is a brown gas). Calculate the longest wavelength required for step (1) at 25°C. [*Hint:* You need to first calculate ΔH and hence ΔU for (1). Next, determine the wavelength for decomposing NO_2 from ΔU.]

20.75 Although the hydroxyl radical (OH) is present only in a trace amount in the troposphere, it plays a central role in its chemistry because it is a strong oxidizing agent and can react with many pollutants as well as some CFC substitutes (see Section 20.3). The hydroxyl radical is formed by the following reactions:

$$O_3 \xrightarrow{\lambda < 320\text{ nm}} O^* + O_2$$

$$O + H_2O \longrightarrow 2OH$$

where O* denotes an electronically excited atom. (a) Explain why the concentration of OH is so small even though the concentrations of O_3 and H_2O are quite large in the troposphere. (b) What property makes OH a strong oxidizing agent? (c) The reaction between OH and NO_2 contributes to acid rain. Write an equation for this process. (d) The hydroxyl radical can oxidize SO_2 to H_2SO_4. The first step is the formation of a neutral HSO_3 species, followed by its reaction with O_2 and H_2O to form H_2SO_4 and the hydroperoxyl radical (HO_2). Write equations for these processes.

20.76 The equilibrium constant (K_P) for the reaction

$$2CO(g) + O_2(g) \rightleftharpoons 2CO_2(g)$$

is 1.4×10^{90} at 25°C. Given this enormous value, why doesn't CO convert totally to CO_2 in the troposphere?

20.77 A person was found dead of carbon monoxide poisoning in a well-insulated cabin. Investigation showed that he had used a blackened bucket to heat water on a butane burner. The burner was found to function properly with no leakage. Explain, with an appropriate equation, the cause of his death.

20.78 The carbon dioxide level in the atmosphere today is often compared with that in preindustrial days. Explain how scientists use tree rings and air trapped in polar ice to arrive at the comparison.

20.79 What is ironic about the following cartoon?

20.80 Calculate the standard enthalpy of formation (ΔH_f°) of ClO from the following bond enthalpies: Cl_2: 243 kJ/mol; O_2: 498 kJ/mol; ClO: 234 kJ/mol.

20.81 Methyl bromide (CH_3Br, b.pt. 3.6°C) is used as a soil fumigant to control insects and weeds. It is also a marine by-product. Photodissociation of the C—Br bond produces Br atoms that can react with ozone similar to Cl, except more effectively. Do you expect CH_3Br to be photolyzed in the troposphere? The bond enthalpy of the C—Br bond is about 293 kJ/mol.

20.82 The effective incoming solar radiation per unit area on Earth is 342 W/m^2. Of this radiation, 6.7 W/m^2 is absorbed by CO_2 at 14,993 nm in the atmosphere. How many photons at this wavelength are absorbed per second in 1 m^2 by CO_2? (1 W = 1 J/s)

20.83 As stated in the chapter, about 50 million tons of sulfur dioxide are released into the atmosphere every year. (a) If 20 percent of the SO_2 is eventually converted to H_2SO_4, calculate the number of 1000-lb marble statues the resulting acid rain can damage. As an estimate, assume that the acid rain only destroys the surface layer of each statue, which is made up of 5 percent of its total mass. (b) What is the other undesirable result of the acid rain damage?

20.84 Peroxyacetyl nitrate (PAN) undergoes thermal decomposition as follows:

$$CH_3(CO)OONO_2 \longrightarrow CH_3(CO)OO + NO_2$$

The rate constant is $3.0 \times 10^{-4}\ s^{-1}$ at 25°C. At the boundary between the troposphere and stratosphere, where the temperature is about −40°C, the rate constant is reduced to $2.6 \times 10^{-7}\ s^{-1}$. (a) Calculate the activation energy for the decomposition of PAN. (b) What is the half-life of the reaction (in minutes) at 25°C?

20.85 How are past temperatures determined from ice cores obtained from the Artic or Antarctica? (*Hint:* Look up the stable isotopes of hydrogen and oxygen. How does energy required for vaporization depend on the masses of H_2O molecules containing different isotopes? How would you determine the age of an ice core?)

20.86 The balance between SO_2 and SO_3 is important in understanding acid rain formation in the troposphere. From the following information at 25°C

$$S(s) + O_2(g) \rightleftharpoons SO_2(g) \qquad K_1 = 4.2 \times 10^{52}$$
$$2S(s) + 3O_2(g) \rightleftharpoons 2SO_3(g) \qquad K_2 = 9.8 \times 10^{128}$$

calculate the equilibrium constant for the reaction

$$2SO_2(g) + O_2(g) \rightleftharpoons 2SO_3(g)$$

20.87 Draw Lewis structures of the species in each step in the conversion of SO_2 to H_2SO_4 discussed in Section 20.6.

20.88 The HO_3 radical was once thought of as a temporary reservoir of atmospheric OH radicals. Draw a Lewis structure of the species.

20.89 What is the difference between weather and climate? In the winter months of 2010, record snows fell on the East Coast from Washington, D.C., to the Deep South. Can this occurrence be taken as evidence against global warming? Can you suggest one way in which global warming might actually be responsible for what happened?

Interpreting, Modeling, & Estimating

20.90 Estimate the annual production of carbon dioxide (in kilograms) by an average passenger car in the United States.

20.91 The following reactions are common in the stratosphere:

(a) $NO_2(g) \longrightarrow NO(g) + O(g)$

(b) $N_2O(g) \longrightarrow N_2(g) + O(g)$

(c) $H_2(g) + O(g) \longrightarrow OH(g) + H(g)$

(d) $CH_4(g) + O(g) \longrightarrow OH(g) + CH_3(g)$

First, without doing any calculations, predict whether each of the reactions is endothermic or exothermic. Then estimate $\Delta H°$ for each reaction, and comment on your predictions and estimates.

Answers to Practice Exercises

20.1 1.12×10^3 nm. **20.2** H_2O. **20.3** 1.2×10^{-6} mol/L.

Answers to Review of Concepts & Facts

20.5.1 (a) and (d).

CHAPTER

21

Metallurgy and the Chemistry of Metals

Purified iron metal is cast into various shapes by pouring the molten liquid into molds and letting it cool to a solid.
yongyut rukkachatsuwa/Shutterstock

CHAPTER OUTLINE

Up to this point we have concentrated mainly on fundamental principles: theories of chemical bonding, intermolecular forces, rates and mechanisms of chemical reactions, equilibrium, the laws of thermodynamics, and electrochemistry. An understanding of these topics is necessary for the study of the properties of representative metallic elements and their compounds.

The use and refinement of metals date back to early human history. For example, archeologists have found evidence that in the first millennium A.D. inhabitants of Sri Lanka used monsoon winds to run iron-smelting furnaces to produce high-carbon steel. Through the years, these furnaces could have been sources of steel for the legendary Damascus swords, known for their sharpness and durability.

In this chapter we will study the methods for extracting, refining, and purifying metals and examine the properties of metals that belong to the representative elements. We will emphasize (1) the occurrence and preparation of metals, (2) the physical and chemical properties of some of their compounds, and (3) their uses in modern society and their roles in biological systems.

21.1 Occurrence of Metals

Learning Objective

- Define key metallurgy terms, including *mineral, ore, alloy,* and *amalgam.*

Most metals come from minerals. A ***mineral*** is *a naturally occurring substance with a range of chemical composition. A mineral deposit concentrated enough to allow economical recovery of a desired metal* is known as ***ore.*** Table 21.1 lists the principal types of minerals, and Figure 21.1 shows a classification of metals according to their minerals.

The most abundant metals, which exist as minerals in Earth's crust, are aluminum, iron, calcium, magnesium, sodium, potassium, titanium, and manganese (see the Chemistry in Action essay "Distribution of Elements on Earth and in Living Systems" in Section 2.4). Seawater is a rich source of some metal ions, including Na^+, Mg^{2+}, and Ca^{2+}. Furthermore, vast areas of the ocean floor are covered with *manganese*

Table 21.1 Principal Types of Minerals

Type	Minerals
Uncombined metals	Ag, Au, Bi, Cu, Pd, Pt
Carbonates	$BaCO_3$ (witherite), $CaCO_3$ (calcite, limestone), $MgCO_3$ (magnesite), $CaCO_3 \cdot MgCO_3$ (dolomite), $PbCO_3$ (cerussite), $ZnCO_3$ (smithsonite)
Halides	CaF_2 (fluorite), NaCl (halite), KCl (sylvite), Na_3AlF_6 (cryolite)
Oxides	$Al_2O_3 \cdot 2H_2O$ (bauxite), Al_2O_3 (corundum), Fe_2O_3 (hematite), Fe_3O_4 (magnetite), Cu_2O (cuprite), MnO_2 (pyrolusite), SnO_2 (cassiterite), TiO_2 (rutile), ZnO (zincite)
Phosphates	$Ca_3(PO_4)_2$ (phosphate rock), $Ca_5(PO_4)_3OH$ (hydroxyapatite)
Silicates	$Be_3Al_2Si_6O_{18}$ (beryl), $ZrSiO_4$ (zircon), $NaAlSi_3O_8$ (albite), $Mg_3(Si_4O_{10})(OH)_2$ (talc)
Sulfides	Ag_2S (argentite), CdS (greenockite), Cu_2S (chalcocite), FeS_2 (pyrite), HgS (cinnabar), PbS (galena), ZnS (sphalerite)
Sulfates	$BaSO_4$ (barite), $CaSO_4$ (anhydrite), $PbSO_4$ (anglesite), $SrSO_4$ (celestite), $MgSO_4 \cdot 7H_2O$ (epsomite)

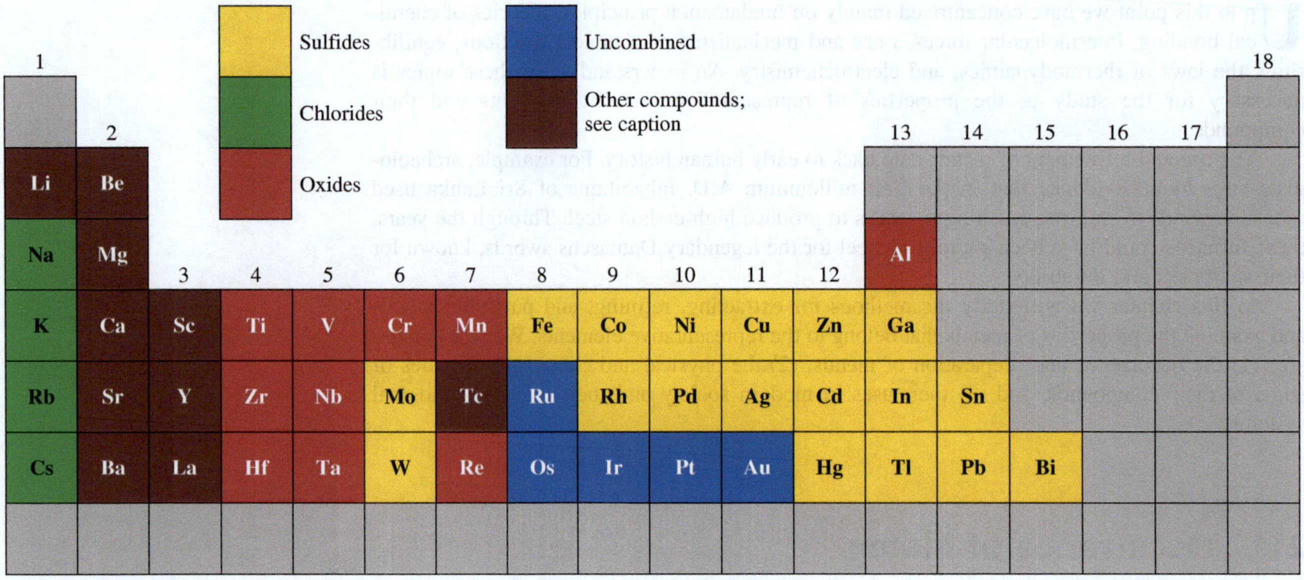

Figure 21.1 *Metals and their best-known minerals. Lithium is found in spodumene ($LiAlSi_2O_6$), and beryllium in beryl (see Table 21.1). The rest of the alkaline earth metals are found in minerals that are carbonates and sulfates. The minerals for Sc, Y, and La are the phosphates. Some metals have more than one type of important mineral. For example, in addition to the sulfide, iron is found as the oxides hematite (Fe_2O_3) and magnetite (Fe_3O_4); and aluminum, in addition to the oxide, is found in beryl ($Be_3Al_2Si_6O_{18}$). Technetium (Tc) is a synthetic element.*

Figure 21.2 *Manganese nodules on the ocean floor.*

Peter Ryan/Science Photo Library/Alamy Stock Photo

nodules, which are made up mostly of manganese, along with iron, nickel, copper, and cobalt in a chemically combined state (Figure 21.2).

Summary of Concepts & Facts

- Depending on their reactivities, metals exist in nature in either the free or the combined state.

21.2 Metallurgical Processes

Learning Objectives

- Describe the three steps involved in obtaining a metal from its ore.
- List the three common methods used for purifying metals.

Metallurgy *is the science and technology of separating metals from their ores and of compounding alloys.* An ***alloy*** *is a solid solution either of two or more metals, or of a metal or metals with one or more nonmetals.*

The three principal steps in the recovery of a metal from its ore are (1) preparation of the ore, (2) production of the metal, and (3) purification of the metal.

Preparation of the Ore

In the preliminary treatment of an ore, the desired mineral is separated from waste materials—usually clay and silicate minerals—which are collectively called the *gangue.* One very useful method for carrying out such a separation is called *flotation.*

In this process, the ore is finely ground and added to water containing oil and detergent. The liquid mixture is then beaten or blown to form a froth. The oil preferentially wets the mineral particles, which are then carried to the top in the froth, while the gangue settles to the bottom. The froth is skimmed off, allowed to collapse, and dried to recover the mineral particles.

Another physical separation process makes use of the magnetic properties of certain minerals. ***Ferromagnetic*** *metals are strongly attracted to magnets.* The mineral magnetite (Fe_3O_4), in particular, can be separated from the gangue by using a strong electromagnet. Cobalt is another ferromagnetic metal.

Mercury forms amalgams with a number of metals. An ***amalgam*** is *an alloy of mercury with another metal or metals.* Mercury can therefore be used to extract metal from ore. Mercury dissolves the silver and gold in an ore to form a liquid amalgam, which is easily separated from the remaining ore. The gold or silver is recovered by distilling off mercury.

Production of Metals

Because metals in their combined forms always have positive oxidation numbers, the production of a free metal is a reduction process. Preliminary operations may be necessary to convert the ore to a chemical state more suitable for reduction. For example, an ore may be *roasted* to drive off volatile impurities and at the same time to convert the carbonates and sulfides to the corresponding oxides, which can be reduced more conveniently to yield the pure metals:

$$CaCO_3(s) \longrightarrow CaO(s) + CO_2(g)$$
$$2PbS(s) + 3O_2(g) \longrightarrow 2PbO(s) + 2SO_2(g)$$

This last equation points out the fact that the conversion of sulfides to oxides is a major source of sulfur dioxide, a notorious air pollutant (see Section 20.6).

How a pure metal is obtained by reduction from its combined form depends on the standard reduction potential of the metal (see Table 18.1). Table 21.2 outlines the reduction processes for several metals. Most major metallurgical processes now in use involve ***pyrometallurgy,*** *procedures carried out at high temperatures.* The reduction in these procedures may be accomplished either chemically or electrolytically.

Chemical Reduction

We can use a more electropositive metal (a more electropositive metal has a more negative standard reduction potential) as a reducing agent to separate a less electropositive metal from its compound at high temperatures:

$$V_2O_5(s) + 5Ca(l) \longrightarrow 2V(l) + 5CaO(s)$$
$$TiCl_4(g) + 2Mg(l) \longrightarrow Ti(s) + 2MgCl_2(l)$$
$$Cr_2O_3(s) + 2Al(s) \longrightarrow 2Cr(l) + Al_2O_3(s)$$
$$3Mn_3O_4(s) + 8Al(s) \longrightarrow 9Mn(l) + 4Al_2O_3(s)$$

Table 21.2 Reduction Processes for Some Common Metals

	Metal	Reduction Process
Decreasing activity of metals ↓	Lithium, sodium, magnesium, calcium	Electrolytic reduction of the molten chloride
	Aluminum	Electrolytic reduction of anhydrous oxide (in molten cryolite)
	Chromium, manganese, titanium, vanadium, iron, zinc	Reduction of the metal oxide with a more electropositive metal, or reduction with coke and carbon monoxide
	Mercury, silver, platinum, copper, gold	These metals occur in the free (uncombined) state or can be obtained by roasting their sulfides

In some cases, even molecular hydrogen can be used as a reducing agent, as in the preparation of tungsten (used as filaments in lightbulbs) from tungsten(VI) oxide:

$$WO_3(s) + 3H_2(g) \longrightarrow W(s) + 3H_2O(g)$$

Electrolytic Reduction

Electrolytic reduction is suitable for very electropositive metals, such as sodium, magnesium, and aluminum. The process is usually carried out on the anhydrous molten oxide or halide of the metal:

$$2MO(l) \longrightarrow 2M \text{ (at cathode)} + O_2 \text{ (at anode)}$$
$$2MCl(l) \longrightarrow 2M \text{ (at cathode)} + Cl_2 \text{ (at anode)}$$

We will describe the specific procedures later in this chapter.

The Metallurgy of Iron

Iron exists in Earth's crust in many different minerals, such as iron pyrite (FeS_2), siderite ($FeCO_3$), hematite (Fe_2O_3), and magnetite (Fe_3O_4, often represented as $FeO \cdot Fe_2O_3$). Of these, hematite and magnetite are particularly suitable for the extraction of iron. The metallurgical processing of iron involves the chemical reduction of the minerals by carbon (in the form of coke) in a blast furnace (Figure 21.3). The concentrated iron ore, limestone ($CaCO_3$), and coke are introduced into the furnace from the top. A blast of hot air is forced up the furnace from the bottom—hence the name *blast furnace*. The oxygen gas reacts with the carbon in the coke to form mostly carbon monoxide and some carbon dioxide. These reactions are highly exothermic, and

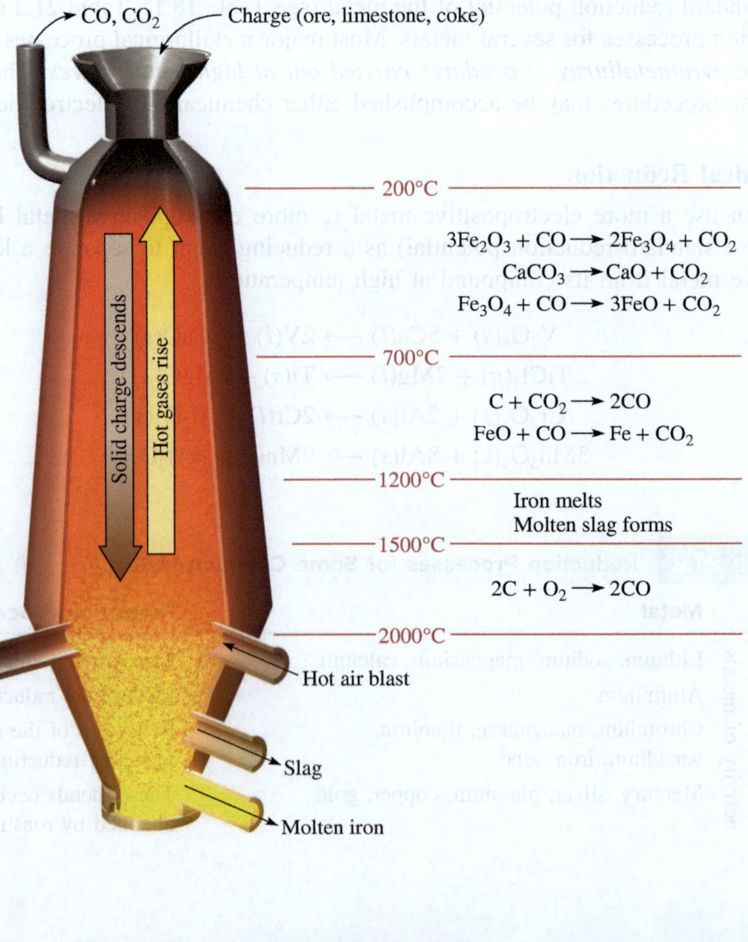

Figure 21.3 *A blast furnace. Iron ore, limestone, and coke are introduced at the top of the furnace. Iron is obtained from the ore by reduction with carbon.*

as the hot CO and CO_2 gases rise they react with the iron oxides in different temperature zones, as shown in Figure 21.3. The key steps in the extraction of iron are

$$3Fe_2O_3(s) + CO(g) \longrightarrow 2Fe_3O_4(s) + CO_2(g)$$
$$Fe_3O_4(s) + CO(g) \longrightarrow 3FeO(s) + CO_2(g)$$
$$FeO(s) + CO(g) \longrightarrow Fe(l) + CO_2(g)$$

The limestone decomposes in the furnace as follows:

$$CaCO_3(s) \longrightarrow CaO(s) + CO_2(g)$$

The calcium oxide then reacts with the impurities in the iron, which are mostly sand (SiO_2) and aluminum oxide (Al_2O_3):

$$CaO(s) + SiO_2(s) \longrightarrow CaSiO_3(l)$$
$$CaO(s) + Al_2O_3(s) \longrightarrow Ca(AlO_2)_2(l)$$

The mixture of calcium silicate and calcium aluminate that remains molten at the furnace temperature is known as *slag.*

By the time the ore works its way down to the bottom of the furnace, most of it has already been reduced to iron. The temperature of the lower part of the furnace is above the melting point of impure iron, and so the molten iron at the lower level can be run off to a receiver. The slag, because it is less dense, forms the top layer above the molten iron and can be run off at that level, as shown in Figure 21.3.

Iron extracted in this way contains many impurities and is called *pig iron;* it may contain up to 5 percent carbon and some silicon, phosphorus, manganese, and sulfur. Some of the impurities stem from the silicate and phosphate minerals, while carbon and sulfur come from coke. Pig iron is granular and brittle. It has a relatively low melting point (about 1180°C), so it can be cast in various forms; for this reason it is also called *cast iron.*

Steelmaking

Steel manufacturing is one of the most important metal industries. In the United States, the annual consumption of steel is well above 100 million tons. Steel is an iron alloy that contains from 0.03 to 1.4 percent carbon plus various amounts of other elements. The wide range of useful mechanical properties associated with steel is primarily a function of chemical composition and heat treatment of a particular type of steel.

Whereas the production of iron is basically a reduction process (converting iron oxides to metallic iron), the conversion of iron to steel is essentially an oxidation process in which the unwanted impurities are removed from the iron by reaction with oxygen gas. One of several methods used in steelmaking is the *basic oxygen process.* Because of its ease of operation and the relatively short time (about 20 minutes) required for each large-scale (hundreds of tons) conversion, the basic oxygen process is by far the most common means of producing steel today.

Figure 21.4 shows the basic oxygen process. Molten iron from the blast furnace is poured into an upright cylindrical vessel. Pressurized oxygen gas is introduced via a water-cooled tube above the molten metal. Under these conditions, manganese, phosphorus, and silicon, as well as excess carbon, react with oxygen to form oxides. These oxides are then reacted with the appropriate fluxes (for example, CaO or SiO_2) to form slag. The type of flux chosen depends on the composition of the iron. If the main impurities are silicon and phosphorus, a basic flux such as CaO is added to the iron:

$$SiO_2(s) + CaO(s) \longrightarrow CaSiO_3(l)$$
$$P_4O_{10}(l) + 6CaO(s) \longrightarrow 2Ca_3(PO_4)_2(l)$$

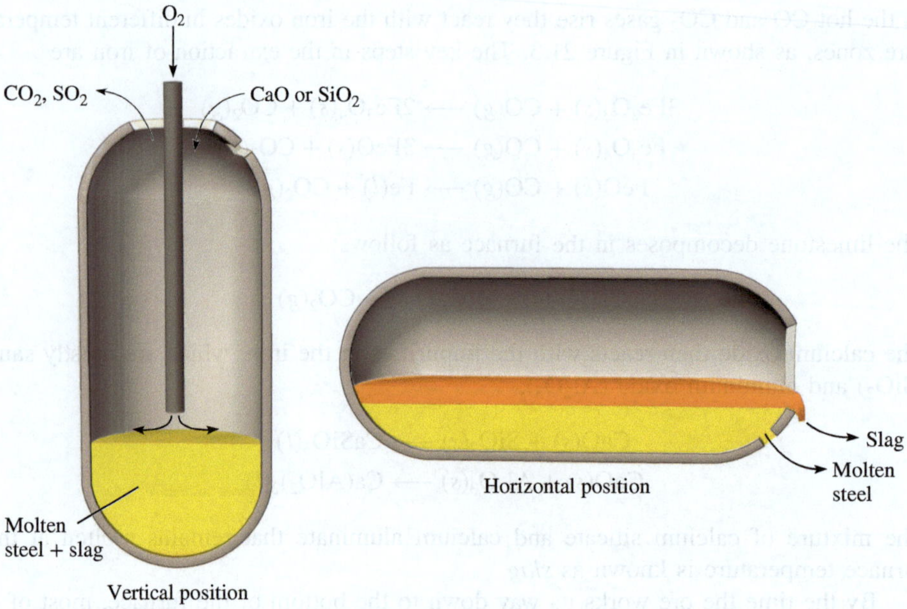

Figure 21.4 *The basic oxygen process of steelmaking. The capacity of a typical vessel is 100 tons of cast iron.*

On the other hand, if manganese is the main impurity, then an acidic flux such as SiO_2 is needed to form the slag:

$$MnO(s) + SiO_2(s) \longrightarrow MnSiO_3(l)$$

The molten steel is sampled at intervals. When the desired blend of carbon and other impurities has been reached, the vessel is rotated to a horizontal position so that the molten steel can be tapped off (Figure 21.5).

The properties of steel depend not only on its chemical composition but also on the heat treatment. At high temperatures, iron and carbon in steel combine to form iron carbide, Fe_3C, called *cementite:*

$$3Fe(s) + C(s) \rightleftharpoons Fe_3C(s)$$

The forward reaction is endothermic, so that the formation of cementite is favored at high temperatures. When steel containing cementite is cooled slowly, the preceding equilibrium shifts to the left, and the carbon separates as small particles of graphite, which give the steel a gray color. (Very slow decomposition of cementite also takes place at room temperature.) If the steel is cooled rapidly, equilibrium is not attained and the carbon remains largely in the form of cementite, Fe_3C. Steel containing cementite is light in color, and it is harder and more brittle than that containing graphite.

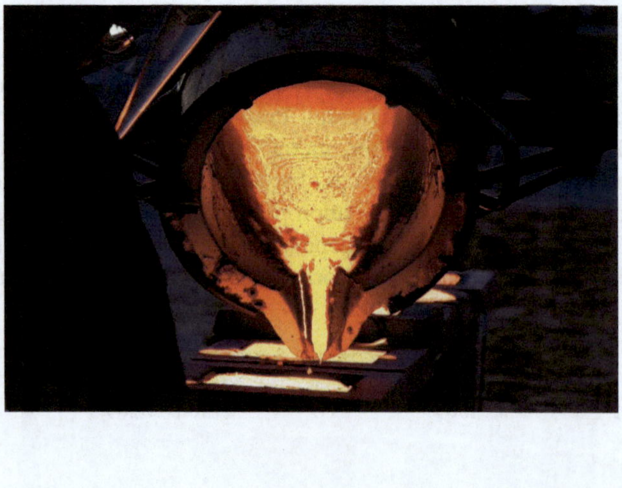

Figure 21.5 *Steelmaking.*
DuffB/iStock/Getty Images

Table 21.3 Types of Steel

	Composition (Percent by Mass)*								
Type	**C**	**Mn**	**P**	**S**	**Si**	**Ni**	**Cr**	**Others**	**Uses**
Plain	1.35	1.65	0.04	0.05	0.06	—	—	Cu (0.2–0.6)	Sheet products, tools
High-strength	0.25	1.65	0.04	0.05	0.15–0.9	0.4–1.0	0.3–1.3	Cu (0.01–0.08)	Construction, steam turbines
Stainless	0.03–1.2	1.0–10	0.04–0.06	0.03	1–3	1–22	4.0–27	—	Kitchen utensils, razor blades

*A single number indicates the maximum amount of the substance present.

Heating the steel to some appropriate temperature for a short time and then cooling it rapidly to give it the desired mechanical properties is known as "tempering." In this way, the ratio of carbon present as graphite and as cementite can be varied within rather wide limits. Table 21.3 shows the composition, properties, and uses of various types of steel.

Purification of Metals

Metals prepared by reduction usually need further treatment to remove impurities. The extent of purification, of course, depends on how the metal will be used. Three common purification procedures are distillation, electrolysis, and zone refining.

Distillation

Metals that have low boiling points, such as mercury, magnesium, and zinc, can be separated from other metals by fractional distillation. One well-known method of fractional distillation is the *Mond*[†] *process* for the purification of nickel. Carbon monoxide gas is passed over the impure nickel metal at about 70°C to form the volatile tetracarbonylnickel (b.p. 43°C), a highly toxic substance, which is separated from the less volatile impurities by distillation:

$$Ni(s) + 4CO(g) \longrightarrow Ni(CO)_4(g)$$

Pure metallic nickel is recovered from $Ni(CO)_4$ by heating the gas at 200°C:

$$Ni(CO)_4(g) \longrightarrow Ni(s) + 4CO(g)$$

The carbon monoxide that is released is recycled back into the process.

Electrolysis

Electrolysis is another important purification technique. The copper metal obtained by roasting copper sulfide usually contains impurities such as zinc, iron, silver, and gold. The more electropositive metals are removed by an electrolysis process in which the impure copper acts as the anode and *pure* copper acts as the cathode in a sulfuric acid solution containing Cu^{2+} ions (Figure 21.6). The reactions are

$$\text{Anode (oxidation):} \quad Cu(s) \longrightarrow Cu^{2+}(aq) + 2e^-$$
$$\text{Cathode (reduction):} \quad Cu^{2+}(aq) + 2e^- \longrightarrow Cu(s)$$

Reactive metals in the copper anode, such as iron and zinc, are also oxidized at the anode and enter the solution as Fe^{2+} and Zn^{2+} ions. They are not reduced at the cathode, however. The less electropositive metals, such as gold and silver, are not oxidized

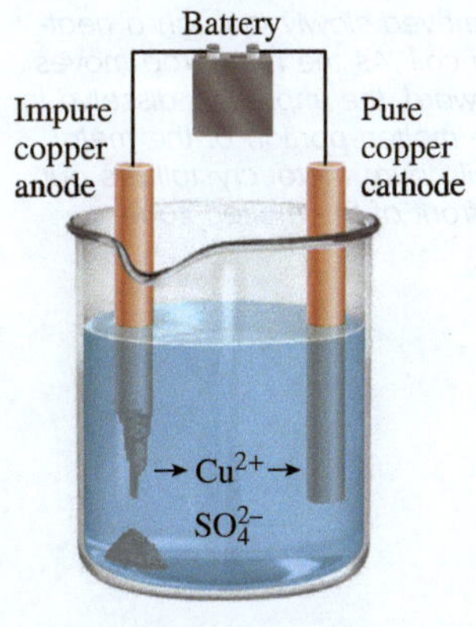

Figure 21.6 *Electrolytic purification of copper.*

[†]Ludwig Mond (1839–1909). British chemist of German origin. Mond made many important contributions to industrial chemistry. His method for purifying nickel by converting it to the volatile $Ni(CO)_4$ compound has been described as having given "wings" to the metal.

Figure 21.7 *Copper cathodes used in the electrorefining process.*
Blaize Pascall/Alamy Stock Photo

at the anode. Eventually, as the copper anode dissolves, these metals fall to the bottom of the cell. Thus, the net result of this electrolysis process is the transfer of copper from the anode to the cathode. Copper prepared this way has a purity greater than 99.5 percent (Figure 21.7).

Zone Refining

Another often-used method of obtaining extremely pure metals is zone refining. In this process, a metal rod containing a few impurities is drawn through an electrical heating coil that melts the metal (Figure 21.8). Most impurities dissolve in the molten metal. As the metal rod emerges from the heating coil, it cools and the pure metal crystallizes, leaving the impurities in the molten metal portion that is still in the heating coil. (This is analogous to the freezing of seawater, in which the solid that separates is mostly pure solvent—water. In zone refining, the liquid metal acts as the solvent and the impurities as the solutes.) When the molten zone carrying the impurities, now at increased concentration, reaches the end of the rod, it is allowed to cool and is then cut off. Repeating this procedure a number of times results in metal with a purity greater than 99.99 percent.

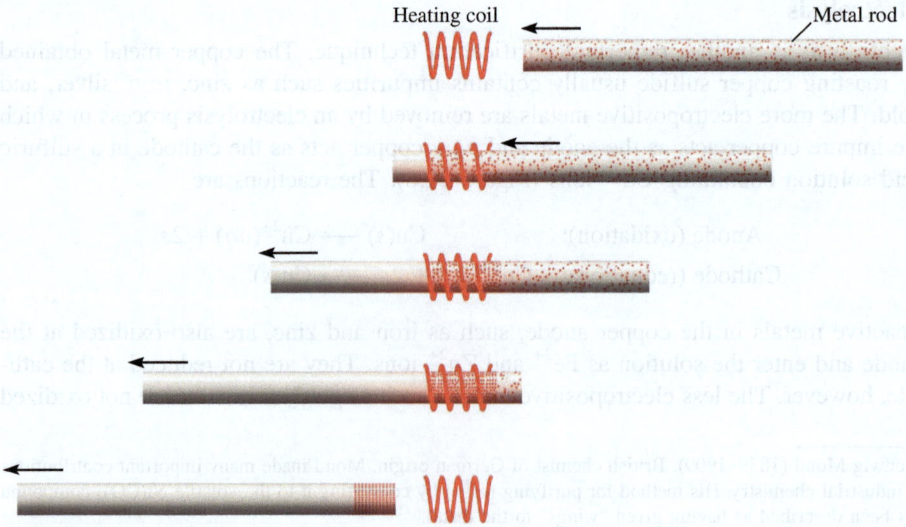

Figure 21.8 *Zone-refining technique for purifying metals. Top to bottom: An impure metal rod is moved slowly through a heating coil. As the metal rod moves forward, the impurities dissolve in the molten portion of the metal while pure metal crystallizes out in front of the molten zone.*

Summary of Concepts & Facts

- Recovering a metal from its ore is a three-stage process. First, the ore must be prepared. The metal is then separated, usually by a reduction process, and finally it is purified.
- The methods commonly used for purifying metals are distillation, electrolysis, and zone refining.

21.3 Band Theory of Electrical Conductivity

Learning Objectives

- Explain the relationship of band theory to metallic bonding.
- Distinguish between conductors and semiconductors.

In Section 11.6 we saw that the ability of metals to conduct heat and electricity can be explained with molecular orbital theory. To gain a better understanding of the conductivity properties of metals we must also apply our knowledge of quantum mechanics. The model we will use to study metallic bonding is ***band theory,*** so called because it states that *delocalized electrons move freely through "bands" formed by overlapping molecular orbitals.* We will also apply band theory to certain elements that are semiconductors.

Conductors

Metals are characterized by high electrical conductivity. Consider magnesium, for example. The electron configuration of Mg is [Ne]$3s^2$, so each atom has two valence electrons in the 3*s* orbital. In a metallic crystal, the atoms are packed closely together, so the energy levels of each magnesium atom are affected by the immediate neighbors of the atom as a result of orbital overlaps. In Chapter 10 we saw that, in terms of molecular orbital theory, the interaction between two atomic orbitals leads to the formation of a bonding and an antibonding molecular orbital. Because the number of atoms in even a small piece of magnesium is enormously large (on the order of 10^{20} atoms), the number of molecular orbitals they form is also very large. These molecular orbitals are so closely spaced on the energy scale that they are more appropriately described as a "band" (Figure 21.9). The closely spaced *filled* energy levels

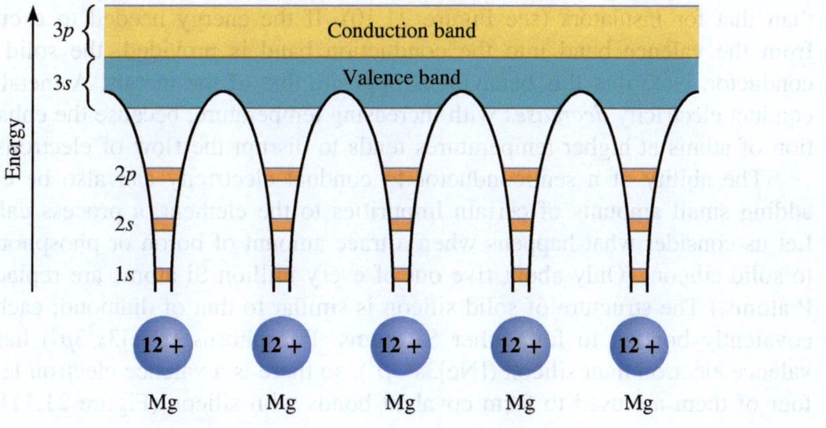

Figure 21.9 *Formation of conduction bands in magnesium. The electrons in the 1s, 2s, and 2p orbitals are localized on each Mg atom. However, the 3s and 3p orbitals overlap to form delocalized molecular orbitals. Electrons in these orbitals can travel throughout the metal, and this accounts for the electrical conductivity of the metal.*

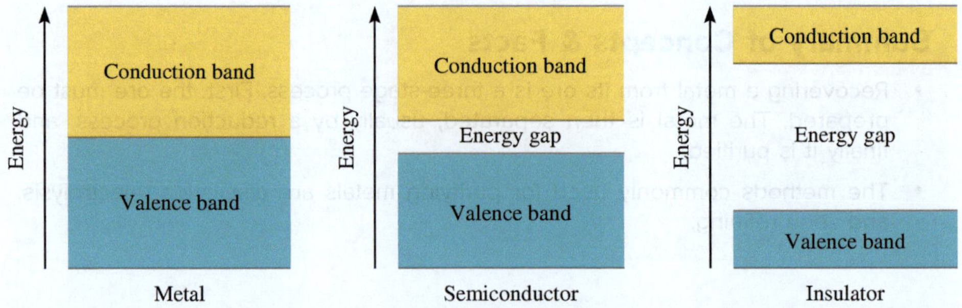

Figure 21.10 *Comparison of the energy gaps between the valence band and the conduction band in a metal, a semiconductor, and an insulator. In a metal, the energy gap is virtually nonexistent; in a semiconductor the energy gap is small; and in an insulator the energy gap is very large, thus making the promotion of an electron from the valence band to the conduction band difficult.*

make up the *valence band.* The upper half of the energy levels corresponds to the empty, delocalized molecular orbitals formed by the overlap of the 3*p* orbitals. This set of closely spaced *empty* levels is called the *conduction band.*

We can imagine a metallic crystal as an array of positive ions immersed in a sea of delocalized valence electrons (see Figure 11.31). The great cohesive force resulting from the delocalization is partly responsible for the strength noted in most metals. Because the valence band and the conduction band are adjacent to each other, the amount of energy needed to promote a valence electron to the conduction band is negligible. There, the electron can travel freely through the metal, because the conduction band is void of electrons. This freedom of movement accounts for the fact that metals are good ***conductors,*** that is, they are *capable of conducting electric current.*

Why don't substances like wood and glass conduct electricity as metals do? Figure 21.10 provides an answer to this question. Basically, the electrical conductivity of a solid depends on the spacing and the state of occupancy of the energy bands. In magnesium and other metals, the valence bands are adjacent to the conduction bands; therefore, these metals readily act as conductors. In wood and glass, on the other hand, the gap between the valence band and the conduction band is considerably greater than that in a metal. Consequently, much more energy is needed to excite an electron into the conduction band. Lacking this energy, electrons cannot move freely. Therefore, glass and wood are ***insulators,*** *ineffective conductors of electricity.*

Semiconductors

A number of elements are ***semiconductors,*** that is, they *normally are not conductors, but will conduct electricity at elevated temperatures or when combined with a small amount of certain other elements.* The Group 14 elements silicon and germanium are especially suited for this purpose. The use of semiconductors in transistors and solar cells, to name two applications, has revolutionized the electronic industry in recent decades, leading to increased miniaturization of electronic equipment.

The energy gap between the filled and empty bands of these solids is much smaller than that for insulators (see Figure 21.10). If the energy needed to excite electrons from the valence band into the conduction band is provided, the solid becomes a conductor. Note that this behavior is opposite that of the metals. A metal's ability to conduct electricity *decreases* with increasing temperature, because the enhanced vibration of atoms at higher temperatures tends to disrupt the flow of electrons.

The ability of a semiconductor to conduct electricity can also be enhanced by adding small amounts of certain impurities to the element, a process called *doping.* Let us consider what happens when a trace amount of boron or phosphorus is added to solid silicon. (Only about five out of every million Si atoms are replaced by B or P atoms.) The structure of solid silicon is similar to that of diamond; each Si atom is covalently bonded to four other Si atoms. Phosphorus ([Ne]$3s^23p^3$) has one more valence electron than silicon ([Ne]$3s^23p^3$), so there is a valence electron left over after four of them are used to form covalent bonds with silicon (Figure 21.11). This extra

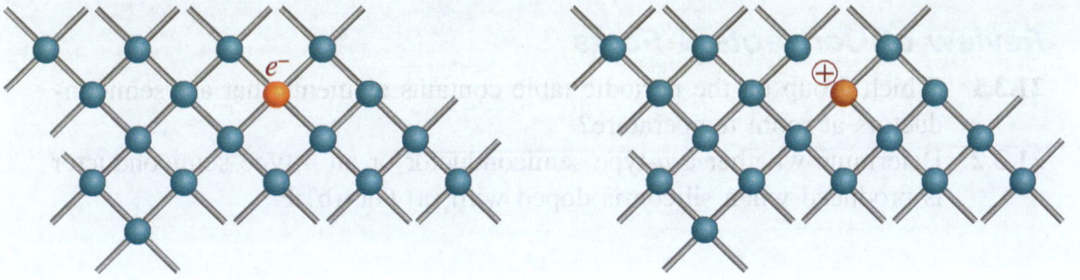

(a) (b)

Figure 21.11 *(a) Silicon crystal doped with phosphorus. (b) Silicon crystal doped with boron. Note the formation of a negative center in (a) and of a positive center in (b).*

electron can be removed from the phosphorus atom by applying a voltage across the solid. The free electron can move through the structure and function as a conduction electron. Impurities of this type are known as ***donor impurities,*** because they *provide conduction electrons. Solids containing donor impurities* are called ***n-type semiconductors,*** where *n* stands for negative (the charge of the "extra" electron).

The opposite effect occurs if boron is added to silicon. A boron atom has three valence electrons ($1s^22s^22p^1$). Thus, for every boron atom in the silicon crystal there is a single *vacancy* in a bonding orbital. It is possible to excite a valence electron from a nearby Si into this vacant orbital. A vacancy created at that Si atom can then be filled by an electron from a neighboring Si atom, and so on. In this manner, electrons can move through the crystal in one direction while the vacancies, or "positive holes," move in the opposite direction, and the solid becomes an electrical conductor. *Impurities that are electron deficient* are called ***acceptor impurities.*** Semiconductors that *contain acceptor impurities* are called ***p-type semiconductors***, where *p* stands for positive.

In both the *p*-type and *n*-type semiconductors, the energy gap between the valence band and the conduction band is effectively reduced, so that only a small amount of energy is needed to excite the electrons. Typically, the conductivity of a semiconductor is increased by a factor of 100,000 or so by the presence of impurity atoms.

The growth of the semiconductor industry since the early 1960s has been truly remarkable. Today semiconductors are essential components of nearly all electronic equipment, ranging from radios and television sets to pocket calculators and computers. One of the main advantages of solid-state devices over vacuum-tube electronics is that the former can be made on a single "chip" of silicon no larger than the cross section of a pencil eraser. Consequently, much more equipment can be packed into a small volume—a point of particular importance in space travel, as well as in handheld calculators and microprocessors (computers-on-a-chip).

Summary of Concepts & Facts

- Metallic bonds can be thought of as the force between positive ions immersed in a sea of electrons. In terms of band theory, the atomic orbitals merge to form energy bands. A substance is a conductor when electrons can be readily promoted to the conduction band, where they are free to move through the substance.
- In insulators, the energy gap between the valence band and the conduction band is so large that electrons cannot be promoted into the conduction band. In semiconductors, electrons can cross the energy gap at higher temperatures, and therefore conductivity increases with increasing temperature as more electrons are able to reach the conduction band.
- *n*-Type semiconductors contain donor impurities and extra electrons. *p*-Type semiconductors contain acceptor impurities and "positive holes."

Review of Concepts & Facts

21.3.1 Which group on the periodic table contains elements that are semiconductors at room temperature?

21.3.2 Determine whether a *p*-type semiconductor or an *n*-type semiconductor is produced when silicon is doped with (a) Ga, (b) P.

21.4 Periodic Trends in Metallic Properties

Learning Objective

- Use the periodic table to predict the metallic character of a series of metals.

Metals are lustrous in appearance, solid at room temperature (with the exception of mercury), good conductors of heat and electricity, malleable (can be hammered flat), and ductile (can be drawn into wire). Figure 21.12 shows the positions of the representative metals and the Group 12 metals in the periodic table. (The transition metals are discussed in Chapter 23.) As we saw in Figure 9.5, the electronegativity of elements increases from left to right across a period and from bottom to top in a group. The metallic character of metals increases in just the opposite directions—that is, from right to left across a period and from top to bottom in a group. Because metals generally have low electronegativities, they tend to form cations and almost always have positive oxidation numbers in their compounds. However, beryllium and magnesium in Group 2 and metals in Group 13 and beyond also form covalent compounds.

In Sections 21.5, 21.6, 21.7, and 21.8 we will study the chemistry of selected metals from Group 1 (the alkali metals), Group 2 (the alkaline earth metals), Group 13 (aluminum), and Groups 8 and 11, respectively.

Summary of Concepts & Facts

- The metallic character of metals increases from right to left across a period and from top to bottom in a group. Metals generally have low electronegativities and tend to form cations.

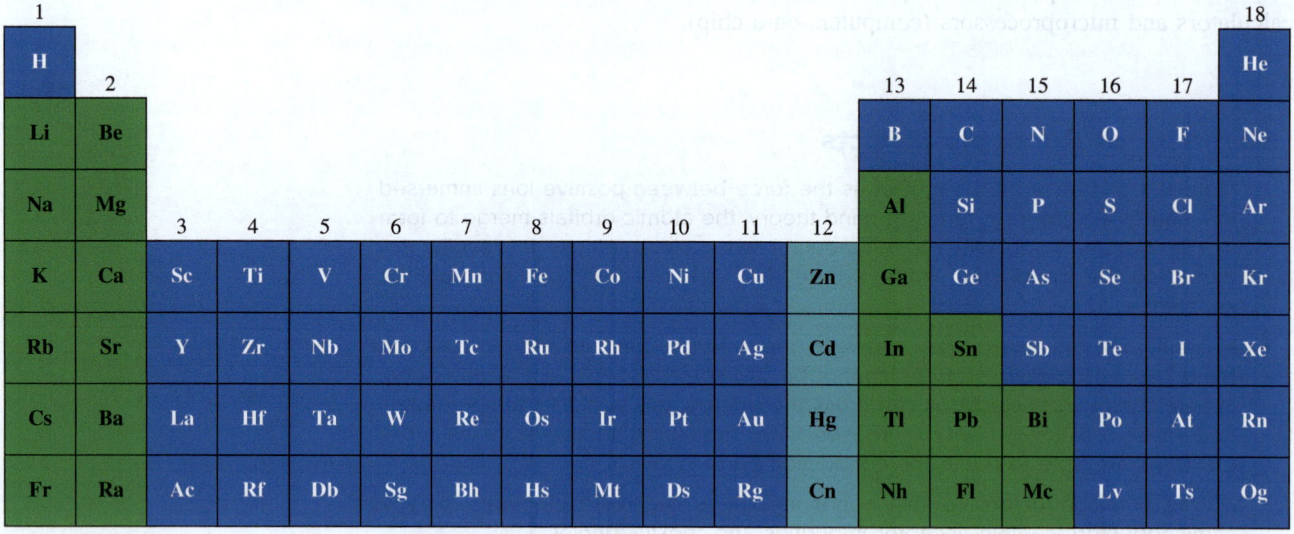

Figure 21.12 *Representative metals (green) and Group 12 metals (light blue) according to their positions in the periodic table.*

21.5 The Alkali Metals

Learning Objective

- Provide some of the common characteristics of the Group 1 metals, including melting point, density, and relative reactivity.

As a group, the alkali metals (the Group 1 elements) are the most electropositive (or the least electronegative) elements known. They exhibit many similar properties, some of which are listed in Table 21.4. From their electron configurations we expect the oxidation number of these elements in their compounds to be +1 because the cations would be isoelectronic with the noble gases. This is indeed the case.

The alkali metals have low melting points and are soft enough to be sliced with a knife (see Figure 8.14). These metals all possess a body-centered crystal structure (see Figure 11.29) with low packing efficiency. This accounts for their low densities among metals. In fact, lithium is the lightest metal known. Because of their great chemical reactivity, the alkali metals never occur naturally in elemental form; they are found combined with halide, sulfate, carbonate, and silicate ions. In this section we will describe the chemistry of two members of Group 1—sodium and potassium. The chemistry of lithium, rubidium, and cesium is less important; all isotopes of francium, the last member of the group, are radioactive.

Sodium and potassium are about equally abundant in nature. They occur in silicate minerals such as albite ($NaAlSi_3O_8$) and orthoclase ($KAlSi_3O_8$). Over long periods of time (on a geologic scale), silicate minerals are slowly decomposed by wind and rain, and their sodium and potassium ions are converted to more soluble compounds. Eventually rain leaches these compounds out of the soil and carries them to the sea. Yet when we look at the composition of seawater, we find that the concentration ratio of sodium to potassium is about 28 to 1. The reason for this uneven distribution is that potassium is essential to plant growth, while sodium is not. Plants take up many of the potassium ions along the way, while sodium ions are free to move on to the sea. Other minerals that contain sodium or potassium are halite (NaCl; shown in Figure 21.13), Chile saltpeter ($NaNO_3$), and sylvite (KCl). Sodium chloride is also obtained from rock salt (see Chemistry in Action essay "Sodium Chloride—A Common and Important Ionic Compound" in Section 9.3).

Metallic sodium is most conveniently obtained from molten sodium chloride by electrolysis in the Downs cell (see Section 18.8). The melting point of sodium chloride is rather high (801°C), and much energy is needed to keep large amounts of the substance molten. Adding a suitable substance, such as $CaCl_2$, lowers the melting point to about 600°C—a more convenient temperature for the electrolysis process.

Figure 21.13 *Halite (NaCl).*
Florea Marius Catalin/iStockphoto/Getty Images

Table 21.4 **Properties of Alkali Metals**

	Li	Na	K	Rb	Cs
Valence electron configuration	$2s^1$	$3s^1$	$4s^1$	$5s^1$	$6s^1$
Density (g/cm^3)	0.534	0.97	0.86	1.53	1.87
Melting point (°C)	179	97.6	63	39	28
Boiling point (°C)	1317	892	770	688	678
Atomic radius (pm)	152	186	227	248	265
Ionic radius (pm)*	78	98	133	148	165
Ionization energy (kJ/mol)	520	496	419	403	375
Electronegativity	1.0	0.9	0.8	0.8	0.8
Standard reduction potential (V)†	−3.04	−2.71	−2.93	−2.93	−2.92

*Refers to the cation M^+ where M denotes an alkali metal atom.

†The half-reaction $M^+(aq) + e^- \longrightarrow M(s)$.

Metallic potassium cannot be easily prepared by the electrolysis of molten KCl because it is too soluble in the molten KCl to float to the top of the cell for collection. Moreover, it vaporizes readily at the operating temperatures, creating hazardous conditions. Instead, it is usually obtained by the distillation of molten KCl in the presence of sodium vapor at 892°C. The reaction that takes place at this temperature is

$$Na(g) + KCl(l) \rightleftharpoons NaCl(l) + K(g)$$

This reaction may seem strange given that potassium is a stronger reducing agent than sodium (see Table 21.4). However, potassium has a lower boiling point (770°C) than sodium (892°C), so it is more volatile at 892°C and distills off more easily. According to Le Châtelier's principle, constant removal of potassium vapor shifts the abovementioned equilibrium from left to right, assuring recovery of metallic potassium.

Sodium and potassium are both extremely reactive, but potassium is the more reactive of the two. Both react with water to form the corresponding hydroxides. In a limited supply of oxygen, sodium burns to form sodium oxide (Na_2O). However, in the presence of excess oxygen, sodium forms the pale yellow peroxide:

$$2Na(s) + O_2(g) \longrightarrow Na_2O_2(s)$$

Sodium peroxide reacts with water to give an alkaline solution and hydrogen peroxide:

$$Na_2O_2(s) + 2H_2O(l) \longrightarrow 2NaOH(aq) + H_2O_2(aq)$$

Like sodium, potassium forms the peroxide. In addition, potassium also forms the superoxide when it burns in air:

$$K(s) + O_2(g) \longrightarrow KO_2(s)$$

When potassium superoxide reacts with water, oxygen gas is evolved:

$$2KO_2(s) + 2H_2O(l) \longrightarrow 2KOH(aq) + O_2(g) + H_2O_2(aq)$$

This reaction is utilized in breathing equipment (Figure 21.14). Exhaled air contains both moisture and carbon dioxide. The moisture reacts with KO_2 in the apparatus to generate oxygen gas as shown in the preceding equation. Furthermore, KO_2 also reacts with exhaled CO_2, which produces more oxygen gas:

$$4KO_2(s) + 2CO_2(g) \longrightarrow 2K_2CO_3(s) + 3O_2(g)$$

Figure 21.14 *Self-contained breathing apparatus.*
Sandro Di Carlo Darsa/PhotoAlto/Getty Images

Thus, a person using the apparatus can continue to breathe oxygen without being exposed to toxic fumes outside.

Sodium and potassium metals dissolve in liquid ammonia to produce a beautiful blue solution:

$$\mathrm{Na} \xrightarrow{\mathrm{NH_3}} \mathrm{Na^+} + e^-$$
$$\mathrm{K} \xrightarrow{\mathrm{NH_3}} \mathrm{K^+} + e^-$$

Both the cation and the electron exist in the solvated form; the solvated electrons are responsible for the characteristic blue color of such solutions. Metal-ammonia solutions are powerful reducing agents (because they contain free electrons); they are useful in synthesizing both organic and inorganic compounds. It was discovered that the hitherto unknown alkali metal anions, M^-, are also formed in such solutions. This means that an ammonia solution of an alkali metal contains ion pairs such as Na^+Na^- and K^+K^-! (Keep in mind that in each case the metal cation exists as a complex ion with *crown ether,* an organic compound with a high affinity for cations.) In fact, these "salts" are so stable that they can be isolated in crystalline form (see the photo at the beginning of Chapter 21). This finding is of considerable theoretical interest, for it shows clearly that the alkali metals can have an oxidation number of −1, although −1 is not found in ordinary compounds.

Sodium and potassium are essential elements of living matter. Sodium ions and potassium ions are present in intracellular and extracellular fluids, and they are essential for osmotic balance and enzyme functions. We now describe the preparations and uses of several of the important compounds of sodium and potassium.

Sodium Chloride

The source, properties, and uses of sodium chloride were discussed in the Chemistry in Action essay "Sodium Chloride—A Common and Important Ionic Compound" in Section 9.3.

Sodium Carbonate

Sodium carbonate (called soda ash) is used in all kinds of industrial processes, including water treatment and the manufacture of soaps, detergents, medicines, and food additives. Today about half of all Na_2CO_3 produced is used in the glass industry (in soda-lime glass; see Section 11.7). Sodium carbonate ranks eleventh among the chemicals produced in the United States (11 million tons in 2010). For many years Na_2CO_3 was produced by the Solvay† process, in which ammonia is first dissolved in a saturated solution of sodium chloride. Bubbling carbon dioxide into the solution results in the precipitation of sodium bicarbonate as follows:

$$\mathrm{NH_3}(aq) + \mathrm{NaCl}(aq) + \mathrm{H_2CO_3}(aq) \longrightarrow \mathrm{NaHCO_3}(s) + \mathrm{NH_4Cl}(aq)$$

Sodium bicarbonate is then separated from the solution and heated to give sodium carbonate:

$$2\mathrm{NaHCO_3}(s) \longrightarrow \mathrm{Na_2CO_3}(s) + \mathrm{CO_2}(g) + \mathrm{H_2O}(g)$$

However, the rising cost of ammonia and the pollution problem resulting from by-products have prompted chemists to look for other sources of sodium carbonate. The last plant using the Solvay process in the United States closed in 1986 though the process is still in use worldwide. One is the mineral *trona* $[Na_5(CO_3)_2(HCO_3) \cdot 2H_2O]$,

†Ernest Solvay (1838–1922). Belgian chemist. Solvay's main contribution to industrial chemistry was the development of the process for the production of sodium carbonate that now bears his name.

large deposits of which have been found in Wyoming. When trona is crushed and heated, it decomposes as follows:

$$2Na_5(CO_3)_2(HCO_3) \cdot 2H_2O(s) \longrightarrow 5Na_2CO_3(s) + CO_2(g) + 3H_2O(g)$$

The sodium carbonate obtained this way is dissolved in water, the solution is filtered to remove the insoluble impurities, and the sodium carbonate is crystallized as $Na_2CO_3 \cdot 10H_2O$. Finally, the hydrate is heated to give pure, anhydrous sodium carbonate.

Sodium Hydroxide and Potassium Hydroxide

The properties of sodium hydroxide and potassium hydroxide are very similar. These hydroxides are prepared by the electrolysis of aqueous NaCl and KCl solutions (see Section 18.8); both hydroxides are strong bases and very soluble in water. Sodium hydroxide is used in the manufacture of soap and many organic and inorganic compounds. Potassium hydroxide is used as an electrolyte in some storage batteries, and aqueous potassium hydroxide is used to remove carbon dioxide and sulfur dioxide from air.

Sodium Nitrate and Potassium Nitrate

Large deposits of sodium nitrate (*Chile saltpeter*) are found in Chile. It decomposes with the evolution of oxygen at about 500°C:

$$2NaNO_3(s) \longrightarrow 2NaNO_2(s) + O_2(g)$$

Potassium nitrate (*saltpeter*) is prepared beginning with the "reaction"

$$KCl(aq) + NaNO_3(aq) \longrightarrow KNO_3(aq) + NaCl(aq)$$

This process is carried out just below 100°C. Because KNO_3 is the least soluble salt at room temperature, it is separated from the solution by fractional crystallization. Like $NaNO_3$, KNO_3 decomposes when heated.

Gunpowder consists of potassium nitrate, wood charcoal, and sulfur in the approximate proportions of 6:1:1 by mass. When gunpowder is heated, the reaction is

$$2KNO_3(s) + S(l) + 3C(s) \longrightarrow K_2S(s) + N_2(g) + 3CO_2(g)$$

The sudden formation of hot expanding gases causes an explosion.

Summary of Concepts & Facts

- The alkali metals are the most reactive of all the metallic elements. They have an oxidation number of +1 in their compounds. Under special conditions, some of them also form uninegative ions.

21.6 The Alkaline Earth Metals

Learning Objective

- Provide some of the common characteristics of the Group 2 metals including melting point, density, and relative reactivity.

The alkaline earth metals are somewhat less electropositive and less reactive than the alkali metals. Except for the first member of the family, beryllium, which resembles aluminum (a Group 13 metal) in some respects, the alkaline earth metals have similar

Table 21.5 Properties of Alkaline Earth Metals

	Be	Mg	Ca	Sr	Ba
Valence electron configuration	$2s^2$	$3s^2$	$4s^2$	$5s^2$	$6s^2$
Density (g/cm^3)	1.86	1.74	1.55	2.6	3.5
Melting point (°C)	1280	650	838	770	714
Boiling point (°C)	2770	1107	1484	1380	1640
Atomic radius (pm)	112	160	197	215	222
Ionic radius (pm)*	34	78	106	127	143
First and second ionization	899	738	590	548	502
energies (kJ/mol)	1757	1450	1145	1058	958
Electronegativity	1.6	1.3	1.0	1.0	0.9
Standard reduction potential (V)†	−1.85	−2.37	−2.87	−2.90	−2.91

*Refers to the cation M^{2+}, where M denotes an alkali earth metal atom.

†The half-reaction is $M^{2+}(aq) + 2e^- \longrightarrow M(s)$.

chemical properties. Because their M^{2+} ions attain the stable electron configuration of the preceding noble gas, the oxidation number of alkaline earth metals in the combined form is almost always +2. Table 21.5 lists some common properties of these metals. Radium is not included in the table because all radium isotopes are radioactive and it is difficult and expensive to study the chemistry of this Group 2 element.

Magnesium

Magnesium (see Figure 8.15) is the sixth most plentiful element in Earth's crust (about 2.5 percent by mass). Among the principal magnesium ores are brucite, $Mg(OH)_2$; dolomite, $CaCO_3 \cdot MgCO_3$ (Figure 21.15); and epsomite, $MgSO_4 \cdot 7H_2O$. Seawater is a good source of magnesium; there are about 1.3 g of magnesium in each kilogram of seawater. As is the case with most alkali and alkaline earth metals, metallic magnesium is obtained by electrolysis, in this case from its molten chloride, $MgCl_2$ (obtained from seawater; see the Chemistry in Action essay "Metal from the Sea" in Section 4.7).

Figure 21.15 *Dolomite ($CaCO_3 \cdot MgCO_3$).*
Brendan Hunter/iStock/Getty Images

The chemistry of magnesium is intermediate between that of beryllium and the heavier Group 2 elements. Magnesium does not react with cold water but does react slowly with steam:

$$Mg(s) + H_2O(g) \longrightarrow MgO(s) + H_2(g)$$

It burns brilliantly in air to produce magnesium oxide and magnesium nitride (see Figure 4.9):

$$2Mg(s) + O_2(g) \longrightarrow 2MgO(s)$$
$$3Mg(s) + N_2(g) \longrightarrow Mg_3N_2(s)$$

This property makes magnesium (in the form of thin ribbons or fibers) useful in flash photography and flares.

Magnesium oxide reacts very slowly with water to form magnesium hydroxide, a white solid suspension called *milk of magnesia* (see Chemistry in Action essay "Antacids and the pH Balance in Your Stomach" in Section 15.3), which is used to treat acid indigestion:

$$MgO(s) + H_2O(l) \longrightarrow Mg(OH)_2(s)$$

Magnesium is a typical alkaline earth metal in that its hydroxide is a strong base. [The only alkaline earth hydroxide that is not a strong base is $Be(OH)_2$, which is amphoteric.]

Figure 21.16 *Fluorite (CaF_2).*
Ken Karp/McGraw Hill

The major uses of magnesium are in lightweight structural alloys, for cathodic protection (see Section 18.7), in organic synthesis, and in batteries. Magnesium is essential to plant and animal life, and Mg^{2+} ions are not toxic. It is estimated that the average adult ingests about 0.3 g of magnesium ions daily. Magnesium plays several important biological roles. It is present in intracellular and extracellular fluids. Magnesium ions are essential for the proper functioning of a number of enzymes. Magnesium is also present in the green plant pigment chlorophyll, which plays an important part in photosynthesis.

Calcium

Earth's crust contains about 3.4 percent calcium (see Figure 8.15) by mass. Calcium occurs in limestone, calcite, chalk, and marble as $CaCO_3$; in dolomite as $CaCO_3 \cdot MgCO_3$ (see Figure 21.15); in gypsum as $CaSO_4 \cdot 2H_2O$; and in fluorite as CaF_2 (Figure 21.16). Metallic calcium is best prepared by the electrolysis of molten calcium chloride ($CaCl_2$).

As we read down Group 2 from beryllium to barium, we note an increase in metallic properties. Unlike beryllium and magnesium, calcium (like strontium and barium) reacts with cold water to yield the corresponding hydroxide, although the rate of reaction is much slower than those involving the alkali metals (see Figure 4.14):

$$Ca(s) + 2H_2O(l) \longrightarrow Ca(OH)_2(aq) + H_2(g)$$

Calcium hydroxide [$Ca(OH)_2$] is commonly known as slaked lime or hydrated lime.

Lime (CaO), which is also referred to as quicklime, is one of the oldest materials known to mankind. Quicklime is produced by the thermal decomposition of calcium carbonate (see Section 17.5):

$$CaCO_3(s) \longrightarrow CaO(s) + CO_2(g)$$

while slaked lime is produced by the reaction between quicklime and water:

$$CaO(s) + H_2O(l) \longrightarrow Ca(OH)_2(aq)$$

Quicklime is used in metallurgy (see Section 21.2) and the removal of SO_2 when fossil fuel is burned (see Section 20.6). Slaked lime is used in water treatment.

For many years, farmers have used lime to lower the acidity of soil for their crops (a process called *liming*). Nowadays lime is also applied to lakes affected by acid rain (see Section 20.6).

Metallic calcium has rather limited uses. It serves mainly as an alloying agent for metals like aluminum and copper and in the preparation of beryllium metal from its compounds. It is also used as a dehydrating agent for organic solvents.

Calcium is an essential element in living matter. It is the major component of bones and teeth; the calcium ion is present in a complex phosphate salt, hydroxyapatite, $Ca_5(PO_4)_3OH$. A characteristic function of Ca^{2+} ions in living systems is the activation of a variety of metabolic processes. Calcium plays a vital role in heart action, blood clotting, muscle contraction, and nerve impulse transmission.

Summary of Concepts & Facts

- The alkaline earth metals are somewhat less reactive than the alkali metals. They almost always have an oxidation number of +2 in their compounds. The properties of the alkaline earth elements become increasingly metallic from top to bottom in their periodic group.

21.7 Aluminum

Learning Objectives

- Provide some of the common characteristics of aluminum, including melting point, density, and relative reactivity.
- Summarize the Hall process.

Aluminum (see Figure 8.16) is the most abundant metal and the third most plentiful element in Earth's crust (7.5 percent by mass). The elemental form does not occur in nature; its principal ore is bauxite ($Al_2O_3 \cdot 2H_2O$). Other minerals containing aluminum are orthoclase ($KAlSi_3O_8$), beryl ($Be_3Al_2Si_6O_{18}$), cryolite (Na_3AlF_6), and corundum (Al_2O_3; Figure 21.17).

Aluminum is usually prepared from bauxite, which is frequently contaminated with silica (SiO_2), iron oxides, and titanium(IV) oxide. The ore is first heated in sodium hydroxide solution to convert the silica into soluble silicates:

$$SiO_2(s) + 2OH^-(aq) \longrightarrow SiO_3^{2-}(aq) + H_2O(l)$$

At the same time, aluminum oxide is converted to the aluminate ion (AlO_2^-):

$$Al_2O_3(s) + 2OH^-(aq) \longrightarrow 2AlO_2^-(aq) + H_2O(l)$$

Iron oxide and titanium oxide are unaffected by this treatment and are filtered off. Next, the solution is treated with acid to precipitate the insoluble aluminum hydroxide:

$$AlO_2^-(aq) + H_3O^+(aq) \longrightarrow Al(OH)_3(s)$$

After filtration, the aluminum hydroxide is heated to obtain aluminum oxide:

$$2Al(OH)_3(s) \longrightarrow Al_2O_3(s) + 3H_2O(g)$$

Anhydrous aluminum oxide, or *corundum,* is reduced to aluminum by the *Hall*† *process*. Figure 21.18 shows a Hall electrolytic cell, which contains a series of carbon

Figure 21.17 *Corundum (Al_2O_3).*
Melissa Carroll/E+/Getty Images

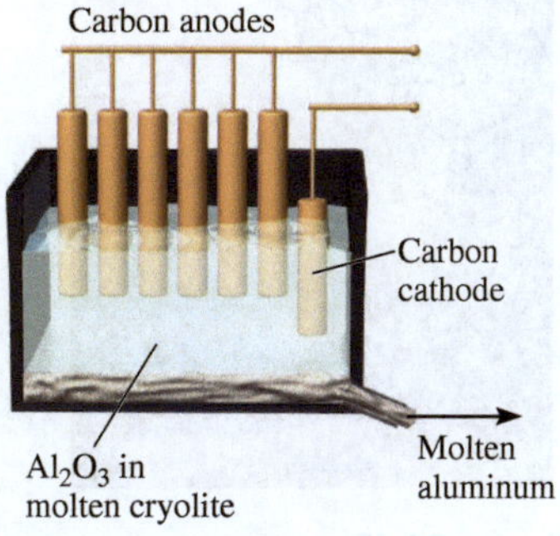

Figure 21.18 *Electrolytic production of aluminum based on the Hall process.*

†Charles Martin Hall (1863–1914). American inventor. While Hall was an undergraduate at Oberlin College, he became interested in finding an inexpensive way to extract aluminum. Shortly after graduation, when he was only 22 years old, Hall succeeded in obtaining aluminum from aluminum oxide in a backyard woodshed. Amazingly, the same discovery was made at almost the same moment in France, by Paul Héroult, another 22-year-old inventor working in a similar makeshift laboratory.

anodes. The cathode is also made of carbon and constitutes the lining inside the cell. The key to the Hall process is the use of cryolite, or Na_3AlF_6 (m.p. 1000°C), as the solvent and conducting medium for aluminum oxide (m.p. 2045°C). The mixture is electrolyzed to produce aluminum and oxygen gas:

$$\begin{array}{rl} \text{Anode (oxidation):} & 3[2O^{2-} \longrightarrow O_2(g) + 4e^-] \\ \text{Cathode (reduction):} & 4[Al^{3+} + 3e^- \longrightarrow Al(l)] \\ \hline \text{Overall:} & 2Al_2O_3 \longrightarrow 4Al(l) + 3O_2(g) \end{array}$$

Video: Aluminum Production

Oxygen gas reacts with the carbon anodes (at elevated temperatures) to form carbon monoxide, which escapes as a gas. The liquid aluminum metal (m.p. 660.2°C) sinks to the bottom of the vessel, from which it can be drained from time to time during the procedure.

Aluminum is one of the most versatile metals known. It has a low density (2.7 g/cm^3) and high tensile strength (that is, it can be stretched or drawn out). Aluminum is malleable, it can be rolled into thin foils, and it is an excellent electrical conductor. Its conductivity is about 65 percent that of copper. However, because aluminum is cheaper and lighter than copper, it is widely used in high-voltage transmission lines. Although aluminum's chief use is in aircraft construction, the pure metal itself is too soft and weak to withstand much strain. Its mechanical properties are greatly improved by alloying it with small amounts of metals such as copper, magnesium, and manganese, as well as silicon. Aluminum is not used by living systems and is generally considered to be nontoxic.

As we read across the periodic table from left to right in a given period, we note a gradual decrease in metallic properties. Thus, although aluminum is considered an active metal, it does not react with water as do sodium and calcium. Aluminum reacts with hydrochloric acid and with strong bases as follows:

$$2Al(s) + 6HCl(aq) \longrightarrow 2AlCl_3(aq) + 3H_2(g)$$
$$2Al(s) + 2NaOH(aq) + 2H_2O(l) \longrightarrow 2NaAlO_2(aq) + 3H_2(g)$$

Aluminum readily forms the oxide Al_2O_3 when exposed to air:

$$4Al(s) + 3O_2(g) \longrightarrow 2Al_2O_3(s)$$

A tenacious film of this oxide protects metallic aluminum from further corrosion and accounts for some of the unexpected inertness of aluminum.

Aluminum oxide has a very large exothermic enthalpy of formation ($\Delta H_f^\circ = -1670$ kJ/mol). This property makes aluminum suitable for use in solid propellants for rockets such as those used for some space shuttles. When a mixture of aluminum and ammonium perchlorate (NH_4ClO_4) is ignited, aluminum is oxidized to Al_2O_3, and the heat liberated in the reaction causes the gases that are formed to expand with great force. This action lifts the rocket.

The great affinity of aluminum for oxygen is illustrated nicely by the reaction of aluminum powder with a variety of metal oxides, particularly the transition metal oxides, to produce the corresponding metals. A typical reaction is

$$2Al(s) + Fe_2O_3(s) \longrightarrow Al_2O_3(s) + 2Fe(l) \qquad \Delta H^\circ = -852 \text{ kJ/mol}$$

which can result in temperatures approaching 3000°C. This reaction, which is used in the welding of steel and iron, is called the *thermite reaction* (Figure 21.19).

Figure 21.19 *The temperature of a thermite reaction can reach 3000°C.*
Charles D. Winters/McGraw Hill

Aluminum chloride exists as a dimer:

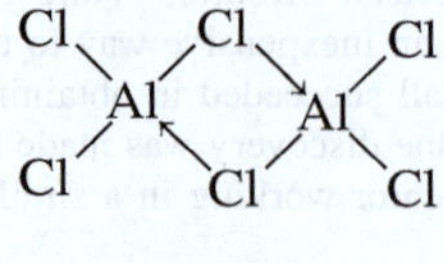

CHEMISTRY *in Action*

Recycling Aluminum

Aluminum beverage cans were virtually unknown in 1960; yet by the early 1970s over 1.3 billion pounds of aluminum had been used for these containers. The reasons for aluminum's popularity in the beverage industry are that it is nontoxic, odorless, tasteless, and lightweight. Furthermore, it is thermally conducting, so the fluid inside the container can be chilled rapidly.

The tremendous increase in the demand for aluminum does have a definite drawback, however. More than 3 billion pounds of the metal cans and foils are discarded in the United States annually. They litter the countryside and clog landfills. The best solution to this environmental problem, and the way to prevent the rapid depletion of a finite resource, is recycling.

What are the economic benefits of recycling aluminum? Let us compare the energy consumed in the production of aluminum from bauxite with that consumed when aluminum is recycled. The overall reaction for the Hall process can be represented as

$$Al_2O_3 \text{ (in molten cryolite)} + 3C(s) \longrightarrow 2Al(l) + 3CO(g)$$

for which $\Delta H° = 1340$ kJ/mol and $\Delta S° = 586$ J/K · mol. At 1000°C, which is the temperature of the process, the standard free-energy change for the reaction is given by

$$\begin{aligned}\Delta G° &= \Delta H° - T\Delta S° \\ &= 1340 \text{ kJ/mol} - (1273 \text{ K})\left(\frac{586 \text{ J}}{\text{K} \cdot \text{mol}}\right)\left(\frac{1 \text{ kJ}}{1000 \text{ J}}\right) \\ &= 594 \text{ kJ/mol}\end{aligned}$$

Equation (18.3) states that $\Delta G° = -nFE°$; therefore, the amount of electrical energy needed to produce 1 mole of Al from bauxite is 594 kJ/2, or 297 kJ.

Recycling aluminum requires only enough energy to heat the metal to its melting point (660°C) plus the heat of fusion (10.7 kJ/mol). The heat change where 1 mole of aluminum is heated from 25°C to 660°C is

$$\begin{aligned}\text{heat input} &= \mathscr{M}s\Delta t \\ &= (27.0 \text{ g})(0.900 \text{ J/g} \cdot °\text{C})(660 - 25)°\text{C} \\ &= 15.4 \text{ kJ}\end{aligned}$$

where $\mathscr{M}$ is the molar mass, s is the specific heat of Al, and Δt is the temperature change. Thus, the total energy needed to recycle 1 mole of Al is given by

$$\begin{aligned}\text{total energy} &= 15.4 \text{ kJ} + 10.7 \text{ kJ} \\ &= 26.1 \text{ kJ}\end{aligned}$$

To compare the energy requirements of the two methods we write

$$\begin{aligned}&\frac{\text{energy needed to recycle 1 mol Al}}{\text{energy needed to produce 1 mol Al by electrolysis}} \\ &= \frac{26.1 \text{ kJ}}{297 \text{ kJ}} \times 100\% \\ &= 8.8\%\end{aligned}$$

Thus, by recycling aluminum cans we can save about 91 percent of the energy required to extract the metal from bauxite. Recycling most of the aluminum cans thrown away each year saves 20 billion kilowatt-hours of electricity—about 1 percent of the electric power used in the United States annually. (Watt is the unit for power, 1 watt = 1 joule per second.)

Left: Collecting aluminum cans for recycling.
Right: Melting and purifying recycled aluminum.
(left): Monty Rakusen/Image Source/Getty Images; (right): Chris Knapton/Alamy Stock Photo

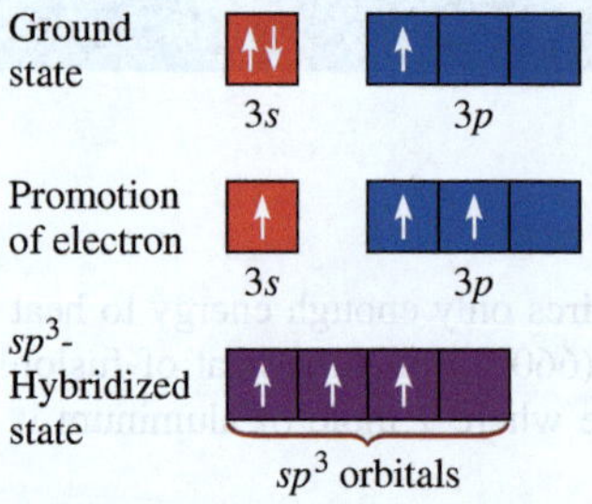

Figure 21.20 *The sp^3 hybridization of an Al atom in Al_2Cl_6. Each Al atom has one vacant sp^3 hybrid orbital that can accept a lone pair from the bridging Cl atom.*

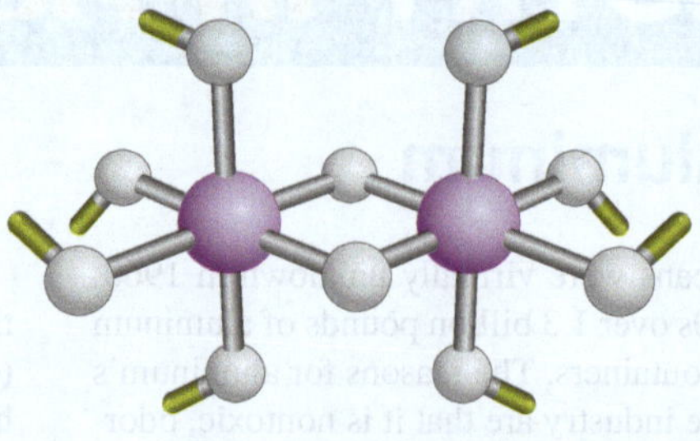

Figure 21.21 *Structure of aluminum hydride. Note that this compound is a polymer. Each Al atom is surrounded octahedrally by six bridging H atoms.*

Each of the bridging chlorine atoms forms a normal covalent bond and a coordinate covalent bond (indicated by $\longrightarrow$) with two aluminum atoms. Each aluminum atom is assumed to be sp^3-hybridized, so the vacant sp^3 hybrid orbital can accept a lone pair from the chlorine atom (Figure 21.20). Aluminum chloride undergoes hydrolysis as follows:

$$AlCl_3(s) + 3H_2O(l) \longrightarrow Al(OH)_3(s) + 3HCl(aq)$$

Aluminum hydroxide, like $Be(OH)_2$, is amphoteric:

$$Al(OH)_3(s) + 3H^+(aq) \longrightarrow Al^{3+}(aq) + 3H_2O(l)$$
$$Al(OH)_3(s) + OH^-(aq) \longrightarrow Al(OH)_4^-(aq)$$

In contrast to the boron hydrides, which are a well-defined series of compounds, aluminum hydride is a polymer in which each aluminum atom is surrounded octahedrally by bridging hydrogen atoms (Figure 21.21).

When an aqueous mixture of aluminum sulfate and potassium sulfate is evaporated slowly, crystals of $KAl(SO_4)_2 \cdot 12H_2O$ are formed. Similar crystals can be formed by substituting Na^+ or NH_4^+ for K^+, and Cr^{3+} or Fe^{3+} for Al^{3+}. These compounds are called *alums,* and they have the general formula

$$M^+M^{3+}(SO_4)_2 \cdot 12H_2O \qquad \begin{array}{l} M^+\text{: } K^+, Na^+, NH_4^+ \\ M^{3+}\text{: } Al^{3+}, Cr^{3+}, Fe^{3+} \end{array}$$

Alums are examples of double salts, that is, salts that contain two different cations.

Summary of Concepts & Facts

- Aluminum does not react with water due to the formation of a protective oxide; its hydroxide is amphoteric.
- The Hall process is used to reduce aluminum oxide to aluminum.

21.8 Iron and Copper

Learning Objective

- Summarize the chemistry of iron and copper.

Figure 21.22 shows the first-row transition metals. In this section, we will briefly survey the chemistry of two of these elements—iron and copper—paying particular attention to their occurrence, preparation, uses, and important compounds.

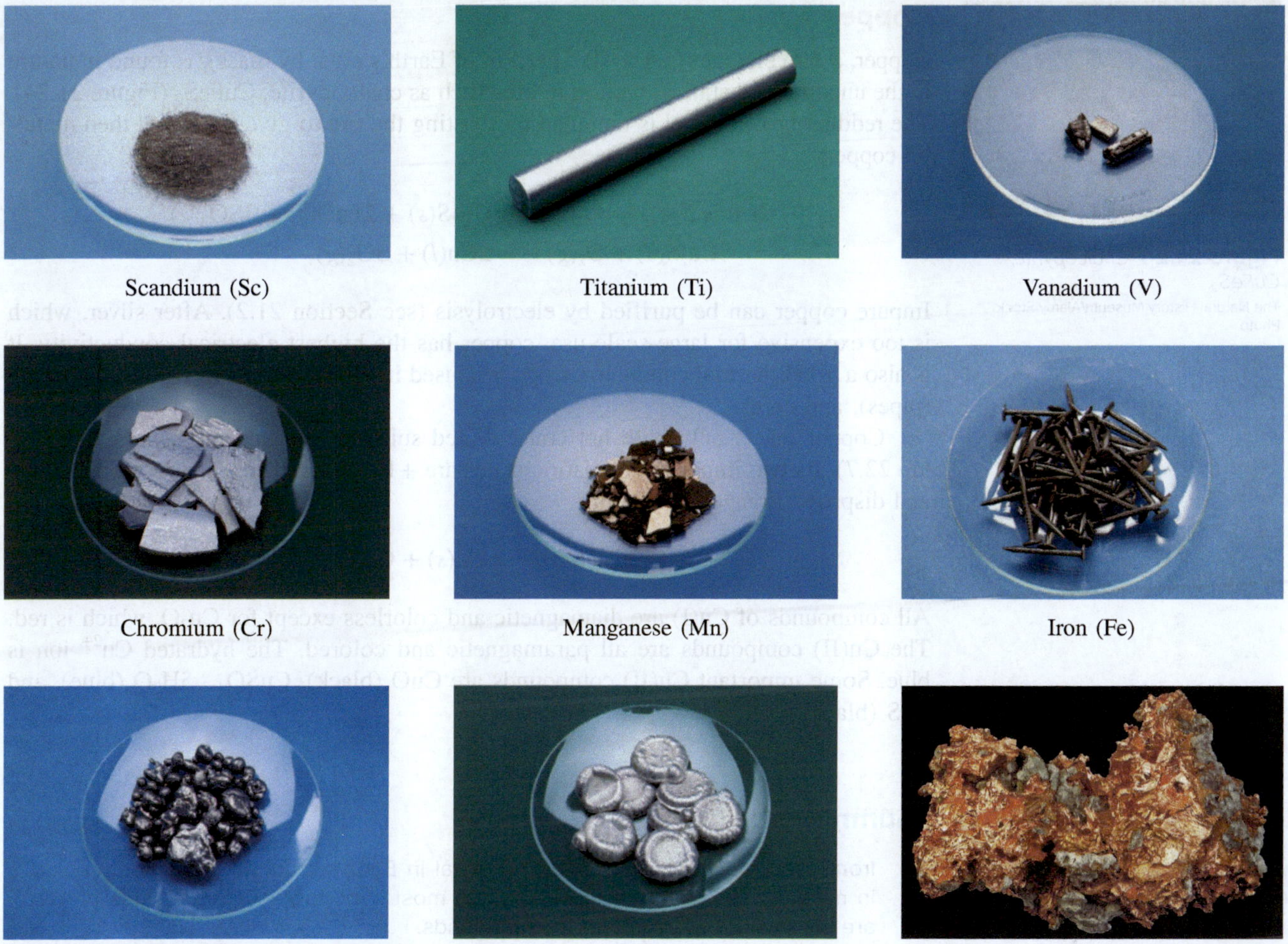

Scandium (Sc) Titanium (Ti) Vanadium (V)

Chromium (Cr) Manganese (Mn) Iron (Fe)

Cobalt (Co) Nickel (Ni) Copper (Cu)

Figure 21.22 *The first-row transition metals.*
(Scandium, Titanium, Vanadium, Chromium, Manganese, Iron, Cobalt, Nickel): Ken Karp/McGraw Hill; (Copper): ScottOrr/E+/Getty Images

Iron

After aluminum, iron is the most abundant metal in Earth's crust (6.2 percent by mass). It is found in many ores; some of the important ones are *hematite,* Fe_2O_3; *siderite,* $FeCO_3$; and *magnetite,* Fe_3O_4 (Figure 21.23).

The preparation of iron in a blast furnace and steelmaking were discussed in Section 21.2. Pure iron is a gray metal and is not particularly hard. It is an essential element in living systems.

Iron reacts with hydrochloric acid to give hydrogen gas:

$$Fe(s) + 2H^+(aq) \longrightarrow Fe^{2+} + H_2(g)$$

Concentrated sulfuric acid oxidizes the metal to Fe^{3+}, but concentrated nitric acid renders the metal "passive" by forming a thin layer of Fe_3O_4 over the surface. One of the best-known reactions of iron is rust formation (see Section 18.7). The two oxidation states of iron are +2 and +3. Iron(II) compounds include FeO (black), $FeSO_4 \cdot 7H_2O$ (green), $FeCl_2$ (yellow), and FeS (black). In the presence of oxygen, Fe^{2+} ions in solution are readily oxidized to Fe^{3+} ions. Iron(III) oxide is reddish brown, and iron(III) chloride is brownish black.

Figure 21.23 *The iron ore magnetite,* Fe_3O_4.
Dirk Wiersma/Science Source

Figure 21.24 *Chalcopyrite, $CuFeS_2$.*
The Natural History Museum/Alamy Stock Photo

Copper

Copper, a rare element (6.8×10^{-3} percent of Earth's crust by mass), is found in nature in the uncombined state as well as in ores such as chalcopyrite, $CuFeS_2$ (Figure 21.24). The reddish-brown metal is obtained by roasting the ore to give Cu_2S and then metallic copper:

$$2CuFeS_2(s) + 4O_2(g) \longrightarrow Cu_2S(s) + 2FeO(s) + 3SO_2(g)$$
$$Cu_2S(s) + O_2(g) \longrightarrow 2Cu(l) + SO_2(g)$$

Impure copper can be purified by electrolysis (see Section 21.2). After silver, which is too expensive for large-scale use, copper has the highest electrical conductivity. It is also a good thermal conductor. Copper is used in alloys, electrical cables, plumbing (pipes), and coins.

Copper reacts only with hot concentrated sulfuric acid and nitric acid (see Figure 22.7). Its two important oxidation states are +1 and +2. The +1 state is less stable and disproportionate in solution:

$$2Cu^+(aq) \longrightarrow Cu(s) + Cu^{2+}(aq)$$

All compounds of Cu(I) are diamagnetic and colorless except for Cu_2O, which is red. The Cu(II) compounds are all paramagnetic and colored. The hydrated Cu^{2+} ion is blue. Some important Cu(II) compounds are CuO (black), $CuSO_4 \cdot 5H_2O$ (blue), and CuS (black).

Summary of Concepts & Facts

- Iron is the most common transition metal in Earth's crust and commonly found in minerals containing oxygen. The two most important oxidation states of iron are +2 and +3 when found in compounds.
- Copper can be found in its free state and as an ore (chalcopyrite). The two most important oxidation states in compounds are +1 and +2.

Chapter Summary

Occurrence of Metals Depending on their reactivities, metals exist in nature in either the free or combined state; more reactive metals are found combined with other elements. Most metals are found in minerals. Minerals with high metal content are called ores. (Section 21.1)

Metallurgy Metallurgy involves recovering metal from ores. The three stages of metal recovery are *preparation, separation,* and *purification.* An alloy is a solid mixture of one or more metals, sometimes also containing one or more *nonmetals.* An amalgam is a mixture of mercury and one or more other metals. The methods commonly used for purifying metals are distillation, electrolysis, and zone refining. Pyrometallurgy refers to metallurgical processes carried out at high temperatures. (Section 21.2)

Band Theory Metallic bonds can be thought of as the force between positive ions immersed in a sea of electrons. In terms of band theory, the atomic orbitals merge to form energy bands. A substance is a conductor when electrons can be readily promoted to the conduction band, where they are free to move through the substance. In an insulator, the energy gap between the valence band and the conduction band is so large that electrons cannot be promoted into the conduction band. Semiconductors are substances that normally are not conductors but will conduct electricity at elevated temperatures or when combined with a small amount of certain other elements. Semiconductors in which an electron-rich impurity is added to enhance conduction are known as *n*-type semiconductors. Semiconductors in which an electron-poor impurity is added to enhance conduction are known as *p*-type semiconductors. (Section 21.3)

Periodic Trends in Metals Metals typically are good conductors and are malleable and ductile. Metallic character increases from top to bottom in a group and decreases from left to right across a period. (Section 21.4)

Group 1, Group 2, and Aluminum The alkali metals are the most reactive of all the metallic elements. They have an oxidation state of +1 in their compounds. Under special conditions, some of them can form anions with an oxidation state of −1. The alkaline earth metals are somewhat less reactive than the alkali metals. They almost always have an oxidation number of +2 in their compounds. The properties of the alkaline earth elements become increasingly metallic from top to bottom in their group. Aluminum ordinarily does not react with water due to a protective coating of aluminum oxide; its hydroxide is amphoteric. The Hall process is used to reduce aluminum oxide to aluminum. (Sections 21.5, 21.6, and 21.7)

Iron and Copper Iron is the most abundant metal in Earth's crust commonly occurring in oxygen-containing ores. The two most important oxidation states of iron are +2 and +3. Copper can be found in an uncombined state or in the common ore chalcopyrite. Impure copper can be refined by electrolysis. Copper in compounds most commonly have oxidation states of +1 and +2. (Section 21.8)

Key Words

Acceptor impurity, p. 945
Alloy, p. 936
Amalgam, p. 937
Band theory, p. 943
Conductor, p. 944
Donor impurity, p. 945
Ferromagnetic, p. 937
Insulator, p. 944
Metallurgy, p. 936
Mineral, p. 935
n-Type semiconductor, p. 945
Ore, p. 935
p-Type semiconductor, p. 945
Pyrometallurgy, p. 936
Semiconductors, p. 944

Questions & Problems

Red numbered problems solved in Student Solutions Manual

21.1 Occurrence of Metals

Review Questions

21.1 Define mineral, ore, and metallurgy.

21.2 List three metals that are usually found in an uncombined state in nature and three metals that are always found in a combined state in nature.

21.3 Write chemical formulas for the following minerals: (a) calcite, (b) dolomite, (c) fluorite, (d) halite, (e) corundum, (f) magnetite, (g) beryl, (h) galena, (i) epsomite, (j) anhydrite.

21.4 Name the following minerals: (a) $MgCO_3$, (b) Na_3AlF_6, (c) Al_2O_3, (d) Ag_2S, (e) HgS, (f) ZnS, (g) $SrSO_4$, (h) $PbCO_3$, (i) MnO_2, (j) TiO_2.

21.2 Metallurgical Processes

Review Questions

21.5 Describe the main steps involved in the preparation of an ore.

21.6 What does roasting mean in metallurgy? Why is roasting a major source of air pollution and acid rain?

21.7 Describe with examples the chemical and electrolytic reduction processes used in the production of metals.

21.8 Describe the main steps used to purify metals.

21.9 Describe the extraction of iron in a blast furnace.

21.10 Briefly discuss the steelmaking process.

Problems

21.11 In the Mond process for the purification of nickel, CO is passed over metallic nickel to give $Ni(CO)_4$:

$$Ni(s) + 4CO(g) \rightleftharpoons Ni(CO)_4(g)$$

Given that the standard energies of formation of CO(g) and $Ni(CO)_4(g)$ are −137.2 kJ/mol and −587.4 kJ/mol, respectively, calculate the equilibrium constant of the reaction at 80°C. (Assume ΔG_f° to be independent of temperature.)

21.12 Copper is purified by electrolysis (see Figure 21.6). A 5.00-kg anode is used in a cell where the current is 37.8 A. How long (in hours) must the current run to dissolve this anode and electroplate it onto the cathode?

21.13 Consider the electrolytic procedure for purifying copper described in Figure 21.6. Suppose that a sample of copper contains the following impurities: Fe, Ag, Zn, Au, Co, Pt, and Pb. Which of the metals will be oxidized and dissolved in solution and which will be unaffected and simply form the sludge that accumulates at the bottom of the cell?

21.14 How would you obtain zinc from sphalerite (ZnS)?

21.15 Starting with rutile (TiO_2), explain how you would obtain pure titanium metal. (*Hint:* First convert TiO_2 to $TiCl_4$. Next, reduce $TiCl_4$ with Mg. Look up physical properties of $TiCl_4$, Mg, and $MgCl_2$ in a chemistry handbook.)

21.16 A certain mine produces 2.0×10^8 kg of copper from chalcopyrite ($CuFeS_2$) each year. The ore contains only 0.80 percent Cu by mass. (a) If the density of the ore is 2.8 g/cm^3, calculate the volume (in cm^3) of ore removed each year. (b) Calculate the mass (in kg) of SO_2 produced by roasting (assume chalcopyrite to be the only source of sulfur).

21.17 Which of the following compounds would require electrolysis to yield the free metals: Ag_2S, $CaCl_2$, NaCl, Fe_2O_3, Al_2O_3, $TiCl_4$?

21.18 Although iron is only about two-thirds as abundant as aluminum in Earth's crust, mass for mass it costs only about one-quarter as much to produce. Why?

21.3 Band Theory of Electrical Conductivity

Review Questions

21.19 Define the following terms: *conductor, insulator, semiconducting elements, donor impurities, acceptor impurities, n-type semiconductors,* and *p-type semiconductors.*

21.20 Briefly discuss the nature of bonding in metals, insulators, and semiconducting elements.

21.21 Describe the general characteristics of *n*-type and *p*-type semiconductors.

21.22 State whether silicon would form *n*-type or *p*-type semiconductors with the following elements: Ga, Sb, Al, As.

21.5 The Alkali Metals

Review Questions

21.23 How is sodium prepared commercially?

21.24 Why is potassium usually not prepared electrolytically from one of its salts?

21.25 Describe the uses of the following compounds: NaCl, Na_2CO_3, NaOH, KOH, KO_2.

21.26 Under what conditions do sodium and potassium form Na^- and K^- ions?

Problems

21.27 Complete and balance the following equations:
(a) $K(s) + H_2O(l) \longrightarrow$
(b) $NaH(s) + H_2O(l) \longrightarrow$
(c) $Na(s) + O_2(g) \longrightarrow$
(d) $K(s) + O_2(g) \longrightarrow$

21.28 Write a balanced equation for each of the following reactions: (a) Sodium reacts with water; (b) an aqueous solution of NaOH reacts with CO_2; (c) solid Na_2CO_3 reacts with a HCl solution; (d) solid $NaHCO_3$ reacts with a HCl solution; (e) solid $NaHCO_3$ is heated; (f) solid Na_2CO_3 is heated.

21.29 Sodium hydride (NaH) can be used as a drying agent for many organic solvents. Explain how it works.

21.30 Calculate the volume of CO_2 at 10.0°C and 746 mmHg pressure obtained by treating 25.0 g of Na_2CO_3 with an excess of hydrochloric acid.

21.6 The Alkaline Earth Metals

Review Questions

21.31 List the common ores of magnesium and calcium.

21.32 How are the metals magnesium and calcium obtained commercially?

Problems

21.33 From the thermodynamic data in Appendix 2, calculate the $\Delta H°$ values for the following decompositions:
(a) $MgCO_3(s) \longrightarrow MgO(s) + CO_2(g)$
(b) $CaCO_3(s) \longrightarrow CaO(s) + CO_2(g)$
Which of the two compounds is more easily decomposed by heat?

21.34 Starting with magnesium and concentrated nitric acid, describe how you would prepare magnesium oxide. [*Hint:* First convert Mg to $Mg(NO_3)_2$. Next, MgO can be obtained by heating $Mg(NO_3)_2$.]

21.35 Describe two ways of preparing magnesium chloride.

21.36 The second ionization energy of magnesium is only about twice as great as the first, but the third ionization energy is 10 times as great. Why does it take so much more energy to remove the third electron?

21.37 List the sulfates of the Group 2 metals in order of increasing solubility in water. Explain the trend. (*Hint:* You need to consult a chemistry handbook.)

21.38 Helium contains the same number of electrons in its outer shell as do the alkaline earth metals. Explain why helium is inert whereas the Group 2 metals are not.

21.39 When exposed to air, calcium first forms calcium oxide, which is then converted to calcium hydroxide, and finally to calcium carbonate. Write a balanced equation for each step.

21.40 Write chemical formulas for (a) quicklime, (b) slaked lime, (c) limewater.

21.7 Aluminum

Review Questions

21.41 Describe the Hall process for preparing aluminum.

21.42 What action renders aluminum inert?

Problems

21.43 Before Hall invented his electrolytic process, aluminum was produced by the reduction of its chloride with an active metal. Which metals would you use for the production of aluminum in that way?

21.44 With the Hall process, how many hours will it take to deposit 664 g of Al at a current of 32.6 A?

21.45 Aluminum forms the complex ions $AlCl_4^-$ and AlF_6^{3-}. Describe the shapes of these ions. $AlCl_6^{3-}$ does not form. Why? (*Hint:* Consider the relative sizes of Al^{3+}, F^-, and Cl^- ions.)

21.46 The overall reaction for the electrolytic production of aluminum by means of the Hall process may be represented as

$$Al_2O_3(s) + 3C(s) \longrightarrow 2Al(l) + 3CO(g)$$

At 1000°C, the standard Gibbs-energy change for this process is 594 kJ/mol. (a) Calculate the minimum voltage required to produce 1 mole of aluminum at this temperature. (b) If the actual voltage applied is exactly three times the ideal value, calculate the energy required to produce 1.00 kg of the metal.

21.47 In basic solution, aluminum metal is a strong reducing agent and is oxidized to AlO_2^-. Give balanced equations for the reaction of Al in basic solution with the following: (a) $NaNO_3$, to give ammonia; (b) water, to give hydrogen; (c) Na_2SnO_3, to give metallic tin.

21.48 Write a balanced equation for the thermal decomposition of aluminum nitrate to form aluminum oxide, nitrogen dioxide, and oxygen gas.

21.49 Describe some of the properties of aluminum that make it one of the most versatile metals known.

21.50 The pressure of gaseous Al_2Cl_6 increases more rapidly with temperature than predicted by the ideal gas equation even though Al_2Cl_6 behaves like an ideal gas. Explain.

21.51 Starting with aluminum, describe with balanced equations how you would prepare (a) Al_2Cl_6, (b) Al_2O_3, (c) $Al_2(SO_4)_3$, (d) $NH_4Al(SO_4)_2 \cdot 12H_2O$.

21.52 Explain the change in bonding when Al_2Cl_6 dissociates to form $AlCl_3$ in the gas phase.

Additional Problems

21.53 In steelmaking, nonmetallic impurities such as P, S, and Si are removed as the corresponding oxides. The inside of the furnace is usually lined with $CaCO_3$ and $MgCO_3$, which decompose at high temperatures to yield CaO and MgO. How do CaO and MgO help in the removal of the nonmetallic oxides?

21.54 When 1.164 g of a certain metal sulfide was roasted in air, 0.972 g of the metal oxide was formed. If the oxidation number of the metal is +2, calculate the molar mass of the metal.

21.55 An early view of metallic bonding assumed that bonding in metals consisted of localized, shared electron–pair bonds between metal atoms. What evidence would help you to argue against this viewpoint?

21.56 Referring to Figure 21.6, would you expect H_2O and H^+ to be reduced at the cathode and H_2O oxidized at the anode?

21.57 A 0.450-g sample of steel contains manganese as an impurity. The sample is dissolved in acidic solution and the manganese is oxidized to the permanganate ion MnO_4^-. The MnO_4^- ion is reduced to Mn^{2+} by reacting with 50.0 mL of 0.0800 *M* $FeSO_4$ solution. The excess Fe^{2+} ions are then oxidized to Fe^{3+} by 22.4 mL of 0.0100 *M* $K_2Cr_2O_7$. Calculate the percent by mass of manganese in the sample.

21.58 Given that $\Delta G_f^\circ(Fe_2O_3) = -742.2$ kJ/mol and that $\Delta G_f^\circ(Al_2O_3) = -1582.3$ kJ/mol, calculate ΔG° for the following reactions at 25°C:

(a) $2Fe_2O_3(s) \longrightarrow 4Fe(s) + 3O_2(g)$

(b) $2Al_2O_3(s) \longrightarrow 4Al(s) + 3O_2(g)$

21.59 Use compounds of aluminum as an example to explain what is meant by amphoterism.

21.60 When an inert atmosphere is needed for a metallurgical process, nitrogen is frequently used. However, in the reduction of $TiCl_4$ by magnesium, helium is used. Explain why nitrogen is not suitable for this process.

21.61 It has been shown that Na_2 species form in the vapor phase. Describe the formation of the "disodium molecule" in terms of a molecular orbital energy-level diagram. Would you expect the alkaline earth metals to exhibit a similar property?

21.62 Explain each of the following statements: (a) An aqueous solution of $AlCl_3$ is acidic. (b) $Al(OH)_3$ is soluble in NaOH solutions but not in NH_3 solution.

21.63 Write balanced equations for the following reactions: (a) the heating of aluminum carbonate; (b) the reaction between $AlCl_3$ and K; (c) the reaction between solutions of Na_2CO_3 and $Ca(OH)_2$.

21.64 Write a balanced equation for the reaction between calcium oxide and dilute HCl solution.

21.65 What is wrong with the following procedure for obtaining magnesium?

$$MgCO_3 \longrightarrow MgO(s) + CO_2(g)$$
$$MgO(s) + CO(g) \longrightarrow Mg(s) + CO_2(g)$$

21.66 Explain why most metals have a flickering appearance.

21.67 Predict the chemical properties of francium, the last member of Group 1.

21.68 Describe a medicinal or health-related application for each of the following compounds: NaF, Li_2CO_3, $Mg(OH)_2$, $CaCO_3$, $BaSO_4$, $Al(OH)_2NaCO_3$. (You will need to do a Web search for some of these compounds.)

21.69 The following are two reaction schemes involving magnesium. *Scheme I:* When magnesium burns in oxygen, a white solid (A) is formed. A dissolves in 1 *M* HCl to give a colorless solution (B). Upon addition of Na_2CO_3 to B, a white precipitate is formed (C). On heating, C decomposes to D and a colorless gas is generated (E). When E is passed through limewater [an aqueous suspension of $Ca(OH)_2$], a white precipitate appears (F). *Scheme II:* Magnesium reacts with 1 *M* H_2SO_4 to produce a colorless solution (G).

Treating G with an excess of NaOH produces a white precipitate (H). H dissolves in 1 *M* HNO_3 to form a colorless solution. When the solution is slowly evaporated, a white solid (I) appears. On heating I, a brown gas is given off. Identify A-I and write equations representing the reactions involved.

21.70 Lithium and magnesium exhibit a diagonal relationship in some chemical properties. How does lithium resemble magnesium in its reaction with (i) oxygen and (ii) nitrogen? Consult a handbook of chemistry and compare the solubilities of (iii) carbonates, fluorides, and phosphates of these metals.

21.71 To prevent the formation of oxides, peroxides, and superoxides, alkali metals are sometimes stored in an inert atmosphere. Which of the following gases should not be used for lithium: Ne, Ar, N_2, Kr? Explain why.

21.72 Which of the following metals is not found in the free state in nature: Ag, Cu, Zn, Au, Pt?

21.73 After heating, a metal surface (such as that of a cooking pan or skillet) develops a color pattern like an oil slick on water. Explain.

21.74 Chemical tests of four metals A, B, C, and D show the following results.

(a) Only B and C react with 0.5 *M* HCl to give H_2 gas.

(b) When B is added to a solution containing the ions of the other metals, metallic A, C, and D are formed.

(c) A reacts with 6 *M* HNO_3 but D does not.

Arrange the metals in the increasing order as reducing agents. Suggest four metals that fit these descriptions.

21.75 The electrical conductance of copper metal decreases with temperature, but that of a $CuSO_4$ solution increases with temperature. Explain.

21.76 As stated in the chapter, potassium superoxide (KO_2) is a useful source of oxygen employed in breathing equipment. Calculate the pressure at which oxygen gas stored at 20°C would have the same density as the oxygen gas provided by KO_2. The density of KO_2 at 20°C is 2.15 g/cm^3.

21.77 A sample of 10.00 g of sodium reacts with oxygen to form 13.83 g of sodium oxide (Na_2O) and sodium peroxide (Na_2O_2). Calculate the percent composition of the mixture.

Answers to Review of Concepts & Facts

21.3.1 14. **21.3.2** (a) *p*-type. (b) *n*-type.

CHAPTER

22 Nonmetallic Elements and Their Compounds

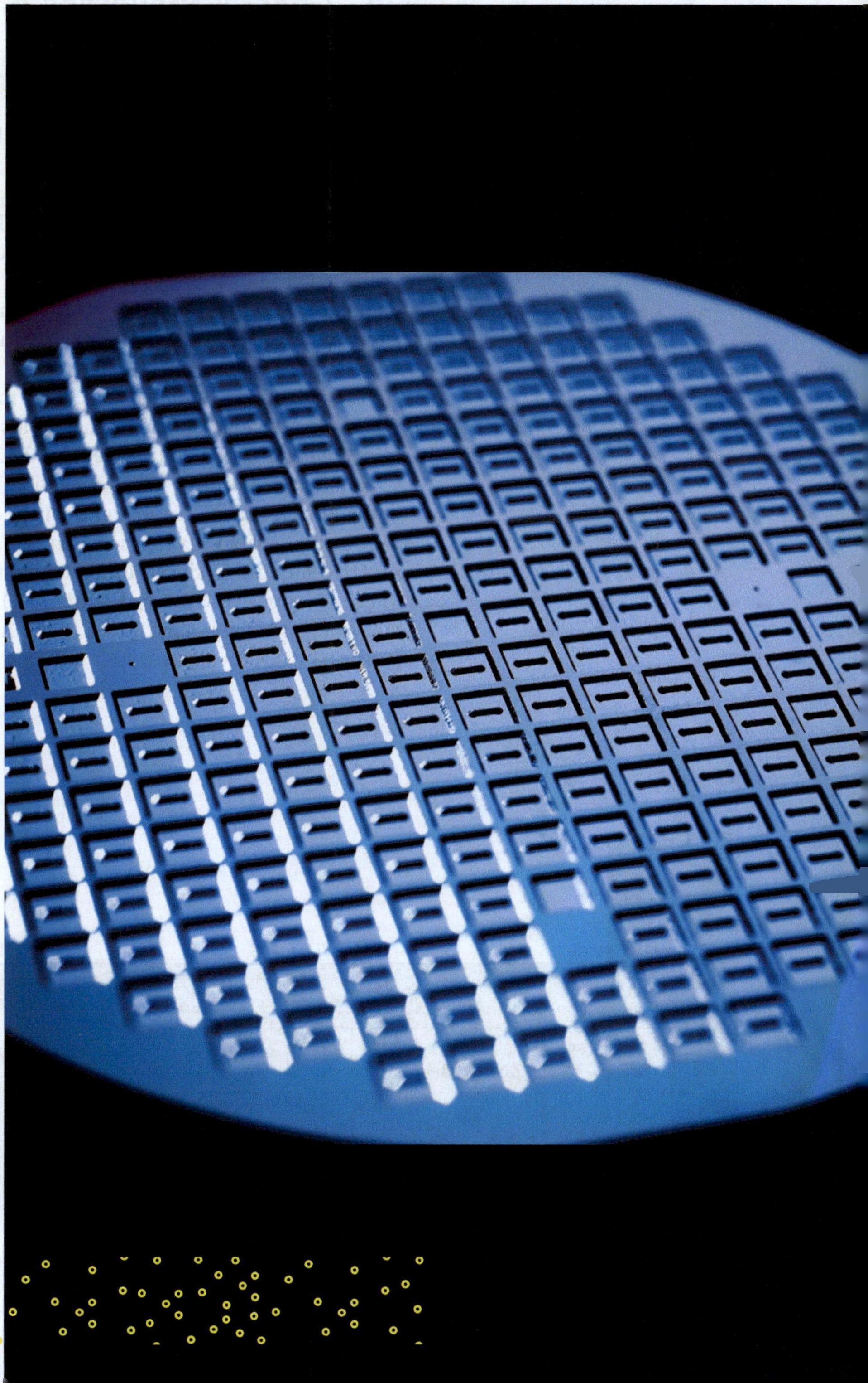

Silicon wafers, thin slices of semiconducting material, are used for the fabrication of computer circuits and the manufacture of solar cells.
Science Photo Library/Getty Images

CHAPTER OUTLINE

Of the 118 elements known, only 25 are nonmetallic elements. Unlike the metals, the chemistry of these elements is diverse. Despite their relatively small number, most of the essential elements in biological systems are nonmetals (H, C, N, P, O, S, Cl, and I). This group of nonmetallic elements also includes the most unreactive of the elements—the noble gases. The unique properties of hydrogen set it aside from the rest of the elements in the periodic table. A whole branch of chemistry—organic chemistry—is based on carbon compounds.

In this chapter, we continue our survey of the elements by concentrating on the nonmetals. The emphasis will again be on important chemical properties and on the roles of nonmetals and their compounds in industrial, chemical, and biological processes.

22.1 General Properties of Nonmetals

Learning Objective

- Identify examples of the uses of isotopes in science and medicine.

Properties of nonmetals are more varied than those of metals. A number of nonmetals are gases in the elemental state: hydrogen, oxygen, nitrogen, fluorine, chlorine, and the noble gases. Only one, bromine, is a liquid. All the remaining nonmetals are solids at room temperature. Unlike metals, nonmetallic elements are poor conductors of heat and electricity; they exhibit both positive and negative oxidation numbers.

A small group of elements, called metalloids, have properties characteristic of both metals and nonmetals. The metalloids boron, silicon, germanium, and arsenic are semiconducting elements (see Section 21.3).

Nonmetals are more electronegative than metals. The electronegativity of elements increases from left to right across any period and from bottom to top in any group in the periodic table (see Figure 9.5). With the exception that hydrogen has no totally suitable position, the nonmetals are concentrated in the upper right-hand corner of the periodic table (Figure 22.1). Compounds formed by a combination of metals and nonmetals tend to be ionic, having a metallic cation and a nonmetallic anion.

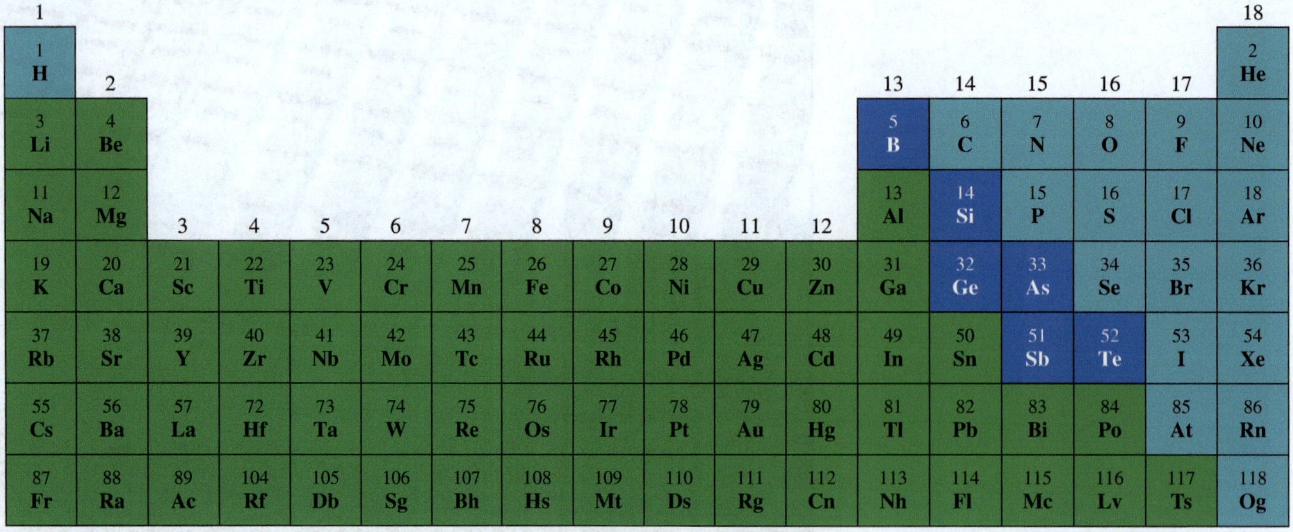

Figure 22.1 *Representative nonmetallic elements (in blue) and metalloids (dark blue).*

In this chapter, we will discuss the chemistry of a number of common and important nonmetallic elements: hydrogen; carbon (Group 14); nitrogen and phosphorus (Group 15); oxygen and sulfur (Group 16); and fluorine, chlorine, bromine, and iodine (Group 17).

Summary of Concepts & Facts

- The nonmetals are concentrated in the top right corner of the periodic table. Nonmetals are generally more electronegative than metals.

22.2 Hydrogen

Learning Objectives

- Compare the two types of hydrides that can form.
- List the three isotopes of hydrogen.
- Explain the relevance of the hydrogen economy.

Hydrogen is the simplest element known—its most common atomic form contains only one proton and one electron. The atomic form of hydrogen exists only at very high temperatures, however. Normally, elemental hydrogen is a diatomic molecule, the product of an exothermic reaction between H atoms:

$$H(g) + H(g) \longrightarrow H_2(g) \qquad \Delta H^\circ = -436 \text{ kJ/mol}$$

Molecular hydrogen is a colorless, odorless, and nonpoisonous gas. At 1 atm, liquid hydrogen has a boiling point of $-252.9°C$ (20.3 K).

Hydrogen is the most abundant element in the universe, accounting for about 70 percent of the universe's total mass. It is the tenth most abundant element in Earth's crust, where it is found in combination with other elements. Unlike Jupiter and Saturn, Earth does not have a strong enough gravitational pull to retain the lightweight H_2 molecules, so hydrogen is not found in our atmosphere.

The ground-state electron configuration of H is $1s^1$. It resembles the alkali metals in that it can be oxidized to the H^+ ion, which exists in aqueous solutions in the hydrated form. On the other hand, hydrogen resembles the halogens in that it forms the uninegative hydride ion (H^-), which is isoelectronic with helium ($1s^1$). Hydrogen is found in a large number of covalent compounds. Hydrogen typically has +1 oxidation number in its compounds, but in ionic hydrides it has a −1 oxidation number. It also has the unique capacity for hydrogen-bond formation (see Section 11.2).

Hydrogen gas plays an important role in industrial processes. About 95 percent of the hydrogen produced is used captively; that is, it is produced at or near the plant where it is used for industrial processes, such as the synthesis of ammonia. The large-scale industrial preparation is the reaction between propane (from natural gas and also as a product of oil refineries) and steam in the presence of a catalyst at 900°C:

$$C_3H_8(g) + 3H_2O(g) \longrightarrow 3CO(g) + 7H_2(g)$$

In another process, steam is passed over a bed of red-hot coke:

$$C(s) + H_2O(g) \longrightarrow CO(g) + H_2(g)$$

Figure 22.2 *Apparatus for the laboratory preparation of hydrogen gas. The gas is collected over water, as is also the case of oxygen gas (see Figure 5.15).*

The mixture of carbon monoxide and hydrogen gas produced in this reaction is commonly known as *water gas.* Because both CO and H_2 burn in air, water gas was used as a fuel for many years. But because CO is poisonous, water gas has been replaced by natural gases, such as methane and propane.

Small quantities of hydrogen gas can be prepared conveniently in the laboratory by reacting zinc with dilute hydrochloric acid (Figure 22.2):

$$Zn(s) + 2HCl(aq) \longrightarrow ZnCl_2(aq) + H_2(g)$$

Hydrogen gas can also be produced by the reaction between an alkali metal or an alkaline earth metal (Ca or Ba) and water (see Section 4.4), but these reactions are too violent to be suitable for the laboratory preparation of hydrogen gas. Very pure hydrogen gas can be obtained by the electrolysis of water, but this method consumes too much energy to be practical on a large scale.

Binary Hydrides

Binary hydrides are compounds containing hydrogen and another element, either a metal or a nonmetal. Depending on structure and properties, these hydrides are broadly divided into three types: (1) ionic hydrides, (2) covalent hydrides, and (3) interstitial hydrides.

Ionic Hydrides

Ionic hydrides are formed when molecular hydrogen combines directly with any alkali metal or with the alkaline earth metals Ca, Sr, or Ba:

$$2Li(s) + H_2(g) \longrightarrow 2LiH(s)$$

$$Ca(s) + H_2(g) \longrightarrow CaH_2(s)$$

All ionic hydrides are solids that have the high melting points characteristic of ionic compounds. The anion in these compounds is the hydride ion, H^-, which is a very strong Brønsted base. It readily accepts a proton from a proton donor such as water:

$$H^-(aq) + H_2O(l) \longrightarrow OH^-(aq) + H_2(g)$$

Due to their high reactivity with water, ionic hydrides are frequently used to remove traces of water from organic solvents.

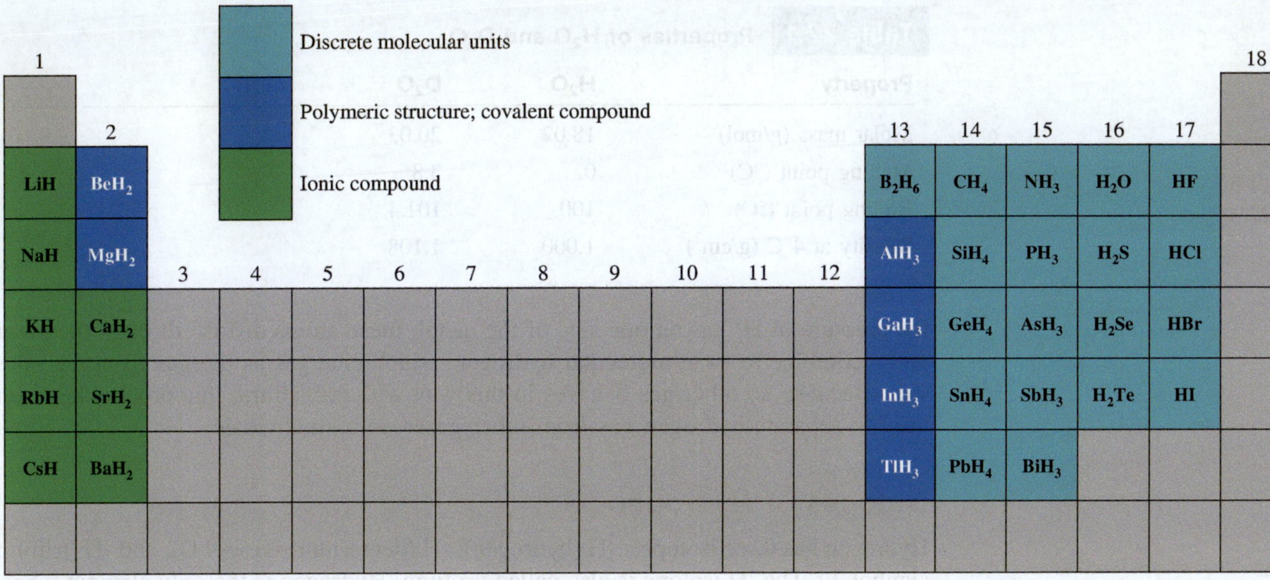

Figure 22.3 *Binary hydrides of the representative elements. In cases in which hydrogen forms more than one compound with the same element, only the formula of the simplest hydride is shown. The properties of many of the transition metal hydrides are not well characterized.*

Covalent Hydrides

In *covalent hydrides,* the hydrogen atom is covalently bonded to the atom of another element. There are two types of covalent hydrides—those containing discrete molecular units, such as CH_4 and NH_3, and those having complex polymeric structures, such as $(BeH_2)_x$ and $(AlH_3)_x$, where x is a very large number (this is an example of the diagonal relationship between Be and Al).

Figure 22.3 shows the binary ionic and covalent hydrides of the representative elements. The physical and chemical properties of these compounds change from ionic to covalent across a given period. Consider, for example, the hydrides of the second-period elements: LiH, BeH_2, B_2H_6, CH_4, NH_3, H_2O, and HF. LiH is an ionic compound with a high melting point (680°C). The structure of BeH_2 (in the solid state) is polymeric; it is a covalent compound. The molecules B_2H_6 and CH_4 are nonpolar. In contrast, NH_3, H_2O, and HF are all polar molecules in which the hydrogen atom is the *positive* end of the polar bond. Of this group of hydrides (NH_3, H_2O, and HF), only HF is acidic in water.

As we move down any group in Figure 22.3, the compounds change from covalent to ionic. In Group 2, for example, BeH_2 and MgH_2 are covalent, but CaH_2, SrH_2, and BaH_2 are ionic.

Interstitial Hydrides

Molecular hydrogen forms a number of hydrides with transition metals. In some of these compounds, the ratio of hydrogen atoms to metal atoms is *not* a constant. Such compounds are called *interstitial hydrides.* For example, depending on conditions, the formula for titanium hydride can vary between $TiH_{1.8}$ and TiH_2. Interstitial compounds are sometimes called nonstoichiometric compounds as they do not obey the law of definite proportions (see Section 2.1).

Many of the interstitial hydrides have metallic properties such as electrical conductivity. Yet it is known that hydrogen is definitely bonded to the metal in these compounds, although the exact nature of the bonding is often not clear.

Molecular hydrogen interacts in a unique way with palladium (Pd). Hydrogen gas is readily adsorbed onto the surface of the palladium metal, where it dissociates into atomic hydrogen. The H atoms then "dissolve" into the metal. On heating and under

Table 22.1 Properties of H_2O and D_2O

Property	H_2O	D_2O
Molar mass (g/mol)	18.02	20.03
Melting point (°C)	0	3.8
Boiling point (°C)	100	101.4
Density at 4°C (g/cm^3)	1.000	1.108

the pressure of H_2 gas on one side of the metal, these atoms diffuse through the metal and recombine to form molecular hydrogen, which emerges as the gas from the other side. Because no other gas behaves in this way with palladium, this process has been used to separate hydrogen gas from other gases on a small scale.

Isotopes of Hydrogen

Hydrogen has three isotopes: ^1_1H (hydrogen), ^2_1H (deuterium, symbol D), and ^3_1H (tritium, symbol T). The ^1_1H isotope is also called protium. Hydrogen is the only element whose isotopes are given different names. The natural abundances of the stable hydrogen isotopes are hydrogen, 99.985 percent; and deuterium, 0.015 percent. Tritium is a radioactive isotope with a half-life of about 12.5 years.

Table 22.1 compares some of the common properties of H_2O with those of D_2O. Deuterium oxide, or heavy water as it is commonly called, is used in some nuclear reactors as a coolant and a moderator of nuclear reactions (see Chapter 19). D_2O can be separated from H_2O by fractional distillation because H_2O boils at a lower temperature, as Table 22.1 shows. Another technique for its separation is electrolysis of water. Because H_2 gas is formed about eight times as fast as D_2 during electrolysis, the water remaining in the electrolytic cell becomes progressively enriched with D_2O. Interestingly, the Dead Sea, which for thousands of years has entrapped water that has no outlet other than through evaporation, has a higher $[D_2O]/[H_2O]$ ratio than water found elsewhere.

Although D_2O chemically resembles H_2O in most respects, it is a toxic substance. The reason is that deuterium is heavier than hydrogen; thus, its compounds often react more slowly than those of the lighter isotope. Regular drinking of D_2O instead of H_2O could prove fatal because of the slower rate of transfer of D^+ compared with that of H^+ in the acid-base reactions involved in enzyme catalysis. This *kinetic isotope effect* also manifests in acid ionization constants. For example, the ionization constant of acetic acid

Student Hot Spot

Student data indicate you may struggle with the nature of hydrogen. Access your eBook for additional Learning Resources on this topic.

$$CH_3COOH(aq) \rightleftharpoons CH_3COO^-(aq) + H^+(aq) \qquad K_a = 1.8 \times 10^{-5}$$

is about three times as large as that of deuterated acetic acid:

$$CH_3COOD(aq) \rightleftharpoons CH_3COO^-(aq) + D^+(aq) \qquad K_a = 6 \times 10^{-6}$$

Hydrogenation

Hydrogenation *is the addition of hydrogen to compounds containing multiple bonds, especially C=C and C≡C bonds.* A simple hydrogenation reaction is the conversion of ethylene to ethane:

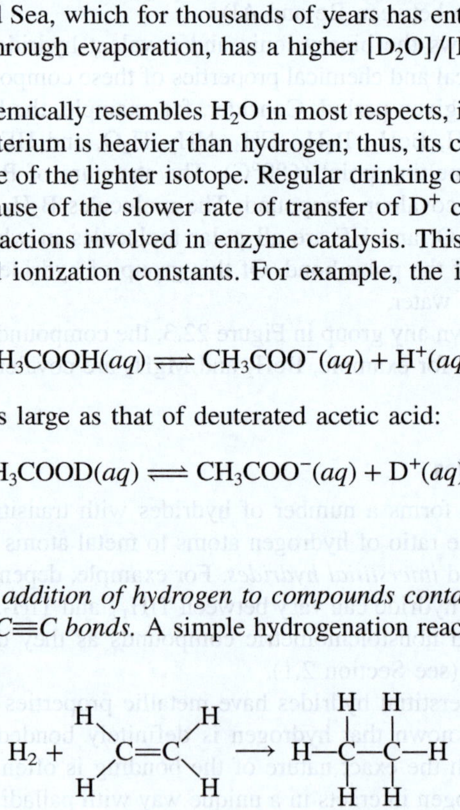

ethylene ethane

This reaction is quite slow under normal conditions, but the rate can be greatly increased by the presence of a catalyst such as nickel or platinum. As in the Haber synthesis of ammonia (see Section 13.6), the main function of the catalyst is to weaken the H—H bond and facilitate the reaction.

Hydrogenation is an important process in the food industry. Vegetable oils have considerable nutritional value, but some oils must be hydrogenated before we can use them because of their unsavory flavor and their inappropriate molecular structures (i.e., there are too many C═C bonds present). Upon exposure to air, these *polyunsaturated* molecules (that is, molecules with many C═C bonds) undergo oxidation to yield unpleasant-tasting products (oil that has oxidized is said to be rancid). In the hydrogenation process, a small amount of nickel (about 0.1 percent by mass) is added to the oil and the mixture is exposed to hydrogen gas at high temperature and pressure. Afterward, the nickel is removed by filtration. Hydrogenation reduces the number of double bonds in the molecule but does not completely eliminate them. If all the double bonds are eliminated, the oil becomes hard and brittle. Under controlled conditions, suitable cooking oils and margarine may be prepared by hydrogenation from vegetable oils extracted from cottonseed, corn, and soybeans.

Platinum catalyst on alumina (Al_2O_3) used in hydrogenation.
Ken Karp/McGraw Hill

The Hydrogen Economy

The world's fossil fuel reserves are being depleted at an alarmingly fast rate. Faced with this dilemma, scientists have made intensive efforts in recent years to develop a method of obtaining hydrogen gas as an alternative energy source. Hydrogen gas could replace gasoline to power automobiles (after considerable modification of the engine, of course) or be used with oxygen gas in fuel cells to generate electricity (see Section 18.6). One major advantage of using hydrogen gas in these ways is that the reactions are essentially free of pollutants; the end product formed in a hydrogen-powered engine or in a fuel cell would be water, just as in the burning of hydrogen gas in air:

$$2H_2(g) + O_2(g) \longrightarrow 2H_2O(l)$$

Of course, success of a hydrogen economy would depend on how cheaply we could produce hydrogen gas and how easily we could store it.

Although electrolysis of water consumes too much energy for large-scale application, if scientists can devise a more practical method of "splitting" water molecules, we could obtain vast amounts of hydrogen from seawater. The total volume of ocean water is about 1×10^{21} L. Thus, the ocean contains an almost inexhaustible supply of hydrogen. One approach that is currently in the early stages of development would use solar energy. In this scheme, a catalyst (a complex molecule containing one or more transition metal atoms, such as ruthenium) absorbs a photon from solar radiation and becomes energetically excited. In its excited state, the catalyst is capable of reducing water to molecular hydrogen.

Some of the interstitial hydrides we have discussed would make suitable storage compounds for hydrogen. The reactions that form these hydrides are usually reversible, so hydrogen gas can be obtained simply by reducing the pressure of the hydrogen gas above the metal. The advantages of using interstitial hydrides are as follows: (1) Many metals have a high capacity to take up hydrogen gas—sometimes up to three times as many hydrogen atoms as there are metal atoms; and (2) because these hydrides are solids, they can be stored and transported more easily than gases or liquids.

The Chemistry in Action essay "Metallic Hydrogen" describes what happens to hydrogen under pressure.

CHEMISTRY *in Action*

Metallic Hydrogen

Scientists have long been interested in how nonmetallic substances, including hydrogen, behave under exceedingly high pressure. It was predicted that when atoms or molecules are compressed, their bonding electrons might be delocalized, producing a metallic state. In 1996, physicists at the Lawrence Livermore Laboratory used a 60-foot-long gun to generate a shock compression onto a thin (0.5 mm) layer of liquid hydrogen. For an instant, at pressures between 0.9 and 1.4 million atm, they were able to measure the electrical conductivity of the hydrogen sample and found that it was comparable to that of cesium metal at 2000 K. (The temperature of the hydrogen sample rose as a result of compression, although it remained in the molecular form.) As the pressure fell rapidly, the metallic state of hydrogen disappeared.

The Livermore experiment suggested that metallic hydrogen, if it can be kept in a stable state, may act as a room-temperature superconductor. The fact that hydrogen becomes metallic at pressures lower than previously thought possible also has provided new insight into planetary science. For many years scientists were puzzled by Jupiter's strong magnetic field, which is 20 times greater than that of Earth. A planet's magnetic field results from the convection motion of electrically conductive fluid in its interior. (For example, Earth's magnetic field is due to the heat-driven motion of liquid iron within its core.) Jupiter is composed of an outer layer of nonmetallic molecular hydrogen that continuously transforms hydrogen within the core to metallic fluid hydrogen. It is now believed that this metallic layer is much closer to the surface (because the pressure needed to convert molecular hydrogen to metallic hydrogen is not as high as previously thought), which would account for Jupiter's unusually strong magnetic field.

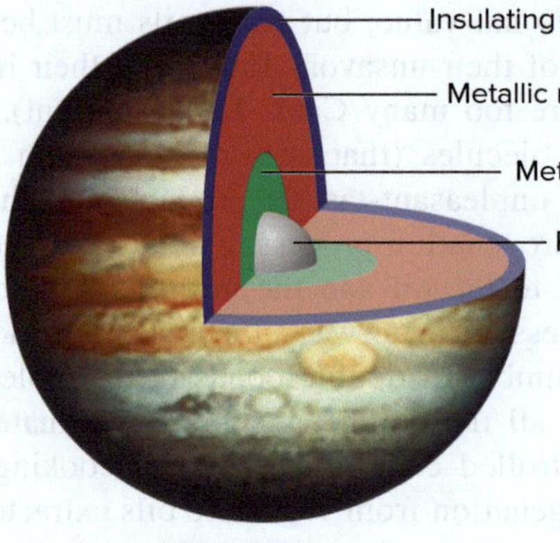

Interior composition of Jupiter.

Summary of Concepts & Facts

- Hydrogen atoms contain one proton and one electron. They are the simplest atoms. Hydrogen combines with many metals and nonmetals to form hydrides; some hydrides are ionic and some are covalent.
- There are three isotopes of hydrogen: ${}^1_1H$, ${}^2_1H$ (deuterium), and ${}^3_1H$ (tritium). Heavy water contains deuterium.

22.3 Carbon

Learning Objective

- Categorize the major compounds that are formed from carbon.

Although it constitutes only about 0.09 percent by mass of Earth's crust, carbon is an essential element of living matter. It is found free in the form of diamond and graphite (see Figure 8.17), and it is also a component of natural gas, petroleum, and coal. (Coal is a natural dark brown to black solid used as a fuel; it is formed from fossilized plants and consists of amorphous carbon with various organic and some inorganic compounds.) Carbon combines with oxygen to form carbon dioxide in the atmosphere and occurs as carbonate in limestone and chalk.

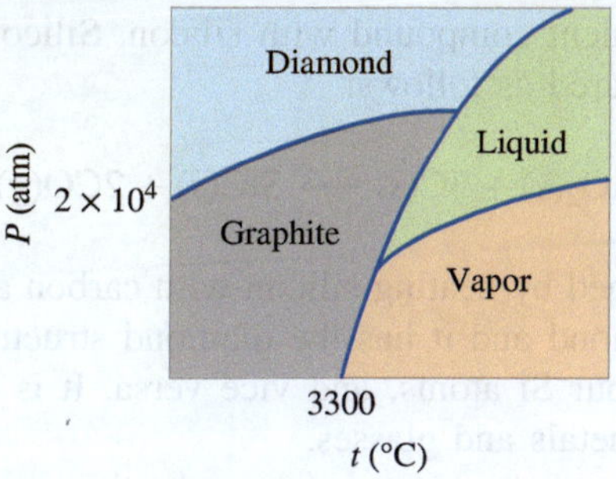

Figure 22.4 *Phase diagram of carbon. Note that under atmospheric conditions, graphite is the stable form of carbon.*

Diamond and graphite are allotropes of carbon. Figure 22.4 shows the phase diagram of carbon. Although graphite is the stable form of carbon at 1 atm and 25°C, owners of diamond jewelry need not be alarmed, for the rate of the spontaneous process

$$C(diamond) \longrightarrow C(graphite) \qquad \Delta G^\circ = -2.87 \text{ kJ/mol}$$

is extremely slow. Millions of years may pass before a diamond turns to graphite.

Synthetic diamond can be prepared from graphite by applying very high pressures and temperatures. Figure 22.5 shows a synthetic diamond and its starting material, graphite. Synthetic diamonds generally lack the optical properties of natural diamonds. They are useful, however, as abrasives and in cutting concrete and many other hard substances, including metals and alloys. The uses of graphite are described in Section 11.6.

Carbon has the unique ability to form long chains (consisting of more than 50 C atoms) and stable rings with five or six members. This phenomenon is called ***catenation,*** *the linking of like atoms.* Carbon's versatility is responsible for the millions of organic compounds (made up of carbon and hydrogen and other elements such as oxygen, nitrogen, and halogens) found on Earth. The chemistry of organic compounds is discussed in Chapter 24.

Carbides and Cyanides

Carbon combines with metals to form ionic compounds called ***carbides,*** such as CaC_2 and Be_2C, in which carbon is in the form of C_2^{2-} or C^{4-} ions. These ions are strong Brønsted bases and react with water as follows:

$$C_2^{2-}(aq) + 2H_2O(l) \longrightarrow 2OH^-(aq) + C_2H_2(g)$$
$$C^{4-}(aq) + 4H_2O(l) \longrightarrow 4OH^-(aq) + CH_4(g)$$

Figure 22.5 *Synthetic diamonds and the starting material—graphite.*
David A. Tietz/Editorial Image, LLC

Carbon also forms a covalent compound with silicon. Silicon carbide, SiC, is called *carborundum* and is prepared as follows:

$$SiO_2(s) + 3C(s) \longrightarrow SiC(s) + 2CO(g)$$

Carborundum is also formed by heating silicon with carbon at 1500°C. Carborundum is almost as hard as diamond and it has the diamond structure; each carbon atom is tetrahedrally bonded to four Si atoms, and vice versa. It is used mainly for cutting, grinding, and polishing metals and glasses.

Another important class of carbon compounds, the ***cyanides,*** *contain the anion group* $:C\equiv N:^-$. Cyanide ions are extremely toxic because they bind almost irreversibly to the Fe(III) ion in cytochrome oxidase, a key enzyme in metabolic processes. Hydrogen cyanide, which has the aroma of bitter almonds, is even more dangerous because of its volatility (b.p. 26°C). A few tenths of 1 percent by volume of HCN in air can cause death within minutes. Hydrogen cyanide can be prepared by treating sodium cyanide or potassium cyanide with acid:

$$NaCN(s) + HCl(aq) \longrightarrow NaCl(aq) + HCN(aq)$$

Because HCN (in solution, called hydrocyanic acid) is a very weak acid ($K_a = 6.2 \times 10^{-10}$), most of the HCN produced in this reaction is in the nonionized form and leaves the solution as hydrogen cyanide gas. For this reason, acids should never be mixed with metal cyanides in the laboratory without proper ventilation.

Cyanide ions are used to extract gold and silver. Although these metals are usually found in the uncombined state in nature, in other metal ores they may be present in relatively small concentrations and are more difficult to extract. In a typical process, the crushed ore is treated with an aqueous cyanide solution in the presence of air to dissolve the gold by forming the soluble complex ion $[Au(CN)_2]^-$:

$$4Au(s) + 8CN^-(aq) + O_2(g) + 2H_2O(l) \longrightarrow 4[Au(CN)_2]^-(aq) + 4OH^-(aq)$$

The complex ion $[Au(CN)_2]^-$ (along with some cation, such as Na^+) is separated from other insoluble materials by filtration and treated with an electropositive metal such as zinc to recover the gold:

$$Zn(s) + 2[Au(CN)_2]^-(aq) \longrightarrow [Zn(CN)_4]^{2-}(aq) + 2Au(s)$$

Figure 22.6 shows an aerial view of cyanide ponds used for the extraction of gold.

Figure 22.6 *Cyanide ponds for extracting gold from metal ore.*
Larry Mayer/Jupiter Images/Creatas/Alamy Stock Photo

CHEMISTRY *in Action*

Synthetic Gas from Coal

The very existence of our technological society depends on an abundant supply of energy. Although the United States has only 5 percent of the world's population, we consume about 20 percent of the world's energy! At present, the two major sources of energy are nuclear fission and fossil fuels (discussed in Chapters 19 and 24, respectively). Coal, oil (which is also known as petroleum), and natural gas (mostly methane) are collectively called fossil fuels because they are the end result of the decomposition of plants and animals over tens or hundreds of millions of years. Oil and natural gas are cleaner-burning and more efficient fuels than coal, so they are preferred for most purposes. However, supplies of oil and natural gas are being depleted at an alarming rate, and research is underway to make coal a more versatile source of energy.

Coal consists of many high-molar-mass carbon compounds that also contain oxygen, hydrogen, and small amounts of nitrogen and sulfur. Coal constitutes about 90 percent of the world's fossil fuel reserves. For centuries coal has been used as a fuel both in homes and in industry. However, underground coal mining is expensive and dangerous, and strip mining (i.e., mining in an open pit after removal of the overlaying earth and rock) is tremendously harmful to the environment. Another problem, this one associated with the burning of coal, is the formation of sulfur dioxide (SO_2) from the sulfur-containing compounds. This process leads to the formation of "acid rain," discussed in Section 20.6.

One of the most promising methods for making coal a more efficient and cleaner fuel involves the conversion of coal to a gaseous form, called *syngas* for "synthetic gas." This process is called *coal gasification.* In the presence of very hot steam and air, coal decomposes and reacts according to the following simplified scheme:

$$C(s) + H_2O(g) \longrightarrow CO(g) + H_2(g)$$
$$C(s) + 2H_2(g) \longrightarrow CH_4(g)$$

The main component of syngas is methane. In addition, the first reaction yields hydrogen and carbon monoxide gases and other useful by-products. Under suitable conditions, CO and H_2 combine to form methanol:

$$CO(g) + 2H_2(g) \longrightarrow CH_3OH(l)$$

Methanol has many uses, for example, as a solvent and a starting material for plastics. Syngas is easier than coal to store and transport. What's more, it is not a major source of air pollution because sulfur is removed in the gasification process.

Underground coal mining.
Monty Rakusen/Cultura/Getty Images

Oxides of Carbon

Of the several oxides of carbon, the most important are carbon monoxide, CO, and carbon dioxide, CO_2. Carbon monoxide is a colorless, odorless gas formed by the incomplete combustion of carbon or carbon-containing compounds:

$$2C(s) + O_2(g) \longrightarrow 2CO(g)$$

Carbon monoxide is used in the metallurgical process for extracting nickel (see Section 21.2), in organic synthesis, and in the production of hydrocarbon fuels with hydrogen. Industrially, it is prepared by passing steam over heated coke. Carbon monoxide burns readily in oxygen to form carbon dioxide:

$$2CO(g) + O_2(g) \longrightarrow 2CO_2(g) \qquad \Delta H° = -566 \text{ kJ/mol}$$

Carbon monoxide is not an acidic oxide (it differs from carbon dioxide in that regard), and it is only slightly soluble in water.

Carbon dioxide is a colorless and odorless gas. Unlike carbon monoxide, CO_2 is nontoxic. It is an acidic oxide (see Section 15.11). Carbon dioxide is used in beverages, in fire extinguishers, and in the manufacture of baking soda, $NaHCO_3$, and soda ash, Na_2CO_3. Solid carbon dioxide, called *dry ice,* is used as a refrigerant (see Figure 11.43).

Summary of Concepts & Facts

- The important inorganic compounds of carbon are the carbides; the cyanides, most of which are extremely toxic; carbon monoxide, also toxic and a major air pollutant; the carbonates and bicarbonates; and carbon dioxide, an end product of metabolism and a component of the global carbon cycle.

22.4 Nitrogen and Phosphorus

Learning Objectives

- Summarize the common compounds of nitrogen.
- Indicate some of the common applications of phosphorus in industry.

Nitrogen

About 78 percent of air by volume is nitrogen. The most important mineral sources of nitrogen are saltpeter (KNO_3) and Chile saltpeter ($NaNO_3$). Nitrogen is an essential element of life; it is a component of proteins and nucleic acids.

Molecular nitrogen is obtained by fractional distillation of air (the boiling points of liquid nitrogen and liquid oxygen are −196°C and −183°C, respectively). In the laboratory, very pure nitrogen gas can be prepared by the thermal decomposition of ammonium nitrite:

$$NH_4NO_2(s) \longrightarrow 2H_2O(g) + N_2(g)$$

The N_2 molecule contains a triple bond and is very stable with respect to dissociation into atomic species. However, nitrogen forms a large number of compounds with hydrogen and oxygen in which the oxidation number of nitrogen varies from −3 to +5 (Table 22.2). Most nitrogen compounds are covalent; however, when heated with certain metals, nitrogen forms ionic nitrides containing the N^{3-} ion:

$$6Li(s) + N_2(g) \longrightarrow 2Li_3N(s)$$

The nitride ion is a strong Brønsted base and reacts with water to produce ammonia and hydroxide ions:

$$N^{3-}(aq) + 3H_2O(l) \longrightarrow NH_3(g) + 3OH^-(aq)$$

Table 22.2 Common Compounds of Nitrogen

Oxidation Number	Compound	Formula	Structure
−3	Ammonia	NH_3	H—N—H (with H bonded below N)
−2	Hydrazine	N_2H_4	H—N—N—H (with H bonded below each N)
−1	Hydroxylamine	NH_2OH	H—N—O—H (with H bonded below N)
0	Nitrogen* (dinitrogen)	N_2	:N≡N:
+1	Nitrous oxide (dinitrogen monoxide)	N_2O	:N≡N—O:
+2	Nitric oxide (nitrogen monoxide)	NO	:N=O
+3	Nitrous acid	HNO_2	O=N—O—H
+4	Nitrogen dioxide	NO_2	:O—N=O
+5	Nitric acid	HNO_3	O=N—O—H (with :O: bonded below N)

*We list the element here as a reference.

Ammonia

Ammonia is one of the best-known nitrogen compounds. It is prepared industrially from nitrogen and hydrogen by the Haber process (see Section 13.6). It can be prepared in the laboratory by treating ammonium chloride with sodium hydroxide:

$$NH_4Cl(aq) + NaOH(aq) \longrightarrow NaCl(aq) + H_2O(l) + NH_3(g)$$

Ammonia is a colorless gas (b.p. −33.4°C) with an irritating odor. About three-quarters of the ammonia produced annually in the United States (about 18 million tons in 2010) is used in fertilizers.

Liquid ammonia, like water, undergoes autoionization:

$$2NH_3(l) \rightleftharpoons NH_4^+ + NH_2^-$$

or simply

$$NH_3(l) \rightleftharpoons H^+ + NH_2^-$$

where NH_2^- is called the *amide ion*. Note that both H^+ and NH_2^- are solvated with the NH_3 molecules. (Here is an example of ion-dipole interaction.) At −50°C, the ion product $[H^+][NH_2^-]$ is about 1×10^{-33}, considerably smaller than 1×10^{-14} for water at 25°C. Nevertheless, liquid ammonia is a suitable solvent for many electrolytes, especially when a more basic medium is required or if the solutes react with water. The ability of liquid ammonia to dissolve alkali metals was discussed in Section 21.5.

Hydrazine

Another important hydride of nitrogen is hydrazine:

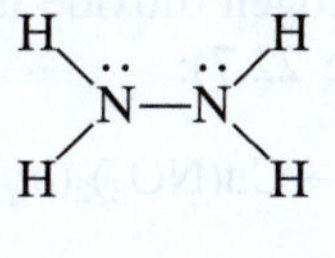

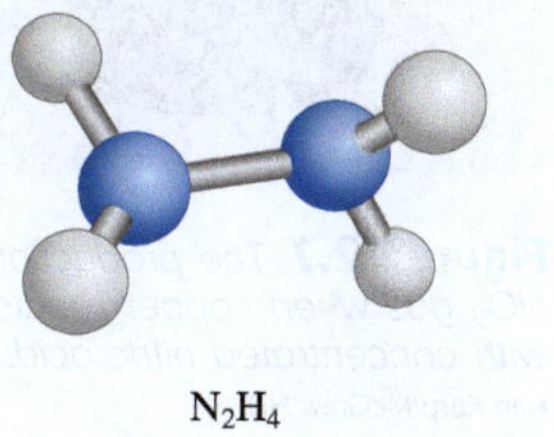

N_2H_4

Each N atom is sp^3-hybridized. Hydrazine is a colorless liquid that smells like ammonia. It melts at 2°C and boils at 114°C.

Hydrazine is a base that can be protonated to give the $N_2H_5^+$ and $N_2H_6^{2+}$ ions. As a reducing agent, it can reduce Fe^{3+} to Fe^{2+}, MnO_4^- to Mn^{2+}, and I_2 to I^-. Its reaction with oxygen is highly exothermic:

$$N_2H_4(l) + O_2(g) \longrightarrow N_2(g) + 2H_2O(l) \qquad \Delta H° = -622.2 \text{ kJ/mol}$$

Hydrazine and its derivative methylhydrazine, $N_2H_3(CH_3)$, together with the oxidizer dinitrogen tetroxide (N_2O_4), are used as rocket fuels. Hydrazine also plays a role in polymer synthesis and in the manufacture of pesticides.

Oxides and Oxoacids of Nitrogen

There are many nitrogen oxides, but the three particularly important ones are nitrous oxide, nitric oxide, and nitrogen dioxide.

Nitrous oxide, N_2O, is a colorless gas with a pleasing odor and sweet taste. It is prepared by heating ammonium nitrate to about 270°C:

$$NH_4NO_3(s) \longrightarrow N_2O(g) + 2H_2O(g)$$

Nitrous oxide resembles molecular oxygen in that it supports combustion. It does so because it decomposes when heated to form molecular nitrogen and molecular oxygen:

$$2N_2O(g) \longrightarrow 2N_2(g) + O_2(g)$$

It is chiefly used as an anesthetic in dental procedures and other minor surgery. Nitrous oxide is also called "laughing gas" because a person inhaling the gas becomes somewhat giddy. No satisfactory explanation has yet been proposed for this unusual physiological response. Nitrous oxide is also used as the propellant in cans of whipped cream due to its high solubility in the whipped cream mixture.

Nitric oxide, NO, is a colorless gas. The reaction of N_2 and O_2 in the atmosphere

$$N_2(g) + O_2(g) \rightleftharpoons 2NO(g) \qquad \Delta G° = 175.2 \text{ kJ/mol}$$

is a form of nitrogen fixation (see Section 20.1). The equilibrium constant for the above reaction is very small at room temperature: K_P is only 4.0×10^{-31} at 25°C, so very little NO will form at that temperature. However, the equilibrium constant increases rapidly with temperature, for example, in a running auto engine. According to Le Châtelier's principle, the forward endothermic reaction is favored by heating. An appreciable amount of nitric oxide is formed in the atmosphere by the action of lightning. In the laboratory, the gas can be prepared by the reduction of dilute nitric acid with copper:

$$3Cu(s) + 8HNO_3(aq) \longrightarrow 3Cu(NO_3)_2(aq) + 4H_2O(l) + 2NO(g)$$

The nitric oxide molecule is paramagnetic, containing one unpaired electron. It can be represented by the following resonance structures:

$$\dot{\text{N}}{=}\ddot{\text{O}} \longleftrightarrow {}^{-}\ddot{\text{N}}{=}\dot{\text{O}}^{+}$$

As we noted in Chapter 9, this molecule does not obey the octet rule. The properties of nitric oxide are discussed in the Chemistry in Action essay "Just Say NO" in Section 9.9.

Unlike nitrous oxide and nitric oxide, nitrogen dioxide is a highly toxic yellow-brown gas with a choking odor. The role of NO_2 in smog formation is discussed in Section 20.7. In the laboratory, nitrogen dioxide is prepared by the action of concentrated nitric acid on copper (Figure 22.7):

$$Cu(s) + 4HNO_3(aq) \longrightarrow Cu(NO_3)_2(aq) + 2H_2O(l) + 2NO_2(g)$$

Figure 22.7 *The production of NO_2 gas when copper reacts with concentrated nitric acid.*
Ken Karp/McGraw Hill

Nitrogen dioxide is paramagnetic. It has a strong tendency to dimerize to dinitrogen tetroxide, which is a diamagnetic molecule:

$$2NO_2 \rightleftharpoons N_2O_4$$

This reaction occurs in both the gas phase and the liquid phase.

Nitrogen dioxide is an acidic oxide; it reacts rapidly with cold water to form both nitrous acid, HNO_2, and nitric acid:

$$2NO_2(g) + H_2O(l) \longrightarrow HNO_2(aq) + HNO_3(aq)$$

(see Section 4.4) in which the oxidation number of nitrogen changes from +4 (in NO_2) to +3 (in HNO_2) and +5 (in HNO_3). Note that this reaction is quite different from that between CO_2 and H_2O, in which only one acid (carbonic acid) is formed.

Nitric acid is one of the most important inorganic acids. It is a liquid (b.p. 82.6°C), but it does not exist as a pure liquid because it decomposes spontaneously to some extent as follows:

$$4HNO_3(l) \longrightarrow 4NO_2(g) + 2H_2O(l) + O_2(g)$$

The major industrial method of producing nitric acid is the Ostwald process, discussed in Section 13.6. The concentrated nitric acid used in the laboratory is 68 percent HNO_3 by mass (density 1.42 g/cm^3), which corresponds to 15.7 *M*. On standing, a concentrated nitric acid solution turns slightly yellow as a result of NO_2 formation.

Nitric acid is a powerful oxidizing agent. The oxidation number of N in HNO_3 is +5. The most common reduction products of nitric acid are NO_2 (oxidation number of N = +4), NO (oxidation number of N = +2), and NH_4^+ (oxidation number of N = −3). Nitric acid can oxidize metals both below and above hydrogen in the activity series (see Figure 4.16). For example, copper is oxidized by concentrated nitric acid, as discussed earlier.

In the presence of a strong reducing agent, such as zinc metal, nitric acid can be reduced all the way to the ammonium ion:

$$4Zn(s) + 10H^+(aq) + NO_3^-(aq) \longrightarrow 4Zn^{2+}(aq) + NH_4^+(aq) + 3H_2O(l)$$

Concentrated nitric acid does not oxidize gold. However, when the acid is added to concentrated hydrochloric acid in a 1:3 ratio by volume (one part HNO_3 to three parts HCl), the resulting solution, called *aqua regia,* can oxidize gold, as follows:

$$Au(s) + 3HNO_3(aq) + 4HCl(aq) \longrightarrow HAuCl_4(aq) + 3H_2O(l) + 3NO_2(g)$$

The oxidation of Au is promoted by the complexing ability of the Cl^- ion (to form the $AuCl_4^-$ ion).

Concentrated nitric acid also oxidizes a number of nonmetals to their corresponding oxoacids:

$$P_4(s) + 20HNO_3(aq) \longrightarrow 4H_3PO_4(aq) + 20NO_2(g) + 4H_2O(l)$$
$$S(s) + 6HNO_3(aq) \longrightarrow H_2SO_4(aq) + 6NO_2(g) + 2H_2O(l)$$

Nitric acid is used in the manufacture of fertilizers, dyes, drugs, and explosives. The Chemistry in Action essay "Ammonium Nitrate—The Explosive Fertilizer" describes a nitrogen-containing fertilizer that can be highly explosive.

Phosphorus

Like nitrogen, phosphorus is a member of the Group 15 family; in some respects the chemistry of phosphorus resembles that of nitrogen. Phosphorus occurs most commonly in nature as *phosphate rocks,* which are mostly calcium phosphate, $Ca_3(PO_4)_2$,

CHEMISTRY *in Action*

Ammonium Nitrate—The Explosive Fertilizer

Ammonium nitrate is the most important fertilizer in the world (see Chemistry in Action essay "Chemical Fertilizers" in Section 3.10). It ranked fifteenth among the industrial chemicals produced in the United States in 2009 (8 million tons). Unfortunately, it is also a powerful explosive. In 1947 an explosion occurred aboard a ship being loaded with the fertilizer in Texas. The fertilizer was in paper bags and apparently blew up after sailors tried to stop a fire in the ship's hold by closing a hatch, thereby creating the compression and heat necessary for an explosion. More than 600 people died as a result of the accident. More recent disasters involving ammonium nitrate took place at the World Trade Center in New York City in 1993 and at the Alfred P. Murrah Federal Building in Oklahoma City in 1995.

A strong oxidizer, ammonium nitrate is stable at room temperature. At 250°C, it begins to decompose as follows:

$$NH_4NO_3(g) \longrightarrow N_2O(g) + 2H_2O(g)$$

At 300°C, different gaseous products and more heat are produced:

$$2NH_4NO_3(g) \longrightarrow 2N_2(g) + 4H_2O(g) + O_2(g)$$

About 1.46 kJ of heat are generated per gram of the compound decomposed. When it is combined with a combustible material, such as fuel oil, the energy released increases almost threefold. Ammonium nitrate can also be mixed with charcoal, flour, sugar, sulfur, rosin, and paraffin to form an explosive. Intense heat from the explosion causes the gases to expand rapidly, generating shock waves that destroy most objects in their path.

Federal law regulates the sale of explosive-grade ammonium nitrate, which is used for 95 percent of all commercial blasting in road construction and mining. However, the wide availability of large quantities of ammonium nitrate and other substances that enhance its explosive power make it possible for anyone who is so inclined to construct a bomb. The bomb that destroyed the federal building in Oklahoma City is estimated to have contained 4000 pounds of ammonium nitrate and fuel oil, which was set off by another small explosive device.

How can the use of ammonium nitrate by terrorists be prevented? The most logical approach is to desensitize or neutralize the compound's ability to act as an explosive, but to date no satisfactory way has been found to do so without diminishing its value as a fertilizer. A more passive method is to add to the fertilizer an agent known as a *taggant,* which would allow law enforcement to trace the source of an ammonium nitrate explosive. A number of European countries now forbid the sale of ammonium nitrate without taggants, although the U.S. Congress has yet to pass such a law.

The Alfred P. Murrah building after a deadly explosion caused by an ammonium nitrate bomb.
Bill Waugh/AP Images

A bag of ammonium nitrate fertilizer, which is labeled as an explosive.
Metropolitan Police/Getty Images

Figure 22.8 *Phosphate mining.*
T. O'Keefe/PhotoLink/Photodisc/Getty Images

and fluoroapatite, $Ca_5(PO_4)_3F$ (Figure 22.8). Elemental phosphorus can be obtained by heating calcium phosphate with coke and silica sand:

$$2Ca_3(PO_4)_2(s) + 10C(s) + 6SiO_2(s) \longrightarrow 6CaSiO_3(s) + 10CO(g) + P_4(s)$$

There are several allotropic forms of phosphorus, but only white phosphorus and red phosphorus (see Figure 8.18) are of importance. White phosphorus consists of discrete tetrahedral P_4 molecules (Figure 22.9). A solid (m.p. 44.2°C), white phosphorus is insoluble in water but quite soluble in carbon disulfide (CS_2) and in organic solvents such as chloroform ($CHCl_3$). White phosphorus is a highly toxic substance. It bursts into flames spontaneously when exposed to air; hence it is used in incendiary bombs and grenades:

$$P_4(s) + 5O_2(g) \longrightarrow P_4O_{10}(s)$$

The high reactivity of white phosphorus is attributed to structural strain: The P—P bonds are compressed in the tetrahedral P_4 molecule. White phosphorus was once used in matches, but because of its toxicity it has been replaced by tetraphosphorus trisulfide, P_4S_3.

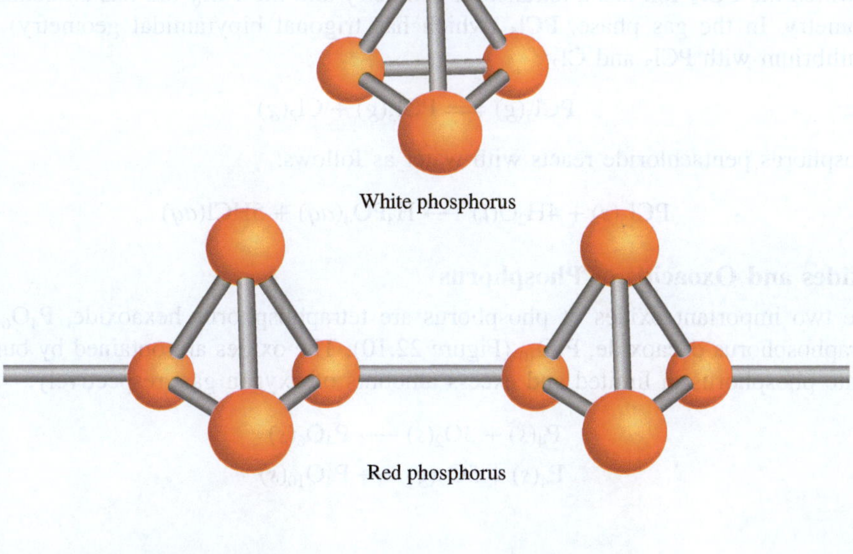

Figure 22.9 *The structures of white and red phosphorus. Red phosphorus is believed to have a chain structure, as shown.*

When heated in the absence of air, white phosphorus is slowly converted to red phosphorus at about 300°C:

$$nP_4(\text{white phosphorus}) \longrightarrow (P_4)_n(\text{red phosphorus})$$

Red phosphorus has a polymeric structure (see Figure 22.9) and is more stable and less volatile than white phosphorus.

Hydride of Phosphorus

The most important hydride of phosphorus is phosphine, PH_3, a colorless, very poisonous gas formed by heating white phosphorus in concentrated sodium hydroxide:

$$P_4(s) + 3NaOH(aq) + 3H_2O(l) \longrightarrow 3NaH_2PO_2(aq) + PH_3(g)$$

Phosphine is moderately soluble in water and more soluble in carbon disulfide and organic solvents. Its aqueous solution is neutral, unlike that of ammonia. In liquid ammonia, phosphine dissolves to give $NH_4^+PH_2^-$. Phosphine is a strong reducing agent; it reduces many metal salts to the corresponding metal. The gas burns in air:

$$PH_3(g) + 2O_2(g) \longrightarrow H_3PO_4(s)$$

Halides of Phosphorus

Phosphorus forms binary compounds with halogens: the trihalides, PX_3, and the pentahalides, PX_5, where X denotes a halogen atom. In contrast, nitrogen can form only trihalides (NX_3). Unlike nitrogen, phosphorus has a 3*d* subshell, which can be used for valence-shell expansion. We can explain the bonding in PCl_5 by assuming that phosphorus undergoes sp^3d hybridization of its 3*s*, 3*p*, and 3*d* orbitals (see Example 10.4). The five sp^3d hybrid orbitals also account for the trigonal bipyramidal geometry of the PCl_5 molecule (see Table 10.4).

Phosphorus trichloride is prepared by heating white phosphorus in chlorine:

$$P_4(l) + 6Cl_2(g) \longrightarrow 4PCl_3(g)$$

A colorless liquid (b.p. 76°C), PCl_3 is hydrolyzed according to the equation:

$$PCl_3(l) + 3H_2O(l) \longrightarrow H_3PO_3(aq) + 3HCl(g)$$

In the presence of an excess of chlorine gas, PCl_3 is converted to phosphorus pentachloride, which is a light yellow solid:

$$PCl_3(l) + Cl_2(g) \longrightarrow PCl_5(s)$$

X-ray studies have shown that solid phosphorus pentachloride exists as $[PCl_4^+][PCl_6^-]$, in which the PCl_4^+ ion has a tetrahedral geometry and the PCl_6^- ion has an octahedral geometry. In the gas phase, PCl_5 (which has trigonal bipyramidal geometry) is in equilibrium with PCl_3 and Cl_2:

$$PCl_5(g) \rightleftharpoons PCl_3(g) + Cl_2(g)$$

Phosphorus pentachloride reacts with water as follows:

$$PCl_5(s) + 4H_2O(l) \longrightarrow H_3PO_4(aq) + 5HCl(aq)$$

Oxides and Oxoacids of Phosphorus

The two important oxides of phosphorus are tetraphosphorus hexaoxide, P_4O_6, and tetraphosphorus decaoxide, P_4O_{10} (Figure 22.10). The oxides are obtained by burning white phosphorus in limited and excess amounts of oxygen gas, respectively:

$$P_4(s) + 3O_2(s) \longrightarrow P_4O_6(s)$$
$$P_4(s) + 5O_2(g) \longrightarrow P_4O_{10}(s)$$

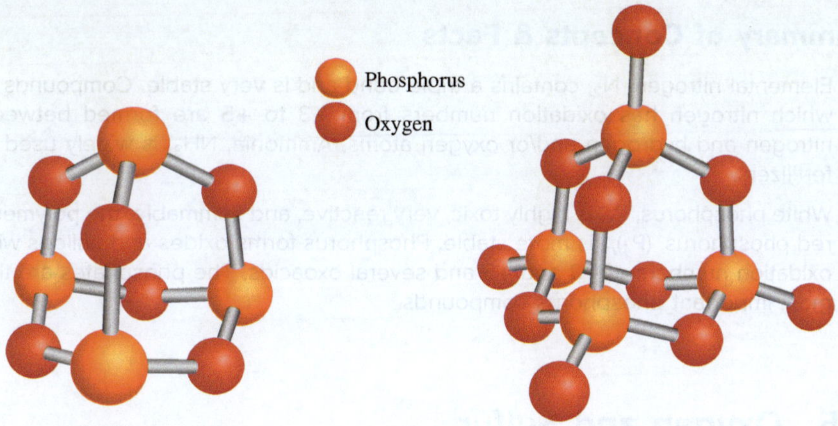

P_2O_6 P_4O_{10}

Figure 22.10 *The structures of P_4O_6 and P_4O_{10}. Note the tetrahedral arrangement of the P atoms in P_4O_{10}.*

Both oxides are acidic; that is, they are converted to acids in water. The compound P_4O_{10} is a white flocculent powder (m.p. 420°C) that has a great affinity for water:

$$P_4O_{10}(s) + 6H_2O(l) \longrightarrow 4H_3PO_4(aq)$$

For this reason, it is often used for drying gases and for removing water from solvents.

There are many oxoacids containing phosphorus. Some examples are phosphorous acid, H_3PO_3; phosphoric acid, H_3PO_4; hypophosphorous acid, H_3PO_2; and triphosphoric acid, $H_5P_3O_{10}$ (Figure 22.11). Phosphoric acid, also called orthophosphoric acid, is a weak triprotic acid (see Section 15.8). It is prepared industrially by the reaction of calcium phosphate with sulfuric acid:

$$Ca_3(PO_4)_2(s) + 3H_2SO_4(aq) \longrightarrow 2H_3PO_4(aq) + 3CaSO_4(s)$$

In the pure form phosphoric acid is a colorless solid (m.p. 42.2°C). The phosphoric acid we use in the laboratory is usually an 82 percent H_3PO_4 solution (by mass). Phosphoric acid and phosphates have many commercial applications in detergents, fertilizers, flame retardants, and toothpastes, and as buffers in carbonated beverages.

Like nitrogen, phosphorus is an element that is essential to life. It constitutes only about 1 percent by mass of the human body, but it is a very important 1 percent. About 23 percent of the human skeleton is mineral matter. The phosphorus content of this mineral matter, calcium phosphate, $Ca_3(PO_4)_2$, is 20 percent. Our teeth are basically $Ca_3(PO_4)_2$ and $Ca_5(PO_4)_3OH$. Phosphates are also important components of the genetic materials deoxyribonucleic acid (DNA) and ribonucleic acid (RNA).

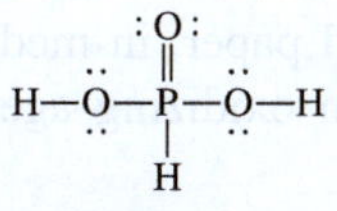

Phosphorous acid (H_3PO_3)

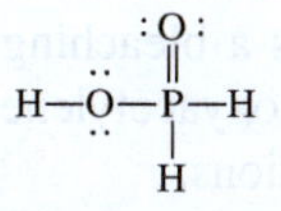

Hypophosphorous acid (H_3PO_2)

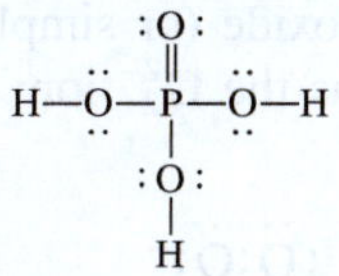

Phosphoric acid (H_3PO_4)

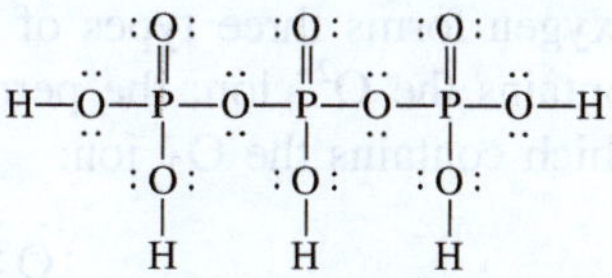

Triphosphoric acid ($H_5P_3O_{10}$)

Figure 22.11 *Structures of some common phosphorus-containing oxoacids.*

Summary of Concepts & Facts

- Elemental nitrogen, N_2, contains a triple bond and is very stable. Compounds in which nitrogen has oxidation numbers from −3 to +5 are formed between nitrogen and hydrogen and/or oxygen atoms. Ammonia, NH_3, is widely used in fertilizers.
- White phosphorus, P_4, is highly toxic, very reactive, and flammable; the polymeric red phosphorus, $(P_4)_n$, is more stable. Phosphorus forms oxides and halides with oxidation numbers of +3 and +5 and several oxoacids. The phosphates are the most important phosphorus compounds.

22.5 Oxygen and Sulfur

Learning Objectives

- Identify the common compounds formed by oxygen.
- Discuss the Frasch process.

Oxygen

Oxygen is by far the most abundant element in Earth's crust, constituting about 46 percent of its mass. In addition, the atmosphere contains about 21 percent molecular oxygen by volume (23 percent by mass). Like nitrogen, oxygen in the free state is a diatomic molecule (O_2). In the laboratory, oxygen gas can be obtained by heating potassium chlorate (see Figure 5.15):

$$2KClO_3(s) \longrightarrow 2KCl(s) + 3O_2(g)$$

The reaction is usually catalyzed by manganese(IV) dioxide, MnO_2. Pure oxygen gas can be prepared by electrolyzing water (see Section 18.8). Industrially, oxygen gas is prepared by the fractional distillation of liquefied air (see Section 12.6). Oxygen gas is colorless and odorless.

Oxygen is a building block of practically all biomolecules, accounting for about a fourth of the atoms in living matter. Molecular oxygen is the essential oxidant in the metabolic breakdown of food molecules. Without it, a human being cannot survive for more than a few minutes.

Properties of Diatomic Oxygen

Although oxygen has two allotropes, O_2 and O_3, when we speak of molecular oxygen, we normally mean O_2. Ozone, O_3, is less stable than O_2. The O_2 molecule is paramagnetic because it contains two unpaired electrons (see Section 10.7).

A strong oxidizing agent, molecular oxygen is one of the most widely used industrial chemicals. Its main uses are in the steel industry (see Section 21.2) and in sewage treatment. Oxygen is also used as a bleaching agent for pulp and paper, in medicine to ease breathing difficulties, in oxyacetylene torches, and as an oxidizing agent in many inorganic and organic reactions.

Oxides, Peroxides, and Superoxides

Oxygen forms three types of oxides: the normal oxide (or simply the oxide), which contains the O^{2-} ion; the peroxide, which contains the O_2^{2-} ion; and the superoxide, which contains the O_2^- ion:

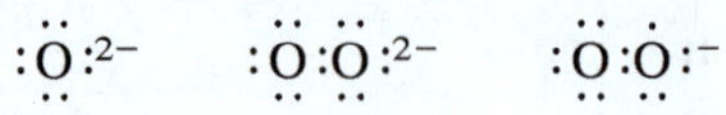

oxide peroxide superoxide

The ions are all strong Brønsted bases and react with water as follows:

$$\text{Oxide:}\quad O^{2-}(aq) + H_2O(l) \longrightarrow 2OH^-(aq)$$
$$\text{Peroxide:}\quad 2O_2^{2-}(aq) + 2H_2O(l) \longrightarrow O_2(g) + 4OH^-(aq)$$
$$\text{Superoxide:}\quad 4O_2^-(aq) + 2H_2O(l) \longrightarrow 3O_2(g) + 4OH^-(aq)$$

Note that the reaction of O^{2-} with water is a hydrolysis reaction, but those involving O_2^{2-} and O_2^- are redox processes.

The nature of bonding in oxides changes across any period in the periodic table (see Figure 15.8). Oxides of elements on the left side of the periodic table, such as those of the alkali metals and alkaline earth metals, are generally ionic solids with high melting points. Oxides of the metalloids and of the metallic elements toward the middle of the periodic table are also solids, but they have much less ionic character. Oxides of nonmetals are covalent compounds that generally exist as liquids or gases at room temperature. The acidic character of the oxides increases from left to right.

Consider the oxides of the third-period elements (see Table 8.4):

$$\underbrace{Na_2O\quad MgO}_{\text{basic}}\quad \underbrace{Al_2O_3}_{\text{amphoteric}}\quad \underbrace{SiO_2\quad P_4O_{10}\quad SO_3\quad Cl_2O_7}_{\text{acidic}}$$

The basicity of the oxides increases as we move down a particular group. MgO does not react with water but reacts with acid as follows:

$$MgO(s) + 2H^+(aq) \longrightarrow Mg^{2+}(aq) + H_2O(l)$$

On the other hand, BaO, which is more basic, undergoes hydrolysis to yield the corresponding hydroxide:

$$BaO(s) + H_2O(l) \longrightarrow Ba(OH)_2(aq)$$

The best-known peroxide is hydrogen peroxide (H_2O_2). It is a colorless, syrupy liquid (m.p. −0.9°C), prepared in the laboratory by the action of cold dilute sulfuric acid on barium peroxide octahydrate:

$$BaO_2 \cdot 8H_2O(s) + H_2SO_4(aq) \longrightarrow BaSO_4(s) + H_2O_2(aq) + 8H_2O(l)$$

The structure of hydrogen peroxide is shown in Figure 22.12. Using the VSEPR method we see that the H—O and O—O bonds are bent around each oxygen atom in a configuration similar to the structure of water. The lone-pair–bonding-pair repulsion is greater in H_2O_2 than in H_2O, so that the HOO angle is only 97° (compared with 104.5° for HOH in H_2O). Hydrogen peroxide is a polar molecule (μ = 2.16 D).

Figure 22.12 *The structure of H_2O_2.*

Hydrogen peroxide readily decomposes when heated or exposed to sunlight or even in the presence of dust particles or certain metals, including iron and copper:

$$2H_2O_2(l) \longrightarrow 2H_2O(l) + O_2(g) \qquad \Delta H° = -196 \text{ kJ/mol}$$

Note that this is a disproportionation reaction. The oxidation number of oxygen changes from −1 to −2 and 0.

Hydrogen peroxide is miscible with water in all proportions due to its ability to hydrogen-bond with water. Dilute hydrogen peroxide solutions (3 percent by mass), available in drugstores, are used as mild antiseptics; more concentrated H_2O_2 solutions are employed as bleaching agents for textiles, fur, and hair. The high heat of decomposition of hydrogen peroxide also makes it a suitable component in rocket fuel.

Hydrogen peroxide is a strong oxidizing agent; it can oxidize Fe^{2+} ions to Fe^{3+} ions in an acidic solution:

$$H_2O_2(aq) + 2Fe^{2+}(aq) + 2H^+(aq) \longrightarrow 2Fe^{3+}(aq) + 2H_2O(l)$$

It also oxidizes SO_3^{2-} ions to SO_4^{2-} ions:

$$H_2O_2(aq) + SO_3^{2-}(aq) \longrightarrow SO_4^{2-}(aq) + H_2O(l)$$

In addition, hydrogen peroxide can act as a reducing agent toward substances that are stronger oxidizing agents than itself. For example, hydrogen peroxide reduces silver oxide to metallic silver:

$$H_2O_2(aq) + Ag_2O(s) \longrightarrow 2Ag(s) + H_2O(l) + O_2(g)$$

and permanganate, MnO_4^-, to manganese(II) in an acidic solution:

$$5H_2O_2(aq) + 2MnO_4^-(aq) + 6H^+(aq) \longrightarrow 2Mn^{2+}(aq) + 5O_2(g) + 8H_2O(l)$$

If we want to determine hydrogen peroxide concentration, this reaction can be carried out as a redox titration, using a standard permanganate solution.

There are relatively few known superoxides, that is, compounds containing the O_2^- ion. In general, only the most reactive alkali metals (K, Rb, and Cs) form superoxides.

We should take note of the fact that both the peroxide ion and the superoxide ion are by-products of metabolism. Because these ions are highly reactive, they can inflict great damage on living cells. Fortunately, our bodies are equipped with the enzymes catalase, peroxidase, and superoxide dismutase, which convert these toxic substances to water and molecular oxygen.

Liquid ozone.
Ken Karp/McGraw Hill

Ozone

Ozone is a rather toxic, light blue gas (b.p. −111.3°C). Its pungent odor is noticeable around sources of significant electrical discharges (such as a subway train). Ozone can be prepared from molecular oxygen, either photochemically or by subjecting O_2 to an electrical discharge (Figure 22.13):

$$3O_2(g) \longrightarrow 2O_3(g) \qquad \Delta G° = 326.4 \text{ kJ/mol}$$

Because the standard free energy of formation of ozone is a large positive quantity [$\Delta G_f^\circ = (326.4/2)$ kJ/mol or 163.2 kJ/mol], ozone is less stable than molecular oxygen. The ozone molecule has a bent structure in which the bond angle is 116.5°:

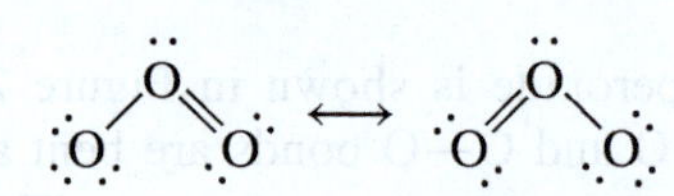

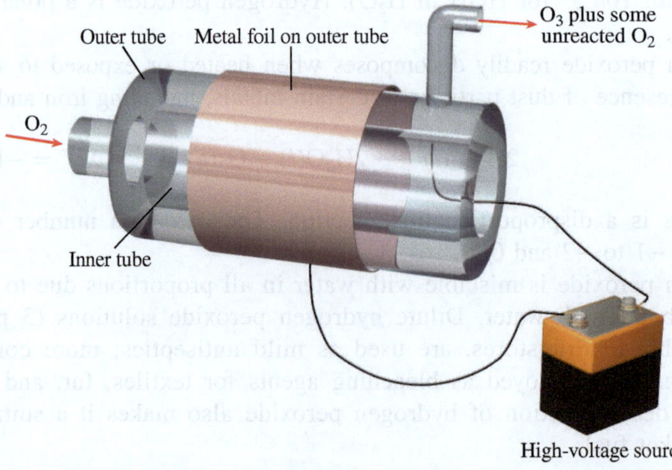

Figure 22.13 *The preparation of O_3 from O_2 by electrical discharge. The outside of the outer tube and the inside of the inner tube are coated with metal foils that are connected to a high-voltage source. (The metal foil on the inside of the inner tube is not shown.) During the electrical discharge, O_2 gas is passed through the tube. The O_3 gas formed exits from the upper right-hand tube, along with some unreacted O_2 gas.*

Figure 22.14 *Pyrite (FeS_2), commonly called "fool's gold" because of its gold luster.*
precismarketing/iStock/Getty Images

Ozone is mainly used to purify drinking water, to deodorize air and sewage gases, and to bleach waxes, oils, and textiles.

Ozone is a very powerful oxidizing agent—its oxidizing power is exceeded only by that of molecular fluorine (see Table 18.1). For example, ozone can oxidize sulfides of many metals to the corresponding sulfates:

$$4O_3(g) + PbS(s) \longrightarrow PbSO_4(s) + 4O_2(g)$$

Ozone oxidizes all the common metals except gold and platinum. In fact, a convenient test for ozone is based on its action on mercury. When exposed to ozone, mercury loses its metallic luster and sticks to glass tubing (instead of flowing freely through it). This behavior is attributed to the change in surface tension caused by the formation of mercury(II) oxide:

$$O_3(g) + 3Hg(l) \longrightarrow 3HgO(s)$$

The beneficial effect of ozone in the stratosphere and its undesirable action in smog formation were discussed in Chapter 20.

Sulfur

Although sulfur is not a very abundant element (it constitutes only about 0.06 percent of Earth's crust by mass), it is readily available because it occurs commonly in nature in the elemental form. The largest known reserves of sulfur are found in sedimentary deposits. In addition, sulfur occurs widely in gypsum ($CaSO_4 \cdot 2H_2O$) and various sulfide minerals such as pyrite (FeS_2) (Figure 22.14). Sulfur is also present in natural gas as H_2S, SO_2, and other sulfur-containing compounds.

Sulfur is extracted from underground deposits by the *Frasch*† *process,* shown in Figure 22.15. In this process, superheated water (liquid water heated to about 160°C under high pressure to prevent it from boiling) is pumped down the outermost pipe to melt the sulfur. Next, compressed air is forced down the innermost pipe. Liquid sulfur mixed with air forms an emulsion that is less dense than water and therefore rises to the surface as it is forced up the middle pipe. Sulfur produced in this manner, which amounts to about 10 million tons per year, has a purity of about 99.5 percent.

There are several allotropic forms of sulfur, the most important being the rhombic and monoclinic forms. Rhombic sulfur is thermodynamically the most stable form; it has a puckered S_8 ring structure:

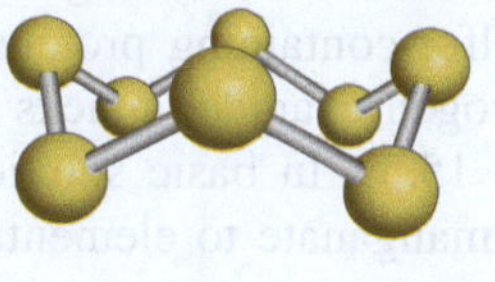

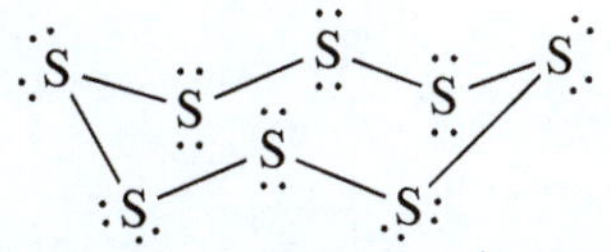

S_8

†Herman Frasch (1851–1914). German chemical engineer. Besides inventing the process for obtaining pure sulfur, Frasch developed methods for refining petroleum.

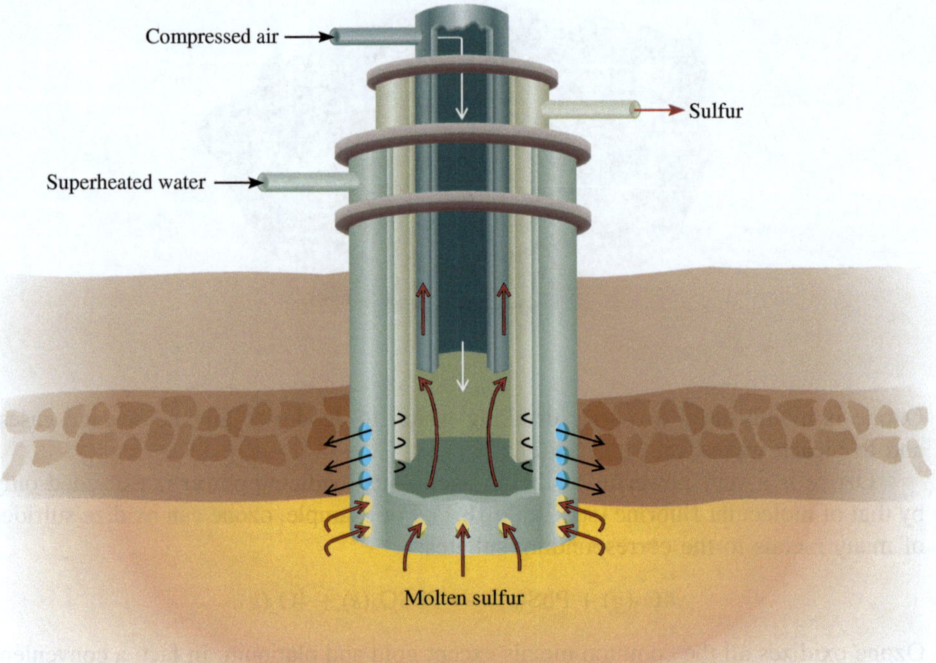

Figure 22.15 *The Frasch process. Three concentric pipes are inserted into a hole drilled down to the sulfur deposit. Superheated water is forced down the outer pipe into the sulfur, causing it to melt. Molten sulfur is then forced up the middle pipe by compressed air.*

It is a yellow, tasteless, and odorless solid (m.p. 112°C) (see Figure 8.19) that is insoluble in water but soluble in carbon disulfide. When heated, it is slowly converted to monoclinic sulfur (m.p. 119°C), which also consists of the S_8 units. When liquid sulfur is heated above 150°C, the rings begin to break up, and the entangling of the sulfur chains results in a sharp increase in the liquid's viscosity. Further heating tends to rupture the chains, and the viscosity decreases.

Like nitrogen, sulfur shows a wide variety of oxidation numbers in its compounds (Table 22.3). The best-known hydrogen compound of sulfur is hydrogen sulfide, which is prepared by the action of an acid on a sulfide; for example,

$$FeS(s) + H_2SO_4(aq) \longrightarrow FeSO_4(aq) + H_2S(g)$$

Today, hydrogen sulfide used in qualitative analysis (see Section 16.8) is prepared by the hydrolysis of thioacetamide:

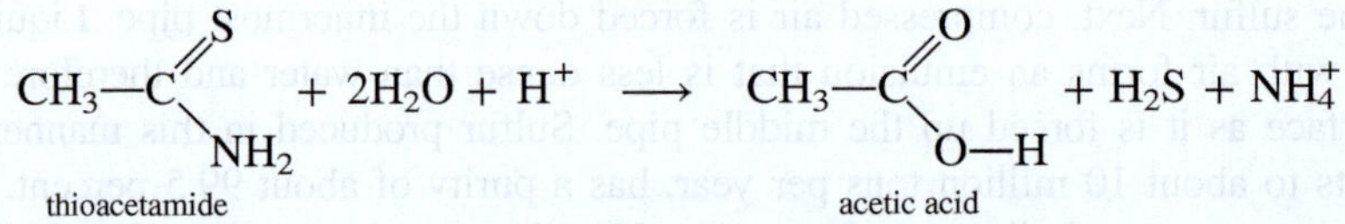

Hydrogen sulfide is a colorless gas (b.p. −60.2°C) that smells like rotten eggs. (The odor of rotten eggs actually comes from hydrogen sulfide, which is formed by the bacterial decomposition of sulfur-containing proteins.) Hydrogen sulfide is a highly toxic substance that, like hydrogen cyanide, attacks respiratory enzymes. It is a very weak diprotic acid (see Table 15.5). In basic solution, H_2S is a reducing agent. For example, it is oxidized by permanganate to elemental sulfur:

$$3H_2S(aq) + 2MnO_4^-(aq) \longrightarrow 3S(s) + 2MnO_2(s) + 2H_2O(l) + 2OH^-(aq)$$

Table 22.3 Common Compounds of Sulfur

Oxidation Number	Compound	Formula	Structure
−2	Hydrogen sulfide	H_2S	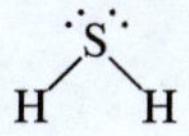
0	Sulfur*	S_8	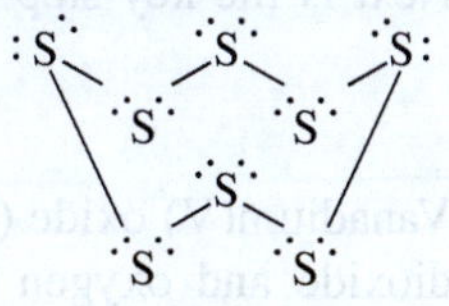
+1	Disulfur dichloride	S_2Cl_2	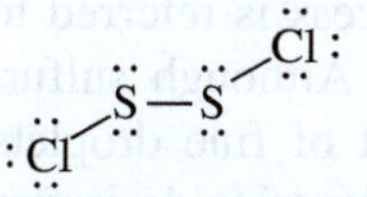
+2	Sulfur dichloride	SCl_2	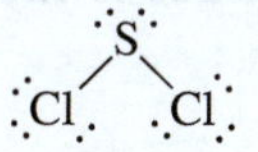
+4	Sulfur dioxide	SO_2	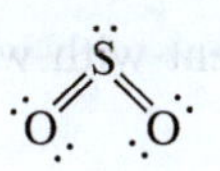
+6	Sulfur trioxide	SO_3	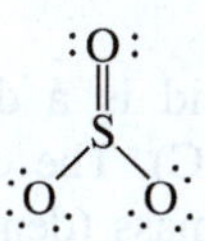

*We list the element here as a reference.

Oxides of Sulfur

Sulfur has two important oxides: sulfur dioxide, SO_2; and sulfur trioxide, SO_3. Sulfur dioxide is formed when sulfur burns in air:

$$S(s) + O_2(g) \longrightarrow SO_2(g)$$

In the laboratory, it can be prepared by the action of an acid on a sulfite; for example,

$$2HCl(aq) + Na_2SO_3(aq) \longrightarrow 2NaCl(aq) + H_2O(l) + SO_2(g)$$

or by the action of concentrated sulfuric acid on copper:

$$Cu(s) + 2H_2SO_4(aq) \longrightarrow CuSO_4(aq) + 2H_2O(l) + SO_2(g)$$

Sulfur dioxide (b.p. −10°C) is a pungent, colorless gas that is quite toxic. As an acidic oxide, it reacts with water as follows:

$$SO_2(g) + H_2O(l) \rightleftharpoons H^+(aq) + HSO_3^-(aq)$$

Sulfur dioxide is slowly oxidized to sulfur trioxide, but the reaction rate can be greatly enhanced by a platinum or vanadium oxide catalyst (see Section 13.6):

$$2SO_2(g) + O_2(g) \longrightarrow 2SO_3(g)$$

Sulfur trioxide dissolves in water to form sulfuric acid:

$$SO_3(g) + H_2O(l) \longrightarrow H_2SO_4(aq)$$

The contributing role of sulfur dioxide to acid rain is discussed in Section 20.6.

Vanadium oxide on alumina (Al_2O_3).
Ken Karp/McGraw Hill

Sulfuric Acid

Sulfuric acid is the world's most important industrial chemical. Globally, more than 266 million tons were produced in 2018. It is prepared industrially by first burning sulfur in air:

$$S(s) + O_2(g) \longrightarrow SO_2(g)$$

Next is the key step of converting sulfur dioxide to sulfur trioxide:

$$2SO_2(g) + O_2(g) \longrightarrow 2SO_3(g)$$

Vanadium(V) oxide (V_2O_5) is the catalyst used for the second step. Because the sulfur dioxide and oxygen molecules react in contact with the surface of solid V_2O_5, the process is referred to as the *contact process*.

Although sulfur trioxide reacts with water to produce sulfuric acid, it forms a mist of fine droplets of H_2SO_4 with water vapor that is hard to condense. Instead, sulfur trioxide is first dissolved in 98 percent sulfuric acid to form *oleum* ($H_2S_2O_7$):

$$SO_3(g) + H_2SO_4(aq) \longrightarrow H_2S_2O_7(aq)$$

On treatment with water, concentrated sulfuric acid can be generated:

$$H_2S_2O_7(aq) + H_2O(l) \longrightarrow 2H_2SO_4(aq)$$

Sulfuric acid is a diprotic acid (see Table 15.5). It is a colorless, viscous liquid (m.p. 10.4°C). The concentrated sulfuric acid we use in the laboratory is 98 percent H_2SO_4 by mass (density: 1.84 g/cm^3), which corresponds to a concentration of 18 *M*. The oxidizing strength of sulfuric acid depends on its temperature and concentration. A cold, dilute sulfuric acid solution reacts with metals above hydrogen in the activity series (see Figure 4.16), thereby liberating molecular hydrogen in a displacement reaction:

$$Mg(s) + H_2SO_4(aq) \longrightarrow MgSO_4(aq) + H_2(g)$$

This is a typical reaction of an active metal with an acid. The strength of sulfuric acid as an oxidizing agent is greatly enhanced when it is both hot and concentrated. In such a solution, the oxidizing agent is actually the sulfate ion rather than the hydrated proton, $H^+(aq)$. Thus, copper reacts with concentrated sulfuric acid as follows:

$$Cu(s) + 2H_2SO_4(aq) \longrightarrow CuSO_4(aq) + SO_2(g) + 2H_2O(l)$$

Depending on the nature of the reducing agents, the sulfate ion may be further reduced to elemental sulfur or the sulfide ion. For example, reduction of H_2SO_4 by HI yields H_2S and I_2:

$$8HI(aq) + H_2SO_4(aq) \longrightarrow H_2S(aq) + 4I_2(s) + 4H_2O(l)$$

Concentrated sulfuric acid oxidizes nonmetals. For example, it oxidizes carbon to carbon dioxide and sulfur to sulfur dioxide:

$$C(s) + 2H_2SO_4(aq) \longrightarrow CO_2(g) + 2SO_2(g) + 2H_2O(l)$$
$$S(s) + 2H_2SO_4(aq) \longrightarrow 3SO_2(g) + 2H_2O(l)$$

Other Compounds of Sulfur

Carbon disulfide, a colorless, flammable liquid (b.p. 46°C), is formed by heating carbon and sulfur to a high temperature:

$$C(s) + 2S(l) \longrightarrow CS_2(l)$$

It is only slightly soluble in water. Carbon disulfide is a good solvent for sulfur, phosphorus, iodine, and nonpolar substances such as waxes and rubber.

Another interesting compound of sulfur is sulfur hexafluoride, SF_6, which is prepared by heating sulfur in an atmosphere of fluorine:

$$S(l) + 3F_2(g) \longrightarrow SF_6(g)$$

Sulfur hexafluoride is a nontoxic, colorless gas (b.p. −63.8°C). It is the most inert of all sulfur compounds; it resists attack even by molten KOH. The structure and bonding of SF_6 were discussed in Chapters 9 and 10 and its critical phenomenon illustrated in Chapter 11 (see Figure 11.38).

Summary of Concepts & Facts

- Elemental oxygen, O_2, is paramagnetic and contains two unpaired electrons. Oxygen forms ozone (O_3), oxides (O^{2-}), peroxides (O_2^{2-}), and superoxides (O_2^-). The most abundant element in Earth's crust, oxygen is essential for life on Earth.
- Sulfur is taken from Earth's crust by the Frasch process as a molten liquid. Sulfur exists in a number of allotropic forms and has a variety of oxidation numbers in its compounds.
- Sulfuric acid is the cornerstone of the chemical industry. It is produced from sulfur via sulfur dioxide and sulfur trioxide by means of the contact process.

22.6 The Halogens

Learning Objectives

- Contrast the chemical behavior of fluorine to that of the other halogens.
- Describe processes used to prepare halogens.
- Assess some of the common uses of halogens.

The halogens—fluorine, chlorine, bromine, and iodine—are reactive nonmetals (see Figure 8.20). Table 22.4 lists some of the properties of these elements. Although all halogens are highly reactive and toxic, the magnitude of reactivity and toxicity generally decreases from fluorine to iodine. The chemistry of fluorine differs from that of the rest of the halogens in the following ways (recall that the first member of a group usually differs in properties from the rest of the members of the group) (see Section 8.6):

1. Fluorine is the most reactive of all the halogens. The difference in reactivity between fluorine and chlorine is greater than that between chlorine and bromine. Table 22.4 shows that the F—F bond is considerably weaker than the Cl—Cl bond. The weak bond in F_2 can be explained in terms of the lone pairs on the F atoms:

$$:\ddot{\underset{..}{F}}-\ddot{\underset{..}{F}}:$$

   The small size of the F atoms (see Table 22.4) allows a close approach of the three lone pairs on each of the F atoms, resulting in a greater repulsion than that found in Cl_2, which consists of larger atoms.
2. Hydrogen fluoride, HF, has a high boiling point (19.5°C) as a result of strong intermolecular hydrogen bonding, whereas all other hydrogen halides have much lower boiling points (see Figure 11.7).
3. Hydrofluoric acid is a weak acid, whereas all other hydrohalic acids (HCl, HBr, and HI) are strong acids.

Table 22.4 Properties of the Halogens

Property	F	Cl	Br	I
Valence electron configuration	$2s^22p^5$	$3s^23p^5$	$4s^24p^5$	$5s^25p^5$
Melting point (°C)	−223	−102	−7	114
Boiling point (°C)	−187	−35	59	183
Appearance*	Pale yellow gas	Yellow-green gas	Red-brown liquid	Dark violet vapor Dark metallic-looking solid
Atomic radius (pm)	72	99	114	133
Ionic radius (pm)†	133	181	195	220
Ionization energy (kJ/mol)	1680	1251	1139	1003
Electronegativity	4.0	3.2	3.0	2.7
Standard reduction potential (V)*	2.87	1.36	1.07	0.54
Bond enthalpy (kJ/mol)*	159	243	194	152

*These values and descriptions apply to the diatomic species X_2, where X represents a halogen atom. The half-reaction is $X_2(g) + 2e^- \longrightarrow 2X^-(aq)$.

†Refers to the anion X^-.

4. Fluorine reacts with cold sodium hydroxide solution to produce oxygen difluoride as follows:

$$2F_2(g) + 2NaOH(aq) \longrightarrow 2NaF(aq) + H_2O(l) + OF_2(g)$$

The same reaction with chlorine or bromine, on the other hand, produces a halide and a hypohalite:

$$X_2(g) + 2NaOH(aq) \longrightarrow NaX(aq) + NaXO(aq) + H_2O(l)$$

where X stands for Cl or Br. Iodine does not react under the same conditions.

5. Silver fluoride, AgF, is soluble. All other silver halides (AgCl, AgBr, and AgI) are insoluble (see Table 4.2).

The element astatine also belongs to the Group 17 family. However, all isotopes of astatine are radioactive; its longest-lived isotope is astatine-210, which has a half-life of 8.3 h. Therefore, it is both difficult and expensive to study astatine in the laboratory.

The halogens form a very large number of compounds. In the elemental state they form diatomic molecules, X_2. In nature, however, because of their high reactivity, halogens are always found combined with other elements. Chlorine, bromine, and iodine occur as halides in seawater, and fluorine occurs in the minerals fluorite (CaF_2) (see Figure 21.16) and cryolite (Na_3AlF_6).

Preparation and General Properties of the Halogens

Because fluorine and chlorine are strong oxidizing agents, they must be prepared by electrolysis rather than by chemical oxidation of the fluoride and chloride ions. Electrolysis does not work for aqueous solutions of fluorides, however, because fluorine is a stronger oxidizing agent than oxygen. From Table 18.1 we find that

$$F_2(g) + 2e^- \longrightarrow 2F^-(aq) \qquad E° = 2.87\text{ V}$$
$$O_2(g) + 4H^+(aq) + 4e^- \longrightarrow 2H_2O(l) \qquad E° = 1.23\text{ V}$$

If F_2 were formed by the electrolysis of an aqueous fluoride solution, it would immediately oxidize water to oxygen. For this reason, fluorine is prepared by electrolyzing

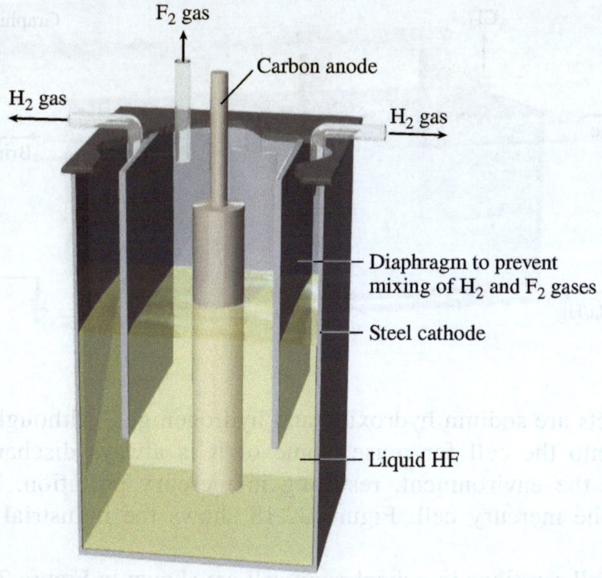

Figure 22.16 *Electrolytic cell for the preparation of fluorine gas. Note that because H_2 and F_2 form an explosive mixture, these gases must be separated from each other.*

liquid hydrogen fluoride containing potassium fluoride to increase its conductivity, at about 70°C (Figure 22.16):

$$\begin{array}{rl} \text{Anode(oxidation):} & 2F^- \longrightarrow F_2(g) + 2e^- \\ \text{Cathode(reduction):} & 2H^+ + 2e^- \longrightarrow H_2(g) \\ \hline \text{Overall reaction:} & 2HF(l) \longrightarrow H_2(g) + F_2(g) \end{array}$$

Chlorine gas, Cl_2, is prepared industrially by the electrolysis of molten NaCl (see Section 18.8) or by the ***chlor-alkali process,*** *the electrolysis of a concentrated aqueous NaCl solution* (called brine). (*Chlor* denotes chlorine and *alkali* denotes an alkali metal, such as sodium.) Two of the common cells used in the chlor-alkali process are the mercury cell and the diaphragm cell. In both cells the overall reaction is

$$2NaCl(aq) + 2H_2O(l) \xrightarrow{\text{electrolysis}} 2NaOH(aq) + H_2(g) + Cl_2(g)$$

As you can see, this reaction yields two useful by-products, NaOH and H_2. The cells are designed to separate the molecular chlorine from the sodium hydroxide solution and the molecular hydrogen to prevent side reactions such as

$$2NaOH(aq) + Cl_2(g) \longrightarrow NaOCl(aq) + NaCl(aq) + H_2O(l)$$
$$H_2(g) + Cl_2(g) \longrightarrow 2HCl(g)$$

These reactions consume the desired products and can be dangerous because a mixture of H_2 and Cl_2 is explosive.

Figure 22.17 shows the mercury cell used in the chlor-alkali process. The cathode is a liquid mercury pool at the bottom of the cell, and the anode is made of either graphite or titanium coated with platinum. Brine is continuously passed through the cell as shown in the diagram. The electrode reactions are

$$\begin{array}{rl} \text{Anode(oxidation):} & 2Cl^-(aq) \longrightarrow Cl_2(g) + 2e^- \\ \text{Cathode(reduction):} & 2Na^+(aq) + 2e^- \xrightarrow{Hg(l)} 2Na/Hg \\ \hline \text{Overall reaction:} & 2NaCl(aq) \longrightarrow 2Na/Hg + Cl_2(g) \end{array}$$

where Na/Hg denotes the formation of sodium amalgam. The chlorine gas generated this way is very pure. The sodium amalgam does not react with the brine solution but decomposes as follows when treated with pure water outside the cell:

$$2Na/Hg + 2H_2O(l) \longrightarrow 2NaOH(aq) + H_2(g) + 2Hg(l)$$

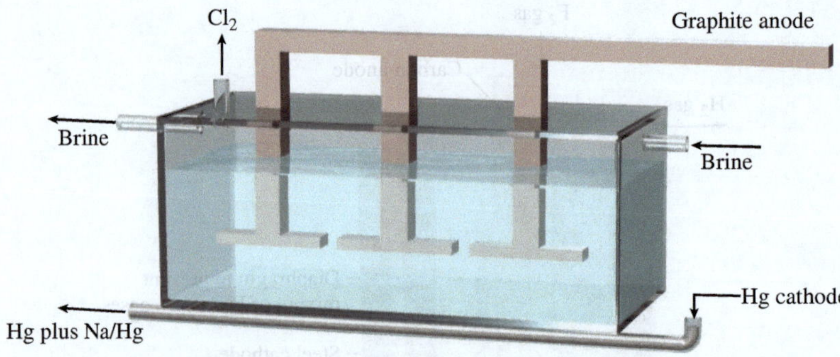

Figure 22.17 *Mercury cell used in the chlor-alkali process. The cathode contains mercury. The sodium-mercury amalgam is treated with water outside the cell to produce sodium hydroxide and hydrogen gas.*

The by-products are sodium hydroxide and hydrogen gas. Although the mercury is cycled back into the cell for reuse, some of it is always discharged with waste solutions into the environment, resulting in mercury pollution. This is a major drawback of the mercury cell. Figure 22.18 shows the industrial manufacture of chlorine gas.

The half-cell reactions in a diaphragm cell are shown in Figure 22.19. The asbestos diaphragm is permeable to the ions but not to the hydrogen and chlorine gases and so prevents the gases from mixing. During electrolysis a positive pressure is applied on the anode side of the compartment to prevent the migration of the OH^- ions from the cathode compartment. Periodically, fresh brine solution is added to the cell and the sodium hydroxide solution is run off as shown. The diaphragm cell presents no pollution problems. Its main disadvantage is that the sodium hydroxide solution is contaminated with unreacted sodium chloride.

The preparation of molecular bromine and iodine from seawater by oxidation with chlorine was discussed in Section 4.4. In the laboratory, chlorine, bromine, and iodine can be prepared by heating the alkali halides (NaCl, KBr, or KI) in concentrated sulfuric acid in the presence of manganese(IV) oxide. A representative reaction is

$$MnO_2(s) + 2H_2SO_4(aq) + 2NaCl(aq) \longrightarrow MnSO_4(aq) + Na_2SO_4(aq) + 2H_2O(l) + Cl_2(g)$$

Compounds of the Halogens

Most of the halides can be classified into two categories. The fluorides and chlorides of many metallic elements, especially those belonging to the alkali metal and alkaline earth metal (except beryllium) families, are ionic compounds. Most of the halides of nonmetals such as sulfur and phosphorus are covalent compounds. As Figure 4.11 shows, the oxidation numbers of the halogens can vary from −1 to +7. The only exception is fluorine. Because it is the most electronegative element, fluorine can have only two oxidation numbers, 0 (as in F_2) and −1, in its compounds.

Figure 22.18 *The industrial manufacture of chlorine gas.*
Liang-Zheng Hong/iStock/Getty Images

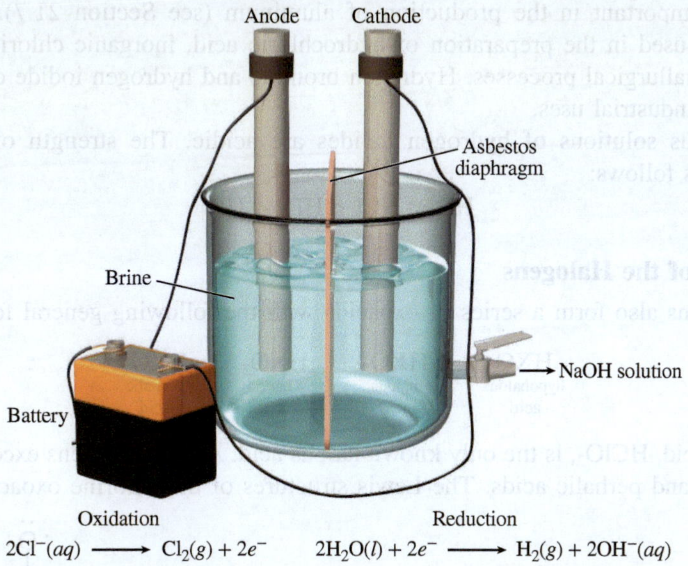

Figure 22.19 *Diaphragm cell used in the chlor-alkali process.*

The Hydrogen Halides

The hydrogen halides, an important class of halogen compounds, can be formed by the direct combination of the elements:

$$H_2(g) + X_2(g) \rightleftharpoons 2HX(g)$$

where X denotes a halogen atom. These reactions (especially the ones involving F_2 and Cl_2) can occur with explosive violence. Industrially, hydrogen chloride is produced as a by-product in the manufacture of chlorinated hydrocarbons:

$$C_2H_6(g) + Cl_2(g) \longrightarrow C_2H_5Cl(g) + HCl(g)$$

In the laboratory, hydrogen fluoride and hydrogen chloride can be prepared by reacting the metal halides with concentrated sulfuric acid:

$$CaF_2(s) + H_2SO_4(aq) \longrightarrow 2HF(g) + CaSO_4(s)$$
$$2NaCl(s) + H_2SO_4(aq) \longrightarrow 2HCl(g) + Na_2SO_4(aq)$$

Hydrogen bromide and hydrogen iodide cannot be prepared this way because they are oxidized to elemental bromine and iodine. For example, the reaction between NaBr and H_2SO_4 is

$$2NaBr(s) + 2H_2SO_4(aq) \longrightarrow Br_2(l) + SO_2(g) + Na_2SO_4(aq) + 2H_2O(l)$$

Instead, hydrogen bromide is prepared by first reacting bromine with phosphorus to form phosphorus tribromide:

$$P_4(s) + 6Br_2(l) \longrightarrow 4PBr_3(l)$$

Next, PBr_3 is treated with water to yield HBr:

$$PBr_3(l) + 3H_2O(l) \longrightarrow 3HBr(g) + H_3PO_3(aq)$$

Hydrogen iodide can be prepared in a similar manner.

The high reactivity of HF is demonstrated by the fact that it attacks silica and silicates:

$$6HF(aq) + SiO_2(s) \longrightarrow H_2SiF_6(aq) + 2H_2O(l)$$

This property makes HF suitable for etching glass and is the reason that hydrogen fluoride must be kept in plastic or inert metal (for example, Pt) containers. Hydrogen fluoride is used in the manufacture of Freons (see Chapter 20); for example,

$$CCl_4(l) + HF(g) \longrightarrow CFCl_3(g) + HCl(g)$$
$$CFCl_3(g) + HF(g) \longrightarrow CF_2Cl_2(g) + HCl(g)$$

It is also important in the production of aluminum (see Section 21.7). Hydrogen chloride is used in the preparation of hydrochloric acid, inorganic chlorides, and in various metallurgical processes. Hydrogen bromide and hydrogen iodide do not have any major industrial uses.

Aqueous solutions of hydrogen halides are acidic. The strength of the acids increases as follows:

$$HF \ll HCl < HBr < HI$$

Oxoacids of the Halogens

The halogens also form a series of oxoacids with the following general formulas:

HXO	HXO_2	HXO_3	HXO_4
hypohalous acid	halous acid	halic acid	perhalic acid

Chlorous acid, $HClO_2$, is the only known halous acid. All the halogens except fluorine form halic and perhalic acids. The Lewis structures of the chlorine oxoacids are

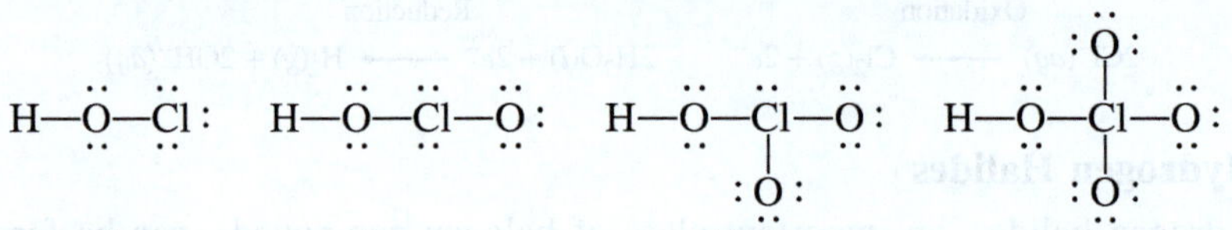

hypochlorous acid   chlorous acid   chloric acid   perchloric acid

For a given halogen, the acid strength decreases from perhalic acid to hypohalous acid; the explanation of this trend is discussed in Section 15.9.

Table 22.5 lists some of the halogen compounds. Periodic acid, HIO_4, does not appear because this compound cannot be isolated in the pure form. Instead the formula H_5IO_6 is often used to represent periodic acid.

Uses of the Halogens

Fluorine

Applications of the halogens and their compounds are widespread in industry, health care, and other areas. One is fluoridation, the practice of adding small quantities of fluorides (about 1 ppm by mass) such as NaF to drinking water to reduce dental caries.

One of the most important inorganic fluorides is uranium hexafluoride, UF_6, which is essential to the gaseous diffusion process for separating isotopes of uranium (U-235 and U-238). Industrially, fluorine is used to produce polytetrafluoroethylene, a polymer better known as Teflon:

$$-\!(CF_2-CF_2)_n\!-$$

where *n* is a large number. Teflon is used in electrical insulators, high-temperature plastics, cooking utensils, and so on.

Table 22.5 Common Compounds of Halogens*

Compound	F	Cl	Br	I
Hydrogen halide	HF (−1)	HCl (−1)	HBr (−1)	HI (−1)
Oxides	OF_2 (−1)	Cl_2O (+1)	Br_2O (+1)	I_2O_5 (+5)
		ClO_2 (+4)	BrO_2 (+4)	
		Cl_2O_7 (+7)		
Oxoacids	HFO (−1)	$HClO$ (+1)	$HBrO$ (+1)	HIO (+1)
		$HClO_2$ (+3)		
		$HClO_3$ (+5)	$HBrO_3$ (+5)	HIO_3 (+5)
		$HClO_4$ (+7)	$HBrO_4$ (+7)	H_5IO_6 (+7)

*The number in parentheses indicates the oxidation number of the halogen.

Chlorine

Chlorine plays an important biological role in the human body, where the chloride ion is the principal anion in intracellular and extracellular fluids. Chlorine is widely used as an industrial bleaching agent for paper and textiles. Ordinary household laundry bleach contains the active ingredient sodium hypochlorite (about 5 percent by mass), which is prepared by reacting chlorine gas with a cold solution of sodium hydroxide:

$$Cl_2(g) + 2NaOH(aq) \longrightarrow NaCl(aq) + NaClO(aq) + H_2O(l)$$

Chlorine is also used to purify water and disinfect swimming pools. When chlorine dissolves in water, it undergoes the following reaction:

$$Cl_2(g) + H_2O(l) \longrightarrow HCl(aq) + HClO(aq)$$

It is thought that the ClO^- ions destroy bacteria by oxidizing life-sustaining compounds within them.

Chlorinated methanes, such as carbon tetrachloride and chloroform, are useful organic solvents. Large quantities of chlorine are used to produce insecticides, such as DDT. However, in view of the damage they inflict on the environment, the use of many of these compounds is either totally banned or greatly restricted in the United States. Chlorine is also used to produce polymers such as poly(vinyl chloride).

Bromine

So far as we know, bromine compounds occur naturally only in some marine organisms. Seawater is about 1×10^{-3} *M* Br^-; therefore, it is the main source of bromine. Bromine is used to prepare ethylene dibromide ($BrCH_2CH_2Br$), which is used as an insecticide and as a scavenger for lead (that is, to combine with lead) in gasoline to keep lead deposits from clogging engines. Studies have shown that ethylene dibromide is a very potent carcinogen.

Bromine combines directly with silver to form silver bromide (AgBr), which is used in photographic films.

Iodine

Iodine is not used as widely as the other halogens. A 50 percent (by mass) alcohol solution of iodine, known as *tincture of iodine,* is used medicinally as an antiseptic. Iodine is an essential constituent of the thyroid hormone thyroxine:

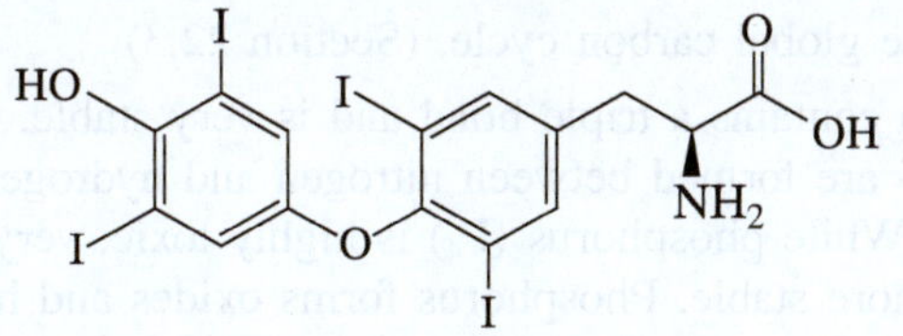

Iodine deficiency in the diet may result in enlargement of the thyroid gland (known as goiter). Iodized table salt sold in the United States usually contains 0.01 percent KI or NaI, which is more than sufficient to satisfy the 1 mg of iodine per week required for the formation of thyroxine in the human body.

A compound of iodine that deserves mention is silver iodide, AgI. It is a pale yellow solid that darkens when exposed to light. In this respect it is similar to silver bromide. Silver iodide is sometimes used in cloud seeding, a process for inducing rainfall on a small scale (Figure 22.20). The advantage of using silver iodide is that enormous numbers of nuclei (that is, small particles of silver iodide on which ice crystals can form) become available. About 10^{15} nuclei are produced from 1 g of AgI by vaporizing an acetone solution of silver iodide in a hot flame. The nuclei are then dispersed into the clouds from an airplane.

Figure 22.20 *Cloud seeding using AgI particles.*
Inga Spence/Alamy Stock Photo

Summary of Concepts & Facts

- The halogens are toxic and reactive elements that are found only in compounds with other elements. Fluorine and chlorine are strong oxidizing agents and are prepared by electrolysis.
- The reactivity, toxicity, and oxidizing ability of the halogens decrease from fluorine to iodine. The halogens all form binary acids **(HX)** and a series of oxoacids.

Chapter Summary

Nonmetals Properties of nonmetals vary more than properties of metals. Nonmetal elements may be solid, liquid, or gaseous and may exhibit both positive and negative oxidation numbers in their compounds. (Section 22.1)

Hydrogen Atoms of hydrogen contain one proton and one electron. They are the simplest atoms. Hydrogen combines with many metals and nonmetals to form hydrides; some hydrides are ionic and some are covalent. There are three isotopes of hydrogen: $^{1}_{1}H$, $^{2}_{1}H$ (deuterium), and $^{3}_{1}H$ (tritium). "Heavy water" (D_2O) contains deuterium. (Section 22.2)

Carbon The important inorganic compounds of carbon are the carbides; the cyanides, most of which are extremely toxic; carbon monoxide, also toxic and a major air pollutant; the carbonates and bicarbonates; and carbon dioxide, an end product of metabolism and a component of the global carbon cycle. (Section 22.3)

Nitrogen and Phosphorus Elemental nitrogen (N_2) contains a triple bond and is very stable. Compounds in which nitrogen has oxidation numbers from −3 to +5 are formed between nitrogen and hydrogen and/or oxygen atoms. Ammonia (NH_3) is widely used in fertilizers. White phosphorus (P_4) is highly toxic, very reactive, and flammable; the polymeric red phosphorus $[(P_4)_n]$ is more stable. Phosphorus forms oxides and halides with oxidation numbers of +3 and +5 and several oxoacids. The phosphates are the most important phosphorus compounds. (Section 22.4)

Oxygen and Sulfur Elemental oxygen (O_2) is paramagnetic and contains two unpaired electrons. Oxygen forms ozone (O_3), oxides (O^{2-}), peroxides (O_2^{2-}), and superoxides (O_2^-). The most abundant element in Earth's crust, oxygen is essential for life on Earth. Sulfur is obtained from Earth's crust as a molten liquid via the Frasch process. Sulfur exists in a number of allotropic forms and has a variety of oxidation numbers in its compounds. Sulfuric acid is the cornerstone of the chemical industry. It is produced from sulfur via sulfur dioxide and sulfur trioxide by means of the contact process. (Section 22.5)

The Halogens The halogens are toxic and reactive elements that are found only in compounds with other elements. Fluorine and chlorine are strong oxidizing agents and are prepared by electrolysis. With the exception of fluorine, the halogens may have both negative and positive oxidation states. The oxidation state of fluorine in its compounds is always −1. (Section 22.6)

Key Words

Carbide, p. 971
Catenation, p. 971
Chlor-alkali process, p. 991
Cyanide, p. 972
Hydrogenation, p. 968

Questions & Problems

Red numbered problems solved in Student Solutions Manual

22.1 General Properties of Nonmetals

Review Questions

22.1 Without referring to Figure 22.1, state whether each of the following elements are metals, metalloids, or nonmetals: (a) Cs, (b) Ge, (c) I, (d) Kr, (e) W, (f) Ga, (g) Te, (h) Bi.

22.2 List two chemical and two physical properties that distinguish a metal from a nonmetal.

22.3 Make a list of physical and chemical properties of chlorine (Cl_2) and magnesium. Comment on their differences with reference to the fact that one is a metal and the other is a nonmetal.

22.4 Carbon is usually classified as a nonmetal. However, the graphite used in "lead" pencils conducts electricity. Look at a pencil and list two nonmetallic properties of graphite.

22.2 Hydrogen

Review Questions

22.5 Explain why hydrogen has a unique position in the periodic table.

22.6 Describe two laboratory and two industrial preparations for hydrogen.

22.7 Hydrogen exhibits three types of bonding in its compounds. Describe each type of bonding with an example.

22.8 What are interstitial hydrides?

22.9 Give the name of (a) an ionic hydride and (b) a covalent hydride. In each case describe the preparation and give the structure of the compound.

22.10 Describe what is meant by the "hydrogen economy."

Problems

22.11 Elements number 17 and 20 form compounds with hydrogen. Write the formulas for these two compounds and compare their chemical behavior in water.

22.12 Give an example of hydrogen as (a) an oxidizing agent and (b) a reducing agent.

22.13 Compare the physical and chemical properties of the hydrides of each of the following elements: Na, Ca, C, N, O, Cl.

22.14 Suggest a physical method that would enable you to separate hydrogen gas from neon gas.

22.15 Write a balanced equation to show the reaction between CaH_2 and H_2O. How many grams of CaH_2 are needed to produce 26.4 L of H_2 gas at 20°C and 746 mmHg?

22.16 How many kilograms of water must be processed to obtain 2.0 L of D_2 at 25°C and 0.90 atm pressure? Assume that deuterium abundance is 0.015 percent and that recovery is 80 percent.

22.17 Predict the outcome of the following reactions:

(a) $CuO(s) + H_2(g) \longrightarrow$

(b) $Na_2O(s) + H_2(g) \longrightarrow$

22.18 Starting with H_2, describe how you would prepare (a) HCl, (b) NH_3, (c) LiOH.

22.3 Carbon

Review Questions

22.19 Give an example of a carbide and a cyanide.

22.20 How are cyanide ions used in metallurgy?

22.21 Briefly discuss the preparation and properties of carbon monoxide and carbon dioxide.

22.22 What is coal?

22.23 Explain what is meant by coal gasification.

22.24 Describe two chemical differences between CO and CO_2.

Problems

22.25 Describe the reaction between CO_2 and OH^- in terms of a Lewis acid-base reaction such as that shown in Section 15.12.

22.26 Draw a Lewis structure for the C_2^{2-} ion.

22.27 Balance the following equations:

(a) $Be_2C(s) + H_2O(l) \longrightarrow$

(b) $CaC_2(s) + H_2O(l) \longrightarrow$

22.28 Unlike $CaCO_3$, Na_2CO_3 does not readily yield CO_2 when heated. On the other hand, $NaHCO_3$ undergoes thermal decomposition to produce CO_2 and Na_2CO_3. (a) Write a balanced equation for the reaction. (b) How would you test for the CO_2 evolved? [*Hint:* Treat the gas with limewater, an aqueous solution of $Ca(OH)_2$.]

22.29 Two solutions are labeled A and B. Solution A contains Na_2CO_3 and solution B contains $NaHCO_3$. Describe how you would distinguish between the two solutions if you were provided with a $MgCl_2$ solution. (*Hint:* You need to know the solubilities of $MgCO_3$ and $MgHCO_3$.)

22.30 Magnesium chloride is dissolved in a solution containing sodium bicarbonate. On heating, a white precipitate is formed. Explain what causes the precipitation.

22.31 A few drops of concentrated ammonia solution added to a calcium bicarbonate solution cause a white precipitate to form. Write a balanced equation for the reaction.

22.32 Sodium hydroxide is hygroscopic—that is, it absorbs moisture when exposed to the atmosphere. A student placed a pellet of NaOH on a watch glass. A few days later, she noticed that the pellet was covered with a white solid. What is the identity of this solid? (*Hint:* Air contains CO_2.)

22.33 A piece of red-hot magnesium ribbon will continue to burn in an atmosphere of CO_2 even though CO_2 does not support combustion. Explain.

22.34 Is carbon monoxide isoelectronic with nitrogen (N_2)?

22.4 Nitrogen and Phosphorus

Review Questions

22.35 Describe a laboratory and an industrial preparation of nitrogen gas.

22.36 What is meant by nitrogen fixation? Describe a process for fixation of nitrogen on an industrial scale.

22.37 Describe an industrial preparation of phosphorus.

22.38 Why is the P_4 molecule unstable?

Problems

22.39 Nitrogen can be obtained by (a) passing ammonia over red-hot copper(II) oxide and (b) heating ammonium dichromate [one of the products is Cr(III) oxide]. Write a balanced equation for each preparation.

22.40 Write balanced equations for the preparation of sodium nitrite by heating (a) sodium nitrate, (b) sodium nitrate with carbon.

22.41 Sodium amide ($NaNH_2$) reacts with water to produce sodium hydroxide and ammonia. Describe this reaction as a Brønsted acid-base reaction.

22.42 Write a balanced equation for the formation of urea, $(NH_2)_2CO$, from carbon dioxide and ammonia. Should the reaction be run at high or low pressure to maximize the yield?

22.43 Some farmers believe that lightning helps produce a better crop. What is the scientific basis for this belief?

22.44 At 620 K the vapor density of ammonium chloride relative to hydrogen (H_2) under the same conditions of temperature and pressure is 14.5, although, according to its formula mass, it should have a vapor density of 26.8. How would you account for this discrepancy?

22.45 Explain, giving one example in each case, why nitrous acid can act as both a reducing agent and an oxidizing agent.

22.46 Explain why nitric acid can be reduced but not oxidized.

22.47 Write a balanced equation for each of the following processes: (a) On heating, ammonium nitrate produces nitrous oxide. (b) On heating, potassium nitrate produces potassium nitrite and oxygen gas. (c) On heating, lead nitrate produces lead(II) oxide, nitrogen dioxide (NO_2), and oxygen gas.

22.48 Explain why, under normal conditions, the reaction of zinc with nitric acid does not produce hydrogen.

22.49 Potassium nitrite can be produced by heating a mixture of potassium nitrate and carbon. Write a balanced equation for this reaction. Calculate the theoretical yield of KNO_2 produced by heating 57.0 g of KNO_3 with an excess of carbon.

22.50 Predict the geometry of nitrous oxide, N_2O, by the VSEPR method and draw resonance structures for the molecule. (*Hint:* The atoms are arranged as NNO.)

22.51 Consider the reaction

$$N_2(g) + O_2(g) \rightleftharpoons 2NO(g)$$

Given that the $\Delta G°$ for the reaction at 298 K is 175.2 kJ/mol, calculate (a) the standard Gibbs energy of formation of NO, (b) K_P for the reaction, (c) K_c for the reaction.

22.52 From the data in Appendix 2, calculate $\Delta H°$ for the synthesis of NO (which is the first step in the manufacture of nitric acid) at 25°C:

$$4NH_3(g) + 5O_2(g) \longrightarrow 4NO(g) + 6H_2O(l)$$

22.53 Explain why two N atoms can form a double bond or a triple bond, whereas two P atoms normally can form only a single bond.

22.54 When 1.645 g of white phosphorus are dissolved in 75.5 g of CS_2, the solution boils at 46.709°C, whereas pure CS_2 boils at 46.300°C. The molal boiling-point elevation constant for CS_2 is 2.34°C/*m*. Calculate the molar mass of white phosphorus and give the molecular formula.

22.55 Starting with elemental phosphorus, P_4, show how you would prepare phosphoric acid.

22.56 Dinitrogen pentoxide is a product of the reaction between P_4O_{10} and HNO_3. Write a balanced equation for this reaction. Calculate the theoretical yield of N_2O_5 if 79.4 g of P_4O_{10} are reacted with an excess of HNO_3. (*Hint:* One of the products is HPO_3.)

22.57 Explain why (a) NH_3 is more basic than PH_3, (b) NH_3 has a higher boiling point than PH_3, (c) PCl_5 exists but NCl_5 does not, (d) N_2 is more inert than P_4.

22.58 What is the hybridization of phosphorus in the phosphonium ion, PH_4^+?

22.5 Oxygen and Sulfur

Review Questions

22.59 Describe one industrial and one laboratory preparation of O_2.

22.60 Give an account of the various kinds of oxides that exist and illustrate each type by two examples.

22.61 Hydrogen peroxide can be prepared by treating barium peroxide with sulfuric acid. Write a balanced equation for this reaction.

22.62 Describe the Frasch process for obtaining sulfur.

22.63 Describe the contact process for the production of sulfuric acid.

22.64 How is hydrogen sulfide generated in the laboratory?

Problems

22.65 Draw molecular orbital energy level diagrams for O_2, O_2^-, and O_2^{2-}.

22.66 One of the steps involved in the depletion of ozone in the stratosphere by nitric oxide may be represented as

$$NO(g) + O_3(g) \longrightarrow NO_2(g) + O_2(g)$$

From the data in Appendix 2 calculate $\Delta G°$, K_P, and K_c for the reaction at 25°C.

22.67 Hydrogen peroxide is unstable and decomposes readily:

$$2H_2O_2(aq) \longrightarrow 2H_2O(l) + O_2(g)$$

This reaction is accelerated by light, heat, or a catalyst. (a) Explain why hydrogen peroxide sold in drugstores comes in dark bottles. (b) The concentrations of aqueous hydrogen peroxide solutions are normally expressed as percent by mass. In the decomposition of hydrogen peroxide, how many liters of oxygen gas can be produced at STP from 15.0 g of a 7.50 percent hydrogen peroxide solution?

22.68 What are the oxidation numbers of O and F in HFO?

22.69 Oxygen forms double bonds in O_2, but sulfur forms single bonds in S_8. Explain.

22.70 In 2008, about 48 million tons of sulfuric acid were produced in the United States. Calculate the amount of sulfur (in grams and moles) used to produce that amount of sulfuric acid.

22.71 Sulfuric acid is a dehydrating agent. Write balanced equations for the reactions between sulfuric acid and the following substances: (a) HCOOH, (b) H_3PO_4, (c) HNO_3, (d) $HClO_3$. (*Hint:* Sulfuric acid is not decomposed by the dehydrating action.)

22.72 Calculate the amount of $CaCO_3$ (in grams) that would be required to react with 50.6 g of SO_2 emitted by a power plant.

22.73 SF_6 exists but OF_6 does not. Explain.

22.74 Explain why SCl_6, SBr_6, and SI_6 cannot be prepared.

22.75 Compare the physical and chemical properties of H_2O and H_2S.

22.76 The bad smell of water containing hydrogen sulfide can be removed by the action of chlorine. The reaction is

$$H_2S(aq) + Cl_2(aq) \longrightarrow 2HCl(aq) + S(s)$$

If the hydrogen sulfide content of contaminated water is 22 ppm by mass, calculate the amount of Cl_2 (in grams) required to remove all the H_2S from 2.0×10^2 gallons of water. (1 gallon = 3.785 L.)

22.77 Describe two reactions in which sulfuric acid acts as an oxidizing agent.

22.78 Concentrated sulfuric acid reacts with sodium iodide to produce molecular iodine, hydrogen sulfide, and sodium hydrogen sulfate. Write a balanced equation for the reaction.

22.6 The Halogens

Review Questions

22.79 Describe an industrial method for preparing each of the halogens.

22.80 Name the major uses of the halogens.

Problems

22.81 Metal chlorides can be prepared in a number of ways: (a) direct combination of metal and molecular chlorine, (b) reaction between metal and hydrochloric acid, (c) acid-base neutralization, (d) metal carbonate treated with hydrochloric acid, (e) precipitation reaction. Give an example for each type of preparation.

22.82 Sulfuric acid is a weaker acid than hydrochloric acid. Yet hydrogen chloride is evolved when concentrated sulfuric acid is added to sodium chloride. Explain.

22.83 Show that chlorine, bromine, and iodine are very much alike by giving an account of their behavior (a) with hydrogen, (b) in producing silver salts, (c) as oxidizing agents, and (d) with sodium hydroxide. (e) In what respects is fluorine not a typical halogen element?

22.84 A 375-gallon tank is filled with water containing 167 g of bromine in the form of Br^- ions. How many liters of Cl_2 gas at 1.00 atm and 20°C will be required to oxidize all the bromide to molecular bromine?

22.85 Draw structures for (a) $(HF)_2$, (b) HF_2^-.

22.86 Hydrogen fluoride can be prepared by the action of sulfuric acid on sodium fluoride. Explain why hydrogen bromide cannot be prepared by the action of the same acid on sodium bromide.

22.87 Aqueous copper(II) sulfate solution is blue. When aqueous potassium fluoride is added to the $CuSO_4$ solution, a green precipitate is formed. If aqueous potassium chloride is added instead, a bright green solution is formed. Explain what happens in each case.

22.88 What volume of bromine (Br_2) vapor measured at 100°C and 700 mmHg pressure would be obtained if 2.00 L of dry chlorine (Cl_2), measured at 15°C and 760 mmHg, were absorbed by a potassium bromide solution?

22.89 Use the VSEPR method to predict the geometries of the following species: (a) I_3^-, (b) $SiCl_4$, (c) PF_5, (d) SF_4.

22.90 Iodine pentoxide, I_2O_5, is sometimes used to remove carbon monoxide from the air by forming carbon dioxide and iodine. Write a balanced equation for this reaction and identify species that are oxidized and reduced.

Additional Problems

22.91 Write a balanced equation for each of the following reactions: (a) Heating phosphorous acid yields phosphoric acid and phosphine (PH_3). (b) Lithium carbide reacts with hydrochloric acid to give lithium chloride and methane. (c) Bubbling HI gas through an aqueous solution of HNO_2 yields molecular iodine and nitric oxide. (d) Hydrogen sulfide is oxidized by chlorine to give HCl and SCl_2.

22.92 (a) Which of the following compounds has the greatest ionic character: PCl_5, $SiCl_4$, CCl_4, BCl_3? (b) Which of the following ions has the smallest ionic radius: F^-, C^{4-}, N^{3-}, O^{2-}? (c) Which of the following atoms has the highest ionization energy: F, Cl, Br, I? (d) Which of the following oxides is most acidic: H_2O, SiO_2, CO_2?

22.93 Both N_2O and O_2 support combustion. Suggest one physical and one chemical test to distinguish between the two gases.

22.94 What is the change in oxidation number for the following reaction?

$$3O_2 \longrightarrow 2O_3$$

22.95 Describe the bonding in the C_2^{2-} ion in terms of the molecular orbital theory.

22.96 Starting with deuterium oxide (D_2O), describe how you would prepare (a) NaOD, (b) DCl, (c) ND_3, (d) C_2D_2, (e) CD_4, (f) D_2SO_4.

22.97 Solid PCl_5 exists as $[PCl_4^+][PCl_6^-]$. Draw Lewis structures for these ions. Describe the hybridization state of the P atoms.

22.98 Consider the Frasch process. (a) How is it possible to heat water well above 100°C without turning it into steam? (b) Why is water sent down the outermost pipe? (c) Why would excavating a mine and digging for sulfur be a dangerous procedure for obtaining the element?

22.99 Predict the physical and chemical properties of astatine, a radioactive element and the last member of Group 17.

22.100 Lubricants used in watches usually consist of long-chain hydrocarbons. Oxidation by air forms solid polymers that eventually destroy the effectiveness of the lubricants. It is believed that one of the initial steps in the oxidation is removal of a hydrogen atom (hydrogen abstraction). By replacing the hydrogen atoms at reactive sites with deuterium atoms, it is possible to substantially slow the overall oxidation rate. Why? (*Hint:* Consider the kinetic isotope effect.)

22.101 How are lightbulbs frosted? (*Hint:* Consider the action of hydrofluoric acid on glass, which is made of silicon dioxide.)

22.102 Life evolves to adapt to its environment. In this respect, explain why life most frequently needs oxygen for survival, rather than the more abundant nitrogen.

22.103 As mentioned in Chapter 3, ammonium nitrate is the most important nitrogen-containing fertilizer in the world. Given only air and water as starting materials and any equipment and catalyst at your disposal, describe how you would prepare ammonium nitrate. State conditions under which you can increase the yield in each step.

22.104 As we saw in Section 21.2, the reduction of iron oxides is accomplished by using carbon monoxide as a reducing agent. Starting with coke in a blast furnace, the following equilibrium plays a key role in the extraction of iron:

$$C(s) + CO_2(g) \rightleftharpoons 2CO(g)$$

Use the data in Appendix 2 to calculate the equilibrium constant at 25°C and 1000°C. Assume $\Delta H°$ and $\Delta S°$ to be independent of temperature.

22.105 Assuming ideal behavior, calculate the density of gaseous HF at its normal boiling point (19.5°C). The experimentally measured density under the same conditions is 3.10 g/L. Account for the discrepancy between your calculated value and the experimental result.

22.106 A 10.0-g sample of white phosphorus was burned in an excess of oxygen. The product was dissolved in enough water to make 500 mL of solution. Calculate the pH of the solution at 25°C.

CHAPTER

23 Coordination Chemistry

Precious gemstones have various colors due to the various transition metal ions in their chemical structures. The presence of transition metal ions leads to the absorbance of different wavelengths of light.
Akkaradet Bangchun/Shutterstock

CHAPTER OUTLINE

The series of elements in the periodic table in which the *d* and *f* subshells are gradually filled are called the transition elements. There are about 50 transition elements, and they have widely varying and fascinating properties. To present even one interesting feature of each transition element is beyond the scope of this book. We will therefore limit our discussion to the transition elements that have incompletely filled *d* subshells and to their most commonly encountered property—the tendency to form complex ions.

23.1 Properties of the Transition Metals

Learning Objective

- Describe some of the properties of transition metals.

Transition metals typically have incompletely filled *d* subshells or readily give rise to ions with incompletely filled *d* subshells (Figure 23.1). (The Group 12 metals—Zn, Cd, and Hg—do not have this characteristic electron configuration and so, although they are sometimes called transition metals, they really do not belong in this category.) This attribute is responsible for several notable properties, including distinctive coloring, formation of paramagnetic compounds, catalytic activity, and especially a great tendency to form complex ions. In this chapter we focus on the first-row elements from scandium to copper, the most common transition metals. Table 23.1 lists some of their properties.

As we read across any period from left to right, atomic numbers increase, electrons are added to the outer shell, and the nuclear charge increases by the addition of protons. In the third-period elements—sodium to argon—the outer electrons weakly shield one another from the extra nuclear charge. Consequently, atomic radii decrease rapidly from sodium to argon, and the electronegativities and ionization energies increase steadily (see Figures 8.5, 8.11, and 9.5).

For the transition metals, the trends are different. Looking at Table 23.1 we see that the nuclear charge, of course, increases from scandium to copper, but electrons are being added to the inner 3*d* subshell. These 3*d* electrons shield the 4*s* electrons from the increasing nuclear charge somewhat more effectively than outer-shell electrons can shield one another, so the atomic radii decrease less rapidly. For the same reason, electronegativities and ionization energies increase only slightly from scandium across to copper compared with the increases from sodium to argon.

1	2	3	4	5	6	7	8	9	10	11	12	13	14	15	16	17	18
1 H																	2 He
3 Li	4 Be											5 B	6 C	7 N	8 O	9 F	10 Ne
11 Na	12 Mg											13 Al	14 Si	15 P	16 S	17 Cl	18 Ar
19 K	20 Ca	21 Sc	22 Ti	23 V	24 Cr	25 Mn	26 Fe	27 Co	28 Ni	29 Cu	30 Zn	31 Ga	32 Ge	33 As	34 Se	35 Br	36 Kr
37 Rb	38 Sr	39 Y	40 Zr	41 Nb	42 Mo	43 Tc	44 Ru	45 Rh	46 Pd	47 Ag	48 Cd	49 In	50 Sn	51 Sb	52 Te	53 I	54 Xe
55 Cs	56 Ba	57 La	72 Hf	73 Ta	74 W	75 Re	76 Os	77 Ir	78 Pt	79 Au	80 Hg	81 Tl	82 Pb	83 Bi	84 Po	85 At	86 Rn
87 Fr	88 Ra	89 Ac	104 Rf	105 Db	106 Sg	107 Bh	108 Hs	109 Mt	110 Ds	111 Rg	112 Cn	113 Nh	114 Fl	115 Mc	116 Lv	117 Ts	118 Og

Figure 23.1 *The transition metals (light blue). Note that although the Group 12 elements (Zn, Cd, Hg) are described as transition metals by some chemists, neither the metals nor their ions possess incompletely filled d subshells.*

Table 23.1 Electron Configurations and Other Properties of the First-Row Transition Metals

	Sc	Ti	V	Cr	Mn	Fe	Co	Ni	Cu
Electron configuration									
M	$4s^23d^1$	$4s^23d^2$	$4s^23d^3$	$4s^13d^5$	$4s^23d^5$	$4s^23d^6$	$4s^23d^7$	$4s^23d^8$	$4s^13d^{10}$
M^{2+}	—	$3d^2$	$3d^3$	$3d^4$	$3d^5$	$3d^6$	$3d^7$	$3d^8$	$3d^9$
M^{3+}	[Ar]	$3d^1$	$3d^2$	$3d^3$	$3d^4$	$3d^5$	$3d^6$	$3d^7$	$3d^8$
Electronegativity	1.4	1.5	1.6	1.7	1.6	1.8	1.9	1.9	1.9
Ionization energy (kJ/mol)									
First	631	658	650	652	717	759	760	736	745
Second	1235	1309	1413	1591	1509	1561	1645	1751	1958
Third	2389	2650	2828	2986	3250	2956	3231	3393	3578
Radius (pm)									
M	162	147	134	130	135	126	125	124	128
M^{2+}	—	90	88	85	91	82	82	78	72
M^{3+}	83	68	74	64	66	67	64	—	—
Standard reduction potential (V)*	−2.08	−1.63	−1.2	−0.74	−1.19	−0.45	−0.28	−0.26	0.34

*The half-reaction is $M^{2+}(aq) + 2e^- \longrightarrow M(s)$ (except for Sc and Cr, where the ions are Sc^{3+} and Cr^{3+}, respectively).

Although the transition metals are less electropositive (or more electronegative) than the alkali and alkaline earth metals, the standard reduction potentials of the first-row transition metals suggest that all of them except copper should react with strong acids such as hydrochloric acid to produce hydrogen gas. However, most transition metals are inert toward acids or react slowly with them because of a protective layer of oxide. A case in point is chromium: Despite a rather negative standard reduction potential, it is quite inert chemically because of the formation on its surfaces of chromium(III) oxide, Cr_2O_3. Consequently, chromium is commonly used as a protective and noncorrosive plating on other metals. On the bumpers and trim of vintage automobiles, chromium plating serves a decorative as well as a functional purpose.

General Physical Properties

Most of the transition metals have a close-packed structure (see Figure 11.30) in which each atom has a coordination number of 12. Furthermore, these elements have relatively small atomic radii. The combined effect of closest packing and small atomic size results in strong metallic bonds. Therefore, transition metals have higher densities, higher melting points and boiling points, and higher heats of fusion and vaporization than the Group 1, 2, and 12 metals (Table 23.2).

Table 23.2 Physical Properties of Elements K to Zn

	1	2	Transition Metals								12
	K	Ca	Sc	Ti	V	Cr	Mn	Fe	Co	Ni	Cu	Zn
Atomic radius (pm)	227	197	162	147	134	130	135	126	125	124	128	138
Melting point (°C)	63.7	838	1539	1668	1900	1875	1245	1536	1495	1453	1083	419.5
Boiling point (°C)	760	1440	2730	3260	3450	2665	2150	3000	2900	2730	2595	906
Density (g/cm^3)	0.86	1.54	3.0	4.51	6.1	7.19	7.43	7.86	8.9	8.9	8.96	7.14

Sc	Ti	V	Cr	Mn	Fe	Co	Ni	Cu
				+7				
			+6	+6	+6			
		+5	+5	+5	+5			
	+4	+4	+4	+4	+4	+4		
+3	+3	+3	+3	+3	+3	+3	+3	+3
	+2	+2	+2	+2	+2	+2	+2	+2
								+1

Figure 23.2 *Oxidation states of the first-row transition metals. The most stable oxidation numbers are shown in color. The zero oxidation state is encountered in some compounds, such as $Ni(CO)_4$ and $Fe(CO)_5$.*

Electron Configurations

The electron configurations of the first-row transition metals were discussed in Section 7.9. Calcium has the electron configuration $[Ar]4s^2$. From scandium across to copper, electrons are added to the $3d$ orbitals. Thus, the outer electron configuration of scandium is $4s^2 3d^1$, that of titanium is $4s^2 3d^2$, and so on. The two exceptions are chromium and copper, whose outer electron configurations are $4s^1 3d^5$ and $4s^1 3d^{10}$, respectively. These irregularities are the result of the extra stability associated with half-filled and completely filled $3d$ subshells.

When the first-row transition metals form cations, electrons are removed first from the $4s$ orbitals and then from the $3d$ orbitals. (This is the opposite of the order in which orbitals are filled in atoms.) For example, the outer electron configuration of Fe^{2+} is $3d^6$, not $4s^2 3d^4$.

Oxidation States

Transition metals exhibit variable oxidation states in their compounds. Figure 23.2 shows the oxidation states from scandium to copper. Note that the common oxidation states for each element include +2, +3, or both. The +3 oxidation states are more stable at the beginning of the series, whereas toward the end the +2 oxidation states are more stable. The reason for this trend can be understood by examining the ionization energy plots in Figure 23.3. In general, the ionization energies increase gradually from left to right. However, the third ionization energy (when an electron is removed from the $3d$ orbital) increases more rapidly than the first and second ionization energies. Because it takes more energy to remove the third electron from the metals near the end of the row than from those near the beginning, the metals near the end tend to form M^{2+} ions rather than M^{3+} ions.

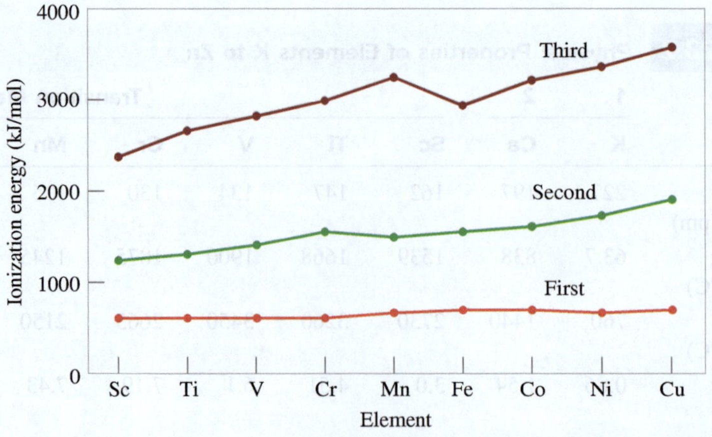

Figure 23.3 *Variation of the first, second, and third ionization energies for the first-row transition metals.*

The highest oxidation state for a transition metal is +7, for manganese ($4s^2 3d^5$). For elements to the right of Mn (Fe to Cu), oxidation numbers are lower. Transition metals usually exhibit their highest oxidation states in compounds with very electronegative elements such as oxygen and fluorine—for example, V_2O_5, CrO_3, and Mn_2O_7. Recall that oxides in which the metal has a high oxidation number are covalent and acidic, whereas those in which the metal has a low oxidation number are ionic and basic (see Section 15.11).

Student Hot Spot

Student data indicate you may struggle with the general trends of transition metals. Access your eBook for additional Learning Resources on this topic.

Summary of Concepts & Facts

- Transition metals usually have incompletely filled *d* subshells and a pronounced tendency to form complexes.
- The first-row transition metals (scandium to copper) are the most common of all the transition metals; their chemistry is characteristic, in many ways, of the entire group.

Review of Concepts & Facts

23.1.1 Identify the following transition metal atoms and ions according to their electron configurations. Atoms: (a) $[Kr]5s^2 4d^5$, (b) $[Xe]6s^2 4f^{14} 5d^4$. Ions: (c) $[Ar]3d^3$ (a +4 ion), (d) $[Xe]4f^{14} 5d^8$ (a +3 ion). (See Table 7.3.)

23.2 Coordination Compounds

Learning Objectives

- Define coordination compound.
- List some common ligands found in transition metal complexes.
- Identify the coordination number of a coordination complex.
- Explain what a chelating agent does.
- Assign names to coordination compounds using nomenclature rules.

Transition metals have a distinct tendency to form complex ions (see Section 16.7). A ***coordination compound*** *typically consists of a complex ion and counter ion.* A complex ion contains a central metal bonded to one or more ions or molecules. Note that some coordination compounds such as $Fe(CO)_5$ do not contain complex ions. Our understanding of the nature of coordination compounds stems from the classic work of Alfred Werner,† who prepared and characterized many coordination compounds. In 1893, at the age of 26, Werner proposed what is now commonly referred to as *Werner's coordination theory*.

Nineteenth-century chemists were puzzled by a certain class of reactions that seemed to violate valence theory. For example, the valences of the elements in cobalt(III) chloride and in ammonia seem to be completely satisfied, and yet these two substances react to form a stable compound having the formula $CoCl_3 \cdot 6NH_3$. To explain this behavior, Werner postulated that most elements exhibit two types of valence: *primary valence* and *secondary valence*. In modern terminology, primary valence corresponds to the oxidation number and secondary valence to the coordination number of the element. In $CoCl_3 \cdot 6NH_3$, according to Werner, cobalt has a primary valence of 3 and a secondary valence of 6.

†Alfred Werner (1866–1919). Swiss chemist. Werner started as an organic chemist but became interested in coordination chemistry. For his theory of coordination compounds, Werner was awarded the Nobel Prize in Chemistry in 1913.

Table 23.3 Some Common Ligands

Name	Structure
	Monodentate ligands
Ammonia	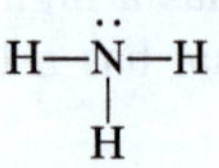
Carbon monoxide	$:C{\equiv}O:$
Chloride ion	$:\ddot{\underset{..}{Cl}}:^-$
Cyanide ion	$[:C{\equiv}N:]^-$
Thiocyanate ion	$[:\ddot{\underset{..}{S}}-C{\equiv}N:]^-$
Water	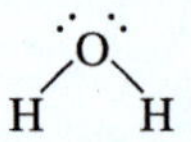
	Bidentate ligands
Ethylenediamine	$H_2\ddot{N}-CH_2-CH_2-\ddot{N}H_2$
Oxalate ion	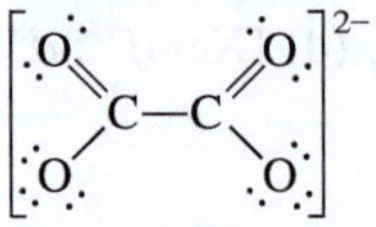
	Polydentate ligand
Ethylenediaminetetraacetate ion (EDTA)	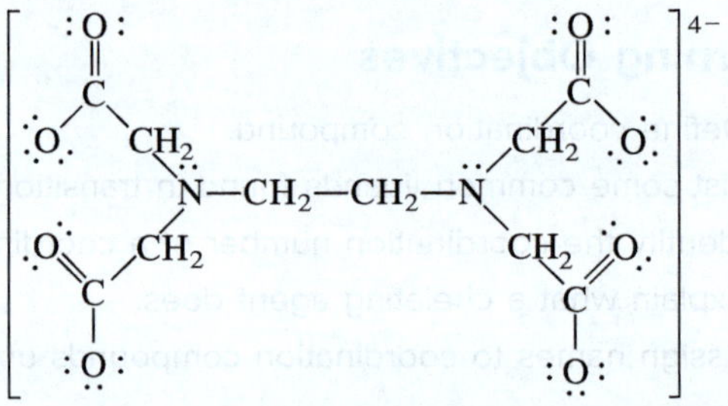

Today we use the formula $[Co(NH_3)_6]Cl_3$ to indicate that the ammonia molecules and the cobalt atom form a complex ion; the chloride ions are not part of the complex but are held to it by ionic forces. Most, but not all, of the metals in coordination compounds are transition metals.

The molecules or ions that surround the metal in a complex ion are called ***ligands*** (Table 23.3). The interactions between a metal atom and the ligands can be thought of as Lewis acid-base reactions. As we saw in Section 15.12, a Lewis base is a substance capable of donating one or more electron pairs. Every ligand has at least one unshared pair of valence electrons; therefore, ligands play the role of Lewis bases. On the other hand, a transition metal atom (in either its neutral or positively charged state) acts as a Lewis acid, accepting (and sharing) pairs of electrons from the Lewis bases. Thus, the metal-ligand bonds are usually coordinate covalent bonds (see Section 9.9).

Student Hot Spot

Student data indicate you may struggle with how ligands bind metals. Access your eBook for additional Learning Resources on this topic.

The atom in a ligand that is bound directly to the metal atom is known as the ***donor atom.*** For example, nitrogen is the donor atom in the $[Cu(NH_3)_4]^{2+}$ complex ion. The ***coordination number*** in coordination compounds is defined as *the number of donor atoms surrounding the central metal atom in a complex ion*. For example, the coordination number of Ag^+ in $[Ag(NH_3)_2]^+$ is 2, that of Cu^{2+} in $[Cu(NH_3)_4]^{2+}$ is 4, and that of Fe^{3+} in $[Fe(CN)_6]^{3-}$ is 6. The most common coordination numbers are 4 and 6, but coordination numbers such as 2 and 5 are also known. In a crystal lattice, the coordination number of an atom (or ion) is defined as the number of atoms (or ions) surrounding the atom (or ion).

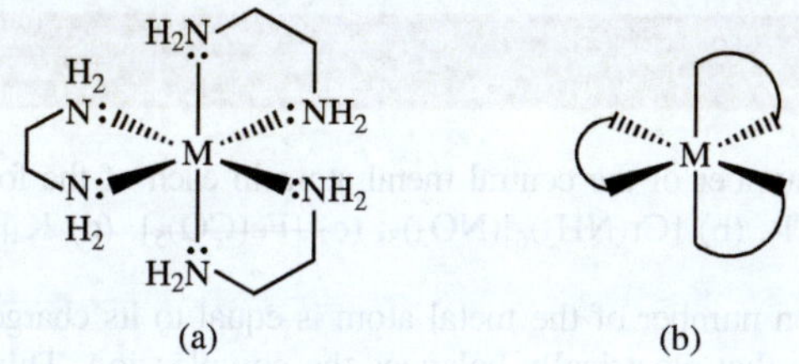

Figure 23.4 *(a) Structure of a metal-ethylenediamine complex cation, such as $[Co(en)_3]^{2+}$. Each ethylenediamine molecule provides two N donor atoms and is therefore a bidentate ligand. (b) Simplified structure of the same complex cation.*

Depending on the number of donor atoms present, ligands are classified as *monodentate, bidentate,* or *polydentate* (see Table 23.3). H_2O and NH_3 are monodentate ligands with only one donor atom each. One bidentate ligand is ethylenediamine (sometimes abbreviated "en"):

$$H_2N—CH_2—CH_2—NH_2$$

The two nitrogen atoms can coordinate with a metal atom, as shown in Figure 23.4. *Bidentate and polydentate ligands* are also called ***chelating agents*** because of *their ability to hold the metal atom like a claw* (from the Greek *chele,* meaning "claw"). One example is ethylenediaminetetraacetate ion (EDTA), a polydentate ligand used to treat metal poisoning (Figure 23.5). Six donor atoms enable EDTA to form a very stable complex ion with lead. In this form, it is removed from the blood and tissues and excreted from the body. EDTA is also used to clean up spills of radioactive metals.

Oxidation Numbers of Metals in Coordination Compounds

Another important property of coordination compounds is the oxidation number of the central metal atom. The term "oxidation state" is commonly used when referring to the oxidation number of the transition metal in coordination compounds. The net charge of a complex ion is the sum of the charges on the central metal atom and its surrounding ligands. In the $[PtCl_6]^{2-}$ ion, for example, each chloride ion has an oxidation number of -1, so the oxidation number of Pt must be +4. If the ligands do not bear net charges, the oxidation number of the metal is equal to the charge of the complex ion. Thus, in $[Cu(NH_3)_4]^{2+}$ each NH_3 is neutral, so the oxidation number of Cu is +2.

Example 23.1 deals with oxidation numbers of metals in coordination compounds.

Student Hot Spot

Student data indicate you may struggle with determining oxidation states of a metal in a coordination complex. Access your eBook for additional Learning Resources on this topic.

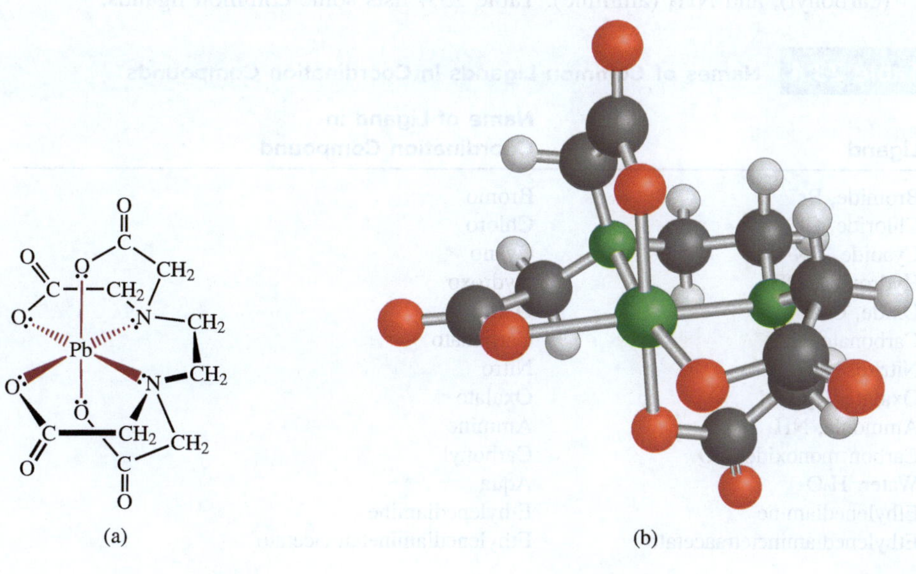

Figure 23.5 *(a) EDTA complex of lead. The complex bears a net charge of 2− because each of the six O donor atoms has a charge of 1− and the lead ion carries a charge of 2+. Only the lone pairs that participate in bonding are shown. Note the octahedral geometry around the Pb^{2+} ion. (b) Molecular model of the Pb^{2+}–EDTA complex. The green sphere is the Pb^{2+} ion.*

Example 23.1

Specify the oxidation number of the central metal atom in each of the following compounds: (a) $[Ru(NH_3)_5(H_2O)]Cl_2$, (b) $[Cr(NH_3)_6](NO_3)_3$, (c) $[Fe(CO)_5]$, (d) $K_4[Fe(CN)_6]$.

Strategy The oxidation number of the metal atom is equal to its charge. First we examine the anion or the cation that electrically balances the complex ion. This step gives us the net charge of the complex ion. Next, from the nature of the ligands (charged or neutral species) we can deduce the net charge of the metal and hence its oxidation number.

Solution

(a) Both NH_3 and H_2O are neutral species. Because each chloride ion carries a −1 charge, and there are two Cl^- ions, the oxidation number of Ru must be +2.

(b) Each nitrate ion has a charge of −1; therefore, the cation must be $[Cr(NH_3)_6]^{3+}$. NH_3 is neutral, so the oxidation number of Cr is +3.

(c) Because the CO species are neutral, the oxidation number of Fe is zero.

(d) Each potassium ion has a charge of +1; therefore, the anion is $[Fe(CN)_6]^{4-}$. Next, we know that each cyanide group bears a charge of −1, so Fe must have an oxidation number of +2.

Practice Exercise Write the oxidation numbers of the metals in the compound $K[Au(OH)_4]$.

Similar problems: 23.13, 23.14.

Naming Coordination Compounds

Now that we have discussed the various types of ligands and the oxidation numbers of metals, our next step is to learn what to call these coordination compounds. The rules for naming coordination compounds are as follows:

1. The cation is named before the anion, as in other ionic compounds. The rule holds regardless of whether the complex ion bears a net positive or a negative charge. For example, in $K_3[Fe(CN)_6]$ and $[Co(NH_3)_4Cl_2]Cl$, we name the K^+ and $[Co(NH_3)_4Cl_2]^+$ cations first, respectively.
2. Within a complex ion the ligands are named first, in alphabetical order, and the metal ion is named last.
3. The names of anionic ligands end with the letter *o*, whereas a neutral ligand is usually called by the name of the molecule. The exceptions are H_2O (aqua), CO (carbonyl), and NH_3 (ammine). Table 23.4 lists some common ligands.

Table 23.4 Names of Common Ligands in Coordination Compounds

Ligand	Name of Ligand in Coordination Compound
Bromide, Br^-	Bromo
Chloride, Cl^-	Chloro
Cyanide, CN^-	Cyano
Hydroxide, OH^-	Hydroxo
Oxide, O^{2-}	Oxo
Carbonate, CO_3^{2-}	Carbonato
Nitrite, NO_2^-	Nitro
Oxalate, $C_2O_4^{2-}$	Oxalato
Ammonia, NH_3	Ammine
Carbon monoxide, CO	Carbonyl
Water, H_2O	Aqua
Ethylenediamine	Ethylenediamine
Ethylenediaminetetraacetate	Ethylenediaminetetraacetato

4. When several ligands of a particular kind are present, we use the Greek prefixes *di-*, *tri-*, *tetra-*, *penta-*, and *hexa-* to name them. Thus, the ligands in the cation $[Co(NH_3)_4Cl_2]^+$ are "tetraamminedichloro." (Note that prefixes are ignored when alphabetizing ligands.) If the ligand itself contains a Greek prefix, we use the prefixes *bis* (2), *tris* (3), and *tetrakis* (4) to indicate the number of ligands present. For example, the ligand ethylenediamine already contains *di;* therefore, if two such ligands are present the name is *bis(ethylenediamine).*
5. The oxidation number of the metal is written in Roman numerals following the name of the metal. For example, the Roman numeral III is used to indicate the +3 oxidation state of chromium in $[Cr(NH_3)_4Cl_2]^+$, which is called tetraamminedichlorochromium(III) ion.
6. If the complex is an anion, its name ends in *-ate*. For example, in $K_4[Fe(CN)_6]$ the anion $[Fe(CN)_6]^{4-}$ is called hexacyanoferrate(II) ion. Note that the Roman numeral II indicates the oxidation state of iron. Table 23.5 gives the names of anions containing metal atoms.

Examples 23.2 and 23.3 deal with the nomenclature of coordination compounds.

Table 23.5

Names of Anions Containing Metal Atoms

Metal	Name of Metal in Anionic Complex
Aluminum	Aluminate
Chromium	Chromate
Cobalt	Cobaltate
Copper	Cuprate
Gold	Aurate
Iron	Ferrate
Lead	Plumbate
Manganese	Manganate
Molybdenum	Molybdate
Nickel	Nickelate
Silver	Argentate
Tin	Stannate
Tungsten	Tungstate
Zinc	Zincate

Student Hot Spot

Student data indicate you may struggle with naming coordination compounds. Access your eBook for additional Learning Resources on this topic.

Example 23.2

Write the systematic names of the following coordination compounds: (a) $Ni(CO)_4$, (b) $NaAuF_4$, (c) $K_3[Fe(CN)_6]$, (d) $[Cr(en)_3]Cl_3$.

Strategy We follow the preceding procedure for naming coordination compounds and refer to Tables 23.4 and 23.5 for names of ligands and anions containing metal atoms.

Solution

(a) The CO ligands are neutral species and therefore the Ni atom bears no net charge. The compound is called tetracarbonylnickel(0), or more commonly, nickel tetracarbonyl.

(b) The sodium cation has a positive charge; therefore, the complex anion has a negative charge (AuF_4^-). Each fluoride ion has a negative charge so the oxidation number of gold must be +3 (to give a net negative charge). The compound is called sodium tetrafluoroaurate(III).

(c) The complex ion is the anion and it bears three negative charges because each potassium ion bears a +1 charge. Looking at $[Fe(CN)_6]^{3-}$, we see that the oxidation number of Fe must be +3 because each cyanide ion bears a −1 charge (−6 total). The compound is potassium hexacyanoferrate(III). This compound is commonly called potassium ferricyanide.

(d) As we noted earlier, en is the abbreviation for the ligand ethylenediamine. Because there are three chloride ions each with a −1 charge, the cation is $[Cr(en)_3]^{3+}$. The en ligands are neutral so the oxidation number of Cr must be +3. Because there are three en groups present and the name of the ligand already contains *di* (rule 4), the compound is called tris(ethylenediamine)chromium(III) chloride.

Practice Exercise What is the systematic name of $[Cr(H_2O)_4Cl_2]Cl$?

Similar problems: 23.15, 23.16.

Example 23.3

Write the formulas for the following compounds: (a) pentaamminechlorocobalt(III) chloride, (b) dichlorobis(ethylenediamine)platinum(IV) nitrate, (c) sodium hexanitrocobaltate(III).

Strategy We follow the preceding procedure and refer to Tables 23.4 and 23.5 for names of ligands and anions containing metal atoms.

(Continued)

Solution

(a) The complex cation contains five NH_3 groups, a Cl^- ion, and a Co ion having a +3 oxidation number. The net charge of the cation must be +2, $[Co(NH_3)_5Cl]^{2+}$. Two chloride anions are needed to balance the positive charges. Therefore, the formula of the compound is $[Co(NH_3)_5Cl]Cl_2$.

(b) There are two chloride ions (−1 each), two en groups (neutral), and a Pt ion with an oxidation number of +4. The net charge on the cation must be +2, $[Pt(en)_2Cl_2]^{2+}$. Two nitrate ions are needed to balance the +2 charge of the complex cation. Therefore, the formula of the compound is $[Pt(en)_2Cl_2](NO_3)_2$.

(c) The complex anion contains six nitro groups (−1 each) and a cobalt ion with an oxidation number of +3. The net charge on the complex anion must be −3, $[Co(NO_2)_6]^{3-}$. Three sodium cations are needed to balance the −3 charge of the complex anion. Therefore, the formula of the compound is $Na_3[Co(NO_2)_6]$.

Practice Exercise Write the formula for the following compound: tris(ethylenediamine) cobalt(III) sulfate.

Similar problems: 23.17, 23.18.

Summary of Concepts & Facts

- Compounds that contain complex ions are called coordination compounds.
- Complex ions consist of a metal ion surrounded by ligands. The donor atoms in the ligands each contribute an electron pair to the central metal ion in a complex.

Review of Concepts & Facts

23.2.1 Determine the oxidation number of the metal center in the following coordination complexes: (a) $[Co(H_2O)_6]Cl_3$, (b) $K_4[MnCl_6]$, (c) $[Cr(en)_3]Br_3$.

23.2.2 What is the difference between these two compounds: $CrCl_3 \cdot 6H_2O$ and $[Cr(H_2O)_6]Cl_3$?

23.2.3 A student writes the name of the compound $[Cr(H_2O)_4Cl_2]Cl$ as dichlorotetraaquachromium chloride. Is this correct? If not, provide a proper systematic name.

23.3 Structure of Coordination Compounds

Learning Objectives

- Predict the structure of a coordination compound.
- Distinguish between geometric and optical isomers.
- Summarize the following stereochemical terms: *chiral, dextorotatory, levorotatory,* and *enantiomer.*

In studying the geometry of coordination compounds, we often find that there is more than one way to arrange ligands around the central atom. Compounds rearranged in this fashion have distinctly different physical and chemical properties. Figure 23.6 shows four different geometric arrangements for metal atoms with monodentate

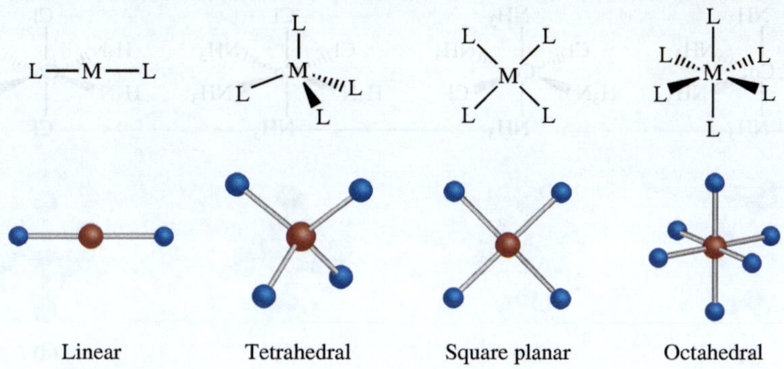

Figure 23.6 *Common geometries of complex ions. In each case, M is a metal and L is a monodentate ligand.*

ligands. In these diagrams, we see that structure and coordination number of the metal atom relate to each other as follows:

Coordination number	Structure
2	Linear
4	Tetrahedral or square planar
6	Octahedral

Stereoisomers are *compounds that are made up of the same types and numbers of atoms bonded together in the same sequence but with different spatial arrangements.* There are two types of stereoisomers: geometric isomers and optical isomers. Coordination compounds may exhibit one or both types of isomerism. Note, however, that many coordination compounds do not have stereoisomers.

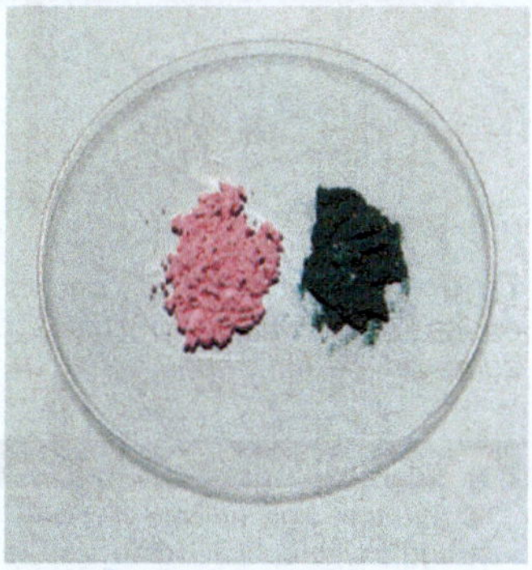

cis-tetraamminedichlorocobalt(III) chloride (left) and trans-tetraamminedichlorocobalt(III) chloride (right).
Ken Karp/McGraw Hill

Geometric Isomers

Geometric isomers are *stereoisomers that cannot be interconverted without breaking a chemical bond.* Geometric isomers usually come in pairs. We use the terms "*cis*" and "*trans*" to distinguish one geometric isomer of a compound from the other. *Cis* means that two particular atoms (or groups of atoms) are adjacent to each other, and *trans* means that the atoms (or groups of atoms) are on opposite sides in the structural formula. The *cis* and *trans* isomers of coordination compounds generally have quite different colors, melting points, dipole moments, and chemical reactivities. Figure 23.7 shows the *cis* and *trans* isomers of diamminedichloroplatinum(II). Note that although the types of bonds are the same in both isomers (two Pt–N and two Pt–Cl bonds), the spatial arrangements are different. Another example is tetraamminedichlorocobalt(III) ion, shown in Figure 23.8.

Student Hot Spot

Student data indicate you may struggle with geometric isomers of coordination complexes. Access your eBook for additional Learning Resources on this topic.

Optical Isomers

Optical isomers are *nonsuperimposable mirror images.* ("Superimposable" means that if one structure is laid over the other, the positions of all the atoms will match.) Like

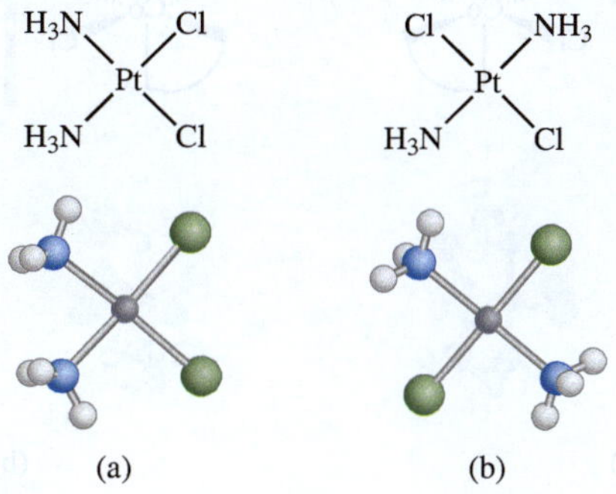

Figure 23.7 *The (a) cis and (b) trans isomers of diamminedichloroplatinum(II). Note that the two Cl atoms are adjacent to each other in the cis isomer and diagonally across from each other in the trans isomer.*

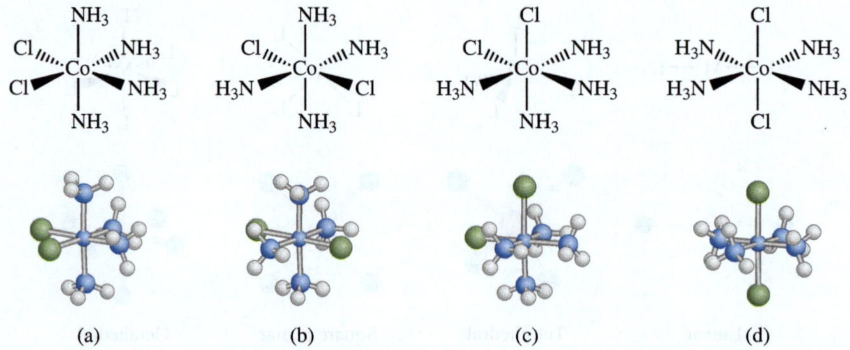

Figure 23.8 *The (a) cis and (b) trans isomers of the tetraamminedichlorocobalt(III) ion, $[Co(NH_3)_4Cl_2]^+$. The structure shown in (c) can be generated by rotating that in (a), and the structure shown in (d) can be generated by rotating that in (b). The ion has only two geometric isomers, (a) [or (c)] and (b) [or (d)].*

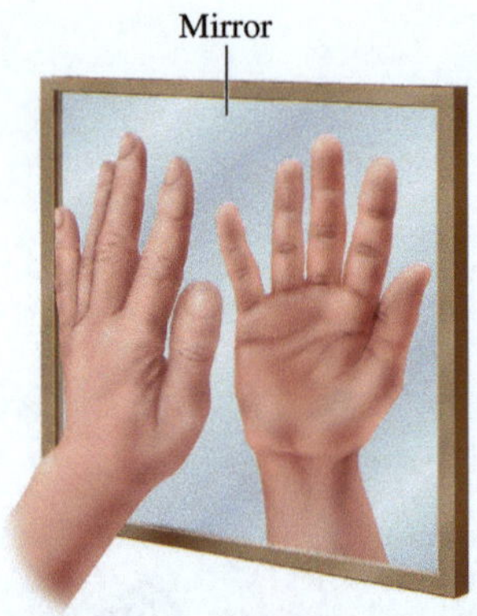

Figure 23.9 *A left hand and its mirror image, which looks the same as the right hand.*

geometric isomers, optical isomers come in pairs. However, the optical isomers of a compound have *identical* physical and chemical properties, such as melting point, boiling point, dipole moment, and chemical reactivity toward molecules that are not optical isomers themselves. Optical isomers differ from each other in their interactions with plane-polarized light, as we will see.

The structural relationship between two optical isomers is analogous to the relationship between your left and right hands. If you place your left hand in front of a mirror, the image you see will look like your right hand (Figure 23.9). We say that your left hand and right hand are mirror images of each other. However, they are nonsuperimposable, because when.you place your left hand over your right hand (with both palms facing down), they do not match.

Figure 23.10 shows the *cis* and *trans* isomers of dichlorobis(ethylenediamine)-cobalt(III) ion and their images. Careful examination reveals that the *trans* isomer and its mirror image are superimposable, but the *cis* isomer and its mirror image are not. Therefore, the *cis* isomer and its mirror image are optical isomers.

Optical isomers are described as ***chiral*** (from the Greek word for "hand") because, like your left and right hands, chiral molecules are nonsuperimposable. Isomers that are superimposable with their mirror images are said to be *achiral.* Chiral molecules play a vital role in enzyme reactions in biological systems. Many drug molecules are chiral. It is interesting to note that frequently only one of a pair of chiral isomers is biologically effective.

Chiral molecules are said to be optically active because of their ability to rotate the plane of polarization of polarized light as it passes through them. Unlike ordinary light, which vibrates in all directions, *plane-polarized light* vibrates only in a single plane. We use a ***polarimeter*** to *measure the rotation of polarized light by optical isomers* (Figure 23.11). A beam of unpolarized light first passes through a Polaroid sheet, called the polarizer, and then through a sample tube containing a solution of an optically active, chiral compound. As the polarized light passes through the sample tube, its plane of polarization is rotated either to the right or to the left. This rotation can be measured directly by turning the analyzer in the appropriate direction until minimal

Student Hot Spot

Student data indicate you may struggle with optical isomers of coordination complexes. Access your eBook for additional Learning Resources on this topic.

Video: Chirality

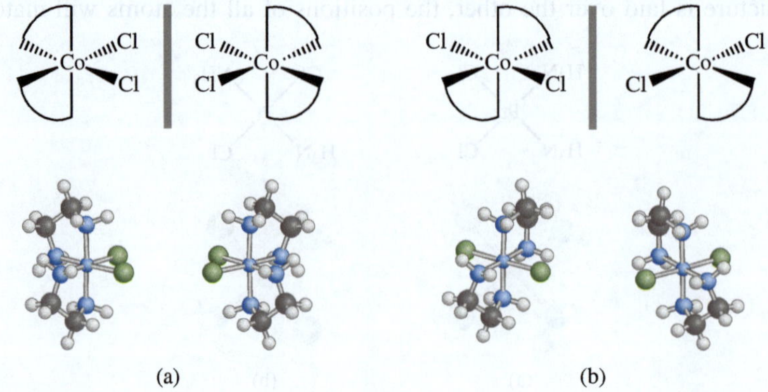

Figure 23.10 *The (a) cis and (b) trans isomers of the dichlorobis(ethylenediamine) cobalt(III) ion and their mirror images. If you could rotate the mirror image in (b) 90° clockwise about the vertical position and place the ion over the trans isomer, you would find that the two are superimposable. No matter how you rotated the cis isomer and its mirror image in (a), however, you could not superimpose one on the other.*

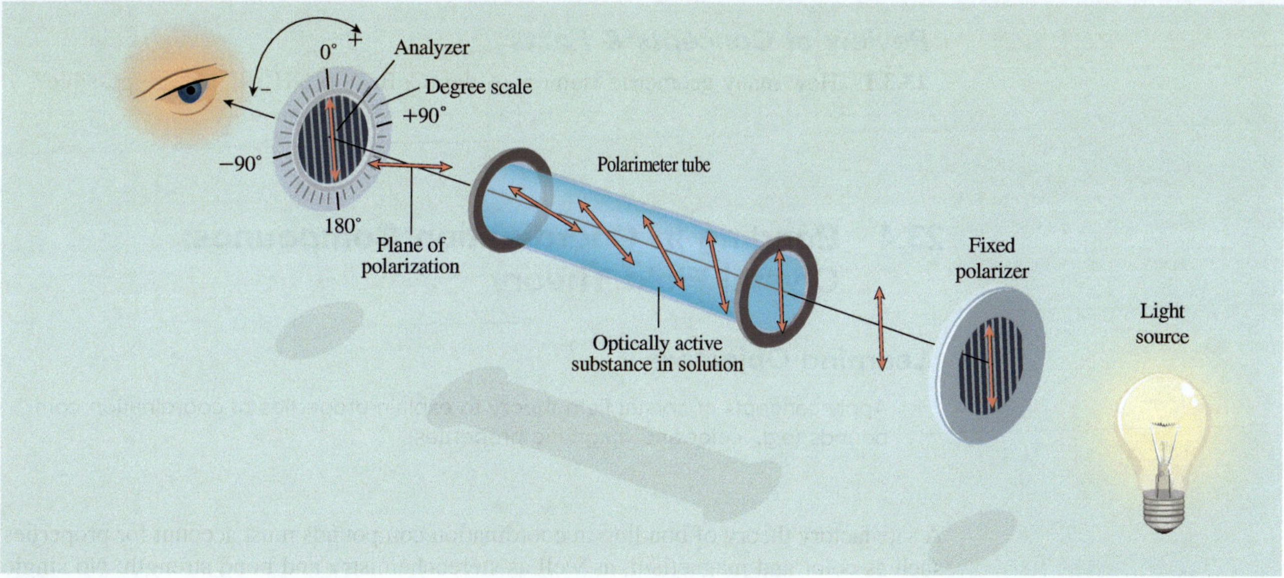

Figure 23.11 *Operation of a polarimeter. Initially, the tube is filled with an achiral compound. The analyzer is rotated so that its plane of polarization is perpendicular to that of the polarizer. Under this condition, no light reaches the observer. Next, a chiral compound is placed in the tube as shown. The plane of polarization of the polarized light is rotated as it travels through the tube so that some light reaches the observer. Rotating the analyzer (either to the left or to the right) until no light reaches the observer again enables the angle of optical rotation to be measured.*

light transmission is achieved (Figure 23.12). If the plane of polarization is rotated to the right, the isomer is said to be *dextrorotatory (d);* it is *levorotatory (l)* if the rotation is to the left. *The d and l isomers of a chiral substance,* called ***enantiomers,*** always rotate the light by the same amount, but in opposite directions. Thus, in *an equimolar mixture of two enantiomers,* called a ***racemic mixture,*** the net rotation is zero.

Summary of Concepts & Facts

- Coordination compounds may display geometric and/or optical isomerism.

Figure 23.12 *With one Polaroid sheet over a picture, light passes through. With a second sheet of Polaroid placed over the first so that the axes of polarization of the sheets are perpendicular, little or no light passes through. If the axes of polarization of the two sheets were parallel, light would pass through.*
David A. Tietz/Editorial Image, LLC

Review of Concepts & Facts

23.3.1 How many geometric isomers of the $[CoBr_2(en)(NH_3)_2]^+$ ion are possible?

23.4 Bonding in Coordination Compounds: Crystal Field Theory

Learning Objective

- Apply concepts of crystal field theory to explain properties of coordination compounds (e.g., color and magnetic properties).

A satisfactory theory of bonding in coordination compounds must account for properties such as color and magnetism, as well as stereochemistry and bond strength. No single theory as yet does all this for us. Rather, several different approaches have been applied to transition metal complexes. We will consider only one of them here—crystal field theory—because it accounts for both the color and magnetic properties of many coordination compounds. The name "crystal field" is associated with the theory used to explain the properties of solid, crystalline materials. The same theory is used to study coordination compounds.

We will begin our discussion of crystal field theory with the most straightforward case, namely, complex ions with octahedral geometry. Then we will see how it is applied to tetrahedral and square-planar complexes.

Crystal Field Splitting in Octahedral Complexes

Crystal field theory explains the bonding in complex ions purely in terms of electrostatic forces. In a complex ion, two types of electrostatic interaction come into play. One is the attraction between the positive metal ion and the negatively charged ligand or the negatively charged end of a polar ligand. This is the force that binds the ligands to the metal. The second type of interaction is electrostatic repulsion between the lone pairs on the ligands and the electrons in the *d* orbitals of the metals.

As we saw in Chapter 7, *d* orbitals have different orientations, but in the absence of external disturbance they all have the same energy. In an octahedral complex, a central metal atom is surrounded by six lone pairs of electrons (on the six ligands), so all five *d* orbitals experience electrostatic repulsion. The magnitude of this repulsion depends on the orientation of the *d* orbital that is involved. Take the $d_{x^2-y^2}$ orbital as an example. In Figure 23.13, we see that the lobes of this orbital point toward corners of the octahedron along the *x* and *y* axes, where the lone-pair electrons are positioned. Thus, an electron residing in this orbital would experience a greater repulsion from the ligands than an electron would in, say, the d_{xy} orbital. For this reason, the energy of the $d_{x^2-y^2}$ orbital is increased relative to the d_{xy}, d_{yz}, and d_{xz} orbitals. The d_{z^2} orbital's energy is also greater, because its lobes are pointed at the ligands along the *z*-axis.

As a result of these metal-ligand interactions, the five *d* orbitals in an octahedral complex are split between two sets of energy levels: a higher level with two orbitals ($d_{x^2-y^2}$ and d_{z^2}) having the same energy and a lower level with three equal-energy orbitals (d_{xy}, d_{yz}, and d_{xz}), as shown in Figure 23.14. The ***crystal field splitting* (Δ)** is *the energy difference between two sets of d orbitals in a metal atom when ligands are present.* The magnitude of Δ depends on the metal and the nature of the ligands; it has a direct effect on the color and magnetic properties of complex ions.

Student Hot Spot

Student data indicate you may struggle with crystal field splitting. Access your eBook for additional Learning Resources on this topic.

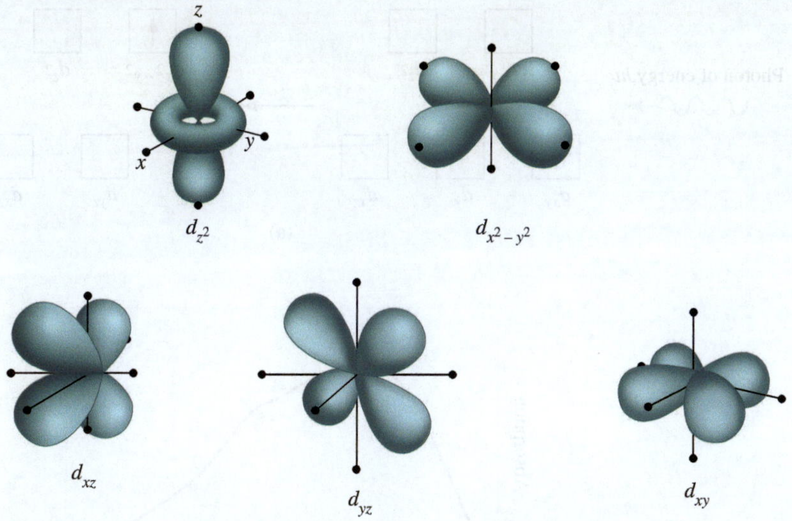

Figure 23.13 *The five d orbitals in an octahedral environment. The metal atom (or ion) is at the center of the octahedron, and the six lone pairs on the donor atoms of the ligands are at the corners.*

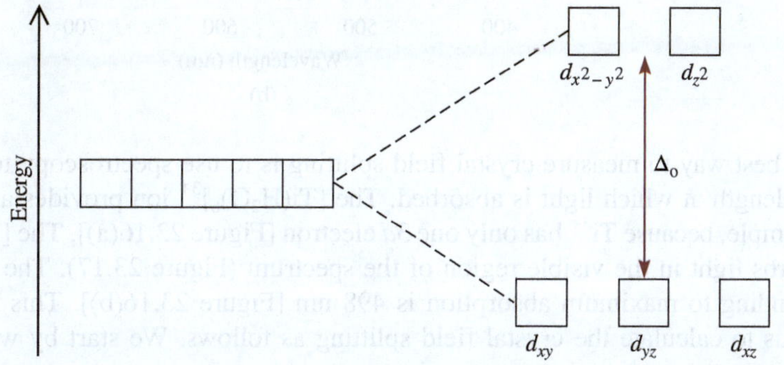

Figure 23.14 *Crystal field splitting between d orbitals in an octahedral complex.*

Color

Video: Absorption of Color

In Chapter 7 we learned that white light, such as sunlight, is a combination of all colors. A substance appears black if it absorbs all the visible light that strikes it. If it absorbs no visible light, it is white or colorless. An object appears green if it absorbs all light but reflects the green component. An object also looks green if it reflects all colors except red, the *complementary* color of green (Figure 23.15).

What has been said of reflected light also applies to transmitted light (that is, the light that passes through the medium, for example, a solution). Consider the hydrated cupric ion, $[Cu(H_2O)_6]^{2+}$, which absorbs light in the orange region of the spectrum so that a solution of $CuSO_4$ appears blue to us. Recall from Chapter 7 that when the energy of a photon is equal to the difference between the ground state and an excited state, absorption occurs as the photon strikes the atom (or ion or compound), and an electron is promoted to a higher level. This knowledge enables us to calculate the energy change involved in the electron transition. The energy of a photon, given by Equation (7.2), is

$$E = h\nu$$

where h represents Planck's constant (6.63×10^{-34} J · s) and ν is the frequency of the radiation, which is 5.00×10^{14}/s for a wavelength of 600 nm. Here $E = \Delta$, so we have

$$\begin{aligned}\Delta &= h\nu \\ &= (6.63 \times 10^{-34}\ \text{J} \cdot \text{s})(5.00 \times 10^{14}/\text{s}) \\ &= 3.32 \times 10^{-19}\ \text{J}\end{aligned}$$

(Note that this is the energy absorbed by *one* ion.) If the wavelength of the photon absorbed by an ion lies outside the visible region, then the transmitted light looks the same (to us) as the incident light—white—and the ion appears colorless. A *d*-to-*d* transition must occur for a transition metal complex to show color. Therefore, ions with d^0 or d^{10} electron configurations are usually colorless.

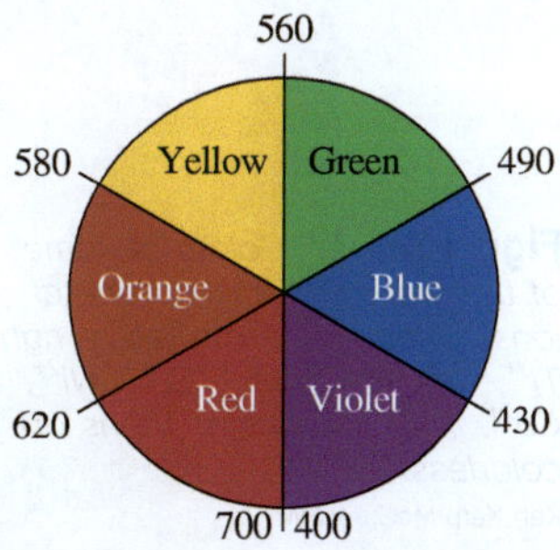

Figure 23.15 *A color wheel with appropriate wavelengths. A compound that absorbs in the green region will appear red, the complementary color of green.*

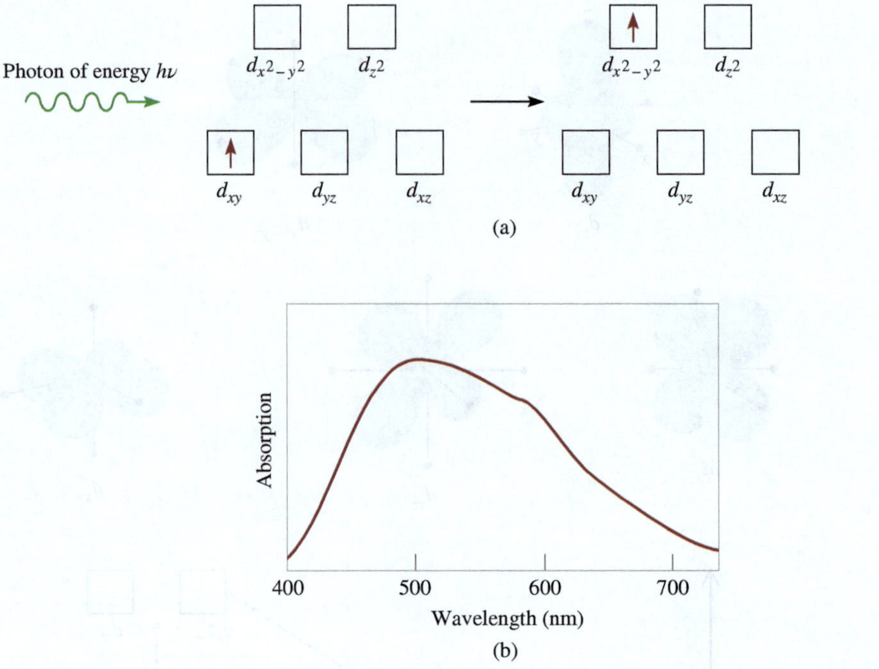

Figure 23.16 *(a) The process of photon absorption and (b) the absorption spectrum of $[Ti(H_2O)_6]^{3+}$. The energy of the incoming photon is equal to the crystal field splitting. The maximum absorption peak in the visible region occurs at 498 nm.*

Student Hot Spot
Student data indicate you may struggle with the colors of coordination complexes. Access your eBook for additional Learning Resources on this topic.

The best way to measure crystal field splitting is to use spectroscopy to determine the wavelength at which light is absorbed. The $[Ti(H_2O)_6]^{3+}$ ion provides a straightforward example, because Ti^{3+} has only one 3*d* electron [Figure 23.16(a)]. The $[Ti(H_2O)_6]^{3+}$ ion absorbs light in the visible region of the spectrum (Figure 23.17). The wavelength corresponding to maximum absorption is 498 nm [Figure 23.16(b)]. This information enables us to calculate the crystal field splitting as follows. We start by writing

$$\Delta = h\nu \tag{23.1}$$

Also

$$\nu = \frac{c}{\lambda}$$

where *c* is the speed of light and λ is the wavelength. Therefore,

$$\Delta = \frac{hc}{\lambda} = \frac{(6.63 \times 10^{-34}\ \text{J}\cdot\text{s})(3.00 \times 10^{8}\ \text{m/s})}{(498\ \text{nm})(1 \times 10^{-9}\ \text{m/1 nm})}$$
$$= 3.99 \times 10^{-19}\ \text{J}$$

This is the energy required to excite *one* $[Ti(H_2O)_6]^{3+}$ ion. To express this energy difference in the more convenient units of kilojoules per mole, we write

$$\Delta = (3.99 \times 10^{-19}\ \text{J/ion})(6.02 \times 10^{23}\ \text{ions/mol})$$
$$= 240{,}000\ \text{J/mol}$$
$$= 240\ \text{kJ/mol}$$

Figure 23.17 *Colors of some of the first-row transition metal ions in solution. From left to right: Ti^{3+}, Cr^{3+}, Mn^{2+}, Fe^{3+}, Co^{2+}, Ni^{2+}, Cu^{2+}. The Sc^{3+} and V^{5+} ions are colorless.*

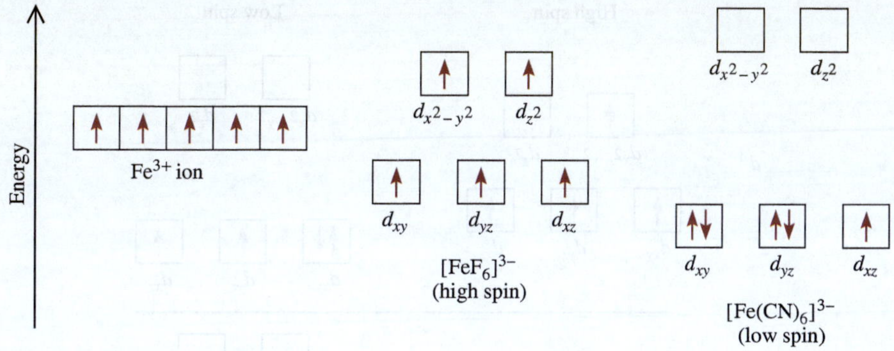

Figure 23.18 *Energy-level diagrams for the Fe^{3+} ion and for the $[FeF_6]^{3-}$ and $[Fe(CN)_6]^{3-}$ complex ions.*

Aided by spectroscopic data for a number of complexes, all having the same metal ion but different ligands, chemists calculated the crystal splitting for each ligand and established a ***spectrochemical series,*** which is *a list of ligands arranged in increasing order of their abilities to split the d orbital energy levels:*

$$I^- < Br^- < S^{2-} < \underline{S}CN^- < Cl^- < F^- < OH^- < H_2O < \underline{N}CS^- < NH_3 < en < CN^- < CO$$

These ligands are arranged in the order of increasing value of Δ. CO and CN^- are called *strong-field ligands,* because they cause a large splitting of the *d* orbital energy levels. The halide ions and hydroxide ion are *weak-field ligands,* because they split the *d* orbitals to a lesser extent. For the thiocyanate ion, when it is bound to a metal ion through the S atom (as shown by the underlined atom in the list), it is a weak-field ligand. When it is bound to a metal ion through the N atom, it has a stronger splitting effect on the *d* orbital energy levels.

The crystal field splitting also depends on the nature of the central metal ion. The listing in approximate order of increasing splitting is:

$$Mn^{2+} < Ni^{2+} < Co^{2+} < Fe^{2+} < V^{2+} < Fe^{3+} < Co^{3+} < Mo^{3+} <$$
$$Rh^{3+} < Ru^{3+} < Pd^{4+} < Ir^{3+} < Pt^{4+}$$

The crystal field splitting increases with increasing oxidation state of the metal ion. The splitting also increases down a group (for example, compare the Group 9 metals Co, Rh, and Ir).

Magnetic Properties

The magnitude of the crystal field splitting also determines the magnetic properties of a complex ion. The $[Ti(H_2O)_6]^{3+}$ ion, having only one *d* electron, is always paramagnetic. However, for an ion with several *d* electrons, the situation is less clear-cut. Consider, for example, the octahedral complexes $[FeF_6]^{3-}$ and $[Fe(CN)_6]^{3-}$ (Figure 23.18). The electron configuration of Fe^{3+} is $[Ar]3d^5$, and there are two possible ways to distribute the five *d* electrons among the *d* orbitals. According to Hund's rule (see Section 7.8), maximum stability is reached when the electrons are placed in five separate orbitals with parallel spins. But this arrangement can be achieved only at a cost; two of the five electrons must be promoted to the higher-energy $d_{x^2-y^2}$ and d_{z^2} orbitals. No such energy investment is needed if all five electrons enter the d_{xy}, d_{yz}, and d_{xz} orbitals. According to Pauli's exclusion principle (Section 7.8), there will be only one unpaired electron present in this case.

Figure 23.19 shows the distribution of electrons among *d* orbitals that results in low- and high-spin complexes. The actual arrangement of the electrons is determined by the amount of stability gained by having maximum parallel spins versus the investment in energy required to promote electrons to higher *d* orbitals. Because F^- is a weak-field ligand, the five *d* electrons enter five separate *d* orbitals with parallel spins to create a high-spin complex (see Figure 23.18). On the other hand, the cyanide ion is a strong-field ligand, so it is energetically preferable for all five electrons to be in the lower orbitals and therefore a low-spin complex is formed. High-spin complexes are more paramagnetic than low-spin complexes.

Student Hot Spot
Student data indicate you may struggle with high-spin and low-spin coordination complexes. Access your eBook for additional Learning Resources on this topic.

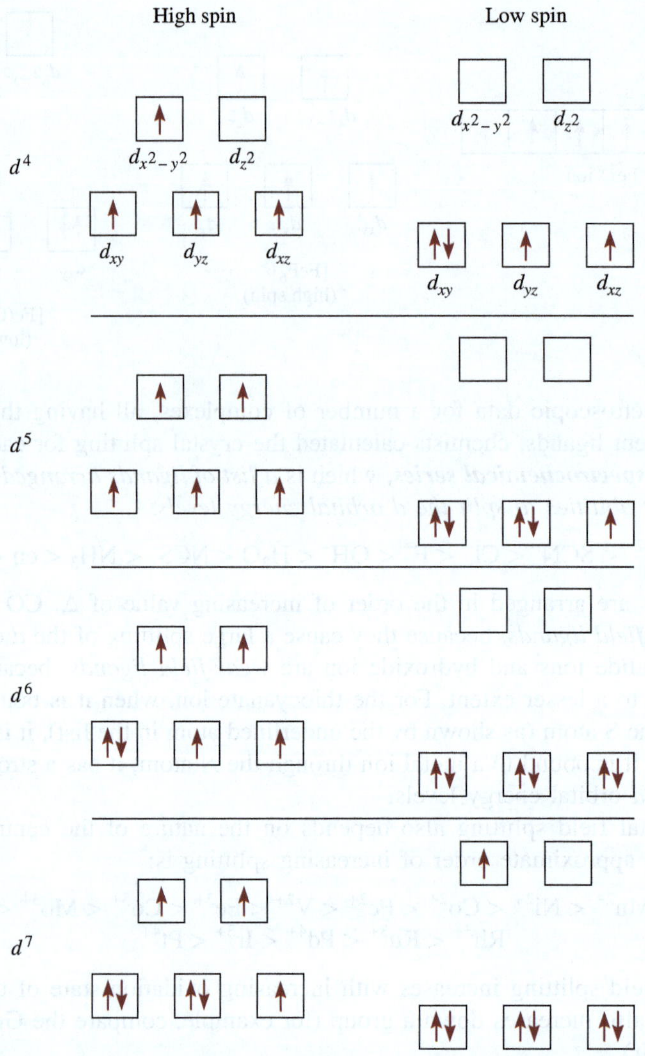

Figure 23.19 *Orbital diagrams for the high-spin and low-spin octahedral complexes corresponding to the electron configurations d^4, d^5, d^6, and d^7. No such distinctions can be made for d^1, d^2, d^3, d^8, d^9, and d^{10}.*

Student Hot Spot

Student data indicate you may struggle with determining the number of unpaired electrons in coordination complexes. Access your eBook for additional Learning Resources on this topic.

The actual number of unpaired electrons (or spins) in a complex ion can be found by magnetic measurements, and in general, experimental findings support predictions based on crystal field splitting. However, a distinction between low- and high-spin complexes can be made only if the metal ion is a d^4, d^5, d^6, or d^7 configuration, as shown in Figure 23.19.

Example 23.4

Predict the number of unpaired electrons in the $[Cr(en)_3]^{2+}$ ion.

Strategy The magnetic properties of a complex ion depend on the strength of the ligands. Strong-field ligands, which cause a high degree of splitting among the *d* orbital energy levels, result in low-spin complexes. Weak-field ligands, which cause a small degree of splitting among the *d* orbital energy levels, result in high-spin complexes.

Solution The electron configuration of Cr^{2+} is $[Ar]3d^4$. Because en is a strong-field ligand, we expect $[Cr(en)_3]^{2+}$ to be a low-spin complex. According to Figure 23.19, all four electrons will be placed in the lower-energy *d* orbitals (d_{xy}, d_{yz}, and d_{xz}) and there will be a total of two unpaired spins.

Practice Exercise How many unpaired electrons are in $[Mn(H_2O)_6]^{2+}$? (H_2O is a weak-field ligand.)

Similar problem: 23.35.

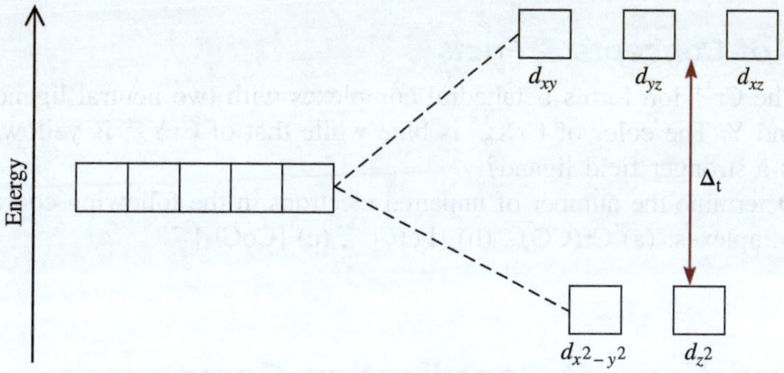

Figure 23.20 *Crystal field splitting between d orbitals in a tetrahedral complex.*

Tetrahedral and Square-Planar Complexes

So far we have concentrated on octahedral complexes. The splitting of the *d* orbital energy levels in two other types of complexes—tetrahedral and square planar—can also be accounted for satisfactorily by the crystal field theory. In fact, the splitting pattern for a tetrahedral ion is just the reverse of that for octahedral complexes. In this case, the d_{xy}, d_{yz}, and d_{xz} orbitals are more closely directed at the ligands and therefore have more energy than the $d_{x^2-y^2}$ and d_{z^2} orbitals (Figure 23.20). Most tetrahedral complexes are high-spin complexes. Presumably, the tetrahedral arrangement reduces the magnitude of metal-ligand interactions, resulting in a smaller Δ value compared to the octahedral case. This is a reasonable assumption because the number of ligands is smaller in a tetrahedral complex.

As Figure 23.21 shows, the splitting pattern for square-planar complexes is the most complicated of the three cases. Clearly, the $d_{x^2-y^2}$ orbital possesses the highest energy (as in the octahedral case), and the d_{xy} orbital the next highest. However, the relative placement of the d_{z^2} and the d_{xz} and d_{yz} orbitals cannot be determined simply by inspection and must be calculated.

Summary of Concepts & Facts

- Crystal field theory explains bonding in complexes in terms of electrostatic interactions. According to crystal field theory, the *d* orbitals are split into two higher-energy and three lower-energy orbitals in an octahedral complex. The energy difference between these two sets of *d* orbitals is the crystal field splitting.
- Strong-field ligands cause a large crystal field splitting, and weak-field ligands cause a small splitting. Electron spins tend to be parallel with weak-field ligands and paired with strong-field ligands, where a greater investment of energy is required to promote electrons into the high-lying *d* orbitals.

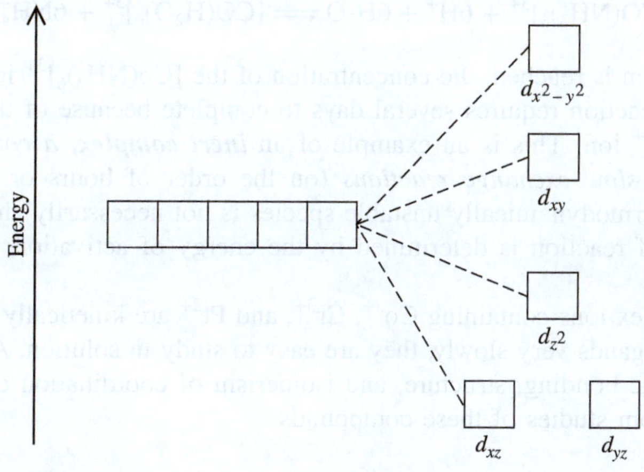

Figure 23.21 *Energy-level diagram for a square-planar complex. Because there are more than two energy levels, we cannot define crystal field splitting as we can for octahedral and tetrahedral complexes.*

Review of Concepts & Facts

23.4.1 The Cr^{3+} ion forms octahedral complexes with two neutral ligands X and Y. The color of CrX_6^{3+} is blue while that of CrY_6^{3+} is yellow. Which is a stronger field ligand?

23.4.2 Determine the number of unpaired electrons in the following coordination complexes: (a) $Cr(CO)_6$, (b) $[FeF_6]^{3-}$, (c) $[CoCl_4]^{2-}$.

23.5 Reactions of Coordination Compounds

Learning Objective

- Discuss the nature of ligand exchange reactions, including an explanation of kinetic lability.

Complex ions undergo ligand exchange (or substitution) reactions in solution. The rates of these reactions vary widely, depending on the nature of the metal ion and the ligands.

In studying ligand exchange reactions, it is often useful to distinguish between the stability of a complex ion and its tendency to react, which we call *kinetic lability*. Stability in this context is a thermodynamic property, which is measured in terms of the species' formation constant K_f (see Section 16.7). For example, we say that the complex ion tetracyanonickelate(II) is stable because it has a large formation constant ($K_f \approx 1 \times 10^{30}$):

$$Ni^{2+} + 4CN^- \rightleftharpoons [Ni(CN)_4]^{2-}$$

By using cyanide ions labeled with the radioactive isotope carbon-14, chemists have shown that $[Ni(CN)_4]^{2-}$ undergoes ligand exchange very rapidly in solution. The following equilibrium is established almost as soon as the species are mixed:

$$[Ni(CN)_4]^{2-} + 4{*CN}^- \rightleftharpoons [Ni({*CN})_4]^{2-} + 4CN^-$$

where the asterisk denotes a ^{14}C atom. Complexes like the tetracyanonickelate(II) ion are termed ***labile complexes*** because they *undergo rapid ligand exchange reactions.* Thus, a thermodynamically stable species (that is, one that has a large formation constant) is not necessarily unreactive. (In Section 13.4 we saw that the smaller the activation energy, the larger the rate constant, and hence the greater the rate.)

A complex that is thermodynamically *unstable* in acidic solution is $[Co(NH_3)_6]^{3+}$. The equilibrium constant for the following reaction is about 1×10^{20}:

$$[CO(NH_3)_6]^{3+} + 6H^+ + 6H_2O \rightleftharpoons [CO(H_2O)_6]^{3+} + 6NH_4^+$$

When equilibrium is reached, the concentration of the $[Co(NH_3)_6]^{3+}$ ion is very low. However, this reaction requires several days to complete because of the inertness of the $[Co(NH_3)_6]^{3+}$ ion. This is an example of an ***inert complex,*** *a complex ion that undergoes very slow exchange reactions* (on the order of hours or even days). It shows that a thermodynamically unstable species is not necessarily chemically reactive. The rate of reaction is determined by the energy of activation, which is high in this case.

Most complex ions containing Co^{3+}, Cr^{3+}, and Pt^{2+} are kinetically inert. Because they exchange ligands very slowly, they are easy to study in solution. As a result, our knowledge of the bonding, structure, and isomerism of coordination compounds has come largely from studies of these compounds.

Summary of Concepts & Facts

- Discuss the nature of ligand exchange reactions, including an explanation of kinetic lability.

23.6 Applications of Coordination Compounds

Learning Objective

- Provide examples of common applications of coordination compounds.

Coordination compounds are found in living systems and have many uses in the home, in industry, and in medicine. We describe a few examples here and in the Chemistry in Action essay, "Coordination Compounds in Living Systems," in Section 23.5.

Metallurgy

The extraction of silver and gold by the formation of cyanide complexes (Section 22.3) and the purification of nickel (Section 21.2) by converting the metal to the gaseous compound $Ni(CO)_4$ are typical examples of the use of coordination compounds in metallurgical processes.

Therapeutic Chelating Agents

Earlier we mentioned that the chelating agent EDTA is used in the treatment of lead poisoning. Certain platinum-containing compounds can effectively inhibit the growth of cancerous cells. A specific case is discussed in the Chemistry in Action essay, "Cisplatin—The Anticancer Drug."

Chemical Analysis

Although EDTA has a great affinity for a large number of metal ions (especially 2+ and 3+ ions), other chelates are more selective in binding. For example, dimethylglyoxime,

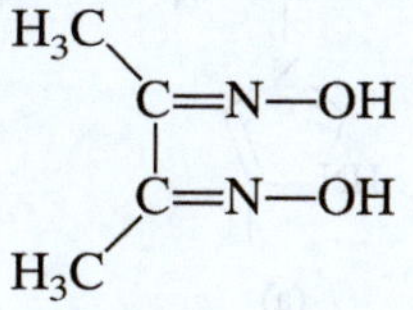

forms an insoluble brick-red solid with Ni^{2+} and an insoluble bright yellow solid with Pd^{2+}. These characteristic colors are used in qualitative analysis to identify nickel and palladium. Further, the quantities of ions present can be determined by gravimetric analysis (see Section 4.6) as follows: To a solution containing Ni^{2+} ions, say, we add an excess of dimethylglyoxime reagent, and a brick-red precipitate forms. The precipitate is then filtered, dried, and weighed. Knowing the formula of the complex (Figure 23.22), we can readily calculate the amount of nickel present in the original solution.

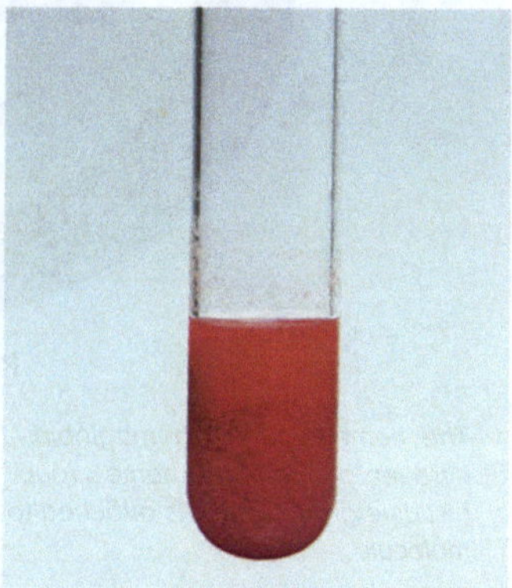

An aqueous suspension of bis(dimethylglyoximato)-nickel(II).
Ken Karp/McGraw Hill

Detergents

The cleansing action of soap in hard water is hampered by the reaction of the Ca^{2+} ions in the water with the soap molecules to form insoluble salts or curds. In the late 1940s the detergent industry introduced a "builder," usually sodium tripolyphosphate, to circumvent this problem. The tripolyphosphate ion is an effective chelating agent

CHEMISTRY *in Action*

Coordination Compounds in Living Systems

Coordination compounds play many important roles in animals and plants. They are essential in the storage and transport of oxygen, as electron transfer agents, as catalysts, and in photosynthesis. Here we focus on coordination compounds containing iron and magnesium.

Because of its central function as an oxygen carrier for metabolic processes, hemoglobin is probably the most studied of all the proteins. The molecule contains four folded long chains called *subunits*. Hemoglobin carries oxygen in the blood from the lungs to the tissues, where it delivers the oxygen molecules to myoglobin. Myoglobin, which is made up of only one subunit, stores oxygen for metabolic processes in muscle.

The porphine molecule forms an important part of the hemoglobin structure. Upon coordination to a metal, the H^+ ions that are bonded to two of the four nitrogen atoms in porphine are displaced. Complexes derived from porphine are called *porphyrins,* and the iron-porphyrin combination is called the *heme* group. The iron in the heme group has an oxidation number of +2; it is coordinated to the four nitrogen atoms in the porphine group and also to a nitrogen donor atom in a ligand that is attached to the protein. The sixth ligand is a water molecule, which binds to the Fe^{2+} ion on the other side of the ring to complete the octahedral complex. This hemoglobin molecule is called *deoxyhemoglobin* and imparts a bluish tinge to venous

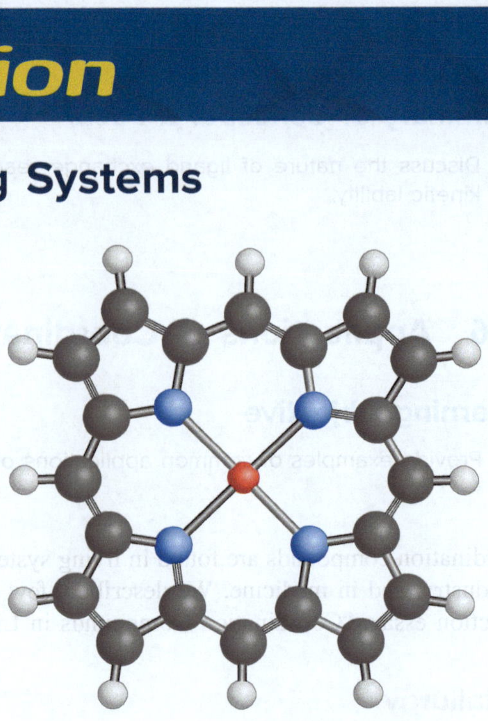

Simplified structures of the porphine molecule and the Fe^{2+}-porphyrin complex.

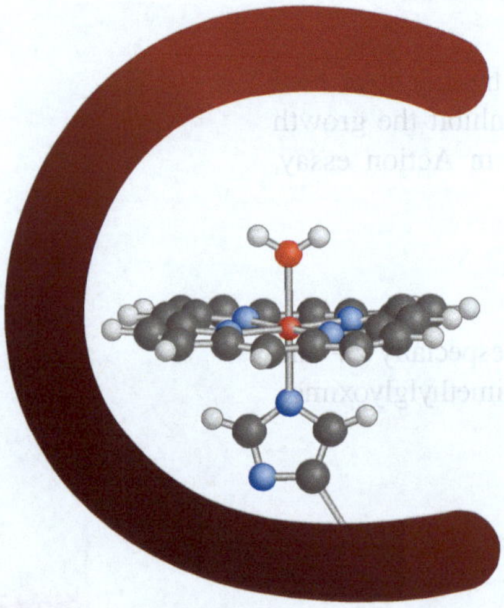

The heme group in hemoglobin. The Fe^{2+} ion is coordinated with the nitrogen atoms of the heme group. The ligand below the porphyrin is the histidine group, which is attached to the protein. The sixth ligand is a water molecule.

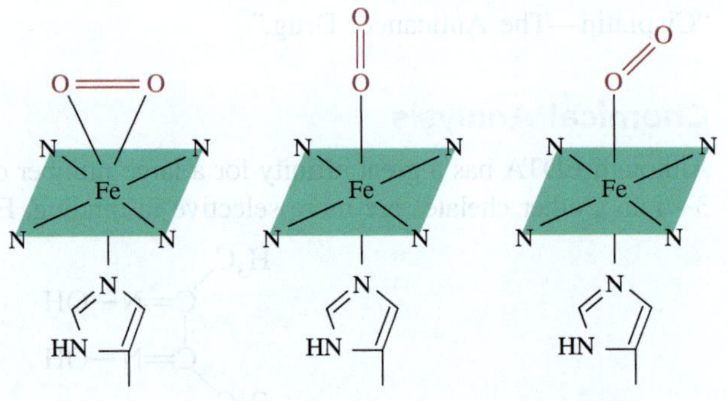

Three possible ways for molecular oxygen to bind to the heme group in hemoglobin. The structure shown in (a) would have a coordination number of 7, which is considered unlikely for Fe(II) complexes. Although the end-on arrangement in (b) seems the most reasonable, evidence points to the structure in (c) as the correct one. The structure shown in (c) is the most plausible.

blood. The water ligand can be replaced readily by molecular oxygen to form red oxyhemoglobin found in arterial blood. Each subunit contains a heme group, so each hemoglobin molecule can bind up to four O_2 molecules.

There are three possible structures for oxyhemoglobin. For a number of years, the exact arrangement of the oxygen molecule relative to the porphyrin group was not clear. Most experimental evidence suggests that the bond between O and Fe is bent relative to the heme group.

The porphyrin group is a very effective chelating agent, and not surprisingly, we find it in a number of biological systems. The iron-heme complex is present in another class of proteins, called the *cytochromes*. The iron forms an octahedral complex in these proteins, but because both the histidine and the methionine groups are firmly bound to the metal ion, they cannot be displaced by oxygen or other ligands. Instead, the cytochromes act as electron carriers, which are essential to metabolic processes. In cytochromes, iron undergoes rapid reversible redox reactions:

$$Fe^{3+} + e^- \rightleftharpoons Fe^{2+}$$

which are coupled to the oxidation of organic molecules such as the carbohydrates.

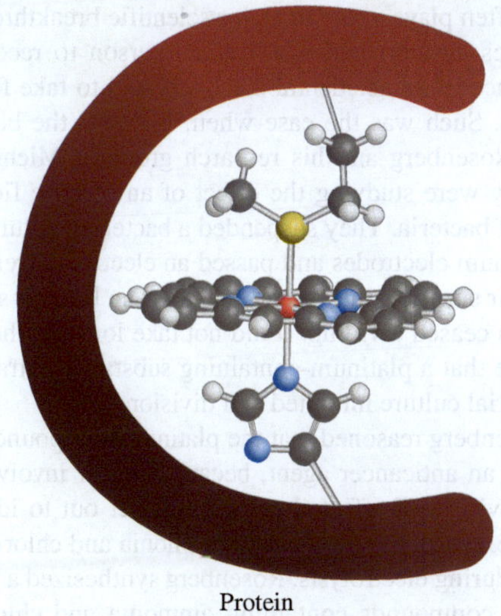

The heme group in cytochrome c. The ligands above and below the porphyrin are the methionine group and histidine group of the protein, respectively.

The chlorophyll molecule, which is necessary for plant photosynthesis, also contains the porphyrin ring, but in this case the metal ion is Mg^{2+} rather than Fe^{2+}.

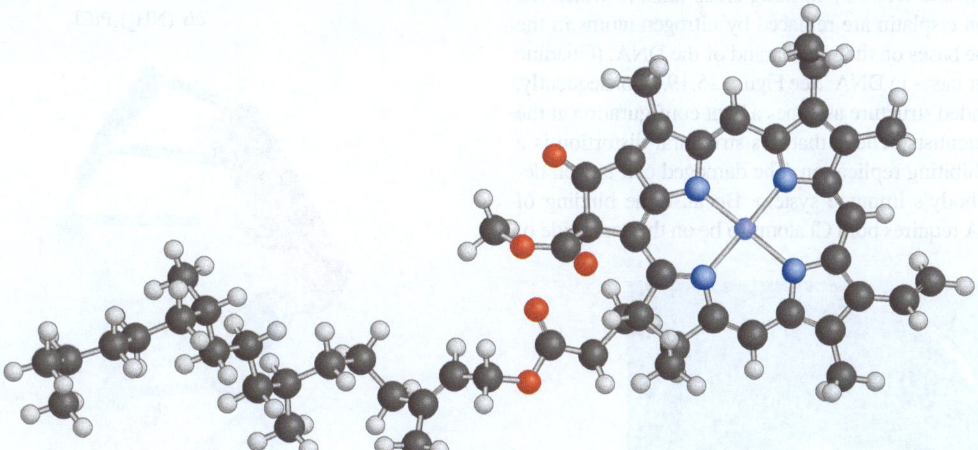

The porphyrin structure in chlorophyll. The dotted lines indicate the coordinate covalent bonds. The electron-delocalized portion of the molecule is shown in color.

CHEMISTRY *in Action*

Cisplatin—The Anticancer Drug

Luck often plays a role in major scientific breakthroughs, but it takes an alert and well-trained person to recognize the significance of an accidental discovery and to take full advantage of it. Such was the case when, in 1964, the biophysicist Barnett Rosenberg and his research group at Michigan State University were studying the effect of an electric field on the growth of bacteria. They suspended a bacterial culture between two platinum electrodes and passed an electric current through it. To their surprise, they found that after an hour or so the bacteria cells ceased dividing. It did not take long for the group to determine that a platinum-containing substance extracted from the bacterial culture inhibited cell division.

Rosenberg reasoned that the platinum compound might be useful as an anticancer agent, because cancer involves uncontrolled division of affected cells, so he set out to identify the substance. Given the presence of ammonia and chloride ions in solution during electrolysis, Rosenberg synthesized a number of platinum compounds containing ammonia and chlorine. The one that proved most effective at inhibiting cell division was *cis*-diamminedichloroplatinum(II) $[Pt(NH_3)_2Cl_2]$, also called cisplatin.

The mechanism for the action of cisplatin is the chelation of DNA (deoxyribonucleic acid), the molecule that contains the genetic code. During cell division, the double-stranded DNA splits into two single strands, which must be accurately copied for the new cells to be identical to their parent cell. X-ray studies show that cisplatin binds to DNA by forming cross-links in which the two chlorides on cisplatin are replaced by nitrogen atoms in the adjacent guanine bases on the *same* strand of the DNA. (Guanine is one of the four bases in DNA. See Figure 25.19.) Consequently, the double-stranded structure assumes a bent configuration at the binding site. Scientists believe that this structural distortion is a key factor in inhibiting replication. The damaged cell is then destroyed by the body's immune system. Because the binding of cisplatin to DNA requires both Cl atoms to be on the same side of the complex, the *trans* isomer of the compound is totally ineffective as an anticancer drug. Unfortunately, cisplatin can cause serious side effects, including severe kidney damage. Therefore, ongoing research efforts are directed toward finding related complexes that destroy cancer cells with less harm to healthy tissues.

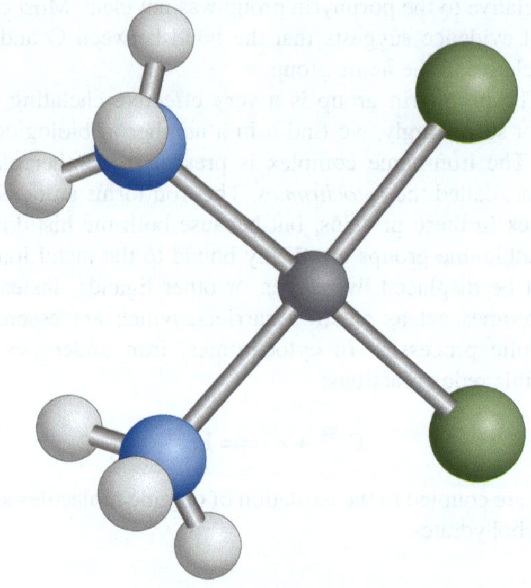

cis-$(NH_3)_2PtCl_2$

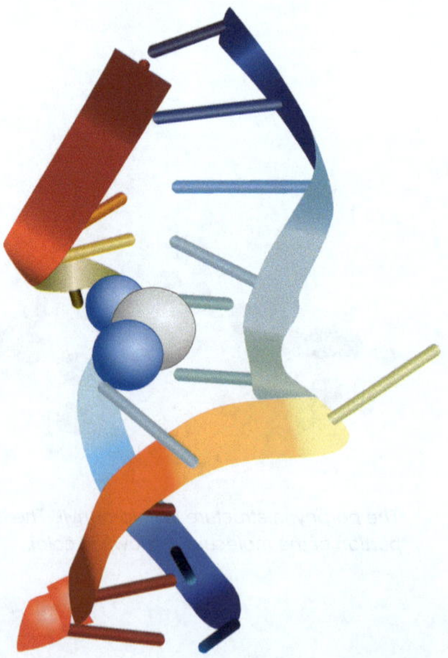

Cisplatin destroys the cancer cells' ability to reproduce by changing the configuration of their DNA. It binds to two sites on a strand of DNA, causing it to bend about 33° away from the rest of the strand. The structure of this DNA adduct was elucidated by Professor Stephen Lippard's group at MIT.

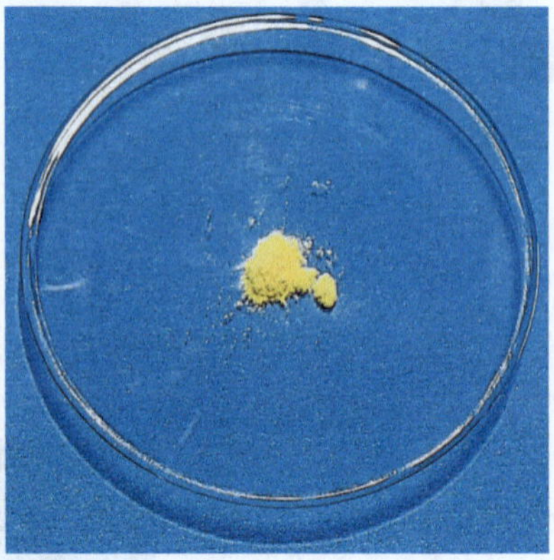

Cisplatin, a bright yellow compound, is administered intravenously to cancer patients.
Courtesy of Raymond Chang

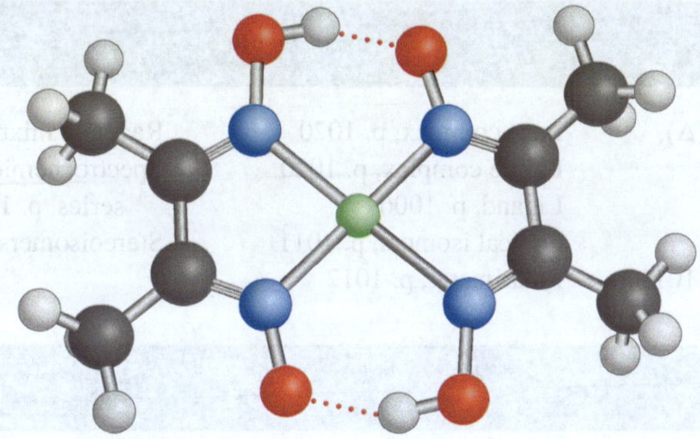

Figure 23.22 *Structure of nickel dimethylglyoxime. Note that the overall structure is stabilized by hydrogen bonds.*

that forms stable, soluble complexes with Ca^{2+} ions. Sodium tripolyphosphate revolutionized the detergent industry. However, because phosphates are plant nutrients, waste waters containing phosphates discharged into rivers and lakes cause algae to grow, resulting in oxygen depletion. Under these conditions most or all aquatic life eventually succumbs. This process is called *eutrophication.* Consequently, many states have banned phosphate detergents since the 1970s, and manufacturers have reformulated their products to eliminate phosphates.

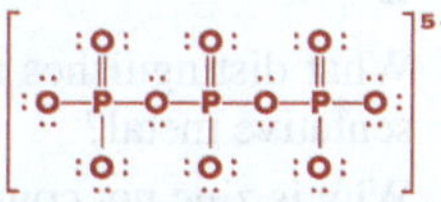

Tripolyphosphate ion.

Summary of Concepts & Facts

- Coordination compounds find application in many different areas, for example, as antidotes for metal poisoning and in chemical analysis.

Chapter Summary

Transition metals Transition metals are those that have incompletely filled *d* subshells—or that *give rise* to ions with incompletely filled *d* subshells. Transition metals exhibit variable oxidation states ranging from +1 to +7. (Section 23.1)

Coordination Compounds A coordination compound contains one or more complex ions in which a small number of molecules or ions surround a central metal atom or ion, usually of the transition metal family. Common geometries of coordination compounds are linear, tetrahedral, square planar, and octahedral. (Sections 23.2 and 23.3)

Bonding in Coordination Compounds Crystal field theory explains the bonding in a complex ion in terms of electrostatic forces. The approach of the ligands toward the metal causes a splitting in energy in the five *d* orbitals. The extent of the splitting, called crystal field splitting, depends on the nature of the ligands. Crystal field theory successfully accounts for the color and magnetic properties of many complex ions. (Section 23.4)

Coordination Complex Reactions Complex ions undergo ligand exchange in solution. The rate at which ligand exchange occurs is a measure of a complex's kinetic lability and does not necessarily correspond directly to the complex's thermodynamic stability. (Section 23.5)

Coordination Compound Applications Coordination chemistry is important in many biological, medical, and industrial processes. (Section 23.6)

Key Equation

$\Delta = h\nu$ (23.1)	Crystal field splitting

Key Words

Chelating agent, p. 1007
Chiral, p. 1012
Coordination compound, p. 1005
Coordination number, p. 1006
Crystal field splitting (Δ), p. 1014
Donor atom, p. 1006
Enantiomers, p. 1013
Geometric isomers, p. 1011
Inert complex, p. 1020
Labile complex, p. 1020
Ligand, p. 1006
Optical isomers, p. 1011
Polarimeter, p. 1012
Racemic mixture, p. 1013
Spectrochemical series, p. 1017
Stereoisomers, p. 1011

Questions & Problems

Red numbered problems solved in Student Solutions Manual

23.1 Properties of the Transition Metals

Review Questions

23.1 What distinguishes a transition metal from a representative metal?

23.2 Why is zinc not considered a transition metal?

23.3 Explain why atomic radii decrease very gradually from scandium to copper.

23.4 Without referring to the text, write the ground-state electron configurations of the first-row transition metals. Explain any irregularities.

23.5 Write the electron configurations of the following ions: V^{5+}, Cr^{3+}, Mn^{2+}, Fe^{3+}, Cu^{2+}, Sc^{3+}, Ti^{4+}.

23.6 Why do transition metals have more oxidation states than other elements?

23.7 Give the highest oxidation states for scandium to copper.

23.8 Why does chromium seem to be less reactive than its standard reduction potential suggests?

23.2 Coordination Compounds

Review Questions

23.9 Define the following terms: *coordination compound, ligand, donor atom, coordination number,* and *chelating agent.*

23.10 Describe the interaction between a donor atom and a metal atom in terms of a Lewis acid-base reaction.

Problems

23.11 Complete the following statements for the complex ion $[Co(en)_2(H_2O)CN]^{2+}$: (a) en is the abbreviation for ____. (b) The oxidation number of Co is ____. (c) The coordination number of Co is ____. (d) ____ is a bidentate ligand.

23.12 Complete the following statements for the complex ion $[Cr(C_2O_4)_2(H_2O)_2]^-$: (a) The oxidation number of Cr is ____. (b) The coordination number of Cr is ____. (c) ____ is a bidentate ligand.

23.13 Give the oxidation numbers of the metals in the following species: (a) $K_3[Fe(CN)_6]$, (b) $K_3[Cr(C_2O_4)_3]$, (c) $[Ni(CN)_4]^{2-}$.

23.14 Give the oxidation numbers of the metals in the following species: (a) Na_2MoO_4, (b) $MgWO_4$, (c) $Fe(CO)_5$.

23.15 What are the systematic names for the following ions and compounds?
(a) $[Co(NH_3)_4Cl_2]^+$
(b) $Cr(NH_3)_3Cl_3$
(c) $[Co(en)_2Br_2]^+$
(d) $[Co(NH_3)_6]Cl_3$

23.16 What are the systematic names for the following ion and compounds?
(a) $[cis\text{-}Co(en)_2Cl_2]^+$
(b) $[Pt(NH_3)_5Cl]Cl_3$
(c) $[Co(NH_3)_5Cl]Cl_2$

23.17 Write the formulas for each of the following ions and compounds: (a) tetrahydroxozincate(II), (b) pentaaquachlorochromium(III) chloride, (c) tetrabromocuprate(II), (d) ethylenediaminetetraacetatoferrate(II).

23.18 Write the formulas for each of the following ions and compounds: (a) bis(ethylenediamine)dichlorochromium(III), (b) pentacarbonyliron(0), (c) potassium tetracyanocuprate(II), (d) tetraammineaquachlorocobalt(III) chloride.

23.3 Structure of Coordination Compounds

Review Questions

23.19 Define the following terms: *stereoisomers, geometric isomers, optical isomers,* and *plane-polarized light.*

23.20 Specify which of the following structures can exhibit geometric isomerism: (a) linear, (b) square planar, (c) tetrahedral, (d) octahedral.

23.21 What determines whether a molecule is chiral? How does a polarimeter measure the chirality of a molecule?

23.22 Explain the following terms: (a) enantiomers, (b) racemic mixtures.

Problems

23.23 The complex ion $[Ni(CN)_2Br_2]^{2-}$ has a square-planar geometry. Draw the structures of the geometric isomers of this complex.

23.24 How many geometric isomers are in the following species: (a) $[Co(NH_3)_2Cl_4]^-$, (b) $[Co(NH_3)_3Cl_3]$?

23.25 Draw structures of all the geometric and optical isomers of each of the following cobalt complexes:
(a) $[Co(NH_3)_6]^{3+}$
(b) $[Co(NH_3)_5Cl]^{2+}$
(c) $[Co(C_2O_4)_3]^{3-}$

23.26 Draw structures of all the geometric and optical isomers of each of the following cobalt complexes: (a) $[Co(NH_3)_4Cl_2]^+$, (b) $[Co(en)_3]^{3+}$.

23.4 Bonding in Coordination Compounds: Crystal Field Theory

Review Questions

23.27 Briefly describe crystal field theory.

23.28 Define the following terms: *crystal field splitting, high-spin complex, low-spin complex,* and *spectrochemical series.*

23.29 What is the origin of color in a coordination compound?

23.30 Compounds containing the Sc^{3+} ion are colorless, whereas those containing the Ti^{3+} ion are colored. Explain.

23.31 What factors determine whether a given complex will be diamagnetic or paramagnetic?

23.32 For the same type of ligands, explain why the crystal field splitting for an octahedral complex is always greater than that for a tetrahedral complex.

Problems

23.33 The $[Ni(CN)_4]^{2-}$ ion, which has a square-planar geometry, is diamagnetic, whereas the $[NiCl_4]^{2-}$ ion, which has a tetrahedral geometry, is paramagnetic. Show the crystal field splitting diagrams for those two complexes.

23.34 Transition metal complexes containing CN^- ligands are often yellow in color, whereas those containing H_2O ligands are often green or blue. Explain.

23.35 Predict the number of unpaired electrons in the following complex ions: (a) $[Cr(CN)_6]^{4-}$, (b) $[Cr(H_2O)_6]^{2+}$.

23.36 The absorption maximum for the complex ion $[Co(NH_3)_6]^{3+}$ occurs at 470 nm. (a) Predict the color of the complex and (b) calculate the crystal field splitting in kJ/mol.

23.37 From each of the following pairs, choose the complex that absorbs light at a longer wavelength: (a) $[Co(NH_3)_6]^{2+}$, $[Co(H_2O)_6]^{2+}$; (b) $[FeF_6]^{3-}$, $[Fe(CN)_6]^{3-}$; (c) $[Cu(NH_3)_4]^{2+}$, $[CuCl_4]^{2-}$.

23.38 A solution made by dissolving 0.875 g of $Co(NH_3)_4Cl_3$ in 25.0 g of water freezes at −0.56°C. Calculate the number of moles of ions produced when 1 mole of $Co(NH_3)_4Cl_3$ is dissolved in water and suggest a structure for the complex ion present in this compound.

23.5 Reactions of Coordination Compounds

Review Questions

23.39 Define the terms (a) labile complex, (b) inert complex.

23.40 Explain why a thermodynamically stable species may be chemically reactive and a thermodynamically unstable species may be unreactive.

Problems

23.41 Oxalic acid, $H_2C_2O_4$, is sometimes used to clean rust stains from sinks and bathtubs. Explain the chemistry underlying this cleaning action.

23.42 The $[Fe(CN)_6]^{3-}$ complex is more labile than the $[Fe(CN)_6]^{4-}$ complex. Suggest an experiment that would prove that $[Fe(CN)_6]^{3-}$ is a labile complex.

23.43 Aqueous copper(II) sulfate solution is blue in color. When aqueous potassium fluoride is added, a green precipitate is formed. When aqueous potassium chloride is added instead, a bright green solution is formed. Explain what is happening in these two cases.

23.44 When aqueous potassium cyanide is added to a solution of copper(II) sulfate, a white precipitate, soluble in an excess of potassium cyanide, is formed. No precipitate is formed when hydrogen sulfide is bubbled through the solution at this point. Explain.

23.45 A concentrated aqueous copper(II) chloride solution is bright green in color. When diluted with water, the solution becomes light blue. Explain.

23.46 In a dilute nitric acid solution, Fe^{3+} reacts with thiocyanate ion (SCN^-) to form a dark red complex:

$$[Fe(H_2O)_6]^{3+} + SCN^- \rightleftharpoons H_2O + [Fe(H_2O)_5NCS]^{2+}$$

The equilibrium concentration of $[Fe(H_2O)_5NCS]^{2+}$ may be determined by how darkly colored the solution is (measured by a spectrometer). In one such experiment, 1.0 mL of 0.20 *M* $Fe(NO_3)_3$ was mixed with 1.0 mL of 1.0×10^{-3} *M* KSCN and 8.0 mL of dilute HNO_3. The color of the solution quantitatively indicated that the $[Fe(H_2O)_5NCS]^{2+}$ concentration was 7.3×10^{-5} *M*. Calculate the formation constant for $[Fe(H_2O)_5NCS]^{2+}$.

Additional Problems

23.47 As we read across the first-row transition metals from left to right, the +2 oxidation state becomes more stable in comparison with the +3 state. Why is this so?

23.48 Which is a stronger oxidizing agent in aqueous solution, Mn^{3+} or Cr^{3+}? Explain your choice.

23.49 Carbon monoxide binds to Fe in hemoglobin some 200 times more strongly than oxygen. This is the reason why CO is a toxic substance. The metal-to-ligand sigma bond is formed by donating a lone pair from the donor atom to an empty sp^3d^2 orbital on Fe. (a) On the basis of electronegativities, would you expect the C or O atom to form the bond to Fe? (b) Draw a diagram illustrating the overlap of the orbitals involved in the bonding.

23.50 What are the oxidation states of Fe and Ti in the ore ilmenite, $FeTiO_3$? (*Hint:* Look up the ionization energies of Fe and Ti in Table 23.1; the fourth ionization energy of Ti is 4180 kJ/mol.)

23.51 A student has prepared a cobalt complex that has one of the following three structures: $[Co(NH_3)_6]Cl_3$, $[Co(NH_3)_5Cl]Cl_2$, or $[Co(NH_3)_4Cl_2]Cl$. Explain how the student would distinguish between these possibilities by an electrical conductance experiment. At the student's disposal are three strong electrolytes—NaCl, $MgCl_2$, and $FeCl_3$—which may be used for comparison purposes.

23.52 Chemical analysis shows that hemoglobin contains 0.34 percent of Fe by mass. What is the minimum possible molar mass of hemoglobin? The actual molar mass of hemoglobin is about 65,000 g. How do you account for the discrepancy between your minimum value and the actual value?

23.53 Explain the following facts: (a) Copper and iron have several oxidation states, whereas zinc has only one. (b) Copper and iron form colored ions, whereas zinc does not.

23.54 A student in 1895 prepared three coordination compounds containing chromium, with the following properties:

Formula	Color	Cl^- Ions in Solution per Formula Unit
(a) $CrCl_3 \cdot 6H_2O$	Violet	3
(b) $CrCl_3 \cdot 6H_2O$	Light green	2
(c) $CrCl_3 \cdot 6H_2O$	Dark green	1

Write modern formulas for these compounds and suggest a method for confirming the number of Cl^- ions present in solution in each case. (*Hint:* Some of the compounds may exist as hydrates and Cr has a coordination number of 6 in all the compounds.)

23.55 The formation constant for the reaction $Ag^+ + 2NH_3 \rightleftharpoons [Ag(NH_3)_2]^+$ is 1.1×10^7 and that for the reaction $Ag^+ + 2CN^- \rightleftharpoons [Ag(CN)_2]^-$ is 1.3×10^{21} at 25°C (see Table 16.4). Calculate the equilibrium constant and $\Delta G°$ at 25°C for the reaction

$$[Ag(NH_3)_2]^+ + 2CN^- \rightleftharpoons [Ag(CN)_2]^- + 2NH_3$$

23.56 From the standard reduction potentials listed in Table 18.1 for Zn/Zn^{2+} and Cu^+/Cu^{2+}, calculate $\Delta G°$ and the equilibrium constant for the reaction

$$Zn(s) + 2Cu^{2+}(aq) \longrightarrow Zn^{2+}(aq) + 2Cu^+(aq)$$

23.57 Using the standard reduction potentials listed in Table 18.1 and the *Handbook of Chemistry and Physics,* show that the following reaction is favorable under standard-state conditions:

$$2Ag(s) + Pt^{2+}(aq) \longrightarrow 2Ag^+(aq) + Pt(s)$$

What is the equilibrium constant of this reaction at 25°C?

23.58 The Co^{2+}-porphyrin complex is more stable than the Fe^{2+}-porphyrin complex. Why, then, is iron the metal ion in hemoglobin (and other heme-containing proteins)?

23.59 What are the differences between geometric isomers and optical isomers?

23.60 Oxyhemoglobin is bright red, whereas deoxyhemoglobin is purple. Show that the difference in color can be accounted for qualitatively on the basis of high-spin and low-spin complexes. (*Hint:* O_2 is a strong-field ligand; see the Chemistry in Action essay, "Coordination Compounds in Living Systems," in Section 23.5.)

23.61 Hydrated Mn^{2+} ions are practically colorless (see Figure 23.20) even though they possess five 3*d* electrons. Explain. (*Hint:* Electronic transitions in which there is a change in the number of unpaired electrons do not occur readily.)

23.62 Which of the following hydrated cations are colorless: $Fe^{2+}(aq)$, $Zn^{2+}(aq)$, $Cu^+(aq)$, $Cu^{2+}(aq)$, $V^{5+}(aq)$, $Ca^{2+}(aq)$, $Co^{2+}(aq)$, $Sc^{3+}(aq)$, $Pb^{2+}(aq)$? Explain your choice.

23.63 Aqueous solutions of $CoCl_2$ are generally either light pink or blue. Low concentrations and low temperatures favor the pink form while high concentrations and high temperatures favor the blue form. Adding hydrochloric acid to a pink solution of $CoCl_2$ causes the solution to turn blue; the pink color is restored by the addition of $HgCl_2$. Account for these observations.

23.64 Suggest a method that would allow you to distinguish between *cis*-$Pt(NH_3)_2Cl_2$ and *trans*-$Pt(NH_3)_2Cl_2$.

23.65 You are given two solutions containing $FeCl_2$ and $FeCl_3$ at the same concentration. One solution is light yellow and the other one is brown. Identify these solutions based only on color.

23.66 The label of a certain brand of mayonnaise lists EDTA as a food preservative. How does EDTA prevent the spoilage of mayonnaise?

23.67 The compound 1,1,1-trifluoroacetylacetone (tfa) is a bidentate ligand:

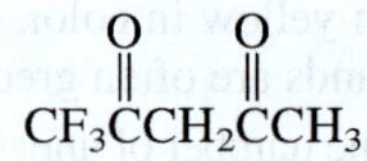

It forms a tetrahedral complex with Be^{2+} and a square-planar complex with Cu^{2+}. Draw structures of these complex ions and identify the type of isomerism exhibited by these ions.

23.68 How many geometric isomers can the following square-planar complex have?

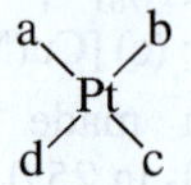

23.69 $[Pt(NH_3)_2Cl_2]$ is found to exist in two geometric isomers designated I and II, which react with oxalic acid as follows:

$$I + H_2C_2O_4 \longrightarrow [Pt(NH_3)_2C_2O_4]$$
$$II + H_2C_2O_4 \longrightarrow [Pt(NH_3)_2(HC_2O_4)_2]$$

Comment on the structures of I and II.

23.70 The K_f for the complex ion formation between Pb^{2+} and $EDTA^{4-}$

$$Pb^{2+} + EDTA^{4-} \rightleftharpoons Pb(EDTA)^{2-}$$

is 1.0×10^{18} at 25°C. Calculate $[Pb^{2+}]$ at equilibrium in a solution containing $1.0 \times 10^{-3}\ M\ Pb^{2+}$ and $2.0 \times 10^{-3}\ M\ EDTA^{4-}$.

23.71 Manganese forms three low-spin complex ions with the cyanide ion with the formulas $[Mn(CN)_6]^{5-}$, $[Mn(CN)_6]^{4-}$, and $[Mn(CN)_6]^{3-}$. For each complex ion, determine the oxidation number of Mn and the number of unpaired *d* electrons present.

23.72 Commercial silver-plating operations frequently use a solution containing the complex $Ag(CN)_2^-$ ion. Because the formation constant (K_f) is quite large, this procedure ensures that the free Ag^+ concentration in solution is low for uniform electrodeposition. In one process, a chemist added 9.0 L of 5.0 *M* NaCN to 90.0 L of 0.20 *M* $AgNO_3$. Calculate the concentration of free Ag^+ ions at equilibrium. See Table 16.4 for K_f value.

23.73 Draw qualitative diagrams for the crystal field splittings in (a) a linear complex ion ML_2, (b) a trigonal-planar complex ion ML_3, (c) a trigonal-bipyramidal complex ion ML_5.

23.74 (a) The free Cu(I) ion is unstable in solution and has a tendency to disproportionate:

$$2Cu^+(aq) \rightleftharpoons Cu^{2+}(aq) + Cu(s)$$

Use the information in Table 18.1 to calculate the equilibrium constant for the reaction. (b) Based on your result in (a), explain why most Cu(I) compounds are insoluble.

23.75 Consider the following two ligand exchange reactions:

$$[Co(H_2O)_6]^{3+} + 6NH_3 \rightleftharpoons [Co(NH_3)_6]^{3+} + 6H_2O$$
$$[Co(H_2O)_6]^{3+} + 3en \rightleftharpoons [Co(en)_3]^{3+} + 6H_2O$$

(a) Which of the reactions should have a larger $\Delta S°$? (b) Given that the Co—N bond strength is approximately the same in both complexes, which reaction will have a larger equilibrium constant? Explain your choices.

23.76 Copper is also known to exist in the +3 oxidation state, which is believed to be involved in some biological electron-transfer reactions. (a) Would you expect this oxidation state of copper to be stable? Explain. (b) Name the compound K_3CuF_6 and predict the geometry of the complex ion and its magnetic properties. (c) Most of the known Cu(III) compounds have square planar geometry. Are these compounds diamagnetic or paramagnetic?

Answers to Practice Exercises

23.1 K: +1; Au: +3. **23.2** Tetraaquadichlorochromium(III) chloride. **23.3** $[Co(en)_3]_2(SO_4)_3$. **23.4** 5.

Answers to Review of Concepts & Facts

23.1.1 (a) Tc, (b) W, (c) Mn^{4+}, (d) Au^{3+}. **23.2.1** (a) +3. (b) +2. (c) +3. **23.2.2** $CrCl_3 \cdot 6H_2O$ is a hydrate compound. The water molecules are associated with the $CrCl_3$ unit. $[Cr(H_2O)_6]Cl_3$ is a coordination compound. The water molecules are ligands that bonded to the Cr^{3+} ion. **23.2.3** No; tetraaquadichlorochromium(III) chloride. **23.3.1** Three geometric isomers. **23.4.1** The yellow color of CrY_6^{3+} means that the compound absorbs in the blue-violet region, which has a larger ligand-field splitting. Thus Y has a stronger field strength. **23.4.2** (a) 0. (b) 5. (c) 3.

CHAPTER

24

Organic Chemistry

The oceans are a fertile source in the search for new and useful natural organic products. Corals, sponges, fish, and even marine microorganisms produce a bounty of biologically potent organic compounds with numerous medicinal properties.
Amith Nag Photography/Moment/Getty Images

CHAPTER OUTLINE

Organic chemistry is the study of carbon compounds. The word "organic" was originally used by eighteenth-century chemists to describe substances obtained from living sources—plants and animals. These chemists believed that nature possessed a certain vital force and that only living things could produce organic compounds. This romantic notion was disproved in 1828 by Friedrich Wohler, a German chemist who prepared urea, an organic compound, from the reaction between inorganic compounds lead cyanate and aqueous ammonia:

$$Pb(OCN)_2 + 2NH_3 + 2H_2O \longrightarrow \underset{\text{urea}}{2(NH_2)_2CO} + Pb(OH)_2$$

Today, well over 20 million synthetic and natural organic compounds are known. This number is significantly greater than the 100,000 or so known inorganic compounds.

24.1 Classes of Organic Compounds

Learning Objective

- List characteristics of carbon that allow it to form so many different compounds.

Carbon can form more compounds than any other element because carbon atoms are able not only to form single, double, and triple carbon-carbon bonds, but also to link up with each other in chains and ring structures. Recall that the linking of like atoms is called catenation. The ability of carbon to catenate is discussed in Section 22.3. *The branch of chemistry that deals with carbon compounds is* ***organic chemistry.***

Classes of organic compounds can be distinguished according to functional groups they contain. A ***functional group*** is *a group of atoms that is largely responsible for the chemical behavior of the parent molecule*. Different molecules containing the same kind of functional group or groups undergo similar reactions. Thus, by learning the characteristic properties of a few functional groups, we can study and understand the properties of many organic compounds. In the second half of this chapter, we will discuss the functional groups known as alcohols, ethers, aldehydes and ketones, carboxylic acids, and amines.

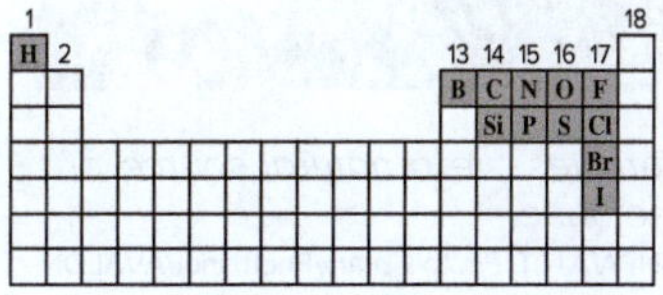

Common elements in organic compounds.

Most organic compounds are derived from a group of compounds known as ***hydrocarbons*** because they are *made up of only hydrogen and carbon*. On the basis of structure, hydrocarbons are divided into two main classes—aliphatic and aromatic. ***Aliphatic hydrocarbons*** *do not contain the benzene group, or the benzene ring,* whereas ***aromatic hydrocarbons*** *contain one or more benzene rings.*

Summary of Concepts & Facts

- Because carbon atoms can link up with other carbon atoms in straight and branched chains, carbon can form more compounds than any other element.
- Organic compounds are derived from two types of hydrocarbons: aliphatic hydrocarbons and aromatic hydrocarbons.

24.2 Aliphatic Hydrocarbons

Learning Objectives

- Name alkanes, alkenes, and alkynes.
- Produce structural isomers of alkanes given the formula.
- Predict chiral molecules based on their structures.
- Contrast *cis* and *trans* isomers of alkenes.

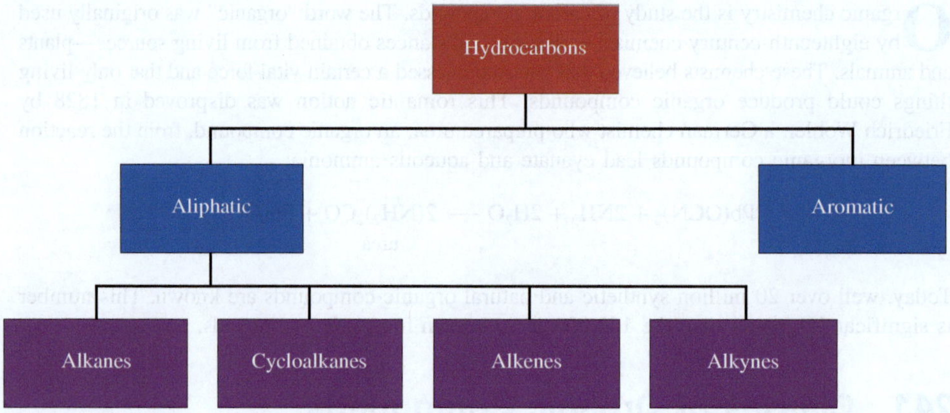

Figure 24.1 *Classification of hydrocarbons.*

Aliphatic hydrocarbons are divided into alkanes, alkenes, and alkynes, discussed next (Figure 24.1).

Alkanes

Termites are a natural source of methane.
NHPA/A.N.T. Photo Library/Photoshot/AVALON

Alkanes** have the general formula C_nH_{2n+2}, where $n = 1, 2, \ldots$.* The essential characteristic of alkane hydrocarbon molecules is that *only single covalent bonds are present.* The alkanes are known as ***saturated hydrocarbons because they *contain the maximum number of hydrogen atoms that can bond with the number of carbon atoms present.*

The simplest alkane (that is, with $n = 1$) is methane CH_4, which is a natural product of the anaerobic bacterial decomposition of vegetable matter under water. Because it was first collected in marshes, methane became known as "marsh gas." A rather improbable but proven source of methane is termites. When these voracious insects consume wood, the microorganisms that inhabit their digestive system break down cellulose (the major component of wood) into methane, carbon dioxide, and other compounds. An estimated 170 million tons of methane are produced annually by termites! It is also produced in some sewage treatment processes. Commercially, methane is obtained from natural gas. The Chemistry in Action essay "Ice That Burns" describes an interesting compound formed by methane and water molecules.

Figure 24.2 shows the structures of the first four alkanes ($n = 1$ to $n = 4$). Natural gas is a mixture of methane, ethane, and a small amount of propane. We discussed the bonding scheme of methane in Chapter 10. Indeed, the carbon atoms in all the alkanes can be assumed to be sp^3-hybridized. The structures of ethane and propane are straightforward, for there is only one way to join the carbon atoms in these molecules. Butane, however, has two possible bonding schemes resulting in the ***structural isomers*** *n*-butane (*n* stands for normal) and isobutane, *molecules that have the same molecular formula, but different structures.* Alkanes such as the structural isomers of butane are described as having the straight-chain or branched-chain structures.

Figure 24.2 *Structures of the first four alkanes. Note that butane can exist in two structurally different forms, called structural isomers.*

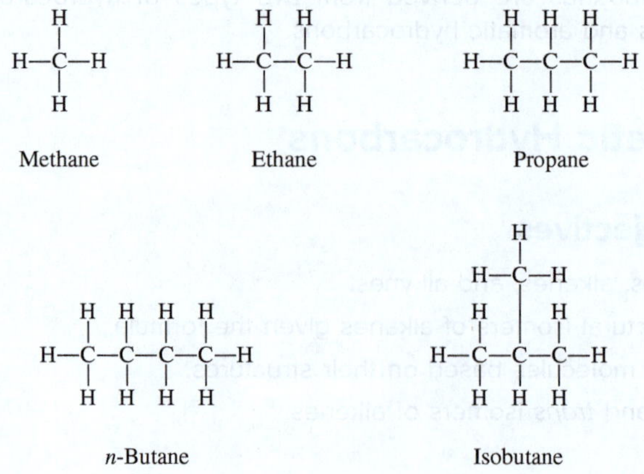

CHEMISTRY *in Action*

Ice That Burns

Ice that burns? Yes, there is such a thing. It is called *methane hydrate*, and there is enough of it to meet America's energy needs for years. But scientists have yet to figure out how to mine it without causing an environmental disaster.

Bacteria in the sediments on the ocean floor consume organic material and generate methane gas. Under high-pressure and low-temperature conditions, methane forms methane hydrate, which consists of single molecules of the natural gas trapped within crystalline cages formed by frozen water molecules. A lump of methane hydrate looks like a gray ice cube, but if one puts a lighted match to it, it will burn.

Oil companies have known about methane hydrate since the 1930s, when they began using high-pressure pipelines to transport natural gas in cold climates. Unless water is carefully removed before the gas enters the pipeline, chunks of methane hydrate will impede the flow of gas.

The total reserve of the methane hydrate in the world's oceans is estimated to be 10^{13} tons of carbon content, about twice the amount of carbon in all the coal, oil, and natural gas on land. However, harvesting the energy stored in methane hydrate presents a tremendous engineering challenge. It is believed that methane hydrate acts as a kind of cement to keep the ocean floor sediments together. Tampering with the hydrate deposits could cause underwater landslides, leading to the discharge of methane into the atmosphere. This event could have serious consequences for the environment because methane is a potent greenhouse gas (see Section 20.5). In fact, scientists have speculated that the abrupt release of methane hydrates may have hastened the end of the last ice age about 10,000 years ago. As the great blanket of continental ice melted, global sea levels swelled by more than 90 m, submerging Arctic regions rich in hydrate deposits. The relatively warm ocean water would have melted the hydrates, unleashing tremendous amounts of methane, which led to global warming.

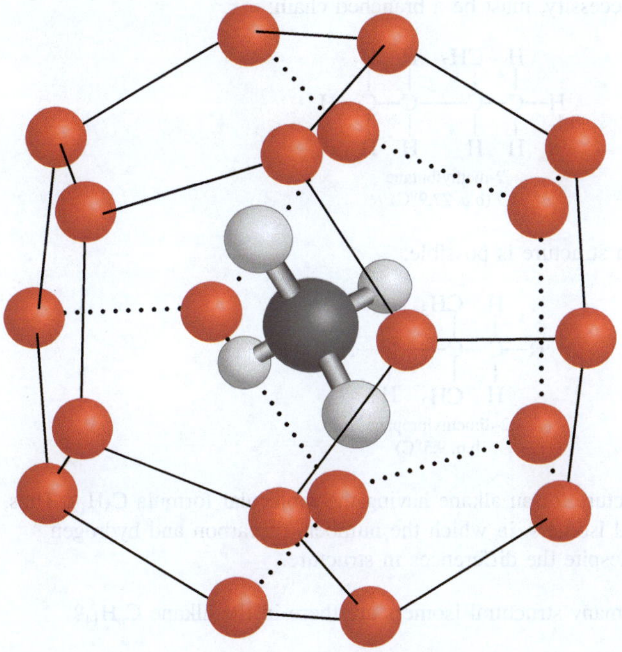

Methane hydrate. The methane molecule is trapped in a cage of frozen water molecules (red spheres) held together by hydrogen bonds.

Methane hydrate burning in air.
L. Stern and J. Pinkston/U.S. Geological Survey

n-Butane is a straight-chain alkane because the carbon atoms are joined along one line. In a branched-chain alkane like isobutane, one or more carbon atoms are bonded to at least three other carbon atoms.

In the alkane series, as the number of carbon atoms increases, the number of structural isomers increases rapidly. For example, butane, C_4H_{10}, has two isomers; decane, $C_{10}H_{22}$, has 75 isomers; and the alkane $C_{30}H_{62}$ has over 400 million, or 4×10^8, possible isomers! Obviously, most of these isomers do not exist in nature nor have they been synthesized. Nevertheless, the numbers help to explain why carbon is found in so many more compounds than any other element.

Example 24.1 deals with the number of structural isomers of an alkane.

Example 24.1

How many structural isomers can be identified for pentane, C_5H_{12}?

Strategy For small hydrocarbon molecules (eight or fewer C atoms), it is relatively easy to determine the number of structural isomers by trial and error.

Solution The first step is to write the straight-chain structure:

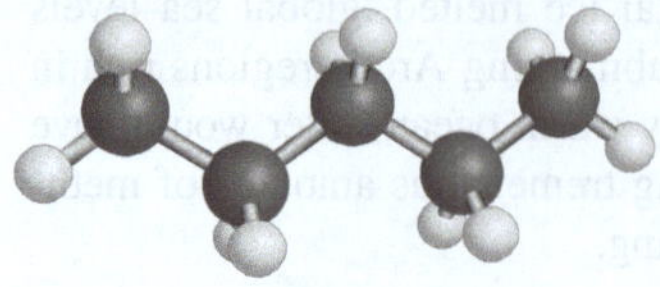

n-pentane

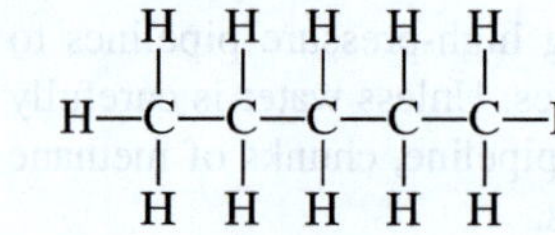

n-pentane
(b.p. 36.1°C)

The second structure, by necessity, must be a branched chain:

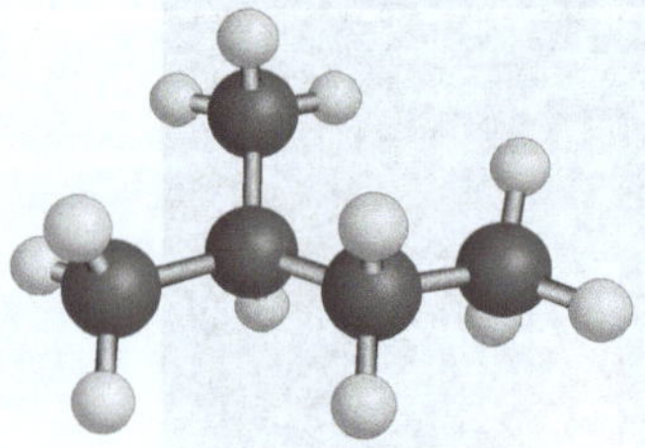

2-methylbutane

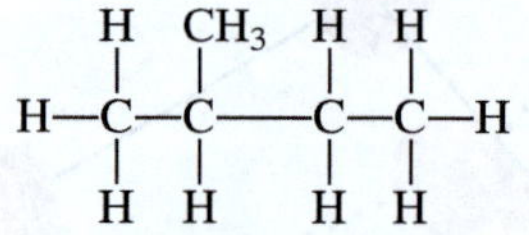

2-methylbutane
(b.p. 27.9°C)

Yet another branched-chain structure is possible:

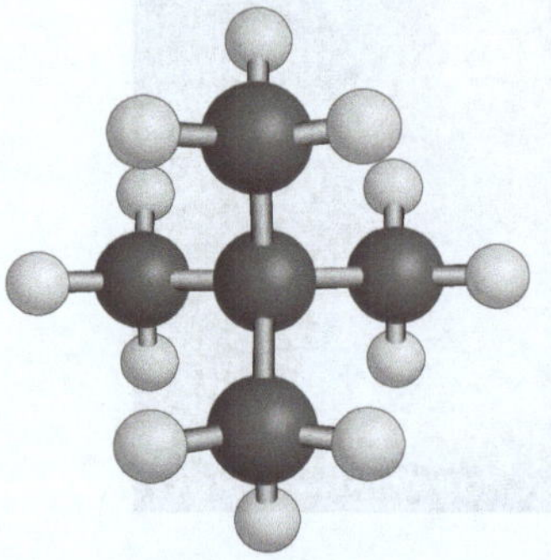

2,2-dimethylpropane

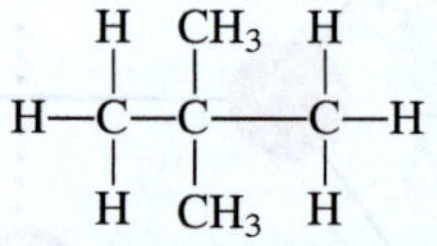

2,2-dimethylpropane
(b.p. 9.5°C)

We can draw no other structure for an alkane having the molecular formula C_5H_{12}. Thus, pentane has three structural isomers, in which the numbers of carbon and hydrogen atoms remain unchanged despite the differences in structure.

Practice Exercise How many structural isomers are there in the alkane C_6H_{14}?

Similar problem: 24.11.

Table 24.1 shows the melting and boiling points of the straight-chain isomers of the first 10 alkanes. The first four are gases at room temperature; and pentane through decane are liquids. As molecular size increases, so does the boiling point, because of the increasing dispersion forces (see Section 11.2).

Alkane Nomenclature

The nomenclature of alkanes and all other organic compounds is based on the recommendations of the International Union of Pure and Applied Chemistry (IUPAC). The first four alkanes (methane, ethane, propane, and butane) have nonsystematic names. As Table 24.1 shows, the number of carbon atoms is reflected in the Greek prefixes for

Table 24.1 **The First 10 Straight-Chain Alkanes**

Name of Hydrocarbon	Molecular Formula	Number of Carbon Atoms	Melting Point (°C)	Boiling Point (°C)
Methane	CH_4	1	−182.5	−161.6
Ethane	$CH_3—CH_3$	2	−183.3	−88.6
Propane	$CH_3—CH_2—CH_3$	3	−189.7	−42.1
Butane	$CH_3—(CH_2)_2—CH_3$	4	−138.3	−0.5
Pentane	$CH_3—(CH_2)_3—CH_3$	5	−129.8	36.1
Hexane	$CH_3—(CH_2)_4—CH_3$	6	−95.3	68.7
Heptane	$CH_3—(CH_2)_5—CH_3$	7	−90.6	98.4
Octane	$CH_3—(CH_2)_6—CH_3$	8	−56.8	125.7
Nonane	$CH_3—(CH_2)_7—CH_3$	9	−53.5	150.8
Decane	$CH_3—(CH_2)_8—CH_3$	10	−29.7	174.0

the alkanes containing 5 to 10 carbons. We now apply the IUPAC rules to the following examples:

1. The parent name of the hydrocarbon is that given to the longest continuous chain of carbon atoms in the molecule. Thus, the parent name of the following compound is heptane because there are seven carbon atoms in the longest chain:

$$\overset{1}{CH_3}—\overset{2}{CH_2}—\overset{3}{CH_2}—\overset{\overset{\displaystyle CH_3}{|}}{\underset{4}{CH}}—\overset{5}{CH_2}—\overset{6}{CH_2}—\overset{7}{CH_3}$$

2. An alkane less one hydrogen atom is an *alkyl* group. For example, when a hydrogen atom is removed from methane, we are left with the CH_3 fragment, which is called a *methyl* group. Similarly, removing a hydrogen atom from the ethane molecule gives an *ethyl* group, or C_2H_5. Table 24.2 lists the names of several common alkyl groups. Any chain branching off the longest chain is named as an alkyl group.

3. When one or more hydrogen atoms are replaced by other groups, the name of the compound must indicate the locations of carbon atoms where replacements are made. The procedure is to number each carbon atom on the longest chain in

Table 24.2 **Common Alkyl Groups**

<table>
<tr><th>Name</th><th>Formula</th></tr>
<tr><td>Methyl</td><td>$—CH_3$</td></tr>
<tr><td>Ethyl</td><td>$—CH_2—CH_3$</td></tr>
<tr><td>n-Propyl</td><td>$—CH_2—CH_2—CH_3$</td></tr>
<tr><td>n-Butyl</td><td>$—CH_2—CH_2—CH_2—CH_3$</td></tr>
<tr><td>Isopropyl</td><td>CH_3
|
—C—H
|
CH_3</td></tr>
<tr><td>t-Butyl*</td><td>CH_3
|
$—C—CH_3$
|
CH_3</td></tr>
</table>

*The letter *t* stands for tertiary.

the direction that gives the smaller numbers for the locations of all branches. Consider the two different systems for the *same* compound shown here:

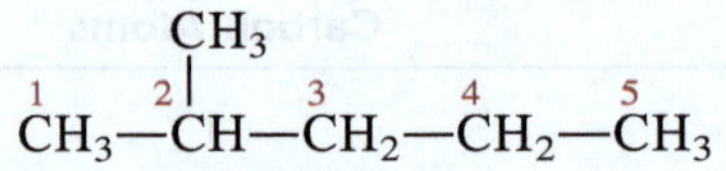
2-methylpentane

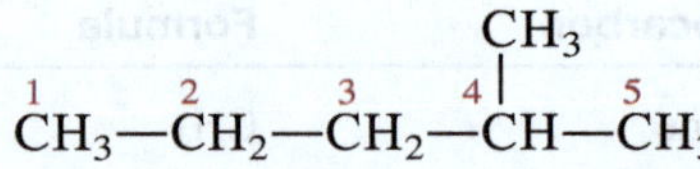
4-methylpentane

The compound on the left is numbered correctly because the methyl group is located at carbon 2 of the pentane chain; in the compound on the right, the methyl group is located at carbon 4. Thus, the name of the compound is 2-methylpentane, not 4-methylpentane. Note that the branch name and the parent name are written as a single word, and a hyphen follows the number.

4. When there is more than one alkyl branch of the same kind present, we use a prefix such as *di-, tri-,* or *tetra-* with the name of the alkyl group. Consider the following examples:

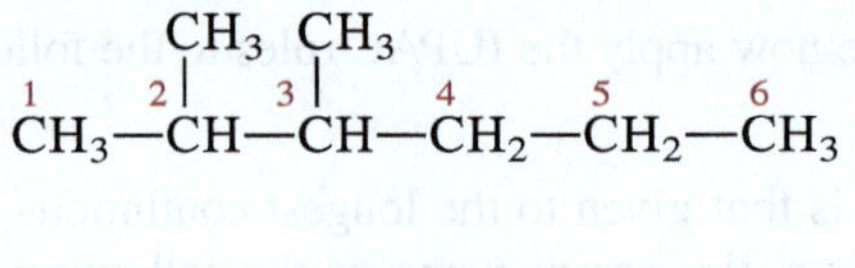
2,3-dimethylhexane

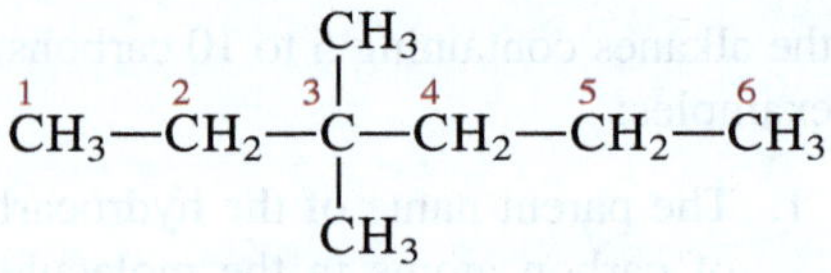
3,3-dimethylhexane

When there are two or more different alkyl groups, the names of the groups are listed alphabetically. For example,

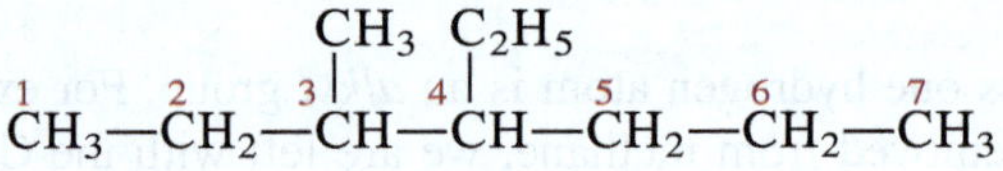
4-ethyl-3-methylheptane

5. Of course, alkanes can have many different types of substituents. Table 24.3 lists the names of some substituents, including nitro and bromo. Thus, the compound

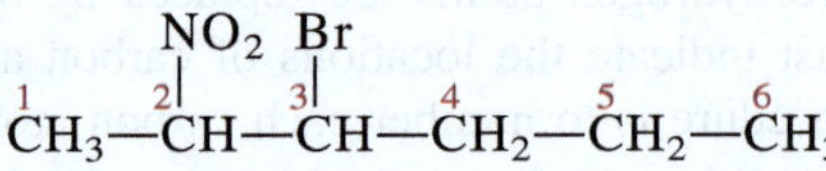

is called 3-bromo-2-nitrohexane. Note that the substituent groups are listed alphabetically in the name, and the chain is numbered in the direction that gives the lowest number to the first substituted carbon atom.

Table 24.3

Names of Common Substituent Groups

Functional Group	Name
$-NH_2$	Amino
$-F$	Fluoro
$-Cl$	Chloro
$-Br$	Bromo
$-I$	Iodo
$-NO_2$	Nitro
$-CH{=}CH_2$	Vinyl

Student Hot Spot

Student data indicate you may struggle with naming alkanes. Access your eBook for additional Learning Resources on this topic.

Example 24.2

Give the IUPAC name of the following compound:

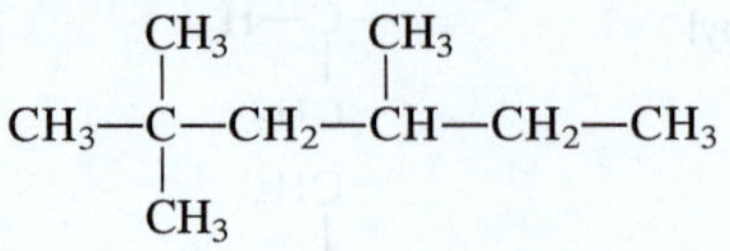

Strategy We follow the IUPAC rules and use the information in Table 24.2 to name the compound. How many C atoms are there in the longest chain?

(Continued)

Solution The longest chain has six C atoms so the parent compound is called hexane. Note that there are two methyl groups attached to carbon number 2 and one methyl group attached to carbon number 4.

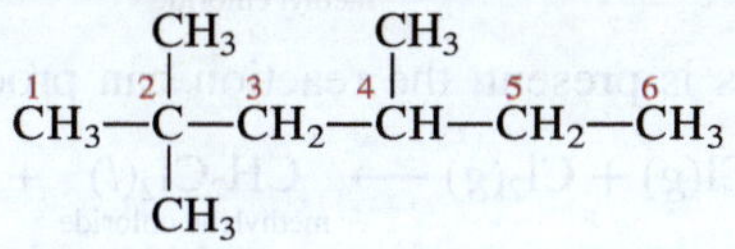

Therefore, we call the compound 2,2,4-trimethylhexane.

Practice Exercise Give the IUPAC name of the following compound:

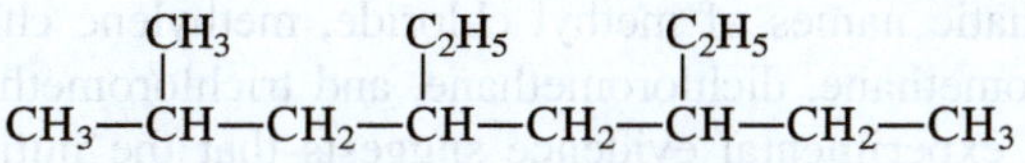

Similar problem: 24.26.

Example 24.3 shows that prefixes such as *di-*, *tri-*, and *tetra-* are used as needed, but are ignored when alphabetizing.

Example 24.3

Write the structural formula of 3-ethyl-2,2-dimethylpentane.

Strategy We follow the preceding procedure and the information in Table 24.2 to write the structural formula of the compound. How many C atoms are there in the longest chain?

Solution The parent compound is pentane, so the longest chain has five C atoms. There are two methyl groups attached to carbon number 2 and one ethyl group attached to carbon number 3. Therefore, the structure of the compound is

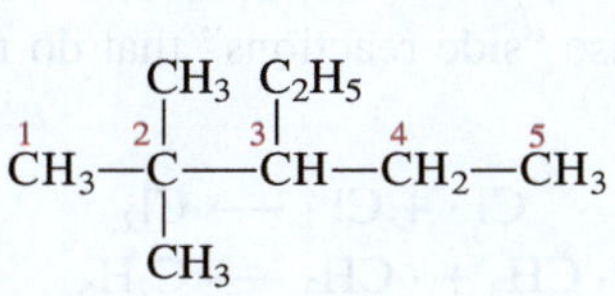

Practice Exercise Write the structural formula of 5-ethyl-2,4,6-trimethyloctane.

Similar problem: 24.27.

Reactions of Alkanes

Alkanes are generally not considered to be very reactive substances. However, under suitable conditions they do react. For example, natural gas, gasoline, and fuel oil are alkanes that undergo highly exothermic combustion reactions:

$$CH_4(g) + 2O_2(g) \longrightarrow CO_2(g) + 2H_2O(l) \qquad \Delta H^\circ = -890.4 \text{ kJ/mol}$$

$$2C_2H_6(g) + 7O_2(g) \longrightarrow 4CO_2(g) + 6H_2O(l) \qquad \Delta H^\circ = -3119 \text{ kJ/mol}$$

These, and similar combustion reactions, have long been utilized in industrial processes and in domestic heating and cooking.

Halogenation of alkanes—that is, the replacement of one or more hydrogen atoms by halogen atoms—is another type of reaction that alkanes undergo. When a mixture of methane and chlorine is heated above 100°C or irradiated with light of a suitable wavelength, methyl chloride is produced:

$$CH_4(g) + Cl_2(g) \longrightarrow \underset{\text{methyl chloride}}{CH_3Cl(g)} + HCl(g)$$

If an excess of chlorine gas is present, the reaction can proceed further:

$$CH_3Cl(g) + Cl_2(g) \longrightarrow \underset{\text{methylene chloride}}{CH_2Cl_2(l)} + HCl(g)$$

$$CH_2Cl_2(l) + Cl_2(g) \longrightarrow \underset{\text{chloroform}}{CHCl_3(l)} + HCl(g)$$

$$CHCl_3(l) + Cl_2(g) \longrightarrow \underset{\text{carbon tetrachloride}}{CCl_4(l)} + HCl(g)$$

Note that the systematic names of methyl chloride, methylene chloride, and chloroform are monochloromethane, dichloromethane, and trichloromethane, respectively.

A great deal of experimental evidence suggests that the initial step of the first halogenation reaction occurs as follows:

$$Cl_2 + \text{energy} \longrightarrow Cl\cdot + Cl\cdot$$

Thus, the covalent bond in Cl_2 breaks and two chlorine atoms form. We know it is the Cl—Cl bond that breaks when the mixture is heated or irradiated because the bond enthalpy of Cl_2 is 242.7 kJ/mol whereas about 414 kJ/mol are needed to break C—H bonds in CH_4.

A chlorine atom is a *radical,* which contains an unpaired electron (shown by a single dot). Chlorine atoms are highly reactive and attack methane molecules according to the equation

$$CH_4 + Cl\cdot \longrightarrow \cdot CH_3 + HCl$$

This reaction produces hydrogen chloride and the methyl radical $\cdot CH_3$. The methyl radical is another reactive species; it combines with molecular chlorine to give methyl chloride and a chlorine atom:

$$\cdot CH_3 + Cl_2 \longrightarrow CH_3Cl + Cl\cdot$$

The production of methylene chloride from methyl chloride and any further reactions can be explained in the same way. The actual mechanism is more complex than the scheme we have shown because "side reactions" that do not lead to the desired products often take place, such as

$$Cl\cdot + Cl\cdot \longrightarrow Cl_2$$

$$\cdot CH_3 + \cdot CH_3 \longrightarrow C_2H_6$$

Alkanes in which one or more hydrogen atoms have been replaced by a halogen atom are called *alkyl halides*. Among the large number of alkyl halides, the best known are chloroform ($CHCl_3$), carbon tetrachloride (CCl_4), methylene chloride (CH_2Cl_2), and the chlorofluorohydrocarbons.

Chloroform is a volatile, sweet-tasting liquid that was used for many years as an anesthetic. However, because of its toxicity (it can severely damage the liver, kidneys, and heart) it has been replaced by other compounds. Carbon tetrachloride, also a toxic substance, serves as a cleaning liquid, for it removes grease stains from clothing. Methylene chloride was used as a solvent to decaffeinate coffee and as a paint remover.

The preparation of chlorofluorocarbons and the effect of these compounds on ozone in the stratosphere were discussed in Chapter 20.

Optical Isomerism of Substituted Alkanes

Video: Chirality

Optical isomers are compounds that are nonsuperimposable mirror images of each other. Figure 24.3 shows perspective drawings of the substituted methanes CH_2ClBr and $CHFClBr$ and their mirror images. The mirror images of CH_2ClBr are superimposable

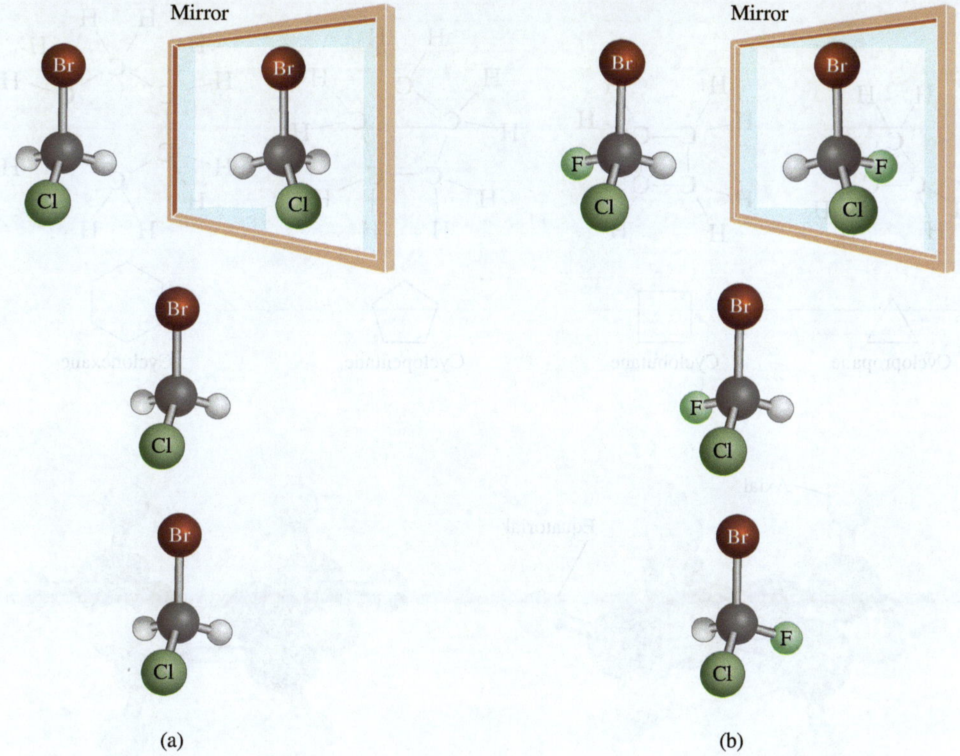

Figure 24.3 *(a) The CH_2ClBr molecule and its mirror image. Because the molecule and its mirror image are superimposable, the molecule is said to be achiral. (b) The CHFClBr molecule and its mirror image. Because the molecule and its mirror image are not superimposable, no matter how we rotate one with respect to the other, the molecule is said to be chiral.*

but those of CHFClBr are not, no matter how we rotate the molecules. Thus, the CHFClBr molecule is chiral. Most simple chiral molecules contain at least one *asymmetric* carbon atom—that is, a carbon atom bonded to four different atoms or groups of atoms.

Student Hot Spot
Student data indicate you may struggle with optical isomers. Access your eBook for additional Learning Resources on this topic.

Example 24.4

Is the following molecule chiral?

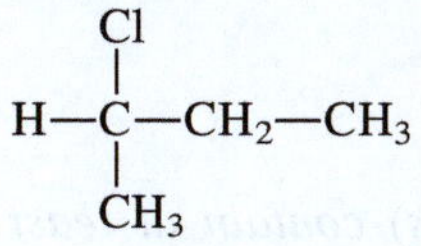

Strategy Recall the condition for chirality. Is the central C atom asymmetric; that is, does it have four different atoms or different groups attached to it?

Solution We note that the central carbon atom is bonded to a hydrogen atom, a chlorine atom, a —CH_3 group, and a —CH_2—CH_3 group. Therefore, the central carbon atom is asymmetric and the molecule is chiral.

Practice Exercise Is the following molecule chiral?

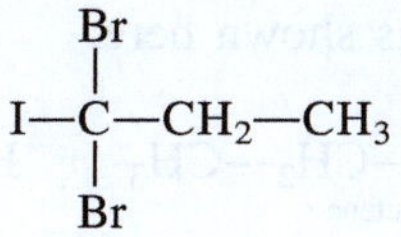

Similar problem: 24.25.

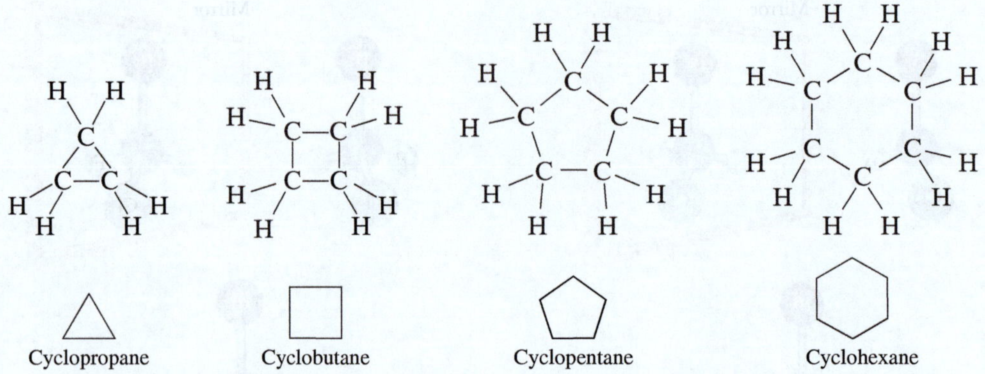

Figure 24.4 *Structures of the first four cycloalkanes and their simplified forms.*

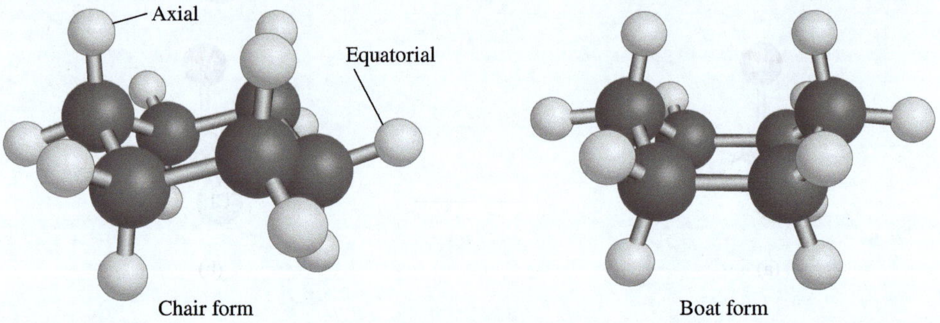

Figure 24.5 *The cyclohexane molecule can exist in various shapes. The most stable shape is the chair form and a less stable one is the boat form. Two types of H atoms are labeled axial and equatorial, respectively.*

Cycloalkanes

Alkanes whose carbon atoms are joined in rings are known as ***cycloalkanes.*** They have the general formula C_nH_{2n}, where n = 3, 4, ... The simplest cycloalkane is cyclopropane, C_3H_6 (Figure 24.4). Many biologically significant substances such as cholesterol, testosterone, and progesterone contain one or more such ring systems. Theoretical analysis shows that cyclohexane can assume two different geometries that are relatively free of strain (Figure 24.5). By "strain" we mean that bonds are compressed, stretched, or twisted out of normal geometric shapes as predicted by sp^3 hybridization. The most stable geometry is the *chair form.*

Alkenes

The ***alkenes*** (also called *olefins*) *contain at least one carbon-carbon double bond. Alkenes have the general formula C_nH_{2n}, where n = 2, 3, ...* The simplest alkene is C_2H_4, ethylene, in which both carbon atoms are sp^2-hybridized and the double bond is made up of a sigma bond and a pi bond (see Section 10.5).

Alkene Nomenclature

In naming alkenes, we indicate the positions of the carbon-carbon double bonds. The names of compounds containing C═C bonds end with *-ene.* As with the alkanes, the name of the parent compound is determined by the number of carbon atoms in the longest chain (see Table 24.1), as shown here:

$$\underset{\text{1-butene}}{CH_2{=}CH{-}CH_2{-}CH_3} \qquad \underset{\text{2-butene}}{H_3C{-}CH{=}CH{-}CH_3}$$

The numbers in the names of alkenes refer to the lowest numbered carbon atom in the chain that is part of the C═C bond of the alkene. The name "butene" means that there are four carbon atoms in the longest chain. Alkene nomenclature must specify whether a given molecule is *cis* or *trans* if it is a geometric isomer, such as

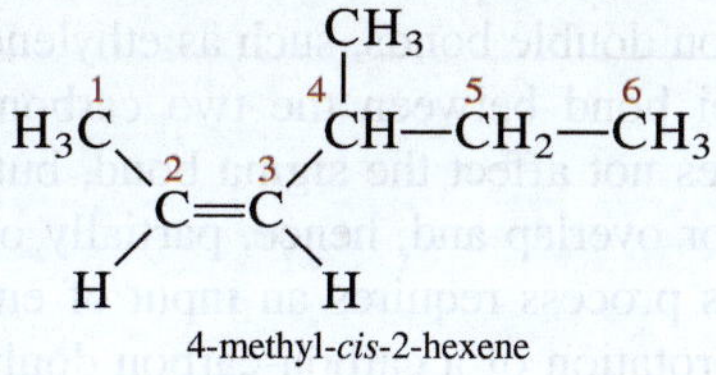

4-methyl-*cis*-2-hexene

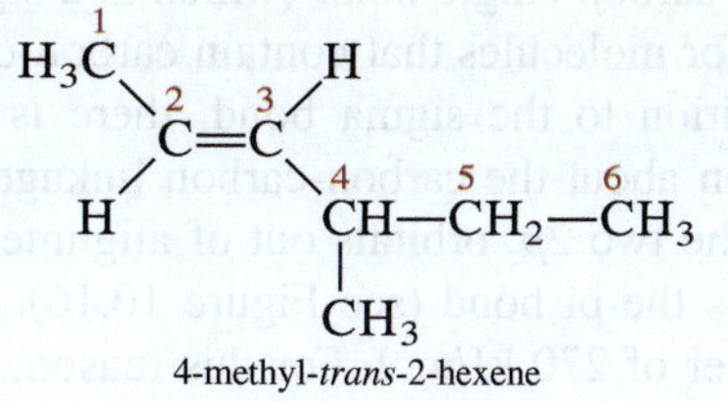

4-methyl-*trans*-2-hexene

In the *cis* isomer, the two H atoms are on the same side of the C=C bond; in the *trans* isomer, the two H atoms are across from each other. Geometric isomerism was introduced in Section 23.3.

Properties and Reactions of Alkenes

Ethylene is an extremely important substance because it is used in large quantities for the manufacture of organic polymers (to be discussed in Chapter 25) and in the preparation of many other organic chemicals. Ethylene is prepared industrially by the *cracking* process, that is, the thermal decomposition of a large hydrocarbon into smaller molecules. When ethane is heated to about 800°C, it undergoes the following reaction:

$$C_2H_6(g) \xrightarrow{\text{Pt catalyst}} CH_2{=}CH_2(g) + H_2(g)$$

Other alkenes can be prepared by cracking the higher members of the alkane family.

Alkenes are classified as ***unsaturated hydrocarbons,*** *compounds with double or triple carbon-carbon bonds that enable them to add hydrogen atoms.* Unsaturated hydrocarbons commonly undergo ***addition reactions,*** in which *one molecule adds to another to form a single product.* Hydrogenation (see Section 22.2) is an example of addition reaction. Other addition reactions to the C=C bond include

$$C_2H_4(g) + HX(g) \longrightarrow CH_3{-}CH_2X(g)$$
$$C_2H_4(g) + X_2(g) \longrightarrow CH_2X{-}CH_2X(g)$$

where X represents a halogen (Cl, Br, or I).

The addition of a hydrogen halide to an unsymmetrical alkene such as propene is more complicated because two products are possible:

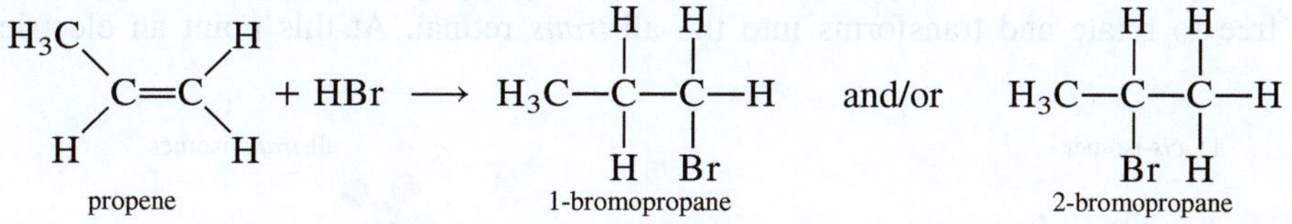

propene 1-bromopropane 2-bromopropane

In reality, however, only 2-bromopropane is formed. This phenomenon was observed in all reactions between unsymmetrical reagents and alkenes. In 1871, Vladimir Markovnikov[†] postulated a generalization that enables us to predict the outcome of such an addition reaction. This generalization, now known as *Markovnikov's rule,* states that in the addition of unsymmetrical (that is, polar) reagents to alkenes, the positive portion of the reagent (usually hydrogen) adds to the carbon atom that already has the most hydrogen atoms.

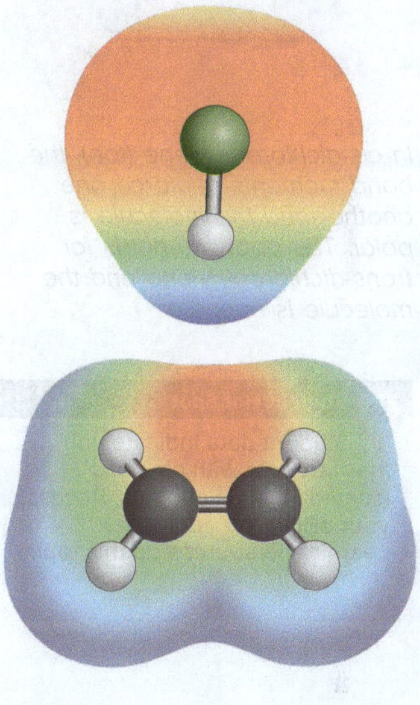

The addition reaction between HCl and ethylene. The initial interaction is between the positive end of HCl (blue) and the electron-rich region of ethylene (red), which is associated with the pi electrons of the C=C bond.

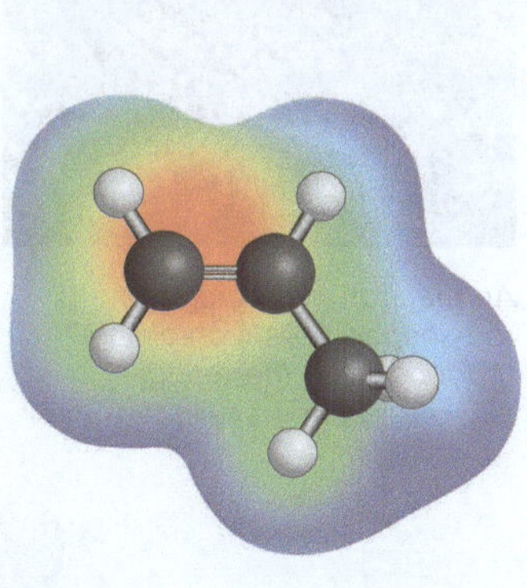

The electron density is higher on the carbon atom in the CH_2 group in propene.

[†]Vladimir W. Markovnikov (1838–1904). Russian chemist. Markovnikov's observations of the addition reactions to alkenes were published a year after his death.

Geometric Isomers of Alkenes

In a compound such as ethane, C_2H_6, the rotation of the two methyl groups about the carbon-carbon single bond (which is a sigma bond) is quite free. The situation is different for molecules that contain carbon-carbon double bonds, such as ethylene, C_2H_4. In addition to the sigma bond, there is a pi bond between the two carbon atoms. Rotation about the carbon-carbon linkage does not affect the sigma bond, but it does move the two $2p_z$ orbitals out of alignment for overlap and, hence, partially or totally destroys the pi bond (see Figure 10.16). This process requires an input of energy on the order of 270 kJ/mol. For this reason, the rotation of a carbon-carbon double bond is considerably restricted, but not impossible. Consequently, molecules containing carbon-carbon double bonds (that is, the alkenes) may have geometric isomers, which cannot be interconverted without breaking a chemical bond.

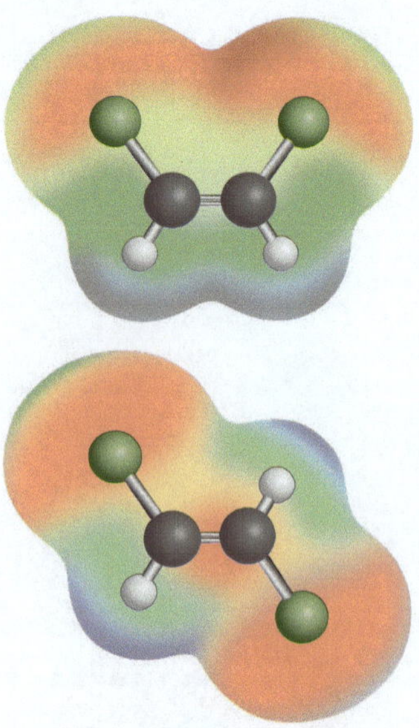

In cis-dichloroethylene (top), the bond moments reinforce one another and the molecule is polar. The opposite holds for trans-dichloroethylene and the molecule is nonpolar.

The molecule dichloroethylene, ClHC=CHCl, can exist as one of the two geometric isomers called *cis*-dichloroethylene and *trans*-dichloroethylene:

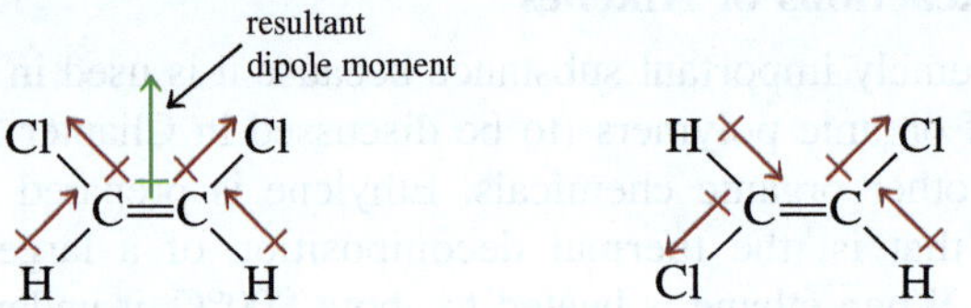

cis-dichloroethylene
$\mu = 1.89$ D
b.p. 60.3°C

trans-dichloroethylene
$\mu = 0$
b.p. 47.5°C

Student Hot Spot
Student data indicate you may struggle with *cis-trans* isomerization in alkenes. Access your eBook for additional Learning Resources on this topic.

where the term *cis* means that two particular atoms (or groups of atoms) are adjacent to each other, and *trans* means that the two atoms (or groups of atoms) are across from each other. Generally, *cis* and *trans* isomers have distinctly different physical and chemical properties. Heat or irradiation with light is commonly used to bring about the conversion of one geometric isomer to another, a process called *cis-trans isomerization,* or geometric isomerization. As the data show, dipole moment measurements can be used to distinguish between geometric isomers. In general, *cis* isomers possess a dipole moment, whereas *trans* isomers do not.

An electron micrograph of rod-shaped cells (containing rhodopsins) in the retina.
Science History Images/Alamy Stock Photo

Cis-Trans Isomerization in the Vision Process

The molecules in the retina that respond to light are rhodopsin, which has two components called 11-*cis* retinal and opsin (Figure 24.6). Retinal is the light-sensitive component and opsin is a protein molecule. Upon receiving a photon in the visible region, the 11-*cis* retinal isomerizes to the all-*trans* retinal by breaking a carbon-carbon pi bond. With the pi bond broken, the remaining carbon-carbon sigma bond is free to rotate and transforms into the all-*trans* retinal. At this point an electrical

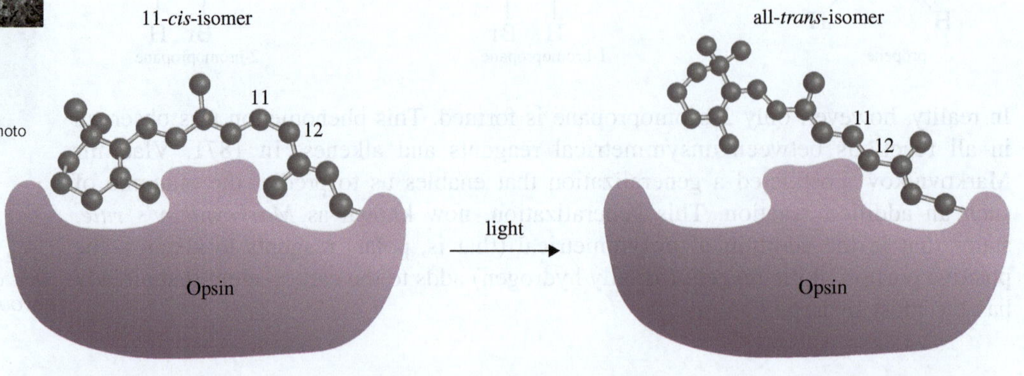

Figure 24.6 *The primary event in the vision process is the conversion of 11-cis retinal to the all-trans isomer on rhodopsin. The double bond at which the isomerization occurs is between carbon-11 and carbon-12. For simplicity, most of the H atoms are omitted. In the absence of light, this transformation takes place about once in a thousand years!*

impulse is generated and transmitted to the brain, which forms a visual image. The all-*trans* retinal does not fit into the binding site on opsin and eventually separates from the protein. In time, the *trans* isomer is converted back to 11-*cis* retinal by an enzyme (in the absence of light) and rhodopsin is regenerated by the binding of the *cis* isomer to opsin and the visual cycle can begin again.

Alkynes

***Alkynes** contain at least one carbon-carbon triple bond.* They have the *general formula* C_nH_{2n-2}, *where* $n = 2, 3, \ldots$.

Alkyne Nomenclature

Names of compounds containing C≡C bonds end with *-yne*. Again, the name of the parent compound is determined by the number of carbon atoms in the longest chain (see Table 24.1 for names of alkane counterparts). As in the case of alkenes, the names of alkynes indicate the position of the carbon-carbon triple bond, as, for example, in

$$\underset{\text{1-butyne}}{HC{\equiv}C{-}CH_2{-}CH_3} \qquad \underset{\text{2-butyne}}{H_3C{-}C{\equiv}C{-}CH_3}$$

Properties and Reactions of Alkynes

The simplest alkyne is ethyne, better known as acetylene (C_2H_2). The structure and bonding of C_2H_2 were discussed in Section 10.5. Acetylene is a colorless gas (b.p. −84°C) prepared by the reaction between calcium carbide and water:

$$CaC_2(s) + 2H_2O(l) \longrightarrow C_2H_2(g) + Ca(OH)_2(aq)$$

The reaction of calcium carbide with water produces acetylene, a flammable gas.
Charles D. Winters/McGraw Hill

Acetylene has many important uses in industry. Because of its high heat of combustion acetylene burned in an "oxyacetylene torch" gives an extremely hot flame (about 3000°C). Thus, oxyacetylene torches are used to weld metals (see Example 6.9).

$$2C_2H_2(g) + 5O_2(g) \longrightarrow 4CO_2(g) + 2H_2O(l) \qquad \Delta H° = -2599.2 \text{ kJ/mol}$$

The standard Gibbs energy of formation of acetylene is positive ($\Delta G_f° = 209.2$ kJ/mol), unlike that of the alkanes. This means that the molecule is unstable (relative to its elements) and has a tendency to decompose:

$$C_2H_2(g) \longrightarrow 2C(s) + H_2(g)$$

In the presence of a suitable catalyst or when the gas is kept under pressure, this reaction can occur with explosive violence. To be transported safely, the gas must be dissolved in an organic solvent such as acetone at moderate pressure. In the liquid state, acetylene is very sensitive to shock and is highly explosive.

Acetylene, an unsaturated hydrocarbon, can be hydrogenated to yield ethylene:

$$C_2H_2(g) + H_2(g) \longrightarrow C_2H_4(g)$$

It undergoes the following addition reactions with hydrogen halides and halogens:

$$C_2H_2(g) + HX(g) \longrightarrow CH_2{=}CHX(g)$$
$$C_2H_2(g) + X_2(g) \longrightarrow CHX{=}CHX(g)$$
$$C_2H_2(g) + 2X_2(g) \longrightarrow CHX_2{-}CHX_2(g)$$

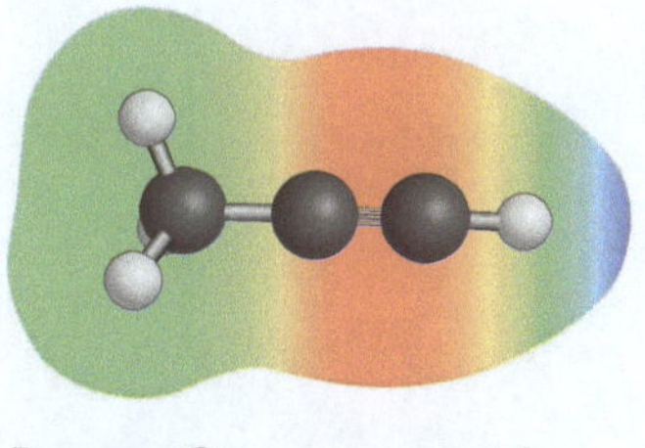

Propyne. Can you account for Markovnikov's rule in this molecule?

Methylacetylene (propyne), $CH_3{-}C{\equiv}H$, is the next member in the alkyne family. It undergoes reactions similar to those of acetylene. The addition reactions of propyne also obey Markovnikov's rule:

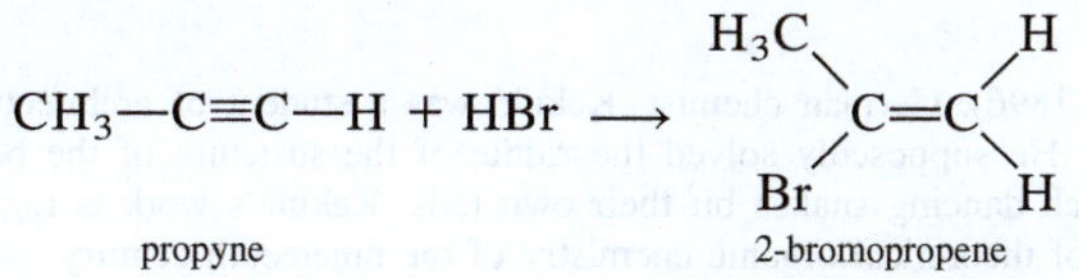

propyne 2-bromopropene

Summary of Concepts & Facts

- Methane, CH_4, is the simplest of the alkanes, a family of hydrocarbons with the general formula C_nH_{2n+2}. Cyclopropane, C_3H_6, is the simplest of the cycloalkanes, a family of alkanes whose carbon atoms are joined in a ring. Alkanes and cycloalkanes are saturated hydrocarbons.
- Ethylene, $CH_2{=}CH_2$, is the simplest of the olefins, or alkenes, a class of hydrocarbons containing carbon-carbon double bonds and having the general formula C_nH_{2n}.
- Acetylene, $CH{\equiv}CH$, is the simplest of the alkynes, which are compounds that have the general formula C_nH_{2n-2} and contain carbon-carbon triple bonds.

Review of Concepts & Facts

24.2.1 What is the formula of the alkane 2,3,4-trimethylpentane?

24.2.2 For which of the following compounds are *cis-trans* isomers possible: (a) $CH_3CH{=}CH_2$, (b) $CH_3CH{=}CHCl$, (c) $CHCl{=}CHBr$?

24.3 Aromatic Hydrocarbons

Learning Objective

- Describe an aromatic compound.

Benzene, the parent compound of this large family of organic substances, was discovered by Michael Faraday in 1826. Over the next 40 years, chemists were preoccupied with determining its molecular structure. Despite the small number of atoms in the molecule, there are quite a few ways to represent the structure of benzene without violating the tetravalency of carbon. However, most proposed structures were rejected because they did not explain the known properties of benzene. Finally, in 1865, August Kekulé[†] deduced that the benzene molecule could be best represented by a ring structure—a cyclic compound consisting of six carbon atoms:

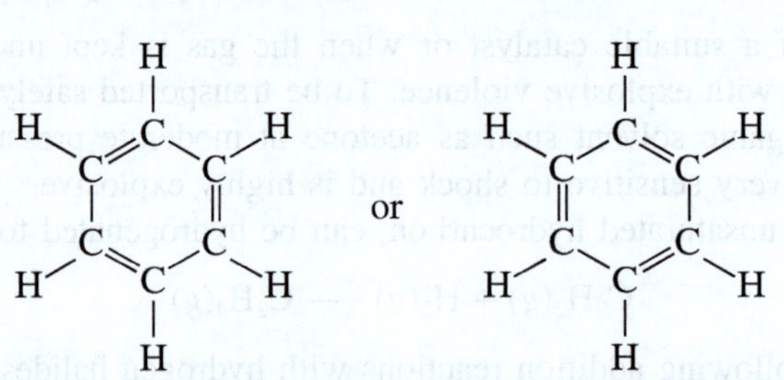

As we saw in Section 9.8, the properties of benzene are best represented by both of these resonance structures. Alternatively, the properties of benzene can be explained in terms of delocalized molecular orbitals (see Section 10.8):

[†]August Kekulé (1829–1896). German chemist. Kekulé was a student of architecture before he became interested in chemistry. He supposedly solved the riddle of the structure of the benzene molecule after having a dream in which dancing snakes bit their own tails. Kekulé's work is regarded by many as the crowning achievement of theoretical organic chemistry of the nineteenth century.

Nomenclature of Aromatic Compounds

The naming of monosubstituted benzenes, that is, benzenes in which one H atom has been replaced by another atom or a group of atoms, is quite straightforward, as shown here:

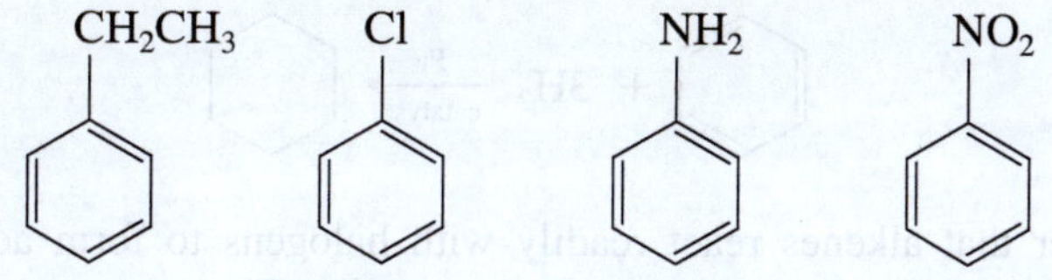

ethylbenzene | chlorobenzene | aminobenzene (aniline) | nitrobenzene

If more than one substituent is present, we must indicate the location of the second group relative to the first. The systematic way to accomplish this is to number the carbon atoms as follows:

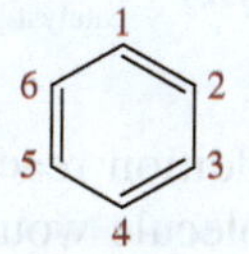

Three different dibromobenzenes are possible:

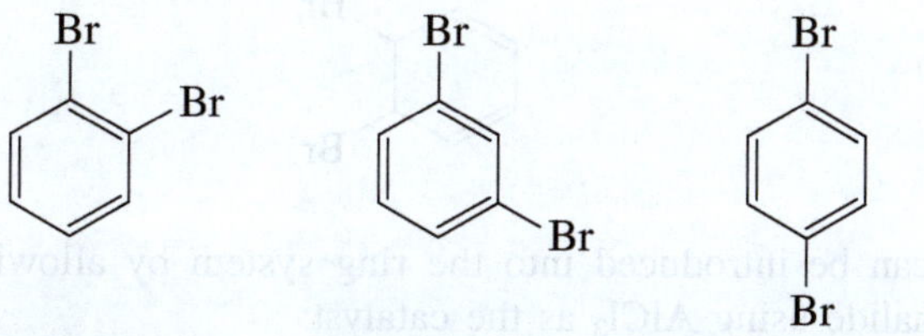

1,2-dibromobenzene (*o*-dibromobenzene) | 1,3-dibromobenzene (*m*-dibromobenzene) | 1,4-dibromobenzene (*p*-dibromobenzene)

The prefixes *o*- (*ortho*-), *m*- (*meta*-), and *p*- (*para*-) are also used to denote the relative positions of the two substituted groups, as just shown for the dibromobenzenes. Compounds in which the two substituted groups are different are named accordingly. Thus,

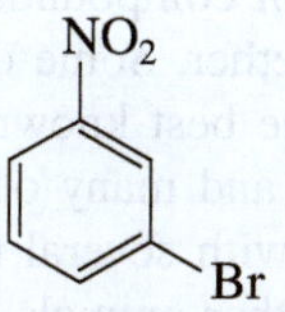

is named 3-bromonitrobenzene, or *m*-bromonitrobenzene.

Finally, we note that the group containing benzene minus a hydrogen atom (C_6H_5) is called the *phenyl* group. Thus, the following molecule is called 2-phenylpropane (also called isopropyl benzene):

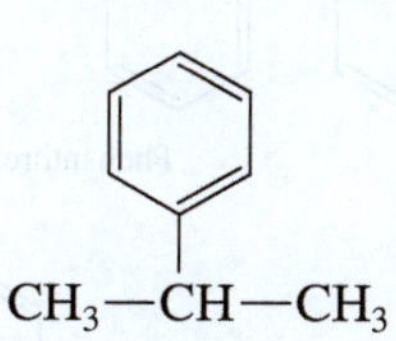

Properties and Reactions of Aromatic Compounds

Benzene is a colorless, flammable liquid obtained chiefly from petroleum and coal tar. Perhaps the most remarkable chemical property of benzene is its relative inertness. Although it has the same empirical formula as acetylene (CH) and a high degree of unsaturation, it is much less reactive than either ethylene or acetylene. The stability of benzene is the result of electron delocalization. In fact, benzene can be hydrogenated, but only with difficulty. The following reaction is carried out at significantly higher temperatures and pressures than are similar reactions for the alkenes:

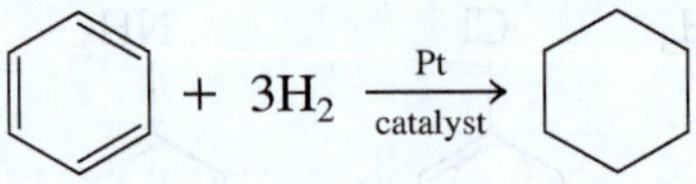

We saw earlier that alkenes react readily with halogens to form addition products, because the pi bond in C=C can be broken easily. The most common reaction of halogens with benzene is the ***substitution reaction,*** in which *an atom or group of atoms replaces an atom or group of atoms in another molecule.* For example,

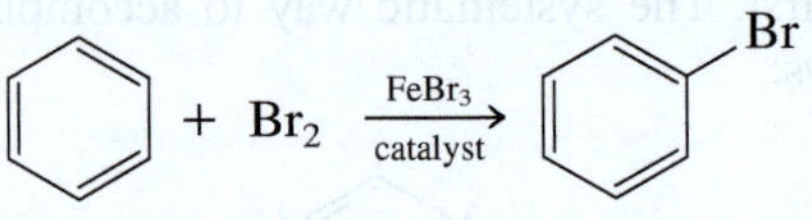

Note that if the reaction were an addition reaction, electron delocalization would be destroyed in the product and the molecule would not have the aromatic characteristic of chemical unreactivity.

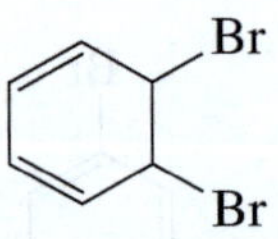

Alkyl groups can be introduced into the ring system by allowing benzene to react with an alkyl halide using $AlCl_3$ as the catalyst:

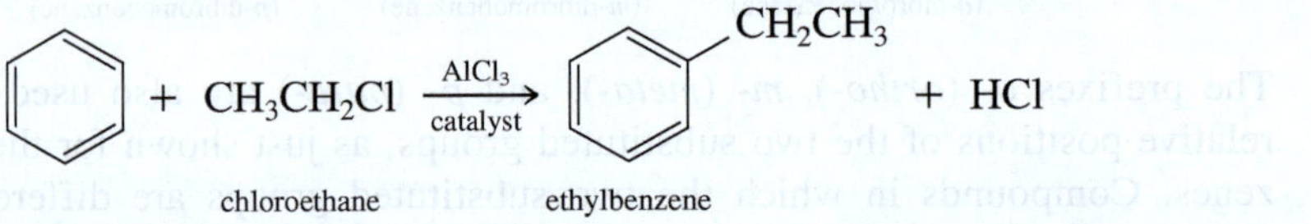

An enormously large number of compounds can be generated from substances in which benzene rings are fused together. Some of these *polycyclic* aromatic hydrocarbons are shown in Figure 24.7. The best known of these compounds is naphthalene, which is used in mothballs. These and many other similar compounds are present in coal tar. Some of the compounds with several rings are powerful carcinogens—they can cause cancer in humans and other animals.

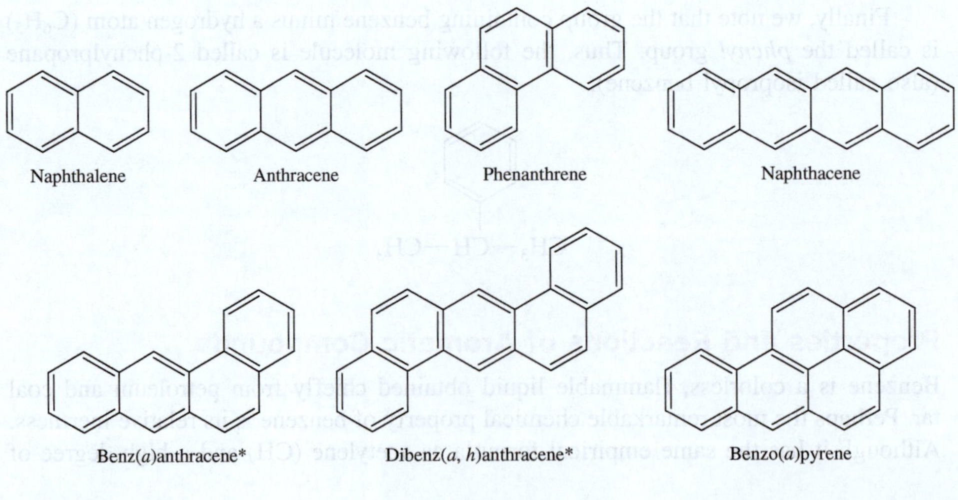

Figure 24.7 *Some polycyclic aromatic hydrocarbons. Compounds denoted by * are potent carcinogens. An enormous number of such compounds exist in nature.*

Summary of Concepts & Facts

- Compounds that contain one or more benzene rings are called aromatic hydrocarbons. These compounds undergo substitution by halogens and alkyl groups.

Review of Concepts & Facts

24.3.1 Name the following aromatic compound.

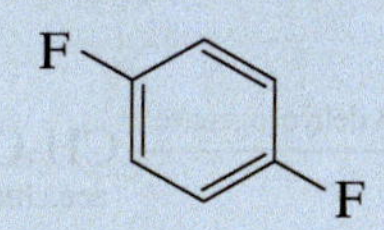

24.4 Chemistry of the Functional Groups

Learning Objective

- Identify the common functional groups found in organic compounds.

We now examine in greater depth some organic functional groups, groups that are responsible for most of the reactions of the parent compounds. In particular, we focus on oxygen-containing and nitrogen-containing compounds.

Alcohols

All ***alcohols*** contain *the hydroxyl functional group, —OH*. Some common alcohols are shown in Figure 24.8. Ethyl alcohol, or ethanol, is by far the best known. It is produced biologically by the fermentation of sugar or starch. In the absence of oxygen, the enzymes present in bacterial cultures or yeast catalyze the reaction

$$C_6H_{12}O_6(aq) \xrightarrow{\text{enzymes}} \underset{\text{ethanol}}{2CH_3CH_2OH(aq)} + 2CO_2(g)$$

This process gives off energy, which microorganisms, in turn, use for growth and other functions.

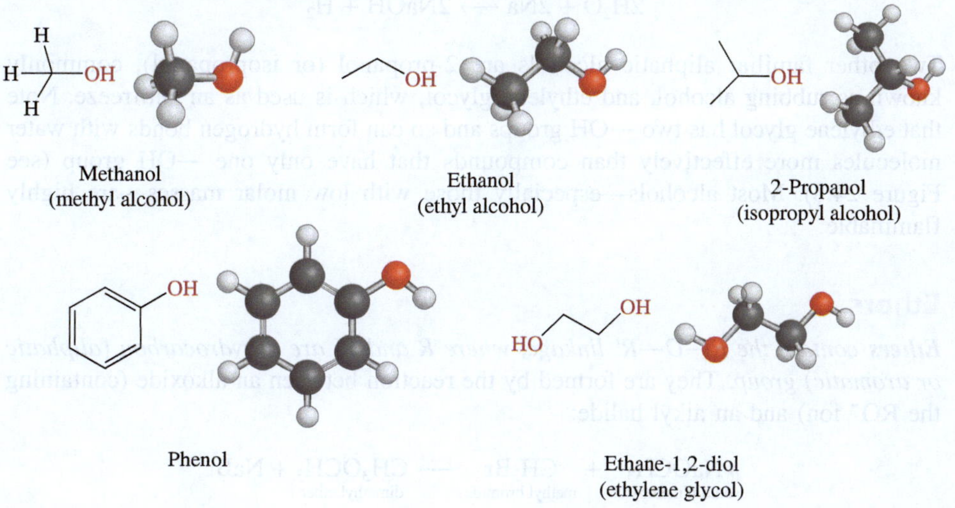

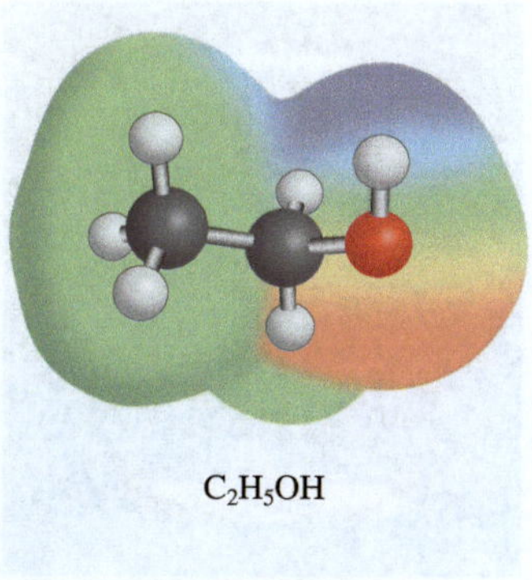

Figure 24.8 *Common alcohols. Note that all the compounds contain the OH group. The properties of phenol are quite different from those of the aliphatic alcohols.*

Commercially, ethanol is prepared by an addition reaction in which water is combined with ethylene at about 280°C and 300 atm:

$$CH_2{=}CH_2(g) + H_2O(g) \xrightarrow{H_2SO_4} CH_3CH_2OH(g)$$

Ethanol has countless applications as a solvent for organic chemicals and as a starting compound for the manufacture of dyes, synthetic drugs, cosmetics, and explosives. It is also a constituent of alcoholic beverages. Ethanol is the only nontoxic (more properly, the least toxic) of the straight-chain alcohols; our bodies produce an enzyme, called *alcohol dehydrogenase,* which helps metabolize ethanol by oxidizing it to acetaldehyde:

$$CH_3CH_2OH \xrightarrow{\text{alcohol dehydrogenase}} \underset{\text{acetaldehyde}}{CH_3CHO} + H_2$$

This equation is a simplified version of what actually takes place; the H atoms are taken up by other molecules, so that no H_2 gas is evolved.

Ethanol can also be oxidized by inorganic oxidizing agents, such as acidified dichromate, to acetaldehyde and acetic acid:

$$CH_3CH_2OH \xrightarrow[H^+]{Cr_2O_7^{2-}} CH_3CHO \xrightarrow[H^+]{Cr_2O_7^{2-}} CH_3COOH$$

Ethanol is called an aliphatic alcohol because it is derived from an alkane (ethane). The simplest aliphatic alcohol is methanol, CH_3OH. Called *wood alcohol,* it was prepared at one time by the dry distillation of wood. It is now synthesized industrially by the reaction of carbon monoxide and molecular hydrogen at high temperatures and pressures:

$$CO(g) + 2H_2(g) \xrightarrow[\text{catalyst}]{Fe_2O_3} \underset{\text{methanol}}{CH_3OH(l)}$$

Alcohols react more slowly with sodium metal than water does.
Ken Karp/McGraw Hill

Methanol is highly toxic. Ingestion of only a few milliliters can cause nausea and blindness. Ethanol intended for industrial use is often mixed with methanol to prevent people from drinking it. Ethanol containing methanol or other toxic substances is called *denatured alcohol.*

The alcohols are very weakly acidic; they do not react with strong bases, such as NaOH. The alkali metals react with alcohols to produce hydrogen:

$$2CH_3OH + 2Na \longrightarrow \underset{\text{sodium methoxide}}{2CH_3ONa} + H_2$$

However, the reaction is much less violent than that between Na and water:

$$2H_2O + 2Na \longrightarrow 2NaOH + H_2$$

Two other familiar aliphatic alcohols are 2-propanol (or isopropanol), commonly known as rubbing alcohol, and ethylene glycol, which is used as an antifreeze. Note that ethylene glycol has two —OH groups and so can form hydrogen bonds with water molecules more effectively than compounds that have only one —OH group (see Figure 24.8). Most alcohols—especially those with low molar masses—are highly flammable.

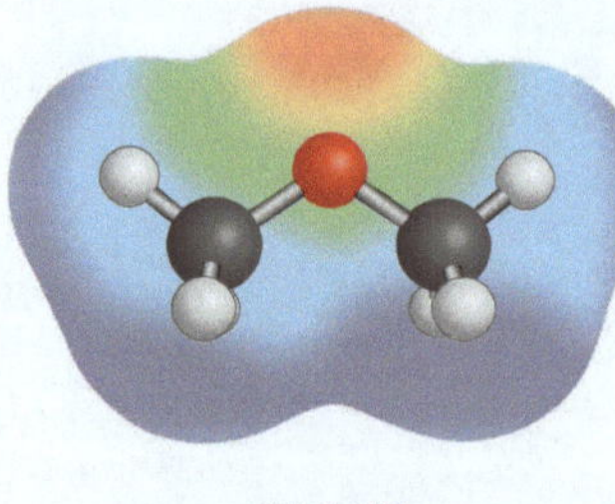

CH_3OCH_3

Ethers

***Ethers** contain the R—O—R′ linkage, where R and R′ are a hydrocarbon (aliphatic or aromatic) group.* They are formed by the reaction between an alkoxide (containing the RO^- ion) and an alkyl halide:

$$\underset{\text{sodium methoxide}}{NaOCH_3} + \underset{\text{methyl bromide}}{CH_3Br} \longrightarrow \underset{\text{dimethyl ether}}{CH_3OCH_3} + NaBr$$

Diethyl ether is prepared on an industrial scale by heating ethanol with sulfuric acid at 140°C:

$$C_2H_5OH + C_2H_5OH \longrightarrow C_2H_5OC_2H_5 + H_2O$$

This reaction is an example of a ***condensation reaction,*** which is characterized by *the joining of two molecules and the elimination of a small molecule, usually water.*

Like alcohols, ethers are extremely flammable. When left standing in air, they have a tendency to slowly form explosive peroxides:

$$\underset{\text{diethyl ether}}{C_2H_5OC_2H_5} + O_2 \longrightarrow \underset{\text{1-ethyoxyethyl hydroperoxide}}{C_2H_5O\text{—}C(CH_3)(H)\text{—}O\text{—}O\text{—}H}$$

Peroxides contain the —O—O linkage; the simplest peroxide is hydrogen peroxide, H_2O_2. Diethyl ether, commonly known as "ether," was used as an anesthetic for many years. It produces unconsciousness by depressing the activity of the central nervous system. The major disadvantages of diethyl ether are its irritating effects on the respiratory system and the occurrence of postanesthetic nausea and vomiting. "Neothyl," or methyl propyl ether, $CH_3OCH_2CH_2CH_3$, is currently favored as an anesthetic because it is relatively free of side effects.

Aldehydes and Ketones

Under mild oxidation conditions, it is possible to convert alcohols to aldehydes and ketones:

$$CH_3OH + \tfrac{1}{2}O_2 \longrightarrow \underset{\text{formaldehyde}}{H_2C{=}O} + H_2O$$

$$C_2H_5OH + \tfrac{1}{2}O_2 \longrightarrow \underset{\text{acetaldehyde}}{(H_3C)(H)C{=}O} + H_2O$$

$$CH_3\text{—}CH(OH)\text{—}CH_3 + \tfrac{1}{2}O_2 \longrightarrow \underset{\text{acetone}}{(H_3C)_2C{=}O} + H_2O$$

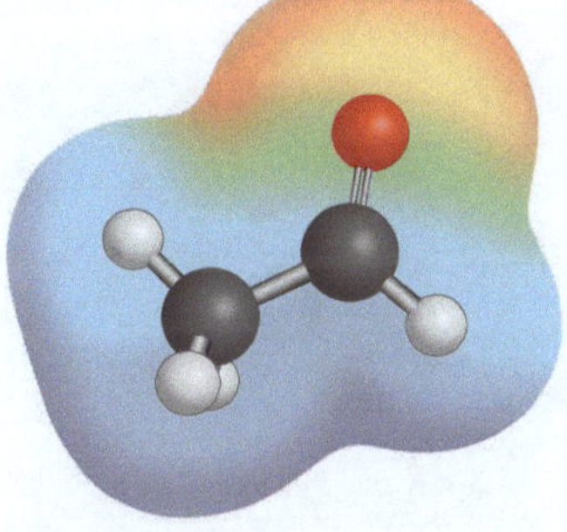

CH_3CHO

The functional group in these compounds is the *carbonyl group,* $\rangle$*C=O*. In an ***aldehyde*** *at least one hydrogen atom is bonded to the carbon in the carbonyl group.* In a ***ketone,*** *the carbon atom in the carbonyl group is bonded to two hydrocarbon groups.*

The simplest aldehyde, formaldehyde ($H_2C{=}O$) has a tendency to *polymerize;* that is, the individual molecules join together to form a compound of high molar mass. This action gives off much heat and is often explosive, so formaldehyde is usually prepared and stored in aqueous solution (to reduce the concentration). This rather disagreeable-smelling liquid is used as a starting material in the polymer industry (see Chapter 25) and in the laboratory as a preservative for animal specimens. Interestingly, the higher molar mass aldehydes, such as cinnamic aldehyde, have a pleasant odor (cinnamic aldehyde gives cinnamon its characteristic aroma) and are used in the manufacture of perfumes.

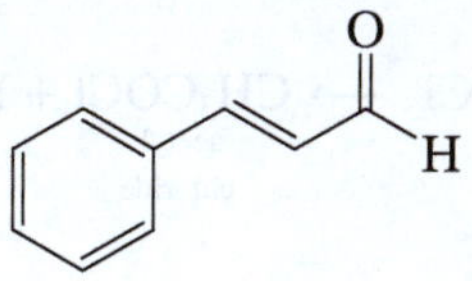

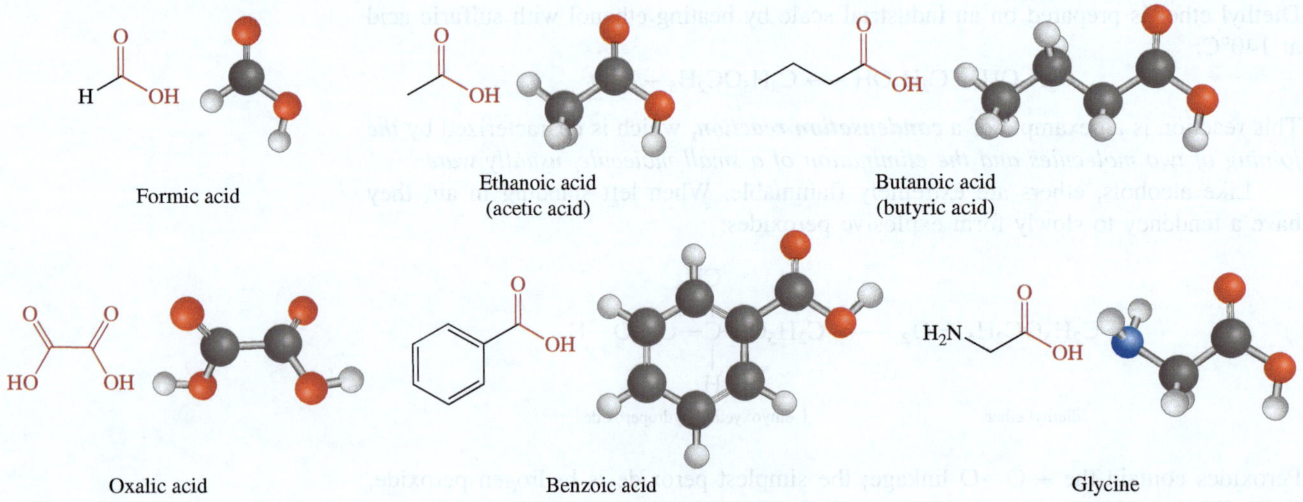

Figure 24.9 *Some common carboxylic acids. Note that they all contain the COOH group. (Glycine is one of the amino acids found in proteins.)*

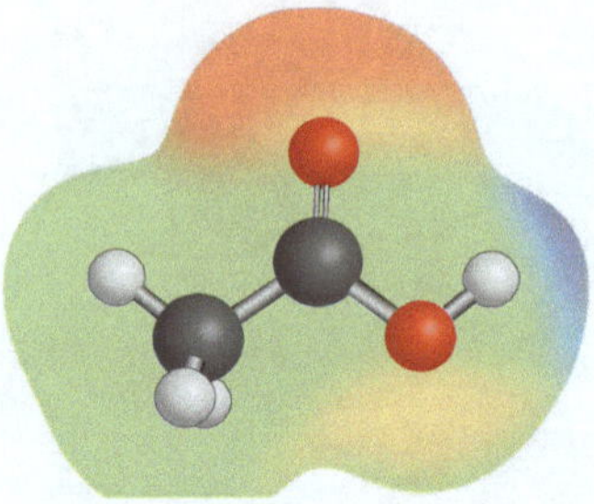

CH_3COOH

Ketones generally are less reactive than aldehydes. The simplest ketone is acetone, a pleasant-smelling liquid that is used mainly as a solvent for organic compounds and nail polish remover.

Carboxylic Acids

Under appropriate conditions both alcohols and aldehydes can be oxidized to ***carboxylic acids,*** *acids that contain the carboxyl group, —COOH:*

$$CH_3CH_2OH + O_2 \longrightarrow CH_3COOH + H_2O$$
$$CH_3CHO + \tfrac{1}{2}O_2 \longrightarrow CH_3COOH$$

These reactions occur so readily, in fact, that wine must be protected from atmospheric oxygen while in storage. Otherwise, it would soon turn to vinegar due to the formation of acetic acid. Figure 24.9 shows the structure of some of the common carboxylic acids.

Carboxylic acids are widely distributed in nature; they are found in both the plant and animal kingdoms. All protein molecules are made of amino acids, a special kind of carboxylic acid containing an amino group ($—NH_2$) and a carboxyl group (—COOH).

Unlike the inorganic acids HCl, HNO_3, and H_2SO_4, carboxylic acids are usually weak. They react with alcohols to form pleasant-smelling esters:

$$\underset{\text{acetic acid}}{CH_3COOH} + \underset{\text{ethanol}}{HOCH_2CH_3} \longrightarrow \underset{\text{ethyl acetate}}{CH_3—\overset{\overset{\displaystyle O}{\|}}{C}—O—CH_2CH_3} + H_2O$$

Other common reactions of carboxylic acids are neutralization

$$CH_3COOH + NaOH \longrightarrow CH_3COONa + H_2O$$

and formation of acid halides, such as acetyl chloride

$$CH_3COOH + PCl_5 \longrightarrow \underset{\substack{\text{acetyl}\\\text{chloride}}}{CH_3COCl} + HCl + \underset{\substack{\text{phosphoryl}\\\text{chloride}}}{POCl_3}$$

Acid halides are reactive compounds used as intermediates in the preparation of many other organic compounds. They hydrolyze in much the same way as many nonmetallic halides, such as $SiCl_4$:

$$CH_3COCl(l) + H_2O(l) \longrightarrow CH_3COOH(aq) + HCl(g)$$

$$SiCl_4(l) + 3H_2O(l) \longrightarrow \underset{\text{silicic acid}}{H_2SiO_3(s)} + 4HCl(g)$$

Esters

***Esters** have the general formula R′COOR, where R′ can be H or a hydrocarbon group and R is a hydrocarbon group.* Esters are used in the manufacture of perfumes and as flavoring agents in the confectionery and soft-drink industries. Many fruits owe their characteristic smell and flavor to the presence of small quantities of esters. For example, bananas contain 3-methylbutyl acetate [$CH_3COOCH_2CH_2CH(CH_3)_2$], oranges contain octyl acetate ($CH_3COOCHCH_3C_6H_{13}$), and apples contain methyl butyrate ($CH_3CH_2CH_2COOCH_3$).

The odor of fruits is mainly due to the ester compounds they contain.
Ken Karp/McGraw Hill

The functional group in esters is the —COOR group. In the presence of an acid catalyst, such as HCl, esters undergo hydrolysis to yield a carboxylic acid and an alcohol. For example, in acid solution, ethyl acetate hydrolyzes as follows:

$$\underset{\text{ethyl acetate}}{CH_3COOC_2H_5} + H_2O \rightleftharpoons \underset{\text{acetic acid}}{CH_3COOH} + \underset{\text{ethanol}}{C_2H_5OH}$$

However, this reaction does not go to completion because the reverse reaction—that is, the formation of an ester from an alcohol and an acid—also occurs to an appreciable extent. On the other hand, when NaOH solution is used in hydrolysis the sodium acetate does not react with ethanol, so this reaction does go to completion from left to right:

$$\underset{\text{ethyl acetate}}{CH_3COOC_2H_5} + NaOH \longrightarrow \underset{\text{sodium acetate}}{CH_3COO^-Na^+} + \underset{\text{ethanol}}{C_2H_5OH}$$

For this reason, ester hydrolysis is usually carried out in basic solutions. Note that NaOH does not act as a catalyst; rather, it is consumed by the reaction. The term ***saponification*** (meaning *soapmaking*) was originally used to describe the alkaline hydrolysis of fatty acid esters to yield soap molecules (sodium stearate):

$$\underset{\text{ethyl stearate}}{C_{17}H_{35}COOC_2H_5} + NaOH \longrightarrow \underset{\text{sodium stearate}}{C_{17}H_{35}COO^-Na^+} + C_2H_5OH$$

Saponification has now become *a general term for alkaline hydrolysis of any type of ester.* Soaps are characterized by a long nonpolar hydrocarbon chain and a polar head (the $—COO^-$ group). The hydrocarbon chain is readily soluble in oily substances, while the ionic carboxylate group ($—COO^-$) remains outside the oily nonpolar surface. Section 12.8 describes the action of soap.

Amines

***Amines** are organic bases having the general formula R_3N, where R may be H or a hydrocarbon group.* As with ammonia, the reaction of amines with water is

$$RNH_2 + H_2O \longrightarrow RNH_3^+ + OH^-$$

Table 24.4 Important Functional Groups and Their Reactions

Functional Group	Name	Typical Reactions
$>C=C<$	Carbon-carbon double bond	Addition reactions with halogens, hydrogen halides, and water; hydrogenation to yield alkanes
$-C\equiv C-$	Carbon-carbon triple bond	Addition reactions with halogens, hydrogen halides; hydrogenation to yield alkenes and alkanes
$-\ddot{X}:$ (X = F, Cl, Br, I)	Halogen	Exchange reactions: $CH_3CH_2Br + KI \longrightarrow CH_3CH_2I + KBr$
$-\ddot{O}-H$	Hydroxyl	Esterification (formation of an ester) with carboxylic acids; oxidation to aldehydes, ketones, and carboxylic acids
$>C=\ddot{O}$	Carbonyl	Reduction to yield alcohols; oxidation of aldehydes to yield carboxylic acids
$-C(=O)-\ddot{O}-H$	Carboxyl	Esterification with alcohols; reaction with phosphorus pentachloride to yield acid chlorides
$-C(=O)-\ddot{O}-R$ (R = hydrocarbon)	Ester	Hydrolysis to yield acids and alcohols
$-\ddot{N}(R)(R)$ (R = H or hydrocarbon)	Amine	Formation of ammonium salts with acids

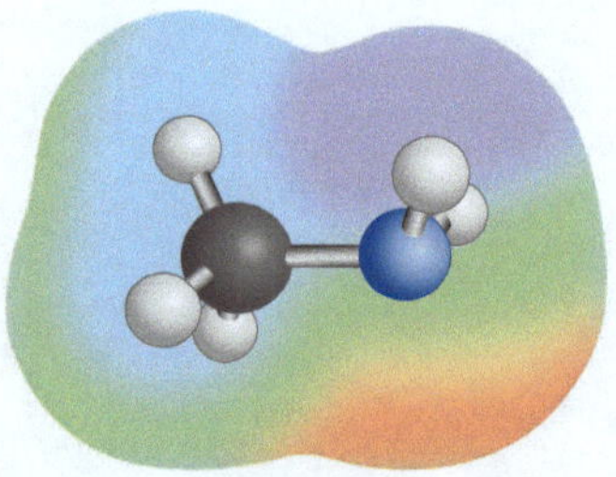

CH_3NH_2

where R represents a hydrocarbon group. Like all bases, the amines form salts when allowed to react with acids:

$$\underset{\text{ethylamine}}{CH_3CH_2NH_2} + HCl \longrightarrow \underset{\text{ethylammonium chloride}}{CH_3CH_2NH_3^+Cl^-}$$

These salts are usually colorless, odorless solids.

Aromatic amines are used mainly in the manufacture of dyes. Aniline, the simplest aromatic amine, itself is a toxic compound; a number of other aromatic amines such as 2-naphthylamine and benzidine are potent carcinogens:

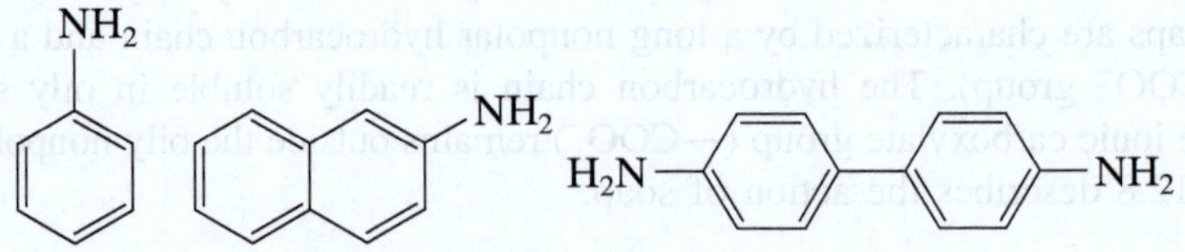

Summary of Functional Groups

Table 24.4 summarizes the common functional groups, including the C═C and C≡C groups. Organic compounds commonly contain more than one functional group. Generally, the reactivity of a compound is determined by the number and types of functional groups in its makeup.

Example 24.5 shows how we can use the functional groups to predict reactions.

Student Hot Spot

Student data indicate you may struggle with identifying functional groups. Access your eBook for additional Learning Resources on this topic.

Example 24.5

Cholesterol is a major component of gallstones, and it is believed that the cholesterol level in the blood is a contributing factor in certain types of heart disease. From the following structure of the compound, predict its reaction with (a) Br_2, (b) H_2 (in the presence of a Pt catalyst), (c) CH_3COOH.

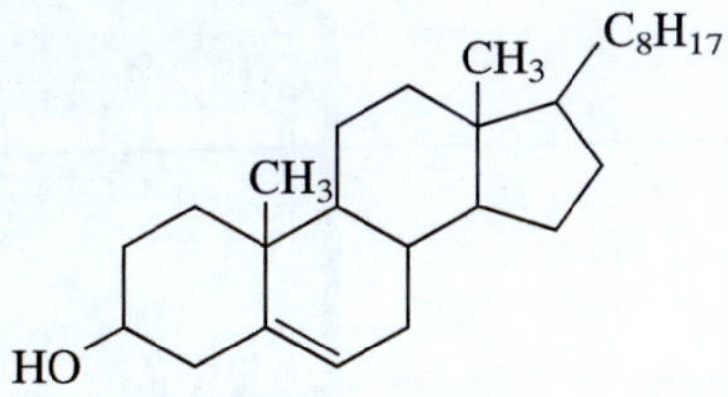

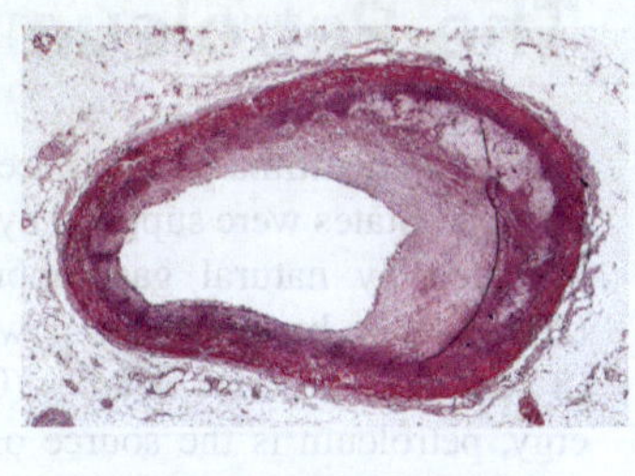

An artery becoming blocked by cholesterol.
Kateryna Kon/Shutterstock

Strategy To predict the type of reactions a molecule may undergo, we must first identify the functional groups present (see Table 24.4).

Solution There are two functional groups in cholesterol: the hydroxyl group and the carbon-carbon double bond.

(a) The reaction with bromine results in the addition of bromine to the double-bonded carbons, which become single bonded.

(b) This is a hydrogenation reaction. Again, the carbon-carbon double bond is converted to a carbon-carbon single bond.

(c) The acid reacts with the hydroxyl group to form an ester and water. Figure 24.10 shows the products of these reactions.

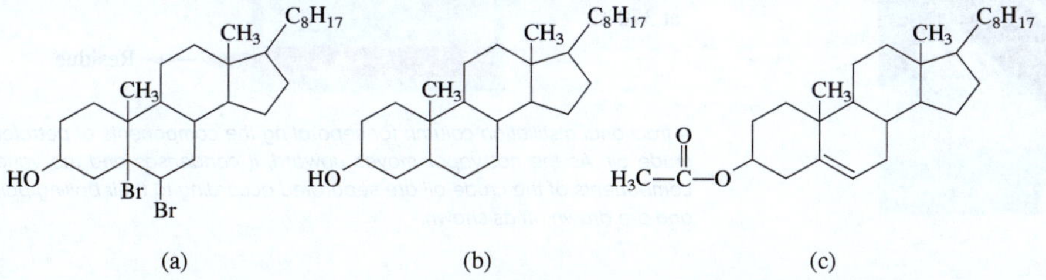

Figure 24.10 *The products formed by the reaction of cholesterol with (a) molecular bromine, (b) molecular hydrogen, and (c) acetic acid.*

Practice Exercise Predict the products of the following reaction:

$$CH_3OH + CH_3CH_2COOH \longrightarrow ?$$

Similar problem: 24.41.

The Chemistry in Action essay "The Petroleum Industry" shows the key organic compounds present in petroleum.

Summary of Concepts & Facts

- Functional groups impart specific types of chemical reactivity to molecules. Classes of compounds characterized by their functional groups include alcohols, ethers, aldehydes and ketones, carboxylic acids and esters, and amines.

CHEMISTRY *in Action*

The Petroleum Industry

In 2010 an estimated 40 percent of the energy needs of the United States were supplied by oil or petroleum. The rest was provided by natural gas (approximately 25 percent), coal (23 percent), hydroelectric power (4 percent), nuclear power (8 percent), and other sources (0.5 percent). In addition to energy, petroleum is the source of numerous organic chemicals used to manufacture drugs, clothing, and many other products.

Unrefined petroleum, a viscous, dark brown liquid, is often called crude oil. A complex mixture of alkanes, alkenes, cycloalkanes, and aromatic compounds, petroleum was formed in Earth's crust over the course of millions of years by the anaerobic decomposition of animal and vegetable matter by bacteria.

Petroleum deposits are widely distributed throughout the world, but they are found mainly in North America, Mexico, Russia, China, Venezuela, and, of course, the Middle East. The actual composition of petroleum varies with location. In the United States, for example, Pennsylvania crude oils are mostly aliphatic hydrocarbons, whereas the major components of western crude oils are aromatic in nature.

Although petroleum contains literally thousands of hydrocarbon compounds, we can classify its components according to the range of their boiling points. These hydrocarbons can be

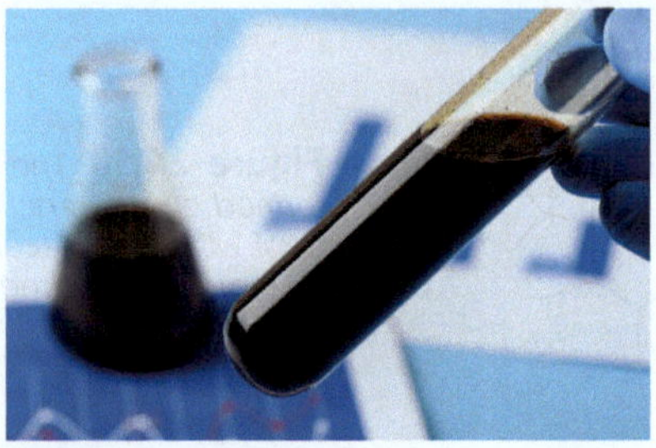

Crude oil.
nevodka/Shutterstock

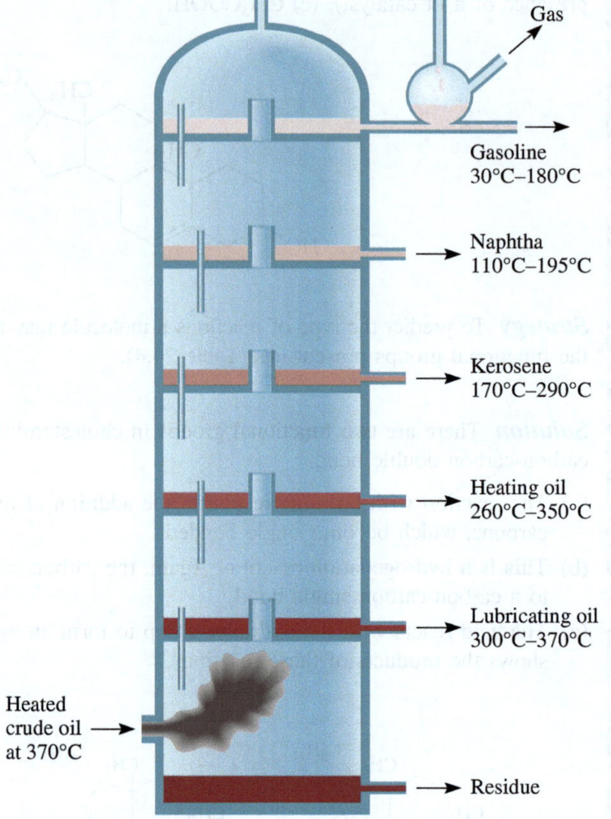

A fractional distillation column for separating the components of petroleum crude oil. As the hot vapor moves upward, it condenses and the various components of the crude oil are separated according to their boiling points and are drawn off as shown.

Major Fractions of Petroleum

Fraction	Carbon Atoms*	Boiling Point Range (°C)	Uses
Natural gas	C_1–C_4	−161 to 20	Fuel and cooking gas
Petroleum ether	C_5–C_6	30–60	Solvent for organic compounds
Ligroin	C_7	20–135	Solvent for organic compounds
Gasoline	C_6–C_{12}	30–180	Automobile fuels
Kerosene	C_{11}–C_{16}	170–290	Rocket and jet engine fuels, domestic heating
Heating fuel oil	C_{14}–C_{18}	260–350	Domestic heating and fuel for electricity production
Lubricating oil	C_{15}–C_{24}	300–370	Lubricants for automobiles and machines

*The entries in this column indicate the numbers of carbon atoms in the compounds involved. For example, C_1–C_4 tells us that in natural gas the compounds contain 1 to 4 carbon atoms, and so on.

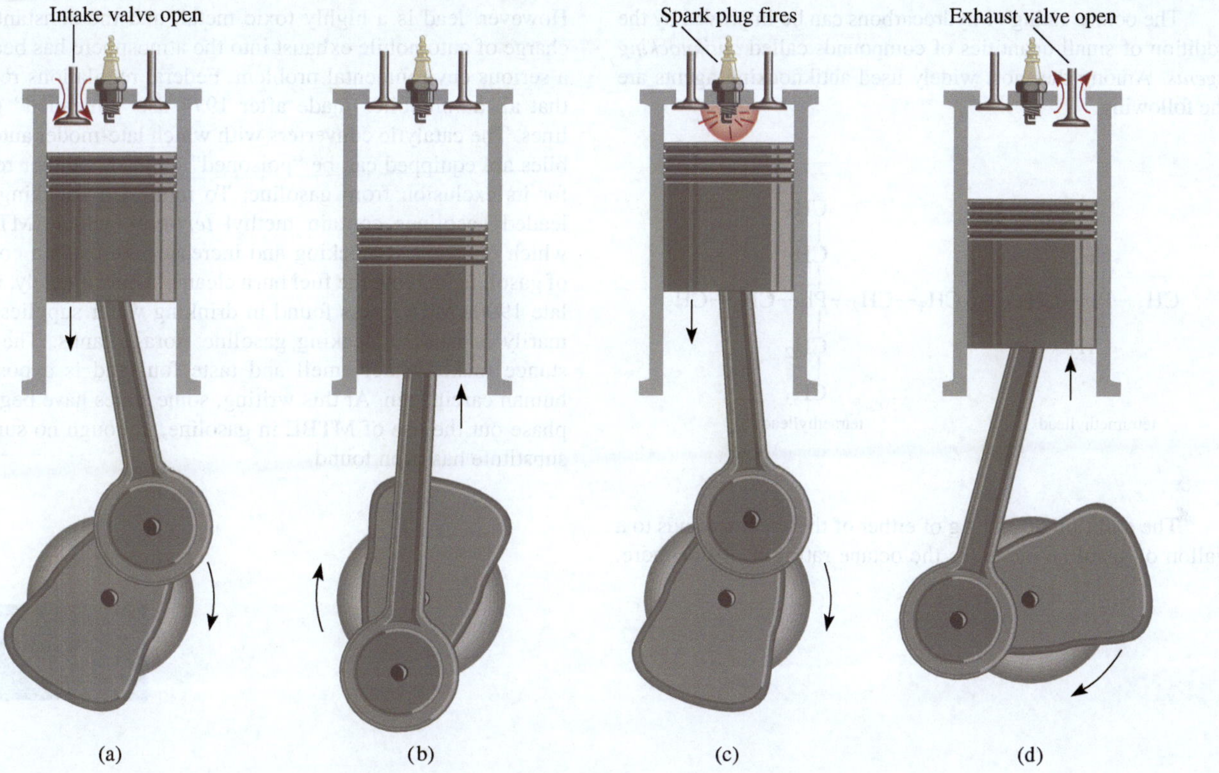

The four stages of operation of an internal combustion engine. This is the type of engine used in practically all automobiles and is described technically as a four-stroke Otto cycle engine. (a) The intake valve opens to let in a gasoline-air mixture. (b) During the compression stage the two valves are closed. (c) The spark plug fires and the piston is pushed outward. (d) Finally, as the piston is pushed downward, the exhaust valve opens to let out the exhaust gas.

separated on the basis of molar mass by fractional distillation. Heating crude oil to about 400°C converts the viscous oil into hot vapor and fluid. In this form it enters the fractionating tower. The vapor rises and condenses on various collecting trays according to the temperatures at which the various components of the vapor liquefy. Some gases are drawn off at the top of the column, and the unvaporized residual oil is collected at the bottom.

Gasoline is probably the best-known petroleum product. A mixture of volatile hydrocarbons, gasoline contains mostly alkanes, cycloalkanes, and a few aromatic hydrocarbons. Some of these compounds are far more suitable for fueling an automobile engine than others, and herein lies the problem of the further treatment and refinement of gasoline.

Most automobiles employ the four-stroke operation of the *Otto cycle* engine. A major engineering concern is to control the burning of the gasoline-air mixture inside each cylinder to obtain a smooth expansion of the gas mixture. If the mixture burns too rapidly, the piston receives a hard jerk rather than a smooth, strong push. This action produces a "knocking" or "pinging" sound, as well as a decrease in efficiency in the conversion of combustion energy to mechanical energy. It turns out that straight-chain hydrocarbons have the greatest tendency to produce knocking, whereas the branched-chain and aromatic hydrocarbons give the desired smooth push.

Gasolines are usually rated according to the *octane number,* a measure of their tendency to cause knocking. On this scale, a branched C_8 compound (2,2,4-trimethylpentane, or isooctane) has been arbitrarily assigned an octane number of 100, and that of *n*-heptane, a straight-chain compound, is zero. The higher the octane number of the hydrocarbon, the better its performance in the internal combustion engine. Aromatic hydrocarbons such as benzene and toluene have high octane numbers (106 and 120, respectively), as do aliphatic hydrocarbons with branched chains.

(Continued)

CHEMISTRY *in Action*

(Continued)

The octane rating of hydrocarbons can be improved by the addition of small quantities of compounds called *antiknocking agents.* Among the most widely used antiknocking agents are the following:

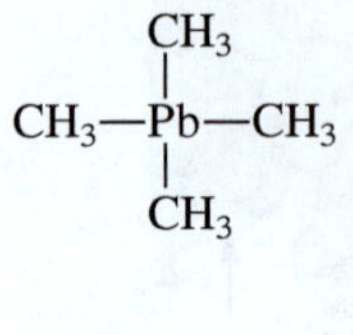

tetramethyllead

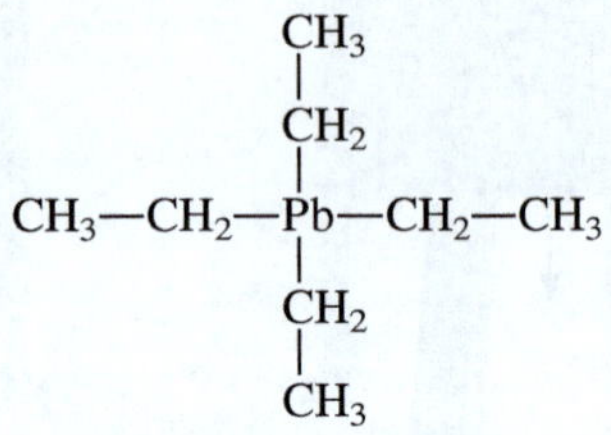

tetraethyllead

The addition of 2 to 4 g of either of these compounds to a gallon of gasoline increases the octane rating by 10 or more. However, lead is a highly toxic metal, and the constant discharge of automobile exhaust into the atmosphere has become a serious environmental problem. Federal regulations require that all automobiles made after 1974 use "unleaded" gasolines. The catalytic converters with which late-model automobiles are equipped can be "poisoned" by lead, another reason for its exclusion from gasoline. To minimize knocking, unleaded gasolines contain methyl *tert*-butyl ether (MTBE), which minimizes knocking and increases the oxygen content of gasoline, making the fuel burn cleaner. Unfortunately, in the late 1990s MTBE was found in drinking water supplies, primarily because of leaking gasoline storage tanks. The substance makes water smell and taste foul and is a possible human carcinogen. At this writing, some states have begun to phase out the use of MTBE in gasoline, although no suitable substitute has been found.

Review of Concepts & Facts

24.4.1 Identify the functional groups present in the following molecule:

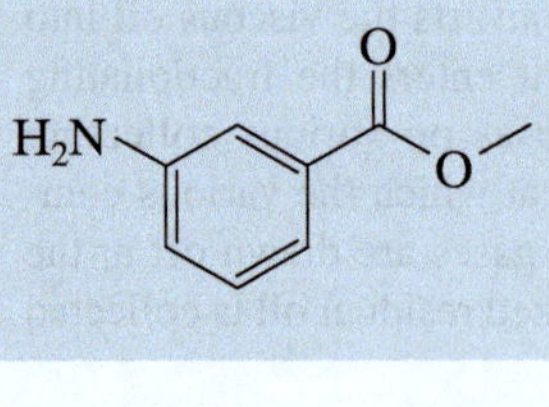

Chapter Summary

Organic Compounds Organic compounds contain primarily carbon and hydrogen atoms, plus nitrogen, oxygen, sulfur, and atoms of other elements. The parent compounds of all organic compounds are the hydrocarbons—the alkanes (containing only single bonds), the alkenes (containing carbon-carbon double bonds), the alkynes (containing carbon-carbon triple bonds), and the aromatic hydrocarbons (containing the benzene ring). (Sections 24.1, 24.2, 24.3)

Functional Groups The reactivity of organic compounds can be reliably predicted by the presence of functional groups, which are groups of atoms that are largely responsible for the chemical behavior of the compounds. (Section 24.4)

Key Words

Addition reactions, p. 1041
Alcohol, p. 1047
Aldehyde, p. 1049
Aliphatic hydrocarbon, p. 1031
Alkane, p. 1032
Alkene, p. 1040
Alkyne, p. 1043
Amine, p. 1051
Aromatic hydrocarbon, p. 1031
Carboxylic acid, p. 1050
Condensation reaction, p. 1049
Cycloalkane, p. 1040
Ester, p. 1051
Ether, p. 1048
Functional group, p. 1031
Hydrocarbon, p. 1031
Ketone, p. 1049
Organic chemistry, p. 1031
Saponification, p. 1051
Saturated hydrocarbon, p. 1032
Structural isomer, p. 1032
Substitution reaction, p. 1046
Unsaturated hydrocarbon, p. 1041

Questions & Problems

Red numbered problems solved in Student Solutions Manual

24.1 Classes of Organic Compounds

Review Questions

24.1 Explain why carbon is able to form so many more compounds than any other element.

24.2 What is the difference between aliphatic and aromatic hydrocarbons?

24.2 Aliphatic Hydrocarbons

Review Questions

24.3 What do "saturated" and "unsaturated" mean when applied to hydrocarbons? Give examples of a saturated hydrocarbon and an unsaturated hydrocarbon.

24.4 Give three sources of methane.

24.5 Alkenes exhibit geometric isomerism because rotation about the C=C bond is restricted. Explain.

24.6 Why is it that alkanes and alkynes, unlike alkenes, have no geometric isomers?

24.7 What is Markovnikov's rule?

24.8 Describe reactions that are characteristic of alkanes, alkenes, and alkynes.

24.9 What factor determines whether a carbon atom in a compound is chiral?

24.10 Give examples of a chiral substituted alkane and an achiral substituted alkane.

Problems

24.11 Draw all possible structural isomers for the following alkane: C_7H_{16}.

24.12 How many distinct chloropentanes, $C_5H_{11}Cl$, could be produced in the direct chlorination of *n*-pentane, $CH_3(CH_2)_3CH_3$? Draw the structure of each molecule.

24.13 Draw all possible isomers for the molecule C_4H_8.

24.14 Draw all possible isomers for the molecule C_3H_5Br.

24.15 The structural isomers of pentane, C_5H_{12}, have quite different boiling points (see Example 24.1). Explain the observed variation in boiling point, in terms of structure.

24.16 Discuss how you can determine which of the following compounds might be alkanes, cycloalkanes, alkenes, or alkynes, without drawing their formulas: (a) C_6H_{12}, (b) C_4H_6, (c) C_5H_{12}, (d) C_7H_{14}, (e) C_3H_4.

24.17 Draw the structures of *cis*-2-butene and *trans*-2-butene. Which of the two compounds would have the higher heat of hydrogenation? Explain.

24.18 Would you expect cyclobutadiene to be a stable molecule? Explain.

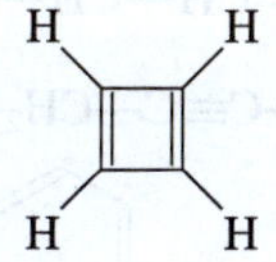

24.19 How many different isomers can be derived from ethylene if two hydrogen atoms are replaced by a fluorine atom and a chlorine atom? Draw their structures and name them. Indicate which are structural isomers and which are geometric isomers.

24.20 Suggest two chemical tests that would help you distinguish between these two compounds:
(a) $CH_3CH_2CH_2CH_2CH_3$
(b) $CH_3CH_2CH_2CH{=}CH_2$

24.21 Sulfuric acid (H_2SO_4) adds to the double bond of alkenes as H^+ and $^-OSO_3H$. Predict the products when sulfuric acid reacts with (a) ethylene and (b) propene.

24.22 Acetylene is an unstable compound. It has a tendency to form benzene as follows:

$$3C_2H_2(g) \longrightarrow C_6H_6(l)$$

Calculate the standard enthalpy change in kilojoules per mole for this reaction at 25°C.

24.23 Predict products when HBr is added to (a) 1-butene and (b) 2-butene.

24.24 Geometric isomers are not restricted to compounds containing the C=C bond. For example, certain disubstituted cycloalkanes can exist in the *cis* and the *trans* forms. Label the following molecules as the *cis* and *trans* isomer, of the same compound:

(a) 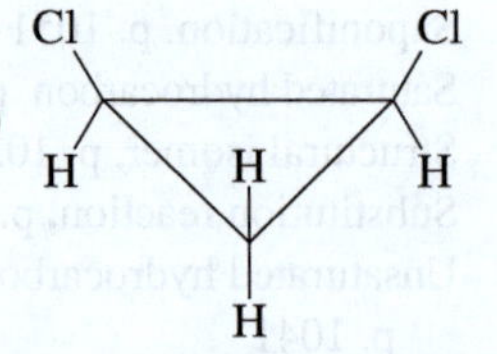 (b) 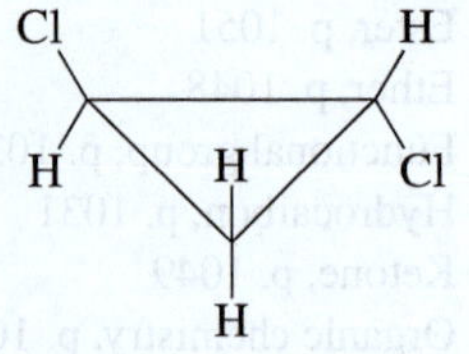

24.25 Which of the following amino acids are chiral: (a) $CH_3CH(NH_2)COOH$, (b) $CH_2(NH_2)COOH$, (c) $CH_2(OH)CH(NH_2)COOH$?

24.26 Name the following compounds:

(a) 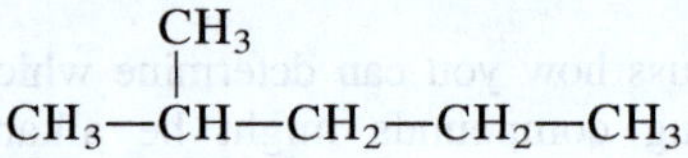

(b) 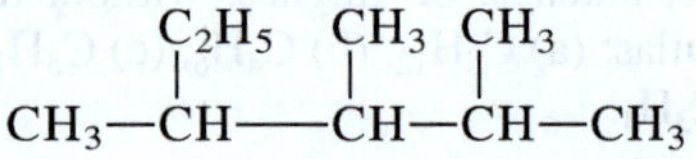

(c) 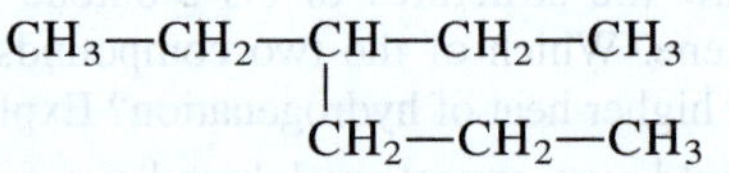

(d) 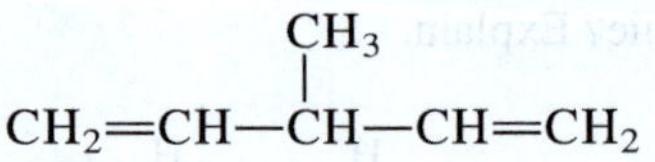

(e) $CH_3—C≡C—CH_2—CH_3$

(f) 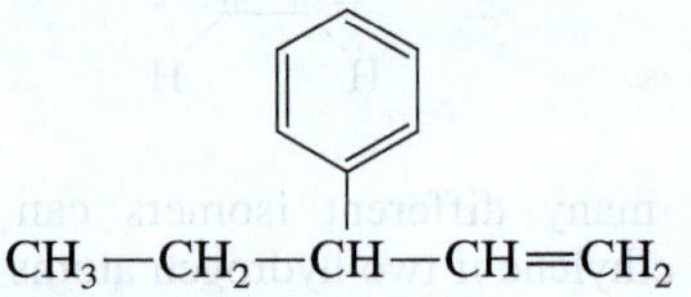

24.27 Write structural formulas for the following organic compounds: (a) 3-methylhexane, (b) 1,3,5-trichlorocyclohexane, (c) 2,3-dimethylpentane, (d) 2-bromo-4-phenylpentane, (e) 3,4,5-trimethyloctane.

24.28 Write structural formulas for the following compounds: (a) *trans*-2-pentene, (b) 2-ethyl-1-butene, (c) 4-ethyl-*trans*-2-heptene, (d) 3-phenyl-1-butyne.

24.3 Aromatic Hydrocarbons

Review Questions

24.29 Comment on the extra stability of benzene compared to ethylene. Why does ethylene undergo addition reactions while benzene usually undergoes substitution reactions?

24.30 Benzene and cyclohexane molecules both contain six-membered rings. Benzene is a planar molecule, and cyclohexane is nonplanar. Explain.

Problems

24.31 Write structures for the following compounds: (a) 1-bromo-3-methylbenzene, (b) 1-chloro-2-propylbenzene, (c) 1,2,4,5-tetramethylbenzene.

24.32 Name the following compounds:

(a) 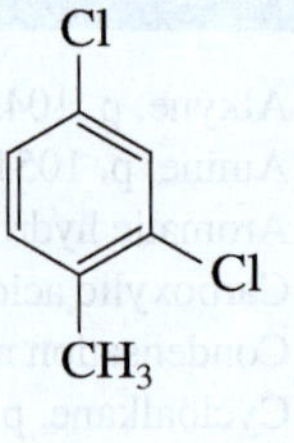 (b) 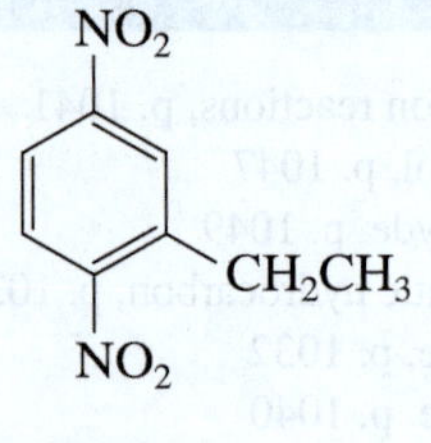

(c) 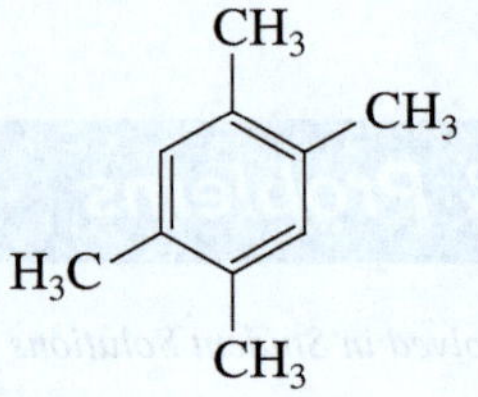

24.4 Chemistry of the Functional Groups

Review Questions

24.33 What are functional groups? Why is it logical and useful to classify organic compounds according to their functional groups?

24.34 Draw the Lewis structure for each of the following functional groups: alcohol, ether, aldehyde, ketone, carboxylic acid, ester, amine.

Problems

24.35 Draw structures for molecules with the following formulas: (a) CH_4O, (b) C_2H_6O, (c) $C_3H_6O_2$, (d) C_3H_8O.

24.36 Classify each of the following molecules as alcohol, aldehyde, ketone, carboxylic acid, amine, or ether:

(a) $CH_3—O—CH_2—CH_3$

(b) $CH_3—CH_2—NH_2$

(c) 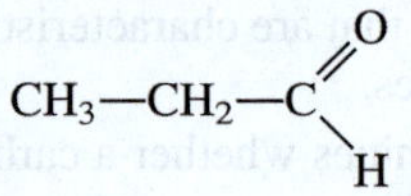

(d) 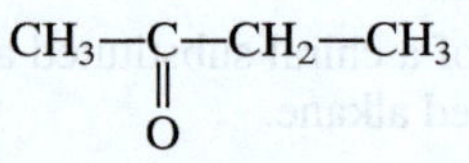

(e) 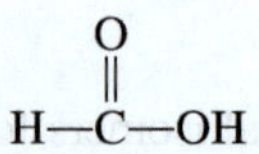

(f) $CH_3—CH_2—CH_2—OH$

(g) 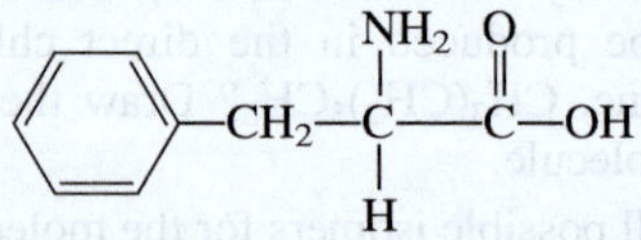

24.37 Generally aldehydes are more susceptible to oxidation in air than are ketones. Use acetaldehyde and acetone as examples and show why ketones such as acetone are more stable than aldehydes in this respect.

24.38 Complete the following equation and identify the products:

$$HCOOH + CH_3OH \longrightarrow$$

24.39 A compound has the empirical formula $C_5H_{12}O$. Upon controlled oxidation, it is converted into a compound of empirical formula $C_5H_{10}O$, which behaves as a ketone. Draw possible structures for the original compound and the final compound.

24.40 A compound having the molecular formula $C_4H_{10}O$ does not react with sodium metal. In the presence of light, the compound reacts with Cl_2 to form three compounds having the formula C_4H_9OCl. Draw a structure for the original compound that is consistent with this information.

24.41 Predict the product or products of each of the following reactions:

(a) $CH_3CH_2OH + HCOOH \longrightarrow$

(b) $H{-}C{\equiv}C{-}CH_3 + H_2 \longrightarrow$

(c) 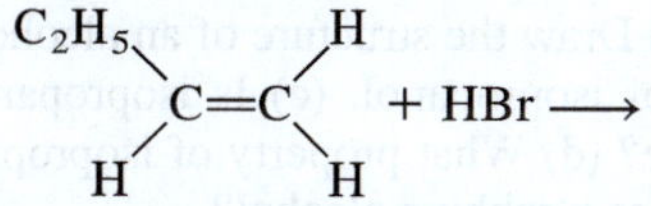

24.42 Identify the functional groups in each of the following molecules:

(a) $CH_3CH_2COCH_2CH_2CH_3$

(b) $CH_3COOC_2H_5$

(c) $CH_3CH_2OCH_2CH_2CH_2CH_3$

Additional Problems

24.43 Draw all the possible structural isomers for the molecule having the formula C_7H_7Cl. The molecule contains one benzene ring.

24.44 Given these data

$$C_2H_4(g) + 3O_2(g) \longrightarrow 2CO_2(g) + 2H_2O(l) \qquad \Delta H° = -1411 \text{ kJ/mol}$$

$$2C_2H_2(g) + 5O_2(g) \longrightarrow 4CO_2(g) + 2H_2O(l) \qquad \Delta H° = -2599 \text{ kJ/mol}$$

$$H_2(g) + \tfrac{1}{2}O_2(g) \longrightarrow H_2O(l) \qquad \Delta H° = -285.8 \text{ kJ/mol}$$

calculate the heat of hydrogenation for acetylene:

$$C_2H_2(g) + H_2(g) \longrightarrow C_2H_4(g)$$

24.45 State which member of each of the following pairs of compounds is the more reactive and explain why: (a) propane and cyclopropane, (b) ethylene and methane, (c) acetaldehyde and acetone.

24.46 State which of the following types of compounds can form hydrogen bonds with water molecules: (a) carboxylic acids, (b) alkenes, (c) ethers, (d) aldehydes, (e) alkanes, (f) amines.

24.47 An organic compound is found to contain 37.5 percent carbon, 3.2 percent hydrogen, and 59.3 percent fluorine by mass. The following pressure and volume data were obtained for 1.00 g of this substance at 90°C:

P (atm)	**V (L)**
2.00	0.332
1.50	0.409
1.00	0.564
0.50	1.028

The molecule is known to have no dipole moment. (a) What is the empirical formula of this substance? (b) Does this substance behave as an ideal gas? (c) What is its molecular formula? (d) Draw the Lewis structure of this molecule and describe its geometry. (e) What is the systematic name of this compound?

24.48 State at least one commercial use for each of the following compounds: (a) 2-propanol (isopropanol), (b) acetic acid, (c) naphthalene, (d) methanol, (e) ethanol, (f) ethylene glycol, (g) methane, (h) ethylene.

24.49 How many liters of air (78 percent N_2, 22 percent O_2 by volume) at 20°C and 1.00 atm are needed for the complete combustion of 1.0 L of octane, C_8H_{18}, a typical gasoline component that has a density of 0.70 g/mL?

24.50 How many carbon-carbon sigma bonds are present in each of the following molecules: (a) 2-butyne, (b) anthracene (see Figure 24.7), (c) 2,3-dimethylpentane?

24.51 How many carbon-carbon sigma bonds are present in each of the following molecules: (a) benzene, (b) cyclobutane, (c) 3-ethyl-2-methylpentane?

24.52 The combustion of 20.63 mg of compound Y, which contains only C, H, and O, with excess oxygen gave 57.94 mg of CO_2 and 11.85 mg of H_2O. (a) Calculate how many milligrams of C, H, and O were present in the original sample of Y. (b) Derive the empirical formula of Y. (c) Suggest a plausible structure for Y if the empirical formula is the same as the molecular formula.

24.53 Draw all the structural isomers of compounds with the formula $C_4H_8Cl_2$. Indicate which isomers are chiral and give them systematic names.

24.54 The combustion of 3.795 mg of liquid B, which contains only C, H, and O, with excess oxygen gave 9.708 mg of CO_2 and 3.969 mg of H_2O. In a molar mass determination, 0.205 g of B vaporized at 1.00 atm and 200°C and occupied a volume of 89.8 mL. Derive the empirical formula, molar mass, and molecular formula of B and draw three plausible structures.

24.55 Beginning with 3-methyl-1-butyne, show how you would prepare the following compounds:

(a) $CH_2{=}C(Br){-}CH(CH_3){-}CH_3$

(b) 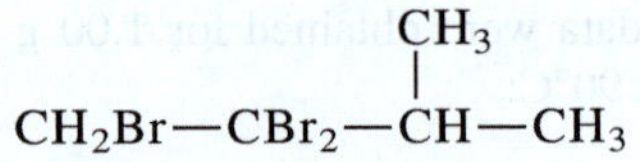

(c) 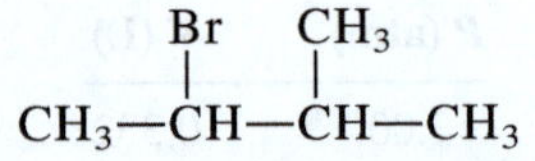

24.56 Indicate the asymmetric carbon atoms in the following compounds:

(a) 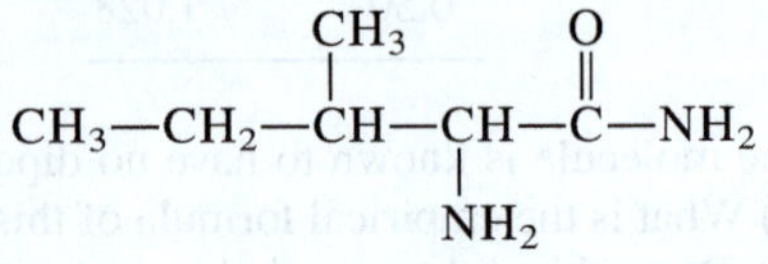

(b) 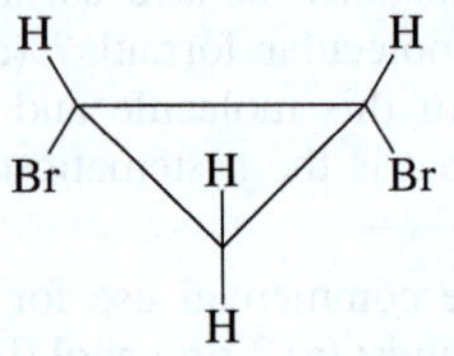

24.57 Suppose benzene contained three distinct single bonds and three distinct double bonds. How many different isomers would there be for dichlorobenzene ($C_6H_4Cl_2$)? Draw all your proposed structures.

24.58 Write the structural formula of an aldehyde that is a structural isomer of acetone.

24.59 Draw structures for the following compounds: (a) cyclopentane, (b) *cis*-2-butene, (c) 2-hexanol, (d) 1,4-dibromobenzene, (e) 2-butyne.

24.60 Name the classes to which the following compounds belong:

(a) C_4H_9OH

(b) $CH_3OC_2H_5$

(c) C_2H_5CHO

(d) C_6H_5COOH

(e) CH_3NH_2

24.61 Ethanol, C_2H_5OH, and dimethyl ether, CH_3OCH_3, are structural isomers. Compare their melting points, boiling points, and solubilities in water.

24.62 Amines are Brønsted bases. The unpleasant smell of fish is due to the presence of certain amines. Explain why cooks often add lemon juice to suppress the odor of fish (in addition to enhancing the flavor).

24.63 You are given two bottles, each containing a colorless liquid. You are told that one liquid is cyclohexane and the other is benzene. Suggest one chemical test that would allow you to distinguish between these two liquids.

24.64 Give the chemical names of the following organic compounds and write their formulas: marsh gas, grain alcohol, wood alcohol, rubbing alcohol, antifreeze, mothballs, chief ingredient of vinegar.

24.65 The compound $CH_3{-}C{\equiv}C{-}CH_3$ is hydrogenated to an alkene using platinum as the catalyst. Predict whether the product is the pure *trans* isomer, the pure *cis* isomer, or a mixture of *cis* and *trans* isomers. Based on your prediction, comment on the mechanism of the heterogeneous catalysis.

24.66 How many asymmetric carbon atoms are present in each of the following compounds?

(a) 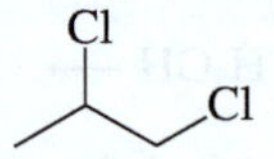

(b) 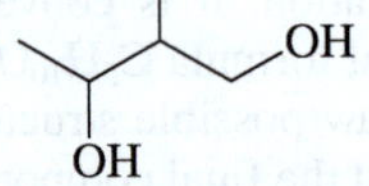

(c) 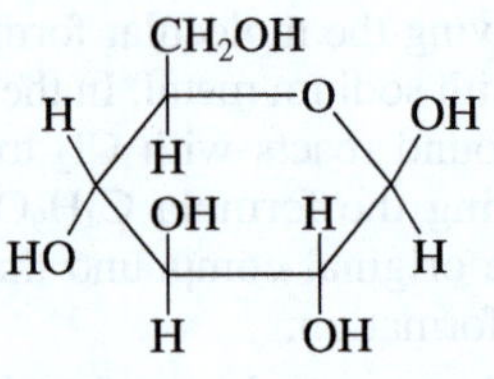

24.67 Isopropanol is prepared by reacting propylene (CH_3CHCH_2) with sulfuric acid, followed by treatment with water. (a) Show the sequence of steps leading to the product. What is the role of sulfuric acid? (b) Draw the structure of an alcohol that is an isomer of isopropanol. (c) Is isopropanol a chiral molecule? (d) What property of isopropanol makes it useful as a rubbing alcohol?

24.68 When a mixture of methane and bromine vapor is exposed to light, the following reaction occurs slowly:

$$CH_4(g) + Br_2(g) \longrightarrow CH_3Br(g) + HBr(g)$$

Suggest a mechanism for this reaction. (*Hint:* Bromine vapor is deep red; methane is colorless.)

24.69 Under conditions of acid catalysis, alkenes react with water to form alcohols. As in the case with hydrogen halides, the addition reaction in the formation of alcohols is also governed by Markovnikov's rule. An alkene of approximate molar mass of 42 g reacts with water and sulfuric acid to produce a compound that reacts with acidic potassium dichromate solution to produce a ketone. Identify all the compounds in the preceding steps.

24.70 2-Butanone can be reduced to 2-butanol by reagents such as lithium aluminum hydride ($LiAlH_4$). (a) Write the formula of the product. Is it chiral? (b) In reality, the product does not exhibit optical activity. Explain.

24.71 Write the structures of three alkenes that yield 2-methylbutane on hydrogenation.

24.72 An alcohol was converted to a carboxylic acid with acidic potassium dichromate. A 4.46-g sample of the acid was added to 50.0 mL of 2.27 *M* NaOH and the excess NaOH required 28.7 mL of 1.86 *M* HCl for neutralization. What is the molecular formula of the alcohol?

24.73 Write the structural formulas of the alcohols with the formula $C_6H_{13}O$ and indicate those that are chiral. Show only the C atoms and the —OH groups.

24.74 Fat and oil are names for the same class of compounds, called triglycerides, which contain three ester groups

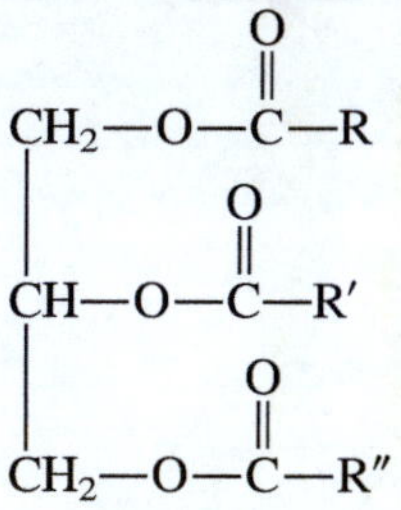

A fat or oil

where R, R′, and R″ represent long hydrocarbon chains. (a) Suggest a reaction that leads to the formation of a triglyceride molecule, starting with glycerol and carboxylic acids (see Section 11.3 for the structure of glycerol). (b) In the old days, soaps were made by hydrolyzing animal fat with lye (a sodium hydroxide solution). Write an equation for this reaction. (c) The difference between fats and oils is that at room temperature, the former are solids and the latter are liquids. Fats are usually produced by animals, whereas oils are commonly found in plants. The melting points of these substances are determined by the number of C=C bonds (or the extent of unsaturation) present—the larger the number of C=C bonds, the lower the melting point and the more likely that the substance is a liquid. Explain. (d) One way to convert liquid oil to solid fat is to hydrogenate the oil, a process by which some or all of the C=C bonds are converted to C—C bonds. This procedure prolongs shelf life of the oil by removing the more reactive C=C group and facilitates packaging. How would you carry out such a process (that is, what reagents and catalyst would you employ)? (e) The degree of unsaturation of oil can be determined by reacting the oil with iodine, which reacts with the C=C bond as follows:

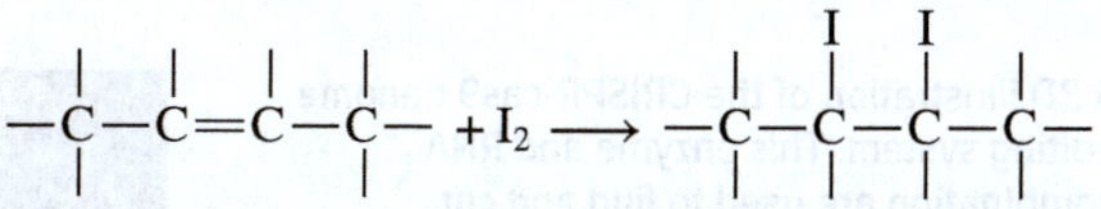

The procedure is to add a known amount of iodine to the oil and allow the reaction to go to completion. The amount of excess (unreacted) iodine is determined by titrating the remaining iodine with a standard sodium thiosulfate ($Na_2S_2O_3$) solution:

$$I_2 + 2Na_2S_2O_3 \longrightarrow Na_2S_4O_6 + 2NaI$$

The number of grams of iodine that react with 100 g of oil is called the *iodine number.* In one case, 43.8 g of I_2 were treated with 35.3 g of corn oil. The excess iodine required 20.6 mL of a 0.142 *M* $Na_2S_2O_3$ for neutralization. Calculate the iodine number of the corn oil.

Answers to Practice Exercises

24.1 5.
24.2 4,6-diethyl-2-methyloctane.
24.3 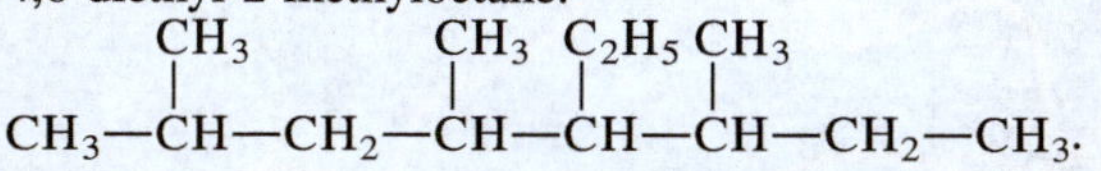
24.4 No.
24.5 $CH_3CH_2COOCH_3$ and H_2O.

Answers to Review of Concepts & Facts

24.2.1 C_8H_{18}. **24.2.2** (b) and (c). **24.3.1** 1,4-difluorobenzene (or *p*-difluorobenzene). **24.4.1** Amine and ester.

CHAPTER

25

Synthetic and Natural Organic Polymers

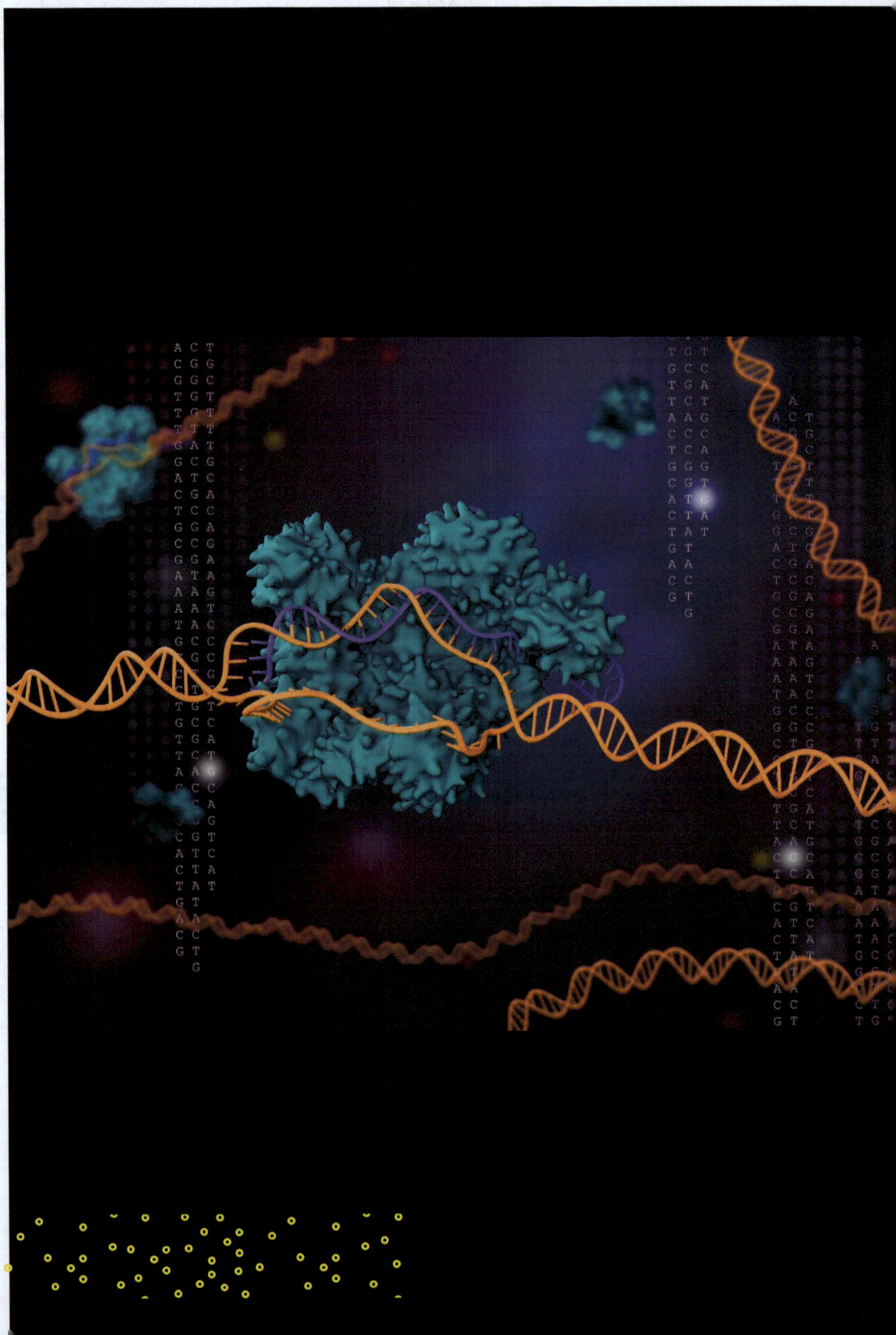

A 3D illustration of the CRISPR-cas9 genome editing system. This enzyme and RNA combination are used to find and cut specific sequences in DNA.

Meletios Verras/Shutterstock

CHAPTER OUTLINE

Polymers are very large molecules containing hundreds or thousands of atoms. People have been using polymers since prehistoric time, and chemists have been synthesizing them for the past century. Natural polymers are the basis of all life processes, and our technological society is largely dependent on synthetic polymers.

This chapter discusses some of the preparation and properties of important synthetic organic polymers in addition to two naturally occurring polymers that are vital to living systems—proteins and nucleic acids.

25.1 Properties of Polymers

Learning Objective

- Define polymer.

A ***polymer*** is *a molecular compound distinguished by a high molar mass, ranging into thousands and millions of grams, and made up of many repeating units.* The physical properties of these so-called macromolecules differ greatly from those of small, ordinary molecules, and special techniques are required to study them.

Naturally occurring polymers include proteins, nucleic acids, cellulose (polysaccharides), and rubber (polyisoprene). Most synthetic polymers are organic compounds. Familiar examples are nylon, poly(hexamethylene adipamide); Dacron, poly(ethylene terephthalate); and Lucite or Plexiglas, poly(methyl methacrylate).

The development of polymer chemistry began in the 1920s with the investigation into a puzzling behavior of certain materials, including wood, gelatin, cotton, and rubber. For example, when rubber, with the known empirical formula of C_5H_8, was dissolved in an organic solvent, the solution displayed several unusual properties—high viscosity, low osmotic pressure, and negligible freezing-point depression. These observations strongly suggested the presence of solutes of very high molar mass, but chemists were not ready at that time to accept the idea that such giant molecules could exist. Instead, they postulated that materials such as rubber consist of aggregates of small molecular units, like C_5H_8 or $C_{10}H_{16}$, held together by intermolecular forces. This misconception persisted for a number of years until Hermann Staudinger† clearly showed that these so-called aggregates are, in fact, enormously large molecules, each of which contains many thousands of atoms held together by covalent bonds.

Once the structures of these macromolecules were understood, the way was open for manufacturing polymers, which now pervade almost every aspect of our daily lives. About 90 percent of today's chemists, including biochemists, work with polymers.

Summary of Concepts & Facts

- Polymers are large molecules made up of small, repeating units called monomers.
- Proteins, nucleic acids, cellulose, and rubber are natural polymers. Nylon, Dacron, and Lucite are examples of synthetic polymers.

†Hermann Staudinger (1881–1963). German chemist. One of the pioneers in polymer chemistry. Staudinger was awarded the Nobel Prize in Chemistry in 1953.

25.2 Synthetic Organic Polymers

Learning Objectives

- Define monomer.
- Describe the mechanism of addition polymerization.
- Distinguish between addition and condensation polymers.
- Differentiate between homopolymers and copolymers.

Because of their size, we might expect molecules containing thousands of carbon and hydrogen atoms to form an enormous number of structural and geometric isomers (if C=C bonds are present). However, these molecules are made up of ***monomers,*** *simple repeating units,* and this type of composition severely restricts the number of possible isomers. Synthetic polymers are created by joining monomers together, one at a time, by means of addition reactions and condensation reactions.

Addition Reactions

Addition reactions involve unsaturated compounds containing double or triple bonds, particularly C=C and C≡C. Hydrogenation and reactions of hydrogen halides and halogens with alkenes and alkynes are examples of addition reactions.

Polyethylene, a very stable polymer used in packaging wraps, is made by joining ethylene monomers via an addition-reaction mechanism. First an *initiator* molecule (R_2) is heated to produce two radicals:

$$R_2 \longrightarrow 2R\cdot$$

The reactive radical attacks an ethylene molecule to generate a new radical:

$$R\cdot + CH_2{=}CH_2 \longrightarrow R{-}CH_2{-}CH_2\cdot$$

which further reacts with another ethylene molecule, and so on:

$$R{-}CH_2{-}CH_2\cdot + CH_2{=}CH_2 \longrightarrow R{-}CH_2{-}CH_2{-}CH_2{-}CH_2\cdot$$

Very quickly a long chain of CH_2 groups are built. Eventually, this process is terminated by the combination of two long-chain radicals to give the polymer called polyethylene:

$$R{\left(CH_2{-}CH_2\right)}_n CH_2CH_2\cdot + R{\left(CH_2{-}CH_2\right)}_n CH_2CH_2\cdot \longrightarrow$$
$$R{\left(CH_2{-}CH_2\right)}_n CH_2CH_2{-}CH_2CH_2{\left(CH_2{-}CH_2\right)}_n R$$

where ${\left(CH_2{-}CH_2\right)}_n$ is a convenient shorthand convention for representing the repeating unit in the polymer. The value of n is understood to be very large, on the order of hundreds.

Common mailing envelopes made of Tyvek.
Ken Karp/McGraw Hill

The individual chains of polyethylene pack together well and so account for the substance's crystalline properties (Figure 25.1). Polyethylene is mainly used in films, frozen food packaging, and other product wrappings. A specially treated type of polyethylene called Tyvek is used for home insulation.

Polyethylene is an example of a ***homopolymer,*** which is *a polymer made up of only one type of monomer.* Other homopolymers that are synthesized by the radical

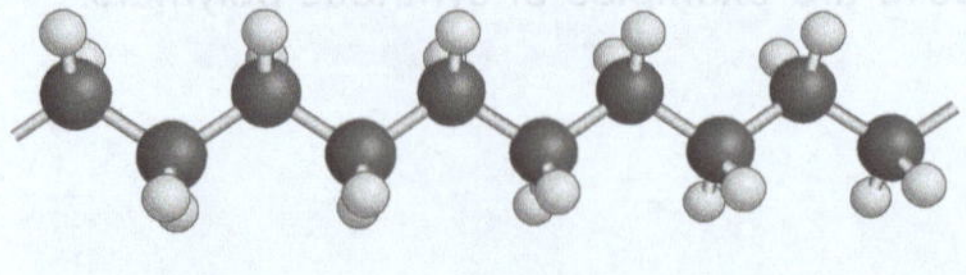

Figure 25.1 *Structure of polyethylene. Each carbon atom is* sp^3*-hybridized.*

mechanism are Teflon, polytetrafluoroethylene (Figure 25.2), and poly(vinyl chloride) (PVC):

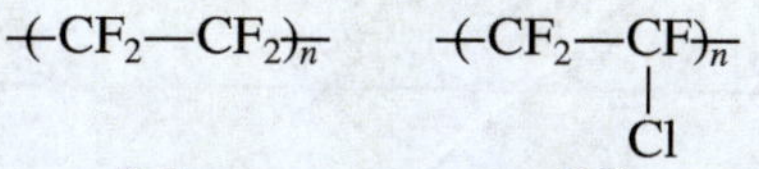

Teflon PVC

The chemistry of polymers is more complex if the starting units are asymmetric:

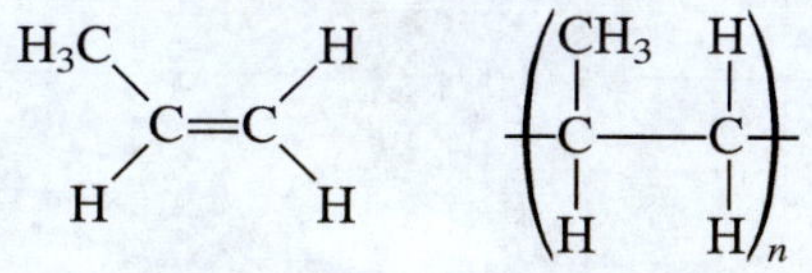

propene polypropene

Figure 25.2 *A cooking utensil coated with Silverstone, which contains polytetrafluoroethylene.*
Ken Karp/McGraw Hill

Several geometric isomers can result from an addition reaction of propenes (Figure 25.3). If the additions occur randomly, we obtain *atactic* polypropenes, which do not pack together well. These polymers are rubbery, amorphous, and relatively weak. Two other possibilities are an *isotactic* structure, in which the R groups are all on the same side of the asymmetric carbon atoms, and a *syndiotactic* form, in which the R groups alternate to the left and right of the asymmetric carbons. Of these, the isotactic isomer has the highest melting point and greatest crystallinity and is endowed with superior mechanical properties.

A major problem that the polymer industry faced in the beginning was how to synthesize either the isotactic or syndiotactic polymer selectively without having it contaminated by other products. The solution came from Giulio Natta† and Karl Ziegler,‡ who demonstrated that certain catalysts, including triethylaluminum [$Al(C_2H_5)_3$] and titanium trichloride ($TiCl_3$), promote the formation only of specific isomers. Using Natta-Ziegler catalysts, chemists can design polymers to suit any purpose.

Rubber is probably the best known organic polymer and the only true hydrocarbon polymer found in nature. It is formed by the radical addition of the monomer isoprene.

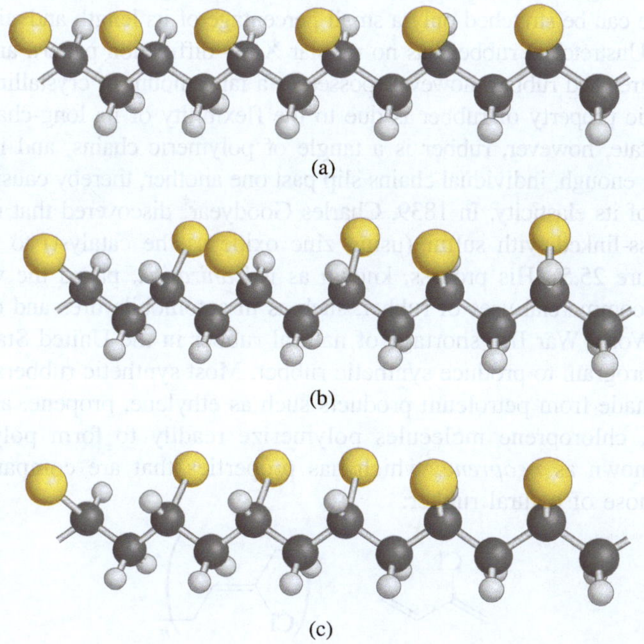

Figure 25.3 *Stereoisomers of polymers. When the R group (yellow sphere) is CH_3, the polymer is polypropene. (a) When the R groups are all on one side of the chain, the polymer is said to be isotactic. (b) When the R groups alternate from side to side, the polymer is said to be syndiotactic. (c) When the R groups are disposed at random, the polymer is atactic.*

†Giulio Natta (1903–1979). Italian chemist. Natta received the Nobel Prize in Chemistry in 1963 for discovering stereospecific catalysts for polymer synthesis.

‡Karl Ziegler (1898–1976). German chemist. Ziegler shared the Nobel Prize in Chemistry in 1963 with Natta for his work in polymer synthesis.

Figure 25.4 *Latex (aqueous suspension of rubber particles) being collected from a rubber tree.*
Suphatthra China/Shutterstock

Actually, polymerization can result in either poly-*cis*-isoprene or poly-*trans*-isoprene—or a mixture of both, depending on reaction conditions:

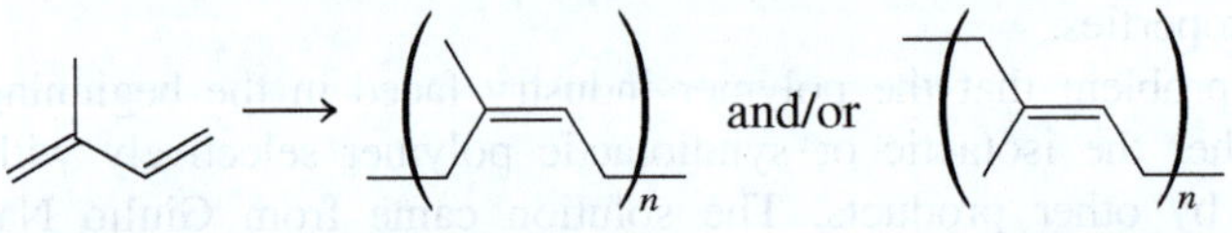

isoprene poly-*cis*-isoprene poly-*trans*-isoprene

Note that in the *cis* isomer the two CH_2 groups are on the same side of the C=C bond, whereas the same groups are across from each other in the *trans* isomer. Natural rubber is poly-*cis*-isoprene, which is extracted from the tree *Hevea brasiliensis* (Figure 25.4).

An unusual and very useful property of rubber is its elasticity. Rubber will stretch up to 10 times its length and, if released, will return to its original size. In contrast, a piece of copper wire can be stretched only a small percentage of its length and still return to its original size. Unstretched rubber has no regular X-ray diffraction pattern and is therefore amorphous. Stretched rubber, however, possesses a fair amount of crystallinity and order.

The elastic property of rubber is due to the flexibility of its long-chain molecules. In the bulk state, however, rubber is a tangle of polymeric chains, and if the external force is strong enough, individual chains slip past one another, thereby causing the rubber to lose most of its elasticity. In 1839, Charles Goodyear[†] discovered that natural rubber could be cross-linked with sulfur (using zinc oxide as the catalyst) to prevent chain slippage (Figure 25.5). His process, known as *vulcanization,* paved the way for many practical and commercial uses of rubber, such as in automobile tires and dentures.

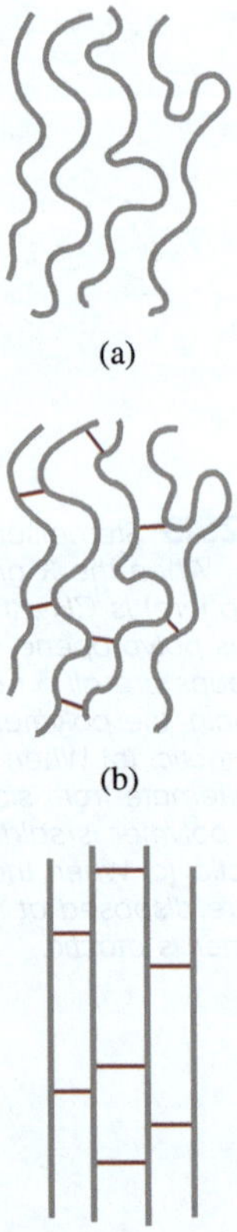

Figure 25.5 *Rubber molecules ordinarily are bent and convoluted. Parts (a) and (b) represent the long chains before and after vulcanization, respectively; (c) shows the alignment of molecules when stretched. Without vulcanization these molecules would slip past one another, and rubber's elastic properties would be gone.*

During World War II a shortage of natural rubber in the United States prompted an intensive program to produce synthetic rubber. Most synthetic rubbers (called *elastomers*) are made from petroleum products such as ethylene, propene, and butadiene. For example, chloroprene molecules polymerize readily to form polychloroprene, commonly known as *neoprene,* which has properties that are comparable or even superior to those of natural rubber:

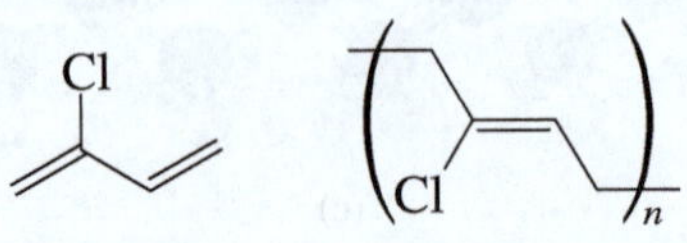

chloroprene polychloroprene

[†]Charles Goodyear (1800–1860). American chemist. Goodyear was the first person to realize the potential of natural rubber. His vulcanization process made rubber usable in countless ways and opened the way for the development of the automobile industry.

Another important synthetic rubber is formed by the addition of butadiene to styrene in a 3:1 ratio to give styrene-butadiene rubber (SBR). Because styrene and butadiene are different monomers, SBR is called a ***copolymer,*** which is *a polymer containing two or more different monomers.* Table 25.1 shows a number of common and familiar homopolymers and one copolymer produced by addition reactions.

Bubble gums contain synthetic styrene-butadiene rubber.
Kwame Zikomo/Purestock/Alamy Stock Photo

Table 25.1 Some Monomers and Their Common Synthetic Polymers

Monomer		**Polymer**	
Formula	**Name**	**Name and Formula**	**Uses**
$H_2C{=}CH_2$	Ethylene	Polyethylene $\lbrace CH_2{-}CH_2 \rbrace_n$	Plastic piping, bottles, electrical insulation, toys
$H_2C{=}C(H)(CH_3)$	Propene	Polypropene $\left(CH{-}CH_2{-}CH{-}CH_2 \right)_n$ with CH_3 on each CH	Packaging film, carpets, crates for soft-drink bottles, lab wares, toys
$H_2C{=}C(H)(Cl)$	Vinyl chloride	Poly(vinyl chloride) (PVC) $\lbrace CH_2{-}CH(Cl) \rbrace_n$	Piping, siding, gutters, floor tile, clothing, toys
$H_2C{=}C(H)(CN)$	Acrylonitrile	Polyacrylonitrile (PAN) $\left(CH_2{-}CH(CN) \right)_n$	Carpets, knitwear
$F_2C{=}CF_2$	Tetrafluoro-ethylene	Polytetrafluoroethylene (Teflon) $\lbrace CF_2{-}CF_2 \rbrace_n$	Coating on cooking utensils, electrical insulation, bearings
$H_2C{=}C(CH_3){-}C({=}O){-}O{-}CH_3$	Methyl methacrylate	Poly(methyl methacrylate) (Plexiglas) $\lbrace CH_2{-}C(COOCH_3)(CH_3) \rbrace_n$	Optical equipment, home furnishings
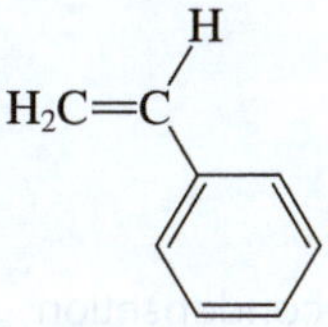	Styrene	Polystyrene 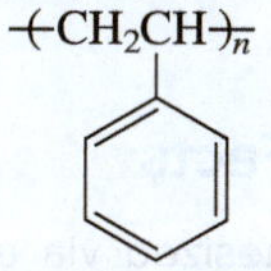	Containers, thermal insulation (ice buckets, water coolers), toys
$H_2C{=}C(H){-}C(H){=}CH_2$	Butadiene	Polybutadiene $\lbrace CH_2CH{=}CHCH_2 \rbrace_n$	Tire tread, coating resin
See above structures	Butadiene and styrene	Styrene-butadiene rubber (SBR) 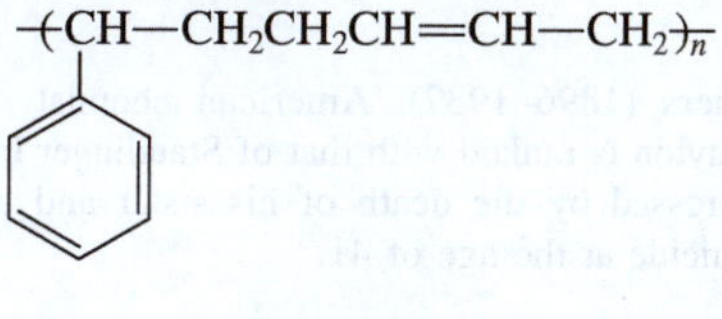	Synthetic rubber

Figure 25.7 *The nylon rope trick. Adding a solution of adipoyl chloride (an adipic acid derivative in which the OH groups have been replaced by Cl groups) in cyclohexane to an aqueous solution of hexamethylenediamine causes nylon to form at the interface of the two solutions, which do not mix. It can then be drawn off.*

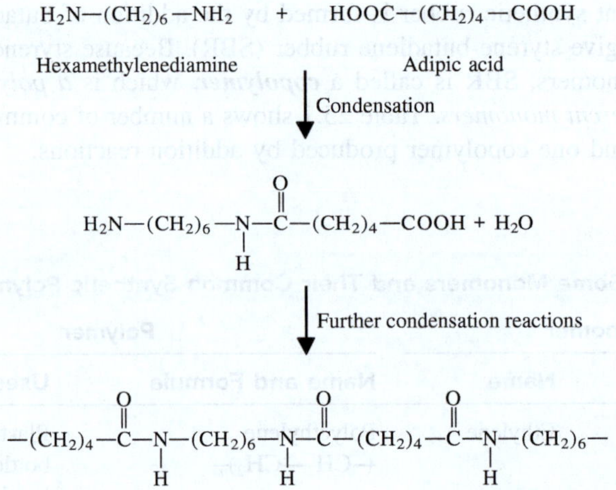

Figure 25.6 *The formation of nylon by the condensation reaction between hexamethylenediamine and adipic acid.*

Condensation Reactions

One of the best-known polymer condensation processes is the reaction between hexamethylenediamine and adipic acid, shown in Figure 25.6. The final product, called nylon 66 (because there are six carbon atoms each in hexamethylenediamine and adipic acid), was first made by Wallace Carothers[†] at Du Pont in 1931. The versatility of nylons is so great that the annual production of nylons and related substances now amounts to several billion pounds. Figure 25.7 shows how nylon 66 is prepared in the laboratory.

Condensation reactions are also used in the manufacture of Dacron (polyester):

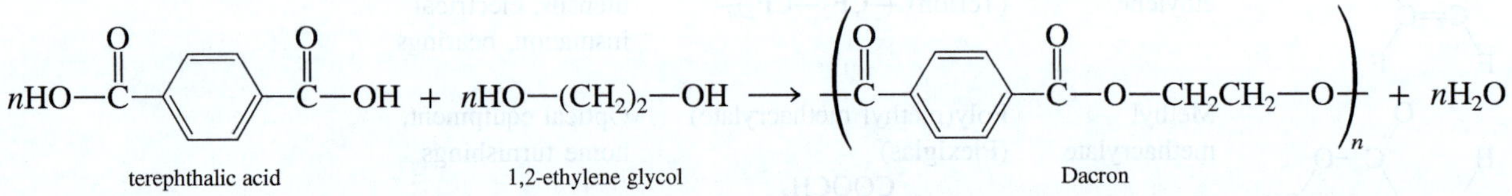

Polyesters are used in fibers, films, and plastic bottles.

Summary of Concepts & Facts

- Organic polymers can be synthesized via addition reactions or condensation reactions.
- Stereoisomers of a polymer made up of asymmetric monomers have different properties, depending on how the starting units are joined together.
- Synthetic rubbers include polychloroprene and styrene-butadiene rubber, which is a copolymer of styrene and butadiene.

[†]Wallace H. Carothers (1896–1937). American chemist. Besides its enormous commercial success, Carothers' work on nylon is ranked with that of Staudinger in clearly elucidating macromolecular structure and properties. Depressed by the death of his sister and convinced that his life's work was a failure, Carothers died by suicide at the age of 41.

25.3 Proteins

Learning Objectives

- Explain what a protein is.
- Summarize peptide bonds.
- Assess the four levels of structure within a protein.

Proteins are *polymers of amino acids;* they play a key role in nearly all biological processes. Enzymes, the catalysts of biochemical reactions, are mostly proteins. Proteins also facilitate a wide range of other functions, such as transport and storage of vital substances, coordinated motion, mechanical support, and protection against diseases. The human body contains an estimated 100,000 different kinds of proteins, each of which has a specific physiological function. As we will see in this section, the chemical composition and structure of these complex natural polymers are the basis of their specificity.

Amino Acids

Proteins have high molar masses, ranging from about 5000 g to 1×10^7 g, and yet the percent composition by mass of the elements in proteins is remarkably constant: carbon, 50 to 55 percent; hydrogen, 7 percent; oxygen, 23 percent; nitrogen, 16 percent; and sulfur, 1 percent.

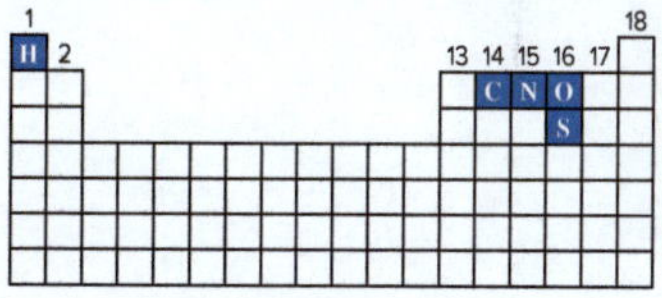

Elements in proteins.

The basic structural units of proteins are *amino acids.* An ***amino acid*** is *a compound that contains at least one amino group (—NH_2) and at least one carboxyl group (—COOH):*

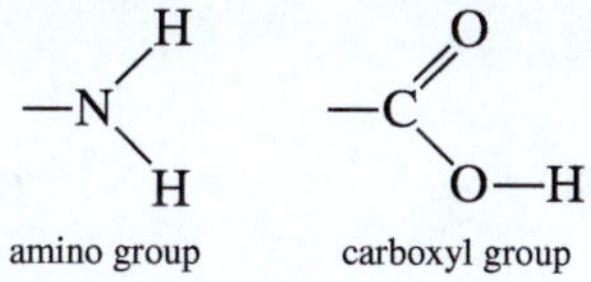

Twenty different amino acids are the building blocks of all the proteins in the human body. Table 25.2 shows the structures of these vital compounds, along with their three-letter abbreviations.

Amino acids in solution at neutral pH exist as *dipolar ions,* meaning that the proton on the carboxyl group has migrated to the amino group. Consider glycine, the simplest amino acid. The un-ionized form and the dipolar ion of glycine are

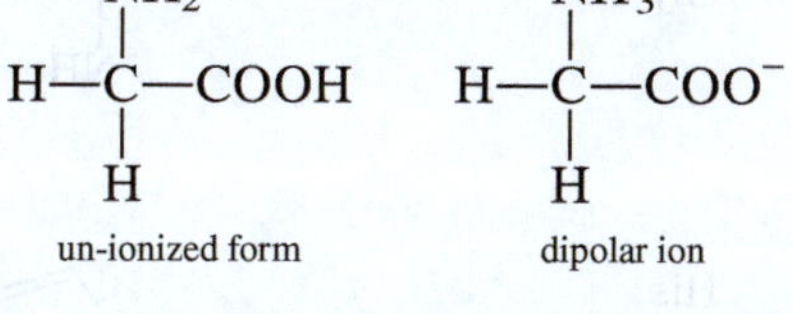

The first step in the synthesis of a protein molecule is a condensation reaction between an amino group on one amino acid and a carboxyl group on another amino acid. The molecule formed from the two amino acids is called a *dipeptide,* and the bond joining them together is a *peptide bond:*

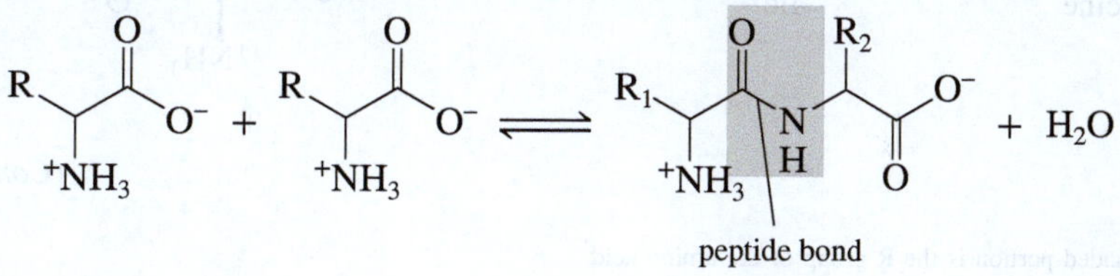

Table 25.2 The 20 Amino Acids Essential to Living Organisms*

Name	Abbreviation	Structure
Alanine	Ala	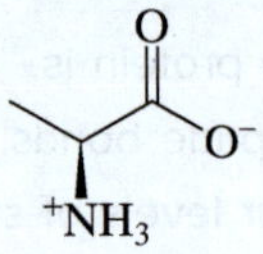
Arginine	Arg	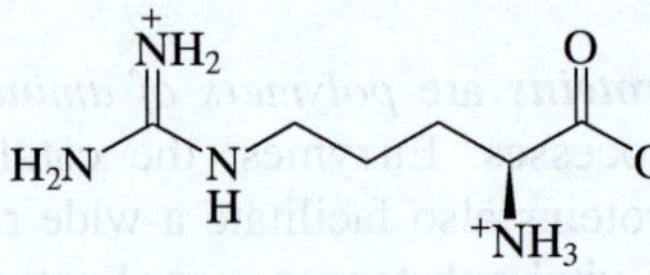
Asparagine	Asn	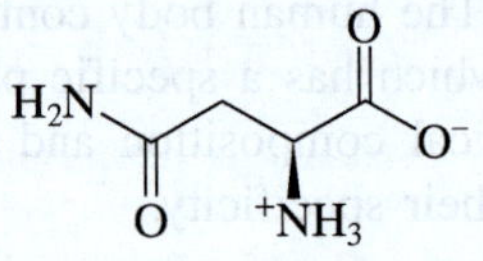
Aspartic acid	Asp	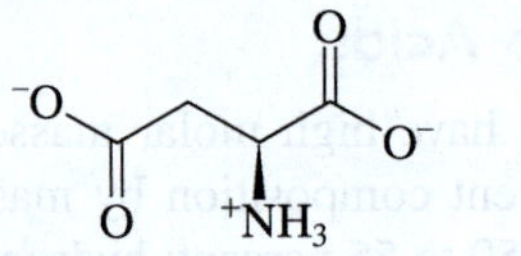
Cysteine	Cys	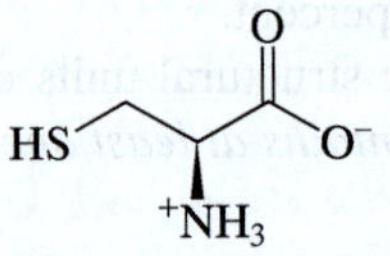
Glutamic acid	Glu	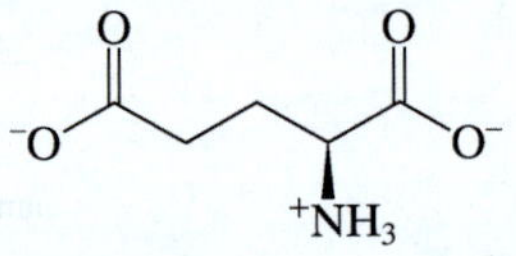
Glutamine	Gln	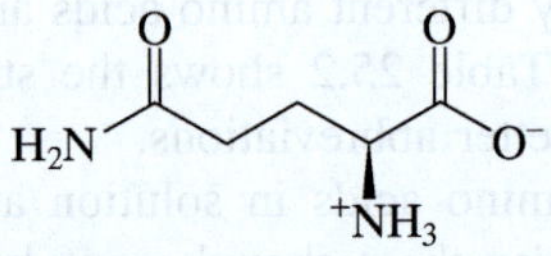
Glycine	Gly	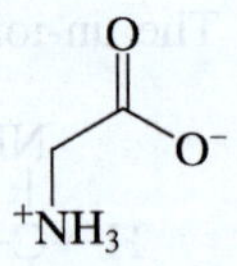
Histidine	His	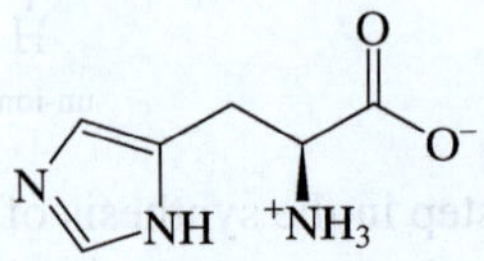
Isoleucine	Ile	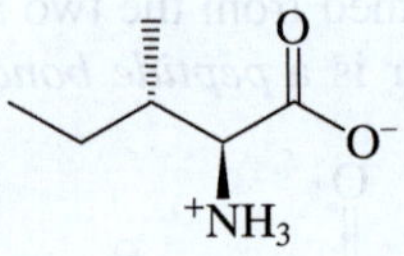

(Continued)

*The shaded portion is the R group of the amino acid.

Table 25.2 The 20 Amino Acids Essential to Living Organisms* (*continued*)

Name	Abbreviation	Structure
Leucine	Leu	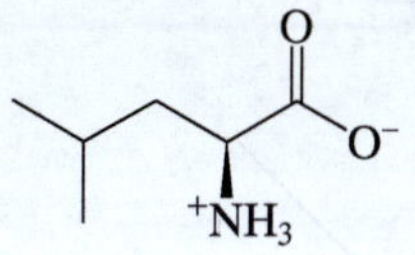
Lysine	Lys	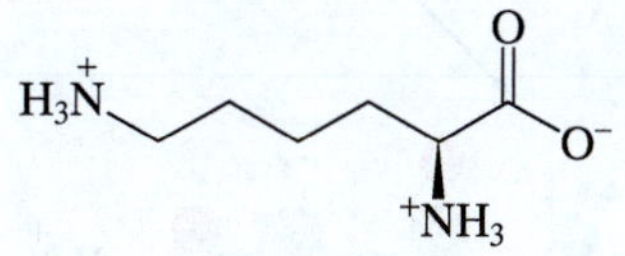
Methionine	Met	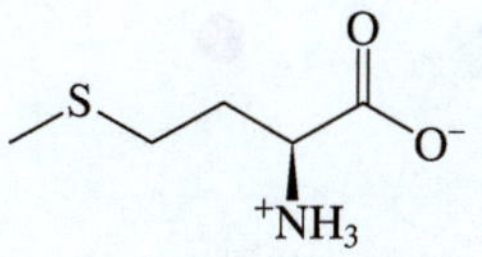
Phenylalanine	Phe	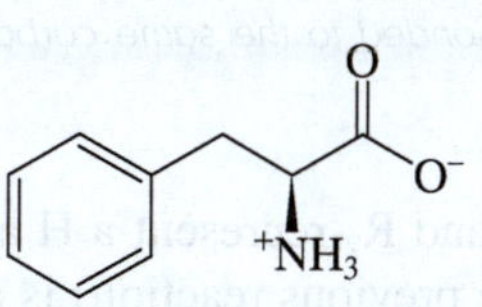
Proline	Pro	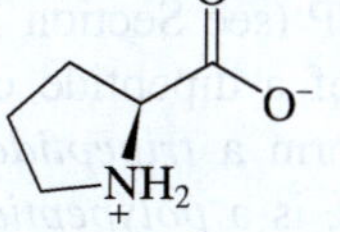
Serine	Ser	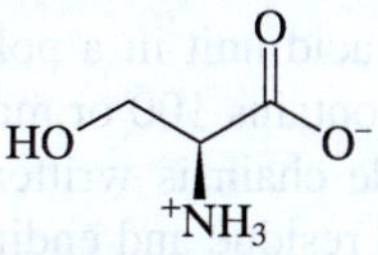
Threonine	Thr	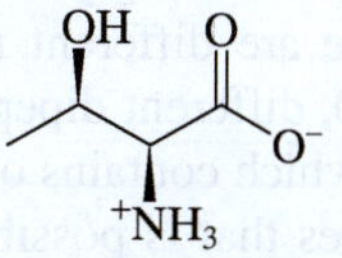
Tryptophan	Trp	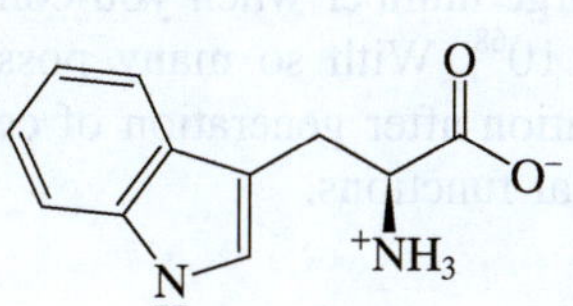
Tyrosine	Tyr	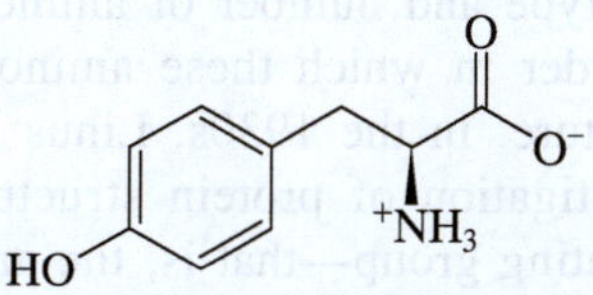
Valine	Val	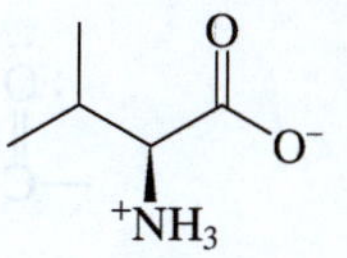

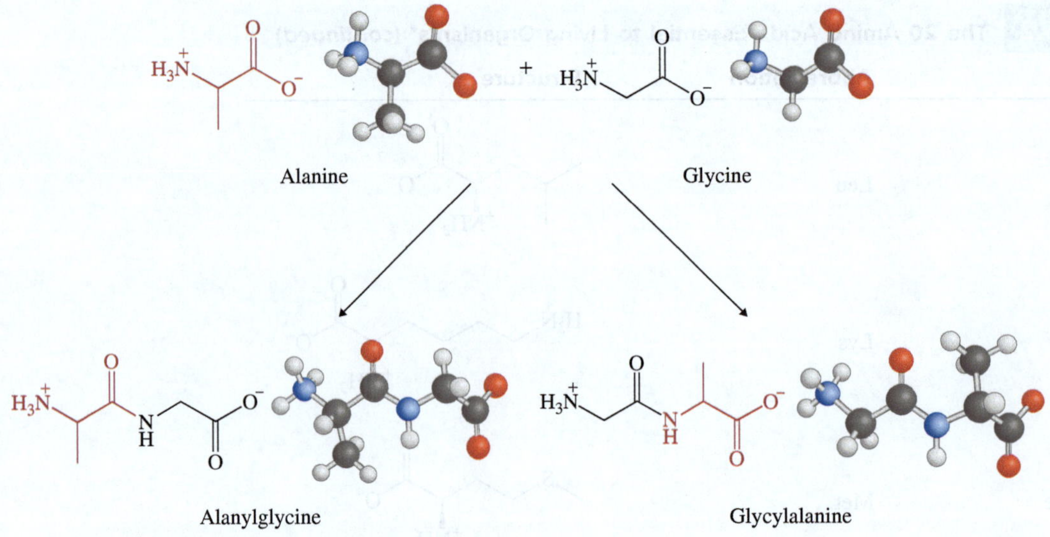

Figure 25.8 *The formation of two dipeptides from two different amino acids. Alanylglycine is different from glycylalanine in that in alanylglycine the amino and methyl groups are bonded to the same carbon atom.*

where R_1 and R_2 represent a H atom or some other group; —CO—NH— (the shaded area in the previous reaction) is also called the *amide group*. Because the equilibrium of the reaction joining two amino acids lies to the left, the process is coupled to the hydrolysis of ATP (see Section 17.6).

Either end of a dipeptide can engage in a condensation reaction with another amino acid to form a *tripeptide,* a *tetrapeptide,* and so on. The final product, the protein molecule, is a *polypeptide;* it can also be thought of as a polymer of amino acids.

An amino acid unit in a polypeptide chain is called a *residue*. Typically, a polypeptide chain contains 100 or more amino acid residues. The sequence of amino acids in a polypeptide chain is written conventionally from left to right, starting with the amino-terminal residue and ending with the carboxyl-terminal residue. Let us consider a dipeptide formed from glycine and alanine. Figure 25.8 shows that alanylglycine and glycylalanine are different molecules. With 20 different amino acids to choose from, 20^2, or 400, different dipeptides can be generated. Even for a very small protein such as insulin, which contains only 50 amino acid residues, the number of chemically different structures that is possible is on the order of 20^{50} or 10^{65}! This is an incredibly large number when you consider that the total number of atoms in our galaxy is about 10^{68}. With so many possibilities for protein synthesis, it is remarkable that generation after generation of cells can produce identical proteins for specific physiological functions.

Protein Structure

The type and number of amino acids in a given protein along with the sequence or order in which these amino acids are joined together determine the protein's structure. In the 1930s, Linus Pauling and his coworkers conducted a systematic investigation of protein structure. First, they studied the geometry of the basic repeating group—that is, the amide group—which is represented by the following resonance structures:

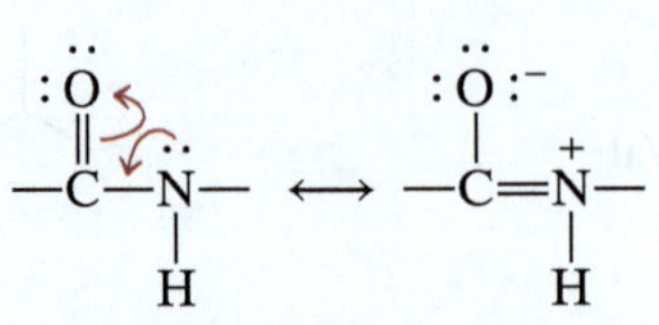

Because it is more difficult (that is, it would take more energy) to twist a double bond than a single bond, the four atoms in the amide group become locked in the same plane (Figure 25.9). Figure 25.10 depicts the repeating amide group in a polypeptide chain.

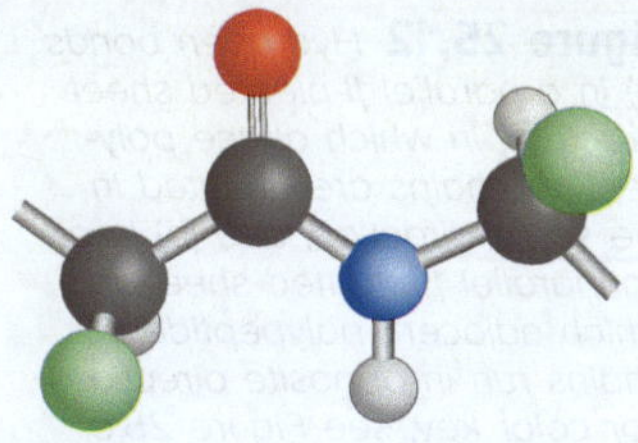

Figure 25.9 *The planar amide group in protein. Rotation about the peptide bond in the amide group is hindered by its double-bond character. The black atoms represent carbon; blue, nitrogen; red, oxygen; green, R group; and gray, hydrogen.*

On the basis of models and X-ray diffraction data, Pauling deduced that there are two common structures for protein molecules, called the *α helix* and the *β-pleated* sheet. The *α*-helical structure of a polypeptide chain is shown in Figure 25.11. The helix is stabilized by *intramolecular* hydrogen bonds between the NH and CO groups of the main chain, giving rise to an overall rodlike shape. The CO group of each amino acid is hydrogen-bonded to the NH group of the amino acid that is four residues away in the sequence. In this manner all the main-chain CO and NH groups take part in hydrogen bonding. X-ray studies have shown that the structure of a number of proteins, including myoglobin and hemoglobin, is to a great extent *α*-helical in nature.

The *β*-pleated structure is markedly different from the *α* helix in that it is like a sheet rather than a rod. The polypeptide chain is almost fully extended, and each chain forms many *intermolecular* hydrogen bonds with adjacent chains. Figure 25.12 shows the two different types of *β*-pleated structures, called *parallel* and *antiparallel*. Silk molecules possess the *β* structure. Because its polypeptide chains are already in extended form, silk lacks elasticity and extensibility, but it is quite strong due to the many intermolecular hydrogen bonds.

It is customary to divide protein structure into four levels of organization. The *primary structure* refers to the unique amino acid sequence of the polypeptide chain. The *secondary structure* includes those parts of the polypeptide chain that are stabilized by a regular pattern of hydrogen bonds between the CO and NH groups

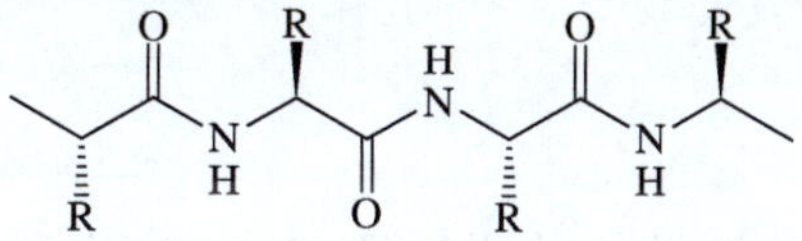

Figure 25.10 *A polypeptide chain. Note the repeating units of the amide group. The symbol R represents part of the structure characteristic of the individual amino acids. For glycine, R is simply a H atom.*

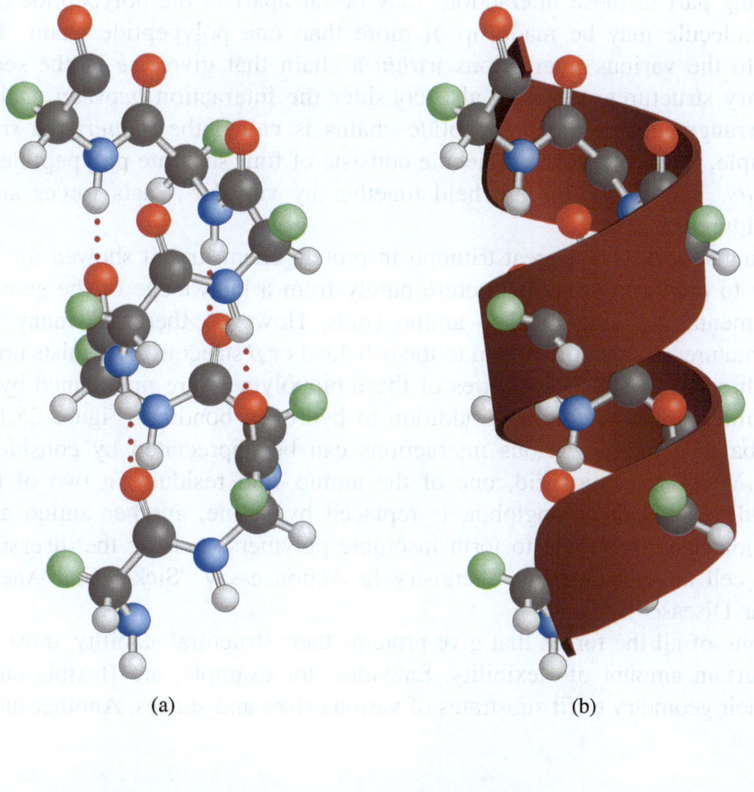

Figure 25.11 *The α-helical structure of a polypeptide chain. The gray spheres are hydrogen atoms. The structure is held in position by intramolecular hydrogen bonds, shown as dotted lines. For color key, see Figure 25.9.*

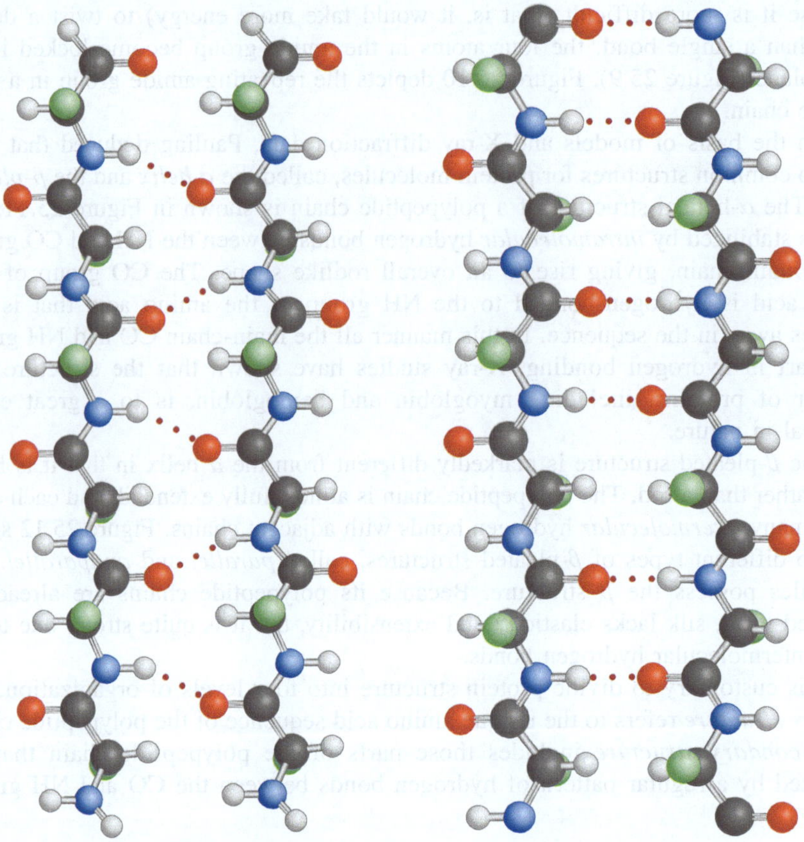

Figure 25.12 *Hydrogen bonds (a) in a parallel β-pleated sheet structure, in which all the polypeptide chains are oriented in the same direction, and (b) in an antiparallel β-pleated sheet, in which adjacent polypeptide chains run in opposite directions. For color key, see Figure 25.9.*

of the backbone, for example, the α helix. The term *tertiary structure* applies to the three-dimensional structure stabilized by dispersion forces, hydrogen bonding, and other intermolecular forces. It differs from secondary structure in that the amino acids taking part in these interactions may be far apart in the polypeptide chain. A protein molecule may be made up of more than one polypeptide chain. Thus, in addition to the various interactions *within* a chain that give rise to the secondary and tertiary structures, we must also consider the interaction *between* chains. The overall arrangement of the polypeptide chains is called the *quaternary structure*. For example, the hemoglobin molecule consists of four separate polypeptide chains, or *subunits*. These subunits are held together by van der Waals forces and ionic forces (Figure 25.13).

Pauling's work was a great triumph in protein chemistry. It showed for the first time how to predict a protein structure purely from a knowledge of the geometry of its fundamental building blocks—amino acids. However, there are many proteins whose structures do not correspond to the α-helical or β structure. Chemists now know that the three-dimensional structures of these biopolymers are maintained by several types of intermolecular forces in addition to hydrogen bonding (Figure 25.14). The delicate balance of the various interactions can be appreciated by considering an example: When glutamic acid, one of the amino acid residues in two of the four polypeptide chains in hemoglobin, is replaced by valine, another amino acid, the protein molecules aggregate to form insoluble polymers, causing the disease known as sickle cell anemia (see the Chemistry in Action essay "Sickle Cell Anemia—A Molecular Disease").

In spite of all the forces that give proteins their structural stability, most proteins have a certain amount of flexibility. Enzymes, for example, are flexible enough to change their geometry to fit substrates of various sizes and shapes. Another interesting

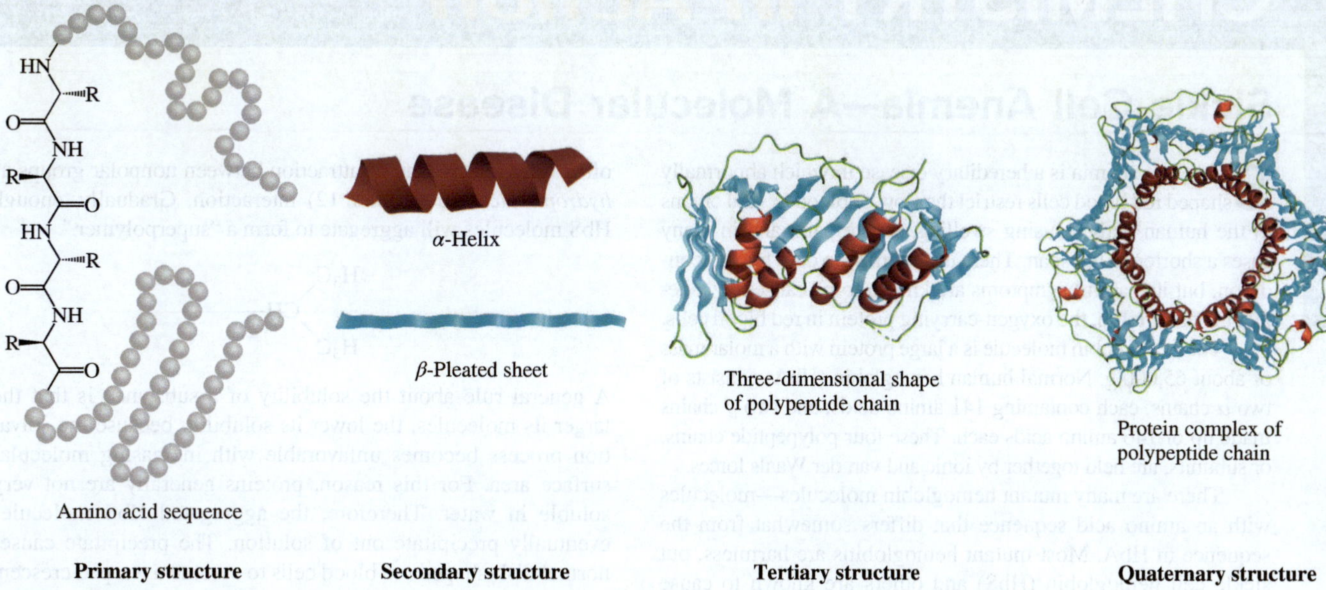

Figure 25.13 *The primary, secondary, tertiary, and quaternary structure of the hemoglobin molecule.*

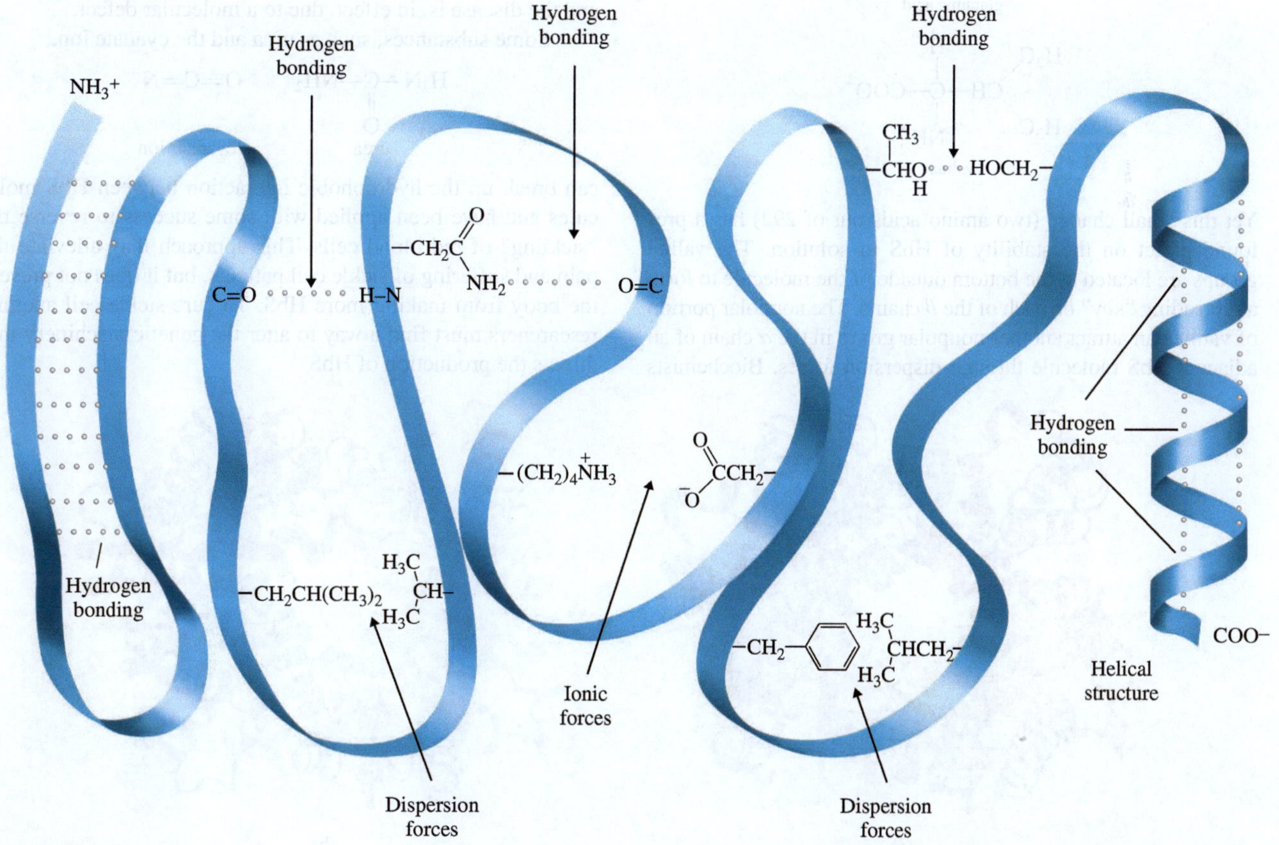

Figure 25.14 *Intermolecular forces in a protein molecule: (a) ionic forces, (b) hydrogen bonding, (c) dispersion forces, and (d) dipole-dipole forces.*

CHEMISTRY *in Action*

Sickle Cell Anemia—A Molecular Disease

Sickle cell anemia is a hereditary disease in which abnormally shaped red blood cells restrict the flow of blood to vital organs in the human body, causing swelling, severe pain, and in many cases a shortened life span. There is currently no cure for this condition, but its painful symptoms are known to be caused by a defect in hemoglobin, the oxygen-carrying protein in red blood cells.

The hemoglobin molecule is a large protein with a molar mass of about 65,000 g. Normal human hemoglobin (HbA) consists of two α chains, each containing 141 amino acids, and two β chains made up of 146 amino acids each. These four polypeptide chains, or subunits, are held together by ionic and van der Waals forces.

There are many mutant hemoglobin molecules—molecules with an amino acid sequence that differs somewhat from the sequence in HbA. Most mutant hemoglobins are harmless, but sickle cell hemoglobin (HbS) and others are known to cause serious diseases. HbS differs from HbA in only one very small detail. A valine molecule replaces a glutamic acid molecule on each of the two β chains:

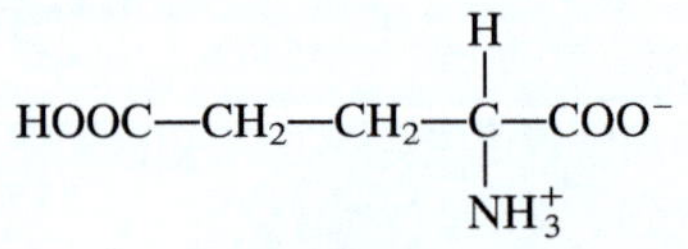

glutamic acid

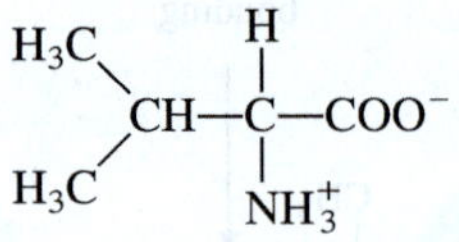

valine

Yet this small change (two amino acids out of 292) has a profound effect on the stability of HbS in solution. The valine groups are located at the bottom outside of the molecule to form a protruding "key" on each of the β chains. The nonpolar portion of valine can attract another nonpolar group in the α chain of an adjacent HbS molecule through dispersion forces. Biochemists often refer to this kind of attraction between nonpolar groups as *hydrophobic* (see Chapter 12) interaction. Gradually, enough HbS molecules will aggregate to form a "superpolymer."

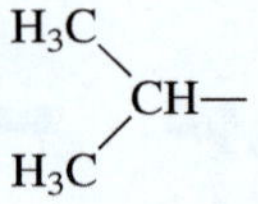

A general rule about the solubility of a substance is that the larger its molecules, the lower its solubility because the solvation process becomes unfavorable with increasing molecular surface area. For this reason, proteins generally are not very soluble in water. Therefore, the aggregated HbS molecules eventually precipitate out of solution. The precipitate causes normal disk-shaped red blood cells to assume a warped crescent or sickle shape (see figure in Chemistry in Action essay "Electron Microscopy" in Section 7.4). These deformed cells clog the narrow capillaries, thereby restricting blood flow to organs of the body. It is the reduced blood flow that gives rise to the symptoms of sickle cell anemia. Sickle cell anemia has been termed a molecular disease by Linus Pauling, who did some of the early important chemical research on the nature of the affliction, because the destructive action occurs at the molecular level and the disease is, in effect, due to a molecular defect.

Some substances, such as urea and the cyanate ion,

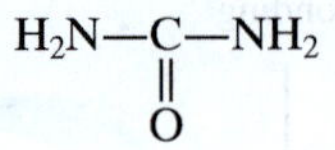

urea

$O{=}C{=}N^-$

cyanate ion

can break up the hydrophobic interaction between HbS molecules and have been applied with some success to reverse the "sickling" of red blood cells. This approach may alleviate the pain and suffering of sickle cell patients, but it does not prevent the body from making more HbS. To cure sickle cell anemia, researchers must find a way to alter the genetic machinery that directs the production of HbS.

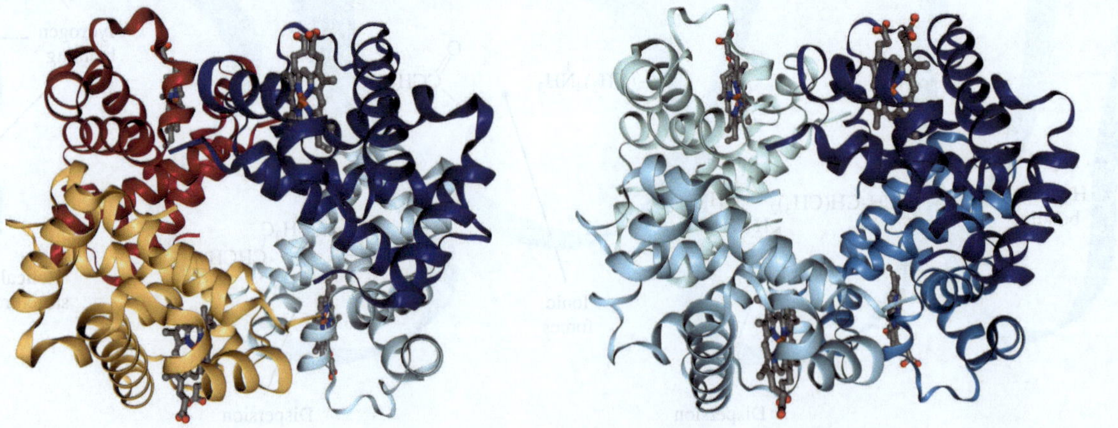

(a) (b)

The overall structure of hemoglobin. Each hemoglobin molecule contains two α chains and two β chains. Each of the four chains is similar to a myoglobin molecule in structure, and each also contains a heme group for binding oxygen. In sickle cell hemoglobin, the defective regions (the valine groups) are located near the ends of the β chains, as indicated by the dots.

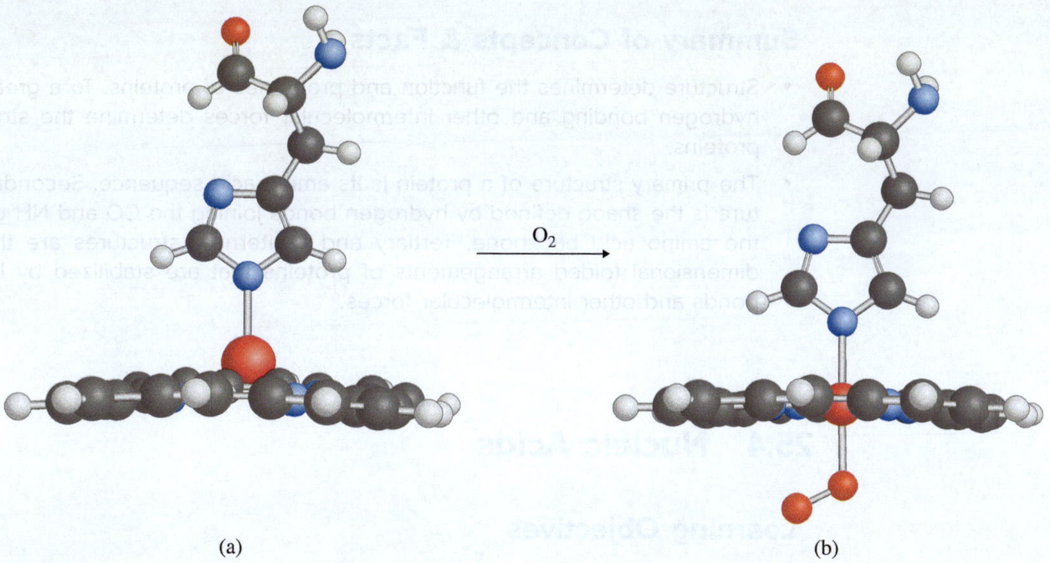

Figure 25.15 *The structural changes that occur when the heme group in hemoglobin binds to an oxygen molecule. (a) The heme group in deoxyhemoglobin. (b) Oxyhemoglobin.*

example of protein flexibility is found in the binding of hemoglobin to oxygen. Each of the four polypeptide chains in hemoglobin contains a heme group that can bind to an oxygen molecule (see Section 23.6). In deoxyhemoglobin, the affinity of each of the heme groups for oxygen is about the same. However, as soon as one of the heme groups becomes oxygenated, the affinity of the other three hemes for oxygen is greatly enhanced. This phenomenon, called *cooperativity,* makes hemoglobin a particularly suitable substance for the uptake of oxygen in the lungs. By the same token, once a fully oxygenated hemoglobin molecule releases an oxygen molecule (to myoglobin in the tissues), the other three oxygen molecules will depart with increasing ease. The cooperative nature of the binding is such that information about the presence (or absence) of oxygen molecules is transmitted from one subunit to another along the polypeptide chains, a process made possible by the flexibility of the three-dimensional structure (Figure 25.15). It is believed that the Fe^{2+} ion has too large a radius to fit into the porphyrin ring of deoxyhemoglobin. When O_2 binds to Fe^{2+}, however, the ion shrinks somewhat so that it can fit into the plane of the ring. As the ion slips into the ring, it pulls the histidine residue toward the ring and thereby sets off a sequence of structural changes from one subunit to another. Although the details of the changes are not clear, biochemists believe that this is how the binding of an oxygen molecule to one heme group affects another heme group. The structural changes drastically affect the affinity of the remaining heme groups for oxygen molecules.

When proteins are heated above body temperature or when they are subjected to unusual acid or base conditions or treated with special reagents called *denaturants,* they lose some or all of their tertiary and secondary structure. For example, hard boiling an egg denatures the proteins in the egg white. Called ***denatured proteins,*** proteins in this state *no longer exhibit normal biological activities.* Figure 25.16 shows the variation of rate with temperature for a typical enzyme-catalyzed reaction. Initially, the rate increases with increasing temperature, as we would expect. Beyond the optimum temperature, however, the enzyme begins to denature and the rate falls rapidly. If a protein is denatured under mild conditions, its original structure can often be regenerated by removing the denaturant or by restoring the temperature to normal conditions. This process is called *reversible denaturation.*

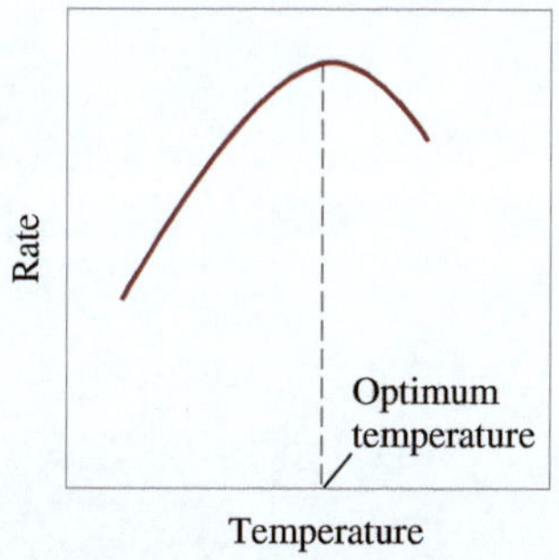

Figure 25.16 *Dependence of the rate of an enzyme-catalyzed reaction on temperature. Above the optimum temperature at which an enzyme is most effective, its activity drops off as a consequence of denaturation.*

Summary of Concepts & Facts

- Structure determines the function and properties of proteins. To a great extent, hydrogen bonding and other intermolecular forces determine the structure of proteins.
- The primary structure of a protein is its amino acid sequence. Secondary structure is the shape defined by hydrogen bonds joining the CO and NH groups of the amino acid backbone. Tertiary and quaternary structures are the three-dimensional folded arrangements of proteins that are stabilized by hydrogen bonds and other intermolecular forces.

25.4 Nucleic Acids

Learning Objectives

- Contrast the two types of nucleic acids.
- Evaluate the role of base pairs in the structure of DNA.

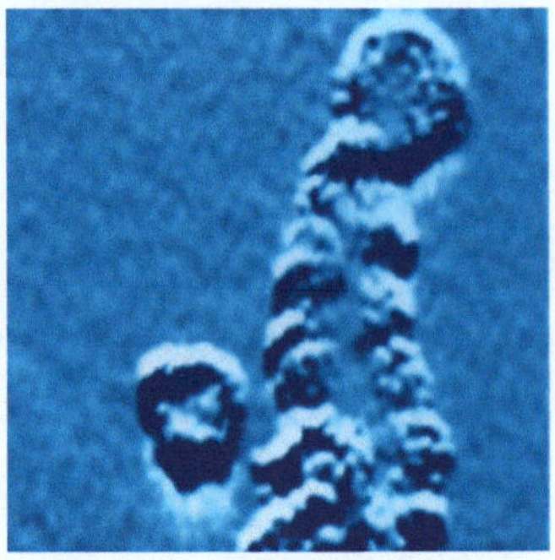

An electron micrograph of a DNA molecule. The double-helical structure is evident. If the DNA molecules from all the cells in a human were stretched and joined end to end, the length would be about 100 times the distance from Earth to the sun!
Lawrence Berkeley National Laboratory

Nucleic acids are *high molar mass polymers that play an essential role in protein synthesis.* ***Deoxyribonucleic acid (DNA)*** and ***ribonucleic acid (RNA)*** are *the two types of nucleic acid.* DNA molecules are among the largest molecules known; they have molar masses of up to tens of billions of grams. On the other hand, RNA molecules vary greatly in size, some having a molar mass of about 25,000 g. Compared with proteins, which are made of up to 20 different amino acids, nucleic acids are fairly simple in composition. A DNA or RNA molecule contains only four types of building blocks: purines, pyrimidines, furanose sugars, and phosphate groups (Figure 25.17). Each purine or pyrimidine is called a *base.*

In the 1940s, Erwin Chargaff[†] studied DNA molecules obtained from various sources and observed certain regularities. *Chargaff's rules,* as his findings are now known, describe these patterns:

1. The amount of adenine (a purine) is equal to that of thymine (a pyrimidine); that is, A = T, or A/T = 1.
2. The amount of cytosine (a pyrimidine) is equal to that of guanine (a purine); that is, C = G, or C/G = 1.
3. The total number of purine bases is equal to the total number of pyrimidine bases; that is, A + G = C + T.

Based on chemical analyses and information obtained from X-ray diffraction measurements, James Watson[‡] and Francis Crick[§] formulated the double-helical structure for the DNA molecule in 1953. Watson and Crick determined that the DNA molecule has two helical strands. Each strand is made up of ***nucleotides,***

[†]Erwin Chargaff (1905–2002). American biochemist of Austrian origin. Chargaff was the first to show that different biological species contain different DNA molecules.

[‡]James Dewey Watson (1928–). American biologist. Watson shared the 1962 Nobel Prize in Physiology or Medicine with Crick and Maurice Wilkins for their work on the DNA structure, which is considered by many to be the most significant development in biology in the twentieth century.

[§]Francis Harry Compton Crick (1916–2004). British biologist. Crick started as a physicist but became interested in biology after reading the book *What Is Life?* by Erwin Schrödinger (see Chapter 7). In addition to elucidating the structure of DNA, for which he was a corecipient of the Nobel Prize in Physiology or Medicine in 1962, Crick made many significant contributions to molecular biology.

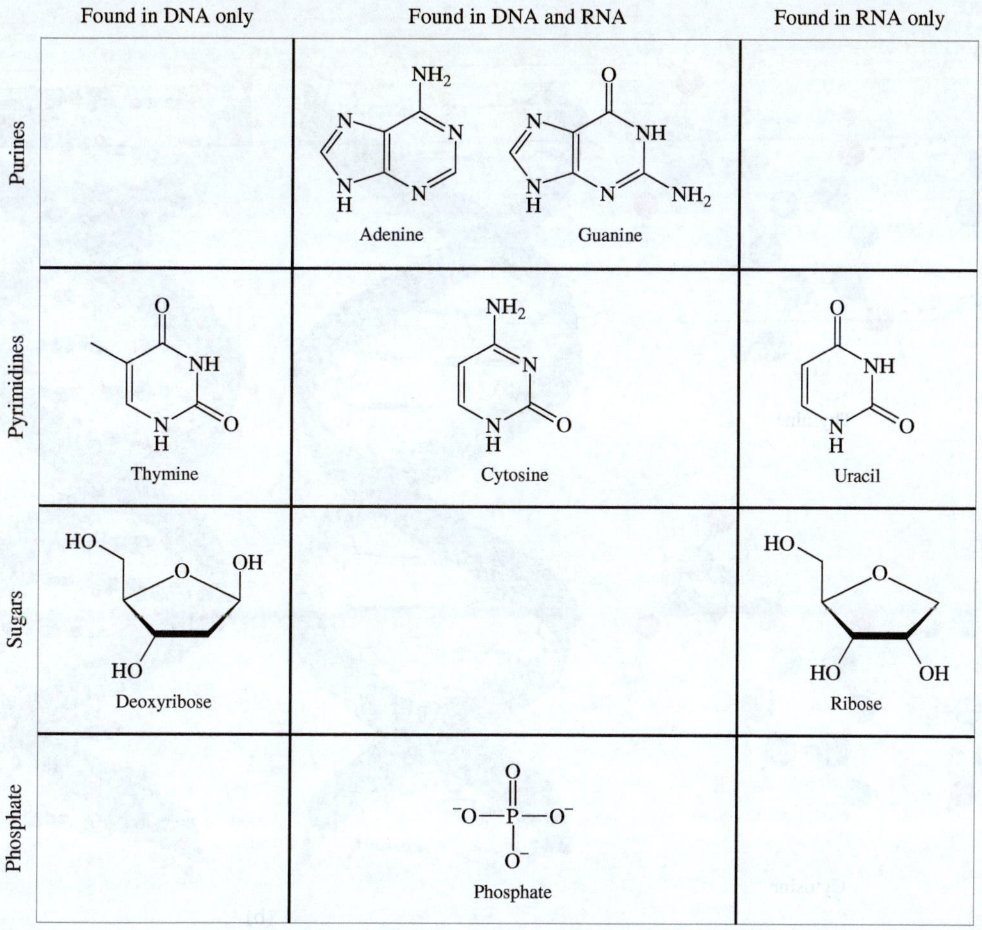

Figure 25.17 *The components of the nucleic acids DNA and RNA.*

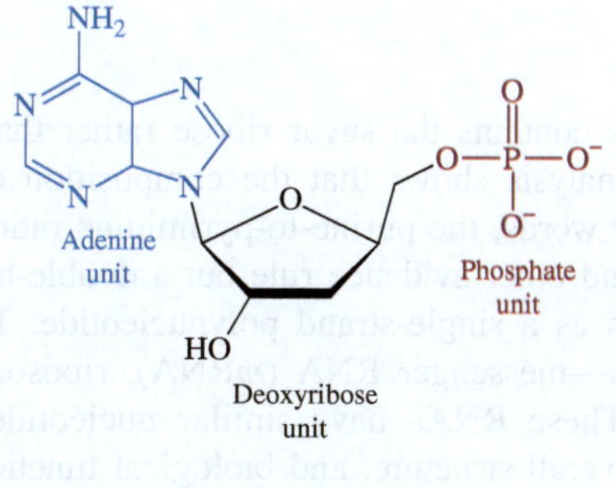

Figure 25.18 *Structure of a nucleotide, one of the repeating units in DNA.*

which *consist of a base, a deoxyribose, and a phosphate group linked together* (Figure 25.18).

The key to the double-helical structure of DNA is the formation of hydrogen bonds between bases in the two strands of a molecule. Although hydrogen bonds can form between any two bases, called *base pairs,* Watson and Crick found that the most favorable couplings are between adenine and thymine and between cytosine and guanine (Figure 25.19). Note that this scheme is consistent with Chargaff's rules, because every purine base is hydrogen-bonded to a pyrimidine base, and vice versa (A + G = C + T). Other attractive forces such as dipole-dipole interactions and van der Waals forces between the base pairs also help to stabilize the double helix.

The structure of RNA differs from that of DNA in several respects. First, as shown in Figure 25.17, the four bases found in RNA molecules are adenine, cytosine, guanine,

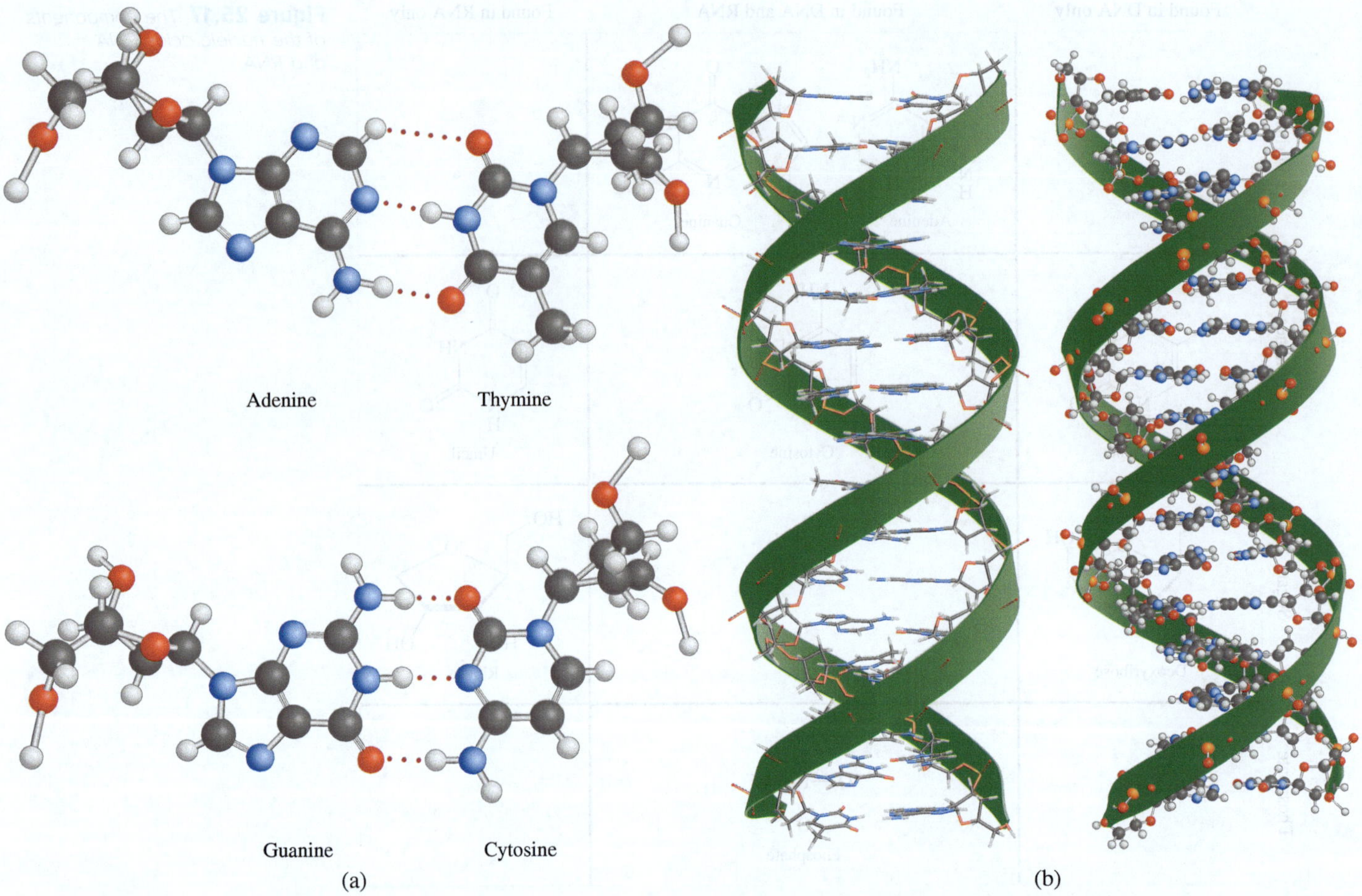

Figure 25.19 *(a) Base-pair formation by adenine and thymine and by cytosine and guanine. (b) The double-helical strand of a DNA molecule held together by hydrogen bonds (and other intermolecular forces) between base pairs A-T and C-G.*

and uracil. Second, RNA contains the sugar ribose rather than the 2-deoxyribose of DNA. Third, chemical analysis shows that the composition of RNA does not obey Chargaff's rules. In other words, the purine-to-pyrimidine ratio is not equal to 1 as in the case of DNA. This and other evidence rule out a double-helical structure. In fact, the RNA molecule exists as a single-strand polynucleotide. There are actually three types of RNA molecules—messenger RNA (*m*RNA), ribosomal RNA (*r*RNA), and transfer RNA (*t*RNA). These RNAs have similar nucleotides but differ from one another in molar mass, overall structure, and biological functions.

DNA and RNA molecules direct the synthesis of proteins in the cell, a subject that is beyond the scope of this book. Introductory texts in biochemistry and molecular biology explain this process.

The Chemistry in Action essay "DNA Fingerprinting" describes a technique in crime investigation that is based on our knowledge of DNA sequence.

Summary of Concepts & Facts

- Nucleic acids—DNA and RNA—are high-molar-mass polymers that carry genetic instructions for protein synthesis in cells. Nucleotides are the building blocks of DNA and RNA. DNA nucleotides each contain a purine or pyrimidine base, a deoxyribose molecule, and a phosphate group. RNA nucleotides are similar but contain different bases and ribose instead of deoxyribose.

CHEMISTRY *in Action*

DNA Fingerprinting

The human genetic makeup, or *genome,* consists of about 3 billion nucleotides. These 3 billion units compose the 23 pairs of chromosomes, which are continuous strands of DNA ranging in length from 50 million to 500 million nucleotides. Encoded in this DNA and stored in units called *genes* are the instructions for protein synthesis. Each of about 100,000 genes is responsible for the synthesis of a particular protein. In addition to instructions for protein synthesis, each gene contains a sequence of bases, repeated several times, that has no known function. What is interesting about these sequences, called *minisatellites,* is that they appear many times in different locations, not just in a particular gene. Furthermore, each person has a unique number of repeats. Only identical twins have the same number of minisatellite sequences.

In 1985 a British chemist named Alec Jeffreys suggested that minisatellite sequences provide a means of identification, much like fingerprints. *DNA fingerprinting* has since gained prominence with law enforcement officials as a way to identify crime suspects.

To make a DNA fingerprint, a chemist needs a sample of any tissue, such as blood or semen; even hair and saliva contain DNA. The DNA is extracted from cell nuclei and cut into fragments by the addition of so-called restriction enzymes. These fragments, which are negatively charged, are separated by an electric field in gel. The smaller fragments move faster than larger ones, so they eventually separate into bands. The bands of DNA fragments are transferred from the gel to a plastic membrane, and their position is thereby fixed. Then a DNA probe—a DNA fragment that has been tagged with a radioactive label—is added. The probe binds to the fragments that have a complementary DNA sequence. An X-ray film is laid directly over the plastic sheet, and bands appear on the exposed film in the positions corresponding to the fragments recognized by the probe. About four different probes are needed to obtain a profile that is unique to just one individual. It is estimated that the probability of finding identical patterns in the DNA of two randomly selected individuals is on the order of 1 in 10 billion.

The first U.S. case in which a person was convicted of a crime with the help of DNA fingerprints was tried in 1987. Today, DNA fingerprinting has become an indispensable tool of law enforcement.

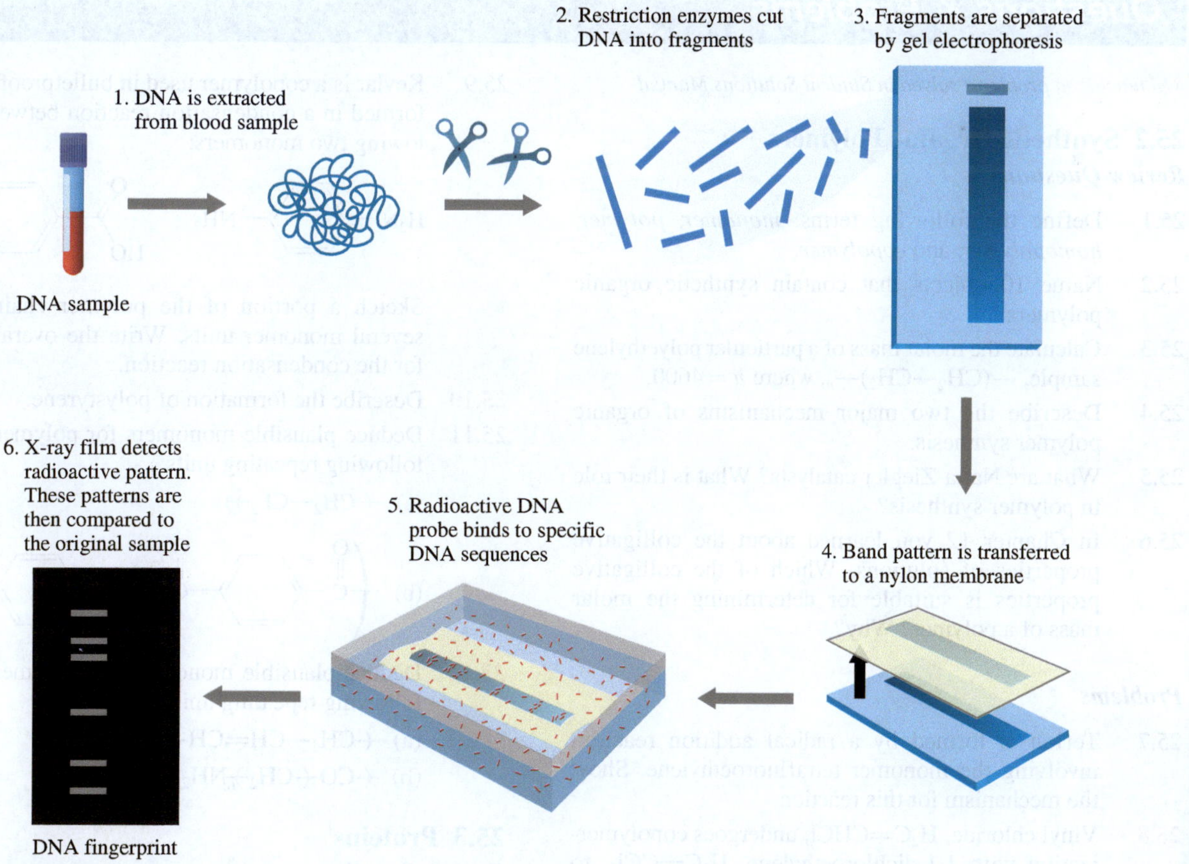

Procedure for obtaining a DNA fingerprint. The developed film shows the DNA fingerprint, which is compared with patterns from known subjects.

Chapter Summary

Synthetic Organic Polymers Many organic polymers have been synthesized by various chemical processes. They mimic and sometimes surpass the properties of those that occur naturally. Nylon is the best known of all synthetic organic polymers. (Sections 25.1 and 25.2)

Proteins Proteins are natural polymers made of amino acids. They perform a host of functions, including catalysis, transport and storage of vital substances, coordinated motion, and protection against diseases. The complex structures of proteins have been analyzed in terms of their primary, secondary, tertiary, and quaternary structures. The three-dimensional integrity of protein molecules is maintained by various intermolecular forces and hydrogen bonding. (Section 25.3)

Nucleic Acids Deoxyribonucleic acid (DNA) carries all the genetic information, and ribonucleic acid (RNA) controls the synthesis of proteins. The elucidation of the double helical structure of DNA was one of the major accomplishments in science in the twentieth century. (Section 25.4)

Key Words

Amino acid, p. 1069
Copolymer, p. 1067
Denatured protein, p. 1077
Deoxyribonucleic acid (DNA), p. 1078
Homopolymer, p. 1064
Monomer, p. 1064
Nucleic acid, p. 1078
Nucleotide, p. 1078
Polymer, p. 1063
Protein, p. 1069
Ribonucleic acid (RNA), p. 1078

Questions & Problems

Red numbered problems solved in Student Solutions Manual

25.2 Synthetic Organic Polymers

Review Questions

25.1 Define the following terms: *monomer, polymer, homopolymer,* and *copolymer.*

25.2 Name 10 objects that contain synthetic organic polymers.

25.3 Calculate the molar mass of a particular polyethylene sample, —$(CH_2—CH_2)—_n$, where $n = 4600$.

25.4 Describe the two major mechanisms of organic polymer synthesis.

25.5 What are Natta-Ziegler catalysts? What is their role in polymer synthesis?

25.6 In Chapter 12 you learned about the colligative properties of solutions. Which of the colligative properties is suitable for determining the molar mass of a polymer? Why?

Problems

25.7 Teflon is formed by a radical addition reaction involving the monomer tetrafluoroethylene. Show the mechanism for this reaction.

25.8 Vinyl chloride, $H_2C{=}CHCl$, undergoes copolymerization with 1,1-dichloroethylene, $H_2C{=}CCl_2$, to form a polymer commercially known as Saran. Draw the structure of the polymer, showing the repeating monomer units.

25.9 Kevlar is a copolymer used in bulletproof vests. It is formed in a condensation reaction between the following two monomers:

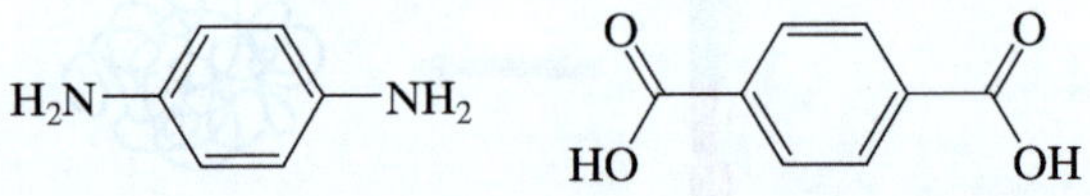

Sketch a portion of the polymer chain showing several monomer units. Write the overall equation for the condensation reaction.

25.10 Describe the formation of polystyrene.

25.11 Deduce plausible monomers for polymers with the following repeating units:

(a) $\left(CH_2—CF_2\right)_n$

(b) 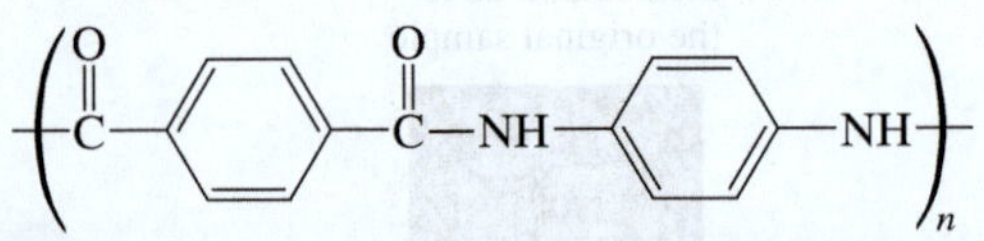

25.12 Deduce plausible monomers for polymers with the following repeating units:

(a) $\left(CH_2—CH{=}CH—CH_2\right)_n$

(b) $\left(CO\left(CH_2\right)_6NH_2\right)_n$

25.3 Proteins

Review Questions

25.13 Discuss the characteristics of an amide group and its importance in protein structure.

25.14 What is the α-helical structure in proteins?

25.15 Describe the β-pleated structure present in some proteins.

25.16 Discuss the main functions of proteins in living systems.

25.17 Briefly explain the phenomenon of cooperativity exhibited by the hemoglobin molecule in binding oxygen.

25.18 Why is sickle cell anemia called a molecular disease?

Problems

25.19 Draw the structures of the dipeptides that can be formed from the reaction between the amino acids glycine and alanine.

25.20 Draw the structures of the dipeptides that can be formed from the reaction between the amino acids glycine and lysine.

25.21 The amino acid glycine can be condensed to form a polymer called polyglycine. Draw the repeating monomer unit.

25.22 The following are data obtained on the rate of product formation of an enzyme-catalyzed reaction:

Temperature (°C)	Rate of Product Formation (*M*/s)
10	0.0025
20	0.0048
30	0.0090
35	0.0086
45	0.0012

Comment on the dependence of rate on temperature. (No calculations are required.)

25.4 Nucleic Acids

Review Questions

25.23 Describe the structure of a nucleotide.

25.24 What is the difference between ribose and deoxyribose?

25.25 What are Chargaff's rules?

25.26 Describe the role of hydrogen bonding in maintaining the double-helical structure of DNA.

Additional Problems

25.27 Discuss the importance of hydrogen bonding in biological systems. Use proteins and nucleic acids as examples.

25.28 Proteins vary widely in structure, whereas nucleic acids have rather uniform structures. How do you account for this major difference?

25.29 If untreated, fevers of 104°F or higher may lead to brain damage. Why?

25.30 The "melting point" of a DNA molecule is the temperature at which the double-helical strand breaks apart. Suppose you are given two DNA samples. One sample contains 45 percent C-G base pairs while the other contains 64 percent C-G base pairs. The total number of bases is the same in each sample. Which of the two samples has a higher melting point? Why?

25.31 When fruits such as apples and pears are cut, the exposed parts begin to turn brown. This is the result of an oxidation reaction catalyzed by enzymes present in the fruit. Often the browning action can be prevented or slowed by adding a few drops of lemon juice to the exposed areas. What is the chemical basis for this treatment?

25.32 "Dark meat" and "white meat" are one's choices when eating a turkey. Explain what causes the meat to assume different colors. (*Hint:* The more active muscles in a turkey have a higher rate of metabolism and need more oxygen.)

25.33 Nylon can be destroyed easily by strong acids. Explain the chemical basis for the destruction. (*Hint:* The products are the starting materials of the polymerization reaction.)

25.34 Despite what you may have read in science fiction novels or seen in horror movies, it is extremely unlikely that insects can ever grow to human size. Why? (*Hint:* Insects do not have hemoglobin molecules in their blood.)

25.35 How many different tripeptides can be formed by lysine and alanine?

25.36 Chemical analysis shows that hemoglobin contains 0.34 percent Fe by mass. What is the minimum possible molar mass of hemoglobin? The actual molar mass of hemoglobin is four times this minimum value. What conclusion can you draw from these data?

25.37 The folding of a polypeptide chain depends not only on its amino acid sequence but also on the nature of the solvent. Discuss the types of interactions that might occur between water molecules and the amino acid residues of the polypeptide chain. Which groups would be exposed on the exterior of the protein in contact with water and which groups would be buried in the interior of the protein?

25.38 What kind of intermolecular forces are responsible for the aggregation of hemoglobin molecules that leads to sickle cell anemia? (*Hint:* See the Chemistry in Action essay "Sickle Cell Anemia—A Molecular Disease" in Section 25.3.)

25.39 Draw structures of the nucleotides containing the following components: (a) deoxyribose and cytosine, (b) ribose and uracil.

25.40 When a nonapeptide (containing nine amino acid residues) isolated from rat brains was hydrolyzed, it gave the following smaller peptides as identifiable products: Gly-Ala-Phe, Ala-Leu-Val, Gly-Ala-Leu, Phe-Glu-His, and His-Gly-Ala. Reconstruct the amino acid sequence in the nonapeptide, giving your reasons. (Remember the convention for writing peptides.)

25.41 At neutral pH amino acids exist as dipolar ions. Using glycine as an example, and given that the pK_a of the carboxyl group is 2.3 and that of the ammonium group is 9.6, predict the predominant form of the molecule at pH 1, 7, and 12. Justify your answers using Equation (16.4).

25.42 In Lewis Carroll's tale "Through the Looking Glass," Alice wonders whether "looking-glass milk" on the other side of the mirror would be fit to drink. Based on your knowledge of chirality and enzyme action, comment on the validity of Alice's concern.

25.43 Nylon was designed to be a synthetic silk. (a) The average molar mass of a batch of nylon 66 is 12,000 g/mol. How many monomer units are there in this sample? (b) Which part of nylon's structure is similar to a polypeptide's structure? (c) How many different tripeptides (made up of three amino acids) can be formed from the amino acids alanine (Ala), glycine (Gly), and serine (Ser), which account for most of the amino acids in silk?

25.44 The enthalpy change in the denaturation of a certain protein is 125 kJ/mol. If the entropy change is 397 J/K · mol, calculate the minimum temperature at which the protein would denature spontaneously.

25.45 When deoxyhemoglobin crystals are exposed to oxygen, they shatter. On the other hand, deoxymyoglobin crystals are unaffected by oxygen. Explain. (Myoglobin is made up of only one of the four subunits, or polypeptide chains, in hemoglobin.)

25.46 The α-helix and β-sheet structures are prevalent in proteins. What is the common feature that they have which makes them suitable for this role?

25.47 In protein synthesis, the selection of a particular amino acid is determined by the so-called genetic code, or a sequence of three bases in DNA. Will a sequence of only two bases unambiguously determine the selection of 20 amino acids found in proteins? Explain.

25.48 Consider the fully protonated amino acid valine:

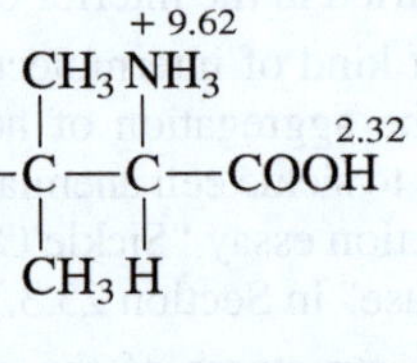

where the numbers denote the pK_a values. (a) Which of the two groups ($-\overset{+}{N}H_3$ or —COOH) is more acidic? (b) Calculate the predominant form of valine at pH 1.0, 7.0, and 12.0. (c) Calculate the isoelectric point of valine. (*Hint:* See Problem 16.137.)

25.49 Consider the formation of a dimeric protein

$$2P \longrightarrow P_2$$

At 25°C, we have $\Delta H° = 17$ kJ/mol and $\Delta S° = 65$ J/K · mol. Is the dimerization favored at this temperature? Comment on the effect of lowering the temperature. Does your result explain why some enzymes lose their activities under cold conditions?

25.50 Molar mass measurements play an important role in characterizing polymer solutions. Number-average molar mass ($\overline{M}_n$) is defined as the total molar mass (given by $\Sigma N_i M_i$) divided by the total number of molecules:

$$\overline{M}_n = \frac{\Sigma N_i M_i}{\Sigma N_i}$$

where N_i is the number of molecules with molar mass M_i. Another important definition is the weight-average molar mass ($\overline{M}_w$) where

$$\overline{M}_w = \frac{\Sigma N_i M_i^2}{\Sigma N_i M_i}$$

The difference between these two definitions is that $\overline{M}_w$ is based on experimental measurements that are affected by the size of molecules. (a) Consider a solution containing five molecules of molar masses 1.0, 3.0, 4.0, 4.0, and 6.0 kg/mol. Calculate both $\overline{M}_n$ and $\overline{M}_w$. (b) $\overline{M}_w$ is always greater than $\overline{M}_n$ because of the square term in the definition. However, if all the molecules have identical molar mass, then we have $\overline{M}_n = \overline{M}_w$. Show that this is the case if we have four molecules having the same molar mass of 5 kg/mol. (c) Explain how a comparison of these two average molar masses gives us information about the distribution of the size of synthetic polymers like polyethylene and poly(vinyl chloride). (d) Proteins like myoglobin and cytochrome *c* have the same $\overline{M}_n$ and $\overline{M}_w$, while this is not the case for hemoglobin. Explain.

25.51 The diagram (left) shows the structure of the enzyme ribonuclease in its native form. The three-dimensional protein structure is maintained in part by the disulfide bonds (—S—S—) between the amino acid residues (each color sphere represents an S atom). Using certain denaturants, the compact structure is destroyed and the disulfide bonds are converted to sulfhydryl groups (—SH) shown on the right of the arrow. (a) Describe the bonding scheme in the disulfide bond in terms of hybridization. (b) Which amino acid in Table 25.2 contains the —SH group? (c) Predict the signs of ΔH and ΔS for the denaturation process. If denaturation is induced by a change in temperature, show why a rise in temperature would favor denaturation. (d) The sulfhydryl groups can be oxidized (that is, removing the H atoms) to form the disulfide bonds. If the formation of the disulfide bonds is totally random between any two —SH groups, what is the fraction of the regenerated protein structures that corresponds to the native form? (e) An effective remedy to deodorize a dog that has been sprayed by a skunk is to rub the affected areas with a solution of an oxidizing agent such as hydrogen peroxide. What is the chemical basis for this action? (*Hint:* An odiferous component of a skunk's secretion is 2-butene-1-thiol, $CH_3CH{=}CHCH_2SH$.)

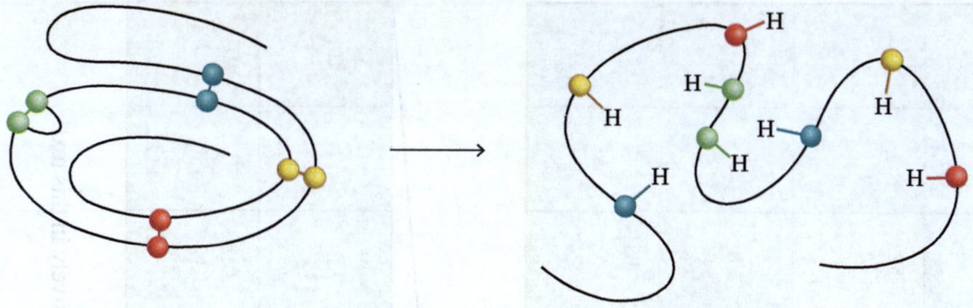

Native form

Denatured form

Interpreting, Modeling, & Estimating

25.52 Assume the energy of hydrogen bonds per base pair to be 10 kJ/mol. Given two complementary strands of DNA containing 10 base pairs each, estimate the ratio of two separate strands to hydrogen-bonded double helix in solution at 300 K. The ratio for a single base pair is given by the formula exp $(-\Delta E/RT)$, where ΔE is the energy of hydrogen bond per base pair, R is the gas constant, and T is the temperature in kelvins.

25.53 The average distance between base pairs measured parallel to the axis of a DNA molecule is 3.4 Å (see Figure 25.19). The average molar mass of a pair of nucleotides is 650 g/mol. Estimate the length in centimeters of a DNA molecule of molar mass 5.0×10^9 g/mol.

Periodic Table

1	2	3	4	5	6	7	8	9	10	11	12	13	14	15	16	17	18
1 **H** Hydrogen 1.008																	2 **He** Helium 4.003
3 **Li** Lithium 6.938	4 **Be** Beryllium 9.012											5 **B** Boron 10.81	6 **C** Carbon 12.01	7 **N** Nitrogen 14.01	8 **O** Oxygen 16.00	9 **F** Fluorine 19.00	10 **Ne** Neon 20.18
11 **Na** Sodium 22.99	12 **Mg** Magnesium 24.30											13 **Al** Aluminum 26.98	14 **Si** Silicon 28.08	15 **P** Phosphorus 30.97	16 **S** Sulfur 32.06	17 **Cl** Chlorine 35.45	18 **Ar** Argon 39.79
19 **K** Potassium 39.10	20 **Ca** Calcium 40.08	21 **Sc** Scandium 44.96	22 **Ti** Titanium 47.87	23 **V** Vanadium 50.94	24 **Cr** Chromium 52.00	25 **Mn** Manganese 54.94	26 **Fe** Iron 55.85	27 **Co** Cobalt 58.93	28 **Ni** Nickel 58.69	29 **Cu** Copper 63.55	30 **Zn** Zinc 65.38	31 **Ga** Gallium 69.72	32 **Ge** Germanium 72.63	33 **As** Arsenic 74.92	34 **Se** Selenium 78.97	35 **Br** Bromine 79.90	36 **Kr** Krypton 83.80
37 **Rb** Rubidium 85.47	38 **Sr** Strontium 87.62	39 **Y** Yttrium 88.91	40 **Zr** Zirconium 91.22	41 **Nb** Niobium 92.91	42 **Mo** Molybdenum 95.95	43 **Tc** Technetium (98)	44 **Ru** Ruthenium 101.1	45 **Rh** Rhodium 102.9	46 **Pd** Palladium 106.4	47 **Ag** Silver 107.9	48 **Cd** Cadmium 112.4	49 **In** Indium 114.8	50 **Sn** Tin 118.7	51 **Sb** Antimony 121.8	52 **Te** Tellurium 127.6	53 **I** Iodine 126.9	54 **Xe** Xenon 131.3
55 **Cs** Cesium 132.9	56 **Ba** Barium 137.3	57 **La** Lanthanum 138.9	72 **Hf** Hafnium 178.5	73 **Ta** Tantalum 180.9	74 **W** Tungsten 183.8	75 **Re** Rhenium 186.2	76 **Os** Osmium 190.2	77 **Ir** Iridium 192.2	78 **Pt** Platinum 195.1	79 **Au** Gold 197.0	80 **Hg** Mercury 200.6	81 **Tl** Thallium 204.4	82 **Pb** Lead 207.2	83 **Bi** Bismuth 209.0	84 **Po** Polonium (210)	85 **At** Astatine (210)	86 **Rn** Radon (222)
87 **Fr** Francium (223)	88 **Ra** Radium (226)	89 **Ac** Actinium (227)	104 **Rf** Rutherfordium (261)	105 **Db** Dubnium (262)	106 **Sg** Seaborgium (266)	107 **Bh** Bohrium (272)	108 **Hs** Hassium (277)	109 **Mt** Meitnerium (276)	110 **Ds** Darmstadtium (281)	111 **Rg** Roentgenium (280)	112 **Cn** Copernicium (285)	113 **Nh** Nihonium (285)	114 **Fl** Flerovium (287)	115 **Mc** Moscovium (289)	116 **Lv** Livermorium (291)	117 **Ts** Tennessine (293)	118 **Og** Oganesson (294)

58 **Ce** Cerium 140.1	59 **Pr** Praseodymium 140.9	60 **Nd** Neodymium 144.2	61 **Pm** Promethium (145)	62 **Sm** Samarium 150.4	63 **Eu** Europium 152.0	64 **Gd** Gadolinium 157.3	65 **Tb** Terbium 158.9	66 **Dy** Dysprosium 162.5	67 **Ho** Holmium 164.9	68 **Er** Erbium 167.3	69 **Tm** Thulium 168.9	70 **Yb** Ytterbium 173.0	71 **Lu** Lutetium 175.0
90 **Th** Thorium 232.0	91 **Pa** Protactinium 231.0	92 **U** Uranium 238.0	93 **Np** Neptunium (237)	94 **Pu** Plutonium (239)	95 **Am** Americium (243)	96 **Cm** Curium (247)	97 **Bk** Berkelium (249)	98 **Cf** Californium (252)	99 **Es** Einsteinium (252)	100 **Fm** Fermium (257)	101 **Md** Mendelevium (258)	102 **No** Nobelium (259)	103 **Lr** Lawrencium (262)

Key: 11 **Na** Sodium 22.99 — Atomic number (11), Atomic mass (22.99)

Metals

Metalloids

Nonmetals

The 1–18 group designation has been recommended by the International Union of Pure and Applied Chemistry (IUPAC) and will be used exclusively in this text. In 2011 IUPAC revised the atomic masses of some elements. The changes are minor and they are not adopted in the present edition of this text.

The Elements with Their Symbols and Atomic Masses*

Element	Symbol	Atomic Number	Atomic Mass†	Element	Symbol	Atomic Number	Atomic Mass†
Actinium	Ac	89	(227)	Mendelevium	Md	101	(257)
Aluminum	Al	13	26.98	Mercury	Hg	80	200.6
Americium	Am	95	(243)	Molybdenum	Mo	42	95.95
Antimony	Sb	51	121.8	Moscovium	Mc	115	(289)
Argon	Ar	18	39.79	Neodymium	Nd	60	144.2
Arsenic	As	33	74.92	Neon	Ne	10	20.18
Astatine	At	85	(210)	Neptunium	Np	93	(237)
Barium	Ba	56	137.3	Nickel	Ni	28	58.69
Berkelium	Bk	97	(249)	Nihonium	Nh	113	(285)
Beryllium	Be	4	9.012	Niobium	Nb	41	92.91
Bismuth	Bi	83	209.0	Nitrogen	N	7	14.01
Bohrium	Bh	107	(262)	Nobelium	No	102	(259)
Boron	B	5	10.81	Oganesson	Og	118	(294)
Bromine	Br	35	79.90	Osmium	Os	76	190.2
Cadmium	Cd	48	112.4	Oxygen	O	8	16.00
Calcium	Ca	20	40.08	Palladium	Pd	46	106.4
Californium	Cf	98	(252)	Phosphorus	P	15	30.97
Carbon	C	6	12.01	Platinum	Pt	78	195.1
Cerium	Ce	58	140.1	Plutonium	Pu	94	(239)
Cesium	Cs	55	132.9	Polonium	Po	84	(210)
Chlorine	Cl	17	35.45	Potassium	K	19	39.10
Chromium	Cr	24	52.00	Praseodymium	Pr	59	140.9
Cobalt	Co	27	58.93	Promethium	Pm	61	(145)
Copernicium	Cn	112	(285)	Protactinium	Pa	91	(231.0)
Copper	Cu	29	63.55	Radium	Ra	88	(226)
Curium	Cm	96	(247)	Radon	Rn	86	(222)
Darmstadtium	Ds	110	(281)	Rhenium	Re	75	186.2
Dubnium	Db	105	(262)	Rhodium	Rh	45	102.9
Dysprosium	Dy	66	162.5	Roentgenium	Rg	111	(280)
Einsteinium	Es	99	(252)	Rubidium	Rb	37	85.47
Erbium	Er	68	167.3	Ruthenium	Ru	44	101.1
Europium	Eu	63	152.0	Rutherfordium	Rf	104	(261)
Fermium	Fm	100	(257)	Samarium	Sm	62	150.4
Flerovium	Fl	114	(287)	Scandium	Sc	21	44.96
Fluorine	F	9	19.00	Seaborgium	Sg	106	(266)
Francium	Fr	87	(223)	Selenium	Se	34	78.97
Gadolinium	Gd	64	157.3	Silicon	Si	14	28.08
Gallium	Ga	31	69.72	Silver	Ag	47	107.9
Germanium	Ge	32	72.63	Sodium	Na	11	22.99
Gold	Au	79	197.0	Strontium	Sr	38	87.62
Hafnium	Hf	72	178.5	Sulfur	S	16	32.06
Hassium	Hs	108	(277)	Tantalum	Ta	73	180.9
Helium	He	2	4.003	Technetium	Tc	43	(98)
Holmium	Ho	67	164.9	Tellurium	Te	52	127.6
Hydrogen	H	1	1.008	Tennessine	Ts	117	(293)
Indium	In	49	114.8	Terbium	Tb	65	158.9
Iodine	I	53	126.9	Thallium	Tl	81	204.4
Iridium	Ir	77	192.2	Thorium	Th	90	232.0
Iron	Fe	26	55.85	Thulium	Tm	69	168.9
Krypton	Kr	36	83.80	Tin	Sn	50	118.7
Lanthanum	La	57	138.9	Titanium	Ti	22	47.87
Lawrencium	Lr	103	(262)	Tungsten	W	74	183.8
Lead	Pb	82	207.2	Uranium	U	92	238.0
Lithium	Li	3	6.938	Vanadium	V	23	50.94
Livermorium	Lv	116	(291)	Xenon	Xe	54	131.3
Lutetium	Lu	71	175.0	Ytterbium	Yb	70	173.0
Magnesium	Mg	12	24.30	Yttrium	Y	39	88.91
Manganese	Mn	25	54.94	Zinc	Zn	30	65.38
Meitnerium	Mt	109	(276)	Zirconium	Zr	40	91.22

*All atomic masses have four significant figures. These values are recommended by the Committee on Teaching of Chemistry, International Union of Pure and Applied Chemistry.

†Approximate values of atomic masses for radioactive elements are given in parentheses.

Appendix 1

Units for the Gas Constant

In this appendix, we will see how the gas constant R can be expressed in units J/K · mol. Our first step is to derive a relationship between atm and pascal. We start with

$$\text{pressure} = \frac{\text{force}}{\text{area}}$$
$$= \frac{\text{mass} \times \text{acceleration}}{\text{area}}$$
$$= \frac{\text{volume} \times \text{density} \times \text{acceleration}}{\text{area}}$$
$$= \text{length} \times \text{density} \times \text{acceleration}$$

By definition, the standard atmosphere is the pressure exerted by a column of mercury exactly 76 cm high of density 13.5951 g/cm^3, in a place where acceleration due to gravity is 980.665 cm/s^2. However, to express pressure in N/m^2 it is necessary to write

$$\text{density of mercury} = 1.35951 \times 10^4 \text{ kg/m}^3$$
$$\text{acceleration due to gravity} = 9.80665 \text{ m/s}^2$$

The standard atmosphere is given by

$$1 \text{ atm} = (0.76 \text{ m Hg})(1.35951 \times 10^4 \text{ kg/m}^3)(9.80665 \text{ m/s}^2)$$
$$= 101{,}325 \text{ kg m/m}^2 \cdot \text{s}^2$$
$$= 101{,}325 \text{ N/m}^2$$
$$= 101{,}325 \text{ Pa}$$

From Section 5.4 we see that the gas constant R is given by 0.082057 L · atm/K · mol. Using the conversion factors

$$1 \text{ L} = 1 \times 10^{-3} \text{ m}^3$$
$$1 \text{ atm} = 101{,}325 \text{ N/m}^2$$

we write

$$R = \left(0.082057 \frac{\text{L} \cdot \text{atm}}{\text{K} \cdot \text{mol}}\right)\left(\frac{1 \times 10^{-3} \text{ m}^3}{1 \text{ L}}\right)\left(\frac{101{,}325 \text{ N/m}^2}{1 \text{ atm}}\right)$$
$$= 8.314 \frac{\text{N} \cdot \text{m}}{\text{K} \cdot \text{mol}}$$
$$= 8.314 \frac{\text{J}}{\text{K} \cdot \text{mol}}$$

and

$$1 \text{ L} \cdot \text{atm} = (1 \times 10^{-3} \text{ m}^3)(101{,}325 \text{ N/m}^2)$$
$$= 101.3 \text{ N} \cdot \text{m}$$
$$= 101.3 \text{ J}$$

Likewise

$$1 \text{ J} = 9.872 \times 10^{-3} \text{ L} \cdot \text{atm}$$

We can express R in units of bar as well. Using the conversion factor 1 atm = 1.01325 bar, we write

$$R = \left(0.082057 \frac{\text{L} \cdot \text{atm}}{\text{K} \cdot \text{mol}}\right)\left(1.01325 \frac{\text{bar}}{\text{atm}}\right)$$

$$= 0.08315 \frac{\text{L} \cdot \text{bar}}{\text{K} \cdot \text{mol}}$$

Additional values of R are summarized in Table A1.1 along with the values derived here. These can be derived by using appropriate unit conversions.

Table A1.1 Additional Values of the Gas Constant, R

R	Units
8.314	$\frac{\text{J}}{\text{K} \cdot \text{mol}}$
0.08206	$\frac{\text{L} \cdot \text{atm}}{\text{K} \cdot \text{mol}}$
0.08315	$\frac{\text{L} \cdot \text{bar}}{\text{K} \cdot \text{mol}}$
8.314	$\frac{\text{L} \cdot \text{kPa}}{\text{K} \cdot \text{mol}}$
62.364	$\frac{\text{L} \cdot \text{torr}}{\text{K} \cdot \text{mol}}$
1.987	$\frac{\text{cal}}{\text{K} \cdot \text{mol}}$

Appendix 2

Thermodynamic Data at 1 atm and 25°C*

Table A2.1 **Inorganic Substances***

Substance	ΔH_f° (kJ/mol)	ΔG_f° (kJ/mol)	S° (J/K · mol)
$Ag(s)$	0	0	42.6
$Ag^+(aq)$	105.6	77.1	72.7
$AgCl(s)$	−127.0	−109.8	96.3
$AgBr(s)$	−100.4	−96.9	107.1
$AgI(s)$	−61.8	−66.2	115.5
$AgNO_3(s)$	−124.4	−33.4	140.9
$Al(s)$	0	0	28.3
$Al^{3+}(aq)$	−531.0	−485.0	−321.7
$AlCl_3(s)$	−704.2	−628.8	109.3
$Al_2O_3(s)$	−1675.7	−1582.3	50.9
$As(s)$	0	0	35.1
$AsO_4^{3-}(aq)$	−888.1	−648.4	−162.8
$AsH_3(g)$	66.4	68.9	222.8
$H_3AsO_4(s)$	−906.3	−766.0	184.0
$Au(s)$	0	0	47.4
$Au_2O_3(s)$	80.8	163.2	125.5
$AuCl(s)$	−34.7		92.9
$AuCl_3(s)$	−117.6	−55.2	147.3
$B(s)$	0	0	5.9
$B_2O_3(s)$	−1273.5	−1194.3	54.0
$H_3BO_3(s)$	−1094.3	−968.9	90.0
$H_3BO_3(aq)$	−1072.8	−963.3	162.4
$Ba(s)$	0	0	62.5
$Ba^{2+}(aq)$	−537.6	−560.8	9.6
$BaO(s)$	−548.0	−520.3	72.1
$BaCl_2(s)$	−855.0	−806.7	123.7
$BaSO_4(s)$	−1473.2	−1362.2	132.2
$BaCO_3(s)$	−1213.0	−1134.4	112.1
$Be(s)$	0	0	9.5
$BeO(s)$	−609.4	−580.1	13.8
$Br_2(l)$	0	0	152.2
$Br_2(g)$	30.9	3.1	245.5
$Br^-(aq)$	−121.6	−104.0	82.4
$HBr(g)$	−36.3	−53.4	198.7
C(graphite)	0	0	5.7
C(diamond)	1.9	2.9	2.4
$CO(g)$	−110.5	−137.2	197.7
$CO_2(g)$	−393.5	−394.4	213.8
$CO_2(aq)$	−413.3	−386.2	119.4
$CO_3^{2-}(aq)$	−677.1	−527.8	−56.9
$HCO_3^-(aq)$	−692.0	−586.8	91.2
$H_2CO_3(aq)$	−698.7	−623.4	191.0
$CS_2(g)$	116.7	67.1	237.8
$CS_2(l)$	89.0	64.6	151.3
$HCN(aq)$	105.4	112.1	128.9

*The thermodynamic quantities of ions are based on the reference states that $\Delta H_f^\circ[H^+(aq)] = 0$, $\Delta G_f^\circ[H^+(aq)] = 0$, and $S^\circ[H^+(aq)] = 0$ (see Section 17.3).

Substance	ΔH_f° (kJ/mol)	ΔG_f° (kJ/mol)	S° (J/K · mol)
$CN^-(aq)$	150.6	172.4	94.1
$(NH_2)_2CO(s)$	−333.1	−197.2	104.3
$(NH_2)_2CO(aq)$	−319.2	−203.8	173.9
$Ca(s)$	0	0	41.6
$Ca^{2+}(aq)$	−542.8	−553.6	−53.1
$CaO(s)$	−634.9	−603.3	38.1
$Ca(OH)_2(s)$	−985.2	−897.5	83.4
$CaF_2(s)$	−1228.0	−1175.6	68.5
$CaCl_2(s)$	−795.4	−748.8	108.4
$CaSO_4(s)$	−1434.5	−1322.0	106.5
$CaCO_3(s)$	−1207.6	−1129.1	91.7
$Cd(s)$	0	0	51.8
$Cd^{2+}(aq)$	−75.9	−77.6	−73.2
$CdO(s)$	−258.4	−228.7	54.8
$CdCl_2(s)$	−391.5	−343.9	115.3
$CdSO_4(s)$	−933.3	−822.7	123.0
$Cl_2(g)$	0	0	223.1
$Cl^-(aq)$	−167.2	−131.2	56.5
$HCl(g)$	−92.3	−95.3	186.9
$Co(s)$	0	0	30.0
$Co^{2+}(aq)$	−58.2	−54.4	−113.0
$CoO(s)$	−237.9	−214.2	53.0
$Cr(s)$	0	0	23.8
$Cr^{2+}(aq)$	−138.9	−165	−100
$Cr_2O_3(s)$	−1139.7	−1058.1	118.7
$CrO_4^{2-}(aq)$	−881.2	−727.8	50.2
$CrO_7^{2-}(aq)$	−1490.3	−1301.1	261.9
$Cs(s)$	0	0	85.2
$Cs^+(aq)$	−258.3	−292.0	133.1
$Cu(s)$	0	0	33.2
$Cu^+(aq)$	71.7	50.0	40.6
$Cu^{2+}(aq)$	64.8	65.5	−99.6
$CuO(s)$	−157.3	−129.7	42.6
$Cu_2O(s)$	−168.6	−146.0	93.1
$CuCl(s)$	−137.2	−119.9	86.2
$CuCl_2(s)$	−220.1	−175.7	108.1
$CuS(s)$	−53.1	−53.6	66.5
$CuSO_4(s)$	−771.4	−662.2	109.2
$F_2(g)$	0	0	202.8
$F^-(aq)$	−332.6	−278.8	−13.8
$HF(g)$	−273.3	−275.4	173.8
$Fe(s)$	0	0	27.3
$Fe^{2+}(aq)$	−89.1	−78.9	−137.7
$Fe^{3+}(aq)$	−48.5	−4.7	−315.9
$FeCl_3(s)$	−399.5	−334.0	142.3
$FeO(s)$	−272.0	−255.2	60.8
$Fe_2O_3(s)$	−824.2	−742.2	87.4
$Fe(OH)_2(s)$	−574.0	−483.6	87.9
$Fe(OH)_3(s)$	−832.6	−696.5	104.6
$H(g)$	218.0	203.3	114.7
$H_2(g)$	0	0	130.7
$H^+(aq)$	0	0	0
$OH^-(aq)$	−230.0	−157.2	−10.8
$H_2O(l)$	−285.8	−237.1	70.0

(Continued)

Substance	ΔH_f° (kJ/mol)	ΔG_f° (kJ/mol)	S° (J/K · mol)
$H_2O(g)$	−241.8	−228.6	188.8
$H_2O_2(l)$	−187.8	−120.4	109.6
$Hg(l)$	0	0	75.9
$Hg^{2+}(aq)$	171.1	164.4	−32.2
$HgO(s)$	−90.8	−58.5	70.3
$HgCl_2(s)$	−224.3	−178.6	146.0
$Hg_2Cl_2(s)$	−265.4	−210.7	191.6
$HgS(s)$	−58.2	−50.6	48.4
$HgSO_4(s)$	−707.5	−595.0	140.0
$Hg_2SO_4(s)$	−743.1	−625.8	200.7
$I_2(s)$	0	0	116.1
$I_2(g)$	62.4	19.3	260.7
$I(g)$	106.8	70.2	180.8
$I^-(aq)$	−55.2	−51.6	111.3
$HI(g)$	26.5	1.7	206.6
$K(s)$	0	0	64.7
$K^+(aq)$	−252.4	−283.3	102.5
$KOH(s)$	−424.6	−379.4	81.2
$KCl(s)$	−436.5	−408.5	82.6
$KClO_3(s)$	−397.7	−296.3	143.1
$KClO_4(s)$	−432.8	−303.1	151.0
$KBr(s)$	−393.8	−380.7	95.9
$KI(s)$	−327.9	−324.9	106.3
$KNO_3(s)$	−494.6	−394.9	133.1
$Li(s)$	0	0	29.1
$Li^+(aq)$	−278.5	−293.3	13.4
$Li_2O(s)$	−597.9	−561.2	37.6
$LiOH(s)$	−487.5	−441.5	42.8
$Mg(s)$	0	0	32.7
$Mg^{2+}(aq)$	−466.9	−454.8	−138.1
$MgO(s)$	−601.6	−569.3	27.0
$Mg(OH)_2(s)$	−924.5	−833.5	63.2
$MgCl_2(s)$	−641.3	−591.8	89.6
$MgSO_4(s)$	−1284.9	−1170.6	91.6
$MgCO_3(s)$	−1095.8	−1012.1	65.7
$Mn(s)$	0	0	32.0
$Mn^{2+}(aq)$	−220.8	−228.1	−73.6
$MnO_2(s)$	−520.0	−465.1	53.1
$N_2(g)$	0	0	191.6
$N_3^-(aq)$	275.1	348.2	107.9
$NH_3(g)$	−45.9	−16.4	192.8
$NH_4^+(aq)$	−132.5	−79.3	113.4
$NH_4Cl(s)$	−314.4	−202.9	94.6
$NH_4NO_3(s)$	−365.6	−183.9	151.1
$NH_3(aq)$	−80.3	−26.5	111.3
$N_2H_4(l)$	50.6	149.3	121.2
$NO(g)$	91.3	87.6	210.8
$NO_2(g)$	33.2	51.3	240.1
$N_2O_4(g)$	11.1	99.8	304.4
$N_2O(g)$	81.6	103.7	220.0
$HNO_2(aq)$	−118.8	−53.6	254.1
$HNO_3(l)$	−174.1	−80.7	155.6
$NO_3^-(aq)$	−207.4	−111.3	146.4
$Na(s)$	0	0	51.3
$Na^+(aq)$	−240.1	−261.9	59.0
$Na_2O(s)$	−414.2	−375.5	75.1

Substance	ΔH_f° (kJ/mol)	ΔG_f° (kJ/mol)	S° (J/K · mol)
NaCl(*s*)	−411.2	−384.1	72.1
NaI(*s*)	−287.8	−286.1	98.5
Na_2SO_4(*s*)	−1387.1	−1270.2	149.6
$NaNO_3$(*s*)	−467.9	−367.0	116.5
Na_2CO_3(*s*)	−1130.7	−1044.4	135.0
$NaHCO_3$(*s*)	−950.8	−851.0	101.7
Ni(*s*)	0	0	29.9
Ni^{2+}(*aq*)	−54.0	−45.6	−128.9
NiO(*s*)	−244.4	−216.3	38.6
$Ni(OH)_2$(*s*)	−529.7	−447.2	88.0
O(*g*)	249.2	231.7	161.1
O_2(*g*)	0	0	205.2
O_3(*aq*)	−12.1	16.3	110.8
O_3(*g*)	142.7	163.2	238.9
P(white)	0	0	41.1
P(red)	−17.6	−12.1	22.8
PO_4^{3-}(*aq*)	−1277.4	−1018.7	−220.5
P_4O_{10}(*s*)	−2984	−2675.2	228.9
PH_3(*g*)	5.4	13.5	210.2
HPO_4^{2-}(*aq*)	−1292.1	−1089.2	−33.5
$H_2PO_4^-$(*aq*)	−1296.3	−1130.2	90.4
Pb(*s*)	0	0	64.8
Pb^{2+}(*aq*)	−1.7	−24.4	10.5
PbO(*s*)	−219.0	−188.9	66.5
PbO_2(*s*)	−277.4	−217.3	68.6
$PbCl_2$(*s*)	−359.4	−314.1	136.0
PbS(*s*)	−100.4	−98.7	91.2
$PbSO_4$(*s*)	−920.0	−813.0	148.5
Pt(*s*)	0	0	41.6
$PtCl_4^{2-}$(*aq*)	−516.3	−384.5	175.7
Rb(*s*)	0	0	76.8
Rb^+(*aq*)	−251.2	−284.0	121.5
S(rhombic)	0	0	32.1
S(monoclinic)	0.3	0.1	32.6
SO_2(*g*)	−296.8	−300.1	248.2
SO_3(*g*)	−395.7	−371.1	256.8
SO_3^{2-}(*aq*)	−635.5	−486.5	−29.0
SO_4^{2-}(*aq*)	−909.3	−744.5	20.1
H_2S(*g*)	−20.6	−33.4	205.8
HSO_3^-(*aq*)	−626.2	−527.7	139.7
HSO_4^-(*aq*)	−887.3	−755.9	131.8
H_2SO_4(*l*)	−814.0	−690.0	156.9
SF_6(*g*)	−1220.5	−1116.5	291.5
Si(*s*)	0	0	18.8
SiO_2(*s*)	−910.7	−856.3	41.5
Sr(*s*)	0	0	55.0
Sr^{2+}(*aq*)	−545.8	−559.5	−32.6
$SrCl_2$(*s*)	−1220.1	−1140.1	97.1
$SrSO_4$(*s*)	−1453.1	−1340.9	117.0
$SrCO_3$(*s*)	−828.9	−781.1	114.9
Zn(*s*)	0	0	41.3
Zn^{2+}(*aq*)	−153.9	−147.1	−112.1
ZnO(*s*)	−350.5	−320.5	43.7
$ZnCl_2$(*s*)	−415.1	−369.4	111.5
ZnS(*s*)	−206.0	−201.3	57.7
$ZnSO_4$(*s*)	−982.8	−871.5	110.5

Table A2.2 Organic Substances

Substance	Formula	ΔH_f° (kJ/mol)	ΔG_f° (kJ/mol)	S° (J/K · mol)
Acetic acid(*l*)	CH_3COOH	−484.3	−389.9	159.8
Acetaldehyde(*g*)	CH_3CHO	−166.2	−133.0	263.8
Acetone(*l*)	CH_3COCH_3	−248.4	−153.6	199.8
Acetylene(*g*)	C_2H_2	227.4	209.9	200.9
Benzene(*l*)	C_6H_6	49.1	124.5	173.4
Butane(*g*)	C_4H_{10}	−125.7	−15.7	310.0
Ethanol(*l*)	C_2H_5OH	−277.6	−174.8	160.7
Ethanol(*g*)	C_2H_5OH	−234.8	−167.9	281.6
Ethane(*g*)	C_2H_6	−84.0	−32.0	229.2
Ethylene(*g*)	C_2H_4	52.4	68.4	219.3
Formic acid(*l*)	$HCOOH$	−425.0	−316.4	129.0
Glucose(*s*)	$C_6H_{12}O_6$	−1273.3	−910.6	212.1
Methane(*g*)	CH_4	−74.6	−50.5	186.3
Methanol(*l*)	CH_3OH	−239.2	−166.6	126.8
Propane(*g*)	C_3H_8	−103.8	−23.4	270.3
Sucrose(*s*)	$C_{12}H_{22}O_{11}$	−2226.1	−1544.3	392.4

Dissociation Constants for Weak Acids and Bases at 25°C

Table A3.1 Dissociation Constants for Weak Acids at 25°C

Name	Formula	K_{a_1}	K_{a_2}	K_{a_3}
Acetic	$HC_2H_3O_2(CH_3COOH)$	1.8×10^{-5}		
Acetylsalicylic	$HC_9H_7O_4$	3.0×10^{-4}		
Ascorbic	$H_2C_6H_6O_6$	9.1×10^{-5}	1.6×10^{-12}	
Benzoic	$HC_7H_5O_2(C_6H_5COOH)$	6.3×10^{-5}		
Carbonic	H_2CO_3	4.5×10^{-7}	4.7×10^{-11}	
Chloroacetic	$HC_2H_2O_2Cl(CH_2ClCOOH)$	1.5×10^{-3}		
Chlorous	$HClO_2$	1.1×10^{-2}		
Citric	$H_3C_6H_5O_7$	7.4×10^{-4}	1.7×10^{-5}	4.0×10^{-7}
Dichloroacetic	$HC_2HO_2Cl_2(CHCl_2COOH)$	3.3×10^{-2}		
Formic	$HCHO_2(HCOOH)$	1.8×10^{-4}		
Hydrocyanic	HCN	6.2×10^{-10}		
Hydrofluoric	HF	6.3×10^{-4}		
Hydrosulfuric	H_2S	8.9×10^{-8}	$\sim 1 \times 10^{-19}$	
Hypobromous	$HOBr$	2.3×10^{-9}		
Hypochlorous	$HOCl$	4.0×10^{-8}		
Hypoiodous	HOI	2.3×10^{-11}		
Nitrous	HNO_2	5.6×10^{-4}		
Oxalic	$H_2C_2O_4$	5.6×10^{-2}	1.5×10^{-4}	
Phenol	C_6H_5OH	1.0×10^{-10}		
Phosphoric	H_3PO_4	6.9×10^{-3}	6.2×10^{-8}	4.8×10^{-13}
Phosphorus	H_3PO_3	5×10^{-2}	2×10^{-7}	
Sulfuric	H_2SO_4	very large	1.0×10^{-2}	
Sulfurous	H_2SO_3	1.3×10^{-2}	6.3×10^{-8}	
Trichloroacetic	$HC_2O_2Cl_3(CCl_3COOH)$	2.0×10^{-1}		
Trifluoroacetic	$HC_2O_2F_3(CF_3COOH)$	3.0×10^{-1}		

Table A3.2 Dissociation Constants for Weak Bases at 25°C

Name	Formula	K_b
Ammonia	NH_3	1.8×10^{-5}
Aniline	$C_6H_5NH_2$	7.4×10^{-10}
Ethylamine	$C_2H_5NH_2$	4.5×10^{-4}
Hydrazine	N_2H_4	1.3×10^{-6}
Methylamine	CH_3NH_2	4.6×10^{-4}
Pyridine	C_5H_5N	1.7×10^{-9}
Triethylamine	$C_6H_{15}N$	5.6×10^{-4}
Urea	H_2NCONH_2	1.3×10^{-14}

Appendix 4

Solubility Product Constants at 25°C

Table A4.1 Solubility Product Constants at 25°C

Compound	Dissolution Equilibrium	K_{sp}
Bromides		
Copper(I) bromide	$CuBr(s) \rightleftharpoons Cu^+(aq) + Br^-(aq)$	6.3×10^{-9}
Lead(II) bromide	$PbBr_2(s) \rightleftharpoons Pb^{2+}(aq) + 2Br^-(aq)$	6.6×10^{-6}
Mercury(I) bromide	$Hg_2Br_2(s) \rightleftharpoons Hg_2^{2+}(aq) + 2Br^-(aq)$	6.4×10^{-23}
Silver bromide	$AgBr(s) \rightleftharpoons Ag^+(aq) + Br^-(aq)$	5.4×10^{-13}
Carbonates		
Barium carbonate	$BaCO_3(s) \rightleftharpoons Ba^{2+}(aq) + CO_3^{2-}(aq)$	2.6×10^{-9}
Calcium carbonate	$CaCO_3(s) \rightleftharpoons Ca^{2+}(aq) + CO_3^{2-}(aq)$	3.4×10^{-9}
Cadmium carbonate	$CdCO_3(s) \rightleftharpoons Cd^{2+}(aq) + CO_3^{2-}(aq)$	1.0×10^{-12}
Iron(II) carbonate	$FeCO_3(s) \rightleftharpoons Fe^{2+}(aq) + CO_3^{2-}(aq)$	7.4×10^{-14}
Lead(II) carbonate	$PbCO_3(s) \rightleftharpoons Pb^{2+}(aq) + CO_3^{2-}(aq)$	7.4×10^{-14}
Magnesium carbonate	$MgCO_3(s) \rightleftharpoons Mg^{2+}(aq) + CO_3^{2-}(aq)$	6.8×10^{-6}
Nickel(II) carbonate	$NiCO_3(s) \rightleftharpoons Ni^{2+}(aq) + CO_3^{2-}(aq)$	1.4×10^{-7}
Silver carbonate	$Ag_2CO_3(s) \rightleftharpoons 2Ag^+(aq) + CO_3^{2-}(aq)$	8.5×10^{-12}
Strontium carbonate	$SrCO_3(s) \rightleftharpoons Sr^{2+}(aq) + CO_3^{2-}(aq)$	5.6×10^{-10}
Zinc carbonate	$ZnCO_3(s) \rightleftharpoons Zn^{2+}(aq) + CO_3^{2-}(aq)$	1.5×10^{-10}
Chlorides		
Lead(II) chloride	$PbCl_2(s) \rightleftharpoons Pb^{2+}(aq) + 2Cl^-(aq)$	1.7×10^{-5}
Mercury(I) chloride	$Hg_2Cl_2(s) \rightleftharpoons Hg_2^{2+}(aq) + 2Cl^-(aq)$	1.4×10^{-18}
Silver chloride	$AgCl(s) \rightleftharpoons Ag^+(aq) + Cl^-(aq)$	1.8×10^{-10}
Chromates		
Barium chromate	$BaCrO_4(s) \rightleftharpoons Ba^{2+}(aq) + CrO_4^{2-}(aq)$	1.2×10^{-10}
Lead(II) chromate	$PbCrO_4(s) \rightleftharpoons Pb^{2+}(aq) + CrO_4^{2-}(aq)$	2.0×10^{-14}
Mercury(I) chromate	$Hg_2CrO_4(s) \rightleftharpoons Hg_2^{2+}(aq) + 2Cl^-(aq)$	2×10^{-9}
Silver chromate	$Ag_2CrO_4(s) \rightleftharpoons 2Ag^+(aq) + CrO_4^{2-}(aq)$	1.1×10^{-12}
Fluorides		
Barium fluoride	$BaF_2(s) \rightleftharpoons Ba^{2+}(aq) + 2F^-(aq)$	1.8×10^{-7}
Calcium fluoride	$CaF_2(s) \rightleftharpoons Ca^{2+}(aq) + 2F^-(aq)$	3.5×10^{-11}
Lead(II) fluoride	$PbF_2(s) \rightleftharpoons Pb^{2+}(aq) + 2F^-(aq)$	3.3×10^{-8}
Hydroxides		
Aluminum hydroxide	$Al(OH)_3(s) \rightleftharpoons Al^{3+}(aq) + 3OH^-(aq)$	1.8×10^{-33}
Calcium hydroxide	$Ca(OH)_2(s) \rightleftharpoons Ca^{2+}(aq) + 2OH^-(aq)$	5.0×10^{-6}
Chromium(III) hydroxide	$Cr(OH)_3(s) \rightleftharpoons Cr^{3+}(aq) + 3OH^-(aq)$	3.0×10^{-29}
Cobalt(II) hydroxide	$Co(OH)_2(s) \rightleftharpoons Co^{2+}(aq) + 2OH^-(aq)$	5.9×10^{-15}
Cobalt(III) hydroxide	$Co(OH)_3(s) \rightleftharpoons Co^{3+}(aq) + 3OH^-(aq)$	3.0×10^{-29}
Copper(II) hydroxide	$Cu(OH)_2(s) \rightleftharpoons Cu^{2+}(aq) + 2OH^-(aq)$	2.2×10^{-20}
Iron(II) hydroxide	$Fe(OH)_2(s) \rightleftharpoons Fe^{2+}(aq) + 2OH^-(aq)$	4.9×10^{-17}
Iron(III) hydroxide	$Fe(OH)_3(s) \rightleftharpoons Fe^{3+}(aq) + 3OH^-(aq)$	2.8×10^{-39}
Magnesium hydroxide	$Mg(OH)_2(s) \rightleftharpoons Mg^{2+}(aq) + 2OH^-(aq)$	5.6×10^{-12}
Strontium hydroxide	$Sr(OH)_2(s) \rightleftharpoons Sr^{2+}(aq) + 2OH^-(aq)$	3.2×10^{-4}
Zinc hydroxide	$Zn(OH)_2(s) \rightleftharpoons Zn^{2+}(aq) + 2OH^-(aq)$	3.0×10^{-17}

Compound	Dissolution Equilibrium	K_{sp}
Iodides		
Copper(I) iodide	$CuI(s) \rightleftharpoons Cu^+(aq) + I^-(aq)$	1.3×10^{-12}
Lead(II) iodide	$PbI_2(s) \rightleftharpoons Pb^{2+}(aq) + 2I^-(aq)$	9.8×10^{-9}
Silver iodide	$AgI(s) \rightleftharpoons Ag^+(aq) + I^-(aq)$	8.5×10^{-17}
Phosphates		
Barium phosphate	$Ba_3PO_4(s) \rightleftharpoons 3Ba^{2+}(aq) + 2PO_4^{3-}(aq)$	6×10^{-39}
Calcium phosphate	$Ca_3(PO_4)_2(s) \rightleftharpoons 3Ca^{2+}(aq) + 2PO_4^{3-}(aq)$	2.1×10^{-33}
Iron(III) phosphate	$FePO_4(s) \rightleftharpoons Fe^{3+}(aq) + PO_4^{3-}(aq)$	1.3×10^{-22}
Lead phosphate	$Pb_3(PO_4)_2(s) \rightleftharpoons 3Pb^{2+}(aq) + 2PO_4^{3-}(aq)$	1×10^{-54}
Silver phosphate	$Ag_3PO_4(s) \rightleftharpoons 3Ag^+(aq) + PO_4^{3-}(aq)$	8.9×10^{-17}
Sulfates		
Barium sulfate	$BaSO_4(s) \rightleftharpoons Ba^{2+}(aq) + SO_4^{2-}(aq)$	1.1×10^{-10}
Calcium sulfate	$CaSO_4(s) \rightleftharpoons Ca^{2+}(aq) + SO_4^{2-}(aq)$	4.9×10^{-5}
Lead(II) sulfate	$PbSO_4(s) \rightleftharpoons Pb^{2+}(aq) + SO_4^{2-}(aq)$	2.5×10^{-8}
Mercury(I) sulfate	$Hg_2SO_4(s) \rightleftharpoons Hg_2^{2+}(aq) + SO_4^{2-}(aq)$	6.5×10^{-7}
Silver sulfate	$Ag_2SO_4(s) \rightleftharpoons 2Ag^+(aq) + SO_4^{2-}(aq)$	1.2×10^{-5}
Strontium sulfate	$SrSO_4(s) \rightleftharpoons Sr^{2+}(aq) + SO_4^{2-}(aq)$	3.4×10^{-7}
Sulfides		
Bismuth sulfide	$Bi_2S_3(s) \rightleftharpoons 2Bi^{3+}(aq) + 3S^{2-}(aq)$	1.6×10^{-72}
Cadmium sulfide	$CdS(s) \rightleftharpoons Cd^{2+}(aq) + S^{2-}(aq)$	8.0×10^{-28}
Cobalt(II) sulfide	$CoS(s) \rightleftharpoons Co^{2+}(aq) + S^{2-}(aq)$	4.0×10^{-21}
Copper(II) sulfide	$CuS(s) \rightleftharpoons Cu^{2+}(aq) + S^{2-}(aq)$	6.0×10^{-37}
Iron(II) sulfide	$FeS(s) \rightleftharpoons Fe^{2+}(aq) + S^{2-}(aq)$	6.0×10^{-19}
Lead(II) sulfide	$PbS(s) \rightleftharpoons Pb^{2+}(aq) + S^{2-}(aq)$	3.4×10^{-28}
Manganese(II) sulfide	$MnS(s) \rightleftharpoons Mn^{2+}(aq) + S^{2-}(aq)$	3.0×10^{-14}
Mercury(II) sulfide	$HgS(s) \rightleftharpoons Hg^{2+}(aq) + S^{2-}(aq)$	4.0×10^{-54}
Nickel(II) sulfide	$NiS(s) \rightleftharpoons Ni^{2+}(aq) + S^{2-}(aq)$	1.4×10^{-24}
Silver sulfide	$Ag_2S(s) \rightleftharpoons 2Ag^+(aq) + S^{2-}(aq)$	6.0×10^{-51}
Tin(II) sulfide	$SnF_2(s) \rightleftharpoons Sn^{2+}(aq) + S^{2-}(aq)$	1.0×10^{-26}
Zinc sulfide	$ZnF_2(s) \rightleftharpoons Zn^{2+}(aq) + S^{2-}(aq)$	3.0×10^{-23}

Appendix 5

Mathematical Operations

Logarithms

Common Logarithms

The concept of the logarithms is an extension of the concept of exponents, which is discussed in Chapter 1. The *common,* or base-10, logarithm of any number is the power to which 10 must be raised to equal the number. The following examples illustrate this relationship:

Logarithm	Exponent
$\log 1 = 0$	$10^0 = 1$
$\log 10 = 1$	$10^1 = 10$
$\log 100 = 2$	$10^2 = 100$
$\log 10^{-1} = -1$	$10^{-1} = 0.1$
$\log 10^{-2} = -2$	$10^{-2} = 0.01$

In each case, the logarithm of the number can be obtained by inspection.

Because the logarithms of numbers are exponents, they have the same properties as exponents. Thus, we have

Logarithm	Exponent
$\log AB = \log A + \log B$	$10^A \times 10^B = 10^{A+B}$
$\log \frac{A}{B} = \log A - \log B$	$\frac{10^A}{10^B} = 10^{A-B}$

Furthermore, $\log A^n = n \log A$.

Now suppose we want to find the common logarithm of 6.7×10^{-4}. On most electronic calculators, the number is entered first and then the log key is pressed. This operation gives us

$$\log 6.7 \times 10^{-4} = -3.17$$

Note that there are as many digits *after* the decimal point as there are significant figures in the original number. The original number has two significant figures and the "17" in −3.17 tells us that the log has two significant figures. The "3" in −3.17 serves only to locate the decimal point in the number 6.7×10^{-4}. Other examples are

Number	Common Logarithm
62	1.79
0.872	−0.0595
1.0×10^{-7}	−7.00

Sometimes (as in the case of pH calculations) it is necessary to obtain the number whose logarithm is known. This procedure is known as taking the antilogarithm; it is simply the reverse of taking the logarithm of a number. Suppose in a certain calculation we have pH = 1.46 and are asked to calculate $[H^+]$. From the definition of pH (pH = $-\log [H^+]$) we can write

$$[H^+] = 10^{-1.46}$$

Many calculators have a key labeled $\log^{-1}$ or INV log to obtain antilogs. Other calculators have a 10^x or y^x key (where x corresponds to −1.46 in our example and y is 10 for base-10 logarithm). Therefore, we find that $[H^+] = 0.035\ M$.

Natural Logarithms

Logarithms taken to the base e instead of 10 are known as natural logarithms (denoted by ln or $\log_e$); e is equal to 2.7183. The relationship between common logarithms and natural logarithms is as follows:

$$\log 10 = 1 \qquad 10^1 = 10$$
$$\ln 10 = 2.303 \qquad e^{2.303} = 10$$

Thus,

$$\ln x = 2.303 \log x$$

To find the natural logarithm of 2.27, say, we first enter the number on the electronic calculator and then press the ln key to get

$$\ln 2.27 = 0.820$$

If no ln key is provided, we can proceed as follows:

$$2.303 \log 2.27 = 2.303 \times 0.356$$
$$= 0.820$$

Sometimes we may be given the natural logarithm and asked to find the number it represents. For example,

$$\ln x = 59.7$$

On many calculators, we simply enter the number and press the e key:

$$x = e^{59.7} = 8 \times 10^{25}$$

The Quadratic Equation

A quadratic equation takes the form

$$ax^2 + bx + c = 0$$

If coefficients a, b, and c are known, then x is given by

$$x = \frac{-b \pm \sqrt{b^2 - 4ac}}{2a}$$

Suppose we have the following quadratic equation:

$$2x^2 + 5x - 12 = 0$$

Solving for x, we write

$$x = \frac{-5 \pm \sqrt{(5)^2 - 4(2)(-12)}}{2(2)}$$
$$= \frac{-5 \pm \sqrt{25 + 96}}{4}$$

Therefore,

$$x = \frac{-5 + 11}{4} = \frac{3}{2}$$

and

$$x = \frac{-5 - 11}{4} = -4$$

Appendix 6

Fundamental Constants, Useful Conversion Factors, and SI Prefixes

Table A6.1 Fundamental Constants

Atomic mass unit (amu)	$1.66052886 \times 10^{-24}$ g
Avogadro's number (N_A)	6.0221413×10^{23}
Bohr radius (a_0)	$5.29177249 \times 10^{-11}$ m
Boltzmann constant (k_B)	$1.38064852 \times 10^{-23}$ J/K
Electron charge (e)	$-1.60217653 \times 10^{-19}$ C
Electron mass	$9.1093826 \times 10^{-28}$ g
Faraday constant (F)	96,485.3383 C/mol e^-
Neutron mass	$1.67492728 \times 10^{-24}$ g
Permittivity of vacuum (ε_0)	$8.854187817 \times 10^{-12}$ C^2 J^{-1} m^{-2}
Planck constant (h)	$6.6260693 \times 10^{-34}$ J · s
Proton charge (p)	$1.60217653 \times 10^{-19}$ C
Proton mass	1.672621×10^{-24} g
Rydberg constant (R_H)	2.179872×10^{-18} J
Speed of light in vacuum (c)	2.99792458×10^{8} m/s

Table A6.2 Useful Conversion Factors and Relationships

Mass	1 lb = 16 oz = 0.45359237 kg (exactly) = 453.6 g
Length	1 in = 2.54 cm (exactly)
	1 ft = 12 in = 0.3048 m (exactly)
	1 mi = 1.609 km
	1 km = 0.62137 mi
	1 pm = 1×10^{-12} m = 1×10^{-10} cm
	1 Å = 10^{-8} cm = 10^{-10} m = 0.1 nm
Volume	1 gal = 4 quarts = 3.785412 L (exactly)
Pressure	1 atm = 760 mmHg = 760 torr = 101,325 N/m^2 = 101,325 Pa = 1.01325 bar
Energy	1 cal = 4.184 J (exactly)
	1 L · atm = 101.325 J
	1 J = 1 C × 1 V
	1 eV = 1.602×10^{-19} J = 96.4853 kJ/mol
Temperature	$?°C = (°F - 32°F) \times \frac{5°C}{9°F}$
	$?°F = \frac{9°F}{5°C} \times (°C) + 32°F$
	$?K = (°C + 273.15°C)\left(\frac{1\ K}{1°C}\right)$

Table A6.3 Some Prefixes Used with SI Units

exa (E)	10^{18}	centi (c)	10^{-2}
peta (P)	10^{15}	milli (m)	10^{-3}
tera (T)	10^{12}	micro (μ)	10^{-6}
giga (G)	10^{9}	nano (n)	10^{-9}
mega (M)	10^{6}	pico (p)	10^{-12}
kilo (k)	10^{3}	femto (f)	10^{-15}
deci (d)	10^{-1}	atto (a)	10^{-18}

Appendix 7

Color Codes for Molecular Models

Glossary

The number in parentheses is the number of the section in which the term first appears.

A

absolute temperature scale A temperature scale that uses the absolute zero of temperature as the lowest temperature. (5.3)

absolute zero Theoretically the lowest attainable temperature. (5.3)

acceptor impurity An impurity that can accept electrons from a semiconductor. (21.3)

accuracy The closeness of a measurement to the true value of the quantity that is measured. (1.4)

acid A substance that yields hydrogen ions (H^+) when dissolved in water. (2.7)

acid ionization constant (K_a) The equilibrium constant for the acid ionization. (15.5)

actinoid series Elements that have incompletely filled 5*f* subshells or readily give rise to cations that have incompletely filled 5*f* subshells. (7.9)

activated complex The species temporarily formed by the reactant molecules as a result of the collision before they form the product. (13.4)

activation energy (E_a) The minimum amount of energy required to initiate a chemical reaction. (13.4)

activity series A summary of the results of many possible displacement reactions. (4.4)

actual yield The amount of product actually obtained in a reaction. (3.10)

addition reaction A reaction in which one molecule adds to another. (24.2)

adhesion Attraction between unlike molecules. (11.3)

alcohol An organic compound containing the hydroxyl group —OH. (24.4)

aldehydes Compounds with a carbonyl functional group and the general formula RCHO, where R is an H atom, an alkyl, or an aromatic group. (24.4)

aliphatic hydrocarbons Hydrocarbons that do not contain the benzene group or the benzene ring. (24.1)

alkali metals The Group 1 elements (Li, Na, K, Rb, Cs, and Fr). (2.4)

alkaline earth metals The Group 2 elements (Be, Mg, Ca, Sr, Ba, and Ra). (2.4)

alkanes Hydrocarbons having the general formula C_nH_{2n+2}, where $n = 1, 2, \ldots$. (24.2)

alkenes Hydrocarbons that contain one or more carbon-carbon double bonds. They have the general formula C_nH_{2n}, where $n = 2, 3, \ldots$. (24.2)

alkynes Hydrocarbons that contain one or more carbon-carbon triple bonds. They have the general formula C_nH_{2n}, where $n = 2, 3, \ldots$. (24.2)

allotropes Two or more forms of the same element that differ significantly in chemical and physical properties. (2.6)

alloy A solid solution composed of two or more metals, or of a metal or metals with one or more nonmetals. (21.2)

alpha (α) rays Helium ions with a positive charge of +2. (2.2)

alpha decay A radioactive process in which an alpha particle is emitted from a nucleus. (19.2)

alpha particles See alpha rays.

amalgam An alloy of mercury with another metal or metals. (21.2)

amines Organic bases that have the functional group $—NR_2$, where R may be H, an alkyl group, or an aromatic group. (24.4)

amino acids A compound that contains at least one amino group and at least one carboxyl group. (25.3)

amorphous solid A solid that lacks a regular three-dimensional arrangement of atoms or molecules. (11.7)

amphoteric oxide An oxide that exhibits both acidic and basic properties. (8.6)

amplitude The vertical distance from the middle of a wave to the peak or trough. (7.1)

anion An ion with a net negative charge. (2.5)

anode The electrode at which oxidation occurs. (18.2)

antibonding molecular orbital A molecular orbital that is of higher energy and lower stability than the atomic orbitals from which it was formed. (10.6)

aqueous solution A solution in which the solvent is water. (4.1)

aromatic hydrocarbon A hydrocarbon that contains one or more benzene rings. (24.1)

atmospheric pressure The pressure exerted by Earth's atmosphere. (5.2)

atom The basic unit of an element that can enter into chemical combination. (2.2)

atom economy A measure of the overall efficiency of converting reactants into useful products. (3.10)

atomic mass The mass of an atom in atomic mass units. (3.1)

atomic mass unit (amu) A mass exactly equal to $\frac{1}{12}$ the mass of one carbon-12 atom. (3.1)

atomic number (Z) The number of protons in the nucleus of an atom. (2.3)

atomic orbital The wave function (Ψ) of an electron in an atom. (7.5)

atomic radius One-half the distance between the two nuclei in two adjacent atoms of the same element in a metal. For elements that exist as diatomic units, the atomic radius is one-half the distance between the nuclei of the two atoms in a particular molecule. (8.3)

Aufbau principle As protons are added one by one to the nucleus to build up the elements, electrons similarly are added to the atomic orbitals. (7.9)

Avogadro's law At constant pressure and temperature, the volume of a gas is directly proportional to the number of moles of the gas present. (5.3)

Avogadro's number (N_A) 6.022×10^{23}; the number of particles in a mole. (3.2)

B

band theory Delocalized electrons move freely through "bands" formed by overlapping molecular orbitals. (21.3)

barometer An instrument that measures atmospheric pressure. (5.2)

base A substance that yields hydroxide ions (OH^+) when dissolved in water. (2.7)

base ionization constant (K_b) The equilibrium constant for the base ionization. (15.6)

battery A galvanic cell, or a series of combined galvanic cells, that can be used as a source of direct electric current at a constant voltage. (18.6)

beta (β) rays Electrons. (2.2)

beta particles See beta rays.

bimolecular reaction An elementary step that involves two molecules. (13.5)

binary compounds Compounds formed from just two elements. (2.7)

boiling point The temperature at which the vapor pressure of a liquid is equal to the external atmospheric pressure. (11.8)

boiling-point elevation (ΔT_b) The boiling point of the solution (T_b) minus the boiling point of the pure solvent (T°_b). (12.6)

bond enthalpy The enthalpy change required to break a bond in 1 mole of gaseous molecules. (9.10)

bond length The distance between the nuclei of two bonded atoms in a molecule. (9.4)

bond order The difference between the numbers of electrons in bonding molecular orbitals and antibonding molecular orbitals, divided by two. (10.7)

bonding molecular orbital A molecular orbital that is of lower energy and greater stability than the atomic orbitals from which it was formed. (10.6)

Born-Haber cycle The cycle that relates lattice energies of ionic compounds to ionization energies, electron affinities, heats of sublimation and formation, and bond enthalpies. (9.3)

boundary surface diagram Diagram of the region containing a substantial amount of the electron density (about 90 percent) in an orbital. (7.7)

Boyle's law The volume of a fixed amount of gas maintained at constant temperature is inversely proportional to the gas pressure. (5.3)

breeder reactor A nuclear reactor that produces more fissionable materials than it uses. (19.5)

Brønsted acid A substance capable of donating a proton. (4.3)
Brønsted base A substance capable of accepting a proton. (4.3)
buffer solution A solution of (a) a weak acid or base and (b) its salt; both components must be present. The solution has the ability to resist changes in pH upon the addition of small amounts of either acid or base. (16.2)

C

calorimetry The measurement of heat changes. (6.5)
carbides Ionic compounds containing the C^{4-} ion. (22.3)
carboxylic acids Acids that contain the carboxyl group —COOH. (24.4)
catalyst A substance that increases the rate of a chemical reaction without itself being consumed. (13.6)
catenation The ability of the atoms of an element to form bonds with one another. (22.3)
cathode The electrode at which reduction occurs. (18.2)
cation An ion with a net positive charge. (2.5)
cell voltage Difference in electrical potential between the anode and the cathode of a galvanic cell. (18.2)
Charles's and Gay-Lussac's law The volume of a fixed amount of gas maintained at constant pressure is directly proportional to the absolute temperature of the gas. (5.3).
Charles's law The volume of a fixed amount of gas maintained at constant pressure is directly proportional to the absolute temperature of the gas. (5.3)
chelating agent A substance that forms complex ions with metal ions in solution. (23.2)
chemical energy Energy stored within the structural units of chemical substances. (6.1)
chemical equation An equation that uses chemical symbols to show what happens during a chemical reaction. (3.7)
chemical equilibrium A state in which the rates of the forward and reverse reactions are equal. (14.1)
chemical formula An expression showing the chemical composition of a compound in terms of the symbols for the atoms of the elements involved. (2.6)
chemical kinetics The area of chemistry concerned with the speeds, or rates, at which chemical reactions occur. (13.1)
chemical property Any property of a substance that cannot be studied without converting the substance into some other substance. (1.9)
chemical reaction A process in which a substance (or substances) is changed into one or more new substances. (3.7)
chemistry The study of matter and the changes it undergoes. (1.1)
chiral Compounds or ions that are not superimposable with their mirror images. (23.3)
chlor-alkali process The production of chlorine gas by the electrolysis of aqueous NaCl solution. (22.6)
closed system A system that enables the exchange of energy (usually in the form of heat) but not mass with its surroundings. (6.2)
closest packing The most efficient arrangements for packing atoms, molecules, or ions in a crystal. (11.4)
cohesion The intermolecular attraction between like molecules. (11.3)
colligative properties Properties of solutions that depend on the number of solute particles in solution and not on the nature of the solute particles. (12.6)
colloid A dispersion of particles of one substance (the dispersed phase) throughout a dispersing medium made of another substance. (12.8)
combination reaction A reaction in which two or more substances combine to form a single product. (4.4)
combustion reaction A reaction in which a substance reacts with oxygen, usually with the release of heat and light, to produce a flame. (4.4)
common ion effect The shift in equilibrium caused by the addition of a compound having an ion in common with the dissolved substances. (16.1)
complex ion An ion containing a central metal cation bonded to one or more molecules or ions. (16.7)
compound A substance composed of atoms of two or more elements chemically united in fixed proportions. (1.7)
concentration of a solution The amount of solute present in a given quantity of solvent or solution. (4.5)
condensation The phenomenon of going from the gaseous state to the liquid state. (11.8)
condensation reaction A reaction in which two smaller molecules combine to form a larger molecule. Water is invariably one of the products of such a reaction. (24.4)
conductor Substance capable of conducting electric current. (21.3)
conjugate acid-base pair An acid and its conjugate base or a base and its conjugate acid. (15.1)
coordinate covalent bond A bond in which the pair of electrons is supplied by one of the two bonded atoms; also called a dative bond. (9.9)
coordination compound A neutral species containing one or more complex ions. (23.2)
coordination number In a crystal lattice it is defined as the number of atoms (or ions) surrounding an atom (or ion) (11.4). In coordination compounds it is defined as the number of donor atoms surrounding the central metal atom in a complex. (23.2)
copolymer A polymer containing two or more different monomers. (25.2)
core electrons All nonvalence electrons in an atom. (8.2)
corrosion The deterioration of metals by an electrochemical process. (18.7)
Coulomb's law The potential energy between two ions is directly proportional to the product of their charges and inversely proportional to the distance between them. (9.3)
covalent bond A bond in which two electrons are shared by two atoms. (9.4)
covalent compounds Compounds containing only covalent bonds. (9.4)
critical mass The minimum mass of fissionable material required to generate a self-sustaining nuclear chain reaction. (19.5)
critical point A point on a phase diagram where a phase boundary ends. (11.9)
critical pressure (P_c) The minimum pressure necessary to bring about liquefaction at the critical temperature. (11.8)
critical temperature (T_c) The temperature above which a gas will not liquefy. (11.8)
crystal field splitting (Δ) The energy difference between two sets of *d* orbitals in a metal atom when ligands are present. (23.4)
crystalline solid A solid that possesses rigid and long-range order; its atoms, molecules, or ions occupy specific positions. (11.4)
crystallization The process in which dissolved solute comes out of solution and forms crystals. (12.1)
cyanides Compounds containing the CN^- ion. (22.3)
cycloalkanes Alkanes whose carbon atoms are joined in rings. (24.2)

D

Dalton's law of partial pressures The total pressure of a mixture of gases is just the sum of the pressures that each gas would exert if it were present alone. (5.6)
decomposition reaction The breakdown of a compound into two or more components. (4.4)
degenerate orbitals Atomic orbitals with the same energy. (7.6)
delocalized molecular orbitals Molecular orbitals that are not confined between two adjacent bonding atoms but actually extend over three or more atoms. (10.8)
denatured protein Protein that does not exhibit normal biological activities. (25.3)
density The mass of a substance divided by its volume. (1.9)
deoxyribonucleic acid (DNA) A type of nucleic acid. (25.4)
deposition The process in which the molecules go directly from the vapor into the solid phase. (11.8)
diagonal relationship Similarities between pairs of elements in different groups and periods of the periodic table. (8.6)
diamagnetic Repelled by a magnet; a diamagnetic substance contains only paired electrons. (7.8)
diatomic molecule A molecule that consists of two atoms. (2.5)
diffusion The gradual mixing of molecules of one gas with the molecules of another by virtue of their kinetic properties. (5.7)

dilution A procedure for preparing a less concentrated solution from a more concentrated solution. (4.5)

dipole moment (μ) The product of charge and the distance between the charges in a molecule. (10.2)

dipole-dipole forces Forces that act between polar molecules. (11.2)

diprotic acid Each unit of the acid yields two hydrogen ions upon ionization. (4.3)

dispersion forces The attractive forces that arise as a result of temporary dipoles induced in the atoms or molecules; also called London forces. (11.2)

displacement reaction An atom or an ion in a compound is replaced by an atom of another element. (4.4)

disproportionation reaction A reaction in which an element in one oxidation state is both oxidized and reduced. (4.4)

donor atom The atom in a ligand that is bonded directly to the metal atom. (23.2)

donor impurity An impurity that provides conduction electrons to a semiconductor. (21.3)

double bond Two atoms are held together by two pairs of electrons. (9.4)

dynamic equilibrium The condition in which the rate of a forward process is exactly balanced by the rate of a reverse process. (11.8)

E

effective nuclear charge (Z_{eff}) The nuclear charge felt by an electron when both the actual charge (*Z*) and the repulsive effect (shielding) of the other electrons are taken into account. (8.3)

effusion A process by which a gas under pressure escapes from one compartment of a container to another by passing through a small opening. (5.7)

electrochemistry The branch of chemistry that deals with the interconversion of electrical energy and chemical energy. (18.1)

electrolysis A process in which electrical energy is used to cause a nonspontaneous chemical reaction to occur. (18.8)

electrolyte A substance that, when dissolved in water, results in a solution that can conduct electricity. (4.1)

electrolytic cell An apparatus for carrying out electrolysis. (18.8)

electromagnetic radiation The emission and transmission of energy in the form of electromagnetic waves. (7.1)

electromagnetic wave A wave that has an electric field component and a mutually perpendicular magnetic field component. (7.1)

electromotive force (emf) (*E*) The voltage difference between electrodes. (18.2)

electron A subatomic particle that has a very low mass and carries a single negative electric charge. (2.2)

electron affinity (*EA*) The negative of the enthalpy change when an electron is accepted by an atom in the gaseous state to form an anion. (8.5)

electron capture A type of radioactive decay in which a nucleus captures an orbital electron, usually a 1*s* electron. (19.2)

electron configuration The distribution of electrons among the various orbitals in an atom or molecule. (7.8)

electron density The probability that an electron will be found at a particular region in an atomic orbital. (7.5)

electronegativity The ability of an atom to attract electrons toward itself in a chemical bond. (9.5)

electrostatic energy Potential energy (measured in joules) that results from conservative coulomb forces and is associated with the configuration of a particular set of charges. (6.1)

element A substance that cannot be separated into simpler substances by chemical means. (1.7)

elementary steps A series of simple reactions that represent the progress of the overall reaction at the molecular level. (13.5)

emission spectra Continuous or line spectra emitted by substances. (7.3)

empirical formula An expression showing the types of elements present and the simplest ratios of the different kinds of atoms. (2.6)

enantiomers Optical isomers, that is, compounds and their nonsuperimposable mirror images. (23.3)

end point The pH at which the indicator changes color. (16.4)

endothermic processes Processes that absorb heat from the surroundings. (6.2)

energy The capacity to do work or to produce change. (6.1)

enthalpy (*H*) A thermodynamic quantity used to describe heat changes taking place at constant pressure. (6.4)

enthalpy of reaction (ΔH_{rxn}) The difference between the enthalpies of the products and the enthalpies of the reactants. (6.4)

enthalpy of solution (ΔH_{soln}) The heat generated or absorbed when a certain amount of solute is dissolved in a certain amount of solvent. (6.7)

entropy (*S*) A measure of how dispersed the energy of a system is among the different ways that system can contain energy. (17.2)

enzyme A biological catalyst. (13.6)

equilibrium constant (*K*) A number equal to the ratio of the equilibrium concentrations of products to the equilibrium concentrations of reactants, each raised to the power of its stoichiometric coefficient. (14.1)

equilibrium vapor pressure The vapor pressure measured under dynamic equilibrium of condensation and evaporation at some temperature. (11.8)

equivalence point The point at which the acid has completely reacted with or been neutralized by the base. (4.7)

esters Compounds that have the general formula R′COOR, where R′ can be H or an alkyl group or an aromatic group and R is an alkyl group or an aromatic group. (24.4)

ether An organic compound containing the R—O—R′ linkage, where R and R′ are alkyl and/or aromatic groups. (24.4)

evaporation The process in which a liquid is transformed into a gas; also called vaporization. (11.8)

excess reactants One or more reactants present in quantities greater than necessary to react with the quantity of the limiting reagent. (3.9)

excited state (or level) A state that has higher energy than the ground state. (7.3)

exothermic processes Processes that give off heat to the surroundings. (6.2)

extensive property A property that depends on how much matter is being considered. (1.9)

F

family The elements in a vertical column of the periodic table. (2.4)

Faraday constant (*F*) Charge contained in 1 mole of electrons, equivalent to 96,485.3 coulombs. (18.4)

ferromagnetic Attracted by a magnet. The unpaired spins in a ferromagnetic substance are aligned in a common direction. (21.2)

first law of thermodynamics Energy can be converted from one form to another, but cannot be created or destroyed. (6.3)

first-order reaction A reaction whose rate depends on reactant concentration raised to the first power. (13.3)

formal charge The difference between the valence electrons in an isolated atom and the number of electrons assigned to that atom in a Lewis structure. (9.7)

formation constant (K_f) The equilibrium constant for the complex ion formation. (16.7)

fractional crystallization The separation of a mixture of substances into pure components on the basis of their different solubilities. (12.4)

fractional distillation A procedure for separating liquid components of a solution that is based on their different boiling points. (12.6)

free energy (*G*) See Gibbs energy.

freezing point The temperature at which the solid and liquid phases of a substance coexist at equilibrium. (11.8)

freezing-point depression (ΔT_f) The freezing point of the pure solvent (T_f°) minus the freezing point of the solution (T_f). (12.6)

frequency (ν) The number of waves that pass through a particular point per unit time. (7.1)

fuel cell A galvanic cell that requires a continuous supply of reactants to keep functioning. (18.6)

functional group That part of a molecule characterized by a special arrangement of atoms that is largely responsible for the chemical behavior of the parent molecule. (24.1)

G

galvanic cell The experimental apparatus for generating electricity through the use of a spontaneous redox reaction. (18.2)
gamma (γ) rays High-energy radiation. (2.2)
gamma emission The type of radioactive decay in which gamma rays are emitted from an excited nucleus. (19.2)
gas constant (*R*) The constant that appears in the ideal gas equation. It is usually expressed as 0.08206 L · atm/K · mol, or 8.314 J/K · mol. (5.4)
geometric isomers Compounds with the same type and number of atoms and the same chemical bonds but different spatial arrangements; such isomers cannot be interconverted without breaking a chemical bond. (23.3)
Gibbs energy The energy available to do useful work. (17.4)
glass The optically transparent fusion product of inorganic materials that has cooled to a rigid state without crystallizing. (11.7)
Graham's law of diffusion Under the same conditions of temperature and pressure, rates of diffusion for gases are inversely proportional to the square roots of their molar masses. (5.7)
gravimetric analysis An experimental procedure that involves the measurement of mass. (4.6)
greenhouse effect Carbon dioxide and other gases' influence on Earth's temperature. (20.5)
ground state (or level) The lowest energy state of a system. (7.3)
group The elements in a vertical column of the periodic table. (2.4)

H

half-cell reactions Oxidation and reduction reactions at the electrodes. (18.2)
half-life ($t_{½}$) The time required for the concentration of a reactant to decrease to half of its initial concentration. (13.3)
half-reaction A reaction that explicitly shows electrons involved in either oxidation or reduction. (4.4)
halogens The nonmetallic elements in Group 17 (F, Cl, Br, I, and At). (2.4)
heat Transfer of energy between two bodies that are at different temperatures. (6.2)
heat capacity (*C*) The amount of heat required to raise the temperature of a given quantity of the substance by one degree Celsius. (6.5)
heat of dilution The heat change associated with the dilution process. (6.7)
heat of hydration (ΔH_{hydr}) The heat change associated with the hydration process. (6.7)
heat of solution See enthalpy of solution.
Heisenberg uncertainty principle It is impossible to know simultaneously both the momentum and the position of a particle with certainty. (7.5)
Henry's law The solubility of a gas in a liquid is proportional to the pressure of the gas over the solution. (12.5)
Hess's law When reactants are converted to products, the change in enthalpy is the same whether the reaction takes place in one step or in a series of steps. (6.6)
heterogeneous equilibrium An equilibrium state in which the reacting species are not all in the same phase. (14.2)
heterogeneous mixture The individual components of a mixture remain physically separated and can be seen as separate components. (1.7)
homogeneous equilibrium An equilibrium state in which all reacting species are in the same phase. (14.2)
homogeneous mixture The composition of a mixture, after sufficient stirring, is the same throughout the solution. (1.7)
homonuclear diatomic molecule A diatomic molecule containing atoms of the same element. (10.7)
homopolymer A polymer that is made from only one type of monomer. (25.2)
Hund's rule The most stable arrangement of electrons in subshells is the one with the greatest number of parallel spins. (7.8)
hybrid orbitals Atomic orbitals obtained when two or more nonequivalent orbitals of the same atom combine. (10.4)
hybridization The process of mixing the atomic orbitals in an atom (usually the central atom) to generate a set of new atomic orbitals. (10.4)
hydrates Compounds that have a specific number of water molecules attached to them. (2.7)
hydration A process in which an ion or a molecule is surrounded by water molecules arranged in a specific manner. (4.1)
hydrocarbons Compounds made up only of carbon and hydrogen. (24.1)
hydrogen bond A special type of dipole-dipole interaction between the hydrogen atom bonded to an atom of a very electronegative element (F, N, O) and another atom of one of the three electronegative elements. (11.2)
hydrogenation The addition of hydrogen, especially to compounds with double and triple carbon-carbon bonds. (22.2)
hydronium ion The hydrated proton, H_3O^+. (4.3)
hydrophilic Water-loving. (12.8)
hydrophobic Water-fearing. (12.8)
hypothesis A tentative explanation for a set of observations. (1.2)

I

ideal gas A hypothetical gas whose pressure-volume-temperature behavior can be completely accounted for by the ideal gas equation. (5.4)
ideal gas equation An equation expressing the relationships among pressure, volume, temperature, and amount of gas ($PV = nRT$, where R is the gas constant). (5.4)
ideal solution Any solution that obeys Raoult's law. (12.6)
indicators Substances that have distinctly different colors in acidic and basic media. (4.7)
induced dipole The separation of positive and negative charges in a neutral atom (or a nonpolar molecule) caused by the proximity of an ion or a polar molecule. (11.2)
inert complex A complex ion that undergoes very slow ligand exchange reactions. (23.5)
inert pair effect The tendency for heavier Group 13 to 16 atoms to form ions with charges two less than their normal valency. (8.2)
inorganic compounds Compounds other than organic compounds. (2.7)
insulator A substance incapable of conducting electricity. (21.3)
intensive property A property that does not depend on how much matter is being considered. (1.9)
intermediate A species that appears in the mechanism of the reaction (that is, the elementary steps) but not in the overall balanced equation. (13.5)
intermolecular forces Attractive forces that exist among molecules. (11.2)
International System of Units (SI) A system of units based on metric units. (1.3)
intramolecular forces Forces that hold atoms together in a molecule. (11.2)
ion An atom or a group of atoms that has a net positive or negative charge. (2.5)
ion pair One or more cations and one or more anions held together by electrostatic forces. (12.7)
ion-dipole forces Forces that operate between an ion and a dipole. (11.2)
ion-product constant Product of hydrogen ion concentration and hydroxide ion concentration (both in molarity) at a particular temperature. (15.2)
ionic bond The electrostatic force that holds ions together in an ionic compound. (9.2)
ionic compound Any neutral compound containing cations and anions. (2.5)
ionic equation An equation that shows dissolved species as free ions. (4.2)
ionic radius The radius of a cation or an anion as measured in an ionic compound. (8.3)
ionization energy (*IE*) The minimum energy required to remove an electron from an isolated atom (or an ion) in its ground state. (8.4)
ionosphere The uppermost layer of the atmosphere. (20.1)
isoelectronic Ions, or atoms and ions, that possess the same number of electrons, and hence the same ground-state electron configuration, are said to be isoelectronic. (8.2)
isolated system A system that does not allow the transfer of either mass or energy to or from its surroundings. (6.2)
isotopes Atoms having the same atomic number but different mass numbers. (2.3)

J

Joule (J) Unit of energy given by newtons × meters. (5.7)

K

kelvin The SI base unit of temperature. (1.3)

Kelvin temperature scale A temperature scale that uses the absolute zero of temperature as the lowest temperature. (5.3).

ketones Compounds with a carbonyl functional group and the general formula RR′CO, where R and R′ are alkyl and/or aromatic groups. (24.4)

kinetic energy (E_k) Energy available because of the motion of an object. (5.7)

kinetic molecular theory of gases Treatment of gas behavior in terms of the random motion of molecules. (5.7)

L

labile complex Complexes that undergo rapid ligand exchange reactions. (23.5)

lanthanoids (rare earth) series Elements that have incompletely filled 4*f* subshells or readily give rise to cations that have incompletely filled 4*f* subshells. (7.9)

lattice A three-dimensional array of cations and anions. (2.6)

lattice energy (*U*) The energy required to completely separate one mole of a solid ionic compound into gaseous ions. (6.7)

law A concise verbal or mathematical statement of a relationship between phenomena that is always the same under the same conditions. (1.2)

law of conservation of energy The total quantity of energy in the universe is constant. (6.1)

law of conservation of mass Hypothesis stating that matter can be neither created nor destroyed. (2.1)

law of definite proportions Different samples of the same compound always contain its constituent elements in the same proportions by mass. (2.1)

law of mass action For reversible reaction at equilibrium and a constant temperature, a certain ratio of reactant and product concentrations has a constant value, *K* (the equilibrium constant). (14.1)

law of multiple proportions If two elements can combine to form more than one type of compound, the masses of one element that combine with a fixed mass of the other element are in ratios of small whole numbers. (2.1)

Le Châtelier's principle If an external stress is applied to a system at equilibrium, the system will adjust itself in such a way as to partially offset the stress as the system reaches a new equilibrium position. (14.5)

Lewis acid A substance that can accept a pair of electrons. (15.12)

Lewis base A substance that can donate a pair of electrons. (15.12)

Lewis dot symbol The symbol of an element with one or more dots that represent the number of valence electrons in an atom of the element. (9.1)

Lewis structure A representation of covalent bonding using Lewis symbols. Shared electron pairs are shown either as lines or as pairs of dots between two atoms, and lone pairs are shown as pairs of dots on individual atoms. (9.4)

ligand A molecule or an ion that is bonded to the metal ion in a complex ion. (23.2)

limiting reagent The reactant used up first in a reaction. (3.9)

line spectra Spectra produced when radiation is absorbed or emitted by substances only at some wavelengths. (7.3)

liter The volume occupied by one cubic decimeter. (1.3)

lone pairs Valence electrons that are not involved in covalent bond formation. (9.4)

M

macroscopic properties Properties that can be measured directly. (1.3)

manometer A device used to measure the pressure of gases. (5.2)

many-electron atoms Atoms that contain two or more electrons. (7.5)

mass A measure of the quantity of matter contained in an object. (1.9)

mass defect The difference between the mass of an atom and the sum of the masses of its protons, neutrons, and electrons. (19.2)

mass number (*A*) The total number of neutrons and protons present in the nucleus of an atom. (2.3)

matter Anything that occupies space and possesses mass. (1.7)

melting point The temperature at which solid and liquid phases coexist in equilibrium. (11.8)

mesosphere A region between the stratosphere and the ionosphere. (20.1)

metalloid An element with properties intermediate between those of metals and nonmetals. (2.4)

metallurgy The science and technology of separating metals from their ores and of compounding alloys. (21.2)

metals Elements that are good conductors of heat and electricity and have the tendency to form positive ions in ionic compounds. (2.4)

metathesis reaction A reaction that involves the exchange of parts between two compounds. (4.2)

microscopic properties Properties that cannot be measured directly without the aid of a microscope or other special instrument. (1.3)

mineral A naturally occurring substance with a range of chemical composition. (21.1)

miscible Two liquids that are completely soluble in each other in all proportions are said to be miscible. (12.2)

mixture A combination of two or more substances in which the substances retain their identity. (1.7)

moderator A substance that can reduce the kinetic energy of neutrons. (19.5)

molality The number of moles of solute dissolved in 1 kg of solvent. (12.3)

molar concentration See molarity.

molar heat of fusion (ΔH_{fus}) The energy (in kilojoules) required to melt 1 mole of a solid. (11.8)

molar heat of sublimation (ΔH_{sub}) The energy (in kilojoules) required to sublime 1 mole of a solid. (11.8)

molar heat of vaporization (ΔH_{vap}) The energy (in kilojoules) required to vaporize 1 mole of a liquid. (11.8)

molar mass ($\mathscr{M}$) The mass (in grams or kilograms) of one mole of atoms, molecules, or other particles. (3.2)

molar solubility The number of moles of solute in 1 liter of a saturated solution (mol/L). (16.5)

molarity (M) The number of moles of solute in 1 liter of solution. (4.5)

mole (mol) The amount of substance that contains as many elementary entities (atoms, molecules, or other particles) as there are atoms in exactly 12 g (or 0.012 kg) of the carbon-12 isotope. (3.2)

mole fraction Ratio of the number of moles of one component of a mixture to the total number of moles of all components in the mixture. (5.6)

mole method An approach for determining the amount of product formed in a reaction. (3.8)

molecular equations Equations in which the formulas of the compounds are written as though all species existed as molecules or whole units. (4.2)

molecular formula An expression showing the exact numbers of atoms of each element in a molecule. (2.6)

molecular mass The sum of the atomic masses (in atomic mass units) present in the molecule. (3.3)

molecular orbital An orbital that results from the interaction of the atomic orbitals of the bonding atoms. (10.6)

molecularity of a reaction The number of molecules reacting in an elementary step. (13.5)

molecule An aggregate of at least two atoms in a definite arrangement held together by special forces. (2.5)

monatomic ion An ion that contains only one atom. (2.5)

monomer The single repeating unit of a polymer. (25.2)

monoprotic acid Each unit of the acid yields one hydrogen ion upon ionization. (4.3)

multiple bonds Bonds formed when two atoms share two or more pairs of electrons. (9.4)

N

***n*-type semiconductors** Semiconductors that contain donor impurities. (21.3)

Nernst equation The relation between the emf of a galvanic cell and the standard emf and the concentrations of the oxidizing and reducing agents. (18.5)

net ionic equation An equation that indicates only the ionic species that actually take part in the reaction. (4.2)

neutralization reaction A reaction between an acid and a base. (4.3)

neutron A subatomic particle that bears no net electric charge. Its mass is slightly greater than a proton's mass. (2.2)

newton (N) The SI unit for force. (5.2)

nitrogen fixation The conversion of molecular nitrogen into nitrogen compounds. (20.1)

noble gas core The electron configuration of the noble gas element that most nearly precedes the element being considered. (7.9)

noble gases Nonmetallic elements in Group 18 (He, Ne, Ar, Kr, Xe, and Rn). (2.4)

node The point at which the amplitude of the wave is zero. (7.1, 7.4)

nonelectrolyte A substance that, when dissolved in water, gives a solution that is not electrically conducting. (4.1)

nonmetals Elements that are usually poor conductors of heat and electricity. (2.4)

nonpolar molecule A molecule that does not possess a dipole moment. (10.2)

nonvolatile Does not have a measurable vapor pressure. (12.6)

nuclear binding energy The energy required to break up a nucleus into its protons and neutrons. (19.2)

nuclear chain reaction A self-sustaining sequence of nuclear fission reactions. (19.5)

nuclear fission A heavy nucleus (mass number > 200) divides to form smaller nuclei of intermediate mass and one or more neutrons. (19.5)

nuclear fusion The combining of small nuclei into larger ones. (19.6)

nuclear transmutation The change undergone by a nucleus as a result of bombardment by neutrons or other particles. (19.1)

nucleic acids High molar mass polymers that play an essential role in protein synthesis. (25.4)

nucleon A general term for the protons and neutrons in a nucleus. (19.2)

nucleotide The repeating unit in each strand of a DNA molecule, which consists of a base-deoxyribose-phosphate linkage. (25.4)

nucleus The central core of an atom. (2.2)

O

octet rule An atom other than hydrogen tends to form bonds until it is surrounded by eight valence electrons. (9.4)

open system A system that can exchange mass and energy (usually in the form of heat) with its surroundings. (6.2)

optical isomers Compounds that are nonsuperimposable mirror images. (23.3)

ore The material of a mineral deposit in a sufficiently concentrated form to allow economical recovery of a desired metal. (21.1)

organic chemistry The branch of chemistry that deals with carbon compounds. (24.1)

organic compounds Compounds that contain carbon, usually in combination with elements such as hydrogen, oxygen, nitrogen, and sulfur. (2.7)

osmosis The net movement of solvent molecules through a semipermeable membrane from a pure solvent or from a dilute solution to a more concentrated solution. (12.6)

osmotic pressure (π) The pressure required to stop osmosis. (12.6)

overvoltage The difference between the electrode potential and the actual voltage required to cause electrolysis. (18.8)

oxidation number The number of charges an atom would have in a molecule if electrons were transferred completely in the direction indicated by the difference in electronegativity. (4.4)

oxidation reaction The half-reaction that involves the loss of electrons. (4.4)

oxidation state See oxidation number.

oxidation-reduction (redox) reaction A reaction that involves the transfer of electron(s) or the change in the oxidation state of reactants. (4.4)

oxidizing agent A substance that can accept electrons from another substance or increase the oxidation numbers in another substance. (4.4)

oxoacid An acid containing hydrogen, oxygen, and another element (the central element). (2.7)

oxoanion An anion derived from an oxoacid. (2.7)

P

***p*-type semiconductors** Semiconductors that contain acceptor impurities. (21.3)

paramagnetic Attracted by a magnet. A paramagnetic substance contains one or more unpaired electrons. (7.8)

partial pressure Pressure of one component in a mixture of gases. (5.6)

pascal (Pa) A pressure of one newton per square meter (1 N/m^2). (5.2)

Pauli exclusion principle No two electrons in an atom can have the same four quantum numbers. (7.8)

percent by mass The ratio of the mass of a solute to the mass of the solution, multiplied by 100 percent. (12.3)

percent composition by mass The percent by mass of each element in a compound. (3.5)

percent ionization Ratio of ionized acid concentration at equilibrium to the initial concentration of acid. (15.5)

percent yield The ratio of actual yield to theoretical yield, multiplied by 100 percent. (3.10)

period A horizontal row of the periodic table. (2.4)

periodic table A tabular arrangement of the elements. (2.4)

pH The negative logarithm of the hydrogen ion concentration. (15.3)

phase A homogeneous part of a system in contact with other parts of the system but separated from them by a well-defined boundary. (11.1)

phase boundary The boundaries separating different phases in a phase diagram. (11.9)

phase change Transformation from one phase to another. (11.8)

phase diagram A diagram showing the conditions at which a substance exists as a solid, liquid, or vapor. (11.9)

photochemical smog Formation of smog by the reactions of automobile exhaust in the presence of sunlight. (20.7)

photoelectric effect A phenomenon in which electrons are ejected from the surface of certain metals exposed to light of at least a certain minimum frequency. (7.2)

photon A particle of light. (7.2)

physical equilibrium An equilibrium in which only physical properties change. (14.1)

physical property Any property of a substance that can be observed without transforming the substance into some other substance. (1.9)

pi (π) bond A covalent bond formed by sideways overlapping orbitals; its electron density is concentrated above and below the plane of the nuclei of the bonding atoms. (10.5)

pi molecular orbital A molecular orbital in which the electron density is concentrated above and below the plane of the two nuclei of the bonding atoms. (10.6)

plasma A gaseous mixture of positive ions and electrons. (19.6)

polar covalent bond In such a bond, the electrons spend more time in the vicinity of one atom than the other. (9.5)

polar molecule A molecule that possesses a dipole moment. (10.2)

polarimeter The instrument for measuring the rotation of polarized light by optical isomers. (23.3)

polyatomic ion An ion that contains more than one atom. (2.5)

polyatomic molecule A molecule that consists of more than two atoms. (2.5)

polymer A compound distinguished by a high molar mass, ranging into thousands and millions of grams, and made up of many repeating units. (25.1)

positron A particle that has the same mass as the electron, but bears a + 1 charge. (19.1)

positron emission A type of radioactive decay in which a positron is emitted from a nucleus. (19.2)

potential energy Energy available by virtue of an object's position. (6.1)

precipitate An insoluble solid that separates from the solution. (4.2)

precipitation reaction A reaction that results in the formation of a precipitate. (4.2)

precision The closeness of agreement of two or more measurements of the same quantity. (1.4)

pressure Force applied per unit area. (5.2)

product The substance formed as a result of a chemical reaction. (3.7)

protein Polymers of amino acids. (25.3)

proton A subatomic particle having a single positive electric charge. The mass of a proton is about 1840 times that of an electron. (2.2)

pyrometallurgy Metallurgical processes that are carried out at high temperatures. (21.2)

Q

qualitative Consisting of general observations about the system. (1.2)
qualitative analysis The determination of the types of ions present in a solution. (16.8)
quantitative Comprising numbers obtained by various measurements of the system. (1.2)
quantitative analysis The determination of the amount of substances present in a sample. (4.5)
quantum The smallest quantity of energy that can be emitted (or absorbed) in the form of electromagnetic radiation. (7.1)
quantum numbers Numbers that describe the distribution of electrons in hydrogen and other atoms. (7.6)

R

racemic mixture An equimolar mixture of the two enantiomers. (23.3)
radiant energy Energy transmitted in the form of waves. (6.1)
radiation The emission and transmission of energy through space in the form of particles and/or waves. (2.2)
radical Any neutral fragment of a molecule containing an unpaired electron. (19.8)
radioactive decay series A sequence of nuclear reactions that ultimately result in the formation of a stable isotope. (19.3)
radioactivity The spontaneous breakdown of an atom by emission of particles and/or radiation. (2.2)
random errors Errors that are not predictable, leading to measured values that vary greatly from the true value. (1.4)
Raoult's law The vapor pressure of the solvent over a solution is given by the product of the vapor pressure of the pure solvent and the mole fraction of the solvent in the solution. (12.6)
rare earth series See lanthanoids (rare earth) series.
rate constant (*k*) Constant of proportionality between the reaction rate and the concentrations of reactants. (13.1)
rate law An expression relating the rate of a reaction to the rate constant and the concentrations of the reactants. (13.2)
rate-determining step The slowest step in the sequence of steps leading to the formation of products. (13.5)
reactant The starting substance in a chemical reaction. (3.7)
reaction mechanism The sequence of elementary steps that leads to product formation. (13.5)
reaction order The sum of the powers to which all reactant concentrations appearing in the rate law are raised. (13.2)
reaction quotient (Q_c) A number equal to the ratio of product concentrations to reactant concentrations, each raised to the power of its stoichiometric coefficient at some point other than equilibrium. (14.4)
reaction rate The change in the concentration of reactant or product with time. (13.1)
redox reaction See oxidation-reduction reaction.
reducing agent A substance that can donate electrons to another substance or decrease the oxidation numbers in another substance. (4.4)
reduction reaction The half-reaction that involves the gain of electrons. (4.4)
representative elements Elements in Groups 1, 2, and 13–17, all of which have incompletely filled *s* or *p* subshells of highest principal quantum number. (8.2)
resonance The use of two or more Lewis structures to represent a particular molecule. (9.8)
resonance structure One of two or more alternative Lewis structures for a molecule that cannot be described fully with a single Lewis structure. (9.8)
reversible reaction A reaction that can occur in both directions. (4.1)
ribonucleic acid (RNA) A form of nucleic acid. (25.4)
root-mean-square (rms) speed (u_{rms}) A measure of the average molecular speed at a given temperature. (5.7)

S

salt An ionic compound made up of a cation other than H^+ and an anion other than OH^- or O^{2-}. (4.3)
salt hydrolysis The reaction of the anion or cation, or both, of a salt with water. (15.10)
saponification Soapmaking. (24.4)
saturated hydrocarbons Hydrocarbons that contain the maximum number of hydrogen atoms that can bond with the number of carbon atoms present. (24.2)
saturated solution At a given temperature, the solution that results when the maximum amount of a substance has dissolved in a solvent. (12.1)
scientific method A systematic approach to research. (1.2)
second law of thermodynamics The entropy of the universe increases in a spontaneous process and remains unchanged in an equilibrium process. (17.3)
second-order reaction A reaction whose rate depends on reactant concentration raised to the second power or on the concentrations of two different reactants, each raised to the first power. (13.3)
semiconductors Elements that normally cannot conduct electricity, but can have their conductivity greatly enhanced either by raising the temperature or by adding certain impurities. (21.3)
semipermeable membrane A membrane that enables solvent molecules to pass through, but blocks the movement of solute molecules. (12.6)
sigma (σ) bond A covalent bond formed by orbitals overlapping end-to-end; its electron density is concentrated between the nuclei of the bonding atoms. (10.5)
sigma molecular orbital A molecular orbital in which the electron density is concentrated around a line between the two nuclei of the bonding atoms. (10.6)
significant figures The number of meaningful digits in a measured or calculated quantity. (1.4)
single bond Two atoms are held together by one electron pair. (9.4)
solubility The maximum amount of solute that can be dissolved in a given quantity of solvent (commonly 1 L) at a specific temperature. (4.2, 16.5)
solubility product (K_{sp}) The product of the molar concentrations of the constituent ions, each raised to the power of its stoichiometric coefficient in the equilibrium equation. (16.5)
solute The substance present in smaller amount in a solution. (4.1)
solution A homogeneous mixture of two or more substances. (4.1)
solvation The process in which an ion or a molecule is surrounded by solvent molecules arranged in a specific manner. (12.2)
solvent The substance present in the larger amount in a solution. (4.1)
specific heat (*c*) The amount of heat energy required to raise the temperature of one gram of a substance by one degree Celsius. (6.5)
spectator ions Ions that are not involved in the overall reaction. (4.2)
spectrochemical series A list of ligands arranged in increasing order of their abilities to split the *d* orbital energy levels. (23.4)
standard atmospheric pressure (1 atm) The pressure that supports a column of mercury exactly 76 cm high at 0°C at sea level. (5.2)
standard emf (E°_{cell}) The difference of the standard reduction potential of the substance that undergoes reduction and the standard reduction potential of the substance that undergoes oxidation. (18.3)
standard entropy of reaction The entropy change when the reaction is carried out under standard-state conditions. (17.3)
standard enthalpy of formation The heat change that results when one mole of a compound is formed from its elements in their standard states. (6.6)
standard enthalpy of reaction The enthalpy change when the reaction is carried out under standard-state conditions. (6.6)
standard Gibbs-energy of formation The Gibbs-energy change when 1 mole of a compound is synthesized from its elements in their standard states. (17.4)
standard Gibbs-energy of reaction The Gibbs-energy change when the reaction is carried out under standard-state conditions. (17.4)
standard reduction potential The voltage measured as a reduction reaction occurs at the electrode when all solutes are 1 *M* and all gases are at 1 atm. (18.3)
standard solution A solution of accurately known concentration. (4.7)
standard state The condition of 1 atm of pressure. (6.6)
standard temperature and pressure (STP) 0°C and 1 atm. (5.4)

state function A property that is determined by the state of the system. (6.3)

state of a system The values of all pertinent macroscopic variables (for example, composition, volume, pressure, and temperature) of a system. (6.3)

stereoisomers Compounds that are made up of the same types and numbers of atoms bonded together in the same sequence but with different spatial arrangements. (23.3)

stoichiometric amounts The exact molar amounts of reactants and products that appear in the balanced chemical equation. (3.9)

stoichiometry The quantitative study of reactants and products in a chemical reaction. (3.8)

stratosphere The region of the atmosphere extending upward from the troposphere to about 50 km from Earth. (20.1)

strong acids Strong electrolytes that are assumed to ionize completely in water. (15.4)

strong bases Strong electrolytes that are assumed to ionize completely in water. (15.4)

strong nuclear force The force that holds particles together in the atomic nucleus. (19.2)

structural formula A chemical formula that shows how atoms are bonded to one another in a molecule. (2.6)

structural isomers Molecules that have the same molecular formula but different structures. (24.2)

sublimation The process in which molecules go directly from the solid into the vapor phase. (11.8)

substance A form of matter that has a definite or constant composition (the number and type of basic units present) and distinct properties. (1.7)

substitution reaction A reaction in which an atom or group of atoms replaces an atom or groups of atoms in another molecule. (24.3)

supercooling Cooling of a liquid below its freezing point without forming the solid. (11.8)

supersaturated solution A solution that contains more of the solute than is present in a saturated solution. (12.1)

surface tension The amount of energy required to stretch or increase the surface of a liquid by a unit area. (11.3)

surroundings The rest of the universe outside a system. (6.2)

system Any specific part of the universe that is of interest to us. (6.2)

systematic errors Errors that occur in a predictable manner, leading to measured values that are constantly off from the true value. (1.4)

T

termolecular reaction An elementary step that involves three molecules. (13.5)

ternary compounds Compounds consisting of three elements. (2.7)

theoretical yield The amount of product predicted by the balanced equation when all of the limiting reagent has reacted. (3.10)

theory A unifying principle that explains a body of facts and the laws that are based on them. (1.2)

thermal energy Energy associated with the random motion of atoms and molecules. (6.1)

thermochemical equation An equation that shows both the mass and enthalpy relations. (6.4)

thermochemistry The study of heat changes in chemical reactions. (6.2)

thermodynamics The scientific study of the interconversion of heat and other forms of energy. (6.3)

thermonuclear reactions Nuclear fusion reactions that occur at very high temperatures. (19.6)

thermosphere The region of the atmosphere in which the temperature increases continuously with altitude. (20.1)

third law of thermodynamics The entropy of a perfect crystalline substance is zero at the absolute zero of temperature. (17.3)

titration The gradual addition of a solution of accurately known concentration to another solution of unknown concentration until the chemical reaction between the two solutions is complete. (4.7)

tracers Isotopes, especially radioactive isotopes, that are used to trace the path of the atoms of an element in a chemical or biological process. (19.7)

transition metals Elements that have incompletely filled *d* subshells or readily give rise to cations that have incompletely filled *d* subshells. (7.9)

transition state See activated complex.

transuranium elements Elements with atomic numbers greater than 92. (19.4)

triple bond Two atoms are held together by three pairs of electrons. (9.4)

triple point The point at which the vapor, liquid, and solid states of a substance are in equilibrium. (11.9)

triprotic acid Each unit of the acid yields three protons (H^+) upon ionization. (4.3)

troposphere The layer of the atmosphere which contains about 80 percent of the total mass of air and practically all of the atmosphere's water vapor. (20.1)

U

unimolecular reaction An elementary step in which only one reacting molecule participates. (13.5)

unit cell The basic repeating unit of the arrangement of atoms, molecules, or ions in a crystalline solid. (11.4)

unsaturated hydrocarbons Hydrocarbons that contain carbon-carbon double bonds or carbon-carbon triple bonds. (24.2)

unsaturated solution A solution that contains less solute than it has the capacity to dissolve. (12.1)

V

valence electrons The outer electrons of an atom, which are those involved in chemical bonding. (8.2)

valence shell The outermost electron-occupied shell of an atom, which holds the electrons that are usually involved in bonding. (10.1)

valence-shell electron-pair repulsion (VSEPR) model A model that accounts for the geometrical arrangements of shared and unshared electron pairs around a central atom in terms of the repulsions between electron pairs. (10.1)

van der Waals equation An equation that describes the *P*, *V*, and *T* of a nonideal gas. (5.8)

van der Waals forces The dipole-dipole, dipole-induced dipole, and dispersion forces. (11.2)

van't Hoff factor (*i*) The ratio of actual number of particles in solution after dissociation to the number of formula units initially dissolved in solution. (12.7)

vaporization The escape of molecules from the surface of a liquid; also called evaporation. (11.8)

viscosity A measure of a fluid's resistance to flow. (11.3)

volatile Has a measurable vapor pressure. (12.6)

volume It is the length cubed. (1.9)

W

wave A vibrating disturbance by which energy is transmitted. (7.1)

wavelength (λ) The distance between identical points on successive waves. (7.1)

weak acids Weak electrolytes that ionize only to a limited extent in water. (15.4)

weak bases Weak electrolytes that ionize only to a limited extent in water. (15.4)

weight The force that gravity exerts on an object. (1.3)

work Directed energy change resulting from a process. (6.1)

work function (Φ) The amount of energy binding an electron in a metal. (7.2)

X

X-ray diffraction The scattering of X rays by the units of a regular crystalline solid. (11.5)

Answers to Even-Numbered Problems

Chapter 1

1.4 (a) Hypothesis. (b) Law. (c) Theory. **1.10** 71.2 g. **1.12** (a) 41°C. (b) 11.3°F. (c) 1.1×10^4°F. (d) 233°C. **1.14** (a) −196°C. (b) −269°C. (c) 328°C. **1.18** (a) 0.0152. (b) 0.0000000778. **1.20** (a) 1.80×10^{-2}. (b) 1.14×10^{10}. (c) -5×10^4. (d) 1.3×10^3. **1.22** (a) One. (b) Three. (c) Three. (d) Four. (e) Two or three. (f) One. (g) One or two. **1.24** (a) 1.28. (b) 3.18×10^{-3} mg. (c) 8.14×10^7 dm. (d) 3.8 m/s. **1.26** Tailor X's measurements are the most precise. Tailor Y's measurements are the least accurate and least precise. Tailor Z's measurements are the most accurate. **1.28** (a) 1.10×10^8 mg. (b) 6.83×10^{-5} m^3. (c) 7.2×10^3 L. (d) 6.24×10^{-8} lb. **1.30** 3.1557×10^7 s. **1.32** (a) 118 in/s. (b) 1.80×10^2 m/min. (c) 10.8 km/h. **1.34** 178 mph. **1.36** 3.7×10^{-3} g Pb. **1.38** (a) 1.5×10^2 lb. (b) 4.4×10^{17} s. (c) 2.3 m. (d) 8.86×10^4 L. **1.40** 6.25×10^{-4} g/cm^3. **1.42** 27 y. **1.44** Float: 8 lb, 9 lb, 10 lb, 11 lb, 12 lb. **1.50** (a) Cs. (b) Ge. (c) Ga. (d) Sr. (e) U. (f) Se. (g) Ne. (h) Cd. **1.52** (a) Homogeneous mixture. (b) Element. (c) Compound. (d) Homogeneous mixture. (e) Heterogeneous mixture. (f) Heterogeneous mixture. (g) Element. **1.58** (a) Physical change. (b) Chemical change. (c) Physical change. (d) Chemical change. (e) Physical change. **1.60** (a) Chemical. (b) Chemical. (c) Physical. (d) Physical. (e) Chemical. **1.62** 2.6 g/cm^3 **1.64** 9.20 cm. **1.66** 767 mph. **1.68** Liquid must be less dense than ice; temperature below 0°C. **1.70** 2.3×10^3 cm^3. **1.72** 6.4¢. **1.74** 73°S. **1.76** (a) 8.6×10^3 L air/day. (b) 0.018 L CO/day. **1.78** 26,700,000 basketballs. **1.80** 7.0×10^{20} L. **1.82** 88 lb; 40 kg. **1.84** O: 4.0×10^4 g; C: 1.1×10^4 g; H: 6.2×10^3 g; N: 2×10^3 g; Ca: 9.9×10^2 g; P: 7.4×10^2 g. **1.86** 4.6×10^2°C; 8.6×10^2°F. **1.88** \$$2.4 \times 10^{12}$. **1.90** 5.4×10^{22} Fe atoms. **1.92** 29 times. **1.94** 1.450×10^{-2} mm. **1.96** 1.3×10^3 mL. **1.98** (a) 11.063 mL. (b) 0.78900 g/mL. (c) 7.140 g/mL. **1.100** 0.88 s. **1.102** (a) 327 L CO. (b) 5.0×10^{-8} g/L. (c) 1.20×10^3 μg/mL. **1.104** 0.85^3 cm. **1.106** 4.97×10^4 g. **1.108** 2.413 g/mL. **1.110** The glass bottle would crack.

Chapter 2

2.8 0.12 mi. **2.14** 145. **2.16** N(7,8,7); S(16,17,16); Cu(29,34,29); Sr(38,46,38); Ba(56,74,56); W(74,112,74); Hg(80,122,80). **2.18** (a) $^{186}_{74}W$. (b) $^{201}_{80}Hg$. **2.24** (a) Metallic character increases down a group. (b) Metallic character decreases from left to right. **2.26** F and Cl; Na and K; P and N. **2.32** (a) Diatomic molecule and compound. (b) Polyatomic molecule and compound. (c) Polyatomic molecule and element. **2.34** (a) H_2 and F_2. (b) HCl and CO. (c) S_8 and P_4. (d) H_2O and $C_{12}H_{22}O_{11}$ (sucrose). **2.36** (protons, electrons): K^+(19,18); Mg^{2+}(12,10); Fe^{3+}(26,23); Br^-(35,36); Mn^{2+}(25,23); C^{4-}(6,10); Cu^{2+}(29,27). **2.38** (a) $^{52}_{25}Mn$. (b) $^{22}_{10}Ne$. (c) $^{107}_{47}Ag$. (d) $^{127}_{53}I$. (e) $^{239}_{94}Pu$. **2.46** (a) CuBr. (b) Mn_2O_3. (c) Hg_2I_2. (d) $Mg_3(PO_4)_2$. **2.48** (a) $AlBr_3$. (b) $NaSO_2$. (c) N_2O_5. (d) $K_2Cr_2O_7$. **2.50** C_2H_6O. **2.52** Ionic: NaBr, BaF_2, CsCl. Molecular: CH_4, CCl_4, ICl, NF_3. **2.60** (a) Potassium hypochlorite. (b) Silver carbonate. (c) Iron(II) chloride. (d) Potassium permanganate. (e) Cesium chlorate. (f) Hypoiodous acid. (g) Iron(II) oxide. (h) Iron(III) oxide. (i) Titanium(IV) chloride. (j) Sodium hydride. (k) Lithium nitride. (l) Sodium oxide. (m) Sodium peroxide. (n) Iron(III) chloride hexahydrate. **2.62** (a) CuCN. (b) $Sr(ClO_2)_2$. (c) $HBrO_4$. (d) HI(*aq*). (e) $Na_2(NH_4)PO_4$. (f) $PbCO_3$. (g) SnF_2. (h) P_4S_{10}. (i) HgO. (j) Hg_2I_2. (k) SeF_6. **2.64** (a) Dinitrogen pentoxide (N_2O_5). (b) Boron trifluoride (BF_3). (c) Dialuminum hexabromide (Al_2Br_6). **2.66** (c) Changing the electrical charge of an atom usually has a major effect on its chemical properties. **2.68** I^-. **2.70** NaCl is an ionic compound. It does not form molecules. **2.72** Element: (b), (c), (e), (f), (g), (j), (k). Molecules but not compounds: (b), (f), (g), (k). Compounds but not molecules: (i), (l). Compounds and molecules: (a), (d), (h). **2.74** (a) Ne: 10 p, 10 n. (b) Cu: 29 p, 34 n. (c) Ag: 47 p, 60 n. (d) W: 74 p, 108 n. (e) Po: 84 p, 119 n. (f) Pu: 94 p, 140 n. **2.76** (a) Cu. (b) P. (c) Kr. (d) Cs. (e) Al. (f) Sb. (g) Cl. (h) Sr. **2.78** (a) The magnitude of a particle scattering depends on the number of protons present. (b) Density of nucleus: 3.25×10^{14} g/cm^3; density of space occupied by electrons: 3.72×10^{-4} g/cm^3. The result supports Rutherford's model. **2.80** The empirical and molecular formulas of acetaminophen are both $C_8H_9NO_2$. **2.82** (a) Tin(IV) chloride. (b) Copper(I) oxide. (c) Cobalt(II) nitrate. (d) Sodium dichromate. **2.84** (a) Ionic compounds formed between metallic and nonmetallic elements. (b) Transition metals, lanthanides, and actinides. **2.86** ^{23}Na. **2.88** Hg and Br_2. **2.90** H_2, N_2, O_2, F_2, Cl_2, He, Ne, Ar, Kr, Xe, Rn. **2.92** Unreactive. He, Ne, and Ar are chemically inert. **2.94** Ra is a radioactive decay product of U-238. **2.96** $^{77}Se^{2-}$. **2.98** (a) NaH, sodium hydride. (b) B_2O_3, diboron trioxide. (c) Na_2S, sodium sulfide. (d) AlF_3, aluminum fluoride. (e) OF_2, oxygen difluoride. (f) $SrCl_2$, strontium chloride. **2.100** NF_3 (nitrogen trifluoride), PBr_5 (phosphorus pentabromide), SCl_2 (sulfur dichloride). **2.102** 1st row: Mg^{2+}, HCO_3^-, $Mg(HCO_3)_2$. 2nd row: Sr^{2+}, Cl^-, strontium chloride. 3rd row: $Fe(NO_2)_3$, iron(III) nitrite. 4th row: Mn^{2+}, ClO_3^-, $Mn(ClO_3)_2$. 5th row: Sn^{4+}, Br^-, tin(IV) bromide. 6th row: $Co_3(PO_4)_2$, cobalt(II) phosphate. 7th row: Hg_2I_2, mercury(I) iodide. 8th row: Cu^+, CO_3^{2-}, copper(I) carbonate. 9th row: Li^+, N^{3-}, Li_3N. 10th row: Al_2S_3, aluminum sulfide. **2.104** 1.91×10^{-8} g. Mass is too small to be detected. **2.106** (a) Volume of a sphere is given by $V = (4/3)\pi r^3$. Volume is also proportional to the number of neutrons and protons present, or the mass number A. Therefore, $r^3 \propto A$ or $r \propto A^{1/3}$. (b) 5.1×10^{-44} m^3. (c) The nucleus occupies only 3.5×10^{-13}% of the atom's volume. The result supports Rutherford's model. **2.108** (a) Yes. (b) Ethane: CH_3 and C_2H_6. Acetylene: CH and C_2H_2. **2.110** Manganese (Mn). **2.112** From left to right: chloric acid, nitrous acid, hydrocyanic acid, and sulfuric acid. **2.114** XY_2. X is likely in Group 4 or Group 14 and Y is likely in Group 16. Examples: titanium(IV) oxide (TiO_2), tin(IV) oxide (SnO_2), and lead(IV) oxide (PbO_2).

Chapter 3

3.6 7.5% and 92.5%. **3.8** 5.1×10^{24} amu. **3.12** 5.8×10^3 light-yr. **3.14** 9.96×10^{-15} mol Co. **3.16** 3.01×10^3 g Au. **3.18** (a) 1.244×10^{-22} g/As atom. (b) 9.746×10^{-23} g/Ni atom. **3.20** 6.0×10^{20} Cu atoms. **3.22** Pb. **3.24** (a) 73.89 g. (b) 76.15 g. (c) 119.37 g. (d) 176.12 g. (e) 101.11 g. (f) 100.95 g. **3.26** 6.69×10^{21} C_2H_6 molecules. **3.28** C: 1.10×10^{26} atoms; S: 5.50×10^{25} atoms; H: 3.30×10^{26} atoms; O: 5.50×10^{25} atoms. **3.30** 8.56×10^{22} molecules. **3.34** 7. **3.38** C: 10.06%; H: 0.8442%; Cl: 89.07%. **3.40** (d) Ammonia, NH_3. **3.42** 5.97 g F. **3.44** (a) CH_2O. (b) KCN. **3.48** C_6H_6. **3.50** $C_2H_3NO_5$. **3.52** $C_5H_8O_4NNa$. **3.58** (a) $2N_2O_5 \longrightarrow 2N_2O_4 + O_2$. (b) $2KNO_3 \longrightarrow 2KNO_2 + O_2$. (c) $NH_4NO_3 \longrightarrow N_2O + 2H_2O$. (d) $NH_4NO_2 \longrightarrow N_2 + 2H_2O$. (e) $2NaHCO_3 \longrightarrow Na_2CO_3 + H_2O + CO_2$. (f) $P_4O_{10} + 6H_2O \longrightarrow 4H_3PO_4$. (g) $2HCl + CaCO_3 \longrightarrow CaCl_2 + H_2O + CO_2$. (h) $2Al + 3H_2SO_4 \longrightarrow Al_2(SO_4)_3 + 3H_2$. (i) $CO_2 + 2KOH \longrightarrow K_2CO_3 + H_2O$. (j) $CH_4 + 2O_2 \longrightarrow CO_2 + 2H_2O$. (k) $Be_2C + 4H_2O \longrightarrow 2Be(OH)_2 + CH_4$. (l) $3Cu + 8HNO_3 \longrightarrow 3Cu(NO_3)_2 + 2NO + 4H_2O$. (m) $S + 6HNO_3 \longrightarrow H_2SO_4 + 6NO_2 + 2H_2O$. (n) $2NH_3 + 3CuO \longrightarrow 3Cu + N_2 + 3H_2O$. **3.62** (d). **3.64** 1.01 mol. **3.66** 20 mol. **3.68** 39.3 g S. **3.70** (a) $2NaHCO_3 \longrightarrow Na_2CO_3 + CO_2 + H_2O$. (b) 78.3 g. **3.72** 255.9 g; 0.324 L. **3.74** 0.294 mol. **3.76** (a) $NH_4NO_3 \longrightarrow N_2O + 2H_2O$. (b) 20 g.

3.78 18.0 g. **3.82** 1 mole H_2 left and 6 moles NH_3 produced. **3.84** (a) $2NH_3 + H_2SO_4 \longrightarrow (NH_4)_2SO_4$. (b) 5.23 g NH_3; 21.0 g H_2SO_4. **3.86** HCl; 23.4 g. **3.90** (a) 7.05 g. (b) 92.9%. **3.92** 3.48×10^3 g. **3.94** 8.55 g; 76.6%. **3.96** 100%. **3.98** ^{85}Rb: 72.1%; ^{87}Rb: 27.9%. **3.100** (b). **3.102** (a) $C_5H_{12} + 8O_2 \longrightarrow 5CO_2 + 6H_2O$. (b) $NaHCO_3 + HCl \longrightarrow CO_2 + NaCl + H_2O$. (c) $6Li + N_2 \longrightarrow 2Li_3N$. (d) $PCl_3 + 3H_2O \longrightarrow H_3PO_3 + 3HCl$. (e) $3CuO + 2NH_3 \longrightarrow 3Cu + N_2 + 3H_2O$. **3.104** Cl_2O_7. **3.106** 18.7 g. **3.108** (a) 0.212 mol. (b) 0.424 mol. **3.110** 18. **3.112** 2.4×10^{23} atoms. **3.114** 65.4 amu; Zn. **3.116** 89.5%. **3.118** CH_2O; $C_6H_{12}O_6$. **3.120** 51.9 g/mol; Cr. **3.122** 29.54%. **3.124** 1.6×10^4 g/mol. **3.126** NaBr: 24.03%; Na_2SO_4: 75.97%. **3.128** $C_3H_8 + 5O_2 \longrightarrow 3CO_2 + 4H_2O$. **3.130** Ca: 38.76%; P: 19.97%; O: 41.27%. **3.132** Yes. **3.134** 2.01×10^{21} molecules. **3.136** 16.00 amu. **3.138** (e). **3.140** $PtCl_2$; $PtCl_4$. **3.142** (a) 12 g; 28 mL. (b) 15 g. **3.144** (a) X: MnO_2; Y: Mn_3O_4. (b) $3MnO_2 \longrightarrow Mn_3O_4 + O_2$. **3.146** 6.1×10^5 tons. **3.148** $C_3H_2ClF_5O$. C: 19.53%; H: 1.093%; Cl: 19.21%; F: 51.49%; O: 8.672%. **3.150** Mg_3N_2 (magnesium nitride). **3.152** PbC_8H_{20}. **3.154** (a) 4.3×10^{22} atoms. (b) 1.6×10^2 pm. **3.156** 28.97 g/mol. **3.158** (a) $Fe_2O_3 + 6HCl \longrightarrow 2FeCl_3 + 3H_2O$. (b) 396 g $FeCl_3$. **3.160** (a) $C_3H_8 + 3H_2O \longrightarrow 3CO + 7H_2$. (b) 9.09×10^2 kg. **3.162** (a) There is only one reactant so the use of "limiting reagent" is unnecessary. (b) The term "limiting reagent" usually applies only to one reactant. **3.164** (a) \$0.47/kg. (b) 0.631 kg K_2O. **3.166** $BaBr_2$. **3.168** NaCl: 32.17%; Na_2SO_4: 20.09%; $NaNO_3$: 47.75%.

Chapter 4

4.8 (c). **4.10** (a) Strong electrolyte. (b) Nonelectrolyte. (c) Weak electrolyte. (d) Strong electrolyte. **4.12** (b) and (c). **4.14** HCl does not ionize in benzene. **4.18** (b). **4.20** (a) Insoluble. (b) Soluble. (c) Soluble. (d) Insoluble. (e) Soluble. **4.22** (a) Ionic: $2Na^+ + S^{2-} + Zn^{2+} + 2Cl^- \longrightarrow ZnS + 2Na^+ + 2Cl^-$. Net ionic: $Zn^{2+} + S^{2-} \longrightarrow ZnS$. (b) Ionic: $6K^+ + 2PO_4^{3-} + 3Sr^{2+} + 6NO_3^- \longrightarrow Sr_3(PO_4)_2 + 6KNO_3$. Net ionic: $3Sr^{2+} + 2PO_4^{3-} \longrightarrow Sr_3(PO_4)_2$. (c) Ionic: $Mg^{2+} + 2NO_3^- + 2Na^+ + 2OH^- \longrightarrow Mg(OH)_2 + 2Na^+ + 2NO_3^-$. Net ionic: $Mg^{2+} + 2OH^- \longrightarrow Mg(OH)_2$. **4.24** (a) Add chloride ions. (b) Add hydroxide ions. (c) Add carbonate ions. (d) Add sulfate ions. **4.32** (a) Brønsted base. (b) Brønsted base. (c) Brønsted acid. (d) Brønsted base and Brønsted acid. **4.34** (a) Ionic: $CH_3COOH + K^+ + OH^- \longrightarrow K^+ + CH_3COO^- + H_2O$; Net Ionic: $CH_3COOH + OH^- \longrightarrow CH_3COO^- + H_2O$. (b) Ionic: $H_2CO_3 + 2Na^+ + 2OH^- \longrightarrow 2Na^+ + CO_3^{2-} + 2H_2O$; Net Ionic: $H_2CO_3 + 2OH^- \longrightarrow CO_3^{2-} + 2H_2O$. (c) Ionic: $2H^+ + 2NO_3^- + Ba^{2+} + 2OH^- \longrightarrow Ba^{2+} + 2NO_3^- + 2H_2O$; Net Ionic: $H^+ + OH^- \longrightarrow H_2O$. **4.44** (a) $Fe \longrightarrow Fe^{3+} + 3e^-$; $O_2 + 4e^- \longrightarrow 2O^{2-}$. Oxidizing agent: O_2; reducing agent: Fe. (b) $2Br^- \longrightarrow Br_2 + 2e^-$; $Cl_2 + 2e^- \longrightarrow 2Cl^-$. Oxidizing agent: Cl_2; reducing agent: Br^-. (c) $Si \longrightarrow Si^{4+} + 4e^-$; $F_2 + 2e^- \longrightarrow 2F^-$. Oxidizing agent: F_2; reducing agent: Si. (d) $H_2 \longrightarrow 2H^+ + 2e^-$; $Cl_2 + 2e^- \longrightarrow 2Cl^-$. Oxidizing agent: Cl_2; reducing agent: H_2. **4.46** (a) +5. (b) +1. (c) +3. (d) +5. (e) +5. (f) +5. **4.48** All are zero. **4.50** (a) −3. (b) −1/2. (c) −1. (d) +4. (e) +3. (f) −2. (g) +3. (h) +6. **4.52** (b) and (d). **4.54** (a) No reaction. (b) No reaction. (c) $Mg + CuSO_4 \longrightarrow MgSO_4 + Cu$. (d) $Cl_2 + 2KBr \longrightarrow Br_2 + 2KCl$. **4.56** (a) Combination. (b) Decomposition. (c) Displacement. (d) Disproportionation. **4.58** O_2^+. **4.64** Dissolve 15.0 g $NaNO^3$ in enough water to make up 250 mL. **4.66** 10.8 g. **4.68** (a) 1.37 *M*. (b) 0.426 *M*. (c) 0.716 *M*. **4.70** (a) 6.50 g CsI. (b) 2.45 g H_2SO_4. (c) 2.65 g Na_2CO_3. (d) 7.36 g $K_2Cr_2O_7$. (e) 3.95 g $KMnO_4$. **4.74** 0.0433 *M*. **4.76** 126 mL. **4.78** 1.09 *M*. **4.82** 35.73%. **4.84** 0.00975 *M*. **4.92** 0.217 *M*. **4.94** (a) 6.00 mL. (b) 8.00 mL. **4.96** 9.44×10^{-3} g. **4.98** 0.06020 *M*. **4.100** 6.15 mL. **4.102** 0.232 mg. **4.104** (i) Only oxygen supports combustion. (ii) Only CO_2 reacts with $Ca(OH)_2(aq)$ to form $CaCO_3$ (white precipitate). **4.106** 1.26 *M*. **4.108** (a) 15.6 g $Al(OH)_3$. (b) $[Al^{3+}] = 0.250\ M$, $[NO_3^-] = 2.25\ M$, $[K^+]1.50\ M$. **4.110** 0.171 *M*. **4.112** 0.115 *M*. **4.114** Ag: 1.25 g; Zn: 2.12 g. **4.116** 0.0721 *M* NaOH. **4.118** 24.0 g/mol; Mg. **4.120** 2. **4.122** 1.72 *M*. **4.124** Only Fe(II) is oxidized by $KMnO_4$ solution and can therefore change the purple color to colorless. **4.126** Ions are removed as the $BaSO_4$ precipitates. **4.128** $FeCl_2 \cdot 4H_2O$. **4.130** (i) Conductivity test. (ii) Only NaCl reacts with $AgNO_3$ to form AgCl precipitate. **4.132** The Cl^- ion cannot accept any electrons. **4.134** Reaction is too violent. **4.136** Use sodium bicarbonate: $HCO_3^- + H^+ \longrightarrow H_2O + CO_2$. NaOH is a caustic substance and unsafe to use in this manner. **4.138** (a) Conductivity. Reaction with $AgNO_3$ to form AgCl. (b) Soluble in water. Nonelectrolyte. (c) Possesses properties of acids. (d) Soluble. Reacts with acids to give CO_2. (e) Soluble, strong electrolyte. Reacts with acids to give CO_2. (f) Weak electrolyte and weak acid. (g) Soluble in water. Reacts with NaOH to produce $Mg(OH)_2$ precipitate. (h) Strong electrolyte and strong base. (i) Characteristic odor. Weak electrolyte and weak base. (j) Insoluble. Reacts with acids. (k) Insoluble. Reacts with acids to produce CO_2. **4.140** NaCl: 44.11%; KCl: 55.89%. **4.142** (a) $AgOH(s) + HNO_3(aq) \longrightarrow AgNO_3(aq) + H_2O(l)$. **4.144** 1.33 g. **4.146** 56.18%. **4.148** (a) 1.40 *M*. (b) 4.96 g. **4.150** (a) $NH_4^+ + OH^- \longrightarrow NH_3 + H_2O$. (b) 97.99%. **4.152** Zero. **4.154** 0.224%. Yes. **4.156** (a) $Zn + H_2SO_4 \longrightarrow ZnSO_4 + H_2$. (b) $2KClO_3 \longrightarrow 2KCl + 3O_2$. (c) $Na_2CO_3 + 2HCl \longrightarrow 2NaCl + CO_2 + H_2O$. (d) $NH_4NO_2 \longrightarrow N_2 + 2H_2O$. **4.158** Yes. **4.160** (a) $8.316 \times 10^{-7}\ M$. (b) 3.286×10^{-5} g. **4.162** $[Fe^{2+}] = 0.0920\ M$, $[Fe^{3+}] = 0.0680\ M$. **4.164** (a) Precipitation: $Mg^{2+} + 2OH^- \longrightarrow Mg(OH)_2$; acid-base: $Mg(OH)_2 + 2HCl \longrightarrow MgCl_2 + 2H_2O$; redox: $MgCl_2 \longrightarrow Mg + Cl_2$. (b) NaOH is more expensive than CaO. (c) Dolomite provides additional Mg. **4.166** D < A < C < B. D = Au, A = Cu, C = Zn, B = Mg. **4.168** (a) $Cu^{2+} + SO_4^{2-} + Ba^{2+} + 2OH^- \longrightarrow Cu(OH)_2 + BaSO_4$. (b) 14.6 g $Cu(OH)_2$, 35.5 g $BaSO_4$. $[Cu^{2+}] = [SO_4^{2-}] = 0.0417\ M$.

Chapter 5

5.14 0.797 atm; 80.8 kPa. **5.18** (1) b. (2) a. (3) c. (4) a. **5.20** 53 atm. **5.22** (a) 0.69 L. (b) 61 atm. **5.24** 1.3×10^2 K. **5.26** ClF_3. **5.32** 6.2 atm. **5.34** 745 K. **5.36** 1.9 atm. **5.38** 0.82 L. **5.40** 45.1 L. **5.42** 6.1×10^{-3} atm. **5.44** 35.1 g/mol. **5.46** N_2: 2.1×10^{22}; O_2: 5.7×10^{21}; Ar: 3×10^{20}. **5.48** 2.98 g/L. **5.50** SF_4. **5.52** F_2: 59.7%; Cl_2: 40.3%. **5.54** 370 L. **5.56** 88.9%. **5.58** $M + 3HCl \longrightarrow (3/2)H_2 + MCl_3$; M_2O_3, $M_2(SO_4)_3$. **5.60** 2.84×10^{-2} mol CO_2; 94.7%. The impurities must not react with HCl to produce CO_2. **5.62** 1.71×10^3 L. **5.64** 86.0%. **5.68** (a) 0.89 atm. (b) 1.4 L. **5.70** 349 mmHg. **5.72** 19.8 g. **5.74** H_2: 650 mmHg; N_2: 217 mmHg. **5.76** (a) Box on right. (b) Box on left. **5.82** N_2: 472 m/s; O_2: 441 m/s; O_3: 360 m/s. **5.84** 2.8 m/s; 2.7 m/s. Squaring favors the larger values. **5.86** 1.0043. **5.88** 4. **5.94** No. **5.96** Ne. **5.98** C_6H_6. **5.100** 445 mL. **5.102** (a) 9.53 atm. (b) $Ni(CO)_4$ decomposes to give CO, which increases the pressure. **5.104** 1.30×10^{22} molecules; CO_2, O_2, N_2, H_2O. **5.106** 5.25×10^{18} kg. **5.108** 0.0701 *M*. **5.110** He: 0.16 atm; Ne: 2.0 atm. **5.112** HCl dissolves in the water, creating a partial vacuum. **5.114** 7. **5.116** (a) 61.2 m/s. (b) 4.58×10^{-4} s. (c) 328 m/s; 366 m/s. The velocity 328 m/s is that of a particular atom and u_{rms} is an average value. **5.118** 2.09×10^4 m; 1.58×10^4 L. **5.120** Higher partial pressure of C_2H_4 inside the paper bag. **5.122** To equalize the pressure as the amount of ink decreases. **5.124** (a) $NH_4NO_3 \longrightarrow N_2O + 2H_2O$. (b) 0.0821 L · atm/K · mol. **5.126** C_6H_6. **5.128** The low atmospheric pressure caused the harmful gases (CO, CO_2, CH_4) to flow out of the mine, and the person suffocated. **5.130** Br_2 (159.8 g/mol; red); SO_3 (80.07 g/mol; yellow); N_2 (28.02 g/mol; green); CH_4 (16.04 g/mol; blue). **5.132** (a) 5×10^{-22} atm. (b) 5×10^{20} L/g H. **5.134** 91%. **5.136** 1.7×10^{12} molecules. **5.138** 4.66 L. **5.140** 3.8×10^{-3} m/s; 1.0×10^{-30} J. **5.142** 2.3×10^3 L. **5.144** 1.8×10^2 mL. **5.146** (a) 1.09×10^{44} molecules. (b) 1.18×10^{22} molecules/breath. (c) 2.60×10^{30} molecules. (d) 2.39×10^{-14}; 3×10^8 molecules. (e) Complete mixing of air; no molecules escaped to the outer atmosphere; no molecules used up during metabolism, nitrogen fixation, etc. **5.148** 3.7 nm; 0.31 nm. **5.150** 0.54 atm.

5.152 H_2: 0.5857; D_2: 0.4143. **5.154** 53.4%. **5.156** CO: 54.4%; CO_2: 45.6%. **5.158** CH_4: 0.789; C_2H_6: 0.211. **5.160** (a) $8(4\pi r^3/3)$. (b) $(16/3)N_A\pi r^3$. The excluded volume is 4 times the volumes of the atoms. **5.162** CH_4. **5.164** NO. **5.166** (b). **5.168** (i) (b) 8.0 atm. (c) 5.3 atm. (ii) P_T = 5.3 atm. P_A = 2.65 atm. P_B = 2.65 atm. **5.170** CH_4: 2.3 atm. C_2H_6: 0.84 atm. C_3H_8: 1.4 atm.

Chapter 6

6.16 (a) 0. (b) −9.5 J. (c) −18 J. **6.18** 48 J. **6.20** -3.1×10^3 J. **6.26** 1.59×10^4 kJ. **6.28** −553.8 kJ/mol. **6.32** 0.237 J/g · °C. **6.34** 3.31 kJ. **6.36** 98.6 g. **6.38** 22.39°C. **6.46** O_2. **6.48** (a) $\Delta H^\circ_f[Br_2, (l)] = 0$; $\Delta H^\circ_f[Br_2, (g)] > 0$. (b) $\Delta H^\circ_f[I_2, (s)] = 0$; $\Delta H^\circ_f[I_2, (g)] > 0$. **6.50** Measure ΔH° for the formation of Ag_2O from Ag and O_2 and of $CaCl_2$ from Ca and Cl_2. **6.52** (a) −167.2 kJ/mol. (b) −55.8 kJ/mol. **6.54** (a) −1411 kJ/mol. (b) −1124 kJ/mol. **6.56** 218 kJ/mol. **6.58** 71.58 kJ/g. **6.60** 2.70×10^2 kJ. **6.62** −84.0 kJ/mol. **6.64** −851.3 kJ/mol. **6.72** 11 kJ. **6.74** -2.90×10^2 kJ/mol. **6.76** (a) −337.0 kJ/mol. (b) NH_3. NH_3 release more energy per kilogram of substance, it would be a better fuel. **6.78** 26.5 kJ/mol. **6.80** 43.6 kJ. **6.82** 0. **6.84** −351 kJ/mol. **6.86** −558.2 kJ/mol. **6.88** 0.492 J/g · °C. **6.90** The first (exothermic) reaction can be used to promote the second (endothermic) reaction. **6.92** 1.09×10^4 L. **6.94** 4.10 L. **6.96** 5.60 kJ/mol. **6.98** (a). **6.100** (a) 0. (b) −9.1 J. (c) 2.4 L; −48 J. **6.102** (a) A more fully packed freezer has a greater mass and hence a larger heat capacity. (b) Tea or coffee has a greater amount of water, which has a higher specific heat than noodles. **6.104** 1.84×10^3 kJ. **6.106** 3.0×10^9. **6.108** 5.35 kJ/°C. **6.110** -5.2×10^6 kJ. **6.112** (a) 3.4×10^5 g. (b) -2.0×10^8 J. **6.114** −86.7 kJ/mol. **6.116** (a) 1.4×10^2 kJ. (b) 3.9×10^2 kJ. **6.118** (a) −52.3 kJ/mol. (b) +3.5 kJ/mol. **6.120** −110.5 kJ/mol. It will form both CO and CO_2. **6.122** (a) 0.50 J. (b) 32 m/s. (c) 0.12°C. **6.124** −277.0 kJ/mol. **6.126** 104 g. **6.128** 296 kJ. **6.130** 9.9×10^8 J; 304°C. **6.132** (a) $CaC_2 + 2H_2O \longrightarrow C_2H_2 + Ca(OH)_2$. (b) 1.51×10^3 kJ. **6.134** ΔU = −5153 kJ/mol; ΔH = −5158 kJ/mol. **6.136** −564.2 kJ/mol. **6.138** 96.21%. **6.140** (a) CH. (b) 49 kJ/mol. **6.142** (a) Heating water at room temperature to its boiling point. (b) Heating water at its boiling point. (c) A chemical reaction taking place in a bomb calorimeter (an isolated system) where there is no heat exchange with the surroundings. **6.144** −101.3 J. Yes, because in a cyclic process, the change in a state function must be zero. **6.146** (a) Exothermic. (b) No clear conclusion. It is a balance between the energy needed to break the ionic bond and the energy released during hydration. (c) No clear conclusion. It is a balance between the energy needed to break the A—B bond and the energy released when the A—C bond is formed. (d) Endothermic.

Chapter 7

7.8 (a) 6.58×10^{14}/s. (b) 1.22×10^8 nm. **7.10** 2.5 min. **7.12** 4.95×10^{14}/s. **7.16** (a) 4.0×10^2 nm. (b) 5.0×10^{-19} J. **7.18** 1.2×10^2 nm (UV). **7.20** (a) 3.70×10^2 nm. (b) UV. (c) 5.38×10^{-19} J. **7.22** 8.16×10^{-19} J. **7.26** Use a prism. **7.28** Compare the emission spectra with those on Earth of known elements. **7.30** 3.027×10^{-19} J. **7.32** 6.17×10^{14}/s. 486 nm. **7.34** 5. **7.40** 1.37×10^{-6} nm. **7.42** 1.7×10^{-23} nm. **7.52** $\ell = 0$: $m_\ell = 0$. $\ell = 1$: $m_\ell = -1, 0, 1$. $\ell = 2$: $m_\ell = -2, -1, 0, 1, 2$. **7.54** (a) $n = 0$; $\ell = 0$; $m_\ell = 0$; $m_s = +1/2$ or $-1/2$. (b) $n = 4$; $\ell = 1$; $m_\ell = -1, 0, 1$; $m_s = +1/2$ or $-1/2$. (c) $n = 3$; $\ell = 2$; $m_\ell = -2, -1, 0, 1, 2$; $m_s = +1/2$ or $-1/2$. **7.56** For $n = 6$, the allowed values of l are 0, 1, 2, 3, 4, and 5[$l = 0$ to $(n - 1)$, integer values]. These l values correspond to the 6*s*, 6*p*, 6*d*, 6*f*, 6*g*, and 6*h* subshells. These subshells each have 1, 3, 5, 7, 9, and 11 orbitals, respectively (number of orbitals = $2l + 1$). **7.62** The three *p* orbitals are identical in size, shape, and energy; they differ from one another only in orientation. **7.64** $2n^2$. **7.66** (a) 3. (b) 6. (c) 0. **7.68** There is no shielding in an H atom. **7.70** (a) $2s < 2p$. (b) $3p < 3d$. (c) $3s < 4s$. (d) $4d < 5f$. **7.76** Al: $1s^22s^22p^63s^23p^1$. B: $1s^22s^22p^1$. F: $1s^22s^22p^5$. **7.78** B(1), Ne(0), P(3), Sc(1), Mn(5), Se(2), Kr(0), Fe(4), Cd(0), I(1), Pb(2). **7.88** $[Kr]5s^24d^5$. **7.90** Ge: $[Ar]4s^23d^{10}4p^2$. Fe: $[Ar]4s^23d^6$. Zn: $[Ar]4s^23d^{10}$. Ni: $[Ar]4s^23d^8$. W: $[Xe]6s^24f^{14}5d^4$. Tl: $[Xe]6s^24f^{14}5d^{10}6p^1$. **7.92** S^+. **7.94** 6.68×10^{16} photons. **7.96** (a) Incorrect. (b) Correct. (c) Incorrect. **7.98** (a) 4*e*: An *e* in a 2*s* and an *e* in each 2*p* orbital. (b) 6*e*: 2*e* each in a 4*p*, a 4*d*, and a 4*f* orbital. (c) 10*e*: 2*e* in each of the five 3*d* orbitals. (d) 1*e*: An *e* in a 2*s* orbital. (e) 2*e*: 2*e* in a 4*f* orbital. **7.100** Wave properties. **7.102** (a) 1.05×10^{-25} nm. (b) 8.86 nm. **7.104** (a) $n = 2$. The possible ℓ values are from 0 to $(n - 1)$ integer values. (b) Possible ℓ values are 0, 1, 2, or 3. Possible m_ℓ values range from $-\ell$ to $+\ell$ integer values. **7.106** (a) 1.20×10^{18} photons. (b) 3.76×10^8 W. **7.108** 419 nm. Yes. **7.110** Ne. **7.112** He^+: 164 nm, 121 nm, 109 nm, 103 nm (all in the UV region). H: 657 nm, 487 nm, 434 nm, 411 nm (all in the visible region). **7.114** 1.2×10^2 photons. **7.116** 2.5×10^{17} photons. **7.118** Yellow light will generate more electrons; blue light will generate electrons with greater kinetic energy. **7.120** (a) He. (b) N. (c) Na. (d) As. (e) Cl. See Table 7.3 for ground-state electron configurations. **7.122** They might have discovered the wave properties of electrons. **7.124** 7.39×10^{-2} nm. **7.126** (a) False. (b) False. (c) True. (d) False. (e) True. **7.128** 2.0×10^{-5} m/s. **7.130** (a) and (f) violate the Pauli exclusion principle; (b), (d), and (e) violate Hund's rule. **7.132** 2.8×10^6 K. **7.134** 2.76×10^{-11} m. **7.136** 17.4 pm. **7.138** 0.929 pm; 3.23×10^{20}/s. **7.140** $n_i = 5$ to $n_f = 3$. **7.142** (a) B: $4 \longrightarrow 2$; C: $5 \longrightarrow 2$. (b) A: 41.1 nm; B: 30.4 nm. (c) 2.18×10^{-18} J. (d) At high values of *n*, the energy levels are very closely spaced, leading to a continuum of lines. **7.144** $n = 1$: 1.96×10^{-17} J; $n = 5$: 7.85×10^{-19} J. 10.6 nm. **7.146** 9.5×10^3 m/s. **7.148** 3.87×10^5 m/s. **7.150** Photosynthesis and vision. **7.152** 1.06 nm. **7.154** (a) 1.12 pm. (b) Smaller than the molecule.

Chapter 8

8.20 (a) $1s^22s^22p^63s^23p^5$. (b) Representative. (c) Paramagnetic. **8.22** (a) and (d); (b) and (e); (c) and (f). **8.24** (a) Group 1. (b) Group 15. (c) Group 18. (d) Group 10. **8.26** Fe. **8.28** (a) [Ne]. (b) [Ne]. (c) [Ar]. (d) [Ar]. (e) [Ar]. (f) $[Ar]3d^6$. (g) $[Ar]3d^9$. (h) $[Ar]3d^{10}$. **8.30** (a) Cr^{3+}. (b) Sc^{3+}. (c) Rh^{3+}. (d) Ir^{3+}. **8.32** Be^{2+} and He; F^- and N^{3-}; Fe^{2+} and Co^{3+}; S^{2-} and Ar. **8.38** Na > Mg > Al > P > Cl. **8.40** F. **8.42** The effective nuclear charge that the outermost electrons feel increases across the period. **8.44** $Mg^{2+} < Na^+ < F^- < O^{2-} < N^{3-}$. **8.46** Te^{2-}. **8.48** H^- is larger. **8.52** K < Ca < P < F < Ne. **8.54** The single 3*p* electron in Al is well shielded by the 1*s*, 2*s*, and 3*s* electrons. **8.56** $1s^22s^22p^6$: 2080 kJ/mol. **8.58** 8.40×10^6 kJ/mol. **8.62** Greatest: Cl; least: He. **8.64** The ns^1 configuration enables them to accept another electron. **8.68** Fr should be the most reactive toward water and oxygen, forming FrOH and Fr_2O, Fr_2O_2, and FrO_2. **8.70** The Group 11 elements have higher ionization energies due to the incomplete shielding of the inner *d* electrons. **8.72** (a) $Li_2O + H_2O \longrightarrow 2LiOH$. (b) $CaO + H_2O \longrightarrow Ca(OH)_2$. (c) $SO_3 + H_2O \longrightarrow H_2SO_4$. **8.74** BaO. **8.76** (a) Bromine. (b) Nitrogen. (c) Rubidium. (d) Magnesium. **8.78** P^{3-}, S^{2-}, Cl^-, K^+, Ca^{2+}, Sc^{3+}, Ti^{4+}, V^{5+}, Cr^{6+}, Mn^{7+}. **8.80** M is K; X is Br. **8.82** N and O^+; Ne and N^{3-}; Ar and S^{2-}; Zn and As^{3+}; Cs^+ and Xe. **8.84** (a) and (d). **8.86** Yellow-green gas: F_2; yellow gas: Cl_2; red liquid: Br_2; dark solid: I_2. **8.88** (a) ΔH = 1532 kJ/mol. (b) ΔH = 12,405 kJ/mol. **8.90** Fluorine. **8.92** H^-. **8.94** Li_2O (basic); BeO (amphoteric); B_2O_3 (acidic); CO_2 (acidic); N_2O_5 (acidic). **8.96** It forms both the H^+ and H^- ions; H^+ is a single proton. **8.98** 0.65. **8.100** 79.9%. **8.102** 418 kJ/mol. Use maximum wavelength. **8.104** 7.28×10^3 kJ/mol. **8.106** X: Sn or Pb; Y: P; Z: alkali metal. **8.108** 495.9 kJ/mol. **8.110** 343 nm (UV). **8.112** 604.3 kJ/mol. **8.114** K_2TiO_4. **8.116** $2K_2MnF_6 + 4SbF_5 \longrightarrow 4KSbF_6 + 2MnF_3 + F_2$. **8.118** N_2O (+1), NO (+2), N_2O_3 (+3), NO_2 and N_2O_4 (+4), N_2O_5 (+5). **8.120** The larger the effective nuclear charge, the more tightly held are the electrons. The atomic radius will be small and the ionization energy will be large. **8.122** m.p.: 6.3°C; b.p.: 74.9°C. **8.124** An alkaline earth metal. **8.126** (a) It was discovered that the periodic table was based on atomic number, not atomic mass. (b) Ar: 39.95 amu; K: 39.10 amu. **8.128** Z = 119; $[Rn]7s^25f^{14}6d^{10}7p^68s^1$. **8.130** Group 13. **8.132** (a) SiH_4, GeH_4, SnH_4, PbH_4. (b) RbH more ionic. (c) $Ra + 2H_2O \longrightarrow Ra(OH)_2 + H_2$. (d) Be. **8.134** Mg^{2+} is the smallest cation and has the largest charge density and is closest to the negative ion. Ba^{2+} is just the opposite. Thus, Mg^{2+} binds the tightest and Ba^{2+} the least. **8.136** See chapter. **8.138** Carbon (diamond). **8.140** 419 nm. **8.142** The first ionization energy of He is less than twice the ionization of H because the radius of He is greater than that of H and the shielding in He makes Z_{eff} less than 2. In He^+, there is no shielding and the greater nuclear attraction makes the second ionization of He greater than twice the ionization energy of H. **8.144** Z_{eff}: Li(1.26); Na(1.84); K(2.26). Z_{eff}/n: Li(0.630); Na(0.613); K(0.565). Z_{eff} increases as *n* increases. Thus, Z_{eff}/n remains fairly constant. **8.146** Go to the recommended website. Click on "Biology" tab above the periodic table and then click on each of the listed elements. A brief summary of the biological role of each element is provided.

Chapter 9

9.16 (a) RbI, rubidium iodide. (b) Cs_2SO_4, cesium sulfate. (c) Sr_3N_2, strontium nitride. (d) Al_2S_3, aluminum sulfide.
9.18 (a) $\cdot Sr \cdot + \cdot \ddot{Se} \cdot \longrightarrow Sr^{2+} : \ddot{Se} :^{2-}$
(b) $\cdot Ca \cdot + 2H \cdot \longrightarrow Ca^{2+}\ 2H :^-$
(c) $3Li \cdot + \cdot \dot{N} \cdot \longrightarrow 3Li^+ : \ddot{N} :^{3-}$
(d) $2 \cdot \dot{Al} \cdot + 3 \cdot \ddot{S} \cdot \longrightarrow 2Al^{3+}\ 3 : \ddot{S} :^{2-}$
9.20 (a) BF_3, covalent. Boron trifluoride. (b) KBr, ionic. Potassium bromide. **9.26** 2195 kJ/mol. **9.36** C—H < Br—H < F—H < Li—Cl < Na—Cl < K—F. **9.38** Cl—Cl < Br—Cl < Si—C < Cs—F. **9.40** (a) Covalent. (b) Polar covalent. (c) Ionic. (d) Polar covalent.

9.44
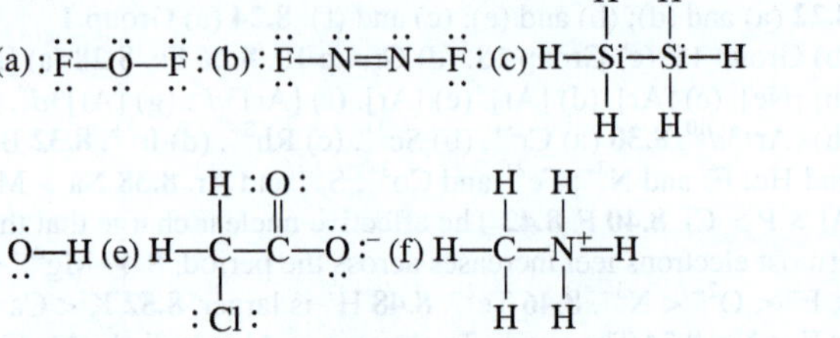

9.46
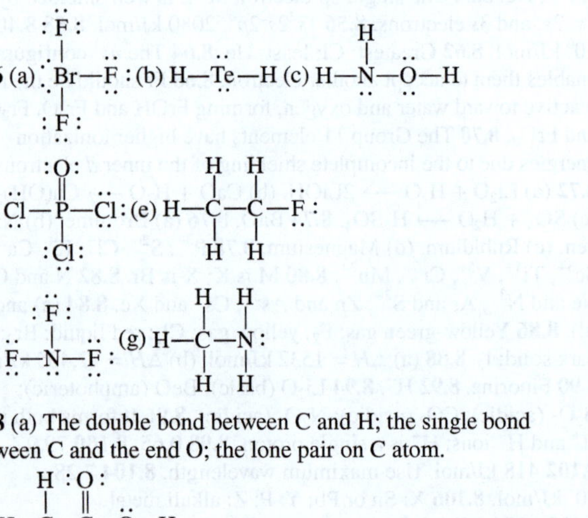

9.48 (a) The double bond between C and H; the single bond between C and the end O; the lone pair on C atom.

(b)
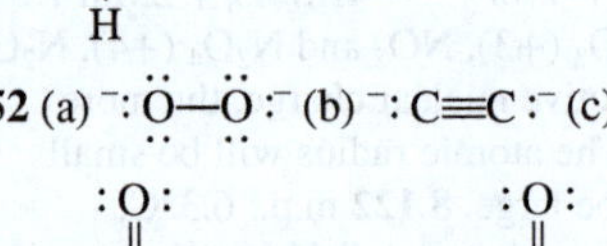

9.52 (a) $^-$:Ö—Ö:$^-$ (b) $^-$:C≡C:$^-$ (c) :N≡O:$^+$

9.56

9.58
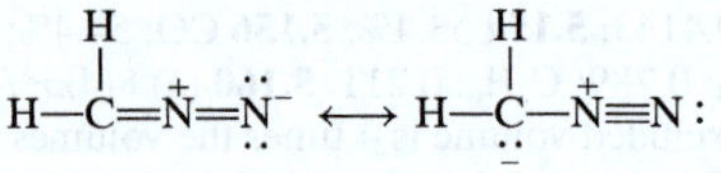

9.60
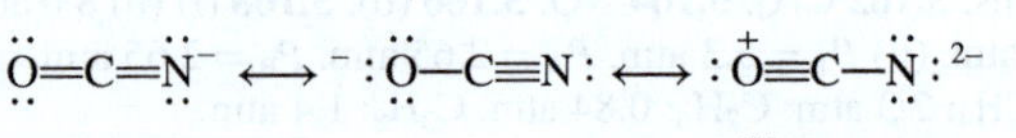

9.66 $^+$Cl=Be=Cl$^+$ Not plausible. **9.68**
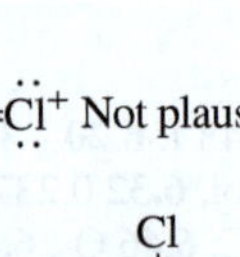
The octet rule is not obeyed. **9.70**
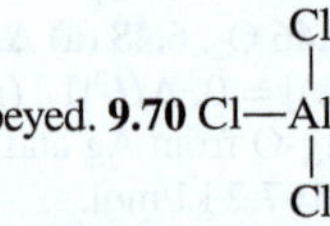
Coordinate covalent bond.

9.74 302 kJ/mol. **9.76** (a) −2810 kJ/mol. (b) −2856.8 kJ/mol. **9.78** Ionic: RbCl, KO_2; covalent: PF_5, BrF_3, CI_4. **9.80** Ionic: NaF, MgF_2, AlF_3; covalent: SiF_4, PF_5, SF_6, ClF_3. **9.82** KF: ionic, high melting point, soluble in water, its melt and solution conduct electricity. C_6H_6: covalent and discrete molecule, low melting point, insoluble in water, does not conduct electricity.

9.84
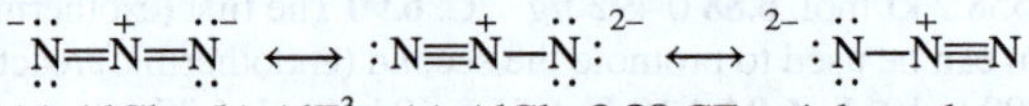

9.86 (a) $AlCl_4^-$. (b) AlF_6^{3-}. (c) $AlCl_3$. **9.88** CF_2: violates the octet rule; LiO_2: lattice energy too low; $CsCl_2$: second ionization energy too high to produce Cs^{2+}; PI_5: I atom too bulky to fit around P. **9.90** (a) False. (b) True. (c) False. (d) False. **9.92** −63 kJ/mol. **9.94** N_2. **9.96** NH_4^+ and CH_4; CO and N_2; $B_3N_3H_6$ and C_6H_6.

9.98
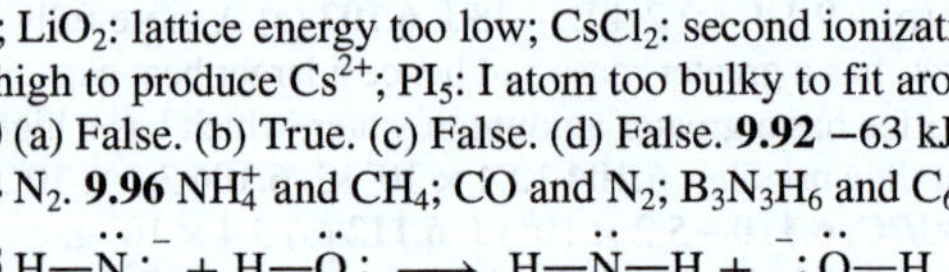

9.100 F_3^- violates the octet rule.
9.102 CH_3—N=C=O ⟷ CH_3—$\overset{+}{N}$≡C—$\overset{-}{O}$:
9.104 (c) No bond between C and O. (d) Large formal charges.

9.106
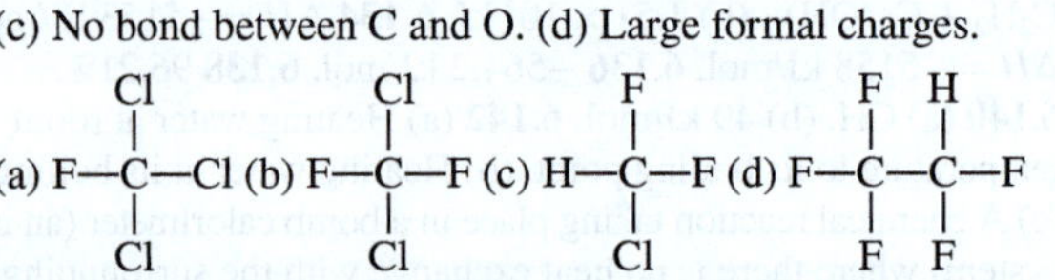

9.108 (a) −8 kJ/mol. (b) −9.4 kJ/mol.
9.110 (a) $^-$:C≡O:$^+$ (b) :N≡O:$^+$ (c) $^-$:C≡N: (d) :N≡N:
9.112 True. **9.114** (a) 121 kJ/mol. (b) Extra electron increases repulsion between F atoms. **9.116** Lone pair on C and negative formal charge on C. **9.118** (a) :N=O ⟷ $\overset{-}{N}$=O$^+$ (b) No.

9.120
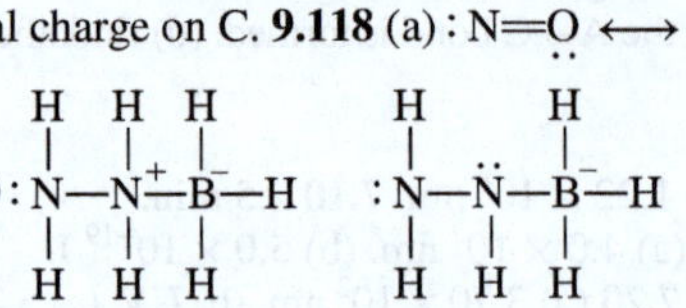

9.122 The OCOO structure leaves a lone pair and a negative charge on C.

9.124
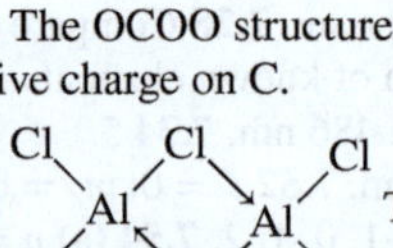
The arrows indicate coordinate covalent bonds. **9.126** 351 kJ/mol. **9.128** From bond enthalpies: −147 kJ/mol; from standard enthalpies of formation: −184 kJ/mol.

9.130
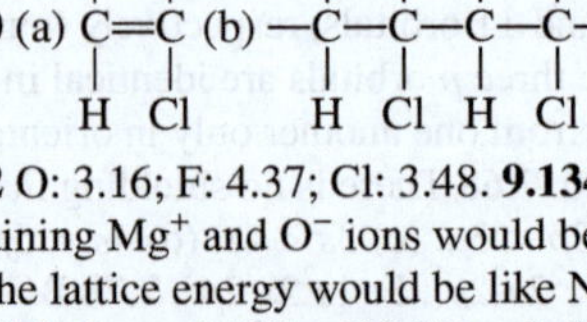
(c) -1.3×10^6 kJ.

9.132 O: 3.16; F: 4.37; Cl: 3.48. **9.134** (1) The MgO solid containing Mg^+ and O^- ions would be paramagnetic. (2) The lattice energy would be like NaCl (too low). **9.136** 71.5 nm. **9.138** −629 kJ/mol. **9.140** 264 nm.

9.142 (a) From bond enthalpies: −1465 kJ/mol; from standard enthalpies of formation: −1413.9 kJ/mol. (b) 162 L. (c) 11.0 atm. **9.144** The repulsion between lone pairs on adjacent atoms weakens the bond. There are two lone pairs on each O atom in H_2O_2. The repulsion is the greatest; it has the smallest bond enthalpy (about 210 kJ/mol). There is one lone pair on each N atom in N_2H_4; it has the intermediate bond enthalpy (about 270 kJ/mol). There are no lone pairs on the C atoms in C_2H_6; it has the greatest bond enthalpy (about 370 kJ/mol). **9.146** 244 kJ/mol.

Chapter 10

10.8 (a) Trigonal planar. (b) Linear. (c) Tetrahedral. **10.10** (a) Tetrahedral. (b) T-shaped. (c) Bent. (d) Trigonal planar. (e) Tetrahedral. **10.12** (a) Tetrahedral. (b) Bent. (c) Trigonal planar. (d) Linear. (e) Square planar. (f) Tetrahedral. (g) Trigonal bipyramidal. (h) Trigonal pyramidal. (i) Tetrahedral. **10.14** $SiCl_4$, CI_4, $CdCl_4^{2-}$. **10.20** Electronegativity decreases from F to I. **10.22** Higher. **10.24** (b) = (d) < (c) < (a). **10.32** sp^3 for both. **10.34** B: sp^2 to sp^3; N: remains at sp^3. **10.36** sp^3d. **10.38** IF_4^-. **10.40** From left to right. (a) sp^3. (b) sp^3, sp^2, sp^2. (c) sp^3, sp, sp, sp^3. (d) sp^3, sp^2. (e) sp^3, sp^2. **10.42** sp. **10.44** 9 pi bonds and 9 sigma bonds. **10.50** Electron spins must be paired in H_2. **10.52** $Li_2^- = Li_2^+ < Li_2$. **10.54** B_2^+. **10.56** MO theory predicts C_2 is paramagnetic. **10.58** $O_2^{2-} < O_2^- < O_2 < O_2^+$. **10.60** B_2 contains a pi bond; C_2 contains 2 pi bonds. **10.62** (1) Atoms far apart. No interaction. (2) The 2*p* orbitals begin to overlap. Attractive forces operating. (3) The system at its most stable state. The potential energy reaches a minimum. (4) As the distance decreases further, nuclear-nuclear and electron-electron repulsion increase. (5) Further decrease in distance leads to instability of F_2 molecule. **10.66** The circle shows electron delocalization.
10.68 (a) 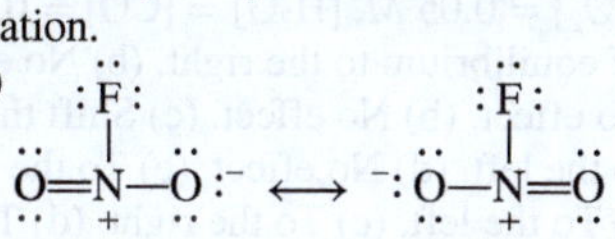
(b) sp^2. (c) N forms sigma bonds with F and O atoms. There is a pi molecular orbital delocalized over N and O atoms. **10.70** sp^2. **10.72** Linear. Dipole moment measurement. **10.74** The large size of Si results in poor sideways overlap of *p* orbitals to form pi bonds. **10.76** (a) $C_8H_{10}N_4O_2$. (b) C atoms in the ring and O are sp^2. C atom in CH_3 group is sp^3. Double-bonded N is sp^2; single-bonded N is sp^3. (c) Geometry about the sp^2 C and N atoms is trigonal planar. Geometry about sp^3 C and N atom is tetrahedral. **10.78** XeF_3^+: T − shaped; XeF_5^-: square pyramidal; SbF_6^-: octahedral. **10.80** (a) 180°. (b) 120°. (c) 109.5°. (d) About 109.5°. (e) 180°. (f) About 120°. (g) About 109.5°. (h) 109.5°. **10.82** sp^3d. **10.84** ICl_2^- and $CdBr_2$. **10.86** (a) sp^2. (b) Molecule on the right. **10.88** The pi bond in *cis*-dichloroethylene prevents rotation. **10.90** O_3, CO, CO_2, NO_2, N_2O, CH_4, $CFCl_3$. **10.92** C: all single-bonded C atoms are sp^3, the double-bonded C atoms are sp^2; N: single-bonded N atoms are sp^3, N atoms that form one double bond are sp^2, N atom that forms two double bonds is *sp*. **10.94** Si has 3*d* orbitals so water can add to Si (valence shell expansion). **10.96** C: sp^2; N: N atom that forms a double bond is sp^2, the others are sp^3. **10.98** (a) Use a conventional oven. (b) No. Polar molecules would absorb microwaves. (c) Water molecules absorb part of microwaves. **10.100** (a) and (b) are polar. **10.102** The small size of F results in a shorter bond and greater lone pair repulsion. **10.104** 43.6%. **10.106** Second and third vibrations. CO, NO_2, N_2O. **10.108** (a) The two 90° rotations will break and make the pi bond and convert *cis*-dichloroethylene to *trans*-dichloroethylene. (b) The pi bond is weaker because of the lesser extent of sideways orbital overlap. (c) 444 nm. **10.110** (a) H_2. The electron is removed from the more stable bonding molecular orbital. (b) N_2. Same as (a). (c) O. The atomic orbital in O is more stable than the antibonding molecular orbital in O_2. (d) The atomic orbital in F is more stable than the antibonding molecular orbital in F_2. **10.112** (a) $[Ne_2](\sigma_{3s})^2(\sigma^*_{3s})^2(\pi_{3py})^2(\pi_{3pz})^2(\sigma_{3px})^2$. (b) 3. (c) Diamagnetic. **10.114** For all the electrons to be paired in O_2 (see Table 10.5), energy is needed to flip the spin of one of the electrons in the antibonding molecular orbitals. This arrangement is less stable according to Hund's rule. **10.116** ClF_3: T-shaped; sp^3d. AsF_5: Trigonal bipyramidal; sp^3d; ClF_2^+: bent; sp^3; AsF_6^-: Octahedral; sp^3d^2. **10.118** (a) Planar and no dipole moment. (b) 20 sigma bonds and 6 pi bonds. **10.120** (a) The negative formal charge is placed on the less electronegative carbon, so there is less charge separation and a smaller dipole moment. (b) Both the Lewis structure and the molecular orbital treatment predict a triple bond. (c) C. **10.122** Yes. O=C=C=C=O. The molecule is linear and nonpolar. **10.124** $NO^{2-} < NO^- < NO = NO^{2+} < NO^+$.

Chapter 11

11.8 Methane. **11.10** (a) Dispersion forces. (b) Dispersion and dipole-dipole forces. (c) Same as (b). (d) Dispersion and ion-ion forces. (e) Same as (a). **11.12** Only (e). **11.14** Only 1-butanol can form hydrogen bonds. **11.16** (a) Xe. (b) CS_2. (c) Cl_2. (d) LiF. (e) NH_3. **11.18** (a) Hydrogen bond and dispersion forces. (b) Dispersion forces. (c) Dispersion forces. (d) Covalent bond. **11.20** The compound on the left can form an intramolecular hydrogen bond, reducing intermolecular hydrogen bonding. **11.32** Between ethanol and glycerol. **11.38** scc: 1; bcc: 2; fcc: 4. **11.40** 6.20×10^{23} Ba atoms/mol. **11.42** 458 pm. **11.44** XY_3. **11.48** 0.220 nm. **11.52** Molecular solid. **11.54** Molecular solids: Se_8, HBr, CO_2, P_4O_6, SiH_4. Covalent solids: Si, C. **11.56** Each C atom in diamond is covalently bonded to four other C atoms. Graphite has delocalized electrons in two dimensions. **11.76** 2.67×10^3 kJ. **11.78** 47.03 kJ/mol. **11.80** Freezing, sublimation. **11.82** When steam condenses at 100°C, it releases heat equal to heat of vaporization. **11.84** 323 mmHg. **11.86** The small amount of liquid nitrogen will evaporate quickly, extracting little heat from the skin. Boiling water will release much more heat to the skin as it cools. Water has a high specific heat. **11.90** Initially ice melts because of the increase in pressure. As the wire sinks into the ice, the water above it refreezes. In this way, the wire moves through the ice without cutting it in half. **11.92** (a) Ice melts. (b) Water vapor condenses to ice. (c) Water boils. **11.94** (d). **11.96** Covalent crystal. **11.98** Orthorhombic. **11.100** 760 mmHg. **11.102** It is the critical point. **11.104** Crystalline SiO_2. **11.106** (c) and (d). **11.108** (a), (b), (d). **11.110** 8.3×10^{-3} atm. **11.112** (a) K_2S. Ionic. (b) Br_2. Dispersion. **11.114** SO_2. It is a polar molecule. **11.116** 62.4 kJ/mol. **11.118** 304°C. **11.120** Small ions have more concentrated charges and are more effective in ion-dipole interaction, resulting in a greater extent of hydration. The distance of separation between cation and anion is also shorter. **11.122** (a) 30.9 kJ/mol. (b) 194 kJ/mol. **11.124** (a) Decreases. (b) No change. (c) No change. **11.126** (a) 1 Cs^+ ion and 1 Cl^- ion. (b) 4 Zn^{2+} ions and 4 S^{2-} ions. (c) 4 Ca^{2+} ions and 8 F^- ions. **11.128** $CaCO_3(s) \longrightarrow CaO(s) + CO_2(g)$. Three phases. **11.130** SiO_2 is a covalent crystal. CO_2 exists as discrete molecules. **11.132** 66.8%. **11.134** scc: 52.4%; bcc: 68.0%; fcc: 74.0%. **11.136** 1.69 g/cm^3. **11.138** (a) Two (diamond/graphite/liquid and graphite/liquid/vapor). (b) Diamond. (c) Apply high pressure at high temperature. **11.140** Molecules in the cane are held together by intermolecular forces. **11.142** When the tungsten filament is heated to a high temperature (ca. 3000°C), it sublimes and condenses on the inside walls. The inert pressurized Ar gas retards sublimation. **11.144** When methane burns in air, it forms CO_2 and water vapor. The latter condenses on the outside of the cold beaker. **11.146** 6.019×10^{23} Fe atoms/mol. **11.148** Na (186 pm and 0.965 g/cm^3). **11.150** (d). **11.152** 0.833 g/L. Hydrogen bonding in the gas phase.

Chapter 12

12.10 Cyclohexane cannot form hydrogen bonds. **12.12** The longer chains become more nonpolar. **12.16** (a) 25.9 g. (b) 1.72×10^3 g. **12.18** (a) 2.68 *m*. (b) 7.82 *m*. **12.20** 0.010 *m*. **12.22** 5.0×10^2 *m*; 18.3 *M*. **12.24** (a) 2.41 *m*. (b) 2.13 *M*. (c) 0.0587 L. **12.28** 45.9 g. **12.36** CO_2 pressure is greater at the bottom of the mine. **12.38** 0.28 L. **12.50** 1.3×10^3 g. **12.52** Ethanol: 30.0 mmHg; 1-propanol: 26.3 mmHg. **12.54** 128 g. **12.56** 0.59 *m*. **12.58** 120 g/mol; $C_4H_8O_4$. **12.60** −8.6°C. **12.62** 4.3×10^2 g/mol; $C_{24}H_{20}P_4$. **12.64** 1.75×10^4 g/mol. **12.66** 343 g/mol. **12.70** Boiling point, vapor pressure, osmotic pressure. **12.72** 0.50 *m* glucose > 0.50 *m* acetic acid > 0.50 *m* HCl. **12.74** 0.9420 *m*. **12.76** 7.6 atm. **12.78** 1.6 atm. **12.82** (c). **12.84** 3.5 atm. **12.86** (a) 104 mmHg. (b) 116 mmHg. **12.88** 2.95×10^3 g/mol. **12.90** 12.6 g. **12.92** No. **12.94** No. $AlCl_3$ dissociates into Al^{3+} and 3 Cl^- ions. **12.96** O_2: 4.7×10^{-6}; N_2: 9.7×10^6. The mole fraction of O_2 compared to the mole fraction of N_2 in water is greater compared to that in air. **12.98** The molar mass in B (248 g/mol) is twice as large as that in A (124 g/mol). A dimerization process. **12.100** (a) Last alcohol. (b) Methanol. (c) Last alcohol. **12.102** I_2-water: weak dipole–induced dipole; I_3^--water: favorable ion-dipole interaction. **12.104** (a) Same NaCl solution on both sides. (b) Only water would move from left to right. (c) Normal osmosis. **12.106** 12.3 *M*. **12.108** 14.2%. **12.110** (a) and (d). **12.112** (a) Decreases with increasing lattice energy. (b) Increases with increasing polarity of solvent. (c) Increases with increasing enthalpy of hydration. **12.114** 1.80 g/mL. 5.0×10^2 *m*. **12.116** 0.815. **12.118** NH_3 can form hydrogen bonds with water. **12.120** 3%. **12.122** 1.2×10^2 g/mol. It forms a dimer in benzene. **12.124** (a) 1.1 *m*. (b) The protein prevents the formation of ice crystals. **12.126** It is due to the precipitated minerals that refract light and create an opaque appearance. **12.128** 1.9 *m*. **12.130** (a) $X_A = 0.524$, $X_B = 0.476$. (b) A: 50 mmHg; B: 20 mmHg. (c) $X_A = 0.71$, $X_B = 0.29$. $P_A = 67$ mmHg. $P_B = 12$ mmHg. **12.132** 2.7×10^{-3}. **12.134** From $n = kP$ and $PV = nRT$, show that $V = kRT$. **12.136** −0.737°C. **12.138** The polar groups (C═O) can bind the K^+ ions. The exterior is nonpolar (due to the $-CH_3$ groups), which enables the molecule to pass through the cell membranes containing nonpolar lipids. **12.140** The string is wetted and laid on top of the ice cube. Salt is shaken onto the top of the ice cube and the moistened string. The presence of salt lowers the freezing point of the ice, resulting in the melting of the ice on the surface. Melting is an endothermic process. The water in the moist string freezes, and the string becomes attached to the ice cube. The ice cube can now be lifted out of the glass.

Chapter 13

13.6 (a) Rate $= -(1/2)\Delta[H_2]/\Delta t = -\Delta[O_2]/\Delta t = (1/2)\Delta[H_2O]/\Delta t$. (b) Rate $= -(1/4)\Delta[NH_3]/\Delta t = -(1/5)\Delta[O_2]/\Delta t = (1/4)\Delta[NO]/\Delta t = (1/6)\Delta[H_2O]/\Delta t$. **13.8** (a) 0.049 *M*/s. (b) 0.025 *M*/s. **13.14** 2.4×10^{-4} *M*/s. **13.16** (a) Third order. (b) 0.38 *M*/s. **13.18** (a) 0.046 s^{-1}. (b) $0.13/M \cdot s$. **13.20** First order. 1.08×10^{-3} s^{-1}. **13.26** (a) 0.0198 s^{-1}. (b) 151 s. **13.28** 3.6 s. **13.30** (a) The relative rates for (i), (ii), and (iii) are 4:3:6. (b) The relative rates would be unaffected, but each of the absolute rates would decrease by 50%. (c) The relative half-lives are 1:1:1. **13.38** 135 kJ/mol. **13.40** 103 kJ/mol. **13.42** 644 K. **13.44** 9.25×10^3 s^{-1}. **13.46** 51.0 kJ/mol. **13.56** (a) Rate $= k[X_2][Y]$. (b) Reaction is zero order in Z. (c) $X_2 + Y \longrightarrow XY + X$ (slow). $X + Z \longrightarrow XZ$ (fast). **13.58** Mechanism I. **13.66** Rate $= (k_1k_2/k_{-1})$ [E] [S]. **13.68** This is a first-order reaction. The rate constant is 0.046 min^{-1}. **13.70** Temperature, energy of activation, concentration of reactants, catalyst. **13.72** 22.6 cm^2; 44.9 cm^2. The large surface area of grain dust can result in a violent explosion. **13.74** (a) Third order. (b) $0.38/M^2 \cdot s$. (c) $H_2 + 2NO \longrightarrow N_2 + H_2O + O$ (slow); $O + H_2 \longrightarrow H_2O$ (fast). **13.76** Water is present in excess so its concentration does not change appreciably. **13.78** $10.7/M \cdot s$. **13.80** 2.63 atm. **13.82** M^{-2} s^{-1}. **13.84** 56.4 min. **13.86** rate $= k[A][B]^2$. **13.88** (b), (d), (e). **13.90** 9.8×10^{-4}. **13.92** (a) Increased. (b) Decreased. (c) Decreased. (d) Increased. **13.94** 0.0896 min^{-1}. **13.96** 1.12×10^3 min. **13.98** (a) I_2 absorbs visible light to form I atoms. (b) UV light is needed to dissociate H_2. **13.100** (a) Rate $= k[X][Y]^2$. (b) $1.9 \times 10^{-2}/M^2 \cdot s$. **13.102** Second order. $2.4 \times 10^7/M \cdot s$. **13.104** Because the engine is relatively cold so the exhaust gases will not fully react with the catalytic converter. **13.106** $H_2(g) + ICl(g) \longrightarrow HCl(g) + HI(g)$ (slow). $HI(g) + ICl(g) \longrightarrow HCl(g) + I_2(g)$ (fast). **13.108** 5.7×10^5 yr. **13.110** (a) Mn^{2+}; Mn^{3+}; first step. (b) Without the catalyst, reaction would be termolecular. (c) Homogeneous. **13.112** 0.45 atm. **13.114** (a) $k_1[A] - k_2[B]$. (b) $[B] = (k_1/k_2)$ [A]. **13.116** (a) 2.47×10^{-2} yr^{-1}. (b) 9.8×10^{-4}. (c) 186 yr. **13.118** (a) 3. (b) 2. (c) $C \longrightarrow D$. (d) Exothermic. **13.120** 1.8×10^3 K. **13.122** (a) 2.5×10^{-5} *M*/s. (b) Same as (a). (c) 8.3×10^{-6} *M*. **13.126** (a) 1.13×10^{-3} *M*/min. (b) 6.83×10^{-4} *M*/min; 8.8×10^{-3} *M*. **13.128** Second order. $0.42/M \cdot min$. **13.130** 60% increase. The result shows the profound effect of an exponential dependence. **13.132** 2.6×10^{-4} *M*/s. **13.134** 404 kJ/mol. **13.136** (a) Rate $= k[NO]^2[O_2]$. (b) Rate $= k_{obs}[NO]^2$. (c) 1.3×10^3 min.

Chapter 14

14.14 (a) $A + C \rightleftharpoons AC$. (b) $A + D \rightleftharpoons AD$. **14.16** 1.08×10^7. **14.18** 3.5×10^{-7}. **14.20** (a) 0.082. (b) 0.29. **14.22** 0.105; 2.05×10^{-3}. **14.24** 7.09×10^{-3}. **14.26** 3.3. **14.28** 0.0353. **14.30** 4.0×10^{-6}. **14.32** 5.6×10^{23}. **14.36** $0.64/M^2 \cdot s$. **14.40** $[NH_3]$ will increase and $[N_2]$ and $[H_2]$ will decrease. **14.42** NO: 0.50 atm; NO_2: 0.020 atm. **14.44** $[I] = 8.58 \times 10^{-4}$ *M*; $[I_2] = 0.0194$ *M*. **14.46** (a) 0.52. (b) $[CO_2] = 0.48$ *M*, $[H_2] = 0.020$ *M*, $[CO] = 0.075$ *M*, $[H_2O] = 0.065$ *M*. **14.48** $[H_2] = [CO_2] = 0.05$ *M*, $[H_2O] = [CO] = 0.11$ *M*. **14.54** (a) Shift position of equilibrium to the right. (b) No effect. (c) No effect. **14.56** (a) No effect. (b) No effect. (c) Shift the position of equilibrium to the left. (d) No effect. (e) To the left. **14.58** (a) To the right. (b) To the left. (c) To the right. (d) To the left. (e) No effect. **14.60** No change. **14.62** (a) More CO_2 will form. (b) No change. (c) No change. (d) Some CO_2 will combine with CaO to form $CaCO_3$. (e) Some CO_2 will react with NaOH so equilibrium will shift to the right. (f) HCl reacts with $CaCO_3$ to produce CO_2. Equilibrium will shift to the left. (g) Equilibrium will shift to the right. **14.64** (a) NO: 0.24 atm; Cl_2: 0.12 atm. (b) 0.017. **14.66** $[A_2] \rightleftharpoons [B_2] = 0.040$ *M*. $[AB] = 0.020$ *M*. **14.68** (a) No effect. (b) More CO_2 and H_2O will form. **14.70** (a) 8×10^{-44}. (b) The reaction has a very large activation energy. **14.72** (a) 1.7. (b) A: 0.69 atm, B: 0.81 atm. **14.74** 1.5×10^5. **14.76** H_2: 0.28 atm, Cl_2: 0.049 atm, HCl: 1.67 atm. **14.78** 5.0×10^1 atm. **14.80** 3.84×10^{-2}. **14.82** 3.13. **14.84** N_2: 0.860 atm; H_2: 0.366 atm; NH_2: 4.40×10^{-3} atm. **14.86** (a) 1.16. (b) 53.7%. **14.88** (a) 0.49 atm. (b) 0.23. (c) 0.037. (d) Greater than 0.037 mol. **14.90** $[H_2] = 0.070$ *M*, $[I_2] = 0.182$ *M*, $[HI] = 0.825$ *M*. **14.92** (c). **14.94** (a) 4.2×10^{-4}. (b) 0.83. (c) 1.1. (d) In (b): 2.3×10^3; in (c): 0.021. **14.96** 0.0231; 9.60×10^4. **14.98** NO_2: 1.2 atm; N_2O_4: 0.12 atm. $K_P = 12$. **14.100** (a) $K_c = 33.3$. (b) $Q_c = 2.8$. Shift to the right. (c) $Q_c = 169$. Shift to the left. **14.102** (a) The equilibrium will shift to the right. (b) To the right. (c) No change. (d) No change. (e) No change. (f) To the left. **14.104** NO_2: 0.100 atm; N_2O_4: 0.09 atm. **14.106** (a) 1.03 atm. (b) 0.39 atm. (c) 1.67 atm. (d) 0.620. **14.108** (a) $K_P = 2.6 \times 10^{-6}$; $K_c = 1.1 \times 10^{-7}$. (b) 22 mg/m^3. Yes. **14.110** Temporary dynamic equilibrium between the melting ice cubes and the freezing of the water between the ice cubes. **14.112** $[NH_3] = 0.042$ *M*, $[N_2] = 0.086$ *M*, $[H_2] = 0.026$ *M*. **14.114** 1.3 atm. **14.116** PCl_5: 0.683 atm; PCl_3: 1.11 atm; Cl_2: 0.211 atm. **14.118** −115 kJ/mol. **14.120** SO_2: 2.71 atm; Cl_2: 2.71 atm; SO_2Cl_2: 3.58 atm. **14.122** 4.0. **14.124** (a) The plot curves toward higher pressure at low values of $1/V$. (b) The plot curves toward higher volume as *T* increases.

Chapter 15

15.4 (a) NO_2^-. (b) HSO_4^-. (c) HS^-. (d) CN^-. (e) $HCOO^-$. **15.6** (a) H_2S. (b) H_2CO_3. (c) HCO_3^-. (d) H_3PO_4. (e) $H_2PO_4^-$. (f) HPO_4^{2-}. (g) H_2SO_4. (h) HSO_4^-. (i) HSO_3^-. **15.8** (a) CH_2ClCOO^-. (b) IO_4^-. (c) $H_2PO_4^-$. (d) HPO_4^{2-}. (e) PO_4^{3-}. (f) HSO_4^-. (g) SO_4^{2-}. (h) IO_3^-. (i) SO_3^{2-}. (j) NH_3. (k) HS^-. (l) S^{2-}. (m) OCl^-. **15.16** 1.6×10^{-14} *M*. **15.18** (a) 10.74. (b) 3.28. **15.20** (a) 6.3×10^{-6} *M*. (b) 1.0×10^{-16} *M*. (c) 2.7×10^{-6} *M*. **15.22** (a) Acidic. (b) Neutral. (c) Basic. **15.24** 1.98×10^{-3} mol. 0.444. **15.26** 0.118. **15.32** (1) (c). (2) (b) and (d). **15.34** (a) Strong. (b) Weak. (c) Weak. (d) Weak. (e) Strong. **15.36** (b) and (c). **15.38** No. **15.44** $[H^+] = [CH_3COO^-] = 5.8 \times 10^{-4}$ *M*, $[CH_3COOH] = 0.0181$ *M*. **15.46** 2.2×10^{-3} *M*. **15.48** (a) 3.2%. (b) 30%. (c) 75%. Percent ionization increases with dilution. **15.50** (a) 3.9%. (b) 0.30%. **15.54** (c) < (a) < (b). **15.56** 7.1×10^{-7}. **15.58** 1.5%. **15.64** HCl: 1.40; H_2SO_4: 1.31. **15.66** $[H^+] = [HCO_3^-] = 1.0 \times 10^{-4}$ *M*, $[CO_3^{2-}] = 5.0 \times 10^{-11}$ *M*. **15.70** (a) $H_2SO_4 > H_2SeO_4$. (b) $H_3PO_4 > H_3AsO_4$. **15.72** The conjugate base of phenol can be stabilized by resonance. **15.78** (a) Neutral. (b) Basic. (c) Acidic. (d) Acidic. **15.80** HZ < HY < HX. **15.82** 4.82. **15.84** Basic. **15.88** (a) $Al_2O_3 < BaO < K_2O$. (b) $CrO_3 < Cr_2O_3 < CrO$. **15.90** $Al(OH)_3 + OH^- \longrightarrow Al(OH)_4^-$. Lewis acid-base reaction. **15.94** $AlCl_3$ is the Lewis acid, Cl^- is the Lewis base. **15.96** CO_2 and BF_3. **15.98** 0.0094 *M*. **15.100** 0.106 L. **15.102** No. **15.104** No, volume is the same. **15.106** CrO is basic and CrO_3 is acidic. **15.108** 3.2×10^{-2}. **15.110** 7.00. **15.112** NH_3. **15.114** (a) 7.43. (b) pD < 7.43. (c) pD + pOD = 14.87. **15.116** 1.79. **15.118** F^- reacts with HF to form HF_2^-, thereby shifting the ionization of HF to the right. **15.120** (b) 6.80. **15.122** $[H^+] = [H_2PO_4^-] = 0.0221$ *M*, $[H_3PO_4] = 0.076$ *M*, $[HPO_4^{2-}] = 6.2 \times 10^{-8}$ *M*, $[PO_4^{3-}] = 1.3 \times 10^{-18}$ *M*. **15.124** Pyrex glass contains 10–25% B_2O_3, an acidic oxide. **15.126** $[Na^+] = 0.200$ *M*, $[CO_3^{2-}] = 0.096$ *M* $[HCO_3^-] = 4.5 \times 10^{-3}$ *M* $[OH^-] = 4.5 \times 10^{-3}$ *M*, $[H_2CO_3] = 2.1 \times 10^{-8}$ *M*, $[H^+] = 2.2 \times 10^{-12}$ *M*. **15.128** The H^+ ions convert CN^- to HCN, which escapes as a gas. **15.130** 0.23 g. **15.132** −0.20. **15.134** (a) Equilibrium will shift to the right. (b) To the left. (c) No effect. (d) To the right. **15.136** The amines are converted to their salts RNH_3^+. **15.138** 1.4×10^{-4}. **15.140** 4.40. **15.142** In a basic medium, the ammonium salt is converted to the pungent-smelling ammonia. **15.144** (c). **15.146** 21 mL. **15.148** HX is the stronger acid. **15.150** Mg. **15.152** 1.60. $[CN^-] = 2.3 \times 10^{-8}$ *M* in 1.00 *M* HF and 2.5×10^{-5} *M* in 1.00 *M* HCN. HF is a stronger acid than HCN. **15.154** 6.02. **15.156** 1.21. **15.158** (a) pH = 7.24. (b) 10,000 H_3O^+ ions for every OH^- ion. **15.160** Both are −55.8 kJ/mol because they have the same net ionic equation.

Chapter 16

16.6 (a) 11.28. (b) 9.08. **16.10** (a), (b), and (c). **16.12** 4.74 for both. (a) is more effective because it has a higher concentration. **16.14** 7.03. **16.16** 10. More effective against the acid. **16.18** (a) 4.82. (b) 4.64. **16.20** HC. **16.22** (1) (a): 5.10. (b): 4.82. (c): 5.22. (d): 5.00. (2) 4.90. (3) 5.22. **16.24** 0.53 mol. **16.28** 90.1 g/mol. **16.30** 0.467 *M*. **16.32** $[H^+] = 3.0 \times 10^{-13}$ *M*, $[OH^-] = 0.0335$ *M*, $[Na^+] = 0.0835$ *M*, $[CH_3COO^-] = 0.0500$ *M*, $[CH_3COOH] = 8.4 \times 10^{-10}$ *M*. **16.34** 8.23. **16.36** (a) 11.36. (b) 9.55. (c) 8.95. (d) 5.19. (e) 1.70. **16.38** (1) (c). (2) (a). (3) (d). (4) (b). pH < 7 at the equivalence point. **16.40** 6.0×10^{-6}. **16.44** CO_2 dissolves in water to form H_2CO_3, which neutralizes NaOH. **16.46** 5.70. **16.54** (a) 7.8×10^{-10}. (b) 1.8×10^{-18}. **16.56** 1.80×10^{-10}. **16.58** 2.2×10^{-4} *M*. **16.60** 2.3×10^{-9}. **16.62** $[Na^+] = 0.045$ *M*, $[NO_3^-] = 0.076$ *M*, $[Sr^{2+}] = 0.016$ *M*, $[F^-] = 1.1 \times 10^{-4}$ *M*. **16.64** pH greater than 2.48 and less than 6.77. **16.70** (a) 0.012 *M*. (b) 1.7×10^{-4} *M*. (c) 5.7×10^{-3} *M*. **16.72** (a) 1.0×10^{-5} *M*. (b) 1.1×10^{-10} *M*. **16.74** (b), (c), (d), and (e). **16.76** (a) 4.9×10^{-4} *M*. (b) 4.9×10^{-9} *M*. **16.78** Yes. **16.80** $[Cd^{2+}] = 1.3 \times 10^{-20}$ *M*, $[Cd(CN)_4^{2-}] = 4.2 \times 10^{-3}$ *M*, $[CN^-] = 0.48$ *M*. **16.82** 3.1×10^{-5} *M*. **16.84** (a) $Cu^{2+} + 4NH_3 \rightleftharpoons Cu(NH_3)_4^{2+}$. (b) $Ag^+ + 2CN^- \rightleftharpoons Ag(CN)_2^-$. (c) $Hg^+ + 4Cl^- \rightleftharpoons HgCl_4^{2-}$. **16.88** 7.6×10^{-4} *M*. **16.90** Use Cl^- ions or flame test. **16.92** From 2.51 to 4.41. **16.94** 1.8×10^2 mL. **16.96** 1.39 *M*. **16.98** $[H^+] = 3.0 \times 10^{-13}$ *M*, $[OH^-] = 0.0335$ *M*, $[HCOO^-] = 0.0500$ *M*, $[HCOOH] = 8.4 \times 10^{-11}$ *M*, $[Na^+] = 0.0835$ *M*. **16.100** 9.97 g. pH = 13.04. **16.102** 1.0×10^4. **16.104** 0.036 g/L. **16.106** (a) 1.37. (b) 5.98. (c) 10.26. **16.108** Original precipitate was HgI_2. In the presence of excess KI, it redissolves as HgI_4^{2-}. **16.110** 7.82 – 10.38. **16.112** (a) 3.60. (b) 9.69. (c) 6.07. **16.114** (a) $MCO_3 + 2HCl \longrightarrow MCl_2 + H_2O + CO_2$. $HCl + NaOH \longrightarrow NaCl + H_2O$. (b) 24.3 g/mol. Mg. **16.116** 2. **16.118** (a) 12.6. (b) 8.8×10^{-6} *M*. **16.120** (a) Sulfate. (b) Sulfide. (c) Iodide. **16.122** They are insoluble. **16.124** The ionized polyphenols have a dark color. The H^+ ions from lemon juice shift the equilibrium to the light color acid. **16.126** Yes. **16.128** (c). **16.130** (a) 6.8×10^{-8} *M*. (b) $MgCO_3$ is more soluble than $CaCO_3$. (c) 12.34. (d) 1.1×10^{-8} *M*. (e) Ca^{2+} because it is present in larger amount. **16.132** pH = 1.0, fully protonated; pH = 7.0, dipolar ion; pH = 12.0, fully ionized. **16.134** (a) 7.7 mL. (b) 12.5 mL. (c) 27.0 mL. **16.136** (a) 4.74 before and after dilution. (b) 2.52 before and 3.02 after dilution. **16.138** 4.75. **16.140** (a) 0.0014 g. (b) 3.1×10^{-8} g. (c) 8.4×10^{-5} g. **16.142** (1) The initial pH of acid (a) is lower. (2) The pH at halfway to the equivalence point is lower for (a). (3) The pH at the equivalence point is lower for acid (a), indicating that (a) forms a weaker conjugate base than (b). Thus, (a) is the stronger acid. **16.144** $[Cu^{2+}] = 1.8 \times 10^{-7}$ *M*. $[OH^-] = 3.6 \times 10^{-7}$ *M*. $[Ba^{2+}] = [SO_4^{2-}] = 1.0 \times 10^{-5}$ *M*.

Chapter 17

17.6 (a) 0.25. (b) 8×10^{-31}. (c) ≈ 0. For a macroscopic system, the probability is practically zero that all the molecules will be found only in one bulb. **17.10** (c) < (d) < (e) < (a) < (b). Solids have smaller entropies than gases. More complex structures have higher entropies. **17.12** (a) 48.7 J/K · mol. (b) −12.9 J/K · mol. (c) −242.9 J/K · mol. **17.14** (a) $\Delta S < 0$. (b) $\Delta S > 0$. (c) $\Delta S > 0$. (d) $\Delta S < 0$. **17.18** (a) −1138.6 kJ/mol. (b) −142.0 kJ/mol. (c) −2936.2 kJ/mol. **17.20** (a) At all temperatures. (b) Below 111 K. **17.24** 8.0×10^1 kJ/mol. **17.26** 4.572×10^2 kJ/mol. 7.2×10^{-81}. **17.28** (a) −24.6 kJ/mol. (b) −1.33 kJ/mol. **17.30** −341 kJ/mol. **17.32** −2.9 kJ/mol. The process has a high activation energy. **17.36** 1×10^3. glucose + ATP $\longrightarrow$ glucose 6 – phosphate + ADP. 1×10^3. **17.38** (a) 0. (b) 4.0×10^4 J/mol. (c) -3.2×10^4 J/mol. (d) 6.4×10^4 J/mol. **17.40** Positive. **17.42** (a) No reaction is possible because $\Delta G > 0$. (b) The reaction has a very large activation energy. (c) Reactants and products already at their equilibrium concentrations. **17.44** In all cases $\Delta H > 0$ and $\Delta S > 0$. $\Delta G < 0$ for (a), = 0 for (b), and > 0 for (c). **17.46** $\Delta S > 0$. **17.48** (a) Most liquids have similar structure so the changes in entropy from liquid to vapor are similar. (b) ΔS_{vap} are larger for ethanol and water because of hydrogen bonding (there are fewer microstates in these liquids). **17.50** (a) $2CO + 2NO \longrightarrow 2CO_2 + N_2$. (b) Oxidizing agent: NO; reducing agent: CO. (c) 3×10^{120}. (d) 1.2×10^{18}. From left to right. (e) No. **17.52** 1.9×10^{-10}. **17.54** 2.9×10^{-9}. **17.56** 981 K. **17.58** $\Delta S < 0$; $\Delta H < 0$. **17.60** 55 J/K · mol. **17.62** Increase in entropy of the surroundings offsets the decrease in entropy of the system. **17.64** 55 J/K. **17.66** 5.2×10^5. **17.68** 1.9×10^{-76} atm. **17.70** (c). **17.72** $C + CuO \rightleftharpoons CO + Cu$. 3.9. **17.74** 575.1 K. **17.76** (a) 6.8×10^{13}. (b) 3.3×10^{-12}. **17.78** (a) A reverse disproportionation reaction. (b) 6.7×10^{15}. Yes, a large *K* makes this an efficient process. (c) Less effective. **17.80** 1.5×10^{70}. Reaction has a large activation energy. **17.82** Heating the ore alone is not a feasible process. −213.9 kJ/mol. **17.84** $K_P = 36.981$ K. No. **17.86** Negative. **17.88** Mole percents: butane = 29%; isobutane = 71%. Yes. **17.90** (a) Na(*l*): 57.9 J/K · mol. (b) $S_2Cl_2(g)$: 331.5 J/K · mol. (c) $FeCl_2(s)$: 118 J/K · mol. **17.92** Mole fractions are: CO = 0.45, CO_2 = 0.55. Use ΔG_f° values at 25°C for 900°C.

17.94 617 J/K. **17.96** 3×10^{-13} s. **17.98** $\Delta S_{sys} = -327$ J/K · mol, $\Delta S_{surr} = 1918$ J/K · mol, $\Delta S_{univ} = 1591$ J/K · mol. **17.100** q, w. **17.102** $\Delta H < 0$, $\Delta S < 0$, $\Delta G < 0$. **17.104** (a) 5.76 J/K · mol. (b) The orientation is not totally random. **17.106** $\Delta H° = 33.8$ kJ/mol; $\Delta S° = 95.8$ J/K · mol; $\Delta G° = 5.3$ kJ/mol. This is an endothermic liquid to vapor process so both $\Delta H°$ and $\Delta S°$ are positive. $\Delta G°$ is positive because the temperature is below the boiling point of benzene (80.1°C). **17.108** $\Delta G° = 62.5$ kJ/mol; $\Delta H° = 157.8$ kJ/mol; $\Delta S° = 109$ J/K · mol. **17.110** Slightly larger than 0.052 atm.

Chapter 18

18.2 (a) $Mn^{2+} + H_2O_2 + 2OH^- \longrightarrow MnO_2 + H_2O$. (b) $2Bi(OH)_3 + 3SnO_2^{2-} \longrightarrow 2Bi + 3H_2O + 3SnO_3^{2-}$. (c) $Cr_2O_7^{2-} + 14H^+ + 3C_2O_4^{2-} \longrightarrow 2Cr^{3+} + 6CO_2 + 7H_2O$. (d) $2Cl^- + 2ClO_3^- + 4H^+ \longrightarrow Cl_2 + 2ClO_2 + 2H_2O$. **18.12** 2.46 V. $Al + 3Ag^+ \longrightarrow 3Ag + Al^{3+}$. **18.14** $Cl_2(g)$ and $MnO_4^-(aq)$ **18.16** Only (a) and (d) are spontaneous. **18.18** (a) Li. (b) H_2. (c) Fe^{2+}. (d) Br^-. **18.20** −1.80 V. **18.24** 0.370 V. **18.26** (a) −432 kJ/mol, 5×10^{75}. (b) −104 kJ/mol, 2×10^{18}. (c) −178 kJ/mol, 1×10^{31}. (d) -1.28×10^3 kJ/mol, 1×10^{224}. **18.28** 0.37 V, −36 kJ/mol, 2×10^6. **18.32** (a) 2.23 V, 2.23 V, −430 kJ/mol. (b) 0.02 V, 0.04 V, −23 kJ/mol. **18.34** 0.083 V. **18.36** 0.010 V. **18.40** 1.09 V. **18.48** (b) 0.64 g. **18.50** (a) $\$2.10 \times 10^3$. (b) $\$2.46 \times 10^3$. (c) $\$4.70 \times 10^3$. **18.52** (a) 0.14 mol. (b) 0.121 mol. (c) 0.10 mol. **18.54** (a) $Ag^+ + e^- \longrightarrow Ag$. (b) $2H_2O \longrightarrow O_2 + 4H^+ + 4e^-$. (c) 6.0×10^2 C. **18.56** (a) 0.589 Cu. (b) 0.133 A. **18.58** 2.3 h. **18.60** 9.66×10^4 C. **18.62** 0.0710 mol. **18.64** (a) Anode: $Cu(s) \longrightarrow Cu^{2+}(aq) + 2e^-$. Cathode: $Cu^{2+}(aq) + 2e^- \longrightarrow Cu(s)$. (b) 2.4×10^2 g. (c) Copper is more easily oxidized than Ag and Au. Copper ions (Cu^{2+}) are more easily reduced than Fe^{2+} and Zn^{2+}. **18.66** 0.0296 V. **18.68** 0.156 M. $Cr_2O_7^{2-} + 6Fe^{2+} + 14H^+ \longrightarrow 2Cr^{3+} + 6Fe^{3+} + 7H_2O$. **18.70** 45.1%. **18.72** (a) $2MnO_4^- + 16H^+ + 5C_2O_4^{2-} \longrightarrow 2Mn^{2+} + 10CO_2 + 8H_2O$. (b) 5.40%. **18.74** 0.231 mg Ca^{2+}/mL blood. **18.76** (a) 0.80 V. (b) $2Ag^+ + H_2 \longrightarrow 2Ag + 2H^+$. (c) (i) 0.92 V. (ii) 1.10 V. (d) The cell operates as a pH meter. **18.78** Fluorine gas reacts with water. **18.80** 2.5×10^2 h. **18.82** Hg_2^{2+}. **18.84** $[Mg^{2+}] = 0.0500\ M$, $[Ag^+] = 7 \times 10^{-55}\ M$, 1.44 g. **18.86** (a) 0.206 L H_2. (b) 6.09×10^{23}/mol e^-. **18.88** (a) −1356.8 kJ/mol. (b) 1.17 V. **18.90** +3. **18.92** 6.8 kJ/mol, 0.064. **18.94** In both cells, the anode is on the left and the cathode is on the right. In the galvanic cell, the anode is negatively charged and the cathode is positively charged. The opposite holds for the electrolytic cell. Electrons flow from the anode in the galvanic cell to the cathode in the electrolytic cell and electrons flow from the anode in the electrolytic cell to the cathode in the galvanic cell. **18.96** 1.4 A. **18.98** +4. **18.100** 1.60×10^{-19} C/e^-. **18.102** A cell made of Li^+/Li and F_2 /F^- gives the maximum voltage of 5.92 V. Reactive oxidizing and reducing agents are hard to handle. **18.104** 0.030 V. **18.106** 2×10^{20}. **18.108** (a) $E°$ for X is negative; $E°$ for Y is positive. (b) 0.59 V. **18.110** (a) The reduction potential of O_2 is insufficient to oxidize gold. (b) Yes. (c) $2Au + 3F_2 \longrightarrow 2AuF_3$. **18.112** $[Fe^{2+}] = 0.0920\ M$, $[Fe^{3+}] = 0.0680\ M$. **18.114** $E° = 1.08$ V. Spontaneous. **18.116** (a) Ni. (b) Pb. (c) Zn. (d) Fe. **18.118** (a) Unchanged. (b) Unchanged. (c) Squared. (d) Doubled. (e) Doubled. **18.120** Stronger. **18.122** 4.4×10^2 atm. **18.124** (a) $Zn \longrightarrow Zn^{2+} + 2e^-$; $(1/2)O_2 + 2e^- \longrightarrow O^{2-}$. 1.66 V (b) 1.64 V. (c) 4.87×10^3 kJ/kg. (d) 62 L. **18.126** −3.04 V. **18.128** 1×10^{-14}. **18.130** (a) 1 A · h = 1 A × 3600 s = 3600 C. (b) 104 A · h. The concentration of H_2SO_4 keeps decreasing. (c) 2.00 V; -3.86×10^2 kJ/mol. **18.132** \$217. **18.134** −0.043 V. **18.136** 2×10^{37}. **18.138** 5 mol ATP. **18.140** 2.89 V.

Chapter 19

19.8 (a) ${}^{0}_{-1}\beta$. (b) ${}^{40}_{20}Ca$. (c) ${}^{4}_{2}\alpha$. (d) ${}^{1}_{0}n$. **19.16** (a) ${}^{9}_{3}Li$. (b) ${}^{25}_{11}Na$. (c) ${}^{48}_{21}Sc$. **19.18** (a) ${}^{17}_{10}Ne$. (b) ${}^{45}_{20}Ca$. (c) ${}^{92}_{43}Tc$. (d) ${}^{195}_{80}Hg$. (e) ${}^{242}_{96}Cm$. **19.20** 6×10^9 kg/s. **19.22** (a) 4.55×10^{-12} J; 1.14×10^{-12} J/nucleon. (b) 2.36×10^{-10} J; 1.28×10^{-12} J/nucleon. **19.26** 0.251 d^{-1}. 2.77 d. **19.28** 2.7 d. **19.30** ${}^{208}_{82}Pb$. **19.32** A: 0; B: 0.25 mole; C: 0; D: 0.75 mole. **19.34** ${}^{224}_{88}Ra$. **19.38** (a) ${}^{80}_{34}Se + {}^{2}_{1}H \longrightarrow {}^{1}_{1}p + {}^{81}_{34}Se$. (b) ${}^{9}_{4}Be + {}^{2}_{1}H \longrightarrow 2{}^{1}_{1}p + {}^{9}_{3}Li$. (c) ${}^{10}_{5}B + {}^{1}_{0}n \longrightarrow {}^{4}_{2}\alpha + {}^{7}_{3}Li$. **19.40** ${}^{198}_{80}Hg + {}^{1}_{0}n \longrightarrow {}^{198}_{79}Au + {}^{1}_{1}p$. **19.52** IO_3^- is only formed from IO_4^-. **19.54** Incorporate Fe-59 into a person's body. After a few days isolate red blood cells and monitor radioactivity from the hemoglobin molecules. **19.56** (a) ${}^{50}_{25}Mn \longrightarrow {}^{50}_{24}Cr + {}^{0}_{+1}\beta$. (b) Three half-lives. **19.58** An analogous Pauli exclusion principle for nucleons. **19.60** (a) 0.343 mCi. (b) ${}^{237}_{93}Np \longrightarrow {}^{4}_{2}\alpha + {}^{233}_{91}Pa$. **19.62** (a) 1.040×10^{-12} J/nucleon. (b) 1.111×10^{-12} J/nucleon. (c) 1.199×10^{-12} J/nucleon. (d) 1.410×10^{-12} J/nucleon. **19.64** ${}^{18}_{7}N \longrightarrow {}^{18}_{8}O + {}^{0}_{-1}\beta$. **19.66** Radioactive dating. **19.68** (a) ${}^{209}_{83}Bi + {}^{4}_{2}\alpha \longrightarrow {}^{211}_{85}At + 2{}^{1}_{0}n$. (b) ${}^{209}_{83}Bi(\alpha, 2n){}^{211}_{85}At$. **19.70** The sun exerts a much greater gravity on the particles. **19.72** 2.77×10^3 yr. **19.74** (a) ${}^{40}_{19}K \longrightarrow {}^{40}_{18}Ar + {}^{0}_{+1}\beta$. (b) 3.0×10^9 yr. **19.76** (a) ${}^{90}Sr$: 5.59×10^{-15} J; ${}^{90}Y$: 2.84×10^{-13} J. (b) 0.024 mole. (c) 4.26×10^6 kJ. **19.78** 2.7×10^{14} I-131 atoms. **19.80** 5.9×10^{23}/mol. **19.82** All except gravitational. **19.84** U-238 and Th-232. Long half-lives. **19.86** 8.3×10^{-4} nm. **19.88** ${}^{3}_{1}H$. **19.90** The reflected neutrons induced a nuclear chain reaction. **19.92** 2.1×10^2 g/mol. **19.94** First step: ${}^{234}_{90}Th \longrightarrow {}^{234}_{91}Pa + {}^{0}_{-1}\beta$. Second step: ${}^{234}_{91}Pa \longrightarrow {}^{234}_{92}U + {}^{0}_{-1}\beta$. Third step: ${}^{234}_{92}U \longrightarrow {}^{230}_{90}Th + {}^{4}_{2}\alpha$. Fourth step: ${}^{230}_{90}Th \longrightarrow {}^{226}_{88}Ra + {}^{4}_{2}\alpha$. Fifth step: ${}^{226}_{88}Ra \longrightarrow {}^{222}_{86}Rn + {}^{4}_{2}\alpha$. **19.96** (a) ${}^{238}_{94}Pu \longrightarrow {}^{4}_{2}\alpha + {}^{234}_{92}U$. (b) $t = 0$: 0.58 mW; $t = 10$ yr: 0.53 mW. **19.98** 0.49 rem. **19.100** The high temperature attained during the chain reaction causes a small-scale nuclear fusion: ${}^{2}_{1}H + {}^{3}_{1}H \longrightarrow {}^{4}_{2}He + {}^{1}_{0}n$. The additional neutrons will result in a more powerful fission bomb. **19.102** 21.5 mL. **19.104** No. According to Equation (19.1), energy and mass are interconvertible. **19.106** (a) 1.69×10^{-12} J. (b) 1.23×10^{-12} J. Because a proton feels the repulsion from other protons, it has a smaller binding energy than a neutron.

Chapter 20

20.6 3.3×10^{-4} atm. **20.8** N_2: 3.96×10^{18} kg; O_2: 1.22×10^{18} kg; CO_2: 2.63×10^{15} kg. **20.12** 3.57×10^{-19} J. **20.22** 5.2×10^6 kg/day. 5.6×10^{14} kJ. **20.24** The wavelength is not short enough. **20.26** 434 nm. Both.

20.28 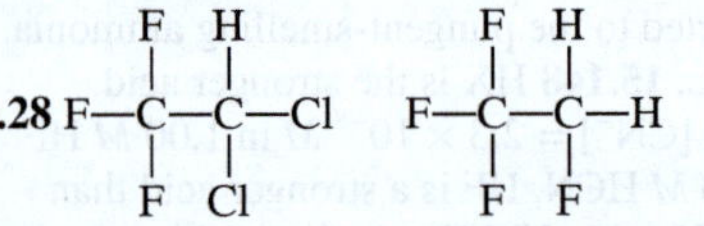

20.40 1.3×10^{10} kg. **20.42** Ethane and propane are greenhouse gases. **20.50** 4.34. **20.58** 1.2×10^{-11} M/s. **20.60** (b). **20.66** 0.12%. **20.68** Endothermic. **20.70** O_2. **20.72** 5.72. **20.74** 394 nm. **20.76** It has a high activation energy. **20.78** Size of tree rings is related to CO_2 content. Age of CO_2 in ice can be determined by radiocarbon dating. **20.80** 137 kJ/mol. **20.82** 5.1×10^{20} photons. **20.84** (a) 62.6 kJ/mol. (b) 38 min. **20.86** 5.6×10^{23}. **20.88** H—Ö—Ö—Ö·

Chapter 21

21.12 111 h. **21.14** Roast the sulfide followed by reduction of the oxide with coke or carbon monoxide. **21.16** (a) 8.9×10^{12} cm^3. (b) 4.0×10^8 kg. **21.18** Iron does not need to be produced electrolytically. **21.28** (a) $2Na + 2H_2O \longrightarrow 2NaOH + H_2$. (b) $2NaOH + CO_2 \longrightarrow Na_2CO_3 + H_2O$. (c) $Na_2CO_3 + 2HCl \longrightarrow 2NaCl + CO_2 + H_2O$. (d) $NaHCO_3 + HCl \longrightarrow NaCl + CO_2 + H_2O$. (e) $2NaHCO_3 \longrightarrow Na_2CO_3 + CO_2 + H_2O$. (f) No reaction. **21.30** 5.59 L. **21.34** First react Mg with HNO_3 to form $Mg(NO_3)_2$. On heating, $2Mg(NO_3)_2 \longrightarrow 2MgO + 4NO_2 + O_2$. **21.36** The third electron is removed from the neon core. **21.38** Helium has a closed-shell noble gas configuration. **21.40** (a) CaO. (b) $Ca(OH)_2$. (c) An aqueous suspension of $Ca(OH)_2$. **21.44** 60.7 h. **21.46** (a) 1.03 V. (b) 3.32×10^4 kJ/mol. **21.48** $4Al(NO_3)_3 \longrightarrow 2Al_2O_3 + 12NO_2 + 3O_2$.

21.50 Because Al_2Cl_6 dissociates to form $AlCl_3$. **21.52** From sp^3 to sp^2. **21.54** 65.4 g/mol. **21.56** No. **21.58** (a) 1484 kJ/mol. (b) 3164.6 kJ/mol. **21.60** Magnesium reacts with nitrogen to form magnesium nitride. **21.62** (a) Al^{3+} hydrolyzes in water to produce H^+ ions. (b) $Al(OH)_3$ dissolves in a strong base to form $Al(OH)_4^-$. **21.64** $CaO + 2HCl \longrightarrow CaCl_2 + H_2O$. **21.66** Electronic transitions (in the visible region) between closely spaced energy levels. **21.68** NaF: toothpaste additive; Li_2CO_3: to treat mental illness; $Mg(OH)_2$: antacid; $CaCO_3$: antacid; $BaSO_4$: for X-ray diagnostic of digestive system; $Al(OH)_2NaCO_3$: antacid. **21.70** (i) Both Li and Mg form oxides. (ii) Like Mg, Li forms nitride. (iii) The carbonates, fluorides, and phosphates of Li and Mg have low solubilities. **21.72** Zn. **21.74** D < A < C < B. **21.76** 727 atm.

Chapter 22

22.12 (a) Hydrogen reacts with alkali metals to form hydrides. (b) Hydrogen reacts with oxygen to form water. **22.14** Use palladium metal to separate hydrogen from other gases. **22.16** 11 kg. **22.18** (a) $H_2 + Cl_2 \longrightarrow 2HCl$. (b) $N_2 + 3H_2 \longrightarrow 2NH_3$. (c) $2Li + H_2 \longrightarrow 2LiH$, $LiH + H_2O \longrightarrow LiOH + H_2$. **22.26** $:C\equiv C:^{2-}$. **22.28** (a) $2NaHCO_3 \longrightarrow Na_2CO_3 + H_2O + CO_2$. (b) CO_2 reacts with $Ca(OH)_2$ solution to form a white precipitate ($CaCO_3$). **22.30** On heating, the bicarbonate ion decomposes: $2HCO_3^- \longrightarrow CO_3^{2-} + H_2O + CO_2$. Mg^{2+} ions combine with CO_3^{2-} ions to form $MgCO_3$. **22.32** First, $2NaOH + CO_2 \longrightarrow Na_2CO_3 + H_2O$. Then, $Na_2CO_3 + CO_2 + H_2O \longrightarrow 2NaHCO_3$. **22.34** Yes. **22.40** (a) $2NaNO_3 \longrightarrow 2NaNO_2 + O_2$. (b) $NaNO_3 + C \longrightarrow NaNO_2 + CO$. **22.42** $2NH_3 + CO_2 \longrightarrow (NH_2)_2CO + H_2O$. At high pressures. **22.44** NH_4Cl decomposes to form NH_3 and HCl. **22.46** N is in its highest oxidation state (+5) in HNO_3. **22.48** Favored reaction: $4Zn + NO_3^- + 10H^+ \longrightarrow 4Zn^{2+} + NH_4^+ + 3H_2O$. **22.50** Linear. **22.52** −1168 kJ/mol. **22.54** P_4. 125 g/mol. **22.56** $P_4O_{10} + 4HNO_3 \longrightarrow 2N_2O_5 + 4HPO_3$. 60.4 g. **22.58** sp^3. **22.66** −199.5 kJ/mol, 6×10^{34}. **22.68** 0; −1. **22.70** 4.4×10^{11} mol; 1.4×10^{13} g. **22.72** 79.1 g. **22.74** Cl, Br, and I atoms are too bulky around the S atom. **22.76** 35 g. **22.78** $9H_2SO_4 + 8NaI \longrightarrow H_2S + 4I_2 + 4H_2O + 8NaHSO_4$. **22.82** $H_2SO_4 + NaCl \longrightarrow HCl + NaHSO_4$. The HCl gas escapes, driving the equilibrium to the right. **22.84** 25.3 L. **22.86** Sulfuric acid oxidizes sodium bromide to molecular bromine. **22.88** 2.81 L. **22.90** $I_2O_5 + 5CO \longrightarrow I_2 + 5CO_2$. C is oxidized; I is reduced. **22.92** (a) $SiCl_4$. (b) F^-. (c) F. (d) CO_2. **22.94** No change. **22.96** (a) $2Na + D_2O \longrightarrow 2NaOD + D_2$. (b) $2D_2O \longrightarrow 2D_2 + O_2$ (electrolysis). $D_2 + Cl_2 \longrightarrow 2DCl$. (c) $Mg_3N_2 + 6D_2O \longrightarrow 3Mg(OD)_2 + 2ND_3$. (d) $CaC_2 + 2D_2O \longrightarrow C_2D_2 + Ca(OD)_2$. (e) $Be_2C + 4D_2O \longrightarrow 2Be(OD)_2 + CD_4$. (f) $SO_3 + D_2O \longrightarrow D_2SO_4$. **22.98** (a) At elevated pressure, water boils above 100°C. (b) So the water is able to melt a larger area of sulfur deposit. (c) Sulfur deposits are structurally weak. Conventional mining would be dangerous. **22.100** The C—D bond breaks at a slower rate. **22.102** Molecular oxygen is a powerful oxidizing agent, reacting with substances such as glucose to release energy for growth and function. Molecular nitrogen (containing the nitrogen-to-nitrogen triple bond) is too unreactive at room temperature to be of any practical use. **22.104** 25°C: 9.61×10^{-22}; 1000°C: 138. High temperature favors the formation of CO. **22.106** 1.20.

Chapter 23

23.12 (a) +3. (b) 6. (c) Oxalate. **23.14** (a) Na: +1, Mo: +6. (b) Mg: +2, W: +6. (c) Fe: 0. **23.16** (a) *cis*-dichlorobis(ethylenediamine) cobalt(III). (b) pentaamminechloroplatinum(IV) chloride. (c) pentaamminechlorocobalt(III) chloride. **23.18** (a) $[Cr(en)_2Cl_2]^+$. (b) $Fe(CO)_5$. (c) $K_2[Cu(CN)_4]$. (d) $[Co(NH_3)_4(H_2O)Cl]Cl_2$. **23.24** (a) 2. (b) 2. **23.26** (a) Two geometric isomers:

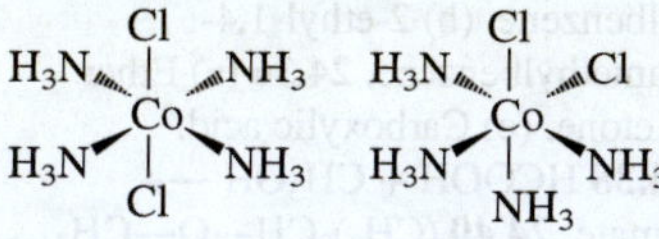

trans *cis*

(b) Two optical isomers:

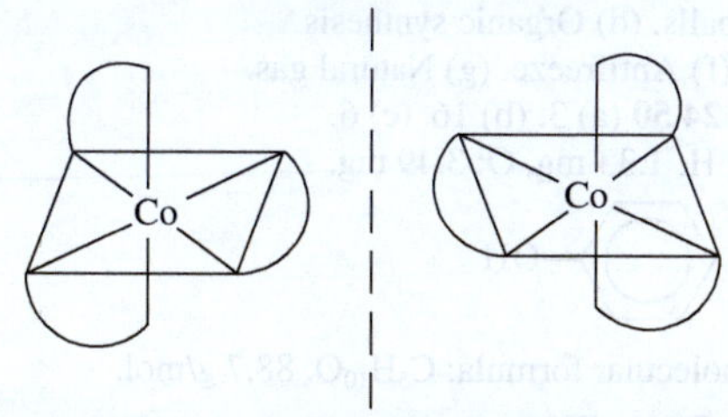

23.34 CN^- is a strong-field ligand. Absorbs near UV (blue) so appears yellow. **23.36** (a) Orange. (b) 255 kJ/mol. **23.38** $[Co(NH_3)_4Cl_2]Cl$. 2 moles. **23.42** Use $^{14}CN^-$ label (in NaCN). **23.44** First $Cu(CN)_2$ (white) is formed. It redissolves as $Cu(CN)_4^{2-}$. **23.46** 1.4×10^2. **23.48** Mn^{3+}. The $3d^3$ electron configuration of Cr^{3+} is stable. **23.50** Ti: +3; Fe: +3. **23.52** Four Fe atoms per hemoglobin molecule. 1.6×10^4 g/mol. **23.54** (a) $[Cr(H_2O)_6]Cl_3$. (b) $[Cr(H_2O)_5Cl]Cl_2 \cdot H_2O$. (c) $[Cr(H_2O)_4Cl_2]Cl \cdot 2H_2O$. Compare electrical conductance with solutions of NaCl, $MgCl_2$, and $FeCl_3$ of the same molar concentration. **23.56** -1.8×10^2 kJ/mol; 6×10^{30}. **23.58** Iron is more abundant. **23.60** Oxyhemoglobin is low spin and therefore absorbs higher energy light. **23.62** All except Fe^{2+}, Cu^{2+}, and Co^{2+}. The colorless ions have electron configurations d^0 and d^{10}. **23.64** Dipole moment measurements. **23.66** EDTA sequesters essential metal ions (Ca^{2+}, Mg^{2+}). **23.68** 3. **23.70** 1.0×10^{-18} *M*. **23.72** 2.2×10^{-20} *M*. **23.74** (a) 2.7×10^6. (b) Cu^+ ions are unstable in solution. **23.76** (a) Cu^{3+} is unstable in solution because it can be easily reduced. (b) Potassium hexafluorocuprate(III). Octahedral. Paramagnetic. (c) Diamagnetic.

Chapter 24

24.12 $CH_3CH_2CH_2CH_2CH_2Cl$. $CH_3CH_2CH_2CHClCH_3$. $CH_3CH_2CHClCH_2CH_3$.

24.14

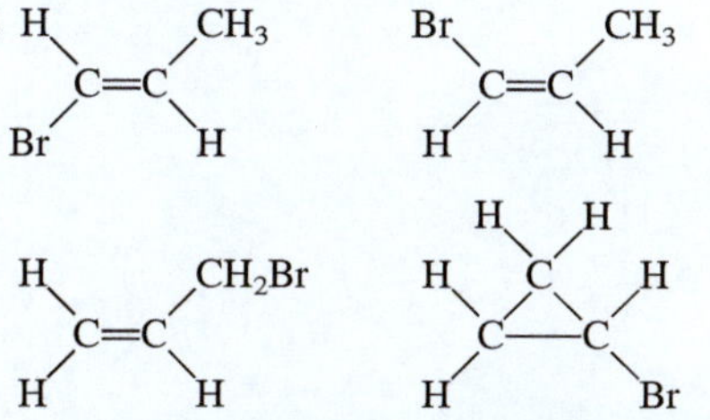

24.16 (a) Alkene or cycloalkane. (b) Alkyne. (c) Alkane. (d) Like (a). (e) Alkyne. **24.18** No, too much strain. **24.20** (a) is alkane and (b) is alkene. Only an alkene reacts with a hydrogen halide and hydrogen. **24.22** −630.8 kJ/mol. **24.24** (a) *cis*-1,2-dichlorocylopropane. (b) *trans*-1,2-dichlorocylopropane. **24.26** (a) 2-methylpentane. (b) 2,3,4-trimethylhexane. (c) 3-ethylhexane. (d) 3-methyl-1,4-pentadiene. (e) 2-pentyne. (f) 3-phenyl-1-pentene.

24.28 (a)

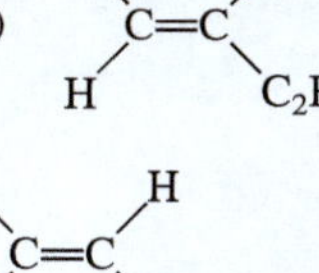

(b)

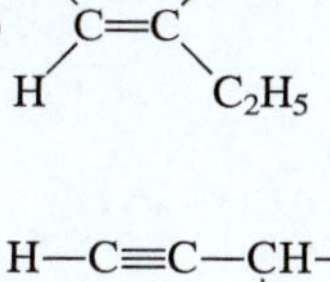

(c)

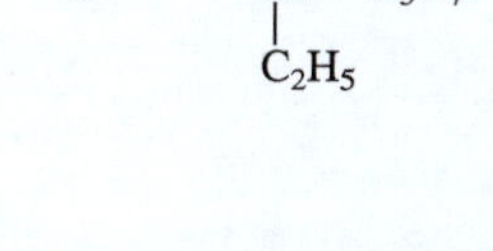

(d)

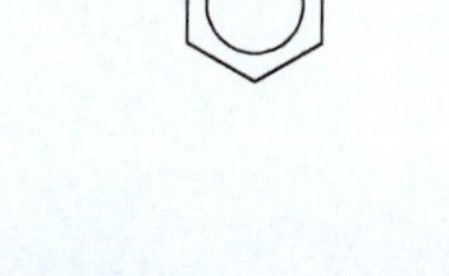

24.32 (a) 2,4-dichloro-1-methylbenzene. (b) 2-ethyl-1,4-dinitrobenzene. (c) 1,2,4,5-tetramethylbenzene. **24.36** (a) Ether. (b) Amine. (c) Aldehyde. (d) Ketone. (e) Carboxylic acid. (f) Alcohol. (g) Amino acid. **24.38** $HCOOH + CH_3OH \longrightarrow HCOOCH_3 + H_2O$. Methyl formate. **24.40** $(CH_3)_2CH—O—CH_3$. **24.42** (a) Ketone. (b) Ester. (c) Ether. **24.44** −174 kJ/mol. **24.46** (a), (c), (d), and (f). **24.48** (a) Rubbing alcohol. (b) Vinegar. (c) Moth balls. (d) Organic synthesis. (e) Organic synthesis. (f) Antifreeze. (g) Natural gas. (h) Synthetic polymer. **24.50** (a) 3. (b) 16. (c) 6. **24.52** (a) C: 15.81 mg, H: 1.33 mg, O: 3.49 mg.

(b) C_6H_6O. (c) Phenol. 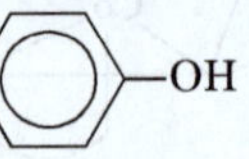

24.54 Empirical and molecular formula: $C_5H_{10}O$. 88.7 g/mol.

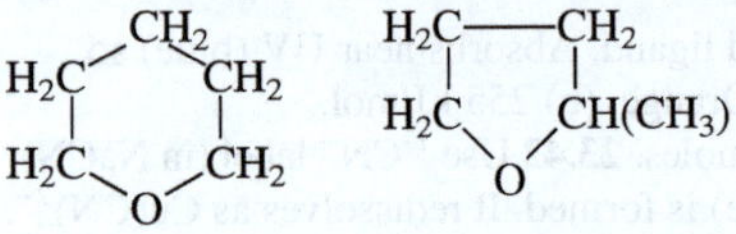

24.56 (a) The C atoms bonded to the methyl group and the amino group and the H atom. (b) The C atoms bonded to Br. **24.58** CH_3CH_2CHO. **24.60** (a) Alcohol. (b) Ether. (c) Aldehyde. (d) Carboxylic acid. (e) Amine. **24.62** The acids in lemon juice convert the amines to the ammonium salts, which have very low vapor pressures. **24.64** Methane (CH_4), ethanol (C_2H_5OH), methanol (CH_3OH), isopropanol (C_3H_7OH), ethylene glycol (CH_2OHCH_2OH), naphthalene ($C_{10}H_8$), acetic acid (CH_3COOH). **24.66** (a) 1. (b) 2. (c) 5. **24.68** Br_2 dissociates into Br atoms, which react with CH_4 to form CH_3Br and HBr.

24.70 (a) $CH_3—\underset{H}{\overset{OH}{C}}—CH_2—CH_3$. The compound is chiral.

(b) The product is a racemic mixture.

24.72 $CH_3CH_2CH_2OH$ or $CH_3—\overset{OH}{CH}—CH_3$.
24.74 (a) Reaction between glycerol and carboxylic acid (formation of an ester). (b) Fat or oil (shown in problem) + $NaOH \longrightarrow$ Glycerol + $3RCOO^-Na^+$(soap). (c) Molecules having more $^0_2$N bonds are harder to pack tightly. Consequently, they have a lower melting point. (d) H_2 gas with a homogeneous or heterogeneous catalyst. (e) 123.

Chapter 25

25.8 $+CH_2—CHCl—CH_2—CCl_2+$. **25.10** By an addition reaction involving styrene monomers. **25.12** (a) $CH_2═CH—CH═CH_2$. (b) $HO_2C(CH_2)_6NH_2$. **25.22** At 35°C the enzyme begins to denature. **25.28** Proteins are made of 20 amino acids. Nucleic acids are made of four building blocks (purines, pyrimidines, sugar, phosphate group) only. **25.30** C-G base pairs have three hydrogen bonds and higher boiling point; A-T base pairs have two hydrogen bonds. **25.32** Leg muscles are active, have a high metabolic rate and hence a high concentration of myoglobin. The iron content in Mb makes the meat look dark. **25.34** Insects have blood that contains no hemoglobin. It is unlikely that a human-sized insect could obtain sufficient oxygen for metabolism by diffusion. **25.36** There are four Fe atoms per hemoglobin molecule. 1.6×10^4 g/mol. **25.38** Mostly dispersion forces. **25.40** Gly-Ala-Phe-Glu-His-Gly-Ala-Leu-Val. **25.42** No. Enzymes only act on one of the two optical isomers of a compound. **25.44** 315 K. **25.46** Hydrogen bonding. **25.48** (a) The —COOH group. (b) pH = 1.0: The valine is in the fully protonated form. pH = 7.0: Only the —COOH group is ionized. pH = 12.0: Both groups are ionized. (c) 5.97. **25.50** (a) $\overline{M}_n$ = 3.6 kg/mol; $\overline{M}_w$ = 4.3 kg/mol. (b) $\overline{M}_n$ = 5 kg/mol; $\overline{M}_w$ = 5 kg/mol. (c) If $\overline{M}_n$ and $\overline{M}_w$ are close in value, that indicates a small spread in the distribution of polymer sizes. (d) The four subunits in hemoglobin molecule dissociate in solution, giving a distribution of molar masses. There are no subunits in myoglobin or cytochrome *c*, so no distribution of molar masses.

Index

Note: Page numbers followed by *f*, indicate figures; *mn*, marginal notes; and *t*, tables.

D

E

N

O

P

Q

R

T

U

V

W

X

Y

Z